Organotransition Metal Chemistry

Organotransition Metal Chemistry

From Bonding to Catalysis

John F. Hartwig

UNIVERSITY OF ILLINOIS
URBANA-CHAMPAIGN

University Science Books
Mill Valley, California

University Science Books
www.uscibooks.com

Production Manager: Jennifer Uhlich at Wilsted and Taylor
Manuscript Editor: John Murdzek
Illustrator: Lineworks, Inc.
Design: Yvonne Tsang at Wilsted and Taylor
Compositor: MPS Limited, a Macmillan Company
Cover Design: Genette Itoko McGrew
Indexer: Grant Hackett
Printer & Binder: Edwards Brothers, Inc.

Library of Congress Cataloging-in-Publication Data

Hartwig, John F., 1964-
  Organotransition metal chemistry : from bonding to catalysis / John Hartwig.
     p. cm.
  Includes bibliographical references and index.
  ISBN 978-1-891389-53-5 (alk. paper)
  1. Organotransition metal compounds. 2. Ligands. I. Title.
  QD411.8.T73H37 2010
  547'.056—dc22
                              2009020537

Printed in the United States of America
10   9   8   7   6   5   4   3   2

Abbreviated Contents

Chapter 1. Structure and Bonding   1

Chapter 2. Dative Ligands   27

Chapter 3. Covalent (X-Type) Ligands Bound Through Metal–Carbon and
Metal–Hydrogen Bonds   85

Chapter 4. Covalent (X-Type) Ligands Bound Through Metal–Heteroatom Bonds   147

Chapter 5. Ligand Substitution Reactions   217

Chapter 6. Oxidative Addition of Nonpolar Reagents   261

Chapter 7. Oxidative Addition of Polar Reagents   301

Chapter 8. Reductive Elimination   321

Chapter 9. Migratory Insertion Reactions   349

Chapter 10. Elimination Reactions   397

Chapter 11. Nucleophilic Attack on Coordinated Ligands   417

Chapter 12. Electrophilic Attack on Coordinated Ligands   453

Chapter 13. Metal–Ligand Multiple Bonds   481

Chapter 14. Principles of Catalysis (Written with Prof. Patrick J. Walsh)   539

Chapter 15. Homogeneous Hydrogenation   575

Chapter 16. Hydrofunctionalization and Oxidative Functionalization of Olefins   667

Chapter 17. Catalytic Carbonylation   745

Chapter 18. Catalytic C–H Functionalization   825

Chapter 19. Transition Metal-Catalyzed Coupling Reactions   877

Chapter 20. Allylic Substitution   967

Chapter 21. Metathesis of Olefins and Alkynes   1015

Chapter 22. Polymerization and Oligomerization of Olefins   1047

Contributor Listing   1101

Index   1103

Contents

Chapter 1. Structure and Bonding 1

1.1. General Properties of the Ligands 1
1.1.1. Classification of Ligands as Dative or Covalent, Neutral or Anionic, Even- or Odd-Electron, L-Type or X-Type 1
1.1.2. Classification by Number of Electrons Donated to the Metal 3
1.1.3. π-Bonded Ligands 4
1.1.4. Combinations of σ- and π-Donors 5
1.1.5. Cationic Ligands 6

1.2. Properties of the Metal 6
1.2.1. Oxidation State 6
1.2.2. The Relationship Between Oxidation State and the Number of *d*-Electrons 7
1.2.3. Trends in the Properties of Transition Metals 8
1.2.3.1. Trends in Ionization Potentials 8
1.2.3.2. Trends in Size 9
1.2.3.3. Trends in Bond Strengths 9

1.3. Metal–Ligand Complexes 10
1.3.1. Electron Counting 10
1.3.2. The 18-Electron Rule 13
1.3.3. Metal–Metal Bonding and Electron Counting in Polynuclear Complexes 13
1.3.4. Geometries of Transition Metal Complexes 14
1.3.5. Isoelectronic and Isolobal Analogies 15
1.3.6. Molecular Orbitals for Transition Metal Complexes 17
1.3.7. π-Bonding in Organotransition Metal Complexes 19
1.3.7.1. π-Bonding of CO and its Analogs 19
1.3.7.2. π-Bonding of Carbene and Carbyne Complexes 20
1.3.7.3. π-Bonding in Olefin Complexes 21
1.3.7.4. π-Bonding with Other Unsaturated Ligands 22
1.3.8. π-Donor Ligands 22
References and Notes 26

Chapter 2. Dative Ligands 27

2.1. Introduction 27
2.2. Carbon Monoxide and Related Ligands 27
2.2.1. Properties of Free Carbon Monoxide 27

2.2.2. Types of Metal Carbonyl Complexes 28
2.2.3. Models for CO Binding: Introduction of Backbonding 29
2.2.4. Evidence for Backbonding in Terminal Carbonyls 30
2.2.5. Infrared and X-Ray Diffraction Data for Complexes with Bridging Carbonyls 31
2.2.6. Thermodynamics of the M–CO Bond 31
2.2.7. Isoelectronic Analogs of CO: Isocyanides and Thiocarbonyls 32

2.3. Dative Phosphorus Ligands and Heavier Congeners 33
2.3.1. Tertiary Phosphines and Related Ligands 33
2.3.2. Chelating Phosphines 34
2.3.3. Properties of Free Phosphines 35
2.3.4. Properties of Phosphine Complexes 36
2.3.4.1. Bonding and Electronic Properties 36
2.3.4.2. Steric Properties 38
2.3.4.3. Effects of Phosphine Steric and Electronic Properties on Structure and Reactivity 39
2.3.5. Pathways for the Decomposition of Phosphorus Ligands 39
2.3.6. NMR Spectroscopic Properties of Phosphines 40
2.3.7. Heavier Congeners of Phosphorus Ligands 41

2.4. Carbenes 41
2.4.1. Classes of Free and Coordinated Carbenes 41
2.4.1.1. Properties of Free Carbenes 41
2.4.1.2. Properties of Carbene Complexes 41
2.4.2. Bonding of Carbenes 44
2.4.3. Spectroscopic Characteristics of Carbene Complexes 45

2.5. Transition Metal Carbyne Complexes 45
2.5.1. Bonding and Structure of Carbyne Complexes 45
2.5.2. Spectroscopic Characteristics of Carbyne Complexes 46

2.6. Organic Ligands Bound Through More than One Atom 47
2.6.1. Olefin Complexes 47
2.6.1.1. Stability of Metal–Olefin Complexes 47

2.6.1.2. Structures of Metal–Olefin Complexes   49
   2.6.1.2.1. Structural Changes Upon Binding   49
   2.6.1.2.2. Orientation of Coordinated Olefins   49
2.6.1.3. Spectral Properties of Metal–Olefin Complexes   51
2.6.2. Alkyne Complexes   51
  2.6.2.1. Structural Characteristics of Alkyne Complexes   51
  2.6.2.2. Physical and Chemical Properties of Alkyne Ligands   52
2.6.3. Complexes of Organic Carbonyl Compounds   53
2.6.4. η⁶-Arene and Related Complexes   53

2.7. Complexes of Ligands Bound Through N, O and S   57
2.7.1. Neutral Nitrogen Donor Ligands   57
  2.7.1.1. Amine Complexes   57
  2.7.1.2. Pyridine and Imine Complexes   58
  2.7.1.3. Dinitrogen Complexes   59
  2.7.1.4. Complexes of Neutral Oxygen Donors   62
  2.7.1.5. Complexes of Neutral Sulfur Donors   63

2.8. Sigma Complexes   64
2.8.1. Overview of Sigma Complexes   64
2.8.2. Dihydrogen Complexes   66
  2.8.2.1. Properties that Lead to Stable H₂ Complexes   67
  2.8.2.2. Spectroscopic Signatures of H₂ Complexes   67
  2.8.2.3. Reactivity of H₂ Complexes   68
2.8.3. Alkane and Silane Complexes   70
  2.8.3.1. Stability Relative to H₂ Complexes   70
  2.8.3.2. Evidence for Alkane Complexes   70
  2.8.3.3. Intramolecular Coordination of Aliphatic C–H Bonds (Agostic Interactions)   71
References and Notes   73

Chapter 3. Covalent (X-Type) Ligands Bound Through Metal–Carbon and Metal–Hydrogen Bonds   85

3.1. Introduction   85
3.2. Transition Metal Hydrocarbyl Ligands   85
3.2.1. Alkyl Ligands (Written with Prof. Jack R. Norton)   86
  3.2.1.1. History of Transition Metal–Alkyl Complexes   86
  3.2.1.2. Thermodynamic Properties of M–Alkyl Bonds   86
  3.2.1.3. Synthesis of Metal–Alkyl Complexes   87
   3.2.1.3.1. Synthesis of Alkyl Complexes by Transmetallation   87

   3.2.1.3.2. Synthesis of Alkyl Complexes by Alkylation   88
   3.2.1.3.3. Synthesis of Alkyl Complexes by Other Methods   89
  3.2.1.4. Selected Reactions of Metal–Alkyl Complexes   90
3.2.2. Aryl, Vinyl, and Alkynyl Complexes (Written with Prof. Jack R. Norton)   92
  3.2.2.1. Synthesis of Complexes Containing Terminal Aryl Ligands   92
  3.2.2.2. Complexes with Bridging Aryl Ligands   94
  3.2.2.3. Properties of Metal–Aryl Complexes   95
3.2.3. Vinyl Complexes (Written with Prof. Jack R. Norton)   96
3.2.4. Alkynyl Complexes   97

3.3. Enolate Complexes (Written with Prof. Erik J. Alexanian)   98
3.3.1. Overview   98
3.3.2. Structure of Enolate Complexes   98
3.3.3. Spectral Features of Enolate Complexes   100
3.3.4. Synthesis of Enolate Complexes   101

3.4. Cyanide Complexes (Written with Prof. Jesse W. Tye)   102
3.4.1. Overview   102
3.4.2. Properties of the Free Molecule   102
3.4.3. Structures and Electron Counting of Metal–Cyanide Complexes   102
3.4.4. Thermodynamics of M–CN Linkages   102
3.4.5. Spectral Features of M–CN Complexes   103
3.4.6. Synthesis of CN⁻ Complexes   103

3.5. Allyl, η³-Benzyl, Pentadienyl, and Trimethylenemethane Ligands (Written with Dr. Mark J. Pouy)   104
3.5.1. Allyl Ligands   104
  3.5.1.1. Overview   104
  3.5.1.2. Structures of Allyl Ligands   104
  3.5.1.3. Dynamics of Metal–Allyl Complexes   106
  3.5.1.4. Synthesis of π-Allyl Complexes   107
  3.5.1.5. Reactions of Allyl Complexes   108
3.5.2. η³-Benzyl Complexes   108
3.5.3. Higher Anionic π-Ligands   109
3.5.4. η⁴-Trimethylenemethane (TMM) Complexes   110

3.6. Cyclopentadienyl and Related Compounds (Written with Prof. Jack R. Norton)   111
3.6.1. Overview   111
3.6.2. Bonding and Thermodynamics of Cp Ligands   111

3.6.3. Synthesis of η^5-Cyclopentadienyl Complexes 111

3.6.4. Examples of Substituted Cyclopentadienyl Ligands 112

3.7. Ansa Metallocenes 113

3.7.1. Types of Cyclopentadienyl Complexes 113

    3.7.1.1. Cp₃M and Their Permethyl Derivatives Cp₃*M 114

    3.7.1.2. Metallocenes Cp₂M and their Permethyl Derivatives Cp₂*M 114

    3.7.1.3. Structures of "Sandwich Complexes" 114

    3.7.1.4. Bent Metallocenes Cp₂MLₓ and Related Compounds 115

    3.7.1.5. "Half-Sandwich" Compounds CpMLᵧ 117

    3.7.1.6. Other Modes of Binding of Cyclopentadienyl Ligands 118

3.7.2. Ligands That Are Electronically Similar to the Cyclopentadienyl Ligand 118

3.7.3. Reactions of Cyclopentadienyl Complexes 120

3.8. Hydride Ligands (Written by Prof. Jack R. Norton) 122

3.8.1. Structural Features 122

    3.8.1.1. Terminal Hydrides 122

    3.8.1.2. Bridging Hydrides 123

    3.8.1.3. Spectroscopic Properties 124

3.8.2. Synthesis of Metal–Hydride Complexes 124

    3.8.2.1. From Hydrogen 124

    3.8.2.2. By Protonation 126

    3.8.2.3. From Main Group Hydrides 127

    3.8.2.4. From Other Reagents 128

3.8.3. Acidities of Hydride Complexes 129

3.8.4. Strength of M–H Bonds 131

3.8.5. Hydricities 133

3.8.6. Hydrogen Bonding 136

References and Notes 137

Chapter 4. Covalent (X-Type) Ligands Bound Through Metal–Heteroatom Bonds 147

4.1. Overview and Scope 147

4.2. Complexes Containing Metal–Nitrogen Bonds 147

4.2.1. Metal–Amido Complexes 147

    4.2.1.1. Late-Metal–Amido Complexes (Written with Prof. Pinjing Zhao) 148

    4.2.1.1.1. Overview of Metal–Amido Complexes of the Late Transition-Metals 148

    4.2.1.1.2. Bonding of Late-Metal–Amido Complexes 148

    4.2.1.1.3. Thermodynamic Properties of Late-Metal–Amido Complexes 149

    4.2.1.1.4. Spectral Properties of Late-Metal–Amido Complexes 150

    4.2.1.1.5. Synthesis of Late-Metal–Amido Complexes 150

    4.2.1.1.6. Reactivity of Late-Metal–Amido Complexes 151

    4.2.1.2. Amido Complexes of the Early Transition Metals (Written with Prof. Seth B. Herzon) 152

    4.2.1.2.1. Overview 152

    4.2.1.2.2. Thermodynamic Properties of Early-Metal–Amido Complexes 153

    4.2.1.2.3. Synthesis of Early-Metal–Amido Complexes 154

    4.2.1.2.4. Reactivity of Early-Metal–Amido Complexes 154

4.2.2. Amidate and Amidinate Complexes of the Early Transition Metals (Written with Prof. Seth B. Herzon) 155

4.2.3. Complexes of Anionic Nitrogen Heterocycles (Written with Prof. Jianrong (Steve) Zhou) 155

    4.2.3.1. Overview 155

    4.2.3.2. Metal–Azolyl Bonding 155

    4.2.3.3. Synthesis of Metal–Azolyl Complexes 156

    4.2.3.4. Reactivity of Metal–Azolyl Complexes 157

4.2.4. Nitrosyl Complexes (Written with Prof. Jesse W. Tye) 158

    4.2.4.1. Overview 158

    4.2.4.2. Properties of the Free Molecule 159

    4.2.4.3. Structures and Electron Counting of Metal–Nitrosyl Complexes 159

    4.2.4.4. Thermodynamics of M–NO linkages 160

    4.2.4.5. Spectral Features of M–NO Complexes 161

    4.2.4.6. Synthesis of NO Complexes 161

    4.2.4.7. Reactivity of Metal–Nitrosyl Complexes 162

4.2.5. Polydentate Nitrogen Donor Ligands 162

    4.2.5.1. Organometallic Porphyrin and Corrin Complexes (Written with Giang Vo) 162

    4.2.5.1.1. Overview 162

    4.2.5.1.2. Structures of Metal–Porphyrin Complexes 163

    4.2.5.1.3. Synthesis of Metal–Porphyrin Complexes 164

    4.2.5.1.4. Reactivity of Metal–Porphyrin Complexes 164

4.2.5.2. Bis-Sulfonamide Complexes (Written with Prof. Patrick J. Walsh) 165
4.2.5.2.1. Bonding in Bis-Sulfonamido Complexes 165
4.2.5.2.2. Synthesis of Bis-Sulfonamide Complexes 166
4.2.5.2.3. Thermodynamics of Metal–Bis-Sulfonamido Bonds 166
4.2.5.3. Pyrazolylborate Ligands (Written with Dr. Jaclyn M. Murphy) 167
4.2.5.3.1. Overview 167
4.2.5.3.2. Bonding of Polypyrazolylborate Ligands 168
4.2.5.3.3. Synthesis of Polypyrazolylborate Ligands and Complexes 169
4.2.5.3.4. Reactions of Polypyrazolylborate Complexes 170
4.2.5.4. β-Diketiminate Complexes 170
4.2.5.4.1. Overview 170
4.2.5.4.2. Structure and Bonding of β-Diketiminate Ligands 170
4.2.5.4.3. Synthesis of β-Diketimines and β-Diketiminate Complexes 171
4.2.5.4.4. Examples of β-Diketiminate Complexes 172

4.3. Transition Metal Complexes with Anionic Oxygen Ligands (Written with Prof. Pinjing Zhao) 173
4.3.1. Transition Metal–Alkoxo Complexes 173
4.3.1.1. Overview 173
4.3.1.2. Alkoxide Complexes of the Early Transition Metals 174
4.3.1.2.1. Overview 174
4.3.1.2.2. Bonding of Early-Metal Alkoxides 174
4.3.1.2.3. Preparation of Early-Metal–Alkoxo Complexes 174
4.3.1.2.4. Reactivity of Early-Metal–Alkoxo Complexes 175
4.3.1.2.5. Early-Metal Alkoxides as Ancillary Ligands 175
4.3.1.2.5.1. Steric and Electronic Properties 175
4.3.1.2.5.2. Catalytic Reactions of Early-Metal–Alkoxo Complexes 176
4.3.1.3. Alkoxide Complexes of the Late Transition Metals 177
4.3.1.3.1. Overview 177

4.3.1.3.2. Bonding of Late-Metal Alkoxides 177
4.3.1.3.3. Thermodynamics of Late-Metal–Alkoxo Bonds 178
4.3.1.3.4. Late-Metal Alkoxides as Ancillary Ligands 180
4.3.1.3.5. Preparation of Late-Metal–Alkoxo Complexes 180
4.3.1.3.6. Reactivity of Late-Metal–Alkoxo Complexes 183
4.3.1.3.7. Catalytic Reactions of Late-Metal–Alkoxo Complexes 185
4.3.2. Metal β-Diketonate Complexes 185

4.4. Transition-Metal–Boryl Complexes (Written with Dr. Jaclyn M. Murphy) 186
4.4.1. Overview 186
4.4.2. Metal–Boryl Bonding 187
4.4.3. Thermodynamics of Metal–Boryl Complexes 188
4.4.4. Synthesis of Metal–Boryl Complexes 188
4.4.5. Reactivity of Metal–Boryl Complexes 190

4.5. Transition-Metal–Phosphido Complexes (Written with Prof. Jack R. Norton) 190
4.5.1. Structures of Phosphido Complexes 191
4.5.2. Dynamics of Phosphido Complexes 192
4.5.3. Thermodynamic Properties of Phosphido Complexes 192
4.5.4. Reactivity of Phosphido Complexes 192

4.6. Transition Metal-Thiolate–Complexes (Written with Dr. Elsa Alvaro) 194
4.6.1. Overview 194
4.6.2. Bonding and Structures of Transition-Metal–Thiolate Complexes 194
4.6.3. Thermodynamics of M–SR Bonds 195
4.6.4. Synthesis of Metal–Thiolate Complexes 196
4.6.5. Reactivity of Thiolate Complexes 197

4.7. Transition-Metal–Silyl Complexes (Written with Dr. Tim A. Boebel) 197
4.7.1. Overview 197
4.7.2. Electronic Properties of Free and Coordinated Silyl Groups 197
4.7.3. Structures of Metal–Silyl Complexes 198
4.7.4. Spectral Properties of Metal–Silyl Complexes 198
4.7.5. Synthesis of Metal–Silyl Complexes 199
4.7.6. Stability and Reactivity of Silyl Complexes 200

4.8. Halide Ligands 200
  4.8.1. Overview 200
  4.8.2. Steric and Electronic Properties 201
  4.8.3. Reactivity of Metal–Halide Complexes 203
References and Notes 204

Chapter 5. Ligand Substitution Reactions 217
5.1. Introduction 217
  5.1.1. Overview of Ligand Substitution 217
  5.1.2. Definitions of Associative, Dissociative, and Interchange 217
  5.1.3. The Basic Factors that Control Ligand Substitution Mechanisms 219
  5.1.4. Scope of the Chapter 220
5.2. Thermochemical Considerations 220
5.3. Mechanisms of Ligand Substitutions 223
  5.3.1. Mechanisms of Ligand Substitutions of 16-Electron and 17-Electron Complexes 223
    5.3.1.1. Associative Substitutions of Square-Planar d^8 Complexes 223
    5.3.1.1.1. Stereochemistry of Associative Substitution and Cis–Trans Isomerization 224
    5.3.1.1.2. The Rate Law for Associative Substitutions 225
    5.3.1.1.3. Dependence of the Rates on the Incoming Ligand, the Departing Ligand, and the Metal Center 226
    5.3.1.1.4. Trans and Cis Effects 228
    5.3.1.2. Associative versus Dissociative Substitutions of Square-Planar Complexes 229
    5.3.1.3. Associative Substitutions of 17-Electron Complexes 231
5.4. Substitution Reactions of 18-Electron Complexes 233
  5.4.1. Dissociative Substitution Reactions 233
    5.4.1.1. General Features of the Kinetics of Dissociative Ligand Substitution 233
    5.4.1.2. Reactions of $Ni(CO)_4$ as Quintessential Examples of Dissociative Substitutions 235
    5.4.1.3. Steric Effects on Dissociative Substitution 235
    5.4.1.4. Stereochemistry of Dissociative Substitution 236
    5.4.1.5. Substitution of Weakly Bound Ligands in 18-Electron Complexes 237
    5.4.1.6. Electronic Effect of Ancillary Ligands on the Rates of Dissociative Substitution Reactions—The Cis Effect 238
    5.4.1.7. Stereochemistry of Substitutions of Octahedral Compounds 240

  5.4.2. Substitutions of 18-Electron Complexes that Deviate from Pure Thermally Induced Dissociative Mechanisms 241
    5.4.2.1. Substitutions of $M(CO)_6$ Complexes Occur with an Associative Term in the Rate Law 241
    5.4.2.2. Catalyzed and Assisted Ligand Substitution Reactions 242
    5.4.2.2.1. Ligand Substitution Catalyzed by Electron Transfer 242
    5.4.2.2.2. Ligand Substitutions by Radical Chains Initiated by Atom Abstractions 243
    5.4.2.2.3. Photoinduced Dissociation of Ligands 244
    5.4.2.2.4. Oxidation of Coordinated CO 246
    5.4.2.2.5. Other Assisted Ligand Substitutions 246
5.5. Substitution Reactions Involving Polyhapto Ligands 247
  5.5.1. Substitutions for Dienes and Trienes 247
  5.5.2. Substitutions for Arenes and Arene Exchange Reactions 248
  5.5.3. Associative Substitution by Pentadienyl Ligand Ring Slip 250
5.6. Ligand Substitutions in Metal–Metal Bonded Bimetallic and Higher Nuclearity Clusters 253
5.7. Summary 255
References and Notes 255

Chapter 6. Oxidative Addition of Nonpolar Reagents 261
6.1. Definitions, Examples, and Trends 261
  6.1.1. Definition of Oxidative Addition 261
  6.1.2. Qualitative Trends for Oxidative Addition 263
  6.1.3. Thermodynamics of Oxidative Addition 264
6.2. Oxidative Addition of Dihydrogen 266
  6.2.1. General Mechanism for the Oxidative Addition of H_2 266
  6.2.2. Examples of Oxidative Addition of H_2 to a Single Metal Center 268
  6.2.3. Oxidative Addition of H_2 to Two Metal Centers 269
6.3. Oxidative Addition of Silanes 270
6.4. Oxidative Addition of C–H Bonds 272
  6.4.1. Early History of C–H Bond Oxidative Addition 272
  6.4.2. Intramolecular C–H Oxidative Addition 273
  6.4.3. Intermolecular Oxidative Addition of C–H Bonds 275

6.4.4. Selectivity of Alkane Oxidative Addition   278

6.4.5. Mechanism of Oxidative Addition of C–H Bonds   279

6.4.6. Examples of Complexes that Oxidatively Add Alkanes   281

6.4.7. Synthetic Applications of C–H Oxidative Addition of Alkyl Groups   282

6.4.8. Dinuclear Activation of Hydrocarbons   282

6.5. Addition of H–H and C–H Bonds to Transition Metal Complexes Without Oxidation and Reduction   283

6.5.1. Sigma-Bond Metathesis Involving d^0 Complexes   283

6.5.2. Potential Sigma-Bond Metatheses Involving Late Transition Metal Complexes   285

6.5.3. [2 + 2] Additions Across Metal–Ligand Multiple Bonds   287

6.6. Oxidative Addition of C–C Bonds   289

6.7. Oxidative Addition of E–E Bonds   291

6.8. Summary   292

References and Notes   292

Chapter 7. Oxidative Addition of Polar Reagents   301

7.1. Introduction   301

7.2. Oxidative Addition by S_N2 Pathways   301

7.3. Oxidative Additions by One-Electron Mechanisms   304

7.3.1. Inner-Sphere Electron Transfer and Caged Radical Pairs   305

7.3.2. Radical Chain Pathways   306

7.3.3. Outer-Sphere Electron-Transfer Mechanisms   308

7.3.4. Atom Abstraction and Combination of the Resulting Radical with a Second Metal   309

7.4. Concerted Oxidative Additions   310

7.4.1. Concerted Oxidative Additions of Reagents with C–X Bonds of Medium Polarity   310

7.4.2. Oxidative Addition of Reagents with H–X Bonds of Medium Polarity   313

7.5. Dinuclear Oxidative Additions of Electrophilic A–B   315

7.6. Summary   317

References and Notes   317

Chapter 8. Reductive Elimination   321

8.1. Overview   321

8.1.1. Changes in Electron Count and Oxidation State   321

8.1.2. Factors that Affect the Rates of Reductive Elimination   322

8.1.2.1. *Effect of Metal Identity and Electron Density   322*

8.1.2.2. *The Effect of Steric Properties   322*

8.1.2.3. *The Effect of Participating Ligands   323*

8.1.2.4. *The Effect of Coordination Number   323*

8.1.2.5. *The Effect of Geometry   324*

8.1.2.6. *The Effect of Light: Photochemically Induced Reductive Elimination   324*

8.2. Reductive Eliminations Organized by Type of Bond Formation   325

8.2.1. Reductive Elimination to Form C–H Bonds   325

8.2.1.1. *Overview and Principles   325*

8.2.1.2. *Examples   326*

8.2.1.3. *Evidence for Intermediate Alkane and Arene Complexes   327*

8.2.1.4. *The Effect of Ancillary Ligands on C–H Bond-Forming Reductive Elimination   329*

8.2.2. Reductive Elimination to Form X–H Bonds   330

8.2.3. Reductive Elimination to Form C–C Bonds   331

8.2.3.1. *Trends and Principles   331*

8.2.3.2. *The Effect of Participating Groups   332*

8.2.3.3. *The Effect of Coordination Number   334*

8.2.3.4. *The Effect of Bite Angle   335*

8.2.3.5. *Survey of Carbon–Carbon Bond-Forming Reductive Eliminations   336*

8.2.4. Reductive Elimination to Form C–X Bonds   338

8.2.4.1. *Mechanisms of Reductive Eliminations to Form C–X Bonds   338*

8.2.4.2. *Survey of Reductive Eliminations to Form C–X Bonds   341*

8.2.4.2.1. Reductive Eliminations to Form C–X Bonds from Aryl and Alkylplatinum(IV) Complexes   341

8.2.4.2.2. Reductive Eliminations to Form C–X Bonds from Arylpalladium(II) Complexes   342

8.2.4.2.3. Reductive Eliminations to Form C–X Bonds from Acyl Complexes   344

8.3. Summary   345

References and Notes   345

Chapter 9. Migratory Insertion Reactions   349

9.1. Overview and Basic Principles   349

9.1.1. Description of Migratory Insertion and Elimination   349

9.1.2. Changes in Geometry and Electron Count During Migratory Insertion and Elimination   350

9.2. Specific Classes of Insertions   350

9.2.1. Insertions of Ligands Bound by a Single Atom   351

9.2.1.1. Insertions of Carbon Monoxide   351

9.2.1.1.1. Examples of CO Insertions into Metal–Hydrocarbyl Complexes   351

9.2.1.1.2. Examples of Insertions of CO into M–X Bonds (X = N, O, and Si)   352

9.2.1.1.3. Kinetics and Mechanism of CO Insertions into Metal–Alkyl Complexes   354

9.2.1.1.3.1. Insertions into 18-Electron Complexes   354

9.2.1.1.3.2. Insertions into 16-Electron d^8 Complexes   355

9.2.1.1.3.3. Stereochemistry at Carbon   356

9.2.1.1.3.4. Stereochemistry at the Metal   357

9.2.1.1.3.5. Structure of the Unsaturated Intermediate   358

9.2.1.1.3.6. Solvent Effects   359

9.2.1.1.4. Migratory Aptitudes of R   360

9.2.1.1.4.1. Thermodynamic Effects on Migratory Aptitudes   360

9.2.1.1.4.2. Kinetic Effects on Migratory Aptitudes   361

9.2.1.1.5. Catalysis of CO Insertion   362

9.2.1.1.5.1. Catalysis by Lewis Acids   362

9.2.1.1.5.2. Redox Acceleration   363

9.2.1.2. Insertions of Other Ligands Bound Through a Single Atom   364

9.2.1.3. Insertions of Carbenes   365

9.2.2. Insertions of Polyhapto Ligands into Metal–Ligand Covalent Bonds   366

9.2.2.1. Insertions into Metal–Hydride Bonds   366

9.2.2.1.1. Insertions of Olefins into Metal–Hydride Bonds   366

9.2.2.1.2. Insertions of Alkynes into Metal–Hydride Bonds   368

9.2.2.1.3. Insertion of Ketones and Imines into Metal–Hydride Bonds   370

9.2.2.2. Insertions of Olefins into Metal–Carbon Bonds   371

9.2.2.2.1. Insertions of Olefins into Metal–Hydrocarbyl σ-Bonds   371

9.2.2.2.2. Insertions of Olefins into Metal–Acyl Bonds   377

9.2.2.2.3. Insertions of Alkynes into Metal–Carbon Bonds   379

9.2.2.2.4. Insertions of Polyenes into Metal–Carbon Bonds   381

9.2.2.2.5. Insertions of Aldehydes and Imines into Metal–Carbon Bonds   381

9.2.2.3. Insertions of Olefins and Acetylenes into Metal–Heteroatom Bonds   383

9.2.2.3.1. Insertion of Olefins into Metal–Oxygen Bonds   383

9.2.2.3.2. Insertions of Olefins into Metal–Nitrogen Bonds   385

9.2.2.3.3. Insertions of Olefins and Acetylenes into Metal–Silicon and Metal–Boron Bonds   388

9.3. Summary   389

References and Notes   390

Chapter 10. Elimination Reactions   397

10.1. Overview of the Chapter   397

10.2. Scope of Organometallic Elimination Chemistry   397

10.3. β-Elimination Processes   398

10.3.1. β-Hydrogen Eliminations   398

10.3.1.1. β-Hydrogen Elimination from Metal–Alkyl Complexes   398

10.3.1.1.1. Effect of Conformation and Coordination Number on the Rate of β-Hydrogen Elimination   399

10.3.1.1.2. Effect of Electronics on the Rate of β-Hydrogen Elimination   400

10.3.1.1.3. Effect of Ancillary Ligands on the Rate of β-Hydrogen Elimination   402

10.3.1.2. β-Hydrogen Elimination from Metal Alkoxides and Amides   402

10.3.1.3. β-Hydrogen Elimination from Metal–Silyl Complexes   405

10.3.2. β-Hydrocarbyl Eliminations   406

10.3.2.1. β-Alkyl Eliminations from Alkyl Complexes   406

10.3.2.2. β-Alkyl and β-Aryl Eliminations from Alkoxido and Amido Complexes   408

10.3.3. β-Halide and Alkoxide Elimination   409

10.4. α-Hydrogen Eliminations and Abstractions   410

10.5. Summary   413

References and Notes   414

Chapter 11. Nucleophilic Attack on Coordinated Ligands   417

11.1. Fundamental Principles   417

11.2. Nucleophilic Attack on Transition Metal Complexes of Carbon Monoxide and Isonitriles   419

11.2.1. General Trends   419

11.2.2. Examples of Nucleophilic Attack on Carbon Monoxide and Isonitriles   420

11.3. Nucleophilic Attack On Carbene and Carbyne Complexes   421

11.4. Nucleophilic Cleavage of Metal–Carbon σ-Bonds   422

11.4.1. General Principles and Trends   422

11.4.2. Examples of Nucleophilic Attack on σ-Bound Ligands   423

11.5. Nucleophilic Attack on η^2-Unsaturated Hydrocarbon Ligands   427

11.5.1. General Trends   427

11.5.2. Nucleophilic Attack on η^2-Olefin Complexes   428

   11.5.2.1. Overview of Nucleophilic Attack on η^2-Olefin Complexes   428

   11.5.2.2. Specific Examples of Nucleophilic Attack on η^2-Olefin Complexes: Reactions of $[CpFe^{II}(CO)_2]^+$, $[CpPd^{II}L]^+$ and Square Planar M^{II} ($M=Pd, Pt$) Olefin Complexes   429

11.5.3. Nucleophilic Attack on Square Planar Pd(II) Diene and Allene Complexes   433

11.5.4. Nucleophilic Attack on η^2-Alkyne Complexes   434

11.5.5. Reactions of η^2-Arene Complexes   435

11.6. Nucleophilic Attack on Imine and Aldehyde Complexes   435

11.7. Nucleophilic Attack on Polyhapto (η^3–η^6) Ligands   436

11.7.1. Nucleophilic Attack on η^3-Allyl Complexes   436

11.7.2. Nucleophilic Attack on η^4-Diene Complexes   439

11.7.3. Nucleophilic Attack on η^5-Dienyl Complexes   441

11.7.4. Nucleophilic Attack on η^6-Arene and Cycloheptatrienyl Complexes   442

   11.7.4.1. Overview of Nucleophilic Attack on η^6-Arene Complexes   442

   11.7.4.2. Examples of Nucleophilic Attack on π-Arene Complexes   444

11.8. Summary   446

References and Notes   447

Chapter 12. Electrophilic Attack on Coordinated Ligands   453

12.1. Overview and Basic Principles   453

12.2. Electrophilic Cleavage of Metal–Carbon and Metal–Hydride σ-Bonds   454

12.2.1. Scope of Electrophilic Cleavage of Metal–Carbon and Metal–Hydride σ-Bonds   454

12.2.2. Mechanism of Electrophilic Attack   457

   12.2.2.1. Mechanism of Attack of Main Group Electrophiles on Alkyl Complexes Possessing d-Electrons   457

   12.2.2.2. Mechanism of Protonolysis of Metal–Carbon Bonds in Complexes Possessing d-Electrons   460

   12.2.2.3. Mechanism of Protonation of Metal–Hydride Bonds in Complexes Containing d-Electrons   461

12.2.3. Mechanism of Electrophilic Attack on Alkyl Complexes that Lack d-Electrons   461

12.3. Electrophilic Insertion Reactions: Sulfur Dioxide, Carbon Dioxide and Related Electrophiles   462

12.4. Electrophilic Modification of Coordinated Ligands   465

12.4.1. Attack at the α-Position   465

   12.4.1.1 Attack at the α-Position of an Alkyl Group   465

   12.4.1.2. Electrophilic Attack on Carbene and Carbyne Complexes   466

12.4.2. Attack at the β-Position   466

12.4.3. Attack at the γ-Position   469

12.5. Attack on Coordinated Olefins and Polyenes   471

12.5.1. Attack of Carbonyl Compounds and Protons on Olefin Complexes   471

12.5.2. Hydride Abstraction by Electrophilic Attack on Diene Complexes   472

12.5.3. Electrophilic Attack on π-Polyenyl Complexes   474

12.5.4. Electrophilic Attack on η^2-Arene and Heteroarene Complexes   475

12.6. Summary   476

References and Notes   477

Chapter 13. Metal–Ligand Multiple Bonds   481

13.1. Introduction to Metal–Ligand Multiple Bonds   481

13.2. Carbene Complexes   482

13.2.1. Classes of Carbene Complexes   482

13.2.2. Origin of the Electronic Properties of Fischer and Schrock Carbenes   483

13.2.3. Synthesis of Carbene Complexes   484

13.2.3.1. *Synthesis of Fischer Carbene Complexes   484*

13.2.3.2. *Synthesis of Vinylidene Complexes   486*

13.2.3.3. *Synthesis of Some Classic Alkylidene Complexes   488*

13.2.3.3.1. Synthesis of the First Schrock Carbene Complexes   488

13.2.3.3.2. Synthesis of the Schrock Alkylidene Catalysts   488

13.2.3.3.3. Synthesis of Tebbe's Reagent   490

13.2.4. Synthesis of *N*-Heterocyclic Carbene Complexes   491

13.2.5. Reactivity of Carbene Complexes   492

13.2.5.1. *Reactivity of Fischer Carbene Complexes   492*

13.2.5.1.1. Reactions with Nucleophiles   493

13.2.5.1.2. Conversion to Carbyne Complexes   493

13.2.5.1.3. Reactions Related to Those of Enolates   494

13.2.5.1.4. Cyclopropanations   495

13.2.5.1.5. Annulations: The Dötz Reaction   496

13.2.5.2. *Reactivity of Vinylidene Complexes   498*

13.2.5.3. *Reactivity of Alkylidene and Alkylidyne Complexes   498*

13.2.5.3.1. Examples of [2+2] Reactions of Alkylidenes and Alkylidynes   499

13.2.5.3.2. Fomal [2+2] Reactions with C–H σ-Bonds   503

13.3. Silylene Complexes   505

13.3.1. Overview of Silylene Complexes   505

13.3.2. Bonding of Silylene Complexes   505

13.3.3. Examples of Isolated Silylene Complexes   506

13.3.4. Reactivity of Silylene Complexes   507

13.4. Metal–Heteroatom Multiple Bonds   508

13.4.1. Scope of the Section   508

13.4.2. Overview   509

13.4.3. Bonding of Oxo and Imido Complexes   510

13.4.4. Synthesis of Metal–Imido and Metal–Oxo Complexes   512

13.4.4.1. *Synthesis of Metal–Imido Complexes   512*

13.4.4.2. *Synthesis of Metal–Oxo Complexes   514*

13.4.5. Reactions of Imido and Oxo Compounds   515

13.4.5.1. *[2+2] and [3+2] Cycloadditions   515*

13.4.5.2. *Atom Transfer of Oxo and Imido Groups to Olefins   518*

13.4.5.3. *Reactions with C–H Bonds   521*

13.4.5.4. *Reactions with Electrophiles   523*

13.4.5.5. *Migrations of Alkyl and Hydride Groups from M to O or N   524*

13.4.5.6. *Catalytic Reactions of Imido and Metal–Oxo Compounds Through Organometallic Intermediates   525*

13.4.6. Nitrido Ligands (Written with Dr. Devon C. Rosenfeld)   527

13.4.6.1. *Overview   527*

13.4.6.2. *Bonding of Nitrido Ligands   527*

13.4.6.3. *Structural and Spectral Features   528*

13.4.6.4. *Synthesis of Metal–Nitrido Complexes   528*

13.4.6.5. *Reactions of Metal–Nitrido Complexes   529*

References and Notes   530

Chapter 14. Principles of Catalysis (Written with Prof. Patrick J. Walsh)   539

14.1. General Principles   539

14.1.1. Definition of a Catalyst   539

14.1.2. Energetics of Catalysis   540

14.1.3. Reaction Coordinate Diagrams of Catalytic Reactions   540

14.1.4. Origins of Transition State Stabilization   542

14.1.5. Terminology of Catalysis   543

14.1.5.1. *The Catalytic Cycle   543*

14.1.5.2. *Catalyst Precursors, Catalyst Deactivation, and Promoters   544*

14.1.5.3. *Quantification of Efficiency   545*

14.1.6. Kinetics of Catalytic Reactions and Resting States   546

14.1.7. Homogeneous vs. Heterogeneous Catalysis   546

14.1.7.1. *Distinguishing Homogeneous from Heterogeneous Catalysts   547*

14.2. Fundamentals of Asymmetric Catalysis   549

14.2.1. Importance of Asymmetric Catalysis   549

14.2.2. Classes of Asymmetric Transformations   550

14.2.3. Nomenclature   551

14.2.3.1. *Description of Stereoselectivity   551*

14.2.3.2. *The Origin of Stereoselection   552*

14.2.4. Energetics of Stereoselectivity   552

14.2.4.1. *Reaction Coordinates of Catalytic Enantioselective Reactions   553*

14.2.4.1.1. Reactions with a Single Enantioselectivity-Determining Step   554

14.2.4.1.2. Reactions with Reversiblity Prior to the Enantioselectivity-Determining Step: The Curtin–Hammett Principle Applied to Asymmetric Catalysis   555

    14.2.4.1.2.1. *Theory*   555

    14.2.4.1.3.2. *Two Examples of Reactions Under Curtin–Hammett Conditions*   556

      14.2.4.1.3.2.1. *Asymmetric Hydrogenation*   556

      14.2.4.1.3.2.2. *Asymmetric Allylic Alkylation*   557

14.2.5. Transmission of Asymmetry   559

    14.2.5.1. *Effect of C₂ Symmetry*   559

    14.2.5.2. *Quadrant Diagrams*   559

    14.2.5.3. *Structures of Ligands Generating Highly Selective Catalysts ("Privileged Ligands")*   561

14.2.6. Alternative Asymmetric Processes: Kinetic Resolutions and Desymmetrizations   563

    14.2.6.1. *Kinetic Resolutions*   563

      14.2.6.1.1. Quantification of Selectivity in Kinetic Resolutions   564

      14.2.6.1.2. Energetics of Selectivity in Kinetic Resolutions   565

      14.2.6.1.3. Examples of Kinetic Resolutions   565

    14.2.6.2. *Dynamic Kinetic Resolution*   567

      14.2.6.2.1. Example of Dynamic Kinetic Resolutions: Dynamic Kinetic Resolution of 1,3-Dicarbonyl Compounds Through Asymmetric Hydrogenation   567

    14.2.6.4. *Dynamic Kinetic Asymmetric Transformations*   568

    14.2.6.5. *Desymmetrization Reactions*   569

      14.2.6.5.1. Two Examples of Desymmetrization   570

        14.2.6.5.1.1. *Desymmetrization of Achiral Dienes via Catalytic Asymmetric Hydrosilylation*   570

        14.2.6.5.1.2. *Desymmetrization via the Palladium-Catalyzed Heck Reaction*   570

14.3. Summary   571

References and Notes   571

Chapter 15. Homogeneous Hydrogenation   575

15.1. Introduction   575

15.2. A Perspective on the Homogeneous Catalytic Hydrogenation of Olefins   576

15.3. Selected Examples of Achiral Homogeneous Hydrogenation Catalysts   578

15.3.1. Rhodium Catalysts for Olefin Hydrogenation   578

    15.3.1.1. *Neutral Rhodium Catalysts*   578

      15.3.1.1.1. Preparation of Wilkinson's Catalyst   578

      15.3.1.1.2. The Reactivity of Wilkinson's Catalyst   579

    15.3.1.2. *Cationic Rhodium Catalysts*   581

15.3.2. Iridium Catalysts: Crabtree's Catalyst   582

15.3.3. Ruthenium Catalysts for Olefin Hydrogenation   583

15.3.4. Lanthanide Catalysts   584

15.4. Directed Hydrogenation   584

15.5. Mechanisms of Homogeneous Olefin and Ketone Hydrogenation   585

15.5.1. Background   585

15.5.2 Overview of the Typical Mechanisms   585

    15.5.2.1. *Mechanisms Occurring by Insertions of Olefins into Dihydride Complexes*   588

      15.5.2.1.1. Hydrogenation by Wilkinson's Catalyst   588

        15.5.2.1.1.1. *Mechanism of the Oxidative Addition Step*   589

        15.5.2.1.1.2. *Mechanism of the Migratory Insertion Step*   590

      15.5.2.1.2. Hydrogenation by Cationic Rhodium Catalysts   590

        15.5.2.1.2.1. *Cationic Rhodium Complexes Containing Aromatic Phosphines*   590

        15.5.2.1.2.2. *Cationic Rhodium Catalysts Containing Alkylphosphines*   592

      15.5.2.1.3. Cationic Iridium Catalysts Containing Alkylphosphines   594

    15.5.2.2. *Catalysts that React by Insertions of Olefins into Monohydride Intermediates*   596

      15.5.2.2.1. Hydrogenation by Rhodium Carbonyl Hydride Catalysts   596

      15.5.2.2.2. Hydrogenation by Ruthenium Catalysts   597

        15.5.2.2.2.1. *Mechanism of Hydrogenation by Ru(PPh₃)₃H(Cl)*   597

      15.5.2.2.3. Mechanism of Hydrogenation of Olefins and Ketones by RuL₂(κ²-OAc)₂ and [RuL₂Cl₂]₂   597

15.5.2.2.4. Monohydride Catalysts Reacting Through Radical Pathways 599

15.5.2.2.5. d^0-Monohydride Catalysts Reacting Through σ-Bond Metathesis Pathways 600

15.5.2.3. *Outer-Sphere Mechanism for the Hydrogenation of Ketones and Imines* 600

15.5.2.4. *Ionic Hydrogenations* 602

15.6. Ligands Used for Asymmetric Hydrogenation 603

15.6.1. Aromatic Bisphosphines 603

15.6.1.1. *Aromatic Bisphosphines Containing Backbone Chirality* 603

15.6.1.1.1. Ligands Containing Axial Chiral Backbones 603

15.6.1.1.2. Compounds Containing Chiral Ferrocenyl Backbones 606

15.6.1.1.3. Ligands Containing Aliphatic Backbones 607

15.6.2. Aliphatic Bisphosphines 608

15.6.3. *P*-Chiral Phosphines 609

15.6.4. P,N Ligands 609

15.6.5. Phosphites and Phosphoramidites 610

15.7. Examples of Asymmetric Hydrogenation and Transfer Hydrogenation 611

15.7.1. Classes of Asymmetric Hydrogenations of Olefins 612

15.7.1.1. *Asymmetric Hydrogenation of Enamides* 612

15.7.1.1.1. Asymmetric Hydrogenation of Dehydro α-Amino Acids [α-(Acylamino)acrylic Acids and Esters] 612

15.7.1.1.2. Asymmetric Hydrogenation of Dehydro β-Amino Acids [β-(Acylamino)acrylic Acids and Esters] 614

15.7.1.1.3. Asymmetric Hydrogenation of Simple Enamides 615

15.7.1.2. *Asymmetric Hydrogenation of α-(Acyloxy)-acrylates* 616

15.7.1.3. *Asymmetric Hydrogenation of Acrylic Acids* 616

15.7.1.4. *Asymmetric Hydrogenation of Unsaturated Alcohols* 618

15.7.1.5. *Asymmetric Hydrogenation of Unfunctionalized Olefins* 618

15.7.1.6. *Asymmetric Hydrogenation of Ketones* 620

15.7.1.6.1. Asymmetric Hydrogenations of Functionalized Ketones 621

15.7.1.6.1.1. *Asymmetric Hydrogenations of α-Keto Esters* 621

15.7.1.6.1.2. *Asymmetric Hydrogenation of β-Keto Esters* 622

15.7.1.6.1.3. *Asymmetric Hydrogenations of β-Diketones* 624

15.7.1.6.1.4. *Asymmetric Hydrogenations of α- and β-Amino and Hydroxy Ketones* 624

15.7.1.6.2. Hydrogenation of Unfunctionalized Ketones 626

15.7.1.7. *Asymmetric Hydrogenation of Imines* 629

15.7.1.7.1. Asymmetric Hydrogenation of Cyclic Imines 629

15.7.1.7.2. Asymmetric Hydrogenation of Acyclic N-Alkyl Imines 630

15.7.1.7.3. Asymmetric Hydrogenation of Acyclic N-Aryl Imines 631

15.7.1.7.4. Asymmetric Hydrogenation of Aroylhydrazones and Phosphinylketimines 632

15.7.2. Asymmetric Transfer Hydrogenation of Ketones and Imines 633

15.7.3. Mechanism of Asymmetric Catalytic Hydrogenation of α-Acetamidocinnamic Acid Esters 636

15.8. Hydrogenation of Alkynes and Conjugated Dienes 640

15.8.1. Rhodium-Catalyzed Hydrogenation of Alkynes and Conjugated Dienes 640

15.8.2. Chromium-Catalyzed Hydrogenation of Alkynes and Conjugated Dienes 642

15.8.3. Palladium-Catalyzed Hydrogenation of Alkynes and Conjugated Dienes 643

15.9. Homogeneous Catalytic Hydrogenation of Arenes and Heteroarenes 644

15.9.1. Homogeneous Catalytic Hydrogenation of Polycyclic Arenes 644

15.9.2. Hydrogenation of Monocyclic Arenes 647

15.9.3. Asymmetric Hydrogenation of Heteroarenes 647

15.9.3.1. *Asymmetric Hydrogenation of Six-Membered Ring Heteroarenes* 648

15.9.3.2. *Asymmetric Hydrogenation of Five-Membered Ring Heteroarenes* 649

15.10. Homogeneous Hydrogenation of Other Functional Groups (Written with Prof. Jing Zhao)   651

15.10.1. Hydrogenation of Esters   651

15.10.2. Hydrogenation of Carboxylic Anhydrides and Imides   653

15.10.3. Hydrogenation of Nitriles   655

15.11. Summary   656

References and Notes   657

Chapter 16. Hydrofunctionalization and Oxidative Functionalization of Olefins   667

16.1. Introduction and Scope   667

16.2. Homogeneous Catalytic Hydrocyanation of Olefins and Alkynes   668

16.2.1. Introduction to Hydrocyanation   668

16.2.2. Examples of Alkene Hydrocyanation   668

16.2.3. Mechanism of Hydrocyanation   670

16.2.3.1. Mechanism of the Hydrocyanation of Alkenes   670

16.2.3.2. Mechanism of Deactivation   673

16.2.4. Hydrocyanation of Dienes   673

16.2.5. Asymmetric Hydrocyanation   674

16.2.6. Hydrocyanation of Alkynes   676

16.2.7. Summary of Catalytic Hydrocyanation   676

16.3. Hydrosilylation and Disilylation   677

16.3.1. Introduction to Hydrosilylation and Disilylation   677

16.3.2. Purpose for Hydrosilylation   677

16.3.3. History and Types of Catalyst   678

16.3.4. Examples of Hydrosilylations   679

16.3.4.1. Hydrosilylation of Olefins with Achiral Catalysts   679

16.3.4.2. Hydrosilylation of Vinylarenes   680

16.3.4.3. Hydrosilylation of Dienes   680

16.3.4.4. Dehydrogenative Silylation of Olefins   681

16.3.4.5. Hydrosilylation of Alkynes   681

16.3.4.6. Asymmetric Hydrosilylation of Olefins   683

16.3.4.7. Hydrosilylation of Ketones and Imines   684

16.3.5. Mechanism of Hydrosilylation   686

16.3.5.1. Induction Periods and Phase of the Reactions Catalyzed by Speier's and Karstedt's Catalysts   686

16.3.5.2. Overall Catalytic Cycles   686

16.3.5.2.1. The Chalk–Harrod Mechanism   688

16.3.5.2.2. Evidence for a Modified Chalk–Harrod Mechanism   688

16.3.5.2.3. Alkene Hydrosilylation by σ-Bond Metathesis   689

16.3.5.2.4. Mechanism of Alkyne Hydrosilylation   690

16.3.6. Disilation   690

16.4. Transition-Metal-Catalyzed Hydroboration, Diboration, Silylboration, and Stannylboration   691

16.4.1. Overview of Hydroboration and Diboration   691

16.4.2. History of Catalytic Hydroboration   691

16.4.3. Examples of Metal-Catalyzed Hydroboration   692

16.4.4. Asymmetric Hydroboration   694

16.4.5. Mechanism of the Hydroboration of Olefins   695

16.4.6. Diboration, Silylboration, and Stannylboration   697

16.4.6.1. Diboration, Silylboration, and Stannylboration of Alkynes   697

16.4.6.2. Diboration of Alkenes   698

16.4.6.3. Mechanism of Diborations   699

16.5. Transition-Metal-Catalyzed Hydroamination of Olefins and Alkynes   700

16.5.1. Introduction and Fundamentals of Hydroamination   700

16.5.2. Scope of Hydroamination   701

16.5.2.1. Hydroamination of Alkenes   701

16.5.2.2. Hydroamination of Vinylarenes   705

16.5.2.3. Hydroamination of Allenes   707

16.5.2.4. Hydroamination of 1,3-Dienes   708

16.5.2.5. Hydroamination of Alkynes   710

16.5.2.5.1. Hydroamination of Alkynes Catalyzed by Group 4 Metal Complexes   710

16.5.2.5.2. Hydroamination of Alkynes Catalyzed by Lanthanide and Actinide Complexes   711

16.5.2.5.3. Hydroamination of Alkynes Catalyzed by Rhodium and Palladium Complexes   711

16.5.3. Mechanisms of Transition-Metal-Catalyzed Hydroamination   712

16.5.3.1. Overview of the Mechanisms of Transition-Metal-Catalyzed Hydroaminations   712

*16.5.3.2. Hydroamination by Attack of Amines on
    π-Complexes   713*
    16.5.3.2.1. Hydroamination by Attack on
        π-Olefin and Alkyne Complexes   713
    16.5.3.2.2. Hydroamination by Attack of
        Amines on π-Allyl and π-Benzyl
        Complexes   713
    16.5.3.2.3. Hydroamination by Attack of
        Amines on π-Arene Complexes   714
*16.5.3.3. Hydroamination by Insertions of Olefins into
    Metal Amides   715*
*16.5.3.4. Hydroamination by [2+2]
    Cycloadditions   716*

16.6. Oxidative Functionalization of Olefins   717

16.6.1. Overview   717

16.6.2. The Wacker Process   718

16.6.2.1. Description of the Process   718

*16.6.2.2. Mechanism of the Wacker Process (Written
    with Prof. Jack R. Norton)   719*

*16.6.2.3. Olefin Oxidations Related to the Wacker
    Process   722*

    16.6.2.3.1. Intermolecular Additions of Alcohols
        and Carboxylates   722
    16.6.2.3.2. Intramolecular Additions of Alcohols
        and Carboxylates   724
    16.6.2.3.3. Wacker-Type Oxidations in Natural
        Products Synthesis   726

16.6.3. Oxidative Aminations of Olefins   728

16.6.3.1. Intermolecular Oxidative Aminations   728

16.6.3.2. Intramolecular Oxidative Amination   730

*16.6.3.3. Palladium-Catalyzed Difunctionalizations of
    Olefins   730*

16.6.4. Mechanistic Studies on Wacker Oxidations
    with Alcohol, Phenol, and Amide
    Nucleophiles   731

16.6.4.1. Overview   731

16.6.4.2. Mechanism of C–X Bond Formation   732

16.6.4.3. Mechanism of Reoxidation   733

16.7. Summary   735

References and Notes   735

Chapter 17. Catalytic Carbonylation   745

17.1. Overview   745

**17.2. Catalytic Carbonylation to form Acetic Acid and
    Acetic Anhydride (Written with Prof. Charles
    P. Casey)   746**

17.2.1. Rhodium-Catalyzed Carbonylation of Methanol:
    Monsanto's Acetic Acid Process   746

17.2.2. Carbonylation of Methyl Acetate: Eastman
    Chemical's Acetic Anhydride Process   748

17.2.3. Iridium-Catalyzed Carbonylation of Methanol:
    BP's Cativa™ Process   749

**17.3. Hydroformylation of Olefins (Written with Prof.
    Charles P. Casey)   751**

17.3.1. Overview   751

17.3.2. Hydroformylation Catalyzed by
    HCo(CO)₄   752

*17.3.2.1. Mechanism of Hydroformylation Catalyzed
    by HCo(CO)₄   752*

*17.3.2.2. Regioselectivity of Hydroformylation
    Catalyzed by HCo(CO)₄   754*

17.3.3. Hydroformylation Catalyzed by
    HCo(CO)₃(PR₃)   754

*17.3.3.1. Comparison of Rate, Selectivity, and
    Mechanism to Hydroformylation Catalyzed
    by HCo(CO)₄   754*

*17.3.3.2. Hydroformylation of Internal Alkenes
    Catalyzed by HCo(CO)₃(PR₃)   755*

17.3.4. Rhodium-Catalyzed Hydroformylation   756

17.3.4.1. Overview   756

*17.3.4.2. Rhodium Catalysts for Hydroformylation
    Containing Triarylphosphine Ligands   756*

    17.3.4.2.1. Discovery and Reactivity of the
        Original Catalyst   756
    17.3.4.2.2. Mechanism of Hydroformylation
        Catalyzed by HRh(CO)₂(PPh₃)₂   757

*17.3.4.3. Water-Soluble Rhodium Hydroformylation
    Catalysts   758*

*17.3.4.4. Rhodium Catalysts Containing Chelating
    Diphosphine Ligands   759*

    17.3.4.4.1. Early Studies with Less Selective
        Catalysts   759
    17.3.4.4.2. Catalysts Containing Wide-Bite-
        Angle Bisphosphines   760
    17.3.4.4.3. Effect of Diphosphine Electronic
        Properties on Regioselectivity   762

*17.3.4.5. Rhodium-Catalyzed Hydroformylation of
    Internal Alkenes   763*

*17.3.4.6. Hydroformylation Catalyzed by Rhodium
    Complexes of Phosphites   763*

*17.3.4.7. Rhodium-Catalyzed Hydroformylation of
    Functionalized Alkenes   764*

17.3.4.8. Enantioselective Hydroformylation   765

17.4. Hydroaminomethylation 769
   17.4.1. History and Overview of Recent Developments 769
   17.4.2. Scope of Hydroaminomethylation 770
   17.4.3. Mechanism of Hydroaminomethylation 774
17.5. Hydrocarboxylation and Hydroesterification of Alkenes and Alkynes 775
   17.5.1. Overview 775
   17.5.2. Synthetic Targets for Hydroesterification and Hydrocarboxylation 775
   17.5.3. Catalysts for the Hydroesterification and Hydrocarboxylation of Olefins and Alkynes 777
   17.5.4. Scope of Hydroesterification and Hydrocarboxylation 778
     17.5.4.1. Hydroesterification and Hydrocarboxylation of Alkenes 778
       17.5.4.1.1. Intermolecular Hydroesterification and Hydrocarboxylation of Alkenes 778
       17.5.4.1.2. Intramolecular Hydroesterification of Olefins 780
     17.5.4.2. Hydroesterification of Alkynes 781
     17.5.4.3. Hydroesterification of Butadiene 782
   17.5.5. Mechanism of Hydroesterification 782
17.6. Carbonylation of Epoxides and Aziridines (Written with Prof. Geoffrey W. Coates) 784
   17.6.1. Ring-Expansion Carbonylation of Epoxides and Aziridines 784
     17.6.1.1. Overview 784
     17.6.1.2. History of Epoxide and Aziridine Carbonylation 785
     17.6.1.3. Types of Catalysts and Scope of Substrates for Epoxide Carbonylation 786
   17.6.2. Carbonylation of Lactones and Epoxides to Succinic Anhydrides 787
   17.6.3. Ring-Opening Epoxide Carbonylation 788
   17.6.4. Types of Catalysts and Scope of Substrates for Aziridine Carbonylation 790
   17.6.5. Mechanism of Epoxide Carbonylation 792
17.7. Carbonylations of Organic Halides 794
   17.7.1. Carbonylations of Organic Halides to form Esters and Amides 795
     17.7.1.1. Discovery and Scope 795
     17.7.1.2. Mechanism of Aryl Halide Esterification and Amidation 797
17.8. Copolymerization of CO and Olefins 798
   17.8.1. Overview of the Process and Polymer Properties 798

17.8.2. Development of Catalysts for the Synthesis of CO/Ethylene Copolymerization 798
17.8.3. Mechanism of the Coplymerization of CO and Ethylene 800
     17.8.3.1. Overall Cycle: The Steps of Chain Propagation 800
     17.8.3.2. Chain Termination and Catalyst Decomposition 802
17.8.4. Copolymerization of CO and α-Olefins 804
     17.8.4.1. Overview of the Copolymerization of CO and α-Olefins 804
     17.8.4.2. Copolymerization of Carbon Monoxide and Styrene 804
       17.8.4.2.1. Overall Mechanism 804
       17.8.4.2.2. Control of Stereochemistry 805
     17.8.4.3. Copolymerization of Carbon Monoxide and Propene 806
       17.8.4.3.1. Regiochemistry of Insertion 807
       17.8.4.3.2. Stereochemistry of Insertion 807
       17.8.4.3.3. Polymer Structure from the Copolymerization of CO and Propene 808
17.9. Pauson–Khand Reactions (Written with Dr. Qilong Shen) 809
   17.9.1. Overview 809
   17.9.2. Origin of the Pauson–Khand Reaction 809
   17.9.3. Effects of Additives 810
   17.9.4. Catalysts Other Than $Co_2(CO)_8$ 810
   17.9.5. Pauson–Khand Reactions with Allenes 811
   17.9.6. Catalytic Asymmetric Pauson–Khand Reactions 812
   17.9.7. Intermolecular Pauson–Khand Reaction 812
   17.9.8. Applications of the PKR 814
   17.9.9. Mechanism of the Pauson–Khand Reaction 814
References and Notes 816

Chapter 18. Catalytic C–H Functionalization 825
18.1. Overview 825
18.2. Platinum-Catalyzed Alkane and Arene Oxidations via Organometallic Intermediates 827
   18.2.1. Early Platinum-Catalyzed C–H Activation Processes 827
   18.2.2. More Practical Platinum Catalysts for Alkane Functionalization 827
   18.2.3. Mechanism of the Pt-Catalyzed Oxidations 829

18.3. Directed Oxidations, Aminations, and Halogenations of Alkanes and Arenes   832

18.4. Carbonylation of Arenes and Alkanes   835

18.4.1. Oxidative Carbonylation of Alkanes and Arenes   835

18.4.2. Alkylative Carbonylation of Alkanes and Arenes   837

18.4.3. Direct Carbonylation to Aldehydes   838

18.5. Dehydrogenation   839

18.5.1. Early Studies   839

18.5.2. Dehydrogenation Catalyzed by Complexes of Pincer Ligands   840

18.5.3. Alkane Metathesis via Dehydrogenation   842

18.5.4. Mechanism of Dehydrogenation   844

18.6. Hydroarylation   846

18.6.1. Directed Hydroarylation of Olefins   846

18.6.1.1. Overview   846

18.6.1.2. Reaction Scope and Catalysts   847

18.6.1.3. Mechanisms of Directed Hydroarylation of Olefins   849

18.6.2. Directed Hydroarylation of Alkynes   850

18.6.3. Undirected Hydroarylation and Oxidative Arylation of Olefins   850

18.7. Functionalization of Alkanes and Arenes with Main Group Reagents   852

18.7.1. Borylation of Alkanes   852

18.7.2. Borylation of Arenes   853

18.7.3. Borylation of Polyolefins   855

18.7.4. Mechanism of the Alkane and Arene Borylation   855

18.7.5. Silylation of Aromatic and Aliphatic C–H Bonds   857

18.8. Hydroacylation   859

18.8.1. Overview   859

18.8.2. Intermolecular Hydroacylation   860

18.8.3. Intramolecular Hydroacylation   860

18.8.4. Mechanism of Hydroacylation   861

18.8.5. Directed Intermolecular Hydroacylation   863

18.9. Functionalization of C–H Bonds by Carbene Insertions   864

18.9.1. Overview   864

18.9.2. Intramolecular C–H Functionalization by Carbene Insertion   865

18.9.3. Intermolecular C–H Functionalization by Carbene Insertion   867

18.10. H/D Exchange   869

References and Notes   870

Chapter 19. Transition Metal-Catalyzed Coupling Reactions   877

19.1. Overview of Cross-Coupling   877

19.2. The Classes of C–C Bond-Forming Coupling Reactions   878

19.2.1. Early Studies on Cross-Coupling: Coupling with Organomagnesium Reagents   878

19.2.2. Coupling of Organozinc Reagents   878

19.2.3. Coupling of Organotin Reagents   879

19.2.4. Coupling of Organosilicon Reagents   879

19.2.5. Coupling of Organoboron Reagents   880

19.2.6. Coupling of Alkynes   880

19.2.7. Coupling of Enolates and Related Reagents   881

19.2.8. Coupling at Aliphatic Electrophiles   882

19.2.9. Coupling of Olefins   883

19.2.10. Coupling of Cyanide   883

19.3. Enantioselective Cross Coupling   884

19.4. The Mechanisms of Cross Coupling   890

19.4.1. Mechanism of the Overall Catalytic Processes   890

19.4.1.1. Mechanism of Palladium-Catalyzed Cross Coupling with Main Group Organometallic Nucleophiles   890

19.4.1.2. Mechanism of Homocoupling   891

19.4.1.3. Mechanism of the Olefination of Aryl Halides (Mizoroki–Heck Reaction)   892

19.4.2. Mechanism of the Individual Steps of the Cross-Coupling Process   893

19.4.2.1. The Oxidative Addition Step   893

19.4.2.2. Mechanism of Transmetallation   895

19.4.2.3. Mechanism of Reductive Elimination   899

19.4.3. Effects of Catalyst Structure on Cross Coupling   899

19.4.3.1. Effect of Chelation   899

19.4.3.2. Effect of Steric Properties   901

19.4.3.3. Effect of Ligand Electronic Properties   902

19.5. Applications of C–C Cross Coupling   903

19.6. Cross-Coupling Reactions that Form Carbon–Heteroatom Bonds   907

19.6.1. Overview   907

19.6.2. Coupling of Aryl Halides with Amines   907

19.6.2.1. Scope of the Reaction   907

19.6.2.2. Catalysts for C–N Coupling   910

19.6.2.3. Mechanism of the C–N Coupling   911

19.7. Carbonylative Coupling Processes   914

19.7.1. Carbonylation of Organic Halides to Form Ketones   914

19.7.2. Mechanism of Carbonylative Coupling to form Ketones   916

19.7.3. Formylation of Organic Halides   917

19.8. Copper-Mediated Cross-Coupling Reactions (Written with Dr. Shashank Shekhar)   918

19.8.1. Copper-Mediated Cross Coupling to Form C(aryl)–N, C(aryl)–O and C(aryl)–S Bonds   920

19.8.1.1. Classes of Copper Catalysts for Carbon–Heteroatom Bond-Forming Coupling Reactions   920

19.8.1.2. Copper-Catalyzed Carbon–Nitrogen Cross-Coupling Reactions   922

19.8.1.2.1. Copper-Catalyzed Coupling of Amines   922

19.8.1.2.1.1. Copper-Catalyzed Coupling of Arylamines   922

19.8.1.2.1.2. Copper-Catalyzed Coupling of Alkylamines   923

19.8.1.2.2. Copper-Catalyzed Coupling of Amides with Aryl Halides   925

19.8.1.2.3. Copper-Catalyzed Reactions of Aryl Halides with Heterocyclic Amines   925

19.8.1.3. Copper-Catalyzed Coupling of Aryl Halides with Alcohols and Thiols   926

19.8.1.3.1. Reactions of Aryl Halides with Phenols   926

19.8.1.3.2. Reactions of Aryl Halides with Aliphatic Alcohols   928

19.8.1.3.3. Reactions of Aryl Halides with Amino Alcohols   929

19.8.1.3.4. Copper-Catalyzed Reactions of Aryl Halides with Thiols   929

19.8.2. Mechanism of Copper-Catalyzed Coupling of Aryl Halides with Amines, Alcohols, and Thiols   930

19.8.3. Reactions of Aryl Boronic Acids with Amines and Alcohols (Chan–Evans–Lam Couplings)   932

19.8.4. Copper-Catalyzed Cross Coupling to Form C–C Bonds   933

19.8.4.1. Cross Coupling to Form C(Alkyl)–C Bonds with Copper   933

19.8.4.1.1. C(sp^3)–C(sp^3) Coupling Mediated by Copper Reagents   933

19.8.4.1.2. Copper-Catalyzed C(sp^3)–C(sp^3) Coupling   934

19.8.4.2. Copper-Catalyzed Cross Coupling to Form Aromatic C–C Bonds   936

19.8.4.2.1. Coupling of β-Diketones, Cyanoesters, and Malonates   936

19.8.4.2.2. Copper-Catalyzed Stille and Suzuki Couplings   937

19.9. Direct Arylation (Written with Dr. Mark E. Scott, Dr. Dino Alberico, and Prof. Mark Lautens)   938

19.9.1. Introduction and Overview   938

19.9.2. Mechanisms of Direct Arylations   938

19.9.3. Transition Metal Catalysts for Direct Arylation   939

19.9.4. Regioselectivity of Direct Arylations   943

19.9.5. General Comments on Reaction Conditions for Direct Arylation   948

19.10. Catalytic Direct Oxidative Cross Couplings (Written with Dr. Mark E. Scott, Dr. Dino Alberico, and Prof. Mark Lautens)   949

19.11. Summary   950

References and Notes   951

Chapter 20. Allylic Substitution   967

20.1. Overview   967

20.2. Early Developments Toward Enantioselective Allylic Substitution   968

20.2.1. Stoichiometric Attack on Palladium Allyl Complexes   968

20.2.2. The First Catalytic Allylic Substitutions   968

20.2.3. The First Catalysts for Allylic Substitutions   969

20.3. Substrate Scope and Catalysts   969

20.3.1. Scope of Electrophile   969

20.3.2. Scope of Nucleophile   972

20.3.3. Metals Used for Allylic Substitutions   973

20.4. Mechanism of Allylic Substitution   974

20.4.1. Mechanism of Palladium-Catalyzed Reactions   974

20.4.2. Mechanism of Reactions Catalyzed by Complexes Other Than Palladium   977

20.5. Regioselectivity of Allylic Substitutions   979

20.5.1. Trends and Origins of Regioselectivity of Palladium-Catalyzed Reactions   979

20.5.1.1. Reactions of Carbon Nucleophiles   979

20.5.1.2. Reactions of Heteroatom Nucleophiles   981

20.5.2. Memory Effect with Palladium 982

20.5.3. Regioselectivity of Reactions Catalyzed by Complexes of Other Metals 983

20.6. Enantioselective Allylic Substitution 984

20.6.1. Overview of Enantioselective Allylic Substitution 984

20.6.1.1. Forms of Enantioselective Allylic Substitution 984

20.6.1.2. Catalysts for Enantioselective Substitutions 985

20.6.2. Enantioselective Allylic Substitution Classified by Electrophile 987

20.6.2.1. Enantioselective Allylic Substitution of Acyclic Electrophiles 987

20.6.2.1.1. Enantioselective Allylic Substitution of Symmetric Acyclic Allylic Esters 987

20.6.2.1.2. Enantioselective Opening of Vinyl Epoxides 987

20.6.2.1.3. Enantioselective Reactions of Unsymmetrical Acyclic Substrates 988

20.6.2.1.3.1. Enantioselective Reactions of Unsymmetrical Acyclic Substrates Catalyzed by Palladium Complexes 988

20.6.2.1.3.2. Enantioselective Reactions of Unsymmetrical Allylic Esters Catalyzed by Molybdenum, Ruthenium, Rhodium, and Iridium 989

20.6.2.2. Enantioselective Substitution of Cyclic Substrates 993

20.6.2.2.1. Enantioselective Substitution of Cyclic Allylic Monoesters 993

20.6.2.2.2. Enantioselective Substitution of Meso Cyclic Diesters 994

20.6.3. Kinetic Resolution 995

20.6.4. Enantioselective Allylation of Prochiral Nucleophiles 996

20.7. Copper-Catalyzed Allylic Substitution (Written with Levi Stanley) 999

20.7.1. Fundamentals 999

20.7.2. Mechanism of Copper-Catalyzed Allylic Substitution 1000

20.7.3. Enantioselective Copper-Catalyzed Allylic Substitution 1001

20.7.3.1. Diorganozinc Reagents as Nucleophiles 1002

20.7.3.2. Grignard Reagents as Nucleophiles 1004

20.7.3.3. Organoaluminum Reagents as Nucleophiles 1006

20.7.4. Miscellaneous Copper-Catalyzed Allylic Substitution Reactions 1007

20.8. Summary 1008

References and Notes 1008

Chapter 21. Metathesis of Olefins and Alkynes 1015

21.1. Introduction 1015

21.1.1. Overview of the Catalytic Metathesis of Carbon–Carbon Multiple Bonds 1015

21.1.2. Overview of the Classes of Metathesis Processes 1015

21.2. Olefin Metathesis 1017

21.2.1. Overview of Catalysts for Olefin Metathesis 1017

21.2.2. History of Olefin Metathesis 1019

21.2.3. Mechanism of Olefin Metathesis 1020

21.2.4. Catalyst Decomposition 1022

21.2.5. Examples of Olefin Metathesis 1023

21.2.5.1. Ring-Closing Olefin Metathesis 1023

21.2.5.2. Olefin Cross Metathesis 1026

21.2.6. Enantioselective Ring-Closing and Ring-Opening Metathesis 1028

21.2.7. Ring-Opening Metathesis Polymerization 1031

21.2.7.1. Utility of Ring-Opening Metathesis Polymerization 1031

21.2.7.2. Mechanism of Ring-Opening Metathesis Polymerization 1033

21.3. Alkyne Metathesis 1034

21.3.1. Examples of Alkyne Metathesis 1034

21.3.2. Mechanism of Alkyne Metathesis 1036

21.3.3. Applications of Alkyne Metathesis 1036

21.3.4. Alkyne Cross Metathesis 1038

21.3.5. Ring-Closing Alkyne Metathesis 1039

21.4. Enyne Metathesis 1040

21.4.1. Examples of Enyne Metathesis 1040

21.4.2. Mechanism of Enyne Metathesis 1041

21.5. Summary 1042

References and Notes 1043

Chapter 22. Polymerization and Oligomerization of Olefins 1047

22.1. Introduction 1047

22.1.1. A Primer on Polyolefin Chemistry (Written with Prof. Geoffrey W. Coates and Prof. Gregory J. Domski) 1048

22.2. Mechanism(s) of Monoene Polymerization and Oligomerization 1050

22.3. Ethylene-Based Polymers (Written with Prof. Geoffrey W. Coates and Prof. Gregory J. Domski) 1051

22.3.1. Catalysts for the Synthesis of HDPE 1052

22.3.2. Catalysts for the Synthesis of LDPE Materials from Only Ethylene 1054

22.3.3. Hyperbranched Polyethylenes from Late Metal Catalysts 1054

22.4. Propylene-Based Polymers (Written with Prof. Geoffrey W. Coates and Prof. Gregory J. Domski) 1057

22.4.1. Mechanism of Stereocontrol in Isotactic Polypropylene Synthesis 1057

22.4.2. Synthesis of Stereodefined Polypropylenes 1060

22.4.2.1. *Synthesis of Isotactic and Syndiotactic Polypropylene 1060*

22.4.2.2. *Synthesis of Hemiisotactic Polypropylene 1062*

22.4.2.3. *Synthesis of Stereoblock Polypropylenes 1062*

22.4.2.3.1. Isotactic–Atactic Stereoblock Polypropylene Generated from Heterogeneous Catalysts 1063

22.4.2.3.2. Isotactic–Atactic Stereoblock Polypropylene Generated from Homogeneous Catalysts 1063

22.4.2.3.3. Stereoblock Copolymers by Alternation of the Ligand Sphere 1065

22.4.2.3.4. Stereoblock Copolymers by Chain Transfer 1065

22.4.2.3.5. Stereoblock Copolymers from Living Catalysts 1065

22.5. Hyperbranched Polypropylenes 1066

22.6. Ethylene–α-Olefin Copolymers (Written with Prof. Geoffrey W. Coates and Prof. Gregory J. Domski) 1067

22.6.1. Alternating Ethylene–Propylene Copolymers 1068

22.6.2. Ethylene–Propylene Block Copolymers 1069

22.7. Single-Site Catalysts for the Polymerization of Styrene (Written with Prof. Geoffrey W. Coates and Prof. Gregory J. Domski) 1070

22.7.1. Synthesis of Syndiotactic Polystyrene 1070

22.7.2. Synthesis of Isotactic Polystyrene 1072

22.8. Further Mechanistic Information on Alkene Polymerization 1072

22.8.1. The Mechanism of the Chain Propagation Step 1073

22.8.2. Mechanism of Chain Transfer and Scope of Chain Transfer Agents 1076

22.8.3. Effect of Catalyst Steric Properties on Chain Transfer 1078

22.9. Oligomerization of Alkenes 1079

22.9.1. Ethylene Oligomerization 1080

22.9.1.1. *The Shell Higher Olefin Process 1080*

22.9.1.2. *Ethylene Oligomerization with Metals Other than Nickel 1082*

22.9.2. Olefin Dimerization by Insertion into Metal–Carbon Bonds 1082

22.9.3. Olefin Oligomerization Through Metallacyclic Intermediates 1084

22.9.3.1. *Dimerization of Alkenes by a Metallacyclic Mechanism 1084*

22.9.3.2. *Trimerization and Tetramerization of Alkenes by a Metallacyclic Mechanism 1084*

22.10. Oligomerization and Polymerization of Conjugated Dienes 1086

22.10.1. Polymerization of 1,3-Dienes 1087

22.10.2. Oligomerization and Telomerization of Conjugated Dienes 1088

22.10.2.1. *Linear Oligomerization of Butadiene 1088*

22.10.2.2. *Cyclooligomerization of 1,3-Dienes 1090*

22.11. Summary 1092

References and Notes 1093

Contributor Listing 1101

Index 1103

Preface

Principles and Applications of Organotransition Metal Chemistry by Jim Collman, Lou Hegedus, Jack Norton, and Rick Finke was published in 1987 during my first year as a graduate student. The way my contemporaries and I think about the bonding, reactivity, and catalysis of organometallic systems was shaped in large part by reading this important text. When I became an assistant professor and taught my own course in organometallic chemistry, I opened that text and created notes that followed this book's organization. Thus, when Jane Ellis and Bruce Armbruster at University Science Books provided me the chance in 2002 to contribute to the next edition of this book, I seized the opportunity to share with others beginning to study the field what I have learned over the past two decades.

Ultimately, it proved an impossible task to squeeze twenty years of advances in organometallic chemistry into the second edition of this text to create an "updated" third edition as originally planned. The current book contains some of the structure of the second edition by Collman, Hegedus, Norton, and Finke, and readers who are familiar with the previous book will recognize chapter headings and may even recognize (in a contemporary form) several of the figures and schemes from the previous text. However, every chapter, section, paragraph, sentence, and, yes, title are revised from the book published in 1987.

Like the book by Collman, Hegedus, Norton, and Finke, this work was written to serve as a textbook for students who are serious about the topic of organometallic chemistry and, now, for the many chemists who practice catalytic chemistry with organometallic systems but have not had formal training in the subject. I also hope that people who are experts and practitioners in certain areas of organometallic chemistry can turn to chapters of this book to learn about new topics and to find seminal references. This book was written with the intention that chapters would begin at a level that is appropriate for a newcomer to the field but would progress to cover examples and concepts sufficiently advanced to be useful for those experienced in organometallic chemistry.

Although the organization of this text parallels many of the chapters and topics of the Collman, Hegedus, Norton, and Finke book, the structure diverts from its past history in several important ways. Like the 1987 book, this new book begins with chapters on bonding and families of ligands that are typically found in organometallic chemistry. The first chapter covers structure and bonding and seeks to convey overarching principles that are used to predict phenomena of organometallic systems. The next three chapters present the classes of ligands that are commonly used in organometallic chemistry. I appreciate that the "ligand chapter" of the prior text provided a challenge for those using the 1987 book to teach a course, and I have now made this challenge even greater by dividing this material into three chapters. However, it seemed appropriate to continue to provide an extensive discussion of the "functional groups" of organometallic chemistry that includes topics any instructor would teach in a course, as well as topics that are best considered reference or instructional material for practitioners. The first two of these chapters cover ligands that are most conventionally considered within the realm of organometallic chemistry, while Chapter 4 presents ligands that are bound to the metal through oxygen, nitrogen, sulfur, phosphorus, silicon, boron, and even a single halide, but that have been shown to undergo much of the reactivity considered to characterize organometallic systems. Thus, people using this text for a course should pick and choose sections that they feel are most important from this portion of the book. For example, sections in Chapter 2 on the structure, bonding, and electronic properties of phosphine, carbonyl and alkene ligands are key elements to any organometallic course, but properties of ether, thioethes, and amine ligands are probably best saved for specialists.

Chapters 5–12 cover the fundamental reactions of organometallic systems. The first of these chapters, Ligand Substitution Reactions, also encompasses many concepts of coordination chemistry, but I have sought in this chapter to emphasize mechanisms that are followed by organometallic systems. For example, this chapter includes mechanisms of substitution reactions of carbonyl complexes and mechanisms of substitution reactions that occur by changes in hapticity of unsaturated organic ligands. The chapter on oxidative addition and reductive elimination has now become three chapters due to the large amount of information that has been gained on C–H bond activation, oxidative additions of carbon–halogen bonds, and reductive eliminations that are part of many catalytic processes developed over the past years. Insertion and elimination reactions are now also divided into two separate chapters because of the many new classes of migratory insertion reactions that have been developed, again, as part of important new catalytic processes. Chapter 13—Metal–Ligand Multiple Bonds—is completely new. The advances in olefin metathesis and the interest in organometallic oxidation processes have made this topic an important one for organometallic chemists and an appropriate one for a full chapter.

Chapters 14–22 cover transformations catalyzed by organometallic complexes and are completely or largely new chapters. The applications to organic synthesis in the book by Collman, Hegedus, Norton, and Finke were organized in many cases by the manner in which intermediates were generated. A majority of the material from this section of the text is now part of a valuable book by Lou Hegedus and Björn Söderberg titled *Transition Metals in the Synthesis of Complex Organic Molecules*. Now, the applications section of this text focuses completely on principles of catalysis and classes of catalytic reactions. Chapter 14 is new and presents fundamental principles of catalysis that apply to all subsequent chapters of the text, and it includes principles of asymmetric catalysis that pertain to discussions of enantioselective reactions. Chapters presenting classes of organometallic reactions that are commonly practiced in industry and in academic laboratories complete the book.

The first of these chapters, Chapter 15, covers hydrogenation. This chapter, now much longer than it was in 1987, describes many classes of hydrogenations, including asymmetric hydrogenations of a series of substituted alkenes, ketones and imines, that have been developed over the past twenty years. Chapter 16, also new, presents hydrofunctionalization and oxidative functionalization of alkenes. Catalytic carbonylation reactions are the topic of Chapter 17. The roots of this chapter lie in Chapter 12 of the 1987 book, but since many new systems for carbonylation and new reactions involving CO have been discovered during the past twenty-two years, this chapter has been expanded, and many sections are new. Chapters 18–21 are all new and present classes of catalytic organometallic reactions—such as cross coupling, C–H bond functionalization, allylic substitution, and olefin metathesis—that have been developed extensively since 1987. Finally, the last chapter on olefin polymerization stems from Chapter 11 of the 1987 text, but this chapter has been completely rewritten due to the many advances made in the field of alkene polymerization. Certainly, one of the success stories of organometallic chemistry over the past 20 years has been the development of single-site catalysts for ethylene and α-olefin polymerization.

I am sure this book contains many biases, errors of judgment, and errors of omission. I am also sure that many researchers' contributions have been mistakenly or inappropriately overlooked. No doubt the text also contains grammatical or typographical errors, and the structures or contents of certain graphics will be incorrect. There are certainly unbalanced equations, molecules with five bonds to carbon, and complexes containing impossible oxidation states that escaped many people's eyes. Any new updates, supplements, sites to share exercises, or errata pertaining to this book can be found on its book page at www.uscibooks.com. Please send comments and corrections to jhartwig@illinois.edu, and appropriate changes will be made for future printings and editions.

The number of students, faculty, and industrial chemists inventing and using fundamental and catalytic organometallic chemistry has exploded during the past twenty years. This text thus includes many connections between fundamental, stoichiometric organometallic chemistry and catalytic reactions used in the synthesis of organic molecules in many different contexts. By including these connections, I hope readers will see why so many chemists are passionate about the structure, bonding, and reaction chemistry of organotransition metal complexes. I hope that people learning organometallic chemistry during this part of the twenty-first century gain as much from this book as I did from *Principles and Applications of Organotransition Metal Chemistry* during the early stages of my career.

John F. Hartwig

Acknowledgments

Extensive input from friends and colleagues in the field of organometallic chemistry made this writing project possible. Many people quickly answered questions and supplied detailed information and references to support their answers. Others provided assistance by drafting certain sections of this text that covered topics on which they are particular experts. A list of these contributors appears on page 1101; when citing material from these sections, please include their names. Prof. Pat Walsh wrote (and rewrote) a draft of a section on asymmetric catalysis that is close to the form you will find as the second half of Chapter 14. Prof. Jack Norton drafted sections within the chapter on compounds containing metal-carbon bonds, and the section on metal hydrides is his alone. He also drafted sections on the mechanism of the Wacker process. Prof. Chuck Casey kindly provided drafts of sections on the synthesis of acetic acid and on hydroformylation. Prof. Geoff Coates supplied crucial text on the carbonylation of small ringed heterocycles, and he and his student Gregory Domsky provided a large amount of material that enabled me to write the chapter on olefin polymerization. Prof. Jing Zhou drafted sections on the hydrogenation of esters, amides, imides, and nitriles, Dr. Shashank Shekhar drafted a section on copper-catalyzed cross coupling, Prof. Mark Lautens and two postdocs from his group, Dr. Mark Scott and Dr. Dino Alberico, provided a draft on direct coupling, and Dr. Levi Stanley provided a draft of the section on copper-catalyzed allylic substitution. Finally, at a time when I could not stomach writing one more word, many graduate students and postdoctoral researchers in my own group teamed up to launch an assault on Chapters 3 and 4 covering X-type ligands, along with a few loose ends of other chapters, and new sections were drafted within weeks. These saviors were Erik Alexanian, Elsa Alvaro, Tim Boebel, Seth Herzon, Jaclyn M. Murphy, Mark Pouy, Devon Rosenfeld, Qilong Shen, Jesse Tye, Giang Vo, Jing Zhao, Pinjing Zhao, and Jianrong (Steve) Zhou.

I am grateful to the many others in the organometallic community who answered my questions over the course of the past six or seven years: Jim Atwood, Guy Bazan, Bob Crabtree, Huw Davies, Scott Denmark, Steve Diver, Odile Eisenstein, Jack Faller, Greg Girolami, John Gladysz, Alan Goldman, Bob Grubbs, Mike Heinekey, Marissa Kazlowski, Ryoichi Kuwano, Janis Louie, Jim Mayer, Tom Rauchfuss, Martin Semmelhack, Matt Sigman, Shannon Stahl, Don Tilley, Pete Wolczanski, and Zhumu Zhang. Numerous organometallic chemists generously reviewed individual chapters, and all of them provided helpful, critical, expert comments. Their reviews were crucial to making the book what it is. Thanks to Christian Amatore, Jim Atwood, Jan Bäckvall, Steve Bergens, Maurice Brookhart, Morris Bullock, Don Darensbourg, Steven Diver, David Glueck, Alan Goldman, Harry Gray, Bob Grubbs, Mike Hall, Mike Heinekey, Greg Hillhouse, Takao Ikariya, Bill Jones, Jay Labinger, Jim Mayer, David Milstein, Ei-Ichi Negishi, Ged Parkin, Andreas Pfaltz, T. V. (Babu) RajanBabu, Melanie Sanford, Martin Semmelhack, Matt Sigman, Don Tilley, and Antonio Togni.

Many others helped in the production of this book. Carole Velleca from Yale University worked tirelessly for many years obtaining literature and transcribing sections of the text, and Nasrin Ghavari took over when I moved to Illinois. Nan Holda worked with me for the past two years, being crucial to bringing this project to completion, and conducted the administrative portions of the production process at Illinois. Jane Ellis deftly oversaw the entire project, and I thank her and Bruce Armbruster for getting me into this mess in the first place. Jennifer Uhlich, who tolerated my prodding emails, oversaw the production of this book with patience and grace. John Murdzek was a quick, precise editor and caught many of my errors; and Thomas Webster, who converted my hand drawings and rough electronic files into attractive figures, schemes, and equations, is responsible for the greatly improved graphics of this book. Carl Liskey, Dale Pahls, Mark Pouy, Cassady Richers, Daniel Robbins, Levi Stanley and Giang Vo proof read many of the graphics at a late stage of production.

I also thank my faculty colleagues with whom I have worked for twenty-two years. I thank my original organometallic mentors Dick Andersen and Bob Bergman at Berkeley, my more recent organometallic colleagues Bob Crabtree and Jack Faller at Yale, and my most recent organometallic colleagues at Illinois, Scott Denmark, Greg Girolami, and Tom Rauchfuss for helping me understand the field. I also thank my father for teaching me to put both a noun and a verb into the same sentence in high school, Bob Bergman for teaching me to put them into a scientific sentence in graduate school, and my mother for attempting to show me when enough is enough.

Finally, I thank my two girls, Amelia and Pauline, for sleeping through the night since they were a few months old, for keeping me company in their bassinets on the patio at five a.m., for being dedicated nappers on weekend afternoons, and for beaming the most charming smiles when nap and worktime were over. I also thank my wife and colleague, Anne Baranger, for listening to my repeated declarations over many years that the project is almost finished and for waiting, many times, just a couple more minutes.

John F. Hartwig

To Anne, Amelia, and Pauline

CHAPTER 1

Structure and Bonding

The principles of structure and bonding provide a foundation for the reaction chemistry presented in this text. Because organometallic complexes comprise a vast array of metals, oxidation states, and ligands, general principles that apply to all or most of these systems are needed. The number of valence electrons, d-electrons, and formal charges of the metal can be used to explain a diverse set of reactions and structural phenomena observed with organometallic systems, and the interactions of metal and ligand orbitals create a foundation to understand the strengths of these bonds and the rates and selectivities of reactions occurring at these sites. Thus, it is important to convey some of the fundamental principles chemists follow to understand trends in structures and bonding of organometallic complexes prior to discussing specific ligands, complexes or reactions of these complexes.

Chapter 1 describes a framework with which organometallic chemists classify ligands, metal–ligand interactions, and properties of metals and ligands. This framework allows one to predict structures, relative reaction rates, and the thermodynamics of metal–ligand complexes. Some of the discussion will focus on ways to classify metals and ligands. Although these formal classifications only partly reflect the true properties of the complexes, some relationship between these formalisms and the true properties does exist and does provide the organometallic chemist with a power to predict structure and reactivity patterns. Chapter 1 also describes the orbital interactions that control the strength and reactivity of metal–ligand bonds.

1.1. General Properties of the Ligands

1.1.1. Classification of Ligands as Dative or Covalent, Neutral or Anionic, Even- or Odd-Electron, L-Type or X-Type

Ligands in organometallic systems are generally classified into two groups. This classification is based on the charges assigned to them by oxidation state formalisms or the number of electrons in the ligand considered to contribute to the metal–ligand bond(s). Several different terms are used to describe the same class of ligand. "Covalent ligands" are considered to form a bond to the transition metal by a sharing of one electron from the ligand and one from the metal. "Dative ligands" (also called dative covalent bonds or coordinate bonds) are considered to form a bond to the transition metal by the donation of two electrons from the ligand and no electrons from the metal. The bond between BF_3 and NH_3 is one classic "dative" bond.

Two systems have been developed for the classification of ligands in transition metal organometallic chemistry, and the choice of system is a matter of preference. By one system, ligands are assigned to be neutral (dative ligands) or charged (covalent ligands). When charged, the ligand is usually assigned a negative charge. Although these ligands are assigned a charge in this classification system, one should realize that the bond between these ligands and the metal are just as covalent as the C–C, C–O, or C–N bonds in alkanes, ethers, or amines. The difference in electronegativity between carbon and platinum is smaller than the difference in electronegativity between carbon and nitrogen. By the

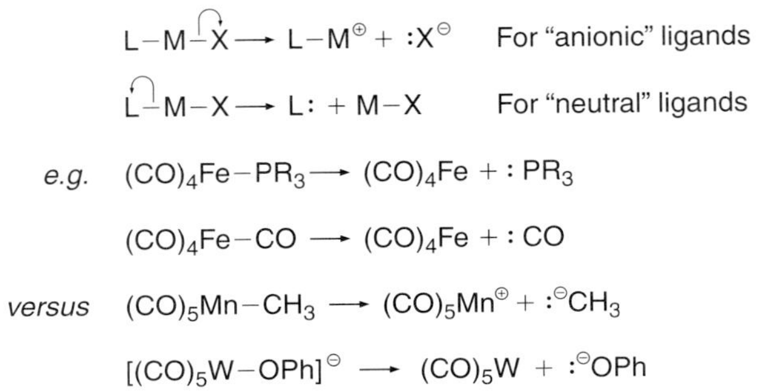

$$L-M-X \longrightarrow L-M^{\oplus} + :X^{\ominus} \qquad \text{For "anionic" ligands}$$

$$L-M-X \longrightarrow L: + M-X \qquad \text{For "neutral" ligands}$$

e.g. $(CO)_4Fe-PR_3 \longrightarrow (CO)_4Fe + :PR_3$

$(CO)_4Fe-CO \longrightarrow (CO)_4Fe + :CO$

versus $(CO)_5Mn-CH_3 \longrightarrow (CO)_5Mn^{\oplus} + :^{\ominus}CH_3$

$[(CO)_5W-OPh]^{\ominus} \longrightarrow (CO)_5W + :^{\ominus}OPh$

Figure 1.1.
Mnemonic for determining if a ligand is anionic or neutral.

second system, all ligands are considered to be neutral. In this case, some ligands donate two electrons and have been termed[1,2] "L-type" ligands, and other ligands donate one electron to the metal center and have been termed "X-type" ligands.[3] Ligands that donate more than two electrons have been termed LX ligands (three-electron donors), L_2 ligands (four-electron donors), L_2X (five-electron donors), etc.

A simple mnemonic allows one to determine in the first classification system if the ligand is neutral or charged. If one formally cleaves the metal–ligand bond by placing both electrons in the bond on the ligand to generate a closed-shell configuration,[4] then the free ligand will lack a formal charge if it is a "neutral" or "dative" ligand, and it will be charged if it is an anionic or cationic ligand. As shown in Figure 1.1, this formal cleavage of a metal–phosphine bond or a metal–carbonyl bond generates a free phosphine, which is a stable neutral molecule, but cleavage of the metal–alkyl linkage in this manner generates a carbanion and cleavage of a metal–alkoxide bond generates an alkoxide anion.

A second mnemonic in Figure 1.2 allows one to determine if the ligand donates one or two electrons in the second classification system that treats all ligands as neutral. In this case, one divides the electrons in the metal–ligand bond to create a neutral organic group. Thus, one would assign the two electrons in the metal–ligand bond to the phosphine, but one would assign one electron of a metal–alkyl bond to the metal and one electron of a metal–alkyl bond to the alkyl group.

$(CO)_4Fe-PR_3 \longrightarrow (CO)_4Fe + :PR_3$ *or* $(CO)_4Fe-CO \longrightarrow (CO)_4Fe + :CO$

$(CO)_5Mn-CH_3 \longrightarrow (CO)_5Mn^{\oplus} + CH_3^{\ominus}$ *or* $[(CO)_5W-OPh]^{\ominus} \longrightarrow (CO)_5W + :OPh^{\ominus}$

Figure 1.2. Mnemonic for determining if a ligand is a one-electron or two-electron donor.

Sometimes the bonds between the metal and the formally neutral and formally anionic ligands are distinguished graphically by drawing the former with arrows, indicating that the bond is akin to a simple Lewis acid–base complex between NH_3 and BH_3, and the latter with lines indicating covalent bonds. When $(Et_3P)_2PtBu_2$ is drawn in this way (Figure 1.3), it is clearer that the complex contains two dative and two formally charged or one-electron ligands, and that the central atom is Pt(II). However, organometallic complexes are not usually depicted in this way because there would be too many arrows. Thus, one must learn to distinguish between neutral and charged ligands to assign oxidation states and to recognize, for example, $(Et_3P)_2PtBu_2$ as a Pt(II) complex when all metal–ligand bonds are drawn with simple lines (Figure 1.3).

X-type ligands often bridge two metals. The most appropriate way to draw these bridging ligands has been a subject of intense debate. One method advocated by Green, which avoids ambiguity in electron counting, involves solid lines, arrows, and half-arrows. Bridging halide, alkoxide, thiolate, or amide ligands bond to one metal in one resonance structure by a typical M–X bond and to the second metal through a typical dative or L-type interaction. An equally important resonance structure would contain the opposite arrangements of the two bonds, as depicted on the left of Figure 1.4. A bridging hydride would bind in one resonance structure to one metal through a typical M–X bond and to the second metal by the donation of electrons from the M–X bond to create a two-electron three-center bond. Again, an equally important resonance structure would contain the opposite arrangements of the two bonds, as depicted on the right of Figure 1.4.

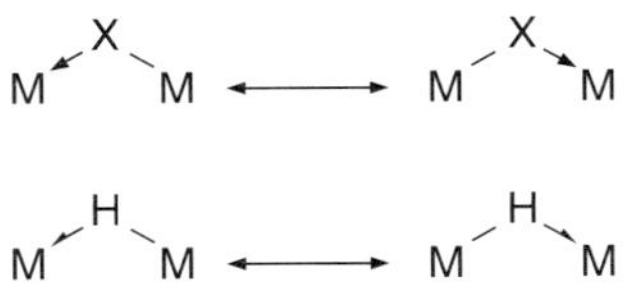

Figure 1.3.
Convention with dative bonds drawn as arrows vs. the more usual convention with all bonds drawn as lines.

Figure 1.4.
Conventions for depicting bridging ligands with lines, arrows, and half-arrows.

1.1.2. Classification by Number of Electrons Donated to the Metal

An important formalism about metal–ligand bonds to address is the number of electrons on the ligand involved in bonding to the metal when the ligands are more complex than a phosphine or an alkyl group. Because these interactions are in large part Lewis acid–base interactions, with the metal acting as a Lewis acid and the ligand as a Lewis base, these electrons from the ligand are often called the electrons "donated" from the ligand to the metal. Thus, a phosphine is considered to be a two-electron donor and is, therefore, a "neutral, two-electron" ligand.

A methyl group is considered to "donate" two electrons and to be an "anionic, two-electron" ligand by the first classification system and is considered to "donate" one electron and to be an X-type ligand by the second classification system.

Table 1.1 summarizes the structures, the formal charges assigned by the first classification system, and the number of electrons donated to the metal center for a series of common ligands. In most cases, ligands coordinated through a single electron pair on a single

Table 1.1. Electron counts and changes of common ligands.

Ligand	Formal charge	Electrons donated		Bonding† mode
		Ionic model	Covalent model	
$H^{\ominus}$	−1	2	1	
Halide (terminal)	−1	2	1	
Halide (bridging)	−1	4	3	μ^2
Akyl, aryl, vinyl, alkynyl, silyl, germyl, stannyl, alkoxo, amido, thiolato, or phosphido	−1	2	1	
	−1	2	1	μ^2
	0	2	2	μ^2
	−1	4	3	η^2
	−1	4	3	μ^2, η^2
	−1	4	3	η^3
	0	4	4	η^4
	−1	6	5	η^5
	0	6	6	η^6

†The symbol η^n or "hapticity" denotes the number (n) of contiguous atoms, typically in a π-system, that are bound to a metal. The symbol κ^n or "denticity" refers to the number (n) of atoms of a ligand that is bound through n non-contiguous atoms. The symbol μ^n refers to the number of metals (n) to which a ligand is bound.

(continued)

Table 1.1. *(continued)*

Ligand	Formal charge	Electrons donated		Bonding mode
		Ionic model	Covalent model	
M–C(=O)R (η^1 acyl)	−1	2	1	η^1
M←C(=O)R (η^2 acyl)	−1	4	3	η^2
M–C(=O) with R, R	0	2	2	η^2
O=C(R)R bound to M (η^1)	0	2	2	η^1
M–C(R)≡	0	2 or 4	2 or 4	η^2
M–C(R)≡C(R)–M	0	4	4	η^2, η^2
M=C(OR)R	0	2	2	
M=C(R)R	−2	4	2	
M–CR_2–M (bridging)	−2	4	2	μ^2
M=O, M=NR	−2	4 or 6	2 or 4	
M–N=O	−1	2	1	
M–N≡O	+1	2	3	

metal center donate two electrons, and ligands coordinated through a single π-system donate two electrons. One can then combine these interactions to determine the number of electrons donated by ligands with more than one atom or π-system bound to the metal or with a combination of σ- and π-bonding to the metal center. The "number of electrons donated" includes the electrons of the ligand that participate in σ-bonds with the metal and those that participate in π-interactions with the metal.

1.1.3. π-Bonded Ligands

In contrast to the metal–ligand bonds shown in Section 1.1, which are much like main group Lewis acid–base interactions, many of the ligands in organometallic chemistry bind to the metal by an interaction of the metal with the π-system of an unsaturated organic group. More details on the interaction between metals and ligands that bond through their π-systems are provided in Chapter 2. For now, for the purpose of classification, ligands containing an even number of carbon atoms bound to the metal are considered to be neutral ligands. Those containing an odd number of atoms bound to the metal are considered to be charged and to donate an even number of electrons by the first classification system, and are considered to be neutral and to donate an odd number of electrons by the second system. The number of electrons donated to the metal is equal to the number of electrons in a neutral π-ligand containing an even number of atoms. For example, ethylene is a neutral, two-electron donor ligand, and it is stable as the free neutral molecule. The number of electrons donated to the metal is equal to the number of electrons in the anionic version of an acyclic π-ligand containing an odd number of atoms (Figure 1.5). For example, an allyl ligand, shown in Figure 1.5, is considered to be a four-electron donor, anionic ligand by the first classification system and a three-electron donor ligand by the second classification system.

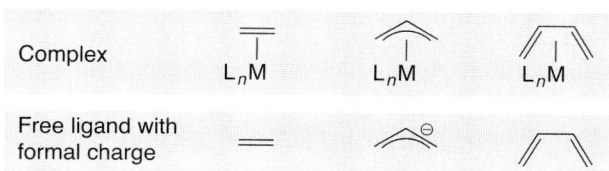

Figure 1.5.
Even-electron π-systems are considered to be neutral ligands and odd-electron π-systems are considered to be charged ligands or odd-electron donors.

The number of electrons donated to the metal of a cyclic π-ligand by the first electron-counting method corresponds to the number of electrons in the most stable aromatic system (Figure 1.6). This method for assigning charges begins to reveal the potential complexity in keeping track of charges and electrons donated by various π-donor ligands. By the second classification (based on neutral ligands), the π-ligands simply donate the number of electrons in the neutral π-system without regard for the most stable aromatic system.

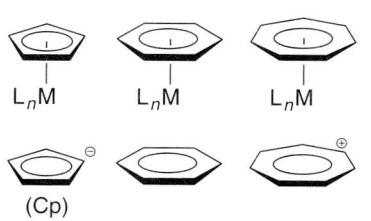

Figure 1.6.
Comparison of the charges on five-, six-, and seven-membered ring unsaturated ligands.

1.1.4. Combinations of σ- and π-Donors

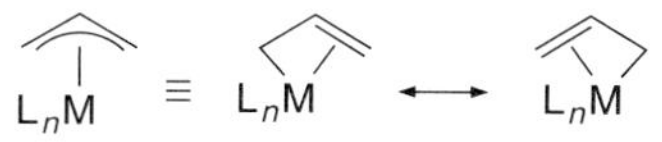

The bonding of the allyl fragment can also be envisioned to result from a combination of resonance structures. As shown in Figure 1.7, the two resonance structures of the allyl ligand each contain one σ-bound ligand and one π-ligand in which an even number of atoms are bound to the metal. Based on this structure, the allyl ligand is anionic and shares four electrons with the metal in the first classification system because an alkyl ligand is treated as an anionic two-electron donor, while an olefin is a neutral two-electron donor. These resonance structures also reveal why the allyl ligand is a three-electron donor by the second classification system: The π-system donates two electrons and the M–C single bond donates one electron. The allyl group is a classic LX-type ligand.

Figure 1.7.
The two resonance structures of an allyl ligand.

Another common anionic ligand is the cyclopentadienyl ligand (Cp). Many derivatives of this ligand, such as the pentamethylcyclopentadienyl ligand (Cp*), have also been prepared and are common ligands. This class of ligand binds to the metal through the π-system in most cases. This π-system contains an odd number of carbon atoms, and the anionic form of the ligand contains six electrons in its π-system. Thus, Cp is considered to be an anionic ligand that donates six electrons to the metal center by the first classification system. This ligand is particularly stable because it is a six-electron, cyclic π-system and is, therefore, aromatic. Although the resonance forms of this ligand break the aromaticity and are, therefore, poor representations of the bonding, five resonance forms containing one σ-bond and two two-electron π-donors can be drawn. These resonance forms underscore the origin of the anionic, six-electron donation of a Cp ligand by the first classification system or the five-electron donation by the second classification system.

1.1.5. Cationic Ligands

A few ligands are considered to be cationic in the first classification system because the free species is most stable in the cationic form. The most common of these ligands is the nitrosyl ligand, NO. NO is easily oxidized to NO^+, and $NO[BF_4]$, for example, is stable enough to be sold commercially. Thus, the charge of the NO ligand is cationic when it is bound in its most common geometry with the M, N, and O atoms lying in a nearly linear arrangement (Figure 1.8). However, when this ligand is "bent" and the M–N–O angle is closer to 120° than to 180°, the electron donation is more closely related to that of an anionic ligand linked to the metal through a single bond and is considered by the first classification scheme to possess a negative charge, as shown in the center of Figure 1.8. In the second "neutral" counting scheme, the "linear" NO ligand is classified as a three-electron donor because of the combination of one σ- and one π-bond, whereas "bent" NO is classified as a one-electron donor because of one single covalent bond to the metal.

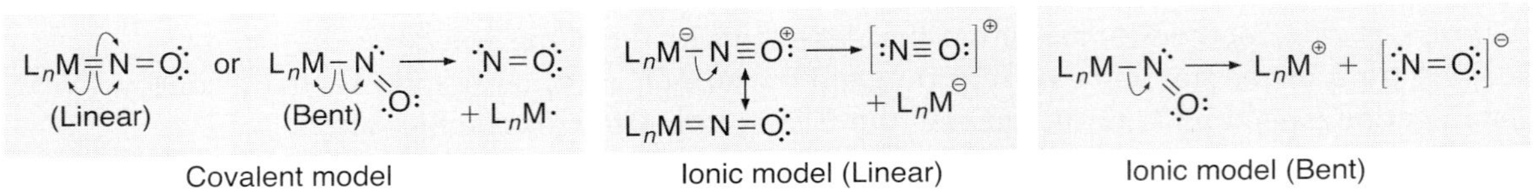

Figure 1.8. Linear and bent nitrosyl ligands.

Another ligand that can be considered cationic is the cycloheptadienyl ligand. Because the cycloheptadienyl anion possesses eight π-electrons and this number of electrons in a cyclic structure generates an antiaromatic π-system, the charge on a fully unsaturated seven-membered ring acting as a ligand is often considered to be cationic, as was shown in Figure 1.6.

1.2. Properties of the Metal

1.2.1. Oxidation State

Oxidation state is one common formalism used to classify the metal centers in organometallic and coordination compounds. Oxidation state is the formal charge on the metal center that balances the overall charge of the complex and the sum of the formal charges on the ligands. For example, the oxidation state of the metal in a neutral complex that has two anionic ligands is +2, while the oxidation state of a complex that has an overall +1 charge and two formally anionic ligands is +3. Because of Pauling's electroneutrality principle, which states that no atom in a complex will have an actual charge greater than ±1,[5] and the high degree of covalent bond character in many organometallic compounds, the value of the oxidation state is far from the true charge on the metal center. However, the oxidation state can be used to determine the number of valence electrons in orbitals with predominantly metal character, and the combination of oxidation state and number of valence electrons can be used to predict structures and trends in reactivity.

The difference between oxidation state and true charge can be appreciated by considering the wide range of formal oxidation states found in organometallic compounds. The formal oxidation state of the iron in $[Fe(CO)_4]^{2-}$ is -2, although the iron atom bears little if any negative charge. Similarly, the formal oxidation states of iridium in $(\eta^5\text{-}C_5Me_5)IrMe_4$, $(\eta^5\text{-}C_5Me_5)IrMe_2(dmso)$, and $(\eta^5\text{-}C_5Me_5)Ir(CO)_2$[6] are V, III, and I, but the three complexes have nearly the same ionization energies. In fact, the complex with the highest ionization energy is the Ir(I) complex. Because oxidation states in organometallic chemistry do not reflect the true electronic properties of the metal centers, formal "oxidations" do not necessarily decrease the electron density at the metal, and formal "reductions" do not necessarily increase the electron density at the metal. More detailed presentations of bonding and d-electron configurations in inorganic complexes,[7a] molecular orbitals of appropriate symmetry,[7b] and bonding in transition metal organometallic complexes can be found in other books and in specialized reviews.[8–13]

1.2.2. The Relationship Between Oxidation State and the Number of d-Electrons

The oxidation state of the metal center is directly related to the number of valence electrons in orbitals of predominantly metal character. Two simple rules allow one to determine quickly the number of electrons in the metal d-orbitals.

First, the electron-counting formalism assigns the metal valence electrons of a transition metal organometallic compound in a way that fills the metal's $(n)d$-shell first and the $(n+1)s$-shell second. The $4s$ orbital of an atom in the gas phase is lower in energy than the $3d$ orbitals, but the $3d$ orbitals of a metal center in almost all transition metal complexes are lower in energy than the $4s$ orbitals. The $4s$ orbital of a neutral atom is lower in energy than the $3d$ orbital, despite the higher principal quantum number of the $4s$ orbital, because it is less shielded from the nuclear charge than the $3d$ orbital. The difference in shielding between s- and d-orbitals is smaller than that in the corresponding neutral atom when the metal possesses a partial positive charge, in part because the positively charged ion is smaller. As a result of the smaller difference in shielding, the principal quantum number determines the relative energies of the orbitals, and the energies of the $4s$ and $3d$ orbitals of the cationic metal center are the opposite of those of the neutral atom. Because most transition metal complexes contain electronegative ligands that lead to polarized metal–ligand bonds, albeit weakly polarized, the metal center possesses a partial positive charge regardless of whether it is in a neutral or positive oxidation state. As a result, the $3d$ orbitals are lower in energy than the $4s$ orbitals in most transition metal complexes.

Organometallic chemists often use the terminology "number of d-electrons." The number of d-electrons generally describes the number of electrons not involved in the primary metal–ligand bonding interactions. The number of d-electrons is simply equal to the number of the column of the metal in the periodic table, which is often called the group number, minus the oxidation state or minus the sum of the overall charge of the molecules and number of X-type ligands. This rule predicts the number of d-electrons because the group numbers equal the number of valence electrons in the neutral atom. Thus, an iron complex with the metal in the $+2$ oxidation state would possess a d^6 metal center because iron is in the eighth column and the oxidation state or sum of the overall charge of the molecules and number of X-type ligands is two. An iron complex in the -2 oxidation state would possess a d^{10} metal center. The homoleptic rhenium hyride complex ReH_9^{2-} possesses no d-electrons because rhenium is in group seven and the oxidation state of the metal in ReH_9^{2-} is $+7$.

Consideration of the group number and column number allows a quick prediction of whether the metal complex will be paramagnetic or whether it can be diamagnetic. Complexes containing a metal atom from an even column that has an even oxidation state and complexes containing a metal from an odd column that has an odd oxidation state will possess an even number of electrons in the valence shell. The complexes containing these metal centers are, therefore, likely to be diamagnetic. Complexes containing a metal from an even column that has an odd oxidation state or complexes containing a metal from an odd column that has an even oxidation state will possess an odd number of electrons in the valence shell and will be paramagnetic.

1.2.3. Trends in the Properties of Transition Metals

The stability, basicity, and *d*-orbital energies vary from left to right and top to bottom in the transition metals in ways that can be different from the trends in stability, basicity, and valence orbital energies of main group elements. These properties have a large effect on reactivity. In most cases, these trends are not continuous throughout the transition series. Thus, it is often best to consider sections of the transition metals when drawing trends, rather than the entire set of these elements. Furthermore, the trends in properties one would draw when comparing the elements are often different from those one would draw when comparing a series of metal centers within complexes. Likewise, the trends in properties of the metals vary with oxidation state. Nevertheless, it is valuable to highlight some trends here, even if there will be exceptions noted later in the text.

1.2.3.1. Trends in Ionization Potentials

The energies of the *d*-orbitals can be estimated from ionization potentials. Figure 1.9 provides a graphical representation of the first and second ionization potentials of the group 4–10 transition metals. Although the trend is not completely continuous, these graphs show a general trend of increasing ionization potential from left to right in the transition series. This trend follows from the rule that the energies of the orbitals of a more electronegative element are lower than the energies of the same orbitals of a less electronegative element, and the trend in electronegativity can be traced to the higher effective nuclear charge from left to right in the transition series. Thus, the more easily oxidized, more basic, and more electron-rich of two metal centers with similar ligand sets and the same oxidation state will generally be the one that lies at left in the transition series or the "earlier metal." Thus, a Zr(II) complex is likely to be more easily oxidized, more basic, and more electron-rich than a Pd(II) complex.

The relative first and second ionization energies of first-, second-, and third-row transition metals vary substantially. For some groups, oxidation of the second-row metal requires less energy than oxidation of the first, while oxidation of the third-row metal requires more energy to oxidize than the second. This trend is reversed for some other groups and can be reversed for the second oxidation. At the same time, some general trends can be stated about the ability of first-, second-, and third-row metals to form high oxidation state complexes. For the middle to late transition metals, the accessibility of oxidation states higher than +3 is greater for the third-row metals than the second, and is greater for the second-row metals than for the first. Thus, Pt(IV) compounds are often stable, while the generation of a Ni(IV) compound would be challenging. Likewise, Os(VIII) compounds are well known (e.g., OsO_4), while an Fe(VIII) compound would be challenging to access, and Ir(V)

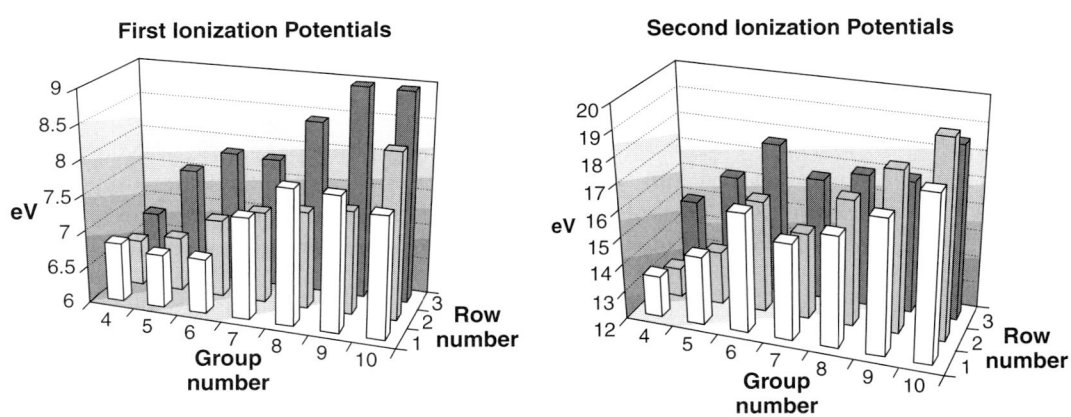

Figure 1.9.
Trends in ionization potentials for the transition metals in groups 4–10. Data from Douglas, B. E.; McDaniel, D. H.; Alexander, J. J. *Concepts and Models of Inorganic Chemistry*, 3rd ed.; Wiley: New York, 1994.

compounds are well known [e.g., Ir(O)Me$_3$], while Co(V) compounds would again be challenging to form.

Although the origins of the trends in acidity and basicity are currently being studied, in many cases the third-row metal complexes are more basic than the second, which are more basic than the first. As discussed in more detail in Chapter 3, Os(CO)$_4$(H)$_2$ is less acidic than Ru(CO)$_4$(H)$_2$, which is less acidic than Fe(CO)$_4$(H)$_2$, and CpW(CO)$_3$H is less acidic than CpMo(CO)$_3$H, which is less acidic than CpCr(CO)$_3$H.

1.2.3.2. Trends in Size

Trends in covalent radii are illustrated in Figure 1.10. These data show that the second-row metals are larger than their first-row congeners, but the third-row metals are about the same size as the second-row elements. The increased size of the second-row metals, relative to the first-row metals, can be attributed to the higher quantum number of the valence orbitals. This same argument can be used to rationalize the larger size of chlorine compared to fluorine or the larger size of phosphorus compared to nitrogen. However, the lanthanide elements fall in between the second- and third-row transition elements. Thus, the nuclear charges of the third-row elements are much higher than those of the second-row elements. This increased charge balances the effect of the increased quantum number and makes the third-row transition metals roughly the same size as the second-row metals. This effect is called the "lanthanide contraction."

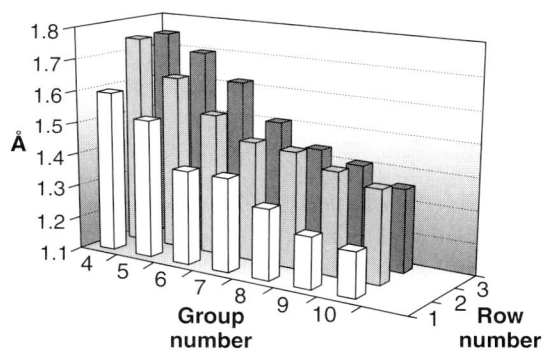

Figure 1.10.
Trends in covalent radii for the transition metals. Data from Cordero, B.; Gómez, V.; Platero-Prats, A. E.; Revés, M.; Echeverría, J.; Cremades, E.; Barragán, F.; Alvarez, S. *J. Chem. Soc., Dalton Trans.* **2008**, 2832.

For low-spin metal complexes of groups 4–7, with the same charge and oxidation state, the size decreases from left to right. This effect results from the greater nuclear charge of the elements on the right side of the periodic table and the lack of an increase in the principal quantum number across the transition series.

1.2.3.3. Trends in Bond Strengths

Metal–ligand bonds tend to be stronger in third-row metal complexes than in second-row metal complexes, and metal–ligand bonds in second-row metal complexes tend to be stronger than those in first-row metal complexes. This trend is the *opposite* of the trend in bond strengths among molecules containing main group elements (e.g., the carbon–iodine bond in methyl iodide is weaker than the carbon–chlorine bond in methyl chloride). The origin of this trend in bond strengths for the transition series is complex because of the large variation in properties of the transition metals. Both ionic and covalent interactions contribute to bond strengths. Greater orbital overlap and a smaller energy difference between overlapping orbitals increase the covalent contribution to the bond strength. The increasing bond strengths from first to second to third row can be attributed, at least in part, to a greater spatial overlap of the metal with the ligand orbitals. This increase can also be attributed to a closer match of the energy of the ligand orbitals with those of the third-row metals than of the second-row metals, as reflected by the respective first ionization potentials. The effect of the orbital energies of the first- and second-row metals on bond strength is less clear because of the variations in relative ionization potentials for the first- and second-row metals across the transition series.

1.3. Metal–Ligand Complexes

1.3.1. Electron Counting

In contrast to the formalisms of oxidation state and ligand charge, which assign electrons and charges to individual atoms in a manner that is largely arbitrary, the number of valence electrons on the metal complex, often termed the "electron count," is a true, measurable quantity. This number of electrons is crucial to understanding and predicting the properties and reactivity of organometallic complexes. By the first ligand classification system, the total number of valence electrons on the complex equals the sum of the d-electron count and the total number of electrons donated by each ligand. By the second classification system, the total number of electrons equals the sum of the number of valence electrons in the neutral metal and the number of electrons donated by all even- and odd-electron ligands. The number of d-electrons predicted by the two electron counting systems, if followed properly, will be the same. The two systems simply divide the same number of total electrons in different ways between the metal and ligand.

Determination of the number of total valence electrons on an organometallic complex is simplest when following the second classification system in which all ligands are considered neutral and will, therefore, be presented first. By this system, the number of total electrons can be determined by the formula:

Total valence electrons = Metal group + Electrons donated by all even-
and odd-electron ligands − Overall charge on the complex

To follow this system, simply determine the number of electrons on the neutral version of each ligand.

The second classification system has the advantage of revealing the oxidation state as part of the determination of the total number of electrons. The reader should realize, however, that oxidation state and total number of electrons are not linked. Both the system with charges and the system without charges lead to the same total number of electrons.

The relationship between the charge of the complex, oxidation state, and assigned charges of ligands can be written as

Charge of the complex = Oxidation state + Charge of the ligands

As noted earlier in this chapter, the relationship between the number of d-electrons and the oxidation state can be written as

Number of d-electrons = Group number − Oxidation state of the metal

By the first ligand classification system that includes charges, the total number of valence electrons is then equal to the number of d-electrons plus the total number of electrons donated by both neutral and charged ligands.

One procedure to determine the total electron count with ligands assigned formal charges includes the following steps:

1. Determine the total charge of the ligand set and the number of electrons donated by the ligands.
2. Determine the oxidation state of the metal.
3. Determine the d-electron count of the metal.
4. Determine the total number of electrons by summing the number of d-electrons and the total number of electrons donated by the ligands.

A few examples of the counting of electrons illustrate these formalisms.

EXAMPLE

$CpFe(CO)_2(C_2H_5)$

By the neutral system, this complex contains
1. A metal from group 8
2. Two two-electron CO ligands
3. A five-electron Cp ligand
4. A one-electron C_2H_5 (ethyl) ligand

The total number of electrons is then $8 + (2 \times 2) + 5 + 1 = 18$.

By the charged system, one can determine the total number of electrons in $CpFe(CO)_2(C_2H_5)$ by the following system:
1. This complex contains the following types of ligands:
   - Two CO ligands: Neutral, two electrons each = four electrons
   - Cp: Anionic, six-electron donor
   - C_2H_5: Anionic, two-electron donor
   - Total: 2– charge, 12 electrons donated
2. Oxidation state: The complex is neutral overall and possesses two anionic ligands. The metal, therefore, has an oxidation state of 2+.
3. d-Electron count: Iron is in group 8. With a 2+ charge, the metal is d^6.
4. Total number of electrons: 12 electrons from the ligands + six d-electrons = 18 total electrons.

EXAMPLE

$[CpMo(CO)_3(H_2O)]^+$

By the neutral system, this complex contains
1. A metal from group 6
2. Three two-electron CO ligands
3. One two-electron OH_2 ligand
4. One five-electron Cp ligand

The total number of electrons is then $6 + (3 \times 2) + 2 + 5 = 19 - 1$ overall positive charge = 18 electrons.

By the charged system, one can determine the total number of electrons in $[CpMo(CO)_3(H_2O)]^+$ by the following system:
1. This complex contains the following types of ligands:
   - Three CO ligands: Neutral, two electrons each = six electrons
   - OH_2: Netural, two-electron ligand
   - Cp: Anionic, six-electron donor
   - Total: 14 electrons donated
2. Oxidation state: The complex is a cation overall and possesses one anionic ligand. The metal, therefore, has an oxidation state of 2+.
3. d-Electron count: Molybdenum is in group 6. With a 2+ charge, the metal is d^4.
4. Total number of electrons: 14 electrons from the ligands + four d-electrons = 18 total electrons.

EXAMPLE

$Ru(NO)(PPh_3)_2Cl_3$

By the neutral system, this complex contains
1. A metal from group 8
2. Two two-electron PPh_3 ligands
3. One three-electron linear NO ligand
4. Three one-electron Cl ligands

The total number of electrons is then $8 + (2 \times 2) + 3 + 3 = 18$ electrons.

By the charged system, one can determine the total number of electrons in $Ru(NO)$ $(PPh_3)_2Cl_3$ by the following system:
1. This complex contains the following types of ligands:
   - Two PPh_3 ligands: Neutral, two-electron ligands
   - One NO ligand: Cationic, two-electron ligand
   - Three Cl ligands: Anionic two-electron donors
   - Total: 12 electrons donated
2. Oxidation state: The complex is overall neutral and possesses three anionic ligands and one cationic ligand. The metal, therefore, has an oxidation state of 2+.
3. d-Electron count: Ruthenium is in group 8. With a 2+ charge, the metal is d^6.
4. Total number of electrons: 12 electrons from the ligand + six d-electrons = 18 total electrons.

EXAMPLE

$(Et_3P)_2Pt(n-Bu)_2$

By the neutral system, this complex contains
1. A metal from group 10
2. Two two-electron PEt_3 ligands
3. Two one-electron Bu ligands

The total number of electrons is then $10 + (2 \times 2) + 2 = 16$ electrons.

By the charged system, one can determine the total number of electrons in $(Et_3P)_2Pt(n-Bu)_2$ by the following system:
1. This complex contains the following types of ligands:
   - Two PEt_3 ligands: Neutral, two-electron ligands
   - Two alkyl groups: Anionic, two-electron ligands
   - Total: eight electrons donated
2. Oxidation state: The complex is overall neutral and possesses two anionic ligands. The metal, therefore, has an oxidation state of 2+.
3. d-Electron count: Platinum is in group 10. With a 2+ charge, the metal is d^8.
4. Total number of electrons: eight electrons from the ligand + eight d-electrons = 16 total electrons.

1.3.2. The 18-Electron Rule

Stable, diamagnetic, mononuclear organotransition metal complexes almost always contain 18 or fewer valence electrons. This trend is known as the 18-electron rule (sometimes called the effective atomic number rule). This rule is largely empirical, and the often-quoted origin of this rule has been scrutinized recently. It has often been stated that transition metal complexes tend to adopt 18-electron configurations because of the number of valence orbitals on the metal, just as the eight-electron rule for organic compounds originates from the number of valence orbitals on carbon (see Figure 1.10). A transition metal has nine valence orbitals—five nd orbitals (n is the principal quantum number), three $(n - 1)p$ orbitals, and one $(n + 1)s$ orbital—while a carbon atom has one s- and three p-orbitals in its valence shell. However, modern theoretical studies have implied that the p-orbitals on the metal do not participate significantly in metal–ligand bonding.[15–19] Thus, we will consider this "rule" to be an empirical trend, until further analysis provides a clear origin.

1.3.3. Metal–Metal Bonding and Electron Counting in Polynuclear Complexes

The maximum number of formal metal–metal bonds in polynuclear complexes has often been determined from the total number of valence electrons. This counting scheme assumes that each metal will adopt at maximum an 18-electron configuration and that a metal–metal bond is a shared pair that contributes to the 18-electron configuration of both metals. As noted below, these restrictions do not apply to metal polynuclear complexes. However, when these restrictions do apply, the number of metal–metal bonds is 18 × (the number of metals) − (the total number of actual valence electrons), and the number of formal metal–metal bonds (shared pairs) is given by Equation 1.1.

$$\text{Number of M--M bonds} = \frac{18 \bullet M - N}{2}$$

M = number of metals
N = total number of valence electrons

(1.1)

This equation does not predict the appropriate number of metal–metal bonds when the compound contains two-electron three-center bonds. Many complexes containing more than one metal contain bridging hydride or alkyl ligands that participate in this type of bond. In these cases, the simple formula in Equation 1.1 typically overestimates the M–M bond orders.

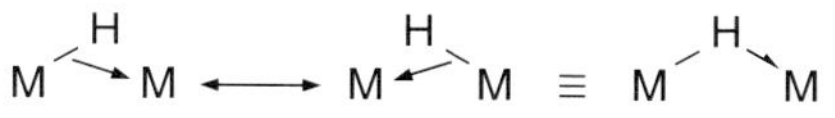

$$M \overset{H}{\diagup\!\!\diagdown} M \longleftrightarrow M \overset{H}{\diagup\!\!\diagdown} M \equiv M \overset{H}{\diagup\!\!\diagdown} M$$

Figure 1.11.
Half-arrow description for a bridging hydride ligand.

For complexes with bridging hydrides or other types of two-electron three-center bonds, the use of the half-arrow notation for bridging ligands shown in Figure 1.11 provides the best correlation between predicted M–M bond order and calculated M–M bond order from quantum mechanics using density functional theory.[14] The half-arrow notation was described by Green[2] and is one way to depict a two-electron three-center bonding situation. This depiction takes into account the two equal resonance structures of a symmetric bridging hydride that donates electron density to both metals. In one resonance structure the hydride is bound as an X-type ligand to one metal and the M–H bond acts as an L-type ligand to the second metal. This combination of electron donation can be considered to cause the hydride to act as a "three-electron" ligand to the overall complex in the neutral counting scheme or a "four-electron" ligand to the overall complex in the ionic counting scheme.

Three examples of counting electrons in complexes containing M–M bonds are shown in Figure 1.12. $Co_2(CO)_8$ possesses eight carbonyl ligands and two cobalt atoms that are formally Co(0) and contribute nine electrons each. The total number of valence electrons is, therefore, 34, and the number of formal Co–Co bonds is one. In $[Re(CO)_4(\mu\text{-Cl})]_2$, which has two bridging chlorides, the total number of valence electrons is 36 (note from Table 1.1 that

$$\frac{18 \cdot 2 - 34}{2} = 1$$

$$\frac{18 \cdot 2 - 36}{2} = 0$$

Neutral, half-arrow method:
[18 · 3 − (3 · 8) − (10 · 2) −
(2 · 3)]/2
= [54 − 24 − 20 −6]/2
= 4/2 = 2 M–M bonds

Figure 1.12. Determination of the number of metal–metal bonds by electron counting.

bridging halides are four-electron donors). Thus, this complex contains no formal Re–Re bond.

The best description of $Os_3(CO)_{10}(\mu\text{-H})_2$ has been the subject of much written and verbal debate.[14] The total number of electrons from the three osmium centers by the neutral, half-arrow counting scheme is 8•3 = 24. The number of electrons donated from the carbonyl ligands is 10•2 = 20. The number of electrons from the bridging hydrides by the "half-arrow" method is 2•3 = 6. Thus, the total number of electrons from the ligands and the metal equals 50, and four electrons remain to be used in Os–Os bonds. This method then leads to the appropriate conclusion that the complex contains two Os–Os bonds and does not contain an osmium–osmium bond between the two metals bridged by the hydrides. If the hydride is treated as a one-electron ligand, the counting scheme of Equation 1.1 predicts an Os=Os bond between the two metals bridged by the hydrides, and the appropriate orbitals for such a multiple bond are not present.[14]

Metal–metal bonds in polynuclear complexes containing more than four metals are not appropriately described as localized bonds. Wade[15] and Mingos[16,17] have developed a scheme (supported by theory[18–23]) for predicting polyhedral shapes by counting skeletal electron pairs and Teo has described a topological electron-counting scheme.[24,25] These schemes are beyond the scope of this book.

1.3.4. Geometries of Transition Metal Complexes

Transition metal complexes can adopt many geometries. The number of valence electrons and the formal d-electron configuration can allow one to predict with frequent success the geometry of organometallic complexes. The possible geometries for three-coordinate to six-coordinate compounds are shown in Figure 1.13. The sterically preferred geometry for a three-coordinate compound is trigonal, for a four-coordinate complex is tetrahedral, for a five-coordinate complex is trigonal bipyramidal, and for a six-coordinate complex is octahedral. These steric preferences can be appreciated by comparing the number of interactions between ligands with the smallest bond angles. This approach mimics the way one would predict the sterically most favorable conformation for a main group compound that lacks electron pairs. Assigning a coordination number to many organometallic compounds is not straightforward because some ligands, such as an allyl or cyclopentadienyl ligand, are bound to the metal by many atoms. IUPAC defines coordination number as the number of atoms directly linked to the specified atom. However, the properties of organometallic compounds are sometimes better described by considering the coordination number as the number of electron pairs donated to the metal (Table 1.1) and not by the number of atoms attached to the metal. By the latter convention, the coordination number of Ru(NO)(PPh$_3$)$_2$Cl, CpFe(CO)$_2$C$_2$H$_5$, and CpCr(NO)(CO)(C$_2$H$_2$) is six, while the coordination number of [CpMo(CO)$_3$(H$_2$O)]$^+$ is seven. In many cases, the geometry favored by steric effects and the geometry favored by electronic effects are different. Often, the electronic effects will override the steric effects, and an arrangement of ligands that is sterically unfavorable

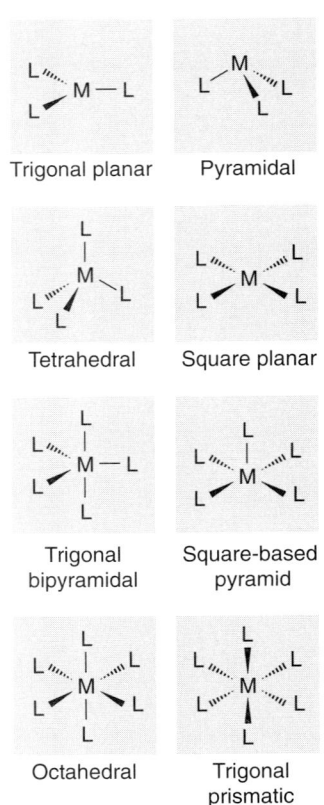

Trigonal planar   Pyramidal

Tetrahedral   Square planar

Trigonal bipyramidal   Square-based pyramid

Octahedral   Trigonal prismatic

Figure 1.13.
Common geometries of transition metal complexes.

is observed. The electronic preference for one geometry over another results from the total energy of the filled valence orbitals (and typically the energy of the highest occupied molecular orbital, HOMO) of one geometry being lower than the total energy (and typically the energy of the HOMO) of these orbitals of other possible geometries.

Thus, one must be able to predict which complexes will adopt the sterically preferred geometry and which will have a strong electronic preference for a geometry that is less favored by steric effects. Many transition metal complexes contain nonbonding electron pairs, but these electron pairs do not affect the geometry like an electron pair in a main group compound. Instead, the geometries of transition metal complexes possessing d-electrons are best predicted by a diagram of the energies of the d-orbitals. This diagram can be generated by the crystal-field analysis commonly used in coordination chemistry or a molecular orbital diagram that can be found in several texts.

Certain d-electron configurations cause transition metal complexes to adopt particular geometries for electronic reasons. One should commit to memory that four-coordinate d^8 complexes of second- and third-row metals are almost always square planar. The origin of this effect should be familiar to those who have studied coordination chemistry: a square planar geometry causes one orbital, the $d_{x^2-y^2}$ orbital, to be high in energy, and the other four d-orbitals to be lower in energy. The eight d-electrons then fill the lower energy orbitals.

The trend that four-coordinate d^8 complexes are square planar is often violated by first-row transition metal centers. The electronic preference for a square planar geometry is smaller for the first-row metals than for second- and third-row metals, and first-row metal centers often adopt a geometry with its four ligands in an arrangement that is closer to tetrahedral. For example, $Ni(PPh_3)_2Cl_2$ adopts a geometry closer to tetrahedral than square planar. However, most organometallic compounds possess some ligands that are strongly electron donating. This strong electron donation creates strong electronic preferences for one geometry over the other. Thus, $Ni(PPh_3)_2Me_2$ is square planar. It is also helpful for understanding the stereochemistry of several reactions discussed in this text that five-coordinate d^8 complexes tend to adopt trigonal bipyramidal geometries, while five-coordinate d^6 complexes tend to adopt square-based pyramidal geometries for electronic reasons.

1.3.5. Isoelectronic and Isolobal Analogies

The models we will use to describe more precisely the bonding of ligands in Chapter 2 draw upon orbital interaction diagrams that include the orbitals of the ligands involved in bonding with the metal and the highest occupied molecular orbital (HOMO) and lowest unoccupied molecular orbital (LUMO) of the metal. This approach is similar to that used in organic chemistry to describe, for example, the reaction of a carbene or boron hydride with an olefin to generate a cyclopropane or alkylborane.

To allow the use of a small number of orbital diagrams to describe a large variety of metal–ligand interactions, we often categorize the metal fragments into groups that have the same structure and number of electrons. The members of these groups are said to be "isoelectronic." Following Hoffmann's lead, organometallic chemists often group compounds into classes of molecules with the same number, symmetry, approximate energy, and shape of the frontier orbitals. Frontier orbitals are the higher occupied and lower unoccupied molecular orbitals. The members of the groups of molecules with these similar orbital properties are said to be "isolobal."

$[V(CO)_6]^-$, $Cr(CO)_6$, and $[Mn(CO)_6]^+$ are "isoelectronic" because they are all d^6 octahedral complexes. Even though the ligands on $Ni(CO)_4$, $Co(NO)(CO)_3$, and $Fe(NO)_2(CO)_2$ are different, these molecules are considered to be isoelectronic because the metal centers possess 10 d-electrons, similar ligands, and similar tetrahedral arrangements of the ligands. Many chemists would also consider $CpMn(CO)_3$ and $[CpRu(CO)_3]^+$ to be "isoelectronic," even though the metals lie in different rows of the periodic table.

The analogy of common fragments of main group elements in organic chemistry to common fragments of transition metal complexes in organometallic chemistry is provided in

Table 1.2. Isoelectronic relationships between organic and transition metal fragments.

Neutral hydrocarbon fragment	CH_4	CH_3	CH_2	CH	C
Charged hydrocarbon fragments related by $\pm$ H^+	CH_3^-	CH_2^-	CH^- CH_3^+	CH_2^+	CH^+
Common isolobal inorganic fragments containing metals of the first transition series	$Cr(CO)_6$ $Fe(CO)_5$ $Ni(CO)_4$	$Mn(CO)_5$ $CpFe(CO)_2$ $Co(CO)_4$	$Cr(CO)_5$ $Fe(CO)_4$ $CpCo(CO)$ $Ni(CO)_3$ $CpCu$ $Ni(CO)_2$	$CpCr(CO)_2$ $Mn(CO)_4$ $CpFe(CO)$ $Co(CO)_3$ $CpNi$	$Cr(CO)_4$ $CpMn(CO)$ $Fe(CO)_3$ $CpCo$

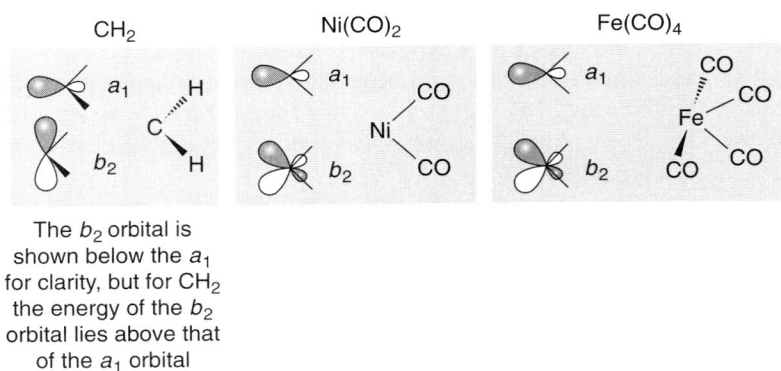

The b_2 orbital is shown below the a_1 for clarity, but for CH_2 the energy of the b_2 orbital lies above that of the a_1 orbital

Figure 1.14. An isolobal analogy of transition metal and organic fragments.

Table 1.2. The table lists only common cyclopentadienyl- and carbonyl-containing fragments of the first-row elements, but the entries in this table can be extended to the second- and third-row congeners in a straightforward manner. As shown graphically for one example in Figure 1.14, the frontier orbitals of $Fe(CO)_4$ and $Ni(CO)_2$ are similar to those of methylene, CH_2. All have one frontier orbital of a_1 symmetry that can participate in a σ-bond and one of b_2 symmetry that can form a π-bond. Although the order of the energies of the orbitals in a carbene and in the metal fragments is reversed, all three fragments are isolobal.

A similar examination of the frontier orbitals of $Mn(CO)_5$, which consist of a single a_1 orbital with a single electron, shows that this fragment is isolobal with $CH_3 \bullet$. Removal of an electron from both species leads to the isolobal fragments $[Mn(CO)_5]^+$ and CH_3^+, while addition of an electron to both species creates the isolobal fragments $[Mn(CO)_5]^-$ and CH_3^-.

An isolobal relationship can also be drawn between metal fragments with apparently distinct ligand sets because the orbitals of seemingly unrelated ligands can be similar. As a particularly important example, the frontier orbitals of Cp^- (the η^5-cyclopentadienyl ligand)

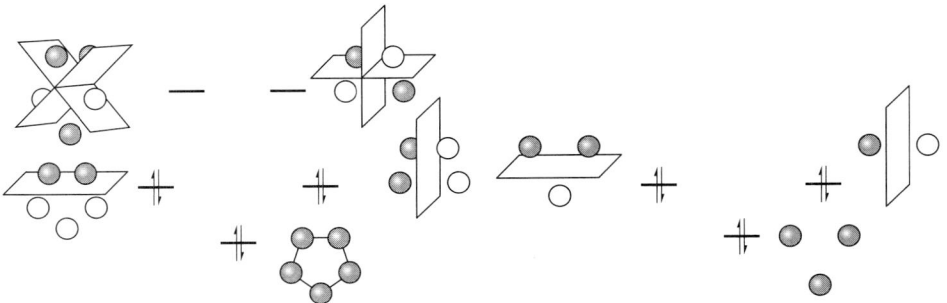

Figure 1.15. The top lobes of the π-orbitals of Cp^- vs. three dative ligands.

are isolobal with three mutually cis carbonyl ligands, as shown in Figure 1.15. Thus, $Mn(CO)_5$ is isolobal with $[CpMn(CO)_2]^-$ and $CpFe(CO)_2$. Similarly, $[CpFe(CO)]^-$ and $CpCo(CO)$ are isolobal with $Fe(CO)_4$ and CH_2, while $CpCr(CO)_2$ is isolobal with $[Cr(CO)_5]^+$ and CH_3^+.

The relationship between two organic fragments, CH and CH^+, and transition metal fragments are considered as the final example. The methylidyne fragment, CH, is isolobal with fragments such as $Co(CO)_3$ and CpNi, and the CH^+ fragment is isolobal with fragments such as $Fe(CO)_3$ and CpCo.

Consideration of these isolobal analogies presages much of the bonding of ligands that is described in Chapter 2. Methyl radical dimerizes to form ethane. Thus, one might expect that the bonding between a methyl group and $Mn(CO)_5$ will involve orbital interactions similar to those in the C–C bond in ethane. Furthermore, carbenes dimerize to form one σ- and one π-bond in ethene. Thus, one might expect carbenes to bind to the metal fragments that are isolobal with methylene by creating one metal–carbon σ-bond and one metal–carbon π-bond. These predictions are correct, and the discussion in Chapter 2 draws upon these relationships to describe the various types of common ligands in organometallic chemistry and how they bind to metal centers.

1.3.6. Molecular Orbitals for Transition Metal Complexes

A molecular orbital diagram for a transition metal complex can be generated from the orbitals of the metal and the symmetry-adapted linear combinations (SALCs) of the orbitals of the ligands. The SALCs are typically illustrated on one side of the diagram, the orbitals of the metal on the other side, and the molecular orbitals that result from combining the

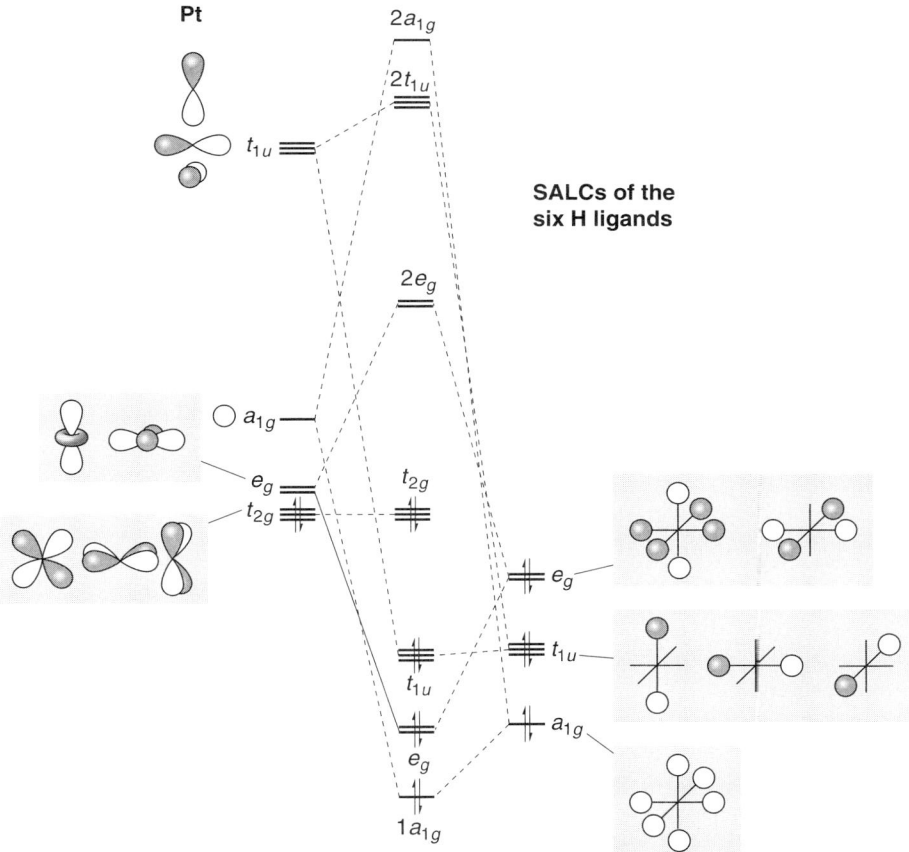

Figure 1.16.
Molecular orbital diagram for PtH_6^{2-}, an octahedral compound with ligands that are pure σ-donors.

orbitals of the SALCs and the metal with the same symmetry are typically displayed in the middle. Figure 1.16 shows a molecular orbital diagram for a simple octahedral compound containing ligands that are pure σ-donors. The SALCs generated from the six σ-bonding orbitals in an octahedral arrangement are shown at the right. The symmetry labels from group theory for these linear combinations are included.[26] The six σ-bonding ligands generate SALCs of e_g, t_{1u}, and a_{1g} symmetries. In an octahedral geometry, the symmetry labels of the metal valence orbitals are t_{1u} for the $(n + 1)p$ orbitals, a_{1g} for the $(n + 1)s$ orbitals, and t_{2g} for three of the nd orbitals, and e_g for the other two metal nd orbitals.

Metal and ligand orbitals of the same symmetry mix to form bonding molecular orbitals that are lower in energy than the isolated metal or ligand orbitals and antibonding molecular orbitals that are higher in energy than the isolated metal or ligand orbitals. For an octahedral complex, ligand–metal σ-interactions generate six bonding and six antibonding molecular orbitals. The order of the molecular orbitals given in Figure 1.16 is that of PtH_6^{2-}.

The nonbonding and antibonding molecular orbitals with high d-orbital character more often control structure and reactivity than the metal–ligand bonding orbitals. No SALC of the ligand orbitals possesses t_{2g} symmetry. Therefore, the metal nd orbitals of t_{2g} symmetry do not generate bonding and antibonding orbitals with any σ-bonding orbitals of the ligand and are nonbonding. The metal nd orbitals of e_g symmetry do form bonding and antibonding orbitals with the ligand group orbitals. Two of the metal–ligand bonding orbitals are generated from the nd e_g orbitals, and two of the most important metal–ligand antibonding orbitals are generated from the nd e_g orbitals. This antibonding orbital is the LUMO of the metal center.

Eighteen electrons fill the bonding molecular orbitals of e_g, a_{1g}, and t_{1u} symmetries, along with the nonbonding t_{2g} orbitals. For such complexes, the HOMO of the complex is the t_{2g} orbital, and the LUMO is the antibonding e_g orbital. Even though the antibonding e_g orbital is a molecular orbital constructed from the ligand and metal orbitals, the t_{2g} and e_g orbitals are generally considered to be the "d-orbitals" of the metal complex. The antibonding e_g orbital bears a higher fraction of d-orbital character than ligand orbital character because the energy of the metal orbitals commonly is higher than that of the ligand orbitals. For a complex of any symmetry, the bonding and (metal-based) nonbonding molecular orbitals can always accommodate exactly 18 electrons, because the nine valence orbitals of the metal give rise to either a bonding or a nonbonding molecular orbital. To accommodate more than 18 electrons, each electron in excess of 18 must occupy an antibonding molecular orbital, and the high energy created by filling the antibonding orbital has classically been considered to be the origin of the 18-electron rule.

Valence bond theory involving hybrid orbitals can also predict geometries of transition metal complexes, and recent work has illustrated how this theory can predict some initially unexpected structures.[27–35] By this theory, the hybrid orbitals are constructed from the s- and d-orbitals. The p-orbitals are *not* used because they are much higher in energy than the s- and d-orbitals, which are close to each other in energy. The hybrid orbital is generated from the s-orbital and $n - 1$ d-orbitals when there are n metal–ligand bonds. Electron pairs occupy purely d-orbitals. When the number of valence orbitals exceeds 12, some M–L bonds are considered as delocalized three-center four-electron bonding units. The L–M–L angles that are electronically preferred for the sd^n hybrid orbitals are provided in Table 1.3.

Two examples illustrate this approach to predicting structure. By this theory, W–Me bonds of the 12-electron compound WMe_6 comprise six sd^5 orbitals with optimal 63° and 117° ligand–metal–ligand (L–M–L) angles (Figure 1.17).[31,33] Therefore, a geometry other than octahedral would be predicted. Two C_{3v} geometries with these L–M–L angles are shown in Figure 1.17, and the observed geometry corresponds to the less sterically constrained structure shown on the left of the figure. In contrast, the 18-electron complex PtH_6^{2-} would be considered to contain three lone pairs of electrons in pure d-orbitals, leaving two d-orbitals to create three sd^2 hybrid orbitals. The optimal L–M–L angle for the sd^2 hybrid orbitals is 90°. Thus, the six M–L bonds would consist of three pairs of three-center two-electron bonds oriented 90° from each other, and the observed octahedral structure would be predicted.[31,33]

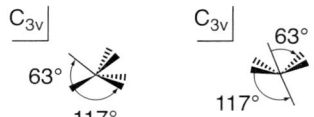

Figure 1.17.
Two structures of WMe_6 predicted using valence bond theory with sd^5 hybrid orbitals.[31,33]

Table 1.3. Favored L−M−L angles predicted for different sd^n hybrid orbitals.

Hybrid orbital	Favored L–M–L angles (°)
sd	90
sd^2	90
sd^3	71, 109
sd^4	66, 114
sd^5	63, 117
d	55, 125

Although most known organometallic compounds are diamagnetic, some possess unpaired electrons and are paramagnetic. Complexes that violate the 18-electron rule are often paramagnetic. Molecular orbital theory predicts that additional electrons in an octahedral complex would occupy the degenerate e_g^* orbitals. Thus, an octahedral 20-electron compound would be paramagnetic, even though it possesses an even number of electrons, and would adopt a triplet ground state. For example, the unusual 20-electron complex nickelocene (Cp_2Ni) is paramagnetic. Complexes with other even numbers of valence electrons can also be paramagnetic if the geometry is appropriate. For example, d^8 tetrahedral or d^6 trigonal bipyramidal complexes possessing 16 valence electrons are paramagnetic. All complexes with odd numbers of valence electrons are paramagnetic.

1.3.7. π-Bonding in Organotransition Metal Complexes

1.3.7.1. π-Bonding of CO and its Analogs

Many ligands in organometallic chemistry and coordination chemistry act not only as σ-donors but also as π-acceptors or π-donors or both. The bonding of ligands that act as π-acceptors is presented first. The π-accepting interaction is extremely important for the stabilization of complexes in low formal oxidation states. The most strongly π-accepting ligands are carbon monoxide and its close relatives linear nitrosyl (NO^+ in Table 1.1) and the isocyanides (:CNR).

Many of the ligands that are involved in π-bonding with transition metals have unoccupied π*-orbitals. These π*-orbitals overlap with filled metal d-orbitals to generate a lower energy filled bonding orbital and a higher energy unoccupied antibonding orbital, as shown for a carbonyl ligand in Figure 1.18. Because the filled bonding molecular orbital contains some carbonyl π* character, the interaction of the metal and the ligand π*-orbital leads to the delocalization of electron density from the metal onto the carbonyl ligand. This delocalization decreases the C–O bond order and more than compensates for the donation of electron density to the metal by the lone pair on carbon.

This donation of electron density from the metal to the π*-orbital is known as "backbonding." There is much physical evidence for backbonding, and some of these data are presented in Chapter 2. In brief, the C–O distance in carbonyl ligands is longer than that

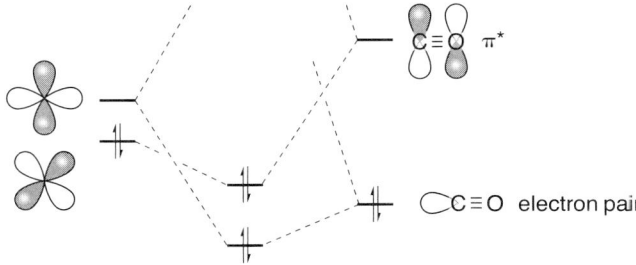

Figure 1.18.
The orbitals involved in the π-bonding of a carbonyl ligand.

in free CO, and the infrared stretching frequency for the carbonyl group is lower than that in free CO. Both of these observations reflect the decreased C–O bond order that results from the backbonding interaction.

1.3.7.2. π-Bonding of Carbene and Carbyne Complexes

From the vantage point of the metal fragment, the symmetry of the frontier orbitals of CO, carbene (CR_2) and carbyne (CR) ligands is related. As shown on the right of the two molecular orbital diagrams in Figure 1.19, the carbene possesses an orbital of σ-symmetry and an orbital of π-symmetry, and the carbyne possesses an orbital of σ-symmetry and two orbitals of π-symmetry. The orbital of σ-symmetry is filled in a singlet carbene, which typically possesses an electronegative group bound to carbon through a heteroatom. The orbital of π-symmetry is unoccupied. Thus, the orbital on the carbene of σ-symmetry can act as a σ-donor, like the orbital corresponding to the electron pair on the carbon of carbon monoxide, while the orbital on the carbene of π-symmetry can act as a π-acceptor, like a π*-orbital of carbon monoxide. The orbital interactions between a metal complex and a carbyne ligand are similar, but two orthogonal *d*-orbitals interact with the two orthogonal *p*-orbitals of the CR unit. The carbyne ligand is often assigned to be trianionic. In this case, the orbitals with σ- and π-symmetry would all be filled, and these orbitals would interact with unoccupied metal *d*-orbitals, as shown on the right in Figure 1.19.

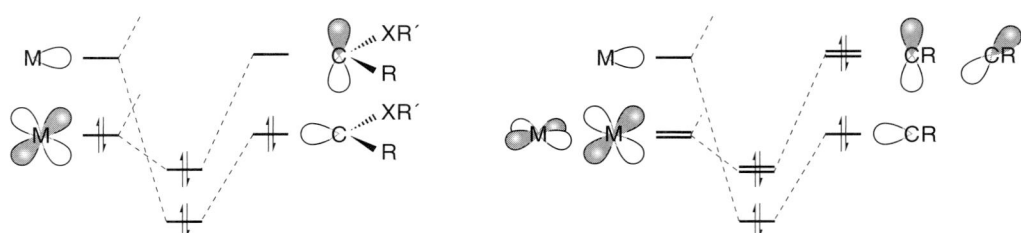

Figure 1.19. π-Bonding in Fischer carbene (left) and carbyne (right) complexes.

The charge on carbene ligands containing alkyl groups or hydrogens at the carbene carbon ("alkylidenes") is considered to be different from the charge on carbene ligands containing electronegative groups at the carbene carbon. For reasons based on reactivity trends discussed in more detail in Chapter 3, alkylidenes are considered to be dianionic, four-electron donors. Thus, both orbitals are filled in the free ligand fragment and they act as σ- and π-donors. Following this convention, Table 1.1 distinguishes between carbene ligands bearing at least one heteroatom or other substituent capable of π-interaction [e.g., C(OMe)Me, C(NMe_2)Me, and CPh_2] and carbene ligands lacking such substituents. By this convention, the two chromium complexes in Figure 1.20 are considered to be Cr(0) complexes, and the tantalum complex is considered to be a Ta(V) complex. This convention can be confusing for those beginning the study of organometallic chemistry, but this convention simply results from an assignment of electrons and charges to atoms. The total number of electrons involved in the metal–carbene bond is the same in all cases, and the true charge on the carbon of an alkylidene is far from −2.

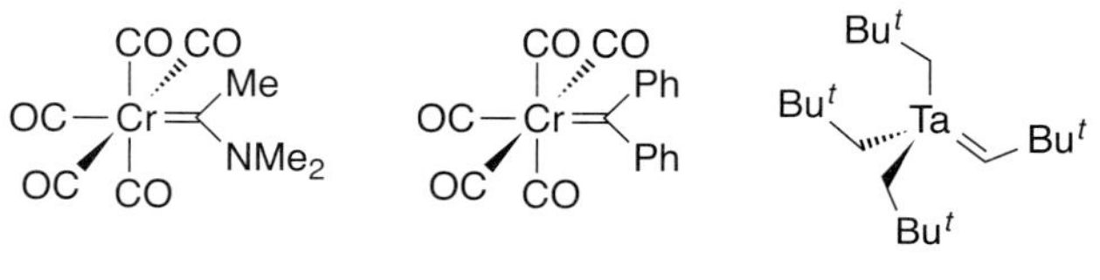

Figure 1.20.
Three carbene complexes of different classes.

1.3.7.3. π-Bonding in Olefin Complexes

Olefins and related unsaturated organic molecules that bind to the metal through their π-bonds constitute another important class of π-acceptor ligands. Recall from organic chemistry that acid-catalyzed additions of water to olefins are initiated by protonation of the olefin at the π-bond, that hydroborations are initiated by coordination of the unoccupied orbital of boron to the π-system, and that brominations of olefins proceed through bromonium ions formed by formal addition of Br^+ to an olefin. In contrast to the high reactivity of the intermediates formed by addition of electrophiles or Lewis acids to the π-system of an olefin, transition metal complexes of olefins are often stable. The stability of transition metal olefin complexes is due in large part to back donation of electron density from the metal d-orbitals to the π*-orbital of the olefin that is absent in the bonding of a proton or simple Lewis acid to an olefin.

The orbital interactions involved in a metal–olefin complex are shown in Figure 1.21. This description of metal–olefin bonding, first developed by Chatt, Dewar, and Duncanson,[12,36,37] involves both σ-donating and π-accepting interactions between the olefin ligand and the metal. One interaction involves donation of electron density from the π-bonding orbital of the olefin to an unoccupied orbital at the metal. Coordination to the face of the olefin creates one σ-bonding orbital because there is no nodal plane in the π-bonding orbital along this bond axis. The second interaction involves overlap of an occupied orbital at the metal with the π-antibonding orbital of the olefin. Instead of overlap of the metal with the π-system of CO through the p-orbital of one atom at the end of the ligand, the overlap occurs between the metal d-orbitals and the p-orbitals of two carbons at the face of the π-system. The face of the π*-orbital of an olefin has one nodal plane, and the orbital formed by interaction of the filled orbital at the metal with the π*-orbital of the olefin possesses a nodal plane along the bond axis. This interaction is, therefore, considered to be a π-bond.

Unlike carbonyl ligands, which are always net π-acceptors in complexes possessing d-electrons, olefin ligands are predominantly σ-donors when bound to some metals and predominantly π-acceptors when bound to others. Olefins are predominantly σ-donors in complexes in which the metal bears a significant positive charge due to an overall cationic charge of the complex or a high oxidation state of the metal. The C–C bond length of the coordinated double bond in such complexes differs little from that in the free olefin, and these types of complexes often react with nucleophiles at the olefin ligand, as presented in Chapter 11.

Other olefin complexes are stabilized considerably by donation of electron density from the metal d-orbital into the unoccupied π*-orbital of the olefin. Olefins are predominantly π-acceptors in complexes that possess electron-rich metal centers, such as those in low formal oxidation states. Like the interaction of the π*-orbital of CO with a filled metal d-orbital, the interaction of the π*-orbital of an olefin with a filled orbital at the metal leads to a flow of electron density into the π*-orbital of the ligand and reduces the bond order of the π-bond in the ligand.

Thus, the carbon–carbon bond in such a complex is longer than that in the free olefin. In addition, the hybridization at the olefinic carbons increases from sp^2 toward sp^3, and the substituents on the olefin are bent away from the metal. Such olefin ligands undergo electrophilic attack, as is discussed in Chapter 12.

In an alternative valence bond language, two resonance forms can be considered to describe metal–olefin complexes, as shown in Figure 1.22. An olefin bound to a mildly electron-rich or electron-poor metal center in which the σ-donation dominates the bonding will correspond predominantly to resonance form **A** in Figure 1.22. One example of such a complex is the cationic iron complex in Figure 1.22. In contrast, olefin complexes of strongly backbonding metals, such as those with d^2 or d^6 electron counts, will correspond

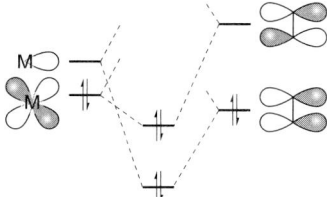

Figure 1.21.
Orbital interactions in the Chatt–Dewar–Duncanson bonding model of olefin binding.

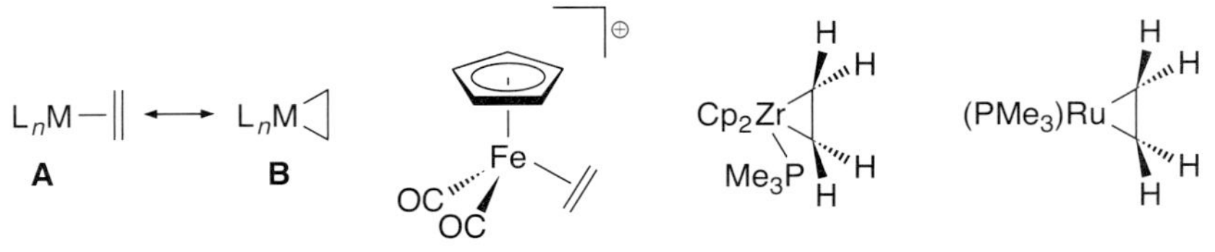

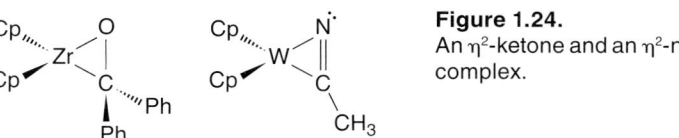

Figure 1.22.
Olefin complex and metallacyclopropane resonance structures and examples of complexes with predominantly olefin character and predominantly metallacyclopropane character.

predominantly to the metallacyclopropane structure **B**. Structure **B** results from strong enough π-backbonding into the π^*-orbital of the olefin to reduce the C–C bond order to nearly one. For example, the olefin complexes $Ru(PMe_3)_4(C_2H_4)$ and $Cp_2Zr(butene)(PMe_3)$ can be described as metallacyclopropanes (Figure 1.22). One can interpret both structure and reactivity with these two resonance forms.

1.3.7.4. π-Bonding with Other Unsaturated Ligands

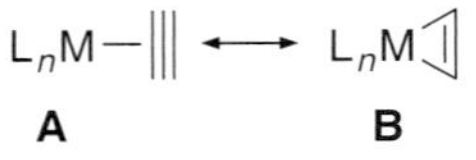

Metal–ligand bonds can be formed between a transition metal and other types of multiple bonds through a similar combination of σ-donor and π-acceptor interactions. For example, alkynes can bind a metal through similar interactions. A complex containing an alkyne that is a sufficiently strong π-acceptor can adopt a structure of a metallacyclopropene (Figure 1.23) with substituents on the alkyne carbon bent away from the metal. The barrier to rotation about the metal–acetylene bond is sometimes high, even when there is little steric effect on the orientation of the ligand.[38] This observation suggests that alkynes can act as strong π-acceptors.

Figure 1.23.
Alkyne complex and metallacyclopropene resonance forms.

Ketones, imines, and nitriles can coordinate to a metal by the same interactions. Ketones bind to hard transition metal complexes through an electron pair at oxygen, and imines and nitriles through the electron pair on nitrogen. However, they can also bind to softer metals that are capable of backbonding through the carbon–oxygen or carbon–nitrogen multiple bond. For example, the benzophenone complex in Figure 1.24 contains an η^2-benzophenone[39,41,49] ligand, and the structure can be described as an oxazirconacyclopropane resulting from strong backbonding from the zirconium(II) into the benzophenone. Likewise, the tungsten nitrile complex in Figure 1.24 contains a strongly backbonding W(II) center, and the nitrile is coordinated through the C–N multiple bond instead of the electron pair on nitrogen.

Figure 1.24.
An η^2-ketone and an η^2-nitrile complex.

1.3.8. π-Donor Ligands

Coordinated polyenes and arenes are a simple, but important, class of π-donor ligand. Although we considered an η^6-arene ligand to be a six-electron donor in Section 1.1, because it contains resonance structures with three double bonds, each of which are two-electron donors, the bonding of such ligands to the metal is more complex. These ligands are more accurately described by a combination of σ- and π-donation from the molecular orbitals of the entire ring. This combination of interactions is illustrated by the molecular orbitals of a sandwich compound, such as bis(benzene)chromium.

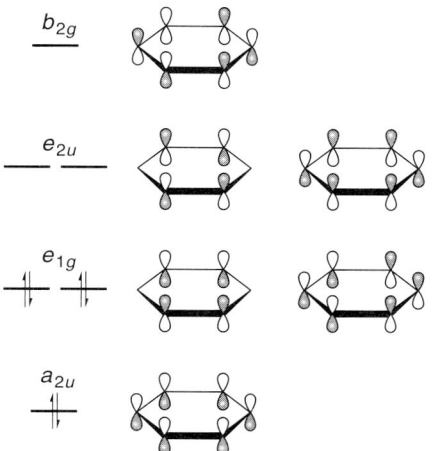

b_{2g}

e_{2u}

e_{1g}

a_{2u}

Figure 1.25.
Molecular orbitals for the π-system of benzene.

The molecular orbitals of the π-system of benzene are shown in Figure 1.25. The orbitals of the two benzene rings combine to generate the set of ligand orbitals shown in Figure 1.26. The lowest energy π-orbitals of the two benzene rings, each of a_{2u} symmetry, generate ligand orbitals of a_{1g} and a_{2u} symmetries. As shown in Figure 1.26, these are the difference and the sum of the individual benzene orbitals of a_{2u} symmetry. Similarly, the e_{1g} orbitals of the two benzene rings give rise to ligand orbitals of e_{1g} and e_{1u} symmetry.

The interactions of these ligand orbitals with the chromium orbitals possessing the appropriate symmetry give rise to the molecular orbitals of the overall complex. The orbitals of the metal and benzene rings that interact are shown in Figure 1.26. For example, the a_{2u} ligand orbital has the appropriate symmetry to interact with the p_z orbital of the metal, and the a_{1g} ligand orbital has the appropriate symmetry to interact with the s- and d_{z^2} orbitals of the metal. Similarly, the e_{1g} ligand orbital shown in Figure 1.26 has the appropriate symmetry to interact with the metal d_{xz} orbital, and the e_{1u} ligand orbital shown in Figure 1.26 has the appropriate symmetry to interact with the metal p_x orbital.

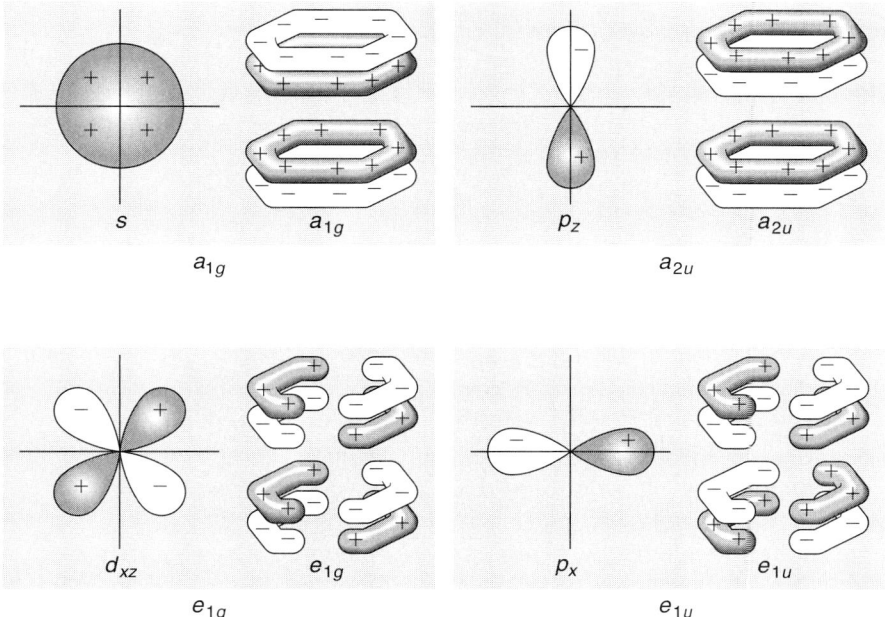

Figure 1.26.
The combinations of metal and ligand orbitals that form the molecular orbitals of bis(benzene)chromium.

The resulting molecular orbital diagram for bis(benzene)chromium is shown in Figure 1.27. The molecular orbital resulting from the combination of the a_{2u} ligand orbital and the Cr p_z orbital has σ-symmetry, and the orbital resulting from the a_{1g} ligand orbital and the Cr s- and d_{z^2} orbitals has σ-symmetry. However, the orbital resulting from the combination of the e_{1g} ligand orbital and the Cr d_{xz} orbital has π-symmetry, and the orbital resulting from the e_{1u} ligand orbital and the Cr p_x orbital also has π-symmetry. Because the six π-electrons of benzene fill the a_{2u} and e_{1g} orbitals shown in Figure 1.26, the e_{1g} and e_{1u} ligand orbitals of each benzene are filled and serve as π-donors to the chromium metal center.

Arenes also act as acceptors of electron density from the metal. The e_{2g} orbitals in Figure 1.27, which are generated from the empty e_{2u} orbitals of the individual benzenes, have the appropriate symmetry to overlap with the d_{xy} and $d_{x^2-y^2}$ orbitals of the metal. Because the e_{2g} orbitals of the ligand are unoccupied and the e_{2u} orbitals of the metal are filled, the e_{2g} orbitals can act as acceptors of electron density from the d_{xy} and $d_{x^2-y^2}$ orbitals. This interaction has δ-symmetry. Orbitals of δ-symmetry contain two nodal planes along the bond axis.

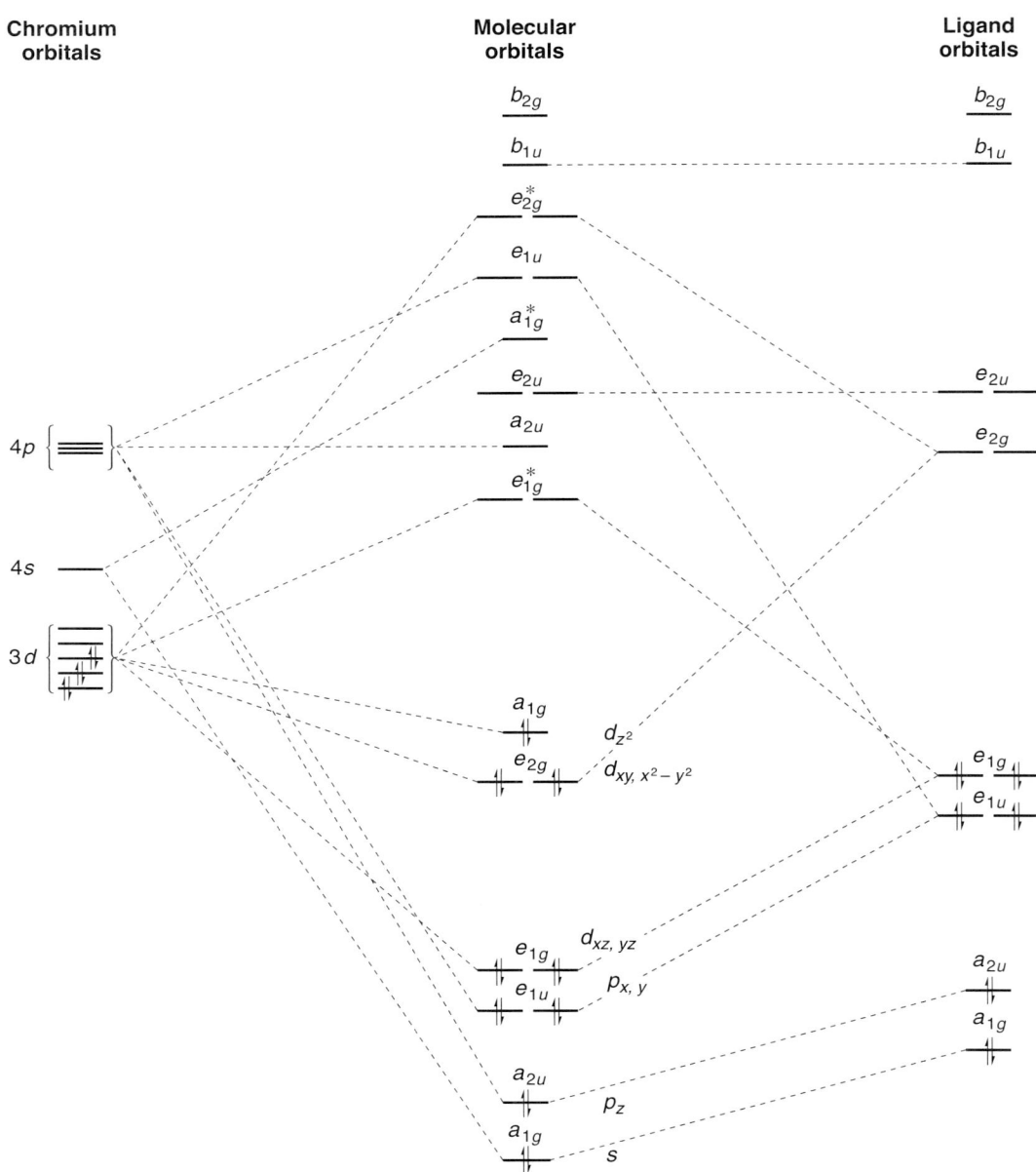

Figure 1.27. Molecular orbital diagram for bis(benzene)chromium.

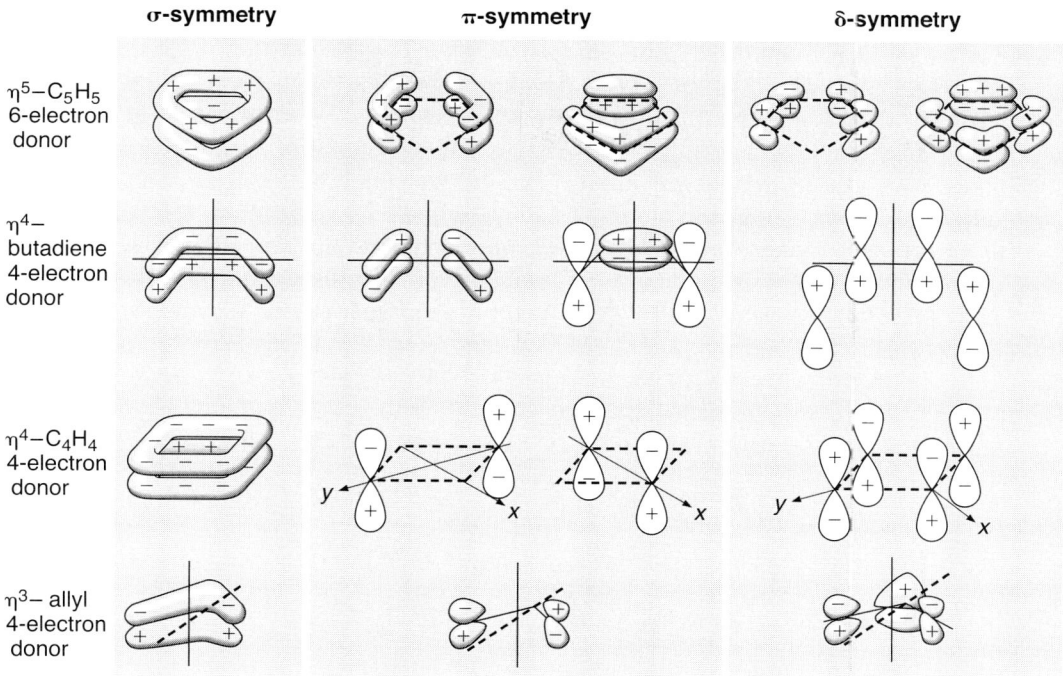

Figure 1.28.
Symmetries of the interactions between metal orbitals and the molecular orbitals of some cyclic and acyclic polyenes.

The π-molecular orbitals of many other cyclic and acyclic polyenes can also participate in σ-, π-, and δ-interactions with orbitals of the metal, as shown in Figure 1.28. In some of these cases, it is artificial to divide these overlaps into "donor" and "acceptor" interactions; the distinction depends on the formal charge we assign the ligands and on whether or not these molecular orbitals are populated in the free ligand as a result.

Some simpler ligands can also serve as π-donors. We saw earlier that acetylenes can act as σ-donors and π-acceptors in the same way as olefins. However, the π-bonding orbitals that are perpendicular to those oriented toward the metal can also serve as π-donors, as shown in Figure 1.29. Acetylene ligands in complexes that possess less than 18 electrons without such π-donation are sometimes considered to be "four-electron donors." This is the origin of the listing of alkynes as 2-electron or 4-electron ligands in Table 1.1.

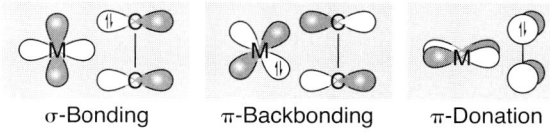

σ-Bonding   π-Backbonding   π-Donation

Figure 1.29.
σ-Donation, π-acceptance, and π-donation in metal–alkyne complexes.

Alkoxides and amides are also common π-donor ligands, as shown in Figure 1.30. As will be presented in more detail in Chapter 4, the M–O–C angle of alkoxide ligands can approach 180°, and dialkylamido ligands are usually planar at nitrogen. A linear geometry of an alkoxide ligand implies that the hybridization of the orbital of oxygen that binds to the metal is *sp* and that the two electron pairs can form π-bonds to the metal. The planarity of the amido ligand implies that the hybridization at nitrogen is sp^2 and that the electron pair can form a π-bond to the metal. Halide, oxide, and nitride ligands, which contain nonbonding electron pairs in the unbound ligand, are also good π-donors to transition metals.

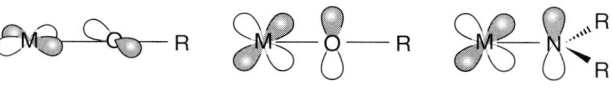

Figure 1.30.
π-Donation n alkoxides and amides.

References and Notes

1. Crabtree, R. H. *The Organometallic Chemistry of the Transition Metals*, 4th ed.; Wiley: New York, 2005.
2. Green, M. L. H. *J. Organomet. Chem.* **1995**, *500*, 127.
3. Some rare ligands, such as boranes, donate zero electrons, and have been termed "Z-type" ligands.
4. Nyholm, R. S. *Pure Appl. Chem.* **1968**, *17*, 1.
5. Pauling, L. *The Nature of the Chemical Bond*, 3rd ed.; Cornell University Press: Ithaca, NY, 1960.
6. Dudeney, N.; Kirchner, O. N.; Green, J. C.; Maitlis, P. M. *J. Chem. Soc., Dalton Trans.* **1984**, 1877.
7. (a) Cotton, F. A.; Wilkinson, G. *Advanced Inorganic Chemistry*, 4th ed.; Wiley-Interscience: New York, 1980; (b) Cotton, F. A. *Chemical Applications of Group Theory*, 2nd ed.; Wiley-Interscience: New York, 1971.
8. Fenske, R. F. *Prog. Inorg. Chem.* **1976**, *21*, 179.
9. Burdett, J. K. *Molecular Shapes: Theoretical Models of Inorganic Stereochemistry*; Wiley-Interscience: New York, 1980.
10. Hoffmann, R. *Science* **1981**, *211*, 995.
11. Hoffmann, R. *Angew. Chem., Int. Ed. Engl.* **1982**, *21*, 711.
12. Mingos, D. M. P. In *Comprehensive Organometallic Chemistry*; Wilkinson, G., Stone, F. G. A., Abel, E. W., Eds.; Pergamon Press: New York, 1982; Vol. 3, p 1.
13. Albright, T. A.; Burdett, J. K.; Whangbo, M.-H. *Orbital Interactions in Chemistry*; Wiley-Interscience: New York, 1985.
14. Baik, M. H.; Friesner, R. A.; Parkin, G. *Polyhedron* **2004**, *23*, 2879.
15. Wade, K. *Adv. Inorg. Chem. Radiochem.* **1976**, *18*, 1.
16. Mingos, D. M. P. *Acc. Chem. Res.* **1984**, *17*, 311.
17. Mingos, D. M. P. *J. Chem. Soc., Chem. Commun.* **1985**, 1352.
18. Stone, A. J. *Inorg. Chem.* **1981**, *20*, 563.
19. Fowler, P. W.; Porterfield, W. W. *Inorg. Chem.* **1985**, *24*, 3511.
20. Woolley, R. G. *Inorg. Chem.* **1985**, *24*, 3519.
21. Woolley, R. G. *Inorg. Chem.* **1985**, *24*, 3525.
22. Lauher, J. W. *J. Am. Chem. Soc.* **1978**, *100*, 5305.
23. Lauher, J. W. *J. Am. Chem. Soc.* **1979**, *101*, 2604.
24. Teo, B. K. *Inorg. Chem.* **1984**, *23*, 1251.
25. Teo, B. K.; Longoni, G.; Chung, F. R. K. *Inorg. Chem.* **1984**, *23*, 1257.
26. For the meaning of the Mulliken symbols, which give the symmetries for each orbital and the derivation of the symmetries for the case under consideration, see any of the standard texts (e.g., [8]) on the applications of group theory to chemistry. In this book, Mulliken symbols will be used only as labels. You need know only that orbitals with different labels have different symmetries and cannot combine to generate new molecular orbitals. Orbitals with *a* and *b* symmetries are nondegenerate and can accommodate only two electrons, whereas orbitals with *e* symmetries are doubly degenerate and can accommodate four electrons, and orbitals with *t* symmetries are triply degenerate and can accommodate six electrons.
27. Weinhold, F.; Landis, C. R. *Valency and Bonding: A Natural Bond Orbital Donor-Acceptor Perspective*; Cambridge University Press: Cambridge, U.K., 2005.
28. Root, D. M.; Landis, C. R.; Cleveland, T. *J. Am. Chem. Soc.* **1993**, *115*, 4201.
29. Landis, C. R.; Cleveland, T.; Firman, T. K. *J. Am. Chem. Soc.* **1995**, *117*, 1859.
30. Cleveland, T.; Landis, C. R. *J. Am. Chem. Soc.* **1996**, *118*, 6020.
31. Firman, T. K.; Landis, C. R. *J. Am. Chem. Soc.* **1998**, *120*, 12650.
32. Landis, C. R.; Cleveland, T.; Firman, T. K. *J. Am. Chem. Soc.* **1998**, *120*, 2641.
33. Landis, C. R.; Firman, T. K.; Root, D. M.; Cleveland, T. *J. Am. Chem. Soc.* **1998**, *120*, 1842.
34. Firman, T. K.; Landis, C. R. *J. Am. Chem. Soc.* **2001**, *123*, 11728.
35. Landis, C. R.; Cleveland, T.; Firman, T. K. *Science* **1996**, 179.
36. Dewar, M. J. S. *Bull. Soc. Chim. Fr.* **1951**, *18*, 79.
37. Chatt, J.; Duncanson, L. A. *J. Chem. Soc.* **1953**, 2939.
38. Miki, K.; Kai, Y.; Kasai, N.; Kurosawa, H. *J. Am. Chem. Soc.* **1983**, *105*, 2482.
39. Erker, G.; Rosenfeldt, F. *J. Organomet. Chem.* **1982**, *224*, 29.
40. Erker, G.; Dorf, U.; Czisch, P.; Petersen, J. L. *Organometallics* **1986**, *5*, 668.
41. Note that Cp_2 8-(O=Cph_2) is a dimer in the solid state.

CHAPTER 2

Dative Ligands

2.1. Introduction

In Chapters 2–4, organotransition metal compounds are surveyed according to the type of ligand. Organometallic complexes are typically named according to the metal and type of ligand that would undergo reaction, such as a "zirconium methyl complex" (**A** in Figure 2.1), a "palladium phosphine complex" (**B**), or a "tungsten carbene complex" (**C**). However, most complexes contain several types of ligands, and one must learn to identify which ligands will serve as the point of reaction and which ligands will modulate the structural and electronic properties of the metal to control reactivity. The number of types of ligands and members of each type is vast and increases with time. Chapters 2–4, therefore, present an illustrative but not a comprehensive summary of the types of complexes in organometallic chemistry. Chapter 2 presents descriptions of steric and electronic properties, methods for preparation when appropriate, and information on the spectroscopic features of neutral ligands. Chapters 3 and 4 present this information on formally anionic ligands.

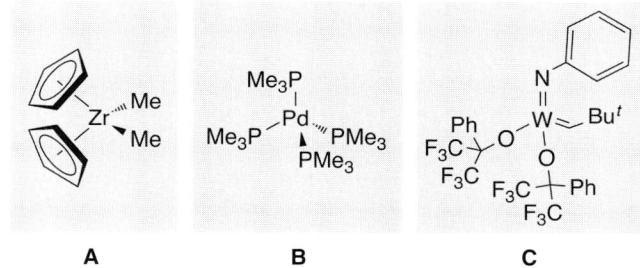

| A | B | C |

Figure 2.1. Three typical complexes used in organometallic chemistry.

By organizing the types of complexes by ligand, rather than by metal, the type of complex is more closely aligned with the classes of reactions presented in later chapters. When combined with the principles of bonding presented in Chapter 1, the information in Chapters 2–4 should allow the reader to classify structure and reactivity by patterns and principles of metal–ligand bonding.

2.2. Carbon Monoxide and Related Ligands

2.2.1. Properties of Free Carbon Monoxide

Carbon monoxide is a stable, inexpensive, toxic gas that is produced in massive quantities in parallel with hydrogen by the reaction of methane or coal with water at high temperatures and pressures. Carbon monoxide has a C–O triple bond and a small dipole moment with the negative end located at the carbon. It produces a strong vibration observed in the infrared spectrum centered at 2143 cm^{-1}. Carbon monoxide binds to transition metals as a neutral ligand and is most commonly bound to the metal through the lone pair of electrons on carbon.[12] Complexes of carbon monoxide are commonly called "carbonyl complexes." In most cases, carbon monoxide binds to one metal, but CO can also bridge two or more metals with M–C–M angles that are typically much less than the 120° angle of an sp^2-hybridized carbon.

2.2.2. Types of Metal Carbonyl Complexes

Metal carbonyl complexes are readily available and serve as starting materials for the preparation of many other organometallic compounds. Numerous reviews and books discuss their preparation and structure.[3–10] The common stable metal carbonyls are given in Table 2.1, and the commercially available ones are marked with an asterisk.[11]

Metal carbonyl complexes can be prepared in some cases from the bulk metal and from the reduction of complexes in higher oxidation states. For example, the highly toxic $Ni(CO)_4$ can be prepared directly from nickel metal and carbon monoxide at one atmosphere pressure,[12,13] and small amounts of $Fe(CO)_5$ are found in carbon monoxide that has been stored under pressure in steel cylinders. However, most metal carbonyls are best made in an ordinary laboratory by reductive carbonylation, as shown in Equations 2.1 and 2.2. The reducing agents used include electropositive metals and aluminum alkyls, as well as carbon monoxide itself.

Homoleptic carbonyl complexes of metals in the second and third rows of the transition series form polynuclear carbonyl complexes more often than they form mononuclear carbonyl complexes. For example, $Fe(CO)_5$ is stable and is the most common iron carbonyl, but $Os(CO)_5$[14] is much less stable; $Os_3(CO)_{12}$ is more stable.

$$Re_2O_7 + 17\ CO \longrightarrow Re_2(CO)_{10} + 7\ CO_2 \qquad (2.1)$$

$$WCl_6 + CO + Et_3Al \longrightarrow W(CO)_6 \qquad (2.2)$$

Some of the CO ligands in polynuclear carbonyl complexes, such as $Fe_2(CO)_9$[15] and $Os_3(CO)_{12}$, bridge two or more metals. The bridging between the two metals is not always symmetric, and structural classifications that are now used commonly[7,8,16,17] are shown in drawings I–VI of Figure 2.2. Type I carbonyl ligands are bound to a single metal with M–C–O angles near 180°. Type II carbonyl ligands bridge two metals with M–C–M angles typically between 77° and 90°.[8,16,18] Type III carbonyls bridge three metals.

Table 2.1. Stable neutral binary metal carbonyls.

$V(CO)_6$*, $Cr(CO)_6$*, $Mn_2(CO)_{10}$*, $Fe(CO)_5$*, $Co_2(CO)_8$*, $Ni(CO)_4$*
$Fe_2(CO)_9$*, $Co_4(CO)_{12}$*
$Fe_3(CO)_{12}$*, $Co_6(CO)_{16}$
$Mo(CO)_{l6}$*, $Tc_2(CO)_{10}$, $Ru(CO)_5$
$Ru_3(CO)_{12}$*, $Rh_4(CO)_{12}$*
$Rh_6(CO)_{16}$*
$W(CO)_6$*, $Re_2(CO)_{10}$*, $Os(CO)_5$
$Os_3(CO)_{12}$*, $Ir_4(CO)_{12}$*
$Ir_6(CO)_{16}$

*Commercially available.

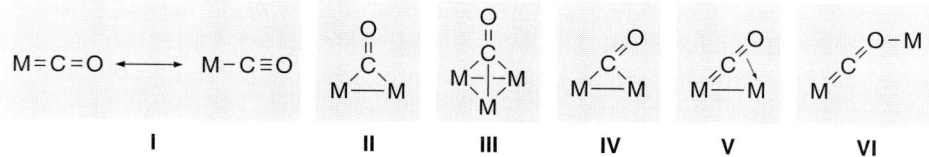

Figure 2.2. Classification of metal carbonyl complexes.

Type IV, V and VI carbonyl ligands bind two metals in an unsymmetrical fashion. Type IV carbonyls bind one metal in a terminal fashion and a second metal through the carbonyl carbon. This binding mode is often called "semibridging." Type V carbonyls bind one metal in a terminal fashion and a second metal through the CO π-system, much like a metal binds an olefin as described in Chapter 1. This type of carbonyl is a four-electron donor resulting from the combination of two interactions that donate two electrons each. A type VI carbonyl binds one metal in a terminal fashion and a second metal through an electron pair at oxygen in a Lewis acid–base interaction. The latter part of this binding mode is also called an "end-on" or isocarbonyl mode.

2.2.3. Models for CO Binding: Introduction of Backbonding

The bonding of carbon monoxide to a transition metal is more complex than the simple combination of an electron pair and an unoccupied orbital in a Lewis acid–base interaction. Carbon monoxide is weakly basic.[19, 20] The formation of HCO^+ does not occur under standard synthetic conditions.[21–23] HCO^+ has been invoked as an intermediate in superacid solution[24,25] and has been identified in interstellar space.[19,20,22–25] Yet CO binds tightly to many transition metals and even binds weakly to simple main group Lewis acids like BH_3 and d^0 transition metals.

CO binds strongly to electron-rich, low-valent metals. Two major factors contribute to this stronger binding to electron-rich, low-valent metals than to electron-poor, high-valent metals, as was presented in Chapter 1. First, hard Lewis acids, such as a proton, trihaloboranes, or d^0 transition metal complexes are a poor match for the soft carbon on CO. Second, the donation of electrons from low-valent, electron-rich metals into the LUMO of CO, which is the π-antibonding (π*) orbital, provides an additional bonding interaction. The overlap of metal orbitals with the σ- and π*-orbitals of CO described in Chapter 1 is repeated in Figure 2.3.[26]

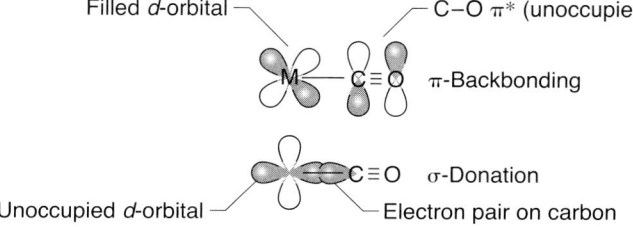

Figure 2.3.
Orbital interactions involved in the binding of CO to a single transition metal.

The bonding of CO to the metal occurs by two types of interactions. One involves overlap of the orbital of CO containing the electron pair with the unoccupied orbital of the metal of σ-symmetry. This interaction is the conventional Lewis acid–base interaction. However, a second interaction makes the bonding of CO distinct from the bonding of simple Lewis bases. A pair of molecular orbitals are generated from the occupied orbitals of the metal of π-symmetry and the unoccupied π*-orbitals of CO. Two electrons from the metal orbital occupy each of these two orbitals. The generation of these orbitals creates two effects. First, the energy of the resulting filled orbital is lower than that of the original occupied metal orbital. The stability of the complex is, therefore, increased. Second, this interaction delocalizes electron density from the metal into the $\pi*$-orbitals of the carbonyl ligand and, therefore, decreases the C–O bond order. This delocalization of electron density from the metal into the $\pi*$-orbital of the ligand is known as backbonding.

Binding of CO to two metals requires rehybridization of the ligand orbitals, and the orbital interactions that lead to binding of CO to two metals in a type II bridging mode are shown in Figure 2.4. The standard symmetric bridging mode is not directly analogous (or can be compared only loosely) to the bonding of the central carbonyl unit in a ketone because the M–C–M angles of 77° to 90° are much more acute than the C–C–C angle of acetone. Instead, the bonding of the bridging carbonyl is highly delocalized and involves the overlap of metal orbitals with the s and p molecular orbitals of CO.[5,31,32] The σ-bonding orbital overlaps with the bonding orbital of the two metals and the $\pi*$ orbital overlaps with filled d-orbitals on the two metals. The resulting CO bond order of a bridging carbonyl ligand is closer to two than is the CO bond order in a terminal carbonyl. A CO ligand can also bridge three metals. The bonding between the three metals and the carbonyl is again highly delocalized, and the angles at carbon are acute.

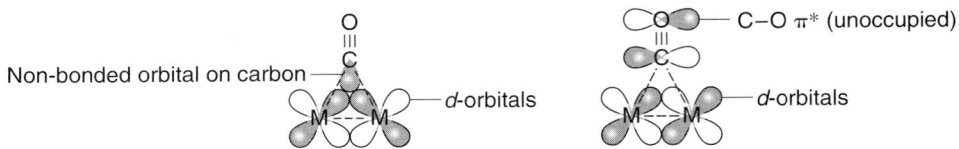

Figure 2.4.
Orbital interactions that create bonding orbitals for the coordination of CO between two transition metals.

2.2.4. Evidence for Backbonding in Terminal Carbonyls

Much physical evidence reveals the importance of backbonding. For example, the C–O distance in carbonyl ligands is longer than that in free CO, and the carbonyl stretching frequency is lower than that in free CO. Both of these observations result from the decrease in C–O bond order that results from backbonding.

The value of v_{CO} for terminal carbonyl ligands on metal fragments with d-electrons is typically between 2125 and 1850 cm^{-1}, while the value for free CO is 2143 cm^{-1}. Unusually electron-rich neutral complexes can display even lower values of v_{CO}. For example, Fe(dmpe)$_2$(CO) has a v_{CO} of 1812 cm^{-1}.[33] A positive charge on a complex decreases backbonding and increases v_{CO}, as can be seen by comparing the v_{CO} value of compounds **A** and **B** in Figure 2.5;[34] on the other hand, a negative charge increases backbonding and decreases v_{CO}, as can be seen by comparing the v_{CO} values for compounds **C**–**E**.[35] Carbonyl groups bound to metal centers lacking d-electrons are rare, but several have been identified[36] and recently isolated.[37–41] The v_{CO} values of these compounds are usually higher than that of free CO. Although an explanation involving a removal of electron density from a $\sigma*$-orbital has been put forth, later studies indicate that the higher stretching frequency results from an electrostatic effect that increases the covalency of the C–O bond, decreases the C–O distance, and increases the C–O stretching force constant.[42]

Comparisons of v_{CO} values between compounds with different numbers of carbonyl ligands or different symmetries are not straightforward. The observed frequencies of vibration

Ph$_3$P, CO Ir Cl, PPh$_3$	Ph$_3$P, ⊕, CO Pt Cl, PPh$_3$	Ni(CO)$_4$	Ce(CO)$_4^-$	Fe(CO)$_4^{2-}$
1944	2120	2057	1886	1785
A	**B**	**C**	**D**	**E**

Figure 2.5.
CO stretching frequencies of five different metal carbonyl compounds.

Table 2.2. Infrared and Raman active normal modes for common carbonyl symmetries.

Molecular point group	Example	Infrared	Raman
C_{2v}	$Ni(CO)_2L_2$	Two: A_1 and B_1''	Two: A_1 and B_1
D_{2h}	$RuCl_2(CO)_2L_2$ (all trans)	One: B_{1u}''	One: A_g
C_{3v}	$Ni(CO)_3L$	Two: A_1 and E	Two: A_1 and E
D_{3h}	$Ru(CO)_3L_2$ (tbp with equatorial carbonyls)	One: E'	Two: A_1' and E'
C_{4v}	$Mn(CO)_5I$	Three: Two A_1 and one E	Four: Two A_1, one B_1, and one E
D_{4h}	trans-$Mo(CO)_4L_2$	One: E_u	Two: A_{1g} and B_{1g}
T_d	$Ni(CO)_4$	One: T_2	Two: A_1 and T_2
O_h	$Cr(CO)_6$	One: T_{1u}	Two: A_{1g} and E_g

are actually due to independent motions called normal modes, which comprise in carbonyl compounds a combination of the vibrations of more than one carbonyl ligand. The number of normal modes in the carbonyl stretching region and the activity of each mode in the infrared and Raman spectra are easily predicted for carbonyl ligands in a particular geometry by group theoretical methods. This information is given in Table 2.2 for the common geometries of homoleptic metal carbonyl complexes. Thus, the IR bands arising from C–O stretching vibrations of terminal carbonyl ligands are strong and sharp (at least in nonpolar solvents), and their pattern is characteristic of a particular geometry. Detailed analyses of representative spectra, and tables of vibrational data for metal carbonyl complexes can be found in two excellent books.[43,44]

2.2.5. Infrared and X-Ray Diffraction Data for Complexes with Bridging Carbonyls

As expected from the more reduced C–O bond order in bridging carbonyl ligands than in terminal carbonyl ligands, the stretching frequencies v_{CO} of bridging carbonyls are lower than those of terminal carbonyls. In neutral complexes containing type II bridging carbonyl ligands, the v_{CO} values are typically between 1850 and 1700 cm^{-1}, although the interaction of several normal modes can lead to values somewhat above 1850 cm^{-1}. The stretching frequencies of type III, triply bridging, carbonyl ligands typically fall between 1675 and 1600 cm^{-1}, although the values of v_{CO} for $Co_6(CO)_{16}$ and $Ir_6(CO)_{16}$ lie near 1800 cm^{-1} for the type III carbonyl ligands.[45] In agreement with a lower C–O bond order for bridging carbonyl ligands than for terminal carbonyl ligands, the C–O distances, usually between 1.17 and 1.22 Å, are somewhat longer than those in terminal carbonyl ligands.

The C–O distance in typical terminal carbonyl ligands is between 1.12 and 1.18 Å. This distance is slightly longer than the 1.128 Å C–O distance in free CO because of a reduction in the C–O bond order from backbonding into the π^*-orbital. For the same reason, the M–C bond length is slightly shorter than that of an M–C single bond because of an increase in the M–C bond order due to backbonding.[46]

2.2.6. Thermodynamics of the M–CO Bond

The dissociation of carbonyl ligands is a key step in many reactions, and accurate values for M–CO bond dissociation energies are useful to estimate whether CO dissociation could occur as part of a reaction pathway. Some representative bond energies are provided

Table 2.3. Selected M–CO bond dissociation energies (BDE) from theory and experiment.

M–CO	Calculated first BDE (kJ/mol)	Experimental first BDE (kJ/mol)	Calculated mean BDE (kJ/mol)[48]	Experimental mean BDE (kJ/mol)
$V(CO)_6^-$	171[48]			
$Cr(CO)_6$	147[48]	162,[53] 155[54]	107	110[55]
$Mo(CO)_6$	119[48]	126,[53] 142[54]	126	151[55]
$W(CO)_6$	142[48]	166,[53] 159[54]	156	179[55]
$Mn(CO)_6^+$	92[48]			
$Rh(PPr_3^i)_2Cl(CO)$	35[56]	36		
$Ir(PPr_3^i)_2Cl(CO)$	84[56]	> 72		
$Ni(CO)_4$	106[48]	104[53]	179	191[55]
$Pd(CO)_4$	27[48]		44	
$Pt(CO)_4$	38[48]		59	

in Table 2.3, and these values demonstrate that metal–carbonyl bond energies vary dramatically with ancillary ligands. The range of bond energies spans from 6–11 kcal/mol for palladium carbonyl[47,48] to 72 kcal/mol for $IrL_2Cl(CO)$ (L = PPr_3^i).[49]

Thus, one cannot use a single value to approximate the bond energies of all metal–carbonyl bonds. Yet, some trends are clear. The value for palladium carbonyl is low enough that the compound is unstable and has been observed only in low temperature matrices.[50–52] In contrast, the value of the iridium–carbonyl bond in $(CO)Ir(PPr_3^i)_2Cl$ is so strong that dissociation would require more than 10^{14} centuries at room temperature and would not be part of any reasonable reaction mechanism. A comparison of the mean bond dissociation energies of the M–CO bonds of the Cr, Mo, and W complexes shows that this value generally increases down a triad. Furthermore, the M–CO bond strengths increase with an increase in backbonding. Thus, the first M–CO bond strength of $Cr(CO)_6$ is higher than that of $Ni(CO)_4$, in part because Cr^0 is a better backbonder than Ni^0 due to the higher energies of the orbitals on the left of the transition series than at the right. The Ir–CO bond strength of $[Ir(PPr_3^i)_2Cl(CO)]$ is particularly high because iridium is a third-row metal and is very electron rich due to the alkylphosphine donors in this complex. In contrast to trends in the strengths of many metal–ligand bonds, the strengths of some metal–carbonyl linkages for homoleptic carbonyl compounds of the first-row metals are stronger than those for the second- and third-row metals. For example, the mean M–CO bond dissociation energy of $Ni(CO)_4$ is higher than that of $Pd(CO)_4$ and the first M–CO bond dissociation energy of $Cr(CO)_6$ is higher than that for $Mo(CO)_6$.

2.2.7. Isoelectronic Analogs of CO: Isocyanides and Thiocarbonyls

Isocyanides (also called isonitriles, C≡NR) and carbon monosulfides (CS) are two common ligands that are isoelectronic with CO. Isocyanides are stable molecules that can be purchased from common suppliers with phenyl, mesityl, *t*-butyl, and cyclohexyl groups at nitrogen. As one would expect from simple arguments based on electronegativity, an isocyanide is a stronger σ-donor and weaker π-acceptor than is carbon monoxide.[57,58] Moreover, the carbon–nitrogen π-bond in an isocyanide is weaker than the carbon–oxygen π-bond of carbon monoxide.[59] This difference in bond strength can change the thermodynamics of a reaction from unfavorable for CO to favorable for RNC. Complexes of isocyanides can be prepared by ligand substitution with free isocyanide as shown in Equation 2.3. Unlike homoleptic carbonyl complexes, homoleptic isocyanide complexes are unavailable commercially, but some of these species have been prepared by reduction of metal halides in the presence of RNC (Equation 2.4).[58,60,61]

$$Fe(CO)_5 + RNC \xrightarrow{\text{CoCl}_2} Fe(CO)_{5-x}(CNR)_x \qquad (2.3)$$

$$(RN \equiv C)_4 RuCl_2 \xrightarrow[\text{Na/Naphthylene}]{\text{RNC}} (RN \equiv C)_6 Ru^{2-} \qquad (2.4)$$

Carbon monosulfide is not a stable molecule, but carbon monosulfide complexes (more commonly called thiocarbonyl complexes) are known for most of the transition metals. Because of the instability of free CS, most CS complexes have been prepared by generation of CS from CS_2, Cl_2CS, or $EtOC(S)Cl$ in the coordination sphere of the metal, as shown in Equations 2.5,[62] 2.6,[63] and 2.7.[64] Thiocarbonyl ligands vibrate between 1160 and 1410 cm[-1]; free CS (in a matrix) vibrates at 1274 cm[-1]. A detailed analysis of force constants of mixed carbonyl and thiocarbonyl complexes indicates that the CS ligand can be a weaker or stronger π-acceptor than CO.[65] Seleno-[64,66] and tellurocarbonyl complexes are also known.[67,68]

$$(2.5)$$

$$Fe(TPP) + Cl_2CS \xrightarrow[\text{2) py}]{\text{1) Fe(O)}} Fe(TPP)(Py)(CS)$$

TPP = tetraphenylporphyrin

$$(2.6)$$

$$(2.7)$$

2.3. Dative Phosphorus Ligands and Heavier Congeners

2.3.1. Tertiary Phosphines and Related Ligands

Phosphines and related trivalent phosphorus ligands are among the most important ancillary ligands in homogeneous catalysis and perhaps in all of organometallic chemistry. All transition metals, particularly late transition metals, form complexes with trivalent compounds of phosphorus. The soft phosphorus donor matches well with soft low-valent metals, and the substituents on the phosphorus atom can dramatically affect the properties and reactivity of the metal center. The number of known transition metal complexes containing tertiary phosphine ligands is immense. Some reviews of transition metal phosphine complexes and phosphorus ligands appeared a number of years ago,[69-77] but the vast array of ligands prepared in the last decade has caused recent reviews to focus on certain classes of phosphorus ligands.[78] Classic tertiary phosphine and related ligands based on group 5B atoms are shown in Figure 2.6. These ligands exhibit a range of steric and electronic effects that are discussed in detail in the next two sections.

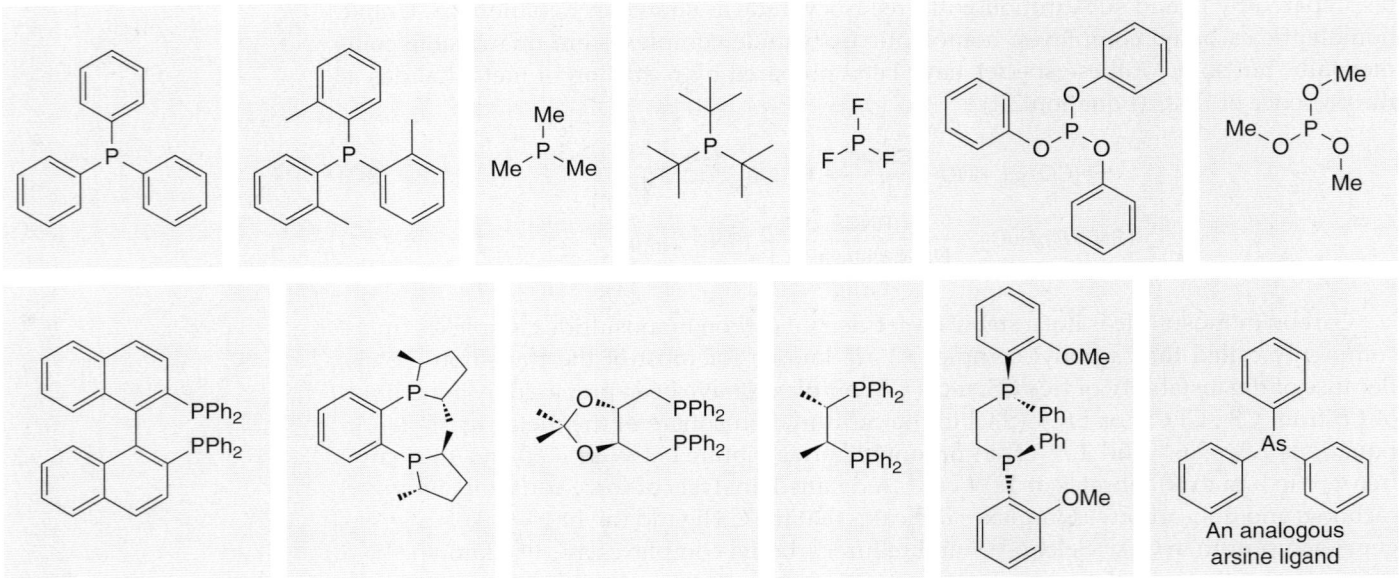

Figure 2.6. Representative common and historically important monophosphines and bisphosphines.

2.3.2. Chelating Phosphines

In contrast to CO, most dative ligands, including phosphines, can be tethered to other donor atoms to create ligands that bind the metal through more than one donor atom. Ligands containing two, three, or four donor atoms are often called bidentate, tridentate, and tetradentate, respectively, for obvious etymological reasons. These ligands bind metals in what is called a "chelating" (pronounced *kelating*) fashion and, as a group, are often called "chelating ligands." This term is derived from the Greek claw or "chela" ("chely") of the lobster and other crustaceans due to the analogy of the ligands to the caliper-like function of the claw.[83] Phosphorus donors are most often combined with another phosphorus donor to create a symmetric bidentate ligand. However, unsymmetric bisphosphine ligands are becoming more common, and many ligands combine a phosphorus donor with another neutral donor, such as a nitrogen heterocycle, to form an unsymmetrical, neutral, bidentate ligand, or with the charged donor ligands discussed in Chapter 3 to create charged polydentate ligands. The tether length of most chelating ligands is short enough to enforce a cis disposition of the two donor atoms. However, certain bidentate phosphines, such as those in Figure 2.7,[84–86] have been designed to ensure or encourage the two phosphine donors to bind in a trans orientation. Others have large enough tethers that they may adopt cis or trans conformations.

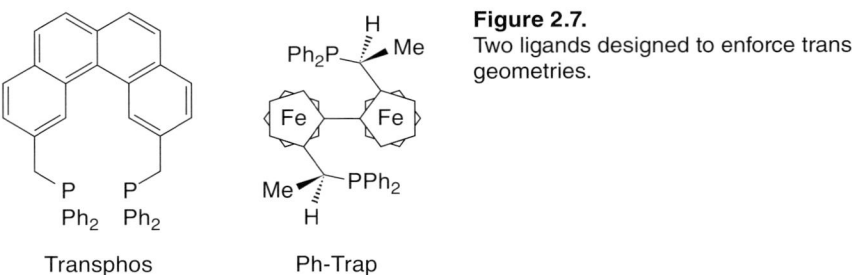

Transphos   Ph-Trap

Figure 2.7.
Two ligands designed to enforce trans geometries.

The P–M–P angle enforced by the backbone of a bidentate ligand can strongly affect the reactivity of a complex containing a chelating ligand in both catalytic and stoichiometric reactions. Casey and Whiteker[89] have defined a range of angles that a series of chelating ligands prefer to adopt and have termed this angle the "natural bite angle." Some ligands prefer to bind to a metal with angles less than or near to 90°. These ligands create complexes that are stable with the two donor atoms bound cis to each other in an octahedral or square planar geometry or bound with one of the donors in the axial and one in the equatorial position of a trigonal bipyramid. Other chelating ligands prefer to bind to a metal with angles closer to 120°. These ligands prefer to bind, for example, to two equatorial positions of a trigonal bipyramidal geometry.[90]

2.3.3. Properties of Free Phosphines

Free trialkylphopshines have similar protic basicities[91–94] (Table 2.4) to the corresponding amines. The pK_a of $HPEt_3$ in DMSO solvent is 9.1, while the pK_a of $HNEt_3$ in this solvent is 9. However, most transition metals are much softer acids than a proton and, therefore, bind more strongly to the soft phosphorus in a phosphine than to the harder nitrogen in an amine.[95–97] Moreover, the larger size of the phosphorus atom makes the M–P distance longer and the steric effects of the substituents in analogous nitrogen and phosphorus ligands less pronounced for the phosphorus ligand. Thus, tertiary phosphines usually bind to transition metals with much higher affinity than tertiary amines.

Phosphines are more susceptible to oxidation than amines because P(V) is a stable oxidation state. Thus, some phosphines are air sensitive and must be handled under an inert atmosphere. Many alkylphosphines are air sensitive, but arylphosphines and phosphites, which are less electron rich than alkylphosphines, are less sensitive or are indefinitely stable to air. Also, more sterically hindered alkylphosphines are less air sensitive than less-hindered alkylphosphines.[98,99]

Table 2.4. Enthalpy of protonation and pK_a values of selected phosphines.

PR_3	ΔH_{HP} (kcal/mol)[a]	pK_a
$(p\text{-}ClC_6H_4)_3P$	17.9 (0.2)[b]	1.03[c]
$(p\text{-}FC_6H_4)_3P$	19.6 (0.2)	1.97[c]
Ph_3P	21.2 (0.1)	2.73[d]
$(o\text{-}MeC_6H_4)_3P$	22.6 (0.2)	3.08[c]
$(p\text{-}MeC_6H_4)_3P$	23.2 (0.3)	3.84[c]
$(p\text{-}MeOC_6H_4)_3P$	24.1 (0.2)	4.57[c]
$MePh_2P$	24.7 (0.0)	4.59[e]
Me_2PhP	28.4 (0.2)	6.50[d]
Me_3P	31.6 (0.2)	8.65[d]
$(c\text{-}C_6H_{11})P$	33.2 (0.4)	9.70[d]
Et_3P	33.7 (0.3)	8.69[d]
Bu^t_3P	36.6 (0.3)	11.4[c]

[a] For protonation with CF_3SO_3H in dichloroethane solvent at 25.0 °C.
[b] Numbers in parentheses are average deviations.
[c] Source: Allman, T.; Goel, R. G. *Can. J. Chem.* **1982**, *60*, 716.
[d] Source: Streuli, C. A. *Anal. Chem.* **1960**, *32*, 985.
[e] Source: Golovin, M. N.; Rahman, M. M.; Belmonte, J. E.; Giering, W. P. *Organometallics* **1985**, *4*, 1981.

The barrier to inversion at phosphorus is much higher than the barrier to inversion at nitrogen and typically ranges from 29–35 kcal/mol (Figure 2.8).[100,101] Thus, an amine containing three different substituents will consist of a racemic mixture of conformers in solution, but most phosphines containing three different substituents can be prepared in optically active form. Such a phosphine can be resolved and stored indefinitely as a nonracemic mixture. A "chiral-at-phosphorus" or "P-chiral" compound, MeP(cyclohexyl)(o-anisyl) in Figure 2.8, was the first ligand that generated an enantioselective catalyst for hydrogenation.[102,103] The development of P-chiral ligands has experienced a renaissance in recent years.[104–106]

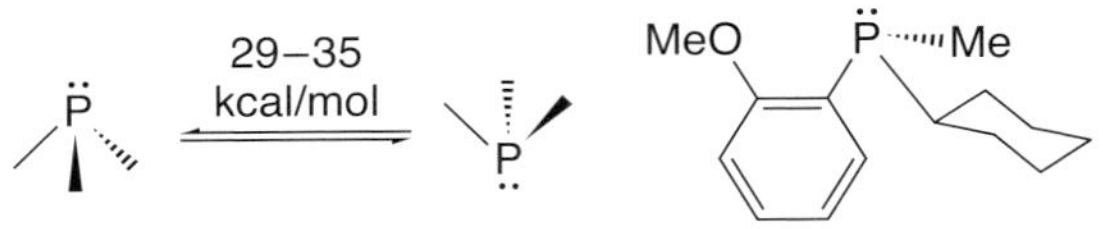

Figure 2.8.
Barrier to interconversion and one of the first ligands for asymmetric hydrogenation.

2.3.4. Properties of Phosphine Complexes

2.3.4.1. Bonding and Electronic Properties

Trialkylphosphine ligands bind to transition metals predominantly by Lewis acid–base interactions. The soft diffuse lone pair serves as a strong Lewis base to the soft transition metal Lewis acids. In contrast to many other dative ligands, monophosphines bind to a single metal center in almost all cases.[107,108] In general, trialkylphosphines are the most electron donating of the dative phosphorus ligands, and arylphosphines are less electron donating. This trend is observed, in part, because the greater s-character of the sp^2-hybridized orbital of the aryl group makes it a weaker electron donor than an alkyl group. Similarly, phosphites, which contain three alkoxy groups at phosphorus, are less electron donating than phosphines because electron donation by the alkoxo groups is weaker than that by alkyl or aryl groups.

Tertiary phosphines and phosphites can also serve as π-acceptors. These ligands were once thought to stabilize transition metal alkyl derivatives through dπ–dπ backbonding in which filled metal d-orbitals overlap with the vacant d-orbitals on phosphorus. However, more recent studies on the potential of phosphines and phosphites to act as π-acceptors have indicated that the acceptor orbital is a hybrid of the P–X σ*-orbitals and the phosphorus d-orbital with the dominant component the P–X σ*-orbitals.[109–112] The orbitals resulting from combining the P–X σ* orbitals and phosphorus d-orbitals and the symmetry of this backbonding interaction are shown in Figure 2.9. The π-accepting ability has been shown

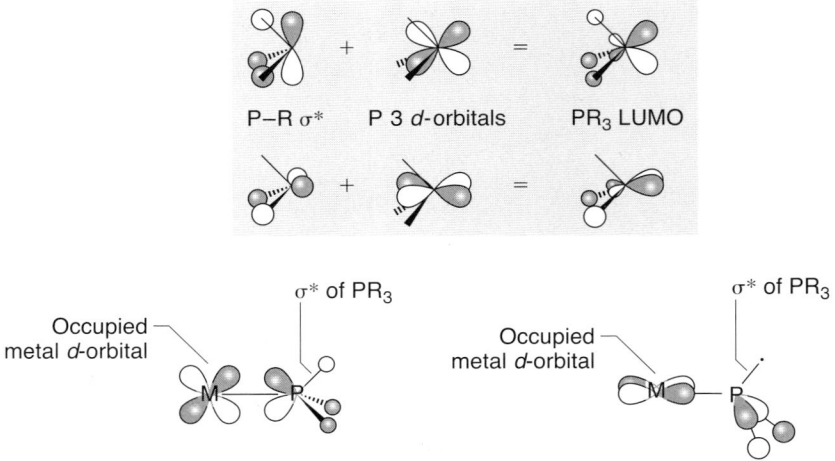

P–R σ*   P 3 d-orbitals   PR₃ LUMO

Occupied metal d-orbital   σ* of PR₃   Occupied metal d-orbital   σ* of PR₃

Figure 2.9.
Mixing of σ*-orbitals in P(III) ligands with the phosphorus d-orbital and the symmetry of orbital interactions in metal–phosphine backbonding.

by structural data on pairs of phosphine and phosphite compounds with the same ligands and geometry but two different oxidation states. If the ligand were a simple σ-donor, then the M–P bond would be shorter in the complex containing the metal in the higher oxidation state. However, the M–P bond was longer and the P–X distance was shorter in the complexes of higher oxidation state. This phenomenon was rationalized by weaker donation of electron density from the metal into the P–X σ*-orbital within the compound in the higher oxidation state.[113,114]

The magnitude of the back-donation depends on the substituents at phosphorus because the electronegativity of the group affects the energy of the P–C σ*-orbitals. One computational paper provides a scale of the π-accepting abilities of different ligands.[115] This scale, resulting from a natural bond orbital analysis, is shown in Figure 2.10 and sets the π-accepting ability of CO at 100. Although these numbers are generated from a somewhat arbitrary weighting of data from calculations of different compounds, they do illustrate the relative π-accepting ability of different phosphorus ligands, relative to CO and amine ligands.

Experimental data on the π-accepting ability of alkylphosphines are ambiguous. Photoelectron spectroscopy indicates that aliphatic phosphines are poor π-acceptors,[116–120] but one study on a complex with an electron-rich d^2 metal center suggested that the relative energies of electronic states could be influenced by π-back-donation into the σ-bonding orbitals of an aliphatic phosphine.[121] The NBO analysis concluded that alkylphosphines are effective π-acids. Thus, further debate is likely to ensue about the π-accepting ability of alkylphosphines, but other phosphorus ligands are clearly π-acceptors.

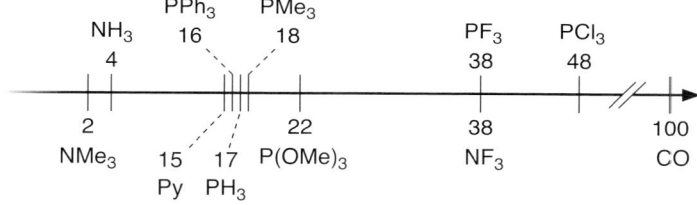

Figure 2.10.
Calculated π-acceptor index for P- and N-based ligands, relative to CO, determined by a natural bond orbital (NBO) analysis. (Taken from Figure 3 of *Organometallics* **2007**, *26*, 2637.)

Several approaches have been followed to reveal the electronic effects of phosphorus ligands. In a classic study, Chadwick Tolman at DuPont measured the value of the sharp, highest-frequency v_{CO} band in the infrared spectra of many [Ni(CO)$_3$L] complexes in which L is a phosphine or phosphite.[122] As presented in the earlier section on the binding of CO, the value of v_{CO} depends on the degree of backbonding. The degree of backbonding depends on the electron density of the metal center, which, in turn, depends on the electron-donating ability of the phosphine in the series of complexes [Ni(CO)$_3$L]. A plot of the representative v_{CO} values is shown in Table 2.5. These data show that PBu$^t_3$ and PCy$_3$ are the strongest donors of the set, that alkylphosphines tend to be stronger donors than aryl phosphines, that phosphites are weaker donors than phosphines, and that PF$_3$ is the weakest donor of this set of phosphorus ligands. In other cases, the electronic effects of phosphines on structure and reactivity have been studied by modifying the para substituents on triarylphosphines.[123] The magnitudes of the changes in rates that result from varying these substituents is typically small, but these studies can reveal whether a reaction is favored by a stronger or weaker donor.

Table 2.5. Representative v_{CO} values for [Ni(CO)$_3$L].[a]

L	v_{CO} (cm^{-1})
PBu$^t_3$	2056
PCy$_3$	2056
PMe$_3$	2064
P(C$_6$H$_4$-4-OMe)$_3$	2066
PPh$_3$	2069
P(OMe)$_3$	2079
P(OPh)$_3$	2085
PF$_3$	2110

[a] Source: Tolman, C. A. *Chem. Rev.* **1977**, *77*, 313.

2.3.4.2. Steric Properties

The steric properties of phosphorus ligands have been exploited many times to control the reactivity of organometallic compounds in both stoichiometric and catalytic chemistry. Thus, much effort has been spent on quantifying the steric properties of phosphine ligands. Tolman can again be credited with an early systematic treatment of the steric properties of phosphorus ligands.[69,70] Tolman devised a structural parameter he called "cone angle," which is the angle defined by the outer edge of the substituents at phosphorus and the metal center of a space-filling model, as shown in Figure 2.11. This work was conducted well before computer modeling, and computational energy minimization procedures. Thus, the angles were measured with space-filling models, wood, and a protractor. Yet, this conceptually simple paper has been cited more than 2500 times because this model, though flawed in details, explains an enormous amount of reactivity in organometallic chemistry.

The cone angles of several common ligands are listed in Table 2.6. Two significant drawbacks to the cone angle parameter should be kept in mind. First, the cone angle depends

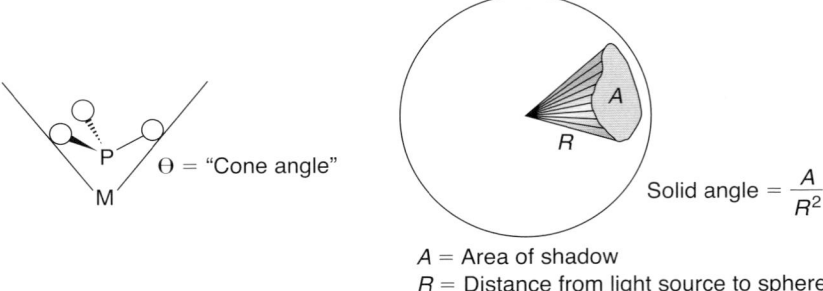

Θ = "Cone angle"

Solid angle = $\dfrac{A}{R^2}$

A = Area of shadow
R = Distance from light source to sphere

Figure 2.11. Definition of phosphine cone angles and solid angles.

Table 2.6. Phosphine ligand cone angles.[a]

Phosphorus ligand	Cone angle (°)	Solid angle (°)[c]
PH_3	87	
$P(OCH_2)_3CR$	101	82
PF_3	104	
$P(OMe)_3$	107	113
PMe_3	118	124
PMe_2Ph	122	126
$Ph_2PCH_2CH_2PPh_2$	123	
$PHPh_2$	128	112
$P(OPh)_3$	128	135
PEt_3	132	143
PPh_3	145	129
$PPh_2(Bu^t)$	157	149
$PPh(Bu^t)_2$	170	168
PCy_3	170 (163–181)[b]	181
$P(Bu^t)_3$	182	
$P(o\text{-tol})_3$	194 (183–198)[b]	142
$P(mesityl)_3$	212	

[a] Cone angle data from reference 122.
[b] Range of cone angles determined from the X-ray diffraction of PCy_3 and $P(o\text{-tol})_3$ complexes.
[c] Solid angle data from references 126–131.

on conformation, and the conformation of the ligand depends on the complex. Because the cone angle is based on the conformation that is the least hindered, the cone angle can vary significantly from one structure to another. Second, the proper definition of the cone angle of an unsymmetrical ligand is less straightforward than the definition of a cone angle of a symmetric ligand. An average of the cone angles for the symmetric phosphines, weighted for the number of substituents, was used originally. However, the conformation of the substituents modulates the steric properties of the ligands. Thus, the highly cited cone angle provides a rough measure of *trends* in steric properties, but this value should not be considered an absolute measure of the steric properties of these ligands. A range of cone angles better describes the steric properties, as shown for PCy_3 and $P(o\text{-tol})_3$ in Table 2.6. Solid angles have also been used to describe the sizes of phosphines, and Table 2.6 includes some of these more modern "solid-angle" values. Solid-angle values are based on a three-dimensional rather than a two-dimensional angle. A solid angle (θ) of a ligand is defined as the normalized area of the shadow cast by the ligand on a sphere encompassing the entire complex with the metal as the point source of light.[126–131] Such a shadow is shown on the right in Figure 2.11.

2.3.4.3. Effects of Phosphine Steric and Electronic Properties on Structure and Reactivity

Phosphines and related P(III) compounds typically serve as ancillary ligands, but the dissociation of these ligands is crucial to the reactivity described in later chapters. Tolman correlated the ligand cone angle with the equilibrium for dissociation from NiL_4 compounds.[132] The extent of ligand dissociation in these nickel complexes and in related palladium complexes increases in the order $PMe_3 < PMe_2Ph < PMePh_2 < PEt_3 < PPh_3 < PPr^i_3 < PCy_3 < PPhBu^t_2$.

The steric and electronic properties of phosphines also affect the overall geometry of a complex. For example, bulky phosphines tend to bind trans to one another. Furthermore, the presence of several bulky phosphine ligands in the same coordination sphere can cause deviations from the idealized coordination geometry. For example, the Wilkinson hydrogenation catalyst, $RhCl(PPh_3)_3$, contains a nonplanar arrangement of the four donor atoms, as determined by X-ray diffraction,[133,134] instead of the square planar arrangement expected for a d^8, Rh(I) complex.

2.3.5. Pathways for the Decomposition of Phosphorus Ligands

Several reactions of phosphines and related ligands are undesirable because they cause decomposition of the ligand. First, the more electron rich of the phosphines are air sensitive, as mentioned previously. When coordinated, the phosphine is much less prone to oxidation, but reversible dissociation of a phosphine can allow oxidation to occur. Second, the P–C bonds can undergo cleavage.[135–138] Organometallic phosphine complexes often undergo exchange of aryl groups at the metal with aryl groups on the phosphine.[136–138] Several mechanisms are possible for this reaction, but the reversible formation of phosphonium ions shown in Equation 2.8 is one likely pathway.[138] The formation of a phosphonium salt from an organometallic phosphine complex has been observed many times.[138–140]

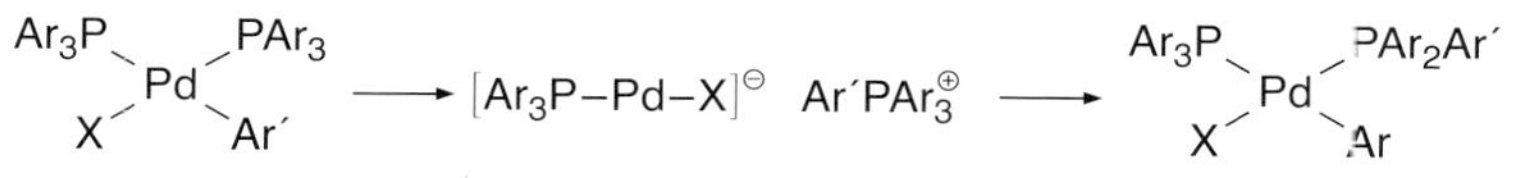

$$\underset{X}{\overset{Ar_3P}{\diagdown}}\underset{Ar'}{\overset{PAr_3}{\diagup}}Pd \longrightarrow [Ar_3P-Pd-X]^{\ominus}\ Ar'PAr_3^{\oplus} \longrightarrow \underset{X}{\overset{Ar_3P}{\diagdown}}\underset{Ar}{\overset{PAr_2Ar'}{\diagup}}Pd \qquad (2.8)$$

The groups on phosphites and phosphoramides can also undergo cleavage, typically with water and alcohols.[141–144] This process has been a concern when phosphites are used as ligands in hydrocyanation reactions that form nitriles[145] and in hydroformylation reactions that form aldehydes with alcohol side products.[146] Phosphites that are hindered and that are constructed from diols are less prone to cleavage.[141] With hindered groups, such as those in the bisphosphines developed at Union Carbide (Figure 2.12), attack is inhibited and cleavage of the P–O bond involving a diolate would require opening of a stable ring.

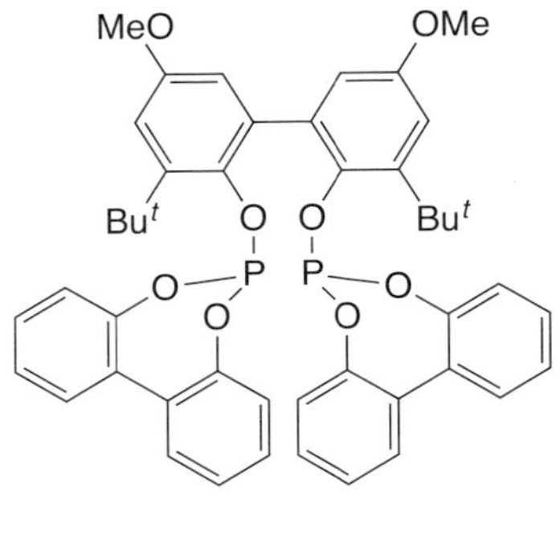

Figure 2.12.
A phosphite based on diolate substituents used in hydroformylation reactions.

The P–H bonds in complexes of the parent phosphine (M–PH$_3$) and in complexes of primary and secondary phosphines (M–PH$_2$R and M–PHR$_2$) are reactive. These ligands can be used to prepare novel phosphide (M–PR$_2$) complexes, but the reactivity of the P–H bond has caused primary and secondary phosphines to be used less commonly as ancillary ligands. A few recent examples, however, demonstrate that they can be valuable in both enantioselective and coupling processes.[149,150]

2.3.6. NMR Spectroscopic Properties of Phosphines

The ^{31}P nucleus has a spin of 1/2 and is 100% abundant. Its receptivity, a measure of the sensitivity toward NMR analysis, is 0.0665, in comparison with 1 for protons and 0.000175 for ^{13}C.[151] Thus, ^{31}P NMR spectroscopy is an important method for the characterization of organometallic complexes. This subject has been extensively reviewed.[152,153]

^{31}P NMR chemical shifts depend on several parameters. Thus, predictions of ^{31}P NMR chemical shifts of complexes of phosphines and related ligands are tenuous. However, empirical ^{31}P NMR chemical shift data are useful as a "fingerprint," and the chemical shifts of phosphites, phosphinites, halophosphines, and phosphines fall in distinct regions. Splitting patterns and coupling constants in the ^{31}P NMR spectra are particularly informative about structure. For example, the complex with a trans geometry on the left in Figure 2.13 possesses four equivalent phosphorus nuclei and generates a singlet resonance in the proton-decoupled ^{31}P NMR spectrum, but the cis isomer contains two pairs of equivalent phosphorus nuclei and generates a spectrum containing two triplet resonances. In addition, the magnitude of the ^{31}P–^{31}P coupling can reveal geometry. For example, the two-bond ^{31}P–^{31}P coupling of phosphines located trans to each other is much larger than that for phosphines located cis to each other. As one might expect, the pattern and magnitude

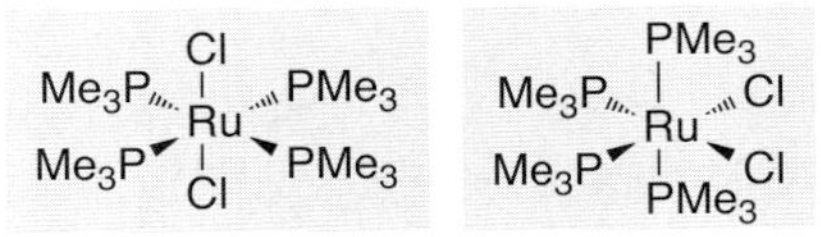

δ −6.63 (s)   δ 9.0 (t)
δ −12.7 (t)
J = 35 Hz

Figure 2.13.
^{31}P NMR data for *cis*- and *trans*-(PMe$_3$)$_4$RuCl$_2$.

of the coupling of a phosphorus nucleus to ^{1}H and ^{13}C nuclei of another ligand can also be used to determine geometry. Couplings between a phosphorus atom and the ^{1}H and ^{13}C nuclei of a ligand located trans to it are typically larger than those to a ligand located cis to it.

Spin coupling between ^{31}P and various metal nuclei, such as ^{195}Pt or ^{183}W, in phosphine complexes can be very large. The magnitude of this coupling is particularly sensitive to the nature of the ligand trans to the phosphorus.[153] The influence of the ligand trans to the phosphine on the magnitude of ^{195}Pt–^{31}P coupling has been studied extensively, and this phenomenon has been termed the "trans influence."[159]

2.3.7. Heavier Congeners of Phosphorus Ligands

Trends among ligands comprised of heavier congeners of phosphorus deserve comment. In general, the strength of the metal–ligand bonding decreases from phosphines to arsines,[154] and presumably from arsines to stibines and bismuthines. Steric effects of the substituents increase in the order Sb < As < P because the M–L bond lengths decrease in the order M–Sb > M–As > M–P. Bismuthines have been little studied. Transition metal reagents or catalysts rarely contain tertiary stibine, or bismuthine ligands, but arsine ligands have been studied more extensively in palladium-catalyzed reactions[154–156] and as a component of nonracemic chiral ligands for enantioselective catalysis[157,158] in recent years. Complexes of mixed-donor chelating ligands have also been prepared.[76] One comparison of the relative rates for P–C and As–C bond cleavage implies that arsine ligands tend to undergo As–C bond cleavage slower than phosphine ligands undergo P–C bond cleavage.[137]

2.4. Carbenes

2.4.1. Classes of Free and Coordinated Carbenes

2.4.1.1. Properties of Free Carbenes

As the reader may recall from introductory organic chemistry, carbenes usually are reactive intermediates containing a divalent carbon atom with one nonbonding orbital of π-symmetry and one of σ-symmetry. Carbenes can be more stable in the singlet or triplet states, depending on the substituents at the carbene carbon. Carbenes with electronegative substituents are often more stable in the singlet spin state,[160–165] while carbenes with only hydrogen or alkyl groups are typically more stable in the triplet spin state.[164]

Although free carbenes are usually reactive intermediates, some stable triplet[166] and singlet carbenes[167] have been prepared recently. The stable singlet carbenes have become important ancillary ligands in organometallic chemistry. In 1991, Arduengo isolated crystalline, sterically hindered, singlet carbenes with two amino substituents in a cyclic structure typically called N-heterocyclic carbenes.[168] The nitrogen acts as a strong π-donor to the unoccupied p-orbital of the carbene carbon. Many of these carbenes possess unsaturated, cyclic, six-electron π-systems. The participation of the p-orbital at carbon in this aromatic system makes it even less available for bonding than that in the acyclic diaminocarbenes.

2.4.1.2. Properties of Carbene Complexes

Reactive organic fragments can be stabilized by coordination to a metal center, and many stable complexes of carbenes have been prepared. Several types of carbene complexes

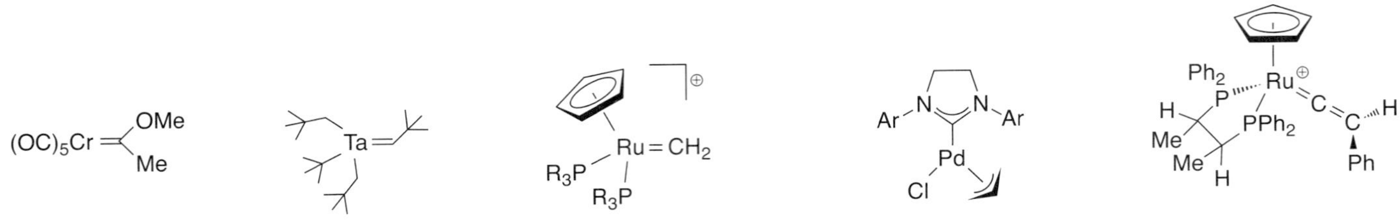

A The first
"Fischer carbene
complex"

B The first
"Schrock carbene
complex"

C Hybrid of a Fischer
and Schrock carbene
complex

D *N*-Heterocyclic
carbene compex

E Vinylidene complex

Figure 2.14. Examples of five classes of carbene complexes.

are shown in Figure 2.14. Some excellent reviews of the preparation and properties of carbene ligands have been published.[169–181] Other reviews focus on the structure of their complexes[182] and on the application of these complexes to organic synthesis.[183,184]

Coordination of a carbene ligand to a metal generates a complex with a formal metal–ligand double bond in most cases. These complexes with metal–ligand multiple bonds have been used commonly as stoichiometric reagents in organic synthesis and are crucial intermediates in important catalytic processes, such as cyclopropanation and olefin metathesis. Although carbenes are well-known organic reactive intermediates, and carbene complexes are common organometallic compounds, carbene complexes neither give rise to, nor are typically generated from, free carbenes. The exceptions to this trend are *N*-heterocyclic carbenes, which are now common ancillary ligands. *N*-heterocyclic carbenes bind to transition metals through metal–ligand single bonds and the complexes are often prepared from the free carbene.

The carbene carbon of a carbene complex can react as a nucleophile or an electrophile. Generally speaking, coordinated carbenes that possess electronegative substituents are electrophilic. These carbenes are typically bound to late metals of groups 6–8 in a low oxidation state bearing π-accepting ligands, such as CO. The metal–carbon bonds in complexes of these metals are not strongly polarized. This property of the metal–ligand bond and the presence of an electronegative substituent at the carbon helps make these complexes electrophilic. These carbene complexes are called "Fischer carbenes" because E. O. Fischer was the first to prepare an example of them (**A** in Figure 2.14).[169] They are typically counted as neutral ligands because of the similarity of the electron donation and backbonding to that in the ligation of CO.

Carbene complexes that possess hydrogen or alkyl groups at the carbene carbon are typically nucleophilic. These carbene ligands are most commonly bound to early transition metals in groups 4–6 in high oxidation states. Organometallic compounds of early metals in high oxidation states have more polar M–C bonds than those of late metals in lower oxidation states. This polarization of the metal carbon bond and the presence of alkyl substituents on the carbene carbon both cause the carbene carbon to be nucleophilic. These carbene complexes are often called alkylidenes or "Schrock carbenes" because Richard Schrock first prepared an example of them (**B** in Figure 2.14). These carbene ligands are generally assigned a dianionic charge because of the polarity of the metal–carbon bond and the nucleophilicity of the carbene carbon.

Two additional classes of common carbene complexes have evolved in the past decade. One reactive class of carbene complex is a hybrid of a Fischer and Schrock carbene (**C** in Figure 2.14). Complexes of this type contain a carbene ligand lacking a heteroatom substituent bound to a late metal in a low oxidation state. These complexes often possess strongly donating ancillary ligands, instead of the carbon monoxide ligands in Fischer carbene complexes, and the carbene ligand often contains an unsaturated hydrocarbyl group such as phenyl or vinyl, which may help to stabilize them. The carbene ligands in these complexes are often considered neutral ligands, even though they lack the alkoxo, amino, or thiolato group at the carbene carbon of a typical Fischer carbene. These carbene complexes have become important precursors and reactive intermediates for olefin metathesis.

$$\text{(2.9a)}$$

$$\text{(2.9b)}$$

$$\text{(2.9c)}$$

$$\text{(2.9d)}$$

A fourth class of carbene complex is comprised of diamino and N-heterocylic carbenes. In 1968, Ofele and Wanzlick independently prepared chromium and mercury N-heterocyclic carbene complexes by reaction of chromium carbonyl hydride anion and mercuric acetate with N,N-dimethyl and N,N-diphenyl imidazolium salts (Equations 2.9a and 2.9b). In 1971 and 1972, Lappert reported that electron-rich Wanzlick diaminoalkenes[185] oxidatively added to a single metal center to form biscarbene complexes or to two low-valent metals to form two diaminocarbene complexes, as shown in Equations 2.9c and 2.9d.[186–188] More thoroughly studied complexes of diaminocarbenes are N-heterocyclic carbene complexes that are stable when free in solution. These free carbenes are often called "Arduengo carbenes" because Arduengo first prepared them.[167–190] Particularly important carbene ligands for catalysis have been the N-aryl, N-heterocyclic carbenes. Hindered mesityl (2,4,6-trimethylphenyl) or 2,6-diisopropylphenyl groups at nitrogen make the free carbenes stable as monomers.[167,168,190] Herrmann first demonstrated the utility of these carbenes for catalysis,[191] and the highly active "second-generation Grubbs catalyst" for olefin metathesis (**D** in Figure 2.14)[192] contains a saturated version of an Arduengo carbene. Nolan has developed efficient preparative methods to generate this carbene complex[193] and has shown the utility of Arduengo carbene complexes of palladium for cross-coupling chemistry.[194–199] Others have shown their utility in C–N coupling,[202] reductive coupling,[203] diene telomerization, and many other reactions discussed later in this text.

A final class of carbene complexes includes vinylidenes (**E** in Figure 2.14).[205,206] Vinylidenes possess one methylene substituent at the carbene carbon instead of two substituents connected by single bonds. Free vinylidene is electrophilic, and the singlet state is calculated to lie below the triplet state by 43.1 kcal/mol.[207–210] The electrophilicity of free vinylidene is similar to that of a dihalocarbene. It is highly reactive toward rearrangement to the much more stable alkyne tautomer.

Many examples of vinylidene complexes have been isolated, and the two tautomers of alkyne and vinylidene complexes shown in Equations 2.10a and 2.10b are similar in stability. As shown by these two equations, either can be more stable, depending on the number and identity of the ligands on the metal fragment. Vinylidene complexes have also been invoked as intermediates in both stoichiometric and catalytic reactions. Vinylidenes are typically electrophilic at the carbene carbon (C_α) and nucleophilic at the metal and C_β. The ^{13}C NMR chemical shift of the carbene carbon typically lies downfield, between 250 and 380 ppm, while the ^{13}C NMR chemical shift of the β-carbon typically lies between 87 and 143 ppm.

$$\text{(2.10a)}$$

$$L = PMe_2Ph$$

$$\text{(2.10b)}$$

$$L = P(OMe)_3$$

2.4.2. Bonding of Carbenes

Carbene complexes possess formal metal–ligand double bonds. As presented in Chapter 1 on bonding in organometallic compounds, the symmetry of the orbitals involved in bonding of singlet carbene fragments to transition metals is closely related to the bonding of carbon monoxide (see Figure 2.15). In a singlet carbene, the orbital of σ-symmetry contains an electron pair, while the orbital of π-symmetry is unoccupied. Thus, the singlet carbenes donate two electrons to the metal in a dative bond and accept d-electrons in a π-backbond.[160] These carbene complexes often have weak π-bonds; the barrier to rotation is typically 8–10 kcal/mol.[211,212] This π-bonding is weak because of a large energy difference between the occupied metal orbital and the carbene p-orbital. The weakness of this bond and the presence of two orthogonal orbitals at the metal with the appropriate symmetry to overlap with the carbene p-orbital make the barrier to rotation small. A 90° rotation generates a new π-bond instead of fully cleaving the π-bond.

The bonding of a triplet carbene is less closely related to the metal–ligand bonds discussed thus far. The carbenes in these complexes are considered dianionic. With this assignment of electrons and charges, the σ- and π-orbitals of this dianionic fragment can be considered to bind the metal by donation of two electron pairs, one into each of the unoccupied orbitals of the metal of σ- and π-symmetry. These carbene ligands have π-bonds that are stronger than those in Fischer carbenes, but much weaker than those in alkenes. The barrier to rotation in these complexes is typically 19 kcal/mol.[213,214]

N-Heterocyclic carbenes are strong σ-donors and weak π-acceptors. These ligands are strong σ-donors because the carbon of these carbenes is soft and less electronegative than most heteroatom Lewis bases. These ligands are weak π-acceptors because the p-orbital of the carbene carbon participates in strong π-bonding with the amino substituents. This strong σ-donation from a soft donor and lack of π-acceptance makes the electronic properties of *N*-heterocyclic carbenes similar to those of alkylphosphines.

Relative to phosphines, however, the carbenes are stronger σ-donors, have stronger M–L bond strengths, and have steric properties that are distinct from those of a ligand with

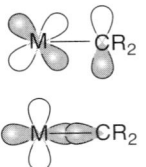

Figure 2.15.
Orbital interactions in the bonding of a carbene ligand.

a central atom bearing three substituents.[215,216] Symmetric *N*-heterocylic carbenes have two-fold rather than three-fold symmetry and tend to leave space unoccupied above and below the N–C–N plane of the ligand. They have been termed "fences" rather than "cones." The hindered *N*-aryl *N*-heterocyclic carbenes bind well to the metal, yet undergo dissociation under mild conditions to generate unsaturated intermediates.

2.4.3. Spectroscopic Characteristics of Carbene Complexes

The ^{1}H and ^{13}C NMR chemical shifts of coordinated carbenes are distinctive.[217] The ^{13}C chemical shifts of the carbene carbon in Fischer carbene complexes with oxygen donors on the carbene resonate between 290 and 365 ppm versus tetramethylsilane. Those with nitrogen donors resonate less far downfield, but generally give rise to signals between 185 and 280 ppm. The ^{13}C chemical shifts of the carbene carbons in Schrock carbene complexes are also far downfield. These carbons typically resonate between 240 and 330 ppm. The ^{1}H NMR chemical shifts of these species are also far downfield and are typically found between 10 and 20 ppm. The infrared vibrations are difficult to locate because the M=C bond vibrates at a low frequency,[218] and these bands are generally not identified.

2.5. Transition Metal Carbyne Complexes

In 1973 E. O. Fischer reported the first complex containing a metal–carbon triple bond.[219] These compounds are named alkylidyne or carbyne complexes. Carbyne complexes are intermediates in alkyne metathesis, as shown by Schrock[220] and developed recently by others.[221–227] Carbyne complexes have been prepared by three general routes[179]: (1) the direct introduction of a carbyne ligand by metathesis of a metal–metal or other metal–ligand triple bond with an alkyne, (2) the transformation of an existing ligand (usually a carbene) into a carbyne ligand within the coordination sphere of the metal, and (3) the modification of existing carbyne complexes to exchange the substituent at the carbyne carbon. Carbyne complexes undergo several reactions, including [2+2] cycloadditions (Chapters 13 and 20) and attack by both nucleophiles (Chapter 11) and electrophiles (Chapter 12).

2.5.1. Bonding and Structure of Carbyne Complexes

Most carbyne complexes have been prepared with group 5–7 metal complexes. The carbyne ligand is typically considered trianionic,[228] and this assignment of charge places the metal in a high oxidation state. This large separation of charge bears no relationship to the true charge at the ligand and at the metal, however, and Roper has recommended that they be considered monocationic ligands to draw analogies to linear nitrosyl ligands.[179] When these ligands are considered trianionic, the formal distribution of electrons in the two fragments is similar to that of a Schrock alkylidene. As you will see later in this text, the carbyne ligand can be susceptible to nucleophilic attack, and no atom with a large true negative charge would be the site of such reactivity.

The bonding interactions of a carbyne ligand are essentially those of a metal carbene complex, but with an additional π-bond (Figure 2.16). One orbital of σ-symmetry and two of π-symmetry overlap with three metal orbitals of appropriate symmetry. When considered trianionic, all three orbitals of the ligand fragment contain two electrons. Theoretical studies of heteroatom-substituted or "Fischer-type" carbyne complexes[229,230] indicate that the HOMO predominantly consists of the metal fragment, and the LUMO consists of one of the π*-orbitals of the metal–carbon bond. This result explains the tendency of nucleophiles to attack the carbyne carbon in carbyne complexes, just as they attack the carbene carbon in Fischer carbene complexes.

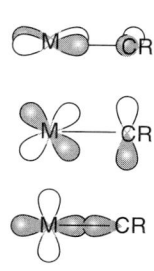

Figure 2.16.
Orbital symmetries involved in the bonding of an alkylidyne ligand.

The electronic character of the carbyne ligand is markedly influenced by the substituent on the carbyne carbon. For example, a π-donor group, such as R_2N, decreases the π-acidity of the carbyne, while an electropositive group, such as R_3Si, increases the π-acidity. A pure σ-donor, such as an alkyl group, increases the energies of all the orbitals. This effect on the orbital energies, along with the electrophilicity of the metals in a high oxidation state, causes the metal–carbon triple bond of an alkyl-substituted carbene complex to be polarized such that the carbyne carbon is nucleophilic. This trend mirrors the high nucleophilicity of Schrock carbenes that results from the high oxidation state and polarization of the metal–carbon bond.

Consistent with the metal–ligand triple bond, structural studies reveal that transition metal carbyne complexes possess metal–carbon distances that are short and metal–carbon–X angles that are close to 180°. The relationship between metal–carbon bond distance and bond order is illustrated by the structure of Schrock's tungsten complex in Figure 2.17, which contains a metal–carbon single, double, and triple bond.[231] The M–C distances between the metal and the alkyl, carbene, and carbyne ligands reflect the M–C bond orders. Metal–carbyne distances are affected little by the type of substituent on the carbyne carbon. The similarity of the M–C distance in the two chromium carbyne complexes in Figure 2.17,[232,233] as well as others,[179] shows this constancy of the metal–carbyne bond length.

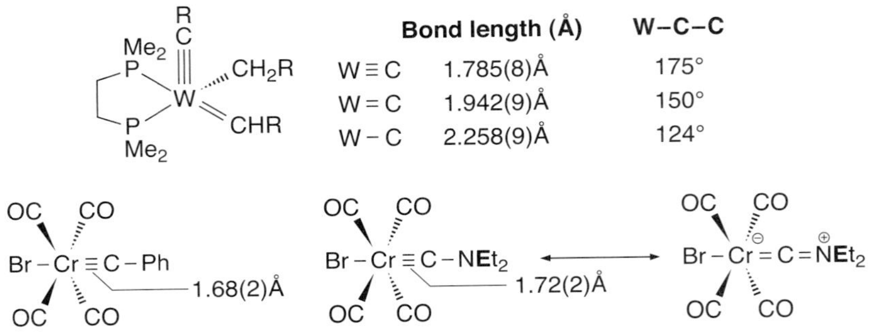

	Bond length (Å)	W–C–C
W≡C	1.785(8)Å	175°
W=C	1.942(9)Å	150°
W–C	2.258(9)Å	124°

Figure 2.17.
X-ray diffraction data of a complex with a carbyne, carbene, and alkyl group and two carbyne complexes with different substituents.

2.5.2. Spectroscopic Characteristics of Carbyne Complexes

The ^{13}C NMR resonances of carbyne complexes fall between 200 and 350 ppm downfield of tetramethylsilane. Detailed IR and Raman studies[179,234,235] show the metal–carbon triple bond vibrates between 1250 and 1400 cm^{-1}, but that the M–C vibration is strongly coupled with vibrational modes of the phenyl ring in aryl-substituted carbyne complexes. The carbyne complex in Figure 2.18 illustrates these spectroscopic features.[179] Neutral carbyne complexes exhibit C–O bands at low frequencies, and this low frequency shows that the carbyne ligands withdraw little electron density from the metal.

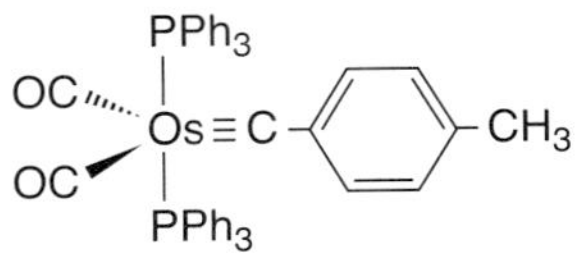

$\nu_{Os≡C}$	: 1375 cm^{-1}
ν_{CO}	: 2010 and 1944 cm^{-1}
^{13}C NMR CO	: δ 183 (t, $^2J_{CP}$ = 9.5 Hz)
Os≡C	: δ 331 (t, $^2J_{CP}$ = 11 Hz)

Figure 2.18. Spectroscopic data of a carbyne complex.

2.6. Organic Ligands Bound Through More than One Atom

2.6.1. Olefin Complexes

Alkenes are common, neutral, two-electron ligands for transition metals. In fact, the oldest transition metal organometallic compound, Zeiss's salt (Figure 2.19), contained coordinated ethylene. Many olefin complexes are commercially available and serve as starting materials for further synthesis. The olefin can be a supporting or spectator ligand, or it can be the point of reaction in either stoichiometric or catalytic transformations. Catalytic processes that are presented in later chapters that involve reactions of coordinated olefins include alkene hydrogenation, dimerization, polymerization, cyclization, hydrocarbonylation, hydrocyanation, isomerization, and certain types of oxidation, such as the "Wacker–Schmidt" process. In recent years, chiral diene ligands (Figure 2.19) and hybrid phosphine-alkene ligands have also been designed that have become valuable for catalytic asymmetric processes.[236–240]

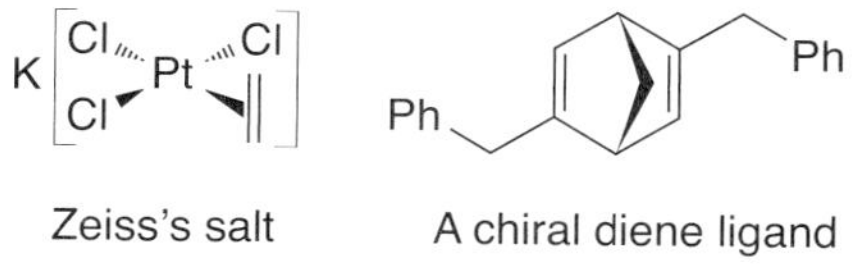

Figure 2.19.
The first olefin complex and a modern chiral diene for asymmetric catalysis.

Zeiss's salt   A chiral diene ligand

As noted in Chapter 1, olefins bind to metals through σ-donation and π-backbonding. This combination of σ-donation and π-backbonding in the coordination of olefins to metals is summarized in Figure 2.20, and has been termed the Chatt–Dewar–Duncanson bonding model.[241,242] This bonding model can be used to explain the stability, structure, spectroscopic features, and reactivity of olefin complexes. The alternative valence bond model, which includes the two resonance forms noted in Figure 1.23 (Section 1.3.7.3), is also included in Figure 2.20. The stability, structure, and spectroscopy of these complexes are presented here. The reactivity of olefin ligands pervades the chapters that describe the reactions of organometallic complexes.

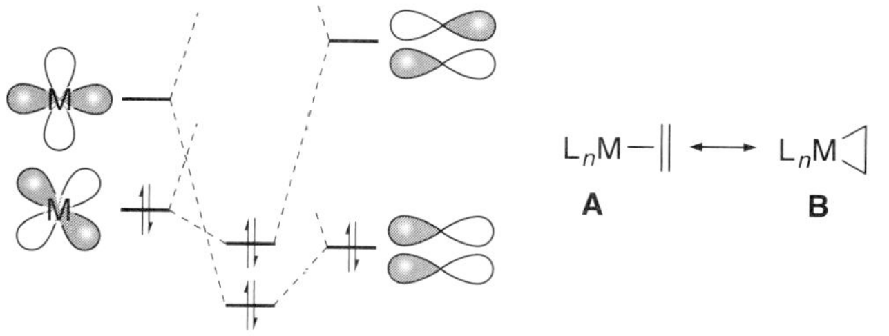

Figure 2.20.
Orbital symmetries and the valence bond description of transition metal olefin complexes.

$$L_nM—\| \longleftrightarrow L_nM◁$$
A   **B**

2.6.1.1. Stability of Metal–Olefin Complexes

The relative magnitude of σ-donation and π-acceptance in a metal–olefin bond depends upon the metal and upon the olefin substituents. Many relative binding constants of various olefins to metal centers possessing different electronic properties have been measured, and the relative stabilities depend on the match of the electronic properties of the olefin and the metal center.[243] For example, electron-rich metal centers, such as Ni(0), bind more strongly to olefins bearing electron-withdrawing substituents, such as cyano or carbalkoxy groups, but electrophilic metal centers, such as Pd(II), bind more weakly to olefins bearing electron-withdrawing substituents. Maleic anhydride binds roughly 10^6 times more strongly to (bipyridine)Ni(0) than does vinyl acetate (Equation 2.11a),[244] but methacrylate binds 10^6 times more weakly to the cationic palladium complex [(bis-imine)Pd(Me)]$^+$ in Equation 2.11b than does ethylene and 10^3 times more weakly than propene.[245] Similarly, styrenes containing electron-withdrawing groups have been shown to bind more weakly to a neutral Pd(II) complex than do styrenes with electron-donating groups.[246]

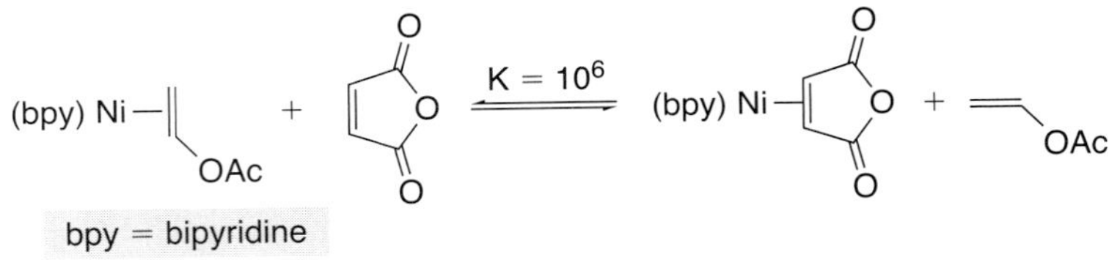

$$(2.11a)$$

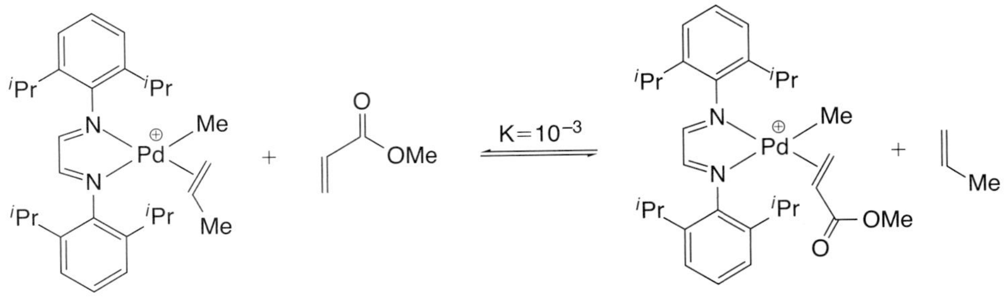

bpy = bipyridine

$$(2.11b)$$

Olefin ligands are less common in complexes of metals in higher oxidation states, with charges greater than +1, and with d^0 electron counts. Yet, olefin complexes of d^0 metals are important intermediates in olefin polymerization. Because of the importance of these d^0 olefin complexes, the direct observation of such complexes has been sought, and such olefin complexes have recently been identified, usually at low temperatures.[247–260] Two examples of d^0 olefin complexes generated from the intermolecular coordination of ethylene are shown in Figure 2.21.

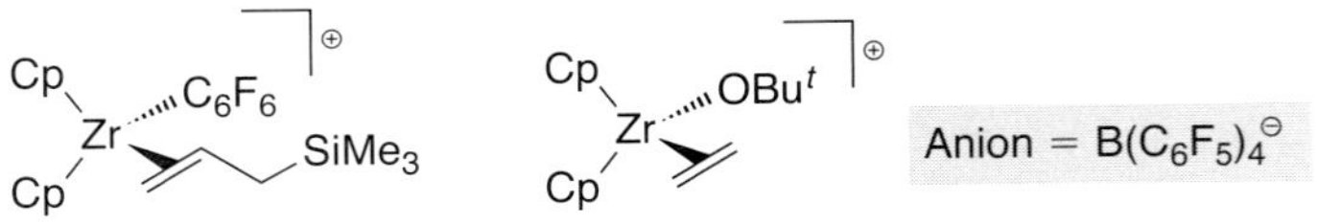

Figure 2.21.
Two examples of d^0 olefin complexes observed at low temperature.

The stability of olefin complexes is also sensitive to steric effects. The binding of ethylene is stronger than that of α-olefins in nearly all cases. However, the magnitude of the steric effect on the binding of an olefin depends on the other ligands. For example, the ethylene complex of bis(amine)PdMe$^+$ in Equation 2.11b is 10-fold more stable than the propylene or hexene complex, while the ethylene complex of PtCl$_3^-$ is only two-fold more stable than the corresponding propene complex.[243]

Dienes that are able to act as bidentate chelating ligands form especially stable transition metal complexes because of the chelate effect. Examples of stable complexes of dienes, such as 1,3-butadiene, norbornadiene, and 1,5-cyclooctadiene, are illustrated in Figure 2.22.

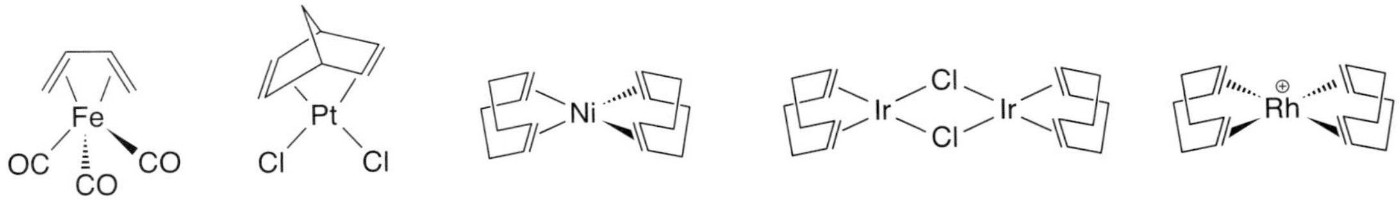

Figure 2.22. Common diene complexes of metals used as starting materials or as catalysts themselves.

2.6.1.2. Structures of Metal–Olefin Complexes

2.6.1.2.1. Structural Changes Upon Binding

Binding of olefins to metals alters the bond lengths and angles from those in the free olefin. The flow of electron density from a filled orbital at the metal into the π^*-orbital of the ligand reduces the order of the π-bond. This reduced bond order lengthens the C–C bond of a coordinated alkene. Greater backbonding occurs in the bonding of olefins to metals in low formal oxidation states, and the alkene C–C bond length changes more upon coordination to these metals than it does upon coordination to metals in higher oxidation states. Data consistent with this trend are shown in Figure 2.23. The length of the C–C bond in the Pt(II) Zeiss's salt (Figure 2.23, left) is similar to that in ethylene, but the length of the C–C bond in the related Pt(0) complex (Figure 2.23, right)[261] is a longer 1.43 Å. The latter distance falls between the length of a typical carbon–carbon double bond (1.34 Å) and a typical carbon–carbon single bond (1.54 Å).

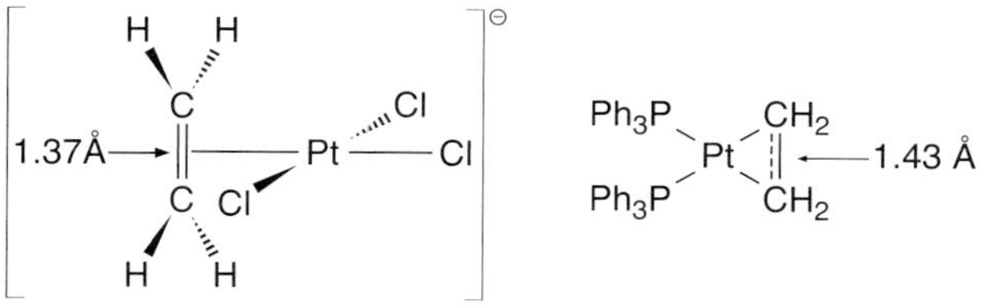

Figure 2.23.
Changes in C–C distances upon binding of ethylene to an electron-poor Pt(II) and electron-rich Pt(0) center.

The change in hybridization of the carbons of the olefin toward sp^3 upon coordination to a transition metal relieves strain in strained olefins. Thus, strained olefins bind more strongly than unstrained olefins if the steric and electronic properties of substituents are similar. This effect is illustrated by the displacement of cis-cyclooctene by trans-cyclooctene (Equation 2.12).[262]

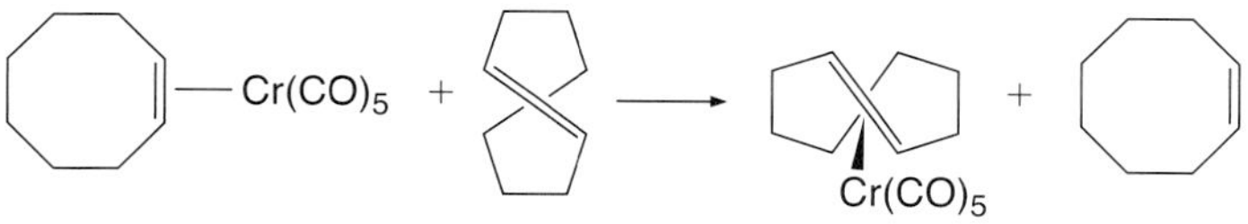

$$(2.12)$$

2.6.1.2.2. Orientation of Coordinated Olefins

A coordinated olefin is often most stable in one orientation, relative to the other ligands at the metal. In many cases, the barrier to rotation about the metal–olefin bond axis can be measured by solution NMR methods.[263–267] The rotational barriers (10–25 kcal/mol) arise from a combination of steric and electronic effects.

A σ-bond lacks a nodal plane along the bond axis. One would expect, therefore, an absence of an electronic preference for orientation about the metal–olefin axis in a complex in which the olefin acts only as a σ-donor. In contrast, one would expect an electronic preference for the orientation of an olefin to a metal center that acts as a σ-donor and a π-acceptor. This preferred orientation would place the olefin in position to maximize overlap between its empty π^* orbital of the olefin and a filled d-orbital on the metal.

Typically, the orientation of the olefin is controlled by the orientation of the metal HOMO that would serve as the electron donor to the π^*-orbital of the olefin. For example, the HOMO of the d^{10} ML$_2$ fragment lies in the plane of the metal and the metal-bound atoms of the two ligands, and as shown in Figure 2.24 the olefinic carbons lie in this plane in the most stable geometry. The HOMO of a sawhorse C_{2v} geometry d^8 fragment lies in the plane of the metal and equatorial ligands, and the olefinic carbons lie in the equatorial plane of a trigonal bipyramidal, five-coordinate d^8 olefin complex. The HOMO of an analogous d^6 complex lies along the z-axis, and the olefin then lies along this axis (Figure 2.24).

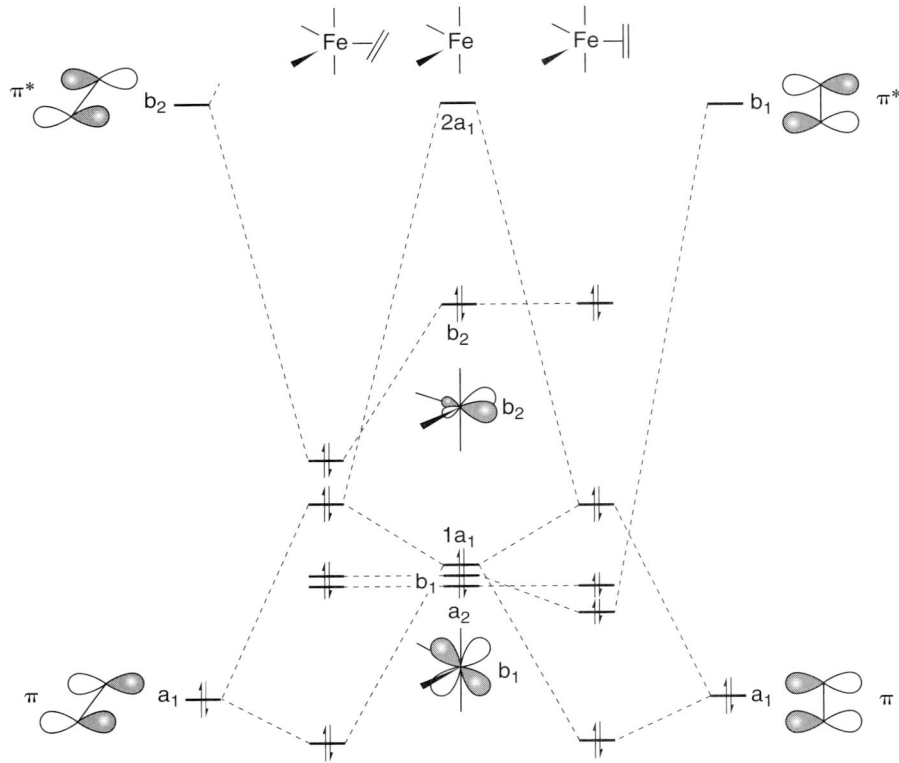

Figure 2.24.
Orientation of olefins in d^{10} trigonal planar, d^8 trigonal bipyramidal, and d^6 trigonal bipyramidal complexes.

The binding of an olefin to a trigonal bipyramidal metal center illustrates the effect of π-bonding on the orientation of an olefin. The orbital interactions between an olefin in a trigonal bipyramidal coordination sphere in two orientations of the trigonal plane are shown in Figure 2.25.

Figure 2.25.
Orbital diagram to predict the conformation of olefins in trigonal bipyramidal complexes. From Figure 19.3 of Albright, T. A.; Burdett, J. K.; Whangbo, M.-H. *Orbital Interactions in Chemistry*; Wiley-Interscience: New York, 1985.

If one arbitrarily considers the axis of the metal–olefin bond to be the z-axis, then the metal d_{xz} and d_{yz} orbitals will be at or near the energy level of the HOMO and can both overlap with the olefin π*-orbital. The interaction of the olefin π*-orbital with the d_{xz} and d_{yz} orbitals of the metal fragment then determines the relative energies of the conformer with the C–C bond of the olefin in the equatorial plane (the left of Figure 2.25) or along the axial direction (the right of Figure 2.25). If the ML_n fragment is axially symmetric about the z-axis, as in a square-pyramidal ML_n fragment, then the d_{xz} and d_{yz} orbitals are degenerate. In this case, there is little electronic origin to the barrier to rotation, and typical values for the barrier to rotation are less than 10 kcal/mol. If the ML_n fragment lacks axial symmetry about the z-axis, but the d_{xz} and d_{yz} orbitals are similar in energy, then the rotational barriers are also low. In these two cases, the origin of the barrier to rotation is largely steric. If the ML_n fragment lacks axial symmetry about the z-axis and the d_{xz} and d_{yz} orbitals are very different in energy [e.g. Pt(alkene)(PPh$_3$)$_2$], then the barrier to rotation can be closer to 25 kcal/mol. These same geometries are found in complexes of other ligands that act as π-acids, such as acetylenes, allenes, azo compounds, and η^2-O_2. For a more thorough analysis, refer to the comprehensive reviews.[261,268,269]

2.6.1.3. Spectral Properties of Metal–Olefin Complexes

The ^1H and ^{13}C NMR resonances of an olefin complex of an electron-poor metal center lie close to those of a free olefin. These metal centers act as weak π-backbonders, and the olefin complexes can be described with resonance structure **A** of Figure 2.20 in this case. For example, the ^1H and ^{13}C NMR resonances of the olefin in Zeiss's salt, K[Pt(C$_2$H$_4$)Cl$_3$], are at 4.76 ppm and 67 ppm, and the ^1H and ^{13}C NMR resonances of the olefin in the neutral (COD)PtMe$_2$ complex are at 4.65 and 98.8 ppm.[270] In contrast, the ^1H and ^{13}C NMR resonances of an olefin complex of an electron-rich metal center lie closer to the region of alkyl resonances than olefinic resonances. These metal centers are strong π-backbonders, and the corresponding olefin complexes can be described by resonance structure **B** in Figure 2.20, which contains sp^3-hybridized carbons. For example, the ^1H and ^{13}C NMR resonances of the Pt(0) complex (PPh$_3$)$_2$Pt(C$_2$H$_4$) lie upfield at 2.15 and 39.6 ppm[270] of the corresponding resonances of the Pt(II) olefin complexes. Tolman[271] has shown that the change in chemical shift upon binding correlates with the equilibrium constant for displacement of a phosphine ligand in NiL$_3$ by olefin. This relationship exists because these Ni(0) complexes are stabilized by strong π-backbonding.

Infrared spectroscopy of olefin complexes is a less useful probe of π-bonding than infrared spectroscopy of CO complexes. Binding of an olefin to an electron-rich metal center does reduce the C–C stretching frequency, as one would expect from the reduction of the C–C bond order due to π-backbonding. However, the C–C stretch of a coordinated olefin is weaker than that of coordinated carbon monoxide because the vibration of the olefin creates a smaller change in the dipole moment. (Recall that symmetric vibrations are not observed in the infrared spectrum because of a lack of change in the dipole moment.) Thus, the olefin stretch is weak and lies at a frequency that overlaps with other bands.

2.6.2. Alkyne Complexes

Metal complexes of alkynes are intermediates in the cyclooligomerization of acetylene to form arenes and cyclooctatetraene (Section 10.5), polymerization of alkynes to polyacetylenes, and the metathesis of alkynes. The bonding of alkynes can be described by three interactions: σ-donation by the one set of π-bonding orbitals, π-bonding by the orthogonal set of π-bonding orbitals, and π-backbonding into a $\pi*$ orbital of the alkyne. These three interactions were described in Chapter 1 and are repeated in Figure 2.26.

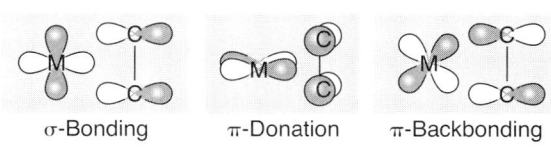

σ-Bonding π-Donation π-Backbonding

Figure 2.26.
The three bonding interactions of metal–alkyne complexes.

2.6.2.1. Structural Characteristics of Alkyne Complexes

Many of the structural changes that occur upon bonding of an olefin to a metal also occur upon bonding of an alkyne to a metal. The acetylenic group is distorted toward the geometry of a cis olefin[272] when coordination occurs. The carbon–carbon bond becomes longer, and the two substituents bend away from the metal center. The degree of distortion varies with the degree of orbital overlap in the metal–alkyne complexes. Often the π-back-donation from a metal into the π*-orbital of an alkyne is strong enough that the complex adopts a structure best described as a "metallacyclopropene." In this structure, the metal–alkyne bond is particularly strong, the length and strength of the C–C bond are closer to the parameters of a double bond than that of a triple bond, and the substituents are bent away from the metal.

Like strained olefins, strained alkynes, some of which are unstable when free in solution, form stable complexes with transition metals. Cycloheptyne,[273] cyclohexyne,[274,275] cyclopentyne,[276] and benzyne complexes have all been prepared. The lengths of the coordinated and uncoordinated C–C bonds of most benzyne complexes are equivalent. Thus, the structures and reactions of the benzyne complexes are more like those of arene dianions[286,287] than those of a fragment containing a strained C–C triple bond. In other words, these complexes are better described by resonance structure **B** in Figure 2.27 than by resonance structure **A**.

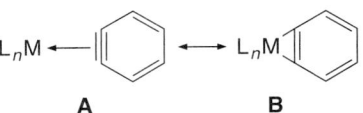

A   B

Figure 2.27.
Resonance structures of benzyne complexes.

The effects of π-donation from the filled alkyne orbital orthogonal to the σ-donating orbital are illustrated by the three structurally characterized Mo(II) diphenylacetylene complexes shown in Figure 2.28. Hoffmann[288] has analyzed the bonding in these complexes. The alkyne ligand in the first complex is adequately described as a two-electron donor, whereas a detailed electronic analysis shows that the alkyne in the porphyrin complex must participate in additional π-donation to the metal. As a result of this difference in bonding, the acetylenic C–C distance increases across the series of compounds, and the distance from the midpoint of the acetylene to the Mo decreases.

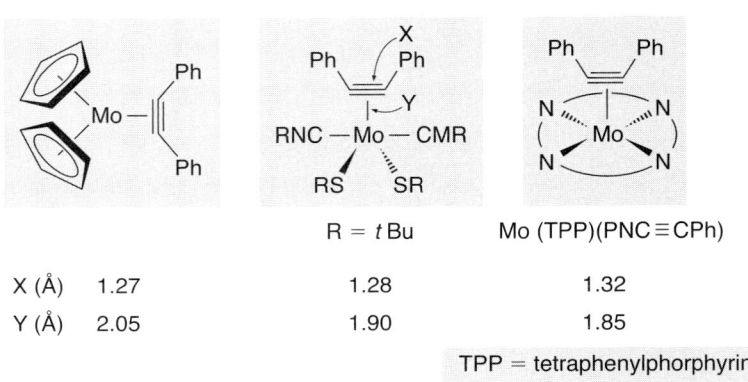

R = tBu   Mo (TPP)(PNC≡CPh)

X (Å)	1.27	1.28	1.32
Y (Å)	2.05	1.90	1.85

TPP = tetraphenylphorphyrin

Figure 2.28.
Effect of acetylene π-donation on the structures of metal–alkyne complexes.

2.6.2.2. Physical and Chemical Properties of Alkyne Ligands

Coordination of an acetylene to a transition metal makes the acetylene stretching frequency infrared "allowed" and shifts the vibration to frequencies that are 150 to 450 cm⁻¹ below that of the Raman band of the free alkyne. Larger changes in the vibrational frequency would be expected when the complex adopts the structure of a "metallacyclopropene" instead of a coordinated alkyne, but this prediction has not been carefully explored. Like alkene complexes, alkyne complexes can possess rotational barriers about the metal–acetylene axis,[268,269] and these rotational barriers depend on the symmetry of the orbitals of the metal fragment.

2.6.3. Complexes of Organic Carbonyl Compounds

Ketones, aldehydes, esters, and amides can bind in an η^1-mode through the carbonyl oxygen or in an η^2-mode that is analogous to the binding of an alkene. Electron-poor metal fragments, such as Lewis acidic early transition metal centers, tend to bind to the oxygen. These metals are harder and are less capable of participating in backbonding into the carbonyl π-system. In contrast, electron-rich metal fragments,[289] such as those of low-valent late metals, often bind aldehydes and ketones in an η^2-fashion and create stable complexes by back-donation into the C=O $\pi*$ orbital. A simple example of an η^2-aldehyde complex is that in Figure 2.29.[289] Some metal fragments, such as the low–valent, cationic, middle transition metal fragment in Figure 2.30, bind with nearly equal energies to the carbonyl oxygen in an η^1-mode and to the π-system in an η^2-mode.[290-294] Carbon dioxide and related hetercumules, such as the isocyanates, can also bind in an η^2-fashion.[295-298]

Figure 2.29.
Two simple η^2-aldehyde and ketone complexes.

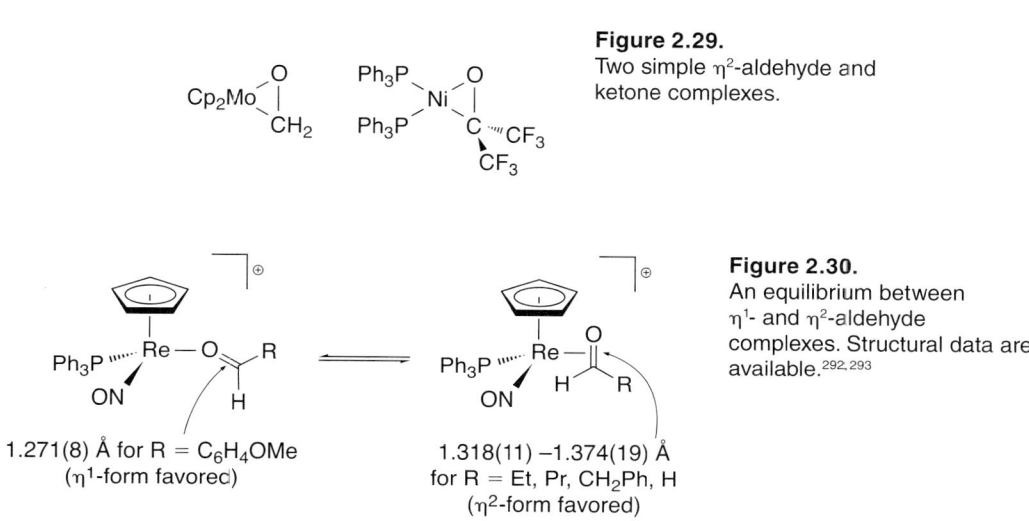

Figure 2.30.
An equilibrium between η^1- and η^2-aldehyde complexes. Structural data are available.[292,293]

1.271(8) Å for R = C_6H_4OMe
(η^1-form favored)

1.318(11) –1.374(19) Å
for R = Et, Pr, CH_2Ph, H
(η^2-form favored)

The C–O bonds of aldehydes and ketones in η^2-complexes are longer than those of aldehydes and ketones in η^1-complexes and those of free aldehydes and ketones. For example, the C–O bond lengths in the propionaldehyde and phenylacetaldehyde bound to rhenium in Figure 2.30,[290,293] are roughly 0.05–0.1 Å longer than the C–O distances in the related η^1-aldehyde complex and 0.10–0.15 Å longer than the C–O distance in a free aldehyde.[299] The carbonyl stretching vibrations of η^2-aldehyde and ketone complexes occur at low energy, frequently 1000–1300 cm^{-1}, and are usually weak. The C–O bond length and vibrational frequency is less perturbed by coordination to a Lewis acid. The carbonyl carbon becomes more electrophilic, but the ground state structure is not severely changed.[300-303]

2.6.4. η^6-Arene and Related Complexes

Arenes form π-complexes with all the transition metals. Many π-arene complexes have been isolated, structurally characterized,[304,305] and studied by a plethora of spectroscopic techniques. The first η^6-arene complexes were prepared by Hein in 1919, but the structures of these compounds were not fully recognized until 1954. The prototype "sandwich" complex, bis(benzene)chromium, Cr(C$_6$H$_6$)$_2$, was prepared by E. O. Fischer in 1955.[306]

The coordinated arene serves many functions in transition metal chemistry. It can be a spectator ligand, a labile ligand that temporarily occupies and then releases vacant coordination sites, a catalyst poison, or a substrate for catalytic and stoichiometric transformations. An arene is a stronger electron donor than three CO ligands. For example, the strong infrared absorption of Mo(CO)$_6$ at 2000 cm^{-1} is higher in frequency than any of the three absoprtions of (mesitylene)Mo(CO)$_3$ at 1970, 1932, and 1865 cm^{-1}.[307] Representative η^6-arene complexes are depicted in Figure 2.31. In such compounds the aromatic ring is typically planar, and the carbon–carbon distances in the arene rings have equal lengths. The parent compound, bis(benzene)chromium, has rigorous D_{6h} symmetry. Electron-releasing substituents on the aromatic ring invariably stabilize the η^6-arene ligand, but steric interactions counteract this electronic effect. The cationic hexamethylbenzene complexes of Tc, Re, Fe, Ru, and Os (Figure 2.31) illustrate the stabilization of cationic arene complexes by electron-releasing substituents. Examples of mixed-ligand η^6-arene compounds are also shown; all are 18-electron complexes. Arene complexes have been extensively reviewed.[308,309]

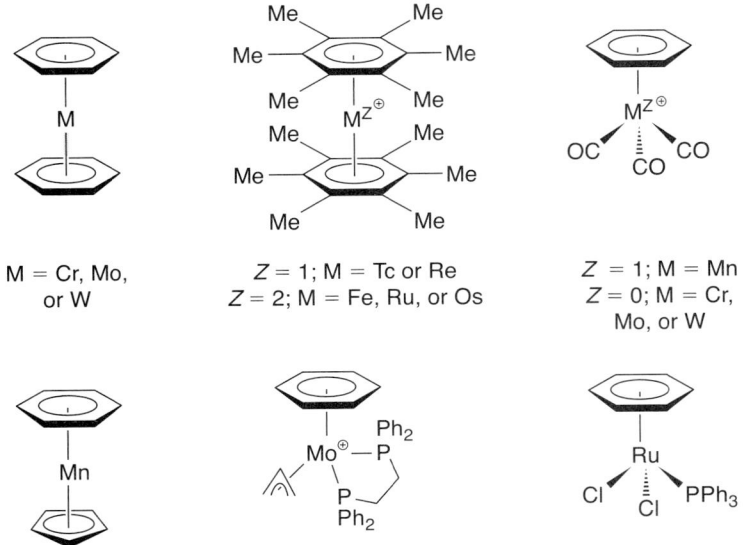

Figure 2.31. Representative η^6-arene complexes.

Arene complexes are usually prepared by the displacement of more weakly bound, monodentate ligands, by oxidation or evolution of CO ligands, or by reducing a metal from a higher oxidation state in the presence of an arene. These methods have been reviewed.[310,311] In some cases, one arene will displace another. For example, hexamethylbenzene replaces *p*-diisopropyl benzene in the tungsten complex of Equation 2.13. This reaction is undoubtedly controlled by steric effects. The mechanisms of these arene exchange reactions have been clarified by Traylor[160] and are discussed in Section 4.55.

$$\text{(2.13)}$$

Arene complexes can also be prepared by metal–atom vapor techniques.[312] This method is most practical with lower-boiling (less-refractory) metals, but chromium arene complexes have been prepared by this method.[313] A metastable iron–benzene complex has been prepared on a 10–20 g scale by this method.[314] Such complexes are often useful inorganic synthons in subsequent reactions.

Arenes bind to metals with orbital interactions like those of other π-ligands, but the orbital interactions are complicated by the larger number of atoms and corresponding larger number of π-orbitals of the arene fragment. An interaction diagram for the orbitals of benene and the orbitals of the CrL_3 fragment was provided in Chapter 1. With the metal–arene bonding axis considered the C_3 axis, a metal orbital of σ-symmetry possesses the appropriate symmetry to overlap with the a_1 orbital of benzene, and two metal orbitals of predominantly d_{xz} and d_{yz} origin have the appropriate symmetry to overlap with the e_{1g} HOMOs of the arene. Back-donation from the metal to the ligand occurs from the orbitals of predominantly d_{xy} and $d_{x^2-y^2}$ origin to the e_{2u} LUMOs of the arene.

Thus, coordination with a transition metal depletes the arene ligand of much of its π-electron density. For example, the η^6-aniline complex $(\eta^6\text{-}C_6H_5NH_2)Cr(CO)_3$ is a far weaker base than free aniline, whereas the η^6-benzoic complex $(\eta^6\text{-}C_6H_5CO_2H)Cr(CO)_3$ is a stronger acid than benzoic acid (Figure 2.32). Even methyl substituents on cationic arene complexes are acidic and can be alkylated in the presence of strong bases.[315] The electron deficiency of coordinated arenes is manifested by many other properties and is especially important in synthetic applications. For example, nucleophilic substitutions of complexed arenes are faster than those of free arenes, while electrophilic substitutions of complexed arenes are slower. Neutral and cationic π-benzene complexes are known, but apparently none have a charge greater than $2+$. Anionic arene complexes are rare.

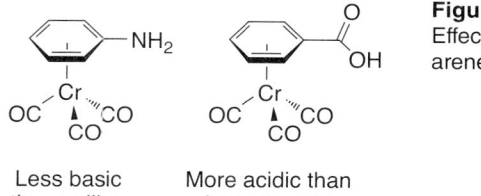

Figure 2.32.
Effect of acid-base properties from arene binding.

Less basic than aniline

More acidic than benzoic acid

In addition to complexes in which arenes are six-electron, η^6-donor ligands, complexes in which arenes are four- and two-electron η^4- and η^2-ligands are known (Figure 2.33). These forms of the ligand are more common as reactive intermediates than as stable complexes, but several examples of η^4- and η^2-arene complexes have been isolated.[316–320] The formulation as an η^6-, η^4- or η^2-arene complex can typically be predicted from the 18-electron rule.

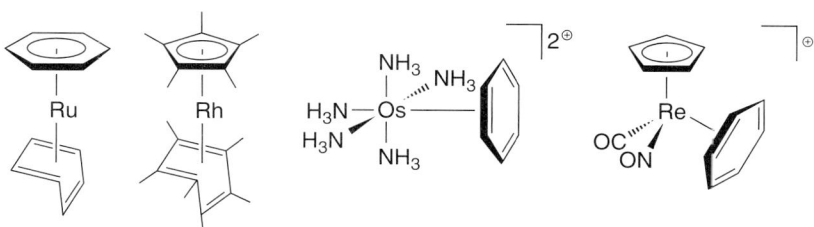

Figure 2.33. Examples of η^4- and η^2-arene complexes.

η^4-Arene complexes display structures with distortions of the ring from planarity and carbon–carbon bond lengths that differ from typical distances.[321–325] The aromaticity is largely lost in these structures, and the arene is bound, at least formally, with the diene portion of a hypothetical "Kekule structure." Structurally well-characterized examples of η^4-arene complexes are the similar ruthenium[321,322] and rhodium[323] complexes in Figure 2.33, which were prepared by reduction of the (η^6-arene) dications. In these complexes the length of the uncoordinated double bond is shorter than that in an arene, whereas the remaining distances exhibit the bond alternation of a diene complex. The dihedral angles between the planes intersecting at the C1–C4 axis are 43° and 48°.

η^2-Arene complexes containing late metals have been known for many years, but the scope and utility of these complexes have increased in recent years. Copper(I) and silver(I) form labile arene complexes of various stoichiometries that are apparently η^2-arene complexes. A few of these complexes have been structurally characterized.[325] More recently, a large number of η^2-arene and heteroarene complexes of osmium,[318] rhenium,[327] molybdenum,[328,329] and tungsten[330] have been prepared for the purpose of dearomatization of the arene or heteroarenes. Two examples are shown in Figure 2.33. This dearomatization creates a diene or vinyl unit that undergoes the organic chemistry of these isolated units,[318,319,331] instead of the chemistry of an arene. η^2-Arene complexes of rhodium[320,332] and platinum[333] have been characterized structurally and studied in the context of their likely intermediacy in the oxidative addition of arene C–H bonds.[334]

Heteroarenes also can form π-complexes with metals, as shown in Figure 2.34. Particularly well-studied examples are complexes of thiophenes, because of the importance of the catalytic hydrodesulfurization of these heteroarenes. Neutral η^4-thiophene complexes are known, as are netural η^2-thiophene complexes with the metal bound to a C–C bond.[335,336] Although most pyridine complexes form by coordination of the nitrogen electron pair, 2-substituted pyridines often bind in an η^6-fashion,[337,338] and fused six-membered heteroarenes, such as quinolines, can bind in an η^6-fashion to either the aryl or pyridyl portion of the heterocycle.[337–339] Main group heterocyclic arenes also form η^6-heteroarene complexes. For example, borazines, which are isoelectronic with arenes, form π-complexes with fragments such as the chromium(0) tricarbonyl shown in Figure 2.34.[340] The neutral iron(II) sandwich complex in this figure[341] is generated from the anion of borabenzene.[342–345] More recently, a number of group 4 and group 5 boratabenzene complexes have been prepared and studied for their ability to polymerize olefins.[346–349] Complexes of silabenzene have also been prepared. Free silabenzene is unstable, but a ruthenium complex of silabenzene has been isolated.[350]

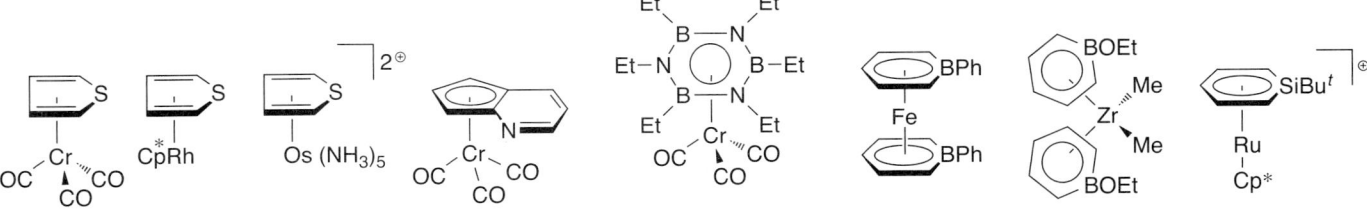

Figure 2.34. π-Heteroarene complexes.

Arenes can also bridge two metals, as shown in Figure 2.35. In one example shown, the arene bridges and sandwiches two palladium atoms. The arene in the dinuclear nickel complex also bridges two metals. The arene in the osmium complex binds symmetrically to each of the three metal atoms on the triangular face of a cluster. This compound provides a molecular view of the chemisorption of benzene on a metal surface.[351]

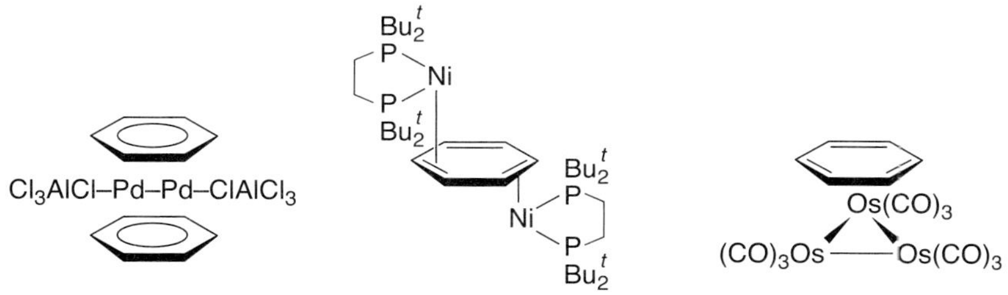

Figure 2.35. Complexes with bridging arenes.

η^2-Arene interactions can be part of a ligand with other points of attachment, as shown in Figure 2.36. η^3-Benzyl complexes formally contain a metal–carbon bond to an anionic benzylic carbon and an η^2-interaction with the C–C bond of the ipso and ortho carbons of the arene ring.[352–359] The structures of these complexes are similar to those of η^3-allyl complexes. Vinylarenes can bind in an η^4-fashion with two of the donor atoms being the ipso and ortho carbons of the arene.

P̂ P = R-tol-BINAP

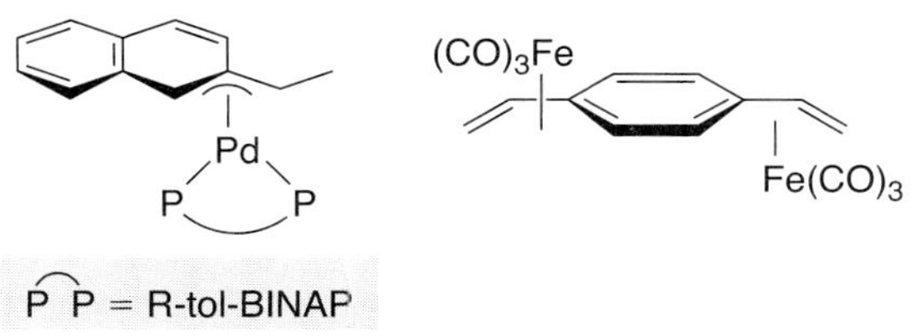

Figure 2.36.
Examples of η^3-benzyl and η^4-vinylarene complexes.

2.7. Complexes of Ligands Bound Through N, O and S

2.7.1. Neutral Nitrogen Donor Ligands

2.7.1.1. Amine Complexes

Ammonia and amines are classic ligands in coordination chemistry, but these Lewis bases are less commonly used as ancillary ligands in organotransition metal compounds. The N–H protons of coordinated amines tend to be reactive, and tertiary amines bind weakly. Because the C–N bond is shorter than the C–P bond and the C–N–C bond angle is larger than the C–P–C bond angle, complexes of tertiary amines tend to be more sterically congested than those of tertiary phosphines. Cone angles for a number of amine ligands have been determined.[126,360] Representative cone angles are provided in Table 2.7. These structural differences between amines and phosphines, along with the hard–soft mismatch with most metals, cause amines to bind to transition metals more weakly than phosphines.

Diamines and polyamines bind more tightly than monoamines, and organometallic complexes containing these ligands are more common than those of monoamines. Three representative complexes are shown in Figure 2.37. (TMEDA)PdMe$_2$[361,362] and (TMEDA)PdPh(I)[363,364] (in which TMEDA = tetramethylethylenediamine) serve as useful starting materials, and the C–H activation and rearrangement chemistry of [(triazacyclononane)Rh[P(OMe)$_3$](R)(H)] and [(trimethyltriazacyclononane)Rh[P(OMe)$_3$](R)(H)] complexes has been studied extensively.[365,366]

Table 2.7. Selected cone angles of amine ligands.

Amine	Cone angle (°)[a]
NH$_3$	94
NH$_2$Et	106
NHEt$_2$	125
NEt$_3$	150
NPr$^i_3$	160
NPh$_3$	166
NPr$^i_3$	220

[a]For comparison, the cone angle of PEt$_3$ is 132° (see Table 2.6).

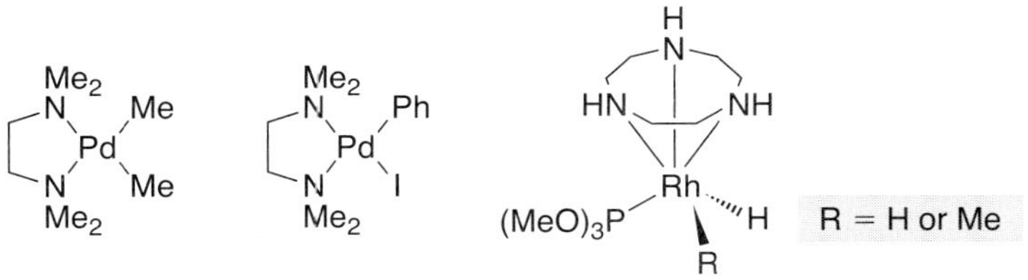

R = H or Me

Figure 2.37.
Representative organometallic complexes with diamine or triamine ligands.

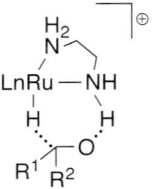

Figure 2.38.
Proposed transition state for the transfer of hydrogen to a ketone from a hydride complex with a coordinated diamine.

Complexes of primary and secondary amines can serve as reactive ligands. For example, 1,2-diamines have acted in a hybrid fashion on catalysts for enantioselective hydrogenation of ketones and imines, serving a role in both controlling structure and delivering the hydrogen to the ketone or imine, as shown schematically in Figure 2.38.[367–369]

2.7.1.2. Pyridine and Imine Complexes

Bipyridines,[370] chelating α-diimines (often called bisimines),[371] bisoxazolines,[372–374] and combinations of imine and oxazoline groups with pyridines[375–377] and phosphines[377] are more common ancillary ligands than amines in organometallic chemistry. Representative examples of these ligands are shown in Figure 2.39. These ligands are more common than amines in organometallic systems because the nitrogen in pyridines, imines, and oxazolines are softer than those in saturated amines, these heteroarenes can act as π-acceptors,[378] and they lack reactive N–H bonds. The major use of bipyridine ligands has been in coordination chemistry, but some catalytic reactions with complexes of bipyridine ligands have been reported.[379–381]

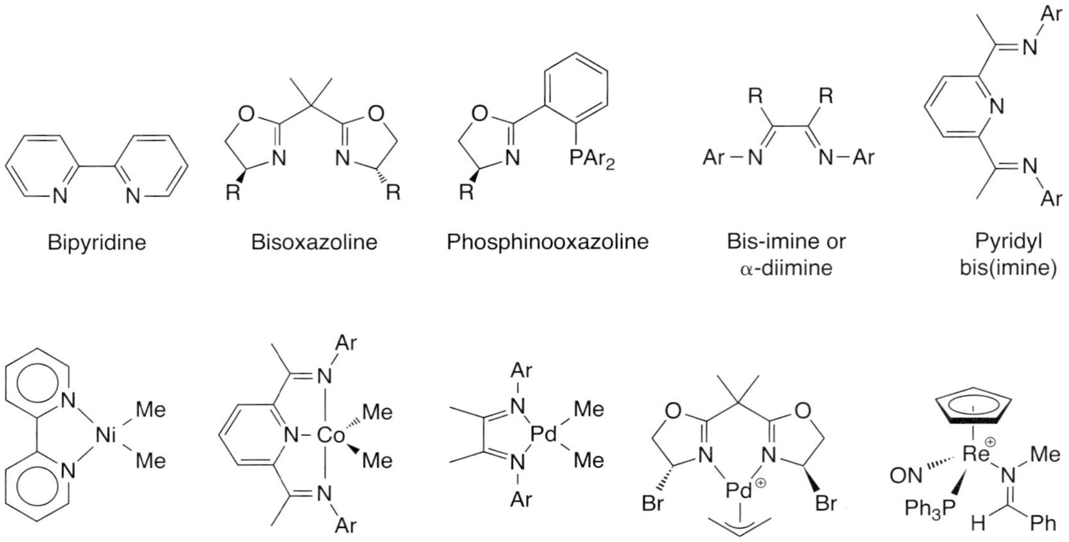

Bipyridine   Bisoxazoline   Phosphinooxazoline   Bis-imine or α-diimine   Pyridyl bis(imine)

Figure 2.39.
Representative pyridine, oxazoline, and imine ligands used in organometallic chemistry and complexes of these classes of ligands.

The use of ligands containing bisimine and oxazoline units has surged in the past decade. These ligands are trivial to prepare, and a rich coordination chemistry had been developed in the past.[371] More recently, middle and late transition complexes containing sterically hindered imine ligands have been shown to be highly active catalysts for olefin polymerization.[382] Less hindered versions act as supporting ligands for related oligomerization catalysts.[383] Complexes of optically active, bisoxazoline ligands and mixed phosphine–oxazoline ligands derived from amino acids are highly selective catalysts for a wide variety of enantioselective transformations, including allylic substitution and hydrogenation.[373]

Monoimines tend to be reactive groups, rather than supporting ligands, in organometallic chemistry. Imines bind more often to transition metals through the lone pair on nitrogen than through the C=N π-system (for these two binding modes, see Figure 2.40). For this reason, imine complexes of transition metals often react with external nucleophiles at the imine carbon as shown, rather than undergoing intramolecular processes that parallel the reactions of η^2-olefin complexes.

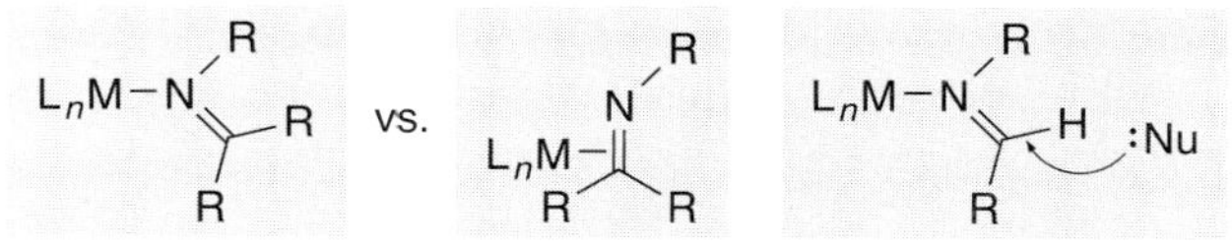

Figure 2.40.
Binding mode of imine ligands.

η^1-Imine complex η^2-Imine complex
More common   Less common

2.7.1.3. Dinitrogen Complexes

The conversion of dinitrogen and hydrogen to ammonia catalyzed by heterogeneous iron catalysts is one of the largest chemical processes, and enzymes containing molybdenum and iron convert dinitrogen, protons, and electrons to ammonia under mild conditions. Thus, many studies of dinitrogen complexes have been conducted with the goal of understanding and mimicking these reactions with small, soluble metal complexes.[384, 385] Dinitrogen complexes are now known with most, if not all, of the transition metals,[385–390] despite the typical inertness of dinitrogen. The first dinitrogen complex, $[Ru(NH_3)_5N_2]^{2+}$, was generated from the reaction of $RuCl_3$ with aqueous hydrazine (Equation 2.14).[391]

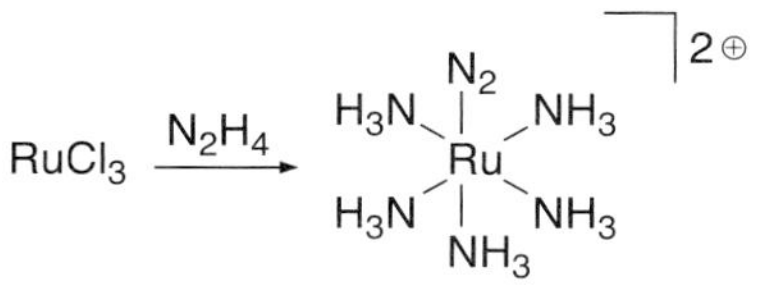

$$(2.14)$$

N_2 is isoelectronic with CO and binds most often as an η^1-ligand. It is less basic and less π-acidic than CO.[392] The π-accepting ability of N_2 is, however, stronger than the σ-donating property, and dinitrogen is generally more electron-accepting than it is electron-donating. Because dinitrogen is such a weak Lewis base, most dinitrogen complexes are formed with metal centers that can act as strong backbonders into the N–N π*-orbital. Thus, most dinitrogen complexes contain electron-rich metal centers. Some of the most stable dinitrogen complexes (see Figure 2.41) have been generated with d^2 metal centers.[34, 393] One bond between the electron-rich d^8 transition metal center $[Ir(PPr_3)_2Cl]$ and dinitrogen was measured to be as high as 36 kcal/mol and calculated by DFT to be 49 kcal/mol.

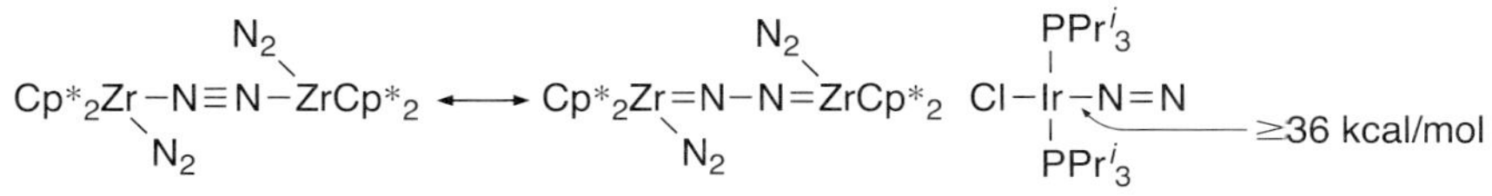

Figure 2.41. Two of the most stable N_2 complexes.

Most dinitrogen complexes are mononuclear and have a collinear arrangement of the metal and the two nitrogen atoms. Structural studies have shown that the N–N bond lengths in most terminal N_2 complexes are within 0.02 Å of that of free N_2. However, the M–N distances are somewhat shorter than the sum of the covalent radii, and this short distance provides evidence for $d\pi$–$p\pi$ bonding. The infrared bands for coordinated N_2 are narrow and intense and, like the bands for CO, are shifted to frequencies that are lower than those of the free molecule (ν_{N_2} Raman = 2331 cm^{-1}) due to backbonding. The IR stretching frequencies for several N_2 complexes are provided in Figure 2.42. Thus, the N–N stretching frequencies reflect the electronic properties of the metal, just as ν_{CO} bands do.

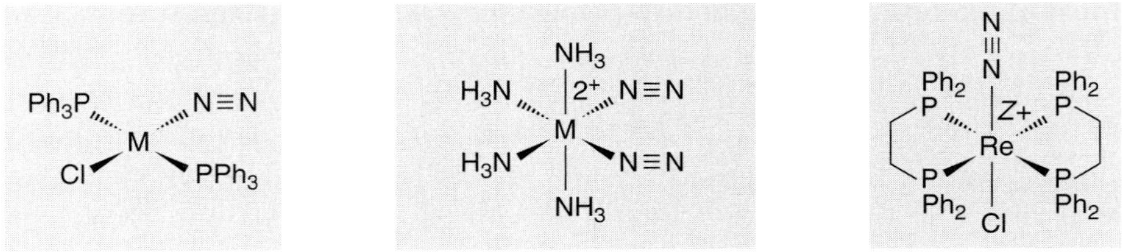

$M = Rh\ \nu_{N_2} = 2152\ cm^{-1}$
$M = Ir\ \nu_{N_2} = 2105\ cm^{-1}$

$M = Ru\ \nu_{N_2} = 2220$ and $2185\ cm^{-1}$
$M = Os\ \nu_{N_2} = 2175$ and $2120\ cm^{-1}$

$Z = O\ \nu_{N_2} = 1980\ cm^{-1}$
$Z = +1\ \nu_{N_2} = 2060\ cm^{-1}$

Figure 2.42. N_2 stretching frequencies for several N_2 complexes.

Although the vast majority of metal dinitrogen complexes are mononuclear with η^1-dinitrogen ligands, a variety of dinuclear dinitrogen complexes have been isolated. Some of the first dinitrogen complexes of early metals were dinuclear, such as the zirconium complex in Figure 2.41, with linear M–N–N–M arrangements.[394–396] Dinuclear dinitrogen complexes of the later metals, such as iron, have also been prepared recently.[397] In both cases, the N–N distance is much longer than that in free dinitrogen. Most dinitrogen complexes of the middle transition metals, such as molybdenum, contain a terminal N_2 ligand, but a molybdenum trisamido fragment has been shown to cleave dinitrogen to form two Mo–nitrido complexes by way of the dinuclear intermediate shown in Scheme 2.1.[398] A derivative of the likely intermediate dinuclear dinitrogen complex has been isolated in this system, and the N–N bond is elongated by roughly 0.1 Å over that in free N_2. The N–N bond cleavage is proposed, based on computational work,[399] to occur through a transition state containing a zig-zag arrangment of the Mo–N–N–Mo atoms.[400]

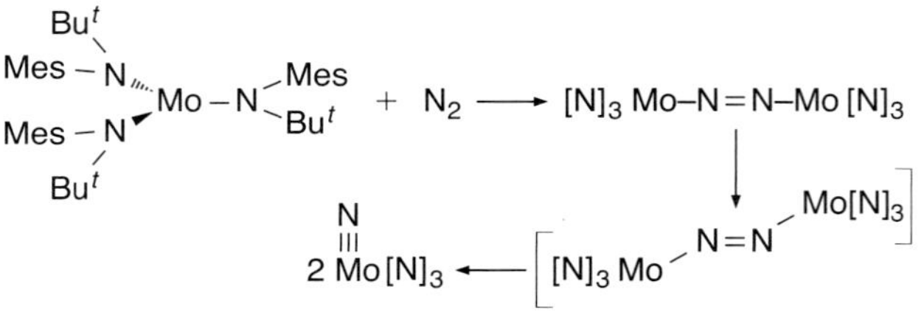

Scheme 2.1

Other dinuclear complexes of dinitrogen contain a side-bound η^2-N_2 ligand (for two examples, see Figure 2.43).[401–408] In these complexes, the N–N bond distance is generally much longer than the N–N distance in free dinitrogen. For example, Fryzuk's original side-bound dinitrogen complex of zirconium contains an N–N distance of 1.55 Å, which is 0.45 Å longer than that in free N_2.

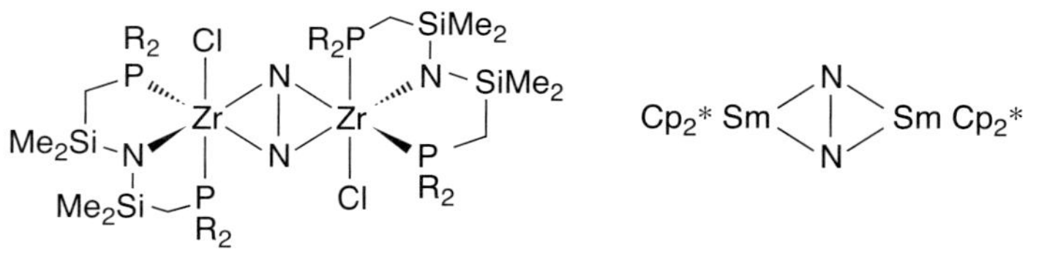

Figure 2.43. Two examples of complexes with side-bound N_2 ligands.

Much effort has been spent to extract reactivity from dinitrogen after coordination to a transition metal.[385,389,390,392] However, this ligand has not displayed the reactivity that is characteristic of unsaturated ligands such as carbon monoxide or ethylene. For example, no examples of migratory insertion and few additions of nucleophiles[409] to coordinated N_2 have been reported.

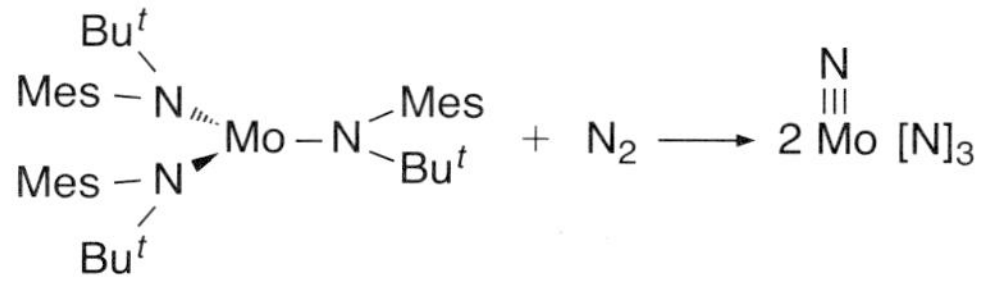

$$(2.15)$$

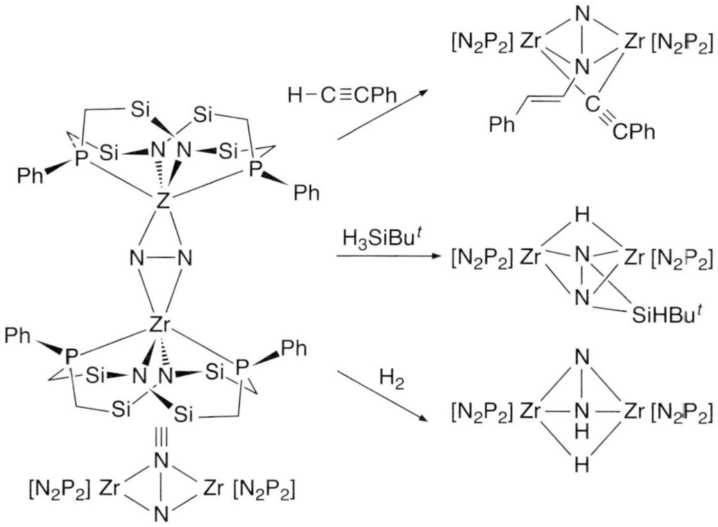

Scheme 2.2

However, reactions of N_2 that are less characteristic of organometallic systems have been observed. In addition to the cleavage of N_2, in Equation 2.15 by the mechanism of Scheme 2.1, reactions of alkynes, silanes, and dihydrogen occur with complexes of zirconium. These reactions occur with complexes containing side-bound dinitrogen ligands, as shown in Scheme 2.2.[404] In one recent case,[407] measurable amounts of ammonia were formed by the reaction of a dinitrogen complex with hydrogen (Equation 2.16). However, the more common chemistry of coordinated dinitrogen involves protonation or alkylation of dinitrogen that is coordinated to highly reduced metal centers, as illustrated in Equation 2.17.[410–412] This reactivity occurs because the N_2 bound to such an electron-rich metal center is reduced enough to behave more like an anionic hydrazido ligand. The product of the protonation process is generally a hydrazine or a hydrazido complex.[385] In one case, this reduction and protonation has been shown to occur catalytically, although the number of turnovers is limited.[413]

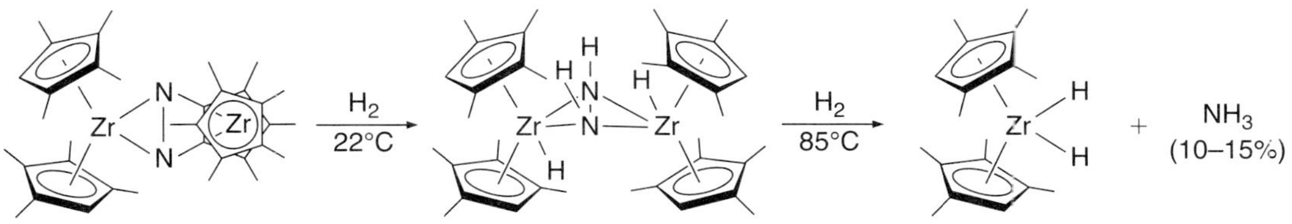

$$(2.16)$$

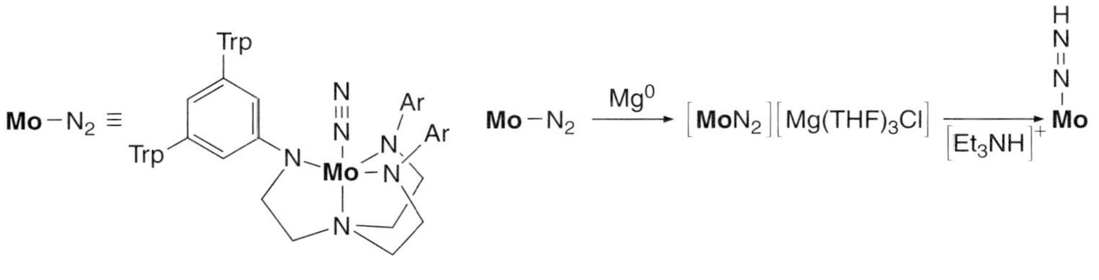

$$(2.17)$$

2.7.1.4. Complexes of Neutral Oxygen Donors

The most common complexes containing dative metal–oxygen bonds in organometallic systems include those with H_2O, MeOH, THF, DME, acetone, or Ar_3PO as the oxygen donor. In addition, DMSO binds to hard metal centers through oxygen.[414] The weakly basic, "hard" oxygen and alcohol ligands bind poorly to low-valent transition metals and even dissociate readily from more oxophilic high-valent metal complexes. Thus, the organometallic chemistry of ether and alcohol complexes is limited. Cyclic ethers, such as THF, bind more strongly than acyclic ethers, and complexes of ethers that can chelate, such as dimethoxyethane (DME), form more stable complexes than simple ethers. Phosphine oxides tend to bind more strongly than ethers. For example, PPh_3O displaces THF from $Cp_2Zr(N-Bu^t)$ (THF) to form the corresponding PPh_3O complex.[415] Despite the weak binding of neutral oxygen donors, the coordination of water, alcohols, ethers, and sulfoxides can influence reaction chemistry, even when stable complexes are not formed. These solvents can bind transiently to unsaturated intermediates to stabilize them or affect their selectivity.

THF, DME, and phosphine oxides are among the more common neutral oxygen ligands found in organotransition complexes. Examples of stable ether complexes are shown in Figure 2.44. In some cases, THF and DME complexes are used as starting materials for further synthesis. For example, the tungsten DME complex is the starting material to prepare Schrock's tungsten alkylidenes, and the DME complex of tungsten is a useful precursor to a variety of organometallic systems, including metallocenes by reaction with cyclopentadienyl anions.[416,417] A variety of $CpRe(CO)_2L$ complexes have been prepared by displacement of THF from $CpRe(CO)_2(THF)$ by softer, more basic dative ligands. In other cases, ether complexes have been used as reagents for synthetic organic transformations. For example, a low-valent niobium DME complex has been used as a reagent for the reductive coupling of imines and aldehydes.[419,420] In other cases, THF, ethers, or phosphine oxides have been used to trap reactive intermediates. In two important examples, the cationic zirconium alkyl complexes that are intermediates in polymerization chemistry were trapped with THF to make an adduct that was characterized by X-ray diffraction.[421] Diethylether adducts of palladium and nickel catalysts for olefin polymerization have been observed directly by low-temperature NMR methods.[422–424] Similarly, two different zirconium imido reactive intermediates[425,426] were identified, in part, by trapping with THF to make isolable THF adducts.

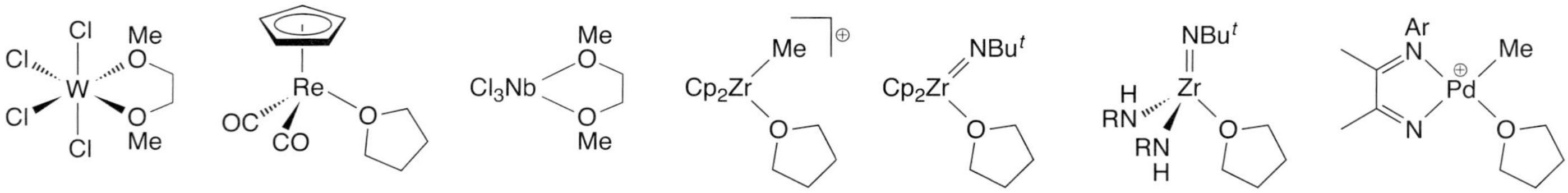

Figure 2.44. Representative ether complexes relevant to organotransition metal chemistry.

In these and related reactions, the ether and alcohol ligands dissociate readily. The coordinated THF dissociates from the zirconium methyl complex to allow olefin polymerization, albeit more slowly than in the absence of THF.[421] The ether dissociates from palladium and nickel to allow olefin to bind to the cationic palladium and nickel species, and alcohol and water are easily displaced from related Pt(II) complexes by hydrocarbons prior to C–H activation processes.[427] Likewise, the THF and phosphine oxide ligands reversibly dissociate from the zirconium imido complex (Equation 2.18) prior to [2+2] additions with alkynes.[415,428]

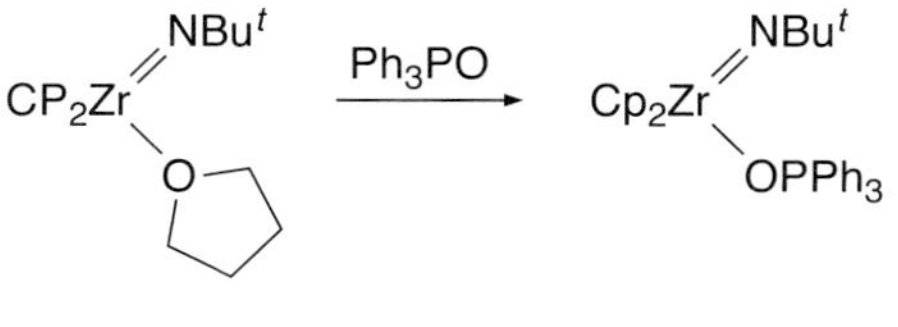

$$(2.18)$$

Several complexes of ligands with a mixture of phosphine and neutral oxygen donors shown in Figure 2.45 have been studied as catalysts for enantioselective transformations. Two well-known ligands are the 2-diphenylphosphino-2′-methoxybinaphthyl (MOP)[429–431] and a related ferrocenylphosphine shown on the left of Figure 2.45.[432–434] Complexes of these ligands catalyze allylic substitution and enantioselective cross-coupling. In these cases, however, the ether group may bind only transiently, if at all, to the soft palladium metal center. Mixed phosphine–phosphine oxide ligands such as BINAP monoxide are now available by a selective oxidation procedure.[435] These mixed phosphine–phosphine oxide ligands, such as DuPhos monoxide, have begun to be useful for enantioselective transformations, such as the alkylation of imines.[436]

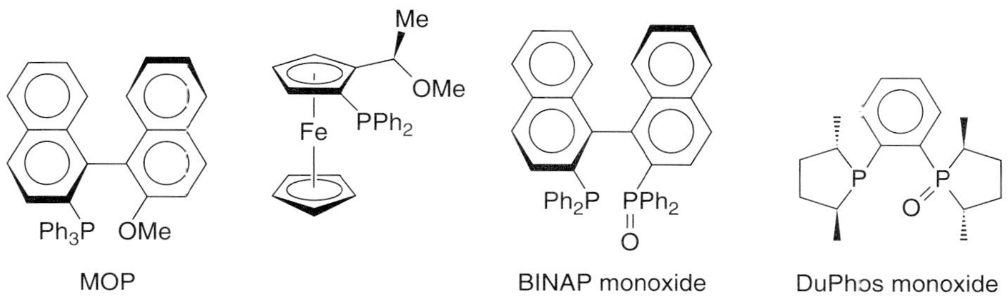

Figure 2.45.
Examples of ligands with ether units that are potential neutral oxygen-donor ligands.

2.7.1.5. Complexes of Neutral Sulfur Donors [437]

Neutral sulfur donors are softer and more polarizable than the neutral oxygen donors of Section 2.7.1.4. Thioethers and sulfoxides are the most common neutral sulfur donors. The trans influence of neutral sulfur donors is greater than that of ethers and is more comparable to that of amines, as based on M–Cl bond lengths. The strength of the metal–thioether bond is usually stronger than that of the metal–amine bond, but weaker than the metal–phosphorus bond. They lie in the spectrochemical series in between chloride and phosphine ligands.

Because the metal–thioether bond is relatively weak, few organometallic complexes contain monodentate thioether ligands. Thus, some complexes contain two thioether donors, such as dimethyl ethanedithiol. However, a majority of the metal–thioether linkages are supported by a second donor. For example, complexes of a number of ligands containing a combination of phosphorus and sulfur donors or carbon and sulfur donors have been prepared. In one case (Figure 2.46, left), a metallacycle with one carbon and two sulfur donors was prepared and studied as a precursor to catalysts for palladium-catalyzed coupling chemistry.[438] In another case (Figure 2.46, right), BINAP monosulfide was used as a ligand for palladium-catalyzed allylic amination.[439]

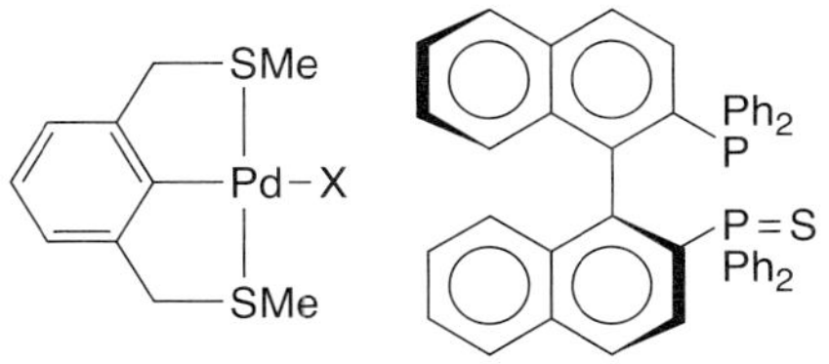

Figure 2.46.
Two sulfur-donor ligands with accompanying carbon or phosphorus donors.

DMSO complexes have also been prepared and studied in contexts related to organometallic chemistry. DMSO binds to low-valent late metals through the electron pair on sulfur, to harder high-valent metals through the electron pairs on oxygen, and to some complexes with a mixture of O- and S-bound forms.[414] Palladium complexes of DMSO and of bidentate sulfoxide ligands[440] have been used as catalysts for the oxidative cyclization of enols and tosylamides,[441–443] and for the allylic oxidation of terminal olefins.[444] Furthermore, DMSO can induce the isomerization of square planar complexes, presumably by coordination and generation of a stereochemically nonrigid five-coordinate intermediate.[445,446] Isolated organometallic DMSO complexes include a few olefin complexes (Figure 2.47, top)[447] and a cyclometallated species related to some catalytic C–H activation (Figure 2.47, middle).[448] The ruthenium complexes at the bottom of Figure 2.47 are classic examples of DMSO complexes.

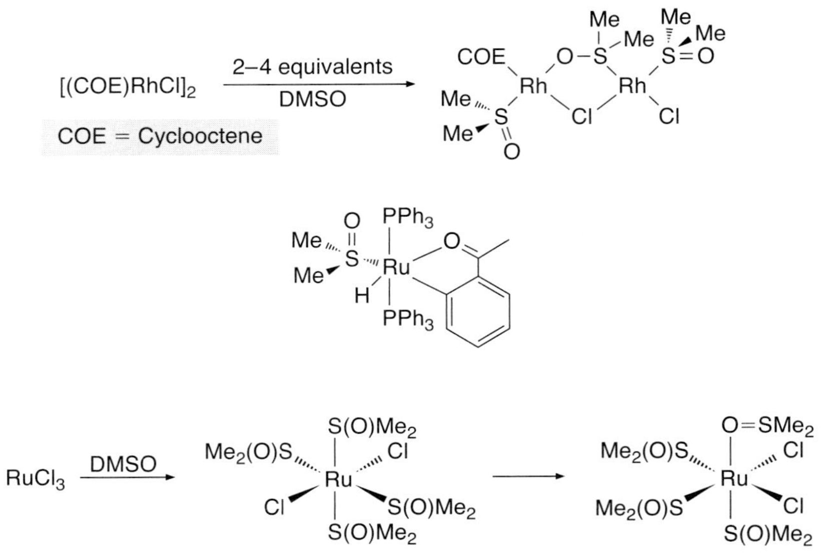

Figure 2.47.
Two DMSO complexes of organometallic systems and a ruthenium DMSO complex with mixed S- and O-donor atoms.

2.8. Sigma Complexes

2.8.1. Overview of Sigma Complexes

A final set of dative ligands, in this case weakly bound dative ligands, includes coordinated dihydrogen, alkanes, silanes, and boranes. These neutral molecules are bound to the metal through H–H, C–H, Si–H, or B–H bonds, respectively.[449] These groups bind to the transition metal by a combination of electron donation from the X–H sigma bonding orbital and back-donation of electron density from the metal into the X–H σ*-orbital (Figure 2.48).[451–454] These interactions resemble σ-donation from the π-bonding orbital and π-backbonding into the π*-orbital in the Chatt–Dewar–Duncanson model. However, the bonding between a metal and an X–H σ-bond is weaker than the bonding between a metal and a π-bond, because the σ-bonding orbital is lower in energy and therefore less basic,

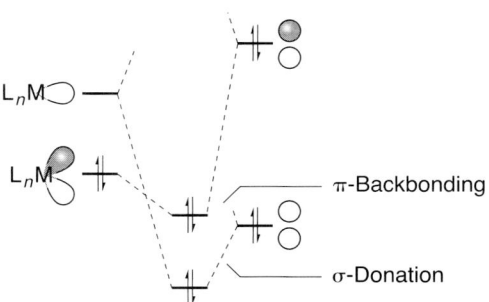

Figure 2.48.
Orbital interactions in a Chatt–Dewar–Duncanson bonding model applied to H_2 complexes.

and the σ* orbital is higher in energy and therefore less π-acidic. One might expect the difference in energies of the σ- and σ*-orbitals to be too large for the metal d-orbitals to interact strongly with both. However, the complexation of the X–H σ-bond leads to lengthening of the X–H bond, and the lengthening of this bond leads to a decrease in the difference in energy between the σ- and σ*-orbitals.[450]

Silane complexes were the first species formed solely by coordination of a metal to a σ-bond.[455–460] The bonding in these complexes was recognized to be unusual and was described as a hydride bridged M–Si bond, instead of the later language of an X–H σ-bond coordinated to a transition metal. The first monomeric example of such a complex was the manganese piano stool complex shown in Figure 2.49.[456,458]

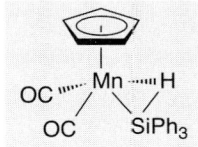

First silane σ-complex   A borane σ-complex

Figure 2.49.
The first silane σ-complex and a borane σ-complex.

$$0.82 \text{ Å by neutron defraction}$$
$$\nu_{H-H} = 2690 \text{ cm}^{-1}$$

(2.19)

For product of addition of H–D J_{H-D} = 33.5 Hz
(versus 43.2 Hz for free H–D)

In 1984, Kubas reported the isolation of [W(PPr$^i_3$)$_2$(CO$_3$)(H$_2$)] in Equation 2.19 from the reaction of dihydrogen with a group 6 metal complex that possessed an open coordination site occupied by an interaction between a ligand C–H bond and the metal (this type of interaction is discussed below).[461] This unsaturated complex bound weak ligands like nitrogen, and unusual IR vibrations were noted for the complex generated by the addition of H$_2$.[462] A series of spectroscopic and structural studies revealed that the complex formed by addition of hydrogen could be described as a dihydrogen complex with H–H bonding. Most striking, the J_{HD} for the complex generated from H–D was 33.5 Hz, and this value is close to the 43.2 Hz value for free H–D. Furthermore, the H–H distance determined by neutron diffraction was 0.82 Å, which is near the H–H distance of 0.74 Å in free H$_2$. The isolation of this complex led to a wide range of studies of dihydrogen complexes.

Considering this result and proposals of alkane complexes in matrix isolation studies,[463,464] both before and after Kubas's discovery, Bergman explained results of H–D exchange between the hydride and cyclohexyl group of the complex in Equation 2.20 and kinetic isotope effects for reductive elimination by invoking an intermediate in which the alkane is bound to the transition metal through a C–H bond.[465] This proposal of an alkane complex is now fully accepted,[466] although the precise structure of such a complex in solution is unknown.

via

(2.20)

Most recently, several examples of borane complexes have been isolated (for one example, see Figure 2.49).[467–471] NMR spectroscopy provided evidence for B–H bonding, in addition to M–H and M–B bonding. In the manganese complex of Figure 2.49, NMR spectral data included B–H scalar coupling. Theoretical studies[468] show that the bonding interactions in these borane complexes are different from those between alkanes, silanes, and dihydrogen and transition metals. Because the LUMO of the borane is the unoccupied *p*-orbital, back-donation occurs into the boron *p*-orbital instead of the X–H σ*-orbital. The lower energy of the *p*-orbital, relative to a σ*-orbital, creates stronger backbonding, and the borane complexes tend to be more stable than alkane or silane complexes.[467]

The interaction between a X–H σ-bond and a transition metal can be different than a hydrogen bonding interaction between a metal center or a metal hydride and a proton donor.[472] One way to consider the distinction between a σ-complex and a complex formed by hydrogen bonding is by the number of electrons involved in the bonding. In a σ-complex, there is a dual flow of electrons from the X–H σ-bonding orbital into an unoccupied metal and from the metal into the X–H σ*-orbital. In a hydrogen bonding interaction, there is a single flow of electrons from the filled metal hydrogen bond acceptor orbital to the proton of the hydrogen bond donor (Figure 2.50). For example, complexes have been prepared in which an acidic X–H bond coordinates to the axial position of a square planar platinum compound (Figure 2.51)[473–479] by an interaction best be described as a hydrogen bond to the metal.[480]

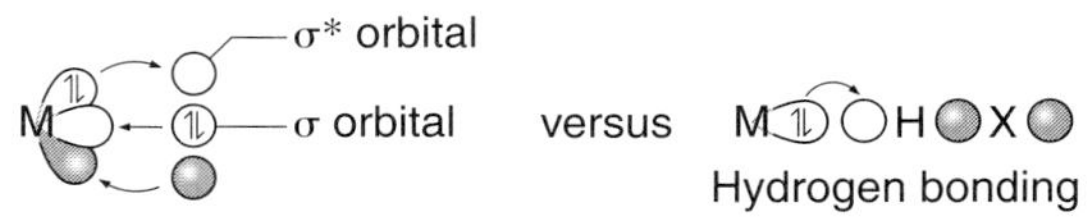

Figure 2.50.
Comparison of electrons involved in a σ-complex and in hydrogen bonding.

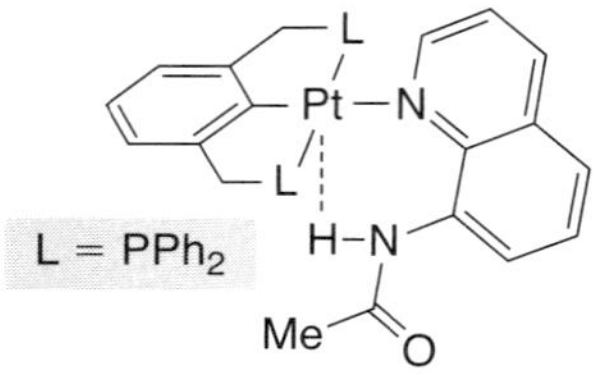

Figure 2.51.
A complex involving hydrogen bonding from an N–H proton to the axial position of a square planar platinum.

Complexes have also been generated in which an acidic X–H bond coordinates intramolecularly or intermolecularly to a metal hydride.[481–489] Three examples of this unusual type of hydrogen bond are shown in Figure 2.52.[484,487] This type of hydrogen bonding is also found between protic materials and boron hydrides.[490]

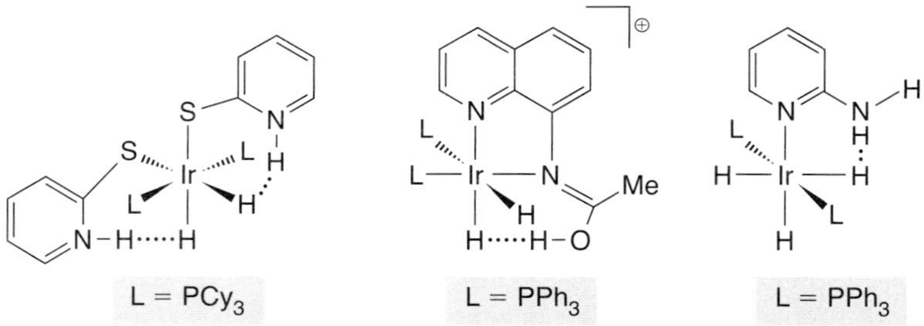

Figure 2.52.
Examples of hydrogen bonding from a protic group to a metal hydride.

2.8.2. Dihydrogen Complexes

The largest amount of information on the structures, dynamics, and reactivity of σ-complexes is derived from dihydrogen complexes.[449,491] These complexes have been studied intensively because they are considered to be intermediates in the oxidative addition of dihydrogen and could serve as a type of interaction in a material that could reversibly store molecular hydrogen. Furthermore, many complexes of dihydrogen have succumbed to isolation, and these complexes posses a variety of properties that depend on the steric and electronic properties of the metal center.

2.8.2.1. Properties that Lead to Stable H₂ Complexes

Dihydrogen complexes are often envisioned as resulting from an arrested oxidative addition of dihydrogen. Dihydrogen complexes are formed from transition metal fragments that backbond with the H_2 σ* orbital, but weakly enough to cleave the H–H bond only partially. Thus, most of the isolated dihydrogen complexes contain transition metal centers in relatively low oxidation states (because of the importance of backbonding to the stability of H_2 complexes). However, many of these low-valent complexes are cationic because a neutral complex that is low valent would likely be electron rich enough to cleave fully the H–H bond. Thus, many neutral dihydrogen complexes contain π-accepting carbonyl groups. Dihydrogen complexes have been calculated as intermediates on the pathway for hydrogenolysis of d^0 metal alkyl complexes,[492] but these complexes are not stable enough to be detected, and no d^0 dihydrogen complex has been isolated.

In many cases, the complexes of second-row metals adopt structures with dihydrogen ligands, while the directly analogous third-row metal complexes adopt structures with two terminal hydrides.[493,494] For example, the ruthenium polyhydride complex $[RuH_2(H_2)(PPh_3)_3]$ contains a dihydrogen ligand, while the analogous osmium complex (containing P(p-Tol)$_3$ as ligand) possesses all terminal hydrides.[493] Similarly, $[Mo(H_2)(CO)(dppe)_2]$ is a dihydrogen complex, but $[WH_2(CO)(dppe)_2]$ is a dihydride complex.[495] The importance of charge and electron density is illustrated by a comparison of the structures of Heinekey's $[Re(H_2)(PPr^i_3)_3(CO)_3]^+$ and Kubas's $[W(H_2)(PPr^i_3)_3(CO)_3]$. The neutral tungsten complex is a mixture of dihydrogen and dihydride species,[496] while the cationic rhenium complex is solely a dihydrogen complex.[497]

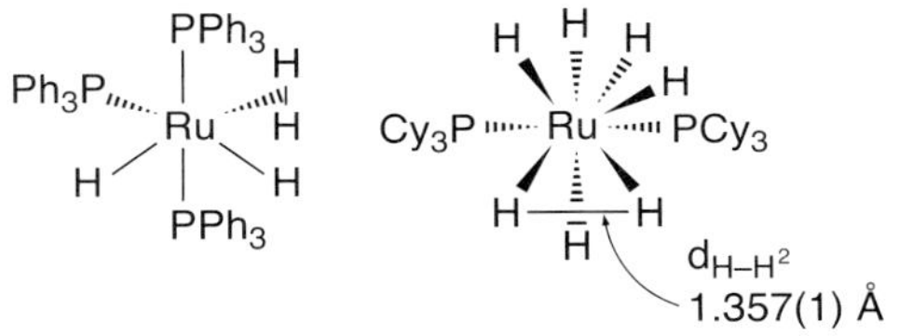

Figure 2.53.
Examples of polyhydrides containing a hydrogen ligand.

Many of the dihydrogen ligands are also part of polyhydride complexes, and several known polyhydrides have now been shown to contain dihydrogen ligands.[481,493,498–503] For two examples, see Figure 2.53.[493,502] To avoid the high oxidation state of a classical penta- or heptahydride complex, the polyhydride compounds contain some H–H bonding between two or more of the hydrogens. With a large number of hydride ligands, little structural change is needed to establish an H–H bond.

2.8.2.2. Spectroscopic Signatures of H₂ Complexes

Methods to distinguish between dihydrogen and dihydride complexes were challenging to develop, but several methods are now well accepted. X-ray diffraction is not an effective method to determine the bonding mode because of the low electron density of a hydride ligand. Instead, neutron diffraction has been used to characterize several metal dihydrogen complexes. In these structures, the H–H bond distance, which is 0.74 Å in free dihydrogen, has ranged from 0.8–1.0 Å in typical dihydrogen complexes to 1.0–1.36 Å in so-called elongated dihydrogen complexes. The H–H distance in a cis dihydride is typically greater than 1.6 Å.

Because of the difficulty in obtaining this type of structural data, spectroscopic methods have also been developed. The H–D coupling constant of a partially labeled complex has been used to provide evidence for H–H bonding in dihydrogen complexes. An H–D coupling constant is 1/6.5 of an H–H coupling due to the difference in the gyromagnetic ratios of H and D. The H–D coupling constant in Kubas's original dihydrogen complex was 34 Hz, while the coupling in free H–D is 43 Hz (see Equation 2.19).[461] The graph in

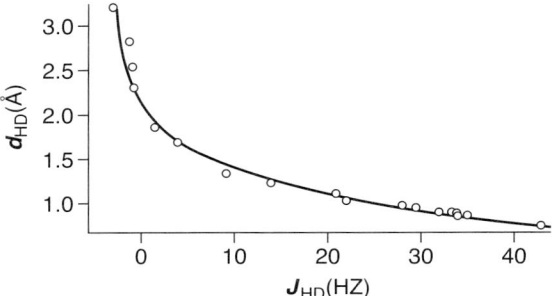

Figure 2.54.
Relationship between H–H distances obtained by neutron diffraction and the magnitude of the H–D coupling constants measured by solid-state NMR spectroscopy. Adapted from reference 505b. The author thanks Prof. Michael D. Heinekey for assistance with this graphic.

Figure 2.54 shows the relationship between H–H distances obtained by neutron diffraction and the magnitude of the H–D coupling constants measured by solid-state NMR spectroscopy.[486,505]

The infrared stretching frequency has also been used to distinguish dihydrogen from dihydride complexes. The Raman frequency of free H_2 is 4300 cm^{-1} and typical ν_{M-H} vibrations are found between 1700 and 2300 cm^{-1}. The dihydrogen ligand vibrates at a frequency between that of free H_2 and those of metal hydrides.[504] Typical values for the H–H stretch in dihydrogen complexes lie between 2400 and 3100 cm^{-1}. For example, Kubas's compound contained a band at 2690 cm^{-1} (see Equation 2.19) for the H–H stretch. Although observed in some cases, the H–H stretch is weak and has not been identified in the majority of H_2 complexes.

An additional, frequently reported method to distinguish between dihydrogen and dihydride complexes is the value termed $T_{1(min)}$. This value is the minimum value of the T_1 for relaxation of the hydride resonance in the ^{1}H NMR spectrum as a function of temperature. The value of this method was first demonstrated[506] and later refined from T_1 to $T_{1(min)}$ by Crabtree.[507] The value of T_1 depends on the H–H distance in cases when the dominant contribution to relaxation is the dipolar relaxation of one hydrogen by the nearest hydrogen. In this case, T_1 is proportional to the sixth power of the H–H distance. Because the T_1 value of hydride ligands depends strongly on temperature and decreases in value before increasing again at the lowest temperatures, one must determine the minimum value of T_1 on a curve of T_1 versus temperature. The T_1 of the hydride in most or all dihydride complexes has been greater than 50 ms at 250 MHz (100 ms at 500 MHz), and the T_1 of most or all dihydrogen complex has been less than 35 ms.[508]

Many parameters have been shown to effect the value of T_1. Halpern conducted an extensive study of the factors that can effect the value of T_1 in transition metal hydride complexes. He noted, among other conclusions, that relaxation can be affected by metals having a high gyromagnetic ratio, such as cobalt, rhenium, and manganese.[512] Despite these issues, and later studies[513] that express reservations about Halpern's conclusions, the T_1 criterion is commonly used and can be powerful if used for metals other than Co, Re, and Mn (or with corrections for the relaxation provided by these metals) and with an understanding of potential pitfalls.

2.8.2.3. Reactivity of H_2 Complexes

The reactivity of dihydrogen ligands is dominated by the dissociation of dihydrogen, the addition of dihydrogen to form a dihydride, and the deprotonation of the hydrogen by external base.

$$\text{[complex]} \xrightarrow{\text{NCPh}} \text{[complex]} + H_2 \tag{2.21}$$

$$(2.22)$$

Examples of the replacement of dihydrogen by conventional dative ligands are shown in Equations 2.21 and 2.22.[514,515] The stability of the dihydrogen ligand depends on the electronic properties of the complex, but some general trends can be drawn.[516] As an anchor point, the enthalpy of binding of H_2 to $[W(CO)_3(PCy_3)_2]$ is estimated as 9.4 kcal/mol by calorimetry.[515,517] Thus, dihydrogen is typically replaced by strongly binding ligands, such as CO or PR_3. It is also typically displaced by nitriles. Dihydrogen is usually a stronger ligand than water or ethers.[517] However, the relative strength of the binding versus H_2 depends on the complex, and entropic terms are surprisingly large.[518]

The rates of the substitution reactions are typically zero order in the incoming ligand, implying that the rate of displacement corresponds to the rate of dissociation of dihydrogen.[514] In this case, the enthalpy of activation for ligand substitution is close to the enthalpy of binding of H_2. The rates of substitution imply that the enthalpy for dissociation of H_2 can be much higher than the enthalpy of binding of H_2 in the Kubas compound. For example, $\Delta H^\ddagger$ is 27–29 kcal/mol for the dissociation of H_2 from $[FeH(H_2)(depe)_2]^+$ in the substitution of H_2 by nitriles (Equation 2.21).[514]

$$(2.23)$$

R = Pr$_3^i$, Cyclohexyl or Cyclopentyl

$$[ReH_4(CO)L_3]^{\oplus} \rightleftharpoons [ReH_2(H_2)(CO)L_3]^{\oplus} \qquad (2.24)$$

$$(2.25)$$

Dihydrogen complexes also undergo oxidative addition to cleave the H–H bond. In many cases, an equilibrium between the dihydrogen complex and the dihydride has been observed.[496,519–533] Three examples of this equilibrium are shown in Equations 2.23–2.25.[496,520,521] The barrier to interconversion of the two forms is believed to result largely from structural changes that must occur to the coordination geometry and ancillary ligands to accommodate two hydride ligands. In some cases, the dihydrogen complex has been observed as a kinetic product at low temperature, followed by cleavage of the H–H bond to form a complex with two terminal hydride ligands.[521]

Finally, many cationic dihydride complexes are acidic, and some dicationic complexes are very strong acids. Dihydrogen has a $pK_a > 49$ in THF,[534] but this acidity can increase up to 52 orders of magnitude by the coordination to a metal and rival that of triflic acid.[535,536] A series of values is provided in Kubas's book,[537] and these values range from 36 for the electron-rich $[RuH_2(H_2)(PPh_3)_3]$[534] to negative values. The pK_a of the monocationic $[Cp*Re(H_2)(CO)(NO)]^+$[538] and $[Cp*Ru(CO)_2(H_2)]^+$ is -2,[522] and the pK_a of the dicationic $[Os(CO)(H_2)(dppp)_2]^{2+}$ is -5.7.[536]

This increase in the acidity of dihydrogen upon coordination has led to the proposal that dihydrogen complexes are intermediates in σ-bond metathesis reactions and that the alkyl group undergoing hydrogenolysis acts as the base to deprotonate the dihydrogen ligand.[492] This type of heterolytic cleavage of H_2 could also occur with late metals and may be involved in the hydrogenolysis of rhodium formate complexes during the hydrogenation of carbon dioxide.[539] In other cases, the dihydrogen ligand simply acts as a Brønsted acid and delivers protons to an added base.

2.8.3. Alkane and Silane Complexes

The properties and reactivity of alkane[450,466] and silane[450,460,540,541] complexes are closely related to those of dihydrogen complexes. However, the thermodynamic stabilities of the complexes are much different. The silane complexes have been studied in less detail than dihydrogen complexes, but, broadly speaking, they are similar in stability to the dihydrogen complexes or only slightly less stable. Alkane complexes have been studied intensively as reactive intermediates,[366,542–550] but detailed structural and reactivity studies have not been conducted because alkane complexes are unstable in solution. The most detailed data have been gained by NMR spectroscopic studies of $CpRe(CO)_2(RH)$ complexes.[551,552] These data imply that the metal binds in an η^2-mode to a single C–H bond of the alkane.

2.8.3.1. Stability Relative to H_2 Complexes

Silane complexes are similar in stability to dihydrogen complexes, despite the greater steric effect, for two reasons.[553] First, the silane Si–H bond is more basic than H_2 or the R–H bond. In addition, the Si–H bond is longer and weaker than that in H_2 or a hydrocarbon C–H bond and has a lower energy σ*-orbital. Thus, there is a smaller difference in energy between the σ- and σ*-orbitals of the silane than between the σ- and σ*-orbitals of hydrogen or an alkane. This smaller difference in energy between the σ- and σ*-orbitals leads to stronger bonding by a Chatt–Dewar–Duncanson model. Alkane complexes are less stable than H_2 or silane complexes because the steric effect is not compensated by this smaller difference in energy between the σ- and σ*-orbitals.

$$(2.26)$$

$$(2.27)$$

2.8.3.2. Evidence for Alkane Complexes

Much kinetic data has been obtained that support alkane complexes as intermediates in the oxidative addition of alkane C–H bonds. Flash photolysis experiments have even provided rates of reactions for alkane complexes (Equation 2.26).[546,548,554] In addition, studies have been conducted to identify alkane complexes generated with metal centers that do not cleave the alkane C–H bond (Equation 2.27). Most studies have focused on studying alkane complexes of the $[M(CO)_5]$ (M = Cr or W) fragment. Photoacoustic calorimetry showed that the strength of the binding of heptane to $[Cr(CO)_5]$ is 9.6 kcal/mol[555] and to $[Mo(CO)_5]$ is 11 kcal/mol,[556] with errors that are roughly 20–25% of the magnitude of the binding energy. Equilibrium studies in the gas phase showed that $[W(CO)_5]$ binds hexane with a binding energy of 11 kcal/mol.[557] These binding energies are remarkable, considering the lack of electron pairs and steric factors. Yet, they are similar to the energy cost of forming one molecule from two ($T\Delta S \approx 30$ eu $\times$ 300 K $\approx$ 9 kcal/mol).

The most direct evidence for alkane complexes has been gained by Ball, who has reported NMR spectral data on alkane σ-complexes generated by the continuous photolysis of $CpRe(CO)_3$ (Scheme 2.3) and $(Pr^i–Cp)Re(CO)_3$ in alkane solvent in an NMR spectrometer probe.[551,552] These experiments were conducted in both cyclopentane and n-pentane. The studies in cyclopentane initially showed that these complexes could be observed directly by ^1H NMR spectroscopy using techniques that suppress signals of the free alkane. The studies conducted in n-pentane showed that this metal center binds with a slight preference for methylene over methyl resonances, after correcting for the number of each type of hydrogen, and that the isomeric alkane complexes interconvert on the time scale of 1–10 s^{-1} at 173 K. The effect of isotopes on chemical shift implies an unsymmetrical binding mode, consistent with the η^2-coordination through one C–H bond predicted by calculations.[558–560]

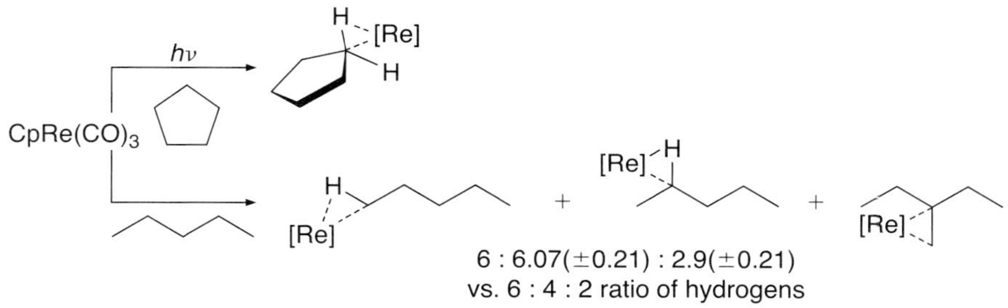

6 : 6.07($\pm$0.21) : 2.9($\pm$0.21)
vs. 6 : 4 : 2 ratio of hydrogens

Scheme 2.3

2.8.3.3. Intramolecular Coordination of Aliphatic C–H Bonds (Agostic Interactions)

Although alkane complexes have not yet been isolated, many complexes have been isolated in which the C–H bond is coordinated to the metal in an intramolecular fashion. Complexes with an intramolecular interaction between a C–H bond and a metal center have been termed "agostic" complexes. To form these complexes, the entropic penalty of joining two molecules need not be paid, and hundreds of complexes have now been prepared in which the C–H bond is coordinated to a metal center through the C–H σ-bond.[561,562] The bonding interactions in these complexes are the same as those of σ-complexes described previously. Like dihydrogen complexes, these agostic complexes can be envisioned as resulting from arrested oxidative additions or β-hydrogen eliminations. Many of the early agostic complexes were generated from cationic metal centers, and some classic examples were generated by the protonation of an olefin complex (Equation 2.28). A classic review was published by Green and Brookhart,[562] and an updated version of this review has appeared.[561]

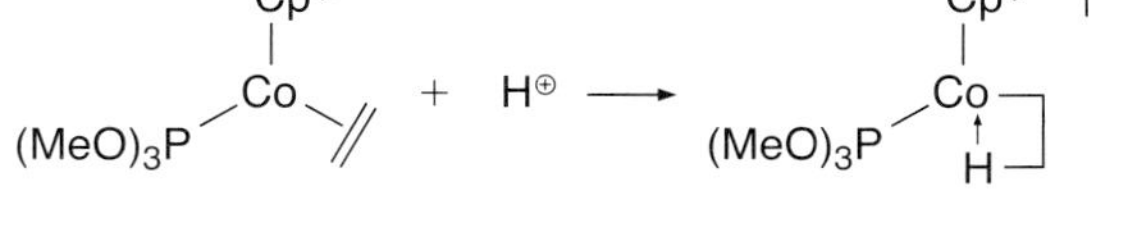

(2.28)

Agostic interactions of many types have been reported, as summarized in Figure 2.55. Most common are "β-agostic interactions," which involve the interaction of the metal with a hydrogen on the carbon β to the metal. These complexes can be considered intermediates on the way to β-hydrogen elimination (Chapter 10). In addition, α-agostic interactions have been found in complexes of electrophilic metal centers.[424,563,564] Carbene complexes of high oxidation state metals have also been identified that possess α-agostic interactions.[565,566]

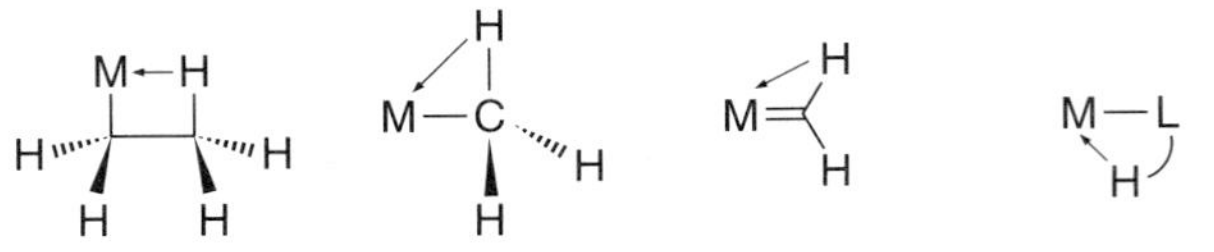

β-Agostic   **α-Agostic**   **α-Agostic**   Agostic interaction with a dative ligand

Figure 2.55. Four classes of agostic interactions.

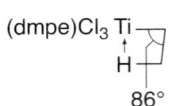

Figure 2.56.
An titanium alkyl with severe distortion from an agostic interaction.

These two interactions can be considered intermediates in α-elimination processes. Agostic interactions between unsaturated metals and ligand C–H bonds are also common, such as in the precursor to Kubas's dihydrogen complex,[461] in $[Ru(PPh_3)_3Cl_2]$,[567] and in a highly unsaturated palladium(II)[568] and Rh(I) species.[569]

Several spectroscopic methods have been used to identify agostic complexes.[561] X-ray diffraction does not reliably identify the position of the hydrogen, but a short M–C distance can indicate an M–H–C interaction. This criterion of the distance between heavy atoms is much like the criterion to support hydrogen bonding interactions in organic systems. However, distortions of heavy atoms to place the hydrogen in the position for an agostic interaction is the most compelling evidence for this interaction.[570] One such example of severe distortion involves the titanium complex in Figure 2.56, which possesses a M–C–C angle of only 86°.[571,572]

Several pieces of spectroscopic data can provide evidence for an agostic interaction. First, a reduced ν_{C-H} stretching frequency due to a reduced C–H bond order can often be observed. Second, upfield shifts of the ^1H and ^{13}C NMR resonances of the carbon and hydrogen atoms involved in the agostic interactions are often observed, because the agostic interaction lies partway to the generation of a metal–hydride and metal–carbon bonds, and both hydrides and carbons bound to transition metals tend to resonate upfield of unligated hydrogens and carbons. Finally, a decrease in the one-bond C–H coupling constant is often observed because of a reduction of the C–H bond order. The chemical shift and coupling constants for the classic cobalt complex with an agostic interaction are shown in Figure 2.57.[573]

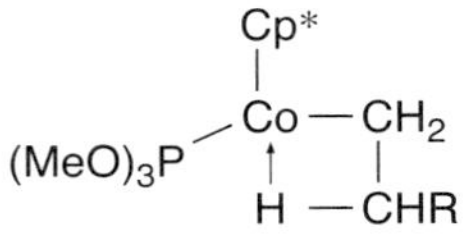

$$R = Me \qquad J_{CH} = 38 \text{ Hz}$$
$$\delta H = -13.2 \text{ ppm}$$

$$R = H \qquad J_{CH} = 61 \text{ Hz}$$
$$\delta^1H = -12.1 \text{ ppm}$$
$$\delta^{13}C = -5.8 \text{ ppm}$$

Figure 2.57.
Low-temperature NMR spectroscopic data for a cobalt complex with a β-agostic interaction.

In most cases, an agostic interaction to one hydrogen is dynamic, and this hydrogen exchanges with other, otherwise equivalent, hydrogens. For example, the three hydrogens of a methyl group involved in a single agostic interaction typically appear equivalent by NMR spectroscopy, even at low temperature.[573,574] Partial deuteration of the methyl group, however, perturbs the equilibrium for binding and favors binding of the C–H bond over binding of the C–D bond. This perturbation of the equilibrium leads to a change in average chemical shift and a J_{C-H} value for the hydrogens of the methyl group. The elegant experiment devised by Martin Saunders for studies of carbocations, termed "isotopic perturbation of degeneracy" or "isotopic perturbation of equilibrium,"[575–577] was first applied to organometallic compounds by John Shapley[578] and has been used to reveal agostic interactions in many types of complexes.[564,574,579]

References and Notes

1. For a likely exception with lanthanide metals, see reference 2.
2. Maron, L.; Perrin, L.; Eisenstein, O.; Andersen, R. A. *J. Am. Chem. Soc.* **2002**, *124*, 5614.
3. Calderazzo, F.; Ercoli, R.; Natta, G. *Organic Syntheses via Metal Carbonyls*; Interscience Publishers: New York, 1968; Vol. 1.
4. Abel, E. W.; Stone, F. G. A. *Q. Rev.* **1969**, *23*, 325.
5. Braterman, P. S. *Struct. Bonding* **1972**, *10*, 57.
6. Braterman, P. S. *Struct. Bonding* **1976**, *26*, 1.
7. Cotton, F. A. *Prog. Inorg. Chem.* **1976**, *21*, 1.
8. Lukehart, C. M. *Fundamental Transition Metal Organometallic Chemistry*; Brooks/Cole: Monterey, CA, 1985.
9. Dyson, P. J.; McIndoe, J. S. *Transition Metal Carbonyl Cluster Chemistry*; Gordon and Breach Science Publishers: Amsterdam, 2000.
10. See the chapter in each year's edition of the internet e-brary resource Organometallic Chemistry.
11. For a reliable preparation of Ru$_3$(CO)$_{12}$, see Eady, C. R.; Jackson, P. F.; Johnson, B. F. G.; Lewis, J.; Malatesta, M. C.; McPartlin, M.; Nelson, W. J. H. *J. Chem. Soc., Dalton Trans.* **1980**, 383.
12. Gilliland, W. L.; Blanchard, A. A. *Inorg. Synth.* **1946**, *2*, 234.
13. (a) Mond, L.; Quinete, F. *J. Chem. Soc., Trans.* **1891**, *59*, 604; (b) Mond, L; Langer, C. *J. Chem. Soc., Trans.* **1891**, *59*, 1090.
14. Rushman, P.; Buuren, G. N. V.; Shiralian, M.; Pomeroy, R. K. *Organometallics* **1983**, *2*, 693.
15. Braye, E. H.; Hubel, W. *Inorg. Synth.* **1966**, *8*, 178.
16. Colton, R.; McCormick, M. J. *Coord. Chem. Rev.* **1980**, *31*, 1.
17. Horwitz, C. P.; Shriver, D. F. *Adv. Organomet. Chem.* **1984**, *23*, 219.
18. Chini, P. *Inorg. Chim. Acta Rev.* **1968**, *2*, 31.
19. Pritchar, H.; Harrison, A. G. *J. Chem. Phys.* **1968**, *48*, 5623.
20. Pritchar, H.; Harrison, A. G. *J. Chem. Phys.* **1968**, *48*, 2827.
21. Illies, A. J.; Jarrold, M. F.; Bowers, M. T. *J. Am. Chem. Soc.* **1983**, *105*, 2562.
22. Nobes, R. H.; Radom, L. *Chem. Phys.* **1981**, *60*, 1.
23. Bruna, P. J.; Peyerimhoff, S. D.; Buenker, R. J. *Chem. Phys.* **1975**, *10*, 323.
24. Olah, G. A.; Laali, K.; Farooq, O. *J. Org. Chem.* **1985**, *50*, 1483.
25. Olah, G. A.; Pelizza, F.; Kobayashi, S.; Olah, J. A. *J. Am. Chem. Soc.* **1976**, *98*, 296.
26. Caulton, K. G.; Fenske, R. F. *Inorg. Chem.* **1968**, *7*, 1273.
27. Baerends, E. J.; Ros, P. *Mol. Phys.* **1975**, *30*, 1735.
28. Bursten, B. E.; Freier, D. G.; Fenske, R. F. *Inorg. Chem.* **1980**, *19*, 1810.
29. Hubbard, J. L.; Lichtenberger, D. L. *J. Am. Chem. Soc.* **1982**, *104*, 2132.
30. Bauschlicher, C. W.; Bagus, P. S. *J. Chem. Phys.* **1984**, *81*, 5889.
31. Kostic, N. M.; Fenske, R. F. *Inorg. Chem.* **1983**, *22*, 666.
32. Jemmis, E. D.; Pinhas, A. R.; Hoffmann, R. *J. Am. Chem. Soc.* **1980**, *102*, 2576.
33. Tolman, C. A.; Ittel, S. D.; English, A. D.; Jesson, J. P. *J. Am. Chem. Soc.* **1978**, *100*, 4080.
34. Roddick, D. M.; Fryzuk, M. D.; Seidler, P. F.; Hillhouse, G. L.; Bercaw, J. E. *Organometallics* **1985**, *4*, 97.
35. Manriquez, J. M.; McAlister, D. R.; Sanner, R. D.; Bercaw, J. E. *J. Am. Chem. Soc.* **1978**, *100*, 2716.
36. Manriquez, J. M.; McAlister, D. R.; Sanner, R. D.; Bercaw, J. E. *J. Am. Chem. Soc.* **1976**, *98*, 6733.
37. Howard, W. A.; Parkin, G.; Rheingold, A. L. *Polyhedron* **1995**, *14*, 25.
38. Hurlburt, P. K.; Anderson, O. P.; Strauss, S. H. *J. Am. Chem. Soc.* **1991**, *113*, 6277.
39. Antonelli, D. M.; Tjaden, E. B.; Stryker, J. M. *Organometallics* **1994**, *13*, 763.
40. Guo, Z. Y.; Swenson, D. C.; Guram, A. S.; Jordan, R. F. *Organometallics* **1994**, *13*, 756.
41. Weber, L. *Angew. Chem., Int. Ed. Engl.* **1994**, *33*, 1077.
42. Goldman, A. S.; Krogh-Jespersen, K. *J. Am. Chem. Soc.* **1996**, *118*, 12159.
43. Braterman, P. S. *Metal Carbonyl Spectra*; Academic Press: New York, 1975.
44. Adams, D. M. *Metal-Ligand and Related Vibrations*; St. Martin's Press: New York, 1968.
45. Chini, P. *Inorg. Chem.* **1969**, *8*, 1206.
46. Churchill, M. R. In *Perspectives in Structural Chemistry*; Dunitz, J. D., Ed.; Wiley: New York, 1970; Vol. 3, p 91.
47. Ehlers, A. W.; Frenking, G. *Organometallics* **1995**, *14*, 423.
48. Ziegler, T.; Tschinke, V.; Ursenbach, C. *J. Am. Chem. Soc.* **1987**, *109*, 4825.
49. Wang, K.; Rosini, G. P.; Nolan, S. P.; Goldman, A. S. *J. Am. Chem. Soc.* **1995**, *117*, 5082.
50. Kundig, E. P.; McIntosh, D.; Moskovit, M.; Ozin, G. A. *J. Am. Chem. Soc.* **1973**, *95*, 7234.
51. Kundig, E. P.; Moskovit, M.; Ozin, G. A. *Can. J. Chem.* **1972**, *50*, 3587.
52. Huber, H.; Moskovit, M.; Kundig, P.; Ozin, G. A. *Nature-Phys. Sci.* **1972**, *235*, 98.

53. Lewis, K. E.; Golden, D. M.; Smith, G. P. *J. Am. Chem. Soc.* **1984**, *106*, 3905.
54. Bernstein, M.; Simon, J. D.; Peters, K. S. *Chem. Phys. Lett.* **1983**, *100*, 241.
55. Connor, J. A. *Top. Curr. Chem.* **1977**, *71*, 71.
56. Rosini, G. P.; Liu, F. C.; Krogh-Jespersen, K.; Goldman, A. S.; Li, C. B.; Nolan, S. P. *J. Am. Chem. Soc.* **1998**, *120*, 9256.
57. For spectroscopic and electrochemical evidence of this assertion, see the references in reference 58.
58. Leach, P. A.; Geib, S. J.; Corella, J. A.; Warnock, G. F.; Cooper, N. J. *J. Am. Chem. Soc.* **1994**, *116*, 8566.
59. Luo, Y. R. *Handbook of Bond Dissociation Energies in Organic Compounds*; CRC Press: Boca Raton, FL, 2003.
60. Leach, P. A.; Corella, J. A.; Geib, S. J.; Cooper, N. J. *Abstr. Papers Am. Chem. Soc.* **1994**, *208*, 23.
61. Corella, J. A.; Thompson, R. L.; Cooper, N. J. *Angew. Chem., Int. Ed. Engl.* **1992**, *31*, 83.
62. Busetto, L.; Belluco, U.; Angelici, R. J. *J. Organomet. Chem.* **1969**, *18*, 213.
63. Mansuy, D.; Battioni, J. P.; Chottard, J. C. *J. Am. Chem. Soc.* **1978**, *100*, 4311.
64. Butler, I. S.; Cozak, D.; Stobart, S. R. *J. Chem. Soc., Chem. Commun.* **1975**, 103.
65. Andrews, M. A. *Inorg. Chem.* **1977**, *16*, 496.
66. Clark, G. R.; Grundy, K. R.; Harris, R. O.; James, S. M.; Roper, W. R. *J. Organomet. Chem.* **1975**, *90*, C37.
67. Clark, G. R.; Marsden, K.; Roper, W. R.; Wright, L. J. *J. Am. Chem. Soc.* **1980**, *102*, 1206.
68. Roper, W. R. *Chem. New Zealand* **1984**, *48*, 149.
69. Tolman, C. A. *Chem. Rev.* **1977**, *77*, 313.
70. Clark, H. C. *Isr. J. Chem.* **1977**, *15*, 210.
71. McAuliffe, C. A. In *Transition Metal Complexes of Phosphorus, Arsenic, and Antimony Ligands*; McAuliffe, C. A., Ed.; Wiley: New York, 1973.
72. McAuliffe, C. A.; Levason, W. *Phosphine, Arsine, and Stibine Complexes of the Transition Elements (Studies in Inorganic Chemistry, Vol. 1)*; Elsevier: Amsterdam, 1978; Vol. 1.
73. Levason, W.; McAuliffe, C. A. *Acc. Chem. Res.* **1978**, *11*, 363.
74. Levason, W.; McAuliffe, C. A. *Adv. Inorg. Chem. Radiochem.* **1972**, *14*, 173.
75. Booth, G. *Adv. Inorg. Chem. Radiochem.* **1964**, *6*, 1.
76. McAuliffe, C. A. In *Transition Metal Complexes of Phosphorus, Arsenic, and Antimony Ligands*; McAuliffe, C. A., Ed.; Wiley: New York, 1973.
77. Stelzer, O. *Topics in Phosphorus Chemistry*; Wiley: New York, 1973; Vol. 9, p 1.
78. Richards, C. J.; Locke, A. J. *Tetrahedron: Asymmetry* **1998**, *9*, 2377.
79. Gavrilov, K. N.; Polosukhin, A. I. *Russ. Chem. Rev.* **2000**, *69*, 661.
80. Freixa, Z.; van Leeuwen, P. W. N. M. *J. Chem. Soc., Dalton Trans.* **2003**, 1890.
81. Valentine, D. H.; Hillhouse, J. H. *Synthesis* **2003**, 2437.
82. Brunel, J. M. *Mini-Rev. Org. Chem.* **2004**, *1*, 249.
83. *Oxford English Dictionary Online*; Oxford University Press: Oxford, U. K., 2004.
84. De Stefano, N. J.; Johnson, D. K.; Lane, R. M.; Venanzi, L. M. *Helv. Chim. Acta* **1976**, *59*, 2674.
85. Gillie, A.; Stille, J. K. *J. Am. Chem. Soc.* **1980**, *102*, 4933.
86. Kuwano, R.; Uemura, T.; Saitoh, M.; Ito, Y. *Tetrahedron Lett.* **1999**, *40*, 1327.
87. Kamer, P. C. J.; van Leeuwen, P. W. N.; Reek, J. N. H. *Acc. Chem. Res.* **2001**, *34*, 895.
88. Yin, J.; Buchwald, S. L. *J. Am. Chem. Soc.* **2002**, *124*, 6043.
89. Casey, C. P.; Whiteker, G. T. *Isr. J. Chem.* **1990**, *30*, 299.
90. Casey, C. P.; Whiteker, G. T.; Melville, M. G.; Petrovich, L. M.; Gavney, J. A., Jr.; Powell, D. R. *J. Am. Chem. Soc.* **1992**, *114*, 5535.
91. Bush, R. C.; Angelici, R. J. *Inorg. Chem.* **1988**, *27*, 681.
92. Allman, T.; Goel, R. G. *Can. J. Chem.* **1982**, *60*, 716.
93. Streuli, C. A. *Anal. Chem.* **1960**, *32*, 985.
94. Streitwieser, A.; McKeown, A. E.; Hasanayn, F.; Davis, N. R. *Org. Lett.* **2005**, *7*, 1259.
95. Pearson, R. G. *J. Am. Chem. Soc.* **1963**, *85*, 3533.
96. Pearson, R. G. *J. Chem. Educ.* **1968**, *45*, 643.
97. Pearson, R. G. *J. Chem. Educ.* **1968**, *45*, 581.
98. Kataoka, N.; Shelby, Q.; Stambuli, J. P.; Hartwig, J. F. *J. Org. Chem.* **2002**, *67*, 5553.
99. Wolfe, J. P.; Tomori, H.; Sadighi, J. P.; Yin, J. J.; Buchwald, S. L. *J. Org. Chem.* **2000**, *65*, 1158.
100. Mislow, K. *Trans. New York Acad. Sci.* **1973**, *35*, 227.
101. Baechler, R. D.; Mislow, K. *J. Am. Chem. Soc.* **1970**, *92*, 3090.
102. Knowles, W. S.; Vineyard, B. D.; Sabacky, M. J. *J. Chem. Soc., Chem. Commun.* **1972**, 10.
103. Knowles, W. S. *Angew. Chem., Int. Ed. Engl.* **2002**, *41*, 1999.
104. Juge, S.; Stephan, M.; Laffitte, J. A.; Genet, J. P. *Tetrahedron Lett.* **1990**, *31*, 6357.
105. Yamanoi, Y.; Imamoto, T. *J. Org. Chem.* **1999**, *64*, 2988.
106. Maienza, F.; Spindler, F.; Thommen, M.; Pugin, B.; Malan, C.; Mezzetti, A. *J. Org. Chem.* **2002**, *67*, 5239.

107. For recently discovered examples of bridging phosphine ligands, see reference 108.
108. Werner, H. *Angew. Chem., Int. Ed. Engl.* **2004**, *43*, 938.
109. Xiao, S. X.; Trogler, W. C.; Ellis, D. E.; Berkovitchyellin, Z. *J. Am. Chem. Soc.* **1983**, *105*, 7033.
110. Marynick, D. S. *J. Am. Chem. Soc.* **1984**, *106*, 4064.
111. Tossell, J. A.; Moore, J. H.; Giordan, J. C. *Inorg. Chem.* **1985**, *24*, 1100.
112. Giordan, J. C.; Moore, J. H.; Tossell, J. A. *Acc. Chem. Res.* **1986**, *19*, 281.
113. Dunne, B. J.; Morris, R. B.; Orpen, A. G. *J. Chem. Soc., Dalton Trans.* **1991**, 653.
114. Orpen, A. G.; Connelly, N. G. *Organometallics* **1990**, *9*, 1206.
115. Leyssens, T.; Peeters, D.; Orpen, A. G.; Harvey, J. N. *Organometallics* **2007**, *26*, 2637.
116. Higginson, B. R.; Lloyd, D. R.; Connor, J. A.; Hillier, I. H. *J. Chem. Soc., Faraday Trans. II* **1974**, *70*, 1418.
117. Yarbrough, L. W.; Hall, M. B. *Inorg. Chem.* **1978**, *17*, 2269.
118. Daamen, H.; Oskam, A.; Stufkens, D. J. *Inorg. Chim. Acta* **1980**, *38*, 71.
119. Gerloch, M.; Woolley, R. G. *Prog. Inorg. Chem.* **1983**, *31*, 371.
120. Bursten, B. E.; Darensbourg, D. J.; Kellogg, G. E.; Lichtenberger, D. L. *Inorg. Chem.* **1984**, *23*, 4361.
121. Morris, R. J.; Girolami, G. S. *Inorg. Chem.* **1990**, *29*, 4167.
122. Tolman, C. A. *Chem. Rev.* **1977**, *77*, 313.
123. Mastryukova, T. A.; Kabachnik, M. I. *Russ. Chem. Rev.* **1969**, *38*, 795.
124. Glueck, D. S.; Newman Winslow, L. J.; Bergman, R. G. *Organometallics* **1991**, *10*, 1462.
125. Thompson, W. H.; Sears, C. T. *Inorg. Chem.* **1977**, *16*, 769.
126. Brown, T. L.; Lee, K. J. *Coord. Chem. Rev.* **1993**, *128*, 89.
127. Coville, N. J.; Loonat, M. S.; White, D.; Carlton, L. *Organometallics* **1992**, *11*, 1082.
128. White, D.; Taverner, B. C.; Leach, P. G. L.; Coville, N. J. *J. Comput. Chem.* **1993**, *14*, 1042.
129. White, D.; Taverner, B. C.; Leach, P. G. L.; Coville, N. J. *J. Organomet. Chem.* **1994**, *478*, 205.
130. White, D.; Coville, N. J. In *Advances in Organometallic Chemistry, Vol. 36*; West, R., Stone, F. G. A., Eds.; Academic Press: New York, 1994; Vol. 36, p 95.
131. White, D.; Taverner, B. C.; Coville, N. J.; Wade, P. W. *J. Organomet. Chem.* **1995**, *495*, 41.
132. Tolman, C. A. *J. Am. Chem. Soc.* **1970**, *92*, 2956.
133. Hitchcock, P. B.; McPartlin, M.; Mason, R. *J. Chem. Soc., Chem. Commun.* **1969**, 1367.
134. Bennett, M. J.; Donaldson, P. B. *Inorg. Chem.* **1977**, *16*, 655.
135. Garrou, P. E. *Chem. Rev.* **1985**, *85*, 171.
136. Kong, K.-C.; Cheng, C.-H. *J. Am. Chem. Soc.* **1991**, *113*, 6313.
137. Morita, D. K.; Stille, J. K.; Norton, J. R. *J. Am. Chem. Soc.* **1995**, *117*, 8576.
138. Goodson, F. E.; Wallow, T. I.; Novak, B. M. *J. Am. Chem. Soc.* **1997**, *119*, 12441.
139. Marcoux, D.; Charette, A. B. *J. Org. Chem.* **2008**, *73*, 590.
140. Rybin, L. V.; Petrovskaya, E. A.; Rubinskaya, M. I.; Kuz'mina, L. G.; Struchkov, Y. T.; Kaverin, V. V.; Koneva, N. Y. *J. Organomet. Chem.* **1985**, *288*, 119.
141. Bryant, D. R.; Nicholson, J. C.; Billig, E. Union Carbide Chemicals & Plastics Technology Corp., USA **1998**, 96-756788 19961126, CAN 129:55722, US, 33.
142. *Rhodium Catalyzed Hydroformylation*; van Leeuwen, P. W. N. M., Claver, C., Eds.; *Catalysis in Metal Complexes Series*; Springer Publishing: New York, 2000; Vol. 22, p 284.
143. Burrows, A. D.; Mahon, M. F.; Palmer, M. T.; Varrone, M. *Inorg. Chem.* **2002**, *41*, 1695.
144. Bedford, R. B.; Hazelwood, S. L.; Limmert, M. E.; Brown, J. M.; Ramdeehul, S.; Cowley, A. R.; Coles, S. J.; Hursthouse, M. B. *Organometallics* **2003**, *22*, 1364.
145. Tolman, C. A.; Seidel, W. C.; Druliner, J. D.; Domaille, P. J. *Organometallics* **1984**, *3*, 33.
146. Nugent, W. A.; McKinney, R. J. *J. Org. Chem.* **1985**, *50*, 5370.
147. Briggs, J. R.; Klosin, J.; Whiteker, G. T. *Org. Lett.* **2005**, *7*, 4795.
148. Briggs, J. R.; Whiteker, G. T. *J. Chem. Soc., Chem. Commun.* **2001**, 2174.
149. Ostermeier, M.; Priess, J.; Helmchen, G. *Angew. Chem., Int. Ed. Engl.* **2002**, *41*, 612.
150. Schnyder, A.; Indolese, A. F.; Studer, M.; Blaser, H. U. *Angew. Chem., Int. Ed. Engl.* **2002**, *41*, 3668.
151. Günther, H. In *NMR Spectroscopy*; Wiley: New York, 1992; p 463.
152. Davies, J. A. In *The Chemistry of the Metal–Carbon Bond*; Hartley, F. R., Patai, S., Eds.; Wiley: New York, 1982; Vol. 1.
153. Pregosin, P. S.; Kunz, R.W. In *NMR Basic Principles and Progress*; Diehl, P., Fluck, E., Kosfeld, R., Eds.; Springer-Verlag: Berlin, 1979; Vol. 16, p 55.
154. Farina, V.; Krishnan, B. *J. Am. Chem. Soc.* **1991**, *113*, 9585.
155. Cheng, C. R.; Leung, P.; Mok, K. F. *Inorg. Chim. Acta* **1997**, *260*, 137.
156. Amatore, C.; Bucaille, A.; Fuxa, A.; Jutand, A.; Meyer, G.; Ntepe, A. N. *Chem. Eur. J.* **2001**, *7*, 2134.
157. Kojima, A.; Boden, C. D. J.; Shibasaki, M. *Tetrahedron Lett.* **1997**, *38*, 3459.
158. Cho, S. Y.; Shibasaki, M. *Tetrahedron Lett.* **1998**, *39*, 1773.
159. Appleton, T. G.; Clark, H. C.; Manzer, L. E. *Coord. Chem. Rev.* **1973**, *10*, 335.

160. Taylor, T. E.; Hall, M. B. *J. Am. Chem. Soc.* **1984**, *106*, 1576.
161. The observation that the singlet or triplet character of the corresponding free carbene correctly predicts the reactivity of a carbene ligand is due to Prof. A. K. Rapp.
162. Nazran, A. S.; Griller, D. *J. Chem. Soc., Chem. Commun.* **1983**, 850.
163. Rondan, N. G.; Houk, K. N.; Moss, R. A. *J. Am. Chem. Soc.* **1980**, *102*, 1770.
164. Harrison, J. F.; Liedtke, R. C.; Liebman, J. F. *J. Am. Chem. Soc.* **1979**, *101*, 7162.
165. Wentrup, C. In *Reactive Molecules: The Neutral Reactive Intermediates in Organic Chemistry*; Wiley: New York, 1984; Chapter 4.
166. Iwamoto, E.; Hirai, K.; Tomioka, H. *J. Am. Chem. Soc.* **2003**, *125*, 14664.
167. Arduengo, A. J. I. *Acc. Chem. Res.* **1999**, *32*, 913.
168. Arduengo, A. J.; Harlow, R. L.; Kline, M. *J. Am. Chem. Soc.* **1991**, *113*, 361.
169. Fischer, E. O. *Adv. Organomet. Chem.* **1976**, *14*, 1.
170. Cardin, D. J.; Cetinkay.B; Lappert, M. F. *Chem. Rev.* **1972**, *72*, 545.
171. Fehlhammer, W. P.; Bartel, K.; Plaia, U.; Volkl, A.; Liu, A. T. *Chem. Ber.* **1985**, *118*, 2235.
172. Fehlhammer, W. P.; Bartel, K.; Weinberger, B.; Plaia, U. *Chem. Ber.* **1985**, *118*, 2220.
173. Cardin, D. J.; Cetinkay, B.; Doyle, M. J.; Lappert, M. F. *Chem. Soc. Rev.* **1973**, *2*, 99.
174. Brown, F. J. *Prog. Inorg. Chem.* **1980**, *27*, 1.
175. Fischer, H. In *The Chemistry of the Metal–Carbon Bond*; Hartley, F. R., Patai, S., Eds.; Wiley: New York, 1982; Vol. 1, Chapter 4.
176. Dotz, K. H.; Fischer, H.; Hoffman, P.; Kreissl, F. R.; Schubert, U.; Weiss, K. *Transition Metal Carbene Complexes*; Verlag Chemie: Deerfield Beach, FL, 1983.
177. Casey, C. P. In *Reactive Intermediates*; Jones, M., Jr., Moss, R. A., Eds.; Wiley: New York, 1981; Vol. 2, p 135.
178. Casey, C. P. In *Reactive Intermediates*; Jones, M., Jr., Moss, R. A., Eds.; Wiley: New York, 1985; Vol. 2, p 150.
179. Gallop, M. A.; Roper, W. R. *Adv. Organomet. Chem.* **1986**, *25*, 121.
180. Dörwald, F. Z. *Metal Carbenes in Organic Synthesis*; Wiley-VCH: New York, 1999.
181. Dotz, K. H.; Jahr, H. C. In *Carbene Chemistry*; Bertrand, G., Ed.; Marcel Dekker: New York, 2002; p 231.
182. Schubert, U. *Coord. Chem. Rev.* **1984**, *55*, 261.
183. Casey, C. P. In *Transition Metal Organometallics in Organic Synthesis*; Alper, H., Ed.; Academic Press: New York, 1976; Vol. 1, Chapter 3.
184. Dotz, K. H. *Angew. Chem., Int. Ed. Engl.* **1984**, *23*, 587.
185. Wanzlick, H. W. *Angew. Chem.* **1962**, *74*, 129.
186. Doyle, M. J.; Lappert, M. F.; Pye, P. L.; Terreros, P. *J. Chem. Soc., Dalton Trans.* **1984**, 2355.
187. Cardin, D. J.; Cetinkay, B.; Lappert, M. F.; Manojlov, L.; Muir, K. W. *J. Chem. Soc., Chem. Commun.* **1971**, 400.
188. Cetinkaya, B.; Dixneuf, P.; Lappert, M. F. *J. Chem. Soc., Chem. Commun.* **1973**, 206.
189. Arduengo, A. J., III; Rasika Dias, H. V.; Harlow, R. L.; Kline, M. *J. Am. Chem. Soc.* **1992**, *112*, 5530.
190. Arduengo, A. J.; Krafczyk, R.; Schmutzler, R. *Tetrahedron* **1999**, *55*, 14523.
191. Herrmann, W. A.; Kocher, C. *Angew. Chem., Int. Ed. Engl.* **1997**, *36*, 2163.
192. Trnka, T. M.; Grubbs, R. H. *Acc. Chem. Res.* **2001**, *34*, 18.
193. Jafarpour, L.; Hillier, A. C.; Nolan, S. P. *Organometallics* **2002**, *21*, 442.
194. Huang, J.; Nolan, S. *J. Am. Chem. Soc.* **1999**, *121*, 9889.
195. Huang, J.; Grasa, G.; Nolan, S. P. *Org. Lett.* **1999**, *1*, 1307.
196. Zhang, C. M.; Huang, J. K.; Trudell, M. L.; Nolan, S. P. *J. Am. Chem. Soc.* **2000**, *122*, 12051.
197. Grasa, G. A.; Nolan, S. P. *Org. Lett.* **2001**, *3*, 119.
198. Grasa, G. A.; Viciu, M. S.; Huang, J. K.; Nolan, S. P. *J. Org. Chem.* **2001**, *66*, 7729.
199. Viciu, M. S.; Kissling, R. M.; Stevens, E. D.; Nolan, S. P. *Org. Lett.* **2002**, *4*, 2229.
200. Viciu, M. S.; Germaneau, R. F.; Nolan, S. P. *Org. Lett.* **2002**, *4*, 4053.
201. Yang, C. L.; Nolan, S. P. *Organometallics* **2002**, *21*, 1020.
202. Stauffer, S. R.; Lee, S.; Stambuli, J. P.; Hauck, S. I.; Hartwig, J. F. *Org. Lett.* **2000**, *2*, 1423.
203. Louie, J.; Gibby, J. E.; Farnworth, M. V.; Tekavec, T. N. *J. Am. Chem. Soc.* **2002**, *124*, 15188.
204. Jackstell, R.; Andreu, M. G.; Frisch, A.; Selvakumar, K.; Zapf, A.; Klein, H.; Spannenberg, A.; Rottger, D.; Briel, O.; Karch, R.; Beller, M. *Angew. Chem., Int. Ed. Engl.* **2002**, *41*, 986.
205. Bruce, M. I. *Chem. Rev.* **1991**, *91*, 197.
206. Bruce, M. I.; Swincer, A. G. *Adv. Organomet. Chem.* **1983**, *22*, 59.
207. Ervin, K. M.; Ho, J.; Lineberger, W. C. *J. Chem. Phys.* **1989**, *91*, 5974.
208. Ervin, K. M.; Gronert, S.; Barlow, S. E.; Gilles, M. K.; Harrison, A. G.; Bierbaum, V. M.; Depuy, C. H.; Lineberger, W. C.; Ellison, G. B. *J. Am. Chem. Soc.* **1990**, *112*, 5750.
209. Hayes, R. L.; Fattal, E.; Govind, N.; Carter, E. A. *Abstr. Papers Am. Chem. Soc.* **1999**, *217*, U334.
210. Hayes, R. L.; Fattal, E.; Govind, N.; Carter, E. A. *J. Am. Chem. Soc.* **2001**, *123*, 641.
211. Brookhart, M.; Tucker, J. R.; Flood, T. C.; Jensen, J. F. *J. Am. Chem. Soc.* **1980**, *102*, 1203.

212. Kegley, S. E.; Brookhart, M.; Husk, G. R. *Organometallics* **1982**, *1*, 760.
213. Kiel, W. A.; Lin, G.-Y.; Constable, A. G.; McCormick, F. B.; Strouse, C. E.; Eisenstein, O. E.; Gladysz, J. A. *J. Am. Chem. Soc.* **1982**, *104*, 4865.
214. Schrock, R. R. *Science* **1983**, *219*, 13.
215. Jafarpour, L.; Nolan, S. P. *Adv. Organomet. Chem.* **2001**, *46*, 181.
216. Huang, J.; Schanz, H.-J.; Stevens, E. D.; Nolan, S. P. *Organometallics* **1999**, *18*, 2370.
217. Dörwald, F. Z.; *Metal Carbenes in Organic Synthesis*; Wiley-VCH: New York, 1999; p 10.
218. Cardin, D. J.; Cetinkay, B.; Lappert, M. F. *J. Organomet. Chem.* **1974**, *72*, 139.
219. Fischer, E. O.; Kreis, G.; Kreiter, C. G.; Muller, J.; Huttner, G.; Lorenz, H. *Angew. Chem., Int. Ed. Engl.* **1973**, *12*, 564.
220. Wengrovius, J. H.; Sancho, J.; Schrock, R. R. *J. Am. Chem. Soc.* **1981**, *103*, 3932.
221. Pschirer, N. G.; Bunz, U. H. F. *Tetrahdron Lett.* **1999**, *40*, 2481.
222. Bunz, U. H. F. *Acc. Chem. Res.* **2001**, *34*, 998.
223. Furstner, A.; Grela, K. *Angew. Chem., Int. Ed. Engl.* **2000**, *39*, 1234.
224. Furstner, A.; Mathes, C. *Org. Lett.* **2001**, *3*, 221.
225. Furstner, A.; Mathes, C.; Lehmann, C. W. *Chem. Eur. J.* **2001**, *7*, 5299.
226. Furstner, A.; Seidel, G. *Angew. Chem., Int. Ed. Engl.* **1998**, *37*, 1734.
227. Zhang, W.; Kraft, S.; Moore, J. S. *J. Am. Chem. Soc.* **2004**, *126*, 329.
228. Holmes, S. J.; Schrock, R. R. *J. Am. Chem. Soc.* **1981**, *103*, 4599.
229. Kostic, N. M.; Fenske, R. F. *J. Am. Chem. Soc.* **1981**, *103*, 4677.
230. Kegley, S. E.; Brookhart, M.; Husk, G. R. *Organometallics* **1982**, *1*, 760.
231. Churchill, M. R.; Youngs, W. J. *Inorg. Chem.* **1979**, *18*, 2454.
232. Huttner, G.; Frank, A.; Fischer, E. O. *Isr. J. Chem.* **1977**, *15*, 133.
233. Fischer, E. O.; Kleine, W.; Kreis, G.; Kreissl, F. R. *Chem. Ber.* **1978**, *111*, 3542.
234. Dao, N. Q.; Fischer, E. O.; Wagner, W. R.; Neugebauer, D. *Chem. Ber.* **1979**, *112*, 2552.
235. Dao, N. Q.; Fischer, E. O.; Kappenstein, C. *New J. Chem.* **1980**, *4*, 85.
236. Hayashi, T.; Ueyama, K.; Tokunaga, N.; Yoshida, K. *J. Am. Chem. Soc.* **2003**, *125*, 11508.
237. Duan, W. L.; Iwamura, H.; Shintani, R.; Hayashi, T. *J. Am. Chem. Soc.* **2007**, *129*, 2130.
238. Paquin, J. F.; Stephenson, C. R. J.; Defieber, C.; Carreira, E. M. *Org. Lett.* **2005**, *7*, 3821.
239. Berthon-Gelloz, G.; Hayashi, T. *J. Org. Chem.* **2006**, *71*, 8957.
240. Lang, F.; Breher, F.; Stein, D.; Grutzmacher, H. *Organometallics* **2005**, *24*, 2997.
241. Dewar, M. J. S. *Bull. Soc. Chim. Fr.* **1951**, *18*, 79.
242. Chatt, J.; Duncanson, L. A. *J. Chem. Soc.* **1953**, 2939.
243. Hartley, F. R. *Chem. Rev.* **1973**, *73*, 163.
244. Yamamoto, T.; Yamamoto, A.; Ikeda, S. *J. Am. Chem. Soc.* **1971**, *93*, 3360.
245. Mecking, S.; Johnson, L. K.; Wang, L.; Brookhart, M. *J. Am. Chem. Soc.* **1998**, *120*, 888.
246. (a) Ban, E. M.; Powell, J.; Hughes, R. P. *J. Chem. Soc., Chem. Commun.* **1973**, 591; (b) Ban, E. M.; Hughes, R. P.; Powell, J. *J. Organomet. Chem.* **1974**, *69*, 455; (c) Kurosawa, H.; Urabe, A.; Emoto, M. *J. Chem. Soc., Dalton Trans.* **1986**, 891.
247. Wu, Z.; Jordan, R. F.; Petersen, J. L. *J. Am. Chem. Soc.* **1995**, *117*, 5867.
248. Witte, P. T.; Meetsma, A.; Hessen, B.; Budzelaar, P. H. M. *J. Am. Chem. Soc.* **1997**, *119*, 10561.
249. Galakhov, M. V.; Heinz, G.; Royo, P. *J. Chem. Soc., Chem. Commun.* **1998**, 17.
250. Carpentier, J. F.; Wu, Z.; Lee, C. W.; Stromberg, S.; Christopher, J. N.; Jordan, R. F. *J. Am. Chem. Soc.*, **2000**, *122*, 7750.
251. Casey, C. P.; Klein, J. F.; Fagan, M. A. *J. Am. Chem. Soc.* **2000**, *122*, 4320.
252. Humphries, M. J.; Douthwaite, R. E.; Green, M. L. H. *J. Chem. Soc., Dalton Trans.* **2000**, 2952.
253. Brandow, C. G.; Mendiratta, A.; Bercaw, J. E. *Organometallics* **2001**, *20*, 4253.
254. Carpentier, J. F.; Maryin, V. P.; Luci, J.; Jordan, R. F. *J. Am. Chem. Soc.* **2001**, *123*, 898.
255. Casey, C. P.; Lee, T. Y.; Tunge, J. A.; Carpenetti, D. W. *J. Am. Chem. Soc.* **2001**, *123*, 10762.
256. Casey, C. P.; Carpenetti, D. W.; Sakurai, H. *Organometallics* **2001**, *20*, 4262.
257. Carpenetti, D. W. *Inorg. Chem. Commun.* **2003**, *6*, 1287.
258. Casey, C. P.; Tunge, J. A.; Lee, T. Y.; Fagan, M. A. *J. Am. Chem. Soc.* **2003**, *125*, 2641.
259. Stoebenau, E. J.; Jordan, R. F. *J. Am. Chem. Soc.* **2003**, *125*, 3222.
260. Stoebenau, E. J.; Jordan, R. F. *J. Am. Chem. Soc.* **2004**, *126*, 11170.
261. Mingos, D. M. P. In *Comprehensive Organometallic Chemistry*; Wilkinson, G., Stone, F. G. A., Abel, E. W., Eds.; Pergamon Press: New York, 1982; Vol. 3, Chapter 19.
262. Grevels, F. W.; Skibbe, V. *J. Chem. Soc., Chem. Commun.* **1984**, 681.
263. Holloway, C. E.; Hulley, G.; Johnson, B. F. G.; Lewis, J. *J. Chem. Soc. A* **1970**, 1653.
264. Segal, J. A.; Johnson, B. F. G. *J. Chem. Soc., Dalton Trans.* **1975**, 677.
265. Alt, H.; Herberho, M.; Kreiter, C. G.; Strack, H. *J. Organomet. Chem.* **1974**, *77*, 353.

266. Kreiter, C. G.; Nist, K.; Alt, H. G. *Chem. Ber.* **1981**, *114*, 1845.

267. Koemm, U.; Kreiter, C. G. *J. Organomet. Chem.* **1982**, *240*, 27.

268. Mann, B. E. In *Comprehensive Organometallic Chemistry*; Wilkinson, G., Stone, F. G. A., Abel, E. W., Eds.; Pergamon Press: New York, 1982; Vol. 3, Chapter 20.

269. Albright, T. A.; Hoffmann, R.; Thibeault, J. C.; Thorn, D. L. *J. Am. Chem. Soc.* **1979**, *101*, 3801.

270. Hall, P. W.; Puddephatt, R. J.; Tipper, C. F. H. *J. Organomet. Chem.* **1974**, *71*, 145.

271. Tolman, C. A.; English, A. D.; Manzer, L. E. *Inorg. Chem.* **1975**, *14*, 2353.

272. Ittel, S. D.; Ibers, J. A. *Adv. Organomet. Chem.* **1976**, *14*, 33.

273. Robertson, G. B.; Whimp, P. O. *J. Organomet. Chem.* **1971**, *32*, C69.

274. Bennett, M. A.; Robertson, G. B.; Whimp, P. O.; Yoshida, T. *J. Am. Chem. Soc.* **1971**, *93*, 3797.

275. Buchwald, S. L.; Nielsen, R. B. *Chem. Rev.* **1988**, *88*, 1047.

276. Buchwald, S. L.; Lum, R. T.; Fisher, R. A.; Davis, W. M. *J. Am. Chem. Soc.* **1989**, *111*, 9113.

277. Hughes, R. P.; Laritchev, R. B.; Williamson, A.; Incarvito, C. D.; Zakharov, L. N.; Rheingold, A. L. *Organometallics* **2003**, *22*, 2134.

278. Retboll, M.; Edwards, A. J.; Rae, A. D.; Willis, A. C.; Bennett, M. A.; Wenger, E. *J. Am. Chem. Soc.* **2002**, *124*, 8348.

279. Jones, W. M.; Klosin, J. *Adv. Organomet. Chem.* **1998**, *42*, 147.

280. Bennett, M. A.; Dirnberger, T.; Hockless, D. C. R.; Wenger, E.; Willis, A. C. *J. Chem. Soc., Dalton Trans.* **1998**, 271.

281. Bennett, M. A.; Wenger, E. *Chem. Ber.* **1997**, *130*, 1029.

282. Cámpora, J.; Llebaría, A.; Moretó, J. M.; Poveda, M. L.; Carmona, E. *Organometallics* **1993**, *12*, 4032.

283. Hartwig, J. F.; Bergman, R. G.; Andersen, R. A. *J. Am. Chem. Soc.* **1991**, *113*, 3404.

284. Bennett, M. A.; Hambley, T. W.; Roberts, N. K.; Robertson, G. B. *Organometallics* **1985**, *4*, 1992.

285. McLain, S. J.; Schrock, R. R.; Sharp, P. R.; Churchill, M. R.; Youngs, W. J. *J. Am. Chem. Soc.* **1979**, *101*, 263.

286. Hartwig, J. F.; Andersen, R. A.; Bergman, R. G. *J. Am. Chem. Soc.* **1989**, *111*, 2717.

287. Hartwig, J. F.; Bergman, R. G.; Andersen, R. A. *J. Am. Chem. Soc.* **1991**, *113*, 3404.

288. Tatsumi, K.; Hoffmann, R.; Templeton, J. L. *Inorg. Chem.* **1982**, *21*, 466.

289. Gambarotta, S.; Floriani, C.; Chiesivilla, A.; Guastini, C. *J. Am. Chem. Soc.* **1985**, *107*, 2985.

290. Garner, C. M.; Mendez, N. Q.; Kowalczyk, J. J.; Fernandez, J. M.; Emerson, K.; Larsen, R. D.; Gladysz, J. A. *J. Am. Chem. Soc.* **1990**, *112*, 5146.

291. Klein, D. P.; Dalton, D. M.; Mendez, N. Q.; Arif, A. M.; Gladysz, J. A. *J. Organomet. Chem.* **1991**, *412*, C7.

292. Klein, D. P.; Mendez, N. Q.; Seyler, J. W.; Arif, A. M.; Gladysz, J. A. *J. Organomet. Chem.* **1993**, *450*, 157.

293. Mendez, N. Q.; Seyler, J. W.; Arif, A. M.; Gladysz, J. A. *J. Am. Chem. Soc.* **1993**, *115*, 2323.

294. Gladysz, J. A.; Boone, B. J. *Angew. Chem., Int. Ed. Engl.* **1997**, *36*, 551.

295. Aresta, M.; Nobile, C. F.; Albano, V. G.; Forni, E.; Manassero, M. *J. Chem. Soc., Chem. Commun* **1975**, 636.

296. Aresta, M.; Nobile, C. F. *J. Chem. Soc., Dalton Trans.* **1977**, 708.

297. Hoberg, H.; Schaefer, D.; Burkhart, G.; Kruger, C.; Romao, M. J. *J. Organomet. Chem.* **1984**, *266*, 203.

298. Gibson, D. H. *Chem. Rev.* **1996**, *96*, 2063.

299. Carey, F. A.; Sundberg, R. J. *Advanced Organic Chemistry*; Plenum Press: New York, 1990; p 12.

300. Faller, J. W.; Ma, Y.; Smart, C. J.; DiVerdi, M. J. *J. Organomet. Chem.* **1991**, *420*, 237.

301. Amendola, M. C.; Stockman, K. E.; Hoic, D. A.; Davis, W. M.; Fu, G. C. *Angew. Chem., Int. Ed. Engl.* **1997**, *36*, 267.

302. Asao, N.; Kii, S.; Hanawa, H.; Maruoka, K. *Tetrahedron Lett.* **1998**, *39*, 3729.

303. Parks, D. J.; Piers, W. E.; Parvez, M.; Atencio, R.; Zaworotko, M. J. *Organometallics* **1998**, *17*, 1369.

304. For a review of the structures of metal arene complexes, see reference 305.

305. Hubig, S. M.; Lindeman, S. V.; Kochi, J. K. *Coord. Chem. Rev.* **2000**, *200–202*, 831.

306. Fischer, E. O.; Hafner, W. *Z. Naturforsch., B: Chem. Sci.* **1955**, *10*, 665.

307. Jones, L. H.; McDowell, R. S.; Goldblatt, M. *Inorg. Chem.* **1969**, *8*, 2349.

308. Zeiss, H.; Wheatley, P. J.; Winkler, J. F. S. *Benzoid-Metal Complexes*; Ronald Press: New York, 1966.

309. Silverthorn, W. E. *Adv. Organomet. Chem.* **1975**, *13*, 47.

310. Marr, G.; Rockett, B. W. In *The Chemistry of the Metal–Carbon Bond*; Hartley, F. R., Patai, S., Eds.; Wiley: New York, 1982; Vol. 1, Chapter 11.

311. Pape, A. R.; Kaliappan, K. P.; Kundig, E. P. *Chem. Rev.* **2000**, *100*, 2917.

312. McGlinchey, M. J. In *The Chemistry of the Metal–Carbon Bond*; Hartley, F. R., Patai, S., Eds.; Wiley: New York, 1982; Vol. 1, Chapter 13.

313. Markle, R. J.; Pettijohn, T. M.; Lagowski, J. J. *Organometallics* **1985**, *4*, 1529.

314. Green, M. L. H.; Wong, L. L. *J. Chem. Soc., Chem. Commun.* **1984**, 1442.

315. Hamon, J.-R.; Saillard, J.-Y.; LeBeuze, A.; McGlinchey, M. J.; Astruc, D. *J. Am. Chem. Soc.* **1982**, *104*, 7549.

316. Bowyer, W. J.; Geiger, W. E. *J. Am. Chem. Soc.* **1985**, *107*, 5657.

317. Bowyer, W. J.; Merkert, J. W.; Geiger, W. E. *Organometallics* **1989**, *8*, 191.

318. Harman, W. D. *Adv. Chem. Ser.* **1997**, *253*, 39.

319. Harman, W. D. *Chem. Rev.* **1997**, *97*, 1953.

320. Iverson, C. N.; Lachicotte, R. J.; Muller, C.; Jones, W. D. *Organometallics* **2002**, *21*, 5320.

321. Huttner, G.; Lange, S. *Acta Crystallogr., Sect. B* **1972**, *B 28*, 2049.

322. Huttner, G.; Lang, S.; Fischer, E. O. *Angew. Chem., Int. Ed. Engl.* **1971**, *10*, 556.

323. Churchill, M. R.; Mason, R. *Proc. R. Soc. London, Ser. A* **1966**, *292*, 61.

324. Thompson, R. L.; Lee, S.; Rheingold, A. L.; Cooper, N. J. *Organometallics* **1991**, *10*, 1657.

325. For a recent study that includes tabulation of such data, see Lindeman S. V.; Rathov, R; Kochi, J. K. *Inorg. Chem.* **2000**, *39*, 5707.

326. (a) Veauthier, J. M.; Chow, A.; Fraenkel, G.; Geib, S. J.; Cooper, N. J. *Organometallics* **2000**, *19*, 661; (b) Sweet, J. R.; Graham, W. A. G. *J. Am. Chem. Soc.* **1983**, *105*, 305.

327. Meiere, S. H.; Valahovic, M. T.; Harman, W. D. *J. Am. Chem. Soc.* **2002**, *124*, 15099.

328. Meiere, S. H.; Keane, J. M.; Gunnoe, T. B.; Sabat, M.; Harman, W. D. *J. Am. Chem. Soc.* **2003**, *125*, 2024.

329. Mocella, C. J.; Delafuente, D. A.; Keane, J. A.; Warner, G. R.; Friedman, L. A.; Sabat, M.; Harman, W. D. *Organometallics* **2004**, *23*, 3772.

330. Graham, P. M.; Meiere, S. H.; Sabat, M.; Harman, W. D. *Organometallics* **2003**, *22*, 4364.

331. Brooks, B. C.; Gunnoe, T. B.; Harman, W. D. *Coord. Chem. Rev.* **2000**, *206*, 3.

332. Chin, R. M.; Dong, L.; Duckett, S. B.; Partridge, M. G.; Jones, W. D.; Perutz, R. N. *J. Am. Chem. Soc.* **1993**, *115*, 7685.

333. Reinartz, S.; White, P. S.; Brookhart, M.; Templeton, J. L. *J. Am. Chem. Soc.* **2001**, *123*, 12724.

334. Davidson, J. M.; Triggs, C. *J. Chem. Soc. A* **1968**, 1324.

335. Rauchfuss, T. B. *Prog. Inorg. Chem.* **1991**, *39*, 259.

336. Spera, M. L.; Harman, W. D. *Organometallics* **1999**, *18*, 2988.

337. Goti, A.; Semmelhack, M. F. *J. Organomet. Chem.* **1994**, *470*, C4.

338. Fish, R. H.; Fong, R. H.; Tran, A.; Baralt, E. *Organometallics* **1991**, *10*, 1209.

339. Zhu, G.; Tanski, J. M.; Churchill, D. G.; Janak, K. E.; Parkin, G. *J. Am. Chem. Soc.* **2002**, *124*, 13658.

340. Lagowski, J. J. *Coord. Chem. Rev.* **1977**, *22*, 185.

341. Ashe, A. J.; Meyers, E.; Shu, P.; Vonlehmann, T.; Bastide, J. *J. Am. Chem. Soc.* **1975**, *97*, 6865.

342. Ashe, A. J.; Shu, P. *J. Am. Chem. Soc.* **1971**, *93*, 1804.

343. Herberich, G. E.; Greiss, G.; Heil, H. F. *Angew. Chem., Int. Ed. Engl.* **1970**, *9*, 805.

344. Herberich, G. E.; Becker, H. J. *Angew. Chem., Int. Ed. Engl.* **1975**, *14*, 184 and references therein.

345. Herberich, G. E.; Ohst, H. *Adv. Organomet. Chem.* **1986**, *25*, 199.

346. Bazan, G. C.; Rodriguez, G.; Ashe, A. J.; AlAhmad, S.; Muller, C. *J. Am. Chem. Soc* **1996**, *118*, 2291.

347. Bazan, G. C.; Rodriguez, G.; Ashe, A. J.; AlAhmad, S.; Kampf, J. W. *Organometallics* **1997**, *16*, 2492.

348. Barnhart, R. W.; Bazan, G. C.; Mourey, T. *J. Am. Chem. Soc.* **1998**, *120*, 1082.

349. Rogers, J. S.; Lachicotte, R. J.; Bazan, G. C. *J. Am. Chem. Soc.* **1999**, *121*, 1288.

350. Dysard, J. M.; Tilley, T. D.; Woo, T. K. *Organometallics* **2001**, *20*, 1195.

351. Gomez-Sad, M. P.; Johnson, B. F. G.; Lewis, J.; Raithby, P. R.; Wright, A. H. *J. Chem. Soc., Chem. Commun.* **1985**, 1682.

352. King, R. B.; Fronzaglia, A. *J. Am. Chem. Soc.* **1965**, 709.

353. Fitton, P.; McKeon, J. E.; Ream, B. C. *J. Chem. Soc., Chem. Commun.* **1969**, 370.

354. Stevens, R. R.; Shier, G. D. *J. Organomet. Chem.* **1970**, *21*, 495.

355. Becker, Y.; Stille, J. K. *J. Am. Chem. Soc.* **1978**, *100*, 845.

356. Galamb, V.; Palyi, G. *J. Chem. Soc., Chem. Commun.* **1982**, 487.

357. Carmona, E.; Marin, J. M.; Paneque, M.; Poveda, M. L. *Organometallics* **1987**, *6*, 1757.

358. Gatti, G.; Lopez, J. A.; Mealli, C.; Musco, A. *J. Organomet. Chem.* **1994**, *483*, 77.

359. Nettekoven, U.; Hartwig, J. F. *J. Am. Chem. Soc.* **2002**, *124*, 1166.

360. Seligson, A. L.; Trogler, W. C. *J. Am. Chem. Soc.* **1991**, *113*, 2520.

361. Diversi, P.; Ingrosso, G.; Lucherini, A.; Lumini, T.; Marchetti, F.; Merlino, S.; Adovasio, V.; Nardelli, M. *J. Chem. Soc., Dalton Trans.* **1988**, 461.

362. Degraaf, W.; Boersma, J.; Smeets, W. J. J.; Spek, A. L.; Vankoten, G. *Organometallics* **1989**, *8*, 2907.

363. Degraaf, W.; Vanwegen, J.; Boersma, J.; Spek, A. L.; Vankoten, G. *Recl. Trav. Chim. Pays-Bas* **1989**, *108*, 275.

364. Markies, B. A.; Canty, A. J.; Degraaf, W.; Boersma, J.; Janssen, M. D.; Hogerheide, M. P.; Smeets, W. J. J.; Spek, A. L.; Vankoten, G. *J. Organomet. Chem.* **1994**, *482*, 191.

365. Wang, C. M.; Ziller, J. W.; Flood, T. C. *J. Am. Chem. Soc.* **1995**, *117*, 1647.

366. Flood, T. C.; Janak, K. E.; Iimura, H.; Zhen, H. *J. Am. Chem. Soc.* **2000**, *122*, 6783.

367. Yamakawa, M.; Ito, H.; Noyori, R. *J. Am. Chem. Soc.* **2000**, *122*, 1466.

368. Noyori, R.; Ohkuma, T. *Angew. Chem., Int. Ed. Engl.* **2001**, *40*, 40.

369. Noyori, R.; Hashiguchi, S. *Acc. Chem. Res.* **1997**, *30*, 97.

370. McWhinnie, W. R.; Miller, J. D. *Adv. Inorg. Chem. Radiochem.* **1969**, *12*, 135.

371. Van Koten, G.; Vrieze, K. *Adv. Inorg. Chem. Radiochem.* **1982**, *21*, 152.

372. Pfaltz, A. *Acc. Chem. Res.* **1993**, *26*, 339.

373. Pfaltz, A. *Acta Chem. Scand.* **1996**, *50*, 189.

374. Ghosh, A. K.; Packiarajan, M.; Cappiello, J. *Tetrahedron: Asymmetry* **1998**, *9*, 1.

375. Nishiyama, H.; Kondo, M.; Nakamura, T.; Itoh, K. *Organometallics* **1991**, *10*, 500.

376. Desimoni, G.; Faita, G.; Quadrelli, P. *Chem. Rev.* **2003**, *103*, 3119.

377. (a) McManus, H. A.; Guiry, P. J. *Chem. Rev.* **2004**, *104*, 4151; (b) Helmchen, G.; Pfaltz, A. *Acc. Chem. Res.* **2000**, *33*, 336.

378. (a) Lindoy, L. F.; Livingstone, S. E. *Coord. Chem. Rev.* **1967**, *2*, 173; (b) Busch, D. H.; Bailar, J. C. *J. Am. Chem. Soc.* **1956**, *78*, 1137; (c) Krumholz, P. *J. Am. Chem. Soc.* **1953**, *75*, 2163.

379. Ishiyama, T.; Takagi, J.; Ishida, K.; Miyaura, N.; Anastasi, N.; Hartwig, J. F. *J. Am. Chem. Soc.* **2002**, *124*, 390.

380. Ishiyama, T.; Takagi, J.; Hartwig, J. F.; Miyaura, N. *Angew. Chem., Int. Ed. Engl.* **2002**, *41*, 3056.

381. Wolfe, J. P.; Buchwald, S. L. *J. Am. Chem. Soc.* **1997**, *119*, 6054.

382. Ittel, S. D.; Johnson, L. K.; Brookhart, M. *Chem. Rev.* **2000**, *100*, 1169.

383. Small, B. L.; Brookhart, M. *J. Am. Chem. Soc.* **1998**, *120*, 7143.

384. MacKay, B. A.; Fryzuk, M. D. *Chem. Rev.* **2004**, *104*, 385.

385. Gambarotta, S. *J. Organomet. Chem.* **1995**, *500*, 117.

386. Chatt, J.; Dilworth, J. R.; Richards, R. L. *Chem. Rev.* **1978**, *78*, 589.

387. Pombeiro, A. J. In *New Trends in the Chemistry of Nitrogen Fixation*; Chatt, J., de Camara Pina, L. M., Richards, R. L., Eds.; Academic Press: London, 1980; p 153.

388. Dilworth, J. R.; Richards, R. L. In *Comprehensive Organometallic Chemistry*; Wilkinson, G., Stone, F. G. A., Abel, E. W., Eds.; Pergamon Press: New York, 1982; Vol. 8.

389. Hidai, M.; Mizobe, Y. *Chem. Rev.* **1995**, *95*, 1115.

390. Fryzuk, M. D.; Johnson, S. A. *Coord. Chem. Rev.* **2000**, *200–202*, 379.

391. Allen, A. D.; Senoff, C. W. *J. Chem. Soc., Chem. Commun.* **1965**, 621.

392. Bazhenova, T. A.; Shilov, A. E. *Coord. Chem. Rev.* **1995**, *144*, 69.

393. Bercaw, J. E. *J. Am. Chem. Soc.* **1974**, *96*, 5087.

394. Manriquez, J. M.; Bercaw, J. E. *J. Am. Chem. Soc.* **1974**, *96*, 6229.

395. Manriquez, J. M.; Sanner, R. D.; Marsh, R. E.; Bercaw, J. E. *J. Am. Chem. Soc.* **1976**, *98*, 3042.

396. Manriquez, J. M.; McAlister, D. R.; Rosenberg, E.; Shiller, A. M.; Williamson, K. L.; Chan, S. I.; Bercaw, J. E. *J. Am. Chem. Soc.* **1978**, *100*, 3078.

397. Smith, J. M.; Lachicotte, R. J.; Pittard, K. A.; Cundari, T. R.; Lukat-Rodgers, G.; Rodgers, K. R.; Holland, P. L. *J. Am. Chem. Soc.* **2001**, *123*, 9222.

398. Laplaza, C. E.; Cummins, C. C. *Science* **1995**, *268*, 861.

399. Cui, Q.; Musaev, D. G.; Svensson, M.; Sieber, S.; Morokuma, K. *J. Am. Chem. Soc.* **1995**, *117*, 12366.

400. Laplaza, C. E.; Johnson, M. J. A.; Peters, J. C.; Odom, A. L.; Kim, E.; Cummins, C. C.; George, G. N.; Pickering, I. J. *J. Am. Chem. Soc.* **1996**, *118*, 8623.

401. MacLachlan, E. A.; Fryzuk, M. D. *Organometallics* **2006**, *25*, 1530.

402. Fryzuk, M. D.; Haddad, T. S.; Mylvaganam, M.; McConville, D. H.; Rettig, S. J. *J. Am. Chem. Soc.* **1993**, *115*, 2782.

403. Cohen, J. D.; Mylvaganam, M.; Fryzuk, M. D.; Loehr, T. M. *J. Am. Chem. Soc.* **1994**, *116*, 9529.

404. Fryzuk, M. D.; Love, J. B.; Rettig, S. J.; Young, V. G. *Science* **1997**, *275*, 1445.

405. Morello, L.; Love, J. B.; Patrick, B. O.; Fryzuk, M. D. *J. Am. Chem. Soc.* **2004**, *126*, 9480.

406. Pool, J. A.; Lobkovsky, E.; Chirik, P. J. *J. Am. Chem. Soc.* **2003**, *125*, 2241.

407. Pool, J. A.; Lobkovsky, E.; Chirik, P. J. *Nature* **2004**, *427*, 527.

408. Evans, W. J.; Lee, D. S.; Ziller, J. W. *J. Am. Chem. Soc.* **2004**, *126*, 454.

409. Fryzuk, M. D.; MacKay, B. A.; Johnson, S. A.; Patrick, B. O. *Angew. Chem., Int. Ed. Engl.* **2002**, *41*, 3709.

410. Ritleng, V.; Yandulov, D. V.; Weare, W. W.; Schrock, R. R.; Hock, A. S.; Davis, W. M. *J. Am. Chem. Soc.* **2004**, *126*, 6150.

411. Yandulov, D. V.; Schrock, R. R. *J. Am. Chem. Soc.* **2002**, *124*, 6252.

412. Kol, M.; Schrock, R. R.; Kempe, R.; Davis, W. M. *J. Am. Chem. Soc.* **1994**, *116*, 4382.

413. Yandulov, D. V.; Schrock, R. R. *Science* **2003**, *301*, 76.

414. Calligaris, M.; Carugo, O. *Coord. Chem. Rev.* **1996**, *153*, 83.

415. Walsh, P. J.; Hollander, F. J.; Bergman, R. G. *Organometallics* **1993**, *12*, 3705.

416. Persson, C.; Andersson, C. *Inorg. Chim. Acta* **1993**, *203*, 235.

417. Persson, C.; Andersson, C. *Organometallics* **1993**, *12*, 2370.

418. Sellmann, D.; Kleinschmidt, E. *Z. Naturforsch., B: Chem. Sci.* **1977**, *32*, 795.

419. Roskamp, E. J.; Pedersen, S. F. *J. Am. Chem. Soc.* **1987**, *109*, 6551.
420. Roskamp, E. J.; Pedersen, S. F. *J. Am. Chem. Soc.* **1987**, *109*, 3152.
421. Jordan, R. F.; Bajgur, C. S.; Willett, R.; Scott, B. *J. Am. Chem. Soc.* **1986**, *108*, 7410.
422. Rix, F. C.; Brookhart, M.; White, P. S. *J. Am. Chem. Soc.* **1996**, *118*, 4746.
423. Tempel, D. J.; Johnson, L. K.; Huff, R. L.; White, P. S.; Brookhart, M. *J. Am. Chem. Soc.* **2000**, *122*, 6686.
424. Shultz, L. H.; Tempel, D. J.; Brookhart, M. *J. Am. Chem. Soc.* **2001**, *123*, 11539.
425. Walsh, P. J.; Hollander, F. J.; Bergman, R. G. *J. Am. Chem. Soc.* **1988**, *110*, 8729.
426. Cummins, C. C.; Baxter, S. M.; Wolczanski, P. D. *J. Am. Chem. Soc.* **1988**, *110*, 8731.
427. Zhong, H. A.; Labinger, J. A.; Bercaw, J. E. *J. Am. Chem. Soc.* **2002**, *124*, 1378.
428. Baranger, A. M.; Walsh, P. J.; Bergman, R. G. *J. Am. Chem. Soc.* **1993**, *115*, 2753.
429. Uozumi, Y.; Hayashi, T. *J. Am. Chem. Soc.* **1991**, *113*, 9887.
430. Uozumi, Y.; Tanahashi, A.; Lee, S.; Hayashi, T. *J. Org. Chem.* **1993**, *58*, 1945.
431. Uozumi, Y.; Suzuki, N.; Ogiwara, A.; Hayashi, T. *Tetrahedron* **1994**, *50*, 4293.
432. Hayashi, T.; Mise, T.; Fukishima, M.; Kagotani, M.; Nagashima, N.; Hamada, Y.; Matsumoto, A.; Kawakami, S.; Konoshi, M.; Yamamoto, K.; Kumada, M. *Bull. Chem. Soc. Jpn.* **1980**, *53*, 1138.
433. Hayashi, T. *Pure Appl. Chem.* **1988**, *60*, 7.
434. Hayashi, T.; Yamamoto, A.; Ito, Y. *Tetrahedron Lett.* **1988**, *29*, 99.
435. Grushin, V. V. *J. Am. Chem. Soc.* **1999**, *121*, 5831.
436. Boezio, A. A.; Pytkowicz, J.; Cote, A.; Charette, A. B. *J. Am. Chem. Soc.* **2003**, *125*, 14260.
437. Murray, S. G.; Hartley, F. R. *Chem. Rev.* **1981**, *81*, 365.
438. (a) Kiewel, K.; Liu, Y.; Bergbreiter, D. E.; Sulikowski, G. A. *Tetrahedron Lett.* **1999**, *40*, 8945; (b) Bergbreiter, D. E.; Osburn, P. L.; Liu, Y. S. *J. Am. Chem. Soc.* **1999**, *121*, 9531.
439. Faller, J. W.; Wilt, J. C. *Org. Lett.* **2005**, *7*, 633.
440. Pettinari, C.; Pellei, M.; Cavicchio, G.; Crucianelli, M.; Panzeri, W.; Colapietro, M.; Cassetta, A. *Organometallics* **1999**, *18*, 555.
441. Larock, R. C.; Hightower, T. R. *J. Org. Chem.* **1993**, *58*, 5298.
442. Larock, R. C.; Hightower, T. R.; Hasvold, L. A.; Peterson, K. P. *J. Org. Chem.* **1996**, *61*, 3584.
443. Steinhoff, B. A.; Fix, S. R.; Stahl, S. S. *J. Am. Chem. Soc.* **2002**, *124*, 766.
444. Chen, M. S.; White, M. C. *J. Am. Chem. Soc.* **2004**, *126*, 1346.
445. Gillie, A.; Stille, J. K. *J. Am. Chem. Soc.* **1980**, *102*, 4933.
446. Moraviskiy, A.; Stille, J. K. *J. Am. Chem. Soc.* **1981**, *103*, 4182.
447. Dorta, R.; Rozenberg, H.; Shimon, L. J. W.; Milstein, D. *Chem. Eur. J.* **2003**, *9*, 5237.
448. Jazzar, R. F. R.; Mahon, M. F.; Whittlesey, M. K. *Organometallics* **2001**, *20*, 3745.
449. (a) Kubas, G. J. *Metal-Dihydrogen and Sigma-Bond Complexes: Structure, Theory, and Reactivity*; Kluwer Academic: New York, 2001; (b) Crabtree, R. H. *Angew. Chem., Int. Ed. Engl.* **1993**, *32*, 789.
450. Maseras, F.; Lledos, A.; Clot, E.; Eisenstein, O. *Chem. Rev.* **2000**, *100*, 601.
451. Saillard, J.-Y.; Hoffmann, R. *J. Am. Chem. Soc.* **1984**, *106*, 2006.
452. Frenking, G.; Frohlich, N. *Chem. Rev.* **2000**, *100*, 717.
453. Tsipis, C. A. *Coord. Chem. Rev.* **1991**, *108*, 163.
454. Jean, Y.; Eisenstein, O.; Volatron, F.; Maouche, B.; Sefta, F. *J. Am. Chem. Soc.* **1986**, *108*, 6587.
455. Jetz, W.; Graham, W. A. G. *J. Am. Chem. Soc.* **1969**, *91*, 3375.
456. Hoyano, J. K.; Elder, M.; Graham, W. A. G. *J. Am. Chem. Soc.* **1969**, *91*, 4568.
457. Hart-Davis, A. J.; Graham, W. A. G. *J. Am. Chem. Soc.* **1971**, *93*, 4388.
458. Jetz, W.; Graham, W. A. G. *Inorg. Chem.* **1971**, *10*, 4.
459. Graham, W. A. G. *J. Organomet. Chem.* **1986**, *300*, 81.
460. Schubert, U. *Adv. Organomet. Chem.* **1990**, *30*, 151.
461. Kubas, G. J.; Ryan, R. R.; Swanson, B. I.; Vergamini, P. J.; Wasserman, H. J. *J. Am. Chem. Soc.* **1984**, *106*, 451.
462. Kubas, G. J. *Metal-Dihydrogen and Sigma-Bond Complexes: Structure, Theory, and Reactivity*; Kluwer Academic: New York, 2001; p 26.
463. Poliakoff, M.; Turner, J. J. *J. Chem. Soc., Dalton Trans.* **1974**, 2276.
464. Church, S. P.; Grevels, F.-W.; Herrmann, H.; Schaffner, K. *J. Chem. Soc., Chem. Commun.* **1985**, 794.
465. Buchanan, J. M.; Stryker, J. M.; Bergman, R. G. *J. Am. Chem. Soc.* **1986**, *108*, 1537.
466. Hall, C.; Perutz, R. N. *Chem. Rev.* **1996**, *96*, 3125.
467. Schlecht, S.; Hartwig, J. F. *J. Am. Chem. Soc.* **2000**, *122*, 9435.
468. Hartwig, J. F.; He, X.; Muhoro, C. N.; Eisenstein, O.; Bosque, R.; Maseras, F. *J. Am. Chem. Soc.* **1996**, *118*, 10936.
469. (a) Muhoro, C. N.; Hartwig, J. F. *Angew. Chem., Int. Ed. Engl.* **1997**, *36*, 1510; (b) Muhoro, C. N.; He, X. M.; Hartwig, J. F. *J. Am. Chem. Soc.* **1999**, *121*, 5033.
470. For a distinct α^2-bozane complex, see Alcaraz, G.; Clot, E.; Helmstedt, U.; Vendien, L.; Sabo-Etienne, S. *J. Am. Chem.* Soc. **2007**, *129*, 8704.

471. Crestani, M. G.; Munoz-Hernandez, M.; Arevalo, A.; Acosta-Ramirez, A.; Garcia, J. J. *J. Am. Chem. Soc.* **2005**, *127*, 18066.

472. Brammer, L.; Zhao, D.; Ladipo, F. T.; Braddock-Wilking, J. *Acta Crystallogr.* **1995**, *B51*, 632.

473. Albinati, A.; Lianza, F.; Pregosin, P. S.; Muller, B. *Inorg. Chem.* **1994**, *33*, 2522.

474. Albinati, A.; Anklin, C. G.; Pregosin, P. S. *Inorg. Chim. Acta* **1984**, *90*, L37.

475. Hedden, D.; Roundhill, D. M.; Fultz, W. C.; Rheingold, A. L. *Organometallics* **1986**, *5*, 336.

476. Albinati, A.; Anklin, C. G.; Ganazzoli, F.; Ruegg, H.; Pregosin, P. S. *Inorg. Chem.* **1987**, *26*, 503.

477. Albinati, A.; Arz, C.; Pregosin, P. S. *Inorg. Chem.* **1987**, *26*, 508.

478. Brammer, L.; Charnock, J. M.; Goggin, P. L.; Goodfellow, R. J.; Orpen, A. G.; Koetzle, T. F. *J. Chem. Soc., Dalton Trans.* **1991**, 1789.

479. Wehmanooyevaar, I. C. M.; Grove, D. M.; Kooijman, H.; Vandersluis, P.; Spek, A. L.; Vankoten, G. *J. Am. Chem. Soc.* **1992**, *114*, 9916.

480. See reference 28 of reference 473.

481. Vandersluys, L. S.; Eckert, J.; Eisenstein, O.; Hall, J. H.; Huffman, J. C.; Jackson, S. A.; Koetzle, T. F.; Kubas, G. J.; Vergamini, P. J.; Caulton, K. G. *J. Am. Chem. Soc.* **1990**, *112*, 4831.

482. Crabtree, R. H.; Siegbahn, P. E. M.; Eisenstein, O.; Rheingold, A. L. *Acc. Chem. Res.* **1996**, *29*, 348.

483. Park, S. H.; Ramachandran, R.; Lough, A. J.; Morris, R. H. *J. Chem. Soc., Chem. Commun.* **1994**, 2201.

484. Lough, A. J.; Park, S.; Ramachandran, R.; Morris, R. H. *J. Am. Chem. Soc.* **1994**, *116*, 8356.

485. Alkorta, I.; Elguero, J.; Foces-Foces, C. *J. Chem. Soc., Chem. Commun.* **1996**, 1633.

486. Gelabert, R.; Moreno, M.; Lluch, J. M.; Lledos, A. *Organometallics* **1997**, *16*, 3805.

487. Lee, J. C.; Peris, E.; Rheingold, A. L.; Crabtree, R. H. *J. Am. Chem. Soc.* **1994**, *116*, 11014.

488. Liu, Q.; Hoffmann, R. *J. Am. Chem. Soc.* **1995**, *117*, 10108.

489. Stevens, R. C.; Bau, R.; Milstein, D.; Blum, O.; Koetzle, T. F. *J. Chem. Soc., Dalton Trans.* **1990**, 1429.

490. Gatling, S. C.; Jackson, J. E. *J. Am. Chem. Soc.* **1999**, *121*, 8655.

491. Heinekey, D. M.; Oldham, W. J. *Chem. Rev.* **1993**, *93*, 913.

492. Ziegler, T.; Folga, E.; Berces, A. *J. Am. Chem. Soc.* **1993**, *115*, 636 and references therein.

493. Crabtree, R. H.; Hamilton, D. G. *J. Am. Chem. Soc.* **1986**, *108*, 3124.

494. Kubas, G. J. *Metal-Dihydrogen and Sigma-Bond Complexes: Structure, Theory, and Reactivity*; Kluwer Academic: New York, 2001; Table 4.7.

495. Ishida, T.; Mizobe, Y.; Tanase, T.; Hidai, M. *J. Organomet. Chem.* **1991**, *409*, 355.

496. Khalsa, G. R. K.; Kubas, G. J.; Unkefer, C. J.; Vandersluys, L. S.; Kubatmartin, K. A. *J. Am. Chem. Soc.* **1990**, *112*, 3855.

497. Heinekey, D. M.; Radzewich, C. E.; Voges, M. H.; Schomber, B. M. *J. Am. Chem. Soc.* **1997**, *119*, 4172.

498. Ashworth, T. V.; Singleton, E. *J. Chem. Soc., Chem. Commun.* **1976**, 705.

499. Morris, R. H.; Sawyer, J. F.; Shiralian, M.; Zubkowski, J. D. *J. Am. Chem. Soc.* **1985**, *107*, 5581.

500. Arliguie, T.; Chaudret, B.; Morris, R. H.; Sella, A. *Inorg. Chem.* **1988**, *27*, 598.

501. Hamilton, D. G.; Crabtree, R. H. *J. Am. Chem. Soc.* **1988**, *110*, 4126.

502. Brammer, L.; Howard, J. A. K.; Johnson, O.; Koetzle, T. F.; Spencer, J. L.; Stringer, A. M. *J. Chem. Soc., Chem. Commun.* **1991**, 241.

503. Borowski, A. F.; Donnadieu, B.; Daran, J. C.; Sabo-Etienne, S.; Chaudret, B. *J. Chem. Soc., Chem. Commun.* **2000**, 543.

504. For an excellent summary of vibrational spectroscopy of dihydrogen complexes, see reference 491.

505. (a) Grundeman, S.; Limbach, H. H.; Buntkowsky, G.; Sabo-Etienne, S.; Chaudret, B. *J. Phys. Chem. A.* **1999**, *103*, 4752; (b) Gelabert, R.; Moreno, M.; Lluch, J. M.; Lledos, A.; Pons, V.; Heinekey, D. M. *J. Am. Chem. Soc.* **2004**, *126*, 8813.

506. (a) Crabtree, R. H.; Lavin, M. *J. Chem. Soc., Chem. Commun.* **1985**, 1661; (b) Crabtree, R. H.; Lavin, M.; Bonneviot, L. *J. Am. Chem. Soc.* **1986**, *108*, 4032.

507. Hamilton, D. G.; Crabtree, R. H. *J. Am. Chem. Soc.* **1988**, *110*, 4126.

508. Crabtree, R. H. *Acc. Chem. Res.* **1990**, *23*, 95.

509. Gusev, D. G.; Vymenits, A. B.; Bakhmutov, V. I. *Inorg. Chem.* **1991**, *30*, 3116.

510. Cotton, F. A.; Luck, R. L.; Root, D. R.; Walton, R. A. *Inorg. Chem.* **1990**, *29*, 43.

511. Bautista, M. T.; Earl, K. A.; Maltby, P. A.; Morris, R. H.; Schweitzer, C. T.; Sella, A. *J. Am. Chem. Soc.* **1988**, *110*, 7031.

512. Desrosiers, P. J.; Cai, L. H.; Lin, Z. R.; Richards, R.; Halpern, J. *J. Am. Chem. Soc.* **1991**, *113*, 4173.

513. Gusev, D. G.; Nietlispach, D.; Vymenits, A. B.; Bakhmutov, V. I.; Berke, H. *Inorg. Chem.* **1993**, *32*, 3270.

514. Helleren, C. A.; Henderson, R. A.; Leigh, G. J. *J. Chem. Soc., Dalton Trans.* **1999**, 1213.

515. Gonzalez, A. A.; Zhang, K.; Nolan, S. P.; Delavega, R. L.; Mukerjee, S. L.; Hoff, C. D.; Kubas, G. J. *Organometallics* **1988**, *7*, 2429.

516. Kubas, G. J. *Metal-Dihydrogen and Sigma-Bond Complexes: Structure, Theory, and Reactivity*; Kluwer Academic: New York, 2001; Table 7.3.

517. Kubas, G. J.; Burns, C. J.; Khalsa, G. R. K.; Vandersluys, L. S.; Kiss, G.; Hoff, C. D. *Organometallics* **1992**, *11*, 3390.

518. Kubas, G. J. *Metal-Dihydrogen and Sigma-Bond Complexes: Structure, Theory, and Reactivity*; Kluwer Academic: New York, 2001; Table 7.5.

519. Kubas, G. J.; Ryan, R. R.; Wrobleski, D. A. *J. Am. Chem. Soc.* **1986**, *108*, 1339.

520. Luo, X. L.; Crabtree, R. H. *J. Am. Chem. Soc.* **1990**, *112*, 6912.

521. Chinn, M. S.; Heinekey, D. M. *J. Am. Chem. Soc.* **1990**, *112*, 5166.

522. Chinn, M. S.; Heinekey, D. M.; Payne, N. G.; Sofield, C. D. *Organometallics* **1989**, *8*, 1824.

523. Kubas, G. J.; Unkefer, C. J.; Swanson, B. I.; Fukushima, E. *J. Am. Chem. Soc.* **1986**, *108*, 7000.

524. Albertin, G.; Antoniutti, S.; Garcia-Fontan, S.; Carballo, R.; Padoan, F. *J. Chem. Soc., Dalton Trans.* **1998**, 2071.

525. Bullock, R. M.; Song, J. S.; Szalda, D. J. *Organometallics* **1996**, *15*, 2504.

526. Gusev, D. G.; Nietlispach, D.; Eremenko, I. L.; Berke, H. *Inorg. Chem.* **1993**, *32*, 3628.

527. Luo, X. L.; Michos, D.; Crabtree, R. H. *Organometallics* **1992**, *11*, 237.

528. Haward, M. T.; George, M. W.; Hamley, P.; Poliakoff, M. *J. Chem. Soc., Chem. Commun.* **1991**, 1101.

529. Esteruelas, M. A.; Oro, L. A.; Valero, C. *Organometallics* **1991**, *10*, 462.

530. Jia, G.; Morris, R. H. *J. Am. Chem. Soc.* **1991**, *113*, 875.

531. Arliguie, T.; Chaudret, B. *J. Chem. Soc., Chem. Commun.* **1989**, 155.

532. Conroy-Lewis, F. M.; Simpson, S. J. *J. Chem. Soc., Chem. Commun.* **1987**, 1675.

533. Chinn, M. S.; Heinekey, D. M. *J. Am. Chem. Soc.* **1987**, *109*, 5865.

534. Morris, R. H. *Can. J. Chem.* **1996**, *31*, 1471.

535. Pons, V.; Heinekey, D. M. *J. Am. Chem. Soc.* **2003**, *125*, 8428.

536. Luther, T. A.; Heinekey, D. M. *Inorg. Chem.* **1998**, *37*, 127.

537. Kubas, G. J. *Metal-Dihydrogen and Sigma-Bond Complexes: Structure, Theory, and Reactivity*; Kluwer Academic: New York, 2001; Table 9.1.

538. Rocchini, E.; Mezzetti, A.; Ruegger, H.; Burckhardt, U.; Gramlich, V.; DelZotto, A.; Martinuzzi, P.; Rigo, P. *Inorg. Chem.* **1997**, *36*, 711.

539. Jessop, P. G.; Ikariya, T.; Noyori, R. *Chem. Rev.* **1995**, *95*, 259.

540. Corey, J. Y.; Braddock-Wilking, J. *Chem. Rev.* **1999**, *99*, 175.

541. Lin, Z. Y. *Chem. Soc. Rev.* **2002**, *31*, 239.

542. Buchanan, J. M.; Stryker, J. M.; Bergman, R. G. *J. Am. Chem. Soc.* **1986**, *108*, 1537.

543. Periana, R. A.; Bergman, R. G. *J. Am. Chem. Soc.* **1986**, *108*, 7346.

544. Weiller, B. H.; Wasserman, E. P.; Bergman, R. G.; Moore, C. B.; Pimentel, G. C. *J. Am. Chem. Soc.* **1989**, *111*, 8288.

545. Wasserman, E. P.; Moore, C. B.; Bergman, R. G. *Science* **1992**, *255*, 315.

546. Bengali, A. A.; Schultz, R. H.; Moore, C. B.; Bergman, R. G. *J. Am. Chem. Soc.* **1994**, *116*, 9585.

547. Lian, T.; Bromberg, S. E.; Yang, H.; Proulx, G.; Bergman, R. G.; Harris, C. B. *J. Am. Chem. Soc.* **1996**, *118*, 3769.

548. McNamara, B. K.; Yeston, J. S.; Bergman, R. G.; Moore, C. B. *J. Am. Chem. Soc.* **1999**, *121*, 6437.

549. Northcutt, T. O.; Wick, D. D.; Vetter, A. J.; Jones, W. D. *J. Am. Chem. Soc.* **2001**, *123*, 7257.

550. Jones, W. D. *Acc. Chem. Res.* **2003**, *36*, 140.

551. Geftakis, S.; Ball, G. E. *J. Am. Chem. Soc.* **1998**, *120*, 9953.

552. Lawes, D. J.; Geftakis, S.; Ball, G. E. *J. Am. Chem. Soc.* **2005**, *127*, 4134.

553. Kubas, G. J. *Metal-Dihydrogen and Sigma-Bond Complexes: Structure, Theory, and Reactivity*; Kluwer Academic: New York, 2001; p 330.

554. Wasserman, E. P.; Moore, C. B.; Bergman, R. G. *Science* **1992**, *255*, 315.

555. Morse, J. M.; Parker, G. H.; Burkey, T. J. *Organometallics* **1989**, *8*, 2471.

556. Jiao, T. J.; Leu, G. L.; Farrell, G. J.; Burkey, T. J. *J. Am. Chem. Soc.* **2001**, *123*, 4960.

557. Brown, C. E.; Ishikawa, Y.; Hackett, P. A.; Rayner, D. M. *J. Am. Chem. Soc.* **1990**, *112*, 2530.

558. Cobar, E. A.; Khaliullin, R. Z.; Bergman, R. G.; Head-Gordon, M. *Proc. Natl. Acad. Sci. U. S. A.* **2007**, *104*, 6963.

559. Bergman, R. G.; Cundari, T. R.; Gillespie, A. M.; Gunnoe, T. B.; Harman, W. D.; Klinckman, T. R.; Temple, M. D.; White, D. P. *Organometallics* **2003**, *22*, 2331.

560. Yang, H.; Asplund, M. C.; Kotz, K. T.; Wilkens, M. J.; Frei, H.; Harris, C. B. *J. Am. Chem. Soc.* **1998**, *120*, 10154.

561. Brookhart, M.; Green, M. L. H.; Wong, L.-L. *Prog. Inorg. Chem.* **1988**, *36*, 1.

562. Brookhart, M.; Green, M. L. H. *J. Organomet. Chem.* **1983**, *250*, 395.

563. Dawoodi, Z.; Green, M. L. H.; Mtetwa, V. S. B.; Prout, K. *J. Chem. Soc., Chem. Commun.* **1982**, 1410.

564. Guo, Z. Y.; Swenson, D. C.; Jordan, R. F. *Organometallics* **1994**, *13*, 1424.

565. Schultz, A. J.; Brown, R. K.; Williams, J. M.; Schrock, R. R. *J. Am. Chem. Soc.* **1981**, *103*, 169.

566. Churchill, M. R.; Wasserman, H. J.; Turner, H. W.; Schrock, R. R. *J. Am. Chem. Soc.* **1982**, *104*, 1710.

567. La Placa, S. J.; Ibers, J. A. *Inorg. Chem.* **1965**, *4*, 778.

568. Stambuli, J. P.; Incarvito, C. D.; Buhl, M.; Hartwig, J. F. *J. Am. Chem. Soc.* **2004**, *126*, 1184.

569. Urtel, H.; Meier, C.; Eisentrager, F.; Rominger, F.; Joschek, J. P.; Hofmann, P. *Angew. Chem., Intl. Ed. Engl.* **2001**, *40*, 781.

570. Huang, D.; Streib, W. E.; Bollinger, J. C.; Caulton, K. G.; Winter, R. F.; Scheiring, T. *J. Am. Chem. Soc.* **1999**, *121*, 8087.

571. Dawoodi, Z.; Green, M. L. H.; Mtetwa, V. S. B.; Prout, K. *J. Chem. Soc., Chem. Commun.* **1982**, 802.

572. Dawoodi, Z.; Green, M. L. H.; Mtetwa, V. S. B.; Prout, K.; Schultz, A. J.; Williams, J. M.; Koetzle, T. F. *J. Chem. Soc., Dalton Trans.* **1986**, 1629.

573. Schmidt, G. F.; Brookhart, M. *J. Am. Chem. Soc.* **1985**, *107*, 1443.

574. Jordan, R. F.; Bradley, P. K.; Baenziger, N. C.; Lapointe, R. E. *J. Am. Chem. Soc.* **1990**, *112*, 1289.

575. Saunders, M.; Telkowski, L.; Kates, M. R. *J. Am. Chem. Soc.* **1977**, *99*, 8070.

576. Saunders, M.; Kates, M. R. *J. Am. Chem. Soc.* **1977**, *99*, 8071.

577. Saunders, M.; Kates, M. R.; Wiberg, K. B.; Pratt, W. *J. Am. Chem. Soc.* **1977**, *99*, 8072.

578. Calvert, R. B.; Shapley, J. R. *J. Am. Chem. Soc.* **1978**, *100*, 7726.

579. Grubbs, R. H.; Coates, G. W. *Acc. Chem. Res.* **1996**, *29*, 85.

CHAPTER 3

Covalent (X-Type) Ligands Bound Through Metal–Carbon and Metal–Hydrogen Bonds

3.1. Introduction

Chapters 3 and 4 provide an overview of the physical properties, bonding interactions, structures, spectral features, thermodynamic properties, syntheses, and reactivities of X-type and XL_n-type ligands. Recall from Chapter 1 that X-type and XL_n-type ligands are assigned negative charges in the ionic bonding formalism or an odd number of electrons in the neutral bonding formalism. Chapter 3 includes descriptions of the classic organometallic X-type and XL_n-type ligands that contain metal–carbon and metal–hydride bonds. The X-type ligands described in this chapter include σ-bonded fragments, such as alkyl, aryl, vinyl, alkynyl, enolate, and cyanide ligands, and the XL_n-type ligands in this chapter include π-bound fragments, such as η^3-allyl, cyclopentadienyl, and cyclopentadienyl analogs. In addition, Chapter 3 surveys the chemistry of metal hydrides. Chapter 4 includes descriptions of X-type and XL_n-type ligands in which the metal is bound to a heteroatom. These include κ^1, σ-bonded ligands, such as amido, alkoxo, and boryl groups, in which the metal is bound to a first-row element; as well as κ^1,σ-bonded ligands, such as phosphido, thiolato, and silyl groups in which the metal is bound to a second-row element. Chapter 4 also includes a discussion of halide ligands. The chemistry of complexes containing metal–ligand multiple bonds, such as metal–carbene, –imido, –oxo, –carbyne, and –nitrido complexes is presented in Chapter 13. Chapter 3 begins with a survey of σ-bound hydrocarbyl ligands before describing the chemistry of π-bound ligands.

3.2. Transition Metal Hydrocarbyl Ligands

Hydrocarbyl ligands, such as those shown in Figure 3.1, lie at the heart of transition metal organometallic chemistry. Virtually all reactions catalyzed by transition metals, such as hydrogenation (Chapter 15), cross-coupling (Chapter 19), hydroformylation (Chapter 16), and olefin polymerization (Chapter 22), involve the formation of such complexes. Stable hydrocarbyl complexes are known for all transition metals, and the early work in organometallic chemistry focused on their synthesis.

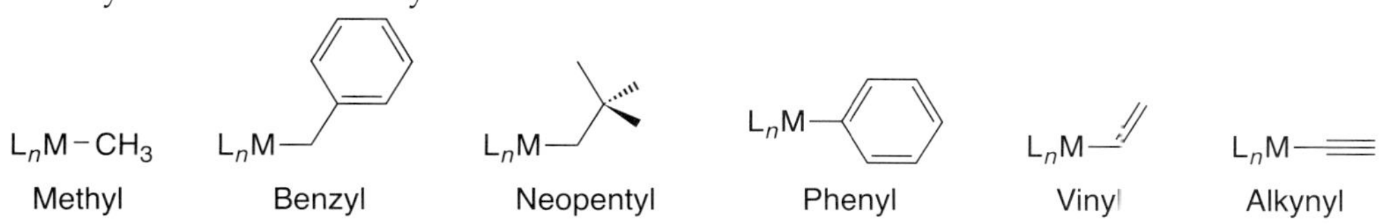

Figure 3.1. Common σ-bonded hydrocarbyl ligands.

3.2.1. Alkyl Ligands (Written with Prof. Jack R. Norton)

Methyl groups are the most common alkyl ligands, followed by benzyl and neopentyl groups. The bond between a late transition metal and an alkyl ligand is, as noted in Chapter 1, largely covalent. A late transition metal is more electropositive than carbon, but the difference in electronegativity between these elements is often smaller than the difference between carbon and nitrogen. The bonds between alkyl ligands and early transition metals are more polar and are more susceptible to protonolysis, but they still contain significant covalent character. For example, the difference in electronegativity between the group 5 metals in their atomic form and carbon is similar to that between carbon and oxygen.

3.2.1.1. History of Transition Metal–Alkyl Complexes

On March 21, 1907, W. J. Pope and S. J. Peachey reported to the Chemical Society in London the discovery of "Me_3PtI" from treating $PtCl_4$ with MeMgI in a mixture of ether and benzene.[1,2] It is now clear that Me_3PtI is a tetramer, held together by bridging iodides.[3] The chairman of the Chemical Society, Sir Henry E. Roscoe, congratulated Pope and Peachey "on having opened out an entirely new branch of investigation." Indeed, Pope (with C. S. Gibson) reported some ethylgold compounds soon afterward,[1] and work (largely by Gibson and co-workers) on organogold compounds continued through the 1930s and 1940s.[4–6] However, alkyl complexes of other transition metals were unknown until later. Krause and von Grosse surveyed the topic in 1937,[7] and Cotton wrote an excellent critical review ("Alkyls and Aryls of Transition Metals") in 1955,[3] in which he mentioned that several preparations of metal methyl complexes MMe_n had proven irreproducible.

Ferrocene[8,9] and a number of methyl complexes with supporting ligands, such as $CpCr(NO)_2Me$,[10] $CpFe(CO)_2Me$,[11,12] Cp_2TiMe_2,[12] and cis-$(Et_3P)_2PtMe_2$, were prepared in the 1950s.[13] Soon after, syntheses of "homoleptic" alkyl complexes MR_n, which contain a single type of ligand, were reported. Examples of homoleptic transition metal–alkyl complexes include $TiMe_4$[14] and $[CrMe_6]$.[3–15] An exceptionally stable and paradigmatic homoleptic alkyl complex is WMe_6, which was reported in 1973.[16–18] The early history of homoleptic alkyl complexes was reviewed in 1974,[19] and the early history of all metal hydrocarbyls was reviewed in 1976.[20] A more recent review has revisited "the decomposition of transition metal alkyls."[21]

3.2.1.2. Thermodynamic Properties of M–Alkyl Bonds

The strengths of transition metal–alkyl bonds vary over a wide range.[22,23] As discussed in more detail in this section, these bonds are typically weaker than C–C bonds and in most cases are weaker than the M–H bonds involving the same metal fragment. Moreover, these bonds, in the absence of electron-withdrawing groups on the alkyl group, are weaker than the corresponding M–aryl bonds. Metal–alkyl bond strengths are significantly influenced by steric factors (more hindered systems tend to have weaker M–C bonds), but are also affected by electronic factors. Substituents that stabilize radicals through typical mesomeric effects weaken metal–carbon bonds, and electron-withdrawing groups on the α-carbon often strengthen metal–carbon bonds through inductive effects.[24,25] This inductive effect presumably results from stabilizing a partial negative charge on the alkyl ligand and increasing the ionic character of the bond. As discussed in Section 3.2.1.4, complexes of primary alkyl ligands lacking aromatic or electron-withdrawing substituents are typically more stable than complexes of the isomeric secondary alkyl ligands due to these steric and electronic factors.

Metal–carbon bonds between first-row metals and benzyl groups are known with particular accuracy because these bonds are weak enough that their homolysis can be followed directly. From such studies, the manganese–benzyl bond in $ArCH_2Mn(CO)_4P(OAr)_3$ has been estimated to be 25–30 kcal/mol[26]; the same bond in $PhCH_2Mn(CO)_5$ was measured to be only slightly stronger, about 31 kcal/mol.[22,27] The chromium–benzyl bond in $[PhCH_2Cr(H_2O)_5]^{2+}$ was found to be about 29 kcal/mol.[24]

Transition metal–methyl bonds tend to be stronger than those of higher alkyl ligands. For example, the U–C bond in $(Cp'')_3U$–CH_3 (Cp''=η^5-$Me_3SiC_5H_4$) is 49 kcal/mol, but the

U–C bond in $(Cp'')_3U–CH_2Si(CH_3)_3$ is only 45 kcal/mol.[28] Calculations comparing the strengths of $M–CH_3$ and $M–C_2H_5$ bonds to second-row metals suggest that the $M–C_2H_5$ bond is always weaker, although the difference in energy between the methyl and ethyl groups decreases to the right in the transition series.[29]

M–CH$_3$ bond strengths increase from the first, to the second, to the third row of a triad, as shown in Figure 3.2.[30] The difference between M–H and M–C bond strengths—which relates to the gain in stability from β-hydrogen elimination from an alkyl complex—increases from left to right in the transition series. This difference in bond energies is small for actinides,[30,31] which has made it straightforward to synthesize tertiary alkyl complexes, such as $(Cp')_3UC(CH_3)_3$.[32]

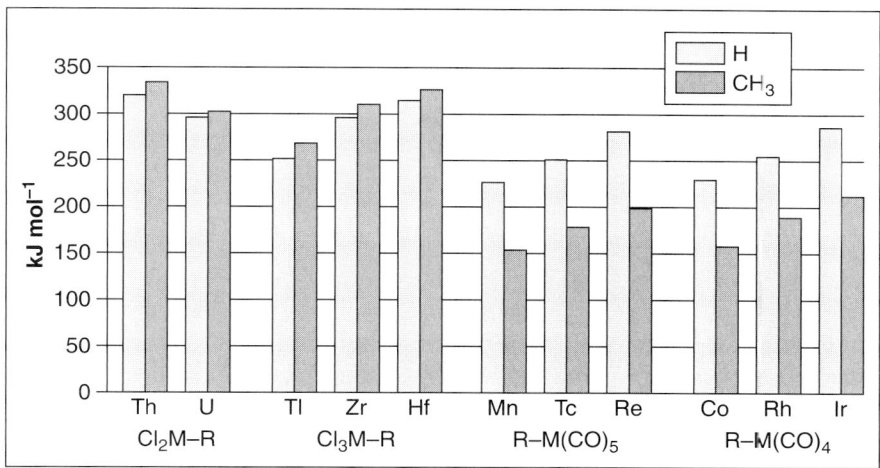

Figure 3.2. Relative strengths of representative M–C bonds. Data from reference 30.

The strength of cobalt–carbon bonds has been extensively investigated because of the presence of a Co–C bond in coenzyme B$_{12}$. Reported cobalt–alkyl bond dissociation energies range from 17 to 37 kcal/mol in complexes of the type Co(N$_4$-macrocycle)R, with and without axial bases.[33,34] Much of the variation in Co–C bond energy appears to result from steric factors.[35] Indeed, structural data reveal a steric influence on the length of the Co–C bond.[36]

The homolysis of the Co–C bond of neopentylcobalamin in aqueous solution has been examined with particular care. These studies showed that $\Delta H^{\ddagger}$ for this homolysis is 28.3(2) kcal/mol[37]; kinetic studies in ethylene glycol and corrections for solvent cage effects indicate that the Co–neopentyl bond strength is 28(2) kcal/mol, in this medium and is probably similar in other media.[38] The Rh–C bond is stronger. The Rh–C bond in (porphyrin)Rh–CH$_3$ is 57 kcal/mol and in (porphyrin)Rh–C$_2$H$_5$[39] is 50 kcal/mol.

3.2.1.3. Synthesis of Metal–Alkyl Complexes

3.2.1.3.1. Synthesis of Alkyl Complexes by Transmetallation

The most common method for the synthesis of metal–alkyl complexes is the reaction between a transition metal halide complex and a main group organometallic species (Equation 3.1). This reaction is typically called a "transmetallation."[‡] The main group organometallic reagents most commonly used in these transmetallation processes are organolithium (RLi), Grignard (RMgX), organozinc (R$_2$Zn), and organoaluminum reagents. All four types of reagents are readily available or easily prepared for a wide variety of alkyl groups, R.[40–47] Li, Mg, Zn, and Al are all highly electropositive, and this property causes the equilibrium in Equation 3.1 to lie far to the right in most cases. The M–X product is

---

[‡]The term transmetallation can also be spelled transmetalation. This term was introduced and defined by Henry Gilman with the spelling transmetalation (*J. Am. Chem. Soc.* **1934**, *56*, 1415). In addition, metalation appears in the Oxford English Dictionary with a spelling containing one "l". The more common spelling with two "l"s is used throughout this text. The author thanks Dr. Scott Denmark for information about this issue.

stabilized by the ionic property of the bond between the more electropositive metal and the electronegative group X.

$$RM \ + \ L_nM'X \ \rightleftharpoons \ L_nRM' \ + \ MX \tag{3.1}$$

M = Li, Mg, Zn, Al
M′ = transition metal

The dependence of the position of this equilibrium on the main group metal was described by Coates, Green, Powell, and Wade[48] and later re-emphasized by Negishi.[49] Most transition elements are less electropositive than Li, Mg, Zn, and Al, but there are exceptions, particularly among the early transition elements. For example, the position of the transmetallation equilibrium between Zr and Al depends on the identity of R and the ancillary ligands (Equations 3.2 and 3.3).[50,51]

$$Cp_2Zr\overset{C_8H_{17}}{\underset{Cl}{<}} \ + \ Al_2Cl_6 \ \longrightarrow \ 2 \ Cp_2ZrCl_2 \ + \ n\text{-}C_8H_{17}AlCl_2 \tag{3.2}$$

$$Cp_2Zr\overset{O}{\underset{Cl}{\overset{}{<}}}\overset{}{\underset{Cl}{ZrCp_2}} \ + \ Al_2Me_6 \ \longrightarrow \ Cp_2Zr\overset{CH_3}{\underset{Cl}{<}} \ + \ (Me_2Al)_2O \tag{3.3}$$

A disadvantage of the synthesis of metal–alkyl complexes by transmetallation is the potential for nucleophilic attack to occur at ancillary ligands. Carbonyl ligands are particularly vulnerable to nucleophilic attack, as shown in Equation 3.4, and described in more detail in Chapter 11.[52] However, the extent to which this process competes with transmetallation depends on the electronic property of the metal.[53,54] Thus, some carbonyl halides can be converted into alkyl carbonyl complexes (Equation 3.5).[55] In addition, electron transfer from electron-rich carbon nucleophiles can limit the yields of transmetallation processes.

$$Re(CO)_5Br \ + \ CH_3Li \ \longrightarrow \ \left[cis\text{-}(CO)_4Re\overset{\overset{O}{\underset{||}{CCH_3}}}{\underset{Br}{<}} \right]^{\ominus} \ + \ CO \tag{3.4}$$

$$Co(PMe_3)_3(CO)Cl \ + \ CH_3Li \ \longrightarrow \ Co(PMe_3)_3(CO)CH_3 \tag{3.5}$$

3.2.1.3.2. Synthesis of Alkyl Complexes by Alkylation

The second most common method to prepare metal–alkyl complexes is the alkylation of nucleophilic metal complexes. Like protonations, these alkylations are formal oxidative additions, and are discussed in that context in Chapter 7, which covers oxidative addition of polar reagents. The reaction between an anionic metal–alkyl complex and an alkyl halide is most common (Equations 3.6 and 3.7),[56] but examples of such reactions between neutral metal–alkyl complexes and alkyl halides are also known (Equation 3.8).[55]

$$[CpFe(CO)_2]^{\ominus} \ + \ EtI \ \longrightarrow \ OC^{\text{\tiny{\textbackslash}}}\underset{OC}{\overset{Cp}{\underset{|}{Fe}}}\diagdown Et \ + \ I^{\ominus} \tag{3.6}$$

$$Li[(\eta^5\text{-}C_5Me_5)Ir(PMe_3)H] \ + \ n\text{-}C_5H_{11}OTf \ \longrightarrow \ Me_3P\overset{Cp^*}{\underset{}{\overset{|}{Ir}}}\overset{\ }{\underset{H}{\diagdown}} \diagup\diagdown\diagup$$

$$+$$

$$LiOTf \tag{3.7}$$

$$(Me_3P)_4Co(CH_3) \ + \ MeBr \ \longrightarrow \ (Me_3P)_3CoMe_2Br \tag{3.8}$$

3.2.1.3.3. Synthesis of Alkyl Complexes by Other Methods

Many additional routes to metal–alkyl complexes other than transmetallation and alkylation are discussed in later chapters of this text. For example, metal–alkyl complexes are generated by insertion of an olefin into a metal–hydride[57] or or metal–hydrocarbyl species. Such insertion reactions are discussed in Chapter 9, but an example of the synthesis of a zirconium alkyl by olefin insertion into a zirconium hydride is shown in Equation 3.9. Metal–alkyl complexes are also generated by nucleophilic attack on coordinated olefins (Equation 3.10)[58] or carbene ligands (Equation 3.11).[59] These reactions are presented in detail in Chapter 11.

$$Cp_2ZrHCl \ + \ \diagup\!\!\!\diagdown\diagup\diagdown\diagup\diagdown\diagup \ \longrightarrow \ Cp_2Zr\!\diagdown\!\diagup\diagdown\diagup\diagdown\diagup\diagdown \qquad (3.9)$$

$$\begin{bmatrix} OC^{\cdots}\overset{\displaystyle Cp}{\underset{\displaystyle OC}{Fe}}\!\diagdown \end{bmatrix}^{\oplus} \!+\ NaBH_4 \ \longrightarrow \ CpFe(CO)_2Bu^t \qquad (3.10)$$

$$\begin{bmatrix} ON^{\cdots}\overset{\displaystyle Cp}{\underset{\displaystyle Ph_3P}{Re}}\!\!=\!CH_2 \end{bmatrix}^{\oplus} \!+\ EtLi \ \longrightarrow \ ON^{\cdots}\overset{\displaystyle Cp}{\underset{\displaystyle Ph_3P}{Re}}\!\diagup\diagdown\diagup \qquad (3.11)$$

Perfluoroalkyl complexes have been studied as stable ancillary ligands, as species that could be intermediates in the formation of fluoroalkylarenes, and as species that could lead to the defluorination of fluorocarbon pollutants.[60] Fluoroalkyl complexes cannot be obtained by the alkylation of anionic metal complexes because perfluoroalkyl halides are not electrophilic. Instead, perfluoroalkyl complexes have been prepared by the oxidative addition of perfluoroalkyl iodides or bromides to low-valent neutral complexes (Equation 3.12),[61] the decarbonylation of perfluoroacyl complexes (Equation 3.13), or transmetallation.[62–65] Each of these methods has limitations. Oxidative addition can be used to prepare fluoroalkyl complexes in higher oxidation states, but is less commonly used as a method to prepare perfluoroalkyl complexes containing metals in lower oxidation states. The decarbonylation method is limited because decarbonylation typically requires high temperatures. Transmetallation is limited by the instability of perfluoroalkyllithium and -magnesium reagents and the toxicity of perfluoroalkylcadmium and -mercury reagents. However, transmetallation to form perfluoroalkyl complexes is a useful route when the transmetallation is performed with perfluoroalkylsilanes, Me_3SiR_F, in combination with metal–fluoride complexes (Equation 3.14) or metal complexes of other halides with added fluoride (Equation 3.15).[62,66]

$$Ir(PPh_3)_2(CO)Cl \ + \ CF_3I \ \longrightarrow \ Ir(PPh_3)_2(CO)(Cl)(I)CF_3 \qquad (3.12)$$

$$[Co(CO)_4]^{\ominus} + \ C_3F_7\overset{\displaystyle O}{\overset{\displaystyle \|}{C}}Cl \ \longrightarrow \ [C_3F_7\overset{\displaystyle O}{\overset{\displaystyle \|}{C}}Co(CO)_4] \ \xrightarrow{-CO} \ C_3F_7Co(CO)_4 \qquad (3.13)$$

$$\begin{array}{c} \diagup\!\!\!\!\diagdown \\ \diagdown\!\!\!\!\diagup\!Rh\!\overset{F}{\diagdown}_{PR_3} \end{array} \ \xrightarrow[-Me_3SiF]{Me_3SiR_F} \ \begin{array}{c} \diagup\!\!\!\!\diagdown \\ \diagdown\!\!\!\!\diagup\!Rh\!\overset{R_F}{\diagdown}_{PR_3} \end{array} \qquad (3.14)$$

$$\begin{array}{c} \overset{Ph_2}{P}\diagdown_{o\text{-}Tol} \\ Pd \\ \overset{P}{\underset{Ph_2}{}}\diagdown_{Br} \end{array} \ \xrightarrow{Me_3SiCF_3,\ TBAT} \ \begin{array}{c} \overset{Ph_2}{P}\diagdown_{o\text{-}Tol} \\ Pd \\ \overset{P}{\underset{Ph_2}{}}\diagdown_{CF_3} \end{array} \qquad (3.15)$$

TBAT = Tetrabutylammonium Triphenyldifluorosilicate

3.2.1.4. Selected Reactions of Metal–Alkyl Complexes

Many complexes that contain alkyl ligands bearing β-hydrogens readily decompose to form olefins and metal–hydride complexes. Such β-hydrogen elimination is the most common process that limits the stability of metal–alkyl complexes, although other elimination processes noted below can occur. Kochi[67] summarized early available information on the mechanism of such β-eliminations, and there is also considerable information in early reviews[68–70] of the chemistry of alkyl ligands. More recent reviews survey all of the decomposition modes available to transition metal alkyl complexes, but emphasize the work of individual authors.[21,71,72]

In order to study the reactions of metal–alkyl complexes in the absence of β-hydrogen elimination, a large fraction of organometallic chemistry has been conducted with complexes containing methyl, benzyl, neopentyl (Me_3CCH_2), and trimethylsilylmethyl ligands that lack a β-hydrogen. Classic homoleptic examples of this type include WMe_6,[17] $Ta(CH_2Ph)_5$,[73] and $Ti(CH_2Ph)_4$.[74] As discussed in Chapter 10 on elimination reactions, alkyl complexes possessing certain properties can be stable to β-hydrogen elimination. For example, coordinatively saturated alkyl complexes containing tightly bound ancillary ligands, such as $[(H_2O)_5CrCH(CH_3)_2]^{2+}$,[75] can be stable to β-hydrogen elimination because this elimination process requires a vacant coordination site,[76] and alkyl complexes, such as 1-adamantyl[77] and 1-norbornyl[78] complexes, which would form strained olefins upon β-hydrogen elimination, tend to be stable. Likewise, metallacyclic or sterically hindered complexes, such as $(PBu_3)_2Pt(CH_2CH_2CH_2CH_2)$[79] and $Cr(Bu^t)_4$,[80] can be stable to β-hydrogen elimination.

Some alkyl ligands decompose by α-hydrogen elimination, yielding a coordinated alkylidene and a hydride ligand. This elimination process is also described in detail in Chapter 10. Such elimination is accelerated by steric congestion. For example, $TaMe_5$ and $Ta(CH_2Ph)_5$ are formed by transmetallation reactions like that shown in Equation 3.16,[81] but efforts to prepare the neopentyl analog $Ta[CH_2C(CH_3)_3]_5$ led to the elimination of neopentane and formation of the first alkylidene complex (Equation 3.17).[82]

$$TaCl_5 \ + \ 1.5 \ Zn(CH_3)_2 \longrightarrow Ta(CH_3)_3Cl_2 \xrightarrow{\ 2 \ CH_3Li\ } Ta(CH_3)_5 \qquad (3.16)$$

$$TaCl_5 \xrightarrow{\ 1.5 \ Zn(CH_2Bu^t)_2\ } Ta(CH_2Bu^t)_3Cl_2 \xrightarrow{\ 2 \ LiCH_2Bu^t\ } (Bu^tCH_2)_3Ta{=}\!\!<\!\!\curlyvee \qquad (3.17)$$

In most cases, complexes of primary alkyl ligands are more stable than the isomeric complexes of secondary or tertiary alkyl ligands. For example, Reger has shown that the secondary butyl iron complex in Equation 3.18 isomerizes to the corresponding primary *n*-butyl complex,[83] and that the isopropyl palladium complex in Equation 3.19 isomerizes to the more stable *n*-propyl isomer.[84,85] Likewise, secondary zirconocene alkyl complexes isomerize to the linear isomers (Equation 3.20), as shown many years ago by Schwartz,[86] and Labinger and Bercaw have recently shown that the *sec*-butyl complex of zirconocene, generated by the hydrozirconation of *cis*-2-butene, isomerizes in several hours to the corresponding *n*-butyl complex.[87]

$$(3.18)$$

$$K = 10 \qquad (3.19)$$

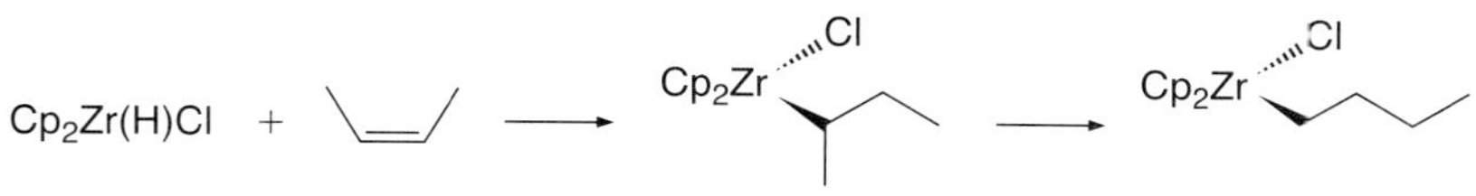

$$(3.20)$$

However, electronic effects can cause a secondary alkyl complex to be more stable than the primary alkyl isomer. For example, an electron-withdrawing substituent can cause the secondary alkyl complex to be more stable than the corresponding primary alkyl complex (Equations 3.21 and 3.22).[84,85,88]

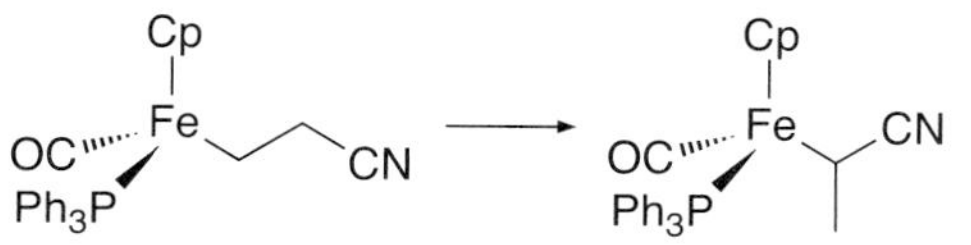

$$(3.21)$$

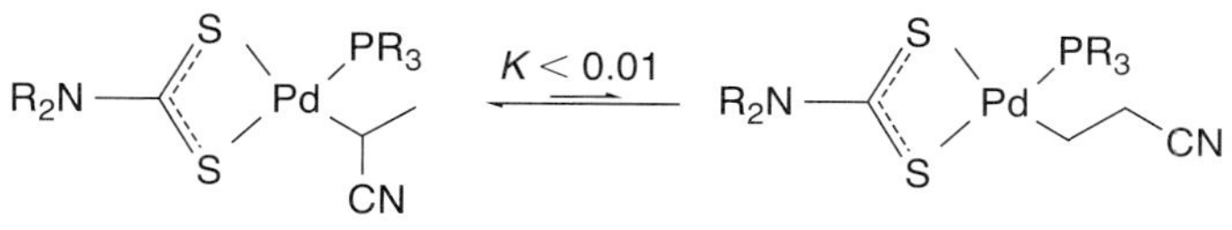

$$(3.22)$$

In addition, Brookhart has reported low-temperature[89] NMR spectroscopic studies of alkyl palladium cations that have shown that the *secondary* alkyl is *more stable* than the primary when the metal is electrophilic and an open coordination site is present cis to the alkyl group (Scheme 3.1). This combination of factors gives rise to a β-agostic interaction (see the discussion of σ-complexes in Chapter 2) that is stronger to the primary C–H bond than to a secondary C–H bond and stabilizes the secondary alkyl complex. Coordination of a fourth ligand reverses the order of stability because of the loss of the agostic interaction, and the primary alkyl complex becomes more stable than the secondary alkyl isomer.

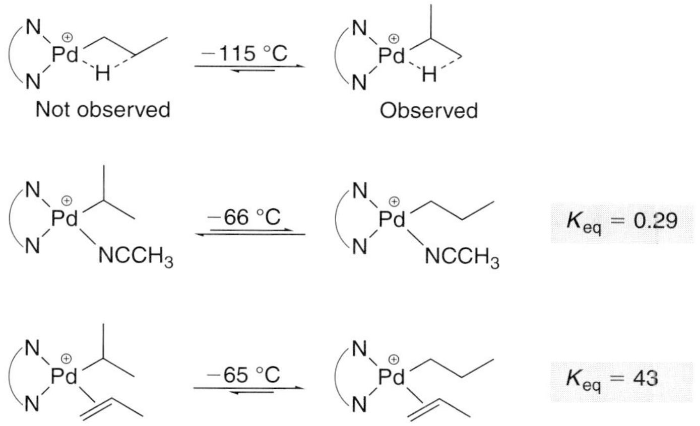

Scheme 3.1

Harvey[90] has examined these issues by density functional calculations, and has concluded that (1) "primary alkyl complexes are usually more stable than secondary and tertiary ones," that (2) "this is an *electronic* effect, due to the partial carbanionic character of the alkyl group," and that (3) "steric effects, . . . usually invoked in the literature . . . play only a minor role in many cases." The electronic effect is proposed to parallel that in alkyl lithium reagents. Because of the partial negative charge on the α-carbon in alkyl lithium and transition metal complexes, the stability of the alkyl complexes parallels the stability of the carbanions. When the charge on the α-carbon is small, as in neutral, late transition metal complexes, other factors, such as steric effects and agostic interactions, can dominate the stability of the isomeric alkyl complexes.

3.2.2. Aryl, Vinyl, and Alkynyl Complexes (Written with Prof. Jack R. Norton)

Some ligands bound through metal–carbon bonds display properties that are similar to those of metal–alkyl complexes. Aryl, vinyl, and alkynyl (often called acetylide) ligands bind to transition metals by bonding interactions that are similar to those between transition metals and alkyl ligands. These ligands are all bound to the transition metal through an M–C σ-bond, although the presence of a π-system can lead to additional interactions between the π-system and the metal. Aryl, vinyl, and alkynyl complexes undergo reactions that are similar to those of alkyl complexes, but the rates of the reactions of these complexes are different from those of alkyl complexes. The following sections provide an overview of the synthesis, properties, and reactivity of aryl, vinyl, and alkynyl complexes.

The stability of aryl, vinyl, and alkynyl complexes follow several trends. First, the strengths of metal–carbon bonds involving aryl and vinyl ligands are greater than those involving alkyl ligands.[91–93] Second, β-hydrogen elimination from vinyl and aryl ligands is rare.[94–96] Abstraction of the β-hydrogen of vinyl and aryl ligands by an accompanying alkyl group can lead, however, to alkyne and benzyne complexes.[97,98] Third, alkynyl ligands are often formed by cleavage of the C–H bond of a terminal alkyne. This process is common because of the substantial acidity of terminal alkynes and the high stability of the acetylide complexes.

3.2.2.1. Synthesis of Complexes Containing Terminal Aryl Ligands

Aryl ligands are X-type ligands that are usually coordinated to a single transition metal through a single σ-bond, although examples of complexes containing bridging aryl ligands are known and are described briefly below. Aryl ligands must not be confused with arene ligands, described in Chapter 2, which are neutral six-electron ligands bound through an arene π-system. Most of the synthetic methods described in Section 3.2.1 that are used to prepare alkyl complexes can be used to prepare aryl complexes.

One of the most common methods used to prepare transition metal aryl complexes is transmetallation from a main group organometallic compound to the transition metal. Reactions of aryllithium, magnesium, zinc, boron, tin, and mercury have all been used extensively. A representative example of transmetallation from lithium is shown in Equation 3.23.[99] Examples of transmetallation from tin, zinc, and boron are shown in Equations 3.24,[100] 3.25,[101] and 3.26.[102] Transmetallation from tin, zinc, and boron are important steps in the catalytic cross coupling reactions described in Chapter 19. Arylmercury compounds, which are readily prepared by mercuration of arenes, have also been used to prepare transition metal aryl complexes,[103] although the toxicity of organomercury compounds has limited their use in more recent times.

(3.23)

(3.24)

(3.25)

$$(3.26)$$

Often the synthesis of transition metal aryl complexes, like the synthesis of transition metal alkyl complexes, benefits from the use of the mildest main group organometallic reagent that will undergo transmetallation. The difference in reactivity of aryl copper, magnesium and lithium reagents can be seen in the synthesis of $CpRe(NO)(CO)C_6H_5$.[104] Nucleophilic attack at the carbonyl ligand of $CpRe(NO)(CO)Cl$ was the exclusive product from reactions with PhLi (Equation 3.27) and the major product from reactions with PhMgBr. In contrast the desired arylrhenium product was formed in 91% yield from the reaction with an arylcopper reagent (Equation 3.28). In a similar vein, the arylrhodium complex in Equation 3.25 was formed in low yield from reaction with p-tolyl lithium, but in high yield from reaction with the milder $Zn(p\text{-tol})_2$.[101,105]

$$(3.27)$$

$$(3.28)$$

Aryl complexes can also be obtained by the reaction of anionic or neutral metal complexes with aryl halides or sulfonates. These reactions are most facile with metal(0) complexes of the nickel triad; a classic example with palladium(0) is shown in Equation 3.29. These reactions are included in Chapter 7, which describes the oxidative addition of polar reagents.[106,107] To prepare aryl complexes from organic halides using species that are relatively unreactive toward aryl halides, an alternative strategy can be followed involving the addition of acyl halides, followed by decarbonylation of the resulting phenacyl complex. An example of this sequence, starting with the manganese carbonyl anion, is shown in Equation 3.30.[108] Such formal oxidative additions are also described in Chapter 7. In other cases, metal–aryl complexes have been formed from arenes. These reactions can occur by oxidative addition at the C–H bond of an arene,[109] by σ-bond metathesis,[110] or by coordination and abstraction of the aryl C–H bond by a basic ligand or external base.[111,112] These processes are summarized in Scheme 3.2, and a full discussion of these classes of processes is presented in Chapter 6.

$$(3.29)$$

$$(3.30)$$

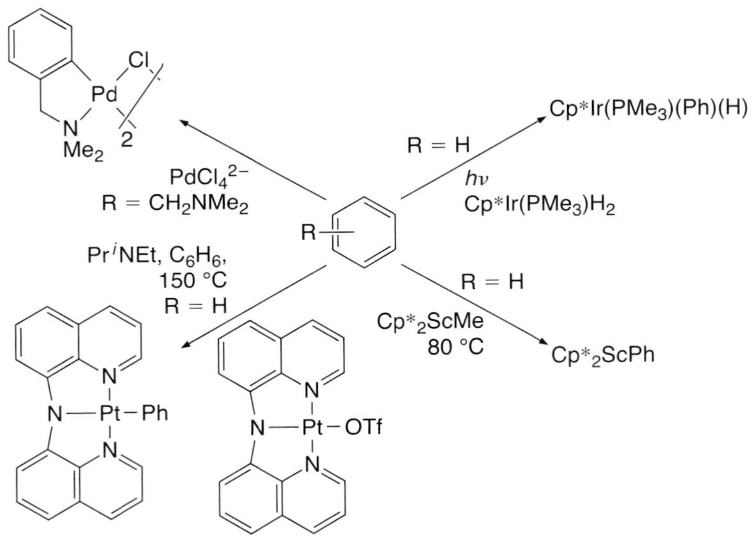

Scheme 3.2

3.2.2.2. *Complexes with Bridging Aryl Ligands*

Complexes containing bridging aryl ligands are also known, and complexes with these structures have been invoked as intermediates in exchanges of aryl groups from one metal to another. Although they are established intermediates, isolated dimeric complexes containing bridging aryl ligands are uncommon. An example of such a complex containing a bridging halogenated aryl group is shown in Equation 3.31. This complex was prepared by transmetallation from silver as part of a study to delineate the factors that lead to complexes containing bridging aryl groups.[113]

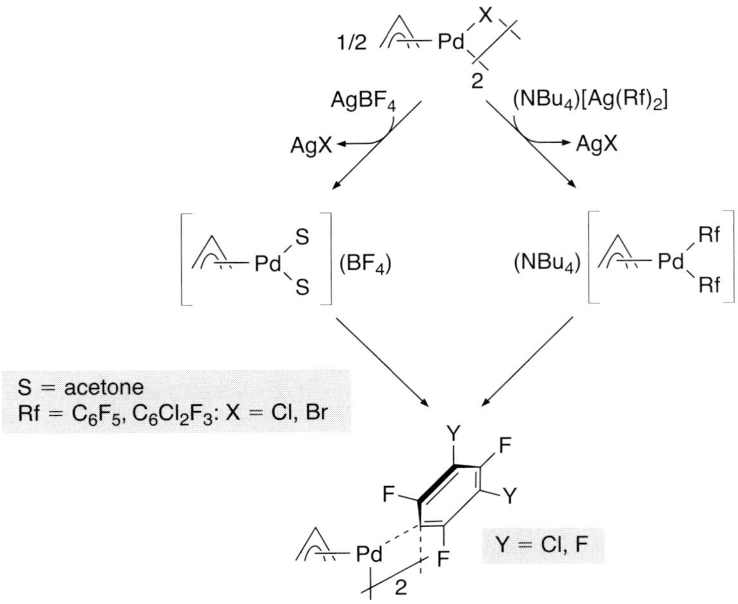

(3.31)

S = acetone
Rf = C_6F_5, $C_6Cl_2F_3$: X = Cl, Br

Y = Cl, F

3.2.2.3. Properties of Metal–Aryl Complexes

In general, aryl–metal bonds are stronger than alkyl–metal bonds.[29, 114] Several examples illustrate this property. In no complex has the metal–alkyl bond been shown to be stronger than the metal–aryl bond containing the same metal–ligand fragment. For example, the Mn–Ph bond in $PhMn(CO)_5$ is 4 kcal/mol stronger than the Mn–Me bond in $MeMn(CO)_5$.[115] The Rh–Ph bond dissociation energy (BDE) in the TpRh complex of Figure 3.3 is 16 kcal/mol greater than the analogous Rh–Me bond,[116] while the Ti–Ph BDE in the titanium complex of Figure 3.3 is 7.5 kcal/mol greater than the analogous Ti–Me bond.[117] The difference seems largest for third-row transition metals. The Ir–Ph BDE in $Cp^*Ir(PMe_3)Ph_2$ (82 kcal/mol) is 26 kcal/mol greater than the Ir–Me bond in $Cp^*Ir(PMe_3)Me_2$ (56 kcal/mol).[118] In the case of a set of lanthanide complexes, the Th–Ph BDE in $(Cp^*)_2ThPh_2$ (91 kcal/mol) is only 9 kcal/mol greater than the Th–Me bond in $(Cp^*)_2ThMe_2$ (82 kcal/mol).[119] Because of this greater bond strength and the lack of available pathways for β-hydrogen elimination, aryl complexes tend to be more easily isolated than alkyl complexes. Consistent with this trend, the equilibrium in Equation 3.32 lies far to the left.[120] Moreover, in several cases, such as Equations 3.33 and 3.34, aryl ligands undergo reactions without affecting the M–Ar bond.[121]

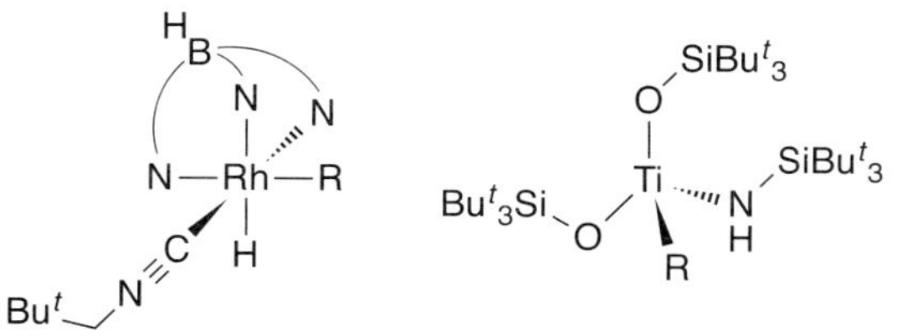

Figure 3.3.
Two complexes in which M–C BDE data are available for R = Ph and R = Me.

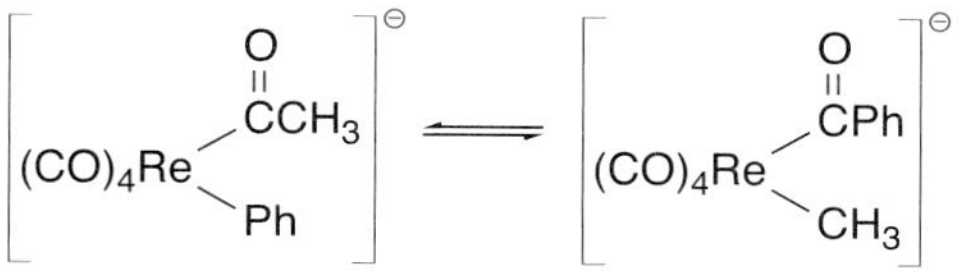

$$(3.32)$$

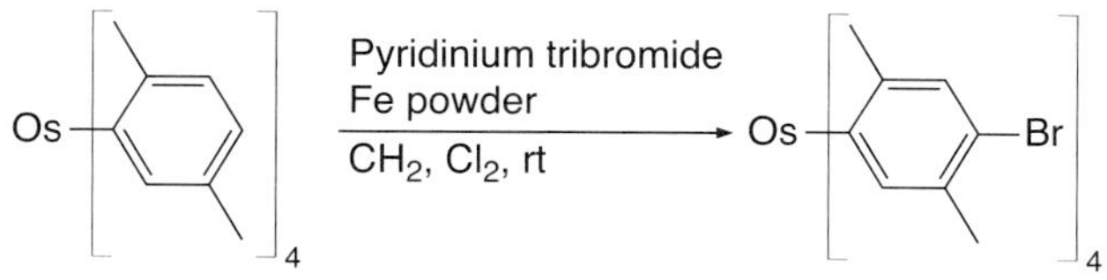

$$(3.33)$$

$$(3.34)$$

Aryl complexes often undergo rotation about the M–C bond with rates that are similar to the NMR time scale. The origin of this barrier is largely steric. Thus, the rotation of the aryl group in some complexes is faster than the NMR time scale, while rotation of the aryl group in others is slower than the NMR time scale, and discrete signals for the two ortho and two meta hydrogens are observed. An example in which slow rotation gives rise to two diastereomers is shown in Figure 3.4. Through-space coupling of the ortho fluorines of a biarylpalladium complex has also allowed an assessment of the barrier to rotation about a metal–aryl bond. The ortho fluorines in the fluoroaryl group of the complex in Figure 3.4 give rise to an AA′XX′ spin system in the ^{19}F NMR spectrum, implying that rotation is slower than the NMR time scale.

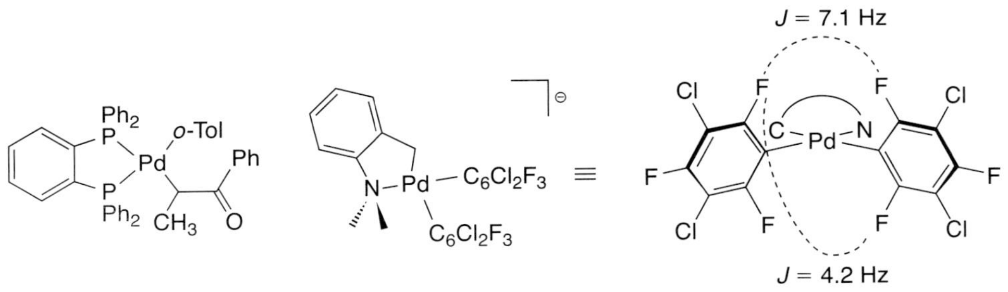

Figure 3.4.
Two complexes with rates of rotation about the M–Ar bond that are slower than the NMR time scale.

3.2.3. Vinyl Complexes (Written with Prof. Jack R. Norton)

Vinyl complexes are typically prepared by the same methods used to prepare aryl complexes. Vinyl mercury compounds, like aryl mercury compounds, are easily prepared (by the mercuration of acetylenes), and are therefore useful for the preparation of vinyl transition metal complexes by transmetallation. The use of vinyl lithium reagents has permitted the synthesis of homoleptic vinyl complexes by transmetallation (Equation 3.35).[122] Reactive low-valent transition metal complexes also form vinyl complexes by the oxidative addition of vinyl halides with retention of stereochemistry about the double bond (Equation 3.36).[123] Vinyl complexes have also been formed by the insertion of alkynes into transition metal hydride bonds (Equation 3.37),[57] by sequential electrophilic and nucleophilic addition to alkynyl ligands (Equation 3.38),[124] and by the addition of nucleophiles to alkyne complexes (Equation 3.39).[125] The insertion of alkynes into transition metal alkyl complexes is presented in Chapter 9 and, when rearrangements are slower than insertion, occurs by syn addition. In contrast, nucleophilic attack on coordinated alkynes, presented in Chapter 11, generates products from anti addition.

$$ZrCl_4 \; + \; 4Li(PhC=CMe_2) \longrightarrow Zr(CPh=CMe_2)_4 \tag{3.35}$$

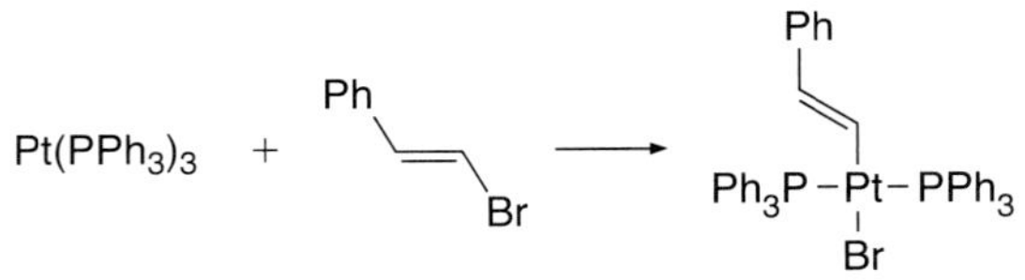

(3.36)

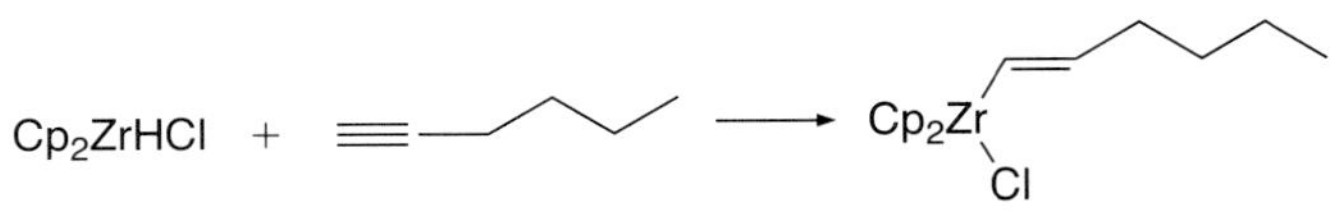

(3.37)

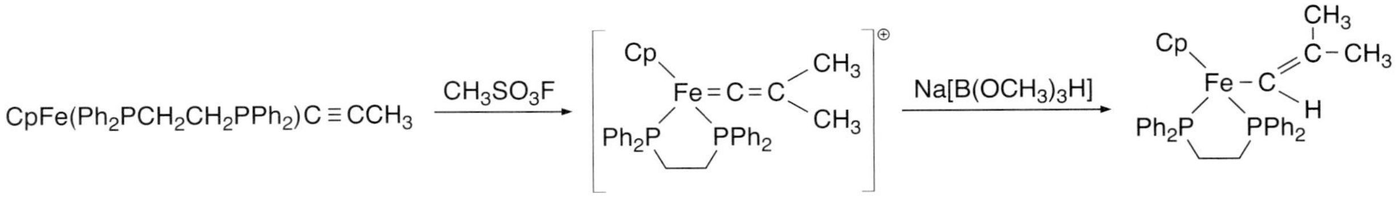

(3.38)

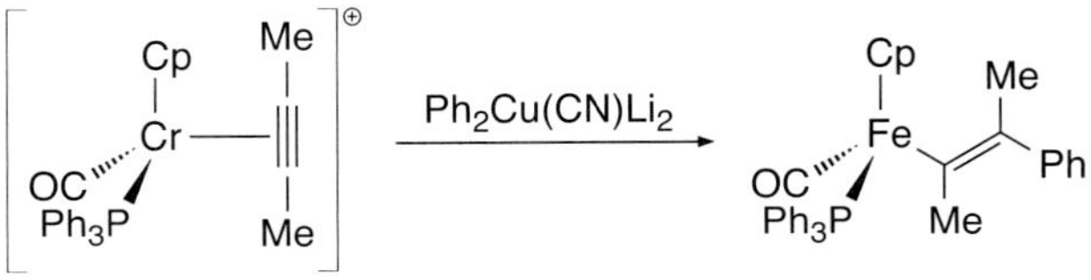

(3.39)

Recent calculations suggest that the M–C bonds in vinyl complexes have approximately the same strength as those in aryl complexes.[114] However, the Ir–vinyl bond in $Cp^*Ir(PMe_3)$ (vinyl)$_2$, for which the Ir–vinyl BDE is 71 kcal/mol, is 11 kcal/mol weaker than the Ir–C bond in $Cp^*Ir(PMe_3)Ph_2$, for which the Ir–Ph BDE is 82 kcal/mol.[118]

Although most vinyl ligands are bound in an η^1 fashion by a single σ-bond, some complexes contain what have been termed η^2-vinyl ligands.[126] Each can be formed by nucleophilic attack on a coordinated alkyne. η^1-Vinyl ligands have been formed, as shown in Equation 3.39,[125] by nucleophilic attack on two-electron-donor alkyne ligands, whereas η^2-vinyl ligands are typically formed, as shown in Equation 3.40,[127,128] by nucleophilic attack on four-electron-donor alkyne ligands. Alternatively, η^2-vinyl complexes have been prepared by the protonation of two-electron-donor alkyne ligands (Equation 3.41). η^2-Vinyl complexes are best described as metallacyclopropenes.[129] These complexes contain one short metal–carbon bond (formally, a double bond) and one long metal–carbon bond (formally, a single bond).

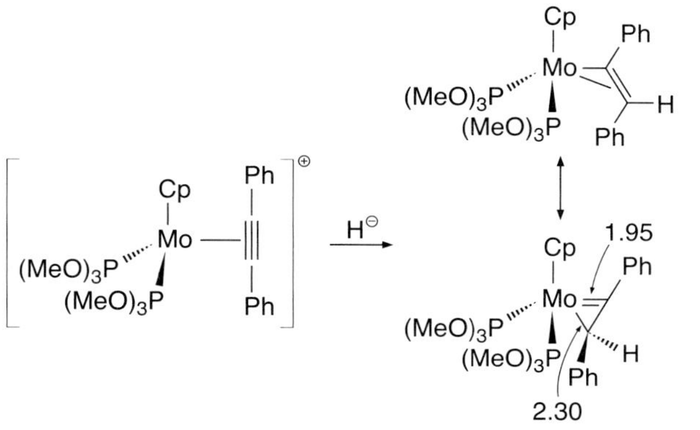

(3.40)

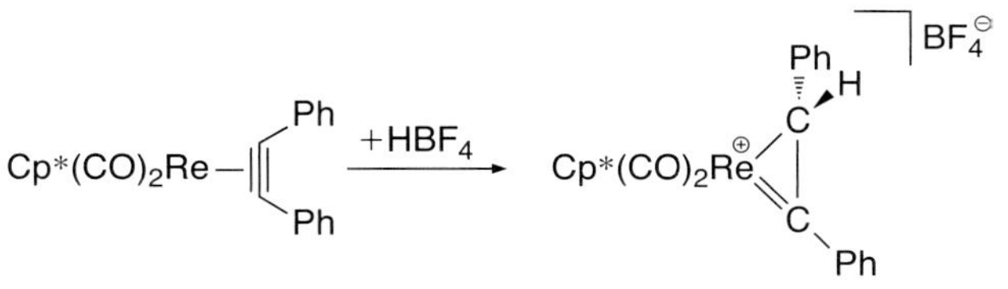

(3.41)

3.2.4. Alkynyl Complexes

Alkynyl complexes contain metal–carbon bonds in which the metal is bound to the sp-hybridized carbon at the terminus of a metal–carbon triple bond. The materials properties of these complexes have been investigated extensively. The properties of these complexes include luminescence, optical nonlinearity, electrical conductivity, and liquid crystallinity.[130–134] These properties derive largely from the extensive overlap of the metal orbitals with the π-orbitals on the alkynyl ligand.[135] The M–C bonds in alkynyl complexes appear to be considerably stronger than those in methyl, phenyl, or vinyl complexes.[29,136] Alkynyl complexes are sometimes prepared from acetylide anions generated from terminal alkynes and lithium bases (e.g., method A in Equation 3.42), but the acidity of alkynyl C–H bonds, particularly after coordination of the alkyne to the transition metal, makes it possible to form alkynyl complexes from alkynes and relatively weak bases (e.g., method B in Equation 3.42). Alkynyl copper complexes are easily prepared and often used to make alkynylnickel, -palladium, or -platinum complexes by transmetallation (Equation 3.43).[107] This reaction is a step in the preparation of Ni, Pd, or Pt alkynyl complexes from an alkyne, base, and a catalytic amount of CuI (Equation 3.44).[107] This protocol for the generation of metal–alkynyl complexes using copper–alkynyl species is also a step of the palladium-catalyzed coupling of aryl halides with alkynes described in Chapter 19.

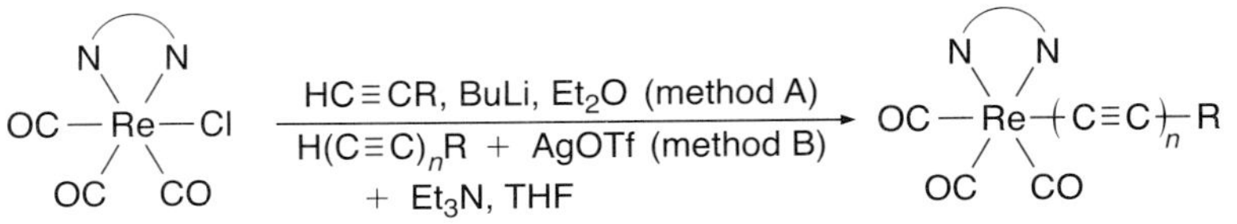

$$\underset{\substack{\text{OC} \quad \text{CO}}}{\overset{\substack{\text{N} \quad \text{N}}}{\text{OC}-\text{Re}-\text{Cl}}} \quad \xrightarrow[\substack{\text{H(C≡C)}_n\text{R} \;+\; \text{AgOTf (method B)}\\ +\; \text{Et}_3\text{N, THF}}]{\text{HC≡CR, BuLi, Et}_2\text{O (method A)}} \quad \underset{\substack{\text{OC} \quad \text{CO}}}{\overset{\substack{\text{N} \quad \text{N}}}{\text{OC}-\text{Re}+\text{C≡C}\!\!\underset{n}{+}\text{R}}} \qquad (3.42)$$

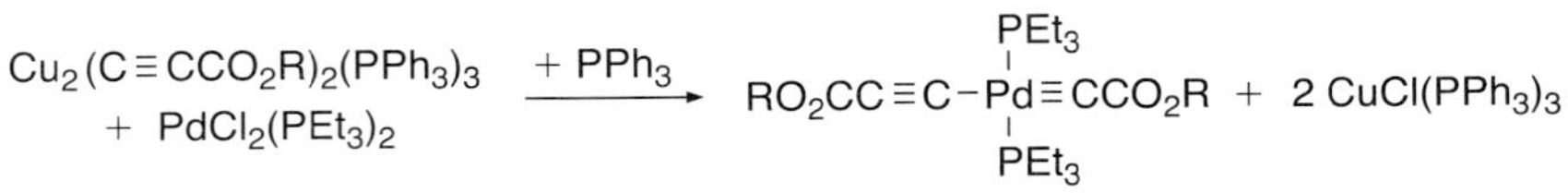

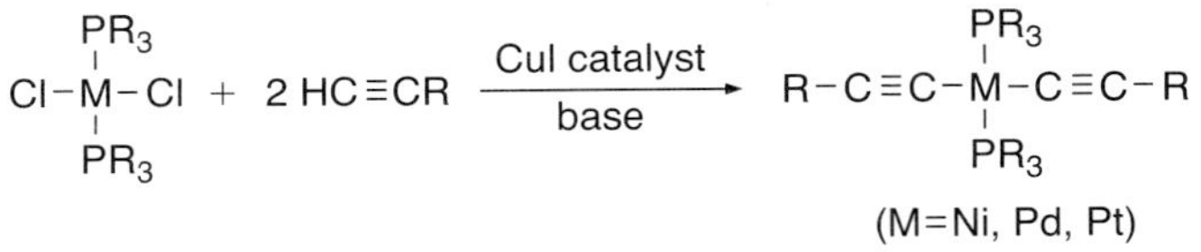

$$\underset{\substack{+ \text{PdCl}_2(\text{PEt}_3)_2}}{\text{Cu}_2(\text{C≡CCO}_2\text{R})_2(\text{PPh}_3)_3} \;\; \overset{+\,\text{PPh}_3}{\longrightarrow} \;\; \underset{\substack{\text{PEt}_3}}{\overset{\substack{\text{PEt}_3}}{\text{RO}_2\text{CC≡C}-\text{Pd}≡\text{CCO}_2\text{R}}} \;+\; 2\,\text{CuCl(PPh}_3)_3 \qquad (3.43)$$

$$\underset{\substack{\text{PR}_3}}{\overset{\substack{\text{PR}_3}}{\text{Cl}-\text{M}-\text{Cl}}} \;+\; 2\,\text{HC≡CR} \;\; \xrightarrow[\text{base}]{\text{CuI catalyst}} \;\; \underset{\substack{\text{PR}_3}}{\overset{\substack{\text{PR}_3}}{\text{R}-\text{C≡C}-\text{M}-\text{C≡C}-\text{R}}} \qquad (3.44)$$

$$(\text{M}=\text{Ni, Pd, Pt})$$

3.3. Enolate Complexes (Written with Prof. Erik J. Alexanian)

3.3.1. Overview

Transition metal enolate complexes have been prepared with most or all transition metals, and the enolate ligands have been shown to adopt a variety of bonding modes. Both early and late transition metal enolate complexes are intermediates in a number of important catalytic processes. Early transition metal enolates are important intermediates in asymmetric aldol reactions that exploit the Lewis acidic character of early metals.[137] Late transition metal enolates are intermediates in aldol and Michael addition processes,[138,139] Saegusa oxidations,[140] Heck-type processes,[141,142] catalytic asymmetric conjugate additions,[143,144] and cross coupling of enolate nucleophiles.[145,146]

This section includes the synthesis and structures of both C-bound and O-bound transition metal enolate complexes. Reviews of reactions catalyzed by Lewis acidic, early metal enolate intermediates have been published.[137] Cross-coupling of enolates is described in Chapter 19. Additional reactions of enolate nucleophiles, such as enolate allylation, are thought to occur by external attack of the enolate nucleophile on π-ligands and are described in Chapters 11 and 20. Late transition metal enolate complexes undergo a number of the classes of stoichiometric reactions presented in Chapters 3–12 of this text, including β-hydride elimination to produce enones,[140,147] insertion of unsaturated π-systems,[148,149] and reductive eliminations.[145,146,150,151]

3.3.2. Structure of Enolate Complexes

Several coordination modes are known for transition metal enolate complexes. These coordination modes are illustrated in Figure 3.5 and include complexes in which the

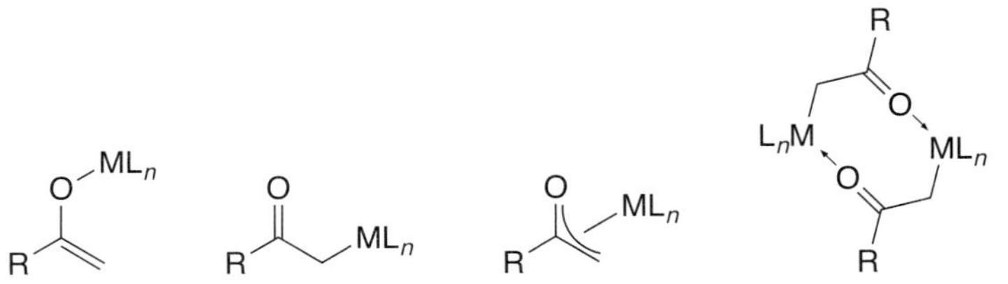

Figure 3.5. Binding modes of transition metal enolate complexes.

enolate is bound to a single metal in an η^1-fashion through the oxygen, in an η^1-fashion through the carbon, or in an η^3-fashion through the oxygen and two carbons of the enolate unit. In other cases, the enolate bridges two metals in a μ_2-C,O fashion. Complexes containing η^3-bound enolate groups are sometimes referred to as π-oxaallyl (or less correctly, oxoallyl) species, in analogy with their π-allyl relatives. The particular bonding mode of the enolate ligand depends on the properties of the metal (hard vs. soft), the steric properties of the metal and the enolate ligand, the ancillary ligands on the metal, and the coordination number of the complex.

Due to the oxophilicity of the early transition metals, enolate complexes of early metals typically exist as η^1-O-bound ligands. Sterically demanding enolate ligands, such as the adamantylidene enolate in Figure 3.6 (left), have been used to stabilize the low-coordinate early metal enolate containing tantalum.[152] Bidentate enolate ligands, such as that in the pyridine–enolate complexes of titanium and zirconium in Figure 3.6 (right) are also known.[153]

Enolate complexes of low-valent, middle transition metal complexes are also known. An 18-electron η^1-C-bound enolate complex of tungsten has been prepared, and is shown in Figure 3.7.[154] A chromium complex of an η^3-bound enolate ligand is also known (Figure 3.7, right).[155] The softer character of these low-valent metals presumably leads to the binding of the metal to the carbon and to the π-system of the enolates in these complexes.

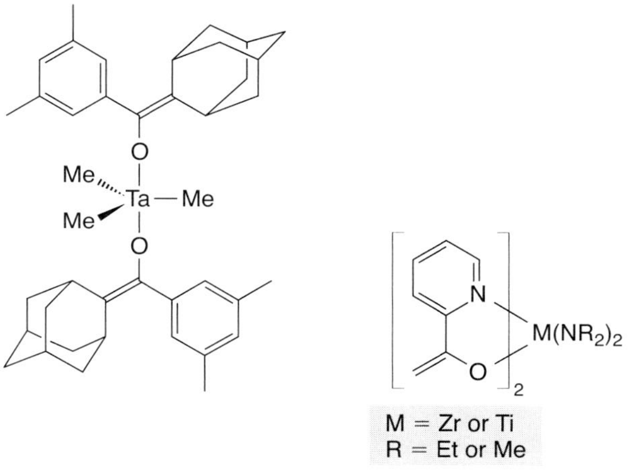

M = Zr or Ti
R = Et or Me

Figure 3.6. Representative early metal enolate complexes.

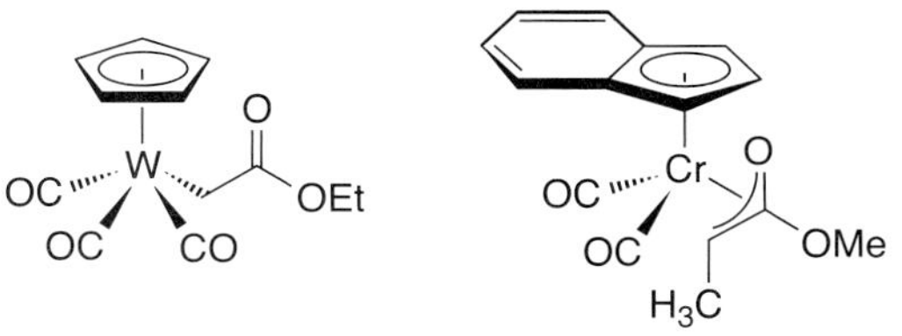

Figure 3.7.
Representative enolate complexes of low-valent middle transition metals.

There is a fine balance between the stability of O- and C-bound forms. One of the first observations of an equilibrium between O- and C-bound forms was made with ruthenium complexes ligated by PMe$_3$ (Equation 3.45).[156] Both O- and C-bound forms have also been observed during mechanistic studies of the reductive elimination from arylpalladium(II) enolate complexes, and some of the factors that control the binding mode of these ligands to late metals were revealed.[157] Enolates that are disubstituted at the α-carbon were always O-bound because of the steric congestion that would result from the binding of a trisubstituted carbon to the metal. However, for less substituted enolates, the C-bound form (e.g, Figure 3.8, left) was favored electronically if the enolate ligand was trans to phosphine, but the O-bound form (Figure 3.8, right) was favored if the enolate ligand was trans to an aryl group.

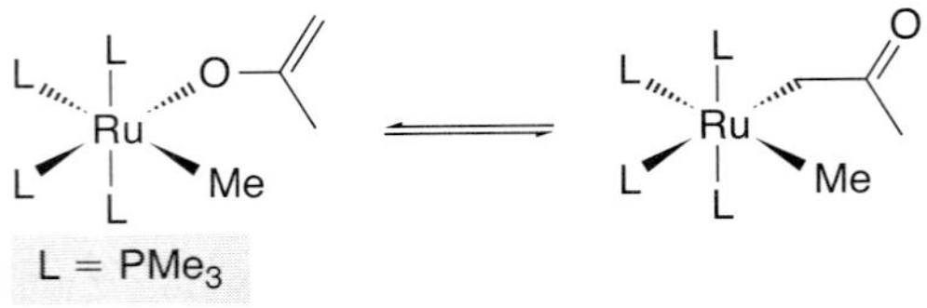

$$(3.45)$$

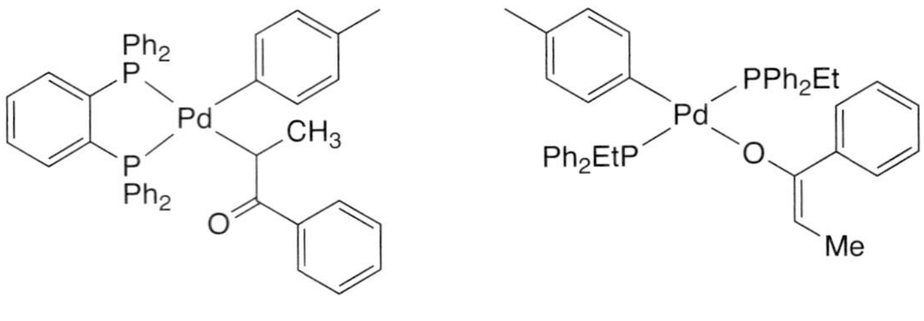

Figure 3.8.
The effect of the trans ligand on *C*- vs. *O*-binding modes of palladium enolates.

Late transition metal enolate complexes in which the enulate is bound in an η^3 fashion have also been observed. One of the first examples of an η^3-enolate complex involved a Rh(I) fragment ligated by phosphines (the reactant in Equation 3.46).[158] The unsaturation at Rh favored the η^3-form. Addition of a strong binding ligand (CO) produced the four-coordinate complex containing an η^1-*O*-enolate (Equation 3.46).

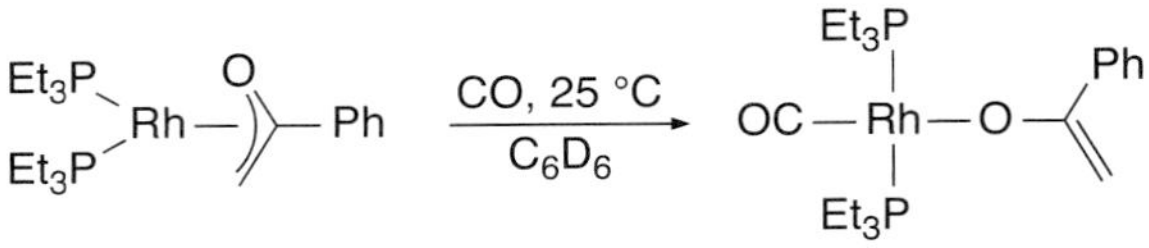

$$(3.46)$$

Related cyanoalkyl complexes are also known, and the cyanoalkyl ligand binds by a number of the same coordination modes as enolates, including η^1-*C*-bound and η^1-*N*-bound forms, as well as a μ_2-*C,N* bridging mode. Palladium–cyanoalkyl complexes displaying each of these coordination modes were prepared during a study of the α-arylation of nitriles (Figure 3.9).[159] The preferred form depended on phosphine ligand hapicity and steric effects. The presence of bidentate ligands at the metal enforced an η^1-binding mode, and the larger bidentate ligands led to the *N*-bound form. The greater lability of monodentate ligands allowed for ligand dissociation and a bridging mode to be observed.

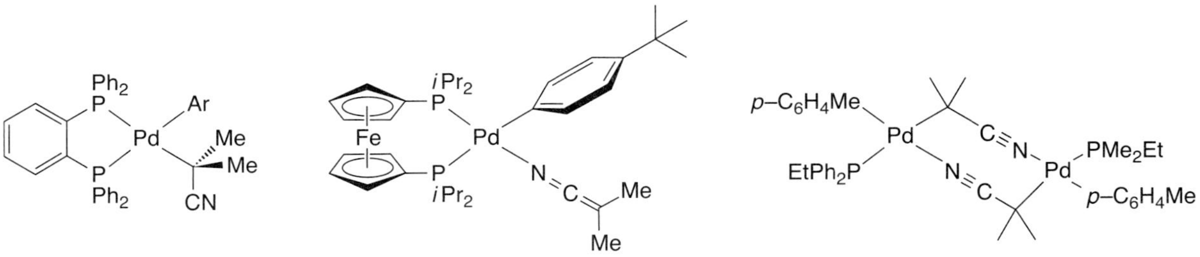

Figure 3.9. Binding modes of cyanoalkyl ligands.

3.3.3. Spectral Features of Enolate Complexes

The different coordination modes of the enolate complex can be distinguished by IR, and ^{1}H, ^{13}C, and ^{31}P NMR spectroscopy. A *C*-bound enolate complex can be identified by the carbonyl stretching frequency of the enolate fragment in the region of 1800–1600 cm^{-1} and the carbonyl resonance in the ^{13}C NMR spectra near 200 ppm. An O-bound enolate complex can be identified by methylene or methine resonances in the ^{1}H NMR spectrum between 4 and 5 ppm and the absence of a carbonyl stretching frequency in the infrared spectrum. An η^3-enolate complex can be identified by ^{1}H NMR resonances for the methylene or methine hydrogens located slightly upfield from the resonance of the η^1-O-bound form due to the increased electron density at the α-carbon of the ligand. For instance, the ^{1}H NMR signals for the methylene protons of the η^3-complex in Figure 3.9 resonated at 3.48 ppm, which is 0.8 ppm upfield of the chemical shift of the corresponding η^1-bound species.[158]

3.3.4. Synthesis of Enolate Complexes

The four most common methods for the synthesis of late transition metal enolates are oxidative addition to halocarbonyl compounds,[160] ligand metathesis with main group enolates[157] or silyl enol ethers, nucleophilic addition of anionic metal complexes to halocarbonyl electrophiles,[154] and insertion of an α,β-unsaturated carbonyl compound into a metal hydride.[161] Examples of these synthetic routes are shown in Equation 3.47–Equation 3.50. Equation 3.47 shows the synthesis of a palladium enolate complex by oxidative addition of ClCH$_2$C(O)CH$_3$ to Pd(PPh$_3$)$_4$, Equation 3.48 shows the synthesis of a palladium enolate complex by the addition of a potassium enolate to an aryl Pd(II) halide complex, and Equation 3.49 shows the synthesis of the C-bound W(II) enolate complex in Figure 3.7 by the addition of Na[(η^5-C$_5$R$_5$)(CO)$_3$W] to the α-halocarbonyl compound. Finally, Equation 3.50 shows the synthesis of a rhodium enolate complex by insertion of but-1-en-3-one into a rhodium hydride. This last route has also been used to prepare enolates as intermediates in reductive aldol processes.[162,163]

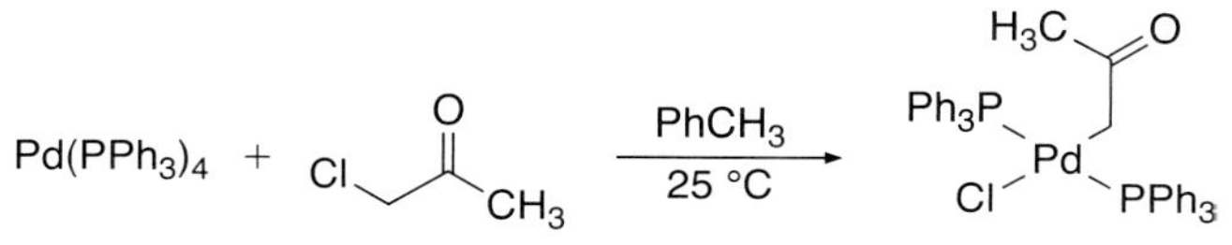

(3.47)

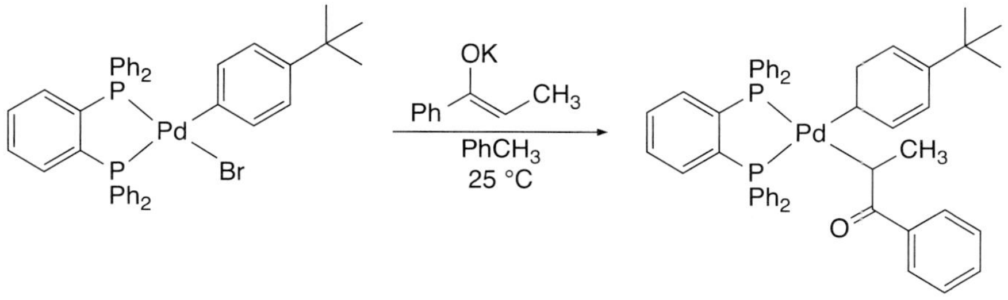

(3.48)

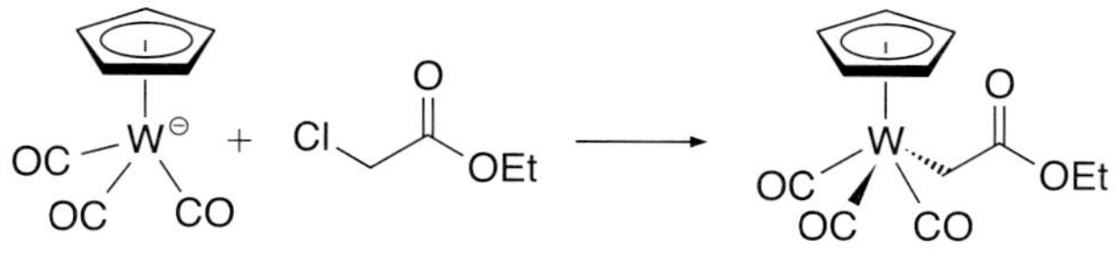

(3.49)

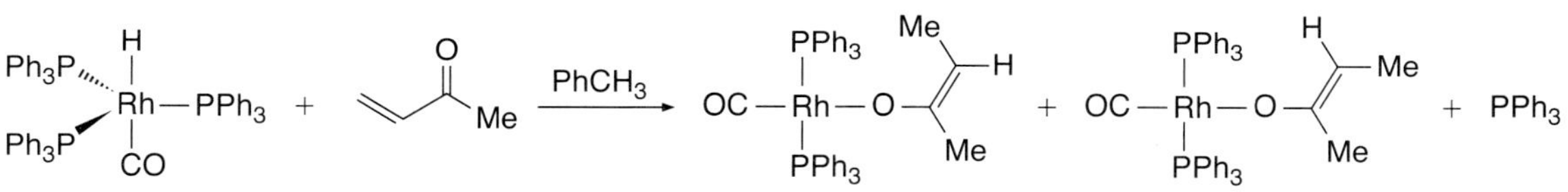

(3.50)

3.4. Cyanide Complexes (Written with Prof. Jesse W. Tye)

3.4.1. Overview

The cyanide ligand (CN^-) is a reactive ligand in important organometallic catalytic reactions, and is an ancillary ligand in coordination and bioinorganic chemistry. The CN^- ligand draws analogies to isoelectronic CO. The hydrocyanation of butadiene is an important step in the synthesis of nylon, and the cyanation of aryl halides generates useful benzonitrile derivatives. The coordination chemistry of CN^- is important in biology due to the high toxicity of CN^- and the presence of CN^- ligands in the hydrogenase enzymes.[164–166] Potassium cyanide is even used in the purification of silver and gold.[167] The chemistry of transition metal cyanide complexes has been reviewed.[168–170]

Cyanide salts and the colorless gaseous HCN are highly toxic and should be handled with care. The primary toxicity of CN^- results from inhibition of cytochrome c oxidase (a key enzyme in the electron transport chain), which catalyzes the four electron reduction of dioxygen to H_2O.[164] The cyanide ligand inhibits the enzyme by occupying the O_2 binding site on one of the enzyme's iron centers.

3.4.2. Properties of the Free Molecule

The cyanide anion is a weak base, but a good nucleophile. The pK_a of HCN is 9.0 and the H–CN bond dissociation energy is 125.5 kcal mol^{-1}.[168,171] CN^- is isoelectronic with N_2 and CO and has a C–N bond order of 3. The molecular orbital diagram of CN^- is qualitatively similar to that of CO or N_2. This similarity in bond order between CN^-, N_2, and CO is supported by infrared and Raman spectroscopic data. The fundamental stretching frequency of a dimethyl-formamide solution of $[Et_4N]^+[CN]^-$ is 2070 cm^{-1}.[172] These values can be compared to the fundamental stretching frequencies of 2143 and 2330 cm^{-1} for CO and N_2, respectively.[173]

3.4.3. Structures and Electron Counting of Metal–Cyanide Complexes

The cyanide ligand is counted as a one-electron donor in the covalent counting method and as an anionic two-electron donor in the ionic formalism. As an anion, the CN^- ligand is isoelectronic with CO. However, the formal negative charge on CN^- and the lower electron density of the nitrogen in the cyanide ligand, versus the oxygen of CO, gives rise to several differences. First, CN^- is a weaker π-acceptor and a stronger σ-donor ligand than CO. Second, the nitrogen of the cyanide ligand is nucleophilic and can be alkylated to form a coordinated isonitrile. Third, the bonding modes by which CN^- and CO bridge two metals are different. CN^- bridges two metals by binding the nitrogen atom to a second metal center to generate a linear M–CN–M' bridge, while CO ligands often bridge metal centers through the carbon atom of CO (see Chapter 2, Section 2.2.2).[169]

3.4.4. Thermodynamics of M–CN Linkages

Metal–cyanide linkages are strong. The M–CN bond strength is higher than would be predicted from correlations of M–X bond energies with H–X bond energies. The M–CN linkage is also more stable toward heterolytic cleavage than would be predicted based on the relatively weak Brønsted basicity of CN^-. The M–CN bond is strong, most likely, because of the combination of π-back-donation, soft donor character, ability to stabilize anionic charge in an ionic bond, and small size.

The exceptional strength of the metal–cyanide linkage makes it challenging to develop catalytic processes that involve excess CN^- as reagent. The Pd-catalyzed cyanation of aryl halides discussed in Chapter 19 to form aryl nitriles initially suffered from the formation of catalytically inactive cyanide complexes containing multiple CN^- ligands[174,175]; the reaction has been improved by the use of $Zn(CN)_2$ or $K_2[Fe(CN)_6]$ as the CN^- source.[174,176–180]

Likewise, the catalysts for the hydrocyanation of olefins presented in Chapter 16 are deactivated by formation of $[Ni(CN)_4]^{2-}$, and the key C–C bond-forming reaction is most often accelerated by the coordination of a Lewis acid to the nitrogen of the CN^- ligand.[181]

3.4.5. Spectral Features of M–CN Complexes

The most distinctive spectral property of the cyanide ligand is the strong infrared v_{CN} band. Cyanide complexes exhibit strong v_{CN} bands over a fairly narrow frequency range—typically 2200–2000 cm^{-1}.[169,173] These values tend to be higher than those at 2070 cm^{-1} for $[Et_4N]^+[CN]^-$ noted previously, but lower than the values of v_{CN} = 2240–2260 cm^{-1} for organic nitriles. Thus, the v_{CN} stretching frequency in metal–cyanide complexes is decreased less by π-backbonding than is the v_{CO} stretching frequency in metal–carbonyl complexes. This difference reflects the weaker π-back-donation in metal–cyanide complexes, relative to metal–carbonyl complexes. The higher v_{CN} stretching frequency in metal–cyanide complexes, relative to the v_{CN} of free cyanide, is similar to the higher v_{CO} stretching frequency in d^0 metal complexes, relative to the v_{CO} of free CO. Like the rationalization for the higher v_{CO} in d^0 metal complexes,[182] the higher v_{CN} stretching frequency in metal–cyanide complexes can be rationalized by the change in polarization of the C–N bond and the change in contributions from resonance structures containing C–N double and triple bonds upon metal binding. The v_{CN} values of bridging CN^- ligands are usually higher than the v_{CN} values for terminal cyanides. This observation also contrasts with the trends for coordinated CO—bridging CO ligands have much lower v_{CO} values than do terminal CO's.

Free and coordinated CN^- can also be detected by ^{13}C NMR spectroscopy. The ^{13}C NMR chemical shift of $[Me_4N][CN]$ in CD_3CN is 167 ppm.[183] A series of homoleptic cyanide complexes were shown to have ^{13}C NMR chemical shifts that range from a value of 84 ppm for $[Pt(CN)_4]^{2-}$ to 177 ppm for $[Fe(CN)_6]^{2-}$.[184] The ^{15}N chemical shifts for KCN is 106 ppm (vs. δ = 0 ppm for nitromethane), but few values for coordinated cyanide have been measured.[185]

3.4.6. Synthesis of M–CN Complexes

Several methods have typically been used to prepare cyanide complexes.[169,170] Representative examples of these processes are shown in Equations 3.51–3.53. As shown in Equations 3.51 and 3.52, the oxidative addition of HCN or cyanogen (NC–CN) to a transition metal forms hydridometal cyanide complexes or dicyano metal complexes. Although less common, the oxidative addition of nitriles has even been shown to form organometal cyanide complexes.[186–190] These reactions have been observed for Ni(0) species, as shown in Equation 3.53.

$$(3.51)$$

$$(3.52)$$

$$(3.53)$$

Alternatively, metal–cyanide complexes have been formed by the reaction of a metal halide with an anionic cyanide. The addition of alkali metal cyanides to aqueous solutions of transition metal halides has been used to form many homoleptic metal cyanides [$M(CN)_n$], as illustrated in Equation 3.54.[191] Alternatively, the reactions of tetraalkyl ammonium cyanides, which are soluble in a range of organic solvents, have been used to prepare low-valent metal–cyanide complexes (Equation 3.55).[192]

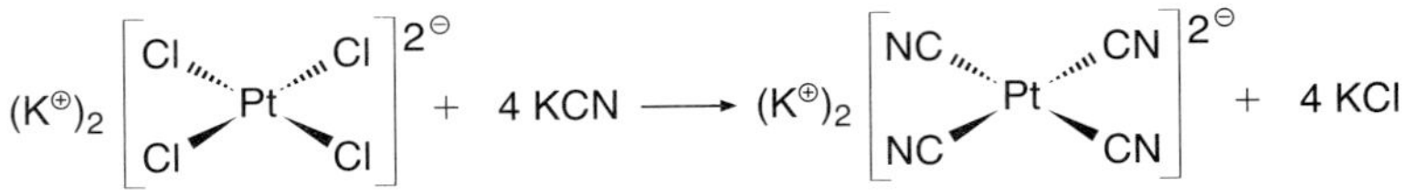

$$(K^\oplus)_2 \begin{bmatrix} Cl_{\prime\prime\prime\prime} & & Cl \\ & Pt & \\ Cl & & Cl \end{bmatrix}^{2\ominus} + 4\ KCN \longrightarrow (K^\oplus)_2 \begin{bmatrix} NC_{\prime\prime\prime\prime} & & CN \\ & Pt & \\ NC & & CN \end{bmatrix}^{2\ominus} + 4\ KCl \qquad (3.54)$$

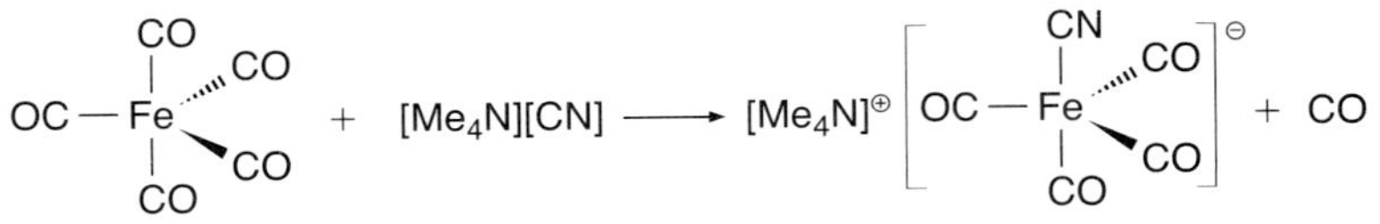

$$OC\!-\!Fe\begin{matrix} CO \\ | \\ \\ | \\ CO \end{matrix}\!\!\overset{CO}{\underset{CO}{<}} + [Me_4N][CN] \longrightarrow [Me_4N]^\oplus \begin{bmatrix} CN \\ | \\ OC\!-\!Fe \\ | \\ CO \end{bmatrix}\overset{CO}{\underset{CO}{<}}^\ominus + CO \qquad (3.55)$$

3.5. Allyl, η^3-Benzyl, Pentadienyl, and Trimethylenemethane Ligands (Written with Dr. Mark J. Pouy)

3.5.1. Allyl Ligands

3.5.1.1. Overview

An η^3-allyl or π-allyl group is a ligand in which the metal binds to the three adjacent carbon atoms of a conjugated system. Transition metal allyl complexes have been studied extensively and have a long history. The structures, the stoichiometric reaction chemistry, and many catalytic processes involving allyl complexes have been studied. The varied chemistry of allyl complexes results, in part, from the varied structures of allyl complexes. The allyl group can adopt a series of binding modes in both homoleptic and mixed ligand complexes (Figure 3.10); many of the early studies on fluxional molecules focused on the dynamics of π-allyl groups and the interconversions between different binding modes.

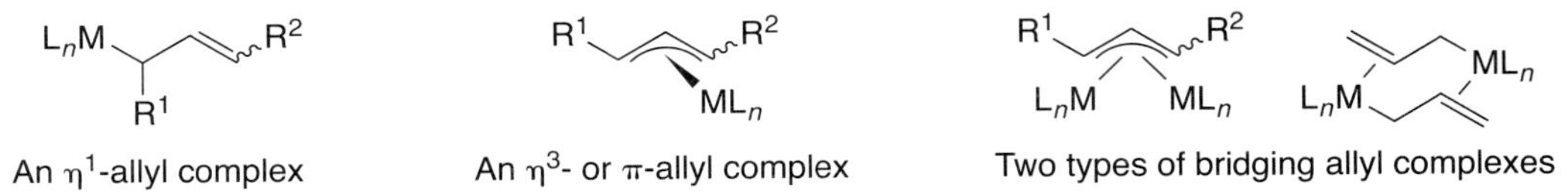

| An η^1-allyl complex | An η^3- or π-allyl complex | Two types of bridging allyl complexes |

Figure 3.10. Typical structures of monomeric and bridging allyl complexes.

Allyl ligands are reactive groups in catalytic processes, such as allylic substitution (Chapter 20) and transition-metal-catalyzed additions of allyl groups to carbonyl compounds (Chapter 12). They are also intermediates in a variety of catalytic processes involving dienes, including the hydroamination (Chapter 16) and telomerization of dienes (Chapter 22). They are also formed by cleavage of the allylic C–H bonds of olefins to form intermediates in allylic oxidation chemistry and as stable species that inhibit the polymerization of α-olefins (Chapter 22). The synthesis, structure, and occurrence of η^3-allyl complexes have been reviewed.[193–197]

3.5.1.2. Structures of Allyl Ligands

Allyl ligands bind to transition metals most commonly by one of the two monomeric structures shown in Figure 3.10. In addition, the π-allyl ligand can bridge two metals in either a symmetrical η^3-mode[198] or the unusual η^1,η^2- or η^1,η^3-modes shown on the right in

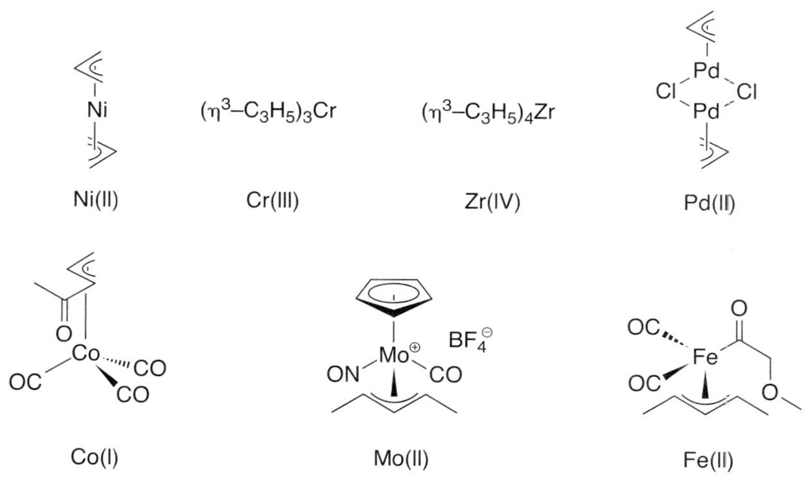

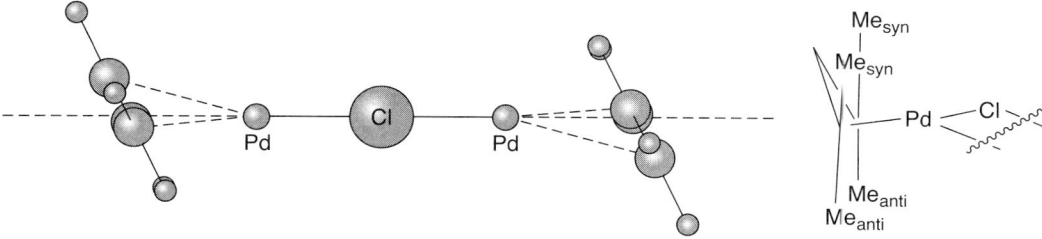

Ni(II)   Cr(III)   Zr(IV)   Pd(II)

$(\eta^3-C_3H_5)_3Cr$ $(\eta^3-C_3H_5)_4Zr$

Co(I)   Mo(II)   Fe(II)

Figure 3.11. Examples of η^3-allyl complexes.

Figure 3.12.
Structures of allylpalladium complexes illustrating the canting of the allyl group and the positions of substituents. Adapted from reference 201.

Figure 3.10.[199,200] The most common binding mode is η^3, and examples of η^3-allyl complexes of early, middle, and late transition metals are shown in Figure 3.11. The orbital interactions in a π-bound ligand, like an allyl group, were presented in Chapter 1. In brief, this interaction involves donation of the π-electrons from the HOMO of the allyl group to the metal and backbonding of metal d-electrons into the LUMO of the allyl group. This interaction requires that the metal bind to the face of the allyl group, although the allyl group is typically tilted in a manner that places the terminal carbons further from the metal than the central carbon.

Typical structural characteristics of an η^3-allyl ligand are illustrated for the common $(\eta^3$-allyl)palladium chloride dimer in Figure 3.12.[201] Because the η^3-allyl ligand is symmetric, the C–C bond lengths are nearly equal (1.357 Å vs. 1.395 Å), and the Pd–C distances are very nearly equal (2.132 Å vs. 2.108 Å vs. 2.121 Å). The C–C–C bond angle is 119.8° and the dihedral angle between the square plane of the palladium and the plane of the allyl group is 111.5°. The terminal substituents on the allyl group do not lie in the plane defined by the three central carbons of the allyl group. This position of the substituents is illustrated by the line drawing in Figure 3.12 of the tetramethylallyl group. The substituent anti to the group on the central carbon is displaced away from the palladium, while the substituent located syn to the group on the central carbon is displaced toward the palladium.[202]

These characteristics contrast greatly with the structure of an η^1-allyl ligand. An η^1-allyl ligand is bound to the metal through a saturated carbon atom, and the π-system of the olefin is not associated with the metal. These features are illustrated by the structure of $trans$-Pt(η^1-C_3H_5)Br(PEt$_3$)$_2$ in Figure 3.13.[203] The coordination sphere of the platinum is square planar. The two alkenyl carbons lie perpendicular to the plane, and there is no evidence for an interaction between the metal and the C=C bond. The Pt–C bond length

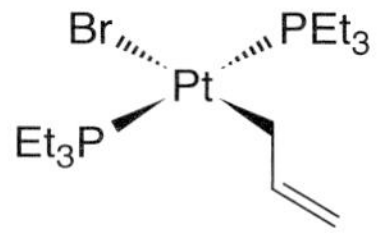

Figure 3.13. An example of an η^1-allyl complex.

is 2.086 Å, which is similar to that of analogous Pt(II) alkyl complexes. The C–C bond distances differ by nearly 0.2 Å (1.486 Å vs.1.298 Å), indicating localization of the carbon–carbon single and double bonds.

For simple η^3-allyl groups that are static on the NMR time scale, the ^{1}H NMR spectrum contains several characteristic features. The anti protons, which are closest to the metal and thus shielded, resonate at the highest field of the different allyl protons, typically between δ 1 and δ 3. The syn protons appear between δ 2 and δ 5, and the central proton between δ 4 and δ 6.5. The coupling constant between the syn and anti protons is often too small to measure, but they do couple more strongly with the central proton, with coupling constants of approximately 7 and 11 Hz, respectively. The relative chemicals shifts of the allyl carbons in the ^{13}C NMR spectra mirror those of the protons. The terminal carbons resonate between δ 80 and δ 90, and the central carbon resonates between δ 110 and δ 130.

3.5.1.3. Dynamics of Metal–Allyl Complexes

Many η^3-allyl groups undergo a number of dynamic processes that complicate their ^{1}H NMR spectra.[204] The rates of these interconversions, which can vary significantly between different metal and ligand systems, are important in determining the regioselectivity and enantioselectivity of allylic substitution reactions. As discussed in Chapter 20, the rate of equilibration of the allyl ligand from η^3- to η^1-forms, relative to the rate of subsequent reations, can affect the regiochemistry and enantioselectivity of metal-catalyzed substitution reactions.

One common dynamic process of metal–allyl complexes is the equilibration of syn and anti substituents by a sequence of $\eta^3-\eta^1-\eta^3$ interconversions. The interconversion of one set of syn and anti protons of an unsubstituted allyl group is shown in Equation 3.56. This η^3- to η^1- (π- to σ-) allyl transformation generates a vacant coordination site on the metal. Thus, the rate of this interconversion can be controlled by addition of Lewis basic additives that can bind to the open coordination site or coordinate to the initial η^3-allyl complex and induce an η^3- to η^1-rearrangement.

$$(3.56)$$

Second, allyl complexes undergo rearrangements that lead to a formal rotation along the metal–allyl axis, as illustrated in Equation 3.57. This process occurs most commonly within stereochemically non-rigid complexes, such as those with coordination numbers 5 or 7 (considering a Cp ligand to occupy three coordination sites). A formal rotation about the metal–allyl axis can also occur by rearrangements of the ancillary ligands. Thus, an apparent rotation of an allyl group in four-coordinate complexes can occur by an associative mechanism, as shown in Equation 3.58,[205] or by a dissociative mechanism, as shown in Equation 3.59.[206]

$$(3.57)$$

$$(3.58)$$

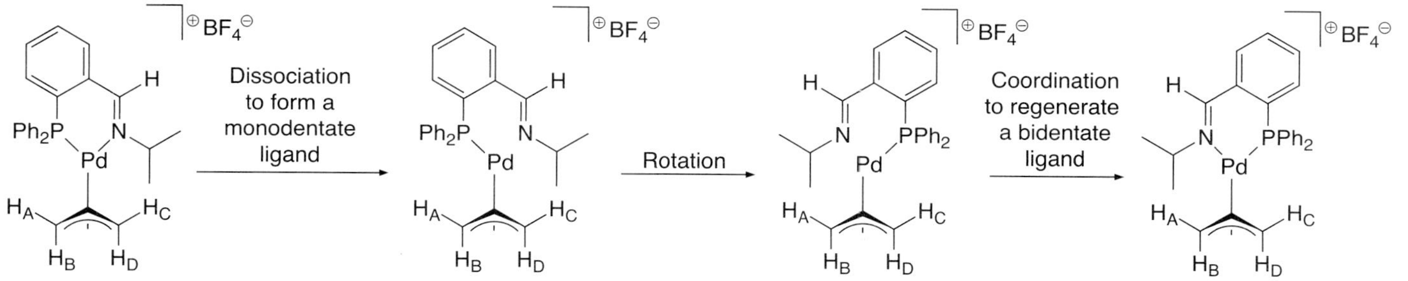

(3.59)

3.5.1.4. Synthesis of π-Allyl Complexes

π-Allyl complexes have been prepared by many methods.[197] Some of the methods involve addition of nucleophilic or electrophilic allyl reagents to a transition metal complex. Other methods involve transformations of coordinated dienes. These routes are illustrated by the examples in Equations 3.60–3.63.

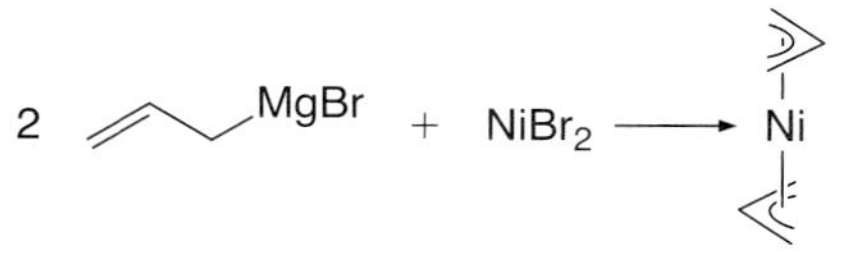

(3.60)

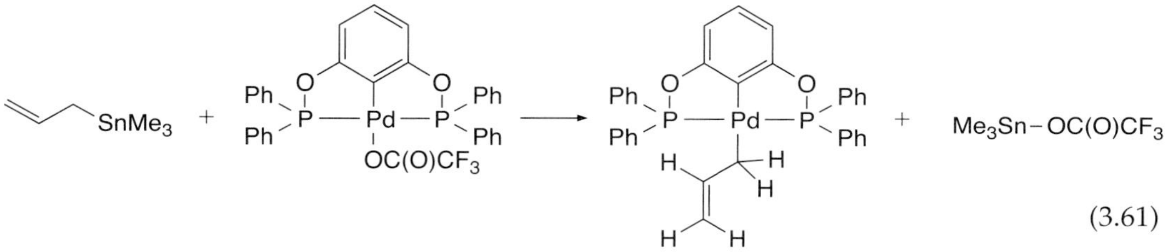

(3.61)

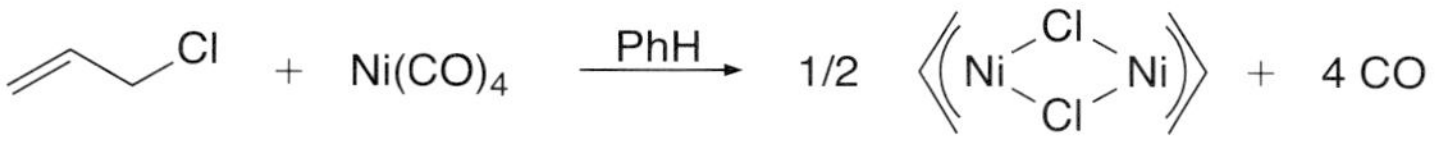

(3.62)

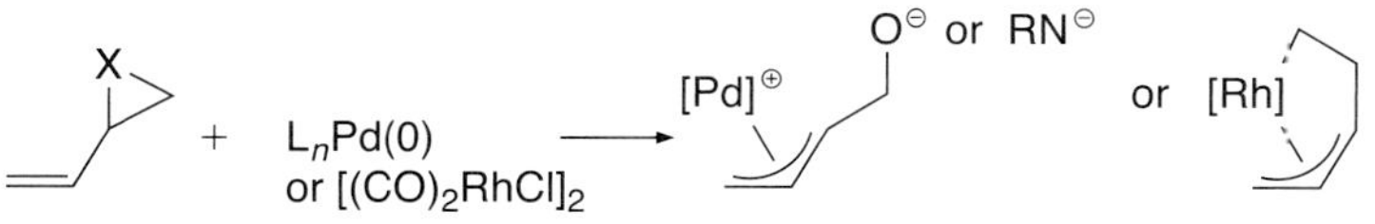

(3.63)

Allyl complexes can be prepared by displacing a good leaving group with a nucleophilic main group metal allyl reagent.[195] For example, the simple π-allyl complex Ni(allyl)$_2$ in Equation 3.60 has been prepared by the reaction of NiBr$_2$ with allyl Grignard. The addition of a less-reactive main group allyl–metal reagent to form an η¹-allyl species is illustrated by Equation 3.61. Alternatively, allyl complexes can be prepared by the oxidative addition of allylic halides or esters to low-valent metal complexes in a process akin to nucleophilic attack of a metal on an electrophilic allyl group (Equation 3.62). (See Chapters 7 and 20.) Similar oxidative additions of allyl regents to form allyl complexes occur between low-valent metals and allylic ethers, amines, alcohols, and even nitro compounds to generate the catalytic intermediates discussed in Chapter 20. Reactions of Pd(0) complexes with vinylepoxides and vinylaziridines and reactions of Rh(I) complexes with vinylcyclopropanes have also been used to generate π-allyl intermediates in catalytic allylic substitutions and catalytic cycloadditions, respectively. These reactions are summarized in Equation 3.63.[207]

$$RCo(CO)_4 \; + \; \diagup\!\!\!\diagdown\!\!\!\diagup \longrightarrow \left[R\diagdown\!\!\!\diagup\!\!\!\diagdown Co(CO)_4 \right] \xrightarrow{-CO} \diagdown\!\!\!\big\langle\!\!\diagup Co(CO)_3 \qquad (3.64)$$

R = H, alkyl, R(O)C

$$(3.65)$$

$$(3.66)$$

$$(3.67)$$

$$R\diagup\!\!\!\diagdown\!\!\!\diagup \xrightarrow[\substack{BQ(2\ equiv.),\ 4\text{–}\text{Å MS} \\ DMSO:AcOH\ (1:1,\ v/v) \\ Air,\ 40\ °C}]{Pd(OAc)_2(10\ mol\ \%)} R\diagup\!\!\!\diagdown\!\!\!\diagup OAc \qquad (3.68)$$

Equations 3.64–3.66 illustrate routes to allyl complexes from dienes, diene complexes, and olefins. Allyl complexes have been prepared by the insertion of a conjugated diene into a metal hydride, alkyl, or acyl linkage, as illustrated for the cobalt complexes in Equation 3.64.[208] Alternatively, allyl complexes have been prepared by nucleophilic or electrophilic attack on a coordinated diene. Equation 3.65 shows the formation of allyl complexes by the addition of carbanions to a cationic diene complex,[209] and Equation 3.66 shows the formation of a cationic diene complex by the protonation of a neutral 1,3-diene complex.[210] Allyl complexes have also been formed by the abstraction of an allylic proton from a metal–olefin complex, either by a base or by the metal itself.[211] This reaction has been proposed as a step in the isomerization of olefins (Equation 3.67)[212] and in the allylic oxidation of olefins (Equation 3.68).[213,214]

3.5.1.5. Reactions of Allyl Complexes

The reactivity of allyl complexes is included in many of the later chapters of this text. Although examples of reductive eliminations and migratory insertions of allyl complexes are known, the dominant reaction chemistry involves the attack by nucleophiles on the allyl ligand and attack of the allyl ligand on external and coordinated electrophiles. The latter reaction leads to catalytic allylic substitution reactions. The reactions of nucleophiles and electrophiles with allyl complexes are described in Chapters 10 and 11, and the catalytic allylic substitution is described in Chapter 20.

3.5.2. η³-Benzyl Complexes

Well-characterized complexes with π-benzyl ligands have been reported. A generic structure of an η³- or π-benzyl complex is shown in Figure 3.14. This structure is analogous

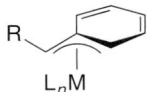

Figure 3.14.
Generic structure of an
η³-benzyl complex.

to that of a π-allyl complex, but the three-carbon π-system consists of the benzylic carbon, the ipso carbon, and one ortho carbon of the aryl ring. π-Benzyl complexes have been shown to be likely intermediates in several stoichiometric and catalytic processes described in later chapters of this text. The intermediacy of such complexes is important because this binding mode often causes insertions of vinylarenes into metal hydrides to form products with the aryl group located on the α-carbon. Thus, the intermediacy of π-benzyl complexes in the metal-catalyzed hydrofunctionalizations of styrenes contributes to the tendency of many additions to vinylarenes to form branched products.

π-Benzyl complexes often exist in equilibrium with η^1-benzyl complexes. Consistent with the scheme for interconversions between η^3- and η^1-allyl complexes, an asymmetric benzyl carbon has been shown to retain its configuration during this process (Equation 3.69).[215]

$$\tag{3.69}$$

π-Benzyl complexes typically form when a benzyl group is present in a coordinatively unsaturated middle or late metal center that can participate in back-donation into the π-benzyl unit. π-Benzyl complexes have been prepared by the addition of benzyl Grignard reagents to unsaturated metal–halide complexes,[216] by the addition of benzyl halides to low-valent metal complexes,[217] or by the insertion of vinylarenes into metal hydrides[218] or alkyls.[219] These three types of reactions are illustrated by the examples in Equations 3.70–3.72.

$$\tag{3.70}$$

$$\tag{3.71}$$

$$\tag{3.72}$$

L_2 = Tetramethylethylenediamine (TMEDA)

3.5.3. Higher Anionic π-Ligands

There are also complexes containing polyhapto ligands with a more extended π-system than that in η^3-allyl or η^3-benzyl groups. Pentadienyl systems are the next higher anionic homolog of an allyl system. Pentadienyl systems are often formed by the addition of protons to, or removal of hydrides from, transition metal polyene complexes. Alternatively, pentadienyl groups have been generated by the addition of nucleophiles to coordinated arenes. Such complexes are often synthetic intermediates in processes described later in this text and elsewhere.[220] Examples of the protonation of a coordinated hexadiene system, hydride abstraction from a coordinated diene, and nucleophilic attack on a coordinated arene to form pentadienyl ligands are given in Equations 3.73–3.75.[221–223]

$$(3.73)$$

$$(3.74)$$

$$(3.75)$$

3.5.4. η^4-Trimethylenemethane (TMM) Complexes

The η^4-trimethylenemethane (TMM) ligand is related to the η^3-allyl ligand, but is less common. The TMM ligand consists of a four-carbon π-system that is well studied in physical organic chemistry as the free neutral species. TMM ligands have an umbrella-like pucker, with the central carbon above the plane formed by the three terminal carbons.

The synthesis, structure, and reactivity of TMM complexes have been reviewed.[224] Stable TMM complexes of a variety of metals are known. They are typically formed by the dehalogenation of dihaloisobutylene derivatives (Equation 3.76),[225] from the thermal extrusion of HX from η^3-methylallyl complexes (Equation 3.77),[226] or from the elimination of Me$_3$SiX from γ-substituted allylsilanes(Equation 3.78).[227,228] Coordinated TMM ligands undergo a variety of reactions, but cycloaddition reactions involving the TMM ligand have been the most studied. For example, the palladium-catalyzed cycloaddition reactions of electrophilic olefins with (2-acetoxymethyl-3-allyl)trimethylsilane are thought to occur through palladium–TMM complexes (Equation 3.78).[227,228] Enantioselective versions of this process have been developed.[229] A number of oxa-, aza-, sila-, and phosphatrimethylene-methane complexes have also been prepared, although catalytic chemistry based on these structures has not been developed.[230–235]

$$(3.76)$$

$$(3.77)$$

$$\text{AcO} \diagup \diagdown \text{TMS} \ + \ L_4Pd \ \longrightarrow \ \left[\ \underset{L}{\overset{}{\underset{}{Pd}}} \diagup \diagdown L \ \right] \ \overset{Y \quad X}{\longrightarrow} \ \overset{Y}{\underset{X}{\diagdown}} \qquad (3.78)$$

3.6. Cyclopentadienyl and Related Compounds (Written with Prof. Jack R. Norton)

3.6.1. Overview

The cyclopentadienyl ligand, generally abbreviated as "Cp," is a common anionic ancillary ligand in organometallic chemistry. Complexes of this ligand or its substituted derivatives are known for all transition metals and for most f-block metals.[236,237] Ferrocene is the classic cyclopentadienyl compound, and the identification of the structure of this species by Woodward, Wilkinson, and co-workers, after initial misassignment,[238,239] is legendary.[240]

Cyclopentadienyl ligands are present on some of the most active catalysts for olefin polymerization,[241] on the first complexes to add alkanes[242,243] and on the first catalysts to functionalize the terminal position of alkanes regiospecifically.[244] They are ancillary ligands on complexes that undergo all of the classes of reactions described in later chapters of this book, and for these reasons many derivatives of Cp ligands have been prepared. Section 3.6 describes the thermodynamic, electronic, and steric properties of cyclopentadienyl ligands, the synthesis of cyclopentadienyl complexes, and some fundamental reactivity of the Cp ligand.

3.6.2. Bonding and Thermodynamics of Cp Ligands

The η^5-cyclopentadienyl ligand is an L_2X-type ligand, and can be considered to occupy three facially oriented coordination sites. The molecular orbitals of the π-system of a cyclopentadienyl ligand were shown in Chapter 1. These orbitals consist of one filled orbital of σ-symmetry and two filled orbitals of π-symmetry, along with two unoccupied orbitals of δ-symmetry that can act as electron acceptors.

Cp ligands tend to bind tighter than arene ligands because of the additional electrostatic attraction generated by the partial anionic and cationic charges of the ligand and metal, respectively. Thus, full dissociation of Cp anion or radical from a metal occurs only in rare instances. The ΔH for homolysis of the Fe–Cp bond in ferrocene (Equation 3.79) is 79 kcal/mol,[245] and the ΔH for homolysis of the M–Cp bond in Equation 3.80 is 70 kcal/mol for M = Ti, 100 kcal/mol for M = Zr, and 101 kcal/mol for M = Hf.[246] These bond strengths are comparable to those of η^5-pyrrolyl ligands[246] and exceed those of η^6-arene ligands.

$$Cp_2Fe \ \rightleftharpoons \ Cp\cdot \ + \ \cdot FeCp \qquad \Delta H = 79 \text{ kcal/mol} \qquad (3.79)$$

$$(\eta^5\text{-Cp})MCl_3 \ \rightleftharpoons \ Cp\cdot \ + \ \cdot MCl_3 \qquad \begin{array}{l} \Delta H = 79 \text{ kcal/mol for Ti} \\ \Delta H = 100 \text{ kcal/mol for Zr} \\ \Delta H = 101 \text{ kcal/mol for Hf} \end{array} \qquad (3.80)$$

3.6.3. Synthesis of η^5-Cyclopentadienyl Complexes

Cyclopentadienyl complexes are usually prepared by taking advantage of the acidity of cyclopentadiene ($pK_a = 16.0$).[247] By one method, Cp compounds are prepared by combining salts of its anion (usually NaCp) with metal halides, as shown for ferrocene in Equation 3.81.[238,239,248,249] Reactions of the toxic, air-stable TlCp have been used to prepare metal–cyclopentadienyl complexes when side reactions such as electron transfer limit the yields of reactions of NaCp.[250–252] Alternatively, reactions of cyclopentadiene or its derivatives with metal–amido complexes lead to Cp complexes and the free amine.[253–260] An example of this process to form Cp complexes of group IV metals is shown in Equation 3.82.[253]

$$FeCl_2 \ + \ NaCp \ \longrightarrow \ Cp_2Fe \ + \ 2\,NaCl \tag{3.81}$$

$$Zr(NMe_2)_4 \ + \ CpH \ \longrightarrow \ Cp_2Zr(NMe_2)_2 \tag{3.82}$$

3.6.4. Examples of Substituted Cyclopentadienyl Ligands

Substituted Cp ligands are common, and the most common of these derivatives is the bulky, strongly electron-donating C_5Me_5 ligand, generally abbreviated as Cp*.[261] Pentamethylcyclopentadiene is commercially available, and $C_5Me_5MgCl(THF)$ is crystalline, storable, and easily weighed. Other commercially available substituted cyclopentadiene derivatives (e.g., C_5Ph_5) are even bulkier than Cp*.[262–264]

Many cyclopentadienyl ligands bearing functional groups have been prepared.[265] A general review of "side-chain-functionalized cyclopentadienyl compounds" has appeared,[266] and there have been more specialized reviews on Cp derivatives with pendant O-,[267] P-, As-, and S-donors.[268] Even pentahalocyclopentadienes have been prepared.[269]

Annulated derivatives of cyclopentadienyl, such as indenyl[270] and fluorenyl[271] (Figure 3.15), as well as acenaphthyl-substituted cyclopentadienyl,[272] and pentalenyl,[273,274] are some of the most important cyclopentadienyl derivatives. They, and their derivatives with interannular bridges, play a major role in controlling stereochemistry during the metal-catalyzed polymerization of olefins, as is described in Chapter 21. A number of permethylindenyl complexes have been reported.[275]

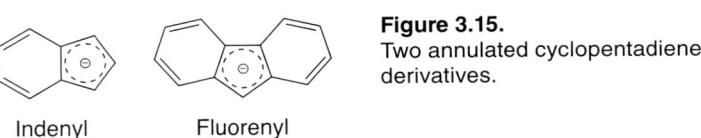

Figure 3.15.
Two annulated cyclopentadiene derivatives.

Indenyl   Fluorenyl

Although an early study of C–O stretching frequencies on Rh suggested that indenyl ligands are weaker electron donors than Cp ligands,[276] electrochemistry, photoelectron spectroscopy, and a study of C–O stretching frequencies on Fe imply that indenyl ligands are slightly better donors.[277,278] DFT calculations imply that η^5-indenyl ligands are more weakly bound than η^5-Cp ligands, but they also show that η^3-indenyl ligands are more strongly bound than η^3-Cp ligands[270]—results that appear to explain how indenyl ligands facilitate associative substitution (Section 5.5.3).

Chiral substituted cyclopentadienyl ligands in enantiopure or enantioenriched form are known, and some of these contain substituents derived from the chiral pool. Examples include **A**,[279] **B**,[280] and **C**[281] in Figure 3.16. There are several reviews of such ligands and their complexes.[282–284]

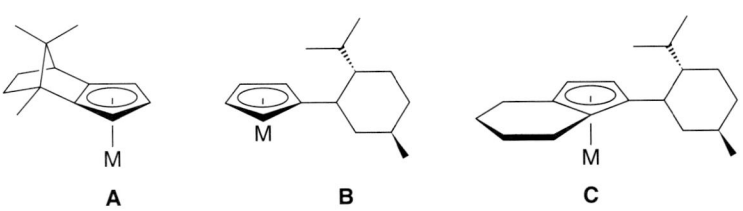

A   B   C

Figure 3.16. Examples of chiral, non-racemic cyclopentadienyl ligands.

In other cases, substituted cyclopentadienyl ligands have enantiotopic faces and form chiral complexes upon coordination to a metal. For example, coordination of the two faces of the 1,3-MePhCp ligand generates enantiomeric complexes, as shown in Figure 3.17.[282, 285–287]

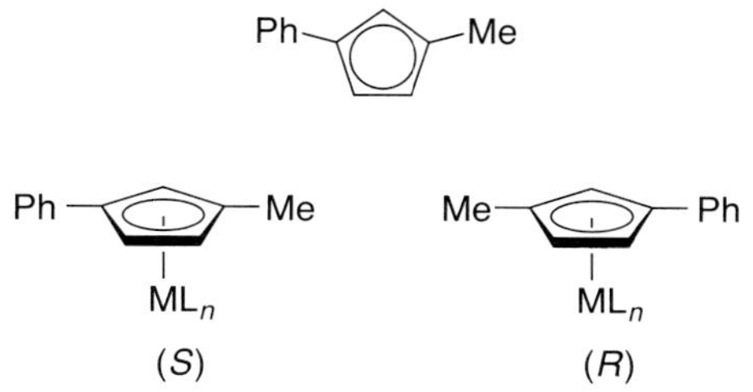

Figure 3.17.
Enantiomers of 1,3-unsymmetrically disubsituted cyclopentadienyl complexes.

(*S*)   (*R*)

3.7. Ansa Metallocenes

Many zirconium and titanium polymerization catalysts contain two cyclopentadienyl derivatives connected by a linking group. This linker has been termed an "*ansa*" bridge.[*,288–291] These linking groups prevent simple rotation of the Cp derivative; as shown for the fragment in Figure 3.18, if this linker bridges two tetrahydroindenyl groups, then the complex is locked into a chiral conformation.[292,293] A more detailed description of ansa metallocenes is provided in Section 3.7.1.4.

Cyclopentadienyl ligands can also be connected to other types of donors, as in the "constrained geometry" complex in Figure 3.19. First reported by Bercaw and co-workers,[294] this amidocyclopentadienyl ligand is contained in commercially important olefin polymerization catalysts (see Chapter 22).[295–297]

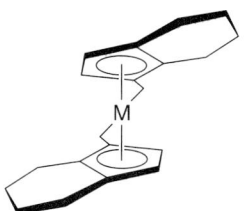

Figure 3.18.
Structure of a *rac*-ansa-metallocene fragment.

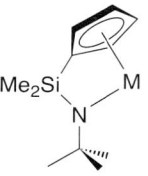

Figure 3.19.
A metal with a "constrained geometry" ligand.

3.7.1. Types of Cyclopentadienyl Complexes

Most mononuclear cyclopentadienyl complexes fall into one of the four categories shown in Figure 3.20. The first (and newest) category, (**A**) Cp_3M, consists of molecules with three cyclopentadienyl ligands. The second (but the first to be discovered), (**B**) Cp_2M, has two cyclopentadienyl ligands, which are generally parallel; such "sandwich" compounds are traditionally called "metallocenes." The third, (**C**) Cp_2ML_x, consists of "bent metallocenes," in which the cyclopentadienyl rings are *not* parallel and the number x of unidentate ligands L (H, R, CO, etc.) varies between one and three. The fourth, (**D**) $CpML_y$, consists of "half-sandwich" compounds, in which the number y of unidentate ligands L varies from one to four. Other, less common, structural types will be mentioned after these categories have been discussed.

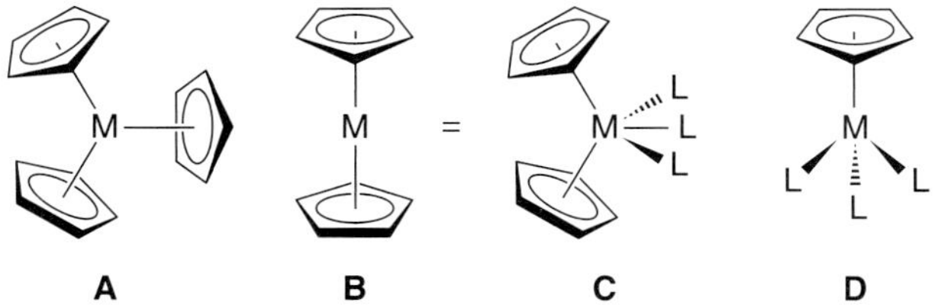

A   B   =   C   D

Figure 3.20. Four classes of Cp complexes.

---

*"Ansa," meaning "handle" in Latin, was first used by Lüttringhaus for bridged aromatic compounds and later applied by him to ferrocenes with interannular bridges; Brintzinger extended the term to titanocenes containing interannular bridges. The history of this usage can be found in footnote 3 of reference 242.

3.7.1.1. Cp₃M and Their Permethyl Derivatives Cp*₃M

With a few exceptions,[251] most tris(cyclopentadienyl) complexes contain lanthanides (Ln) or actinides as the central metal. Cp_3Ln have polymeric structures, with chains of (η^5-Cp)₂Ln units linked by cyclopentadienides that are bound in an η^1-fashion toward one or both of the metals they bridge.[236] Cp^*_3Ln and Cp^*_3U are monomeric. The Sm(III) complex has been prepared from Sm(II) and Cp^*_2Pb (Equation 3.83); the U(III) complex was prepared from the U(III) hydride and tetramethylfulvene (Equation 3.84).[237]

$$2 \quad \text{Sm—O} \quad + \quad \text{Pb} \quad \xrightarrow[\text{−2 Et}_2\text{O}]{\text{−Pb}} \quad 2 \quad \text{Sm} \qquad (3.83)$$

$$(C_5Me_5)_2UH(dmpe) \quad + \qquad \xrightarrow{\text{−dmpe}} \qquad \text{U} \qquad (3.84)$$

3.7.1.2. Metallocenes Cp₂M and their Permethyl Derivatives Cp*₂M [251]

Many people consider the discovery in 1951[8] and subsequent characterization of Cp_2Fe to be the beginning of modern organotransition metal chemistry.[9] Certainly, the binding of an aromatic group to a metal through the face of the ligand was a new bonding mode, and thousands of papers, and even commercial asymmetric catalytic processes,[298] are based upon derivatives of ferrocene.

3.7.1.3. Structures of "Sandwich Complexes"

In the gas phase, the eclipsed conformation (D_{5h}) of metallocenes Cp_2M is lower in energy than the staggered conformation (D_{5d})[299] (see Figure 3.21). However, the difference in energy is small, and the barrier to rotation in metallocenes is low.

The frontier orbitals of metallocenes (Figure 3.22) are similar to those of bis(benzene) chromium discussed in Chapter 1. One E_1'' pair is a bonding combination of the metal d_{xz} and d_{yz} orbitals with ligand orbitals, and the other E_1'' pair is the antibonding combination of these orbitals, which is denoted $E_1''^*$. Three frontier orbitals are largely metal-centered and nonbonding: a degenerate E_2' orbital pair ($d_{x^2-y^2}$ and d_{xy}) and an A_1' (d_{z^2}) orbital. These orbitals in ferrocene are filled and are the HOMOs.

Some metallocenes contain less than 18 valence electrons, and some contain more than 18 valence electrons. Metallocenes containing metals to the left of iron, such as vanadocene, chromocene, and manganocene, possess fewer than 18 valence electrons. The E_2' and A_1' orbitals of these complexes are partially occupied. Thus, they are paramagnetic and often adopt more complex structures than simple monomers. For example, manganocene adopts a distorted chain structure in the solid state (Figure 3.23) in which the bridging Cp ligands bind in an η^2 fashion toward one Mn and in an η^3 fashion toward another.[300] Metallocenes containing metals to the right of iron, such as cobaltocene and nickelocene, possess more than 18 valence electrons. In these cases, the valence electrons occupy the antibonding $E_1''^*$ orbitals (Figure 3.22). The number of unpaired

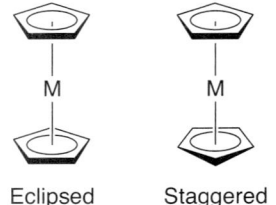

Eclipsed   Staggered

Figure 3.21.
Eclipsed and staggered conformations of metallocenes.

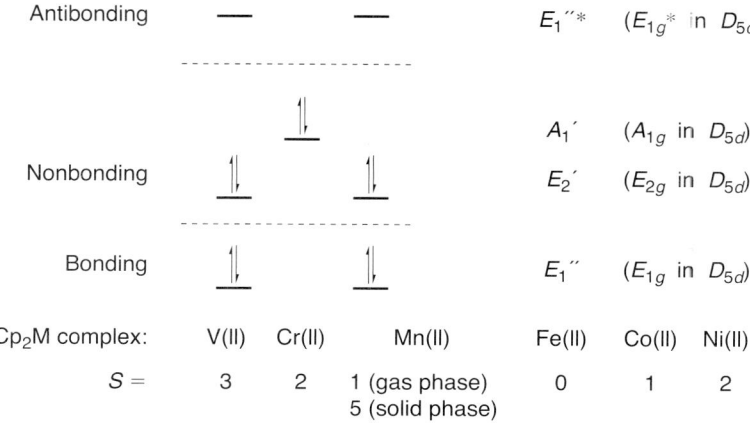

Figure 3.22.
Frontier orbitals of metallocenes containing parallel Cp ligands (the electron occupancy shown is that of ferrocene); the spin states of several metallocenes.

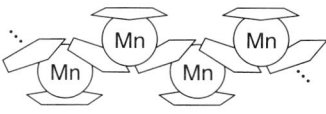

Figure 3.23.
Polymeric structure of manganocene.

electrons in the metallocenes containing different metals are given at the bottom of Figure 3.22. Related metallocenes of the second- and third-row metals are highly reactive. For example, Cp_2Re can only be prepared in a matrix at 20 K[301] and the osmocinium ion exists as a diamagnetic dimer, $[Cp_2Os]_2(PF_6)_2$, connected by a 3.04 Å Os–Os bond (Figure 3.24).[302, 303]

Metallocenes containing Cp^* ligands are also common. These metallocenes are more soluble, and often more stable than their counterparts containing Cp ligands. The bulky Cp^* ligand inhibits intermolecular interactions; thus, Cp^*_2Re does not adopt the dimeric structure of the isoelectronic $[Cp_2Os]_2(PF_6)_2$,[304] and Cp^*_2Mn does not form the chain structure of Cp_2Mn. Morever, Cp^*_2Mn and Cp^*_2Re are low spin with one unpaired electron.[304] Cp^* metallocenes are also more easily oxidized and more easily protonated than their Cp analogs. For example, the diamagnetic d^6 dication $Cp^*_2Ni^{2+}$ is stable,[305] whereas the products from oxidation of Cp_2Ni decompose, and Cp^*_2Os undergoes protonation under milder conditions than Cp_2Os.[306, 307]

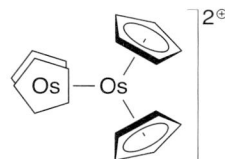

Figure 3.24.
Dimeric structure of the osmocinium ion.

3.7.1.4. Bent Metallocenes Cp_2ML_x and Related Compounds

In bent metallocenes,[308] the angle between vectors normal to each ring is less than 180°. These complexes contain one (**A**), two (**B**), or three (**C**) additional ligands, as shown in Figure 3.25.

An understanding of the orbitals of bent metallocenes is essential for understanding their structures and reactivity. The empty orbital in 16-electron complexes, such as Cp_2ZrCl_2 in Figure 3.25 is the "d_{x^2}" orbital shown in Figure 3.26. Complexes in which this orbital is partially or fully occupied have smaller X–M–X angles. For example, the angle in Cp_2ZrCl_2 is 97.1°, whereas this angle in Cp_2NbCl_2 is is 85.6° and in Cp_2MoCl_2 is 82.0°.[309] The orbital can serve as a π-acceptor, explaining the perpendicular orientation of the THF in $[Cp_2Zr(CH_3)(THF)]^+$ with respect to the plane defined by CH_3, Zr, and the THF oxygen (Figure 3.27).[310]

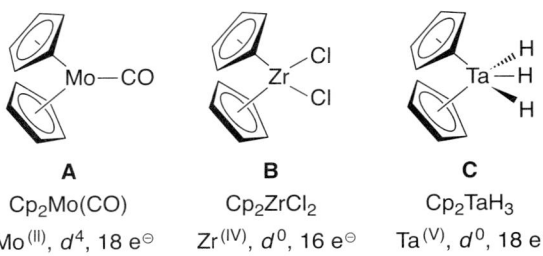

Figure 3.25.
Bent metallocenes containing one, two, and three additional ligands.

A	**B**	**C**
$Cp_2Mo(CO)$	Cp_2ZrCl_2	Cp_2TaH_3
$Mo^{(II)}$, d^4, 18 $e^{\ominus}$	$Zr^{(IV)}$, d^0, 16 $e^{\ominus}$	$Ta^{(V)}$, d^0, 18 $e^{\ominus}$

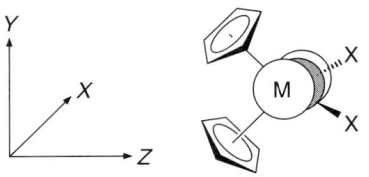

Figure 3.26.
Unoccupied, partially or fully occupied orbital in $Cp_2Mo(CO)$, Cp_2ZrCl_2, and Cp_2TaH_3.

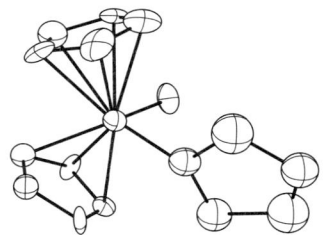

Figure 3.27.
Structure of $[Cp_2Zr(CH_3)(THF)]^+$ showing the orientation of the THF ligand. From reference 310.

Metallocenes can be forced to adopt bent conformations by ansa bridges. Two common precursors to ansa metallocenes, ethylenebis(indenyl) (EBI) and ethylenebis-(tetrahydroindenyl) (EBTHI)″, are shown in Figure 3.28. The ansa-bridged EBI is a weaker electron donor to Zr than two Cp ligands, but the partially hydrogenated derivative EBTHI is a *stronger* donor to Zr. A single-atom ansa bridge (particularly Me_2Si) between cyclopentadienyl ligands has an electron-withdrawing effect; single-atom ansa-bridged Cp analogs are less electron donating than two Cp ligands on Zr.[291]

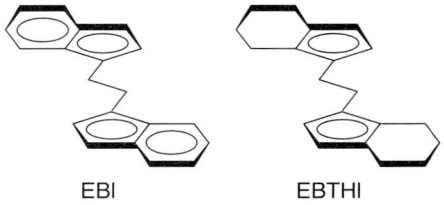

EBI   EBTHI

Figure 3.28.
Two common ligands that lead to ansa metallocenes.

The ansa metallocenes of Ti and Zr, with structures of the fragments shown in Figure 3.18 have become extremely important in the stereoselective catalysis of olefin polymerization and other reactions. While the first ansa titanocene was prepared by Katz in 1970,[311] and the first enantiopure ansa titanocene was isolated in 1982 when Brintzinger resolved (EBTHI)TiCl$_2$ (Figure 3.18 with M = Ti) using a binaphtholate derivative,[312] the stereoselective synthesis of the *rac* form of the ansa metallocenes has been challenging. More recently, Jordan has found that the addition of bis(indenyl) dianions to the bis(amide) Zr complex in Equation 3.85 leads *exclusively* to the *rac* diastereomer.[313] He has also shown that a zirconium complex containing a chiral, non-racemic diamido ligand (Equation 3.86) leads to the ansa-metallocene in enantio-enriched form.[314]

$$(3.85)$$

$$X = SiMe_2 \text{ or } CH_2CH_2$$

$$(3.86)$$

3.7.1.5. "Half-Sandwich" Compounds CpML$_y$

Many half-sandwich complexes, which contain one Cp ligand or one Cp derivative and a set of additional ligands, have been prepared and studied intensively. Complexes of this type containing one, two, three, and four additional ligands are known, and representative examples are shown in Figure 3.29. The most common contain 2–4 additional ligands.

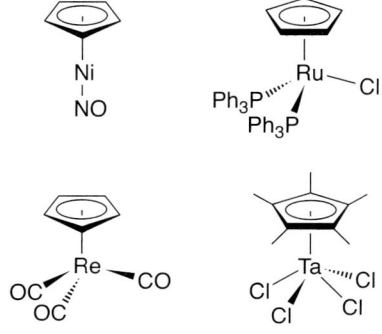

Figure 3.29.
Half-sandwich Cp complexes with one, two, three, and four additional ligands.

A general review of the half-sandwich cyclopentadienyl complexes of Ru and Os was published in 1987,[315] and a review of CpRu(PPh$_3$)$_2$Cl and its derivatives was published in 1990.[316] Other reviews of half-sandwich complexes have been more specialized: chiral derivatives of CpRu(PPh$_3$)$_2$Cl,[317] complexes with a single-ring substituent,[318] V analogs of Cp*TaCl$_4$,[319] complexes with stereogenic metal atoms,[320] coordinatively unsaturated Fe and Ru complexes,[321,322] alkylidene complexes,[323] complexes with carbene ligands,[324] and complexes used in supramolecular chemistry.[325]

3.7.1.6. Other Modes of Binding of Cyclopentadienyl Ligands

Cyclopentadienyl ligands occasionally bridge two metals. Palladium complexes containing such bridges are known. These complexes, such as the one in Figure 3.30, contain a Pd–Pd bond. The bridging "cyclopentadienyl" ligand in this complex is really more similar to a bridging allyl group and an uncoordinated double bond.[326] As noted previously, the Cp ligands in manganocene in the solid state bond η^2 to one Mn and η^3 to another. Cyclopentadienyl ligands can bind to a single metal in an η^1- or η^5-fashion. The rearrangements that η^1-cyclopentadienyl ligands undergo are discussed in Section 3.7.3.

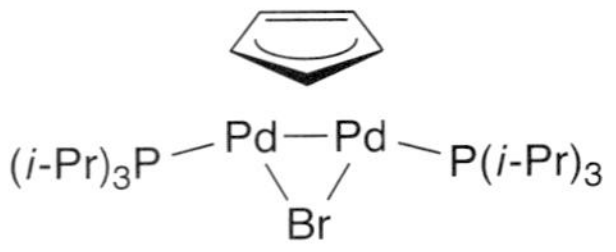

Figure 3.30.
Example of a bridging Cp ligand.

3.7.2. Ligands That Are Electronically Similar to the Cyclopentadienyl Ligand

Several ligands possess frontier orbitals that share the symmetry properties of cyclopentadienyl ligands. For example, the HOMOs on the open C$_2$B$_3$ face of the carboranyl dianion [$nido$-C$_2$B$_9$H$_{11}$]$^{2-}$,[327] and on the pentagonal C$_2$B$_3$ face of [$nido$-C$_2$B$_4$H$_6$]$^{2-}$,[328] are similar to those of a cyclopentadienyl monoanion. These carboranyl ligands can then serve as dianionic analogs of a Cp ligand, and complexes,[329] such as those in Figure 3.31, can be prepared that are related to cyclopentadienyl complexes.

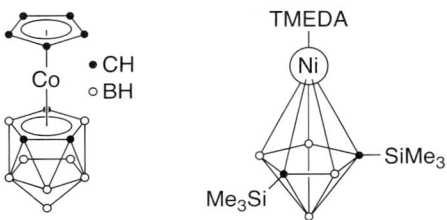

Figure 3.31.
Examples of carboranyl complexes.

Planar (or almost planar) boron analogs to cyclopentadienyl ligands include borollides,[330] 1,2-azaborolyls,[331] azaborindenyls,[332] and boratabenzenes (Figure 3.32).[333] These analogs of cylopentadienyl ligands containing boron have been reviewed by Grimes.[334]

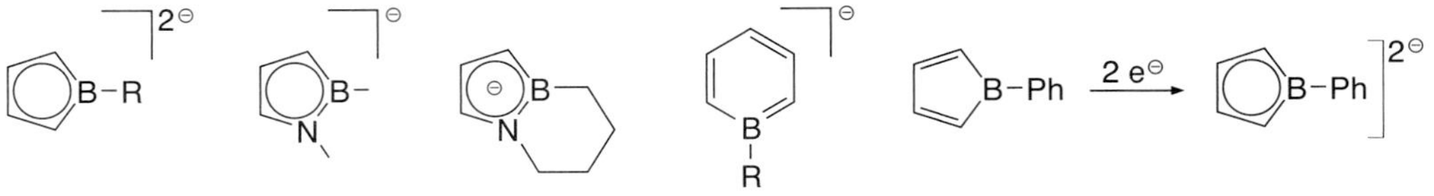

Figure 3.32. Structures of borollides, 1,2-azaborolyls, azaborindenyls, and boratabenzenes.

Most closely analogous to cyclopentadienyls are borollides, the dianions of boroles. Borollide complexes are often prepared by a procedure like that in Scheme 3.3, beginning with the dehydrogenation of a borolene. This reaction with a rhodium(I) species formed a "triple-decker sandwich."[330, 333, 335]

One 1,2-azaborolyl ligand has been prepared by dehydrohalogenation of the β-chloroboracycle that is available from the transmetallation of the stannacycle in Scheme 3.4. Another has been prepared by ring-closing metathesis from a vinyl aminoborane (Scheme 3.5).

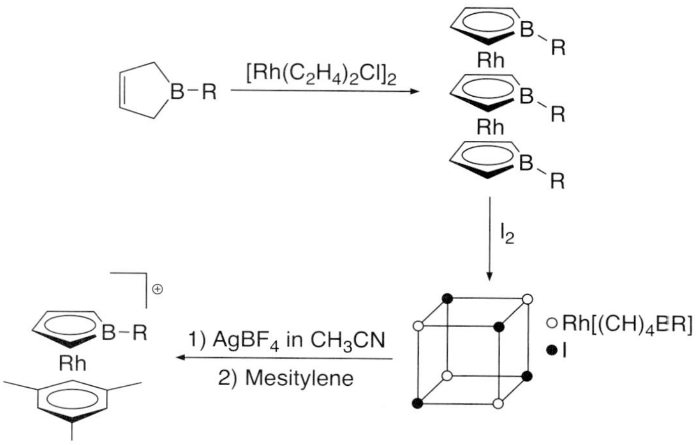

Scheme 3.3

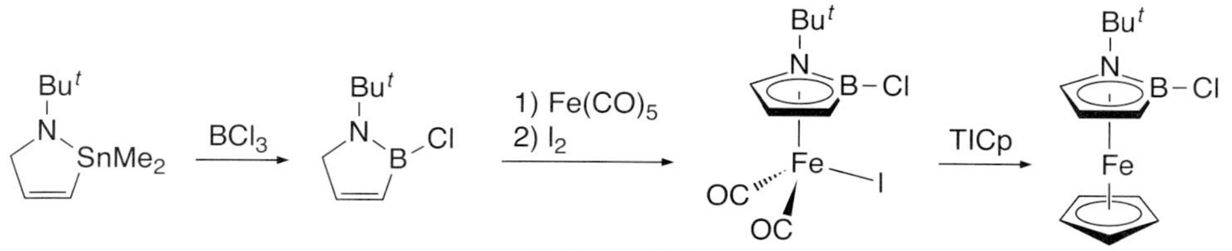

Scheme 3.4

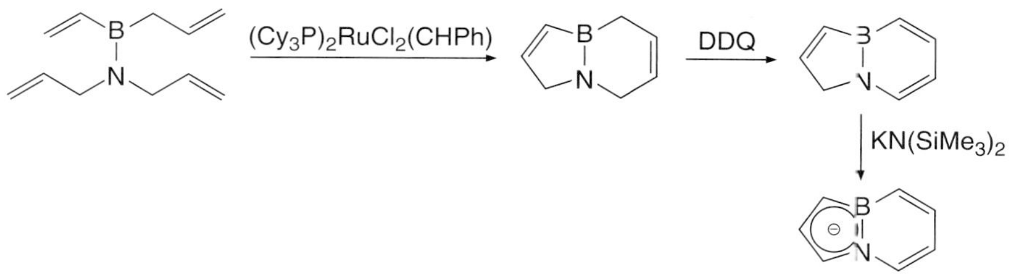

Scheme 3.5

Boratabenzene anions have been called "more akin [than any other anion] to cyclopentadienyl," although they are less basic and less nucleophilic than the cyclopentadienyl anion.[333] Several reviews on boratabenzene complexes have been published.[335–337] The boratabenzene Li(C_5H_5BMe) can be prepared efficiently by the method in Scheme 3.6.[338] Its complexes are often prepared from a transition metal halide and a Si or Sn derivative, as shown in Equation 3.87.[339] Some aminoboratabenzene complexes (Figure 3.33) are catalysts for ethylene polymerization,[340,341] as are some complexes of boratabenzene analogs of ansa metallocenes[342] and boratabenzene analogs of constrained geometry complexes.[343]

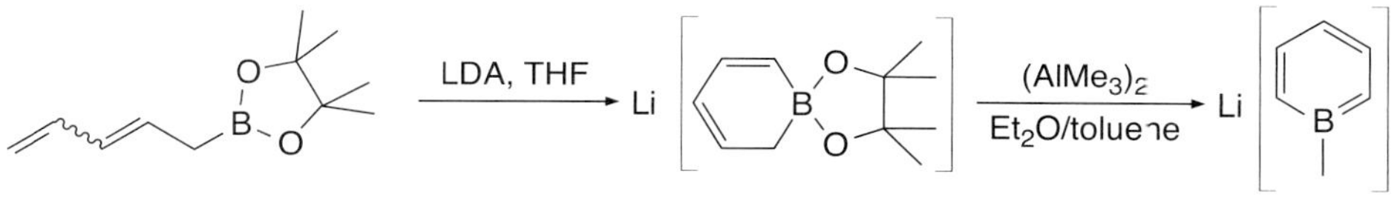

Scheme 3.6

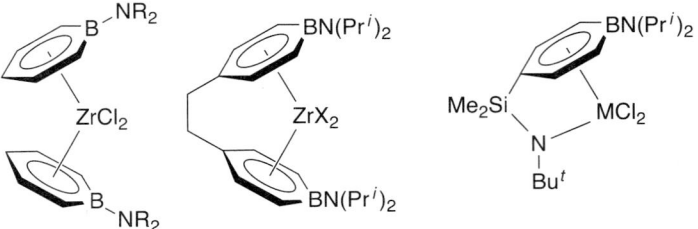

$$(3.87)$$

$$E = Si \text{ or } Sn$$

Figure 3.33. Three types of aminoboratabenzene complexes.

3.7.3. Reactions of Cyclopentadienyl Complexes

η^1-Cyclopentadienyl ligands often rearrange by 1,2-shifts,[344] and this rearrangement makes all five carbons and hydrogens appear equivalent by NMR spectroscopy. One classic example of this rearrangment is the η^1-cyclopentadienyl ligand in $(\eta^5\text{-}C_5H_5)Fe(CO)_2(\eta^1\text{-}C_5H_5)$. In some complexes the η^5- and η^1-rings exchange. This exchange occurs within $(\eta^5\text{-}C_5H_5)_2Ti(\eta^1\text{-}C_5H_5)_2$.[345]

Associative substitution reactions of dative ligands in cyclopentadienyl complexes can occur by a decrease in the hapticity of the Cp ligand from η^5 to η^3. These "ring-slip" reactions, which occur more rapidly for indenyl complexes than for cyclopentadienyl complexes, are discussed in Section 5.5.3. This increase in rate is called the "indenyl effect."[270]

Electrophilic attack on coordinated cyclopentadienyl rings, particularly those in ferrocene, is well established. This process occurs in a similar fashion to electrophilic attack on arenes and was used to establish the η^5 binding mode of the Cp ligand in ferrocene (see Equation 3.88). These reactions of the Cp ligand with electrophiles are described in more detail in Chapter 12, which covers electrophilic attack on coordinated ligands. Friedel–Crafts acylation, formylation, aminomethylation, and mercuration are all known.[251,346]

However, electrophilic aromatic substitution on ferrocene presents two complications: (1) its susceptibility to one-electron oxidation, which precludes halogenation, nitration, or direct (H_2SO_4) sulfonation; and (2) competition between the Fe and the ring as the initial site of electrophilic attack. It is now clear from ^{1}H NMR spectroscopy of the product of the protonation of ferrocene with a superacid at –122 °C that protonation of ferrocene initially occurs *endo*. The detailed structure of the protonation product has not been unambiguously established, but calculations imply that the proton is associated at least as much with the cyclopentadienyl carbons as with the iron.[347]

$$(3.88)$$

Ferrocenes greatly stabilize positive charge α to a cyclopentadienyl ring.[348] This stabilization of positive charge allows the ferrocenylmethyl acetate in Equation 3.89 to serve as a precursor for stereoselective exchanges of the acetate by phosphines at the pseudo-benzylic position. This reaction generates an important set of ligands for asymmetric catalysis and cross-coupling. The positive charge can be stabilized by the resonance form shown in the equation.[349–351]

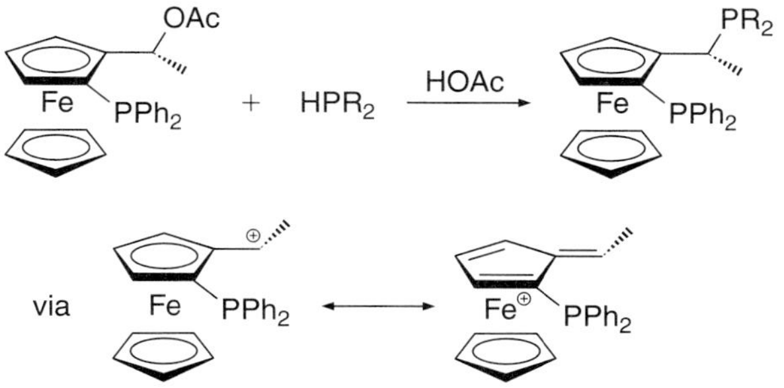

(3.89)

Cyclopentadienyl complexes also react with strong bases to undergo net deprotonation of the Cp ring. Ferrocene can be lithiated on a single ring with modest selectivity or on separate rings (Equation 3.90).[352] The ortholithiation of substituted ferrocenes occurs with high diastereoselectivity when the ring is attached to certain chiral substituents, such as aminoalkyl groups and oxazolines (Equation 3.91).[353] This diastereoselective ortholithiation also occurs with sulfoxides, and the enantiopure sulfoxide in Equation 3.92 allows the preparation of enantiopure 1,2-disubstituted ferrocenes.[354,355]

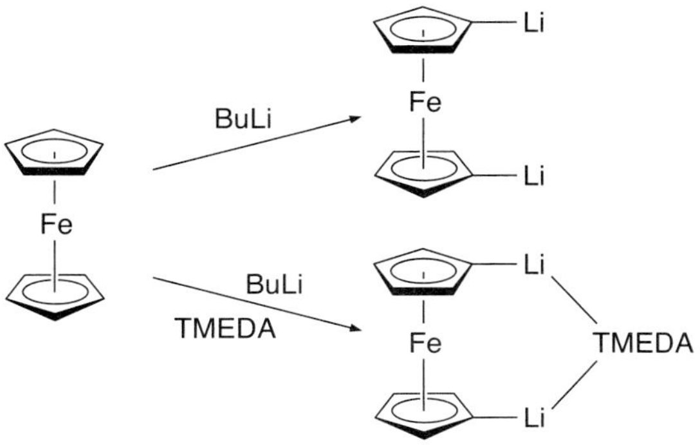

(3.90)

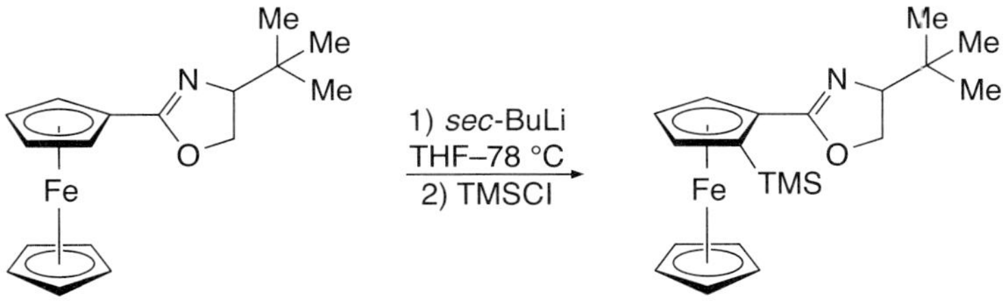

(3.91)

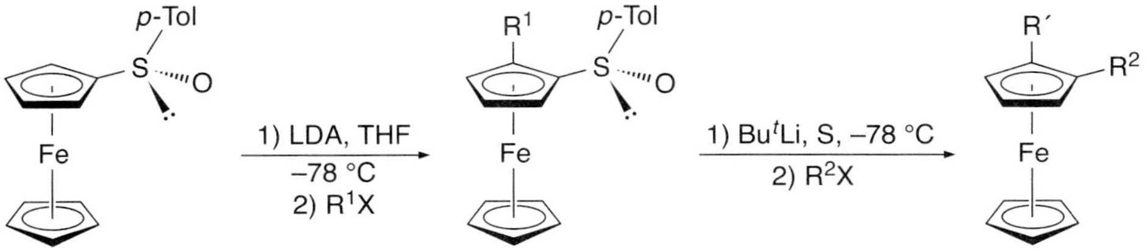

(3.92)

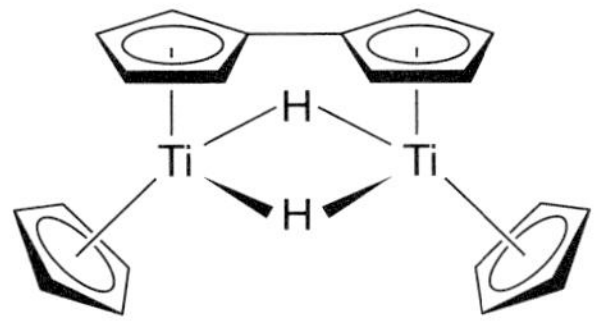

Figure 3.34.
Dimer resulting from the cleavage of a cyclopentadienyl C–H bond.

Cp complexes and their derivatives also undergo insertion of the metal into a ligand C–H bond. For example, the dimer in Figure 3.34 forms from the reduction of Cp_2TiCl_2.[356,357] Cleavage of the methyl C–H bonds of coordinated Cp* also occurs frequently. An example of the activation of the Cp* methyl groups to form dinuclear products is shown in Equation 3.93,[358] and an example of the intramolecular activation of a Cp* ligand is shown in Equation 3.94.[359]

$$\text{(3.93)}$$

$$\text{Cp*WCl}_4 \; + \; 3.1 \; \text{Na/Hg(0.8\% W/W)} \quad \xrightarrow[\substack{C_6H_6 \\ -3 \text{ NaCl}}]{65\,°C,\ 16\ h} \quad \text{(3.94)}$$

Bioorganometallic chemistry of ferrocene has also been developed. This chemistry is beyond the scope of this text, but the reader is referred to a recent review.[360] Considerable earlier literature is cited therein.

3.8. Hydride Ligands (Written by Prof. Jack R. Norton)

Hydrogen atoms bound to transition metals are called "hydride ligands," and hydrides are among the most common ligands in transition metal organometallic chemistry. Hydride ligands react with many organic substrates (e.g., alkenes, as in Equation 3.95, and alkynes, as in Equation 3.96), and serve as precursors for a variety of organic ligands. Although these ligands are called hydrides, they need not have the properties expected of an H^- group. In fact, "hydride" complexes are often acidic, as described in Section 3.8.3. Although most hydride complexes contain ancillary ligands, a few complexes are known in which the entire coordination sphere consists of hydride ligands. Numerous books and reviews have been published on hydride complexes.[361–378]

$$M–H \; + \; {>}C{=}C{<} \quad \longrightarrow \quad M \cdots C \cdots C \cdots H \qquad \text{(3.95)}$$

$$M–H \; + \; –C{\equiv}C– \quad \longrightarrow \quad {>}C{=}C{<} \qquad \text{(3.96)}$$

3.8.1. Structural Features

3.8.1.1. Terminal Hydrides

Because of the small amount of electron density associated with a hydride ligand, the positions of hydride ligands are difficult to determine accurately by X-ray diffraction. The most accurate structural data on hydride complexes have been obtained by neutron diffraction, although electron diffraction has also been used to gain structural data on volatile hydride complexes.[365]

In general, hydride ligands are "stereochemically active;" that is, they occupy a normal coordination site. The structure of $HMn(CO)_5$ reported by Ibers and co-workers in 1969 is often cited as evidence that a hydride ligand occupies a coordination site.[379] Although the carbonyl ligands cis to the hydride bend slightly toward it [the CO(axial)–CO(equatorial) angle is about 95°], the structure as a whole is clearly pseudo-octahedral (Figure 3.35). The structures of many complexes of the general formula $M(PR_3)_xH_y$, including $H_5Re(PMePh_2)_3$ (Figure 3.35),[363,369,374,380] further illustrate the stereochemical activity of hydride ligands. There are complexes, however, such as $HRh(PPh_3)_4$ (Figure 3.35),[381] in which the other ligands are sufficiently bulky that the "coordination site" of the hydride is masked.

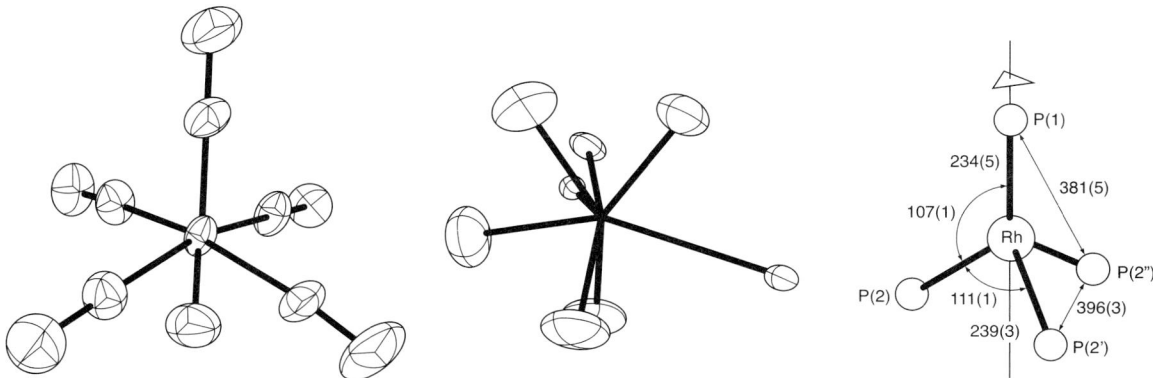

Figure 3.35.
Structures of $HMn(CO)_5$, $H_5Re(PMePh_2)_3$, and $HRh(PPh_3)_4$. [From LaPlaca, S. J.; Hamilton, W. C.; Ibers, J. A.; Davidson, A. *Inorg. Chem.* **1969**, *8*, 1928; Emge, T. J.; Koetzle, T. F.; Bruno, J. W.; Caulton, K. G. *Inorg. Chem.* **1984**, *23*, 4012; and Baker, R. W.; Pauling, P. *J. Chem. Soc. D* **1969**, 1435. Reproduced with permission.]

3.8.1.2. Bridging Hydrides

In addition to serving as terminal ligands on a single metal, hydrides can bridge two or three metals (Figure 3.36). As many as four hydride bridges [in $H_8Re_2(PEt_2Ph)_4$[382]] have been observed between the same pair of metal centers. Hydrides can even occupy interstitial sites, in which they fit inside a cage of metal atoms. The Ru_6 complex on the right in Figure 3.36 is one example.[365,383,384]

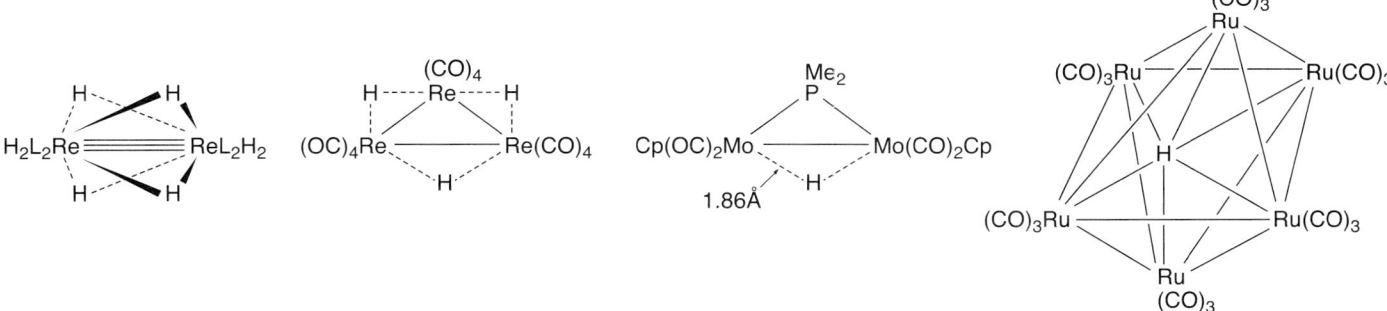

Figure 3.36. Some polynuclear complexes with bridging hydrides.

Hydrides associated with two or more metals are even more difficult than terminal hydrides to locate by X-ray diffraction. For example, the hydrides in $(Cp^*)_2Co_2(\mu\text{-H})_3$ were missed entirely in the first X-ray investigation of its structure.[385] The presence of three hydrides was confirmed by the observation of a parent ion at m/z 391 in the mass spectrum.[386]

The M–H–M angle in an $M\mu(\mu\text{-H})M$ linkage is always less than 180°.[383,387] This bent geometry implies a closed three-center two-electron bond like that in diborane, with an M–M distance that is longer than that in a simple two-center two-electron metal–metal bond. For example, the Cr–Cr distance increases upon protonation of $[Cr_2(CO)_{10}]^{2-}$ (Figure 3.37).

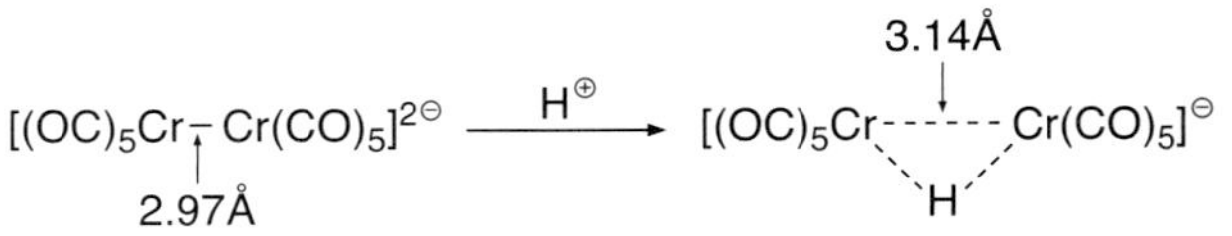

$$[(OC)_5Cr\!-\!Cr(CO)_5]^{2\ominus} \xrightarrow{H^{\oplus}} [(OC)_5Cr\cdots\cdots Cr(CO)_5]^{\ominus}$$

2.97Å

3.14Å

H

Figure 3.37.
Change in Cr–Cr distance upon protonation.

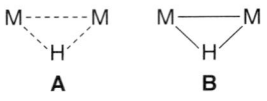

Figure 3.38.
Line description of the two-electron three-center **C** bonding of bridging hydrides.

These three-center two-electron bonds are often illustrated by the use of three dotted lines (**A**), as shown in Figure 3.38, or with three solid lines (**B**) for simplicity. For electron counting, the half-arrow notation **C** for a bridging hydride discussed in Chapter 1 should be used.

3.8.1.3. Spectroscopic Properties

The term "hydride" complex arose because early researchers noted that the ^1H NMR chemical shifts of these ligands were frequently upfield of TMS, and incorrectly attributed these upfield shifts to hydridic character of the ligand. For example, the ^1H NMR resonance of $HCo(CO)_4$ occurs at δ −10.7, and the ^1H NMR resonance of $HMn(CO)_5$ occurs at δ −7.5, despite the fact that both "hydrides" are acidic. Instead, the upfield shifts arise from the mixing of low-lying electronic excited states into the ground state when the magnetic field is applied.[388] Thus, these upfield shifts are more often observed for compounds with d^1–d^9 electron configurations, which generally possess low-lying excited states. The ^1H NMR chemical shifts of hydride ligands bound to d^0 and d^{10} metal centers are not, as a rule, upfield of TMS. For example, the ^1H NMR chemical shift of the hydride in $(C_5Me_5)_2ZrH_2$ is δ 7.46,[389] that of the hydride in $[HCuP(p\text{-tolyl})_3]_6$ is δ 3.50,[390] and that of the hydride in $[HCu(carbene)]_2$ [in which "carbene" = 1,3-bis(2,6-diisopropylphenyl)imidazol-2-ylidene] is δ 2.67.[391] The ^1H NMR resonances of bridging hydrides usually lie at higher fields than those of terminal hydrides,[364,367] but there are exceptions to this rule.

Infrared spectroscopy is also useful for identifying the presence of a hydride ligand. Terminal hydride ligands typically vibrate at 2200–1600 cm^{-1}, although the intensities of these bands can vary from weak to medium.[364,367,392] Hydrides that bridge two metals give rise to IR bands that are broader, weaker, and at lower energy (1600–800 cm^{-1}) than the corresponding bands of terminal hydrides. The M–H stretching vibrations of both terminal and bridging hydrides are usually more intense in the Raman spectrum.

3.8.2. Synthesis of Metal–Hydride Complexes

3.8.2.1. From Hydrogen

Some hydride complexes (see Equations 3.97–3.99) can be prepared directly from hydrogen gas, by oxidative addition (see Chapters 6 and 7). Formation of dihydride complexes by the oxidative addition of H_2 to coordinatively unsaturated complexes (Equation 3.97)[393] is usually faster than their formation by oxidative addition to coordinatively saturated complexes (Equation 3.98)[394] because ligand dissociation is required in the latter case. In some cases (Equation 3.99),[395] the open coordination site for the oxidative addition of H_2 is generated by the reduction of metal halides.

$$IrCl(PPh_3)_3 \;+\; H_2 \xrightarrow[\text{1 atm}]{\text{25 °C}} H_2IrCl(PPh_3)_3 \tag{3.97}$$

$$Os(CO)_4(PPh_3) \;+\; H_2 \xrightarrow[\text{100 °C}]{\text{80 atm}} H_2Os(CO)_3(PPh_3) \;+\; CO \tag{3.98}$$

$$W(PMe_3)_3Cl_4 \xrightarrow[\substack{-78\ °C\ THF \\ +PMe_3}]{H_2,\ Na/Hg} WH_4(PMe_3)_4 \qquad (3.99)$$

Some hydride complexes (Equation 3.100) are produced by the heterolytic cleavage of H_2, a reaction that probably proceeds through the deprotonation of an intermediate dihydrogen complex.[369] Other hydride complexes (Equation 3.101) are produced by the hydrogenolysis of metal–alkyl bonds.[396]

$$[(COD)Ir(PPh_3)_2]^\oplus \xrightarrow{H_2,\ Et_3N} IrH_3(PPh_3)_2 \qquad (3.100)$$

$$(3.101)$$

The addition of H_2 to $Cp^*_2Ta(H)=CH_2$ occurs, not by path **B** in Equation 3.102 (addition across the Ta=C double bond), but by path **A** [addition to an unsaturated Ta(III) intermediate].[397]

$$(3.102)$$

Some important hydride complexes are available by a reaction that appears to be the hydrogenolysis of a metal–metal single bond (Equation 3.103). However, this reaction occurs only when the metal–metal bond is weak, and probably proceeds via the third-order reaction with H_2 of the two metalloradicals [i.e., $\bullet Co(CO)_4$] formed when the metal–metal bond dissociates.[398,399] More recent observations of the reaction of monomeric metalloradicals with H_2 are shown in Equations 3.104 and 3.105.[400,401]

$$CO_2(CO)_8 + H_2 \longrightarrow 2\ HCo(CO)_4 \qquad (3.103)$$
$$\text{via} \searrow 2\ \bullet Co(CO)_4 \nearrow H_2$$

$$2 \; (TMP)Rh\cdot \; + \; H_2 \longrightarrow 2 \; HRh(TMP)$$

(3.104)

TMP = Tetramesitylporphyrinato

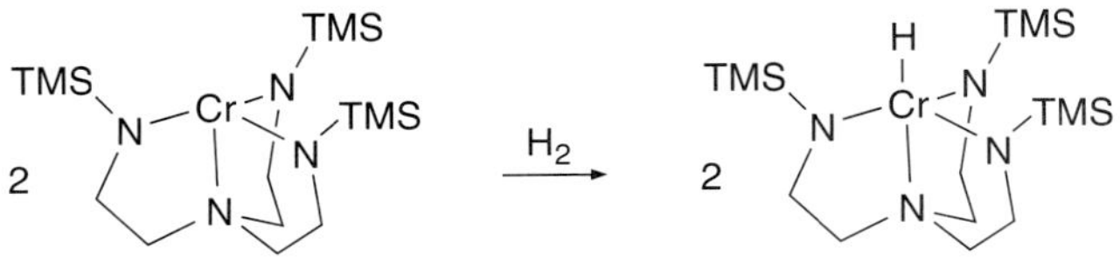

(3.105)

A dinuclear complex with a single hydride ligand per metal can be prepared by the addition of hydrogen to a metal–metal double bond (Equation 3.106),[402] although the reaction is undoubtedly not concerted.

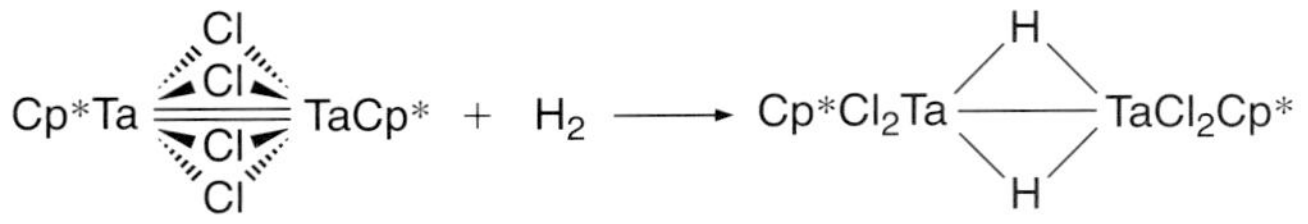

(3.106)

3.8.2.2. By Protonation

Acids often generate hydride complexes by oxidative addition. The starting material in Equation 3.107, $Mn(CO)_5{}^-$, is Mn(–I), d^8, whereas the product, $HMn(CO)_5$, is Mn(I), d^6.

$$Mn(CO)_5{}^\ominus \xrightarrow{\; H^\oplus \;} HMn(CO)_5$$

(3.107)

When the starting materal is cationic and the acid is HX, X$^-$ may coordinate before protonation. The mechanism shown in Equation 3.108 was deduced from kinetic data[403] and confirmed by isolation of the neutral intermediate.[404]

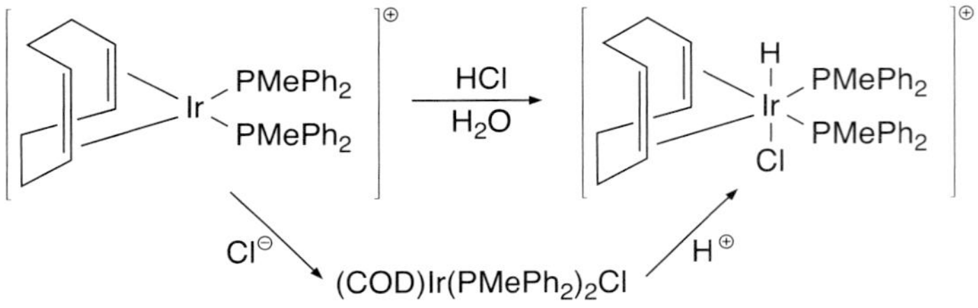

(3.108)

Other hydride complexes are generated by protonation without initial X$^-$ coordination. Interstitial hydrides, such as the $[HRu_6(CO)_{18}]^-$[405] in Figure 3.36, are generally prepared in this way.

Some metal–hydrides react with acidic reagents. For example, the $HMn(CO)_5$ formed by the protonation of $Mn(CO)_5{}^-$ evolves H_2 when dissolved in pure CF_3SO_3H (in Equation 3.109).[406] Hydride ligands often undergo protonation faster than the metal center.[407]

$$Mn(CO)_5 \xrightarrow[\text{THF}]{CF_3SO_3H} HMn(CO)_5 \xrightarrow{CF_3SO_3H} [Mn(CO)_5][O_3SCF_3]$$
$$+ \; H_2$$

(3.109)

One reagent that is effective for the clean monoprotonation of reactive anions is *tert*-butyl chloride (Equation 3.110),[408] which is not acidic but undergoes a facile E2 elimination when attacked by strong bases.

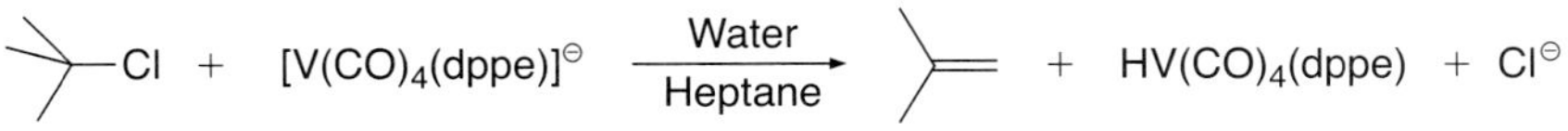

(3.110)

Many neutral complexes also react with acids (Equations 3.111 and 3.112) to yield cationic hydrides. In Equation 3.111,[409] the metal is formally oxidized from Os(0), d^8, to Os(II), d^6. Ammonium hexafluorophosphate (Equation 3.112) is a convenient reagent for the addition of one proton to a basic metal center {stronger acids result in loss of H_2 and formation of the dication, $[CpCo(PMe_3)_3]^{2+}$}.[410]

$$Os(CO)_3(PPh_3)_2 \; + \; HClO_4 \longrightarrow \; [Os(CO)_3(PPh_3)_2H] \, [ClO_4] \qquad (3.111)$$

$$CpCo(PMe_3)_2 \xrightarrow{NH_4PF_6} \; [CpCo(PMe_3)_2H] \, [PF_6] \qquad (3.112)$$

The kinetically favored site of protonation is frequently a basic oxygen or nitrogen atom rather than a metal (see the discussion of hydride kinetic acidities below). In Equation 3.113, $Ru_3(CO)_{10}NO^-$ undergoes kinetic protonation on the nitrosyl oxygen, although the Ru–Ru bond is the thermodynamically favored site of protonation.[411] In Equation 3.114, the kinetic site of protonation is the piperidine nitrogen, although the thermodynamic site is the iridium.[412]

$$(3.113)$$

$$(3.114)$$

3.8.2.3. From Main Group Hydrides

Hydride complexes, particularly those of the early transition metals, are often prepared from main group hydrides (Equations 3.115,[413] 3.116,[414] 3.117,[415] 3.118,[416] and 3.119[417]). Borohydride[418] and aluminohydride[419] complexes are known and have been observed as intermediates in a few reactions that form transition metal hydrides.

$$WCl_6 \; + \; NaBH_4 \; + \; NaCp \longrightarrow \; Cp_2WH_2 \qquad (3.115)$$

$$(3.116)$$

$$\text{(3.117)}$$

$$\text{(3.118)}$$

$$(2-\text{MeSBI})Y-N(\text{SiHMe}_2)_2 \xrightarrow{\text{AlHBu}^i_2} [(2-\text{MeSBI})YH]_2 \qquad \text{(3.119)}$$

$$2-\text{MeSBI} = \text{Me}_2\text{Si}(2-\text{Me-Indenyl})_2$$

3.8.2.4. From Other Reagents

A convenient and widely used method for the preparation of late transition metal hydride complexes involves the generation of coordinated alkoxide from base and an alcohol, followed by elimination of an aldehyde or ketone. In one case,[420] the alkoxide intermediate has been observed, while use of the labeled alcohol in Equation 3.120[421] has confirmed that the α-hydrogen becomes the hydride ligand. This reaction has even been used to produce a Co(III) hydride from a Co(III) aquo complex (Equation 3.121).[422] Only primary and secondary alcohols can be used.

$$K_2\text{IrCl}_6 \xrightarrow[\text{PPh}_3]{\text{CH}_3\text{CD}_2\text{OH/H}_2\text{O}} \text{IrDCl}_2(\text{PPh}_3)_3$$

$$\text{(3.120)}$$

$$H_3C-\underset{D}{\overset{D}{C}}-O-Ir(III) \longrightarrow H_3C-\overset{O}{\overset{\|}{C}}D + D-Ir(III)$$

$$[\text{Co(tmen)}_2(\text{OH}_2)_2]^{3\oplus} + \text{CH}_3\text{CH}_2\text{OH} \longrightarrow [\text{Co(tmen)}_2(\text{OH}_2)\text{H}]^{2\oplus} + \text{CH}_3\text{CHO} + \text{H}_3\text{O}^{\oplus} \qquad \text{(3.121)}$$

Other ligands from which hydrides are often generated are formate (Equation 3.122),[423] hydroxycarbonyl (Equation 3.123),[424] and t-butyl (Equation 3.124).[425] The hydroxycarbonyl ligand is generated by the addition of OH⁻ to a carbonyl ligand; its stability toward decarboxylation varies and may depend upon the availability of a vacant coordination site.[371] Alkyl ligands that contain β-hydrogens readily eliminate olefins to form hydride ligands; a secondary or tertiary organolithium, organoaluminum, or organomagnesium reagent is, like the one in Equation 3.124, most likely to yield a hydride complex.

$$[\text{HCO}_2\text{Cr(CO)}_5]^{\ominus} \xrightarrow{25\,°\text{C}} [\text{HCr(CO)}_5]^{\ominus} + \text{CO}_2 \qquad \text{(3.122)}$$

$$\text{(3.123)}$$

$$\text{Cp}_2\text{ZrCl}_2 + \text{Me}_3\text{CMgCl} \longrightarrow \text{Cp}_2\text{ZrHCl} \qquad \text{(3.124)}$$

3.8.3. Acidities of Hydride Complexes

The pK_a values are now available for many hydride complexes.[370,371,426–429] Extensive tables have been compiled recently by Bullock[430] and by Tilset.[431] The rate of proton transfer to and from transition metals is rather slow (see below), so it is often possible to detect separate NMR signals for M–H and M⁻, and thus to determine the position of proton transfer equilibria between hydride complexes (M–H) and bases (B), or metal bases (M⁻) and organic acids (HA). The pK_a values in Table 3.1 have been obtained in acetonitrile, an excellent solvent for acid–base chemistry because it solvates cations well enough to minimize ion pair formation; it is both a weak acid and a weak base, with a very low autoprotolysis constant (ion product).[‡][432,433]

---

‡ The autoprotolysis constant of CH_3CN is at most 10^{-33} and may be as low as 10^{-44}.

Table 3.1. The pK_a values for carbonyl, cyclopentadienyl, and bis(diphosphine) hydrides in acetonitrile, the oxidation potentials (E°) for their conjugate bases, and the bond dissociation energies (BDE) of the hydrides. Many entries are taken from reference 431.

Metal hydride M–H	pK_a(M–H)	E° (M⁻/M)[a] (V) vs. FcH⁺/FcH	BDE (M–H) (kcal mol⁻¹)	References[b]
HV(CO)$_4$(dppe)[c]	17.4	−1.12	57.5	434
CpCr(CO)$_3$H	13.3	−0.69	61	435,436
CpCr(CO)$_2$(PPh$_3$)H	21.8	−1.29	60	437,438
Cp*Cr(CO)$_3$H	17.0	−0.83	63.6 62.3[d]	434,438,439
CpMo(CO)$_3$H	13.9	−0.50[e]	69	435,436
Cp*Mo(CO)$_3$H	17.1	−0.71[e]	69	435,436
TpMo(CO)$_3$H	10.7	−0.52	62	440
CpW(CO)$_3$H	16.1	−0.49[e]	72	435,436
CpW(CO)$_2$(PMe$_3$)H	26.6	−1.23[e]	70	435,436
[CpW(CO)$_2$(PMe$_3$)H$_2$]⁺	5.6			441
TpW(CO)$_3$H	14.4	−0.58	66	440
Mn(CO)$_5$H	14.1	−0.56[e]	68	435,436
Mn(CO)$_4$(PPh$_3$)H	20.4	−0.87[e]	68	435,436
Re(CO)$_5$H	21.1	−0.69[e]	75	435,436
CpFe(CO)$_2$H	19.4	−1.35[e]	57	435,436
Fe(CO)$_4$H$_2$	11.4	−0.40[e]	68	435,436
CpRu(CO)$_2$H	20.2	−1.06[e]	65	435,436
Cp*$_2$OsH⁺	9.9	−0.06	71	442
H$_2$Os$_2$(CO)$_8$	20.4			427
[Rh$_{13}$(CO)$_{24}$H$_3$]²⁻	11.0			443
[Rh$_{13}$(CO)$_{24}$H$_2$]³⁻	16.5			443
HNi(depp)$_2$⁺[c]	23.3	−1.34	55[f]	444
HNi(depe)$_2$⁺[c]	23.8	−1.29	62	445

(continued)

Table 3.1. *continued*

Metal hydride M–H	pK_a (M–H)	$E°$ (M⁻/M)[a] (V) vs. FcH⁺/FcH⁺	BDE (M–H) (kcal mol⁻¹)	References[b]
HPd(depp)$_2$⁺[c]	22.9	−1.22	62[f]	446
HPd(depx)$_2$⁺[c]	19.8	−1.02	62[f]	446
HPd(depPE)$_2$⁺[c]	18.3	−0.92	62[f]	446
HPd(EtXanthphos)$_2$⁺[c]	18.5	−0.94	62[f]	446
HPt(EtXanthphos)$_2$⁺[c]	27.3	−0.97	74[f]	447

[a] $E°$ data are from reversible voltammograms unless otherwise noted.

[b] References given are for the $E°$ and BDE data, which generally cite the source of the pK_a data. References for the pK_a data are given *only* when the pK_a data are not available in the $E°$/BDE references, or when no $E°$/BDE data are available.

[c] The ligands depp = Et$_2$PCH$_2$CH$_2$CH$_2$PEt$_2$, depx = 1,2-(Et$_2$PCH$_2$)$_2$C$_6$H$_4$, depe = Et$_2$PCH$_2$CH$_2$PEt$_2$, dppe = Ph$_2$PCH$_2$CH$_2$PPh$_2$, depPE = bis[2-(diethylphosphino)phenyl]ether, and EtXanthphos = 9,9-dimethyl-4,5-bis(diethylphosphino)xanthene.

[d] Calorimetric measurement from the reaction of Cp*Cr(CO)$_3$• with H$_2$.

[e] $E°$ data from irreversible voltammograms. Kinetic potential shifts were applied before use in Equation 3.126.

[f] A $T\Delta S°$ term of 5 kcal mol⁻¹ has been added to the ΔG(dissociation) data provided in the reference.

The pK_a values in Table 3.1 for hydride complexes in acetonitrile rely on the pK_a values of neutral organic acids HA and protonated organic bases BH⁺.[448] The established pK_a values for neutral acids HA have now been placed on a single scale,[449] spanning 24 orders of magnitude and anchored to the traditional value of 11.00 for picric acid.[450] The established pK_a values for the conjugate acids BH⁺ of neutral bases B have also been placed on a single scale,[451] spanning 28 orders of magnitude and anchored to a revised value of 12.53 for pyridine. The hydride pK_a values in Table 3.1 have *not* been recalculated on the basis of the new scales for HA and BH⁺—that is, the values of the hydride pK_a's in Table 3.1 are those published in the references given. Their *relative* values are reasonably accurate, but there is some uncertainty in their *absolute* values because of uncertainty in the absolute values of the HA and BH⁺ pK_a's.*[428,448,450] Ab initio calculations of hydride pK_a values in CH$_3$CN have recently been published.[452]

In less polar solvents it is impractical to determine pK_a values, either because the solvent is not readily protonated (CH$_2$Cl$_2$) or because the solvent permits extensive ion pair formation and its protonated form is unstable (THF). However, the positions of proton transfer equilibria in such solvents (e.g., Equation 3.125) have been used *to estimate relative aqueous* pK_a's for hydride complexes that are insoluble in (or unstable in) water. Cy$_3$PH⁺, whose aqueous pK_a is 9.7, has often been used as an anchor—that is, the acidities of other R$_3$PH⁺ and hydride complexes have been measured relative to Cy$_3$PH⁺, leading to the "pseudo-aqueous" pK_a values in Table 3.2.[431] Morris has published acidity scales in CH$_2$Cl$_2$[453,454] and in THF.[455,456] However, complications due to ion pairing make such pK_a values less reliable than direct measurements in CH$_3$CN (or water).[457] Morris has linked his THF acidity scale with acidities in DMSO, a solvent that minimizes ion pair formation, but is incompatible with many organometallic complexes.[456]

$$\text{EtPh}_2\text{PH}^\oplus + \text{Fe(CO)}_3(\text{PCy}_3)_2 \xrightleftharpoons[]{\text{CD}_2\text{Cl}_2} \text{EtPh}_2\text{P} + \text{Fe(CO)}_3(\text{PCy}_3)_2\text{H}^\oplus \qquad (3.125)$$

The pK_a of HCo(CO)$_4$ has been reported as 8.4 in CH$_3$CN,[428] but should be considered uncertain, as the bond strength (67 kcal mol⁻¹) that results[435,436] from the use of that pK_a in a thermodynamic cycle is too high to be consistent with the reactivity of HCo(CO)$_4$ in H•

---

* Fundamental to many of the basic measurements is the belief that perchloric is a strong acid in CH$_3$CN. This belief is based on early work (see reference 448 and references therein) with "anhydrous" solutions prepared by adding Ac$_2$O to CH$_3$CN solutions of 70% aqueous perchloric.

Table 3.2. Pseudo-aqueous pK_a values for cationic hydride complexes.

Cationic hydride, MH$^+$	pK_a (MH$^+$)	Reference
Fe(CO)$_3$(PCy$_3$)$_2$H$^+$	4.4	454
Cp*Ru(dppm)H$_2^{+a}$	8.8	458
Cp*Ru(dppm)(η^2-H$_2$)$^{+a}$	9.2	458
trans-CpOs(dppe)H$_2^{+b}$	11.8	459
cis-CpOs(dppe)H$_2^{+b}$	9.9	459
TpRu(PPh$_3$)$_2$(η^2-H$_2$)$^+$	7.6	459
TpOs(PPh$_3$)$_2$(η^2-H$_2$)$^+$	8.9	459
trans-HRu(dppe)$_2$(η^2-H$_2$)$^{+b}$	15.0	460
trans-HOs(dppe)$_2$(η^2-H$_2$)$^{+b}$	13.6	460

^{a}The ligand dppm = Ph$_2$PCH$_2$PPh$_2$.
^{b}The ligand dppe = Ph$_2$PCH$_2$CH$_2$PPh$_2$.

transfer reactions. Klingler and Rathke have determined the H–Co(CO)$_4$ bond to be 59 kcal mol^{-1} as a result of ^{13}C NMR and magnetic susceptibility studies that imply a Co–Co bond strength of 19 kcal mol^{-1} in Co$_2$(CO)$_8$.[461]

The most acidic hydrides in Table 3.2 have pK_a values comparable to that of HCl (which has a pK_a between 8 and 9 in CH$_3$CN),[428] while the least acidic have pK_a values exceeding that of phenol.[448] Electron-donating substituents (i.e., the replacement of a carbonyl by a phosphine or phosphite) predictably increase the pK_a of hydride complexes, as does descending a column of the periodic table (Cr < Mo < W, Fe < Ru < Os, and Mn < Re).

The rates at which protons can be removed from transition metal hydrides (their "kinetic acidities") generally parallel their thermodynamic acidities (pK_a values). However, the removal of a proton from a metal is much slower than the removal of a proton from an electronegative atom like nitrogen or oxygen. The reverse is also true: the protonation of a metal (to form a hydride) is slower than the protonation of a nitrogen or an oxygen; examples are shown in Equations 3.113 and 3.114. The low kinetic acidity of transition metal hydrides is much like that of carbon acids in organic chemistry.[370,371,426,427,462]

One-electron oxidation of transition metal hydride complexes significantly increases their acidity—their pK_a declines by about 20 units.[431] For example, the pK_a of CpW(CO)$_2$(PMe$_3$)H is 26.6 in CH$_3$CN, whereas that of [CpW(CO)$_2$(PMe$_3$)H]$^{+\bullet}$ is only 5.1 in CH$_3$CN.[463]

3.8.4. Strength of M–H Bonds

With a thermodynamic cycle (shown in Scheme 3.7), the strengths of an M–H bond can be deduced from the pK_a of the hydride complex and the oxidation potential (against the SHE in the same solvent) of its conjugate base. The resulting expression is Equation 3.126 in acetonitrile.[435,436]

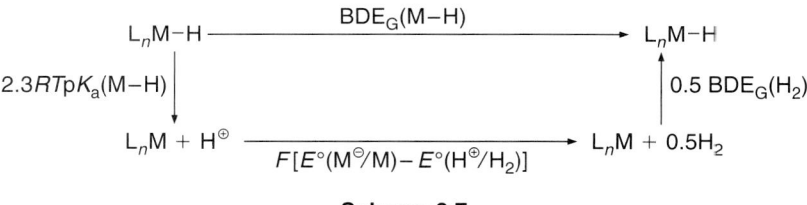

Scheme 3.7

$$BDE = 2.3RTpK_a(M-H, CH_3CN) + FE° (M^{\ominus}/M, solv) + 59.5 \text{ kcal/mol}^{\ominus} \qquad (3.126)$$

Some results from the use of Equation 3.126 were given previously in Table 3.1. Such cycles, used for C–H bonds by Breslow[464–466] and Bordwell,[467] involve thermodynamic parameters (e.g., the enthalpy of solvation of H•) that must be estimated for the solvent being used (in this case CH₃CN), so the *absolute* accuracy of the results is not guaranteed;[†] the *relative* values of the M–H bond strengths calculated from pK_a and $E°$ data should, however, be reliable.

Bond dissociation energies have also been obtained (Table 3.3)[431] from calorimetric measurements of the heat of protonation of neutral complexes with CF_3SO_3H in dichloroethane. These heats correlate linearly with differences in pK_a (even differences in pK_a in another solvent, CH₃CN) and thus can be used along with measured electrochemical potentials to estimate bond dissociation energies (Equation 3.127).

$$BDE(M-H) \text{ in } MH^{\oplus} = -\Delta H_{HM} + 23.06 E_{1/2}(M) + 33.3 \text{ kcal/mol} \qquad (3.127)$$

M–H bond strengths increase from first to second to third row in the periodic table: first < second < third row. The introduction of electron-donating substituents (i.e., the replacement of a carbonyl by a phosphine or phosphite) has little effect on the M–H bond strength, as the effect on the pK_a largely cancels that on the oxidation potential.[429,431]

The sum of the total bond strengths for a bridging hydride can be greater than the strength of a terminal M–H bond. Vites and Fehlner[469] have demonstrated that the interaction energy of the three atoms in the Fe(μ-H)Fe unit is about 83 kcal/mol.

Comparison of M–H bond strengths with M–C bond strengths allows one to predict the energetics of β-hydrogen elimination and thus the stability of alkyl ligands. BDE(M–H)

[†] Potential sources of error in the determination of M–H bond strengths by this method have been reviewed in Eisenberg, D. C.; Norton, J. R. *Isr. J. Chem.* **1991**, *31*, 55, and the relative and absolute uncertainties in such bond strengths have been discussed in Choi, J.; Pulling, M. E.; Smith, D. M.; Norton, J. R. *J. Am. Chem. Soc.* **2008**, *130*, 4250.

Table 3.3. Heats of protonation and reversible oxidation potentials for neutral complexes in 1,2-dichloroethane and BDEs of M–H in the resulting cationic hydride complexes.[468]

Metal hydride, M–H⁺	$-\Delta H_{MH}^+$ (kcal mol⁻¹)	$E°$ (M/M⁺) (V) vs. SCE	BDE (M–H⁺) (kcal mol⁻¹)
$HW(CO)_3(PEtPh_2)_3^+$	17	0.45	61
$HW(CO)_3(PEt_2Ph)_3^+$	18	0.41	61
$HW(CO)_3(PEt_3)_3^+$	25	0.28	65
$HFe(CO)_3(PPh_3)_2^+$	14	0.55	60
$HFe(CO)_3(PMePh_2)_2^+$	18	0.49	62
$HFe(CO)_3(PMe_2Ph)_2^+$	21	0.45	65
$HFe(CO)_3(PMe_3)_2^+$	23	0.41	66
$Cp^*_2RuH^+$	19	0.68	68
$Cp^*_2OsH^+$	27	0.51	72
$CpOs(PPh_3)_2(Br)H^+$	16	0.59	63
$CpOs(PPh_3)_2(Cl)H^+$	20	0.58	66
$CpOs(PPh_3)_2H_2^+$	38	0.13	74

is almost always greater than BDE(M–C).[30,470–472] BDE(M–H) – BDE(M–C) is smaller for actinides than for middle and late transition metals, which "explain[s] why the tendency to β-hydride elimination appears to be least for actinide alkyl compounds."[31]

The strength of M–H bonds has considerable influence on the rate of their H• transfer reactions, although steric effects are substantial. Table 3.4 lists the rate constants, $k_{H•}$, for H• transfer from various hydride complexes to $Ar_3C•$ (Ar = p-$tert$-butylphenyl) at 25 °C.[473] Replacement of Cp by Cp*, which the data in Table 3.1 show to have little effect on the Mo–H bond strength, decreases $k_{H•}$ by a factor of 40.

Table 3.4. H• transfer rate constants k_H [to tris (p-$tert$-butylphenyl) methyl radical at 25°C in toluene].

Metal hydride, M–H	$k_{H•}$ $(M^{-1}s^{-1})$
$HCr(CO)_3Cp^*$	335 (2)
$HRu(CO)_2Cp$	1030 (30)
$HMo(CO)_3Cp^*$	13.9 (5)
$HMo(CO)_3Cp$	514 (2)
$HW(CO)_3Cp$	91 (1)

H• transfer to organic radicals is a common reaction of transition metal hydride complexes. The reaction in Equation 3.128, which is parallel to R_3SnH reductions in organic chemistry, is frequently used for the preparation of stable derivatives and for quantifying the amount of hydride present (by measuring the amount of $CHCl_3$ formed). The mechanism is related to that of the radical chain substitution reactions discussed in Chapter 4.

$$M–H + CCl_4 \longrightarrow M–Cl + CHCl_3 \qquad (3.128)$$

M–H bonds are sufficiently weak that hydrides can also transfer H• to C–C double bonds. Examples of H• acceptors include methyl methacrylate,[474] α-methylstyrene,[475] and the dinuclear organometallic complex in Figure 3.39.[476] The resulting radicals must be stabilized in order for H• transfer to proceed at an appreciable rate.

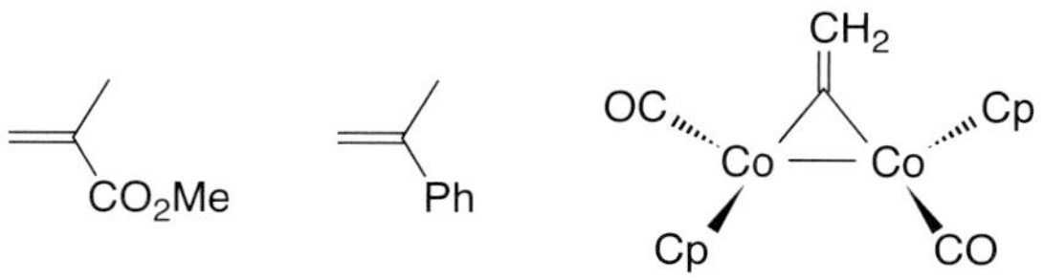

Figure 3.39. Double bonds to which H• can be transferred.

3.8.5. Hydricities

A different thermodynamic cycle (Scheme 3.8 and Equation 3.129) can give the free energy of H^- dissociation from a hydride complex. In some cases the results (Table 3.5)[430,431] have been confirmed by measuring the equilibrium constant for heterolytic cleavage of H_2 by L_nM^{2+} and a base B.

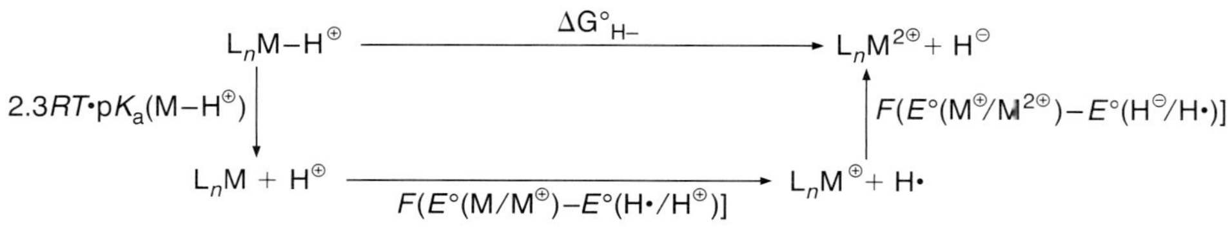

Scheme 3.8

Table 3.5. The pK_a values, conjugate base oxidation potentials, and hydricities of some hydride complexes in acetonitrile.

Metal hydride, M–H$^+$	pK_a (M–H$^+$)	$E°$(M/M$^+$) (V) vs. Fc	$E°$(M$^+$/M^{2+}) (V) vs. Fc	$E°$(M/M^{2+}) (V) vs. Fc	$\Delta G°_{H^-}$ (kcal mol^{-1})	Reference
HCo(dppe)$_2$	38.1	–2.02	–1.56		49	477
HRh(dppb)$_2$^{a}	35.0^b			–2.02	34^b	478
HNi(dppe)$_2$$^+$	14.2	–0.88	–0.70		63	445
HPd(depe)$_2$$^+$	23.2	–1.24^c		–1.48	43	446
HPd(depp)$_2$$^+$	22.9	–1.22		–1.22	55	446
HPd(depx)$_2$$^+$	19.8	–1.02	–0.92		61	446
HPd(depPE)$_2$$^+$	18.3	–0.92	–0.73		66	446
HPd(EtXanthphos)$_2$$^+$	18.5	–0.94	–0.55		71	446
HPt(dppe)$_2$$^+$	22.2			–1.24	53	445
HPt(depe)$_2$$^+$	29.7^d			–1.63	44	444

^{a}The ligand dppb = (Ph$_2$P)$_2$C$_6$H$_4$. The others are as in Table 3.1.
^{b}pK_a value determined from $\Delta G°_{H^-}$ and $E°$ data.
^{c}Estimated from $E°$ vs. pK_a correlation data.
^{d}Originally reported pK_a values^{445} have been revised upward.

$$\Delta G°_{H^-} = 2.3RT\mathrm{p}K_a(\text{M–H}^\oplus, \text{solv}) + F[E°(\text{M/M}^\oplus, \text{solv}) + E°(\text{M}^\oplus/\text{M}^{2\oplus}, \text{solv})] + 79.6 \text{ kcal/mol}^{-1}$$
$$= 2.3RT\mathrm{p}K_a(\text{M–H}^\oplus, \text{solv}) + 2FE°(\text{M/M}^{2\oplus}, \text{solv}) + 79.6 \text{ kcal/mol}^{-1} \qquad (3.129)$$

However, hydride removal creates a vacant coordination site, so $\Delta G°$H$^-$ includes the ΔG of solvation for L$_n$M^{2+}. The interpretation of such data is most straightforward if the coordinatively unsaturated M^{2+} that is generated by abstraction of the hydride does not become solvated—for example, when M^{2+} is a planar 16-electron complex. Many of the results given in Table 3.5 are thus for the removal of H$^-$ from a five-coordinate cation.

Not surprisingly, the free energy of H$^-$ dissociation increases with charge. If we define the "hydricity" as $-\Delta G°_{H^-}$ (the energy liberated by hydride loss), the hydricity increases as the chelating phosphines become better donors—that is, HPt(dppe)$_2$$^+$ < HPt(depe)$_2$$^+$. The hydricity also increases as the size of the chelate ring decreases, so HPt(depp)$_2$$^+$ < HPd(depe)$_2$$^+$. Larger chelate rings raise the energy of the 16-electron product M(P–P)$_2$$^{2+}$ by distorting it from a stable planar geometry toward a tetrahedral one.

The thermodynamic hydricity of an M–H bond is related to the ionicity of that bond, which can be calculated from the ^{2}D quadrupole coupling constant (available from the ^{2}D NMR spectrum of the M–D analog).[416,479] Such data can be compared to *rate constants* for H$^-$ transfer, or *kinetic* hydricities. The rate constants k_{H^-} for transfer of the hydride in a series of complexes to trityl cation in CH$_2$Cl$_2$ (Equation 3.130 and Table 3.6),[430,480] and from a series of CpRu(P–P)H complexes to the iminium cation in Equation 3.131 (Table 3.7) have been measured.[481]

$$\text{Ph}_3\text{C}^\oplus \text{ BF}_4^\ominus + \text{MH} \xrightarrow[\text{CH}_2\text{Cl}_2]{k_{H^-}} \text{Ph}_3\text{C–H} + \text{M–F–BF}_3 \qquad (3.130)$$

Table 3.6. Rate constants (CH_2Cl_2, 25 °C) for H^- transfer from various hydride complexes to $[Ph_3C]BF_4$.[431,433]

Metal hydride, M–H	k_{H^-} (M^{-1}s^{-1})
$(CO)_5MnH$	5.0×10^1
$Cp^*(CO)_3CrH$	5.7×10^1
$Cp(CO)_3WH$	7.6×10^1
$HSiEt_3$	1.5×10^2
$Cp(CO)_3MoH$	3.8×10^2
$(CO)_5ReH$	2.0×10^3
$Cp^*(CO)_3MoH$	6.5×10^3
$Cp(NO)_2WH$	1.9×10^4
$trans$-$Cp(CO)_2(PPh_3)MoH$	5.7×10^5
$trans$-$Cp(CO)_2(PMe_3)MoH$	4.6×10^6

$$CpRu(P-P)H + \text{[Iminium cation]} \xrightarrow[CD_2Cl_2]{\substack{k_{H-}\\ \text{4 equiv. } CH_3CN}} [CpRu(P-P)(CH_3CN)]BF_4 + \text{[product]} \quad (3.131)$$

Iminium cation

Table 3.7. Rate constants (CD_2Cl_2, 300 K) for hydride transfer from CpRu(P–P)H to the iminium cation in Equation 3.131.[432]

Chelating ligand (P–P)	k_{H^-} (M^{-1}s^{-1})
dppm	6.4×10^{-1}
dppe	2.0×10^{-2}
dppp	2.8×10^{-3}
dppb	No reaction ($< 4 \times 10^{-6}$)

With each acceptor the "kinetic hydricities" cover a range of about 10^5. The data in Table 3.6 show that many hydride complexes are better H^- donors than Et_3SiH, that second- and third-row hydride complexes are faster H^- donors than first-row ones, and that electronic factors are more important than steric factors (the substitution of Cp^* for Cp or PR_3 for a CO increases the rate). The data in Table 3.7 show that the rate of H^- transfer from CpRu(P–P)H increases as the size of the chelate ring decreases. (The effect is similar to that of chelate ring size on the *thermodynamic* hydricity of the five-coordinate hydrides in Table 3.5.) Smaller rings make CpRu(P–P)$^+$ more pyramidal and raise the energy of its vacant hydride acceptor orbital, making hydride donation from CpRu(P–P)H more favorable.[481]

The hydride complexes containing metals on the left of the periodic table (i.e., those of Zr, Ti, and Nb) have the most "hydridic" character, as expected by the electropositive character of these metals, but the difference has proven difficult to quantify.[434,435] The relative rates at which such hydrides react with acetone[482] or methanol[483] have been offered as evidence. However, it is possible that acetone and methanol react only after coordination. In this case the observed rates may really be the rates of ligand dissociation or coordination.[483]

The *M–H bond* behaves as a nucleophile in a number of reactions related to hydride transfer, such as Equations 3.132[484] and 3.133.[366] However, the range of nucleophilicities observed in these reactions is not very large (about 10^2), and steric effects are much more important than in true H⁻ transfer reactions; for example, CpW(CO)$_2$(PMe$_3$)H is slightly less nucleophilic than CpW(CO)$_3$H in Equation 3.132.

Sometimes it is difficult to distinguish an H⁻ transfer from a single-electron transfer followed by an H• transfer. There is clear evidence [i.e., broadening of the ^{1}H NMR resonances of Cp*Ru(dppf)H, in which dppf = 1,1'-bis(diphenylphosphino)ferrocene, by electron exchange with the corresponding radical cation] for electron transfer in the reaction of Cp*Ru(dppf)H with the phenyl-substituted aziridinium cation in Equation 3.134.[485]

$$[EtC(O)](CH_3CN)Re(CO)_4 \ + \ HRe(CO)_5 \longrightarrow EtCHO \ + \ (CH_3CN)Re_2(CO)_9 \quad (3.132)$$

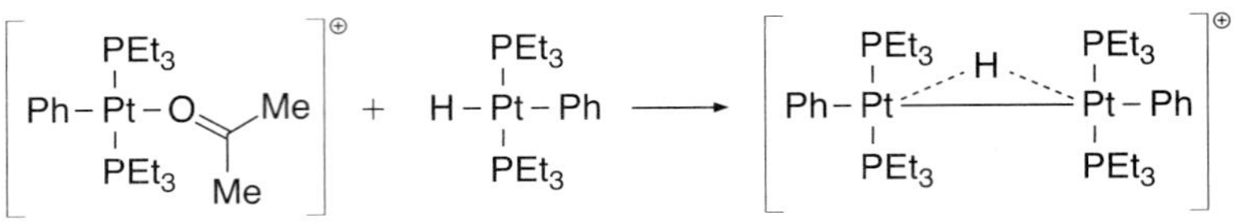

$$(3.133)$$

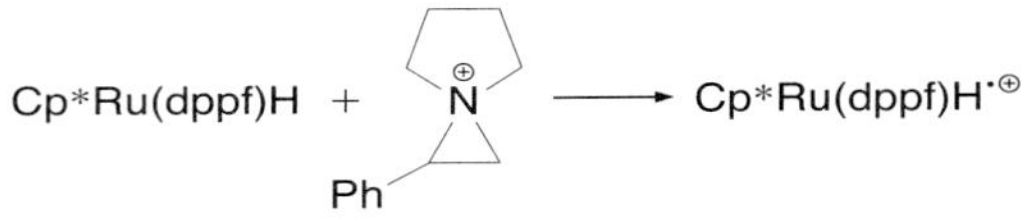

$$(3.134)$$

3.8.6. Hydrogen Bonding[486–489]

In view of the acidity of hydride complexes, one might expect them to form hydrogen-bonded intermediates during proton-transfer reactions. However, it is relatively uncommon for *neutral* hydride complexes to form hydrogen bonds M–H•••A because most M–H bonds are polarized as M(δ+)–H(δ–), not as M(δ–)–H(δ+). (The need to reverse the observed polarization during deprotonation is a major cause of the low kinetic acidity of transition metal hydrides, mentioned previously.) The first M–H•••A hydrogen bond from a neutral hydride has just been reported: CpM(CO)$_3$H (M = Mo and W) serves as a hydrogen bond donor to (octyl)$_3$P=O and even to pyridine, apparently because there is M(δ–)–H(δ+) polarization in its M–H bond.[490]

It is more common for *cationic* hydride complexes to serve as hydrogen bond donors, presumably because the positive charge polarizes the M–H bond away from the hydride ligand. Examples are (Cp*)$_2$OsH⁺ and WH$_3$(dppe)$_2$⁺ in Figure 3.40.[306,491]

Figure 3.40.
Examples of metal–hydride ligands as hydrogen bond donors.

Transition metals, and even hydride ligands, can also serve as hydrogen bond *acceptors*. A transition metal that serves as an acceptor must have a filled d-orbital that is suitably oriented, such as the d_{z^2} orbital in Figure 3.41.[492]

Hydride ligands serve as acceptors when approached from the side by acids HA, giving the nonlinear geometry in Figure 3.42.[493]

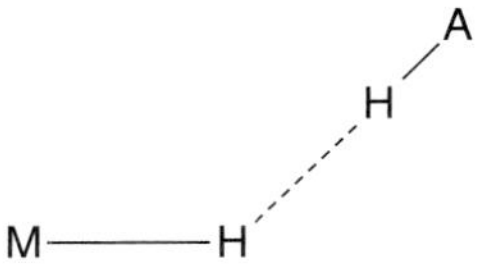

Examples of this interaction, sometimes called a "dihydrogen bond," are shown in Figure 3.43.[493,494]

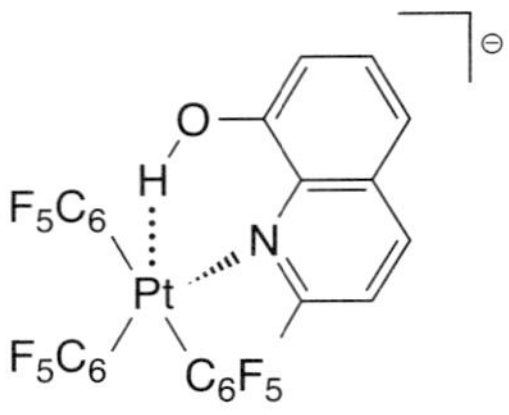

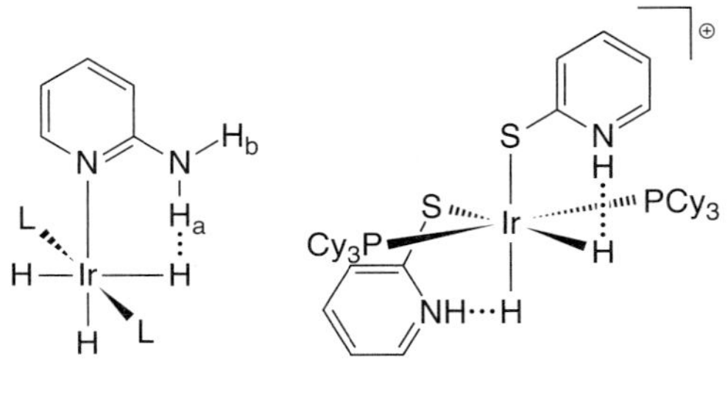

Figure 3.41.
Example of a transition metal acting as a hydrogen bond acceptor.

Figure 3.42.
General depiction of hydrogen bonding to a metal hydride.

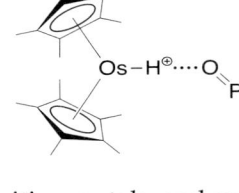

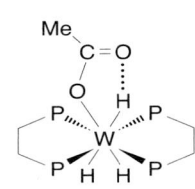

Figure 3.43. Examples of compounds with hydrogen bonds to metal hydrides.

References and Notes

1. Pope, W. J.; Gibson, C. S. *J. Chem. Soc., Trans.* **1907**, *91*, 2061.
2. Pope, W. J.; Peachey, S. J. *J. Chem. Soc., Trans.* **1909**, *95*, 571.
3. Cotton, F. A. *Chem. Rev.* **1955**, *55*, 551.
4. Gibson, C. S.; Simonsen, J. L. *J. Chem. Soc.* **1930**, 2531.
5. Brain, F. H.; Gibson, C. S. *J. Chem. Soc.* **1939**, 762.
6. Foss, M. E.; Gibson, C. S. *J. Chem. Soc.* **1949**, 3063.
7. Krause, E.; von Grosse, A. *Die Chemie der Metall-Organischen Verbindungen*; Verlag von Gebrüder Borntraeger: Berlin, 1937.
8. Kealy, T. J.; Pauson, P. L. *Nature* **1951**, *168*, 1039.
9. Adams, R. D. *J. Organomet. Chem.* **2001**, *637*, 1.
10. Piper, T. S.; Wilkinson, G. *Chem. Ind.* **1955**, 1296.
11. Piper, T. S.; Lemal, D.; Wilkinson, G. *Naturwissenschaften* **1956**, *43*, 129.
12. Piper, T. S.; Wilkinson, G. *J. Inorg. Nucl. Chem.* **1956**, *3*, 104.
13. Chatt, J.; Shaw, B. L. *J. Chem. Soc.* **1959**, 705.
14. Berthold, H. J.; Groh, G. *Z. Anorg. Allg. Chem.* **1963**, *319*, 230.
15. Kurras, E.; Otto, J. *J. Organomet. Chem.* **1965**, *4*, 114.
16. Shortland, A. J.; Wilkinson, G. *J. Chem. Soc., Dalton Trans.* **1973**, 872.
17. Galyer, A. L.; Wilkinson, G. *J. Chem. Soc., Dalton Trans.* **1976**, 2235.
18. Kleinhenz, S.; Pfennig, V.; Seppelt, K. *Chem. Eur. J.* **1998**, *4*, 1687.
19. Davidson, P. J.; Lappert, M. F.; Pearce, R. *Acc. Chem. Res.* **1974**, *7*, 209.
20. Davidson, P. J.; Lappert, M. F.; Pearce, R. *Chem. Rev.* **1976**, *76*, 219.
21. Eisch, J. J.; Adeosun, A. A.; Dutta, S.; Fregene, P. O. *Eur. J. Org. Chem.* **2005**, 2657.
22. Simoes, J. A. M.; Beauchamp, J. L. *Chem. Rev.* **1990**, *90*, 629.
23. *Bonding Energetics in Organometallic Compounds*; Marks, T. J., Ed.; American Chemical Society: Washington, DC, 1990; Vol. 428.
24. Blau, R. J.; Espenson, J. H.; Bakac, A. *Inorg. Chem.* **1984**, *23*, 3526.
25. Kirker, G. W.; Bakac, A.; Espenson, J. H. *J. Am. Chem. Soc.* **1982**, *104*, 1249.
26. Nappa, M. J.; Santi, R.; Halpern, J. *Organometallics* **1985**, *4*, 34.
27. Simoes, J. A. M.; Schultz, J. C.; Beauchamp, J. L. *Organometallics* **1985**, *4*, 1238.
28. Schock, L. E.; Seyam, A. M.; Sabat, M.; Marks, T. J. *Polyhedron* **1988**, *7*, 1517.
29. Siegbahn, P. E. M. *J. Phys. Chem.* **1995**, *99*, 12723.
30. Ziegler, T.; Tschinke, V. In *Bonding Energies in Organometallic Compounds*; Marks, T. J., Ed.; American Chemical Society: Washington, DC, 1990; Vol. 428, p 279.
31. Jemine, X.; Goffart, J.; Berthet, J. C.; Ephritikhine, M. *J. Chem. Soc., Dalton Trans.* **1992**, 2439.
32. Weydert, M.; Brennan, J. G.; Andersen, R. A.; Bergman, R. G. *Organometallics* **1995**, *14*, 3942.
33. See reference 34, and references therein.
34. Andruniow, T.; Zgierski, M. Z.; Kozlowski, P. M. *J. Am. Chem. Soc.* **2001**, *123*, 2679.
35. Halpern, J. *Acc. Chem. Res.* **1982**, *15*, 238.
36. Brescianipahor, N.; Forcolin, M.; Marzilli, L. G.; Randaccio, L.; Summers, M. F.; Toscano, P. J. *Coord. Chem. Rev.* **1985**, *63*, 1.
37. Brown, K. L.; Evans, D. R. *Inorg. Chem.* **1994**, *33*, 6380.
38. Waddington, M. D.; Finke, R. G. *J. Am. Chem. Soc.* **1993**, *115*, 4629.
39. Cui, W. H.; Wayland, B. B. *J. Am. Chem. Soc.* **2004**, *126*, 8266.
40. Wakefield, B. J. *The Chemistry of Organolithium Compounds*; Pergamon Press: New York, 1974.
41. Wardell, J. L. In *Comprehensive Organometallic Chemistry*; Wilkinson, G., Ed.; Pergamon Press: New York, 1982; Vol. 1.
42. Lindsell, W. E. In *Comprehensive Organometallic Chemistry*; Wilkinson, G., Ed.; Pergamon Press: New York, 1982; Vol. 1.
43. *Handbook of Grignard Reagents*; Silverman, G. S.; Rakita, P. E., Eds.; Marcel Dekker, Inc.: New York, 1996.
44. *Organozinc Reagents: A Practical Approach*; Knochel, P.; Jones, P., Eds.; Oxford University Press: New York, 1999.
45. Mole, T.; Jeffery, E. A. *Organoaluminum Compounds*; Elsevier: New York, 1972.
46. Eisch, J. J. In *Comprehensive Organometallic Chemistry*; Wilkinson, G., Ed.; Pergamon Press: New York, 1982; Vol. 1.
47. Eisch, J. J. In *Comprehensive Organometallic Chemistry II: A Review of the Literature 1982–1994*; Housecroft, C. E., Ed.; Pergamon Press: Oxford, U.K., 1995; Vol. 1, p 431.

48. Coates, G. E.; Green, M. L.; Powell, P.; Wade, K. *Principles of Organometallic Chemistry*; Methuen: London, 1971.

49. Negishi, E.-i. *Organometallics in Organic Synthesis*; Wiley: New York, 1980.

50. Carr, D. B.; Schwartz, J. *J. Am. Chem. Soc.* **1979**, *101*, 3521.

51. Wailes, P. C.; Weigold, H.; Bell, A. P. *J. Organomet. Chem.* **1971**, *33*, 181.

52. Darst, K. P.; Lenhert, P. G.; Lukehart, C. M.; Warfield, L. T. *J. Organomet. Chem.* **1980**, *195*, 317.

53. Carbonyl ligands with stretching frequencies below 2000 cm^{-1} are not usually subject to attack by these reagents. However, accurate prediction of the susceptibility of carbonyl ligands to nucleophilic attack requires knowledge of their C–O stretching force constants rather than of v_{CO}.

54. (a) Darensbourg, D.; Darensbourg, M. *Inorg. Chem.* **1970**, *9*, 1691; (b) Darensbourg, M.; Conder, H. L.; Darensbourg, D.; Hasday, C. *J. Am. Chem. Soc.* **1973**, *95*, 5919.

55. Klein, H. F.; Karsch, H. H. *Chem. Ber. Rec.* **1975**, *108*, 944.

56. Gilbert, T. M.; Bergman, R. G. *J. Am. Chem. Soc.* **1985**, *107*, 3502.

57. Schwartz, J.; Labinger, J. A. *Angew. Chem. Int. Ed.* **1976**, *15*, 333.

58. Giering, W. P.; Rosenblum, M. *J. Organomet. Chem.* **1970**, *25*, C71.

59. Kiel, W. A.; Lin, G. Y.; Bodner, G. S.; Gladysz, J. A. *J. Am. Chem. Soc.* **1983**, *105*, 4958.

60. Garratt, S. A.; Hughes, R. P.; Kovacik, I.; Ward, A. J.; Willemsen, S.; Zhang, D. H. *J. Am. Chem. Soc.* **2005**, *127*, 15585, and references therein.

61. Greene, T. R.; Roper, W. R. *J. Organomet. Chem.* **1986**, *299*, 245.

62. Vicente, J.; Gil-Rubio, J.; Guerrero-Leal, J.; Bautista, D. *Organometallics* **2004**, *23*, 4871.

63. Brothers, P. J.; Roper, W. R. *Chem. Rev.* **1988**, *88*, 1293.

64. Bruce, M. I.; Stone, F. G. A. In *Preparative Inorganic Reactions*; Jolly, W. L., Ed.; Wiley-Interscience: Bristol, U.K., 1968; Vol. 4, p 177.

65. Morrison, J. A. In *Advances in Organometallic Chemistry*; Stone, F. G. A., West, R., Eds.; Academic Press: New York, 1993; Vol. 35, p 211.

66. Culkin, D. A.; Hartwig, J. F. *Organometallics* **2004**, *23*, 3398.

67. Kochi, J. K. *Organometallic Mechanisms and Catalysis*; Academic Press: New York, 1978.

68. Davidson, P. J.; Lappert, M. F.; Pearce, R. *Chem. Rev.* **1976**, *76*, 219.

69. Schrock, R. R.; Parshall, G. W. *Chem. Rev.* **1976**, *76*, 243.

70. Holton, J.; Lappert, M. F.; Pearce, R.; Yarrow, P. I. W. *Chem. Rev.* **1983**, *83*, 135.

71. Yamamoto, A. *J. Organomet. Chem.* **1986**, *300*, 347.

72. Moss, J. R. *J. Mol. Catal. A* **1996**, *107*, 169.

73. Groysman, S.; Goldberg, I.; Kol, M.; Goldschmidt, Z. *Organometallics* **2003**, *22*, 3793.

74. Zucchini, U.; Albizzati, E.; Giannini, U. *J. Organomet. Chem.* **1971**, *26*, 357.

75. Leslie, J. P.; Espenson, J. H. *J. Am. Chem. Soc.* **1976**, *98*, 4839.

76. Reger, D. L.; Culbertson, E. C. *J. Am. Chem. Soc.* **1976**, *98*, 2789.

77. Bochmann, M.; Wilkinson, G.; Young, G. B. *J. Chem. Soc., Dalton Trans.* **1980**, 1879.

78. Byrne, E. K.; Theopold, K. H. *J. Am. Chem. Soc.* **1989**, *111*, 3887.

79. McDermott, J. X.; White, J. F.; Whitesides, G. M. *J. Am. Chem. Soc.* **1976**, *98*, 6521.

80. Kruse, W. *J. Organomet. Chem.* **1972**, *42*, C39.

81. Schrock, R. R. *J. Organomet. Chem.* **1976**, *122*, 209.

82. Schrock, R. R.; Fellmann, J. D. *J. Am. Chem. Soc.* **1978**, *100*, 3359.

83. Reger, D. L.; Culbertson, E. C. *Inorg. Chem.* **1977**, *16*, 3104.

84. Reger, D. L.; Garza, D. G.; Lebioda, L. *Organometallics* **1992**, *11*, 4285.

85. Reger, D. L.; Garza, D. G.; Lebioda, L. *Organometallics* **1991**, *10*, 902.

86. Schwartz, J.; Labinger, J. A. *Angew. Chem. Int. Ed.* **1976**, *15*, 333.

87. Chirik, P. J.; Day, M. W.; Labinger, J. A.; Bercaw, J. E. *J. Am. Chem. Soc.* **1999**, *121*, 10308.

88. Reger, D. L.; McElligott, P. J. *J. Organomet. Chem.* **1981**, *216*, C12.

89. Tempel, D. J.; Johnson, L. K.; Huff, R. L.; White, P. S.; Brookhart, M. *J. Am. Chem. Soc.* **2000**, *122*, 6686.

90. Harvey, J. N. *Organometallics* **2001**, *20*, 4887.

91. Buchanan, J. M.; Stryker, J. M.; Bergman, R. G. *J. Am. Chem. Soc.* **1986**, *108*, 1537.

92. Schock, L. E.; Marks, T. J. *J. Am. Chem. Soc.* **1988**, *110*, 7701.

93. Bennett, J. L.; Wolczanski, P. T. *J. Am. Chem. Soc.* **1997**, *119*, 10696.

94. Limmert, M. E.; Roy, A. H.; Hartwig, J. F. *J. Org. Chem.* **2005**, *70*, 9364.

95. Hansen, A. L.; Ebran, J.-P.; Ahlquist, E. M.; Norrby, P.-O.; Skrydstrup, T. *Angew. Chem. Int. Ed.* **2006**, *45*, 3349.

96. Ebran, J.-P.; Hansen, A. L.; Gogsig, T. M.; Skrydstrup, T. *J. Am. Chem. Soc.* **2007**, *129*, 6931.

97. Buchwald, S. L.; Nielsen, R. B. *Chem. Rev.* **1988**, *88*, 1047.

98. Hartwig, J. F.; Bergman, R. G.; Andersen, R. A. *J. Am. Chem. Soc.* **1991**, *113*, 3404.

99. Krug, C.; Hartwig, J. F. *J. Am. Chem. Soc.* **2004**, *126*, 2694.

100. Brune, H. A.; Schmidtberg, G.; Weisemann, C. *J. Organomet. Chem.* **1989**, *371*, 121.
101. Krug, C.; Hartwig, J. F. *J. Am. Chem. Soc.* **2002**, *124*, 1674.
102. Osakada, K.; Onodera, H.; Nishihara, Y. *Organometallics* **2005**, *24*, 190.
103. Clark, A. M.; Rickard, C. E. F.; Roper, W. R.; Wright, L. J. *J. Organomet. Chem.* **2000**, *598*, 262.
104. Sweet, J. R.; Graham, W. A. G. *J. Organomet. Chem.* **1983**, *241*, 45.
105. Krug, C.; Hartwig, J. F. *Organometallics* **2004**, *23*, 4594.
106. Amatore, C.; Carre, E.; Jutand, A.; Mbarki, M. A. *Organometallics* **1995**, *14*, 1818.
107. Osakada, K.; Yamamoto, T. *Coord. Chem. Rev.* **2000**, *198*, 379.
108. Stewart, R. P.; Treichel, P. M. *J. Am. Chem. Soc.* **1970**, *92*, 2710.
109. Janowicz, A. H.; Bergman, R. G. *J. Am. Chem. Soc.* **1983**, *105*, 3929.
110. Thompson, M. E.; Baxter, S. M.; Bulls, A. R.; Burger, B. J.; Nolan, M. C.; Santarsiero, B. D.; Schaefer, W. P.; Bercaw, J. E. *J. Am. Chem. Soc.* **1987**, *109*, 203.
111. Cope, A. C.; Kliegman, J. M.; Friedrich, E. C. *J. Am. Chem. Soc.* **1967**, *89*, 287.
112. Harkins, S. B.; Peters, J. C. *Organometallics* **2002**, *21*, 1753.
113. Albéniz, A. C.; Espinet, P.; Lopez-Cimas, O.; Martin-Ruiz, B. *Chem. Eur. J.* **2004**, *11*, 242.
114. Clot, E.; Mégret, C.; Eisenstein, O.; Perutz, R. N. *J. Am. Chem. Soc.* **2006**, *128*, 8350.
115. Connor, J. A.; Zafaranimoattar, M. T.; Bickerton, J.; Elsaied, N. I.; Suradi, S.; Carson, R.; Altakhin, G.; Skinner, H. A. *Organometallics* **1982**, *1*, 1166.
116. Wick, D. D.; Jones, W. D. *Organometallics* **1999**, *18*, 495.
117. Bennett, J. L.; Wolczanski, P. T. *J. Am. Chem. Soc.* **1997**, *119*, 10696.
118. Stoutland, P. O.; Bergman, R. G.; Nolan, S. P.; Hoff, C. D. *Polyhedron* **1988**, *7*, 1429.
119. Bruno, J. W.; Marks, T. J.; Morss, L. R. *J. Am. Chem. Soc.* **1983**, *105*, 6824.
120. Casey, C. P.; Scheck, D. M. *J. Am. Chem. Soc.* **1980**, *102*, 2723.
121. Lau, M. K.; Zhang, Q. F.; Chim, J. L. C.; Wong, W. T.; Leung, W. H. *Chem. Commun.* **2001**, 1478.
122. Cardin, C. J.; Cardin, D. J.; Kelly, J. M.; Norton, R. J.; Roy, A. *J. Organomet. Chem.* **1977**, *132*, C23.
123. Rajaram, J.; Pearson, R. G.; Ibers, J. A. *J. Am. Chem. Soc.* **1974**, *96*, 2103.
124. Davison, A.; Selegue, J. P. *J. Am. Chem. Soc.* **1980**, *102*, 2455.
125. Reger, D. L.; Belmore, K. A.; Mintz, E.; Charles, N. G.; Griffith, E. A. H.; Amma, E. L. *Organometallics* **1983**, *2*, 101.
126. Frohnapfel, D. S.; Templeton, J. L. *Coord. Chem. Rev.* **2000**, *206*, 199.
127. Davidson, J. L.; Shiralian, M.; Manojlovicmuir, L.; Muir, K. W. *J. Chem. Soc., Dalton Trans.* **1984**, 2167.
128. Allen, S. R.; Beevor, R. G.; Green, M.; Norman, N. C.; Orpen, A. G.; Williams, I. D. *J. Chem. Soc., Dalton Trans.* **1985**, 435.
129. Casey, C. P.; Brady, J. T.; Boller, T. M.; Weinhold, F.; Hayashi, R. K. *J. Am. Chem. Soc.* **1998**, *120*, 12500.
130. Yam, V. W. W. *J. Organomet. Chem.* **2004**, *689*, 1393 and references therein.
131. Szafert, S.; Gladysz, J. A. *Chem. Rev.* **2003**, *103*, 4175.
132. Szafert, S.; Gladysz, J. A. *Chem. Rev.* **2006**, *106*, 1.
133. Long, N. J.; Charlotte K. Williams, D. *Angew. Chem. Int. Ed.* **2003**, *42*, 2586.
134. Yam, V. W. W. *Acc. Chem. Res.* **2002**, *35*, 555.
135. Manna, J.; John, K. D.; Hopkins, M. D. *Adv. Organomet. Chem.* **1995**, *38*, 79.
136. Bulls, A. R.; Bercaw, J. E.; Manriquez, J. M.; Thompson, M. E. *Polyhedron* **1988**, *7*, 1409.
137. *Modern Aldol Reactions*; Mahrwald, R., Ed.; Wiley-VCH: Weinheim, 2004.
138. Hamashima, Y.; Sodeoka, M. *Chem. Rec.* **2004**, *4*, 231.
139. Slough, G. A.; Bergman, R. G.; Heathcock, C. H. *J. Am. Chem. Soc.* **1989**, *111*, 938.
140. Tsuji, J. *Palladium Reagents and Catalysts: New Perspectives for the 21st Century*; Wiley: New York, 2004.
141. Yang, S. C.; Tseng, F. H.; Hung, C. W. *J. Chin. Chem. Soc.* **1999**, *46*, 211.
142. Glorius, F. *Tetrahedron Lett.* **2003**, *44*, 5751.
143. Fagnou, K.; Lautens, M. *Chem. Rev.* **2003**, *103*, 169.
144. Takaya, Y.; Ogasawara, M.; Hayashi, T.; Sakai, M.; Miyaura, N. *J. Am. Chem. Soc.* **1998**, *120*, 5579.
145. Culkin, D. A.; Hartwig, J. F. *Acc. Chem. Res.* **2003**, *36*, 234.
146. Palucki, M.; Buchwald, S. L. *J. Am. Chem. Soc.* **1997**, *119*, 11108.
147. Ito, Y.; Hirao, T.; Saegusa, T. *J. Org. Chem.* **1978**, *43*, 1011.
148. Mori, M.; Kubo, Y.; Ban, Y. *Tetrahedron* **1988**, *44*, 4321.
149. Ogoshi, S.; Morimoto, T.; Nishio, K.; Ohe, K.; Murai, S. *J. Org. Chem.* **1993**, *58*, 9.
150. Tsuda, T.; Chujo, Y.; Nishi, S.; Tawara, K.; Saegusa, T. *J. Am. Chem. Soc.* **1980**, *102*, 6381.
151. Tsuji, J.; Minami, I. *Acc. Chem. Res.* **1987**, *20*, 140.
152. Sen Soo, H.; Diaconescu, P. L.; Cummins, C. C. *Organometallics* **2004**, *23*, 498.
153. Joung, U. G.; Kim, T. H.; Joe, D. J.; Lee, B. Y.; Shin, D. M.; Chung, Y. K. *Polyhedron* **2004**, *23*, 1587.
154. Burkhardt, E. R.; Doney, J. J.; Slough, G. A.; Stack, J. M.; Heathcock, C. H.; Bergman, R. G. *Pure Appl. Chem.* **1988**, *60*, 1.

155. Kündig, E. P.; Bernardinelli, G.; Kondratenko, M.; Robvieux, F.; Romanens, P. *Helv. Chim. Acta* **2003**, *86*, 4169.
156. Hartwig, J. F.; Andersen, R. A.; Bergman, R. G. *J. Am. Chem. Soc.* **1990**, *112*, 5670.
157. Culkin, D. A.; Hartwig, J. F. *J. Am. Chem. Soc.* **2001**, *123*, 5816.
158. Slough, G. A.; Hayashi, R.; Ashbaugh, J. R.; Shamblin, S. L.; Aukamp, A. M. *Organometallics* **1994**, *13*, 890.
159. Culkin, D. A.; Hartwig, J. F. *J. Am. Chem. Soc.* **2002**, *124*, 9330.
160. Albeniz, A. C.; Catalina, N. M.; Espinet, P.; Redon, R. *Organometallics* **1999**, *18*, 5571.
161. Walczuk, E. B.; Kamer, P. C. J.; van Leeuwen, P. *Angew. Chem., Int. Ed.* **2003**, *42*, 4665.
162. Taylor, S. J.; Morken, J. P. *J. Am. Chem. Soc.* **1999**, *121*, 12202.
163. Taylor, S. J.; Duffey, M. O.; Morken, J. P. *J. Am. Chem. Soc.* **2000**, *122*, 4528.
164. Piantadosi, C. A.; Sylvia, A. L.; Jöbsis, F. F. *J. Clin. Invest.* **1983**, *72*, 1224.
165. Liu, X.; Ibrahim, S. K.; Tard, C.; Pickett, C. J. *Coord. Chem. Rev.* **2005**, *249*, 1641.
166. Volbeda, A.; Fontecilla-Camps, J. C. *Coord. Chem. Rev.* **2005**, *249*, 1609.
167. de Andrade Lima, L. R. P.; Hodouin, D. *Int. J. Min. Process.* **2006**, *80*, 15.
168. Corain, B. *Coord. Chem. Rev.* **1982**, *47*, 165.
169. Griffith, W. P. *Coord. Chem. Rev.* **1975**, *17*, 177.
170. Rigo, P.; Turco, A. *Coord. Chem. Rev.* **1974**, *13*, 133.
171. Dunford, H. B.; Hewson, W. D. *J. Phys. Chem.* **1979**, *83*, 3307.
172. Loupy, A.; Corset, J. *J. Solution Chem.* **1976**, *5*, 817.
173. Nakamoto, K. *Infrared and Raman Spectra of Inorganic and Coordination Compounds*; Wiley: New York, 1978.
174. Marcantonio, K. M.; Frey, L. F.; Liu, Y.; Chen, Y.; Strine, J.; Phenix, B.; Wallace, D. J.; Chen, C. Y. *Org. Lett.* **2004**, *6*, 3723.
175. Dobbs, K. D.; Marshall, W. J.; Grushin, V. V. *J. Am. Chem. Soc.* **2007**, *129*, 30.
176. Schareina, T.; Zapf, A.; Beller, M. *J. Chem. Soc., Chem. Commun.* **2004**, 1388.
177. Schareina, T.; Zapf, A.; Beller, M. *J. Organomet. Chem.* **2004**, *689*, 4576.
178. Weissman, S. A.; Zewge, D.; Chen, C. *J. Org. Chem.* **2005**, *70*, 1508.
179. Kubota, H.; Rice, K. C. *Tetrahedron Lett.* **1998**, *39*, 2907.
180. Jin, F.; Confalone, P. N. *Tetrahedron Lett.* **2000**, *41*, 3271.
181. Huang, J. K.; Haar, C. M.; Nolan, S. P.; Marcone, J. E.; Moloy, K. G. *Organometallics* **1999**, *18*, 297.
182. Goldman, A. S.; Krogh-Jespersen, K. *J. Am. Chem. Soc.* **1996**, *118*, 12159.
183. Sun, H.; DiMagno, S. G. *J. Am. Chem. Soc.* **2005**, *127*, 2050.
184. Pesek, J. J.; Mason, W. R. *Inorg. Chem.* **1979**, *18*, 924.
185. Witanowski, M. *Tetrahedron* **1967**, *23*, 4299.
186. Favero, G.; Morvillo, A. *J. Organomet. Chem.* **1984**, *260*, 363.
187. Abla, M.; Yamamoto, T. *J. Organomet. Chem.* **1997**, *532*, 267.
188. Garcia, J. J.; Jones, W. D. *Organometallics* **2000**, *19*, 5544.
189. Garcia, J. J.; Jones, W. D. *Organometallics* **2000**, *19*, 5544.
190. Garcia, J. J.; Arevalo, A.; Brunkan, N. M.; Jones, W. D. *Organometallics* **2004**, *23*, 3997.
191. Pesek, J. J.; Mason, W. R. *Inorg. Chem.* **1983**, *22*, 2958.
192. Ruff, J. K. *Inorg. Chem.* **1969**, *8*, 86.
193. Clarke, H. L. *J. Organomet. Chem.* **1974**, *80*, 155.
194. Green, M. L. H.; Nagy, P. L. O. *Adv. Organomet. Chem.* **1964**, *2*, 325.
195. Wilke, G.; Bogdanovic, B.; Hardt, P.; Heimbach, P.; Keim, W.; Kröner, M.; Oberkirch, W.; Tanaka, K.; Steinrücke, E.; Walter, D.; Zimmermann, H. *Angew. Chem. Int. Ed.* **1966**, *5*, 151.
196. *Comprehensive Organometallic Chemistry*; Wilkinson, G.; Stone, F. G. A.; Abel, E. W., Eds.; Pergamon Press: New York, 1982; Vol. 6.
197. Beckhaus, R. In *Synthetic Methods of Organometallic and Inorganic Chemistry*; Herrmann, W. A., Ed.; Georg Thieme Verlag: New York, 2000; Vol. 9, p 1.
198. Werner, H.; Kuhn, A. *Angew. Chem. Int. Ed.* **1977**, *16*, 412.
199. Raper, G.; McDonald, W. S. *J. Chem. Soc., Dalton Trans.* **1972**, 265
200. Bandoli, G.; Dolmella, A.; Fanizzi, F. P.; Di Masi, N. G.; Maresca, L.; Natile, G. *Organometallics* **2001**, *20*, 805.
201. Smith, A. E. *Acta Crystallogr.* **1965**, *18*, 331.
202. Mason, R.; Wheeler, A. G. *J. Chem. Soc. A* **1968**, 2543.
203. Huffman, J. C.; Laurent, M. P.; Kochi, J. K. *Inorg. Chem.* **1977**, *16*, 2639.
204. Vrieze, K. In *Dynamic Nuclear Magnetic Resonance Spectroscopy*; Jackman, L. M., Cotton, F. A., Eds.; Academic Press: New York, 1975, p 441.

205. Hansson, S.; Norrby, P. O.; Sjogren, M. P. T.; Akermark, B.; Cucciolito, M. E.; Giordano, F.; Vitagliano, A. *Organometallics* **1993**, *12*, 4940.

206. Faller, J. W.; Stokes-Huby, H. L.; Albrizzio, M. A. *Helv. Chim. Acta* **2001**, *84*, 3031.

207. Yu, Z. X.; Wender, P. A.; Houk, K. N. *J. Am. Chem. Soc.* **2004**, *126*, 9154.

208. Hegedus, L. S.; Inoue, Y. *J. Am. Chem. Soc.* **1982**, *104*, 4917.

209. Pearson, A. J.; Khan, M. N. I. *J. Am. Chem. Soc.* **1984**, *106*, 1872.

210. Johnson, B. F. G.; Lewis, J.; Yarrow, D. J. *J. Chem. Soc., Dalton Trans.* **1972**, 2084.

211. Trost, B. M.; Strege, P. E.; Weber, L.; Fullerton, T. J.; Dietsche, T. J. *J. Am. Chem. Soc.* **1978**, *100*, 3407.

212. Tulip, T. H.; Ibers, J. A. *J. Am. Chem. Soc.* **1979**, *101*, 4201.

213. Chen, M. S.; Prabagaran, N.; Labenz, N. A.; White, M. C. *J. Am. Chem. Soc.* **2005**, *127*, 6970.

214. Mitsudome, T.; Umetani, T.; Nosaka, N.; Mori, K.; Mizugaki, T.; Ebitani, K.; Kaneda, K. *Angew. Chem. Int. Ed.* **2006**, *45*, 481.

215. Becker, Y.; Stille, J. K. *J. Am. Chem. Soc.* **1978**, *100*, 845.

216. Werner, H.; Schafer, M.; Nurnberg, O.; Wolf, J. *Chem. Ber.* **1994**, *127*, 27.

217. Roberts, J. S.; Klabunde, K. J. *J. Am. Chem. Soc.* **1977**, *99*, 2509.

218. Nettekoven, U.; Hartwig, J. F. *J. Am. Chem. Soc.* **2002**, *124*, 1166.

219. Gatti, G.; Lopez, J. A.; Mealli, C.; Musco, A. *J. Organomet. Chem.* **1994**, *483*, 77.

220. Hegedus, L. S. *Transition Metals in the Synthesis of Complex Organic Molecules,* 2nd ed.; University Science Books: Sausalito, CA, 1999.

221. Brookhart, M.; Karel, K. J.; Nance, L. E. *J. Organomet. Chem.* **1977**, *140*, 203.

222. Pearson, A. J. *Acc. Chem. Res.* **1980**, *13*, 463.

223. Semmelhack, M. F.; Hall, H. T.; Farina, R.; Yoshifuji, M.; Clark, G.; Bargar, T.; Hirotsu, K.; Clardy, J. *J. Am. Chem. Soc.* **1979**, *101*, 3535.

224. Jones, M. D.; Kemmitt, R. D. W. In *Advances in Organometallic Chemistry*; Stone, F. G. A., West, R., Eds.; Academic Press: New York, 1989; Vol. 27, p 279.

225. Ward, J. S.; Pettit, R. *J. Chem. Soc., Chem. Commun.* **1970**, 1419.

226. Ehrlich, K.; Emerson, G. F. *J. Am. Chem. Soc.* **1972**, *94*, 2464.

227. Trost, B. M.; Chan, D. M. T. *J. Am. Chem. Soc.* **1983**, *105*, 2326.

228. Trost, B. M.; Chan, D. M. T. *J. Am. Chem. Soc.* **1983**, *105*, 2315.

229. Trost, B. M.; Stambuli, J. P.; Silverman, S. M.; Schworer, U. *J. Am. Chem. Soc.* **2006**, *128*, 13328.

230. Hartwig, J. F.; Bergman, R. G.; Andersen, R. A. *J. Am. Chem. Soc.* **1990**, *112*, 5670.

231. Ohe, K.; Ishihara, T.; Chatani, N.; Murai, S. *J. Am. Chem. Soc.* **1990**, *112*, 9646.

232. Ando, W.; Yamamoto, T.; Saso, H.; Kabe, Y. *J. Am. Chem. Soc.* **1991**, *113*, 2791.

233. Kemmitt, R. D. W.; Moore, M. R. *Transition Met. Chem.* **1993**, *18*, 348.

234. Kabe, Y.; Yamamoto, T.; Ando, W. *Organometallics* **1994**, *13*, 4606.

235. Huy, N. H. T.; Marinetti, A.; Ricard, L.; Mathey, F. *Organometallics* **2001**, *20*, 593.

236. Schumann, H.; Meese-Marktscheffel, J. A.; Esser, L. *Chem. Rev.* **1995**, *95*, 865.

237. Evans, W. J.; Davis, B. L. *Chem. Rev.* **2002**, *102*, 2119.

238. Kealy, T. J.; Pauson, P. L. *Nature* **1951**, *168*, 1039.

239. Miller, S. A.; Tebboth, J. A.; Tremaine, J. F. *J. Chem. Soc.* **1952**, 632.

240. Laszlo, P.; Hoffman, R. *Angew. Chem. Int. Ed.* **2000**, *39*, 123.

241. Brintzinger, H. H.; Fischer, D.; Mulhaupt, R.; Rieger, B.; Waymouth, R. M. *Angew. Chem. Int. Ed.* **1995**, *34*, 1143.

242. Hoyano, J. K.; Graham, W. A. G. *J. Am. Chem. Soc.* **1982**, *104*, 3723.

243. Janowicz, A. H.; Bergman, R. G. *J. Am. Chem. Soc.* **1982**, *104*, 352.

244. Chen, H.; Schlecht, S.; Semple, T. C.; Hartwig, J. F. *Science* **2000**, *287*, 1995.

245. Ryan, M. F.; Eyler, J. R.; Richardson, D. E. *J. Am. Chem. Soc.* **1992**, *114*, 8611.

246. Dias, A. R.; Veiros, L. F. *J. Organomet. Chem.* **2005**, *690*, 1840.

247. Streitwieser, A.; Nebenzahl, L. L. *J. Am. Chem. Soc.* **1976**, *98*, 2188.

248. Jolly, W. L. *Inorg. Synth.* **1968**, *11*, 120.

249. Wilkinson, G.; Rosenblum, M.; Whiting, M. C.; Woodward, R. B. *J. Am. Chem. Soc.* **1952**, *74*, 2125

250. Marr, G.; Rockett, B. W. In *The Chemistry of the Metal-Carbon Bond*; Hartley, F. R., Patai, S., Eds.; Wiley: New York, 1982; Vol. 1, Chapter 10.

251. Long, N. J. *Metallocenes: An Introduction to Sandwich Complexes*; Blackwell Science: London, 1998.

252. Esteruelas, M. A.; Lopez, A. M.; Ruiz, N.; Tolosa, J. I. *Organometallics* **1997**, *16*, 4657.

253. Chandra, G.; Lappert, M. F. *J. Chem. Soc. A* **1968**, 1940.

254. Lappert, M. F.; Power, P. P.; Sanger, A. R.; Srivastava, R. C. *Metal and Metalloid Amides*; Ellis Horwood: Chichester, U. K., 1980.

255. Lappert, M. F.; Power, P. P.; Sanger, A. R.; Srivatava, R. C. In *Metal and Metalloid Amides: Syntheses, Structures and Physical and Chemical Properties*; Halsted Press: New York, 1979, p 616.

256. Diamond, G. M.; Jordan, R. F.; Petersen, J. L. *Organometallics* **1996**, *15*, 4030.

257. Diamond, G. M.; Jordan, R. F.; Petersen, J. L. *Organometallics* **1996**, *15*, 4045.

258. Diamond, G. M.; Jordan, R. F.; Petersen, J. L. *J. Am. Chem. Soc.* **1996**, *118*, 8024.

259. LoCoco, M. D.; Zhang, X. W.; Jordan, R. F. *J. Am. Chem. Soc.* **2004**, *126*, 15231.

260. LoCoco, M. D.; Jordan, R. F. *J. Am. Chem. Soc.* **2004**, *126*, 13918.

261. Wolczanski, P. T.; Bercaw, J. E. *Acc. Chem. Res.* **1980**, *13*, 121.

262. Janiak, C.; Schumann, H. *Adv. Organomet. Chem.* **1991**, *33*, 291.

263. Okuda, J. *Top. Curr. Chem.* **1992**, *160*, 97.

264. Morales, D.; Poli, R.; Andrieu, J. *Inorg. Chim. Acta* **2000**, *300*, 709.

265. Macomber, D. W.; Hart, W. P.; Rausch, M. D. *Adv. Organomet. Chem.* **1982**, *21*, 1.

266. Muller, C.; Vos, D.; Jutzi, P. *J. Organomet. Chem.* **2000**, *600*, 127.

267. Siemeling, U. *Chem. Rev.* **2000**, *100*, 1495.

268. Butenschön, H. *Chem. Rev.* **2000**, *100*, 1527.

269. Sunkel, K. *Chem. Ber. Rec.* **1997**, *130*, 1721.

270. Calhorda, M. J.; Romao, C. C.; Veiros, L. F. *Chem. Eur. J.* **2002**, *8*, 868.

271. Alt, H. G.; Samuel, E. *Chem. Soc. Rev.* **1998**, *27*, 323.

272. Repo, T.; Jany, G.; Hakala, K.; Klinga, M.; Polamo, M.; Leskela, M.; Rieger, B. *J. Organomet. Chem.* **1997**, *549*, 177.

273. Katz, T. J.; Acton, N. *J. Am. Chem. Soc.* **1972**, *94*, 3281.

274. Cloke, F. G. N. *Pure Appl. Chem.* **2001**, *73*, 233.

275. Trnka, T. M.; Bonanno, J. B.; Bridgewater, B. M.; Parkin, G. *Organometallics* **2001**, *20*, 3255.

276. Kakkar, A. K.; Taylor, N. J.; Marder, T. B.; Shen, J. K.; Hallinan, N.; Basolo, F. *Inorg. Chim. Acta* **1992**, *200*, 219.

277. Frankcom, T. M.; Green, J. C.; Nagy, A.; Kakkar, A. K.; Marder, T. B. *Organometallics* **1993**, *12*, 3688.

278. Zargarian, D. *Coord. Chem. Rev.* **2002**, *233*, 157.

279. Halterman, R. L.; Vollhardt, K. P. C. *Tetrahedron Lett.* **1986**, *27*, 1461.

280. Cesarotti, E.; Kagan, H. B.; Goddard, R.; Kruger, C. *J. Organomet. Chem.* **1978**, *162*, 297.

281. Knickmeier, M.; Erker, G.; Fox, T. *J. Am. Chem. Soc.* **1996**, *118*, 9623.

282. Halterman, R. L. *Chem. Rev.* **1992**, *92*, 965.

283. Halterman, R. L. In *Metallocenes: Synthesis—Reactivity—Applications*; Togni, A., Halterman, R. L., Eds.; Wiley-VCH: Weinheim, 1998; Vol. 1, p 455.

284. Paley, R. S. *Chem. Rev.* **2002**, *102*, 1493.

285. Attig, T. G.; Teller, R. G.; Wu, S. M.; Bau, R.; Wojcicki, A. *J. Am. Chem. Soc.* **1979**, *101*, 619.

286. Flood, T. C.; Miles, D. L. *J. Organomet. Chem.* **1977**, *127*, 33.

287. Brunner, H. *Adv. Organomet. Chem.* **1980**, *18*, 151.

288. Luttringhaus, A.; Gralheer, H. *Annal. Chem.* **1942**, *550*, 67.

289. Luttringhaus, A.; Kullick, W. *Angew. Chem.* **1958**, *70*, 438.

290. Smith, J. A.; Vonseyerl, J.; Huttner, G.; Brintzinger, H. H. *J. Organomet. Chem.* **1979**, *173*, 175.

291. Zachmanoglou, C. E.; Docrat, A.; Bridgewater, B. M.; Parkin, G.; Brandow, C. G.; Bercaw, J. E.; Jardine, C. N.; Lyall, M.; Green, J. C.; Keister, J. B. *J. Am. Chem. Soc.* **2002**, *124*, 9525.

292. Shapiro, P. J. *Coord. Chem. Rev.* **2002**, *231*, 67.

293. Prashar, S.; Antiñolo, A.; Otero, A. *Coord. Chem. Rev.* **2006**, *250*, 133.

294. Shapiro, P. J.; Bunel, E.; Schaefer, W. P.; Bercaw, J. E. *Organometallics* **1990**, *9*, 867.

295. McKnight, A. L.; Waymouth, R. M. *Chem. Rev.* **1998**, *98*, 2587.

296. Okuda, J.; Eberle, T. In *Metallocenes: Synthesis—Reactivity—Applications*; Togni, A., Halterman, R. L., Eds.; Wiley-VCH: Weinheim, 1998; Vol. 1, p 415.

297. Okuda, J. *J. Chem. Soc., Dalton Trans.* **2003**, 2367.

298. Blaser, H.; Spindler, F. *Chimia* **1997**, *51*, 297.

299. Swart, M. *Inorg. Chim. Acta* **2007**, *360*, 179.

300. Heise, H.; Köhler, F. H.; Xie, X. L. *J. Magn. Reson.* **2001**, *150*, 198.

301. Chetwynd-Talbot, J.; Grebenik, P.; Perutz, R. N.; Powell, M. H. A. *Inorg. Chem.* **1983**, *22*, 1675.

302. Droege, M. W.; Harman, W. D.; Taube, H. *Inorg. Chem.* **1987**, *26*, 1309.

303. Watanabe, M.; Sato, M.; Nagasawa, A.; Kai, M.; Motoyama, I.; Takayama, T. *Bull. Chem. Soc. Jpn.* **1999**, *72*, 715.

304. Bandy, J. A.; Cloke, F. G. N.; Cooper, G.; Day, J. P.; Girling, R. B.; Graham, R. G.; Green, J. C.; Grinter, R.; Perutz, R. N. *J. Am. Chem. Soc.* **1988**, *110*, 5039.

305. Robbins, J. L.; Edelstein, N.; Spencer, B.; Smart, J. C. *J. Am. Chem. Soc.* **1982**, *104*, 1882.

306. Epstein, L. M.; Shubina, E. S.; Krylov, A. N.; Kreindlin, A. Z.; Rybinskaya, M. I. *J. Organomet. Chem.* **1993**, *447*, 277.
307. Kamyshova, A. A.; Kreindlin, A. Z.; Rybinskaya, M. I.; Petrovskii, P. V.; Kruglova, N. V.; Borisov, Y. A. *Russ. Chem. Bull.* **2003**, *52*, 2424.
308. *Metallocenes: Synthesis—Reactivity—Applications*; Togni, A.; Halterman, R. L., Eds.; Wiley-VCH: Weinheim, 1998; Vols. 1 and 2.
309. Green, J. C. *Chem. Soc. Rev.* **1998**, *27*, 263.
310. Jordan, R. F.; Bajgur, C. S.; Willett, R.; Scott, B. *J. Am. Chem. Soc.* **1986**, *108*, 7410.
311. Katz, T. J.; Acton, N. *Tetrahedron Lett.* **1970**, 2497.
312. Wild, F.; Zsolnai, L.; Huttner, G.; Brintzinger, H. H. *J. Organomet. Chem.* **1982**, *232*, 233.
313. LoCoco, M. D.; Zhang, X. W.; Jordan, R. F. *J. Am. Chem. Soc.* **2004**, *126*, 15231.
314. LoCoco, M. D.; Jordan, R. F. *J. Am. Chem. Soc.* **2004**, *126*, 13918.
315. Albers, M. O.; Robinson, D. J.; Singleton, E. *Coord. Chem. Rev.* **1987**, *79*, 1.
316. Davies, S. G.; McNally, J. P.; Smallridge, A. J. *Adv. Organomet. Chem.* **1990**, *30*, 1.
317. Consiglio, G.; Morandini, F. *Chem. Rev.* **1987**, *87*, 761.
318. Coville, N. J.; Duplooy, K. E.; Pickl, W. *Coord. Chem. Rev.* **1992**, *116*, 1.
319. Herberhold, M.; Dietel, A. M.; Goller, A.; Milius, W. *Z. Anorg. Allg. Chem.* **2003**, *629*, 871.
320. Ganter, C. *Chem. Soc. Rev.* **2003**, *32*, 130.
321. Sapunov, V. N.; Schmid, R.; Kirchner, K.; Nagashima, H. *Coord. Chem. Rev.* **2003**. *238*, 363.
322. Jimenez-Tenorio, M.; Puerta, M. C.; Valerga, P. *Eur. J. Org. Chem.* **2004**, 17.
323. Cadierno, V.; Gamasa, M. P.; Gimeno, J. *Coord. Chem. Rev.* **2004**, *248*, 1627.
324. Werner, H. *Organometallics* **2005**, *24*, 1036.
325. Severin, K. *J. Chem. Soc., Chem. Commun.* **2006**, 3859.
326. Murahashi, T.; Kurosawa, H. *Coord. Chem. Rev.* **2002**, *231*, 207.
327. Hawthorn, M. F.; Young, D. C.; Wegner, P. A. *J. Am. Chem. Soc.* **1965**, *87*, 1818.
328. Hosmane, N. S.; Maguire, J. A. *Pure Appl. Chem.* **2006**, *78*, 1333.
329. Zhang, H. M.; Wang, Y.; Saxena, A. K.; Oki, A. R.; Maguire, J. A.; Hosmane, N. S. *Organometallics* **1993**, *12*, 3933.
330. Herberich, G. E.; Eckenrath, H. J.; Englert, U. *Organometallics* **1998**, *17*, 519.
331. Liu, S. Y.; Lo, M. M. C.; Fu, G. C. *Angew. Chem., Int. Ed.* **2002**, *41*, 174.
332. Ashe, A. J.; Yang, H.; Fang, X. D.; Kampf, J. W. *Organometallics* **2002**, *21*, 4578.
333. Herberich, G. E.; Englert, U.; Schmitz, A. *Organometallics* **1997**, *16*, 3751.
334. Grimes, R. N. In *Comprehensive Organometallic Chemistry III*; Housecroft, C. E., Ed.; Elsevier: Amsterdam, 2006; Vol. 3, p 1.
335. Herberich, G. E.; Ohst, H. *Adv. Organomet. Chem.* **1986**, *25*, 199.
336. Herberich, G. E. In *Comprehensive Organometallic Chemistry II*; Housecroft, C. E.. Ed.; Pergamon Press: Oxford, U.K., 1995; Vol. 1, p 197.
337. Fu, G. C. In *Advances in Organometallic Chemistry*; West, R., Hill, A. F., Eds., Academic Press: New York, 2001; Vol. 47, p 101.
338. Herberich, G. E.; Schmidt, B.; Englert, U. *Organometallics* **1995**, *14*, 471.
339. (a) Herberich, G. E.; Rosenplanter, J.; Schmidt, B.; Englert, U. *Organometallics* **1997**. *16*, 926; (b) Herberich, G. E.; Englert, U.; Schmite, A. *Organometallics*, **1997**, *16*, 3751.
340. Bazan, G. C.; Rodriguez, G.; Ashe, A. J.; Al-Ahmad, S.; Muller, C. *J. Am. Chem. Soc.* **1996**, *118*, 2291.
341. Komon, Z. J. A.; Rogers, J. S.; Bazan, G. C. *Organometallics* **2002**, *21*, 3189.
342. Ashe, A. J.; Al-Ahmad, S.; Fang, X. G.; Kampf, J. W. *Organometallics* **1998**, *17*, 3883.
343. Ashe, A. J.; Fang, X. G.; Kampf, J. W. *Organometallics* **1999**, *18*, 1363.
344. Cotton, F. A.; Marks, T. J. *J. Am. Chem. Soc.* **1969**, *91*, 7523.
345. Calderon, J. L.; Cotton, F. A.; Takats, J. *J. Am. Chem. Soc.* **1971**, *93*, 3587.
346. *Ferrocenes*; Togni, A.; Hayashi, T., Eds.; Wiley-VCH: Weinheim, 1995.
347. Karlsson, A.; Broo, A.; Ahlberg, P. *Can. J. Chem.* **1999**, *77*, 628.
348. Wagner, G.; Herrmann, R. In *Ferrocenes*; Togni, A., Hayashi, T., Eds.; Wiley-VCH: Weinhem, 1995; p 173.
349. Togni, A.; Breutel, C.; Soares, M. C.; Zanetti, N.; Gerfin, T.; Gramlich, V.; Spindler, F.; Rihs, G. *Inorg. Chim. Acta* **1994**, *222*, 213.
350. Togni, A.; Breutel, C.; Schnyder, A.; Spindler, F.; Landert, H.; Tijani, A. *J. Am. Chem. Soc.* **1994**, *116*, 4062.
351. Blaser, H. U.; Buser, H. P.; Hausel, R.; Jalett, H. P.; Spindler, F. *J. Organomet. Chem.* **2001**, *621*, 34.
352. Sander, R.; Mueller-Wersterhoff, U. T. *J. Organomet. Chem.* **1996**, *512*, 219.
353. Sammakia, T.; Latham, H. A. *J. Org. Chem.* **1995**, *60*, 6002.
354. Riant, O.; Argouarch, G.; Guillaneux, D.; Samuel, O.; Kagan, H. B. *J. Org. Chem.* **1998**, *63*, 3511.

355. Ferber, B.; Kagan, H. B. *Adv. Synth. Catal.* **2007**, *349*, 493.

356. Brintzinger, H. H.; Bercaw, J. E. *J. Am. Chem. Soc.* **1970**, *92*, 6182.

357. Davison, A.; Wreford, S. S. *J. Am. Chem. Soc.* **1974**, *96*, 3017.

358. Bernskoetter, W. H.; Lobkovsky, E.; Chirik, P. J. *J. Am. Chem. Soc.* **2005**, *127*, 14051.

359. Mork, B. V.; Tilley, T. D. *J. Am. Chem. Soc.* **2004**, *126*, 4375.

360. van Staveren, D. R.; Metzler-Nolte, N. *Chem. Rev.* **2004**, *104*, 5931.

361. Schunn, R. A. In *Transition Metal Hydrides*, 1st ed.; Muetterties, E. L., Ed.; Marcel Dekker: New York, 1971; Vol. 1, p 203.

362. Geoffroy, G. L.; Lehman, J. R. In *Advances in Inorganic Chemistry and Radiochemistry*; Eméleus, H. J., Sharpe, A. G., Eds.; Academic Press: New York, 1977; Vol. 20, p 189.

363. Bau, R.; Carroll, W. E.; Hart, D. W.; Teller, R. G. In *Transition Metal Hydrides*; Bau, R., Ed.; American Chemical Society: Washington DC, 1978; Vol. 167, p 73.

364. Humphries, A. P.; Kaesz, H. D. In *Progress in Inorganic Chemistry*; Lippard, S. J., Ed.; Wiley: New York, 1979; Vol. 25, p 145.

365. Teller, R. G.; Bau, R. *Struct. Bonding* **1981**, *44*, 1.

366. Venanzi, L. M. *Coord. Chem. Rev.* **1982**, *43*, 251.

367. Moore, D. S.; Robinson, S. D. *Chem. Soc. Rev.* **1983**, *12*, 415.

368. James, B. R. In *Comprehensive Organometallic Chemistry*; Wilkinson, G., Ed.; Pergamon Press: Oxford, U.K., 1982; Vol. 8, p 290.

369. Hlatky, G. G.; Crabtree, R. H. *Coord. Chem. Rev.* **1985**, *65*, 1.

370. Pearson, R. G. *Chem. Rev.* **1985**, *85*, 41.

371. Norton, J. R. In *Inorganic Reactions and Methods: The Formation of Bonds to Hydrogen (Part 2)*; Zuckerman, J. J., Ed.; Wiley-VCH: Weinheim, 1988; Vol. 2, p 204.

372. Darensbourg, M. Y.; Ash, C. E. *Adv. Organomet. Chem.* **1987**, *27*, 1.

373. Berke, H.; Burger, P. *Comments Inorg. Chem.* **1994**, *16*, 279.

374. Maseras, F.; Lledós, A.; Clot, E.; Eisenstein, O. *Chem. Rev.* **2000**, *100*, 601.

375. Albinati, A.; Venanzi, L. M. *Coord. Chem. Rev.* **2000**, *200*, 687.

376. King, R. B. *Coord. Chem. Rev.* **2000**, *200*, 813.

377. Puddephatt, R. J. *Coord. Chem. Rev.* **2001**, *219*, 157.

378. Hoskin, A. J.; Stephan, D. W. *Coord. Chem. Rev.* **2002**, *233*, 107.

379. Laplaca, S. J.; Hamilton, W. C.; Ibers, J. A.; Davison, A. *Inorg. Chem.* **1969**, *8*, 1928.

380. Emge, T. J.; Koetzle, T. F.; Bruno, J. W.; Caulton, K. G. *Inorg. Chem.* **1984**, *23*, 4012.

381. Baker, R. W.; Pauling, P. *J. Chem. Soc., Chem. Commun.* **1969**, 1495.

382. Bau, R.; Carroll, W. E.; Teller, R. G.; Koetzle, T. F. *J. Am. Chem. Soc.* **1977**, *99*, 3872.

383. Dahl, L. F. *Ann. N. Y. Acad. Sci.* **1983**, *415*, 1.

384. Jackson, P. F.; Johnson, B. F. G.; Lewis, J.; Raithby, P. R.; McPartlin, M.; Nelson, W. J. H.; Rouse, K. D.; Allibon, J.; Mason, S. A. *J. Chem. Soc., Chem. Commun.* **1980**, 295.

385. Schneider, J. J.; Goddard, R.; Werner, S.; Kruger, C. *Angew. Chem. Int. Ed.* **1991**, *30*, 1124.

386. Kersten, J. L.; Rheingold, A. L.; Theopold, K. H.; Casey, C. P.; Widenhoefer, R. A.; Hop, C. *Angew. Chem. Int. Ed.* **1992**, *31*, 1341.

387. Churchill, M. R. In *Transition Metal Hydrides*; Bau, R., Ed.; American Chemical Society: Washington DC, 1978; Vol. 167, p 36.

388. Buckingham, A. D.; Stephens, P. F. *J. Chem. Soc.* **1964**, 4583.

389. Manriquez, J. M.; McAlister, D. R.; Sanner, R. D.; Bercaw, J. E. *J. Am. Chem. Soc.* **1978**, *100*, 2716.

390. Goeden, G. V.; Caulton, K. G. *J. Am. Chem. Soc.* **1981**, *103*, 7354.

391. Mankad, N. P.; Laitar, D. S.; Sadighi, J. P. *Organometallics* **2004**, *23*, 3369.

392. Cooper, C. B., III; Shriver, D. F.; Onaka, S. In *Transition Metal Hydrides*; Bau, R., Ed.; American Chemical Society: Washington DC, 1978; Vol. 167, p 232.

393. Bennett, M. A.; Milner, D. L. *J. Am. Chem. Soc.* **1969**, *91*, 6983.

394. L'Eplattenier, F.; Calderazzo, F. *Inorg. Chem.* **1968**, *7*, 1290.

395. Lyons, D.; Wilkinson, G. *J. Chem. Soc., Dalton Trans.* **1985**, 587.

396. Love, J. B.; Clark, H. C. S.; Cloke, F. G. N.; Green, J. C.; Hitchcock, P. B. *J. Am. Chem. Soc.* **1999**, *121*, 6843.

397. Bregel, D. C.; Oldham, S. M.; Eisenberg, R. *J. Am. Chem. Soc.* **2002**, *124*, 13827.

398. Hoff, C. D. *Coord. Chem. Rev.* **2000**, *206*, 451.

399. Capps, K. B.; Bauer, A.; Kiss, G.; Hoff, C. D. *J. Organomet. Chem.* **1999**, *586*, 23.

400. Wayland, B. B.; Ba, S. J.; Sherry, A. E. *Inorg. Chem.* **1992**, *31*, 148.

401. Filippou, A. C.; Schneider, S.; Schnakenburg, G. *Angew. Chem. Int. Ed.* **2003**, *42*, 4486.

402. Ting, C.; Baenziger, N. C.; Messerle, L. *J. Chem. Soc., Chem. Commun.* **1988**, *1988*, 1133.

403. Ashworth, T. V.; Singleton, J. E.; Dewaal, D. J. A.; Louw, W. J.; Singleton, E.; Vanderstok, E. *J. Chem. Soc., Dalton Trans.* **1978**, 340.

404. Crabtree, R. H.; Quirk, J. M.; Fillebeenkhan, T.; Morris, G. E. *J. Organomet. Chem.* **1979**, *181*, 203.

405. Eady, C. R.; Jackson, P. F.; Johnson, B. F. G.; Lewis, J.; Malatesta, M. C.; McPartlin, M.; Nelson, W. J. H. *J. Chem. Soc., Dalton Trans.* **1980**, 383.

406. Trogler, W. C. *J. Am. Chem. Soc.* **1979**, *101*, 6459.

407. Papish, E. T.; Magee, M. P.; Norton, J. R. In *Recent Advances in Hydride Chemistry*; Peruzzini, M., Poli, R., Eds.; Elsevier: New York, 2001; p 39.

408. Davison, A.; Ellis, J. E. *J. Organomet. Chem.* **1972**, *36*, 131.

409. Laing, K. R.; Roper, W. R. *J. Chem. Soc. A* **1969**, 1889.

410. Werner, H.; Hofmann, W. *Chem. Ber.* **1977**, *110*, 3481.

411. Stevens, R. E.; Gladfelter, W. L. *J. Am. Chem. Soc.* **1982**, *104*, 6454.

412. Abad, M. M.; Atheaux, I.; Maisonnat, A.; Chaudret, B. *J. Chem. Soc., Chem. Commun.* **1999**, 381.

413. Green, M. L. H.; Knowles, P. J. *J. Chem. Soc., Perkins Trans. 1* **1973**, 989.

414. Reinartz, S.; White, P. S.; Brookhart, M.; Templeton, J. L. *Organometallics* **2000**, *19*, 3748.

415. Cheng, T. Y.; Szalda, D. J.; Zhang, J.; Bullock, R. M. *Inorg. Chem.* **2006**, *45*, 4712.

416. Cugny, J.; Schmalle, H. W.; Fox, T.; Blacque, O.; Alfonso, M.; Berke, H. *Eur. J. Inorg. Chem.* **2006**, 540.

417. Klimpel, M. G.; Sirsch, P.; Scherer, W.; Anwander, R. *Angew. Chem. Int. Ed.* **2003**, *42*, 574.

418. Marks, T. J.; Kolb, J. R. *Chem. Rev.* **1977**, *77*, 263.

419. Girolami, G. S.; Howard, C. G.; Wilkinson, G.; Dawes, H. M.; Thorntonpett, M.; Motevalli, M.; Hursthouse, M. B. *J. Chem. Soc., Dalton Trans.* **1985**, 921.

420. Arnold, D. P.; Bennett, M. A. *Inorg. Chem.* **1984**, *23*, 2110.

421. Vaska, L.; Diluzio, J. W. *J. Am. Chem. Soc.* **1962**, *84*, 4989.

422. Rahman, A. K. F.; Jackson, W. G.; Willis, A. C.; Rae, A. D. *J. Chem. Soc., Chem. Commun.* **2003**, 2748.

423. Darensbourg, D. J.; Kudaroski, R. A. *Adv. Organomet. Chem.* **1983**, *22*, 129.

424. Grice, N.; Kao, S. C.; Pettit, R. *J. Am. Chem. Soc.* **1979**, *101*, 1627.

425. Negishi, E.; Miller, J. A.; Yoshida, T. *Tetrahedron Lett.* **1984**, *25*, 3407.

426. Walker, H. W.; Pearson, R. G.; Ford, P. C. *J. Am. Chem. Soc.* **1983**, *105*, 1179.

427. Jordan, R. F.; Norton, J. R. *J. Am. Chem. Soc.* **1982**, *104*, 1255.

428. Moore, E. J.; Sullivan, J. M.; Norton, J. R. *J. Am. Chem. Soc.* **1986**, *108*, 2257.

429. Kristjánsdóttir, S. S.; Norton, J. R. In *Transition Metal Hydrides*; Dedieu, A., Ed.; Wiley-VCH: New York, 1991; p 309.

430. Bullock, R. M. In *Handbook of Homogeneous Hydrogenation*; de Vries, J. G., Elsevier, C. J., Eds.; Wiley-VCH: Weinheim, 2007; Vol. 1, Chapter 7.

431. Tilset, M. In *Comprehensive Organometallic Chemistry III, Introduction: Fundamentals*, 1st ed.; Parkin, G., Ed.; Elsevier: New York, 2007; Vol. 1, p 279.

432. Kolthoff, I. M.; Chantooni, M. K. *J. Phys. Chem.* **1968**, *72*, 2270.

433. Schwesinger, R.; Schlemper, H. *Angew. Chem. Int. Ed.* **1987**, *26*, 1167.

434. Choi, J.; Pulling, M. E.; Smith, D. M.; Norton, J. R. *J. Am. Chem. Soc.* **2008**, *130*, 4250.

435. Tilset, M.; Parker, V. D. *J. Am. Chem. Soc.* **1989**, *111*, 6711.

436. Tilset, M. *J. Am. Chem. Soc.* **1990**, *112*, 2843.

437. Parker, V. D.; Handoo, K. L.; Roness, F.; Tilset, M. *J. Am. Chem. Soc.* **1991**, *113*, 7493.

438. Tilset, M. *J. Am. Chem. Soc.* **1992**, *114*, 2740.

439. Kiss, G.; Zhang, K.; Mukerjee, S. L.; Hoff, C. D.; Roper, G. C. *J. Am. Chem. Soc.* **1990**, *112*, 5657.

440. Skagestad, V.; Tilset, M. *J. Am. Chem. Soc.* **1993**, *115*, 5077.

441. Papish, E. T.; Rix, F. C.; Spetseris, N.; Norton, J. R.; Williams, R. D. *J. Am. Chem. Soc.* **2000**, *122*, 12235.

442. Pedersen, A.; Skagestad, V.; Tilset, M. *Acta Chem. Scand.* **1995**, *49*, 632.

443. Weberg, R. T.; Norton, J. R. *J. Am. Chem. Soc.* **1990**, *112*, 1105.

444. Curtis, C. J.; Miedaner, A.; Ellis, W. W.; Dubois, D. L. *J. Am. Chem. Soc.* **2002**, *124*, 1918.

445. Berning, D. E.; Noll, B. C.; Dubois, D. L. *J. Am. Chem. Soc.* **1999**, *121*, 11432.

446. Raebiger, J. W.; Miedaner, A.; Curtis, C. J.; Miller, S. M.; Anderson, O. P.; Dubois, D. L. *J. Am. Chem. Soc.* **2004**, *126*, 5502.

447. Miedaner, A.; Raebiger, J. W.; Curtis, C. J.; Miller, S. M.; DuBois, D. L. *Organometallics* **2004**, *23*, 2670.

448. Coetzee, J. F. *Prog. Phys. Org. Chem.* **1967**, *4*, 45.

449. Kütt, A.; Leito, I.; Kaljurand, I.; Soováli, L.; Vlasov, V. M.; Yagupolskii, L. M.; Koppel, M. A. *J. Org. Chem.* **2006**, *71*, 2829.

450. Kolthoff, I. M.; Chantooni, M. K. *J. Am. Chem. Soc.* **1965**, *87*, 4428.

451. Kaljurand, I.; Kütt, A.; Soováli, L.; Rodima, T.; Mäemets, V.; Leito, I.; Koppel, I. A. *J. Org. Chem.* **2005**, *70*, 1019.

452. Qi, X. J.; Liu, L.; Fu, Y.; Guo, Q. X. *Organometallics* **2006**, *25*, 5879.

453. Li, T. S.; Lough, A. J.; Zuccaccia, C.; Macchioni, A.; Morris, R. H. *Can. J. Chem.* **2006**, *84*, 164.

454. Li, T.; Lough, A. J.; Morris, R. H. *Chem. Eur. J.* **2007**, *13*, 3796.

455. Abdur-Rashid, K.; Fong, T. P.; Greaves, B.; Gusev, D. G.; Hinman, J. G.; Landau, S. E.; Lough, A. J.; Morris, R. H. *J. Am. Chem. Soc.* **2000**, *122*, 9155.

456. Hinman, J. G.; Lough, A. J.; Morris, R. H. *Inorg. Chem.* **2007**, *46*, 4392

457. Streitwieser, A.; McKeown, A. E.; Hasanayn, F.; Davis, N. R. *Org. Lett.* **2005**, *7*, 1259.

458. Chin, B.; Lough, A. J.; Morris, R. H.; Schweitzer, C. T.; Dagostino, C. *Inorg. Chem.* **1994**, *33*, 6278.

459. Ng, W. S.; Jia, G. C.; Huang, M. Y.; Lau, C. P.; Wong, K. Y.; Wen, L. B. *Organometallics* **1998**, *17*, 4556.

460. Cappellani, E. P.; Drouin, S. D.; Jia, G. C.; Maltby, P. A.; Morris, R. H.; Schweitzer, C. T. *J. Am. Chem. Soc.* **1994**, *116*, 3375.

461. Klingler, R. J.; Rathke, J. W. *J. Am. Chem. Soc.* **1994**, *116*, 4772.

462. Walker, H. W.; Kresge, C. T.; Ford, P. C.; Pearson, R. G. *J. Am. Chem. Soc.* **1979**, *101*, 7428.

463. Ryan, O. B.; Tilset, M.; Parker, V. D. *J. Am. Chem. Soc.* **1990**, *112*, 2618.

464. Breslow, R.; Balasubramanian, K. *J. Am. Chem. Soc.* **1969**, *91*, 5182.

465. Breslow, R.; Chu, W. *J. Am. Chem. Soc.* **1973**, *95*, 411.

466. Jaun, B.; Schwarz, J.; Breslow, R. *J. Am. Chem. Soc.* **1980**, *102*, 5741.

467. Bordwell, F. G.; Cheng, J. P.; Harrelson, J. A. *J. Am. Chem. Soc.* **1988**, *110*, 1229.

468. Wang, D. M.; Angelici, R. J. *J. Am. Chem. Soc.* **1996**, *118*, 935.

469. Vites, J.; Fehlner, T. P. *Organometallics* **1984**, *3*, 491.

470. Schock, L. E.; Marks, T. J. *J. Am. Chem. Soc.* **1988**, *110*, 7701.

471. Ziegler, T.; Tschinke, V.; Becke, A. *J. Am. Chem. Soc.* **1987**, *109*, 1351.

472. Lichtenberger, D. L.; Darsey, G. P.; Kellogg, G. E.; Sanner, R. D.; Young, V. G.; Clark, J. R. *J. Am. Chem. Soc.* **1989**, *111*, 5019.

473. Eisenberg, D. C.; Lawrie, C. J. C.; Moody, A. E.; Norton, J. R. *J. Am. Chem. Soc.* **1991**, *113*, 4888.

474. Choi, J.; Tang, L. H.; Norton, J. R. *J. Am. Chem. Soc.* **2007**, *129*, 234.

475. Sweany, R. L.; Halpern, J. *J. Am. Chem. Soc.* **1977**, *99*, 8335.

476. Jacobsen, E. N.; Bergman, R. G. *J. Am. Chem. Soc.* **1985**, *107*, 2023.

477. Ciancanelli, R.; Noll, B. C.; Dubois, D. L.; Dubois, M. R. *J. Am. Chem. Soc.*, **2002**, *124*, 2984.

478. Price, A. J.; Ciancanelli, R.; Noll, B. C.; Curtis, C. J.; Dubois, D. L.; Dubois, M. R. *Organometallics* **2002**, *21*, 4833.

479. Jacobsen, H.; Berke, H. In *Recent Advances in Hydride Chemistry*, 1st ed.; Poli, R., Peruzzini, M., Eds.; Elsevier: Amsterdam, 2001; p 89.

480. Cheng, T. Y.; Brunschwig, B. S.; Bullock, R. M. *J. Am. Chem. Soc.* **1998**, *120*, 13121.

481. Guan, H. R.; Iimura, M.; Magee, M. P.; Norton, J. R.; Janak, K. E. *Organometallics* **2003**, *22*, 4084.

482. Labinger, J. A.; Komadina, K. H. *J. Organomet. Chem.* **1978**, *155*, C25.

483. Mayer, J. M.; Bercaw, J. E. *J. Am. Chem. Soc.* **1982**, *104*, 2157.

484. Martin, B. D.; Warner, K. E.; Norton, J. R. *J. Am. Chem. Soc.* **1986**, *108*, 33.

485. Guan, H. R.; Saddoughi, S. A.; Shaw, A. P.; Norton, J. R. *Organometallics* **2005**, *24*, 6358.

486. Crabtree, R. H. *Science* **1998**, *282*, 2000.

487. Epstein, L. M.; Shubina, E. S. *Coord. Chem. Rev.* **2002**, *231*, 165.

488. Brammer, L. *J. Chem. Soc., Dalton Trans.* **2003**, 3145.

489. Belkova, N. V.; Shubina, E. S.; Epstein, L. M. *Acc. Chem. Res.* **2005**, *38*, 624.

490. Belkova, N. V.; Gutsul, E. I.; Filippov, O. A.; Levina, V. A.; Valyaev, D. A.; Epstein, L. M.; Lledos, A.; Shubina, E. S. *J. Am. Chem. Soc.* **2006**, *128*, 3486.

491. Fairhurst, S. A.; Henderson, R. A.; Hughes, D. L.; Ibrahim, S. K.; Pickett, C. J. *J. Chem. Soc., Chem. Commun.* **1995**, 1569.

492. Casas, J. M.; Falvello, L. R.; Fornies, J.; Martin, A.; Welch, A. J. *Inorg. Chem.* **1996**, *35*, 6009.

493. Clot, E.; Eisenstein, O.; Lee, D.-H.; Crabtree, R. In *Recent Advances in Hydride Chemistry*; Peruzzini, M., Poli, R., Eds.; Elsevier: New York, 2001; Chapter 3.

494. Lough, A. J.; Park, S.; Ramachandran, R.; Morris, R. H. *J. Am. Chem. Soc.* **1994**, *116*, 8356.

CHAPTER 4

Covalent (X-Type) Ligands Bound Through Metal–Heteroatom Bonds

4.1. Overview and Scope

Many anionic ligands are bound to transition metals through a metal–silicon, –boron, –nitrogen, –oxygen, –sulfur, or –phosphorus bond, and these ligands can display reactivity that resembles that of ligands bound to transition metals through carbon. The use of amide and alkoxo groups as ancillary ligands in organometallic systems and the participation of silyl, boryl, amido, and alkoxo ligands in the types of reactions presented in later chapters of this book have increased significantly in the past decades. The later chapters of this book include reactions of complexes of these types of ligands. For this reason, descriptions of these ligands have been provided in a full chapter devoted to the properties of complexes containing metal–heteroatom bonds to X-type ligands.

Like Chapter 3, Chapter 4 will present the physical properties, structural classes, bonding interactions, spectral features, stability, synthesis, and reactivity of common ligands. First, Chapter 4 will survey complexes of σ-bonded ligands in which the metal is attached to nitrogen. These complexes include metal amides, with separate discussions for the much different early and late metal–amido complexes. This section of the chapter also includes complexes of σ-bonded nitrogen heterocycles, as well as the important cationic, ancillary nitrosyl ligand. Second, Chapter 4 will survey the chemistry of polydentate nitrogen ligands, such as porphyrins, tris(pyrazolylborate), and β-diketiminates, which are now important ancillary ligands in organometallic chemistry. Third, Chapter 4 will survey the chemistry of anionic oxygen ligands, such as alkoxides, carboxylates, and β-diketonates. Fourth, Chapter 4 will survey the chemistry of the heavier analogs of amido and alkoxo ligands—namely, metal–phosphides and –thiolates. Finally, Chapter 4 will survey the chemistry of main group silyl and boryl ligands.

4.2. Complexes Containing Metal–Nitrogen Bonds

4.2.1. Metal–Amido Complexes

Metal–amido complexes have been known for decades,[1] but the organometallic chemistry of these ligands has been largely developed in the past few decades. As discussed in Chapters 15–22 of this text covering catalysis, olefin polymerization catalysts containing amido ligands have been some of the most commercially important organometallic catalysts developed in the late 20th and early 21st century, and the coupling of aryl and amido ligands to form arylamines has become an important reaction in synthetic organic chemistry. In addition, emerging catalytic processes, such as certain hydroamination and oxidative amination

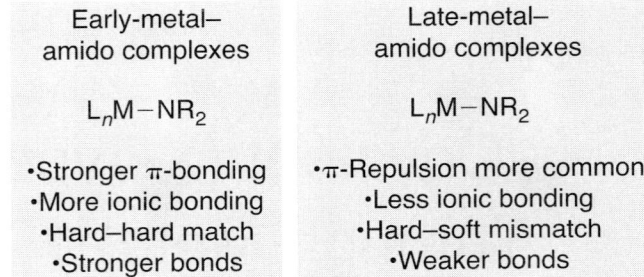

Figure 4.1.
A comparison of typical properties of early- and late-metal amides.

processes, occur through amido ligands. The bonds to early metals and late metals have different degrees of π-bonding, different degrees of ionic bonding, and different matching of a hard ligand with a hard or soft metal. Thus, they have different thermodynamic strengths, as summarized in Figure 4.1. These two classes of complexes also display much different reactivities, and the chemistry of transition metal amido complexes presented here in Chapter 4 is divided into a section on late-metal amides and one on early-metal amides.

4.2.1.1. Late-Metal–Amido Complexes (Written with Prof. Pinjing Zhao)

4.2.1.1.1. Overview of Metal–Amido Complexes of the Late Transition Metals

The reaction chemistry of late-transition-metal–amido complexes resembles that of organometallic complexes more than that of early-transition-metal amides. Thus, the chemistry of this class of amido complex is presented first. Several reviews of the chemistry of late-metal amido complexes have been published.[2]

For many years, transition metal–amido compounds containing low-valent transition metal centers were rare. Since about 1990, the chemistry of this class of compounds has developed significantly. Prior to this time, late-transition-metal–amido complexes were thought to be unstable toward decomposition pathways, such as β-hydrogen elimination. This instability was thought to originate from the mismatch of a soft metal with a hard ligand and the destabilization from π-donation of the nitrogen nonbonded electron pair to a saturated metal center. Indeed, the number of isolated late-transition-metal–amido complexes containing β-hydrogens is small, and the number of dialkylamido complexes containing β-hydrogens is particularly small. At the same time, such complexes have now been shown to be intermediates in some important catalytic processes, including the cross-coupling of aryl halides and amines to form arylamines (Chapter 19). Certain oxidative amination processes have also been proposed to occur through secondary alkyamido complexes (Chapter 16).

In addition to serving as reactive ligands in organometallic compounds, late-transition-metal–amido linkages have begun to be used as ancillary ligands. The set of amido groups that have been used as ancillary ligands for late transition metals is much more limited than that for early metals, but several types of pincer ligands contain nitrogen as the central atom. For example, the chemistry of complexes containing silylamido ligands developed by Fryzuk[3–7] and more recently, diarylamido ligands developed by Ozerov[8–11] and Liang[12] is being developed.

4.2.1.1.2. Bonding of Late-Metal–Amido Complexes

The properties of late-transition-metal–amido complexes result, in part, from the mismatch of a hard ligand with a soft metal. This pairing of ligand and metal leads to a degree of ionic bonding and to π-repulsion between the *d*-electron of the metal and the electron pair on nitrogen. This ionic character, along with the π-repulsion, causes late-transition-metal–amido complexes to be more basic than early-transition-metal–amido complexes.

As noted again later in the section on late-transition-metal–alkoxo complexes, the role of hard/soft and *d*π–*p*π interactions in controlling the properties of these compounds has been debated. In some cases, the π-repulsion has been cited as a factor that leads to the instability of these complexes, but in other cases arguments have been made that the reactivity of these complexes can be explained without invoking π-donation. In other cases, π-donation from an amido group into an unoccupied orbital of a 16-electron late transition metal complex has been proposed to account for the structures of such species.[13]

Certain saturated, late-transition-metal–amido complexes have been shown to be highly basic. For example, 18-electron ruthenium– and iron–amido complexes have been shown to deprotonate hydrocarbons including the benzylic C–H bond of toluene.[14–18] At the same time, the basicity of the amido group, relative to that of an alkoxo group, can be tempered by the slightly softer character of the amido ligand, relative to that of the alkoxide ligand. A set of equilibria involving exchange between the combination of palladium–hydroxo compounds and amines and the combination of palladium–amido complexes and water have revealed that this equilibrium lies in the direction of palladium–amido complexes.[19,20] This acid–base chemistry contrasts with that of alkali metal amides and alkoxides and presumably results from the formation of the strong O–H bond and the less favorable match of the palladium(II) center with the hard alkoxo ligand. Moreover, exchange studies conducted by Bryndza and Bercaw that encompass arylamines indicate that the ruthenium– and platinum–nitrogen bond strengths fall in line with those expected in the absence of overriding π-effects.[21]

4.2.1.1.3. Thermodynamic Properties of Late-Metal–Amido Complexes

Although there are few if any absolute M–N bond energies for late-metal–amido complexes, studies of the relative strengths of metal–ligand bonds have shown that the strengths of the M–N bond in late-metal–amido complexes lie between those of the M–C and M–O bonds of late-metal–alkyl and –alkoxo complexes. Studies on the exchanges between monomeric metal alkoxides and both aryl- and diarylamines showed that the equilibrium for exchange was relatively close to 1.[21] Thus, the strength of the M–N versus M–O bond tracked with the difference between the H–N versus H–O bonds in amines and alcohols. From this finding, one can conclude that the M–N bond in late-metal–amido complexes is not strongly destabilized by the *d*π–*p*π repulsion.

The relative strengths of M–N bonds when the amido group bridges two metals can be different than those for monomeric complexes. Because the amido group is more basic than an alkoxide, it serves as a more strongly bridging ligand. In other words, the sum of the two M–N bonds in a bridging amido group can be greater than the sum of the M–O bonds in a bridging alkoxo complex. As one manifestation of this phenomenon, the equilibrium shown in Equation 4.1 lies toward the side of the amido ligand and free water.[20]

$$(4.1)$$

4.2.1.1.4. Spectral Properties of Late-Metal–Amido Complexes

The spectral properties of amido ligands include upfield ^{1}H NMR chemical shifts and high-frequency infrared vibrations of weak intensity corresponding to the N–H stretches in the amido ligands. Although the ^{1}H NMR chemical shift of the amido ligand varies, this proton resonance often overlaps with the alkyl region for terminal amido complexes,[22,23] and can be found at even higher fields for bridging amido species.[20,24] The infrared stretching frequencies of the N–H bond in an amido group vibrate between 3300 and 3400 cm^{-1};[20,24] however, the intensities of these bands are weak and sometimes unreported. Insufficient data on ^{15}N NMR chemical shifts are available to draw trends between structural types and ^{15}N chemical shifts, but ^{15}N-labeled amido groups have been used to determine the stereochemistry of the complex by measuring the coupling between this nitrogen and the cis or trans phosphine.[25]

4.2.1.1.5. Synthesis of Late-Metal–Amido Complexes

Late-transition-metal–amido complexes have been prepared by metathetical substitution reactions, σ-bonded ligand exchange, deprotonation of amine complexes, and oxidative addition of N–H bonds. Metathetical substitution is the most common route to late-metal–alkylamido complexes, whereas metathetical substitution and σ-bonded ligand exchange have both been used commonly to prepare arylamido compounds.

$$(4.2)$$

$$(4.3)$$

$$(4.4)$$

Representative examples of the synthesis of primary and secondary amides containing β-hydrogens by metathetical substitution are shown in Equation 4.2. In this case, the classic Vaska's complex reacts with a lithium amide to form primary or secondary alkylamido complex containing β-hydrogens.[26] Examples of arylamido complexes formed by σ-bonded ligand exchange are shown in Equations 4.3 and 4.4. The first example was used to measure relative M–O and M–N bond strengths, whereas the second example generates an amido complex that undergoes reductive eliminations (discussed in Chapter 8) and is an intermediate in the catalytic chemistry discussed in Chapter 19. This is an example of a reaction displaying the unusual proton transfer behavior described above.

$$(4.5)$$

$$(4.6)$$

The deprotonation of coordinated amines has also been used to prepare late-transition-metal compounds of the parent amido group ($-NH_2$). This route was first used to prepare a ruthenium parent amido complex, as shown in Equation 4.5[27], and subsequently other metal–amido complexes.[28] In addition, the deprotonation of amine complexes is thought to be the mechanism by which arylpalladium–amido intermediates are generated during the mechanism of the cross-coupling to form amines presented in Chapter 19.[29–33] This step was observed as a stoichiometric process during studies on the intermediates of the original system for this type of cross coupling (Equation 4.6).[34]

$$(4.7)$$

$$(4.8)$$

Finally, late-metal–amido complexes have been prepared by the oxidative addition of amines. These oxidative additions are included in Chapter 7. In brief, the oxidative addition of aniline is favored for several different types of metals. This reaction of electron-poor pentafluoroaniline to Pt(0) was reported many years ago by Stone (Equation 4.7),[35] and the reaction of the parent aniline with Ir(PEt$_3$)$_3$Cl was reported later by Casalnuovo and Milstein.[36] More recently, the oxidative addition of ammonia to an iridium(I) complex containing an electron-rich pincer ligand was observed (Equation 4.8).[28]

4.2.1.1.6. Reactivity of Late-Metal–Amido Complexes

Much of the organometallic reactivity of late-metal–amido complexes is presented in later chapters of this text. In general, these complexes are reactive toward many classic organometallic processes, such as reductive elimination, migratory insertion, and β-hydrogen elimination. Although it was once thought that late-transition-metal–amido complexes containing β-hydrogens would be very unstable,[37] it has been shown more recently that β-hydrogen elimination from an iridium–amido complex is much slower than β-hydrogen elimination from the analogous alkyl species (Equation 4.9).[26] Thus, amido complexes containing β-hydrogen can be isolated and can serve as intermediates in productive catalytic chemistry.

$$(4.9)$$

$$(4.10)$$

$$(4.11)$$

$$(4.12)$$

Examples of the reductive elimination and migratory insertion reactions of representative late-metal–amido complexes are shown in Equations 4.10–4.12. Many organopalladium–amido complexes that are either isolated or intermediates in catalytic coupling reactions undergo reductive elimination; one example is shown in Equation 4.10. Although less common than reductive elimination, examples of the insertions of unsaturated organic reagents into late-metal–amido complexes are beginning to emerge, and examples of reactions that occur by the insertions of α-olefins into metal amides have been reported.[38] The insertion of CO into a metal–amido group is shown in Equation 4.11,[27,39] and the insertion of an α-olefin into a rhodium–arylamido complex is shown in Equation 4.12. In one unusual process involving carbon monoxide and an amido complex, insertion of CO into the N–H bond of a parent amido ligand occurred.[40] These insertion reactions are presented in more detail in Chapter 9.

Amido complexes have also been shown to undergo proton transfers, as discussed above. These proton transfers are observed because the nitrogen atoms in these complexes are often highly basic.[14,15,18]

4.2.1.2. Amido Complexes of the Early Transition Metals (Written with Prof. Seth B. Herzon)[1,41,42]

4.2.1.2.1. Overview

As noted in the introductory section to amido complexes, early-transition-metal–amido complexes have been used as catalysts for olefin polymerization,[43–50] and they have been used as catalysts or precatalysts for alkyne and olefin hydroamination,[51–55] olefin metathesis,[56,57]

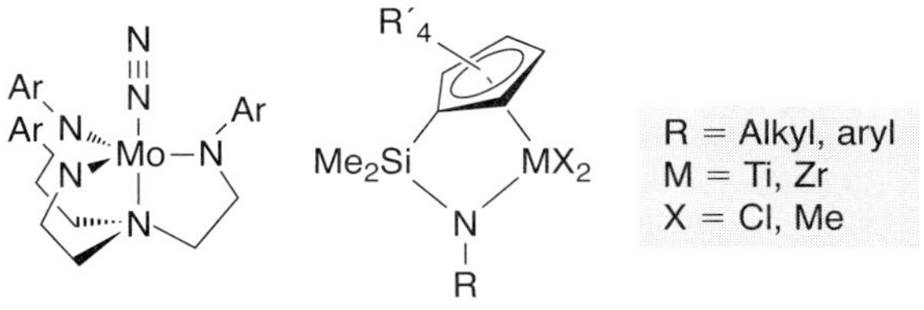

Figure 4.2.
Two types of early-metal
complexes containing
chelating amido ligands.

and transamidation.[58,59] They have also served as precursors to species that activate alkane C–H bonds,[60–64] as important precursors to metallocene complexes,[49,65] and even as precursors to metal–nitride materials by organometallic chemical vapor deposition.[66]

These complexes are highly basic and have strong M–N bonds, but they do undergo β-hydrogen abstraction processes to generate η^2-imine complexes.[67–71] In addition to the role of amido groups as reactive ligands, much chemistry has been described with complexes in which bulky monodentate[72] and polydentate amides[73,74] serve as ancillary ligands. A particularly thoroughly studied set of complexes are triamidoamine complexes.[75,76]

Amido groups are also part of many polydentate ligands containing one or more amide donors and another covalent or dative donor. Complexes such as the species on the left of Figure 4.2, which contains bulky, tetradentate triamidoamine ligands, have been developed for the catalytic reduction of dinitrogen to ammonia.[77] Complexes such as the species on the right of Figure 4.2, which contains an amido ligand tethered to a cyclopentadienyl derivative, are so-called "constrained-geometry" catalysts for the polymerization and copolymerization of α-olefins.[43–46]

4.2.1.2.2. Thermodynamic Properties of Early-Metal–Amido Complexes

Early-transition-metal amides are extremely reactive toward protic acids and are therefore useful precursors to a variety of M–X complexes (in which X is an anionic ligand less basic than an amide) by reaction of the amido complex with H–X (Equation 4.13).[78–81] Common exchange processes that favor elimination of amine involve the reaction with alcohols, amides, amidines, and guanidines. However, weakly acidic hydrocarbons, such as cyclopentadiene derivatives, also react to form Cp complexes (Equation 4.14).[82]

$$[M]-NR_2 \ + \ H-X \ \longrightarrow \ [M]-X \ + \ H-NR_2 \tag{4.13}$$

$$Zr[N(CH_3)_2]_4 \ + \ 2\,Cp-H \ \xrightarrow[\text{Reflux, 2h}]{\text{Benzene}} \ Cp_2Zr[N(CH_3)_2]_2 \ + \ 2\,HN(CH_3)_2 \tag{4.14}$$
$$54\%$$

The bond dissociation energies of the M–N bonds in early-transition-metal–amido complexes are high. For example, the M–N bond dissociation energies in the complexes $M[N(CH_2CH_3)_2]_4$ are 91, 90, and 95 kcal/mol for M = Ti, Zr, and Hf, respectively.[83] Amide ligands are considered to be hard,[84] and there is a significant ionic contribution to bonding when they interact with highly oxidized, electropositive metal centers. This ionic property contributes to the high homolytic bond strengths. The order of bond strengths to early metals is typically M–O > M–N > M–C,[83] and this trend reflects the relative hardness of alkoxide, amide, and alkyl ligands and the favorable pairing of a hard metal with a hard ligand.

In addition, a body of data indicates that these early-metal M–N bonds are strengthened by $p_N{\to}d_M$ π-bonding. The nitrogen lone pairs interact with unoccupied d-orbitals of the high-valent early metals to create this stabilizing interaction. Structural studies of homoleptic amide complexes indicate that the nitrogen atom is nearly planar, and this planarity has often been attributed to this $p_N \rightarrow d_M$ π-bonding.[85] In hexakis(dimethylamino) tungsten, $W[N(CH_3)_2]_6$, such π-bonding is estimated to provide a stabilization energy of ~0.8–1.2 eV (~15–22 kcal/mol).[85] Thus, some combination of the favorable hard–hard interaction, ionic character to the M–N bond, and stabilization from π-bonding explains why amido complexes of high-valent early transition metals are more abundant and more stable than amido complexes of low-valent late transition metals.

4.2.1.2.3. Synthesis of Early-Metal–Amido Complexes

Synthetic routes to homoleptic early-metal–amido complexes are discussed in detail in the review by Lappert.[1] The most convenient route to homoleptic early-metal–amido complexes was developed by Bradley and co-workers and involves treatment of the appropriate metal chloride with excess lithium amide (Equation 4.15).[86,87,88]

$$MCl_n + n\,LiNR_2 \longrightarrow M(NR_2)_n + n\,LiCl \qquad (4.15)$$

An alternative route to early-metal–amido complexes involves treatment of the metal halide complex with excess amine to form the metal–amido complex and amine hydrohalide (Equation 4.16). While somewhat milder than the transmetalation described above, this process does not usually form homoleptic amido species; mixed halo–amido complexes are the more common products of this type of reaction.

$$MCl_n + HNR_2\ (excess) \longrightarrow M(NR_2)_n + n\,HNR_2 \cdot HCl \qquad (4.16)$$

Finally, transamination of dimethylamide complexes with larger alkyl amines has been reported. Removal of the more volatile amine may help to drive the reaction. In this case, steric factors often prevent complete transamination.

4.2.1.2.4. Reactivity of Early-Metal–Amido Complexes

Insertion and α-elimination reactions of early-metal–amido complexes are presented in Chapters 9 and 10 of this book. In addition to these processes, amido complexes undergo several important reactions that cause elimination of amine or formation of imido complexes.

Much of the reactivity of amido ligands involves proton exchange processes that eliminate amine. The exchange is believed to occur by an associative mechanism; consequently, the rate of reaction decreases for sterically congested metal complexes. For example, treatment of $V[N(CH_3)_2]_5$ with *tert*-butanol at room temperature forms $V(O\text{-}t\text{-}Bu)_4$ in good yield,[89] while at this temperature the more encumbered complex $W[N(CH_3)_2]_6$ reacts only slowly with methanol and ethanol, and not at all with *tert*-butanol.[90] The use of bulky N-alkyl substituents allows for the isolation of low-coordinate complexes, such as $Cr[N(i\text{-}Pr)_2]_3$.[72,91] Amido complexes derived from primary amines can also serve as precursors to imido complexes. In many cases, amido halide complexes form imido complexes by loss of hydrogen chloride in the presence of a base (Equation 4.17).[92]

4.2.2. Amidate and Amidinate Complexes of the Early Transition Metals (Written with Prof. Seth B. Herzon)

Amidate and amidinate[93] complexes of early metals are also known. In these complexes, the π-bonding interaction is reduced because of the adjacent π-system. These ligands may bind in a κ[1]- or κ[2]-fashion, or they can bind in an η[3]-fashion with the metal interacting with the entire π-system of the amidate, as shown in Figure 4.3. The synthetic routes used to prepare these complexes are similar to those used to prepare early-metal–amido complexes. The reactivity of these ligands is limited—most often they serve as ancillary ligands that modulate the steric and electronic properties of the central metal. As such, amidate complexes have been employed as catalysts for the hydroamination of terminal alkenes[54,55] and alkynes[53] that is described in Chapter 14.

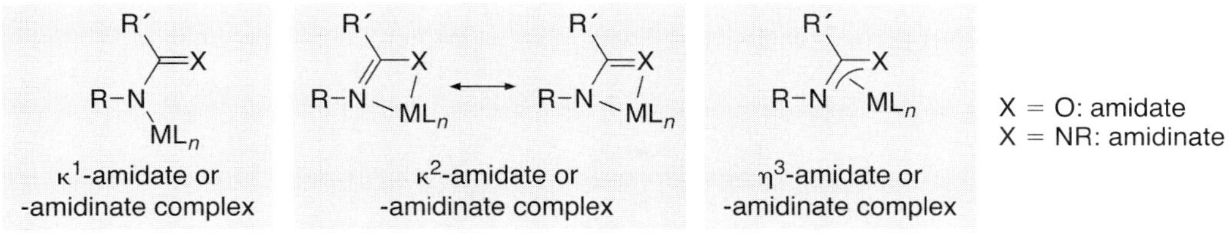

$$\kappa^1\text{-amidate or -amidinate complex} \qquad \kappa^2\text{-amidate or -amidinate complex} \qquad \eta^3\text{-amidate or -amidinate complex} \qquad X = O: \text{amidate} \qquad X = NR: \text{amidinate}$$

Figure 4.3. Generic structures of κ[1]-, κ[2]-, and η[3]-amidate and -amidinate complexes.

4.2.3. Complexes of Anionic Nitrogen Heterocycles (Written with Prof. Jianrong (Steve) Zhou)

4.2.3.1. Overview

Transition metal complexes ligated to the nitrogen of anionic heterocyclic ligands constitute a subclass of compounds containing metal–nitrogen bonds, but they can display distinct structural, bonding, and reactivity patterns from common alkyl and amido complexes.[94–98] Typical heterocycles in this class include pyrroles, indoles, and carbazoles. Some of these complexes have been shown or implied to be reactive intermediates in palladium-[99–103] and copper-catalyzed[104–108] C–N bond formation between azoles and aryl halides or sulfonates and in copper-mediated, oxidative couplings of azoles with arylboronic acids,[109–113] arylsiloxanes,[114] and arylstannanes.[115] In addition, the azolyl group has been used as an ancillary ligand, and complexes of these ligands have been used in asymmetric nucleophilic catalysis.

4.2.3.2. Metal–Azolyl Bonding

Similar to simple amido ligands derived from aliphatic and aromatic amines, anionic N-heterocycles can bind in a monodentate manner through a σ-bond to the metal (Figure 4.4, left). However, they can also bind through the π-system, either on the azoyl group (Figure 4.4, center) or in a fused aromatic group (Figure 4.4, right). A majority of azolyl complexes are bound η[1], but the η[5]-binding mode involving five-membered heterocyclic rings is also relatively common. This η[5]-binding mode is isoelectronic and isostructural to that of η[5]-cyclopentadienyl ligands, and azaferrocene (Equation 4.18)[116,117] is the classic complex containing an η[5]-pyrrolyl ligand.

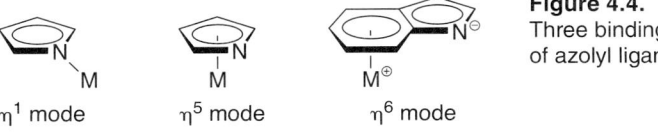

η[1] mode   η[5] mode   η[6] mode

Figure 4.4.
Three binding modes
of azolyl ligands.

Several factors control the binding mode of azolyl ligands. Coordinative unsaturation and electron deficiency on the metal center favor the η^5 mode, while bulky and π-donating ancillary ligands favor the η^1-binding mode. Azoles containing bulky substituents surrounding the nitrogen tend to bind in an η^5 fashion.[118–120] Complexes of benzo-fused heterocycles can bind in an η^6-coordination mode through the arene rings, especially when the heterocyclic moieties contain alkyl groups that sterically hinder binding.[121] In fact, all three coordination modes of the indolyl ligand have been observed on the molybdenum–phosphine fragment in Equation 4.19, depending on available coordination sites and reaction conditions.[122]

$$(4.18)$$

$$(4.19)$$

Because it is less basic than an amido group, the azolyl nitrogen is not as strong a σ-donor as the nitrogen in an alkylamido group. Due to delocalization of the nitrogen lone pair electrons into the π-systems, π-donation from the nitrogen is not as efficient as dialkylamido ligands. The degree of this π-donation in group 4 metallocene–pyrroyl complexes has been the subject of several papers. The most reasonable conclusion seems to be that the π-donation is not sufficiently strong to significantly shorten the M–N bond or to override steric effects on the conformation of the pyrroyl group.[123–125]

Variation of the substituents at nitrogen allows for modulation of the $p_N{\rightarrow}d_M$ π-bonding interaction. For example, titanium complexes containing chelating bis(pyrrolyl) ligands have been synthesized by Odom and co-workers by exchange of the pyrrole ligand with the amino groups of a titanium–dimethylamido complex (Equation 4.20).[52] Because the nitrogen lone pairs are delocalized into the pyrrole rings, $p_N{\rightarrow}d_M$ π-bonding is significantly reduced, and the Lewis acidity of the metal centers is enhanced. These complexes have been shown to be active catalysts for the hydroamination of alkynes with primary amines.[52]

$$(4.20)$$

4.2.3.3. Synthesis of Metal–Azolyl Complexes

Anion displacement of halides with azolyl anions is one common route to azolyl complexes. One example of such a synthesis is shown in Equation 4.21.[126] In other cases, azolyl complexes have been prepared by proton transfer between the free azole and a metal alkoxide[127] or hydroxide.[20] An example involving the synthesis of palladium–azolyl complexes is shown in Equation 4.22. In some rare cases, reactions of pyrrole and d^0 early metal alkyls also lead to the formation of a metal-nitrogen bond via σ-bond metathesis, as shown in Equation 4.23.[128] Finally, several late-transition-metal–azolyl complexes possessing accompanying hydride ligands have been prepared by N–H activation of pyrrole and other azoles.

Examples of reactions with low-valent group 8,[129,130] 9,[131,132] and 10[35,133–135] complexes are shown in Equations 4.24–4.26.

$$(4.21)$$

$$(4.22)$$

$$(4.23)$$

$$(4.24)$$

$$(4.25)$$

$$(4.26)$$

4.2.3.4. Reactivity of Metal–Azolyl Complexes

The reactivity of azolyl complexes are controlled by the electronic properties of these groups and by the η^1- versus η^5-binding mode. Like free pyrrole, an η^1-bound pyrrolyl ligand does not have a basic or nucleophilic nitrogen. Thus, η^1-pyrrolyl complexes tend to undergo protonation and acetylation[136] at the C-2 position. The protonated intermediate has been isolated and was shown to subsequently rearrange to carbon-ligated isomers, as shown in Equation 4.27.[137] Consistent with the attenuated nucleophilicity, the pyrrolyl ligand was shown to be much less reactive toward reductive elimination from palladium(II) than amido ligands or even methyleneamido ligands that are also bound to the metal through an sp^2-hybridized nitrogen, as summarized in Equation 4.28.[99]

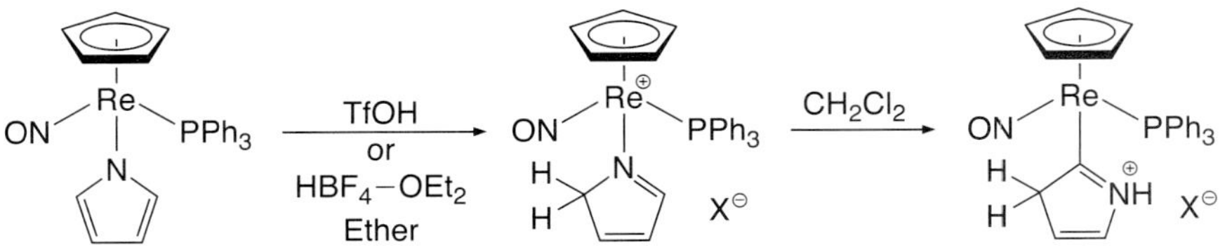

(4.27)

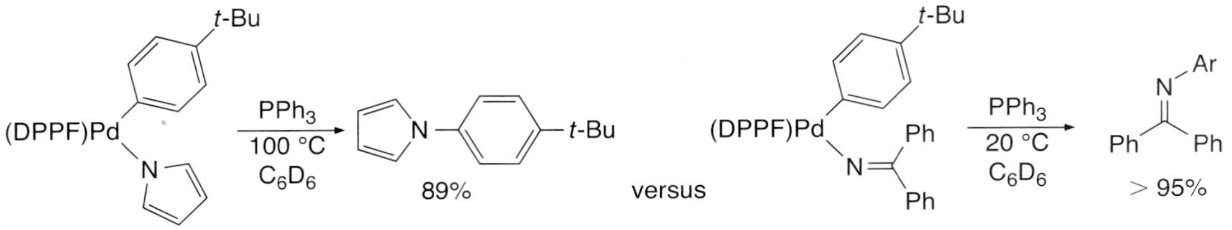

(4.28)

η^5-Pyrrolyl complexes are stable due to the polyhapto binding mode, but they are less stable than η^5-cyclopentadienyl compounds. It remains controversial whether this difference in stability stems from a thermodynamic or kinetic effect.[138,139] Nevertheless, the high stability of η^5-pyrrolyl complexes allows several chemical modifications to be performed directly on the five-membered ring.[140,141] Lithiation and alkylation of the resulting anion is shown in Equation 4.29.[141]

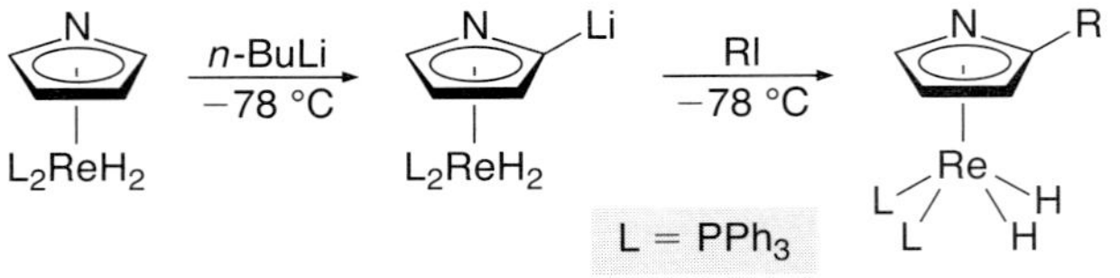

(4.29)

The N_{sp2} lone pair in the η^5-pyrrolyl ligands is the most basic site in η^5-azolyl complexes, and is the site of protonation and electrophilic attack.[120,142] Novel planar chiral catalysts have been prepared, based on the azaferrocene unit, and these complexes have been used for asymmetric nucleophilic catalysis.[143] One example of such a complex is shown on the left in Figure 4.5.[144] Some linked versions of azaferrocenes possessing C_2 symmetry create an unconventional chiral environment. The ligand on the right in Figure 4.5 has been used for several asymmetric copper-catalyzed transformations.[145]

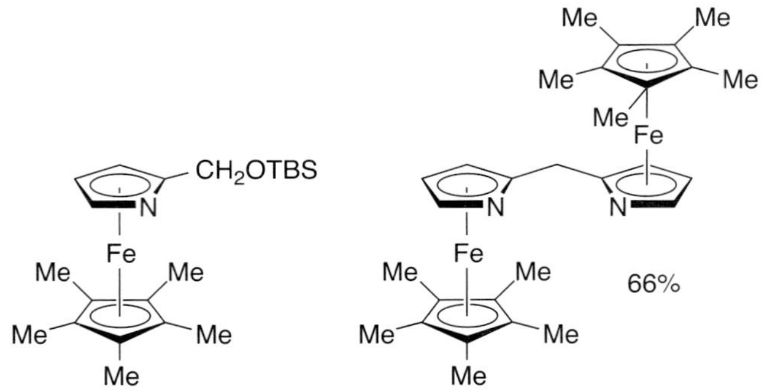

Figure 4.5.
Examples of azaferrocenes used in asymmetric catalysis.

4.2.4. Nitrosyl Complexes (Written with Prof. Jesse W. Tye)

4.2.4.1. Overview

Nitric oxide (NO) is a common ancillary ligand, and NO complexes are known for all the transition metals except those of the Sc and Ti triads. In its cationic form, NO^+, the NO ligand is isoelectronic with CO. The reaction chemistry of coordinated NO is more limited than that of CO, but insertions of NO into metal–alkyl linkages and nucleophilic attack on coordinated NO (classic reactions of CO are discussed later in this text) are known. The coordination chemistry of NO has become particularly important in biology because an NO ligand is found in the nitrile hydratase enzyme[146] and in complexes that are part of biological signaling pathways.[147] The chemistry of transition metal NO complexes has been extensively reviewed.[148–157]

4.2.4.2. Properties of the Free Molecule

Nitric oxide is a simple neutral paramagnetic molecule, and the open shell influences the properties and reactivity of NO. For example, one-electron oxidation of NO to yield the diamagnetic nitrosonium cation (NO^+) is much more facile than the oxidation of the closed-shell analog CO. In fact, the nitrosonium cation is stable, and crystalline salts such as $[NO]^+[BF_4]^-$ are commercially available.

The dimeric form of NO, $(NO)_2$, lies uphill of the monomer in free energy at room temperature. Formation of the NO dimer is unfavorable in the gas phase[158] [$K_{eq} < 3 \times 10^{-4}$ M^{-1} at 306 K for the reaction $2NO \rightleftarrows (NO)_2$], but is naturally more favorable at lower temperatures and in the condensed phase. The enthalpy for dimerization is between -1 and -3 kcal/mol in the gas phase.[159–162]

NO is isoelectronic with N_2^-, CO^- and O_2^+. The molecular orbital diagram of NO is qualitatively similar to that of CO or N_2. The additional electron of NO lies in an antibonding π^*-orbital, which causes NO to have an N–O bond order of 2.5. Oxidation of NO to NO^+ generates a molecular cation that lacks this electron in the antibonding π^*-orbital. This cation is isoelectronic with N_2 and CO and contains an N–O bond order of 3. This difference in bond order between NO• and NO^+ is supported by infrared and Raman spectroscopic data. The fundamental stretching frequencies of NO and NO^+ are 1876 and 2377 cm^{-1}, respectively. These values can be compared to the fundamental stretching frequencies of 1860, 2143, and 2330 cm^{-1} for O_2^+, CO, and N_2, respectively.[156, 163]

4.2.4.3. Structures and Electron Counting of Metal–Nitrosyl Complexes

The structures of metal–nitrosyl complexes fall into two broad categories: complexes with "linear" M–N–O angles ($\angle$ M–N–O ≈ 160–$180°$) and those with bent M–N–O angles ($\angle$ M–N–O ≈ 120–$140°$). Representative complexes containing linear and bent nitrosyl ligands are shown in Figure 4.6. A majority of NO complexes contain a relatively linear M–N–O linkage. The NO ligand bound in a linear fashion is generally considered to be a poorer σ-donor and a better π-acceptor than CO.[164–167]

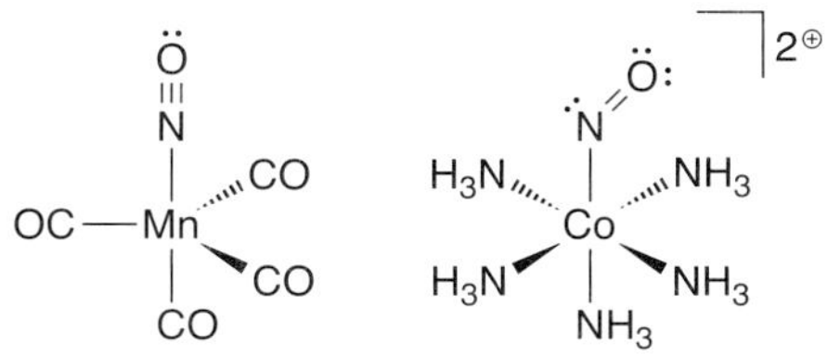

Figure 4.6.
Examples of complexes containing linear and bent NO ligands.

A given NO complex may exist solely as the linear or bent form or may exist as an equilibrium mixture of the two forms. In other words, the linear and bent structures of NO ligands can be discrete energy minima for the same complex. The two forms are often separated by such a low kinetic barrier that they equilibrate, and this equilibration leads to crystallographic disorder in the positions of the NO ligands. Thus, the precise value of the M–N–O bond angle is often difficult to determine, even when the crystallographic R-factor is low.[168]

The charge and electron donation by linear and bent metal–nitrosyl complexes are often considered distinct. The binding of NO to a transition metal center to form a linear metal–nitrosyl linkage has been considered to involve a formal one-electron reduction of the metal center. In the ionic bonding formalism, the bent NO ligand is typically assigned as NO^-, while the linear NO ligand is typically assigned as NO^+. These charges affect the oxidation state and formal d-electron count. Binding of NO in a linear fashion would decrease the oxidation state by one, while binding of NO in a bent fashion would increase the oxidation state by one.

Although this assignment of charge and oxidation state is often found in the literature, the assignment of oxidation states to the metal and NO ligand is often controversial. Enemark and Feltham have stressed the covalency of the M–N bond in metal–nitrosyl complexes and suggest that the electron count of the M–NO unit is the most important factor controlling the M–N–O angle.[168] In fact, the counting of electrons in metal–nitrosyl complexes is more conveniently conducted with the covalent model. Using this method, bent NO is assigned as a one-electron donor and linear NO is assigned as a three-electron donor because of the additional electron donation from the nitrogen electron pair.

To clarify the assignments of electrons and formal charges, examples of electron counting in the bent nitrosyl complex $Mn(NO)(CO)_4$ and the linear nitrosyl complex $[Co(NO)(NH_3)_5]^{2+}$ are given below using both the covalent and ionic formalisms.

EXAMPLE

$Mn(NO)(CO)_4$ containing a linear metal–nitrosyl group ($\sphericalangle$Mn–N–O = 180°)

Covalent electron-counting method:
  Mn = 7 e⁻
  One linear NO ligand = 3 e⁻
  Four CO ligands = 8 e⁻
  Total = 18 e⁻

Ionic electron-counting method:
  Mn⁻ = 8 d e⁻
  Linear NO ⟹ NO⁺ ligand = 2 e⁻
  Four CO ligands = 8 e⁻
  Total = 18 e⁻

EXAMPLE

$[Co(NO)(NH_3)_5]^{2+}$ containing a bent metal–nitrosyl group ($\sphericalangle$Co–N–O = 119°)

Covalent electron-counting method:
  Co = 9 e⁻
  One bent NO ligand = 1 e⁻
  Five NH₃ ligands = 10 e⁻
  2+ total charge = –2 e⁻
  Total = 18 e⁻

Ionic electron counting-method:
  Co³⁺ = 6 e⁻
  Bent NO ⟹ NO⁻ ligand = 2 e⁻
  Five NH₃ ligands = 10 e⁻
  Total = 18 e⁻

4.2.4.4. Thermodynamics of M–NO linkages

In general, the metal–nitrogen bond orders of metal–nitrosyl complexes are larger (i.e., the M–N bonds are stronger) than the metal–carbon bond orders for the corresponding metal carbonyl complexes.[169,170] The nitrogen–oxygen bond orders of metal–nitrosyl complexes are smaller than the carbon–oxygen bond orders of the corresponding carbonyl complexes, based on calculations of bond force constants.[169,170] These differences are thought to result from the stronger π-accepting ability of the nitrosyl ligand.[171]

4.2.4.5. Spectral Features of M–NO Complexes

The most distinctive spectral property of the NO ligand is the strong infrared ν_{NO} band. When NO binds to a transition metal, its ν_{NO} increases in some cases and decreases in others.[156] Terminal mononitrosyl complexes exhibit strong ν_{NO} bands over a very wide frequency range: $\nu_{NO} = 1950–1450$ cm^{-1}. The ν_{NO} values of complexes containing a bridging NO ligand are usually lower ($\nu_{NO} = 1650–1300$ cm^{-1}). These frequencies, which reflect varying degrees of π-backbonding, can be compared to the value for free neutral NO of 1876 cm^{-1} and the value for NO$^+$ salts of 2200–2300 cm^{-1}.

Since the range of ν_{NO} values for linear and bent NO ligands overlap, the ν_{NO} frequency cannot be used as a reliable criterion to distinguish between linear and bent terminal NO groups.[156] A better diagnostic criterion for bent nitrosyl groups is the ^{14}N or ^{15}N NMR chemical shift.[172] The range of ^{15}N NMR chemical shifts for linear NO complexes is −100 to +200 ppm (relative to liquid nitromethane = 0 ppm), while the range of ^{15}N NMR chemical shifts for bent NO complexes is 300–1000 ppm.

4.2.4.6. Synthesis of NO Complexes

Many methods have been used to prepare NO complexes. The reactions of transition metal complexes with gaseous NO or solutions of the dissolved gas have been used to form NO complexes. One example of such a reaction to form a homoleptic nitrosyl complex is shown in Equation 4.30.[173] Alternatively, the reactions of metal complexes with nitrosonium salts form metal–nitrosyl complexes. (Note: Alcohol solvents should be used with caution, because they convert NO$^+$ to alkyl nitrites.[174] In addition, tetrahydrofuran should not be used as solvent because it can be polymerized by NO$^+$.[175]) Two examples of such processes are shown in Equations 4.31[176] and 4.32.[177] A range of organic nitroso compounds release "NO" under appropriate conditions[178,179] and have been used to introduce NO into transition metal complexes. The generation of a metal–nitrosyl complex from F$_3$CNO is shown in Equation 4.33.[180] Inorganic nitrites have also been used to prepare NO complexes,[151,181,182] as shown in Equation 4.34.[182]

$$\text{(4.30)}$$

$$\text{(4.31)}$$

$$\text{(4.32)}$$

$$\text{(4.33)}$$

$$\text{OC} - \underset{\underset{\text{CO}}{|}}{\overset{\overset{\text{CO}}{|}}{\text{Fe}}} \overset{\text{,,,,CO}}{\underset{\text{CO}}{}} + \text{NO}_2^{\ominus} \longrightarrow \left[\text{OC} - \underset{\text{CO}}{\overset{\overset{\text{NO}}{|}}{\text{Fe}}} \overset{\text{,,,,CO}}{} \right]^{\ominus} + \text{CO}_2 \qquad (4.34)$$

4.2.4.7. Reactivity of Metal–Nitrosyl Complexes

Nitrosyl complexes undergo reactions at the metal that are influenced by the properties of the NO ligand, and they undergo reactions at the nitrosyl ligand itself. Most often, the nitrosyl ligand is an ancillary ligand that affects the electronic properties of the metal. However, the ability of the ligand to interconvert between linear and bent forms can impact the rates of ligand substitution reactions.

Ligand exchange reactions of NO complexes are often faster than those of the CO analogs, as described in more detail in Chapter 5.[183] This effect results from the ability of NO to undergo associative ligand substitution processes by a concomitant change in bonding of the NO ligand from linear to bent to decrease the electron count of the metal center and allow binding of an additional ligand. The NO ligands themselves are much more resistant to ligand substitution than are CO groups.

$$\underset{\text{ON}}{\overset{}{}}\overset{\text{Co}}{\underset{\text{CH}_3}{}} + \text{PPh}_3 \longrightarrow \overset{}{O=}\underset{\underset{\text{CH}_3}{|}}{N}\overset{\text{Co}}{\underset{\text{PPh}_3}{}} \qquad (4.35)$$

Nitrosyl complexes have also been shown to undergo migratory insertion processes with alkyl ligands to form metal–nitroso complexes—that is, M–[N(O)R]—as shown in Equation 4.35 and described in more detail in Chapter 9. This chemistry parallels the insertion of CO into metal–alkyl bonds.

4.2.5. Polydentate Nitrogen Donor Ligands

In contrast to amido ligands that are often the reactive site in transition metal complexes, polydentate nitrogen ligands are more often ancillary ligands. Although less common than phosphines in catalytic organometallic chemistry, many types of polydentate nitrogen donor ligands have been used in stoichiometric organometallic chemistry, and the use of such ligands in catalysis has been growing rapidly. Polydentate nitrogen ligands include porphyrins and related tetradentate dianionic ligands, bis-sulfonamide ligands that have been important in asymmetric catalysis with early metal complexes, bis- and trispyrazolylborate ligands, and β-diketiminate ligands. The following sections describe the properties of these types of ligands, the synthesis of complexes of these ligands, and selected reactions of complexes containing these ancillary ligands. The reader is urged to consult the more comprehensive reviews cited in each section. The compounds discussed in the following subsections were selected because they relate most closely to the subject matter of the subsequent chapters of the text.

Many polydentate nitrogen ligands are useful supporting ligands in organometallic chemistry. These ligands can serve as 4-, 6-, or 8-electron donors possessing varying electronic properties. The next few sections surveys these classes of nitrogen ligands that are typically assigned with negative charges in the ionic electron-counting formalism.

4.2.5.1. Organometallic Porphyrin and Corrin Complexes[184,185] (Written with Giang Vo)

4.2.5.1.1. Overview

Porphyrin ligands are dianionic ligands containing four nitrogen donors. The core of the porphyrin ring, shown in Figure 4.7, consists of four pyrrole groups linked by methene

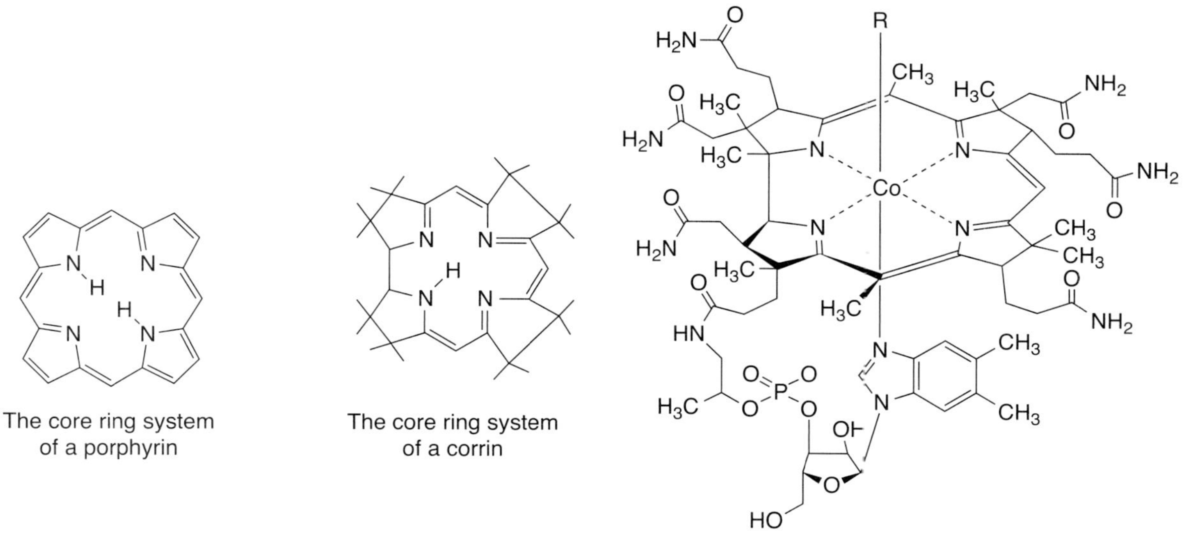

Figure 4.7. Structure of porphyrin and corin ligands, and of vitamin B_{12}.

The core ring system of a porphyrin

The core ring system of a corrin

groups. The corrin ring system, which is closely related to the porphyrin ring, contains one set of pyrrole groups directly linked. Corrins are monoprotic and are typically assigned as monoanionic ligands. Porphyrin and corrin ligands originate from metallobiochemistry, but have been used to prepare organometallic compounds. Some organometallic porphyrin complexes are discussed further in the sections on C–H bond cleavage, insertions of olefins, and carbenoid chemistry. In addition, a cobalt–corrin unit is the core of vitamin B_{12}, which is a rare example of an organometallic species in biology.

4.2.5.1.2. Structures of Metal–Porphyrin Complexes

The structures of metal–porphyrin complexes depend on the radius of the metal. Complexes of the larger early metals typically contain a metal that lies outside of the plane of the porphyrin ring, while complexes of the smaller late metals contain the metal within the plane of the porphyrin.[186] For example, [(OEP)ZrCl$_2$] (where OEP = octaethylpoyrphyrin) contains a metal that lies 0.89 Å[187] outside of the plane, while the rhodium–porphyrin complex contains the metal in the porphyrin plane. This geometry controls the relative orientation of the additional ligands. For example, the out-of-plane geometry directs the additional ligands on the metal to be cis to each other, while the in-plane geometry forces them to be trans (Figure 4.8).

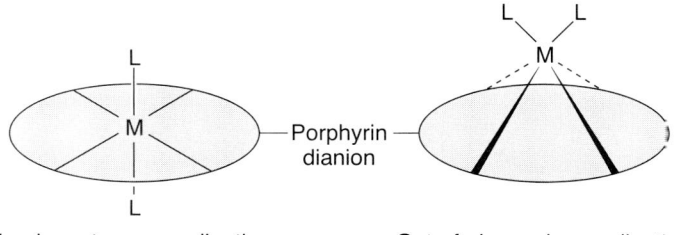

In-plane, trans coordination   Out-of-plane, cis coordination

Figure 4.8.
Effect of metal size on coordination mode and disposition of the ancillary ligands.

The steric properties of the porphyrin ring help control the nuclearity of the species. For example, the rhodium–tetraphenylporphyrin (TPP) complex in the equilibrium of Equation 4.36 is predominantly dimeric, while the tetramesitylporphyrin (TMP) complex in Equation 4.37 is predominantly monomeric.[188,189] Generally, the bond dissociation energies of the Rh–Rh bond in dimeric rhodium–porphyrin complexes lie in the range of 8–25 kcal/mol.[188,190,191]

$$(TPP)Rh-Rh(TPP) \rightleftharpoons 2\ (TPP)Rh\cdot \qquad (4.36)$$

$$(TMP)Rh-Rh(TMP) \rightleftharpoons 2\ (TMP)Rh\cdot \qquad (4.37)$$

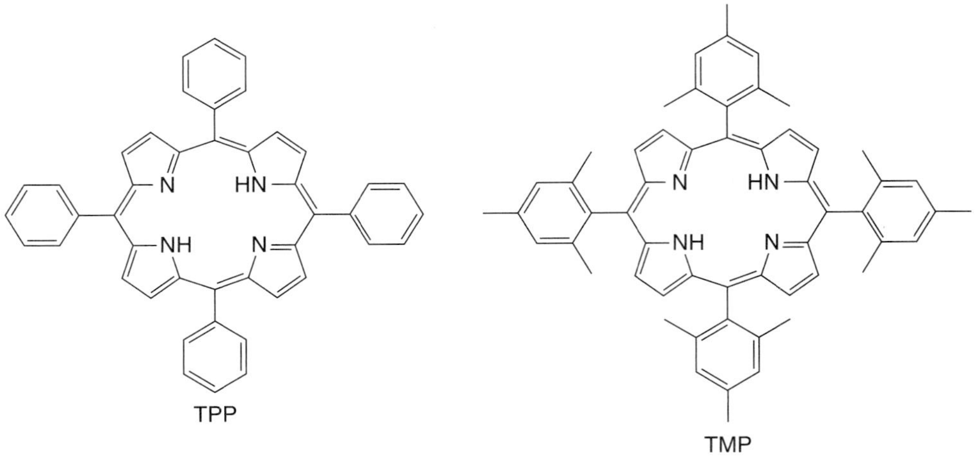

TPP

TMP

4.2.5.1.3. Synthesis of Metal–Porphyrin Complexes

Transition-metal–porphyrin complexes are typically prepared by the reaction of metal halides with the alkali metal salts of the porphyrin ligand.[186,187,192,193] For example, [Fe(OEP)Cl] is prepared by the addition of the dilithium salt to FeCl$_3$, as shown in Equation 4.38,[194] while [Zr(OEP)(OAc)$_2$] has been prepared from the same porphyrin reagent and Zr(acetylacetonate)$_2$, followed by workup with acetic acid and pyridine (Equation 4.38).[195] However, alternative routes include the reaction of the protonated porphyrin with the metal halide.[189,196] After formation of the metal–porphyrin complex, reactions of these complexes with alkylating agents forms organometallic derivatives, much like reactions of compounds containing softer ancillary ligands. Carbenoid and nitrene intermediates of metal–porphyrin species have been generated by the reaction of the porphyrin complex with diazoalkanes and the combination of sulfonamide and PhI(OAc)$_2$.[197–199]

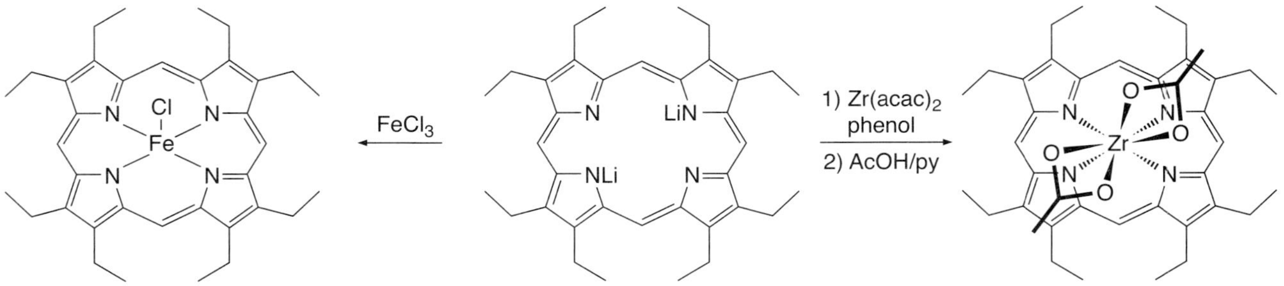

$$(4.38)$$

4.2.5.1.4. Reactivity of Metal–Porphyrin Complexes

The reactions of organometallic porphyrin complexes depend on the geometry. For example, late-metal–porphyrin complexes often react by radical pathways because there is no open coordinate site cis to the reactive ligands. These radical reactions include polymerizations and C–H activations. Cobalt–porphyrin complexes, such as Co(TMP), catalyze polymerizations by the atom-transfer radical polymerization mechanism shown in Scheme 4.1.[200, 201] C–H activation by rhodium–porphyrin complexes[202] is described in Chapter 6. In contrast, early-metal–porphyrin complexes that contain two cis ligands undergo reactions that require a cis geometry, such as migratory insertion into metal–alkyl bonds and concerted reductive eliminations.

$$(TMP)Co^{III}-R \; \rightleftharpoons \; (TMP)Co^{II\bullet} + {}^{\bullet}R$$

$$n\,CH_2{=}CH(CO_2R') + {}^{\bullet}R \longrightarrow {}^{\bullet}P$$

$$(TMP)Co^{II\bullet} + {}^{\bullet}P \; \rightleftharpoons \; (TMP)Co^{III}-P$$

P = polymer

Scheme 4.1.
Mechanism of atom-transfer radical–radical polymerization.

4.2.5.2. Bis-Sulfonamide Complexes (Written with Prof. Patrick J. Walsh)

Bis-sulfonamides have become important ligands in asymmetric catalysis. These ligands, which are deprotonated to give bis-sulfonamido complexes, have predominantly been used for early-metal-catalyzed processes, such as the enantioselective additions of alkylzinc reagents to aldehydes[203–205] and ketones.[206,207] They have also been used successfully in the asymmetric cyclopropanation of allylic alcohols with a zinc bis-sulfonamide-based catalyst,[208] as well as a variety of processes catalyzed by Lewis acids.[209–211] The pK_a values of the conjugate acids of sulfonamido ligands are much different from those of amido ligands. The pK_a of the sulfonamide N–H is around 10,[212] whereas that of a dialkyl amine is well over 30. The decrease in donor ability of sulfonamido groups makes bis-sulfonamido complexes Lewis acidic.

4.2.5.2.1. Bonding in Bis-Sulfonamido Complexes

Bis-sulfonamido ligands can adopt several binding modes, and the one adopted depends on the nature of the metal center. For example, the ligand can bind through the nitrogens alone or through a combination of the nitrogen and the sulfonyl oxygens in an intermolecular[213, 214] or intramolecular fashion.[215,216] Some coordination modes of bis-sulfonamido ligands found in the solid state are illustrated in Figure 4.9. In general, late-metal[214,2:7] and main group metals interact with the sulfonamido nitrogens and do not coordinate the sulfonyl oxygens

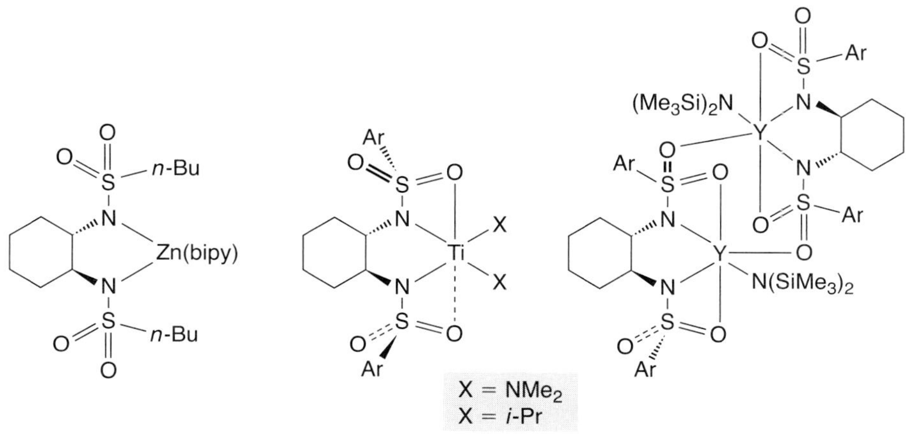

Figure 4.9. Solid state structures of some bis-sulfonamido complexes.

intramolecularly. They can bind them intermolecularly, however, to form dimers[218] or oligomers.[214] In contrast, highly oxophilic early transition metals often coordinate the sulfonyl oxygens intramolecularly in the solid state.[213,215,216,219,220] If strongly donating ancillary ligands are present, such as amido ligands (X = NMe$_2$), the Lewis acidity of the metal is reduced, and only one or two of the four sulfonyl oxygens binds to the metal. If less electron-donating ligands, such as alkoxides, are present, then high-valent early metal centers tend to bind one oxygen of each of the sulfonamido groups.

The bonding of the sulfonamido group is distinct from that of an amido group (NR$_2$). The Ti–N distances of sulfonamido complexes range from 2.048(3) to 2.103(2) Å, while the Ti–N distances of amido complexes are significantly shorter, ranging from 1.858(3) to 1.896(3) Å.[215] In contrast to the nitrogens in early-metal–amido complexes, which tend to be planar, the nitrogens of sulfonamido ligands are often pyramidal. The large difference in the Ti–N distances is due to two factors, the delocalization of the sulfonamido nitrogen lone pairs into the electron-withdrawing sulfonyl groups in the sulfonamido complexes, and delocalization of the lone pair of the amido ligand into an empty orbital on the titanium center.

Bis-sulfonamido ligands are useful in asymmetric catalysis, in part because coordination of a sulfonyl oxygen renders the sulfurs stereogenic. This increases the number of stereogenic centers in the bis-sulfonamido ligand to four and leads to a stereogenic center close to the metal center. Evidence supporting the C$_2$ symmetry of the ligand during catalytic asymmetric additions of organozinc reagents to aldehydes suggests that sulfonyl coordination may be an important factor in the transmittance of asymmetry to the substrate.[221]

4.2.5.2.2. Synthesis of Bis-Sulfonamide Complexes

Because of the high acidity of the sulfonamide N–H, acid–base reactions are used most commonly to prepare metal–sulfonamido complexes. Some examples of such acid–base reactions are illustrated in Equations 4.39 and 4.40. Reaction of Ti(NMe$_2$)$_4$ with one equivalent of a bis-sulfonamide gives a (bis-sulfonamido)Ti(NMe$_2$)$_2$ complex, which can be isolated or can react with an additional equivalent of ligand to generate an unusual eight-coordinate titanium complex.[222] Reaction of the platinum carbonate with the bis-sulfonamide results in elimination of CO$_2$ and water to give the (bis-sulfonamido)Pt(dppe) complex in Equation 4.40.[217]

$$\text{(4.39)}$$

$$\text{(4.40)}$$

4.2.5.2.3. Thermodynamics of Metal–Bis-Sulfonamido Bonds

Although thermodynamic data on bis-sulfonamido complexes are scarce, some useful comparisons of bond strengths have been made. The exchange equilibrium between a bis-sulfonamide and the alkoxosulfonamido ligand in the platinum complex of Equation 4.41

results in formation of the bis-sulfonamido complex ($K_{eq} > 360$ for both R = CF$_3$ and 4-C$_6$H$_4$-CMe$_3$). Moreover, the less basic bis-sulfonamido containing the electron-withdrawing CF$_3$ groups on sulfur is preferred over the more basic sulfonamido complex containing aryl groups on sulfur (Equation 4.42). Both of these results are consistent with the preference of Pt(II) for the less basic ligands (i.e., NSO$_2$CF$_3$ > NSO$_2$Ar > OR).[217]

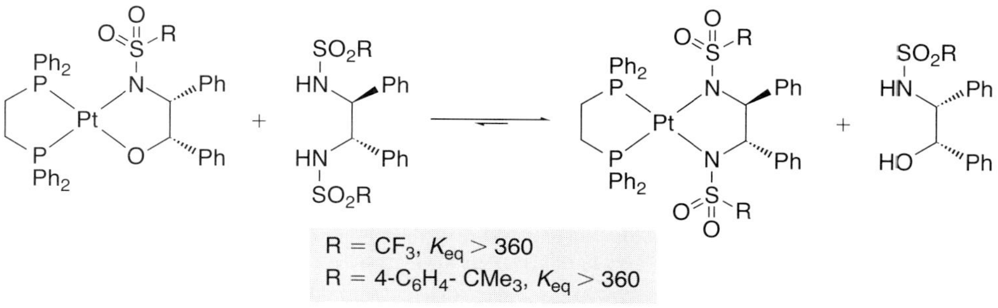

$$R = CF_3, K_{eq} > 360$$
$$R = 4\text{-}C_6H_4\text{-} CMe_3, K_{eq} > 360$$

(4.41)

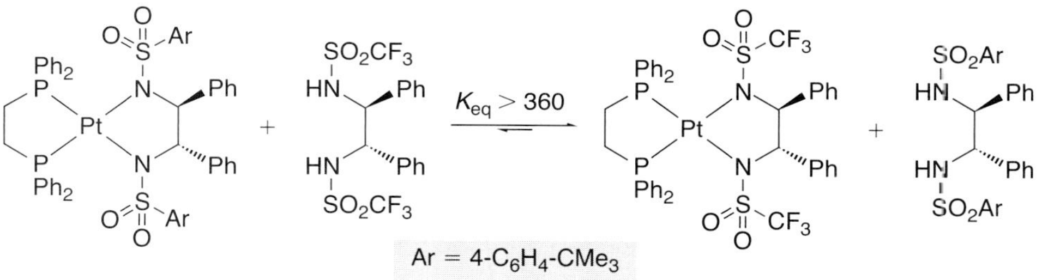

$$Ar = 4\text{-}C_6H_4\text{-}CMe_3$$

(4.42)

The relative binding energies of alkoxo and sulfonamido groups to early metals are different from those to platinum and other late transition metals. High-valent early metals form stronger bonds to oxygen than nitrogen. This equilibrium is important when generating sulfonamido-ligated catalysts in situ from early metal alkoxides. Initially, it was assumed that the reaction of the bis-sulfonamide ligand with a group 4 alkoxide complex would generate the free alcohol and the bis-sulfonamide complex. This equilibrium, however, lies far toward the free bis-sulfonamide ligand and Ti(O-i-Pr)$_4$. Thus, isolated group 4 complexes of these ligands have been prepared by reaction of Ti(O-i-Pr)$_2$(NMe$_2$)$_2$ with the bis-sulfonamide ligand to extrude the free amine, as shown in Equation 4.43.[216]

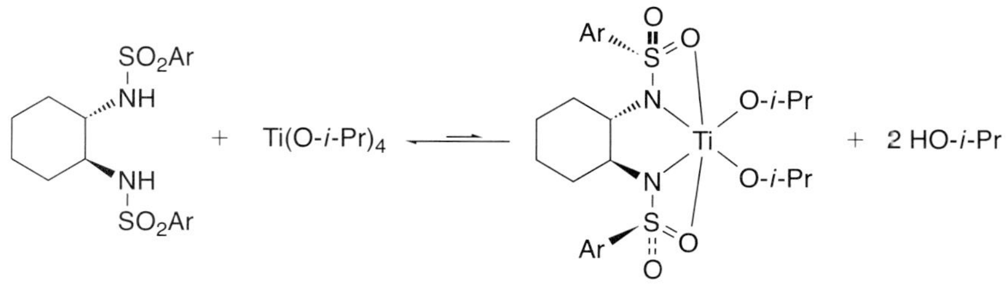

(4.43)

4.2.5.3. Pyrazolylborate Ligands (Written with Dr. Jaclyn M. Murphy)

4.2.5.3.1. Overview

Polypyrazolylborate ligands are now common ancillary ligands in organometallic chemistry, and complexes of the most commonly used trispyrazolylborate ligands are known for all of the transition metals. Pyrazolylborate ligands have been known for many decades. Bis- and trispyrazolylborate ligands and complexes of them were first reported by Trofimenko in 1966.[223] Hundreds if not thousands of compounds have now been prepared containing pyrazolylborate ligands that possess a variety of substitution patterns on the pyrazolylborate ligand. Pyrazolylborate complexes have been reviewed many times.[224–226] Often the trispyrazolylborate ligand is abbreviated Tp, and substituted Tp ligands (vide infra) are abbreviated Tp′ or Tp*. Bispyrazolylborate ligands are usually abbreviated Bp.

4.2.5.3.2. Bonding of Polypyrazolylborate Ligands

Trispyrazolylborate ligands are formally anionic, and this charge results from the four-co-ordinate boron linking the pyrazolyl groups. Neutral silicon analogs of these ligands have also been prepared, but this chemistry has been studied less extensively. Likewise, bispyrazolylborate ligands are considered to be anionic. The ligands bind to the transition metal through one nitrogen electron pair on each pyrazolyl group. Thus, the trispyrazolylborates are considered to be six-electron donors and the bispyrazolylborates to be four-electron donors.

The similarity of the formal charge and number of electrons of the trispyrazolylborate ligands and cyclopentadienyl ligands have led people to consider the relationship between the properties of these two classes of ligands. In fact, the Tp abbreviation results from the similarities between trispyrazolylborate ligands and cyclopentadienyl (Cp) ligands. Comparsions of the electron-donating ability of Cp versus Tp ligands have shown that this property depends on the metal center.[227] For example, Cp* was found to be more electron donating than the Tp* ligand in Cp*Ir(III) and Tp*Ir(III) complexes,[228] but Tp was found to be a stronger electron donor than Cp in the vanadium complexes VL(Cl)$_2$(= N^{t}Bu), in which L = Tp or Cp.[229,230] The electron donating ability of the various members of the class of Tp ligands depends, however, on the substituents bound to the pyrazolyl group.

The difference in steric properties of the Tp and Cp ligand is more striking. The orientation of substituents on a Cp and Tp ligand, relative to the metal, is much different. As shown in Figure 4.10, the substituents on the Tp ligand are directed more toward the metal than are those on a Cp ligand. Thus, Tp ligands containing methyl and *tert*-butyl groups at the 3- and 5-positions are much more hindered than Cp ligands containing alkyl substituents.

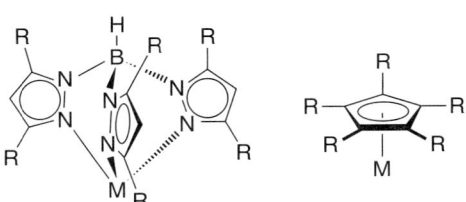

Figure 4.10.
A comparison of the orientation of substituents in Tp and Cp derivatives. This figure is adapted from reference 227.

Complexes of the Tp ligand tend to adopt certain geometries because of the relative positioning of the pyrazolyl groups. The three pyrazolyl groups must bond to the metal in a facial manner. Thus, octahedral and tetrahedral complexes are common. Sterically hindered Tp ligands that tend to generate complexes of low coordination number have even been nicknamed "tetrahedral enforcers."[231] Because many Tp complexes of this geometry possess more than four electrons, complexes of sterically hindered Tp ligands are often paramagnetic.

$$\kappa^3 \text{ - Tp ligand} \qquad \qquad \kappa^2 \text{ - Tp ligand}$$

M = Rh (I), Ir (I) d^8 M = Rh (I), Ir (I) d^8

(4.44)

In other cases, Tp ligands change their binding mode from κ^3 to κ^2, with one pyrazolyl group free from the metal (Equation 4.44). This binding mode has been common for d^8 complexes containing two additional ligands. Rather than adopt an electronically disfavored five-coordinate structure with a d^8 metal center, the complex adopts a square planar structure with the Tp ligand in a κ^2 mode.[232] The relationship between this change in binding mode and the mechanism of C–H activation by rhodium[233] and platinum complexes[234] is discussed in Chapter 6.

4.2.5.3.3. Synthesis of Polypyrazolylborate Ligands and Complexes

Many polypyrazolylborate ligands are prepared by the reaction of borohydride with the free pyrazole. More specifically, Tp' ligands are most commonly prepared by slowly heating potassium borohydride and a pyrazole together in a melt (Equation 4.45).[235] The stoichiometry of borohydride to pyrazole determines whether a hydridotrispyrazolyl borate ligand or a dihydridobispyrazolyl borate ligand is obtained. Trofimenko and co-workers synthesized a series of 3,5-substituted trispyrazolylborate ligands in which the substituents are alkyl or aryl groups.[224,236,237] Alternatively, trispyrazolylborate ligands are prepared from an alkyl or aryl dihaloborane and the alkali metal pyrazolyl group. Related ligands, such as $PhB(CH_2PPh_2)_3$, are likewise prepared by the addition of anions to haloboranes. $PhB(CH_2PPh_2)_3$ was reported by Tilley and co-workers by the addition of $[Li(TMEDA)][CH_2PPh_2]$ to $PhBCl_2$.[238]

$$(4.45)$$

The synthesis of Tp' ligands that contain different substituents on the 3- and 5-positions on the pyrazole have been reported, For example, Graham and co-workers reported the synthesis of tris[3-(trifluoromethyl)-5-methyl]pyrazole borate.[239] However, the most commonly used polypyrazolylborate ligands are substituted symmetrically at the 3- and 5-position. The unsymmetrical systems can undergo shifts of boron between the two nitrogens, and this shift scrambles the 3- and 5- positions.[224,240,241]

$$(4.46)$$

Complexes of these pyrazolyl groups are typically prepared by reaction of the metal halide with the anionic pyrazolyl ligand. For example, bis-olefin complexes of group 9 metals, such as rhodium and iridium, are prepared from complexes of the general type $[M(olefin)_2Cl]_2$, in which M = Ir and Rh, and olefin = ethylene, COE, COD, norbornadiene, etc., upon the addition of Tp salts (Equation 4.46).[242,243] Although a majority of Tp complexes are obtained from the reaction of a metal halide with a salt of the anionic Tp ligand, some Tp complexes have been synthesized by alternate routes. For example, Tp*VO(acac) was synthesized from VO(acac)$_2$[244] and TpMoF[=N–N(p-tolyl)$_2$], which is obtained from the reaction between TlTp and {CpMo[=N–N(p-tolyl)$_2$](PPh$_3$)}[BF$_4$].[245]

4.2.5.3.4. Reactions of Polypyrazolylborate Complexes

Because Tp ligands and their analogs are typically bound tightly to the metal, they are more often ancillary ligands than reactive ligands. Examples of reactions of complexes in which the Tp ligand is an ancillary ligand are found in many of the subsequent chapters of this book. However, much of the organometallic chemistry of Tp ligands has focused on the study of C–H activation and on the study of paramagnetic organometallic species. As noted above, one common reaction of Tp ligands is the dissociation of one nitrogen donor to open a coordination site on the metal or to accommodate the electronic changes resulting from reactions that involve redox processes. Tp ligands can also decompose by exchange between the anionic ligand at the metal and the pyrazolyl group. An example of exchange between a benzyl and a pyrazolyl group is shown in Equation 4.47.[246]

$$\text{Tp}'\text{Zr(CH}_2\text{Ph)}_3 \xrightarrow[\substack{\text{CH}_2\text{Cl}_2 \\ -60\ ^\circ\text{C}}]{\text{Ph}_3\text{C}^{\oplus}} \qquad \xrightarrow{0\ ^\circ\text{C}} \qquad \tag{4.47}$$

4.2.5.4. β-Diketiminate Complexes

4.2.5.4.1. Overview

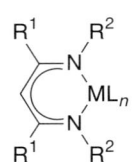

Figure 4.11.
A generic structure of a β-diketiminate complex.

β-Diketiminate ligands are anionic nitrogen donor ligands that have recently been studied intensively as ancillary ligands in organometallic chemistry. A generic structure of a β-diketiminate complex is shown in Figure 4.11. The β-diketiminate ligand can adopt several binding modes that donate varying numbers of electrons. These ligands have been nicknamed "NacNac" ligands because they are analogous to acetyl acetonate ligands (acac), but with nitrogen instead of oxygen donors. Much organometallic chemistry of NacNac complexes has been developed in the late 1990s and first decade of the twenty-first century. The classic chemistry of these β-diketiminate complexes was reviewed by Lappert in 2002.[247a] An older review focusing more on homoleptic complexes was published by Holm in 1966.[247b]

Complexes of these ligands have not yet been developed into commonly used catalysts, although early metal β-diketiminate complexes with co-catalysts do catalyze the polymerization of ethylene. More striking, these ligands have led to the synthesis of a number of unusual low-coordinate complexes. Examples of such complexes are described later in this section. Other examples are shown in later chapters of this text.[198, 199]

4.2.5.4.2. Structure and Bonding of β-Diketiminate Ligands

β-Diketiminate ligands are generally considered to be anionic when using the ionic bonding model. Although anionic in most or all cases, they can donate two, four, or six electrons to a single metal center. In the most common binding mode, they coordinate to a single metal in a κ^2 fashion through the two nitrogen atoms in either a planar or nonplanar ring. In this case, they would be considered a four-electron donor in the absence of additional π-donation. Alternatively, they have been shown to bind in a κ¹ fashion through the central carbon, which causes them to be a two-electron donor ligand, or in an η⁵ fashion through the π-system, in which case they would donate six electrons in the ionic bonding model. Although less common, they have also been shown to bridge two metals with the second metal bound to the central carbon or the π-system.

The steric, as well as electronic, properties of these ligands can be tuned by the choice of substituent at nitrogen. The renaissance of β-diketiminate chemistry resulted from the synthesis of β-diketiminate complexes in which the substituents on nitrogen were sterically demanding.[248] This steric effect led to the predominant formation of monomeric complexes and to the formation of unsaturated early and late metal complexes. The earlier β-diketiminate complexes containing smaller substituents at nitrogen tended to be stable and saturated.

4.2.5.4.3. Synthesis of β-Diketimines and β-Diketiminate Complexes

β-Diketimines are prepared by the modification of classical organic reactions. In one case, the diketimines are formed from the diketone (Equation 4.48).[249] The first step involves simple formation of an imine from the ketone and the amine. The second ketone is typically converted to an imine by alkylation of the oxygen to form a vinyl ether, followed by reaction of the resulting vinyl ether with the second equivalent of amine. Alternatively, these ligands can be formed from the reaction of an arylamine, acid, and the β-diacetal of the diketone (Equation 4.49).[250] Or, they can be prepared by addition of an alkyllithium or alkylpotassium reagent to two equivalents of nitrile, as shown in Equation 4.50.[251]

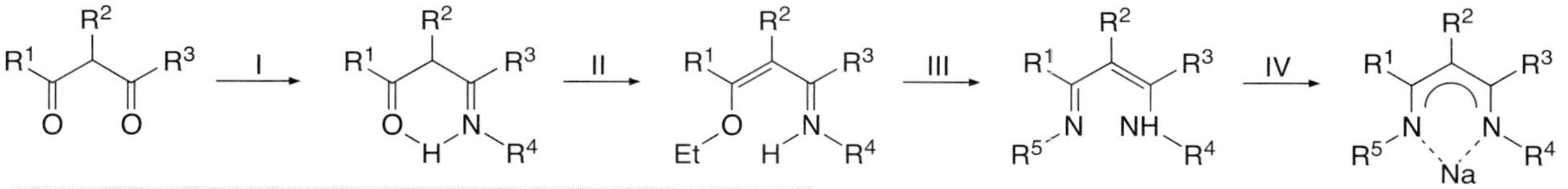

(4.48)

(I) NH₂R⁴, C₆H₆, azeotropic distillation; (II) [Et₃O][BF₄], Et₂O, 278 °C; (III) NH₂R⁵, Et₃O; (IV) NaOMe, MeOH.

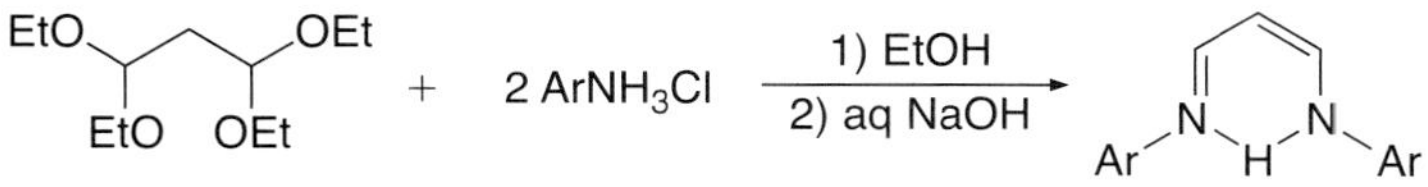

(4.49)

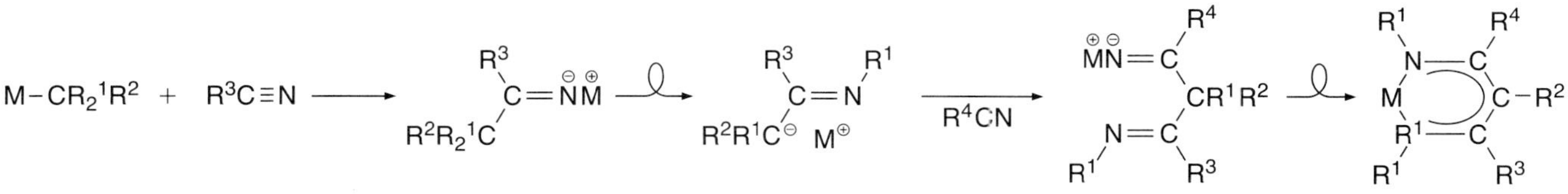

(4.50)

Transition metal β-diketiminate complexes are typically prepared by one of three routes. In one, these complexes are prepared by the reaction of a metal halide with an alkali metal β-diketiminate generated from the reaction of the β-diketiminate with an alkali metal base. An example of this synthesis is shown for the scandium system in Equation 4.51.[252] In a second method, these complexes are prepared by the reaction of a transition metal complex containing a basic ligand, such as an alkyl or amido group, with the neutral β-diketimine. Two examples of this route for zirconium systems are shown in Equations 4.52[253] and 4.53.[253]

$$(4.51)$$

$$(4.52)$$

$$R^1 = Ph, C_6H_4CF_3-4$$

$$(4.53)$$

4.2.5.4.4. Examples of β-Diketiminate Complexes

As noted in the overview paragraphs on β-diketiminate complexes, many of the organometallic β-diketiminate complexes are unsaturated. In addition, the steric demands of this ligand can lead to unusual geometries. Several examples of early metal β-diketiminate complexes were shown in Equations 4.51–4.53. These early metal dichloro complexes containing β-diketiminate ligands have been converted to organometallic complexes by standard alkylation with organometallic reagents of alkali metals. One example is shown for the scandium system in Equation 4.54. The benzyl complexes of this system are coordinatively unsaturated and contain the metal outside of the plane of the five atoms of the ligand.

Many middle and late transition metal complexes containing β-diketiminate ligands have also been prepared. For example, the highly unsaturated iron– and cobalt–methyl compounds have been prepared from the β-diketiminate ligand containing *tert*-butyl groups on the carbon alpha to nitrogen and 2,6-diisopropylphenyl groups attached to nitrogen (Equation 4.55).[254] A five-coordinate Pt(IV) complex has been prepared by reaction of the β-diketiminate containing methyl groups on the carbon alpha to nitrogen and 2,6-diisopropylphenyl groups attached to nitrogen (Equation 4.56).[255]

$$(4.54)$$

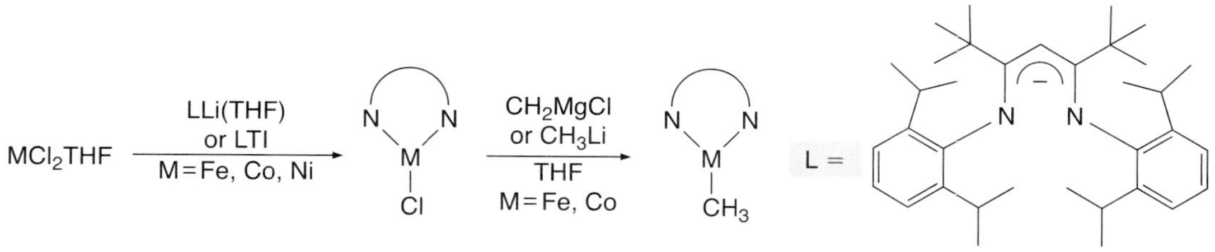

(4.55)

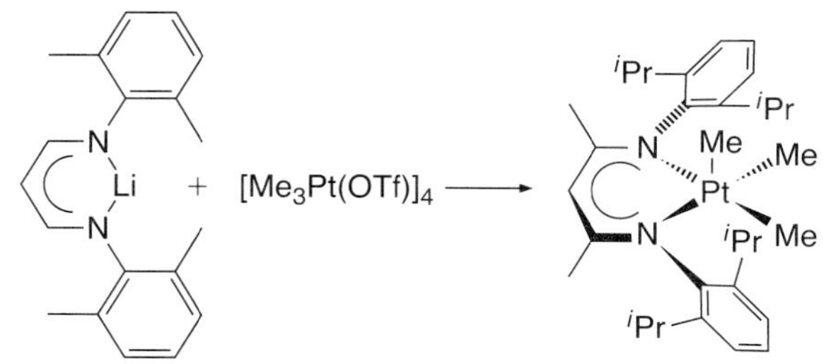

(4.56)

4.3. Transition Metal Complexes with Anionic Oxygen Ligands (Written with Prof. Pinjing Zhao)

Many transition metal complexes contain anionic oxygen donor ligands, and many of these complexes display structures and reactivity that resemble that of more conventional organometallic compounds containing metal–carbon bonds. In some cases the alkoxo ligand is the site of reaction, and in other cases the alkoxide is an ancillary ligand. This section will focus on three main classes of compounds: alkoxides (including aryloxides and the parent hydroxides), carboxylates, and β-diketonates.

4.3.1. Transition Metal–Alkoxo Complexes

4.3.1.1. Overview

Hydroxo, alkoxo, and aryloxo complexes are known for all transition metal elements. Early-transition-metal–alkoxo bonds are strong, and the M–O linkage tends to be less reactive than other types of anionic ligands bound to early metals. Thus, alkoxo groups often act as ancillary ligands in early metal complexes in both coordination and organometallic chemistry.[257] In contrast, late-metal–alkoxo bonds are weaker, and the M–O linkage tends to be more reactive than that in early metal complexes. Thus, late-transition-metal–alkoxo complexes undergo a number of elementary organometallic reactions discussed in later chapters of this text. They have also been shown to be reactive intermediates in many important catalytic processes.[258,259]

Free alkoxide and aryloxide anions are Brønsted bases with pK_a values of the corresponding alcohols ranging from 5 to 20 in water.[260,261] The basicity is highly dependent on the electronic properties of the alkyl or aryl moieties. For example, the pK_a value of hexafluoro-*tert*-butanol, $(CF_3)_2MeCOH$, is 9.6, which is considerably lower than the pK_a value of *tert*-butanol (19.2), but roughly the same as that of phenol (9.9). Such differences in electronic, as well as steric, environments often leads to the different structures and reactivity patterns for compounds containing similar ancillary ligands, but different alkoxides or aryloxides.

A common structural feature of transition metal hydroxide, alkoxide, and aryloxide ligands is their tendency to bridge two or more metals. The extent of such aggregation is reduced when the complexes contain more sterically hindered alkoxides, such as OBu^t, $OCBu^t_3$, bulky trialkylsiloxides,[262] and 2,6-di-*tert*-butylphenoxide.[263,264]

4.3.1.2. Alkoxide Complexes of the Early Transition Metals

4.3.1.2.1. Overview

Many early-transition-metal–alkoxo complexes are known, and the structures and reactions of these complexes range from organometallic chemistry to coordination chemistry and material science.[257,265] As noted in the previous section, the reactivity of early-metal–alkoxo linkages toward the types of reactions presented later in this text is limited because the early transition metal–oxygen bond is strong. At the same time, the strength of this metal–oxygen bond has allowed alkoxo groups to serve as ancillary ligands for organometallic reactivity at other sites of early metal complexes. The thermodynamics of the metal–oxygen bond, some of the properties of these complexes, and their role as ancillary ligands are presented in the following sections.

4.3.1.2.2. Bonding of Early-Metal Alkoxides

Many factors contribute to the strength of early-metal–oxygen bonds.[266,267] Alkoxo ligands and high-valent, early transition metal centers are both considered to be hard.[268–270] Therefore, the strength of the early-metal–oxygen bond can be attributed, in part, to the favorable match of a hard ligand with a hard metal center. In other terms, the large difference in electronegativity between the electropositive early metals and the electronegative alkoxo ligand creates a metal–ligand bond that has significant ionic character.[271] This ionic character causes the metal–oxygen bond to have a high homolytic bond dissociation energy. This polarization of the M–O bond causes the metal center to be Lewis acidic. In addition to these affects on the metal–ligand σ-bond, π-donation from the electron pairs on oxygen to the early metal center is thought to contribute to the strength of early-metal–oxygen bonds, much like π-donation from nitrogen to the transition metal strengthens early-metal–amido bonds.[272]

The strength of several types of early-metal–oxygen bonds has been measured by theoretical methods[267,271] as well as reaction calorimetry.[266,273–278] The bond dissociation energies are remarkably high. For example, the strength of the zirconium–hydroxo bond in $Cp^*_2Zr(OH)Ph$ is 115 kcal/mol.[266] The analogous zirconium–phenoxide bond is weaker due to the stability of the phenoxy radical, but this bond is still 91 kcal/mol. These bond strengths are roughly 30–50 kcal/mol stronger than bonds between late transition metals and hydroxo or phenoxo ligands (vide infra).[267,271,279–281]

4.3.1.2.3. Preparation of Early-Metal–Alkoxo Complexes

Several common synthetic strategies for the formation of transition metal–heteroatom bonds can be used for the preparation of early transition metal alkoxides, as shown in Equation 4.57. These methods include: (1) metathesis between metal halides and alkali metal salts of the corresponding alcohols; (2) direct reaction of the alcohols with metal halides in the presence of a base; and (3) protonolysis of metal alkyls, amides, or alkoxides with more acidic alcohols.

$$L_nM-X \xrightarrow{\text{NaOR}} L_nM-OR \xleftarrow{\text{HOR}} L_nM-R' \text{ or } L_nM-NR'_2 \text{ or } L_nM-OR'$$

$$L_nM-X \xrightarrow[\text{Base}]{\text{HOR}} L_nM-OR \tag{4.57}$$

In addition, some less common reactions can be used to form early metal alkoxides. For example, the formation of metal alkoxides by insertions of a ketone into a metal–alkyl linkage is rare for late-metal–alkyl complexes, but occurs with various early-metal–alkyl complexes.[282–285] The reaction in Equation 4.58 is one example.[285]

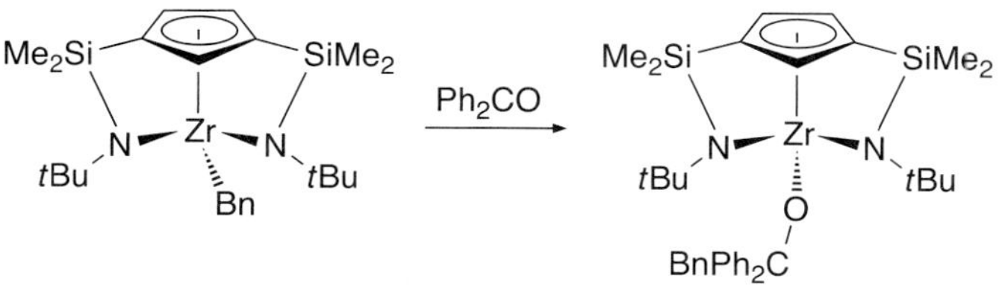

$$(4.58)$$

Various cyclopentadienyl-ligated early metal and actinide alkoxides were prepared by reactions of the alcohols with the metal hydride precursors to eliminate hydrogen gas.[266,275,277,278] Early metal alkoxides, as well as their high-valent late-metal analogs, can also be formed by attack of an alkyl group at an oxo ligand, where the alkyl could be added as a carbocation, a radical, or a carbanion (Equation 4.59).[286]

$$L_nM^{n+} = O + R^+ \rightleftharpoons [L_mM^{n+} - OR]^{\oplus}$$

$$L_nM^{n+} = O + R^{\bullet} \rightleftharpoons [L_mM^{(n-1)+} - OR]$$

$$L_nM^{n+} = O + R^- \rightleftharpoons [L_mM^{(n-2)+} - OR]^{\ominus}$$

$$(4.59)$$

4.3.1.2.4. Reactivity of Early-Metal–Alkoxo Complexes

The reactivity of the alkoxide ligand in early transition-metal–alkoxo complexes is relatively limited. For example, early-transition-metal–alkoxo ligands tend to be stable toward β-hydrogen elimination. This stability is observed, even though many alkoxo complexes contain an open coordination site. Labeling studies have demonstrated the absence of β-hydrogen elimination processes: H/D exchange occurs preferentially in the methyl position over the methylene position of an ethoxide ligand.[287]

Early-transition-metal–alkoxo complexes are also stable toward other typical organometallic processes, such as reductive elimination. Hydridometal–alkoxo, alkylmetal–alkoxo, and arylmetal–alkoxo complexes of the early metals are known, but none of them undergoes reductive elimination to form a carbon–oxygen bond.[266] Early-metal–alkoxo complexes also do not tend to undergo migratory insertion processes. In fact, this lack of insertion was exploited by Jordan to illustrate coordination of alkenes to d^0 metal centers.[288–290]

4.3.1.2.5. Early-Metal Alkoxides as Ancillary Ligands

4.3.1.2.5.1. Steric and Electronic Properties

This lack of reactivity of early-metal–alkoxo complexes has been used to create alkoxo ligands that are stable and serve as ancillary ligands. These ligands have been used to mimic the electronic properties of a cyclopentadienyl ligand, while providing an alternative means to control steric properties. Sterically demanding alkoxide and aryloxide[291,292] ligands have stabilized early metal complexes with low coordination numbers and other novel structural features.[263,264,293] Two such ligands are the tri-*tert*-butyl methoxide ligand studied by Rothwell, termed tritox, and the tributyl siloxide ligand studied by Wolczanski, termed silox.[262,264] Representative complexes of these ligands are shown in Figure 4.12.[294,295]

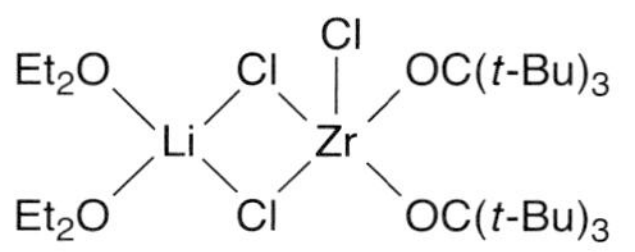

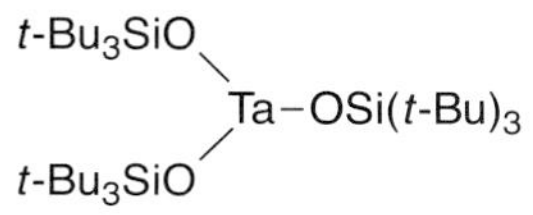

Figure 4.12.
Examples of sterically hindered early-metal–alkoxo complexes.

The electronic properties of these ligands mimic the properties of a cyclopentadienyl ligand because they can act as anionic six-electron donors (Figure 4.13).[294] The donation of six electrons results from the σ-bond along with two π-bonds created by donation of two electron pairs to the metal. The steric bulk of these ligands exceeds the steric bulk of cylcopentadienyl ligands because of the large *tert*-butyl groups.

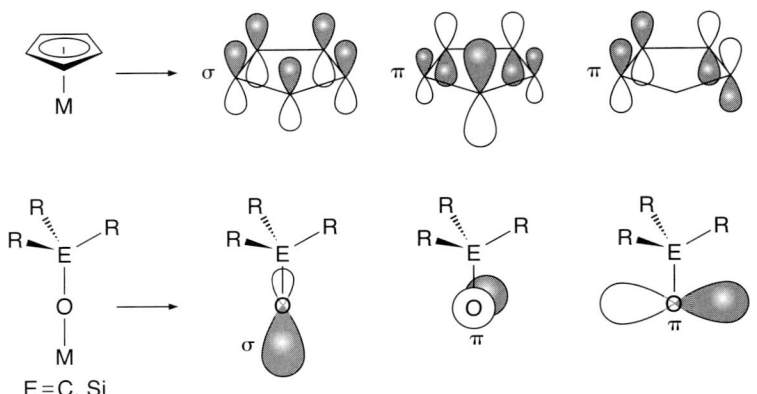

Figure 4.13. Orbital analogy between a Cp and an alkoxo ligand.

4.3.1.2.5.2. Catalytic Reactions of Early-Metal–Alkoxo Complexes

Because of the stability of early-metal–alkoxo ligands, much of the catalytic reactivity of metal–alkoxo complexes is based on their properties as ancillary ligands. The effect of alkoxides on the activity of high-valent group 6 olefin metathesis catalysts discussed in Chapter 21 is significant. Much work has been conducted to tune the steric and electronic properties of these ligands, and this work has been reviewed.[293] In brief, alkoxides have been shown to generate highly active catalysts, and substituted, optically active binolate ligands have been shown to form catalysts that react with high enantioselectivity.[296]

Early-metal–alkoxide complexes also serve as precursors to Lewis acid catalysts. For example, titanium tetraisopropoxide is a common starting material for the generation of Lewis acid catalysts by proton transfers with binaphthols, sulfonamides, and related ligands.[297,298] In most of these processes, no intermediate containing a metal–carbon bond is generated. However, mechanistic studies on the addition of alkylzinc reagents to aldehydes catalyzed by titanium complexes with BINOLate ligands imply that this addition occurs by formation of a titanium–alkyl complex.[297] Two possible mechanisms to transfer the alkyl group from titanium to the aldehyde are shown in Figure 4.14.

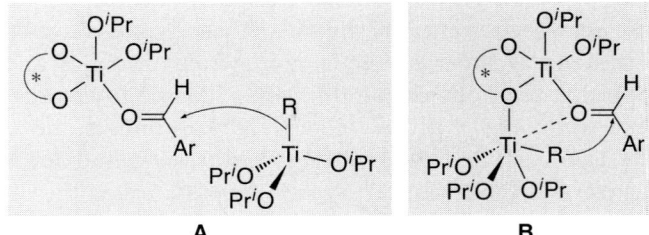

Figure 4.14.
Proposed mechanism for the transfer of alkyl groups to aldehydes with group 4 metal–alkoxo complexes.

In parallel with the catalytic reactivity of early-transition-metal–alkoxo complexes, the catalytic reactivity of lanthanide–alkoxo and –phenoxo complexes has been developed. The homogenous hydrogenation of a variety of arene substrates can be catalyzed by niobium and tantalum hydride catalysts that are stabilized by bulky ancillary aryloxide ligands.[299] Lanthanide catalysts containing ancillary alkoxo ligands derived from chiral diols are particularly useful in asymmetric catalysis. For example, chiral lanthanide biphenolate and binaphtholate complexes have emerged as an important class of non-metallocene rare earth catalysts for enantioselective hydroamination of aminoalkenes.[300] Other applications of chiral lanthanide alkoxide catalysts include catalytic enantioselective cyanide transfer reactions, such as cyanosilylation of aldehydes and ketones and Strecker reactions of ketimines.[301] In these reactions, the gadolinium–alkoxo complexes were proposed to act as bifunctional catalysts containing Lewis acid and Lewis basic units (Figure 4.15).

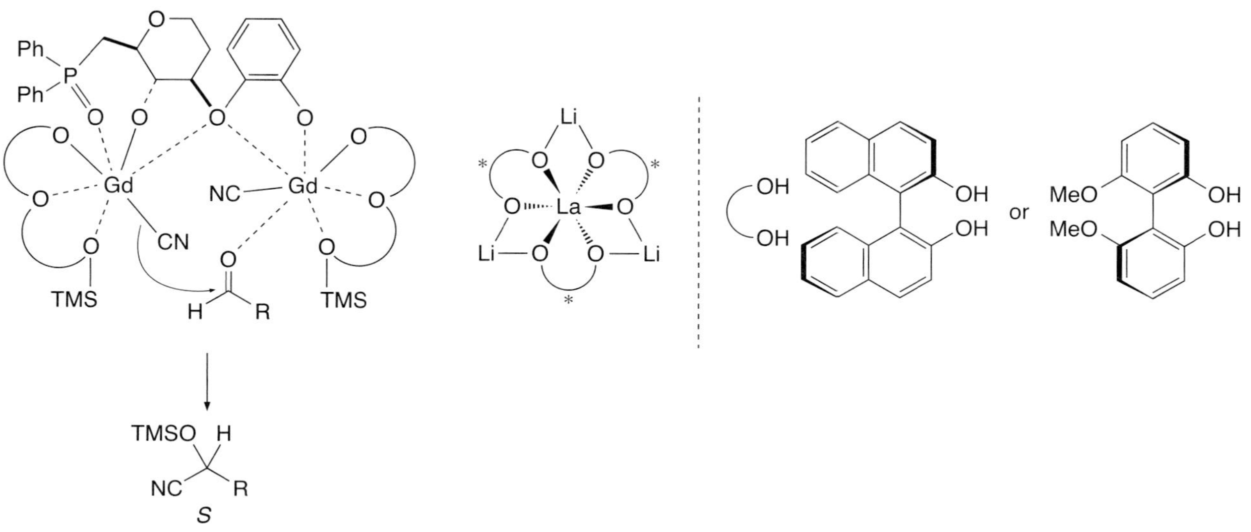

Figure 4.15.
Bifunctional Lewis acidic–Lewis basic alkoxo complex for asymmetric processes including Strecker and nitroaldol reactions.

One of the more reactive and selective catalysts of this type involves a bifunctional catalyst containing an alkali metal cation and an anionic lanthanide complex resulting from addition of excess binolate with lanthanide halides. Such catalysts have been used in asymmetric nitroaldol (Henry) reactions of ketones. Heterobimetallic Li–La alkoxo complexes (Figure 4.15) catalyzed these reactions with particularly high enantioselectivity.[302]

4.3.1.3. Alkoxide Complexes of the Late Transition Metals

4.3.1.3.1. Overview

Late transition metal alkoxo complexes contain weaker metal–oxygen bonds than early metal alkoxides, but these bonds are not any weaker than late-metal–carbon bonds. Instead of being triggered by weak metal–oxygen bonds, the chemistry of these complexes is controlled more by the significant polarity of these bonds and the basicity of the alkoxo group. The theory of bonding between alkoxides, the role of alkoxides as ancillary ligands in late metal compounds, the synthetic routes to late metal alkoxides, and the reactivity of late metal alkoxides are described in the following several sections.

4.3.1.3.2. Bonding of Late-Metal Alkoxides

The strength of late-transition-metal–oxygen bonds can be rationalized by several theories.[258,259] According to the hard soft acid base (HSAB) principle,[303] the M–O bond should be weaker than an early-metal–alkoxo bond. This prediction is consistent with experimental data. However, this theory also predicts that the late-metal–alkoxo bond will be weaker than the late-metal–alkyl bond, and this prediction does not fit with observed relative bond strengths.[21]

Like late-transition-metal–amido complexes, $p\pi$–$d\pi$ interactions between the electron pair on oxygen and the filled d-orbitals on the metal can affect the thermodynamic stability and the reactivity of alkoxo complexes. Naturally, this effect in metal–alkoxo complexes is less pronounced than in metal–amido complexes because of the lower basicity of an alkoxide. At the same time, the presence of two electron pairs on oxygen causes this effect in alkoxo complexes to depend less on geometry than in amido complexes. Such π-interactions have been studied in detail by Caulton,[272] and have been used to rationalize the geometries, nucleophilicity, and basicity of late metal alkoxides and amides (Figure 4.16).[259]

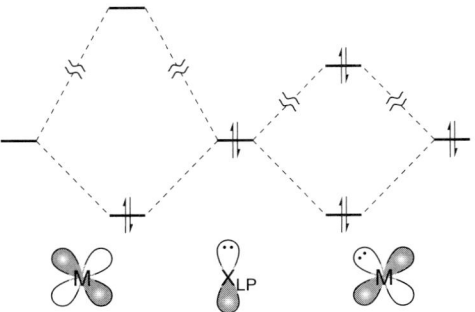

Figure 4.16.
Interaction between an electron pair of π-symmetry and the d-orbitals of a transition metal.

As noted in the previous section on late-metal–amido complexes, the origin of the strengths of late-metal–alkoxide and –amide bonds remains unresolved. The control of these bond strengths has been attributed to a combination of σ- and π-effects,[304,305] as well as to effects through σ-bonds alone.[306] In either case, the ionic character of M–O bonds is significant; this ionic character is likely to contribute to both the homolytic strength of the M–O bond[259] and the significant reactivity of the alkoxo ligand in late metal complexes.

The electronic character of the M–O bond in late metal alkoxides can been assessed by the value of the CO stretching frequencies of ancillary carbonyl ligands. Vaska[307] measured the CO stretching frequency of hydroxo and phenoxo complexes of the square planar *trans*-X-M(PPh$_3$)$_2$(CO) complexes (where M = Rh and Ir, and X = anionic ligands) and compared these values to those of complexes with less electron-donating X groups, such as acetates, thiocyanates, and cyanide. The ν_{CO} value of the phenoxide complex was close to that of the fluoride complex. More recently, Caulton has evaluated the overall donating ability of anionic ligands in the series of square pyramidal complexes RuH(X)(CO)(PBu$^t_2$Me)$_2$ and RuH(X)(CO)(pyridine)(PBu$^t_2$Me)$_2$ by the CO stretching frequencies. The alkoxo ligands were stronger donors than halides, and the detailed ranking of donor ability depended on the steric and electronic characters of various alkoxo ligands (Figure 4.17).[308]

OEt > OCPh$_3$ > OSiMe$_3$ > OSiMe$_2$Ph > OSiPh$_3$ >
F ≈ OCH$_2$CF$_3$ > OH ≈ OB(Mesityl)$_2$ >
NHPh ≈ OPh > SPh ≈ Cl > C$_2$Ph ≈ Br > I

Figure 4.17.
Relative electron-donating ability of various ligands bound through M–X bonds.

4.3.1.3.3. Thermodynamics of Late-Metal–Alkoxo Bonds

The relative M–X bond strengths in late metal amides and alkoxides have been measured by the studies on reversible M–X/H–X exchange (Equations 4.60 and 4.61).[279] These essentially thermoneutral processes suggested a roughly 1:1 correlation of H–X and L$_n$M–X bond strengths, which provide an estimation of relative L$_n$M–X homolytic bond strengths

(Equation 4.62) using known H–X bond energies. Importantly, these results suggested that M–O bond strengths for alkoxides are higher than M–N bond strengths and comparable to M–C(sp^3) bond strengths. Deviations from this 1:1 correlation were observed in the series of Ni(II) complexes shown in Equation 4.63. This deviation was explained by the strong ground state polarization of late metal M–X bonds.[309] Therefore, M–O bond strengths depend on a combination of π-donation or repulsion, depending on the degree of unsaturation and the geometry of the metal center, as well as the bond polarity created by the different ligands on the metal and the substituents on the alkoxo group.

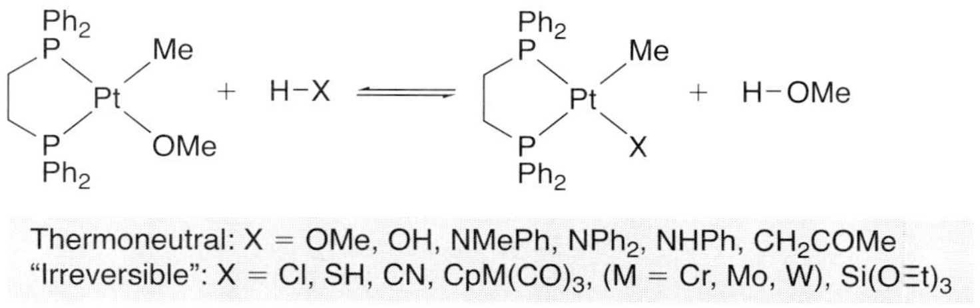
Ph2
P
Me
Pt
OMe
P
Ph2
+ H–X
Ph2
P
Me
Pt
X
P
Ph2
+ H–OMe
Thermoneutral: X = OMe, OH, NMePh, NPh2, NHPh, CH2COMe
"Irreversible": X = Cl, SH, CN, CpM(CO)3, (M = Cr, Mo, W), Si(OEt)3

(4.60)

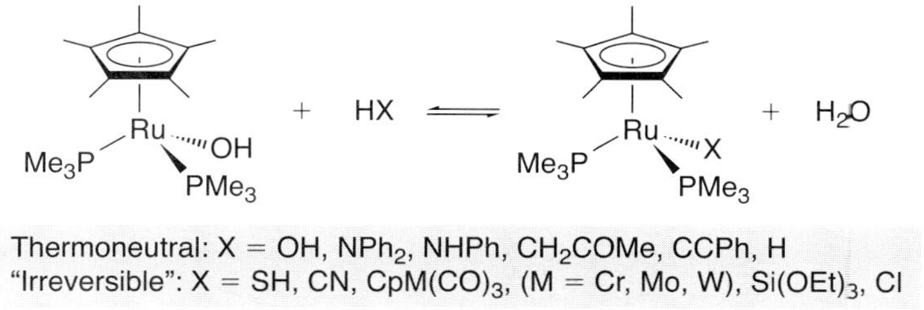
Ru
OH
Me3P
PMe3
+ HX
Ru
X
Me3P
PMe3
+ H2O
Thermoneutral: X = OH, NPh2, NHPh, CH2COMe, CCPh, H
"Irreversible": X = SH, CN, CpM(CO)3, (M = Cr, Mo, W), Si(OEt)3, Cl

(4.61)

$$L_nM\text{–}X \;+\; H\text{–}Y \;\xrightleftharpoons{\Delta H \approx 0}\; L_nM\text{–}Y \;+\; M\text{–}X \qquad (4.62)$$
$$\text{If } \Delta H \approx 0, \text{ then } D(H\text{–}Y) - D(H\text{–}X) \approx D(M\text{–}Y) - D(M\text{–}X)$$

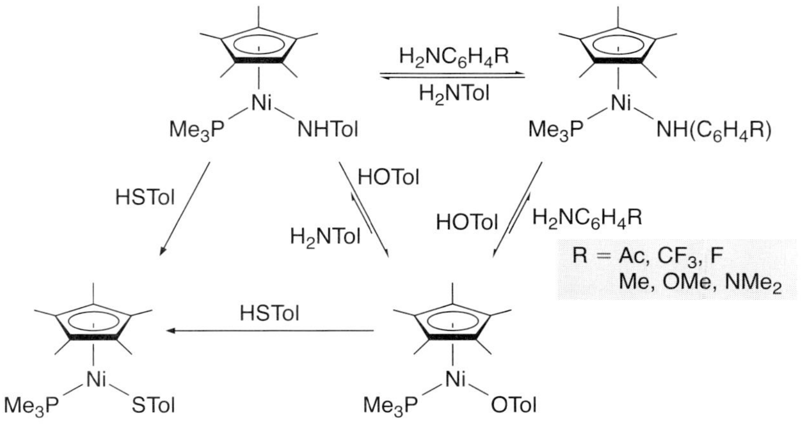
Ni
Me3P
NHTol
H2NC6H4R
H2NTol
Ni
Me3P
NH(C6H4R)
HSTol
HOTol
H2NTol
HOTol
H2NC6H4R
R = Ac, CF3, F
Me, OMe, NMe2
HSTol
Ni
Me3P
STol
Ni
Me3P
OTol

(4.63)

Measurements of the strengths of late-metal–alkoxo bonds by theoretical methods[267,271] and reaction calorimetry[280] have provided a few benchmarks. These results provide some clearer comparisons between the strengths of M–O bonds in early and late metal alkoxides. For example, the Co–O bond strength in $Co(CO)_4(OH)$ was calculated to be 55 kcal/mol, and the Rh–O bond dissociation energy (BDE) in several octaethylporphyrin rhodium alkoxide complexes was estimated to be ~50–55 kcal/mol by thermodynamic measurements and reactivity studies.[280] These values are much lower than the M–O bond strengths noted in the previous section on early metal alkoxides.

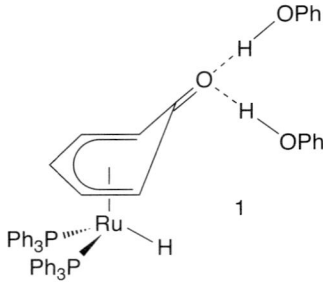

Figure 4.18.
Structure of an η^5-phenoxide.

4.3.1.3.4. Late-Metal Alkoxides as Ancillary Ligands

The weaker M–O bonds in late metal alkoxides make alkoxo ligands less suitable as ancillary ligands on late metals than on early metals. Nevertheless, sterically demanding alkoxide and aryloxide ligands can stabilize late metal complexes with low coordination numbers or other novel structural features.[264] For example, Grubbs has reported four-coordinate ruthenium alkylidene complexes possessing trigonal pyramidal geometries that contain tertiary alkoxo ligands (Equations 4.64 and 4.65).[310] In some cases, such as the η^5 complex shown in Figure 4.18, phenoxides serve as π-ligands by coordination of the metal to the aryl group. Structure of an η^5-phenoxide was reported by Wilkinson.[311]

$$(4.64)$$

$$(4.65)$$

$$R = C(CF_3)_2(CH_3) \text{ and } C(CF_3)_3$$

4.3.1.3.5. Preparation of Late-Metal–Alkoxo Complexes

Many of the typical reactions that form transition metal–heteroatom bonds can be used to prepare late transition metal alkoxides.[258,259] For example, metathetical exchange between late metal halides and alkali metal salts of the corresponding alcohols often forms late metal alkoxides. Complexes that are more reactive than metal halides, such as metal acetates or triflates, are sometimes used when exchanges with halides are slow or reversible (Equation 4.66).[312] Late metal fluorides can also be used for these exchanges, as shown in Equation 4.67.[313]

$$(4.66)$$

$$(4.67)$$

In chemistry related to the reactions of alkali metal alkoxides with late metal halide complexes, late metal alkoxides have been formed by the reaction of alcohols with metal halides in the presence of a base. This process has been used to form arylpalladium alkoxides during the coupling of aryl halides with alcohols discussed in Chapter 19 (cross-coupling).[314,315] In addition, Milstein used an intramolecular version of this process to prepare the metallaoxetane in Equation 4.68.[316]

$$(4.68)$$

Protonolysis of metal alkyls, amides,[317] or alkoxides with more acidic alcohols has also been used to generate alkoxides. Yamamoto has reported the preparation of a series of Ni(II) and Pd(II) aryloxides by alcoholysis of the corresponding metal–methyl complexes to release methane (Equation 4.69).[318] Metal–silylamido complexes are also useful precursors to late metal alkoxides, presumably due to release of steric congestion.[319,320]

$$(4.69)$$

The third general synthetic route for late metal alkoxides involves O–H oxidative additions of alcohols to a low-valent metal center to form a hydrido alkoxo complex. An early study by Green showed that oxidative additions of phenols occurred to electron-rich Pt(0) complexes (Equation 4.70).[321] Oxidative additions of water to late metals has also been reported by Otsuka and Milstein with platinum(0) and iridium(I) systems, respectively (Equations 4.71 and 4.72).[322–325]

$$Pt(PCy_3)_2 \ + \ HOPh \ \longrightarrow \ \underset{\underset{PCy_3}{|}}{\overset{\overset{PCy_3}{|}}{H-Pt-OPh}} \qquad (4.70)$$

$$Pt[P(i\text{-}Pr)_3]_n \ + \ H_2O \ \rightleftharpoons \ \underset{\underset{P(i\text{-}Pr)_3}{|}}{\overset{\overset{P(i\text{-}Pr)_3}{|}}{H-Pt-OH}} \qquad (4.71)$$

$$n = 2, 3$$

$$(4.72)$$

Oxidative additions to form late-metal alkoxides by C–O bond cleavage are also known, but are less common than oxidative additions of O–H bonds. Graziani reported insertions of palladium(0) complexes into C–O bonds in activated epoxides (Equation 4.73).[326] The resulting metallaoxetanes were stabilized by the cyano substituents. More common is the oxidative addition of esters to form acylmetal–alkoxo complexes. Yamamoto reported such oxidative additions to cobalt and rhodium systems (Equations 4.74 and 4.75, respectively).[327]

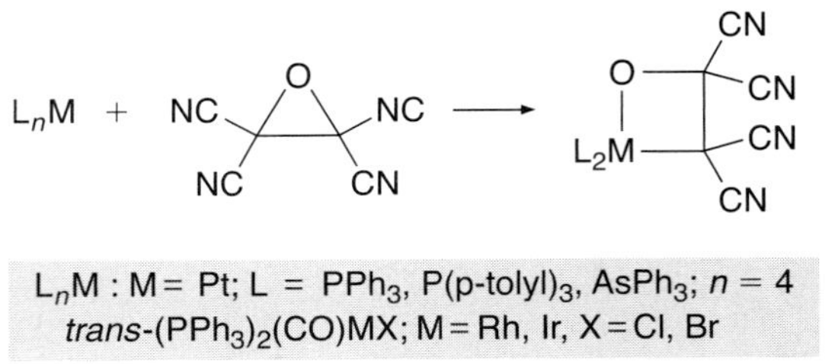

$$L_nM \ + \ \text{(structure)} \longrightarrow \text{(structure)} \qquad (4.73)$$

$$L_nM : M = Pt; \ L = PPh_3, \ P(p\text{-tolyl})_3, \ AsPh_3; \ n = 4$$
$$trans\text{-}(PPh_3)_2(CO)MX; \ M = Rh, \ Ir, \ X = Cl, \ Br$$

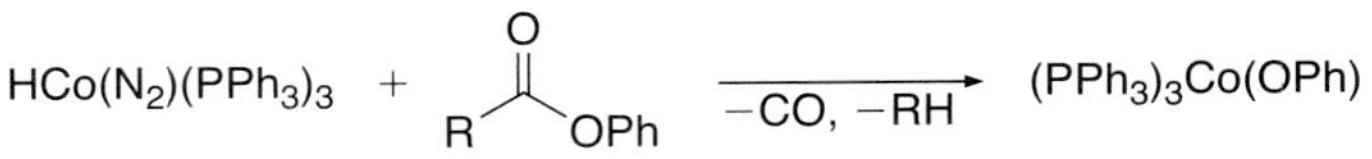

$$HCo(N_2)(PPh_3)_3 \ + \ \text{(structure)} \xrightarrow[-CO, \ -RH]{} (PPh_3)_3Co(OPh) \qquad (4.74)$$

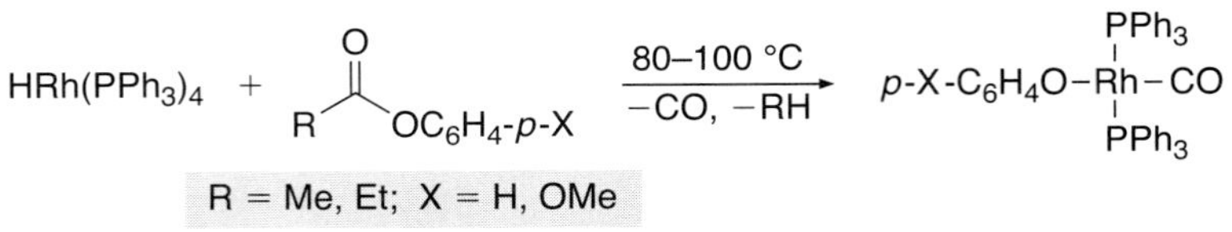

$$HRh(PPh_3)_4 \ + \ \text{(structure)} \xrightarrow[-CO, \ -RH]{80-100 \ °C} p\text{-}X\text{-}C_6H_4O\text{-}Rh\text{-}CO \qquad (4.75)$$
$$R = Me, \ Et; \ X = H, \ OMe$$

More recently, the first direct observation of C–O oxidative addition of an aromatic ether to a transition metal complex has been reported by Kakiuchi (Equation 4.76).[328] This reaction occurs by initial chelate-assisted C–H bond cleavage, followed by isomerization to form the final Ru(II) aryloxide complex.

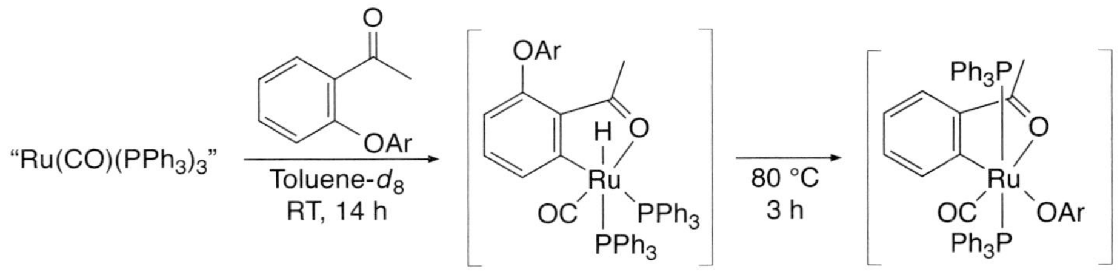

$$\text{"Ru(CO)(PPh}_3)_3\text{"} \xrightarrow[\text{RT, 14 h}]{\text{Toluene-}d_8} \text{(structure)} \xrightarrow[\text{3 h}]{80 \ °C} \text{(structure)} \qquad (4.76)$$

Finally, a variety of oxidation methods have been used to prepare late metal alkoxides. For example, an Ir(III) hydroxo complex, [Ir(OH)(NO)(PPh_3)_3]BF_4, was prepared by the oxidation of [Ir(H)(NO)(PPh_3)_3]BF_4 with O_2.[329] Hillhouse reported the preparation of a series of Ni(II) complexes with chelating alkoxo alkyl ligands by oxidations of the chelating dialkyl precursors (Equation 4.77).[330,331] Finally, several methods have been developed to form metallaoxetanes by olefin oxidations,[332,333] such as the oxidation of coordinated olefins by hydrogen peroxide to form Rh(III) rhodaoxetanes (Equation 4.78).[332] The formation of metallaoxetanes by reaction of olefins with a bridging platinum–oxo complex has also been reported.[334,335]

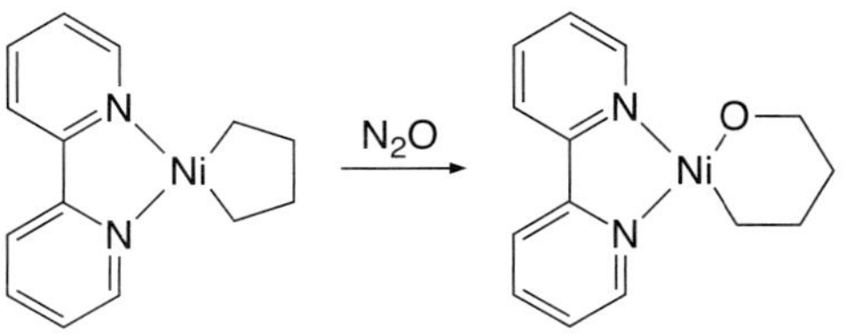

$$\text{(structure)} \xrightarrow{N_2O} \text{(structure)} \qquad (4.77)$$

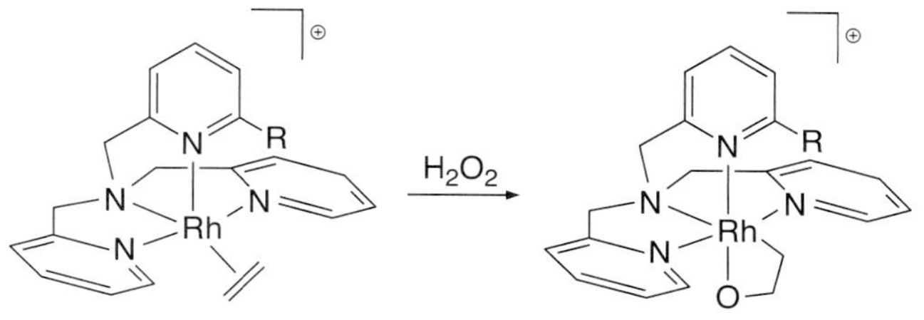

(4.78)

4.3.1.3.6. Reactivity of Late-Metal–Alkoxo Complexes

Because the M–O bonds in late metal alkoxides are more akin to the M–C bonds in late-metal–alkyl complexes than the strong M–O bonds in early metal alkoxides, late-metal alkoxides undergo a diverse set of reactions. Many of these reactions are contained in later chapters of the text. For example, alkoxo complexes undergo reductive eliminations to form O–H and C–O bonds; they undergo some insertions of CO and olefins; and they undergo β-hydrogen elimination to form metal hydrides[336,337] and β-carbon elimination to form metal–hydrocarbyl complexes.[319,338,339] In addition, they undergo proton transfer processes that are have been used, in part, to evaluate M–O bond strengths.

Because these reactions are presented in more detail in later chapters, only representative examples are shown here. Many late metal alkoxides undergo β-hydrogen elimination, and this is a common route to the decomposition of such complexes. This β-hydrogen elimination is a common route to metal hydrides and to the reduction of higher-valent late metal complexes to lower-valent species. One complex that undergoes β-hydrogen elimination is shown in Equation 4.79.[340] This complex undergoes both reductive elimination of alcohol and β-hydrogen elimination from the alkoxide. The reductive elimination is reversible, and the final product results from β-hydrogen elimination. Reductive eliminations of late metal alkoxides to form carbon–oxygen bonds (Equation 4.80) have also been studied.[331,341–343] The rates of these reactions tend to be slower than the rates of reductive eliminations to form C–C and C–N bonds.

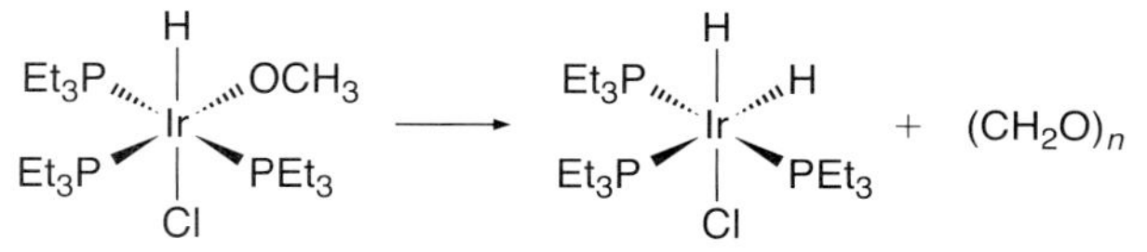

(4.79)

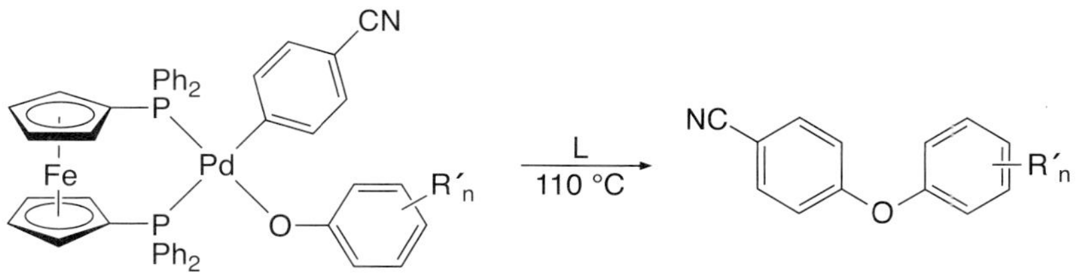

(4.80)

Late metal alkoxides can also undergo decomposition by pathways other than β-hydrogen elimination. For example, several rhodium–aryloxide complexes, $(PPh_3)_3Rh(OAr)$, undergo decomposition by cyclometallation to eliminate free alcohols and form metallacycles (Equation 4.81).[344]

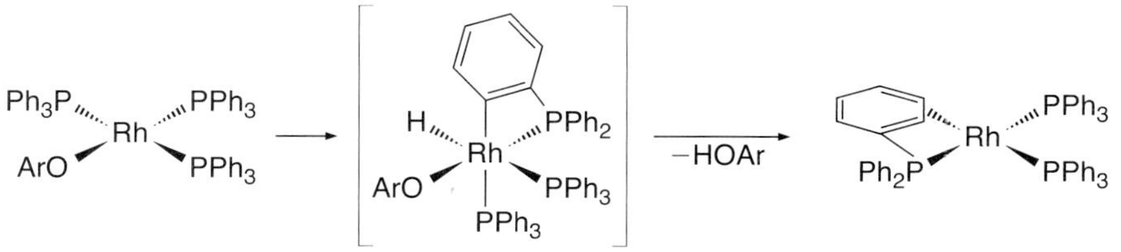

(4.81)

Migratory insertions of unsaturated substrates into late metal alkoxides have been demonstrated. For example, migratory insertions of CO into late-metal–oxygen bonds have been reported,[258] and olefin insertions into alkoxides have been deduced to be one of the steps of palladium-catalyzed oxidations.[345–348] The insertion of an olefin into a rhodium–alkoxo bond has been observed directly with activated olefins[349] or intramolecularly with a rhodium alkoxide containing a pendant olefin (Equation 4.82).[320] In other cases, a formal insertion of olefin occurs by nucleophilic attack of a metal alkoxide on an olefin coordinated to a second metal.[350,351] A variety of late metal alkoxides have also been shown to insert CO_2 and other heterocumulenes by direct attack of the alkoxide at the electrophilic reagent, rather than by migratory insertion (Equation 4.83).[352,353] Late metal alkoxides also react as nucleophiles towards other organic electrophiles, including esters (Equation 4.84), acyl halides, carboxylic esters, carboxylic anhydrides,[258] and aldehydes.[354] These reactivities are analogous to those of main group alkoxides.[353]

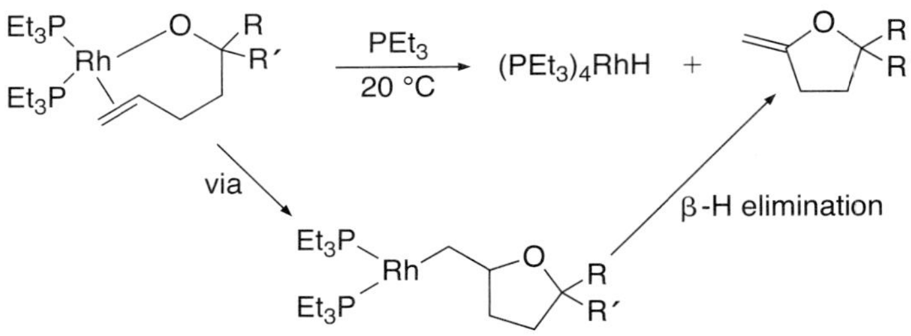

(4.82)

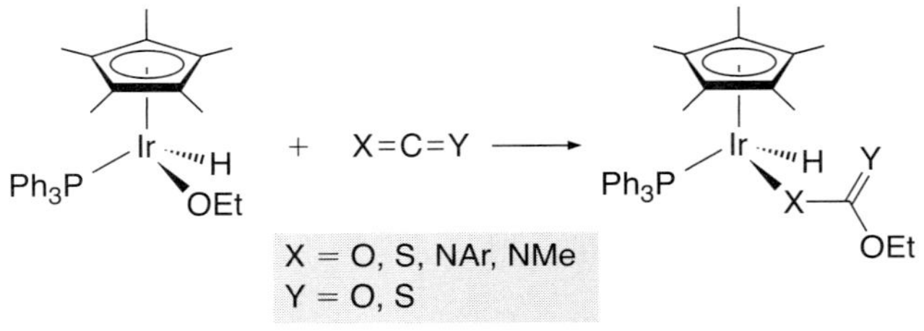

X = O, S, NAr, NMe
Y = O, S

(4.83)

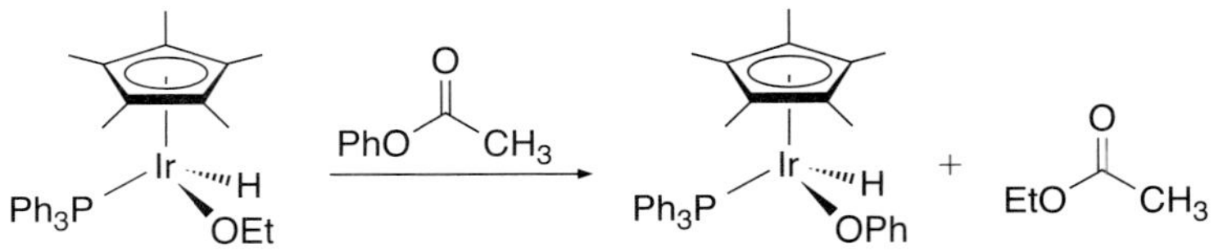

(4.84)

Late metal alkoxides also react as bases to deprotonate organic acids. As mentioned in the bonding section, late metal alkoxides undergo ligand exchange with other compounds containing O–H, N–H, and C–H bonds. The reactions with alcohols form new metal alkoxides, usually by proton transfer with alcohols that are more acidic than the alcohol eliminated. Mechanistic investigations of the reactions of alkoxides with alcohols have revealed the formation of hydrogen-bonded intermediates (Equation 4.85).[355] Such hydrogen bonding of alcohols to late metal alkoxides is common,[356] and the strengths of these hydrogen bonds are indicative of a polarized M–O bond.[357]

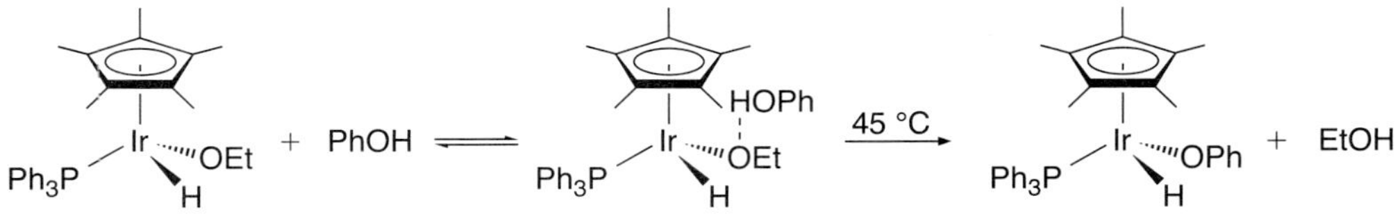

(4.85)

4.3.1.3.7. Catalytic Reactions of Late-Metal–Alkoxo Complexes

Late-metal–alkoxo complexes have been proposed as reactive intermediates in several common catalytic processes. For example, late transition metal–hydroxide complexes are common hydrolysis intermediates in metal-catalyzed reactions carried out in aqueous media.[256,358] Palladium hydroxides and alkoxides are thought to undergo migratory insertions of olefins during classic olefin oxidations.[345–348,359] Metal alkoxides are also potential intermediates in some metal-catalyzed hydrogenations of aldehydes, ketones, and epoxides, as well as the mechanistically related catalytic arylations of aldehydes and ketones.[354,358,360] The recently developed catalytic etherifications of aryl halides and pseudo-halides most likely involve carbon–oxygen bond-forming reductive elimination from arylpalladium(II) alkoxide intermediates.[342] In addition, the generation of hydrides are proposed to occur by β-hydride eliminations from a metal alkoxide during the catalytic transfer hydrogenations of ketones.[361] When β-hydrogens are unavailable at the alkoxo group, β-carbon eliminations can occur, and some catalytic reactions occurring through carbon–carbon bond cleavage of metal alkoxides have been reported.[339] Finally, 2-metallaoxetane complexes are potential intermediates in the catalytic formation of aldehydes and ketones by epoxide isomerizations.[362]

4.3.2. Metal β-Diketonate Complexes

The anions of β-diketones form very stable, six-membered-ring chelate structures with many transition metals. In these complexes, both oxygen donors bind to the metal center in the six-membered ring (Figure 4.19).[363] The chemistry of β-diketoneate complexes is illustrated here using acetylacetonate (abbreviated as acac), the most common β-diketonenolate ligand.

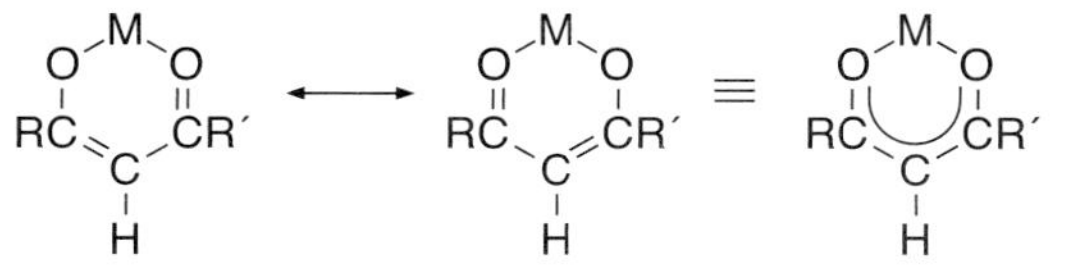

Figure 4.19.
Resonance structures of O,O-κ^2, β-diketonate complexes.

Homoleptic, transition metal acac complexes usually adopt the stoichiometries of $M(acac)_3$ and $M(acac)_2$, such as $Mn(acac)_3$, $Fe(acac)_3$, $Co(acac)_3$, $Ni(acac)_2$, and $Cu(acac)_2$. Structures of $M(acac)_3$ complexes usually display D_3 symmetry with six oxygen donor atoms in an octahedral geometry. $M(acac)_2$ complexes are often isolated as oligomers, but mononuclear $M(acac)_2$ species are favored when the β-diketonate contains more sterically demanding substituents than those in an acac ligand.[364] Alternatively, monomeric complexes are often generated when the complex contains additional ligands, as shown in Equation 4.86.[365] The acetylacetonate anion can also bind to the metal center through the central carbon.[366,367] Complexes with this binding mode are observed when only one binding site is available, so the κ^2 structure is less favorable. For example, the reaction of dimethylmalonate with the hydroxopalladium complex in Equation 4.87 gives the κ^1 complex of the dimethylmalonate anion.[368]

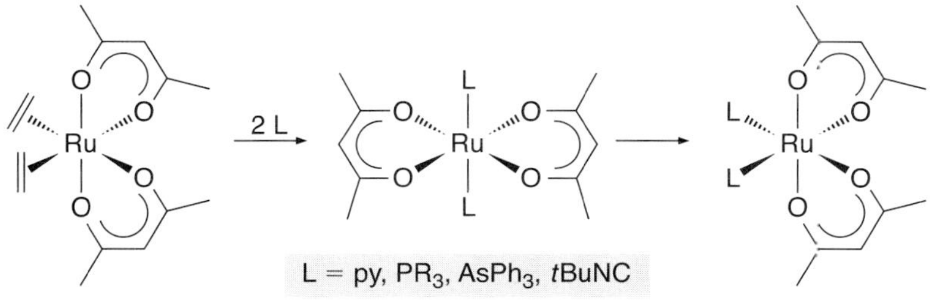

$$L = \text{py, } PR_3\text{, AsPh}_3\text{, } t\text{BuNC}$$

(4.86)

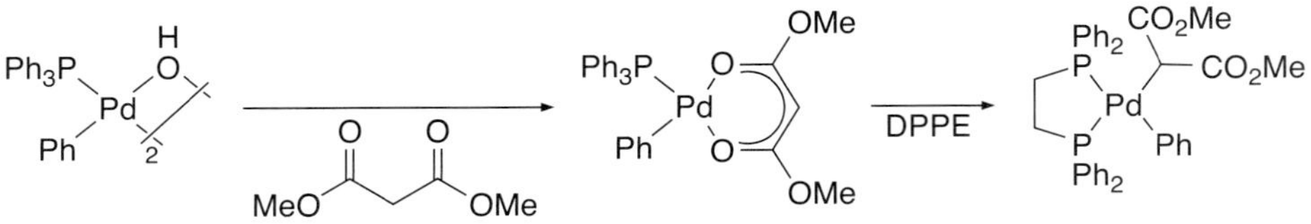

(4.87)

Compared with related metal–halide complexes, transition metal β-diketonate complexes are often more soluble in common organic solvents. Therefore, they are widely used as catalyst precursors and starting materials for the synthesis of other organometallic complexes. For example, the commercially available Cu(acac)$_2$ catalyzes a variety of coupling and carbene transfer reactions involving diazo compounds.[369] The synthetic utility of Au(I) and Au(III) as reagents for other gold complexes has been summarized in a review by Vicente.[370] More recently, bis(acac)-ligated Ir(III) complexes have shown high reactivity towards alkane and arene C–H activations[371,372] (Equation 4.88) and were applied to related catalytic processes.[373]

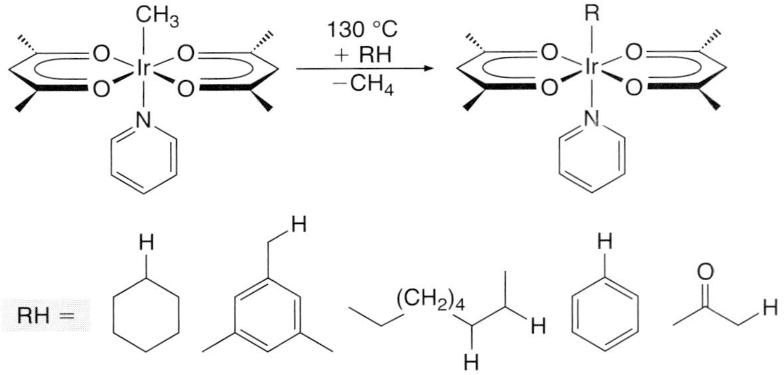

(4.88)

4.4. Transition-Metal–Boryl Complexes (Written with Dr. Jaclyn M. Murphy)

4.4.1. Overview

Transition-metal–boryl complexes[374] contain a covalent bond between a metal and three-coordinate boron. Boryl complexes are a subset of the variety of ligands containing a single boron atom that would encompass borane, boryl, and borene ligands. Of this group, boryl complexes are the most abundant.

Boryl complexes participate in a variety of catalytic processes, including transition-metal-catalyzed hydroboration and diboration of olefin and acetylene C–C π-bonds (Chapter 16) and the functionalization of alkane and arene C–H bonds (Chapter 17). The chemistry of these complexes has been developed primarily since 1990. A body of literature on metal–boryl complexes was published in the 1960s,[375,376] but the structures of these compounds were apparently misassigned in the absence of modern X-ray crystallographic and NMR methods. Boryl complexes of all transition metals except for group 3 (Sc, Y, and La) and group 4 (Ti, Zr, Hf) metals[377,378] have been isolated. A majority of transition-metal boryl–complexes contain late metals.

Free from a transition metal, the boryl group is highly basic. The first fully characterized alkali metal boryl anion was reported in 2006 by Yamashita, Nozaki and co-workers.[379] Previous studies on the generation of boryl anions have not been reproduced.[380] Theoretical studies imply that much of the charge on a free boryl anion would actually reside on the substituents, and that the boron would bear positive charge.[381]

Braunscheweig and co-workers have developed a classification system of compounds containing transition-metal–boron bonds based upon the binding mode and coordination number of the boron atom (Figure 4.20). Borane complexes contain a boron fragment that possesses three substituents other than the metal and is neutral as the free molecule. Borylene ligands are bound to one or two metals and contain a single substituent at boron other than the metals. Boryl ligands contain two substituents and are counted as monoanionic in the ionic bonding formalism. Most boryl ligands are bound to a single metal center, although metal complexes with bridging boryl ligands are known.[378,382–389] Boryl ligands can contain alkyl, aryl, halogen, amido, or alkoxo substituents. The majority of the isolated complexes contain tethered substituents in the form of a diolate derived from catechol or pinacol.

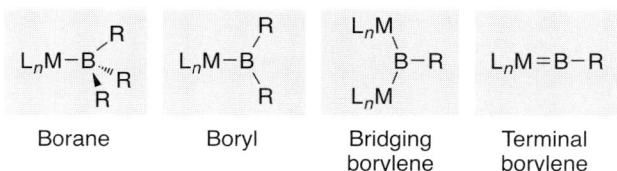

| Borane | Boryl | Bridging borylene | Terminal borylene |

Figure 4.20.
Classification system for boron ligands containing a single boron atom.

4.4.2. Metal–Boryl Bonding

The bonding in metal–boryl complexes has been the topic of many computational studies.[390–394] The metal–boryl bond is similar to the bonds between boron and nonmetals, and is strong (vide infra). Figure 4.21 shows the orbitals that can participate in the different metal–boron interactions. The metal–boryl fragment contains a σ-bond between the metal and the boron atom. In addition, the empty p-orbital at boron creates the potential for π-bonding interactions between this orbital and d-orbitals of appropriate symmetry on the metal center. Several boryl ligands have been shown to possess a geometry that allows for π-bonding between the HOMO of the metal center and the p-orbital at boron.[378] The computational studies have implied that a weak, but significant, π-bond connects the metal to boron,[390] and the absence of metal-to-ligand π-donation in d^0 complexes might contribute to the scarcity of d^0 boryl complexes.

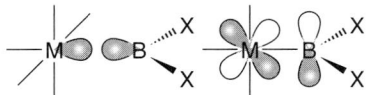

Figure 4.21.
Bonding in metal–boryl complexes.

The electron deficiency of the boron in a boryl complex is often attenuated by the substituents on the boron atom; most metal–boryl complexes contain substituents that are capable of π-donation into the boron p-orbital. The electron deficiency can also be attenuated by metal-to-ligand backbonding. Thus, few complexes in which Lewis bases are bound to boron have been isolated. These examples were some of the first boryl compounds prepared and contain hydride substituents at boron that are incapable of π-donation into the p-orbitals. The compounds include $(CO)_4Co\text{-}BH_2\cdot X$, where $X = $ THF, NEt_3, or SMe_2.[395]

Several studies have experimentally probed the electronic properties of boryl ligands. The CO stretching frequencies of $[CpRu(CO)_2(X)]$, in which $X = $ H and Me, and a series of boryl groups have been measured.[396] The electron donation, according to these ν_{CO} values, follows the trend $BR_2 > $ Bpin $ > $ Bcat (where R = alkyl, pin = pinacolate, and cat = catecholate). These data imply that the largest factor controlling the electronic properties is the σ-donation by the boryl group. Moreover, these studies imply that the overall electron-donating abilities of these boryl groups are similar to those of alkyl or hydride groups. Studies have also probed the trans influence of boryl groups.[397] These studies show that the boryl group imparts a strong trans influence. Much of this property was attributed to strong σ-donation, enhanced by π-acceptance.

4.4.3. Thermodynamics of Metal–Boryl Complexes

Few data on metal–boryl bond strengths are available, but the existing data does provide some benchmarks. An iridium–boryl bond has been measured by calorimetry, in combination with calculations of B–H bond strengths, to be 66 kcal/mol.[398,399] This value is much higher than the metal–methyl bond strength of about 35 kcal/mol between a methyl group and the same iridium fragment, and it is similar to the metal–hydride bond strength of about 60 kcal/mol involving the same iridium fragment.[400] Thus, the metal–boryl bond is strong in the context of organometallic systems. A boron–boron bond is approximately 104 kcal/mol and a typical boron–hydrogen bond is 110 kcal/mol.[400] Thus, the oxidative addition of either a diboron reagent (B_2R_4) or a borane (HBR_2) to a metal center is enthalpically favored by approximately 10 to 15 kcal/mol.

4.4.4. Synthesis of Metal–Boryl Complexes

Most metal–boryl complexes have been synthesized by salt elimination or oxidative addition. Examples of these reactions are shown below. Other less common routes to metal–boryl complexes include transmetallation of a boryl group from one transition metal to another[401] and reactions of boranes with metal–olefin complexes.[402,403] Boryl complexes have also been generated as intermediates in catalytic processes by "transmetallation" of a metal halide with a diboron reagent.[404]

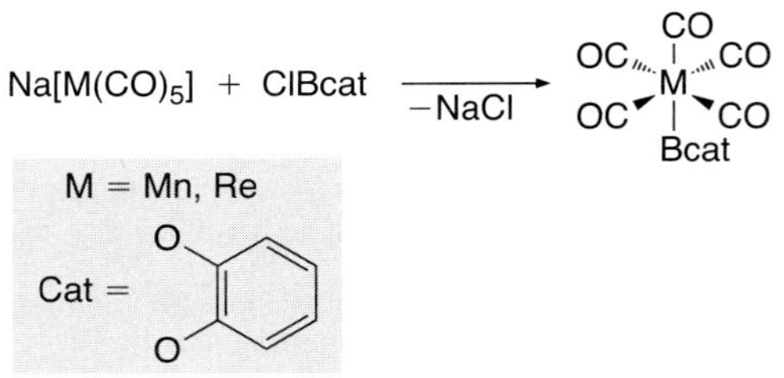

(4.89)

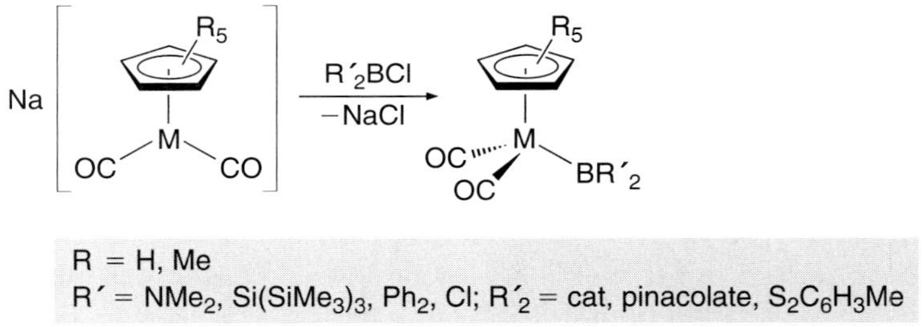

(4.90)

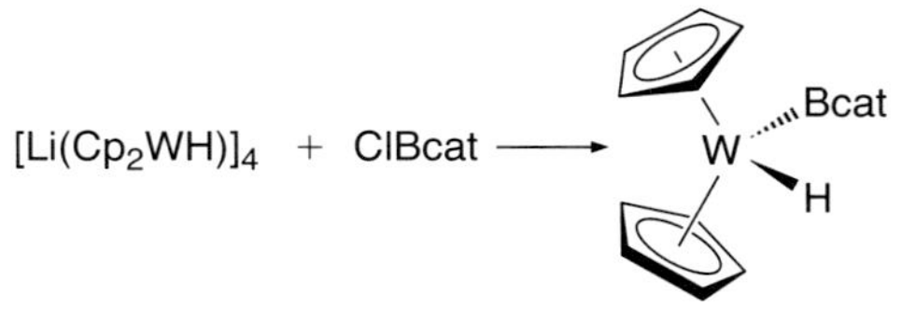

R = H, Me
R′ = NMe_2, $Si(SiMe_3)_3$, Ph_2, Cl; R′$_2$ = cat, pinacolate, $S_2C_6H_3Me$

(4.91)

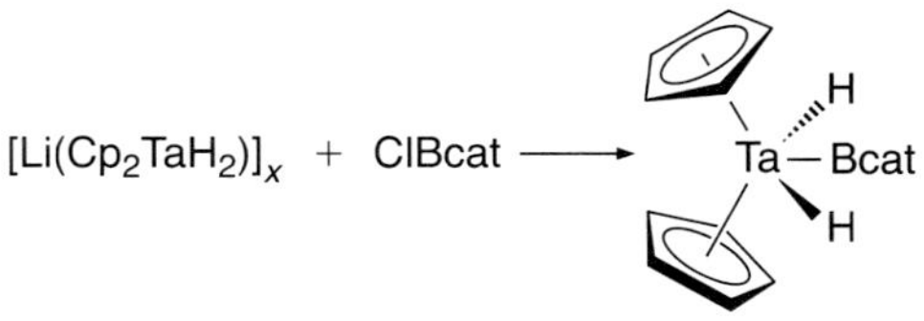

(4.92)

Examples of boryl complexes prepared by salt eliminations are shown in Equations 4.89–4.92. The salt elimination chemistry is conducted with transition metal anions and haloboranes because the availability of boryl anions is limited. Thus, the scope of this approach is limited by the availability of *metal* anions, and most of the compounds prepared by this method have been generated from the anions of metal compounds containing supporting CO ligands. Nevertheless, a series of boryl complexes from $[CpM(CO)_n]^-$ or $[Cp^*M(CO)_n]^-$ and amino chloroboranes, dialkyl chloroboranes, dithiocatechol chloroboranes, catechol chloroboranes, and pinacol chloroboranes have been prepared.[405-410] The reaction of ClBcat with the anions of metallocenes have also produced boryl compounds. For example, the reaction of ClBcat with $Li[Cp_2WH]$ yielded the tungsten boryl hydride complex in Equation 4.91.[411,412] The reaction of $Li[Cp_2TaH_2]$ in Equation 4.92 generated a product[413] that appears to be the closest to a bonafide d^0 metal–boryl hydride complex,[411] although these compounds are likely stabilized by the weak B–H interaction.[414] In addition, dianionic metal complexes can be used to prepare bisboryl complexes. For example, Collman's reagent, $Na_2[Fe(CO)_4]$, has been used as the precursor for the synthesis of $Fe(CO)_4(Bcat)_2$ (in which cat = catecholate) (Equation 4.93).[412]

$$Na_2[Fe(CO)_4] \ + \ 2\ ClBcat^* \quad \xrightarrow{-2NaCl} \quad \underset{OC}{\overset{OC}{\longrightarrow}} Fe \underset{Bcat^*}{\overset{Bcat^*}{\longleftarrow}} \tag{4.93}$$

Examples of complexes prepared by oxidative addition of boranes or diboron reagents are shown in Equations 4.94–4.99. Oxidative addition reactions are the subject of Chapters 6 and 7. A few illustrative examples of oxidative additions to form metal–boryl complexes are provided here. Diborane reagents undergo oxidative addition to late transition metals, such as Co, Rh, Ir, and Pt,[415-417] after dissociation of phosphines or other labile ligands. For example, B_2pin_2 adds to $[Pt(PPh_3)_4]$ to yield cis-$[(pinB)_2Pt(PPh_3)_2]$ (Equation 4.94)[417] and B_2cat_2 adds to $[Co(PMe_3)_4]$ to form $[(catB)_2Co(PMe_3)_3]$.[418] A series of Rh– and Ir–bisboryl complexes have also been obtained from the reaction of $[MCl(PPh_3)_3]$ and diboron reagents (Equation 4.95).[377,416,419,420] In other cases, the open coordination site is generated photochemically, as shown for the iron–carbonyl complex in Equation 4.96.[421]

$$[(PPh_3)_2Pt(C_2H_4)] \ + \quad \text{(diboron reagent)} \quad \xrightarrow{-C_2H_4} \quad Ph_3P-Pt-PPh_3 \tag{4.94}$$

$$[(Ph_3P)_3MCl] \ + \ R_2B-BR_2 \quad \xrightarrow{-PPh_3} \quad \underset{Ph_3P}{\overset{Ph_3P}{\longrightarrow}} M \underset{Cl}{\overset{BR_2}{\underset{}{\longleftarrow}}} BR_2 \tag{4.95}$$

M = Rh and Ir

$$Fe(CO)_5 \ + \ B_2cat^*_2 \quad \xrightarrow[-CO]{h\nu} \quad \underset{cat^*B}{\overset{cat^*B}{\longrightarrow}} Fe \underset{CO}{\overset{CO}{\underset{}{\longleftarrow}}} CO \tag{4.96}$$

Boranes (HBR$_2$) also undergo oxidative additions with transition metals to form boryl complexes, as shown generically in Equation 4.97. The first fully characterized boryl complexes were prepared by this route. In 1990, Baker, Marder, and co-workers reported the full characterization of *fac*-[IrH$_2$(PMe$_3$)$_3$(BC$_8$H$_{14}$)] and *fac*-[IrH(PMe$_3$)$_3$(η^2-CH$_2$BHC$_8$H$_{14}$)] from the reaction of [IrMe(PPh$_3$)$_4$] and HBR$_2$ (Equation 4.98).[422] Also in 1990, Merola and Knorr described the isolation and characterization of [(PMe$_3$)$_3$IrCl(H)(Bcat)] from the reaction of Ir(COE)(PMe$_3$)$_3$Cl and HBcat (Equation 4.99).[423]

$$L_nM \ + \ HBR_2 \ \longrightarrow \ L_nM\begin{smallmatrix} H \\ \diagdown \\ BR_2 \end{smallmatrix} \tag{4.97}$$

$$[IrH(PMe_3)_4] \ + \ HBR_2 \ \longrightarrow \ \tag{4.98}$$

$$[Ir(COE)(PMe_3)_3Cl] \ + \ HBcat \ \longrightarrow \ \tag{4.99}$$

4.4.5. Reactivity of Metal–Boryl Complexes

Metal–boryl complexes have been observed to undergo several types of reactions at the M–B bond, at the metal, and at the boron without involvement of the metal center. These topics have been reviewed.[374] Boryl ligands have been shown to react with protic reagents, such as water, alcohols, and amines, to cleave the M–B bond.[408,412] In addition, a few complexes react with these protic reagents to exchange groups at the boron. For example, the reaction of water at the dichloroboryl ligand in [OsCl(CO)(PPh$_3$)$_2$(BCl$_2$)] produced {OsCl(CO)(PPh$_3$)$_2$(L)[B(OH)$_2$]}.[424] Reaction of the same complex with methanol and ethanol yielded the alkoxyboryl derivatives (Equation 4.100).[424]

$$\tag{4.100}$$

4.5. Transition-Metal–Phosphido Complexes
(Written with Prof. Jack R. Norton)

Phosphide ligands are related to amide ligands because both contain two substituents and an electron pair, but these two kinds of ligands differ in many respects from each other. Phosphido ligands are softer then amide ligands, and the phosphorus is more polarizable and nucleophilic than the nitrogen of an amido group. In addition, the geometric parameters differ from those of amido groups, and phosphido groups tend to be stronger bridging ligands than amido groups are. Thus, the structure and reaction chemistry of phosphido complexes differ greatly from that of amido complexes.

4.5.1. Structures of Phosphido Complexes

In contrast to amido ligands, which are typically planar at nitrogen, phosphide ligands are often pyramidal. This structural difference parallels the more acute angles in phosphines, relative to those in amines. More planar geometries in phosphide ligands are observed when donation from the p-orbital on phosphorus occurs into a vacant metal orbital, as in resonance structure **A** in Figure 4.22. More pyramidal geometries, as shown in structure **B**, are observed when the metal lacks a vacant orbital to act as a π-acceptor. Conversion of a pyramidal phosphide into a planar phosphide has been observed after one-electron oxidation of the complex to make the metal a better π-acceptor.[425]

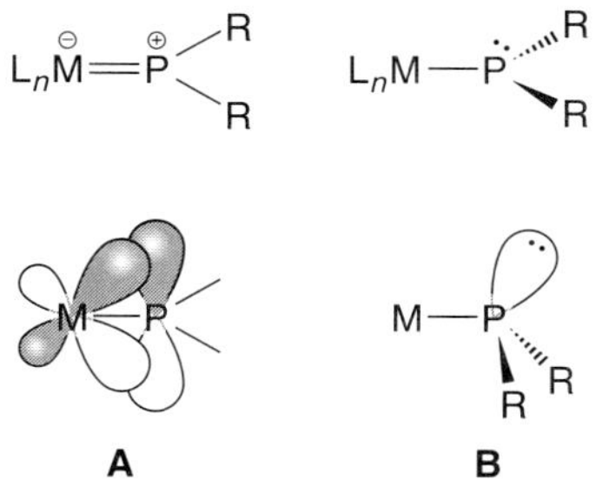

Figure 4.22.
Differences in π-bonding of planar and pyramidal phosphido complexes.

Phosphido complexes that contain a strong contribution from resonance structure **A** in Figure 4.22 tend to possess short M–P bonds and large M–P–R angles. For example, the W–PHPh bond in the tungsten phosphide of Figure 4.23 is 0.26 Å shorter than the W–PEt$_3$ bond in the same complex, and the W–P–C angle is 140°.[426] Likewise, the Ru–PCy$_2$ bond in the ruthenium phosphide of Figure 4.23 is 0.11 Å shorter than the Ru–PPh$_3$ bond in the same complex, and the Ru–P–C angles average 127°.[427]

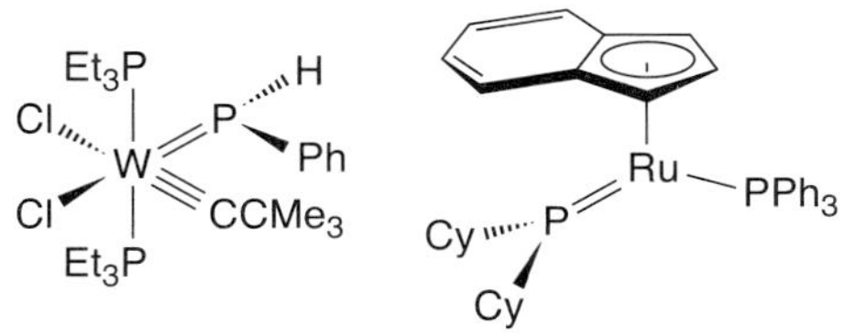

Figure 4.23.
Two planar phosphido complexes in which the phosphido and phosphine geometric parameters can be compared.

In contrast, phosphido complexes that are pyramidal tend to have long M–P bonds and compressed M–P–R angles. For example, the Os–PHPh bond in the osmium–phosphido complex in Figure 4.24 is 0.1 Å longer than the Os–PPh$_3$ bond, and the Os–P–C angle is only 113°.[428] Likewise, the Ru–P(Me)Ph bond in the ruthenium–phosphido complex of Figure 4.24 is 0.17 Å longer than the related 16-electron cationic phosphido complex, [(dmpe)$_2$Ru(PH(Me)Ph)H]$^+$, and the angles around the phosphido P average 109°.[429]

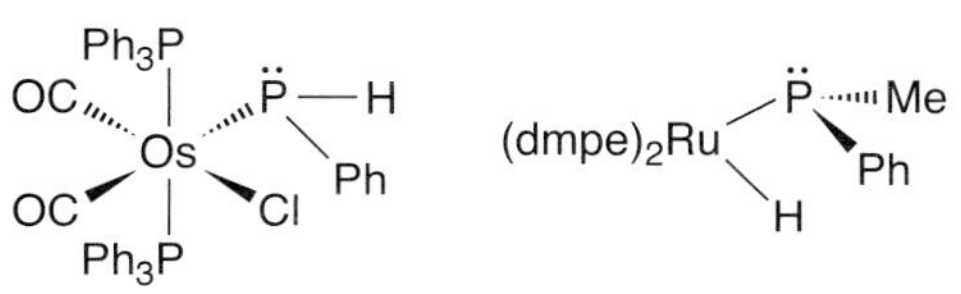

Figure 4.24.
Two pyramidal phosphido complexes in which the phosphido and phosphine geometric parameters can be compared.

4.5.2. Dynamics of Phosphido Complexes

As described in Chapter 2, phosphines have sufficiently large barriers to inversion that phosphines with three different substituents can be resolved. In contrast, the inversion of configuration at phosphorus in metal–phosphido complexes tends to occur near room temperature. In unsaturated transition metal complexes, a vacant acceptor orbital stabilizes the planar transition state. This effect is shown by the interconversion between pyramidal structures through a planar transition state, like the one shown in Equation 4.101[430] and formed during inversion of related compounds.[431,432] In saturated middle and late transition metal complexes, inductive effects explained below destabilize the ground state and lead to lower barriers to inversion.[433] The presence of an ancillary planar phosphide[431,432] or amide[430] ligand can contribute to a low barrier by accepting the lone pair from a pyramidal phosphide, as in Equation 4.101.

$$(4.101)$$

4.5.3. Thermodynamic Properties of Phosphido Complexes

A few studies have revealed the bond energies of metal–phosphido complexes. The BDE of $(Cp^*)_2Sm–PR_2$ has been measured as 33(2) kcal/mol.[434] Glueck and Nolan reported relative Pt–P bond energies for a series of phosphido complexes, and data on Pt–P versus Pt–S bond energies.[435] They found that the Pt–P bond energies followed the order Pt–PHPh > Pt–PHMes > Pt–PHMes* > Pt–PPh₂ > Pt–PMes₂. Thus, steric factors appear to have a large effect on the relative Pt–P bond strengths. In addition, comparisons of the strength of the Pt–P to Pt–S bonds showed the Pt–S bond in Pt(dppe)(Me)(SMes*) is stronger than the analogous Pt–P bonds in Pt(dppe)(Me)(PHMes*) and Pt(dppe)(Me)(PHPh).

Additional studies have shown that the P–H bond is much more acidic than its N–H counterpart: the pK_a of PH_3 in DMSO is 24, 16 units lower than that of NH_3; the pK_a of R_2PH in DMSO is 34, while that of Ar_2PH is 22.[436] Coordination increases the acidity of primary and secondary phosphines; alkoxides deprotonate the coordinated secondary phosphines but not the free phosphines.[429,437]

4.5.4. Reactivity of Phosphido Complexes

The lone pair of a pyramidal phosphide of a coordinatively saturated complex is extremely nucleophilic. This high nucleophilicity arises from the interaction between a filled metal d-orbital [like that of the $CpRe(NO)PPh_3^+$ fragment illustrated in Figure 4.25] and the lone pair on the phosphorus in the phosphide ligand.[438] The interaction is similar to that responsible for the gauche effect in organic chemistry (i.e., the preference for lone pairs on adjacent atoms to adopt a gauche conformation),[439,440] and the α-effect (i.e., enhancement of the nucleophilicity of lone pairs on adjacent atoms).[441] This nucleophilicity allows for the alkylation described in the next paragraph and the fast reductive eliminations discussed in Chapter 8.

The enhanced nucleophilicity of a pyramidal phosphide P has been used by Bergman and Toste[429] and by Glueck[437] to develop the catalytic enantioselective synthesis of P-stereogenic phosphines depicted in Equation 4.102. The system reported by Bergman and Toste involves [(R,R)-MeBPE]₂Ru(H)(PMePh) as the phosphido intermediate and Na(OCMe₂Et) as the base;[429] the system reported by Glueck involves Pt[(R,R)-MeDuphos](Ph)(PMeAr) as the phosphido intermediate and Na(OSiMe₃) as the base.[437]

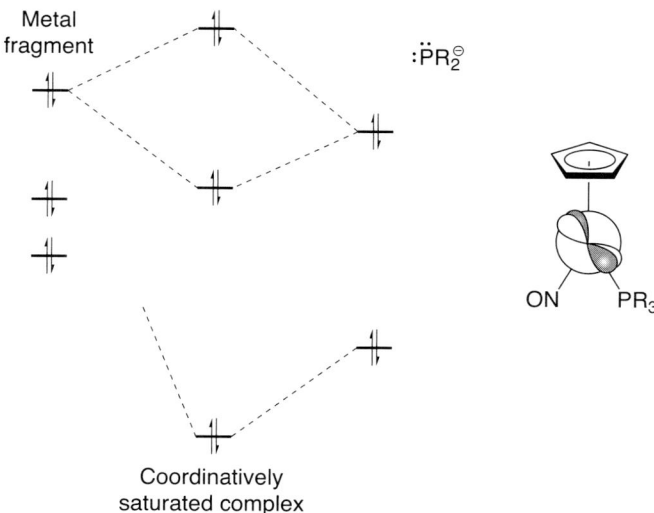

Figure 4.25.
HOMO of CpRe(NO)PPh$_3^+$ and the MO diagram resulting from its interaction with a phosphido ligand.

The mechanism of this process, shown generically in Equation 4.102, occurs by generation of a phoshido complex by deprotonation of a cationic complex of a secondary phosphine. If the phosphido complex contains a chiral ligand, then two diastereomeric phosphido complexes are generated, and these complexes are related by the stereochemistry at phosphorus. For this process to occur catalytically, the electrophile must react with the phosphide ligand, but not with its secondary phosphine precursor PH(Me)Ar. If the inversion at phosphorus is faster than the alkylation, then the stereoselectivity will result from the combination of the relative energies of the diastereomeric phosphido complexes and the relative rates for reaction of these two diastereomers. This scenario is an example of a "Curtin–Hammett [442] situation," and more details of this control of enantioselectivity are contained in Chapter 14, which describes the principles of catalysis, including asymmetric catalysis.

$$R-X \ + \ \underset{\text{Racemic}}{PH(Me)Ar} \ \xrightarrow{L^*M} \ :P \overset{R}{\underset{Ar}{\overset{\text{\tiny\textbackslash}}{\cdots}}} Me$$

via

(4.102)

4.6. Transition Metal-Thiolate–Complexes (Written with Dr. Elsa Alvaro)

4.6.1. Overview

Transition-metal–thiolate complexes are intermediates in synthetically useful processes, such as C–S coupling to form aryl thioethers,[443-448] hydrothiolation,[449-454] and hydrodesulfurization.[455] In addition, thiolates are common ligands in coordination chemistry and in bioinorganic systems, such as ferredoxins and thioneins.[456] Thiolates tend to form stable linkages to many types of transition metals. Thus, thiolate groups have been known to poison or to reduce the activity of homogeneous catalysts, but this deactivation by thiolates can be controlled and does not have the indiscriminate poisoning effect that it is typically thought to have on heterogeneous catalysts.

4.6.2. Bonding and Structures of Transition-Metal–Thiolate Complexes

Several properties of thiolate ligands cause the transition metal–sulfur bond to be a particularly stable linkage in transition metal chemistry. When considering the bonding of thiolate ligands, an overarching property is the softness of the ligand, relative to the first-row alkoxide congeners. Because many transition metals are softer than main group metals, this property contributes to the stability of the M–S bond. In addition, the weakness of C=S double bonds makes β-hydrogen elimination unfavorable, and M–S π-bonding can contribute to the stability of metal–thiolate complexes.

A thiolate ligand has a σ-donor orbital and a filled nonbonding orbital with mainly sulfur p-orbital character. The latter orbital possesses the appropriate symmetry for a π-interaction with a metal d-orbital (Figure 4.26). Therefore, thiolates can stabilize electron deficient and high oxidation state metal centers by π-donation. For example, the Mo–S distance in the six-coordinate Mo(IV) complex on the left in Figure 4.27 (235.5 ppm) is shorter than that in the seven-coordinate species on the right in the figure (253.3 ppm). While coordination number and steric effects may contribute to changes in bond distances, this difference in bond length may imply that the six-coordinate species contains M–S π-bonding.[457]

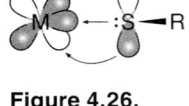

Figure 4.26.
A qualitative description of the bonding in a metal–thiolate complex.

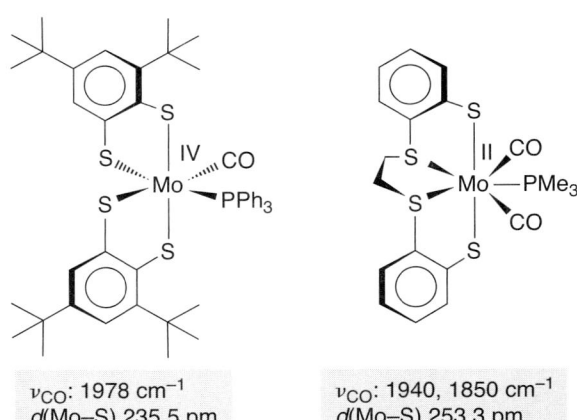

ν_{CO}: 1978 cm^{-1}
d(Mo–S) 235.5 pm

ν_{CO}: 1940, 1850 cm^{-1}
d(Mo–S) 253.3 pm

Figure 4.27.
Structural features of two transition-metal–thiolate complexes.

Thiolate complexes often adopt structures containing bridging thiolate ligands. Most of these compounds contain thiolate ligands that bridge between two metals, and the sulfur in the thiolate ligands that bridges these two metals is usually pyramidal.[458] At the same time, complexes in which the thiolate bridges three and even four metal centers are known.[459] This propensity for forming bridges results from several factors besides the presence of nonbonded electron pairs. It results from the greater propensity of second-row elements (E) to form two single bonds than to form one double bond,[460] the propensity of second-row elements to form structures with small X–E–X angles,[461] and the large size and high polarizability of sulfur. Bridging structures can be discouraged by steric bulk on the thiolate ligand. For example, the reaction of Cp_3Yb with HSR (R = Pr or Bu) afforded $[Cp_2Yb(\mu\text{-}SR)_2]_2$ (Equation 4.103),[462] whereas the reaction with the more sterically encumbered 2,4,6-tris(trifluoromethyl)thiophenol in THF solvent gave the related mononuclear complex containing a THF ligand (Equation 4.104).[463]

$$2\,Cp_3Yb \;+\; 2\,HSR \;\longrightarrow\; [Cp_2Yb(\mu\text{-}SR)]_2 \;+\; 2\,C_5H_6 \tag{4.103}$$

$$\text{(4.104)}$$

This trend in structures is important because bridging thiolates are less reactive toward the types of organometallic transformations described in this text than are terminal thiolates. For example, bridging and terminal thiolate complexes of palladium complexes containing similar ancillary ligands are known, and the bridging complexes do not undergo the same reductive elimination reactions that the terminal thiolate complexes undergo.[464]

4.6.3. Thermodynamics of M–SR Bonds

Several revealing studies on M–S bond strengths have been reported. When considering the thermodynamics of M–S bonds, keep in mind that bonds to second-row elements tend to be weaker than those to first-row elements, and the H–SR bond in a thiol is weaker than the H–OR bond in an alcohol containing the same R group, because of the greater stability of the thiyl radical.

Bond enthalpies reported by Dias and Martinho Simoes[465,466] for a series of $MCp_2(XR)_2$ complexes (where X = O and S) show that the M–S bonds are 24–29 kcal/mol weaker than the corresponding M–O bonds. For instance, the M–X bond enthalpy in $Ti(Cp)_2(SC_6H_5)_2$ is 86 kcal/mol, whereas the M–X bond enthalpy in the related alkoxo complex, $Ti(Cp)_2(OC_6H_5)_2$, is 110 kcal/mol. Thus, the early-metal–thiolate bond is weaker (or the early-metal–alkoxide bond is stronger) than would be expected based on the H–X bond strengths.

The relative strengths of late-metal–thiolate and late-metal–alkoxide bonds are different than are the relative strengths of early-metal–thiolate and early-metal–alkoxide bonds. In this case, the late-metal–thiolate bond is stronger than would be expected based on H–SR and H–OR bond energies.[467] For example, the reaction in Equation 4.105 lies fully to the right and occurs with $\Delta H = -14$ kcal/mol,[468] and the M–S bonds in (DPPE)MePtSR and $Cp^*(PMe)_3RuSR$ are stronger than would be expected, based on S–H bond strengths.[21]

$$Cp^*Ni(PEt_3)(OTol) \;+\; TolSH \;\longrightarrow\; Cp^*Ni(PEt_3)(STol) \;+\; TolOH \tag{4.105}$$

This greater strength of the late-metal–thiolate bond compared to the early-metal–thiolate bond or the late-metal–alkoxide bond has been rationalized in two ways. First, the late-metal–thiolate bond involves the favorable match of a soft ligand with a soft metal.[21] Second, this bond involves a larger covalent component than the bond of a late metal to an alkoxide or an early metal to a thiolate.[468]

Metal–thiolate complexes also tend to be more kinetically stable toward decomposition pathways that are common for organometallic complexes. For example, the thiolate ligands in metal–thiolate complexes tend to be more stable than alkyl or alkoxo ligands toward β-hydrogen elimination. In addition to the strength of the M–S bond, the weakness of the C=S double bond noted earlier makes the thermodynamics for this type of elimination process unfavorable.

4.6.4. Synthesis of Metal–Thiolate Complexes

Several approaches have been used to prepare transition-metal–thiolate complexes. The most common synthetic route involves the metathesis reaction of a metal halide with an alkali metal thiolate salt. For instance, [(DPPE)Pd(Ar)(S-*t*-Bu)] was obtained by treatment of the (DPPE) palladium aryl iodide complex with sodium *tert*-butyl thiolate (Equation 4.106).[469] Additionally, these compounds have been formed by proton exchange reactions of thiols with M–C (Equation 4.107),[470] M–N,[471] and M–O,[472] bonds (Equation 4.108). Finally, the oxidative addition of H_2S,[473,474] RSH,[475] or alkyl and aryl disulfides[476,477] is a useful way to obtain transition metal thiolates. Examples of these reactions are provided in Equations 4.109–4.111.

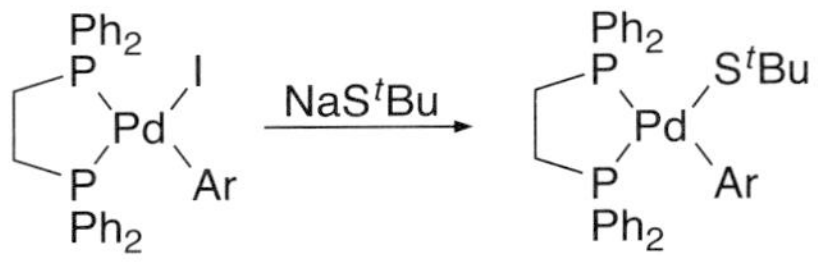

(4.106)

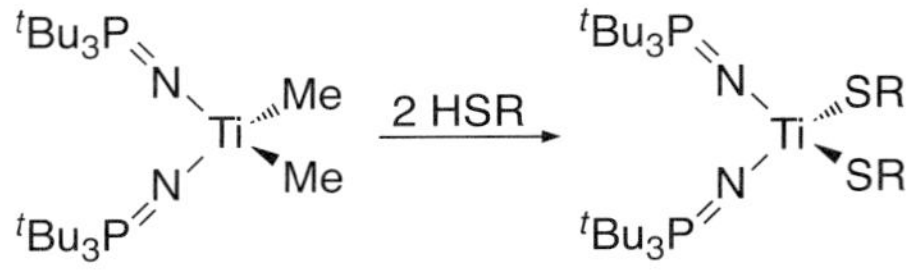

(4.107)

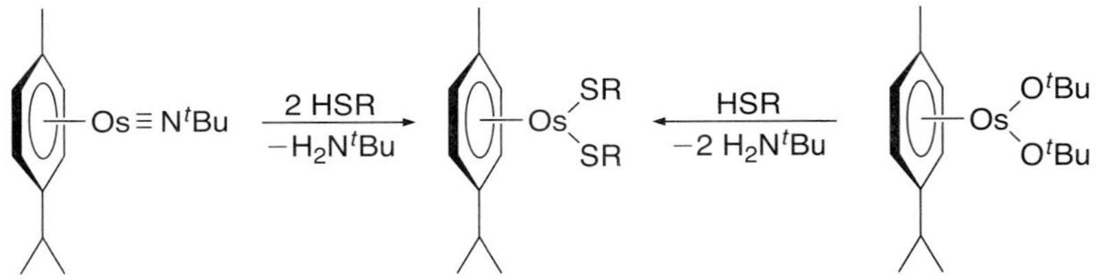

(4.108)

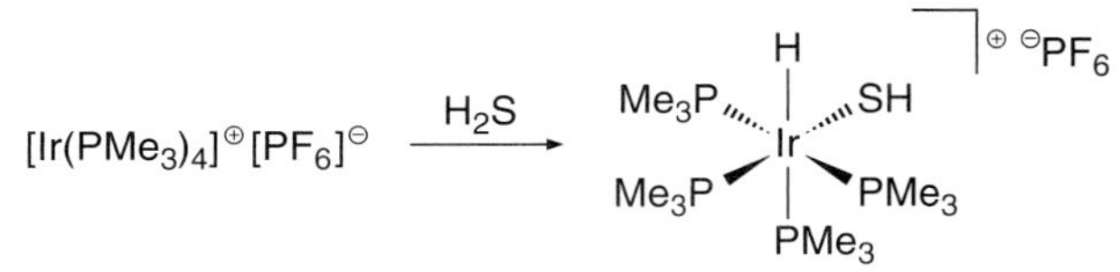

(4.109)

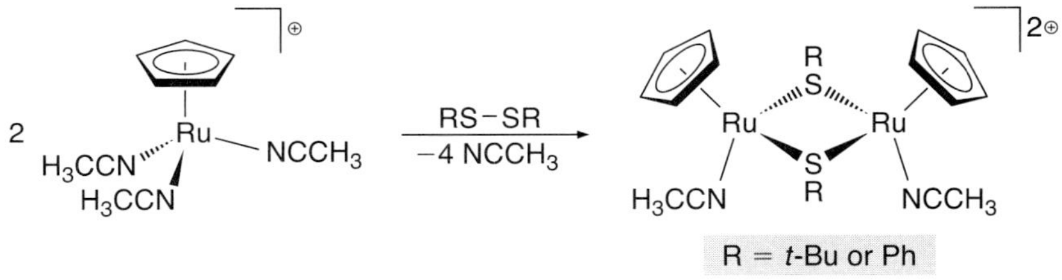

R = *t*-Bu or Ph

(4.110)

$$Cp_2Ti(CO)_2 + RSSR \longrightarrow Cp_2Ti(SR)_2 + 2\,CO$$

(4.111)

4.6.5. Reactivity of Thiolate Complexes

Complexes containing terminal thiolate ligands undergo a wide range of reactions and much of this reactivity is described in later chapters of the text. These reactions include reductive eliminations to form aryl sulfides, migratory insertions of alkenes and alkynes to form C–S bonds, and nucleophilic attack on electrophiles, such as α,β-unsaturated carbonyl compounds.[478] As a result, thiolates are not common spectator ligands in organometallic chemistry. It is more common that thiolates are reactive ligands in organometallic species. This contrasts with the common role of thiolates as spectator ligands in coordination and bioinorganic chemistry. As just one example, thiolate ligands are spectators in the oxygen transfer reactions of the two Re(V) dithiolate complexes in Figure 4.28.[479]

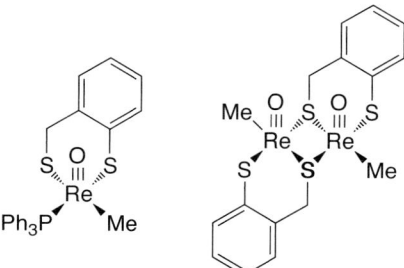

Figure 4.28.
Two thiolate complexes involved in oxygen atom transfer reactions.

Hydrosulfido complexes, which contain M–SH groups,[480] can react by an additional pathway unavailable to metal–thiolate complexes. Hydrosulfido complexes can react by cleavage of the S–H group to generate metal–sulfido complexes. The synthesis and reactivity of sulfido complexes are described in Chapter 13 (metal–ligand multiple bonds).

4.7. Transition-Metal–Silyl Complexes (Written with Dr. Tim A. Boebel)

4.7.1. Overview

The study of transition-metal–silyl derivatives began with Wilkinson's preparation of CpFe(CO)$_2$SiMe$_3$ in 1956.[481] Further examples of metal–silyl complexes were initially slow to appear, but in the ensuing 50 years this class of compounds has expanded to include most, if not all, transition metals.[482] In addition to a large number of isolated examples of metal–silyl complexes, a large number of metal–silyl complexes have been proposed as intermediates in catalytic processes. These processes include hydrosilylation (Chapter 16), dehydrogenative silylation (Chapter 16), dehydrogenative coupling of silanes to form polysilanes, and the large-scale copper-catalyzed reaction of elemental silicon with methyl chloride to generate methylchlorosilanes (known as the "direct process.")

4.7.2. Electronic Properties of Free and Coordinated Silyl Groups

Figure 4.29 summarizes many of the electronic properties of silyl ligands and silyl complexes. Silyl ligands are treated as two-electron, anionic ligands in the ionic bonding model and one-electron ligands in the neutral counting method. Little is known of the acid–base chemistry of free silyl anions, although it is presumed that they should be less basic than free alkyl anions; predicted values fall into a general range of 30–50 pK_a units in DMSO solvent.[483] The pK_a values of two hydrosilanes have been experimentally determined: (Me$_3$Si)$_3$SiH with a pK_a of 29.4 in diethyl ether,[484] and Ph$_3$SiH with a pK_a of 35.1 in tetrahydrofuran.[485,486]

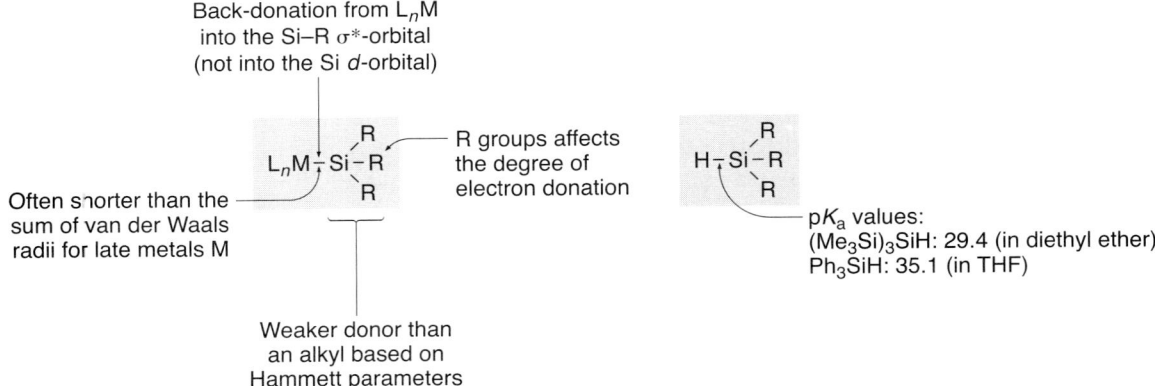

Figure 4.29. Summary of several of the electronic properties of silyl ligands and silyl complexes.

Silyl groups as substituents also tend to be less electron donating than alkyl groups. The Hammett σ-parameter for the trimethylsilyl group is roughly 0.17 units less donating than that of a methyl group, although the trimethylsilylmethyl group is thought to stabilize carbocations.[487] As a ligand for transition metals, the silyl group possesses some propensity for π-acceptance into the Si–X σ* orbitals.[488] Halogen-, alkoxy-, and aryl-substituted silyl ligands can act as π-acceptors, at least to some extent.[489] This strong σ-donating ability and partial π-accepting property of silyl ligands gives them a pronounced trans effect.[490]

The bond dissociation energy of M–Si bonds involving later transition metals should be higher than those of earlier transition metals, and the bonds should be stronger as one descends a column of transition metals. For example, the M–Si bond strengths in the series of complexes $Cl_3M–SiH_3$, in which M = Ti, Zr, and Hf, were calculated to be 50.4, 57.2, and 65.0 kcal/mol, respectively.[491] Moreover, the introduction of electron-withdrawing substituents onto silyl ligands strengthens the M–Si bond. For example, the BDE of the Co–Si bond in $Co(CO)_4SiH_3$ (50.6 kcal/mol)[491] is lower than that in $Co(CO)_4SiCl_3$ (75.9 kcal/mol).[492] Unfortunately, there is relatively little experimental data to corroborate this computational work. Two experimentally measured metal–silicon bond strengths are those of Cp_3USiPh_3 (37.3 kcal/mol)[493] and $PtIMe_2(SiMe_3)(bpy)$ (55.6 kcal/mol).[494]

4.7.3. Structures of Metal–Silyl Complexes

Metal–silicon bonds of low-valent or late transition metal derivatives are frequently shorter than the sum of the van der Waals radii of the two atoms. For example, the iron-silicon bond length in $CpFe(CO)_2SiCl_3$ of 2.21 Å is considerably shorter than the sum of the van der Waals radii of Fe and Si (2.43 Å).[492] The origin of the short M–Si bond is a matter of some debate, but has been attributed in some studies to backbonding into the Si–X σ* ligands. The ability of silanes to serve as strong σ-donors could also be a contributing factor to the shortening of the metal–silicon bond. Conversely, d^0 metal–silyl complexes, which should not exhibit $d\pi–d\pi$ or $d\pi–\sigma^*$ backbonding, often display metal–silyl bond lengths equal to, or greater than, predicted values.[495] An extreme example is found in the zirconium complex $Cp_2Zr(SiPh_3)Cl$, in which the metal–silicon bond length is 2.813 Å. This distance is much greater than the value predicted by the sum of the van der Waals radii (2.63 Å).[496]

4.7.4. Spectral Properties of Metal–Silyl Complexes

Informative spectral features of metal–silyl complexes include the CO stretching frequencies of ancillary CO ligands and ^{29}Si NMR chemical shifts of the silicon in the silyl group. Because silyl groups tend to be weaker donors than alkyl groups, the CO stretching frequencies of ancillary CO ligands in metal–silyl complexes tend to be higher than the CO stretching frequencies of the CO ligands in analogous metal–alkyl complexes. Moreover, the magnitude of the trans effect increases with electron density on silicon. For example, the Ir–CO stretching frequency in the iridium complexes $IrCl(H)(SiR_3)(CO)(PEt_3)_2$ decreases in the order $SiF_3 \approx SiCl_3 > SiCl_2Me > Si(OEt)_3 >> SiClMe_2$.[497] The Ir–CO stretching frequency of the $SiCl_2Me$ and $Si(OEt)_3$ complexes were similar to those of the analogous hydride complex, and that in the $SiClMe_2$ complex was similar to that in the analogous acyl complex.

A standard method used to characterize silyl complexes is ^{29}Si NMR spectroscopy. These complexes exhibit a wide range of ^{29}Si chemical shifts (~550 ppm), although shift values are typically in the range of –30 to 70 ppm. For example, the manganese complex $Mn(CO)_5SiMe_3$ has a ^{29}Si resonance 18.0 ppm, while the corresponding rhenium species $Re(CO)_5SiMe_3$ has an observed shift at –14.0 ppm.[498] Although the prediction of ^{29}Si chemical shifts is not straightforward, the magnitude of a given ^{29}Si resonance tends to decrease in a homologous series of compounds from the first to the second to the third row.[499] The opposite trend seems to hold for d^0 complexes and for the main group silyls $M[Si(SiMe_3)_3]_3$ (M = Zn, Cd, and Hg).[500] In addition to chemical shift values, the coupling constants between silicon and NMR-active metal nuclei can provide valuable information on the nature of the M–Si bond. Although, the ^{29}Si nucleus suffers from rather low sensitivity, indirect detection methods to overcome this difficulty have been developed.[501]

4.7.5. Synthesis of Metal–Silyl Complexes

Metal–silyl complexes are typically prepared by salt displacements, oxidative additions of an Si–H bond (Chapter 6), and σ bond metatheses involving Si–H bonds. The reactions in Equations 4.112–4.117 illustrate examples of these processes.

$$\text{(4.112)}$$

$$\text{(4.113)}$$

As shown in Equations 4.112 and 4.113, salt displacements to form metal–silyl complexes can be conducted by displacement of the halogen of halosilanes with metal anions or by the displacement of transition metal halides with silyl anions. For example, $CpFe(CO)_2SiMe_3$ was prepared by the reaction of $Na[CpFe(CO)_2]$ with Me_3SiCl (Equation 4.112).[481] Equally common is the synthesis of metal–silyl complexes by the reactions of silyl anions, typically as the silyllithium reagents, to metal centers bearing an appropriate leaving group. The preparation of $(t\text{-}BuCH_2)_3ZrSi(SiMe_3)_3$ by the reaction of tris(neopentyl)zirconium chloride with tris(trimethylsilyl)silyl lithium is one example (Equation 4.113).[502]

$$\text{(4.114)}$$

$$\text{(4.115)}$$

Transition-metal–silyl complexes are also formed by the reactions of metal–alkyl complexes with silanes to form free alkane and a metal–silyl complex. Two examples are shown in Equations 4.114 and 4.115.[503,504] The synthesis of silyl complexes by this method has been accomplished with both early and late transition metal complexes. The formation of metal–silyl complexes from late-metal–alkyl complexes resembles the hydrogenolysis of metal–alkyl complexes to form metal hydrides and an alkane. The mechanisms of these reactions are discussed in Chapter 6. In brief, these reactions with late transition metal complexes to form silyl complexes typically occur by a sequence of oxidative addition of the silane, followed by reductive elimination of alkane. An example of this is shown in the coupling of 1,2-bis-dimethylsilyl benzene with a dimethyl platinum(II) complex (Equation 4.114).[503] Similar reactions occur with d^0 early metal complexes by a σ-bond metathesis process that avoids these redox events. For example, the reaction of Cp*$_2$ScPh with MesSiH$_3$ has been shown to proceed through this pathway (Equation 4.115).[504]

Finally, metal–silyl complexes are commonly prepared by insertion of a transition metal fragment into the Si–H or Si–X bond of a silane or halosilane. These oxidative addition processes are presented in detail in Chapter 6. This process forms the silyl complexes that are the intermediates in transformations of silanes catalyzed by late transition metal complexes. Two examples of this route to silyl complexes, one by addition of an Si–H bond and one by addition of an Si–X bond, are shown in Equations 4.116 and 4.117.[505,506]

$$\text{Cp*Ru(PPr}^i{}_3)\text{Cl} \xrightarrow{\text{PhSiH}_3} \text{Cp*Ru(PPr}^i{}_3)(\text{SiH}_2\text{Ph})(\text{H})\text{Cl} \qquad (4.116)$$

$$\text{Pt(PEt}_3)_4 \xrightarrow[\text{X = Br, I}]{\text{XSiMe}_3} \textit{trans}\text{-Pt(PEt}_3)_2(\text{SiMe}_3)\text{X} \qquad (4.117)$$

4.7.6. Stability and Reactivity of Silyl Complexes

Many of the reactions of transition metal silyl derivatives are similar to those of the analogous alkyl compounds. However, there are some important differences that affect their stability. As described in the section of this chapter on metal–alkyl complexes, the most common pathway for decomposition of metal–alkyl complexes is β-hydrogen elimination. Moreover, secondary and tertiary alkyl complexes often isomerize to the primary alkyl complex by a sequence involving β-hydrogen elimination. As discussed in more detail in Chapter 10, metal–silyl complexes are less prone to β-hydrogen elimination because the Si=C bond that is formed by the β-hydrogen elimination of a metal–silyl complex is less stable than the C=C bond formed by the β-hydrogen elimination of an alkyl complex. Moreover, the strength of M–Si bonds tends to disfavor reactions of metal–silyl complexes by homolysis to form silyl radicals.

4.8. Halide Ligands

4.8.1. Overview

Halides are common ligands in all types of transition metal chemistry. In organometallic chemistry they can be reactive units in ligand substitutions (Chapter 6) or ancillary ligands that influence reactivity in ways that are beginning to be delineated.[507,508] Halides are present in many of the commercially available starting materials because reactions of mineral acids are often used to generate transition metal complexes from bulk metal or to recover the metal components from chemical mixtures. They are also increasingly identified as ligands that affect the rates and selectivities in catalytic organometallic chemistry.[507]

The steric and electronic effects of halide ligands are subtle and have been the subject of detailed recent studies. It is now accepted that halides participate in π-bonding and that both σ- and π-bonding affect the overall electronic properties of the metal.[508] It is also well established that the hard or soft character of the different halides has a pronounced effect on the stability of the metal–halogen linkage. The properties of the metal–fluoride ligands often lie at the far extremes of the trends in properties of the halogens when acting as ligands.

4.8.2. Steric and Electronic Properties

The steric properties of halide ligands follow the trends taught in any introductory chemistry course—namely, fluoride is the smallest and iodide is the largest of these ligands. As shown in Table 4.1, the cone angles of these ligands ranges from 92° for fluoride to 107° for iodide.[509] The large size of the iodide ligand is partially attenuated by the longer metal–iodide bond. Therefore, the cone angle of iodide is much less than twice the cone angle of fluoride, even though the inoic radius of iodice is almost twice that of fluoride.

Table 4.1. Steric properties of halides and halide ligands.

	F	Cl	Br	I
Ionic radius (Å)[510]	1.36	1.81	1.95	2.16
Cone angle (°)[509]	92	102	105	107

The electronic properties of the halides are less intuitive. Many studies have shown that the fluoride ligand is a stronger electron donor than the iodide ligand,[511–517] despite the greater electronegativity of a lighter halogen, relative to that of a heavier halogen. This greater electron-donating ability has been attributed to greater π-donation by the lighter halogen, which is thought to be particularly strong in a fluoride ligand.

Evidence that the halides have an effect on the electron density at the metal center has been achieved by measuring the enthalpies of protonation of metal complexes containing the four different halogens.[514,515] As shown in Equation 4.118, these studies show that the absolute value of the enthalpy of protonation of the osmium–fluoride complex at the metal center is much larger than that of the protonation of the chloride complex, which is larger than that of the protonation of the bromide and iodide complexes.

$-\Delta H_{HM}$	(kcal mol^{-1})
I	14.1
Br	15.3
Cl	19.7
F	37.3

$$(4.118)$$

π-Donation can stabilize or destabilize metal–ligand bonds, as presented in Chapter 1. π-Donation into the unoccupied orbital of an unsaturated metal center can lead to particularly strong metal–halogen bonds. However, π-electrons on a ligand can also lead to destabilizing interactions if the interaction occurs with a filled metal orbital possessing π-symmetry. A combination of a π-donor ligand trans to a π-acceptor ligand can also favor binding of the more strongly π-donating halogen. These effects are illustrated by the data in Table 4.2 and Figure 4.30.

Table 4.2. Zr–Cl and Zr–I bond strengths.

$Cp^*_2ZrCl_2$	Zr–Cl$_{ave}$ 115 kcal/mol
$Cp^*_2ZrI_2$	Zr–I$_{ave}$ = 80.4 kcal/mol

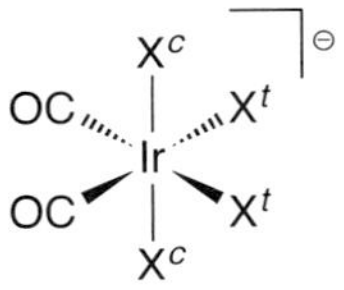

Relative binding affinities:
X^t = Cl > Br > I
X^c = I > Br > Cl

Figure 4.30.
Relative binding affinities of three halides to the cis and trans positions of an Ir(III) complex, illustrating π-replusion, soft–soft interactions for the ligands cis to CO, and the push–pull interaction for the halides trans to CO.

The stabilizing influence of hard–hard interactions and π-donation into an unsaturated metal center is illustrated by the metal–halogen bond energies in Table 4.2. The Zr–Cl bond in $Cp^*_2ZrCl_2$ is much stronger (115 kcal/mol) than the analogous Zr–I bond (80 kcal/mol).[518] The relative binding affinities of the different halides to the cis and trans positions in the iridium complex of Figure 4.30 illustrate the effects of π-replusion, soft–soft interactions, and push–pull interactions. The iodide is thought to bind strongest to the positions cis to CO because the softer character and weaker π-donating ability of the iodide favors binding to an 18-electron, low-valent late transition metal center. The opposite order of affinities to the position trans to CO is thought to result from a stabilizing push–pull interaction between the stronger π-donating halide and the strong π-accepting CO ligand trans to the halide.

At the same time, these π-effects alone cannot account for some of the trends in binding affinities. The binding of halides trans to the phenyl group in [Pd(PPh$_3$)$_2$PhX] complexes is stronger for the lighter halogens than for the heavier halogens, and the phenyl ligand is surely a weaker π-acceptor than CO.[519] Thus, the higher basicity of smaller halides in aprotic media is likely to account for the strength of the interactions between transition metals and the lighter halogens.

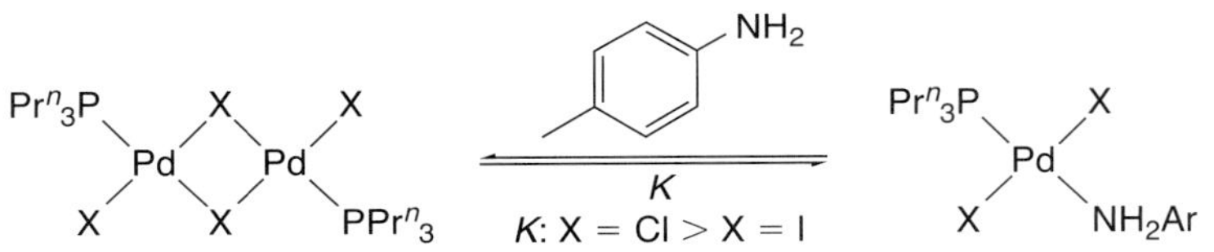

(4.119)

K: X = Cl > X = I

The stability of the different halides toward bridging binding modes can also affect the reactivity of organometallic species. The strength of the bridging interactions follow the trend I > Br > Cl > F. As early as 1957, Chatt illustrated that an aromatic amine would fully cleave the bridging chloride interaction in [Pd(PPr$^n_3$)Cl$_2$]$_2$, but it would only partially cleave this interaction in the analogous [Pd(PPr$^n_3$)I$_2$]$_2$ complex (Equation 4.119).[520] Similar data have been obtained more recently with P(o-tol)$_3$ as ligand.[521]

The different halides also have different abilities to act as hydrogen-bond acceptors. Crabtree has shown that the hydrogen-bonding interaction of a pyridylamine group to fluoride is substantially stronger than to chloride and bromide and that the interaction to iodide is weaker still (Figure 4.31).[522]

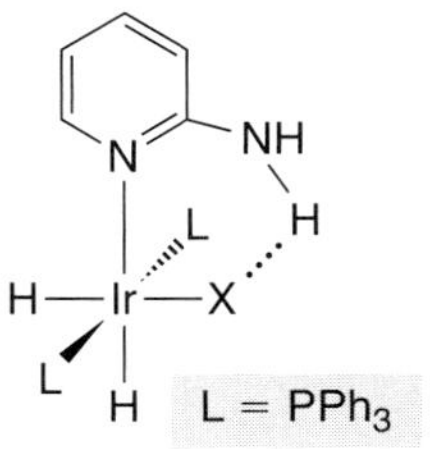

X	Ir X–HN (kcal mol^{-1})
I	< 1.3
Br	1.8
Cl	2.1
F	5.2

Figure 4.31.
Strengths of hydrogen bonding to the different halides bound to Ir(III).

4.8.3. Reactivity of Metal–Halide Complexes

The most common reaction of metal halides is substitution by nucleophiles to generate metal compounds containing new metal X-type ligand bonds or displacement of halide by a strong L-type ligand to form a cationic species (Chapter 6). Abstraction of halide by silver salts to generate insoluble silver halides is also a common transformation of the metal–halide complexes described in Chapter 12. In polar media, halides are also known to undergo heterolytic dissociation to form solvated species.

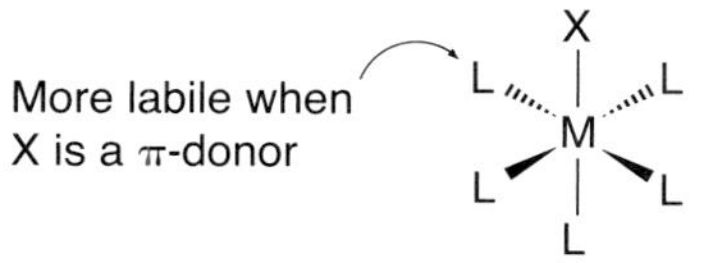

More labile when X is a π-donor

Figure 4.32.
Generic description of the "cis effect."

Metal–halide complexes also undergo a variety of reactions in which the halide affects reaction chemistry as an ancillary ligand. These include an effect on the rates of ligand dissociation from octahedral complexes (Chapter 6) by a phenomenon called the "cis effect." This effect (Figure 4.32) is presented in Chapter 6, and involves the stabilization of the intermediate formed by ligand dissociation through π-bonding of the halide electron pairs. These effects also influence the rates of oxidative additions. As described in more detail in Chapter 6 and summarized in Equation 4.120, the oxidative additions of dihydrogen to Vaska-type complexes [Ir(PPh$_3$)$_2$(CO)(X)] containing different halogens are calculated to be more favored enthalpically to the complexes containing the more weakly π-donating iodide ligand than to the complex containing the more strongly π-donating fluoride ligand,[523,524] even though ν_{CO} values indicate that the fluoride complex is more electron rich. The π-donation is expected to stabilize the starting material by donation into the trans carbonyl ligand and to destabilize the 18-electron trigonal bipyramidal intermediate. Finally, the effect of halides on the rates of isomerization processes can be illustrated by the rearrangements of π-allyl complexes. As noted earlier in Chapter 3 and in more detail in Chapter 20, the coordination of halide can induce η^3–η^1 interconversions by the mechanism shown in Equation 4.121.[525]

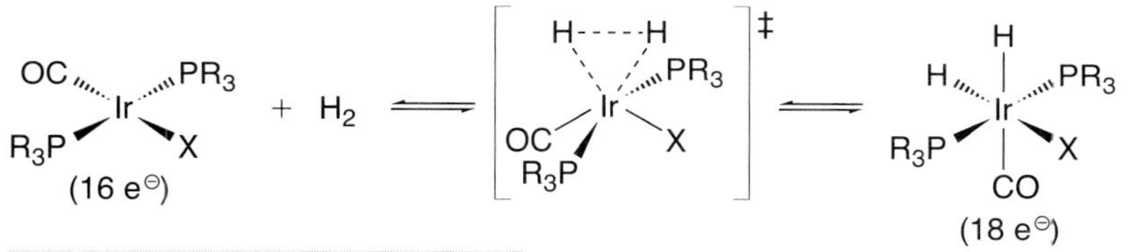

$\tilde{\nu}_{CO}$ [cm^{-1}]		ΔH_{exp} (kcal mol^{-1})
1957	F	> −10
1965	Cl	−14
1966	Br	−17
1967	I	−19

$$(4.120)$$

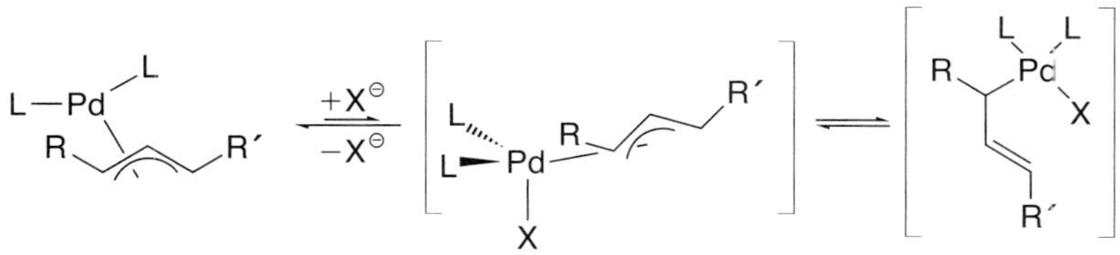

$$(4.121)$$

The affect of halogens on catalytic chemistry is also beginning to be developed. Olefin metathesis is the subject of Chapter 21, and one of the most studied and most active olefin metathesis catalysts contains chlorides as ancillary ligands. The complexes containing chloride as ancillary ligands are more active than catalysts containing iodide. As shown in Equation 4.122, the origin of this effect can be traced to the relative reactivity of the two intermediates with olefin to form product and phosphine to regenerate the starting species.[526]

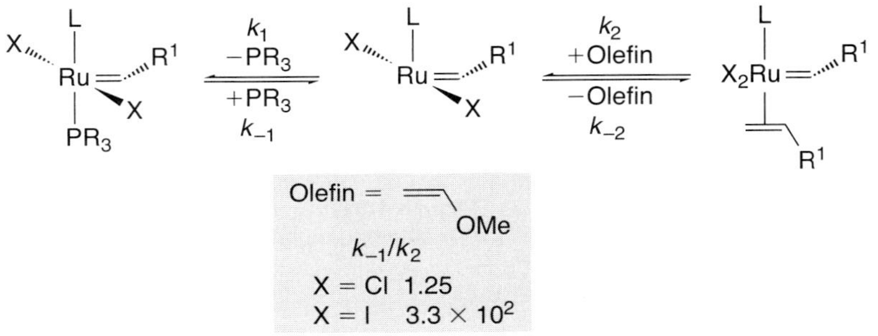

$$\text{Olefin} = \underset{\text{OMe}}{=\!\!=\!\!=}$$

$$k_{-1}/k_2$$

$$X = Cl \quad 1.25$$

$$X = I \quad 3.3 \times 10^2$$

(4.122)

In other cases, the properties of fluoride ligands have been exploited to generate more active catalysts than are obtained in the absence of added fluoride or with other halogen ligands. As described in Chapter 16, added fluoride accelerates the rate of the iridium-catalyzed hydroamination of norbornene with aniline.[527] The origin of this effect has not been revealed.

The effects of halides on enantioselective processes have also been studied. The effect of fluoride as a ligand is illustrated by enhanced rates and selectivities of titanium-[519,528–533] and copper-catalyzed enantioselective aldol[534,535] and allylation[536] processes. The effect on the titanium-catalyzed process is believed to be due to greater Lewis acidity of the titanium containing a fluoride ligand, whereas the effect of the fluoride on the copper-catalyzed chemistry is believed to result from faster reactions of the copper fluoride with the silicon enolate[534,535,537] or allylsilane reagents.[536] The identity of added halide also has a large impact on the rate and enantioselectivity of the asymmetric ring-opening of [2.2.1]oxabicylic alkenes. As shown in Equation 4.123, the effect of halide is the opposite of that observed in the titanium and copper chemistry. The yields of these reactions are highest in the presence of added iodide, and no product was observed in the presence of added fluoride.[538]

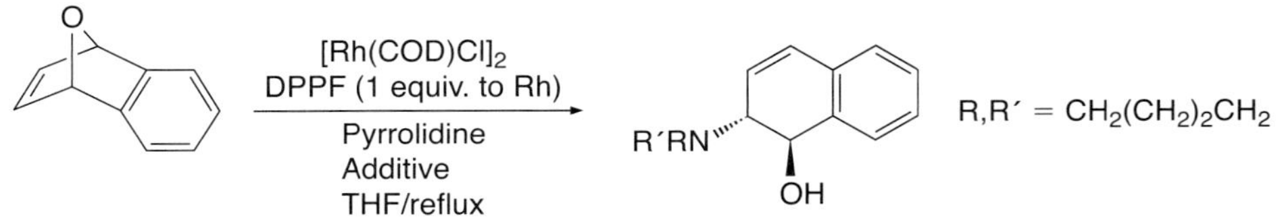

(4.123)

Additive	Yield
Bu$_4$NF (5 equiv.)	NR
Bu$_4$NCl (5 equiv.)	17
Bu$_4$NBr (5 equiv.)	60
Bu$_4$NI (5 equiv.)	85

References and Notes

1. Lappert, M. F.; Power, P. P.; Sanger, A. R.; Srivastava, R. C. *Metal and Metalloid Amides*; Wiley: New York, 1980.
2. Hartwig, J. F. *Inorg. Chem.* **2007**, *46*, 1936.
3. Fryzuk, M. D.; MacNeil, P. A. *J. Am. Chem. Soc.* **1981**, *103*, 3592.
4. Fryzuk, M. D. *Inorg. Chem.* **1982**, *21*, 2134.
5. Fryzuk, M. D.; MacNeil, P. A.; Rettig, S. J.; Secco, A. S.; Trotter, J. *Organometallics* **1982**, *1*, 918.
6. Fryzuk, M. D.; Gao, X.; Joshi, K.; MacNeil, P. A.; Massey, R. L. *J. Am. Chem. Soc.* **1993**, *115*, 10581.
7. Fryzuk, M. D.; Gao, X.; Rettig, S. J. *J. Am. Chem. Soc.* **1995**, *117*, 3106.
8. Fan, L.; Foxman, B. M.; Ozerov, O. V. *Organometallics* **2004**, *23*, 326.

9. Fan, L.; Parkin, S.; Ozerov, O. V. *J. Am. Chem. Soc.* **2005**, *127*, 16772.

10. Gatard, S.; Celenligil-Cetin, R.; Guo, C. Y.; Foxman, B. M.; Ozerov, O. V. *J. Am. Chem. Soc.* **2006**, *128*, 2808.

11. Weng, W.; Guo, C.; Celenligil-Cetin, R.; Foxman, B. M.; Ozerov, O. V. *J. Chem. Soc., Chem. Commun.* **2006**, 197.

12. (a) Liang, L.-C.; Chien, P.-S.; Huang, Y.-L. *J. Am. Chem. Soc.* **2006**, *128*, 15562; (b) Liang, L.-C.; Chien, P.-S.; Lin, J.-M.; Huang, M.-H.; Huang, Y.-L.; Liao, J.-H. *Organometallics* **2006**, *25*, 1399.

13. Riehl, J.F.; Jean, Y.; Eisenstein, O.; Pelissier, M. *Organometallics* **1992**, *11*, 729.

14. Fulton, J. R.; Bouwkamp, M. W.; Bergman, R. G. *J. Am. Chem. Soc.* **2000**, *122*, 8799.

15. Fulton, J. R.; Sklenak, S.; Bouwkamp, M. W.; Bergman, R. G. *J. Am. Chem. Soc.* **2002**, *124*, 4722.

16. Holland, A. W.; Bergman, R. G. *J. Am. Chem. Soc.* **2002**, *124*, 14684.

17. Rais, D.; Bergman, R. G. *Chem. Eur. J.* **2004**, *10*, 3970.

18. Fox, D. J.; Bergman, R. G. *Organometallics* **2004**, *23*, 1656.

19. Driver, M. S.; Hartwig, J. F. *J. Am. Chem. Soc.* **1996**, *118*, 4206.

20. Driver, M. S.; Hartwig, J. F. *Organometallics* **1997**, *16*, 5706.

21. Bryndza, H. E.; Fong, L. K.; Paciello, R. A.; Tam, W.; Bercaw, J. E. *J. Am. Chem. Soc.* **1987**, *109*, 1444.

22. Cowan, R. L.; Trogler, W. C. *Organometallics* **1987**, *6*, 2451.

23. Park, S.; Rheingold, A. L.; Roundhill, D. M. *Organometallics* **1991**, *10*, 615.

24. Driver, M. S.; Hartwig, J. F. *J. Am. Chem. Soc.* **1997**, *119*, 8232.

25. Yamashita, M.; Cuevas Vicario, J. V.; Hartwig, J. F. *J. Am. Chem. Soc.* **2003**, *125*, 16347.

26. Hartwig, J. F. *J. Am. Chem. Soc.* **1996**, *118*, 7010.

27. Joslin, F. L.; Johnson, M. P.; Mague, J. T.; Roundhill, D. M. *Organometallics* **1991**, *10*, 2781.

28. Zhao, J.; Goldman, A. S.; Hartwig, J. F. *Science* **2005**, *307*, 1080.

29. Hartwig, J. F. *Angew. Chem. Int. Ed.* **1998**, *37*, 2046.

30. Wolfe, J. P.; Wagaw, S.; Marcoux, J.-F.; Buchwald, S. L. *Acc. Chem. Res.* **1998**, *31*, 805.

31. Hartwig, J. F. In *Modern Amination Methods*; Ricci, A., Ed.; Wiley-VCH: Weinheim, 2000; p 195.

32. Hartwig, J. F. In *Handbook of Organopalladium Chemistry for Organic Synthesis*; Negishi, E.-i., Ed.; Wiley-Interscience: New York, 2002; Vol. 1, p 1051.

33. Muci, A. R.; Buchwald, S. L. *Top. Curr. Chem.* **2002**, *219*, 131.

34. Louie, J.; Paul, F.; Hartwig, J. F. *Organometallics* **1996**, *15*, 2794.

35. Fornies, T.; Green, M.; Spencer, J. L.; Stone, F. G. A. *J. Chem. Soc., Dalton Trans.* **1977**, 1006.

36. Casalnuovo, A. L.; Calabrese, J. C.; Milstein, D. *J. Am. Chem. Soc.* **1988**, *110*, 6738.

37. Fryzuk, M. D.; Montgomery, C. D. *Coord. Chem. Rev.* **1989**, *95*, 1.

38. Zhao, P. J.; Krug, C.; Hartwig, J. F. *J. Am. Chem. Soc.* **2005**, *127*, 12066.

39. Bryndza, H. E.; Fultz, W. C.; Tam, W. *Organometallics* **1985**, *4*, 939.

40. Fox, D. J.; Bergman, R. G. *J. Am. Chem. Soc.* **2003**, *125*, 8984.

41. Bradley, D. C.; Chisholm, M. H. *Acc. Chem. Res.* **1976**, *9*, 273.

42. Rhett, K. *Angew. Chem. Int. Ed.* **2000**, *39*, 468.

43. Scollard, J. D.; McConville, D. H. *J. Am. Chem. Soc.* **1996**, *118*, 10008.

44. Scollard, J. D.; McConville, D. H.; Payne, N. C.; Vittal, J. J. *Macromolecules* **1996**, *29*, 5241.

45. Liang, L. C.; Schrock, R. R.; Davis, W. M.; McConville, D. H. *J. Am. Chem. Soc.* **1999**, *121*, 5797.

46. Giesbrecht, G. R.; Shafir, A.; Arnold, J. *Inorg. Chem.* **2001**, *40*, 6069.

47. Stevens, J. C. *Stud. Surf. Sci. Catal.* **1996**, *101*, 11.

48. McKnight, A. L.; Waymouth, R. M. *Chem. Rev.* **1998**, *98*, 2587.

49. Zhang, X.; Zhu, Q.; Guzei, I. A.; Jordan, R. F. *J. Am. Chem. Soc.* **2000**, *122*, 8093.

50. Gibson, V. C.; Spitzmesser, S. K. *Chem. Rev.* **2003**, *103*, 283.

51. Kim, H.; Lee, P. H.; Livinghouse, T. *J. Chem. Soc., Chem. Commun.* **2005**, 5205.

52. Odom, A. L. *J. Chem. Soc., Dalton Trans.* **2005**, 225.

53. Zhang, Z.; Leitch, D. C.; Lu, M.; Patrick, B. O.; Schafer, L. L. *Chem. Eur. J.* **2001**, *13*, 2012.

54. Watson, D. A.; Chiu, M.; Bergman, R. G. *Organometallics* **2006**, *25*, 4731.

55. Wood, M. C.; Leitch, D. C.; Yeung, C. S.; Kozak, J. A.; Schafer, L. L. *Angew. Chem., Int. Ed. Engl.* **2007**, *46*, 354.

56. Schrock, R. R.; Hoveyda, A. H. *Angew. Chem., Int. Ed. Engl.* **2003**, *42*, 4592.

57. Sinha, A.; Schrock, R. R.; Muller, P.; Hoveyda, A. H. *Organometallics* **2006**, *25*, 4621.

58. Eldred, S. E.; Stone, D. A.; Gellman, S. H.; Stahl, S. S. *J. Am. Chem. Soc.* **2003**, *125*, 3422.

59. Hoerter, J. M.; Otte, K. M.; Gellman, S. H.; Stahl, S. S. *J. Am. Chem. Soc.* **2006**, *128*, 5177.

60. Cummins, C. C.; Baxter, S. M.; Wolczanski, P. T. *J. Am. Chem. Soc.* **1988**, *110*, 8731.

61. De With, J.; Horton, A. D. *Angew. Chem., Int. Ed. Engl.* **1993**, *32*, 903.

62. Schaller, C. P.; Cummins, C. C.; Wolczanski, P. T. *J. Am. Chem. Soc.* **1996**, *118*, 591.

63. Bennett, J. L.; Wolczanski, P. T. *J. Am. Chem. Soc.* **1997**, *119*, 10696.

64. Hoyt, H. M.; Michael, F. E.; Bergman, R. G. *J. Am. Chem. Soc.* **2004**, *126*, 1018.

65. LoCoco, M. D.; Zhang, X.; Jordan, R. F. *J. Am. Chem. Soc.* **2004**, *126*, 15231.

66. Girolami, G. S.; Gozum, J. E. *Mater. Res. Soc. Symp. Proc.* **1990**, *168*, 319.

67. Clerici, M. G.; Maspero, F. *Synthesis* **1980**, 305.

68. Nugent, W. A.; Ovenall, D. W.; Holmes, S. J. *Organometallics* **1983**, *2*, 161.

69. Buchwald, S. L.; Watson, B. T.; Wannamaker, M. W.; Dewan, J. C. *J. Am. Chem. Soc.* **1989**, *111*, 4486.

70. Barluenga, J.; Rodriguez, F.; Alvarez-Rodrigo, L.; Fananas, F. J. *Chem. Soc. Rev.* **2005**, *34*, 762.

71. Cummings, S. A.; Tunge, J. A.; Norton, J. R. *Top. Organomet. Chem.* **2005**, *10*, 1.

72. Cummins, C. C. *J. Chem. Soc., Chem. Commun. (Camb)* **1998**, 1777.

73. Spencer, L. P.; Winston, S.; Fryzuk, M. D. *Organometallics* **2004**, *23*, 3372.

74. Spencer, L. P.; Beddie, C.; Hall, M. B.; Fryzuk, M. D. *J. Am. Chem. Soc.* **2006**, *128*, 12531.

75. Plass, W.; Verkade, J. G. *Inorg. Chem.* **1993**, *32*, 3762.

76. Schrock, R. R. *Acc. Chem. Res.* **1997**, *30*, 9.

77. Schrock, R. R. *Acc. Chem. Res.* **2005**, *38*, 955.

78. Lavanant, L.; Toupet, L.; Lehmann, C. W.; Carpentier, J.-F. *Organometallics* **2005**, *24*, 5620.

79. Rubio, R. J.; Andavan, G. T. S.; Bauer, E. B.; Hollis, T. K.; Cho, J.; Tham, F. S.; Donnadieu, B. *J. Organomet. Chem.* **2005**, *690*, 5353.

80. Sebe, E.; Guzei, I. A.; Heeg, M. J.; Liable-Sands, L. M.; Rheingold, A. L.; Winter, C. H. *Eur. J. Inorg. Chem.* **2005**, 3955.

81. Gendler, S.; Segal, S.; Goldberg, I.; Goldschmidt, Z.; Kol, M. *Inorg. Chem.* **2006**, *45*, 4783.

82. Chandra, G.; Lappert, M. F. *J. Chem. Soc. A* **1968**, 1940.

83. Lappert, M. F.; Patil, D. S.; Pedley, J. B. *J. Chem. Soc., Chem. Commun.* **1975**, 830.

84. Pearson, R. G. *Coord. Chem. Rev.* **1990**, *100*, 403.

85. Andersen, R. A.; Beach, D. B.; Jolly, W. L. *Inorg. Chem.* **1985**, *24*, 4741.

86. Attempts to prepare Ta[N(CH$_3$)$_2$]$_5$ by this method have sometimes resulted in explosions, and references 87 and 88 report a modified procedure.

87. Chestnut, R. W.; Rothwell, I. P.; Holl, M. B.; Wolczanski, P. T. *Chem. Eng. News* **1990**, *68*, 2.

88. Riley, P. N.; Parker, J. R.; Fanwick, P. E.; Rothwell, I. P. *Organometallics* **1999**, *18*, 3579.

89. Thomas, I. M. *Can. J. Chem.* **1961**, *39*, 1386.

90. Bradley, D. C.; Chisholm, M. H.; Extine, M. W.; Stager, M. E. *Inorg. Chem.* **1977**, *16*, 1794.

91. Alyea, E. C.; Basi, J. S.; Bradley, D. C.; Chisholm, M. H. *J. Chem. Soc., Chem. Commun.* **1968**, 495.

92. Gavenonis, J.; Tilley, T. D. *Organometallics* **2004**, *23*, 31.

93. Barker, J.; Kilner, M. *Coord. Chem. Rev.* **1994**, *133*, 219.

94. Sadimenko, A. P. *Adv. Heterocyc. Chem.* **2001**, *79*, 115.

95. Sadimenko, A. P.; Garnovskii, A. D.; Retta, N. *Coord. Chem. Rev.* **1993**, *126*, 237.

96. Sadimenko, A. P.; Basson, S. S. *Coord. Chem. Rev.* **1996**, *147*, 247.

97. Pannell, K. H.; Kalsotra, B. L.; Parkanyi, C. *J. Heterocycl. Chem.* **1978**, *15*, 1057.

98. Nief, F. O. *Eur. J. Inorg. Chem.* **2001**, *2001*, 891.

99. Mann, G.; Hartwig, J. F.; Driver, M. S.; Fernandez-Rivas, C. *J. Am. Chem. Soc.* **1998**, *120*, 827.

100. Hartwig, J. F.; Kawatsura, M.; Hauck, S. I.; Shaughnessy, K. H.; Alcazar-Roman, L. M. *J. Org. Chem.* **1999**, *64*, 5575.

101. Old, D. W.; Harris, M. C.; Buchwald, S. L. *Org. Lett.* **2000**, *2*, 1403.

102. Movassaghi, M.; Ondrus, A. E. *J. Org. Chem.* **2005**, *70*, 8638.

103. Lebedev, A. Y.; Izmer, V. V.; Kazyul'kin, D. N.; Beletskaya, I. P.; Voskoboynikov, A. Z. *Org. Lett.* **2002**, 623.

104. Klapars, A.; Antilla, J. C.; Huang, X.; Buchwald, S. L. *J. Am. Chem. Soc.* **2001**, *123*, 7727.

105. Antilla, J. C.; Klapars, A.; Buchwald, S. L. *J. Am. Chem. Soc.* **2002**, *124*, 11684.

106. Antilla, J. C.; Baskin, J. M.; Barder, T. E.; Buchwald, S. L. *J. Org. Chem.* **2004**, *69*, 5578.

107. Cristau, H.-J.; Cellier, P. P.; Spindler, J.-F.; Taillefer, M. *Eur. J. Org. Chem.* **2004**, *2004*, 695.

108. Cristau, H.-J.; Cellier, P. P.; Spindler, J.-F.; Taillefer, M. *Chem. Eur. J.* **2004**, *10*, 5607.

109. Collman, J. P.; Zhong, M.; Zhang, C.; Costanzo, S. *J. Org. Chem.* **2001**, *66*, 7892.

110. Collman, J. P.; Zhong, M. *Org. Lett.* **2000**, 1233.

111. Yu, S.; Saenz, J.; Srirangam, J. K. *J. Org. Chem.* **2002**, *67*, 1699.

112. Lam, P. Y. S.; Clark, C. G.; Saubern, S.; Adams, J.; Winters, M. P.; Chan, D. M. T.; Combs, A. *Tetrahedron Lett.* **1998**, *39*, 2941.

113. Lam, P. Y. S.; Vincent, G.; Clark, C. G.; Deudon, S.; Jadhav, P. K. *Tetrahedron Lett.* **2001**, *42*, 3415.

114. Lam, P. S.; Deudon, S.; Averill, K. M.; Li, R.; He, M. Y.; DeShong, P.; Clark, C. G. *J. Am. Chem. Soc.* **2000**, *122*, 7600.

115. Allred, G. D.; Liebeskind, L. S. *J. Am. Chem. Soc.* **1996**, *118*, 2748.

116. Joshi, K. K.; Pauson, P. L.; Quazi, A. R.; Stubbs, W. H. *J. Organomet. Chem.* **1964**, *1*, 471.

117. Zakrzewski, J.; Giannotti, C. *J. Organomet. Chem.* **1990**, *388*, 175.

118. Zakrzewski, J.; Giannotti, C. *J. Organomet. Chem.* **1990**, *388*, 175.

119. Efraty, A.; Jubran, N. *Inorg. Chim. Acta* **1980**, *44*, L191.

120. Joshi, K. K.; Pauson, P. L.; Qazi, A. R.; Stubbs, W. H. *J. Organomet. Chem.* **1964**, *1*, 471.

121. Chen, S.; Carperos, V.; Noll, B.; Swope, R. J.; DuBois, M. R. *Organometallics* **1995**, *14*, 1221.

122. Zhu, G.; Tanski, J. M.; Churchill, D. G.; Janak, K. E.; Parkin, G. *J. Am. Chem. Soc.* **2002**, *124*, 13658.

123. Lee, H.; Bonanno, J. B.; Bridgewater, B. M.; Churchill, D. G.; Parkin, G. *Polyhedron* **2005**, *24*, 1356.

124. Bynum, R. V.; Hunter, W. E.; Rogers, R. D.; Atwood, J. L. *Inorg. Chem.* **1980**, *19*, 2368.

125. Bynum, R. V.; Hunter, W. E.; Atwood, J. L. *Can. J. Chem.* **1986**, *64*, 1254.

126. Driver, M. J.; Hartwig, J. F. *Organometallics* **1998**, *17*, 1134.

127. Simpson, R. D.; Bergman, R. G. *Organometallics* **1992**, 3980.

128. Arndt, S.; Spaniol, T. P.; Okuda, J. *Eur. J. Inorg. Chem.* **2001**, *2001*, 73.

129. Hsu, G. C.; Kosar, W. P.; Jones, W. D. *Organometallics* **1994**, *13*, 385.

130. Morikita, T.; Hirano, M.; Sasaki, A.; Komiya, S. *Inorg. Chim. Acta* **1999**, *291*, 341.

131. Ladipo, F. T.; Merola, J. S. *Inorg. Chem.* **1990**, *29*, 4172.

132. Jones, W. D.; Dong, L.; Myers, A. W. *Organometallics* **1995**, *14*, 855.

133. Jonas, K.; Wilke, G. *Angew. Chem., Int. Ed. Eng.* **1969**, *8*, 519.

134. López, C.; Baron, G.; Arevalo, A.; Munoz-Hernandez, M. A.; Garcia, J. J. *J. Organomet. Chem.* **2002**, *664*, 170.

135. Chantson, J. T.; Lotz, S. *J. Organomet. Chem.* **2004**, *689*, 1315.

136. Pyshnograeva, N. I.; Setkina, V. N.; Andrianov, V. G.; Struchkov, Y. T.; Kursancv, D. N. *J. Organomet. Chem.* **1977**, *128*, 381.

137. Johnson, T. J.; Arif, A. M.; Gladysz, J. A. *Organometallics* **1993**, *12*, 4728.

138. Kershner, D. L.; Basolo, F. *Coord. Chem. Rev.* **1987**, *79*, 279.

139. Pignataro, S.; Cataliotti, R.; Foffani, A. *Inorg. Chem.* **1970**, *9*, 2594.

140. Zakrzewski, J. *J. Organomet. Chem.* **1987**, *333*, 71.

141. Zakrzewski, J. *J. Organomet. Chem.* **1989**, *362*, C31.

142. Pyshnograeva, N. I.; Batsanov, A. S.; Struchkov, Y. T.; Ginzburg, A. G.; Setkina, V. N. *J. Organomet. Chem.* **1985**, *297*, 69.

143. Fu, G. C. *Acc. Chem. Res.* **2000**, *33*, 412.

144. Hodous, B. L.; Ruble, J. C.; Fu, G. C. *J. Am. Chem. Soc.* **1999**, *121*, 2637.

145. Fu, G. C. *Acc. Chem. Res.* **2006**, *39*, 853.

146. Artaud, I.; Chatel, S.; Chauvin, A. S.; Bonnet, D.; Kopf, M. A.; Leduc, P. *Coord. Chem. Rev.* **1999**, *190–192*, 577.

147. Davis, K. L.; Martin, E.; Turko, I. V.; Murad, F. *Ann. Rev. Pharmacol. Toxicol.* **2001**, *41*, 203.

148. Caulton, K. G. *Coord. Chem. Rev.* **1975**, *14*, 317.

149. Feltham, R. D.; Enemark, J. H. In *Topics in Inorganic and Organometallic Stereochemistry*; Wiley: New York, 1981; Vol. 12, p 155.

150. Frenz, B. A.; Ibers, J. A. In *MTP International Review of Science, Physical Chemistry Series One*; Emeleus, H. J., Tobe, M. L., Eds.; Butterworths: London, 1972; p 33.

151. Gladfelter, W. L. *Adv. Organomet. Chem.* **1985**, *24*, 41.

152. Hayton, T. W.; Legzdins, P.; Sharp, W. B. *Chem. Rev.* **2002**, *102*, 935.

153. McCleverty, J. *Chem. Rev.* **1979**, *79*, 53.

154. Mingos, D. M. P.; Sherman, D. J. *Adv. Inorg. Chem.* **1989**, *34*, 292.

155. Richter-Addo, G. B.; Legzdins, P. *Chem. Rev.* **1988**, *88*, 991.

156. Richter-Addo, G. B.; Legzdins, P. *Metal Nitrosyls*; Oxford University Press: New York, 1992.

157. Wescott, B. L.; Enemark, J. H. In *Inorganic Electronic Spectroscopy*; Lever, A. B. P., Solomon, E. I., Eds.; Wiley: New York, 1999; Vol. 2.

158. Lim, M. D.; Lorkovic, I. M.; Ford, P. C. *Inorg. Chem.* **2002**, *41*, 1026.

159. Forte, E.; Vandenbergh, H. *Chem. Phys.* **1978**, *30*, 325.

160. Hetzler, J. R.; Casassa, M. P.; King, D. S. *J. Phys. Chem.* **1991**, *95*, 8086.

161. Wade, E. A.; Cline, J. I.; Lorenz, K. T.; Hayden, C.; Chandler, D. W. *J. Chem. Phys.* **2002**, *116*, 4755.

162. Zhao, Y. L.; Bartberger, M. D.; Goto, K.; Shimada, K.; Kawashima, T.; Houk, K. N. *J. Am. Chem. Soc.* **2005**, *127*, 7964.

163. Nakamoto, K. *Infrared and Raman Spectra of Inorganic and Coordination Compounds*; Wiley: New York, 1978.

164. Bursten, B. E.; Gatter, M. G. *J. Am. Chem. Soc.* **1984**, *106*, 2554.

165. Bursten, B. E.; Gatter, M. G.; Goldberg, K. I. *Polyhedron* **1990**, *9*, 2001.

166. Bursten, B. E.; Jensen, J. R.; Gordon, D. J.; Treichel, P. M.; Fenske, R. F. *J. Am. Chem. Soc.* **1981**, *103*, 5226.

167. Fenske, R. F.; Jensen, J. R. *J. Chem. Phys.* **1979**, *71*, 3374.

168. Enemark, J. H.; Feltham, R. D. *Coord. Chem. Rev.* **1974**, *13*, 339.

169. Avanzino, S. C.; Bakke, A. A.; Chen, H. W.; Donahue, C. J.; Jolly, W. L.; Lee, T. H.; Ricco, A. J. *Inorg. Chem.* **1980**, *19*, 1931.

170. Hedberg, L.; Hedberg, K.; Satija, S. K.; Swanson, B. I. *Inorg. Chem.* **1985**, *24*, 2766.

171. Avanzino, S. C.; Bakke, A. A.; Chen, H.-W.; Donahue, C. J.; Jolly, W. L.; Lee, T. H.; Ricco, A. J. *J. Inorg. Chem.* **1980**, *19*, 1931.

172. Mason, J.; Larkworthy, L. F.; Moore, E. A. *Chem. Rev.* **2002**, *102*, 913.

173. Satija, S. K.; Swanson, B. I. *Inorg. Synth.* **1976**, *16*, 1.

174. Grossi, L; Strazzari, S. *J. Org. Chem.* **1999**, *64*, 8076.

175. (a) Eckstein, Y.; Dreyfuss, P. *J. Polym. Sci., Polym. Chem. Ed.* **1979**, *17*, 4115; (b) Seung, S. L. N.; Dreyfuss, P.; Fetters, L. J. *Polym. Bull.* **1985**, *13*, 337.

176. Hodgson, D. J.; Payne, N. C.; McGinnety, J. A.; Pearson, R. G.; Ibers, J. A. *J. Am. Chem. Soc.* **1968**, *90*, 4486.

177. Connelly, N. G.; Dahl, L. F. *J. Chem. Soc. D* **1970**, 880.

178. Lee, J.; Chen, L.; West, A. H.; Richter-Addo, G. B. *Chem. Rev.* **2002**, *102*, 1019.

179. Wang, P. G.; Xian, M.; Tang, X. P.; Wu, X. J.; Wen, Z.; Cai, T. W.; Janczuk, A. J. *Chem. Rev.* **2002**, *102*, 1091.

180. Green, M.; Osborn, R. B. L.; Rest, A. J.; Stone, F. G. A. *J. Chem. Soc.* **1968**, 2525.

181. Stevens, R. E.; Fjare, D. E.; Gladfelter, W. L. *J. Organomet. Chem.* **1988**, *347*, 373.

182. Stevens, R. E.; Yanta, T. J.; Gladfelter, W. L. *J. Am. Chem. Soc.* **1981**, *103*, 4981.

183. Niu, S. Q.; Hall, M. B. *Chem. Rev.* **2000**, *100*, 353.

184. Brothers, P. J.; Collman, J. P. *Acc. Chem. Res.* **1986**, *19*, 209.

185. Brothers, P. J. *Adv. Organomet. Chem.* **2001**, *46*, 223.

186. Brand, H.; Arnold, J. *Organometallics* **1993**, *12*, 3655.

187. Brand, H.; Arnold, J. *Coord. Chem. Rev.* **1995**, *140*, 137.

188. Zhang, X.-X.; Wayland, B. B. *J. Am. Chem. Soc.* **1994**, *116*, 7897.

189. Fu, X.; Li, S.; Wayland, B. B. *Inorg. Chem.* **2006**, *45*, 9884.

190. Wayland, B. B.; Coffin, V. L.; Farnos, M. D. *Inorg. Chem.* **1988**, *27*, 2745.

191. Wayland, B. B. *Polyhedron* **1988**, *7*, 1545.

192. Brothers, P. J.; Collman, J. P. *Acc. Chem. Res.* **1986**, *19*, 209.

193. Guilard, R.; Kadish, K. M. *Chem. Rev.* **1988**, *88*, 1121.

194. *The Porphyrins*; Dolphin, D., Ed.; Academic Press: New York, 1978; Vol. 1, Part A.

195. Buchler, J. W.; Eikelman, G.; Puppe, L.; Rohbock, K.; Schneehage, H. H.; Weck, D. *Ann. Chem.-J. Liebig* **1971**, *745*, 135.

196. Cui, W.; Wayland, B. B. *J. Am. Chem. Soc.* **2004**, *126*, 8266.

197. Maxwell, J. L.; Brown, K. C.; Bartley, D. W.; Kodadek, T. *Science* **1992**, *256*, 1544.

198. Cenini, S.; Gallo, E.; Penoni, A.; Ragaini, F.; Tollari, S. *J. Chem. Soc., Chem. Commun.* **2000**, 2265.

199. Liang, J. L.; Yuan, S. X.; Huang, J. S.; Yu, W. Y.; Che, C. M. *Angew. Chem., Int. Ed. Engl.* **2002**, *41*, 3465.

200. Wayland, B. B.; Basickes, L.; Mukerjee, S.; Wei, M.; Fryd, M. *Macromolecules* **1997**, *30*, 8109.

201. Lu, Z.; Fryd, M.; Wayland, B. B. *Macromolecules* **2004**, *37*, 2686.

202. Wayland, B. B.; Ba, S.; Sherry, A. E. *J. Am. Chem. Soc.* **1991**, *113*, 5305.

203. Takahashi, H.; Kawakita, T.; Ohno, M.; Yoshioka, M.; Kobayashi, S. *Tetrahedron* **1992**, *48*, 5691.

204. Takahashi, H.; Kawakita, T.; Yoshioka, M.; Kobayashi, S.; Ohno, M. *Tetrahedron Lett.* **1989**, *30*, 7095.

205. Knochel, P.; Vettel, S.; Eisenberg, C. *Appl. Organomet. Chem.* **1995**, *9*, 175.

206. García, C.; LaRochelle, L. K.; Walsh, P. J. *J. Am. Chem. Soc.* **2002**, *124*, 10970.

207. Betancort, J. M.; García, C.; Walsh, P. J. *Synlett* **2004**, 749.

208. Takahashi, H.; Yoshioka, M.; Shibasaki, M.; Ohno, M.; Imai, N.; Kobayashi, S. *Tetrahedron* **1995**, *51*, 12013.

209. Corey, E. J.; Imwinkelreid, R. Pikul, S.; Xiang, Y. B. *J. Am. Chem. Soc.* **1989**, *111*, 5493.

210. Corey, E. J.; Yu, C.-M.; Kim, S. S. *J. Am. Chem. Soc.* **1989**, *111*, 5495.

211. Evans, D. A.; Nelson, S. G. *J. Am. Chem. Soc.* **1997**, *119*, 6452.

212. Gordon, A. J.; Ford, R. A. *The Chemist's Companion*; Wiley: New York, 1972.

213. Görlitzer, H. W.; Spiegler, M.; Anwander, R. *Eur. J. Inorg. Chem.* **1998**, 1009.

214. Nanthakumar, A.; Miura, J.; Diltz, S. I.; Lee, C.-K.; Aguirre, G.; Ortega, F.; Ziller, J. W.; Walsh, P. J. *Inorg. Chem.* **1999**, *38*, 3010.

215. Pritchett, S.; Gantzel, P.; Walsh, P. J. *Organometallics* **1999**, *18*, 823.

216. Pritchett, S.; Woodmansee, D. H.; Gantzel, P.; Walsh, P. J. *J. Am. Chem. Soc.* **1998**, *120*, 6423.

217. Becker, J. J.; White, P. S.; Gagné, M. R. *Inorg. Chem.* **1999**, *38*, 798.

218. Corey, E. J.; Sarshar, S.; Lee, D.-H. *J. Am. Chem. Soc.* **1994**, *116*, 12089.

219. Pritchett, S.; Gantzel, P.; Walsh, P. J. *Organometallics* **1997**, *16*, 5130.

220. Armistead, L. T.; White, P. S.; Gagné, M. R. *Organometallics* **1998**, *17*, 216.

221. Balsells, J.; Betancort, J. M.; Walsh, P. J. *Angew. Chem., Int. Ed. Engl.* **2000**, *39*, 3428.

222. Royo, E.; Betancort, J. M.; Davis, T. J.; Walsh, P. J. *Organometallics* **2000**, *19*, 4840.

223. Trofimenko, S. *J. Am. Chem. Soc.* **1966**, *88*, 1842.

224. Trofimenko, S. *Chem. Rev.* **1993**, *93*, 943.

225. Trofimenko, S. *Prog. Inorg. Chem.* **1986**, *34*, 115.

226. Niedenzu, K.; Trofimenko, S. *Top. Curr. Chem.* **1986**, *131*, 1.

227. Tellers, D. M.; Skoog, S. J.; Bergman, R. G.; Gunnoe, T. B.; Harman, W. D. *Organometallics* **2000**, *19*, 2428.

228. Tellers, D. M.; Bergman, R. G. *J. Am. Chem. Soc.* **2000**, *122*, 954.

229. Scheuer, S.; Fischer, J.; Kress, J. *Organometallics* **1995**, *14*, 2627.

230. Preuss, F.; Becker, H.; Hausler, H. J. *Z. Naturforsch* **1987**, *42B*, 881.

231. Trofimenko, S.; Calabrese, J. C.; Thompson, J. S. *Inorg. Chem.* **1987**, *26*, 1507.

232. Paneque, M.; Sirol, S.; Trujillo, M.; Gutierrez-Puebla, E.; Monge, M. A.; Carmona, E. *Angew. Chem., Int. Ed. Engl.* **2000**, *39*, 218.

233. Bromberg, S. E.; Yang, H.; Asplund, M. C.; Lian, T.; McNamara, B. K.; Kotz, K. T.; Yeston, J. S.; Wilkens, M.; Frei, H.; Bergman, R. G.; Harris, C. B. *Science* **1997**, *278*, 260.

234. Wick, D. D.; Goldberg, K. I. *J. Am. Chem. Soc.* **1997**, *119*, 10235.

235. Trofimenko, S. *Inorg. Synth.* **1970**, *12*, 99.

236. Trofimenko, S. *Prog. Inorg. Chem.* **1986**, *34*, 115.

237. Niedenzu, K.; Trofimenko, S. *Top. Curr. Chem.* **1986**, *131*, 131.

238. Peters, J.; Feldman, J.; Tilley, T. *J. Am. Chem. Soc.* **1999**, *121*, 9871.

239. Ghosh, C. K.; Hoyano, J. K.; Krentz, R.; Graham, W. A. G. *J. Am. Chem. Soc.* **1989**, *111*, 5480.

240. Albinati, A.; Bovens, M.; Riegger, H.; Venanzi, L. M. *Inorg. Chem.* **1997**, *36*, 5991.

241. Slugovc, C.; Padilla-Martinez, I.; Sirol, S.; Carmona, E. *Coord. Chem. Rev.* **2001**, *213*, 129.

242. Trofimenko, S. *J. Am. Chem. Soc.* **1969**, *91*, 588.

243. Slugovc, C.; Padilla-Martinez, I.; Sirol, S.; Carmona, E. *Coord. Chem. Rev.* **2001**, *213*, 129.

244. Collison, D.; Mabbs, F. E.; Passand, M. A.; Rigby, K.; Cleland, W. E. *Polyhedron* **1989**, *8*, 1827.

245. Ferguson, G.; Ruhl, B. L.; Lalor, F. J.; Deane, M. E. *J. Organomet. Chem.* **1985**, *282*, 75.

246. Lee, H.; Jordan, R. F. *J. Am. Chem. Soc.* **2005**, *127*, 9384.

247. (a) Bourget-Merle, L.; Lappert, M. F.; Severn, J. R. *Chem. Rev.* **2002**, *102*, 3031; (b) Holin, R. H., Eventh, G. W., Jr.; Chatravarty, A. *Prog. Inorg. Chem.* **1966**, *7*, 53.

248. Feldman, J.; McLain, S. J.; Parthasarathy, A.; Marshall, W. J.; Calabrese, J. C.; Arthur, S. D. *Organometallics* **1997**, *16*, 1514.

249. McGeachin, S. G. *Can. J. Chem.* **1968**, *46*, 1903.

250. Barry, W. J.; Finar, I. L.; Mooney, E. F. *Spectrochim. Acta* **1965**, *21*, 1095.

251. Hitchcock, P. B.; Lappert, M. F.; Liu, D. S. *J. Chem. Soc., Chem. Commun.* **1994**, 1699.

252. Hayes, P. G.; Piers, W. E.; Lee, L. W. M.; Knight, L. K.; Parvez, M.; Elsegood, M. R. J.; Clegg, W. *Organometallics* **2001**, *20*, 2533.

253. Rahim, M.; Taylor, N. J.; Xin, S. X.; Collins, S. *Organometallics* **1998**, *17*, 1315.

254. Holland, P. L.; Cundari, T. R.; Perez, L. L.; Eckert, N. A.; Lachicotte, R. J. *J. Am. Chem. Soc.* **2002**, *124*, 14416.

255. Fekl, U.; Kaminsky, W.; Goldberg, K. I. *J. Am. Chem. Soc.* **2001**, *123*, 6423.

256. Roesky, H. W.; Singh, S.; Yusuff, K. K. M.; Maguire, J. A.; Hosmane, N. S. *Chem. Rev.* **2006**, *106*, 3813.

257. Chisholm, M. H. *Chemtracts: Inorg. Chem.* **1992**, *4*, 273.

258. Bryndza, H. E.; Tam, W. *Chem. Rev.* **1988**, *88*, 1163.

259. Fulton, J. R.; Holland, A. W.; Fox, D. J.; Bergman, R. G. *Acc. Chem. Res.* **2002**, *35*, 44.

260. *The Chemistry of the Hydroxyl Group, Part. 2*; Patai, S., Ed.; Wiley-Interscience: New York, 1971.

261. Willis, C. J. *Coord. Chem. Rev.* **1988**, *88*, 133.

262. Marciniec, B.; Maciejewski, H. *Coord. Chem. Rev.* **2001**, *223*, 301.

263. Wolczanski, P. T. *Polyhedron* **1995**, *14*, 3335.

264. Power, P. P. *J. Organomet. Chem.* **2004**, *689*, 3904.

265. Van der Sluys, W. G.; Sattelberger, A. P. *Chem. Rev.* **1990**, *90*, 1027.

266. Schock, L. E.; Marks, T. J. *J. Am. Chem. Soc.* **1988**, *110*, 7701.

267. Uddin, J.; Morales, C. M.; Maynard, J. H.; Landis, C. R. *Organometallics* **2006**, *25*, 5566.

268. Pearson, R. G. *J. Am. Chem. Soc.* **1963**, *85*, 3533.

269. Pearson, R. G. *J. Chem. Educ.* **1968**, *45*, 581.

270. Pearson, R. G. *J. Chem. Educ.* **1968**, *45*, 643.

271. Ziegler, T.; Tschinke, V.; Versluis, L.; Baerends, E. J.; Ravenek, W. *Polyhedron* **1988**, *7*, 1625.

272. Caulton, K. G. *New J. Chem.* **1994**, *18*, 25.

273. Lappert, M. F.; Patil, D. S.; Pedley, J. B. *J. Chem. Soc., Chem. Commun.* **1975**, 830.

274. Dias, A. R.; Salema, M. S.; Martinho Simoes, J. A. *J. Organomet. Chem.* **1981**, *222*, 69.

275. Bruno, J. W.; Marks, T. J.; Morss, L. R. *J. Am. Chem. Soc.* **1983**, *105*, 6824.

276. Calhorda, M. J.; Dias, A. R.; Salema, M. S.; Simoes, J. A. M. *J. Organomet. Chem.* **1983**, *255*, 81.

277. Bruno, J. W.; Stecher, H. A.; Morss, L. R.; Sonnenberger, D. C.; Marks, T. J. *J. Am. Chem. Soc.* **1986**, *108*, 7275.

278. Nolan, S. P.; Stern, D.; Marks, T. J. *J. Am. Chem. Soc.* **1989**, *111*, 7844.

279. Bryndza, H. E.; Fong, L. K.; Paciello, R. A.; Tam, W.; Bercaw, J. E. *J. Am. Chem. Soc.* **1987**, *109*, 1444.

280. Wayland, B. B. *Polyhedron* **1988**, *7*, 1545.

281. Hartwig, J. F.; Andersen, R. A.; Bergman, R. G. *Organometallics* **1991**, *10*, 1875.

282. Weidmann, B.; Seebach, D. *Angew. Chem.* **1983**, *95*, 12.

283. Jordan, R. F.; Dasher, W. E.; Echols, S. F. *J. Am. Chem. Soc.* **1986**, *108*, 1718.

284. Dormond, A.; Aaliti, A.; Moise, C. *J. Org. Chem.* **1988**, *53*, 1034.

285. Cano, J. S.; Sudupe, M. A.; Royo, P.; Mosquera, M. E. G. *Organometallics* **2005**, *24*, 2424.

286. Mayer, J. M. *Polyhedron* **1995**, *14*, 3273.

287. Nugent, W. A.; Ovenall, D. W.; Holmes, S. J. *Organometallics* **1983**, *2*, 161.

288. Wu, Z.; Jordan, R. F.; Petersen, J. L. *J. Am. Chem. Soc.* **1995**, *117*, 5867.

289. Carpentier, J.-F.; Wu, Z.; Lee, C. W.; Stroemberg, S.; Christopher, J. N.; Jordan, R. F. *J. Am. Chem. Soc.* **2000**, *122*, 7750.

290. Stoebenau, E. J., III; Jordan, R. F. *J. Am. Chem. Soc.* **2006**, *128*, 8162.

291. Chamberlain, L.; Huffman, J. C.; Keddington, J.; Rothwell, I. P. *J. Chem. Soc., Chem. Commun.* **1982**, 805.

292. Chamberlain, L.; Keddington, J.; Rothwell, I. P.; Huffman, J. C. *Organometallics* **1982**, *1*, 1538.

293. Schrock, R. R. *Polyhedron* **1995**, *14*, 3177.

294. Lubben, T. V.; Wolczanski, P. T.; Van Duyne, G. D. *Organometallics* **1984**, *3*, 977.

295. Neithamer, D. R.; LaPointe, R. E.; Wheeler, R. A.; Richeson, D. S.; Van Duyne, G. D.; Wolczanski, P. T. *J. Am. Chem. Soc.* **1989**, *111*, 9056.

296. Schrock, R. R.; Hoveyda, A. H. *Angew. Chem., Int. Ed. Engl.* **2003**, *42*, 4592.

297. Walsh, P. J. *Acc. Chem. Res.* **2003**, *36*, 739.

298. Ellman, J. A.; Owens, T. D.; Tang, T. P. *Acc. Chem. Res.* **2002**, *35*, 984.

299. Rothwell, I. P. *J. Chem. Soc., Chem. Commun.* **1997**, 1331.

300. Hultzsch, K. C.; Gribkov, D. V.; Hampel, F. *J. Organomet. Chem.* **2005**, *690*, 4441.

301. Kanai, M.; Kato, N.; Ichikawa, E.; Shibasaki, M. *Synlett* **2005**, 1491.

302. Tosaki, S.-Y.; Hara, K.; Gnanadesikan, V.; Morimoto, H.; Harada, S.; Sugita, M.; Yamagiwa, N.; Matsunaga, S.; Shibasaki, M. *J. Am. Chem. Soc.* **2006**, *128*, 11776.

303. Pearson, R. G. *Chemical Hardness: Applications from Molecules to Solids*; Wiley-VCH: New York, 1997.

304. Drago, R. S.; Dadmun, A. P. *J. Am. Chem. Soc.* **1994**, *116*, 1792.

305. Drago, R. S. *Polyhedron* **1994**, *13*, 2017.

306. Holland, P. L.; Andersen, R. A.; Bergman, R. G. *Comments Inorg. Chem.* **1999**, *21*, 115.

307. Vaska, L.; Peone, J., Jr. *J. Chem. Soc., Chem. Commun.* **1971**, 418.

308. Poulton, J. T.; Sigalas, M. P.; Folting, K.; Streib, W. E.; Eisenstein, O.; Caulton, K. G. *Inorg. Chem.* **1994**, *33*, 1476.

309. Holland, P. L.; Andersen, R. A.; Bergman, R. G.; Huang, J.; Nolan, S. P. *J. Am. Chem. Soc.* **1997**, *119*, 12800.

310. Sanford, M. S.; Henling, L. M.; Day, M. W.; Grubbs, R. H. *Angew. Chem,. Int. Ed. Engl.* **2000**, *39*, 3451.

311. Cole-Hamilton, D. J.; Young, R. J.; Wilkinson, G. *J. Chem. Soc., Dalton Trans.* **1976**, 1995.

312. Glueck, D. S.; Bergman, R. G. *Organometallics* **1991**, *10*, 1479.

313. Campora, J.; Matas, I.; Palma, P.; Graiff, C.; Tiripicchio, A. *Organometallics* **2005**, *24*, 2827.

314. Palucki, M.; Wolfe, J. P.; Buchwald, S. L. *J. Am. Chem. Soc.* **1996**, *118*, 10333.

315. Palucki, M.; Wolfe, J. P.; Buchwald, S. L. *J. Am. Chem. Soc.* **1997**, *119*, 3395.

316. Zlota, A. A.; Frolow, F.; Milstein, D. *J. Am. Chem. Soc.* **1990**, *112*, 6411.

317. Bartlett, R. A.; Ellison, J. J.; Power, P. P.; Shoner, S. C. *Inorg. Chem.* **1991**, *30*, 2888.

318. Komiya, S.; Akai, Y.; Tanaka, K.; Yamamoto, T.; Yamamoto, A. *Organometallics* **1985**, *4*, 1130.

319. Zhao, P.; Incarvito, C. D.; Hartwig, J. F. *J. Am. Chem. Soc.* **2006**, *128*, 3124.

320. Zhao, P.; Incarvito, C. D.; Hartwig, J. F. *J. Am. Chem. Soc.* **2006**, *128*, 9642.

321. Fornies, J.; Green, M.; Spencer, J. L.; Stone, F. G. A. *J. Chem. Soc., Dalton Trans.* **1977**, 1006.

322. Yoshida, T.; Matsuda, T.; Okano, T.; Kitani, T.; Otsuka, S. *J. Am. Chem. Soc.* **1979**, *101*, 2027.

323. Milstein, D.; Calabrese, J. C.; Williams, I. D. *J. Am. Chem. Soc.* **1986**, *108*, 6387.

324. Dorta, R.; Rozenberg, H.; Shimon, L. J. W.; Milstein, D. *J. Am. Chem. Soc.* **2002**, *124*, 188.

325. Blum, O.; Milstein, D. *J. Am. Chem. Soc.* **2002**, *124*, 11456.

326. Schlodder, R.; Ibers, J. A.; Lenarda, M.; Graziani, M. *J. Am. Chem. Soc.* **1974**, *96*, 6893.

327. Hayashi, Y.; Yamamoto, T.; Yamamoto, A.; Komiya, S.; Kushi, Y. *J. Am. Chem. Soc.* **1986**, *108*, 385.

328. Ueno, S.; Mizushima, E.; Chatani, N.; Kakiuchi, F. *J. Am. Chem. Soc.* **2006**, *128*, 16516.

329. Reed, C. A.; Roper, W. R. *J. Chem. Soc., Dalton Trans.* **1973**, 1014.

330. Matsunaga, P. T.; Hillhouse, G. L.; Rheingold, A. L. *J. Am. Chem. Soc.* **1993**, *115*, 2075.

331. Han, R.; Hillhouse, G. L. *J. Am. Chem. Soc.* **1997**, *119*, 8135.

332. De Bruin, B.; Boerakker, M. J.; Donners, J. J. J. M.; Christiaans, B. E. C.; Schlebos, P. P. J.; De Gelder, R.; Smits, J. M. M.; Spek, A. L.; Gal, A. W. *Angew. Chem,. Int. Ed. Engl.* **1997**, *36*, 2064.

333. Cinellu, M. A.; Minghetti, G.; Cocco, F.; Stoccoro, S.; Zucca, A.; Manassero, M. *Angew. Chem., Int. Ed. Engl.* **2005**, *44*, 6892.

334. Szuromi, E.; Shan, H.; Sharp, P. R. *J. Am. Chem. Soc.* **2003**, *125*, 10522.

335. Szuromi, E.; Wu, J.; Sharp, P. R. *J. Am. Chem. Soc.* **2006**, *128*, 12088.

336. Zhao, J.; Hesslink, H.; Hartwig, J. F. *J. Am. Chem. Soc.* **2001**, *123*, 7220.

337. Ng, S. M.; Zhao, C.; Lin, Z. *J. Organomet. Chem.* **2002**, *662*, 120.

338. Hartwig, J. F.; Bergman, R. G.; Andersen, R. A. *J. Am. Chem. Soc.* **1990**, *112*, 3234.

339. Satoh, T.; Miura, M. *Top. Organomet. Chem.* **2005**, *14*, 1.

340. Blum, O.; Milstein, D. *Angew. Chem., Int. Ed. Engl.* **1995**, *34*, 229.

341. Bernard, K. A.; Churchill, M. R.; Janik, T. S.; Atwood, J. D. *Organometallics* **1990**, *9*, 12.

342. Hartwig, J. F. *Acc. Chem. Res.* **1998**, *31*, 852.

343. Williams, B. S.; Goldberg, K. I. *J. Am. Chem. Soc.* **2001**, *123*, 2576.

344. Kuznetsov, V. F.; Yap, G. P. A.; Bensimon, C.; Alper, H. *Inorg. Chim. Acta* **1998**, *280*, 172.

345. Baeckvall, J. E.; Akermark, B.; Ljunggren, S. O. *J. Am. Chem. Soc.* **1979**, *101*, 2411.

346. Henry, P. M. In *Handbook of Organopalladium Chemistry for Organic Synthesis*; Negishi, E.-i., Ed.; Wiley: New York, 2002; Vol. 2, p 2119.

347. Hayashi, T.; Yamasaki, K.; Mimura, M.; Uozumi, Y. *J. Am. Chem. Soc.* **2004**, *126*, 3036.

348. Trend, R. M.; Ramtohul, Y. K.; Stoltz, B. M. *J. Am. Chem. Soc.* **2005**, *127*, 17778.

349. Bryndza, H. E. *Organometallics* **1985**, *4*, 406.

350. Ritter, J. C. M.; Bergman, R. G. *J. Am. Chem. Soc.* **1997**, *119*, 2580.

351. Tellers, D. M.; Ritter, J. C. M.; Bergman, R. G. *Inorg. Chem.* **1999**, *38*, 4810.

352. Yin, X.; Moss, J. R. *Coord. Chem. Rev.* **1999**, *181*, 27.

353. Glueck, D. S.; Winslow, L. J. N.; Bergman, R. G. *Organometallics* **1991**, *10*, 1462.

354. Krug, C.; Hartwig, J. F. *J. Am. Chem. Soc.* **2002**, *124*, 1674.

355. Newman, L. J.; Bergman, R. G. *J. Am. Chem. Soc.* **1985**, *107*, 5314.

356. Braga, D.; Sabatino, P.; Di Bugno, C.; Leoni, P.; Pasquali, M. *J. Organomet. Chem.* **1987**, *334*, C46.

357. Kegley, S. E.; Schaverien, C. J.; Freudenberger, J. H.; Bergman, R. G.; Nolan, S. P.; Hoff, C. D. *J. Am. Chem. Soc.* **1987**, *109*, 6563.

358. Fagnou, K.; Lautens, M. *Chem. Rev.* **2003**, *103*, 169.

359. Farnworth, M. V.; Cross, M. J.; Louie, J. *Tetrahedron Lett.* **2004**, *45*, 7441.

360. Krug, C.; Hartwig, J. F. *Organometallics* **2004**, *23*, 4594.

361. Noyori, R.; Hashiguchi, S. *Acc. Chem. Res.* **1997**, *30*, 97.

362. Calhorda, M. J.; Galvao, A. M.; Unaleroglu, C.; Zlota, A. A.; Frolow, F.; Milstein, D. *Organometallics* **1993**, *12*, 3316.

363. Mehrotra, R. C.; Bohra, R.; Gaur, D. P. *Metal Beta-Diketonates and Allied Derivatives*; Academic Press, New York, 1978.

364. Baker, G. J.; Raynor, J. B.; Smits, J. M. M.; Beurskens, P. T.; Vergoossen, H.; Keijzers, C. P. *J. Chem. Soc., Dalton Trans.* **1986**, 2655.

365. Bennett, M. A.; Byrnes, M. J.; Kovacik, I. *J. Organomet. Chem.* **2004**, *689*, 4463.

366. Gibson, D. *Coord. Chem. Rev.* **1969**, *4*, 225.

367. Okey, S.; Kawaguchi, S. *Inorg. Chem.* **1977**, *16*, 1730.

368. Wolkowski, J. P.; Hartwig, J. F. *Angew. Chem., Int. Ed. Engl.* **2002**, *41*, 4289.

369. Doyle, M. P.; McKervey, M. A.; Ye, T. *Modern Catalytic Methods for Organic Synthesis with Diazo Compounds: From Cyclopropanes to Ylides*; Wiley: New York, 1998.

370. Vicente, J.; Chicote, M. T. *Coord. Chem. Rev.* **1999**, *193–195*, 1143.

371. Wong-Foy, A. G.; Bhalla, G.; Liu, X. Y.; Periana, R. A. *J. Am. Chem. Soc.* **2003**, *125*, 14292.

372. Bhalla, G.; Liu, X. Y.; Oxgaard, J.; Goddard, W. A., III; Periana, R. A. *J. Am. Chem. Soc.* **2005**, *127*, 11372.

373. Bhalla, G.; Oxgaard, J.; Goddard, W. A., II; Periana, R. A. *Organometallics* **2005**, *24*, 5499.

374. Irvine, G. J.; Lesley, M. J. G.; Marder, T. B.; Norman, N. C.; Rice, C. R.; Robins, E. G.; Whittell, G. R.; Wright, L. *J. Chem. Rev.* **1998**, *98*, 2685.

375. Noth, H.; Schmidt, G. *Angew. Chem., Int. Ed. Engl.* **1963**, *2*, 623.

376. Schmid, G. *Angew. Chem., Int. Ed. Engl.* **1970**, *9*, 819.

377. Braunschweig, H.; Colling, M. *Coord. Chem. Rev.* **2001**, *223*, 1.

378. Irvine, G. J.; Lesley, M. J. G.; Marder, T. B.; Norman, N. C.; Rice, C. R.; Robins, E. G.; Roper, W. R.; Whittell, G. R.; Wright, L. *J. Chem. Rev.* **1998**, *98*, 2685.

379. Segawa, Y.; Yamashita, M.; Nozaki, K. *Science* **2006**, *314*, 113.

380. Auten, R. W.; Kraus, C. A. *J. Am. Chem. Soc.* **1952**, *74*, 3398.

381. Wagner, M.; Hommes, N. J. R. v. E.; Nöth, H.; Schleyer, P. v. R. *Inorg. Chem.* **1995**, *34*, 607.

382. Curtis, D.; Lesley, M. J. G.; Norman, N. C.; Orpen, A. G.; Starbuck, J. *J. Chem. Soc., Dalton Trans.* **1999**, 1687.

383. Westcott, S. A.; Marder, T. B.; Baker, R. T.; Harlow, R. L.; Calabrese, J. C.; Lam, K. C.; Lin, Z. *Polyhedron* **2004**, *23*, 2665.

384. Braunschweig, H.; Radacki, K.; Rais, D.; Whittell, G. R. *Angew. Chem., Int. Ed. Engl.* **2005**, *44*, 1192.

385. Murphy, J. M.; Lawrence, J. D.; Kawamura, K.; Incarvito, C.; Hartwig, J. F. *J. Am. Chem. Soc.* **2006**, *128*, 13684.

386. Braunschweig, H.; Kollann, C.; Rais, D. *Angew. Chem., Int. Ed. Engl.* **2006**, *45*, 5254.

387. Aldridge, S.; Coombs, D. L.; Jones, C. *J. Chem. Soc., Chem. Commun.* **2002**, 856.

388. Braunschweig, H.; Radacki, K.; Rais, D.; Seeler, F.; Uttinger, K. *J. Am. Chem. Soc.* **2005**, *127*, 1386.

389. Braunschweig, H.; Forster, M.; Radacki, K. *Angew. Chem., Int. Ed. Engl.* **2006**, *45*, 2132.

390. Cundari, T. R.; Zhao, Y. *Inorg. Chim. Acta* **2003**, *345*, 70.

391. Sivignon, G.; Fleurat-Lessard, P.; Onno, J. M.; Volatron, F. *Inorg. Chem.* **2002**, *41*, 6656.

392. Dickinson, A. A.; Willock, D. J.; Calder, R. J.; Aldridge, S. *Organometallics* **2002**, *21*, 1146.

393. Giju, K. T.; Bickelhaupt, F. M.; Frenking, G. *Inorg. Chem.* **2000**, *39*, 4776.

394. Sakaki, S.; Kai, S.; Sugimoto, M. *Organometallics* **1999**, *18*, 4825.

395. Basil, J. D.; Aradi, A. A.; Nripendra, K. B.; Nigam, P. R.; Eigenbrot, C.; Fehlner, T. P. *Inorg. Chem.* **1990**, *29*, 1260.

396. Waltz, K. M.; Hartwig, J. F. *J. Am. Chem. Soc.* **2000**, *122*, 11358.

397. Zhu, J.; Lin, Z.; Marder, T. B. *Inorg. Chem.* **2005**, *44*, 9284.

398. Rablen, P. R.; Hartwig, J. F.; Nolan, S. P. *J. Am. Chem. Soc.* **1994**, *116*, 4121.

399. Sakaki, S.; Kikuno, T. *Inorg. Chem.* **1997**, *36*, 226.

400. Rablen, P.; Hartwig, J. F. *J. Am. Chem. Soc.* **1996**, *118*, 4648.

401. Schmid, G.; Noth, H. *Chem. Ber.* **1967**, *100*, 2899.

402. Smith, M. R. I.; Motry, D. H. *J. Am. Chem. Soc.* **1995**, *117*, 6615.

403. Motry, D. H.; Brazil, A. G.; Smith, M. R., III *J. Am. Chem. Soc.* **1997**, *119*, 2743.

404. Sumimoto, M.; Iwane, N.; Takahama, T.; Sakaki, S. *J. Am. Chem. Soc.* **2004**, *126*, 10457.

405. Hartwig, J. F.; Waltz, K. M.; Muhoro, C. N. In *Advances in Boron Chemistry*; Siebert, W., Ed.; The Royal Society of Chemistry: Cambridge, 1997; p 373.

406. Waltz, K. M.; He, X.; Muhoro, C. N.; Hartwig, J. F. *J. Am. Chem. Soc.* **1995**, *117*, 11357.

407. Waltz, K. M.; Hartwig, J. F. *Science* **1997**, *277*, 211.

408. Hartwig, J. F.; Huber, S. *J. Am. Chem. Soc.* **1993**, *115*, 4908.

409. Braunschweig, H.; Kollann, C.; Englert, U. *Eur. J. Inorg. Chem* **1998**, 465.

410. Braunschweig, H.; Kollann, C.; Klinkhammer, K. W. *Eur. J. Inorg. Chem.* **1999**, 1523.

411. Hartwig, J. F.; Huber, S. *J. Am. Chem. Soc.* **1994**, *116*, 3661.

412. Hartwig, J. F.; He, X. *Organometallics* **1996**, *15*, 5350.

413. Lantero, D. R.; Motry, D. H.; Ward, D. L.; Smith, M. R., III *J. Am. Chem. Soc.* **1994**, *116*, 10811.

414. Muhoro, C. N.; Hartwig, J. F. *Angew. Chem., Int. Ed. Engl.* **1997**, *36*, 1510.

415. Dai, C. Y.; Stringer, G.; Corrigan, J. F.; Taylor, N. J.; Marder, T. B.; Norman, N. C. *J. Organomet. Chem.* **1996**, *513*, 273.

416. Nguyen, P.; Lesley, G.; Taylor, N. J.; Marder, T. B.; Pickett, N. L.; Clegg, W.; Elsegood, M. R. J.; Norman, N. C. *Inorg. Chem.* **1994**, *33*, 4623.

417. Ishiyama, T.; Matsuda, N.; Miyaura, N.; Suzuki, A. *J. Am. Chem. Soc.* **1993**, *115*, 11018.

418. Dai, C.; Stringer, G.; Corrigan, J. F.; Taylor, N. J.; Marder, T. B.; Norman, N. C. *J. Organomet. Chem.* **1996**, *513*, 273.

419. Clegg, W.; Lawlor, F. J.; Marder, T. B.; Nguyen, P.; Norman, N. C.; Orpen, A. G.; Quayle, M. J.; Rice, C. R.; Robins, E. G.; Scott, A. J.; Souza, F. E. S.; Stringer, G.; Whittell, G. R. *J. Chem. Soc., Dalton Trans.* **1998**, 301.

420. Dai, C. Y.; Stringer, G.; Marder, T. B.; Baker, R. T.; Scott, A. J.; Clegg, W.; Norman, N. C. *Can. J. Chem.* **1996**, *74*, 2026.

421. He, X.; Hartwig, J. F. *Organometallics* **1995**, *15*, 400.
422. Baker, R. T.; Ovenall, D. W.; Calabrese, J. C.; Westcott, S. A.; Taylor, N. J.; Williams, I. D.; Marder, T. B. *J. Am. Chem. Soc.* **1990**, *112*, 9399.
423. Knorr, J. R.; Merola, J. S. *Organometallics* **1990**, *9*, 3008.
424. Clark, G. R.; Irvine, G. J.; Roper, W. R.; Wright, L. J. *J. Organomet. Chem.* **2003**, *680*, 81.
425. Melenkivitz, R.; Mindiola, D. J.; Hillhouse, G. L. *J. Am. Chem. Soc.* **2002**, *124*, 3846.
426. Rocklage, S. M.; Schrock, R. R.; Churchill, M. R.; Wasserman, H. J. *Organometallics* **1982**, *1*, 1332.
427. Derrah, E. J.; Pantazis, D. A.; McDonald, R.; Rosenberg, L. *Organometallics* **2007**, *26*, 1473.
428. Bohle, D. S.; Jones, T. C.; Rickard, C. E. F.; Roper, W. R. *Organometallics* **1986**, *5*, 1612.
429. Chan, V. S.; Stewart, I. C.; Bergman, R. G.; Toste, F. D. *J. Am. Chem. Soc.* **2006**, *128*, 2786.
430. Fryzuk, M. D.; Joshi, K. *Organometallics* **1989**, *8*, 722.
431. Baker, R. T.; Krusic, P. J.; Tulip, T. H.; Calabrese, J. C.; Wreford, S. S. *J. Am. Chem. Soc.* **1983**, *105*, 6763.
432. Baker, R. T.; Whitney, J. F.; Wreford, S. S. *Organometallics* **1983**, *2*, 1049.
433. Rogers, J. R.; Wagner, T. P. S.; Marynick, D. S. *Inorg. Chem.* **1994**, *33*, 3104.
434. Nolan, S. P.; Stern, D.; Marks, T. J. *J. Am. Chem. Soc.* **1989**, *111*, 7844.
435. Wicht, D. K.; Paisner, S. N.; Lew, B. M.; Glueck, D. S.; Yap, G. P. A.; Liable-Sands, L. M.; Rheingold, A. L.; Haar, C. M.; Nolan, S. P. *Organometallics* **1998**, *17*, 652.
436. Li, J. N.; Liu, L.; Fu, Y.; Guo, Q. X. *Tetrahedron* **2006**, *62*, 4453.
437. Scriban, C.; Glueck, D. S. *J. Am. Chem. Soc.* **2006**, *128*, 2788.
438. Buhro, W. E.; Zwick, B. D.; Georgiou, S.; Hutchinson, J. P.; Gladysz, J. A. *J. Am. Chem. Soc.* **1988**, *110*, 2427.
439. Eliel, E. L.; Wilen, S. H. *Stereochemistry of Organic Compunds*; Wiley: New York, 1994.
440. Smith, M. B.; March, J. *March's Advanced Organic Chemistry: Reactions, Mechanisms and Structure*, 5th ed.; Wiley: New York, 2001.
441. Ren, Y.; Yamataka, H. *J. Org. Chem.* **2007**, *72*, 566.
442. Carey, F. A.; Sundberg, R. J. *Advanced Organic Chemistry. Part A: Structure and Mechanisms*, 4th ed.; Kluwer Academic/Plenum Publishers: New York, 2000.
443. Fernandez-Rodriguez, M. A.; Shen, Q.; Hartwig, J. F. *J. Am. Chem. Soc.* **2006**, *128*, 2180.
444. Ley, S. N.; Thomas, A. W. *Angew. Chem., Int. Ed. Engl.* **2003**, *42*, 5400.
445. Hartwig, J. F. In *Handbook of Organopalladium Chemistry for Organic Synthesis*; Negishi, E.-i., Ed.; Wiley-Interscience: New York, 2002; Vol. 1, p 1097.
446. Barañano, D.; Mann, G.; Hartwig, J. F. *Curr. Org. Chem.* **1997**, *1*, 287.
447. Kosugi, M.; Shimizu, K.; Ohtani, A.; Migita, T. *Chem. Lett.* **1981**, 829.
448. Murahashi, S. I.; Yamamura, M.; Yanagisawa, K.; Mita, N.; Kondo, K. *J. Org. Chem.* **1979**, *44*, 2408.
449. Ogawa, A.; Ikeda, T.; Kimura, K.; Hirao, T. *J. Am. Chem. Soc.* **1999**, *121*, 5108.
450. Kondo, T.; Mitsudo, T. *Chem. Rev.* **2000**, *100*, 3205.
451. Ogawa, A. *J. Organomet. Chem.* **2000**, *611*, 463.
452. Alonso, F.; Beletskaya, I. P.; Yus, M. *Chem. Rev.* **2004**, *104*, 3079.
453. Cao, C. S.; Fraser, L. R.; Love, J. A. *J. Am. Chem. Soc.* **2005**, *127*, 17614.
454. Misumi, Y.; Seino, H.; Mizobe, Y. *J. Organomet. Chem.* **2006**, *691*, 3157.
455. Bianchini, C.; Meli, A. *Acc. Chem. Res.* **1998**, *31*, 109.
456. Krebs, B.; Henkel, G. *Angew. Chem., Int. Ed. Engl.* **1991**, *30*, 769.
457. Sellmann, D.; Grasser, F.; Knoch, F.; Moll, M. *Angew. Chem., Int. Ed. Engl.* **1991**, *30*, 1311.
458. Killops, S. D.; Knox, S. A. R. *J. Chem. Soc., Dalton Trans.* **1978**, 1260.
459. Dance, I. G. *Polyhedron* **1986**, *5*, 1037.
460. Huheey, J. E.; Keiter, E. A.; Keiter, R. L. *Inorganic Chemistry,* 4th ed.; HarperCollins: New York, 1993; p 863.
461. Huheey, J. E.; Keiter, E. A.; Keiter, R. L. *Inorganic Chemistry*, 4th ed.; HarperCollins: New York, 1993; p 226.
462. Wu, Z.-Z.; Huang, Z.-E.; Cai, R.-F.; Zhou, X.-G.; Xu, Z.; You, X.-Z.; Huang, X.-Y. *J. Organomet. Chem.* **1996**, *506*, 25.
463. Poremba, P.; Noltemeyer, M.; Schmidt, H.-G.; Edelmann, F. T. *J. Organomet. Chem.* **1995**, *501*, 315.
464. Louie, J.; Hartwig, J. F. *J. Am. Chem. Soc.* **1995**, *117*, 11598.
465. Calhorda, M. J.; Carrondo, M. A. A. F. d. C. T.; Dias, A. R.; Domingos, A. M. T. S.; Simoes, J. A. M.; Teixeira, C. *Organometallics* **1986**, *5*, 660.
466. Calhorda, M. J.; Carrondo, M. A. A. F. d. C. T.; Dias, A. R.; Frazao, C. F.; Hursthouse, M. B.; Simoes, J. A. M.; Teixeira, C. *Inorg. Chem.* **1988**, *27*, 2513.
467. Luo, Y. R. *Handbook of Bond Dissociation Energies in Organic Compounds*; CRS Press: Boca Raton, 2003.
468. Holland, P. L.; Andersen, R. A.; Bergman, R. G.; Huang, J. K.; Nolan, S. P. *J. Am. Chem. Soc.* **1997**, *119*, 12800.

469. Baranano, D.; Hartwig, J. F. *J. Am. Chem. Soc.* **1995**, *117*, 2937.
470. Ong, C.; Kickham, J.; Clemens, S.; Guerin, F.; Stephan, D. W. *Organometallics* **2002**, *21*, 1646.
471. Michelman, R. I.; Andersen, R. A.; Bergman, R. G. *J. Am. Chem. Soc.* **1991**, *113*, 5100.
472. Michelman, R. I.; Ball, G. E.; Bergman, R. G.; Andersen, R. A. *Organometallics* **1994**, *13*, 869.
473. Milstein, D.; Calabrese, J. C.; Williams, I. D. *J. Am. Chem. Soc.* **1986**, *108*, 6387.
474. Coto, A.; Jimenez Tenorio, M.; Puerta, M. C.; Valerga, P. *Organometallics* **1998**, *17*, 4392.
475. Zhao, X.; Hsiao, Y. M.; Lai, C. H.; Reibenspies, J. H.; Darensbourg, M. Y. *Inorg. Chem.* **2002**, *41*, 699.
476. Becker, E.; Mereiter, K.; Schmid, R.; Kirchner, K. *Organometallics* **2004**, *23*, 2876.
477. Fachinetti, G.; Floriani, C. *J. Chem. Soc., Dalton Trans.* **1974**, 2433.
478. Kaneko, Y.; Suzuki, N.; Nishiyama, A.; Suzuki, T.; Isobe, K. *Organometallics* **1998**, *17*, 4875.
479. Espenson, J. H. *Coord. Chem. Rev.* **2005**, *249*, 329.
480. Kuwata, S.; Hidai, M. *Coord. Chem. Rev.* **2001**, *213*, 211.
481. Piper, T. S.; Lemal, D.; Wilkinson, G. *Naturwissenschaften* **1956**, *43*, 129.
482. Eisen, M. S. In *The Chemistry of Organic Silicon Compounds*; Rappoport, Z., Apeloig, Y., Eds.; Wiley: New York, 1998; p 2037.
483. Fu, Y.; Liu, L.; Li, R.-Q.; Liu, R.; Guo, Q.-X. *J. Am. Chem. Soc.* **2004**, *126*, 814.
484. Korogodsky, G.; Bendikov, M.; Bravo-Zhivotovskii, D.; Apeloig, Y. *Organometallics* **2002**, *21*, 3157.
485. Buncel, E.; Venkatachalam, T. K. *J. Organomet. Chem.* **2000**, *604*, 208.
486. Notably, the measured pK_a of Ph$_3$SiH (35.1) is greater than that of Ph$_3$CH (31.4). This is presumably due to poorer resonance stabilization of Ph$_3$Si$^-$ as compared with Ph$_3$C$^-$.
487. Siehl, H.-U.; Müller, T. In *The Chemistry of Organic Silicon Compounds*; Rappoport, Z., Apeloig, Y., Eds.; Wiley: New York, 1998; Vol. 2, p 595.
488. The participation of *d*-orbitals in such interactions is no longer considered significant.
489. Pannell, K. H.; Wu, C. C.; Long, G. J. *J. Organomet. Chem.* **1980**, *186*, 85.
490. Chatt, J.; Eaborn, C.; Ibekwe, S. D.; Kapoor, P. N. *J. Chem. Soc. A* **1970**, 1343.
491. Ziegler, T.; Tschinke, V.; Versluis, L.; Baerends, E. J.; Ravenek, W. *Polyhedron* **1988**, *7*, 1625.
492. Novak, I.; Huang, W.; Luo, L.; Huang, H. H.; Ang, H. G.; Zybill, C. E. *Organometallics* **1997**, *16*, 1567.
493. Nolan, S. P.; Porchia, M.; Marks, T. J. *Organometallics* **1991**, *10*, 1450.
494. Levy, C. J.; Puddephatt, R. J. *Organometallics* **1997**, *16*, 4115.
495. Tilley, T. D. In *The Silicon–Heteroatom Bond*; Patai, S., Rappoport, Z., Eds.; Wiley: New York, 1991; p 245.
496. Muir, K. W. *J. Chem. Soc. A* **1971**, 2663.
497. Haszeldine, R. N.; Parish, R. V.; Setchfield, J. H. *J. Organomet. Chem.* **1973**, *57*, 279.
498. Li, S.; Johnson, D. L.; Gladysz, J. A.; Servis, K. L. *J. Organomet. Chem.* **1979**, *166*, 317.
499. Krentz, R.; Pomeroy, R. K. *Inorg. Chem.* **1985**, *24*, 2976.
500. Arnold, J.; Tilley, T. D.; Rheingold, A. L.; Geib, S. J. *Inorg. Chem.* **1987**, *26*, 2106.
501. Blinka, T. A.; Helmer, B. J.; West, R. *Adv. Organomet. Chem.* **1984**, *23*, 193.
502. Xue, Z. L.; Li, L. T.; Hoyt, L. K.; Diminnie, J. B.; Pollitte, J. L. *J. Am. Chem. Soc.* **1994**, *116*, 2169.
503. Pfeiffer, J.; Kickelbick, G.; Schubert, U. *Organometallics* **2000**, *19*, 62.
504. Sadow, A. D.; Tilley, T. D. *J. Am. Chem. Soc.* **2005**, *127*, 643.
505. Campion, B. K.; Heyn, R. H.; Tilley, T. D. *J. Chem. Soc., Chem. Commun.* **1988**, 278.
506. Yamashita, H.; Hayashi, T.; Kobayashi, T.; Tanaka, M.; Goto, M. *J. Am. Chem. Soc.* **1988**, *110*, 4417.
507. Fagnou, K.; Lautens, M. *Angew. Chem., Int. Ed. Engl.* **2002**, *41*, 26.
508. Caulton, K. G. *New J. Chem.* **1994**, *18*, 25.
509. Tolman, C. A. *Chem. Rev.* **1977**, *77*, 313.
510. Huheey, J. E.; Keiter, E. A.; Keiter, R. L. *Inorganic Chemistry*, 4th ed.; Harper Collins: New York, 1993; p 848.
511. Poulton, J. T.; Folting, K.; Streib, W. E.; Caulton, K. G. *Inorg. Chem.* **1992**, *31*, 3190.
512. Doherty, N. M.; Hoffman, N. W. *Chem. Rev.* **1991**, *91*, 553.
513. Agbossou, S. K.; Roger, C.; Igau, A.; Gladysz, J. A. *Inorg. Chem.* **1992**, *31*, 419.
514. Rottink, M. K.; Angelici, R. J. *J. Am. Chem. Soc.* **1992**, *114*, 8296.
515. Rottink, M. K.; Angelici, R. J. *J. Am. Chem. Soc.* **1993**, *115*, 7267.
516. Procopio, L. J.; Carroll, P. J.; Berry, D. H. *J. Am. Chem. Soc.* **1994**, *116*, 177.
517. Lukens, W. W.; Smith, M. R.; Andersen, R. A. *J. Am. Chem. Soc.* **1996**, *118*, 1719.
518. Schock, L. E.; Marks, T. J. *J. Am. Chem. Soc.* **1988**, *110*, 7701.
519. Flemming, J. P.; Pilon, M. C.; Borbulevitch, O. Y.; Antipin, M. Y.; Grushin, V. V. *Inorg. Chim. Acta* **1998**, *280*, 87.
520. Chatt, J.; Venanzi, M. *J. Chem. Soc.* **1957**, 2445.
521. Widenhoefer, R. A.; Zhong, H. A.; Buchwald, S. L. *Organometallics* **1996**, *15*, 2745.
522. Peris, E.; Lee, J. C.; Rambo, J. R.; Eisenstein, O.; Crabtree, R. H. *J. Am. Chem. Soc.* **1995**, *117*, 3485.

523. Abuhasanayn, F.; Kroghjespersen, K.; Goldman, A. S. *Inorg. Chem.* **1993**, *32*, 495.

524. Abuhasanayn, F.; Goldman, A. S.; Kroghjespersen, K. *Inorg. Chem.* **1994**, *33*, 5122.

525. Hansson, S.; Norrby, P. O.; Sjogren, M. P. T.; Akermark, B.; Cucciolito, M. E.; Giordano, F.; Vitagliano, A. *Organometallics* **1993**, *12*, 4940.

526. Sanford, M. S.; Love, J. A.; Grubbs, R. H. *J. Am. Chem. Soc.* **2001**, *123*, 6543.

527. Dorta, R.; Egli, P.; Zurcher, F.; Togni, A. *J. Am. Chem. Soc.* **1997**, *119*, 10857.

528. Duthaler, R. O.; Hafner, A. *Angew. Chem., Int. Ed. Engl.* **1997**, *36*, 43.

529. Gauthier, D. R.; Carreira, E. M. *Angew. Chem., Int. Ed. Engl.* **1996**, *35*, 2363.

530. Singer, R. A.; Carreira, E. M. *J. Am. Chem. Soc.* **1995**, *117*, 12360.

531. Kruger, J.; Carreira, E. M. *J. Am. Chem. Soc.* **1998**, *120*, 837.

532. Kim, Y.; Singer, R. A.; Carreira, E. M. *Angew. Chem., Int. Ed. Engl.* **1998**, *37*, 1261.

533. Pagenkopf, B. L.; Kruger, J.; Stojanovic, A.; Carreira, E. M. *Angew. Chem., Int. Ed. Engl.* **1998**, *37*, 3124.

534. Suto, Y.; Kumagai, N.; Matsunaga, S.; Kanai, M.; Shibasaki, M. *Org. Lett.* **2003**, *5*, 3147.

535. Oisaki, K.; Suto, Y.; Kanai, M.; Shibasaki, M. *J. Am. Chem. Soc.* **2003**, *125*, 5644.

536. Wada, R.; Shibuguchi, T.; Makino, S.; Oisaki, K.; Kanai, M.; Shibasaki, M. *J. Am. Chem. Soc.* **2006**, *128*, 7687.

537. Tomita, D.; Wada, R.; Kanai, M.; Shibasaki, M. *J. Am. Chem. Soc.* **2005**, *127*, 4138.

538. Lautens, M.; Fagnou, K. *J. Am. Chem. Soc.* **2001**, *123*, 7170.

CHAPTER 5

Ligand Substitution Reactions

5.1. Introduction

5.1.1. Overview of Ligand Substitution

A reaction in which a free ligand replaces a coordinated ligand is called ligand substitution (Equation 5.1).[1-24] Several types of mechanisms for ligand substitution have been established. One class of mechanism occurs by initial dissociation of one ligand from the metal, a second class by initial association of the incoming ligand to the metal, and other classes by apparent simultaneous dissociation of the bound ligand and association of the free ligand. These mechanisms are typically called dissociative, associative, and interchange, respectively. In addition to these common mechanisms, other pathways for ligand substitution have been identified. Several of the most important for organometallic systems involve initial oxidation of the metal, initial change in hapticity of a ligand prior to the substitution process, photochemical extrusion of a ligand, and oxidation of a coordinated ligand.

Ligand substitution is a common and important step in the synthesis of most organometallic compounds. It is difficult to imagine the synthesis of an organometallic or coordination compound from simple metal starting materials without the involvement of a ligand substitution process. Moreover, ligand dissociation, the first step of dissociative substitution processes, is the first step of many stoichiometric and catalytic reactions of transition metal organometallic complexes, as shown generically in Equation 5.2. For example, replacement of a coordinated phosphine or solvent by an incoming olefin is an important step of olefin hydrogenation. It has even been stated that "a site of coordinative unsaturation is perhaps the single most important property of a homogeneous catalyst."[25]

$$ML_x + n\,L' \longrightarrow ML_{x-n}L_n' + n\,L \tag{5.1}$$

$$ML_x \rightleftharpoons \underset{\text{Reactive intermediate}}{ML_{x-1}} + L \tag{5.2}$$

5.1.2. Definitions of Associative, Dissociative, and Interchange

The dissociative (D) and associative (A) mechanisms of substitutions at transition metal complexes are closely related to the S_N1 and S_N2 mechanisms of organic nucleophilic substitution reactions, as summarized in Scheme 5.1. Dissociative substitution occurs by initial cleavage of a metal–ligand bond. The initiation of the reaction by a dissociative event is similar to the initial heterolytic cleavage of a bond between carbon and an anionic leaving group that begins many S_N1 reactions. An associative substitution reaction occurs by initial bonding of an incoming ligand to the metal center. This initial bond formation draws some parallels to the events of an S_N2 reaction, which occurs by partial bonding of an incoming nucleophile to the carbon center. Although the associative and S_N2 mechanisms are related,[11–14,26,27] the species formed by association of a ligand to a transition metal is typically a minimum on the energy surface and a reaction intermediate, while the species formed by reaction of a nucleophile and an electrophile in an organic S_N2 reaction is a maximum on the energy surface and is, therefore, a transition state. There is perhaps a stronger parallel between an associative substitution at a transition metal and the generic base-catalyzed transesterification shown in Scheme 5.1. In such a transesterification, addition of an alkoxide leads to an intermediate possessing an increased coordination number at the central atom, and dissociation of the alkoxide leads to the final product with a "three-coordinate" carbon in the carbonyl group.

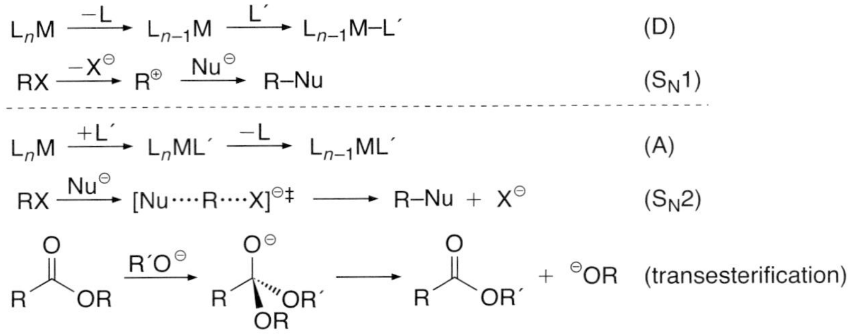

$$L_nM \xrightarrow{-L} L_{n-1}M \xrightarrow{L'} L_{n-1}M{-}L' \qquad\qquad (D)$$

$$RX \xrightarrow{-X^\ominus} R^\oplus \xrightarrow{Nu^\ominus} R{-}Nu \qquad\qquad (S_N1)$$

$$L_nM \xrightarrow{+L'} L_nML' \xrightarrow{-L} L_{n-1}ML' \qquad\qquad (A)$$

$$RX \xrightarrow{Nu^\ominus} [Nu{\cdots}R{\cdots}X]^{\ominus\ddagger} \longrightarrow R{-}Nu\ +\ X^\ominus \qquad\qquad (S_N2)$$

(transesterification)

Scheme 5.1

Many examples of associative and dissociative mechanisms are presented in later sections. One classic example of ligand substitution that exhibits the hallmarks of an associative substitution process is the replacement of chloride in a 16-electron complex, such as $Pt(PEt_3)_2Cl_2$. One classic example of a ligand substitution that exhibits the hallmarks of a dissociative mechanism is the substitution of phosphine for carbonyl in $Ni(CO)_4$.

The terms "associative substitution" and "dissociative substitution" refer to mechanisms at the extremes of a mechanistic continuum. An interchange mechanism (I) can be defined by considering the degree of bond making and bond breaking during the first step of the reaction mechanism. A purely associative mechanism occurs by the formation of a reaction intermediate of increased coordination number by exclusive bond making. A purely dissociative mechanism occurs by formation of an initial intermediate of decreased coordination number by exclusive bond breaking.[12] In between these mechanistic extremes lie pathways in which more bond making (associative interchange) or bond breaking (dissociative interchange) occurs in the transition state of the substitution reaction, but the bond making and bond breaking is not exclusive. In these cases, an intermediate with full bonding of the incoming ligand and departing ligand would not be present in the associative mechanism, and an intermediate with full bond cleavage would not be present in the dissociative mechanism. Instead, a reaction in which an intermediate or transition state is formed predominantly by establishing a bond between the metal and an incoming ligand, but with some cleavage of the bond of the departing ligand, is called associative interchange (I_A). A reaction in which an intermediate or transition state that is formed predominantly by cleavage of the bond between the metal and the departing ligand, but with some formation of a bond between the incoming ligand and the metal, is called a dissociative interchange (I_D).

These interchange mechanisms are particularly common in coordination chemistry when charged complexes react with incoming ligands of opposite charge. In these cases, an "encounter complex" is formed by association of the two charged species (Equation 5.3). In the usual case, a cationic metal complex and an anionic ligand associate to form a caged ion pair. Within this encounter complex, the associating and dissociating ligands would change positions, and the products would then dissociate. Because organometallic reactions often involve neutral complexes, nonpolar organic solvents, and neutral incoming ligands like olefins and CO, interchange mechanisms through encounter complexes are less common in organometallic chemistry than in coordination chemistry. Nevertheless, be aware that organometallic systems with the appropriate properties can react through encounter complexes, and dissociative interchange mechanisms do occur in many cases with organometallic species.

$$ML_n + L' \longrightarrow L_nM \bullet\bullet\bullet\bullet L' \longrightarrow L'L_{n-1}M \bullet\bullet\bullet\bullet L \longrightarrow ML_{n-1}L' + L$$

"Encounter complex"
I_A if this is the transition state of the rate-determining step

I_D if this is the transition state of the rate-determining step

(5.3)

5.1.3. The Basic Factors that Control Ligand Substitution Mechanisms

The factors that control the mechanism for ligand substitution and the characteristics of the associative and dissociative pathways are summarized in Table 5.1. The degree of saturation or unsaturation is often the principal factor that controls whether an associative or dissociative ligand substitution reaction will occur. In general, coordinatively saturated, 18-electron complexes tend to react by dissociative pathways, whereas coordinatively unsaturated, 16-electron complexes tend to react by associative, two-electron pathways. These trends follow from the 18-electron rule. Association of a ligand to an 18-electron complex would generate a 20-electron species, which is typically too unstable to be an intermediate in facile ligand substitution reactions. However, association of a ligand to a 16-electron complex would generate a relatively stable 18-electron intermediate. Although

Table 5.1. Trends for associative and dissociative substitution reactions.

	Associative mechanisms	Dissociative mechanisms
Type of complex	Occurs with 16-e⁻ and 17-e⁻ complexes	Occurs with 18-e⁻ complexes
Rate law	First order in entering ligand	Typically zero order in entering ligand
Activation parameters	Large negative $\Delta S^{\ddagger}$ Large negative $\Delta V^{\ddagger}$	Small positive $\Delta S^{\ddagger}$ Small positive $\Delta V^{\ddagger}$
Electronic effects	Ligand: Favored for more basic entering ligands Metal: Favored for more electrophilic metal centers	Can be favored for more electron-rich or more electron-poor metal centers
Effect of departing ligand	Affected only slightly by the departing ligand	Affected strongly by the strength of the bond to the departing ligand
Steric effects	Favored for sterically accessible metal centers	Favored for sterically hindered metal centers
Additional factors		Cationic complexes can react through "encounter" complexes Reduction weakens M–L and accelerates dissociation of a ligand

Note: In Type of complex, the superscripts should read 16-e⁻ and 17-e⁻ (i.e., $16\text{-}e^-$ and $17\text{-}e^-$), and $18\text{-}e^-$.

one might expect that a 17-electron complex would react by a dissociative pathway to avoid formation of a 19-electron intermediate, experimental data described below strongly imply that 17-electron intermediates undergo ligand substitution by associative mechanisms.

Ligand substitutions by associative and dissociative pathways are typically distinguished by kinetic methods. Although many different kinetic behaviors have been observed, substitution by an associative pathway generally occurs with a rate law that contains at least one term that is second order, whereas substitution by a dissociative pathway most often occurs with a rate law that is first order overall. Substitution by an associative pathway occurs with a large and negative $\Delta S^{\ddagger}$ because two molecules are assembled in the transition state, while substitution by a dissociative pathway occurs with a positive and smaller $\Delta S^{\ddagger}$ because the transition state involves bond lengthening but its molecularity is the same as that of the reactant. Likewise, substitution by an associative pathway occurs with a negative volume of activation ($\Delta V^{\ddagger}$).[28]

5.1.4. Scope of the Chapter

The following sections provide detailed information on the pathways for ligand substitution reactions. These sections describe how thermodynamic properties of the compounds have an effect on the rates of reaction through these pathways, how the properties of ancillary ligands trans and cis to the departing ligand affect the rates of ligand substitutions, and how redox events, changes in hapticity of ancillary ligands, photochemistry, and oxidation allow substitutions to occur on 18-electron complexes. These principles arise many times in subsequent chapters of the text.

5.2. Thermochemical Considerations

Because ligand substitution often involves the simple cleavage of one metal–ligand bond and the formation of a new metal–ligand bond, bond dissociation energies (BDE) dictate whether a ligand substitution process is thermodynamically favorable (see Equation 5.4). In general, the stability of the departing ligand controls the rate of a ligand substitution, just as the stability of a leaving group controls the rate of organic nucleophilic substitution reactions. Ligand substitution reactions do not usually displace a highly basic charged ligand. For example, replacement of a hydride or alkyl group through ligand substitution rarely occurs because hydride and alkyl anions are unstable.[29] Instead, ligand substitution reactions of organometallic compounds typically occur to replace a dative ligand or a ligand that is particularly stable as an anion, such as a halide.

$$L_nM-X + Y \longrightarrow L_nM-Y + X$$

BDE of M−Y must be
greater than the BDE
of M−X

(5.4)

Information on the trends that control bond dissociation energies was contained in Chapter 1–3. Benchmark values for dative ligands that are relevant to ligand substitution reactions are provided in Tables 5.2 and 5.3. Recall that the "bond strength" or BDE for bonds to formally charged ligands is defined as the formation of two radicals, but the BDE for bonds to neutral dative ligands is defined as cleavage to form the unsaturated metal fragment and the free Lewis base. A variety of relative M–L BDEs of dative ligands have been measured.[30–42] Some reliable gas-phase M–L BDE values, which have been selected for the purposes of this chapter, are presented in Table 5.2. Some relative BDE values for equilibria in solution phase that reveal the effects of ligand steric and electronic properties are also listed in Table 5.2, along with Table 5.3. From these data, the following five points can be deduced:

Table 5.2. Selected M–L bond dissociation energy estimates (gas-phase values unless indicated; otherwise, D = BDE in kcal/mol).

Compound process	References
$Ni(CO)_4 \longrightarrow Ni(CO)_3 + CO$ $D = 25 \pm 2$	a, b
$Fe(CO)_5 \longrightarrow Fe(CO)_4 + CO$ $D = 41 \pm 2$ (singlet) $D = 41$ (triplet)	b, c
$M(CO)_6 \longrightarrow M(CO)_5 + CO$ $D(Cr) = 37 \pm 2$ $D(Mo) = 40 \pm 2$ $D(W) = 46 \pm 2$	b
$IrCl(PPr^i_3)_2CO \xrightarrow{D > 72} IrCl(PPr^i_3)_2 + CO$	d
$Mn_2(CO)_{10} \xrightarrow{D = 36 \pm 2} Mn_2(CO)_9 + CO$	e, f
$Mn_2(CO)_{10} \xrightarrow[D = 38 \pm 5]{hexane} 2\,{}^\bullet Mn(CO)_5$	g, h
$CpNi^\oplus(ethylene) \xrightarrow{D = 38 \pm 5} CpNi^\oplus + ethylene$	i
$Cp_2Co(III)^\oplus \xrightarrow[D_1 = 118 \pm 10]{-Cp\bullet} CpCo(II)^\oplus \xrightarrow[D_2 = 85 \pm 10]{-Cp\bullet} Co(I)^\oplus$	j
$[W(PCy_3)_2(CO)_3L] \longrightarrow [W(PCy_3)_2(CO)_3] + L$ L = H₂, 9.9; N₂, 13.5; NCCH₃, 15.1; py, 18.9; P(OMe)₃, 26.5; CO, 30.4	k

(a) Stevens, A. E.; Feigerle, G. S.; Lineberger, W. C. *J. Am. Chem. Soc.* **1982**, *104*, 5026; (b) Lewis, K. E.; Golden, D. M.; Smith, G. P. *J. Am. Chem. Soc.* **1984**, *106*, 3905; (c) Engelking, P. C.; Lineberger, W. C. *J. Am. Chem. Soc.* **1979**, *101*, 5569; (d) Rosini, G. P.; Liu, F. C.; Krogh-Jespersen, K.; Goldman, A. S.; Li, C. B.; Nolan, S. P. *J. Am. Chem. Soc.* **1998**, *120*, 9256; (e) Smith, G. P. Unpublished results cited in reference f; (f) Coville, N. J.; Stolzenberg, A. M.; Muetterties, E. L. *J. Am. Chem. Soc.* **1983**, *105*, 2499; (g) This is the value from recent photoacoustic calorimetry studies: Goodman, J. L.; Peters, K. S.; Vaida, V. *Organometallics* **1986**, *5*, 816; (h) Values ranging from 19 to 41 kcal/mol have been reported for $D(Mn–Mn)$; see Table I in Simoes, J. A. M.; Schultz, J. L.; Beauchamp, J. L. *Organometallics* **1985**, *4*, 1238; (i) Tolbert, M. A.; Beauchamp, J. L. *J. Am. Chem. Soc.* **1984**, *106*, 8117; (j) Jacobsen, D. B.; Freiser, B. S. *J. Am. Chem. Soc.* **1985**, *107*, 7400; (k) Gonzalez, A. A.; Zhang, K.; Nolan, S. P.; Delavega, R. L.; Mukerjee, S. L.; Hoff, C. D.; Kubas, G. J. *Organometallics* **1988**, *7*, 2429.

1) In most cases, the bonds between the metal and two-electron donors such as CO are less than half as strong as typical bonds in organic chemistry. For example, the M–CO BDE values in Table 5.2, which are typical for metal-carbonyl compounds, range from 25 to 46 kcal/mol, whereas the CH_3–CH_3 and CH_3–H BDE values are 88 kcal/mol and 104 kcal/mol, respectively. Such M–CO ligands undergo dissociative ligand substitution at temperatures between 50 °C and 150 °C. Homolysis of CH_3–CH_3 to form $2CH_3{}^\bullet$ requires much higher temperatures than these. At the same time, these bond energies vary with the metal and ancillary ligand set, and the M–CO bond in $[Ir(PPr^i_3)_2(CO)(Cl)]$ has been shown to exceed 72 kcal/mol.[32,36]

Table 5.3. Relative enthalpies for coordination of various phosphorus ligands to ruthenium and osmium dimers.

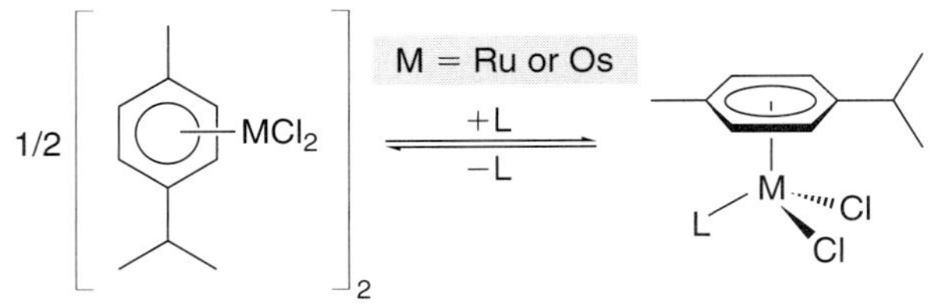

Reaction enthalpy (kcal/mol)

L	Os	Ru
PPh_3	43.1	36.3
$P(OMe)_3$	49.8	39.0
PMe_3	61.6	55.3
PEt_3	54.9	51.3
PCy_3	47.9	34.4

2) Polyhapto ligands—those bound η^3 or higher such as η^5-cyclopentadienyl (Cp_2Co^+, Table 4.2) or η^6 arenes (Equation 5.5)—bind with significantly higher BDE values than do ligands bound through a single atom. It is unlikely, therefore, that low-temperature reactions involving polyhapto ligands will involve their complete dissociation in a single step. However, multiple equivalents of a monodentate ligands can displace polyhapto ligands, such as arene and even cyclopentadienyl ligands.[43] For example, the ΔH for the replacement of toluene in (toluene)$Mo(CO)_3$ by the weak donor THF (Equation 5.5) is favorable by 9.3 kcal/mol,[44] and the displacement of toluene by pyridine is favorable by 30.9 kcal/mol (Equation. 5.5).[42]

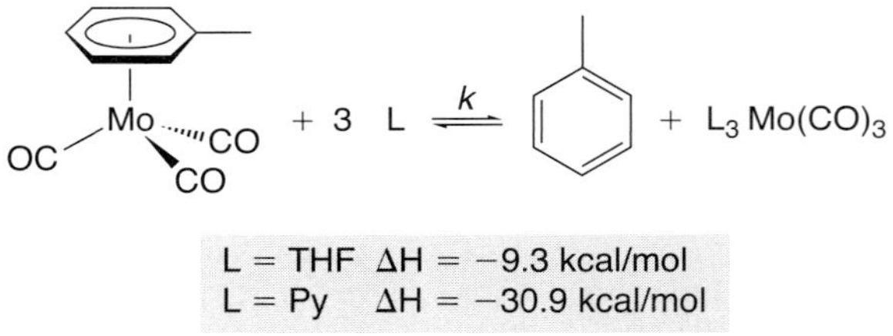

$$L = THF \quad \Delta H = -9.3 \text{ kcal/mol}$$
$$L = Py \quad \Delta H = -30.9 \text{ kcal/mol}$$

(5.5)

3) M–L bond strengths generally increase in the order first-row transition metals < second row < third row. For example, the mean M–CO bond energies of $Cr(CO)_6$, $Mo(CO)_6$, and $W(CO)_6$ are 37 ± 2, 40 ± 2, and 46 ± 2 kcal/mol, respectively. This trend is not always followed, however, and the magnitudes of single and average BDE values can follow different trends. Computational and experimental studies indicate that the first M–CO bond energies follow the trend $Cr(CO)_6 > Mo(CO)_6 < W(CO)_6$, and this order tracks with the relative rates of ligand substitution starting with these complexes. Thus, the bonds to second row metals are often, but not always, greater than those to analogous first-row metals; the bonds to third-row metals are almost always stronger than those to second-row metals.

4) An approximate working order of selected M–L BDEs can be constructed from the data in Table 5.2 and other literature.[45–49] This series of energies applies to low-valent metals able to participate in backbonding into π-accepting ligands:[50] $Cp > Me_xC_6H_{6-x} > C_6H_6 > CO \approx PMe_3 \approx$ ethylene $> PPh_3 > py > CH_3CN > N_2 > H_2 >$ THF, acetone, and ethanol.

5) Steric properties of the ligand affect the M–L bond strength among a series of ligands. The data in Table 5.3 show that more hindered ligands bind more weakly than less hindered ligands and that these steric effects can override the higher bond energy that would result from one ligand being a stronger Lewis base than another.

5.3. Mechanisms of Ligand Substitutions

The remainder of this chapter presents the established mechanisms of ligand substitution reactions of complexes with 16, 17, and 18 valence electrons. These are the first mechanistic data presented in this text. Keep in mind that a mechanism is a theory deduced from the available experimental data.[51,52] Logically, a mechanism can only be proven to be invalid and is, therefore, impossible to "prove" to be true; the accepted mechanism explains the available data and allows additional predictions that can be tested experimentally. Modifications and refinements of the mechanisms presented in this chapter and in the rest of the text are likely to be made in the future.

5.3.1. Mechanisms of Ligand Substitutions of 16-Electron and 17-Electron Complexes

Ligand substitutions at d^8, square-planar, 16-electron Pt(II) complexes constitute the most extensively investigated class of ligand substitution reaction.[11–14,16–24,26,27] Aspects of the mechanisms of these processes are described in this section. These reactions encompass the quintessential examples of associative substitution reactions involving coordinatively unsaturated complexes. In rare cases, dissociative processes involving 16-electron complexes have been deduced. Although dissociative substitutions of 16-electron complexes are clearly the exception, a few examples of these processes are described in this section. Most 17-electron complexes also undergo ligand substitution by associative pathways, and the data in support of this mechanism are also described here.

5.3.1.1. Associative Substitutions of Square-Planar d^8 Complexes

Associative substitution pathways are typically followed by square-planar, 16-electron compounds. Thus, closely related mechanisms for ligand substitution have been deduced for reactions of square-planar d^8 complexes of Pd(II), Ni(II), Au(III), Rh(I), and Ir(I) metal centers. This associative ligand substitution can occur without the assistance of solvent or with the initial replacement of the departing ligand by solvent and subsequent replacement of solvent by the incoming ligand. The participation of solvent depends on its coordinating ability. Often, the pathway involving solvent assistance occurs in parallel with the pathway lacking solvent assistance.

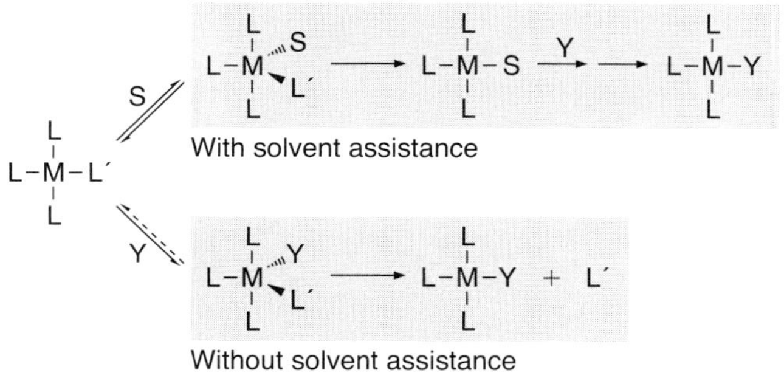

Scheme 5.2

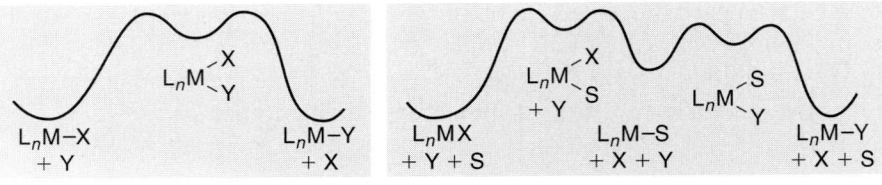

Figure 5.1.
Reaction coordinate diagram for associative ligand substitution reactions with and without solvent assistance.

The mechanism for associative ligand substitution of a square-planar complex with and without the assistance of solvent is shown in Scheme 5.2. Reaction coordinate diagrams for the pathways with and without solvent assistance are provided in Figure 5.1. The mechanism without solvent assistance involves initial binding of the incoming ligand to the platinum, followed by dissociation of the departing ligand. The solvent-assisted pathway is similar and simply occurs by two successive associative substitutions, the first by solvent and the second by incoming ligand. The stereochemistry of these substitutions and the factors that control the rate of the substitution have been studied extensively.

5.3.1.1.1. Stereochemistry of Associative Substitution and Cis–Trans Isomerization

Associative substitution reactions generally, although not always,[16] occur with retention of stereochemistry at the metal. The principle of microscopic reversibility, which states that the mechanism of a forward and reverse reaction must occur by the same mechanism,[51] leads to retention of configuration. As shown in Scheme 5.3, the incoming and departing ligands are present in the equatorial plane of the trigonal bipyramidal intermediate. If the formation of the bond between the metal and the incoming ligand occurred in the equatorial plane, then the cleavage of the bond must occur in the equatorial plane, and this requirement causes the reaction to occur with retention of stereochemistry.

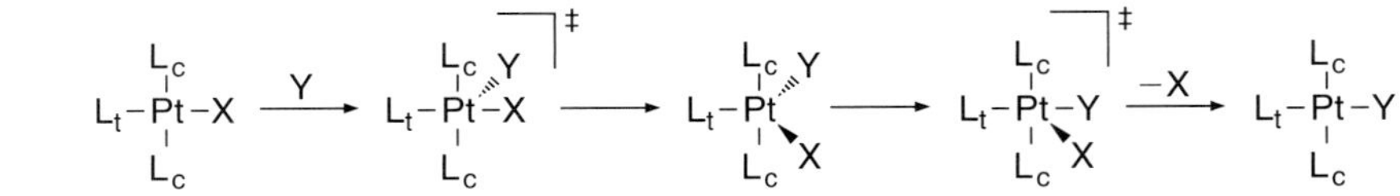

Scheme 5.3

If the intermediate is long lived, however, Berry pseudorotation can occur, and the geometry can be scrambled before the departing ligand is released from the equatorial plane. Five-coordinate complexes with d^8 electron counts are known to be highly fluxional by pseudorotation (Equation 5.6), as well as other processes.[53,54]

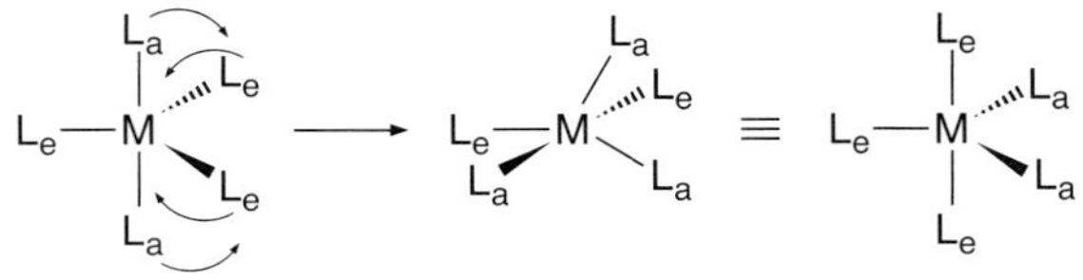

$$(5.6)$$

Thus, the five-coordinate intermediates formed by association of a ligand to a square-planar complex can also serve as an intermediate in cis–trans isomerization reactions[18–20] of d^8 square-planar complexes.[55] Because dissociation of a ligand from square-planar complexes is typically less favorable than association of ligand, isomerization of square-planar metal centers more often occurs by this association of added ligand, by coordination of a small amount of free ligand impurity, or by coordination of solvent[16,55,56] than by dissociation of ligand to generate a nonrigid three-coordinate intermediate. A series of Berry pseudorotations that can isomerize a square-planar complex are shown in Equation 5.7.

$$(5.7)$$

5.3.1.1.2. The Rate Law for Associative Substitutions

The rate behavior of associative ligand substitutions of square-planar complexes reflects the two pathways that often occur in parallel with and without solvent assistance. Thus, the corresponding rate law provided in Equation 5.8 contains two terms, one of which typically corresponds to the rate constant for associative substitution with solvent assistance and one of which typically corresponds to the rate constant for associative substitution without solvent assistance. Often a plot of k_{obs} versus [Y] appears like the plot below in Equation 5.8; in this case, the y-intercept corresponds to the rate constant for the solvent-assisted substitution path.

$$(5.8)$$

Rate $= k_{obs}[L_3ML']$ $k_{obs} = k_1 + k_2[Y]$

For example, the rate law for the replacement of chloride in [Pt(dien)Cl] by thiourea, thiocyanate, iodide, azide, pyridine, nitrite, or Cl⁻ in aqueous solution (Equation 5.9a) is $-d[Pt(dien)Cl]/dt = k_{obs}[Pt(dien)Cl]$, in which $k_{obs} = k_1 + k_2[Y]$. The reaction of ¹⁴C-labeled $NHEt_2$ to displace the unlabeled amine in $PtLCl_2(NHEt_2)$ in Equation 5.9b follows the same type of two-term rate law, $-d[PtLCl_2(NHEt_2)]/dt = k_{obs}[PtLCl_2(NHEt_2)]$, in which $k_{obs} = k_1 + k_2[*NHEt_2]$.[57-59] The first terms of these equations lack a dependence on the concentration of incoming amine because they correspond to the solvent-assisted pathway in which the replacement of the ligand in the starting complex by solvent is rate limiting. The second terms of these equations do contain the concentration of incoming ligand because they correspond to the second pathway in which association of the amine ligand occurs prior to the rate-limiting step. A more organometallic example of this ligand substitution mechanism is shown in Equation 5.10. In this case, exchange of thiocyanate and chloride ligands is fast on the NMR time scale in DMSO solvent.[60]

$$(5.9a)$$

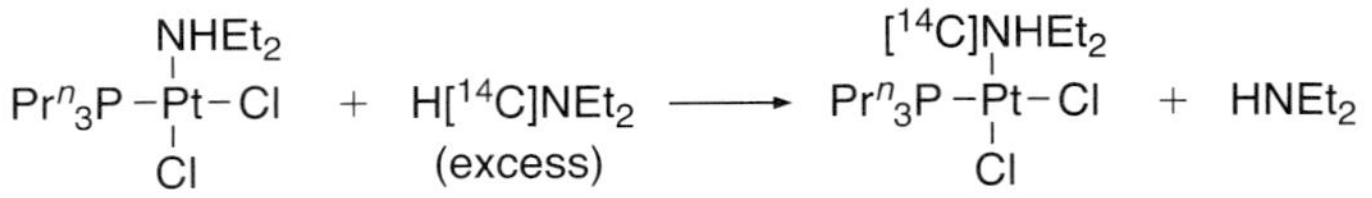

$$Pr^n_3P-\underset{\underset{Cl}{|}}{\overset{\overset{NHEt_2}{|}}{Pt}}-Cl \quad + \quad \underset{(excess)}{H[^{14}C]NEt_2} \quad \longrightarrow \quad Pr^n_3P-\underset{\underset{Cl}{|}}{\overset{\overset{[^{14}C]NHEt_2}{|}}{Pt}}-Cl \quad + \quad HNEt_2 \qquad (5.9b)$$

$$P\overbrace{\qquad}P \; = \; Ph_2P\overbrace{\qquad}PPh_2$$

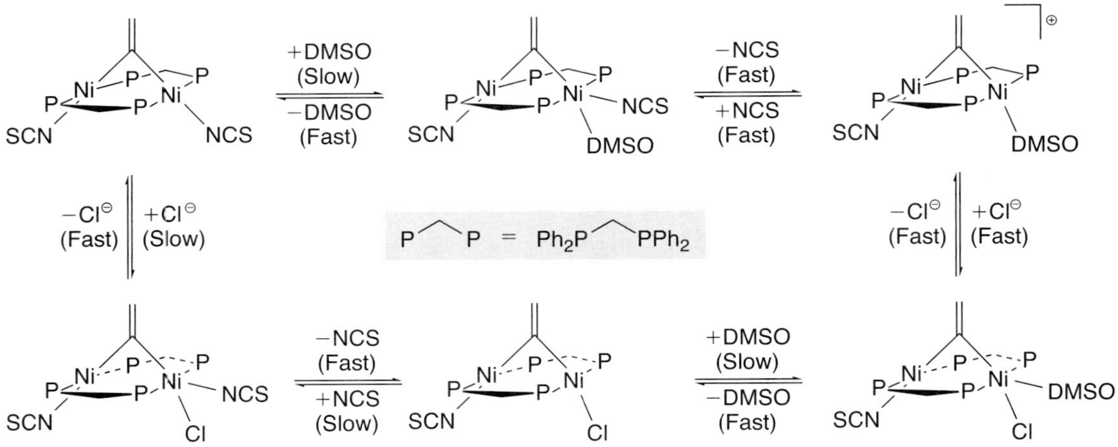

(5.10)

5.3.1.1.3. Dependence of the Rates on the Incoming Ligand, the Departing Ligand, and the Metal Center

The incoming nucleophile, the leaving group, the central metal, and the ancillary ligands trans and cis to the leaving group all affect the rate of these associative substitution reactions. The dependence of the observed rate constants on the identity of the incoming nucleophile can span a range of about 10^5.[26] For example, the platinum center in many of the classic ligand substitutions is relatively soft (e.g., that in Equations 5.10 and 5.11), and the reactions of the most basic, soft nucleophiles are, therefore, fastest. The dependence of the rate constant on the incoming ligand for the reaction of $[Pt(pyridine)_2Cl_2]$ in Equation 5.11 to replace the chloride ligand follows the trend $PPh_3 >> SCN^- >> I^-, Br^- > N_3^- > Cl^- \approx NO_2^- \approx NH_3$.

$$trans\text{-}PtPy_2Cl_2 \; + \; Y^n \; \longrightarrow \; trans\text{-}PtPy_2ClY^{(n+1)\oplus} \; + \; Cl^{\ominus}$$

Y	$k(M^{-1} s^{-1}) \times 10^3$
Cl^-	0.45
NH_3	0.47
NO_2^-	0.68
N_3^-	1.55
Br^-	3.7
I^-	10.7
SCN^-	180
PPh_3	249,000

(5.11)

The dependence of the observed rate constant on the identity of the departing ligand tends to be less than the dependence on the property of the incoming nucleophile, but is large enough to be easily detected. This dependence correlates with the strength of the metal–ligand bond. For the reaction of $Pt(dien)X^+$ with pyridine in Table 5.4, this dependence follows the trend $X = NO_3^- > H_2O > Cl^- > Br^- > I^- > N_3^- > SCN^- > NO_2^- > CN^-$.[61,62] Since these substitutions are associative, this trend reflects the extent to which the M–X bond is weakened in the transition state of the substitution process and the ability of these ligands to increase the electrophilicity of the metal center.

Table 5.4. Effect of the leaving group on the rate of ligand substitution of a platinum coordination complex.

X	Rate constant $k_{ds} \times 10^6 \ s^{-1}$
NO_3^-	Fast
H_2O	1900
Cl^-	35
Br^-	23
I^-	10
N_3^-	0.83
SCN^-	0.3
NO_2^-	0.05
CN^-	0.017

The dependence of the observed rate constants on metal ion can be large and has been shown in some cases to span a range of 10^6 (Tables 5.5 and 5.6). The rates of exchange of free cyanide with the coordinated cyanide of $M(CN)_4^{2-}$ in Table 5.5 and the rate of replacement of chloride by pyridine in Table 5.6 has been studied for Ni, Pd, and Pt complexes.[63,64] These data clearly show that the rates follow the trend Ni > Pd > Pt. The rate of exchange of cyanide is more than 10^3 times faster with $Ni(CN)_4^{2-}$ than with $Pd(CN)_4^{2-}$, and the exchange of free cyanide with $Pd(CN)_4^{2-}$ is about five times faster than the exchange with $Pt(CN)_4^{2-}$. Likewise, the replacement of pyridine for the chloride of $[M(PEt_3)_2(o\text{-tol})Cl]$ is about 10^2 faster for the nickel complex than for the palladium complex, and the reaction of the palladium complex is about 10^5 faster than the reaction of the platinum analog.

These relative rates fit the general trend that reactions of first-row metal complexes tend to be faster than reactions of their second-row analogs, which tend to be faster than reactions of their third-row congeners, but the precise origin of this effect in associative ligand substitutions has been rationalized in more than one way. One text has attributed this trend for associative ligand substitution as reflecting the relative propensities of the metals to form five-coordinate, 18-electron complexes.[17,64–66] In a few cases, five-coordinate

Table 5.5. Effect of the metal on the rate of ligand substitution—the rate constant for exchange of free and coordinated cyanide.

Complex	$k(M^{-1}s^{-1})$
$Ni(CN)_4^{2-}$	$> 5 \times 10^5$
$Pd(CN)_4^{2-}$	120
$Pt(CN)_4^{2-}$	26

Table 5.6. Effect of the metal on the rate of ligand substitution of pyridine for chloride.

M	$k_1(min^{-1})$
Ni	2000
Pd	35
Pt	4×10^{-4}

intermediates have been detected and even structurally characterized.[67] A second text has attributed this trend to the strength of the M–X bond being cleaved, but this argument would apply only if significant M–X bond cleavage occurs in the transition state for attack by the incoming ligand or if the irreversible step of the mechanism were M–X bond cleavage after reversible association of the entering group.[68,2]

5.3.1.1.4. Trans and Cis Effects

The ancillary ligand in d^8 square-planar complexes located trans to the ligand that undergoes substitution has a large effect on the rate of ligand substitution. This phenomenon is called the "trans effect."[69] The trans effect can span several orders of magnitude in rate constants. The cis ligand can also affect the rates of ligand substitution reactions, and this effect is called the "cis effect." The effect of a cis ligand on the rate of ligand substitution in square-planar complexes are typically smaller than that of a trans ligand.

A separate phenomenon refers to a thermodynamic, rather than a kinetic, effect and has been called the "trans influence." This phenomenon is attributed solely to effects on the ground state. The trans influence has been defined most generally as "the extent to which a ligand weakens the bond trans to itself." For example, one ligand can influence the magnitude of the M–P coupling constant or the metal–ligand bond length involving the ligand trans to it. Detailed information on trans effects and trans influences are treated in standard texts on coordination chemistry.[21–24]

The examples in Tables 5.7 and 5.8 illustrate the magnitude of the trans effect and cis effect on the ligand substitution reaction of typical platinum(II) coordination and

Table 5.7. Example of the trans effect on the rate of ligand substitution on organometallic platinum complexes.[a]

$$\textit{trans-}Pt(PEt_3)_2(X)Cl + py \xrightleftharpoons{\text{EtOH}} \textit{trans-}Pt(PEt_3)_2(X)(py)^{\oplus} + Cl^{\ominus}$$

X	$k_{obs}(s^{-1})$[b]	t (°C)
H^-	4.7×10^{-2}	0
Me^-	6.0×10^{-4}	25
$C_6H_5^-$	1.2×10^{-4}	25
Cl^-	3.5×10^{-6}	25

[a]Data from p. 25 of reference 11; [py] = 0.006 M.
[b]For the approach to equilibrium.

Table 5.8. Example of the cis effect on the rate of ligand substitution on organometallic platinum complexes.[a]

$$\textit{cis-}Pt(PEt_3)_2(X)Cl + py \xrightleftharpoons{\text{MeOH}} \textit{cis-}Pt(PEt_3)_2(X)(py)^{\oplus} + Cl^{\ominus}$$

X	$k_{obs}(s^{-1})$[b]	t (°C)
Me^-	11.4	25
$C_6H_5^-$	7.92	25
Cl^-	4.17	25

[a]Data from p. 30 of reference 11; [py] = 0.006 M.
[b]For the approach to equilibrium.

organometallic complexes.[64] These data show that the cis effect is much smaller than the trans effect. The trans effect in these examples spans a factor of 10^4, while the cis effect spans a factor of only 3. These values show that the hydride has among the largest trans effects, but that alkyl and aryl groups also have a large accelerating effect on the rate of ligand substitution. More extensive lists of the relative trans influence of different ligands are available.[21-24,69] The general ordering of trans effects and trans influences are only approximate because they depend upon the identity of both mutually trans ligands.[69,70]

A common approximate trans influence series is $R_3Si^- > H^- \sim CH_3^- \approx CN^- \approx$ olefins, $CO > PR_3 \approx NO_2^- \approx I^- \approx SCN^- > Br^- > Cl^- > RNH_2 \approx NH_3 > OH^- > NO_3^- \approx H_2O$. The presence of hydride and alkyl ligands at the end of the series with a large effect shows that strong σ-donating ability leads to a large trans effect. However, the presence of CO and olefin ligands at the same position in the series shows that strong π-accepting ability also leads to a large effect. Some efforts have been made to separate the trans and cis influences into steric and electronic components.[71]

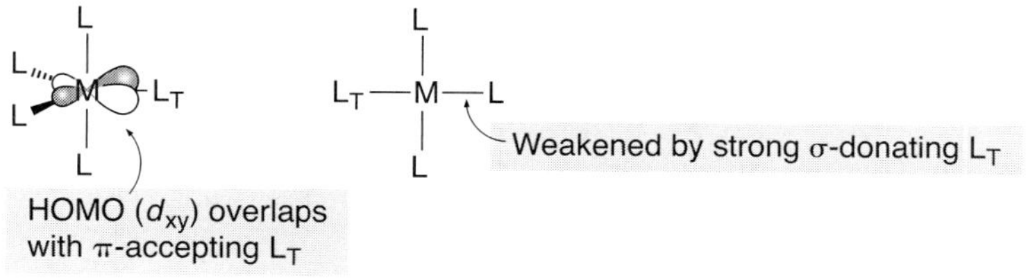

Figure 5.2. Two factors that control the magnitude of the trans effect.

A brief assessment of the factors that control the magnitude of the trans effect can be made based on the mechanism of associative ligand substitution. Association of the incoming ligand generates a five-coordinate intermediate with the original trans ligand in the equatorial plane. Thus, factors that stabilize the five-coordinate intermediate, relative to the starting square-planar complex, control the trans effect. The effect of strong π-acceptors is often attributed to their ability to stabilize the five-coordinate intermediate. As shown in Figure 5.2, a π-acceptor ligand in the equatorial plane would overlap with the HOMO of an 18-electron, d^8, trigonal bipyramidal complex and would, therefore, stabilize this intermediate more than it would stabilize the four-coordinate reactant. In contrast, the effect of strong σ-donors on the trans effect is typically attributed to changes in the bonding of the starting four-coordinate complex. A strong σ-donor should destabilize the bond trans to it in a square-planar complex more than it destabilizes the bonds of the ancillary ligands in a trigonal bipyramidal species because there is direct overlap of the same orbital with two trans ligands in the starting square-planar complex. Thus, both a strong σ-donor and a strong π-acceptor reduce the difference in energy between the starting complex and the reactive intermediate, and both types of ligands reduce the barrier for associative ligand substitution and give rise to a large trans effect.

5.3.1.2. Associative versus Dissociative Substitutions of Square-Planar Complexes

Because most, but not all, ligand substitutions of square-planar complexes occur by an associative mechanism, it is worth considering what conditions lead to a dissociative substitution mechanism for this class of complex. Two combinations of conditions have been shown to lead to dissociative ligand substitutions from square-planar complexes. First, dissociative substitution occurs when a weakly bound ligand is located trans to a ligand with a strong trans influence and the unsaturated intermediate can be stabilized by a polydentate binding mode of the remaining ligands. For example, substitution of the DMSO ligand in cis-$PtPh_2(Me_2SO)_2$ occurs by a dissociative mechanism (Equation 5.12). In this complex, the relatively weakly Lewis basic DMSO is bound trans to a phenyl ligand, which has a high trans influence. The DMSO exchanges with DMSO-d_6 by a predominantly dissociative pathway.[72] The dissociation has also been suggested to occur because an intermediate in which the remaining DMSO ligand binds in an η^2 fashion can form, and this binding mode would stabilize the highly unsaturated intermediate that would form from dissociation of the second DMSO ligand.[20] However, several reports of dissociative exchange with closely related compounds have now been published, and the one in Equation 5.13 cannot be assisted by a change in hapticity of an ancillary ligand.[73,74] Thus, it seems that the combination of weakly bound ligand and strong trans influence ligand alone can induce a dissociative substitution mechanism.

$$\text{(5.12)}$$

Proposed

$$\text{(5.13)}$$

Dissociative substitution of square-planar complexes can also occur when attack at the axial positions of the square plane is disfavored by severe steric hindrance. Two examples of dissociative ligand substitutions of olefins occur with square-planar iridium complexes containing ancillary ligands with large groups positioned above and below the square plane. As part of studies on the oxidative addition of ammonia, the complex [Ir(PCP)(propene)] in Equation 5.14 has been shown to react with ethylene to form [Ir(PCP)(ethylene)] by initial dissociation of propene to generate the reactive intermediate [Ir(PCP)].[75] Likewise, exchange of free olefin with bound olefin in [Ir(POCOP)(olefin)] (Equation 5.15) occurs at elevated temperatures on the NMR time scale by a dissociative process.[76] The large *tert*-butyl groups above and below the square plane and the high trans influence of the aryl group trans to the olefin, presumably, make dissociative substitution faster than associative substitution in these systems.

$$\text{(5.14)}$$

Exchanges with free cyclooctene at 70–110 °C   via

$$\text{(5.15)}$$

In addition, dissociative mechanisms for substitution of d^8 complexes can occur when the geometry is distorted from square planar or if the coordination number is higher than four. For example, replacement of the monophosphine in the highly distorted [*cis*-Pt(SiMePh$_2$)$_2$(PMe$_2$Ph)$_2$] with a bisphosphine has been shown recently to occur by dissociative substitution.[77] The dissociation is favored in this case because of the strong trans influence of the silyl group and because the complex is distorted from a standard square-planar geometry. Likewise, the six-coordinate, formally 16-electron Mo(II) alkyne complex $Mo(CO)(RC\equiv CR')(S_2CNMe_2)_2$ adds a second $RC\equiv CR'$ to form $Mo(RC\equiv CR')_2(S_2CNMe_2)_2$ by a mechanism that occurs by initial dissociation of CO.[78] This dissociative mechanism is favorable because of the ability of the alkyne to act as a four-electron donor ligand (see Chapter 2) and to stabilize the unsaturated intermediate formed by dissociation of CO.

The typically faster rate for associative substitution does not preclude dissociation of ligand occurring as the first step of reactions other than ligand substitutions (Equation 5.10). For example, dissociation of ligands from square-planar, d^8 complexes occurs as the first step of reductive eliminations (Chapter 8). Thus, dissociation of ligands from this class of compound can occur, but association to form a five-coordinate intermediate is typically faster.

To illustrate the propensity of these complexes to react by associative processes, consider the exchange of alkanes and arenes in complexes of the type $[PtL_2(R)(R–H)]^+$ by associative or dissociative mechanisms. Although it is difficult to conceive of a more weakly bound ligand than the alkane in a σ-complex, the rate of associative substitution of acetonitrile for the bound alkanes is faster than simple dissociation of the alkane (Equation 5.16).[79,80]

(5.16)

5.3.1.3. Associative Substitutions of 17-Electron Complexes

Most 17-electron transition metal complexes undergo ligand substitution faster than their 18-electron analogs.[81–83] For example, the rate of ligand substitution of the 17-electron complex $\bullet V(CO)_6$ is about 10^{10} times faster than the rate of ligand substitution of the 18-electron complex $Cr(CO)_6$.[84] In a similar vein, $\bullet V(CO)_6$ undergoes substitution of PPh_3 for CO at −70 °C in 90 minutes,[85] whereas the 18-electron analogue $V(CO)_6^-$ is inert to molten PPh_3.[86] Ratios of rate constants for ligand substitutions by 17-electron and 18-electron complexes of 10^3 to 10^7 are common.[87–90]

Considering the 18-electron rule, it was initially thought[81,91–93] that 17-electron species, such as $\bullet V(CO)_6$ or the transient $\bullet Mn(CO)_5$, would undergo ligand substitution by the same dissociative pathways of coordinatively saturated complexes. In this case, the complexes would dissociate CO to form the 15-electron intermediates $\bullet V(CO)_5$ and $\bullet Mn(CO)_4$. In contrast to this initial expectation, these complexes undergo associative ligand substitutions.

A molecular orbital scheme for the interaction of $\bullet Mn(CO)_5$ with L[94] (Figure 5.3) shows, to a first approximation, that half of an M–L bond is gained by formation of the 19-electron species. This partial bond is formed because two electrons (the odd electron of the starting complex and one electron of the entering ligand) fill a bonding molecular orbital and only one electron fills an antibonding orbital.[95] Several stable complexes with formal 19-electron configurations have even been isolated.[94,96–98] The unpaired electron in these stable complexes typically lies in an orbital that is predominantly ligand based; hence, these complexes are best considered 18-electron species that are bound by a radical-anion ligand. A review by Tyler summarizes ligand substitutions and other reactions of complexes possessing 17- and 19-electron configurations.[94]

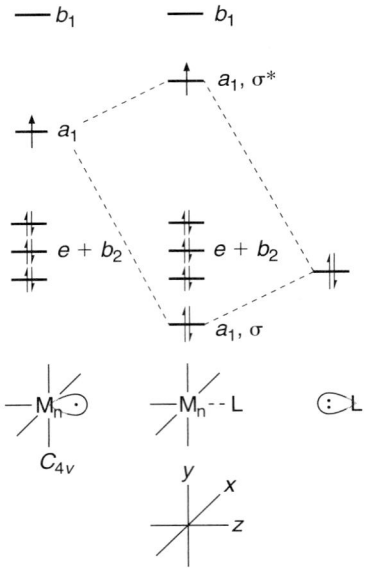

Figure 5.3.
Molecular orbital scheme for the interaction of the •Mn(CO)$_5$ fragment with a Lewis base.

Seventeen-electron intermediates have been generated by photolysis and oxidation, among other routes.[99] For example, photolysis of M$_2$(CO)$_{10}$ (M = Mn and Re) cleaves the M–M bond to generate two equivalents of •M(CO)$_5$.[99–101] One-electron oxidation of the 18-electron complex (MeCp)Mn(CO)$_2$L forms [(MeCp)Mn(CO)$_2$L]$^+$,[87] and one-electron oxidation of (arene)Cr(CO)$_3$ forms [(arene)M(CO)$_3$]$^+$.[83] Each of these complexes undergoes fast substitution after generation of the 17-electron intermediate, as shown for (MeCp)Mn(CO)$_2$L in Equation 5.17.

$$(5.17)$$

Poë provided solid evidence that 17-electron complexes undergo ligand substitution by associative mechanisms. He studied the reactions of •Re(CO)$_5$,[99–101] which was generated by photolysis of Re$_2$(CO)$_{10}$ in the presence of both CCl$_4$ and PPh$_3$. The reaction of •Re(CO)$_5$ with CCl$_4$ forms Re(CO)$_5$Cl, and the reaction with PPh$_3$ forms •Re(CO)$_4$PPh$_3$, which subsequently reacts with CCl_4 to form $Re(CO)_4(PPh_3)Cl$ (Equation 5.18). Added CO had no effect on the ratio of $Re(CO)_4(PPh_3)Cl$ to $Re(CO)_5Cl$, and these data showed that dissociation of CO does not contribute to the rate of the reaction. Brown and co-workers[101] showed that the controversial case[91–93,100] of ligand substitution of PR_3 and even the poorer nucleophile AsR_3 for CO in the $Mn(CO)_5\bullet$ intermediate occurs by an associative mechanism.

$$(CO)_5Mn-Mn(CO)_5 \xrightarrow{h\nu} (CO)_5Mn\bullet \xrightarrow{CCl_4} (CO)_5MnCl$$

$$\downarrow PPh_3$$

$$(PPh_3)(CO)_5Mn\bullet \xrightarrow{-CO} (PPh_3)(CO)_4Mn\bullet \xrightarrow{CCl_4} (PPh_3)(CO)_4MnCl$$

$$(CO)_5MnCl/(PPh_3)(CO)_4MnCl \text{ was independent of [CO]}$$

(5.18)

A thorough study of ligand substitutions of an isolated 17-electron complex, $\bullet V(CO)_6$, by Trogler, Basolo, and co-workers[84] (Equation 5.19) further supported ligand substitution by an associative mechanism. A second-order rate law and a negative $\Delta S^\ddagger$ of -28 ± 2 eu were observed for the reaction with PPh_3. Furthermore, a structure-reactivity study involving seven different phosphines showed that the basicity and nucleophilicity of the phosphine affected the rate. All of these data supported an associative mechanism for ligand substitution of the 17-electron compound. Studies of ligand substitution of the long-lived $\bullet Mn(CO)_3L_2$ [L = $P(n-Bu)_3$ or $P(i-Bu)_3$][102] and of $MeCpMn(CO)_2L^+$ generated electrochemically[87,89,90,103] also support an associative mechanism for the ligand substitution of these 17-electron complexes.

$$\bullet V(CO)_6 + PR_3 \xrightarrow[\text{Hexane}]{25\ °C} \bullet V(CO)_5(PR_3) + CO$$

Rate is proportional to $[V(CO)_5][PR_3]$
Rate is dependent on the R of PR_3
$\Delta S^\ddagger = -28$ eu

(5.19)

5.4. Substitution Reactions of 18-Electron Complexes

5.4.1. Dissociative Substitution Reactions

5.4.1.1. General Features of the Kinetics of Dissociative Ligand Substitution

Ligand substitutions at 18-electron, coordinatively saturated complexes typically occur by dissociative substitution mechanisms. Because this reaction mechanism begins with bond cleavage, these reactions are often slower than the associative substitutions of 16-electron complexes. For example, the 16-electron complex $Rh(acac)(ethylene)_2$ reacts with ethylene (1 atm) by an associative mechanism with a rate constant of 10^4 s^{-1} at 25 °C,[104] but the related 18-electron complex $CpRh(ethylene)_2$ reacts by dissociative substitution with a rate constant that is about 10^{14} times slower (about 4×10^{-10} s^{-1} at 25 °C)[105].

The mechanism and reaction coordinate diagram for dissociative substitution reactions are shown in Scheme 5.4 and Figure 5.4, respectively. This pathway occurs by initial cleavage of the metal–ligand bond to form an unsaturated 16-electron intermediate. This intermediate then undergoes competitive addition of the departing or incoming ligand. If the reaction thermodynamics favor the substitution process, then addition of the incoming ligand is most often faster than re-addition of the departing ligand. However, the differences in rate constants for reactions of five-coordinate, 16-electron intermediates for different Lewis bases can be small.[106,107] Thus, high concentrations of the incoming ligand can make the forward reaction of the intermediate much faster than reversion to the starting complex.

$$L_nM-X \underset{k_{-1}}{\overset{k_1}{\rightleftharpoons}} L_nM \ + \ X \ \underset{Y}{\overset{k_2}{\longrightarrow}} \ L_nM-Y$$

Scheme 5.4.

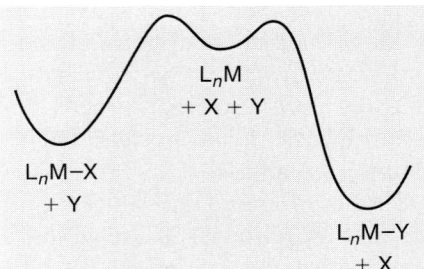

Figure 5.4.
Reaction coordinate diagram for a dissociative substitution reaction.

The rate equation for this dissociative mechanism is shown in Equation 5.20. Depending on the relative magnitudes of the two terms in the denominator, the reaction will be zero order, first order, or an intermediate order in the incoming ligand. If the $k_2[Y]$ term is much larger than the $k_{-1}[X]$ term, the sum of the two denominator terms is approximately equal to $k_2[Y]$, and the overall equation simplifies to $k_1[ML_nX]$. Under these conditions, a dissociative ligand substitution is independent of the concentration of the incoming ligand and first order overall.

$$\text{Rate} = -\frac{d[L_nMX]}{dt} = \frac{k_1 k_2[L_nMX][Y]}{k_{-1}[X] + k_2[Y]}$$

Usually $k_2[Y] >> k_{-1}[X]$

$$\text{and rate} = -\frac{d[L_nMX]}{dt} = k_1[L_nMX]$$

(5.20)

Because the rate-limiting step for a dissociative process is typically the cleavage of the metal–ligand bond in the starting complex, the metal–ligand bond energy correlates closely with the reaction rate. As shown in Figure 5.5, the enthalpy of activation is usually close to the enthalpy for dissociation of ligand. In all cases, for a purely dissociative substitution, the enthalpy of activation must be larger than the bond dissociation enthalpy. Typically, there is a small energy difference between these two values because ligands rapidly associate to an unsaturated metal center. Thus, a ligand substitution that occurs under mild conditions and, therefore, has an enthalpy of activation of 20–25 kcal/mol (the entropy of activation of a dissociative process is usually small, see Section 5.4.1.2) cannot occur by initial cleavage of a metal–ligand bond with a BDE exceeding 20–25 kcal/mol.[108]

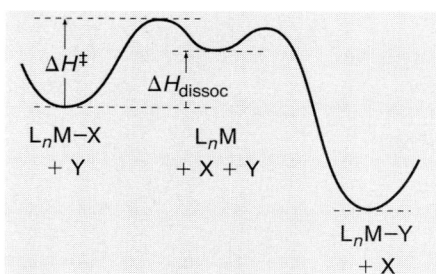

Figure 5.5.
Relationship between the enthalpy of activation and enthalpy of bond cleavage in a dissociative substitution reaction.

As a result, activation enthalpies of many dissociative substitution reactions of 18-electron complexes are relatively high and close to the M–L BDE. The weak M–CO bond energy of $Ni(CO)_4$ (25 ± 2 kcal/mol) allows this complex to react rapidly at room temperature, but the higher M–L bond energies of the olefin in $CpRh(ethylene)_2$ and CO in $Cr(CO)_6$ (31 kcal/mol and 37 ± 2 kcal/mol, respectively) cause the reactions of these complexes to require temperatures of > 100 °C[105] and 80–140 °C,[109] respectively, to occur at reasonable rates.

5.4.1.2. Reactions of $Ni(CO)_4$ as Quintessential Examples of Dissociative Substitutions

Ligand substitution on the d^{10}, 18-electron, tetrahedral complex $Ni(CO)_4$ is a classic example of a dissociative ligand substitution.[3,5,110] Kinetic studies show that the rate is first order in nickel and independent of [L]. These data are consistent with rate-determining dissociation of CO (Equation 5.21).

$$\begin{array}{cc} & (5.21) \end{array}$$

As expected, the activation enthalpy, $\Delta H^{\ddagger}$, is close to the $(CO)_3Ni{-}CO$ BDE of 25 ± 2 kcal/mol. Also as predicted for a dissociative mechanism, the activation entropy, $\Delta S^{\ddagger}$, is positive (8 ± 1 eu in toluene and 13 ± 1 eu in hexane). The positive $\Delta S^{\ddagger}$ lies in the range expected for a reaction in which the M–L bond is stretched to two or three times the length of the bond in the ground state[111] and experiences additional rotational freedom in the transition state.[111, 112] Because the starting complex and the transition state are both single particles, the contribution of translational entropy to $\Delta S^{\ddagger}$ is zero,[112] and the absolute value of the entropy term is smaller than that for an associative process.

5.4.1.3. Steric Effects on Dissociative Substitution

The equilibrium constant for dissociation of bulky ligands is usually larger than for dissociation of less-hindered ligands, and this trend is particularly well followed for ligands in a series with similar electronic properties. These more favorable thermodynamics usually give rise to faster rates of ligand substitution, and complexes containing sterically bulky ligands tend to undergo dissociative substitution faster than complexes containing less-hindered ligands. The stable form of palladium phosphine complexes and of nickel phosphite complexes are classic examples of the effect of steric bulk on the equilibrium constant and rate constant for dissociation.

For example, palladium–trimethylphosphine complexes are most stable as $[Pd(PMe_3)_4]$,[113] but the combination of palladium and tri-*tert*-butylphosphine is most stable as the bis-ligated complex $[Pd(PBu^t_3)_2]$.[114,115] The coordinative unsaturation of the latter complex is important to achieve fast reactions of this complex as described in Chapter 19 on catalytic cross-coupling. Likewise, the $Ni[P(OR)_3]_n$ complexes that catalyze hydrocyanation (Chapter 16) illustrate the importance of steric effects on ligand dissociation. The size of the phosphite ligand changes the equilibrium constant for dissociation by more than 10^{10} (Equation 5.22 and Table 5.9), and the complexes containing the more hindered phosphites are generally more reactive catalysts for this process than those containing the less hindered phosphites.[116]

$$L_{\cdots}Ni(L)(L)(L) \underset{PhH,\ 25\ ^\circ C}{\overset{K_D}{\rightleftharpoons}} L_2Ni-L + L \qquad (5.22)$$

Table 5.9. K_D in Equation 5.22 as a function of the size of L.

L	$P(OEt)_3$	$P(O\text{-}p\text{-tolyl})_3$	$P(O\text{-}i\text{-Pr})_3$	$P(O\text{-}o\text{-tolyl})_3$	PPh_3
Tolman cone angle	109°	128°	130°	141°	145°
K_D (M)	Too small to measure	6×10^{-10}	2.7×10^{-5}	4.0×10^{-2}	No NiL_4 detected

The electronic properties of the ligand also affect this equilibrium constant. For example, the K_D for dissociation of the phosphite $P(O\text{-}o\text{-tolyl})_3$ from the corresponding $Ni[P(OR)_3]_n$ complex is smaller than the K_D for dissociation of PPh_3, despite the similar cone angles of the two ligands. This difference in equilibrium constants was ascribed to the ability of the phosphite ligand to stabilize the low oxidation state of Ni(0) by acting as a π-acceptor.[116] $P(O\text{-}o\text{-tolyl})_3$ is the ligand that was initially used in the commercial hydrocyanation process.

5.4.1.4. Stereochemistry of Dissociative Substitution

Pseudo-four-coordinate, 18-electron "piano stool" complexes can possess a stereogenic metal center, and these chiral complexes have been resolved. These complexes undergo racemization by dissociative ligand substitution mechanisms[117] (Equation 5.23). Two steps control the barrier to racemization. First, reversible dissociation generates a pyramidal 16-electron intermediate. Second, this pyramidal intermediate undergoes an inversion or "umbrella flip" that is similar to the inversion at the nitrogen in an amine. Depending on the ligands at the metal, this inversion can be much slower than the inversion at nitrogen in an amine and contribute significantly to the barrier for racemization. Re-coordination of the ligand would then generate the two products. Work involving chiral metal complexes has been reviewed by Brunner.[117]

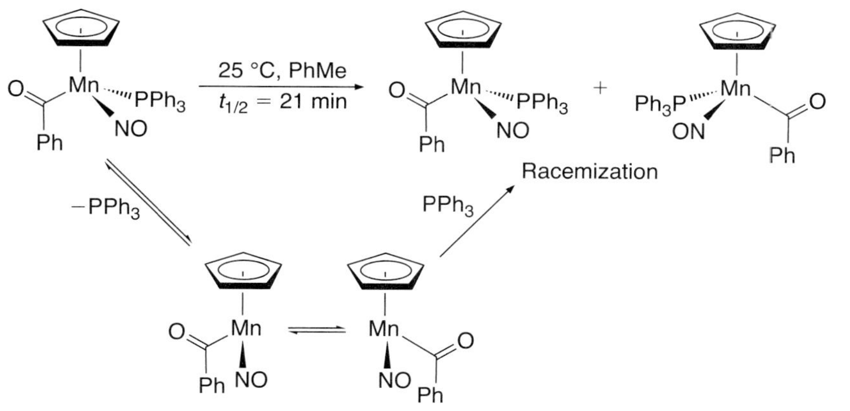

(5.23)

5.4.1.5. Substitution of Weakly Bound Ligands in 18-Electron Complexes

Dissociative substitution reactions, even in alkane solvents, occur through intermediates that typically contain coordinated solvent. For this reason and because of the intermediacy of alkane complexes in C–H activation, a number of studies have been conducted on the mechanism of the displacement of alkanes and even noble gases from the coordination sphere of a metal. A few of these studies are summarized in Scheme 5.5.

$$M(CO)_5 \bullet solvent \xrightarrow[k_1]{\Delta H^\ddagger} M(CO)_5 + solvent \xrightarrow{CO} M(CO)_6$$

M	Solvent[a]	$\Delta H^\ddagger$
Cr		5 kcal/mol
W		<4 kcal/mol
Cr	Heptane	10 kcal/mol

[a] k_1 is 10^2 times faster when the solvent = C_6F_{12}.

Scheme 5.5

Kinetic data indicate that the reaction of $Cr(CO)_5 \bullet$ heptane with an incoming ligand L to give $Cr(CO)_5L$ and heptane[118,119] occurs by a dissociative pathway. Dissociation of solvent, followed by capture of the unsaturated species by ligand generates the substitution product.[120] Flash photolysis of $Cr(CO)_6$ in cyclohexane produces $Cr(CO)_5 \bullet$(cyclohexane), and capture of this intermediate by CO occurs with a $\Delta H^\ddagger = 5 \pm 1$ kcal/mol.[121] A similar value (≤ 4 kcal/mol) has been reported for the $\Delta H^\ddagger$ for dissociation of cyclohexane from $W(CO)_5$ (cyclohexane).[120] A larger value (about 10 kcal/mol) has been reported for the $\Delta H^\ddagger$ for dissociation of heptane from $Cr(CO)_5$(heptane).[118,119] For comparison, the binding energy of H_2 to $Cr(CO)_5$ and $W(CO)_5$ is strong enough that the complexes can be observed directly by low-temperature NMR spectroscopy.[122]

Ligand substitutions of bound fluoroalkanes and noble gases have also been studied. In perfluorocyclohexane, the reaction of $Cr(CO)_5 \bullet (C_6F_{12})$ with CO yields $Cr(CO)_6$ with a rate constant that is 10^2 times larger than that for recombination in cyclohexane.[123,124] Similar fast kinetic studies have been conducted on 16-electron complexes bound by noble gases. The reactions of Cp*Rh(CO)(Kr) with alkanes forms Cp*Rh(CO)(alkane) prior to oxidative addition of the alkane. Kinetic studies on these processes have suggested that the replacement of krypton by alkane (Equation 5.24) occurs in this complex by a mechanism other than a purely dissociative process.[125] Substitutions of 18-electron complexes that deviate from purely dissociative mechanisms are described in detail in the next section.

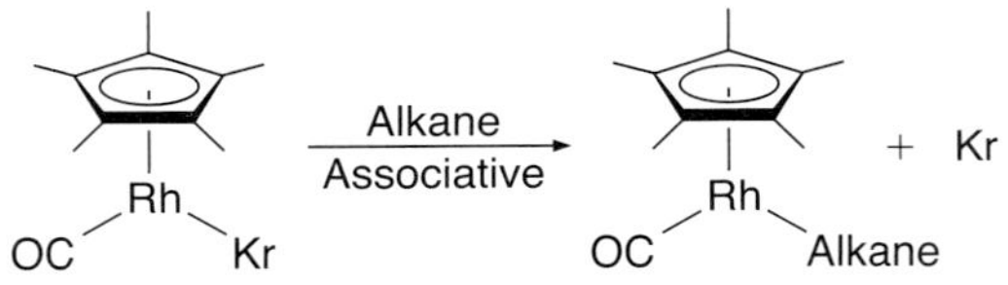

$$\text{(5.24)}$$

5.4.1.6. Electronic Effect of Ancillary Ligands on the Rates of Dissociative Substitution Reactions—The Cis Effect

Both the rates and the stereochemical outcome of ligand replacement reactions are influenced by the ancillary ligands.[6,26] In addition to the effects of the steric properties of ancillary ligands noted earlier in this chapter, the electronic properties of ancillary ligands can affect the rate and site of ligand dissociation. In many cases, a ligand will affect the rate of dissociation of the ligands cis to it in an octahedral or pseudo-octahedral complex (Scheme 5.6). This "cis effect" in dissociative substitutions of octahedral compounds is much larger than the cis effect on the associative substitutions of square-planar compounds.

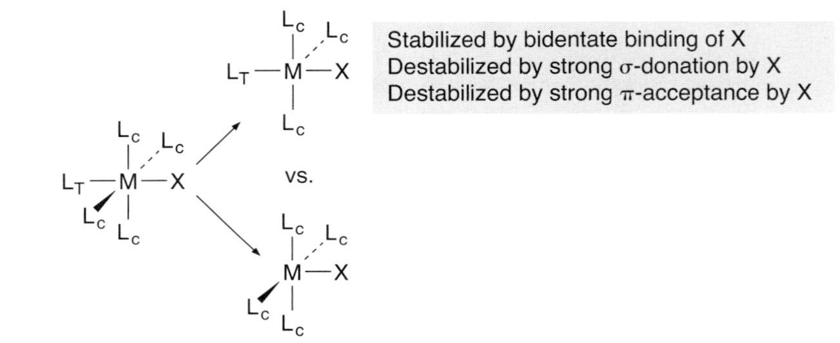

Stabilized by bidentate binding of X
Destabilized by strong σ-donation by X
Destabilized by strong π-acceptance by X

Scheme 5.6.
Cis effect and the origin of the cis effect on the ligand substitutions of octahedral compounds.

The order of the cis labilizing ability of selected ligands[26] is $NO_3^- > OAc^-, HCO_2^-,$[6] $RC(O)^- > NHR > SH^-$[126] $> OR > Cl^- > Br^- > I^- >$ carbene $> PPh_3 > H^-, CO$. The origin of this trend in cis effect is electronic, but many electronic properties can affect the energy of the reactive intermediate, relative to the energy of the starting complex. The more the cis ligand reduces this difference in energy, the more favorable the equilibrium will be for dissociation and the more likely the dissociation will be faster.

Chelation and π-donation can affect the stability of the intermediate. A ligand that can change from monodentate to bidentate when a coordination site is open will stabilize the unsaturated intermediate. The presence of NO_3^-, OAc^- and HCO_2^- on the high end of the cis effect series results from the ability of these ligands to adopt a bidentate binding mode.

The presence of σ-donors and π-acceptors can also affect the stability of the unsaturated intermediate. Extended Hückel calculations have shown that d^6, 16-electron, trigonal bipyramidal complexes are more stable when they contain weak σ-donors and poor π-acceptors in the basal plane.[127,128] Hence, the higher rate of dissociation of a ligand cis to a halide than trans to a halide can result from the weak σ-donation of the halide, and the higher rate of dissociation of a ligand trans to CO than cis to CO likely results from the strong π-accepting property of CO that destabilizes a geometry in which the CO lies in the basal plane.

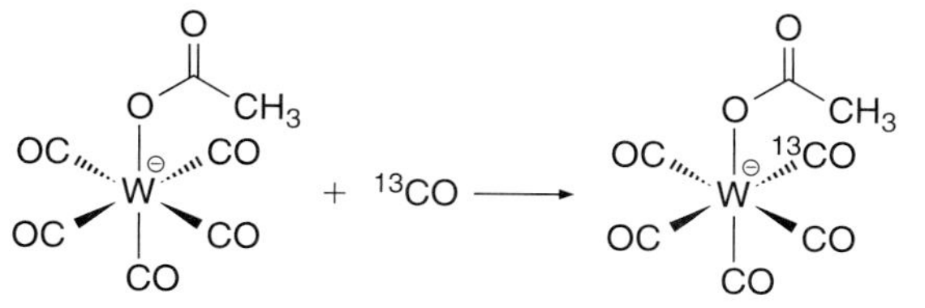

$$\text{(5.25)}$$

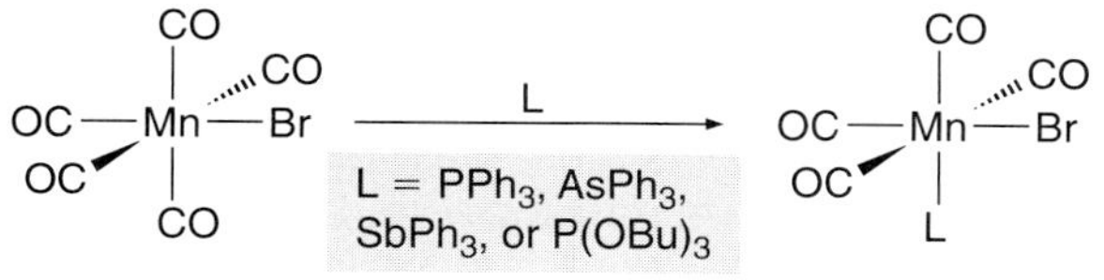

$$\text{(5.26)}$$

The high cis effect of a ligand that can bind in a bidentate mode is illustrated by the reactions of $W(CO)_5(OAc)^-$. A ^{13}C NMR labeling study (Equation 5.25)[e,129,130] showed that the carbonyl cis to the acetate is replaced when $W(CO)_5(OAc)^-$ is treated with labeled CO. This exchange process is much faster than typical ligand substitutions involving $W(CO)_6$. An example of the cis effect of a halide is demonstrated by the site of substitution in $Mn(CO)_5Br$ (Equation 5.26). Infrared studies have shown that dissociation of the CO ligand in $Mn(CO)_5Br$ located cis to the bromide occurs at least 10 times faster than dissociation of the CO ligand trans to the bromide.[26,131,132]

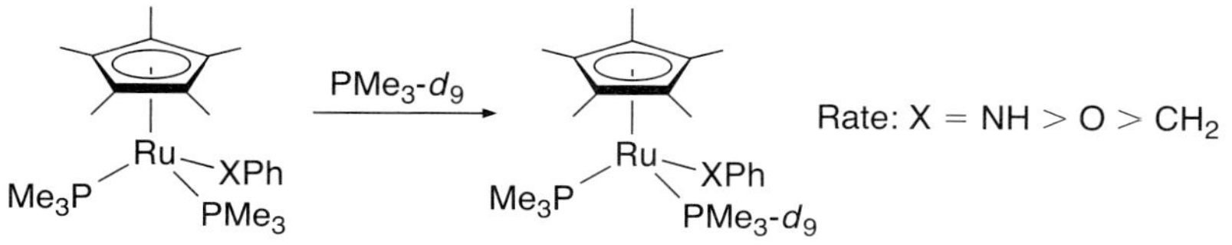

Rate: X = NH > O > CH$_2$

$$\text{(5.27)}$$

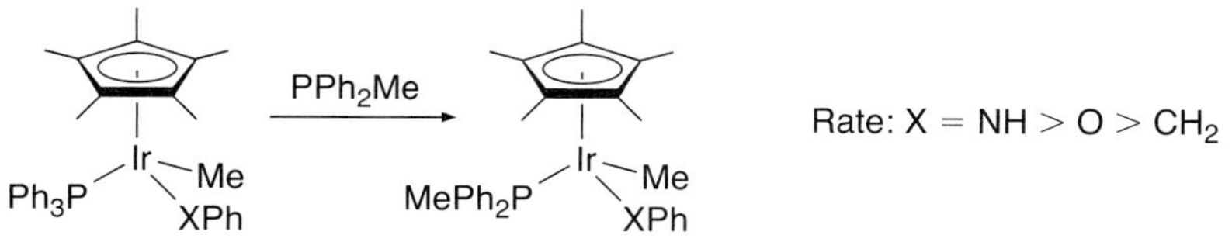

Rate: X = NH > O > CH$_2$

$$\text{(5.28)}$$

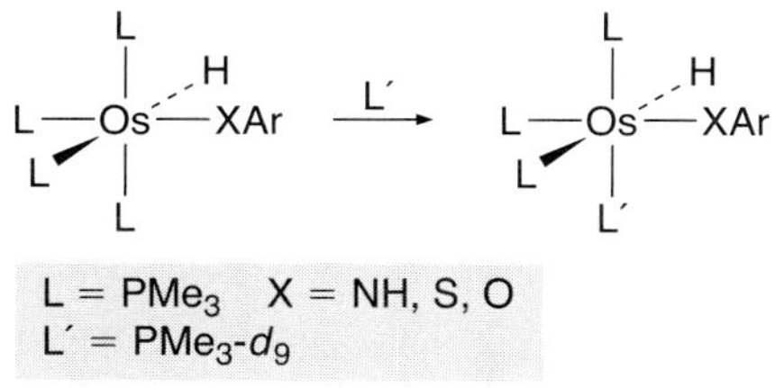

$$\text{(5.29)}$$

L = PMe$_3$ X = NH, S, O
L′ = PMe$_3$-d_9

π-Donating ligands also appear to promote dissociation of ligands from piano-stool and pseudo-octahedral compounds. For example, the rate of exchange of PMe$_3$-d_9 into a series of three-legged piano stool ruthenium complexes (Equation 5.27) followed the trend $Cp^*Ru(PMe_3)_2(NHPh) > Cp^*Ru(PMe_3)_2(OPh) > Cp^*Ru(PMe_3)_2(CH_2Ph)$.[133] This order was rationalized by the stronger π-donating ability of the anilide group than the phenoxide group and the absence of π-donation by the benzyl group in the 16-electron intermediate. The rate of ligand substitution in $Cp^*Ir(PPh_3)(Me)(X)$ (X = Me, Cl, OAc, CH$_2$Ph, NHPh, or OPh) (Equation 5.28) was also consistent with stabilization of an unsaturated intermediate by a π-donor.[134] Arguments have also been made that π-donation can affect the rates of ligand substitution of pseudo-octahedral compounds. Studies on the exchange of PMe_3-d_9 into $[Os(PMe_3)_4(H)(XAr)]$ (X=CH_2, NH, or S) (Equation 5.27)[135] suggested that the ligands cis to the anilide were labilized more than those trans to the anilide. The effect of substituents on the anilide group implied that the labilizing effect resulted from π-donation. This effect should not arise from π-donation into the unoccupied orbital of a square pyramidal complex, however, because this orbital has improper symmetry to overlap with a π-orbital on a cis ligand.

There are many other ways, other than the cis effect, that auxiliary ligands can affect the rates of ligand-replacement reactions. Many of these effects are not well understood, however. For example, the normally stable Co(III) center in vitamin B_{12} undergoes fast substitutions that are influenced by the axial 5,6-dimethylbenzimidazole base appended to the tetradentate corrin macrocyclic ligand.[136–139] Replacement of an axial H_2O by CN^- in $(H_2O)B_{12}$ occurs on a millisecond time scale,[136–139] and these substitutions are even faster than those of the well-studied B_{12} model complexes.[139–141]

5.4.1.7. Stereochemistry of Substitutions of Octahedral Compounds

The stereochemistry of degenerate ligand exchanges must obey the principle of microscopic reversibility, first introduced in Section 5.3.1.1.1. In a degenerate exchange involving an octahedral compound, microscopic reversibility dictates that the site from which the ligand dissociates must be the site at which the incoming ligand approaches. For example, the exclusive product of a single substitution of ^{13}CO for the CO cis to the bromide in $[Mn(CO)_5Br]$ is thought to be cis-$Mn(CO)_4(^{13}CO)Br$ (Equation 5.30).[142] Other isomers form after multiple substitutions of ^{13}CO occur.

However, the substitution of ^{13}CO for a phosphine in $trans$-$Cr(CO)_4(PPh_3)_2$ (Equation 5.31) forms a product in which the labeled CO is located cis to the remaining phosphine in the product.[6,143] These data illustrate that the five-coordinate intermediate formed by dissociation of ligand can undergo scrambling of the stereochemistry faster than it reacts with the incoming ligand. In the case of Equation 5.31, the CO can approach a coordination site other than the one from which PPh_3 departed because the product is different from the reactant, and microscopic reversibility does not apply. The coordinating ability of solvent can affect the stereochemistry of these reactions because coordination of solvent generates an octahedral intermediate that has rigid stereochemistry. A particularly thorough treatment of the effect of the properties of the ligand on the stereochemical stability of the five-coordinate intermediate was reported by Flood when studying ligand substitution of $[Os(PMe_3)_4(H)(XAr)]$, in which X = CH_2, NH, and O.[135]

$$(5.30)$$

$$(5.31)$$

5.4.2. Substitutions of 18-Electron Complexes that Deviate from Pure Thermally Induced Dissociative Mechanisms

Ligand substitution of coordinatively saturated complexes can be accelerated in a variety of ways. Irradiation, initial electron transfer, catalysis by phosphine oxides, catalysis by acid or base, and initiation of radical-chain mechanisms all allow ligand substitution involving 18-electron complexes to occur by processes that are faster than would occur by a purely dissociative substitution mechanism. Small changes in the reaction conditions can cause these alternative reaction pathways to occur. For example, the reaction of PBu_3 with $HRe(CO)_5$ occurs within 20 minutes with photochemical irradiation, but no reaction is observed for 60 days at 25 °C if light is rigorously excluded and extremely pure reagents are used (Equation 5.32).[92,144] Similarly, $Fe(CO)_5$ does not undergo substitution at < 90 °C if carefully purified,[145] but substitutions at 60 °C have been reported with less carefully purified material.[146] The following sections present these various assisted reaction pathways for ligand substitution.

$$HRe(CO)_5 \begin{cases} \xrightarrow[\substack{h\nu \\ 20\ min}]{PBu_3} HRe(CO)_4(PBu_3) + CO \\ \\ \xrightarrow[\substack{PBu_3 \\ 60\ days,\ 75\ °C}]{\substack{No\ light \\ Pure\ reagents}} No\ reaction \end{cases} \qquad (5.32)$$

5.4.2.1. Substitutions of $M(CO)_6$ Complexes Occur with an Associative Term in the Rate Law

Particularly well-investigated cases of assisted substitutions include the reactions of $M(CO)_6$ and $M(CO)_5$(amine) complexes (M=Cr, Mo, W)[5,6,26,147,148] (Equation 5.33). Two terms are observed in the rate law for substitution by $[Cr(CO)_6]$: $-d[Cr(CO)_6]/dt = \{k_1 + k_2[L]\}[Cr(CO)_6]$ and in the rate law for substitution reactions of other complexes in this class, such as the substitutions of PR_3 for a CO in $Mo(CO)_6$ and $W(CO)_6$,[5,6,26,147] and the substitutions for the amine ligand in $Mo(CO)_5$(amine).[148] The first term of this rate law corresponds to reaction by a dissociative path, while the second term corresponds to reaction by a pathway in which some bonding to the incoming ligand occurs prior to full dissociation of CO or amine.

$$M(CO)_5L \begin{cases} \xrightarrow{Decalin} (CO)_5M-S + L \xrightarrow{L'} (CO)_5M-L' \quad (\text{Zero-order in L}) \\ \\ \xrightarrow[-L]{+L'} M(CO)_5L \cdot L' \longrightarrow (CO)_5M-L' + L \quad (\text{First-order in L}) \end{cases} \qquad (5.33)$$

M = Cr, Mo, W
L = CO, amine
L' = PBu_3, PPh_3, $P(OEt)_3$

$$\text{Overall} = -\frac{d[M(CO)_6]}{dt} = k_1[M(CO)_6] + k_2[M(CO)_6][L]$$

Partial reaction by a "dissociative interchange" mechanism $(I_d)^{6,26,148}$ has been suggested to account for the second-order term. This I_d mechanism would occur by formation of an intermediate "$Cr(CO)_6 \bullet PR_3$." Because formation of a complex in which two neutral materials would associate by a noncovalent interaction is unlikely, and attack at the metal to form a weakly bound seven-coordinate complex or addition of the incoming ligand to the carbonyl carbon (see Section 5.5) seem equally unlikely, a mechanism that accounts for the second-order term is difficult to pinpoint.

In contrast, the activation enthalpy of the k_1 term of the rate equation for ligand substitution by $M(CO)_6$ correlates well with the M–CO BDE values in Table 2.3 of Chapter 2 and is consistent with a dissociative substitution mechanism. The $\Delta H^\ddagger$ (40.2 $\pm$ 0.6 kcal/mol) for reaction of $Cr(CO)_6$ is within experimental error of the BDE of $Cr(CO)_5$–CO (37 $\pm$ 2 kcal/mol, Table 5.2), and the $\Delta H^\ddagger$ terms[149–153] for the reaction of $Mo(CO)_6$ (31.7 kcal/mol) and $W(CO)_6$ (39.9 kcal/mol) are similar to the BDE values for dissociation of the first CO ligand in $(CO)_5M$–CO (30–34 kcal/mol for M = Mo; 38–40 kcal/mol for M = W).[147] The small positive entropies of activation[149–153] [6.7 eu for $Mo(CO)_6$; 13.8 eu for $W(CO)_6$][147] are also consistent with a dissociative event.[154–160]

5.4.2.2. Catalyzed and Assisted Ligand Substitution Reactions

Because dissociation of CO from many metal–carbonyl complexes is slow, methods for the catalytic or stoichiometric assistance of ligand substitution reactions have received considerable attention.[5,8–10,161] These methods are presented in the following sections.

5.4.2.2.1. Ligand Substitution Catalyzed by Electron Transfer

Catalysis by single-electron transfer is well known in both organic[162–165] and organometallic chemistry and has been reviewed several times.[161–165] The initial single electron transfer event can be initiated either electrochemically or by an added reagent. Depending upon whether the substitution reaction is oxidatively (O) or reductively (R) activated by a nucleophile (N) or an electrophile (E), the resulting pathways have been labeled $S_{ON}2$, $S_{OE}1$, $S_{RE}2$, and $S_{RN}1$.[165] The general form of these catalytic cycles is shown in Scheme 5.7.

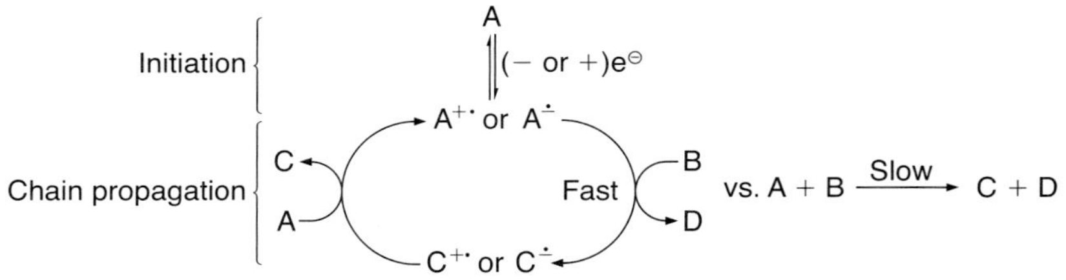

Scheme 5.7.
A general scheme for ligand substitution initiated by electron transfer.

The reaction of $(\eta^5\text{-}C_5H_4Me)Mn(CO)_2(NCCH_3)$ with PPh_3 by an $S_{ON}2$ pathway revealed by Kochi's work illustrates ligand substitution catalyzed by a single-electron transfer (Scheme 5.8).[166–168] In the absence of oxidants, displacement of the CH_3CN ligand in $(\eta^5\text{-}C_5H_4Me)Mn(CO)_2(NCCH_3)$ by PPh_3 occurs much more slowly than it does with catalysis by electron transfer. The electron transfer was induced with a platinum electrode set at a constant potential that generates a small concentration of the 17-electron intermediate, $[(\eta^5\text{-}C_5H_4Me)Mn(CO)_2(NCCH_3)]^{\bullet+}$. This labile intermediate undergoes rapid substitution with PPh_3 to generate $[(\eta^5\text{-}C_5H_4Me)Mn(CO)_2(PPh_3)]^{\bullet+}$. Because the $[(\eta^5\text{-}C_5H_4Me)Mn(CO)_2PPh_3]/[(\eta^5\text{-}C_5H_4Me)Mn(CO)_2PPh_3]^+$ redox couple occurs at a more positive potential than the $[(\eta^5\text{-}C_5H_4Me)Mn(CO)_2(NCCH_3)]/[(\eta^5\text{-}C_5H_4Me)Mn(CO)_2(NCCH_3)]^+$ couple, electron transfer from $[(\eta^5\text{-}C_5H_4Me)Mn(CO)_2(NCCH_3)]$ to $[(\eta^5\text{-}C_5H_4Me)Mn(CO)_2(PPh_3)]^+$ occurs essentially irreversibly to generate the final neutral 18-electron substitution product

$[(\eta^5\text{-}C_5H_4Me)Mn(CO)_2(PPh_3)]$ and to propagate the chain by regenerating the 17-electron species $[(\eta^5\text{-}C_5H_4Me)Mn(CO)_2(NCCH_3)]^{\bullet+}$. Chain termination occurs when $[(\eta^5\text{-}C_5H_4Me)Mn(CO)_2PPh_3]^{\bullet+}$ is reduced by the ca. -0.06 V electrode. At the potential of -0.06 V, the substitution reaction was complete in 10 minutes at 22 °C and 1,013 moles of product were formed for each mole of electrons.

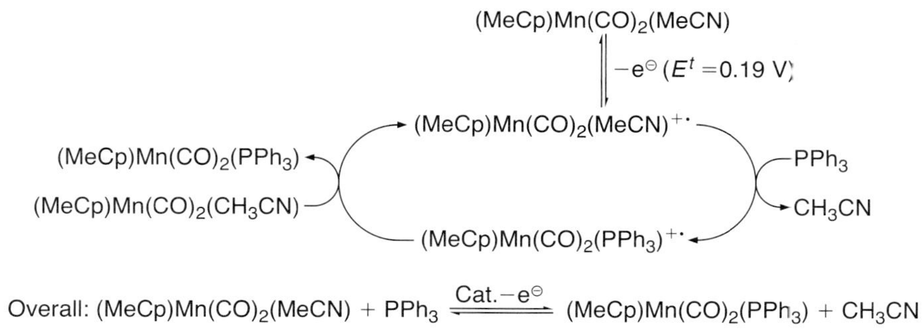

$$(MeCp)Mn(CO)_2(MeCN)$$
$$\big\downarrow\, -e^\ominus\ (E^t = 0.19\ \text{V})$$
$$(MeCp)Mn(CO)_2(MeCN)^{+\bullet}$$

$(MeCp)Mn(CO)_2(PPh_3)$ ←

$(MeCp)Mn(CO)_2(CH_3CN)$ —

—PPh$_3$

→ CH$_3$CN

$(MeCp)Mn(CO)_2(PPh_3)^{+\bullet}$ ←

Overall: $(MeCp)Mn(CO)_2(MeCN) + PPh_3 \underset{}{\overset{Cat.-e^\ominus}{\rightleftharpoons}} (MeCp)Mn(CO)_2(PPh_3) + CH_3CN$

Scheme 5.8

Ligand substitutions of saturated complexes can also be accelerated by reduction. This reductive activation of ligand substitution is most common in metal–carbonyl clusters.[161, 169–171] Presumably, a radical anion is formed, and dissociation of ligand is rapid from this 19-electron species. The ketyl radical anion, $Na^+Ph_2CO^\bullet$,[172–174] has often been used as a reducing agent to induce ligand substitution. A prototypical example is the acceleration of the reaction of $Fe(CO)_5$ with L by the ketyl radical anion to give $Fe(CO)_4L$ [L = PPh_3, $P(OPh)_3$, or $P(OMe)_3$].[172] Photochemically initiated electron-transfer catalysis is also known in organometallic chemistry.[163, 175]

5.4.2.2.2. Ligand Substitutions by Radical Chains Initiated by Atom Abstractions

Ligand substitutions of 18-electron complexes can also occur by radical-chain processes initiated by atom abstraction. These radical chains occur through 17-electron intermediates that undergo facile associative substitutions. Substitutions of metal carbonyl hydrides, halides, and stannyl complexes by this mechanism are all known. These reactions are particularly prevalent in first-row metal hydrides because the M–H bond is weaker than the M–H bond in second- and third-row metal complexes, and hydrogen atom abstraction is one step of the radical chain. However, they have also been proposed to occur with third-row metal–hydride complexes

$$HRe(CO)_5 + PBu_3 \xrightarrow{-CO} HRe(CO)_4L + HRe(CO)_3L_2$$
$$L = PBu_3 \tag{5.34}$$

Byers and Brown uncovered a radical chain pathway for ligand substitution in $HRe(CO)_5$.[92, 176] This reaction is shown in Equation 5.34, and a modification of the originally proposed pathway is shown in Equation 5.35.[177] The reaction is initiated by abstraction of the hydride ligand as a hydrogen atom. The initial hydrogen atom abstraction likely occurs by a species generated by trace oxygen. The resulting 17-electron species then undergoes associative substitution, and the organometallic product of the substitution abstracts a hydrogen atom from the starting complex to form the final substitution product and to regenerate the 17-electron intermediate. Dimerization of the intermediate radicals terminates the chain.

$$In\bullet\ +\ HRe(CO)_5\ \longrightarrow\ In\text{–}H\ +\ \bullet Re(CO)_5$$

$$\bullet Re(CO)_5\ +\ L\ \longrightarrow\ \bullet Re(CO)_4L\ +\ CO$$

$$\bullet Re(CO)_4L\ +\ L\ \longrightarrow\ \bullet Re(CO)_3L_2\ +\ CO \tag{5.35}$$

$$\bullet Re(CO)_xL_{5-x}\ +\ HRe(CO)_5\ \longrightarrow\ HRe(CO)_xL_{5-x}\ +\ \bullet Re(CO)_5$$

$$2\bullet Re(CO)_xL_{5-x}\ \longrightarrow\ (CO)_xL_{5-x}Re\text{–}Re(CO)_xL_{5-x}$$

Several observations implied that this substitution process occurred by a chain mechanism: (1) no reaction occurred over 60 days at 25 °C when light was excluded and ultrapure reagents were used; (2) the rates were not reproducible; (3) the reactions occurred with an induction period and unusual product versus time curves;[92] (4) the reactions were accelerated by $\bullet Re(CO)_5$, which was generated from photolysis of $Re_2(CO)_{10}$; and (5) the reaction was accelerated by small amounts of O_2, but was inhibited by large amounts of O_2.

5.4.2.2.3. Photoinduced Dissociation of Ligands

Although additives to induce radical chemistry have allowed ligand substitutions of 18-electron complexes to be conducted under mild conditions, photochemical reactions provide a common and practical alternative. Photochemically induced dissociation of carbonyl ligands is most common, but photochemical dissociations of other dative ligands are known. Several examples are shown in Equations 5.36–5.40. These examples illustrate the dissociation of CO from homoleptic carbonyl compounds of iron[178,179] and chromium,[180,181] the dissociation of CO from piano-stool carbonyl compounds,[182–184] the dissociation of N_2,[185] and the dissociation of a carbodiimide to generate an intermediate that coordinates and cleaves the C–H bonds of alkanes.[186] In some cases, like the formation of the two THF complexes, the products of the photochemical process are not isolated; instead, they are treated in situ with a ligand, such as a phosphine, to form monosubstitution products selectively.

$$Fe(CO)_5\ \xrightarrow[PPh_3]{h\nu}\ Fe(CO)_3(PPh_3)_2\ +\ 2\ CO$$

$$PEt_3\ \diagdown\ versus \tag{5.36}$$

$$Fe(CO)_4(PEt_3)\ +\ CO$$

$$(5.37)$$

$$(5.38)$$

$$(5.39)$$

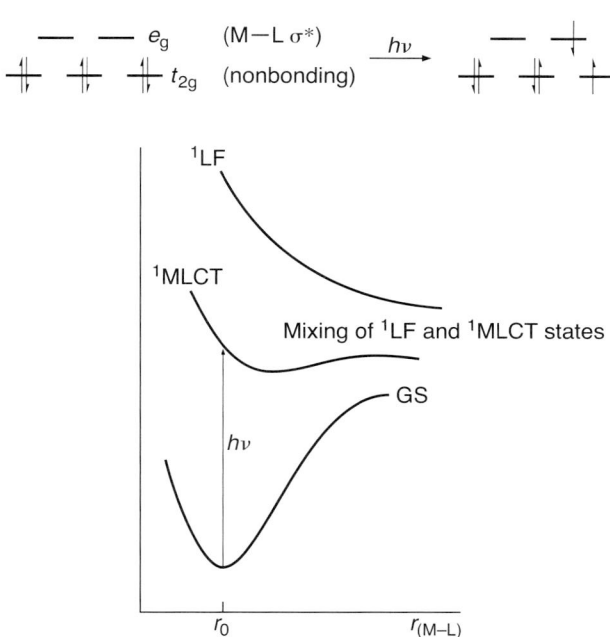

$$(5.40)$$

Photolysis was for many years thought to induce ligand substitution by transferring an electron from a nonbonding orbital to a metal–ligand antibonding orbital. In a d^6 octahedral complex, such as $Cr(CO)_6$, photolysis was thought to transfer an electron from a t_{2g} orbital to an e_g orbital, as shown in Figure 5.6 in a ligand field transition. Because the e_g orbital has M–L antibonding character, this promotion of an electron would reduce the metal–ligand bond order and make thermal dissociation fast.[187]

Figure 5.6.
Top: Electron promotion initially thought to occur upon photolysis of an octahedral d^6 complex. Bottom: Mixing of excited states that are now thought to more accurately reflect the events that occur after photolysis of metal–carbonyl compounds, such as $M(CO)_6$ (M = Cr, Mo, and W).

However, more recent calculations have shown that the events leading to photochemical dissociation are more complex. Photolysis of $M(CO)_6$ compounds (M = Cr, Mo, and W) is now thought to cause a metal-to-ligand charge-transfer (MLCT) transition from the metal d-orbitals to the π^*-orbitals involving the CO ligands. The excited state generated by this process then undergoes a Jahn–Teller distortion from an octahedral geometry to a C_{4v} geometry formed by lengthening of one M–CO bond. This geometric distortion lowers the energy of the excited state that would be formed by a ligand field transition, and this state mixes with the MLCT state. This mixing of states, in turn, populates the M–CO antibonding orbitals and makes the excited state potential energy surface dissociative.

These photochemically induced ligand dissociations have been studied by ultrafast spectroscopic methods. The dissociation of CO occurs on the time scale of tens of picoseconds (1 ps = 10^{-12} s). In some cases, the product from the initial CO dissociation event has been assigned to have a triplet spin state (Equation 5.41).[188] The incoming reagents and final products have singlet spin states. Therefore, the initially formed triplet intermediate undergoes intersystem crossing to the singlet spin state prior to

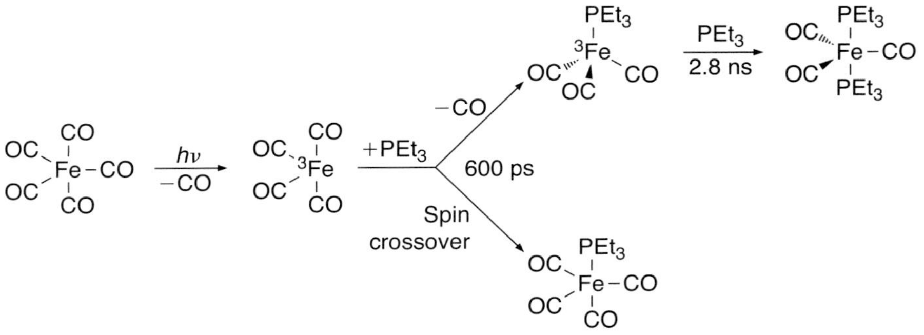

(5.41)

coordination of the phosphine to form one of the two reaction products. In parallel, this intermediate undergoes a spin-conserving associative substitution, followed by reaction with a second incoming ligand to form a second product by a process involving spin crossover. The triplet intermediate $^3Fe(CO)_4$ reacts on the time scale of hundreds of picoseconds.

5.4.2.2.4. Oxidation of Coordinated CO

Many examples of ligand substitution of stable metal carbonyl compounds have been induced by oxidation of coordinated CO by amine oxides.[189] The prototypical examples are conducted with stoichiometric amounts of $Me_3NO \cdot 2H_2O$.[190,191] For example, the reaction of $Fe(CO)_5$ with Me_3NO (Equation 5.42) forms $Fe(CO)_4(NMe_3)$ at -30 °C, and the reaction of $M(CO)_6$ (M = Cr, Mo, and W) with Me_3NO forms $M(CO)_5(NMe_3)$.[192,193] Most often, this reaction is conducted in the presence of a ligand, such as a phosphine, that displaces the amine from the initial product. The use of Me_3NO to extrude CO has been reviewed.[161]

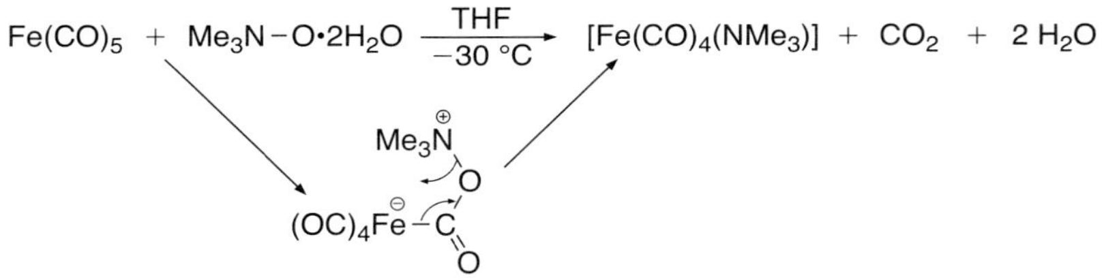

(5.42)

This oxidation occurs by nucleophilic attack at the coordinated CO and fragmentation to form CO_2 and the complex of NMe_3. If the overall reaction leads to the replacement of CO by a more strongly donating ligand, like a phosphine, then the product is more electron rich than the reactant and is less susceptible to attack by Me_3NO. In this case, selective oxidation of the CO ligand in the reagent occurs, and the reaction forms a product from substitution of a single CO ligand.

The ability of Me_3NO to remove a CO ligand can be correlated with the ν_{CO} value. This correlation exists because complexes with higher values of ν_{CO} are more electrophilic, as noted in Chapter 2. As a result, complexes with higher ν_{CO} values are more susceptible to attack by the Me_3NO and oxidation of the CO ligand to CO_2.

5.4.2.2.5. Other Assisted Ligand Substitutions

Many other reagents and reaction conditions accelerate ligand substitution, but are less defined than the examples described above. A variety of nucleophilic reagents, such as phosphine oxides, accelerate ligand dissociation from stable octahedral carbonyl complexes, but the mechanism that accounts for this acceleration is not well understood.[161,194–196]

The site of initial attack by the phosphine oxide is unknown;[194,195,197] although attack at the carbonyl carbon is possible, the phosphine oxide and coordinated carbonyl ligand do not form CO_2 in the same manner as do amine oxide and coordinated CO.

Acids and bases can also accelerate ligand substitution. Acids accelerate ligand substitution of $Fe(CO)_5$,[198] although later work revealed a much smaller effect than did the original work.[199] Bases catalyze the hydrolysis of $[Co(NH_3)_5Cl]^{2+}$ in aqueous solution. This base-catalyzed process is a classic mechanism of ligand substitution in coordination chemistry.[66] Heterogeneous metal catalysts, such as PtO_2, PdO_2, Pd/carbon, $Pd/CaCO_3$, supported Pt, Pd, Ru, or Rh, and $CoCl_2$ and $NiCl_2$ have been shown to accelerate the dissociation of CO.[200] As might be expected for a heterogeneous reaction of this type, the mechanism of this accelerating affect is not well defined. Finally, sonochemical methods lead to ligand dissociation and even cause displacement of the typically stable Cp ligand during a conversion of Cp_2Co and CO to $CpCo(CO)_2$.[201]

5.5. Substitution Reactions Involving Polyhapto Ligands

5.5.1. Substitutions for Dienes and Trienes

Substitution for polyhapto ligands occur in many cases by mechanisms that are similar to those for replacement of monodentate ligands. Thus, substitutions for polyhapto ligands in 18-electron octahedral complexes, like substitutions for CO in the 18-electron octahedral complexes presented in Section 5.4.2.1, often occur by competing dissociative and associative pathways with a two-term rate expression, $k_{obs} = k_1 + k_2[L]$.[5,202]

$$(5.43)$$

$Cr(CO)_4$(butadiene) reacts with relatively strong dative ligands like $P(OMe)_3$ to form $Cr(CO)_4[P(OMe)_3]_2$ and free diene, as shown in Equation 5.43, and the rates of these reactions fit a two-term rate law consisting of one term for a first-order process in which the rate constant k_1 corresponds to dissociation of one half of the diene ligand and a second, measurable $k_2[P(OMe)_3]$ term reflecting a second order process with some associative component.[202] In contrast, the reaction of $Cr(CO)_4$(butadiene) with the more weakly nucleophilic 1,5-COD forms $Cr(CO)_4$(1,5-COD) and free butadiene[202] by a single first-order pathway initiated by dissociation of half of the diene.

If the number of olefins in the polyene is greater than two, then the number of steps in the overall displacement reaction is simply larger. For example, $M(CO)_3$(1,3,5-cycloheptatriene) (wherein which M = Cr, Mo, and W) reacts with three equivalents of PhCN to form $Mo(CO)_3(NCPh)_3$ and 1,3,5-cycloheptatriene by a stepwise process.[203,204] The rate law for reaction of $Cr(CO)_3$(1,3,5-cycloheptatriene) is first order in complex and first order in benzonitrile. However, the reaction of $Mo(CO)_3$(1,3,5-cycloheptatriene) is first order in complex and second order in benzonitrile. The kinetic data for reaction of the molybdenum complex have been rationalized by a stepwise pathway in which the first benzonitrile ligand reversibly adds to the molybdenum complex, and a second benzonitrile adds by an associative pathway (Scheme 5.9).[203, 204]

5.5.2. Substitutions for Arenes and Arene Exchange Reactions

Arene displacements (Equation 5.44) and arene exchange reactions (Equation 5.45) have been known since 1961.[205–211, 212] Strohmeier first studied arene exchange,[207–209] and Traylor has contributed significantly to our understanding of these processes.[205, 213, 214] The arene exchange process is of particular interest because coordinated arenes undergo a variety of reactions that are distinct from reactions of free arenes. Selected transformations of coordinated arenes are presented in Chapter 11. If arene exchange could be made more facile, then these stoichiometric transformations could be developed into catalytic processes.[215–217] Such catalytic chemistry has been achieved in recent years, but in isolated cases.[218, 219] In addition, substituents on the ancillary ligands that accelerate arene exchange have been reported.[220, 221]

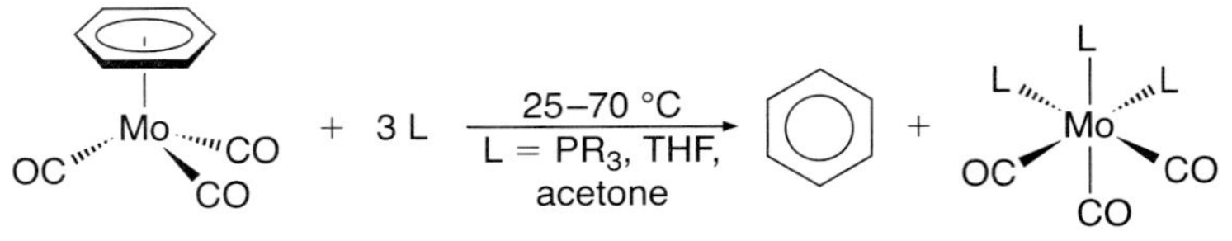

$$(5.44)$$

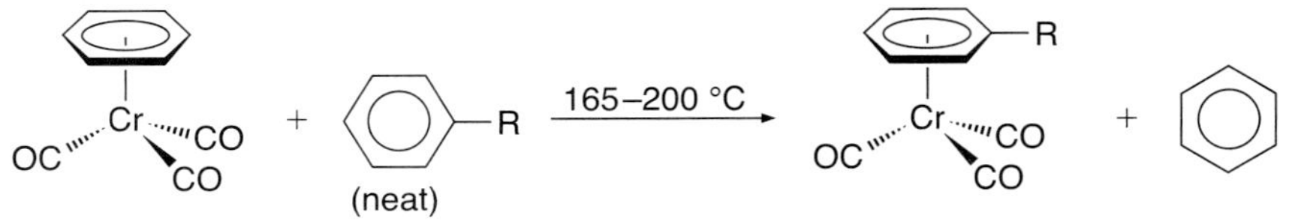

$$(5.45)$$

Donor solvents and weakly binding ligands significantly increase the rates of arene exchange.[205, 211] For example, ketones, nitriles, dienes, THF, glymes, pyridine, and $Ph_3P=O$ all catalyze arene exchanges, like that in Equation 5.45. $Cr(CO)_3(C_6Me_6)$ even appears to accelerate arene exchange by intermolecular coordination of the oxygen of the carbonyl group.[210, 211] The effectiveness of these various species as a catalyst for the arene exchange follows the order cyclohexanone (1,300) > PhCN (600) > THF (30) ≈ $Cr(CO)_3(C_6Me_6)$ (30) > uncatalyzed (1).

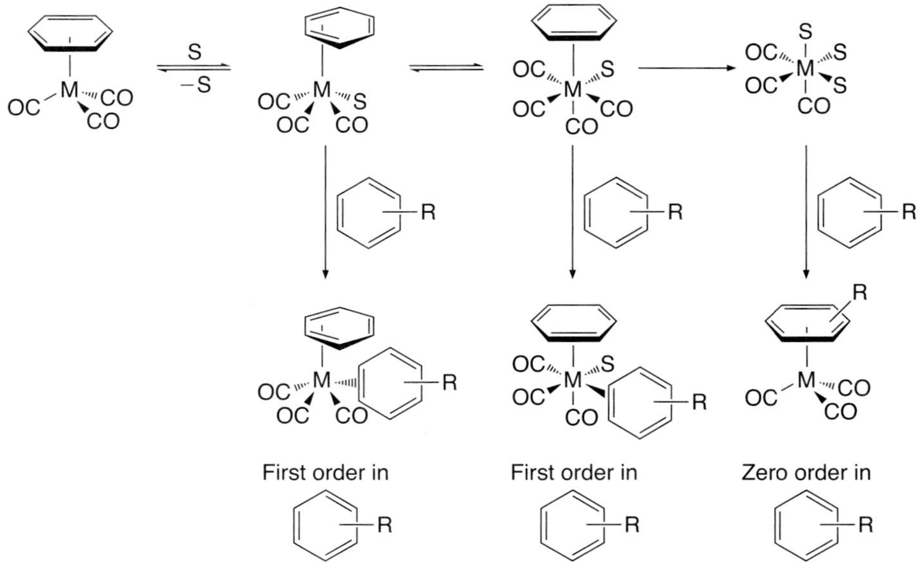

Scheme 5.9

The arene exchanges occur by stepwise conversion from the η^6 form to the η^4 form to an η^2 form to the free ligand. As shown in Scheme 5.9, the entering arene can attack the intermediate containing an η^4-arene, the intermediate containing an η^2-arene, or a complex formed by full dissociation of the arene. Thus, the arene exchange can be zero or first order in the incoming arene, depending on whether the incoming arene binds after the original arene is released completely or before it is released and is bound in an η^4 or η^2 fashion.

The coordination sites generated by the change in hapticity of the ligand become occupied by the coordinating solvent. As a result, the binding ability of the solvent can affect the rate law. In noncoordinating hydrocarbon solvents, a rate law that is first order in complex and first order in free arene is observed. In coordinating solvents that adequately fill the coordination sphere as the departing arene is replaced,[211] a rate law that is zero order in arene is observed.

The solvated intermediates have been observed in some cases. Ir(1,5-COD)(acetone)$_x^+$ has been observed during the exchange of free arene with Ir(η^6-arene)(η^4-1,5-COD),[222] and the solvated complex of a partially displaced arene, Cr(CO)$_3$(η^4-naphthalene) (THF), has been detected at low temperatures[223] during the arene exchange reactions of Cr(CO)$_3$(η^6-naphthalene) in THF. Taken together, these species suggest a stepwise η^6–η^4–η^2 "unzipping" of the arene during its replacement.

A series of methods to accelerate arene exchange have been developed. Irradiation[224–226] can lead to displacement of the arene ligand.[224–227] In addition, oxidation of the metal by I$_2$ is commonly used to release an arene after it has been modified by stoichiometric chemistry; this oxidation generates labile 17-electron complexes such as Cr(CO)$_3$(arene)$^+$I$^-$.[216] Catalysis of arene substitution by reductive electron-transfer is also known, as in the case of CpFe(arene)$^+$.[217]

$$(5.46)$$

$$(5.47)$$

Although few catalytic cycles involving arene exchange have been developed, several examples of systems that exchange arenes under mild conditions have been reported. The ruthenium complex in Equation 5.46 containing a carbonyl ligand and a chelating disilyl ligand exchanges arenes at room temperature.[228] This exchange is currently the mildest reported. In addition, the ruthenium complex in Equation 5.47 containing a PCP ligand exchanges arenes under milder conditions than most previous examples of arene exchange.[219] This exchange is part of a catalytic hydroamination of vinylarenes.[229] Finally, Semmelhack has designed ligands, summarized in Table 5.10, that contain substituents that will encourage exchange of arenes with [Cr(arene)(CO)$_2$(PPyr$_2$Pyr′)] (in which Pyr = a pyrrolyl

Table 5.10. Effect of ancillary ligand on arene ring exchange.

Complex	Half-life
L = SMe	8.7 h at 70 °C
L = SCF$_3$	< 5% conversion/12 h at 70 °C
L = SPh	30.6 h at 70 °C
L = CH$_2$SMe	28.7 h at 70 °C
L = CH$_2$SPh	Very slow (decomp.) at 70 °C
L = CONMe$_2$	9 h at 22 °C
L = S-t-Bu	9.9 h at 70 °C
L = CO$_2$Me	0.5 h (~115 h at 23 °C)
L = 2-Py	8 h at 22 °C[a]

[a] Too fast to measure at 70 °C

group and Pyr′ = a substituted pyrrolyl group). PPyr$_3$ is a π-accepting phosphine that mimics the properties of CO. Pyrroyl groups possessing *ortho*-acyl and amide groups led to the greatest increase in rate versus the complex containing the unsubstituted PPyr$_3$ ligand. Complexes of the ligand containing a 2-thiomethyl group also exchanged arenes much faster than complexes of the PPyr$_3$ ligand lacking any substituents.

5.5.3. Associative Substitution by Pentadienyl Ligand Ring Slip

Coordinatively saturated complexes ligated by η^5-dienyl ligands, such as η^5-fluorenyl, η^5-indenyl, η^5-cyclopentadienyl, and η^5-pentadienyl, can undergo substitution of accompanying monodentate ligands by associative processes.[4] During such associative processes the 18-electron complex reacts by a change in hapticity of the dienyl ligand from η^5 to η^3, and this change in hapticity is accompanied by attack of the incoming ligand. This change in hapticity of a cyclic π-ligand is often called "ring slip."

(5.48)

$$k_2 [PMePh_2] \text{ is approximately equal to } k_{-1}.$$

(5.49)

Equation 5.48[230–232] illustrates a classic ligand substitution reaction that occurs by a ring slip, and Equation 5.49 depicts a more recent related study. In the process of Equation 5.48, the 18-electron (η^5-indenyl)Rh(CO)$_2$ undergoes a change in hapticity from η^5 to η^3 prior to or during attack of PPh$_3$. The mechanism of this reaction was revealed by several observations. First, the rate of the reaction was second order (first order in the rhodium complex and first order in PPh$_3$). Second, the entropy of activation was negative, $\Delta S^{\ddagger} \approx -20$ eu. These data are consistent with an associative attack of the ligand accompanied by ring slip (Path A) or fully reversible ring slip followed by attack of the incoming ligand (Path B).

The ligand substitution reaction of PPh$_2$Me with the Ir(III) complex Cp*Ir(PPh$_3$)(Me)-(NHPh) also appeared to occur by a change in hapticity of the cyclopentadienyl-type ligand from η^5 to η^3,[134] but the rate data were slightly different. In this case, the kinetic expression implied that the reaction occurred in discrete steps: reversible ring slip, followed by attack of the incoming ligand on the 16-electron intermediate and dissociation of the PPh$_3$.

Table 5.11. Effect of η^5 ligands on the rate of ligand substitution in (η^5-Cp′)Co(CO)$_2$.

——	3.8×10^8	6.1×10^5	1.2×10^4	1.1×10^2	1	2.2×10^{-2}

The effect of the identity of the η^5-dienyl ligand can be large. Table 5.11 shows the relative rates of reactions of complexes containing the series of η^5-ligands toward the ligand substitution in Equation 5.48. These data show that the rates of substitution span a factor of 10^{10}. The relative rates are best rationalized by the relative energies of both the ground states and ring-slipped intermediates, and these relative energies have been attributed to a series of effects. They have been proposed to result from the changes in aromaticity experienced upon ring slip, the ability of the η^5 ligand to stabilize the more localized, anionic charge of the η^3 intermediate, the steric demands and greater electron donation of the methyl groups, and the differences in bond-orders between the metal and Cp-type ligand in the starting and ring-slipped species. The fastest reactions are observed with fluorenyl and indenyl groups, and this has been attributed to the aromatic unit in these ligands remaining intact upon formation of the η^3 structure,[4,233,234] but being disrupted upon formation of an η^3-Cp ligand.

However, recent studies have led to alternative arguments. Calculations reported by Calhorda[235] indicate that the distortions of free Cp and indenyl anions from a planar structure to the folded structure in η^3-Cp and -indenyl complexes (vide infra) leads to similar changes in energy for the two ligands. This similarity argues against the rationalizations based on changes in aromaticity upon folding the two types of ligands. Instead, Calhorda argues that the bond orders are larger for an η^5-Cp ligand than for an η^5-indenyl ligand, but smaller for an η^3-Cp ligand than for an η^3-indenyl ligand. These differences are attributed to the differences in orbital symmetry of the indenyl and cyclopentadienyl HOMO and

LUMO. These orbitals are distributed symmetrically in the Cp ligand, whereas they have lower symmetry in the five-membered ring of the indenyl ligand due to the ring juncture. In any case, binding of η^5-indenyl ligands is often weaker than binding of η^5-Cp ligands, and the ground-state structures of η^5-indenyl complexes are often distorted toward those of η^3 structures.[236]

Substituents on cyclopentadienyl ligands affect the rates of ring slip for different reasons. The reactions of complexes containing an electron-withdrawing substituent are faster than those lacking such a group because the electron-withdrawing group stabilizes the more localized negative charge in the intermediate containing the η^3 ligand. Likewise the presence of electron-donating methyl groups in the Cp* ligand destabilizes the more localized negative charge on the η^3 ligand. Slippage of acyclic pentadienyl ligands during associative substitutions has also been documented.[237]

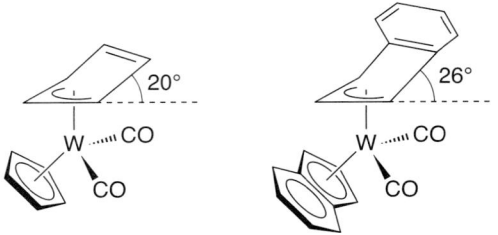

Figure 5.7.
Geometry of η^3-dienyl ligands resulting from ring slip.

The geometry of the dienyl ligand changes upon ring slip.[238] Crystal structures show that the η^3-C_5H_5 ligand in $(\eta^5$-$C_5H_5)(\eta^3$-$C_5H_5)W(CO)_2$ (Figure 5.7) is bent by 20°,[239] the η^3-indenyl group in $(\eta^5$-indenyl)$(\eta^3$-indenyl)$W(CO)_2$ (Figure 5.7) is bent by 26°,[240] and the η^3-indenyl group in $(\eta^5$-indenyl)$Ir(PPhMe_2)_3$ is bent by 28°.[234] The orientation of the η^3 groups in these complexes is similar to that in η^3-allyl complexes. The metal is about 0.02 Å closer to the central carbon than it is to the terminal carbons of the allyl unit. In contrast to the changes in geometry upon ring slip of Cp or indenyl ligands, the more rigid fluorenyl ligand is not bent appreciably in η^3 complexes, such as $(\eta^5$-fluorenyl)$(\eta^3$-fluorenyl)$ZrCl_2$.[241]

Complexes containing η^1-Cp ligands have also been characterized, and Cp and related ligands have even been completely displaced from the metal in some cases. The addition of PMe_3 to $[CpRe(NO)(PMe_3)Me]$ causes reversible formation of the trisphosphine complex containing an η^1-Cp ligand, as shown in Equation 5.50.[43,233] Heating this system leads to complete displacement of the Cp⁻. A complex that contains an η^1-C_5H_5 ligand that is isostructural with the one in Equation 5.50 has been characterized by X-ray diffraction.[43] Indenyl complexes of rhenium and manganese, $Re(\eta^5$-$C_9H_7)(CO)_3$[233] and $Mn(\eta^5$-$C_9H_7)$ $(CO)_3$,[242] undergo η^5 to η^1 slippage, and an η^1-fluorenyl complex has been characterized.[242]

(5.50)

A change in binding mode of other ligands can allow associative ligand substitutions to occur with formally 18-electron complexes. For example, bending of a nitrosyl ligand permits associative substitution of coordinatively saturated nitrosyl complexes. The reaction of the 18-electron complex $Fe(NO)_2(CO)_2$ exhibits a two-term rate law in which one term corresponds to association of solvent and one corresponds to association of the entering ligand (solvent =MeOH or THF).[243] Likewise, the rate equation for substitution of CO in the 18-electron omplex $V(CO)_5(NO)$ contains both an associative $k_2[L]$ term, and a dissociative k_1 term.[244]

5.6. Ligand Substitutions in Metal–Metal Bonded Bimetallic and Higher Nuclearity Clusters

Ligand substitutions of bimetallic complexes and higher-nuclearity clusters occur by the dissociative, associative, and catalyzed pathways that have already been presented. However, they also react by pathways that are available because of the presence of a metal–metal bond. This section discusses pathways that involve the metal–metal bond. Reviews of ligand substitutions in metal–metal bonded systems are available.[245–247]

Reversible homolysis of relatively weak first-row M–M bonds generates 17-electron metal fragments. As stated previously in this chapter, 17-electron metal-centered radicals undergo fast associative ligand substitution. Thus, substitutions of complexes containing metal–metal bonds can occur by initial dissociation into two 17-electron fragments.

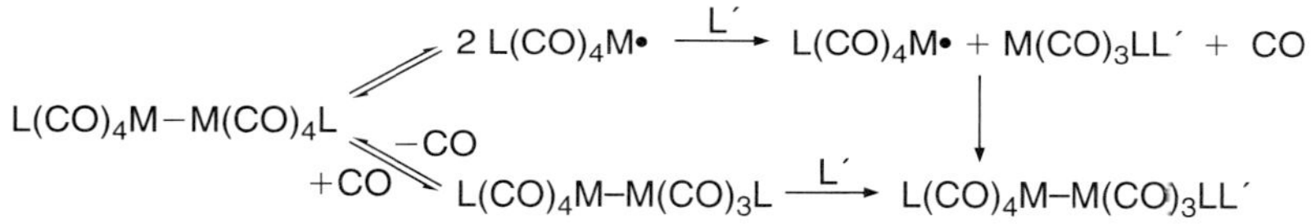

Scheme 5.10

The pathways for ligand substitution by dissociation of CO and by dissociation of the M–M bond are depicted in Scheme 5.10. The relative strength of the M–CO and M–M bonds dictates whether a dinuclear complex undergoes ligand substitution by initial homolysis or by initial dissociation of ligand. The enthalpy of activation for the homolysis of the M–M bond is related to the M–M BDE, just as the enthalpy of activation for a dissociative ligand substitution is related to the M–L BDE. Steric effects are a significant factor in weakening these M–M bonds. The enthalpies of activation for homolysis of complexes of the type $Mn_2(CO)_8L_2$ follow the trend L = $PEt_3 > PPh_3 > PCy_3$.[248]

Kinetic evidence obtained by Poë and co-workers implied that the reaction of $P(OPh)_3$ with $[Mn(CO)_4(PPh_3)]_2$ at 30–50 °C occured by initial homolysis of the Mn–Mn bond (top path of Scheme 5.10).[95] The reaction of $Mn_2(CO)_8(PPh_3)_2$ with $Mn_2(CO)_8(PCy_3)_2$ to give two equivalents of $Mn_2(CO)_8(PPh_3)(PCy_3)$[248,249] is consistent with cleavage of the M–M bond. The rate constant for recombination of two •$Mn(CO)_4(PPh_3)$ radicals is only about 10^2 less than that of a diffusion-controlled process.

$$Mn_2(CO)_{10} + Mn_2(^{13}CO)_{10} \xrightarrow[80\ °C,\ C_5H_6]{CNBu^t} Mn_2(CO)_9(CNBu^t) + Mn_2(^{13}CO)_9(CNBu^t)$$

$$\text{No } Mn_2(CO)_4(^{13}CO)_5(CNBu^t)$$

$$\text{or } Mn_2(CO)_5(^{13}CO)_4(CNBu^t)$$

(5.51)

Wawersik and Basolo earlier had proposed[250] that ligand replacements in the homoleptic $Mn_2(CO)_{10}$, which contains a stronger Mn–Mn bond than the analogs possessing phosphine ligands, occurred by CO dissociation rather than Mn–Mn homolysis. Studies reported since that time have been controversial,[160, 251–253] but the most recent experiments corroborate the original mechanism in which the M–M bond remains intact. This conclusion was drawn mainly from Muetterties' observation that the substitution shown in Equation 5.51 occurred with the mixture of $Mn_2(^{12}CO)_{10}$ and $Mn_2(^{13}CO)_{10}$[254] without crossover to form $Mn_2(^{12}CO)_4(^{13}CO)_5$ or $Mn_2(^{13}CO)_5(^{13}CO)_4$, and substitution of ^{13}CO for coordinated CO in the mixture of $^{185}Re_2(CO)_{10}$ and $^{187}Re_2(CO)_{10}$[255] occurred without crossover to form $^{185}Re^{187}Re(CO)_{10}$.[251, 256, 257] This lack of crossover implies that M–M bond cleavage did not occur during these reactions and rules out substitution after homolysis of the M–M bond as the dominant pathway.[254, 255, 257] Instead, the results support reaction by the bottom pathway of Scheme 5.10, which is initiated by dissociation of CO. This conclusion is consistent with estimates of the Mn–Mn and Mn–CO bond energies in Table 5.2 (38 ± 5 and 36 ± 2 kcal/mol, respectively), indicating that the M–CO bond is weaker than the M–M bond.

Metal carbonyl clusters, $M_x(CO)_y$, in many cases do undergo cleavage into fragments of lower nuclearity when treated with strong nucleophiles, such as PBu_3, or with high pressures of CO. An example of competitive ligand substitutions of clusters and cleavage of the clusters is shown in Equation 5.52.[258, 259] At high temperatures and low concentrations of phosphine, the products from ligand substitution at $Os_3(CO)_{12}$ are formed. However, at 70 °C and a high concentration of PBu_3, a bimolecular, associative pathway for ligand substitution is favored, and mononuclear fragmentation products, $Os(CO)_{5-x}L_x$, are formed.

$$Os_3(CO)_{12} \begin{array}{l} \xrightarrow[\text{[PBu}_3] \leq 0.1\text{ M}]{170\ ^\circ\text{C}} Os_3(CO)_{11}L \ + \ Os_3(CO)_{10}L_2 \ + \ Os_3(CO)_9L_3 \\[2em] \xrightarrow[\text{[PBu}_3] \geq 1\text{ M}]{70\ ^\circ\text{C}} Os(CO)_4L \ + \ Os(CO)_3L_2 \ + \ CO \end{array}$$

$$(5.52)$$

Many reactions of metal cluster complexes occurring by electron-transfer catalysis at low temperatures lead to substitutions without fragmentation to monomeric compounds. For example, derivatives of $Ru_3(CO)_{12}$ and $H_4Ru_4(CO)_{12}$ have been prepared using $Na^+Ph_2CO^-$ as catalyst without detectable byproducts from cluster fragmentation.[173] Successive substitutions on intact clusters often proceed with increasing facility. For example, $Ru_3(CO)_{12}$ reacts with less than 0.01 M PBu_3 at ≥ 60 °C to form the trisubstituted complex $Ru_3(CO)_9(PBu_3)_3$ as the first observable product.[260]

Successive sequential substitutions of cluster complexes can also occur by different mechanisms. In a classic study of the substitution reactions between $Ir_4(CO)_{12}$ and PPh_3 (Equation 5.53),[261] the initial substitution was shown to occur by an associative mechanism, but the second substitution was shown to occur by a dissociative mechanism. Under the conditions employed, the relative rates for the first, second, and third substitutions were 1:30:920. This trend has been ascribed, in part, to the presence of bridging carbonyl ligands[262–264] in the substitution products. Steric effects[265] also may be important.

$$(5.53)$$

5.7. Summary

Ligand substitution reactions occur by many different pathways. The simplest pathways are dissociative and associative. Dissociative pathways occur most commonly with electronically saturated metal complexes, and associative pathways with electronically unsaturated complexes. These two pathways occur through an intermediate of decreased or increased coordination number, respectively. However, interchange pathways that lie between these two extremes on a mechanistic continuum are also well documented. In addition to these classic mechanisms for ligand substitution, organometallic compounds undergo ligand substitutions by photochemical processes, after oxidation or reduction to form odd-electron intermediates, after oxidation of a coordinated CO by amine oxide, or by catalysis with nucleophiles. Complexes of polyhapto ligands have been shown to undergo ligand substitution by several different pathways, including those occurring after dissociation of one portion of the π-ligand and by ring slip of cyclic π-ligands. Finally, polynuclear complexes can undergo substitution by pathways involving cleavage of metal–metal bonds, and the importance of this type of pathway depends on the relative strengths of the M–M and M–L bonds.

References and Notes

1. Richens, D. T. *Chem. Rev.* **2005**, *105*, 1961.
2. Atwood, J. D. *Inorganic and Organometallic Reaction Mechanisms*, 2nd ed.; Wiley-VCH: Monterey, CA, 1997.
3. Basolo, F. *Inorg. Chim. Acta* **1981**, *50*, 65.
4. Basolo, F. *Inorg. Chim. Acta* **1985**, *100*, 33.
5. Howell, J. A. S.; Burkinshaw, P. M. *Chem. Rev.* **1983**, *83*, 557.
6. Darensbourg, D. J. *Adv. Organomet. Chem.* **1982**, *21*, 113.
7. Atwood, J. D.; Wovkulich, M. J.; Sonnenberger, D. C. *Acc. Chem. Res.* **1983**, *16*, 350.
8. *Mechanisms of Inorganic and Organometallic Reactions*; Twigg, M. V., Ed.; Plenum: New York, 1983; Vol. 1, Part 3.
9. *Mechanisms of Inorganic and Organometallic Reactions*; Twigg, M. V., Ed.; Plenum: New York, 1984; Vol. 2, Part 3.
10. *Mechanisms of Inorganic and Organometallic Reactions*; Twigg, M. V., Ed.; Plenum: New York, 1985; Vol. 3, Part 3.
11. Langford, C. H.; Gray, H. B. *Ligand Substitution Processes*; W. A. Benjamin: New York, 1965.
12. Basolo, F.; Pearson, R. G. *Mechanisms of Inorganic Reactions*, 2nd ed.; Wiley: New York, 1967.
13. Wilkins, R. G. *The Study of Kinetics and Mechanism of Reactions of Transition Metal Complexes*; Allyn & Bacon: Boston, MA, 1974.
14. Tobe, M. L. *Inorganic Reaction Mechanisms*; Nelson: London, 1972.
15. *Mechanistic Aspects of Inorganic Reactions*; Rorabacher, D. B., Endicott, J. F., Eds.; American Chemical Society: Washington, DC, 1982; Vol. 198.
16. Cross, R. J. *Chem. Soc. Rev.* **1985**, 197.
17. Peloso, A. *Coord. Chem. Rev.* **1973**, *10*, 123.
18. *Mechanisms of Inorganic and Organometallic Reactions*; Twigg, M. V., Ed.; Plenum: New York, 1983; Vol. 1, Part 2.
19. *Mechanisms of Inorganic and Organometallic Reactions*; Twigg, M. V., Ed.; Plenum: New York, 1984; Vol. 2, Part 2.
20. *Mechanisms of Inorganic and Organometallic Reactions*; Twigg, M. V., Ed.; Plenum: New York, 1985; Vol. 3, Part 2.
21. Huheey, J. E. *Inorganic Chemistry*, 3rd ed.; Harper & Row: New York, 1983.
22. Douglas, B.; McDaniel, D. H.; Alexander, J. J. *Concepts and Models of Inorganic Chemistry*, 2nd ed.; Wiley: New York, 1983.
23. Purcell, K. F.; Kotz, J. C. *Inorganic Chemistry*; Saunders: Philadelphia, 1977.
24. Cotton, F. A.; Wilkinson, G. *Advanced Inorganic Chemistry*, 4th ed.; Wiley: New York, 1980.
25. Collman, J. P. *Acc. Chem. Res.* **1968**, *1*, 136.
26. Atwood, J. D. *Inorganic and Organometallic Reaction Mechanisms*; Brooks/Cole: Monterey, CA, 1985.

27. Jordan, R. B. *Reaction Mechanisms of Inorganic and Organometallic Systems*, 3rd ed.; Oxford University Press: Oxford, 2007.

28. The volume of activation is interpreted, according to transition state theory, "as the difference between the partial molar volumes of the transition state (V) and the sums of the partial volumes of the reactants at the same temperature and pressure." IUPAC Compendium of Chemical Terminology, 2nd ed., 1997; http://goldbook.iupac.org/index.html.

29. Alkyl groups do transfer from one metal to another, but ligand substitutions typically generate the free ligand, and these groups are not eliminated as the unstable free ligand.

30. Li, C.; Stevens, E. D.; Nolan, S. P. *Organometallics* **1995**, *14*, 3791.

31. Luo, L.; Li, C.; Cucullu, M. E.; Nolan, S. P. *Organometallics* **1995**, *14*, 1333.

32. Wang, K.; Rosini, G. P.; Nolan, S. P.; Goldman, A. S. *J. Am. Chem. Soc.* **1995**, *117*, 5082.

33. Wang, K.; Goldman, A. S.; Nolan, S. P. *Organometallics* **1995**, *14*, 4010.

34. Serron, S. A.; Nolan, S. P. *Inorg. Chim. Acta* **1996**, *252*, 107.

35. Serron, S. A.; Luo, L.; Stevens, E. D.; Nolan, S. P.; Jones, N. L.; Fagan, P. J. *Organometallics* **1996**, *15*, 5209.

36. Rosini, G. P.; Liu, F. C.; Krogh-Jespersen, K.; Goldman, A. S.; Li, C. B.; Nolan, S. P. *J. Am. Chem. Soc.* **1998**, *120*, 9256.

37. Huang, J. K.; Schanz, H. J.; Stevens, E. D.; Nolan, S. P. *Organometallics* **1999**, *18*, 2370.

38. Jafarpour, L.; Nolan, S. P. *Adv. Organomet. Chem.* **2001**, *46*, 181.

39. Zhang, K.; Gonzalez, A. A.; Mukerjee, S. L.; Chou, S. J.; Hoff, C. D.; Kubatmartin, K. A.; Barnhart, D.; Kubas, G. J. *J. Am. Chem. Soc.* **1991**, *113*, 9170.

40. Dorta, R.; Stevens, E. D.; Scott, N. M.; Costabile, C.; Cavallo, L.; Hoff, C. D.; Nolan, S. P. *J. Am. Chem. Soc.* **2005**, *127*, 2485.

41. Dorta, R.; Stevens, E. D.; Hoff, C. D.; Nolan, S. P. *J. Am. Chem. Soc.* **2003**, *125*, 10490.

42. Mukerjee, S. L.; Lang, R. F.; Ju, T.; Kiss, G.; Hoff, C. D.; Nolan, S. P. *Inorg. Chem.* **1992**, *31*, 4885.

43. Casey, C. P.; O'Connor, J. M.; Haller, K. J. *J. Am. Chem. Soc.* **1985**, *107*, 1241.

44. Hoff, C. D. *J. Organomet. Chem.* **1985**, *282*, 201.

45. Skinner, H. A.; Connor, J. A. *Pure. Appl. Chem.* **1985**, *57*, 79.

46. Pilcher, G.; Skinner, H. A. In *The Chemistry of the Metal–Carbon Bond*; Hartley, F. R., Patai, S., Eds.; Wiley: New York, 1982; Vol. 1, Chapter 2.

47. Connor, J. A. *Top. Curr. Chem.* **1977**, *71*, 71.

48. Mondal, J. U.; Blake, D. M. *Coord. Chem. Rev.* **1982**, *47*, 205.

49. Nolan, S. P.; Hoff, C. D. *J. Organomet. Chem.* **1985**, *290*, 365.

50. A clear exception to this trend is the M–CO bond strength in $Pd(CO)_4$ and $Pt(CO)_4$, as discussed in Chapter 2.

51. Lowry, T. H.; Richardson, K. S. *Mechanism and Theory in Organic Chemistry*, 3rd ed.; Harper & Row: New York, 1987.

52. Carpenter, B. K. *Determination of Organic Reaction Mechanisms*; Wiley-Interscience: New York, 1984.

53. Berry, R. S. *J. Chem. Phys.* **1960**, *32*, 933.

54. Shapley, J. R.; Osborn, J. A. *Acc. Chem. Res.* **1973**, *6*, 305.

55. Yamashita, F.; Kuniyasu, H.; Terao, J.; Kambe, N. *Inorg. Chem.* **2006**, *45*, 1399.

56. Cooper, M. K.; Downes, J. M. *J. Chem. Soc., Chem. Commun.* **1981**, 381.

57. Odell, A. L.; Raethel, H. A. *J. Chem. Soc., Chem. Commun.* **1968**, 1323.

58. Cheeseman, T. P.; Odell, A. L.; Raethel, H. A. *J. Chem. Soc., Chem. Commun.* **1968**, 1496.

59. Odell, L. A.; Raethel, H. A. *J. Chem. Soc., Chem. Commun.* **1969**, 87.

60. Heise, J. D.; Raftery, D.; Breedlove, B. K.; Washington, J.; Kubiak, C. P. *Organometallics* **1998**, *17*, 4461.

61. Gray, H. B.; Olcott, R. J. *Inorg. Chem.* **1962**, *1*, 481.

62. Basolo, F.; Gray, H. B.; Pearson, R. G. *J. Am. Chem. Soc.* **1960**, *82*, 4200.

63. Pesek, J. J.; Mason, W. R. *Inorg. Chem.* **1983**, *22*, 2958.

64. Basolo, F.; Pearson, R. G.; Chatt, J.; Gray, H. B.; Shaw, B. L. *J. Chem. Soc.* **1961**, 2207.

65. See page 49 of reference 70.

66. Langford, C. H.; Gray, H. B. *Ligand Substitution Processes*; W.A. Benjamin: Reading, MA, 1966.

67. *Mechanisms of Inorganic and Organometallic Reactions*; Twigg, M. V., Ed.; Plenum: New York, 1984; Vol. 2, p 119.

68. See page 67 of reference 2 and page 62 of reference 26.

69. Appleton, T. G.; Clark, H. C.; Manzer, L. E. *Coord. Chem. Rev.* **1973**, *10*, 335.

70. McWeeny, R.; Mason, R.; Towl, A. D. C. *Discuss. Faraday Soc.* **1969**, 20.

71. Brescianipahor, N.; Forcolin, M.; Marzilli, L. G.; Randaccio, L.; Summers, M. F.; Toscano, P. J. *Coord. Chem. Rev.* **1985**, *63*, 1.

72. Lanza, S.; Minniti, D.; Romeo, R.; Moore, P.; Sachinidis, J.; Tobe, M. L. *J. Chem. Soc., Chem. Commun.* **1984**, 542.

73. Frey, U.; Helm, L.; Merbach, A. E.; Romeo, R. *J. Am. Chem. Soc.* **1989**, *111*, 8161.

74. Romeo, R.; Grassi, A.; Scolaro, L. M. *Inorg. Chem.* **1992**, *31*, 4383.

75. Zhao, J.; Goldman, A. S.; Hartwig, J. F. *Science* **2005**, *307*, 1080.

76. Gottker-Schnetmann, I.; Brookhart, M. *J. Am. Chem. Soc.* **2004**, *126*, 9330.

77. Wendt, O. F.; Deeth, R. J.; Elding, L. I. *Inorg. Chem.* **2000**, *39*, 5271.

78. Herrick, R. S.; Leazer, D. M.; Templeton, J. L. *Organometallics* **1983**, *2*, 834.

79. Johansson, L.; Tilset, M. *J. Am. Chem. Soc.* **2001**, *123*, 739.

80. Zhong, H. A.; Labinger, J. A.; Bercaw, J. E. *J. Am. Chem. Soc.* **2002**, *124*, 1378.

81. Brown, T. L. *Ann. N. Y. Acad. Sci.* **1980**, *333*, 80.

82. Kochi, J. K. *Organometallic Mechanisms and Catalysis*; Academic: New York, 1978.

83. Doxsee, K. M.; Grubbs, R. H.; Anson, F. C. *J. Am. Chem. Soc.* **1984**, *106*, 7819.

84. Shi, Q. Z.; Richmond, T. G.; Trogler, W. C.; Basolo, F. *J. Am. Chem. Soc.* **1984**, *106*, 71.

85. Ellis, J. E.; Faltynek, R. A.; Rochfort, G. L.; Stevens, R. E.; Zank, G. A. *Inorg. Chem.* **1980**, *19*, 1082.

86. Davison, A.; Ellis, J. E. *J. Organomet. Chem.* **1971**, *31*, 239.

87. Non-labile 17-electron compounds are known, such as Mn(η^4-butadiene)$_2$L and Co[P(OMe)$_3$]$_4$. See reference 89, 90.

88. Zizelman, P. M.; Amatore, C.; Kochi, J. K. *J. Am. Chem. Soc.* **1984**, *106*, 3771.

89. Harlow, R. L.; Krusic, P. J.; McKinney, R. J.; Wreford, S. S. *Organometallics* **1982**, *1*, 1506.

90. Muetterties, E. L.; Bleeke, J. R.; Yang, Z. Y.; Day, V. W. *J. Am. Chem. Soc.* **1982**, *104*, 2940.

91. Kidd, D. R.; Brown, T. L. *J. Am. Chem. Soc.* **1978**, *100*, 4095.

92. Byers, B. H.; Brown, T. L. *J. Am. Chem. Soc.* **1977**, *99*, 2527.

93. Wegman, R. W.; Olsen, R. J.; Gard, D. R.; Faulkner, L. R.; Brown, T. L. *J. Am. Chem. Soc.* **1981**, *103*, 6089.

94. Stiegman, A. E.; Tyler, D. R. *Comments Inorg. Chem.* **1986**, *5*, 215.

95. Fawcett, J. P.; Jackson, R. A.; Poë, A. *J. Chem. Soc., Chem. Commun.* **1975**, 733.

96. Kaim, W. *Inorg. Chem.* **1984**, *23*, 504.

97. Fenske, D. *Chem. Ber.* **1979**, *112*, 363.

98. Creber, K. A. M.; Wan, J. K. S. *J. Am. Chem. Soc.* **1981**, *103*, 2101.

99. Meckstroth, W. K.; Walters, R. T.; Waltz, W. L.; Wojcicki, A.; Dorfman, L. M. *J. Am. Chem. Soc.* **1982**, *104*, 1842.

100. Fox, A.; Malito, J.; Poë, A. *J. Chem. Soc., Chem. Commun.* **1981**, 1052.

101. Herrinton, T. R.; Brown, T. L. *J. Am. Chem. Soc.* **1985**, *107*, 5700.

102. McCullen, S. B.; Walker, H. W.; Brown, T. L. *J. Am. Chem. Soc.* **1982**, *104*, 4007.

103. For reviews of 17-electron complexes, see (a) Baird, M. *Chem. Rev.* **1988**, *88*, 1217; (b) Astruc, D. *Acc. Chem. Res.* **1991**, *24*, 36; (c) Steigman, A. E.; Tyler, D. R. *Comments Inorg. Chem.* **1986**, *5*, 215.

104. Cramer, R. *J. Am. Chem. Soc.* **1964**, *86*, 217.

105. Cramer, R. *J. Am. Chem. Soc.* **1972**, *94*, 5681.

106. Hyde, C. L.; Darensbourg, D. J. *Inorg. Chem.* **1973**, *12*, 1286.

107. Covey, W. D.; Brown, T. L. *Inorg. Chem.* **1973**, *12*, 2820.

108. Because the dissociation will have a positive entropy of activation, the value of the enthalpy of bond cleavage can be higher than the free energy of activation.

109. Wovkulich, M. J.; Atwood, J. D. *J. Organomet. Chem.* **1980**, *184*, 77.

110. Day, J. P.; Basolo, F.; Pearson, R. G. *J. Am. Chem. Soc.* **1968**, *90*, 6927.

111. Benson, S. W. *Thermochemical Kinetics*, 2nd ed.; Wiley-Interscience: New York, 1976; p 89.

112. Benson, S. W. *Thermochemical Kinetics*, 2nd ed.; Wiley-Interscience: New York, 1976; Chapter 3.

113. Kuran, W.; Musco, A. *Inorg. Chim. Acta* **1975**, *12*, 187.

114. Otsuka, S.; Yoshida, T.; Matsumoto, M.; Nakatsu, K. *J. Am. Chem. Soc.* **1976**, *98*, 5850.

115. Yoshida, T.; Otsuka, S. *Inorg. Synth.* **1985**, *28*, 113.

116. Tolman, C. A. *Chem. Rev.* **1977**, *77*, 313.

117. Brunner, H. *Adv. Organomet. Chem.* **1980**, *18*, 151.

118. Yang, G. K.; Peters, K. S.; Vaida, V. *Chem. Phys. Lett.* **1986**, *125*, 566.

119. Simon, J. D.; Peters, K. S. *Chem. Phys. Lett.* **1983**, *98*, 53.

120. Lees, A. J.; Adamson, A. W. *Inorg. Chem.* **1981**, *20*, 4381.

121. Church, S. P.; Grevels, F. W.; Hermann, H.; Schaffner, K. *Inorg. Chem.* **1985**, *24*, 418.

122. Matthews, S. L.; Pons, V.; Heinekey, D. M. *J. Am. Chem. Soc.* **2005**, *127*, 850.

123. Kelly, J. M.; Gustorf, E. K. V. *J. Chem. Soc., Chem. Commun.* **1973**, 105.

124. Von Gustorf, E. K.; Leenders, L. H. G.; Fischler, I.; Perutz, R. N. *Adv. Inorg. Chem. Radiochem.* **1976**, *19*, 65.

125. Weiller, B. H.; Wasserman, E. P.; Bergman, R. G.; Moore, C. B.; Pimentel, G. C. *J. Am. Chem. Soc.* **1989**, *111*, 8288.
126. Darensbourg, D. J.; Rokicki, A.; Kudaroski, R. *Organometallics* **1982**, *1*, 1161.
127. Elian, M.; Hoffmann, R. *Inorg. Chem.* **1975**, *14*, 1058.
128. Burdett, J. K. *J. Chem. Soc., Faraday Trans.* **1974**, *70*, 1599.
129. Cotton, F. A.; Darensbourg, D. J.; Kolthammer, B. W. S. *J. Am. Chem. Soc.* **1981**, *103*, 398.
130. Cotton, F. A.; Darensbourg, D. J.; Kolthammer, B. W. S.; Kudaroski, R. *Inorg. Chem.* **1982**, *21*, 1656.
131. Atwood, J. D.; Brown, T. L. *J. Am. Chem. Soc.* **1975**, *97*, 3380.
132. Atwood, J. D.; Brown, T. L. *J. Am. Chem. Soc.* **1976**, *98*, 3155.
133. Bryndza, H. E.; Domaille, P. J.; Paciello, R. A.; Bercaw, J. E. *Organometallics* **1989**, *8*, 379.
134. Glueck, D. S.; Bergman, R. G. *Organometallics* **1991**, *10*, 1479.
135. Flood, T. C.; Kim, J. K.; Deming, M. A.; Keung, W. *Organometallics* **2000**, *19*, 1166.
136. Randall, W. C.; Alberty, R. A. *Biochemistry* **1966**, *5*, 3189.
137. Randall, W. C.; Alberty, R. A. *Biochemistry* **1967**, *6*, 1520.
138. Thusius, D. *J. Am. Chem. Soc.* **1971**, *93*, 2629.
139. Hague, D. N.; Halpern, J. *Inorg. Chem.* **1967**, *6*, 2059.
140. Bresciani-Pahor, N.; Forcolin, M.; Marzilli, L. G.; Randaccio, L.; Summers, M. F.; Toscano, P. J. *Coord. Chem. Rev.* **1985**, *63*, 1.
141. Toscano, P. J.; Marzilli, L. G. *Prog. Inorg. Chem.* **1983**, *31*, 105.
142. Atwood, J. D.; Brown, T. L. *J. Am. Chem. Soc.* **1975**, *97*, 3380.
143. Darensbourg, D. J.; Kudaroski, R.; Schenk, W. *Inorg. Chem.* **1982**, *21*, 2488.
144. Hoffman, N. W.; Brown, T. L. *Inorg. Chem.* **1978**, *17*, 613.
145. Siefert, E. E.; Angelici, R. J. *J. Organomet. Chem.* **1967**, *8*, 374.
146. Hieber, W.; Vonpigenot, D. *Chem. Ber.* **1956**, *89*, 193.
147. Graham, J. R.; Angelici, R. J. *Inorg. Chem.* **1967**, *6*, 2082.
148. Covey, W. D.; Brown, T. L. *Inorg. Chem.* **1973**, *12*, 2820.
149. Franck, T.; Rabinowitsch, E. *Trans. Faraday Soc.* **1934**, *30*, 120.
150. Noyes, R. M. *J. Chem. Phys.* **1954**, *22*, 1349.
151. Noyes, R. M. *J. Am. Chem. Soc.* **1955**, *77*, 2042.
152. Koenig, T. In *Free Radicals*; Kochi, J., Ed.; Wiley: New York, 1973; Vol. 1.
153. Koenig, T. In *Free Radicals*; Pryor, W. A., Ed.; American Chemical Society: Washington, DC, 1978, Chapter 9.
154. Kerr, J. A. *Chem. Rev.* **1966**, *66*, 465.
155. Benson, S. W. In *Free Radicals*; Kochi, J., Ed.; Wiley: New York, 1973; Vol. 2, p 293.
156. Hay, B. P.; Finke, R. G. *J. Am. Chem. Soc.* **1986**, *108*, 4820.
157. Edwards, J. O.; Monacelli, F; Ortaggi, G. *Inorg. Chim. Acta* **1974**, *11*, 47.
158. Benson, S. W. *The Foundations of Chemical Kinetics*; McGraw-Hill: New York, 1960; p 91.
159. Benson, S. W. *Thermochemical Kinetics,* 2nd ed.; Wiley-Interscience: New York, 1976.
160. Atwood, J. D. *Inorg. Chem.* **1981**, *20*, 4031.
161. Albers, M. O.; Coville, N. J. *Coord. Chem. Rev.* **1984**, *53*, 227.
162. Chanon, M. *Bull. Soc. Chim. Fr.* **1982**, 197.
163. Julliard, M.; Chanon, M. *Chem. Rev.* **1983**, *83*, 425.
164. Eberson, L. *J. Mol. Catal.* **1983**, *20*, 27.
165. Alder, R. W. *J. Chem. Soc., Chem. Commun.* **1980**, 1184.
166. Hershberger, J. W.; Kochi, J. K. *J. Chem. Soc., Chem. Commun.* **1982**, 212.
167. Hershberger, J. W.; Klingler, R. J.; Kochi, J. K. *J. Am. Chem. Soc.* **1982**, *104*, 3034.
168. Hershberger, J. W.; Klingler, R. J.; Kochi, J. K. *J. Am. Chem. Soc.* **1983**, *105*, 61.
169. Darchen, A.; Mahe, C.; Patin, H. *J. Chem. Soc., Chem. Commun.* **1982**, 243.
170. Jensen, S.; Robinson, B. H.; Simpson, J. *J. Chem. Soc., Chem. Commun.* **1983**, 1081.
171. Cunninghame, R. G.; Downard, A. J.; Hanton, L. R.; Jensen, S. D.; Robinson, B. H.; Simpson, J. *Organometallics* **1984**, *3*, 180.
172. Butts, S. B.; Shriver, D. F. *J. Organomet. Chem.* **1979**, *169*, 191.
173. Bruce, M. I.; Matisons, J. G.; Nicholson, B. K. *J. Organomet. Chem.* **1983**, *247*, 321.
174. Horwitz, C. P.; Holt, E. M.; Shriver, D. F. *Organometallics* **1985**, *4*, 1117.
175. Goldman, A. S.; Tyler, D. R. *Inorg. Chim. Acta* **1985**, *98*, L47.
176. Byers, B. H.; Brown, T. L. *J. Am. Chem. Soc.* **1975**, *97*, 947.
177. The original mechanism was modified to include associative substitutions of the 17-electron intermediates. Associative mechanisms for substitution of 17-electron intermediates were shown to occur after this initial study that illustrated a radical chain mechanism.
178. Strohmeier, W.; Muller, F. J. *Chem. Ber.* **1969**, *102*, 3613.

179. Therien, M. J.; Trogler, W. C. *Inorg. Synth.* **1990**, *28*, 173.
180. Strohmeier, W.; Gerlach, K. *Chem. Ber.* **1961**, *94*, 398.
181. Schubert, U.; Friedrich, P.; Orama, O. *J. Organomet. Chem.* **1978**, *144*, 175.
182. Herberhold, M.; Schmidkonz, B. *J. Organomet. Chem.* **1986**, *308*, 35.
183. Herrmann, W. A.; Hecht, C.; Herdtweck, E.; Kneuper, H. J. *Angew. Chem., Int. Ed Engl.* **1987**, *26*, 132.
184. Casey, C. P.; Sakaba, H.; Hazin, P. N.; Powell, D. R. *J. Am. Chem. Soc.* **1991**, *113*, 8165.
185. Zhuang, J. M.; Sutton, D. *Organometallics* **1991**, *10*, 1516.
186. Jones, W. D.; Hessell, E. T. *J. Am. Chem. Soc.* **1993**, *115*, 554.
187. Geoffroy, G. L.; Wrighton, M. S. *Organometallic Photochemistry*; Academic: New York, 1979.
188. Snee, P. T.; Payne, C. K.; Mebane, S. D.; Kotz, K. T.; Harris, C. B. *J. Am. Chem. Soc.* **2001**, *123*, 6909.
189. Shvo, Y.; Hazum, E. *J. Chem. Soc., Chem. Commun.* **1975**, 829.
190. Shapley, J. R.; Sievert, A. C.; Churchill, M. R.; Wasserman, H. J. *J. Am. Chem. Soc.* **1981**, *103*, 6975.
191. Lawson, R. J.; Shapley, J. R. *J. Am. Chem. Soc.* **1976**, *98*, 7433.
192. Koelle, U. *J. Organomet. Chem.* **1977**, *133*, 53.
193. Maher, J. M.; Beatty, R. P.; Cooper, N. J. *Organometallics* **1985**, *4*, 1354.
194. Darensbourg, D. J.; Walker, N.; Darensbourg, M. Y. *J. Am. Chem. Soc.* **1980**, *102*, 1213.
195. Darensbourg, D. J.; Darensbourg, M. Y.; Walker, N. *Inorg. Chem.* **1981**, *20*, 1918.
196. Darensbourg, D. J.; Ewen, J. A. *Inorg. Chem.* **1981**, *20*, 4168.
197. Webb, S. L.; Giandomenico, C. M.; Halpern, J. *J. Am. Chem. Soc.* **1986**, *108*, 345.
198. Basolo, F.; Brault, A. T.; Poe, J. *J. Chem. Soc.* **1964**, 676.
199. Noack, K.; Ruch, M. *J. Organomet. Chem.* **1969**, *17*, 309.
200. Albers, M.; Coville, N. J.; Singleton, E. *J. Chem. Soc., Chem. Commun* **1982**, 96.
201. Suslick, K. S. In *Modern Synthetic Methods*; Scheffold, R., Ed.; Springer-Verlag: Heidelberg, 1986; Vol. 4, p 1.
202. Dixon, D. T.; Burkinshaw, P. M.; Howell, J. A. S. *J. Chem. Soc., Dalton Trans.* **1980**, 2237.
203. Gower, M.; Kanemaguire, L. A. P. *Inorg. Chim. Acta* **1979**, *37*, 79.
204. Al-Kathumi, K. M.; Kane-Maguire, L. A. P. *J. Chem. Soc., Dalton Trans.* **1974**, 428.
205. Muetterties, E. L.; Bleeke, J. R.; Sievert, A. C. *J. Organomet. Chem.* **1979**, *178*, 197.
206. Zingales, F.; Chiesa, A.; Basolo, F. *J. Am. Chem. Soc.* **1966**, *88*, 2707.
207. Strohmeier, W.; Mittnacht, H. Z. *Z. Phys. Chem. (Frankfurt)* **1961**, *29*, 339.
208. Strohmeier, W.; Staricco, E. H. *Z. Phys. Chem. (Frankfurt)* **1963**, *38*, 315.
209. Strohmeier, W.; Muller, R. *Z. Phys. Chem. (Frankfurt)* **1964**, *40*, 85.
210. Traylor, T. G.; Stewart, K. *Organometallics* **1984**, *3*, 325.
211. Traylor, T. G.; Stewart, K. J.; Goldberg, M. J. *J. Am. Chem. Soc.* **1984**, *106*, 4445.
212. See also Table 3 of Mutteries, E. L.; Bleeke, J. R.; Sievert, A. C. *J. Organomet. Chem.* **1979**, *178*, 197.
213. Albright, T. A.; Carpenter, B. K. *Inorg. Chem.* **1980**, *19*, 3092.
214. Muetterties, E. L.; Bleeke, J. R.; Wucherer, E. J.; Albright, T. A. *Chem. Rev.* **1982**, *82*, 499.
215. Howell, J. A. S.; Dixon, D. T.; Kola, J. C.; Ashford, N. F. *J. Organomet. Chem.* **1985**, *294*, C1.
216. Harrison, J. J. *J. Am. Chem. Soc.* **1984**, *106*, 1487.
217. Darchen, A. *J. Chem. Soc., Chem. Commun.* **1983**, 768.
218. Houghton, R. P.; Voyle, M. *J. Chem. Soc., Perkin Trans. I* **1984**, 925.
219. Takaya, J.; Hartwig, J. F. *J. Am. Chem. Soc.* **2005**, *127*, 5756.
220. Semmelhack, M. F.; Chlenov, A.; Wu, L. Y.; Ho, D. *J. Am. Chem. Soc.* **2001**, *123*, 8438.
221. Semmelhack, M. F.; Chlenov, A.; Ho, D. M. *J. Am. Chem. Soc.* **2005**, *127*, 7759.
222. Sievert, A. C.; Muetterties, E. L. *Inorg. Chem.* **1981**, *20*, 489.
223. Dabard, R.; Jaouen, G.; Simonneaux, G.; Cais, M.; Kohn, D. H.; Lapid, A.; Tatarsky, D. *J. Organomet. Chem.* **1980**, *184*, 91.
224. Gill, T. P.; Mann, K. R. *Inorg. Chem.* **1980**, *19*, 3007.
225. Gill, T. P.; Mann, K. R. *J. Organomet. Chem.* **1981**, *216*, 65.
226. Lee, C. C.; Iqbal, M.; Gill, U. S.; Sutherland, R. G. *J. Organomet. Chem.* **1985**, *288*, 89.
227. Hartwig, J. F.; Cook, K. S.; Hapke, M.; Incarvito, C.; Fan, Y.; Webster, C. E.; Hall, M. B. *J. Am. Chem. Soc.* **2005**, *127*, 2538.
228. Tobita, H.; Hasegawa, K.; Minglana, J. J. G.; Luh, L. S.; Okazaki, M.; Ogino, H. *Organometallics* **1999**, *18*, 2058.
229. Utsunomiya, M.; Hartwig, J. F. *J. Am. Chem. Soc.* **2004**, *126*, 2702.
230. Schuster-Woldan, H. G.; Basolo, F. *J. Am. Chem. Soc.* **1966**, *88*, 1657.
231. Rerek, M. E.; Basolo, F. *J. Am. Chem. Soc.* **1984**, *106*, 5908.
232. Cramer, R.; Seiwell, L. P. *J. Organomet. Chem.* **1975**, *92*, 245.
233. Casey, C. P.; O'Connor, J. M. *Organometallics* **1985**, *4*, 384.
234. Merola, J. S.; Kacmarcik, R. T.; Vanengen, D. *J. Am. Chem. Soc.* **1986**, *108*, 329.

235. Calhorda, M. J.; Romao, C. C.; Veiros, L. F. *Chem. Eur. J.* **2002**, *8*, 868.

236. Zargarian, D. *Coord. Chem. Rev.* **2002**, *233–234*, 157.

237. Paz-Sandoval, M. A.; Powell, P.; Drew, M. G. B.; Perutz, R. N. *Organometallics* **1984**, *3*, 1026.

238. Cotton, F. A. *Discuss. Faraday Soc.* **1969**, 79.

239. Huttner, G.; Brintzinger, H. H.; Bell, L. G.; Friedrich, P.; Bejenke, V.; Neugebauer, D. *J. Organomet. Chem.* **1978**, *145*, 329.

240. Nesmeyanov, A. N.; Ustynyuk, N. A.; Makarova, L. G.; Andrianov, V. G.; Struchkov, Y. T.; Andrae, S.; Ustynyuk, Y. A.; Malyugina, S. G. *J. Organomet. Chem.* **1978**, *159*, 189.

241. Kowala, C.; Wunderlich, J. A. *Acta Crystallogr., Sect. B* **1976**, *32*, 820.

242. Ji, L. N.; Rerek, M. E.; Basolo, F. *Organometallics* **1984**, *3*, 740.

243. Morris, D. E.; Basolo, F. *J. Am. Chem. Soc.* **1968**, *90*, 2531.

244. Shi, Q. Z.; Richmond, T. G.; Trogler, W. C.; Basolo, F. *Inorg. Chem.* **1984**, *23*, 957.

245. Muetterties, E. L.; Burch, R. R.; Stolzenberg, A. M. *Ann. Rev. Phys. Chem.* **1982**, *33*, 89.

246. Poë, A. *Chem. in Br.* **1983**, *19*, 997.

247. Chisholm, M. H.; Rothwell, I. P. *Prog. Inorg. Chem.* **1982**, *29*, 1.

248. Jackson, R. A.; Poë, A. *Inorg. Chem.* **1979**, *18*, 3331.

249. Poë, A.; Sekhar, C. V. *J. Am. Chem. Soc.* **1985**, *107*, 4874.

250. Wawersik, H.; Basolo, F. *Inorg. Chim. Acta* **1969**, *3*, 113.

251. Sonnenberger, D.; Atwood, J. D. *J. Am. Chem. Soc.* **1980**, *102*, 3484.

252. Poë, A. *Inorg. Chem.* **1981**, *20*, 4029.

253. Poë, A. *Inorg. Chem.* **1981**, *20*, 4032.

254. Coville, N. J.; Stolzenberg, A. M.; Muetterties, E. L. *J. Am. Chem. Soc.* **1983**, *105*, 2499.

255. Stolzenberg, A. M.; Muetterties, E. L. *J. Am. Chem. Soc.* **1983**, *105*, 822.

256. For related experiments, see references 251 and 257.

257. Schmidt, S. P.; Trogler, W. C.; Basolo, F. *Inorg. Chem.* **1982**, *21*, 1698.

258. Taube, H. *Chem. Rev.* **1952**, *50*, 69.

259. Douglas, B.; McDaniel, D. H.; Alexander, J. J. *Concepts and Models of Inorganic Chemistry*, 2nd ed.; Wiley: New York, 1983.

260. Brodie, N.; Poë, A.; Sekhar, V. *J. Chem. Soc., Chem. Commun.* **1985**, 1090.

261. Karel, K. J.; Norton, J. R. *J. Am. Chem. Soc.* **1974**, *96*, 6812.

262. Sonnenberger, D.; Atwood, J. D. *Inorg. Chem.* **1981**, *20*, 3243.

263. Sonnenberger, D. C.; Atwood, J. D. *J. Am. Chem. Soc.* **1982**, *104*, 2113.

264. Sonnenberger, D. C.; Atwood, J. D. *Organometallics* **1982**, *1*, 694.

265. Darensbourg, D. J.; Baldwinzuschke, B. J. *J. Am. Chem. Soc.* **1982**, *104*, 3906.

CHAPTER 6

Oxidative Addition of Nonpolar Reagents

6.1. Definitions, Examples, and Trends

6.1.1. Definition of Oxidative Addition

Oxidative addition reactions[1–11] form products containing new metal–ligand bonds by cleavage of a bond in an organic or main group reagent or tethered fragment. By this overall process, the metal is formally oxidized. For this reason, these reactions are termed "oxidative additions." More specifically, oxidative additions of mononuclear compounds lead to an increase in the oxidation state of the metal reactant by two, and they cleave a bond in the organic reagent to form a product containing two additional ligands. Oxidative addition reactions constitute an important step in many of the catalytic and stoichiometric applications of transition metal reagents to organic synthesis. They begin the process of creating or changing the composition of functional groups in organic molecules.

The reader may be familiar with a few reactions in organic chemistry that parallel oxidative additions to transition metal complexes (Equations 6.2a and 6.2b). The formation of a Grignard reagent from an alkyl halide and Mg^0, during which Mg metal is transformed into a reagent containing Mg^{2+}, is an oxidative addition process. Few reactions of purely organic systems can be considered oxidative additions, though, because the chemistry of carbon is dominated by a single oxidation state. Nevertheless, there are a few exceptions, such as the insertion of carbenes into C–H or O–H bonds or the additions of carbenes to olefins in which the divalent carbon of a carbene is transformed into a tetravalent carbon in the insertion or addition product. Thus, the difference between the common occurrence of oxidative addition in transition metal chemistry and the rare occurrence of this process in organic chemistry contributes to the distinction between the reactivity of transition metal complexes and molecules comprised of carbon, hydrogen, oxygen, nitrogen, and sulfur.

$$L_nM + A-B \longrightarrow L_nM^{n+2} \overset{A}{\underset{B}{\diagup\diagdown}} \quad \text{Oxidative addition}$$

X valence electrons at M	$X + 2$ valence electrons at M
m d-electrons	$m - $ d-electrons
n oxidation state of M	$n + 2$ oxidation state of M

$$\tag{6.1}$$

$$L_nM^{n+2} \overset{A}{\underset{B}{\diagup\diagdown}} \longrightarrow L_nM^n + A-B \quad \text{Reductive elimination}$$

Examples from introductory chemistry:

$$RX + Mg \longrightarrow RMgX \quad \text{(Formation of a Grignard reagent)} \tag{6.2a}$$

$$:CH_2 + ROH \longrightarrow ROCH_3 \quad \text{(Carbene insertion)} \tag{6.2b}$$

Oxidative additions occur with both mononuclear and multinuclear complexes. As a result, oxidative additions can involve two-electron (Equation 6.1) changes at a single metal or one-electron (Equation 6.3) changes at two metal centers. In the former, the oxidation state of the single-metal center increases from n to $n + 2$, and the number of electrons in its d-electron configuration decreases by two. In the latter, the oxidation state of each metal increases from n to $n + 1$, and the number of electrons in the d-electron configuration of each metal decreases by one. The coordination number of the metal increases in both cases.

Some oxidative additions occur by the simple coordination of an electrophile, such as a proton, or ligation of a strongly π-accepting ligand that forms π-complexes, such as O_2. In these reactions, the electrophile becomes formally monoanionic (in the case of a proton) or dianionic (in the case of dioxygen) in the counting scheme of the ionic model, and the metal is formally oxidized. To distinguish these reactions from those that cleave bonds to form the addition products, these reactions involving coordination of electrophiles or strongly electrophilic π-ligands can be considered "oxidative ligations," as noted in Figure 6.1.

$$L_nM^n - M^nL_n + A - B \longrightarrow L_nM^{n+1} - A + L_nM^{n+1} - B$$
$$2\,L_nM^n + A - A \longrightarrow 2\,L_nM^{n+1} - A$$

Dinuclear oxidative additions

(6.3)

X valence electrons at M	$X + 1$ valence electrons at M
m d-electrons	$m - 1$ d-electrons
n oxidation state of M	$n + 1$ oxidation state of M

$$L_nM^z + H^\oplus \longrightarrow [L_nM^{z+2} - H]^\oplus$$

Cationic   Anionic

$$L_nM^z + O_2 \longrightarrow L_nM^{z+2} \overset{O}{\underset{O}{<}|}$$

Neutral   Dianionic

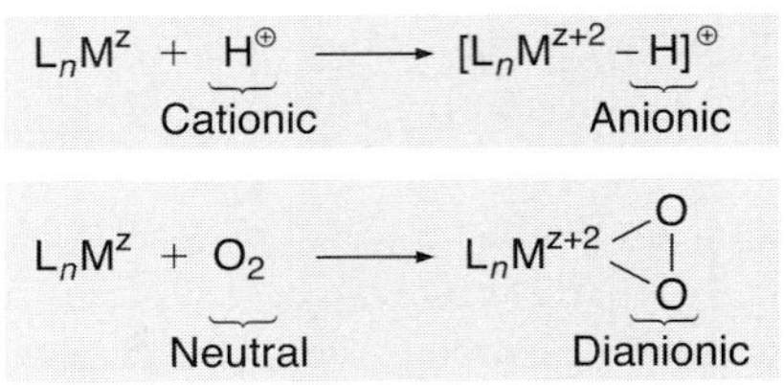

Figure 6.1. Examples of oxidative ligation.

The reagents A–B that undergo oxidative addition can be divided into three broad groups: species that are nonpolar or have low polarity, such as H_2, alkanes, and silanes; reagents that are highly polar, such as alkyl halides, alkyl sulfonates, and strong acids; and reagents that are intermediate in polarity, such as amines, alcohols, thiols, and haloarenes. Examples of these classes of reagents are listed in Table 6.1. The mechanisms of the oxidative additions of nonpolar reagents have been studied in detail and are involved in a variety of commercial and developing catalytic processes. Yet, they are less varied than the mechanisms involving additions of the more polar reagents and tend to occur by two-electron processes. Thus, this first chapter on oxidative addition presents examples and mechanisms of the oxidative addition of relatively nonpolar reagents, and the second chapter presents the more varied examples and mechanisms of the oxidative additions of more polar reagents.

Table 6.1. Examples of reagents with low polarity, high polarity, and intermediate polarity that undergo oxidative addition to transition metal complexes.

Reagents that are nonpolar or have low polarity:

H_2, RH, ArH, R_3SiH, R_3SnH, $R_3Sn–SnR_3$, R_2B-H, R_2BBR_2, RSSR, NC–CN, and Ph-PPh_2

Reagents that are highly polar:

HX, X_2, RCO_2H, RX, ROTs, RC(O)X, RSO_2X, $ROSO_2X$, RC(O)–OPh, HgX_2, and $SnCl_4$, $GeCl_4$, R_3SnCl, R_3PbCl

Reagents with medium polarity:

RSH, ROH, RNH_2, ArX, ArCN, CCl_4, and $CHCl_3$

6.1.2. Qualitative Trends for Oxidative Addition

A series of trends predict the reactivity of metal complexes toward oxidative addition. These trends stem from the increase in charge and the coordination number of the metal that typically occur during oxidative addition. These trends only serve as a guide, however, because there are exceptions to such general statements.

1. *Oxidative addition to more electron-rich metal centers tends to be more favorable than oxidative addition to more electron-poor metal centers.*[12] Although the actual charges at the metal change much less than the oxidation states during oxidative addition, the tendency of oxidative addition to occur more favorably to complexes possessing more electron-rich metal centers results from the increase in partial positive charge at the metal center during most oxidative addition reactions. Therefore, faster addition to more electron-rich metal centers originates from a ground state or thermodynamic effect.

2. *Oxidative addition to less-hindered metal centers tends to be more favorable than oxidative addition to more-hindered metal centers.*[12–14] This trend can be understood by the changes in coordination number that occur during oxidative addition. The more hindered the metal center, the more destabilized are the higher coordinate products from an oxidative addition. Thus, this steric effect, too, originates from ground state or thermodynamic properties.

3. *Oxidative addition of nonpolar reagents requires a site of unsaturation and a d-electron count of 16 or less.* Because ligands are added to the metal center during an oxidative addition process, an open coordination site is needed to accommodate the incoming reagent. In addition, the two electrons of the bond of the nonpolar reagent are contained in the product of the oxidative addition. This increase in valence electron count requires that the metal complex in the elementary oxidative addition step possess a *d*-electron count of 16 or less to abide by the 18-electron rule. The site of unsaturation and an electron configuration that contains 16 or fewer valence electrons may be present in the starting material. Alternatively, the site of unsaturation may be generated by the dissociation of a ligand from the metal center prior to the elementary step of oxidative addition. In this case, oxidative addition can occur to an 18-electron complex, but the process must be initiated by ligand dissociation, and the free ligand must be one of the products of the reaction.

    Oxidative addition of polar electrophiles, such MeX or HX (X = halide, sulfonate, or related stable anion) need not occur to a 16-electron metal center if a cationic metal center is generated with an uncoordinated anion. In this "oxidative ligation" mentioned previously and shown in Figure 6.1, the electrons of the Me–X or H–X bond do not add to the electron count of the metal because they are contained in the free anion of the product.

4. *Rates and equilibrium constants for ligand dissociation or association that occur prior to oxidative addition affect the rates of the overall addition processes.* Because oxidative addition is often preceded by ligand dissociation, the propensity of the starting complex to undergo dissociation of ligand affects the overall rate of oxidative addition.[15] For example, complexes containing tightly chelating bidentate ligands often undergo oxidative addition more slowly than analogous complexes containing monodentate ligands, and complexes containing strongly bound monodentate ligands are often less reactive toward addition than complexes containing weakly bound ligands.[16,17]

SUMMARY

General trends for oxidative addition and reductive elimination:

1. Oxidative addition tends to be more favorable to electron-rich metal centers, and reductive elimination more favorable from electron-poor metal centers.
2. Oxidative addition tends to be favored by less-hindered metal centers, and reductive elimination favored by more-hindered metal centers.

3. Oxidative addition of nonpolar reagents requires a site of unsaturation and a d-electron count of 16 or less.

4. Rates and equilibrium constants for ligand dissociation or association that occur prior to oxidative addition affect the rates of the addition processes.

6.1.3. Thermodynamics of Oxidative Addition

For oxidative addition to be enthalpically favorable, the strengths of the two new metal–ligand bonds must exceed the strength of the bond in the reagent that is cleaved (plus the strength of the bond to the dissociated ligand, if dissociation precedes addition). Table 6.2 provides thermodynamic data on the oxidative addition of H_2 and MeI to an analog of Vaska's complex that contains trimethylphosphine.[18] This table also provides estimates of the thermodynamics for the oxidative addition of the C–H bond of methane and the C–C bond of ethane. This analysis shows that the enthalpy of oxidative addition of dihydrogen is more favorable than the enthalpy of oxidative addition of methane or ethane because the metal–hydride bond is stronger than the metal–methyl bond.[19–21] This analysis also shows that the oxidative addition of methyl iodide is more favorable than the addition of any of the three nonpolar reagents because the carbon–iodine bond is weaker

Table 6.2. Estimates of the enthalpy and free energy for oxidative addition of several nonpolar and polar reagents.

Reaction	ΔH (estimated)[a] (kcal/mol)	ΔG (estimated)[a] (kcal/mol)
	-15^b	-6^b
	-2	$+3$
	-4	$+6$
	-35^c	-25

[a] Computed using the following approximate bond dissociation energy (BDE) values: Ir–H (60 kcal/mol), Ir–CH₃ (46 kcal/mol), Ir–I (45 kcal/mol): Mondal, J. U.; Blake, D. M. *Coord. Chem. Rev.* **1982**, *47*, 205 (particularly the data in Table 9), and references therein. The approximate $-T\Delta S_{298}$ of +10 kcal/mol (standard state = 1 M) term has been calculated at 298 K using S_{298}(translation) of −35 eu: Page, M. I. *Angew. Chem., Int. Ed. Engl.* **1977**, *16*, 449.
[b] The experimental values for the PPh₃ complex, IrCl(CO)(PPh₃)₂, are ΔH = −15 kcal/mol and ΔG = −6 kcal/mol, from Table 2 of Mondal, J. U.; Blake, D. M. *Coord. Chem. Rev.* **1982**, *47*, 205, and from Vaska, L. *Acc. Chem. Res.* **1968**, *1*, 335; Vaska, L.; Werneke, M. F. *Ann. N. Y. Acad. Sci.* **1971**, *172*, 546; Strohmeier, W.; Mueller, F. J. *Z. Naturforsch. B.* **1969**, *24*, 931; and Strohmeier, W. *J. Organomet. Chem.* **1971**, *32*, 137.
[c] The experimental value for the PMe₃ complex, IrCl(CO)(PMe₃)₂, is ΔH = 28 kcal/mol.

than the H–H, C–H, or C–C bond, and the metal–iodide bond is similar in strength to the metal–methyl bond.[18, 21]

Finally, this analysis of the data in Table 6.2 shows that the oxidative addition of the C–H bond of methane and the C–C bond of ethane to this iridium compound is thermodynamically unfavorable. Although the reaction is predicted to be slightly favored enthalpically, the free energy is positive because of the large positive $T\Delta S$ term for a process that generates one product from two reactants. The entropy for a reaction of this type is usually dominated by translational entropy, and this value is typically about –35 eu. Near room temperature, $T\Delta S$ will then equal about –10 kcal/mol, and the free energy for addition of methane and ethane to this Ir(I) complex is positive.

However, a change in ancillary ligands can change the thermodynamics for oxidative addition significantly. Table 6.3 provides values for the iridium–methyl, –hydride, and –iodide bond energies in a series of complexes, including the products from oxidative addition to Vaska-type complexes and the Ir(III) complexes $Cp^*Ir(PMe_3)X_2$ (X = Me, H, Cl, Br, and I). The values for the metal–ligand bond strengths in $Cp^*Ir(PMe_3)X_2$ are much higher than those for the same metal–ligand bonds in Vaska-type complexes.[21] As shown in Table 6.4, the greater strength of these iridium–hydride and –alkyl bonds make oxidative addition of C–C and C–H bonds of hydrocarbons thermodynamically

Table 6.3. Metal–ligand bond energies for two different Ir(III) systems.

X	$Cp^*Ir(PMe_3)X_2$[a]	$L_2Cl(CO)IrX_2$[b]
H	74.2	60
Cl	90.3	71
Br	76.0	53
I	63.8	35
CH_3		35.4
C_6H_{11}	50.8[c]	
C_6H_5	80.6[c]	

[a] Average uncertainties in absolute bond strength values are on the order of ±5 kcal/mol. For experimental errors on specific measurements see the source.
[b] Yoneda, G.; Blake, D. M. *Inorg. Chem.* **1981**, *20*, 67 and references therein.
[c] These bond strength estimates are based on the enthalpies of reaction of $Cp^*Ir(PMe_3)(R)(H)$ complexes.

Table 6.4. Enthapy for oxidative addition of various reagents to $Cp^*Ir(PMe_3)$.

$Cp^*(PMe_3)Ir + A–B \rightarrow Cp^*(PMe_3)Ir(A)(B)$

A–B	$-\Delta H^a$	A–B	$-\Delta H^b$	A–B	$-\Delta H^a$
$C_6H_{11}–C_6H_{11}$	22	Cl_2	123	HCl	61
$C_6H_{11}–H$	30	Br_2	106	HBr	63
H–H	44[b]	I_2	91	HI	57
$C_6H_3–H$	45				
$C_6H_5–C_6H_5$	49				

[a] All enthalpies of reaction are in kcal/mol and are calculated by using bond strength data taken from Table 6.3 and accepted bond strength values.
[b] Estimated bond strength based on the assumption that the Ir–R bond strength is the same for both Ir(R)(H) and Ir(R)_2.

favorable. Although the addition of the C–C bond of ethane is more favorable thermodynamically than the addition of the C–H bond, addition of the C–H bond is more favorable kinetically. An example of intermolecular oxidative addition of the C–C bond of an alkane that is faster than intermolecular addition of the C–H bond of the alkane has not yet been observed.

The remainder of this chapter presents the reactions of nonpolar reagents with a range of metals. The reactions of dihydrogen are presented first; the reactions of silanes are presented second because they often react like dihydrogen. The additions of C–H bonds are presented third; examples of intramolecular additions of C–H bonds are presented before intermolecular examples. This chapter closes with a discussion of the oxidative addition of C–C bonds.

6.2. Oxidative Addition of Dihydrogen

The addition of dihydrogen to a transition metal complex[22] occurs during many catalytic processes, including the catalytic hydrogenation of olefins, alkynes, and arenes (Chapter 13), and the hydroformylation of alkenes (Chapter 15). Because dihydrogen is the simplest nonpolar substrate, and oxidative addition of dihydrogen is synthetically important, reactions of this species have been studied in detail. Information on this reaction helps explain the reactivity of other reagents of low polarity, such as silanes and alkanes. Several examples of the oxidative addition of dihydrogen are provided in Equations 6.4–6.6.

$$\begin{array}{c} \underset{Ph_3P}{\overset{Ph_3P}{\diagdown}} Rh \underset{Cl}{\overset{PPh_3}{\diagup}} + H_2 \longrightarrow \underset{Ph_3P}{\overset{Ph_3P}{\diagdown}} \underset{Cl}{\overset{H}{\underset{|}{Rh}}} \underset{PPh_3}{\overset{H}{\diagup}} \end{array} \qquad (6.4)$$

$$\underset{Ph_3P}{\overset{OC}{\diagdown}} Ir \underset{Cl}{\overset{PPh_3}{\diagup}} \xrightarrow{H_2} \underset{Ph_3P}{\overset{H}{\underset{CO}{Ir}}} \underset{Cl}{\overset{PPh_3}{\diagup}} \quad via \quad \underset{OC}{\overset{H}{\underset{|}{Ph_3P}}} \underset{Cl}{\overset{H}{Ir}} PPh_3 \qquad (6.5)$$

$$Pt(PCy_3)_2 + H_2 \longrightarrow \underset{Cy_3P}{\overset{Cy_3P}{\diagdown}} Pt \underset{H}{\overset{H}{\diagup}} \qquad (6.6)$$

6.2.1. General Mechanism for the Oxidative Addition of H₂

Many theoretical studies of the oxidative addition of H_2 have been conducted with a variety of metal complexes.[23–31] General views of the reaction coordinate and orbital interactions are shown in Figure 6.2. All of the studies have shown that (1) a side-on approach of H_2 is preferred over an end-on approach; (2) cis addition of H_2 has a low barrier that is generally less than 10 kcal/mol, whereas trans addition of H_2 is "forbidden" and has a high barrier; (3) the transition state is "early" with little stretching of the H–H bond; and (4) the transition state is formed by two different types of electron flow illustrated in Figure 6.2, one from the σ-bond of H_2 to an unoccupied orbital of the metal of σ-symmetry and one from a metal orbital of π-symmetry into the σ*-orbital of dihydrogen.

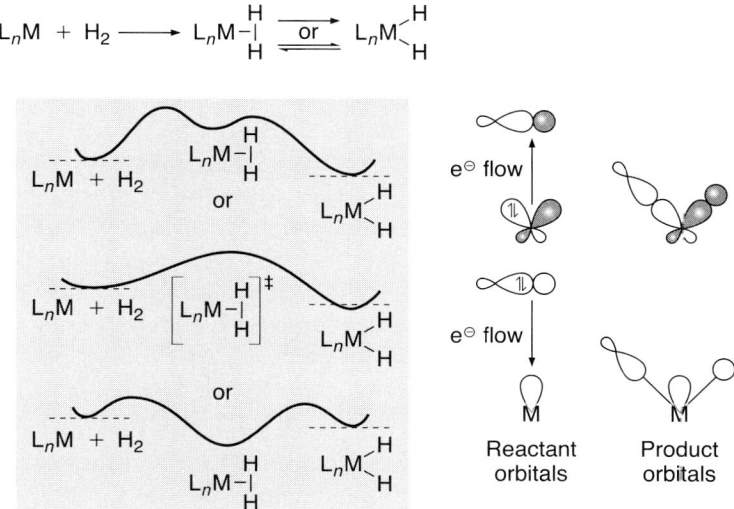

Figure 6.2.
Reaction coordinate diagrams and orbital interactions that occur during the oxidative addition of H_2 to transition metal complexes. (Orbital interactions from Hall, M. B. et al. *Chem. Rev.* **2000**, *100*, 353.)

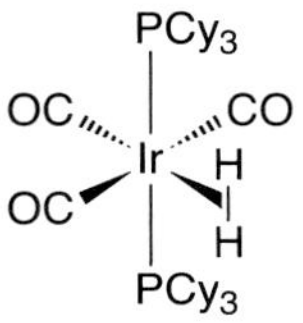

In some cases, coordination of dihydrogen to form a dihydrogen complex appears to precede the actual H–H bond cleavage event.[32] These dihydrogen complexes are illustrated as part of the reaction coordinate in Figure 6.2. The first dihydrogen complex, shown in Figure 6.3, was discovered[33,34] by Kubas at Los Alamos National Laboratory, and these complexes were described in Chapter 2. Information on the presence or absence of such intermediates in the oxidative addition of dihydrogen has been gained, in part, from computational studies. Computational results on some systems suggested that their reactions with H_2 proceed through a dihydrogen intermediate,[32] while computational results on others suggested that they react with H_2 without the intermediacy of a dihydrogen complex.[35,36] Thus, species formed from association of intact H_2 to the metal center are thought to lie along the reaction coordinate in most oxidative additions of H_2. These complexes are thought to be minima on the reaction coordinate for such oxidative additions in many cases, but not all.

Information on dihydrogen complexes that are related to the oxidative addition of dihydrogen have also been obtained experimentally. Either the dihydrogen complex or the dihydride complex can be most stable, and examples of equilibrium mixtures of the two have been reported.[37–49] Even complexes that display structural data that lie in between those of classic dihydride complexes and dihydrogen complexes, known as "stretched" dihydrogen complexes, have been reported.[50–58] Thus, the electronic properties of the metal center control the position at which the reaction stops along the oxidative addition reaction coordinate. For example, an electron-poor metal complex will donate less electron density into the σ*-orbital and will leave some H–H bonding intact in the final product, while a more electron-rich metal complex will donate enough electron density into the σ*-orbital to cleave the H–H bond and generate a dihydride complex.

The stereochemistry of the metal center resulting from the addition of H_2 has been studied,[59,60] and the resulting data have revealed several principles. First, the kinetic product from addition of H_2 has a cis disposition of the two hydride ligands, and the trans products are formed by isomerization of the initial cis products. Second, the addition of hydrogen to a square-planar complex can occur with high selectivity along one of the two bond axes due to electronic effects (Equation 6.7).[61–63]

Figure 6.3.
Kubas' first dihydrogen complex.

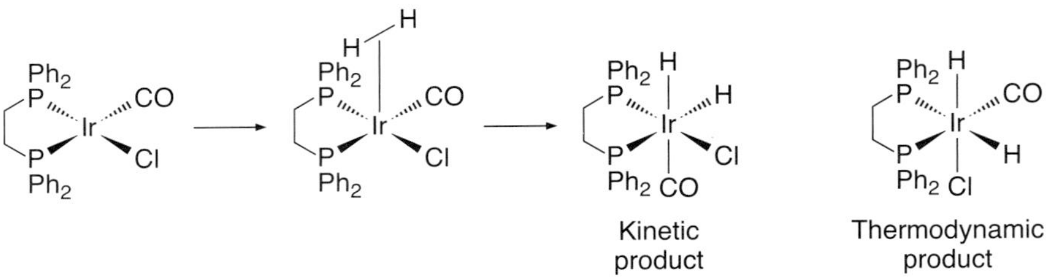

$$(6.7)$$

Kinetic product   Thermodynamic product

6.2.2. Examples of Oxidative Addition of H_2 to a Single Metal Center

One of the most significant oxidative additions of dihydrogen occurs to the complex $RhCl(PPh_3)_3$ (Equation 6.4)[64,65] which is "Wilkinson's hydrogenation catalyst."[66-68] Kinetic studies[64,65,69,70] of this oxidative addition reaction show that addition of H_2 can occur to the starting complex $RhCl(PPh_3)_3$, or the complex $RhCl(PPh_3)_2$ formed by dissociation of PPh_3 (Scheme 6.1). The complex $RhCl(PPh_3)_2$ was estimated to be roughly 10^4 times more reactive than $RhCl(PPh_3)_3$. Thus, the relative contributions of the reactions of these two species depend on their relative concentrations, and these relative concentrations depend on the concentration of free PPh_3. In the absence of a large concentration of added PPh_3, the reaction occurs predominantly through the 14-electron species, even though it is present in low concentrations.

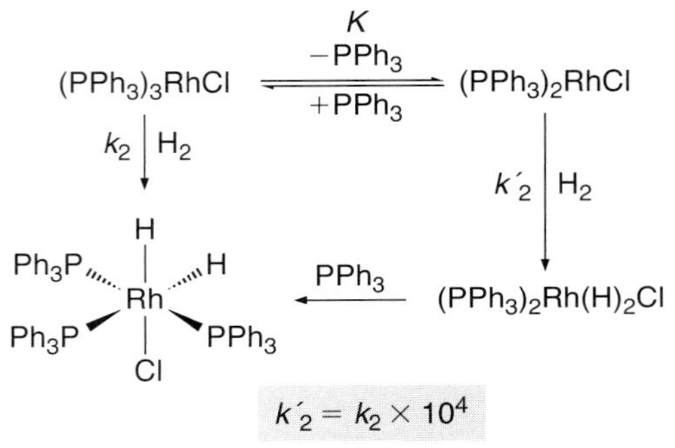

Scheme 6.1

Rate constants for the oxidative addition to the 14-electron intermediate were obtained directly from flash photolysis experiments.[71] Flash photolysis of $RhCl(CO)(PPh_3)_2$ causes selective dissociation of CO to form the 14-electron intermediate $RhCl(PPh_3)_2$ or a derivative of it containing coordinated benzene (Scheme 6.2). Oxidative addition of dihydrogen to this intermediate was observed directly by time-resolved UV−vis spectroscopy and was shown to be fast ($k = 1.0 \times 10^5\,M^{-1}\,s^{-1}$ and $\Delta G^{\ddagger} = 10$ kcal/mol). A small kinetic isotope effect was measured, $k_H/k_D = 1.4$, and this value is consistent with an early, three-centered transition state in which little H–H bond breaking has occurred.

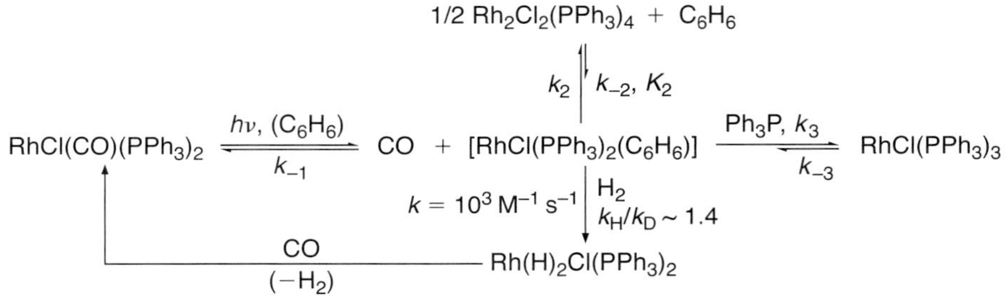

Scheme 6.2

A second classic oxidative addition is the reaction of H_2 with Vaska's complex shown in Equation 6.5. This reaction is reversible. The dihydride complex has been isolated, but when dissolved under vacuum, dihydrogen eliminates, and the starting Ir(I) complex is regenerated.[72] The effect of ancillary ligands on the rate of oxidative addition to Vaska-type complexes $[Ir(PPh_3)_2(CO)X]$ has been measured. The relative rate constants for addition follow the trend, X = I > Br > Cl, and the relative magnitudes of the rate constants are 100:14:1.[73] In this case, the electron density at the metal center has been shown to follow the trend Cl > Br > I, and the relative rates for oxidative addition of MeI follow the trend Cl > Br > I, as described in Chapter 7.[73a] Thus, the relative rates for addition of H_2 cannot be attributed to the degree of electron density at the metal center. Instead, they have been attributed to the degree of halide-to-Ir π-donation, with the faster additions occurring to the complexes containing the weaker π-donation and, thereby, the greater ability to interact with H_2.[74]

A third example of the oxidative addition of H_2 is provided in Equation 6.6. This addition is also reversible. In this case, addition occurs to a d^{10} metal center to form a product containing a d^8 metal center.[75,76]

6.2.3. Oxidative Addition of H_2 to Two Metal Centers

In addition to oxidative addition of dihydrogen to a single metal center by a concerted mechanism, oxidative addition of dihydrogen to two metal centers to generate two metal hydride products is known, each with an oxidation state that has increased by one. Several examples of this type of oxidative addition are shown in Equations 6.8–6.10c. This type of oxidative addition has been observed most commonly with metal complexes of the cobalt triad that contain cyanide or macrocyclic nitrogen ligands. These reactions can occur with mononuclear compounds or with dinuclear starting complexes if the metal–metal bond of the dinuclear species is weak enough to generate a reactive mononuclear species. The sum of the energies of the two new M–H bonds must exceed the sum of the 104 kcal/mol bond of H_2 and the strength of any M–M bond in the metal reactant.

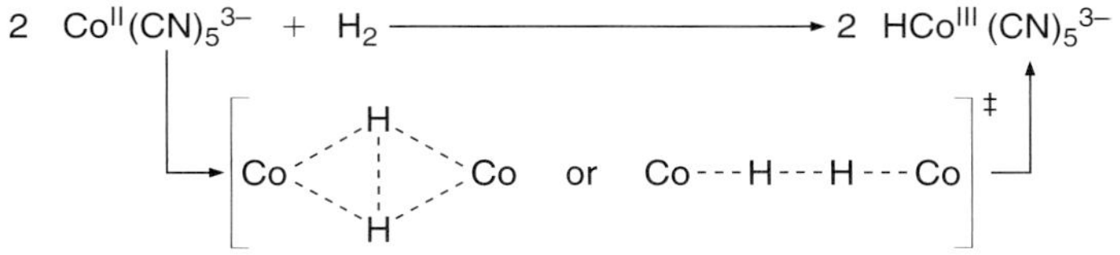

(6.8)

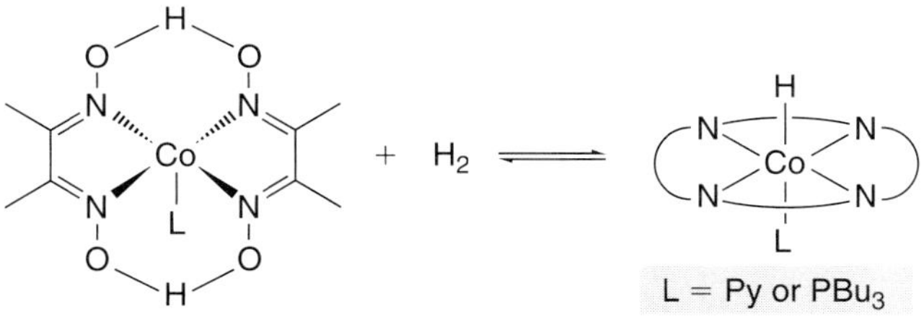

(6.9)

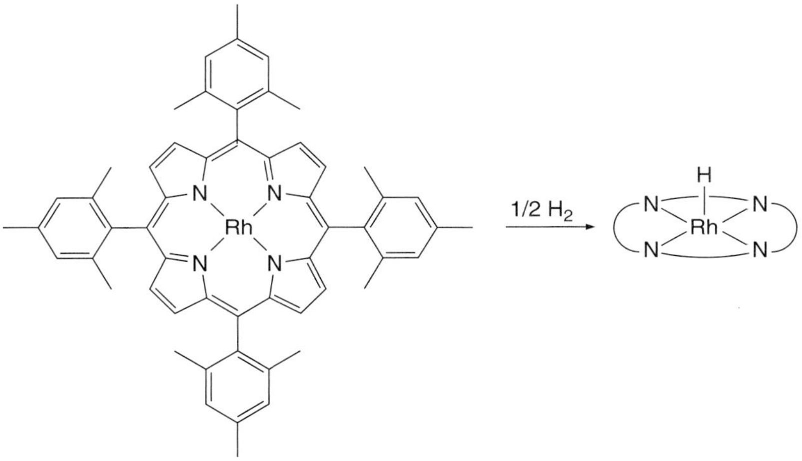

L = Py or PBu_3

(6.10a)

$$[Rh(TMTAA)]_2 \ + \ H_2 \longrightarrow \qquad\qquad\qquad\qquad\qquad (6.10b)$$

$$\text{Rh} - O(CH_2)_6O - \text{Rh} \ + \ H_2 \ \rightleftharpoons \ \overset{H}{\text{Rh}} - O(CH_2)_6O - \overset{H}{\text{Rh}} \qquad (6.10c)$$

The reactions in Equations 6.8–6.10c of the d^7 KCo^{II}(CN)$_5^{3-}$, KCo^{II}(dmgH)$_2$, Rh(TMP),[77] Rh(TMTAA)[78] and dinuclear analogs[77] of Rh(TMP) complexes are among the best-studied examples (dmgH = the monoanion of dimethylglyoximate, TMP = tetramesityl porphyrinate, and TMTAA = dibenzotetramethylazaannulene dianion). Additional examples can be found in a review by James.[79] Abstraction of a hydrogen atom from H_2 to generate a new M–H bond and H• is prohibitively endothermic and cannot be the first step of these reactions of dihydrogen. Instead, the reaction appears to occur through a transition state formed by the combination of hydrogen and two metal centers. Consistent with this assertion, the rates for these reactions of the monomeric metal complexes have been shown to be third order overall: first order in dihydrogen and second order in the monomeric metal complexes. For example, the rate of the reaction of KCo^{II}(CN)$_5^{3-}$ with dihydrogen is proportional to $[H_2][Co(CN)_5^{3-}]^2$,[80,81] the reaction of KCo(dmgH)$_2$ (Equation 6.9)[82,83] is proportional to $[H_2][Co(dmgH)_2^{2-}]^2$, and the reaction of (tetramesitylporphrinato)rhodium(II) is proportional to $[H_2][Rh(TMP)]^2$.[84]

The rates for these reactions can be slow because of the termolecularity of the reaction and the need to assemble two complexes with the same charge or hindered units into a single transition-state structure. Thus, an appropriate tether between the two metal centers can preassemble a dinuclear species containing two metalloradicals. A dirhodium complex in which the two porphyrins ligands are attached by a hexanediolate unit leads to a more reactive system for homolytic cleavage of H_2 and related nonpolar reactants (Equation 6.10c).[77]

Several well-documented examples of the oxidative addition of dihydrogen to bimetallic complexes to form dinuclear products are also known (e.g., Equation 6.11[85]), but little definitive mechanistic work on these reactions has appeared. In each case, the addition of dihydrogen is thought to occur at a single metal site, such as the ruthenium in Equation 6.11,[85] rather than simultaneously across two metals. One piece of evidence in support of this assertion is the absence of reactivity of dihydrogen toward bimetallic complexes that possess a metal–metal bond, but lack sites of coordinative unsaturation. For instance, $(OC)_4Fe(\mu^2\text{-}AsMe_2)Mn(CO)_4$ does not react with even 1,000 psi H_2 at 140 °C.[86]

$$\text{(6.11)}$$

6.3. Oxidative Addition of Silanes

The oxidative addition of the Si–H bond in silanes to transition metal complexes draws many parallels with the oxidative addition of dihydrogen. It occurs readily and is a step in the most common mechanism of the hydrosilylation of olefins, alkynes, and ketones by late transition metal complexes. Moreover, the mechanism is similar to that of the oxidative addition of dihydrogen. The addition often stops at the formation of a silane complex, much like the addition of dihydrogen often stops at the formation of a dihydrogen complex. In fact, the isolation of silane complexes occurred well before the isolation of dihydrogen complexes.[87−89]

Oxidative addition of a silicon–hydrogen bond is one of the most common methods to form transition metal complexes containing silyl ligands.[90−99] Several examples of this reaction[95−98] are shown in Equations 6.12–6.15. Oxidative addition occurs to Wilkinson's catalyst,[100−103] Vaska's complex,[104,105] and platinum(0) species.[106] In addition, silanes add to Ir(III) complexes to form Ir(V) silyl hydrides[107−110] and even Rh(V) silyl hydrides.[97,109,111,112] In many cases, reactions of metal–alkyl complexes with silanes generate alkane and metal–silyl complexes, and one example is shown in Equation 6.16.[113,114] These reactions most often occur by oxidative addition of the silane and subsequent reductive elimination of the alkane.[114]

$$(6.12)$$

$$(6.13)$$

$$(6.14)$$

$$(6.15)$$

$$Cp*L_2RuCH_2SiMe_3 + (x + 2y)\ HSiPh_2Cl \xrightarrow[L\ =\ PMe_3]{-SiMe_4} x\,Cp*L_2RuSiPh_2Cl + y\,Cp*L_2Ru(SiPh_2Cl)_2H \qquad (6.16)$$

In other cases, low-valent metal centers react with silanes to form η^2-silane complexes. This process occurs with less electron-rich metal centers and with metal centers that are less able to expand their coordination number. For example, photolysis of $(MeCp)Mn(CO)_3$ in the presence of a silane leads to dissociation of one CO ligand and coordination of the silane to form $(MeCp)Mn(CO)_2(\eta^2\text{-}HSiPh_3)$, as shown in Equation 6.17.[37−89]

$$(6.17)$$

$$\text{(6.18)}$$

The mechanism of the oxidative addition of silanes has been studied in detail. Like the addition of H_2, the reaction of silanes occurs by donation of electron density in the H–Si σ-orbital into a metal orbital of σ-symmetry and back-donation of electron density from a metal orbital of π-symmetry into the H–Si σ*-orbital. Reactions of optically active silanes occur with retention of configuration;[106] in the product, the metal occupies the position that the hydrogen occupied in the starting silane (Equation 6.18). Moreover, the oxidative addition has been shown by femtosecond flash photolytic and infrared spectroscopic methods to occur with prior formation of a silane complex,[115–118] perhaps in parallel with the formation of complexes from slipping of the Cp ring or those resulting from coordination of the triethylsilane through ethyl C–H bonds.

6.4. Oxidative Addition of C–H Bonds

The reactions of C–H bonds in hydrocarbons with transition metal complexes to form products containing metal–carbon bonds, often termed "C–H activation," has been a topic of active study for decades because of the long-term goal of conducting selective synthesis with unfunctionalized reactants or unfunctionalized portions of more complex molecules.[119–125] Transformations of alkanes catalyzed by heterogeneous catalysts motivated the search for soluble species that could give rise to similar reactivity but with higher selectivity and at lower temperatures.

Extensive studies of stoichiometric, as well as catalytic, reactions involving the oxidative addition of the C–H bonds of hydrocarbons to transition metal complexes have been conducted. Considering the low polarity of C–H bonds, one might expect the mechanisms of oxidative addition of the C–H bonds of alkanes and arenes to parallel those for oxidative addition of dihydrogen. Indeed, oxidative additions of alkanes to a single metal center and to two metal centers have both been observed, although oxidative addition of an alkane to a single metal center by non-radical processes and through alkane σ-complexes is most common. Most 16- and 14-electron transition metal fragments are isolobal with CH_2, and the singlet form of CH_2 is known to insert into C–H bonds by a non-radical mechanism (Equation 6.19).[126–128] Thus, parallels can be drawn between the chemistry of a carbene and the reactivity of 16- and some 14-electron transition metal fragments.

$$:CH_2 \ + \quad \longrightarrow \quad CH_3 \tag{6.19}$$

6.4.1. Early History of C–H Bond Oxidative Addition

Early reports on the reactions of transition metal complexes with alkanes were published by Shilov. Shilov and coworkers observed H/D exchanges, oxidations, and halogenations of alkanes with platinum complexes (Equation 6.20).[129,130] Work by Crabtree and Felkin, summarized in Equations 6.21a and 6.21b, showed that soluble complexes could be used to conduct the dehydrogenation of alkanes.[131–134] This body of results suggested that oxidative addition of alkanes could occur. This section of this chapter focuses on well-defined examples of the oxidative addition of C–H bonds to soluble transition metal complexes. Chapter 18 presents studies of catalytic C–H bond functionalization invoving organometallic complexes.

$$(6.20)$$

$$(6.21a)$$

$$(6.21b)$$

For many years, only intramolecular C–H additions[135,136] were observed because this type of reaction is favored kinetically and thermodynamically. Intermolecular additions of arenes were later observed, and arenes are more reactive than alkanes toward oxidative addition to all single-metal centers. In 1982, the isolation of an alkyl hydride complex from the oxidative addition of an alkane was first reported.[137,138] Since that time, many complexes have been reported that undergo oxidative additions of alkanes. Many of these complexes do not provide stable alkyl–hydride products, but these complexes can be induced in some cases to undergo productive transformations. The following sections describe the development of intramolecular and intermolecular oxidative addition of the C–H bonds of alkyl groups, aryl groups, alkanes, and arenes.

6.4.2. Intramolecular C–H Oxidative Addition

Intramolecular C–H oxidative additions have been common for many years. These reactions are known as cyclometallations[135,139] (or orthometallations when the metal undergoes addition of an ortho C–H bond of the aromatic group of a ligand). Examples are provided in Equations 6.22–6.25.[140–144] Cyclometallations of azobenzene are common and constitute some of the earliest examples of C–H activation. Two examples with $Co_2(CO)_8$ and Cp_2Ni are shown in Equation 6.22;[140–142] the product from the reaction of Cp_2Ni was the first isolated complex that resulted from cyclometallation.[140–142] Examples more closely related to compounds that undergo intermolecular reactions are shown in Equations 6.23–6.25. Metallation of both aryl[143] and alkylphosphines (Equations 6.23 and 6.24)[145–153] are known, and cyclometallation of bis-neopentyl complexes, such as those in Equation 6.25, with concomitant extrusion of neopentane, are also common. Three-membered and four-membered rings containing a transition metal are much less strained than cyclopropanes and cyclobutanes and form with more favorable rates.

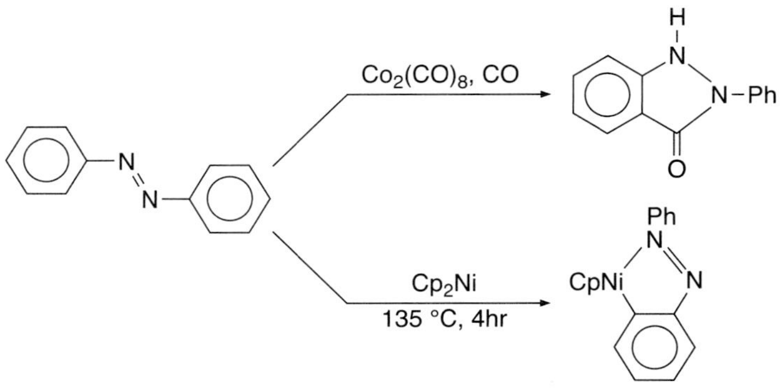

(6.22)

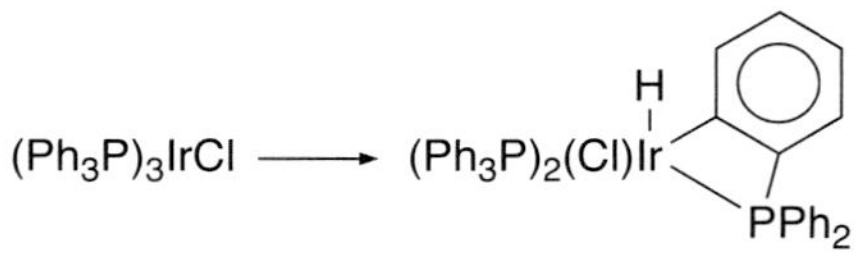

(6.23)

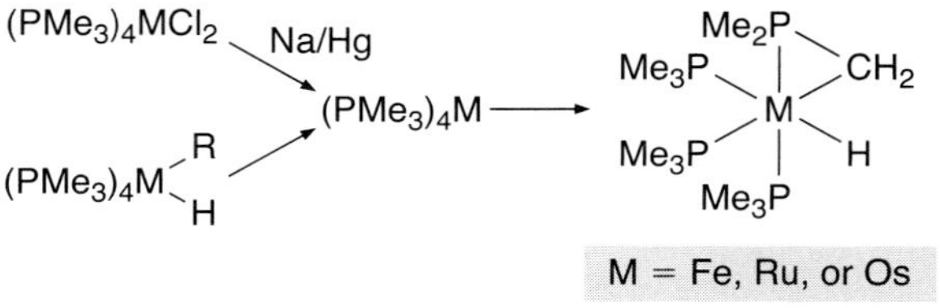

M = Fe, Ru, or Os

(6.24)

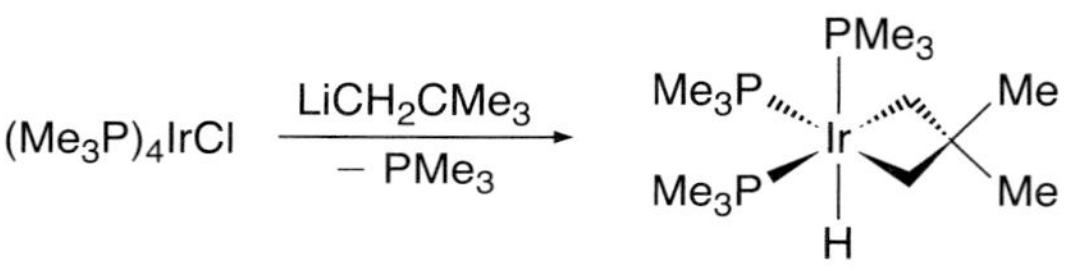

(6.25)

   Cyclometallation of phosphine ligands is a pathway that can deactivate catalysts. Addition of aryl C–H bonds is most favorable, and addition of the primary C–H bonds in methylphosphines is more favorable than addition of the methylene C–H bonds in higher alkylphosphines. Catalytic reactions initiated by compounds containing cyclometallated phosphorus ligands as stable and convenient precursors to active catalysts have been reported, some of which are shown in Figure 6.4.[154–158] Less common are catalysts that become activated by cyclometallation; one example that catalyzes allylic substitution is shown in Figure 6.5.[159]

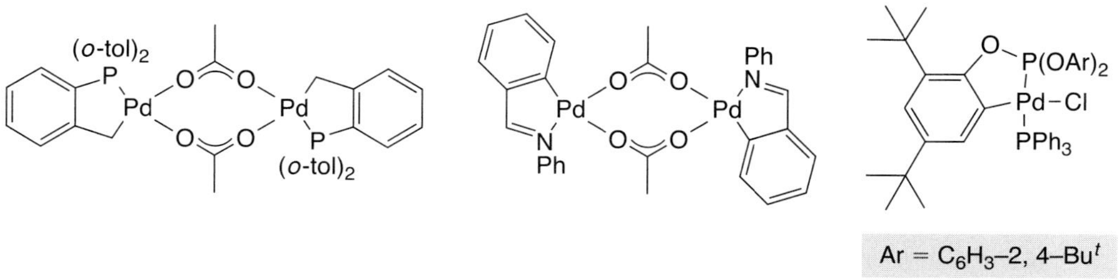

Ar = C_6H_3–2, 4–Bu^t

Figure 6.4. Examples of palladium catalyst precursors synthesized by cyclometallation.

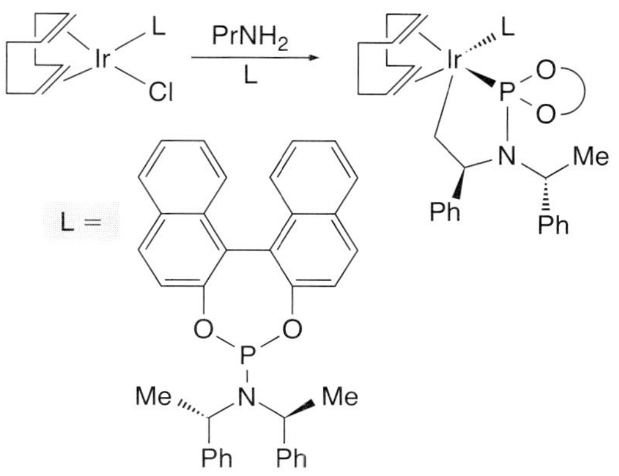

Figure 6.5.
Example of a catalyst for allylic substitution activated by cyclometallation.

Intramolecular C–H bond cleavage of a covalent ligand in L_2PtR_2 has been studied particularly carefully by Whitesides (Scheme 6.3).[144] Relief of steric strain in the starting material provides a large part of the thermodynamic driving force. The Pt–P and Pt–C bonds in the product are shorter than those in the reactant (by 0.037 and 0.035 Å, respectively), and the twisting between the P–Pt–P and C–Pt–C planes in the product is less than in the reactant. A further driving force is gained from the entropic advantage of the formation of two particles from one.

Detailed kinetic studies provide evidence for the stepwise mechanism shown in Scheme 6.3. In this mechanism, dissociation of phosphine precedes oxidative addition. The dissociation of ligand provides a site for coordination of the C–H bond in the square plane prior to oxidative addition and follows the trend noted earlier in this chapter for Wilkinson's catalyst that addition and elimination from three-coordinate species is often faster than addition and elimination from four-coordinate species.

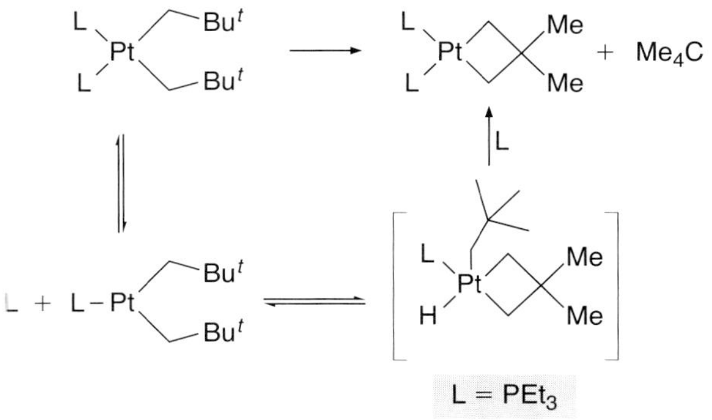

Scheme 6.3

6.4.3. Intermolecular Oxidative Addition of C–H Bonds

Complexes that undergo intermolecular addition of arene C–H bonds were reported by Chatt and Davidson[160] soon after the observation of intramolecular C–H activation of arenes (Equation 6.26). Both kinetic and thermodynamic factors favor the addition of arene C–H bonds over alkane C–H bonds, even though the arene C–H bonds are stronger than the alkane C–H bonds. The origin of the kinetic preference for the addition of arene C–H bonds is continuing to be debated, but contributing factors include coordination of the arene to the metal center as an η^2-arene complex, greater steric accessibility of an arene C–H bond, and lower directionality of sp^2- versus sp^3-hybridized orbitals. The greater thermodynamic driving force for addition of arene C–H bonds results from a difference between metal–aryl and metal–alkyl bond strengths that is greater than the difference between aryl C–H and alkyl C–H bond strengths.[21,161-163]

$$(6.26)$$

The first direct observation of oxidative addition of the C–H bond of a saturated hydrocarbon to a transition metal center was reported in 1982 by Janowicz and Berman[137,164] and by Hoyano and Graham[138] (Equation 6.27). Jones reported the oxidative addition of alkane C–H bonds to the analogous rhodium complexes at nearly the same time (Equation 6.28).[165] These reports have provided the foundation for hundreds of subsequent reports of C–H bond cleavage by late transition metal complexes and studies of the mechanism by which a metal can cleave an alkane C–H bond under mild conditions. In fact, cleavage of the strong C–H bond in methane[166,167] (104 kcal/mol) occurs even at 12 K in a CH_4 matrix.[168]

$$(6.27)$$

Only regioisomer

$$(6.28)$$

Examples of the oxidative addition of alkane C–H bonds to form directly observed transition-metal alkyl–hydride complexes are now known for group 7–10 metals. Rhenium (Equation 6.29),[169,170] iron (Equation 6.30),[171] rhodium,[165,172] iridium,[137,138,164,173] and platinum (Equation 6.31)[174,175] complexes all insert into the C–H bonds of alkanes by oxidative addition to form alkylmetal–hydride complexes that have been observed directly. The largest number of examples of these reactions has been reported with rhodium and iridium. The majority of examples of these additions to rhodium and iridium complexes occur to complexes containing substituted cyclopentadienyl ligands and to complexes containing anionic tridentate ligands, such as substituted tris-pyrazolylborates[176-178] and PCP-pincer ligands (Equations 6.32 and 6.33).[179-181] Additions to group 8 and 10 metals have been observed with complexes of chelating alkylphosphines, which are less susceptible to cyclometallation than complexes of monophosphines and constrain the metal center into an unstable bent conformation.[171,182] The greater strength of metal–carbon bonds of third-row transition metals makes oxidative addition to rhenium, iridium, and platinum most common.

$$\text{(6.29)}$$

$$\text{(6.30)}$$

Cy = cyclohexyl

$$\text{(6.31)}$$

$$\text{(6.32)}$$

$$\text{(6.33)}$$

In many other cases, oxidative additions of alkanes occur readily to transition-metal–alkyl complexes to generate hydride dialkyl intermediates that subsequently eliminate alkane and form a new metal–alkyl complex. For example, cations related to the alkyl hydrides of iridium formed by oxidative addition undergo reaction with alkanes at or below room temperature to generate new alkyl complexes (Equation 6.34).[183,184] Cationic platinum complexes undergo similar reactions with substrates containing aromatic and aliphatic C–H bonds (Equation 6.35).[185–189] The C–H activation of the platinum complexes has been studied, in part, to understand and to develop systems related to the ones reported by Shilov that lead to H/D exchange, and oxidation and halogenation of alkanes.

$$\text{(6.34)}$$

$$\text{(6.35)}$$

Compounds containing ligands bound to the metal through a main group element undergo related C–H bond cleavage reactions. Complexes containing boryl ligands react with alkanes to cleave in one transformation the terminal C–H bonds in alkanes or alkyl chains and the metal–boron bond (Equations 6.36 and 6.37).[190–193] In this process, a metal hydride and a free functionalized product are generated. This release of free functionalized product allowed for the development of catalytic functionalization of alkanes[194] and arenes[195–199] with boron reagents presented in Chapter 18. Metal–silyl complexes also undergo related reactions with arenes, but the yields of arylsilanes from isolated silyl complexes have been lower than the yields of alkyl- and arylboronates.[200] Nevertheless, the catalytic silylation of arenes has been reported.[200–202]

$$\text{(6.36)}$$

$$\text{(6.37)}$$

6.4.4. Selectivity of Alkane Oxidative Addition

The high selectivity for reaction at primary C–H bonds is one of the most striking features of the oxidative addition of alkanes. This selectivity contrasts with that of hydrogen atom abstraction by radical species, which occur fastest at tertiary C–H bonds and slowest at primary C–H bonds.[203] It also contrasts with the chemistry of metal–oxo[204–206] or metal–carbene[207,208] complexes, which tend to insert the oxo or carbene group into the weakest, sterically accessible C–H bond. This selectivity of oxidative addition for reaction at a primary C–H bond implies that alkanes can be converted selectively to products with functional groups in the terminal position.

With few exceptions,[209,210] transition metal complexes of terminal alkyl groups are more stable than those of isomeric secondary or tertiary alkyl groups.[172,211] Consistent with this trend, the terminal alkylmetal hydride complexes of systems that undergo oxidative addition of C–H bonds are thermodynamically more stable than the branched alkylmetal hydride complexes.[161,162] The oxidative addition of alkanes by the Cp*Ir(PMe₃) fragment is reversible, and this reversibility allowed for an evaluation of the thermodynamic stability of alkyl hydride complexes formed from a series of alkanes, as well as from arenes.[173] These data showed that the alkyl hydride complex formed from oxidative addition at the primary C–H bond leads to more stable products than those formed from oxidative addition of the secondary C–H bonds like those in cyclohexane (Equation 6.38). The aryl hydride complexes are much more stable than any of the corresponding alkyl hydride complexes.

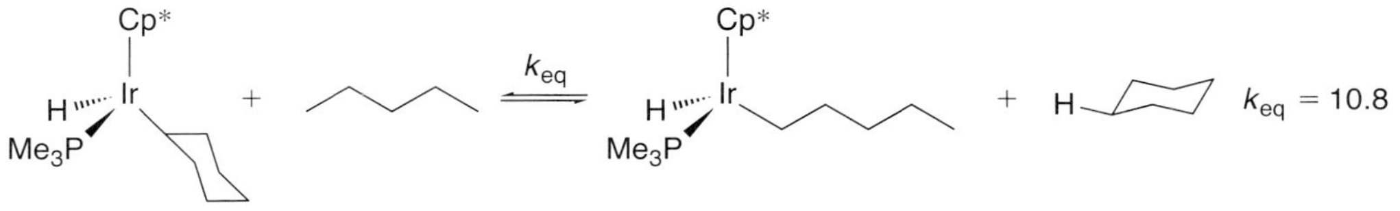

$$H\cdots Ir(Cp^*)(Me_3P)(cyclohexyl) + \xrightarrow{k_{eq}} H\cdots Ir(Cp^*)(Me_3P)(hexyl) + H-cyclohexane \quad k_{eq} = 10.8 \quad (6.38)$$

6.4.5. Mechanism of Oxidative Addition of C–H Bonds

The mechanism of the oxidative addition of alkanes by iridium and rhodium complexes has been studied by many methods, including isotopic labeling,[121] conventional kinetic studies, and kinetic studies on the femtosecond time scale in condensed noble gases.[212–214] The mechanism for the oxidative addition of alkanes, summarized in Scheme 6.4, parallels the mechanism described for the oxidative addition of dihydrogen. An unsaturated intermediate first coordinates the alkane to form an η^2-alkane complex in nearly all cases in which detailed mechanistic information has been gained.[34,215] This alkane complex undergoes oxidative addition to generate the alkyl hydride product. The alkane complex is generated by the same orbital interactions that generate dihydrogen complexes. Donation of electron density from the alkane C–H bond into a metal orbital of σ-symmetry, and back-donation of electron density into the σ^*-orbital of the C–H bond, leads to the formation of the alkane complexes. The oxidative addition process occurs by a further flow of electrons from the C–H σ-bond to the metal and from the metal π-orbital into the alkane C–H σ^*-orbital. Current experimental data gained by NMR spectroscopy of $CpRe(CO)_2$(cyclopentane) and computational studies of this interaction[216] imply that the metal is coordinated to a single C–H bond in the most stable alkane complex.[217] However, experimental data on more than one system is needed to conclude that this binding mode is generally the most favored.

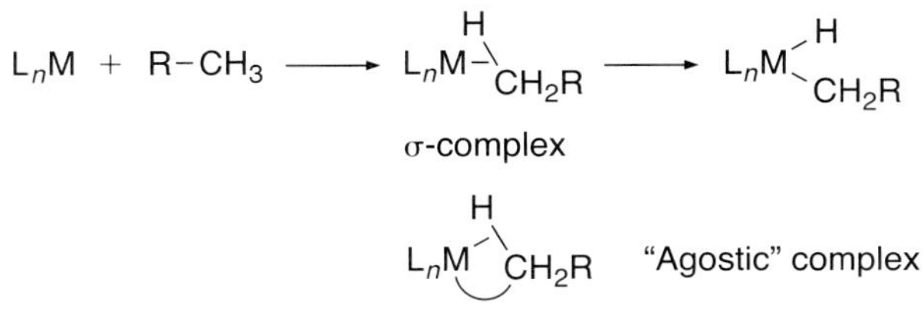

$$L_nM + R-CH_3 \longrightarrow L_nM\overset{H}{\underset{CH_2R}{\diagdown}} \longrightarrow L_nM\overset{H}{\underset{CH_2R}{\diagdown}}$$

σ-complex

$$L_nM\overset{H}{\underset{CH_2R}{\diagdown}} \quad \text{"Agostic" complex}$$

Scheme 6.4

The larger steric properties of the alkane make alkane complexes much less stable than dihydrogen complexes, and complexes that contain coordinated alkanes in solution[218] or in structures determined in the solid state by X-ray diffraction[219] have been difficult to obtain. However, many complexes in which a C–H bond is coordinated intramolecularly to a transition metal have been observed directly by spectroscopic methods in solution and by X-ray and neutron diffraction in the solid state. These complexes were presented in Chapter 3 and were termed "agostic" complexes by Brookhart and Green;[220,221] a general structure of such a complex is given in Scheme 6.4. Crabtree and co-workers[222] have used structural data on many complexes containing agostic interactions to construct the trajectory in Figure 6.6 for the approach of a C–H bond to a metal and its eventual oxidative addition. Calculations of the reaction coordinate provided similar results.[29]

The presence of arene complexes as intermediates formed prior to the oxidative addition of the arenes and alkenes is now equally well accepted. However, the mode of

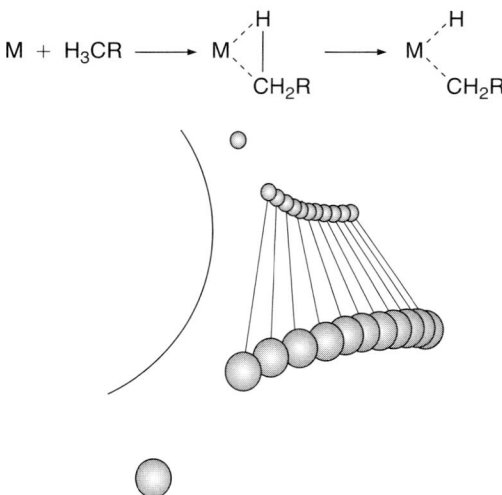

Figure 6.6.
Trajectory for oxidative addition of a C–H bond as determined by Crabtree from X-ray data of a series of agostic complexes. Figure adapted from Crabtree, R. H.; Holt, E. M.; Lavin, M.; Morehouse, S. M. *Inorg. Chem.* **1985**, *24*, 1986.

coordination of the arene and alkane that leads to oxidative addition is not firmly established. An η^2-arene π-complex is most commonly invoked as the species that precedes oxidative addition, and η^2-arene complexes have now been isolated in systems that subsequently react with the arene C–H bonds by oxidative addition.[223–225] However, recent computational data suggests that arene C–H σ-complexes lie on the path to addition, perhaps after initial formation of η^2-arene π-complexes.[226]

$$\text{(6.39)}$$

Ethylene can also coordinate through a C–H bond as a σ-complex or as a more stable η^2 π-complex. η^2-Complexes of ethylene are intermediates in C–H oxidative addition reactions in some cases but not in others. As depicted in Equation 6.39, the complex Cp*Ir(PMe$_3$)(ethylene) is more stable than the vinyl hydride isomer Cp*Ir(PMe$_3$)(vinyl)(H), but the vinyl hydride complex is the kinetically preferred product.[227,228] Thus, the ethylene complex cannot be an intermediate in the oxidative addition of the C–H bond of ethylene in this case. However, the olefin complex may form prior to oxidative addition of the vinyl C–H bond in other cases.[229]

$$\text{(6.40)}$$

e.g. Cp*Ir(PMe$_3$)(D)(C$_6$H$_{11}$), Cp$_2'$W(H)(CD$_3$),[Cp*$_2$Re(H)(CD$_3$)]$^{\oplus}$

Because the alkane complexes are less stable than dihydrogen or olefin complexes, the data that support their intermediacy in the oxidative addition of C–H bonds is less direct.[215] Some of these data were presented in Chapter 2. Many lines of data support their intermediacy. First, alkyl hydride complexes containing deuterium in the hydride position undergo rearrangement to form complexes containing deuterium in the α-position of the alkane faster than they undergo reductive elimination (Equation 6.40). For example, exchange of the deuteride with the α-hydrogen of the cyclohexyl group in Cp*Ir(PMe$_3$)(C$_6$H$_{11}$)(D) has been observed,[173] and related exchanges have been observed many times with other alkyl hydride complexes.[226,230–232] Second, the isotope effects for reductive eliminations (Chapter 8) are often less than one,[174,233,234] and, for reasons discussed in Chapter 8, suggest reversible formation of an alkane complex. Third, fast kinetic studies on C–H activation by Tp*Rh(CO) monitored by infrared spectroscopy have provided data to support the presence of an alkane complex.[212,214] Fourth, the rate behavior of reactions conducted in krypton and xenon solvent requires the presence of an intermediate that is almost certainly the alkane complex.[239–242] Finally, the alkane complex Cp*Re(CO)$_2$(cyclopentane) has been detected directly by low-temperature NMR spectroscopy during continuous irradition of Cp*Re(CO)$_3$ in cyclopentane solvent (Equation 6.41).[218]

$$ (6.41) $$

The origin of the kinetic selectivity for reaction at primary C–H bonds is unclear. Although Cp*Ir(PMe$_3$) does not show high kinetic selectivity for reaction with primary C–H bonds over secondary C–H bonds, the intermediates Cp*Re(PMe$_3$)$_2$[169,170] and Cp*Rh(PMe$_3$)[137,164,243] react with exquisite selectivity for the primary C–H bonds. Recent spectroscopic data on alkane σ-complexes imply that σ-complexes involving primary and secondary C–H bonds are similar in energy.[217] Thus, the greater thermodynamic stability of the primary alkyl complex versus a secondary alkyl complex appears at this time to lead to the high selectivity for oxidative addition of primary C–H bonds in most systems. A comparison of the rate constant for the elementary step of oxidative addition of a primary and secondary C–H bond in an alkane has recently been deduced by Jones and coworkers from reactions of the Tp'Rh(CNCH$_2$^{t}Bu) fragment with a mixture of n-alkanes and cycloalkanes.[244] In these reactions, formation of the primary alkyl hydride product was observed over the cycloalkyl hydride product by factors of 4.5:1 to 6.5:1. These data, along with data on the stability and rates of isomerizations of alkane σ-complexes implied that cleavage of the C–H bond in a σ-complex to the primary C–H bond occurs about 64 times faster than cleavage of the C–H bond in a σ-complex to the secondary C–H bond.

6.4.6. Examples of Complexes that Oxidatively Add Alkanes

Methods to generate the unsaturated intermediate that adds alkane include photolysis of metal–carbonyl or –dihydride complexes, thermolysis of alkyl hydride complexes, and abstraction of halide with a cationic reagent or protonation of an alkyl group with an acid containing a weakly coordinating anion. For example, the photolysis of Cp*Ir(CO)$_2$ or Cp*Ir(PMe$_3$)H$_2$ generates Cp*Ir(L) (L = CO or PMe$_3$),[137,138,164] and photolysis of the related complexes Tp*Ir(CO)$_2$ or Tp*Rh(CO)$_2$ generates the closely related Tp*M(CO).[212] These intermediates have also been generated by thermolysis of the alkyl hydride complexes of the Cp*Ir(L) or Tp*Ir(CO) fragments.[173] Kinetic evidence for the formation of (C$_5$Me$_5$)Ir(PMe$_3$) in these processes include observed rate constants for the thermolysis of (η^5-C$_5$Me$_5$)Ir(PMe$_3$)(H)(C$_6$H$_{11}$) in mixtures of cyclohexane and benzene to form (η^5-C$_5$Me$_5$)Ir(PMe$_3$)(H)(Ph) (Equation 6.42) that are inversely dependent on the ratio of cyclohexane to benzene and independent of the concentration of PMe$_3$. (PCP)Ir(R)(H) complexes also react by reductive elimination of arene or alkane, in this case to form (PCP)Ir. (PCP) Ir(Ph)(H) exchanges with benzene on the NMR time scale by reductive elimination and oxidative addition of benzene and extracts dihydrogen from cyclooctene by reductive elimination of benzene, oxidative addition of cyclooctene, and extrusion of olefin within 30 minutes at room temperature (Equation 6.43).[245] Addition of AgB(Ar$_f$)$_4$ (where Ar$_f$ = 3,5-bis-trifluoromethylphenyl) to Cp*Ir(PMe$_3$)Me(Cl) leads to formation of the cation

$[Cp*Ir(PMe_3)Me(CH_2Cl_2)]^+$ that dissociates methylene chloride to generate the reactive intermediate in one of the C–H activation reactions noted previously in this chapter.[184] Likewise, addition of $[H(OEt_2)_2]B(Ar_f)_4$ to complexes of the type (bis-imine)$PtMe_2$ in $HOCH_2CF_3$ has been shown to generate the intermediates $[(bis-imine)PtMe(HOCH_2CF_3)]^+$ that add arenes.[186−188]

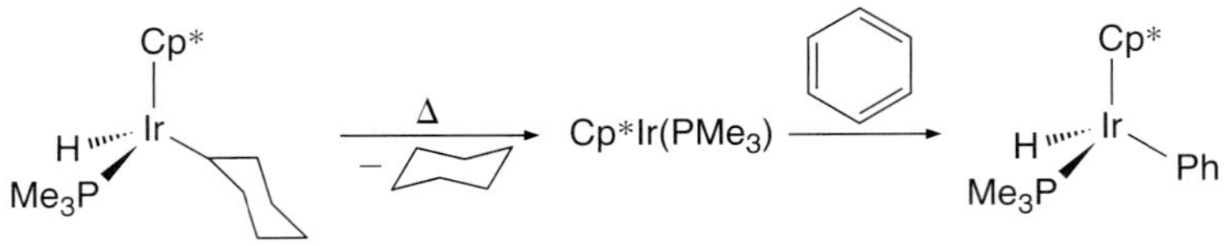

(6.42)

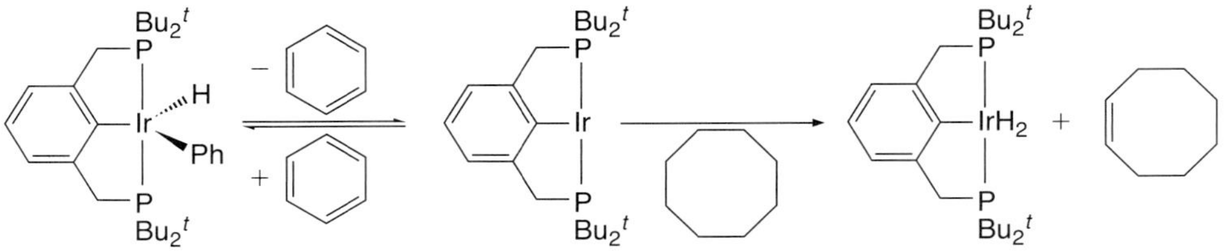

(6.43)

6.4.7. Synthetic Applications of C–H Oxidative Addition of Alkyl Groups

The oxidative addition of the C–H bonds of alkyl groups in more complex molecules than alkanes has begun to be investigated more intensively with the goal of conducting selective conversions of C–H bonds in molecules with existing functionality. For example, an intermediate toward the synthesis of rhazinilam shown in Equation 6.44 was prepared with an N-pyridyl imine to act as a ligand for platinum.[246,247] Treatment of this material with $Me_2Pt(SMe_2)_2$ led to selective cleavage of a C–H bond of one of the ethyl groups and subsequent β-hydrogen elimination to generate the olefin. Likewise, coordination of pyridine nitrogens, ketone oxygens, or imine nitrogens can lead to selective oxidative addition of a metal into the ortho-aromatic C–H bond in a reaction akin to the first orthometallation, and the resulting aryl hydride intermediate can then react with a second reagent in a subsequent step of a catalytic process.[248−255]

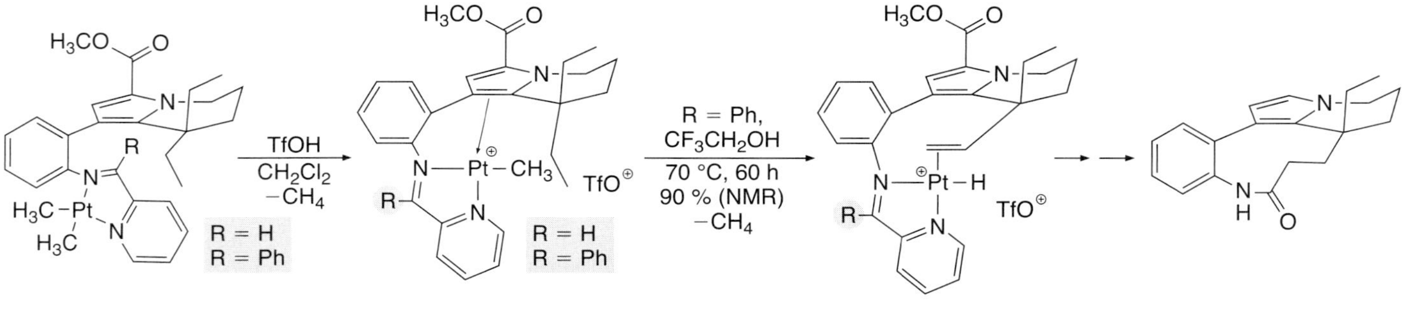

(6.44)

6.4.8. Dinuclear Activation of Hydrocarbons

Examples of oxidative additions of alkanes to two metal centers are rare but can occur by a process that parallels the addition of H_2 to two metal centers that was shown in Equations 6.10a–6.10c. Wayland showed that the Rh(II) complex containing tetramesityl porphyrin cleaves the C–H bond of methane by a termolecular process to generate a rhodium(III) hydride and a rhodium(III) methyl complex (Equation 6.45).[256] The Rh(II) porphyrin complex also reacts with toluene at the methyl C–H bonds. No reaction is observed with benzene. This selectivity is the opposite of that observed for oxidative additions of alkanes and arenes to a single metal center. The reaction of methane with the tethered dinuclear system in Equation 6.46[257] is faster than to two mononuclear species, just as appropriately tethered rhodium centers were noted to react faster with H_2 in Section 6.2.2. This reaction also occurs with highly electrophilic Rh(II) complexes of perfluorophorphyrins.[258]

$$2 \quad \text{[(TMP)Rh]} + CH_4 \longrightarrow (TMP)Rh-Me + (TMP)Rh-H \tag{6.45}$$

$$\tag{6.46}$$

6.5. Addition of H–H and C–H Bonds to Transition Metal Complexes Without Oxidation and Reduction

Transition metal complexes containing metal centers with d^0 electron counts cannot undergo oxidative addition, but they can react with dihydrogen and hydrocarbons by alternative reaction pathways. The most common of these alternative pathways include σ-bond metatheses of d^0 metal–alkyl and –aryl complexes and 2+2 reactions of metal carbene and imido compexes with substrates containing H–H, Si–H, and and C–H bonds.

6.5.1. Sigma-Bond Metathesis Involving d^0 Complexes

"Sigma-bond metathesis"[259] is shown generally in Equation 6.47. This reaction involves addition of an H–H or C–H bond across an M–R bond of a transition metal or f-element complex to generate a product containing a new M–C or M–H bond and a product containing a new C–H bond. An exchange of alkyl groups between a d^0 transition-metal–alkyl complex and an alkane would be an example of a σ-bond metathesis. σ-Bond metathesis reactions are common among d^0 transition metal, lanthanide, and actinide complexes, but main group compounds also undergo this type of reaction. For example, alkyl lithium reagents react with dihydrogen, albeit under forcing conditions, to generate lithium hydride and an alkane.[260]

$$L_nM-R + R'-H \longrightarrow L_nM-R' + R-H \tag{6.47}$$

Like oxidative addition, σ-bond metathesis occurs by a multi-step sequence that requires a site of unsaturation and a complex possessing 16 or fewer valence electrons

(Equation 6.48). Several computational studies[261–265] suggest that the first step of σ-bond metathesis is coordination of the dihydrogen, alkane, or arene to the transition metal center. The metal–ligand bond in this σ-complex is weaker than that in the late metal complexes because the metal center lacks the d-electrons that stabilize such complexes by back-donation into the σ*-orbital. Nevertheless, these complexes can be enthalpically more stable than the free reactant and the 16-electron complex.[263] These σ-complexes then react through a four-centered transition state to generate a new σ-complex containing the coordinated product. This new σ-complex then dissociates the alkane or arene to generate the final separate reaction products. The σ-complex formed by coordination of dihydrogen or the C—H bond of a hydrocarbon increases dramatically the acidity of the H–H or C–H bond. Thus, this mechanism can be considered a heterolytic activation of the H–H or C–H bond by which the nonpolar bond is made acidic by coordination,[266] and the basic X-type ligand of a d^0 complex deprotonates the coordinated reagent.

$$L_nM-R \ + \ R'-H \ \longrightarrow \ \underset{\underset{\underset{\text{Acidic}}{R'-H}}{|}}{L_nM-R} \ \longrightarrow \ L_nM \underset{R'}{\overset{R}{\cdots}} H \ \longrightarrow \ \underset{\underset{L_nM-R'}{|}}{R-H} \ \longrightarrow \ L_nM-R' \ + \ R-H \tag{6.48}$$

The relative reactivities of dihydrogen, arenes, and alkanes toward σ-bond metathesis parallels the relative reactivities of these reagents toward oxidative addition. σ-Bond metathesis between dihydrogen and zirconocene alkyl complexes occurs (Equation 6.49),[267,268] but metathesis between alkanes or arenes and these complexes does not. Examples of σ-bond metatheses with arenes are less common than those with dihydrogen, but they are more common and occur faster than those with alkanes. For example, Cp*$_2$ScMe reacts with benzene to form Cp*$_2$ScPh[259] (Equation 6.50) faster than Cp*$_2$ScMe reacts with labeled methane.

$$\tag{6.49}$$

$$Cp^*_2ScMe \ + \ R-H \ \longrightarrow \ Cp^*_2ScR \ + \ CH_4$$
$$\text{Rate: } R = Ph > {}^{13}CH_3 \tag{6.50}$$

$$Cp^*_2Lu-Me \ + \ {}^{13}CH_3 \ \longrightarrow \ Cp^*_2Lu-{}^{13}CH_3 \ + \ CH_4 \tag{6.51}$$

$$\tag{6.52}$$

Some of the first reactions of soluble metal complexes with methane occurred by σ-bond metathesis. Like the first examples of oxidative addition of alkyl C–H bonds, the first examples of σ-bond metathesis with alkyl C–H bonds were intramolecular. Yet, the lutetium– and yttrium–methyl complexes, Cp*$_2$MMe (M = Lu and Y) were shown by Watson to react intermolecularly with ^{13}C-labeled methane to form the labeled methyl complexes and unlabeled methane at 70 °C (Equation 6.51).[269] Related scandium compounds have now been shown to undergo similar reactions with alkanes,[259] and a thoracyclobutane complex reacts with methane to open the metallacycle and form a methylthorium compound (Equation 6.52).[270]

6.5.2. Potential Sigma-Bond Metatheses Involving Late Transition Metal Complexes

The reaction of a late-transition-metal–alkyl complex containing d-electrons with hydrogen or a hydrocarbon to extrude alkane or arene and generate a new transition-metal–alkyl, –aryl, or –hydride complex usually occurs by a sequence of oxidative addition and reductive elimination. However, this overall transformation could also occur by a σ-bond metathesis (Scheme 6.5). It is difficult to determine by experiment if the reaction of a late-transition-metal–alkyl complex occurs by the sequence of oxidative addition and reductive elimination or by σ-bond metathesis. Therefore, theoretical studies have addressed this issue. In most cases, even when the process would involve a relatively high oxidation state, the process has been calculated to occur by oxidative addition and reductive elimination.[271,272]

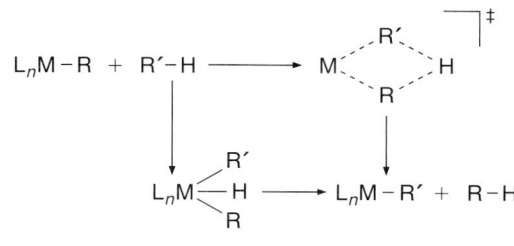

Scheme 6.5

Nevertheless, experimental and computational studies on the reactions of some late transition metal complexes have recently suggested the occurrence of mechanisms that are closely related to the σ-bond metathesis pathway of early-metal systems.[273–275] Heterolytic activation of dihydrogen by late transition metal complexes has been established in some cases, and is more common than heterolytic activation of alkanes or arenes. These reactions are more common because dihydrogen complexes of late transition metals can be relatively stable and acidic.[276] Therefore, even weakly basic covalent ligands are reactive enough to deprotonate the coordinated H_2 and cause reaction by a heterolytic or "sigma-bond metathesis" pathway.[34,278,279] Examples of the heterolytic activation of alkanes and arenes by late transition metal complexes are less well established, but strong evidence for these mechanisms has been gained by computational studies.[271,274,280,281]

Thus, a continuum between two mechanistic extremes is probably the best way to view the mechanistic dichotomy between the σ-bond metathesis and the sequence of oxidative addition and reductive elimination. The distinction between the two classes of mechanism would then be made by the amount of M–H bonding in a transition state (more M–H bonding would indicate an oxidative addition/reductive elimination sequence) and the presence or absence of a minimum on the energy surface that would correspond to the dialkyl intermediate of the oxidative addition/reductive elimination pathway.

More specifically, calculations have suggested that the ruthenium-alkyl complex[274] in Equation 6.53 reacts with arene to exchange covalent ligands by a process closely related to a σ-bond metathesis mechanism. Computational studies of the reactions of a simple iridium–alkyl and alkoxo complex with alkanes to generate new metal–alkyl complexes have also suggested that a mechanism is followed that involves many of the characteristics of a classic σ-bond metathesis transition state.[275] However, calculations of the mechanisms of these two processes imply that the transition state contains some degree of M–H bonding.

Such bonding is thought to be absent (or at least much weaker) in the transition state for σ-bond metathesis with a d^0 metal complex. Thus, to distinguish between these two types of transition state, the mechanism calculated for the complexes containing d-electrons has been called an "oxidative hydrogen migration." In contrast to these results, computational studies have suggested that the reaction of the Ir(III) cationic intermediates [Cp*Ir(PMe$_3$) R]$^+$ with alkanes, shown in Equation 6.34, occurs by the mechanism shown in Equation 6.54 involving oxidative addition to form an Ir(V) intermediate.[271]

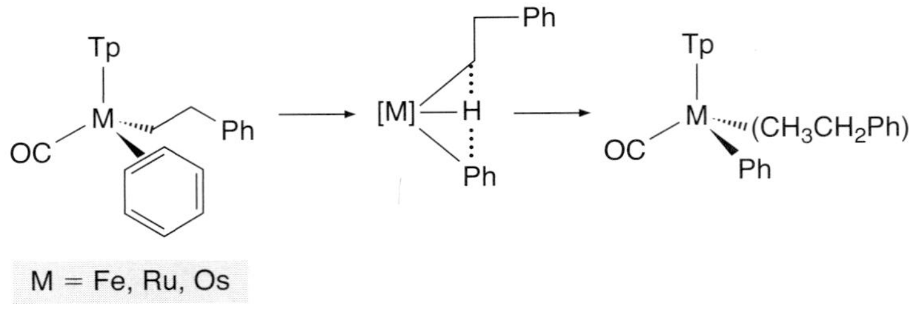

(6.53)

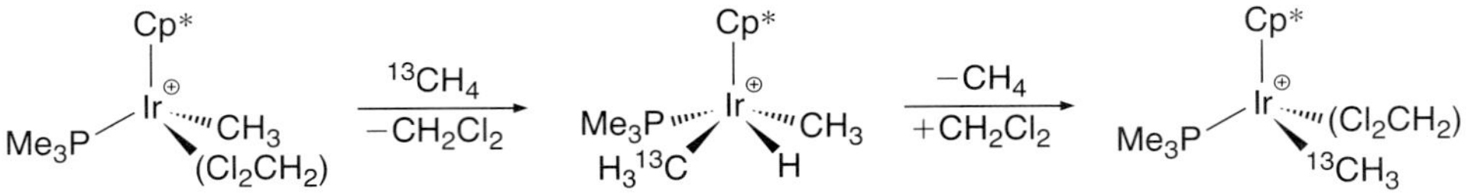

M = Fe, Ru, Os

(6.54)

Likewise, theoretical studies suggest that the reaction of a transition metal boryl complex with an alkane to generate an alkylboronate ester and a transition metal hydride complex occurs in some cases by a mechanism in which the transition state possesses features of a σ-bond metathesis process (Scheme 6.6).[273] Computational studies of this reaction imply that the alkane first coordinates to an open site on the metal, and that the hydrogen of the coordinated C–H bond migrates to the boryl ligand to form an intermediate containing a metal–alkyl ligand and a coordinated borane. This process occurs without an increase in the oxidation state of the metal center. This intermediate then undergoes a second reaction to generate the boron–carbon and metal–hydride bonds in the final products. Calculations[282] of arene borylation by a bipyridine-ligated Ir(III) trisboryl intermediate[195−197] suggest that oxidative addition to generate an Ir(V) intermediate occurs. Thus, different pathways for C–H bond cleavage by metal-boryl complexes appear to be similar in energy.

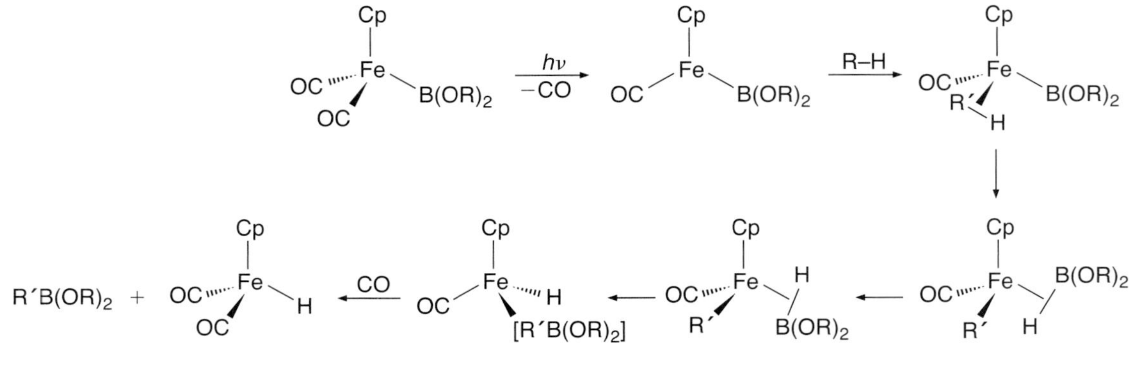

Scheme 6.6

6.5.3. [2+2] Additions Across Metal–Ligand Multiple Bonds

Many transition metal oxo complexes[283] and metalloproteins[284–288] react with the C–H bonds of alkanes to form alcohols and ketones. However, these reactions are not typically thought to occur by formation of a metal–carbon bond.[204,289–292] In contrast, transition metal amido and carbene complexes have been prepared containing a ligand set that leads to addition of the C–H bond of a hydrocarbon across the M=N or M=C bond to generate a transition metal hydrocarbyl complex. Certain titanium–,[293–296] zirconium–,[297–301] tantalum–,[302] and tungsten–imido[303] complexes undergo this type of [2+2] reaction, and a tungsten–carbene complex has been shown to undergo an analogous process.

$$Cp_2Zr\begin{smallmatrix}NHBu^t\\\\Me\end{smallmatrix} \xrightarrow[-CH_4]{\Delta} Cp_2Zr=NBu^t \xrightarrow{} Cp_2Zr\begin{smallmatrix}NHBu^t\\\\Ph\end{smallmatrix} \qquad (6.55)$$

$$\begin{matrix}Zr(NHSiBu_3^t)_4\\\\[Bu_3^tSiN(H)]_3Zr\end{matrix}\!\!\searrow\!\!\nearrow [Bu_3^tSiN(H)]_2Zr=NSiBu_3^t \xrightarrow[R=CH_3\text{ and }Ph]{RH} (Bu_3^tSiNH)_3Zr-R \qquad (6.56)$$

Wolczanski and Bergman concurrently showed that the addition of hydrocarbon C–H bonds occurs to zirconium(IV) imido complexes, as shown in Equations 6.55 and 6.56.[299,304] The silylamidozircononium–imido complex reacted with both alkanes and arenes, whereas the zirconocene–imido complex reacted with arenes. Kinetic studies showed that the generation of the unsaturated zirconium–imido complex is rate limiting and that the unsaturated zirconocene compound is generated reversibly by dissociation of THF prior to the C–H bond cleavage process. Most likely, this reaction occurs by formation of an alkane or arene complex and subsequent reaction of the alkane or arene complex to generate the amido alkyl or amido aryl complexes.[295,303]

Extensive studies on potential intermediates, kinetic isotope effects, and selectivities have been conducted on the silylamido compounds. Studies with tungsten alkyl amido complexes have shown that formation of an alkane complex precedes C–H bond cleavage by these d^0 imido complexes,[303] just as it precedes oxidative addition and σ-bond metathesis.[295,297]

The kinetic and thermodynamic selectivity for reactions of a titanium–imido complex with different types of C–H bonds has been determined.[294] Reactions with substrates that possess primary and secondary C–H bonds occur selectively at the primary C–H bond. In addition, reactions with mixtures of alkanes and arenes occur selectively at the arene C–H bond. Like the stabilities of most low-valent, late metal complexes, the primary alkyl complex is thermodynamically more stable than the secondary alkyl complex, and the aryl complexes are more stable than the alkyl complexes. Activation of olefins at the allylic position occurs more slowly than reaction at the vinyl position, but when it does occur, the reaction generates a stable η³-allyl complex.

Carbene complexes can also react by addition of alkane C–H bonds across the M=C bond to generate a product containing two metal–carbon single bonds or materials derived from such products. This reaction is the reverse of the α-elimination, which is a common reaction that forms Schrock-type carbene complexes, as noted in Chapters 3, 10, and 13. Such reactions of carbene complexes with the C–H bonds of alkyl groups were first observed to occur intramolecularly, particularly after generation of cationic carbene complexes.[305–309] However, intermolecular additions of C–H bonds of arenes[310–312] and alkanes[312] are also known.

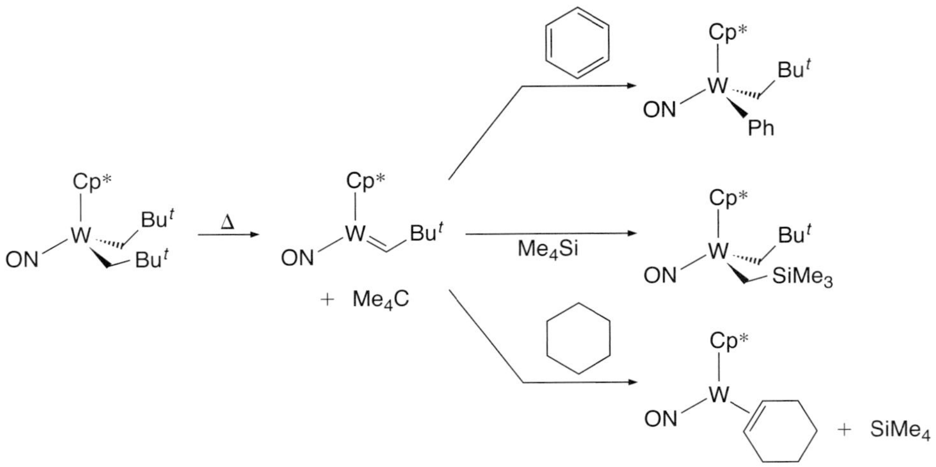

Scheme 6.7

In particular, Legzdins[313,314] uncovered a set of tungsten–carbene complexes that react intermolecularly with alkanes and arenes to generate products possessing two metal–carbon single bonds by addition of the hydrocarbon C–H bond across the M=C bond (Scheme 6.7). Heating of $Cp^*W(NO)(CH_2CMe_3)$ generates the intermediate $Cp^*(NO)$ $W=CHBu^t$ and free CH_3Bu^t. This alkylidene complex reacts with arenes and tetramethylsilane to form alkyl aryl and alkyl silylalkyl complexes. Moreover, this carbene complex reacts with cyclohexane to generate a cyclohexene complex and with ethylcyclohexane to form an allyl complex by further C–H bond cleavage of the alkyl intermediates. The benzyne complex generated from thermolysis of the tungsten diphenyl (Equation 6.57) and the allenyl complex generated by heating of the alkyl vinyl complex also react with aromatic and saturated C–H bonds. Computational data and isotopic labeling studies support the formation of alkane and arene complexes prior to the C–H bond cleavage step of these reactions.[315]

Related C–H activations by 1,2-additions across metal–ligand bonds have also been reported with unusual group 4 alkylidyne complexes.[316,317] As shown in Equation 6.58, a titanium(IV) trimethylsilylmethyl benzylidene complex eliminates tetramethylsilane to generate a Ti(IV) alkylidyne complex. This complex then reacts with benzene to add the arene C–H bond across the metal-alkylidyne unit to form an aryl benzylidene complex.

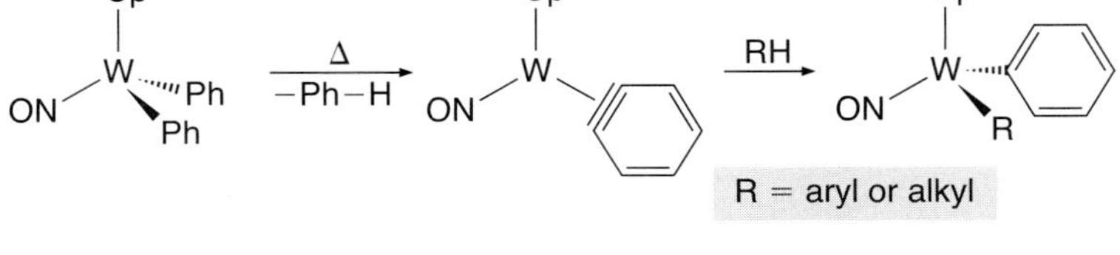

(6.57)

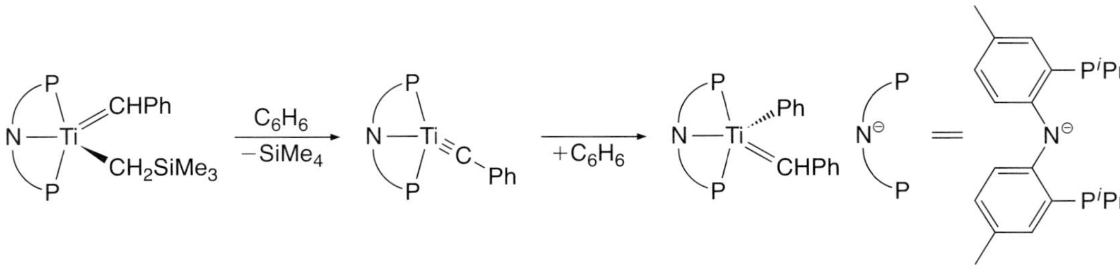

(6.58)

$$(6.59)$$

Certain dinuclear Rh(II) carbene complexes react with alkanes to generate products from insertion of the carbene unit into the alkane C–H bond with high diastereo- and enantioselectivity (Equation 6.59).[208,318] These reactions occur by mechanisms distinct from those of the reactions of C–H bonds with the tungsten alkylidene and alkylidyne complexes just described. The reactions of the dinuclear Rh(II) carbene complexes appear to occur by a mechanism that involves direct reaction of the carbene at the C–H bond without coordination of the alkane and addition across the M=C bond of the carbene.[208,319] Such rhodium carbene complexes have not been isolated, but the absence of an open coordination site cis to the carbene ligand in the accepted carbene intermediate is thought to preclude initial reaction of the substrate at the metal center to form a new metal–carbon bond. The catalytic chemistry that occurs via these carbene complexes is presented in more detail in Chapter 18 (catalytic C–H bond functionalization).

6.6. Oxidative Addition of C–C Bonds[320,321]

As noted in the introduction to this chapter, products from oxidative addition reactions that cleave the C–C bond of an unstrained alkane have not yet been observed. However, many systems have been studied that react to cleave C–C bonds in catalytic processes or to form isolable oxidative addition products from cleavage of strained C–C bonds or C–C bonds in substrates containing ligating atoms. In addition, evidence for mild C–C bond cleavage of simple alkanes by group 4 and 5 metals on surfaces has been gained in recent years.[322–326]

The oxidative addition or σ-bond metathesis of unstrained C–C bonds in alkyl groups or alkylarenes by soluble transition metal complexes has required a docking group for the metal or a highly strained ring system. The need for prior coordination is likely due to the greater steric effects on insertion of a metal fragment into a C–C bond than on insertion into a C–H bond. Thus, oxidative addition can be thermodynamically feasible, but C–H addition is typically faster. Once coordination or C–H addition of the substrate has occurred, intramolecular oxidative addition of a C–C bond can occur selectively and under mild conditions. One exception to this general trend is the oxidative addition of the C–C single bond in aryl nitriles (Equation 6.60).[327–330] C–C bond cleavage can also occur by a β-alkyl elimination; this reaction is presented in Chapter 10.

$$(6.60)$$

A series of studies of the oxidative addition of C–C bonds in substrates containing a group that coordinates the metal center has been reported. Addition of acyl carbon bonds after such coordination is particularly favorable. Suggs reported the C–C oxidative addition of an unstrained C–C bond in an acyl quinoline (Equation 6.61).[331–334] Murakami reported a catalytic reaction involving insertion of an alkene into a cyclobutanone that appears to occur by C(acyl)–C(alkyl) oxidative addition even without an accompanying ligating group.[335]

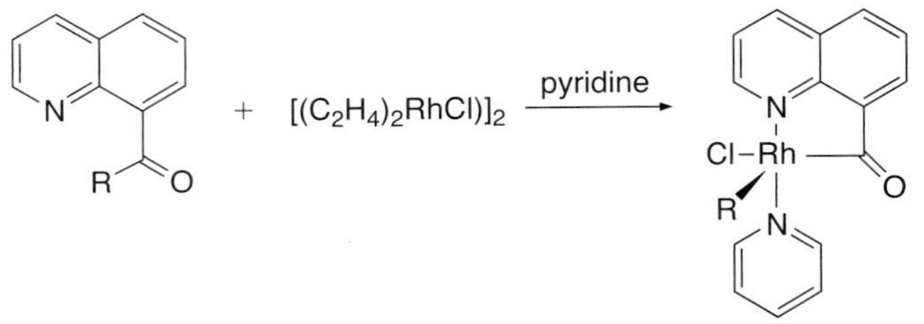

(6.61)

Milstein has reported a series of reactions of 2-methyl- and 1,3-bis-phosphinomethyl-arenes to form rhodium– and iridium–methyl complexes containing PCP, PCO, or PCN pincer ligands (Equations 6.62 and 6.63).[336–344] These studies have revealed several principles. First, oxidative addition of the strong aryl–methyl C–C bond can be favorable thermodynamically and occur under mild conditions. Second, reactions that form products from C–H and C–C activation can occur in parallel, and their relative rates depend on the specific ligand. In some cases, the product of C–H activation is formed reversibly and converts to the more stable product from C–C activation, while in other cases the metal is optimally oriented towards the C–C bond, and either C–H activation is not observed[342] or products from both C–H and C–C activation are formed at similar rates, and the product from C–H activation slowly converts to the product from C–C bond activation.[339] Second, the product from cleavage of the stronger $C(sp^2)$–$C(sp^3)$ bond is more stable than the product from cleavage of the weaker $C(sp^3)$–$C(sp^3)$ bond because of the high stability of an M–$C(sp^2)$ bond. Third, a product possessing an agostic interaction between the metal and a C–C bond is generated when the geometry and thermodynamics are appropriate. The existence of this complex suggests that the oxidative addition of a C–C bond is preceded by formation of a C–C σ-complex (Equation 6.63), just as the addition of H_2 and C–H bonds is preceded by coordination of this σ-bond to the metal.

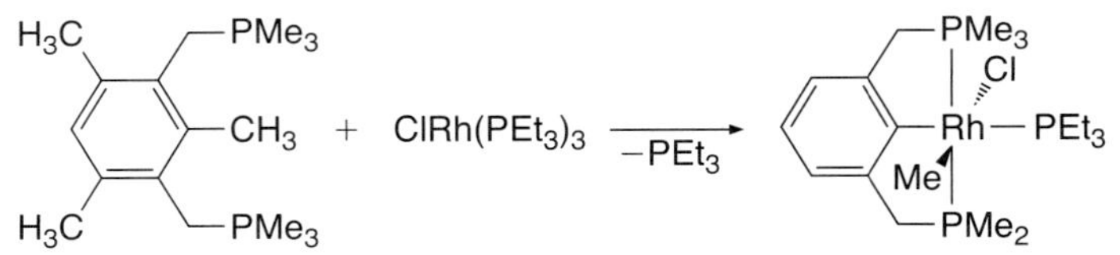

(6.62)

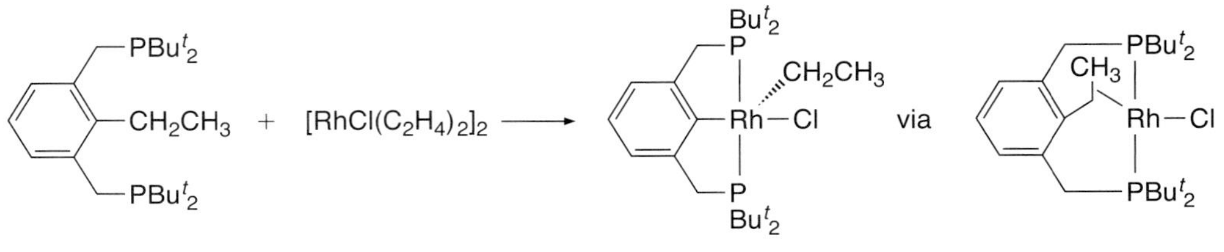

(6.63)

Other cleavages of the C–C bonds of substrates that lack a dative ligand have occurred after addition of the C–H bond. Crabtree and co-workers reported the oxidative addition of an unstrained C–C bond (Equation 6.64),[345] most likely after a C–H bond cleavage process and just prior to establishment of the cyclopentadienyl-type ligand. The reaction of cyclopropane with $Cp*Rh(PMe_3)$ to form a metallacyclobutane occurs by initial C–H bond addition, followed by insertion of the metal center into the C–C bond to form the product that is more stable thermodynamically (Equation 6.65).[346,347]

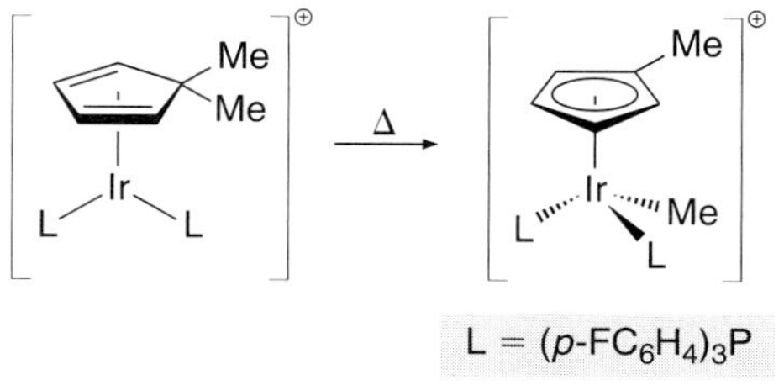

(6.64)

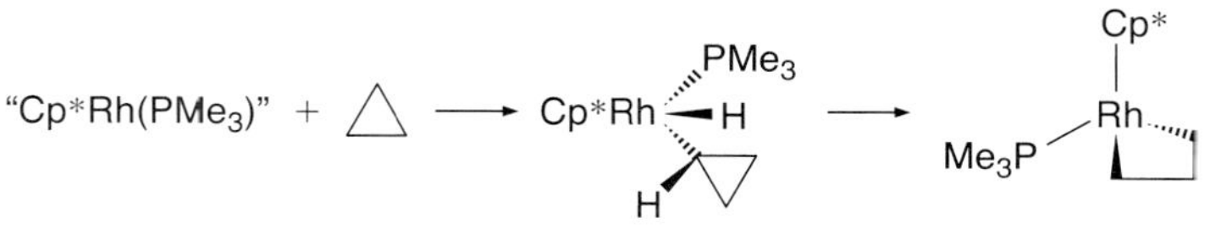

$L = (p\text{-}FC_6H_4)_3P$

(6.65)

C–C cleavage of strained rings and ketones has been used to develop useful catalytic reactions. For example, vinylcyclopropanes and vinylcyclobutanes react with alkynes (Equation 6.66) to generate products from 5+2 and 6+2 addition processes that form seven- and eight-membered ring products by overall transformations that are homologs of the Diels–Alder reaction.[348–354] The mechanism of these catalytic reactions continues to be studied, but these reactions most likely occur by coordination of the olefin to rhodium and insertion of the metal into the cyclopropene or cyclobutane. Decarbonylation of dialkyl ketones,[355] including relatively unstrained cyclic ketones, has been reported and most likely occurs by oxidative addition into the acyl–alkyl C–C bond,[355–357] subsequent de-insertion of CO, and C–C reductive elimination.

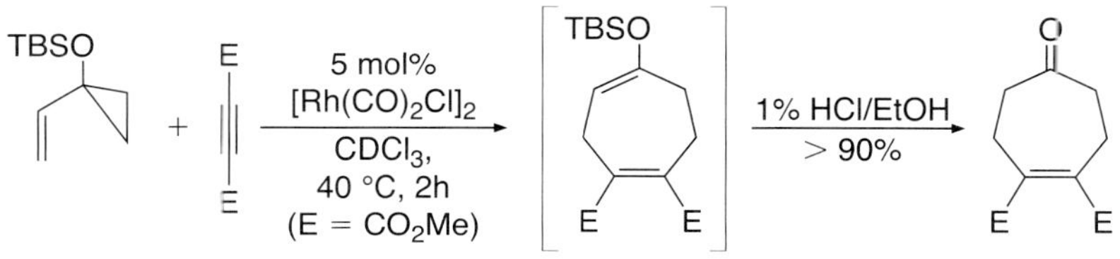

(6.66)

6.7. Oxidative Addition of E–E Bonds

In contrast to the rare occurrence of the oxidative addition of C–C bonds, the oxidative addition of the nonpolar bonds between two main group atoms, such as boron and silicon, can be facile. The oxidative addition of Si–Si bonds in disilanes and the B–B bond in diborane(4) reagents is likely to be a step in a variety of catalytic reactions, including the additions of disilanes,[358,359] diborane(4) reagents, and silaboranes[358,360–363] across olefins and alkynes, the silylation[364] and borylation of arene C–H bonds,[195,197,139,365] the borylation of alkane C–H bonds,[194] and the conversion of aryl halides to arylsilanes[366–368] and arylboronate esters.[369–372]

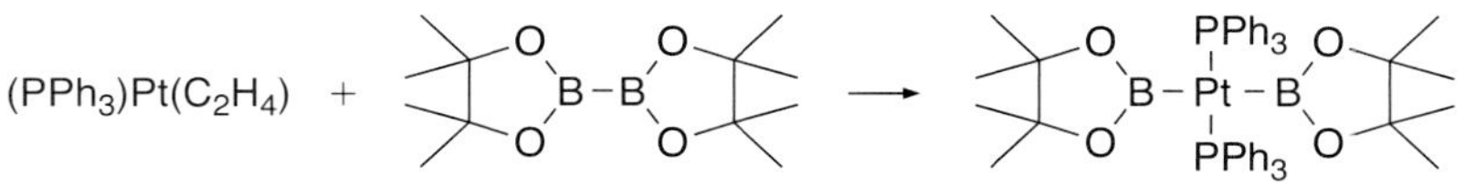

(6.67)

The oxidative addition of disilanes occurs to palladium complexes of isonitrile ligands and platinum complexes of trialkylphosphine ligands as part of the catalytic silylation of alkynes and aryl halides.[366–368,373] The addition of stannylboranes to Pd(0) complexes has also been reported,[374] and the addition of diboron compounds to many metal systems, such as Pt(0) complexes (Equation 6.67), is now common.[375–381] These reactions all occur with metal complexes that do not undergo intermolecular reactions with alkane C–H bonds, let alone C–C bonds. Thus, the Lewis acidic character of these reagents must accelerate the coordination of substrate and cleavage of the E–E bonds.

6.8. Summary

Work during the past 40–50 years has revealed many examples of the insertion of an unsaturated metal fragment into a nonpolar H–X, C–X, and E–E bonds by processes called "oxidative addition" and examples of related non-oxidative processes often called "sigma-bond metathesis." In the oxidative addition processes, the metal is formally oxidized, and the substrate is formally reduced as new metal–ligand bonds are formed. Many complexes add dihydrogen and silanes, and many complexes also add the C–H bonds of arenes. Fewer complexes insert into the C–H bonds of alkanes, but a series of complexes that undergo this process are now known, and the mechanism of the reactions of these complexes has become well established. The mechanism of these reactions occurs in most cases by initial coordination of the H–H, C–H, or other nonpolar H–X bond to the metal center, followed by either full insertion of the metal into the H–X bond to complete an oxidative addition or deprotonation of the coordinated ligand by an accompanying covalent ligand to complete a σ-bond metathesis process. Even fewer examples of the oxidative addition of C–C bonds are known, but intramolecular additions, additions into strained C–C bonds, and insertions into C–C(sp^2) or C–C(sp^3) bonds are known. Insertions into Si–Si, Si–B, and B–B bonds are more facile. These bond cleavage processes are part of catalytic hydrogenations, hydroformylations, hydrosilations, disilations, silylborations, diborations, and alkane or arene functionalizations discussed in Chapters 15–19.

References and Notes

1. Collman, J. P.; Roper, W. R. *Adv. Organomet. Chem.* **1968**, *7*, 53.
2. Collman, J. P. *Acc. Chem. Res.* **1968**, *1*, 136.
3. Halpern, J. *Acc. Chem. Res.* **1970**, *3*, 386.
4. Lappert, M. F.; Lednor, P. W. *Adv. Organomet. Chem.* **1976**, *14*, 345.
5. Osborn, J. A. *Organotransition-Metal Chemistry*; Plenum Press: New York, 1975.
6. Vaska, L. *Acc. Chem. Res.* **1968**, *1*, 335.
7. Ugo, R. *Coord. Chem. Rev.* **1968**, *3*, 319.
8. Cramer, R. *Acc. Chem. Res.* **1968**, *1*, 186.
9. Parshall, G. W. *Acc. Chem. Res.* **1970**, *3*, 139.
10. Stille, J. K.; Lau, K. S. Y. *Acc. Chem. Res.* **1977**, *10*, 434.
11. Stille, J. K. *The Chemistry of the Metal–Carbon Bond, Vol. II. The Nature and Cleavage of Metal–Carbon Bonds*; Wiley: New York, 1985.
12. Mann, G.; Shelby, Q.; Roy, A. H.; Hartwig, J. F. *Organometallics* **2003**, *22*, 2775.
13. Jones, W. D.; Kuykendall, V. L. *Inorg. Chem.* **1991**, *30*, 2615.
14. Yamashita, M.; Hartwig, J. F. *J. Am. Chem. Soc.* **2004**, *126*, 5344.
15. Alcazar-Roman, L. M.; Hartwig, J. F.; Rheingold, A. L.; Liable-Sands, L. M.; Guzei, I. A. *J. Am. Chem. Soc.* **2000**, *122*, 4618.
16. Crumpton-Bregel, D. M.; Goldberg, K. I. *J. Am. Chem. Soc.* **2003**, *125*, 9442.
17. Bartlett, K. L.; Goldberg, K. I.; Borden, W. T. *Organometallics* **2001**, *20*, 2669.
18. Yoneda, G.; Blake, D. M. *Inorg. Chem.* **1981**, *20*, 67 and references therein.
19. Bryndza, H. E.; Fong, L. K.; Paciello, R. A.; Tam, W.; Bercaw, J. E. *J. Am. Chem. Soc.* **1987**, *109*, 1444.

20. Schock, L. E.; Marks, T. J. *J. Am. Chem. Soc.* **1988**, *110*, 7701.
21. Nolan, S. P.; Hoff, C. D.; Stoutland, P. O.; Newman, L. J.; Buchanan, J. M.; Bergman, R. G.; Yang, G. K.; Peters, K. S. *J. Am. Chem. Soc.* **1987**, *109*, 3143.
22. James, B. R. In *Comprehensive Organometallic Chemistry*; Wilkinson, G., Stone, F. G. A., Abel, E. W., Eds.; Pergamon Press: Oxford, 1982; Vol. 8, p 285.
23. Niu, S.; Hall, M. B. *Chem. Rev.* **2000**, *100*, 353.
24. Low, J. J.; Goddard, W. A. *J. Am. Chem. Soc.* **1984**, *106*, 6928.
25. Low, J. J.; Goddard, W. A. *J. Am. Chem. Soc.* **1984**, *106*, 8321.
26. Obara, S.; Kitaura, K.; Morokuma, K. *J. Am. Chem. Soc.* **1984**, *106*, 7482.
27. Noell, J. O.; Hay, P. J. *J. Am. Chem. Soc.* **1982**, *104*, 4578.
28. Balazs, A. C.; Johnson, K. H.; Whitesides, G. M. *Inorg. Chem.* **1982**, *21*, 2162.
29. Saillard, J. Y.; Hoffmann, R. *J. Am. Chem. Soc.* **1984**, *106*, 2006.
30. Nakatsuji, H.; Hada, M. *J. Am. Chem. Soc.* **1985**, *107*, 8264.
31. Siegbahn, P. E. M.; Blomberg, M. R. A.; Bauschlicher, C. W. *J. Chem. Phys.* **1984**, *81*, 1373.
32. Wang, W.; Weitz, E. *J. Phys. Chem. A* **1997**, *101*, 2358.
33. Kubas, G. J.; Ryan, R. R.; Swanson, B. I.; Vergamini, P. J.; Wasserman, H. J. *J. Am. Chem. Soc.* **1984**, *106*, 451.
34. Kubas, G. J. *Metal-Dihydrogen and Sigma-Bond Complexes: Structure, Theory, and Reactivity*; Kluwer Academic: New York, 2001.
35. Matsubara, T.; Maseras, F.; Koga, N.; Morokuma, K. *J. Phys. Chem.* **1996**, *100*, 2573.
36. Musaev, D. G.; Morokuma, K. *J. Organomet. Chem.* **1995**, *504*, 93.
37. Khalsa, G. R. K.; Kubas, G. J.; Unkefer, C. J.; Vandersluys, L. S.; Kubatmartin, K. A. *J. Am. Chem. Soc.* **1990**, *112*, 3855.
38. Kubas, G. J.; Ryan, R. R.; Unkefer, C. J. *J. Am. Chem. Soc.* **1987**, *109*, 8113.
39. Kubas, G. J.; Ryan, R. R.; Wrobleski, D. A. *J. Am. Chem. Soc.* **1986**, *108*, 1339.
40. Kubas, G. J.; Unkefer, C. J.; Swanson, B. I.; Fukushima, E. *J. Am. Chem. Soc.* **1986**, *108*, 7000.
41. Luo, X. L.; Crabtree, R. H. *J. Am. Chem. Soc.* **1990**, *112*, 6912.
42. Luo, X. L.; Michos, D.; Crabtree, R. H. *Organometallics* **1992**, *11*, 237.
43. Luo, X. L.; Kubas, G. J.; Burns, C. J.; Eckert, J. *Inorg. Chem.* **1994**, *33*, 5219.
44. Chinn, M. S.; Heinekey, D. M. *J. Am. Chem. Soc.* **1990**, *112*, 5166.
45. Chinn, M. S.; Heinekey, D. M.; Payne, N. G.; Sofield, C. D. *Organometallics* **1989**, *8*, 1824.
46. Chinn, M. S.; Heinekey, D. M. *J. Am. Chem. Soc.* **1987**, *109*, 5865.
47. Conroy-Lewis, F. M.; Simpson, S. J. *J. Chem. Soc., Chem. Commun.* **1987**, 1675.
48. Cappellani, E. P.; Maltby, P. A.; Morris, R. H.; Schweitzer, C. T.; Steele, M. R. *Inorg. Chem.* **1989**, *28*, 4437.
49. Arliguie, T.; Chaudret, B. *J. Chem. Soc., Chem. Commun.* **1989**, 155.
50. Esteruelas, M. A.; Garcia-Yebra, C.; Olivan, M.; Onate, E.; Tajada, M. A. *Organometallics* **2002**, *21*, 1311.
51. Klooster, W. T.; Koetzle, T. F.; Jia, G. C.; Fong, T. P.; Morris, R. H.; Albinati, A. *J. Am. Chem. Soc.* **1994**, *116*, 7677.
52. Michos, D.; Luo, X. L.; Howard, J. A. K.; Crabtree, R. H. *Inorg. Chem.* **1992**, *31*, 3914.
53. Luo, X. L.; Crabtree, R. H. *J. Chem. Soc., Dalton Trans.* **1991**, 587.
54. Brammer, L.; Howard, J. A. K.; Johnson, O.; Koetzle, T. F.; Spencer, J. L.; Stringer, A. M. *J. Chem. Soc., Chem. Commun.* **1991**, 241.
55. Kim, Y. Y.; Deng, H.; Meek, D. W.; Wojcicki, A. *J. Am. Chem. Soc.* **1990**, *112*, 2798.
56. Pons, V.; Heinekey, D. M. *J. Am. Chem. Soc.* **2003**, *125*, 8428.
57. Law, J. K.; Mellows, H.; Heinekey, D. M. *J. Am. Chem. Soc.* **2002**, *124*, 1024.
58. Law, J. K.; Mellows, H.; Heinekey, D. M. *J. Am. Chem. Soc.* **2001**, *123*, 2085.
59. Kunin, A. J.; Farid, R.; Johnson, C. E.; Eisenberg, R. *J. Am. Chem. Soc.* **1985**, *107*, 5315.
60. Shin, J. H.; Parkin, G. *J. Am. Chem. Soc.* **2002**, *124*, 7652.
61. Johnson, C. E.; Eisenberg, R. *J. Am. Chem. Soc.* **1985**, *107*, 3148.
62. Burk, M. J.; McGrath, M. P.; Wheeler, R.; Crabtree, R. H. *J. Am. Chem. Soc.* **1988**, *110*, 5034.
63. Deutsch, P. P.; Eisenberg, R. *Chem. Rev.* **1988**, *88*, 1147.
64. Halpern, J.; Wong, C. S. *J. Chem. Soc., Chem. Commun.* **1973**, 629.
65. Tolman, C. A.; Meakin, P. Z.; Lindner, D. L.; Jesson, J. P. *J. Am. Chem. Soc.* **1974**, *96*, 2762.
66. Jardine, F. H.; Osborn, J. A.; Wilkinson, G.; Young, J. F. *Chem. Ind.* **1965**, 560.
67. Osborn, J. A.; Jardine, F. H.; Young, J. F.; Wilkinson, G. *J. Chem. Soc. A* **1966**, 1711.
68. Jardine, F. H. *Prog. Inorg. Chem.* **1981**, *28*, 63.
69. Ohtani, Y.; Fujimoto, M.; Yamagishi, A. *Bull. Chem. Soc. Jpn.* **1977**, *50*, 1453.
70. Daniel, C.; Koga, N.; Han, J.; Fu, X. Y.; Morokuma, K. *J. Am. Chem. Soc.* **1988**, *110*, 3773.
71. Wink, D.; Ford, P. C. *J. Am. Chem. Soc.* **1985**, *107*, 1794.
72. Vaska, L.; DiLuzio, J. W. *J. Am. Chem. Soc.* **1962**, *84*, 679.

73. (a) Chock, P. B.; Halpern, J. *J. Am. Chem. Soc.* **1966**, *88*, 3511; (b) Strohmeier, W. *J. Organomet. Chem.* **1971**, *32*, 137.

74. Abu-Hasanayn, F.; Goldman, A. S.; Krogh-Jespersen, K. *J. Phys. Chem.* **1993**, *97*, 5890.

75. Yoshida, T.; Otsuka, S. *J. Am. Chem. Soc.* **1977**, *99*, 2134.

76. Paonessa, R. S.; Trogler, W. C. *J. Am. Chem. Soc.* **1982**, *104*, 1138.

77. Zhang, X. X.; Wayland, B. B. *J. Am. Chem. Soc.* **1994**, *116*, 7897.

78. Van Voorhees, S. L.; Wayland, B. B. *Organometallics* **1987**, *6*, 204.

79. James, B. R. In *Comprehensive Organometallic Chemistry*; Wilkinson, G., Stone, F. G. A., Abel, E. W., Ed.; Pergamon Press: New York, 1982; Vol. 8, p 285.

80. Halpern, J. *Inorg. Chim. Acta* **1983**, *77*, L105.

81. Halpern, J. *Inorg. Chim. Acta* **1982**, *62*, 31.

82. Simandi, L. I.; Budozahonyi, E.; Szeverenyi, Z.; Nemeth, S. *J. Chem. Soc.* **1980**, 276.

83. Chao, T. H.; Espenson, J. H. *J. Am. Chem. Soc.* **1978**, *100*, 129.

84. Wayland, B. B.; Ba, S.; Sherry, A. E. *Inorg. Chem.* **1992**, *31*, 148.

85. Chaudret, B.; Dahan, F.; Sabo, S. *Organometallics* **1985**, *4*, 1490.

86. Casey, C. P.; Bullock, R. M. *J. Mol. Catal.* **1982**, *14*, 283.

87. Jetz, W.; Graham, W. A. G. *Inorg. Chem.* **1971**, *10*, 4.

88. Schubert, U.; Scholz, G.; Muller, J.; Ackermann, K.; Worle, B.; Stansfield, R. F. D. *J. Organomet. Chem.* **1986**, *306*, 303.

89. Schubert, U. *Adv. Organomet. Chem.* **1990**, *30*, 151.

90. Kramer, A. V.; Osborn, J. A. *J. Am. Chem. Soc.* **1974**, *96*, 7832.

91. Kramer, A. V.; Labinger, J. A.; Bradley, J. S.; Osborn, J. A. *J. Am. Chem. Soc.* **1974**, *96*, 7145.

92. Speier, J. L. *Adv. Organomet. Chem.* **1979**, *17*, 407.

93. Cundy, C. S.; Kingston, B. M.; Lappert, M. F. *Adv. Organomet. Chem.* **1973**, *11*, 253.

94. Mackay, K. M.; Nicholson, B. K. *Comprehensive Organometallic Chemistry*; Pergamon Press: New York, 1982; Vol. 6.

95. Kunin, A. J.; Farid, R.; Johnson, C. E.; Eisenberg, R. *J. Am. Chem. Soc.* **1985**, *107*, 5315.

96. Fernandez, M. J.; Bailey, P. M.; Bentz, P. O.; Ricci, J. S.; Koetzle, T. F.; Maitlis, P. M. *J. Am. Chem. Soc.* **1984**, *106*, 5458.

97. Fernandez, M. J.; Maitlis, P. M. *J. Chem. Soc.* **1982**, 310.

98. Carre, F.; Colomer, E.; Corriu, R. J. P.; Vioux, A. *Organometallics* **1984**, *3*, 1272.

99. Schubert, U.; Ackermann, K.; Worle, B. *J. Am. Chem. Soc.* **1982**, *104*, 7378.

100. Muir, K. W.; Ibers, J. A. *Inorg. Chem.* **1970**, *9*, 440.

101. Haszeldine, R. N.; Parish, R. V.; Parry, D. J. *J. Chem. Soc. A* **1969**, 683.

102. DeCharentenay, F.; Osborn, J. A.; Wilkinson, G. *J. Chem. Soc. A* **1968**, 787.

103. Haszeldine, R. N.; Parish, R. V.; Parry, D. J. *J. Organomet. Chem.* **1967**, *9*, 13.

104. Chalk, A. J.; Harrod, J. F. *J. Am. Chem. Soc.* **1965**, *87*, 16.

105. Bennett, M. A.; Romeo, C.; Fraser, P. J. *Aust. J. Chem.* **1977**, *30*, 1201.

106. Eaborn, C.; Tune, D. J.; Walton, D. R. M. *J. Chem. Soc., Dalton Trans.* **1973**, 2255.

107. Ricci, J. S.; Koetzle, T. F.; Fernandez, M. J.; Maitlis, P. M.; Green, J. C. *J. Organomet. Chem.* **1986**, *299*, 383.

108. Ricci, J. S.; Koetzle, T. F.; Fernandez, M.; Maitlis, P. M. *Acta Crystallogr., Sec. A* **1984**, *40*, C301.

109. Fernandez, M. J.; Maitlis, P. M. *J. Chem. Soc., Dalton Trans.* **1984**, 2063.

110. Fernandez, M. J.; Maitlis, P. M. *Organometallics* **1983**, *2*, 164.

111. Fernandez, M. J.; Bailey, P. M.; Bentz, P. O.; Ricci, J. S.; Koetzle, T. F.; Maitlis, P. M. *J. Am. Chem. Soc.* **1984**, *106*, 5458.

112. Ruiz, J.; Mann, B. E.; Spencer, C. M.; Taylor, B. F.; Maitlis, P. M. *J. Chem. Soc., Dalton Trans.* **1987**, 1963.

113. Straus, D. A.; Tilley, T. D.; Rheingold, A. L.; Geib, S. J. *J. Am. Chem. Soc.* **1987**, *109*, 5872.

114. Tilley, T. D. In *The Chemistry of Organic Silicon Compounds*; Patai, S., Rappoport, Z., Eds.; Wiley: New York, 1989; p 1415.

115. Yang, H.; Kotz, K. T.; Asplund, M. C.; Harris, C. B. *J. Am. Chem. Soc.* **1997**, *119*, 9564.

116. Yang, H.; Asplund, M. C.; Kotz, K. T.; Wilkens, M. J.; Frei, H.; Harris, C. B. *J. Am. Chem. Soc.* **1998**, *120*, 10154.

117. Snee, P. T.; Yang, H.; Kotz, K. T.; Payne, C. K.; Harris, C. B. *J. Phys. Chem.* **1999**, *103*, 10426.

118. Snee, P. T.; Payne, C. K.; Kotz, K. T.; Yang, H.; Harris, C. B. *J. Am. Chem. Soc.* **2001**, *123*, 2255.

119. Sen, A. *J. Chem. Soc., Chem. Commun.* **1999**, 550.

120. Labinger, J. A.; Bercaw, J. E. *Nature* **2002**, *417*, 507.

121. Jones, W. D. *Acc. Chem. Res.* **2003**, *36*, 140.

122. Crabtree, R. H. *Chem. Rev.* **1995**, *95*, 987.

123. Arndtsen, B. A.; Bergman, R. G.; Mobley, T. A.; Peterson, T. H. *Acc. Chem. Res.* **1995**, *28*, 154.

124. Crabtree, R. H. *Chem. Rev.* **1985**, *85*, 245.

125. Bergman, R. G. *Science* **1984**, *223*, 902.

126. Doering, W. V.; Prinzbach, H. *Tetrahedron* **1959**, *6*, 24.

127. Baron, W. J.; DeCamp, M. R.; Hendrick, M. E.; Jones, M. J. In *Carbenes*; Jones, M. J., Moss, R. A., Eds.; Wiley: New York, 1973; Vol. 1, p 8.

128. Kirmse, W. *Carbene Chemistry*; Academic Press: New York, 1964; p 8.

129. Shilov, A. E.; Shteinman, A. A. *Coord. Chem. Rev.* **1977**, *24*, 97.

130. Shilov, A. E.; Shul'pin, G. B. *Chem. Rev.* **1997**, *97*, 2879.

131. Crabtree, R. H.; Mihelcic, J. M.; Quirk, J. M. *J. Am. Chem. Soc.* **1979**, *101*, 7738.

132. Crabtree, R. H.; Mellea, M. F.; Mihelcic, J. M.; Quirk, J. M. *J. Am. Chem. Soc.* **1982**, *104*, 107.

133. Baudry, D.; Ephritikhine, M.; Felkin, H.; Holmes-Smith, R. *J. Chem. Soc., Chem. Commun.* **1983**, 788.

134. Felkin, H.; Fillebeen-Khan, T.; Holmes-Smith, R.; Yingrui, L. *Tetrahedron Lett.* **1985**, *26*, 1999.

135. Ryabov, A. D. *Chem. Rev.* **1990**, *90*, 403.

136. Bruce, M. I. *Angew. Chem., Int. Ed. Engl.* **1977**, *16*, 73.

137. Janowicz, A. H.; Bergman, R. G. *J. Am. Chem. Soc.* **1982**, *104*, 352.

138. Hoyano, J. K.; Graham, W. A. G. *J. Am. Chem. Soc.* **1982**, *104*, 3723.

139. Constable, E. C. *Polyhedron* **1984**, *3*, 1037.

140. Horie, S.; Murahashi, S. *Bull. Chem. Soc. Jpn.* **1960**, *33*, 247.

141. Kleiman, J. P.; Dubeck, M. *J. Am. Chem. Soc.* **1963**, *85*, 1544.

142. Bennett, M. A.; Milner, D. L. *J. Am. Chem. Soc.* **1969**, *91*, 6983.

143. Tulip, T. H.; Thorn, D. L. *J. Am. Chem. Soc.* **1981**, *103*, 2448.

144. Ibers, J. A.; Dicosimo, R.; Whitesides, G. M. *Organometallics* **1982**, *1*, 13.

145. Karsch, H. H.; Klein, H.-F.; Schmidbaur, H. *Angew. Chem., Int. Ed. Engl.* **1975**, *14*, 637.

146. Rathke, J. W.; Muetterties, E. L. *J. Am. Chem. Soc.* **1975**, *97*, 3272.

147. Harris, T. V.; Rathke, J. W.; Muetterties, E. L. *J. Am. Chem. Soc.* **1978**, *100*, 6966.

148. Werner, H.; Werner, R. *J. Organomet. Chem.* **1981**, *209*, C60.

149. Werner, H.; Gotzig, J. *Organometallics* **1983**, *2*, 547.

150. Desrosiers, P. J.; Shinomoto, R. S.; Flood, T. C. *J. Am. Chem. Soc.* **1986**, *108*, 1346.

151. Shinomoto, R. S.; Desrosiers, P. J.; Gregory, T.; Harper, P.; Flood, T. C. *J. Am. Chem. Soc.* **1990**, *112*, 704.

152. Zeiher, E. H. K.; DeWit, D. G.; Caulton, K. G. *J. Am. Chem. Soc.* **1984**, *106*, 7006.

153. Hartwig, J. F.; Andersen, R. A.; Bergman, R. G. *J. Am. Chem. Soc.* **1991**, *113*, 6492.

154. Beller, M.; Fischer, H.; Herrmann, W. A.; Öfele, K.; Brossmer, C. *Angew. Chem., Int. Ed Engl.* **1995**, *34*, 1848.

155. Bedford, R. B. *Chem. Commun.* **2003**, 1787.

156. Ryabov, A. D. *Synthesis* **1985**, 233.

157. Rosner, T.; Pfaltz, A.; Blackmond, D. G. *J. Am. Chem. Soc.* **2001**, *123*, 4621.

158. Rosner, T.; Le Bars, J.; Pfaltz, A.; Blackmond, D. G. *J. Am. Chem. Soc.* **2001**, *123*, 1848.

159. Kiener, C. A.; Shu, C.; Incarvito, C.; Hartwig, J. F. *J. Am. Chem. Soc.* **2003**, *125*, 14272.

160. Chatt, J.; Davidson, J. M. *J. Chem. Soc.* **1965**, 843.

161. Janowicz, A. H.; Periana, R. A.; Buchanan, J. M.; Kovac, C. A.; Stryker, J. M.; Wax, M. J.; Bergman, R. G. *Pure. Appl. Chem.* **1984**, *56*, 13.

162. Buchanan, J. M.; Stryker, J. M.; Bergman, R. G. *J. Am. Chem. Soc.* **1986**, *108*, 1537.

163. Jones, W. D.; Feher, F. J. *J. Am. Chem. Soc.* **1984**, *106*, 1650.

164. Janowicz, A. H.; Bergman, R. G. *J. Am. Chem. Soc.* **1983**, *105*, 3929.

165. Jones, W. D.; Feher, F. J. *Organometallics* **1983**, *2*, 562.

166. Wax, M. J.; Stryker, J. M.; Buchanan, J. M.; Kovac, C. A.; Bergman, R. G. *J. Am. Chem Soc.* **1984**, *106*, 1121.

167. Hoyano, J. K.; McMaster, A. D.; Graham, W. A. G. *J. Am. Chem. Soc.* **1983**, *105*, 7190.

168. Rest, A. J.; Whitwell, I.; Graham, W. A. G.; Hoyano, J. K.; McMaster, A. D. *J. Chem. Soc., Chem. Commun.* **1984**, 624.

169. Bergman, R. G.; Seidler, P. F.; Wenzel, T. T. *J. Am. Chem. Soc.* **1985**, *107*, 4358.

170. Wenzel, T. T.; Bergman, R. G. *J. Am. Chem. Soc.* **1986**, *108*, 4856.

171. Baker, M. V.; Field, L. D. *J. Am. Chem. Soc.* **1987**, *109*, 2825.

172. Flood, T. C.; Janak, K. E.; Iimura, H.; Zhen, H. *J. Am. Chem. Soc.* **2000**, *122*, 6783.

173. Buchanan, J. M.; Stryker, J. M.; Bergman, R. G. *J. Am. Chem. Soc.* **1986**, *108*, 1537.

174. Hackett, M.; Ibers, J. A.; Whitesides, G. *J. Am. Chem. Soc.* **1988**, *110*, 1436.

175. Hackett, M.; Whitesides, G. M. *J. Am. Chem. Soc.* **1988**, *110*, 1449.

176. Jones, W. D.; Hessell, E. T. *J. Am. Chem. Soc.* **1993**, *115*, 554.

177. Wick, D. D.; Goldberg, K. I. *J. Am. Chem. Soc.* **1997**, *119*, 10235.

178. Jensen, M. P.; Wick, D. D.; Reinartz, S.; White, P. S.; Templeton, J. L.; Goldberg, K. I. *J. Am. Chem. Soc.* **2003**, *125*, 8614.

179. Gottker-Schnetmann, I.; White, P.; Brookhart, M. *J. Am. Chem. Soc.* **2004**, *126*, 1804.

180. Renkema, K. B.; Kissin, Y. V.; Goldman, A. S. *J. Am. Chem. Soc.* **2003**, *125*, 7770.

181. Jensen, C. M. *Chem. Commun.* **1999**, 2443.

182. Hackett, M.; Whitesides, G. M. *J. Am. Chem. Soc.* **1988**, *110*, 1449.

183. Tellers, D. M.; Yung, C. M.; Arndtsen, B. A.; Adamson, D. R.; Bergman, R. G. *J. Am. Chem. Soc.* **2002**, *124*, 1400.

184. Arndtsen, B. A.; Bergman, R. G. *Science* **1995**, *270*, 1970.

185. Brainard, R. L.; Nutt, W. R.; Lee, T. R.; Whitesides, G. M. *Organometallics* **1988**, *7*, 2379.

186. Johansson, L.; Ryan, O. B.; Tilset, M. *J. Am. Chem. Soc.* **1999**, *121*, 1974.

187. Johansson, L.; Ryan, O. B.; Rømming, C.; Tilset, M. *J. Am. Chem. Soc.* **2001**, *123*, 6579.

188. Zhong, H. A.; Labinger, J. A.; Bercaw, J. E. *J. Am. Chem. Soc.* **2002**, *124*, 1378.

189. Iverson, C. N.; Carter, C. A. G.; Baker, R. T.; Scollard, J. D.; Labinger, J. A.; Bercaw, J. E. *J. Am. Chem. Soc.* **2003**, *125*, 12674.

190. Waltz, K. M.; He, X.; Muhoro, C. N.; Hartwig, J. F. *J. Am. Chem. Soc.* **1995**, *117*, 11357.

191. Waltz, K. M.; Hartwig, J. F. *Science* **1997**, *277*, 211.

192. Waltz, K. M.; Muhoro, C. N.; Hartwig, J. F. *Organometallics* **1999**, *18*, 3383.

193. Kawamura, K.; Hartwig, J. F. *J. Am. Chem. Soc.* **2001**, *123*, 8422.

194. Chen, H.; Schlecht, S.; Semple, T. C.; Hartwig, J. F. *Science* **2000**, *287*, 1995.

195. Ishiyama, T.; Takagi, J.; Ishida, K.; Miyaura, N.; Anastasi, N.; Hartwig, J. F. *J. Am. Chem. Soc.* **2002**, *124*, 390.

196. Ishiyama, T.; Takagi, J.; Hartwig, J. F.; Miyaura, N. *Angew. Chem. Int. Ed.* **2002**, *41*, 3056.

197. Ishiyama, T.; Nobuta, Y.; Hartwig, J. F.; Miyaura, N. *Chem. Commun.* **2003**, 2924.

198. Iverson, C. N.; Smith, M. R., III *J. Am. Chem. Soc.* **1999**, *121*, 7696.

199. Cho, J. Y.; Tse, M. K.; Holmes, D.; Maleczka, R. E.; Smith, M. R. *Science* **2002**, *295*, 305.

200. Gustavson, W. A.; Epstein, P. S.; Curtis, M. D. *Organometallics* **1982**, *1*, 884.

201. Ezbiansky, K.; Djurovich, P. I.; Laforest, M.; Sinning, D. J.; Zayes, R.; Berry, D. H. *Organometallics* **1998**, *17*, 1455.

202. Kakiuchi, F.; Matsumoto, M.; Tsuchiya, K.; Igi, K.; Hayamizu, T.; Chatani, N.; Murai, S. *J. Organomet. Chem.* **2003**, *686*, 134.

203. March, J. *Advanced Organic Chemistry*; Wiley: New York, 1985; p 620.

204. Mayer, J. M. *Acc. Chem. Res.* **1998**, *31*, 441.

205. Nugent, W. A.; Mayer, J. M. *Metal–Ligand Multiple Bonds*; Wiley: New York, 1988.

206. Sheldon, R. A.; Kochi, J. K. *Metal Catalyzed Oxidations of Organic Compounds: Mechanistic Principles and Synthetic Methodology Including Biochemical Processes*; Academic Press: New York, 1981.

207. Davies, H. M. L.; Antoulinakis, E. G. *J. Organomet. Chem.* **2001**, *617*, 47.

208. Davies, H. M. L.; Beckwith, R. E. *J. Chem. Rev.* **2003**, *103*, 2861.

209. Shultz, L. H.; Tempel, D. J.; Brookhart, M. *J. Am. Chem. Soc.* **2001**, *123*, 11539.

210. Tempel, D. J.; Johnson, L. K.; Huff, R. L.; White, P. S.; Brookhart, M. *J. Am. Chem. Soc.* **2000**, *122*, 6686.

211. Northcutt, T. O.; Wick, D. D.; Vetter, A. J.; Jones, W. D. *J. Am. Chem. Soc.* **2001**, *123*, 7257.

212. Lian, T.; Bromberg, S. E.; Yang, H.; Proulx, G.; Bergman, R. G.; Harris, C. B. *J. Am. Chem. Soc.* **1996**, *118*, 3769.

213. Asplund, M. C.; Snee, P. T.; Yeston, J. S.; Wilkens, M. J.; Payne, C. K.; Yang, H.; Kotz, K. T.; Frei, H.; Bergman, R. G.; Harris, C. B. *J. Am. Chem. Soc.* **2002**, *124*, 10605.

214. Bromberg, S. E.; Yang, H.; Asplund, M. C.; Lian, T.; McNamara, B. K.; Kotz, K. T.; Yeston, J. S.; Wilkens, M.; Frei, H.; Bergman, R. G.; Harris, C. B. *Science* **1997**, *278*, 260.

215. Hall, C.; Perutz, R. N. *Chem. Rev.* **1996**, *96*, 3125.

216. Cobar, E. A.; Khaliullin, R. Z.; Bergman, R. G.; Head-Gordon, M. *Proc. Natl. Acad. Sci. U. S. A.* **2007**, *104*, 6963.

217. Lawes, D. J.; Geftakis, S.; Ball, G. E. *J. Am. Chem. Soc.* **2005**, *127*, 4134.

218. Geftakis, S.; Ball, G. E. *J. Am. Chem. Soc.* **1998**, *120*, 9953.

219. Castro-Rodriguez, I.; Nakai, H.; Gantzel, P.; Zakharov, L. N.; Rheingold, A. L.; Meyer, K. *J. Am. Chem. Soc.* **2003**, *125*, 15734.

220. Brookhart, M.; Green, M. L. H.; Wong, L.-L. *Prog. Inorg. Chem.* **1988**, *36*, 1.

221. Brookhart, M.; Green, M. L. H. *J. Organomet. Chem.* **1983**, *250*, 395.

222. Crabtree, R. H.; Holt, E. M.; Lavin, M.; Morehouse, S. M. *Inorg. Chem.* **1985**, *24*, 1986.

223. Iverson, C. N.; Lachicotte, R. J.; Muller, C.; Jones, W. D. *Organometallics* **2002**, *21*, 5320.

224. Reinartz, S.; White, P. S.; Brookhart, M.; Templeton, J. L. *J. Am. Chem. Soc.* **2001**, *123*, 12724.

225. Chin, R. M.; Dong, L.; Duckett, S. B.; Partridge, M. G.; Jones, W. D.; Perutz, R. N. *J. Am. Chem. Soc.* **1993**, *115*, 7685.

226. Churchill, D. G.; Janak, K. E.; Wittenberg, J. S.; Parkin, G. *J. Am. Chem. Soc.* **2003**, *125*, 1403.

227. Stoutland, P. O.; Bergman, R. G. *J. Am. Chem. Soc.* **1988**, *110*, 5732.

228. Stoutland, P. O.; Bergman, R. G. *J. Am. Chem. Soc.* **1985**, *107*, 4581.

229. Paneque, M.; Perez, P. J.; Pizzano, A.; Poveda, M. L.; Toboada, S.; Trujillo, M.; Carmona, E. *Organometallics* **1999**, *18*, 4304.

230. Bullock, R. M.; Headford, C. E. L.; Kegley, S. E.; Norton, J. R. *J. Am. Chem. Soc.* **1985**, *107*, 727.

231. Bullock, R. M.; Headford, C. E. L.; Hennessy, K. M.; Kegley, S. E.; Norton, J. R. *J. Am. Chem. Soc.* **1989**, *111*, 3897.

232. Gould, G. L.; Heinekey, D. M. *J. Am. Chem. Soc.* **1989**, *111*, 5502.

233. Abis, L.; Sen, A.; Halpern, J. *J. Am. Chem. Soc.* **1978**, *100*, 2915.

234. Michelin, R. A.; Faglia, S.; Uguagliati, P. *Inorg. Chem.* **1983**, *22*, 1831.

235. Periana, R. A.; Bergman, R. G. *J. Am. Chem. Soc.* **1986**, *108*, 7332.

236. Wang, C. M.; Ziller, J. W.; Flood, T. C. *J. Am. Chem. Soc.* **1995**, *117*, 1647.

237. Stahl, S. S.; Labinger, J. A.; Bercaw, J. E. *J. Am. Chem. Soc.* **1996**, *118*, 5961.

238. Jensen, M. P.; Wick, D. D.; Reinartz, S.; White, P. S.; Templeton, J. L.; Goldberg, K. I. *J. Am. Chem. Soc.* **2003**, *125*, 8614.

239. Weiller, B. H.; Wasserman, E. P.; Bergman, R. G.; Moore, C. B.; Pimentel, G. C. *J. Am. Chem. Soc.* **1989**, *111*, 8288.

240. Wasserman, E. P.; Moore, C. B.; Bergman, R. G. *Science* **1992**, *255*, 315.

241. McNamara, B. K.; Yeston, J. S.; Bergman, R. G.; Moore, C. B. *J. Am. Chem. Soc.* **1999**, *121*, 6437.

242. Bengali, A. A.; Schultz, R. H.; Moore, C. B.; Bergman, R. G. *J. Am. Chem. Soc.* **1994**, *116*, 9585.

243. Periana, R. A.; Bergman, R. G. *Organometallics* **1984**, *3*, 508.

244. Vetter, A. J.; Flaschenriem, C.; Jones, W. D. *J. Am. Chem. Soc.* **2005**, *127*, 12315.

245. Kanzelberger, M.; Singh, B.; Czerw, M.; Krogh-Jespersen, K.; Goldman, A. S. *J. Am. Chem. Soc.* **2000**, *122*, 11017.

246. Johnson, J. A.; Sames, D. *J. Am. Chem. Soc.* **2000**, *122*, 6321.

247. Johnson, J. A.; Li, N.; Sames, D. *J. Am. Chem. Soc.* **2002**, *124*, 6900.

248. Murai, S.; Kakiuchi, F.; Sekine, S.; Tanaka, Y.; Kamatani, A.; Sonoda, M.; Chatani, N. *Nature* **1993**, *366*, 529.

249. Kakiuchi, F.; Kan, S.; Igi, K.; Chatani, N.; Murai, S. *J. Am. Chem. Soc.* **2003**, *125*, 1698.

250. Jun, C.-H.; Moon, C. W.; Hong, J.-B.; Lim, S.-G.; Chung, K.-Y.; Kim, Y.-H. *Chem. Eur. J.* **2002**, 485.

251. Tan, K. L.; Bergman, R. G.; Ellman, J. A. *J. Am. Chem. Soc.* **2002**, *124*, 13964.

252. Chatani, N.; Asaumi, T.; Yorimitsu, S.; Ikeda, T.; Kakiuchi, F.; Murai, S. *J. Am. Chem. Soc.* **2001**, *123*, 10935.

253. Matsubara, T.; Koga, N.; Musaev, D.; Morokuma, K. *Organometallics* **2000**, *19*, 2318.

254. Chatani, N.; Ie, Y.; Kakiuchi, F.; Murai, S. *J. Am. Chem. Soc.* **1999**, *121*, 8645.

255. Matsubara, T.; Koga, N.; Musaev, D. G.; Morokuma, K. *J. Am. Chem. Soc.* **1998**, *120*, 12692.

256. (a) Sherry, A. E.; Wayland, B. B. *J. Am. Chem. Soc.* **1990**, *112*, 1259; (b) Wayland, B. B.; Ba, S.; Sherry, A. E. *J. Am. Chem. Soc.* **1991**, *113*, 5305.

257. (a) Cui, W. H.; Zhang, X. P.; Wayland, B. B. *J. Am. Chem. Soc.* **2003**, *125*, 4994; (b) Cui, W.; Wayland, B. B. *J. Am. Chem. Soc.* **2004**, *126*, 8266.

258. Nelson, A. P.; DiMagno, S. G. *J. Am. Chem. Soc.* **2000**, *122*, 8569.

259. Thompson, M. E.; Baxter, S. M.; Bulls, A. R.; Burger, B. J.; Nolan, M. C.; Santarsiero, B. D.; Schaefer, W. P.; Bercaw, J. E. *J. Am. Chem. Soc.* **1987**, *109*, 203.

260. Vitale, A. A.; San Filippo, J. *J. Am. Chem. Soc.* **1982**, *104*, 7341.

261. Steigerwald, M. L.; Goddard, W. A. *J. Am. Chem. Soc.* **1984**, *106*, 308.

262. Maron, L.; Eisenstein, O. *J. Am. Chem. Soc.* **2001**, *123*, 1036.

263. Ziegler, T.; Folga, E.; Berces, A. *J. Am. Chem. Soc.* **1993**, *115*, 636.

264. Rappé, A. K. *Organometallics* **1990**, *9*, 466.

265. Rabaa, H.; Saillard, J. Y.; Hoffmann, R. *J. Am. Chem. Soc.* **1986**, *108*, 4327.

266. Heinekey, D. M.; Oldham, W. J. *Chem. Rev.* **1993**, *93*, 913.

267. Gell, K. I.; Posin, B.; Schwartz, J.; Williams, G. M. *J. Am. Chem. Soc.* **1982**, *104*, 1846.

268. Wochner, F.; Brintzinger, H. H. *J. Organomet. Chem.* **1986**, *309*, 65.

269. Watson, P. L. *J. Am. Chem. Soc.* **1983**, *105*, 6491.

270. Fendrick, C. M.; Marks, T. J. *J. Am. Chem. Soc.* **1984**, *106*, 2214.

271. Niu, S. Q.; Hall, M. B. *J. Am. Chem. Soc.* **1998**, *120*, 6169.

272. Gilbert, T. M.; Hristov, I.; Ziegler, T. *Organometallics* **2001**, *20*, 1183.

273. Webster, C. E.; Fan, Y.; Hall, M. B.; Kunz, D.; Hartwig, J. F. *J. Am. Chem. Soc.* **2002**, *124*, 858.

274. Oxgaard, J.; Goddard, W. A. *J. Am. Chem. Soc.* **2004**, *126*, 442.

275. (a) Oxgaard, J.; Muller, R. P.; Goddard, W. A.; Periana, R. A. *J. Am. Chem. Soc.* **2004**, *126*, 352; (b) Tenn, W. J.; Young, K. J. H.; Bhalla, G.; Oxgaard, J.; Goddard, W. A.; Periana, R. A. *J. Am. Chem. Soc.* **2005**, *127*, 14172.

276. See the discussion of dihydrogen ligands in Chapter 2, along with references 277–279 here in Chapter 6.

277. Morris, R. H. *Can. J. Chem.* **1996**, *31*, 1471.

278. Heinekey, D. M.; Radzewich, C. E.; Voges, M. H.; Schomber, B. M. *J. Am. Chem. Soc.* **1997**, *119*, 4172.

279. Jessop, P. G.; Morris, R. H. *Coord. Chem. Rev.* **1992**, *121*, 155.
280. Oxgaard, J.; Periana, R. A.; William A. Goddard, I. *J. Am. Chem. Soc.* **2004**, *126*, 11658
281. Lail, M.; Bell, C. M.; Conner, D.; Cundari, T. R.; Gunnoe, T. B.; Petersen, J. L. *Organometallics* **2004**, *23*, 5007.
282. Tamura, H.; Yamazaki, H.; Sato, H.; Sakaki, S. *J. Am. Chem. Soc.* **2003**, *125*, 16114.
283. Costas, M.; Mehn, M. P.; Jensen, M. P.; Que, L. *Chem. Rev.* **2004**, *104*, 939.
284. Wallar, B. J.; Lipscomb, J. D. *Chem. Rev.* **1996**, *96*, 2625.
285. Merkx, M.; Kopp, D. A.; Sazinsky, M. H.; Blazyk, J. L.; Muller, J.; Lippard, S. J. *Angew. Chem., Int. Ed. Engl.* **2001**, *40*, 2782.
286. Groves, J. T.; McClusky, G. A. *Biochem. Biophys. Res. Commun.* **1978**, *81*, 154.
287. Peters, M. W.; Meinhold, P.; Glieder, A.; Arnold, F. H. *J. Am. Chem. Soc.* **2003**, *125*, 13442.
288. Atkinson, J. K.; Hollenberg, P. F.; Ingold, K. U.; Johnson, C. C.; Le Tadic, M.; Newcomb, M.; Putt, D. A. *Biochemistry* **1994**, *33*, 10630.
289. Barton, D. H. R.; Doller, D. *Acc. Chem. Res.* **1992**, *25*, 504.
290. Barton, D. H. R. *Chem. Soc. Rev.* **1996**, *25*, 237.
291. Stavropoulos, P.; Celenligil-Cetin, R.; Tapper, A. E. *Acc. Chem. Res.* **2001**, *34*, 745.
292. Gardner, K. A.; Mayer, J. M. *Science* **1995**, *269*, 1849.
293. Cummins, C. C.; Schaller, C. P.; Vanduyne, G. D.; Wolczanski, P. T.; Chan, A. W. E.; Hoffmann, R. *J. Am. Chem. Soc.* **1991**, *113*, 2985.
294. Bennett, J. L.; Wolczanski, P. T. *J. Am. Chem. Soc.* **1997**, *119*, 10696.
295. Slaughter, L. M.; Wolczanski, P. T.; Klinckman, T. R.; Cundari, T. R. *J. Am. Chem. Soc.* **2000**, *122*, 7953.
296. Cundari, T. R.; Klinckman, T. R.; Wolczanski, P. T. *J. Am. Chem. Soc.* **2002**, 1481.
297. Schaller, C. P.; Bonanno, J. B.; Wolczanski, P. T. *J. Am. Chem. Soc.* **1994**, *116*, 4133.
298. Schaller, C. P.; Cummins, C. C.; Wolczanski, P. T. *J. Am. Chem. Soc.* **1996**, *118*, 591.
299. Walsh, P. J.; Hollander, F. J.; Bergman, R. G. *J. Am. Chem. Soc.* **1988**, *110*, 8729.
300. Walsh, P. J.; Hollander, F. J.; Bergman, R. G. *Organometallics* **1993**, *12*, 3705.
301. Hoyt, H. M.; Michael, F. E.; Bergman, R. G. *J. Am. Chem. Soc.* **2004**, *126*, 1018.
302. Schaller, C. P.; Wolczanski, P. T. *Inorg. Chem.* **1993**, *32*, 131.
303. Schafer, D. F.; Wolczanski, P. T. *J. Am. Chem. Soc.* **1998**, *120*, 4881.
304. Cummins, C. C.; Baxter, S. M.; Wolczanski, P. D. *J. Am. Chem. Soc.* **1988**, *110*, 8731.
305. Ishii, S.; Zhao, S.; Helquist, P. *J. Am. Chem. Soc.* **2000**, *122*, 5897.
306. Zhao, S. K.; Knors, C.; Helquist, P. *J. Am. Chem. Soc.* **1989**, *111*, 8527.
307. Ishii, S.; Helquist, P. *Synlett* **1997**, 508.
308. Barluenga, J.; Rodriguez, F.; Vadecard, J.; Bendix, M.; Fananas, F. J.; Lopez-Ortiz, F. *J. Am. Chem. Soc.* **1996**, *118*, 6090.
309. Guerchais, V.; Sinbandhit, S. *J. Chem. Soc.* **1990**, 1550.
310. van der Heijden, H.; Hessen, B. *J. Chem. Soc., Chem. Commun.* **1995**, 145.
311. Coles, M. P.; Gibson, V. C.; Clegg, W.; Elsegood, M. R. J.; Porrelli, P. A. *J. Chem. Soc., Chem. Commun.* **1996**, 1963.
312. Cheon, J.; Rogers, D. M.; Girolami, G. S. *J. Am. Chem. Soc.* **1997**, *119*, 6804.
313. Pamplin, C. B.; Legzdins, P. *Acc. Chem. Res.* **2003**, *36*, 223.
314. Blackmore, I. J.; Jin, X.; Legzdins, P. *Organometallics* **2005**, *24*, 4088.
315. Adams, C. S.; Legzdins, P.; McNeil, W. S. *Organometallics* **2001**, *20*, 4939.
316. Bailey, B. C.; Fan, H.; Baum, E. W.; Huffman, J. C.; Baik, M.-H.; Mindiola, D. J. *J. Am. Chem. Soc.* **2005**, *127*, 16016.
317. Bailey, B. C.; Fan, H.; Huffman, J. C.; Baik, M.-H.; Mindiola, D. J. *J. Am. Chem. Soc.* **2007**, *129*, 8781.
318. Davies, H. M. L.; Hansen, T.; Churchill, M. R. *J. Am. Chem. Soc.* **2000**, *122*, 3063.
319. Nowlan, D. T., III; Gregg, T. M.; Davies, H. M. L.; Singleton, D. A. *J. Am. Chem. Soc.* **2003**, *125*, 15902.
320. Murakami, M.; Ito, Y. *Top. Organomet. Chem.* **1999**, *3*, 97.
321. Rybtchinski, B.; Milstein, D. *Angew. Chem., Int. Ed. Engl.* **1999**, *38*, 870.
322. Maury, O.; Lefort, L.; Vidal, V.; Thivolle-Cazat, J.; Basset, J. M. *Angew. Chem., Int. Ed. Engl.* **1999**, *38*, 1952.
323. Lefebvre, F.; Thivolle-Cazat, J.; Dufaud, V.; Niccolai, G. P.; Basset, J. M. *Appl. Catal., A* **1999**, *182*, 1.
324. Rosier, C.; Niccolai, G. P.; Basset, J. M. *J. Am. Chem. Soc.* **1997**, *119*, 12408.
325. Vidal, V.; Theolier, A.; Thivolle-Cazat, J.; Basset, J. M. *Science* **1997**, *276*, 99.
326. Corker, J.; Lefebvre, F.; Lecuyer, C.; Dufaud, V.; Quignard, F.; Choplin, A.; Evans, J.; Basset, J. M. *Science* **1996**, *271*, 966.
327. Morvillo, A.; Turco, A. *J. Organomet. Chem.* **1981**, *208*, 103.
328. Favero, G.; Morvillo, A. *J. Organomet. Chem.* **1984**, *260*, 363.
329. Garcia, J. J.; Brunkan, N. M.; Jones, W. D. *J. Am. Chem. Soc.* **2002**, *124*, 9547.
330. Garcia, J. J.; Jones, W. D. *Organometallics* **2000**, *19*, 5544.

331. Suggs, J. W.; Jun, C. H. *J. Am. Chem. Soc.* **1986**, *108*, 4679.
332. Suggs, J. W.; Wovkulich, M. J.; Williard, P. G.; Lee, K. S. *J. Organomet. Chem.* **1986**, *307*, 71.
333. Suggs, J. W.; Wovkulich, M. J.; Lee, K. S. *J. Am. Chem. Soc.* **1985**, *107*, 5546.
334. Suggs, J. W.; Jun, C. H. *J. Am. Chem. Soc.* **1984**, *106*, 3054.
335. Murakami, M.; Itahashi, T.; Ito, Y. *J. Am. Chem. Soc.* **2002**, *124*, 13976.
336. Gozin, M.; Weisman, A.; Bendavid, Y.; Milstein, D. *Nature* **1993**, *364*, 699.
337. Liou, S. Y.; Gozin, M.; Milstein, D. *J. Chem. Soc., Chem. Commun.* **1995**, 1965.
338. Liou, S. Y.; Gozin, M.; Milstein, D. *J. Am. Chem. Soc.* **1995**, *117*, 9774.
339. Rybtchinski, B.; Vigalok, A.; Ben-David, Y.; Milstein, D. *J. Am. Chem. Soc.* **1996**, *118*, 12406.
340. Liou, S.-Y.; van der Boom, M. E.; Milstein, D. *Chem. Commun.* **1998**, 687.
341. van der Boom, M. E.; Liou, S. Y.; Ben-David, Y.; Gozin, M.; Milstein, D. *J. Am. Chem. Soc.* **1998**, *120*, 13415.
342. Gandelman, M.; Vigalok, A.; Konstantinovski, L.; Milstein, D. *J. Am. Chem. Soc.* **2000**, *122*, 9848.
343. Sundermann, A.; Uzan, O.; Milstein, D.; Martin, J. M. L. *J. Am. Chem. Soc.* **2000**, *122*, 7095.
344. Rybtchinski, B.; Oevers, S.; Montag, M.; Vigalok, A.; Rozenberg, H.; Martin, J. M. L.; Milstein, D. *J. Am. Chem. Soc.* **2001**, *123*, 9064.
345. Crabtree, R. H.; Dion, R. P. *J. Chem. Soc.* **1984**, 1260.
346. Periana, R. A.; Bergman, R. G. *J. Am. Chem. Soc.* **1984**, *106*, 7272.
347. Periana, R. A.; Bergman, R. G. *J. Am. Chem. Soc.* **1986**, *108*, 7346.
348. Wender, P. A.; Pedersen, T. M.; Scanio, M. J. C. *J. Am. Chem. Soc.* **2002**, *124*, 15154.
349. Wender, P. A.; Gamber, G. G.; Hubbard, R. D.; Zhang, L. *J. Am. Chem. Soc.* **2002**, *124*, 2876.
350. Wender, P. A.; Correa, A. G.; Sato, Y.; Sun, R. *J. Am. Chem. Soc.* **2000**, *122*, 7815.
351. Wender, P. A.; Dyckman, A. J.; Husfeld, C. O.; Kadereit, D.; Love, J. A.; Rieck, H. *J. Am. Chem. Soc.* **1999**, *121*, 10442.
352. Wender, P. A.; Glorius, F.; Husfeld, C. O.; Langkopf, E.; Love, J. A. *J. Am. Chem. Soc.* **1999**, *121*, 5348.
353. Wender, P. A.; Rieck, H.; Fuji, M. *J. Am. Chem. Soc.* **1998**, *120*, 10976.
354. Wender, P. A.; Takahashi, H.; Witulski, B. *J. Am. Chem. Soc.* **1995**, *117*, 4720.
355. Murakami, M.; Amii, H.; Shigeto, K.; Ito, Y. *J. Am. Chem. Soc.* **1996**, *118*, 8285.
356. Murakami, M.; Takahashi, K.; Amii, H.; Ito, Y. *J. Am. Chem. Soc.* **1997**, *119*, 9307.
357. Murakami, M.; Miyamoto, Y.; Ito, Y. *J. Am. Chem. Soc.* **2001**, *123*, 6441.
358. Suginome, M.; Nakamura, H.; Ito, Y. *Tetrahedron Lett.* **1997**, *38*, 555.
359. Ito, Y.; Suginome, M.; Murakami, M. *J. Org. Chem.* **1991**, *56*, 1948.
360. Suginome, M.; Hiroshi, N.; Ito, Y. *Chem. Commun.* **1996**, 2777.
361. Suginome, M.; Nakamura, H.; Ito, Y. *Angew. Chem., Int. Ed. Engl.* **1997**, *36*, 2516.
362. Suginome, M.; Matsuda, T.; Yoshimoto, T.; Ito, Y. *Org. Lett.* **1999**, *1*, 1567.
363. Suginome, M.; Ohmori, Y.; Ito, Y. *J. Organomet. Chem.* **2000**, *611*, 403.
364. Ishiyama, T.; Sato, K.; Nishio, Y.; Miyaura, N. *Angew. Chem. Int. Ed.* **2003**, *42*, 5346.
365. Takagi, J.; Sato, K.; Hartwig, J. F.; Ishiyama, T.; Miyaura, N. *Tetrahedron Lett.* **2002**, *43*, 5649.
366. Goossen, L. J.; Ferwanah, A.-R. S. *Synlett* **2000**, 1801.
367. Bottoni, A.; Higueruelo, A. P.; Miscione, G. P. *J. Am. Chem. Soc.* **2002**, *124*, 5506.
368. For oxidative addition of disilanes to Pt(0) complexes, see Gilges, H.; Kickelbick, G.; Schabert, U. *J. Organomet. Chem.* **1997**, *548*, 57.
369. Ishiyama, T.; Miyaura, N. *J. Organomet. Chem.* **2000**, *611*, 392.
370. Murata, M.; Watanabe, S.; Masuda, Y. *J. Org. Chem.* **1997**, *62*, 6458.
371. Murata, M.; Oyama, T.; Watanabe, S.; Masuda, Y. *J. Org. Chem.* **2000**, *65*, 164.
372. Ishiyama, T.; Itoh, Y.; Kitano, T.; Miyaura, N. *Tetrahedron Lett.* **1997**, *38*, 3447.
373. Yamashita, H.; Kobayashi, T.-a.; Hayashi, T.; Tanaka, M. *Chem. Lett.* **1990**, 1447.
374. Onozawa, S.; Hatanaka, Y.; Sakakura, T.; Shimada, S.; Tanaka, M. *Organometallics* **1996**, *15*, 5450.
375. Sakaki, S.; Kikuno, T. *Inorg. Chem.* **1997**, *36*, 226.
376. Dai, C.; Stringer, G.; Marder, T. B. *Inorg. Chem.* **1997**, *36*, 272.
377. Curtis, D.; Lesley, M. J. G.; Norman, N. C.; Orpen, A. G.; Starbuck, J. *J. Chem. Soc.* **1999**, 1687.
378. Marder, T. B.; Norman, N. C.; Rice, C. R.; Robins, E. G. *Chem. Commun.* **1997**, 53.
379. Ishiyama, T.; Matsuda, N.; Murata, M.; Ozawa, F.; Suzuki, A.; Miyaura, N. *Organometallics* **1996**, *15*, 713.
380. Curtis, D.; Lesley, M. J. G.; Norman, N. C.; Orpen, A. G.; Starbuck, J. *J. Chem. Soc., Dalton Trans.* **1999**, 1687.
381. Hartwig, J. F.; He, X. *Angew. Chem., Int. Ed. Engl.* **1996**, 35, 315.

CHAPTER 7

Oxidative Addition of Polar Reagents

7.1. Introduction

Chapter 6 presented the basic principles of oxidative addition. Recall that the term "oxidative addition" refers to reactions that lead to an increase in the oxidation state of the metal center by two units, an increase in the number of valence electrons on the metal center by two, and an increase in the coordination number of the complex by one or two. This term does *not* refer to a particular mechanism by which this transformation occurs.

Chapter 7 presents the oxidative addition of reagents containing more polar A–B bonds than those in the reagents discussed in Chapter 6. Such oxidative additions of polar substrates are common methods to generate stable organometallic complexes, as well as intermediates in reactions catalyzed by organometallic species. This type of oxidative addition is an important step of many valuable and commercially important catalytic reactions, such as the Monsanto acetic acid process and cross-coupling reactions catalyzed by various transition metal complexes. Representative examples of the oxidative addition of polar substrates were given in Table 6.1. Early examples of this reaction have been extensively reviewed.[1–13]

Polar substrates undergo oxidative addition by a variety of mechanisms including those involving ionic or radical intermediates, and those occurring by a series of concerted steps. The mechanisms of the oxidative additions of polar substrates depend on the metal complex, the identity of the polar reagent, and the reaction conditions. Thus, this chapter presents a variety of examples of oxidative additions of substrates containing polar A–B bonds and discusses the variety of pathways by which these reactions occur. The mechanisms of these reactions can be divided into three categories that will be used to organize this chapter: S_N2 pathways, one-electron transfer pathways, and three-centered concerted pathways analogous to those for the oxidative addition of nonpolar substrates. Some guidelines are provided in the conclusions section to predict which of the pathways discussed is likely to be followed for the reaction between a particular combination of metal complex and organic reagent.

$$M^n: \quad \overset{\cdots}{C}-X \longrightarrow \left[M\text{---}\overset{|}{C}\text{---}X \right]^{\ddagger} \longrightarrow \left[M^{(n+2)}-C\blacktriangleleft \right]^{\oplus} X^{\ominus} \longrightarrow X-M-C\blacktriangleleft \qquad (7.1)$$

7.2. Oxidative Addition by S_N2 Pathways

Oxidative additions of alkyl halides and pseudohalides can occur by an S_N2 mechanism in which the metal acts as a nucleophile (Equation 7.1). The first step of this reaction is analogous to the alkylation of an amine. The data in support of an S_N2 mechanism for the oxidative addition of alkyl halides by a transition metal complex are the same as the data that are typically considered to support an S_N2 mechanism for the substitution of alkyl halides by organic nucleophiles—namely, the reaction is faster in more polar solvents,[14,15] the electrophilic carbon undergoes inversion of configuration,[16–18] the reaction kinetics are second order, and the relative rates of alkyl halides are Me > primary > secondary >> tertiary[19] and I > Br > Cl >> F.

$$\text{(7.2)}$$

The oxidative addition of methyl iodide to Vaska's complex, shown in Equation 7.2, is a classic example of the oxidative addition of alkyl halides by an S_N2 mechanism. Strong electrophiles that are sterically accessible, such as methyl iodide, benzyl bromide, allyl halides, and chloromethyl ethers, react with $L_2(CO)IrX$ species by this pathway.[20] A series of data supports addition of these electrophiles by an S_N2 mechanism. For example, the trans stereochemistry of the kinetic product from addition of methyl iodide[21] is inconsistent with a concerted three-centered mechanism; radical traps do not affect the rate or products of the reaction;[22] the reaction rates are faster in more polar solvents;[14,15] and the reactions are first order in both metal and electrophile.[15] Higher alkyl halides add by more complex mechanisms presented below.

$$\text{(7.3)}$$

Palladium(0) complexes also add alkyl halides by S_N2 mechanisms. Oxidative additions of MeI to Pd(0) complexes were some of the early examples of this reaction.[23] Stille showed that oxidative addition of benzyl bromide-d_1 occurs with predominant inversion of configuration,[16,25] and a more recent study demonstrated that oxidative addition of an alkyl tosylate[24] and a higher alkyl bromide occurs by an S_N2 path.[25,26] Equation 7.3 shows the stereochemical evidence for an S_N2 pathway. This equation shows the individual steps that occur during the catalytic addition of an arylborane to a stereochemically defined alkyl tosylate. The product forms with overall inversion of configuration. As is noted in Chapter 8, the final step, reductive elimination to form a C–C bond, occurs with retention of configuration. Thus, the first oxidative addition step must occur with inversion of configuration, and this inversion of configuration signals an S_N2 reaction.

Oxidative addition can also occur between anionic metal complexes and alkyl electrophiles. In this case, the halide becomes a byproduct, rather than a ligand, in the final metal complex. Two well-studied examples of oxidative addition to an anionic metal complex by an S_N2 pathway involves the d^8 and d^{10} "supernucleophiles" Co(I) cobaloximes and Fe(CO)$_4^{2-}$. These complexes react rapidly with most organic halides, including hindered halides containing strong C–X bonds, such as neopentyl chloride.[27] The products from multistep reactions of this complex are consistent with 100% inversion in the oxidative addition step. This and the other evidence in Figure 7.1 support reaction by a classic S_N2 pathway.

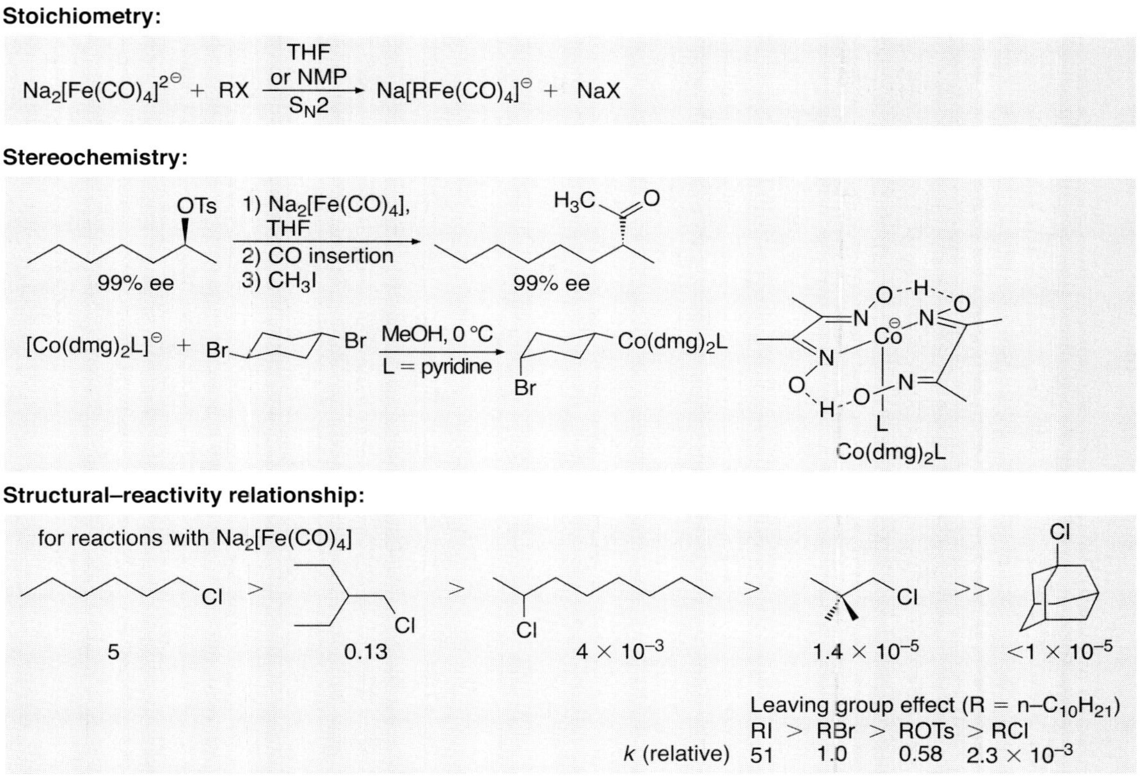
Stoichiometry:
Na2[Fe(CO)4]2⊖ + RX
THF or NMP
SN2
Na[RFe(CO)4]⊖ + NaX
Stereochemistry:
OTs
99% ee
1) Na2[Fe(CO)4], THF
2) CO insertion
3) CH3I
H3C
O
99% ee
[Co(dmg)2L]⊖ + Br
Br
MeOH, 0 °C
L = pyridine
Co(dmg)2L
Br
Co(dmg)2L
Structural–reactivity relationship:
for reactions with Na2[Fe(CO)4]
Cl
5
0.13
4 × 10-3
1.4 × 10-5
<1 × 10-5
Leaving group effect (R = n–C10H21)
RI > RBr > ROTs > RCl
k (relative) 51 1.0 0.58 2.3 × 10-3

Figure 7.1. Data on the oxidative addition of alkyl halides and tosylates by an S_N2 pathway.

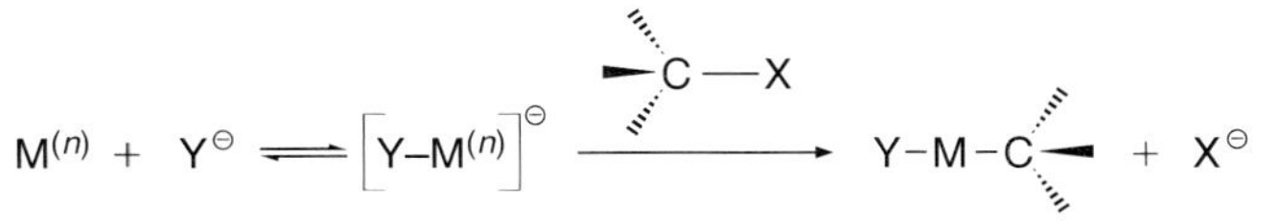
M(n) + Y⊖
[Y–M(n)]⊖
C—X
Y–M–C + X⊖

(7.4)

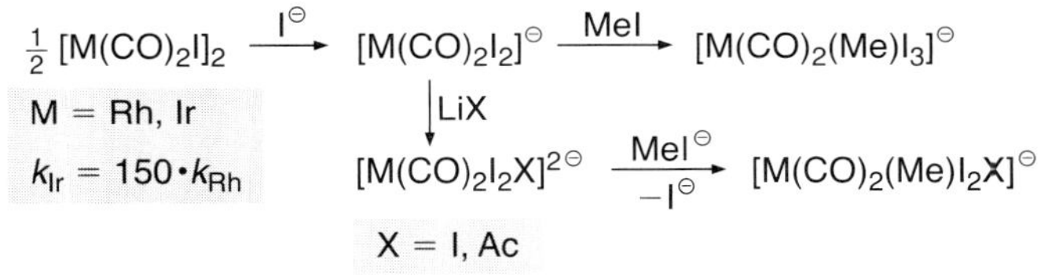
½ [M(CO)2I]2
I⊖
[M(CO)2I2]⊖
MeI
[M(CO)2(Me)I3]⊖
M = Rh, Ir
kIr = 150•kRh
LiX
[M(CO)2I2X]2⊖
MeI⊖
–I⊖
[M(CO)2(Me)I2X]⊖
X = I, Ac

Scheme 7.1

In some cases, oxidative addition occurs by coordination of an anion to a neutral metal complex to generate an anionic complex prior to the C–X bond cleavage step by the S_N2 pathway. This sequence is shown generically in Equation 7.4. A commercially important example of this reaction is the oxidative addition of methyl iodide to diiodorhodium and diiodoiridium–carbonyl complexes $[M(CO)_2I_2]^-$ in the presence of added iodide (Scheme 7.1). This oxidative addition occurs during the carbonylation of methanol to acetic acid. The origin of the effect of iodide has been studied extensively because it improved the synthesis of acetic acid with rhodium catalysts by allowing the use of less water in the medium.[28] Yet, this effect is not fully understood. It is clear that the presence of iodide generates negatively charged rhodium carbonyl compounds, such as $[Rh(CO)_2I_2]^-$, and that this anionic complex is more reactive toward oxidative addition of MeI than its neutral counterpart $[Rh(CO)_2I]_2$. However, reaction of the anionic species $[Rh(CO)_2I_2]^-$ is faster in the presence of additional lithium iodide and lithium acetate, and the origin of the effect of these additional anions is unclear.[19,29,30] The acceleration due to these anions seems to result, in part, from an increase in the polarity of the medium,[30] but reaction through the five-coordinate dianionic species $[Rh(CO)_2I_2X]^{2-}$ (in which X = I and OAc) shown in Scheme 7.1 has also been proposed.[29, 31]

The analogous anionic iridium complex reacts with methyl iodide 150 times faster than the rhodium complex.[19] The iridium complex also reacts 140–200 times faster than the rhodium analog with higher alkyl iodides,[19] but competing radical mechanisms appear to occur during the addition of the higher alkyl iodides. More details on the mechanism of rhodium and iridium-catalyzed carbonylation of methanol are provided in Chapter 17.

Several other anionic transition metal complexes undergo oxidative addition of organic electrophiles, some by S_N2 pathways and others by radical pathways. The reactions by radical pathways parallel the reactions of carbanions with alkyl halides by radical mechanisms.[32–35] An example of a complex that reacts by both pathways is shown in Scheme 7.2. $CpFe(CO)_2$ reacts with alkyl bromides and sulfonates by S_N2 pathways. It reacts with cyclopropylmethyl bromide with little of the ring opening of the cyclopropyl group that would occur if a cyclopropylmethyl radical were generated,[36] and it reacts with an alkyl sulfonate with > 95% inversion of configuration.[17] However, the same iron anion reacts with cyclopropylmethyl iodide to give substantial amounts of product from a cyclopropylmethyl radical intermediate.[36]

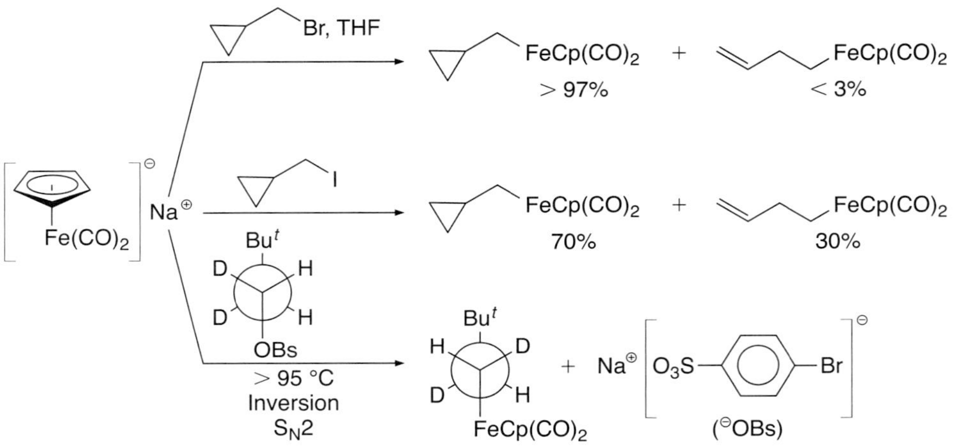

Scheme 7.2

7.3. Oxidative Additions by One-Electron Mechanisms

Many oxidative additions of alkyl iodides other than methyl iodide and many oxidative additions of alkyl bromides to neutral transition metal complexes occur by radical pathways. The factors that lead to reaction by a radical pathway over an S_N2 or concerted pathway are subtle and depend on the identity of the metal and the substrate. Broadly speaking, reactions of weaker and more-hindered electrophiles occur by radical pathways more often than reactions of stronger and less-hindered electrophiles, and reactions of alkyl sulfonates almost always occur by non-radical pathways. Also, reactions of coordinatively saturated metal centers tend to occur by radical pathways more than reactions of coordinatively unsaturated metal centers. Coordinatively saturated complexes tend to react by outer-sphere processes because oxidative additions of organic halides by inner-sphere pathways require an open coordination site. Finally, metal complexes that are prone to undergo one-electron oxidations tend to undergo oxidative addition by radical pathways that involve two metal complexes. These one-electron oxidations by two metal centers are most common with transition metal complexes of the first row.

7.3.1. Inner-Sphere Electron Transfer and Caged Radical Pairs

Oxidative addition by an inner-sphere electron-transfer mechanism is depicted in Scheme 7.3. This process occurs by initial coordination of the organic halide, followed by electron transfer to generate a solvent-caged radical pair consisting of R• and an odd-electron metal–halide complex. Collapse of the radical pair forms the product of overall oxidative addition. Escape of R• from the cage creates the free radical R•, which forms the products typically formed by the same free radical generated by other processes. The oxidative addition product $[R–M^{(n+2)}–X]$ formed from this pathway is the same as that formed from an S_N2 pathway, except that racemization (instead of inversion) of the stereochemistry is usually observed.[37-40] This pathway requires a coordinatively unsaturated metal center to bind the substrate in the first step and an electrochemical potential high enough to reduce the organic electrophile in the second step.

$$M^{(n)} + RX \rightleftharpoons M^{(n)} \leftarrow X-R \rightleftharpoons [\cdot M^{(n+1)}-X + R\cdot] \text{ cage}$$

Collapse / \ Escape

$$R-M^{(n+2)}-X \qquad R\cdot + M^{(n+1)}-X$$

$$M^{(n)}$$

$$R-R, \text{ alkenes, alkanes} \qquad M^{(n+1)}-R$$

$$M^{(n)} = \text{Metal in oxidation state } n$$

Scheme 7.3

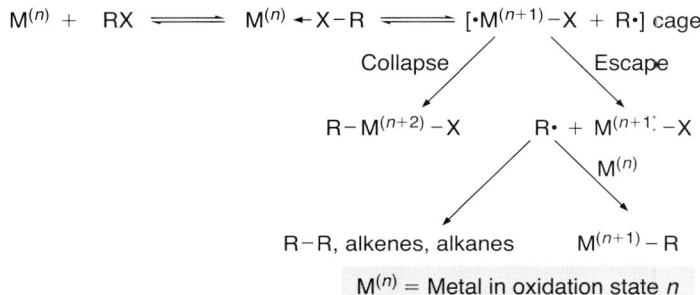

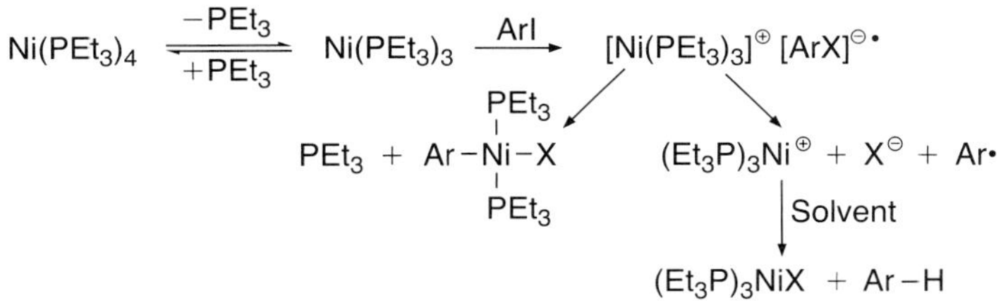

Scheme 7.4

Few examples of this mechanism have been clearly demonstrated because of the difficulty in establishing that this path occurs from experimental data. The most well-established examples are reactions of nickel complexes with aryl halides studied by Tsou and Kochi.[41] The rate of the reaction of $Ni(PEt_3)_3$ with aryl halides was shown to be first order in nickel and in ArX and retarded by added PEt_3. Ortho-methyl substituents had little effect on the rate. Because of the lack of steric effect, electron transfer was proposed to occur after formation of a π-complex between $Ni(PEt_3)_3$ and ArX, rather than by direct insertion of the metal into the carbon–halogen bond by a three-centered mechanism. Moreover, the products of the reaction included the Ni(I) species L_3NiX and arene. These products are likely to result from the pathway in Scheme 7.4, involving electron transfer from Ni(0) to the aryl halide and escape of the aryl radical from the solvent cage. Other studies of oxidative additions of aryl halides and sulfonates to Ni(0) complexes have been reported.[42-46]

Initiation

$$In\cdot \quad X-R \longrightarrow In-X + R\cdot$$

Propagation

$$R\cdot + M^{(n)} \longrightarrow R-M^{(n+1)}$$

$$R-M^{(n+1)} + X-R \longrightarrow R-M^{(n+2)}-X + R\cdot$$

Termination

$$2\,R\cdot$$
$$R\cdot + M\cdot \Bigg\} \longrightarrow \text{Side products}$$

Scheme 7.5

7.3.2. Radical Chain Pathways

Other oxidative additions of organic halides through radical intermediates have been shown to occur by radical chains (Scheme 7.5). Radical chain mechanisms require a coordinatively unsaturated metal center and a source (often unknown) of radicals to act as the initiator, In•. This pathway is signaled by a combination of initiation or inhibition of the reaction by radical initiators or radical traps and by the observation of products from alkyl radicals R•, such as those resulting from racemization at carbon, from coupling of two R• to form R–R, from disproportionation, from trapping of the radical, and from rearrangements of R•. Possible sources of initiators include trace impurities, trace O_2 (which can serve as an initiator or an inhibitor), the RX substrate after an atom abstraction process, and radicals generated from light. In the radical pathway shown in Scheme 7.5, the chain propagation involves the addition of an alkyl radical to a metal center, followed by halide abstraction from an alkyl halide to generate the oxidative addition product and reform the alkyl radical.

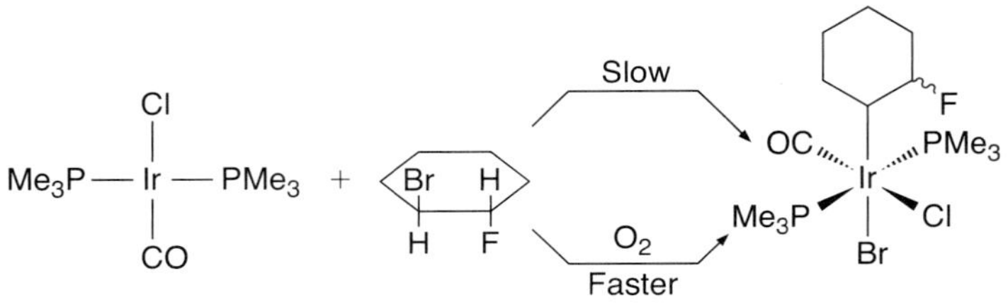

$$(7.5)$$

α-Halo esters and higher alkyl halides than methyl iodide often react by this mechanism. For example, the trimethylphosphine analog of Vaska's complex reacts with 2-bromoethylbenzene and trans–1-bromo-2-fluorocyclohexane (Equation 7.5) by a mechanism that is likely to involve radical chains.[20,22] The reaction rates are faster in the presence of O_2 or the radical initiator azoisobutyronitrile (AIBN). The reactions are also inhibited by hydroquinones. Furthermore, the products from these reactions result from a process that includes racemization at carbon.

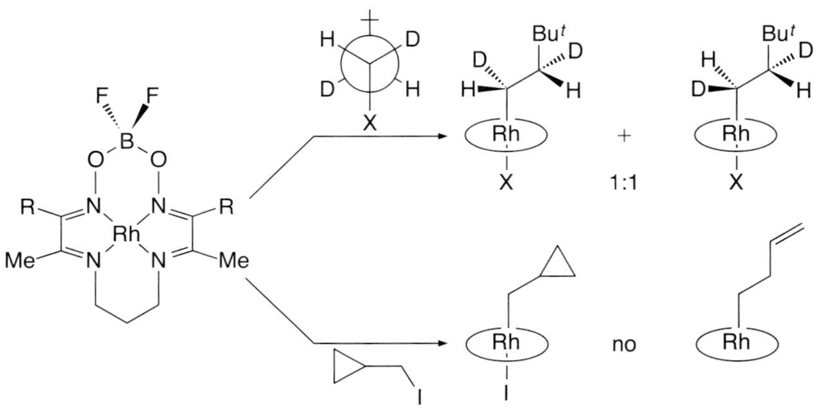

$$(7.6)$$

Oxidative additions of alkyl halides to the 16-electron Rh(I) macrocycle in Equation 7.6 might also occur by radical pathways.[47] This complex is about 10^4 more reactive than Vaska's Ir(I) complex. Although its reaction with cyclopropylcarbinyl iodide (Equation 6.57) does not form $Rh–CH_2CH_2CH=CH_2$ by ring opening of a free cyclopropylcarbinyl radical, reactions of d_2-neohexyl iodide occur with loss of stereochemistry at carbon. Studies on the oxidative additions of alkyl halides to several macrocyclic Co(I) complexes that are analogous to the Rh(I) system, including $[Co(I)(dmgH)_2(PR)_3]^-$, Vitamin B_{12}, Co(tetraphenylporphyrin)$^-$, Co(salen)$^-$, and Co(phthalocyanin)$^-$,[48–50] have also provided evidence for an electron-transfer step.[48]

A quantitative study of the reaction of a d^8 square planar Pt(II) by a free-radical chain mechanism was reported by Hill and Puddephatt and is summarized in Scheme 7.6.[51] The radical chain was initiated by photochemical MLCT (metal-to-ligand charge transfer) to form $[Me_2Pt^+(phen)^-]^*$, which has primarily triplet character. This intermediate then abstracts an iodine atom to generate an alkyl radical. Consistent with the more common occurrence of radical mechanisms with alkyl iodides than bromides, the reaction of isopropyl bromide was unaffected by light, by benzoquinone that would scavenge $R\bullet$, or by the presence or absence of oxygen.[52]

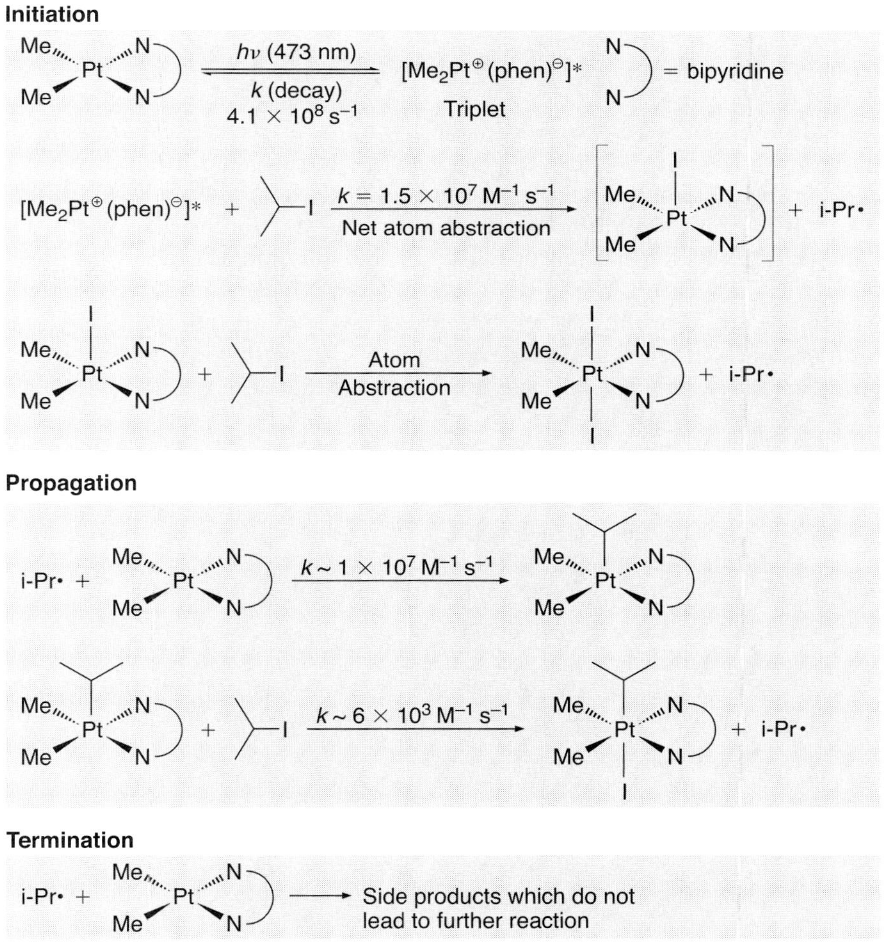

Scheme 7.6

7.3.3. Outer-Sphere Electron-Transfer Mechanisms

Oxidative addition of alkyl halides can also occur in certain cases by an outer-sphere electron transfer mechanism involving a coordinatively saturated metal center and an alkyl halide. This pathway is shown in Scheme 7.7. Oxidative addition by this initial outer-sphere electron transfer pathway tends to occur instead of an S_N2 pathway when the electrophile is particularly susceptible to electron transfer, when the electrophile possesses some steric hindrance, when the electrophile possesses a weak C–X bond, and when the metal lacks an available coordination site. Because of the lack of a coordination site at the metal, the initial electron transfer occurs without prior coordination of the electrophile to the metal. This initial step parallels the electron transfer and subsequent radical chemistry that occurs when some carbanions are treated with alkyl halides.[53]

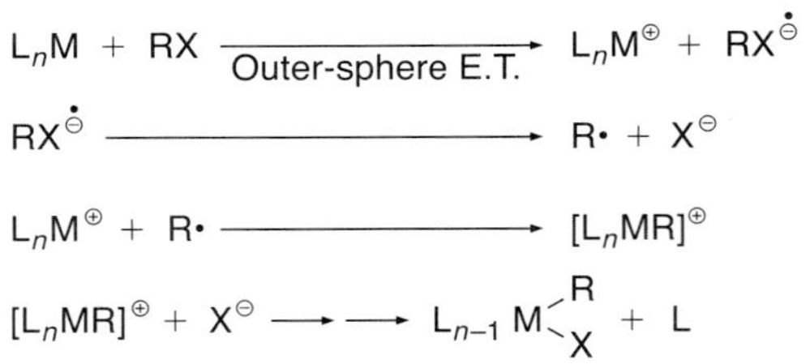

$$L_nM \ + \ RX \xrightarrow{\text{Outer-sphere E.T.}} L_nM^{\oplus} \ + \ RX^{\overset{\bullet}{\ominus}}$$

$$RX^{\overset{\bullet}{\ominus}} \longrightarrow R\cdot \ + \ X^{\ominus}$$

$$L_nM^{\oplus} \ + \ R\cdot \longrightarrow [L_nMR]^{\oplus}$$

$$[L_nMR]^{\oplus} \ + \ X^{\ominus} \longrightarrow L_{n-1}M{\overset{R}{\underset{X}{<}}} \ + \ L$$

Scheme 7.7

The initial outer-sphere electron transfer generates a free radical anion and a metal cation that undergo a series of steps to complete the oxidative addition process. The radical anion $RX^-\bullet$ rapidly fragments to generate a neutral radical $R\bullet$ and a halide anion. The radical $R\bullet$ then combines with the metal cation to give a complex that can be the final or penultimate product. Coordination of halide to this alkylmetal cation often leads to the final product. The principal side products result from coupling and disproportion of $R\bullet$. In many cases, the final product from these reactions contains an alkyl group and a free halide counterion. In other cases, a ligand in the alkylmetal product dissociates, and the final product contains the halide directly bound to the metal.

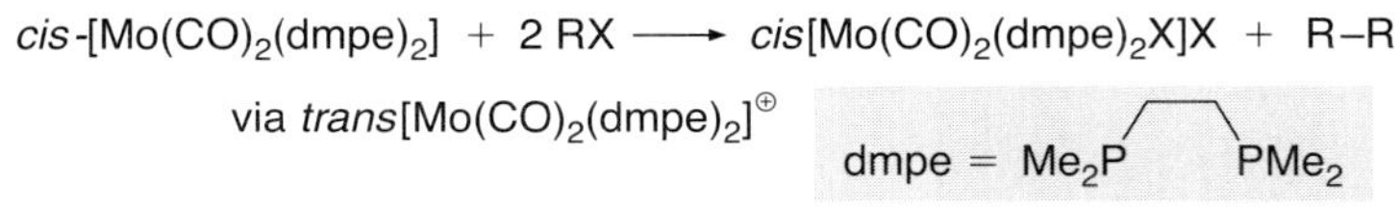

$$cis\text{-}[Mo(CO)_2(dmpe)_2] \ + \ 2\,RX \longrightarrow cis[Mo(CO)_2(dmpe)_2X]X \ + \ R\text{–}R$$

$$\text{via } trans[Mo(CO)_2(dmpe)_2]^{\oplus}$$

$$dmpe = Me_2P\frown PMe_2 \tag{7.7}$$

Complexes such as $CpFe(CO)_2^-$ or MoL_6 react by this mechanism[36,54,55] because they are 18-electron metal centers containing ligands that do not readily dissociate. As noted in Scheme 7.1, the reaction of alkyl iodides with $CpFe(CO)_2^-$ occurs through radical intermediates, most likely formed by initial outer-sphere electron transfer, but the reactions of alkyl bromides and sulfonates occur by S_N2 pathways. Reactions of neutral, coordinatively saturated d^6 complexes containing tightly bound ligands, such as cis-$Mo(CO)_2(dmpe)_2$,[54,55] also occur by an outer-sphere electron transfer mechanism (Equation 7.7).

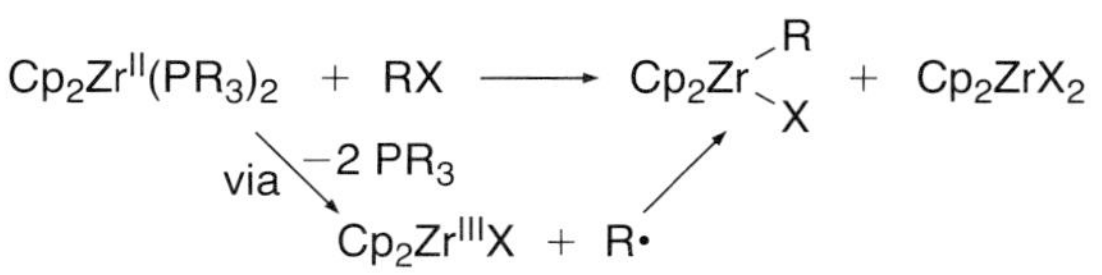

$$Cp_2Zr^{II}(PR_3)_2 \ + \ RX \longrightarrow Cp_2Zr{\overset{R}{\underset{X}{<}}} \ + \ Cp_2ZrX_2$$

$$\text{via} \diagdown {-2\,PR_3} \qquad \diagup$$

$$Cp_2Zr^{III}X \ + \ R\cdot \tag{7.8}$$

Several lines of evidence indicate that the oxidative addition of RX to the electron-rich early-metal complex $Cp_2Zr(II)L_2$ also occurs by a radical pathway initiated by electron transfer (Equation 7.8).[56,57] Zr(III) intermediates were observed by EPR, and evidence for effects of CIDNP (Chemically Induced Dynamic Nuclear Polarization) were observed in the NMR spectra of the reactions. Furthermore, optically active alkyl halides reacted to form products that resulted from racemization, alkyl halides with pendant olefins formed products from the cyclization of radical intermediates, and the reactions of alkyl halides formed both Cp_2ZrX_2 and $Cp_2ZrX(R)$. Finally, the rates of reaction of primary, secondary, and tertiary alkyl halides suggested that electron transfer occurred by an outer-sphere process to generate the R• intermediate. In this case, ligand dissociation from the higher oxidation state intermediate or product allows for coordination of the halide.

$$M^{(n)} + X-R \xrightarrow{\text{rate-determining step}} X-M^{(n+1)} + R\bullet$$

$$R\bullet + M^{(n)} \xrightarrow{\text{Fast}} R-M^{(n+1)}$$

$$\text{net: } 2\, M^{(n)} + X-R \longrightarrow X-M^{(n+1)} + R-M^{(n+1)}$$

$M^{(n)}$ = complex with the metal in oxidation state n

Scheme 7.8

7.3.4. Atom Abstraction and Combination of the Resulting Radical with a Second Metal

Oxidative addition of an alkyl halide to two metal centers to form two products from one-electron oxidations are also known (Scheme 7.8). This overall transformation generates the two products $M^{n+1}-R$ and $M^{n+1}-X$ from two metal complexes M^n. Thus, this reaction parallels the addition of dihydrogen and alkanes to two metal centers, but it is believed to proceed by successive one-electron processes instead of a mechanism through a single transition state containing the reactant and two metal centers. This dinuclear process occurs when the metal complex undergoes facile one-electron oxidation, but slow two-electron oxidation.

Experimental evidence for this pathway is diverse. For example, the reactions occur with a 2:1 stoichiometry, but the rate law is second order (proportional to [M•] and [RX]). Products are generated with racemization at carbon, and the relative reactivity of alkyl halides follows the trend tertiary > secondary > primary > Me. These relative reactivities parallel the stabilities of R• instead of the reactivity as an electrophile in a nucleophilic substitution process. Furthermore the alkyl radical intermediate in this mechanism has been trapped, has undergone rearrangements, and has been detected directly by EPR.

$$Co(CN)_5^{3-} + RX \xrightarrow{\text{Rate-limiting}} XCo(CN)_5^{3-} + R\bullet$$

$$R\bullet + Co(CN)_5^{3-} \xrightarrow{\text{Fast}} RCo(CN)_5^{3-} \tag{7.9}$$

$$2\, Co(CN)_5^{3-} + RX \longrightarrow XCo(CN)_5^{3-} + RCo(CN)_5^{3-}$$

This reaction often occurs with d^7 Co(II) or Rh(II) systems, which are excellent one-electron reductants and poor two-electron reductants. For example, the 17-electron, d^7, Co(II) complex $KCo(CN)_5^{3-}$ reacts with alkyl halides by this path (Equation 7.9).[58] These reactions display second-order kinetics, with the rate decreasing in the order RI > RBr >> RCl. Only activated halides such as $X-CH_2CO_2Me$ react. Alkyl tosylates do not.[59]

The rates of the atom abstraction are influenced by the steric properties of the ligands. The rate of reactions of $Co(dmgH)_2L$ complexes are significantly retarded by increased steric bulk of the coordinated phosphine.[60] Likewise, atom abstraction by 17-electron manganese–carbonyl radicals are slower for more hindered complexes and follow the trend $KMn(CO)_5 > KMn(CO)_4L > KMn(CO)_3L_2$ in which L is $P(i\text{-}Pr)_3$.[61] The reactions of other 17-electron complexes, such as $KRe(CO)_5$, $KM(Cp)(CO)_3$ (in which M = Cr, Mo, and W)[61] and the rhodoxime, $KRh(dmgH)_2PPh_3$,[62] as well as the 15-electron, d^3 Cp_2V,[63] have also been shown to give products from atom abstraction.

The same process occurs with actinides and lanthanides for which only one-electron oxidation is possible.[64] An example of such an oxidative addition is shown in Equation 7.10. Dissociation of an ether precedes the coordination of RX and halide abstraction.

$$Cp^*_2U(THF)Cl \underset{+THF}{\overset{-THF}{\rightleftharpoons}} Cp^*_2UCl \xrightarrow{RX} R\cdot + Cp^*_2UClX$$

$$R\cdot + Cp^*_2U(THF)Cl \longrightarrow Cp^*_2U(R)(Cl) + THF \tag{7.10}$$

$$\text{---}$$

$$2\,Cp^*_2U(THF)Cl + RX \longrightarrow Cp^*_2U(R)(Cl) + Cp^*_2U(X)(Cl)$$

7.4. Concerted Oxidative Additions

7.4.1. Concerted Oxidative Additions of Reagents with C–X Bonds of Medium Polarity

Oxidative addition of substrates possessing C–X or H–X bonds of medium polarity and of substrates possessing Ar–X bonds that cannot undergo S_N2 pathways often occur by concerted pathways involving three-centered transition states more like those of the oxidative additions of nonpolar substrates. The clearest cases in which reactions occur by concerted pathways are the oxidative additions of aryl halides and sulfonates to palladium(0) complexes.[65,66] These reactions have been studied extensively because they are the first step of transition-metal-catalyzed nucleophilic aromatic substitution reactions called cross couplings. The oxidative additions of the O–H and N–H bonds in water, alcohols, and amines also appear to occur by concerted three-centered transition states in many cases.

The mechanism of the reactions of aryl halides cannot occur by the common S_N2 path for the oxidative addition of methyl halides, and most aryl halides lack substituents that would make them sufficiently electrophilic to react by nucleophilic aromatic substitution pathways. As presented in the section on radical pathways for oxidative addition, aryl halides react with metal complexes that readily undergo one-electron oxidation by radical mechanisms. However, metal complexes that do not readily undergo one-electron processes tend to react by two-electron mechanisms. Thus, aryl halides typically react with d^{10} palladium(0) complexes by concerted pathways through three-centered transition states. No strong data for a radical pathway has been gained during the many studies on the oxidative addition of aryl halides to Pd(0). In contrast, evidence that oxidative addition of aryl halides to d^8, iridium, Vaska-type complexes occurs by a radical pathway has been published.[22]

$$L_2Pd(0) + ArX \longrightarrow L_2Pd(ArX) \longrightarrow L_nPd\overset{Ar}{\underset{X}{\diagdown}}$$

$$14e^{\ominus} \qquad L_2Pd(ArX) = \tag{7.11}$$

$$16e^{\ominus}$$

Just as the oxidative addition of H_2 and alkanes involves initial coordination of H_2 or alkane, the oxidative addition of aryl halides by non-radical pathways occurs by initial coordination of the arene. This coordination is then followed by insertion of the metal into the carbon–halogen bond (Equation 7.11). The coordination mode of the aryl halide prior to oxidative addition is difficult to determine precisely. Complexes of aryl halides are known in which the metal is bound to the halogen,[67] but computational studies suggest that the complexes in which the arene is bound in an η^2-fashion are the species that lie on the pathway for oxidative addition.[68,69]

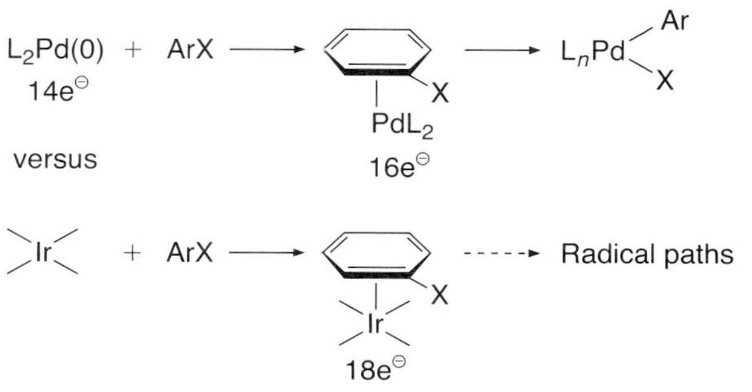

Figure 7.2.
Osborn's proposal for the propensity of Pd(0) complexes to react by concerted mechanisms and Ir(I) species to react by radical pathways.

The oxidative addition of aryl halides to palladium(0) complexes occurs fastest with highly unsaturated metal centers such as those in 12- and 14-electron L_2Pd and LPd species.[70–72] As summarized in Figure 7.2, Osborn suggested[20,22] that the difference in mechanism for reactions of Vaska-type Ir(I) complexes and Pd(0) or Pt(0) complexes results from a difference in the availability of valence orbitals after coordination of the aryl halide. The facile generation of a 14-electron intermediate with palladium and the presence of an orbital available for oxidative addition after coordination of the aryl halide allows this metal to react by a non-radical path.[22] Osborn proposed that the lack of a valence orbital after coordination of an aryl halide to a d^8, 16-electron species leads to slow reactions by non-radical mechanisms and, therefore, reactions by radical paths.

$$(7.12)$$

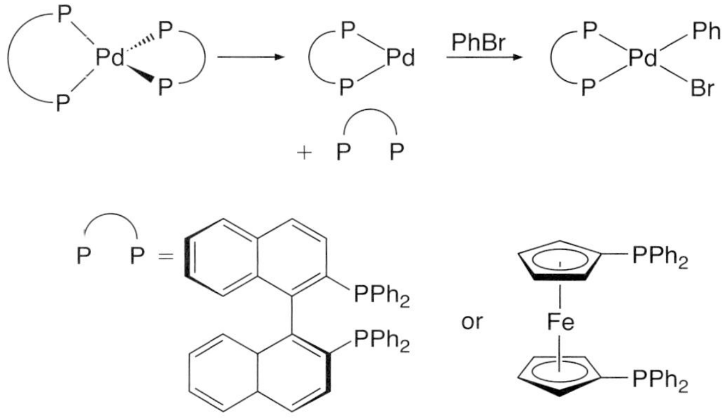

$$(7.13)$$

Many kinetic studies have been conducted on the oxidative addition of aryl halides to palladium(0) complexes, and these studies have revealed several principles that have been used to design more effective catalysts. First, the oxidative addition typically occurs to a 14-electron $L_2Pd(0)$ complex. Oxidative addition to $(PPh_3)_3Pd$ [which is generated upon dissolution of $(PPh_3)_4Pd$] (Equation 7.12) was shown by kinetic studies to occur by initial, reversible dissociation of a phosphine ligand to generate $(PPh_3)_2Pd$.[71–73] Oxidative addition to palladium(0) complexes coordinated by chelating ligands (Equation 7.13) has also been shown to occur to the $L_2Pd(0)$ complex.[74,75] In this case, the $L_2Pd(0)$ is bent into a less stable conformation than the linear $L_2Pd(0)$ fragment, and the frontier orbitals are more available for coordination and addition.[76] Thus, the bent (chelate)Pd(0) intermediate tends to be more reactive than the linear $L_2Pd(0)$ species containing monodentate ligands L.

$$R_3P-Pd-PR_3 \underset{k_{-1}}{\overset{k_1}{\rightleftharpoons}} \underset{+\ L}{R_3P-Pd} \overset{ArBr}{\underset{k_2}{\longrightarrow}} R_3P-Pd\overset{Ar}{\underset{Br}{\big\langle}} \overset{\text{For R = }}{\underset{o\text{-tol}}{\longrightarrow}} \underset{Ar}{\overset{L}{\big\rangle}}Pd\overset{Br}{\underset{Br}{\big\langle}}Pd\overset{Ar}{\underset{L}{\big\langle}}$$

R = o-tol or Bu^t

$k_{-1}[L] \geq k_2$ for R = o-tol
$k_{-1}[L] \leq k_2$ for R = Bu^t

(7.14)

Second, complexes containing extremely hindered phosphine ligands, such as tri(o-tolyl)phosphine and tri-$tert$-butylphosphine, are isolated as the $L_2Pd(0)$ species,[77–79] but they undergo oxidative addition of aryl halides after replacement of a phosphine ligand by the aryl halide to generate a complex containing an aryl halide and a single phosphine ligand (Equation 7.14).[70,80] The replacement of phosphine by the aryl halide could occur by an associative attack of the aryl halide, or when the ligand is P(o-tol)$_3$, by dissociation of the ligand to generate a 12-electron fragment. However, the reaction of phenyl chloride clearly occurs by rate-limiting dissociation of phosphine when the ligand is P(t-Bu)$_3$. The highly unsaturated Pd(0) species might be stabilized by an M–H–C interaction. Because the hindered ligands more readily generate the unsaturated Pd(0) species, complexes of these ligands tend to undergo oxidative addition faster than those that contain smaller phosphine ligands.

As described in Chapter 3, the P–M–P angle contained in the ring system generated by coordination of a chelating ligand is often called the "bite angle." The effect of this angle on the rate of oxidative addition has been studied because complexes containing chelating ligands are often effective catalysts for the reactions of aryl halides.[81,82] The effect of this angle on the rate of oxidative addition is not simple. Complexes containing a small bite angle undergo the C–X bond cleavage step faster than complexes containing a large bite angle. However, the small bite angle makes these complexes less hindered, and this reduced steric hindrance makes the concentration of reactive intermediate resulting from dissociation of other ligands lower.[83] Thus, complexes containing larger bite angles can undergo the overall multistep oxidative addition process faster than those containing a small bite angle. For example, the photochemically generated [bis(di-$tert$-butylphosphanyl)methane]Pt(0) intermediate[84] that contains a small bite angle inserts into the C–F bonds of hexafluorobenzene at room temperature,[85] but oxidative addition of aryl tosylates, which are typically unreactive toward Pd(0) complexes, occurs at room temperature to the complex in Equation 7.15 that contains a bisphosphine possessing a larger bite angle.[86]

$$\underset{Fe}{\overset{Bu^t}{\underset{Cy_2}{P}}} Pd-P(o\text{-tol})_3 \ + \ PhOTs \ \longrightarrow \ \overset{P}{\underset{P}{\big(}}Pd\overset{Ph}{\underset{OTs}{\big\langle}}$$

(7.15)

Third, complexes containing alkylphosphines undergo oxidative addition of less reactive carbon–chlorine bonds more readily than those containing aryl phosphines.[87,88] Aryl chlorides possessing electron-withdrawing substituents add to some complexes of arylphosphines, but aryl chlorides that lack these substituents react much more readily with palladium complexes bearing more electron-donating phosphine ligands.[89] The fastest additions of aryl chlorides occur to complexes containing sterically hindered alkylphosphines or related carbenes, as demonstrated by the catalytic chemistry of such substrates at room temperature.[87,90–93]

Fourth, reactions of aryl triflates and tosylates occur by similar mechanisms,[94–97] although the relative rates of these additions compared to the rates of addition of the various aryl halides depend on the identity of the metal center. In most cases, the rates of reactions of aryl triflates lie between those of aryl bromides and iodides, and the rates of reactions of aryl tosylates[86,98–100] are similar or slower than those of the reactions of aryl chlorides.

$$(PPh_3)_3Pd \rightleftharpoons (PPh_3)_2Pd \xrightarrow[\text{ArI}]{k_2} (PPh_3)_2Pd(Ar)(I)$$
$$+ PPh_3$$

$X^{\ominus}$ | Polar medium $-X^{\ominus}$

$$[(PPh_3)_2PdX]^{\ominus} \xrightarrow[\text{ArI}]{k_2{'}} [(PPh_3)_2Pd(X)(I)Ar]^{\ominus}$$
$$+ PPh_3$$

$k_2 > k_2{'}$, but $[(PPh_3)_2PdX]^{\ominus} > [(PPh_3)_2Pd]$ in polar media

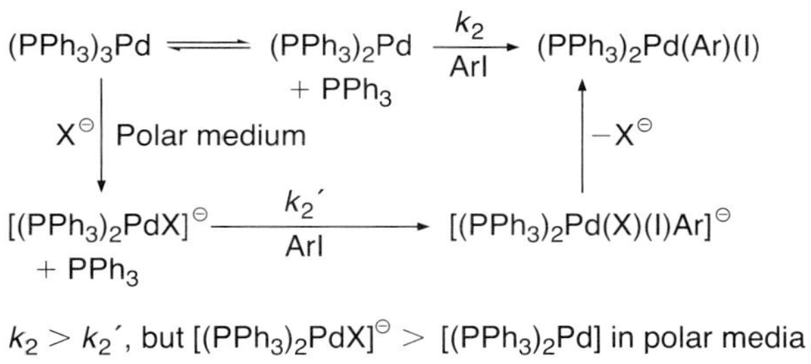

Scheme 7.9

Finally, added anions can influence the rates of the oxidative addition reactions.[66] In some cases, the anions can simply generate a more polar medium,[101] but in other cases the added halogen can coordinate directly to the metal center and alter the mechanism.[101–103] As depicted in Scheme 7.9, the added halogen can change the starting Pd(0) complex from a neutral species to an anionic Pd(0) complex bound by halide when the solvent is polar.[102,103] Although the neutral two-coordinate species is more reactive than the anionic three-coordinate species, the concentration of the anionic three-coordinate species is much higher, and the majority of the reaction can occur through this anion. In other cases, such as an addition of aryl triflates to $Pd[P(o\text{-}tol)_3]_2$, the added halogen can lead to a two-coordinate anionic species that is more reactive than the two-coordinate neutral bisphosphine complex.[101]

Because analogous platinum complexes are less reactive catalysts for cross-coupling chemistry, the oxidative addition of aryl halides to platinum(0) has received less scrutiny than the oxidative addition of aryl halides to palladium(0). Faster rates for the oxidative addition of aryl halides to $(PPh_3)_3Pt$ in the presence of AIBN under photolysis suggested that the reaction occurs, at least in part, by a radical-chain mechanism.[104] However, the small difference between the rates in the presence and absence of this additive in the dark does not make a compelling case for a radical path.

7.4.2. Oxidative Addition of Reagents with H–X Bonds of Medium Polarity

Oxidative additions of substrates containing H–X bonds that are more polar than those in a hydrocarbon, silane, or borane have been studied recently. These reactions can occur through a pathway involving a three-centered transition state, or they can occur by protonation of a basic metal center. The latter pathway requires a highly acidic reagent when the reactions are conducted in nonpolar media that do not stabilize charged intermediates.

Many years ago, Fornies[105] showed that bis-tricyclohexylphosphine platinum(0) reacts with substrates that contain acidic X–H bonds, such as pentafluoroaniline, and Otsuka showed that platinum(0) and palladium(0) complexes containing hindered trialkyl and mixed alkyl aryl phosphine ligands react with acidic substrates, such as carboxylic acids (Equation 7.16), to form products of oxidative addition.[106]

$$L_2M \ + \ HX \ \longrightarrow \ L-\overset{\overset{\displaystyle H}{|}}{\underset{\underset{\displaystyle X}{|}}{M}}-L$$

$$
\begin{array}{ll}
L = PCy_3, \ M = Pt & X = NH \\
L = PBu^t{}_3, \ M = Pt \ or \ Pd & OAc \\
L = PBu^t{}_2Ph, \ M = Pt \ or \ Pd & Cl
\end{array}
\qquad F_5
$$

(7.16)

However, others have shown that oxidative additions of much less acidic substrates, such as water,[107–109] alcohols,[107,108] unactivated anilines[110] and azole heterocycles,[111–113] occur to low-valent transition metal complexes. For example, iridium(I) complexes containing tri-ethylphosphine as ligand undergo oxidative addition of water rapidly at room temperature (Equation 7.17).[107,108] Both [Ir(PEt$_3$)$_4$]$^+$ and [Ir(PEt$_3$)$_3$Cl] also add aniline rapidly at room temperature (Equation 7.17).[110] The latter reaction is a step in the catalytic hydroamination of norbornene.[110] These substrates react readily with early metal complexes, but these reactions do not occur by oxidative addition.[114]

$$[Ir(PEt_3)_3Cl] \ + \ HX \ \longrightarrow \quad
\begin{array}{c}
H \\
Et_3P_{\text{\tiny{,,,}}} \ | \ _{\text{\tiny{,,,}}}PEt_3 \\
Ir \\
Et_3P \ | \ X \\
Cl
\end{array}
$$

X = OH or NHPh

(7.17)

Even ammonia undergoes reaction with similar Ir(I) complexes.[115,116] In some cases, these reactions of ammonia generate dimeric complexes containing bridging amido groups by a pathway thought to involve the oxidative addition of ammonia (Equation 7.18).[117] Presumably, this bridging interaction provides the thermodynamic driving force for this reaction. More recently,[118] the oxidative addition of ammonia to a particularly electron-rich and sterically hindered iridium complex to form a monomeric amido hydride complex was reported. This reaction and its mechanism are shown in Equation 7.19. Kinetic studies showed that the olefin ligand dissociates to generate a 14-electron Ir(I) intermediate. This 14-electron intermediate then adds ammonia, perhaps by initial formation of an ammine complex.

$$[Ir(C_2H_4)(PEt_3)_2Cl] \ + \ NH_3 \ \longrightarrow \quad
\begin{array}{c}
NH_3 \quad\quad H \\
Et_3P \ | \ NH_2 \ | \ PEt_3 \\
Ir \quad\quad Ir \\
Et_3P \ | \ NH_2 \ | \ PEt_3 \\
H \quad\quad\quad NH_3
\end{array}
\ + \ 2 \ C_2H_4
$$

(7.18)

(7.19)

Oxidative addition of the P–H and S–H bonds in phosphines and thiols are less well studied but are more favorable than oxidative aditions of N–H and O–H bonds for thermodynamic and kinetic reasons. S–H and P–H bonds in thiols and phosphines are weaker[119] and more acidic[120–123] than the O–H and N–H bonds in alcohols and amines. Moreover, the resulting products from additions to late transition metal complexes possess proportionately stronger metal–sulfur and metal–phosphorus bonds.[124–128]

The reported oxidative additions of weakly acidic X–H bonds are unlikely to occur by protonation of the metal center and collapse of the anion and cation. Most of these reactions were conducted in nonpolar solvents in which substrates like water, alcohols, and amines have very high pK_a values.[123] Instead, these reactions are more likely to occur by initial coordination of the reactants to the metal center to generate a transient aqua, alcohol, or amine complex, which subsequently undergoes insertion of the metal into the X–H bond. The acidity of the X–H bond does appear to promote the reaction by this pathway. Thus, the oxidative addition of aniline[110] is more common than the oxidative addition of alkylamines and ammonia.[115,117,118,129] In many cases, the product from the coordination of amine is more stable than the oxidative addition product because amines are basic and are common ligands for transition metals. Thus, the product from the coordination of a basic amine or ammonia can be more stable than the amido hydride complex that would form by oxidative addition (Equation 7.20).[129]

(7.20)

7.5. Dinuclear Oxidative Additions of Electrophilic A–B

The oxidative addition of a reagent across two metal centers in a single compound can occur to generate two mononuclear products or to generate a single product in which part of the reagent A–B is bonded to one metal and the other part is bonded to the second metal. Additions of A–B to a single metal center in a dinuclear or polynuclear complex can also occur, but the mechanisms of these processes are closely related to those for the oxidative additions to a mononuclear complex. In addition to reactions of homonuclear bimetallic compounds, reactions of heterobinuclear complexes have been studied with the objective of combining the favorable properties of an early metal complex with those of a late metal complex to generate new stoichiometric and catalytic reactivity. Studies of so-called "early–late heterobimetallic" complexes have also been motivated by the need to understand the chemistry of late transition metals supported on early metal oxides.

$$(CO)_5Mn–Mn(CO)_5 \xrightarrow{Br_2} 2\ (CO)_5MnBr$$

(7.21)

Classic examples of oxidative additions across the M–M bond of homonuclear dimeric species include the reactions of halogens with group 7 metal–carbonyl complexes (Equation 7.21). Many other examples of the oxidative addition of halogens to dinuclear complexes are known,[130–133] including additions to heterobinuclear complexes.[130] This reaction is often used to generate starting materials for further syntheses.

$$(CO)_5Mn–Mn(CO)_5 \xrightarrow{MeI} (CO)_5Mn–Me\ +\ (CO)_5Mn–I$$

(7.22)

(7.23)

$$[Bu_4N]_2[Fe_2(CO)_8] + CH_2I_2 \xrightarrow[0\ °C]{acetone} \text{(structure)} \quad (7.24)$$

$$2 \text{(structure)} + \text{(structure)} \longrightarrow CpCo\text{==}CpCo + \text{(structure)} + 2\,I^{\ominus} \quad (7.25)$$

Alkyl halides also add to dinuclear compounds. Additions can occur across the metal–metal bond of homodinuclear complexes, as shown in Equation 7.22,[130–137] or across the metal–metal bond of heterodinuclear compounds, as shown in Equation 7.23.[130] Oxidative additions of dihaloalkanes across two metal centers to generate bridging methylene complexes[138,139] have been reported for complexes of iron (Equation 7.24)[140] and osmium.[138] Addition of a higher alkyl dihalide to an odd-electron, dinuclear cobalt system (Equation 7.25)[141–143] has also been reported. In this case, the stereochemical outcome suggests that the first substitution process occurs by a radical mechanism.

$$Cp_2Zr\text{---}IrCp^* + HX \longrightarrow Cp_2Zr\text{(structure)}IrCp^* \quad (7.26)$$

$$X = H,\ SiHMePh,\ OBu^t,\ CH_2C(O)Me,\ \text{or}\ SPh$$

$$Cp^*_2Th\text{–}Ru(CO)_2Cp + CH_3CN \longrightarrow CpRu(CO)_2H + Cp_2Th\text{(structure)} \quad (7.27)$$

$$\text{(structure)} + \text{(structure)} \longrightarrow [N_3]ZrOEt + \text{(structure)} \quad (7.28)$$

The oxidative addition of polar and nonpolar H–X bonds can also occur to cleave metal–metal bonds, although few examples of this reaction have been reported. Baranger and Bergman showed that the three-membered ring of the complex in Equation 7.26 undergoes reaction with dihydrogen, phenyl methylsilane, aniline, tert-butanol, acetone, and arene thiols to form products containing one bridging hydride and one terminal hydride, silyl, anilide, alkoxide, enolate, or thiolate group, respectively, on the zirconium.[144] Marks reported that a heterodinuclear complex containing a thorium–ruthenium bond cleaves mildly acidic C–H bonds. These reactions generate a ruthenium–hydride product and complex thorium products that result from the combination of one cyanoalkyl group and two equivalents of nitrile (Equation 7.27) or one enolate of acetone and a second equivalent of the ketone.[145] Gade has also reported a series of reactions of early–late heterobimetallic compounds, in one case to generate products resulting from an oxidative addition that cleaves the C–O bond in an ester (Equation 7.28).[146]

Oxidative addition can also occur to two metal centers of dinuclear complexes that lack a metal–metal bond. For example, the dinuclear d^{10} Au(I) complex containing phosphorus ylide ligands in Equation 7.29 undergoes oxidative addition of CH_3I in a transannular fashion.[147–151] The reactivity of CH_3X, $PhCH_2Br$, and EtBr, but not neopentyl or adamantyl bromide, suggests that this reaction occurs by an S_N2 mechanism.

7.6. Summary

This overview of the oxidative addition of substrates containing polar C–X and H–X bonds shows that these reactions can occur by a diverse set of pathways. The following generalizations can help to predict the mechanism and products of the reaction, although there will certainly be exceptions to these trends. Oxidative addition of MeI tends to occur by an S_N2 pathway. Higher alkyl halides tend to undergo oxidative addition by radical pathways, although some highly unsaturated complexes, such as $Pd(PBu_2Me)_2$, which do not readily undergo one-electron processes, add higher alkyl halides by an S_N2 path. Alkyl sulfonates rarely react by radical pathways. The radical reactions occur by radical chain and non-chain pathways. These are typically initiated by outer-sphere electron transfer when the metal center is coordinatively saturated and by inner-sphere electron transfer when the metal center is unsaturated. Many reactions of aryl halides occur by radical pathways, but additions of aryl halides to Pd(0) complexes occur by concerted pathways. The additions of H–X bonds can occur by initial protonation of the metal when the reactant containing an H–X bond is very acidic, but more often occur by coordination and migration of the proton from X to the metal when the reactant is not very acidic and the solvent is nonpolar. These types of oxidative addition reactions will be seen many times again in the later chapters of this book that present catalytic processes, including the carbonylation and coupling of organic halides, allylic substitution, and reactions of alcohols and amines with olefins.

References and Notes

1. Collman, J. P.; Roper, W. R. *Adv. Organomet. Chem.* **1968**, *7*, 53.
2. Collman, J. P. *Acc. Chem. Res.* **1968**, *1*, 136.
3. Halpern, J. *Acc. Chem. Res.* **1970**, *3*, 386.
4. Lappert, M. F.; Lednor, P. W. *Adv. Organomet. Chem.* **1976**, *14*, 345.
5. Osborn, J. A. *Organotransition-Metal Chemistry*; Plenum Press: New York, 1975.
6. Vaska, L. *Acc. Chem. Res.* **1968**, *1*, 335.
7. Ugo, R. *Coord. Chem. Rev.* **1968**, *3*, 319.
8. Cramer, R. *Acc. Chem. Res.* **1968**, *1*, 186.
9. Parshall, G. W. *Acc. Chem. Res.* **1970**, *3*, 139.
10. Deeming, A. J. *MTP International Review of Science, Inorganic Chemistry Series One, Reaction Mechanisms in Organic Chemistry*; Butterworths: London, 1972.
11. Stille, J. K.; Lau, K. S. Y. *Acc. Chem. Res.* **1977**, *10*, 434.
12. Stille, J. K. *The Chemistry of the Metal-Carbon Bond, Vol. II. The Nature and Cleavage of Metal-Carbon Bonds*; Wiley: New York, 1985.
13. Kochi, J. K. *Organometallic Mechanisms and Catalysis*; Academic: New York, 1978.
14. Stieger, H.; Kelm, H. *J. Phys. Chem.* **1973**, *77*, 290.

15. Chock, P. B.; Halpern, J. *J. Am. Chem. Soc.* **1966**, *88*, 3511.

16. Lau, K. S. Y.; Wong, P. K.; Stille, J. K. *J. Am. Chem. Soc.* **1976**, *98*, 5832.

17. Bock, P. L.; Boschetto, D. J.; Rasmussen, J. R.; Demers, J. P.; Whitesides G. M . *J. Am. Chem. Soc.* **1974**, *96*, 2814.

18. Flood, T. C. *Topics in Inorganic and Organometallic Stereochemistry*; Wiley: New York, 1981; Vol. 12.

19. Ellis, P. R.; Pearson, J. M.; Haynes, A.; Adams, H.; Bailey, N. A.; Maitlis, P. M. *Organometallics* **1994**, *13*, 3215.

20. Labinger, J. A.; Osborn, J. A. *Inorg. Chem.* **1980**, *19*, 3230.

21. Collman, J. P.; Sears, C. T. *Inorg. Chem.* **1968**, *7*, 27.

22. Labinger, J. A.; Osborn, J. A.; Coville, N. J. *Inorg. Chem.* **1980**, *19*, 3236.

23. Fitton, P.; Johnson, M. P.; McKeon, J. E. *J. Chem. Soc., Chem. Commun.* **1968**, 1968.

24. Netherton, M. R.; Fu, G. C. *Angew. Chem. Int. Ed.* **2002**, *41*, 3910.

25. Wong, P. K.; Lau, K. S. Y.; Stille, J. K. *J. Am. Chem. Soc.* **1974**, *96*, 5956.

26. (a) Hills, I. D.; Netherton, M. R.; Fu, G. C. *Angew. Chem. Int. Ed.* **2003**, *42*, 5749. (b) Kirchhoff, J.; Netherton, M. R.; Hills, I.; Fu, G. *J. Am. Chem. Soc.* **2002**, *124*, 13662.

27. (a) Schrauzer, G. N.; Deutsch, E. *J. Am. Chem. Soc.* **1969**, *91*, 3341; (b) Collman, J. P.; Finke, R. G.; Cawse, J. N.; Brauman, J. I. *J. Am. Chem. Soc.* **1977**, *99*, 2515.

28. Celanese Chemical Company, 1985, E.P. 85303127.

29. Murphy, M. A.; Smith, B. L.; Torrence, G. P.; Aguilo, A. *Inorg. Chim. Acta* **1985**, *101*, L47.

30. Forster, D. *J. Am. Chem. Soc.* **1975**, *97*, 951.

31. Murphy, M. A.; Smith, B. L.; Torrence, G. P.; Aguilo, A. *J. Organomet. Chem.* **1986**, *303*, 257.

32. Russell, G. A.; Lamson, D. W. *J. Am. Chem. Soc.* **1969**, *91*, 3967.

33. Fischer, H. *J. Phys. Chem.* **1969**, *73*, 3834.

34. Ward, H. R. *Acc. Chem. Res.* **1972**, *5*, 18.

35. Garst, J. F. In *Free Radicals*; Kochi, J. K., Ed.; Wiley: New York, 1973; Vol. 1.

36. Krusic, P. J.; Fagan, P. J.; Sanfilippo, J. *J. Am. Chem. Soc.* **1977**, *99*, 250.

37. It is possible that cage effects and fast collapse from within the solvent cage can prevent full racemization. For examples, see the following three references.

38. Engstrom, J. P.; Greene, F. D. *J. Org. Chem.* **1972**, *37*, 968.

39. Koenig, T.; Owens, J. M. *J. Am. Chem. Soc.* **1973**, *95*, 8484.

40. Koenig, T.; Owens, J. M. *J. Am. Chem. Soc.* **1974**, *96*, 4052.

41. Tsou, T. T.; Kochi, J. K. *J. Am. Chem. Soc.* **1979**, *101*, 6319.

42. Ishizu, J.; Yamamoto, T.; Yamamoto, A. *Chem. Lett.* **1976**, 1091.

43. Tanaka, M. *Synthesis* **1981**, 47.

44. Basato, M.; Corain, B.; Favero, G.; Rosano, P. *J. Chem. Soc., Dalton Trans.* **1984**, 2513.

45. Eisch, J. J.; Piotrowski, A. M.; Han, K. I.; Kruger, C.; Tsay, Y. H. *Organometallics* **1985**, *4*, 224.

46. Eisch, J. J. *Pure Appl. Chem.* **1984**, *56*, 35.

47. Madonik, A. M. Ph.D. Thesis, Stanford University, 1981.

48. Eberson, L. *Acta Chem. Scan.* **1982**, *36*, 533.

49. Toscano, P. J.; Marzilli, L. G. *Prog. Inorg. Chem.* **1983**, *31*, 105.

50. Cabaret, D.; Maigrot, N.; Welvart, Z.; Vanduong, K. N.; Gaudemer, A. *J. Am. Chem. Soc.* **1984**, *106*, 2870.

51. Hill, R. H.; Puddephatt, R. J. *J. Am. Chem. Soc.* **1985**, *107*, 1218.

52. Ferguson, G.; Monaghan, P. K.; Parvez, M.; Puddephatt, R. J. *Organometallics* **1985**, *4*, 1669.

53. March, J. *Advanced Organic Chemistry*; Wiley: New York, 1985; p 620.

54. Connor, J. A.; Riley, P. I. *J. Chem. Soc., Chem. Commun.* **1976**, 634.

55. Connor, J. A.; Riley, P. I. *J. Chem. Soc., Dalton Trans.* **1979**, 1318.

56. Williams, G. M.; Schwartz, J. *J. Am. Chem. Soc.* **1982**, *104*, 1122.

57. Williams, G. M.; Gell, K. I.; Schwartz, J. *J. Am. Chem. Soc.* **1980**, *102*, 3660.

58. Halpern, J.; Maher, J. P. *J. Am. Chem. Soc.* **1965**, *87*, 5361.

59. Pearson, R. G.; Figdore, P. E. *J. Am. Chem. Soc.* **1980**, *102*, 1541.

60. Blaser, H. U.; Halpern, J. *J. Am. Chem. Soc.* **1980**, *102*, 1684.

61. (a) Herrick, R. S.; Herrinton, T. R.; Walker, H. W.; Brown, T. L. *Organometallics* **1985**, *4*, 42; (b) Biddulph, M. A.; Davis, R.; Wells, C. H. J.; Wilson, F. I. C. *J. Chem. Soc., Chem. Commun.* **1985**, 1287.

62. Espenson, J. H.; Tinner, U. *J. Organomet. Chem.* **1981**, *212*, C43.

63. Cooper, T. A. *J. Am. Chem. Soc.* **1973**, *95*, 4158.

64. (a) Finke, R. G.; Hirose, Y.; Gaughan, G. *J. Chem. Soc., Chem. Commun.* **1981**, 233; (b) Finke, R. G.; Schiraldi, D. A.; Hirose, Y. *J. Am. Chem. Soc.* **1981**, *103*, 1875; (c) Finke, R. G; Kennan, S. R.; Schiraldi, D. A.; Watson, P. L. *Organometallics* **1986**, *5*, 598.

65. Jutand, A. *J. Organomet. Chem.* **1999**, *576*, 254.

66. Amatore, C.; Jutand, A. *Acc. Chem. Res.* **2000**, *33*, 314.

67. Kulawiec, R. J.; Crabtree, R. H. *Coord. Chem. Rev.* **1990**, *99*, 89.

68. Sundermann, A.; Uzan, O.; Martin, J. M. L. *Chem. Eur. J.* **2001**, *7*, 1703.

69. Senn, H. M.; Ziegler, T. *Organometallics* **2004**, *23*, 2980.

70. Hartwig, J. F.; Paul, F. *J. Am. Chem. Soc.* **1995**, *117*, 5373.

71. Fauvarque, J.-F.; Pflüger, F. *J. Organomet. Chem.* **1981**, *208*, 419.

72. Amatore, C.; Pfluger, F. *Organometallics* **1990**, *9*, 2276.

73. Amatore, C.; Jutand, A.; Khalil, F.; Mbarki, M. A.; Mottier, L. *Organometallics* **1993**, *12*, 3168.

74. Alcazar-Roman, L. M.; Hartwig, J. F.; Rheingold, A. L.; Liable-Sands, L. M.; Guzei, I. A. *J. Am. Chem. Soc.* **2000**, *122*, 4618.

75. Amatore, C.; Broeker, G.; Jutand, A.; Khalil, F. *J. Am. Chem. Soc.* **1997**, *119*, 5176.

76. Hofmann, P.; Heiss, H.; Muller, G. *Z. Naturforsch.* **1987**, *B 42*, 395.

77. Otsuka, S.; Yoshida, T.; Matsumoto, M.; Nakatsu, K. *J. Am. Chem. Soc.* **1976**, *98*, 5850.

78. Yoshida, T.; Otsuka, S. *Inorg. Synth.* **1985**, *28*, 113.

79. Paul, F.; Patt, J.; Hartwig, J. F. *Organometallics* **1995**, *14*, 3030.

80. Barrios-Landeros, F.; Carrow, B. P.; Hartwig, J. F. *J. Am. Chem. Soc.* **2009**, *131*, 8141.

81. Kamer, P. C. J.; van Leeuwen, P. W. N.; Reek, J. N. H. *Acc. Chem. Res.* **2001**, *34*, 895.

82. Dierkes, P.; van Leeuwen, P. W. N. M. *J. Chem. Soc., Dalton Trans.* **1999**, 1519.

83. Freixa, Z.; van Leeuwen, P. W. N. M. *J. Chem. Soc., Dalton Trans.* **2003**, 1890.

84. Hofmann, P.; Heiss, H.; Neiteler, P.; Muller, G.; Lachmann, J. *Angew. Chem. Int. Ed.* **1990**, *29*, 880.

85. Hofmann, P.; Unfried, G. *Chem. Ber.* **1992**, *125*, 659.

86. Roy, A. H.; Hartwig, J. F. *J. Am. Chem. Soc.* **2003**, *125*, 8704.

87. Littke, A. F.; Fu, G. C. *Angew. Chem. Int. Ed.* **2002**, *41*, 4176.

88. Portnoy, M.; Milstein, D. *Organometallics* **1993**, *12*, 1665.

89. Liu, S. Y.; Choi, M. J.; Fu, G. C. *Chem. Commun.* **2001**, 2408.

90. Stambuli, J. P.; Kuwano, R.; Hartwig, J. F. *Angew. Chem. Int. Ed.* **2002**, *41*, 4746.

91. Old, D. W.; Wolfe, J. P.; Buchwald, S. L. *J. Am. Chem. Soc.* **1998**, *120*, 9722.

92. Wolfe, J. P.; Buchwald, S. L. *Angew. Chem. Int. Ed.* **1999**, *38*, 2413.

93. Navarro, O.; Kelly, R. A.; Nolan, S. P. *J. Am. Chem. Soc.* **2003**, *125*, 16194.

94. Alcazar-Roman, L. M.; Hartwig, J. F. *Organometallics* **2002**, *21*, 491.

95. Jutand, A.; Mosleh, A. *Organometallics* **1995**, *14*, 1810.

96. Scott, W. J.; Stille, J. K. *J. Am. Chem. Soc.* **1986**, *108*, 3033.

97. Jutand, A.; Negri, S. *Organometallics* **2003**, *22*, 4229.

98. Huang, X. H.; Anderson, K. W.; Zim, D.; Jiang, L.; Klapars, A.; Buchwald, S. L. *J. Am. Chem. Soc.* **2003**, *125*, 6653.

99. Furstner, A.; Leitner, A. *Angew. Chem. Int. Ed.* **2002**, *41*, 609.

100. Hamann, B. C.; Hartwig, J. F. *J. Am. Chem. Soc.* **1998**, *120*, 7369.

101. Roy, A. H.; Hartwig, J. F. *Organometallics* **2004**, *23*, 194.

102. Amatore, C.; Jutand, A.; Suarez, A. *J. Am. Chem. Soc.* **1993**, *115*, 9531.

103. Amatore, C.; Carré, E.; Jutand, A.; M'Barki, M.; Meyer, G. *Organometallics* **1995**, *14*, 5605.

104. Hall, T. L.; Lappert, M. F.; Lednor, P. W. *J. Chem. Soc., Dalton Trans.* **1980**, 1448.

105. Fornies, T.; Green, M.; Spencer, J. L.; Stone, F. G. A. *J. Chem. Soc., Dalton Trans.* **1977**, 1006.

106. Yoshida, T.; Otsuka, S. *J. Am. Chem. Soc.* **1977**, *99*, 2134.

107. Milstein, D.; Calabrese, J. C.; Williams, I. D. *J. Am. Chem. Soc.* **1986**, *108*, 6387.

108. Blum, O.; Milstein, D. *J. Am. Chem. Soc.* **2002**, *124*, 11456.

109. Yoon, M.; Tyler, D. R. *J. Chem. Soc., Chem. Commun.* **1997**, 639.

110. Casalnuovo, A. L.; Calabrese, J. C.; Milstein, D. *J. Am. Chem. Soc.* **1988**, *110*, 6738

111. Jones, W. D.; Dong, L.; Myers, A. W. *Organometallics* **1995**, *14*, 855.

112. Ardizzoia, G. A.; Brenna, S.; LaMonica, G.; Maspero, A.; Masciocchi, N.; Moret, M. *Inorg. Chem.* **2002**, *41*, 610.

113. López, C.; Baron, G.; Arevalo, A.; Munoz-Hernandez, M. A.; Garcia, J. J. *J. Organomet. Chem.* **2002**, *664*, 170.

114. Hillhouse, G. L.; Bercaw, J. E. *J. Am. Chem. Soc.* **1984**, *106*, 5472.

115. Koelliker, R.; Milstein, D. *Angew. Chem., Int. Ed. Engl.* **1991**, *30*, 707.

116. Koelliker, R.; Milstein, D. *J. Am. Chem. Soc.* **1991**, *113*, 8524.

117. Casalnuovo, A. L.; Calabrese, J. C.; Milstein, D. *Inorg. Chem.* **1987**, *26*, 971.

118. Zhao, J.; Goldman, A. S.; Hartwig, J. F. *Science* **2005**, *307*, 1080.

119. Luo, Y. R. *Handbook of Bond Dissociation Energies in Organic Compounds*; CRC Press: Boca Raton, FL, 2003.

120. Olmstead, W. N.; Margolin, Z.; Bordwell, F. G. *J. Org. Chem.* **1980**, *45*, 3295.

121. Bordwell, F. G.; Hughes, D. L. *J. Org. Chem.* **1982**, *47*, 3224.

122. Bordwell, F. G.; McCallum, R. J.; Olmstead, W. N. *J. Org. Chem.* **1984**, *49*, 1424.

123. Bordwell, F. G. *Acc. Chem. Res.* **1988**, *21*, 456.

124. Pearson, R. G. *J. Am. Chem. Soc.* **1963**, *85*, 3533.

125. Pearson, R. G. *J. Chem. Ed.* **1968**, *45*, 643.

126. Pearson, R. G. *J. Chem. Ed.* **1968**, *45*, 581.

127. Bryndza, H. E.; Fong, L. K.; Paciello, R. A.; Tam, W.; Bercaw, J. E. *J. Am. Chem. Soc.* **1987**, *109*, 1444.

128. Wicht, D. K.; Paisner, S. N.; Lew, B. M.; Glueck, D. S.; Yap, G. P. A.; Liable-Sands, L. M.; Rheingold, A. L.; Haar, C. M.; Nolan, S. P. *Organometallics* **1998**, *17*, 652.

129. Kanzelberger, M.; Zhang, X. W.; Emge, T. J.; Goldman, A. S.; Zhao, J.; Incarvito, C.; Hartwig, J. F. *J. Am. Chem. Soc.* **2003**, *125*, 13644.

130. Finke, R. G.; Gaughan, G.; Pierpont, C.; Noordik, J. H. *Organometallics* **1983**, *2*, 1481.

131. Chisholm, M. H.; Rothwell, I. P. *Prog. Inorg. Chem.* **1982**, *29*, 1.

132. Roberts, D. A.; Geoffroy, G. L. *Comprehensive Organometallic Chemistry*; Pergamon Press: New York, 1982; Vol. 6.

133. Halpern, J. *Inorg. Chim. Acta.* **1982**, *62*, 31.

134. Collman, J. P.; Rothrock, R. K.; Finke, R. G.; Moore, E. J.; Rosemunch, F. *Inorg. Chem.* **1982**, *21*, 146.

135. Collman, J. P.; Rothrock, R. K.; Finke, R. G.; Rosemunch, F. *J. Am. Chem. Soc.* **1977**, *99*, 7381.

136. Shyu, S. G.; Wojcicki, A. *Organometallics* **1985**, *4*, 1457.

137. Yu, Y. F.; Gallucci, J.; Wojcicki, A. *J. Am. Chem. Soc.* **1983**, *105*, 4826.

138. Motyl, K. M.; Norton, J. R.; Schauer, C. K.; Anderson, O. P. *J. Am. Chem. Soc.* **1982**, *104*, 7325.

139. Elamane, M.; Maisonnat, A.; Dahan, F.; Pince, R.; Poilblanc, R. *Organometallics* **1985**, *4*, 773.

140. Sumner, C. E.; Riley, P. E.; Davis, R. E.; Pettit, R. *J. Am. Chem. Soc.* **1980**, *102*, 1752.

141. Yang, G. K.; Bergman, R. G. *J. Am. Chem. Soc.* **1983**, *105*, 6045.

142. Theopold, K. H.; Bergman, R. G. *J. Am. Chem. Soc.* **1980**, *102*, 5694.

143. Krause, M. J.; Bergman, R. G. *J. Am. Chem. Soc.* **1985**, *107*, 2972.

144. Baranger, A. M.; Bergman, R. G. *J. Am. Chem. Soc.* **1994**, *116*, 3822.

145. Sternal, R. S.; Sabat, M.; Marks, T. J. *J. Am. Chem. Soc.* **1987**, *109*, 7920.

146. Gade, L. H.; Memmler, H.; Kauper, U.; Schneider, A.; Fabre, S.; Bezougli, I.; Lutz, M.; Galka, C.; Scowen, I. J.; McPartlin, M. *Chem. Eur. J.* **2000**, *6*, 692.

147. Schmidbaur, H.; Franke, R. *Inorg. Chim. Acta.* **1975**, *13*, 85.

148. Fackler, J. P.; Basil, J. D. *Organometallics* **1982**, *1*, 871.

149. Basil, J. D.; Murray, H. H.; Fackler, J. P.; Tocher, J.; Mazany, A. M.; Trzcinska-Bancroft, B.; Knachel, H.; Dudis, D.; Delord, T. J.; Marler, D. O. *J. Am. Chem. Soc.* **1985**, *107*, 6908.

150. Murray, H. H.; Mazany, A. M.; Fackler, J. P. *Organometallics* **1985**, *4*, 154.

151. Fackler, J. P.; Murray, H. H.; Basil, J. D. *Organometallics* **1984**, *3*, 821.

CHAPTER 8

Reductive Elimination

8.1. Overview

Reductive elimination is the reverse of oxidative addition. This class of reaction forms products from the coupling of two covalent ligands at a single transition metal center (Equation 8.1) or two ligands from two different metal centers (Equations 8.2 and 8.3). This reaction is the product-forming step of many catalytic processes. Because oxidative addition and reductive elimination are the same reaction occurring in opposite directions, the formation of products from oxidative addition or reductive elimination depends on the thermodynamics of the two processes. In some cases, equilibria between the two reactions have been observed directly, but more often the thermodynamics favor addition or elimination to a sufficient extent that high yields of either the addition or elimination products are obtained. Information on the thermodynamics for several types of oxidative addition and reductive elimination were provided in Table 6.2.

Because oxidative addition and reductive elimination are the same process run in reverse directions, the intermediates and transition states are the same for the two classes of reactions. Thus, reductive eliminations occur through three-centered transition states, as well as by stepwise mechanisms through cationic or radical intermediates. The mechanism depends on the metal center and the ligands in the complexes undergoing reductive elimination.

$$L_nM\begin{smallmatrix}A\\B\end{smallmatrix} \longrightarrow L_nM + A{-}B \qquad (8.1)$$

$$\underset{L_nM{-}ML_n}{\overset{A\ \ B}{|\ \ |}} \longrightarrow L_nM{=}ML_n + A{-}B \qquad (8.2)$$

$$2\,L_nM{-}A \longrightarrow L_nM{-}M_nL + A{-}A \qquad (8.3)$$

8.1.1. Changes in Electron Count and Oxidation State

Reductive elimination can occur by two-electron (Equation 8.1) changes at a single metal or one-electron (Equations 8.2 and 8.3) changes at each of two metal centers. During a reaction of a mononuclear species, the oxidation state of the single metal center decreases from $n + 2$ to n, and the number of electrons in its d-electron configuration increases by two. During a reaction of a dinuclear species, the oxidation state of each metal decreases from $n + 1$ to n, and the number of electrons in the d-electron configuration of each metal increases by one. These changes to the electron count and oxidation state that result from reductive elimination are the opposite of those that result from oxidative addition.

8.1.2. Factors that Affect the Rates of Reductive Elimination

Many factors control the rates of reductive elimination reactions (Equation 8.4). In many cases, the effects of the steric and electronic properties of the metal complex on the rate of reductive elimination are the opposite of the effects of these properties on the rate of oxidative addition because the effects originate from thermodynamic factors. The steric and electronic properties of the metal that thermodynamically favor oxidative addition must thermodynamically disfavor the opposite reductive elimination reaction.

$$L_nM{\overset{A}{\underset{B}{\diagdown}}} \longrightarrow L_nM + A-B$$

For M: •First row complexes react faster than second row complexes, which react faster than third row complexes.
•More electron-poor complexes react faster than more electron-rich complexes.

For L: •More sterically hindered complexes react faster than less sterically hindered complexes.

For A and B: H reacts faster than R.

For n: complexes in which $n = 1$ or 3 react faster than complexes in which $n = 2$ or 4.

(8.4)

8.1.2.1. Effect of Metal Identity and Electron Density

Reductive eliminations tend to be faster from complexes of first-row metals than from complexes of second-row metals, which in turn tend to be faster than those from complexes of third-row metals. Reductive eliminations from complexes of second-row metals are more favorable thermodynamically than reductive eliminations from third-row metals because the metal–ligand bonds in the second-row reactant are weaker.

Reductive eliminations also tend to occur faster from electron-poor metal centers than from electron-rich metal centers. This trend also results from thermodynamic effects. Reductive eliminations from electron-poor metal centers tend to be faster than those from electron-rich metal centers because the overall reductive elimination process usually creates a product containing more electron density at the metal center. Consistent with this trend, oxidation of the complex, which renders the metal center more electron deficient, often induces reductive elimination.

At the same time, reductive eliminations from d^0 early metal centers to form products containing metals possessing d^2 electron configurations are less common than reductive eliminations from late transition metal d^6 and d^8 metal centers. An example of reductive elimination from Zr(IV) is described later in this chapter, but such examples are unusual. The rarity of reductive eliminations from d^0 metal complexes can be attributed to the high-energy orbitals of an early-metal center and the resulting highly reducing property of the two d-electrons in a d^2 metal center.

8.1.2.2. The Effect of Steric Properties

Steric properties of the metal center tend to affect the rates of oxidative addition and reductive elimination in opposite ways, and these trends, again, result from thermodynamic effects. Reductive elimination tends to be faster from complexes possessing more sterically hindered ancillary ligands than from complexes possessing less sterically hindered ancillary ligands. Similarly, with some exceptions,[1] reductive eliminations also tend to occur faster from complexes containing more hindered covalent ligands participating in the bond-forming process than from complexes containing less hindered covalent ligands participating in the bond-forming process. A more sterically hindered metal center undergoes reductive elimination more rapidly than a less sterically hindered metal center, as long as the appropriate geometry can be adopted, because the steric congestion is relieved when the product is released from the metal.

8.1.2.3. The Effect of Participating Ligands

The properties of the reacting ligands tend to have the same influence on the rate of reductive elimination as on the rate of oxidative addition. These parallel trends result from a larger effect of these properties on the stability of the transition state than on the stability of the ground state. For example, reductive eliminations that form bonds to hydrogen tend to be faster than those that form bonds between two heavy atoms. Thus, reductive elimination to form the C–H bond in an alkane is typically faster than reductive elimination to form the C–C bond in an alkane. This trend parallels the relative rates for oxidative addition to cleave C–H and C–C bonds noted in Chapter 6. The faster rate of reductive eliminations to form X–H bonds results from effects of orbital overlap. The s-orbital of a hydride ligand is less directional than the sp^n-hybridized orbital of a hydrocarbyl group. Thus, the overlap in the transition state is greater for reductive eliminations forming bonds to hydride ligands than for those forming bonds to ligands bound through heavier atoms.[2,3]

8.1.2.4. The Effect of Coordination Number

Complexes of three- or five-coordinate metal centers tend to undergo reductive elimination faster than those of four- and six-coordinate metal centers. This effect results from changes in the energies of frontier orbitals, as discussed in more detail in the next paragraph. This trend also parallels the faster rates for oxidative additions to complexes that are three- and five-coordinate than those that are four- and six-coordinate. The difference in rates between C–C or C–X bond-forming reductive elimination from complexes possessing odd and even coordination numbers tends to be larger than the differences in rates between C–H or H–H bond-forming reductive elimination from complexes possessing odd and even coordination numbers.

The origin of the relative rates for reductive elimination from complexes possessing odd and even coordination numbers can be traced to the orbital of the metal fragment of the product that accepts the two additional d-electrons of the lower-valent product.[4,5] These orbital interactions are summarized in Figure 8.1. One might imagine that reductive elimination is preceded by ligand dissociation, because ligand dissociation would reduce the electron density at the metal center. However, the reverse reaction, oxidative addition, is also faster to complexes possessing odd coordination numbers. Thus, the origin of this effect is due to changes in the transition state energies and reactive intermediates, not due to changes in the electron density at the metal in the ground state.

As shown in Figure 8.1, reductive elimination from four- and six-coordinate complexes leads to occupation of an orbital in the immediate product that is strongly metal–ligand antibonding. In contrast, reductive elimination from trigonal bipyramidal or trigonal complexes leads to occupation of a molecular orbital that is nonbonding. This orbital is nonbonding because the dative ligand of the product, which was coplanar with the covalent ligands, is located in its nodal plane.

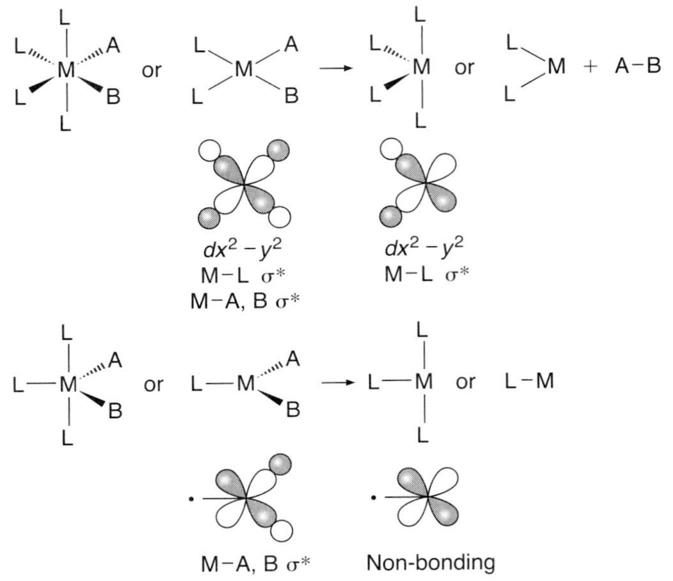

Figure 8.1.
The origin of the relative rates for reductive elimination from
complexes with odd and even coordination numbers.

8.1.2.5. The Effect of Geometry

The geometry of the metal complex also affects the rate and mechanism of reductive elimination, as summarized in Equation 8.5. A cis orientation of the two ligands undergoing reductive elimination is required for this reaction to occur by a three-centered, concerted mechanism. Complexes in which the two ligands undergoing reductive elimination are located trans to each other undergo this reaction by stepwise mechanisms or by prior isomerization of the complex to its cis isomer.

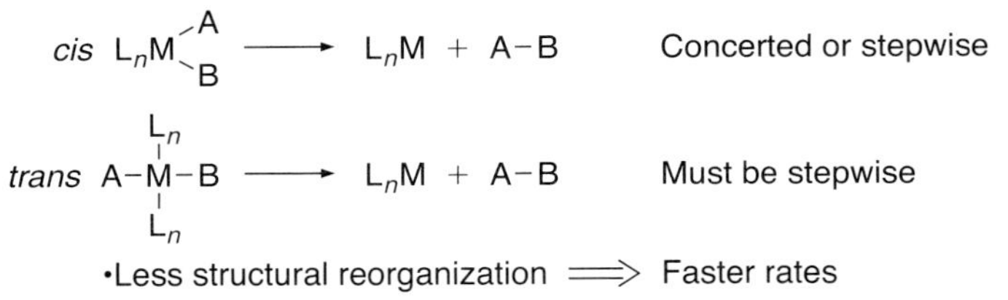

In addition, complexes whose structures resemble the reduced product tend to undergo reductive elimination faster than those that generate this product in an unstable conformation. For example, complexes that can generate a stable linear $L_2M(0)$ fragment (in which M = Pd or Pt) eliminate faster than those that generate a bent (chelate) M(0) fragment.

8.1.2.6. The Effect of Light: Photochemically Induced Reductive Elimination

Most reductive elimination reactions are induced thermally, but some important reductive elimination reactions are induced photochemically. Most examples of such reductive eliminations form dihydrogen from metal–dihydride complexes. Two examples of the photochemical reductive elimination of dihydrogen to generate intermediates that add alkanes are shown in Equations 8.6 and 8.7. In these examples,[6–8] the photochemical irradiation provides the energy to induce the reductive elimination process.

Photochemically induced reductive elimination can also result from initial photochemical extrusion of a dative ligand. For example, irradiation of $[IrH_2Cl(CO)(PPh_3)_2]$ leads to dissociation of CO, and the intermediate resulting from this ligand dissociation undergoes rapid thermal reductive elimination (Equation 8.8).[9] Consistent with the general trend that reductive eliminations from five-coordinate complexes are faster than those from analogous six-coordinate complexes, the reductive elimination of dihydrogen from [IrH$_2$Cl(CO)(PPh$_3$)$_2$] after dissociation of CO was shown in this case to occur about 10^9 times faster than the reductive elimination of dihydrogen from this complex without dissociation of ligand.

$$\text{(8.6)}$$

$$\text{(8.7)}$$

$$\text{IrH}_2\text{Cl(CO)(PPh}_3)_2 \xrightarrow[h\nu]{-\text{CO}} [\text{IrH}_2\text{Cl(PPh}_3)_2] \xrightarrow{-\text{H}_2} [\text{IrCl(PPh}_3)_2] \xrightarrow{+\text{CO}} \text{IrCl(CO)(PPh}_3)_2 \qquad \text{(8.8)}$$

8.2. Reductive Eliminations Organized by Type of Bond Formation

8.2.1. Reductive Elimination to Form C–H Bonds

8.2.1.1. Overview and Principles

Because of the favorable rates for reductive eliminations involving a hydride ligand noted in the introductory section, examples of complexes that undergo reductive elimination to form the C–H bonds in alkanes and arenes span the transition series. Thermally induced reductive eliminations to form dihydrogen (or dihydrogen complexes) from dihydride complexes can also be rapid, but this reaction occurs less frequently because the oxidative addition of dihydrogen is typically favored thermodynamically. Reductive elimination to form a C–H bond is the last step of many catalytic reactions, such as the hydrogenation and hydroformylation of olefins.

Much information has been gained on the mechanism of C–H bond-forming reductive elimination (see Equation 8.9). In addition to creating an understanding of C–H bond formation, this information has been used to understand the mechanism of the opposite reaction, the oxidative addition of C–H bonds. Because reductive eliminations of alkanes are faster from three- and five-coordinate species than from four- and six-coordinate species,[10,11] square planar and octahedral complexes often dissociate or associate a dative ligand prior to reductive elimination. However, elimination to form a C–H bond from a four- or six-coordinate complex can also be fast enough that it occurs directly from the alkylmetal–hydride complexes prior to ligand dissociation.

$$\text{(8.9)}$$

As shown in Equation 8.10, reductive eliminations of alkanes generate σ-complexes as the first-formed product, and the alkane dissociates or is displaced by an incoming ligand by an associative process to form the final products. This σ-complex is often formed reversibly; in these cases, dissociation or displacement of the coordinated alkane is the irreversible step.[12–14] Kinetic isotope effects and isotopic exchanges have revealed the intermediacy of σ-complexes during the reductive eliminations and provided evidence for the reversible formation of σ-complexes.[15]

$$L_nM \overset{H}{\underset{R}{\Big\langle}} \;\rightleftharpoons\; L_nM-\overset{H}{\underset{R}{\Big|}} \quad \overset{L'}{\underset{}{\Big\langle}} \begin{array}{l} L_nM-L' \;+\; R-H \\ or \\ L_nM \;+\; R-H \end{array} \qquad (8.10)$$

8.2.1.2. Examples

Reductive eliminations of methane, toluene, and cyclohexane (Equations 8.11–8.13) occur from $(PPh_3)_2Pt(Me)(H)$,[16,17] $(PMe_3)_4Ru(H)(CH_2Ph)$,[18] and $Cp^*Ir(PMe)_3(C_6H_{11})H$[19] without prior dissociation of ligand. The absence of ligand dissociation prior to reductive elimination was shown in these cases by a zero-order dependence of the reaction rates on the concentration of added phosphine. The rates for these elimination reactions vary dramatically. The platinum complex reacts at -25 °C, the ruthenium complex at 85 °C, and the iridium complex at 135 °C. Although one could rationalize these relative rates in many ways, the platinum compound likely reacts faster than the iridium complex because it is less electron rich, while the ruthenium complex likely reacts faster than the iridium complex because it contains a second-row metal center.

$$(PPh_3)_2Pt\overset{Me}{\underset{H}{\Big\langle}} \quad \xrightarrow[Ph\,\overline{\underline{\quad\quad}}\,Ph]{-25\,°C} \quad Me-H \;+\; (PPh_3)_2Pt(PhC\equiv CPh) \qquad (8.11)$$

$$(PMe_3)_4Ru\overset{CH_2Ph}{\underset{H}{\Big\langle}} \quad \xrightarrow{85\,°C} \quad H-CH_2Ph \;+\; \begin{array}{c} Me_2P \\ Me_3P\diagdown\;\Big|\;\diagup CH_2 \\ Ru \\ Me_3P\diagup\;\Big|\;\diagdown H \\ Me_3P \end{array} \qquad (8.12)$$

$$(8.13)$$

Other reductive eliminations that form C–H bonds do occur after initial dissociation of a ligand. For example, the elimination of carborane in Equation 8.14,[20,21] the elimination of ketone in Equation 8.15,[22,23] and the elimination of aldehyde in Equation 8.16[22,23] all occur after dissociation of a phosphine ligand. In contrast, the reductive elimination of alkane from the zirconocene alkyl hydride complex in Equation 8.17 occurs after association of ligand.[24]

$$H(carb)Ir(PPh_3)_2(CO)Cl \; \underset{+PPh_3}{\overset{-PPh_3}{\rightleftharpoons}} \; H(carb)Ir(PPh_3)(CO)Cl \; \xrightarrow{H-carb} \; Ir(PPh_3)_2(CO)Cl \; \xrightarrow{PPh_3} \; Ir(PPh_3)_2(CO)Cl \qquad (8.14)$$

$$(carb = 7-Ph-1, 7-C_2B_{10}H_{10})$$

$$(8.15)$$

$$L = PMe_3$$

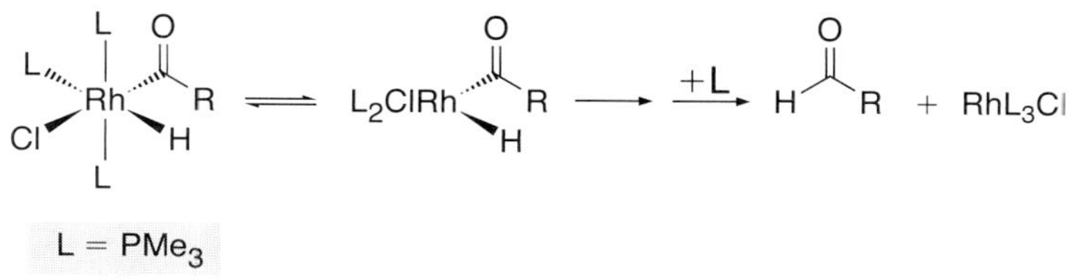

$$\text{(8.16)}$$

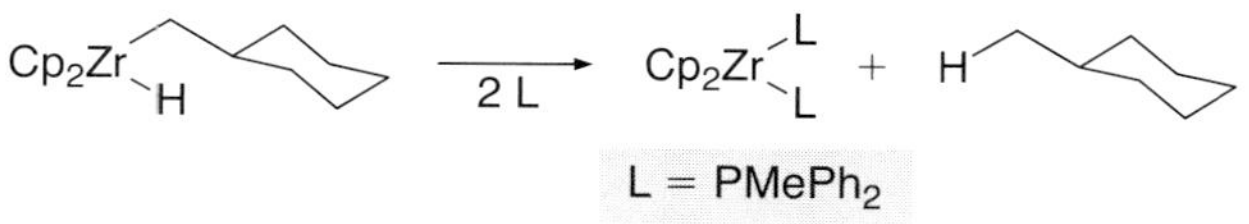

L = PMe$_3$

$$\text{(8.17)}$$

L = PMePh$_2$

8.2.1.3. Evidence for Intermediate Alkane and Arene Complexes

Some of the most compelling evidence for σ-complexes as intermediates in the oxidative addition and reductive elimination of alkanes has been obtained by conducting isotopic labeling studies on the reductive eliminations of alkanes. Many alkyl hydride complexes labeled with deuterium in the hydride or alkyl position undergo scrambling of the label into the other position faster than they undergo reductive elimination. These isotopic interconversions have been rationalized by a reductive coupling of the hydride and alkyl groups to generate a coordinated alkane[19] and a reversion of this step to regenerate an alkyl hydride complex that is faster than dissociation or displacement of the free alkane.

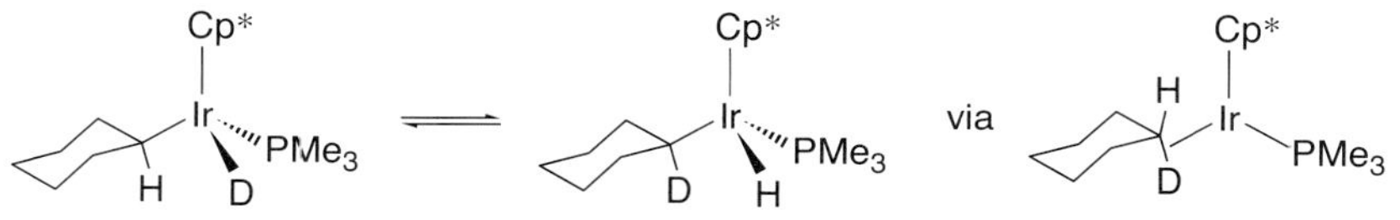

$$\text{via}\quad\text{(8.18)}$$

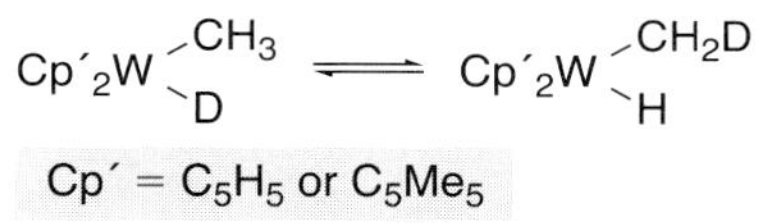

$$\text{(8.19)}$$

Cp′ = C$_5$H$_5$ or C$_5$Me$_5$

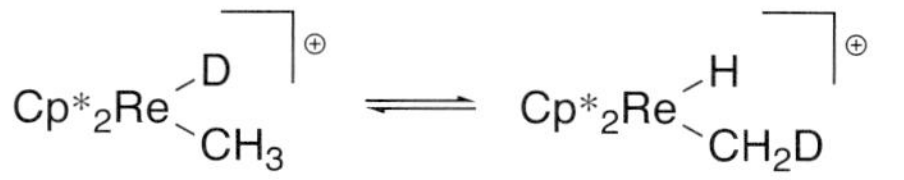

$$\text{(8.20)}$$

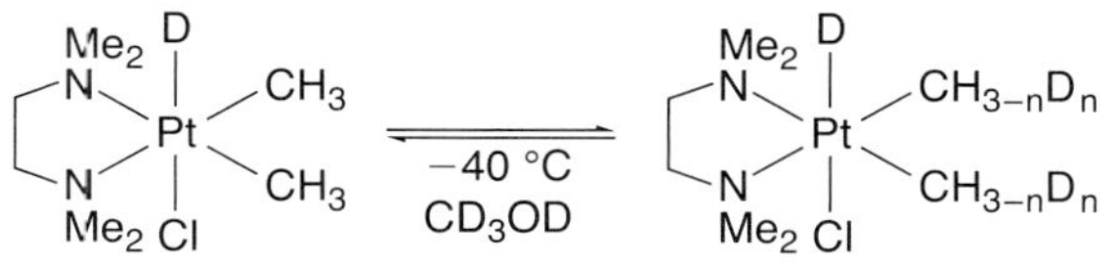

$$\text{(8.21)}$$

Four examples of such isotopic labeling studies are shown in Equations 8.18–8.21. Heating of the iridium cyclohexyl hydride complex in Equation 8.18[19] and related rhodium complexes[25] containing deuterium at the hydride position led to the incorporation of deuterium into the cyclohexyl group in the alpha position. This observation by Bergman's group was the first reported observation of such H/D scrambling and led to their proposal that the scrambling occurred through alkane complexes. This type of isotopic scrambling of alkyl and aryl hydride complexes[26,27] has now been observed in many different systems, ranging from the tungstenocene[28–30] and rhenocene[31] complexes in the middle of the transition series (Equations 8.19 and 8.20) to platinum(IV) complexes (Equation 8.21)[32,33] at the right of the transition series.

Jones[34] and Flood[35] and their co-workers have studied in detail the rates of scrambling between hydrides and long-chain alkyl groups of rhodium complexes. The results of this scrambling are shown in Equations 8.22 and 8.23. They have shown that deuterium migrates from one end of the alkyl group to the other. Thus, the complex of the alkane is relatively stable, and re-addition of the alkane is much faster than dissociation of the alkane.

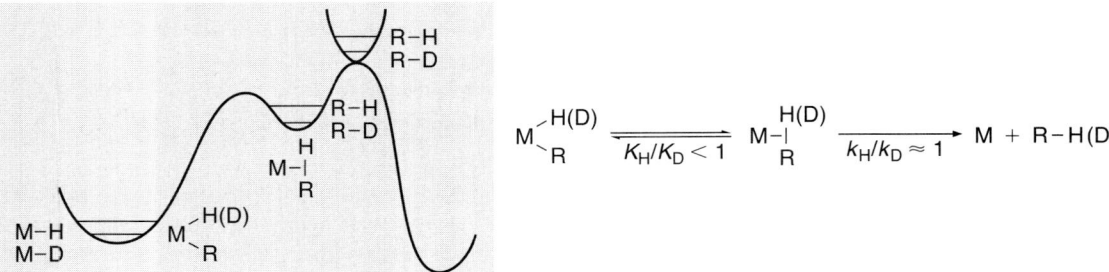

(8.22)

(8.23)

The isotopic labeling studies that show scrambling of the hydride with the hydrogens on the alkyl and aryl groups indicate that alkane and arene complexes are generated, but unusual kinetic isotope effects provided experimental evidence that the alkane complexes lie directly on the pathway to the C–H bond-forming reductive elimination of alkanes. Inverse kinetic isotope effects have been observed for catalytic hydrogenation by an iridium dihydride,[36] for reductive elimination of arenes from arylrhodium hydrides, and for reductive elimination of alkanes from alkyliridium and alkylrhodium hydrides.[19,25,37,38] Some recent reviews summarize these and related isotope effects.[39,40]

Figure 8.2. Proposed origin of the inverse isotope effect on the reductive elimination of alkanes.

This inverse kinetic isotope effect on the reductive elimination of alkanes was rationalized by a reversible formation of an alkane complex that would be followed by irreversible dissociation or displacement of the alkane.[19] This pathway is depicted on an energy diagram in Figure 8.2, which also illustrates the origin of the inverse isotope effect. This mechanism would give rise to an apparent kinetic isotope effect that would comprise an equilibrium isotope effect for C–H bond formation and a kinetic isotope effect for dissocia-tion of the alkane. Because the C–H stretching frequency is higher than the M–H stretching frequency, the equilibrium isotope effect for the reversible formation of an alkane complex should be less than one. Because the C–H bond is already present in the alkane that would dissociate, the kinetic isotope effect for the irreversible dissociation of the alkane is expected to be small. Thus, the product of these two isotope effects would be less than one.

The reductive elimination of arenes from aryl hydride complexes draws many paral-lels with the reductive elimination from alkyl hydrides. For example, arene complexes are believed to form from the reductive elimination prior to release of the free arene. Studies by Feher and Jones provided compelling evidence for the presence of arene complexes,[38, 41] and complementary synthetic studies have led to the isolation of closely related η^2-arene complexes.[42–44] As shown in Equation 8.24, the arylrhodium deuteride complex scrambles deuterium into the aryl ring, just as alkylmetal deuteride complexes scramble deuterium into the alkyl group. This isotopic exchange presumably occurs by formation of the arene complex shown at the center of Equation 8.24.

$$(8.24)$$

Theoretical studies on tungstenocene complexes suggest the intermediacy of arene C–H σ-complexes on the pathway of the oxidative addition and reductive elimination reactions that cleave and form arene C–H bonds.[30] Thus, the η^2-arene complexes likely rearrange to C–H σ-complexes prior to C–H bond cleavage, as depicted in Equation 8.25.

$$(8.25)$$

8.2.1.4. The Effect of Ancillary Ligands on C–H Bond-Forming Reductive Elimination

The electronic effects of ancillary ligands on the reductive elimination to form carbon–hydrogen bonds are significant. Jones has shown that complexes of the type Cp*Rh(PR$_3$)(Ph)(H) possessing more electron-donating phosphines undergo reductive elimination more slowly than those possessing less electron-donating phosphines.[45] As shown in Table 8.1, reductive elimination of benzene from the complex containing PMe$_2$Ph as ancillary ligand was faster than from the complex containing the more electron-donating PMe$_3$ as ligand, even though the cone angles are similar (122° vs. 118°). Likewise, reductive elimination of benzene from the complex containing PMePh$_2$ was faster than that containing the more electron-donating PMe$_2$Bu^t, even though the cone angles are similar (136° vs. 139°).

Table 8.1. The effect of ancillary ligand electronic effects on the rate of reductive elimination of benzene from Cp*Rh(PR$_3$)(Ph)(H).

Entry	PR$_3$	Cone angle (deg)	T (°C)	k (s^{-1})	$\Delta G^{\ddagger}$ (kcal/mol)
1	PMe$_3$	118	23	$3.35(17) \times 10^{-7}$	26.1
2	PMe$_2$Ph	122	23	$1.08(5) \times 10^{-6}$	25.4
3	PMe$_2$Bu^t	139	24.5	6.6×10^{-5}	24.5
4	PMePh$_2$	136	24.5	$1.11(6) \times 10^{-5}$	24.2

Table 8.2. The effect of ancillary ligand steric properties on the rate of reductive elimination of benzene from Cp*Rh(PR$_3$)(Ph)(H).

Entry	PR$_3$	Cone angle (deg)	T (°C)	k (s^{-1})	ΔG‡ (kcal/mol)
1	PMe$_3$	118	23	$3.4(2) \times 10^{-7}$	26.1
2	P(n-Bu)$_3$	132	24.5	$2.4(1) \times 10^{-6}$	25.1
3	PMe$_2$Bu^t	139	24.5	6.6×10^{-5}	24.5

The steric effects of the ancillary ligands on reductive elimination are also significant, even when the elimination involves a small hydride ligand. As shown in Table 8.2, the rate constant for reductive elimination of benzene from Cp*Rh(PR$_3$)(Ph)(H) increased with increasing cone angle of the phosphine ligand. More information on the steric properties of ancillary ligands on the rates of reductive eliminations to form C–C, C–N, and C–O bonds is presented in subsequent sections of this chapter.

Finally, the geometry of the ancillary ligand, which affects the ability of the product to adopt the most stable confirmation, influences the rates of reductive eliminations to form C–H bonds. As is described in detail in later sections, complexes containing phosphines that constrain the product complex into a confirmation that is unstable undergo reductive elimination more slowly than complexes that will contain less geometric constraints in the product. For example, Hoffman has shown that alkyl hydride complexes of platinum ligated by the chelating bis(dialkylphosphino)methane ligands[46] are more stable than the platinum alkyl hydride complexes ligated by the monodentate PMe$_3$ studied by Halpern (Equation 8.26).[16] Similarly, reductive elimination from a metallocene fragment that can adopt a parallel ferrocene-like conformation in the product occurs faster than reductive elimination from a complex in which two cyclopentadiene ligands are bridged in a manner that prevents the parallel orientation of the cyclopentadienyl rings in the product (Equation 8.27).[47,48]

8.2.2. Reductive Elimination to Form X–H Bonds

Reductive eliminations to form O–H and N–H bonds, as well as reductive eliminations to form products such as acetic acid or HCl in the presence of base, occur by mechanisms that are often distinct from those of reductive eliminations to form C–H bonds. Reductive eliminations to form H–X bonds occur upon addition of a dative ligand in most cases. Less well-defined chemistry occurs in the absence of added ligand. One might imagine that reductive elimination to form the N–H and O–H bonds in amines and alcohols might occur by dissociation of anion and subsequent deprotonation of the metal hydride by the free anion. However, this mechanism would imply that the forward oxidative addition occurs by protonation of the metal and subsequent trapping of the cation by the resulting anion. The metal center is typically not basic enough to deprotonate alcohols and amines in nonpolar aprotic organic solvents. Thus, this mechanism for reductive elimination does not usually occur.

Bergman reported a detailed mechanistic study on the reductive elimination of alcohols and amines from iridium complexes that are analogous to alkyl hydride complexes that eliminate alkanes (Equation 8.28).[49] The hydridoiridium alkoxide and amide complexes underwent reductive elimination only upon addition of a dative ligand, such as a triarylphosphine. Several lines of mechanistic data suggested that slippage of the η^5-Cp* ligand to an η^3-Cp* ligand and coordination of the added dative ligand precedes reductive elimination of alcohol or amine from these complexes. In related work, reductive elimination of aniline and phenol was shown to occur upon addition of olefins to platinum hydrido phenoxide and anilide complexes (Equation 8.29). The mechanism of this reaction is not well defined, however.[50] Reductive elimination of pyrrole from an Ir(III) pyrrolyl complex was shown to occur from the octahedral starting complex by a mechanism that appears to involve rate-limiting dissociation of phosphine (Equation 8.30).[51]

$$(8.28)$$

$$(8.29)$$

$$(8.30)$$

8.2.3. Reductive Elimination to Form C–C Bonds[52]

8.2.3.1. Trends and Principles

Reductive eliminations to form carbon–carbon bonds occur from fewer transition metal complexes than reductive eliminations to form carbon–hydrogen bonds. The origin of this effect results from a higher-energy transition state for the C–C bond forming reductive elimination, not from a weaker thermodynamic driving force for C–C bond-forming reductive elimination. Reductive elimination to form the carbon–carbon bond between two saturated carbon centers requires canting of the alkyl groups, and this distortion creates steric hindrance between the substituents on the alkyl groups and the ancillary ligands at the metal.[2,3]

Reductive eliminations to form carbon–carbon bonds occur by concerted mechanisms in most cases. Therefore, a cis coordination geometry is required for this type of reductive elimination. The requirement for cis coordination in concerted reductive eliminations was demonstrated in a classic comparison of reductive elimination from dimethylpalladium complexes possessing ligands that enforce cis and trans geometries. Elimination from the cis complex (Equation 8.31) occurred much faster than elimination from the trans complex (Equation 8.32). Reductive elimination from a concerted pathway also implies that the reaction will occur with retention of configuration of a stereocenter directly attached to the metal. Although the stereochemistry of reductive eliminations to form carbon–carbon bonds is difficult to analyze, a body of indirect evidence supports reaction with retention of configuration.[53] As shown in Equation 8.33, the conversion of the enantioenriched benzylpalladium complex to α-deuterio ethylbenzene occurs with substantial retention of configuration. A related study of the stereochemistry of carbon–phosphorus bond-forming reductive elimination indicated that the reaction occurs with retention of stereochemistry at phosphorus (Equation 8.34).[54]

$$(8.31)$$

$$(8.32)$$

$$(8.33)$$

$$(8.34)$$

[Pd] = Pd[(*S,S*)-Chiraphos]
o-An = 2-MeOC$_6$H$_4$

8.2.3.2. The Effect of Participating Groups

The rates for reductive elimination of groups bound to the metal by sp^2-hybridized carbons can be much different than those for reductive elimination of groups bound by sp^3-hybridized carbons. In general, C–C bond-forming reductive eliminations from aryl and vinyl complexes of nickel, palladium, and platinum occur more rapidly than related reductive eliminations from alkyl complexes of these metals. Because transition metal aryl complexes are more stable thermodynamically than transition metal alkyl complexes, the origin of this effect results from a difference in transition state energies for the two reactions that is larger than the difference in thermodynamic driving forces. The lower transition state energy for reductive elimination involving sp^2-hybridized carbon atoms can be attributed to several effects. First, the two-dimensionality of the aryl or vinyl group reduces the steric hindrance that occurs upon canting of the ligands to form the carbon–carbon bond. Second, an sp^2-hybridized orbital is less directional than an sp^3-hybridized orbital because of the increase in *s*-character, and this lower directionality creates a greater overlap and multi-center bonding in the three-centered transition state.[3]

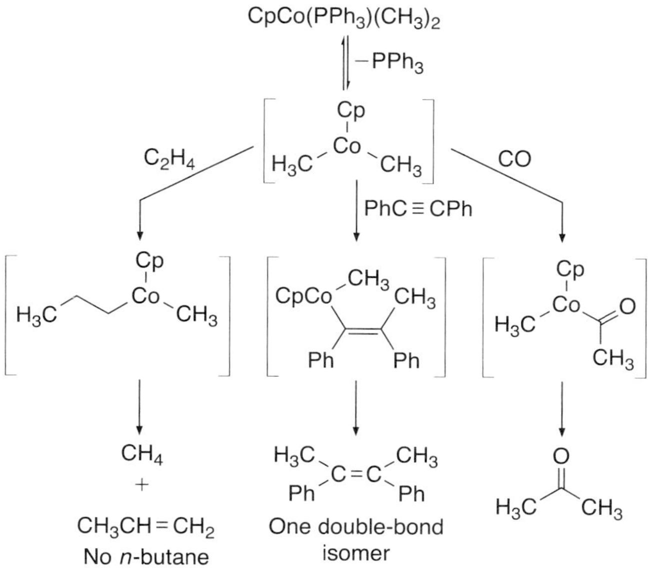

Scheme 8.1

In addition to orbital hybridization, the strength of the coordination of the product to the metal has been proposed to affect the rate of reductive elimination. This coordination can then be present in the transition state and would be expected to lower the barrier for reductive elimination. For example, the π-system of an aryl or vinyl group or the oxygen of an acyl group can interact with the metal during reductive elimination.

The set of reactions of the cobalt compounds in Scheme 8.1 illustrate the potential effect of this type of coordination on the rate of reductive elimination.[55] The dimethyl-cobalt complex is stable, and the ethyl propyl complex undergoes β-hydrogen elimination (for a discussion of β-hydrogen elimination, see Chapter 10) faster than it undergoes C–C bond-forming reductive elimination. However, the vinyl and acyl complexes formed by insertion of alkyne and carbon monoxide, respectively, undergo reductive elimination. The faster reductive eliminations from these vinyl and acyl complexes parallel the faster rates for elimination from vinyl and acyl complexes of other metal centers. These faster rates were attributed to coordination of the olefin and ketone products because coordination of these products would be expected to be stronger than coordination of an alkane product. If the initial product contains a coordinated olefin or ketone, then some degree of coordination of these developing products would be present in the transition state for the reductive elimination, and the energy of this transition state would be lower than it would be in the absence of this coordination.

In contrast to the typical relative rates for reductive elimination of aryl and alkyl groups from most metals, the rate of carbon–carbon bond-forming reductive elimination from alkyl complexes of Pt(IV) is often faster than that from aryl complexes of Pt(IV). For example, reductive elimination from the Pt(IV) complex $[PtI(Me)_2(C_6H_4\text{-}4\text{-}Me)(PMe_2Ph)_2]$ generates ethane and $trans$-$[PtI(C_6H_4\text{-}4\text{-}Me)(PMe_2Ph)_2]$ (Equation 8.35).[56,57] A more recent study on a Pt(IV) intermediate containing a pincer ligand (Equation 8.36) showed that the rate of reductive elimination to form alkyl–alkyl and alkyl–aryl bonds can be almost identical. However, the relative rates of reductive eliminations from Pt(IV) complexes containing acyl and alkyl groups parallel those for reductive elimination from complexes of other metals containing acyl and alkyl groups.[58,59] For example, the dialkyl acyl complex in Equation 8.37 forms acetone and no ethane.[60] These and other studies reveal that the relative rates for reductive elimination from Pt(IV) follow the trend acyl > methyl > aryl.

$$\text{(8.35)}$$

$$\text{(8.36)}$$

$$\text{(8.37)}$$

The rates of reductive elimination to form carbon–carbon bonds are also influenced by the electronic properties of the alkyl, aryl, vinyl, or alkynyl groups. In general, complexes containing less electron-rich reacting groups of a particular class undergo reductive elimination slower than complexes containing more electron-rich reacting groups.[3,4,60–64] This trend has been predicted by theoretical studies and results from a more stable ground state when electron-withdrawing substituents are present on the hydrocarbyl group. These reactions are also accelerated by the pairing of electron-rich and electron-poor groups bound to the metal. For example, reductive elimination of alkylarenes is faster from complexes containing more electron-poor aryl groups and more electron-donating alkyl groups (Equation 8.38).[62,63] Moreover, reductive elimination of biaryls from biarylplatinum complexes is faster from complexes containing one electron-rich and one electron-poor aryl group than from complexes containing two electron-rich or two electron-poor aryl groups (Equation 8.39).[64]

$$\text{(8.38)}$$

Relative rates:
$X = p\text{-CN} > p\text{-CF}_3 > p\text{-H} > p\text{-Me} > p\text{-OMe}$
$Y = \text{H} > \text{Ph} > \text{C(O)R} > \text{CF}_3 > \text{CN}$

$$\text{(8.39)}$$

$R^1 = \text{OMe}, R^2 = \text{CF}_3 > R^1 = R^2 = \text{OMe} > R^1 = R^2 = \text{CF}_3$

8.2.3.3. The Effect of Coordination Number

Reductive elimination to form carbon–carbon bonds can occur directly from the starting complex, from a species formed after dissociation of ligand, or from a species formed by association of ligand. Several examples of such reductive eliminations are presented in Section 8.2.3.4, but a few illustrative examples are shown in Equations 8.40–8.42. If the ancillary ligands are chelated to the metal center, then reductive elimination will often occur from the starting complex. For example, reductive elimination of biaryl from the DPPF-ligated biarylplatinum complex [DPPF =1,1′-bis(diphenylphosphino)ferrocene] in Equation 8.40 occurs directly from the four-coordinate biaryl complex.[64] In contrast, reductive elimination of dimethylcyclopropane from the platinacyclobutane in Equation 8.41 containing monodentate phosphines occurs after dissociation of phosphine,[65] and reductive elimination from the arylnickel methyl complex in Equation 8.42 containing a chelating phosphine undergoes reductive elimination after coordination of PEt_3.[66,67]

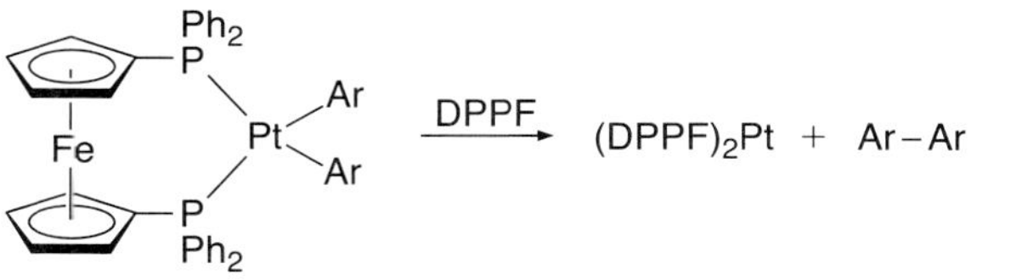

$$(8.40)$$

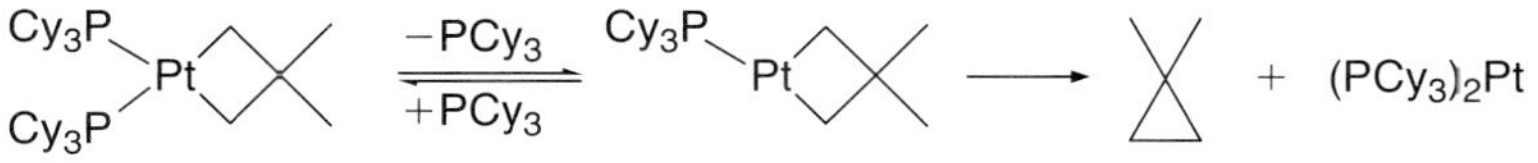

$$(8.41)$$

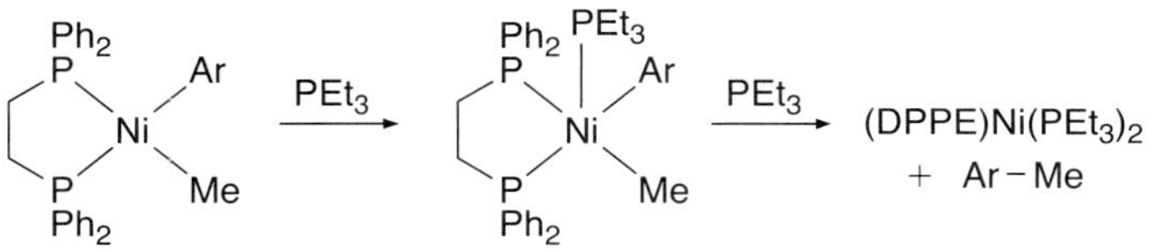

$$(8.42)$$

8.2.3.4. The Effect of Bite Angle

Finally, the rate of reductive elimination can be influenced by the ability of the remaining dative ligands to adopt the stable structure of the reduced product. The reactions in Scheme 8.2, reported by Brown, show the effect of the L–M–L angle or "bite angle" on the rate of reductive elimination.[68] The preferred geometry of the Pd(0) product is linear. Thus, complexes containing a larger bite angle occur faster, in part because the Pd(0) product is more stable when the complex contains a P–Pd–P angle that is closer to 180°. Moreover, the distance between the two ligands that undergo reductive elimination can be shorter in a complex containing an ancillary ligand containing a wider bite angle. This shorter bond would place the starting material further along the reaction coordinate and might be expected to cause reductive elimination to be faster.

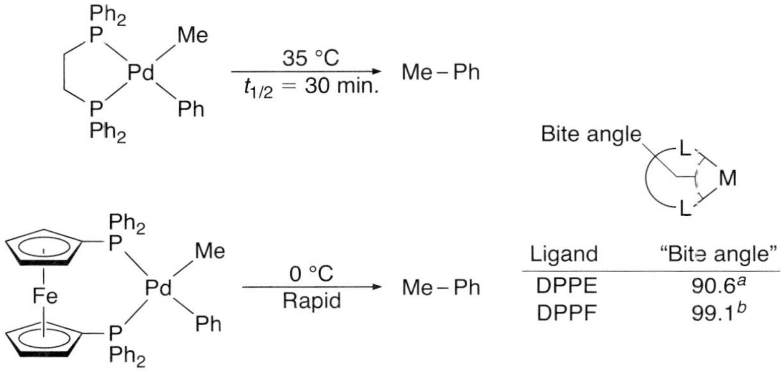

Ligand	"Bite angle"
DPPE	90.6[a]
DPPF	99.1[b]

Scheme 8.2

[a]Steffen, W. L., Palenik, G. J. *Inorg. Chem.* **1976**, 15, 2432.
[b]Hayashi, T., Konishi, M., Kobori, Y., Kumada, M., Higuchi, T., Hirotsu, K. *J. Am. Chem. Soc.* **1984**, 106, 158.

8.2.3.5. Survey of Carbon–Carbon Bond-Forming Reductive Eliminations

Reductive elimination to form carbon–carbon bonds is most common from complexes of group 10 metals. However, complexes of other metals containing covalent ligands that most readily participate in reductive elimination also undergo this C–C bond-forming reaction. Examples of catalytic reactions in which carbon–carbon bond-forming reductive elimination occurs from palladium complexes are legion.[69–71] Catalytic reactions in which carbon–carbon bond-forming reductive eliminations from cobalt[72–75] and iron[76–89] intermediates are also known, and iron-catalyzed couplings were among the first examples of cross-coupling. Reductive eliminations that form C–C bonds have also been proposed to occur during ruthenium-catalyzed C–H bond functionalization processes.[90,91]

$$(8.43)$$

$$(8.44)$$

Many reductive eliminations to form carbon–carbon bonds have been observed directly from Ni(II), Pd(II),[68,92–96] Pd(IV), and Pt(IV) complexes.[97] Examples of directly observed reductive elimination from nickel complexes are shown in Equations 8.43 and 8.44.[98–102] The first example shows, in particular, that addition of an electron-poor olefin can induce the elimination process. The electron-poor olefin might induce the reductive elimination by forming a charge-transfer complex that is more electron poor than the starting four-coordinate complex, or it might induce elimination by generating a complex possessing a coordination number that is more favorable for reductive elimination, or both.[98] The second example shows that addition of electron-donating ligands also induces reductive elimination from nickel complexes, most likely because a complex with an odd coordination number is formed as an intermediate in this case (vide supra).

$$(8.45)$$

$$(8.46)$$

		σ	k_{rel}
R =	CH$_3$	0	> 600
	CH$_2$Ph	0.22	> 250
	CH$_2$C(O)Ar	0.60	31
	CH$_2$CF$_3$	0.92	1.7
	CH$_2$CN	1.30	1
	CF$_3$	2.60	No reaction

$$(8.47)$$

Two examples of the many reductive eliminations to form C–C bonds from Pd(II) complexes are shown in Equations 8.45 and 8.46. These results reveal several of the trends noted in the above sections. Stille and Yamamoto showed that the reductive eliminations from L_2PdMe_2 complexes in Equation 8.45 occur after dissociation of ligand.[103,104] Reaction from the complex containing the DIPHOS ligand in Equation 8.46 occurs more slowly because dissociation of the ligand is disfavored by chelation.

The examples summarized in Equation 8.47 show that the rate of reductive elimination from Pd(II) is faster from complexes containing more electron-donating alkyl groups than from complexes containing less electron-donating alkyl groups. These data show that Taft parameters can be used to predict the relative rates for the reaction of complexes containing different functionalized alkyl groups.[105]

Reductive eliminations to form C–C bonds from platinum complexes also include those containing both Pt(II) and Pt(IV) centers. Reductive elimination of biaryls and cyclopropanes from Pt(II) complexes were shown in Equations 8.39–8.41. Reductive eliminations from Pt(IV) were shown in Equations 8.35 and 8.37.[106]

Few reductive eliminations to form C–C bonds from isolated complexes of other transition metals have been observed directly. However, some classic work on reductive elimination from square planar gold compounds was reported by Kochi (Scheme 8.3). Kochi[107] and Tobias[108] have examined elimination from two types of planar Au(III) complexes, R_3AuL and $[R_2AuL_2]^+$; in both cases, reductive elimination occurred from three-coordinate intermediates formed by dissociation of phosphine. The R_3Au intermediate formed from R_3AuL is thought to be T-shaped. However, because the T-shaped intermediate isomerizes more rapidly than it eliminates in solvents that give rise to purely intramolecular eliminations, the same ratio of CD_3–CH_3 to CH_3–CH_3 is obtained from cis- or trans-CD_3Me_2AuL.[107] R–R elimination from $[R_2AuL_2]^+$ is accelerated by bulky ligands L and by alkyl groups other than methyl, presumably because these factors encourage dissociation of L.[108]

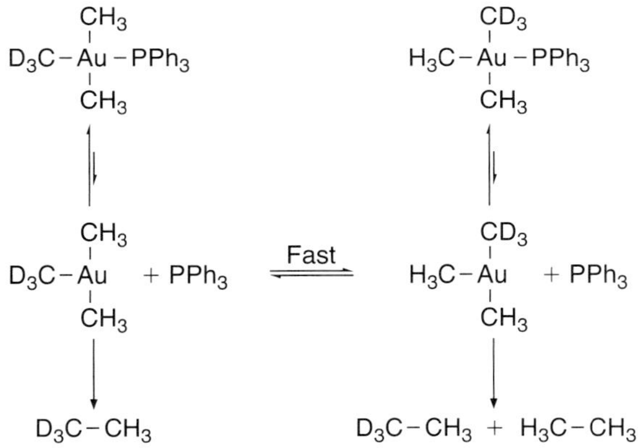

Scheme 8.3

Kochi also showed that $Et_2Fe(bpy)_2$[109] reductively eliminates butane after oxidation to the corresponding dication (Scheme 8.4). In contrast, thermolysis of the neutral species leads to products from β-hydrogen elimination, and reaction of the monocation leads to products from homolysis of the Fe–C bond. Reductive eliminations to form carbon–carbon bonds from cobalt complexes bonded to sp^2-hybridized carbons were shown in Scheme 8.1.[110]

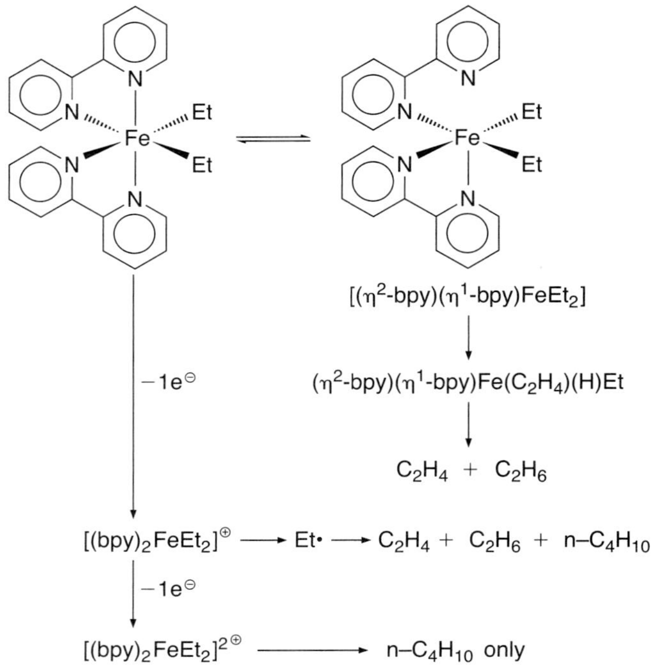

$[(\eta^2\text{-bpy})(\eta^1\text{-bpy})FeEt_2]$

$-1e^\ominus$

$(\eta^2\text{-bpy})(\eta^1\text{-bpy})Fe(C_2H_4)(H)Et$

$C_2H_4 \ + \ C_2H_6$

$[(bpy)_2FeEt_2]^\oplus \longrightarrow Et\cdot \longrightarrow C_2H_4 \ + \ C_2H_6 \ + \ n\text{--}C_4H_{10}$

$-1e^\ominus$

$[(bpy)_2FeEt_2]^{2\oplus} \longrightarrow n\text{--}C_4H_{10}$ only

Scheme 8.4

8.2.4. Reductive Elimination to Form C–X Bonds

Reductive elimination reactions that form carbon–halogen, carbon–oxygen, carbon–nitrogen, carbon–phosphorus, and carbon–sulfur bonds have been investigated for a shorter period of time than reductive eliminations that form nonpolar carbon–hydrogen and carbon–carbon bonds. Most of these types of reductive eliminations occur from Pt(IV), Ni(II), and Pd(II) complexes. However, a particularly important exception includes reductive elimination of acyl halides from rhodium(III). The rates of carbon–heteroatom bond-forming reductive elimination vary widely and depend on the identity of both the carbon and heteroatom ligand. For example, reductive eliminations of esters and amides from acylmetal phenoxide or amide complexes[111,112] occur much faster than reductive eliminations of ethers from alkyl- or arylmetal phenoxide or amide complexes, and reductive eliminations from amido and alkoxo complexes depend strongly on the electronic properties of the heteroatom ligand.[1,113–116]

8.2.4.1. Mechanisms of Reductive Eliminations to Form C–X Bonds[117]

Reductive eliminations to form carbon–heteroatom bonds occur by several mechanisms. As would be expected from microscopic reversibility, these mechanisms parallel those for the oxidative addition of carbon–heteroatom bonds. Recall that oxidative additions to form Pt(IV) alkyl halide complexes and oxidative additions to form Pd(II) alkyl and aryl halide complexes occur by different pathways. The former occurs by a stepwise mechanism involving charged intermediates, and the latter occurs by concerted pathways.

Reductive eliminations to form alkyl carbon–heteroatom bonds from Pt(IV) occur by initial dissociation of the heteroatom ligand as an anion, followed by attack of this anion on the alkyl ligand. Most reductive eliminations to form carbon–heteroatom bonds from Pt(IV) alkyl complexes involve platinum(IV) methyl complexes,[116,118–120] and examples of methylplatinum(IV) complexes that undergo reductive elimination are shown in Equation 8.48. These reactions occur by a pathway involving dissociation of the anion to form a cationic Pt(IV) intermediate and subsequent nucleophilic attack of the anion on the methyl group of this cation. As a result, methylplatinum(IV) complexes containing ligands that do not undergo dissociation are stable to reductive elimination (Equation 8.49). Reductive eliminations to form carbon–heteroatom bonds from arylplatinum(IV) complexes[121] are rare and are typically slower. One example that is likely to occur as part of the oxidation of aryl C–H bonds is shown in Equation 8.50.[122] Oxidation of a Pd(II) complex containing two cyclometallated 2-arylpyridine ligands with iodosylbenzene dibenzoate [PhI(O₂CPh)₂] generates a stable Pt(IV) species that undergoes reductive elimination of 2-pyridylphenyl-benzoate upon heating in the presence of pyridine. The Pt(II) species that results from the reductive elimination is trapped by added pyridine.

$$\text{(8.48)}$$

X = I, OAc, or OAr   Faster for more stable X$^{\ominus}$

$$\text{(8.49)}$$

$$\text{(8.50)}$$

The relative rates for reductive eliminations to form carbon–heteroatom bonds from alkyl and aryl complexes of palladium(II) are different from those for reductive eliminations from alkyl and aryl complexes of platinum(IV) in several ways. First, reductive eliminations that form carbon–heteroatom bonds from arylpalladium(II) complexes are faster than those that form carbon–heteroatom bonds from alkylpalladium(II) complexes. This trend was shown most clearly by studies of reductive elimination from a set of aryl-, vinyl-, and alkylpalladium thiolate complexes, summarized in Table 8.3.[1] Second, reductive eliminations from palladium(II) complexes occur faster from complexes possessing more electron-donating heteroatom ligands, while the examples of reductive elimination from platinum(IV) just discussed occur faster from complexes possessing less electron-donating heteroatom ligands. The contrasting trends in reactivity of alkylplatinum(IV) complexes and arylpalladium(II) amido complexes is summarized in Equations 8.51 and 8.52.[113,114,117,123]

Table 8.3. Rates of reductive elimination from [Pd(DPPE)(R)(S^{t}Bu)].

R	temp, °C	$t_{1/2}$, min
methyl	95	580
phenyl	50	48
alkenyl	50	17

Faster for R = Me than for R = Ph

Faster for less electron-donating $X^{\ominus}$

$$\hspace{10cm} (8.51)$$

Faster for more electron-poor Ar

Faster for more electron-rich amido groups (less stable $^{\ominus}NRR'$)

$$ArNRR' + (DPPF)_2Pd + (PPh_3)_nPd \hspace{3cm} (8.52)$$

These opposite trends for reductive elimination to form carbon–heteroatom bonds from Pt(IV) and Pd(II) result from the differences in the mechanisms for reductive elimination from the two types of complexes. Reductive elimination from Pt(IV) relies on dissociation of X^- and, therefore, is favored by complexes containing X^- ligands that are stable as the free anion. In contrast, reductive elimination to form carbon–heteroatom bonds from Pd(II) occurs by a concerted three-centered mechanism. As a result, the stability of the anion is less important, and reactions from Pd(II) are faster from complexes containing more covalently bound heteroatom ligands (Equation 8.52).[124]

Reductive eliminations from nickel(II) complexes to form carbon–heteroatom bonds in amines and ethers have also been reported.[125–128] Like the mechanisms for oxidative additions to Ni(0) and Pd(0) that cleave carbon–heteroatom bonds, the mechanisms for reductive elimination from nickel(II) and palladium(II) complexes to form carbon–heteroatom bonds are different from each other. Most reductive eliminations from Ni(II) to form carbon–nitrogen bonds occur after oxidation of the Ni(II) to Ni(III) with ferrocenium, oxygen, or iodine (Equations 8.53 and 8.54).[127,128] Reductive eliminations from Ni(II) to form carbon–oxygen bonds in ethers also requires oxidation of the Ni(II) to Ni(III) (Equation 8.55).[125] In contrast, reductive eliminations from Ni(II) to form the ester group of a lactone occurred after a proposed insertion of CO into the nickel–carbon bond of an oxametallacycle without oxidation.[129] Reductive eliminations from isolated arylnickel complexes to form amines and ethers have not been reported.

$$ (8.53) $$

$$ (8.54) $$

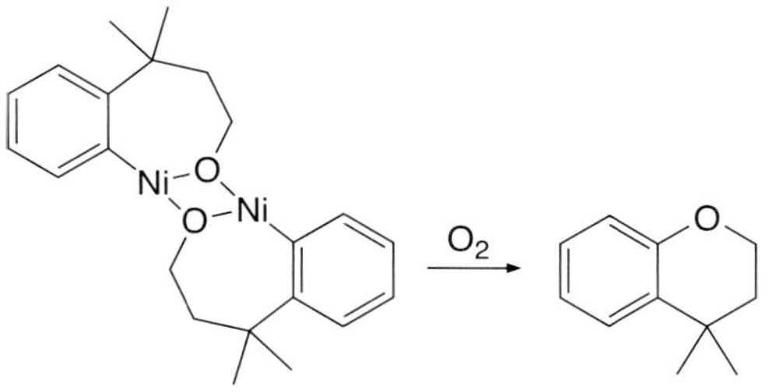

(8.55)

8.2.4.2. *Survey of Reductive Eliminations to Form C–X Bonds*

8.2.4.2.1. Reductive Eliminations to Form C–X Bonds from Aryl
and Alkylplatinum(IV) Complexes

Equations 8.56–8.58 display several examples of reductive eliminations to form carbon–heteroatom bonds from Pt(IV). Reductive elimination of phenyl iodide occurs from a Pt(IV) diphenyl diiodide complex (Equation 8.56).[121,130] This reaction is preceded by dissociation of iodide to generate a cationic Pt(IV) intermediate, as shown by inhibition of the reaction rate by added iodide. Elimination after dissociation of the iodide is favored by the five-coordinate geometry and by the reduced electron density on the cationic metal center.

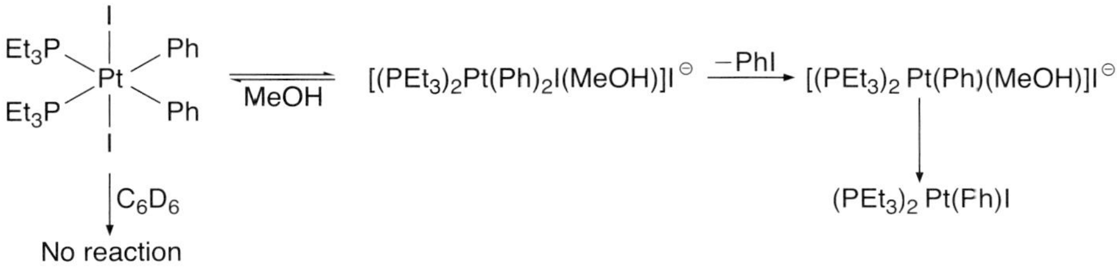

(8.56)

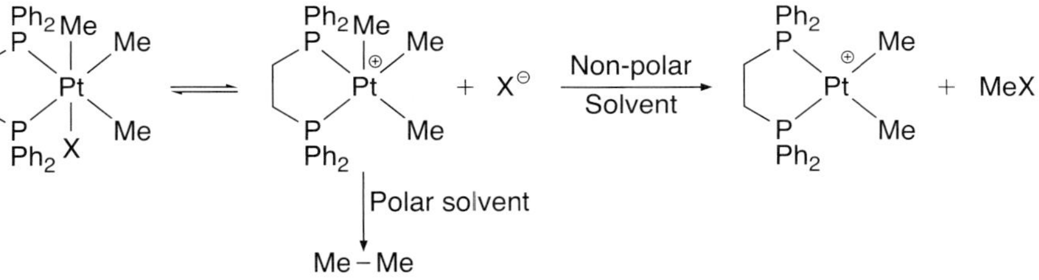

(8.57)

X = I, OAc, or OAr

P⌒P = 1, 2-bis(diphenylphosphine)ethane or 1, 2-bis(diphenylphosphino) benzene

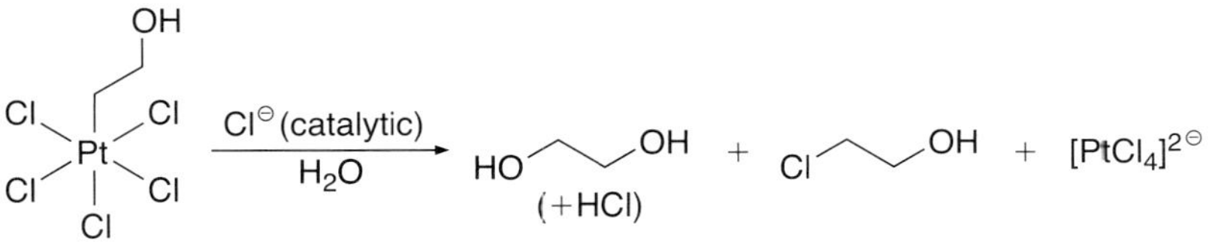

(8.58)

Studies of the reaction in Equation 8.57 show that reductive elimination of methyl iodide from alkylplatinum iodide complexes also occurs after dissociation of halide.[106,119,120] These reductive eliminations occur between groups originally located in mutually trans positions. This result is consistent with the stereochemistry of the Pt(IV) product from oxidative addition by an S_N2 path. The selectivity of the cationic intermediate for reductive elimination to form a new carbon–carbon or carbon–heteroatom bond depends on solvent.[116] The C–C bond-forming reductive elimination is favored in polar protic solvents that stabilize ion pairs, while the C–X bond-forming reductive elimination is favored in less polar solvents that favor collapse of the ion pair.

Additional examples of C–X bond-forming reductive elimination from Pt(IV) form alcohols and alkylchlorides. Bercaw has reported that methylplatinum(IV) complexes undergo attack by chloride or water to generate chloromethane or methanol and that hydroxyethyl complexes generate ethylene glycol and 2-chloroethanol (Equation 8.58).[131] Such backside attack by water on an alkylplatinum(IV) complex is thought to constitute the product-forming step of the catalytic conversion of methane to methanol reported by Shilov. More details about this nucleophilic attack on alkyl ligands are presented in Chapter 11 (nucleophilic attack on coordinated ligands).

8.2.4.2.2. Reductive Eliminations to Form C–X Bonds from Arylpalladium(II) Complexes

A few examples of reductive eliminations to form carbon–heteroatom bonds from Pd(II) are shown in Equations 8.59–8.65. The steric and electronic effects on these reductive eliminations have been studied in detail.[117] Studies of complexes containing various ligands bound to the metal center by a heteroatom have been conducted, including those bearing various substituents on the aromatic rings of thiophenoxide (Equation 8.59)[1] and anilide (Equation 8.60)[132] ligands. These studies have shown that complexes containing heteroatom ligands that are softer and more electron donating undergo reductive elimination faster than those containing hetereoatom ligands that are harder and less electron donating. A similar conclusion was drawn from studies on the reductive elimination of ethers from palladium alkoxides.[115] Consistent with this trend, reductive eliminations to form P–C bonds from phosphido complexes are facile,[54,133–135] including the one in Equation 8.61, which is closely related to the eliminations of amines and thioethers. Similar studies of complexes possessing para and meta substituents on the palladium-bound aryl ring have been conducted.[1,113,115] These studies have shown that complexes containing more electrophilic aryl groups on the metal undergo faster reductive elimination. In addition, these studies have shown that resonance effects of these substituents dominate over inductive effects.

$$\text{ArSR} + \text{Pd(DPPF)}_2 + \text{Pd(PPh}_3)_n \tag{8.59}$$

R = Bu^t or Ar

$$\xrightarrow[0°-65°]{\text{PPh}_3} \text{ArNRR}' + \text{Pd(DPPF)}_2 + \text{Pd(PPh}_3)_n \tag{8.60}$$

R = R′ = p-tol
R = H, R′ = Ph
R = H, R′ = Bu^i

$$\xrightarrow{\text{DPPE}} \text{MePH(Mes)}* + \text{Pd(DPPE)}_2 \tag{8.61}$$

$$(8.62)$$

Reductive elimination from thiolate complexes containing methyl, aryl, vinyl, and alkynyl ligands was studied to reveal the effect of the type of carbon-bound ligand on the rate of carbon–heteroatom bond-forming reductive elimination. As noted briefly above, the rate of reductive elimination depended strongly on the type of hydrocarbyl group (Equation 8.62).[1] Reductive elimination from vinyl complexes occurred faster than from aryl complexes, which occurred faster than from alkynyl complexes. Reductive elimination from the methyl complexes was by far the slowest of the series. Although the origin of this effect has not been studied computationally and is not firmly established, these results are consistent with faster rates of reductive elimination from complexes possessing a greater ability of the hydrocarbyl group to coordinate to the metal center during this reductive elimination process.

$$(8.63)$$

Like the rates of reductive eliminations to form C–H and C–C bonds, the rates of reductive eliminations to form carbon–heteroatom bonds depend on the coordination number of the metal. The reductive elimination to form C–N bonds from Pd(0) has been shown to occur faster from three-coordinate complexes than from four-coordinate complexes.[113,136] The reaction of the triphenylphosphine complex in Equation 8.63 to form triarylamine and Pd(0) was conducted with varying concentrations of added ligand. The rate of the reaction was slower when conducted with higher concentrations of added PPh$_3$. A detailed study of the dependence of the rate of reaction on the concentration of added PPh$_3$ revealed two pathways for reductive elimination of amine—one from a four-coordinate cis complex and one from a three-coordinate complex formed by dissociation of phosphine. Although the relative rates of these two pathways depend on the concentration of added ligand, reaction through the three-coordinate intermediate was the major pathway at the concentrations of free ligand that would be present in most reactions.

$$(8.64)$$

The steric and electronic properties of the phosphine ligand have a large effect on the rate of this type of reductive elimination. The effect of these properties on the rate of reductive elimination to form carbon–oxygen bonds in biaryl ethers and carbon–nitrogen bonds in triarylamines has been studied in detail.[114,137] These results demonstrate that complexes containing more hindered phosphines, such as those bearing *tert*-butyl groups at phosphorus, react much faster than those containing less hindered phosphines. In fact, ligands containing substituents the size of *tert*-butyl groups were required to observe reductive elimination of biaryl ethers from aryl palladium complexes lacking strongly electron-withdrawing substituents on the palladium-bound aryl ring (Equation 8.64).[114] Similarly, arylpalladium diarylamido complexes ligated by P(*t*-Bu)$_3$ underwent reductive elimination much faster than those ligated by arylphosphines (Equation 8.65).[137] Additionally, in the absence of steric effects, arylpalladium phenoxide complexes of phosphine ligands that are less electron donating underwent reductive elimination of biaryl ethers faster than those of phosphine ligands that are more electron donating.[114]

(8.65)

(The analogous complex ligated by two PPh$_3$ ligands does not react at this temperature.)

8.2.4.2.3. Reductive Eliminations to Form C–X Bonds from Acyl Complexes

Reductive elimination to form carbon–heteroatom bonds from acyl complexes occurs more readily than it does from alkyl or aryl complexes. This fast rate is consistent with the trends noted previously—namely, reductive elimination is faster from complexes containing groups that are sp^2 hybridized, that are electrophilic, and that can coordinate to the metal center during the reductive elimination. Reductive elimination of acid halides has been proposed as a step in the Monsanto acetic acid process, and experimental evidence has been obtained for this type of reductive elimination from acetylrhodium halide complexes generated in solution (Equation 8.66).[138,139] A closely related incorporation of ^{13}CO into acid halides catalyzed by Wilkinson's catalyst provides further evidence for reversible addition and elimination of acid halides involving Rh(I) and Rh(III), respectively.[140,141]

In addition, reductive elimination of palladium and nickel complexes to form esters (Equations 8.67 and 8.68), amides, and thioesters has been reported.[93,111,142–144] The reductive eliminations of esters and amides were observed during mechanistic studies on the palladium-catalyzed formation of esters and amides from aryl halides, carbon monoxide, and alcohols or amines.[145] This catalytic process is presented in Chapter 17 (carbonylation processes). The reductive elimination of thioesters from nickel complexes were studied, in part, to understand the C–S bond-forming process of acetyl coenzyme A synthase.[144] Prior to this work, an iron-mediated synthesis of β-lactams had been reported that appears to occur by reductive elimination to form the amide C–N bond of the lactam.[146,147]

(8.66)

$$(8.67)$$

$$(8.68)$$

8.3. Summary

Reductive elimination is the product-forming step in some of the most important catalytic cycles, including hydrogenation, the Monsanto acetic acid process, and various types of cross-couplings. For this reason, detailed studies of this process have been conducted. These studies have revealed examples of reductive eliminations to form H–H and C–H bonds, as well as reductive eliminations to form C–C and C–X bonds (in which X = halide, amide, alkoxide, thiolate, and phosphide). The mechanisms of these processes include the same pathways as have been deduced for oxidative addition (i.e., concerted, ionic, and radical), because reductive elimination is the same as oxidative addition, but in the reverse direction.

In general, reductive eliminations are more favorable from complexes containing electron-poor metal centers, from complexes containing sterically hindered reacting or ancillary ligands, and from complexes containing odd coordination numbers. Reductive eliminations by concerted pathways are usually faster from arylmetal, vinylmetal, and acylmetal complexes than from alkylmetal complexes, but reductive eliminations to form alkyl-heteroatom bonds can be facile from alkylmetal complexes through ionic and radical pathways. Reductive eliminations to form C–H and H–H bonds tend to be fastest because of the favorable overlap with the hydride ligand, as long as they are thermodynamically favored, while reductive eliminations to form C–C and C–X bonds are slower. Furthermore, concerted reductive eliminations to form C–C bonds tend to be faster than concerted reductive eliminations to form C–X bonds. Concerted reductive eliminations to form C–X bonds are faster from complexes containing more electron-donating X groups bound by more covalent M–X bonds, while reductive eliminations to form C–X bonds through ionic intermediates are faster from complexes containing less electron-donating X groups that form more stable free anions and undergo heterolytic C–X bond cleavage. Reductive elimination reactions are known for complexes of all transition metals, although reductive eliminations from rhodium, iridium, nickel, platinum, and palladium have been studied most extensively.

References and Notes

1. Mann, G.; Barañano, D.; Hartwig, J. F.; Rheingold, A. L.; Guzei, I. A. *J. Am. Chem. Soc.* **1998**, *120*, 9205.
2. Low, J. J.; Goddard, W. A. I. *J. Am. Chem. Soc.* **1984**, *106*, 8321.
3. Low, J. J.; Goddard, W. A. I. *J. Am. Chem. Soc.* **1986**, *108*, 6115.
4. Tatsumi, K.; Hoffman, R.; Yamamoto, A.; Stille, J. K. *Bull. Chem. Soc. Jpn.* **1981**, *54*, 1857.
5. Albright, T. A.; Burdett, J. K.; Whangbo, M. H. *Orbital Interactions in Chemistry*; Wiley and Sons, Inc.: New York, 1985; p 339.

6. Janowicz, A. H.; Bergman, R. G. *J. Am. Chem. Soc.* **1982**, *104*, 352.

7. Janowicz, A. H.; Bergman, R. G. *J. Am. Chem. Soc.* **1983**, *105*, 3929.

8. Baker, M. V.; Field, L. D. *J. Am. Chem. Soc.* **1987**, *109*, 2825.

9. Wink, D. A.; Ford, P. C. *J. Am. Chem. Soc.* **1986**, *108*, 4838.

10. Bartlett, K. L.; Goldberg, K. I.; Borden, W. T. *J. Am. Chem. Soc.* **2000**, *122*, 1456.

11. Bartlett, K. L.; Goldberg, K. I.; Borden, W. T. *Organometallics* **2001**, *20*, 2669.

12. Weiller, B. H.; Wasserman, E. P.; Bergman, R. G.; Moore, C. B.; Pimentel, G. C. *J. Am. Chem. Soc.* **1989**, *111*, 8288.

13. Zhong, H. A.; Labinger, J. A.; Bercaw, J. E. *J. Am. Chem. Soc.* **2002**, *124*, 1378.

14. Wik, B. J.; Lersch, M.; Tilset, M. *J. Am. Chem. Soc.* **2002**, *124*, 12116.

15. Buchanan, J. M.; Stryker, J. M.; Bergman, R. G. *J. Am. Chem. Soc.* **1986**, *108*, 1537.

16. Abis, L.; Sen, A.; Halpern, J. *J. Am. Chem. Soc.* **1978**, *100*, 2915.

17. Abis, L.; Santi, R.; Halpern, J. *J. Organomet. Chem.* **1981**, *215*, 263.

18. Hartwig, J. F.; Andersen, R. A.; Bergman, R. G. *J. Am. Chem. Soc.* **1991**, *113*, 6492.

19. Buchanan, J. M.; Stryker, J. M.; Bergman, R. G. *J. Am. Chem. Soc.* **1986**, *108*, 1537.

20. Basato, M.; Morandini, F.; Longato, B.; Bresadola, S. *Inorg. Chem.* **1984**, *23*, 649.

21. Basato, M.; Longato, B.; Morandini, F.; Bresadola, S. *Inorg. Chem.* **1984**, *23*, 3972.

22. Milstein, D. *Acc. Chem. Res.* **1984**, *17*, 221.

23. Milstein, D.; Fultz, W. C.; Calabrese, J. C. *J. Am. Chem. Soc.* **1986**, *108*, 1336.

24. Gell, K. I.; Schwartz, J. *J. Am. Chem. Soc.* **1981**, *103*, 2687.

25. Periana, R. A.; Bergman, R. G. *J. Am. Chem. Soc.* **1986**, *108*, 7332.

26. For an example of H/D exchange thought to involve a benzyne intermediate, see reference 27.

27. Sanner, R. D.; Miller, F. D. *Organometallics* **1988**, *7*, 818.

28. Bullock, R. M.; Headford, C. E. L.; Hennessy, K. M.; Kegley, S. E.; Norton, J. R. *J. Am. Chem. Soc.* **1989**, *111*, 3897.

29. Parkin, G.; Bercaw, J. E. *Organometallics* **1989**, *8*, 1172.

30. Churchill, D. G.; Janak, K. E.; Wittenberg, J. S.; Parkin, G. *J. Am. Chem. Soc.* **2003**, *125*, 1403.

31. Gould, G. L.; Heinekey, D. M. *J. Am. Chem. Soc.* **1989**, *111*, 5502.

32. Stahl, S. S.; Labinger, J. A.; Bercaw, J. E. *J. Am. Chem. Soc.* **1996**, *118*, 5961.

33. Johansson, L.; Ryan, O. B.; Tilset, M. *J. Am. Chem. Soc.* **1999**, *121*, 1974.

34. Northcutt, T. O.; Wick, D. D.; Vetter, A. J.; Jones, W. D. *J. Am. Chem. Soc.* **2001**, *123*, 7257.

35. Flood, T. C.; Janak, K. E.; Iimura, H.; Zhen, H. *J. Am. Chem. Soc.* **2000**, *122*, 6783.

36. Howarth, O. W.; McAteer, C. H.; Moore, P.; Morris, G. E. *J. Chem. Soc., Dalton Trans.* **1984**, 1171.

37. Jones, W. D.; Feher, F. J. *J. Am. Chem. Soc.* **1985**, *107*, 620.

38. Jones, W. D.; Feher, F. J. *J. Am. Chem. Soc.* **1986**, *108*, 4814.

39. Bullock, R. M.; Bender, B. R. In *Encyclopedia of Catalysis*; Horváth, I. T., Ed.; Wiley: New York, 2003.

40. Jones, W. D. *Acc. Chem. Res.* **2003**, *36*, 140.

41. Jones, W. D.; Feher, F. J. *J. Am. Chem. Soc.* **1985**, *106*, 1650.

42. Iverson, C. N.; Lachicotte, R. J.; Muller, C.; Jones, W. D. *Organometallics* **2002**, *21*, 5320.

43. Reinartz, S.; White, P. S.; Brookhart, M.; Templeton, J. L. *J. Am. Chem. Soc.* **2001**, *123*, 12724.

44. Chin, R. M.; Dong, L.; Duckett, S. B.; Partridge, M. G.; Jones, W. D.; Perutz, R. N. *J. Am. Chem. Soc.* **1993**, *115*, 7685.

45. Jones, W. D.; Kuykendall, V. L. *Inorg. Chem.* **1991**, *30*, 2615.

46. Hofmann, P.; Heiss, H.; Neiteler, P.; Muller, G.; Lachmann, J. *Angew. Chem., Int. Ed. Engl.* **1990**, *29*, 880.

47. Labella, L.; Chernega, A.; Green, M. L. H. *J. Chem. Soc., Dalton Trans.* **1995**, 395.

48. Green, J. C.; Jardine, C. N. *J. Chem. Soc., Dalton Trans.* **1998**, 1057.

49. Glueck, D. S.; Newman Winslow, L. J.; Bergman, R. G. *Organometallics* **1991**, *10*, 1462.

50. Cowan, R. L.; Trogler, W. C. *J. Am. Chem. Soc.* **1989**, *111*, 4750.

51. Driver, M. J.; Hartwig, J. F. *Organometallics* **1998**, *17*, 1134.

52. Brown, J. M.; Cooley, N. A. *Chem. Rev.* **1988**, *88*, 1031.

53. Milstein, D.; Stille, J. K. *J. Am. Chem. Soc.* **1979**, *101*, 4981.

54. Moncarz, J. R.; Brunker, T. J.; Glueck, D. S.; Sommer, R. D.; Rheingold, A. L. *J. Am. Chem. Soc.* **2003**, *125*, 1180.

55. Evitt, E. R.; Bergman, R. G. *J. Am. Chem. Soc.* **1980**, *102*, 7003.

56. Jawad, J. K.; Puddephatt, R. J.; Stalteri, M. A. *Inorg. Chem.* **1982**, *21*, 332.

57. Jawad, J. K.; Puddephatt, R. J. *J. Chem. Soc., Chem. Commun.* **1977**, 892.

58. Suggs, J. W.; Wovkulich, M. J.; Cox, S. D. *Organometallics* **1985**, *4*, 1101.

59. Saunders, D. R.; Mawby, R. J. *J. Chem. Soc., Dalton Trans.* **1984**, 2133.

60. Brown, M. P.; Puddephatt, R. J.; Upton, C. E. E.; Lavington, S.W. *J. Chem. Soc., Dalton Trans.* **1974**, 1613.

61. Braterman, P. S.; Cross, R. J.; Young, G. B. *J. Chem. Soc., Dalton Trans.* **1977**, 1892.

62. Culkin, D. A.; Hartwig, J. F. *J. Am. Chem. Soc.* **2001**, *123*, 5816.

63. Culkin, D. A.; Hartwig, J. F. *J. Am Chem. Soc.* **2002**, *124*, 234.

64. Shekhar, S.; Hartwig, J. F. *J. Am. Chem. Soc.* **2004**, *126*, 13016.

65. Dicosimo, R.; Whitesides, G. M. *J. Am. Chem. Soc.* **1982**, *104*, 3601.

66. Komiya, S.; Abe, Y.; Yamamoto, A.; Yamamoto, T. *Organometallics* **1983**, *2*, 1466.

67. Tatsumi, K.; Nakamura, A.; Komiya, S.; Yamamoto, A.; Yamamoto, T. *J. Am. Chem. Soc.* **1984**, *106*, 8181.

68. Brown, J. M.; Guiry, P. J. *Inorg. Chim. Acta* **1994**, *220*, 249.

69. *Handbook of Organopalladium Chemistry for Organic Synthesis*; Negishi, E.-i., Ed ; Wiley-Interscience: New York, 2002.

70. *Metal-Catalyzed Cross-Coupling Reactions*; Diederich, F.; Stang, P. J., Eds.; Wiley-VCH: Weinheim, 1998.

71. *Metal-Catalyzed Cross-Coupling Reactions*; de Meijere, A.; Diederich, F., Eds.; Wiley-VCH: Weinheim, 2004.

72. Avedissian, H.; Berillon, L.; Cahiez, G.; Knochel, P. *Tetrahedron Lett.* **1998**, *39*, 6163.

73. Cahiez, G.; Avedissian, H. *Tetrahedron Lett.* **1998**, *39*, 6159.

74. Le Gall, E.; Gosmini, C.; Nedelec, J. Y.; Perichon, J. *Tetrahedron Lett.* **2001**, *42*, 267.

75. Gomes, P.; Fillon, H.; Gosmini, C.; Labbe, E.; Perichon, J. *Tetrahedron* **2002**, *58*, 8417.

76. Tamura, M.; Kochi, J. *J. Organomet. Chem.* **1971**, *31*, 289.

77. Tamura, M.; Kochi, J. *J. Am. Chem. Soc.* **1971**, *93*, 1487.

78. Kochi, J. K. *Acc. Chem. Res.* **1974**, *7*, 351.

79. Smith, R. S.; Kochi, J. K. *J. Org. Chem.* **1976**, *41*, 502.

80. Molander, G. A.; Rahn, B. J.; Shubert, D. C.; Bonde, S. E. *Tetrahedron Lett.* **1983**, *24*, 5449.

81. Cahiez, G.; Marquais, S. *Tetrahedron Lett.* **1996**, *37*, 1773.

82. Cahiez, G.; Avedissian, H. *Synthesis* **1998**, 1199.

83. Dohle, W.; Kopp, F.; Cahiez, G.; Knochel, P. *Synlett* **2001**, 1901.

84. Furstner, A.; Leitner, A.; Mendez, M.; Krause, H. *J. Am. Chem. Soc.* **2002**, *124*, 13856.

85. Quintin, J.; Franck, X.; Hocquemiller, R.; Figadere, B. *Tetrahedron Lett.* **2002**, *43*, 3547.

86. Furstner, A.; Mendez, M. *Angew. Chem. Int. Ed.* **2003**, *42*, 5355.

87. Duplais, C.; Bures, F.; Sapountzis, I.; Korn, T. J.; Cahiez, G.; Knochel, P. *Angew. Chem. Int. Ed.* **2004**, *43*, 2968.

88. Martin, R.; Furstner, A. *Angew. Chem. Int. Ed.* **2004**, *43*, 3955.

89. Scheiper, B.; Bonnekessel, M.; Krause, H.; Furstner, A. *J. Org. Chem.* **2004**, *69*, 3943.

90. Matsubara, T.; Koga, N.; Musaev, D.; Morokuma, K. *Organometallics* **2000**, *19*, 2318.

91. Matsubara, T.; Koga, N.; Musaev, D. G.; Morokuma, K. *J. Am. Chem. Soc.* **1998**, *120*, 12692.

92. Gillie, A.; Stille, J. K. *J. Am. Chem. Soc.* **1980**, *102*, 4933.

93. Komiya, S.; Yamamoto, A.; Yamamoto, T. *Chem. Lett.* **1981**, 193.

94. Moraviskiy, A.; Stille, J. K. *J. Am. Chem. Soc.* **1981**, *103*, 4182.

95. Ozawa, F.; Ito, T.; Nakamura, Y.; Yamamoto, A. *Bull. Chem. Soc. Jpn.* **1981**, *54*, 1868.

96. Ozawa, F.; Yamamoto, A. *Nippon Kagaku Kaishi* **1987**, *5*, 773.

97. Canty, A. J. *Acc. Chem. Res.* **1992**, *25*, 83.

98. Yamamoto, T.; Yamamoto, A.; Ikeda, S. *J. Am. Chem. Soc.* **1971**, *93*, 3350.

99. Binger, P.; Doyle, M. J. *J. Organomet. Chem.* **1978**, *162*, 195.

100. Goliaszewski, A.; Schwartz, J. *Organometallics* **1985**, *4*, 417.

101. Kurosawa, H.; Emoto, M.; Urabe, A.; Miki, K.; Kasai, N. *J. Am. Chem. Soc.* **1985**, *107*, 8253.

102. Komiya, S.; Akai, Y.; Tanaka, K.; Yamamoto, T.; Yamamoto, A. *Organometallics* **1985**, *4*, 1130.

103. Gillie, A.; Stille, J. K. *J. Am. Chem. Soc.* **1980**, *102*, 4933.

104. Ozawa, F.; Ito, T.; Nakamura, Y.; Yamamoto, A. *Bull. Chem. Soc. Jpn.* **1981**, *54*, 1868.

105. Culkin, D. A.; Hartwig, J. F. *Organometallics* **2004**, *23*, 3398.

106. Brown, M. P.; Puddephatt, R. J.; Upton, C. E. E. *J. Chem. Soc.* **1974**, 2457.

107. Komiya, S.; Albright, T. A.; Hoffmann, R.; Kochi, J. K. *J. Am. Chem. Soc.* **1976**, *98*, 7255.

108. Kuch, P. L.; Tobias, R. S. *J. Organomet. Chem.* **1976**, *122*, 429.

109. Lau, W.; Huffman, J. C.; Kochi, J. K. *Organometallics* **1982**, *1*, 155.

110. Evitt, E. R.; Bergman, R. G. *J. Am. Chem. Soc.* **1980**, *102*, 7003.

111. Kohara, T.; Komiya, S.; Yamamoto, T.; Yamamoto, A. *Chem. Lett.* **1979**, 1513.

112. Srivastava, R. S.; Singh, G.; Nakano, M.; Osakada, K.; Ozawa, F.; Yamamoto, Y. *J. Organomet. Chem.* **1993**, *451*, 221.

113. Driver, M. S.; Hartwig, J. F. *J. Am. Chem. Soc.* **1997**, *119*, 8232.

114. Mann, G.; Shelby, Q.; Roy, A. H.; Hartwig, J. F. *Organometallics* **2003**, 2775.

115. Widenhoefer, R. A.; Buchwald, S. L. *J. Am. Chem. Soc.* **1998**, *120*, 6504.

116. Williams, B. S.; Goldberg, K. I. *J. Am. Chem. Soc.* **2001**, *123*, 2576.

117. (a) Hartwig, J. F. *Acc. Chem. Res.* **1998**, *31*, 852; (b) Hartwig, J. F. *Inorg. Chem,* **2007**, *46*, 1936.

118. Williams, B. S.; Holland, A. W.; Goldberg, K. I. *J. Am. Chem. Soc.* **1999**, *121*, 252.

119. Goldberg, K. I.; Yan, J. Y.; Breitung, E. M. *J. Am. Chem. Soc.* **1995**, *117*, 6889.

120. Goldberg, K. I.; Yan, J. Y.; Winter, E. L. *J. Am. Chem. Soc.* **1994**, *116*, 1573.

121. Ettorre, R. *Inorg. Nucl. Chem. Lett.* **1969**, *5*, 45.

122. Dick, A. R.; Kampf, J. W.; Sanford, M. S. *J. Am. Chem. Soc.* **2005**, *127*, 12790.

123. Mann, G.; Hartwig, J. F.; Driver, M. S.; Fernandez-Rivas, C. *J. Am. Chem. Soc.* **1998**, *120*, 827.

124. Macgregor, S. A.; Neave, G. W.; Smith, C. *Faraday Disc.* **2003**, *124*, 111.

125. Han, R.; Hillhouse, G. L. *J. Am. Chem. Soc.* **1997**, *119*, 8135.

126. Koo, K.; Hillhouse, G. L. *Organometallics* **1995**, *14*, 4421.

127. Koo, K.; Hillhouse, G. L. *Organometallics* **1996**, *15*, 2669.

128. Lin, B. L.; Clough, C. R.; Hillhouse, G. L. *J. Am. Chem. Soc.* **2002**, *124*, 2890.

129. Matsunaga, P. T.; Hillhouse, G. L.; Rheingold, A. L. *J. Am. Chem. Soc.* **1993**, *115*, 2075.

130. Hall, P. W.; Puddephatt, R. J.; Tipper, C. F. H. *J. Organomet. Chem.* **1975**, *84*, 407.

131. Luinstra, G. A.; Wang, L.; Stahl, S. S.; Labinger, J. A.; Bercaw, J. E. *J. Organomet. Chem.* **1995**, *504*, 75.

132. Yamashita, M.; Hartwig, J. F. **2003**, unpublished results.

133. Wicht, D. K.; Kourkine, I. V.; Lew, B. M.; Nthenge, J. M.; Glueck, D. S. *J. Am. Chem. Soc.* **1997**, *119*, 5039.

134. Fryzuk, M. D.; Joshi, K.; Chadha, R. K.; Rettig, S. J. *J. Am. Chem. Soc.* **1991**, *113*, 8724.

135. Gaumont, A. C.; Hursthouse, M. B.; Coles, S. J.; Brown, J. M. *Chem. Commun.* **1999**, 63.

136. Driver, M. S.; Hartwig, J. F. *J. Am. Chem. Soc.* **1995**, *117*, 4708.

137. (a) Yamashita, M.; Hartwig, J. F. *J. Am. Chem. Soc.* **2004**, *126*, 5344; (b) Yamashita, M.; Hartwig, J. F. *J. Am. Chem. Soc.* **2003**, *125*, 16347.

138. Forster, D. *J. Am. Chem. Soc.* **1976**, *98*, 846.

139. Baird, M. C.; Mague, J. T.; Osborn, J. A.; Wilkinson, G. *J. Chem. Soc. A* **1967**, 1347.

140. Kampmeier, J. A.; Mahalingam, S.; Liu, T. Z. *Organometallics* **1986**, *5*, 823.

141. Kampmeier, J. A.; Mahalingam, S. *Organometallics* **1984**, *3*, 489.

142. Komiya, S.; Akai, Y.; Tanaka, K.; Yamamoto, T.; Yamamoto, A. *Organometallics* **1985**, *4*, 1130.

143. Matsunaga, P. T.; Hillhouse, G. L. *Angew. Chem., Int. Ed. Engl.* **1994**, *33*, 1748.

144. Tucci, G. C.; Holm, R. H. *J. Am. Chem. Soc.* **1995**, *117*, 6489.

145. Lin, Y.-S.; Yamamoto, A. *Organometallics* **1998**, *17*, 3466.

146. Wong, P. K.; Madhavarao, M.; Marten, D. F.; Rosenblum, M. *J. Am. Chem. Soc.* **1977**, *99*, 2823.

147. Berryhill, S. R.; Price, T.; Rosenblum, M. *J. Org. Chem.* **1983**, *48*, 158.

CHAPTER 9

Migratory Insertion Reactions

9.1. Overview and Basic Principles

9.1.1. Description of Migratory Insertion and Elimination

Many organometallic complexes react by the insertion of an unsaturated ligand Y into an adjacent metal–ligand bond, M–X. After the insertion step, a Lewis base, L, often binds to the vacant coordination site that is created by the insertion process to generate the final reaction product, as depicted in Equation 9.1. The unsaturated ligand Y can be carbon monoxide, an olefin, an alkyne, a ketone, an aldehyde, an imine, carbon dioxide, or related unsaturated species. This general class of insertion reaction is known for all of the transition metals, and such insertions of a coordinated ligand are typically called "migratory insertions."

Many organometallic complexes also react by the microscopic reverse of these steps: dissociation of ligand, followed by de-insertion of fragment Y to generate a product containing an unsaturated ligand Y and a new M–X bond. This reverse reaction has several names, depending on the groups Y and X. One common class of this reverse reaction occurs when Y = CO, and this reaction is typically called a "de-insertion of CO" or "decarbonylation." In many cases, X = H, and the reaction is called "β-hydrogen elimination." Related reactions when X = alkyl or aryl are known, and these reactions are called "β-alkyl elimination" or "β-aryl elimination," respectively. The term "β-elimination" is used because the group undergoing the elimination process starts the reaction attached to the atom located β to the metal center. These de-insertion or elimination reactions are presented in Chapter 10.

The migration of one ligand to another within the coordination sphere of the metal is distinct from the mechanism of reactions that form products containing the same connectivity (but with different stereochemistry) by intermolecular attack of a nucleophile (Equation 9.2) or electrophile onto the coordinated ligand M–Y. The reactions occurring by nucleophilic or electrophilic attack onto M–Y are discussed in Chapters 11 and 12.

$$
\begin{array}{ccc}
\overset{\displaystyle X}{\underset{\displaystyle d^n}{\overset{|}{M}}-Y} \; \rightleftharpoons & [M-Y-X] & \xrightarrow{\;\;L\;\;} \; \overset{\displaystyle L}{\underset{\displaystyle d^n}{\overset{|}{M}}-Y-X} \\
 & d^n &
\end{array}
$$

L = Lewis base

$$
Y = CO, \quad \text{C=C}, \quad -C\equiv C-, \quad =C\overset{}{\underset{}{}}
$$

$$
X = H, R, Ar, \quad \overset{O}{\underset{R}{\|}}\!\!\!\!, \quad \text{etc.}
$$

(9.1)

$$
M-Y \;\; + \;\; \ddot{X} \longrightarrow M-Y-X
$$

(9.2)

Migratory insertions are one step of many different types of catalytic processes, several of which are conducted on large industrial scales and are presented in later chapters of this text. For example, the mechanism of carbonylation processes, such as hydroformylation, includes the insertion of CO into a metal–carbon bond. Likewise, catalytic hydrogenation occurs by insertion of an olefin into a metal–hydride bond, and olefin polymerizations and couplings of olefins with haloarenes occur by insertions of olefins into metal–carbon bonds. The reverse of these reactions, β-hydride, β-alkyl, and β-aryl eliminations, are principal pathways for the decomposition of metal–alkyl complexes.

9.1.2. Changes in Geometry and Electron Count During Migratory Insertion and Elimination

Like oxidative addition and reductive elimination reactions, migratory insertion reactions lead to changes in the electron count and coordination number of the metal center. However, migratory insertion does not change the oxidation state of the metal. Furthermore, one can imagine migration of one group into the M–X bond or migration of X onto Y, various stereochemical outcomes from the insertion, geometries that would allow or prevent migratory insertion, and additives that could affect the rates of these reactions. The following list summarizes the effect of the coordination number, oxidation state, and overall electron count of the starting organometallic complex on insertion and elimination reactions, and provides a few general statements on factors that influence the rate of insertion.

1. Migratory insertion does not lead to a change in formal oxidation state, unless the inserting ligand Y is an alkylidene, alkylidyne, or isoelectronic ligand bound by a metal–ligand multiple bond.
2. The groups undergoing the migratory insertion process must be coordinated cis to each other within the coordination sphere of the metal.
3. A vacant coordination site is created during the forward insertion reaction of Equation 9.1, and a vacant coordination site must be present for the reverse reaction to occur. This mechanism makes it necessary for coordinatively saturated (18-electron) metal–alkyl complexes to dissociate a ligand prior to β-hydrogen and β-hydrocarbyl elimination reactions.
4. Migratory insertion occurs by a concerted process and, therefore, proceeds with retention of configuration at the α-carbon of a migrating alkyl group.
5. Ligand X can migrate onto Y, creating an open site on the metal at the original position of ligand X, or the unsaturated group Y can insert into the M–X bond, creating an open site on the metal at the original position of group Y.
6. The position of the insertion equilibrium depends upon the strengths of the M–X, M–Y, M–(YX) bonds and M(Y–X) bonds.
7. One-electron oxidation, coordination of Lewis acids, and attack by Lewis bases can accelerate the insertion processes.

9.2. Specific Classes of Insertions

The migratory insertions described in this chapter are divided into two broad classes: insertions of ligands bound through a single atom (e.g., carbon monoxide) and insertions of ligands bound through two atoms (e.g., olefins) or more (e.g., dienes). The following sections will describe the insertions of each type of ligand. The reactions of ligands bound to the metal by one atom are presented first, with a particular emphasis on insertions of carbon monoxide. The reactions of ligands bound to the metal by two or more atoms are presented second, including the insertions of olefins, alkynes, imines, ketones, and aldehydes.

9.2.1. Insertions of Ligands Bound by a Single Atom

9.2.1.1. Insertions of Carbon Monoxide

The transformation of an alkylmetal carbonyl complex into a metal–acyl complex is one of the most common types of migratory insertion reactions (Equation 9.3). Examples of CO insertion into metal–alkyl complexes are known for all of the transition elements. This reaction class has been the subject of review articles.[1] These reactions occur by a family of diverse, delicately balanced reaction pathways; the dominant mechanism depends on the reaction conditions, especially the solvent. Although these pathways are now understood in considerable detail, the precise identities of the intermediates in some of these reaction pathways are unknown.

$$L_nM \overset{R}{\underset{CO}{\big\langle}} \quad \xrightarrow{\;L\;} \quad L_nM \overset{L}{\underset{R}{\big\langle}}{=}O \tag{9.3}$$

9.2.1.1.1. Examples of CO Insertions into Metal–Hydrocarbyl Complexes

As noted in the previous paragraph, insertions of CO into metal–alkyl complexes are known for all of the transition metals. Two examples of these reactions are shown in Equations 9.4 and 9.5. The insertion of CO into the d^6 alkylmanganese carbonyl complex in Equation 9.4 is one of earliest CO insertion processes to be identified.[6] Insertion of CO involving the classic "Fp" alkyl complex in Equation 9.5, which also contains a d^6 metal center, is a second well-studied example of this class of reaction.[7]

$$\text{(Equation 9.4 structures)} \tag{9.4}$$

$$L = NH_2,Cy \text{ or } PPh_3$$

$$\text{(Equation 9.5 structures)} \tag{9.5}$$

These insertions are thought to occur by a migratory insertion mechanism for reasons presented in detail in the following sections. However, one initial piece of evidence involves the apparent need for the carbonyl and alkyl group to be cis to each other. This stereochemical requirement is illustrated by slow reactivity of complexes containing planar tetradentate macrocyclic ligands. The alkyl and acyl derivatives shown in Equation 9.6 undergo neither insertion nor de-insertion. The acetonitrile dissociates readily, but the trans disposition of the alkyl group and the open coordination site severely retards the rates of these insertions and de-insertions.[8]

$$(9.6)$$

Many types of Lewis bases can serve as the incoming ligand, L, in migratory insertions. When L is CO, isotopic labeling studies have repeatedly shown that the labeled, external ^{13}CO is not contained in the acyl ligand unless extensive exchange between free and coordinated CO occurs before the insertion takes place. The insertion process has also been conducted with phosphines, amines, halides, and even metal anions as the ligand that occupies the open coordination site in the initial insertion product. As described in more detail in later sections of this chapter, the rate of the overall reaction depends on the incoming ligand if the CO insertion step of Equation 9.1 is reversible, but the rate is independent of the incoming ligand if the CO insertion of this sequence is irreversible.

Insertion reactions have been observed with metal complexes possessing various d-electron counts. The insertion of CO into the d^8, cobalt(I) carbonyl alkyl complexes occurs during hydroformylation, and a model of this insertion process is shown in Equation 9.7.[9,10] One classic example of CO insertions into a square-planar, d^8 metal–aryl complex[11–14] is shown in Equation 9.8.[11] This step is involved in the synthesis of ketones, esters, and amides from aryl halides, CO and main group organometallic reagents, alcohols, or amines. This work showed that CO displaces the phosphine, not the halide, prior to the insertion event and that the complex containing the more labile PPh_3 as the dative ligand reacts faster than the complex containing the less hindered, more donating and less labile PEt_3 as the dative ligand. Equation 9.9 shows the insertion of CO into the electrophilic zirconocene dimethyl to form an alkyl acyl complex containing an η^2-acyl ligand in which the oxygen of the CO is strongly bound to the hard, electrophilic zirconium(IV).[15]

$$(9.7)$$

M = Pd or Pt
X = Cl, Br, I

$$(9.8)$$

$$(9.9)$$

9.2.1.1.2. Examples of Insertions of CO into M–X Bonds (X = N, O, and Si)

The insertion of CO into metal–alkoxide, –amide, and –silyl bonds is much less common than the insertion of CO into metal–alkyl and metal–aryl bonds. Selected insertions of CO into metal alkoxides are shown in Equations 9.10 and 9.11. The insertion of CO into the platinum alkoxide is believed to occur by a migratory insertion pathway;[16] the insertion of CO into the iridium alkoxide is believed to occur by associative displacement of the alkoxide, coordination of CO, and nucleophilic attack of the alkoxide on the coordinated CO.[17] Theoretical studies are consistent with a migratory insertion pathway for reaction of the platinum alkoxide.[18]

$$(dppe)Pt\underset{Me}{\overset{OMe}{<}} \xrightarrow{CO} (dppe)Pt\underset{Me}{\overset{O}{\underset{}{\overset{\|}{C}}}} OMe \qquad (9.10)$$

$$trans\text{-}(MeO)Ir(CO)(PPh_3)_2 + 2\ CO \longrightarrow [Ir(CO)_3(PPh_3)_2]^{\oplus}\ OMe^{\ominus} \longrightarrow [MeOC(O)]Ir(CO)_2(PPh_3)_2 \qquad (9.11)$$

A few examples of the insertion of CO into transition metal amido complexes are also known.[19,20] Two such examples[20] are shown in Equations 9.12 and 9.13. The mechanism for reaction of the ruthenium complex in Equation 9.12 is unknown and is particularly unclear because the starting complex is an 18-electron species possessing a bidentate ligand. Perhaps ring slip allows coordination of CO. The mechanism of the insertion process in Equation 9.13 also has not been studied, but migratory insertion likely occurs after cleavage of the starting dimer with CO. In addition to these reactions of transition metal amido complexes with CO, the reaction of a thorium–amido complex with CO to generate a κ^2-carbamoyl complex has been reported (Equation 9.14). Insertion of CO into the N–H bond of a ruthenium–amido complex, instead of the Ru–N bond, is also known.[21,22]

$$(9.12)$$

$$\{Cp^*Ni[NH(p\text{-tol})]\}_2 \xrightarrow{2\ CO} 2\ Cp^*Ni\underset{O}{\overset{CO}{<}} NH(p\text{-tol}) \qquad (9.13)$$

$$\xrightarrow[\substack{Toulene \\ M = Th, U}]{0\ ^\circ C,\ 2\ h} \qquad (9.14)$$

The insertion of CO into the M–E bond of more electropositive elements has also been reported. For example, CO insertion into d^0 metal–silyl complexes has been observed (Equations 9.15[23] and 9.16[24]).[25–29] These reactions are driven, at least thermodynamically, by formation of a κ^2 ligand possessing a strong M–O interaction. Insertions of CO into late transition metal silyl complexes that would lack this stabilizing effect of the κ^2 structure are not known.[23] In fact, the 18-electron rhenium κ^1-"acylsilane" complex fac-{Re(CO)$_3$(diphos)[C(O)SiPh$_3$]} undergoes de-insertion of CO upon heating to form fac-[Re(CO)$_3$(diphos)(SiPh$_3$)], and treatment of the silyl complex with 300 atm of CO does not regenerate the acylsilane species.[30]

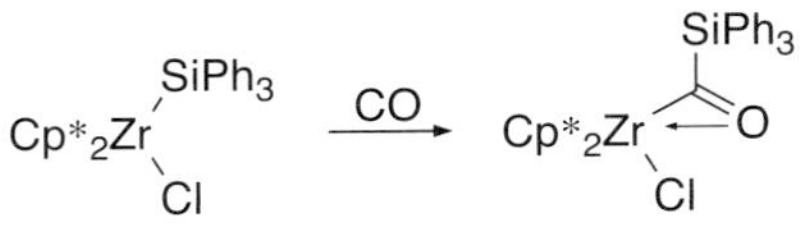

$$\text{Cp*}_2\text{Zr} \overset{\text{SiPh}_3}{\underset{\text{Cl}}{\diagup}} \quad \xrightarrow{\text{CO}} \quad \text{Cp*}_2\text{Zr} \overset{\overset{\text{SiPh}_3}{\|}}{\underset{\text{Cl}}{\diagdown}} \text{O} \tag{9.15}$$

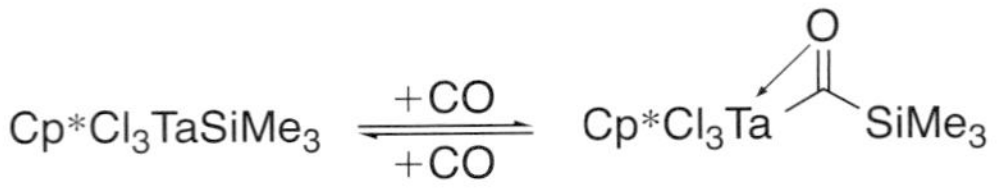

$$\text{Cp*Cl}_3\text{TaSiMe}_3 \underset{+\text{CO}}{\overset{+\text{CO}}{\rightleftharpoons}} \text{Cp*Cl}_3\text{Ta} \overset{\overset{\text{O}}{\|}}{\diagdown} \text{SiMe}_3 \tag{9.16}$$

9.2.1.1.3. Kinetics and Mechanism of CO Insertions into Metal–Alkyl Complexes

9.2.1.1.3.1. Insertions into 18-Electron Complexes

Some of the first kinetic studies on the insertion of CO were conducted on the insertion reactions of the manganese and iron complexes in Equations 9.4 and 9.5.[5,7] These data are consistent with the two-step mechanism in Scheme 9.1. The extent of solvation and the coordination mode (η^1 or η^2) of the acyl ligand in the 16-electron product of the insertion step are unspecified. Experiments that address these issues are presented later in Sections 9.2.1.1.3.5 and 9.2.1.1.3.6.

$$M(CO)R \underset{k_{-1}}{\overset{k_1}{\rightleftharpoons}} M-C(O)R$$

$$M-C(O)R + L \xrightarrow{k_2} M(L)[C(O)R]$$

$$\frac{d[M(L)(CO)R]}{dt} = \frac{k_1 k_2 [M(CO)R][L]}{k_{-1} + k_2 [L]} = k_{obs}[M(CO)R]$$

Scheme 9.1

$$k_{obs} = \frac{k_1 k_2 [L]}{k_{-1} + k_2 [L]} \quad \text{and} \quad \frac{1}{k_{obs}} = \frac{k_{-1}}{k_1 k_2} \cdot \frac{1}{[L]} + \frac{1}{k_1} \tag{9.17}$$

Application of the steady-state approximation to the sequence in Scheme 9.1 leads to the rate law in Equation 9.17. This rate law is common in organometallic chemistry and describes a process in which an intramolecular event leads to an intermediate that reacts with an external reagent, which is the added ligand in this case. The observed reaction order can be first, second, or mixed order, depending on the relative magnitudes of k_{-1} and $k_2[L]$. For example, if k_{-1} is much larger than $k_2[L]$, then the observed rates are overall second order (first order in metal complex and first order in CO). If k_{-1} is much smaller than $k_2[L]$, making the insertion process irreversible, then the reaction is overall first order (first order in metal complex and zero order in CO). If k_{-1} and $k_2[L]$ are similar in magnitude, then the observed reaction order is between first and second. A reaction order between first and second is observed for the reaction of $RMn(CO)_5$ with low pressures of CO to form $[R(O)C]Mn(CO)_5$. First-order kinetics are observed for the reaction of $RMn(CO)_5$ with high pressures of CO. A second-order rate law was observed for the reaction in Equation 9.18 over a wide range of conditions.[31]

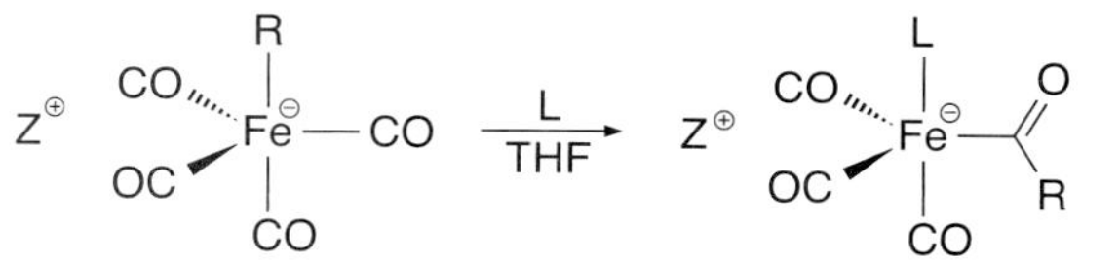

$$Z^{\oplus} \quad \begin{array}{c} CO_{\prime\prime\prime\prime} \\ OC \end{array} \overset{R}{\underset{CO}{\overset{|}{Fe}}} - CO \quad \xrightarrow[THF]{L} \quad Z^{\oplus} \quad \begin{array}{c} CO_{\prime\prime\prime\prime} \\ OC \end{array} \overset{L}{\underset{CO}{\overset{|}{Fe}}} \overset{O}{\underset{R}{\parallel}}$$

(9.18)

$$Z^{\oplus} = Li^{\oplus} > Na^{\oplus} > [(Ph_3P)_2N]^{\oplus}$$
$$L = Me_3P > PhMe_2P > Ph_2MeP > CO$$
$$R = n\text{-alkyl} > PhCH_2$$

In some cases, a one-step mechanism appears to occur in parallel with the two-step mechanism in Scheme 9.1. These parallel mechanisms are revealed by a two-term rate law containing one first-order and one second-order term. An example of a reaction that occurs with such a two-term rate law is shown in Equation 9.19.[32] The rate constant for one of the two terms is insensitive to the polarity or ligating ability of the solvent. Therefore, the insertion process in Equation 9.19 is thought to occur, in part, by a pathway that is initiated by a change in hapticity of the Cp ring (ring slip).

$$CpMo(CO)_3Me + PMePh_2 \longrightarrow CpMo(CO)_2(PMePh_2)[(C(O)Me]$$

(9.19)

9.2.1.1.3.2. Insertions into 16-Electron d^8 Complexes

Carbonylation of four-coordinate d^8 complexes of the type $M(X)(R)(L)_2$ (M = Pt, Pd, and Ni) has been studied extensively.[33] These reactions can occur by several different pathways, three of which are shown in Scheme 9.2. The lowest energy pathway for reactions of complexes containing monodentate phosphines in nonpolar solvents occurs by initial displacement of a phosphine in the starting complex by carbon monoxide to form the square-planar CO complex. Although this complex contains an alkyl group cis to the CO, it rearranges to an isomer containing a phosphine (L) trans to the alkyl ligand prior to insertion. The complex containing a phosphine trans to the alkyl group is thought to be more reactive because the trans effect of the phosphine labilizes the Pt–alkyl bond, and the alkyl group migrates onto CO. This insertion process generates a T-shaped, 14-electron, d^8 intermediate, which coordinates phosphine to form the final product. The ligand displacements and isomerizations in this mechanism presumably take place via the formation of trigonal bipyramidal intermediates and pseudorotation.[34,35]

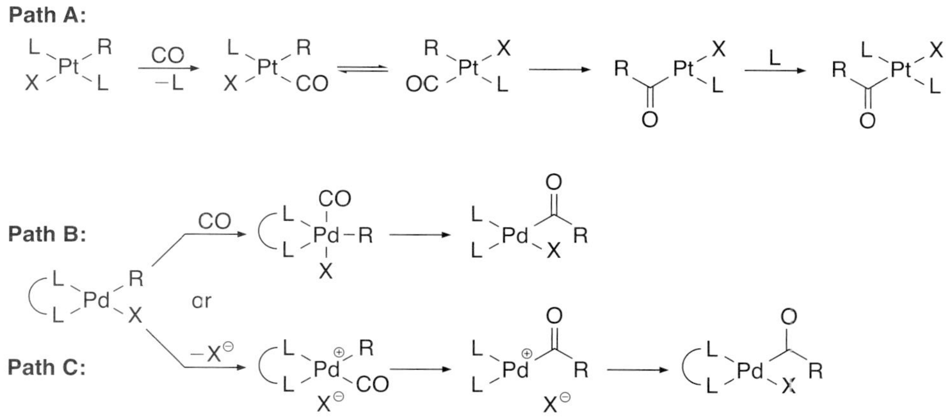

Path A:

Path B:

Path C:

Scheme 9.2

Related complexes containing chelating phosphines undergo insertion more slowly because CO displaces a phosphine in a chelating ligand less readily than it displaces a monodentate phosphine. However, palladium complexes containing bisphosphines do catalyze carbonylation reactions that involve insertion of CO into a metal–alkyl and metal–aryl bond. The CO insertion in these processes follows either Path B or C in Scheme 9.2, involving a five-coordinate or a cationic intermediate, depending on the polarity of the solvent and the coordinating ability of the anionic ligand X. If X is a weakly coordinating ligand, then insertion occurs by displacement of this ligand by CO to form a cationic complex that inserts CO. An example of the direct observation of CO insertion by a related cationic alkyl carbonyl complex is shown in Equation 9.20.[36] If the anion is bound more tightly, then insertion is likely to occur through the five-coordinate intermediate, but at higher temperatures.

$$(phen)Pd\begin{smallmatrix}CO\\CH_3\end{smallmatrix}^{\oplus} \xrightarrow[-66\ °C]{CO,\ CD_2Cl_2} (phen)Pd\begin{smallmatrix}C(O)CH_3\\CO\end{smallmatrix}^{\oplus} \tag{9.20}$$

9.2.1.1.3.3. Stereochemistry at Carbon

Many experiments have shown that CO insertion typically occurs with retention of configuration at the α-carbon.[4] Racemization has been observed in a few cases that occur by radical paths, but inversion has never been observed. The first unequivocal demonstration of the stereochemistry of the α-carbon during insertion[37–39] is illustrated in Equation 9.21. The threo alkyl complex afforts the threo acyl complex.

$$\text{(threo alkyl complex)}\ \begin{smallmatrix}Bu^t\\CpFe(CO)_2\end{smallmatrix} + PPh_3 \xrightarrow{THF} \begin{smallmatrix}Bu^t\\Cp(CO)(PPh_3)Fe\end{smallmatrix} \tag{9.21}$$

The stereochemistry at the α-carbon during carbonylation is often deduced from a two-step sequence, such as oxidative addition followed by carbonylation. Because oxidative addition of nonpolar substrates occurs with retention of configuration, as presented in Chapter 6, the insertion step is considered to proceed with retention of configuration when the overall process occurs with retention of configuration.

$$\xrightarrow[L=PPh_3]{RhClL_3}\ L\ \text{or}\ \text{(product, for X = Me / for X = OMe)} + RhCl(CO)(PPh_3)_2 \tag{9.22}$$

The stereochemistry of CO insertion has also been revealed by studying decarbonylation. Decarbonylation reactions sometimes form products in which racemization or epimerization has occurred at the carbon located α to the carbonyl group; when this stereochemical outcome is observed, a mechanism involving a radical pair is thought to be followed. An example of a system that reacts with epimerization is shown in Equation 9.22. When the substituent X is Me, a high degree of retention of configuration is observed; but when X is OMe, almost complete epimerization occurs.[40]

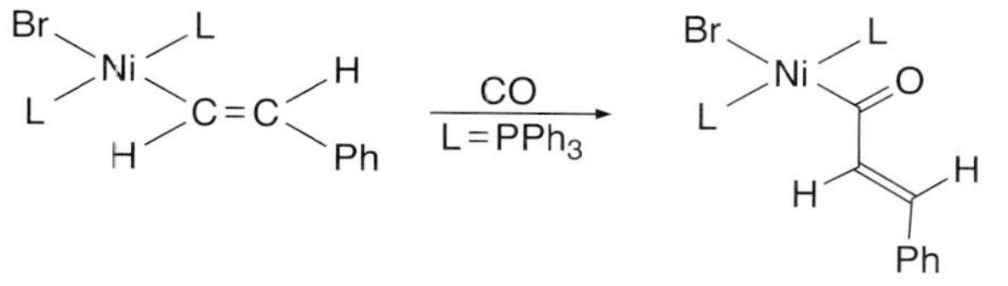

(9.23)

CO insertion into alkenyl–metal bonds also occurs stereospecifically in many cases. An example of insertion of CO into an alkenyl complex is shown in Equation 9.23. Complexes containing either *cis-* or *trans-*CH=CHPh groups insert CO with retention of configuration.[41] Similar stereochemical results have been observed for decarbonylation. For example, the reaction of RhCl(PPh$_3$)$_3$ with *E*-PhCH=C(Et)CHO induces decarbonylation with retention of the double-bond stereochemistry.[40]

9.2.1.1.3.4. Stereochemistry at the Metal

Studies on the stereochemistry of the metal during CO insertion have led to a set of fundamental conclusions about the mechanism of the insertion process. These studies have shown that the alkyl and carbonyl groups that form the acyl ligand are located cis to each other and that the incoming ligand occupies the open coordination site created in the position cis to the newly formed acyl group. These studies have also shown that the alkyl group can migrate to the adjacent CO or the CO can insert into the M–C bond, but that migration of the alkyl group to the CO ligand is more common.

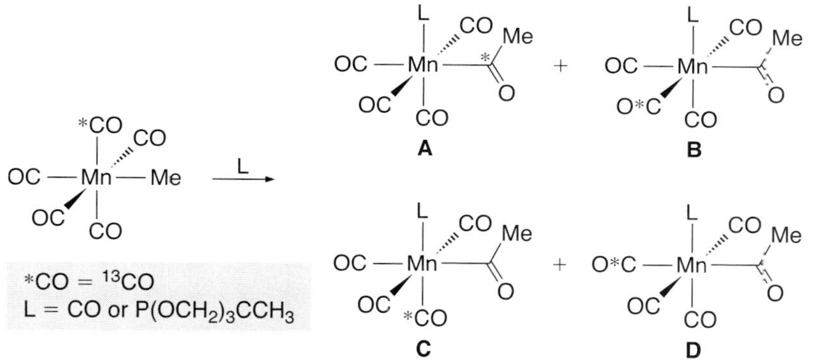

(9.24)

Calderazzo[2,42] first studied the stereochemistry of manganese when MeMn(CO)$_5$ undergoes insertion of CO. He studied the de-insertion of ^{13}CO-labeled complexes and used infrared spectroscopy to determine the position of the ^{13}CO. Flood[4,43] has confirmed Calderazzo's conclusions using ^{13}C NMR spectroscopy. Many permutations of the CO insertion into MeMn(CO)$_5$, and de-insertion of CO from [MeC(O)]Mn(CO)$_5$ have been studied. The reaction in Equation 9.24 illustrates the main points. Methyl migration to the adjacent CO, followed by coordination of L at the position vacated by the migrating methyl group cannot form product **C**; it would form a 1:2:1 ratio of **A:B:D**. In contrast, a formal insertion of CO into the metal–carbon bond, followed by coordination of L at the site previously occupied by CO, would not form **D**; it would form a 1:2:1 ratio of **A:B:C**. With L = CO in acetone/THF, or L = P(OCH$_2$)$_3$CCH$_3$ in either THF or HMPA, the ratio of products is consistent with migration of the methyl group to the adjacent CO. In addition to revealing the ligand that migrates, these results imply that the square pyramidal intermediate scrambles its stereochemistry more slowly than it coordinates the incoming ligand.

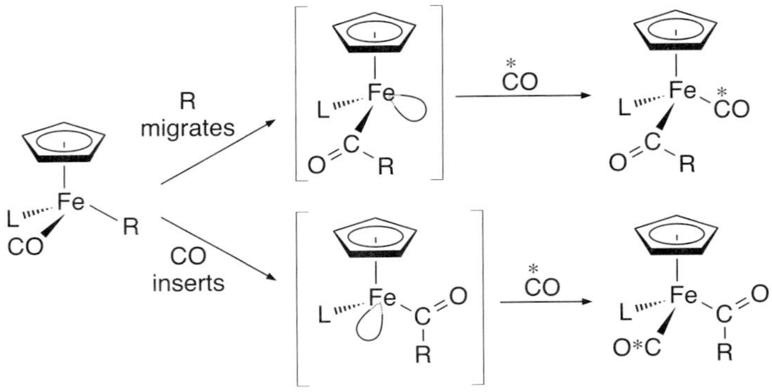

Scheme 9.3

Reactions of complexes containing a stereogenic metal center provide a second means to assess whether the alkyl group migrates to the adjacent CO or CO inserts into the M–C bond.[2,4,5] Scheme 9.3 illustrates the two pathways. When the chiral, pseudotetrahedral reactant forms the acyl product, the pathway occurring by migration of an alkyl group would give the acyl complex (top product) with overall inversion of configuration at iron if the 16-electron intermediate coordinates the incoming ligand faster than it undergoes inversion. In contrast, the pathway involving CO insertion into the M–C bond would form the acyl complex (bottom stereoisomer) with overall retention of configuration at iron if the 16-electron intermediate coordinates the incoming ligand faster than it undergoes inversion. When conducted in nitroethane, the reaction occurred with inversion of configuration at iron, which is consistent with migration of the alkyl group to CO; however, when conducted in HMPA, it formed both stereoisomers, and the isomer resulting from CO insertion predominated.[44,45]

The similar, optically active complex CpFe(CO)(Me)L (in which L = a chiral phosphine) forms an acyl complex at low temperatures when conducted in the presence of CO and added BF_3 to accelerate the rate (see below for the effects of Lewis acids on the rate of insertion). Again the reaction proceeds stereospecifically, and the configuration of the product corresponds to that resulting from migration of the methyl group.[46] The photochemical decarbonylation of a similar iron complex was shown by Davison[47] and then by Wojcicki[48] also to proceed predominantly by formal migration of the alkyl group.

Studies of alkyl carbonyl insertions with pseudooctahedral complexes of Fe,[49–53] Ru,[54] Rh,[55,56] and Ir[57,58] have shown that the stereochemical outcome depends strongly on reaction conditions, that the stereochemical outcome can be affected by the ability of one ligand to affect dissociation of a second ligand, and that ligand rearrangements often obscure the origin of the stereochemical outcome.[2] Thus, the experiments on pseudotetrahedral complexes have been more revealing of the reaction pathway than experiments on related pseudooctahedral complexes.

9.2.1.1.3.5. Structure of the Unsaturated Intermediate

The two-step reaction path in Scheme 9.1 contains an unsaturated, apparently 16-electron, intermediate. For most carbonylation reactions, this intermediate cannot be observed directly. In the absence of strongly coordinating solvents, the oxygen of its acyl ligand could be envisioned to occupy the vacant coordination site in an η^2-acyl complex.[4,45,59] Indeed, examples of such complexes are known,[60,61] including the two in Figure 9.1.[62,63] However, other data indicate that the acyl ligand intermediate is bound η^1. For example, an infrared study of the unsaturated intermediates CpFe(CO)[C(O)Me] and Mn(CO)$_4$[C(O)Me] in methane matrices at 12 K indicated that the acetyl group is bound in an η^1 fashion, rather than an η^2 fashion.[64] Moreover, theoretical studies on the insertion of

CO into $(CO)_4CoMe$ have computed an η^1-acetyl structure in the 16-electron intermediate $MeC(O)Co(CO)_3$ containing an agostic interaction to the acetyl methyl group.[65,66] Finally, stable, coordinatively unsaturated η^1-acyl complexes of rhodium, iridium, and platinum are known.[55,67,68]

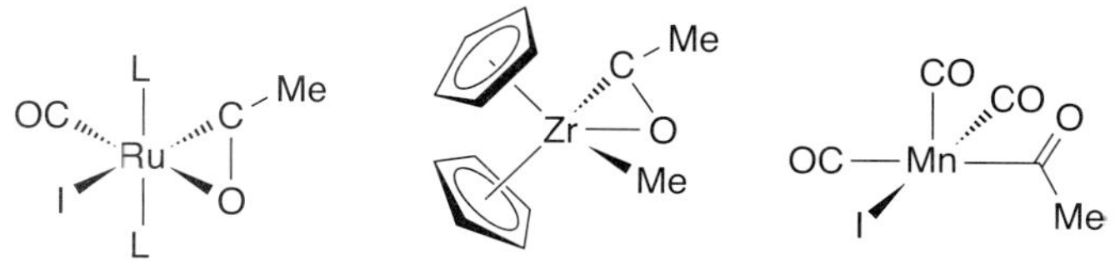

Figure 9.1.
Two η^2-acyl complexes and the structure of a CO insertion product, as determined by low-temperature matrix isolation.

9.2.1.1.3.6. Solvent Effects

The rates of migratory insertion reactions are often strongly affected by the solvent. Insertions of CO conducted in polar solvents are usually faster than those conducted in non-polar solvents. For example, the rate of insertion of CO into the M–C bond of $RMn(CO)_5$ induced by addition of cyclohexylamine is 10^4 times faster in DMF than it is in mesitylene.[69]

The origin of the effect of solvent polarity on this reaction is difficult to pinpoint because the reaction would not be expected to accumulate charge in the transition state. However, several studies suggest that the nucleophilicity of the solvent controls the rate. Bergman[32] studied the kinetic behavior of the reaction in Equation 9.19 in substituted tetrahydrofuran (THF) solvents. This insertion reaction occurs by two competing mechanisms: the normal two-step process shown in Scheme 9.1 and a one-step process by which the added $PMePh_2$ directly attacks the starting complex. Moreover, the rate constants, k_1, for the formation of $M[C(O)R]$ from $M(CO)R$ in THF and its methyl-substituted derivatives decrease in the order THF ≈ 3-methyl THF > 2-methyl THF > 2,5-dimethyl THF. The observed variations are consistent with the relative nucleophilicities of these solvents. The change in k_1 between THF and 2,5-dimethyl THF is a factor of 33.

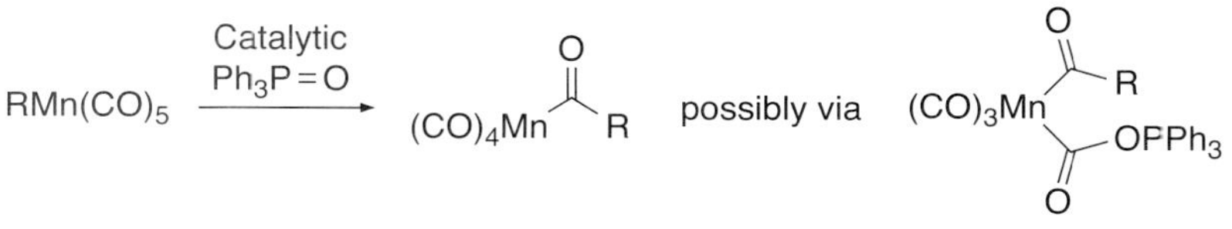

$$RMn(CO)_5 \xrightarrow[\text{Ph}_3\text{P}=\text{O}]{\text{Catalytic}} (CO)_4Mn \overset{O}{\diagup} R \quad \text{possibly via} \quad (CO)_3Mn \tag{9.25}$$

Halpern and co-workers have obtained results that suggest that polar solvents increase the rate constant k_1 in Scheme 9.1 for the formation of $M[C(O)R]$.[70] This effect of polar solvents may be related to the effect of phosphine oxides on migratory insertions and ligand substitutions of coordinated CO. Phosphine oxides have been shown to accelerate the migratory insertions of manganese alkyls (and presumably other alkyls). Halpern's kinetic analysis of the reaction in Equation 9.25 demonstrated that phosphine oxides and some related nucleophiles catalyze the formation of an intermediate that gives rise to the coordinatively unsaturated $RC(O)Mn(CO)_4$. The rate constant for the overall process in Equation 9.25 is faster than the k_1 that corresponds to the unassisted formation of $RC(O)$ $Mn(CO)_4$ from $RMn(CO)_5$. Although the effect of the added ligand is well documented, the nature of the interaction between the phosphine oxide and the starting alkyl complex is unknown (although one is suggested in Equation 9.25).

In some cases, solvent can act as the trapping ligand. For example, reactions of $CpFe(CO)_2R$ with a dative ligand first generate the DMSO complex, which reacts in a second step with the phosphine. In this case, the DMSO complex was directly detected by NMR and IR spectroscopic methods.[71,72]

$$CpFe(CO)_2R \xrightarrow[DMSO]{L} \{CpFe(CO)[C(O)R](DMSO)\} \xrightarrow{L} \{CpFe(CO)L[C(O)R]\} \tag{9.26}$$

Polar solvents can retard migratory insertions by affecting ion pairing. Solvents that decrease the extent of ion pairing decrease the rate of Equation 9.18, and it is thought that this effect is observed because tight ion pairing accelerates the rate of insertion.[31]

9.2.1.1.4. Migratory Aptitudes of R

The migratory aptitudes of the metal-bound group R in carbonyl insertion reactions vary widely. The insertion of CO into some groups is thermodynamically unfavorable, and the migratory aptitudes result from thermodynamic effects, rather than kinetic effects. In other cases, the insertions of CO into two different M–R bonds in the same complex or the insertion of CO into the M–R bonds in two different metal complexes are all favored thermodynamically, and the relative rates are controlled by kinetic effects. Examples of measurements of the relative rates for insertion in these two situations are described in this section.

9.2.1.1.4.1. Thermodynamic Effects on Migratory Aptitudes

In many cases, CO insertion reactions have been shown to be thermodynamically unfavorable. In these cases, de-insertion of CO occurs. For example, acetylmetal complexes do not undergo insertion to form stable α-dicarbonyl compounds. Instead, α-dicarbonyl compounds de-insert CO to form metal–acyl complexes and free CO (Equation 9.27).[73,74] Likewise, the carbonylation of perfluoroalkyl transition metal complexes is usually uphill thermodynamically. Decarbonylation of perfluoroacetyl complexes is one method to prepare perfluoroalkyl complexes (Equation 9.28).[75-77]

$$\text{(9.27)}$$

$$\text{(9.28)}$$

$$\text{(9.29)}$$

For similar reasons, CO rarely inserts into metal hydrides. The M–H bond energy is usually high enough that insertion to form a formyl derivative is thermodynamically uphill. Therefore, decarbonylation of formyl ligands is more typical (Equation 9.29),[78] and formyl complexes of the transition metals are usually prepared by indirect means.[79]

A few exceptions to the unfavorable thermodynamics for insertion of CO into transition metal hydrides have been observed. One exception is shown in Equation 9.30.[80-82] The favorable thermodynamics must result from an unusually strong bond between the formyl carbon and rhodium. A second exception is shown in Equation 9.31, wherein an organoactinide hydride inserts CO to form an η^2-formyl complex. The thermodynamics for this insertion[83] are favorable because of the strong M–O interaction in the η^2-formyl complex to this oxophilic metal.

$$H-Rh(OEP) \; + \; CO \; \longrightarrow \; H \overset{\overset{\displaystyle O}{\|}}{\diagup} Rh(OEP) \tag{9.30}$$
OEP – octaethylporphrin

$$(\eta^5-C_5Me_5)_2ThH(OR) \; \xrightarrow[-78\,°C]{CO} \; (\eta^5-C_5Me_5)_2Th(OR) \tag{9.31}$$

9.2.1.1.4.2. Kinetic Effects on Migratory Aptitudes

When the thermodynamics for insertion are favorable, the relative rates for insertion and de-insertion of CO into different alkyl, aryl, and benzyl groups is controlled by kinetic factors. Several trends have been observed.

$$CpMo(CO)_3CH_2C_6H_4X \; + \; PPh_3 \; \longrightarrow \; CpMo(CO)_2(PPh_3)[C(O)CH_2C_6H_4X] \tag{9.32}$$

First, electron-withdrawing groups on an alkyl or aryl group retard the rate of insertion, and electron-donating groups accelerate insertion. For example, the rates of carbonylation of $RMn(CO)_5$ have been determined for a number of R groups. The results show that the rates of migratory insertion are accelerated by electron-releasing and retarded by electron-withdrawing substituents of the migrating alkyl groups.[84] Similar substituent effects have been found for carbonylation of the benzyl molybdenum complex shown in Equation 9.32[85,86] and for carbonylation of the aryl palladium complexes containing electron-withdrawing substituents shown in Equation 9.8.[11] The rates of the reactions in Equation 9.32 correlated with the electron-releasing character of the para substituents X on the benzyl group.[85,86] Because electron-withdrawing groups tend to strengthen metal–carbon bonds more than they strengthen carbon–carbon bonds, this effect may reflect the relative thermodynamic driving forces for the insertion process.

Second, benzyl groups undergo migratory insertion more slowly than methyl groups. Although a benzyl group might be expected to migrate faster because it can better support an accumulation of charge than a methyl group, the opposite trend is observed. Therefore, the observed trend is best rationalized by slower migration due to the presence of an electron-withdrawing group on the α-carbon.

Third, more sterically demanding alkyl groups undergo insertion faster than less sterically demanding groups. For example, measurements of the relative values of the k_1 step (Scheme 9.1) for CO insertion of $CpFe(CO)_2R$ in DMSO solvent (which acts as the trapping ligand) show that the alkyl complexes react in the following order: $(Me_3Si)_2CH \gg Me_3CCH_2 > sec\text{-}Bu > i\text{-}Pr > Me_3SiCH_2 > CyCH_2 > i\text{-}Bu > Et > Pr > Me$.[71] This trend would also follow the relative driving forces for the insertion process.

Fourth, aryl groups have been shown to migrate more slowly than alkyl groups, although these relative rates have not been measured in many cases. The palladium–methyl and –benzyl complexes in Equation 9.8 underwent carbonylation faster than the corresponding phenyl complex.[11] Moreover, the methyl group migrated faster than the phenyl group from the bis(acyl)rhenium complex shown in Scheme 9.4[87]. This study showed that the phenyl acyl product is more stable than the methyl phenacyl isomer. K_{eq} for

Re(Ph)[C(O)Me]/Re(Me)[C(O)Ph] is 50. However, migration of the methyl group occurs about 30 times faster than migration of the phenyl group.

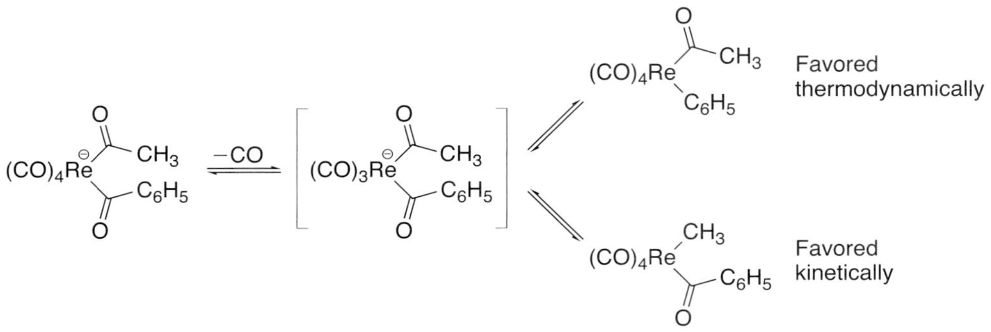

Scheme 9.4

Fifth, the rate of carbonyl insertion as a function of the metal center usually follows the trend $3d > 4d > 5d$ when the complexes are isostructural. Alkyl complexes of the first-row metals are almost always more reactive than the analogs of the third row, and they are usually more reactive than the analogs of the second row. For example, $CpRu(CO)_2Me$ requires higher temperatures to insert CO in the presence of added phosphine than does the analogous iron complex,[88,89] and the osmium analogue fails to react. Similar trends have been observed for other alkyl complexes of metal–carbonyl fragments. For $CpM(CO)_3R$ complexes, the relative rates of insertion are M = Cr and Mo > W, and for $RM(CO)_5$, the relative rates are M = Mn > Re.[57,67,90–97]

Finally, the relative migratory aptitudes of alkyl, aryl, alkoxo, and amido groups have been studied, and the limited data currently available suggest the trend alkoxide > alkyl > aryloxide and arylamide. These trends are revealed by the insertion of CO into the M–O bond of $Pt(DPPE)(Me)(OMe)$[16] and $Pd(DPPE)(Me)[OCH(CF_3)_2]$,[98] by the insertion of CO into the metal–carbon bond of $[Ni(bpy)(Me)(OAr)]$,[99] and by the insertion of CO into the metal–carbon bond of orthometallated ruthenium phenoxo and arylamido complexes.[100] A theoretical study has been conducted on the insertion of CO into the M–O bond of $[Pt(L_2)(Me)(OMe)]$, $[Ni(L_2)(Me)(OMe)]$, and $[Ni(L_2)(Me)(OPh)]$ in which $L_2 = H_2PCH_2CH_2PH_2$ and an α-diimine to model DPPE and bpy. This study suggested that the palladium complexes react by associative displacement of one phosphorus donor of the bisphosphine by CO, and insertion from the resulting four-coordinate complex, while the nickel complex undergoes insertion from the five-coordinate complex generated by coordination of CO.[18]

9.2.1.1.5. Catalysis of CO Insertion

CO insertion into M–C bonds has been shown to be accelerated by coordination of Lewis acids to the carbonyl group and by oxidation of 18-electron complexes to generate 17-electron species. The origin of these effects and examples of such acceleration by additives are described in the following two sections.

9.2.1.1.5.1. Catalysis by Lewis Acids

Several studies show that Lewis acids accelerate the insertion of CO into alkyl groups. This effect has been shown to result from two different reactivities of Lewis acids. In one case, this acceleration is likely related to stronger coordination of Lewis acids to the oxygen of the acyl product than to the oxygen of the CO ligand. In a second case, the acceleration is due to an affect on ion pairing.

Two examples of the effects of Lewis and protic acids on the rate of migratory insertion due to coordination of the acid to the carbonyl oxygen are shown in Equations 9.33 and 9.34. As shown in Equation 9.33, the strong Lewis acids AlX_3 (in which X = Cl or Br) induce the insertion of CO into the alkyl group of $Mn(CO)_5Me$ with a rate acceleration of 10^8.[101, 102] In this case, one halide on aluminum fills the vacant coordination site generated by the migratory insertion, and the aluminum is coordinated to the acyl oxygen. The Lewis acid is incorporated into the product and is, therefore, a reagent, not a catalyst. Protic acids can also accelerate migratory insertion. For example, protic acids accelerate migration of the methyl group in $MeMn(CO)_5$[103] and migration of the ethyl group in the more complex transformation in Equation 9.34.[104]

$$(9.33)$$

$$(9.34)$$

Examples of the acceleration by Lewis acids due to an affect on ion pairing were shown in Equation 9.18. Insertions of CO into the Fe–R bond of the anionic $[RFe(CO)_4]^-$ was faster when a more Lewis acidic counterion was present. Strong ion pairing also increased the rate of the migratory insertion in Equation 9.35.[105] Because the crown ether in the product of Equation 9.35 causes this complex to bind Na^+ more strongly than the reactant, alkali and alkaline earth salts promote the reaction in Equation 9.35. Amphoteric ligands such as $R_2PNRAlR_2$ have also been used to facilitate CO insertions by providing a Lewis acid site (Al) and a basic ligand (P) within the same molecule.[106, 107] Lewis acids have also been shown to accelerate decarbonylation. In this case, the decarbonylation is accelerated by the creation of a vacant coordination site through abstraction of halide from the metal by silver cation.[108]

$$(9.35)$$

9.2.1.1.5.2. Redox Acceleration

Oxidation of the starting alkyl complex can also dramatically accelerate migratory insertion of CO. Oxidation accelerates ligand substitution for CO in 18-electron complexes by formation of a 17-electron species, but it is unclear why oxidation increases the rate of CO insertion. This acceleration is illustrated by the reaction in Equation 9.36. The forward rate constant of this equilibrium increases by at least 10^{11} upon oxidation of the metal from iron(II) to iron(III). As a result K_{eq} is increased by about 10^{11}.[109]

A stoichiometric amount of oxidant can thus convert $CpFe(CO)(L)Me$ to $[CpFe(L)(MeCN)(COMe)]^+$. The similar reaction shown in Equation 9.37 is catalyzed by small amounts of oxidizing agents. Enormous rate enhancements resulted from addition of a few mole percent of oxidizing agents, such as silver ion or ferricinium ion.[110] CO insertions can also be induced by reduction of the metal–alkyl substrate.[111] Redox-promoted and -catalyzed processes may be more prevalent than is generally recognized. Unrecognized adventitious oxidants (such as air) or reductants (such as an organolithium reagent) may accelerate these reactions.

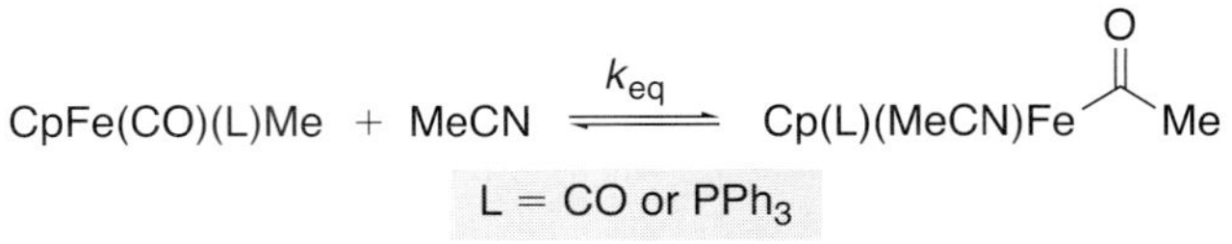

$$CpFe(CO)(L)Me \; + \; MeCN \; \underset{}{\overset{k_{eq}}{\rightleftharpoons}} \; Cp(L)(MeCN)Fe \overset{O}{\underset{}{\diagup\!\!\diagdown}} Me \qquad (9.36)$$

$$L = CO \text{ or } PPh_3$$

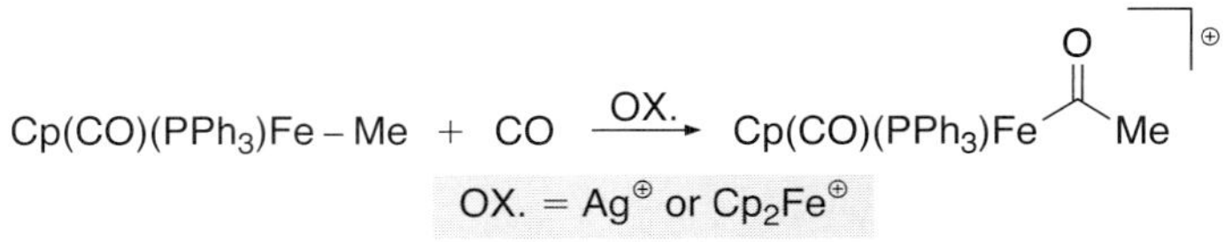

$$Cp(CO)(PPh_3)Fe-Me \; + \; CO \; \xrightarrow{OX.} \; \left[Cp(CO)(PPh_3)Fe \overset{O}{\underset{}{\diagup\!\!\diagdown}} Me \right]^{\oplus} \qquad (9.37)$$

$$OX. = Ag^{\oplus} \text{ or } Cp_2Fe^{\oplus}$$

Cyclic voltammetric studies implied that the mechanism in Scheme 9.5 accounts for the acceleration of CO insertion by oxidation. This process can be catalytic when the 17-electron acyl product is a strong enough oxidant to oxidize the starting alkyl complex. Since an acyl ligand is a weaker electron donor than an alkyl ligand, this condition is often met.

$$[Fe]-Me \; \xrightarrow{-e^{\ominus}} \; [Fe]^{\oplus}\!\!-Me$$

$$[Fe]^{\oplus}\!\!-Me \; + \; CO \; \longrightarrow \; [Fe]^{\oplus}\!\!-C(O)Me$$

$$[Fe]^{\oplus}\!\!-C(O)Me \; + \; [Fe]-Me \; \longrightarrow \; [Fe]^{\oplus}\!\!-Me \; + \; [Fe]-C(O)Me$$

Scheme 9.5

9.2.1.2. Insertions of Other Ligands Bound Through a Single Atom

Isonitrile, thiocarbonyl, and nitrosyl ligands undergo insertion into metal–alkyl complexes, although the number of these types of insertions is much smaller than the number of CO insertions. Isonitrile ligands tend to undergo insertion faster than carbonyl ligands. For example, the alkyl group of the iron complex in Equation 9.38 inserts the isocyanide over the carbonyl ligand.[112] Two factors are likely to make the rate of insertion of isocyanides faster than the rate of insertion of CO. First, the product from insertion can form an η^2-iminoacyl ligand that is more stable than an η^2-acyl ligand. The formation of an η^2-iminoacyl product from insertion is shown in Equation 9.39. Second, the C–N triple bond is weaker than the C–O triple bond. Thus, the insertion of an isonitrile is more favorable thermodynamically than is insertion of CO. In fact, nickel complexes catalyze the sequential insertion of isocyanides to form polyisocyanides, which have helical structures.[113, 114]

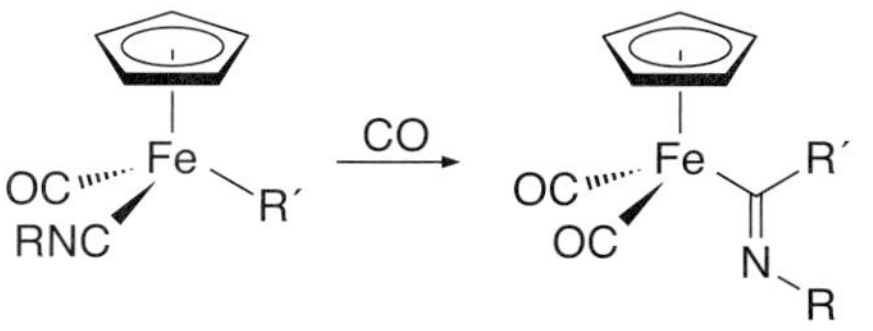

$$(9.38)$$

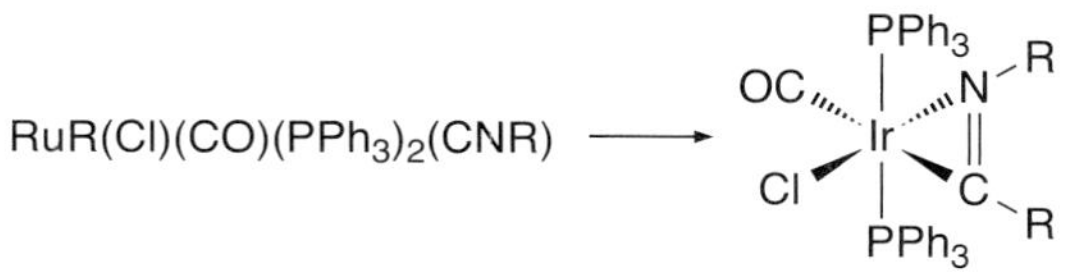

$$RuR(Cl)(CO)(PPh_3)_2(CNR) \longrightarrow \tag{9.39}$$

Insertions of thiocarbonyl ligands are uncommon because of the rarity of the thiocarbonyl ligand. However, insertion appears to occur readily to form κ^2-thioacyl complexes, as illustrated in Equation 9.40.[115]

$$OsAr(X)(CS)(CO)L_2 \underset{\Delta}{\overset{\Delta}{\rightleftharpoons}} Os\,(\kappa^2 - CSAr)X(CO)L_2$$

Ar = $p-MeC_6H_4$ etc.
X = Cl, Br or I   (9.40)
L = PPh_3

Although linear nitrosyl complexes are formally isoelectronic with carbonyl complexes, few examples of migratory insertions of nitrosyl ligands have been reported.[116–118] One example of such an insertion is shown in Scheme 9.6.[117] In the presence of phosphines, intramolecular migratory insertion occurs to form the stable nitrosoalkane complexes at the right of the scheme. A mechanism similar to that for carbonyl insertion in Scheme 9.1 appears to be followed, except that k_1 is the rate-determining step (i.e., $k_2[L] \gg k_1$).

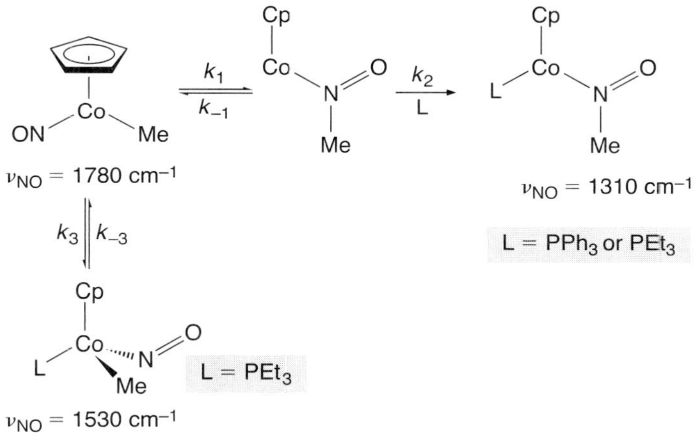

Scheme 9.6

Reactions conducted with the more basic phosphine, PEt_3, occur differently and afford the bent nitrosyl complex at the bottom left of Scheme 9.6. The reactions of this bent nitrosyl complex are inhibited by added phosphine, and this inhibition suggests that the bent NO is unreactive toward insertion. The linear nitrosyl in the starting complex appears to be more reactive. Some data suggest that the insertion of NO is slower than the insertion of CO. For example, $MeFe(NO)(CO)_3$ reacts with CO to form the acyl complex $[MeC(O)]Fe(NO)(CO)_3$.[119]

9.2.1.3. *Insertions of Carbenes*

Carbene ligands are also isolobal with CO, and insertions of carbene units into metal–alkyl and –aryl groups, shown generically in Equation 9.41, have been reported. In many cases, this insertion process is part of a multistep sequence. However, one example that is well defined is shown in Equation 9.42.[120] In this process, abstraction of a hydrogen atom from the starting methyl complex generates an aryltungstenocene carbene intermediate. The aryl group of this complex then migrates to the carbene carbon to generate a benzyltungstenocene complex. A second example that is part of a well-defined multistep sequence is shown in Equation 9.43.[121] In this case, hydride abstraction generates a methyl carbene complex in which the methyl group migrates to the carbene carbon to generate an ethyl group, which then undergoes β-hydrogen elimination (see Chapter 10) to form an olefin hydride complex as product. Migrations of hydride, halide, and alkoxo groups to carbenes can be reversible,[122] and the eliminations of hydride groups from alkyl ligands to form hydridometal carbene complexes are described later in Chapter 10.

$$M = \overset{\displaystyle |}{\underset{\displaystyle X}{CH_2}} \quad \rightleftharpoons \quad M-CH_2-X \tag{9.41}$$

$$(9.42)$$

$$(9.43)$$

9.2.2. Insertions of Polyhapto Ligands into Metal–Ligand Covalent Bonds

Unsaturated ligands bound through more than one atom (polyhapto ligands) also undergo insertion into metal–ligand bonds, and the insertions of alkenes into metal–hydride, metal–alkyl, and metal–aryl bonds are among the most common reactions of organometallic systems. Examples of these reactions in which the insertion processes have been observed directly constitute a small subset of the total number of examples of such insertions. Nevertheless, sufficient data on the insertion process has been gained by a variety of experiments to deduce the mechanism of the insertions of olefins into metal–hydrogen and metal–carbon bonds in detail. Similar reactions of alkynes are also common, although the mechanisms of alkyne insertions are more varied. Like the insertions of CO into metal–carbon bonds, the insertions of olefins into metal–carbon bonds are best described as an intramolecular migratory insertion. The insertions of olefins take place from a complex in which the olefin and hydride, alkyl, or aryl group are coordinated cis to each other.

9.2.2.1. Insertions into Metal–Hydride Bonds

9.2.2.1.1. Insertions of Olefins into Metal–Hydride Bonds

The intramolecular insertion of a hydride into a coordinated olefin is a crucial step in olefin hydrogenation catalyzed by late transition metal complexes, such as those of iridium, rhodium, and ruthenium (Chapter 15), in hydroformylation reactions catalyzed by cobalt, rhodium, and platinum complexes (Chapter 16), and in many other reactions, including the initiation of some olefin polymerizations. The microscopic reverse, β-hydride elimination, is the most common pathway for the decomposition of metal–alkyl complexes and is a mechanism for olefin isomerizations.

A general representation of the insertion of an olefin into a metal hydride and for the reverse reaction, β-hydride elimination, is shown in Equation 9.44. This reaction is fastest when the hydride, the metal, and the two carbon atoms of the olefin are coplanar. As a result, β-hydrogen elimination occurs in a syn fashion through the four-centered transition state shown in this equation. Just as σ-complexes precede the oxidative addition of C–H and H–H bonds, β-agostic interactions (see Chapter 2) are thought to precede the cleavage of a β C–H bond by the β-hydrogen elimination reactions discussed in Chapter 10. Thus, the structures of β-agostic complexes are closely related to the transition states for β-hydrogen elimination and the opposite olefin insertion into a hydride.[123]

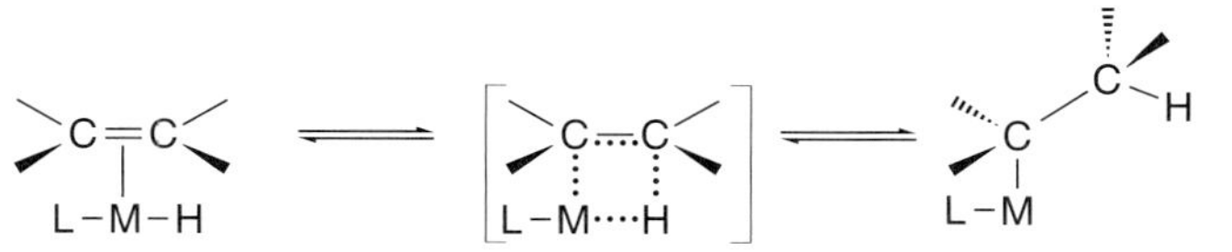

$$\begin{array}{ccc} \ce{>C=C<} & \rightleftharpoons & \left[\ce{>C\cdots C<}\right] & \rightleftharpoons & \ce{C-C-H} \\ \ce{L-M-H} & & \ce{L-M\cdots H} & & \ce{L-M} \end{array} \tag{9.44}$$

The intramolecular migration of a hydride to an olefin is fast enough that few complexes contain an olefin and a hydride ligand cis to each other. Some olefin hydride complexes, such as the iridium[124] and platinum[125] examples below, are stable because they contain hydride and olefin ligands that are mutually trans. These complexes must isomerize before insertion can take place.

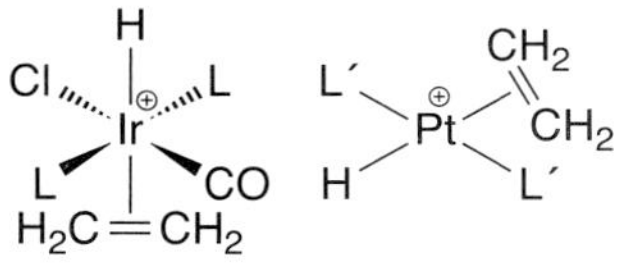

Figure 9.2.
Two stable, trans-hydridometal ethylene complexes.

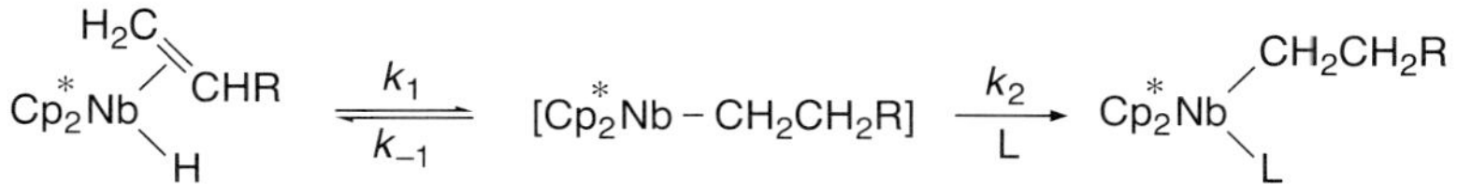

$$\ce{Cp^*_2Nb(H)(H2C=CHR)} \underset{k_{-1}}{\overset{k_1}{\rightleftharpoons}} [\ce{Cp^*_2Nb-CH2CH2R}] \overset{k_2}{\underset{L}{\longrightarrow}} \ce{Cp^*_2Nb(L)(CH2CH2R)}$$

$\ce{Cp^*} = \ce{C5Me5}$, $\ce{L} = \ce{CO}$ or $\ce{MeNC}$, $\ce{R} = \ce{H, Me, Ar}$

$$\text{Rate} = k_{obs}[\text{Nb}] = \frac{k_1 k_2 [\text{L}]}{k_{-1} + k_2 [\text{L}]} [\text{Nb}]$$

Scheme 9.7

Reversible intramolecular insertion of an olefin into the M–H bond of a cis hydride ligand was observed with the niobium complexes in Scheme 9.7.[126] The insertions of olefins into the hydride in this complex were studied in detail. A variety of data imply that the reactivity of this complex follows the mechanism depicted in Scheme 9.7. Addition of CO or MeNC traps the 16-electron intermediate as the saturated alkyl product. The rate of formation of the alkyl complex exhibits a first-order dependence on MeNC (overall second-order behavior). These kinetic data are consistent with a rapid preequilibrium involving insertion and β-H elimination, followed by slow, rate-limiting, bimolecular trapping. This general scheme is the same as that for insertion of CO into metal alkyls (Scheme 9.1). The proposed unsaturated alkyl intermediate in Scheme 9.7 was not observed directly, but its presence was inferred from the exchange of the hydride only with the hydrogens on the substituted end of the olefin. Although the reaction rates (k_1) were insensitive to changes in solvent polarity, the variations in rate as a function of the substituent on the C=C bond (R = Me > Ph > H) were interpreted as indicating the development of a partial positive charge at the β-carbon and migration of the hydride ligand as H⁻.

Another set of insertions of olefins into metal hydrides that has been studied in depth are those involving Rh(III) hydrides that are relevant to the hydrogenation of olefins by Wilkinson's catalyst. An insertion of cyclohexene into such a complex is shown in Equation 9.45a.[127] The cyclohexyl hydride complex is not observed directly because reductive elimination to form cyclohexane is fast. However, the related insertion of ethylene into the PPh₃-ligated hydridorhodium dichloride forms a stable ethyl complex (Equation 9.45b).[128]

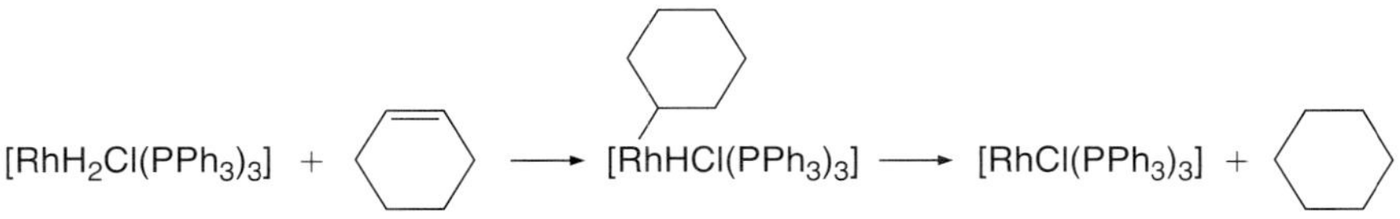

$$\ce{[RhH2Cl(PPh3)3]} + \text{(cyclohexene)} \longrightarrow \ce{[RhHCl(PPh3)3]} \longrightarrow \ce{[RhCl(PPh3)3]} + \text{(cyclohexane)} \tag{9.45a}$$

$$\ce{(PPh3)3Rh(H)Cl2.CH2Cl2} + \text{} \rightleftharpoons \longrightarrow \ce{(PPh3)3Rh(Et)Cl2} \tag{9.45b}$$

The insertion of cyclohexene in Equation 9.45a occurs with an inverse kinetic order in PPh_3. These data imply that a phosphine in the 18-electron starting complex is replaced by olefin and that the insertion process involves a hydride and a coordinated olefin. A study of the reaction between substituted styrenes and $RhH_2Cl(PPh_3)_3$ showed little variation in overall rate.[129] However, this similarity in rate resulted from stronger binding of electron-rich olefins, but slower rates for insertion of these olefins.[127,129]

The insertions of enamides into rhodium hydrides have also been studied in detail (Equation 9.46).[130–132] In this case, oxidative addition of dihydrogen occurs to the olefin complex to generate the intermediate containing cis olefin and hydride ligands. This reaction controls the stereochemical outcome of enantioselective hydrogenations of enamides to form α-amino acids presented in Chapter 15. As noted in this chapter, the oxidative addition forms two diastereomeric complexes, and the less abundant diastereomer reacts much faster and forms the more abundant enantiomeric amino acid product.

$$(9.46)$$

A fourth illustrative example of the insertion of olefins into M–H bonds is the reaction of $Cp_2Zr(H)Cl$, or "Schwartz's reagent"[133] (Equation 9.47).[134] This complex reacts with alkenes with high selectivity for the formation of linear alkyl complexes. The products of these reactions can be oxidized to alcohols or used in coupling processes, much like organoboranes.[135–137] Thus, this addition process has been termed "hydrozirconation." Unlike hydroboration, this addition process generates terminal alkyl complexes from even internal olefins. The mechanism by which this chain-walking process takes place is presented in Chapter 10 where β-hydrogen elimination is discussed. The insertion process has been studied closely to detect the initially formed insertion products. The reaction of Schwartz's reagent with ^{13}C-labeled cis-2-heptene is shown in Equation 9.47.[138]

$$(9.47)$$

9.2.2.1.2. Insertions of Alkynes into Metal–Hydride Bonds

The insertion of alkynes into metal–hydride bonds[139] occurs during a number of catalytic processes, including alkyne hydrogenation, hydrosilylation, silylformylation, hydroesterification and dimerization. This insertion chemistry is more complex mechanistically than the insertions of olefins into metal hydrides. In some cases, cis addition products have been formed, and the reactions appear to occur by a migratory insertion process that is characteristic of olefin insertions. However, trans addition products have been observed in other cases. Some reactions that form trans addition products appear to occur by a combination of cis addition, followed by isomerization of the resulting vinyl complex through an η^2-vinyl intermediate. In some cases, however, the trans addition product is the kinetic product. In these cases, electron-transfer mechanisms and vinyl radical intermediates have been proposed to account for the stereochemical outcome. Although clear trends are difficult to extract from the existing data, it appears that reactions of relatively electron-neutral alkynes tend to give cis addition products by migratory insertion pathways, and strongly electron-deficient olefins tend to give trans addition products by either isomerization or electron-transfer pathways.

The simplest pathway for the insertion of alkynes into metal–alkyl bonds involves a migratory insertion through a four-centered transition state akin to that for the insertion of olefins into metal hydrides.[140–146] Selected examples of this insertion process are shown in Equations 9.48–9.50.[140–146] The formation of cis products from insertion into iridium and rhodium hydrides is shown in Equations 9.48–9.49. In these cases, the open coordination site for binding of an alkyne prior to the actual insertion step is thought to be generated by indenyl ring slip. A case in which the isomerization of the initially formed cis olefin to the trans isomer was observed directly is shown in Equation 9.50.[140,146]

$$R = R' = H$$
$$R = H, R' = t\text{-Bu}$$
$$R = R' = Me$$
$$R = R' = Et$$

$$(9.48)$$

$$(9.49)$$

$$(9.50)$$

In some of these migratory insertions of alkynes, the cis addition is thought to be followed by isomerization of the resulting vinyl complex to its isomeric trans product.[143,147] Many mechanisms for isomerization of the cis vinyl complexes to what would be a trans addition product have been proposed. These include a concerted process,[147] formation of a Zwitterionic carbene intermediate that can rotate about the C–C bond when the vinyl ligand contains an electron-withdrawing group to stabilize anionic charge (Equation 9.51), and formation of an η^2-vinyl intermediate, such as the one shown in Equation 9.52. In the case of the photochemically induced reaction of the tungsten hydride in Equation 9.52, the η^2-vinyl complex is the final product because CO is eliminated from the system.[142] The formation of trans addition products has been observed in catalytic systems, particularly in the hydrosilylation of alkynes.[148–151] This regiochemistry can be synthetically valuable[150, 152–154] because of the ability to oxidize the vinylsilane products.

$$(9.51)$$

$$(9.52)$$

In other cases, the vinyl complex formed from trans addition of the metal hydride across the alkyne is the kinetic product.[155,156] In the reaction shown in Equation 9.53, the product from trans addition isomerizes to the product that would result from cis addition, and this result shows that the trans addition product must be the kinetic one.[157] This stereochemistry is rare for reactions of electron-neutral to slightly electron-poor alkynes like acetylene, alkylacetylenes, arylacetylenes, and diarylacetylenes. However, this stereochemistry has been observed in several systems from reactions of electron-poor alkynes, such as alkyl propiolates and dialkylacetylene dicarboxylates. Clark obtained evidence that this trans product is formed by a mechanism involving electron transfer and radical recombination.[156] The example studied and the proposed mechanism are shown in Equation 9.54. Rate-limiting, outer-sphere electron transfer was proposed to form a geminate radical pair (contained within a solvent cage). Transfer of the metal hydride as a proton would form a Pt(I) hydride and a vinyl radical, which would be more stable in the E-configuration. Recombination of these radicals would form the product from a formal trans addition. A final route to trans addition products involves isomerization of the alkyne complex to a vinylidene and migration of the hydride to the central carbon to generate a vinyl complex as product.[158]

$$(9.53)$$

$$(9.54)$$

9.2.2.1.3. Insertion of Ketones and Imines into Metal–Hydride Bonds

The insertions of ketones and imines into metal hydrides have also been studied, although they have been studied less extensively than the insertion of olefins into metal hydrides. This elementary reaction is relevant to one of the mechanisms for the hydrogenation of ketones and imines.[159] An example of the insertion of a ketone into a ruthenium hydride is shown in Equation 9.55. The alkoxide product is stable toward reductive elimination and β-hydrogen elimination because of the trifluoroalkyl groups. A reversible insertion of imines into dinuclear rhodium hydrides to form a bridging amido complex is shown in Equation 9.56. The amide product is more stable than an amide that would be formed by insertion to form a terminal late-metal–amido complex because it adopts a bridging structure containing two metal–nitrogen bonds.[160]

(9.55)

(9.56)

R′	R
C_6H_5	Pr^i
C_6H_5	OPr^i
CH_3	Pr^i
CH_3	OPr^i
CH_2Ph	Pr^i
CH_2Ph	OPr^i

9.2.2.2. Insertions of Olefins into Metal–Carbon Bonds

The insertion of olefins into metal–alkyl linkages is the cornerstone of the preparation of polyolefins and α-olefins, and the insertions of olefins into metal–aryl bonds is one step of a common class of palladium-catalyzed coupling reactions (the Mizoroki–Heck reaction). The insertions of olefins into early-metal–alkyl bonds is part of the so-called Cossee mechanism[161] of Ziegler–Natta olefin polymerization and of Cramer's mechanism for olefin dimerization.[162] The following sections present examples of these insertion reactions and information on the mechanisms of this type of C–C bond formation.

9.2.2.2.1. Insertions of Olefins into Metal–Hydrocarbyl σ-Bonds

Many examples of insertions of alkenes into metal–alkyl bonds of both early and late transition metal complexes have been observed. In general, the insertion process occurs more readily into metal–alkyl bonds of complexes containing electron-poor metal centers than into metal–alkyl bonds of complexes containing electron-rich metal centers.

C–H and C–C bond strengths, along with the substantial difference between metal–alkyl and metal–hydride bond strengths, imply that migratory insertion of an olefin into a metal–alkyl bond has a larger thermodynamic driving force than insertion of an olefin into a metal–hydride bond. However, the kinetic barrier to insertion of an olefin into a metal–alkyl bond tends to be higher than that for insertion of an olefin into a metal–hydride bond. This difference in the rate of insertion is due, most likely, to lesser steric interactions during formation of a C–H bond than during formation of the C–C bond and to greater overlap involving the less directional orbital of a metal–hydride bond than that involving the more directional orbital of a metal–alkyl bond. The lower barrier for insertions into metal-hydride bonds can also be rationalized by the formation of a β-agostic interaction during olefin insertion into a metal hydride, which would be more stable than the analogous interaction formed during insertion of olefin into a metal–carbon bond.

The chemistry of the cobalt complexes in Figure 9.3 illustrates the different rates for these two types of insertions. The hydride complex undergoes rapid intramolecular insertion of the olefin into the hydride, as revealed by isotopic labeling, but the methyl complex does not undergo the analogous insertion of olefin into the metal–methyl bond.[163] The chemistry of the rhodium complexes in Equations 9.57 and 9.58 provides a more quantitative comparison of the relative rates for insertion of olefins into metal–hydride and

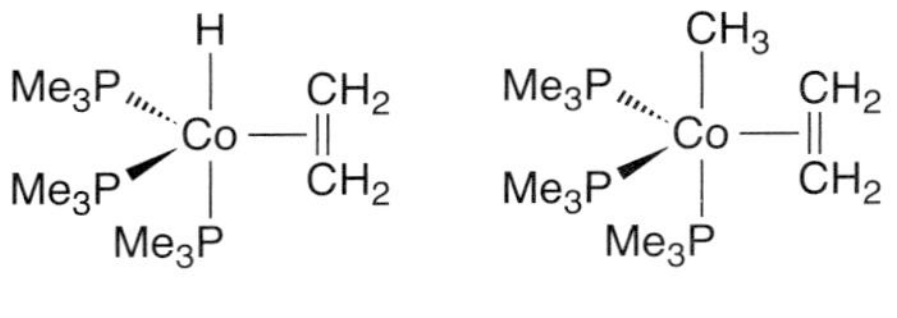

Figure 9.3.
A comparison of the rates of insertion of ethylene into metal–hydride and metal–alkyl bonds. The cobalt–hydride complex inserts ethylene, but the cobalt–alkyl complex does not.

metal–alkyl bonds.[164] The free energy barrier for insertion of ethylene into the rhodium hydride is only 12–15 kcal/mol, while the barrier for insertion of ethylene into the rhodium–ethyl complex is 20–25 kcal/mol.[165]

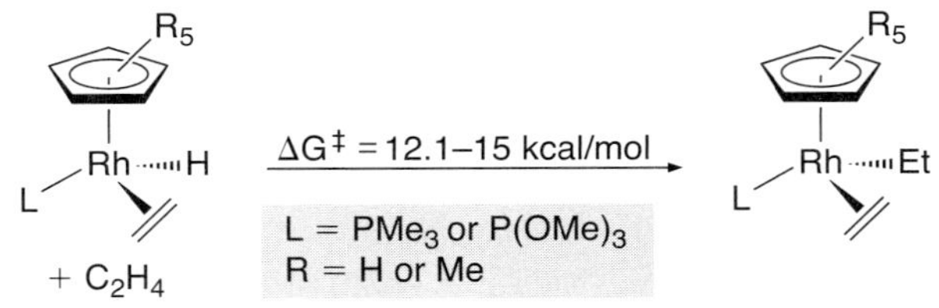

$$\Delta G^{\ddagger} = 12.1\text{–}15 \text{ kcal/mol}$$

L = PMe$_3$ or P(OMe)$_3$
R = H or Me

(9.57)

$$\Delta G^{\ddagger} = 24.4\text{–}24.7 \text{ kcal/mol}$$

L = PMe$_3$ or P(OMe)$_3$
R = H or Me

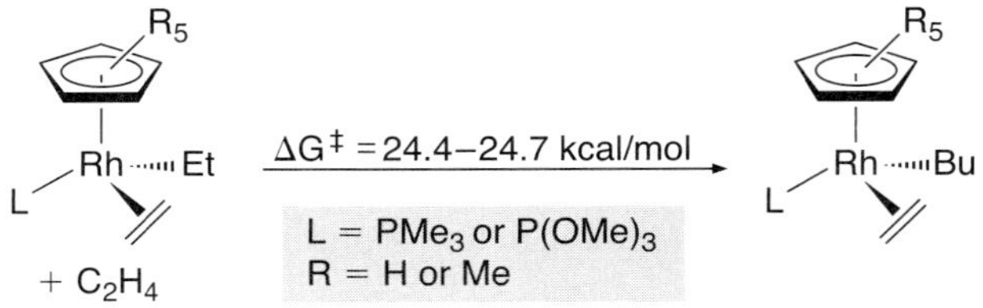

(9.58)

The small number of complexes that have been characterized possessing cis alkyl and alkene ligands that undergo migratory insertion[166] contrasts with the large number of complexes possessing cis alkyl and CO ligands that undergo migratory insertion. However, the strong interest in developing olefin polymerization catalysts has led to an increased number of alkyl olefin complexes that undergo this type of insertion. In 1982 Watson reported one of the first examples of an alkyl complex that inserts propylene, the lutetium–methyl complex Cp*$_2$LuMe.[167] In 1984, Flood showed by labeling experiments that late-metal–alkyl complexes also insert olefins.[168] More recently, Jordan and Casey have studied d^0 scandium and zirconium alkyl olefin complexes that undergo insertion, and Landis has used ^{13}C-labled olefins to monitor directly olefin insertion processes during polymerizations of alkenes catalyzed by zirconium cations.[169] In parallel, Brookhart has studied in detail the insertion reactions of cationic nickel and palladium alkyl olefin complexes that are intermediates in the polymerization and oligomerization of olefins catalyzed by these metals.[170] These studies are described in more detail in Chapter 22.

Like the insertion of CO into metal–alkyl bonds, the insertion of olefins into metal–alkyl bonds occurs by a concerted migratory insertion pathway (Equation 9.59). The alkyl group migrates with retention of configuration at carbon. Syn addition of the metal and the alkyl group occurs across the olefin.

The regiochemistry of metal–alkyl addition can be either Markovnikov (the upper path in Equation 9.59) or anti-Markovnikov (the lower path in Equation 9.59). Several factors control the regiochemistry of this addition. In most cases, olefins containing electron-withdrawing groups, such as carboalkoxy groups or aryl groups, insert with regiochemistry that places the electron-withdrawing group α to the metal. This mode of insertion is often called "2,1-insertion," and several factors cause olefins containing an electron-withdrawing group to insert with this regiochemistry. First, electron-withdrawing groups on an α-carbon tend to strengthen metal–carbon bonds by creating a more polar bond and by stabilizing the partial negative charge on the alkyl group. Second, aryl-substituted olefins often undergo 2,1-insertion because the initial product is stabilized by the adoption of an η^3-benzyl structure.

The insertion of alkenes is often less regioselective than the insertion of olefins possessing an electron-withdrawing group. Insertion of alkenes into Schwartz's reagent Cp$_2$ZrHCl forms the linear alkyl complex with high regioselectivity (vide supra), but insertions into metal–alkyl linkages involving the later metals form mixtures of 1,2- and 2,1-insertion products. The regiochemistry of alkene insertion has been studied extensively in the context of olefin polymerization. These studies have shown that most catalysts react by 1,2-insertions, but that there are exceptions. For example, 2,1-insertions occur during the polymerization of polypropylene catalyzed by titanium(IV) complexes containing salicylaldiminato ligands (Equation 9.59)[171–174] and by iron catalysts containing diiminylpyridine ligands (Figure 9.4).[175] One computational paper on the titanium system suggests that there is a greater electronic preference for titanium containing this ligand set to undergo 2,1-insertion than there is for other types of metal complexes.[176] The regiochemistry of insertion during palladium-catalyzed polymerization of α-olefins has a dramatic affect on polymer structure. This affect is summarized in Scheme 9.8. By a series of insertion and elimination ("chain-running") steps, the product of a 2,1-insertion of hexene can rearrange to incorporate the side chain into the main chain ("chain straightening"), while the product from 1,2-insertion, followed by chain running, leads to the formation of a butyl or methyl branch.

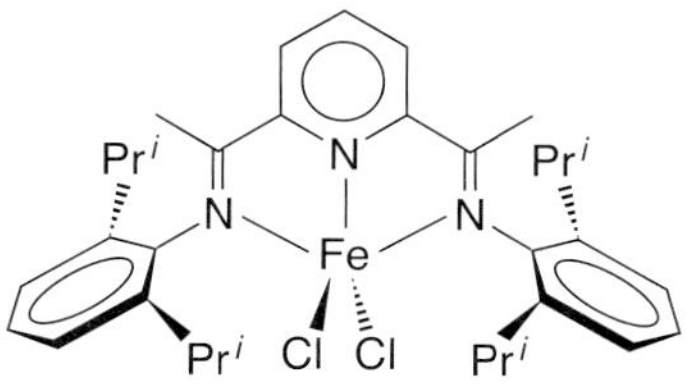

Figure 9.4.
A bisiminylpyridine complex of iron that is a catalyst precursor for polymerizations that occur by 2,1-insertions.

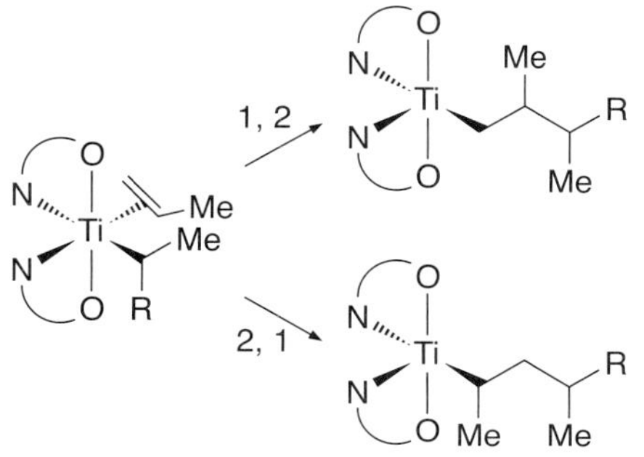

(9.59)

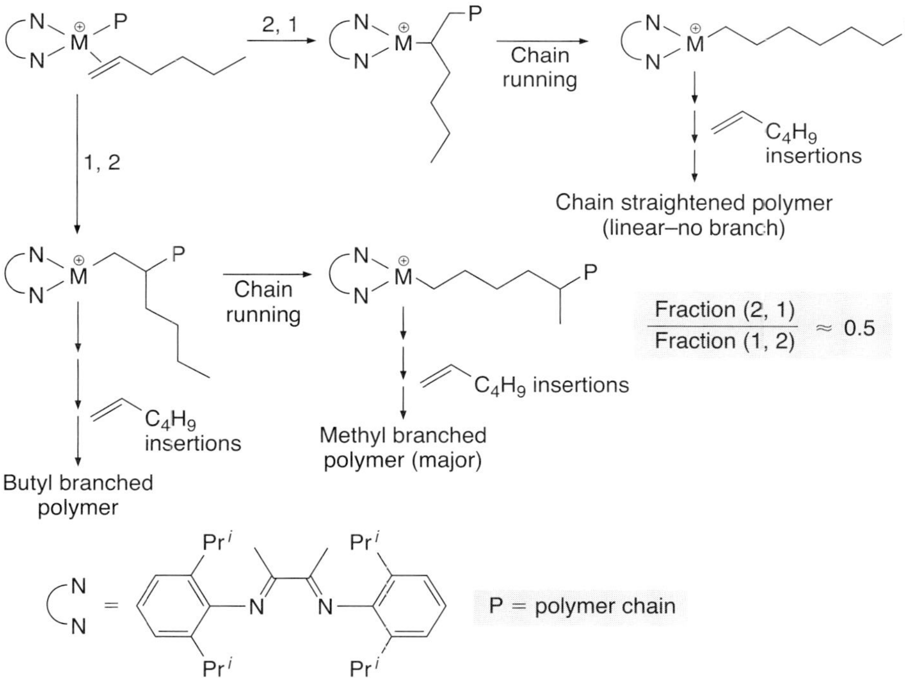

Scheme 9.8

Several examples in which the migratory insertion of olefins into metal–alkyl complexes has been observed directly are shown in Equations 9.60–9.65.[168,177–179] Equations 9.60 and 9.61 show the insertions of propene into lutetium–alkyl bonds reported by Watson. Equation 9.62 depicts the labeling experiment reported by Flood, which revealed the insertion of an olefin into a cationic late-metal–alkyl complex. Equations 9.63 and 9.64 depict low-temperature studies by

Brookhart on the insertion of a single ethylene into a branched alkyl complex of nickel and an example of the 2,1-insertion of vinylacetate into the palladium–methyl complex to form a stable chelated product. Equation 9.65 shows the insertion of a vinylarene into a palladium–aryl complex that is an intermediate in the vinylation of aryl halides known as the Heck reaction.

$$(9.60)$$

$$(9.61)$$

$$(9.62)$$

$$(9.63)$$

$$0 \text{ °C, } k = 1.64(2) \times 10^{-3} \text{ s}^{-1} \quad \Delta G^{\ddagger} = 19.4 \text{ kcal/mol} \qquad (9.64)$$

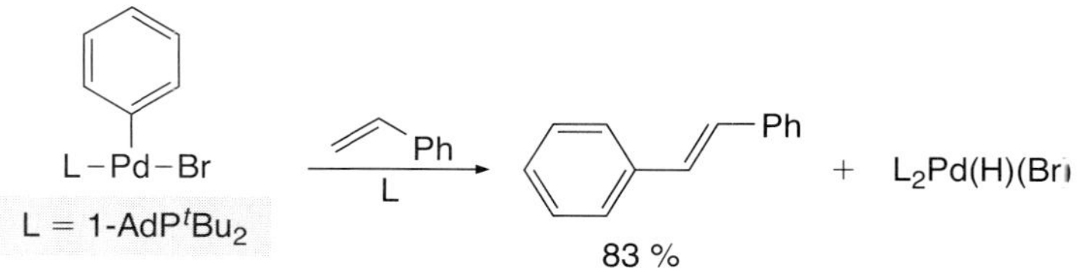

$$(9.65)$$

The identity of the chain-propagation step that forms the C–C bond during metal-catalyzed olefin polymerizations was debated, but subsequent observations of the insertion of alkenes into metal–alkyl bonds makes clear that olefin polymerization occurs by migratory insertion. Prior to this work, however, the top two pathways shown in Scheme 9.9 were both considered. The four-center "migratory" insertion mechanism was proposed by Cossee and Arlman.[161,180–182] The second, more circuitous mechanism, was proposed by Green and Rooney.[183,184] This second mechanism involves an oxidative α-hydrogen migration from an alkyl group to form an alkylidene, followed by a 2+2 insertion and C–H reductive elimination. This mechanism involving α-elimination requires the resulting alkylidene to serve as a dative, L-type ligand, which is now known to be a poor description of an alkylidene ligand on an early transition metal center. However, this "Green–Rooney mechanism" was proposed because of the preponderance of alkyl olefin complexes that did not undergo migratory insertion.

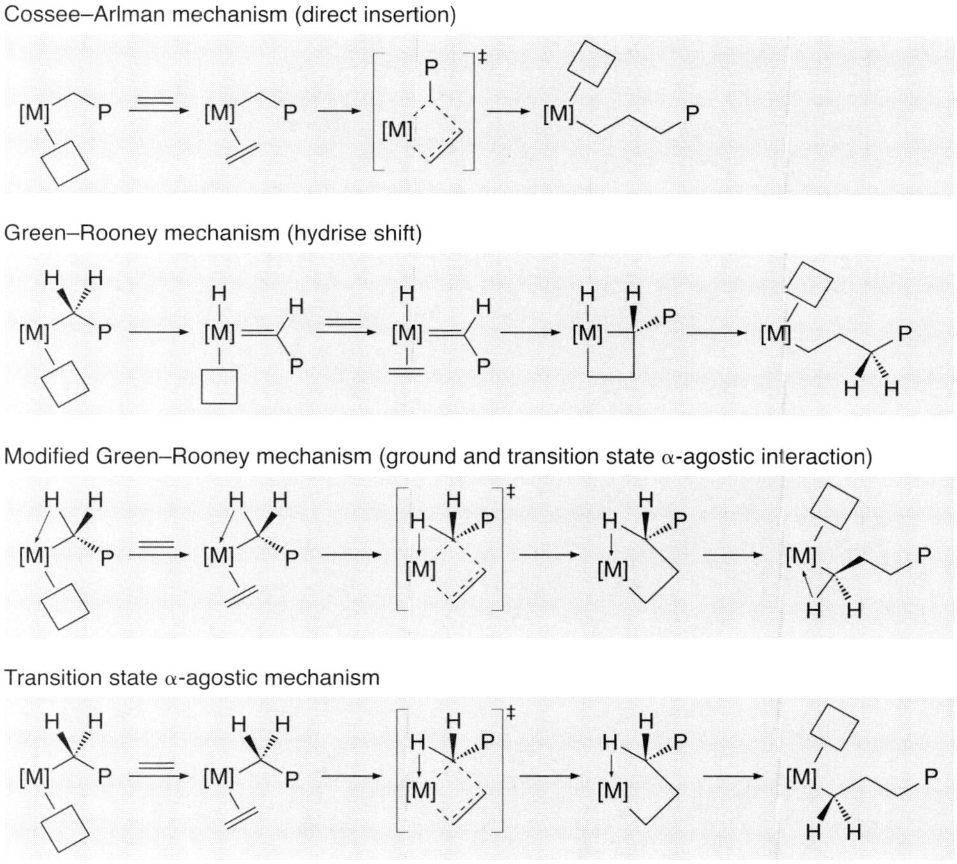

Cossee–Arlman mechanism (direct insertion)

Green–Rooney mechanism (hydrise shift)

Modified Green–Rooney mechanism (ground and transition state α-agostic interaction)

Transition state α-agostic mechanism

Scheme 9.9

Besides the historical interest of this mechanistic discussion, the Green–Rooney mechanism is presented here because aspects of this proposal have rung true. Many experiments have now been conducted to test whether the rate of insertion is accelerated and the stereoselectivity of insertion enhanced[185,186] by a *partial* cleavage of the α-C–H bond in a structure termed an α-agostic complex. This "modified Green–Rooney" mechanism is shown in Scheme 9.9. A further modification of this proposal has been made to include an α-agostic interaction in the transition state, but not the ground state ("transition state α-agostic mechanism" in Scheme 9.9).

Many experiments based on isotope effects have now been conducted to test whether an α-agostic interaction controls the rate and selectivity of olefin insertions into d^0 alkyl complexes. These results often seem to contradict each other, but a review by Grubbs and Coates provides some order to these disparate results.[187] Grubbs measured a 1:1 ratio of stereoisomers for the reaction of the alkyltitanocene complex in Equation 9.66 after abstraction of Cl⁻ by EtAlCl₂.[188] This reaction would have produced an unequal ratio of diastereomers if a difference in the degree of α-agostic interaction was present between the ground state and the transition state. Subsequent experiments by Bercaw and Piers[189] and by Brintzinger and Krauledat[190] showed that the isotope in the α-position *can* affect the stereochemistry of the product when the intramolecular cyclization involves a less reactive neutral group 3 metal complex or when the reaction occurs intermolecularly with a cationic group 4 complex (Equations 9.67 and 9.68).

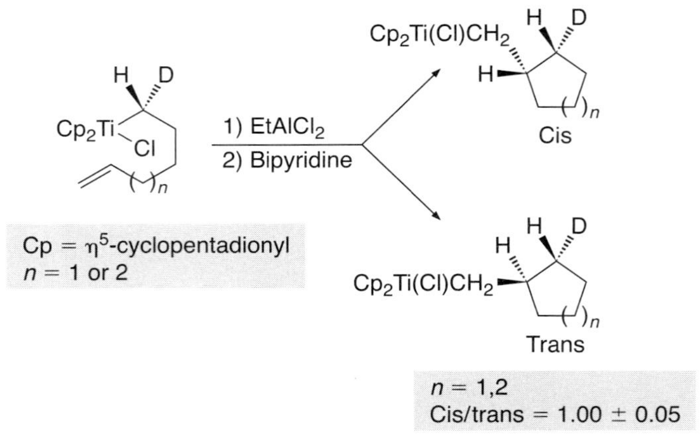

(9.66)

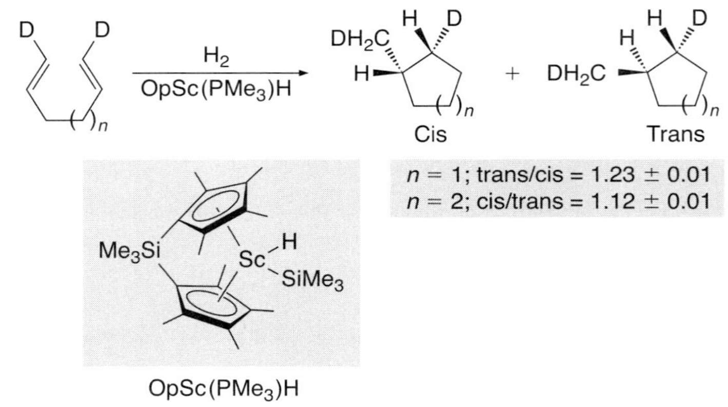

(9.67)

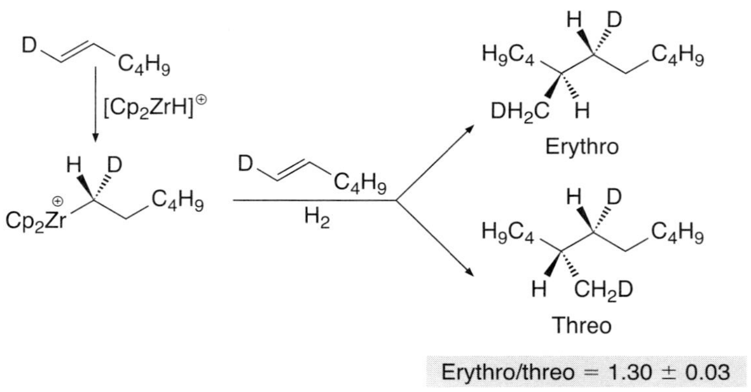

(9.68)

Table 9.1. Summary of α-kinetic isotope effects for insertion reactions of metallocene complexes and catalysts. Table from reference 187.

Entry	Eq	Reaction type	Catalyst system	Insertion[a]	Qualitative rate[b]	KIE
1	1	Cyclization	Ti/AlEtCl$_2$	Intramolecular	Very fast[c]	1.00
2	2	Cyclization	Sc	Intramolecular	Slow[d]	1.12–1.23
3	3	Dimerization	Zr/MAO	Intermolecular	Slow[e]	1.30
4	3	Dimerization	Sc	Intermolecular	Slow[f]	1.27
5	4	Cyclization	Zr/MAO	Intramolecular	Very fast[g]	1.01
6	5	Alkyne insertion	Sc	Intermolecular	Slow[h]	0.98
7	6	Cyclization	Ti/MgX$_2$	Intramolecular	Slow[i]	1.25
8	6	Cyclization	Ti/MAO	Intramolecular	Fast[j]	0.91
9	7	Polymerization	Zr/MAO	Intermolecular	Slow[k]	1.30

[a]Intra, Intramolecular; Inter, intermolecular.

[b]There is a lack of kinetic data for many of these reactions. However, the productivities of these systems (turnovers of substrate per mole of catalyst per hour/initial substrate concentration) can be used to establish the lower limits for the rates of olefin insertion.

[c]Cyclization of a titanium compound in the presence of ethylene at 0 °C yields cyclopentyl-capped polyethylene oligomers.[78] Thus, olefin cyclization occurs in less than the time required for ethylene insertion, which is 95 ms (38 000 h^{-1}) in this system.

[d]Productivity = 135 h^{-1} M^{-1}.[48]

[e]Productivity = 14 h^{-1} M^{-1}.[51]

[f]In 1 h at 25 °C, the catalyst converted 45 equivalents of 1-pentene selectively to the dimer.[41]

[g]There is not sufficient experimental data to estimate the rate of cyclization in this system. However, zirconocene-catalyzed copolymerization of 1,5-hexadiene and ethylene yields soluble polymers with cyclopentane rings in the main chain. Calculations reveal that cyclization in this system is extremely rapid, occurring in less than 0.38 ms (9.4 × 10^6 h^{-1}).[79]

[h]The rate of insertion is presumably very slow, since this reaction can be monitored by NMR.[49]

[i]Slow relative to entry 8.

[j]Fast relative to entry 7.

[k]Productivities for this system range from 364 to 340 h^{-1} M^{-1}.[80,81]

Grubbs and Coates argue that the more active catalysts tend to react with a smaller isotope effect of the α-hydrogen (see Table 9.1) and that the identity of the rate-limiting step of the coordination and migration sequence and the position of the transition state on the reaction coordinate lead to the presence or absence of an isotope effect. When binding of the olefin is rate limiting, an isotope effect is not observed, but when insertion is rate limiting, an isotope effect is observed. Thus, Grubbs' cyclization involving a highly reactive cationic group 4 complex could occur with rate-limiting binding of the olefin. Cyclization in the less-reactive group 3 complex and intermolecular insertion of the cationic group 4 complex could occur with rate-limiting insertion. In addition, the more reactive systems would be likely to react with earlier transition states, and an early transition state could involve less participation of the α-agostic interaction.

9.2.2.2.2. Insertions of Olefins into Metal–Acyl Bonds

The insertion of ethylene and α-olefins into acyl groups is one step of the remarkably selective copolymerization of alkenes and CO to form an alternating copolymer.[191] This process was developed at Shell Chemicals and is discussed in Chapter 17. As depicted in Scheme 9.10, the relative rates for insertion of an alkene into an alkyl group and an acyl group are one factor that controls the selectivity. For high selectivity, the insertion of ethylene into the acyl group must be faster than insertion of ethylene into an alkyl group.[192]

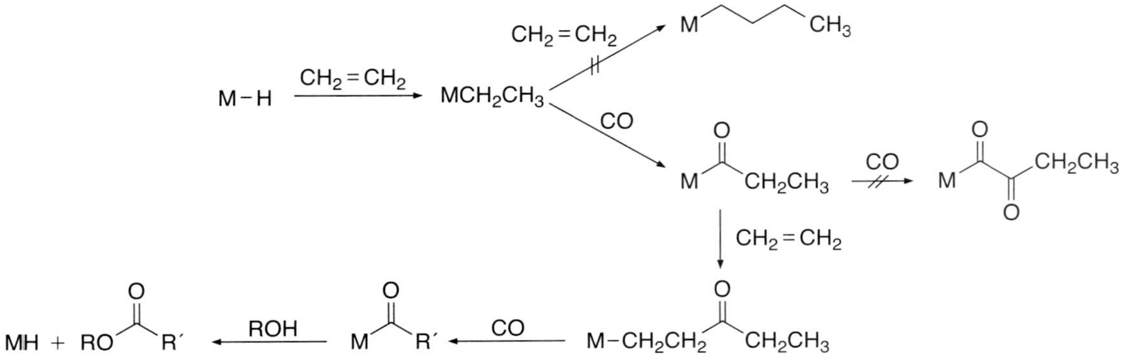

Scheme 9.10

Table 9.2. Comparison of calculated and experimental activation barriers for insertions. Table from reference 194.

Entry	Complex[a]	$\Delta H^{\ddagger}$(exptl)[b]	$\Delta G^{\ddagger}$(exptl)[b]	$\Delta S^{\ddagger}$(exptl)[c]
1	$(N–N)Pd(CH_3)(CH_2=CH_2)^+$		18.5(1)(−25 °C)	
2	$(N–N)Pd(C_2H_5)(CH_2=CH_2)^+$	18.5(6)	19.4(2)(−25 °C)	−3.7(2.0)
3	$(N–N)Pd(acyl)(CH_2=CH_2)^+$		16.6(1)(−46 °C)	
4	$(N–N)Pd(CH_3)(CO)^+$		15.4(1)(−66 °C)	
5	$(P–P)Pd(CH_3)(CH_2=CH_2)^+$	15.2(7)	16.6(1)(−45.6 °C)	−6.2(2.9)
6	$(P–P)Pd(C_2H_5)(CH_2=CH_2)^+$	15.9(8)	16.3(1)(−33.7 °C)	−1.6(3.7)
7	$(P–P)Pd(acyl)(CH_2=CH_2)^+$		12.3(1)(−103.1 °C)	
8	$(P–P)Pd(CH_3)(CO)^+$	14.8(7)	14.8(1)(−81.7 °C)	0.1(3.5)
9	$(P–P)Pd(C_2H_5)(CO)^+$		13.4(1)(−94.2 °C)	

[a]N–N is 1,10-phenanthroline. P–P is $Ph_2P(CH_2)_3PPh_2$.
[b]In kcal·mol^{-1}.
[c]In cal·mol^{-1}·K^{-1}.

Indeed, direct measurements of the rates of insertion of CO and ethylene into alkyl-metal olefin and acylmetal olefin complexes show that the insertion of ethylene into the metal–acyl linkage is faster than the insertion of ethylene into the metal–alkyl linkage. Comparisons of these rates for insertions into cationic palladium complexes containing phenanthroline and bis-diphenylphosphinopropane as ancillary ligand have been made by Brookhart and co-workers. These reactions are shown in Equations 9.69 and 9.70. A summary of the barriers for insertion is provided in Table 9.2. The rate of insertion of ethylene into the metal–acyl bond is orders of magnitude faster than the rate of insertion of ethylene into the metal–alkyl bond.[36, 193, 194]

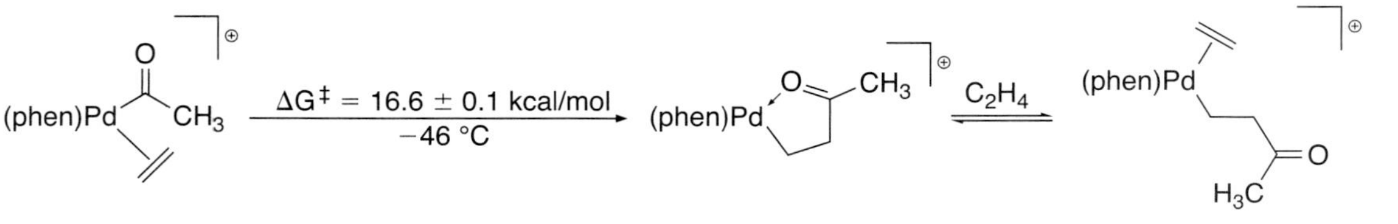

(9.69)

(9.70)

9.2.2.2.3. Insertions of Alkynes into Metal–Carbon Bonds

The insertions of alkynes into metal–carbon σ-bonds are less common than either the insertions of olefins into metal–carbon bonds or the insertions of alkynes into metal–hydride bonds. Nevertheless, several examples of this reaction have been studied, and many examples are part of catalytic processes. Most of the insertions of alkynes into metal–carbon bonds occur by concerted migratory insertion pathways and provide products from cis addition of the metal and hydrocarbyl group across the carbon–carbon multiple bond, as predicted on theoretical grounds by Thorn and Hoffmann.[195] In some cases, the products from trans addition are observed, but these kinetic products are thought to result from isomerization of the vinyl group in reaction intermediates formed by cis addition.

The insertions of alkynes into metal–carbon bonds are thermodynamically more favored than the insertions of olefins into metal–carbon bonds because the cleavage of one carbon–carbon π-bond in an alkyne requires less energy than the cleavage of the C–C π-bond in an olefin[196] and the sp^2-C–M bond in the product of alkyne insertion is stronger than the sp^3-C–M bond in the product of alkene insertion. The insertions of alkynes into the vinyl complexes that result from alkyne insertion are also favored thermodynamically. Thus, multiple insertions of alkynes to form polyacetylenes, just like the multiple insertions of alkenes to form polyolefins, are known. Because of the conducting properties of polyacetylenes, the transition-metal-catalyzed polymerization of alkynes to form polyacetylenes has been studied.[197–205]

Few direct comparisons between the rates for insertions of alkenes and alkynes have been made. However, one comparison indicates that the insertions of alkenes are slower than the insertions of alkynes. The insertion of acetylene into the cationic palladium–alkyl complex in Equation 9.71[206] is directly analogous to the insertions of ethylene into cationic palladium alkyl complexes. This insertion of acetylene is faster than the insertion of ethylene into the same palladium–methyl complex. The insertion of 1-hexyne is much slower than the insertion of acetylene and gives a mixture of the two vinyl complexes that result from 1,2- and 2,1-insertion.

(9.71)

Despite the more favorable thermodynamics and kinetics for insertions of alkynes than for insertions of alkenes, the insertions of alkynes into metal–carbon bonds do not necessarily occur in as high or higher yields. The C–H bond in a terminal acetylene can undergo reactions that compete with migratory insertion, such as oxidative addition, σ-bond metathesis, and protonolysis of a metal–carbon bond to form an acetylide complex.[207] Therefore, many of the insertion reactions of alkynes have been observed with internal alkynes. Some complexes undergo insertions only with alkynes containing strong electron-withdrawing groups, such as dialkyl acetylene dicarboxylates.

An example of an alkyne insertion involving an early metal complex is shown in Equation 9.72. The insertion of dimethylacetylene into the permethylscandocene–alkyl complex occurs in a manner similar to that for the insertion of olefins into d^0 metal–alkyl complexes.[208] This reaction gave the product of a cis addition. The competition experiment shows that there is no measurable isotope effect of the α-hydrogen, implying that the modified Green–Rooney mechanism is not followed in this case.

$$H_3/D_3 = 0.98 \pm 0.05 \tag{9.72}$$

An example of an alkyne insertion into a late, first-row transition metal–alkyl complex is shown in Equation 9.73.[209,210] In this case, the kinetic product is formed by trans addition of the nickel and methyl group across the carbon–carbon bond (right of Equation 9.73). Heating the product leads to isomerization of the vinyl complex resulting from trans addition to products containing a mixture of geometries. To account for the experimental results, the authors proposed that the unsaturated product resulting from the migratory insertion undergoes isomerization prior to formation of the final product. To account for the lack of an effect of added ligand on the product ratios, the authors propose that the isomerization occurs by attack of the phosphorus at the β-carbon of the vinyl group.

$$\tag{9.73}$$

Intramolecular insertions of alkynes have also been observed, and one example of a well-characterized system is shown in Equation 9.74. Kinetic studies of this insertion process have shown that exchange of the alkyne for a coordinated phosphine occurs and that the product is formed by a migratory insertion of the coordinated alkyne into the metal–acyl bond.[211]

$$\tag{9.74}$$

9.2.2.2.4. Insertions of Polyenes into Metal–Carbon Bonds

Migratory insertions of conjugated dienes and other extended olefinic systems into metal–alkyl bonds is common and occurs during the metal–catalyzed polymerizations of dienes. The insertions of dienes into metal–alkyl complexes are often faster than the insertions of simple olefins. The facility of the insertions of dienes likely arises from the formation of stable π-allyl products.[212] Evidence that these reactions occur by a migratory insertion pathway has been gained from the stereochemistry of the product. For example, migratory insertion into an η^4-1,3-diene complex proceeds with syn stereochemistry to afford an anti-η^3-allyl complex (Equation 9.75). (The "anti" term refers to the position of the terminal group relative to the hydrogen at the center of the allyl.) In addition, the neutral manganese alkyl arene complex in Equation 9.76 reacts with added phosphine to yield the endo methylcyclohexadienylmanganese dicarbonyl triphenylphosphine complex.[213] Formation of an acylmanganese derivative is a competing side reaction.

$$(9.75)$$

$$(9.76)$$

9.2.2.2.5. Insertions of Aldehydes and Imines into Metal–Carbon Bonds

The reaction of a Grignard or organolithium reagent with a carbonyl compound or an imine to form a main group metal alkoxide or amide is a formal insertion of the carbonyl or imine group into a metal–carbon bond. Early transition metal complexes possessing polar metal–carbon bonds undergo similar processes. Because of the greater covalency of the late transition metal–carbon bond, the reactions of aldehydes and imines with late transition metal organometallic complexes is more closely related to the migratory insertions of alkenes into the metal–hydride, –alkyl, –alkoxo, and –imido complexes presented above. However, insertions of imines, aldehydes, and ketones into late transition metal–carbon or –hydrogen linkages to form metal amides or alkoxides are less common than such insertions into early metal analogs, and direct observation of this elementary reaction with late transition metal systems is particularly rare.[214–218]

Such insertions of aldehydes and imines into late transition metal–carbon bonds has become important as a step in the additions of aryl and vinyl groups to aldehydes and imines catalyzed by rhodium[219,220] and other late metals.[221–244] The insertion of aldehydes into late transition metal–carbon bonds has been observed directly in a few cases that are likely to be driven by the release of ring strain. One example (Equation 9.77) involved insertion of benzaldehyde into the metal–carbon bond of a ruthenium benzyne complex[217] and another example (Equation 9.78) involved insertion of formaldehyde into the metal–alkyl bond of a Ni metallacycle.[218] Subsequently, examples of the insertion of aldehydes into isolated rhodium–aryl complexes that lack ring strain have been reported (Equation 9.79). In this case, insertion leads to a rhodium–alkoxo complex that undergoes subsequent β-hydrogen elimination to form a ketone and the corresponding rhodium hydride. The rhodium hydride is unstable to formation of the dimeric complex containing a bridging carbonyl shown.[214,215]

$$(9.77)$$

$$(9.78)$$

$$(9.79)$$

$$(9.80)$$

The insertions of imines into late transition metal–carbon bonds are even less common. In one case, the insertion of an imine into Ni– and Pd–acyl bonds occurs with 2,1-regiochemistry to form an aminoalkyl product (Equation 9.80).[266–269] This reaction is likely to occur through a polar transition state formed by attack of a nucleophilic nitrogen at the electrophilic acyl carbon. One set of examples of 1,2-insertions of imines into late metal–carbon bonds have been reported. This example involves insertion of N-aryl aldimines into rhodium–aryl complexes containing a labile pyridine ligand (Equation 9.81). The rates of these reactions were inverse order in added pyridine, suggesting that the reaction occurs by an intramolecular migratory insertion mechanism after replacement of the coordinated pyridine by the imine.

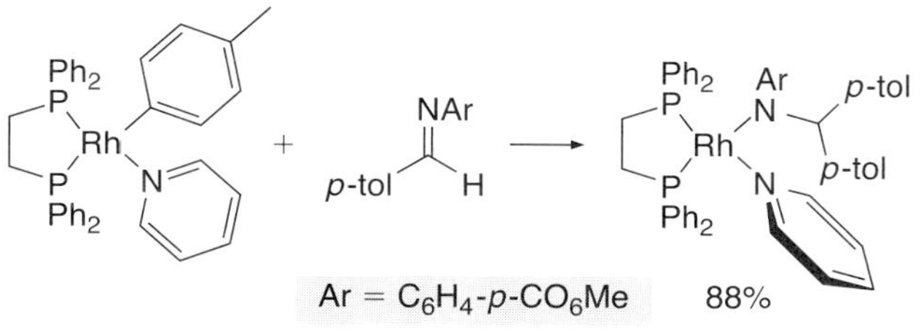

$$Ar = C_6H_4\text{-}p\text{-}CO_6Me \qquad 88\%$$

(9.81)

9.2.2.3. Insertions of Olefins and Acetylenes into Metal–Heteroatom Bonds

9.2.2.3.1. Insertion of Olefins into Metal–Oxygen Bonds

The insertion of olefins into the metal–oxygen bonds of isolated alkoxo, phenoxo, or hydroxo compounds has been observed directly in a few cases. As will be noted in Chapter 11, a hydroxy or alkoxyalkyl group can be formed by insertion of an olefin into a metal–oxygen bond or by attack of hydroxide or an alkoxide on a coordinated olefin. Many studies described in Chapter 11 imply that this type of compound is formed by nucleophilic attack on a coordinatively saturated olefin complex, and this reaction has been proposed as the C–O bond-forming step during oxidations of olefins catalyzed by palladium complexes. However, Henry provided some of the first evidence that the C–O bond forms by insertion of olefins into metal–hydroxo and –alkoxo complexes under certain reaction conditions.[270]

The only isolated metal–hydroxo complex that has been shown to react with olefins to form products from transfer of hydroxide is $Cp*Ir(PMe_3)(Ph)(OH)$. However, this formal insertion process does not occur by a migratory insertion mechanism. Instead, the reaction with olefin is catalyzed by trace amounts of $Cp*Ir(PMe_3)(Ph)(OTf)$ and appears to involve replacement of triflate with ethylene to generate the cationic $[Cp*Ir(PMe_3)(Ph)(C_2H_4)]^+$, which undergoes attack by the separate iridium hydroxido complex, as shown in Scheme 9.11.

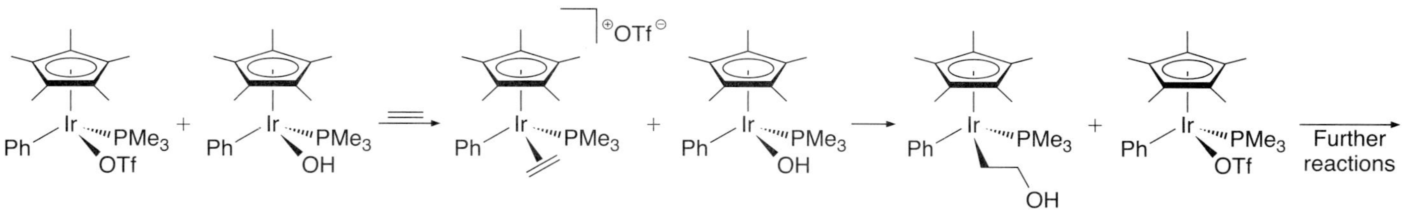

Scheme 9.11

Nevertheless, recent catalytic reactions have suggested that syn addition of a metal–alkoxo linkage across an olefin can occur, and this stereochemistry implies that the reactions occur by a migratory insertion pathway. Henry has shown that the stereochemical outcome of oxypalladations during halohydrin additions to olefins is different at high and low concentrations of chloride. These results imply that migratory insertion occurs under one set of conditions and external attack on coordinated olefin occurs under the other conditions.[270,271] Labeling studies by Hayashi[272] imply that intramolecular syn "oxypalladation" of a phenoxide with a pendant olefin occurs in the absence of added chloride during the reaction depicted in Scheme 9.12. Anti oxypalladation appears to occur in the presence of added chloride. As shown at the bottom of the scheme, syn oxypalladation of the labeled substrate places the metal and the deuterium trans to each other across the ring. As is presented in Chapter 10, elimination of the β-hydrogen requires a cis arrangement across the ring. Thus, the formation of products containing deuterium implies that the oxypalladation occurs in a syn fashion.

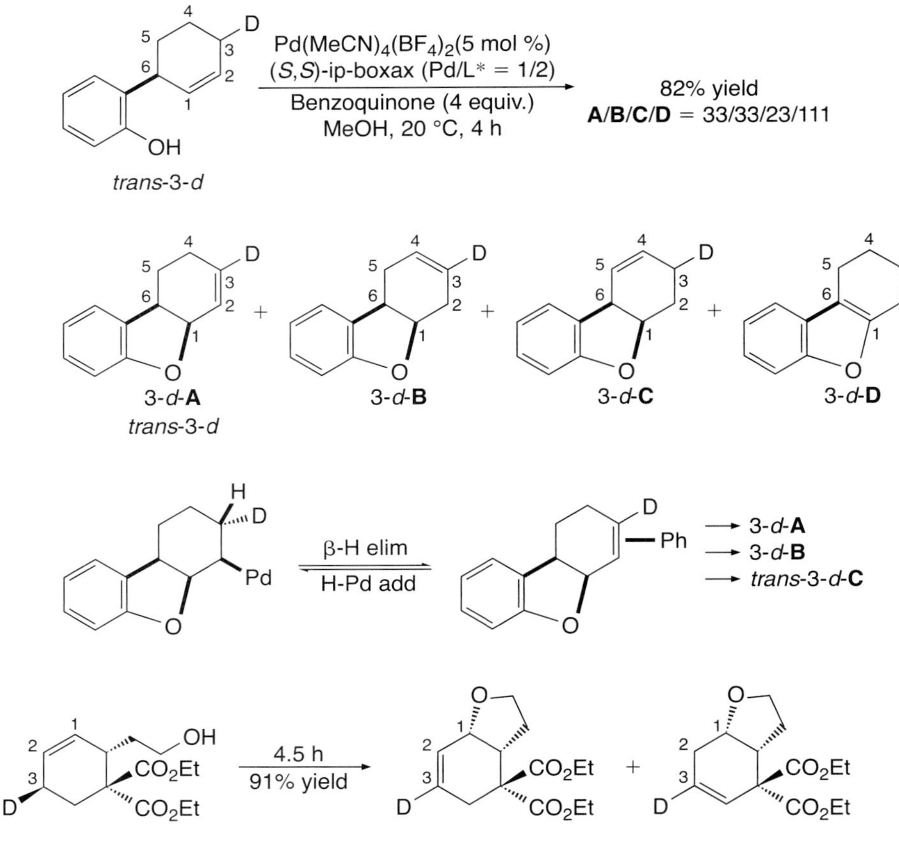

Scheme 9.12

A similar experiment with an unsaturated alcohol reported by Stoltz also implies that syn oxypalladation occurs.[273] Again, the observation of deuterium in the product implies that the alkoxyalkylpalladium intermediate contains the palladium cis to the alkoxo group and trans to the deuterium label. The stereochemical outcome of catalytic cyclizations reported by Wolfe also imply that cis oxypalladation occurs during certain transformations, as shown in Equation 9.82.

Finally, reaction of the discrete rhodium–alkoxo complex containing a tethered olefin shown in Equation 9.83 has been shown to undergo insertion of the alkene into the metal–alkoxo bond, followed by β-hydrogen elimination. This mechanism has also been shown to occur by a stereochemical labeling study in which the stereochemistry of the olefinic product reveals the mode of addition. The stereochemistry of the product from reaction of the alkoxo olefin complex containing deuterium in the trans position of the coordinated alkene shows that a syn addition of the rhodium and oxygen occurred across the carbon–carbon bond.[274]

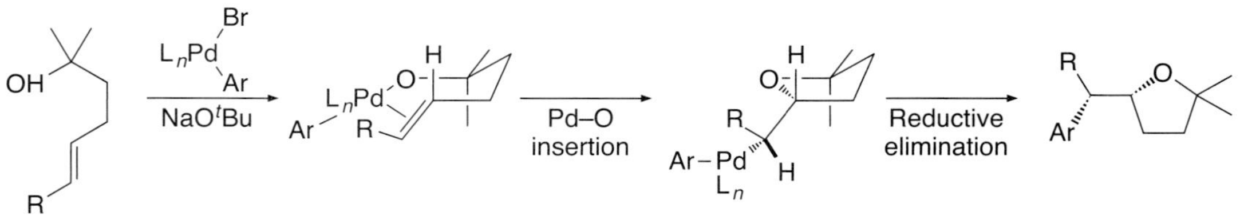

$$(9.82)$$

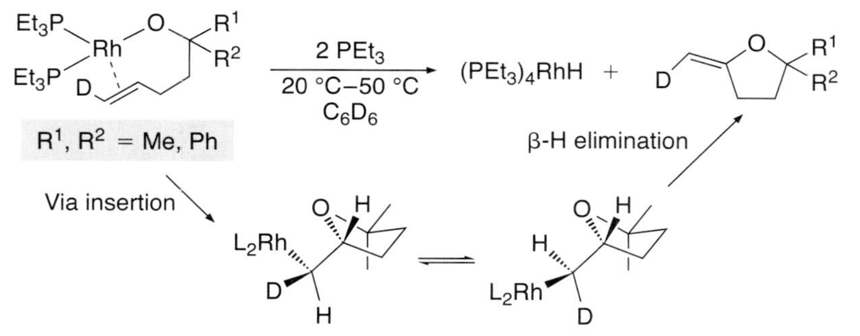

(9.83)

9.2.2.3.2. Insertions of Olefins into Metal–Nitrogen Bonds

Examples of the insertions of alkenes or alkynes into metal–amido bonds are also rare. Examples of the insertions of alkenes into the M–N bonds of isolated amido complexes include the reaction of a rhodium anilide complex with alkenes to form imines with kinetic behavior that is consistent with migratory insertion,[275] and the formal insertion of the strongly electrophilic acrylonitrile into a platinum anilide.[276] Additional examples include reactions of a lanthanide–amido complex generated in situ,[277] a catalytic carboamination process in which the stereochemistry implies insertions of olefins into amides,[278] and a catalytic hydroamination that appears to occur through an aminoalkyl complex generated by syn addition of the iridium and amido groups across the C=C bond of norbornene.

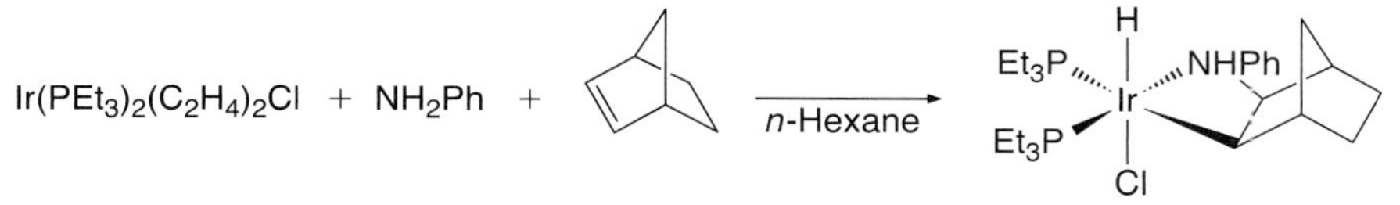

(9.84)

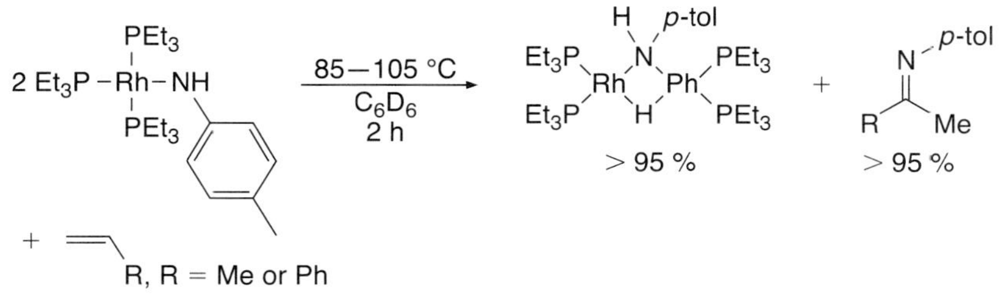

(9.85)

via

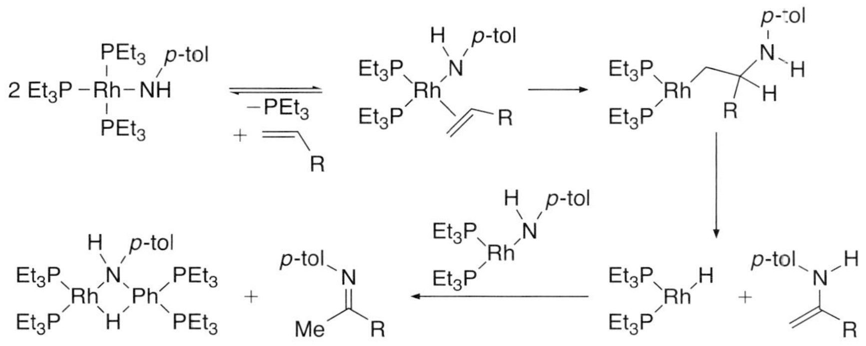

(9.86)

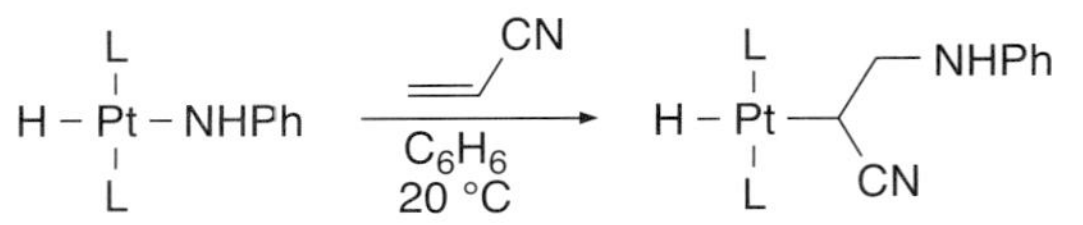

$$(9.87)$$

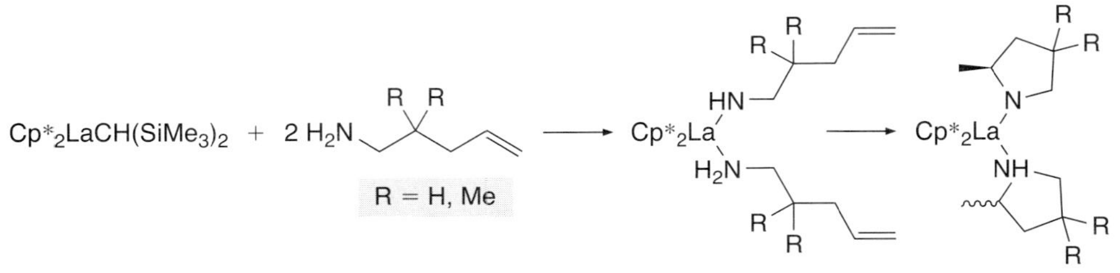

R = H, Me

$$(9.88)$$

via

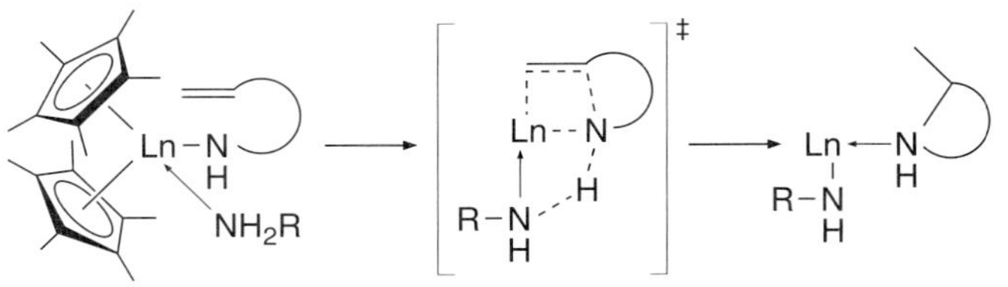

Equation 9.84 depicts the reaction of aniline, norbornene, and [Ir(PEt$_3$)$_2$(C$_2$H$_4$)$_2$Cl] to form an iridium–aminoalkyl complex.[281] This process is thought to occur by oxidative addition of the aniline to form an amido hydride complex, followed by insertion of the strained alkene into the iridium–nitrogen bond. Products from oxidative addition of aniline to the same iridium species were shown to form in the absence of the olefin. The syn stereochemistry of the aminoalkyl product indicated that a migratory insertion pathway was followed.

Equation 9.85 shows the reaction of a similar rhodium–amido complex with unstrained olefins. In this case, the product is a free imine. The inverse order of the reaction in added phosphine and the zero order of the reaction in any free amine imply that the imine is formed by the sequence shown in Equation 9.86, involving olefin insertion, β-hydrogen elimination (see Chapter 10), and tautomerization of the resulting enamine to the imine.

Equation 9.87 involves insertion of acrylonitrile, which is electrophilic enough that the reaction is likely to occur by direct attack of the nitrogen electron pair at the olefin, not by a migratory insertion.

The most direct observation of the insertion of an unactivated olefin into the metal–nitrogen bond of an amido complex involves an intramolecular reaction of a lanthanide–amido complex.[277] As shown in Equation 9.88, the reaction of a lanthanum alkyl with an aminoalkene forms an amido complex containing a tethered olefin. This complex also contains a coordinated aminoalkene. This complex then undergoes insertion of both amino olefins to form the product containing one amido group and one amino group, presumably via initial formation of the aminoalkyl–lanthanide complex and proton transfer to convert the alkyl to the more stable amido complex. The kinetics of the catalytic hydroamination were consistent with two amino groups in the coordination sphere of the metal during insertion. The mechanism in Equation 9.88 involving hydrogen bonding of the amino group to the amido group during insertion was proposed to account for these data. This insertion process is the C–N bond-forming step of a variety of intramolecular olefin hydroaminations catalyzed by lanthanide complexes.[282]

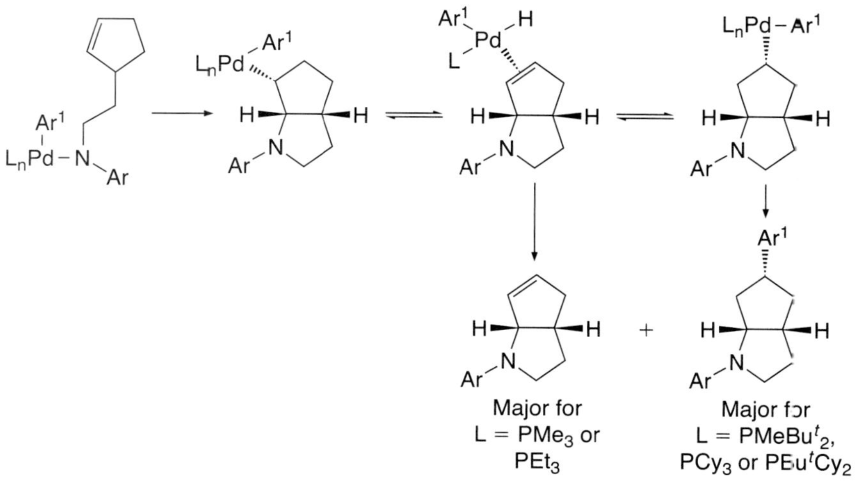

(9.89a)

Finally, several palladium-catalyzed processes appear to occur by insertion of alkenes into palladium amides. Cyclizations of aminoalkenes in the presence of aryl halides to form pyrrolidines appear to occur by intramolecular insertion of alkenes into palladium amides. One example of the cyclizations of aminoalkenes in the presence of aryl halides is shown in Equation 9.89a. The relative stereochemistry of the aryl group and the nitrogen in the ring structure of the product indicates that syn addition of palladium and nitrogen occurred across the alkene.

Oxidative aminations of olefins with sulfonamides also appear to occur by insertions of alkenes into Pd–N bonds in some cases. An example of the use of stereochemistry to probe for syn or anti addition of palladium and nitrogen across an alkene during oxidative amination is shown in Equation 9.89b. The stereochemistry of the alkene unit indicates that the reactions occur by insertions of olefins into Pd–N bonds. As shown in Equation 9.89b, the E-olefin containing a pendant sulfonamide would form the E-product if syn addition of the palladium and sulfonamido group occurs during the catalytic process, but it would form the Z-product if anti addition occurred. The product from syn addition is formed, and such syn stereochemistry is consistent with migratory insertion of the olefin into a palladium–sulfonamido intermediate.[280]

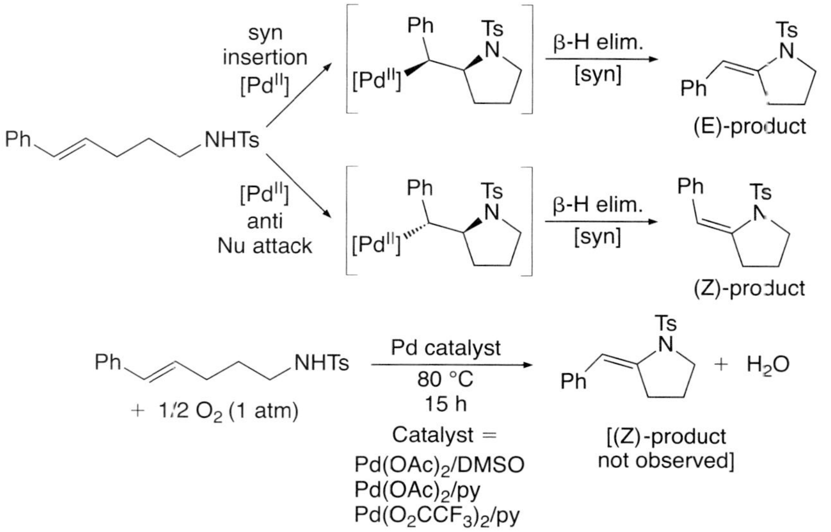

(9.89b)

A few final comments should be made on the insertions of substrates containing C–C multiple bonds into the bonds between a transition metal and an electronegative heteroatom. First, insertions of olefins into related thiolate and phosphide complexes are as rare as insertions into alkoxo and amido complexes. Reactions of acrylonitrile into the metal–phosphorus bonds of palladium– and platinum–phosphido complexes to give products from formal insertions have been observed, and one example is shown in Equation 9.90. However, these reactions are more likely to occur by direct attack of the phosphorus on the electrophilic carbon of acrylonitrile than by migratory insertion. Second, the insertions of alkynes into metal–oxygen or metal–nitrogen covalent bonds are rare,[283] even though the C–C π-bond in an alkyne is weaker than the π-bond in an alkene.

$$(DPPE)Pd\begin{smallmatrix} P(Mes)_2 \\ \\ Me \end{smallmatrix} \; + \; \underset{CN}{=\!\!\!=} \; \longrightarrow \; (DPPE)Pd\begin{smallmatrix} NC \overset{}{\diagup}\!\!\diagdown PMes_2 \\ \\ Me \end{smallmatrix} \qquad (9.90)$$

9.2.2.3.3. Insertions of Olefins and Acetylenes into Metal–Silicon and Metal–Boron Bonds

The insertions of olefins into metal–silyl complexes is an important step in the hydrosilylation of olefins, and the insertions of olefins and alkynes into metal–boron bonds is likely to be part of the mechanism of the diborations[284-298] and silaborations[299,300] of substrates containing C–C multiple bonds. Other reactions, such as the dehydrogenative silylation[301] of olefins can also involve this step. Several studies imply that the rhodium–catalyzed hydrosilylations of olefins occur by insertion of olefins into rhodium–silicon bonds, while side products from palladium- and platinum-catalyzed hydrosilylations are thought to form by insertion of olefins into the metal–silicon bonds. In particular, vinylsilanes are thought to form by a sequence involving olefin insertion into the metal–silicon bond, followed by β-hydrogen elimination (Chapter 10) to form the metal–hydride and vinylsilane products.

A few studies of isolated metal–silyl complexes and the computational study of rhodium–silyl complexes illustrate the insertion of olefins into metal–silicon bonds. Wrighton studied the photochemical reaction of iron–silyl complexes with ethylene (Scheme 9.13). Photolysis of $Cp^*Fe(CO)_2(SiMe_3)$ in the presence of ethylene forms $Cp^*Fe(CO)(C_2H_4)(SiMe_3)$. This complex appears to insert ethylene, but the 16-electron insertion product is unstable and forms the corresponding vinylsilane and iron hydride complexes as products. Photolysis of $Cp^*Fe(CO)_2(SiMe_3)$ in the presence of ethylene *and* CO forms the β-silylalkyl complex containing two CO ligands.

Direct observation of the migratory insertion of olefins into metal–silicon bonds of platinum, palladium, and rhodium are rare because the resulting β-silylalkyl complexes typically undergo subsequent reactions like C–H bond-forming reductive elimination. However, Sakaki has computed barriers for the insertion of olefins into the rhodium–silicon bond of the hypothetical complex $[Rh(PH_3)_2Cl(H)(C_2H_4)(SiMe_3)]$ that would result from oxidative addition of $HSiMe_3$ to an analog of Wilkinson's catalyst, followed by coordination of ethylene.[302] The activation energy for the migratory insertion process is calculated to be only 13.5 kcal/mol.

As noted in the introduction to this section, the insertions of alkynes and alkenes into metal–boron bonds appear to be involved in catalytic additions of borane reagents to alkenes and alkynes. One direct observation of the insertion of an alkyne into a metal–boron bond has been reported, and this reaction is shown in Equation 9.91. In another case, the reaction of an olefin with a metal–boryl complex generated products consistent with the

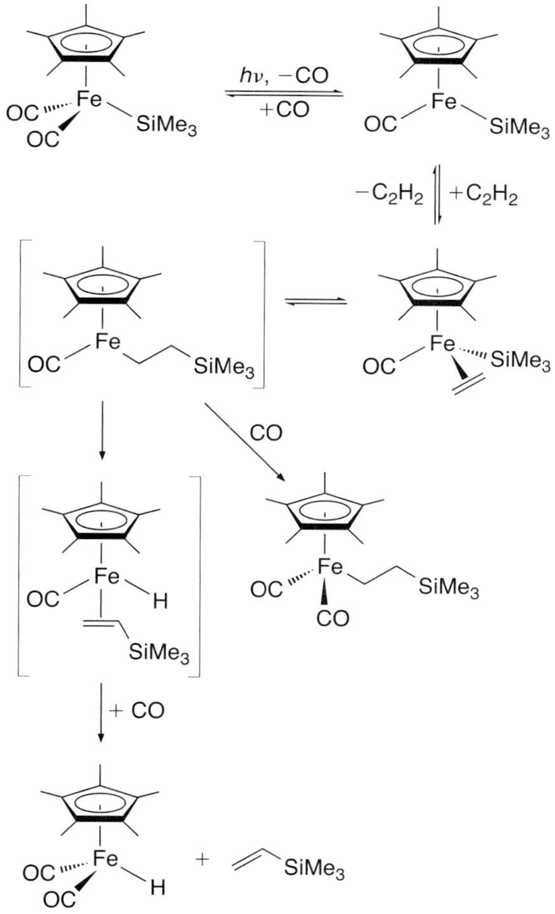

Scheme 9.13

insertion of the olefin into the metal–boron bond, followed by a combination of reductive elimination and β-hydrogen elimination.[303]

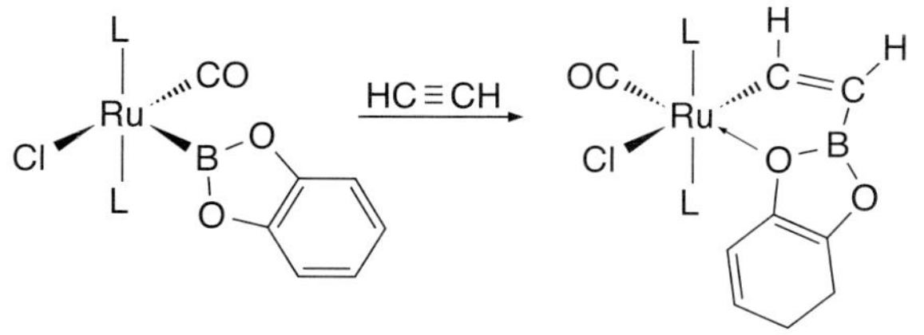

$$(9.91)$$

9.3. Summary

Migratory insertions occur between many types of unsaturated ligands and many types of covalent, σ-bound ligands. Except for insertions of alkylidene and alkylidyne ligands, these reactions do not change the oxidation state of the metal. However, they reduce the coordination number by one and decrease the valence electron count by two. Insertions of alkylidene and alkylidyne ligands into σ-bound ligands decreases the oxidation state of the metal by two and decreases the number of valence electrons by two.

The most common insertion reactions involve the insertions of carbon monoxide into metal–hydrocarbyl ligands and the insertions of olefins and alkynes into metal–hydride and metal–hydrocarbyl ligands. Few insertions of CO into metal hydrides are known because this process is typically unfavorable thermodynamically. CO and alkenes are also now known to insert into many metal–heteroatom bonds involving heteroatoms that are both more electronegative and more electropositive than carbon, and alkynes are known to insert into metal–heteroatom bonds involving silicon and boron.

The insertions of CO and alkenes occur by an intramolecular, migratory insertion mechanism. During the insertion of CO, the coordinated CO can either migrate into the metal–alkyl bond or, more commonly, the hydrocarbyl group can migrate onto the coordinated CO. The added CO or another added dative ligand then occupies the open coordination site generated by the migratory insertion. The irreversible step can be the insertion or trapping step, depending on the complex, added ligand and conditions.

When thermodynamically favored, insertions into metal–hydride bonds are faster than insertions into metal–hydrocarbyl bonds. Studies on migratory aptitudes have shown in some cases that migrations of alkyl groups are faster than migration of aryl groups, even when less favored thermodynamically. Insertions into metal–alkyl bonds are typically faster than insertion into metal–benzyl bonds. Insertions have been shown in certain cases to be accelerated by Lewis or protic acids, nucleophiles, oxidants and reductants. Catalytic processes occurring by migratory insertions are legion and are described in later chapters of the book.

References and Notes

1. Wojcicki, A. *Adv. Organomet. Chem.* **1973**, *11*, 87.
2. Calderazzo, F. *Angew. Chem., Int. Ed. Engl.* **1977**, *16*, 299.
3. Kuhlman, E. J.; Alexander, J. J. *Coord. Chem. Rev.* **1980**, *33*, 195.
4. Flood, T. C. In *Topics in Organic and Organometallic Stereochemistry*; Geoffroy, G. L., Ed.; Wiley: New York, 1981; Vol. 12, p 83.
5. Alexander, J. J. In *The Chemistry of the Metal–Carbon Bond*; Hartley, F. R., Ed.; Wiley: New York, 1985; Vol. 2, Chapter 5.
6. Mawby, R. J.; Basolo, F.; Pearson, R. G. *J. Am. Chem. Soc.* **1964**, *86*, 3994.
7. Butler, I. S.; Basolo, F.; Pearson, R. G. *Inorg. Chem.* **1967**, *6*, 2074.
8. MacLaury, M. Ph. D. Thesis, Stanford University, 1974.
9. Piacenti, F.; Bianchi, M.; Benedetti, E. *Chim. Ind.* **1967**, *49*, 245.
10. Martin, J. T.; Baird, M. C. *Organometallics* **1983**, *2*, 1073.
11. Garrou, P. E.; Heck, R. F. *J. Am. Chem. Soc.* **1976**, *98*, 4115.
12. Bumagin, N. A.; Gulevich, Y. V.; Beletskaya, I. P. *J. Organomet. Chem.* **1985**, *285*, 415.
13. Moser, W. R.; Wang, A. W.; Kildahl, N. K. *J. Am. Chem. Soc.* **1988**, *110*, 2816.
14. Huser, M.; Youinou, M. T.; Osborn, J. A. *Angew. Chem., Int. Ed. Engl.* **1989**, *28*, 1427.
15. Fachinetti, G.; Floriani, C.; Marchetti, F.; Merlino, S. *J. Chem. Soc., Chem. Commun.* **1976**, 522.
16. Bryndza, H. E. *Organometallics* **1985**, *4*, 1686.
17. Rees, W. M.; Churchill, M. R.; Fettinger, J. C.; Atwood, J. D. *Organometallics* **1985**, *4*, 2179.
18. Macgregor, S. A.; Neave, G. W. *Organometallics* **2004**, *23*, 891.
19. Bryndza, H. E.; Fultz, W. C.; Tam, W. *Organometallics* **1985**, *4*, 939.
20. Joslin, F. L.; Johnson, M. P.; Mague, J. T.; Roundhill, D. M. *Organometallics* **1991**, *10*, 2781.
21. Fox, D. J.; Bergman, R. G. *J. Am. Chem. Soc.* **2003**, *125*, 11772.
22. Fox, D. J.; Bergman, R. G. *Organometallics* **2004**, *23*, 1656.
23. Woo, H. G.; Freeman, W. P.; Tilley, T. D. *Organometallics* **1992**, *11*, 2198.
24. Arnold, J.; Tilley, T. D.; Rheingold, A. L.; Geib, S. J.; Arif, A. M. *J. Am. Chem. Soc.* **1989**, *111*, 149.
25. Campion, B. K.; Falk, J.; Tilley, T. D. *J. Am. Chem. Soc.* **1987**, *109*, 2049.
26. Elsner, F. H.; Tilley, T. D.; Rheingold, A. L.; Geib, S. J. *J. Organomet. Chem.* **1988**, *358*, 169.
27. Arnold, J.; Tilley, T. D.; Rheingold, A. L.; Geib, S. J.; Arif, A. M. *J. Am. Chem. Soc.* **1989**, *111*, 149.
28. Roddick, D. M.; Heyn, R. H.; Tilley, T. D. *Organometallics* **1989**, *8*, 324.
29. Campion, B. K.; Heyn, R. H.; Tilley, T. D. *J. Am. Chem. Soc.* **1990**, *112*, 2011.
30. Anglin, J. R.; Calhoun, H. P.; Graham, W. A. G. *Inorg. Chem.* **1977**, *16*, 2281.
31. Collman, J. P.; Finke, R. G.; Cawse, J. N.; Brauman, J. I. *J. Am. Chem. Soc.* **1978**, *100*, 4766.

32. Wax, M. J.; Bergman, R. G. *J. Am. Chem. Soc.* **1981**, *103*, 7028.
33. Anderson, G. K.; Cross, R. J. *Acc. Chem. Res.* **1984**, *17*, 67.
34. Anderson, G. K.; Cross, R. J. *Chem. Soc. Rev.* **1980**, *9*, 185.
35. Anderson, G. K.; Clark, H. C.; Davies, J. A. *Organometallics* **1982**, *1*, 64.
36. Rix, F. C.; Brookhart, M. *J. Am. Chem. Soc.* **1995**, *117*, 1137.
37. Bock, P. L.; Boschetto, D. J.; Rasmussen, J. R.; Demers, J. P.; Whitesides, G. M. *J. Am. Chem. Soc.* **1974**, *96*, 2814.
38. Whitesides, G. M.; Boschetto, D. M. *J. Am. Chem. Soc.* **1969**, *91*, 4313.
39. Whitesides, G. M.; Boschetto, D. J. *J. Am. Chem. Soc.* **1971**, *93*, 1529.
40. Walborsky, H. M.; Allen, L. E. *J. Am. Chem. Soc.* **1971**, *93*, 5465.
41. Cassar, L.; Giarusso, A. *Gazz. Chim. Ital.* **1973**, *103*, 793.
42. Noack, K.; Calderazzo, F. *J. Organomet. Chem.* **1967**, *10*, 101.
43. Flood, T. C.; Jensen, J. E.; Statler, J. A. *J. Am. Chem. Soc.* **1981**, *103*, 4410.
44. Flood, T. C.; Campbell, K. D.; Downs, H. H.; Nakanishi, S. *Organometallics* **1983**, *2*, 1590.
45. Flood, T. C.; Campbell, K. D. *J. Am. Chem. Soc.* **1984**, *106*, 2853.
46. Brunner, H.; Hammer, B.; Bernal, I.; Draux, M. *Organometallics* **1983**, *2*, 1595.
47. Davison, A.; Martinez, N. *J. Organomet. Chem.* **1974**, *74*, C17.
48. Attig, T. G.; Teller, R. G.; Wu, S. M.; Bau, R.; Wojcicki, A. *J. Am. Chem. Soc.* **1979**, *101*, 619.
49. Pankowski, M.; Bigorgne, M. *J. Organomet. Chem.* **1971**, *30*, 227.
50. Pankowski, M.; Samuel, E.; Bigorgne, M. *J. Organomet. Chem.* **1975**, *97*, 105.
51. Pankowski, M.; Bigorgne, M. In *The 13th International Conference on Organometallic Chemistry*; Kyoto, Japan, 1977.
52. Jablonski, C. R.; Wang, Y. P. *Inorg. Chem.* **1982**, *21*, 4037.
53. Wright, S. C.; Baird, M. C. *J. Am. Chem. Soc.* **1985**, *107*, 6899.
54. Barnard, C. F. J.; Daniels, J. A.; Mawby, R. J. *J. Chem. Soc., Dalton Trans.* **1979**, 1331.
55. Egglestone, D. L.; Baird, M. C.; Lock, C. J. L.; Turner, G. *J. Chem. Soc., Dalton Trans.* **1977**, 1576.
56. Bennett, M. A.; Jeffery, J. C.; Robertson, G. B. *Inorg. Chem.* **1981**, *20*, 323.
57. Glyde, R. W.; Mawby, R. J. *Inorg. Chim. Acta* **1971**, *5*, 317.
58. Bennett, M. A.; Jeffery, J. C. *Inorg. Chem.* **1980**, *19*, 3763.
59. Brunner, H.; Vogt, H. *Chem. Ber.* **1981**, *114*, 2186.
60. Curtis, M. D.; Shiu, K. B.; Butler, W. M. *J. Am. Chem. Soc.* **1986**, *108*, 1550.
61. Arnold, J.; Tilley, T. D.; Rheingold, A. L. *J. Am. Chem. Soc.* **1986**, *108*, 5355.
62. Roper, W. R.; Taylor, G. E.; Waters, J. M.; Wright, L. J. *J. Organomet. Chem.* **1979**, *182*, C46.
63. Fachinetti, G.; Fochi, G.; Floriani, C. *J. Chem. Soc., Dalton Trans.* **1977**, 1946.
64. Hitam, R. B.; Narayanaswamy, R.; Rest, A. J. *J. Chem. Soc., Dalton Trans.* **1983**, 615.
65. Versluis, L.; Ziegler, T.; Baerends, E. J.; Ravenek, W. *J. Am. Chem. Soc.* **1989**, *111*, 2018.
66. Goh, S. K.; Marynick, D. S. *Organometallics* **2002**, *21*, 2262.
67. Lau, K. S. Y.; Becker, Y.; Huang, F.; Baenziger, N.; Stille, J. K. *J. Am. Chem. Soc.* **1977**, *99*, 5664.
68. Kubota, M.; Blake, D. M.; Smith, S. A. *Inorg. Chem.* **1971**, *10*, 1430.
69. Mawby, R. J.; Pearson, R. G.; Basolo, F. *J. Am. Chem. Soc.* **1964**, *86*, 3994.
70. Webb, S. L.; Giandomenico, C. M.; Halpern, J. *J. Am. Chem. Soc.* **1986**, *108*, 345.
71. Cotton, J. D.; Crisp, G. T.; Latif, L. *Inorg. Chim. Acta* **1981**, *47*, 171.
72. Nicholas, K.; Raghu, S.; Rosenblum, M. *J. Organomet. Chem.* **1974**, *78*, 133.
73. Casey, C. P.; Bunnell, C. A. *J. Am. Chem. Soc.* **1971**, *93*, 4077.
74. Casey, C. P.; Bunnell, C. A.; Calabrese, J. C. *J. Am. Chem. Soc.* **1976**, *98*, 1166.
75. Treichel, P. M.; Stone, F. G. A. *Adv. Organomet. Chem.* **1964**, *1*, 143.
76. For a system in which the thermodynamics seem to favor insertion, see reference 77.
77. Jablonski, C. R. *Inorg. Chem.* **1981**, *20*, 3940.
78. Collman, J. P.; Winter, S. R. *J. Am. Chem. Soc.* **1973**, *95*, 4089.
79. Gladysz, J. A. *Adv. Organomet. Chem.* **1982**, *20*, 1.
80. Wayland, B. B.; Woods, B. A. *J. Chem. Soc., Chem. Commun.* **1981**, 700.
81. Halpern has shown that this carbonylation occurs by a radical chain process, instead of migratory insertion. See reference 82.
82. Paonessa, R. S.; Thomas, N. C.; Halpern, J. *J. Am. Chem. Soc.* **1985**, *107*, 4333.
83. Fagan, P. J.; Moloy, K. G.; Marks, T. J. *J. Am. Chem. Soc.* **1981**, *103*, 6959.
84. These data are collected in Table 1 of Alexander, J. J. In *The Chemistry of the Metal-Carbon Bond*; Hartley, F. R., Ed.; Wiley: New York, 1985; Vol. 2, Chapter 5.
85. Cotton, J. D.; Dunstan, P. R. *Inorg. Chim. Acta* **1984**, *88*, 223.
86. Cotton, J. D.; Crisp, G. T.; Daly, V. A. *Inorg. Chim. Acta* **1981**, *47*, 165.
87. Casey, C. P.; Scheck, D. M. *J. Am. Chem. Soc.* **1980**, *102*, 2723.

88. Lin, Y. C.; Milstein, D.; Wreford, S. S. *Organometallics* **1983**, *2*, 1461.
89. Coffield, T. H.; Kozikowski, J.; Closson, R. D. *Spec. Publ. — R. Chem. Soc.* **1959**, *13*, 126.
90. Barnett, K. W.; Beach, D. L.; Gaydos, S. P.; Pollmann, T. G. *J. Organomet. Chem.* **1974**, *69*, 121.
91. Glyde, R. W.; Mawby, R. J. *Inorg. Chim. Acta* **1970**, *4*, 331.
92. Egglestone, D.; Baird, M. C. *J. Organomet. Chem.* **1976**, *113*, C25.
93. Shaw, B. L.; Singleton, E. *J. Chem. Soc. A* **1967**, 1683.
94. Hart-Davis, A. J.; Mawby, R. J. *J. Chem. Soc. A* **1969**, 2403.
95. Klein, H. F.; Karsch, H. H. *Chem. Ber.* **1976**, *109*, 2524.
96. Huttner, G.; Orama, O.; Bejenke, V. *Chem. Ber.* **1976**, *109*, 2533.
97. Sacconi, L.; Dapporto, P.; Stoppioni, P. *J. Organomet. Chem.* **1976**, *116*, C33.
98. Kim, Y.; Osakada, K.; Sugita, K.; Yamamoto, T.; Yamamoto, A. *Organometallics* **1988**, *7*, 2182.
99. Komiya, S.; Akai, Y.; Tanaka, K.; Yamamoto, T.; Yamamoto, A. *Organometallics* **1985**, *4*, 1130.
100. Hartwig, J. F.; Bergman, R. G.; Andersen, R. A. *J. Am. Chem. Soc.* **1991**, *113*, 6499.
101. Butts, S. B.; Strauss, S. H.; Holt, E. M.; Stimson, R. E.; Alcock, N. W.; Shriver, D. F. *J. Am. Chem. Soc.* **1980**, *102*, 5093.
102. Richmond, T. G.; Basolo, F.; Shriver, D. F. *Inorg. Chem.* **1982**, *21*, 1272.
103. Butts, S. B.; Richmond, T. G.; Shriver, D. F. *Inorg. Chem.* **1981**, *20*, 278.
104. Grundy, K. R.; Roper, W. R. *J. Organomet. Chem.* **1981**, *216*, 255.
105. McLain, S. J. *J. Am. Chem. Soc.* **1983**, *105*, 6355.
106. Labinger, J. A.; Bonfiglio, J. N.; Grimmett, D. L.; Masuo, S. T.; Shearin, E.; Miller, J. S. *Organometallics* **1983**, *2*, 733.
107. Grimmett, D. L.; Labinger, J. A.; Bonfiglio, J. N.; Masuo, S. T.; Shearin, E.; Miller, J. S. *Organometallics* **1983**, *2*, 1325.
108. Kubota, M.; Rothrock, R. K.; Geibel, J. *J. Chem. Soc., Dalton Trans.* **1973**, 1267.
109. Magnuson, R. H.; Meirowitz, R.; Zulu, S.; Giering, W. P. *J. Am. Chem. Soc.* **1982**, *104*, 5790.
110. Magnuson, R. H.; Meirowitz, R.; Zulu, S. J.; Giering, W. P. *Organometallics* **1983**, *2*, 460.
111. Miholova, D.; Vlcek, A. A. *J. Organomet. Chem.* **1982**, *240*, 413.
112. Yamamoto, Y.; Yamazaki, H. *Inorg. Chem.* **1974**, *13*, 2145.
113. Nolte, R. J. M. *Chem. Soc. Rev.* **1994**, 11.
114. Yamamoto, M.; Onitsuka, K.; Takahashi, S. *Organometallics* **2000**, *19*, 4669.
115. Clark, G. R.; Collins, T. J.; Marsden, K.; Roper, W. R. *J. Organomet. Chem.* **1983**, *259*, 215.
116. Seidler, M. D.; Bergman, R. G. *Organometallics* **1983**, *2*,
117. Weiner, W. P.; Bergman, R. G. *J. Am. Chem. Soc.* **1983**, *105*, 3922.
118. Seidler, M. D.; Bergman, R. G. *J. Am. Chem. Soc.* **1984**, *106*, 6110.
119. Chaudhari, F. M.; Konx, G. R.; Pauson, P. L. *J. Chem. Soc. C* **1967**, 2255.
120. (a) Jernakoff, P.; Cooper, N. J. *J. Am. Chem. Soc.* **1984**, *106*, 3026; (b) Jernakoff, P.; Cooper, N. J. *Organometallics* **1986**, *5*, 747; (c) Jernakoff, P.; Cooper, N. J. *J. Am. Chem. Soc.* **1987**, *109*, 2173; (d) Jernakoff, P.; Cooper, N. J. *J. Am. Chem. Soc.* **1989**, *111*, 7424.
121. (a) Werner, H.; Kletzin, H.; Hohn, A.; Paul, W.; Knaup, W.; Ziegler, M. L.; Serhadli, O. *J. Organomet. Chem.* **1986**, *306*, 227; (b) Kletzin, H.; Werner, H.; Serhadli, O.; Ziegler, M. L. *Angew. Chem., Int. Ed. Engl.* **1983**, *22*, 46.
122. (a) Lisko, J. R.; Jones, W. M. *Organometallics* **1985**, *4*, 944; (b) Stenstrom, Y.; Jones, W. M. *Organometallics* **1986**, *5*, 178; (c) Stenstrom, Y.; Klauck, G.; Koziol, A.; Palenik, G. J.; Jones, W. M. *Organometallics* **1986**, *5*, 2155.
123. Brookhart, M.; Green, M. L. H. *J. Organomet. Chem.* **1983**, *250*, 395.
124. Olgemoller, B.; Beck, W. *Angew. Chem., Int. Ed. Engl.* **1980**, *19*, 834.
125. Deeming, A. J.; Johnson, B. F. G.; Lewis, J. *J. Chem. Soc., Dalton Trans.* **1973**, 1848.
126. Doherty, N. M.; Bercaw, J. E. *J. Am. Chem. Soc.* **1985**, *107*, 2670.
127. Halpern, J.; Okamoto, T.; Zakhariev, A. *J. Mol. Catal.* **1977**, *2*, 65.
128. Baird, M. C.; Mague, J. T.; Osborn, J. A.; Wilkinson, G. *J. Chem. Soc. A* **1967**, 1347.
129. Halpern, J.; Okamoto, T. *Inorg. Chim. Acta* **1984**, *89*, L53.
130. Chan, A. S. C.; Pluth, J. J.; Halpern, J. *J. Am. Chem. Soc.* **1980**, *102*, 5952.
131. Halpern, J. *Inorg. Chim. Acta* **1981**, *50*, 11.
132. Halpern, J. *Science* **1982**, *217*, 401.
133. Schwartz, J.; Labinger, J. A. *Angew. Chem., Int. Ed. Engl.* **1976**, *15*, 333.
134. Hart, D. W.; Schwartz, J. *J. Am. Chem. Soc.* **1974**, *96*, 8115.
135. Negishi, E.; Takahashi, T. *Synthesis* **1988**, 1.
136. Wipf, P.; Takahashi, H.; Zhuang, N. *Pure App. Chem.* **1998**, *70*, 1077.
137. Labinger, J. A. In *Comprehensive Organic Synthesis*; Trost, B. M., Fleming, I., Eds.; Pergamon Press: Oxford, 1991; Chapter 3, Section 3.9.

138. Chirik, P.; Day, M.; Labinger, J.; Bercaw, J. *J. Am. Chem. Soc.* **1999**, *121*, 10308.

139. Otsuka, S.; Nakamura, A. *Adv. Organomet. Chem.* **1976**, *14*, 245.

140. Foo, T.; Bergman, R. G. *Organometallics* **1992**, *11*, 1811.

141. Bassetti, M.; Casellato, P.; Gamasa, M. P.; Gimeno, J.; Gonzalez Bernardo, C.; Martin Vaca, B. *Organometallics* **1997**, *16*, 5470.

142. Frohnapfel, D. S.; White, P. S.; Templeton, J. L. *Organometallics* **2000**, *19*, 1497.

143. Selmeczy, A. D.; Jones, W. D. *Inorg. Chim. Acta* **2000**, *300*, 138.

144. Bassetti, M.; Marini, S.; Diaz, J.; Gamasa, M. P.; Gimeno, J.; Rodriguez-Alvarez, Y.; Garcia-Granda, S. *Organometallics* **2002**, *21*, 4815.

145. Gao, Y.; Jennings, M. C.; Puddephatt, R. J. *Dalton Trans.* **2003**, 261.

146. Navarro, J.; Sola, E.; Martin, M.; Dobrinovitch, I. T.; Lahoz, F. J.; Oro, L. A. *Organometallics* **2004**, *23*, 1908.

147. Nakamura, A.; Otsuka, S. *J. Am. Chem. Soc.* **1972**, *94*, 1886.

148. Tanke, R. S.; Crabtree, R. H. *J. Am. Chem. Soc.* **1990**, *112*, 7984.

149. Ojima, I.; Clos, N.; Donovan, R. J.; Ingallina, P. *Organometallics* **1990**, *9*, 3127.

150. Trost, B. M.; Ball, Z. T.; Joge, T. *J. Am. Chem. Soc.* **2002**, *124*, 7922.

151. Trost, B. M.; Ball, Z. T. *J. Am. Chem. Soc.* **2005**, *127*, 17644.

152. Chung, L. W.; Wu, Y. D.; Trost, B. M.; Ball, Z. T. *J. Am. Chem. Soc.* **2003**, *125*, 11578.

153. Trost, B. M.; Ball, Z. T. *J. Am. Chem. Soc.* **2003**, *125*, 30.

154. Trost, B. M.; Ball, Z. T.; Laemmerhold, K. M. *J. Am. Chem. Soc.* **2005**, *127*, 10028.

155. Herberich, G. E.; Barlage, W. *Organometallics* **1987**, *6*, 1924.

156. Clark, H. C.; Ferguson, G.; Goel, A. B.; Janzen, E. G.; Ruegger, H.; Siew, P. Y.; Wong, C. S. *J. Am. Chem. Soc.* **1986**, *108*, 6961.

157. This stereochemical assignment is a revision of a previous assignment.

158. Li, X.; Vogel, T.; Incarvito, C. D.; Crabtree, R. H. *Organometallics* **2005**, *24*, 62.

159. One mechanism now established for reactions catalyzed by some complexes occurs without a migratory insertion step. See Chapter 15.

160. Fryzuk, M. D.; Piers, W. E. *Organometallics* **1990**, *9*, 986.

161. Cossee, P. *J. Catal.* **1964**, *3*, 80.

162. Cramer, R. *Acc. Chem. Res.* **1968**, *1*, 186.

163. Klein, H. F.; Hammer, R.; Gross, J.; Schubert, U. *Angew. Chem., Int. Ed. Engl.* **1980**, *19*, 809.

164. Brookhart, M.; Hauptman, E.; Lincoln, D. M. *J. Am. Chem. Soc.* **1992**, *114*, 10394.

165. The butyl product of the second ethylene insertion undergoes β-hydrogen elimination (see Chapter 10) to generate free butene and the starting rhodium hydride. Thus, the rhodium complex catalyzes the dimerization of ethylene to form butene.

166. Flood, T. C.; Bitler, S. P. *J. Am. Chem. Soc.* **1984**, *106*, 6076.

167. Watson, P. L.; Roe, D. C. *J. Am. Chem. Soc.* **1982**, *104*, 6471.

168. Flood, T. C.; Bitler, S. P. *J. Am. Chem. Soc.* **1984**, *106*, 6076.

169. (a) Wu, Z.; Jordan, R. F.; Petersen, J. L. *J. Am. Chem. Soc.* **1995**, *117*, 5867; (b) Carpentier, J. F.; Wu, Z.; Lee, C. W.; Stromberg, S.; Christopher, J. N.; Jordan, R. F. *J. Am. Chem. Soc.* **2000**, *122*, 7750; (c) Carpentier, J. F.; Maryin, V. P.; Luci, J.; Jordan, R. F. *J. Am. Chem. Soc.* **2001**, *123*, 898; (d) Stoebenau, E. J.; Jordan, R. F. *J. Am. Chem. Soc.* **2003**, *125*, 3222; (e) Stoebenau, E. J.; Jordan, R. F. *J. Am. Chem. Soc.* **2004**, *126*, 11170; (f) Casey, C. P.; Lee, T. Y.; Tunge, J. A.; Carpenetti, D. W. *J. Am. Chem. Soc.* **2001**, *123*, 10762; (g) Casey, C. P.; Klein, J. F.; Fagan, M. A. *J. Am. Chem. Soc.* **2000**, *122*, 4320; (h) Casey, C. P.; Fagan, M. A.; Hallenbeck, S. L. *Organometallics* **1998**, *17*, 287; (i) Casey, C. P.; Hallenbeck, S. L.; Wright, J. M.; Landis, C. R. *J. Am. Chem. Soc.* **1997**, *119*, 9680; (j) Casey, C. P.; Hallenbeck, S. L.; Pollock, D. W.; Landis, C. R. *J. Am. Chem. Soc.* **1995**, *117*, 9770; (k) Casey, C. P.; Tunge, J. A.; Lee, T. Y.; Fagan, M. A. *J. Am. Chem. Soc.* **2003**, *125*, 2641; (l) Landis, C. R.; Rosaaen, K. A.; Sillars, D. R. *J. Am. Chem. Soc.* **2003**, *125*, 1710.

170. Ittel, S. D.; Johnson, L. K.; Brookhart, M. *Chem. Rev.* **2000**, *100*, 1169.

171. Matsui, S.; Mitani, M.; Saito, J.; Tohi, Y.; Makio, H.; Matsukawa, N.; Takagi, Y.; Tsuru, K.; Nitabaru, M.; Nakano, T.; Tanaka, H.; Kashiwa, N.; Fujita, T. *J. Am. Chem. Soc.* **2001**, *123*, 6847.

172. (a) Tian, J.; Hustad, P. D.; Coates, G. W. *J. Am. Chem. Soc.* **2001**, *123*, 5134; (b) Hustad, P. D.; Coates, G. W. *J. Am. Chem. Soc.* **2002**, *124*, 11578; (c) Hustad, P. D.; Tian, J.; Coates, G. W. *J. Am. Chem. Soc.* **2002**, *124*, 3614.

173. Lamberti, M.; Pappalardo, D.; Zambelli, A.; Pellecchia, C. *Macromolecules* **2002**, *35*, 658.

174. Makio, H.; Kashiwa, N.; Fujita, T. *Adv. Synth. Catal.* **2002**, *344*, 477.

175. Small, B. L.; Brookhart, M. *Macromolecules* **1999**, *32*, 2120.

176. Talarico, G.; Busico, V.; Cavallo, L. *J. Am. Chem. Soc.* **2003**, *125*, 7172.

177. Watson, P. L. *J. Am. Chem. Soc.* **1982**, *104*, 337.

178. Leatherman, M. D.; Svejda, S. A.; Johnson, L. K.; Brookhart, M. *J. Am. Chem. Soc.* **2003**, *125*, 3068.

179. Williams, B. S.; Leatherman, M. D.; White, P. S.; Brookhart, M. *J. Am. Chem. Soc.* **2005**, *127*, 5132.
180. Cossee, P. *Tetrahedron Lett.* **1960**, 12.
181. Cossee, P. *Tetrahedron Lett.* **1960**, 17.
182. Arlman, E. J.; Cossee, P. *J. Catal.* **1964**, *3*, 99.
183. Green, M. L. H. *Pure Appl. Chem.* **1978**, *50*, 27.
184. Ivin, K. J.; Rooney, J. J.; Stewart, C. D.; Green, M. L. H.; Mahtab, R. *J. Chem. Soc., Chem. Commun.* **1978**, 604.
185. Prosenc, M. H.; Janiak, C.; Brintzinger, H. H. *Organometallics* **1992**, *11*, 4036.
186. Roll, W.; Brintzinger, H. H.; Rieger, B.; Zolk, R. *Angew. Chem., Int. Ed. Engl.* **1990**, *29*, 279.
187. Grubbs, R. H.; Coates, G. W. *Acc. Chem. Res.* **1996**, *29*, 85.
188. Clawson, L.; Soto, J.; Buchwald, S. L.; Steigerwald, M. L.; Grubbs, R. H. *J. Am. Chem. Soc.* **1985**, *107*, 3377.
189. Piers, W. E.; Bercaw, J. E. *J. Am. Chem. Soc.* **1990**, *112*, 9406.
190. Krauledat, H.; Brintzinger, H. H. *Angew. Chem., Int. Ed. Engl.* **1990**, *29*, 1412.
191. Drent, E.; Budzelaar, P. H. M. *Chem. Rev.* **1996**, *96*, 663.
192. Insertion of CO into the metal–acyl intermediate must also be slow. As noted previously in this chapter, this reaction is thermodynamically unfavorable.
193. Rix, F. C.; Brookhart, M.; White, P. S. *J. Am. Chem. Soc.* **1996**, *118*, 4746.
194. Shultz, C. S.; Ledford, J.; DeSimone, J. M.; Brookhart, M. *J. Am. Chem. Soc.* **2000**, *122*, 6351.
195. Thorn, D. L.; Hoffman, R. A. *J. Am. Chem. Soc.* **1978**, *101*, 2079.
196. Lowry, T. H.; Richardson, T. H. In *Mechanism and Theory in Organic Chemistry*, 3rd ed.; Harper & Row: New York, 1987; p 162.
197. For references to numerous examples of alkyne polymerization by insertion mechanisms involving d^0 early transition metal catalysts, see references 198–200.
198. Gibson, H. W.; Porchan, J. M. In *Encyclopedia of Polymer Science*; Kvoschwitz, J. I., Ed.; Wiley: New York, 1984; Vol. 1, p 87.
199. Simionescu, C. I.; Percec, V. *Prog. Polym. Sci.* **1982**, *8*, 133.
200. Shirakawa, H.; Masuda, T.; Takeda, K. In *The Chemistry of Triple-Bonded Functional Groups*; Patai, S., Ed.; Wiley: Chichester, 1994; Vol. Suppl. C2, Chapter 17.
201. For examples of acetylene polymerizations catalyzed by rhodium and palladium, see references 202–205.
202. Furlani, A.; Napoletano, C.; Russo, M. V.; Camus, A.; Marsich, N. *J. Poly. Sci. Part A: Polym. Chem.* **1989**, *27*, 75.
203. Furlani, A.; Licoccia, S.; Russo, M. V.; Camus, A.; Marsich, N. *J. Polym. Sci. Part A: Polym. Chem.* **1986**, *24*, 991.
204. Goldberg, Y.; Alper, H. *J. Chem. Soc., Chem. Commun.* **1994**, 1209.
205. Tabata, M.; Yang, W.; Yokota, K. *J. Polym. Sci. Part A: Polym. Chem.* **1994**, *32*, 1113.
206. LaPointe, A. M.; Brookhart, M. *Organometallics* **1998**, *17*, 1530.
207. Rourke, J. P.; Stringer, G.; Chow, P.; Deeth, R. J.; Yufit, D. S.; Howard, J. A. K.; Marder, T. B. *Organometallics* **2002**, *21*, 429.
208. Cotter, W. D.; Bercaw, J. E. *J. Organomet. Chem.* **1991**, *417*, C1.
209. Huggins, J. M.; Bergman, R. G. *J. Am. Chem. Soc.* **1979**, *101*, 4410.
210. Huggins, J. M.; Bergman, R. G. *J. Am. Chem. Soc.* **1981**, *103*, 3002.
211. Samsel, E. G.; Norton, J. R. *J. Am. Chem. Soc.* **1984**, *106*, 5505.
212. Lobach, M. I.; Kormer, V. A. *Russ. Chem. Rev.* **1979**, *48*, 759.
213. Brookhart, M.; Pinhas, A. R.; Lukacs, A. *Organometallics* **1982**, *1*, 1730.
214. Krug, C.; Hartwig, J. F. *J. Am. Chem. Soc.* **2002**, *124*, 1674.
215. Krug, C.; Hartwig, J. F. *Organometallics* **2004**, *23*, 4594.
216. Krug, C.; Hartwig, J. F. *J. Am. Chem. Soc.* **2004**, *126*, 2694.
217. Hartwig, J. F.; Andersen, R. A.; Bergman, R. G. *J. Am. Chem. Soc.* **1989**, *111*, 2717.
218. Carmona, E.; Gutierrezpuebla, E.; Marin, J. M.; Monge, A.; Paneque, M.; Poveda, M. L.; Ruiz, C. *J. Am. Chem. Soc.* **1989**, *111*, 2883.
219. Fagnou, K.; Lautens, M. *Chem. Rev.* **2003**, *103*, 169.
220. Ishiyama, T.; Hartwig, J. F. *J. Am. Chem. Soc.* **2000**, *122*, 12043.
221. Ezoe, A.; Kimura, M.; Inoue, T.; Mori, M.; Tamaru, Y. *Angew. Chem. Int. Ed.* **2002**, *41*, 2784.
222. Kimura, M.; Ezoe, A.; Shibata, K.; Tamaru, Y. *J. Am. Chem. Soc.* **1998**, *120*, 4033.
223. Kimura, M.; Fujimatsu, H.; Ezoe, A.; Shibata, K.; Shimizu, M.; Matsumoto, S.; Tamaru, Y. *Angew. Chem. Int. Ed.* **1999**, *38*, 397.
224. Shibata, K.; Kimura, M.; Shimizu, M.; Tamaru, Y. *Org. Lett.* **2001**, *3*, 2181.
225. Shibata, K.; Kimura, M.; Kojima, K.; Tanaka, S.; Tamaru, Y. *J. Organomet. Chem.* **2001**, *624*, 348.
226. Kimura, M.; Ezoe, A.; Tanaka, S.; Tamaru, Y. *Angew. Chem. Int. Ed.* **2001**, *40*, 3600.

227. Sato, Y.; Saito, N.; Mori, M. *J. Org. Chem.* **2002**, *67*, 9310.

228. Sato, Y.; Takimoto, M.; Hayashi, K.; Katsuhara, T.; Takagi, K.; Mori, M. *J. Am. Chem. Soc.* **1994**, *116*, 9771.

229. Sato, Y.; Takimoto, M.; Mori, M. *Tetrahedron Lett.* **1996**, *37*, 887.

230. Sato, Y.; Saito, N.; Mori, M. *Tetrahedron Lett.* **1997**, *38*, 3931.

231. Takimoto, M.; Hiraga, Y.; Sato, Y.; Mori, M. *Tetrahedron Lett.* **1998**, *39*, 4543.

232. Sato, Y.; Takimoto, M.; Mori, M. *J. Am. Chem. Soc.* **2000**, *122*, 1624.

233. Sato, Y.; Sawaki, R.; Mori, M. *Organometallics* **2001**, *20*, 5510.

234. Sato, Y.; Saito, N.; Mori, M. *Chem. Lett.* **2002**, 18.

235. Sato, Y.; Sawaki, R.; Saito, N.; Mori, M. *J. Org. Chem.* **2002**, *67*, 656.

236. Chowdhury, S. K.; Amarasinghe, K. K. D.; Heeg, M. J.; Montgomery, J. *J. Am. Chem. Soc.* **2000**, *122*, 6775.

237. Tang, X.-Q.; Montgomery, J. *J. Am. Chem. Soc.* **1999**, *121*, 6098.

238. Montgomery, J.; Oblinger, E.; Savchenko, A. V. *J. Am. Chem. Soc.* **1997**, *119*, 4911.

239. Huang, Y. C.; Majumdar, K. K.; Cheng, C. H. *J. Org. Chem.* **2002**, *67*, 1682.

240. Cheng, C.-H.; Majumdar, K. K. *Org. Lett.* **2000**, *2*, 2295.

241. Hu, Y.; Wang, J. X.; Li, W. *Chem. Lett.* **2001**, 174.

242. Wang, J. X.; Fu, Y.; Hu, Y. *Angew. Chem. Int. Ed.* **2002**, *41*, 2757.

243. Ichiyanagi, T.; Kuniyama, S.; Shimizu, M.; Fujisawa, T. *Chem. Lett.* **1998**, 1033.

244. Nakamura, H.; Bao, M.; Yamamoto, Y. *Angew. Chem. Int. Ed.* **2001**, *40*, 3208.

245. Nakamura, H.; Aoyagi, K.; Shim, J. G.; Yamamoto, Y. *J. Am. Chem. Soc.* **2001**, *123*, 372.

246. Nakamura, H.; Ohtaka, M.; Yamamoto, Y. *Tetrahedron Lett.* **2002**, *43*, 7631.

247. Bao, M.; Nakamura, H.; Inoue, A.; Yamamoto, Y. *Chem. Lett.* **2002**, 158.

248. Fernandes, R. A.; Stimac, A.; Yamamoto, Y. *J. Am. Chem. Soc.* **2003**, *125*, 14133.

249. Takeda, A.; Kamijo, S.; Yamamoto, Y. *J. Am. Chem. Soc.* **2000**, *122*, 5662.

250. Nakamura, H.; Nakamura, K.; Yamamoto, Y. *J. Am. Chem. Soc.* **1998**, *120*, 4242.

251. Nakamura, H.; Iwama, H.; Yamamoto, Y. *J. Am. Chem. Soc.* **1996**, *118*, 6641.

252. Nakamura, H.; Sadayori, N.; Sekido, M.; Yamamoto, Y. *Chem. Commun.* **1994**, 2581.

253. Gevorgyan, V.; Quan, L. G.; Yamamoto, Y. *Tetrahedron Lett.* **1999**, *40*, 4089.

254. Quan, L. G.; Lamrani, M.; Yamamoto, Y. *J. Am. Chem. Soc.* **2000**, *122*, 4827.

255. Quan, L. G.; Gevorgyan, V.; Yamamoto, Y. *J. Am. Chem. Soc.* **1999**, *121*, 3545.

256. Nuss, J. M.; Rennels, R. A. *Chem. Lett.* **1993**, 197.

257. Zhao, L.; Lu, X. *Angew. Chem. Int. Ed.* **2002**, *41*, 4343.

258. Larock, R. C.; Doty, M. J.; Cacchi, S. *J. Org. Chem.* **1993**, *58*, 4579.

259. Roesch, K. R.; Larock, R. C. *J. Org. Chem.* **2001**, *66*, 412.

260. Roesch, K. R.; Larock, R. C. *Org. Lett.* **1999**, *1*, 1551.

261. Rayabarapu, D. K.; Cheng, C. H. *Chem. Commun.* **2002**, 942.

262. Rayabarapu, D. K.; Yang, C. H.; Cheng, C. H. *J. Org. Chem.* **2003**, *68*, 6726.

263. Chang, K. J.; Rayabarapu, D. K.; Cheng, C. H. *Org. Lett.* **2003**, *5*, 3963.

264. Li, C. J.; Wei, C. *Chem. Commun.* **2002**, 268.

265. Wei, C.; Li, C. J. *Green Chem.* **2002**, *4*, 39.

266. Kacker, S.; Kim, J. S.; Sen, A. *Angew. Chem. Int. Ed.* **1998**, *37*, 1251.

267. Dghaym, R. D.; Yaccato, K. J.; Arndtsen, B. A. *Organometallics* **1998**, *17*, 4.

268. Davis, J. L.; Arndtsen, B. A. *Organometallics* **2000**, *19*, 4657.

269. Dghaym, R. D.; Dhawan, R.; Arndtsen, B. A. *Angew. Chem. Int. Ed.* **2001**, *40*, 3228.

270. Henry, P. M. In *Handbook of Organopalladium Chemistry for Organic Synthesis*; Negishi, E-i., Ed.; Wiley-Interscience: New York, 2002.

271. Hamed, O.; Thompson, C.; Henry, P. M. *J. Org. Chem.* **1997**, *62*, 7082.

272. Hayashi, T.; Yamasaki, K.; Mimura, M.; Uozumi, Y. *J. Am. Chem. Soc.* **2004**, *126*, 3036.

273. Trend, R. M.; Ramtohul, Y. K.; Stoltz, B. M. *J. Am. Chem. Soc.* **2005**, *127*, 17778.

274. Zhao, P.; Incarvito, C. D.; Hartwig, J. F. *J. Am. Chem. Soc.* **2006**, *128*, 9642.

275. Zhao, P. J.; Krug, C.; Hartwig, J. F. *J. Am. Chem. Soc.* **2005**, *127*, 12066.

276. Cowan, R. L.; Trogler, W. C. *Organometallics* **1987**, *6*, 2451.

277. Gagne, M. R.; Stern, C. L.; Marks, T. J. *J. Am. Chem. Soc.* **1992**, *114*, 275.

278. Lira, R.; Wolfe, J. P. *J. Am. Chem. Soc.* **2004**, *126*, 13906.

279. Ney, J. E.; Wolfe, J. P. *J. Am. Chem. Soc.* **2005**, *127*, 8644.

280. Liu, G.; Stahl, S. S. *J. Am. Chem. Soc.* **2007**, *129*, 6328.

281. Casalnuovo, A. L.; Calabrese, J. C.; Milstein, D. *J. Am. Chem. Soc.* **1988**, *110*, 6738.

282. Hong, S.; Marks, T. J. *Acc. Chem. Res.* **2004**, *37*, 673.

283. Katayev, E.; Li, Y. H.; Odom, A. L. *Chem. Commun.* **2002**, 838.

284. Marder, T. B.; Norman, N. C. *Top. Catal.* **1998**, *5*, 63.

285. Ishiyama, T.; Matsuda, N.; Murata, M.; Ozawa, F.; Suzuki, A.; Miyaura, N. *Organometallics* **1996**, *15*, 713.
286. Lesley, G.; Nguyen, P.; Taylor, N. J.; Marder, T. B.; Scott, A. J.; Clegg, W.; Norman, N. C. *Organometallics* **1996**, *15*, 5137.
287. Ishiyama, T.; Yamamoto, M.; Miyaura, N. *Chem. Commun.* **1997**, 689.
288. Iverson, C. N.; Smith, M. R., III *Organometallics* **1997**, *16*, 2757.
289. Lawson, Y. G.; Lesley, M. J. G.; Marder, T. B.; Norman, N. C.; Rice, C. R. *Chem. Commun.* **1997**, 2051.
290. Clegg, W.; Johann, T. R. F.; Marder, T. B.; Norman, N. C.; Orpen, A. G.; Peakman, T. M.; Quayle, M. J.; Rice, C. R.; Scott, A. J. *J. Chem. Soc.* **1998**, 1431.
291. Ishiyama, T.; Kitano, T.; Miyaura, N. *Tetrahedron Lett.* **1998**, *39*, 2357.
292. Marder, T. B.; Norman, N. C.; Rice, C. R. *Tetrahedron Lett.* **1998**, *39*, 155.
293. Ishiyama, T.; Momota, S.; Miyaura, N. *Synlett* **1999**, *11*, 1790.
294. Ishiyama, T.; Miyaura, N. *J. Organomet. Chem.* **2000**, *611*, 392.
295. Thomas, R. L.; Souza, F. E. S.; Marder, T. B. *J. Chem. Soc., Dalton Trans.* **2001**, 1650.
296. Abu Ali, H.; El Aziz Al Quntar, A.; Goldberg, I.; Srebnik, M. *Organometallics* **2002**, *21*, 4533.
297. Miller, S. P.; Morgan, J. B.; Nepveux, F. J.; Morken, J. P. *Org. Lett.* **2004**, *6*, 131.
298. Pelz, N. F.; Woodward, A. R.; Burks, H. E.; Sieber, J. D.; Morken, J. P. *J. Am. Chem. Soc.* **2004**, *126*, 16328.
299. Suginome, M.; Nakamura, H.; Ito, Y. *Angew. Chem., Int. Ed. Engl.* **1997**, *36*, 2516.
300. Suginome, M.; Nakamura, H.; Matsuda, T.; Ito, Y. *J. Am. Chem. Soc.* **1998**, *120*, 4248.
301. Hayashi, T.; Kawamoto, A. M.; Konayashi, T.; Tanaka, M. *J. Chem. Soc., Chem. Commun.* **1990**, 563.
302. Sakaki, S.; Sumimoto, M.; Fukuhara, M.; Sugimoto, M.; Fujimoto, H.; Matsuzaki, S. *Organometallics* **2002**, *21*, 3788.
303. Baker, R. T.; Calabrese, J. C.; Westcott, S. A.; Nguyen, P.; Marder, T. B. *J. Am. Chem. Soc.* **1993**, *115*, 4367.

CHAPTER 10

Elimination Reactions

10.1. Overview of the Chapter

The insertion reactions presented in Chapter 9 often occur in the reverse direction, and these processes are typically called elimination or de-insertion reactions. As noted in the introduction to Chapter 9, de-insertion of CO or decarbonylation is a common reaction, and β-hydrogen elimination from σ-bonded ligands is one of the most common reactions of organometallic complexes. α-Hydrogen eliminations are also common reactions that are typically observed when β-hydrogens are absent. These "de-insertion" and "elimination" reactions are the microscopic reverse of the insertion processes presented in Chapter 9. Thus, the mechanisms for these processes must be the same as those for the insertion processes, and for this reason they are presented in less detail than were the mechanisms of the insertion processes. However, the scope of β-elimination reactions extends well beyond β-hydrogen elimination. β-Eliminations of hydrogen, alkyl, aryl, alkoxide, and halide groups, as well as α-eliminations of hydrogen, are known, and these processes are presented in this chapter.

10.2. Scope of Organometallic Elimination Chemistry

Elimination reactions can occur by transfer of a group from the α-, β-, or γ-carbon atom of a covalent ligand, and the group can transfer to the metal center or to an incoming reagent. Elimination reactions in which the group transfers to the metal center are presented in this chapter. Reactions in which the group is transferred to an external nucleophile or electrophile are included in Chapters 11 and 12 on nucleophilic and electrophilic attack on coordinated ligands, respectively.

β-Eliminations (Equation 10.1) are the most common type of elimination reaction from transition metal complexes, and β-hydrogen eliminations from metal–alkyl complexes are the most common type of β-elimination reactions. β-Hydrogen elimination from alkoxo and amido complexes has also been observed in a few cases. β-Alkyl elimination, β-aryl elimination, β-alkoxide elimination, and β-chloride elimination have also been observed and have been studied carefully because of their importance as side reactions in catalytic chemistry. Although β-hydrogen elimination from metal–alkyl complexes occurs almost exclusively by migratory de-insertion pathways, β-hydrogen elimination from alkoxides has been shown to occur by several different pathways.

397

β-Elimination processes

$$X {=}\hspace{-0.3em}\langle \genfrac{}{}{0pt}{}{R^1}{H}$$

$$\begin{array}{c} R^1 \\ X {-}\hspace{-0.3em}\langle \\ L_nM \quad R^2 \end{array} \longrightarrow \begin{array}{c} X {=}\hspace{-0.3em}\langle \genfrac{}{}{0pt}{}{R^1}{H} \\ + \\ L_nM{-}R^2 \end{array}$$

(10.1)

Relative rates: $R^2 = H > R^2 =$ alkyl or aryl
For $X = CH_2$, studied for $R^2 = H$, alkyl, Ar, OR and Cl
For $X = O$ or NR, studied for $R^2 = H$ and aryl

α-Elimination processes

$$L_nM{-}XH \longrightarrow L_n\overset{H}{\overset{|}{M}}{=}X$$

(10.2)

$X = O$, NR or CR_2

α-Hydrogen elimination reactions (Equation 10.2) have also been observed in many cases, most often as a route to metal–carbene complexes. This type of α-elimination process is most often part of a multistep sequence, but a number of systems have now been studied in which the α-elimination has been observed directly. In most cases, α-hydrogen elimination processes are slower than β-hydrogen eliminations, and α-eliminations are observed when β-hydrogen eliminations are inhibited or β-hydrogens are absent. However, certain complexes have now been shown to undergo α-hydrogen elimination faster than they undergo β-hydrogen elimination. γ-Elimination reactions typically require reaction of an external reagent or are simply an intramolecular C–H activation process. Thus, γ-elimination reactions are not included in this chapter.

10.3. β-Elimination Processes

10.3.1. β-Hydrogen Eliminations

10.3.1.1. β-Hydrogen Elimination from Metal–Alkyl Complexes

β-Hydrogen elimination from metal–alkyl complexes is the microscopic reverse of migratory insertion of an olefin into a metal–hydride bond. Thus, β-hydrogen elimination occurs by the migratory de-insertion mechanism shown in Equation 10.3. This mechanism is the reverse of the migratory insertion mechanism presented in Chapter 9. Because the β-hydrogen elimination process forms a hydride and a coordinated olefin from a single alkyl group, β-hydrogen elimination requires the presence of an open coordination site at the metal prior to the C–H bond cleavage step. In addition, β-hydrogen elimination occurs most readily from metal–alkyl complexes that can adopt a syn coplanar arrangement of the metal and the hydride groups. These mechanistic requirements explain the kinetic stability of transition metal–alkyl complexes that are coordinatively saturated, of transition metal–alkyl complexes lacking vacant cis coordination sites (such as those containing porphyrin ligands), and of metallacycles that cannot adopt the most favored geometry for a β-hydrogen elimination process. Examples of β-hydrogen eliminations and a discussion of the mechanism of this reaction have been reviewed.[1,2]

(10.3)

10.3.1.1.1. Effect of Conformation and Coordination Number on the Rate of β-Hydrogen Elimination

The requirement of a syn coplanar conformation for β-hydrogen elimination is illustrated by the stereochemical and regiochemical outcome of several types of stoichiometric and catalytic processes. Hundreds of examples of this phenomenon are known, three of which are shown in Equations 10.4–10.6. These examples show that the less stable isomer of the olefinic product can be formed because of the geometric requirement of the β-hydrogen elimination step. Equation 10.4 shows that decarbonylation of a rhodium–acyl complex with erythro stereochemistry on the alkyl chain forms the E-isomer of the olefin after β-hydrogen elimination of the resulting alkyl intermediate,[3] and the analogous reaction of the threo stereoisomer forms the Z-isomer of the olefin after β-hydrogen elimination. The Newman projections of the intermediates in this reaction reveal that the two aryl groups must be eclipsed to create the syn coplanar geometry for the elimination reaction, and this geometry forms the Z-olefin. Similarly, the reactions of Heck's arylpalladium complex generated from arylmercury acetate and palladium acetate reacts with Z-β-methyl styrene by the sequence of olefin insertion, followed by β-hydrogen elimination with the stereochemistry shown in Equation 10.5 to form the Z-olefin product.[4] Finally, Equation 10.6 shows that insertion of a cyclic olefin generates an intermediate in which the β-hydrogen that lies on the carbon containing the migrated group is located trans to the metal. Therefore, β-hydrogen elimination does not occur with this hydrogen. Instead, it occurs to form the 2,5-dihydrofuran.[5]

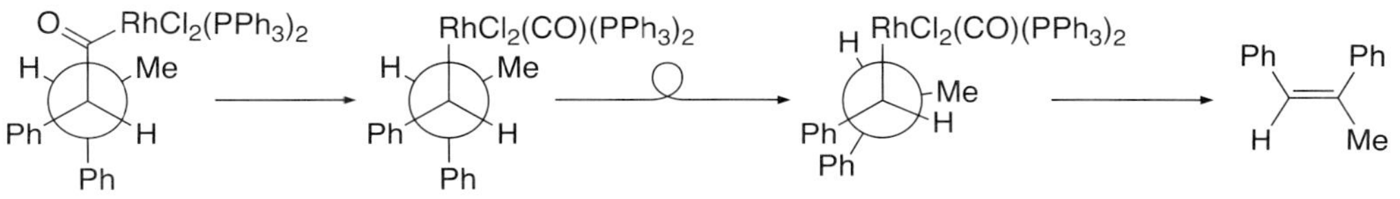

(10.4)

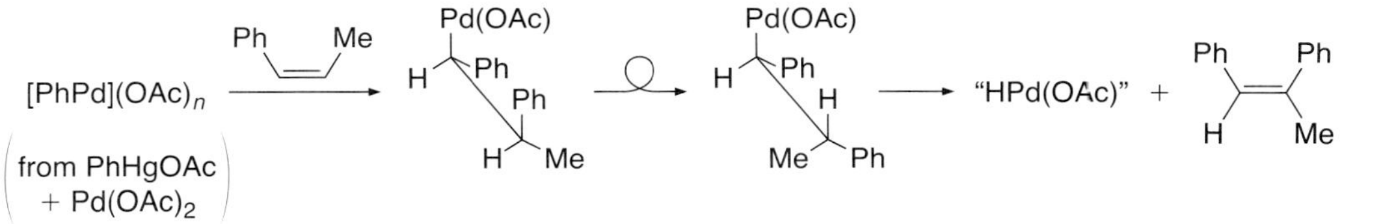

(10.5)

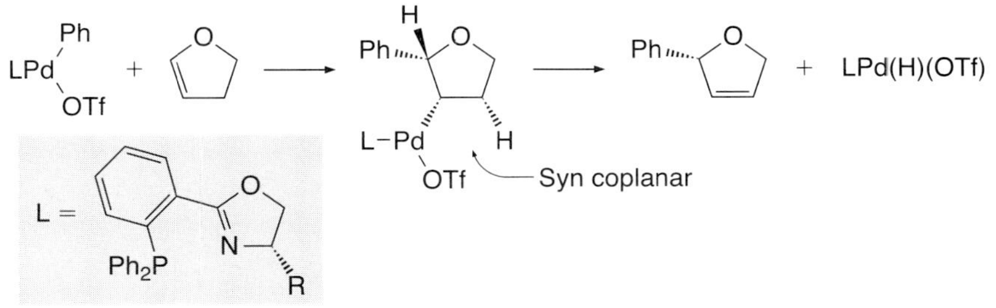

(10.6)

Additional mechanistic information on β-hydrogen elimination from metal–alkyl complexes has been derived from Whiteside's and Yamamoto's work on the thermal decomposition of cis-PtR$_2$(PPh$_3$)$_2$ complexes and the related metallacyclic compound cis-Pt(CH$_2$)$_4$(PPh$_3$)$_2$,[6-11] and from Yamamoto's work on the thermal decomposition of cis-PdR$_2$(PR$_3$)$_2$ complexes.[12,13] Typical reactions are shown in Equations 10.7a, 10.7b, and 10.8. Detailed kinetic studies on these systems have shown: 1) β-hydrogen elimination from square-planar palladium and platinum complexes typically occurs by two pathways, one that occurs after generation of an open coordination site cis to the alkyl group and one that occurs from the four-coordinate, square-planar species; 2) β-hydrogen elimination from these platinum complexes in the absence of added ligand occurred principally by a three-coordinate complex containing an open coordination site in the square plane of the d^8 system; 3) β-hydrogen elimination from the analogous palladium complex containing PMe_2Ph as ligand in the absence of added ligand occurred predominantly without dissociation of ligand; and 4) β-hydrogen elimination from square-planar metallacyclic complexes is slower than that from acyclic alkyl complexes because the metallacycles are less able to adopt the syn coplanar geometry required for this elimination process. Metallacyclic complexes that possess geometries other than square planar undergo β-hydrogen elimination under milder conditions because the open coordination site is more geometrically accessible, as rationalized by recent computational studies.[14]

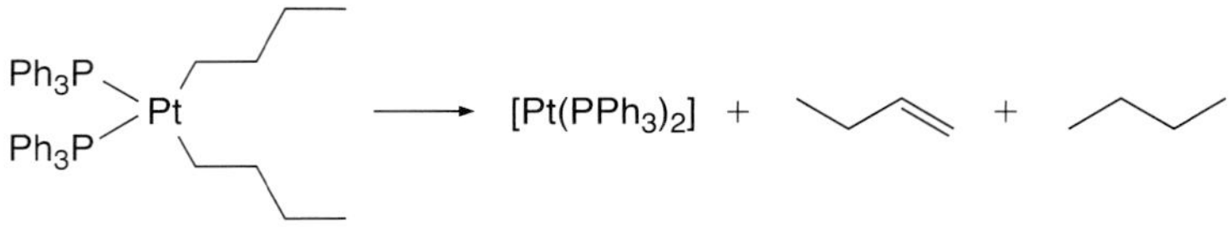

(10.7a)

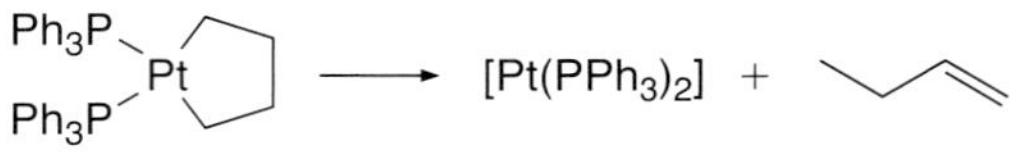

(10.7b)

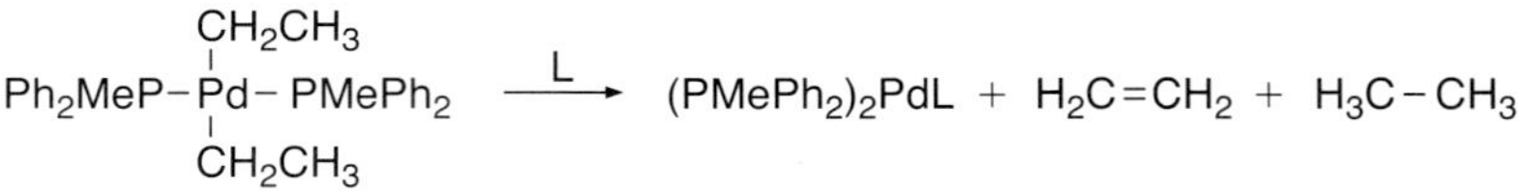

(10.8)

Specifically, the acylic dialkyl complex in Equation 10.7a decomposes in the presence of modest concentrations of added ligand with rates that are inhibited by added triphenylphosphine. The inhibition by phosphine shows that dissociation of phosphine occurs prior to β-hydrogen elimination. A small kinetic isotope effect implies that the β-hydrogen elimination step is fully or partially reversible. In this case, the final irreversible step of the overall reaction would be the reductive elimination of the hydride and alkyl groups to form *n*-butane.

Metallacyclic complexes such as the reactant in Equation 10.7b are much more stable thermally than their acylic analogues. This metallacycle decomposes 10^4 times more slowly than the dibutyl compound, and it decomposes by pathways that do not include β-hydrogen elimination.[15] Additional metal–alkyl complexes that are stable to β-hydrogen elimination because of their metallacyclic structure have been isolated as part of studies on complex molecule synthesis. One example of such a complex is shown in Equation 10.9.[16]

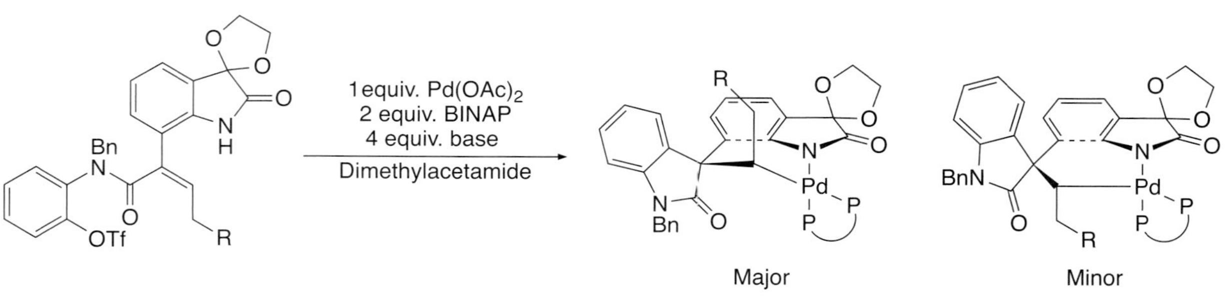

(10.9)

10.3.1.1.2. Effect of Electronics on the Rate of β-Hydrogen Elimination

Although some alkyl complexes are resistant to β-hydride elimination for steric reasons and conformational constraints, some alkyl complexes are particularly susceptible or resistant to β-hydrogen elimination because of the electronic properties. Two examples in which a substituent on the β-carbon appears to promote β-hydrogen elimination are shown in Equations 10.10a and 10.10b. Indirect evidence suggests that β-hydroxyethyl groups readily undergo β-hydrogen elimination during the Wacker oxidation of olefins. As part of this oxidation process[17] a β-hydroxyethylpalladium complex is thought to rearrange rapidly to an α-hydroxyethyl complex by β-hydrogen elimination and re-insertion (Equation 10.10a). A related reaction occurs during the isomerization of epoxides to ketones by rhodium and iridium complexes. As shown in Equation 10.10b,[18] oxidative addition of a low-valent iridium complex to a terminal epoxide at the least-hindered C–O bond gives the intermediate shown. This complex is then proposed to undergo rapid β-hydrogen elimination.

$$Cl_2LPd\!\!-\!\!\diagup\!\!\diagdown OH \longrightarrow Cl_2LPd\diagdown{}_H^{\diagup OH} \longrightarrow Cl_2LPd\diagdown{}^{OH} \qquad (10.10a)$$

$$PhCH\!\!-\!\!CD_2 + (COE)IrL_3Cl \longrightarrow \left[\begin{array}{c} Ph-CH-CD_2 \\ \ominus O \qquad IrL_3Cl \\ \oplus \end{array} \right] \longrightarrow \begin{array}{c} L_{\prime\prime\prime\prime}\!\!\diagup\!\!\overset{H}{\underset{|}{Ir}}\!\!\diagdown^{\cdots CD_2-\overset{\overset{O}{\|}}{C}-Ph} \\ Cl \diagup | \diagdown L \\ L \end{array} \qquad (10.10b)$$

L = PMe₃
COE = cyclooctene

Relative $k_{β\text{-}H}$: X = OCH₃ > CH₃ > H > CF₃ (10.11)

Systematic studies of the effect of substituents on the β-carbon on the rate of β-hydrogen elimination were conduced with zirconocene[19] and scandocene[20] β-arylethyl complexes. Complexes containing varied substituents on the aryl group were studied, as shown in Equation 10.11 for the zirconocene system. β-Hydrogen elimination from these complexes occurred with a negative Hammett parameter (ρ = –1.8 for Zr and –1.9 for Sc), which indicates that electron-withdrawing groups retard β-hydrogen elimination. This electronic effect could result from thermodynamic stabilization of the metal–alkyl bond in the reactant by electron-withdrawing substituents on the aryl ring, from transfer of the hydrogen from the alkyl group to the metal as a hydride, or both.

Substituents on the α-carbon also have a large effect on the rate of β-hydrogen elimination. Typical palladium(II) alkyl complexes containing ligands that dissociate from the metal undergo rapid β-hydrogen elimination, but the complexes in Equation 10.12 that contain strong electron-withdrawing groups on the α-carbon are stable near room temperature. A comparison of the relative rates of β-hydrogen elimination from the complexes in this equation shows that the rate decreases with increasing electron-withdrawing power of the substituent on the α-carbon.[21]

R	Temperature	$t_{1/2}$ for β-elimination
Ph	–20 °C	< 15 min
CO₂^{t}Bu	20 °C	< 6 min
CN	23 °C	25 h

(10.12)

10.3.1.1.3. Effect of Ancillary Ligands on the Rate of β-Hydrogen Elimination

Ancillary ligands also affect the rate of β-hydrogen elimination. In general, an increase in steric hindrance decreases the rate of β-hydrogen elimination[22] and greater electron-donation increases the rate of β-hydrogen elimination.[21] The increased steric hindrance reduces the rate because it disfavors the structure containing the syn coplanar arrangement of the metal, β-hydrogen, and two linking carbon atoms. Second, the β-hydrogen elimination generates two ligands from one ligand, and this increase in coordination number can increase steric hindrance, even though one ligand is a small hydride. Little systematic data are available on how electron density at the metal controls the rate of the C–H bond cleavage step of β-hydrogen elimination (as opposed to controlling the rate or equilibrium for the formation of the open coordination site). However, comparison of the rate of β-hydrogen elimination from the cyanoalkyl complex in Equation 10.12 and from the phosphine analog provides one dramatic example of the effect of the donating ability of the ancillary ligand on the rate of this reaction. Addition of PPh₃ to the stable cyanoalkyl complex containing DMSO as ligand induces rapid β-hydrogen elimination. This β-hydrogen elimination is presumably triggered by substitution of DMSO by PPh_3.[21]

10.3.1.2. β-Hydrogen Elimination from Metal Alkoxides and Amides

β-Hydrogen elimination from alkoxides and amides is a step in many syntheses of metal hydrides, as well as the reductions of metal halides like $RuCl_3$ and $IrCl_3$ to lower valent compounds that are useful for organometallic synthesis.[23,24] β-Hydrogen elimination is also part of the oxidation of alcohols by palladium catalysts[25,26] that has been developed into an enantioselective process.[27–34] β-Hydrogen elimination from alkoxides and amides have also been observed directly in some cases, but only a subset of these reactions appears to occur by a mechanism that parallels the migratory de-insertion mechanism for β-hydrogen elimination from alkyl complexes. β-Hydrogen eliminations from thiolate complexes and phosphide complexes are rare, in part because the carbon–sulfur and carbon–phosphorus double bonds in the products are so weak.

β-Hydrogen elimination from the square-planar iridium(I) alkoxo complexes shown at the top of Scheme 10.1[35,36] and β-hydrogen elimination from the iridium(I) amido complexes shown in Scheme 10.2[37] have been reported. Extensive mechanistic data on the elimination

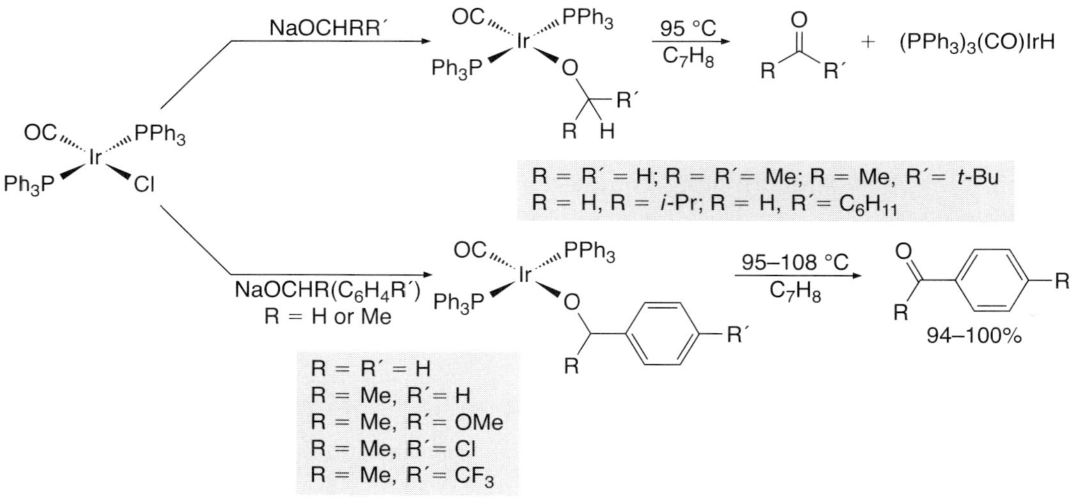

Scheme 10.1

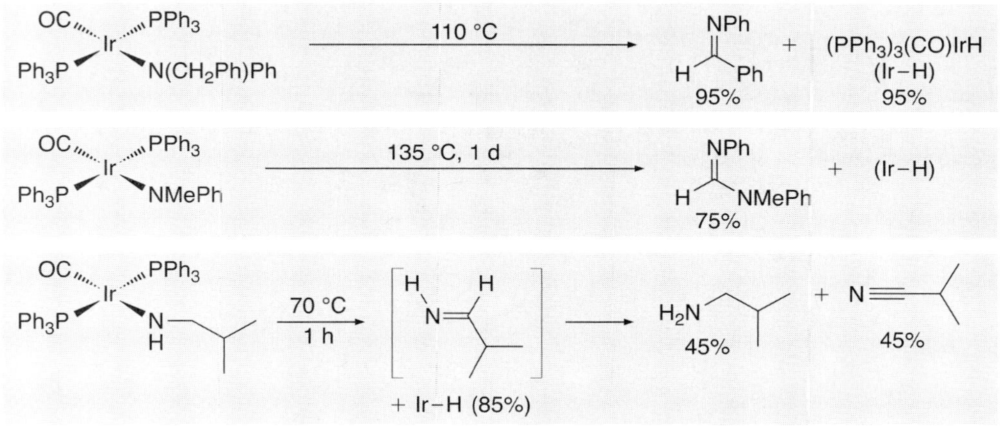

Scheme 10.2

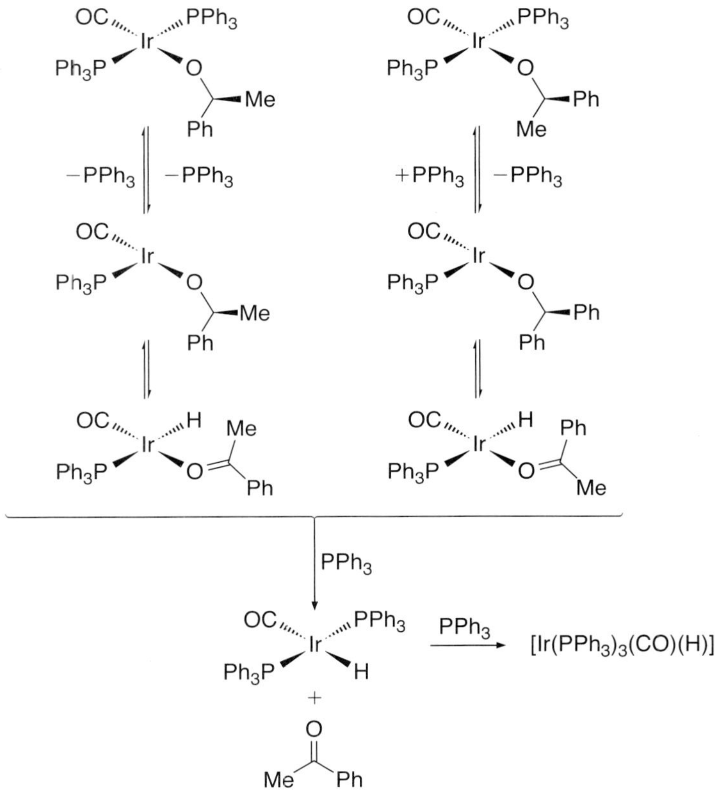

Scheme 10.3

of the alkoxo and amido complexes have revealed detailed mechanistic features, which are summarized in Scheme 10.3. In general, these β-hydrogen eliminations appear to occur by migratory de-insertion. More specifically, studies showed that 1) ligand dissociation to generate a 14-electron intermediate is necessary for both β-hydrogen elimination processes to occur; 2) C–H bond cleavage is the rate-determining step of β-hydrogen elimination from the amido complexes; and 3) C–H bond cleavage can be reversible or irreversible during β-hydrogen elimination from the iridium–alkoxo complexes, depending on the concentration of the added ligand that associatively displaces the ketone product.

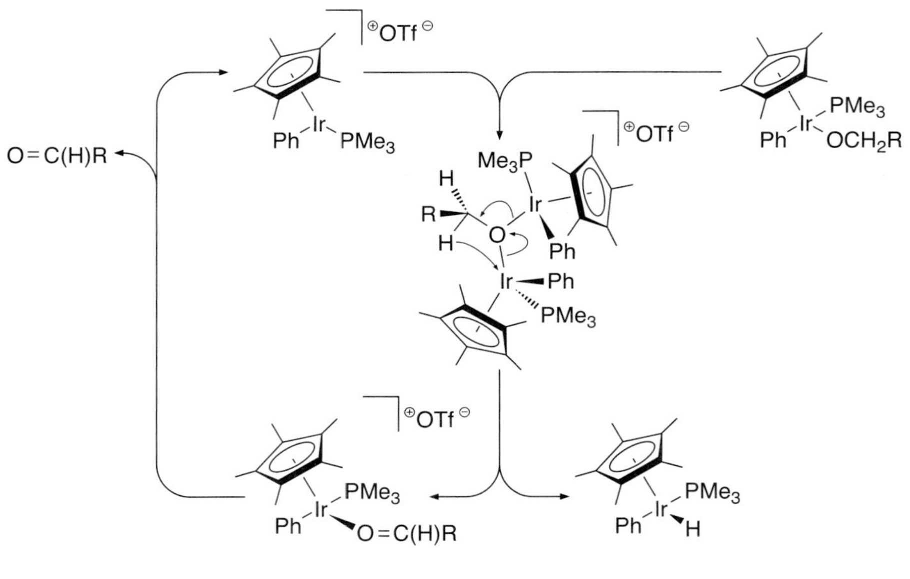

(10.13)

Several examples of β-hydrogen elimination from iridium(III) alkoxo complexes (Equations 10.13–10.15) have been reported, and these reactions occur by pathways other than simple migratory de-insertion. In the reaction shown in Equation 10.13, β-hydrogen elimination occurs from an 18-electron alkoxide complex containing tightly bound dative ligands. Therefore, an open coordination site cis to the alkoxo group is unavailable, and the β-hydrogen elimination occurs by initial abstraction of the β-hydrogen by an iridium cation.[38]

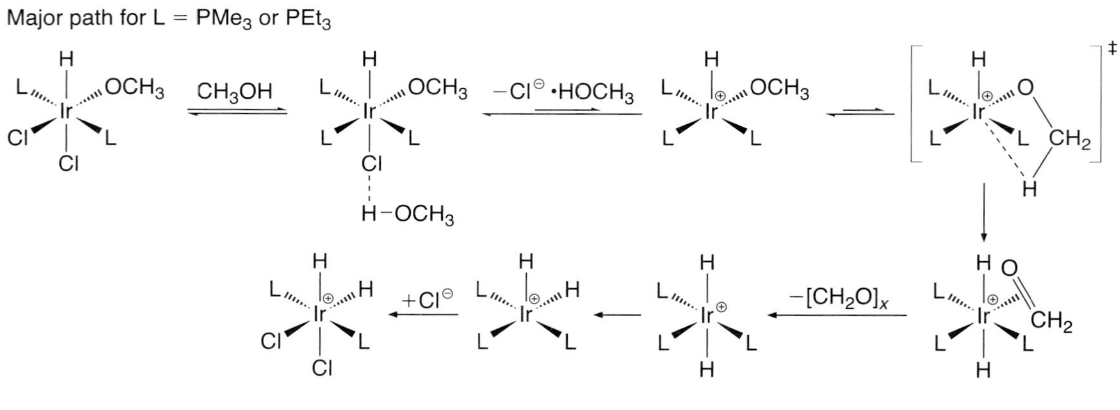

Major path for L = PMe$_3$ or PEt$_3$

(10.14)

In the reaction shown in Equation 10.14,[39] β-hydrogen elimination from a pseudo-octahedral iridium(III) alkoxide complex appears to occur after initial dissociation of chloride, which is triggered by hydrogen bonding of the anion to the methanol solvent. In a third case (Equation 10.15),[40] β-hydrogen elimination from an 18-electron, pseudo-octahedral iridium(III) alkoxide complex that lacks a chloride ligand occurs. This reaction appears to occur by dissociation of the alkoxide. Again, this dissociation is proposed to be assisted by hydrogen bonding to added alcohol. Abstraction of the β-hydrogen from the dissociated alkoxide by the resulting cationic iridium species is proposed to complete the reaction.

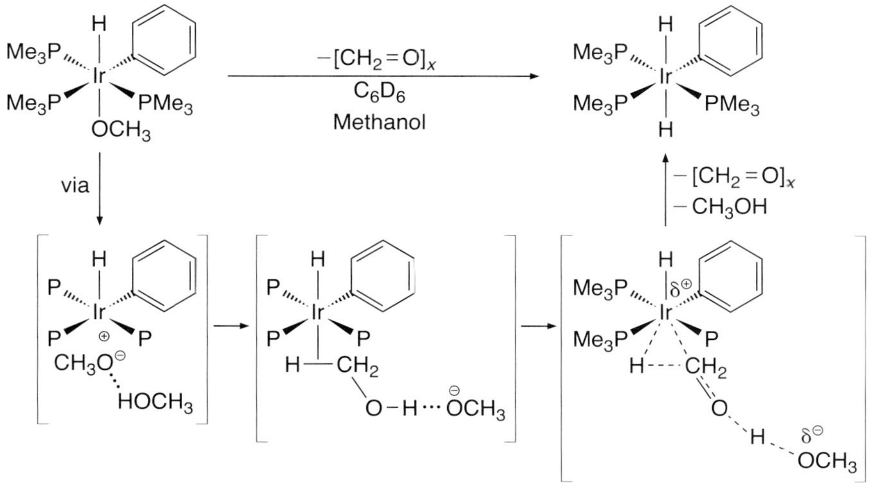

(10.15)

In addition to β-hydrogen elimination from formally anionic amido and alkoxo groups, β-hydrogen elimination from neutral amine ligands to form iminium products has been observed. This reaction can be part of the pathway for the reduction of metal halides, such as palladium(II) halides, in the presence of tertiary amines.[41–43] Such a β-hydrogen elimination was observed directly as a reversible process from the tertiary amine complex $[Os(NH_3)_4(MeOH)(NH_2Pr)]^{2+}$, as shown in Equation 10.16.[44]

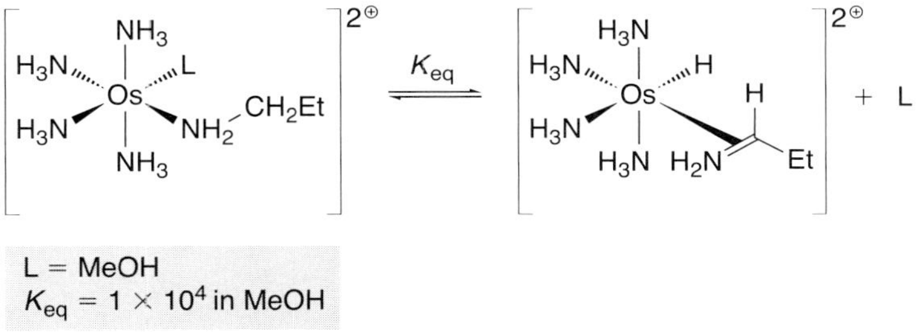

L = MeOH
$K_{eq} = 1 \times 10^4$ in MeOH

(10.16)

10.3.1.3. β-Hydrogen Elimination from Metal–Silyl Complexes

β-Hydrogen elimination from silyl complexes is also known, but is uncommon. β-Hydrogen elimination to form a free silene ($R_2C=SiR_2$) and a metal–hydride complex has not been observed directly and is likely to be thermodynamically unfavorable because of the weak Si–C π-bond. However, β-hydrogen elimination from silyl complexes to form products containing a hydride and a coordinated silene is known and has been proposed to occur during catalytic reactions, such as the dehydrogenative coupling of methylsilanes to form carbosilanes.[45–47] A well-characterized example of such β-hydrogen elimination is shown in Equation 10.17.[48] Sequestering of the dissociated phosphine by BPh$_3$ leads to an intermediate containing an open coordination site, and this intermediate undergoes β-hydrogen elimination to form a silene hydride complex as product. As shown in this equation, the overall process is rapid and generates an equilibrium between the silene hydride complex and the 16-electron silyl complex. Related β-hydrogen eliminations from β-silylalkyl complexes to form silene complexes are also known,[49–56] including the example in Equation 10.18.[49,50]

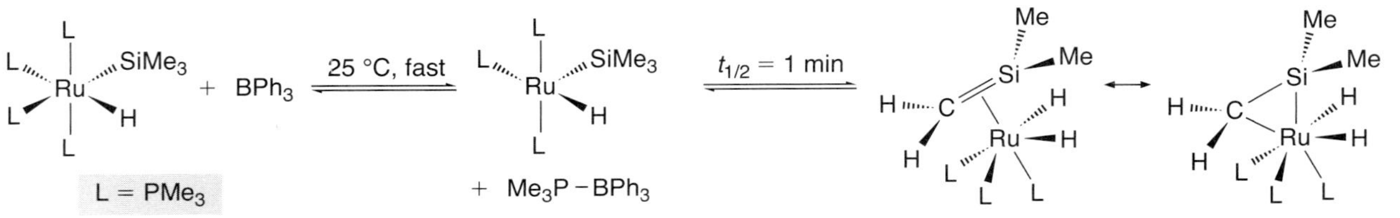

(10.17)

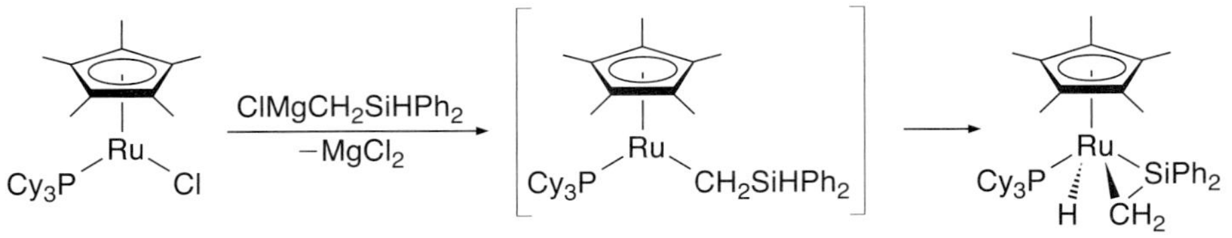

(10.18)

10.3.2. β-Hydrocarbyl Eliminations

Carbon–carbon bond cleavage by β-alkyl or β-aryl elimination has now been observed in a number of different types of transition metal complexes. This reaction has been studied intensively in the past decade because it has been shown to be one pathway for chain transfer during olefin polymerization by group 4 metal complexes.[57–68] In general, this type of β-elimination reaction appears to be fastest from highly electrophilic early transition metal complexes, but an increasing number of examples of β-alkyl eliminations from late transition metal–alkyl complexes have also been observed. In most cases, these β-alkyl eliminations from late transition metal complexes involve the cleavage of a strained ring.

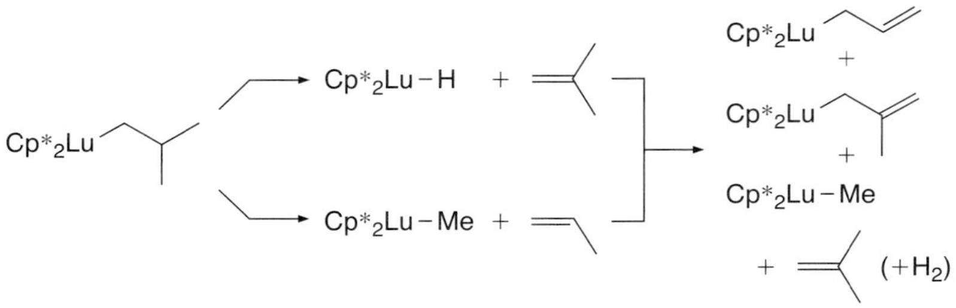

Scheme 10.4

10.3.2.1. β-Alkyl Eliminations from Alkyl Complexes

Several examples of mild carbon–carbon bond cleavage by β-alkyl elimination from d^0 metal complexes have been observed directly. Watson reported one of the first sets of products resulting from β-alkyl elimination from an isolated alkyl complex.[69] In this case, the lutetium–isobutyl complex in Scheme 10.4 converted to a series of lutetium complexes and organic products that were rationalized by a combination of parallel β-hydrogen elimination and β-alkyl elimination.

More recently, scandocene–alkyl complexes were shown to undergo mild cleavage of the C–C bonds to the β-carbon.[58] As shown in Equation 10.19, the scandocene–isobutyl complex bound by tethered Cp ligands and a coordinated phosphine undergoes de-insertion of propene to generate the analogous scandocene–methyl complex. This reaction occurs in parallel with a faster β-hydrogen elimination. Like the transformation of the lutetium complex, the initial products formed from β-alkyl elimination from the scandocene compound undergo a series of transformations and generate a complex set of products. The methyl complex, propene, and some free phosphine generates the final set of products.

(10.19)

(10.20)

M	$\Delta G^{\ddagger}_{\beta-Me}$ (0 °C) (kcal/mol)	$\Delta H^{\ddagger}_{\beta-Me}$ (kcal/mol)	$\Delta S^{\ddagger}_{\beta-Me}$ (kcal/mol·K)
Zr	21.2(0.2)	22.5(0.9)	4.3(3.3)
Hf	20.7(0.2)	17.3(0.9)	−11.9(3.3)

A more straightforward example of β-alkyl elimination is illustrated by the cationic neopentyl zirconocene and hafnocene complexes in Equation 10.20.[57,70] This β-alkyl elimination occurs in high yield to form the corresponding methyl complex and free isobutylene, even at −15 °C. A comparative study of the rates of reaction of the Zr and Hf complexes show that the two complexes react with similar rates and that the differences in the enthalpic and entropic parameters offset each other.

In addition to these reactions of discrete alkyl complexes of d^0 metals, much evidence for β-alkyl elimination has been gained from end-group analysis of polymers generated from olefin polymerization catalyzed by d^0 metal complexes. These data imply that β-alkyl elimination could be a major pathway for chain transfer in the absence of any added chain-transfer agents, such as hydrogen.[59]

Cleavage of the C–C bond to the β-carbon of late-metal–alkyl complexes is slower and less common than that of d^0 early metal complexes. However, a few examples of β-alkyl elimination from late metal complexes are known. The reversible intramolecular insertion of an olefin into the platinum–alkyl complex described in Chapter 9 involves an early example of β-alkyl elimination from a late-metal–alkyl complex.[71] In addition, mild β-alkyl elimination has been reported to occur from a ruthenacylobutane complex. In this case the product is a stable alkyl allyl species.[72,73]

Like β-hydrogen eliminations, β-alkyl eliminations require an open coordination site. This site is generated in the scandocene system by dissociation of PMe$_3$[58] and in the zirconocene and hafnocene complexes by dissociation of the borate from the zwitterionic species.[57,70] The open coordination site is generated in the platinum system by abstraction of chloride[71] and is generated in the ruthenium complex by dissociation of the monodentate phosphine.[72,73] The mild conditions for β-methyl elimination from the ruthenium metallacycle is surprising, considering that it would seem to require the propellane-type transition state shown in Equation 10.21.

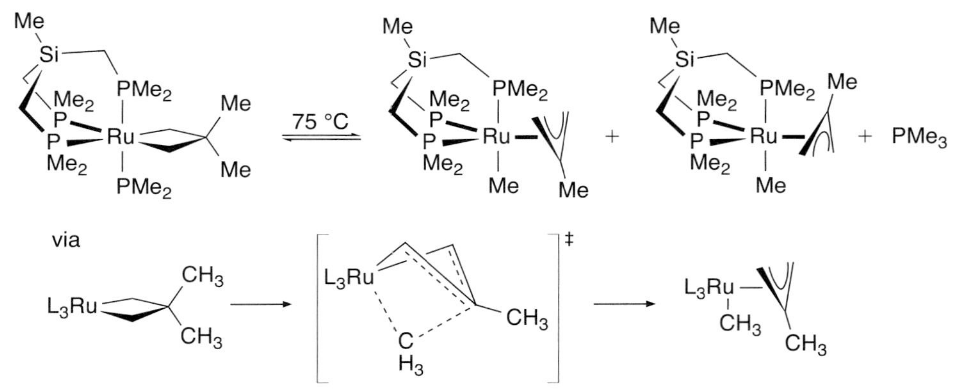

(10.21)

10.3.2.2. β-Alkyl and β-Aryl Eliminations from Alkoxido and Amido Complexes

β-Alkyl and β-aryl eliminations from alkoxide complexes have also been shown to occur during various catalytic and stoichiometric processes.[74–78] β-Methyl elimination occurs under mild conditions from the oxametallacyclobutane in Equation 10.22 that is similar to the ruthenacyclobutane described in Equation 10.21.[79,80] In addition, β-aryl elimination has been reported to occur from triethylphosphine-ligated rhodium–arylmethoxide complexes (Equation 10.23),[81] and from related rhodium β-aryl iminyl complexes.[82] Like β-alkyl or β-aryl eliminations from metal–alkyl complexes, such eliminations from late metal alkoxides appear to require an open coordination site. The analogous oxametallacycle that contains bidentate, rather than monodentate, ligands does not undergo the β-alkyl elimination, and the rhodium complexes undergo the β-aryl elimination from unsaturated alkoxo and iminyl complexes generated by dissociation of a phosphine ligand. An interaction of an alkoxide β-aryl group with the open coordination site of a rhodium arylmethoxide complex was observed by X-ray diffraction (Equation 10.23), and this structure is likely to directly precede the transition state for this β-aryl elimination. As a result, these reactions are currently considered to be migratory de-insertions.

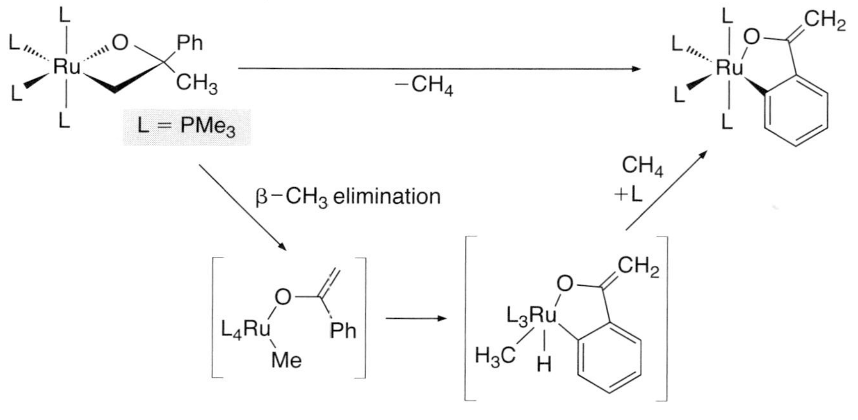

(10.22)

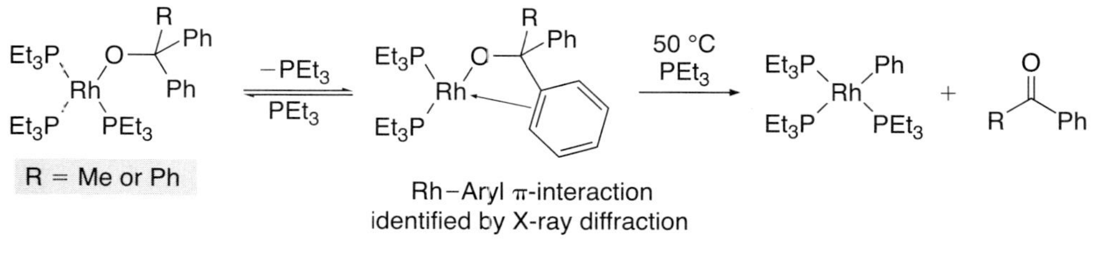

(10.23)

10.3.3. *β*-Halide and Alkoxide Elimination

In addition to the β-elimination of hydrogen, alkyl, and aryl groups, the β-elimination of more electronegative groups has been observed. β-Chloride elimination has been studied because it interferes with the ability to develop catalysts for polymerization of vinyl chloride by a coordination and insertion mechanism. Polymerization of vinyl chloride and vinyl acetate by this mechanism would allow control of the stereochemistry of the resulting polymer by the catalyst.[83] Jordan has shown that β-chloride elimination is particularly fast from β-chloroalkyl complexes of early transition metals, but it also occurs readily from β-chloroalkyl complexes of nickel and palladium. β-Elimination of fluoride and alkoxide groups is also known when these groups are present on the β-carbons of the alkyl ligands in hard, d^0 metal–alkyl complexes.

Several examples of β-chloride elimination are shown in Equations 10.24–10.29. Reaction of vinyl chloride with cationic zirconocene–alkyl complexes (Equations 10.24a and 10.24b) forms propylene and the corresponding zirconocene–chloride complex as the initial products.[84] The final products result from polymerization of the propylene by the starting zirconocene–alkyl species, and generation of a dinuclear cationic chloride species from the resulting cationic chloride complex and a zirconocene dichloride by a less-defined pathway. The propylene is formed by the process shown in Equation 10.24b. Insertion of vinyl chloride into the zirconocene methyl, followed by β-chloride elimination from the β-chloroalkyl intermediate generates propylene and a cationic chloride complex.

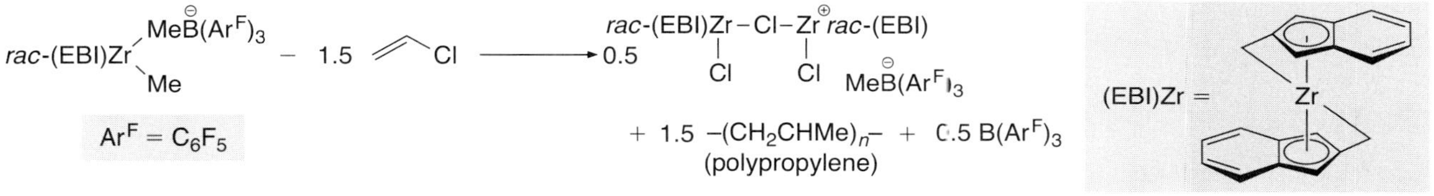

(10.24a)

via

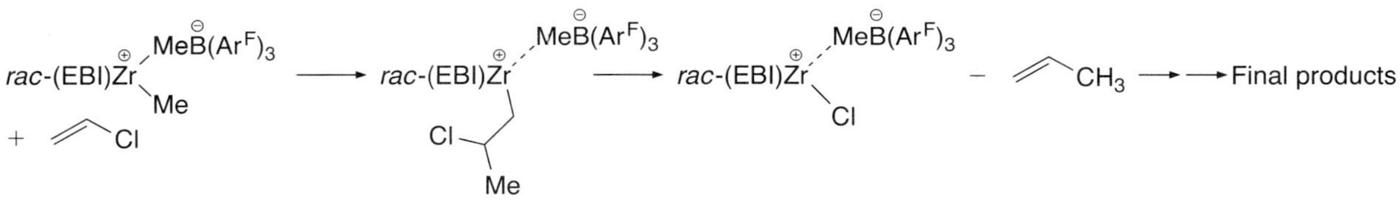

(10.24b)

Simpler β-halide eliminations occur from late transition metal catalysts for olefin polymerization (Equations 10.25 and 10.26).[85] Reactions of the cationic palladium–alkyl complexes occur in a similar fashion to the reactions of the cationic group 4 complexes, despite the softer nature of these species. In this case, propylene and the metal chloride are formed. Even a neutral nickel–hydrocarbyl complex (the salicaldimine complex in Equation 10.26) undergoes reactions with vinyl chloride that involve insertion followed by β-chloride elimination.

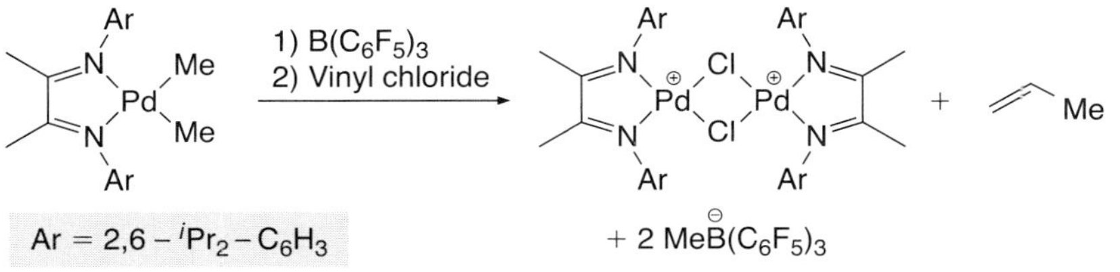

(10.25)

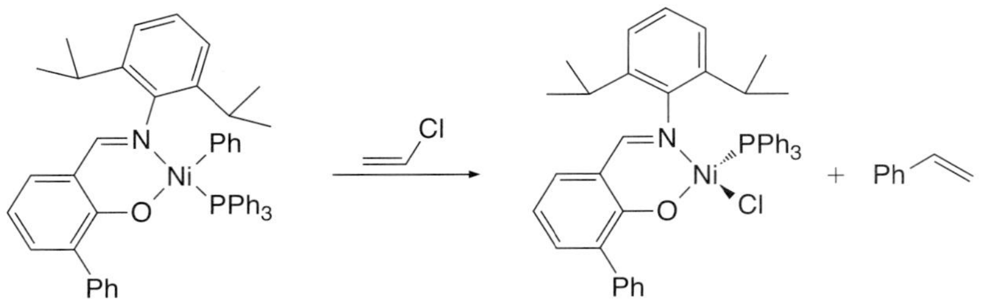

(10.26)

β-Eliminations of fluoride and alkoxide are also known from the alkyl groups of d^0 group 4 metal complexes. Insertion of vinyl fluoride into Schwartz's reagent leads to a β-fluoroalkyl complex that eliminates ethylene and forms the corresponding metal fluoride (Equation 10.27).[86] A similar reaction of Schwartz's reagent with ethyl vinyl ether (Equation 10.28) leads to the corresponding alkoxyalkyl complex which undergoes fast β-alkoxide elimination to form the zirconocene–alkoxide complex and ethylene.[87] The ethylene then reacts with the starting zirconocene hydride to form the corresponding zirconocene–ethyl complex.

The propensity of group 4 metallocenes to undergo β-elimination of halogens and alkoxides is not unique to these metal-ligand sytems. Wolczanski has studied reactions of vinyl halides and vinyl ethers with Ta(V) hydrides, as shown in Equation 10.29. This work shows that the alkoxyalkyl and haloalkyl complexes formed from insertions of vinyl ethers and vinyl halides into nonmetallocene Ta(V) hydrides undergo β-elimination of the alkoxo group or halogen to form the Ta(V) alkoxide and halide complex and olefin. The extruded olefin then re-inserts into the remaining hydride to form some of the observed tantalum–alkyl species.

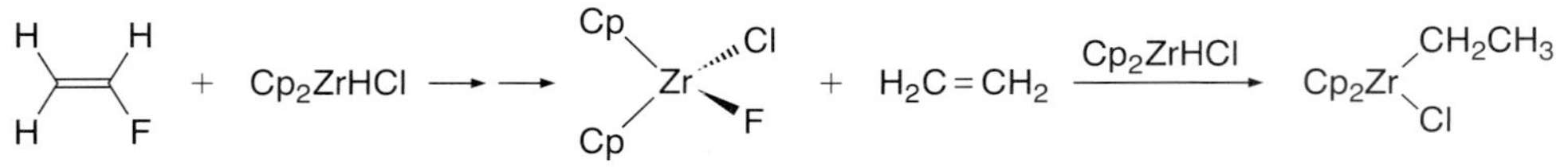

(10.27)

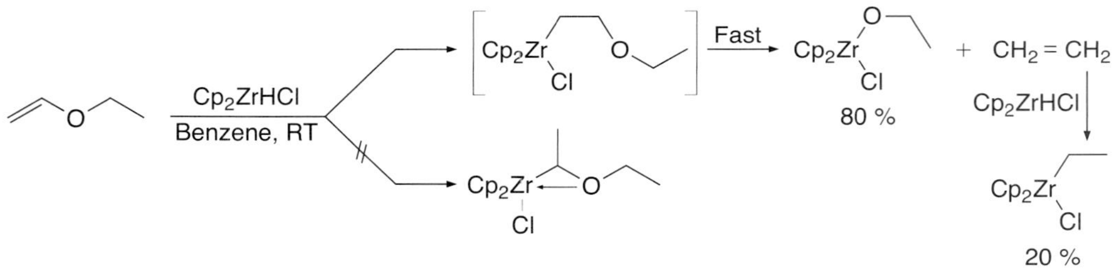

(10.28)

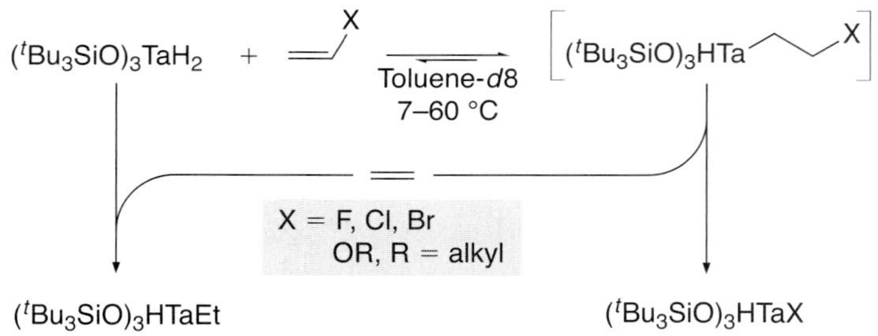

(10.29)

10.4. α-Hydrogen Eliminations and Abstractions

α-Hydrogen elimination is a process that generates a carbene and a hydride ligand from an alkyl ligand (Equation 10.30) or that generates a carbene ligand and alkane from two alkyl ligands (Equation 10.31). α-Hydrogen elimination is the most common method to form Schrock-type alkylidene complexes described in Chapters 2 and 13. α-Hydrogen eliminations are typically slower than β-hydride eliminations. Thus, α-hydrogen elimination occurs most commonly from benzyl, neopentyl, methyl, and other alkyl complexes that cannot undergo β-hydrogen elimination. α-Hydrogen elimination occurs most often among complexes of the early transition metals and typically with d^0 complexes of group 4 and 6 metals. These complexes can access the high oxidation state required for the formation of an alkylidene hydride product, and the absence of d-electrons does not favor formation of a stable olefin complex that would result from β-hydrogen elimination. Steric hindrance can facilitate α-hydrogen eliminations because the M–C–C angle increases as an alkyl group is converted into an alkylidene ligand. This relief of steric hindrance is particularly significant during the process of Equation 10.31 because an alkyl group is also eliminated.

$$L_nM\!-\!\overset{\diagup}{\underset{R}{}} \longrightarrow L_nM\overset{\diagup R}{\underset{H}{=}} \tag{10.30}$$

$$L_nM\overset{\diagup R}{\underset{\diagdown R}{}} \longrightarrow L_nM\overset{R}{\underset{H}{=}} + H\!-\!CH_2R \tag{10.31}$$

$$via\quad L_nM\overset{\diagup R}{\underset{\diagdown R}{\overset{\Vert}{-}}H}\quad or\quad \left[L_nM\overset{R}{\underset{R}{\diamondsuit}}H \right]^{\ddagger}$$

In most cases, the immediate product of α-hydrogen elimination is not observed. Instead, reductive elimination of a hydride and an alkyl group to form an alkane frequently occurs (Equation 10.30). In other cases, a carbene forms from a dialkyl complex of a d^0 metal that cannot accommodate the additional valency required to form an alkylidene and a hydride ligand.[88] In these cases, an alternative four-center pathway involving the transition state shown in Equation 10.31 that does not involve formation of a metal–hydride complex is followed. When the valency of the metal allows either pathway to occur, it is difficult to distinguish between the α-elimination and α-hydrogen abstraction pathways.[1,89]

$$\left(\!\!\diagup\!\!\diagup\!\!\right)_3 TaCl_2 + 2\, LiCH_2CMe_3 \xrightarrow{\text{Pentane}} \left(\!\!\diagup\!\!\diagup\!\!\right)_3 Ta = \diagup\!\!\diagup + 2\, LiCl + CMe_4 \tag{10.32}$$

$$\left(\!\!\diagup\!\!\diagup\!\!\right)_2 TaCl_3 + TlCp \longrightarrow \underset{Cl}{\overset{Cp}{\underset{Cl}{Ta}}} = \!\!\diagup\!\!\diagup + TlCl + \diagup\!\!\diagup H \tag{10.33}$$

Specific examples of α-hydrogen elimination are shown in Equations 10.32–10.38.[90,91] α-Hydrogen elimination was first observed by Cooper and Green[92–94] with tungsten–methyl derivatives. Many subsequent examples were more extensively studied by Schrock using tantalum derivatives (Equations 10.32 and 10.33).[91,95] The synthesis of the first Schrock-type alkylidene complex, shown in Equation 10.32,[88] involves cleavage of the α-hydrogen and elimination of alkane. The species that undergoes the α-hydrogen elimination was not observed directly until later studies by Xue on α-elimination from $Ta(CH_2Bu^t)_5$ and the related, more stable analog $Ta(CH_2SiMe_3)_5$.[96] A more direct, early observation of an α-hydrogen elimination is shown in Equation 10.33.[97] Like many α-hydrogen elimination processes, this reaction was induced by a ligand exchange process (Cp⁻ for chloride). The addition of alkyl phosphines to early transition metal–dialkyl complexes also has induced α-eliminations. The mechanism of such α-elimination chemistry is not always clear, but one example of such an α-hydrogen elimination process is shown in Equation 10.34.[98]

$$(10.34)$$

In some cases, reversible α-hydrogen elimination has been detected by dynamic NMR methods or by isotopic labeling. A clear-cut example of a reversible α-hydride elimination was revealed by dynamic NMR spectroscopy (Equation 10.35).[99] The starting alkylidene hydride complex equilibrates with the alkyl complex, which is not generated in high enough concentration to detect directly. Magnetization-transfer ^{1}H NMR spectroscopy,[100–102] a technique commonly used to study reversible rearrangement reactions of organometallic complexes,[103] shows that the hydride of the starting tantalum complex undergoes intramolecular exchange with the α-hydrogen of the accompanying neopentylidene ligand. More recently, isotopic labeling showed that α-hydrogen elimination can be faster than β-hydrogen elimination.[104] The cyclopentyltungsten(IV) alkyl complex in Equation 10.36 forms the more stable carbene hydride complex reversibly. At higher temperatures a series of β-hydrogen eliminations, the first of which is shown in the equation, occurs to scramble deuterium into all positions of the alkyl group.

$$(10.35)$$

L = PMe$_3$

$$(10.36)$$

Insertion   β-H-elimination   α-H-elimination

N$_3$N = N(CH$_2$CH$_2$NSiMe$_3$)$_3$

α-Hydrogen elimination has also been observed with later transition metal complexes. Two examples are shown in Equations 10.37 and 10.38.[105,106] In one case, the five-coordinate iridium(III) complex bound by a pincer ligand eliminates H$_2$ upon heating under vacuum to form the square-planar carbene complex shown. The reaction is reversed by addition of hydrogen under ambient conditions. It is likely that a hydrido carbene complex is formed as a reaction intermediate. Alternatively, this reaction could proceed via a four-centered transition state or intermediate of the type shown in Equation 10.31. In a more recent example of α-hydrogen elimination from a late-metal–alkyl complex, the hydrido carbene complex was observed by spectroscopy during the reaction. This reversible interconversion between an alkyl complex and a hydrido carbene complex is shown in Equation 10.38.

$$(10.37)$$

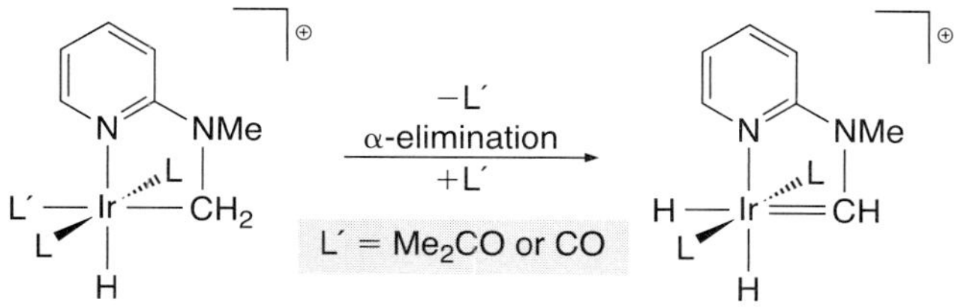

$$(10.38)$$

Carbene complexes have also been shown to undergo α-hydrogen elimination, in this case to form alkylidyne complexes. This reaction is much less common than the reaction of alkyl complexes to form alkylidenes, but a few examples are well documented. Two examples of this transformation are shown in Equations 10.39 and 10.40. Schrock reported the first direct observation of this transformation (Equation 10.39).[107] In this first example, an isolated carbene complex converted to an alkylidyne hydride complex upon abstraction of a chloride ligand with trimethylaluminum. In a second example, a double C–H activation process by two sequential α-hydrogen elimination reactions converts the starting tungsten–methyl complex in Equation 10.40 into a methylidyne complex.[108]

$$Ta(=CHCMe_3)(dmpe)_2Cl \ + \ AlMe_3 \ \rightleftharpoons \ [Ta(\equiv CCMe_3)(H)(dmpe)_2](ClAlMe_3) \qquad (10.39)$$

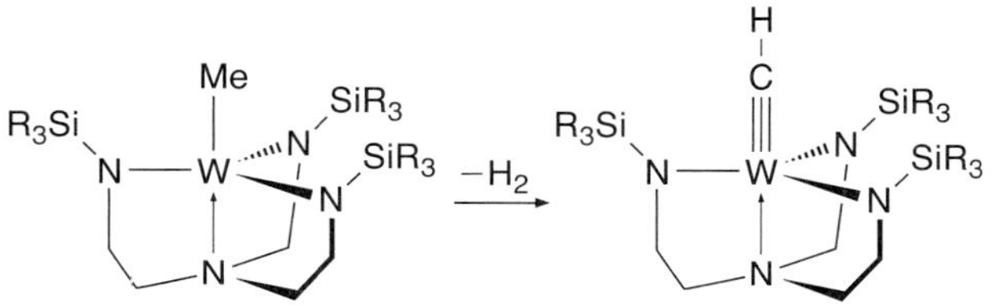

$$(10.40)$$

This section closes with examples of α-hydrogen elimination from amido complexes to form imido complexes. This reaction is described in detail in Chapter 13 (metal–ligand multiple bonds). To illustrate two forms of this reaction, Equations 10.41 and 10.42 are shown here. In these reactions, elimination of arene and elimination of amine occur to generate imido complexes that have been trapped by dative ligands or allowed to undergo [2+2] cycloadditions with alkynes.[109,110]

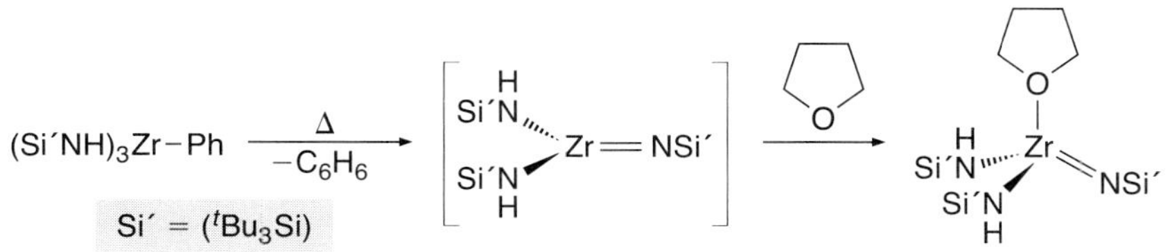

$$(10.41)$$

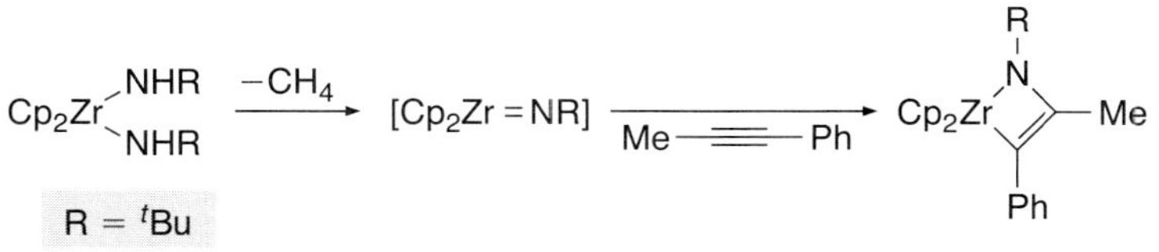

$$(10.42)$$

10.5. Summary

Elimination reactions occur from complexes containing all of the transition metals. The class and rate of elimination reaction depends on the type of reactive ligand, the electronic properties of the ancillary ligands, and the conformational flexibility of the complex.

Many of the elimination reactions discussed in this chapter are the opposite of the migratory insertion reactions presented in the preceding chapter and, therefore, follow related mechanisms.

β-Eliminations from alkyl complexes are the most common class of elimination reaction, and β-hydrogen elimination is the most common class of β-elimination process. β-Elimination reactions increase the coordination number of the complex; therefore, they require an open coordination site at the metal center. These reactions do not change the oxidation state of the metal. This β-elimination chemistry occurs as part of many catalytic processes or as side reactions that limit yields or molecular weights of products formed by catalytic processes.

β-Eliminations of alkyl and aryl groups are now known from both early and late transition metal complexes. β-Hydrogen elimination occurs faster from complexes that can adopt a syn coplanar arrangment of the β-hydrogen, the metal, and the two connecting atoms. Studies on the electronic effects on β-hydrogen elimination imply that this reaction is faster from complexes containing more electron-rich alkyl groups, and, at least in some cases, from complexes containing more electron-donating ancillary ligands.

β-Hydrogen eliminations and β-aryl eliminations from alkoxo and amido complexes are also known. Such eliminations have been shown to occur by migratory de-insertion pathways, as well as alternative β-hydride abstraction mechanisms. β-Hydrogen eliminations from metal–silyl complexes are rare because the silicon–carbon double bond in the product is weak. For similar reasons, β-hydrogen eliminations from metal–thiolate complexes are rare.

In some cases, rapid α-elimination reactions have been observed. These reactions most often occur with early metal complexes and form metal–alkylidene complexes. However, examples of this elimination process from complexes of later transition metals are now known. α-Eliminations from carbene complexes to form carbyne complexes and from amide complexes to form imido complexes are also now well established. Although α-eliminations typically occur with complexes that cannot undergo β-hydrogen elimination, complexes are now known that undergo faster α-hydrogen elimination than β-hydrogen elimination. Such α-elimination reactions give rise to the metal–alkylidene complexes that catalyze the olefin metathesis chemistry described in Chapter 21.

References and Notes

1. Cross, R. J. In *The Chemistry of the Metal–Carbon Bond*; Hartley, F. R., Patai, S., Eds.; Wiley: New York, 1985; Vol. 2, Chapter 8.
2. Bullock, R. M. In *Transition Metal Hydrides*; Dedieu, A., Ed.; VCH: New York, 1992; p 263.
3. Stille, J. K.; Huang, F.; Regan, M. T. *J. Am. Chem. Soc.* **1974**, *96*, 1518.
4. Heck, R. F. *J. Am. Chem. Soc.* **1969**, *91*, 6707.
5. Loiseleur, O.; Hayashi, M.; Schmees, N.; Pfaltz, A. *Synthesis* **1997**, 1338.
6. Whitesides, G. M.; Gaasch, J. F.; Stedronsky, E. R. *J. Am. Chem. Soc.* **1972**, *94*, 5258.
7. McDermott, J. X.; White, J. F.; Whitesides, G. M. *J. Am. Chem. Soc.* **1976**, *98*, 6521.
8. McCarthy, T. J.; Nuzzo, R. G.; Whitesides, G. M. *J. Am. Chem. Soc.* **1981**, *103*, 3396.
9. Nuzzo, R. G.; McCarthy, T. J.; Whitesides, G. M. *J. Am. Chem. Soc.* **1981**, *103*, 3404.
10. Miller, T. M.; Whitesides, G. M. *Organometallics* **1986**, *5*, 1473.
11. Komiya, S.; Morimoto, Y.; Yamamoto, A.; Yamamoto, T. *Organometallics* **1982**, *1*, 1528.
12. Ozawa, F.; Ito, T.; Yamamoto, A. *J. Am. Chem. Soc.* **1980**, *102*, 6457.
13. Ozawa, F.; Ito, T.; Nakamura, Y.; Yamamoto, A. *Bull. Chem. Soc. Jpn.* **1981**, *54*, 1868.
14. Huang, X.; Zhu, J.; Lin, Z. Y. *Organometallics* **2004**, *23*, 4154.
15. Miller, T. M.; Whitesides, G. M. *Organometallics* **1986**, *5*, 1473.
16. Burke, B. J.; Overman, L. E. *J. Am. Chem. Soc.* **2004**, *126*, 16820.
17. Cross, R. J. In *Catalysis; Specialist Periodical Report*; Bond, G.C., Webb, G., Eds.; Royal Society of Chemistry, Oxford, 1982; Vol 5, p 366.
18. Milstein, D. *Acc. Chem. Res.* **1984**, *17*, 221.

19. Chirik, P. J.; Bercaw, J. E. *Organometallics* **2005**, *24*, 5407
20. Burger, B. J.; Thompson, M. E.; Cotter, W. D.; Bercaw, J. E. *J. Am. Chem. Soc.* **1990**, *112*, 1566.
21. Tanaka, D.; Romeril, S. P.; Myers, A. G. *J. Am. Chem. Soc.* **2005**, *127*, 10323.
22. Hartwig, J. F.; Richards, S.; Barañano, D.; Paul, F. *J. Am. Chem. Soc.* **1996**, *118*, 3626.
23. Hallman, P. S.; Stephenson, T. A.; Wilkinson, G. *Inorg. Synth.* **1970**, *12*, 237.
24. See the various preparations in *Inorg. Synth.* **1990**, *28*, 77–92.
25. Steinhoff, B. A.; Fix, S. R.; Stahl, S. S. *J. Am. Chem. Soc.* **2002**, *124*, 766.
26. Schultz, M. J.; Park, C. C.; Sigman, M. S. *Chem. Commun.* **2002**, 3034.
27. Jensen, D. R.; Pugsley, J. S.; Sigman, M. S. *J. Am. Chem. Soc.* **2001**, *123*, 7475.
28. Mueller, J. A.; Jensen, D. R.; Sigman, M. S. *J. Am. Chem. Soc.* **2002**, *124*, 8202.
29. Mueller, J. A.; Sigman, M. S. *J. Am. Chem. Soc.* **2003**, *125*, 7005.
30. Mueller, J. A.; Cowell, A.; Chandler, B. D.; Sigman, M. S. *J. Am. Chem. Soc.* **2005**, *127*, 14817.
31. Ferreira, E. M.; Stoltz, B. M. *J. Am. Chem. Soc.* **2001**, *123*, 7725.
32. Trend, R. M.; Stoltz, B. M. *J. Am. Chem. Soc.* **2004**, *126*, 4482.
33. Nielsen, R. J.; Keith, J. M.; Stoltz, B. M.; Goddard, W. A. *J. Am. Chem. Soc.* **2004**, *126*, 7967.
34. Bagdanoff, J. T.; Stoltz, B. M. *Angew. Chem. Int. Ed.* **2004**, *43*, 353.
35. Bernard, K. A.; Rees, W. M.; Atwood, J. D. *Organometallics* **1986**, *5*, 390.
36. Zhao, J.; Hesslink, H.; Hartwig, J. F. *J. Am. Chem. Soc.* **2001**, *123*, 7220.
37. Hartwig, J. F. *J. Am. Chem. Soc.* **1996**, *118*, 7010.
38. Ritter, J. C. M.; Bergman, R. G. *J. Am. Chem. Soc.* **1998**, *120*, 6826.
39. Blum, O.; Milstein, D. *J. Am. Chem. Soc.* **1995**, *117*, 4582.
40. Blum, O.; Milstein, D. *J. Organomet. Chem.* **2000**, *593–594*, 479.
41. Beletskaya, I., P.; Cheprakov, A. V. *Chem. Rev.* **2000**, *100*, 3009.
42. For one well documented example and related citations, see reference 43.
43. Trzeciak, A. M.; Ciunik, Z.; Ziolkowski, J. J. *Organometallics* **2002**, *21*, 132.
44. Barrera, J.; Orth, S. D.; Harman, W. D. *J. Am. Chem. Soc.* **1992**, *114*, 7316.
45. Procopio, L. J.; Berry, D. H. *J. Am. Chem. Soc.* **1991**, *113*, 4039.
46. Djurovich, P. I.; Dolich, A. R.; Berry, D. H. *J. Chem. Soc., Chem. Commun.* **1994**, 1897.
47. Ezbiansky, K.; Djurovich, P. I.; Laforest, M.; Sinning, D. J.; Zayes, R.; Berry, D. H. *Organometallics* **1998**, *17*, 1455.
48. Dioumaev, V. K.; Plossl, K.; Carroll, P. J.; Berry, D. H. *J. Am. Chem. Soc.* **1999**, *121*, 8391.
49. Campion, B. K.; Heyn, R. H.; Tilley, T. D. *J. Am. Chem. Soc.* **1988**, *110*, 7558.
50. Campion, B. K.; Heyn, R. H.; Tilley, T. D.; Rheingold, A. L. *J. Am. Chem. Soc.* **1993**, *115*, 5527.
51. Pannell, K. H. *J. Organomet. Chem.* **1970**, *21*, P17.
52. Cundy, C. S.; Lappert, M. F.; Pearce, R. *J. Organomet. Chem.* **1973**, *59*, 161.
53. Lewis, C.; Wrighton, M. S. *J. Am. Chem. Soc.* **1983**, *105*, 7768.
54. Randolph, C. L.; Wrighton, M. S. *Organometallics* **1987**, *6*, 365.
55. Campion, B. K.; Heyn, R. H.; Tilley, T. D. *J. Am. Chem. Soc.* **1990**, *112*, 4079.
56. Campion, B. K.; Heyn, R. H.; Tilley, T. D. *J. Chem. Soc., Chem. Commun.* **1992**, 1201.
57. Beswick, C. L.; Marks, T. J. *J. Am. Chem. Soc.* **2000**, *122*, 10358.
58. Hajela, S.; Bercaw, J. E. *Organometallics* **1994**, *13*, 1147.
59. Yang, P.; Baird, M. C. *Organometallics* **2005**, *24*, 6013.
60. Resconi, L.; Camurati, I.; Sudmeijer, O. *Top. Catal.* **1999**, *7*, 145.
61. Shaffer, T. D.; Canich, J. A. M.; Squire, K. R. *Macromolecules* **1998**, *31*, 5145.
62. Horton, A. D. *Organometallics* **1996**, *15*, 2675.
63. Guo, Z. Y.; Swenson, D. C.; Jordan, R. F. *Organometallics* **1994**, *13*, 1424.
64. Resconi, L.; Piemontesi, F.; Franciscono, G.; Abis, L.; Fiorani, T. *J. Am. Chem. Soc.* **1992**, *114*, 1025.
65. Eshuis, J. J. W.; Tan, Y. Y.; Meetsma, A.; Teuben, J. H.; Renkema, J.; Evens, G. G. *Organometallics* **1992**, *11*, 362.
66. Kesti, M. R.; Waymouth, R. M. *J. Am. Chem. Soc.* **1992**, *114*, 3565.
67. Mise, T.; Kageyama, A.; Miya, S.; Yamazaki, H. *Chem. Lett.* **1991**, 1525.
68. Eshuis, J. J. W.; Tan, Y. Y.; Teuben, J. H.; Renkema, J. *J. Mol. Catal.* **1990**, *62*, 277.
69. Watson, P. L.; Roe, D. C. *J. Am. Chem. Soc.* **1982**, *104*, 6471.
70. Yang, P.; Baird, M. C. *Organometallics* **2005**, *24*, 6005.
71. Flood, T. C.; Bitler, S. P. *J. Am. Chem. Soc.* **1984**, *106*, 6076.
72. McNeill, K.; Andersen, R. A.; Bergman, R. G. *J. Am. Chem. Soc.* **1995**, *117*, 3625.
73. McNeill, K.; Andersen, R. A.; Bergman, R. G. *J. Am. Chem. Soc.* **1997**, *119*, 11244.
74. Satoh, T.; Miura, M. *Top. Organomet. Chem.* **2005**, *14*, 1.
75. Nishimura, T.; Uemura, S. *Synlett* **2004**, 201 and the references therein.

76. Kondo, T.; Kodoi, K.; Nishinaga, E.; Okada, T.; Morisaki, Y.; Watanabe, Y.; Mitsudo, T.-a. *J. Am. Chem. Soc.* **1998**, *120*, 5587.

77. Matsuda, T.; Makino, M.; Murakami, M. *Angew. Chem., Int. Ed. Engl.* **2005**, *44*, 4608.

78. Funayama, A.; Satoh, T.; Miura, M. *J. Am. Chem. Soc.* **2005**, *127*, 15354.

79. Hartwig, J. F.; Bergman, R. G.; Andersen, R. A. *J. Am. Chem. Soc.* **1990**, *112*, 3432.

80. Hartwig, J. F.; Bergman, R. G.; Andersen, R. A. *Organometallics* **1991**, *10*, 3326.

81. Zhao, P. J.; Incarvito, C. D.; Hartwig, J. F. *J. Am. Chem. Soc.* **2006**, *128*, 3124.

82. Zhao, P. J.; Hartwig, J. F. *J. Am. Chem. Soc.* **2005**, *127*, 11618.

83. Boone, H. W.; Athey, P. S.; Mullins, M. J.; Philipp, D.; Muller, R.; Goddard, W. A. *J. Am. Chem. Soc.* **2002**, *124*, 8790.

84. Stockland, R. A.; Foley, S. R.; Jordan, R. F. *J. Am. Chem. Soc.* **2003**, *125*, 796.

85. Foley, S. R.; Stockland, R. A.; Shen, H.; Jordan, R. F. *J. Am. Chem. Soc.* **2003**, *125*, 4350.

86. Watson, L. A.; Yandulov, D. V.; Caulton, K. G. *J. Am. Chem. Soc.* **2001**, *123*, 603.

87. Buchwald, S. L.; Nielsen, R. B.; Dewan, J. C. *Organometallics* **1988**, *7*, 2324.

88. Schrock, R. R.; Feldman, J. D. *J. Am. Chem. Soc.* **1978**, *100*, 3359.

89. McDade, C.; Green, J. C.; Bercaw, J. E. *Organometallics* **1982**, *1*, 1629.

90. For other examples, see reference 91.

91. Turner, H. W.; Schrock, R. R.; Fellmann, J. D.; Holmes, S. J. *J. Am. Chem. Soc.* **1983**, *105*, 4942.

92. Cooper, N. J.; Green, M. L. H. *J. Chem. Soc., Chem. Commun.* **1974**, 761.

93. Cooper, N. J.; Green, M. L. H. *J. Chem. Soc., Chem. Commun.* **1974**, 208.

94. Cooper, N. J.; Green, M. L. H. *J. Chem. Soc., Dalton Trans.* **1979**, 1121.

95. Schrock, R. R. *Acc. Chem. Res.* **1979**, *12*, 98.

96. Li, L.; Hung, M.; Xue, Z. *J. Am Chem. Soc.* **1995**, *117*, 12746.

97. Wood, C. D.; McLain, S. J.; Schrock, R. R. *J. Am. Chem. Soc.* **1979**, *101*, 3210.

98. Hessen, B.; Buijink, J. K. F.; Meetsma, A.; Teuben, J. H.; Helgesson, G.; Hakansson, M.; Jagner, S.; Spek, A. L. *Organometallics* **1993**, *12*, 2268.

99. Turner, H. W.; Schrock, R. R.; Fellmann, J. D.; Holmes, S. J. *J. Am. Chem. Soc.* **1983**, *105*, 4942.

100. Forsen, S.; Hoffman, R. A. *J. Chem. Phys.* **1964**, *40*, 1189.

101. Forsen, S.; Hoffman, R. A. *J. Chem. Phys.* **1963**, *39*, 2892.

102. Morris, G. A.; Freeman, R. *J. Magn. Res.* **1978**, *29*, 433.

103. Threlkel, R. S.; Bercaw, J. E. *J. Am. Chem. Soc.* **1981**, *103*, 2650.

104. Schrock, R. R.; Shih, K.-Y.; Dobbs, D. A.; Davis, W. M. *J. Am. Chem. Soc.* **1995**, *117*, 6609.

105. Empsall, H. D.; Hyde, E. M.; Markham, R.; McDonald, W. S.; Norton, M. C.; Shaw, B. L.; Weeks, B. *J. Chem. Soc., Chem. Commun.* **1977**, 589.

106. Clot, E.; Chen, J. Y.; Lee, D. H.; Sung, S. Y.; Appelhans, L. N.; Faller, J. W.; Crabtree, R. H.; Eisenstein, O. *J. Am. Chem. Soc.* **2004**, *126*, 8795.

107. Churchill, M. R.; Wasserman, H. J.; Turner, H. W.; Schrock, R. R. *J. Am. Chem. Soc.* **1982**, *104*, 1710.

108. Shih, K. Y.; Totland, K.; Seidel, S. W.; Schrock, R. R. *J. Am. Chem. Soc.* **1994**, *116*, 12103.

109. Cummins, C. C.; Baxter, S. M.; Wolczanski, P. D. *J. Am. Chem. Soc.* **1988**, *110*, 8731.

110. Walsh, P. J.; Hollander, F. J.; Bergman, R. G. *J. Am. Chem. Soc.* **1988**, *110*, 8729.

Nucleophilic Attack on Coordinated Ligands

The reactions of nucleophiles with organometallic complexes can be divided into two groups: reactions occurring by nucleophilic attack at the metal center and reactions occurring by nucleophilic attack at a ligand coordinated to the metal. These two reaction manifolds are summarized in Scheme 11.1. The former process occurs during the ligand substitution reactions presented in Chapter 5. The latter reaction changes the composition of a ligand in the organometallic reactant and is presented in this chapter. Nucleophilic attack on a coordinated ligand is a common step toward the synthesis of organometallic complexes and in the mechanism of important catalytic processes. This chapter presents principles, examples, and several applications of nucleophilic attack on coordinated ligands. Details of the utility of this process in organic synthesis can be found in the text by Hegedus and Soderberg.[1]

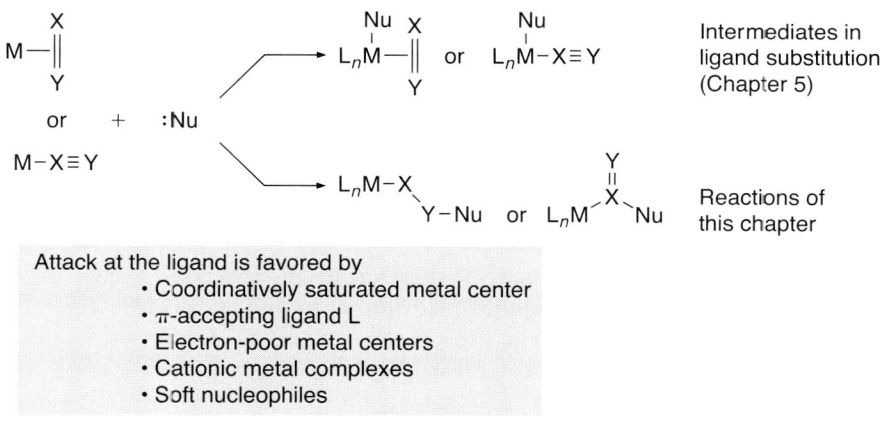

Scheme 11.1

11.1. Fundamental Principles

The coordination of an unsaturated molecule to a transition metal reverses its reactivity from that of a weak nucleophile to that of an electrophile in many cases. Ligands that are most commonly subject to nucleophilic attack include carbon monoxide, isonitriles, olefins, dienes, arenes, and allyl groups. These ligands are nucleophilic when free from the metal and, therefore, react faster with electrophiles than with nucleophiles in the absence of a metal. For example, free carbon monoxide[2,3] reacts with only the strongest of nucleophiles.[3] Even compounds with somewhat activated $C=C$ bonds, such as those in dienes and vinyl-arenes, react with only strong nucleophiles.[4–6] In contrast, CO, olefins, dienes, arenes, and related molecules react readily with nucleophiles of moderate strength when coordinated to certain transition metal centers. This change in reactivity can be rationalized by the net flow of electron density from the ligand to the metal upon coordination.

417

This effect accurately predicts several trends that favor nucleophilic attack at a ligand over the metal (Scheme 11.1). This effect predicts that the unsaturated ligands of more electron-deficient metal centers are more reactive toward nucleophilic attack than the unsaturated ligands of more electron-rich metal centers. Thus, ligands in cationic complexes are more susceptible to nucleophilic attack than are ligands in neutral complexes, and ligands in neutral complexes are more susceptible to attack than are the ligands in anionic complexes. Additional factors that facilitate nucleophilic attack at the ligand include high reactivity of the nucleophile and the presence of spectator ligands on the metal that are π-acceptors and accommodate the increase in charge at the metal center resulting from nucleophilic attack. In addition to considering the overall rate of attack, one must consider the selectivity for attack of the nucleophile at the ligand or at the metal center. Nucleophiles react faster at both the ligand and the metal in more electrophilic complexes, but attack at the ligand is generally favored for complexes that are coordinatively saturated or cannot accommodate an additional ligand for steric or electronic reasons.

Additions of nucleophiles to coordinated ligands do not change the total number of valence electrons at the metal center, but such additions of nucleophiles at the ligand can occur with or without a change in oxidation state, as shown in Scheme 11.2. Nucleophilic attack at a neutral ligand does not lead to a change in oxidation state, but nucleophilic attack at a formally anionic ligand leads to a reduction in the oxidation state of the metal. For example, the attack of an anionic nucleophile on an olefin, which is a neutral, two-electron ligand, leads to an anionic complex containing a formally anionic, two-electron alkyl group. Thus, the oxidation state and overall number of valence electrons does not change. Similarly, attack of an anionic nucleophile on coordinated carbon monoxide generates an anionic complex containing an anionic acyl ligand. Attack of a neutral nucleophile, such as an amine, leads to a zwitterionic intermediate in which the metal contains a formal negative charge and the original nucleophilic moiety contains a formal positive charge. In contrast, attack of a nucleophile on a coordinated allyl ligand reduces the metal oxidation state by two, decreases the number of electrons donated by the ligand by two, and decreases the hapticity of the ligand. This attack on the four-electron, anionic allyl ligand generates a two-electron, neutral olefin ligand. This reaction, therefore, leads to a formal reduction in the oxidation state of the metal by two and can be considered to be a reductive elimination. However, the total number of valence electrons on the metal does not change because the decrease in number of electrons donated by the ligand is compensated for by the change in oxidation state that increases the number of d-electrons by two.

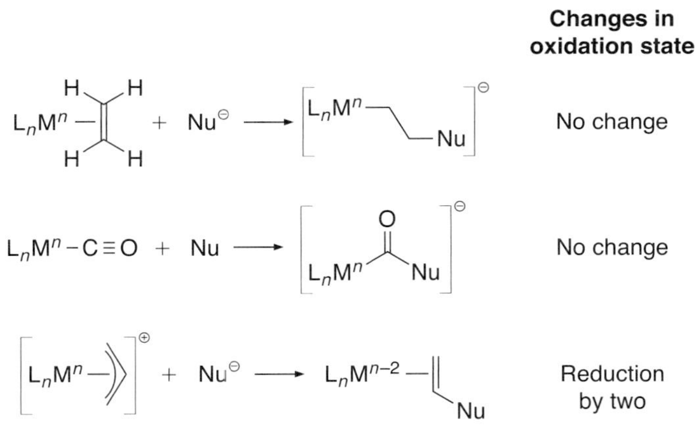

Scheme 11.2

11.2. Nucleophilic Attack on Transition Metal Complexes of Carbon Monoxide and Isonitriles

11.2.1. General Trends

In contrast to reactions of free CO with strong nucleophiles that generate unstable products, reactions of metal–carbonyl and metal–isonitrile complexes with nucleophiles generate stable products, and many relatively weak nucleophiles add to coordinated carbon monoxide. The isonitrile ligand, RNC, is similar to carbon monoxide in its chemical behavior, and, when coordinated to appropriate transition metals, reacts with nucleophiles to give relatively stable metal–carbene complexes. Because of the lower electrophilicity of isonitriles, however, stronger nucleophiles are required for attack on this ligand. A generic scheme for nucleophilic attack on CO and isonitriles is shown in Equation 11.1. Most of these reactions of nucleophiles occur by direct nucleophilic attack on the coordinated ligand without coordination of the nucleophile to the metal prior to the bond-forming process. Certain hydride reagents, organolithium and Grignard reagents, alkoxides, hydroxide, water, amines, and amide anions react with carbonyl complexes to generate formyl, acyl, alkoxycarbonyl, and aminocarbonyl derivatives.

$$L_nM-CO + Nu^{\ominus} \longrightarrow$$
$$L_nM-CNR + Nu^{\ominus} \longrightarrow \quad \overset{X}{\underset{L_n\overset{\ominus}{M}}{\parallel}}{\diagdown}Nu$$
$$X = O \text{ or } NR$$

$$(11.1)$$

These reactions are steps in the synthesis of organometallic complexes, in transformations for organic synthesis, and in the catalytic chemistry of CO. For example, the first synthesis of a carbene complex was achieved by a two-step sequence initiated by nucleophilic attack on a coordinated CO ligand.[7] In addition, the production of CO_2 and H_2 from CO and H_2O—the water-gas shift reaction—occurs by attack of water on a CO ligand that is coordinated to a heterogeneous iron catalyst[8–11] in the commercial processes and to a soluble complex in many homogeneous systems.[12–16]

The susceptibility of metal–carbonyl ligands to nucleophilic attack, summarized in Figure 11.1, generally depends on the amount of backbonding. The more π-acceptance by the CO ligand, the less susceptible the ligand is to nucleophilic attack. Thus, the rate of nucleophilic attack on a cationic metal–carbonyl compound is faster than that of attack on analogous neutral carbonyl compounds. The rate of nucleophilic attack is also decreased by the presence of electron-donating ligands, such as phosphines, at the metal. For example, a metal hexacarbonyl compound $M(CO)_6^{n+}$ reacts with nucleophiles faster than $M(CO)_5(PR_3)^{n+}$, which reacts faster than $M(CO)_4(PR_3)_2^{n+}$. When two carbonyl ligands in the same compound are inequivalent, the carbonyl group possessing the larger stretching force constant and experiencing weaker back-donation undergoes faster attack by nucleophiles.[17] As a guideline, compounds with ν_{co} values below ~2000 cm^{-1} are relatively inert to attack by nucleophiles.[18,19] Thus, water is sufficiently nucleophilic to attack the carbonyl ligand in electron-poor, cationic metal carbonyl complexes, whereas hydroxide is required to attack the carbonyl ligand in most neutral metal carbonyl complexes.

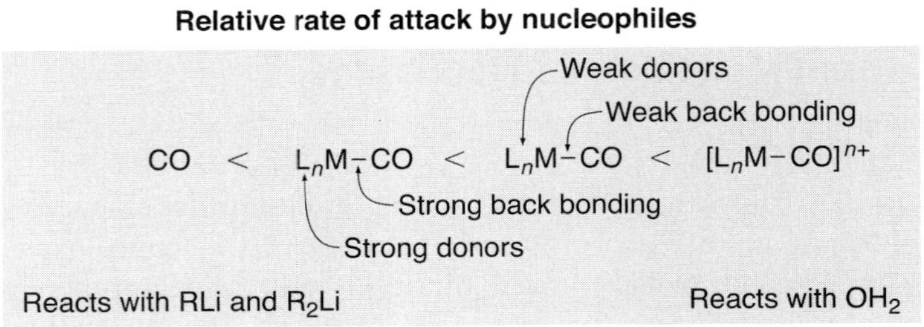

Relative rate of attack by nucleophiles

$$CO \quad < \quad L_nM\text{--}CO \quad < \quad L_nM\text{--}CO \quad < \quad [L_nM\text{--}CO]^{n+}$$

Weak donors
Weak back bonding
Strong back bonding
Strong donors

Reacts with RLi and R_2Li Reacts with OH_2

Figure 11.1.
Effect of coordination, ancillary ligands, and charge on the rate of nucleophilic attack on CO.

11.2.2. Examples of Nucleophilic Attack on Carbon Monoxide and Isonitriles

A hydride is one of the simplest nucleophiles, and Casey[20,21] and Gladysz[22,23] have prepared kinetically stable formyl complexes by the direct attack of hydride on a number of neutral chromium–, molybdenum–, and iron–carbonyl complexes (Equation 11.2). Although these complexes are relatively electron rich, because they possess zero-valent metal centers, the negative charge in the product can be stabilized by the remaining π-accepting CO ligands.

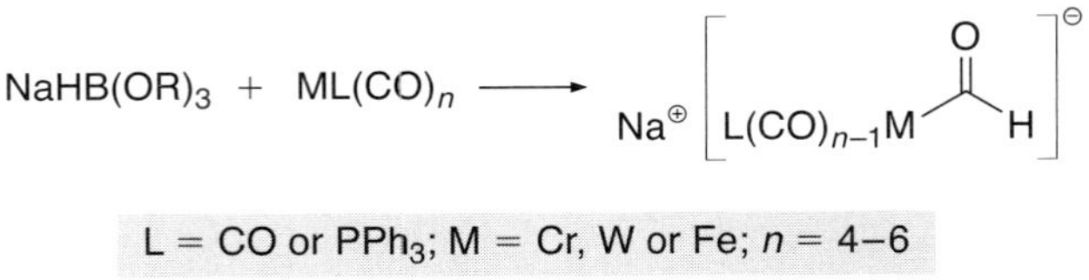

$$NaHB(OR)_3 \;+\; ML(CO)_n \longrightarrow Na^{\oplus} \left[L(CO)_{n-1}M \overset{O}{\underset{H}{\diagdown}} \right]^{\ominus}$$

(11.2)

L = CO or PPh_3; M = Cr, W or Fe; n = 4–6

Alkyl- and aryllithium reagents also attack metal-bound CO groups to generate anionic acyl complexes (Equation 11.3). The negative charge in these complexes is distributed over the metal–acyl system. In parallel with the reactions of related organic enolates, reactions of the anionic acyl complexes with electrophiles can occur at the metal, leading ultimately to products from coupling of the electrophile with the ligand derived from the carbonyl group, or at oxygen, leading to carbene complexes. Organic halides and tosylates tend to react at the metal center,[24,25] while harder and more reactive alkylating reagents, such as Meerwein's reagent ($R_3O^+BF_4^-$) or $ROSO_2F$, react at oxygen to generate carbene complexes.[26]

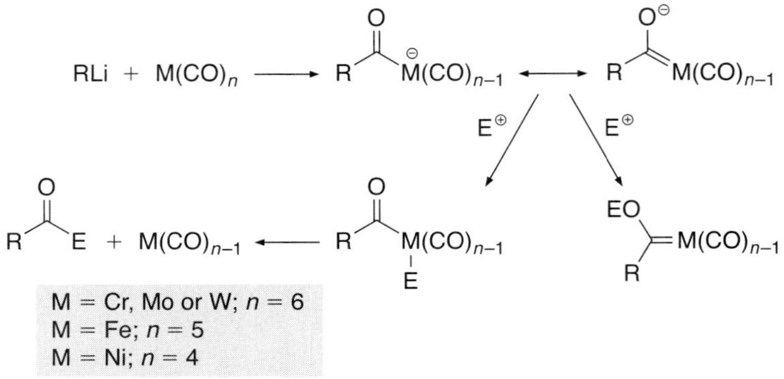

$$RLi \;+\; M(CO)_n \longrightarrow R\overset{O}{\diagup}M(CO)_{n-1}^{\ominus} \longleftrightarrow R\overset{O^{\ominus}}{\diagup}M(CO)_{n-1}$$

(11.3)

M = Cr, Mo or W; n = 6
M = Fe; n = 5
M = Ni; n = 4

Hydroxide ion attacks metal–carbonyl complexes to form hydroxycarbonyl complexes.[27,28] These complexes are generally unstable and extrude CO_2 to produce metal–hydride complexes (Equation 11.4). When the anionic hydride is sufficiently acidic to be deprotonated by the hydroxide, an overall two-electron reduction of the metal results. Thus, CO and aqueous base can reduce a metal complex. The reaction of hydroxide on coordinated CO is important in "Reppe-type" catalysis and in the water gas shift reaction.

$$M-CO + {}^{\ominus}OH \longrightarrow \underset{M}{\overset{O}{\|}}_{OH} \xrightarrow{-CO_2} \overset{\ominus}{M}-H \xrightarrow{OH^{\ominus}} M^{2\ominus} + H_2O \qquad (11.4)$$
$$d^n \qquad\qquad\qquad\qquad\qquad\qquad d^{n+2}$$

Alkoxides and amines also attack coordinated carbon monoxide to give alkoxycarbonyl and aminocarbonyl complexes, respectively[29] (Equations 11.5 and 11.6). Stable complexes of both types have been isolated and thoroughly characterized. Less stable alkoxycarbonyl and carbamoyl complexes are intermediates in a number of synthetically important carbonylation processes discussed in Chapters 17 and 19.

$$[Fe(NO)(CO)_2(PPh_3)_2]^{\oplus} + MeO^{\ominus} \longrightarrow (NO)(CO)(PPh_3)_2Fe\overset{O}{\overset{\|}{\diagdown}}_{OMe} \qquad (11.5)$$

$$L_n\overset{\oplus}{M}-CO + 2\,R_2NH \longrightarrow L_nM\overset{O}{\overset{\|}{\diagdown}}_{NR_2} + R_2\overset{\oplus}{N}H_2 \qquad (11.6)$$

Although many more examples of attack on coordinated CO have been published than examples of attack on isonitriles, the attack on isonitrile ligands is well established. The reaction in Equation 11.7 illustrates one example of such a nucleophilic attack on a coordinated isonitrile.[30] The kinetics of this reaction are consistent with the direct nucleophilic attack of the amine on the coordinated isonitrile to form a stable intermediate, which undergoes a subsequent proton transfer to form the observed carbene complex. The resulting electron-rich carbene complex resists further nucleophilic attack.

$$(11.7)$$

11.3. Nucleophilic Attack On Carbene and Carbyne Complexes

Many carbene complexes undergo nucleophilic attack at the carbene carbon. This chemistry is presented in more detail in Chapter 13 on metal–ligand multiple bonds. In brief, cationic carbene complexes tend to undergo simple addition processes to generate neutral products (Equation 11.8),[31] whereas neutral complexes can react by either addition (Equation 11.9)[32] or by a sequence of addition and elimination reactions (Equation 11.10).[26] Fischer[33] has shown that the aminolysis of alkoxycarbene complexes occurs by initial attack at the carbene carbon by a mechanism similar to that for the aminolysis of organic esters. Although less studied than reactions of carbenes with nucleophiles, reactions of carbyne complexes with nucleophiles are also known, and these reactions generate carbene complexes.[34, 35]

Nucleophilic additions to carbenes are restricted to cationic Fischer carbene complexes or Fischer carbene complexes that possess ancillary ligands that stabilize negative charge. The high-valent, early metal "Schrock-type" carbene complexes[36] are generally nucleophilic, rather than electrophilic, at the carbene carbon, so they react with nucleo-philes and Lewis bases at the metal center.

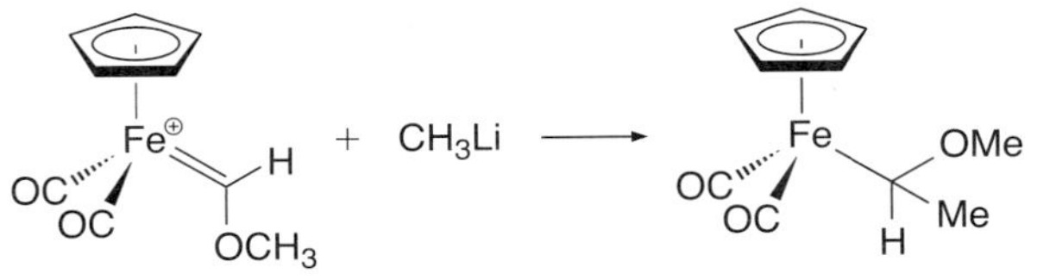

$$(11.8)$$

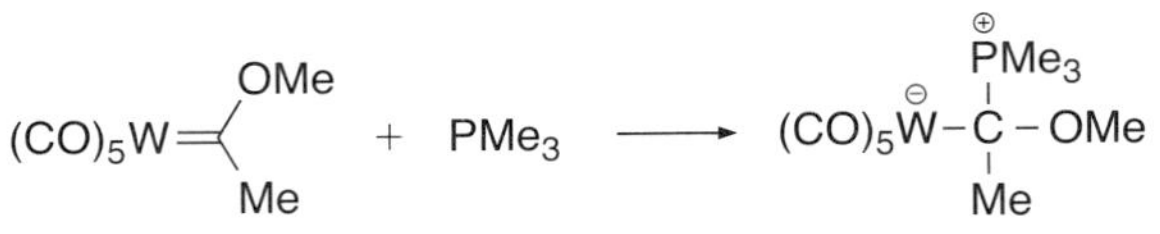

$$(11.9)$$

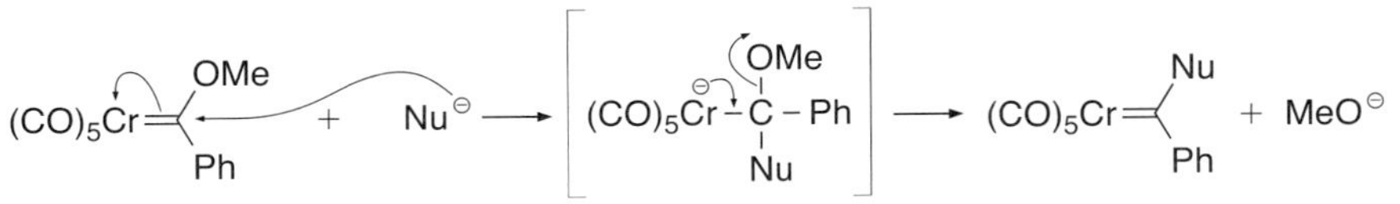

$$(11.10)$$

Nu = RS$^{\ominus}$, R$_2$NH, or PhLi

11.4. Nucleophilic Cleavage of Metal–Carbon σ-Bonds

11.4.1. General Principles and Trends

Trends in the reactivity of nucleophiles with complexes containing metal–carbon bonds with X-type ligands are summarized in Figure 11.2. Nucleophilic attack on σ-bound alkyl ligands is slower than attack on unsaturated ligands bound through a single atom, like CO or Fischer carbenes, or on unsaturated ligands bound through multiple atoms, like allyls, olefins, or polyenes. The slower rate of attack on an alkyl ligand can be attributed to the lack of a low-lying LUMO at the alkyl ligand and the partial negative charge that typically lies at the carbon bound to a transition metal. Yet, electron-poor complexes of σ-alkyl com-plexes can undergo attack by nucleophiles. This attack on a coordinated alkyl group occurs by a mechanism akin to an S$_N$2 reaction in which the metal fragment is a leaving group. The rate of this reaction depends on the amount of electron density at, or the electrophilicity of, the alkyl carbon and on the stability of the metal fragment that would be serve as the leaving group. Such attack on an alkyl complex has been observed most commonly in stoichiometric transformations. Catalytic chemistry occurring by nucleophilic attack on an alkyl group is uncommon, but strong evidence for this reaction in some catalytic reactions that functionalize alkanes has been gained.

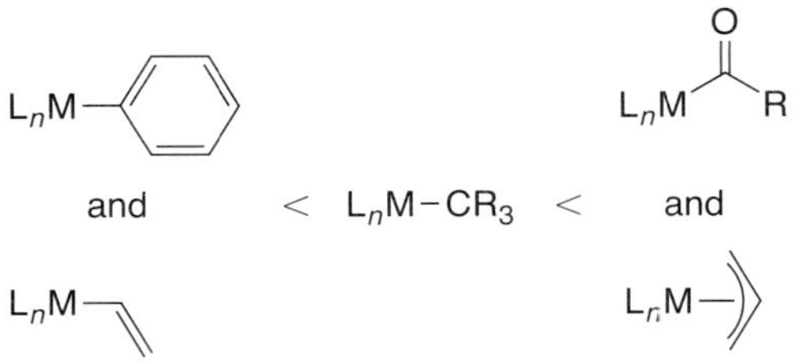

Figure 11.2.
General trends in the relative reactivity of nucleophiles toward various anionic organic ligands.

Nucleophilic attack on unsaturated σ-bound ligands, such as acyl groups, is faster than attack on saturated alkyl ligands. Because the metal-bound carbon of these ligands can possess a partial negative charge, acyl, alkoxycarbonyl, and aminocarbonyl complexes react more slowly with nucleophiles than do organic ketones and esters. Yet, they are more reactive toward nucleophiles than metal alkyls. Just as aryl and vinyl groups are poor electrophiles for simple uncatalyzed substitution reactions, σ-aryl and σ-vinyl groups bound to transition metals are poor electrophiles for external attack by nucleophiles. Attack on aryl and vinyl ligands is much slower than attack on alkyl ligands and is rare or unknown. (Aryl and vinyl groups must not be confused with π-bound, neutral arene and alkene groups, which often react readily with nucleophiles, as described below.)

In contrast to these slow reactions of σ-bound ligands, many complexes of anionic π-bound ligands react readily with nucleophiles. Reactions of π-allyl complexes with nucleophiles are most common, and these reactions have become a staple for the development of catalytic transformations of allylic alcohol derivatives, dienes, and allenes. Attack on electron-poor, often cationic, allyl complexes with relatively weak nucleophiles can occur rapidly below room temperature.[37]

11.4.2. Examples of Nucleophilic Attack on σ-Bound Ligands

Nucleophilic attack of transition metal centers on σ-bound ligands occurs as readily as reaction with the common nucleophiles of organic chemistry. Degenerate exchange of alkyl groups from one metal to another, as exemplified by the reactions in Equations 11.11 and 11.12, is proposed to occur by nucleophilic attack. This exchange of alkyl groups has been proposed as a mechanism to racemize chiral alkylmetal complexes.[38,39]

$$(11.11)$$

$$(11.12)$$

$$(11.13)$$

Favored in
more polar solvents

Faster in
less polar solvents

The attack of a nucleophile, such as an amine, alkoxide, thiolate, or halide on an alkyl group is most common when the alkyl group is bound to an electron-poor, often high-valent, metal center. These reactions occur by mechanisms that are similar to the S_N2 process of organic electrophiles. The reactions are sensitive to the polarity of the solvent, are sensitive to the steric effects at the point of attack, and lead to inversion of configuration at the carbon at the metal that is subject to the attack.

This reaction is part of the mechanism for reductive eliminations of alkyl halides from Pt(IV). As noted in Chapter 8 (reductive elimination), this reaction occurs by initial dissociation of iodide to generate a cationic Pt(IV) methyl complex and subsequent attack of iodide on the platinum(IV) methyl group to generate methyl iodide (Equation 11.13).[40,41] The reductive elimination of methyl aryl ethers, methyl acetate, and methyl trifluoroacetate also occurs by this pathway with phenoxide, acetate, or trifluoroacetate as the nucleophile that attacks the cationic platinum alkyl.[42,43] These reactions are sensitive to solvent in a manner that is counterintuitive. Heterolytic cleavage to generate the cationic platinum and halide anion is more favorable in a nonpolar solvent, but collapse of this charge-separated intermediate to generate neutral products is favored in a polar solvent. Thus, the actual S_N2 process, which is the second step of Equation 11.13, is faster in a nonpolar solvent.[40,41]

The attack of a nucleophile on a methyl group is also a likely step in the oxidation of alkanes catalyzed by platinum complexes.[44–48] The bond-forming process that occurs during these reactions is likely to be attack of water or hydroxide or of halogen onto a platinum(IV) methyl complex.[44] A series of experimental data suggests that platinum(IV) methyl complexes are generated in these oxidation processes by C–H activation of methane, followed by oxidation of a Pt(II) alkyl complex to a platinum(IV) alkyl complex (Equation 11.14). Attack of nucleophiles on such a Pt(IV) alkyl complex has been observed directly by spectroscopic methods in the recent studies by Bercaw and co-workers summarized in Scheme 11.3. In this study, a platinum(IV) methyl complex was generated in aqueous solution and allowed to react with hydroxide to generate methanol or with chloride to generate methyl chloride.[49] Nucleophilic attack of chloride on the diastereomerically enriched platinum(IV) complex in Scheme 11.3 showed that the reaction occurred with inversion of configuration at the platinum-bound carbon.

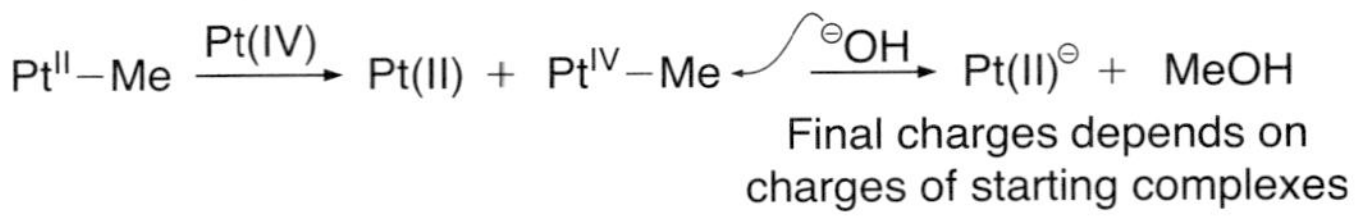

$$Pt^{II}\!-\!Me \xrightarrow{\ Pt(IV)\ } Pt(II) \ + \ Pt^{IV}\!-\!Me \xrightarrow{\ ^{\ominus}OH\ } Pt(II)^{\ominus} \ + \ MeOH \tag{11.14}$$

Final charges depends on
charges of starting complexes

$$[Pt^{IV}\!-\!Me]^{\oplus} \xrightarrow{\ ^{\ominus}OH\ } Pt^{II} \ + \ MeOH$$

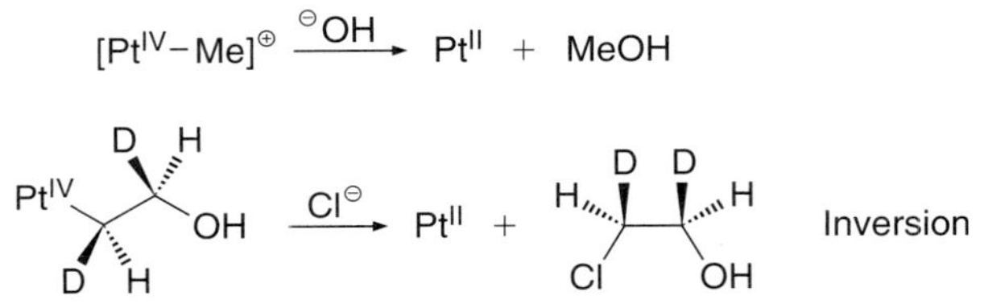

Scheme 11.3

Nucleophilic attack of organic nucleophiles on the alkyl groups in metal–porphyrin complexes has also been observed in several systems. DiMagno has developed a sequence of reactions, shown in Equation 11.15, that leads to the functionalization of methane by rhodium complexes containing fluorinated phorphyrin ligands.[50] Methane activation generates a Rh(III) methyl complex. The electron-poor porphyrin ligand renders the methyl group more like a methyl electrophile than a methyl nucleophile. Attack by PPh_3 generates $MePPh_3^+$ and a rhodium product that can be converted back to the methylrhodium species. Groves and Sanford have generated porphyrin-ligated rhodium–alkyl complexes in which the alkyl group contains a pendant nucleophile.[51] Insertion of the corresponding olefin into a rhodium(III) hydride, as shown in Equation 11.16, generates the alkyl complex that undergoes cyclization after the medium is changed to a polar solvent and the pronucleophile is deprotonated. The hydride can then be regenerated by protonation of the resulting anion.

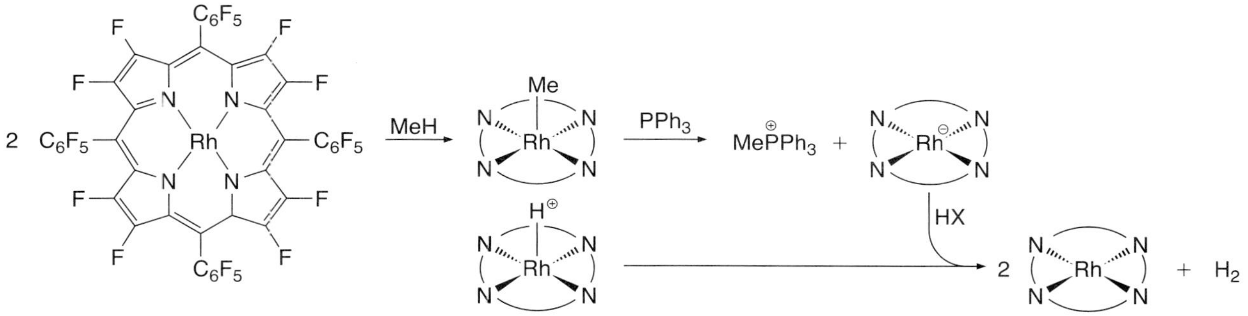

$$(11.15)$$

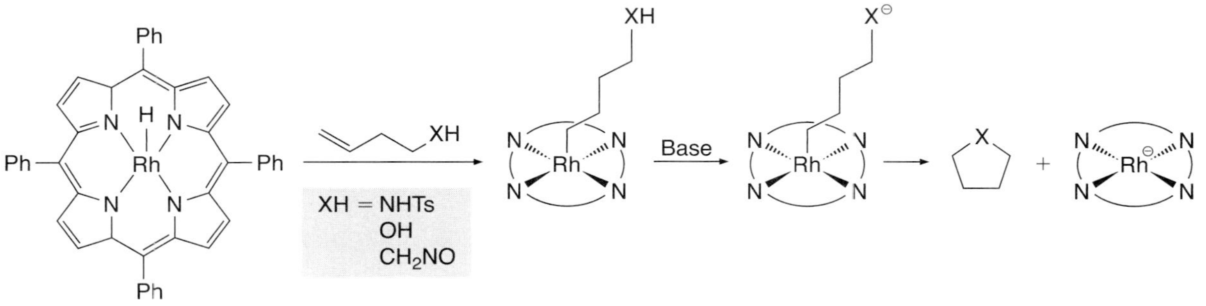

$$(11.16)$$

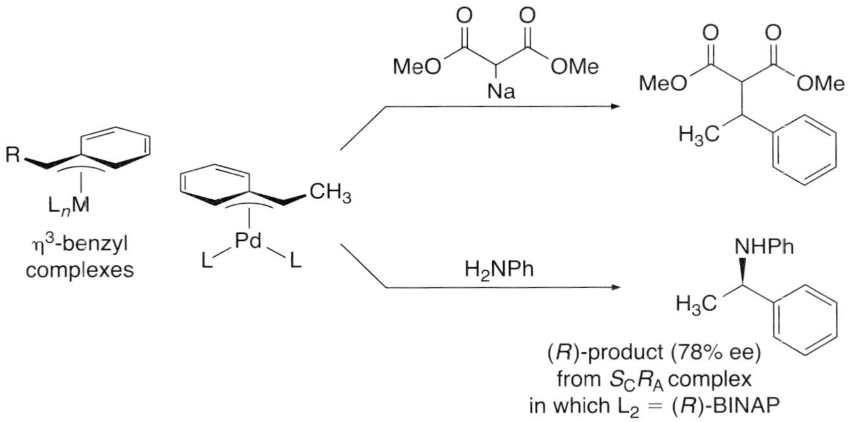

$$(11.17)$$

(R)-product (78% ee)
from $S_C R_A$ complex
in which L_2 = (R)-BINAP

Transition metal–benzyl complexes also react readily with nucleophiles, but the origin of this enhanced reactivity results from effects other than the typical rapid rate of nucleophilic attack at an electrophilic benzylic carbon. As shown in Equation 11.17, benzyl groups can be bound in an η^3 fashion, much like an allyl group.[52–58] As presented later in this chapter, cationic η^3-allyl complexes react with a variety of nucleophiles. η^3-Benzyl groups are common in the chemistry of palladium(II),[54,55,58] and these η^3-benzyl and phenethyl complexes react with a variety of nucleophiles. For example, these complexes react with malonate anions,[59–61] and they have been shown to react with amines[62–65] during some recently developed hydroamination processes. These reactions occur with predominant inversion of configuration.[62]

$$(11.18)$$

Metal–acyl complexes react readily with nucleophiles to produce carboxylic acid derivatives. Because most σ-alkyl complexes insert carbon monoxide to form σ-acyl complexes, the reaction of acyl complexes with nucleophiles occurs during a large number of transition-metal-mediated carbonylation reactions.[66–75] Again, the most rapid reactions occur with acyl groups bound to transition metal fragments that are electrophilic and that are good leaving groups.

Palladium-catalyzed carbonylations that occur by nucleophilic attack on an acyl group are common. Palladium(0) complexes catalyze the carbonylation of aryl halides[76,77] to generate ketones,[78] esters,[79] and amides,[79] by steps shown in Equation 11.18. An acyl complex is generated during these reactions by oxidative addition of the aryl halide and insertion of carbon monoxide. This acyl complex undergoes reaction with a main group organometallic reagent, such as an organotin, organozinc, or organoboron reagent, an alkoxide, or an amine to generate the ketone, ester, or amide.[80] Yamamoto has shown that the reactions to form esters and amides most likely occur by direct attack of the alcohol or amine at the acyl carbon.[81] The reactions to form ketones more likely occur through acylpalladium–alkyl or –aryl intermediates than by direct attack of the organometallic carbon nucleophiles at the acyl ligand.

Allyl groups undergo more rapid nucleophilic attack than any of these σ-bonded groups. As noted in Chapter 3, most allyl groups are η³-ligands bound through the π-system, and the reactions of these coordinated ligands are presented in the next section. However, one could envision reactions of nucleophiles with η¹-allyl complexes. η¹-Allyl groups are much less reactive toward nucleophiles[82] than η³-allyl groups. Nevertheless, differences in reactivity in the presence and absence of added ligand during an early study on the reactivity of η¹- and η³-allyl complexes were explained by reactions of η¹-allyl complexes with nucleophiles. The proposed affect of the added ligand and change in coordination mode is shown in Scheme 11.4. A difference in regioselectivity was observed upon reaction of a crotylpalladium species in the presence and absence of added ligand, and the authors suggested that this difference in reactivity resulted from formation of an η¹-allyl species that generates the branched amine product by an S_N2' pathway.[83] A postulate of reaction through an S_N2' pathway has also been used to explain the regioselectivity of rhodium-catalyzed allylic alkylation.[84] Although η¹-allyl complexes might react with nucleophiles in some cases, η¹-allyl groups react more commonly with electrophiles[85] in catalytic allylations of aldehydes and aldimines.[86–89]

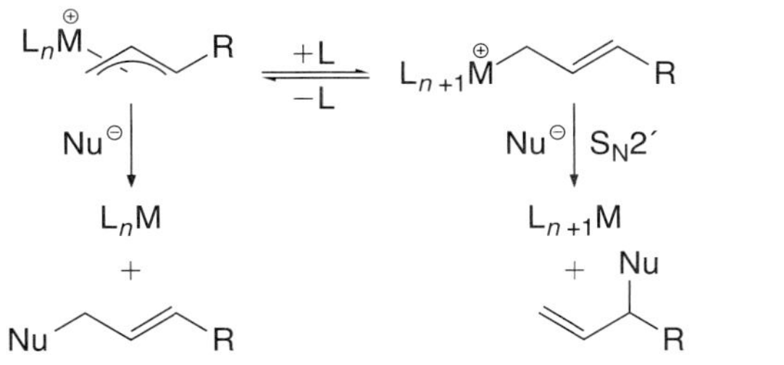

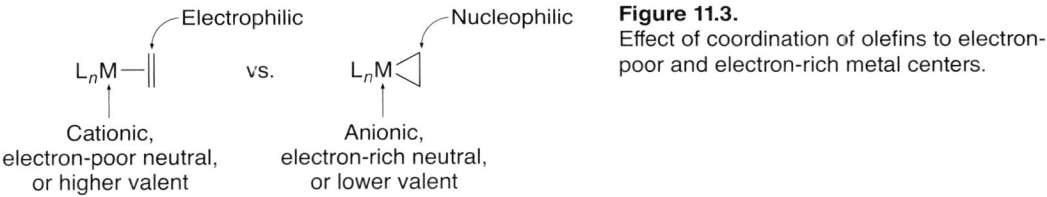

Scheme 11.4

11.5. Nucleophilic Attack on η²-Unsaturated Hydrocarbon Ligands

11.5.1. General Trends

As noted in the introduction of this chapter, free unsaturated hydrocarbons, such as olefins, polyenes, and arenes, typically react with electrophiles, not nucleophiles. However, coordination to an appropriate electron-deficient metal center alters their electronic properties, as shown in Figure 11.3. Coordination of an olefin to an electron-poor transition metal center causes the coordinated olefin to react with nucleophiles because this coordination leads to a net flow of electron density from the unsaturated hydrocarbon to the metal.

Electrophilic ⌐   Nucleophilic ⌐

L_nM—‖   vs. L_nM◁

Cationic,   Anionic,
electron-poor neutral,   electron-rich neutral,
or higher valent   or lower valent

Figure 11.3.
Effect of coordination of olefins to electron-poor and electron-rich metal centers.

The introduction of this chapter noted the changes in oxidation state and charge distribution that occur during nucleophilic attack on coordinated ligands. Attack on a neutral olefin leads to an increase in charge at the metal center. Thus, properties of the metal center that stabilize this negative charge promote nucleophilic attack. Nucleophilic attack, therefore, occurs most readily on unsaturated hydrocarbons that are bound to metal centers in relatively high oxidation states (II, III, or IV) possessing either a full, formal positive charge or a number of electron-withdrawing ancillary ligands, such as carbon monoxide, or both.

In contrast to the increase in reactivity of hydrocarbons with nucleophiles after coordination to electron-accepting metal centers, an increase in reactivity of unsaturated hydrocarbons with electrophiles is observed upon coordination to particularly electron-rich metal centers. This contrasting reactivity is shown schematically in Figure 11.3. Olefin complexes of very electron-rich metal centers are best described as metallacyclopropane complexes, as noted in Chapters 1 and 2. As such, the olefin ligands in these complexes contain a large degree of M–C σ-bond character and react with electrophiles. Reactions of electrophiles with coordinated ligands are described in Chapter 12.

Nucleophilic additions to 18-electron, cationic transition metal complexes of unsaturated hydrocarbons have been particularly well studied. Davies, Green, and Mingos have developed guidelines to predict the most kinetically favorable position for nucleophilic

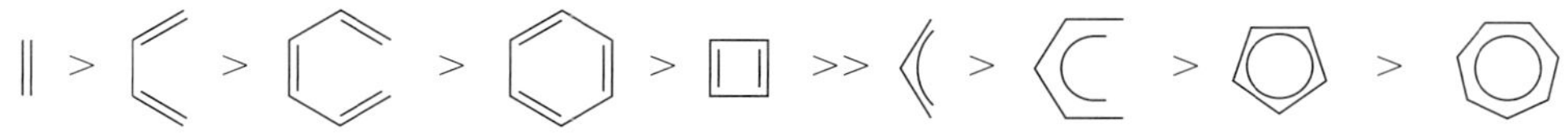

Figure 11.4. Order of reactivity of π-ligands according to the Davies–Green–Mingos rules.

attack on coordinated, unsaturated hydrocarbons in such 18-electron, cationic complexes.[90] These guidelines classify unsaturated hydrocarbon ligands as even or odd, depending on the number of atoms bound to the metal center (η^n), and closed or open, depending on whether they are cyclic or acyclic. The guidelines state that "even systems" are more reactive than "odd systems," and open systems are more reactive than closed systems. These trends lead to the order of reactivity in Figure 11.4. Furthermore, attack of nucleophiles on open polyenes tends to occur at a terminal carbon atom, although there are exceptions noted later in this chapter, and nucleophilic attack on odd open polyenyls typically occurs at a terminal carbon when the metal fragment is cationic or electron poor. As one illustrative example of the relative rates for reaction with different ligands, the starting complex in Equation 11.19, which contains three different unsaturated hydrocarbon ligands, is attacked exclusively at the η^4-diene ligand, and this attack occurs at a terminal position.[91]

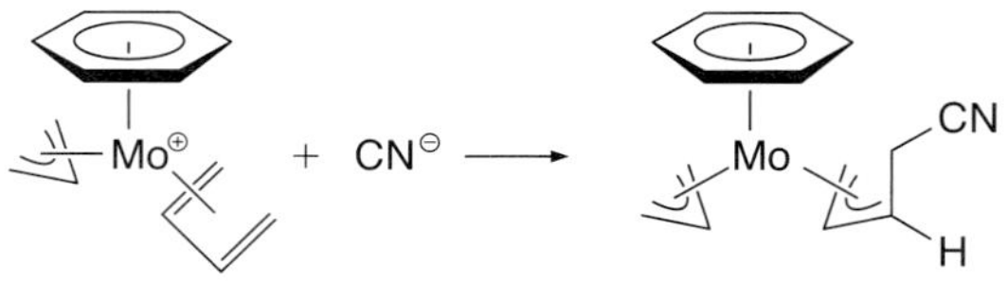

$$(11.19)$$

11.5.2. Nucleophilic Attack on η^2-Olefin Complexes

11.5.2.1. Overview of Nucleophilic Attack on η^2-Olefin Complexes

Nucleophiles add to many olefin complexes, particularly those of Pd(II), Pt(II), and Fe(II) to produce (σ-alkyl)metal complexes. Several modes of attack, summarized in Equation 11.20, are known and are distinguished experimentally by the stereochemistry and regiochemistry of the products. In most cases, nucleophilic attack occurs onto the face of the olefin that is opposite the metal without coordination of the nucleophile to the metal prior to this attack. Reaction by this mechanism generates products from anti addition of the metal and the nucleophile across the olefin.

The changes in structure during this attack are depicted in Equation 11.21. Eisenstein and Hoffmann concluded many years ago that a symmetric metal–olefin complex is deactivated toward nucleophilic attack, but slippage of the olefin creates a LUMO that can interact with the incoming nucleophile. Thus, in the transition state for external attack, "the organometallic olefinic complex cannot be near its equilibrium η^2 structure."[92]

The external attack of a nucleophile occurs predominantly at the more-substituted terminus of substituted olefins. Addition tends to occur at the carbon bearing a greater positive charge, and the more substituted carbon contains a higher positive charge. This larger positive charge can be attributed to several factors. First, the more substituted carbon would be expected to better support positive charge in an olefin complex than a less-substituted carbon for the same reasons it better supports a positive charge in a purely organic system. Second, the position of nucleophilic attack can be traced to the difference in M–C bond distances. The M–C distance is longer to the more-substituted carbon, and the degree of positive charge has been shown to be greater at the carbon bound with a longer M–C distance.

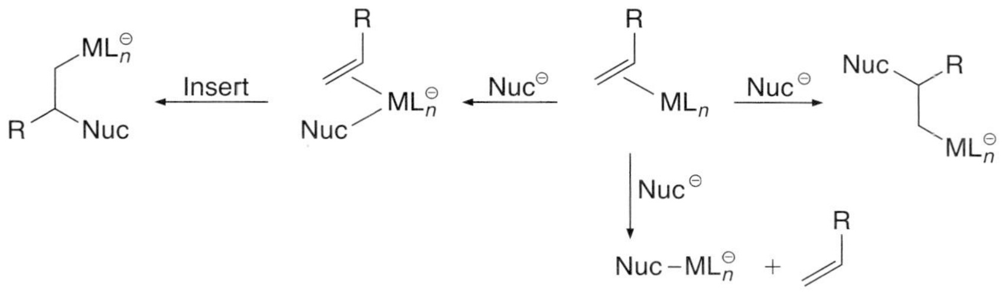

$$(11.20)$$

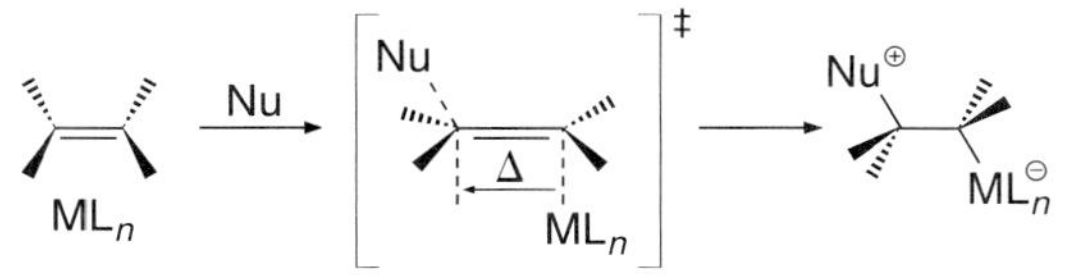

$$(11.21)$$

In some cases, the nucleophile appears to attack the metal before formation of the bond between the nucleophile and the coordinated olefin. In this case, attack of the nucleophile at the metal is followed by insertion of the olefin into the metal-nucleophile bond. Reaction by this mechanism generates products from syn addition of the metal and the nucleophile across the olefin.

Reactions of both types are generally restricted to mono- or disubstituted olefins. More highly substituted olefins coordinate only weakly to most metal centers. Displacement of the more-substituted olefins from the metal by the nucleophile is a major competing reaction, particularly if the nucleophile is also a good ligand for the metal (Equation 11.20, bottom).

11.5.2.2. Specific Examples of Nucleophilic Attack on η^2-Olefin Complexes: Reactions of [CpFe^{II}(CO)$_2$]$^+$, [CpPd^{II}L]$^+$ and Square Planar M^{II} (M=Pd, Pt) Olefin Complexes

Some of the classic studies of nucleophilic attack on coordinated olefins were conducted with iron(II) species. Rosenblum reported the reactions of (η^5-cyclopentadienyl) iron–olefin complexes with a wide range of carbanion and enamine nucleophiles.[93] These reactions produce stable σ-alkyliron complexes (Equation 11.22). The stereochemistry is cleanly trans. However, the regioselectivity of reactions of complexes of unsymmetrical olefins depended on the nucleophile.

Kurosawa has shown that the related palladium(II) olefin complex $[(\eta^5\text{-}C_5H_5)Pd(Ph_3P)$ (CH$_2$=CH$_2$)]$^+$ undergoes clean trans attack by both methoxide[94] and the anion of acetylacetone.[95] These examples of additions to iron– and palladium–olefin complexes occur by attack only at the ligand and lead to stable σ-alkyl products because the metal center is coordinatively saturated. This coordinative saturation disfavors attack at the metal that would lead to either products from syn addition or displacement of the olefin. The σ-alkyl complexes that are products of these examples of nucleophilic attack are stable because the most common mode for decomposition of σ-alkyl complexes, β-hydrogen elimination, requires the presence of a vacant coordination site cis to the alkyl group, as noted in Chapters 8 and 9. Such a site is not present in these σ-alkyl products.

$$(11.22)$$

$$Nu^{\ominus} \text{ or } Nu = MeO^{\ominus}, \, ^{t}BuS^{\ominus}, \, Ph_3P, \, (EtO)_3P, \, R_2NH, \, ^{\ominus}CH_2\,NO_2, \, ^{\ominus}CH(COMe)(CO_2Me)$$
$$^{\ominus}CH(CO_2Me)_2, \, ^{\ominus}CH(CN)(CO_2Me), \, R_2N-CH=C\!\!\!<\, \text{ and } LiCuMe_2$$

Stoichiometric and catalytic reactions that result from attack of nucleophiles onto square-planar palladium(II) and platinum(II) olefin complexes are legion.[96] The σ-alkyl products are stable in some cases, but β-hydrogen elimination to generate products from functionalization of the C–H bond of an olefin occurs in other cases. Nucleophilic attack on palladium(II) and platinum(II) complexes of diolefins often generate stable σ-alkyl products because the resulting alkyl group is stabilized by chelation.

$$(11.23)$$

$$M = Pd \text{ or } Pt, \, Nu = RO^{\ominus}, \, RNH_2, \, N_3^{\ominus}, \, ^{\ominus}CH(CO_2R)_2, \text{ and } ^{\ominus}CH(COR)_2$$

$$(11.24)$$

$$Nu = MeO^{\ominus} \text{ or } CH^{\ominus}(CO_2Et)_2$$

$$(11.25)$$

Thus, norbornadiene and 1,5-cyclooctadiene complexes of both palladium and platinum undergo reactions with alkoxides,[97] amines,[98,99] azide,[100] and stabilized carbanions[101] to generate stable (σ-alkyl) metal complexes. The accumulation of electron density at the metal by conversion of an olefin ligand to a σ-alkyl ligand causes a second attack at the remaining coordinated olefin to be unfavorable (Equation 11.23). Attack occurs to the face of the olefin opposite the metal, and displacement of the olefin is generally not observed. Holton has studied the related reactions of coordinated allylamines and homoallylamines with nucleophiles. Allylamines undergo attack at the more-substituted end of the olefin to give a five-membered palladacycle containing the σ-alkyl ligand stabilized by coordination of the amino group (Equation 11.24).[102, 103] In contrast, homoallylamines undergo attack at the less-substituted carbon to give a five-membered metallacycle containing a σ-alkyl and an amine donor (Equation 11.25).[104] The five-membered palladacycles are more stable than the corresponding four- or six-membered metallacycles. Presumably for this reason, longer chain amines possessing pendant terminal olefins do not undergo similar reactions.

In contrast to complexes of palladium(II) complexes ligated by chelating olefins, palladium(II) complexes ligated by simple monoolefins generate products that are less stable toward β-hydrogen elimination and more often react with nucleophiles to simply displace the olefin from the metal center. Yet, examples of nucleophilic attack on ethylene have been observed and were studied in relation to the mechanism of the important catalytic processes known as the Wacker oxidation of olefins.[105,106] These Wacker processes convert ethylene to acetaldehyde and substituted olefins to ketones. The key bond-forming process occurs by attack of water on the coordinated olefin.[107-109] Related reactions also convert olefins to functionalized products by reaction with carboxylic acids[110-112] and some nitrogen nucleophiles.[113,114]

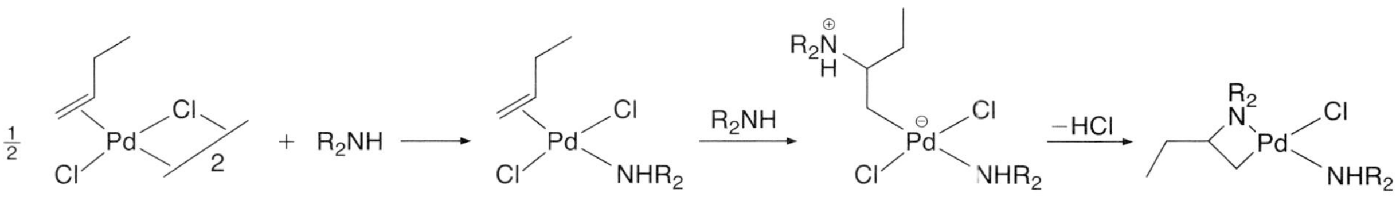

(11.26)

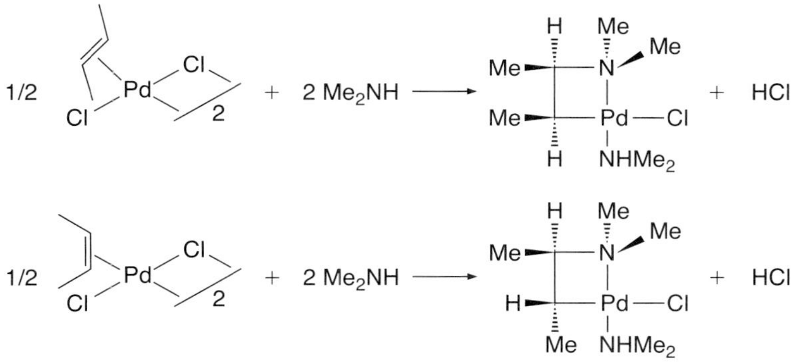

Scheme 11.5

Many reactions of amines with palladium– and platinum–olefin complexes have been reported. Åkermark showed that nucleophiles add to palladium–olefin complexes to generate aminoalkyl complexes, as shown by the example in Equation 11.26. In this case, reactions of a bis-olefin dichloropalladium complex with amines occurs by splitting of the chloro-bridged dimer by the first equivalent of amine to give a neutral olefin-ligated palladium–amine complex that undergoes attack of the coordinated alkene by a second equivalent of amine.[115] The stereochemistry of the amination is cleanly trans. Åkermark and Zetterberg[116] isolated and characterized by ¹³C NMR spectroscopy the σ-alkyl complexes formed by the amination of both *cis-* and *trans-*2-butene, and the stereochemistry of the product alkyl complexes results from external attack by amines, as shown in Scheme 11.5.

Hegedus developed an olefin amination process that generates indoles by intramolecular nucleophilic attack of substituted amines on allylarenes (Equation 11.27).[117] In this process, an *N*-tosyl arylamine attacks the coordinated olefin of an allyl group in the ortho position of the aniline, and β-hydrogen elimination generates a tautomer of indole and a reduced palladium complex. This reduced complex is then re-oxidized by quinone to regenerate the starting palladium(II), and the product tautomerizes to the final indole product. This cycle was one of the earliest catalytic olefin aminations.

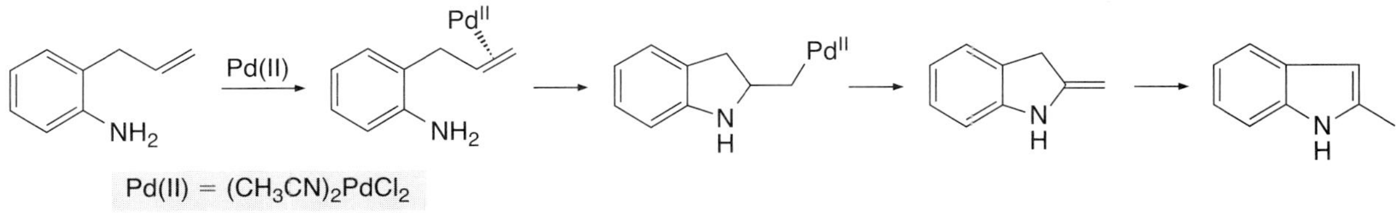

(11.27)

Pd(II) = (CH₃CN)₂PdCl₂

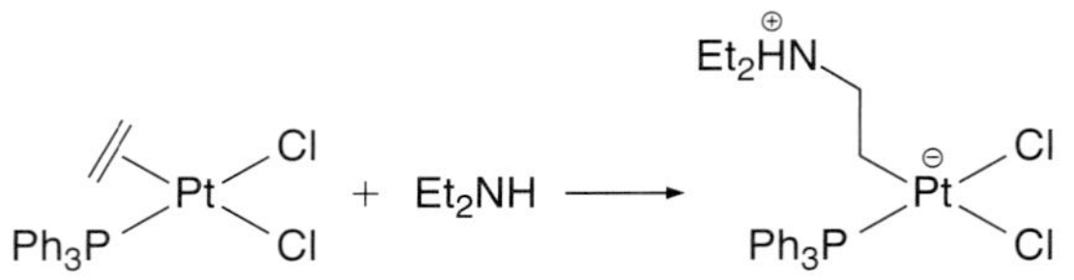

$$\text{(11.28)}$$

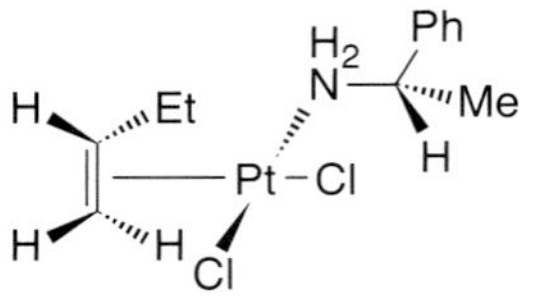

Figure 11.5.
Diastereomerically pure alkene complex used to show trans attack on a coordinated olefin.

Although some palladium σ-alkyl complexes formed from additions of amines are stable by virtue of chelation of the aminoalkyl ligand in the product, σ-alkyl platinum complexes lacking chelation are more stable than the palladium analogs. For example, treatment of *cis*-dichloro(ethylene)triphenylphosphine platinum(II) with diethylamine produces the isolable σ-alkylplatinum complex in Equation 11.28.[99] The mode of attack of amine on a platinum–olefin complex was unequivocally shown to be trans by Panunzi. He showed that the single diastereoisomeric platinum complex, (+)-*cis*-dichloro[(S)1--butene][(S)-α-methylbenzylamine]platinum(II) in Figure 11.5, which was generated from the prochiral olefin 1-butene, reacted with an amine to generate a product resulting from trans addition.[118]

Oxygen nucleophiles, such as water, alcohols, and carboxylic acids, also attack coordinated olefins in simple (monoolefin)palladium(II) complexes. These nucleophiles do not displace the olefin from the metal because they are poor ligands for the soft palladium(II).

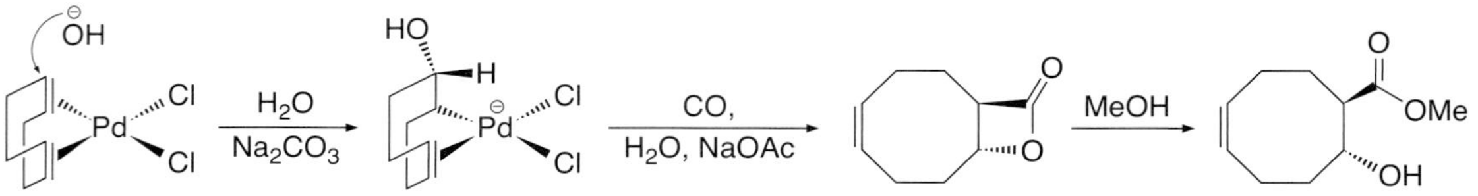

$$\text{(11.29)}$$

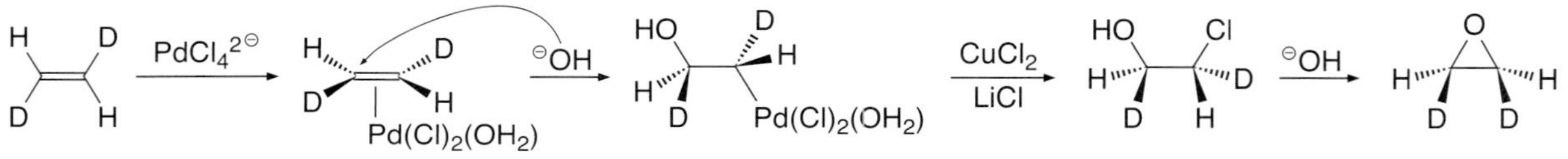

$$\text{(11.30)}$$

A series of studies were conducted on the addition of oxygen nucleophiles to palladium–olefin complexes, and several experiments on the stereochemistry of the process showed that the nucleophilic attack occurred to the face of the olefin opposite the metal center. For example, Stille showed that the cyclooctadiene complex in Equation 11.29 reacts with either methanol or water to give products from external attack.[119,120] Moreover, Bäckvall and Åkermark showed that attack of water on the palladium complex of *trans*-dideuterioethylene during chlorohydrin formation occurred by the external attack in Equation 11.30.[121] Stille showed that attack of water on (*cis*-CHD=CHD)$_2$PdCl$_2$ occurs with inversion of configuration and thereby trans attack to form the *trans*-dideuteriolactone after addition of CO.[122] Although this body of work provides strong evidence for external attack of oxygen and nitrogen nucleophiles on coordinated olefins, a series of studies by Henry on the effect of chloride on the stereochemistry of reactions of chiral, non-racemic allylic alcohols suggests that the mode of attack depends on the concentration of chloride. These studies have provided evidence that products from syn addition form from reactions at low concentrations of chloride and that products from anti addition form from reactions at high concentrations of chloride.[123–126]

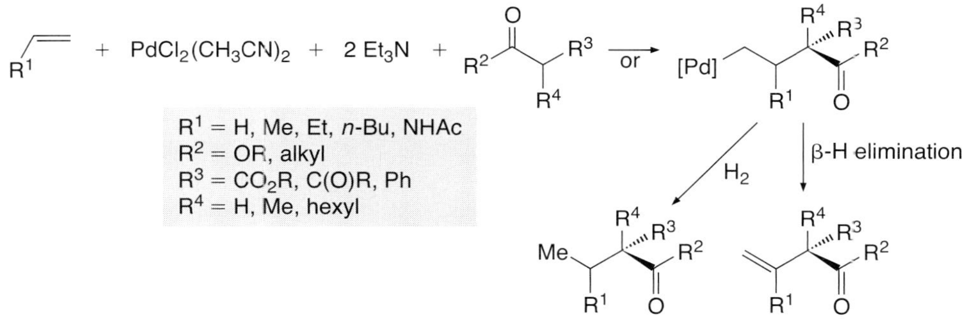

(11.31)

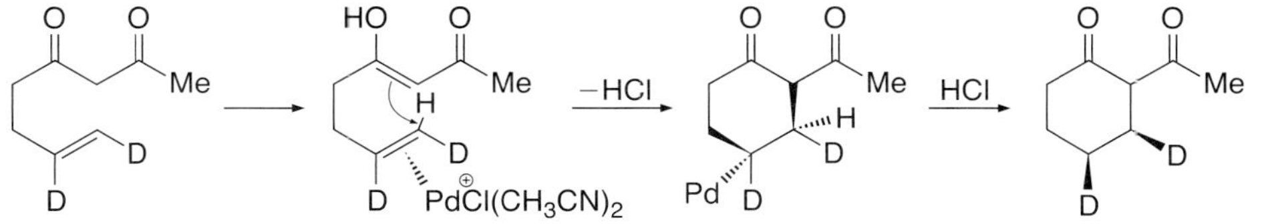

(11.32)

Nucleophilic attack of stabilized carbon nucleophiles on coordinated olefins is also known. Hegedus developed the alkylation of olefins shown in Equation 11.31.[127] The (olefin)palladium(II) chloride complexes did not react with malonate nucleophiles, but the triethylamine adduct does react with this carbon nucleophile to provide the alkylation product. This reaction has recently been incorporated into a catalytic alkylation of olefins by Widenhoefer.[128,129] Intramolecular reaction of the 1,3-dicarbonyl compounds with pendant olefins in the presence of $(CH_3CN)_2PdCl_2$ occurs to generate cyclic products containing a new C–C bond (Equation 11.32).[128–132] Some intermolecular reactions with ethylene and propylene have also been developed by this group.[133] Deuterium labeling studies (Equation 11.32) have shown that the addition occurs by external attack on the coordinated olefin.[131]

11.5.3. Nucleophilic Attack on Square Planar Pd(II) Diene and Allene Complexes

Dienes can bind to metals in an η² or η⁴ fashion. Many metal complexes contain η⁴-diene ligands, and reactions of η⁴-diene complexes are described in Section 11.7.2. However, other metals, particularly Pd(II), bind dienes and allenes in an η² fashion.[134] The reactivity of these species has been exploited to develop useful synthetic methods and is described in this section. Bäckvall[135,136] has developed chemistry based on nucleophilic addition to palladium–diene complexes to generate allyl products, which subsequently react with a second nucleophile to generate free organic products from 1,4-addition of two nucleophiles across the diene. The palladium(0) byproduct is then re-oxidized to Pd(II) with quinone.

Catalytic and stoichiometric additions of oxygen nucleophiles to coordinated dienes are summarized in Equations 11.33 and 11.34. Early studies involved 1,4-additions of two acetoxy or alkoxy groups across a diene.[135] More recently, intermolecular additions of two different nucleophiles have been developed.[137] The stereochemistry for additions across cyclic dienes makes this procedure particularly valuable. Conditions for either cis or trans additions have been developed. Cis addition is typically observed in the presence of added chloride, and trans addition occurs in the absence of chloride.[138,139] Both intermolecular and intramolecular[140–143] 1,4 additions to dienes have been developed, and reactions of nitrogen[144] and carbon[145] nucleophiles have also been reported. More details on these processes are reported in Chapter 16.

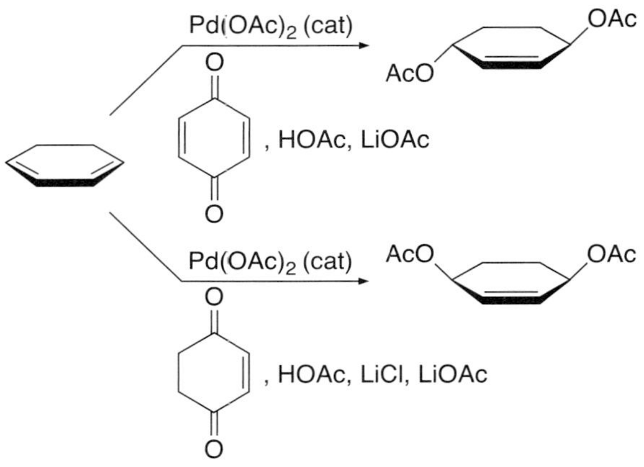

(11.33)

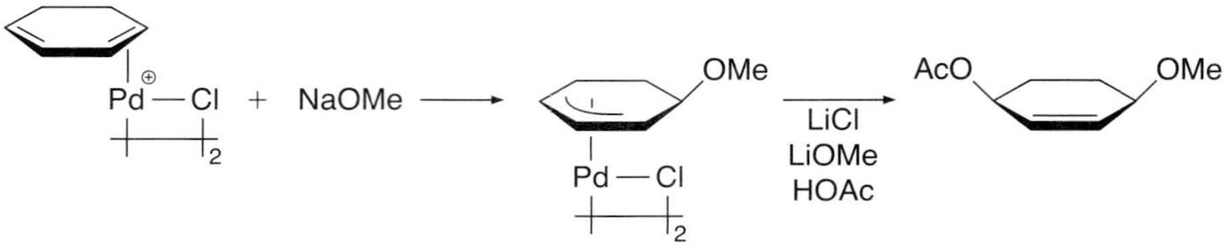

(11.34)

11.5.4. Nucleophilic Attack on η^2-Alkyne Complexes

Transition metal alkyne complexes also react with nucleophiles, in this case to generate σ-vinyl complexes. There are fewer stable alkyne complexes of higher oxidation state or cationic metals than olefin complexes. Because these types of alkyne complexes are most susceptible to nucleophilic attack, less information is available on this reaction than on nucleophilic attack on coordinated alkenes. Nevertheless, reactions of several cationic alkyne complexes with nucleophiles have been reported, and a few examples are presented here.

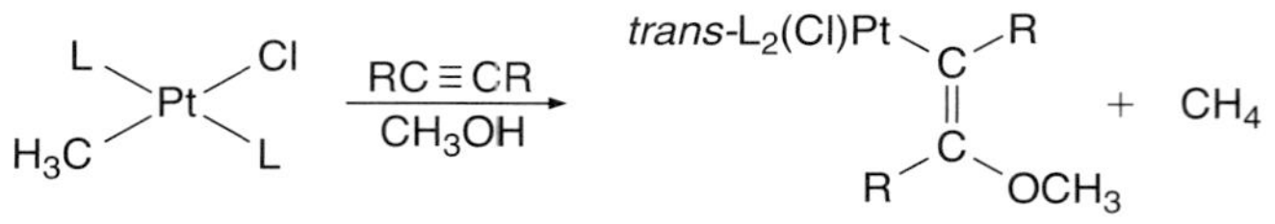

(11.35)

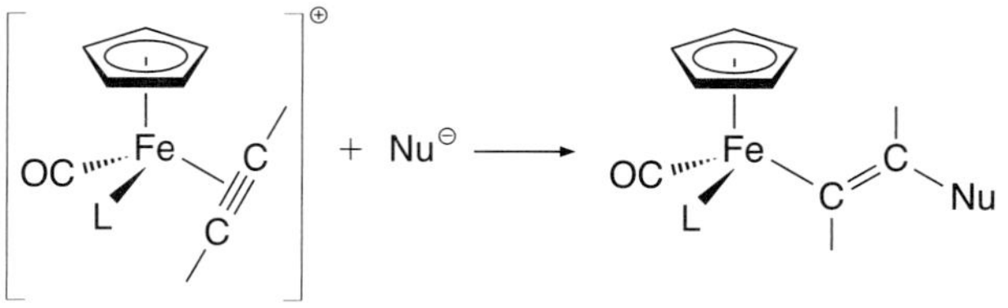

(11.36)

$Nu^\ominus = PhS, {}^\ominus CN, {}^\ominus CH^\ominus(CO_2Et)_2, Ph^\ominus, Me^\ominus, H_2C=CH^\ominus$, or $MeC\equiv C^\ominus$ from $R_2Cu(CN)Li_2$
$L = PPh_3$ or $P(OPh)_3$

Cationic platinum(II) acetylene complexes react with a variety of nucleophiles. Chisholm and Clark[146] showed that the reaction of methanol with coordinated disubstituted acetylenes (generated in situ) affords trans σ-bonded vinyl ether complexes (Equation 11.35). The trans stereochemistry of the σ-vinyl group of the product suggests that the nucleophilic attack occurs external to the metal and does not involve prior coordination of methanol. Reger has shown that stable cationic iron–alkyne complexes undergo reaction with a wide variety of nucleophiles to give stable σ-alkenyl complexes resulting from trans attack of the nucleophile on the coordinated acetylene (Equation 11.36).[147] A variety of hydroaminations of alkynes have been reported to be catalyzed by late metal complexes in the past several years.[148–151] Attack on a coordinated alkyne is likely to be a step in the reaction of at least one if not several of the catalysts.

11.5.5. Reactions of η^2-Arene Complexes

η^2-Arene complexes have been developed as reagents for organic synthesis. A vast majority of this work has been reported by Harmann.[152,153] η^2-Arene and heteroarene complexes have been generated on $(NH_3)_5Os^{2+}$ and more recently on TpM(CO)(L) in which M = Re, Mo, and W, and L = CO or NO.[154] The latter chiral-at-metal system has been resolved.[155] This η^2 coordination leads to de-aromatization of the arene, in effect isolating the diene unit for reactions, as depicted in Figure 11.6. The chemistry of η^2-arene complexes is therefore more akin to the chemistry of dienes, and the chemistry of η^2-furans is more akin of that of vinyl ethers than it is to the chemistry of electron-poor η^5-arenes. Harmann has developed Diels–Alder reactions of the diene unit and additions of electrophiles to the vinyl ether unit.[156,157] In one case, coordination of furan also led to the ability to conduct nucleophilic substitution on the coordinated olefin.[158]

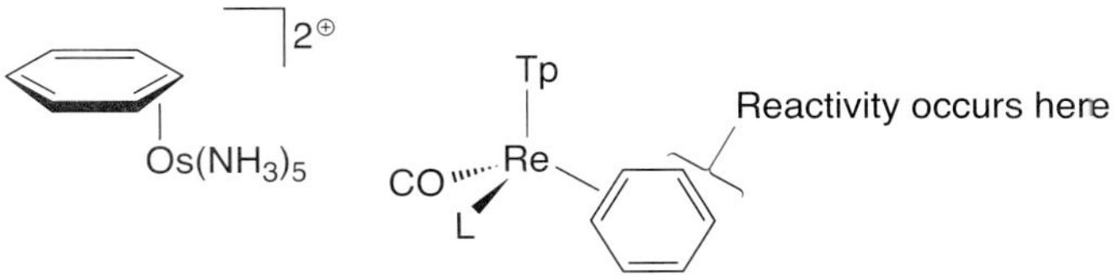

Figure 11.6.
Effect of η^2 coordination on the reactivity of complexed arenes.

11.6. Nucleophilic Attack on Imine and Aldehyde Complexes

Coordination of imines and aldehydes to Lewis acids enhances their reactivity with nucleophiles, and a wide variety of metals have been used to catalyze these types of additions. Gladysz and Templeton have studied the stoichiometric additions of main group organometallic reagents to aldehydes and imines. As part of these studies, Gladysz has shown that aldehydes bind to $CpRe(NO)(PPh_3)^+$ to generate a mixture of η^2- and η^1-aldehyde complexes.[159–161] Kinetic studies show that rearrangement of the η^2-aldehyde complex to the η^1 form precedes attack of cyanide at the aldehydic carbon, as outlined in Equation 11.37.[160]

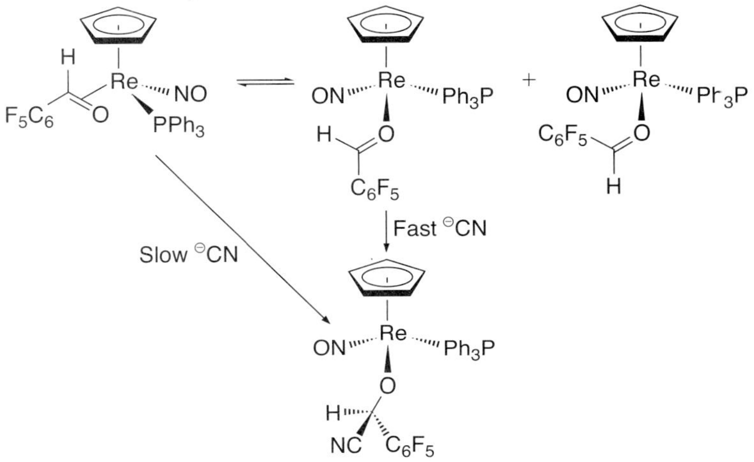

(11.37)

Templeton has studied the additions in Equation 11.38 of hydrides to the complexes formed from a chiral tungsten metal center and imines derived from 2-butanone.[162] These hydride additions occur with high stereoselectivity for addition to one face of the aldehyde over the other to form an amido complex as the major product. Deprotonation at the N–H position forms an iminyl or azavinylidene complex as the minor product. This stereoselectivity for addition to one of two faces is remarkable, considering that these faces are distinguished only by the relative positions of ethyl and methyl groups. Several enantioselective alkylations of imines by organozinc reagents catalyzed by copper complexes containing mixed P,O or P,N donor ligands have been reported recently, and these reactions might occur by attack of the main group organometallic reagent on a coordinated imine.[163,164]

(11.38)

11.7. Nucleophilic Attack on Polyhapto (η^3–η^6) Ligands

11.7.1. Nucleophilic Attack on η^3-Allyl Complexes

A wide range of transition metal–allyl complexes are known to react with many types of nucleophiles. In most cases, these reactions occur between cationic allyl complexes and amines or stabilized, anionic carbon nucleophiles. The reaction typically occurs between the nucleophile and the η^3 form of the allyl complex, and attack usually occurs at the face of the allyl ligand opposite the metal. However, there are exceptions to these trends. For example, several experiments suggest that unstabilized carbon nucleophiles react first at the metal center, and C–C bond formation occurs between the alkyl and the allyl group by reductive elimination.[165,166] In addition, a recent study has shown through deuterium labeling that attack of malonate anion on a molybdenum–allyl complex occurs with retention of configuration.[167]

A variety of (η^3-allyl)metal complexes undergo facile nucleophilic attack at the η^3-allyl ligand. By far the most extensive studies of nucleophilic attack on metal–allyl complexes have been conducted with (η^3-allyl)palladium complexes. These complexes are intermediates in several catalytic reactions, including catalytic allylic substitution processes.[168] However, late transition metal complexes of many metals catalyze reactions that occur through allyl intermediates. Molybdenum,[169–175] tungsten,[174,176,177] ruthenium,[178–183] rhodium[184–186] and iridium[187–197] complexes all catalyze synthetically valuable allylic substitution processes, and nickel[198,199] and palladium[200–205] complexes catalyze useful telomerization reactions that combine a nucleophile and two or more dienes to make products containing longer chains. Although many stereoselective catalytic processes have been developed with a variety of metals in the past decade, and information on the allyl intermediates can be surmised from selectivities in these reactions, fewer studies have focused on the chemistry of isolated allyl complexes. This chapter will focus on the reactions of isolated allyl complexes, whereas Chapter 20 includes catalytic asymmetric allylic substitution reactions.

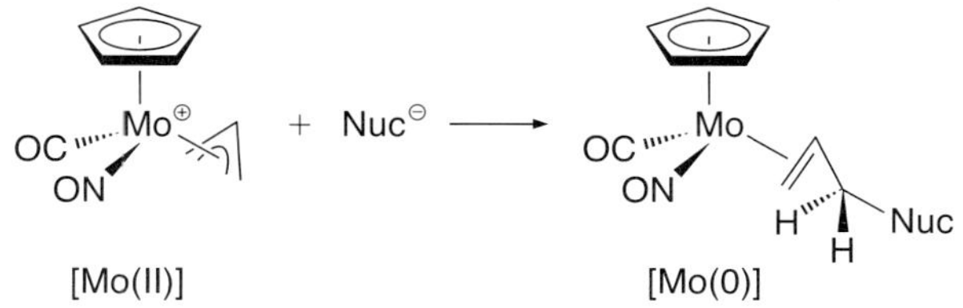

$$(11.39)$$

Nucleophilic attack on allyl complexes of molybdenum, ruthenium, palladium, and nickel has been studied. Faller's studies on the reaction of CpMo(CO)(NO)(η^3-allyl)$^+$ with nucleophiles[206,207] illustrate a number of principles that are general for allylic substitutions (Equation 11.39). The nucleophiles react most readily with electron-poor allyl complexes, and the cationic charge and electron-withdrawing CO and NO ligands of this Mo complex make it susceptible to nucleophilic attack. Furthermore, the ligands located trans to the two termini of an allyl group affect the rate of attack. As a result, the regiochemistry of the attack can be controlled by the electronic properties of these two ligands. In many cases, the attack occurs trans to the ligand that is more donating or has the higher trans influence. This effect is rationalized by a lengthening of the M–C bond trans to this ligand and a greater resulting positive charge or electrophilicity of the carbon more weakly bound to the metal.[208] In the case of the Mo complex, attack occurs exclusively at the allyl terminus trans to the carbonyl ligand.[209,210] Nucleophilic attack occurs from the face of the allyl group opposite the metal to produce a neutral η^2-olefin complex in which the metal has been reduced by two electrons, relative to the starting complex. Attack on an allyl group leads to an olefin complex as the initial product, and the olefin complex was identified as the reaction product from attack on this Mo–allyl species. As the Davies–Green–Mingos "guidelines" state, attack occurs at the terminal carbon of the allyl group and occurs at the open allyl group rather than at the closed η^5-cyclopentadienyl group.

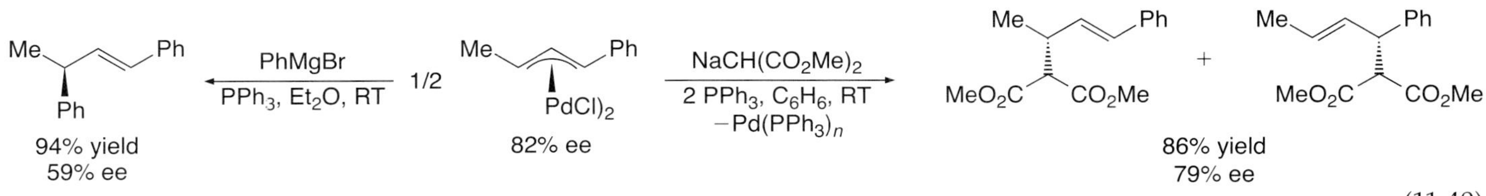

$$(11.40)$$

Many studies of the addition of nucleophiles to palladium–allyl complexes have been conducted. Hayashi has shown that the additions of stabilized anions, such as malonate anions or amine nucleophiles, to chiral, non-racemic allyl complexes occur with inversion of configuration.[165,211] Addition of excess phosphine and either diethyl malonate or dimethylamine to a chiral, non-racemic allyl complex results in nucleophilic attack with nearly complete inversion.[165,211] The reaction with sodium dimethylmalonate is shown at the right of Equation 11.40. In contrast, nonstabilized carbanions such as allyl or phenyl magnesium chloride react with the same η^3-allylpalladium complex with retention of configuration as shown at the left of Equation 11.40.[165] The stereochemistry from reaction of the Grignard reagents likely results from nucleophilic attack at the metal, followed by reductive elimination.

Palladium–allyl complexes tend to form the products from attack at the less substituted terminus of the allyl group.[212,213] However, the regiochemistry can be directed to the more substituted position by electronic effects on the allyl group imparted by the steric properties, bite angle, and array of donor atoms of the ancillary ligands. The influence of the ancillary ligands on the regioselectivity of the reactions of isolated allyl complexes has been analyzed,[213–215] and some conditions cause formation of the branched addition products. Among those studied, complexes generated from electron-accepting ligands, such as alkynes, formed the most branched product. Åkermark showed many years ago that the same regioselectivity is obtained from both catalytic and stoichiometric reactions and from both neutral and cationic η^3-allylpalladium intermediates.[215] Electronic effects of the ancillary ligands have led to the formation of branched substitution products in certain cases, including several types of catalytic allylic substitutions described in more detail in Chapter 20.[208,216–219]

Several recent studies have revealed the dichotomous relationship between the regioselectivity of reactions of unsymmetrical allyl complexes of palladium and that of reactions of unsymmetrical allyl complexes of other metals. An isolated molybdenum–allyl complex containing an allyl group derived from methyl cinnamyl carbonate (PhCH=CHCH$_2$OCO$_2$Me) reacts with the sodium salt of dimethyl malonate in the presence of a source of CO to generate the branched substitution product (Equation 11.41).[167,175] The addition of nucleophile to this allyl complex has been shown to occur by retention of configuration, perhaps by initial attack at the metal, rather than the more common external attack at the allyl group.[167]

$$(11.41)$$

$$(11.42)$$

$$(11.43)$$

In a few exceptional cases, nucleophilic attack at the central carbon of a η^3-allyl group has been observed. Green observed that hydride reagents and methyllithium attack the cationic complex [Cp$_2$Mo(η^3-allyl)]$^+$ and its tungsten analog at the central carbon to generate a stable, isolable metallacycle (Equation 11.42).[220] Molecular orbital analysis by Curtis and Eisenstein rationalized this anomalous behavior by analyzing the symmetry of the LUMO of the allyl complex.[210] In most cases, the LUMO predominantly consists of the combination of a metal d-orbital and the non-bonded orbital of the allyl group, which contains a nodal plane at the central carbon. Complexes with this type of LUMO, therefore, react with nucleophiles at the termini of the allyl group. In other less common cases, the LUMO predominantly consists of the combination of a metal d-orbital and the π^*-orbital of the allyl fragment, which contains a large contribution of the p-orbital on the central carbon. One of the most representative examples of this combination of orbitals is that in

[Cp$_2$Mo(η^3-allyl)]. Nucleophilic addition to this complex occurs at the central carbon of the allyl ligand. Nucleophilic addition to the central carbon in this case is also favored by the presence of significant positive charge at this site. These calculations led to the prediction that "the family of complexes CpLM(η^3-C$_3$H$_5$)$^+$ (L = CO or PR$_3$; M = Co, Rh, and Ir) might be suitable substrates for metallacycle formation." Bergman has indeed observed that (η^5-C$_5$Me$_5$)(PMe$_3$)Rh(η^3-C$_3$H$_5$)$^+$ reacts with hydride donors at the central carbon to generate a metallacyclic product,[221] and Stryker has reported additions of enolates to the cationic iridium–allyl complex in Equation 11.43.[222]

11.7.2. Nucleophilic Attack on η^4-Diene Complexes

According to the guidelines of Davies, Green, and Mingos for the relative reactivity of ligands bound through π-systems, cationic η^4-diene complexes should be among the most reactive substrates toward nucleophilic attack and should undergo preferential attack at a terminal position. Studies on this reaction have been conducted with cationic molybdenum, iron, and palladium. In most cases, attack occurs at the terminal position of the diene, although kinetically favored addition to an internal carbon has been observed in other cases. The η^4-diene complexes are often generated in situ by abstraction of a hydride or alkoxide from the position alpha to an allyl group. Thus, a series of sequential additions of nucleophiles and abstractions can lead to multiply substituted products from simple dienes.

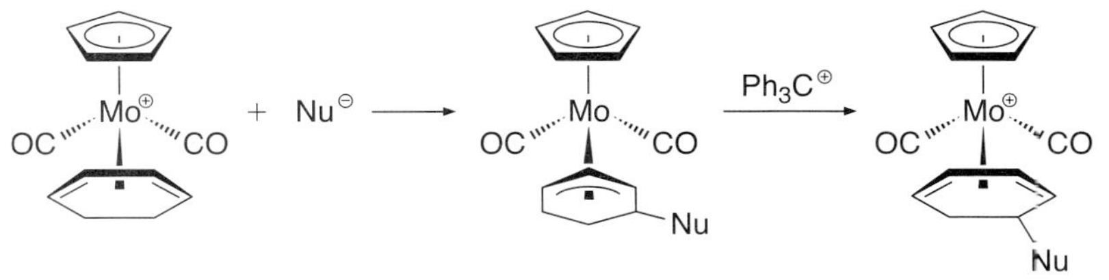

$$(11.44)$$

The chemistry of dienes coordinated to the cationic CpMo(CO)$_2$ fragment has been exploited many times for complex molecule synthesis. Originally, Faller showed that the cationic molybdenum complex in Equation 11.44 undergoes nucleophilic attack by hydride, deuteride, methyl lithium, and enamines to produce the η^3-allyl complex.[223] As expected, attack of the nucleophiles occurs at a terminal position and exclusively from the face opposite the metal. Trityl cation abstracts a hydride from this allyl product from the face opposite the metal to regenerate a diene complex. Pearson has used this sequence of nucleophilic attack and hydride abstraction to synthesize substituted cyclohexenes with control of stereochemistry as shown in Scheme 11.6.[224–226]

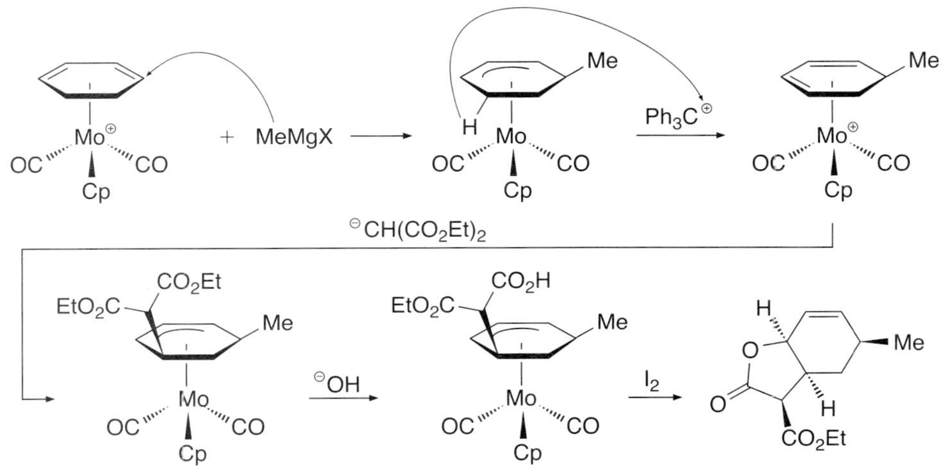

Scheme 11.6

Liebeskind and co-workers have used this type of sequence of addition and abstraction to prepare a number of heterocylic products, in this case by abstraction of alkoxides located alpha to both an allyl unit coordinated to $CpMo(CO)_2$ or $TpMo(CO)_2$ and an oxygen or nitrogen in a heterocyclic ring.[227-230] The cationic dienes that result from this abstraction have been isolated in some cases and used in situ in others. A sequence of abstraction to generate the cationic diene and nucleophilic attack has been used, as shown in Equations 11.45 and 11.46, to generate α,α'-substituted pyrans and piperidines with control of relative and, in some cases, absolute[231] stereochemistry. Use of a chiral auxiliary at nitrogen allows for diastereoselective attack on the diene and generation of optically active products after cleavage of the auxiliary.[227,228]

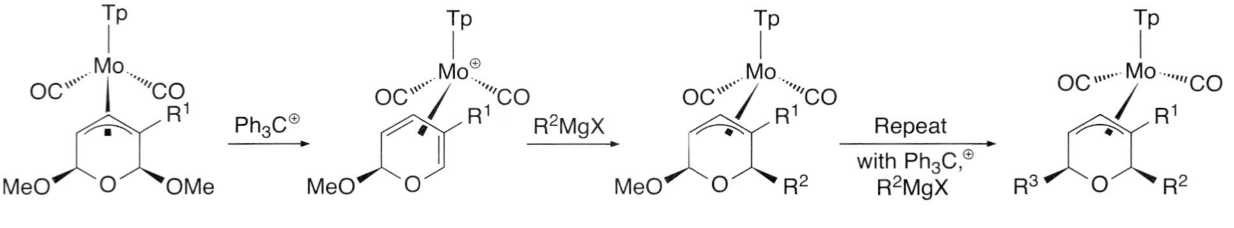

(11.45)

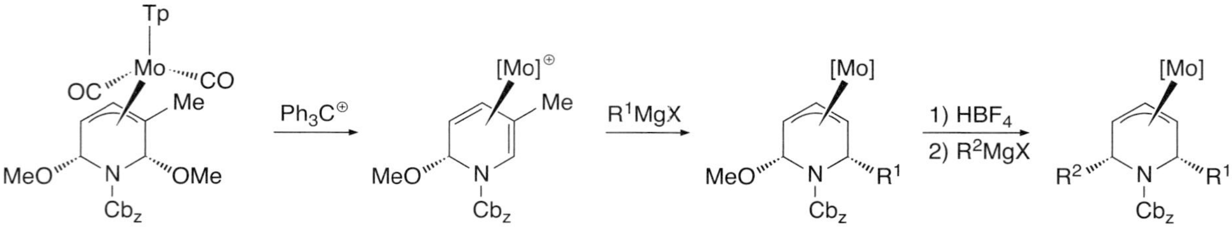

(11.46)

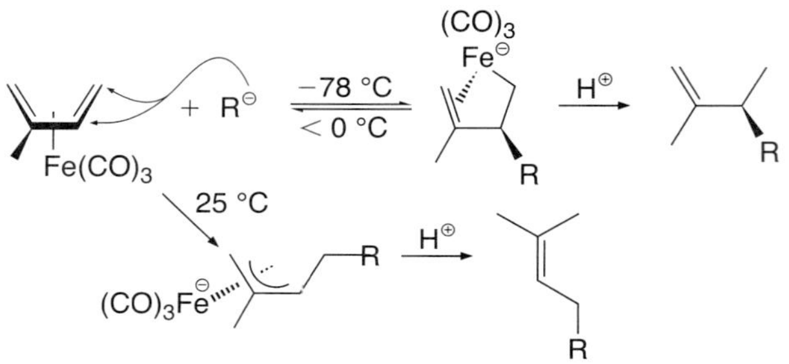

(11.47)

Nucleophilic attack on neutral complexes of η^4-diene ligands is less common than attack on cationic complexes of these ligands but is known. In one case, Semmelhack reported the reactions of iron complexes of acyclic η^4-dienes with reactive carbanions such as $LiCMe_2CN$ and $LiCHPh_2$ (Equation 11.47).[232] At -78 °C, the reaction is rapid, and kinetically controlled. Attack occurs at an unsubstituted, internal position to give an unstable σ-alkyl η^2-olefin complex. This addition is reversible below 0 °C, and the more stable product is then generated from nucleophilic addition at a terminal position of the diene to give the thermodynamically more stable η^3-allyl complex. Cyclohexadiene complexes are similarly alkylated by a range of carbanions.[233]

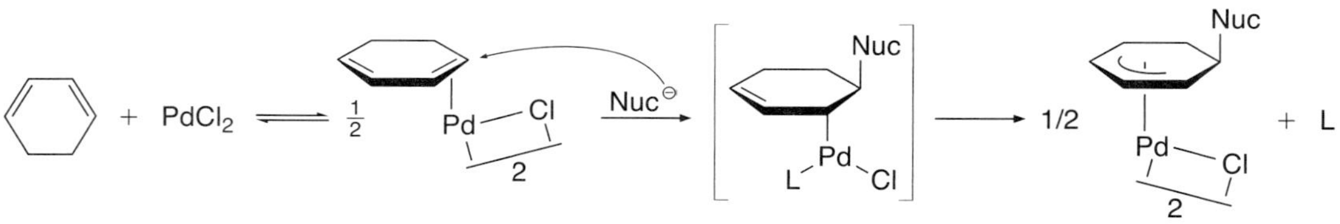

(11.48)

Palladium(II) complexes of 1,3-dienes undergo attack by nucleophiles to give stable η^3-allyl complexes (Equation 11.48). These reactions are likely to occur by nucleophilic additions to η^2-diene complexes, rather than to η^4-diene complexes. For this reason, additions of nucleophiles to palladium diene and allene complexes can be understood with the principles presented in Section 11.5.2 covering the reactions of η^2-olefin complexes.

11.7.3. Nucleophilic Attack on η^5-Dienyl Complexes

By far the most common η^5-dienyl ligand in organometallic chemistry is the cyclopentadienyl ligand, $C_5H_5^-$(Cp). One of the reasons this group is commonly used as an ancillary ligand is its relative inertness toward most reagents, including nucleophiles. However, several instances of attack on this ligand by strong nucleophiles, such as the example in Equation 11.49, are known.[234] Reactions of electrophiles at cyclopentadienyl ligands are more common.[235–237]

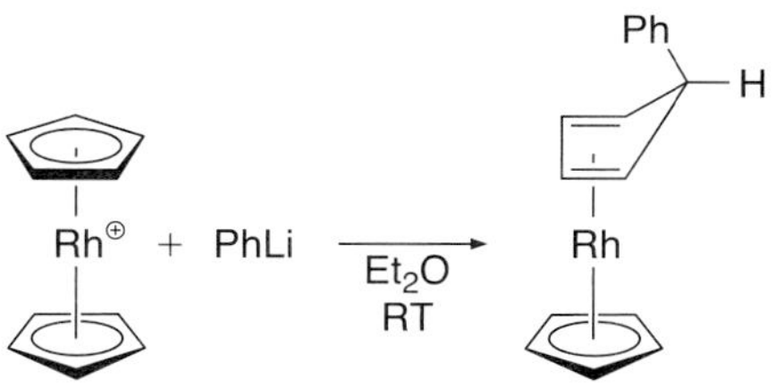

(11.49)

The reactions of cationic (η^5-dienyl)iron–tricarbonyl complexes with nucleophiles have been studied extensively by Birch[238] and Pearson.[239] These complexes are prepared by abstraction of an allylic hydride from η^4-diene complexes, which are readily available from the reaction of 1,3- or 1,4-dienes with $Fe_2(CO)_9$. The parent pentadienyl complex reacts with a variety of nucleophiles exclusively at the terminal carbon, and this reaction generates a new η^4-diene complex (Equation 11.50).[240]

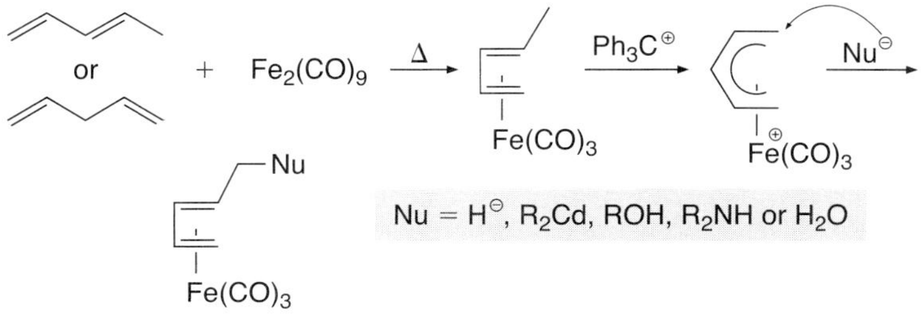

Nu = H$^\ominus$, R$_2$Cd, ROH, R$_2$NH or H$_2$O

(11.50)

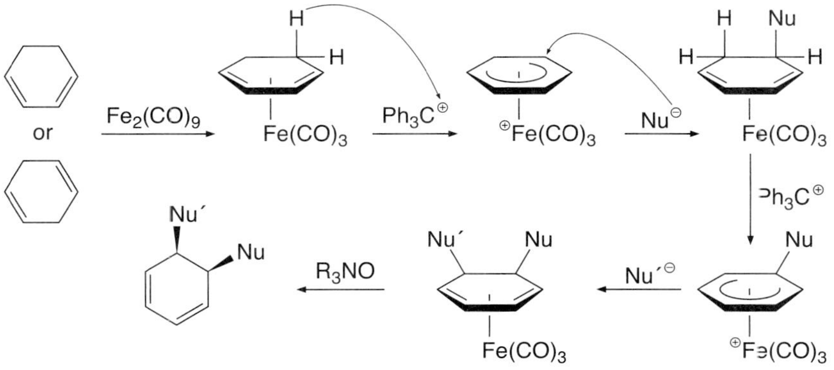

(11.51)

Because the reactions of related η^5-cyclohexadienyl complexes are synthetically valuable, the reactions of this ligand have been studied extensively. An outline of how this chemistry can be conducted on the $Fe(CO)_3$ fragment is shown in Equation 11.51. A variety of cyclohexadienes are readily available from Birch reduction of substituted aromatics. Coordination and abstraction of a hydride, typically by trityl cation, leads to cationic cyclohexadienyl complexes. These cyclohexadienyl complexes are reactive toward organolithium, -copper, -cadmium, and -zinc reagents, ketone enolates, nitroalkyl anions, amines, phthalimide, and even nucleophilic aromatic compounds such as indole and trimethoxybenzene.[241] Attack occurs exclusively from the face opposite the metal, and exclusively at a terminal position of the dienyl system. This combination of hydride abstraction and nucleophilic addition has been repeated to generate cyclohexadiene complexes containing two cis vicinal substituents. The free cyclohexadiene is then released from the metal by oxidation with amine oxides.[242]

$$\text{(11.52)}$$

$R' = CN, SPh, CH_3, nBu, Ph$ or R_2CuLi
$R^\ominus = CH(CO_2Me)_2, CH(COMe)(CO_2Me), CH(CN)(CO_2Me),$ or $CN(SO_2Ph)(CO_2Me)$

The position of substituted cyclohexadienyl ligands at which nucleophilic attack occurs is controlled primarily by electronic factors, although severe steric hindrance can also affect regioselectivity. The regioselectivity of attack on cationic cycloheptadienyliron complexes is less straightforward. Reactions of nucleophiles on the heptadienyl complexes in Equation 11.52 in which L = PPh_3 or $P(OPh)_3$ generates addition products in high yields. Attack at the terminal or internal position depends on the type of nucleophile. Stabilized nucleophiles attack at C-1 to give diene complexes, whereas unstabilized nucleophiles attack at C-2 to give a complex containing an alkyl and an η^3-allyl unit derived from the ring system.[243] Similar reactivity of a diene complex with a nucleophile at the internal position was noted in Section 11.7.2.

11.7.4. Nucleophilic Attack on η^6-Arene and Cycloheptatrienyl Complexes

11.7.4.1. Overview of Nucleophilic Attack on η^6-Arene Complexes

The complexation of an arene to a transition metal has a profound influence on its chemistry and stabilizes anions, cations, and radicals formed at the coordinated arene. Figure 11.7, which is adapted from reviews of π-arene chemistry[244, 245] and texts[246] on synthetic applications of transition metal chemistry, depicts the reactivity of η^6-arene complexes. In general, coordination to electron-withdrawing metal centers makes the arene overall more electron poor and converts the arene from a nucleophile to an electrophile. Consistent with this change in electronic properties, coordination of arenes decreases the pK_a for deprotonation of the arene. This coordination increases the acidity of both the arene hydrogens and benzylic hydrogens.[247–251] Also consistent with this change in electronic properties, coordination makes unsaturated groups conjugated with the arene more electrophilic. At the same time, benzylic carbocations and benzylic radicals of coordinated arenes are more stable than benzylic carbocations and radicals of free arenes.[252]

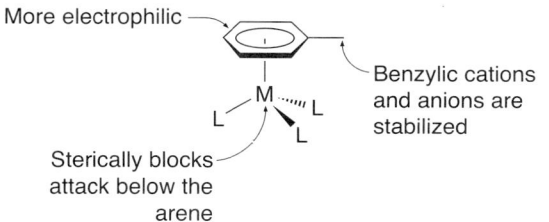

Figure 11.7.
The effects of coordination of an arene to a transition metal fragment.

The greater electrophilicity of coordinated arenes allows nucleophilic attack to occur on the arene unit. The steric effect of the metal fragment leads to attack on the side of the arene opposite to the metal, and this steric effect is used to control the relative stereochemistry of the substituents in the product of sequential reactions. Additions to the face of the arene bound to the metal can also be accomplished indirectly. In this case, the group is introduced by initial addition of a carbon electrophile, rather than nucleophile. The electrophile adds to the metal center, and this addition is generally followed by insertion of a coordinated CO and migration of the acyl group to the endo face of the arene.

The factors that control the site of attack of nucleophiles on arene complexes are complex, and the position of attack is controlled by the electronic properties of the substituents. These properties dictate whether attack will occur meta or para to the substituent. Steric effects discourage attack ortho to the substituent. In many cases, the attack of nucleophiles on coordinated arenes is reversible. Thus, the final productive reaction may result from addition at a site of the arene that is not the kinetic site for attack. For example, nucleophilic attack onto the *ortho*, *meta*, or *para* positions of the phenyl chloride coordinated to Cr(CO)$_3$ in Equation 11.53 is reversible, and it is only attack at the ipso position that leads to productive substitution chemistry.[253,254]

$$(11.53)$$

The reactions of (arene)Cr(CO)$_3$, (arene)Mn(CO)$_3$ and [(arene)MCp]$^+$ (M = Fe or Ru) complexes have been studied extensively.[244] Many of these arene compounds have been prepared by heating the arene with Cr(CO)$_6$, but because dissociations of amines, nitriles or arenes from Cr(0) fragments are typically faster than dissociation of CO (see Chapter 5), reactions of arenes with Cr(CO)$_3$(L)$_3$ (L = NH$_3$ or NCMe$_3$)[255,256] or with (naphthalene) Cr(CO)$_3$[257] are milder than reactions with Cr(CO)$_6$. Manganese–arene complexes [(arene) Mn(CO)$_3$]$^{+258}$ are typically generated from (CO)$_5$MnBr, AgBF$_4$ and arene,[259] while the iron– and ruthenium–arene complexes [(arene)MCp]$^+$ are typically generated by addition of arenes to [CpM(NCMe)$_3$]$^+$(M = Fe or Ru).[260–264]

These arene complexes have been used in a variety of synthetic sequences that depend on the reversibility of the nucleophilic attack. Reversible additions can lead to substitution processes, as noted above. Irreversible reactions allow the addition of a nucleophile, followed by an electrophile, to prepare cis-substituted dienes. Unstabilized nucleophiles, such as organolithium or Grignard reagents, tend to add irreversibly, while more stabilized nucleophiles, such as malonates, amines, and alkoxides are more likely to add reversibly.

After manipulation of the arene, the metal is released by one of several procedures. Free arenes are generally released from the metal by oxidation, which makes the metal poorer at backbonding. Free dienes are typically released by protonation of an intermediate pentadienyl complex. In principle, this chemistry could be catalytic if arene exchange occurred after nucleophilic attack. Semmelhack has made progress in developing ligands that cause displacement of the arene to be faster,[265] although catalytic additions or substitutions of π-arene complexes remain undeveloped. The following paragraphs describe specific examples of the reactions of coordinated arenes.

11.7.4.2. Examples of Nucleophilic Attack on π-Arene Complexes

Semmelhack developed nucleophilic aromatic substitution of neutral arene chromium–tricarbonyl complexes with carbanions.[266,267] This chemistry is summarized in Figure 11.8, and this figure shows the range of carbanions that add to the π-bound arene. In one case, the intermediate pentadienyl complex that is generated prior to release of the leaving group has been characterized by X-ray crystallography. Reactions of substituted aromatic ligands display regioselectivity that is characteristic of the electronic effects of substituents on electrophilic and nucleophilic aromatic substitution. Thus, the order of reactivity toward carbanions is PhCl > PhCH$_3$ > PhOMe. The methoxy and methyl groups direct the nucleophile to the meta position, although the influence of the methyl group is smaller than the influence of the methoxy group. The chlorine directs substitution to the meta and ortho positions over the para position. Trimethylsilyl and trifluoromethyl groups lead to reactions at the para position. Semmelhack has explained this regiochemistry on the basis of the frontier molecular orbitals of the arene.[268]

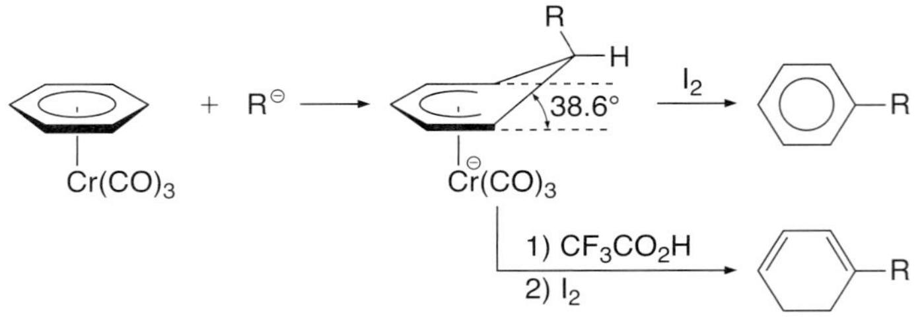

Unreactive: LiCH(CO$_2$R)$_2$, LiCH$_2$COR, MeMgBr, t-BuMgBr, Me$_2$CuLi and LiCH=CH$_2$

Reactive: LiCH$_2$CO$_2$R. LiCH$_2$CN, KCH$_2$CO$_2t$Bu, LiCH(CN)(OR), LiCH$_2$SPh,
Li-1,3-dithiane, LiPh, LiC≡CR, LiCH$_2$CH=CH$_2$, LitBu

Figure 11.8.
Examples of nucleophilic aromatic substitution of (arene)Cr(CO)$_3$ complexes.

Complexes of haloarenes undergo nucleophilic substitution of the halogen by stabilized carbanions to generate the substituted arene complex.[253,254] This reaction occurs with nucleophiles that add reversibly to the arene ring because the kinetically preferred site of attack is ortho and meta to the halogen. As noted above, extrusion of the halogen and a productive substitution reaction occurs upon addition to the less-favored carbon located ipso to the halogen (Equation 11.53).

Many reactions of arenes bound to CpFe$^+$ and Cp*Ru$^+$ fragments have been reported. As predicted by the Davies–Green–Mingos rules, attack of nucleophiles on cationic CpFe(arene) complexes occurs on the complexed arene, not on the cyclopentadienyl ligand. Phenyllithium, methyllithium, and ethyllithium react with these arene complexes to form cyclohexadienyl complexes (Equation 11.54). Reactions of other nucleophiles, such as alkoxides, amides, and some Grignard reagents, are complicated by parallel electron transfer processes. Hydride and carbanionic reagents attack ortho to the halogen of coordinated chlorobenzene, whereas reactions of N, O, and S nucleophiles lead to replacement of the halogen by nucleophilic aromatic substitution processes that are reminiscent of those observed with neutral chromium–carbonyl complexes of chlorobenzene (Equation 11.53). The side reactions of the iron compounds are often eliminated by conducting reactions with the corresponding Cp*Ru(arene) cations. The arene complexes are simple to prepare, and these compounds undergo mild aromatic substitutions with a variety of nucleophiles.

$$(11.54)$$

R = Ph, Me, Et

Although the reactions of π-arene complexes are stoichiometric, not catalytic, this addition chemistry has been used in a variety of synthetic applications. Considering the low cost of many of the metals involved in this chemistry, the labor involved in the preparation of substrates used in the synthetic applications, the small scale of most modern syntheses of biologically active materials, and the short syntheses created by the chemistry of these compounds, greater consideration of the synthetic value of these transformations should be given.

Two examples of the use of these compounds in recent synthetic contexts are shown in Equations 11.55 and 11.56. The formation of biaryl ethers has been a major challenge in the synthesis of natural products of the isodityrosine family, such as Vancomycin. Pearson has prepared cyclic biaryl ether structures of several isodityrosines by intramolecular addition of phenoxides to [Cp*Ru(haloarene)]$^+$ complexes, as shown in Equation 11.55.[261,269–271] Kundig has reported the synthesis of the tricyclic structure of acetoxytubipofuran starting from a complexed imine of benzaldehyde, as shown in Equation 11.56.[272] Addition of vinyl lithium, followed by addition of MeI to the resulting anion leads to methylation at the metal, insertion of CO, and migration of the acyl group to the endo face. Subsequent addition of methyl iodide to the resulting material generates the appropriately substituted diene to carry forward to complete the synthesis.

$$(11.55)$$

(11.56)

Libraries of products from reactions of π-arenes

(11.57)

In addition, several groups have bound π-arene complexes to solid supports and constructed libraries of products derived from π-arenes by the chemistry described in this section (Equation 11.57).[273,274] Oxidation of the final product releases the arene from the metal. The arene–chromium was bound to the support by a pendant phosphine or isocyanide. The isocyanide provides the benefit of maintaining the π-accepting property of the three ancillary ligands.

Finally, methods to control the stereochemistry of the additions have been developed by a number of strategies. For example, chiral auxiliaries have been placed on the arene system,[275–277] chiral nucleophiles, such as Evans' enolates[278,279] or amino acids, have been added, or chiral ligands have been added to the organolithium reagent.[280,281] Planar chiral chromium compounds have also been resolved and then used as a single enantiomer.[275,282–285] No general solution has been developed, but acceptable diastereoselectivities have been obtained in many cases.

11.8. Summary

Coordination of a ligand to a transition metal can reduce the electron density at the ligand, and this flow of electrons from the ligand to the metal often renders the ligand electrophilic. Thus, many ligands on transition metal complexes react with nucleophiles, and the reactions of many classes of nucleophiles with a range of transition metal complexes have been described in this chapter. Reactions at neutral ligands are well documented, and these reactions do not change the overall electron count or oxidation state of the metal. Reactions of several classes of formally anionic ligands, such as coordinated alkyl, allyl or higher formally anionic unsaturated ligands are also well known. These reactions decrease the oxidation state of the metal and can be considered a formal reductive elimination.

Nucleophilic attack typically occurs at coordinated ligands instead of the metal when the metal center is sterically or electronically saturated. In these cases, reactions occur at σ- or π-bound unsaturated ligands, and sometimes at σ-bound, saturated alkyl ligands. The most common examples of this process are nucleophilic attack on coordinated carbon monoxide, nucleophilic attack on coordinated olefins and arenes, and nucleophilic attack on the allyl ligand of cationic allyl complexes. The stereochemistry of these reactions signals direct attack without initial coordination at the metal center. The products from attack at alkyl and allyl ligands typically result from inversion of configuration, the products from attack on alkene ligands generate diastereomers that result from attack anti to the metal center, the products from attack on alkyne ligands give rise to trans vinyl complexes, and the products from nucleophilic additions to dienes and arenes result from attack anti to the metal center. The relative rates of attack on hydrocarbyl ligands follow several trends. Attack at open (acylic) ligands tends to occur faster than at closed ligands possessing the same charge, and attack at neutral ligands tends to occur faster than attack at anionic ligands. Furthermore, attack at open ligands tends to occur at the terminus of the ligand, although many exceptions to the last rule have been observed. These nucleophilic additions to coordinated ligands have been used in stoichiometric reactions that lead to products containing stereochemistry controlled by the position of the metal, and they are part of catalytic processes, including carbonylations, the water-gas shift reaction, and Wacker oxidation processes discussed in later chapters of the text.

References and Notes

1. Hegedus, L. S. *Transition Metals in the Synthesis of Complex Organic Molecules*, 2nd ed.; University Science Books: Mill Valley, CA, 1999.
2. Rautenstrauch, V.; Joyeux, M. *Angew. Chem., Int. Ed. Engl.* **1979**, *18*, 85.
3. Trzupek, L. S.; Newirth, T. L.; Kelly, E. G.; Sbarbati, N. E.; Whitesides, G. M. *J. Am. Chem. Soc.* **1973**, *95*, 8118.
4. Wegler, R.; Pieper, G. *Chem. Ber.* **1950**, *83*, 1.
5. Beller, M.; Breindl, C.; Riermeier, T. H.; Eichberger, M.; Trauthwein, H. *Angew. Chem., Int. Ed. Engl.* **1998**, *37*, 3389.
6. Seayad, J.; Tillack, A.; Hartung, C. G.; Beller, M. *Adv. Synth. Catal.* **2002**, *344*, 795.
7. Fischer, E. O.; Maasböl, A. *Angew. Chem., Int. Ed. Engl.* **1964**, *3*, 580.
8. Sneeden, R. P. A. In *Comprehensive Organometallic Chemistry*; Wilkinson, G., Stone, F. G. A., Abel, E. W., Eds.; Pergamon: Oxford, U. K.,1982; Vol. 8, p 9.
9. Lund, C. R. F.; Kubsh, J. E.; Dumesic, J. A. *ACS Symp. Ser.* **1985**, *279*, 313.
10. Rethwisch, D. G.; Dumesic, J. A. *J. Catal.* **1986**, *101*, 35.
11. Tinkle, M.; Dumesic, J. A. *J. Catal.* **1987**, *103*, 65.
12. King, A. D.; King, R. B.; Yang, D. B. *J. Am. Chem. Soc.* **1980**, *102*, 1028.
13. King, A. D.; King, R. B.; Yang, D. B. *J. Am. Chem. Soc.* **1981**, *103*, 2699.
14. Pearson, R. G.; Mauermann, H. *J. Am. Chem. Soc.* **1982**, *104*, 500.
15. Pac, C.; Miyake, K.; Matsuo, T.; Yanagida, S.; Sakakurai, H. *J. Chem. Soc., Chem. Commun.* **1986**, 1115.
16. Ziessel, R. *Angew. Chem., Int. Ed. Engl.* **1991**, *30*, 844.
17. Darensbourg, D. J.; Baldwin, B. J.; Froelich, J. A. *J. Am. Chem. Soc.* **1980**, *102*, 4688.
18. Darensboourg, D. J.; Darensbourg, M. Y. *Inorg. Chem.* **1970**, *9*, 1691.
19. Darensbourg, M. Y.; Conder, H. L.; Darensbourg, D. J.; Hasday, C. *J. Am. Chem. Soc.* **1973**, *95*, 5919.
20. Casey, C. P.; Neumann, S. M. *J. Am. Chem. Soc.* **1976**, *98*, 5395.
21. Casey, C. P.; Andrews, M. A.; Rinz, J. E. *J. Am. Chem. Soc.* **1979**, *101*, 741.
22. Gladysz, J. A.; Tam, W. *J. Am. Chem. Soc.* **1978**, *100*, 2545.
23. Tam, W.; Wong, W. K.; Gladysz, J. A. *J. Am. Chem. Soc.* **1979**, *101*, 1589.
24. Collman, J. P. *Acc. Chem. Res.* **1975**, *8*, 342.
25. Semmelhack, M. F.; Tamura, R. *J. Am. Chem. Soc.* **1983**, *105*, 4099.
26. Dotz, K. H. *Angew. Chem., Int. Ed. Engl.* **1984**, *23*, 587.
27. Ungermann, C.; Landis, V.; Moya, S. A.; Cohen, H.; Walker, H.; Pearson, R. G.; Rinker, R. G.; Ford, P. C. *J. Am. Chem. Soc.* **1979**, *101*, 5922.
28. Catellani, M.; Halpern, J. *Inorg. Chem.* **1980**, *19*, 566.
29. Angelici, R. J. *Acc. Chem. Res.* **1972**, *5*, 335.
30. Crociani, B.; Uguagliati, P.; Belluco, U. *J. Organomet. Chem.* **1976**, *117*, 189.
31. Casey, C. P.; Miles, W. H. *J. Organomet. Chem.* **1983**, *254*, 333.
32. Kreissl, F. R.; Fischer, E. O.; Kreiter, C. G.; Fischer, H. *Chem. Ber.* **1973**, *106*, 1262.
33. Werner, H.; Fischer, E. O.; Heckl, B.; Kreiter, C. G. *J. Organomet. Chem.* **1971**, *28*, 367.
34. Fischer, E. O.; Richter, K. *Angew. Chem., Int. Ed. Engl.* **1975**, *14*, 345.
35. Kirtley, S. W. In *Comprehensive Organometallic Chemistry*; Wilkinson, G., Stone, F. G. A., Abel, E. W., Eds.; Pergamon Press: Oxford, U. K. 1982; Vol. 3, p 888.
36. Schrock, R. R. *Acc. Chem. Res.* **1979**, *12*, 98.
37. Kuhn, O.; Mayr, H. *Angew. Chem. Int. Ed.* **1999**, *38*, 343.
38. Lau, K. S. Y.; Fries, R. W.; Stille, J. K. *J. Am. Chem. Soc.* **1974**, *96*, 4983.

39. Lau, K. S. Y.; Wong, P. K.; Stille, J. K. *J. Am. Chem. Soc.* **1976**, *98*, 5832.
40. Goldberg, K. I.; Yan, J. Y.; Winter, E. L. *J. Am. Chem. Soc.* **1994**, *116*, 1573.
41. Goldberg, K. I.; Yan, J. Y.; Breitung, E. M. *J. Am. Chem. Soc.* **1995**, *117*, 6889.
42. Williams, B. S.; Holland, A. W.; Goldberg, K. I. *J. Am. Chem. Soc.* **1999**, *121*, 252.
43. Williams, B. S.; Goldberg, K. I. *J. Am. Chem. Soc.* **2001**, *123*, 2526.
44. Stahl, S. S.; Labinger, J. A.; Bercaw, J. E. *Angew. Chem. Int. Ed.* **1998**, *37*, 2181.
45. Shilov, A. E.; Shteinman, A. A. *Coord. Chem. Rev.* **1977**, *24*, 97.
46. Periana, R. A.; Taube, D. J.; Gamble, S.; Taube, H.; Satoh, T.; Fujii, H. *Science* **1998**, *280*, 560.
47. Lin, M.; Shen, C.; Garcia-Zayas, E. A.; Sen, A. *J. Am. Chem. Soc.* **2001**, *123*, 1000.
48. Sen, A. *Acc. Chem. Res.* **1998**, *31*, 550.
49. Luinstra, G. A.; Wang, L.; Stahl, S. S.; Labinger, J. A.; Bercaw, J. E. *J. Organomet. Chem.* **1995**, *504*, 75.
50. Nelson, A. P.; DiMagno, S. G. *J. Am. Chem. Soc.* **2000**, *122*, 8569.
51. Sanford, M. S.; Groves, J. T. *Angew. Chem. Int. Ed.* **2004**, *43*, 588.
52. King, R. B.; Fronzaglia, A. *J. Am. Chem. Soc.* **1965**, 709.
53. Cotton, F. A.; Marks, T. J. *J. Am. Chem. Soc.* **1968**, 1339.
54. Stevens, R. R.; Shier, G. D. *J. Organomet. Chem.* **1970**, *21*, 495.
55. Becker, Y.; Stille, J. K. *J. Am. Chem. Soc.* **1978**, 845.
56. Galamb, V.; Palyi, G. *J. Chem.Soc., Chem. Commun.* **1982**, 487.
57. Mann, B. E.; Shaw, S. D. *J. Organomet. Chem.* **1987**, *326*, C13.
58. Gatti, G.; Lopez, J. A.; Mealli, C.; Musco, A. *J. Organomet. Chem.* **1994**, *483*, 77.
59. Legros, J. Y.; Boutros, A.; Fiaud, J. C.; Toffano, M. *J. Mol. Catal. A* **2003**, *196*, 21.
60. Legros, J. Y.; Toffano, M.; Fiaud, J. C. *Tetrahedron: Asymmetry* **1995**, *6*, 1899.
61. Legros, J. Y.; Toffano, M.; Fiaud, J. C. *Tetrahedron* **1995**, *51*, 3235.
62. Nettekoven, U.; Hartwig, J. F. *J. Am. Chem. Soc.* **2002**, *124*, 1166.
63. Utsunomiya, M.; Hartwig, J. F. *J. Am. Chem. Soc.* **2003**, *125*, 14286.
64. Johns, A. M.; Utsunomiya, M.; Incarvito, C. D.; Hartwig, J. F. *J. Am. Chem. Soc.* **2006**, *128*, 1828.
65. Johns, A. M.; Tye, J. W.; Hartwig, J. F. *J. Am. Chem. Soc.* **2006**, *128*, 16010.
66. Magerlein, W.; Indolese, A. F.; Beller, M. *J. Organomet. Chem.* **2002**, *641*, 30.
67. Kubota, Y.; Nakada, S.; Sugi, Y. *Synlett* **1998**, 183.
68. Schnyder, A.; Beller, M.; Mehltretter, G.; Nsenda, T.; Studer, M.; Indolese, A. *J. Org. Chem.* **2001**, *66*, 4311.
69. Hosoi, K.; Nozaki, K.; Hiyama, T. *Org. Lett.* **2002**, *4*, 2849.
70. Perry, R. J.; Wilson, B. D. *Macromolecules* **1993**, *26*, 1503.
71. El Ali, B.; Alper, H. In *Handbook of Organopalladium Chemistry for Organic Synthesis*; Negishi, E.-i., Ed.; Wiley-Interscience: New York, 2002; Vol. II, p 2333.
72. Pugh, R. I.; Drent, E. *Adv. Synth. Catal.* **2002**, *344*, 837.
73. del Rio, I.; Claver, C.; van Leeuwen, P. *Eur. J. Inorg. Chem.* **2001**, 2719.
74. Pugh, R. I.; Drent, E.; Pringle, P. G. *Chem. Commun.* **2001**, 1476.
75. Drent, E.; Arnoldy, P.; Budzelaar, P. H. M. *J. Organomet. Chem.* **1994**, *475*, 57.
76. Schoenberg, A.; Bartoletti, I.; Heck, R. F. *J. Org. Chem.* **1974**, *1974*, 3318.
77. Heck, R. F. *J. Am. Chem. Soc.* **1968**, *90*, 5546.
78. Tamaru, Y.; Kimura, M. In *Handbook of Organopalladium Chemistry for Organic Synthesis*; Negishi, E.-i., Ed.; Wiley-Interscience: New York, 2002; Vol. II, p 2425.
79. Mori, M. In *Handbook of Organopalladium Chemistry for Organic Synthesis*; Negishi, E.-i., Ed.; Wiley-Interscience: New York, 2002; Vol. II, p 2313.
80. Heck, R. F. *Pure. Appl. Chem.* **1978**, *50*, 691.
81. Lin, Y.-S.; Yamamoto, A. *Organometallics* **1998**, *17*, 3466.
82. Cantat, T.; Genin, E.; Giroud, C.; Meyer, G.; Jutand, A. *J. Organomet. Chem.* **2003**, *687*, 365.
83. Åkermark, B.; Akermark, G.; Hegedus, L. S.; Zetterberg, K. *J. Am. Chem. Soc.* **1981**, *103*, 3037.
84. Evans, P. A.; Nelson, J. D. *J. Am. Chem. Soc.* **1998**, *120*, 5581.
85. Tamaru, Y. In *Handbook of Organopalladium Chemistry for Organic Synthesis*; Negishi, E.-i., Ed.; Wiley-Interscience: New York, 2002; vol II, p 1931.
86. Nakamura, H.; Aoyagi, K.; Shim, J. G.; Yamamoto, Y. *J. Am. Chem. Soc.* **2001**, *123*, 372.
87. Nakamura, H.; Nakamura, K.; Yamamoto, Y. *J. Am. Chem. Soc.* **1998**, *120*, 4242.
88. Nakamura, H.; Iwama, H.; Yamamoto, Y. *J. Am. Chem. Soc.* **1996**, *118*, 6641.
89. Nakamura, H.; Iwama, H.; Yamamoto, Y. *Chem. Commun.* **1996**, 1459.
90. Davies, S. G.; Green, M. L. H.; Mingos, D. M. P. *Tetrahedron* **1978**, *34*, 3047.
91. Green, M. L. H.; Mitchard, L. C.; Silverthorn, W. E. *J. Chem. Soc., Dalton Trans.* **1973**, 1952.
92. Eisenstein, O.; Hoffmann, R. *J. Am. Chem. Soc.* **1981**, *103*, 4308.
93. Lennon, P.; Rosan, A. M.; Rosenblum, M. *J. Am. Chem. Soc.* **1977**, *99*, 8426.

94. Majima, T.; Kurosawa, H. *J. Chem. Soc., Chem. Commun.* **1977**, 610.

95. Kurosawa, H.; Asada, N. *Tetrahedron Lett.* **1979**, 255.

96. *Handbook of Organopalladium Chemistry for Organic Synthesis*; Negishi, E.-i., Ed.; Wiley-Interscience: New York, 2002.

97. Chatt, J.; Vallarino, L. M.; Venanzi, L. M. *J. Chem. Soc.* **1957**, 2496.

98. Palumbo, R.; Derenzi, A.; Panunzi, A.; Paiaro, G. *J. Am. Chem. Soc.* **1969**, *91*, 3874

99. Panunzi, A.; Derenzi, A.; Palumbo, R.; Paiaro, G. *J. Am. Chem. Soc.* **1969**, *91*, 3879

100. Tada, M.; Kuroda, Y.; Sato, T. *Tetrahedron Lett.* **1969**, 2871.

101. Tsuji, J.; Takahash.H *J. Am. Chem. Soc.* **1965**, *87*, 3275.

102. Cope, A. C.; Kliegman, J. M.; Friedrich, E. C. *J. Am. Chem. Soc.* **1967**, *89*, 287.

103. Holton, R. A.; Kjonaas, R. A. *J. Am. Chem. Soc.* **1977**, *99*, 4177.

104. Holton, R. A.; Kjonaas, R. A. *J. Organomet. Chem.* **1977**, *142*, C15.

105. Jira, R. In *Applied Homogeneous Catalysis with Organometallic Compounds: A Comprehensive Handbook in Two Volumes*, 2nd ed.; Cornils, B., Herrmann, W. A., Eds.; Wiley-VCH: Weinheim, 2002; p 386.

106. Jira, R. In *Applied Homogeneous Catalysis with Organometallic Compounds: A Comprehensive Handbook in Two Volumes*; Cornils, B., Herrmann, W. A., Eds.; VCH: New York, 1996; p 374.

107. Smidt, J.; Hafner, W.; Jira, R.; Sieber, R.; Sedlmeier, J.; Sabel, A. *Angew. Chem., Int Ed. Engl.* **1962**, *1*, 80.

108. Hosokawa, T.; Murahashi, S.-I. *Acc. Chem. Res.* **1990**, *23*, 49.

109. Henry, P. M. In *Handbook of Organopalladium Chemistry for Organic Synthesis*; Negishi, E.-i., Ed.; Wiley-Interscience: New York, 2002.

110. Moiseev, I. I.; Vargaftik, M. N.; Syrkin, J. K. *Dok. Akad. Nauk* **1960**, *133*, 377.

111. Corp., M. K. *CHEMTECH* **1988**, 759.

112. Jira, R. In *Applied Homogeneous Catalysis with Organometallic Compounds: A Comprehensive Handbook in Two Volumes*; Cornils, B., Herrmann, W. A., Eds.; VCH: New York, 1996; p 394.

113. Hosokawa, T.; Takano, M.; Kuroki, Y.; Murahashi, S.-I. *Tetrahedron Lett.* **1992**, *33*, 6643.

114. Timokhin, V. I.; Anastasi, N. R.; Stahl, S. S. *J. Am. Chem. Soc.* **2003**, *125*, 12996.

115. Hegedus, L. S.; Akermark, B.; Zetterberg, K.; Olsson, L. F. *J. Am. Chem. Soc.* **1984**, *106*, 7122.

116. Akermark, B.; Zetterberg, K. *J. Am. Chem. Soc.* **1984**, *106*, 5560.

117. Hegedus, L. S.; Aller, G. F.; Bozell, J. J.; Waterman, E. L. *J. Am. Chem. Soc.* **1978**, *100*, 5800.

118. Panunzi, A.; Derenzi, A.; Paiaro, G. *J. Am. Chem. Soc.* **1970**, *92*, 3488.

119. Stille, J. K.; James, D. E. *J. Am. Chem. Soc.* **1975**, *97*, 674.

120. Stille, J. K.; James, D. E. *J. Organomet. Chem.* **1976**, *108*, 401.

121. Bäckvall, J. E.; Åkermark, B.; Ljunggren, S. O. *J. Am. Chem. Soc.* **1979**, *101*, 2411.

122. Stille, J. K.; Divakaruni, R. *J. Organomet. Chem.* **1979**, *169*, 239.

123. Francis, J. W.; Henry, P. M. *Organometallics* **1991**, *10*, 3498.

124. Hamed, O.; Thompson, C.; Henry, P. M. *J. Org. Chem.* **1997**, *62*, 7082.

125. Francis, J. W.; Henry, P. M. *Organometallics* **1992**, *11*, 2832.

126. Hayashi, T.; Yamasaki, K.; Mimura, M.; Uozumi, Y. *J. Am. Chem. Soc.* **2004**, *126*, 3036.

127. Hegedus, L. S.; Williams, R. E.; McGuire, M. A.; Hayashi, T. *J. Am. Chem. Soc.* **1980**, *102*, 4973.

128. Liu, C.; Han, X. Q.; Wang, X.; Widenhoefer, R. A. *J. Am. Chem. Soc.* **2004**, *126*, 3700.

129. Pei, T.; Widenhoefer, R. A. *J. Am. Chem. Soc.* **2001**, *123*, 11290.

130. Wang, X.; Pei, T.; Han, X. Q.; Widenhoefer, R. A. *Org. Lett.* **2003**, *5*, 2699.

131. Qian, H.; Widenhoefer, R. A. *J. Am. Chem. Soc.* **2003**, *125*, 2056.

132. Pei, T.; Wang, X.; Widenhoefer, F. A. *J. Am. Chem. Soc.* **2003**, *125*, 648.

133. Wang, X.; Widenhoefer, R. A. *Chem. Commun.* **2004**, 660.

134. Benn, R.; Jolly, P. W.; Joswig, T.; Mynott, R.; Schick, K.-P. *Z. Naturforsch.* **1986**, *41b*, 680.

135. Bäckvall, J.-E. In *Metal-Catalyzed Cross-Coupling Reactions*; Stang, P. J., Diederich, F., Eds.; Wiley-VCH: Weinheim, 1998; p 339.

136. Andersson, P. G.; Bäckvall, J. E. In *Handbook of Organopalladium Chemistry for Organic Synthesis*; Negishi, E.-i., Ed.; Wiley-Interscience: New York, 2002; p 1859.

137. Aranyos, A.; Szabo, K. J.; Bäckvall, J. E. *J. Org. Chem.* **1998**, *63*, 2523.

138. Bäckvall, J. E.; Nordberg, R. E. *J. Am. Chem. Soc.* **1981**, *103*, 4959.

139. Bäckvall, J. E.; Nordberg, R. E.; Wilhelm, D. *J. Am. Chem. Soc.* **1985**, *107*, 6892.

140. Koroleva, E. B.; Bäckvall, J. E.; Andersson, P. G. *Tetrahedron Lett.* **1995**, *36*, 5397.

141. Andersson, P. G.; Nilsson, Y. I. M.; Bäckvall, J. E. *Tetrahedron* **1994**, *50*, 559.

142. Bäckvall, J. E.; Granberg, K. L.; Andersson, P. G.; Gatti, R.; Gogoll, A. *J. Org. Chem.* **1993**, *58*, 5445.

143. Bäckvall, J. E.; Andersson, P. G. *J. Am. Chem. Soc.* **1992**, *114*, 6374.

144. Bäckvall, J. E.; Andersson, P. G. *J. Am. Chem. Soc.* **1990**, *112*, 3683.

145. Ronn, M.; Andersson, P. G.; Bäckvall, J. E. *Tetrahedron Lett.* **1997**, *38*, 3603.

146. Chisholm, M. H.; Clark, H. C. *Acc. Chem. Res.* **1973**, *6*, 202.

147. Reger, D. L.; Belmore, K. A.; Mintz, E.; McElligott, P. J. *Organometallics* **1984**, *3*, 134.

148. Mizushima, E.; Hayashi, T.; Tanaka, M. *Org. Lett.* **2003**, *5*, 3349.

149. Field, L. D.; Messerle, B. A.; Wren, S. L. *Organometallics* **2003**, *22*, 4393.

150. (a) Beller, M.; Breindl, C.; Eichberger, M.; Hartung, C. G.; Seayad, J.; Thiel, O. R.; Tillack, A.; Trauthwein, H. *Synlett* **2002**, 1579; (b) Hartung, C. G.; Tillack, A.; Trauthwein, H.; Beller, M. *J. Org. Chem.* **2001**, *66*, 6339.

151. (a) Zhang, J.; Yang, C.-G.; He, C. *J. Am. Chem Soc.* **2006**, *128*, 1798; (b) Kennedy-Smith, J. J.; Staben, S.T.; Toste, F. D. *J. Am. Chem. Soc.* **2004**, *126*, 4526; (c) Hashmi, A. S. K.; Weyrauch, J. P.; Frey, W., Bats, J. W. *Org. Lett.* **2004**, *6*, 4391.

152. Harman, W. D. *Chem. Rev.* **1997**, *97*, 1953.

153. Brooks, B. C.; Gunnoe, T. B.; Harman, W. D. *Coord. Chem. Rev.* **2000**, *206*, 3.

154. Keane, J. M.; Chordia, M. D.; Mocella, C. J.; Sabat, M.; Trindle, C. O.; Harman, W. D. *J. Am. Chem. Soc.* **2004**, *126*, 6806.

155. Meiere, S. H.; Valahovic, M. T.; Harman, W. D. *J. Am. Chem. Soc.* **2002**, *124*, 15099.

156. Chen, H. Y.; Liu, R. G.; Myers, W. H.; Harman, W. D. *J. Am. Chem. Soc.* **1998**, *120*, 509.

157. Chen, H. Y.; Hodges, L. M.; Liu, R. G.; Stevens, W. C.; Sabat, M.; Harman, W. D. *J. Am. Chem. Soc.* **1994**, *116*, 5499.

158. Chen, H. Y.; Harman, W. D. *J. Am. Chem. Soc.* **1996**, *118*, 5672.

159. Klein, D. P.; Mendez, N. Q.; Seyler, J. W.; Arif, A. M.; Gladysz, J. A. *J. Organomet. Chem.* **1993**, *450*, 157.

160. Klein, D. P.; Gladysz, J. A. *J. Am. Chem. Soc.* **1992**, *114*, 8710.

161. Klein, D. P.; Dalton, D. M.; Mendez, N. Q.; Arif, A. M.; Gladysz, J. A. *J. Organomet. Chem.* **1991**, *412*, C7.

162. Vogeley, N. J.; White, P. S.; Templeton, J. L. *J. Am. Chem. Soc.* **2003**, *125*, 12422.

163. Fujihara, H.; Nagai, K.; Tomioka, K. *J. Am. Chem. Soc.* **2000**, *122*, 12055.

164. Boezio, A. A.; Charette, A. B. *J. Am. Chem. Soc.* **2003**, *125*, 1692.

165. Hayashi, T.; Konishi, M.; M., K. *J. Chem. Soc. Chem. Commun.* **1984**, 107.

166. Matsushita, H.; Negishi, E.-i. *J. Chem. Soc, Chem. Commun.* **1982**, 160.

167. Lloyd-Jones, G. C.; Krska, S. W.; Hughes, D. L.; Gouriou, L.; Bonnet, V. D.; Jack, K.; Sun, Y.; Reamer, R. A. *J. Am. Chem. Soc.* **2004**, *126*, 702.

168. In *Handbook of Organopalladium Chemistry for Organic Synthesis*; Negishi, E.-i., Ed.; Wiley-Interscience: New York, 2002; p 1669.

169. Trost, B. M.; Dogra, K.; Hachiya, I.; Emura, T.; Hughes, D. L.; Krska, S.; Reamer, R. A.; Palucki, M.; Yasuda, N.; Reider, P. J. *Angew. Chem. Int. Ed.* **2002**, *41*, 1929.

170. Trost, B. M.; Hachiya, I. *J. Am. Chem. Soc.* **1998**, *120*, 1104.

171. Glorius, F.; Pfaltz, A. *Org. Lett.* **1999**, *1*, 141.

172. Glorius, F.; Neuburger, M.; Pfalz, A. *Helv. Chim. Acta* **2001**, *84*, 3178.

173. Palucki, M.; Um, J. M.; Conlon, D. A.; Yasuda, N.; Hughes, D. L.; Mao, B.; Wang, J.; Reider, P. J. *Adv. Synth. Catal.* **2001**, *343*, 46.

174. Malkov, A. V.; Baxendale, I. R.; Dvorak, D.; Mansfield, D. J.; Kocovsky, P. *J. Org. Chem.* **1999**, *64*, 2737.

175. Krska, S. W.; Hughes, D. L.; Reamer, R. A.; Mathre, D. J.; Sun, Y.; Trost, B. M. *J. Am. Chem. Soc.* **2002**, *124*, 12656.

176. Lehmann, J.; Lloyd-Jones, G. C. *Tetrahedron* **1995**, *51*, 8863.

177. Lloyd-Jones, G. C.; Pfaltz, A. *Angew. Chem., Int. Ed. Engl.* **1995**, *34*, 462.

178. Zhang, S. W.; Mitsudo, T.; Kondo, T.; Watanabe, Y. *J. Organomet. Chem.* **1993**, *450*, 197.

179. Morisaki, Y.; Kondo, T.; Mitsudo, T. A. *Organometallics* **1999**, *18*, 4742.

180. Matsushima, Y.; Onitsuka, K.; Kondo, T.; Mitsudo, T.; Takahashi, S. *J. Am. Chem. Soc.* **2001**, *123*, 10405.

181. Trost, B. M.; Fraisse, P. L.; Ball, Z. T. *Angew. Chem., Int. Ed. Eng.* **2002**, *41*, 1059.

182. Mbaye, M. D.; Demerseman, B.; Renaud, J.-L.; Toupet, L. c.; Bruneau, C. *Angew. Chem. Int. Ed.* **2003**, *42*, 5066.

183. Renaud, J.-L.; Bruneau, C.; Demerseman, B. *Synlett* **2003**, 408.

184. Evans, P. A.; Nelson, J. D. *Tetrahedron Lett.* **1998**, *39*, 1725.

185. Evans, P. A.; Robinson, J. E.; Nelson, J. D. *J. Am. Chem. Soc.* **1999**, *121*, 6761.

186. Evans, P. A.; Leahy, D. K. *J. Am. Chem. Soc.* **2002**, *124*, 7882.

187. Janssen, J. P.; Helmchen, G. *Tetrahedron Lett.* **1997**, *38*, 8025.

188. Takeuchi, R. *Polyhedron* **2000**, *19*, 557.

189. Takeuchi, R.; Ue, N.; Tanabe, K.; Yamashita, K.; Shiga, N. *J. Am. Chem. Soc.* **2001**, *123*, 9525.

190. Ohmura, T.; Hartwig, J. F. *J. Am. Chem. Soc.* **2002**, *124*, 15164.

191. Lopez, F.; Ohmura, T.; Hartwig, J. F. *J. Am. Chem. Soc.* **2003**, *125*, 3426.

192. Kiener, C. A.; Shu, C.; Incarvito, C.; Hartwig, J. F. *J. Am. Chem. Soc.* **2003**, *125*, 14272.

193. Kanayama, T.; Yoshida, K.; Miyabe, H.; Kimachi, T.; Takemoto, Y. *J. Org. Chem.* **2003**, *68*, 6197.

194. Kanayama, T.; Yoshida, K.; Miyabe, H.; Takemoto, Y. *Angew. Chem. Int. Ed.* **2003**, *42*, 2054.

195. Miyabe, H.; Yoshida, K.; Matsumura, A.; Yamauchi, M.; Takemoto, Y. *Synlett* **2003**, 567.

196. Fischer, C.; Defieber, C.; Suzuki, T.; Carreira, E. M. *J. Am. Chem. Soc.* **2004**, *126*, 1628.

197. Lipowsky, G.; Helmchen, G. *Chem. Commun.* **2004**, 116.

198. Baker, R.; Cook, A. H.; Halliday, D. E.; Smith, T. N. *J. Chem. Soc. Perkin Trans. 2* **1974**, 1511.

199. Beger, V. J.; Duschek, C.; Fullbier, H.; Gaube, W. *J. Prakt. Chem.* **1974**, *1*, 26.

200. Jackstell, R.; Andreu, M. G.; Frisch, A.; Selvakumar, K.; Zapf, A.; Klein, H.; Spannenberg, A.; Rottger, D.; Briel, O.; Karch, R.; Beller, M. *Angew. Chem.* **2002**, *41*, 986.

201. Maddock, S. M.; Finn, M. G. *Organometallics* **2000**, *19*, 2684.

202. Petrushkina, E. A.; Zakharkin, L. I. *Russ. Chem Bull.* **1992**, *41*, 1392.

203. Zakharkin, M. I.; Guseva, V. V.; Sulaimankulova, D. D.; Petrushkina, E. A. *Zh. Org. Khim.* **1987**, *23*, 1654.

204. Keim, W.; Roper, M.; Schieren, M. *J. Mol. Catal.* **1983**, *20*, 139.

205. Takahashi, S.; Shibano, T.; Hagihara, N. *Bull. Chem. Soc. J.* **1968**, *41*, 454.

206. Schilling, B. E. R.; Hoffmann, R.; Faller, J. W. *J. Am. Chem. Soc.* **1979**, *101*, 592.

207. Adams, R. D.; Chodosh, D. F.; Faller, J. W.; Rosan, A. M. *J. Am. Chem. Soc.* **1979**, *101*, 2570.

208. Hayashi, T.; Kawatsura, M.; Uozumi, Y. *Chem. Commun.* **1997**, 561.

209. Faller, J. W.; Chao, K. H. *J. Am. Chem. Soc.* **1983**, *105*, 3893.

210. Curtis, M. D.; Eisenstein, O. *Organometallics* **1984**, *3*, 887.

211. Hayashi, T.; Hagihara, T.; Konishi, M.; Kumada, M. *J. Am. Chem. Soc.* **1983**, *105*, 7767.

212. van Haaren, R. J.; Keeven, P. H.; van der Veen, L. A.; Goubitz, K.; van Strijdonck, G. P. F.; Oevering, H.; Reek, J. N. H.; Kamer, P. C. J.; van Leeuwen, P. *Inorg. Chim. Acta* **2002**, *327*, 108.

213. van Haaren, R. J.; Goubitz, K.; Fraanje, J.; van Stijdonck, G. P. F.; Oevering, H.; Coussens, B.; Reek, J. N. H.; Kamer, P. C. J.; van Leeuwen, P. W. N. M. *Inorg. Chem.* **2001**, *40*, 3363.

214. For early papers discussing a potential S_N2^1 mechanism, See Magid, R.M. *Tetrahedron* **1980**, *36*, 1901; Fiaud, J.C.; Malleron, J.L. *Tetrahedron Lett.* **1981** *22*, 1399; Trast, B.M.; Schmuff, N.R. *Tetrahedron Lett.* **1981**, *22*, 2999.

215. Åkermark, B.; Hansson, S.; Krakenberger, B.; Vitagliano, A.; Zetterberg, K. *Organometallics* **1984**, *3*, 679.

216. Trost, B. M.; Toste, F. D. *J. Am. Chem. Soc.* **1998**, *120*, 9074.

217. Trost, B. M.; Toste, F. D. *J. Am. Chem. Soc.* **1999**, *121*, 4545.

218. Hayashi, T.; Kawatsura, M.; Uozumi, Y. *J. Am. Chem. Soc.* **1998**, *120*, 1681.

219. You, S. L.; Zhu, X. Z.; Luo, Y. M.; Hou, X. L.; Dai, L. X. *J. Am. Chem. Soc.* **2001**, *123*, 7471.

220. Ephritikhine, M.; Francis, B. R.; Green, M. L. H.; Mackenzie, R. E.; Smith, M. J. *J. Chem. Soc., Dalton Trans.* **1977**, 1131.

221. Periana, R. A.; Bergman, R. G. *J. Am. Chem. Soc.* **1984**, *106*, 7272.

222. Wakefield, J. B.; Stryker, J. M. *J. Am. Chem. Soc.* **1991**, *113*, 7057.

223. Faller, J. W.; Murray, H. H.; White, D. L.; Chao, K. H. *Organometallics* **1983**, *2*, 400.

224. Pearson, A. J.; Khan, M. N. I.; Clardy, J. C.; Cunheng, H. *J. Am. Chem. Soc.* **1985**, *107*, 2748.

225. Pearson, A. J.; Khan, M. N. I. *Tetrahedron Lett.* **1984**, *25*, 3507.

226. Pearson, A. J.; Khan, M. N. I. *J. Am. Chem. Soc.* **1984**, *106*, 1872.

227. Shu, C. T.; Liebeskind, L. S. *J. Am. Chem. Soc.* **2003**, *125*, 2878.

228. Shu, C. T.; Alcudia, A.; Yin, J. J.; Liebeskind, L. S. *J. Am. Chem. Soc.* **2001**, *123*, 12477.

229. Moretto, A. F.; Liebeskind, L. S. *J. Org. Chem.* **2000**, *65*, 7445.

230. Hansson, S.; Miller, J. F.; Liebeskind, L. S. *J. Am. Chem. Soc.* **1990**, *112*, 9660.

231. Yin, J. J.; Llorente, I.; Villanueva, L. A.; Liebeskind, L. S. *J. Am. Chem. Soc.* **2000**, *122*, 10458.

232. Semmelhack, M. F.; Le, H. T. M. *J. Am. Chem. Soc.* **1984**, *106*, 2715.

233. Semmelhack, M. F.; Herndon, J. W. *Organometallics* **1983**, *2*, 363.

234. Angelici, R. J.; Fischer, E. O. *J. Am. Chem. Soc.* **1963**, *85*, 3733.

235. Braunschweig, H.; Wagner, T. *Chem. Ber.* **1994**, *127*, 1613.

236. Hartwig, J. F.; He, X. *Organometallics* **1996**, *15*, 5350.

237. Ruf, W.; Fueller, M.; Siebert, W. *J. Organomet. Chem.* **1974**, *64*, C45.

238. Birch, A. J.; Jenkins, I. D. In *Transition Metal Organometallics in Organic Synthesis*; Alper, H., Ed.; Academic Press: New York, 1976; Chapter 1.

239. Pearson, A. J. *Acc. Chem. Res.* **1980**, *13*, 463.

240. Bayoud, R. S.; Biehl, E. R.; Reeves, P. C. *J. Organomet. Chem.* **1979**, *174*, 297.

241. Hegedus, L. S. In *The Chemistry of the Metal-Carbon Bond*; Hartley, F. R., Ed.; Wiley-Interscience: New York, 1985; Vol. 2, p 482.

242. Pearson, A. J.; Ong, C. W. *J. Org. Chem.* **1982**, *47*, 3780.

243. Pearson, A. J.; Kole, S. L.; Ray, T. *J. Am. Chem. Soc.* **1984**, *106*, 6060.

244. Pape, A. R.; Kaliappan, K. P.; Kundig, E. P. *Chem. Rev.* **2000**, *100*, 2917.
245. Semmelhack, M. F. In *Arene-Metal Complexes in Organic Synthesis*; Seyferth, D., Ed.; Elsevier: New York, 1976, p 361.
246. Hegedus, L. S. In *Transition Metals in the Synthesis of Complex Organic Molecules*, 2nd ed.; University Science Books: Mill Valey, CA, 1999; p 289.
247. Wu, A.; Reeves, P. C.; Biehl, E. R. *J. Chem. Soc., Perkin Trans. 2* **1972**, 449.
248. Ashraf, M.; Jackson, W. R. *J. Chem. Soc., Perkin Trans. 2* **1972**, 103.
249. Nicholls, B.; Whiting, M. C. *J. Chem. Soc.* **1959**, 551.
250. Fischer, E. O.; Ofele, K.; Essler, H.; Frohlich, W.; Mortensen, J. P.; Semmlinger, W. *Chem. Ber.* **1958**, *91*, 2763.
251. Semmelhack, M. F. In *Comprehensive Organometallic Chemistry II*; Wilkinson, G., Stone, F. G. A., Abel, E. W., Eds.; Elsevier.: New York, 1995; Vol. 12, p 1017.
252. Merlic, C. A.; Walsh, J. C.; Tantillo, D. J.; Houk, K. N. *J. Am. Chem. Soc.* **1999**, *121*, 3596.
253. Semmelhack, M. F.; Hall, H. T. *J. Am. Chem. Soc.* **1974**, *96*, 7092.
254. Semmelhack, M. F.; Hall, H. T. *J. Am. Chem. Soc.* **1974**, *96*, 7091.
255. Razuvaev, G. A.; Artemov, A. N.; Aladjin, A. A.; Sirotkin, N. I. *J. Organomet. Chem.* **1976**, *111*, 131.
256. Morley, J. A.; Woolsey, N. F. *J. Org. Chem.* **1992**, *57*, 6487.
257. Kundig, E. P.; Perret, C.; Spichiger, S.; Bernardinelli, G. J. *J. Organomet. Chem.* **1985**, *286*, 183.
258. Kane-Maguire, L. A. P.; Honig, E. D.; Sweigart, D. A. *Chem. Rev.* **1984**, *84*, 525.
259. Pike, R. D.; Sweigart, D. A. *Synlett* **1990**, 565.
260. Kundig, E. P.; Jeger, P.; Bernardinelli, G. *Inorg. Chim. Acta* **2004**, *357*, 1909.
261. Pearson, A. J.; Park, J. G. *J. Org. Chem.* **1992**, *57*, 1744.
262. Gill, T. P.; Mann, K. R. *Organometallics* **1982**, *1*, 485.
263. Robertson, I. W.; Stephenson, T. A.; Tocher, D. A. *J. Organomet. Chem.* **1982**, *228*, 171.
264. Zelonka, R. A.; Baird, M. C. *J. Organomet. Chem.* **1971**, *335*, C43.
265. Semmelhack, M. F.; Chlenov, A.; Wu, L. Y.; Ho, D. *J. Am. Chem. Soc.* **2001**, *123*, 8438.
266. Semmelhack, M. F. *Ann. N.Y. Acad. Sci.* **1977**, *295*, 36.
267. Semmelhack, M. F.; Clark, G. R.; Garcia, J. L.; Harrison, J. J.; Thebtaranonth, Y.; Wulff, W.; Yamashita, A. *Tetrahedron* **1981**, *37*, 3957.
268. Semmelhack, M. F.; Garcia, J. L.; Cortes, D.; Farina, R.; Hong, R.; Carpenter, B. K. *Organometallics* **1983**, *2*, 467.
269. Pearson, A. J.; Zhang, P.; Lee, K. *J. Org. Chem.* **1996**, *61*, 6581.
270. Pearson, A. J.; Bignan, G. *Tetrahedron Lett.* **1996**, *37*, 735.
271. Pearson, A. J.; Lee, K. S. *J. Org. Chem.* **1994**, *59*, 2304.
272. Kundig, E. P.; Cannas, R.; Laxmisha, M.; Liu, R. G.; Tchertchian, S. *J. Am. Chem. Soc.* **2003**, *125*, 5642.
273. Baldoli, C.; Maiorana, S.; Licandro, E.; Casiraghi, L.; Zinzalla, G.; Seneci, P.; De Magistris, E.; Paio, A.; Marchioro, C. *J. Comb. Chem.* **2003**, *5*, 809.
274. Maiorana, S.; Baldoli, C.; Licandro, E.; Casiraghi, L.; de Magistris, E.; Paio, A.; Provera, S.; Seneci, P. *Tetrahedron Lett.* **2000**, *41*, 7271.
275. Kundig, E. P.; Quattropani, A. *Tetrahedron Lett.* **1994**, *35*, 3497.
276. Kundig, E. P.; Amurrio, D.; Anderson, G.; Beruben, D.; Khan, K.; Ripa, A.; Ronggang, L. *Pure. Appl. Chem.* **1997**, *69*, 543.
277. Pearson, A. J.; Gontcharov, A. V. *J. Org. Chem.* **1998**, *63*, 152.
278. Miles, W. H.; Brinkman, H. R. *Tetrahedron Lett.* **1992**, *33*, 589.
279. Miles, W. H.; Smiley, P. M.; Brinkman, H. R. *J. Chem. Soc., Chem. Commun.* **1989**, 1897.
280. Tomioka, K.; Shindo, M.; Koga, K. *J. Am. Chem. Soc.* **1989**, *111*, 8266.
281. Amurrio, D.; Khan, K.; Kundig, E. P. *J. Org. Chem.* **1996**, *61*, 2258.
282. Price, D. A.; Simpkins, N. S.; Macleod, A. M.; Watt, A. P. *J. Org. Chem.* **1994**, *59*, 1961.
283. Price, D. A.; Simpkins, N. S.; Macleod, A. M.; Watt, A. P. *Tetrahedron Lett.* **1994**, *35*, 6159.
284. Uemura, M.; Nishimura, H.; Yamada, S.; Hayashi, Y.; Nakamura, K.; Ishihara, K.; Ohno, A. *Tetrahedron: Asymmetry* **1994**, *5*, 1673.
285. Ewin, R. A.; MacLeod, A. M.; Price, D. A.; Simpkins, N. S.; Watt, A. P. *J. Chem. Soc,. Perkin Trans. 1* **1997**, 401.

CHAPTER 12

Electrophilic Attack on Coordinated Ligands

12.1. Overview and Basic Principles

Electrophiles are known to react with transition metal complexes at the metal in some cases and at the ligand in others. Electrophilic attack at the metal causes a formal increase in the oxidation state by two units and is most often considered an oxidative addition or an oxidative ligation. This reaction was presented, albeit briefly, in Chapter 7. This current chapter describes reactions of electrophiles at the ligands of organometallic complexes to modify these coordinated units. Some of these reactions occur by direct attack of the electrophile at the ligand, and others occur by initial attack at the metal, prior to transformation of the ligand. The scope of electrophile and transition metal complex that undergo this class of reaction, the types of organometallic products formed by reactions with electrophiles, guidelines to predict when an electrophile will react at a ligand over the metal, and applications of these reactions for the synthesis of organic molecules are presented in this chapter.

The electrophile that modifies a coordinated ligand can be a proton, a strong Lewis acid, or an unsaturated electrophile. Examples of Lewis acids include trityl cation or perfluoroarylboranes, and examples of unsaturated electrophiles include CO_2, SO_2, isocyanates, aldehydes, ketones, and related compounds. Reactions of these electrophiles can lead to the formation of cationic metal complexes by abstraction of a hydride or hydrocarbyl group by the Lewis acid, or they can lead to products from insertion of the unsaturated electrophile into the metal–carbon bond. These reactions are shown generically in Equations 12.1–12.3.

$$L_nM-R + E^{\oplus} \longrightarrow L_nM^{\oplus} + R-E \tag{12.1}$$

$$L_nM-R + E \longrightarrow L_nM^{\oplus} + R-E^{\ominus} \tag{12.2}$$

$$L_nM-R + X=Y=Z \longrightarrow L_nM\underset{X}{\overset{Z}{\diagdown}}\overset{\parallel}{Y}\diagdown R \tag{12.3}$$

The factors that control whether electrophilic attack occurs at the metal or the ligand are often subtle. Because electrophiles typically interact with the HOMO, the relative contribution of the metal and the ligand orbitals to the HOMO will often determine the site of electrophilic attack. Although the HOMO of the metal complex typically contains a large contribution from the metal orbitals because metal atoms have relatively high-energy orbitals, the kinetic product sometimes results from direct attack of the electrophile at the ligand. This attack occurs at the ligand because most ligand atoms are more sterically accessible than is the metal center. Furthermore, protonation or addition of electrophiles at the lone pair of an oxygen or nitrogen atom requires less rearrangement than protonation or addition of electrophiles at a metal center or a metal–carbon bond. The smaller amount of rearrangement often makes attack at nitrogen and oxygen favored kinetically.[1–4]

12.2. Electrophilic Cleavage of Metal–Carbon and Metal–Hydride σ-Bonds

12.2.1. Scope of Electrophilic Cleavage of Metal–Carbon and Metal–Hydride σ-Bonds

A proton is the electrophile most commonly used to remove an organic ligand from a transition metal. These protonolyses typically occur by mechanisms that retain the stereochemistry of the organic ligand.[5] For example, the deuterium in the product of the reaction in Equation 12.4 occupies the same position about the double bond as did the ruthenium in the initial vinyl complex.[6] Olefin stereochemistry was also shown to be retained during cleavage of a vinyl–palladium bond with D[+].[7] Cleavage of a metal–alkyl bond has also been shown to occur with retention of stereochemistry (Equation 12.5).[8]

$$(12.4)$$

$$(12.5)$$

Metal–carbon bonds are not only cleaved by protons but by halogens or mercuric halides as the electrophile. These reactions typically lead to the formation of alkyl halides or alkylmercury products (Equation 12.6). In contrast to the more common reactions with protons, these reactions generate functionalized organic products.

$$(12.6)$$

In addition to the cleavage of metal–carbon bonds by additions of electrophilic reagents across M–C bonds, Lewis acidic reagents can abstract alkyl or hydride groups from the metal. In some cases, these reactions have been used to generate cationic organometallic complexes, and in other cases these reactions have been used to generate main group organometallic reagents or reactive intermediates. For example, abstraction of alkyl groups by main group Lewis acids has become a commonly used route to cationic group 4 metal complexes that are intermediates in the polymerization of olefins. These catalysts were first generated by reaction of the alkyl complex with methylaluminoxane, which is generated by partial hydrolysis of trimethylaluminum. Although reaction of these complexes with a large excess of this reagent remains the most common method to activate group 4 alkyl complexes for polymerization, the abstraction of alkyl groups by neutral boranes to form cationic complexes containing borate anions is a more defined, and in some cases a more effective, method to generate these catalysts. Thus, this abstraction process has been studied in some detail.

The thermodynamic parameters for the abstraction process in Equation 12.7 have been measured by Marks.[9,10] The transition metal products of these reactions are zwitterionic species containing a bridging alkyl group. Representative enthalpies for this abstraction

$$ (12.7) $$

Table 12.1. Enthalpies for Equation 12.7 in toluene solution at 25 °C.

M	R	$\Delta H_{form}{}^{a}$ (kcal/mol)
Zr	Me	$-24.6(0.3)^{b}$
Zr	CH$_2$TMS	$-22.6(1.0)$
Zr	CH(TMS)$_2$	$-59.2(1.4)$
Hf	Me	$-20.8(0.5)^{b}$
Hf	CH$_2$TMS	$-31.1(1.6)$

^{a} Values are determined by titration calorimetry.
^{b} From Deck, P. A.; Beswick, C. L.; Marks, T. J. *J. Am. Chem. Soc.* **1998**, *120*, 1772.

process are listed in Table 12.1. These values show that abstraction of alkyl groups by the highly electrophilic tris-perfluorophenylborane is strongly favored enthalpically. The trends for the thermodynamic parameters can be explained by relief of steric congestion and stabilization of the resulting cationic charge by the larger alkyl groups.

Abstraction of hydrides by Lewis acids has also been studied extensively. Several groups have measured the thermodynamics for hydride abstraction by hydride acceptors, such as triarylmethyl cation and 1-benzyl-1,4-dihydronicotinamide (BzNADH). Examples of these hydride abstraction reactions are given in Equations 12.8–12.9.[11–13]

$$ (12.8) $$

$$ (12.9) $$

DuBois has studied many examples of these reactions and from these data and related experiments has generated the "hydricity scale" shown in Table 12.2.[11] Particularly electron-rich complexes and third-row metal complexes are stronger hydride donors than less-electron-rich complexes and first-row analogs.

Table 12.2. Hydricities of selected compounds in acetonitrile.

Compound	ΔG_{H^-} (kcal/mol)	Reference
$[HRh(dppb)_2]^+$	34 ± 2	Price, A. J.; Ciancanelli, R.; Noll, B. C.; Curtis, C. J.; DuBois, D. L.; DuBois, M. R. *Organometallics* **2002**, *21*, 4833.
$[HW(CO)_4(PPh_3)]^-$	36 ± 2	Ellis,W. W.; Ciancanelli, R.; Miller, S. M.; DuBois, M. R.; DuBois, D. L. *J. Am. Chem. Soc.* **2003**, *125*, 12230.
$Cp^*Re(PMe_3)(NO)(CHO)$	42 ± 2	Ellis, W. W.; Miedaner, A.; Curtis, C. J.; Gibson, D. H.; DuBois, D. L. *J. Am. Chem. Soc.* **2002**, *124*, 1926.
$[HNi(dmpe)_2]^+$	51 ± 2	Berning, D. E.; Noll, B. C.; DuBois, D. L. *J. Am. Chem. Soc.* **1999**, *121*, 11432.
$CpRe(CO)(NO)(CHO)$	55 ± 2	Ellis, W. W.; Miedaner, A.; Curtis, C. J.; Gibson, D. H.; DuBois, D. L. *J. Am. Chem. Soc.* **2002**, *124*, 1926.
$CpMo(PMe_3)(CO)_2H$	55 ± 3	Ellis, W. W.; Raebiger, J. W.; Curtis, C. J.; Bruno, J. W.; DuBois, D. L. *J. Am. Chem. Soc.* **2004**, *126*, 2738.
$Cp^*Mo(PMe_3)(CO)_2H$	58 ± 2	Ellis, W. W.; Raibiger, J. W.; Curtis, C. J.; Bruno, J. W.; DuBois, D. L. *J. Am. Chem. Soc.* **2004**, *126*, 2738.
X = C(O)NH₂	59 ± 2	Ellis, W. W.; Raibiger, J. W.; Curtis, C. J.; Bruno, J. W.; DuBois, D. L. *J. Am. Chem. Soc.* **2004**, *126*, 2738.
X = C(O)CH₃	60 ± 2	Ellis, W. W.; Raibiger, J. W.; Curtis, C. J.; Bruno, J. W.; DuBois, D. L. *J. Am. Chem. Soc.* **2004**, *126*, 2738.
$[HNi(dmpp)_2]^+$	60 ± 2	Berning, D. E.; Noll, B. C.; DuBois, D. L. *J. Am. Chem. Soc.* **1999**, *121*, 11432.
X = CN	63 ± 2	Ellis, W. W.; Raibiger, J. W.; Curtis, C. J.; Bruno, J. W.; DuBois, D. L. *J. Am. Chem. Soc.* **2004**, *126*, 2738.
$[HNi(depp)_2]^+$	67 ± 2	Berning, D. E.; Miedaner, A.; Curtis, C. J.; Noll, B. C.; DuBois, M. R.; DuBois, D. L. *Organometallics* **2001**, *20*, 1832.
	70 ± 2	Ellis, W. W.; Raibiger, J. W.; Curtis, C. J.; Bruno, J. W.; DuBois, D. L. *J. Am. Chem. Soc.* **2004**, *126*, 2738.
$p\text{-}(Me_2NC_6H_4)_3CH$	74	Zhang, X.-M.; Bruno, J. W.; Enyinnaya, E. *J. Org. Chem.* **1998**, *63*, 4671.
Ph_3CH	99	Zhang, X.-M.; Bruno, J. W.; Enyinnaya, E. *J. Org. Chem.* **1998**, *63*, 4671.

[a] Value calculated from the ΔG_{H^-} value for BzNADH and from equilibrium data given in references 17 and 18 for hydride transfer reactions between BzNADH and the respective NAD⁺ model compounds.

Addition of main group halide reagents, such as aryl dihaloboranes and diaryl diha-lostannanes, can also lead to abstraction of alkyl groups from d^0 metal complexes to form main group heterocycles. The zirconacyclopentadienes in Equation 12.10 are easily gener-ated by reductive coupling, and these complexes react with phosphorus, asenic, antimony and bismuth halides to transfer the organic portion of the metallacycles and form the cor-responding main group heterocycles.[14,15] In some cases, the products are unstable to Diels–Alder dimerization. The products from reaction of the metallacycles with the main group halides were proposed to occur by electrophilic attack on the M–C bond.

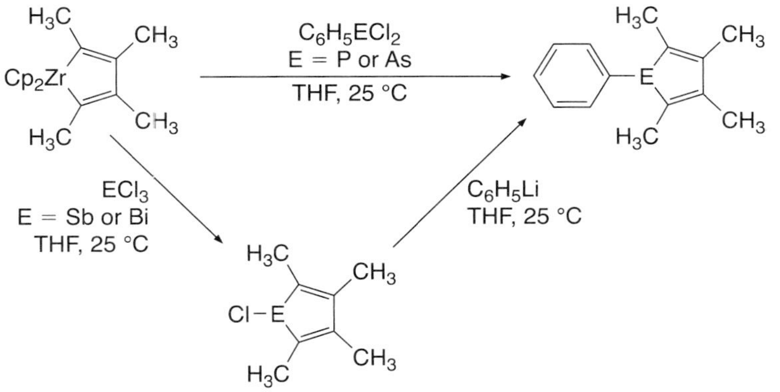

$$(12.10)$$

12.2.2. Mechanism of Electrophilic Attack

12.2.2.1. Mechanism of Attack of Main Group Electrophiles on Alkyl Complexes Possessing d-Electrons

Reactions of electrophiles with transition metal organometallic compounds can occur by initial attack at the metal or direct attack at the M–C bond, as shown in Scheme 12.1. Because the HOMO of most metal complexes possessing d-electrons largely consists of the metal d-orbitals, most reactions of main group electrophiles with metal complexes containing d-electrons cleave the metal–carbon bonds by initial reaction at the metal center, as shown at the top of Scheme 12.1. After initial attack at the metal center, reductive elimination or nucleo-philic attack of an anion on the resulting metal–alkyl complex leads to the final product of M–C bond cleavage. In contrast, protonation of metal–hydride complexes often occurs at the hydride ligand directly to form a metal–dihydrogen complex. This difference in reactivity results from differences between the reorganization energy for protonation of the hydride and the reorganization energy for protonation of the metal, as well as from differences in the avail-ability of the electrons in the M–H and M–C bond, as discussed in more detail below.

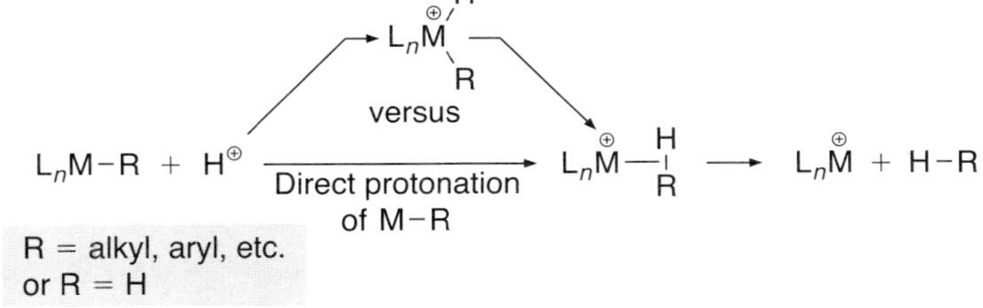

Scheme 12.1

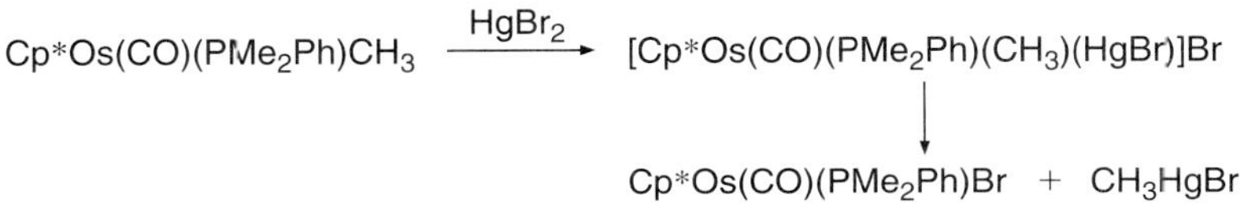

$$(12.11)$$

$$(12.12)$$

Direct evidence for reactions of electrophiles to cleave a metal–carbon bond by initial attack at the metal center was obtained from the reaction of $Cp^*Os(CO)(PMe_2Ph)Me$ with $HgBr_2$.[16] The intermediate shown in Equation 12.11 that results from the reaction of $HgBr_2$ at the metal center in $Cp^*Os(CO)(PMe_2Ph)(Me)$ was isolated from this reaction. Indirect evidence for attack at the metal center during reactions of several other complexes has been obtained by assessing the stereochemistry of the metal complex at partial conversion. A number of studies[17–21] have shown that epimerization at iron has occurred in the starting material recovered from partial reaction of chiral iron–alkyl complexes with electrophiles (Equation 12.12). This epimerization provides evidence for electrophilic attack at the metal center because the epimerization is likely to occur by addition of the electrophile to the metal to generate a complex possessing a four-legged piano stool geometry that is known to be stereochemically nonrigid (Equation 12.13).[22] Equation 12.14 illustrates how the SS–RR enantiomeric pair[‡] (the SS stereoisomer is drawn) is converted into the SR–RS enantiomeric pair more rapidly than the methyl ligand is cleaved.

$$(12.13)$$

$$(12.14)$$

[‡]The first chirality symbol gives the configuration of the substituted cyclopentadienyl ligand and the second gives the configuration about the iron.[18] The Cahn–Ingold–Prelog sequence rules can be extended to chiral four-coordinate metals by considering polyhapto ligands as pseudo-atoms of atomic number equal to the sum of the atomic numbers of all of the atoms bonded to the metal atom. (This principle was first formulated by Stanley and Baird[23] in terms of atomic weights.) A detailed review of stereochemical nomenclature for organometallic and inorganic compounds is available.[24]

The rate expression (Equation 12.15) for electrophilic cleavage of the metal–carbon bond in $CpFe(CO)_2R$ (Equation 12.12) is consistent with a mechanism in which the equilibrium in Equation 12.14 precedes removal of the alkyl ligand from the iron. The rate of the reaction increases with ionic strength and solvent polarity, and these data are consistent with an ionic intermediate. The second-order dependence of the rate on HgX_2 shows that the coordinated electrophile E^+ in the intermediate is likely to be HgX^+ with an HgX_3^- counterion.[25]

$$-\frac{d[CpFe(CO)_2R]}{dt} = k[CpFe(CO)_2R][HgX_2]^2 \qquad (12.15)$$

Information on the detailed mechanism by which the metal–carbon bond is cleaved in these systems can be gained by studying the stereochemistry of the carbon originally attached to the metal. The reaction of $CpFe(CO)_2R$ with HgX_2 or halogen gives $CpFe(CO)_2X$ and the alkylmercury or alkyl halide product with all three possible stereochemical outcomes: retention of stereochemistry at carbon, inversion of stereochemistry at carbon, or loss of stereochemistry at carbon. The different stereochemical outcomes have been explained by the intermediacy of an ionic species. As shown in the general mechanism for the electrophilic cleavage of alkyl ligands from such cyclopentadienyl complexes (Scheme 12.2), the cationic intermediate formed in the initial equilibrium can follow three pathways.

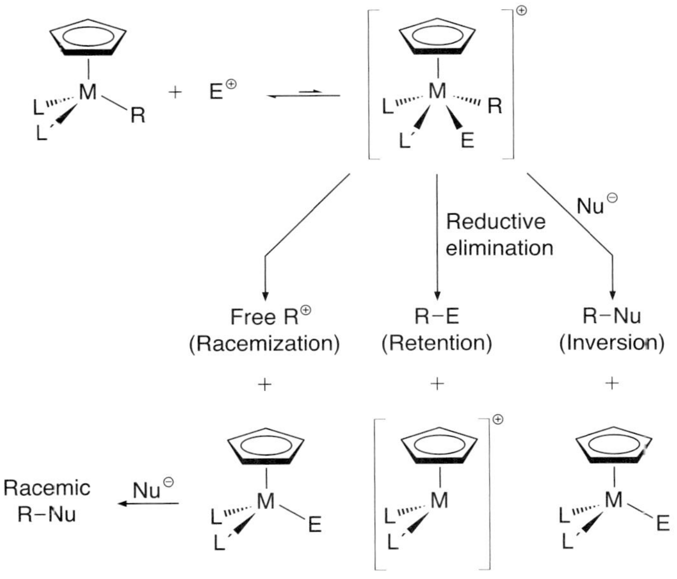

Scheme 12.2

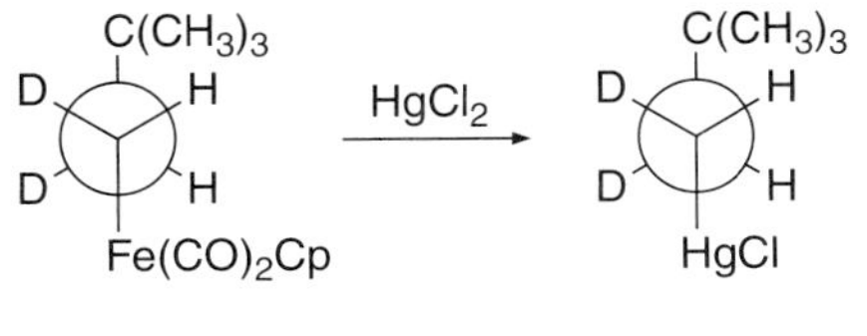

$$(12.16)$$

By one pathway, reductive elimination of R and the electrophile gives R–E with retention of stereochemistry at the metal-bound carbon (recall from Chapter 8 that concerted reductive elimination occurs with retention of configuration at this site). For example, the *threo*-3,3-dimethylbutyl-1,2-d_2 iron complex shown in Equation 12.16 reacts with mercuric chloride to form *threo*-3,3-dimethylbutyl-1,2-d_2 mercuric chloride.[26] Similar results have been obtained for the reactions of *threo*-PhCHDCHDFe(CO)$_2$Cp and *trans-(threo*-PhCHDCHD)

$W(CO)_2(PEt_3)Cp$ with HgX_2,[27] as well as for the reactions of organometallic complexes with halogens to form alkyl halides [e.g., the reaction of *threo*-$PhCHDCHDW(CO)_3Cp$ with iodine[28]]. The metal cation that is formed by reductive elimination of R–E typically undergoes stereochemical inversion slowly.[21,29,30] Thus, the metal product often retains the stereochemistry of the starting material. For example, the reaction shown in Equation 12.12 occurs with predominant retention of configuration at iron.[18]

By a second pathway, the reaction occurs with inversion of configuration at carbon. This pathway occurs when a strong nucleophile is generated during the reaction. For example, halide that is generated by cleavage of the metal–carbon bond by electrophilic halogen can react at the metal-bound carbon by an S_N2 process. This sequence leads to inversion of configuration at the metal-bound carbon. For example, the reaction of bromine with the *erythro* 3,3-dimethylbutyl-1,2-d_2 complex in Equation 12.17 forms *threo*-3,3-dimethylbutyl-1,2-d_2 bromide.[31]

$$\text{(12.17)}$$

In some cases, reactions of electrophiles with metal–alkyl complexes possessing d-electrons appear to occur by S_E2 mechanisms.[27,32–34] In this process, the electrophile can attack the frontside or backside of the alkyl group. Although retention of stereochemistry at the metal-bound carbon from frontside attack at the metal–carbon bond is often observed, inversion of stereochemistry from backside attack at this carbon has also been observed.

$$(+)-PhCH(Me)Fe(CO)_2Cp \xrightarrow{\text{HgCl}_2} (rac)-PhCH(Me)Cl + ClHgFe(CO)_2Cp \quad \text{(12.18)}$$

By a third mechanism, the reaction occurs with loss of stereochemistry at the metal-bound carbon. This stereochemistry is observed if reaction with the electrophile leads to a relatively stable carbocation. For example, the reaction of HgX_2 in Equation 12.18 forms racemic α-methylbenzyl chloride from the optically active α-methylbenzyl complex,[25] presumably due to the stability of the benzylic cation.

12.2.2.2. Mechanism of Protonolysis of Metal–Carbon Bonds in Complexes Possessing d-Electrons

Most reactions of d^n ($n \geq 2$) organometallic complexes with acid to cleave the metal-carbon bond occur by initial protonation at the metal, followed by C–H bond-forming reductive elimination, as shown in the general Scheme 12.2. De Luca and Wojcicki[35] established that the cleavage of $CpFe(CO)_2R$ with CF_3CO_2H occurs by initial formation of $[CpFe(CO)_2(R)H][(O_2CCF_3)_2]$, followed by reductive elimination of R–H. Likewise, cleavage of the platinum(II)–alkyl bond, as shown in Equation 12.19, is thought to occur after protonation at platinum.[36] These reactions occur with retention of configuration at the alkyl carbon, and this stereochemistry is consistent with the mechanism involving protonation at the metal followed by reductive elimination to form the C–H bond (Scheme 12.1, top pathway). Gladysz and co-workers have found that the reaction of $CpRe(NO)(PPh_3)CH_3$ with HX to form methane and $CpRe(NO)(PPh_3)X$ occurs with retention of configuration at rhenium.[30] Less common are mechanisms involving direct protonation of the M–C bond, but arguments have been made that certain electronic properties, such as a strong trans donor ligand, can cause direct protonation to occur at a metal–alkyl bond.[37,38]

$$\text{trans-PtH}[(CH_2)_nCN](PPh_3)_2 \xrightarrow[\text{(1 equiv.)}]{+HCl} \text{trans-PtHCl}(PPh_3)_2 + H(CH_2)_nCN \quad \text{(12.19)}$$
$$(n = 1 \text{ or } 3)$$

12.2.2.3. Mechanism of Protonation of Metal–Hydride Bonds in Complexes Containing d-Electrons

In contrast to the reactions of protons with metal–alkyl complexes, which typically occur by protonation at the metal center, several examples of the reactions of metal–hydride complexes are now well documented in which protonation occurs at the hydride ligand. This protonation of a hydride ligand to generate a dihydrogen complex or to generate the combination of H_2 and a metal cation appears to be faster than protonation of the metal, and therefore to be faster than protonation at a metal–alkyl bond. The faster protonation of a hydride ligand than of an alkyl ligand is likely to result from the relative steric inaccessibility of the electron density of a metal–alkyl bond and the greater directionality of the metal–carbon bond.

Heinekey and Morris showed that protonation of $Cp^*Ru(PR_3)_2H$ complexes at low temperature form the corresponding dihydrogen complexes $[Cp^*Ru(PR_3)_2(H_2)]^+$, which form the corresponding dihydride complexes upon warming (Equation 12.20).[39,40] Likewise, protonation of $Re(NO)(CO)L_2H_2$ with CF_3COOH at low temperature formed a dihydrogen complex prior to loss of H_2.[41] These sequences of events in which the dihydrogen complex is the kinetic product require that protonation of the metal to form the dihydride complex is slower than protonation of the hydride ligand to form the less stable dihydrogen complex. Heinekey also showed by NMR line-shape analysis that the dihydrogen ligand in $[Cp(dmpe)Ru(H_2)]^+$ is more rapidly deprotonated by NEt_3 than are the hydride ligands in the tautomeric $[Cp(dmpe)RuH_2]^+$.[39] In a similar vein, Norton has shown that H/D exchange of the hydride ligand in $[CpW(PMe_3)(CO)_2H]$ with D^+ occurs faster than protonation of the metal to form the corresponding cationic dihydride complex.[42]

$$Cp^*Ru(PR_3)_2H \ + \ H^\oplus \xrightarrow[\text{of M–H}]{\text{Direct protonation}} [Ru]^{\oplus}\!\!-\!\!\overset{H}{\underset{H}{\mid}} \longrightarrow [Cp^*Ru(PR_3)_2(H)_2]^{\oplus} \qquad (12.20)$$

12.2.3. Mechanism of Electrophilic Attack on Alkyl Complexes that Lack d-Electrons

The reactions of electrophiles with metal complexes lacking d-electrons do not occur by initial addition of the electrophile to the metal center. The d-orbitals of a d^0 complex do not make a substantial contribution to the highest occupied molecular orbital (HOMO). Therefore, electrophiles do not react at the metal center in such complexes.

Cleavage of metal–alkyl bonds in d^0 metal complexes, therefore, occur by an S_E2 mechanism involving direct attack on the metal–alkyl bond. The product of this process is usually formed with retention of stereochemistry at carbon. This stereochemistry implies that the electrophile reacts by frontside attack at the M–C bond, rather than backside attack at the α carbon. Equation 12.21 shows an example of electrophilic attack on a d^0 metal complex that occurs with retention of configuration at the metal–carbon bond.[43] The coordinative unsaturation of the 16-electron Zr(IV) complex may facilitate reaction with retention of configuration because it allows coordination of the incipient Br^- during the reaction, as depicted in Figure 12.1.

$$(12.21)$$

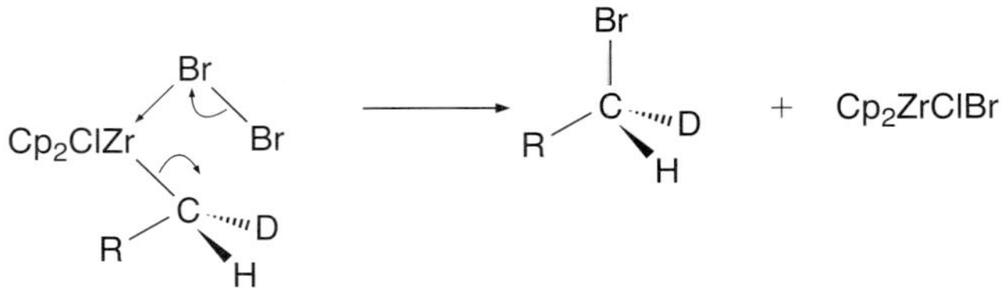

Figure 12.1.
Likely origin of the retention of configuration upon reaction of Br_2 with a Zr(IV) alkyl.

A second example of halogenation of a metal–carbon bond with retention of configuration is shown in Equation 12.22.[44] At –50 °C, the reaction of iodine with the titanacyclobutane leads to cleavage of both metal–carbon bonds. Protonolysis of the intermediate resulting from cleavage of the Ti–C bond containing the label leads to the diastereomer of the 2-phenyl-1-deuterio iodopropane product shown. This diastereomer results from initial cleavage of the M–C bond by iodine with retention of configuration.

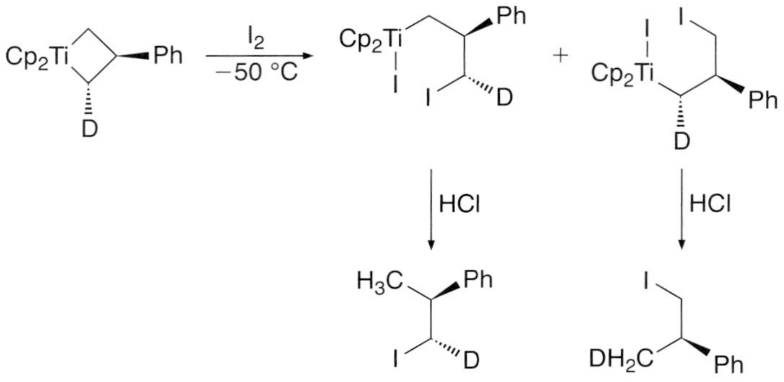

(12.22)

12.3. Electrophilic Insertion Reactions: Sulfur Dioxide, Carbon Dioxide and Related Electrophiles

As noted in Chapter 9 and in the introduction to this chapter, insertion reactions of electrophilic heterocumulenes can occur by a mechanism involving electrophilic attack on a ligand, rather than a mechanism involving migratory insertion. A quintessential example of such a process is the insertion of SO_2 into metal–alkyl bonds by initial abstraction of the alkyl group by SO_2. Some insertions of CO_2 also occur by electrophilic attack on a coordinated ligand. Sulfur dioxide sometimes coordinates to metals possessing d-electrons, but there is no evidence that this coordination occurs during the insertion process. Thus, insertions of these unsaturated electrophiles into metal–ligand bonds are presented in this chapter covering electrophilic attack on coordinated ligands, rather than Chapter 9 covering migratory insertion.

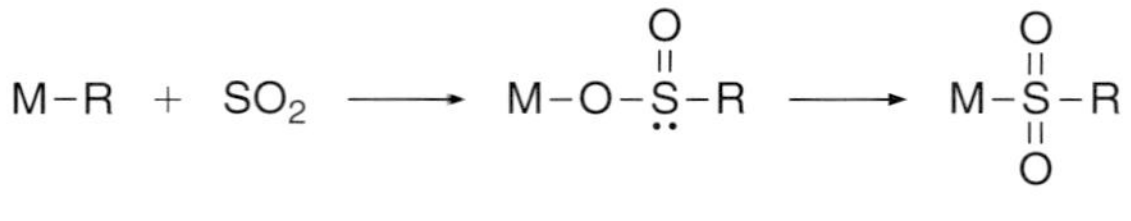

(12.23)

The insertion of SO_2 can lead to two different products. In most cases, the kinetic product of SO_2 insertion is an O-bound sulfinate, whereas the thermodynamic product is an S-bound sulfinate (Equation 12.23). O-bound sulfinates have been observed spectroscopically as intermediates in the reactions of SO_2 with $CpFe(CO)_2R$, $CpMo(CO)_3R$, $Mn(CO)_5R$, and $Re(CO)_5R$.[45,46] In other cases of the insertion of SO_2 into related middle-to-late transition metal–carbon bonds (Equations 12.24[28] and 12.25[47]), only the S-bound sulfinate was observed. In contrast, reactions of SO_2 with hard, early transition metal–alkyl complexes, which are more oxophilic than the softer late transition metal–alkyl complexes, tend to form the O-bound sulfinate as both the kinetic and thermodynamic products of the insertion of SO_2 (Equation 12.26).[48-50]

$$Cp(CO)_3W-CH_2CH_2Ph \ + \ SO_2 \ \longrightarrow \ Cp(CO)_3W-\overset{\overset{O}{\|}}{\underset{\underset{O}{\|}}{S}}-CH_2CH_2Ph \qquad (12.24)$$

$$Cp(Ph_3P)(CO)Ru-CH_3 \ + \ SO_2 \ \longrightarrow \ Cp(Ph_3P)(CO)Ru-\overset{\overset{O}{\|}}{\underset{\underset{O}{\|}}{S}}-CH_3 \qquad (12.25)$$

$$(12.26)$$

There are a number of significant differences between the mechanism of SO_2 insertion reactions[51] and the examples of electrophilic attack on coordinated ligands described in the preceding sections. Most striking, SO_2 insertion generally occurs by inversion of stereochemistry at the carbon bound to the metal. This inversion of configuration implies that the reaction occurs by an S_E2 mechanism involving backside attack at the metal-bound alkyl group, rather than electrophilic attack on the H–X bond. Reaction with inversion of stereochemistry at carbon has been established by many different experiments,[5] including those represented in Equations 12.27[28] and 12.28.[52] Retention of stereochemistry at the metal has been established by X-ray diffraction for the the reaction of SO_2 with the iron complex in Equation 12.29.[53]

$$(12.27)$$

$$[Co] = Cobaloxime \qquad (12.28)$$

$$(12.29)$$

These and other results are best explained, at least when the reaction is carried out in liquid SO_2,[51] by the mechanism shown in Figure 12.2. Initial S_E2 attack by the electrophilic sulfur of SO_2 inverts the configuration at carbon and forms an ion pair. Collapse of the ion pair to form an O-bound sulfinate occurs reversibly, and the more stable S-bound sulfinate is eventually formed. The cation $[CpM(L)(L')]^+$ is slow to invert, leading to retention of stereochemistry at the metal. Sulfur electrophiles that are similar to SO_2, such as N-sulfinylsulfonamides $(RSO_2)NSO$ and sulfur bis(sulfonylimide)s $(RSO_2N)_2S$, also insert into transition metal–carbon σ-bonds with inversion of stereochemistry at carbon,[54,55] probably by a mechanism analogous to that shown in Figure 12.2.

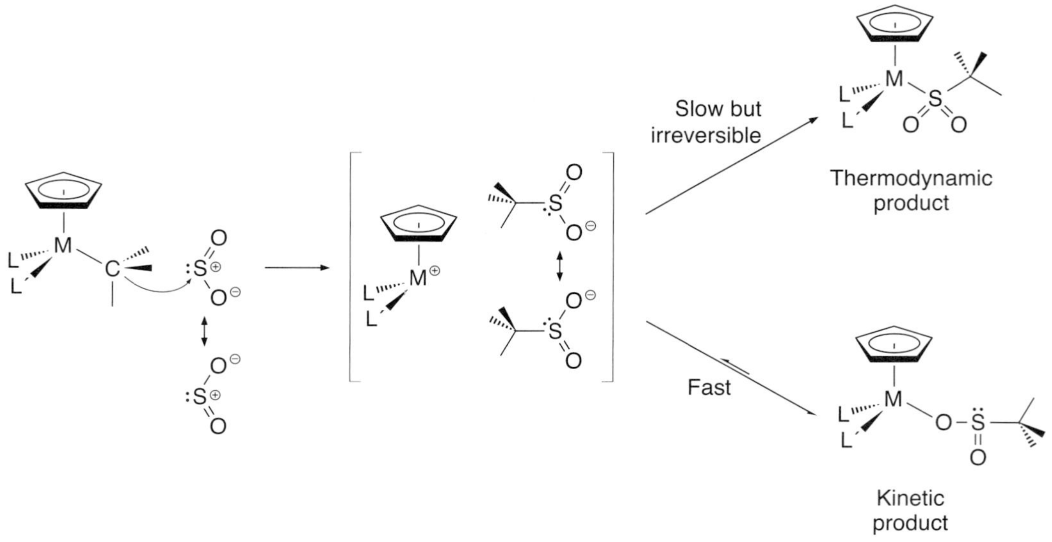

Figure 12.2. Overall mechanism for the insertion of SO_2.

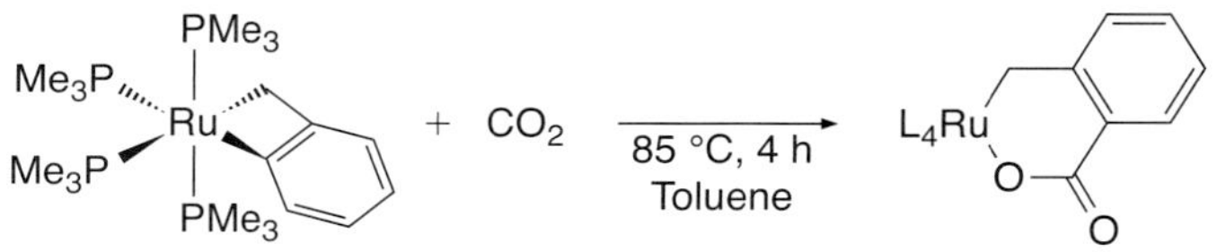

(12.30)

The insertion of CO_2 into transition metal alkyls can occur by migratory insertion or electrophilic attack on a coordinated ligand. Kinetic data show that dissociation of a ligand is necessary for the insertion of CO_2 to occur into the ruthenacycle in Equation 12.30, and these data suggest that reaction occurs by a migratory insertion pathway.[55] However, the closely related amido complex reacts directly with CO_2 to form the product of CO_2 insertion without dissociation of ligand.[55] This difference in mechanism was rationalized by the basicity of the nitrogen electron pair allowing direct attack of the electrophilic CO_2 on the amide ligand, as shown in Equation 12.31.

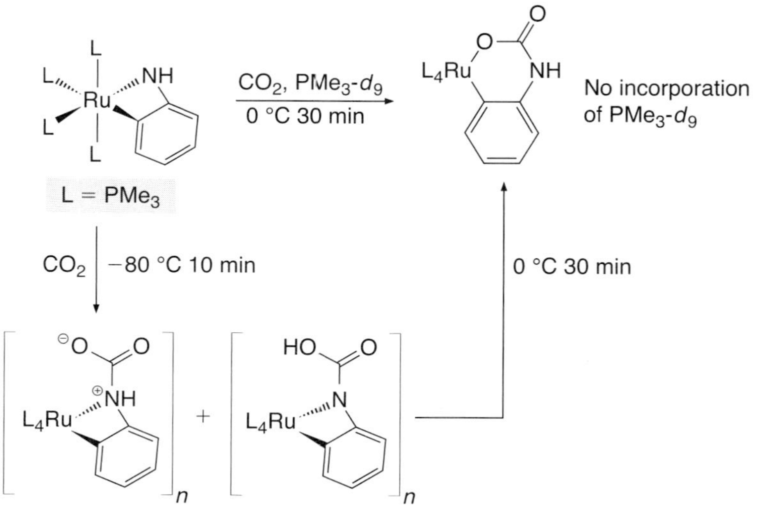

(12.31)

In other cases, reactions of coordinatively saturated alkyl complexes have been proposed to occur by attack of CO_2 on the metal or the alkyl group without dissociation of ligand. Reaction of the anionic tungsten carbonyl alkyl complex with CO_2 forms the product from CO_2 insertion (Equation 12.32), and the mechanism of this reaction has been studied in particular detail.[56] This reaction occurs without dissociation of any of the carbonyl ligands. Because the starting complex is a saturated 18-electron complex, this observation precludes a conventional migratory insertion mechanism. In contrast to the reactions of related metal–alkyl complexes with SO_2, this reaction with CO_2 occurs with retention of configuration at the α-carbon. Thus, migration of the alkyl group to CO_2 is proposed to occur after electrophilic attack of CO_2 at the metal center, as shown on the right of Equation 12.32.

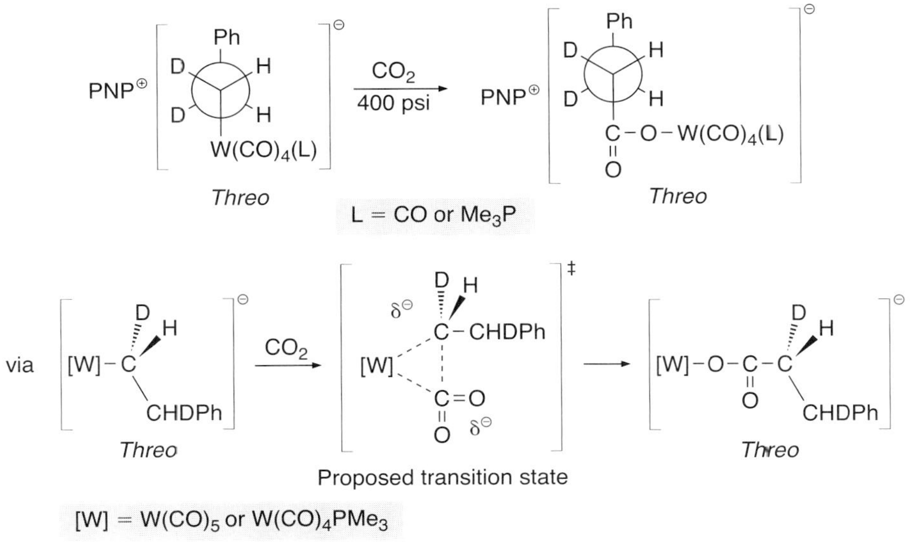

(12.32)

12.4. Electrophilic Modification of Coordinated Ligands

Electrophiles sometimes attack coordinated ligands without breaking the metal–carbon bond. Electrophilic attack can lead to abstraction of a hydride from the α-position of an alkyl group or from the β-position of an alkyl or acyl group. Addition of electrophiles is also known to occur at the γ-position of an allyl or related ligand to release this group from the metal, or to occur at the metal-bound carbon of nucleophilic carbene or carbyne ligands to form an adduct. In all of these reactions a π-bond is present at the beginning of the reaction or is formed by the reaction. Such additions of electrophiles at the α-, β-, or γ-positions of methyl, ethyl, and allyl complexes are shown in a general fashion in Equations 12.33–12.35, and coordination of an electrophile to the carbon of a carbene complex is shown in Equation 12.36.

$$M-CH_3 \ + \ E^{\oplus} \ \longrightarrow \ M=CH_2^{\oplus} \ + \ E-H \tag{12.33}$$

$$[M]-CH_2CH_3 \ + \ E^{\oplus} \ \longrightarrow \ [M]^{\oplus}-\| \ + \ E-H \tag{12.34}$$

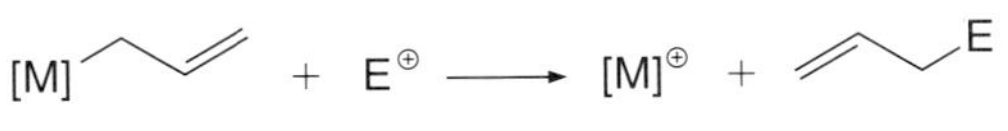

$$[M] \diagup\!\!\diagup \ + \ E^{\oplus} \ \longrightarrow \ [M]^{\oplus} \ + \ \diagup\!\!\diagdown^E \tag{12.35}$$

$$M=CH_2 \ + \ E^{\oplus} \ \longrightarrow \ [M-CH_2E]^{\oplus} \tag{12.36}$$

12.4.1. Attack at the α-Position

12.4.1.1 Attack at the α-Position of an Alkyl Group

In some cases, electrophiles react at the carbon of an alkyl group attached to the metal to abstract a group bound to this carbon. This process typically forms a metal–carbene complex. Often trityl cation (Ph_3C^+) is used as the electrophile. An example of a reaction at the α-position of an alkyl complex to abstract a hydride and generate an alkylidene complex is shown in Equation 12.37.[57,58] This class of reaction has been used to generate a variety of cationic late metal–alkylidene complexes as either stable species[59] or reactive intermediates.[60–62] Strong evidence that this α-hydride abstraction occurs in some cases by a mechanism involving initial electron transfer, followed by H• transfer, has been gained.[59,62]

Alternative routes to the same or related carbene complexes also involve electrophilic abstractions at the α-carbon. As shown in Equations 12.38[57,63] and 12.39,[64] trimethylsilyl triflate can be used to abstract alkoxo or hydroxo groups from the α-carbon to generate cationic carbene complexes.

$$ (12.37) $$

$$ (12.38) $$

$$ (12.39) $$

12.4.1.2. Electrophilic Attack on Carbene and Carbyne Complexes

Electrophilic reagents react not only at the α-carbon of alkyl groups, but at the α-carbon of carbene and carbyne complexes. In these cases, the electrophile can either form a stable adduct by coordination to the nucleophilic carbon, or it can abstract a labile group. As noted in the section of Chapter 13 on carbene complexes, nucleophilic early metal alkylidene complexes, such as Cp_2TiCH_2, coordinate Lewis acids at the carbene carbon. Cp_2TiCH_2 coordinates Me_2AlCl to form Tebbe's reagent.

Examples of the electrophilic modification of α-carbons also include electrophilic attack at carbyne complexes. Calculations suggest that carbyne carbons bear a significant negative charge,[65] but reports of electrophilic attack upon these carbons are rare. In Equation 12.40, protonation of $Cl(Me_3P)_4W\equiv CR$ gives a distorted alkylidene complex;[66] the C–H bond interacts with the coordinatively unsaturated metal in an "agostic" fashion.

$$ (12.40) $$

12.4.2. Attack at the β-Position

Additions of electrophiles to metal–alkyl complexes can also induce abstraction of a hydride from the β-position. Reactions of electrophiles at the β-carbons of alkyl, vinyl, and alkynyl groups have all been reported. Abstraction of a hydride at the β-position by an electrophile generates an olefin complex. Two examples of this reaction are shown in Equations 12.41[67] and 12.42.[68] The factors determining selectivity for α- versus β-hydrogen abstraction from alkyl complexes are not well established.

$$\text{(12.41)}$$

$$\text{(12.42)}$$

$$\text{(12.43)}$$

$$\text{(12.44)}$$

Electrophilic attack at the β-carbon of vinyl ligands can generate carbene complexes. Recall from the earlier sections of this chapter that protonation of a vinyl complex typically occurs at the metal, followed by subsequent reductive elimination. However, reactions of electrophiles other than acids can occur at the ligand instead of the metal. Attack at the β-carbon of a vinyl ligand generates an alkylidene complex, as shown in Equations 12.43 and 12.44. An α-alkoxy substituent helps to induce alkylation of the β-carbon in a neutral vinyl complex.[69] In this case, the reaction closely resembles the alkylation of a vinyl ether (Equation 12.45). In a related transformation, electrophiles react at the oxygen of acyl complexes to generate alkoxycarbene complexes. This reaction is shown in general in Equation 12.46. Specific examples of this process include alkylation of the anionic acyl complexes of Equations 12.47 and 12.48 and the neutral metal–acyl complex in Equation 12.49.[57,70]

$$\text{(12.45)}$$

$$\text{(12.46)}$$

$$\text{(12.47)}$$

$$(CO)_4Fe \overset{\ominus}{\underset{}{\diagdown}} \overset{O}{\overset{\|}{C}} Ph \quad \longleftrightarrow \quad (CO)_4Fe \diagdown \overset{O^{\ominus}}{\underset{}{C}} Ph \qquad \xrightarrow[\text{HMPA}]{\text{EtOSO}_2\text{F}} \quad (CO)_4Fe = \overset{OEt}{\underset{Ph}{C}} \tag{12.48}$$

$$\underset{\underset{R}{L}}{\overset{\text{Cp}}{CO^{\text{\tiny{''''}}}}}Fe \overset{O}{\diagdown} \xrightarrow[25\,°C]{\text{CH}_3\text{OSO}_2\text{CF}_3/\text{CH}_2\text{Cl}_2} \quad \underset{\underset{R}{L}}{\overset{\text{Cp}}{CO^{\text{\tiny{''''}}}}}\overset{\oplus}{Fe} \diagdown \overset{OCH_3}{} \quad {}^{\ominus}OSO_2CF_3 \tag{12.49}$$

L = P(C₆H₅)₃ or CO
R = CH₃, CH₂CH₃ or CH(CH₃)₂

Electrophilic attack on the β-carbon of alkynyl ligands (Equation 12.50) is common and is a route to vinylidene complexes introduced in Chapter 3. Examples of protonation and electrophilic alkylation of an anionic acetylide complex at the β-carbon are shown in Equation 12.51. Attack of two protons on an anionic acetylide complex generates a new carbyne complex, as shown in Equation 12.52.[70] This reaction, presumably, occurs by initial formation of a vinylidene complex.

$$M-C\equiv C- \quad + \quad E^{\oplus} \quad \longrightarrow \quad \overset{\oplus}{M}=C=C\diagdown^{}_{E} \tag{12.50}$$

$$\tag{12.51}$$

$$[Et_4N][(CO)_5WC\equiv CR] \xrightarrow{CF_3SO_3H/I^{\ominus}} I(CO)_4W\equiv CCH_2R \tag{12.52}$$

$$M \diagdown \diagup + \quad E^{\oplus} \quad \longrightarrow \quad \overset{\oplus}{M} \leftarrow \| \diagdown_E \tag{12.53}$$

$$M \diagdown \diagup + \quad A-B \quad \longrightarrow \quad \overset{\oplus}{M} \leftarrow \| \overset{B^{\ominus}}{\diagdown_A} \quad \longrightarrow \quad M \diagdown \overset{B}{\underset{A}{\diagup}} \tag{12.54}$$

12.4.3. Attack at the γ-Position

Electrophilic attack has also been shown to occur at the γ-position of several types of organometallic ligands. Electrophilic attack at the γ-position of η^1-allyls may be the most common. These reactions form substituted olefin complexes, as shown generally in Equations 12.53 and 12.54. Although fewer examples of the reactions of electrophiles with allyl complexes have been reported than the reactions of nucleophiles with allyl complexes, a number of catalytic processes have been developed that are likely to occur by electrophilic attack at the γ-position of allyl complexes, and a few examples of the reactions of electrophiles with isolated η^1-allyl complexes have been reported.[71–78]

$$R^1 \diagup \!\!\!\diagdown^{R^2} \!\!\! SnBu_3 + R^3CHO \xrightarrow[\text{THF}]{\substack{PdCl_2(PPh_3)_2 \\ \text{or } PtCl_2(PPh_3)_2}} \quad (12.55)$$

$$R^1 \diagup \!\!\!\diagdown^{R^2} \!\!\! SnBu_3 + \overset{R^4}{\underset{R^3}{N}}\!\!\!\diagdown_H \xrightarrow[\text{THF}]{\substack{PdCl_2(PPh_3)_2 \\ \text{or } PtCl_2(PPh_3)_2}} \quad (12.56)$$

Catalytic processes that are likely to occur by electrophilic attack at the γ-position of an η^1-allyl ligand include the additions of allylstannanes (Equation 12.55 and 12.56)[73,79] and allyltrifluoroborates[80] to aldehydes and imines catalyzed by palladium complexes. These reactions are believed to occur by formation of an η^1-allyl complex, followed by attack of the electrophile at the γ-carbon. These reactions were first conducted with complexes of phosphine ligands or with simple π-allylpalladium chloride complexes.[73,79,81–85] Subsequently, complexes containing a non-racemic chiral allyl group as the ancillary ligand have been shown to undergo enantioselective additions of the accompanying allyl group to aldehydes and imines, as shown in Equation 12.57.[79,82,84,86]

$$\underset{R_1}{\overset{R_2}{N}}\!\!\!\diagdown_H + \diagup\!\!\!\diagdown SnBu_3 \xrightarrow[\text{1 equiv. } H_2O, \text{THF, 0 °C}]{\text{Catalytic}} \quad 50–91\ \% \quad (12.57)$$

Information on the mechanism of the reactions of electrophiles at η^1-allyl groups has been gained from the reactions of allylpalladium complexes containing pincer ligands. Considering the tridentate coordination of pincer ligands, the allylpalladium intermediate in the catalytic cycle would be expected to contain an η^1-allyl ligand. These pincer-ligated complexes are catalytically active, and this activity implies, in general, that the reactions of electrophiles with organometallic allyl reagents catalyzed by palladium occur through η^1-allyl complexes formed from the allylmetal nucleophile.[87–92] Computational work on these systems suggests that the reactions of η^1-allyl complexes occur at the γ-carbon through a closed, six-membered Zimmerman–Traxler transition state, such as the one shown in Equation 12.58.[93]

$$(PCP)Pd\diagup\!\!\!\diagdown \longrightarrow (PCP)Pd\diagdown_{O}\!\!=\!\!\diagup_R \longrightarrow \left[(PCP)Pd\diagdown_{O}\!\!=\!\!\diagup_R\right]^{\ddagger} \longrightarrow (PCP)Pd\diagdown_{O}\!\!\diagup_R \quad (12.58)$$

The reactions of η^3-allyl complexes with electrophiles are less common, but examples of the reactions of isolated η^3-allyl complexes with electrophiles, such as aldehydes, are known (Equations 12.59 and 12.60).[74–78] Although these reactions are initiated with η^3-allyl complexes, they are proposed to occur by coordination of the aldehyde to generate a nucleophilic η^1-allyl complex that adds to the aldehyde through a chair transition state. This mechanism is depicted in Equation 12.60.

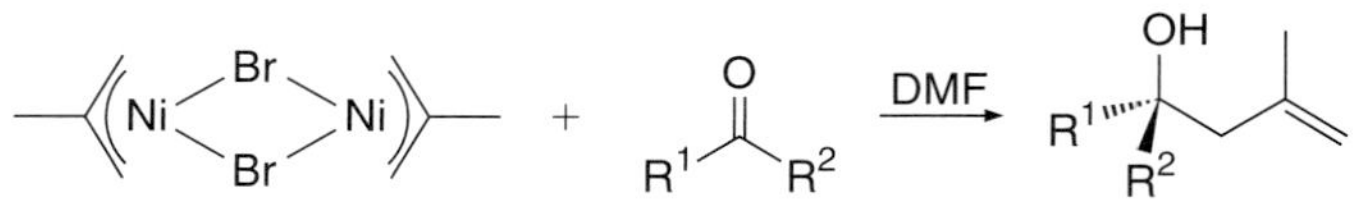

(12.59)

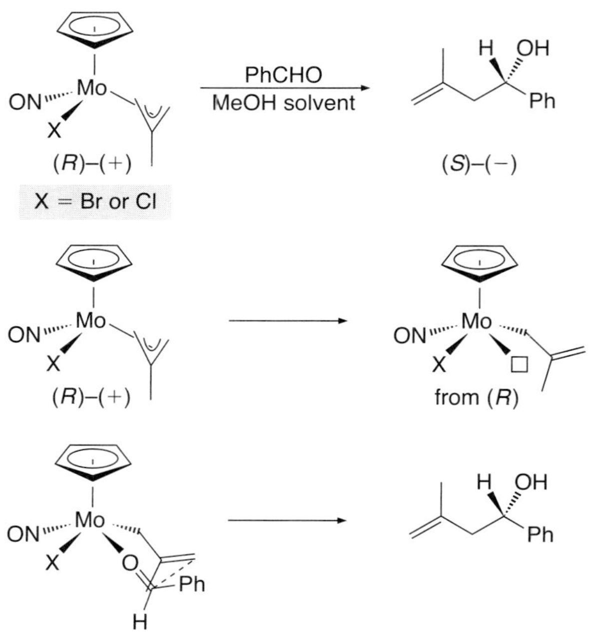

X = Br or Cl

(12.60)

Electrophilic attack also occurs at the γ-position of propargyl ligands, and this reaction converts propargyl complexes to allene complexes. An example of this protonation process is shown in Equation 12.61.[94,95] This reaction has been used to initiate cycloaddition reactions.[96] An example of this type of cycloaddition is the [3+2] reaction of p-toluenesulfonyl isocyanate with the propargyl complex in Equation 12.62.[97] Propargyl and η^1-allyl ligands are about equally reactive toward such cycloadditions.

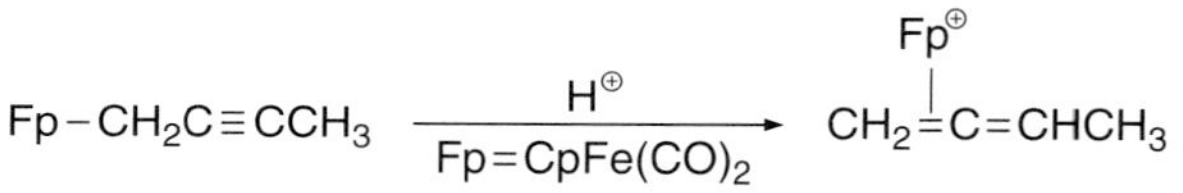

$$Fp-CH_2C\equiv CCH_3 \xrightarrow[Fp=CpFe(CO)_2]{H^\oplus} \overset{Fp^\oplus}{CH_2{=}C{=}CHCH_3}$$

(12.61)

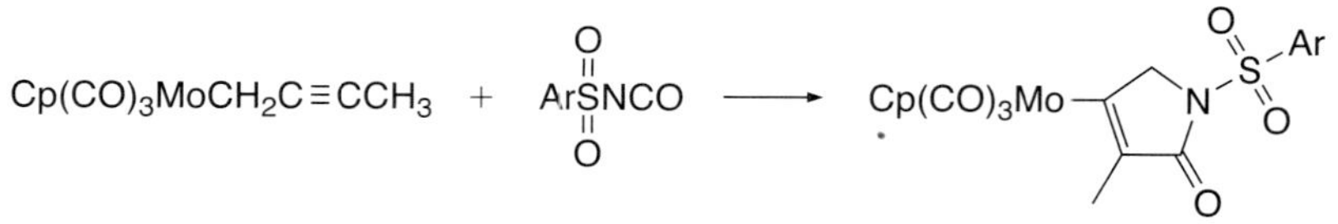

(12.62)

Abstraction of a hydride has been shown to occur by reaction of electrophiles at the γ-position of ligands containing a heteroatom bound to the metal. For example, abstraction of the hydrogen γ to the oxygen of the niobocene–ketene complex in Equation 12.63 by trityl cation generates a vinylacyl complex.[98,99]

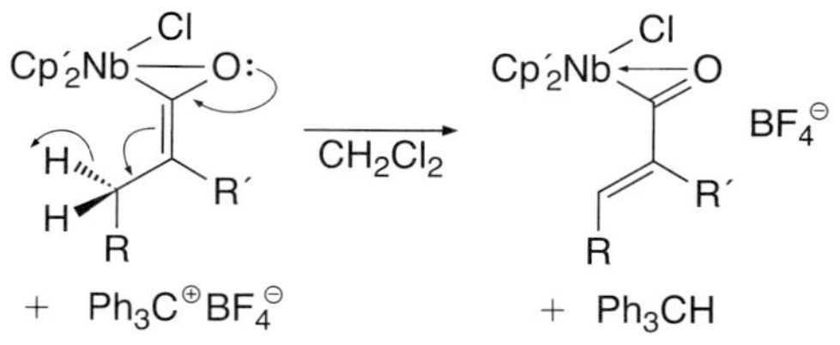

(12.63)

12.5. Attack on Coordinated Olefins and Polyenes

12.5.1. Attack of Carbonyl Compounds and Protons on Olefin Complexes

As discussed in Chapter 3, olefins and dienes bind to electron-poor metal centers by a flow of electrons from the olefin π-system to the metal and from the metal to the olefin π^*-system. Thus, olefins bound to electron-rich and strongly backbonding metal centers react with protons and electrophiles directly at the metal–carbon bond. However, olefins and dienes coordinated to electron-poor metal centers are less reactive toward electrophiles than those bound to electron-rich metal centers or even free olefins and dienes. However, electron-poor olefin and diene complexes do undergo reactions with electrophiles at the coordinated ligand by an indirect pathway. This indirect pathway occurs by insertion of the olefin or diene into the bond formed by attack of the electrophile at the metal.

The protonation of the η^4-cyclohexa-1,3-diene complex in Equation 12.64[100] illustrates a mechanism for the reaction of an electrophile at a coordinated diene via initial protonation of the metal and shows the stereochemistry that results from this pathway. The initially formed cationic hydride complex transfers the hydride onto the endo face of the cyclohexadiene ring by insertion of the C=C double bond into the Rh–H bond. The resulting Rh–C single bond is then part of the η^3-allyl ligand in the product of the migration. Migration of the endo proton at the other end of the η^3-allyl ligand to the metal regenerates a cyclohexadiene complex. As a result of these steps, exchange of the Rh–H with both of the original two endo hydrogens occurs, but exchange does not occur with the other hydrogens on the cyclohexadiene ligand.

$$(12.64)$$

Electrophilic attack on olefin ligands coordinated to electron-rich, strongly backbonding metals is illustrated by the reactions of d^2 group 4 olefin and alkyne complexes, as well as some electron-rich d^6 olefin complexes. Zirconocene– and and hafnocene–olefin complexes generated by reaction of zirconocene dichloride with two equivalents of alkyl lithium[101–103] and isolated upon addition of a phosphine ligand[104] react with carbonyl compounds and weak protic acids to form insertion products and alkyl complexes. Several examples of the reactions of these complexes with electrophiles are shown in Equations 12.65–12.66. Zirconocene–alkyne complexes prepared by thermolysis of vinyl alkyl complexes[104] and titanium–alkyne complexes generated by the reduction of Ti(OPr^i)$_4$ also react with electrophiles, such as aldehydes and acid, to form products from insertion into the M–C bond and protonation of the M–C bond respectively.

$$(12.65a)$$

$$(12.65b)$$

$$(12.65c)$$

$$ \text{(12.66)} $$

The reactions of benzyne complexes of zirconium[105–108] also occur by electrophilic attack at an M–C bond. The isolated phosphine adduct of a zironocene–benzyne complex reacts with ketones to undergo insertion into one of the M–C bonds and with alcohol to make an aryl alkoxo complex, as shown in Equation 12.67.[109] An electron-rich ruthenium–benzyne complex also reacts with electrophiles, such as benzaldehyde or carbon dioxide, to form products from insertion, as shown at the top of Equation 12.68. It also reacts with weak acids, such as aniline, to form products from formal protonation at the Ru–C bond, as shown at the bottom of Equation 12.68.[110, 111] This reaction with aniline could occur by initial protonation at the metal, followed by C–H bond-forming reductive elimination, or by direct protonation of the M–C bond. Initial protonation of the metal center was proposed.

$$ \text{(12.67)} $$

$$ \text{(12.68)} $$

12.5.2. Hydride Abstraction by Electrophilic Attack on Diene Complexes

Electrophilic attack by trityl cation on olefin complexes can also occur to abstract a hydride from the allylic position. This process converts a neutral olefin complex into a cationic allyl complex and has been conducted with complexes that are less strongly backbonding than the complexes described in the preceding two paragraphs. One example of such an allylic hydride abstraction from a rhenium–propene complex is shown in Equation 12.69.[112] An early example to form a homoconjugated system is shown in Equation 12.70.[113]

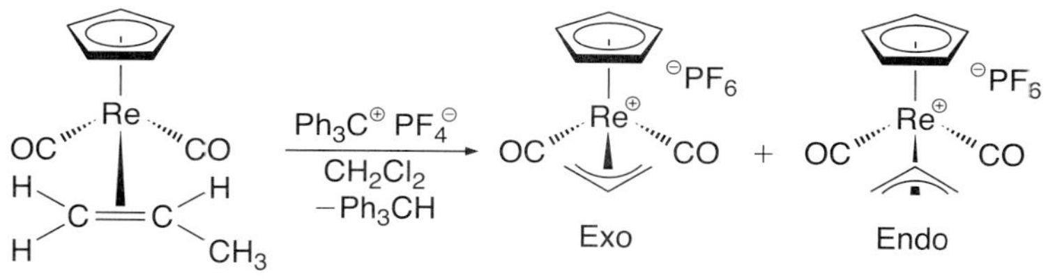

(12.69)

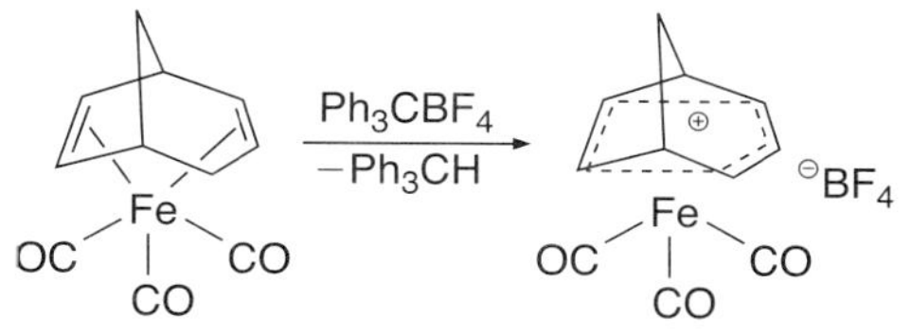

(12.70)

These reactions of electrophiles with alkene complexes to abstract a hydride from the saturated carbon adjoining the coordinated π-system can create a sequence for the functionalization of dienes that are derived from the Birch reduction of arenes. A commonly used[114] example of this reaction is the abstraction of a hydride from (cyclohexadiene)Fe(CO)$_3$ (Equation 12.71) by trityl cation as described by Fischer.[115] Such hydride abstractions have been shown to occur by initial electron transfer, followed by H• transfer in some cases.[68]

The combination of Birch reduction and electrophilic abstraction, followed by nucleophilic attack on the resulting cation, generates a modified coordinated diene complex, as shown in Equation 12.72.[114] Decomplexation and hydrolysis yields the enone product. Alternatively, additions of nucleophiles and electrophiles have been conducted in reverse, and one such sequence is shown in Equation 12.73.[116] In this case, nucleophilic attack on a coordinated polyene leads to a polyenyl system that is susceptible to hydride abstraction. Abstraction of the hydride restores the polyene system and allows for a second nucleophilic attack. Hydrolysis, lactonization of the half acid, and oxidative decomplexation gives the final free lactone. The product of this sequence contains the two nucleophiles cis to each other because they are delivered to the dienyl fragment on the side anti to the metal center.

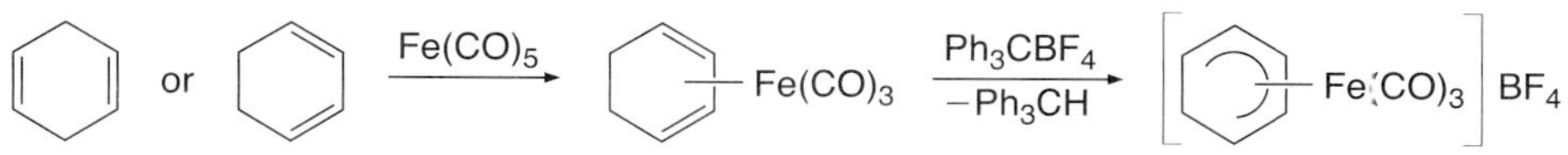

(12.71)

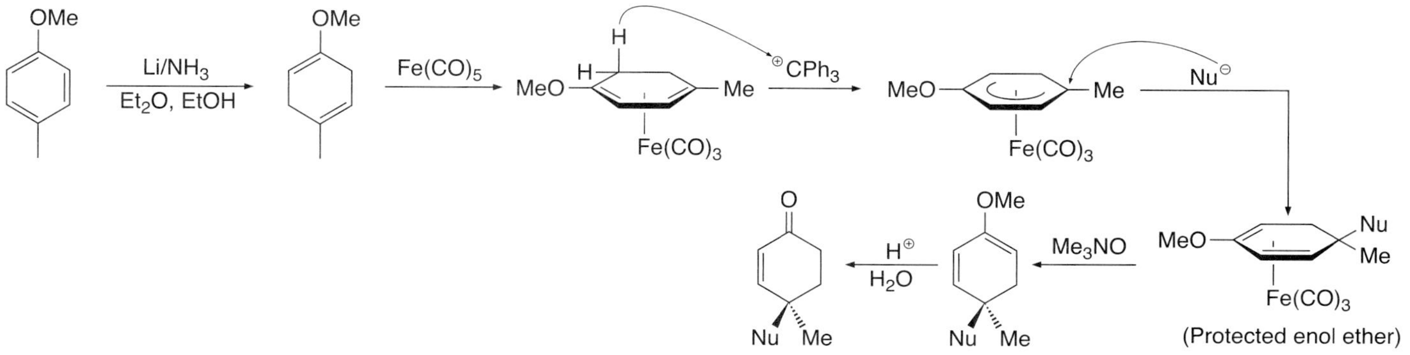

(12.72)

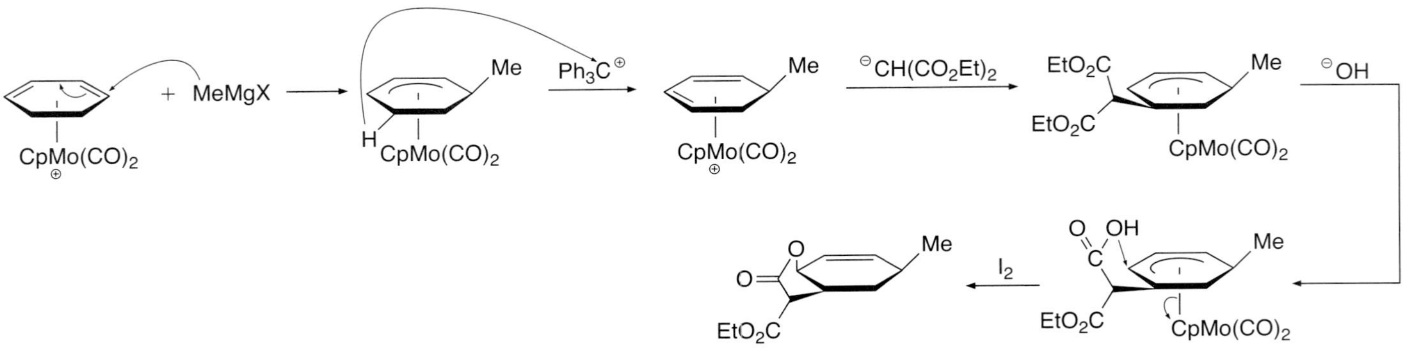

(12.73)

Direct electrophilic attack on the exo side of a coordinated diene is also possible, even to complexes of modestly electron-rich metal centers. For example, Friedel–Crafts acetylation of (cyclohexa-1,3-diene)Fe(CO)$_3$ gives principally the exo isomer.[117] However, such direct exo attack occurs more commonly on the uncoordinated portion of a partially coordinated polyene or polyenyl ligand (recall that uncoordinated olefins and dienes are generally more nucleophilic than coordinated ones). An example of such exo attack on the uncoordinated portion of a pentadienyl ligand is shown in Equation 12.74.[118] In this case, coordination of a CO ligand is thought to displace one or more carbons of the dienyl π-system from the iron center to trigger exo acylation. Similar results have been reported for electrophilic attack on the uncoordinated portion of the polyene system in $[(C_7H_7)Fe(CO)_3]^-$.[119]

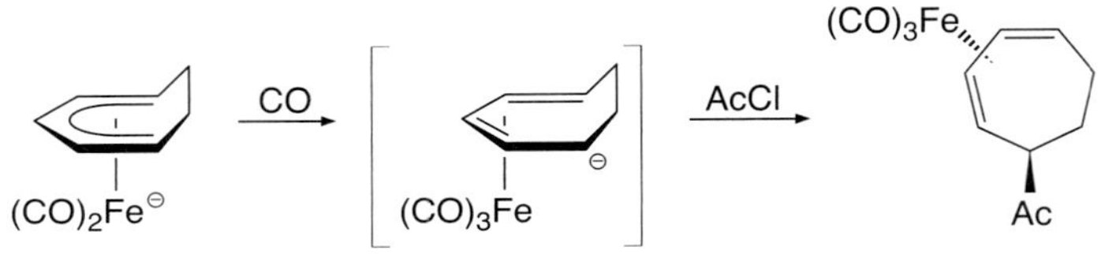

(12.74)

12.5.3. Electrophilic Attack on π-Polyenyl Complexes

Electrophilic attack on coordinated polyenyl ligands is uncommon, but electrophilic attack on the η^5-cyclopentadienyl ligands in certain complexes has become synthetically valuable. For example, the acetylation of ferrocene (Equation 12.75) is a reliable Friedel–Crafts process[120–123] and generates derivatives used to make many ferrocenyl ligands.[124–126] Similar electrophilic aromatic substitution reactions of haloboranes on ferrocene (shown in Equation 12.76) and on the anion generated from deprotonation of tungstenocene dihydride (Equation 12.77) leads to borylcyclopentadienyl complexes.[127–129]

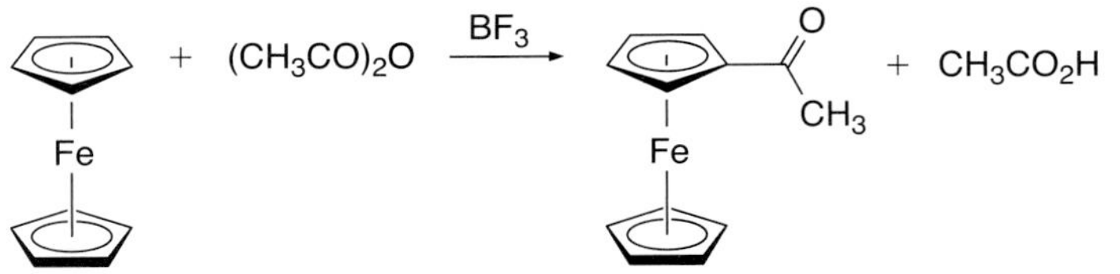

(12.75)

$$(12.76)$$

$$(12.77)$$

12.5.4. Electrophilic Attack on η^2-Arene and Heteroarene Complexes

As noted in Chapters 2 and 11, a series of η^2-arene complexes of osmium have been prepared, and the reactivity of these species has been studied extensively by Harman. The reactions of η^2-arene complexes of Os(II) illustrate how strong backbonding can cause the uncoordinated portion of an aromatic system to be more susceptible to electrophilic attack than the corresponding free arene.[130] Osmium(II) pentamine complexes of phenols,[131] anilines,[132] acetanilides,[132] and anisoles[133] react with electrophiles at the uncoordinated portion of the ring. For example, the simple phenol complex in Equation 12.78 reacts with Michael acceptors at the 4-position of the coordinated phenol in the presence of a mild tertiary amine base. This reactivity and selectivity for reaction at the 4-position is greater than the reactivity of free phenol. The reactions of electrophiles with aniline derivatives occur in a similar fashion and lead to products from alkylation of the aromatic ring predominantaly at the 4-position (Equation 12.79). Related reactions occur with η^2 complexes of electron-rich five-membered pyrrole and furan heterocycles. Examples of electrophilic attack on η^2-pyrrole complexes of Os(II) are shown in Equation 12.80.[132]

$$(12.78)$$

$$(12.79)$$

(12.80)

12.6. Summary

Many types of electrophilic attack on coordinated ligands are known, and these reactions have been used in diverse ways to prepare organometallic species. Electrophilic attack occurs on metal–alkyl complexes and hydride complexes to abstract these groups and form cationic metal complexes. Electrophilic abstraction of a hydride from the α position of an alkyl group forms carbene complexes, and abstraction from the β-position of an alkyl group forms olefin complexes. Electrophilic attack can also occur to abstract hydrides from acyl complexes. Electrophiles also add to the γ-position of allyl complexes, two classes of which occur as part of the catalytic allylation of aldehydes and imines. Attack of unsaturated electrophiles, such as CO_2 and SO_2, at the metal–carbon bond of an alkyl complex leads to overall insertion into the M–C bond. Related electrophilic attack can also occur on vinyl complexes to form carbene complexes. Electrophilic attack on carbene and carbyne complexes at the α-carbon is also known, particularly for early transition metal–carbene complexes. In processes that have been used in synthetic organic contexts, electrophiles commonly abstract hydride from diene complexes and add to dienyl complexes; they can also add to dienes and dienyl groups by initial attack at the metal and migration to the endo face of the π-ligand. In all cases, the electrophile adds to the complex without adding electrons to the valence shell, as one would expect for a reaction that occurs between electrons in the metal complex and an unoccupied orbital in the electrophilic reagent.

References and Notes

1. It can be difficult to distinguish between electrophilic attack by two-electron processes from electrophilic attack by one-electron oxidation. The rates of both, for example, depend similarly upon the energy of the HOMO. For a discussion of this issue, see reference 2.

2. Kochi, J. K. *Organometallic Mechanisms and Catalysis*; Academic Press: New York, 1978.

3. For a general review of electrophilic attack on organometallics, see reference 4.

4. Johnson, M. D. In *The Chemistry of the Metal–Carbon Bond*; Hartley, F. R., Patai, S., Eds.; Wiley: New York, 1985; Vol. 2, Chapter 7.

5. Flood, T. C. In *Topics in Inorganic and Organometallic Stereochemistry*; Geoffroy, G. L., Ed.; Wiley: New York, 1981; Vol. 12, p 37.

6. Komiya, S.; Ito, T.; Cowie, M.; Yamamoto, A.; Ibers, J. A. *J. Am. Chem. Soc.* **1976**, *98*, 3874.

7. Fryzuk, M. D.; Bosnich, B. *J. Am. Chem. Soc.* **1979**, *101*, 3043.

8. Rogers, W. N.; Baird, M. C. *J. Organomet. Chem.* **1979**, *182*, C65.

9. Beswick, C. L.; Marks, T. J. *J. Am. Chem. Soc.* **2000**, *122*, 10358.

10. Deck, P. A.; Beswick, C. L.; Marks, T. J. *J. Am. Chem. Soc.* **1998**, *120*, 1772.

11. Ellis, W. W.; Raebiger, J. W.; Curtis, C. J.; Bruno, J. W.; DuBois, D. L. *J. Am. Chem. Soc.* **2004**, *126*, 2738.

12. Sarker, N.; Bruno, J. W. *J. Am. Chem. Soc.* **1999**, *121*, 2174.

13. Sarker, N.; Bruno, J. W. *Organometallics* **2001**, *20*, 55.

14. Fagan, P. J.; Nugent, W. A. *J. Am. Chem. Soc.* **1988**, *110*, 2310.

15. Fagan, P. J.; Nugent, W. A.; Calabrese, J. C. *J. Am. Chem. Soc.* **1994**, *116*, 1880.

16. Sanderson, L. J.; Baird, M. C. *J. Organomet. Chem.* **1986**, *307*, C1.

17. Attig, T. G.; Wojcicki, A. *J. Organomet. Chem.* **1974**, *82*, 397.

18. Attig, T. G.; Teller, R. G.; Wu, S. M.; Bau, R.; Wojcicki, A. *J. Am. Chem. Soc.* **1979**, *101*, 619.

19. Flood, T. C.; Miles, D. L. *J. Organomet. Chem.* **1977**, *127*, 33.

20. Brunner, H.; Wallner, G. *Chem. Ber.* **1976**, *109*, 1053.

21. Brunner, H. *Adv. Organomet. Chem.* **1980**, *18*, 151.

22. Faller, J. W.; Anderson, A. S. *J. Am. Chem. Soc.* **1970**, *92*, 5852.

23. Stanley, K.; Baird, M. C. *J. Am. Chem. Soc.* **1975**, *97*, 6598.

24. Sloan, T. E. In *Topics in Inorganic and Organometallic Stereochemistry*; Geoffroy, G. L., Ed.; Wiley: New York, 1981; Vol. 12, p 1.

25. Dizikes, L. J.; Wojcicki, A. *J. Am. Chem. Soc.* **1977**, *99*, 5295.

26. Bock, P. L.; Whitesides, G. M. *J. Am. Chem. Soc.* **1974**, *96*, 2826.

27. Dong, D.; Slack, D. A.; Baird, M. C. *Inorg. Chem.* **1979**, *18*, 188.

28. Su, S. C. H.; Wojcicki, A. *Organometallics* **1983**, *2*, 1296.

29. Hofmann, P. *Angew. Chem., Int. Ed. Engl.* **1977**, *16*, 536.

30. Merrifield, J. H.; Fernandez, J. M.; Buhro, W. E.; Gladysz, J. A. *Inorg. Chem.* **1984**, *23*, 4022.

31. Bock, P. L.; Boschetto, D. J.; Rasmussen, J. R.; Demers, J. P.; Whitesides, G. M. *J. Am. Chem. Soc.* **1974**, *96*, 2814.

32. Fritz, H. L.; Espenson, J. H.; Williams, D. A.; Molander, G. A. *J. Am. Chem. Soc.* **1974**, *96*, 2378.

33. Shinozaki, H.; Ogawa, H.; Tada, M. *Bull. Chem. Soc. Jpn.* **1976**, *49*, 775.

34. Dong, D.; Baird, M. C. *J. Organomet. Chem.* **1979**, *172*, 467.

35. Deluca, N.; Wojcicki, A. *J. Organomet. Chem.* **1980**, *193*, 359.

36. Ros, R.; Michelin, R. A.; Bataillard, R.; Roulet, R. *J. Organomet. Chem.* **1978**, *161*, 75.

37. Romeo, R.; Minniti, D.; Lanza, S. *J. Organomet. Chem.* **1979**, *165*, C36.

38. Romeo, R.; D'Amico, G. *Organometallics* **2006**, *25*, 3435.

39. Chinn, M. S.; Heinekey, D. M. *J. Am. Chem. Soc.* **1990**, *112*, 5166.

40. Jia, G.; Morris, R. H. *J. Am. Chem. Soc.* **1991**, *113*, 875.

41. Feracin, S.; Bürgi, T.; Bakhmutov, V.; Eremenko, I.; Vorontsov, E. V.; Vimenitis, A B.; Berke, H. *Organometallics* **1994**, *13*, 4194.

42. Papish, E. T.; Rix, F. C.; Spetseris, N.; Norton, J. R.; Williams, R. D. *J. Am. Chem. Soc.* **2000**, *122*, 12235.

43. Labinger, J. A.; Hart, D. W.; Seibert, W. E.; Schwartz, J. *J. Am. Chem. Soc.* **1975**, *97*, 3851.

44. Ho, S. C. H.; Straus, D. A.; Grubbs, R. H. *J. Am. Chem. Soc.* **1984**, *106*, 1533.

45. Jacobson, S. E.; Reich-Rohrwig, P.; Wojcicki, A. *Inorg. Chem.* **1973**, *12*, 717.

46. Severson, R. G.; Wojcicki, A. *J. Am. Chem. Soc.* **1979**, *101*, 877.

47. Joseph, M. F.; Baird, M. C. *Inorg. Chim. Acta* **1985**, *96*, 229.

48. Dormond, A.; Dahchour, A.; Tirouflet, J. *J. Organomet. Chem.* **1981**, *216*, 49.

49. Dormond, A.; Moise, C.; Dahchour, A.; Leblanc, J. C.; Tirouflet, J. *J. Organomet. Chem.* **1979**, *177*, 191.

50. Dormond, A.; Moise, C.; Dahchour, A.; Tirouflet, J. *J. Organomet. Chem.* **1979**, *177*, 181.

51. Wojcicki, A. *J. Organomet. Chem.* **1974**, *12*, 31.

52. Cotton, J. D.; Crisp, G. T. *J. Organomet. Chem.* **1980**, *186*, 137.

53. Miles, S. L.; Miles, D. L.; Bau, R.; Flood, T. C. *J. Am. Chem. Soc.* **1978**, *100*, 7278.

54. (a) Severson, R. G.; Leung, T. W.; Wojcicki, A. *Inorg. Chem.* **1980**, *19*, 915; (b) Leung, T. W.; Christoph, G. G.; Gallucci, J.; Wojcicki, A. *Organometallics* **1986**, *5*, 846.

55. Hartwig, J. F.; Bergman, R. G.; Andersen, R. A. *J. Am. Chem. Soc.* **1991**, *113*, 6499.

56. Darensbourg, D. J.; Grotsch, G. *J. Am. Chem. Soc.* **1985**, *107*, 7473.

57. Brookhart, M.; Tucker, J. R.; Husk, G. R. *J. Am. Chem. Soc.* **1983**, *105*, 258.

58. Studabaker, W. B.; Brookhart, M. *J. Organomet. Chem.* **1986**, *310*, C39.

59. Guerchais, V.; Lapinte, C. *J. Chem. Soc., Chem. Commun.* **1986**, 663.

60. Hayes, J. C.; Pearson, G. D. N.; Cooper, N. J. *J. Am. Chem. Soc.* **1981**, *103*, 4648.

61. Hayes, J. C.; Pearson, G. D. N.; Cooper, N. J. *Abs. Pap. Am. Chem. Soc.* **1981**, *181*, 64.

62. Hayes, J. C.; Cooper, N. J. *J. Am. Chem. Soc.* **1982**, *104*, 5570.

63. Brookhart, M.; Tucker, J. R.; Husk, G. R. *J. Am. Chem. Soc.* **1981**, *103*, 979.

64. Guerchais, V.; Lapinte, C. *J. Chem. Soc., Chem. Commun.* **1986**, 663.

65. Kostic, N. M.; Fenske, R. F. *J. Am. Chem. Soc.* **1981**, *103*, 4677.

66. Holmes, S. J.; Clark, D. N.; Turner, H. W.; Schrock, R. R. *J. Am. Chem. Soc.* **1982**, *104*, 6322.

67. Laycock, D. E.; Hartgerink, J.; Baird, M. C. *J. Org. Chem.* **1980**, *45*, 291.

68. Mandon, D.; Toupet, L.; Astruc, D. *J. Am. Chem. Soc.* **1986**, *108*, 1320.

69. Smith, D. E.; Gladysz, J. A. *Organometallics* **1985**, *4*, 1480.

70. (a) Fischer, E. O.; Kreiter, C. G.; Kollmeier, H. J.; Müller, J.; Fischer, R. D. *J. Organomet. Chem.* **1971**, *28*, 237; (b) Semmelhack, M. F.; Tamura, R. *J. Am. Chem. Soc.* **1983**, *105*, 4099.

71. Mayr, A.; Schaefer, K. C.; Huang, E. Y. *J. Am. Chem. Soc.* **1984**, *106*, 1517.

72. Hegedus, L. S.; Wagner, S. D.; Waterman, E. L.; Siiralahansen, K. *J. Org. Chem.* **1975**, *40*, 593.

73. (a) Nakamura, H.; Asao, N.; Yamamoto, Y. *J. Chem. Soc., Chem. Commun.* **1995**, 1273; (b) Nakamura, H.; Iwama, H.; Yamamoto, Y. *J. Am. Chem. Soc.* **1996**, *118*, 6641.

74. Faller, J. W.; Nguyen, J. T.; Ellis, W.; Mazzieri, M. R. *Organometallics* **1993**, *12*, 1434.

75. Faller, J. W.; Diverdi, M. J.; John, J. A. *Tetrahedron Lett.* **1991**, *32*, 1271.

76. Collins, S.; Kuntz, B. A.; Hoong, Y. *J. Org. Chem.* **1989**, *54*, 4154.

77. Faller, J. W.; Linebarrier, D. L. *J. Am. Chem. Soc.* **1989**, *111*, 1937.

78. Sato, F.; Iijima, S.; Sato, M. *Tetrahedron Lett.* **1981**, *22*, 243.

79. Nakamura, K.; Nakamura, H.; Yamamoto, Y. *J. Org. Chem.* **1999**, *64*, 2614.

80. Solin, N.; Wallner, O. A.; Szabo, K. J. *Org. Lett.* **2005**, *7*, 689.

81. Nakamura, H.; Shim, J. G.; Yamamoto, Y. *J. Am. Chem. Soc.* **1997**, *119*, 8113.

82. Nakamura, H.; Nakamura, K.; Yamamoto, Y. *J. Am. Chem. Soc.* **1998**, *120*, 4242.

83. Nakamura, H.; Aoyagi, K.; Shim, J. G.; Yamamoto, Y. *J. Am. Chem. Soc.* **2001**, *123*, 372.

84. Fernandes, R. A.; Stimac, A.; Yamamoto, Y. *J. Am. Chem. Soc.* **2003**, *125*, 14133.

85. Kiji, J.; Okano, T.; Nomura, T.; Saiki, K.; Sai, T.; Tsuji, J. *Bull. Chem. Soc. Jpn.* **2001**, *74*, 1939.

86. Bao, M.; Nakamura, H.; Yamamoto, Y. *Tetrahedron Lett.* **2000**, *41*, 131.

87. Szabo, K. J. *Chem. Eur. J.* **2000**, *6*, 4413.

88. Solin, N.; Narayan, S.; Szabo, K. J. *Org. Lett.* **2001**, *3*, 909.

89. Wallner, O. A.; Szabo, K. J. *Org. Lett.* **2002**, *4*, 1563.

90. Wallner, O. A.; Szabo, K. J. *Chem. Eur. J.* **2003**, *9*, 4025.

91. Wallner, O. A.; Szabo, K. J. *J. Org. Chem.* **2003**, *68*, 2934.

92. Szabo, K. J. *Chem. Eur. J.* **2004**, *10*, 5269.

93. Solin, N.; Kjellgren, J.; Szabo, K. J. *J. Am. Chem. Soc.* **2004**, *126*, 7026.

94. Benaim, J.; Merour, J. Y.; Roustan, J. L. *C. R. Hebd. Seances Acad. Sci.* **1971**, *272*, 789.

95. Raghu, S.; Rosenblum, M. *J. Am. Chem. Soc.* **1973**, *95*, 3060.

96. Bucheister, A.; Klemarczyk, P.; Rosenblum, M. *Organometallics* **1982**, *1*, 1679.

97. Bell, P. B.; Wojcicki, A. *Inorg. Chem.* **1981**, *20*, 1585.

98. Hneihen, A. S.; Fermin, M. C.; Maas, J. J.; Bruno, J. W. *J. Organomet. Chem.* **1992**, *429*, C33.

99. Kerr, M. E.; Sarker, N.; Hneihen, A. S.; Schulte, G. K.; Bruno, J. W. *Organometallics* **2000**, *19*, 901.

100. Johnson, B. F. G.; Lewis, J.; Yarrow, D. J. *J. Chem. Soc., Dalton Trans.* **1972**, 2084.

101. Negishi, E.; Cederbaum, F. E.; Takahashi, T. *Tetrahedron Lett.* **1986**, *27*, 2829.

102. Takahashi, T.; Swanson, D. R.; Negishi, E. *Chem. Lett.* **1987**, 623.

103. Negishi, E.; Holmes, S. J.; Tour, J. M.; Miller, J. A.; Cederbaum, F. E.; Swanson, D. R.; Takahashi, T. *J. Am. Chem. Soc.* **1989**, *111*, 3336.

104. Buchwald, S. L.; Watson, B. T.; Huffman, J. C. *J. Am. Chem. Soc.* **1987**, *109*, 2544.

105. Erker, G. *J. Organomet. Chem.* **1977**, *134*, 189.

106. Erker, G.; Kropp, K. *J. Am. Chem. Soc.* **1979**, *101*, 3659.

107. Kropp, K.; Erker, G. *Organometallics* **1982**, *1*, 1246.

108. Buchwald, S. L.; Nielsen, R. B. *Chem. Rev.* **1988**, *88*, 1047.

109. Buchwald, S. L.; Watson, B. T.; Huffman, J. C. *J. Am. Chem. Soc.* **1986**, *108*, 7411.

110. Hartwig, J. F.; Andersen, R. A.; Bergman, R. G. *J. Am. Chem. Soc.* **1989**, *111*, 2717.

111. Hartwig, J. F.; Bergman, R. G.; Andersen, R. A. *J. Am. Chem. Soc.* **1991**, *113*, 3404.

112. Casey, C. P.; Yi, C. S. *Organometallics* **1990**, *9*, 2413.

113. Margulis, T. N.; Schiff, L.; Rosenblum, M. *J. Am. Chem. Soc.* **1965**, *87*, 3269.

114. Pearson, A. J. *Acc. Chem. Res.* **1980**, *13*, 463.

115. Fischer, E. O.; Fischer, R. D. *Angew. Chem., Int. Ed. Engl.* **1960**, *72*, 919.

116. (a) Pearson, A. J.; Khan, M. N. I. *J. Am. Chem. Soc.* **1984**, *106*, 1872; (b) Pearson, A. J.; Khan, M. N. I. *Tetrahedron Lett.* **1984**, *25*, 3507; (c) Pearson, A. J.; Khan, M. N. I.; Clardy, J. C.; Cin-Heng, H. *J. Am. Chem. Soc.* **1985**, *107*, 2748.

117. Johnson, B. F. G.; Lewis, J.; Parker, D. G. *J. Organomet. Chem.* **1977**, *141*, 319.

118. Williams, G. M.; Rudisill, D. E. *J. Am. Chem. Soc.* **1985**, *107*, 3357.

119. Moll, M.; Wurstl, P.; Behrens, H.; Merbach, P. *Z. Naturforsch., B: Chem. Sci.* **1978**, *33*, 1304.

120. Pauson, P. L. *Q. Rev.* **1955**, *9*, 391.

121. Weinmayr, V. *J. Am. Chem. Soc.* **1955**, *77*, 3009.

122. Nesmeyanov, A. N.; Perevalova, E. G.; Golovnga, R. V.; Nesmeyanova, A. *Dok. Akad. Nauk* **1954**, *97*, 459.

123. Perevalova, E. G.; Nikitina, T. V. In *Organometallic Reactions*; Becker, E. I., Tsursui, M., Eds.; Wiley-Interscience: New York, 1972; Vol. 4, p 163.

124. Dai, L. X.; Tu, T.; You, S. L.; Deng, W. P.; Hou, X. L. *Acc. Chem. Res.* **2003**, *36*, 659.

125. Guiry, P. J.; Saunders, C. P. *Adv. Synth. Catal.* **2004**, *346*, 497.

126. Bueno, A.; Moreno, R. M.; Moyano, A. *Tetrahedron: Asymmetry* **2005**, *16*, 1763.

127. Hartwig, J. F.; He, X. *Organometallics* **1996**, *15*, 5350.

128. Ruf, W.; Fueller, M.; Siebert, W. *J. Organomet. Chem.* **1974**, *64*, C45.

129. Braunschweig, H.; Wagner, T. *Chem. Ber.* **1994**, *127*, 1613.

130. Harman, W. D. *Chem. Rev.* **1997**, *97*, 1953.

131. Kopach, M. E.; Harman, W. D. *J. Am. Chem. Soc.* **1994**, *116*, 6581.

132. Kolis, S. P.; Gonzalez, J.; Bright, L. M.; Harman, W. D. *Organometallics* **1996**, *15*, 245.

133. Kopach, M. E.; Harman, W. D. *J. Org. Chem.* **1994**, *59*, 6506.

134. Myers, W. H.; Koontz, J. F.; Harman, W. D. *J. Am. Chem. Soc.* **1992**, *114*, 5684.

CHAPTER 13

Metal–Ligand Multiple Bonds

13.1. Introduction to Metal–Ligand Multiple Bonds

Thus far, this text has focused on the synthesis, properties, and reactions of transition metal organometallic compounds containing metal–carbon single bonds or bonds between a transition metal and the π-system of an unsaturated organic compound. This chapter presents the synthesis, properties, and reactions of organometallic complexes containing multiple bonds between a transition metal and carbon, nitrogen, or oxygen. This type of metal–ligand interaction was introduced in Chapter 2. Complexes containing a metal–carbon multiple bond to a CR_2 unit are called carbene complexes, and those containing a metal–carbon multiple bond to a CR unit are called carbyne complexes. Related complexes possessing metal–nitrogen multiple bonds to an NR (nitrene) unit are usually called imido complexes. Complexes possessing metal–oxygen multiple bonds to an oxygen atom are called oxo complexes, and those containing multiple bonds to nitrogen or carbon atoms are typically called nitride and carbide complexes. This chapter focuses on the chemistry of carbene, imido, and oxo complexes, along with some additional information on carbyne and nitrido complexes. The chemistry of carbene complexes only dates back to the 1960s, and the use of carbenes as both reactive and dative ligands in catalytic transformations has seen rapid development in the past decade. The organic transformations of metal–oxo and –imido complexes have also been studied intensively in recent years. As shown in Figure 13.1, a majority of the complexes containing metal–ligand multiple bonds that are stable enough to isolate and characterize crystallographically contain metals in the middle of the transition series.

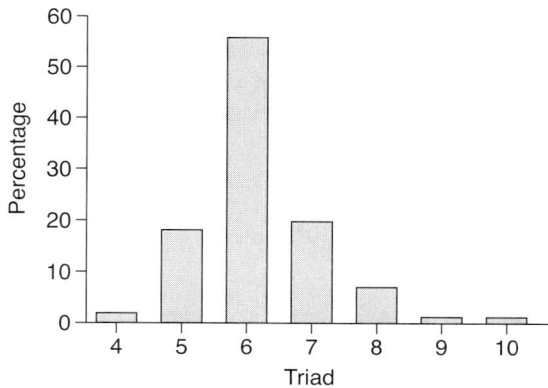

Figure 13.1.
Occurrence of complexes with metal–ligand multiple bonds by triad. Data from the Cambridge Structural Database in 2000. Adapted with permission from Cundari, T. R. *Chem. Rev.* **2000**, *100*, 807.

13.2. Carbene Complexes

13.2.1. Classes of Carbene Complexes

As stated in Chapter 3, carbene complexes can be divided into the five classes illustrated in Figure 13.2. One class of carbene complex encompasses the "Fischer carbenes" that were first prepared in the laboratory of E. O. Fischer.[2] These complexes were the first transition metal carbene complexes prepared, and they contain a π-donating group on the carbene carbon. Complexes of these carbenes are typically electrophilic at the carbene carbon. A second class of carbene complex was first prepared by Richard Schrock.[3] These complexes contain alkyl groups or hydrogens on the carbene carbon and are called alkylidene complexes or often "Schrock carbenes." Complexes of these carbenes are typically electrophilic at the metal and nucleophilic at the carbene carbon.

Acceptor-substituted carbene complexes or "carbenoids" encompass a third class of carbene complex that are rarely isolated but are important intermediates in catalytic transformations. These complexes consist of a carbene containing a carbonyl substituent, such as a carboxyl or acyl group, bound to the carbene carbon. These carbenoid ligands are commonly generated on dinuclear rhodium frameworks, ruthenium centers containing macrocyclic ligands, and copper bound by nitrogen donor ligands.[4] These complexes are intermediates in the cyclopropanations of olefins[5,6] and the functionalization of C–H bonds by insertion of the carbene unit into the C–H bond.[7] A fourth class of carbene complex comprises vinylidenes. Vinylidenes are bound to the metal by a heterocumulene structure in which the metal is attached at one end of the two cumulated double bonds. These complexes are typically electrophilic at the carbene carbon.[8] The fifth class of carbene complex comprises carbenes that are typically unreactive themselves, but serve as ancillary ligands to control the properties and reactivity of other ligands.[9] These carbenes in their free form were originally studied by Wanzlick and usually contain two nitrogen atoms bound to the carbene carbon. Related stable carbenes containing one amino group and one alkyl or aryl group and lacking the cyclic structure have also been reported.[10] The most commonly used version of these carbenes in organometallic chemistry contain two nitrogens bound to the metal and a cyclic structure and are often called N-heterocyclic carbenes or NHCs.[11]

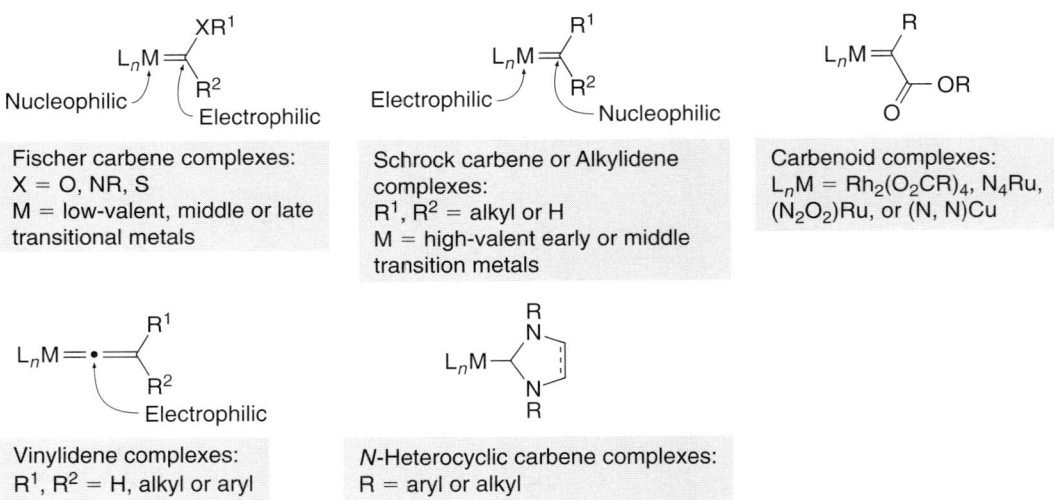

Figure 13.2. Five classes of carbene complexes.

13.2.2. Origin of the Electronic Properties of Fischer and Schrock Carbenes

The diagrams in Figure 13.3 depict qualitatively the orbital interactions that are involved in the bonding of Fischer and Schrock carbene complexes. This information can be used to rationalize the difference in electronic properties between the Fischer-type and the Schrock-type carbene complexes.[4] This molecular orbital diagram indicates that the LUMO of a Fischer carbene contains more ligand than metal character because it is closer in energy to the carbene orbital, while the LUMO of a Schrock carbene contains more metal than ligand character because it is closer in energy to the metal orbital. These orbital properties make the Fischer carbene complexes electrophilic at the ligand, and the Schrock carbene complexes electrophilic at the metal. For similar reasons, the HOMO of a Fischer-type carbene complex contains more metal than ligand character and the HOMO of a Schrock-type carbene complex contains more ligand than metal character. As a result of the character of these orbitals, the Fischer-type carbene complexes are nucleophilic at the metal, while the Schrock-type carbene complexes are nucleophilic at the ligand.

Two differences in the molecular orbital diagrams give rise to these different properties of the HOMO and LUMO. First, the energies of the filled metal d-orbitals of a low valent, middle or late transition metal tend to be lower than the energies of the orbitals of a high valent, early transition metal. Moreover, the difference in energies of the HOMO and LUMO of the heteroatom-substituted carbene in a Fischer carbene complex, which is a singlet carbene in the free state, is larger than that of the the alkylidene unit in a Schrock carbene complex, which is a triplet in the free state. Thus, the electron density in the σ-bond of a Fischer carbene complex is polarized toward the carbon, and the electron density in the π-interaction is primarily localized on the metal. In contrast, the electron density in both the σ- and π-bonds of Schrock carbene complexes is polarized toward the carbene carbon. The MO diagram in Figure 13.3 illustrates how variations in the orbital energies of the metal (early, high valent vs. late, low valent) and carbene units (triplet vs. singlet corresponding to small and large HOMO-LUMO gaps) lead to changes in bond polarity and to nucleophilic or electrophilic properties of the metal center and carbene carbon.

This analysis extends to vinylidene complexes. When free and uncoordinated from the metal center, the singlet state of vinylidene is more stable than the triplet state by 42 kcal/mol. (Of course, either form of the parent vinylidene [:CCH₂] is much less stable than its acetylene tautomer. The difference in energy between acetylene and the singlet vinylidene is estimated to be 45 kcal/mol.[8]) The oxidation state of the metal center in most

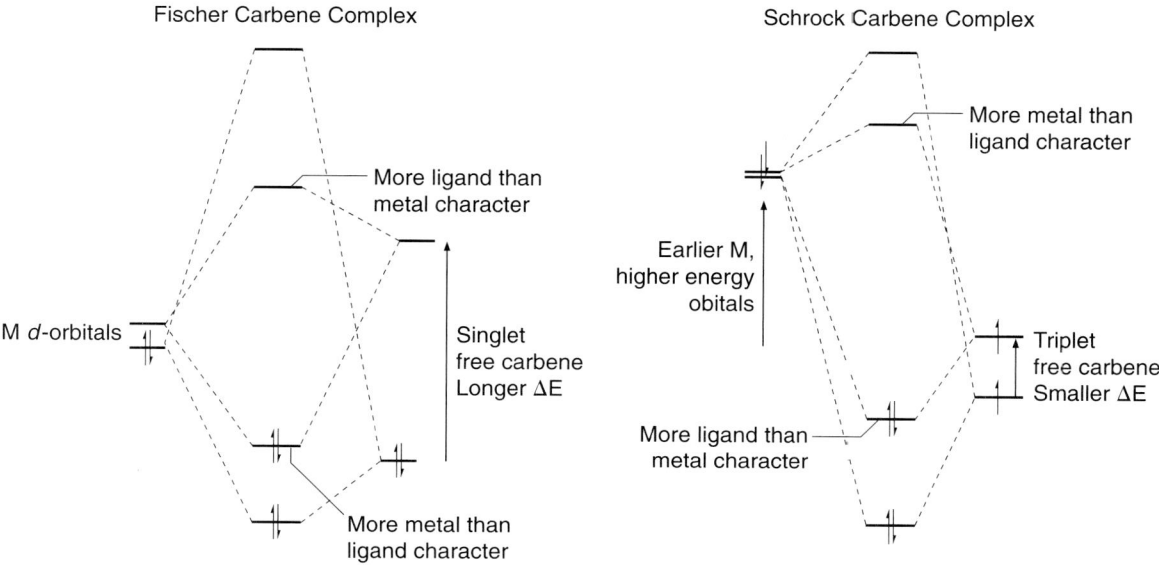

Figure 13.3. Molecular orbital diagram for Fischer and Schrock carbene complexes.

vinylidene complexes is low. This low oxidation state and the singlet spin state of the free vinylidene parallel the electronic properties of Fischer carbene complexes and cause the α-carbon of a vinylidene complex to be electrophilic.

13.2.3. Synthesis of Carbene Complexes

13.2.3.1. Synthesis of Fischer Carbene Complexes

The original synthesis of Fischer carbenes was presented in Chapter 11, the chapter on nucleophilic addition to coordinated ligands, and is summarized again in Equation 13.1.[2] Fischer carbene complexes are generally prepared by addition of a strong nucleophile, such as an alkyllithium reagent, Grignard reagent, or alkali metal amide, to coordinated carbon monoxide. This reaction generates an anionic acyl or carbamoyl ligand that has resonance structures analogous to those of organic enolates. Trapping of this intermediate by an alkyl electrophile, such as trimethyl oxonium salts or methyl triflate, generates a Fischer carbene complex possessing an alkoxy substituent at the carbene carbon. Many variations of this synthetic sequence have been reported, and further manipulation of the substituents at the carbene carbon have been used to generate complexes containing a variety of substitution patterns at the carbene carbon.

$$W(CO)_6 + MeLi \longrightarrow Li\left[(CO)_5W - \overset{\displaystyle O}{\underset{\displaystyle Me}{<}} \quad \longleftrightarrow \quad (CO)_5W = \overset{\displaystyle O^{\ominus}}{\underset{\displaystyle Me}{<}}\right] \xrightarrow{Me_3O^{\oplus} BF_4^{\ominus}} (CO)_5W = \overset{\displaystyle OMe}{\underset{\displaystyle Me}{<}} \qquad (13.1)$$

In addition to the classic synthesis of this type of heteroatom-substituted carbene complex by Fischer, a method to prepare these complexes from metal dianions has been reported and is shown in Equation 13.2.[12,13] Reaction of metal hexacarbonyl complexes with C_8K generates the dianionic metal pentacarbonyl fragment. Reaction of this dianion with an acid chloride generates the same anionic intermediate as is generated from attack of organolithium reagents on carbonyl complexes. Thus, the product from reaction of the hexacarbonyl with C_8K can be trapped with a chloroformate and then with an alkyl electrophile to generate alkoxy-substituted Fischer carbene complexes.[12] Alternatively, reaction of this dianion with an amide and abstraction of the oxygen from the tetrahedral intermediate with trimethylsilyl chloride generates an amino-substituted carbene.[13]

$$W(CO)_6 + C_8K \longrightarrow K_2[W(CO)_5]$$

with branches:
- Upper: $\overset{\displaystyle O}{\underset{\displaystyle}{Cl - C - OR^1}} \xrightarrow{R^2X} (CO)_5W = \overset{\displaystyle OR^2}{\underset{\displaystyle R^1}{<}}$
- Lower: $\overset{\displaystyle O}{\underset{\displaystyle}{R^1 - C - NR^2_2}} \longrightarrow (CO)_5W \overset{O^{\ominus}}{\underset{R^1}{+}} NR^2_2 \xrightarrow{Me_3SiCl} (CO)_5W = \overset{\displaystyle NR^2_2}{\underset{\displaystyle R^1}{<}}$

$$(13.2)$$

Amino-substituted Fischer carbene complexes can also be prepared from isocyanide complexes. This synthesis is less common than the synthesis of alkoxy-substituted carbenes because there are fewer appropriate isocyanide complexes to use as reactants than there are carbonyl complexes, and isocyanide ligands are less electrophilic. Thus, this reaction sequence has often been conducted intramolecularly to promote attack on the isocyanide ligand instead of an ancillary carbonyl group. An intramolecular example of the synthesis of an amino-substituted carbene by attack of an isocyanide ligand is shown in Equation 13.3.[14]

$$(PPh_3)(CO)_4W - C\equiv N \overset{HO}{\underset{}{\bigcirc}} \xrightarrow[\text{2. MeI}]{\text{1. KOBu}^t} (PPh_3)(CO)_4W = \overset{O}{\underset{\underset{Me}{N}}{\bigcirc}} \qquad (13.3)$$

One could imagine generating carbene complexes from typical precursors to carbenes in organic chemistry, such as diazoalkanes. Fewer carbene complexes have been isolated by addition of diazoalkanes to transition metal complexes than by the more indirect methods presented previously, but many carbene or carbenoid complexes have been generated by this method as reactive intermediates. Two examples of carbene complexes that have been isolated from the reaction of a diazoalkane are shown in Equations 13.4a and 13.4b.

$$\text{(13.4a)}$$

$$\text{(13.4b)}$$

Reaction of $CpMn(CO)_2(THF)$ with diphenyldiazomethane and copper powder generated the stable manganese diphenyl carbene complex in Equation 13.4a. This reaction was one of the first uses of diazoalkanes to generate metal–carbene complexes.[15a] The related tungsten pentacarbonyl fragment ligated by THF also reacts with diphenyldiazomethane to generate a diphenylcarbene complex.[15b–d] Reaction of $(Ph_3P)_3RuCl_2$ with benzylidene diazomethane generated the ruthenium benzylidene complex in Equation 13.4b.[16] This complex is a precursor to many carbene complexes studied for olefin metathesis.

More often, the diazoalkanes generate highly reactive carbene complexes. For example, reactions of alkyl diazoacetates catalyzed by dirhodium tetraacetate and related compounds have been proposed to generate a rhodium–carbene complex, as shown generically in Equation 13.5, in which the carbene is bound at the end of the dinuclear fragment. A model of such a compound formed by reaction of an *N*-heterocyclic carbene has been isolated.[17] The rhodium–carbene complexes from diazoacetates have not been observed directly, but computational work has been conducted to outline their structures and reactivity.[17–21] Similarly, reaction of alkyl diazoacetates with the Cu(I) fragment bound by a dative nitrogen-donor ligand has been proposed to generate a copper–carbene intermediate. In a few cases discrete copper–carbene complexes have been isolated from reactions of diazoalkanes.[22] Prior to this work, computational studies[23,24] were used to understand the reactivity of these copper–carbenoid complexes.

$$\text{(13.5)}$$

A series of benzylidene and low-valent alkylidene complexes have also been prepared from sulfur ylides.[25,26] Deprotonation of benzyl or alkyl diphenylsulfonium salts with a strong base, such as $KN(SiMe_3)_2$, $KNPr^i_2$ or $LiNPr^i_2$ generates the corresponding sulfur ylide. Reaction of this ylide with low-valent ruthenium, osmium, rhodium, or iridium complexes containing a labile ligand leads to transfer of the carbene unit from sulfur to the metal. Examples of this sequence to form rhodium and ruthenium benzylidene and alkylidene complexes are shown in Scheme 13.1.

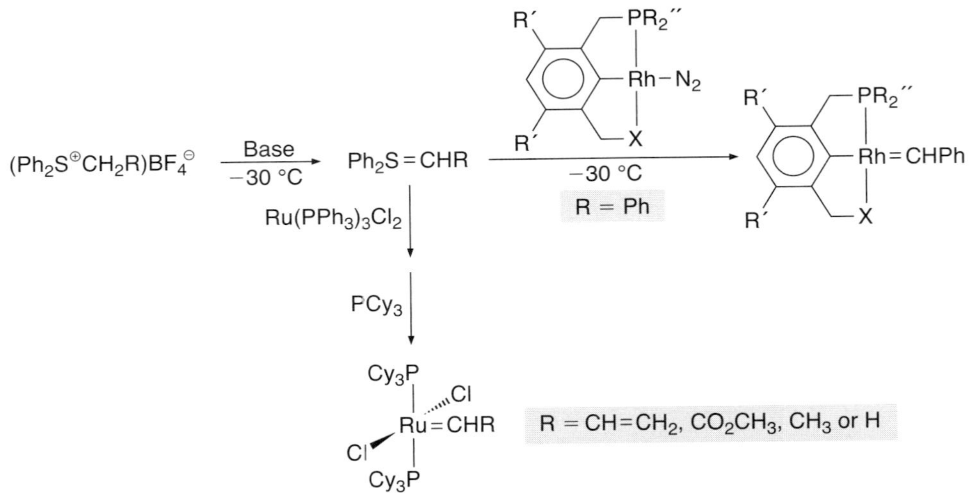

Scheme 13.1

13.2.3.2. Synthesis of Vinylidene Complexes

Several hundred examples of vinylidene complexes have been prepared.[8] Vinylidene complexes have been prepared by rearrangement of alkyne complexes, additions of acid or base to acetylide complexes, by deprotonation of carbyne complexes, by dehydration of acyl complexes, and by α-hydrogen shifts from vinyl complexes. Syntheses from alkyne and from acetylide complexes are most common. A complex of a terminal alkyne and a transition metal can exist as an alkyne complex or as a vinylidene complex. Although the free vinylidene is much higher in energy than the free alkyne, the vinylidene complex is often more stable than the alkyne complex. Vinylidene complexes are most often obtained with late transition metals because this tautomer possesses less repulsion between the filled d-orbitals of the metal and the filled π-orbitals of the ligand.

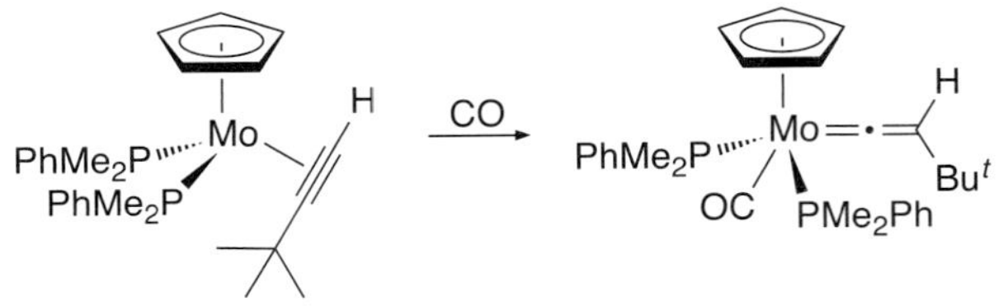

(13.6)

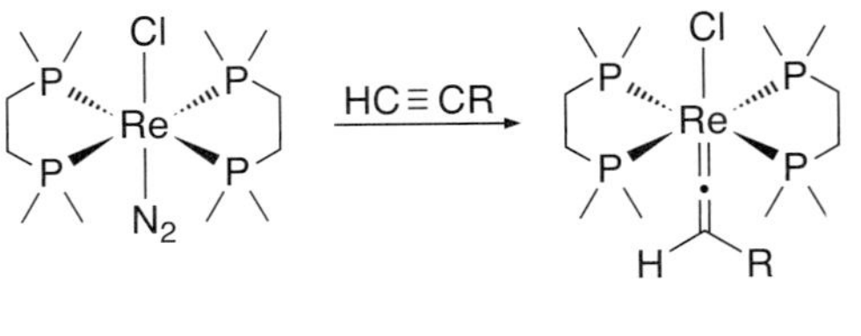

(13.7)

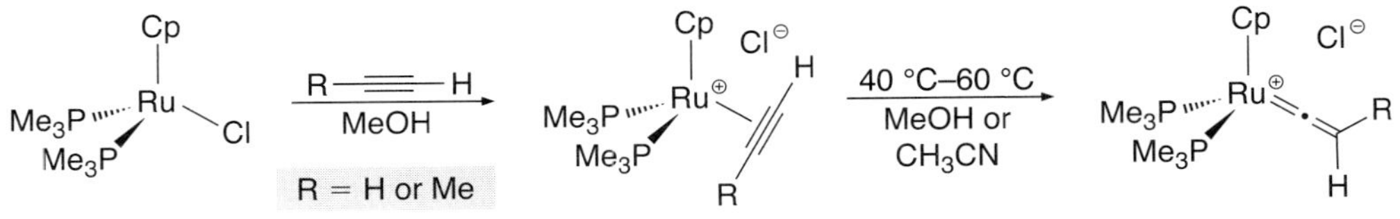

(13.8)

$$\text{(13.9)}$$

Equations 13.6–13.9 show four examples of the synthesis of vinylidene complexes from alkyne complexes. The reaction of the 16-electron molybdenum–alkyne complex in Equation 13.6 with carbon monoxide leads to the rearrangement of the alkyne complex to the vinylidene isomer. In this case, the alkyne complex is more stable than the vinylidene tautomer in the starting material because the alkyne can serve as a four-electron donor to create an 18-electron complex. After coordination of CO, the alkyne would act as a two-electron donor, and the vinylidene tautomer is more stable on this relatively electron-rich, low-valent metal fragment.[27] The rhenium–dinitrogen complex in Equation 13.7 dissociates dinitrogen, and the resulting fragment reacts with a terminal acetylene to generate the rhenium vinylidene.[28, 29] The ruthenium–chloride in Equation 13.8 reacts with a terminal acetylene in a polar protic solvent, such as methanol, to generate the corresponding ruthenium–alkyne complex.[30] Warming of this alkyne complex in methanol or acetonitrile leads to rearrangement of the coordinated alkyne to a vinylidene unit. The rhodium–alkyne complex in Equation 13.9 undergoes a similar rearrangement upon warming to generate the rhodium vinylidene complex.[31, 32] In all cases, if a vinylidene is considered a neutral ligand, these vinylidene complexes contain transition metals in relatively low oxidation states.

$$\text{(13.10)}$$

$$\text{(13.11)}$$

$$\text{(13.12)}$$

Equations 13.10–13.12 show three examples of the synthesis of vinylidene complexes by reactions of metal–acetylide complexes with acid or base. The molybdenum(II) acetylide complex in Equation 13.10 reacts with acid to protonate the β-carbon and generate a cationic vinylidene complex.[27] In this case, the vinylidene complex is thermodynamically unstable. Warming to 0 °C leads to rearrangement of this species to the tautomeric alkyne complex. In contrast, the more electron-rich molybdenum–acetylide complex in Equation 13.11 containing three phosphite donors generates a vinylidene complex upon addition of a proton from alumina to the β-carbon of the acetylide.[27] The vinylidene form of the complex is apparently more stable than the alkyne complex in this case.

Anionic vinylidene complexes can also be prepared by deprotonation of acetylide complexes because acetylides are slightly acidic at the carbon γ to the metal. For example, the reaction of the the molybdenum–carbyne complex in Equation 13.12 with butyllithium generates an anionic vinylidene complex.[33]

13.2.3.3. Synthesis of Some Classic Alkylidene Complexes

13.2.3.3.1.   Synthesis of the First Schrock Carbene Complexes

Schrock carbenes are typically generated by α-hydrogen elimination from a high-valent, early metal, dialkyl complex concomitant with or followed by elimination of alkane. The first synthesis of an alkylidene by Schrock is summarized in Equation 13.13. Addition of neopentyl zinc to tantalum pentachloride generated tris(neopentyl)dichlorotantalum. Subsequent addition of neopentyllithium did not generate the homoleptic neopentyltantalum compound. Instead, this reaction generated the first alkylidene complex, tris(neopentyl) tantalum neopentylidene.[3,34]

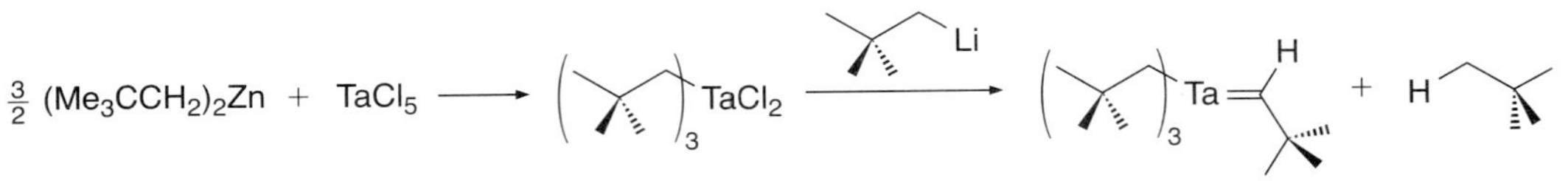

$$\tfrac{3}{2}\ (Me_3CCH_2)_2Zn\ +\ TaCl_5\ \longrightarrow\ \left(\!\!\begin{array}{c}\end{array}\!\!\right)_3 TaCl_2\ \longrightarrow\ \left(\!\!\begin{array}{c}\end{array}\!\!\right)_3 Ta\!=\!\!\begin{array}{c}H\\\end{array}\ +\ H\!\!\begin{array}{c}\end{array}\quad (13.13)$$

Addition of other nucleophiles or Lewis bases to neopentyltantalum complexes also generated alkylidene complexes, as shown in Scheme 13.2. Addition of slightly less than one molar equivalent of bis(neopentyl)zinc to tantalum pentachloride generated bis(neopentyl) tantalum trichloride,[34] and addition of thalium cyclopentadienide to this intermediate generated the Cp-ligated tantalum alkylidene.[35] Addition of dative ligands, such as THF, and monodentate or bidentate phosphines to the bis(neopentyl)tantalum complex generated the octahedral alkylidenes in Scheme 13.2.[36]

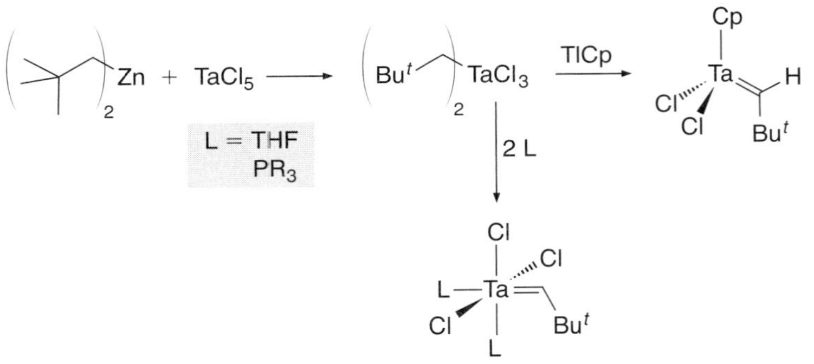

Scheme 13.2

13.2.3.3.2. Synthesis of the Schrock Alkylidene Catalysts

The synthesis of two additional alkylidene complexes by Schrock are shown in Schemes 13.3 and 13.4.[37] These complexes are some of the most active, well-defined catalysts for olefin metathesis. The synthesis of the tungsten complex begins with tris(methoxy)trichlorotungsten(VI). Reaction of this complex with six equivalents of neopentylmagnesium chloride generates the tungsten neopentylidyne complex containing three ancillary neopentyl groups. Addition of hydrochloric acid and dimethoxyethane generates the octahedral neopentylidyne complex, which forms the anilide–alkylidyne complex upon reaction with trimethylsilylaniline. Proton transfer from nitrogen to the carbyne carbon is slow in the absence of catalyst, but addition of triethylamine induces proton transfer[38] to generate a tungsten complex containing one imido group and one neopentylidene ligand. Finally, addition of a lithium alkoxide generates the base-free tetrahedral imido–alkylidene complex. Various alkoxides, including fluorinated alkoxides have been used in this step. The alkylidene ligand can adopt a conformation in which the *tert*-butyl group is anti or syn to the two alkoxy ligands. The isomer in which the *tert*-butyl group is located anti to the alkoxide and syn to the imide is more stable because the alkylidene hydrogen interacts most favorably with the tungsten in the position trans to the imide.

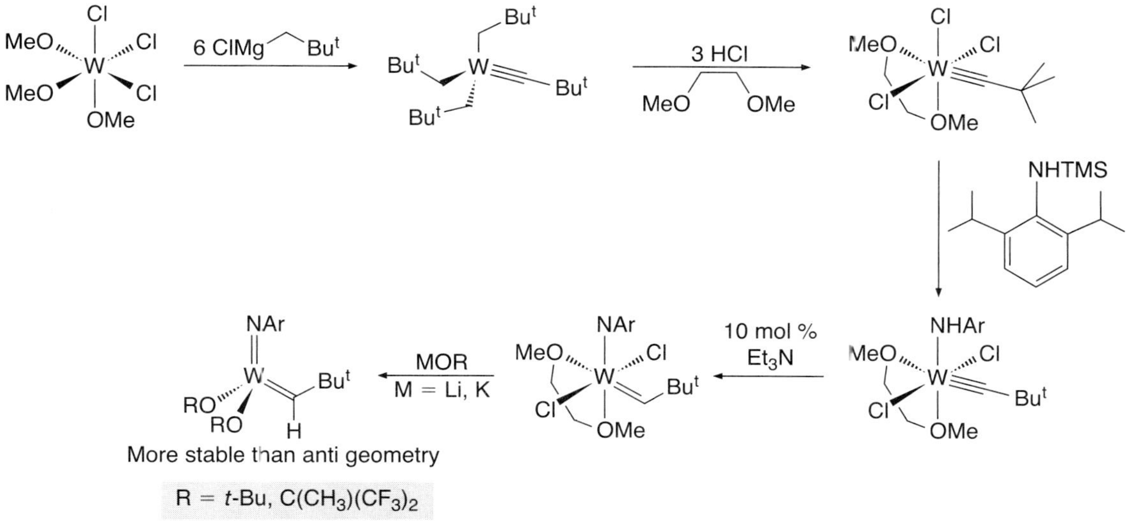

Scheme 13.3

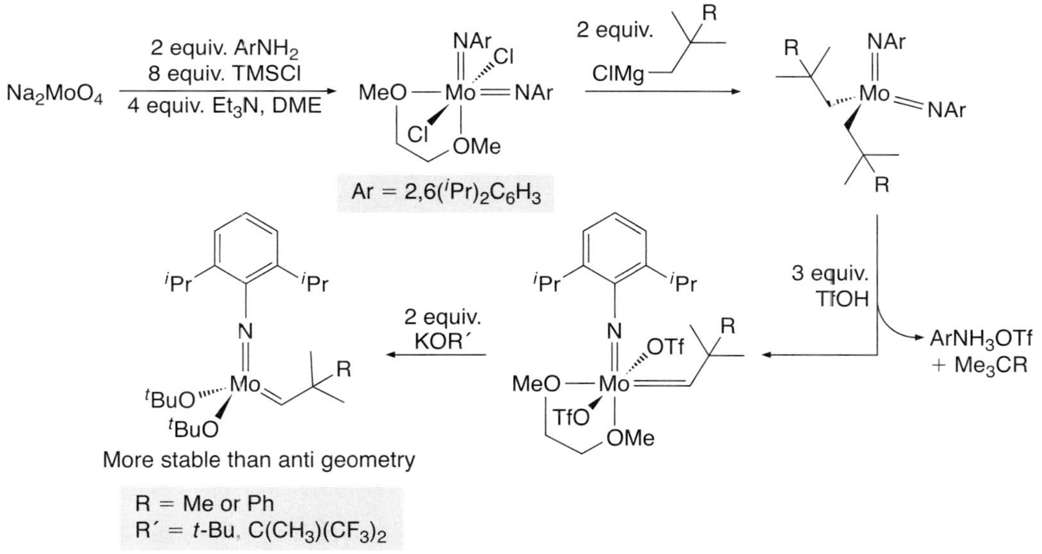

Scheme 13.4

Molybdenum analogs of the tungsten complex tend to be even more active for olefin metathesis. The synthesis of an example of such compounds is summarized in Scheme 13.4[37] and begins with sodium molybdate. Reaction of this material with an aromatic amine, trimethylsilyl chloride, and triethylamine in dimethoxyethane generates a species containing two imido groups. Treatment of this complex with two equivalents of neopentylmagnesium bromide or neophylmagnesium chloride ($ClMgCH_2CMe_2Ph$) leads to a molybdenum complex containing two imido groups and two alkyl groups. One of the imido ligands is then removed with triflic acid in dimethoxyethane to generate an octahedral species containing one chelating ether, two triflates, one imido group, and one alkylidene. Reaction of this species with a hindered alkoxide, such as potassium *tert*-butoxide, leads to the formation of the base-free molybdenum complex containing two alkoxide ligands, one imido group, and one alkylidene. Again, a variety of alkoxides have been used in the final step.

13.2.3.3.3. Synthesis of Tebbe's Reagent

A titanium–methylidene complex was one of the first carbenes used to convert carbonyl compounds to olefins in the fashion of a Wittig reagent and to provide mechanistic information on olefin metathesis. This methylidene was generated from the reaction of titanocene dichloride with trimethylaluminum, as shown at the top of Scheme 13.5.[39] This aluminum adduct was prepared by Fred Tebbe at DuPont and is commonly called "Tebbe's reagent." The methylidene complex is bound to the dimethylaluminum chloride byproduct. As shown at the bottom of Scheme 13.5, the chlorodimethylaluminum dissociates to generate a free titanium methylidene prior to reaction with carbonyl compounds and olefins. This reagent is more reactive than a phosphorus ylide, and readily converts ketones, esters, and amides to the corresponding olefins.

Synthesis of Tebbe's reagent and conversion to a titanacyclobutane:

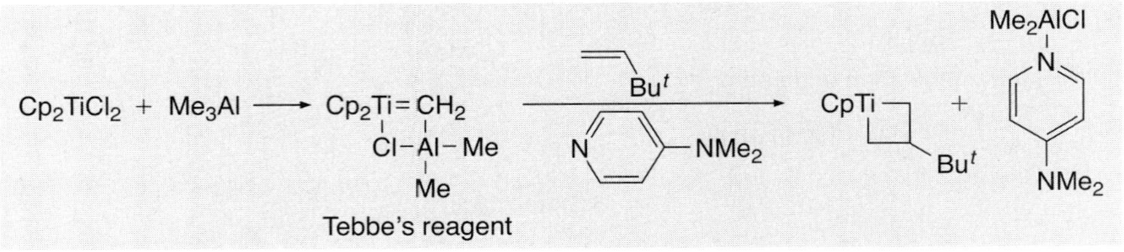

Three methods to generate the titanocene alkylidene:

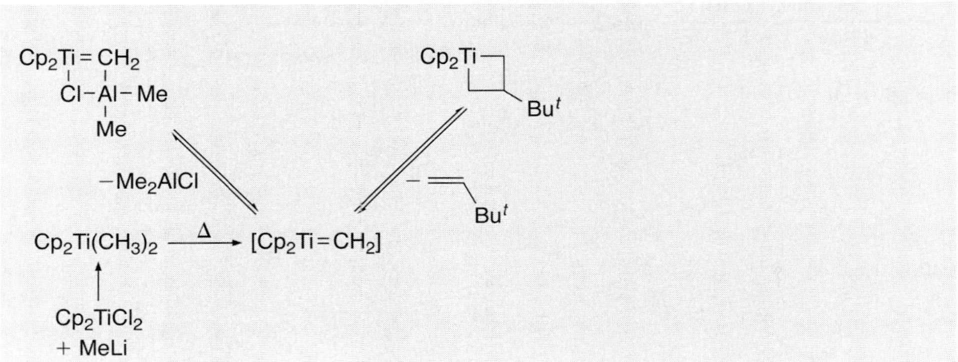

Scheme 13.5

A metallacyclobutane complex that generates the same titanocene–methylidene fragment in the absence of an alkylaluminum species was studied by Grubbs.[40] This complex was formed by reaction of Tebbe's reagent with *tert*-butyl ethylene, as shown at the top of Scheme 13.5. This metallacycle undergoes extrusion of *tert*-butyl ethylene and generates the same methylidene that is generated by dissociation of Me₂AlCl from Tebbe's reagent.

More recently, Petasis has shown that titanocene dimethyl,[41–44] which is easily generated from titanocene dichloride and methyllithium,[45] also generates the same titanium methylidene upon thermolysis[46] (Scheme 13.5, bottom). Thermolysis of titanocene dimethyl is a convenient way to generate the titanocene–methylidene complex because titanocene dimethyl is formed in high yield and is stable to water,[45,47] although it is sensitive to light and can undergo autocatalytic decomposition.[48] This complex has been used in synthesis on a large scale by the process group at Merck, and methods to prevent the autocatalytic decomposition of the titanocene dimethyl have been published.[48]

13.2.4. Synthesis of *N*-Heterocyclic Carbene Complexes

N-Heterocyclic carbene complexes[51] can be generated from the stable, free carbenes or by reaction of carbene precursors. Three examples of the synthesis of *N*-heterocyclic carbene complexes from carbene precursors are shown in Equations 13.14–13.16. The reaction of an imidazolium salt with a complex containing a basic ligand forms *N*-heterocyclic carbene complexes (Equation 13.14), as does transfer of the carbene from a silver–carbene complex (Equation 13.15),[52–54] or reaction of an imidazolium-2-carboxylate, with accompanying decarboxylation (Equation 13.16).[55]

$$(13.14)$$

(cis/trans)

$$(13.15)$$

$$(13.16)$$

Routes to synthesize the free carbenes are shown in Scheme 13.6 The free, unsaturated carbenes that have six-electron π-systems are typically formed by condensation of glyoxal or an α-diketone with an amine to generate an α-diimine.[56] Reaction of this imine with paraformaldehyde and acid generates an imidazolium salt, which is the protonated form of the *N*-heterocyclic carbene.[50,57] Treatment of these salts with a strong base, such as sodium amide in liquid ammonia or an alkoxide generates the free carbene. Alternatively, reaction of the α-diimine with triethylorthoformate generates the alkoxy-substituted imidazolium salt, as shown in Scheme 13.6. Reaction of this adduct with acid generates the imidazolium salt, which is deprotonated by strong base to form the free carbene.

Dihydroimidazolium carbene precursors or "saturated *N*-heterocyclic carbene" precursors are generated by a related sequence that includes reaction of a diamine, often generated by reduction of the glyoxaldimine, with triethylorthoformate, in the presence of acid.[58] This reaction generates the dihydroimidazolium salts, which can be deprotonated to form the free carbene or used directly in the synthesis of carbene complexes. Finally, the acid-catalyzed reaction of the diamine with formaldehyde, followed by iodine and base, generates the free carbene. The silver complexes are typically generated by reaction of the protonated carbenes with Ag₂O.[59]

As noted in Chapter 3, these carbenes are stable monomeric species when the substituents at nitrogen are large enough to prevent dimerization.[50] Typical substituents at nitrogen include *tert*-butyl groups, adamantyl groups, or di-*ortho*-substituted aryl groups, such as a mesityl group or a 2,6-diisopropylphenyl group.

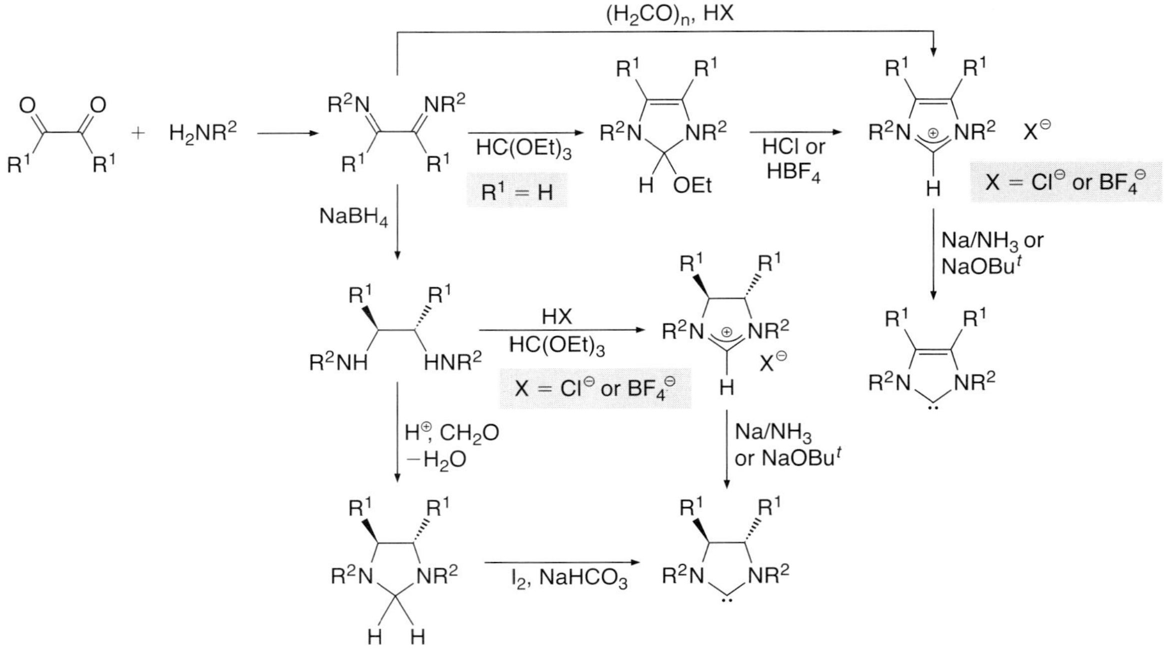

Scheme 13.6

One example of the formation of a carbene complex generates an important catalyst for olefin metathesis and provides an interesting comparison of the reactivity of *N*-heterocyclic carbenes and either an alkylidene or benzylidene. Treatment of the "first-generation Grubbs catalyst" in Equation 13.17 with the free, saturated *N*-heterocyclic carbene leads to replacement of one of the tricyclohexylphosphine ligands and generates the complex containing one *N*-heterocyclic carbene and one benzylidene ligand.[60–66] In this complex, the *N*-heterocyclic carbene acts as a dative ligand, and the benzylidene serves as the site of reactivity. An alternative synthesis of this species in Equation 13.17 that does not require isolation of the neutral carbene involves reaction of the imidazolium salt with potassium *tert*-amyloxide in the presence of the ruthenium complex.[67]

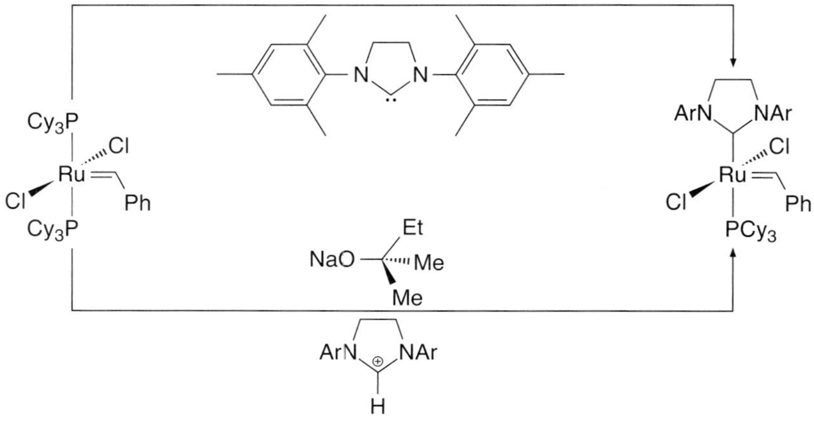

(13.17)

13.2.5. Reactivity of Carbene Complexes

13.2.5.1. Reactivity of Fischer Carbene Complexes [68,69]

Much of the reaction chemistry of Fischer carbene complexes is closely related to the reaction chemistry of organic esters. Like esters, these carbene complexes are electrophilic at the carbene carbon. In particular, they undergo attack by nucleophiles in reactions akin to transesterifications.[70] They also can coordinate Lewis bases to form stable adducts.

13.2.5.1.1. Reactions with Nucleophiles

Fischer carbene complexes undergo reactions with many types of nucleophiles at the carbene carbon, and representative examples of these reactions are summarized in Scheme 13.7. For example, the reaction of aryllithium reagents with the classic Fischer carbene complex of tungsten in Scheme 13.7 (for R = Ph) leads to the tungsten–diphenylcarbene complex.[71] Reactions with alcohols,[72] thiols, or amines also replace the groups at the Fischer carbene to generate, in these cases, carbene complexes containing new alkoxy, thiolate, or amino substituents at the carbene carbon.[73]

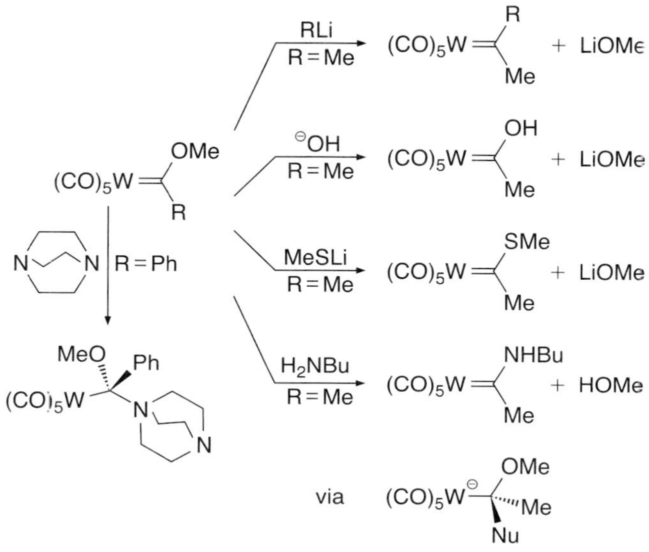

Scheme 13.7

These exchange reactions of Fischer carbene complexes occur through tetrahedral intermediates, just like the transesterification of esters.[70,73] Reaction of a Fischer carbene complex with an alcohol in the presence of base generates the same type of anionic intermediate that an ester does. This intermediate breaks down like an ester to form a new alkoxy-substituted carbene complex. Reactions of amines with Fischer carbene complexes occur similarly,[73] but the high basicity of amines and the high electrophilicity of the carbene complexes alleviates the need for any additional base.

Evidence for the tetrahedral intermediate has been gained in some cases by the reaction of the Fischer carbene complex with a Lewis base. Some of these reactions generate Lewis acid–base adducts in which the the carbene carbon acts as the Lewis acid. For example, addition of quinuclidine to the classic Fischer carbene complexes [M(CO)$_5$=CPh(OMe)] (M = Cr or W) generated stable adducts, as shown for the tungsten example in Scheme 13.7.[74]

13.2.5.1.2. Conversion to Carbyne Complexes

Alternatively, the alkoxy groups of Fischer carbenes can be abstracted with a Lewis acid, as shown in Scheme 13.8.[75,76] Abstraction of an alkoxy group from the Fischer carbene ligand of a neutral complex generates a cationic carbyne complex. The carbyne complex is more electrophilic than the starting carbene complex, and reaction of this carbyne complex with an alcohol regenerates an alkoxy-substituted carbene.

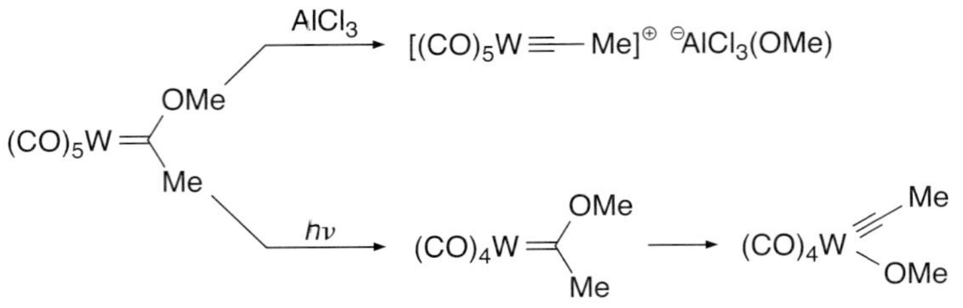

Scheme 13.8

Carbyne complexes have also been generated by migration of a substituent at the carbene to an open coordination site at the metal. For example, dissociation of carbon monoxide from a Fischer carbene complex has been induced photochemically, and this generation of an open coordination site at the metal led to migration of the alkoxy substituent on the carbene to the metal to form a neutral carbyne complex (Scheme 13.8, bottom).[77,78]

13.2.5.1.3. Reactions Related to Those of Enolates

Fischer carbene complexes react like enolates when treated with base. For example, reaction of butyllithium with $(OC)_5Cr=CMe(OMe)$ in Equation 13.18 generates an anion that is resonance-stabilized in a manner similar to the resonance stabilization of an enolate. Like an enolate, these deprotonated Fischer carbene complexes react with electrophiles. For example, deprotonation of the amino-substituted chromium–carbene complex in Equation 13.19 leads to an anion that adds to Michael acceptors to form homologated carbenes.[79] Reaction of the Fischer carbene in Equation 13.20 with benzaldehyde in the presence of a base and trimethylsilyl chloride induces a reaction that is analogous to a Knoevenagel condensation.[80] In this case, the condensation generates the new β,γ-unsaturated carbene complex in Equation 13.20. Sequential reactions of Fischer carbene complexes with electrophiles have also been conducted. Reaction of the anion of the carbene in Equation 13.21 with an epoxide generates a new alkoxide anion, which undergoes attack at the Fischer carbene to displace the methoxy group and to generate the complex of a cyclic carbene.[81]

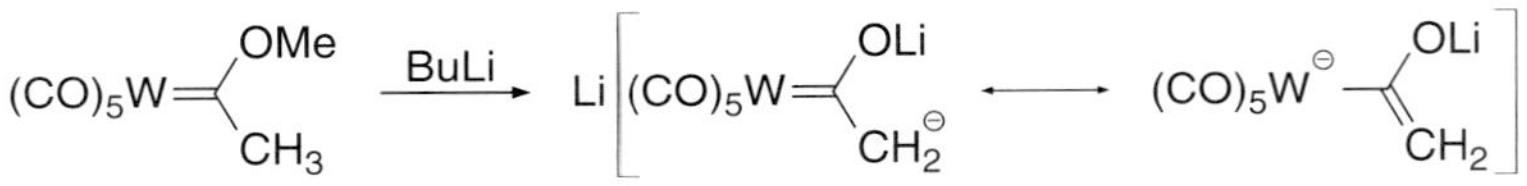

(13.18)

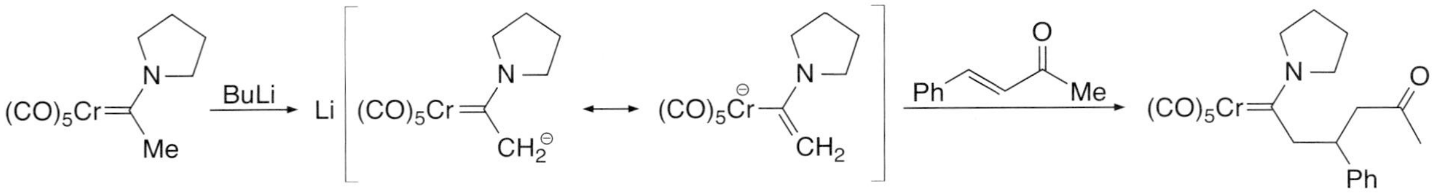

(13.19)

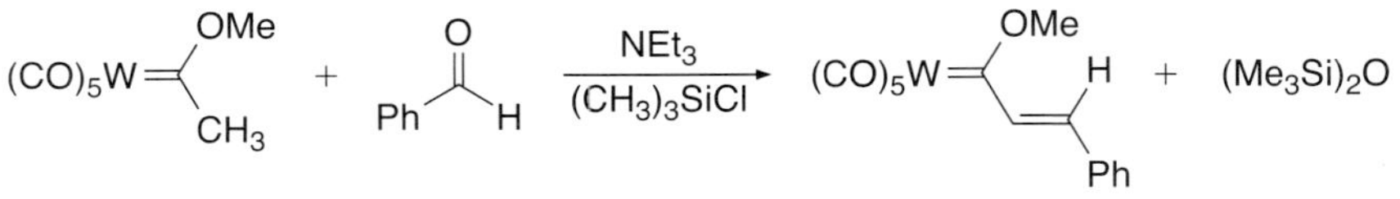

(13.20)

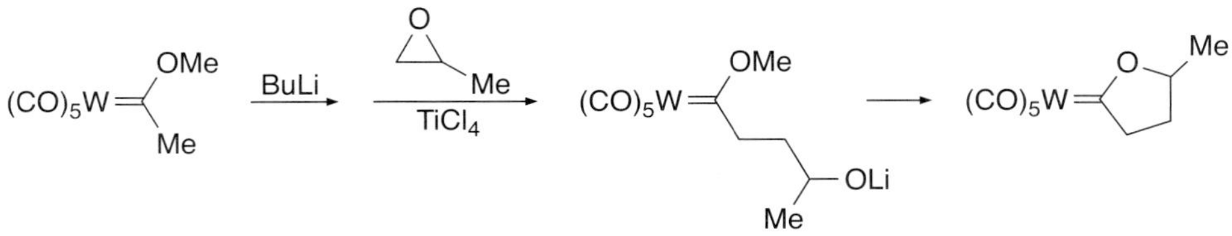

(13.21)

13.2.5.1.4. Cyclopropanations[82]

Many carbene or carbenoid complexes are likely intermediates in cyclopropanation processes, and reactions of discrete, isolated carbene complexes with olefins to generate cyclopropanes have been reported. Several examples of such cyclopropanations are shown in Equations 13.22–13.24.[83–85] The rate of cyclopropanation by the tungsten Fischer carbene complexes is slow, but the rate of cyclopropanation by the cationic iron complex that lacks π-donating groups on the carbene ligand occurs faster. In many cases, including those shown, isomers are often formed, and radical scavengers are needed to observe high yields in some cases.

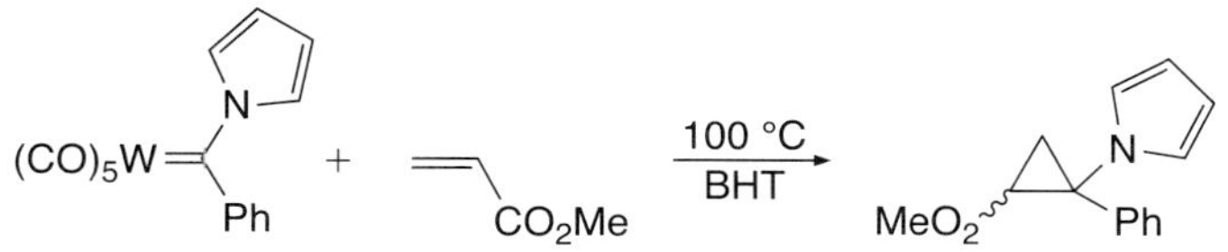

(13.22)

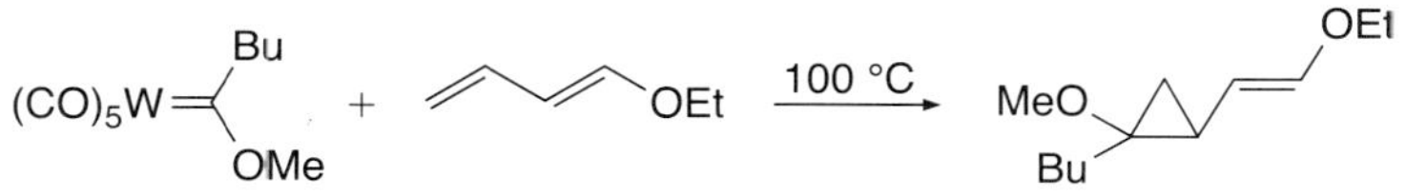

(13.23)

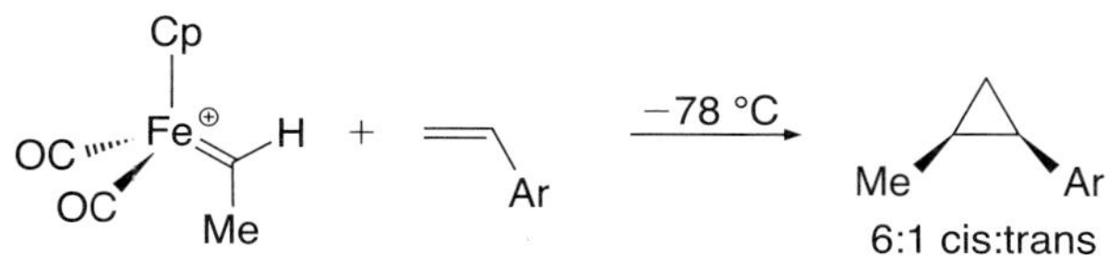

(13.24)

Cyclopropanation can occur by the two pathways in Scheme 13.9. First, a [2+2] process can occur between the metal–carbene unit and the olefin π-bond to generate a metallacyclobutane ring system. Reductive elimination of this metallacycle would generate a reduced metal fragment and a cyclopropane. Alternatively, the olefin could act as a nucleophile that attacks the electrophilic carbene carbon. This interaction would initiate formation of the cyclopropane without the intermediacy of a metallacycle.

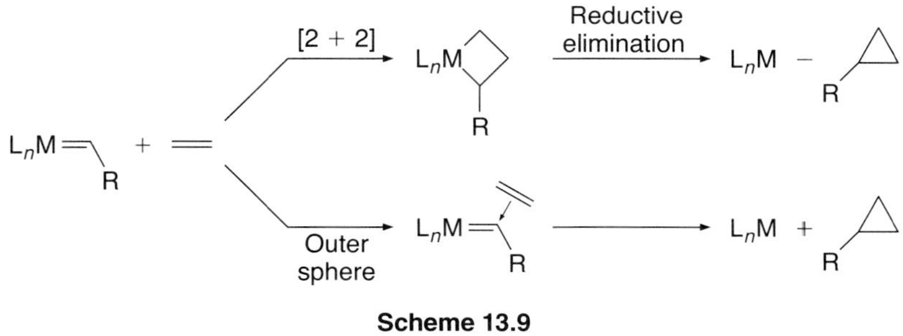

Scheme 13.9

Some studies of the reactivity of carbenes bound to low-valent metals containing carbonyl ligands suggested the intermediacy of carbene complexes in both cyclopropanation and olefin metathesis. For example, reactions of olefins with the pentacarbonyltungsten–diphenylcarbene complex in Scheme 13.10, as well as related benzylidene complexes, generated products from two reaction channels.[86–88] One channel produced a substituted cyclopropane. The other produced 1,1-diphenylethene and an ethoxy-substituted Fischer carbene. Indirect evidence for the formation of cyclopropanes from metallacycle intermediates was obtained more recently by Hillhouse. Cyclopropane products were generated from diphenyldiazomethane, olefin, and a phosphine-ligated Ni(0) complex. The cyclopropane was proposed to form by reductive elimination from a metallacyclobutane containing the L_2Ni unit.[89]

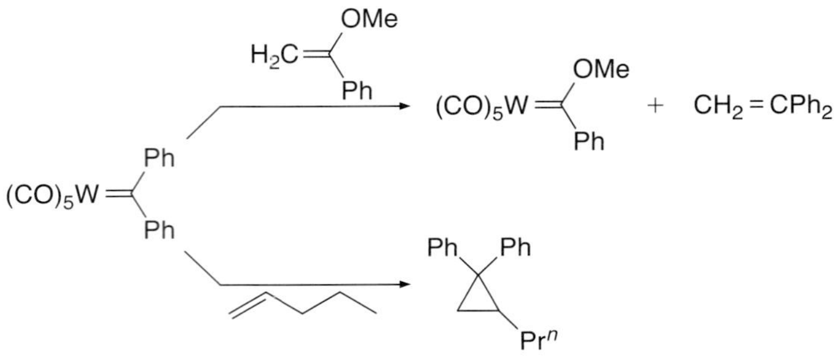

Scheme 13.10

However, reaction by two different pathways without a common metallacyclobutane intermediate seems likely in most reactions of Fischer carbenes.[82,87,90] As shown in Scheme 13.11, the olefin may react with the electrophilic carbene to generate a zwitterionic intermediate. This intermediate could then form the cyclopropane without the intermediacy of a metallacycle.

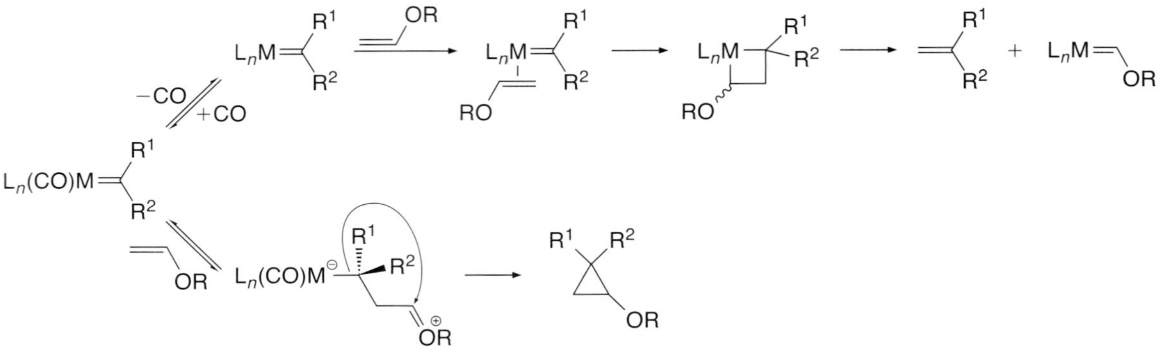

Scheme 13.11

13.2.5.1.5. Annulations: The Dötz Reaction

A final classic reaction of Fischer carbene complexes is the Dötz reaction shown in Equation 13.25. This reaction generates naphthols from a chromium Fischer carbene unit, a carbonyl ligand bound to the metal of the Fischer carbene complex, and an aromatic alkyne. The accepted mechanism of the Dötz reaction is shown in Scheme 13.12.[91-93] This reaction is thought to begin by dissociation of CO and coordination of the alkyne. Intermediates closely related to the unsaturated alkyne complex have been isolated.[94-96] The alkyne complex is then believed to undergo a [2+2] reaction to generate a metallacyclobutene. A retro-[2+2] reaction would then generate a β,γ-unsaturated carbene complex. Coupling of the carbene unit with a carbonyl ligand would then generate a chromium–ketene complex bound in an η^4 fashion. Electrophilic attack of the ketene carbon on the pendant aromatic unit would generate a tautomer of the napthol that would rearrange to the final product. The Dötz reaction has been employed in the synthesis of a number of vitamins,[97] steroids,[99] and antibiotics.[100] One noteworthy example is the synthesis of fredericamycin, as shown in Scheme 13.13.[101]

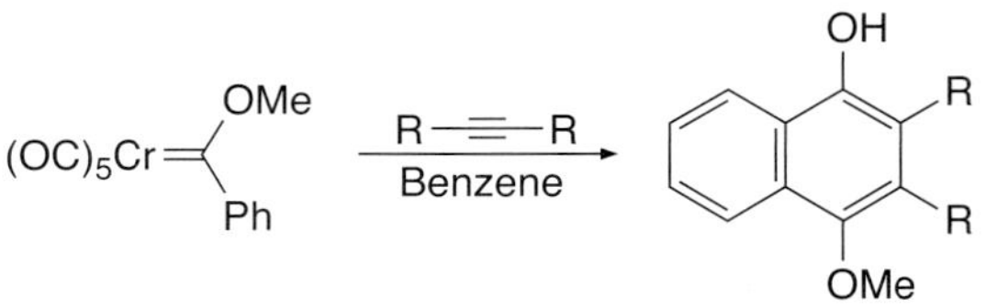

(13.25)

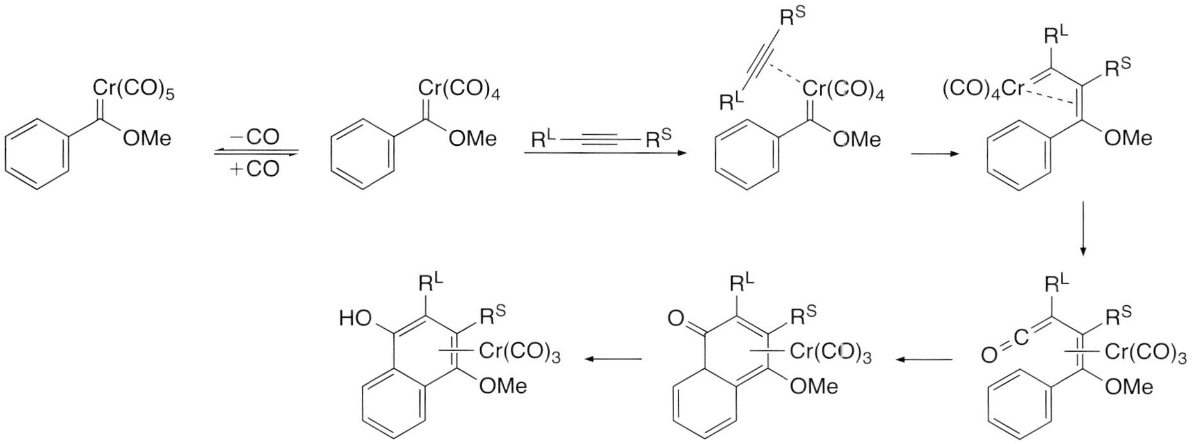

Scheme 13.12

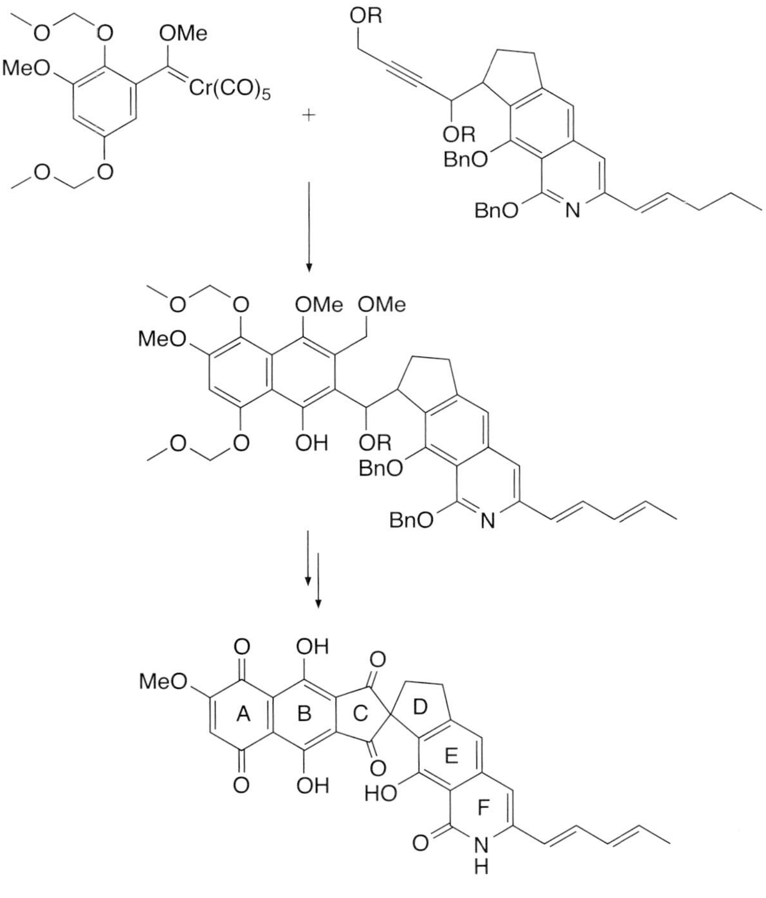

Scheme 13.13

13.2.5.2. Reactivity of Vinylidene Complexes

Vinylidene complexes are electrophilic at the α-carbon. Therefore, they display reactivity more similar to that of Fischer-type carbene complexes than to that of Schrock-type carbene complexes. This trend fits with the greater stability of the singlet spin state (and therefore higher HOMO-LUMO gap) of the free carbene in most Fischer carbene complexes and in vinylidenes. This trend also fits with the low oxidation state of the metal in most Fischer carbene complexes and vinylidene complexes. At the same time, some reactivity of vinylidine complexes is unique to those containing the cumulated unsaturation of a vinylidene unit. For example, the β-carbon of a vinylidene ligand is nucleophilic and basic.

The basicity at the β-carbon is illustrated by the reactions in Equations 13.26 and 13.27. Reaction of the octahedral rhenium vinylidene in Equation 13.26 with HBF_4 generates the cationic carbyne complex from addition of a proton to the basic β-carbon.[102] Because low-valent metals are often basic, addition of a proton to the vinylidene β-carbon is likely to occur by a multi-step process initiated by protonation of the metal center. This initial protonation at the metal center would then be followed by migration of the proton to the β-carbon. The reaction of acid with the iridium vinylidene in Equation 13.27 illustrates this mechanism. In this case, protonation first generates an iridium–hydride complex. The hydride in this complex then migrates to the β-carbon to generate an alkylidyne complex.[103]

$$(13.26)$$

$$(13.27)$$

In some cases, vinylidene complexes undergo [2+2] reactions that are characteristic of Fischer and Schrock carbene complexes. However, these [2+2] reactions involving vinylidene complexes can result from nucleophilic addition at the central carbon, rather than a concerted [2+2] process. For example, the reaction of an imine with the iron–vinylidene complex in Equation 13.28 leads to formation the product of a [2+2] reaction between the carbon–nitrogen double bond and the carbon–carbon double bond.[104,105] This reaction is believed to occur by nucleophilic attack of the nitrogen at the central carbon, followed by ring closure at the β-carbon, instead of a concerted [2+2] process.

$$(13.28)$$

13.2.5.3. Reactivity of Alkylidene and Alkylidyne Complexes

As mentioned in earlier sections of this chapter, alkylidene complexes are crucial intermediates in olefin metathesis, and this process occurs by sequential [2+2] reactions of olefins or alkynes with the metal–carbene complex. Catalytic metathesis of olefins and alkynes is presented in Chapter 21; studies on the elementary [2+2] reaction that forms and cleaves the C–C bond in these metatheses is presented here. In addition, alkylidene complexes undergo formal [2+2] reactions with nonpolar C–H and H–H σ-bonds. These reactions were mentioned in Chapter 6 and are described in more detail in this section.

13.2.5.3.1. Examples of [2+2] Reactions of Alkylidenes and Alkylidynes

A particularly well-defined example of a [2+2] cycloaddition between a carbene complex and an alkene is shown in Equation 13.29. The Schrock tungsten catalyst containing hexafluoro *tert*-butoxide groups reacts with ethylene by sequential [2+2] reactions.[106] The first [2+2] reaction generates the *tert*-butyl-substituted metallacyclobutane. A subsequent retro-[2+2] reaction releases *tert*-butyl ethylene and generates the tungsten–methylidene complex. This methylidene complex then undergoes a rapid [2+2] reaction with ethylene to generate the final unsubstituted tungsten metallacyclobutane.

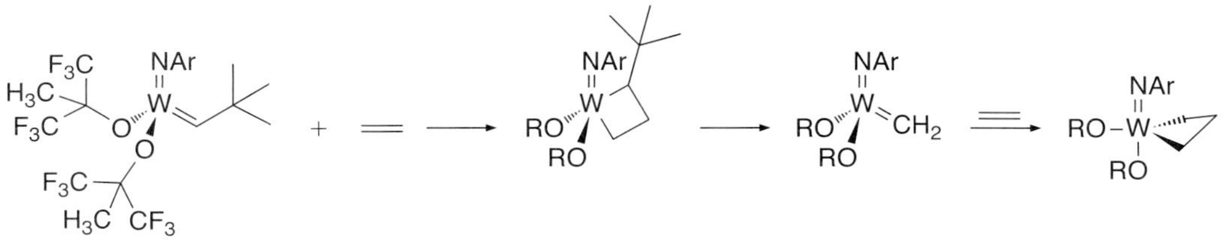

$$(13.29)$$

The [2+2] reactions of the original tantalum(V) alkylidene complex have been investigated with a variety of different unsaturated organic compounds (Scheme 13.14). This alkylidene complex reacts with carbonyl compounds to generate *tert*-butyl-substituted olefins and a tantalum–oxo byproduct. Reactions with nitriles generate the tantalum–imido complex possessing a vinyl substituent, and reaction with an acid chloride generates an enolate complex.[107] Analogous reaction chemistry has also been reported by Tebbe with his titanocene–carbene precursor.[39]

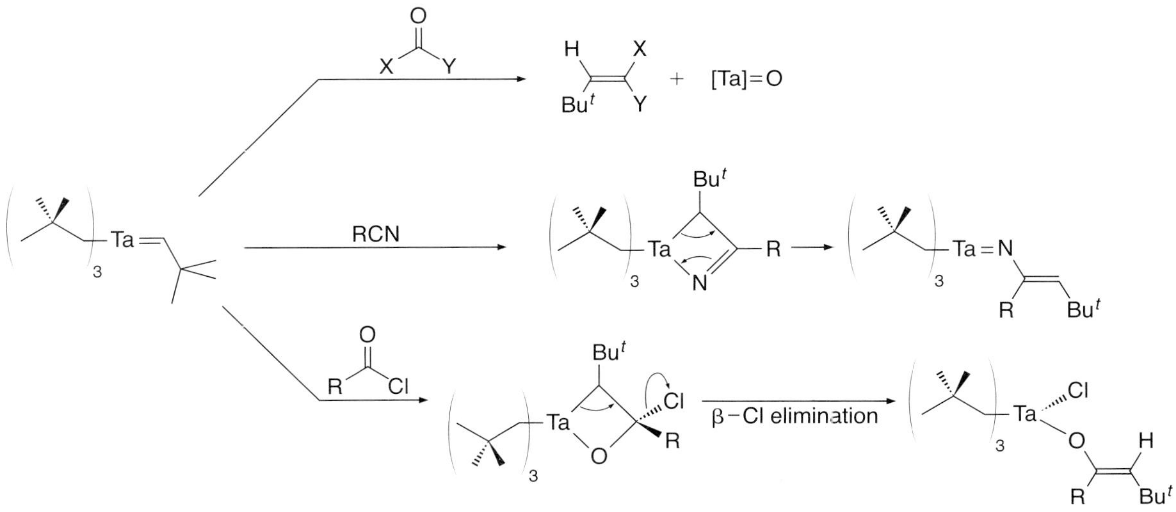

Scheme 13.14

Alkylidyne complexes also undergo [2+2] reactions. The tris(aryloxy)tungsten neopentylidyne complex in Scheme 13.30 undergoes a [2+2] reaction with 3-hexyne to generate a symmetric metallacyclobutadiene complex.[108] This product is a resonance hybrid of the two metallacyclobutadienes shown in the equation.

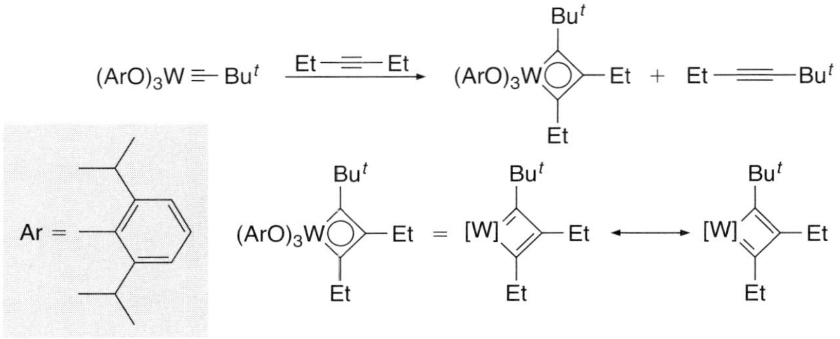

$$(13.30)$$

The mechanism of the [2+2] reactions has been studied in detail with the Schrock-type[109] and Grubbs-type[110] carbene complexes. The results of studies of the mechanism of these [2+2] reactions are summarized in Scheme 13.15. In general, the olefin coordinates to the metal center prior to the [2+2] process. This olefin complex then proceeds through the transition state for the [2+2] reaction and generates a metallacyclobutane product.

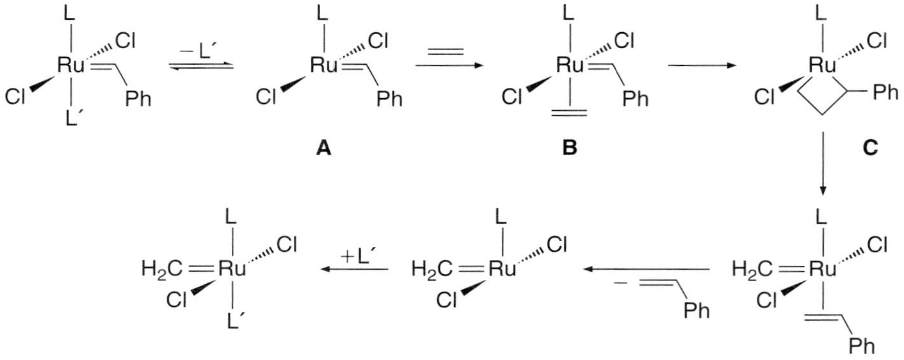

Scheme 13.15

This requirement for dissociation of ligand is shown most clearly with the Grubbs-type ruthenium metathesis catalyst in Scheme 13.15. The [2+2] chemistry of the square pyramidal ruthenium–carbene complex was studied in detail.[110, 111] This complex contains a carbene in the apical position. A ligand cis to the carbene dissociates[111] to generate intermediate **A** in Scheme 13.15 that coordinates olefin to generate intermediate **B**, which undergoes the [2+2] reaction. The resulting metallacyclobutane **C** then undergoes a retro-[2+2] reaction to generate a new carbene complex and a new coordinated olefin. Dissociation of olefin and re-coordination of the ligand completes the process.

Because this sequence of [2+2] and retro-[2+2] reactions is reversible and can be degenerate, one must consider the principle of microscopic reversibility[112–114] when proposing a mechanism for the reaction. This principle states that the mechanism of a reaction in the forward and reverse directions must be the same. A simple [2+2] and retro-[2+2] process starting with complex **B** (Scheme 13.15) proceeding through square pyramidal complexes (Equation 13.31) would not occur by the same intermediates and transition states in the forward and reverse directions. Thus, the metallacyclobutane intermediate must undergo a scrambling of the stereochemistry[110] or adopt the trigonal bipyramidal structure of complex **C**[115] to obey microscopic reversibility.

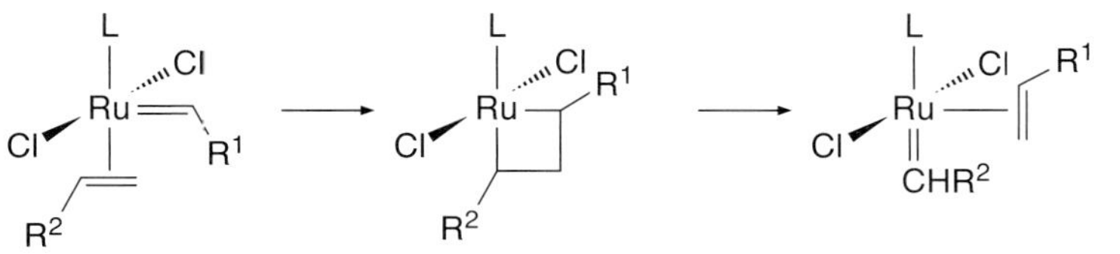

$$(13.31)$$

[2 + 2] sequence that violates microscopic reversibility if $R^1 = R^2$

Related to this issue of microscopic reversibility and the geometry of the olefin complex that leads to the metallacycle intermediate, several computational and experimental studies have led to the conclusion that the species containing cis chlorides and an olefin bound cis to the carbene ligand is similar in energy to the species containing trans chlorides and an olefin bound trans to the carbene ligand (Figure 13.4). Calculations by Cavallo on complexes in the gas phase and in solution showed that the compound containing an olefin trans to the carbene is the more stable species in the gas phase, but that the complex containing the olefin cis to the carbene is more stable in solution.[116] Grubbs has characterized the species shown in Figure 13.5 that is most stable with the olefin cis to the carbene. This compound is stable because a [2+2] cycloaddition would form a strained tricyclic system.[117]

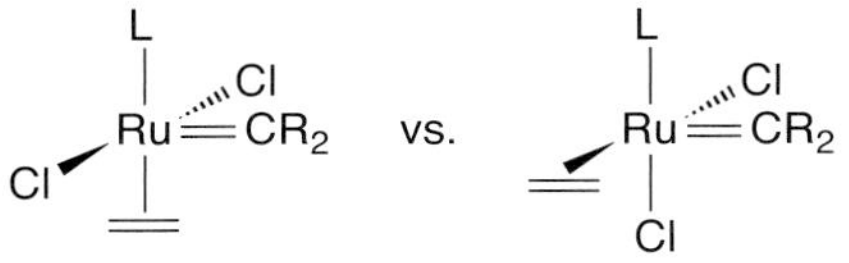

Figure 13.4.
Two potential isomers of carbene–olefin complexes that precede [2+2] processes.

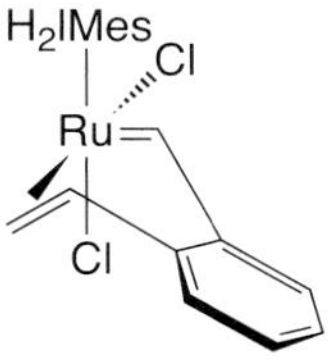

Figure 13.5.
Structure of an isolated ruthenium alkene olefin complex.

Although the identity of the precursor is uncertain and could depend on the medium and ancillary ligands, several studies have led to direct detection of the ruthenacyclobutane intermediate and allowed for particularly detailed assessment of the dynamics and reaction chemistry of this intermediate. Piers and co-workers developed a route to the metallacycle shown in Scheme 13.16.[118,119] Treatment of a ruthenium carbide first isolated by Heppert[120] with acid and phosphine generates the phosphonium salt shown in Scheme 13.16. Because this compound contains an open coordination site, it reacts rapidly with olefins. As shown in Scheme 13.16, the reaction with ethylene at −50 °C generates a vinylphosphinium salt and the parent metallacyclobutane (R = H in the scheme) in high yield. This process is sufficiently fast at this low temperature to generate the metallacycle quantitiatively without decomposition. At higher temperatures, the metallacycle decomposes to propene and unidentified ruthenium products. The use of an unsymmetrical carbene ligand and propene as the olefin further cemented the structural assignment as a trigonal bipyramidal species.[121]

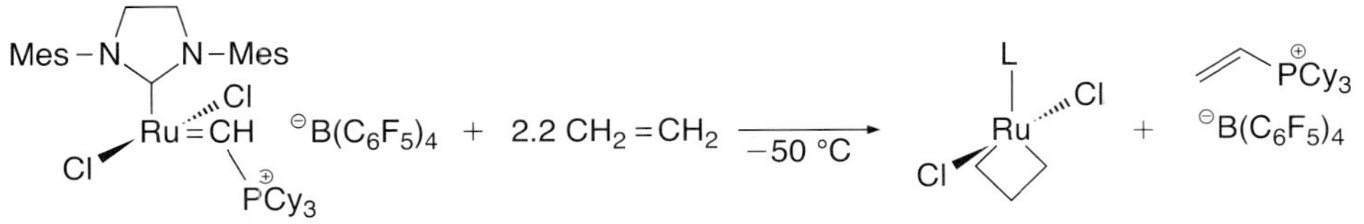

Scheme 13.16

The ability to generate this species has led to studies that allowed an assessment of the rate of the [2+2] process. Isotopic labeling of one carbon of the ethylene led to metallacycle in which the labeled carbon exchanged from the α- to the β-position on the NMR time scale, even at −50 °C (Scheme 13.17). This exchange occurred without dissociation of the olefin into solution. A rapid equilibrium involving a retro-[2+2] reaction to generate an olefin–carbene complex, followed by rotation and a forward [2+2] process would exchange the α- and β-carbons. Ligand substitution of ethylene by an incoming olefin, which would be the next step of the metathesis process, was shown to have a higher barrier and to occur by an associative pathway.

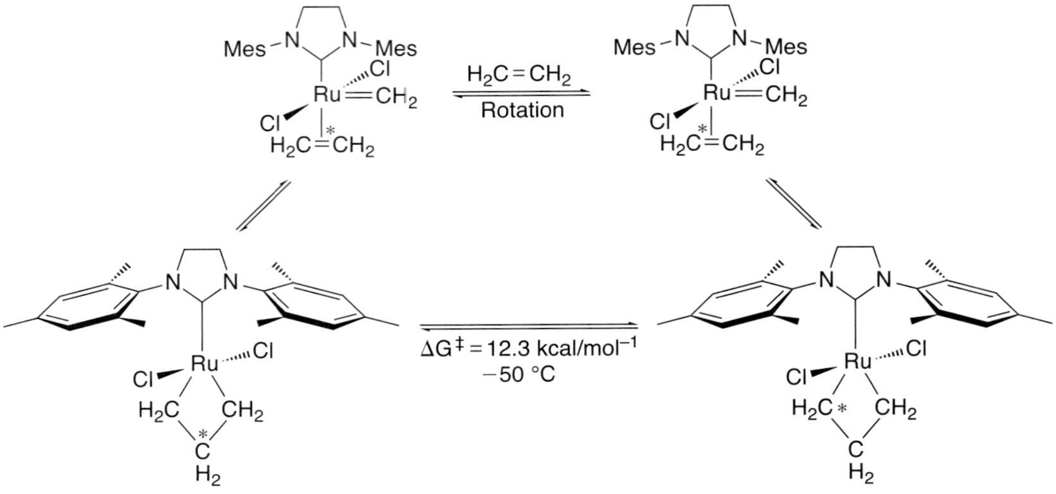

Scheme 13.17

The Schrock-type carbene complexes also undergo the [2+2] reactions shown in Equation 13.32. The mechanism of the [2+2] reaction with molybdenum has been studied most carefully with the bis(hexafluoro-*tert*-butoxide) complex.[37] These [2+2] reactions most likely occur by initial coordination of olefin. The inactivity of five-coordinate molybdenum and tungsten complexes containing a dative Lewis base toward olefin metathesis suggests that a site is needed for coordination of the olefin. Pseudo-tetrahedral molybdenum and tungsten complexes that are formed by removal of the Lewis base or that are prepared in the absence of Lewis base are active for olefin metathesis. The syn isomer of this tetrahedral complex with the carbene hydrogen interacting with the metal at the site trans to the imido nitrogen is more stable than the anti isomer that lacks this interaction. Because this stabilization by coordination of hydrogen is lost as the olefin coordinates and the [2+2] process occurs, this greater stability of the syn isomer makes it less reactive for the [2+2] reaction than the anti isomer.[109]

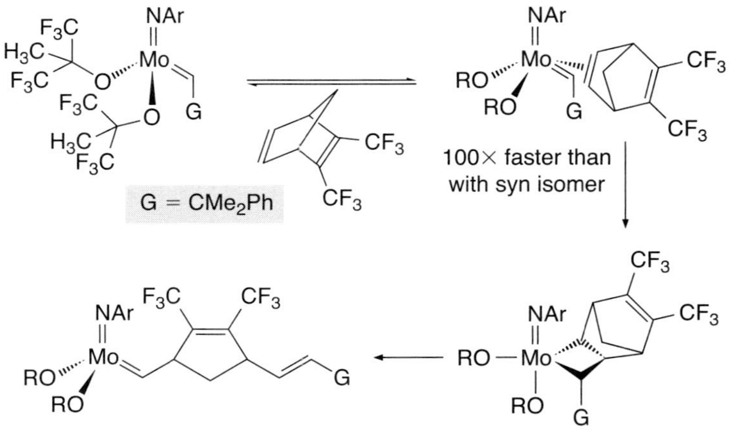

(13.32)

Studies have been conducted on the factors that accelerate the rate of the sequential [2+2] and retro-[2+2] reactions that constitute the olefin metathesis process. The ruthenium system of Grubbs has been amenable to kinetic analysis. The starting complex is a 16-electron species, but the 14-electron species formed by ligand dissociation reacts with olefin in the [2+2] process. Therefore, faster reactions will occur with systems in which dissociation of an ancillary ligand is favored and in which the remaining ancillary ligand on the 14-electron intermediate promotes the [2+2] addition, relative to reassociation of the dissociated ligand.

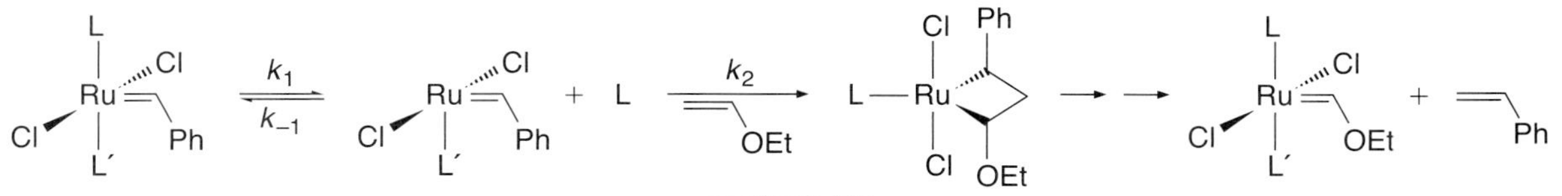

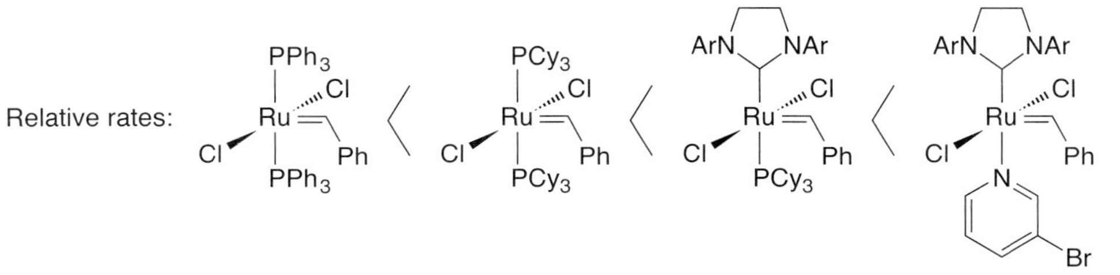

Scheme 13.18

Weakly donating and sterically hindered ligands dissociate most readily. As outlined in Scheme 13.18, pyridine[122] and arylphosphine ligands[123] dissociate more readily from these carbene complexes than do N-heterocyclic carbene or trialkylphosphine ligands.

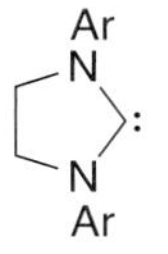

However, the ancillary ligand located trans to the dissociating ligand also influences the dissociation process and will affect the rate of the subsequent [2+2] process. As shown by the rate constants in Scheme 13.18,[110] the presence of an N-heterocyclic carbene on the 14-electron fragment favors the elementary [2+2] reaction over re-coordination of ligand. In contrast, dissociation of tricyclohexyl phosphine is favored by about two orders of magnitude when the ligand trans to it is tricyclohexyl phosphine than when the ligand trans to it is an N-heterocyclic carbene. Thus, the accelerating effect of N-heterocyclic carbene ligands on olefin metathesis results from a favorable partitioning of the 14-electron fragment containing the N-heterocyclic carbene ligand toward a [2+2] reaction over reassociation of ligand. The rate constant for the [2+2] process versus reassociation of ligand is roughly four orders of magnitude more favorable when the dative ligand on ruthenium is an N-heterocyclic carbene than it is when the dative ligand is tricyclohexylphosphine or triphenylphosphine.

13.2.5.3.2. Fomal [2+2] Reactions with C–H σ-Bonds

The addition of C–H bonds in alkanes and arenes across the M=C bond of alkylidene complexes is the opposite of the α-eliminations that form alkylidene complexes. Therefore, one might expect additions of C–H bonds across alkylidenes to be common. Yet, they are rare[124–126] and have been developed only recently.[127–129] Related additions of C–H bonds across alkylidyne complexes are even less common.[130,131] A few examples of such reactions of carbene and carbyne complexes were described in Chapter 6. An example of this reaction reported by Hessen,[124] and two examples of reactions reported by Legzdins are shown in Scheme 13.19.[127,128] Heating of titanocene bis(benzyl) or a nitrosyl-ligated tungsten bis(neopentyl) compound generates unsaturated alkylidene complexes as reactive intermediates. These species add the C–H bonds of arenes to generate aryl alkyl products. Addition of tetramethylsilane to the tungsten complex generates the neopentyl trimethylsilylmethyl complex, which reacts further with tetramethylsilane at higher temperature to form a trimethylsilyl-substituted carbene that adds a second equivalent of tetramethylsilane. Reaction of the same intermediate with methylcyclohexane generates an allyl hydride product. This reaction occurs by a combination of initial C–H activation to form a cyclohexylmethyl complex that undergoes β-hydrogen elimination, extrusion of neopentane, and a second C–H activation to form the allyl hydride product.

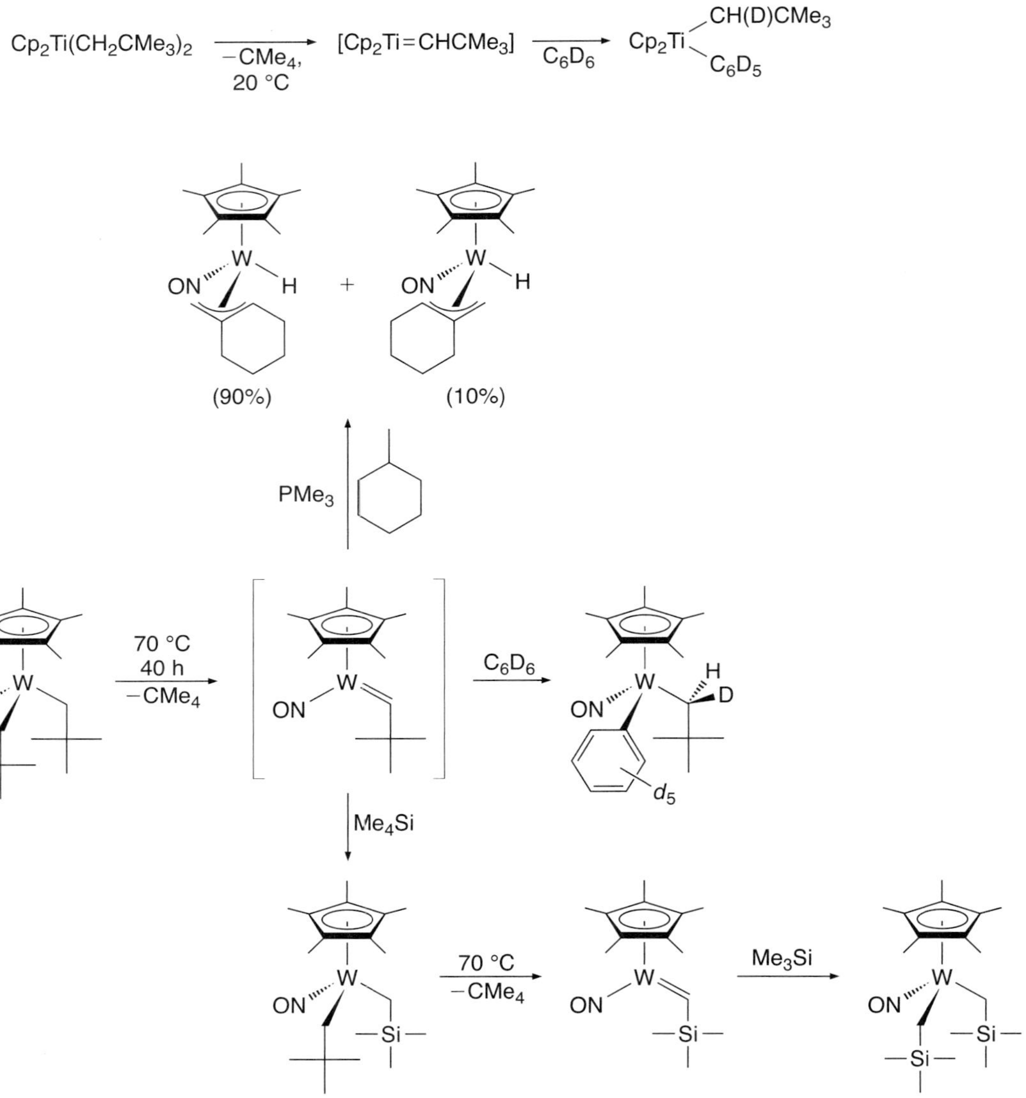

Scheme 13.19

13.3. Silylene Complexes[132–134]

13.3.1. Overview of Silylene Complexes

Transition metal–silylene complexes are the silicon analog of transition metal–carbene complexes. In general, multiple bonds to silicon and other second-row main group elements are less stable than multiple bonds to carbon, oxygen, and nitrogen. Just as the silicon–oxygen double bond is unstable toward oligomerization to form polysiloxane, the transition-metal–silicon double bond is unstable toward formation of bridged structures, and for many years monomeric compounds containing metal–silicon double bonds had not been isolated in pure form.

However, several catalytic reactions had been proposed to occur by a mechanism involving metal–silicon double bonds.[135,136] Kumada[137–139] reported oligomerization of disilanes (Equation 13.33), and Ojima[140] reported the scrambling and oligomerization of silanes in the presence of Wilkinson's catalyst (Equation 13.34), and both processes were proposed to occur by silylene intermediates. The formation of a silylene complex of iron tetracarbonyl was claimed in 1977 (Equation 13.35),[141] but this compound was never isolated. Since the first structural data reported in 1987,[142,143] a variety of metal–silylene complexes have been prepared that are stabilized by coordination of a Lewis base to the silicon of the silylene complex, as in Equation 13.35. The first base-free silylene complex was reported by Tilley in 1993.[144,145]

$$Si_2Me_5H \xrightarrow[90\ °C,\ 18\ h]{trans\text{-}PtCl_2(PEt_3)_2} Si_nMe_{2n+1}H \atop n = 1\text{-}6 \qquad (13.33)$$

$$H_2SiPhMe \xrightarrow[70\ °C,\ 1\ h]{RhCl(PPh_3)_3} SiPh_3Me \ + \ HSiPh_2Me$$

$$+ \ H-\underset{\underset{Me}{|}}{\overset{\overset{Ph}{|}}{Si}}-\underset{\underset{Me}{|}}{\overset{\overset{Ph}{|}}{Si}}-H \ + \ H-\underset{\underset{Me}{|}}{\overset{\overset{Ph}{|}}{Si}}-\underset{\underset{Me}{|}}{\overset{\overset{Ph}{|}}{Si}}-\underset{\underset{Me}{|}}{\overset{\overset{Ph}{|}}{Si}}-H \qquad (13.34)$$

$$Fe(CO)_5 \ + \ HSiMe_2(NEt_2) \xrightarrow[-CO]{h\nu} \overset{\overset{\displaystyle NEt_2}{\diagup}}{(OC)_4Fe=SiMe_2} \qquad (13.35)$$

13.3.2. Bonding of Silylene Complexes

The orbitals involved in the formation of a metal–silicon multiple bond are shown in Figure 13.6. These orbital interactions are the same as those involved in the formation of a metal–carbon double bond in a carbene complex. However, back-donation from the metal d-orbital into the silicon p-orbital is much weaker in a metal–silylene complex than it is in a carbene complex. Therefore, the silicon remains Lewis acidic, and the silylene complexes are often stabilized by Lewis bases.

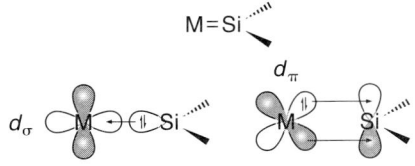

Figure 13.6.
Orbital interactions in metal–silylene complexes.

13.3.3. Examples of Isolated Silylene Complexes

The first well-characterized silylene complexes stabilized by Lewis bases were reported by Zybill (Equation 13.36)[142] and Tilley (Equation 13.37).[143] In one case, the silylene complex was formed by elimination of a salt and in another case by abstraction of triflate in the presence of Lewis base. The first base-free silylene complexes were prepared by using soft π-donating thiolate substituents on silicon (Equation 13.38). One thiolate group from a tris(thiolatosilyl) complex was abstracted with TMSOTf, and exchange of a less coordinating tetraphenylborate anion for triflate formed a stable cationic silylene product lacking coordinated base.[144,145]

$$(^tBuO)_2SiCl_2 \ + \ Na_2Fe(CO)_4 \ \xrightarrow[-\ 2NaCl]{+\ base} \ (OC)_4Fe=Si\overset{\displaystyle base}{\underset{\displaystyle O^tBu}{\diagdown O^tBu}} \tag{13.36}$$

Base = HMPA or THF

$$Cp^*(Me_3P)_2RuSiPh_2(OTf) \ + \ NaBPh_4 \ \xrightarrow[-NaOTf]{+MeCN} \ \left[\ \right]^{\oplus} BPh_4^{\ominus} \tag{13.37}$$

$$Pt(PCy_3)_2 \ + \ HSi(SEt)_3 \ \longrightarrow \ \underset{Cy_3P}{\overset{Cy_3P}{\diagdown}}Pt\underset{Si(SEt)_3}{\overset{H}{\diagup}}$$

$$\Big\downarrow \ \begin{array}{l} +\ Me_3SiOTf \\ -\ Me_3SiSEt \end{array} \tag{13.38}$$

$$\left[\underset{Cy_3P}{\overset{H}{\diagdown}}Pt\underset{Si}{\overset{PCy_3}{\diagup}}\cdots\!SEt \atop SEt \right]^{\oplus} BPh_4^{\ominus} \ \xrightarrow[\ CH_2Cl_2\ ]{\begin{array}{c} +NaBPh_4 \\ -NaOTf \end{array}} \ \underset{Cy_3P}{\overset{H}{\diagdown}}Pt\underset{Si(SEt)_2OTf}{\overset{PCy_3}{\diagup}}$$

Silylene complexes have been prepared by coordination of free silylene, extrusion of dihydrogen from a dihydrosilane, α-hydrogen elimination from a hydrosilyl complex, extraction of hydride, halide, or pseudo-halide from a metal–silyl complex, and photo-extrusion of silylene from a disilanyl complex. Examples of these reactions are shown in Equations 13.39–13.45.

$$2 \ \begin{array}{c} Bu^t \\ \end{array} Si\colon \ \xrightarrow[-2\ CO]{Ni(CO)_4} \ \left(\begin{array}{c} Bu^t \\ \end{array} Si=Ni(CO)_2 \right)_2 \tag{13.39}$$

$$Pt(PCy_3)_2 \ + \ (Me_3Si)_2SiMes_2 \ \xrightarrow[-\ (Me_3Si)_2]{h\nu} \ Pt(Cy_3P)_2=SiMes_2 \tag{13.40}$$

$$Ph_2SiH_2 \ + \ Fe(CO)_5 \ \xrightarrow[\begin{array}{c} +\ HMPA \\ -CO,\ -H_2 \end{array}]{h\nu} \ (OC)_4Fe=SiPh_2 \overset{\displaystyle HMPA}{\diagup} \tag{13.41}$$

$$(13.42)$$

$$(13.43)$$

$$(13.44)$$

$$(13.45)$$

The reaction of nickel carbonyl with a free N-heterocyclic silylene forms the bis(silylene) complex in Equation 13.39.[146,147] Alternatively, an unsaturated metal fragment can trap a silylene complex generated thermally or photochemically, as shown in Equation 13.40.[148] Just as Fischer carbene complexes have been generated by double C–H activation, a base-stabilized silylene complex has been generated under photochemical conditions by two sequential Si–H activation steps and elimination of dihydrogen (Equation 13.41).[149–151] Also analogous to a common method used to generate carbene complexes is the formation of a silylene complex by α-hydrogen elimination, as shown in Equation 13.42.[152] One of the first structurally characterized metal–silylene complexes was formed by abstraction of triflate from a ruthenium–silyl complex (vide supra)[143] and a second related complex was generated by abstraction of hydride, as shown in Equation 13.43.[153] Two groups[154,155] reported evidence that photochemical dissociation of carbon monoxide from the disilanyliron complex in Equation 13.44 leads to silyl migration to generate the silyl silylene complex in which the silylene is replaced during reassociation of carbon monoxide to form the silyl complex. An interesting methoxide-stabilized disilylene complex was also formed by photochemically induced dissociation of carbon monoxide and silicon–silicon bond cleavage (Equation 13.45).[156,157]

13.3.4. Reactivity of Silylene Complexes

The reactivity of metal–silylene complexes is more limited than the reactivity of carbene complexes. The cationic base-stabilized ruthenium–silylene complex in Equation 13.37 does not react with olefins or alkynes to undergo [2+2] addition reactions. However, a related complex did undergo [2+2] addition reactions with isocyanates, as shown in Equation 13.46.[158] Other reactions of silylene complexes are distinct from those of carbene complexes or those of other conventional organometallic compounds. For example, the reaction of the silylene hydride with an acetylene generates a β-silylvinylarene complex

(Equation 13.47), and reaction with a nitrile generates a product resulting from formal migration of the phenyl to the silylene unit, C–C cleavage of the nitrile, and insertion of the cyano group into an M–Si bond (Equation 13.48).[159] Related to the process in Equation 13.47 is the migration of hydrocarbyl groups from the metal to a silylene intermediate. For example, the phenyl group on the reactant of Equation 13.49 migrates to silicon, presumably via the generation of an aryliridium silylene intermediate. The product of migration would then be trapped by the added acetonitrile.[160]

(13.46)

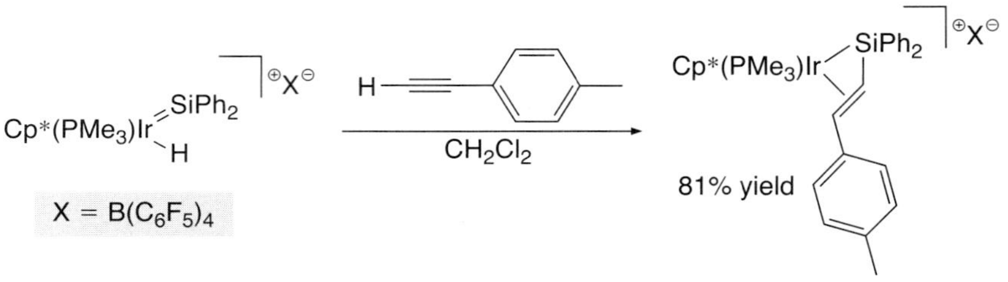

R = Me or Ph

(13.47)

X = B(C₆F₅)₄

81% yield

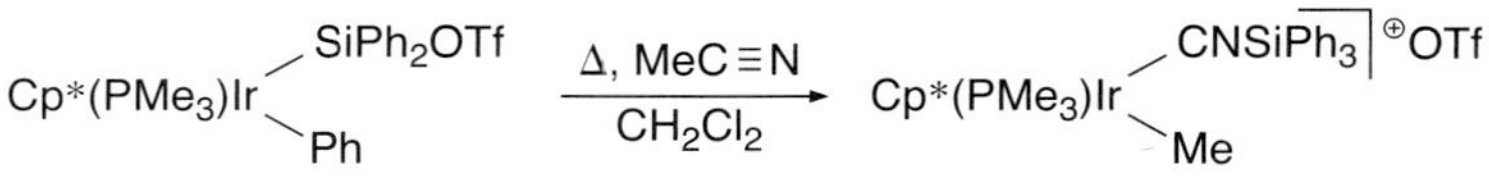

(13.48)

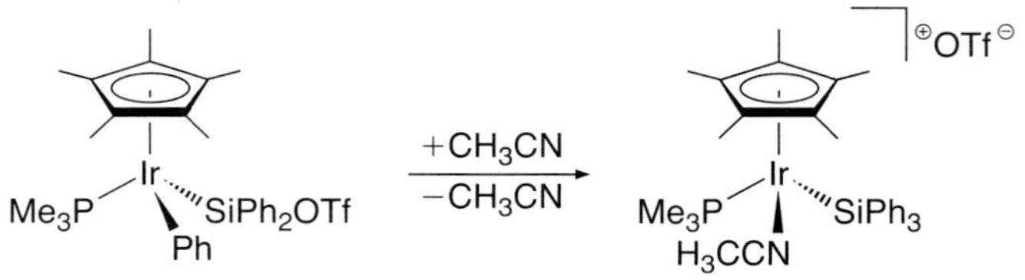

(13.49)

13.4. Metal–Heteroatom Multiple Bonds

13.4.1. Scope of the Section

Oxo, imido,[161,162] sulfido,[163–165] phosphinidine,[166–183] borylene,[184–190] and nitrido[191–195] ligands bond to transition metals through multiple bonds. Some of the bonding and reaction chemistry of these ligands closely parallels that of carbene ligands, but much of it is distinct because of the differences in electronegativity of the elements bound to the metal and because of the presence of nonbonded orbitals on the heteroatom. Oxygen and nitrogen are more electronegative than carbon, and oxo, imido, sulfido, and phosphinidine ligands contain electron pairs that can donate into unoccupied orbitals of the metal center to create a dative metal–ligand π-bond.[196] Boron is less electronegative than carbon, and borylene ligands (B–R) possess an unoccupied *p*-orbital into which the metal can donate electron density to create a weak π-interaction.[197,198] Depending on the electronic properties of the metal center and the extent of electron donation that involves the nonbonded orbitals, oxo, imido, sulfide, and phosphinidine ligands can be nucleophilic or electrophilic.

This section of this chapter primarily focuses on the chemistry of oxo and imido compounds because these compounds are more common as intermediates in homogeneous catalytic transformations than sulfido, phosphinidine, and borylene ligands. Transition metal–oxo complexes are intermediates in many oxidation processes,[199] including reactions that convert alkanes to alcohols and olefins to epoxides.[200–206] Transition metal–imido complexes are intermediates in related reactions, such as intramolecular nitrene insertions into the C–H bonds of alkyl groups[207–211] and aziridinations of olefins.[212–219] Metal–imido complexes are also intermediates in hydroamination reactions catalyzed by titanium and zirconium complexes.[220–241] Many of these reactions of oxo and imido complexes with the C–H bonds of hydrocarbons and with olefins occur by mechanisms that do not generate a metal–carbon bond.[203,204,242,243] However, some reactions of oxo and imido complexes dovetail more closely with classical organometallic chemistry and mimic the reactivity of carbene complexes. This section of the chapter focuses on the reactivity of oxo and imido complexes that is most closely aligned with organometallic compounds, but some representative reactions that do not create a metal–carbon bond are also presented.

13.4.2. Overview

Most transition metal complexes containing terminal oxo and imido ligands possess metals from the middle of the transition series (recall Figure 13.1).[244,245] Group 4 metal–oxo compounds are common, but these materials often contain bridging oxo and imido groups. In the extreme case, these compounds are solid state materials. For example, titanium oxide is common and is a component of white paint pigment. Metal–oxo complexes, such as osmium tetroxide, potassium permanganate, and potassium chromate are used in organic chemistry. An iron enzyme, cytochrome P_{450}, oxidizes organic molecules in the liver through metal–oxo intermediates as part of the body's detoxification system.[246–248]

In transition metal organometallic chemistry, oxo and imido ligands can serve as the point of reaction or as ancillary ligands to control reactivity at other sites on the metal complex. Complexes containing bridging imido and oxo groups are more common than those containing terminal imido and oxo groups, and the bridging imido and oxo complexes tend to be less reactive than the terminal species. Bridging oxo and imido groups more commonly serve as ancillary ligands or structural linkers than the point of reaction. Therefore, this chapter focuses on the reactions of terminal imido and alkoxo complexes with organic substrates. Several classic organometallic compounds containing terminal oxo and imido groups include Hermann's organyl trioxorhenium species ($RReO_3$, in which R = Cp, Me, etc.),[249–251] which reacts at the oxo group by non-radical pathways,[252] and Schrock's metathesis catalyst,[37,253] which contains an imido group as an ancillary ligand. The most reactive of the metal complexes containing terminal amido and oxo complexes are found in the early and late transition metals. For example, some zirconium–imido and –oxo compounds are highly reactive. Chemistry of these complexes is described in this chapter and in Chapters 18 and 6 that include C–H bond activations. Few low-valent, late-transition-metal–imido complexes have been isolated in pure form because they are so unstable. Two examples that have been isolated are the two-coordinate iridium–imido complex and the three-coordinate nickel–imido complex in Figure 13.7, reported by Bergman and Hillhouse, respectively.[254–256]

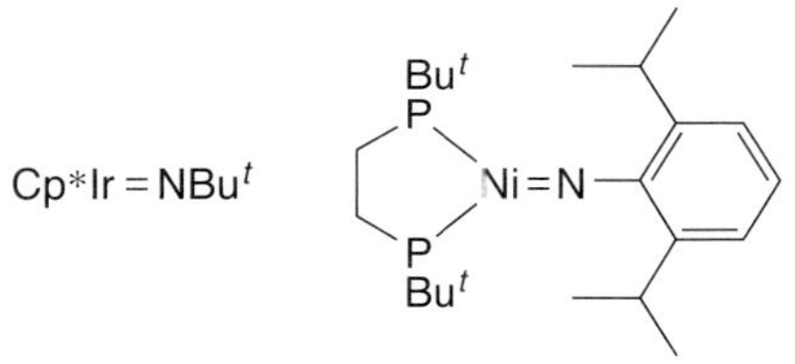

Figure 13.7.
Two low-valent, late-transition-metal–imido complexes.

This chapter does not include detailed information on metal–sulfido, –phosphinidene, and –borylene complexes, but the reader can find information on these types of complexes in a number of references and reviews. In brief, sulfido ligands[163,164] are part of minerals, bioinorganic systems, and catalysts for hydrodesulfurization in the petrochemical industry. The sulfido ligand often bridges two metals; terminal sulfido ligands are less common than bridging sulfido complexes, and the organometallic chemistry of complexes with terminal sulfido ligands is undeveloped.[257–259] Similarly, complexes containing bridging phosphinidine ligands are more common than those containing terminal phosphinidines.[167] A few terminal phosphinidine complexes have been generated as stable species[168] or reactive intermediates.[172–174] Although organometallic chemistry of phosphinidine ligands has begun to be developed recently, this topic remains less investigated than the chemistry of oxo and imido compounds. Transition metal complexes of borylene ligands have also been generated in recent years.[184,187–190,260] The borylene group in these complexes binds much like carbon monoxide, instead of a ligand bound through metal–ligand covalent multiple bonds.[197,198,260] In other words, the binding of the borylene ligand occurs by donation of an electron pair to the metal and π-acceptance into the two orthogonal p-orbitals at the boron atom. This π-backbonding is weaker than the ligand-to-metal donation in a typical metal–ligand multiple bond, such as that in an oxo or imido complex.

13.4.3. Bonding of Oxo and Imido Complexes[1]

A summary of the formal charges on imido and oxo ligands and the localized bonding is presented in Figure 13.8. Imido[162,261] and oxo ligands are typically considered in the ionic bonding model to be dianionic.[262] They are considered to be dianionic, even though some related carbenes are considered neutral, because oxygen and nitrogen are more electronegative than carbon, and more negative charge resides at the heteroatom of an oxo or imido ligand than at the carbon of a carbene ligand on the same metal fragment. In most cases, oxo and imido ligands are considered to be six-electron donors.[263] The metal–ligand covalent (non-dative) double bond involves four of these six electrons. A third bond, which results from π-donation from the ligand to an unoccupied orbital at the metal, accounts for the final two electrons. The degree of this dative π-bonding depends on the electron count at the metal. Because this π-bonding returns electron density to the metal, this π-bonding can influence whether an oxo or imido ligand reacts as a nucleophile or an electrophile.

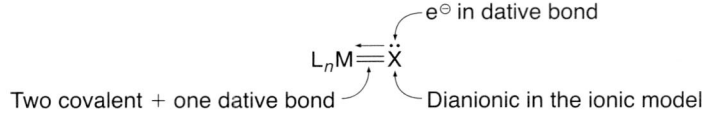

Figure 13.8.
A valence bond description of metal–imido and metal–oxo complexes.

Qualitative molecular orbital diagrams for octahedral, square pyramidal, and tetrahedral complexes containing one oxo, nitrido, or linear imido (six-electron donor) ligand are shown in Figure 13.9. These diagrams show that in the C_{4v} symmetry of the pseudo-octahedral and square pyramidal complexes, the π-bonding of the oxygen or nitrogen with the metal leads to splitting of the degenerate e_g and t_{2g} orbitals of octahedral symmetry. They also show that this π-bonding in a C_{3v} compound leads to splitting of the degenerate t_{2g} orbitals that are typically found in a tetrahedron. The d-orbitals in pseudo-octahedral compounds that contain a single oxo, nitride, or linear imido ligand have d_{xz} and d_{yz} orbitals that are π-antibonding in character and substantially higher in energy than the d_{xy} orbital. Thus, oxo complexes possessing a d^0, d^1, or d^2 electron count are much more stable than octahedral oxo complexes possessing higher d-electron counts that will have partially or fully filled d_{xz} and d_{yz} orbitals. In an alternative valence bond language, octahedral complexes containing a single oxo or imido ligand and two or fewer *d*-electrons can house these two electrons in a *d*-orbital that does not overlap with the filled π-orbitals of the M–L bonding, but complexes possessing more than two *d*-electrons must contain these electrons in a *d̄*-orbital that overlaps with the filled π-orbitals on the oxo, nitride, or linear imido ligand.

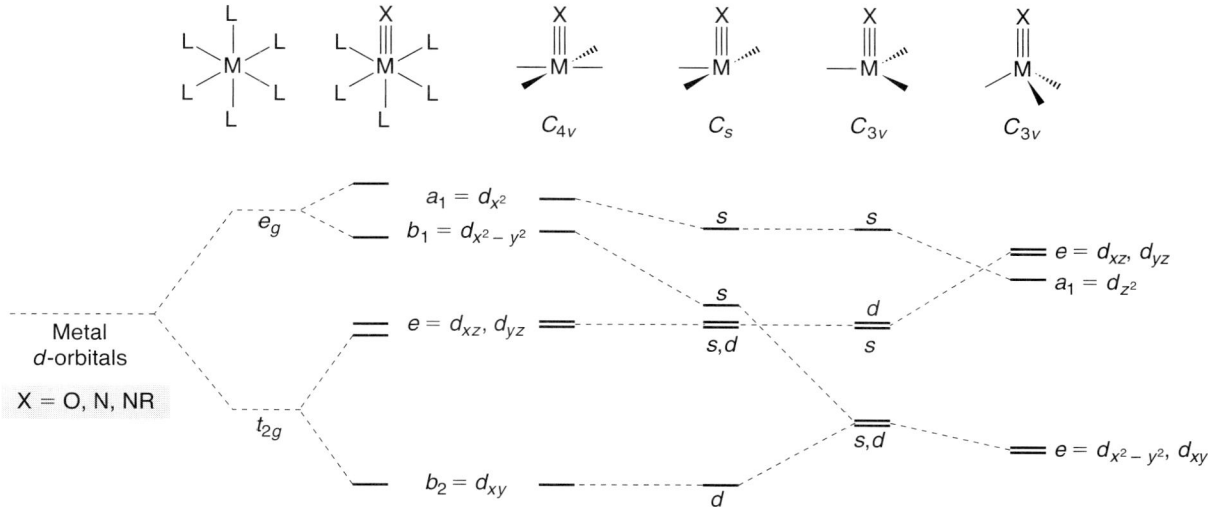

Figure 13.9.
Qualitative molecular orbital diagrams of oxo, nitrido, and linear imido complexes of various geometries. Adapted with permission from Figure 4 of Mayer, J. M., Thorn, D. L., Tulip, T. H. *J. Am. Chem. Soc.* **1985,** *107,* 7454.

At times, the extent of π-donation in imido complexes has been assessed by the bond angle between the metal, the nitrogen, and the carbon bound to nitrogen.[263] This analysis implies that linear imido compounds would have a larger degree of π-donation than bent imido compounds. However, fewer bent imido compounds[264–268] have been prepared than linear amido complexes, in part because the potential energy surface for bending of the group on an imido ligand is relatively flat and the linear structure is sterically favored. Nevertheless, an imido complex containing a single imido ligand has a stronger dative π-interaction than a complex containing multiple imido groups, and bent imido ligands are usually observed in complexes containing more than one imido group.[266] The degree of dative π-bonding in metal–oxo complexes follows a similar trend. Although the lack of a substituent on the oxo ligand prevents assessment of π-bonding by the linearity of the M=XR angle, early theoretical studies have noted changes in bond order upon changes in the number of oxo ligands on the same metal complex.[269, 270]

The small number of bent imido complexes might suggest that imido and oxo compounds are relatively unstable in the absence of this π-bonding. However, many imido complexes that would possess an excess of 18-electrons in the presence of an M–L dative π-bond still have relatively linear M–N–C linkages. For example, the vanadium–[271, 272] and tantalum–phenylimido complexes in Figure 13.10 have relatively linear M–N–C angles.[273] Thus, this linear geometry typically reflects some combination of a low energy for rehybridization of the filled orbital at nitrogen (soft potential for M–N–R bending),[274] steric repulsion between the substituent at nitrogen and the

178.2(6)°   179.7(5)°

Cp$_2$*Nb≡N–Ph   Cp$_2$*Ta≡N–Ph
     1.730(4)Å   1.707(6)Å

Figure 13.10.
Two linear imido complexes possessing 17-electron counts in the absence of a ligand-to-metal dative π-bond.

ancillary ligands at the metal, and π-donation into an orbital typically used for bonding with one of the ancillary ligands.[272]

In some complexes containing more than one oxo or imido ligand, the extent of π-bonding is constrained by the symmetry of the complex. In one particularly illustrative example, the orbital symmetry of the osmium–tris(amido) complex in Figure 13.11 prevents the existence of three full π-bonds.[275] Although group theory provides a more rigorous explanation for the bonding in this compound, one can appreciate without group theory that the *d*-orbitals in the trigonal plane have four-fold symmetry, but the combination of the three imido groups has three-fold symmetry. Thus, three full π-bonds cannot be formed, and each π-interaction with the metal is better considered as containing two-thirds of a π-bond.

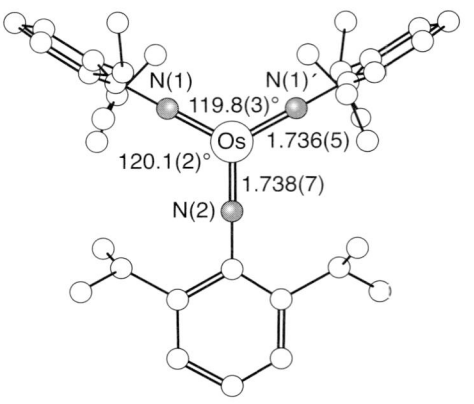

Figure 13.11.
A trisimido complex possessing partial ligand-to-metal π-bonding due to symmetry considerations. Adapted with permission from Schrock, R. R. *J. Am. Chem. Soc.* **1990**, *112*, 1642.

13.4.4. Synthesis of Metal–Imido and Metal–Oxo Complexes

13.4.4.1. Synthesis of Metal–Imido Complexes

Many routes to oxo and imido compounds have been developed. Transition metal–oxo complexes are commonplace and, in many cases, are the most stable form of an element in air.[276] Some imido compounds, including the first imido complex,[277] have been prepared from metal–oxo complexes (Equation 13.50). The reactions of some of the later metal–oxo compounds with a primary amine generate an imido species by extrusion of water.[277–281] In other cases, metal–imido complexes are prepared from oxo complexes by the reaction of phosphinimines (Equation 13.51)[281–283] or from oxo or halo complexes by the reactions of silylamines (Equations 13.52 and 13.53).[284–288] The driving force for these reactions is the formation of phosphine oxide, silyl ethers, or silyl halides.

$$OsO_4 \; + \; H_2N^tBu \xrightarrow{\; -H_2O \;} \underset{O}{\overset{O}{\underset{\|}{O=Os}}}{=}NBu^t \tag{13.50}$$

$$OsO_4 \; + \; 2\,Ph_3P{=}NR \xrightarrow{\; -\,2\,O{=}PPh_3 \;} \underset{O}{\overset{O}{\underset{\|}{O=Os}}}{\overset{NR}{\underset{NR}{<}}} \tag{13.51}$$

$$OsO_4 \; + \; Bu^tNHSiMe_3 \longrightarrow \; \underset{O}{\overset{O}{\underset{\|}{O=Os}}}{=}NBu^t \;\; + \;\; \underset{NBu^t}{\overset{O}{\underset{\|}{O=Os}}}{=}NBu^t \;\; + \;\; \underset{Bu^tN}{\overset{O}{\underset{\|}{O=Os}}}{\overset{NBu^t}{\underset{NBu^t}{<}}} \tag{13.52}$$

$$
\begin{array}{c}
\underset{\underset{\displaystyle \text{PPh}_3}{|}}{\overset{\overset{\displaystyle \text{PPh}_3}{|}}{\text{Cl}\diagdown \underset{\diagup}{\overset{\diagup}{\text{Mo}}}\diagup \text{Cl}}} \quad\xrightarrow{\ (\text{Me}_3\text{Si})\text{NHNMe}_2\ }\quad \left[\ \underset{\underset{\displaystyle \text{PPh}_3}{|}}{\overset{\overset{\displaystyle \text{PPh}_3}{|}}{\text{Cl}-\text{Mo}\!\!\diagdown^{\displaystyle \text{NNMe}_2}_{\displaystyle \text{NNMe}_2}}}\ \right]^{+}\ \text{Cl}^-
\end{array}
\qquad (13.53)
$$

However, the most common route to metal–imido compounds is some type of α-elimination. For example, imido complexes have been prepared by the addition of amine or alkali metal amides to a metal halide (Equations 13.54 and 13.55). This reaction most likely occurs through an α-elimination from an amido halide intermediate.[289] In addition, the first low-valent, late-transition-metal–imido complex was prepared by the simple reaction of [Cp*IrCl$_2$]$_2$ with hindered lithium amides (Equation 13.56).[254,255]

$$
\text{Cp*TaMe}_3\text{Cl} \ + \ \text{LiNHBu}^t \ \longrightarrow \ [\text{Cp*TaMe}_3(\text{NHBu}^t)] \ \longrightarrow \ \text{Cp*Ta(NBu}^t)\text{Me}_2 \ + \ \text{CH}_4 \qquad (13.54)
$$

$$
\text{WCl}_6 \ + \ \text{excess Bu}^t\text{NH}_2 \ \longrightarrow \ \underset{\text{Bu}^t\text{HN}}{\overset{\text{Bu}^t\text{HN}}{\diagdown}}\text{W}\!\!\underset{\text{NBu}^t}{\overset{\text{NBu}^t}{\diagup}} \ + \ 6\ \text{Bu}^t\text{NH}_3\text{Cl} \qquad (13.55)
$$

$$
\tfrac{1}{2}\,[\text{Cp*IrCl}_2]_2 \ + \ 2\ \text{LiNHBu}^t \ \longrightarrow \ \text{Cp*Ir}{=}\text{NBu}^t \ + \ \text{H}_2\text{NBu}^t \qquad (13.56)
$$

The elimination of an alkane from an amido alkyl complex has also formed imido complexes in many cases (Equations 13.57 and 13.58),[290,291] and reactions of a metal polyhalide with a silylamine has generated metal–imido complexes. Deprotonation of a cationic amido complex has been used as a route to a monomeric, late transition metal terminal imido complex.[256]

$$
(\text{Bu}_3{}^t\text{SiHN})_3\text{ZrPh} \ \xrightarrow[\text{THF}]{\Delta}\ \underset{\underset{\displaystyle \text{H}}{|}}{\overset{\overset{\displaystyle \text{H}}{|}}{\underset{\text{Bu}_3{}^t\text{SiN}}{\overset{\text{Bu}_3{}^t\text{SiN}}{}}}}\!\!\!\text{Zr}\!\!\underset{\displaystyle \text{O}}{\overset{\displaystyle \text{NSiBu}_3{}^t}{\diagup}} \qquad (13.57)
$$

$$
\text{Cp}_2\text{Zr}\!\!\underset{\text{NHBu}^t}{\overset{\text{Me}}{\diagup}} \ \xrightarrow[\Delta]{\text{THF}}\ \text{Cp}_2\text{Zr}\!\!\underset{\displaystyle \text{O}}{\overset{\displaystyle \text{NBu}^t}{\diagup}} \qquad (13.58)
$$

Metathesis reactions have also been used to prepare imido compounds. In one such metathesis reaction, an oxo complex reacts with an isocyanate to form an imido complex (Equation 13.59).[292] In another case, a carbene complex reacts with an organic imine to form the corresponding olefin and a transition metal–imido complex (Equation 13.60).[293,294]

$$
\text{ReO}_3(\text{OSiMe}_3) \ \xrightarrow[\text{2) 3 pyHCl/CH}_2\text{Cl}_2]{\text{1) 3 ArNCO/toluene reflux}}\ \underset{\text{ArN}}{\overset{\text{ArN}}{\diagdown\!\!\diagup}}\text{Re}\!\!\underset{\underset{\text{Cl}}{|}}{\overset{\overset{\text{Cl}}{|}}{\diagup\text{Cl}}}\text{py} \qquad (13.59)
$$

$$
\underset{\text{THF}}{\overset{\text{THF}}{}}\!\!\text{Ta}\!\!\underset{\text{CHBu}^t}{\overset{\text{Cl, Cl}}{}} \ \xrightarrow[-\text{Bu}^t\text{CH}=\text{CHPh}]{\text{PhCH}=\text{NR}}\ \underset{\text{THF}}{\overset{\text{THF}}{}}\!\!\text{Ta}\!\!\underset{\text{NR}}{\overset{\text{Cl, Cl}}{}} \qquad (13.60)
$$

$$(13.61)$$

$$(13.62)$$

Imido complexes can also be formed from metal–nitrido species by alkylation of the nitrogen (Equation 13.61).[217] Electron-rich nitrido species undergo alkylation with organic electrophiles, such as alkyl halides and trimethyloxonium reagents.[295,296] Electron-poor nitrido species undergo alkylation with carbon nucleophiles, and an example of the alkylation of a metal–nitride by a Grignard reagent is shown in Equation 13.62.[297]

$$Cp_2V \ + \ Me_3SiN_3 \longrightarrow Cp_2V{=}NSiMe_3 \qquad (13.63a)$$

$$Cp^*_2V \ + \ PhN_3 \longrightarrow Cp^*_2V{=}NPh \qquad (13.63b)$$

A final route to imido compounds has been developed by transfer of an imido group from a reagent such as an organic azide.[271,298,299] Reaction of an unsaturated transition metal complex with an organic azide can form, initially, a complex with the azide, and this complex can then undergo extrusion of dinitrogen to form the imido complex. Two related examples of the generation of imido complexes by transfer of a nitrene group from an azide are shown in Equations 13.63a and 13.63b. In a similar vein, reaction of an unsaturated metal center with an iminoiodoarene has been shown in some cases to transfer a nitrene group to form an imido complex.[300,301]

13.4.4.2. Synthesis of Metal–Oxo Complexes

A number of routes have been developed to prepare transition metal–oxo complexes in a controlled fashion. Many of these reactions parallel those used to prepare imido complexes, and therefore, the presentation here is brief. In most cases, oxo complexes have been generated by oxidation of a lower-valent metal complex. For example, the reaction of Cp_2NbCl with oxygen generates the oxo complex in Equation 13.64.[302] Alternatively, a variety of oxidants that can act as atom transfer reagents, such as NaOCl (bleach), H_2O_2, $ArCO_3H$, N_2O, and iodosyl arenes (PhI=O and more soluble analogs[303]) have been used to generate metal–oxo complexes.[304,305] Metal–oxo compounds have also been generated by metatheses, but these often generate insoluble, polymeric oxo products. For example, the reaction of Tebbe's reagent or Schrock's catalyst with a ketone (Equation 13.65) generates the olefin and the corresponding titanium–[39] or molybdenum–oxo[306] complex. Metal–oxo compounds have also been generated by α-eliminations. For example, reaction of a zirconocene hydroxo triflate complex with base leads to deprotonation of the hydroxo species, extrusion of triflate, and generation of the $Cp^*_2Zr{=}O$ intermediate (Equation 13.66).[258,259] Finally, metal–oxo compounds have been formed in one study by oxidative addition of the C=O bond in ketones and CO_2, as shown in Equation 13.67.[307–309]

$$Cp_2NbCl + O_2 \longrightarrow Cp_2NbCl(O) \tag{13.64}$$

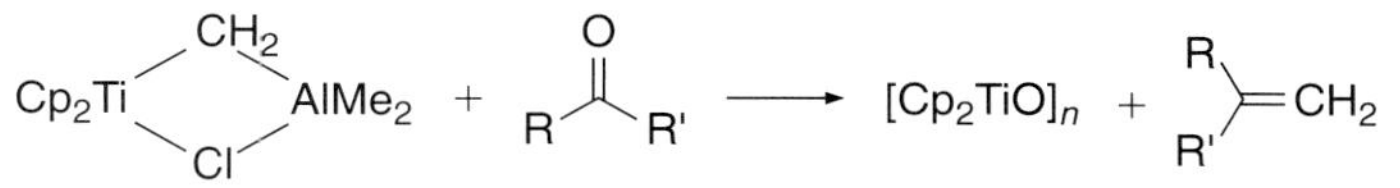

$$(13.65)$$

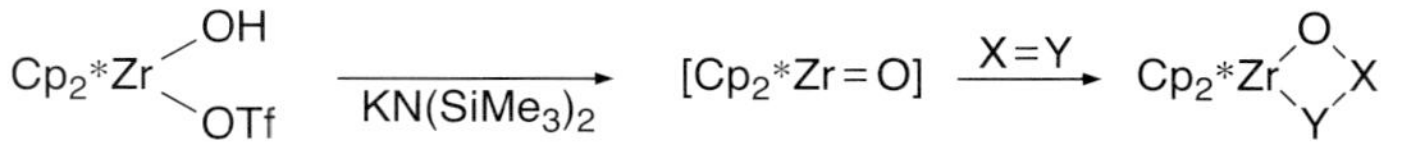

$$(13.66)$$

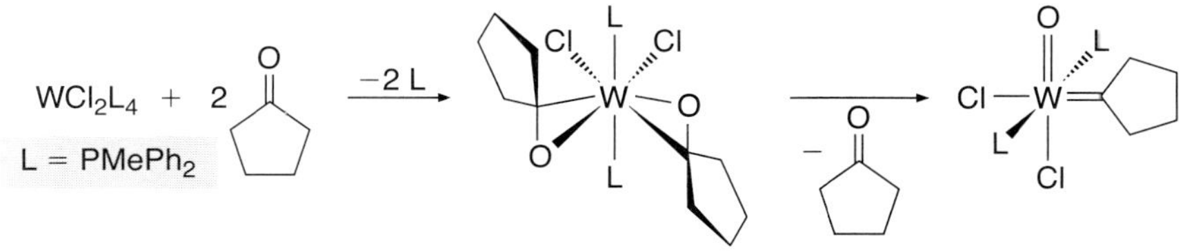

$$(13.67)$$

13.4.5. Reactions of Imido and Oxo Compounds

Imido and oxo complexes undergo a variety of reactions with organic substrates. These reactions can be classified into cycloadditions, alkylations, atom transfers (e.g. to olefins), hydrogen atom abstractions, and migrations of hydride and alkyl groups from the metal to the heteroatom. Generic examples of these reactions are shown in Scheme 13.20 and are described more specifically in the following paragraphs.

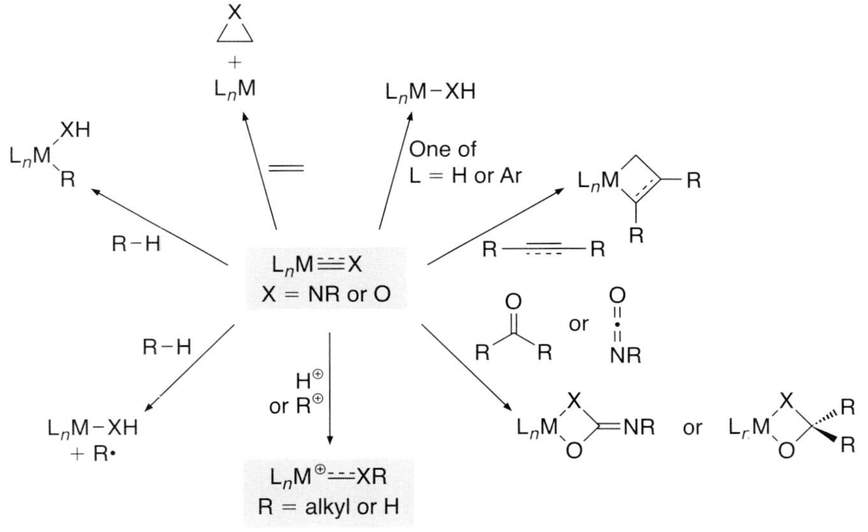

Scheme 13.20

13.4.5.1. [2+2] and [3+2] Cycloadditions

Cycloadditions are among the more common reactions of imido and oxo groups in catalytic transformations. Both [2+2] cycloadditions and [3+2] cycloadditions are well established. The [2+2] cycloaddition chemistry occurs during hydroaminations of alkynes, allenes, and strained olefins catalyzed by imido compounds.[310,311] The [3+2] cycloaddition reactions appear to occur during dihydroxylation and aminohydroxylation of olefins catalyzed by osmium complexes.[312–315]

Examples of the [2+2] cycloaddition of oxo and imido complexes with olefins and alkynes are shown in Equations 13.68–13.73. As shown in Equations 13.68 and 13.69, zirconium– and titanium–imido compounds react with alkynes to generate metallacyclobutenes.[291,311,316] Several early-metal–imido complexes have been shown to react reversibly with ethylene to generate metallaazacyclobutanes (Equations 13.68, 13.70, 13.71, and 13.72).[310,317,318] Although few examples have been published to date, early-metal–oxo complexes can also react with substrates containing carbon–carbon multiple bonds by [2+2] pathways to generate oxametallacycles. One example is shown in Equation 13.73.

$$(13.68)$$

$$(13.69)$$

$$(13.70)$$

$$(13.71)$$

$$(13.72)$$

$$\text{(13.73)}$$

The thermodynamics for the addition of alkenes are much less favorable than those for the addition of alkynes. In several cases, the [2+2] additions of alkynes are thermodynamically favorable enough to occur to completion, but related additions of alkenes occur reversibly. This trend was shown by comparing the two reactions in Equation 13.68. Therefore, the metallacycles generated by [2+2] additions of alkenes to oxo and imido complexes have rarely been isolated.[319,311]

A nickel–imido complex and a ruthenium–imido complex bound by an ancillary porphyrin ligand also react with olefins, in this case to generate aziridine products (Equations 13.74 and 13.75).[300,320,321] A pathway for the formation of an aziridine that occurs by a [2+2] addition of the olefin across the metal imido unit, followed by reductive elimination of the aziridine, was proposed for the reaction of the nickel complex,[320] although the azametallacycle was not observed directly. Because of the lack of coordination site cis to the imido group in the ruthenium–porphyrin system, the transfer of the imido group likely occurs by direct reaction of the imido group with the olefin.[300,321]

$$\text{(13.74)}$$

$$\text{(13.75)}$$

The [2+2] reactions of the zirconium–imido compounds with alkynes and alkenes occurs by a mechanism similar to that for the [2+2] reactions of carbenes with alkynes and alkenes. The alkene or alkyne first binds to an open coordination site at the metal, and this coordination is followed by conversion of the alkyne or alkene complex to the metallacylic product (e.g. Equation 13.76). Thus, the [2+2] reaction requires a 16-electron intermediate to bind the olefin or alkyne, even though the metallacylic product and the imido complex have the same overall electron count. In support of the coordination of alkyne or alkene, albeit weakly, to the d^0 metal center, the rate of the reaction of alkynes with the 18-electron zirconocene–imido compound containing bound pyridine-N-oxide was inhibited by added pyridine-N-oxide (Equation 13.76).[311]

$$\text{(13.76)}$$

The mechanism of the reactions of complexes containing more than one oxo group with olefins has been studied for many years in the context of the catalytic dihydroxylation and aminohydroxylation of olefins.[314] Both the combination of a [2+2] cycloaddition followed by rearrangement to the final [3+2] addition product[322] and direct [3+2] reactions of the olefin with the osmium species have been proposed (Scheme 13.21).[323] Computational studies predict that the barrier for the [2+2] addition is much higher than the barrier for the [3+2] addition and that the barrier for rearrangement of the [2+2] addition product to the final [3+2] addition product also has a high barrier.[324–326] The enthalpic barrier for the [3+2] pathway with model compounds (OsO_4 with and without coordinated NH_3) has been calculated to be less than 2 kcal/mol, while the enthalpic barrier for the [2+2] addition exceeds 40 kcal/mol. In addition, studies that measure ^{13}C isotope effects have shown similar magnitudes for both carbons of the olefin.[327,328] These data imply that a concerted [3+2] cycloaddition occurs directly.

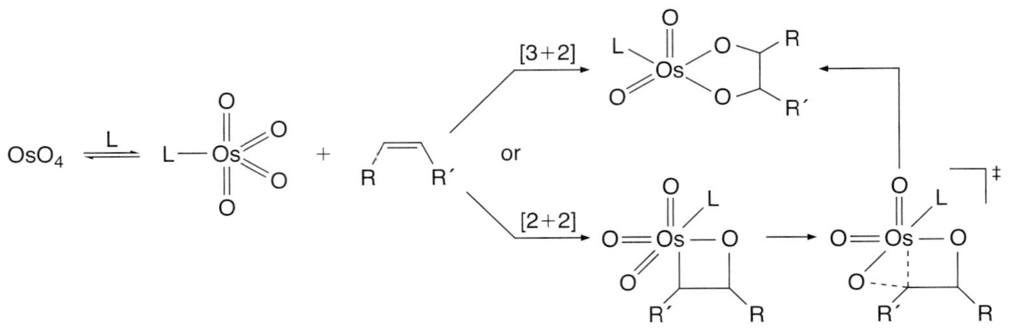

Scheme 13.21

Evidence in favor of the [2+2] mechanism is circumstantial, but it does include several types of studies. This evidence includes nonlinear free energy relationships between the substituent parameters on vinylarenes and the rates of the dihydroxylation,[329] and it includes temperature effects on selectivity.[330] It also includes the results of studies on the cleavage of Cp*Re(O)(diolate) complexes to Cp*ReO₃ and free olefin.[331] The electronic effects, enthalpy of activation versus the strain of the olefin, and secondary deuterium isotope effects obtained from these studies on the rhenium complex support a stepwise cleavage process that could occur by initial formation of an oxametallacyclobutane intermediate. In the end, however, the combination of computational data and isotope-effect measurements seem to have led the community to accept that osmium tetroxide reacts by the direct [3+2] pathway. The mechanism of the reaction of O₃OsNR with olefins during aminohydroxylation presumably follows the same type of [3+2] pathway.

13.4.5.2. Atom Transfer of Oxo and Imido Groups to Olefins

Imido and oxo compounds are intermediates in many of the transfers of oxygen atoms and nitrene units to olefins to form epoxides and aziridines, and they are intermediates in many of the insertions of oxygen atoms and nitrene units into the C–H bonds of hydrocarbons to form alcohols and amine derivatives. The enantioselective epoxidation of allylic alcohols (Scheme 13.22)[332–334] is the most widely used epoxidation process, and the discovery and development of this process was one of the sets of chemistry that led K. Barry Sharpless to receive the Nobel Prize in Chemistry in 2001.[335,336] The mechanism of this process is not well established, despite the long time since its discovery and development. Nevertheless, most people accept that transfer of the oxygen atom occurs from a titanium–peroxo complex[337,338] rather than from an oxo complex. Jacobsen's[339–341] and Katsuki's[342,343] manganese–salen catalysts for the enantioselective epoxidations of unfunctionalized olefins,[344,345] which were based on Kochi's achiral chromium– and manganese–salen complexes,[346–349] are a second set of remarkable epoxidation catalysts (Scheme 13.23). Reaction of these catalysts occurs by transfer of oxygen from an oxo complex to an olefin.[350,351] Likewise, chiral porphyrin catalysts for enantioselective epoxidations are likely to react through metal–oxo intermediates.[352-355] Epoxidations of alkenes with dioxygen as the oxidant remain scarce.[356]

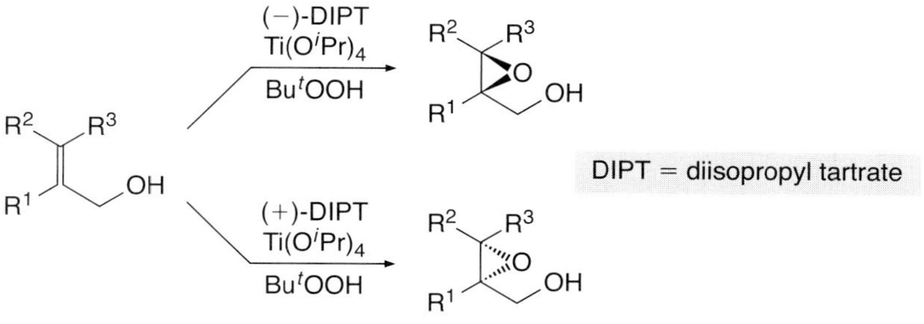

Scheme 13.22

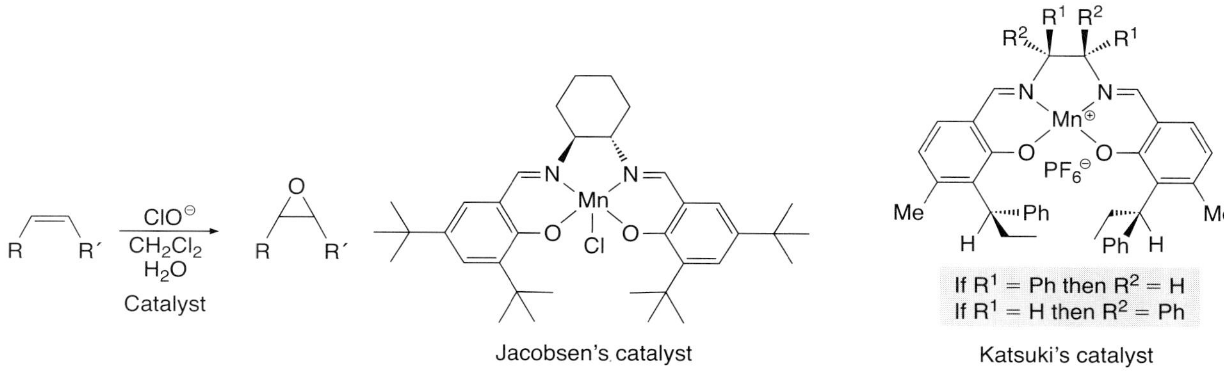

DIPT = diisopropyl tartrate

Jacobsen's catalyst

Katsuki's catalyst

If R¹ = Ph then R² = H
If R¹ = H then R² = Ph

Scheme 13.23

The mechanism of the transfer of an oxygen atom in a metal–oxo complex to an olefin to form an epoxide includes a vast literature, which has been summarized in books on biomimetic oxidations.[357] Thus, it is impossible to even touch on most of the work in a few paragraphs. As a result, this section focuses on examples of well-defined metal–oxo complexes that react with olefins to form epoxides.

Examples of well-characterized metal–oxo complexes that react with olefins are shown in Scheme 13.24 and Equations 13.77 and 13.78. Manganese complexes containing porphyrin ligands are involved in many biological and biomimetic oxidation processes, and Mn(V) oxo complexes have been invoked as intermediates.[357] Groves characterized a porphyrin-ligated Mn(V) oxo complex that reacts with olefins to form epoxides (Scheme 13.24).[358] High-valent iron–oxo complexes have been invoked as intermediates in the oxidations catalyzed by both heme and non-heme iron enzymes.[359-367] A particularly unusual Fe(IV) oxo complex that mimics the structure and reactivity of the proposed intermediates in oxidations by non-heme iron was isolated recently by Que and coworkers and crystallographically characterized (Equation 13.77).[368] Manganese–oxo complexes that are intermediates in the Jacobsen–Katsuki epoxidations of unfunctionalized olefins have also been generated and characterized by mass spectrometry (Equation 13.78).[350,351] Many theoretical studies of the reactions of these complexes have been reported.[369-375]

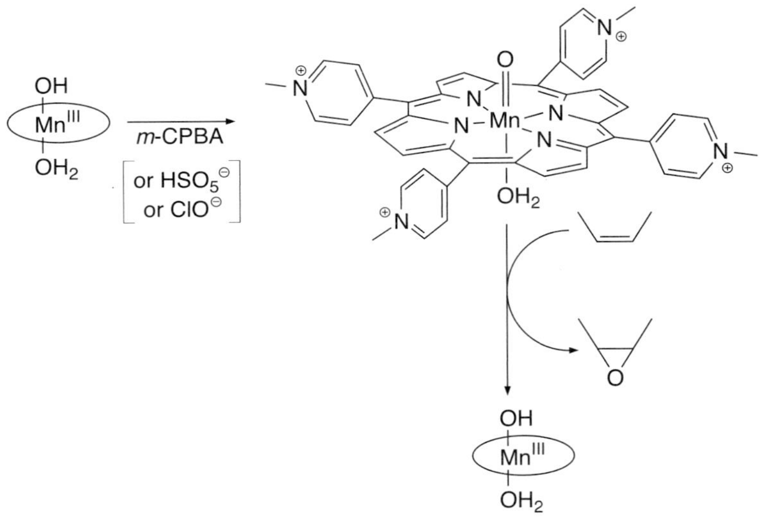

Scheme 13.24

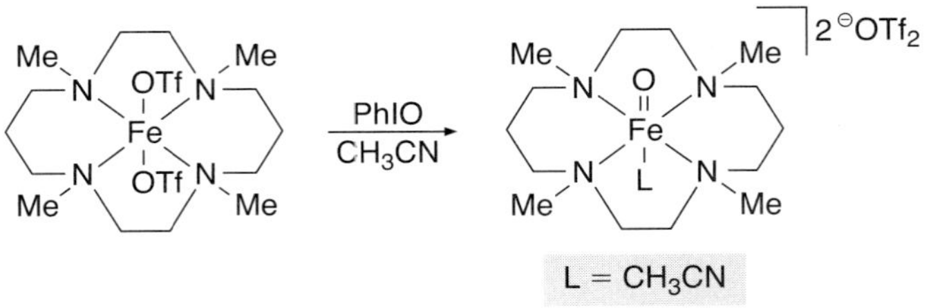

L = CH₃CN

(13.77)

X = ClO₄
L = NCMe

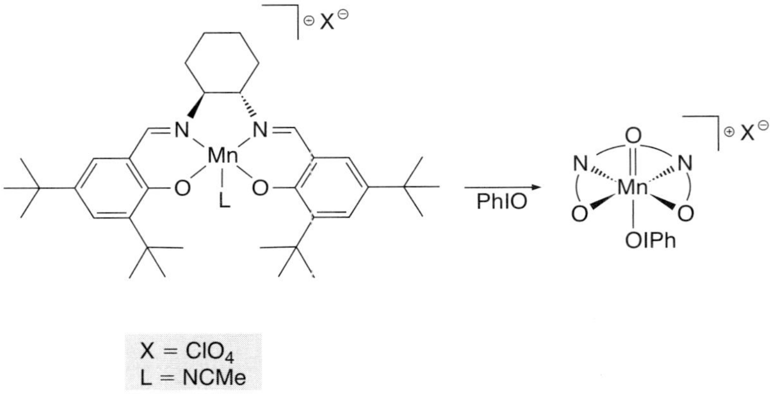

(13.78)

The mechanisms of the epoxidations of olefins have been studied intensively.[339–341,376–378] In general, the reactions of metal–oxo complexes with olefins to form epoxides do not involve intermediates containing metal–carbon bonds. The oxo group tends to act as an electrophile and interact with the HOMO of the olefin during the transfer of the oxo to the olefin. After this initial interaction, the epoxide may form by a non-radical concerted process or by a stepwise process involving radical or cationic intermediates.[372–375]

The epoxidation of olefins does not appear at this time to involve radical intermediates in most systems, but epoxidation through two different spin states of the same complex has gained acceptance.[372–375] In some cases, side products imply that intermediates are present, but these intermediates have been concluded by Bruice to be carbocationic instead of radical. The carbocation would allow for Z/E isomerization and formation of oxidation products other than epoxides.[380–382] Studies on the epoxidations of an olefin attached to ultrafast radical clocks showed than the Mn–salen system of Jacobsen catalyzes the epoxidations without radical intermediates (Equation 13.79).[379]

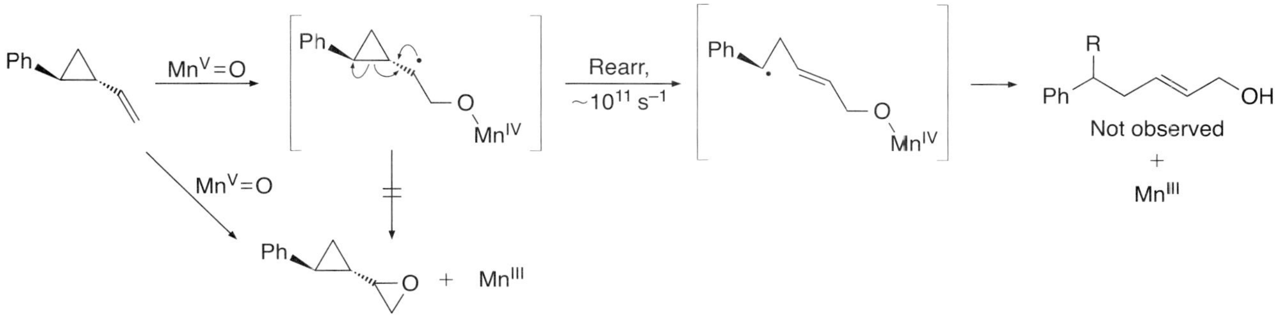

(13.79)

The aziridination of olefins has also been studied, but fewer complexes catalyze this reaction as efficiently as iron and manganese complexes catalyze the epoxidation of olefins. Nevertheless, the aziridinations of olefins catalyzed by copper,[213–215] ruthenium,[383] and rhodium[384] complexes have been reported. The source of nitrogen is usually [N-(p-toluenesulfonyl)imino]phenyliodinane (PhI=NTs) or a precursor to a related iodoarylimine. The aziridine is likely generated from these copper- and rhodium-catalyzed reactions by an outer-sphere process in which the olefin interacts with the LUMO of the complex, which is located at the nitrogen. This mechanism is more likely to be followed by these catalysts than a [2+2] process, followed by reductive elimination.

13.4.5.3. Reactions with C–H Bonds

Imido compounds also undergo [2+2] reactions with C–H σ-bonds. This chemistry was described in Chapter 6. For example, several zirconium–, titanium–, and tungsten–imido compounds undergo [2+2] reactions with the C–H bonds of alkanes and arenes to generate organozirconium–, titanium–, and tungsten–amido products (Equation 13.80).[290,291,317,385–394] Ruthenium–porphyrin[321,395,396] and dirhodium–tetracarboxylate complexes[210,397] also catalyze the insertion of nitrenes into C–H bonds (Equations 13.81–13.83). The lack of an open coordination site cis to the imido group in the proposed intermediate implies that these reactions occur by direct interaction of the imido group with the C–H bond. This mechanism contrasts that for reactions of d^0 imido complexes with alkanes and arenes. The amination of C–H bonds by ruthenium is believed to occur through an alkyl radical intermediate, as deduced from substituent effects on reaction rates.[300]

$X = Cp$ or $Bu^t_3Si(H)N$
$M = Ti$, Zr, or W
$R = Bu^t$, Ar, or $SiBu^t_3$
$R' = $ alkyl or aryl

(13.80)

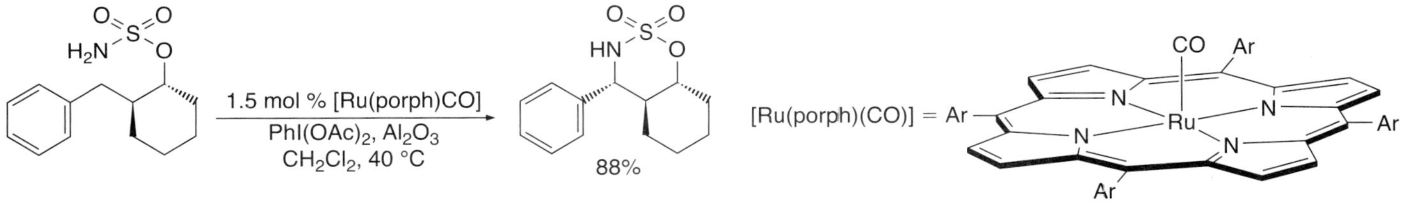

(13.81)

$$\text{(13.82)}$$

53%, 81% ee

$$\text{(13.83)}$$

The oxidation of alkanes through oxo complexes is a process that has potential importance for the selective transformations of alkanes. Thus, the mechanism of these reactions has been studied in detail. Much of this work is beyond the scope of this text on organometallic chemistry because these oxidation reactions are unlikely to generate intermediates containing metal–carbon bonds. Instead, the oxo group interacts with the hydrocarbon C–H bond to transfer the oxygen atom by an outer-sphere mechanism.

The mechanism of the catalytic hydroxylation of alkanes via metal–oxo compounds has been studied intensively. Reactions that are concerted, that occur through alkyl radicals, and that occur through carbocations have all been proposed. A mechanism in which a hydrogen atom transfers to a metal oxo occurs in some cases. The alkyl radical can be trapped by halogen or can undergo reaction with the resulting hydroxide to generate alcohol. This later reaction has been called a "rebound mechanism" (Equation 13.84).[246,398–405]

$$L_nM{=}O \ + \ R{-}H \ \longrightarrow \ L_nM{-}OH \ + \ R{\bullet} \ \longrightarrow \ L_nM \ + \ ROH \qquad \text{(13.84)}$$

Figure 13.12.
Rate constants for hydrogen atom abstraction from dihydroanthracene vs. the strength of the O–H bond formed by oxygen radicals and manganese complexes. Adapted with permission from Mayer, J. M. *Acc. Chem. Res.* **1998**, *31*, 441.

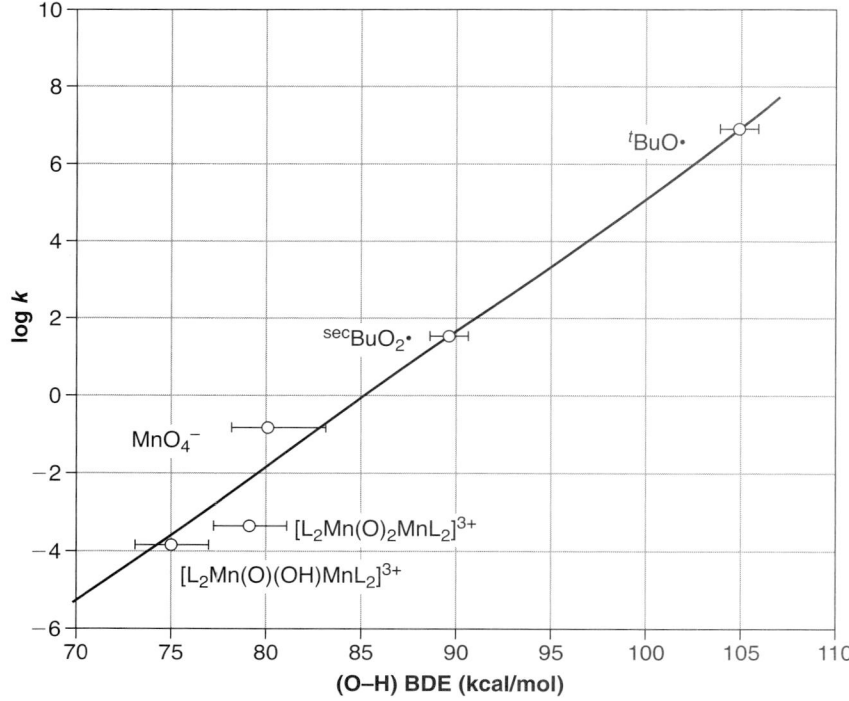

An analogy between the reactivity of a metal–oxo species and an organic alkoxy radical has been noted by Mayer.[243] The rates and selectivities of the hydrogen atom abstraction by some metal–oxo compounds parallel those of the hydrogen atom abstraction by alkoxo radicals. In particular, metal–oxo complexes fit with alkoxo radicals in a correlation between the strength of an O–H bond and the rate of hydrogen atom abstraction (Figure 13.12). This correlation does not require that the metal–oxo compound be a radical species or have radical character at the oxygen.

13.4.5.4. Reactions with Electrophiles

Imido and oxo species react with a variety of electrophiles that include protons, alkyl halides, carbonyl compounds, and heterocumulenes. Representative reactions of oxo and imido complexes with acid and methyl bromide are shown in Equations 13.85–13.89.

In some cases, reactions of oxo compounds with mineral acids generate water and the metal halide (Equation 13.85).[406] Similarly, reactions of imido complexes with mineral acids often generate amine and the metal dihalide (Equation 13.86).[283] The rate of protonation of metal–oxo complexes has been studied. The rate of protonation of bridging oxo compounds can be slow,[407] but the rate of protonation of a terminal oxo species is presumably faster.

$$(13.85)$$

$$(13.86)$$

Imido complexes also react with organic electrophiles. For example, the molybdenum-imido complex in Equation 13.87 reacts with methyl bromide to generate the metal dibromide and trialkylanilinium bromide.[408] Simple alkylations of an oxo group to generate an alkoxide product are uncommon, but one example of alkylation of a zironocene–oxo complex is shown in Equation 13.88.[409] One example of a related silylation to generate a siloxide complex is shown in Equation 13.89.

$$(13.87)$$

$$(13.88)$$

$$AgReO_4 \ + \ Me_3SiCl \ \longrightarrow \ Me_3SiOReO_3 \qquad\qquad (13.89)$$

Oxo and imido compounds also react as nucleophiles with compounds containing polar C=X bonds, such as aldehydes, ketones, imines, and heterocumulenes. These reactions generally occur by a formal [2+2] cycloaddition process to generate a metalacycle that breaks down to the more thermodynamically stable combination of products containing metal–ligand and carbon–heteroatom multiple bonds. Three examples of such reactions are shown in Equations 13.90–13.92.[410–412] In the first two cases, the four-membered metallacycle was isolated or characterized by low-temperature NMR spectroscopy.[410,411]

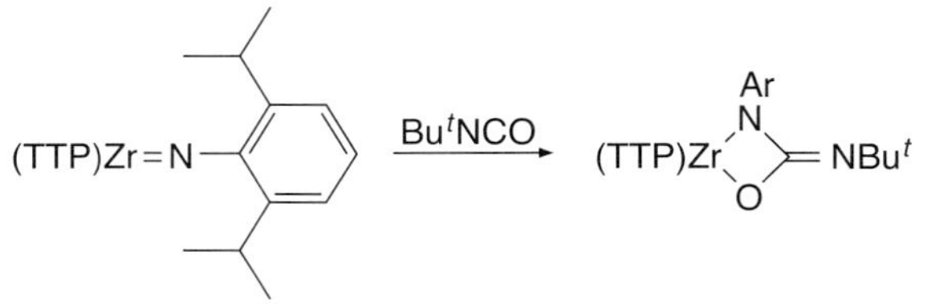

(13.90)

TTP = meso-5,10,15,20-tetra-p-tolyporphyrinato dianion

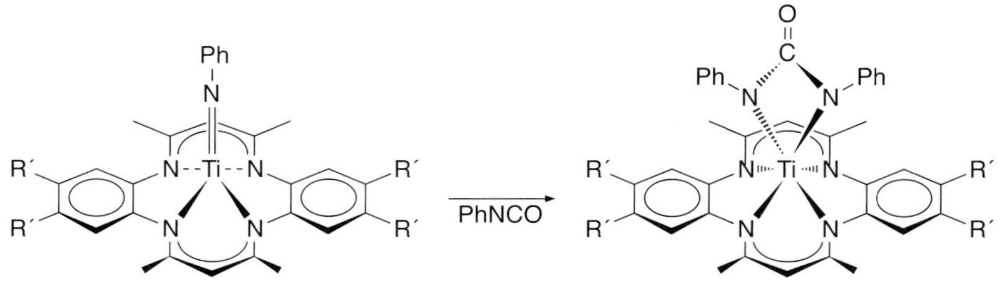

(13.91)

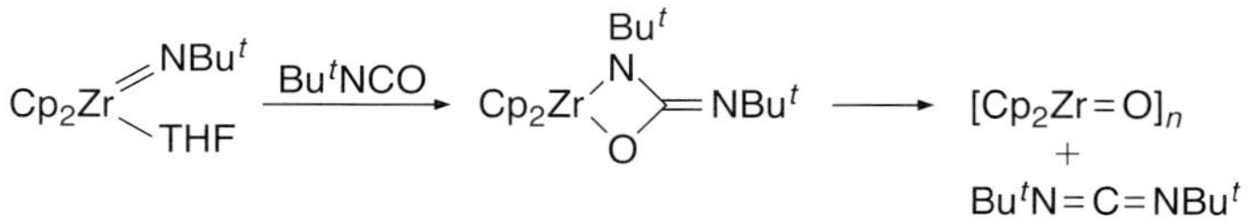

(13.92)

13.4.5.5. Migrations of Alkyl and Hydride Groups from M to O or N

The migration of an alkyl group from the metal center to the oxo or imido ligand is an elementary reaction that is important for achieving the goal of selective oxidation by organometallic C–H activation chemistry. Thus far, only a few examples of this process have been observed. Reversible transfers of protons from the metal to an oxo group are more common. For example, a rhenium–hydroxo compound rearranges to an oxo–hydride compound (Equation 13.93).[413]

Circumstantial evidence for the transfer of an aryl group from the metal to an oxo ligand was gained many years ago.[414,415] The reactions of Ph_2Hg with $OVCl_3$ (Equation 13.94) and Ph_2Zn with $Cr(NBu^i)_2(OR)_2$ generated phenol and an arylamine after addition of acid, but the arylmetal intermediates were not observed. In contrast, the migrations reported by Brown and Mayer shown in Equations 13.95–13.97 were initiated with a defined metal–aryl complex.[416,417] In one case, the migration of the aryl group from the metal to the oxo ligand occurs thermally after generation of the dioxo complex by DMSO or pyridine-*N*-oxide. Both metal–phenolate and metal–catecholate products are formed. In a second system, the migration of a phenyl group, which had been generated by C–H activation of benzene, occurs photochemically. In a third process, the migration of an ethyl group occurs photochemically. In contrast to the migration of aryl groups, the product distribution from reaction of the ethyl complex and the effect of thiophenol on this product distribution implies that the ethyl migration occurs by a radical process.

$$(13.93)$$

$$(13.94)$$

$$(13.95)$$

$$(13.96)$$

$$(13.97)$$

13.4.5.6. Catalytic Reactions of Imido and Metal–Oxo Compounds Through Organometallic Intermediates

Two catalytic processes involving organometallic imido complexes have been studied in depth. Other catalytic reactions of imido and oxo complexes that occur without organometallic intermediates, such as the hydroxylation and amination of C–H bonds, the epoxidations and aziridination of olefins, and the dihydroxylation and aminohydroxylation of olefins, were mentioned along with leading references in the previous sections. In one catalytic process—olefin metathesis—the imido group of an organometallic imido complex serves as an ancillary ligand, while in a second process—alkene and alkyne hydroamination—the imido group serves as the point of reaction. The scope and mechanism of the catalytic olefin metathesis by complexes with ancillary imido ligands is not presented here because it is discussed in Chapter 21 on olefin metathesis. However, the ability of the imido ligand to promote the [2+2] reaction of an accompanying carbene ligand with an olefin has been proposed to be crucial to the high reactivity of these catalysts.[37] The [2+2] reaction with the carbene is particularly favorable because the C=N bond strengthens as the carbene is converted into two metal–carbon single bonds in the product (Equation 13.98).

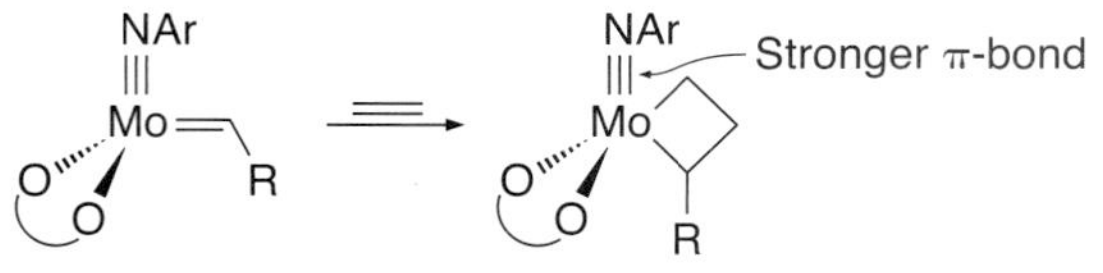

$$(13.98)$$

Hydroaminations of alkynes and alkenes occur through transition metal imido intermediates, and the mechanism of this process has been studied in depth. The mechanism of this catalytic process is shown in Scheme 13.25. A zirconocene–imido complex is generated from α-elimination of a bis(amide).[221,222] The zirconocene–imido compound then undergoes a [2+2] cycloaddition with the alkyne or alkene to generate an azametallacyclobutene or an azametallacyclobutane intermediate. These metallacycles undergo protonation by amines to regenerate a bis(amide) intermediate, and this bis(amide) intermediate undergoes α-elimination to generate the organic product and regenerate the imido complex.

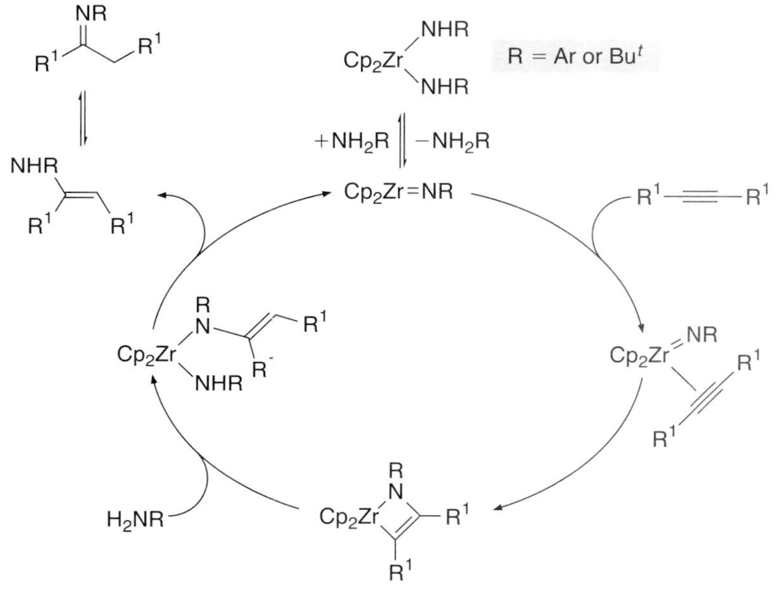

Scheme 13.25

The first hydroaminations by this mechanism were reported by Bergman with zirconocene complexes[221,222] and by Livinghouse[418] with monocyclopentadienyl titanium and zirconium complexes. Bergman reported the intermolecular addition of a hindered aniline to an alkyne. The hindrance of the aniline was important to prevent formation of stable dimeric complexes containing bridging imido groups. Livinghouse reported intramolecular reactions that occurred at lower temperatures over shorter times. The intramolecularity of this process allows the [2+2] cycloaddition of the imido complex with the alkyne to be faster than the dimerization.

Since that time, several types of related complexes have been studied for the hydroamination of alkynes and alkenes. Titanocene complexes have also been used as catalyst precursors. Doye has reported reactions catalyzed by titanocene dimethyl as a precursor.[224] Bergman has shown that one of the cyclopentadienyl ligands in the titanocene catalysts can be eliminated by amine, and the reactions catalyzed by the resulting monocyclopentadienyl complex are faster than those through the bis(cyclopentadienyl) compounds.[419a] Complexes of several anionic multidentate ligands have been tested for these reactions. For example, Odom has reported that complexes of dipyrrolylamine ligands catalyze the hydroamination of alkynes[225] and Bergman has reported reactions catalyzed by bis(sulfonamido) ligands.[419b] Schafer has reported reactions catalyzed by complexes of amidate ligands,[238] and even intramolecular reactions catalyzed by the simple compound $Ti(NMe_2)_4$.[420]

13.4.6. Nitrido Ligands (Written with Dr. Devon C. Rosenfeld)

13.4.6.1. Overview

This chapter closes with a brief discussion of monomeric metal–nitrido complexes. Metal–nitrido complexes contain a metal–nitrogen triple bond and a nonbonded electron pair at nitrogen. The nitride ligand is uncommon in organometallic chemistry, but nitrido complexes have been shown to serve as precursors to imido complexes by reaction with electrophiles and to act as a valuable supporting ligand. In addition, nitrido complexes are postulated intermediates in important chemical transformations such as the Haber–Bosch process.[421] Most metal–nitrido complexes contain early to middle transition metals in the second and third rows of the transition series. Metal–nitrido complexes have been the subject of several reviews.[191–195] Although monomeric and polynuclear metal–nitride complexes are known, this section focuses on the monomeric nitrido complexes that have been most closely related to organometallic processes.

13.4.6.2. Bonding of Nitrido Ligands

The nitrido ligand is typically considered trianionic in the ionic counting scheme. In this case, six of the eight electrons on nitrogen interact with three unoccupied orbitals at the metal, as noted earlier in the section of this chapter on the bonding in metal–heteroatom multiple bonds, and as shown more explicitly in Figure 13.13. As a result, the nitrido–metal bond possesses a formal metal–ligand triple bond, consisting of one σ- and two π-bonds. The σ-bond is formed by overlap of an sp-hybridized orbital at nitrogen and an unoccupied frontier orbital at the metal of σ-symmetry. The π-bonding involves two orthogonal occupied p-orbitals on nitrogen and two d-orbitals of π-symmetry at the metal. The electron pair at nitrogen then lies in the nonbonding sp-hybridized orbital. The formal metal–nitrogen triple bond causes a nitride ligand to be a strong π-donor ligand.

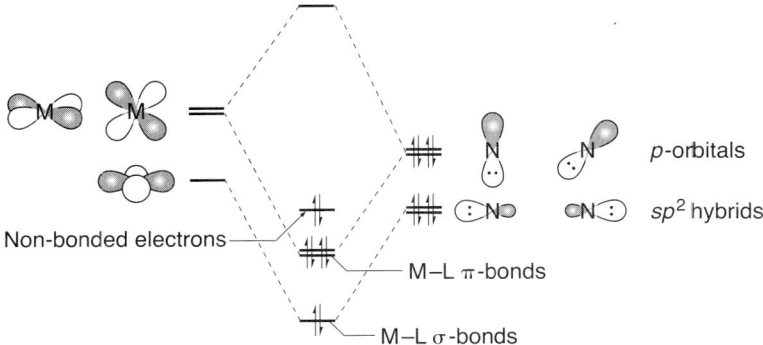

Figure 13.13.
Qualitative orbital diagram of the interaction between a transition metal fragment and a nitrido ligand.

13.4.6.3. Structural and Spectral Features

Many metal–nitrido complexes have been characterized by X-ray crystallography. The metal–nitride linkage exhibits short metal–nitrogen bond lengths that typically range from 1.54–1.67 Å.[194] The specific length depends upon the metal and its oxidation state. Metal–nitride complexes have also been characterized by infrared spectroscopy. The M–N stretching mode for terminal nitrides ranges from 950–1150 cm^{-1} and shifts upon the replacement of ^{14}N with ^{15}N.[194] The substitution of ^{14}N with ^{15}N also allows for characterization of the nitride complex by ^{15}N NMR spectroscopy. The chemical shifts for ^{15}N-labeled nitrido complexes range from +40 to +260 ppm.[193]

13.4.6.4. Synthesis of Metal–Nitrido Complexes

Metal–nitride complexes have been synthesized by a number of methods. The reaction of nitrogen trichloride with a homoleptic metal–carbonyl complex, followed by heating, has been shown to form metal–nitride complexes (Equation 13.99).[422,423] Several ruthenium– and osmium–nitride complexes have been prepared by the reaction of a metal–oxo precursor with sodium azide in acidic media (Equation 13.100),[424,425] or by reaction of a metal chloride with NaN_3 or Me_3SiN_3.[426] The thermolysis or photolysis of azido complexes can also result in the formation of metal–nitride complexes. For example, the thermolysis of the tetraazide pyridine complex of molybdenum formed from reaction of $MoCl_4(py)_2$ with four equivalents of silyl azide forms the pyridine-ligated, triazido nitrido complex in Equation 13.101.[427] In related chemistry, the reaction of hydrazine with the bis(trialkylphosphine) rheniumoxotrichloride complex formed from reaction of $NaReO_4$ with the phosphine forms the bis(trialkylphosphine)rhenium nitrido dichloride complex in Equation 13.102.[428]

$$2 \, Mo(CO)_6 \; + \; 4 \, NCl_3 \; \longrightarrow \qquad\qquad \xrightarrow{100 \,°C} \; Mo(\equiv N)Cl_3 \; + \; Cl_2 \tag{13.99}$$

$$+ \; 12 \, CO \; + \; N_2 \; + \; Cl_2$$

$$\textit{trans-}[Ru(O)_2X_4]^{2-} \; + \; HX \; + \; NaN_3 \; + \; CsX \; \longrightarrow \; Cs_2[Ru(\equiv N)X_5] \tag{13.100}$$

$$(X = Cl \; or \; Br)$$

$$MoCl_4(py)_2 \; + \; 4 \, (CH_3)_3SiN_3 \; \longrightarrow \; Mo(\equiv N)(N_3)_3py \; + \; N_2 \; + \; py \; + \; 3 \, (CH_3)_3SiCl \tag{13.101}$$

$$NaReO_4 \; + \; PR_3 \; + \; N_2H_4 \cdot 2 \, HCl \; \xrightarrow{EtOH} \; [Re(\equiv N)Cl_2(PR_3)_n] \tag{13.102}$$

$$PR_3 = PMe_2Ph$$

Nitrido complexes have also been formed by metathesis and atom transfer processes. The reaction of dinitrogen with a molybdenum(III) species forms a molybdenum–nitrido complex, as shown in Equation 13.103.[429] and described in more detail in the section of Chapter 5 on dinitrogen complexes. In a metathesis process involving related complexes, the reaction of a metal–alkylidyne complex with a nitrile extrudes an alkyne to form a metal nitride that adopts a dimeric structure (Equation 13.104).[430a,b] Related nitrido complexes have been formed from an azabicyclic compound that eliminates anthracene after forming the M–N bond (Equation 13.105).[430c]

$$2 \, ArNR - Mo(NRAr)_2 \; \xrightarrow[\text{1 atm}]{N_2} \; 2 \quad N\equiv Mo(NRAr)_3 \tag{13.103}$$

$$ (13.104) $$

R = 2,6-^{i}Pr$_2$C$_6$H$_3$
R′, R″ = aryl

$$ (13.105) $$

13.4.6.5. Reactions of Metal–Nitrido Complexes

The stability of the metal–nitride triple bonds makes its reaction chemistry limited, but several classes of reactions that metal–nitrido complexes do undergo are summarized here. The uncoordinated lone pair on nitrogen can act as a nucleophile, as a Lewis base, or as a bridge to form μ-nitrido complexes. Some reactions of nitrido complexes form imido complexes, and these reactions were described earlier in this chapter. For example, the addition of alkyl halides or alkyl trifluoromethanesulfonates to [Os(N)(CH$_2$SiMe$_3$)$_4$] forms osmium imido complexes (Equation 13.61).[431] The nitrido group can also be electrophilic and react with phosphines,[432,433] amines,[434] and Grignard reagents.[435,436] The addition of a Grignard reagent to form an imido complex was shown in Equation 13.62.

Nitrido complexes also undergo cycloadditions. A nitrido complex has been shown to react with 3-hexyne to form a molybdenum–carbyne complex in the reverse of Equation 13.105. In addition, a [4+1] cycloaddition reaction between the osmium(VI) nitride TpOs(N)Cl$_2$ and cyclohexadiene forms a bicyclic amido complex TpOs(NC$_6$H$_8$)Cl$_2$ (Equation 13.106).[437] The cationic osmium complex, cis-[(terpy)Os(N)Cl$_2$]PF$_6$, reacts with trans-stilbene and other conjugated alkenes to cleave the olefinic C–C double bond and form an azametallacyclopropane, as shown in Equation 13.107.[438,439] Thus, the stoichiometric reactions of metal–nitrido complexes are becoming increasingly diverse. One target for the chemistry of this ligand is to incorporate the reactions into catalytic processes in which the nitrido ligand is derived from dinitrogen. Perhaps future studies will provide the basis for the conversion of dinitrogen to ammonia or organic products under mild conditions.

$$ (13.106) $$

(13.107)

References and Notes

1. Cundari, T. R. *Chem. Rev.* **2000**, *100*, 807.
2. Fischer, E. O.; Maasböl, A. *Angew. Chem., Int. Ed. Engl.* **1964**, *3*, 580.
3. Schrock, R. R. *J. Am. Chem. Soc.* **1974**, *96*, 6796.
4. *Metal Carbenes in Organic Synthesis*; Dörwald, F. Z., Ed. Wiley-VCH: Weinheim, 1999.
5. Doyle, M. P.; Forbes, D. C. *Chem. Rev.* **1998**, *98*, 911.
6. Doyle, M. P. *Chem. Rev.* **1986**, *86*, 919.
7. Davies, H. M. L.; Beckwith, R. E. J. *Chem. Rev.* **2003**, *103*, 2861.
8. Bruce, M. I. *Chem. Rev.* **1991**, *91*, 197.
9. Jafarpour, L.; Nolan, S. P. *Adv. Organomet. Chem.* **2001**, *46*, 181.
10. Canac, Y.; Soleilhavoup, M.; Conejero, S.; Bertrand, G. *J. Organomet. Chem.* **2004**, *689*, 3857.
11. Arduengo, A. J.; Krafczyk, R.; Schmutzler, R. *Tetrahedron* **1999**, *55*, 14523.
12. Semmelhack, M. F.; Lee, G. R. *Organometallics* **1987**, *6*, 1839.
13. Imwinkelried, R.; Hegedus, L. S. *Organometallics* **1988**, *7*, 702.
14. Hahn, F. E.; Tamm, M. *Organometallics* **1995**, *14*, 2597.
15. (a) Herrmann, W. A. *Chem. Ber.* **1975**, *108*, 486; (b) Casey, C. P.; Burkhardt, T. J. *J. Am. Chem. Soc.* **1973**, *95*, 5883; (c) Casey, C. P.; Burkhardt, T. J. *J. Am. Chem. Soc.* **1974**, *96*, 7808; (d) Casey, C. P.; Burkhardt, T. J.; Bunnell, C. A.; Calabrese, J. C. *J. Am. Chem. Soc.* **1977**, *99*, 2127.
16. Schwab, P.; Grubbs, R. H.; Ziller, J. W. *J. Am. Chem. Soc.* **1996**, *118*, 100.
17. Snyder, J. P.; Padwa, A.; Stengel, T.; Arduengo, A. J.; Jockisch, A.; Kim, H. J. *J. Am. Chem. Soc.* **2001**, *123*, 11318.
18. Sheehan, S. M.; Padwa, A.; Snyder, J. P. *Tetrahedron Lett.* **1998**, *39*, 949.
19. Padwa, A.; Snyder, J. P.; Curtis, E. A.; Sheehan, S. M.; Worsencroft, K. J.; Kappe, C. O. *J. Am. Chem. Soc.* **2000**, *122*, 8155.
20. Nakamura, E.; Yoshikai, N.; Yamanaka, M. *J. Am. Chem. Soc.* **2002**, *124*, 7181.
21. Nowlan, D. T.; Gregg, T. M.; Davies, H. M. L.; Singleton, D. A. *J. Am. Chem. Soc.* **2003**, *125*, 15902.
22. Dai, X. L.; Warren, T. H. *J. Am. Chem. Soc.* **2004**, *126*, 10085.
23. Fraile, J. M.; Garcia, J. I.; Martinez-Merino, V.; Mayoral, J. A.; Salvatella, L. *J. Am. Chem. Soc.* **2001**, *123*, 7616.
24. Buhl, M.; Terstegen, F.; Loffler, F.; Meynhardt, B.; Kierse, S.; Muller, M.; Nather, C.; Luning, U. *Eur. J. Org. Chem.* **2001**, 2151.
25. Gandelman, M.; Rybtchinski, B.; Ashkenazi, N.; Gauvin, R. M.; Milstein, D. *J. Am. Chem. Soc.* **2001**, *123*, 5372.
26. Gandelman, M.; Naing, K. M.; Rybtchinski, B.; Poverenov, E.; Ben-David, Y.; Ashkenazi, N.; Gauvin, R. M.; Milstein, D. *J. Am. Chem. Soc.* **2005**, *127*, 15265.
27. Nickias, P. N.; Selegue, J. P.; Young, B. A. *Organometallics* **1988**, *7*, 2248.
28. Pombeiro, A. J. L.; Almeida, S.; Silva, M.; Jeffrey, J. C.; Richards, R. L. *J. Chem. Soc., Dalton Trans.* **1989**, 2381.
29. Pombeiro, A. J. L.; Jeffery, J. C.; Pickett, C. J.; Richards, R. L. *J. Organomet. Chem.* **1984**, *277*, C7.
30. Bullock, R. M. *J. Am. Chem. Soc.* **1987**, *109*, 8087.
31. Werner, H.; Alonso, F. J. G.; Otto, H.; Wolf, J. *Z. Naturforsch., B: Chem. Sci.* **1988**, *43*, 722.
32. Grotjahn, D. B.; Zeng, X.; Cooksy, A. L. *J. Am. Chem. Soc.* **2006**, *128*, 2798.
33. Beevor, R. G.; Freeman, M. J.; Green, M.; Morton, C. E.; Orpen, A. G. *J. Chem. Soc., Chem. Commun.* **1985**, 68.
34. Schrock, R. R.; Fellmann, J. D. *J. Am. Chem. Soc.* **1978**, *100*, 3359.
35. Wood, C. D.; McLain, S. J.; Schrock, R. R. *J. Am. Chem. Soc.* **1979**, *101*, 3210.
36. Rupprecht, G. A.; Messerle, L. W.; Fellmann, J. D.; Schrock, R. R. *J. Am. Chem. Soc.* **1980**, *102*, 6236.

37. Schrock, R. R.; Hoveyda, A. H. *Angew. Chem. Int. Ed.* **2003**, *42*, 4592.

38. Rocklage, S. M.; Schrock, R. R.; Churchill, M. R.; Wasserman, H. J. *Organometallics* **1982**, *1*, 1332.

39. Tebbe, F. N.; Parshall. G. W.; Reddy, G. S. *J. Am. Chem. Soc.* **1978**, *100*, 3611.

40. Howard, T. R.; Lee, J. R.; Grubbs, R. H. *J. Am. Chem. Soc.* **1980**, *102*, 6876.

41. Petasis, N. A.; Lu, S.-P.; Bzowej, E. I.; Fu, D.-K.; Staszewski, J. P.; Akritopoulou-Zanze, I.; Patane, M. A.; Hu, Y.-H. *Pure Appl. Chem.* **1996**, *68*, 667.

42. Petasis, N. A.; Bzowej, E. I. *J. Am. Chem. Soc.* **1990**, *112*, 6392.

43. Petasis, N. A.; Akritopoulou, I. *Synlett* **1992**, 665.

44. Petasis, N. A.; Bzowej, E. I. *J. Org. Chem.* **1992**, *57*, 1327.

45. Claus, K. B., H. *Ann.* **1962**, *654*, 8.

46. Hughes, D. L.; Payack, J. F.; Cai, D. W.; Verhoeven, T. R.; Reider, P. J. *Organometallics* **1996**, *15*, 663.

47. Payack, J. F.; Hughes, D. L.; Cai, D.; Cottrell, I. F.; Verhoeven, T. R. *Oppi Briefs* **1995**, *27*, 707.

48. Payack, J. F.; Huffman, M. A.; Cai, D. W.; Hughes, D. L.; Collins, P. C.; Johnson, B. K.; Cottrell, I. F.; Tuma, L. D. *Org. Process Res. Dev.* **2004**, *8*, 256.

49. Igau, A.; Baceiredo, A.; Trinquier, G.; Bertrand, G. *Angew. Chem., Int. Ed. Engl.* **1989**, *28*, 621.

50. Arduengo, A. J. I. *Acc. Chem. Res.* **1999**, *32*, 913.

51. Herrmann, W. A.; Weskamp, T.; Bohm, V. P. W. In *Advances in Organometallic Chemistry, Vol. 48*; West, R., Hill, A. F., Eds.; Academic Press: San Diego, CA, 2001; p 1.

52. Wang, H. M. J.; Lin, I. J. B. *Organometallics* **1998**, *17*, 972.

53. Lin, I. J. B.; Vasam, C. S. *Comments Inorg. Chem.* **2004**, *25*, 75.

54. de Frémont, P.; Scott, N. M.; Stevens, E. D.; Ramnial, T.; Lightbody, O. C.; Macdonald, C. L. B.; Clyburne, J. A. C.; Abernethy, C. D.; Nolan, S. P. *Organometallics* **2005**, *24*, 6301.

55. Voutchkova, A. M.; Appelhans, L. N.; Chianese, A. R.; Crabtree, R. H. *J. Am. Chem. Soc.* **2005**, *127*, 17624.

56. Herrmann, W. A. *Angew. Chem. Int. Ed.* **2002**, *41*, 1290.

57. Arduengo, A. J. I. U.S. Patent #5 077 414.

58. Grasa, G. A.; Viciu, M. S.; Huang, J. K.; Nolan, S. P. *J. Org. Chem.* **2001**, *66*, 7729.

59. Wang, H. M. J.; Lin, I. J. B. *Organometallics* **1988**, *17*, 972.

60. Trnka, T. M.; Morgan, J. P.; Sanford, M. S.; Wilhelm, T. E.; Scholl, M.; Choi, T. L.; Ding, S.; Day, M. W.; Grubbs, R. H. *J. Am. Chem. Soc.* **2003**, *125*, 2546.

61. Scholl, M.; Trnka, T. M.; Morgan, J. P.; Grubbs, R. H. *Tetrahedron Lett.* **1999**, *40*, 2247.

62. Scholl, M.; Ding, S.; Lee, C. W.; Grubbs, R. H. *Org. Lett.* **1999**, *1*, 953.

63. Weskamp, T.; Kohl, F. J.; Hieringer, W.; Gleich, D.; Herrmann, W. A. *Angew. Chem. Int. Ed.* **1999**, *38*, 2416.

64. Weskamp, T.; Kohl, F. J.; Herrmann, W. A. *J. Organomet. Chem.* **1999**, *582*, 362.

65. Ackermann, L.; Furstner, A.; Weskamp, T.; Kohl, F. J.; Herrmann, W. A. *Tetrahedron Lett.* **1999**, *40*, 4787.

66. Huang, J.; Stevens, E. D.; Nolan, S. P.; Petersen, J. L. *J. Am. Chem. Soc.* **1999**, *121*, 2674.

67. Jafarpour, L.; Hillier, A. C.; Nolan, S. P. *Organometallics* **2002**, *21*, 442.

68. Barluenga, J.; Fernandez-Rodriguez, M. A.; Aguilar, E. *J. Organomet. Chem.* **2005**, *690*, 539.

69. de Meijere, A.; Schirmer, H.; Duetsch, M. *Angew. Chem. Int. Ed.* **2000**, *39*, 3964.

70. Bernasconi, C. F. *Adv. Phys. Org. Chem.* **2002**, *37*, 137.

71. Casey, C. P.; Burkhardt, T. J.; Bunnell, C. A.; Calabrese, J. C. *J. Am. Chem. Soc.* **1977**, *99*, 2127.

72. Casey, C. P.; Shusterman, A. J. *J. Mol. Catal.* **1980**, *8*, 1.

73. Werner, H.; Fischer, E. O.; Heckl, B.; Kreiter, C. G. *J. Organomet. Chem.* **1971**, *28*, 367.

74. (a) Kreissl, F. R.; Fischer, E. O. *Chem. Ber.* **1974**, *107*, 183; (b) Kreissl, F. R.; Fischer, E. O.; Kreiter, C. G.; Weiss, K. *Angew. Chem., Int. Ed. Engl.* **1973**, *12*, 563.

75. Schubert, U. In *The Chemistry of the Metal–Carbon Bond*; Hartley, F. R., Patai, S., Eds.; Wiley-Interscience: New York, 1982; p 233.

76. Gallop, M. A.; Roper, W. R. *Adv. Organomet. Chem.* **1986**, *25*, 121.

77. Fischer, H.; Motsch, A.; Markl, R.; Ackermann, K. *Organometallics* **1985**, *4*, 726.

78. Fischer, H.; Motsch, A.; Kleine, W. *Angew. Chem., Int. Ed. Engl.* **1978**, *17*, 842.

79. Anderson, B. A.; Wulff, W. D.; Rahm, A. *J. Am. Chem. Soc.* **1993**, *115*, 4602.

80. (a) Shin, G. C.; Hwang, J.; Yang, K.; Koo, I. S.; Lee, I., *Bull. Korean Chem. Soc.* **2005**, *26*, 1981. For related reactions, see (b) Wulff, W. D.; Gilbertson, S. R. *J. Am. Chem. Soc.* **1985**, *107*, 503; (c) Casey, C. P., Boggs, R. A.; Anderson, R. L. *J. Am. Chem. Soc.* **1972**, *94*, 8947.

81. Lattuada, L.; Licardro, E.; Maiorana, S.; Molinari, H.; Papagni, A. *Organometallics* **1991**, *10*, 807.

82. Brookhart, M.; Studabaker, W. B. *Chem. Rev.* **1987**, *87*, 411.

83. Merino, I.; Hegedus, L. S. *Organometallics* **1995**, *14*, 2522.

84. Harvey, D. F.; Lund, K. P. *J. Am. Chem. Soc.* **1991**, *113*, 8916.

85. Kegley, S. E.; Brookhart, M.; Husk, G. R. *Organometallics* **1982**, *1*, 760.

86. Casey, C. P.; Burkhardt, C. P. *J. Am. Chem. Soc.* **1974**, *96*, 7808.

87. Casey, C. P.; Polichnowski, S. W.; Shusterman, A. J.; Jones, C. R. *J. Am. Chem. Soc.* **1979**, *101*, 7282.

88. Casey, C. P.; Polichnowski, S. W. *J. Am. Chem. Soc.* **1977**, *99*, 6097.

89. Waterman, R.; Hillhouse, G. L. *J. Am. Chem. Soc.* **2003**, *125*, 13350.

90. Jaeger, M.; Prosenc, M. H.; Sontag, C.; Fischer, H. *New J. Chem.* **1995**, *19*, 911.

91. Dötz, K. H.; Tomuschat, P. *Chem. Soc. Rev.* **1999**, *28*, 187.

92. Dötz, K. H.; Jahr, H. C. In *Carbene Chemistry*; Bertrand, G., Ed.; Marcel Dekker: New York, 2002; p 231.

93. Dötz, K. H. *Angew. Chem., Int. Ed. Engl.* **1975**, *14*, 644.

94. Barluenga, J.; Aznar, F.; Gutierrez, I.; Martin, A.; Garcia-Granda, S.; Llorca-Baragano, M. A. *J. Am. Chem. Soc.* **2000**, *122*, 1314.

95. Barluenga, J.; Aznar, F.; Martin, A.; Garciagranda, S.; Perezcarreno, E. *J. Am. Chem. Soc.* **1994**, *116*, 11191.

96. Dötz, K. H.; Schafer, T.; Kroll, F.; Harms, K. *Angew. Chem., Int. Ed. Engl.* **1992**, *31*, 1236.

97. Dotz, K. H.; Kuhn, W. *Angew. Chem., Int. Ed. Engl.* **1983**, *22*, 732.

98. Dotz, K. H.; Pruskil, I. *J. Organomet. Chem.* **1981**, *209*, C4.

99. Bao, J. M.; Wulff, W. D.; Dragisich, V.; Wenglowsky, S.; Ball, R. G. *J. Am. Chem. Soc.* **1994**, *116*, 7616.

100. Semmelhack, M. F.; Bozell, J. J.; Sato, T.; Wulff, W.; Spiess, E.; Zask, A. *J. Am. Chem. Soc.* **1982**, *104*, 5850.

101. Boger, D. L.; Huter, O.; Mbiya, K.; Zhang, M. S. *J. Am. Chem. Soc.* **1995**, *117*, 11839.

102. Pombeiro, A. J. L.; Hills, A.; Hughes, D. L.; Richards, R. L. *J. Organomet. Chem.* **1988**, *352*, C5.

103. Höhn, A.; Werner, H. *Angew. Chem., Int. Ed. Engl.* **1986**, *25*, 737.

104. Barrett, A. G. M.; Sturgess, M. A. *J. Org. Chem.* **1987**, *52*, 3940.

105. Barrett, A. G. M.; Sturgess, M. A. *Tetrahedron Lett.* **1986**, *27*, 3811.

106. Schrock, R. R.; Depue, R. T.; Feldman, J.; Schaverien, C. J.; Dewan, J. C.; Liu, A. H. *J. Am. Chem. Soc.* **1988**, *110*, 1423.

107. Schrock, R. R.; Fellmann, J. D. *J. Am. Chem. Soc.* **1978**, *100*, 3359.

108. Churchill, M. R.; Ziller, J. W.; Freudenberger, J. H.; Schrock, R. R. *Organometallics* **1984**, *3*, 1554.

109. Oskam, J. H.; Schrock, R. R. *J. Am. Chem. Soc.* **1993**, *115*, 11831.

110. Sanford, M. S.; Love, J. A.; Grubbs, R. H. *J. Am. Chem. Soc.* **2001**, *123*, 6543.

111. Sanford, M. S.; Ulman, M.; Grubbs, R. H. *J. Am. Chem. Soc.* **2000**, *123*, 749.

112. Tolman, R. C. *Proc. Natl. Acad. Sci. U.S.A.* **1925**, *11*, 436.

113. Lowry, T. H.; Richardson, T. H. *Mechanism and Theory in Organic Chemistry*, 3rd ed.; Harper & Row: New York, 1987.

114. Anslyn, E. V.; Dougherty, D. A. *Modern Physical Organic Chemistry*; University Science Books: Sausalito, CA, 2006.

115. Adlhart, C.; Chen, P. *J. Am. Chem. Soc.* **2004**, *126*, 3496.

116. Correa, A.; Cavallo, L. *J. Am. Chem. Soc.* **2006**, *128*, 13352.

117. Anderson, D. R.; Hickstein, D. D.; O'Leary, D. J.; Grubbs, R. H. *J. Am. Chem. Soc.* **2006**, *128*, 8386.

118. Romero, P. E.; Piers, W. E. *J. Am. Chem. Soc.* **2005**, *127*, 5032.

119. Romero, P. E.; Piers, W. E. *J. Am. Chem. Soc.* **2007**, *129*, 1698.

120. Carlson, R. G.; Gile, M. A.; Heppert, J. A.; Mason, M. H.; Powell, D. R.; Vander Velde, D.; Vilain, J. M. *J. Am. Chem. Soc.* **2002**, *124*, 1580.

121. Wenzel, A. G.; Grubbs, R. H. *J. Am. Chem. Soc.* **2006**, *128*, 16048.

122. Love, J. A.; Morgan, J. P.; Trnka, T. M.; Grubbs, R. H. *Angew. Chem. Int. Ed.* **2002**, *41*, 4035.

123. Love, J. A.; Sanford, M. S.; Day, M. W.; Grubbs, R. H. *J. Am. Chem. Soc.* **2003**, *125*, 10103.

124. van der Heijden, H.; Hessen, B. *J. Chem. Soc., Chem. Commun.* **1995**, 145.

125. Coles, M. P.; Gibson, V. C.; Clegg, W.; Elsegood, M. R. J.; Porrelli, P. A. *Chem. Commun.* **1996**, 1963.

126. Cheon, J.; Rogers, D. M.; Girolami, G. S. *J. Am. Chem. Soc.* **1997**, *119*, 6804.

127. Tran, E.; Legzdins, P. *J. Am. Chem. Soc.* **1997**, *119*, 5071.

128. Adams, C. S.; Legzdins, P.; Tran, E. *J. Am. Chem. Soc.* **2001**, *123*, 612.

129. Pamplin, C. B.; Legzdins, P. *Acc. Chem. Res.* **2003**, *36*, 223.

130. Bailey, B. C.; Fan, H.; Baum, E. W.; Huffman, J. C.; Baik, M.-H.; Mindiola, D. J. *J. Am. Chem. Soc.* **2005**, *127*, 16016.

131. Bailey, B. C.; Fan, H.; Huffman, J. C.; Baik, M.-H.; Mindiola, D. J. *J. Am. Chem. Soc.* **2007**, *129*, 8781.

132. Zybill, C. *Top. Curr. Chem.* **1992**, *160*, 1.

133. Zybill, C.; Handwerker, H.; Friedrich, H. In *Advances in Organometallic Chemistry, Vol. 36*; West, R., Stone, F. G. A., Eds.; Academic Press: San Diego, CA, 1994; p 229.

134. Zybill, C. E.; Liu, C. Y. *Synlett* **1995**, 687.

135. Tilley, T. D. In *The Chemistry of Organic Silicon Compounds*; Patai, S., Rappoport, Z., Eds.; Wiley: New York, 1989; p 1415.

136. Tilley, T. D. In *The Chemistry of Organic Silicon Compounds*; Patai, S., Rappoport, Z., Eds.; Wiley: New York, 1991; p 245.

137. Yamamoto, K.; Okinoshi, H.; Kumada, M. *J. Organomet. Chem.* **1970**, *23*, C7.

138. Yamamoto, K.; Okinoshi, H.; Kumada, M. *J. Organomet. Chem.* **1971**, *27*, C31.

139. Okinoshima, H.; Yamamoto, K.; Kumada, M. *J. Am. Chem. Soc.* **1972**, *94*, 9263.

140. Ojima, I.; Inaba, S.; Kogure, T.; Nagai, Y. *J. Organomet. Chem.* **1973**, *55*, C7.

141. Schmid, G.; Welz, E. *Angew. Chem., Int. Ed. Engl.* **1977**, *16*, 785.

142. Zybill, C.; Muller, G. *Angew. Chem., Int. Ed. Engl.* **1987**, *26*, 669.

143. Straus, D. A.; Tilley, T. D.; Rheingold, A. L.; Geib, S. J. *J. Am. Chem. Soc.* **1987**, *109*, 5872.

144. Grumbine, S. D.; Tilley, T. D.; Arnold, F. P.; Rheingold, A. L. *J. Am. Chem. Soc.* **1993**, *115*, 7884.

145. Grumbine, S. K.; Tilley, T. D.; Arnold, F. P.; Rheingold, A. L. *J. Am. Chem. Soc.* **1994**, *116*, 5495.

146. Denk, M.; Hayashi, R. K.; West, R. *J. Chem. Soc., Chem. Commun.* **1994**, 33.

147. Gehrhus, B.; Hitchcock, P. B.; Lappert, M. F.; Maciejewski, H. *Organometallics* **1998**, *17*, 5599.

148. Feldman, J. D.; Mitchell, G. P.; Nolte, J. O.; Tilley, T. D. *J. Am. Chem. Soc.* **1998**, *120*, 11184.

149. Corriu, R. J. P.; Lanneau, G. F.; Chauhan, B. P. S. *Organometallics* **1993**, *12*, 2001.

150. Corriu, R. J. P.; Chauhan, B. P. S.; Lanneau, G. F. *Organometallics* **1995**, *14*, 1646.

151. Chauhan, B. P. S.; Corriu, R. J. P.; Lanneau, G. F.; Priou, C.; Auner, N.; Handwerker, H.; Herdtweck, E. *Organometallics* **1995**, *14*, 1657.

152. Mitchell, G. P.; Tilley, T. D. *Angew. Chem. Int. Ed.* **1998**, *37*, 2524.

153. Zybill, C.; Muller, G. *Angew. Chem., Int. Ed. Engl.* **1987**, *26*, 669.

154. Pannell, K. H.; Cervantes, J.; Hernandez, C.; Cassias, J.; Vincenti, S. *Organometallics* **1986**, *5*, 1056.

155. Tobita, H.; Ueno, K.; Ogino, H. *Chem. Lett.* **1986**, 1777.

156. Ueno, K.; Tobita, H.; Shimoi, M.; Ogino, H. *J. Am. Chem. Soc.* **1988**, *110*, 4092.

157. Tobita, H.; Ueno, K.; Shimoi, M.; Ogino, H. *J. Am. Chem. Soc.* **1990**, *112*, 3415.

158. Mitchell, G. P.; Tilley, T. D. *J. Am. Chem. Soc.* **1997**, *119*, 11236.

159. Klei, S. R.; Tilley, T. D.; Bergman, R. G. *Organometallics* **2002**, *21*, 4648.

160. Klei, S. R.; Tilley, T. D.; Bergman, R. G. *J. Am. Chem. Soc.* **2000**, *122*, 1816.

161. Nugent, W. A.; Haymore, B. L. *Coord. Chem. Rev.* **1980**, *31*, 123.

162. Wigley, D. E. In *Progress in Inorganic Chemistry, Vol. 42*; Karlin, K. D., Ed.; Wiley-Interscience: New York, 1994; p 239.

163. Stiefel, E. I. In *Transition Metal Sulfur Chemistry*; Stiefel, E. I., Matsumoto, K., Eds.; American Chemical Society: Washington, DC, 1996; p 2.

164. Eichhorn, B. W. In *Progress in Inorganic Chemistry, Vol. 42*; Karlin, K. D., Ed.; Wiley-Interscience: New York, 1994; p 139.

165. Parkin, G. In *Progress in Inorganic Chemistry, Vol. 47*; Karlin, K. D., Ed.; Wiley-Interscience: New York, 1998; p 1.

166. Huttner, G.; Evertz, K. *Acc. Chem. Res.* **1986**, *19*, 406.

167. Cowley, A. H.; Barron, A. R. *Acc. Chem. Res.* **1988**, *21*, 81.

168. Cowley, A. H. *Acc. Chem. Res.* **1997**, *30*, 445.

169. Lammertsma, K.; Vlaar, M. J. M. *Eur. J. Org. Chem.* **2002**, 1127.

170. Mathey, F. *Angew. Chem. Int. Ed.* **2003**, *42*, 1578.

171. Lammertsma, K. In *New Aspects in Phosphorus Chemistry III*; Chivers, T., Crépy, K. V. L., Cristau, H.-J., Gulea, M., Eds.; Springer-Verlag: New York, 2003; Vol. 229, p 95.

172. Hou, Z. M.; Stephan, D. W. *J. Am. Chem. Soc.* **1992**, *114*, 10088.

173. Hou, Z. M.; Breen, T. L.; Stephan, D. W. *Organometallics* **1993**, *12*, 3158.

174. Breen, T. L.; Stephan, D. W. *J. Am. Chem. Soc.* **1995**, *117*, 11914.

175. Melenkivitz, R.; Mindiola, D. J.; Hillhouse, G. L. *J. Am. Chem. Soc.* **2002**, *124*, 3846.

176. Sanchez-Nieves, J.; Sterenberg, B. T.; Udachin, K. A.; Carty, A. J. *J. Am. Chem. Soc.* **2003**, *125*, 2404.

177. Waterman, R.; Hillhouse, G. L. *Organometallics* **2003**, *22*, 5182.

178. Basuli, F.; Bailey, B. C.; Huffman, J. C.; Baik, M. H.; Mindiola, D. J. *J. Am. Chem. Soc.* **2004**, *126*, 1924.

179. Johnson, B. P.; Balazs, G.; Scheer, M. In *New Aspects in Phosphorus Chemistry IV*, 2004; Vol. 232, p 1.

180. Driess, M.; Aust, J.; Merz, K. *Eur. J. Inorg. Chem.* **2005**, 866.

181. Graham, T. W.; Cariou, R. P. Y.; Sanchez-Nieves, J.; Allen, A. E.; Udachin, K. A.; Regragui, R.; Carty, A. J. *Organometallics* **2005**, *24*, 2023.

182. Shaver, M. P.; Fryzuk, M. D. *Organometallics* **2005**, *24*, 1419.

183. Bailey, B. C.; Huffman, J. C.; Mindiola, D. J.; Weng, W.; Ozerov, O. V. *Organometallics* **2005**, *24*, 1390.

184. Braunschweig, H.; Colling, M. *Eur. J. Inorg. Chem.* **2003**, 393.

185. Braunschweig, H.; Colling, M. *J. Organometallic Chem.* **2000**, *614-615*, 18.
186. Braunschweig, H.; Colling, M.; Hu, C. H.; Radacki, K. *Angew. Chem. Int. Ed.* **2003**, *42*, 205.
187. Braunschweig, H.; Colling, M.; Kollann, C.; Merz, K.; Radacki, K. *Angew. Chem. Int. Ed.* **2001**, *40*, 4198.
188. Braunschweig, H.; Kollann, C.; Englert, U. *Angew. Chem. Int. Ed.* **1998**, *37*, 3179.
189. Cowley, A. H.; Lomeli, V.; Voigt, A. *J. Am. Chem. Soc.* **1998**, *120*, 6401.
190. Coombs, D. L.; Aldridge, S.; Rossin, A.; Jones, C.; Willock, D. J. *Organometallics* **2004**, *23*, 2911.
191. Griffith, W. P. *Coord. Chem. Rev.* **1972**, *8*, 369.
192. Dehnicke, K. S., J. *Angew. Chem., Int. Ed. Engl.* **1981**, *20*, 413.
193. Dehnicke, K. S., J. *Angew. Chem., Int. Ed. Engl.* **1992**, *31*, 955.
194. Nugent, W. A.; Mayer, J. M. *Metal–Ligand Multiple Bonds*; Wiley: New York, 1988.
195. Che, C.-M. *Pure Appl. Chem.* **1995**, *67*, 225.
196. Mayer, J. M. *Comments Inorg. Chem.* **1988**, *8*, 125.
197. Uddin, J.; Boehme, C.; Frenking, G. *Organometallics* **2000**, *19*, 571.
198. Uddin, J.; Frenking, G. *J. Am. Chem. Soc.* **2001**, *123*, 1683.
199. Sheldon, R. A.; Kochi, J. K. *Metal Catalyzed Oxidations of Organic Compounds: Mechanistic Principles and Synthetic Methodology Including Biochemical Processes*; Academic Press: New York, 1981.
200. Crabtree, R. H. *Chem. Rev.* **1985**, *85*, 245.
201. Crabtree, R. H. *Chem. Rev.* **1995**, *95*, 987.
202. Yoshizawa, K. *Coord. Chem. Rev.* **2002**, *226*, 251.
203. Feig, A. L.; Lippard, S. J. *Chem. Rev.* **1994**, *94*, 759.
204. Wallar, B. J.; Lipscomb, J. D. *Chem. Rev.* **1996**, *96*, 2625.
205. Joo, H.; Lin, Z.; Arnold, F. H. *Nature* **1999**, *399*, 670.
206. Peters, M. W.; Meinhold, P.; Glieder, A.; Arnold, F. H. *J. Am. Chem. Soc.* **2003**, *125*, 13442.
207. Breslow, R.; Gellman, S. H. *J. Chem. Soc., Chem. Commun.* **1982**, 1400.
208. Breslow, R.; Gellman, S. H. *J. Am. Chem. Soc.* **1983**, *105*, 6728.
209. Svastits, E.; Dawson, J. H.; Breslow, R.; Gellman, S. H. *Biochemistry* **1985**, *24*, 3380.
210. Espino, C. G.; Du Bois, J. *Angew. Chem. Int. Ed.* **2001**, 598.
211. Espino, C. G.; Wehn, P. M.; Chow, J.; Du Bois, J. *J. Am. Chem. Soc.* **2001**, *123*, 6935.
212. Groves, J. T.; Takahashi, T. *J. Am. Chem. Soc.* **1983**, *105*, 2073.
213. Evans, D. A.; Faul, M. M.; Bilodeau, M. T. *J. Org. Chem.* **1991**, *56*, 6744.
214. Evans, D. A.; Faul, M. M.; Bilodeau, M. T.; Anderson, B. A.; Barnes, D. M. *J. Am. Chem. Soc.* **1993**, *115*, 5328.
215. Li, Z.; Conser, K. R.; Jacobsen, E. N. *J. Am. Chem. Soc.* **1993**, *115*, 5326.
216. Li, Z.; Quan, R. W.; Jacobsen, E. N. *J. Am. Chem. Soc.* **1995**, *117*, 5889.
217. DuBois, J.; Tomooka, C. S.; Hong, J.; Carreira, E. M. *Acc. Chem. Res.* **1997**, *30*, 364.
218. Guthikonda, K.; Du Bois, J. *J. Am. Chem. Soc.* **2002**, *124*, 13672.
219. Diaz-Requejo, M. M.; Belderrain, T. R.; Nicasio, M. C.; Trofimenko, S.; Perez, P. J. *J. Am. Chem. Soc.* **2003**, *125*, 12078.
220. Muller, T. E.; Beller, M. *Chem. Rev.* **1998**, *98*, 675.
221. Walsh, P. J.; Baranger, A. M.; Bergman, R. G. *J. Am. Chem. Soc.* **1992**, *114*, 1708.
222. Baranger, A. M.; Walsh, P. J.; Bergman, R. G. *J. Am. Chem. Soc.* **1993**, *115*, 2753.
223. Knight, P. D.; Munslow, I.; O'Shaughnessy, P. N.; Scott, P. *Chem. Commun.* **2004**, 894.
224. Haak, E.; Bytschkov, I.; Doye, S. *Angew. Chem. Int. Ed.* **1999**, *38*, 3389.
225. Cao, C. S.; Ciszewski, J. T.; Odom, A. L. *Organometallics* **2001**, *20*, 5011.
226. Pohlki, F.; Doye, S. *Angew. Chem. Int. Ed.* **2001**, *40*, 2305.
227. Straub, B. F.; Bergman, R. G. *Angew. Chem. Int. Ed.* **2001**, *40*, 4632.
228. Ackermann, L.; Bergman, R. G. *Org. Lett.* **2002**, *4*, 1475.
229. Cao, C.; Shi, Y.; Odom, A. L. *Org. Lett.* **2002**, *4*, 2853.
230. Heutling, A.; Doye, S. *J. Org. Chem.* **2002**, *67*, 1961.
231. Ong, T.-G.; Yap, G. P. A.; Richeson, D. S. *Organometallics* **2002**, *21*, 2839.
232. Pohlki, F.; Heutling, A.; Bytschkov, I.; Hotopp, T.; Doye, S. *Synlett* **2002**, *5*, 799.
233. Tillack, A.; Castro, I. G.; Hartung, C. G.; Beller, M. *Angew. Chem. Int. Ed.* **2002**, *41*, 2541.
234. Bytschkov, I.; Siebeneicher, H.; Doye, S. *Eur. J. Org. Chem.* **2003**, 2888.
235. Cao, C. S.; Shi, Y. H.; Odom, A. L. *J. Am. Chem. Soc.* **2003**, *125*, 2880.
236. Castro, I. G.; Tillack, A.; Hartung, C. G.; Beller, M. *Tetrahedron Lett.* **2003**, *44*, 3217.
237. Siebeneicher, H.; Bytschkov, I.; Doye, S. *Angew. Chem. Int. Ed.* **2003**, *42*, 3042.
238. Zhang, Z.; Schafer, L. L. *Org. Lett.* **2003**, *5*, 4733.
239. Ackermann, L.; Kaspar, L. T.; Gschrei, C. J. *Org. Lett.* **2004**, *6*, 2515.
240. Gribkov, D. V.; Hultzsch, K. C. *Angew. Chem. Int. Ed.* **2004**, *43*, 5542.

241. Li, Y. H.; Shi, Y. H.; Odom, A. L. *J. Am. Chem. Soc.* **2004**, *126*, 1794.

242. Shilov, A. E.; Shteinman, A. A. *Acc. Chem. Res.* **1999**, *32*, 763.

243. Mayer, J. M. *Acc. Chem. Res.* **1998**, *31*, 441.

244. Nugent, W. A.; Haymore, B. L. *Coord. Chem. Rev.* **1980**, *31*, 123.

245. For a classic review of metal–ligand multiple bonds, see reference 244.

246. Ortiz de Montellano, P. R., Ed. *Cytochrome P450: Structure, Mechanism, and Biochemistry*, 3rd ed.; Kluwer Academic/Plenum Publishers: New York, 2005.

247. Denisov, I. G.; Makris, T. M.; Sligar, S. G.; Schlichting, I. *Chem. Rev.* **2005**, *105*, 2253.

248. Shaik, S.; Kumar, D.; de Visser, S. P.; Altun, A.; Thiel, W. *Chem. Rev.* **2005**, *105*, 2279.

249. Herrmann, W. A. *Angew. Chem., Int. Ed. Engl.* **1988**, *27*, 1297.

250. Romao, C. C.; Kuhn, F. E.; Herrmann, W. A. *Chem. Rev.* **1997**, *97*, 3197.

251. Kuhn, F. E.; Herrmann, W. A. In *Metal-Oxo and Metal-Peroxo Species in Catalytic Oxidations*; Meunier, B., Ed.; Springer-Verlag: New York, 2000; Vol. 97, p 213.

252. Espenson, J. H. *Chem. Commun.* **1999**, 479.

253. Schrock, R. R. *Chem. Rev.* **2002**, *102*, 145.

254. Glueck, D. S.; Hollander, F. J.; Bergman, R. G. *J. Am. Chem. Soc.* **1989**, *111*, 2719.

255. Glueck, D. S.; Wu, J.-X.; Hollander, F. J.; Bergman, R. G. *J. Am. Chem. Soc.* **1991**, *113*, 2041.

256. Mindiola, D. J.; Hillhouse, G. L. *J. Am. Chem. Soc.* **2001**, *123*, 4623.

257. Vicic, D. A.; Jones, W. D. *J. Am. Chem. Soc.* **1999**, *121*, 4070.

258. Carney, M. J.; Walsh, P. J.; Bergman, R. G. *J. Am. Chem. Soc.* **1990**, *112*, 6426.

259. Carney, M. J.; Walsh, P. J.; Hollander, F. J.; Bergman, R. G. *Organometallics* **1992**, *11*, 761.

260. Macdonald, C. L. B.; Cowley, A. H. *J. Am. Chem. Soc.* **1999**, *121*, 12113.

261. Cundari, T. R. *J. Am. Chem. Soc.* **1992**, *114*, 7879.

262. Nugent, W. A.; Mayer, J. M. *Metal–Ligand Multiple Bonds*, Wiley: New York, 1988; p 21.

263. Nugent, W. A.; Mayer, J. M. *Metal–Ligand Multiple Bonds*, Wiley: New York, 1988; p 22.

264. Haymore, B. L.; Maatta, E. A.; Wentworth, R. A. D. *J. Am. Chem. Soc.* **1979**, *101*, 2063.

265. Gountchev, T. I.; Tilley, T. D. *J. Am. Chem. Soc.* **1997**, *119*, 12831.

266. Ciszewski, J. T.; Harrison, J. F.; Odom, A. L. *Inorg. Chem.* **2004**, *43*, 3605.

267. Barrie, P.; Coffey, T. A.; Forster, G. D.; Hogarth, G. *J. Chem. Soc., Dalton Trans.* **1999**, 4519.

268. Collier, P. E.; Pugh, S. M.; Clark, H. S. C.; Love, J. B.; Blake, A. J.; Cloke, F. G. N.; Mountford, P. *Inorg. Chem.* **2000**, *39*, 2001.

269. Rappe, A. K.; Goddard, W. A. *J. Am. Chem. Soc.* **1982**, *104*, 3287.

270. Rappe, A. K.; Goddard, W. A. *J. Am. Chem. Soc.* **1982**, *104*, 448.

271. Gambarotta, S.; Chiesivilla, A.; Guastini, C. *J. Organomet. Chem.* **1984**, *270*, C49.

272. Osborne, J. H.; Rheingold, A. L.; Trogler, W. C. *J. Am. Chem. Soc.* **1985**, *107*, 7945.

273. Parkin, G.; Vanasselt, A.; Leahy, D. J.; Whinnery, L.; Hua, N. G.; Quan, R. W.; Henling, L. M.; Schaefer, W. P.; Santarsiero, B. D.; Bercaw, J. E. *Inorg. Chem.* **1992**, *31*, 82.

274. Rankin, D. W. H.; Robertson, H. E.; Danopoulos, A. A.; Lyne, P. D.; Mingos, D. M. P.; Wilkinson, G. *J. Chem. Soc., Dalton Trans.* **1994**, 1563.

275. Schrock, R. R. *J. Am. Chem. Soc.* **1990**, *112*, 1642.

276. Greenwood, N. N.; Earnshaw, A. *Chemistry of the Elements*, 2nd ed.; Butterworth-Heinemann: Oxford, U.K., 1997; p 640.

277. Clifford, A. F.; Kobayashi, C. S. *Inorg. Synth.* **1960**, *6*, 204.

278. Chatt, J.; Rowe, G. A. *J. Chem. Soc.* **1962**, 4019.

279. Chatt, J.; Rowe, G. A.; Garforth, J. D.; Johnson, N. P. *J. Chem. Soc.* **1964**, 1013.

280. Sharpless, K. B.; Patrick, D. W.; Truesdale, L. K.; Biller, S. A. *J. Am. Chem. Soc.* **1975**, *97*, 2305.

281. Chong, A. O.; Oshima, K.; Sharpless, K. B. *J. Am. Chem. Soc.* **1977**, *99*, 3420.

282. Chatt, J.; Dilworth, J. R. *J. Chem. Soc., Chem. Commun.* **1972**, 549.

283. Maatta, E. A.; Haymore, B. L.; Wentworth, R. A. D. *Inorg. Chem.* **1980**, *19*, 1055.

284. Chatt, J.; Crichton, B. A. L.; Dilworth, J. R.; Dahlstrom, P.; Gutkoska, R.; Zubieta, J. *Inorg. Chem.* **1982**, *21*, 2383.

285. Godemeyer, T.; Berg, A.; Grass, H. D.; Muller, U.; Dehnicke, K. *Z. Naturforsch., B: Chem. Sci.* **1985**, *40*, 999.

286. Becker, F. *J. Organomet. Chem.* **1973**, *51*, C9.

287. Wiberg, N.; Haring, H. W.; Schieda, O. *Angew. Chem., Int. Ed. Engl.* **1976**, *15*, 386.

288. Muniz, K. *New J. Chem.* **2005**, *29*, 1371.

289. Nugent, W. A.; Harlow, R. L. *Inorg. Chem.* **1980**, *19*, 777.

290. Cummins, C. C.; Baxter, S. M.; Wolczanski, P. T. *J. Am. Chem. Soc.* **1988**, *110*, 8731.

291. Walsh, P. J.; Hollander, F. J.; Bergman, R. G. *J. Am. Chem. Soc.* **1988**, *110*, 8729.

292. Horton, A. D.; Schrock, R. R.; Freudenberger, J. H. *Organometallics* **1987**, *6*, 893.

293. Rocklage, S. M.; Schrock, R. R. *J. Am. Chem. Soc.* **1980**, *102*, 7808.

294. Rocklage, S. M.; Schrock, R. R. *J. Am. Chem. Soc.* **1982**, *104*, 3077.

295. Shapley, P. A.; Own, Z. Y. *J. Organomet. Chem.* **1987**, *335*, 269.

296. Shapley, P. A. B.; Own, Z. Y.; Huffman, J. C. *Organometallics* **1986**, *5*, 1269.

297. Crevier, T. J.; Bennett, B. K.; Soper, J. D.; Bowman, J. A.; Dehestani, A.; Hrovat, D. A.; Lovell, S.; Kaminsky, W.; Mayer, J. M. *J. Am. Chem. Soc.* **2001**, *123*, 1059.

298. Wiberg, N.; Haring, H. W.; Schubert, U. *Z. Naturforsch., B: Chem. Sci.* **1980**, *35*, 599.

299. Osborne, J. H.; Rheingold, A. L.; Trogler, W. C. *J. Am. Chem. Soc.* **1985**, *107*, 7945.

300. Au, S.-M.; Huang, J.-S.; W.-Y. Yu; Fung, W.-H.; Che, C.-M. *J. Am. Chem. Soc.* **1999**, *121*, 9120.

301. The nomenclature of imido and nitrene complexes is usually interchangeable; nitrene complexes tend to refer to complexes that transfer the NR group.

302. Lemenovskii, D. A.; Baukova, T. V.; Fedin, V. P. *J. Organomet. Chem.* **1977**, *132*, C14.

303. Macikenas, D.; Skrzypszak-Jankun, E.; Protasiewicz, J. D. *J. Am. Chem. Soc.* **1999**, *121*, 7164.

304. Holm, R. H. *Chem. Rev.* **1987**, *87*, 1401.

305. Holm, R. H.; Donahue, J. P. *Polyhedron* **1993**, *12*, 571.

306. Fu, G. C.; Grubbs, R. H. *J. Am. Chem. Soc.* **1993**, *115*, 3800.

307. Bryan, J. C.; Geib, S. J.; Rheingold, A. L.; Mayer, J. M. *J. Am. Chem. Soc.* **1987**, *109*, 2826.

308. Bryan, J. C.; Mayer, J. M. *J. Am. Chem. Soc.* **1987**, *109*, 7213.

309. Bryan, J. C.; Mayer, J. M. *J. Am. Chem. Soc.* **1990**, *112*, 2298.

310. Polse, J. L.; Andersen, R. A.; Bergmann, R. G. *J. Am. Chem. Soc.* **1998**, *120*, 13405.

311. Walsh, P. J.; Hollander, F. J.; Bergman, R. G. *Organometallics* **1993**, *12*, 3705.

312. Kolb, H. C.; VanNieuwenhze, M. S.; Sharpless, K. B. *Chem. Rev.* **1994**, *94*, 2483.

313. Bodkin, J. A.; McLeod, M. D. *J. Chem. Soc., Perkin Trans. I* **2002**, 2733.

314. Deubel, D. V.; Frenking, G. *Acc. Chem. Res.* **2003**, *36*, 645.

315. Muniz, K. *Chem. Soc. Rev.* **2004**, *33*, 166.

316. Ward, B. D.; Maisse-Francois, A.; Mountford, P.; Gade, L. H. *Chem. Commun.* **2004**, 704.

317. Bennett, J. L.; Wolczanski, P. T. *J. Am. Chem. Soc.* **1994**, *116*, 2179.

318. Dewith, J.; Horton, A. D.; Orpen, A. G. *Organometallics* **1993**, *12*, 1493.

319. The one example known at the time of publication involves addition of the strained norbornene. See reference 311.

320. Waterman, R.; Hillhouse, G. L. *J. Am. Chem. Soc.* **2003**, *125*, 13350.

321. Liang, J. L.; Huang, J. S.; Yu, X. Q.; Zhu, N. Y.; Che, C. M. *Chem. Eur. J.* **2002**, *8*, 1563.

322. Sharpless, K. B.; Teranishi, A. Y.; Bäckvall, J. E. *J. Am. Chem. Soc.* **1977**, *99*, 3120.

323. Jorgensen, K. A.; Hoffmann, R. *J. Am. Chem. Soc.* **1986**, *108*, 1867.

324. Dapprich, S.; Ujaque, G.; Maseras, F.; Lledos, A.; Musaev, D. G.; Morokuma, K. *J. Am. Chem. Soc.* **1996**, *118*, 11660.

325. Pidun, U.; Boehme, C.; Frenking, G. *Angew. Chem., Int. Ed. Engl.* **1996**, *35*, 2817.

326. Torrent, M.; Deng, L. Q.; Duran, M.; Sola, M.; Ziegler, T. *Organometallics* **1997**, *16*, 13.

327. DelMonte, A. J.; Haller, J.; Houk, K. N.; Sharpless, K. B.; Singleton, D. A.; Strassner, T.; Thomas, A. A. *J. Am. Chem. Soc.* **1997**, *119*, 9907.

328. Corey, E. J.; Noe, M. C.; Grogan, M. J. *Tetrahedron Lett.* **1996**, *37*, 4899.

329. Nelson, D. W.; Gypser, A.; Ho, P. T.; Kolb, H. C.; Kondo, T.; Kwong, H. L.; McGrath, D. V.; Rubin, A. E.; Norrby, P. O.; Gable, K. P.; Sharpless, K. B. *J. Am. Chem. Soc.* **1997**, *119*, 1840.

330. Gobel, T.; Sharpless, K. B. *Angew. Chem., Int. Ed. Engl.* **1993**, *32*, 1329.

331. Gable, K. P.; Juliette, J. J. J. *J. Am. Chem. Soc.* **1996**, *118*, 2625.

332. Katsuki, T.; Sharpless, K. B. *J. Am. Chem. Soc.* **1980**, *102*, 5974.

333. Martin, V. S.; Woodard, S. S.; Katsuki, T.; Yamada, Y.; Ikeda, M.; Sharpless, K. B. *J. Am. Chem. Soc.* **1981**, *103*, 6237.

334. Katsuki, T. In *Comprehensive Asymmetric Catalysis*; Jacobsen, E. N., Pfaltz, A., Yamamoto, H., Eds.; Springer Publishing: Berlin, 1999; Vol. 2, p 621.

335. Sharpless, K. B. *CHEMTECH* **1985**, *15*, 692.

336. Sharpless, K. B. *Angew. Chem. Int. Ed.* **2002**, *41*, 2024.

337. Williams, I. D.; Pedersen, S. F.; Sharpless, K. B.; Lippard, S. J. *J. Am. Chem. Soc.* **1984**, *106*, 6430.

338. Pedersen, S. F.; Dewan, J. C.; Eckman, R. R.; Sharpless, K. B. *J. Am. Chem. Soc.* **1987**, *109*, 1279.

339. Zhang, W.; Loebach, J. L.; Wilson, S. R.; Jacobsen, E. N. *J. Am. Chem. Soc.* **1990**, *112*, 2801.

340. Jacobsen, E. N.; Zhang, W.; Muci, A. R.; Ecker, J. R.; Deng, L. *J. Am. Chem. Soc.* **1991**, *113*, 7063.

341. Zhang, W.; Jacobsen, E. N. *J. Org. Chem.* **1991**, *56*, 2296.

342. Irie, R.; Noda, K.; Ito, Y.; Katsuki, T. *Tetrahedron Lett.* **1991**, *32*, 1055.

343. Irie, R.; Noda, K.; Ito, Y.; Matsumoto, N.; Katsuki, T. *Tetrahedron: Asymmetry* **1991**, *2*, 481.

344. Jacobsen, E. N. In *Catalytic Asymmetric Synthesis*; Ojima, I., Ed.; Wiley-VCH: Weinheim, 1993; p 159.

345. Wu, M. H.; Jacobsen, E. N. In *Comprehensive Asymmetric Catalysis*; Jacobsen, E. N., Pfaltz, A., Yamamoto, H., Eds.; Springer Publishing: Berlin, 1999; Vol. 2, p 649.

346. Samsel, E. G.; Srinivasan, K.; Kochi, J. K. *J. Am. Chem. Soc.* **1985**, *107*, 7606.

347. Srinivasan, K.; Kochi, J. K. *Inorg. Chem.* **1985**, *24*, 4671.

348. Srinivasan, K.; Perrier, S.; Kochi, J. K. *J. Mol. Catal.* **1986**, *36*, 297.

349. Srinivasan, K.; Michaud, P.; Kochi, J. K. *J. Am. Chem. Soc.* **1986**, *108*, 2309.

350. Feichtinger, D.; Plattner, D. A. *Angew. Chem., Int. Ed. Engl.* **1997**, *36*, 1718.

351. Feichtinger, D.; Plattner, D. A. *Chem. Eur. J.* **2001**, *7*, 591.

352. Groves, J. T.; Viski, P. *J. Org. Chem.* **1990**, *55*, 3628.

353. Collman, J. P.; Zhang, X. M.; Lee, V. J.; Brauman, J. I. *J. Chem. Soc., Chem. Commun.* **1992**, 1647.

354. Collman, J. P.; Lee, V. J.; Zhang, X. M.; Ibers, J. A.; Brauman, J. I. *J. Am. Chem. Soc.* **1993**, *115*, 3834.

355. Collman, J. P.; Zhang, X. M.; Lee, V. J.; Uffelman, E. S.; Brauman, J. I. *Science* **1993**, *261*, 1404.

356. Groves, J. T.; Quinn, R. *J. Am. Chem. Soc.* **1985**, *107*, 5790.

357. Meunier, B., Ed. *Biomimetic Oxidations Catalyzed by Transition Metal Complexes*; Imperial College Press: London, 2000.

358. Groves, J. T.; Lee, J. B.; Marla, S. S. *J. Am. Chem. Soc.* **1997**, *119*, 6269.

359. Chin, D. H.; Lamar, G. N.; Balch, A. L. *J. Am. Chem. Soc.* **1980**, *102*, 5945.

360. Che, C. M.; Wong, K. Y.; Mak, T. C. W. *J. Chem. Soc., Chem. Commun.* **1985**, 546.

361. Groves, J. T.; Han, Y.-Z. *Cytochrome P450: Structure, Mechanism, and Biochemistry*; Plenum Press: New York, 1995.

362. Lange, S. J.; Miyake, H.; Que, L. *J. Am. Chem. Soc.* **1999**, *121*, 6330.

363. Grapperhaus, C. A.; Mienert, B.; Bill, E.; Weyhermuller, T.; Wieghardt, K. *Inorg. Chem.* **2000**, *39*, 5306.

364. Schlichting, I.; Berendzen, J.; Chu, K.; Stock, A. M.; Maves, S. A.; Benson, D. E.; Sweet, B. M.; Ringe, D.; Petsko, G. A.; Sligar, S. G. *Science* **2000**, *287*, 1615.

365. Watanabe, Y.; Fujii, H. In *Metal-Oxo and Metal-Peroxo Species in Catalytic Oxidations*; Meunier, B., Ed.; Springer-Verlag: New York, 2000; Vol. 97, p 61.

366. Miyake, H.; Chen, K.; Lange, S. J.; Que, L. *Inorg. Chem.* **2001**, *40*, 3534.

367. Kellner, D. G.; Hung, S. C.; Weiss, K. E.; Sligar, S. G. *J. Biol. Chem.* **2002**, *277*, 9641.

368. Rohde, J. U.; In, J. H.; Lim, M. H.; Brennessel, W. W.; Bukowski, M. R.; Stubna, A.; Munck, E.; Nam, W.; Que, L. *Science* **2003**, *299*, 1037.

369. Khavrutskii, I. V.; Musaev, D. G.; Morokuma, K. *J. Am. Chem. Soc.* **2003**, *125*, 13879.

370. Khavrutskii, I. V.; Musaev, D. G.; Morokuma, K. *Proc. Nat. Acad. Sci. U.S.A.* **2004**, *101*, 5743.

371. Khavrutskii, I. V.; Rahim, R. R.; Musaev, D. G.; Morokuma, K. *J. Phys. Chem. B* **2004**, *108*, 3845.

372. de Visser, S. P.; Ogliaro, F.; Harris, N.; Shaik, S. *J. Am. Chem. Soc.* **2001**, *123*, 3037.

373. Shaik, S.; de Visser, S. P.; Kumar, D. *J. Biol. Inorg. Chem.* **2004**, *9*, 661.

374. Shaik, S.; de Visser, S. P.; Ogliaro, F.; Schwarz, H.; Schroder, D. *Curr. Opin. Chem. Biol.* **2002**, *6*, 556.

375. de Visser, S. P.; Ogliaro, F.; Sharma, P. K.; Shaik, S. *J. Am. Chem. Soc.* **2002**, *124*, 11809.

376. Groves, J. T.; Watanabe, Y. *J. Am. Chem. Soc.* **1986**, *108*, 507.

377. Jacobsen, E. N.; Zhang, W.; Guler, M. L. *J. Am. Chem. Soc.* **1991**, *113*, 6703.

378. Finney, N. S.; Pospisil, P. J.; Chang, S.; Palucki, M.; Konsler, R. G.; Hansen, K. B.; Jacobsen, E. N. *Angew. Chem., Int. Ed. Engl.* **1997**, *36*, 1720.

379. Hong, F.; Look, G. C.; Wei, Z.; Jacobsen, E. N.; Wong, C. H. *J. Org. Chem.* **1991**, *56*, 6497.

380. Castellino, A. J.; Bruice, T. C. *J. Am. Chem. Soc.* **1988**, *110*, 158.

381. Castellino, A. J.; Bruice, T. C. *J. Am. Chem. Soc.* **1988**, *110*, 7513.

382. Castellino, A. J.; Bruice, T. C. *J. Am. Chem. Soc.* **1988**, *110*, 1313.

383. Au, S. M.; Huang, J. S.; Yu, W. Y.; Fung, W. H.; Che, C. M. *J. Am. Chem. Soc.* **1999**, *121*, 9120.

384. Guthikonda, K.; Du Bois, J. *J. Am. Chem. Soc.* **2002**, *124*, 13672.

385. Cundari, T. R.; Klinckman, T. R.; Wolczanski, P. T. *J. Am. Chem. Soc.* **2002**, *124*, 1481.

386. Slaughter, L. M.; Wolczanski, P. T.; Klinckman, T. R.; Cundari, T. R. *J. Am. Chem. Soc.* **2000**, *122*, 7953.

387. Schafer, D. F.; Wolczanski, P. T. *J. Am. Chem. Soc.* **1998**, *120*, 4881.

388. Bennett, J. L.; Wolczanski, P. T. *J. Am. Chem. Soc.* **1997**, *119*, 10696.

389. Schaller, C. P.; Cummins, C. C.; Wolczanski, P. T. *J. Am. Chem. Soc.* **1996**, *118*, 591.

390. Schaller, C. P.; Bonanno, J. B.; Wolczanski, P. T. *J. Am. Chem. Soc.* **1994**, *116*, 4133.

391. Schaller, C. P.; Wolczanski, P. T. *Inorg. Chem.* **1993**, *32*, 131.

392. Cummins, C. C.; Schaller, C. P.; Vanduyne, G. D.; Wolczanski, P. T.; Chan, A. W. E.; Hoffmann, R. *J. Am. Chem. Soc.* **1991**, *113*, 2985.

393. Walsh, P. J.; Hollander, F. J.; Bergman, R. G. *Organometallics* **1993**, *12*, 3705.

394. Hoyt, H. M.; Michael, F. E.; Bergman, R. G. *J. Am. Chem. Soc.* **2004**, *126*, 1018.

395. Au, S. M.; Zhang, S. B.; Fung, W. H.; Yu, W. Y.; Che, C. M.; Cheung, K. K. *Chem. Commun.* **1998**, 2677.

396. Liang, J. L.; Yuan, S. X.; Huang, J. S.; Yu, W. Y.; Che, C. M. *Angew. Chem. Int. Ed.* **2002**, *41*, 3465.

397. Fleming, J. J.; Fiori, K. W.; Du Bois, J. *J. Am. Chem. Soc.* **2003**, *125*, 2028.

398. Groves, J. T.; McClusky, G. A. *Biochem. Biophys. Res. Commun.* **1978**, *81*, 154.

399. Groves, J. T.; Kruper, W. J. *J. Am. Chem. Soc.* **1979**, *101*, 7613.

400. Groves, J. T. *J. Chem. Ed.* **1985**, *62*, 928.

401. For studies with radical clocks, see reference 402.

402. Newcomb, M.; Toy, P. H. *Acc. Chem. Res.* **2000**, *33*, 449.

403. For a discussion of controversy about the rebound mechanism, see references 404–405.

404. Shaik, S.; Cohen, S.; de Visser, S. P.; Sharma, P. K.; Kumar, D.; Kozuch, S.; Ogliaro, F.; Danovich, D. *Eur. J. Inorg. Chem.* **2004**, 207.

405. Shaik, S.; de Visser, S. P.; Ogliaro, F.; Schwarz, H.; Schroder, D. *Curr. Opin. Chem. Biol.* **2002**, *6*, 556.

406. Legzdins, P.; Phillips, E. C.; Sanchez, L. *Organometallics* **1989**, *8*, 940.

407. Carroll, J. M.; Norton, J. R. *J. Am. Chem. Soc.* **1992**, *114*, 8744.

408. Maatta, E. A.; Wentworth, R. A. D. *Inorg. Chem.* **1979**, *18*, 2409.

409. Howard, W. A.; Waters, M.; Parkin, G. *J. Am. Chem. Soc.* **1993**, *115*, 4917.

410. Thorman, J. L.; Guzei, I. A.; Young, V. G.; Woo, L. K. *Inorg. Chem.* **2000**, *39*, 2344.

411. Blake, A. J.; McInnes, J. M.; Mountford, P.; Nikonov, G. I.; Swallow, D.; Watkin, D. J. *J. Chem. Soc., Dalton Trans.* **1999**, 379.

412. Zuckerman, R. L.; Bergman, R. G. *Organometallics* **2000**, *19*, 4795.

413. Tahmassebi, S. K.; Conry, R. R.; Mayer, J. M. *J. Am. Chem. Soc.* **1993**, *115*, 7553.

414. Nugent, W. A.; Harlow, R. L. *J. Am. Chem. Soc.* **1980**, *102*, 1759.

415. Reichle, W.; Carrick, W. L. *J. Organomet. Chem.* **1970**, 419.

416. (a) Brown, S. N.; Mayer, J. M. *J. Am. Chem. Soc.* **1994**, *116*, 2219; (b) Brown, S. N.; Mayer, J. M. *J. Am. Chem. Soc.* **1996**, *118*, 12119.

417. Brown, S. N.; Mayer, J. H. *Organometallics* **1995**, *14*, 2951.

418. McGrane, P. L.; Jensen, M.; Livinghouse, T. *J. Am. Chem. Soc.* **1992**, *114*, 5459.

419. (a) Johnson, J. S.; Bergman, R. G. *J. Am. Chem. Soc.* **2001**, *123*, 2923; (b) Ackermann, L.; Bergman, R. G.; Loy, R. N. *J. Am. Chem. Soc.* **2003**, *125*, 11956.

420. Bexrud, J. A.; Beard, J. D.; Leitch, D. C.; Schafer, L. L. *Org. Lett.* **2005**, *7*, 1959.

421. Mehn, M. P. P., J. C. *J. Inorg. Biochem.* **2006**, *100*, 634.

422. Frankenau, A. D., K. *Z. Naturforsch., B: Chem. Sci.* **1989**, *44*, 493.

423. Gorge, A. D., K.; Fenske, D. *Z. Naturforsch., B: Chem. Sci.* **1988**, *43*, 677.

424. Griffith, W. P. P., D. *J. Chem. Soc., Dalton Trans.* **1973**, 1315.

425. Williams, D. S.; Coia, G. M.; Meyer, T. J. *Inorg. Chem.* **1995**, *34*, 586.

426. Walstrom, A.; Pink, M.; Yang, X. F.; Tomaszewski, J.; Baik, M. H.; Caulton, K. G. *J. Am. Chem. Soc.* **2005**, *127*, 5330.

427. Schweda, E. S., J. *Z. Naturforsch., B: Chem. Sci.* **1981**, *36*, 662.

428. Chatt, J. F., C. D.; Leigh, G. J.; Poske, R. J. *J. Chem. Soc. A* **1969**, 2288.

429. (a) Laplaza, C. E.; Cummins, C. C. *Science* **1995**, *268*, 861. (b) Laplaza, C. E.; Johnson, M. J. A.; Peters, J. C.; Odom, A. L.; Kim, E.; Cummins, C. C.; George, G. N.; Pickering, I. *J. Am. Chem. Soc.* **1996**, *118*, 8623.

430. (a) Freudenberger, J. H.; Schrock, R. R. *Organometallics* **1986**, *5*, 398. (b) Pollagi, T. P.; Manna, J.; Geib, S. J.; Hopkins, M. D. *Inorg. Chim. Acta.* **1996**, 177. (c) Mindeola, D. J.; Cummins, C. C. *Agnew. Chem., Intl. Ed. Engl.* **1998**, *37*, 945.

431. Shapley, P. A. B.; Own, Z. Y.; Huffman, J. C. *Organometallics* **1986**, *5*, 1269.

432. Griffith, W. P.; Pawson, D. *J. Chem. Soc., Dalton Trans.* **1975**, 417.

433. Demadis, K. D. B., M.; Klesczewski, B. G.; Williams, D. S.; White. P. S.; Meyer, T. J. *Inorg. Chim. Acta.* **1998**, *270*, 511.

434. Huynh, M. H. V.; El-Samanody, E. S.; Demandis, K. D.; Meyer, T. J.; White, P. S. *J. Am. Chem. Soc.* **1999**, *121*, 1403.

435. Crevier, T. J.; Mayer, J. M. *J. Am. Chem. Soc.* **1998**, *120*, 5595.

436. Crevier, T. J.; Bennett, B. K.; Soper, J. D.; Bowman, J. A.; Dehestani, A.; Hrovat, D. A.; Lovell, S.; Kaminsky, W.; Mayer, J. M. *J. Am. Chem. Soc.* **2001**, *123*, 1059.

437. Maestri, A. G.; Cherry, K. S.; Toboni, J. J; Brown, S. N. *J. Am. Chem. Soc.* **2001**, *123*, 7459.

438. Brown, S. N. *J. Am. Chem. Soc.* **1999**, *121*, 9752.

439. Maestri, A. G. T., S. D.; Schuck, S. M.; Brown, S. N. *Organometallics* **2004**, *23*, 1932.

CHAPTER 14

Principles of Catalysis

(Written with Prof. Patrick J. Walsh)

The remaining chapters of this text describe catalytic reactions. Each of these catalytic reactions comprises stoichiometric reactions that have been presented so far in this text. Catalytic reactions are distinct from the stoichiometric reactions because the metal complexes in catalytic reactions are present in substoichiometric amounts. Chapter 14 presents some of the principles that apply to catalytic reactions, including enantioselective catalytic reactions. The content of this chapter applies to all reactions contained in the subsequent chapters of this text, as well as to catalytic reactions that lie outside of the scope of this text. The latter sections of Chapter 14 present principles that are specific to enantioselective catalysis. Catalytic enantioselective reactions form non-racemic, chiral products by virtue of the chirality of the catalyst.

14.1. General Principles

14.1.1. Definition of a Catalyst

The term "catalyst" was first introduced in 1836 by Berzelius and was defined in 1894 by Ostwald. Ostwald stated that a catalyst is a substance that increases the rate of a chemical reaction without itself being consumed. This basic definition continues to be commonly used. During a catalytic reaction, the catalyst undergoes a series of transformations to generate product, and this series of reactions must regenerate the starting catalyst. As a result, the catalyst can be used in substoichiometric amounts relative to the reagents.

The mechanisms of a variety of catalytic processes are described in detail in the subsequent chapters of this text. However, for the purposes of illustrating how a series of steps can regenerate the starting catalyst, the steps of one of the simplest catalytic organometallic reactions, the hydrogenation of ethylene, is shown here. The mechanism of a catalytic reaction is typically drawn as a cycle (vide infra), but the linear sequence of steps shown in Scheme 14.1 illustrates that the starting rhodium complex is regenerated after conversion of one equivalent of ethylene and hydrogen to ethane.

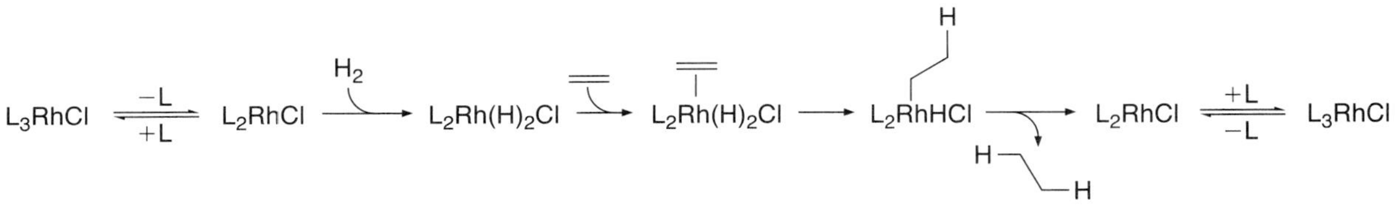

Scheme 14.1. Regeneration of the starting catalyst in a simple catalytic process.

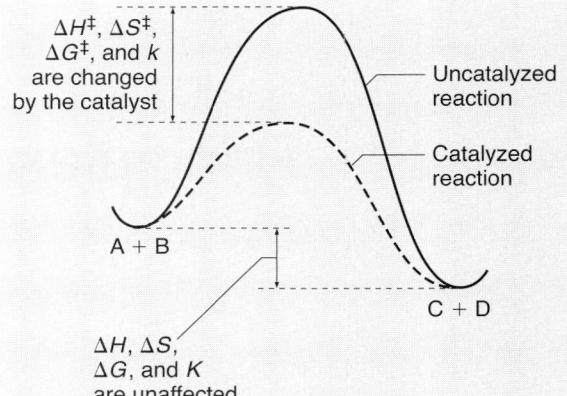

$\Delta H^{\ddagger}$, $\Delta S^{\ddagger}$, $\Delta G^{\ddagger}$, and k are changed by the catalyst

Uncatalyzed reaction

Catalyzed reaction

A + B

C + D

ΔH, ΔS, ΔG, and K are unaffected

Figure 14.1.
Simple depiction of the reaction coordinate for a catalyzed and uncatalyzed reaction.

14.1.2. Energetics of Catalysis

Figure 14.1 shows the simplest depiction of the reaction coordinate for a catalyzed and uncatalyzed reaction. More detailed diagrams are provided later in this section. As shown by this figure, catalysts reduce the free energy of the highest energy transition states and thereby increase the rate of the reaction. Catalysts change only the energies of the transition states of reactions. They do not change the energies of the reactants or products. Therefore, catalysts change only the rates of reaction and do not change the thermodynamics and equilibrium constants that define the molar ratios of the reactants and products. One can envision many hypothetical, synthetically valuable reactions that lie too far uphill in thermodynamics to accumulate appreciable amounts of product. These reactions will not accumulate product at the same temperature in the presence of even the most active catalysts.

However, catalysts can change our ability to obtain higher concentrations of product for more subtle reasons. Equilibria are shifted by changes in conditions, such as temperature, and Le Châtelier's principle predicts that an exothermic reaction will form higher equilibrium concentrations of product at lower temperatures than at higher temperatures. Thus, a catalyst can create the ability to observe an increase in the concentration of products by allowing one to conduct an exothermic reaction at a lower temperature in the presence of a catalyst than in the absence of a catalyst.

Efforts to develop catalysts for the synthesis of ammonia from nitrogen and hydrogen at low temperatures illustrate this effect. The Haber–Bosch process is the well-known reaction of nitrogen with hydrogen at high temperature and pressure to form ammonia in the presence of heterogeneous iron catalysts supported on alumina, as shown in Equation 14.1. The yield of the reaction is limited by thermodynamics. Because this reaction is exothermic (but entropically unfavorable), the yield could be higher if a catalyst were developed that would allow the reaction to be conducted at lower temperatures.

$$3\,H_2\ +\ N_2\ \xrightarrow[\substack{250\ \text{atm} \\ 450\text{–}500\ ^{\circ}\text{C}}]{\substack{\text{Fe/Al}_2\text{O}_3 \\ \text{additives}}}\ \underset{10\text{–}20\%:}{2\,NH_3}\quad (\Delta H = -92.4\ \text{kJ/mol}) \tag{14.1}$$

14.1.3. Reaction Coordinate Diagrams of Catalytic Reactions

A catalyst decreases the activation energy of a reaction, relative to that of the uncatalyzed reaction, by stabilizing the transition state. Because the transition state is an unstable species and is fleeting, the catalyst cannot affect the transition state alone. Instead, the catalyst binds one or more of the reactants and remains bound through the transition state

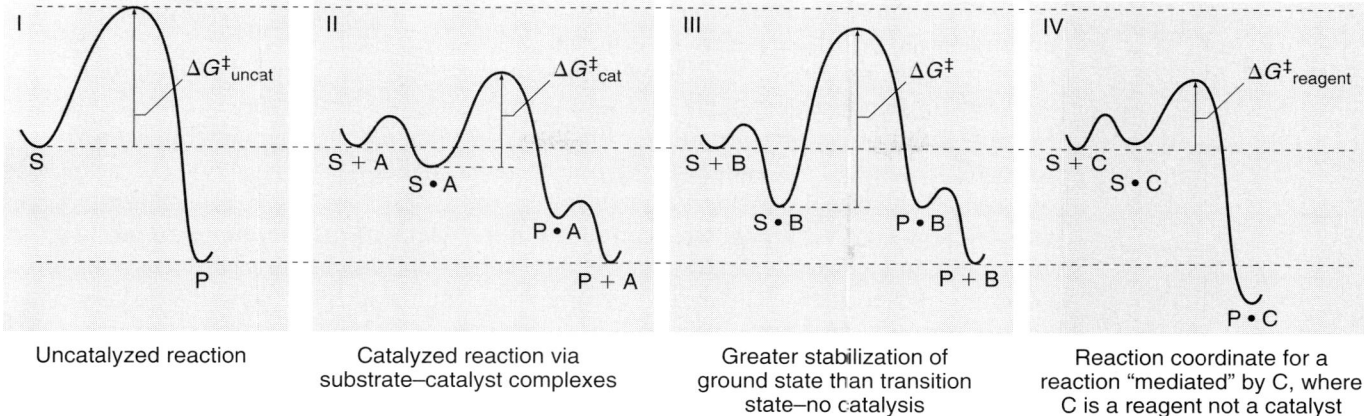

Figure 14.2.
Reaction coordinate diagram for an uncatalyzed reaction, a reaction catalyzed by A, a reaction promoted by an additive B, and a reaction involving a reagent C.

of the catalytic process. Dissociation of the product either regenerates the starting catalyst directly or generates a species that will be converted to the starting catalyst.

More precise reaction coordinate diagrams than the one in Figure 14.1 that reflect the interactions of the catalyst are illustrated in Figure 14.2. These curves illustrate the importance of comparing transition-state energies with ground-state energies and the distinction between a catalyst and a reagent. The first reaction coordinate I corresponds to the thermal, uncatalyzed reaction of substrate (S) to generate product (P). The second, third, and fourth reaction coordinates II–IV correspond to the analogous reactions in the presence of A, B, and C. All three reaction coordinates show the formation of an adduct (S•A, S•B, or S•C) between the substrate and the additives that, for the purpose of illustration, is lower in energy than the reactants. This lower energy is unnecessary for catalysis, and the same principles could be illustrated with a reaction coordinate in which these adducts lie uphill of the reactants.

Most important, three scenarios are possible starting from this complex of the additive with the substrate. In reaction coordinate II, species A stabilizes the transition state more than it stabilizes the ground state S (by forming S•A). In addition, species A releases the free product in this reaction coordinate in a manner that reforms A. This scenario constitutes a catalytic process, and A would be the catalyst. In reaction coordinate III, the complex S•B formed between additive B and substrate S also forms product and regenerates B. However, the overall energy landscape in this reaction coordinate does not lead to catalysis. In this reaction coordinate, additive B stabilizes S (by formation of S•B) more than it stabilizes the transition state. As a result, the activation energy in this scenario is higher than that for the reaction without additive B. S would then convert to P by the lower energy uncatalyzed pathway in the presence of substoichiometric amounts of B. In reaction coordinate IV, a species C forms an adduct with the substrate and the product of the reaction. In this scenario, the reaction coordinate ends with P•C, the complex of the product with C. A second process is then conducted on P•C by addition of another reagent (such as water, acid, or an oxidant) to release the organic product (not shown). Although reaction in the presence of C will occur faster than in the absence of C and could affect selectivity, compound C is changed by the reaction. Thus, compound C is a reagent in this scenario, instead of a catalyst, and must be used in stoichiometric quantities.

The many examples of catalytic reactions described in the following chapters illustrate the first scenario in which A acts as a catalyst, and one example was shown in Scheme 14.1 here. Few examples of the second scenario have been observed because these reactions do not occur with rate acceleration and are typically unreported. However, it is instructive to compare catalytic reactions with those known to follow the third scenario in which the additive is a reagent, rather than a catalyst.

The many reactions of π-arene complexes described in Chapter 11 are examples of reactions in which a metal complex acts as a reagent, rather than a catalyst. One goal in the chemistry of π-arene complexes has been to translate their stoichiometric reactivity into catalytic processes. The classic chemistry of Pt(II) olefin complexes also illustrates the difference between the metal acting as reagent and as catalyst. In the sequence in Equation 14.2, the starting Pt(II) olefin complex has been shown to add amine to form a stable aminoalkyl intermediate. Treatment of this complex with a reducing agent was shown to generate the product of olefin hydroamination. However, this process was not catalytic because the reaction of the aminoallyl complex with $NaBH_4$ to release the organic product formed Pt(0), and the system did not contain a reagent to reform the starting Pt(II) species. Chapter 16, which describes the hydrofunctionalizations of olefins, illustrates how nucleophilic attack on a coordinated olefin has more recently been developed into several types of olefin aminations.

$$\text{(14.2)}$$

$$+ \; R_2NH \cdot HCl$$

14.1.4. Origins of Transition State Stabilization

Catalysts can lower the energy of the highest energy transition state by interacting with and stabilizing a structure that is similar to that of the uncatalyzed reaction, or it can create a completely new reaction pathway. In the reaction coordinate diagram in Figure 14.1, the mechanism of the reaction is closely related to the mechanism of the uncatalyzed reaction. Both occur by single steps (after formation of the adduct) and can occur through a transition state consisting of an arrangement of atoms that is close to that of the uncatalyzed reaction. This situation often occurs in enzyme-catalyzed processes.

This stabilization of a transition state related to the uncatalyzed process can also occur in reactions catalyzed by transition metal complexes, such as those catalyzed by Lewis acids. For example, Diels–Alder reactions catalyzed by transition metal complexes sometimes occur by mechanisms related to the concerted [4+2] mechanism of the uncatalyzed process. In this case, the catalyst changes the electronic properties of the substrate bound to the Lewis acid in a fashion that reduces the barrier for the [4+2] cycloaddition. Figure 14.3 shows the transition state proposed for enantioselective Diels–Alder reactions catalyzed by copper complexes. The transition state structure is proposed on the basis of the calculated structure of the Lewis acid complex formed between the copper–bisoxazoline fragment and the acrylate.

More often in organometallic chemistry, the catalytic reaction occurs by a mechanism that is completely different from the mechanism of the uncatalyzed process. In this case, the reaction typically occurs by more steps, but the activation energy of each of the individual steps is lower than the activation energy of the uncatalyzed process. The overall barrier is then lower than that of the uncatalyzed reaction. A comparison of the uncatalyzed and catalyzed hydroboration of alkenes with a dialkoxyborane $(RO)_2BH$, such as catecholborane (see Chapter 16), illustrates this scenario. Qualitative reaction coordinates for the uncatalyzed and rhodium-catalyzed process are shown in Figure 14.4. In the absence of a catalyst, the B–H bond adds across the alkene through a concerted four-center transition state, albeit at elevated temperatures in neat alkene. In contrast, late transition metal-catalyzed hydroborations first cleave the B–H bond by oxidative addition. Coordination

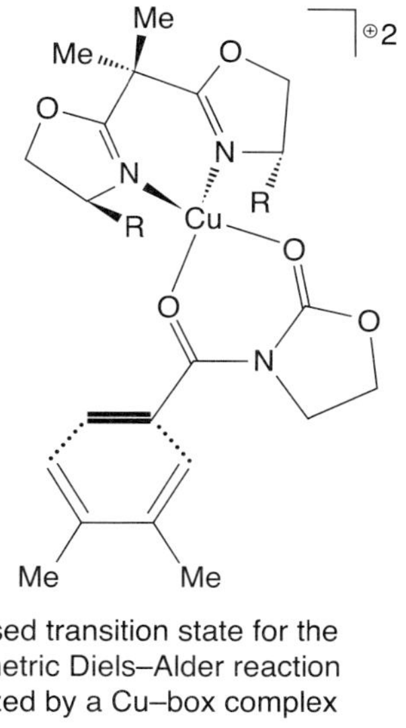

Proposed transition state for the
asymmetric Diels–Alder reaction
catalyzed by a Cu–box complex

Figure 14.3.
Transition state proposed for enantioselective Diels–Alder reactions catalyzed by a copper–bisoxazoline complex. The orientation of the diene and acrylate in this transition state resembles that in the transition state for an uncatalyzed [4+2] cycloaddition. Adapted from Evans, D. A.; Barnes, D. M.; Johnson, J. S.; Leckta, T.; von Matt, P.; Miller, S. J.; Murry, J. A.; Norcross, R. D.; Shaughnessy, E. A.; Campos, K. R. *J. Am. Chem. Soc.* **1999**, *121*, 7582.

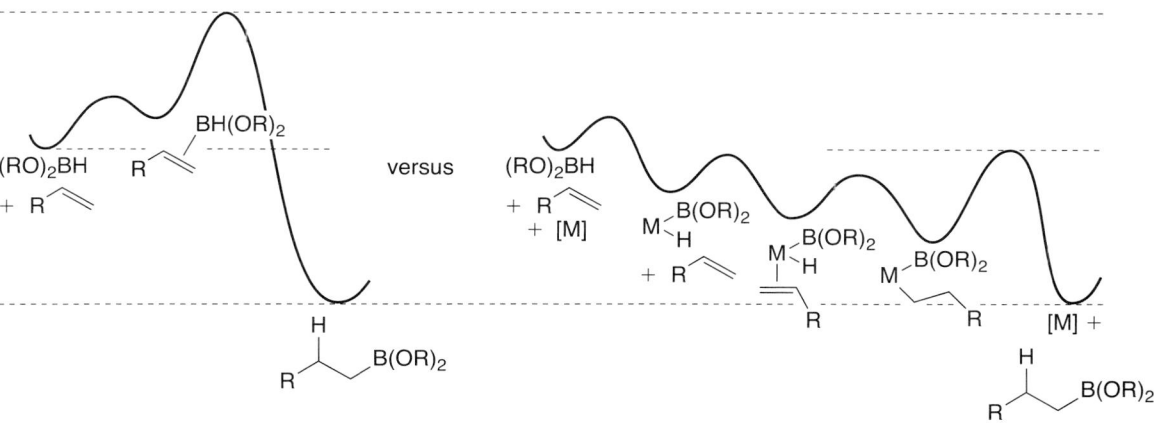

Figure 14.4.
Comparison of the reaction coordinates for an uncatalyzed and late-metal-catalyzed hydroboration of an alkene.

and migratory insertion of the olefin then forms the C–H bond, and reductive elimination forms the B–C bond in the alkylboronate product and regenerates the starting metal complex.

14.1.5. Terminology of Catalysis

14.1.5.1. The Catalytic Cycle

The combination of steps of a catalytic process is often called a *catalytic cycle*. This term refers to the series of steps by which the reagents are consumed, the products are formed, and the catalyst is regenerated. As shown generically in Figure 14.5, this series of reactions is commonly written in the form of a cycle because the starting point of the

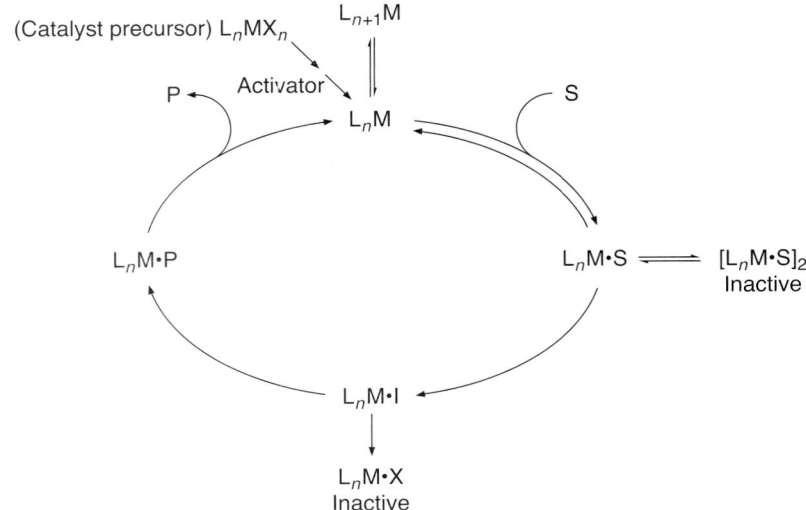

Figure 14.5.
Example of a catalytic cycle with catalyst precursors and both the reversible
and irreversible formation of inactive species.

catalytic process is also the ending point of the reaction. In the case illustrated below, the catalyst precursor, L_nMX_n, connects to the active catalyst L_nM. This active catalyst then binds the substrate to give the catalyst-substrate adduct, $L_nM \bullet S$. The bound substrate undergoes some type of transformation on the metal to form intermediate $L_nM \bullet I$, which is subsequently converted to $L_nM \bullet P$. Dissociation of the product P regenerates the catalyst, which can reversibly exit the catalytic cycle by coordination of the ligand L to form $L_{n+1}M$ or bind another substrate molecule to restart the cycle. For a catalytic cycle comprising multiple steps, the concentration of the *active catalyst* refers to the sum of the concentrations of all catalytic species that lie within the cycle. Maximum efficiency is achieved when all of the catalyst is "active" and lies within the catalytic cycle.

14.1.5.2. Catalyst Precursors, Catalyst Deactivation, and Promoters

Often the material added to the reaction flask to increase the rate is not the true catalyst but is a complex or a mixture of compounds that gives rise to a complex that lies on the catalytic reaction pathway. In this case, the catalytic process does not regenerate the starting complex, and the starting complex would be called a *catalyst precursor*. Because the catalyst precursor leads to the true catalyst, the catalyst precursor is typically drawn at a position off of this catalytic cycle (see Figure 14.5).

The distinction between a catalyst precursor and a true catalyst can be illustrated by the catalytic hydrogenations and cross couplings presented in Chapters 15 and 19. In many cases, hydrogenations are initiated by a complex of the general formula $[Rh(COD)L_2]^+$ (COD = cyclooctadiene), $[RuL_2(OAc)_2]$, or $[RuL_2(H)(COT)]^+$ (COT = cyclooctatriene). As illustrated for the ruthenium complex in Scheme 14.2, the cyclooctatriene ligand must first undergo hydrogenation to lead to the species that reacts with the olefin substrate. Likewise, the active catalyst in many cross-coupling reactions between haloarenes and main group organometallic nucleophiles are generated by a combination of $Pd(OAc)_2$ and a phosphine, which first generates $L_2Pd(OAc)_2$. By mechanisms that depend on the particular reaction conditions, the Pd(II) species generated by this precursor and ligand form a Pd(0) complex of the phosphine.[1] Generation of the active Pd(0) catalyst by reduction of the $L_2Pd(OAc)_2$ with the nucleophile is shown in Scheme 14.3. In some cases, the active catalyst is formed in high yield, but in other cases, the efficiency of the catalytic process is limited by the rate and yield for formation of the active catalyst from the catalyst precursor.[2-4]

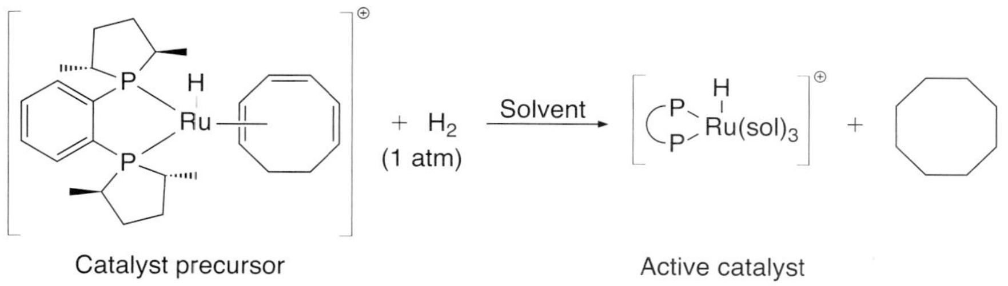

Scheme 14.2

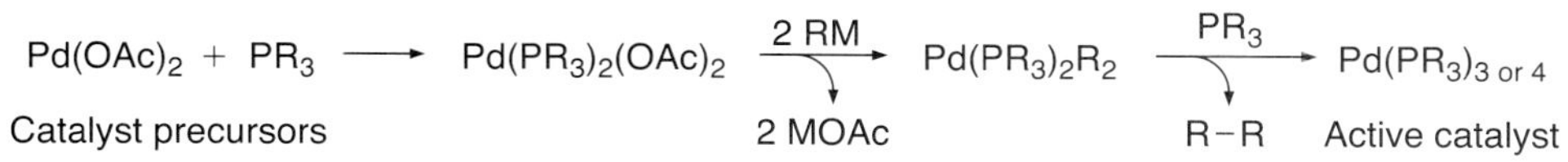

Scheme 14.3

Because the set of reactions that constitute the catalytic cycle must regenerate the active catalyst, irreversible decomposition of a catalyst must occur by a step that is external to the catalytic cycle. Such a step is shown arbitrarily in Figure 14.5 by the irreversible reaction of $L_nM\bullet I$ to form $L_nM\bullet X$. Examples of such routes to catalyst deactivation include P–C bond cleavage or cyclometallation of a phosphine ligand, or dissociation of a ligand and precipitation of the metallic form of the central metal.

Alternatively, reactions external to the cycle can retard the catalytic process by siphoning the catalyst away from the productive steps. These reactions, such as the reversible formation of $L_{n+1}M$ or the formation of dimeric $L_nM\bullet S$ (see Figure 14.5), are drawn at points external to the catalytic cycle. Often dissociation of a dative ligand must occur for stable complexes to enter the catalytic cycle, and this event is depicted by the equilibrium between $L_{n+1}M$ and L_nM in Figure 14.5. One common example of such an equilibrium that lies external to the catalytic cycle is the dissociation of phosphine from a Pd(0) species to generate the reactive intermediate that undergoes oxidative addition. A second example is the reversible formation of dimer as part of the complex system involved in the hydrogenation of olefins catalyzed by $[RhCl(PPh_3)_3]$ presented in Chapter 15. In the former case, the ligand dissociation is typically faster than the overall catalytic cycle.[5,6] In the latter case, the equilibrium between monomer and dimer occurs more slowly than the catalytic cycle.[7]

Often the rates and selectivities of catalytic reactions are enhanced by additional reaction components that are added in small amounts. These added materials are often called *promoters* or *co-catalysts*. Protic acids or Lewis acids are common promoters. As is presented in Chapter 16, triarylboranes are Lewis acid promoters in hydrocyanations catalyzed by transition metal complexes.

14.1.5.3. Quantification of Efficiency

Several quantities are used to define the efficiency of a catalytic process. One of these quantities, which refers to the output of the catalyst, is the *turnover number*. The turnover number is the number of moles of product per mole of catalyst (Equation 14.3). Commodity-scale reactions such as rhodium-catalyzed hydroformylation occur with hundreds of thousands of turnover numbers, whereas other reactions that are less developed might occur with 10 to 20 turnovers. A second quantity is *turnover frequency*. Turnover frequency is the number of moles of product per mole of catalyst per unit time, in other words, turnover numbers per unit time (Equation 14.4). Because the rates of reactions change

$$\text{Turnover number (TON)} = \text{moles of product/moles of catalyst} \qquad (14.3)$$

$$\text{Turnover frequency (TOF)} = \text{TON/time} \qquad (14.4)$$

over time, a reported turnover frequency often implicitly refers to an average turnover frequency over the course of the reaction, to an initial turnover frequency, or to a turnover frequency under conditions in which a gaseous reagent is held at constant pressure. Because the actual concentrations of active catalyst species are generally unknown, most reports of catalytic reaction rates are apparent turnover frequencies for which it is assumed that all of the catalyst precursor has been converted into the active catalyst.

14.1.6. Kinetics of Catalytic Reactions and Resting States

The *rate constants* of different steps of a catalytic cycle are different, but the net *rates* of each step of a catalytic process are identical once the system has reached steady state. The rate of each step is proportional to the rate constant and the concentrations of reagents and species that lie on the catalytic cycle (e.g., for the step, cat_a + $reagent_a \xrightarrow{k_a} cat_b$, the net rate is given by $k_a[cat_a][reagent_a]$). For a reversible step, the net rate is equal to the forward rate minus the reverse rate. The concentrations of species that lie on the catalytic cycle vary to make the rate of each step of the catalytic cycle identical. The step with the smallest pseudo-first order rate constant (given by $k_a[reagent_a]$ in the example above) is called the "turnover-limiting step." This step controls the efficiency of the catalytic process because the rate of a catalytic reaction cannot exceed $k_{tls} \cdot [catalyst][reagents]$. In this situation, essentially all of the catalyst in the system consists of the species preceding the turnover-limiting step. Although this step is often called the rate-limiting step, the use of this term is incorrect because all steps in a catalytic cycle proceed at the same rate.

The term *"catalyst resting state"* refers to the catalytic species that is present in the highest concentration. The resting state can be a species that lies off of the cycle and reversibly generates the catalyst, or it can be one of the species that lies on the catalytic cycle. For example, $L_3Pd(0)$ and $L_4Pd(0)$ complexes are the resting state of many palladium-catalyzed processes that occur by turnover-limiting oxidative addition.

Species that lie off the productive catalytic cycle, but reversibly enter back into the cycle, are sometimes called *dormant states*. Such a dormant state was shown as $L_{n+1}M$ in Figure 14.5, and the $L_3Pd(0)$ or $L_4Pd(0)$ resting states noted in the previous paragraph would be considered to be *dormant states*. If the resting state is one of the species within the catalytic cycle, then it is the species that undergoes the turnover-limiting step. This is the condition of maximum catalytic efficiency.

14.1.7. Homogeneous vs. Heterogeneous Catalysis

Catalysts are often subdivided into groups called *homogenous catalysts* and *heterogeneous catalysts*. A homogeneous catalyst resides in the same phase as the reactants, and this medium is usually the liquid phase. Often, homogeneous catalysts are discrete, molecular species. Heterogeneous catalysts reside in a different phase from the reagents. Often the reactants reside in the gas phase, and the heterogeneous catalyst is solid; in other cases, the reactants reside in the liquid phase (liquid reagents or solids dissolved in solution), and the catalyst is an insoluble solid.

Each class of catalyst has its own advantages and disadvantages. One advantage of a homogeneous catalyst is rapid rates for the diffusion of reagents and heat in the solution phase. A second advantage is the ability to initiate the reaction with a discrete compound that can be characterized by solution-phase methods, such as NMR spectroscopy. A third advantage is the ability to control a three-dimensional architecture created by the geometry of the ligands. Because of this ability to control the structure of homogeneous catalysts, these catalysts are often more reactive and more selective than solid heterogeneous catalysts. The reactions of homogeneous catalysts tend to occur with higher regioselectivity, diastereoselectivity, and enantioselectivity than reactions of heterogeneous catalysts.

At the same time, homogeneous catalysts have disadvantages. These disadvantages include an inability to regenerate the catalyst after decomposition. Usually, the catalyst decomposition occurs by an irreversible reaction of a ligand or by precipitation of the metal. For this reason, regeneration of the catalyst from these decomposition products is difficult. These disadvantages sometimes include the difficulty of separating the catalyst from the reaction products. Products are often separated from the catalyst by distillation at temperatures that would lead to catalyst decomposition. In other cases, small amounts of the catalyst crystallize with the product. This contamination by the catalyst is a particular problem when using a catalyst to prepare a pharmaceutical ingredient.

The advantages of heterogeneous catalysts include low cost, ease of separation from products, high stability, and an ability to be recycled. Often these catalysts are generated from a mixture of metal salts without organic ligands. Because the ligand is often the most expensive part of a homogeneous catalyst and cannot be recycled, the absence of ligands on the heterogeneous catalysts makes them less expensive. Furthermore, a heterogeneous catalyst can simply be removed from the products by filtration when it is a solid and the reactants are liquids, and separation is even simpler when the products are gasses. Heterogeneous catalysts often are stable at high temperature. Thus, reactions that are slow, even in the presence of the catalyst, can be conducted in a practical manner with heterogeneous catalysts by running the catalytic reaction at high temperatures. Finally, heterogeneous catalysts often catalyze oxidation and reduction reactions. In these cases, oxidation or reduction of the deactivated catalyst can regenerate the active catalyst prior to addition of a second batch or stream of reagents.

At the same time, heterogeneous catalysts have disadvantages. These disadvantages include slow rates that result from a low effective concentration. The effective concentration is low because the catalyst can only react with reagents at the interface of the two phases and the active catalyst is often present at sites of defects in the bulk solid. For this reason, much effort has been spent to generate heterogeneous catalysts that have high surface areas. Disadvantages may also include the lack of precise three-dimensional architecture and non-uniformity in the local structures and properties of catalytic sites. This lack of defined, monodisperse, and controllable catalyst structure generally causes the selectivities of reactions with heterogeneous catalysts to be lower than those of reactions with homogeneous catalysts and hinders synthesis of specialized catalyst environments, such as the chiral structures necessary for enantioselective catalysis. For this reason, effort is being spent to attach highly active and selective catalysts to solid supports to combine the advantages of a solid catalyst with the uniformity and sophisticated architecture of a homogeneous catalyst.

14.1.7.1. Distinguishing Homogeneous from Heterogeneous Catalysts

It is often difficult to definitively demonstrate that a catalyst is either homogeneous or heterogeneous. In some cases, reactions catalyzed by soluble precursors are actually catalyzed by nanoparticulate or colloidal metal particles. In other cases, catalysts attached to solid supports actually operate after cleaving from the support and becoming a soluble, homogeneous catalyst.[8–12] One might think that the reactions catalyzed by homogeneous or heterogeneous species could be distinguished by the observation of solids in the reaction solution. However, one cannot determine in the absence of additional data whether the reaction is catalyzed by the material that is dissolved or by the material that is a solid. In other cases, the solutions appear to contain material that is fully dissolved, but the solution actually contains nanoparticulate or colloidal species. But again, if these species are identified by methods described in this section, one cannot determine in the absence of futher data whether the reaction occurred by the fully dissolved material or by the nanoparticulate material. This section will describe how solids are identified in solution and the experiments that have been used to reveal whether a dissolved or solid species is the true catalyst in the system. Several reviews that address this issue in detail have been published.[13–16]

To identify the presence of nanoclusters in solutions that appear to be homogeneous, one typically uses light scattering and tunneling electron microscopy (TEM). Finke argues that TEM is the most definitive method to assess the presence or absence of nanoparticulate

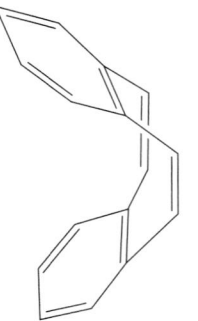

Figure 14.6.
Structure of dibenzo[a,e]
cyclooctatetraene.

material. Because these solids contain metals in low—often the M(0)—oxidation state, catalytic reactions conducted under reducing conditions such as hydrogenation,[15,16] hydrosilylation,[17] and coupling reactions with alcohols or amines containing hydrogens α to the oxygen or nitrogen, or with strongly reducing metal–alkyl reagents,[18,19] have been considered to be the best candidates for generating nanoparticulate material.[14]

Many methods have been used to assess whether the dissolved or solid material is responsible for the catalysis, ranging from effects of additives, the "three phase test," and kinetic profiles. Two of the earliest proposed tests to address the homogeneous vs. heterogeneous catalysis issue are to determine the effect of added mercury and the effect of dibenzo[a,e]cyclooctatetraene (Figure 14.6). Elemental mercury has been shown to poison heterogeneous catalysts by blocking the pores of high-surface-area solids.[20] Thus, reactions catalyzed by heterogeneous catalysts are typically retarded by added mercury, whereas reactions catalyzed by homogeneous systems are unaffected by added mercury. However, the effect of mercury on colloidal systems is not well established and soluble complexes can react and be changed by elemental mercury. Thus, this experiment provides suggestive, but not definitive, data. Conversely, dibenzo[a,e]cyclooctatetraene binds to soluble complexes to poison their catalytic activity, but does not bind as well to metal surfaces.[21] Thus, reactions catalyzed by homogeneous catalysts are typically unaffected by this additive, whereas reactions catalyzed by homogeneous systems are poisoned by it. Thus, the effect of dibenzo[a,e]cyclooctatetraene on the reaction has been use to distinguish homogeneous from heterogeneous catalysts. However, this diene does not bind to all homogeneous catalysts, and these data are again suggestive.

To address these limitations, additional tests have been designed. In one experiment called the "three-phase test,"[21] a substrate is attached to an insoluble support, such as a polymer. If the catalyst is also a solid, then the supported substrate reacts much more slowly than the analogous soluble substrate, but if the catalyst is dissolved, then the two substrates will react with more similar rates. The polymer-bonded substrate must be swollen in a solvent which is compatible with the homogeneous catalyst. Applications of this test for homogeneity are illustrated in Equation 10.54.[21] The virtue of this test is that it is based on the catalytic process itself, rather than on detection of hypothetical catalysts. A variation of this theme involves the use of polymeric catalyst poisons, such as polythiols. These were found to have no effect on heterogeneous catalysts but to retard homogeneous catalysts.[20]

$$\text{P}-\text{C}_6\text{H}_4-\text{NO}_2 + \text{H}_2 \xrightarrow{\text{RhCl}_2(\text{BH}_4)(\text{DMF})(\text{py})_2} \text{P}-\text{C}_6\text{H}_4-\text{NH}_2 \qquad (14.5)$$

$$\text{P}-\text{C}_6\text{H}_4-\underset{\text{CH}_2}{\overset{\text{CH}_3}{\text{C}}} \xrightarrow{\text{cat., H}_2} \text{P}-\text{C}_6\text{H}_4-\underset{\text{CH}_3}{\overset{\text{CH}_3}{\text{CH}}}$$

cat.	yield
RhClL$_3$	100%
Ni(O$_2$CR)$_2$ + NaBH$_4$	–
[Rh(NBD)L$_2$]ClO$_4$	90%
Pd/C	–
[Ir(COD)(PPr$^i_3$)(py)] PF$_6$	100%

(14.6)

Several homogeneous catalysts, such as McQuillin's catalyst (Equation 14.5) and Crabtree's catalyst (last entry of Equation 14.6), that displayed unusual reactivity for homogeneous systems were shown to be truly homogeneous by this three-phase test. In one instance has this test implied the presence of a homogeneous arene hydrogenation catalyst.[21] In other cases, soluble precatalysts were shown to generate heterogeneous catalysts. [Rh(C$_5$Me$_5$)Cl$_2$]$_2$ did not catalyze the hydrogenation of the arene units in cross-linked polystyrene beads or in a hydrated butadiene-styrene copolymer (Equation 14.7), but the olefin groups in a cross-linked polymer were hydrogenated by the same system (Equation 14.8). Apparently this precursor generates an active homogeneous component capable of catalyzing the hydrogenation of olefins, as well as a heterogeneous catalyst from decomposition of the precursor that catalyzes the hydrogenation of arenes.

$$\text{—(CH}_2\text{—CH—)(—CH}_2\text{CH}_2\text{CH—CH}_2\text{—)—} \xrightarrow[\substack{\text{i-PrOH} \\ \text{H}_2 \text{ Et}_3\text{N}}]{[\text{Rh(C}_5\text{Me}_5)\text{Cl}_2]_2} \text{no reaction} \tag{14.7}$$

$$\text{(P)—}\bigcirc\text{—CH=CH}_2 \xrightarrow[\substack{\text{CH}_2\text{Cl}_2,\ \text{Et}_3\text{N} \\ \text{H}_2}]{[\text{Rh(C}_5\text{Me}_5)\text{Cl}_2]_2} \text{(P)—}\bigcirc\text{—CH}_2\text{CH}_3 \tag{14.8}$$

Finally, Finke has argued that kinetic measurements provide the most definitive data to distinguish between homogeneous and heterogeneous catalysts when particulate material is observed visually, with light scattering, or with TEM.[14,15] First, if the observed solid or nanoclusters catalyze the reaction, then the reaction will occur with a sigmoidal profile in which the rate increases as the solid or nanoclusters forms. This reaction profile is shown in Figure 14.7. If the solid material is the catalyst decomposition product of an active soluble catalyst, then the rates will slow more than they would in the absence of this decomposition. Second, it is important to identify the complexes dissolved in solution and to determine independently if these species are competent to be intermediates in the catalytic process.

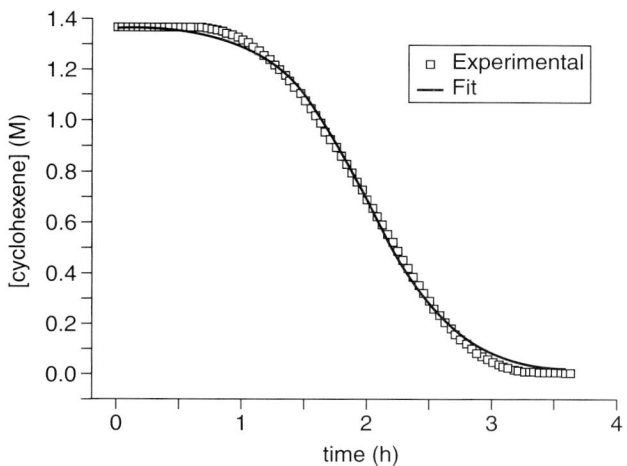

Figure 14.7.
Sigmoidal profile of the cyclohexene concentration vs. time for hydrogenation starting with $[\text{Bu}_4\text{N}]_5\text{Na}_3[(1,5\text{-COD})\text{Ir·P}_2\text{·W}_{15}\text{Nb}_3\text{O}_{62}]$. This profile is characteristic of a reaction in which a nanocluster that catalyzes the reaction forms in an induction period. Adapted with permission from Widegren, J. A.; Finke, R. G. *J. Mol. Catal. A* **2003**, *198*, 317.

14.2. Fundamentals of Asymmetric Catalysis[22]

14.2.1. Importance of Asymmetric Catalysis

Many of the molecules that constitute living organisms, such as amino acids, enzymes, proteins, and DNA, are chiral and present in only one of the two enantiomeric forms. When small, chiral organic molecules interact with the homochiral molecules of biological systems, each enantiomer interacts differently and elicits a different response from the organism. Therefore, to target biological processes more specifically, and to reduce side effects caused by the undesired enantiomer, new pharmaceutical ingredients are generally synthesized as single enantiomers. This need to generate single enantiomers efficiently has spawned the field of asymmetric catalysis. In a catalytic asymmetric reaction, a sub-stoichiometric amount of a non-racemic, chiral catalyst generates a chiral product in non-racemic form.

14.2.2. Classes of Asymmetric Transformations

The most common type of asymmetric transformation involves the action of non-racemic, chiral catalysts on prochiral substrates.[22] A position in a molecule is considered to be prochiral if addition of an achiral substituent (that is different than the other sub-stituents on the prochiral center) to one face or replacement of one group by a new group generates a chiral center. For example, the two faces of an aldehyde RCHO are prochiral if R ≠ H, and the two hydrogens of a methylene group attached to two different groups are prochiral. As shown in Figure 14.8A, addition of a methyl group to the carbonyl unit of benzaldehyde generates a stereogenic center, and, as shown in Figure 14.8B, reaction of N-bromosuccinimide (NBS) with propiophenone under acidic conditions furnishes the chiral, racemic halogenated product. The faces of benzaldehyde and the α-hydrogens of propiophenone are also called enantiotopic.

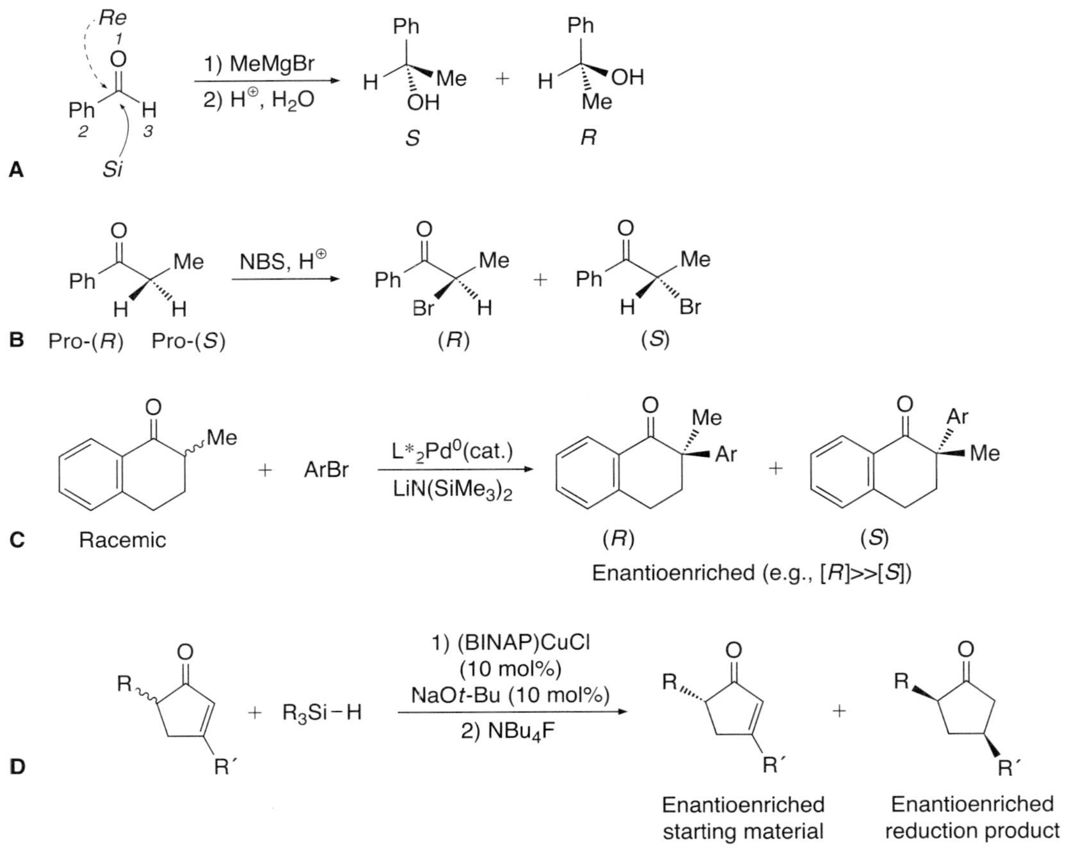

Figure 14.8. Four classes of reactions that form chiral products. **A** illustrates prochiral faces and **B** enantiotopic atoms. **C** and **D** are examples of a dynamic kinetic resolution and kinetic resolution, respectively.

In addition to molecules containing prochiral sites, racemic molecules can participate in catalytic asymmetric transformations. In some transformations, the stereocenter is destroyed in the course of the reaction, and equilibrating prochiral intermediates are formed. An example of such a process is the asymmetric arylation of ketones (Figure 14.8C). In other cases, one of the enantiomers of the substrate reacts with the asymmetric catalyst significantly faster than the other enantiomer. In this case, an enantioenriched product is formed, and the opposite enantiomer of the reactant remains. This last process is called a kinetic resolution and is illustrated for the conjugate reduction of enones via hydrosilylation (Figure 14.8D).[23] In this case, the top and bottom faces of the C–C double bond are diastereotopic because reaction at each face of the enone generates diastereomeric products. This section of the chapter first presents the principles that relate to reactions at prochiral centers of achiral substrates and then presents the principles that relate to reactions of racemic or meso compounds.

14.2.3. Nomenclature

To describe the relationship between two prochiral faces, the substituents of the reacting prochiral center are assigned priority using the Cahn–Ingold–Prelog convention.[24,25] If these groups are arranged with increasing priority in a clockwise direction, the prochiral face is the *Re* face. If they are arranged with increasing priority in a counter-clockwise direction, the prochiral face is the *Si* face. The top face of benzaldehyde in Figure 14.8A is the *Si* face, and the bottom is the *Re* face. In this case, methyl addition to the *Si* face affords the (S)-alcohol, while addition to the *Re* face generates the (R)-product. Similarly, enantiotopic groups are designated pro-(S) or pro-(R) by temporally making one of the identical groups higher priority than the other, without changing the rank ordering of the other two inequivalent atoms or groups. In the case of propiophenone, deuterium could be substituted for one α-hydrogen, but bromine could not, because bromine would then become the highest priority group. If the hypothetical configuration of the stereocenter is (S), then the group made higher priority is the pro-(S) group. If it is (R), then the group made higher priority is the pro-(R) group. There is no correlation between reaction at the *Si* or *Re* face and the (S)- or (R)-configuration of the product, because the product configuration depends on the priority of the new substituent relative to those originally present in the substrate. There is no direct correlation between reaction at a pro-(S) or pro-(R) group and the (S)- or (R)-configuration of the product.

14.2.3.1. Description of Stereoselectivity

The ratio of the product enantiomers can be described in several ways. Before the advent of modern instrumentation, optical rotation was used to determine the ratio of enantiomers and was described as optical purity (Equation 14.9). This method required knowledge of the optical rotation of the enantiomerically pure material, which was not easily obtained. Optical rotations are rarely used today to quantify the ratio of enantiomers. The most common measure is percentage enantiomeric excess, ee (%), which is related to optical purity (Equation 14.10). It is understood that ee is given as percentage and the percentage symbol is often omitted. Another measure, enantiomeric ratio or er, is being used increasingly because it reflects the raw data collected by modern methods and is more directly related to $\Delta\Delta G^{\ddagger}$, the energy difference between the diastereomeric transition states (Equation 14.11). The er is described as a ratio (95:5) rather than a number. Product ee's or er's can be measured using a number of analytical methods, the most common being GC or HPLC equipped with columns packed with chiral stationary phase material.

$$\text{Optical purity (\%)} = \frac{\text{Optical rotation of mixture}}{\text{Optical rotation of single enantiomer}} \times 100 = \text{ee (\%)} \qquad (14.9)$$

$$\text{Enantomeric excess (\%)} = \frac{[\text{Major enantiomer}] - [\text{Minor enantiomer}]}{[\text{Major enantiomer}] + [\text{Minor enantiomer}]} \times 100 = \text{ee (\%)} \qquad (14.10)$$

$$\text{Enantomeric ratio} = \frac{[\text{Major enantiomer}]}{[\text{Minor enantiomer}]} = e^{-(\Delta\Delta G^{\ddagger}/RT)} \qquad (14.11)$$

14.2.3.2. The Origin of Stereoselection

Achiral reagents do not discriminate between the two faces of a prochiral substrate in the absence of a catalyst. Thus, addition of the achiral reagents to the *Re* and *Si* faces occurs with the same energy and the same rate, and equal amounts of (*R*)- and (*S*)-products are formed. In contrast, the reaction of a prochiral substrate with achiral reagents in the presence of an asymmetric catalyst occurs preferentially at one prochiral face, at one of two enantiotopic groups, or with one of two enantiomers in a racemic mixture. This selectivity can occur because the action of a chiral catalyst on a prochiral substrate gives rise to diastereomeric transition states. Like diastereomeric compounds, which have different arrangements of the atoms in space and distinct physical properties, diastereomeric transition states have different spatial relationships between the substrate, reagent, and catalyst. As a result, diastereomeric transition states have unequal energies. The larger the differences in energies of the diastereomeric transition states, the larger the differences in rates for production of the two enantiomeric products. Enantioselective catalysts affect the rates of formation of two enantiomers through irreversible enantioselectivity-determining steps. Equilibration of the product with the reactants would racemize an enantioenriched product.

14.2.4. Energetics of Stereoselectivity

Equation 14.11 illustrates important relationships in asymmetric catalysis between enantioselectivity, temperature, and the difference in energy between the diastereomeric transition states ($\Delta\Delta G^{\ddagger}$). Energy differences and the corresponding enantioselectivities at constant temperature are plotted in Figure 14.9. This plot shows that the difference in ee between a reaction with a $\Delta\Delta G^{\ddagger}$ of 0 kcal/mol and a reaction with a $\Delta\Delta G^{\ddagger}$ of 1 kcal/mol is larger than the difference in ee between a reaction with a $\Delta\Delta G^{\ddagger}$ of 2 kcal/mol and one with a $\Delta\Delta G^{\ddagger}$ of 3 kcal/mol. Much of the difference between more and less synthetically useful catalytic asymmetric processes depends on differences between selectivities in the range of 80–95% ee. As a rule of thumb, a 1.38 (best remembered as 1.4) kcal/mol difference in activation energy at room temperature gives rise to a 10:1 ratio of products, and this ratio of products corresponds to about 80% ee. A 2 kcal/mol difference in activation energies at room temperature gives rise to a product mixture with about 90% ee.

A plot of enantioselectivity versus temperature (°C) at $\Delta\Delta G^{\ddagger}$ values of 0.2, 1.0, 1.4, and 2.6 kcal/mol is illustrated in Figure 14.10. This plot illustrates that high enantioselectivity is achieved with smaller $\Delta\Delta G^{\ddagger}$ values at lower temperatures. For example, an energy difference between the diastereomeric transition states ($\Delta\Delta G^{\ddagger}$) of only 2.6 kcal/mol is needed at 25 °C to generate products with 98% ee (an er of 99:1), but a $\Delta\Delta G^{\ddagger}$ of only 1.8 kcal/mol is needed at −78 °C to achieve the same level of enantioselectivity. Be aware, however, that the correlation between temperature and enantioselectivity is not always as simple as these curves imply. Some systems exhibit a more complex relationship between temperature

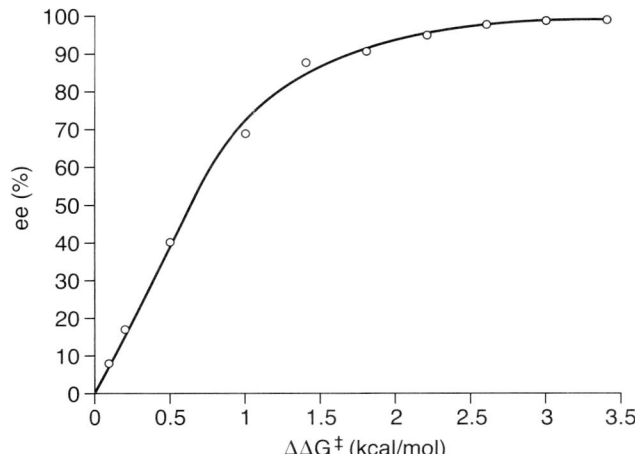

Figure 14.9.
Enantioselectivity as a function of the energy differences between the diastereomeric transition states ($\Delta\Delta G^{\ddagger}$) at 25 °C.

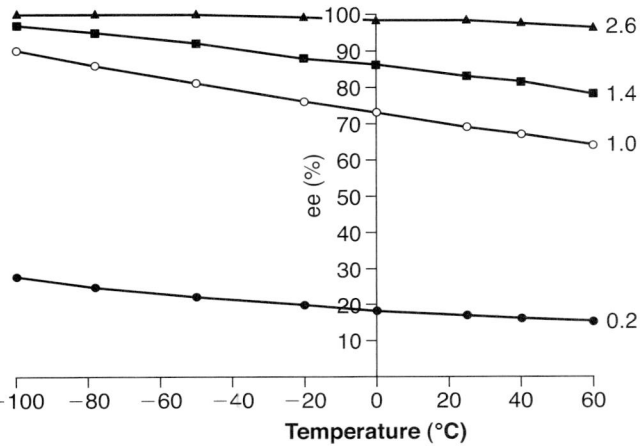

Figure 14.10.
Enantioselectivity as a function of temperature with $\Delta\Delta G^{\ddagger}$ values of 0.2, 1.0, 1.4, and 2.6 kcal/mol.

and enantioselectivity because of a change in the enantioselectivity-determining step or catalyst structure with temperature.

Compared to the overall barriers of the reactions and the strengths of the individual bonds being formed and broken in the reactions, these energy differences between the diastereomeric transition states are small. If such little energy differences are necessary to produce nearly enantiopure products, why is it challenging to obtain high enantioselectivity? The answer is that it is hard to identify the interactions that lead to the differences in the energies of the diastereomeric transition states. Nevertheless, extensive studies have revealed a number of important concepts to help understand the origins of enantioselectivities in asymmetric catalysis. These concepts, as well as several classes of enantioselective reactions, are outlined in the following sections.

14.2.4.1. Reaction Coordinates of Catalytic Enantioselective Reactions

Although reaction mechanisms in asymmetric catalysis vary greatly, each must contain a step that controls the enantioselectivity, often called the enantioselectivity-determining step. This enantioselection occurs in the first irreversible step that takes place through diastereomeric transition states. Later steps in the reaction mechanism have no impact on the enantioselectivity of the process.

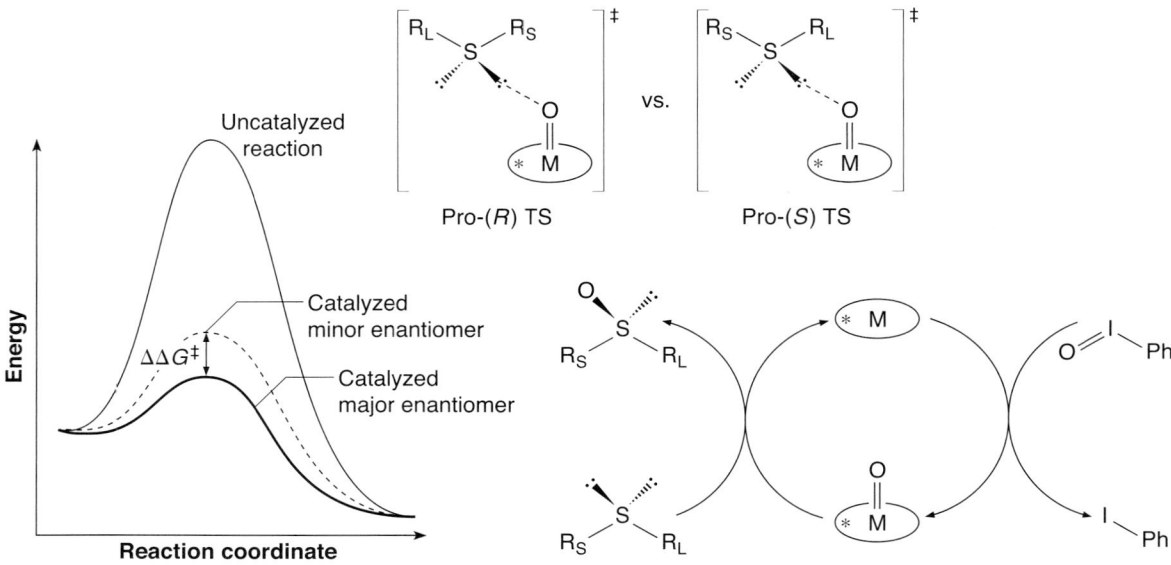

Figure 14.11. Reaction coordinate of the oxidation step in the asymmetric sulfoxidation reaction.

14.2.4.1.1. Reactions with a Single Enantioselectivity-Determining Step

The simplest example of asymmetric induction involves direct reaction of a chiral catalyst with a prochiral substrate without coordination of the substrate to the catalyst before the enantioselectivity-determining step. A reaction coordinate diagram for this process, illustrated in Figure 14.11, consists of two diastereomeric transition states. In this example, direct uncatalyzed reaction of the substrate with the reagent (PhI=O) has a much higher barrier than the catalytic reaction and does not impact the enantioselectivity (i.e., there is no "background reaction").

Asymmetric reactions that can exhibit this type of behavior include atom and group transfer reactions,[26,27] such as the asymmetric oxidation of sulfides,[28] some asymmetric epoxidations of olefins,[29,30] asymmetric aziridination of olefins,[31,32] and asymmetric cyclopropanation of olefins.[33] In the asymmetric oxidation of sulfides, a non-racemic, chiral, low-valent metal complex is oxidized, in this case by iodosobenzene, to generate a highly reactive oxo intermediate. The oxo is then transferred directly to the sulfur to form the sulfoxide in the enantioselectivity-determining step. A representative example is illustrated in Equation 14.12 that involves a chiral salen-based catalyst.[34]

(14.12)

14.2.4.1.2. Reactions with Reversiblity Prior to the Enantioselectivity-Determining Step: The Curtin–Hammett Principle Applied to Asymmetric Catalysis

14.2.4.1.2.1. Theory

The origin of enantioselection becomes more complicated when prochiral substrates bind to chiral catalysts in a separate step from the enantioselectivity-determining step. Such an interaction leads to diastereomeric complexes possessing different energies. In these cases, the relationship between the barrier to interconversion of the diastereomeric complexes and their barriers to reaction to form enantiomeric products can give rise to several scenarios.

The first scenario is relatively simple and relates to the previous example of group transfer. If interconversion of the diastereomeric olefin complexes **I** and **I′** is slow, relative to the rates of conversion of the olefin complexes to product, the enantioselectivity-determining step is binding the prochiral olefin faces to the metal (Figure 14.12A). In contrast, if interconversion of the diastereomeric olefin complexes is significantly faster than their reaction to form product, the enantioselectivity-determining step will be the reaction to form product (Figure 14.12B). This latter scenario is stated to meet "Curtin–Hammett conditions."

The Curtin–Hammett principle[35,36] states that when competing reaction pathways begin from rapidly interconverting isomers, the product ratio is determined by the relative heights of the highest energy barriers leading to the two different products ($G_I^{\ddagger}-G_{I'}^{\ddagger}$, in Figure 14.12B). This principle was first applied to the effect of conformational equilibria and was quantified by Winstein and Holness[37] and by Eliel and co-workers.[38–40] This analysis can be applied more generally, and is now used to describe a variety of processes involving interconverting isomers that form two different products. The simple equation now most commonly applied to a reaction under Curtin–Hammett conditions is shown in Equation 14.13. In this equation, K_{eq} is the equilibrium constant between the interconverting intermediates **I** and **I′**, and k_i and $k_{i'}$ are the rate constants for the reaction of **I** and **I′**, respectively, to form the two products.

$$\frac{[R]}{[S]} = K_{eq}\left(\frac{k_i}{k_{i'}}\right) \tag{14.13}$$

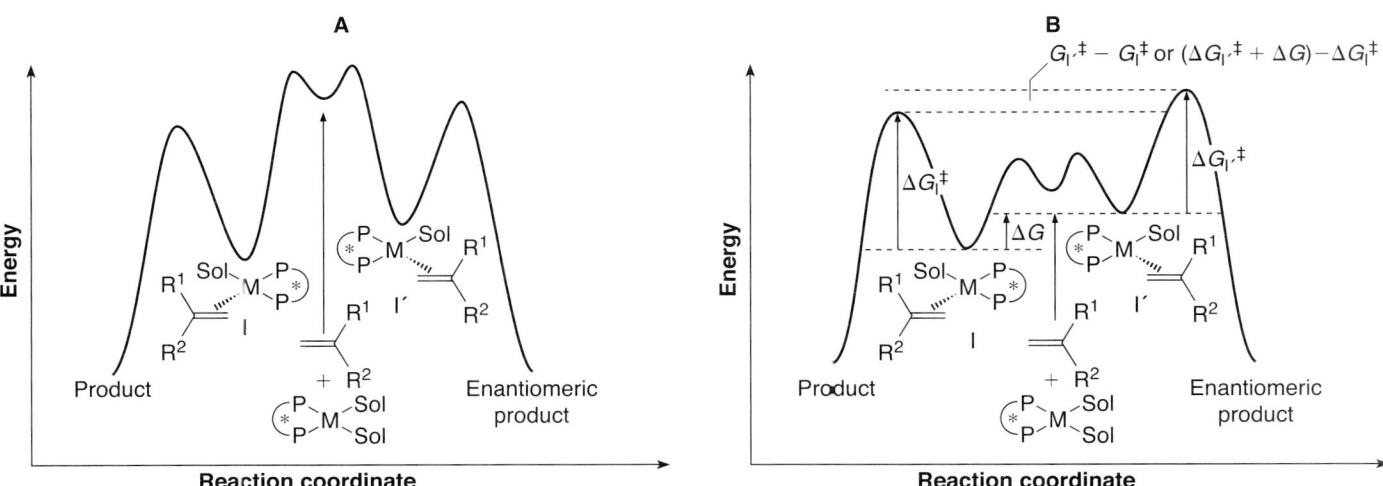

Figure 14.12.
Reaction coordinate diagrams illustrating reactions of diastereomeric olefin complexes. In scenario A, olefin binding is enantio-determining. In B, the diastereomeric olefin complexes are in rapid equilibrium and enantio-determination is the conversion of the olefin adducts to products. B is an example of Curtin–Hammett conditions.

In the context of an asymmetric catalytic process occurring through interconverting diastereomeric intermediates, the Curtin–Hammett principle emphasizes that the enantioselectivity is controlled by the relative energies of the two diastereomeric transition states, rather than the stabilities of the two diastereomeric intermediates. For example, the major enantiomer of the product of certain enantioselective processes has been shown to form from the higher energy, less stable diastereomeric intermediate (see below). Thus, Equation 14.13 illustrates that the equilibrium constant K_{eq} between the diastereomeric intermediates **I** and **I′** and the ratio of rate constants $k_i/k_{i'}$ for reaction of diastereomeric intermediates **I** and **I′** both contribute to the relative rates for formation of the two enantiomeric products. In practice, the ratio of rate constants $k_i/k_{i'}$ is sometimes larger and sometimes smaller than the equilibrium constant for the two diastereomeric intermediates (K_{eq}).

14.2.4.1.3.2. Two Examples of Reactions Under Curtin–Hammett Conditions

14.2.4.1.3.2.1. Asymmetric Hydrogenation

The classic example of the Curtin–Hammett principle in asymmetric catalysis is the asymmetric hydrogenation of ethyl (Z)-α-acetamidocinnamate to provide α-amino acid derivatives (Equation 14.14).

(14.14)

Under the conditions of the catalytic asymmetric hydrogenation, the enamide substrate binds to the rhodium catalyst to generate an equilibrium ratio of diastereomeric four-coordinate intermediates **I** and **I′** that interconvert by substrate dissociation and re-coordination (Figure 14.13). The ratio **I** : **I′** was measured spectroscopically and found to be > 95 : 5. Surprisingly, the more abundant and lower energy intermediate **I** (> 95%) led to the *minor* product enantiomer. At room temperature under 1 atmosphere of hydrogen, the turnover-limiting and enantioselectivity-determining step of this catalytic hydrogenation was stated to be the oxidative addition (OA) of hydrogen.[41–45] If so, then the reaction of the minor diastereomeric intermediate (**I′**) with hydrogen must be much faster than the reaction of the major diastereomer (i.e., $k'_{OA}[\mathbf{I'}][H_2] \gg k_{OA}[\mathbf{I}][H_2]$). With an equilibrium ratio of the diastereomers greater than 95:5, the reactivity of the minor diastereomer must be 1000 times greater than that of the major diastereomer to achieve the observed enantioselectivity! In addition to the importance of mechanistic studies of this process to the development of our understanding of asymmetric catalysis,[46–48] this asymmetric hydrogenation was a step in the industrial-scale manufacture of several pharmaceutical agents, including L-Dopa (used to treat Parkinson's disease).[48] Related asymmetric hydrogenations are used to prepare a number of important pharmaceuticals, including Naproxen (Alleve®, an analgesic, anti-inflammatory drug). More details on the mechanism of this process and the various types of hydrogenation are described in Chapter 15.

An energy diagram for the hydrogenation is illustrated in Figure 14.14. The steps after the turnover-limiting oxidative addition do not impact the enantioselectivity (their energies are approximated in this figure). As seen in this reaction coordinate diagram, the selectivity in this process is controlled by the relative energies of the two diastereomeric transition states.

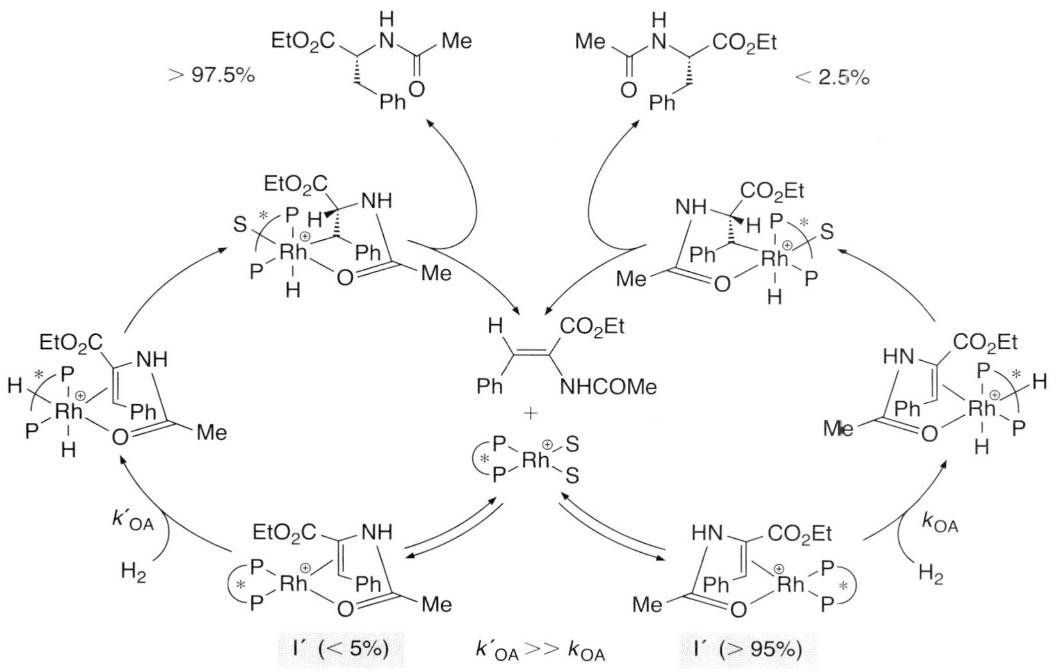

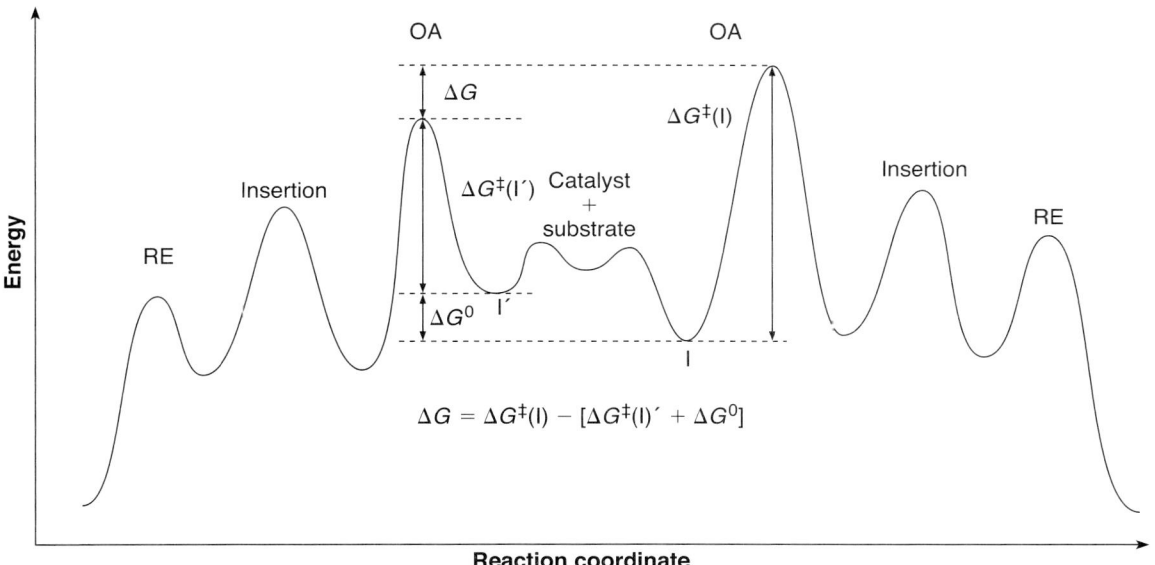

Figure 14.13.
Mechanism of the asymmetric hydrogenation, illustrating a reaction meeting the Curtin–Hammett conditions.

Figure 14.14.
Energy diagram for the asymmetric hydrogenation reaction under Curtin–Hammett conditions. The enantioselectivity-determining step is the oxidative addition (OA) of hydrogen.

14.2.4.1.3.2.2. Asymmetric Allylic Alkylation

A second example of a common asymmetric process discussed later in this text that occurs through interconverting diastereomeric intermediates is palladium-catalyzed allylic alkylation. In this case, interconversion of the diastereomeric intermediates occurs within the coordination sphere of the metal center, rather than by dissociation of the reactant to regenerate the starting catalyst and the free reactant. This interconversion of diastereomeric intermediates in allyic alkylation is illustrated in Figure 14.15. The cationic palladium is bound to a chiral bidentate ligand, L*, which is often a bisphosphine, and the isomerization of the diastereomeric complexes proceeds by reversible generation of the 14-electron η^1-allyl species in a mechanism often described as a π-σ-π interconversion.

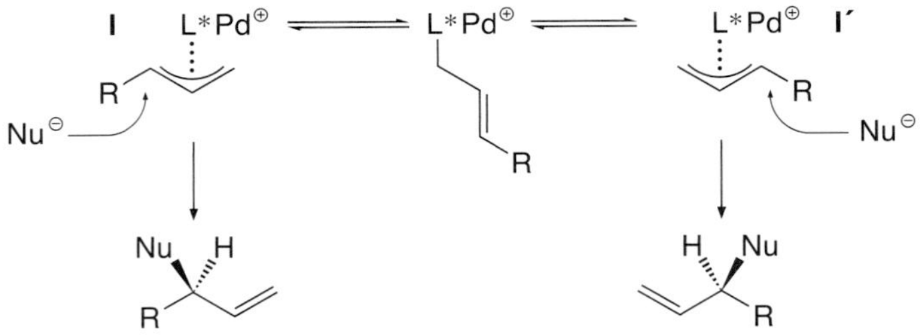

Figure 14.15.
Interconversion of the diastereomeric π-allyls **I** and **I'** occurs via an η^1-allyl. The enantioselectivity-determining step depends on the relative rates of π-σ-π isomerization and nucleophilic attack.

The identity of the enantioselectivity-determining step of this mechanism depends on the relative rates of π-σ-π isomerization and nucleophilic attack on the π-allyl intermediate. When the π-σ-π isomerization through an η^1-allyl complex has a higher barrier than reaction of the η^3-allyl complexes with the nucleophile, the enantioselectivity-determining step is the formation of the diastereomeric η^3-allyl complexes **I** and **I'**. In this scenario, the reaction in Figure 14.15 would not be under Curtin–Hammett control. However, when the π-σ-π isomerization through an η^1-allyl complex has a significantly lower barrier than reaction of the η^3-allyl complexes with the nucleophile, the stereochemistry-determining step is nucleophilic attack on the η^3-allyl complexes **I** and **I'**, and the reaction is under Curtin–Hammett control.

Several experimental variables impact the relative rates of isomerization and nucleophilic attack in the allylic substitution reaction in Figure 14.15. For example, an increase in the rate of interconversion of the π-allyl intermediates, such that the rate of this process is faster than nucleophilic attack, will change the enantioselectivity-determining step, because the diastereomeric complexes now equilibrate rapidly. Because isomerization of the diastereomeric π-allyl complexes is a unimolecular process, but nucleophilic attack on the π-allyl intermediates is a bimolecular process, reactions under dilute conditions will decrease the rate of nucleophilic attack, relative to the unimolecular π-σ-π isomerization, and will help to achieve Curtin–Hammett conditions.

Alternatively, additives can affect the rate of interconversions of diastereomeric intermediates, relative to the rate of formation of product from these intermediates. This affect of additives has been exploited to achieve Curtin–Hammett conditions in allylic alkylation. Halide ions are known to catalyze the π-σ-π isomerization[49] by the mechanism in Equation 14.15. Thus, the concentration of additives can affect enantioselectivity by shifting the enantioselectivity-determining step from formation of the allyl intermediate to reaction of the allyl intermediates with the nucleophile. In fact, this affect of additives on the identity of the enantioselectivity-determining step can even cause reactions in the presence of additives to form the enantiomer that is the opposite of the one formed in the absence of additives.[50] Even variation in temperature can cause a change in the enantioselectivity-determining step of a catalytic process and formation of opposite product enantiomers at low and high temperature. This effect has been observed in the asymmetric hydroformylation processes described in Chapter 17.[51]

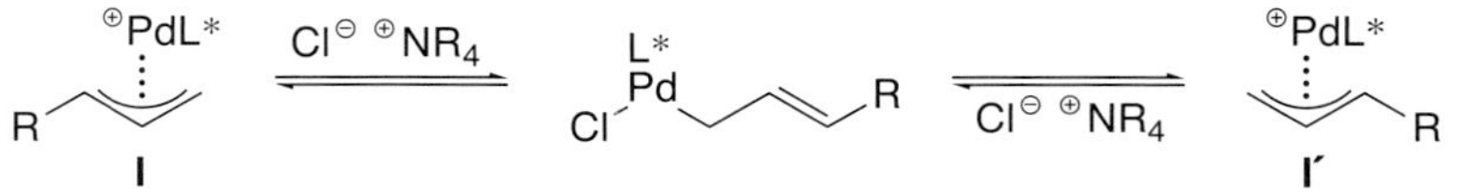

(14.15)

14.2.5. Transmission of Asymmetry

In some cases, asymmetric catalysts bind the substrate and react preferentially with one of the prochiral faces of the substrate. In other cases, asymmetric catalysts bind the substrate, shield one of the prochiral faces, and prevent reaction at that face. Despite the simplicity of these strategies, the mechanics of the transmission of asymmetry from the catalyst to the substrate are complex and not well understood in most systems. Furthermore, there are many classes of chiral ligands and catalysts, and the nature of the transmission of asymmetry from the catalyst to the substrate varies greatly. Thus, this section first introduces some of the means by which the asymmetry of the catalyst is transferred to the substrate and then illustrates these principles by describing reactions of a few classes of catalysts containing chiral ligands that have been extensively researched.

The most common method to transfer asymmetry from a catalyst to the substrate relies on steric biasing. Other catalyst–substrate interactions, such as π-interactions between aromatic groups on the catalyst and substrate or hydrogen bonding between the catalyst and substrate, can also play important roles and may be used in combination with steric biasing.

14.2.5.1. Effect of C_2 Symmetry

In the early days of asymmetric catalysis, it was often observed that catalysts containing C_2-symmetric ligands were the most selective. Kagan proposed that this selectivity resulted from the smaller number of metal-substrate adducts and transition states available to these catalysts than are available to catalysts containing less symmetric ligands. This principle is illustrated in the context of the asymmetric allylation reaction.

A catalyst containing a C_2-symmetric bisphosphine forms two diastereomeric π-allyl intermediates (Figure 14.15). In contrast, a catalyst containing a less symmetric P–N ligand gives rise to two additional diastereomeric intermediates and diastereomeric transition states for the nucleophilic attack. These four diastereomeric allyl intermediates formed by complexes containing the P–N ligand are shown in Figure 14.16 (the chiral ligand backbone is omitted for clarity). Although many catalysts lacking C_2 symmetry exhibit high levels of enantioselectivity, those containing C_2 symmetric ligands remain one of the most important and selective classes of catalysts.

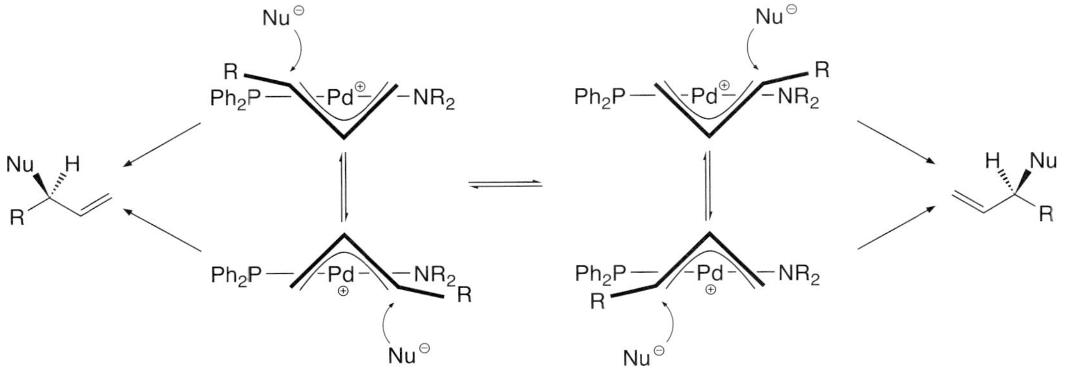

Figure 14.16.
The four diastereomeric π-allyl complexes containing an unsymmetical P–N ligand. These four diastereomeric complexes can be compared to the two diastereomeric π-allyl complexes containing a C_2-symmetric ligand (Figure 14.15).

14.2.5.2. Quadrant Diagrams

A generic model for steric biasing of chiral metal–ligand adducts has been advanced to facilitate the prediction of the facial stereoselectivity in catalyst–substrate complexes and transition states. In this model, the environment around the metal is divided into quadrants in which the horizontal dividing line is congruent with a plane or pseudo-plane in

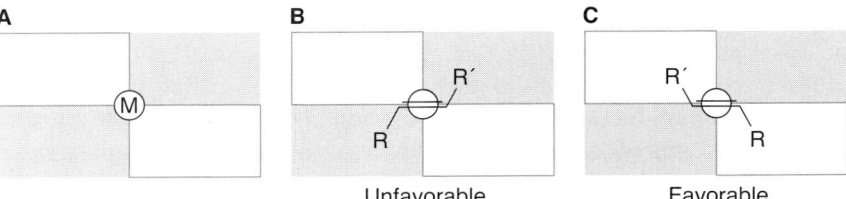

Figure 14.17.
A. Quadrant diagrams for a C_2-symmetric catalyst. **B.** Unfavorable binding and
C. favorable binding of an olefin in diastereomeric olefin complexes.

the catalyst. For simplicity, the quadrant diagram is given for a C_2-symmetric catalyst (Figure 14.17A). The two shaded diagonal quadrants represent space that is occupied by substituents on the ligands that extend forward, whereas the unshaded quadrants correspond to less-hindered space. Binding of the prochiral faces of an olefin to a metal, for example, would give rise to diastereomers in which the more stable diastereomer contains the R and R′ substituents positioned in the open, unshaded quadrants. This difference in stability of these two diastereomers is illustrated in Figure 14.17B and Figure 14.17C.

The means by which metal complexes of chiral ligands block quadrants depends on the nature of the ligand and the metal–ligand adduct. In some cases, the stereogenic centers of the ligands exist in close proximity to the metal. In other cases, the stereogenic centers are located so far from the metal that it is not obvious how the effect of these distant stereocenters can be transmitted to the site of reaction.

Examples of ligands that possess stereogenic centers in close proximity to the metal include bisoxazoline ligands, such as PyBox (Figure 14.18A). On binding to metals, a rigid (PyBox)M adduct is formed. The isopropyl groups then block the lower left and upper right quadrants, yielding the quadrant diagram in Figure 14.17A.

Examples of ligands that possess stereogenic centers more distant from the metal include Chiraphos (Figure 14.18B). The nature of transmission of asymmetry is subtle in this case, and the stereocenters must indirectly influence events at the metal. The conformation of a Chiraphos complex is more stable with the methyl groups occupying pseudo-equatorial positions in the resulting metallacycle. This conformational preference causes the metallacycle to adopt a skewed conformation that is illustrated in an exaggerated form in Figure 14.18C. The phosphorus-bound phenyl groups are then disposed in pseudo-axial (a) and pseudo-equatorial (e) positions. To minimize steric interactions between these phenyl groups on a given phosphorus center, they adopt an edge–face relationship. The faces of the pseudo-equatorial P-bound phenyl groups then project forward and the edges of the pseudo-axial phenyl groups project forward, although to a lesser extent (see below). The faces of the pseudo-equatorial P-bound phenyl groups are more sterically demanding than the edges of the pseudo-axial phenyl groups and block the lower left and upper right quadrants. This arrangement of P-bound phenyl groups then corresponds to the quadrant diagram in Figure 14.17A.

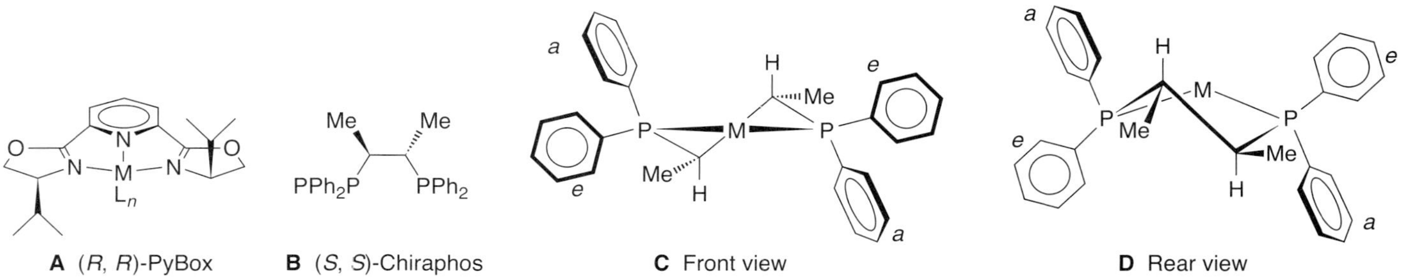

Figure 14.18.
PyBox and Chiraphos, illustrating the positioning of key substituents contributing to the chiral environment of the catalyst bearing these ligands.

14.2.5.3. Structures of Ligands Generating Highly Selective Catalysts ("Privileged Ligands")

Of the myriad of enantioenriched ligands that have been employed in asymmetric catalysis, a few types of ligands consistently generate catalysts that are highly enantioselective for a wide range of reactions. In analogy to the substructures of drugs that are active toward a wide range of biological targets and have been called "privileged structures,"[52] these ligands have been referred to as "privileged ligands."[53] Selected examples of these ligands are shown in Figure 14.19. Both BINAP and BINOL are chiral due to the skewed conformation of the binaphthyl rings. The barrier to rotation about the central C–C bond is very high, because the PPh_2 moieties and the hydrogens at the 8- and 8'-positions would have to pass each other at close distances during racemization. As a result, both BINAP and BINOL are stable to racemization at high temperatures.

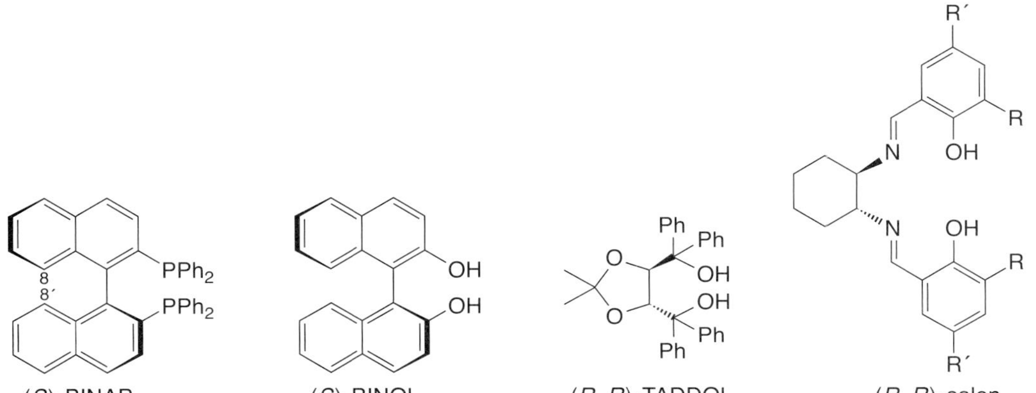

Figure 14.19. Structures of important ligands in asymmetric catalysis.

Although BINAP[54–56] forms rigid metallacycles on binding to metals, this ligand accommodates metals of different radii by rotating about the central aryl–aryl bond and the 2,2' P–C bonds. The efficiency with which BINAP-based catalysts transmit asymmetry to substrates results from an effect of the axial chirality of the binaphthyl backbone on the positions and orientations of the phenyl substituents. Like Chiraphos (Figure 14.18C), the skewed conformation of the seven-membered metalacycle causes the BINAP phenyl groups to assume pseudo-axial and pseudo-equatorial positions, as can be seen in the a ruthenium-BINAP portion of the crystal structure of [(S)-BINAP)] $Ru(O_2C\text{-}t\mathrm{Bu})_2$ shown in Figure 14.20.[57] A stereoview of this structure is provided in Figure 14.21. The pseudo-equatorial phenyl rings extend forward past the metal center, and the pseudo-axial phenyl groups are oriented behind the metal in the orientation shown.

The complexes of many chiral bidentate phosphines (such as the Chiraphos example presented previously) adopt similar conformations, but most do not exhibit such pronounced differences in the positions of the pseudo-equatorial and pseudo-axial phenyl groups. It is the combination of the high degree of protrusion of the equatorial phenyl groups toward the metal binding site and the axial phenyl groups oriented away from this binding site that is thought to be responsible for the exquisite enantiocontrol by complexes of this ligand.[58] Gearing of the phosphorus-bound aryl groups also contributes to the differentiation of the pseudo-axial and pseudo-equatorial phenyl groups, as can be seen from the structure in Figure 14.20.[59, 60]

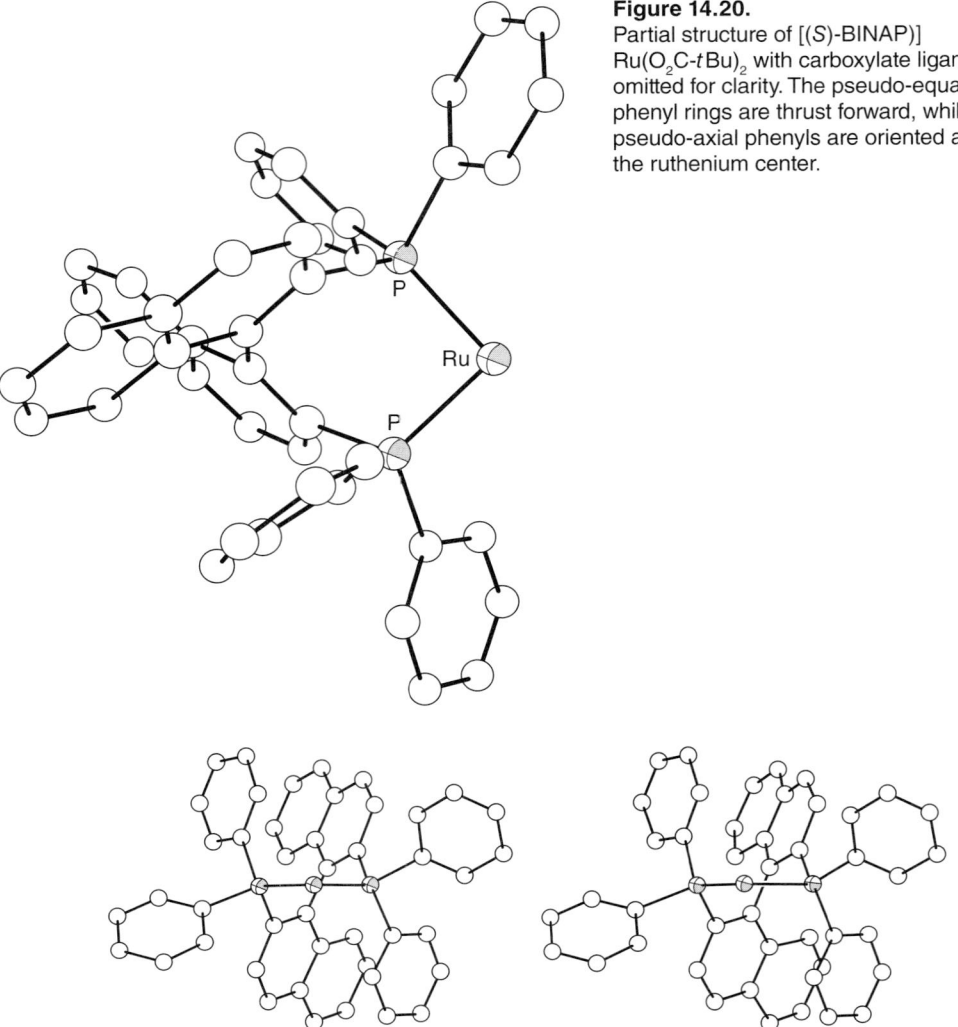

Figure 14.20.
Partial structure of [(S)-BINAP)]
Ru(O$_2$C-tBu)$_2$ with carboxylate ligands
omitted for clarity. The pseudo-equatorial
phenyl rings are thrust forward, while the
pseudo-axial phenyls are oriented away from
the ruthenium center.

Figure 14.21.
Stereoview of the metal–BINAP core, illustrating the chiral environment of the ligand.

One particularly important type of complex between a prochiral substrate and a BINAP–metal fragment is the complex formed between the (BINAP)Ru fragment and β-ketoesters. These complexes are intermediates in one of the most selective asymmetric hydrogenations, shown in Equation 14.16,[61–63] and presented in more detail in Chapter 15. As shown in Figure 14.22, the ketone carbonyl is thought to coordinate in a π-fashion to the ruthenium while the ester binds through an oxygen lone pair.[61, 64] Due to the η^2-binding of the ketone unit, this portion of the substrate places greater steric demands on the ruthenium–BINAP system and occupies the least-hindered site on the catalyst. The interaction of the ketone with the protruding equatorial phenyl group of the BINAP, drawn in bold, makes the diastereomer on the left of Figure 14.22 less stable, and the placement of the ketone carbonyl in the quadrant with the axial P-bound phenyl group, which is directed away from the substrate, makes the diastereomer at the right of the figure more stable. These basic concepts in transmission of asymmetry are applicable to many other systems in asymmetric catalysis.[22]

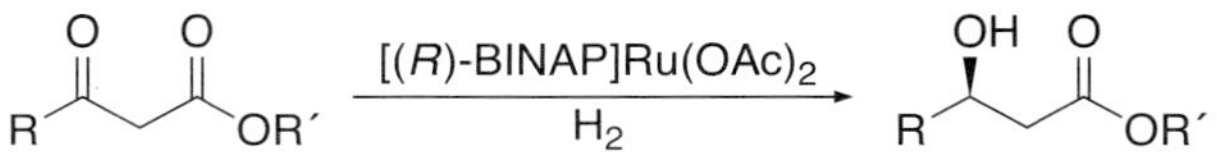

$$
\underset{R}{\overset{O\quad O}{\bigwedge}}\!\!\!OR' \quad \xrightarrow[\text{H}_2]{\text{[(}R\text{)-BINAP]Ru(OAc)}_2} \quad \underset{R}{\overset{OH\quad O}{\bigwedge}}\!\!\!OR'
$$

(14.16)

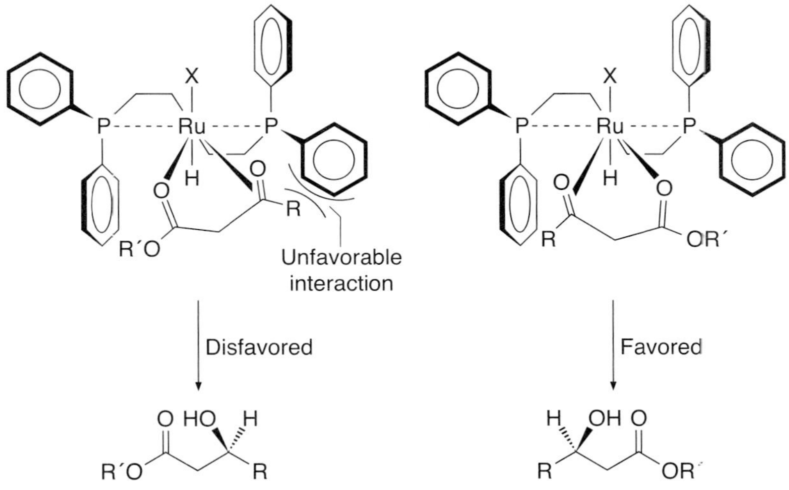

Figure 14.22.
Proposed chelation of β-keto esters to the [(R)-BINAP]Ru moiety in the asymmetric hydrogenation shown in Equation 14.16. The equatorial phenyl groups of the BINAP ligand are shown in bold. The unfavorable interaction of the ketone with the equatorial phenyl on the left destabilizes this diastereomer.

14.2.6. Alternative Asymmetric Processes: Kinetic Resolutions and Desymmetrizations

The asymmetric processes discussed so far in this chapter have focused on reactions that create non-racemic, chiral products from achiral reagents by selective reaction at one prochiral face or position over the other. However, these principles can also be applied to reactions that separate enantiomers of an existing racemic mixture, channel both enantiomers of such a mixture to a single enantiomeric product, or that select between reaction at one of two diastereotopic functional groups in an achiral substrate. These reactions are also synthetically valuable and are called kinetic resolutions, dynamic kinetic resolutions, and desymmetrizations. An understanding of these reactions draws from the principles established so far in this chapter, but they also require some additional principles to be established that apply in a specific way to these classes of asymmetric transformations. Thus, the remainder of Chapter 14 introduces the fundamentals of these classes of asymmetric catalysis.

14.2.6.1. Kinetic Resolutions

Kinetic resolutions (KR) are reactions that occur at different rates with the two enantiomers of a chiral substrate. A kinetic resolution generates two sets of chiral, non-racemic materials. It generates a non-racemic mixture of the substrate enantiomers enriched in the less reactive of the two enantiomers, and it generates a non-racemic mixture of the product resulting from the predominant reaction of the more reactive enantiomer of the substrate. Because the reactant and product have different physical properties, they can be separated more easily than two enantiomers.

Thus, kinetic resolutions are distinct from enantioselective reactions that occur with prochiral substrates and create new chiral elements. Kinetic resolutions do not usually generate additional stereochemistry (an exception is shown in Figure 14.8D). Rather, they distinguish one enantiomer from another by enantioselectively creating new functionality. Kinetic resolutions are also distinct from classical resolutions. Classical resolutions are conducted with stoichiometric amounts of chiral resolving agents, but kinetic resolutions are typically conducted with stoichiometric amounts of achiral reagents and substoichiometric amounts of a chiral, non-racemic catalyst.

Because kinetic resolutions provide a means to separate the two enantiomers of a racemic mixture, the maximum yield of enantiopure product from a kinetic resolution is 50%. For this reason, an increased emphasis has been placed on enantioselective catalysis over kinetic resolution (enantioselective catalysis can generate 100% yield of enantioenriched product, at least in principle). However, the continued use of KR and development of synthetically valuable new kinetic resolutions attests to the utility of these classes of reactions. KR is often the best option when the racemate is inexpensive, when no practical enantioselective route to the single enantiomer is available, or when classical resolution does not provide the desired material with high ee.[65–68]

14.2.6.1.1. Quantification of Selectivity in Kinetic Resolutions

The efficiency of a KR is given by the relative rates of reaction of the substrate enantiomers with the chiral catalyst to generate the enantiomeric products. These two processes are shown for substrates S_R and S_S (the R- and S-enantiomers of the substrate S) with the catalyst Cat_R to form enantiomeric products P_R and P_S in Figure 14.23. The relative rate constants k_{rel} (k_{fast}/k_{slow}) and the selectivity factor, s, described in Figure 14.23 are used interchangeably as a measure of the discrimination of a particular KR toward two enantiomers. The k_{rel} is calculated using Equation 14.17 by experimentally determining the conversion and the ee [not ee (%)!] of the starting material at a given time.

$$S_S + \text{Reagent} \xrightarrow[k_S\,(=\,k_{fast})]{\text{Cat}_R} P_S$$

$$S_R + \text{Reagent} \xrightarrow[k_R\,(=\,k_{slow})]{\text{Cat}_R} P_R$$

$$s = k_{rel} = k_{fast}/k_{slow}$$

Figure 14.23.
Kinetic resolution involving the reaction of a racemic substrate, achiral reagent, and a resolved catalyst. The relative rate of reaction of the enantiomers determines the efficiency of the KR.

$$s = k_{rel} = \frac{\ln[(1 - c)(1 - ee)]}{\ln[(1 - c)(1 + ee)]}$$

c = Conversion of starting material
ee = ee of starting material

(14.17)

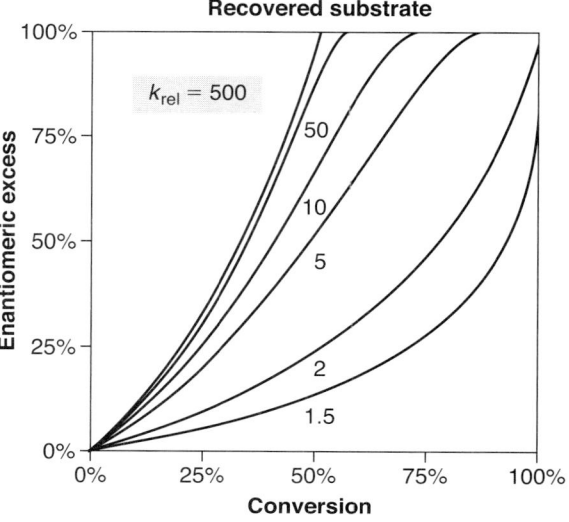

Figure 14.24.
Graph representing the ee (%) as a function of conversion (%) for different k_{rel} values.

The relationship between the ee of the recovered substrate and the conversion for reactions with different k_{rel} values is shown in Figure 14.24. Most obvious from this graph is that the larger values of k_{rel} lead to higher enantioselectivity at 50% conversion. In addition, the ee of the starting material increases as the reaction progresses. Thus, high substrate ee can be obtained, even when k_{rel} is moderate, by running the reaction beyond 50% conversion. The higher conversion used to increase substrate ee, however, causes the yield of recovered resolved substrate to be lower. In general, systems with $k_{rel} > 10$ generate useful quantities of recovered substrate with high ee.

14.2.6.1.2. Energetics of Selectivity in Kinetic Resolutions

Just as the enantiomeric excess of an enantioselective catalytic reaction is controlled by the difference in free energies of activation $\Delta\Delta G^{\ddagger}$, the selectivity factor in a kinetic resolution is controlled by the difference in free energy between the selectivity-determining diastereomeric transition states. This relationship and illustrative reaction coordinate diagrams are shown in Figure 14.25. These diagrams are closely related, but distinct from the reaction coordinate diagram for an enantioselective reaction with an achiral reagent. Although the energies of S_S and S_R are identical because they are enantiomers, two separate reaction coordinate diagrams are shown in Figure 14.18 because a kinetic resolution involves two separate reactions, one with each substrate enantiomer, in the same medium.

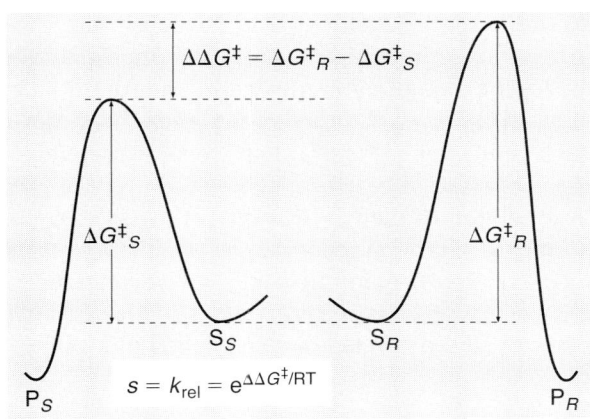

Figure 14.25.
Relationship between s or k_{rel} and free energy, and illustrative reaction coordinates for a kinetic resolution.

14.2.6.1.3. Examples of Kinetic Resolutions

Kinetic resolutions have been an important method to prepare enantioenriched organic compounds. One of the most important KRs since the 1980s has been the resolution of racemic allylic alcohols by the Sharpless asymmetric epoxidation process.[69] An important process developed more recently is the Jacobsen hydrolytic kinetic resolution (HKR) of terminal epoxides.[70] Although valuable synthetically, these reactions lie outside the realm of the organometallic chemistry described in this text. Representative examples of KR involving organometallic systems are illustrated in the figures that follow. The three examples presented include 1) reactions that form a bond between the reagent and the stereogenic center (Figure 14.26), 2) reactions that eliminate the stereogenic center of one enantiomer (Figure 14.27), and 3) reactions that form a new carbon–carbon double bond at a site adjacent to the stereogenic center (Figure 14.28).

The first example involves the kinetic resolution of allylic acetates. In the asymmetric synthesis of (+)-cyclophellitol, the two enantiomers of the tetraacetate in Figure 14.26 were resolved by substitution of a pivalate group for one acetate. The palladium catalyst bearing the (R,R)-L* ligand shown reacted faster with the (R)-allylic acetate.[71]

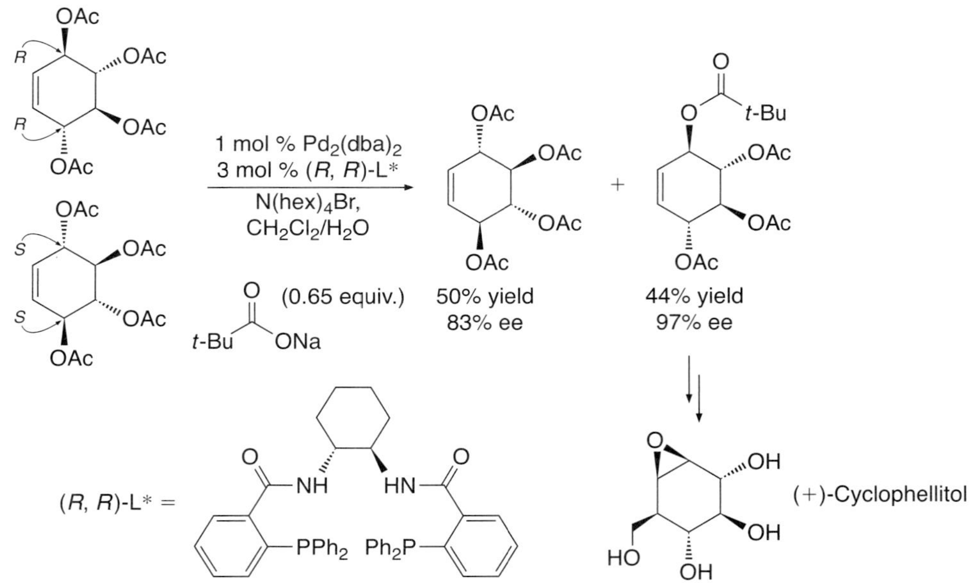

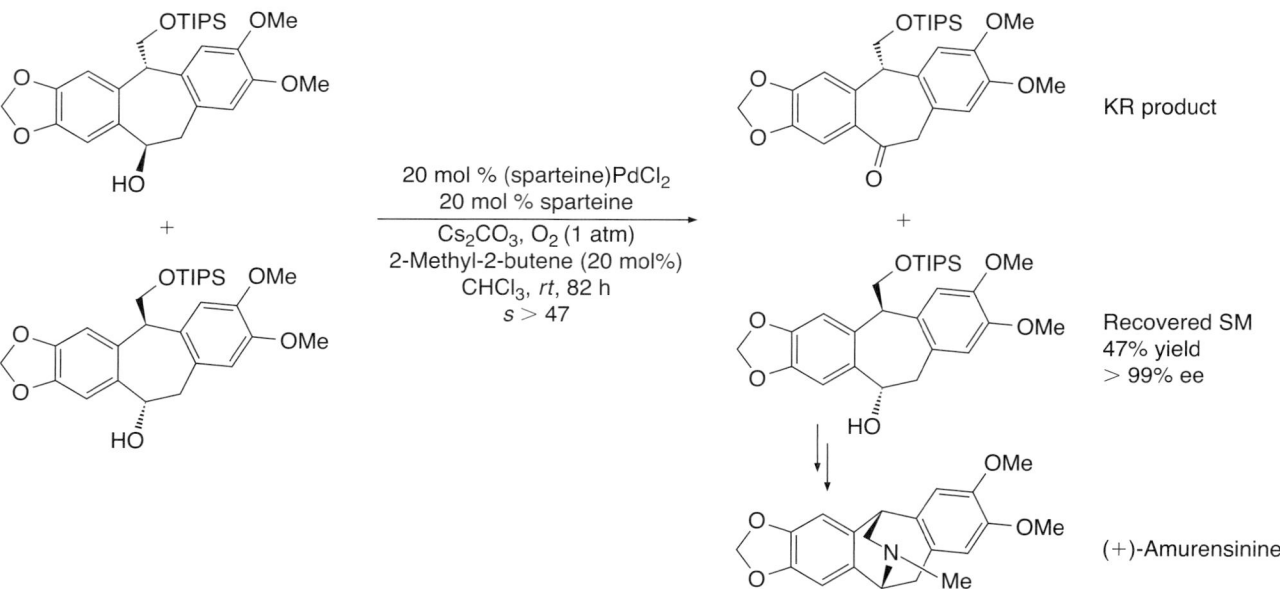

Figure 14.26. Kinetic resolution in the asymmetric allylic substitution.

The second example involves the kinetic resolution of racemic secondary alcohols, a process that also has been used in a total synthesis.[72–74] As illustrated in Figure 14.27, the naturally occurring tricyclic diamine sparteine, in combination with palladium dichloride and oxygen as the terminal oxidant, catalyzes the oxidation of one enantiomer of the racemic benzylic alcohol to the ketone faster than it oxidizes the other enantiomer. The desired unreacted alcohol was isolated in 47% yield with 99% ee ($s > 47$) and was subsequently transformed to (+)-amurensinine.[75]

A third example does not involve manipulation of the functionality at the stereogenic center. This example involves the selective ring closure of one enantiomer of a diene over the other. As described in more detail in Chapter 21, a molybdenum-based catalyst containing a chiral, non-racemic biphenoxide selectively reacts with certain chiral dienes. The example in Figure 14.28 illustrates the kinetic resolution of a silyl-protected dieneol by this catalyst.[76]

Figure 14.27. Oxidative KR in the synthesis of (+)-amurensinine.

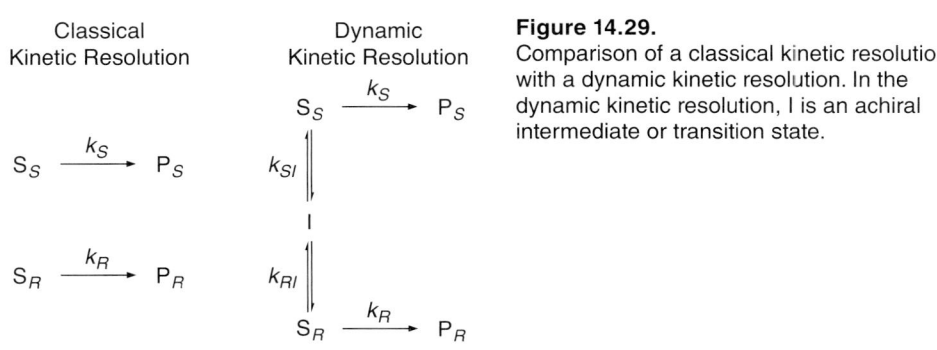

Figure 14.28. Kinetic resolution in the asymmetric ring-closing metathesis.

14.2.6.2. Dynamic Kinetic Resolution

The limitation on the theoretical yield of a KR led to the development of methods to conduct kinetic resolutions in a fashion that allows the conversion of both enantiomers of the reactant into a single enantiomer of the product. Dynamic kinetic resolution (DKR) is one method that has accomplished this goal.[62, 77–81]

Classical Kinetic Resolution	Dynamic Kinetic Resolution

$$S_S \xrightarrow{k_S} P_S$$

$$S_S \xrightarrow{k_S} P_S$$
$$k_{SI} \downarrow$$
$$I$$
$$k_{RI} \downarrow$$
$$S_R \xrightarrow{k_R} P_R$$

$$S_R \xrightarrow{k_R} P_R$$

Figure 14.29.
Comparison of a classical kinetic resolution with a dynamic kinetic resolution. In the dynamic kinetic resolution, I is an achiral intermediate or transition state.

The relationship between kinetic resolutions and dynamic kinetic resolutions is illustrated in Figure 14.29. The DKR couples a kinetic resolution with a rapid in situ racemization of the chiral substrate through an achiral intermediate (I, Figure 14.29) or transition state. While the fast-reacting enantiomer of the substrate is converted to product, the equilibrium between substrate enantiomers is constantly re-established by racemization. If this racemization is significantly faster than the conversion of substrate to product, the ee of the substrate is nearly zero throughout the reaction. Thus, a typical DKR involves a racemization process that proceeds at an equal or greater rate than the catalytic asymmetric processes—that is, $k_{rac} \geq k_{fast}$. If the two enantiomers of the substrate are fully equilibrating, k_{fast}/k_{slow} is about 20, and the two reactions occur independently without product incorporation into the catalyst, then the ee of the product will be 90%. A full mathematical treatment of DKR processes must address the situation when the rate of racemization is not much faster than the rate of reaction of the substrate and is complicated, but it has been reported.[82, 83]

14.2.6.2.1. Example of Dynamic Kinetic Resolutions: Dynamic Kinetic Resolution of 1,3-Dicarbonyl Compounds Through Asymmetric Hydrogenation

A common entry into the DKR manifold involves racemization of stereocenters α to carbonyl groups.[61, 62] The achiral intermediate formed through base-catalyzed enolization allows for facile racemization of the substrate. α-Substituted β-dicarbonyl compounds racemize under particularly mild conditions and undergo highly enantioselective dynamic kinetic resolutions by hydrogenation of the ketone unit to form α-substituted β-hydroxy carbonyl compounds (Figure 14.30). This type of DKR is used to prepare azetidinone, which is a key intermediate in the synthesis of carbapenem antibiotics on an industrial scale (i.e., 120 ton/year).[62]

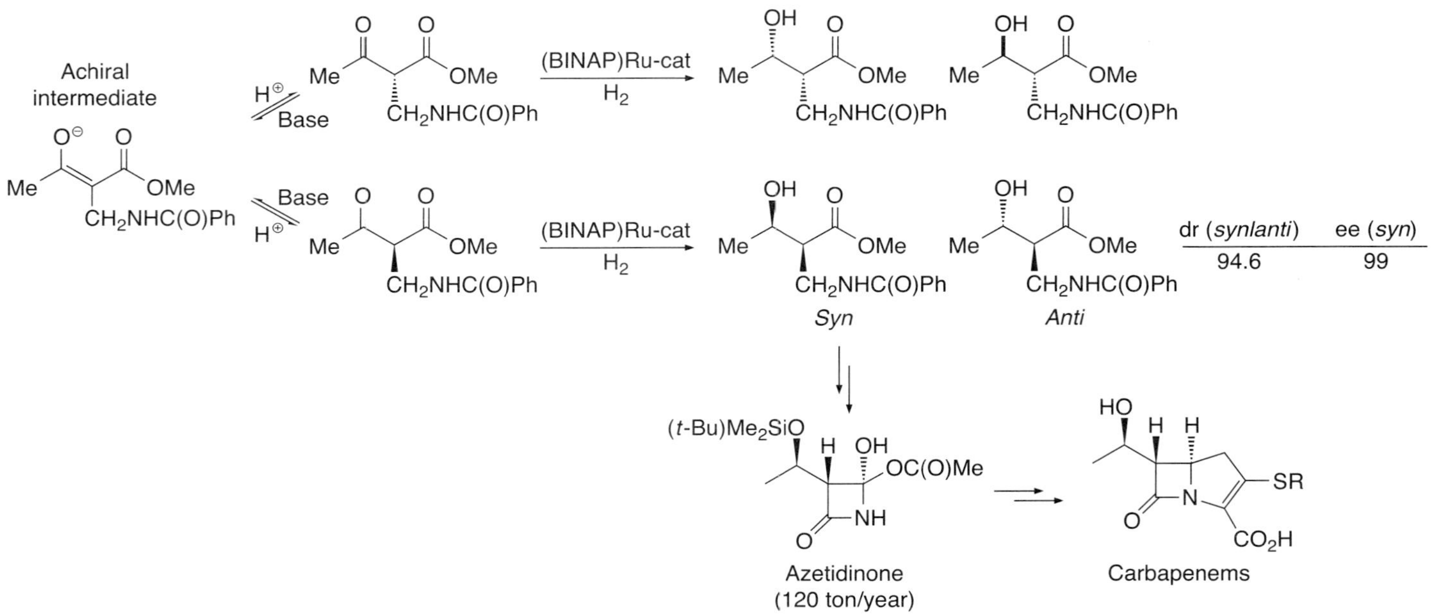

Figure 14.30.
DKR in the asymmetric reduction of α-substituted β-keto esters, illustrating the substrate racemization and the four possible diastereomeric products.

Although reduction of α-substituted β-keto esters in Figure 14.30 can give rise to four stereoisomers, conditions can often be found that furnish a single stereoisomer with high selectivity.[62] The configuration of the catalyst has been shown to determine which face of the ketone is reduced, while the configuration of the α-carbon depends on the identity of the substituent between the carbonyls and the group flanking the ketone. For example, hydrogenation of the substrate shown in Figure 14.30 in the presence of the catalyst containing (R)-BINAP gives the *syn* diastereomer with 94:6 diaselectivity and the (R)-configuration at the β-carbon in 99% ee. This intermediate is then carried forward to the carbapenem antibiotics as shown.

14.2.6.4. Dynamic Kinetic Asymmetric Transformations

The DKR processes just outlined can be contrasted with a related method called a dynamic kinetic asymmetric transformation (DyKAT).[84] The mechanism of stereochemical interconversions distinguishes a DKR and a DyKAT. As was described in Figure 14.29, an achiral intermediate or transition state interconverts the starting enantiomers in a DKR. In general, the catalyst that promotes racemization during a DKR is achiral and unrelated to the catalyst that performs the resolution step. In a DyKAT, the interconversion of the substrate stereochemistry occurs on the asymmetric catalyst (Figure 14.31). Since the catalyst is chiral, inversion of the configuration of the substrate generates diastereomeric complexes of the substrate–catalyst complex and is, therefore, an epimerization.[85]

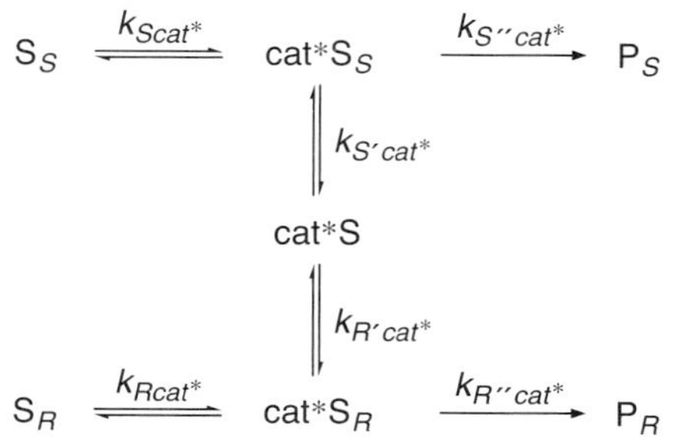

Figure 14.31.
DyKAT reaction pathways involving epimerization. The (R)- and (S)-substrates and products are S_R, S_S, P_R, and P_S. The chiral catalyst is cat*.

One example of a DyKAT is the asymmetric allylic alkylation outlined in Equation 14.18, in which the two diastereomeric π-allyl complexes epimerize via the σ-allyl intermediate, as shown previously in Figure 14.15.

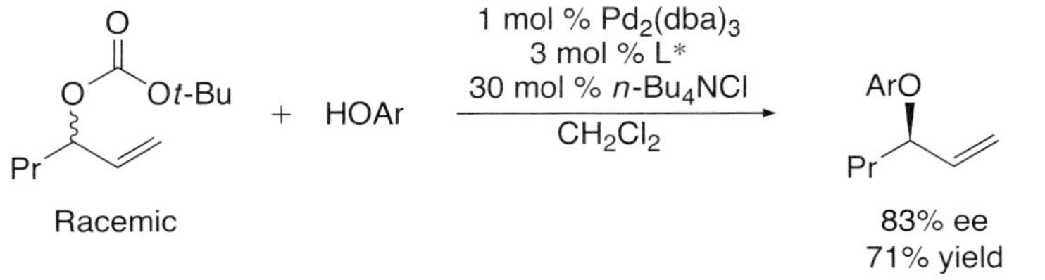

$$(14.18)$$

A second, less obvious example of a DyKAT is the enantioselective phosphination of an aryl halide. This process involves the reaction of a racemic secondary phosphine with an aryl halide to form a tertiary phosphine with excellent conversion and good enantioselectivity (Figure 14.32). The starting secondary phosphine is racemic because of the high inversion barrier of phosphines. However, the phosphorus center undergoes epimerization when it is part of an intermediate phosphido ligand. Thus, the catalytic process occurs by the reaction of an L*Pd(Ph)I intermediate with the racemic secondary phosphine HPArMe and base to form diastereomeric phosphido complexes (Figure 14.33). These diastereomers interconvert via lone pair inversion at phosphorus faster than they undergo reductive elimination to establish the configuration at the phosphorus center.[86] One selective catalyst for this reaction is [(R,R)-(Me-DUPHOS)Pd(Ph)I].

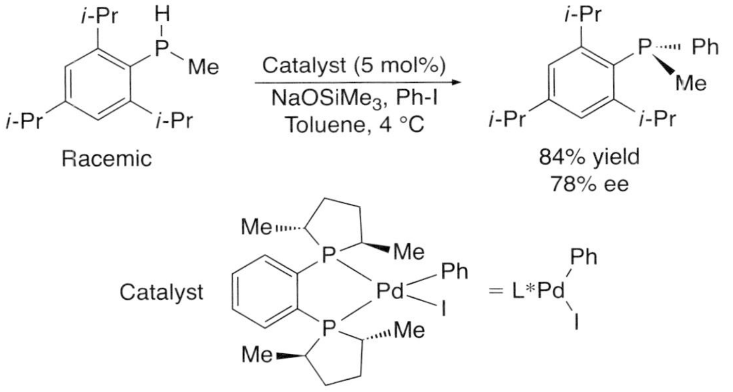

Figure 14.32.
DyKAT of racemic phosphine to generate a tertiary *P*-chirogenic phosphine.

14.2.6.5. Desymmetrization Reactions

Desymmetrization of meso and centrosymmetric substrates by asymmetric catalysts is synthetically valuable because such symmetry-breaking reactions can set several stereogenic centers simultaneously.[87] This strategy generally involves the differential reactivity of enantiotopic atoms or functional groups of a substrate with a chiral reagent or catalyst. As will be illustrated below, the desymmetrization is often similar to a kinetic resolution; instead of differentiating between enantiomers of a racemic substrate, the catalyst differentiates between enantiotopic groups within a single substrate. Thus, catalysts that are reactive and selective in a KR are often reactive and selective for desymmetrization of meso substrates whose enantiotopic functional groups resemble the arrangement of atoms in the substrates of the KR. In contrast to kinetic resolution, the theoretical yield of the enantioenriched product of a desymmetrization is 100%.

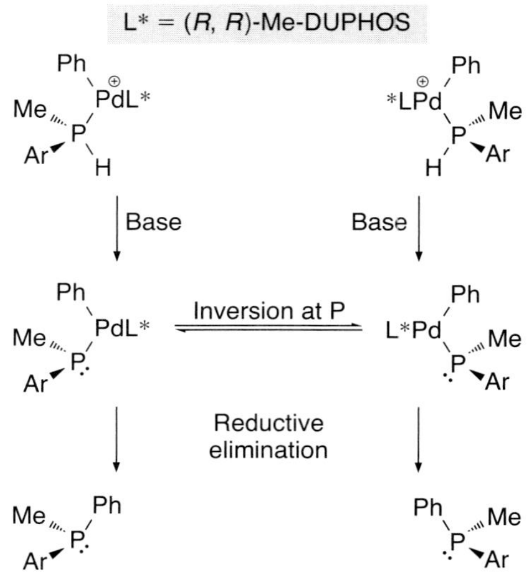

L* = (R, R)-Me-DUPHOS

Figure 14.33.
Proposed mechanism of the DyKAT, where inversion of phosphorus is faster than reductive elimination.

14.2.6.5.1. Two Examples of Desymmetrization

14.2.6.5.1.1. Desymmetrization of Achiral Dienes via Catalytic Asymmetric Hydrosilylation

The addition of an Si–H bond across a C–C double bond is called hydrosilylation, and this reaction is presented in Chapter 16.[88] The alkylsilane products of hydrosilylation can be converted to alcohols upon oxidation of the newly formed Si–C bond.[89,90] Thus, desymmetrization of divinyl carbinols by hydrosilylation can generate enantioenriched 1,3-diols.[91] In the presence of a (R,R)-DIOP-based rhodium catalyst, intramolecular hydrosilylation of a 3,5-dimethylphenyl-substituted silane derivative formed the cyclic product in Figure 14.34 in 93% ee.[91]

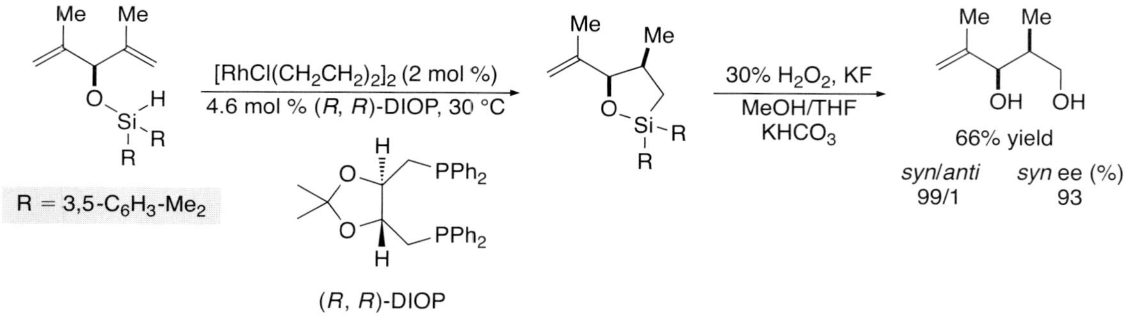

Figure 14.34.
Desymmetrization of dienes by catalytic asymmetric hydrosilylation. Oxidation of the product provides a valuable 1,3-diol.

14.2.6.5.1.2. Desymmetrization via the Palladium-Catalyzed Heck Reaction

Another reaction that has been successfully employed as an enantioselective desymmetrization processes is the Heck reaction. The Heck reaction, which is described in more detail in Chapter 19, couples an aryl or vinyl electrophile with an olefin. As shown in Equation 14.19, this reaction can be run as a desymmetrization in which the catalyst reacts preferentially at one of the enantiotopic olefins over the other to generate the intermediate enol that undergoes isomerization to the ketone product. This ketone as formed in 76% yield and 86% enantioselectivity and was an intermediate in the synthesis of vernolepin.[92]

$$\text{(14.19)}$$

76% yield
36% ee

Vernolepin

14.3. Summary

The concepts and reactions outlined in this chapter are intended to convey the essential features of catalysis involving the organometallic reactions outlined in this book. However, these fundamentals apply to catalysis in general. Further reading on the fundamentals of asymmetric catalysis[22, 58] and more comprehensive treatises[93, 94] are listed in the references and notes.

References and Notes

1. Amatore, C.; Carre, E.; Jutand, A.; M'Barki, M. A. *Organometallics* **1995**, *14*, 1818.
2. Cobley, C. J.; Lennon, I. C.; McCague, R.; Ramsden, J. A.; Zanotti-Gerosa, A. *Tetrahedron Lett.* **2001**, *42*, 7481.
3. Dobbs, D. A.; Vanhessche, K. P. M.; Brazi, E.; Rautenstrauch, V.; Lenoir, J. Y.; Genet, J. P.; Wiles, J.; Bergens, S. H. *Angew. Chem. Int. Ed.* **2000**, *39*, 1992.
4. Wiles, J. A.; Bergens, S. H.; Vanhessche, K. P. M.; Dobbs, D. A.; Rautenstrauch, V. *Angew. Chem. Int. Ed.* **2001**, *40*, 914.
5. Alcazar-Roman, L. M.; Hartwig, J. F. *J. Am. Chem. Soc.* **2001**, *123*, 12905.
6. Shekhar, S.; Ryberg, P.; Hartwig, J. F.; Mathew, J. S.; Blackmond, D. G.; Strieter, E. R.; Buchwald, S. L. *J. Am. Chem. Soc.* **2006**, *128*, 3584.
7. Meakin, P.; Jesson, J. P.; Tolman, C. A. *J. Am. Chem. Soc.* **1972**, *94*, 3240.
8. Davies, I. W.; Matty, L.; Hughes, D. L.; Reider, P. J. *J. Am. Chem. Soc.* **2001**, *123*, 10139.
9. Weck, M.; Jones, C. W. *Inorg. Chem.* **2007**, *46*, 1865.
10. Phan, N. T. S.; Van Der Sluys, M.; Jones, C. W. *Adv. Synth. Catal.* **2006**, *348*, 609.
11. Bergbreiter, D. E.; Osburn, P. L.; Frels, J. D. *Adv. Synth. Catal.* **2005**, *347*, 172.
12. Yu, K. Q.; Sommer, W.; Richardson, J. M.; Weck, M.; Jones, C. W. *Adv. Synth. Catal.* **2005**, *347*, 161.
13. Hamlin, J. E.; Hirai, K.; Millan, A.; Maitlis, P. M. *J. Mol. Catal.* **1980**, *7*, 543.
14. Widegren, J. A.; Finke, R. G. *J. Mol. Catal. A* **2003**, *198*, 317.
15. Widegren, J. A.; Finke, R. G. *J. Mol. Catal. A* **2003**, *191*, 187.
16. Dyson, P. J. *Dalton Trans.* **2003**, 2964.
17. Lewis, L. N.; Colborn, R. E.; Grade, H.; Bryant, G. L.; Sumpter, C. A.; Scott, R. A. *Organometallics* **1995**, *14*, 2202.
18. Reetz, M.; Westermann, E.; Lohmer, R.; Lohmer, G. *Tetrahedron Lett.* **1998**, 8449
19. Reetz, M. T.; Westermann, E. *Angew. Chem., Int. Ed. Engl.* **2000**, *39*, 165.
20. Whitesides, G. M.; Hackett, M.; Brainard, R. L.; Lavalleye, J.-P. P. M.; Sowinski, A. F.; Izumi, A. N.; Moore, S. S.; Brown, D. W.; Staudt, E. M. *Organometallics* **1985**, *4*, 1819.
21. Collman, J. P.; Hegedus, L. S.; Cooke, M. P.; Norton, J. R.; Dolcetti, G.; Marquardt, D. N. *J. Am. Chem. Soc.* **1972**, *94*, 1789.
22. Walsh, P. J.; Kozlowski, M. C. *Fundamentals of Asymmetric Catalysis*; University Science Books: Sausalito, CA, 2009.
23. Jurkauskas, V.; Buchwald, S. L. *J. Am. Chem. Soc.* **2002**, *124*, 2892.
24. Cahn, R. S.; Ingold, C.; Prelog, V. *Angew. Chem., Int. Ed. Engl.* **1966**, *5*, 385.
25. Prelog, V.; Helmchen, G. *Angew. Chem., Int. Ed. Engl.* **1982**, *21*, 567.
26. Nugent, W. A.; Mayer, J. M. *Metal–Ligand Multiple Bonds*; Wiley: New York, 1988.
27. Katsuki, T. In *Comprehensive Coordination Chemistry II*; Ward, M. D., Ed.; Elsevier: Oxford, U.K., 2004; Vol. 9, p 207.

28. Kagan, H. B. In *Catalytic Asymmetric Synthesis*; Ojima, I., Ed.; Wiley-VCH: New York, 2000, p 327.
29. Dalton, C. T.; Ryan, K. M.; Wall, V. M.; Bousquet, C.; Gilheany, D. G. *Top. Catal.* **1998**, *5*, 75.
30. Jacobsen, E. N.; Wu, M. H. In *Comprehensive Asymmetric Catalysis*; Jacobsen, E. N., Pfaltz, A., Yamamoto, H., Eds.; Springer: Berlin, 1999; Vol. 2, p 649.
31. Müller, P.; Fruit, C. *Chem. Rev.* **2003**, *103*, 2905.
32. Halfen, J. *Curr. Org. Chem.* **2005**, *9*, 657.
33. Doyle, M. P.; McKervey, A. M.; Ye, T. *Modern Catalytic Methods for Organic Synthesis with Diazo Compounds*; Wiley and Sons, Inc: New York, 1998.
34. Katsuki, T. *Synlett* **2003**, 281.
35. Curtin, D. Y. *Rec. Chem. Prog.* **1954**, *15*, 111.
36. Seeman, J. I. *Chem. Rev.* **1983**, *83*, 83.
37. Winstein, S.; Holness, N. J. *J. Am. Chem. Soc.* **1955**, *77*, 5562.
38. Eliel, E. L. *Experientia* **1953**, *9*, 91.
39. Eliel, E. L.; Ro, R. S. *Chem. Ind. (London)* **1956**, 251.
40. Eliel, E. L.; Lukach, C. A. *J. Am. Chem. Soc.* **1957**, *79*, 5986.
41. Halpern, J. *Science* **1982**, *217*, 401.
42. For a more recent investigation of this and related systems, see references 43–45.
43. Giovannetti, J. S.; Kelly, C. M.; Landis, C. R. *J. Am. Chem. Soc.* **1993**, *115*, 4040.
44. Landis, C. R.; Hilfenhaus, P.; Feldgus, S. *J. Am. Chem. Soc.* **1999**, *121*, 8741.
45. Feldgus, S.; Landis, C. R. *J. Am. Chem. Soc.* **2000**, *122*, 12714.
46. Halpern, J. *Science* **1982**, *217*, 401.
47. Landis, C. R.; Halpern, J. *J. Am. Chem. Soc.* **1987**, *109*, 1746.
48. Knowles, W. S. *Acc. Chem. Res.* **1983**, *16*, 106.
49. Sjogren, M. P. T.; Hansson, S.; Åkermark, B.; Vitagliano, A. *Organometallics* **1994**, *13*, 1963.
50. Trost, B. M.; Toste, F. D. *J. Am. Chem. Soc.* **1999**, *121*, 4545.
51. Casey, C. P.; Martins, S. C.; Fagan, M. A. *J. Am. Chem. Soc.* **2004**, *126*, 5585.
52. Evans, B. E.; Rittle, K. E.; Bock, M. G.; DiPardo, R. M.; Freidinger, R. M.; Whitter, W. L.; Lundell, G. F.; Veber, D. F.; Anderson, P. S.; Chang, R. S. L.; Lotti, V. J.; Cerino, D. J.; Chen, T. B.; Kling, P. J.; Kunkel, K. A.; Springer, J. P.; Hirshfieldt, J. *J. Med. Chem.* **1988**, *31*, 2235.
53. Yoon, T. P.; Jacobsen, E. N. *Science* **2003**, *299*, 1691.
54. Miyashita, A.; Yasuda, A.; Takaya, H.; Toriumi, T.; Ito, K.; Souchi, T.; Noyori, R. *J. Am. Chem. Soc.* **1980**, *102*, 7932.
55. Noyori, R.; Ohkuma, T. *Angew. Chem. Int. Ed.* **2001**, *40*, 40.
56. Noyori, R.; Yamakawa, M.; Hashiguchi, S. *J. Org. Chem.* **2001**, *66*, 7931.
57. Ohta, T.; Takaya, H.; Noyori, R. *Inorg. Chem.* **1988**, *27*, 566.
58. Noyori, R. *Asymmetric Catalysis in Organic Synthesis*; Wiley: New York, 1994.
59. Morton, D. A. V.; Orpen, A. G. *J. Chem. Soc., Dalton Trans.* **1992**, 641.
60. Brunner, H.; Winter, A.; Breu, J. *J. Organomet. Chem.* **1998**, *553*, 285.
61. Noyori, R.; Takaya, H. *Acc. Chem. Res.* **1990**, *23*, 345.
62. Noyori, R.; Tokunaga, M.; Kitamura, M. *Bull. Chem. Soc. Jpn.* **1995**, *68*, 36.
63. Ashby, M. T.; Halpern, J. *J. Am. Chem. Soc.* **1991**, *113*, 589.
64. Kitamura, M.; Ohkuma, T.; Inoue, S.; Sayo, N.; Kumobayashi, H.; Akutagawa, S.; Ohta, T.; Takaya, H.; Noyori, R. *J. Am. Chem. Soc.* **1988**, *110*, 629.
65. Kagan, H. B.; Fiaud, J. C. *Top. Stereochem.* **1988**, *18*, 249.
66. Keith, J. M.; Larrow, J. F.; Jacobsen, E. N. *Adv. Synth. Catal.* **2001**, *1*, 5.
67. Cook, G. R. *Curr. Org. Chem.* **2000**, *4*, 869.
68. Hoveyda, A. H.; Didiuk, M. T. *Curr. Org. Chem.* **1998**, *2*, 489.
69. Martín, V. S.; Woodard, S. S.; Katsuki, T.; Yamada, Y.; Ikeda, M.; Sharpless, K. B. *J. Am. Chem. Soc.* **1981**, *103*, 6237.
70. Jacobsen, E. N. *Acc. Chem. Res.* **2000**, *33*, 421.
71. Trost, B. M.; Hembre, E. J. *Tetrahedron Lett.* **1999**, *40*, 219.
72. Ferreira, E. M.; Stoltz, B. M. *J. Am. Chem. Soc.* **2001**, *123*, 7725.
73. Jensen, D. R.; Pugsley, J. S.; Sigman, M. S. *J. Am. Chem. Soc.* **2001**, *123*, 7475.
74. Sigman, M. S.; Jensen, D. R. *Acc. Chem. Res.* **2006**, *39*, 221.
75. Tambar, U. K.; Ebner, D. C.; Stoltz, B. M. *J. Am. Chem. Soc.* **2006**, *128*, 11752.
76. Zhu, S. S.; Cefalo, D. R.; La, D. S.; Jamieson, J. Y.; Davis, W. M.; Hoveyda, A. H.; Schrock, R. R. *J. Am. Chem. Soc.* **1999**, *121*, 8251.
77. Ward, R. S. *Tetrahedron: Asymmetry* **1995**, *6*, 1475.
78. El Gihani, M. T.; Williams, J. M. J. *Curr. Opin. Chem. Biol.* **1999**, *3*, 11.
79. Huerta, F. F.; Minidis, A. B. E.; Bäckvall, J.-E. *Chem. Soc. Rev.* **2001**, *30*, 321.

80. Caddick, S.; Jenkins, K. *Chem. Soc. Rev.* **1996**, *25*, 447.
81. Pamies, O.; Bäckvall, J.-E. *Chem. Rev.* **2003**, *103*, 3247.
82. Kitamura, M.; Tokunaga, M.; Noyori, R. *Tetrahedron* **1993**, *49*, 1853.
83. Kitamura, M.; Tokunaga, M.; Noyori, R. *J. Am. Chem. Soc.* **1993**, *115*, 144.
84. Faber, K. *Chem. Eur. J.* **2001**, *7*, 5004.
85. Eliel, E. L.; Wilen, S. H. *Stereochemistry of Organic Compounds*; Wiley: New York; 1994.
86. Moncarz, J. R.; Laritcheva, N. F.; Glueck, D. S. *J. Am. Chem. Soc.* **2002**, *124*, 13356.
87. Willis, M. C. *J. Chem. Soc., Perkin Trans. 1* **1999**, 1765.
88. Hishiyama, H.; Itoh, K. In *Catalytic Asymmetric Synthesis*; Ojima, I., Ed.; Wiley-VCH: New York; 2000, p 111.
89. Tamao, K.; Ishida, N.; Tanaka, T.; Kumada, M. *Organometallics* **1983**, *2*, 1694.
90. Tamao, K.; Kakui, T.; Kumada, M. *J. Am. Chem. Soc.* **1978**, *100*, 2268.
91. Tamao, K.; Tohma, T.; Inui, N.; Nakayama, O.; Ito, Y. *Tetrahedron Lett.* **1990**, *31*, 7333.
92. (a) Sato, Y.; Honda, T.; Shibasaki, M. *Tetrahedron Lett.* **1992**, *33*, 2593; (b) Kondo, K.; Sodeoka, M.; Mori, M.; Shibasaki, M. *Tetrahedron Lett.* **1993**, *34*, 4219.
93. Jacobsen, E. N.; Pfaltz, A.; Yamamoto, H. *Comprehensive Asymmetric Catalysis*; Springer: Berlin, 1999; Vol. 1–3.
94. Ojima, I., Ed. *Catalytic Asymmetric Synthesis*; 2nd ed.; Wiley-VCH: New York, 2000.

CHAPTER 15

Homogeneous Hydrogenation

15.1. Introduction

The hydrogenation of organic substrates is one of the most important classes of reaction for the synthesis of both commodity and fine chemicals. Alkene hydrogenations catalyzed by solid or supported palladium and platinum, such as palladium on carbon or Adam's catalyst derived from platinum oxide, are well known examples. As noted in Chapter 14, these solid catalysts present many advantages for separation and reuse. However, the diversity of selective hydrogenations that occur with homogeneous catalysts is unmatched by that of heterogeneous systems. In many cases, the focus of work on homogeneous catalysts has been to control stereoselectivity, including diastereoselectivity and enantioselectivity. Tremendous advances have been made in stereoselective, homogeneous hydrogenation, and William S. Knowles and Ryoji Noyori were awarded the 2001 Nobel Prize in Chemistry for their contributions to enantioselective hydrogenation. Enantioselective hydrogenation is now the most utilized industrial method to set the stereochemistry of products containing tertiary stereocenters. However, homogeneous catalysts also create the ability to conduct directed hydrogenations, hydrogenations of one type of olefin over another, or selective hydrogenation of the ketone or olefin portion of an enone based on the type of catalyst. Moreover, homogeneous systems catalyze the hydrogenation of many other functional groups, such as acetylenes, aldehydes, ketones, esters, nitroarenes, arenes, and heteroarenes. [1-6]

A vast number of soluble transition metal complexes have been shown to serve as "precatalysts" for the hydrogenation of olefins and other unsaturated substrates. These catalysts include, inter alia, neutral and cationic rhodium, ruthenium, and iridium catalysts, as well as lanthanides. These catalysts encompass the widely known Wilkinson's catalyst and the related $[Rh(CO)(PPh_3)_3H]$ that reduces aldehydes produced by hydroformylation (discussed in Chapter 17), chiral, non-racemic rhodium, iridium, and ruthenium catalysts now widely used for the enantioselective reduction of intermediates for pharmaceuticals, insecticides, fragrances and flavorings, catalysts containing open coordination spheres or labile ligands that allow for the coordination of auxiliary functional groups to direct the stereoselectivity and regioselectivity of hydrogenation, and catalysts under development for the reduction of arenes and heteroarenes. During the past few decades, most research has focused on the enantioselective hydrogenation of C=C, C=O, and C=N bonds.

A complement to hydrogenation that has practical uses is "transfer hydrogenation." In transfer hydrogenation, the hydrogen equivalent is derived from sources other than H_2. These catalysts transfer hydrogen from alcohols, formic acid, metal hydrides (such as $NaBH_4$), or the hydrolysis of coordinated CO (via the water gas shift process discussed in Chapter 11). Since these catalysts exhibit features similar to homogeneous hydrogenation catalysts, they are also discussed in this chapter, albeit briefly. Catalytic hydrosilations also lead to the reduction of ketones and imines, and these reactions are discussed in Chapter 16.

15.2. A Perspective on the Homogeneous Catalytic Hydrogenation of Olefins

In 1938, Melvin Calvin was the first to use the term "homogeneous hydrogenation."[7,8] The reactions published by Calvin involved the reduction of p-benzoquinone by catalysts generated from copper acetate and quinoline. Iguchi reported the first homogeneous hydrogenation with rhodium, using rhodium ammine (NH_3) complexes in aqueous solution,[9] and later homogeneous hydrogenation with cyanocobaltate complexes.[10,11] In the 1960s, Wilkinson, Bennett, and Coffey prepared $[Rh(PPh_3)_3Cl]$.[12–14] Wilkinson[12,15–18] and Coffey[14] showed that this complex catalyzed the hydrogenation of alkenes, and this complex became known as "Wilkinson's catalyst." This catalyst is routinely used for synthetic organic applications because it is reliable, selective, and efficient. The mechanism of olefin hydrogenation by Wilkinson's catalyst was elucidated by Halpern.[19,20] Roughly 20 years later, Crabtree revealed a much more reactive cationic complex for the hydrogenation of alkenes.[21,22] This catalyst, $[Ir(acetone)_2(PCy_3)(pyridine)]$, was particularly remarkable because it contained a third-row metal. Prior to this discovery, most researchers considered complexes of the third row to best serve as models of catalysts containing second-row metals, rather than to be a source for more reactive versions of them. Crabtree's catalyst and related cationic rhodium catalysts of Brown have been used widely for hydrogenations directed by auxiliary functionality and as a stimulus for the design of chiral, non-racemic catalysts for enantioselective hydrogenation.

Soon after the publication of Wilkinson's catalysts, Horner and co-workers in Mainz[23] and Knowles and co-workers at Monsanto[24] published work on chiral, non-racemic rhodium catalysts containing monophosphines in which three different substituents were bound to phosphorus. As noted in Chapter 2, such phosphines can be resolved and are called P-chiral ligands. Measurable enantioselectivities (but less than 30%) were observed for the hydrogenation of dehydroamino acids. Kagan then revealed three important principles for ligand design through his studies showing that rhodium complexes containing the bisphosphine DIOP [(4R,5R)-(−)-O-isopropylidene-2,3-dihydroxy-1,4-bis(diphenylphosphino)butane, see Figure 15.4] as ligand catalyze enantioselective hydrogenation with higher enantioselectivities than those obtained with the P-chiral monophosphines.[25] First, Kagan's studies showed that enantioselectivities were higher for catalysts containing bisphosphines than for those containing monophosphines. The chelation of a bisphosphine significantly rigidifies the ligand and decreases its degrees of freedom. Second, these studies showed that ligands possessing C_2 symmetry reduced the number of isomers that would be formed by coordination of a prochiral olefin and that this reduction in the number of isomers could reduce the number of species that react with competing enantioselectivity. It is now known that C_2 symmetry of a ligand is by no means a requirement to generate a catalyst that reacts with high enantioselectivity (vide infra),[26] but these design principles were incorporated by Knowles and co-workers into the second-generation catalysts containing P-chiral ligands. The DIPAMP [(R,R)-(−)-1,2-bis[(o-methoxy-phenyl)(phenyl)phosphino]ethane, see Figure 15.6] ligand developed at Monsanto is a C_2-symmetric P-chiral bisphosphine, and rhodium complexes of this ligand catalyze the hydrogenation of dehydroamino acids with high enantioselectivity. This chemistry was quickly developed into a technical process to prepare L-DOPA (3,4-dihydroxy-L-phenylalanine), a drug for Parkinson's disease.[27] Third, these studies showed that the center of chirality need not be the phosphorus bound to the metal. Substantial enantioselectivities were obtained from catalysts containing the DIOP ligand, which contains its stereocenters in the linker or "backbone" between the two phosphorus atoms. As noted in Chapter 14, the stereochemistry of the backbone leads to preferred conformations of the phenyl groups bound to phosphorus. This conformational preference transmits the stereochemistry of the backbone to the binding site of the metal and leads to differentiation of the prochiral faces of the substrate by the metal center.

In the 1980s, Noyori showed that the chirality of the backbone does not need to be present in the form of a typical carbon stereocenter. Instead, two aryl groups that do not rotate about the linking C–C bond and, thereby, adopt enantiomeric atropisomeric conformations can create backbone chirality. Moreover, extensive studies by Noyori on complexes of BINAP [2,2′-bis(diphenylphosphino)-1,1′-binaphthyl][28] have shown that the atropisomeric biaryl backbone is one of the most efficient structural motifs for the transfer of chirality. Ikariya and Noyori showed that many of the most active and selective catalysts containing BINAP as ligand contain ruthenium, rather than rhodium, as the central metal. Hydrogenations of many types of olefins are now conducted with ruthenium complexes, such as [Ru(BINAP)(OAc)$_2$]. The dihalide analog [Ru(BINAP)Cl$_2$]$_n$, originally used by Ikariya for the hydrogenation of olefins,[29] was subsequently shown to be highly active and selective as a catalyst for the hydrogenation of functionalized ketones when used in combination with a base.[30] Finally, ruthenium–BINAP complexes, in combination with amine ligands, were shown by Noyori to be highly active and selective for the hydrogenation of ketones by a new mechanism.[31,32] Due to the success of BINAP as a ligand to generate enantioselective catalysts for many types of olefins, ketones, and imines, many related atropisomeric ligands have been developed for enantioselective catalysis.[33]

More recently, several other classes of ligands have been developed that have significantly increased the capabilities of asymmetric hydrogenation. Ligands containing alkylphosphine, rather than arylphosphine, units were developed by Burk[34,35] at DuPont in order to increase the rate of oxidative addition of dihydrogen. The combination of these DuPhos ligands and rhodium precursors led to the hydrogenation of many olefins and imine substrates that did not react with fast rates and selectivities with rhodium catalysts containing arylphosphine ligands. Iridium catalysts bound by highly unsymmetrical "P,N ligands" containing one phosphino substituent and one dative nitrogen donor were developed by Pfaltz[36–38] to mimic the coordination sphere of Crabtree's catalyst, but with a chiral ligand. Finally, rhodium catalysts containing monodentate ligands have been revisited, and rhodium catalysts with phosphoramidite or phosphite ligands possessing a chiral diolate are highly selective for many classes of enantioselective hydrogenation.[39,40]

The hydrogenation of other unsaturated functionalities can also be valuable, but many of these reactions are currently under development or have largely been conducted with heterogeneous catalysts. The hydrogenation of arenes to cycloalkanes is conducted with heterogeneous catalysts on a massive commodity chemical scale as part of petroleum refining,[41] and it is this reduction that makes cyclohexane a useful starting material for the industrial production of C$_6$ materials, such as adipic acid. The development of homogeneous catalysts for arene hydrogenation could lead to catalysts for the asymmetric hydrogenation of arenes, but homogeneous catalysts for mild hydrogenations of arenes have been challenging to identify. Yet, catalysts for the hydrogenation of heteroarenes, including highly enantioselective examples, have been discovered. Homogeneous catalysts for the hydrogenation of arenes and hetereoarenes are described in Section 15.6.

The hydrogenation of functionalities in the carboxylic acid oxidation state can also be useful for small- or large-scale syntheses. For example, the hydrogenation of adiponitrile generates hexamethylenediamine that is one of the two monomers in the production of nylon.[42] This reaction is conducted with a heterogeneous catalyst, but homogeneous catalysts for the reduction of nitriles to amines would be convenient for the conversion of nitriles to amines on a laboratory scale. The hydrogenation of esters to aldehydes would shorten the typical sequence consisting of reduction of an ester to an alcohol, followed by oxidation of the alcohol to the aldehyde, and the hydrogenation of esters to alcohols would provide a cleaner method to conduct this transformation than typical procedures conducted with metal–hydride reagents. Examples of these reactions are described in Section 15.10.

15.3. Selected Examples of Achiral Homogeneous Hydrogenation Catalysts

15.3.1. Rhodium Catalysts for Olefin Hydrogenation

15.3.1.1. Neutral Rhodium Catalysts

15.3.1.1.1. Preparation of Wilkinson's Catalyst

As noted in Section 15.1, one of the most important developments in hydrogenation was the discovery that $[Rh(PPh_3)_3Cl]$ (Wilkinson's catalyst) catalyzed the hydrogenation of alkenes. The reactions of Wilkinson's catalyst with H_2 were presented in Chapter 6. This complex is mildly sensitive to oxygen. When it reacts with oxygen, some of the triphenylphosphine is converted to triphenylphosphine oxide, and the resulting rhodium complex is presumably the dinuclear species $[Rh(PPh_3)_2Cl]_2$, which contains a bridging chloride. In some cases, reactions catalyzed by this "aged Wilkinson's catalyst" occur with different rates and generate products with different regioselectivities or stereoselectivities than those conducted with freshly prepared Wilkinson's catalyst.

Two methods are routinely used to prepare Wilkinson's catalyst. The first is very simple, but is limited to the parent complex containing triphenylphosphine and a few other triarylphosphine ligands. By this method, rhodium(III) chloride is heated with PPh_3 in ethanol. (Equation 15.1).[16,43] The burgundy-red crystalline catalyst is formed, along with the phosphine oxide, which apparently results from the reduction of rhodium(III) by a mole of triphenylphosphine and hydrolysis of the resulting Ph_3PCl_2. The chloride ligand in Wilkinson's catalyst can be exchanged by bromide or iodide by a typical ligand substitution of four-coordinate d^8 complexes (Chapter 4).

$$RhCl_3(H_2O)_x \ + \ PPh_3 \ \xrightarrow{CH_3CH_2OH} \ RhCl(PPh_3)_3 \ + \ Ph_3PCl_2 \xrightarrow{H_2O} Ph_3PO \ + \ 2\,HCl \quad (15.1)$$

$$\tfrac{1}{2} \left[\begin{array}{c} \\ Rh \\ \end{array} \begin{array}{c} Cl \\ \end{array} \right]_2 \xrightarrow[\;n \leq 3\;]{nL} L_nRhCl \quad (L = PPh_3 \ or \ R'R_2P) \quad (15.2)$$

A second method to prepare $[Rh(PPh_3)_3Cl]$ and many other rhodium–phosphine chloride complexes is more versatile (Equation 15.2). In this procedure, the olefin ligands in $[Rh(ethylene)_2Cl]_2$ or the more soluble $[Rh(cyclooctene)_2Cl]_2$ are displaced by PPh_3 or virtually any other phosphine ligand. The phosphine-to-rhodium ratio can be varied by this method, and hydrogenation catalysts having varying degrees of reactivity can thus be generated in situ.

15.3.1.1.2. The Reactivity of Wilkinson's Catalyst

Unconjugated olefins and acetylenes are rapidly hydrogenated in a cis manner in the presence of Wilkinson's catalyst. These reactions typically occur at ambient hydrogen pressure and temperature in benzene when the reaction is conducted with a polar co-solvent, such as ethanol. These polar co-solvents might facilitate migratory insertion, which is the turnover-limiting step of the catalytic cycle presented later in this chapter.

The relative rates for the reaction of different classes of alkenes are shown in Figure 15.1 and cover a range of about 50-fold. These differences in rates can be accounted for by the steric effects on olefin binding constants discussed in Chapter 2. One manifestation of this selectivity can be seen in the reduction of the disubstituted, cis double bond of the reactant in Equation 15.3. However, some olefins that bind particularly strongly, such as ethylene and 1,3-butadiene, undergo hydrogenation slowly under ambient conditions. In this case, the alkenes bind tightly enough to siphon the catalyst from the cycle and thus alter the mechanism.

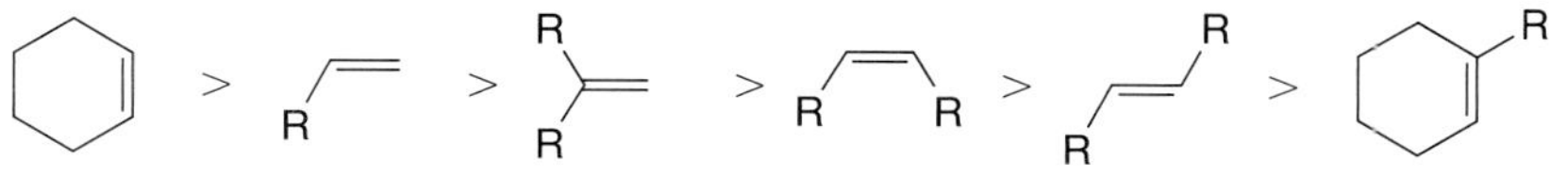

Figure 15.1.
Relative reactivities of alkenes for hydrogenation by Wilkinson's catalyst.

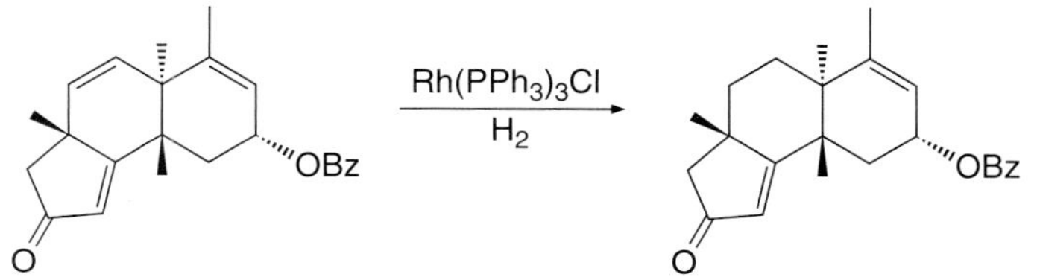

$$(15.3)$$

Wilkinson's catalyst undergoes the hydrogenation of alkenes without isotopic scrambling between H_2 and D_2 and without isotopic scrambling of D_2 and protons in the solvent or on the olefin substrate.[1] Under carefully controlled conditions, remarkably clean cis addition of hydrogen (or deuterium) is achieved. This lack of scrambling implies that the mechanism involves a dihydride intermediate in which both hydrides are transferred to the same alkene and that the mechanism involves a migratory insertion in which the metal and the hydride add in a cis fashion across the alkene, as discussed in Chapter 9.

Wilkinson's catalyst is frequently used to hydrogenate olefins in the presence of additional functional groups. The selectivity for olefin reduction in the presence of esters, ketones, and nitroarenes, and for reduction in the absence of olefin isomerization make this catalyst superior in many cases to heterogeneous catalysts. A few of the early applications of this catalyst are presented here.

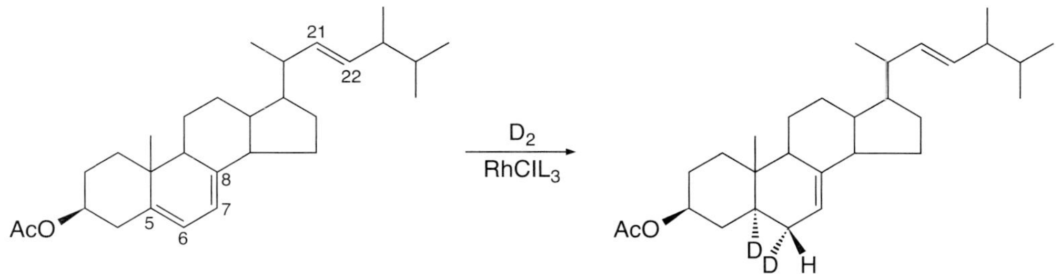

$$(15.4)$$

$$(15.5)$$

Addition of deuterium to ergosterol acetate was shown[1] to form the $5\alpha,6\alpha$-dideuterio reduction product under mild conditions (Equation 15.4). This result illustrates a number of features of the reactivity of Wilkinson's catalyst. First, the conjugated diene is reduced in preference to the C21–C22 double bond. Second, the diene is reduced from the less-crowded α-face of the steroid. Neither the remaining trisubstituted Δ^7 double bond nor the trans disubstituted Δ^{21} double bond are affected. Finally, no isotopic exchange takes place during this reaction. This selectivity has been used to specifically label prostaglandins with tritium (Equation 15.5).[44] The sensitive β-keto alcohol is unperturbed.

Birch[1] also illustrated the difference in selectivity between Wilkinson's catalyst and typical heterogeneous catalysts. Hydrogenation of 1,4-dihydrobenzenes (derived from the "Birch reductions" of arenes) in the presence of Wilkinson's catalyst forms the tetra-substituted alkene product, whereas hydrogenation with classic platinum and palladium heterogeneous catalysts causes disproportionation to form arene byproducts (Equation 15.6). Like the β-keto alcohol in Equation 15.5, the ester in Equation 15.6 is not reduced.

$$(15.6)$$

$$(15.7)$$

$$(15.8)$$

The selectivity for alkene reduction in the presence of other functionalities is also illustrated by selective hydrogenation of the carbon–carbon double bond in nitro olefins in the presence of Wilkinson's catalyst (Equation 15.7). Most heterogeneous catalysts react with the nitro group. In a similar sense, the alkoxy group in vinyl acetals is cleaved by Raney nickel, but the olefin undergoes selective hydrogenation in the presence of Wilkinson's catalyst (Equation 15.8). Olefins in both soluble and crosslinked polymers are hydrogenated in the presence of Wilkinson's catalyst (Equation 15.9), but are less readily reduced by heterogeneous catalysts.[45]

$$(15.9)$$

15.3.1.2. *Cationic Rhodium Catalysts*

Cationic rhodium complexes also catalyze the hydrogenation of olefins, and these catalysts ultimately led to the most common type of rhodium catalyst used for the enantioselective hydrogenation of olefins. These cationic rhodium dihydride catalysts are described by the general formula $[RhL_2(H)_2(S)_2]^+$, in which S represents a polar solvent molecule, such as THF or MeOH, and L_2 represents two tertiary phosphine ligands or, more commonly, one chelating bisphosphine. Some of the early members of this family of catalysts are the one containing the chelating bidentate ligand DIPHOS, and the one containing the mono-phosphine PPh_3. This family of catalysts was first studied by Osborn and Schrock,[46–48] who isolated and studied the reaction chemistry of several complexes of this general formula.

Many catalysts with the general structure $[Rh(diene)(PR_3)_2]^+$ have been used for hydrogenation. One set of experiments[46] summarized in Table 15.1 showed that cationic rhodium complexes containing more electron-donating ligands generate more active catalysts for alkene hydrogenation than complexes of less electron-donating ligands. Subsequent studies have shown that catalysts containing bidentate ligands are more reactive than those containing monodentate ligands, and a particularly active catalyst is [Rh(norbornadiene)(DPPB)]$^+$, in which DPPB is 1,4-bis(diphenylphosphino)butane.[49,50]

$$(15.10)$$

The studies by Osborn showed several fundamental aspects of the hydrogenation by these compounds.[46,51] First, they showed that the active catalyst is generated from the $[Rh(alkene)_2L_2]^+$ species by hydrogenation of the coordinated alkene in an initiation step (Equation 15.10). Second, they showed that the reaction of $[Rh(alkene)_2L_2]^+$ complexes with hydrogen formed cationic dihydride complexes $[RhL_2H_2S_2]^+$. Third, they showed that the hydrides were acidic and the presence of a basic group or added trialkylamine led to neutral species that, presumably, reacted by the same pathways as followed by Wilkinson's catalyst. The cationic Osborn catalysts and the neutral analogs give different products. The cationic manifold generates products with little isomerization of the olefin, while the neutral catalysts containing two phosphines lead to products resulting from isomerization of the olefin during the hydrogenation process. Thus, the reactions of cationic hydrogenation catalysts are typically conducted in the presence of a protic solvent or additive. As discussed in Section 15.3.2, the achiral, cationic catalysts are useful for diastereoselective and regioselective hydrogenations directed by auxiliary functional groups.

Table 15.1. Initial rate constants for hydrogenation of 1-hexene catalyzed by cationic rhodium complexes containing PPh_3, PPh_2Me, and $PhMe_2$ ligands.

Complex	Substrate concentration (mM)	k_{init} x 10^4 (s^{-1})
$[Rh(NBD)(PPh_3)_2]^+$	5.3	0.1
$[Rh(NBD)(PPh_2Me)_2]^+$	3.7	3.0
$[Rh(NBD)(PPhMe_2)_3]^+$	3.5	6.0

15.3.2. Iridium Catalysts: Crabtree's Catalyst

Crabtree and Felkin developed an unsaturated, iridium precatalyst for the reduction of hindered olefins with remarkably high rates.[21,52] This catalyst has become known as "Crabtree's catalyst." This "catalyst," $[Ir(COD)(Py)PCy_3]^+$, is actually a precatalyst in which hydrogen reduces the COD ligand to form cationic Ir(III) hydride complexes. In contrast to most homogeneous catalysts, particularly neutral species, Crabtree's catalyst is unaffected by oxygen. Because it is cationic, it binds polar functionalities, and its activity relies on the use of relatively noncoordinating solvents, such as methylene chloride. Pfaltz has shown that the activity of Crabtree's catalyst and chiral versions of it are significantly improved by the use of bulky, extremely weakly coordinating anions, such as tetrakis(3,5-bis(trifluoromethyl)phenyl) borate (BArF), for the iridium cation.[53,54] The propensity of this cation to bind polar functionalities led to its use for the directed hydrogenations described in the next section.

Suggs and Crabtree reported the hydrogenation of hindered olefins in steroids (Equation 15.11) with high selectivity for addition to the α-face, and this application, along with others in this seminal paper, drew attention to the virtues of Crabtree's catalyst for synthetic applications.[55] Crabtree's catalyst reduces olefins in the presence of many other functional groups. This is illustrated in one case by the reduction of an olefin in the presence of cyclopropyl and gem dibromo groups (Equation 15.12),[55] both of which would react with heterogeneous catalysts.

$$(15.11)$$

$$(15.12)$$

Table 15.2. A comparison of turnover frequencies for the homogeneous hydrogenation of different alkenes catalyzed by rhodium, ruthenium, and iridium catalysts.

Catalyst precursor	Temperature (°C)	Solvent	Turnover frequency (h⁻¹)		
			1-Hexene	Cyclohexene	Tetramethylethylene
$[Ir(COD)(PCy_3)(Py)]^+$	0	CH_2Cl_2	6,400	4,500	4,000
$[Ir(COD)(PMePh_2)_2]^+$	0	CH_2Cl_2	5,100	3,800	50
$[Ir(COD)(PMePh_2)_2]^+$	0	$CH_2C(O)CH_3$	10	0	0
$[Rh(COD)(PPh_3)_2]^+$	25	CH_2Cl_2	4,000	10	0
$[Ru(H)Cl(PPh_3)_3]$	25	C_6H_6	9,000	7	0
$[RhCl(PPh_3)_3]$	25	$C_6H_6/EtOH$	650	700	0
$[RhCl(PPh_3)_3]$	0	$C_6H_6/EtOH$	60	70	0

Table 15.2 compares the rates of several homogeneous catalysts for the hydrogenation of three olefins possessing varying degrees of steric bulk. These data show that Crabtree's catalyst is much more reactive than the neutral, more-hindered rhodium and ruthenium catalysts toward more substituted alkenes. Most striking, these data also show the unusually high reactivity of the cationic iridium complexes for hydrogenation of a tetra-substituted alkene. These data (entries 2 and 3) also show the importance of the use of weakly coordinating solvents on the activity of the cationic iridium catalysts.

15.3.3. Ruthenium Catalysts for Olefin Hydrogenation

In 1961, Halpern, Herod, and James reported the hydrogenation of carboxylic acids in water in the presence of chlororuthenium(II) catalysts.[56] As discussed later in this chapter, the catalytic, asymmetric reduction of the C=C bond in acrylate derivatives are some of the most successful applications of asymmetric homogeneous hydrogenations. Chatt and Hayter then began to report metal–hydride complexes of ruthenium with the general structure $[Ru(PR_3)_n(H)(Cl)]$ containing trialkylphosphines.[57] Just after the first papers on Wilkinson's rhodium catalyst for hydrogenation, Wilkinson showed that the reaction of $Ru(PPh_3)_3Cl_2$ with hydrogen and base gave $Ru(PPh_3)_3(H)(Cl)$, and that this product was a highly active catalyst for the hydrogenation of terminal alkenes (Equation 15.13).[15] Related PPh_3-ligated ruthenium complexes catalyze the hydrogenation of polycyclic arenes, esters, and nitriles, as discussed later in this chapter.[58] This complex was one of the first unsaturated, five-coordinate, d^6 complexes. Ruthenium catalysts for olefin hydrogenation are now some of the most widely used hydrogenation catalysts, particularly for asymmetric hydrogenation.

$$RCH=CH_2 \xrightarrow[\text{HRuCl(PPh}_3)_3]{H_2} RCH_2CH_3 \qquad (15.13)$$

Like Wilkinson's catalyst, the ruthenium–hydride complex $RuH(Cl)(PPh_3)_3$ selectively catalyzes the hydrogenation of terminal alkenes over internal alkenes. This catalyst reacts roughly 1000 times faster with terminal olefins than with internal olefins.[1] The lack of hydrogenation of the more-substituted alkenes and the lack of a binding site for the docking of other functionalities has limited the use of this catalyst.

Nevertheless, the reactivity of ruthenium for homogeneous hydrogenation and the mechanism of the reaction catalyzed by this complex led to the less-saturated catalysts containing chiral ligands, such as $[Ru(BINAP)(OAc)_2]$ and $[Ru(BINAP)Cl_2]_2$, which have been used widely in commercial settings for the asymmetric hydrogenation of olefins and ketones. Specific examples of these reactions are presented later in this chapter, but one example of enantioselective hydrogenation of a β-keto ester is shown here in Equation 15.14.[59]

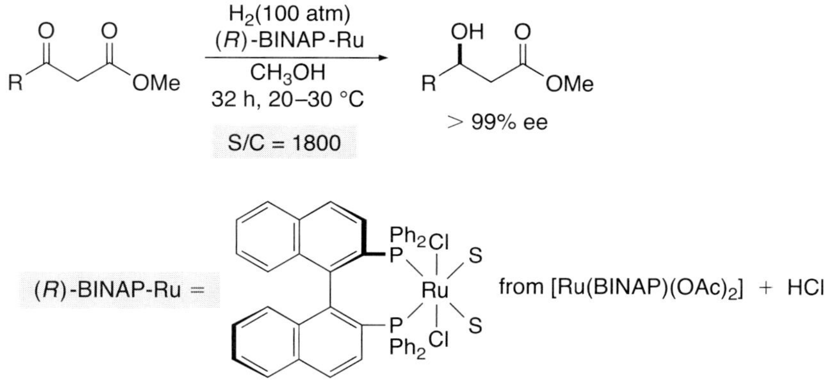

(15.14)

15.3.4. Lanthanide Catalysts

The final system in this section on the different classes of catalysts is discussed only briefly because it has not yet experienced the widespread use that the rhodium, iridium, and ruthenium catalysts have. These catalysts are based on lanthanides, typically containing pentamethylcyclopentadienyl ligands. These catalysts have not been used often in synthetic applications because of their sensitivity to air and moisture and their high reactivity toward auxiliary functional groups, such as esters, nitriles, and amides. However, Marks has shown that certain lanthanide catalysts are spectacularly reactive for the hydrogenation of terminal alkenes in the absence of these other functionalities.[60] Turnover frequencies exceeding 120,000 h^{-1} at 25 °C and 1 atm of H$_2$ for the reduction of 1-hexene have been reported! Related group 4 catalysts have been used for the selective reduction of dienes to monoenes. Chiral lanthanides have been designed for asymmetric hydrogenation,[61] but these catalysts have not led to selectivities that rival those of late metal catalysts.

15.4. Directed Hydrogenation

Crabtree's catalyst and related cationic rhodium catalysts, such as [Rh(norbornadiene)(DPPB)]$^+$, contain weakly bound solvent molecules after reduction of the diene ligands. The lability of the solvent ligands allows the substrate to coordinate to the metal through the carbon–carbon double bond and an additional functional group to direct delivery of hydrogen to the face of the olefin containing the second functional group. Crabtree[62] and Stork[63] published at similar times such "directed hydrogenations." Developments in directed hydrogenation have been reviewed by Brown,[49] and developments in the broader context of directed reactions have been reviewed by Evans and co-workers.[64]

The presence of functionality in the substrate can allow the catalyst to control which C=C bond in a diene or polyene undergoes hydrogenation. However, most applications have exploited substrate binding for the diastereoselective reduction of an olefin in a cyclic or acyclic unit. The cationic charges on the rhodium and iridium catalysts used for directed hydrogenation promote binding of polar functional groups, including alcohols, ethers, esters, ketones, and amides. In general, hydrogen adds to the same face of a cyclic molecule that contains these functional groups. This stereoselectivity often contrasts with the stereoselectivity from the reduction of the olefin by palladium on charcoal, which typically results from addition of hydrogen to the least-hindered face.

Three of many examples of directed hydrogenation are shown in Equations 15.15–15.17. Equation 15.15 shows the reduction of a homoallylic alcohol, which was one of the substrates first used to demonstrate this effect.[63] Equation 15.16 shows a more complex substrate in which the diastereoselective reduction by Crabtree's catalyst is directed by the amide function as part of the synthesis of pulmitoxins.[65] Equation 15.17 shows that the addition of hydrogen can be directed to a hindered face of a bicyclic system. In this case, the cationic rhodium system of Brown, as well as Crabtree's catalyst, led to high selectivity.[50] Many other reactions occur with high selectivity in the presence of Brown's cationic rhodium system. Diastereoselective additions to acyclic systems, along with a rationalization for the selectivity in these types of substrates, can be found in the review by Evans.[64]

$$\text{20\% [Ir(COD)py(PCy}_3\text{)]PF}_6$$

$$\text{H}_2 \text{ (1 atm), room temperature}$$

$$\text{CH}_2\text{Cl}_2$$

R = H   > 99 : 1
R = Me   33 : 1

(15.15)

$$(15.16)$$

$$> 99:1$$

$$(15.17)$$

$$70:1$$

15.5. Mechanisms of Homogeneous Olefin and Ketone Hydrogenation

15.5.1. Background

Some of the earliest mechanistic studies on homogeneous catalysis in organic solvents were reported by Halpern on the hydrogenation of olefins with Wilkinson's catalyst and on the mechanism of the asymmetric hydrogenation of dehydroamino acids by Knowles' rhodium–DIPAMP system. Two important general conclusions were drawn from these studies.[19,20,66–69] First, as noted in Chapter 14, the species in a catalytic system that accumulate in sufficient concentration to be identified spectroscopically may or may not lie directly on the catalytic cycle. In some cases, these species are connected to the catalytic cycle by equilibria, while in other cases they are unproductive species formed irreversibly. By associating particular complexes with the kinetic behavior of the catalytic reaction, one can assess whether the observed complex contributes positively or negatively to the rate of the catalytic process.

Second, to deduce the mechanism of a multistep catalytic reaction, the rate laws of individual steps should be determined independently whenever possible and correlated with the rate law of the overall catalytic process. When the rate and equilibrium parameters for these steps are assembled and shown to account quantitatively for the overall catalytic behavior, the proposed mechanism can be considered to describe the catalytic system. Studies that simply determine the effect of numerous variables on the overall kinetic behavior of a multistep catalytic reaction can be misleading. Such experiments do not generate data that can be used to deduce the mechanism because there are usually too many variables to specify a particular path. The authors of the previous version of this text stated, "A critical reader of the chemical literature will notice that these two lessons are often ignored."

15.5.2 Overview of the Typical Mechanisms

Many different mechanisms are followed by the commonly used systems for hydrogenation. Schemes 15.1–15.5 show a selection of these mechanisms. The mechanisms depicted in Schemes 15.1 and 15.2 involve oxidative additions of dihydrogen and insertions of alkenes into dihydride intermediates, and the mechanism in Scheme 15.3 involves heterolytic splitting of hydrogen and insertion of the alkene into a monohydride intermediate. The mechanism in Scheme 15.1 is typically followed by neutral rhodium catalysts and is often called the "hydride" or "hydrogen-first" mechanism. The mechanism in Scheme 15.2 is typically followed by cationic rhodium catalysts and is often called the "alkene" or "alkene-first" mechanism. The mechanism in Scheme 15.3, shown as originally proposed by Halpern and modified later by Morris, is followed by many neutral ruthenium catalysts for the hydrogenation of unsaturated carboxylic acids. A fourth mechanism (Scheme 15.4) is followed by catalysts for the hydrogenation of ketones and imines that contain a mixture of phosphine and amine or sulfonamidate ligands. This mechanism involves the transfer of the reducing equivalent by an "outer-sphere" mechanism in which hydrogen is transferred as a hydride and proton simultaneously without any oxidative addition of hydrogen or direct coordination of the unsaturated substrate to the metal prior to hydride transfer. The details of the transition state structure and the identity of the initially formed products are currently being studied and are discussed later in this chapter. The final mechanism (Scheme 15.5) is followed by group 4 and lanthanide catalysts for the hydrogenation of alkenes and imines and involves σ-bond metathesis steps. These five mechanisms and some of the data supporting them are described in the next few sections.

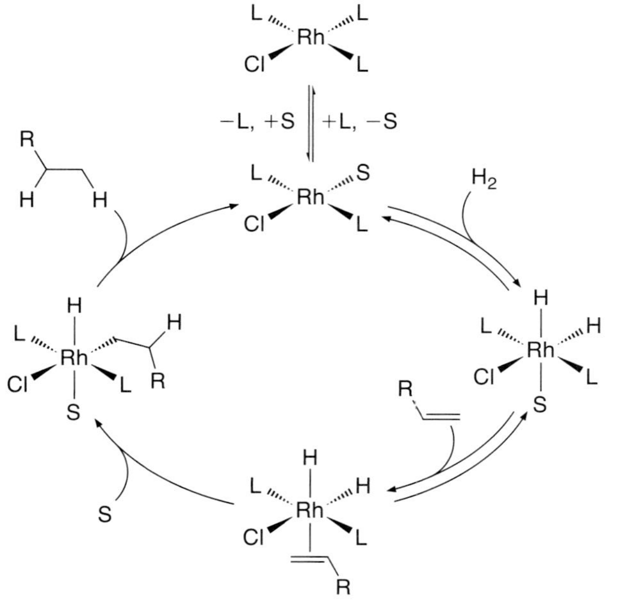

Scheme 15.1.
Core of the mechanism for hydrogenation by Wilkinson's catalyst.

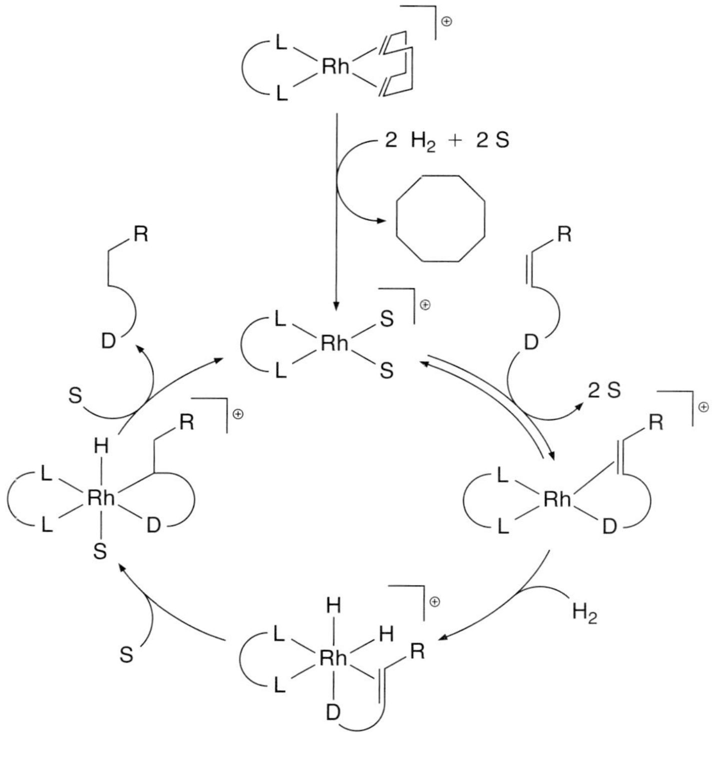

Scheme 15.2.
Core of the mechanism for hydrogenation of olefins containing a donor group D by cationic rhodium complexes.

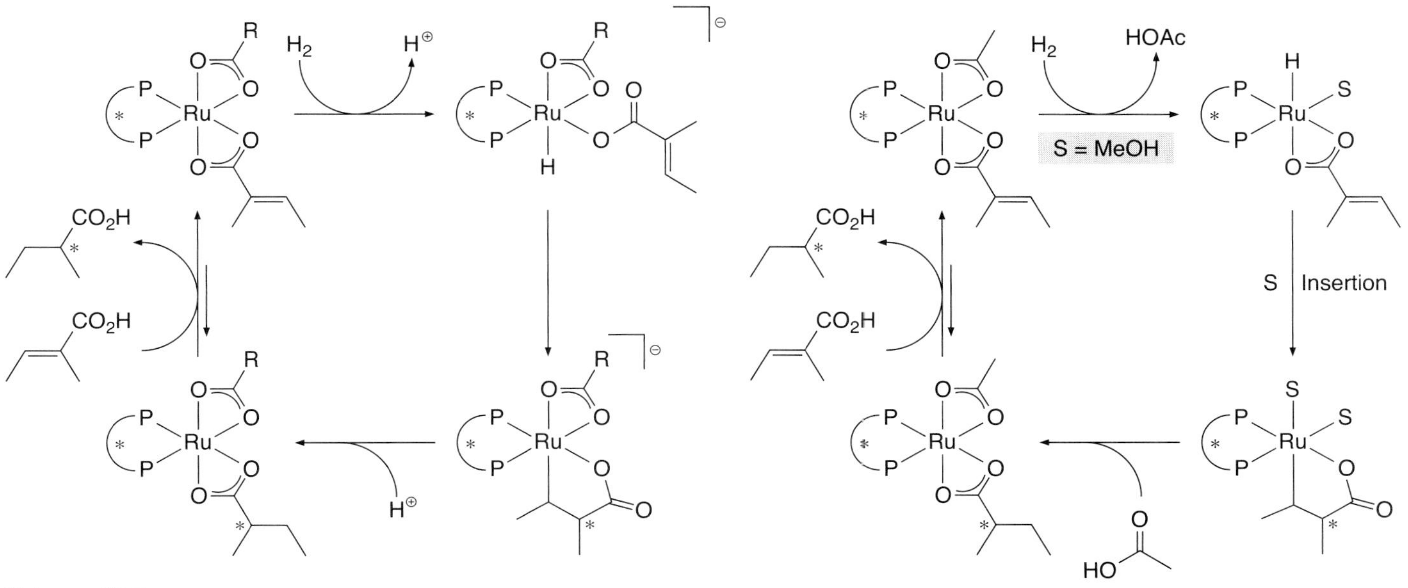

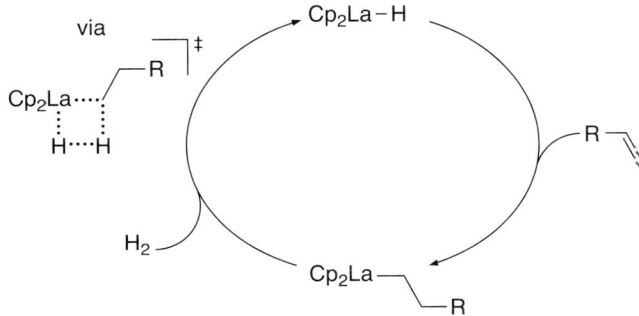

Scheme 15.3.
Core of the mechanism of hydrogenation of acrylic acids by bisphosphine ruthenium diacetates.
Left: Original proposal by Halpern; right: modified scheme proposed by Morris. See Section 15.5.2.2.3.

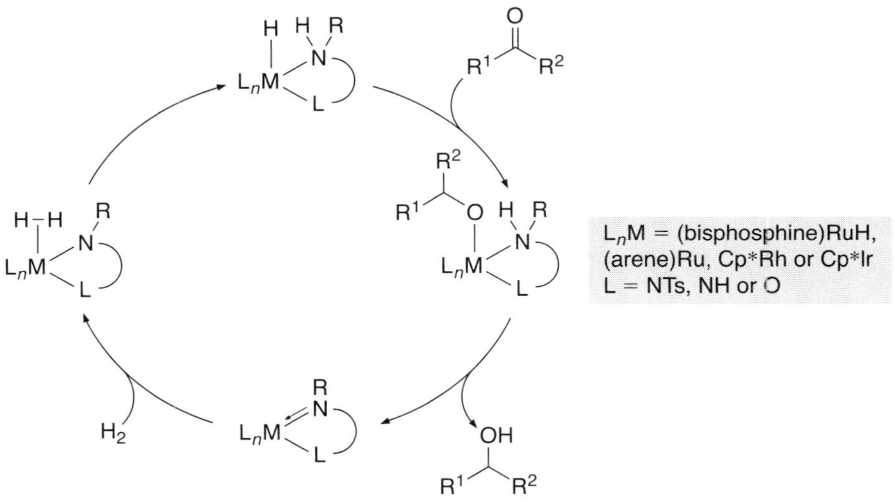

L_nM = (bisphosphine)RuH, (arene)Ru, Cp*Rh or Cp*Ir
L = NTs, NH or O

Scheme 15.4.
Core of the mechanism of hydrogenation of ketones by metal–ligand bifunctional catalysts.

Scheme 15.5.
Core of the mechanism of hydrogenation by lanthanide catalysts through σ-bond metathesis steps.

15.5.2.1. Mechanisms Occurring by Insertions of Olefins into Dihydride Complexes

15.5.2.1.1. Hydrogenation by Wilkinson's Catalyst

Rhodium complexes containing phosphine ligands catalyze the hydrogenation of olefins by mechanisms occurring through dihydride intermediates. The hydrogen-first pathway can be illustrated by studies on Wilkinson's catalyst, and the olefin-first pathway can be illustrated by studies on cationic phosphine-ligated rhodium catalysts. Isotopic labeling studies by Schrock and Osborn showed many years ago that one can switch between these two reaction manifolds by the presence of base or acid in the reaction medium.[46] More recent studies have shown that complexes of alkylphosphines can make the metal in cationic complexes sufficiently electron rich to react by the hydrogen-first mechanism.[70,71] These issues are revisited at the end of this part of the mechanistic discussion, but are noted here to illustrate the small differences in energy between these various pathways and their dependence on ligand and reaction medium. The factors that lead to control of stereochemistry within these pathways by chiral, non-racemic catalysts are presented after this discussion of the elementary steps of the principal classes of hydrogenation mechanisms.

Halpern studied the rate behavior of the overall catalytic hydrogenation of cyclohexene by Wilkinson's catalyst, as well as the rate behavior of the individual steps of the proposed catalytic cycle.[19,20] The core of the catalytic cycle was shown in Scheme 15.1 to illustrate the hydrogen-first pathway, but a more complete view of the pathway deduced from Halpern's studies is shown in Scheme 15.6. This overall scheme was deduced by isolating and determining the rate behavior of each individual stage of the catalytic cycle, demonstrating that these steps can account for the kinetic behavior of the overall catalytic reaction, and demonstrating that several previously isolated compounds were not kinetically competent to be intermediates in the catalytic process.

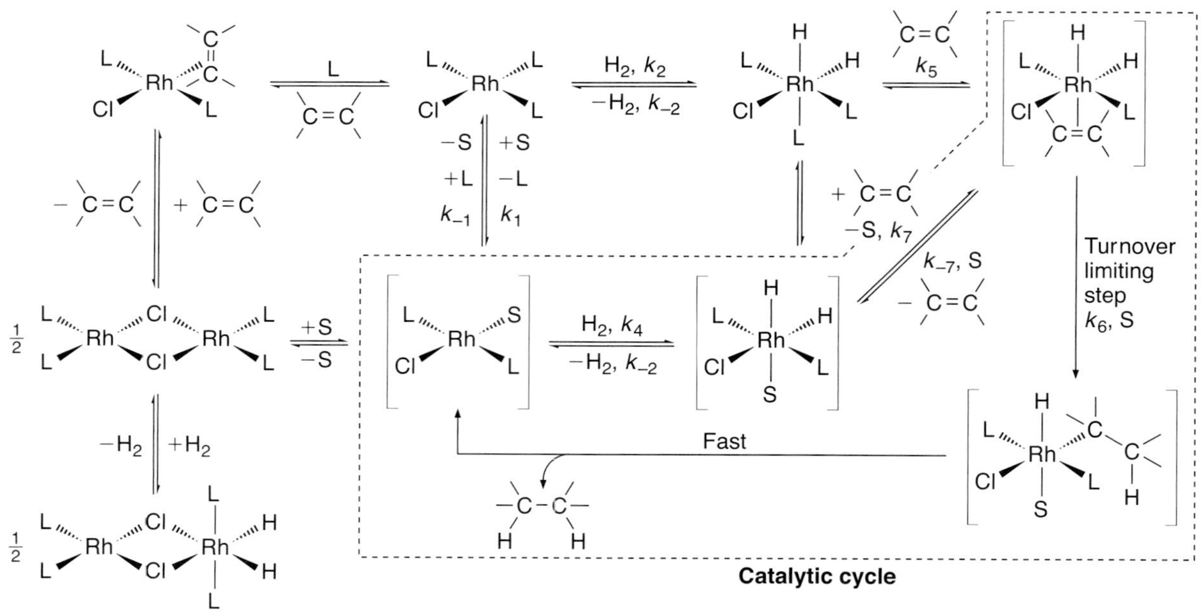

Scheme 15.6

In this system, all of the compounds that have been detected in or isolated from solutions of catalytic reactions lie off of the major catalytic pathway. These species are shown outside of the dotted enclosure in Scheme 15.6. The substances within the dotted enclosure are the proposed catalytic intermediates. The accumulation of the complexes outside the dotted enclosure in Scheme 15.6 reduces the rate of the catalytic reaction. This phenomenon should not be assumed to be the case for *all* catalytic systems; species lying directly on the catalytic cycle *have* been observed directly in many other systems. However, this study did show that the identification of a detectable species in a catalytic system without kinetic data to assess the connection between the observed species and the catalytic cycle can lead to incorrect interpretations of the reaction mechanism. As stated by the authors of the previous version of this text, "Only when kinetic and thermodynamic measurements define the role of complexes along the actual reaction path can the mechanism be defined."

The overall mechanism for olefin hydrogenation by Wilkinson's catalyst can be divided into three parts: the addition of hydrogen to $RhClL_3$, the reaction of RhH_2ClL_3 with the olefin by migratory insertion, and the reductive elimination of the reduced product. Because the hydrogen adds *before* the olefin, the mechanism shown in Schemes 15.1 and 15.6 is referred to as the hydride or hydrogen-first path. Mechanistic information on the first two parts of the catalytic cycle that control the rates of these reactions are discussed in the following two sections.

15.5.2.1.1.1. Mechanism of the Oxidative Addition Step

As discussed in Chapter 6, the addition of hydrogen to Wilkinson's catalyst is inhibited by added PPh_3,[72] and detailed analysis of these results indicated that the reaction occurs by two competing pathways: direct addition to the starting trisphosphine complex and addition to the bisphosphine species formed by dissociation of ligand from the starting complex. The major pathway occurs by the dissociation of a phosphine from $RhClL_3$, thus affording the 14-electron complex $RhClL_2$ (reaction 1 of Scheme 15.6). The equilibrium for dissociation of ligand lies far toward the starting trisphosphine complex ($K_1 = 10^{-5}$ M). After dissociation of phosphine, fast addition of hydrogen to $RhClL_2$ occurs to form the unsaturated dihydride $RhClL_2H_2$, which binds free triphenylphosphine to give the observed saturated dihydride $RhClL_3H_2$. This circuitous route from $RhClL_3$ to $RhClL_3H_2$ (both observable complexes) is faster than the direct addition of hydrogen to $RhClL_3$ at most concentrations of free phosphine. Independent generation of the bisphosphine complex by photolysis of $RhClL_2(CO)$ showed that $RhClL_2$ reacted with H_2 at least 10^4 times faster than $RhClL_3$.[73,74] This addition of H_2 has been shown to be reversible, and the stereochemistry of the addition has been analyzed by detailed NMR spectroscopic studies.[75,76]

The 14-electron bisphosphine complex, $RhClL_2$, has a pronounced tendency to dimerize (Scheme 15.6, $K_{eq} = 10^6$ M^{-1}). The product from the oxidative addition of hydrogen to the resulting dimer shown in Scheme 15.6 was characterized by Tolman et al.,[76,77] but this complex reacts with olefins too slowly to be an intermediate in the overall catalytic process. Moreover, the equilibrium between the $[RhL_2(S)Cl]$ complex and $[Rh_2L_4Cl_2]$ is not established in the catalytic system containing H_2, because H_2 intercepts the bisphosphine complex $RhL_2(S)Cl$ before it forms the inactive dimer.

In the absence of H_2, olefin complexes, such as $[RhL_2(olefin)Cl]$, are formed. With moderate binding olefins, such as cyclohexene, the formation of $[RhL_2(olefin)Cl]$ complexes is an unimportant side reaction during olefin hydrogenation. However, olefins that bind with relatively high equilibrium constants, such as ethylene, act as competitive inhibitors of the dihydride path by forming stable $[RhL_2(olefin)Cl]$ complexes.

15.5.2.1.1.2. Mechanism of the Migratory Insertion Step

A separate kinetic study of the reaction of cyclohexene with the dihydride complex [Rh(PPh$_3$)$_3$H$_2$Cl] was conducted,[20,78] and this study was first discussed in Chapter 9. This study showed that the reaction of cyclohexene with [Rh(PPh$_3$)$_3$H$_2$Cl] obeys the rate law shown in Equation 15.18. The reciprocal of Equation 15.18 predicts that plots of $1/k_{obs}$ versus [L]/[C=C] should be linear, with a slope that corresponds to $1/k_5k_6$ and an intercept that corresponds to $1/k_6$. The migratory insertion step, k_6, proved to be the slowest step of the second stage of the catalytic cycle and the overall catalytic hydrogenation mechanism. A small isotope effect [k_{obs}(H)/k_{obs}(D) = 1.15] was observed for the reaction between cyclohexene and [Rh(PPh$_3$)$_3$H$_2$Cl]. The subsequent product-forming step (the reductive elimination from the alkylrhodium hydride complex—reaction 8 in Scheme 15.6) is sufficiently rapid that the reverse of reaction 6 is kinetically insignificant. This result accounts for the lack of isotopic exchange between D$_2$ and the olefin reagent. The open coordination sites in the unsaturated catalytic intermediates in Scheme 15.6 may be occupied by the polar co-solvent, as shown. With the exception of the degree to which solvent binds, the stoichiometry of the reactive intermediates is based firmly on kinetic and equilibrium studies. However, the stereochemistry of the intermediates is not determined by these experiments. Based on NMR spin-transfer studies and the relative reactivities of related complexes, Brown[75,79] has advanced arguments that the phosphines are cis in the reactive forms of the olefin dihydride and alkyl hydride intermediates of Scheme 15.6.

$$-d[RhH_2ClL_3]/dt = k_{obs}\,[RhH_2ClL_3]$$

$$\text{in which}\quad k_{obs} = \frac{k_5k_6[C=C]}{[L] + k_5[C=C]} \quad \text{and} \quad \frac{1}{k_{obs}} = \left(\frac{1}{k_5k_6}\right)\left(\frac{[L]}{[C=C]}\right) + \frac{1}{k_6} \tag{15.18}$$

The rate law in Equation 15.18 quantitatively accounts for the observed rate behavior for the hydrogenation of cyclohexene. Since the migratory insertion reaction (k_6) is the slowest step in the catalytic cycle, under conditions of constant H$_2$ pressure, high olefin concentrations (> 1 M), and no added phosphine, the rate reaches a limiting value determined by the rate of the migratory insertion step.

Although the mechanism shown in Scheme 15.6 is well established for the hydrogenation of cyclohexene by Wilkinson's catalyst, slight modifications to the substrate can alter the dominant mechanism. For example, styrene (which binds with a higher equilibrium constant than cyclohexene) reacts with rate behavior that requires a parallel path involving an intermediate containing two bound styrenes.[80] Wilkinson's studies also indicate that the hydrogenation of other strongly binding olefins, such as ethylene or 1,3-butadiene, occur with different kinetic behavior, suggesting that binding of these more strongly coordinating olefins occurs *prior* to addition of H$_2$.

15.5.2.1.2. Hydrogenation by Cationic Rhodium Catalysts

15.5.2.1.2.1. Cationic Rhodium Complexes Containing Aromatic Phosphines

The cationic rhodium complexes described earlier in this chapter catalyze the hydrogenation of olefins through a manifold that is different from that followed by neutral rhodium complexes. The synthetic studies of Osborn show that complexes of the formula [RhL$_2$S$_2$]$^+$ undergo oxidative addition of hydrogen to form octahedral dihydride complexes,[46] but kinetic studies by Halpern on the catalytic reactions indicate that the olefin binds *before* hydrogen adds.

The mechanism deduced from the kinetic behavior of the hydrogenation of olefins by the DIPHOS catalyst[66,68,69] is outlined in Scheme 15.7. This mechanism contrasts with that described in Scheme 15.6 for reactions catalyzed by the neutral Wilkinson's catalyst. The dominant mechanism for the hydrogenation of olefins by $[Rh(DIPHOS)S_2]^+$ follows the olefin-first pathway of Scheme 15.2. This pathway is followed for two reasons. First, the oxidative addition of hydrogen to the cationic species is slower and less favored thermodynamically than it is to the neutral species. Second, the olefin can more readily displace the solvent that is weakly bound to the cationic complexes than it can displace the third phosphine bound to the neutral catalysts. In most cases, the cationic catalysts are used for the hydrogenation of olefins that chelate the metal through the olefin unit and a second functional group. Thus, hydrogenation of such olefins by these complexes occurs by 1) initial displacement of the solvent by olefin, 2) oxidative addition of H_2, 3) insertion of the olefin into one of the two rhodium–hydride bonds, and 4) reductive elimination of the product from the resulting alkylrhodium hydride complex. Three of the complexes that were deduced as intermediates from the kinetic data—namely, [Rh(DIPHOS)S_2]$^+$, [Rh(DIPHOS)(olefin)]$^+$, and the product from olefin insertion into the hydride—have been detected and characterized spectroscopically.[66,68,81] The observation of these species is consistent with the assertion that true intermediates in catalytic processes can be observed directly. Contrary to what is often expressed, such true intermediates were even isolated by Halpern's own group when studying olefin hydrogenation.

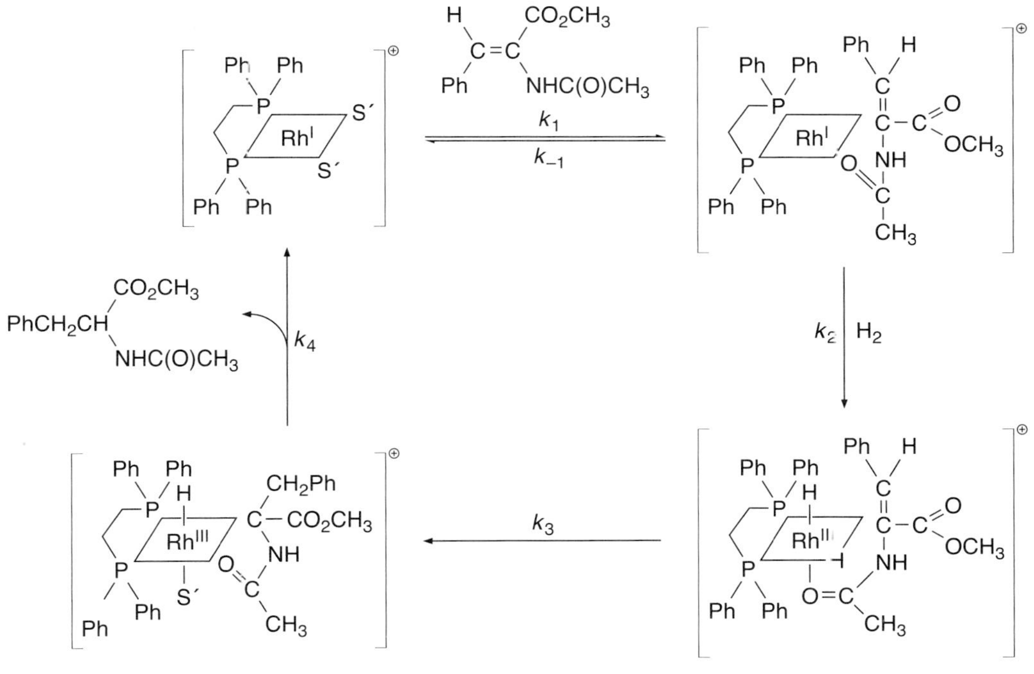

Scheme 15.7

$$[Rh(DIPHOS)]^+ \; + \; \text{unsat.} \; \underset{}{\overset{K_{eq}}{\rightleftharpoons}} \; [Rh(DIPHOS)(\text{unsat.})]^+$$

Unsaturated substrate	$K_{eq}(M^{-1})$
Benzene	18
Toluene	97
o-, m-, or p-Xylene	500
1-Hexene	2
Styrene	20
Methyl acrylate	3

(15.19)

In this olefin-first mechanism, the olefin complex is in equilibrium with the solvated complex. The equilibrium constants for the binding of solvent and olefins are summarized in Equation 15.19. These data show that aromatic compounds bind to [Rh(DIPHOS)S$_2$]$^+$ as well as simple alkenes. Because the arenes are not reduced, they are competitive inhibitors of olefin hydrogenation that tie up the catalyst in an inactive form. The origin of the large association constant between [Rh(DIPHOS)S$_2$]$^+$ and methyl (Z)-α-acetamidocinnamate (MAC) was revealed by the crystal structure of the adduct, [Rh(DIPHOS)(MAC)]$^+$, which is schematically depicted in Scheme 15.7.[81] The MAC substrate chelates to the rhodium atom through the carbonyl oxygen of the amide group, as well as through normal (η^2) coordination to the olefinic unit. The solution structure of [Rh(DIPHOS)(MAC)]$^+$, established by ^{31}P, ^{13}C, and ^{1}H NMR spectroscopy, appears to be the same as that observed in the solid state.[66,82,83]

The alkyl hydride complex resulting from the migratory insertion of the olefin into the rhodium hydride is the dominant species in solution at lower temperatures (−40 °C).[66] Characterization of this complex by multinuclear NMR spectroscopy showed the regiochemistry of the migratory insertion, which generates a five-membered metallacycle involving the partially reduced substrate. To form this ring, the hydride has become attached to the β-carbon, and the rhodium to the α-carbon of the olefin. The olefin dihydride complex in Scheme 15.7 has not been directly characterized, but has been characterized in related rhodium systems.

Thus, the overall catalytic cycle for the reduction of (Z)-α-acetamidocinnamate (MAC) involves an initial equilibrium between the combination of MAC and [Rh(DIPHOS)(S)$_2$]$^+$ and the olefin complex [Rh(DIPHOS)(MAC)]$^+$ in Scheme 15.7. This equilibrium is established faster than the subsequent oxidative addition of H$_2$ and lies well toward the olefin adduct [Rh(DIPHOS)(MAC)]$^+$, even at moderate concentrations of MAC. At 25 °C, and 1 atm of hydrogen, the oxidative addition of H$_2$ to the olefin adduct [Rh(DIPHOS)(MAC)]$^+$ was deduced from the kinetic data to be rate limiting in the overall hydrogenation cycle. This oxidative addition is followed by migratory insertion of the coordinated olefin to form the alkyl hydride complex that was observed at low temperature. At lower temperatures (−40 °C), the reductive elimination step (k_4 in Scheme 15.7) is turnover limiting, allowing direct observation of the alkyl hydride species at this temperature.[66,68,81] The turnover-limiting step changes because the enthalpy of the oxidative addition step, k_2, is lower (6 kcal mol^{-1}) than that of the reductive elimination step, k_4 (17 kcal mol^{-1}). Therefore, the reductive elimination step depends more strongly on temperature and is more significantly retarded at lower temperature. Reductive elimination then regenerates the starting [RhL$_2$S$_2$]$^+$ species. These relative rates contrast those for hydrogenation by Wilkinson's catalyst, in which the dihydride complex accumulated and the insertion of olefin was rate limiting.

15.5.2.1.2.2. Cationic Rhodium Catalysts Containing Alkylphosphines

As noted briefly in Section 15.1 and as is described in Section 15.7 on enantioselective catalysis, a major advance in the asymmetric hydrogenation of olefins was the development of chiral, non-racemic alkylphosphine ligands. One principle behind the design of these ligands was an expectation that the stronger electron-donating properties of the alkylphosphine ligands versus the arylphosphine ligands would increase the rates of catalytic hydrogenation by increasing the rate of the turnover-limiting oxidative addition of H$_2$ to the cationic olefin complexes in the olefin-first mechanism. (Note that this electronic property of the phosphine would be expected to retard the rate of reductive elimination to form the product in the last step, but this reaction has not been found to be turnover limiting in rhodium- and ruthenium-catalyzed hydrogenations under operating conditions.) In fact, catalysts containing these alkylphosphine ligands are sufficiently electron rich, despite being cationic, that they appear to react by a *hydrogen*-first mechanism, rather than the *alkene*-first mechanism.

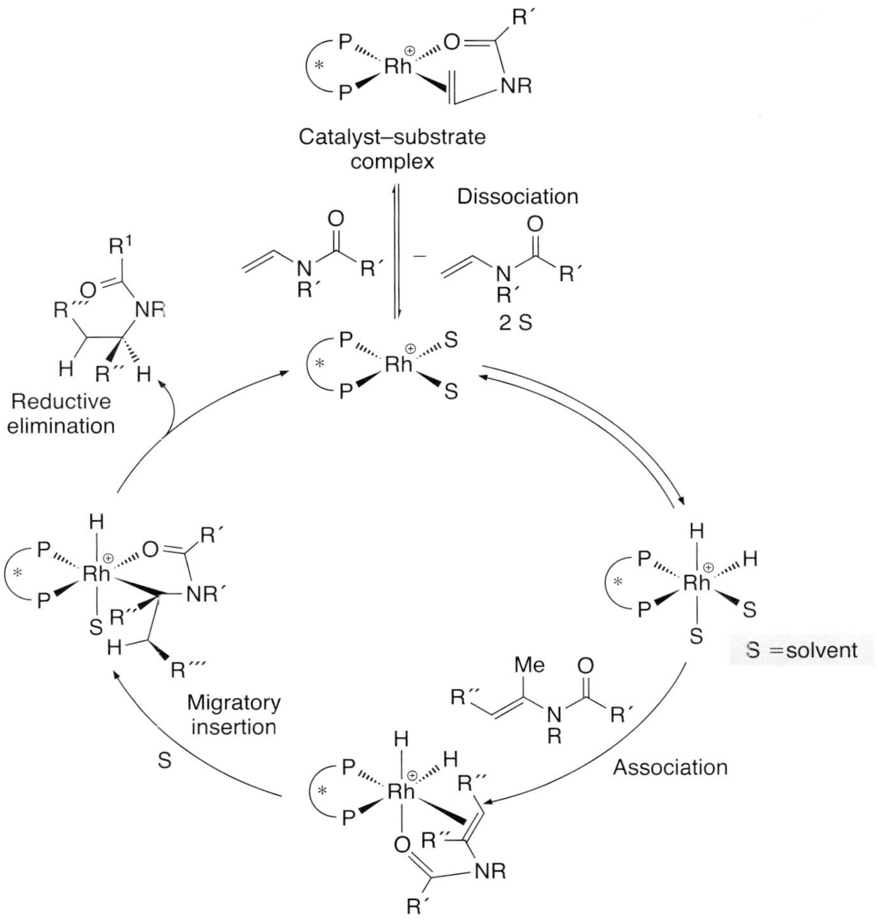

Scheme 15.8

Some of the most detailed studies on this process were conducted by Gridnev and Imamoto on reactions catalyzed by bidentate, *P*-chiral ligands containing alkyl substituents on phosphorus.[70,71,84-86] Two pieces of data together imply that this catalyst reacts by the hydrogen-first mechanism in Scheme 15.8, instead of the alkene-first mechanism. First, the solvated bisphosphine dihydride complex $[Rh(L_2)(S)_2(H)_2]^+$ is formed at $-90\ ^\circ C$, and this complex reacts with MAC at this temperature to form $[Rh(L_2)(MAC)(H)_2]^+$. Thus, this complex is capable of reacting with H_2 prior to the coordination of olefin. Second, the enantioselectivity of the product of warming a related complex formed from an α,β-unsaturated phosphonic acid ester, $\{Rh(L_2)[RCH=CRP(O)(OR)_2](H)_2\}^+$, is higher and closer to that of the catalytic reaction than the product formed from the addition of hydrogen to $\{Rh(L_2)[RCH=CRP(O)(OR)_2]\}^+$. The latter reaction, which occurs in low yield, mimics the portion of the olefin-first mechanism that sets the product stereochemistry. Thus, the hydrogen-first mechanism appears to be the dominant pathway for the reaction of this catalyst, although it might not be the exclusive one. Further studies would be needed to determine if this observation applies to other chiral, cationic rhodium systems ligated by alkylphosphines discussed later in this chapter.

15.5.2.1.3. Cationic Iridium Catalysts Containing Alkylphosphines

Kinetic studies on the hydrogenations of simple alkenes by Crabtree's catalyst have been challenging to conduct because the rates are sufficiently fast that mass transport of hydrogen can be rate limiting. However, many studies on the reactions of Crabtree's catalyst with hydrogen and with alkenes separately have been conducted. Two features of this system are distinct from those of the rhodium systems described so far in this chapter. First, this catalyst readily deactivates by the formation of iridium clusters, and such complexes have been isolated after the oxidative addition of H_2 to Crabtree's catalyst in the absence of alkene.[87–89] Second, olefin dihydride complexes have been observed directly from the oxidative addition of hydrogen.[22] Although debate continues, some researchers have acquired computational data in support of a mechanism for hydrogenation by a cationic iridium complex through Ir(III) and Ir(V) oxidation states,[90,91] rather than Ir(I) and Ir(III).[92] These two mechanistic possibilities are shown in parts A and B of Scheme 15.9.

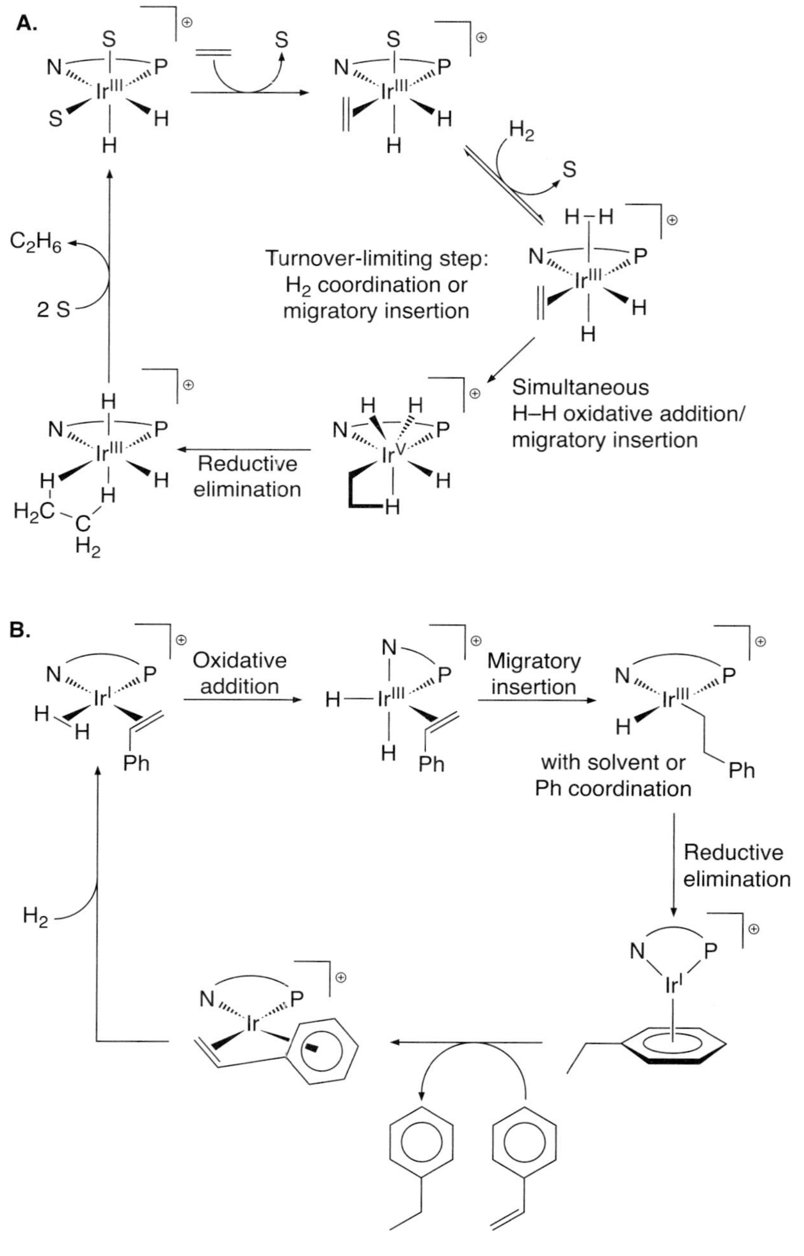

Scheme 15.9

Crabtree has studied the reactions of the cationic complexes [Ir(diene)LL']+ with hydrogen and Lewis bases. First, these complexes coordinate many Lewis basic groups, including arenes. Thus, reactions catalyzed by these complexes are best conducted in weakly coordinating solvents, such as methylene chloride. Second, reactions of these complexes with hydrogen (Equations 15.20 and 15.21) form the bridging polyhydrides when both ligands are phosphines and the trinuclear cluster containing a μ^3-hydride when one ligand is PCy_3 and one is pyridine, as in the most-used Crabtree catalyst. These polynuclear complexes are inactive catalysts. Third, Crabtree has shown that the reactions of the cationic complex [Ir(COD)(PPh$_3$)$_2$]+ with hydrogen forms the corresponding dihydride complex, and that the cis dihydride complex reduces the coordinated ligand (Equation 15.22) while the trans isomer is stable.[93] As noted in Chapter 2, dihydrogen binds to cationic complexes, and iridium–dihydrogen complexes of the type [IrL$_2$H$_2$(H$_2$)$_2$]+ form at low temperatures.[94] Finally, Crabtree has shown that olefins containing pendant coordinating groups form chelates, such as the norbornenol adduct in Equation 15.23.[62,95] The formation of such chelates accounts for the directing effect noted in Section 15.4 on hydrogenations of enols catalyzed by this system. These studies have led to the conclusion that the resting states (Chapter 14) of the iridium in the hydrogenations of alkenes catalyzed by cationic species are [Ir(PR$_3$)LL'L"H$_2$] complexes in which L is a second phosphine or a pyridine, and L' and L" are olefin, solvent, or coordinated dihydrogen, depending on the conditions.

$$2 \text{ [Ir(COD)L}_2\text{]BF}_4 \xrightarrow[\;7\text{ H}_2\;]{-2\text{ C}_8\text{H}_{16}} \cdots \text{BF}_4 + \text{HBF}_4 \qquad (15.20)$$

$$3 \text{ [Ir(COD)LL']BF}_4 \xrightarrow[\;10\text{ H}_2\;]{-3\text{ C}_8\text{H}_{16}} \cdots \text{BF}_4 + \text{HBF}_4 \qquad (15.21)$$

$$(15.22)$$

$$(15.23)$$

More recently, chiral, non-racemic versions of these catalysts have been studied.[36,96,97] Kinetic experiments on reactions catalyzed by iridium complexes of phosphino–oxazoline ligands showed that the reaction is first order in dihydrogen and zero order in substrate.[91] Because hydrogen adds rapidly at low temperatures to the cationic complexes [IrLL'(olefin)] and the resting state of the catalyst contains dihydrogen, it seems likely that the transition state of the turnover-limiting step contains at least two equivalents of dihydrogen in the form of hydrides and coordinated H$_2$. These data and DFT calculations led to the conclusion that alkene insertion occurs into a hydride of a dihydride dihydrogen complex, and that the resulting complex is an Ir(V) alkyl trihydride cation. Reductive elimination of the product would then regenerate the starting cationic Ir(III) dihydride complex. Further studies are needed, however, to define clearly the mechanism of hydrogenation by iridium complexes containing phosphine and nitrogen donors.

15.5.2.2. Catalysts that React by Insertions of Olefins into Monohydride Intermediates

15.5.2.2.1. Hydrogenation by Rhodium Carbonyl Hydride Catalysts

One of the classic catalysts that reacts through a pathway involving the insertion of olefins into monohydride intermediates is $[Rh(H)(PPh_3)_3(CO)]$. The reactions of terminal olefins occur under mild conditions (Equation 15.24). The pronounced substrate selectivity of this catalyst is illustrated by its lack of reactivity with cyclohexene.[98] This complex catalyzes the isomerization of internal olefins, but it does not catalyze their hydrogenation. Other unsaturated functionalities, such as aldehydes, nitriles, and esters, are stable to this catalyst under these mild conditions.

$$RCH=CH_2 \xrightarrow[\text{HRh(CO)(PPh}_3)_3]{25\ °C,\ <1\ atm,\ H_2} RCH_2CH_3 \qquad (15.24)$$

A mechanism for the hydrogenation of olefins catalyzed by this complex is shown in Scheme 15.10.[99,100] This process falls into the category of reactions that follow a "monohydride mechanism." This mechanism begins with the coordination and insertion of olefin into the rhodium hydride after reversible dissociation of PPh_3. The resulting alkyl complex undergoes hydrogenolysis by a sequence of oxidative addition of H_2, followed by reductive elimination of alkane to regenerate the starting hydride complex. This mechanism is supported by an inverse dependence of the rate on the concentration of added PPh_3 and the known insertion of olefin into this hydride. This complex has been studied most extensively as a catalyst for hydroformylation (discussed in Chapter 16). Many rhodium and cobalt complexes that catalyze olefin hydroformylation also catalyze olefin hydrogenation through monohydride intermediates. The hydrogenation of aldehydes to alcohols catalyzed by $[Rh(H)(PPh_3)_3(CO)]$ is slower than the hydroformylation of alkenes, but it is, nevertheless, a competing side reaction under the high temperatures and pressures of hydroformylation processes.

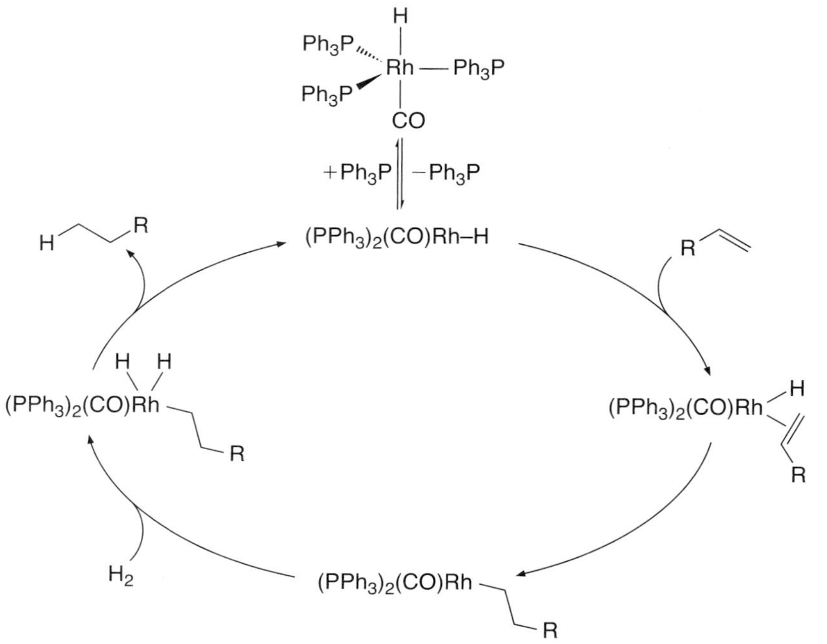

Scheme 15.10

15.5.2.2.2. Hydrogenation by Ruthenium Catalysts

Ruthenium catalysts are now widely used for olefin hydrogenation, and many examples of enantioselective ruthenium-catalyzed hydrogenation are discussed in Section 15.7. Here, before addressing the issues of stereoselectivity, the elementary steps of ruthenium-catalyzed hydrogenation are discussed. These catalysts react through monohydride species containing a second anionic ligand.

15.5.2.2.2.1. Mechanism of Hydrogenation by Ru(PPh₃)₃H(Cl)

Like [Rh(PPh$_3$)$_3$(CO)(H)], the ruthenium monohydride complex [Ru(PPh$_3$)$_3$H(Cl)] catalyzes the hydrogenation of olefins with high selectively for terminal olefins over internal olefins. [Ru(PPh$_3$)$_3$H(Cl)] enters the catalytic cycle by dissociation of a phosphine. Although [Ru(PPh$_3$)$_3$H(Cl)] is an unsaturated, 16-electron complex (stabilized by an agostic interaction with an aryl C–H bond), the rate of the catalytic process is inhibited by added phosphine, and this inhibition by phosphine shows that Ru(PPh$_3$)$_3$H(Cl) first dissociates PPh$_3$.[101] Thus, the hydrogenation of olefins catalyzed by this complex follows the mechanism in Scheme 15.11, which is similar to the mechanism for hydrogenation by [Rh(H)(PPh$_3$)$_3$(CO)]. Olefin insertion into the ruthenium hydride initiated by reversible dissociation of PPh$_3$ is followed by hydrogenolysis of the resulting alkyl. This hydrogenolysis step could involve the oxidative addition of hydrogen to form a ruthenium(IV) intermediate. However, it more likely occurs by formation of a dihydrogen complex and proton transfer from the acidic dihydrogen ligand to the alkyl group.

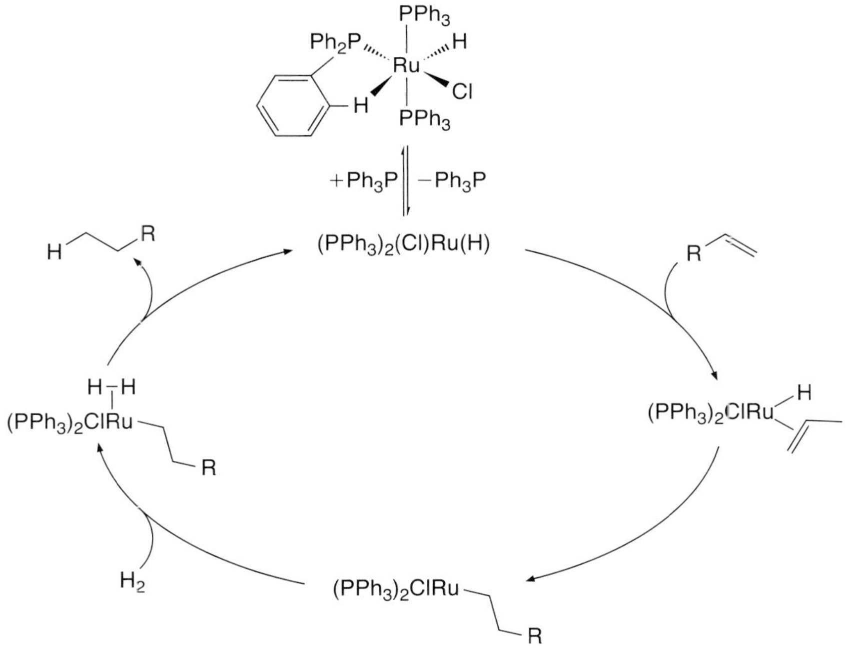

Scheme 15.11

15.5.2.2.3. Mechanism of Hydrogenation of Olefins and Ketones by RuL₂(κ²-OAc)₂ and [RuL₂Cl₂]₂

Ruthenium catalysts became widely used for hydrogenation when Noyori showed that complexes of the formula RuL$_2$(κ²-OAc)$_2$, in which L$_2$ = BINAP, catalyze the asymmetric hydrogenation of a variety of functionalized alkenes, and many related complexes have now been studied as catalysts for asymmetric hydrogenation. Halpern conducted detailed studies on the mechanism of the hydrogenation of acrylic acid derivatives by this complex.[102] The currently accepted mechanism for this process was outlined in Scheme 15.3 and is shown in more detail with deuterium labeling in Scheme 15.12. This pathway involves only ruthenium(II) species. Halpern proposed that the reaction with hydrogen generates an anionic ruthenium hydride, but Morris has modified this proposal in a review article to include neutral ruthenium complexes.[103] By the modified pathway, the diacetate precatalyst converts to the monohydride complex $[RuL_2(S)(\kappa^2\text{-OAc})(H)]$ by coordination of H_2, deprotonation of the coordinated H_2 by the acetate ligand, and displacement of the resulting carboxylate by methanol. Coordination of the olefin substrate, followed by insertion of the olefin into the hydride, forms a ruthenium–alkyl species. This alkyl species converts to the starting acetate complex by reaction with acetic acid. This mechanism for release of the product by protonolysis of the Ru–C bond was demonstrated by observing deuterium in the β-position of the carboxylic acid when the reaction was run in deuterated methanol solvent (the deuterium in the methanol exchanges with the acidic proton of acetic acid to supply the deuterium).

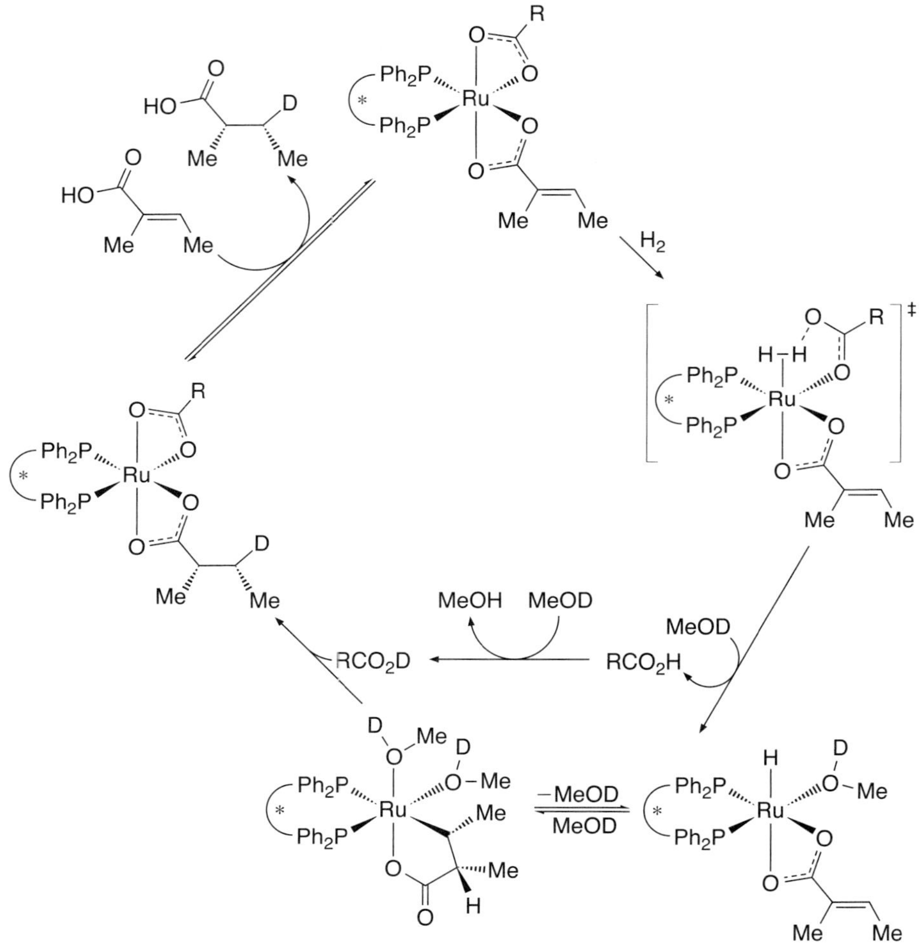

Scheme 15.12

As noted in the perspective on hydrogenation (Section 15.2), the ruthenium diacetate complexes were reactive toward the hydrogenation of olefins, but the combination of the corresponding dichloride $[RuL_2Cl_2]_2$, base, and alcohol (as solvent or additive) generated a practical catalyst for the hydrogenation of functionalized ketones. The currently accepted mechanism for the hydrogenation of functionalized ketones by this catalyst is shown in Scheme 15.13.[104]

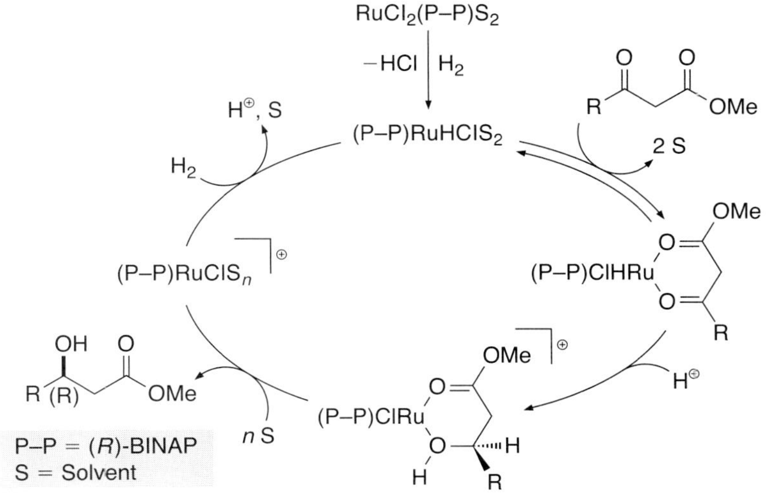

Scheme 15.13

Fewer detailed studies have been conducted on the hydrogenation of ketones with these systems than on the hydrogenation of olefins with the diacetate analogs, but several crucial pieces of data have been obtained. The starting dichloride was shown to convert to the corresponding solvated hydrido chloride complex.[105] This complex binds the β-keto ester and undergoes insertion of the keto functionality into the metal hydride. The resulting alkoxo intermediate has been isolated from reactions of a vicinal tricarbonyl substrate,[106] and an accelerating effect of acid was noted.[107,108] The acceleration by acid is thought to arise from an acceleration of the protonolysis of the alkoxo insertion product. A combination of displacement of the substrate with solvent and heterolytic cleavage of hydrogen then regenerates the starting solvated hydrido chloride complex. Like the hydrogenation of unsaturated carboxylic acids, this cycle is thought to occur with exclusively Ru(II) complexes.

15.5.2.2.4. Monohydride Catalysts Reacting Through Radical Pathways

A final set of catalysts that has largely historical importance consists of monohydride complexes that react by radical pathways. Insertion of the olefin into the hydride occurs through radical intermediates. Many of these catalysts have been shown to reduce dienes, acrylic acid derivatives, and fumarates. As discussed in Chapter 9, migratory insertion requires an open coordination site cis to the hydride for initial binding of the olefin. Thus, catalysts based on metal carbonyl compounds that dissociate CO slowly, or porphyrin ligands that occupy the sites cis to a hydride, react through radical pathways or direct transfer of a true "hydride." The basic mechanism for such direct formal insertions by radical processes is shown in Scheme 15.14.

$$L_nM\text{–}H + PhCH{=}CH_2 \longrightarrow Ph\dot{C}H\text{–}CH_3 + L_nM\cdot$$

$$Ph\dot{C}H\text{–}CH_3 + L_nM\cdot \longrightarrow PhCH(ML_n)\text{–}CH_3$$

Scheme 15.14

The stability of a conjugated radical accounts for the greater reactivity of vinylarenes and conjugated dienes versus that of unconjugated monoenes. A well-documented example of hydrogen atom addition to olefins by this radical pathway is the hydrogenation of α-methyl styrene by $HMn(CO)_5$; Halpern[109] used CIDNP (chemically induced dynamic nuclear polarization) to establish this radical reaction path.

15.5.2.2.5. d^0-Monohydride Catalysts Reacting Through σ-Bond Metathesis Pathways

The final system discussed in this section on hydrogenations through monohydride intermediates involves catalysts that cannot undergo the oxidative addition and reductive elimination steps that are characteristic of late transition metal hydrogenation catalysts. The mechanism of the hydrogenations of alkenes by d^0 complexes can be summarized by the cycle depicted in Scheme 15.5 in Section 15.5.1. The catalyst is typically generated by the hydrogenolysis of a lanthanocene alkyl to generate a lanthanocene hydride. This hydrogenolysis occurs by the σ-bond metathesis process discussed in Chapter 6. This hydride complex then reacts with an alkene to form a lanthanide alkyl by a migratory insertion pathway. Hydrogenolysis of this alkyl by σ-bond metathesis regenerates the starting alkyl complex. For hydrogenations of dienes, the insertion process generates a metal–allyl intermediate that undergoes hydrogenolysis to form the reduced product and regenerate the starting hydride.

15.5.2.3. Outer-Sphere Mechanism for the Hydrogenation of Ketones and Imines

As noted in Section 15.2 and described in more detail in the sections on the asymmetric hydrogenation of ketones, remarkably active catalysts for the hydrogenation of unfunctionalized ketones have been developed. These catalysts contain both phosphine and primary amine or sulfonamide ligands. The details of the mechanism for the hydrogenation of ketones and imines by these catalysts continues to emerge; however, the current data appear to demonstrate clearly that these catalysts react by a pathway that does not involve the oxidative addition, reductive elimination, or migratory insertion that is typical for most hydrogenation catalysts containing late transition metals. Instead, these catalysts react by heterolytic cleavage of hydrogen and by simultaneous interaction of a ketone with a proton and a hydride in the transition state for hydrogen transfer. Noyori has called this mechanism "metal–ligand bifunctional catalysis."

One of the first systems for ketone hydrogenation that reacts by an outer-sphere pathway is a ruthenium catalyst first reported by Shvo. This catalyst contains a hydride ligand and a hydroxy group on the cyclopentadienyl ligand. This catalyst was studied extensively by Bäckvall for the racemization of alcohols in combination with lipases for a dynamic kinetic resolution of alcohols.[110–115] A related system has been studied for the same processes by Kim.[116, 117] Chiral, non-racemic catalysts containing amine ligands were discovered by Noyori,[31, 32, 118–121] and many analogs have been studied by others.[122–132] Many of these complexes containing phosphine ligands catalyze the hydrogenation of ketones and imines with high activity and selectivity, and many complexes containing a diamine, aminosulfonamide, or aminoalcohol as the only chiral ligand, catalyze the transfer hydrogenation of ketones and imines with high enantioselectivities.

A mechanism for reactions catalyzed by the $[Ru(BINAP)(diamine)(H)_2]$ complex is shown in Scheme 15.15.[133–136] Related mechanisms have been proposed for transfer hydrogenations catalyzed by $[Ru(arene)(NH_2CHRCHRNHTs)H]$ complexes and by Shvo's catalyst, except for the portion of the catalytic cycle involving the regeneration of the dihydride. The mechanism deduced by many experiments and calculations involves reaction of the hydrido chloride complex with the ketone by the six-membered ring transition state shown in Scheme 15.15, in which the hydride and proton are transferred simultaneously and the alcohol is hydrogen bonded to the resulting amide ligand and then released into solution.[112,133,136–148] The simultaneous transfer of these two groups was supported by the observation of a kinetic isotope effect for both the hydride and N–H positions and an overall isotope effect that is approximately the product of the two isotope effects for the reaction of [(arene)Ru(NH$_2$CHRCHRNHTs)H] with acetone[143] and by many calculations of the mechanism.[146,147,149–151] Moreover, it is supported by the lack of catalytic activity by complexes that do not contain a proton on the coordinated nitrogen.

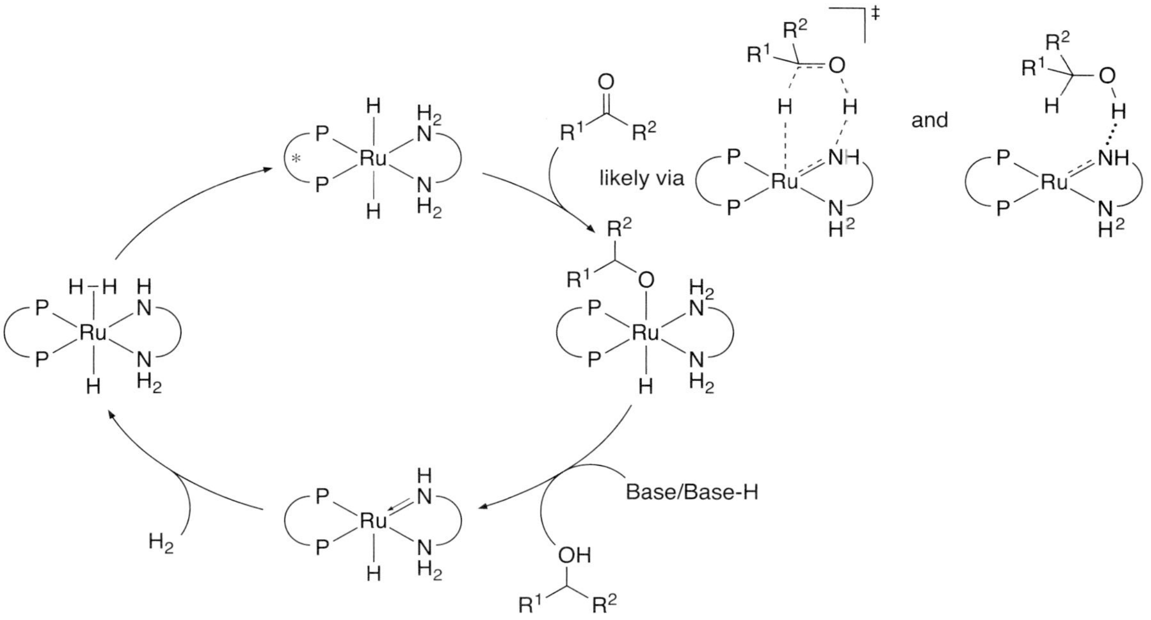

Scheme 15.15

The immediate product from the reaction of the hydride complex with ketone has been a subject of debate. Some have proposed that an alkoxide complex is formed and that proton transfer between the coordinated amine and the alkoxide then forms the alcohol that is ultimately released.[112] Others have supported a direct transfer to form free alcohol (or amine) and then coordination of this species to the open site.[142,144,145] To accommodate the isotope effect data and the absence of an open coordination site for coordination of the ketone, the formation of the alkoxide from the hydride has been proposed to occur by hydride transfer assisted by hydrogen bonding of the amine in the case of the reactions with [Ru(BINAP)(diamine)(H)$_2$], or by ring slip to allow coordination of the ketone (or imine) in the case of the reactions with the Shvo catalyst.[112,152,153]

A recent study by Bergens provided important data on this issue.[133] This study showed that the reaction of a ketone with [Ru(BINAP)(diamine)(H)$_2$] forms an alkoxide complex at low temperature. In the presence of a different alcohol than that formed by the reduction, the initial alkoxide complex is the one that results from the reduction of the ketone. These data imply that the transfer of the hydride occurs by a mechanism that forms alkoxide with assistance by hydrogen bonding of the ketone to the amine, without coordination of the ketone to the metal, and without releasing free alcohol.

These data can be accommodated by a simultaneous addition of the hydride and proton to form an alcohol that is initially hydrogen bonded to the amine and forms the alkoxide without dissociation of free alcohol or by a transition state with a structure resembling those proposed for the direct, simultaneous transfer of the hydride and proton, but leading to the alkoxide complex directly. After this transfer of the hydride to form an alkoxide, a base-assisted elimination of the alkoxide to form the amide complex is proposed to occur. This step parallels the classic conjugate-base mechanism for dissociative ligand substitution discussed in Chapter 5.

The resulting unsaturated amido complex then reacts with H_2 in the hydrogenation reactions to regenerate the starting monohydride species. This process is thought to occur by heterolytic cleavage of hydrogen via initial formation of a dihydrogen complex, followed by deprotonation of the coordinated dihydrogen by the basic amido ligand to regenerate the hydrido amine complex.[135,136] This species reacts with alcohol or formic acid to regenerate the hydride in transfer hydrogenation reactions. The mechanism of the reaction with alcohol is the microscopic reverse of the reaction of the ketone. The mechanism of the reaction with formic acid occurs by protonation of the amide to generate a formate ligand that de-inserts CO_2 (Chapters 9 and 10).[154]

15.5.2.4. Ionic Hydrogenations

A pathway for hydrogenation of ketones that is related to the simultaneous transfer of a hydride and a proton from a hydride and acidic N–H bond is the stepwise addition of a proton and a hydride to a ketone from coordinated H_2. By this mechanism, the hydrogenation of ketones can occur by a dihydrogen complex that lacks an open coordination site. This pathway for hydrogenation has been termed "ionic hydrogenation," and a general scheme for this process is shown in Scheme 15.16.[155] By this pathway, a cationic dihydrogen complex undergoes proton transfer to the oxygen of a ketone, and the resulting oxonium ion undergoes rapid abstraction of the hydride remaining at the metal. Dissociation of the resulting alcohol and coordination of H_2 regenerates the acidic dihydrogen complex. Catalysts that react by this pathway remain under development, but one example of a complex that catalyzes the hydrogenation of ketones through this pathway is $CpW(CO)_2(PPh_3)$ $(O=CEt_2)^+$.[156,157]

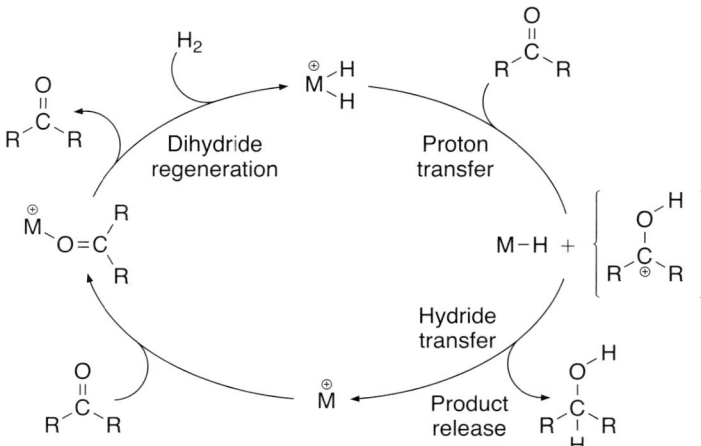

Scheme 15.16

15.6. Ligands Used for Asymmetric Hydrogenation

One of the most significant developments in organotransition metal chemistry is the discovery of soluble complexes that catalyze the asymmetric hydrogenation of prochiral olefins, ketones, and imines.[27,33,158–164] In many examples, optical yields have been achieved that approach 100% enantiomeric excess (ee). This result rivals the stereoselectivity of enzymes, and the scope of homogeneous asymmetric hydrogenations occurring with high enantiomeric excess vastly exceeds that achieved with heterogeneous catalysts.[165]

As discussed later in this chapter, olefins possessing a diverse array of substitution patterns now undergo hydrogenation with fast rates, high turnover numbers, and high enantioselectivities. Two decades ago, such a diverse array of enantioselective hydrogenations was hard to fathom based on the single success with the asymmetric reduction of dehydro amino acids and the unexpected lack of a lock-and-key model for enantioselectivity (vide infra). The authors of the predecessor to this book, *Principles and Applications of Organotransition Metal Chemistry*, even stated, "Only a restricted class of olefinic substrates like MAC are likely to afford high stereoselectivity, and a systematic search for other asymmetric hydrogenation catalysts seems hopeless."

Sections 15.6 and 15.7 present a diverse array of catalysts and asymmetric hydrogenation processes. These data illustrate the tremendous success in catalyst development, although it is true that the identification of catalysts or new hydrogenation processes involves a combination of empiricism and design. Section 15.6 describes the major classes of ligands that have been used to generate catalysts based on rhodium, ruthenium, and iridium for the asymmetric hydrogenation of olefins, ketones, and imines. Section 15.7 shows how these catalysts have been used in these different classes of hydrogenation.

15.6.1. Aromatic Bisphosphines

15.6.1.1. Aromatic Bisphosphines Containing Backbone Chirality

15.6.1.1.1. Ligands Containing Axial Chiral Backbones

As stated in Section 15.1, the first enantioselective hydrogenations were achieved with *P*-chiral bisphosphines. However, the scope of hydrogenation was dramatically increased by the advent of aromatic phosphines containing axial chiral backbones. A selection of such ligands is shown in Figure 15.2. BINAP, the first example of such a ligand, was reported by Noyori,[28] and one synthesis of this ligand is shown in Equation 15.25. The synthesis of BINAP was challenging at first. The conversion of the 2,2'-bi-1,1'-naphthol to the phosphine was not straightforward. Originally, this diol was converted to the dibromide at high temperature. The racemic dibromide was then converted to the bisphosphine oxide, which was resolved and reduced to form enantiopure BINAP.[166] The earliest synthesis involved resolution of racemic BINAP by making diastereomeric palladium complexes.[28] Related routes involving resolution of the bisphophshine oxide are used to prepare many of the derivatives in Figure 15.2. An improved route to BINAP from the resolved binaphthol has now been developed. In this route, the resolved binaphthol[167,168] is converted to the bistriflate and then coupled with a secondary phosphine to form the optically active ligand.[169] This route is shown in Equation 15.26.

(S)-BINAP: R = Ph
(S)-TolBINAP: R = 4-MeC$_6$H$_4$
(S)-XylBINAP: R = 3,5-(Me)$_2$C$_6$H$_3$

(S)-BICHEP: R^1 = Cy; R^2 = CH$_3$
(S)-BIPHEMP: R^1 = Ph; R^2 = CH$_3$
(S)-BIPHEP: R^1 = Ph; R^2 = OCH$_3$

(S)-bisbenzodioxanPhos (SYNPHOS)

(S)-SEGPHOS

(S)-Difluorphos

(S)-H8-BINAP

(S)-P–Phos: Ar = Ph
(S)-Tol–P–Phos: Ar = 4-MeC$_6$H$_4$
(S)-Xyl–P–Phos: Ar = 3,5-(Me)$_2$C$_6$H$_3$

(S)-C$_n$-TunePhos
n = 1–6

Figure 15.2.
A selection of axial chiral bisphosphine ligands. The original BINAP ligand is shown at the top left of the figure.

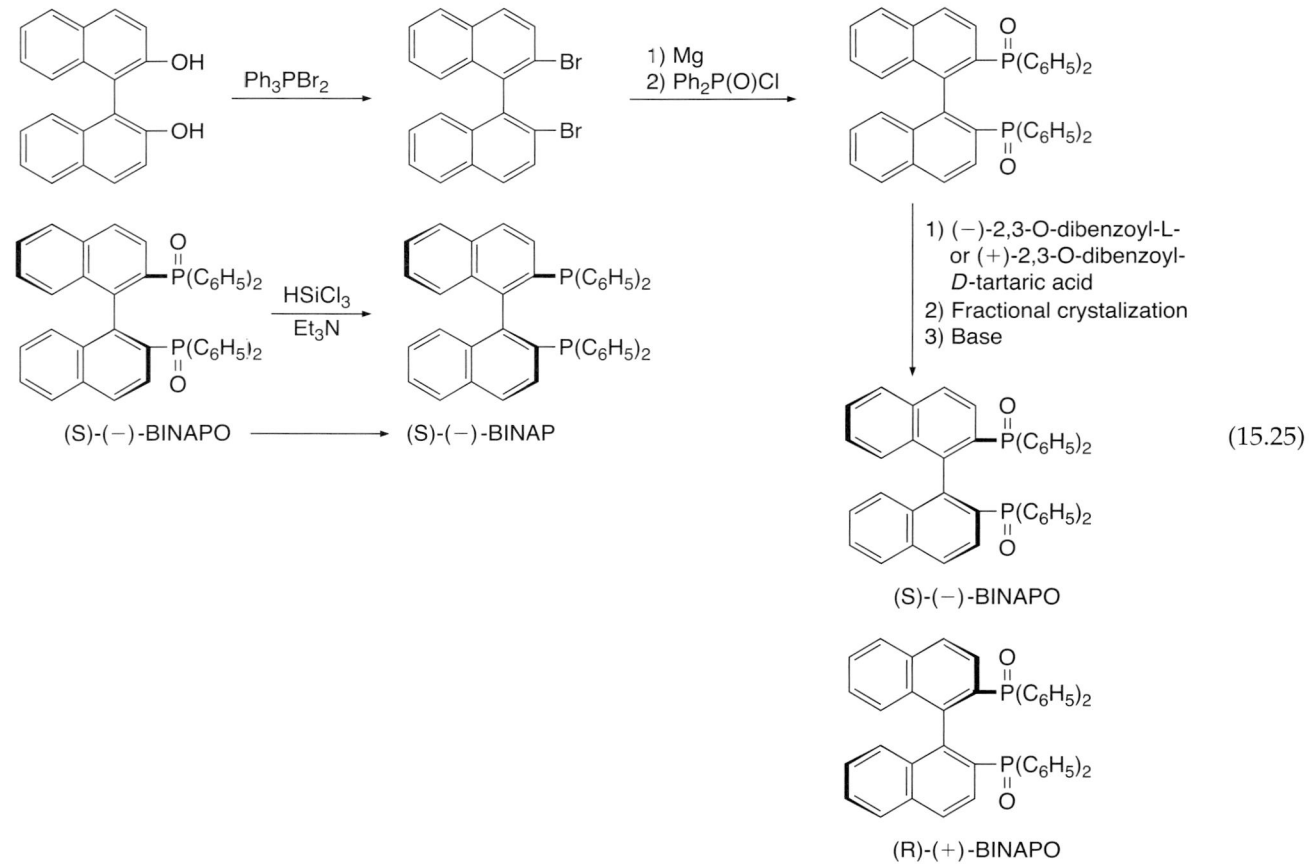

(S)-(−)-BINAPO ⟶ (S)-(−)-BINAP

1) Mg
2) Ph$_2$P(O)Cl

HSiCl$_3$
Et$_3$N

Ph$_3$PBr$_2$

1) (−)-2,3-O-dibenzoyl-L-
   or (+)-2,3-O-dibenzoyl-
   D-tartaric acid
2) Fractional crystalization
3) Base

(S)-(−)-BINAPO

(R)-(+)-BINAPO

(15.25)

(15.26)

The high reactivity and selectivity of catalysts containing BINAP as ligand led to the development of many related ligands containing atropisomeric biaryl backbones. Many of these were developed to circumvent patents on the BINAP structure, but many of them have led to catalysts that react with some combination of faster rates, higher turnover numbers, and higher enantioselectivities for particular reactions. Some conceptual information has emerged from these studies. First, the dihedral angle of the backbone of the ligand can influence enantioselectivity. The dihedral angles of several of these ligands are listed in Table 15.3. The Segphos ligand, developed at Takasago International Corp., possesses a smaller dihedral angle than BINAP, and the hydrogenation of a variety of carbonyl compounds occurs with higher enantioselectivity in the presence of ruthenium catalysts containing this ligand than containing BINAP.[170,171] To provide a systematic means to control the dihedral angle of biaryl ligands on enantioselectivity, Zhang has prepared a series of ligands containing alkyl linkers of different lengths, called TunePhos.[172]

Second, substituents on the P-bound aryl groups, particularly in the 3- and 5-positions of these aryl groups, can increase the enantioselectivity by enhancing the gearing of these groups and limiting the degrees of freedom.[173] For example, the hydrogenation of ketones by [Ru(3,5-xylyl-BINAP)(diamine)H$_2$] typically occurs with much higher enantioselectivity than the same reaction catalyzed by [Ru(BINAP)(diamine)H$_2$].[174]

Finally, the electronic properties of the backbone can affect selectivities. Chan has studied a dipyridylphosphine analog of BINAP called P-Phos. Ruthenium-catalyzed asymmetric hydrogenation of unsaturated carboxylic acids and unfunctionalized ketones occurs in high ee with this ligand.[175–178] Takaya has reported a more electron-donating analog of BINAP containing a partially hydrogenated backbone (H8-BINAP) and has shown that the hydrogenation of unsaturated carboxylic acids occurs with higher ee with ruthenium catalysts containing this ligand than with those containing BINAP.[179,180]

Table 15.3. Backbone dihedral angles of three atropisomeric bisphosphine ligands.

	Ligand	Backbone dihedral angle[a]
	(R)-BINAP	86°
(R)-Segphos =		67°
(R)-MeOBIPHEP =		72°

[a] Values from Jeulin, S.; de Paule, S. D.; Ratovelomanana-Vidal, V.; Genet, J. P.; Champion, N.; Dellis, P. *Proc. Natl. Acad. Sci. U.S.A.* **2004,** *101,* 5799.

15.6.1.1.2. Compounds Containing Chiral Ferrocenyl Backbones

One of the most impressive applications of asymmetric hydrogenation on a bulk scale is the enantioselective hydrogenation of an imine for the synthesis of the herbicide metolachlor, and this process was enabled by the development of chiral bisphosphine ligands possessing a ferrocene unit in the backbone. This family of ligands has generated highly active and selective catalysts for many enantioselective processes, including many types of hydrogenations. These ligands contain two substituents on the same cyclopentadienyl group of the ferrocene. Other chiral bisphosphine ligands contain ferrocene linkers in which one phosphino group is present on each of the two cyclopentadienyl rings. Many of these ligands are based on the diastereoselective lithiation of ferrocenylethylamine, known as the Ugi-amine.[181]

Togni and Spindler reported a class of bisphosphines shown at the top left of Figure 15.3[182] called "Josiphos" ligands after the experimentalist who first prepared them. One of the most remarkable features of catalysts containing these ligands at the time they were developed is their high enantioselectivity without the C_2 symmetry of the P-chiral ligands of Knowles or the BINAP ligand of Noyori. Instead, these ligands contain an unsymmetrical backbone. This backbone contains two types of chiral elements: a classic carbon stereocenter and "planar chirality." A ferrocene unit containing two different substituents on the same Cp ligand is chiral, and this type of topology has been called "planar chirality."

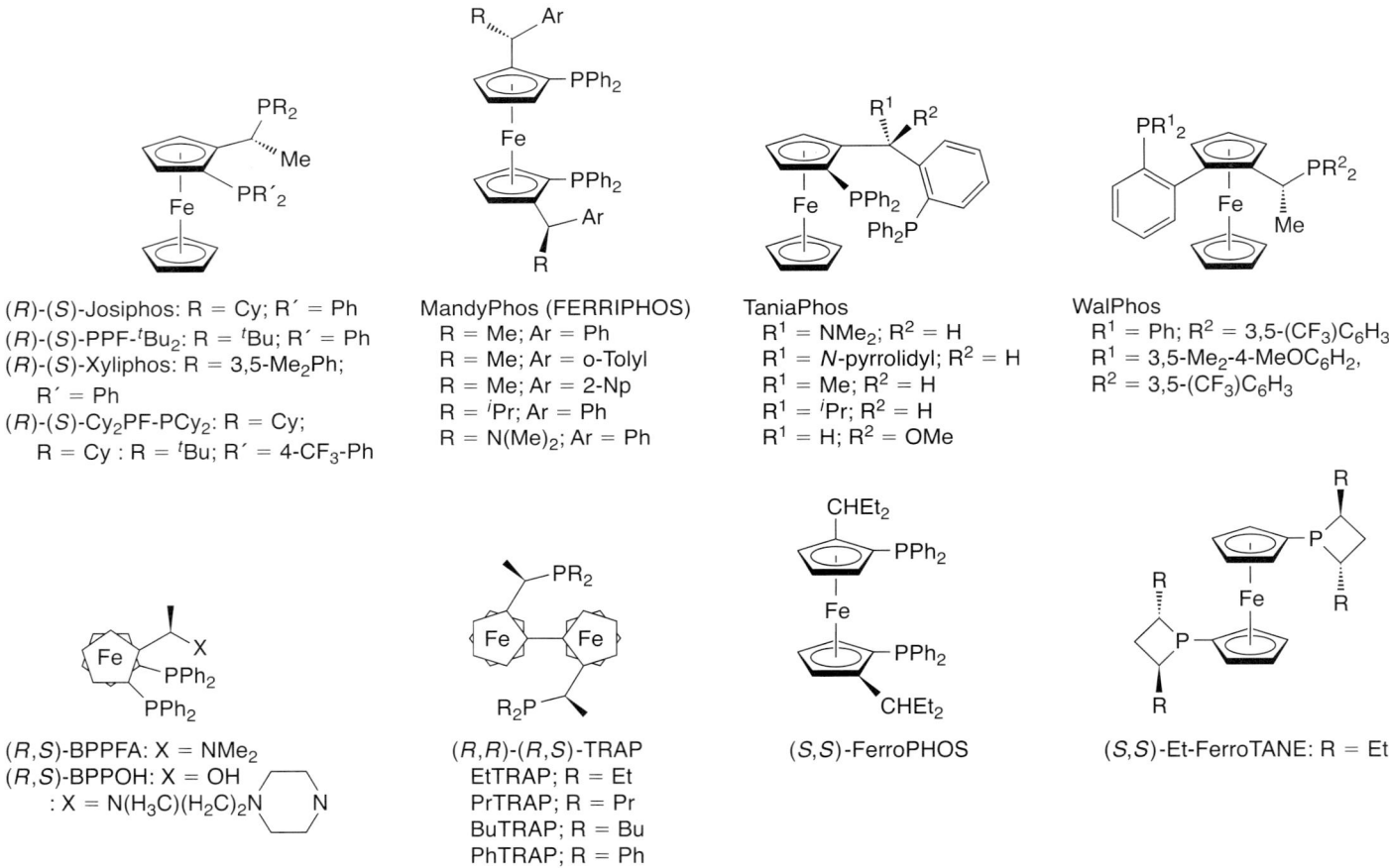

Figure 15.3. A selection of chiral bisphosphines based on a ferrocene backbone.

These Josiphos ligands generate highly active and selective rhodium catalysts for the hydrogenation of α-acetamidocinnamate, dimethylitaconate, and β-keto esters. As noted later in this chapter, the PPF-*tert*-Bu version of the Josiphos-type ligand has been used for a commercial synthesis of (+)-biotin,[183] and the XyliPhos derivative for the synthesis of (S)-metolachlor.[184,185]

Additional examples of chiral ligands in which ferrocene is part of the bridging unit are shown in Figure 15.3. These ligands include those that bind in a trans fashion and are called

"TRAP" ligands for *trans* phosphines. These ligands are discussed in Section 15.9 on heteroarene hydrogenation.[186] These examples also include aryl- and alkyl-substituted bisphosphines containing one phosphino group on each of the two cyclopentadienyl rings. The chirality of the ligand can be derived from the ferrocenyl unit (e.g., MandyPhos) or from the substituents on the phosphorus (e.g., FerroTANE).[187] Et-FerroTANE has been shown by Burk to generate selective rhodium catalysts for the hydrogenation of itaconates[187] and (*E*)-(β-acylamino)acrylates.[188]

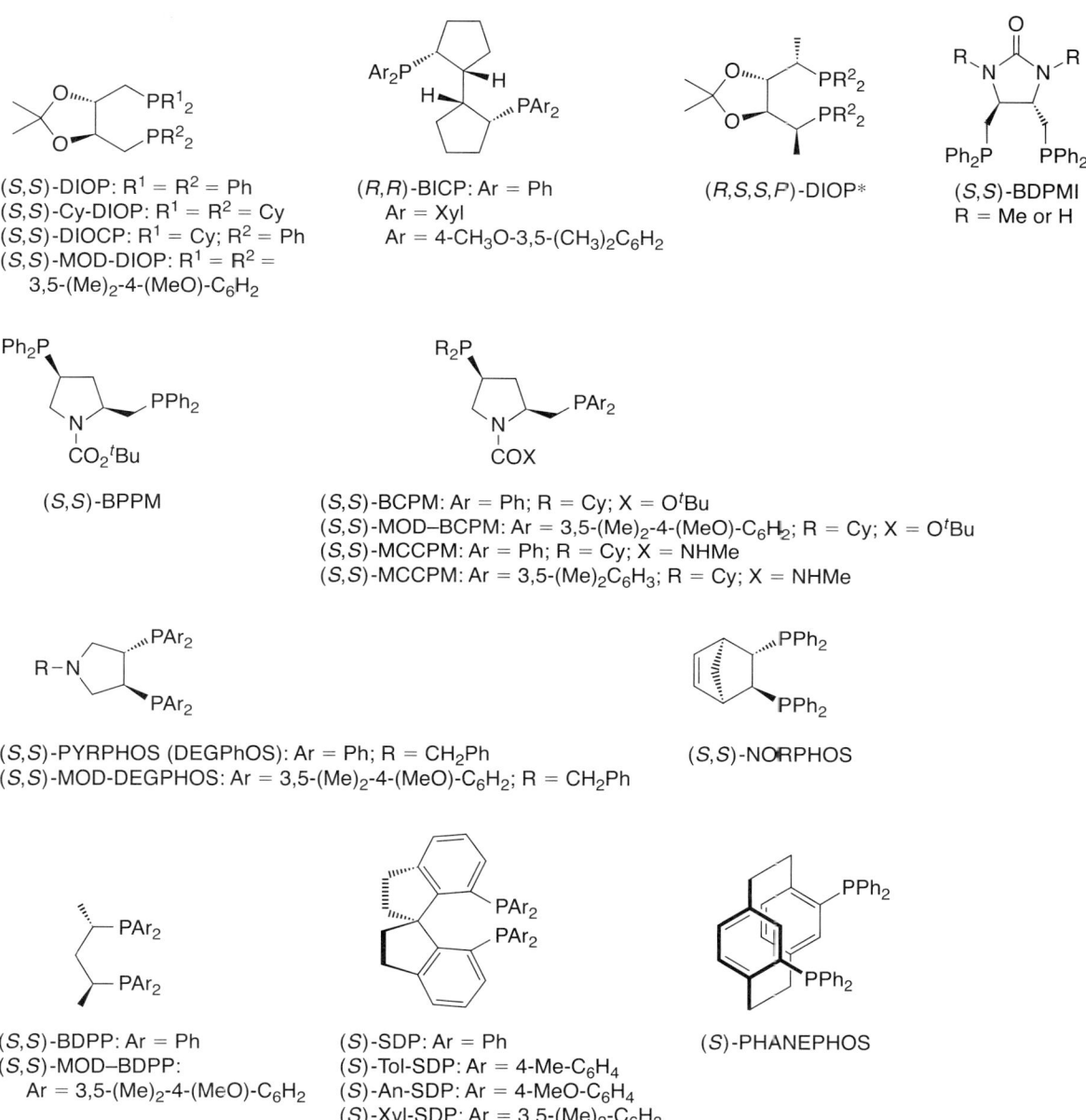

Figure 15.4. A selection of ligands containing a variety of chiral backbones.

15.6.1.1.3. Ligands Containing Aliphatic Backbones

After Kagan's pioneering work on the development of a chiral bisphosphine in which the stereochemistry originates from a chiral backbone, many backbones derived from the chiral pool have been used to generate bisphosphines. Some of them are designed to rigidify the backbone of the original diop ligand and others are built from other accessible and relatively rigid structures. A selection of such ligands is shown in Figure 15.4. The diop* ligand was prepared to enhance the preference for a single conformation of the ring containing the metal and ligand. Others, like BICP and BPPM, contain backbones that rigidify a relatively long linker between the two phosphorus atoms. Others, like DEGPHOS and Norphos, have two-carbon backbones that are expected to be more rigid than that of Chiraphos [(S,S)-2,3-bis(diphenylphopsphino)butane.[189,190] This set of ligands also contains two ligands possessing unusual backbones that have generated highly active and selective catalysts for hydrogenation, among other reactions. One ligand contains a spirocyclic backbone (SDP)[191] and the other (Phanephos)[192] contains a cyclophane that places the two phosphines in positions similar to those of the phosphino groups in BINAP.

15.6.2. Aliphatic Bisphosphines

As noted previously, the development of aliphatic bisphosphines containing the phospholane unit was one of the major advances in ligand design after BINAP. A selection of the original ligands (two top left general structures) and subsequent derivatives prepared later are shown in Figure 15.5. Until this time, the bisphosphine ligands contained chiral backbones or a center of chirality at phosphorus. Moreover, most of these ligands contained aromatic groups on phosphorus.[193] In contrast, the DuPhos and BPE ligands designed by Burk contained chirality on the substituents at phosphorus, and these substituents were aliphatic, rather than aromatic.[35,194–196] With these ligands, the asymmetric hydrogenation of carboxylic acids and itaconic acids catalyzed by rhodium complexes occurred with extremely high efficiencies. These ligands have been used in a commercial setting by several companies for the synthesis of α-amino acids.

Two subsequent versions of these ligands prepared by Zhang have particularly noteworthy activities. BINAPHANE ligands generate rhodium catalysts for the hydrogenation of enamides.[197,198] In particular, they reduce E/Z isomeric mixtures of dehydro β-amino acids with enantioselectivity up to 99%. In addition, rhodium complexes of PennPhos catalyze the hydrogenation of simple dialkyl and alkyl aryl ketones with high enantioselectivity without a diamine co-ligand.[199,200]

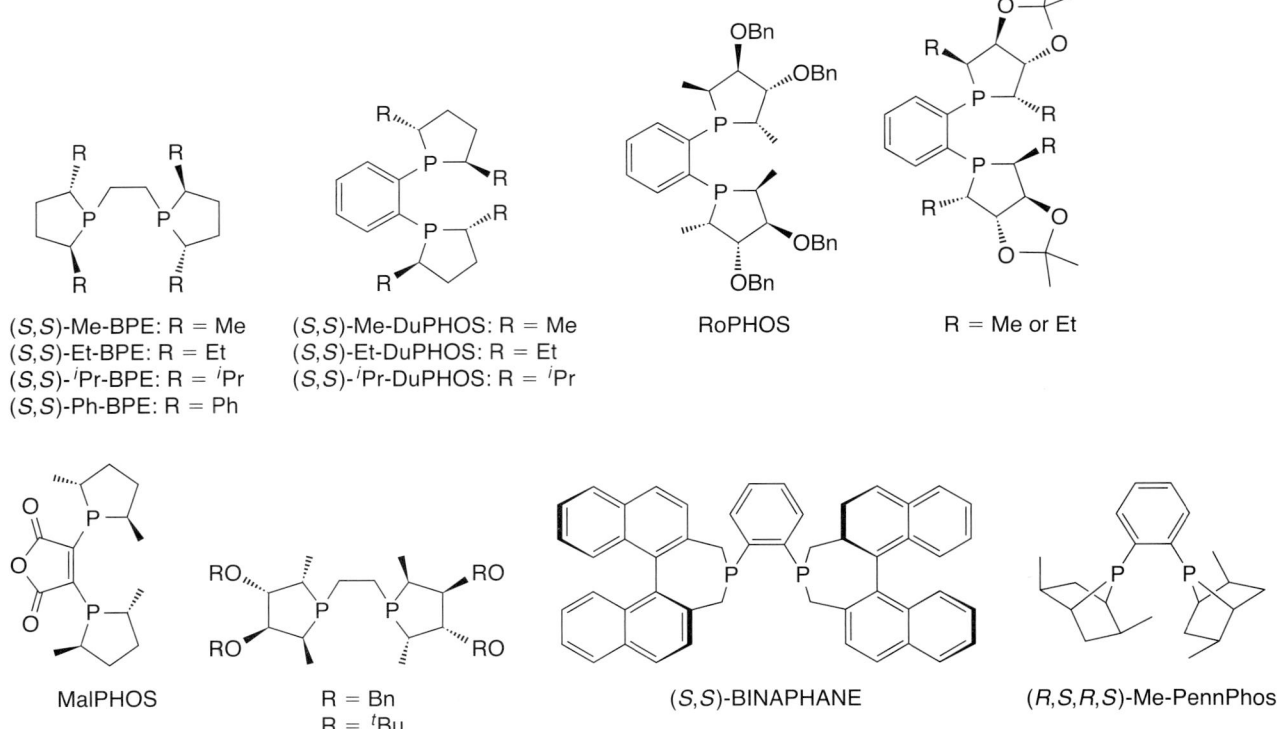

(S,S)-Me-BPE: R = Me
(S,S)-Et-BPE: R = Et
(S,S)-^{i}Pr-BPE: R = ^{i}Pr
(S,S)-Ph-BPE: R = Ph

(S,S)-Me-DuPHOS: R = Me
(S,S)-Et-DuPHOS: R = Et
(S,S)-^{i}Pr-DuPHOS: R = ^{i}Pr

RoPHOS

R = Me or Et

MalPHOS

R = Bn
R = ^{t}Bu

(S,S)-BINAPHANE

(R,S,R,S)-Me-PennPhos

Figure 15.5. A selection of bis-phospholane ligands.

15.6.3. *P*-Chiral Phosphines

As noted several times in this chapter, asymmetric hydrogenation was shown to be practical when Knowles used the C_2-symmetric chelating bisphosphine ligand DIPAMP for the rhodium-catalyzed hydrogenation of dehydro α-amino acids.[201,202] However, the development of additional *P*-chiral bisphosphines was slow, partly because these ligands were difficult to prepare. However, diastereoselective lithiation of tertiary phosphine boranes containing two methyl groups[203] and diastereoselective additions to phosphines containing ephedrin[204] allowed these ligands to be prepared more easily. Using the diastereoselective lithiation, Imamoto prepared a series of *P*-chiral ligands, such as BisP*, that led to the further development of *P*-chiral phosphorus ligands (Figure 15.6).[70,86,205,206] Rhodium catalysts containing the BisP* ligands are highly active and enantioselective for the hydrogenation of many classes of substrates.[70,86,207–209] As noted in the mechanistic discussion, studies on asymmetric hydrogenation by rhodium catalysts of *tert*-Bu-BisP* have provided evidence that these reactions occur through the "hydrogen-first" pathway even though the complexes are cationic.[210,211]

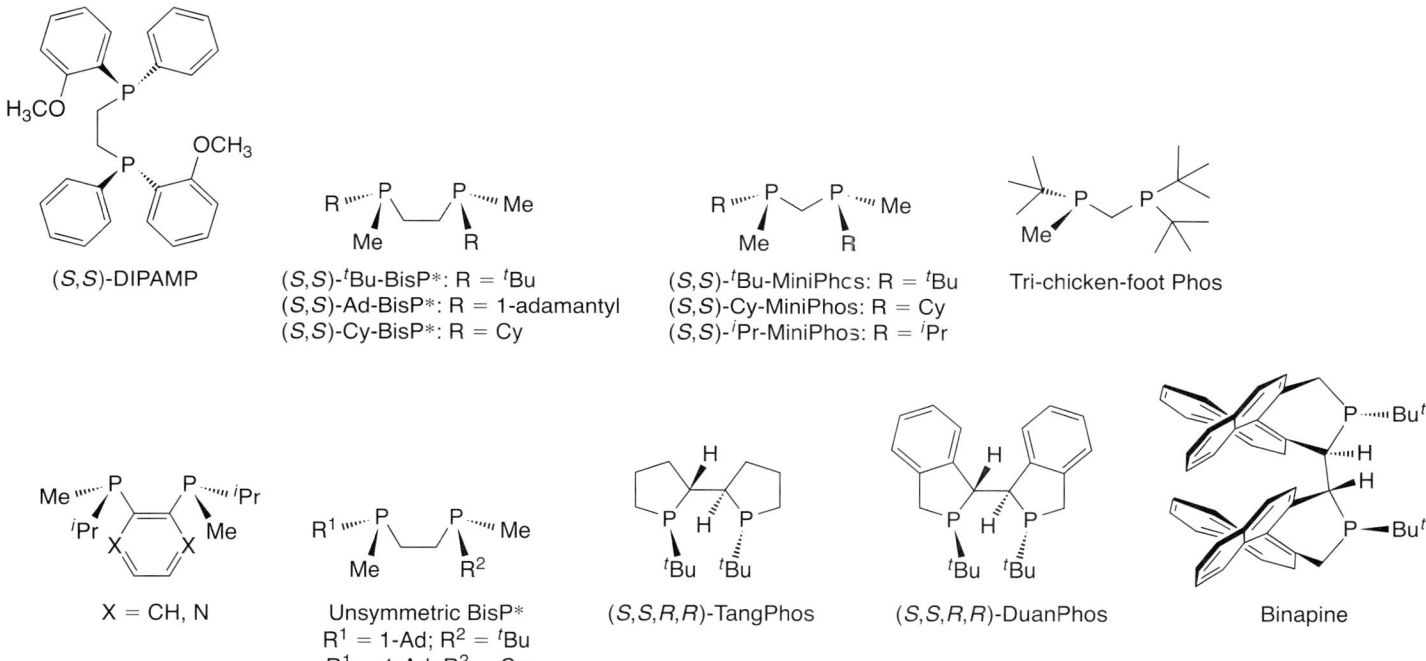

Figure 15.6. A selection of *P*-chiral phosphine ligands.

15.6.4. P,N Ligands

As described in Section 15.3.2, Crabtree's highly active iridium catalyst contains one phosphorus and one nitrogen donor ligand. To generate catalysts based on this structure for asymmetric hydrogenation, chelating P,N ligands that were first used for palladium-catalyzed allylic substitution[212–214] were applied to the problem of asymmetric hydrogenation. A selection of such P,N ligands is shown in Figure 15.7. The first of these ligands contains one phosphino group and an oxazoline unit linked by an aryl ring.[215] This original ligand was called Phox for *ph*osphino *ox*azoline. Later, many other linkers were used to generate related P,N ligands, and iridium catalysts containing these ligands catalyze the hydrogenation of unfunctionalized alkenes with remarkable activities and selectivities.[36–38,216–221] Burgess has also reported a series of related ligands, one of which is called JM-Phos, for the asymmetric hydrogenation of unfunctionalized olefins.[97,222]

In addition, P,N ligands in which the nitrogen bears an acidic proton have been developed for the hydrogenation of polarized functional groups, such as those in imides, esters, and even epoxides. Several examples of such ligands are shown in Figure 15.8. These P,N ligands have been prepared only recently and include achiral derivatives, as well as a chiral, non-racemic derivative prepared from proline. The use of catalysts containing the ligands in Figure 15.8 for the hydrogenation of these less reactive polar functional groups is described in Section 15.10.

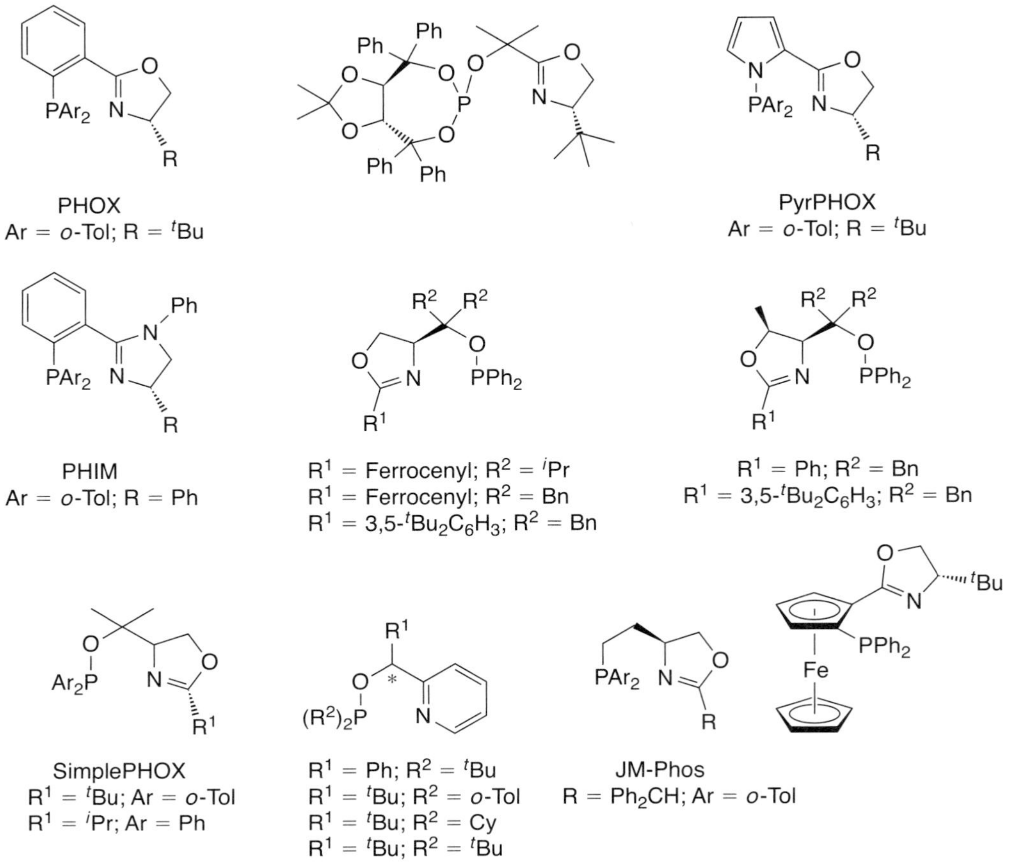

PHOX
Ar = o-Tol; R = ^{t}Bu

PyrPHOX
Ar = o-Tol; R = ^{t}Bu

PHIM
Ar = o-Tol; R = Ph

R^1 = Ferrocenyl; R^2 = ^{i}Pr
R^1 = Ferrocenyl; R^2 = Bn
R^1 = 3,5-^{t}Bu$_2$C$_6$H$_3$; R^2 = Bn

R^1 = Ph; R^2 = Bn
R^1 = 3,5-^{t}Bu$_2$C$_6$H$_3$; R^2 = Bn

SimplePHOX
R^1 = ^{t}Bu; Ar = o-Tol
R^1 = ^{i}Pr; Ar = Ph

R^1 = Ph; R^2 = ^{t}Bu
R^1 = ^{t}Bu; R^2 = o-Tol
R^1 = ^{t}Bu; R^2 = Cy
R^1 = ^{t}Bu; R^2 = ^{t}Bu

JM-Phos
R = Ph$_2$CH; Ar = o-Tol

Figure 15.7. A selection of P,N ligands used for the asymmetric hydrogenation of unfunctionalized olefins.

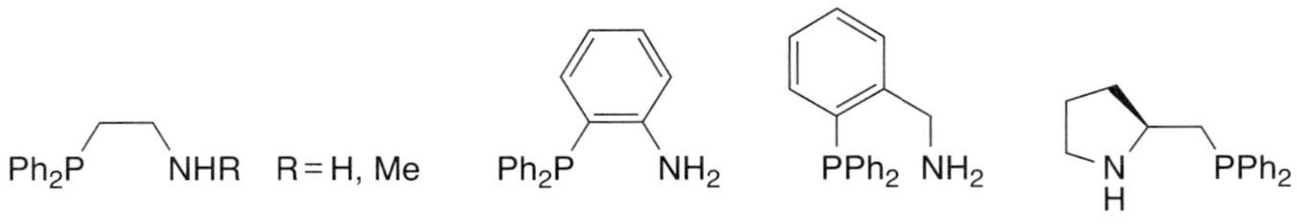

Ph$_2$P NHR   R = H, Me   Ph$_2$P NH$_2$ PPh$_2$ NH$_2$ PPh$_2$

Figure 15.8.
P,N ligands used to generate metal–ligand bifunctional catalysts for the hydrogenation of esters, imides, and epoxides.

15.6.5. Phosphites and Phosphoramidites

The final major class of ligand developed for asymmetric hydrogenation described in this chapter results from a return to studies of monodentate phosphorus ligands. As noted previously, the first asymmetric hydrogenation of an olefin was conducted with a rhodium complex containing a monodentate phosphine. Although the enantioselectivity was measurable, it was low. Thus, it is remarkable that a family of monodentate ligands has recently generated rhodium catalysts that react with high enantioselectivities for the hydrogenation of a wide variety of olefins. A selection of such ligands is shown in Figure 15.9. A particularly valuable aspect of the phosphite, phosphinite, and phosphoramidite ligands is their simple synthesis. These ligands are readily generated by sequential reactions of diols, alcohols, and amines to PCl_3. This simple synthesis has made it possible to prepare large libraries of these monodentate ligands to screen for activity and selectivity.[223,224] The monodentate property of these ligands has also has allowed for the mixing of two different ligands to optimize selectivity.[225,226] At the same time, the monodentate property of these ligands makes the catalysts less stable, and the turnover numbers are generally lower for these catalysts than for those containing the chelating ligands described in the previous sections of this part of the chapter. Nevertheless, up to 40,000 turnovers with 96.9% ee has been achieved for the hydrogenation of dimethyl itaconate.[227]

Reetz, and De Vries and Feringa, reported hydrogenations with these types of ligands nearly simultaneously. Reetz developed a series of monophosphite ligands, such as those shown at the top left of Figure 15.9, which generate rhodium catalysts that are selective for the hydrogenation of dimethyl itaconate.[228,229] De Vries and Feringa developed phosphoramidite ligands, such as those shown on the right of Figure 15.9, and found that rhodium catalysts containing the simple phosphoramidite ligand bearing a dimethylamino group on phosphorus, called MonoPhos, catalyzes the asymmetric hydrogenation of dehydroamino acid derivatives with > 99% ee.[230] Such catalysts are also effective for the hydrogenation of α-aryl enamides.[231] Catalysts containing two similar phosphoramidite ligands are highly enantioselective for the hydrogenation of (β-acylamino)acrylates.[232]

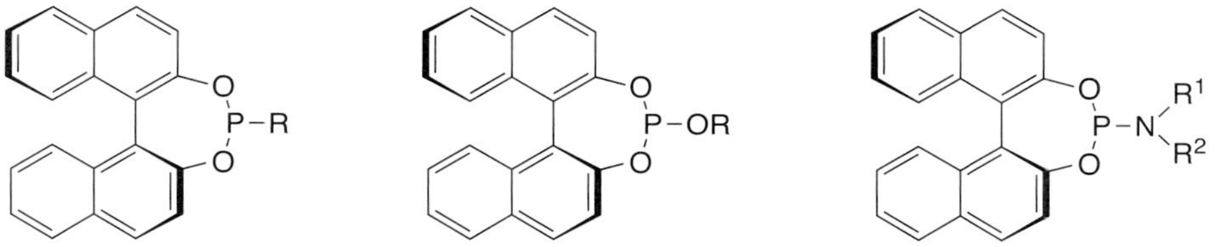

R = ^{t}Bu
R = *o*-Br–C_6H_4

R = ^{i}Pr
R = Ph
R = (*R*)-CH(Me)Ph
R = (*S*)-CH(CH_2OMe)Ph

(*S*)-MonoPhos: R^1 = R^2 = Me
R^1 = Bn; R^2 = Me
R^1 = (*R*)-CH(Me)Ph, R^2 = H
R^1 = R^2 = Et

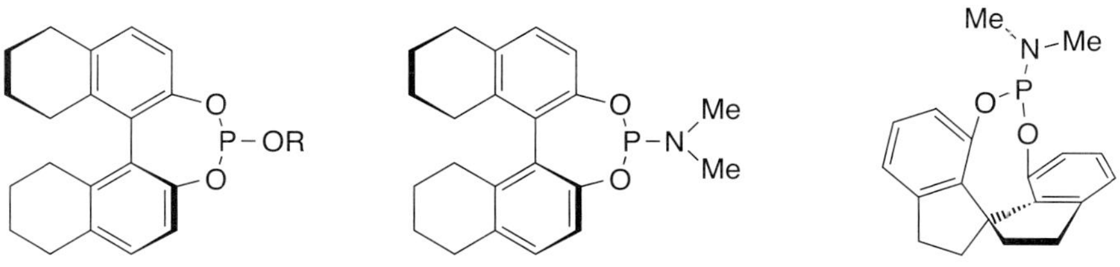

R = (*R* or *S*)-CH(CH_3)Ph
R = CH$(CH_3)_2$

(*S*)-H8–Monophos

(*S*)-SIPHOS

Figure 15.9.
A selection of monodentate phosphite and phosphoramidite ligands used for asymmetric hydrogenation.

15.7. Examples of Asymmetric Hydrogenation and Transfer Hydrogenation

Hydrogenation is now the most utilized homogeneous catalytic reaction in process chemistry for the synthesis of enantioenriched material. The scope of highly enantioselective hydrogenation is spectacular. It ranges from olefins and ketones functionalized with ligating substituents to unfunctionalized alkenes and ketones. Examples of the asymmetric hydrogenations of various classes of substrates that occur by the mechanisms presented so far with catalysts containing the chiral ligands discussed in the last section are presented here in Section 15.7. The hydrogenations of olefins are presented first, ketones second, and imines third. Asymmetric hydrogenations of heteroaromatic substrates are presented fourth. When applicable, cases of the industrial production of chiral products by these hydrogenations are included. Many of the examples of these applications were selected from excellent reviews and monographs by Blaser, Schmidt, Spindler, and Thommen at Solvias.[233,234] Interested readers are urged to consult these articles for further examples, perspectives, and details.

15.7.1. Classes of Asymmetric Hydrogenations of Olefins

15.7.1.1. Asymmetric Hydrogenation of Enamides

As noted several times so far in this chapter, the hydrogenation of the precursor to L-DOPA, an α-amino acid, was the first hydrogenation conducted with high enantioselectivity. As described in Section 15.5 on hydrogenation mechanisms, these olefins bind through the C=C bond and the carbonyl oxygen atom of the amide group. Such two-point binding can lead to selective reactions of substrates beyond precursors to α-amino acids. This section describes the hydrogenation of dehydro α-amino acids, dehydro β-amino acids, and simple enamides lacking the carboxy function of the amino acid precursors.

15.7.1.1.1. Asymmetric Hydrogenation of Dehydro α-Amino Acids [α-(Acylamino)acrylic Acids and Esters]

Although the C_2-symmetric, P-chiral bisphosphine DIPAMP first led to hydrogenations of dehydroamino acids with high yields and enantioselectivities, many catalysts have now been found that catalyze these reactions with even faster rates, higher turnover numbers, and equal or greater enantioselectivities (Equation 15.27). Most of these catalysts contain rhodium and phosphine ligands. This hydrogenation has become one of the test reactions for research groups preparing new ligands for asymmetric catalysis. A review article published in 2004 includes a description of 57 different ligands that generate rhodium catalysts for enantioselective hydrogenation of α-amino acid precursors with enantioselectivities exceeding 96%.[122] These rhodium-catalyzed reactions are typically conducted with the amino acid precursor in the form of an ester. The ligands that generate rhodium catalysts reacting with high enantioselectivities to form α-amino acids include members of all of the classes of ligands presented in Section 15.6.

$$R^1 \diagup CO_2R^2 \xrightarrow{\text{Chiral Rh catalyst}} R^1 \diagup CO_2R^2$$
$$NHC(O)CH_3 \qquad\qquad NHC(O)CH_3$$

A: $R^1 = H$, $R^2 = H$
B: $R^1 = H$, $R^2 = CH_3$
C: $R^1 = Ph$, $R^2 = H$
D: $R^1 = Ph$, $R^2 = CH_3$

(15.27)

Rhodium catalysts containing 57 different ligands generate products with > 96% ee.

Despite the plethora of catalysts that reduce α-acetamidocinnamic acid esters to phenylalanine derivatives, there are a series of challenges for the development of catalysts for the synthesis of other classes of α-amino acids by asymmetric hydrogenation. First, as will be mentioned several times throughout this section, the synthesis of substrates can be as challenging as the hydrogenation itself. For example, the hydrogenation of precursors to aliphatic α-amino acids is challenging because of the difficulty in preparing geometrically pure aliphatic, tri-substituted alkenes. Second, substrates containing heteroaromatic groups with basic nitrogens, such as precursors to pyridyl amino acids, can be difficult to reduce with high activity because the catalyst can bind the pyridine and be deactivated. Third, tetra-substituted substrates that form β-branched α-amino acids are particularly challenging to reduce with high activity because of the weaker binding of the olefin. Fourth, a commercial synthesis requires that the reactions can be conducted with a high ratio of substrate to catalyst (high turnover numbers) and with fast rates, preferably with modest pressures of hydrogen; hydrogenations of α-acetamidocinnamic acid esters by many catalysts do meet these criteria.

The use of geometrically pure olefins is often required to obtain products with high enantiomeric excess because the hydrogenation of E and Z isomers can lead to enantiomeric products and low overall enantioselectivities. Because of the challenge of preparing geometrically pure olefins efficiently on a large scale, a catalyst that converts both isomers to the same enantiomer is often desired. As shown in Equations 15.28 and 15.29, the rhodium–DuPhos system reduces both isomers of the aliphatic dehydro α-amino acid with high enantioselectivity to form the same enantiomer.[235] The challenge of conducting hydrogenations of substrates containing heteroarenes has also been addressed with Rh-Duphos systems by running the reaction in the presence of acid that protonates the pyridyl nitrogen atom.[236] Finally, the highly reactive rhodium–DuPhos system catalyzes the hydrogenation of even tetra-substituted α-acyl enamides with relatively low catalyst loadings and high selectivities (Equation 15.30).[237]

(15.28)

99.6% ee

(15.29)

99.4% ee

(15.30)

Ligand	% ee of product (configuration)
(S,S)-Me-DuPhos	96.0 (S)
(R,R)-Me-BPE	98.2 (R)

Examples of commercial syntheses of α-amino acids by hydrogenation catalyzed by rhodium–DuPhos complexes, including data on the activities and selectivities, are shown in Figure 15.10. Notice that several of these processes occur with high enantioselectivities and turnover numbers ranging from 1000 to 50,000.[238–240] In general, turnover numbers of 1000 or 5000 are needed for the cost of the catalyst to become a minor contributor to the cost of the overall process.

The hydrogenations in Equations 15.31 and 15.32 illustrate the reduction of more complex substrates leading to α-amino acid amides. The first example demonstrates a diastereoselective hydrogenation conducted by the fine chemicals company Lonza as part of the synthesis of the vitamin biotin.[241–243] The second example shows an α-amino acid product produced in quantities greater than 200 tons by Lonza using a rhodium catalyst containing the Josiphos ligand shown.[241] This product is an intermediate in the synthesis of Indinavir (Crixivan), which was used in combination with a reverse transcriptase inhibitor to create the first effective treatment for AIDS.[241,244,245]

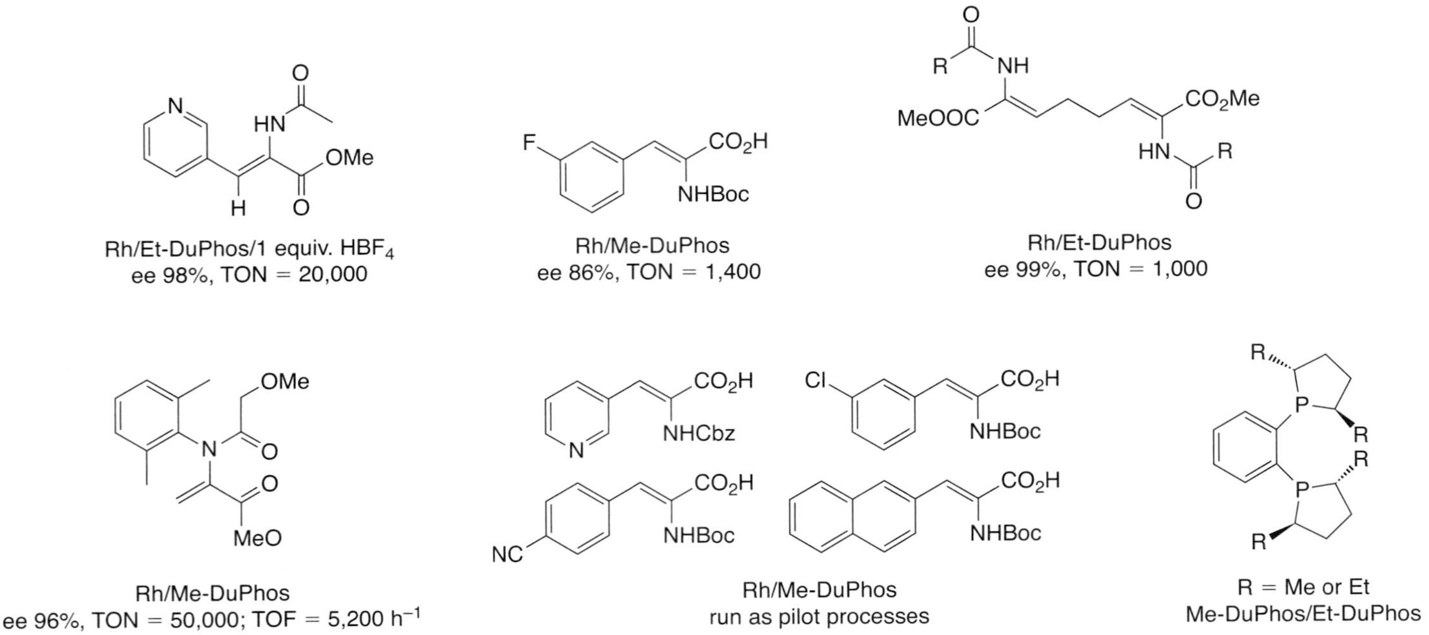

Rh/Et-DuPhos/1 equiv. HBF$_4$
ee 98%, TON = 20,000

Rh/Me-DuPhos
ee 86%, TON = 1,400

Rh/Et-DuPhos
ee 99%, TON = 1,000

Rh/Me-DuPhos
ee 96%, TON = 50,000; TOF = 5,200 h^{-1}

Rh/Me-DuPhos
run as pilot processes

R = Me or Et
Me-DuPhos/Et-DuPhos

Figure 15.10.
Examples of α-amino acid precursors that have been reduced by asymmetric hydrogenation with a rhodium–DuPhos catalyst in commercial processes. Data from references 224–226.

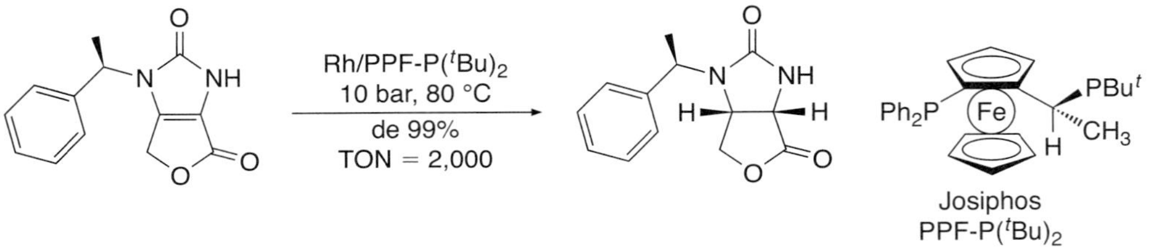

Rh/PPF-P(^{t}Bu)$_2$
10 bar, 80 °C
de 99%
TON = 2,000

Josiphos
PPF-P(^{t}Bu)$_2$

(15.31)

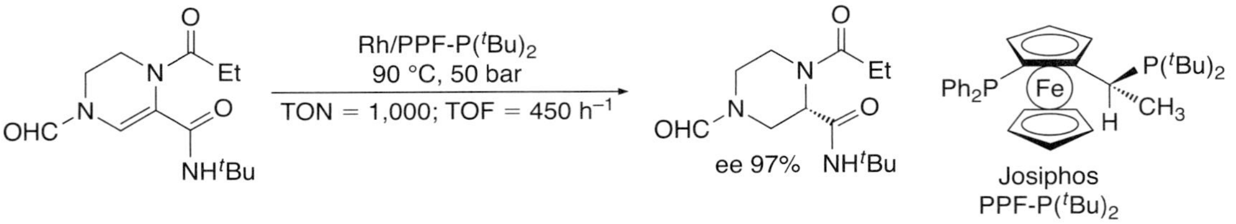

Rh/PPF-P(^{t}Bu)$_2$
90 °C, 50 bar
TON = 1,000; TOF = 450 h^{-1}

ee 97%

Josiphos
PPF-P(^{t}Bu)$_2$

(15.32)

15.7.1.1.2. Asymmetric Hydrogenation of Dehydro β-Amino Acids [β-(Acylamino)acrylic Acids and Esters]

Dehydro β-amino acids can also bind through the olefin and the carbonyl oxygen atom of the amide. Thus, enantioselective hydrogenation of these substrates is also typically run with rhodium catalysts and can be conducted with high enantioselectivities with many different types of ligands. The review by Zhang[122] includes 41 different ligands that form the reduced product with > 92% ee.

For the practical reduction of these substrates, one must again confront the problem of olefin geometry. Most synthetic methods form a mixture of (Z)- and (E)-β-acetamido acrylic acid esters. Thus, catalysts that reduce both geometries of the olefins are important, and catalysts that reduce both geometries of the olefin to form the same enantiomer are most practical. Chiral phosphorus ligands, such as BINAP,[246] DuPhos,[247] BICP,[248] BDPMI,[249] o-Ph–HexaMeO–BIPHEP,[250] DuanPhos,[251] tert-Bu–BisP*,[207] TangPhos,[252] and monophosphoramidites,[183,232] among other ligands, generate rhodium and ruthenium catalysts that reduce (E)-3-alkyl-3-(acetamido)acrylates with high enantioselectivities. However, only a few chiral ligands reduce the mixture of E and Z isomers shown in Equation 15.33. These ligands include BDPMI,[249] TangPhos,[252] (R)-tri-chicken-foot phosphine,[253] and the monophosphonite[254] in this equation containing the spiro diolate.[184,232] Mechanistic studies of the hydrogenation of 3-alkyl-3-(acylamino)acrylates catalyzed by a closely related rhodium–BisP* complex showed that the reaction proceeds through a monohydride intermediate with the β-carbon atom of the insertion product bound to rhodium.[207]

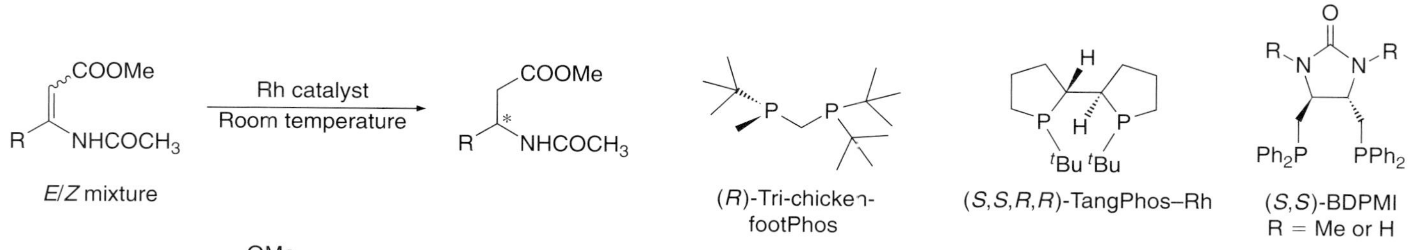

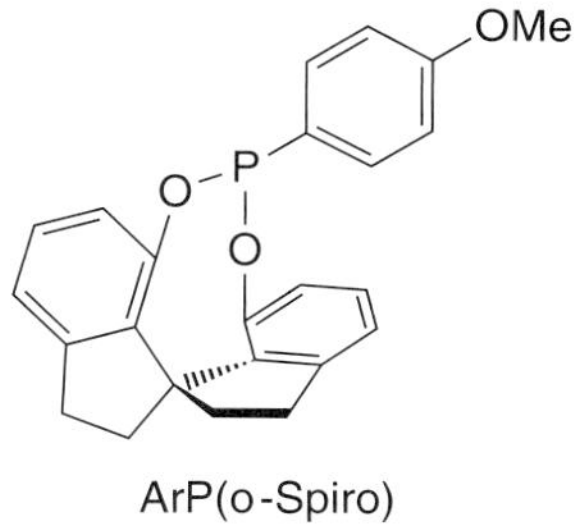

Ligand	Substrate	Conditions	ee (%)
(S,S,R,R)-TangPhos–Rh	R = CH$_3$	THF, 20 psi H$_2$	99.5 (R)
(R)-BDPMI–Rh	R = CH$_3$	THF, 20 psi H$_2$	98 (R)
Tri-chicken-footPhos–Rh	R = p-MeO–Ph	CH$_2$Cl$_2$,100 bar H$_2$	98 (R)
Tri-chicken-footPhos–Rh	R = p-MeO–Ph	CH$_2$Cl$_2$,100 bar H$_2$	95 (R)
ArP(o-Spiro)	R = Ph	Toluene, 100 bar H$_2$	93 (R)

(15.33)

15.7.1.1.3. Asymmetric Hydrogenation of Simple Enamides

The asymmetric hydrogenation of simple enamides provides a route to protected amines containing a chiral center α to nitrogen. Catalysts have been developed that hydrogenate these substrates with extremely high enantioselectivities. Reactions of 1-monosubstituted enamides form α-methylamines, and the reaction in Equation 15.34 illustrates an example that occurs with exceptionally high enantioselectivity. Even the tetra-substituted enamide in Equation 15.35 reacts with high enantioselectivity, in this case with a rhodium catalyst containing the P-chiral (R,R)-tBu-BISP* ligand.[70]

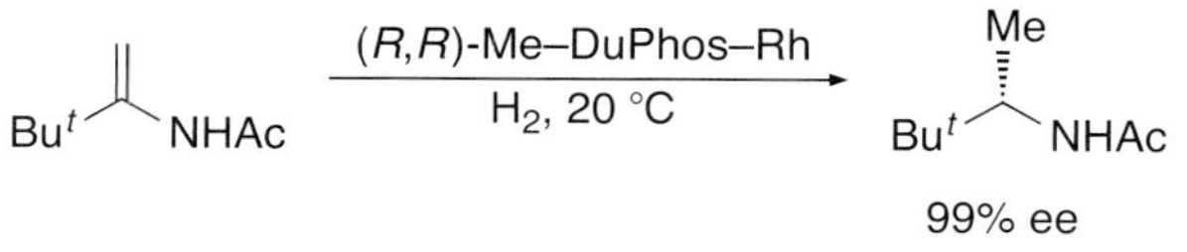

(15.34)

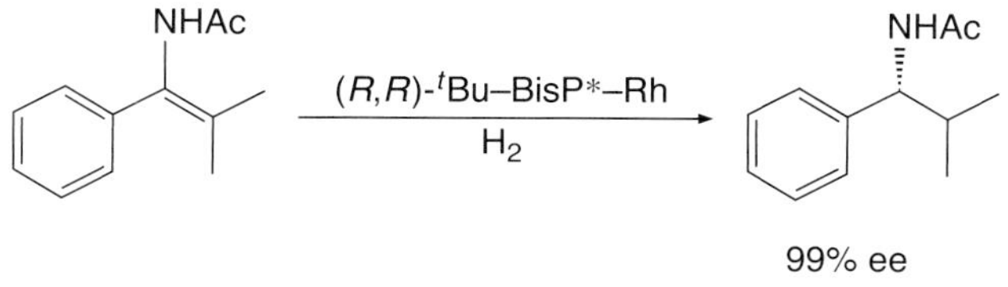

99% ee

(15.35)

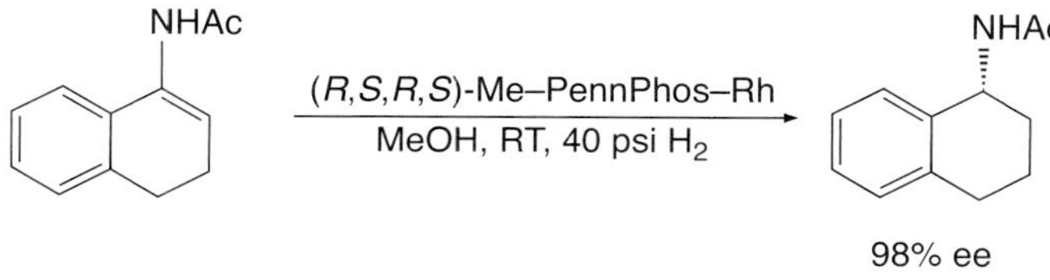

99% ee

(15.36)

98% ee

$$\text{(15.37)}$$

Like reactions of many other prochiral olefins, the geometry of the enamine can affect selectivity. Thus, cyclic enamides that contain one geometry react with high selectivities (Equation 15.36). Alternatively, catalysts that react with both isomers with high selectivity for the same enantiomer can be developed. Equation 15.37 shows that the rhodium–DuanPhos complex catalyzes the hydrogenation of an E/Z mixture of isomeric enamides with high enantioselectivity.[251]

15.7.1.2. Asymmetric Hydrogenation of α-(Acyloxy)acrylates

Although enol esters have a structure that is similar to that of enamides and can bind to the metal through a combination of the C=C bond and the carbonyl oxygen atom, the oxygen of the ester is less basic than that of an amide and binds less tightly to the metal. Thus, the hydrogenation of α-(acyloxy)acrylates has been more challenging to develop than the hydrogenation of α-(acylamino)acrylates. The products of these reductions, however, are valuable α-hydroxy carboxylic acid derivatives.

A few catalysts that hydrogenate α-(acyloxy)acrylates have been reported. Rhodium and ruthenium complexes of DIPAMP,[255,256] DuPhos,[34,257] BINAP,[256] FERRIPHOS,[258] and TaniaPhos[259] catalyze the hydrogenation of 2-(acyloxy)acrylates with high enantioselectivities. Data for reactions catalyzed by the rhodium complex of DuPhos and by the ruthenium complex of BINAP are shown in Equation 15.38. A wide range of 2-(acyloxy)acrylates undergo hydrogenation in the presence of the Et-DuPhos–Rh catalyst with high enantioselectivities.[257] High selectivities for the hydrogenation of E/Z isomeric mixtures of β-substituted α-(acyloxy)acrylates were also obtained with this catalyst.

Catalyst	R	Geometry	Reaction conditions	Percent ee of product (configuration)
(R,R)-Et–DuPhos–Rh	H		MeOH, RT, 2 atm H$_2$	> 99 (R)
(R)-BINAP–Ru	^{i}Pr	E/Z	MeOH, 50 °C, 50 atm H$_2$	98 (S)
(R,R)-Et–DuPhos–Rh	Ph	E/Z	MeOH, RT, 3 atm H$_2$	95.6 (R)

$$\text{(15.38)}$$

15.7.1.3. Asymmetric Hydrogenation of Acrylic Acids

One of the major developments in the asymmetric hydrogenation of olefins has been the discovery of ruthenium catalysts for the hydrogenation of α,β-unsaturated carboxylic acids. As noted in the mechanistic section, these carboxylic acids bind to ruthenium through the carboxylate group as an anionic ligand. Thus, [Ru(BINAP)(OAc)$_2$] is a particularly reactive and selective precatalyst for the asymmetric hydrogenation of these substrates.[260] Many other ligands have been investigated for this process. Other atropisomeric ligands containing biaryl backbones, such as H8-BINAP,[261,262] MeO–BIPHEP,[263] BIPHEMP,[263] and P-Phos,[175–178] among others, catalyze these reactions with high enantioselectivities. For the hydrogenation of tiglic acid (Equation 15.39), the ruthenium catalyst containing H8-BINAP reacted with particularly fast rates and high enantioselectivities. In the case of the hydrogenation of 2-arylpropionic acids, high hydrogenation pressure is required to achieve good enantioselectivities (Equation 15.40). Most recently, an iridium catalyst containing a spirocyclic P,N ligand was shown by Zhou and co-workers to allow these reductions to occur with lower pressures of hydrogen and higher turnover numbers.[264]

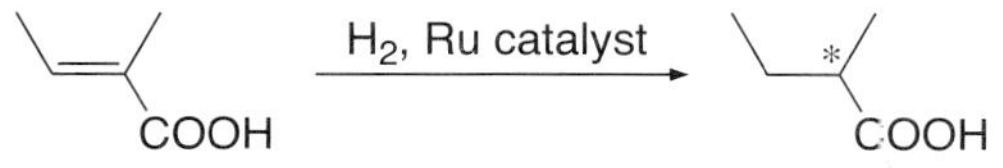

Catalyst	S/C ratio	Reaction conditions	Percent ee of product (configuration)
Ru(OAc)$_2$[(R)-BINAP]	100	MeOH, 15–30 °C, 4 atm H$_2$	91 (R)
Ru(OAc)$_2$[(S)-H8–BINAP]	200	MeOH, 10–25 °C, 1.5 atm H$_2$	97 (S)

$$(15.39)$$

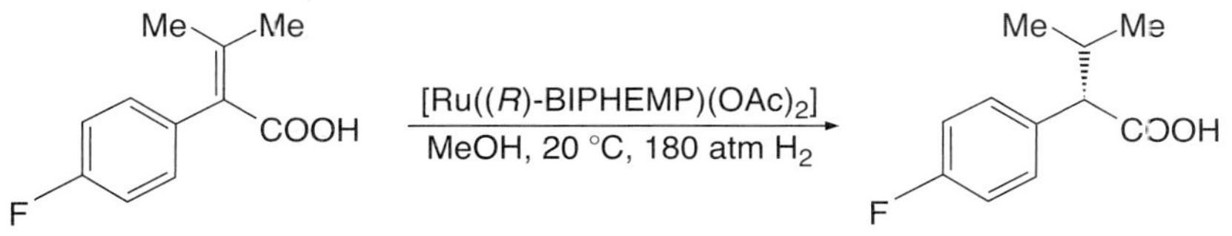

$$(15.40)$$

94% ee

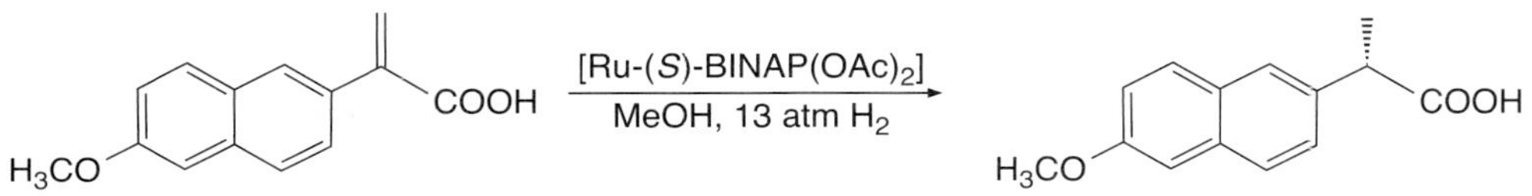

97% ee

$$(15.41)$$

(S)-Naproxen

By these procedures, the anti-inflammatory drugs (S)-ibuprofen and (S)-naproxen (Equation 15.41) can be synthesized by asymmetric hydrogenation. Again, one challenging aspect of this route to naproxen is the synthesis of the substrate, and for this reason, asymmetric hydrogenation has not become the major commercial route to this compound.

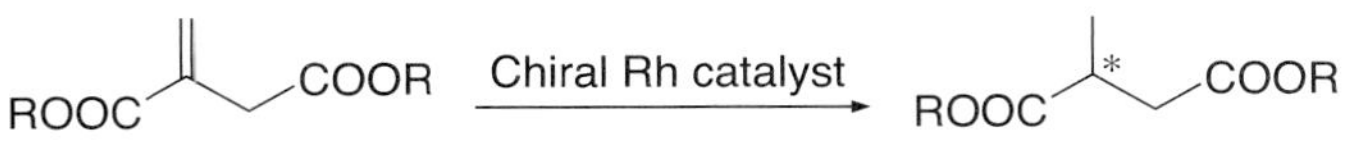

$$(15.42)$$

26 catalysts shown to react with over 95% ee.

Many rhodium and ruthenium catalysts containing chiral phosphorus ligands have also been shown to reduce γ,δ-unsaturated itaconic acids or esters with high enantioselectivities (Equation 15.42). The review by Zhang[122] describes 26 catalysts that reduce this substrate with over 95% ee. The ligands that generate active catalysts include electron-rich phosphines, such as BICHEP,[265] Et-DuPhos,[266] Ph-BPE,[267] TangPhos,[268] and DuanPhos.[251] They also include electron-deficient bisphosphite or phosphonite ligands.[269–271] Even monophosphorus ligands, such as the phosphoramidite MonoPhos[231,272] and some simple phosphites,[227,228] generate catalysts for the hydrogenation of itaconic acid derivatives with fast rates, high turnover numbers, and high enantiomeric excess. Even iridium catalysts containing P,N ligands catalyze the hydrogenation of acrylic acid esters with high enantioselectivity.[273]

Although the substrate contains a 1,2-disubstituted olefin, rather than the 1,1-disubstituted olefin of itaconic acid derivatives, one commercial application of the reduction of a γ,δ-unsaturated ester is shown in Equation 15.43.[274–276] The product is a compound used in the fragrance industry, and multiple tons of this product are produced each year by this hydrogenation. This product has been reported to be produced by the company Firmenich by hydrogenation with ruthenium complexes of electron-rich bisphosphine ligands, such as the Josiphos ligands or Me-DuPhos.[275–277]

$$ (15.43) $$

15.7.1.4. Asymmetric Hydrogenation of Unsaturated Alcohols

Like the hydrogenation of unsaturated carboxylic acids, the asymmetric hydrogenation of unsaturated alcohols has become a useful process due to the development of ruthenium catalysts. The reduction of allylic alcohols occurs with particularly impressive efficiency and is conducted on a commercial scale for the synthesis of fragrances and vitamins. Although the mechanism of these reactions has not been studied in detail, the alcohol presumably binds to create a complex possessing two-point binding of the substrate, like the binding of unsaturated amides and esters. These reactions were first shown to occur with high selectivities in the presence of Ru(BINAP)(OAc)$_2$ as catalyst.[278,279] They have also now been shown to occur with high selectivities with iridium catalysts containing P,N ligands.[36,280]

As shown in Equations 15.44 and 15.45, non-racemic, chiral allylic alcohols are formed on a production scale with Ru–BINAP and Ru–Biphep catalysts with remarkable turn-over numbers and enantioselectivities. Geraniol undergoes hydrogenation to give (S)- or (R)-citronellol in nearly quantitative yield with 96–99% ee.[278,279] The double bond at the C-6 and C-7 positions of the substrate is not reduced. A high hydrogen pressure is required to obtain high enantioselectivities; racemization of geraniol occurs when the hydrogen pressure is lower.[281,282] Other atropisomeric ligands, such as MeO–BIPHEP, also form highly active and selective catalysts for this type of reduction.[281–284] A Ru–Biphep catalyst has been used on a pilot scale for the production of the vitamin E side chain by the hydrogenation shown in Equation 15.45.[283–285]

Geraniol   Citronellol

$$ (15.44) $$

Precursor to the side chain of
Vitamin E (Tocopherol)

$$ (15.45) $$

15.7.1.5. Asymmetric Hydrogenation of Unfunctionalized Olefins[273]

All of the olefins discussed so far contain a functional group, other than the C=C bond, that binds to the metal to create a defined structure. The asymmetric hydrogenation of olefins that lack this second functional group has been a major challenge. Few complexes of any type catalyze the hydrogenation of tri-substituted and tetra-substituted olefins, let alone catalyze asymmetric hydrogenation of these olefins. Recall from Section 15.3 on achiral catalysts for olefin hydrogenation that Wilkinson's catalyst and ruthenium–hydride complexes display little reactivity for the reduction of tri-substituted alkenes, and no reactivity for reduction of tetra-substituted alkenes. Thus, early studies on the asymmetric hydrogenation of "unfunctionalized olefins" were first conducted with group 4 and lanthanide metal complexes. Later studies focused on the development of chiral versions of Crabtree's catalyst.

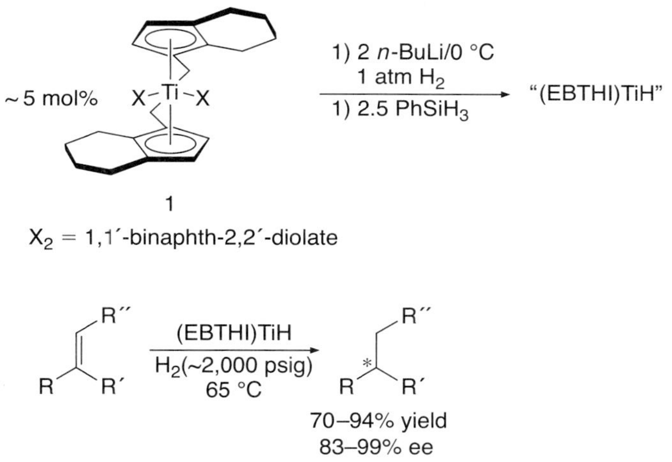

~5 mol% $X-Ti-X$

1

$X_2 = 1,1'$-binaphth-2,2'-diolate

$$\xrightarrow[\text{1) 2.5 PhSiH}_3]{\substack{\text{1) 2 }n\text{-BuLi/0 °C} \\ \text{1 atm H}_2}} \text{"(EBTHI)TiH"}$$

(15.46)

$$\xrightarrow[\substack{\text{H}_2(\sim2,000\text{ psig}) \\ 65\text{ °C}}]{\text{(EBTHI)TiH}}$$

70–94% yield
83–99% ee

Chiral metallocene catalysts containing titanium[286] or zirconium[287] catalyze the asymmetric hydrogenation of unfunctionalized tri- or tetra-substituted olefins. As discussed in Chapter 3, metallocene complexes are chiral when two indenyl or tetrahydroindenyl ligands contain a tether to lock the rotation of these rings. One catalyst developed for olefin polymerization, which is discussed in Chapter 22, is the ethylene bis(tetrahydroindenyl) ligand (EBTHI). (EBTHI)TiH and (EBTHI)ZrH$_2$ have been shown to catalyze the asymmetric hydrogenation of a series of tri- and tetra-substituted aryl olefins with excellent enantioselectivities.[286, 287] The generation of the titanium catalyst and the conditions for hydrogenation are shown in Equation 15.46. Chiral lanthanocene complexes have also been studied in which the chirality is derived from a neomenthyl substituent (Figure 15.11). These complexes catalyzed the hydrogenation of 2-phenyl-1-butene at −80 °C with an ee of 96%.[61] The use of these catalysts has been limited by the high cost or lengthy synthesis of them, the difficulty in preparing derivatives of them, and the difficulty in handling them in many synthetic settings.

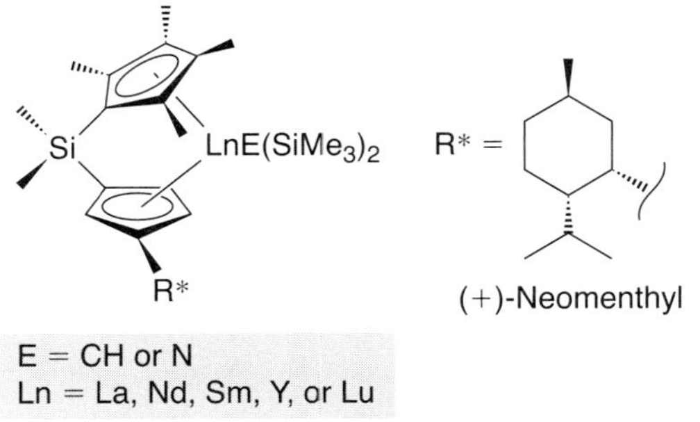

LnE(SiMe$_3$)$_2$ R* =

R*   (+)-Neomenthyl

E = CH or N
Ln = La, Nd, Sm, Y, or Lu

Figure 15.11.
Chiral lanthanide catalysts studied for the asymmetric hydrogenation of unfunctionalized olefins.

More recent effort has focused on developing chiral cationic iridium catalysts containing P,N-ligands. Selected examples of such hydrogenations that occur with high enantioselectivities are shown in Equations 15.47–15.51. A series of para-substituted (E)-methylstilbenes has been hydrogenated with a member of the family of Phox–Ir catalysts (Figure 15.12) with excellent enantioselectivities.[37,38] Hydrogenation of E and Z tri-substituted olefins generate opposite enantiomers of the product. Thus, the use of geometrically pure olefins is important to obtain high enantioselectivity. Hydrogenation of a tetra-substituted olefin has even been conducted with >90% ee.[288] A disubstituted terminal olefin, 2-arylbutene,

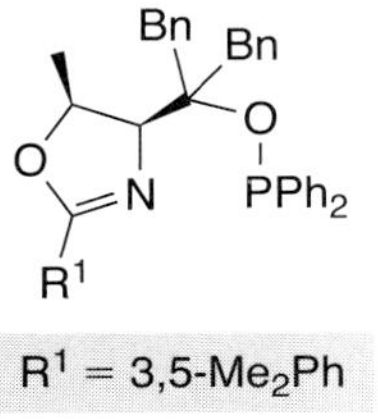

$R^1 = 3,5$-Me$_2$Ph

Figure 15.12.
Ligand for iridium-catalyzed asymmetric hydrogenation of unfunctionalized olefins.

has also been hydrogenated to form the chiral reduced product in 89% ee (Equation 15.50). One of the most impressive examples of stereoselective hydrogenation is the diastereoselective hydrogenation of the multiply unsaturated precursor to the vitamin E side chain.[289]

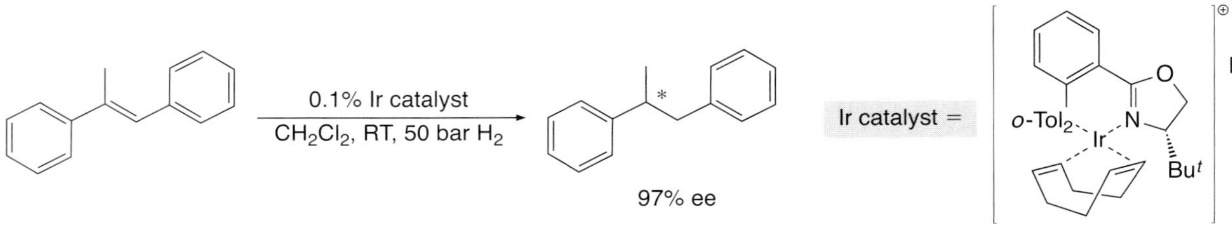

(15.47)

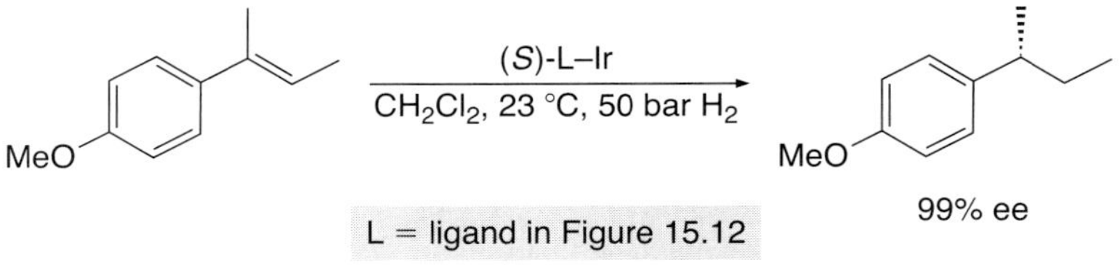

(15.48)

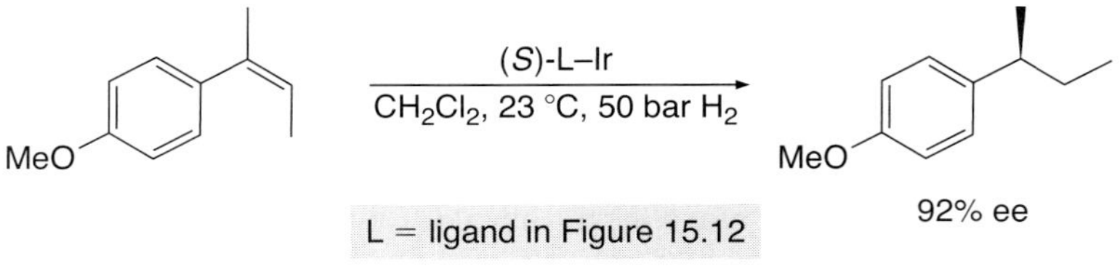

(15.49)

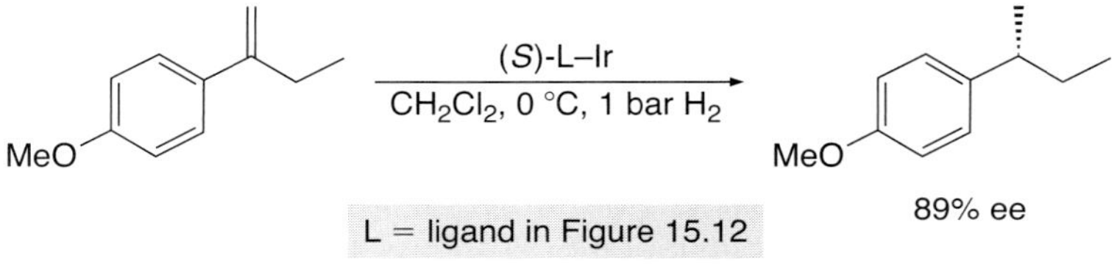

(15.50)

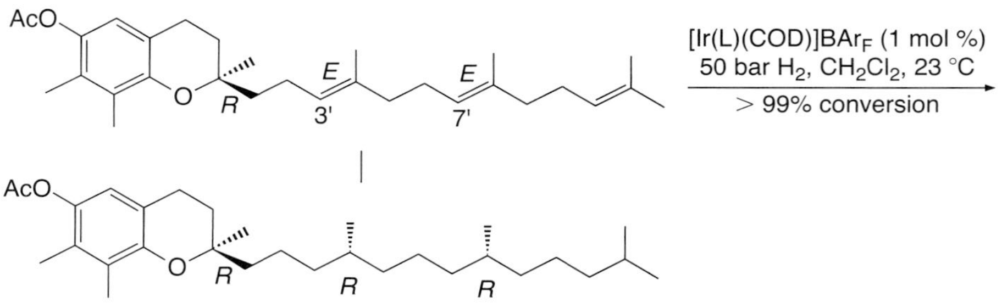

(15.51)

> 98% RRR (< 0.5% RRS; < 0.5% RSR; < 0.5% RSS)

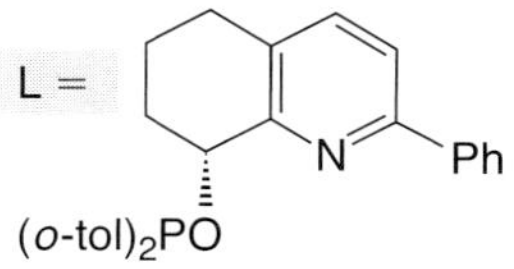

15.7.1.6. Asymmetric Hydrogenation of Ketones

As noted previously, one of the most dramatic advances in asymmetric hydrogenation over the past two decades has been the development of ruthenium catalysts for the hydrogenation of ketones. These hydrogenations can be divided into the hydrogenation of "functionalized" ketones and "unfunctionalized" ketones. "Functionalized" ketones contain another ligating functional group, and include α- or β-ketoesters, α-aminoketones, and α-hydroxyketones. Remarkable selectivities have been achieved with rhodium and ruthenium complexes of bisphosphines, although most of the selective catalysts are based on ruthenium. More recently, catalysts have also been developed for the asymmetric hydrogenation of "unfunctionalized" ketones, such as alkyl aryl ketones and diaryl ketones. These hydrogenations of "unfunctionalized" ketones have been conducted in almost all cases with catalysts containing a bisphosphine and a diamine ligand. Transfer hydrogenations of such ketones have been conducted with catalysts containing a tosylamide, or an amino alcohol without a phosphine. The mechanisms of such hydrogenations and transfer hydrogenations were discussed in Section 15.5 on the mechanism of hydrogenations and transfer hydrogenations.

15.7.1.6.1. Asymmetric Hydrogenations of Functionalized Ketones

15.7.1.6.1.1. Asymmetric Hydrogenations of α-Keto Esters

Ruthenium chloride complexes containing BINAP were shown by Noyori to catalyze the hydrogenation of α-ketoesters, and this finding led to studies with catalysts containing many other atropisomeric bisphosphines. Three sets of results are shown in Equation 15.52. Reactions with ruthenium–Segphos and –BICHEP complexes (see Figure 15.2 for structures) led to the hydrogenation of an alkyl-substituted α-ketoester with very high enantioselectivity. Some neutral rhodium complexes also catalyze this transformation.[290, 291]

$$\text{(15.52)}$$

Ligand	R	XR^1	S/C ratio	Conditions	ee
(R)-SEGPHOS–Ru	^{t}Bu	OEt	1,000	EtOH, 70 °C, 50 atm H$_2$	98.5 (R)
(R)-BICHEP–Ru	Ph	OMe	100	EtOH, 25 °C, 5 atm H$_2$	> 99 (S)
(S)-BINAP–Ru	4-MePh	OMe	150	MeOH, 30 °C, 100 atm H$_2$	93 (S)

One application of this hydrogenation with the structurally related 4,4-dimethyl-2, 3-furandione is shown in Equation 15.53. This α-ketoester undergoes hydrogenation with a neutral rhodium catalyst containing the bpm ligand shown in this equation with spectacularly high turnover numbers and good enantioselectivity.[283] This product is used by Roche in the synthesis of pantothenic acid, a B-vitamin used for the synthesis of coenzyme A.[283]

$$\text{(15.53)}$$

15.7.1.6.1.2. Asymmetric Hydrogenation of β-Keto Esters [292]

Many catalysts for the hydrogenation of β-keto esters have been discovered (Equation 15.54). Hydrogenation of these substrates occurs selectively at the more reactive keto functionality. As noted in Section 15.2 describing the major advances in hydrogenation, a switch in anionic ligand on the ruthenium–BINAP catalysts from acetate to chloride led to species that catalyzes the hydrogenation of functionalized ketones with high activities and selectivities.[59] The reactions of β-keto esters were conducted first with ruthenium–BINAP catalysts. Now, many ligands have been studied for the asymmetric hydrogenation of this type of substrate. The review by Zhang[122] describes 20 ligands, such as BPE,[293] BisP*,[294] and PHANEPHOS,[295] which lead to catalysts for the hydrogenation of aliphatic β-keto esters in greater than 95% ee. However, the original catalyst containing BINAP, as well as its cousin Segphos, are among the most effective. The BINAP catalyst is not particularly selective for the hydrogenation of aryl-substituted β-keto esters. Instead, the anionic ruthenium–Segphos precursor shown in Equation 15.55 leads to a catalyst for the reduction of such substrates with high turnover numbers and excellent enantioselectivity.[170]

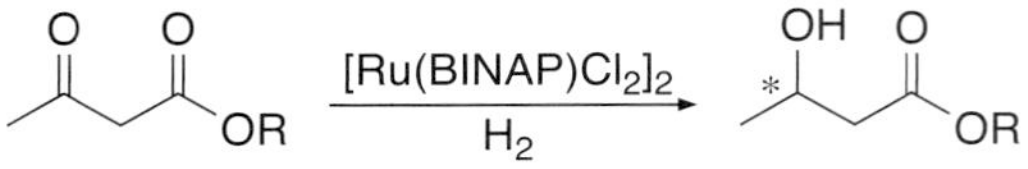

$$\text{(15.54)}$$

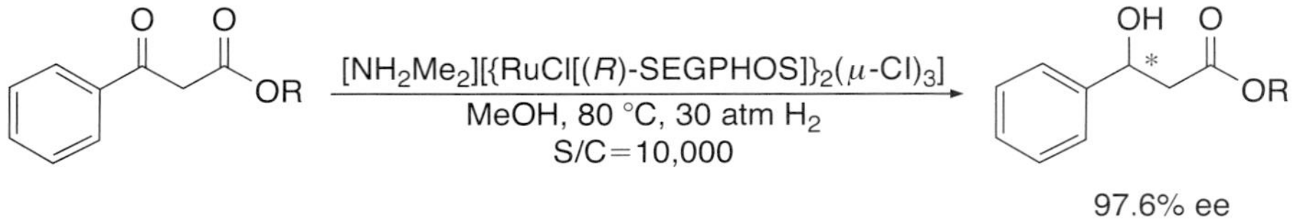

$$\text{(15.55)}$$

97.6% ee

One of the most useful aspects of the hydrogenation of β-keto esters is the ability to use a racemic substrate containing a single substituent on the carbon between the two carbonyl groups. Several examples of such hydrogenations are shown in Equations 15.56 and 15.57.[296, 297] Binding of the 1,3-dicabonyl compound to the metal makes the central hydrogen particularly acidic, and the substrate racemizes faster than the hydrogenation process. One of the two enantiomers of the reactants then undergoes hydrogenation with the stereochemistry controlled by the catalyst. In most cases, the enantiomer that reacts is the one that forms the anti diastereomer. Because the product is much less acidic than the reactant, the product does not undergo racemization, and the product is formed enantioselectively and diastereoselectively. This position can bear an alkyl group, but in more synthetically useful forms of the process, it can be bound to a heteroatom. The chloro- and amino-substituted substrates in Equations 15.56 and 15.57, as well many others, undergo hydrogenation to form essentially one diastereomer with very high enantioselectivity.[170, 298–300]

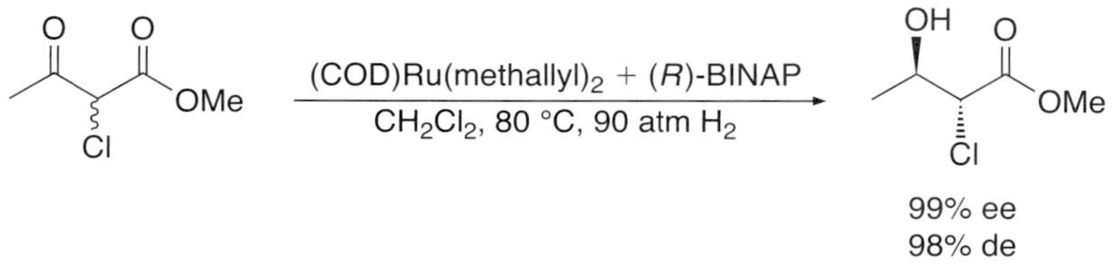

$$\text{(15.56)}$$

99% ee
98% de

$$(15.57)$$

$$(15.58)$$

This type of dynamic kinetic resolution has been conducted on a production scale by the Takasago Company.[170,300,301] [Ru(cymene)I$_2$]$_2$, in combination with Tol-BINAP, generates a catalyst for the hydrogenation of a β-keto ester containing an amidomethyl group on the central carbon (Equation 15.58). The product of this reaction is used for the synthesis of carbapanen antibiotics. This hydrogenation has been reported to be conducted on the scale of 50–120 tons per year.

Other hydrogenations of β-keto esters are also conducted on production scales. MSC Technologies and Lanxess[302,303] conduct these reactions with BINAP and a BIPHEP derivative as ligand. The reactions with BINAP typically occur with turnover numbers of 10,000–20,000 and frequencies of 12,000 h^{-1}.

$$(15.59)$$

$$(15.60)$$

Such dynamic kinetic resolutions can also be conducted on cyclic β-keto esters. Two examples are shown in Equations 15.59 and 15.60. Such cyclic substrates contain a stereocenter at the carbon between the two carbonyl groups. Again, a dynamic kinetic resolution of these substrates by hydrogenation occurs selectively to form predominantly a single stereoisomer. This reaction occurs to form a 99:1 ratio of diastereomers and 93% enantioselectivity of the major diastereomer in the presence of a ruthenium–BINAP catalyst.[85] The positions of the keto and ester functionalities can also be reversed. Reduction of the cyclic β-keto ester in Equation 15.60 generates, in this case, the cis diastereomer with high diastereoselectivity and enantioselectivity.[304]

15.7.1.6.1.3. Asymmetric Hydrogenations of β-Diketones

Hydrogenations of β-diketones form diol products. An example of the hydrogenation of acetylacetone to generate the anti diol containing two different ligands is shown in Equation 15.61. Many ligands have been shown to generate catalysts that are active and enantioselective for this transformation. Again, catalysts containing BINAP and its relative Biphemp are particularly effective for this transformation and form essentially a single enantiomer and diastereomer of the product. These reactions, however, need to be conducted with high pressures of hydrogen.[210,305]

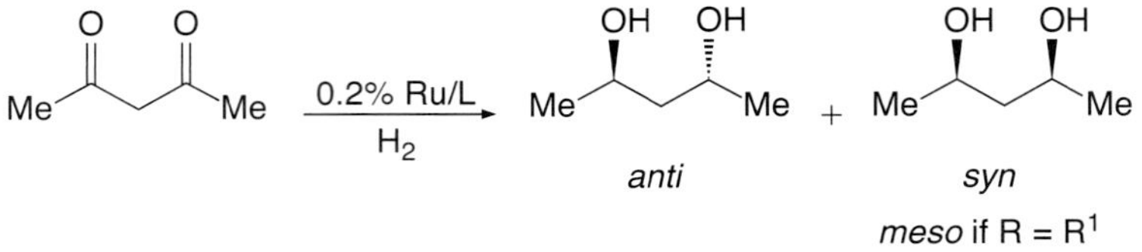

(15.61)

Ligand	Conditions	de (%)
(R)-BINAP	MeOH, RT, 72 atm H$_2$	100 (R,R)
(R)-BIPHEMP	EtOH, 50 °C, 100 atm H$_2$	> 99.9 (R,R)

15.7.1.6.1.4. Asymmetric Hydrogenations of α- and β-Amino and Hydroxy Ketones

The synthesis of chiral, non-racemic amino alcohols can also be conducted by the hydrogenation of ketones. The hydrogenation of α-amino ketones occurs with particularly high selectivity. In this case, the amino group is likely to act as a ligand to allow two-point binding of the substrate. Thus, these reactions have been conducted with ruthenium catalysts ligated by phosphines.[304] For example, reactions of the alkyl-substituted α-amino ketone in Equation 15.62 occurred with very high enantioselectivities in the presence of a ruthenium catalyst containing BINAP as the sole chiral ligand. However, they can also be reduced without binding of the amino group by ruthenium catalysts containing a combination of bisphosphine and diamine discussed in the next section on the asymmetric hydrogenation of unfunctionalized ketones.[306] Hydrogenations of the α-amino acetophenones in Equation 15.63 were reported with Noyori's bifunctional catalyst containing xylyl–BINAP and the daipen diamine ligand. Hydrogenations of α-amino ketones have been used for the syntheses of several drugs containing amino alcohols. For example, this reaction has been used to generate the precursor to denopamine hydrochloride used for the treatment of congestive heart failure. Reduction of the α-amino ketone in Equation 15.64 generates an amino alcohol derivative that is then carried forward in a straightforward fashion to this drug.[306] Additional applications of the hydrogenations of α-amino ketones include the syntheses of precursors to adrenalin and phenyl ephedrin (Equation 15.65).[307]

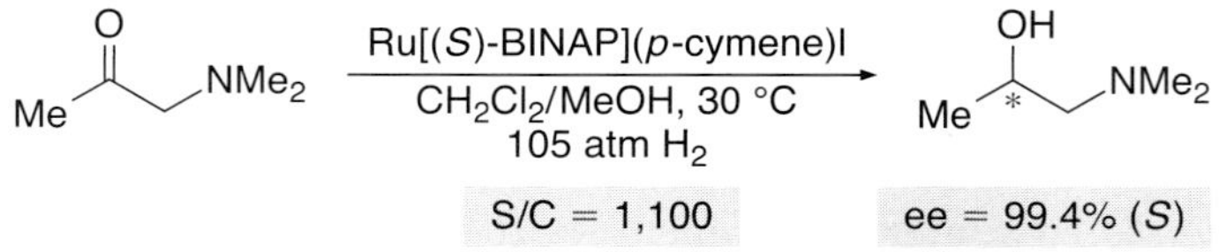

(15.62)

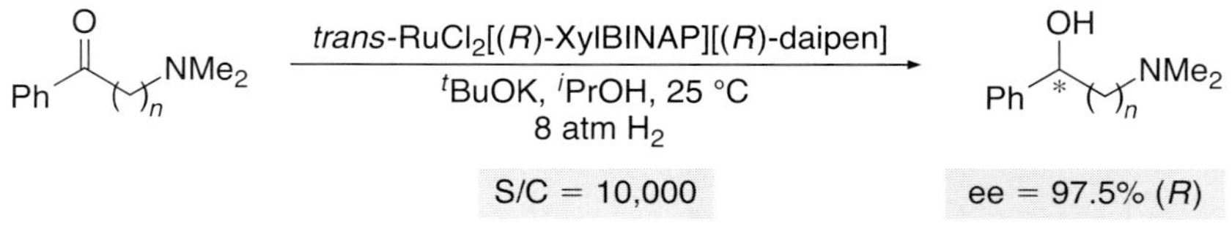

(15.63)

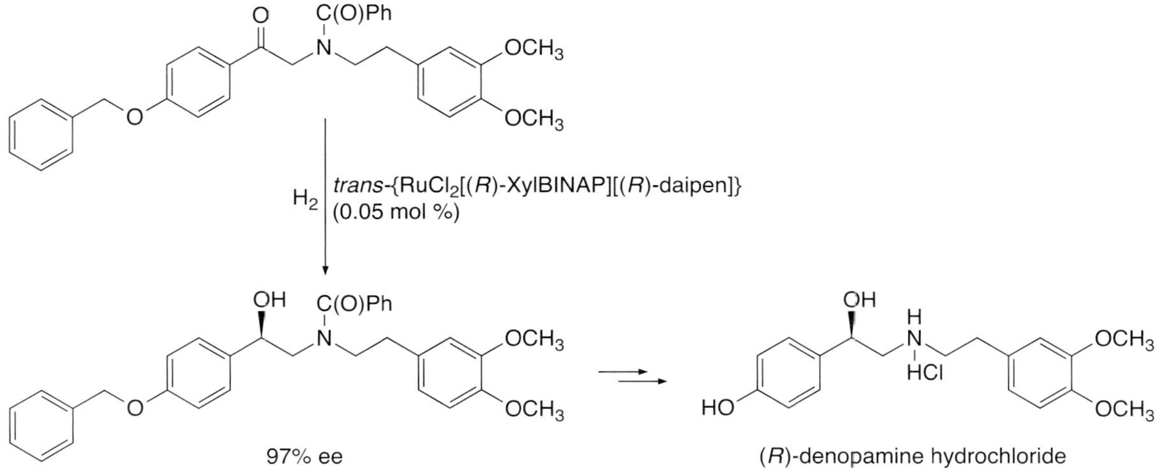
C(O)Ph
OCH3
OCH3
H2
trans-{RuCl2[(R)-XylBINAP][(R)-daipen]}
(0.05 mol %)
97% ee
HCl
(R)-denopamine hydrochloride

(15.64)

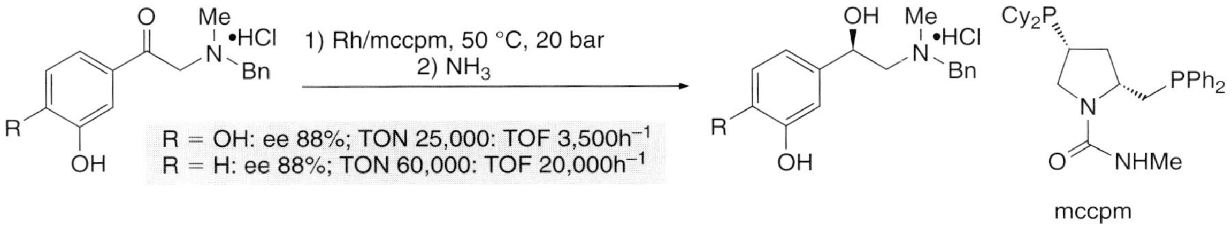
1) Rh/mccpm, 50 °C, 20 bar
2) NH3
R = OH: ee 88%; TON 25,000: TOF 3,500h−1
R = H: ee 88%; TON 60,000: TOF 20,000h−1
mccpm

(15.65)

The asymmetric hydrogenation of β-aminoketones to form optically active amino alcohols is highly desirable because the resulting amino alcohols are precursors to some of the most commonly used antidepressants. It has been more challenging to identify catalysts for this reaction than for the reduction of α-amino ketones, but some success has been achieved. In one study summarized in Equation 15.66, a rhodium complex containing DuanPhos as the ligand catalyzed the hydrogenation of the precursor to (*S*)-fluoxitine (Prozac) in high enantioselectivity.

The hydrogenation of α-hydroxy ketones can also be conducted with high enantioselectivities. In one commercial process shown in Equation 15.67, optically active 1,2-propane diol is prepared in 94% ee by the hydrogenation of α-hydroxy acetone in the presence of a ruthenium catalyst ligated by Tol-BINAP. This process is reported to be conducted on the scale of 50 tons per year for the synthesis of (*S*)-oxafloxazin, a bactericide that is now being sold in enantioenriched form.[170,299]

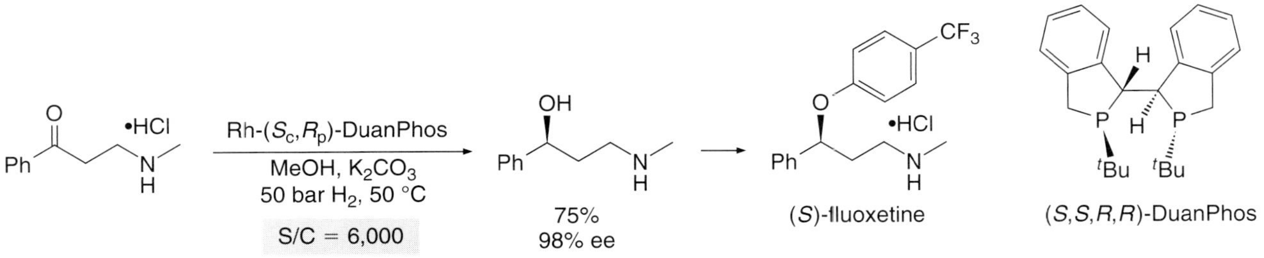
Rh-(Sc,Rp)-DuanPhos
MeOH, K2CO3
50 bar H2, 50 °C
S/C = 6,000
75%
98% ee
(S)-fluoxetine
(S,S,R,R)-DuanPhos

(15.66)

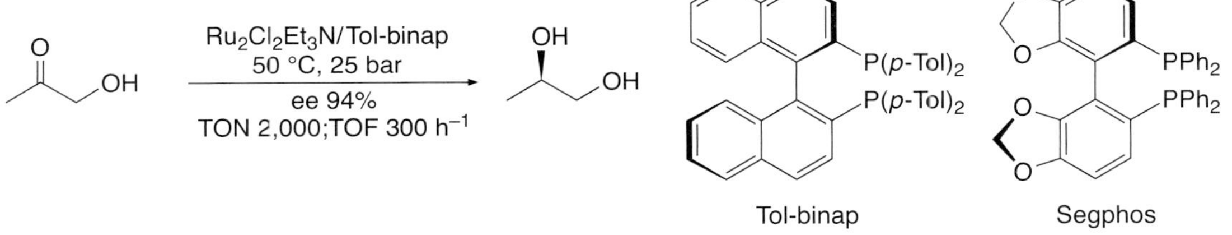
Ru2Cl2Et3N/Tol-binap
50 °C, 25 bar
ee 94%
TON 2,000;TOF 300 h−1
P(p-Tol)2
P(p-Tol)2
PPh2
PPh2
Tol-binap
Segphos

(15.67)

15.7.1.6.2. Hydrogenation of Unfunctionalized Ketones[308]

The hydrogenation of "unfunctionalized" ketones is challenging, just as the asymmetric hydrogenation of unfunctionalized olefins has been challenging, because these ketones lack a neighboring functional group to create a structure with a rigid two-point binding of the substrate to the catalyst. However, practical catalysts for the asymmetric hydrogenation of this class of ketone have been developed and are now used in commercial processes. As described previously, the asymmetric hydrogenation of ketones has become a practical reaction with the discovery by Noyori of catalysts containing the combination of a bisphosphine and a diamine. Zhang also achieved the hydrogenation of unfunctionalized ketones with a rhodium catalyst containing the aliphatic bisphosphine PennPhos. This reaction requires 2,6-lutidine and KBr as additives, and the mechanism of this process is currently unknown. However, high enantioselectivities were observed for alkyl aryl ketones, as well as some dialkyl ketones (Equations 15.68 and 15.69).

$$\underset{Ph}{\overset{O}{\|}}\underset{}{C}Et \quad \xrightarrow[\text{MeOH, RT, 30 atm } H_2, 88 \text{ h}]{\substack{[RhCl(COD)]_2\text{-}(R,S,R,S)\text{-}Me\text{-}PennPhos \\ 2,6\text{-lutidine} + KBr}} \quad \underset{Ph}{\overset{OH}{\underset{*}{\|}}}Et \qquad (15.68)$$

95% ee

$$\underset{^nBu}{\overset{O}{\|}}\underset{}{C}Me \quad \xrightarrow[\text{MeOH, RT, 30 atm } H_2, 88 \text{ h}]{\substack{[RhCl(COD)]_2\text{-}(R,S,R,S)\text{-}Me\text{-}PennPhos \\ 2,6\text{-lutidine} + KBr}} \quad \underset{^nBu}{\overset{OH}{\underset{*}{\|}}}Me \qquad (15.69)$$

75% ee

The first generation of bisphosphine diamine catalysts for the hydrogenation of unfunctionalized ketones was generated from $trans$-{RuCl$_2$[(S)-tol–BINAP][(S)-DPEN]}, shown at the top of Figure 15.13. This dichloride catalyst was converted in situ to the dihydride catalyst by the combination of BuOK in PrOH solvent. The hydrido chloride complex was not active in the absence of added base.[308] As shown for the simple example in Equation 15.70, this complex is remarkably active and selective for the asymmetric hydrogenation of alkyl aryl ketones.[31] Comparable or higher activities have been achieved with catalysts containing other bisphosphines, such as xylyl versions of Phanephos[309] and P-phos.[310] Now, more than 20 catalysts form the product from the reduction of acetophenone with greater than 93% ee. The conditions for two commercial processes are shown in Equation 15.71.[311,312]

Original catalyst: **Optimized catalyst:**

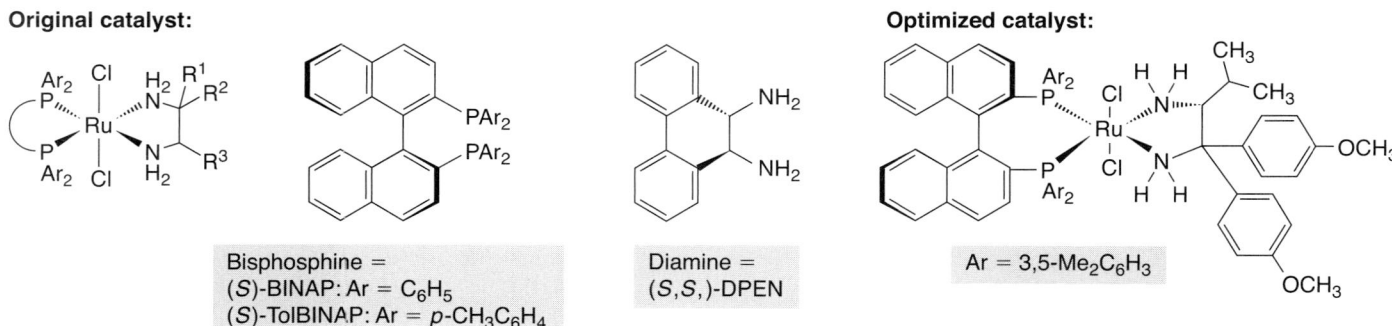

Bisphosphine =
(S)-BINAP: Ar = C$_6$H$_5$
(S)-TolBINAP: Ar = p-CH$_3$C$_6$H$_4$

Diamine =
(S,S,)-DPEN

Ar = 3,5-Me$_2$C$_6$H$_3$

Figure 15.13. Noyori's catalyst for the hydrogenation of unfunctionalized ketones.

$$(15.70)$$

$$(15.71)$$

In some cases, the use of base can be incompatible with substrates or can add an undesirable variable to running a process. This base can be avoided by starting with the complexes $[Ru(bisphosphine)(diamine)(H)(BH_4)]$.[313]

Many combinations of bisphosphines and diamines have now been used for the asymmetric hydrogenation of ketones, and different combinations are optimal for different substrates. One of the most general catalysts for the hydrogenation of unfunctionalized ketones, however, is trans-{RuCl$_2$[(S)-Xyl–BINAP][(S)-DAIPEN]}. This catalyst resulted from optimization of both the bisphosphine and the diamine portion of the catalyst. The Xyl–BINAP ligand contains methyl groups in the 3- and 5-positions of the P-bound aryl groups, which restrict the conformations of this ligand, and the diamine is unusual in that it lacks C_2 symmetry. This catalyst expanded the scope of ketone hydrogenations to include the reduction of α,β-unsaturated ketones to allylic alcohols by chemoselective and enantioselective reduction of the carbonyl group instead of the olefin (Equation 15.72).[174] A particularly impressive example of stereoselective reduction by distinguishing a methyl from vinyl group is shown in Equation 15.73. This reaction was conducted with the base-free borohydride catalyst.[313]

$$(15.72)$$

$$(15.73)$$

The hydrogenation of diaryl ketones or aryl heteroaryl ketones can even be conducted enantioselectively, if the ketones contain different steric or electronic properties. One impressive example of such a hydrogenation is shown in Equation 15.74. In this process, the aryl thazolyl ketone is reduced with the optimized Noyori catalyst with exceptionally high enantioselectivity.[314] This product is then carried forward for the synthesis of an FDE-IV inhibitor.

$$(15.74)$$

The asymmetric hydrogenation of dialkyl ketones is significantly more challenging than the asymmetric hydrogenation of aryl alkyl ketones or enones. The asymmetric hydrogenation of dialkyl ketones requires the catalyst to distinguish between the two faces of a ketone based on the different steric properties of the two alkyl groups. This has not been achieved in a general way just yet, but methyl cycloalkyl and methyl *tert*-butyl ketones can be reduced with good enantioselectivities. In these cases, the ruthenium catalysts of Noyori and the rhodium–PennPhos catalyst of Zhang are currently the most effective. Three examples of such asymmetric hydrogenations are summarized in Equation 15.75.[174, 199]

Catalyst	R	S/C[a]	Conditions	Yield (%)	ee
trans-RuCl$_2$[(S)-XylBINAP][(S)-daipen] + ^{t}BuOK	Cyclohexyl	11,000	^{i}PrOH, 28 °C, 8 atm H$_2$, 20 h	99	85 (R)
trans-RuCl$_2$[(S)-XylBINAP][(S)-daipen] + ^{t}BuOK	CycloC$_3$H$_5$	11,000	^{i}PrOH, 28 °C, 8 atm H$_2$, 12 h	96	95 (R)
[RhCl(cod)]$_2$-(R,S,R,S)-Me–PennPhos + 2,6-lutidine + KBr	^{i}Bu	100	MeOH, RT, 30 atm H$_2$, 75 h	66	85 (S)

$$(15.75)$$

[a]s/c: substrate to catalyst ratio

The ability to catalyze the asymmetric hydrogenation of aliphatic ketones provides an opportunity to extend the scope of these reactions to the dynamic kinetic resolution of ketones that contain a stereocenter in the enolizable α-position. The catalysts containing bisphosphine and diamine ligands are activated by base, and the base used to activate the metal catalyst can also catalyze the racemization of the ketone. An example of such a process is shown in Equation 15.76.[315]

$$(15.76)$$

15.7.1.7. Asymmetric Hydrogenation of Imines

In contrast to the remarkable advances in the asymmetric hydrogenation of various ketones, the asymmetric hydrogenation of imines remains under development.[316] Several factors make the hydrogenation of imines more challenging than the hydrogenation of ketones. First, the imine and the amine product can act as ligands for the metal and either poison the catalyst or serve as competitive inhibitors. Second, a ketimine is more sterically hindered than a ketone because stable prochiral imines contain three substituents on the C=N bond versus two on the ketone. Imines containing a hydrogen on the nitrogen are not stable enough to use as substrates unless generated in situ and consumed rapidly. Third, imines can adopt two geometric isomeric forms. Hydrogenation of these two isomers can lead to the opposite enantiomers and low enantioselectivity for the overall process. Finally, the C=N bond of an imine is less polarized than the C=O bond of a ketone, and this decreased polarity could affect the rates of the insertion step or the external delivery of a hydride and proton by an outer-sphere mechanism followed by "bifunctional" catalysts. In many cases, the desired product is the primary amine. Thus, the substituent at nitrogen must be chosen such that it can be removed after hydrogenation. Some of these types of imines are included in the examples discussed in this section. The transfer hydrogenation of imines has also been studied and is described in Section 15.7.2. Although a general solution to the hydrogenation of imines has not yet been found, one of the earliest and most celebrated cases of the application of asymmetric hydrogenation to polar C=X bonds involves an imine hydrogenation. This hydrogenation was developed at Ciba-Geigy for the synthesis of the herbicide metolachlor on a large scale.

Three classes of catalysts have been studied for the asymmetric hydrogenation of imines. One class of catalyst is generated from late transition metal precursors and bisphosphines. These catalysts have typically been generated from rhodium and iridium precursors. A second class of catalyst is based on the chiral titanocene[317–319] and zirconocene systems[320] presented in the previous section on the asymmetric hydrogenation of unfunctionalized olefins. The third class of catalyst is used for the transfer hydrogenation of imines and consists of ruthenium or rhodium complexes containing diamine, amino tosylamide, or amino alcohol ligands.[321–323]

15.7.1.7.1. Asymmetric Hydrogenation of Cyclic Imines

Some of the greatest successes in the asymmetric hydrogenation of imines have been gained with cyclic imines because of their fixed geometries. Such reactions can generate valuable saturated heterocycles with the control of stereochemistry α to nitrogen. One simple example to generate a 2-aryl-substituted piperidine is shown in Equation 15.77. A particularly valuable set of examples for natural product synthesis includes the imine hydrogenation shown in Equation 15.78.[324] These hydrogenations generate intermediates used by Beyerman and Rice[325–327] for the synthesis of morphine, dextromethoriphan, and related alkaloids (Equation 15.79).[328] An (S)-Tol-BINAP–Ir complex, in combination with a protic amine, such as benzylamine, catalyzes the hydrogenation of 2-phenyl-3,4,5,6-tetrahydropyridine with enantioselectivities in the range of 90% ee.[329]

$$\text{(15.77)}$$

$$(15.78)$$

85-93% ee

$$(15.79)$$

	R^1	R^2	Yield	ee
Beyerman intermediate	CH$_2$C$_6$H$_5$	OCH$_2$C$_6$H$_5$	86%	97%
Rice intermediate	H	H	95%	99%

These precursors to the morphine alkaloids can also be obtained by the hydrogenation of the related enamide structures in which the R group is joined to the ring system by a C=C bond.[330] The preferred route depends largely on the substrate synthesis. The synthesis of the enamides can be more flexible, and Noyori has published the synthesis of the optically active tetrahydroquinolines from the enamide (Equation 15.80).[331]

$$(15.80)$$

> 99.5% ee

15.7.1.7.2. Asymmetric Hydrogenation of Acyclic N-Alkyl Imines

The hydrogenation of acyclic imines has been challenging because of the flexible structures of acyclic substrates and the interconversion of E/Z geometries about the C=N bond. The chiral titanocene system is among the most active and selective for the hydrogenation of N-alkyl imines.[317–319,332] However, the issues of cost and air sensitivity noted previously have limited its use in synthesis. Thus, the majority of effort has focused on developing catalysts based on late transition metals. The highest enantioselectivity (94% ee) for the hydrogenation of acetophenone N-benzylimine was obtained with a neutral monosulfonated (S,S)-BDPP–Rh complex in a mixed solvent system consisting of EtOAc and H$_2$O (Equation 15.81).[333]

$$(15.81)$$

94% ee   (S,S)-BDPP: Ar = Ph

However, the asymmetric hydrogenation of *N*-aryl imines occurs in high enantioselectivities with many different catalysts. The selectivity of the reactions of these imines is high for several reasons. First, the *N*-aryl ketimines derived from acetophenones exist in one major geometric form. Second, two trans aryl groups on a π-system help to distinguish the Re and Si faces. This effect was seen in the asymmetric hydrogenation of unfunctionalized olefins. The catalyst containing Et-DuPhos and (*R*,*R*)-*trans*-diaminocyclohexane as ligands (Equation 15.82) is among the most selective for the hydrogenation of *N*-aryl imines, although this system requires long reaction times at elevated temperatures. An iridium catalyst generated from f-binaphane with iodine promoter is more active and reacts with similar enantioselectivity (Equation 15.83).[334] This catalyst is closely related to the system described in the next paragraph on the hydrogenation to form the precursor to metolachlor.

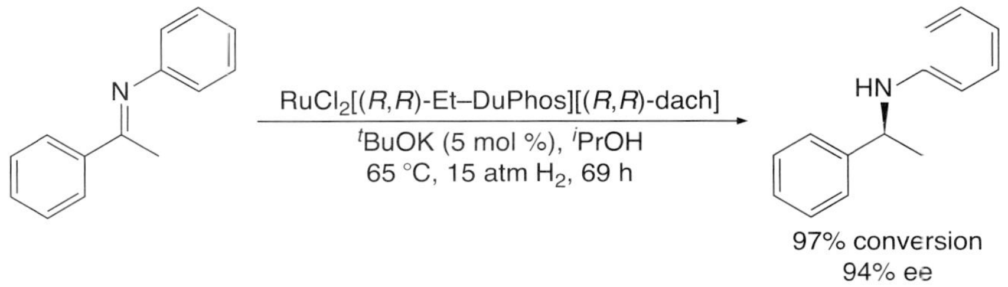

(15.82)

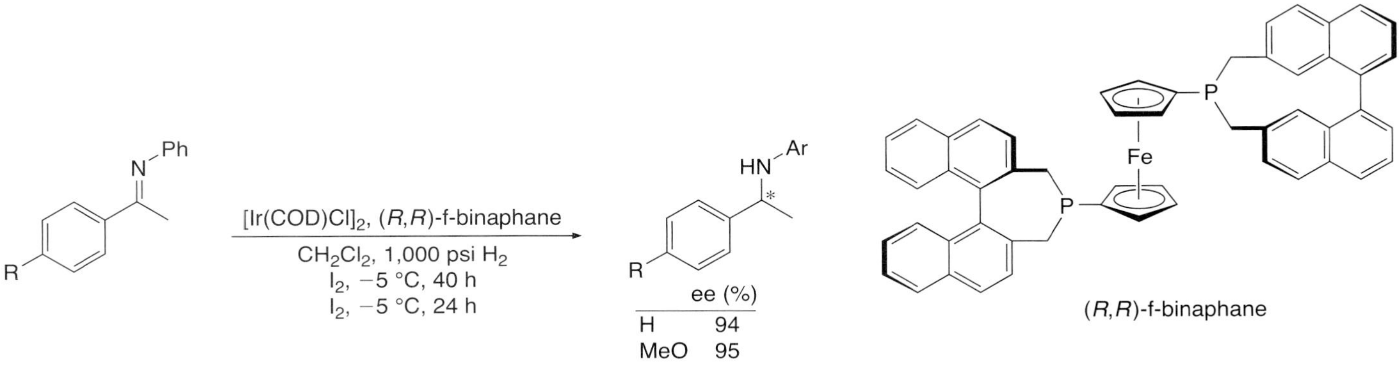

97% conversion
94% ee

	ee (%)
H	94
MeO	95

(*R*,*R*)-f-binaphane

(15.83)

15.7.1.7.3. Asymmetric Hydrogenation of Acyclic *N*-Aryl Imines

A dramatic example of the hydrogenation of an imine is the reaction depicted in Equation 15.84. This process generates the precursor to metolachlor (Figure 15.14) and was developed over the course of roughly a decade. It involved several significant advances. First, it involved an asymmetric catalytic process using iridium, which had not previously been the focus of enantioselective catalysis. Second, it was conducted with the Josiphos ferrocenyl bisphosphines that were a new class of ligand. Third, the rate of hydrogenation far surpassed that of any hydrogenation of a C=X bond (for which X = NR or O). Accounts of the development of this process have been published.[184, 185, 335]

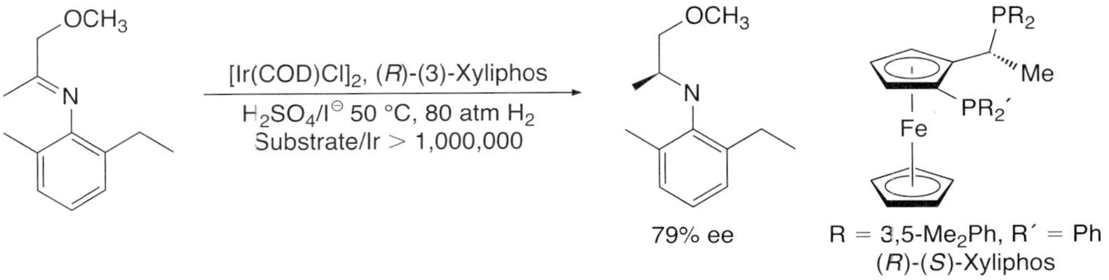

79% ee

R = 3,5-Me$_2$Ph, R′ = Ph
(*R*)-(*S*)-Xyliphos

(15.84)

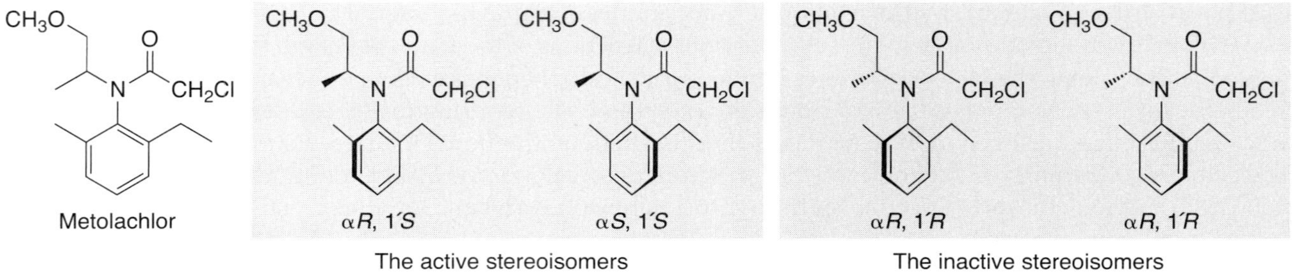

Metolachlor $\alpha R, 1'S$ $\alpha S, 1'S$ $\alpha R, 1'R$ $\alpha R, 1'R$

The active stereoisomers   The inactive stereoisomers

Figure 15.14.
The structure of metolachlor and the importance of stereochemistry in the activity of this herbicide.[335]

This reaction is the largest scale asymmetric process and is conducted on a scale of 10,000 tons/year. The reaction occurs under optimized conditions with 1,000,000 turnovers and with an initial turnover frequency of 1,800,000 h⁻¹. The product generated from the opposite enantiomer of the amine is inactive and nontoxic. Thus, the enantioselectivity of 79% was sufficiently high for production, and the use of enantioenriched material reduces the amount of product that must be applied to the field. Other complexes catalyze this reaction with high enantioselectivities, but without the same activity.

The mechanism of this hydrogenation has not been revealed in detail, but several aspects are noteworthy in the context of the mechanistic discussion earlier in this chapter. First, this reaction almost certainly occurs by the insertion of an imine into a metal hydride (see Chapter 9), rather than by the transfer of a hydride and a proton by a metal–ligand bifunctional system. This iridium catalyst does not contain a protic ligand. Second, acid and iodide are needed as promoters to obtain the fast rates. The iodide is thought to help stabilize an iridium(III) species and the acid is thought to help in the release of the amine product from the metal.

15.7.1.7.4. Asymmetric Hydrogenation of Aroylhydrazones and Phosphinylketimines

The asymmetric hydrogenation of imine derivatives that contain a second ligating functionality to create two-point binding of the substrate have been investigated. Reactions of two types of such substrates are shown in Equations 15.85 and 15.86. In the first example, an aroyl hydrazone undergoes selective hydrogenation to form the enantioenriched N-alkyl hydrazone.[336] Oxidative cleavage of the N–N bond then leads to the primary amine.

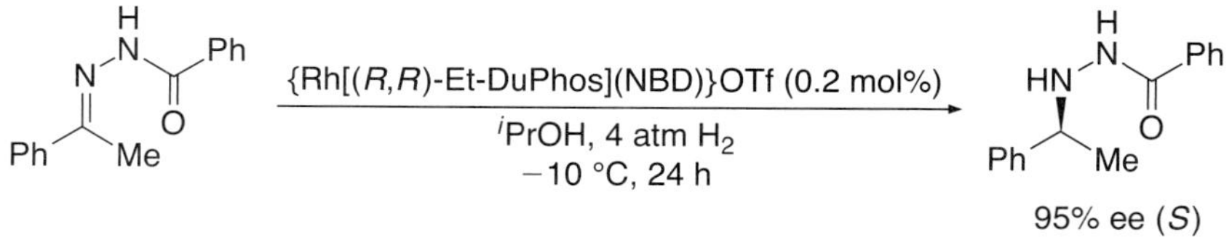

$$\{Rh[(R,R)\text{-Et-DuPhos}](NBD)\}OTf \ (0.2 \ mol\%)$$
$$^iPrOH, \ 4 \ atm \ H_2$$
$$-10\ °C, \ 24 \ h$$

95% ee (S)

(15.85)

Cleavage of a diphenylphosphinylamine to the primary amine occurs under hydrolytic conditions. Thus, the asymmetric hydrogenation of diphenylphosphinylimines can be a synthetically valuable route to optically active primary amines. The presence of the oxygen on phosphorus, like the amide on the aroylhydrazone, can create two-point binding of the substrate to the catalyst. As shown in Equation 15.86, asymmetric hydrogenation of this class of amine can occur with high enantioselectivities in the presence of a rhodium catalyst containing one of the Josiphos ligands.[337]

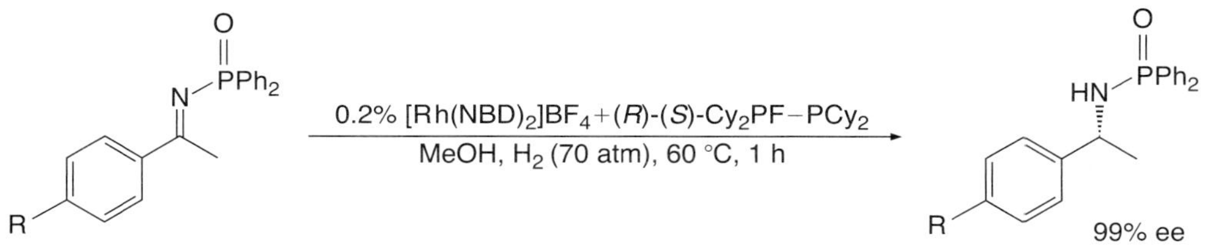

$$0.2\% \; [Rh(NBD)_2]BF_4 + (R)\text{-}(S)\text{-}Cy_2PF\text{-}PCy_2$$
$$MeOH, H_2 \; (70 \; atm), 60 \;°C, 1 \; h$$

(15.86)

99% ee

Two final examples of the asymmetric hydrogenation of an imine involve the synthesis of a pharmaceutical intermediate.[338] Investigation of this process began with the expectation that asymmetric hydrogenation of an unprotected enamine might occur. As shown in Equation 15.87, hydrogenation of the enamine shown generates the product amine with high enantioselectivity. This process is run with a rhodium catalyst containing a Josiphos ligand. However, addition of deuterium does not lead to the product containing deuterium in the two olefinic positions. Instead, deuterium is found only α to nitrogen. This isotopic labeling indicates that the intermediate *NH*-imine is actually the species that undergoes hydrogenation. Generation of this imine in situ by addition of ammonia to the β-keto amide could provide an alternative means to the substrate for hydrogenation. Although not published, researchers at Takasago have disclosed the asymmetric reductive amination of the β-keto amide with high selectivity using ammonium acetate and the corresponding β-keto ester to generate the imine in situ (Equation 15.88).[339] In this case, the reaction was conducted with a ruthenium complex of a 3,5-disubstituted version of their Segphos ligand.

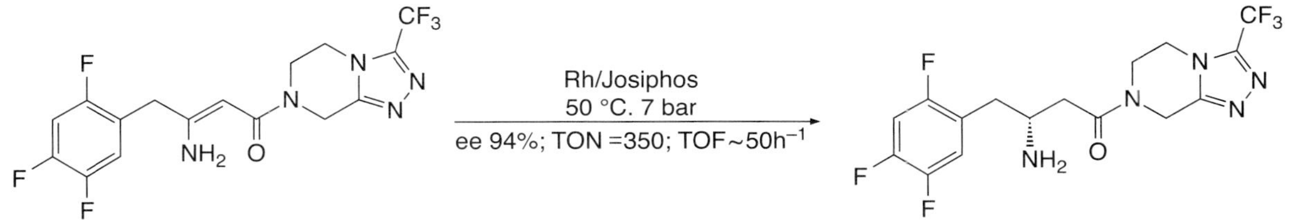

Rh/Josiphos
50 °C. 7 bar
ee 94%; TON = 350; TOF ~50h^{-1}

(15.87)

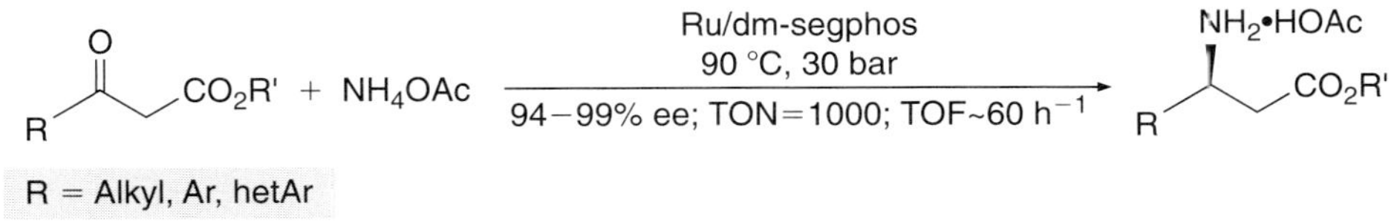

Ru/dm-segphos
90 °C, 30 bar
94 – 99% ee; TON = 1000; TOF ~60 h^{-1}

(15.88)

R = Alkyl, Ar, hetAr

15.7.2. Asymmetric Transfer Hydrogenation of Ketones and Imines[120,123–125,127–132,340,341]

A classic reduction of ketones is the Meerwein–Pondorf–Verlay reaction, in which a ketone is reduced by an alcohol in an equilibrium process catalyzed by aluminum oxides. A modern version of this overall transformation that occurs by a different mechanism is the "transfer hydrogenation" of ketones and imines using alcohols as reagent. For years, little progress on this reaction had been made with transition metal complexes. However, a breakthrough was made when Noyori discovered that ruthenium complexes of amino sulfonamides (Figure 15.15) and amino alcohols catalyze the enantioselective transfer hydrogenations of ketones in alcohol solvents in high yields.[32, 342] These reactions must be conducted carefully because they are equilibrium processes. The product eventually racemizes because of the reversibility of the process; material with lower enantiomeric excess is obtained at longer times. To alleviate this racemization, large excess of the alcohol must be used as a donor solvent.

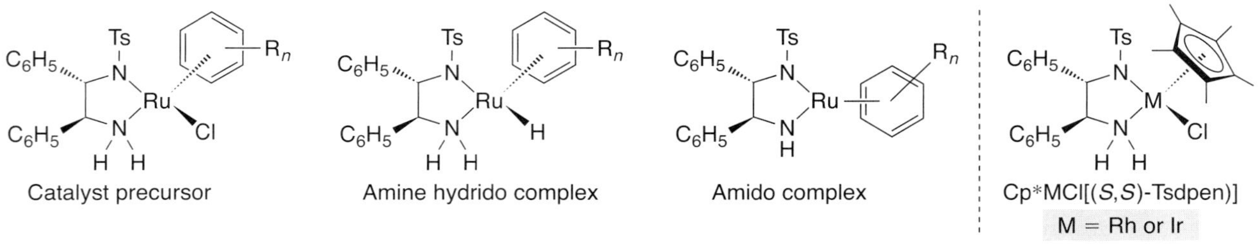

Figure 15.15.
Left: Noyori's original catalyst for transfer hydrogenations in its form as the precursor, the reduced 18-electron reducing species, and the oxidized 16-electron species after transfer of H_2. Right: The Cp*Rh and Cp*Ir analogs of the original Noyori catalysts. Derivatives of the Cp*M fragment ligated by chiral, non-racemic amino alcohols have been developed at Avecia.

Transfer hydrogenation became much more practical when Noyori reported the reductions of ketones using a mixture of formic acid and triethylamine as the source of hydrogen.[118] Related catalysts based on the Cp*Rh and Cp*Ir fragments, combined with a monosulfonated diamine, were reported by several groups (Figure 15.15).[343-345] The ruthenium catalysts have been used commercially by Lanxess, and the rhodium and iridium catalysts have been used by Avecia for commercial transfer hydrogenations. The latter catalysts have been tradenamed CATHy catalysts for *c*atalytic *a*symmetric *t*ransfer *hy*drogenations.

One advantage of transfer hydrogenations is that they occur without the need for hydrogen pressures. A second advantage is that the catalysts are simple to generate because they contain only an amino alcohol, along with the arene or cyclopentadienyl ligand. However, hydrogen is a cheap and clean reagent to use when the appropriate equipment is available, and turnover numbers are often high enough that the catalyst cost is minimal. Thus, most commercial ketone reductions are conducted as hydrogenations, rather than transfer hydrogenations. Nevertheless, transfer hydrogenations are convenient to run on a laboratory scale, little investment needs to be made in catalyst synthesis, and Avecia has run such reactions on a commercial scale. The mechanism of the transfer hydrogenation of ketones and imines by these catalysts was discussed in Section 15.5 on the mechanism of hydrogenations. These reactions are thought to occur by a mechanism involving the outer-sphere bifunctional catalysis summarized in Scheme 15.17. In this scheme, the reducing equivalents are transferred to the catalyst from alcohol solvent by the same metal–ligand bifunctional transition state or from formic acid by protonation of the amide and de-insertion of CO_2 from a formate ligand (Section 15.5.2.3).[154]

Figure 15.16 summarizes a series of optically active alcohols and amines reported (in a review by Blacker at Avecia[346]) to be produced by the CATHy catalysts. These include the types of products described for the different classes of hydrogenations described in the previous sections. These products include those from the reduction of alkyl aryl ketones, α-ketoesters, aliphatic ketones, α,β-unsaturated ketones, cyclic ketones, and α-hydroxy ketones. In addition, transfer hydrogenation allows for the asymmetric formation of amines by the reduction of N-diphenylphosphinylimines. These transfer hydrogenations

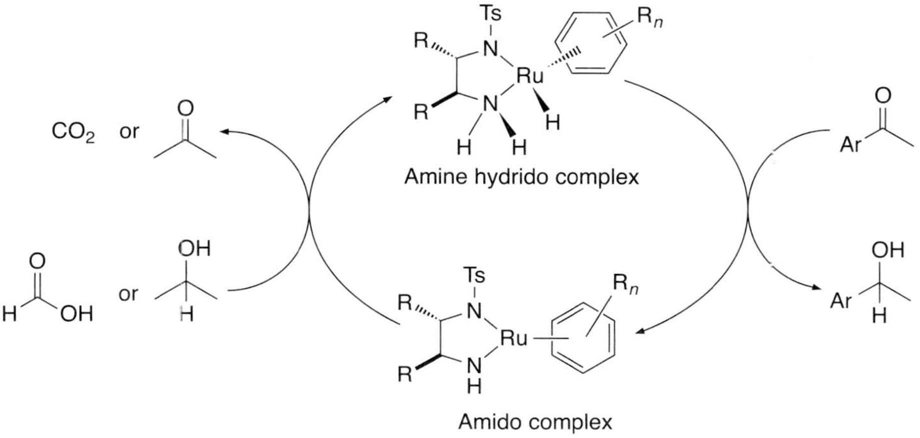

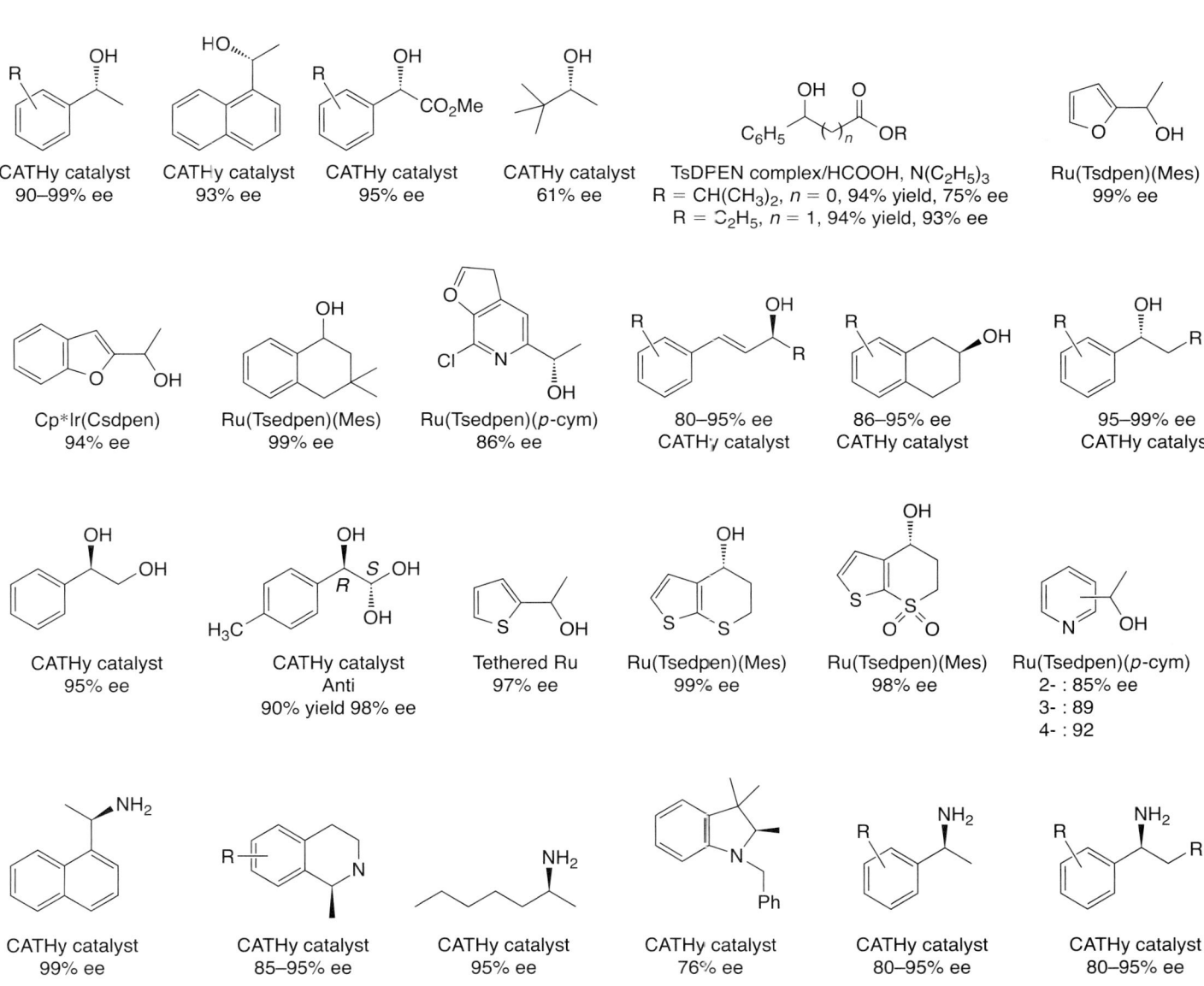

Scheme 15.17

Figure 15.16. Examples of optically active alcohols and amines produced by transfer hydrogenation.

include the reductions of ketimines derived from alkyl aryl ketones, cyclic ketones, and even simple aliphatic ketones.

Optically active amines can also be generated directly from ketones with ammonium formate in a catalytic version of the Leuckart–Wallach reaction[347] (Equation 15.89).[348] The imine is derived from the ammonia generated from the ammonium formate, and the reducing equivalent is derived from the formate. In the presence of the [Ru(BINAP)Cl$_2$]$_2$ catalyst, the amine and formamide were formed as the major products. After hydrolysis of the formamide, the amine was isolated in high yield and ee.

$$
\begin{array}{ccc}
\text{O} & \text{NH}_2 & \text{OH} & \text{NHCHO} \\
\text{Ph} & \xrightarrow[\text{[((R)-tol-binap)RuCl}_2]}{\underset{\text{1 mol \%}}{\text{HCOONH}_4}} & \text{Ph} + \text{Ph} + \text{Ph} \\
& & 75\% & & 19\%
\end{array}
$$

75%
92% after hydrolysis of the formamide
95% ee   (15.89)

15.7.3. Mechanism of Asymmetric Catalytic Hydrogenation of α-Acetamidocinnamic Acid Esters

As described previously, the first commercial application of asymmetric hydrogenation was the Monsanto process for the manufacture of L-Dopa, developed by Knowles (Equation 10.23).[202] L-Dopa is used to treat Parkinson's disease. For these reasons, Halpern and Brown studied the mechanism of this enantioselective process, and the results of these studies were particularly enlightening about how physical organic principles apply to asymmetric catalysis. This process was used as a case study in Chapter 14 to present how enantioselectivity is controlled. The findings are reiterated briefly here.

Prochiral olefins such as the (Z)-α-acetamidocinnamic acid derivatives, MAC and EAC, bind to the rhodium–DIPAMP catalyst to form the diastereomeric olefin complexes A and A′ shown in Scheme 15.18.[69] Oxidative addition of dihydrogen to each of the diastereomers forms diastereomeric hydride complexes B and B′, each of which undergoes migratory insertion to form alkyl hydride complexes C and C′. These complexes then undergo reductive elimination to form the amino acid ester products. The stereochemistry of the organic products can be predicted for the two parallel paths, assuming that the addition of H$_2$ occurs to the face of the olefin coordinated to Rh. The N-acetyl-(R)-phenylalanine ester would be produced from the olefin adduct A and the (S)-product from A′.

Multinuclear NMR studies revealed the presence of only a single diastereomer (A or A′) for reactions of EAC with the rhodium complex of the chiral ligand (S,S)-CHIRAPHOS. One might assume that the product chirality is determined by this preferred mode of substrate binding. This would be a form of "lock-and-key" mechanism that has its origins in Fischer's early concepts of enzymatic stereospecificity. This assumption, however, is incorrect. The predominant enantiomer of the product is actually formed from the minor diastereomer of the catalyst—namely, olefin adduct A.

Two sets of data showed that the minor diastereomeric olefin complex generated the major enantiomer of the product. First, the structure of the major diastereomer of [Rh(S,S-CHIRAPHOS)(EAC)]$^+$ was determined by single-crystal X-ray diffraction. The complex in this structure would form N-acetyl-(S)-phenylalanine ethyl ester, but the major enantiomer of the product was the (R)-isomer. Second, both diastereomers of the olefin complex were observed during the hydrogenation of MAC by the rhodium complex of (R,R)-DIPAMP.[83,349,350] At temperatures low enough that these diastereomers do not interconvert, NMR studies showed that the minor diastereomer reacted much more rapidly with H$_2$ to form the reduced product than the major diastereomer.

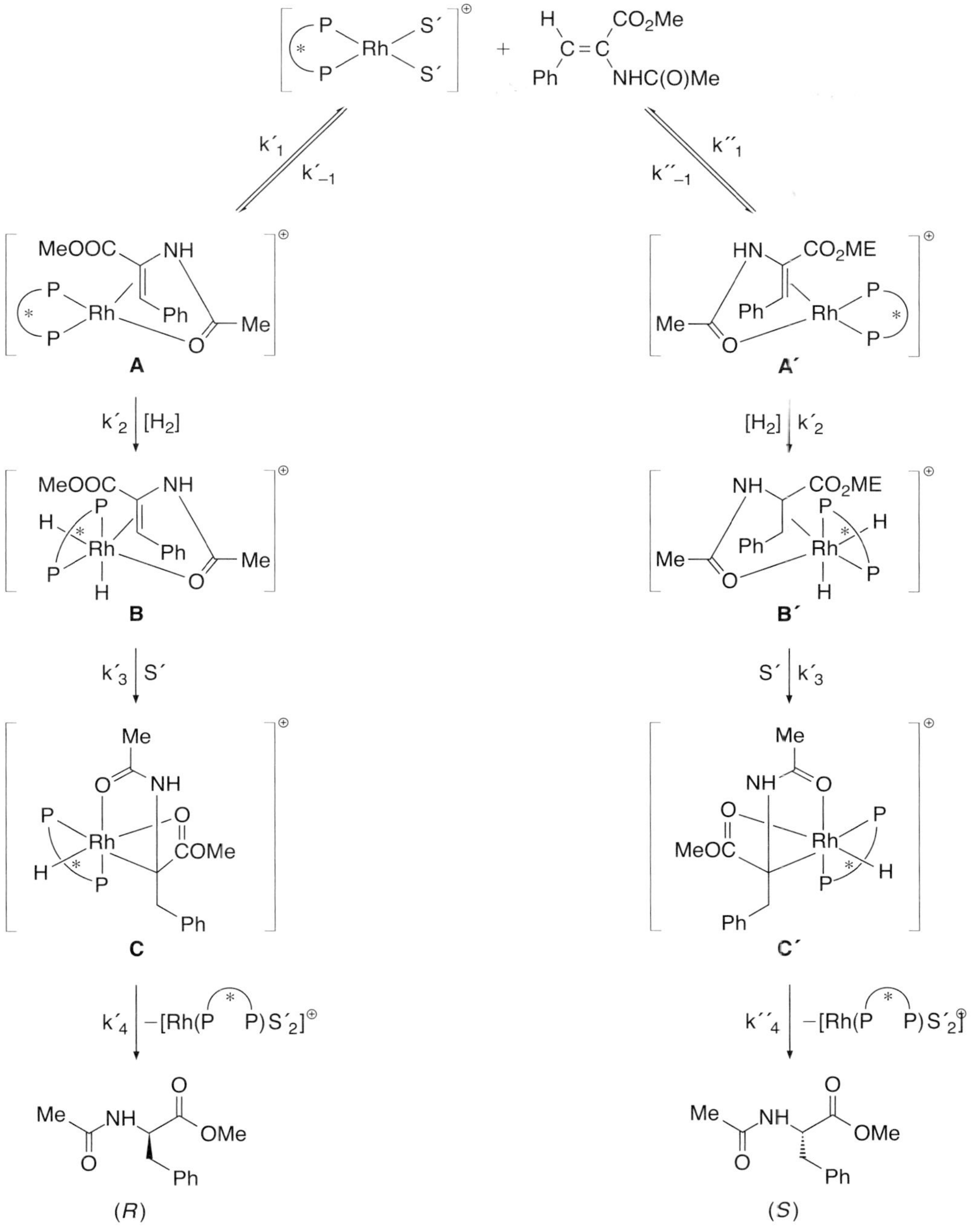

Scheme 15.18

These data imply that there is a large difference between the rate of reaction of the two diastereomeric olefin complexes toward the oxidative addition of H_2. For the MAC–DIPAMP complex A (Figure 10.4), the minor diastereomer must be 10^3 times more reactive than the major diastereomer at 25 °C. Furthermore, the differences in reactivity expressed by the free energies of activation must be greater than the differences in thermodynamic stabilities of the two diastereomeric olefin complexes. These differences in energy are schematically represented by the free-energy diagrams in Figure 15.17.

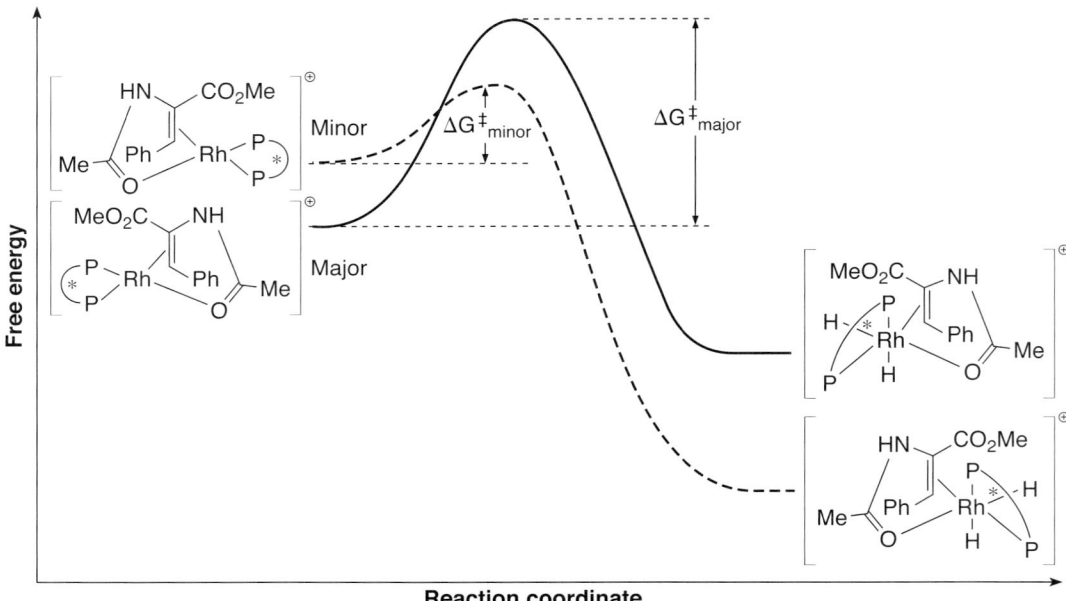

Figure 15.17.
Free-energy diagram for the addition of hydrogen to the two different diastereomeric olefin adducts.

As noted in Chapter 14, the Curtin–Hammett principle applies to this process, and high enantioselectivities are obtained under conditions when the two diastereomeric olefin complexes equilibrate faster than the addition of H_2. Because the relative rates for equilibration versus oxidative addition of H_2 depend on the concentration of hydrogen, enantioselectivities often depend on the pressure of hydrogen. Enantioselectivities of reactions of MAC have been shown in some cases to be higher at lower P_{H_2}.[351,352] In addition, because the equilibration of the diastereomeric olefin complexes is retarded by decreasing the temperature more than addition of H_2 is, hydrogenation at low temperature forms product with *lower* enantioselectivities in some cases.

Many other catalysts for the asymmetric hydrogenation of dehydro α-amino acids appear to occur with relative rates that cause the major enantiomer to be formed from a minor diastereoisomer of the olefin adducts. There are simple experiments that can be used to test for this phenomenon without carrying out a complete kinetic and structural analysis. One criterion pertains to the dependence of the optical yield on the hydrogen pressure. Halpern's kinetic analysis[69] predicts that at higher hydrogen pressure the enantioselectivity should decrease, and in certain cases the predominant product chirality should reverse. This has been observed.[351,352] High hydrogen pressure accelerates hydrogen addition but does not affect the rate of interconversion of the diastereomers A and A'.[351–353] This counterintuitive phenomenon has been used to identify other systems in which the minor diastereomer forms the major enantiomer.

The origins of the enantioselectivity for other types of hydrogenations or for hydrogenations with other ligands have not always led to the same conclusion about the relative rates for reactions of major and minor diastereomers. Evans reported hydrogenations catalyzed by rhodium complexes of a P,S ligand in which the substrate complex was characterized as a single isomer with stereochemistry that would lead to the major enantiomer of the product.[354] A similar observation of the major diastereomer leading to the major enantiomer of the product was reported by Heller for asymmetric hydrogenation of β-acylaminoacrylates by DIPAMP complexes of rhodium.[355,356]

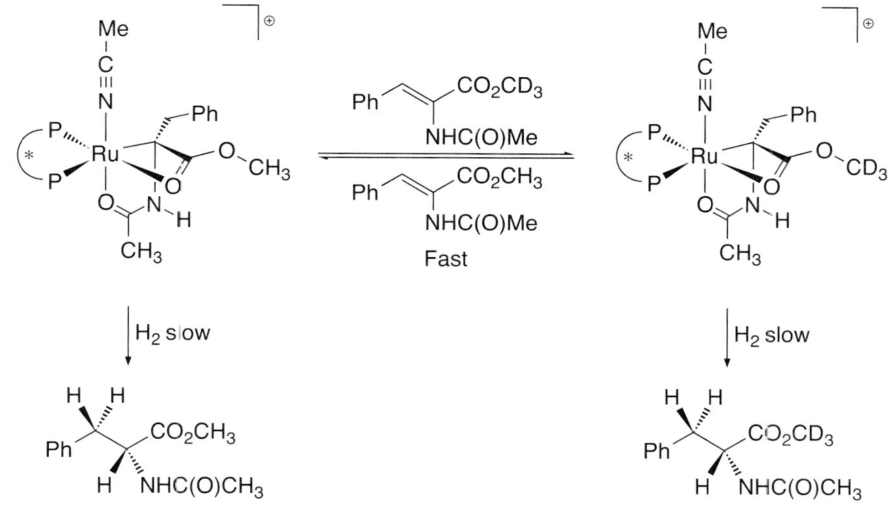

Scheme 15.19

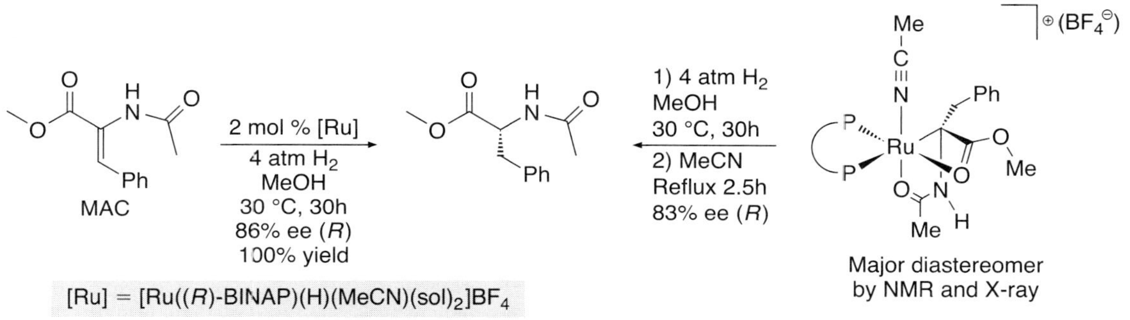

$$(15.90)$$

[Ru] = [Ru((R)-BINAP)(H)(MeCN)(sol)$_2$]BF$_4$

A particularly detailed study on hydrogenation by a ruthenium–BINAP system[357,358] involving isotope labeling, determination of relative rates, and X-ray diffraction of a reaction intermediate showed that the relative rates of the steps in the mechanism of hydrogenation for this catalyst system are likely different than those in the rhodium-catalyzed processes. The insertion of the olefin into the ruthenium hydride appears to be rapid and reversible for this system, and the major diastereomer resulting from insertion was shown from structural studies to correspond to the major enantiomer of the product. Several labeling studies implied that insertion of MAC into the hydride of [Ru(BINAP)(H)(MAC)(NCMe)]⁺ is reversible. One experiment is depicted in Scheme 15.19. This experiment showed that release of the hydrogenation product by reaction of the insertion product with hydrogen is slower than de-insertion and re-insertion of the free substrate.[358] Deuterium incorporation into the β-position of the product from the reaction of the alkyl complex with D$_2$, and the formation of product in only 90% ee from a single diastereomeric alkyl complex, further implied that insertion was reversible. In addition, the major diastereomeric alkyl complex was isolated and characterized by X-ray diffraction.[357] Assuming that hydrogenolysis of the Ru–C bond occurs with retention of configuration, the major diastereomer leads to the major enantiomer of the overall hydrogenation process (Equation 15.90). Thus, Halpern's observation of the formation of the major enantiomer from the minor diastereomeric Rh-MAC complex emphasizes the importance of the Curtin–Hammett principle when considering the origins of enantioselection, but it should not be considered a "rule" for the relationships between the stereochemistry of intermediates and products.

15.8. Hydrogenation of Alkynes and Conjugated Dienes

The hydrogenation of alkynes generates Z-alkenes, and these reactions are typically conducted with heterogeneous catalysts, such as the Lindlar catalyst (palladium on calcium carbonate deactivated with lead).[359] Considering the wide applicability of this catalyst and the lack of chirality in the product of alkyne hydrogenation, the development of the homogeneous hydrogenation of alkynes has received less attention than the homogeneous hydrogenation of olefins, ketones, and imines. However, homogeneous catalysts could lead to fewer side products from Z-to-E isomerization, and isomerization of the position of the C=C bond.

The hydrogenation of conjugated dienes also generates alkenes. In this case, the resulting alkene can possess a stereocenter in the allylic or homoallylic position. However, only recently has the asymmetric hydrogenation of conjugated dienes been studied at all, and no general solution to this synthetic problem has yet been published.[360–363] However, several sets of studies during the early development of homogeneous catalysts focused on understanding the factors controlling the rates of the hydrogenations of dienes, and chromium catalysts are valuable for the selective 1,4-hydrogenation of dienes.

The development of catalysts for the hydrogenation of alkynes and dienes must address several challenges. Most obvious, the catalyst must be much more reactive toward the hydrogenation of the alkyne or diene reagent than the alkene product. Second, the catalyst cannot lead to isomerization of the product by processes such as insertion of the alkene into an intermediate hydride, followed by β-hydrogen elimination, or generation of an allyl hydride intermediate. Third, dienes can undergo hydrogenation to form products of 1,2- and 1,4-additions, and mixtures of such products are undesirable in most cases. Fourth, terminal acetylenes contain reactive C–H bonds, and protonolysis of a ligand by the alkyne C–H bond or oxidative addition of the alkynyl C–H bond can lead to inactive acetylide complexes. Fifth, dienes and alkynes bind more tightly to transition metals than alkenes, and this binding can poison the catalyst. At the same time, the more favorable binding of dienes and alkynes can allow for selective reactions with these functional groups over alkenes.

15.8.1. Rhodium-Catalyzed Hydrogenation of Alkynes and Conjugated Dienes

Because Wilkinson's catalyst reacts much faster with terminal than internal alkenes, it can be used for the selective hydrogenation of internal alkynes, but not terminal alkynes. However, the cationic rhodium catalysts of Osborn-containing achiral phosphine ligands are among the most utilized homogeneous catalysts for the semi-hydrogenation of alkynes.[47] These catalysts have high activities for the hydrogenation of internal alkynes, and low activities for the hydrogenation of the internal alkene products. As noted in Section 15.3.1.2 on the hydrogenation of olefins with these catalysts, added acid to maintain the cationic charge is necessary to prevent isomerization of the olefin. This catalyst is decomposed by terminal alkynes.

These cationic rhodium catalysts have also been studied for the hydrogenation of conjugated dienes. Depending on the substrate and conditions, these reductions of dienes can occur with high turnover frequencies (140–330 turnovers h^{-1}). However, a mixture of 1,2- and 1,4-addition products is formed.[48,364,365]

The mechanism for the hydrogenation of alkynes (Scheme 15.20) is thought to occur by the dihydride route that is analogous to the hydrogenation of olefins discussed earlier in this chapter. Oxidative addition of hydrogen to the $[RhL_2(S)_2]^+$ species is followed by coordination and insertion of the olefin into one of the rhodium–hydride bonds to form a vinyl hydride complex as an intermediate. Reductive elimination from this complex regenerates the starting $[RhL_2(S)_2]^+$ species.

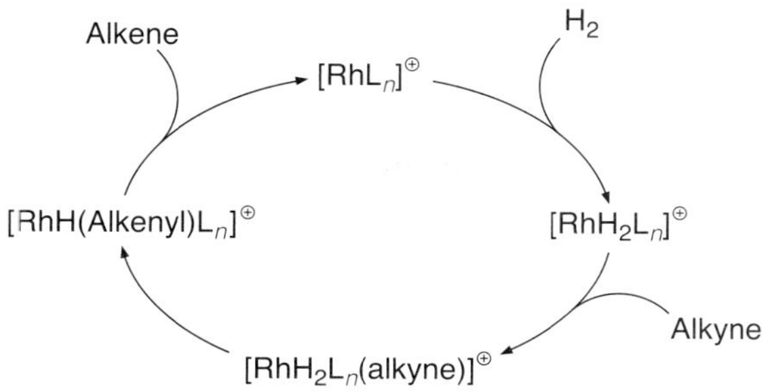

Scheme 15.20

In contrast, the mechanism for the hydrogenation of dienes (Scheme 15.21) appears to occur by initial coordination of the diene, followed by oxidative addition of hydrogen to the diene complex. This order of events results from the high binding affinity of the diene for the $[RhL_2]^+$ fragment. Migratory insertion of the diene into one of the rhodium–hydride bonds then generates an allyl hydride intermediate. Reductive elimination from the allyl hydride complex leads to the monoene products. The allyl hydride intermediate in this process has not been observed, but binding of the allyl group in an η^3-fashion or migration of the metal from one end of the allyl unit to the other through an η^3-intermediate is likely to occur. Either binding as an η^3-complex or rapid migration of the metal will lead to formation of mixtures of 1,2- and 1,4-addition products, as shown in Scheme 15.21. These mixtures of products have been shown to form during the hydrogenation of terpene substrates, such as myrcene, catalyzed ruthenium, rhodium, or iridium catalysts (Equation 15.91).[366]

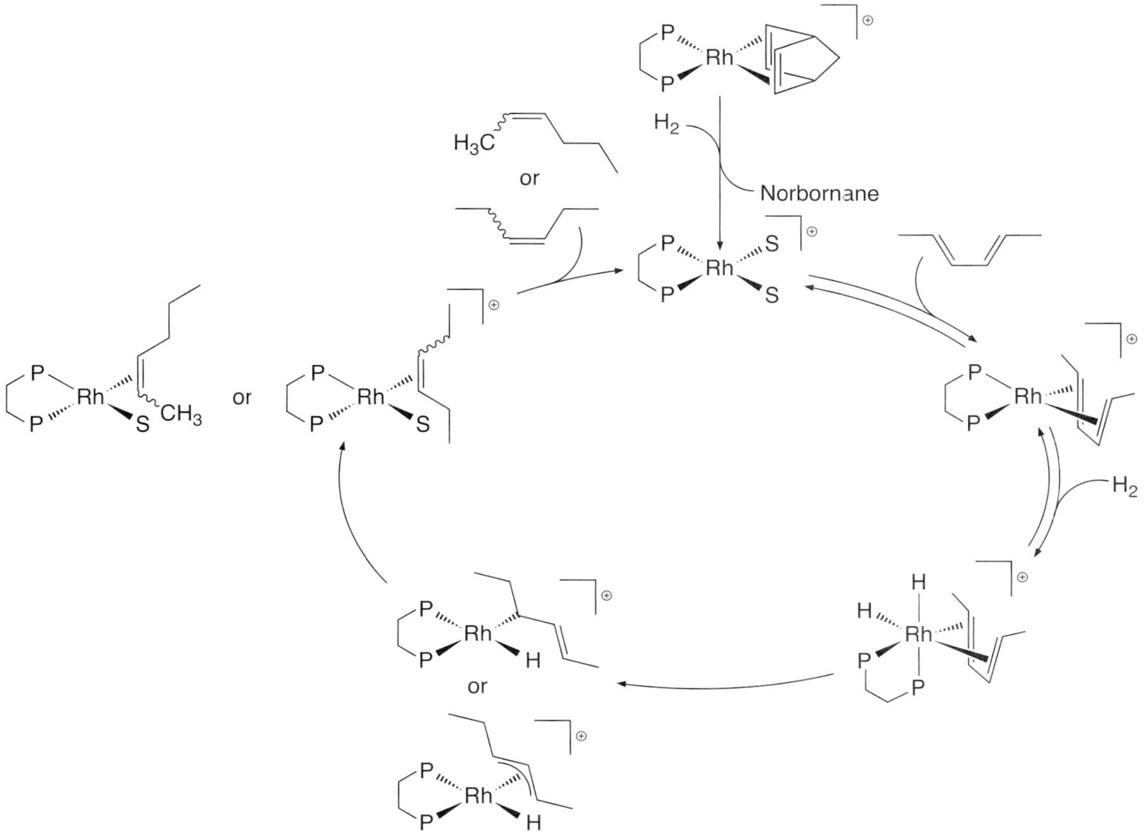

Scheme 15.21

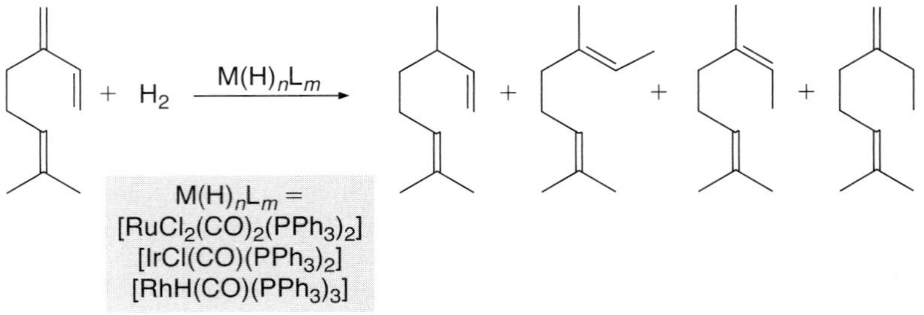

$$M(H)_nL_m =$$
$$[RuCl_2(CO)_2(PPh_3)_2]$$
$$[IrCl(CO)(PPh_3)_2]$$
$$[RhH(CO)(PPh_3)_3]$$

$$(15.91)$$

15.8.2. Chromium-Catalyzed Hydrogenation of Alkynes and Conjugated Dienes

Chromium catalysts of the general form $[Cr(arene)(CO)_3]$ also catalyzed the hydrogenation of alkynes and dienes to form alkenes selectively. These catalysts were reported in 1968 by Cais[367] and by Frankel,[368] about the time of the development of Wilkinson's hydrogenation catalyst.[101] An example of such a hydrogenation is shown in Equation 15.92. This catalyst is exquisitely selective for the hydrogenation of alkynes because it is completely unreactive toward alkenes. However, the hydrogenations of alkynes by these catalysts require very high pressures of hydrogen. Thus, these systems are not frequently used for the semi-reduction of alkynes.

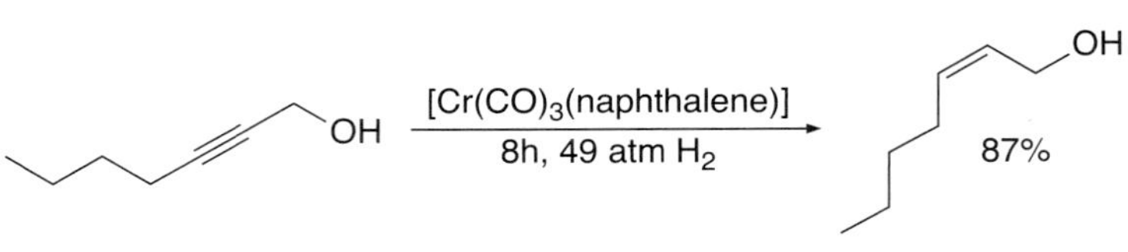

$$(15.92)$$

However, these catalysts have been used for the hydrogenation of dienes because they selectively form Z-monoenes[369–371] by 1,4-additions. This regioselectivity is revealed by the addition of D_2 to the diene in Equation 15.93.[371] Cis monoenes are also produced from 1,4-dienes ("skipped" dienes).[372] In this case, the chromium complex cleaves the activated double allylic C–H bond to induce isomerization to the conjugated diene. Addition of hydrogen then produces cis alkenes. This process has been investigated as a route to partially hydrogenated vegetable oils without the formation of product containing Z-alkene units ("trans fats") that are produced by heterogeneous hydrogenation catalysts.

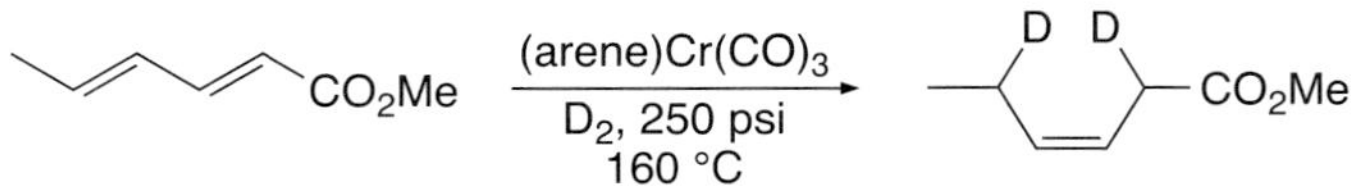

$$(15.93)$$

The best-studied precatalysts for this reaction are of the type $L_3Cr(CO)_3$, in which L_3 is an arene,[372,373] $(CH_3CN)_3$, or $(CO)_3$.[369,370] The arene catalysts are thought to generate $Cr(CO)_3(S)_3$ (in which S = solvent) after thermal or photochemical activation[369,370] of the coordinatively saturated precatalyst. The most active catalyst of this type is the naphthalene complex $(\eta^6\text{-naphthalene})Cr(CO)_3$.[372,373]

A proposed mechanism for this diene hydrogenation is shown in Scheme 15.22. After generation of the $Cr(CO)_3(S)_3$ fragment, this hydrogenation is proposed to occur by initial binding of the alkyne or diene or by initial oxidative addition of H_2. In the case of the diene

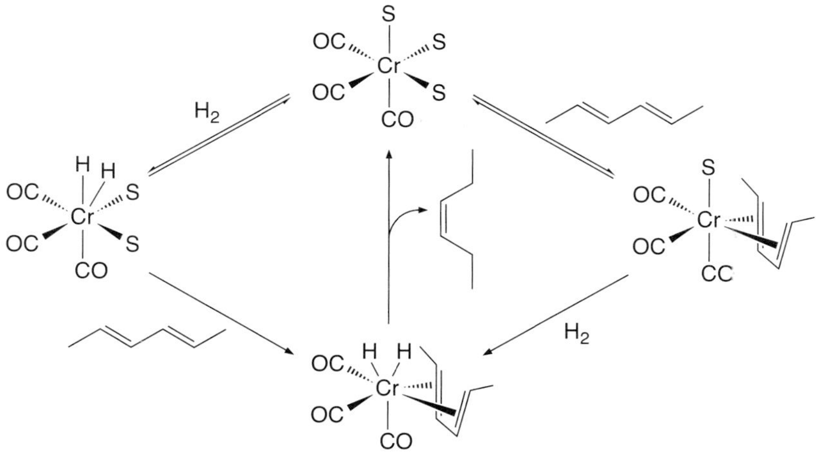

Scheme 15.22

hydrogenation shown in Scheme 15.22, the pathways involving initial coordination of the diene and initial coordination of H_2 are thought to occur in parallel. The origin of the 1,4-regioselectivity is unclear. Studies have shown that the diene must be able to adopt an S-cis geometry for binding to the metal. Addition of the two hydrides to this geometry of the diene by some mechanism gives rise to the 1,4-addition product. An example of this high 1,4-selectivity in the context of a synthetic application is shown in Equation 15.94.[374,375] In this example, Sodeoka and Shibasaki used the chromium-catalyzed diene hydrogenation for the stereoselective preparation of geometrically defined prostacyclins and their analogs.[376]

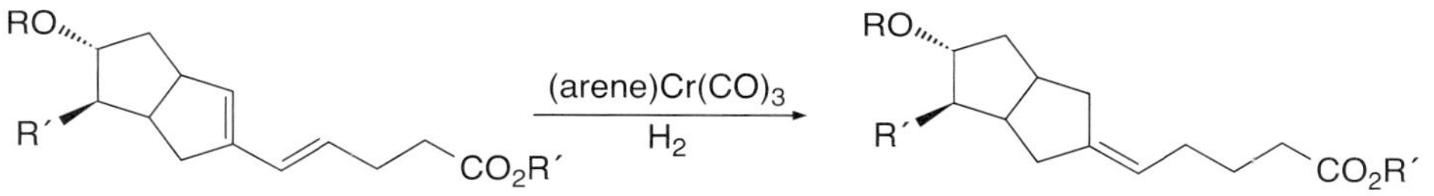

(15.94)

15.8.3. Palladium-Catalyzed Hydrogenation of Alkynes and Conjugated Dienes

Although soluble palladium catalysts for the hydrogenation of alkenes typically undergo reduction and generate palladium colloids, Elsevier has reported a very active and selective catalyst for the cis, semi-hydrogenation of terminal and internal alkynes using either hydrogen[377] or ammonium formate as the hydrogen source.[378] This catalyst for hydrogenation (Equation 15.95) contains an α-diimine ligand that was used to generate palladium catalysts for the polymerization of olefins (discussed in Chapter 22). The catalyst for the transfer hydrogenation with formic acid (Equation 15.95) contains the IMes N-heterocyclic carbene ligand bearing mesityl groups on nitrogen (discussed in Chapter 2). Under the conditions of hydrogenation, internal aliphatic and aromatic alkynes were reduced to form the Z-alkene in > 90% yields. A maximum of 6% E-alkene was formed, and only the hydrogenation of diphenylacetylene was observed. Similar selectivities were obtained for transfer hydrogenations.

$$Ph\text{—}\!\!\equiv\!\!\text{—}Me \xrightarrow[\substack{HCO_2H/NEt_3(5\ equiv.) \\ or\ hydrogen\ 1\ atm \\ THF\ or\ MeCN,\ N_2,\ 20\ °C}]{1\ mol\ \%\ Pd\ catalyst} $$

Ph / Me (cis)   +   Ph / Me

92–95%   2–4%   (15.95)

Pd catalyst
for hydrogenation

Pd catalyst
for transfer
hydrogenation

15.9. Homogeneous Catalytic Hydrogenation of Arenes and Heteroarenes

Many soluble transition metal complexes serve as precatalysts for the hydrogenation of arenes.[164,379–384] These catalysts can be divided into two categories: systems that promote the hydrogenation of all but one ring in polycyclic aromatic substrates, such as naphthalene or anthracene, and systems that promote the hydrogenation of simple monocyclic arenes.[382–384] Studies of reaction mechanisms indicate that at least some of the catalysts in the first category are homogeneous, but all of the current soluble catalysts for the hydrogenation of monocyclic arenes are thought to generate colloidal metal that acts as a heterogenous catalyst.[382–384] Some of the catalysts generated from soluble precursors are very active for arene hydrogenation, particularly those containing a combination of rhodium and palladium.[385–391] More recently, studies to develop homogeneous catalysts for the hydrogenation of heteroaromatic substrates have gained increasing attention with the goal of enantioselective heteroarene hydrogenation. These substrates are more reactive than arenes, and bonafide homogeneous catalysts for the hydrogenation of heteroarenes are known. Some of these catalysts generate chiral, non-racemic saturated heterocycles with high enantioselectivities.[392,393] This section on arene and heteroarene hydrogenation is divided into the hydrogenation of polycyclic arenes, monocyclic arenes, six-membered heteroarenes (pyridines, quinolines, isoquinolines, etc.), and five-membered heteroarenes (pyrroles, indoles, furans, etc).

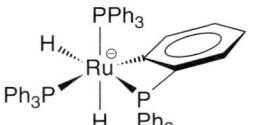

Figure 15.18.
Anionic precatalyst for arene hydrogenation.

15.9.1. Homogeneous Catalytic Hydrogenation of Polycyclic Arenes

A particularly active catalyst for the hydrogenation of polycylic arenes has been generated from the anionic hydridoruthenium precatalyst in Figure 15.18, discovered by Grey and Pez.[58] This system catalyzes the hydrogenation of anthracene to 1,2,3,4-tetrahydroanthracene under relatively mild conditions with a turnover frequency of $> 50\ h^{-1}$ and a turnover number of > 100 (Equation 15.96). Both anthracene and naphthalene are partially hydrogenated in the presence of this catalyst, but monocyclic arenes do not react.

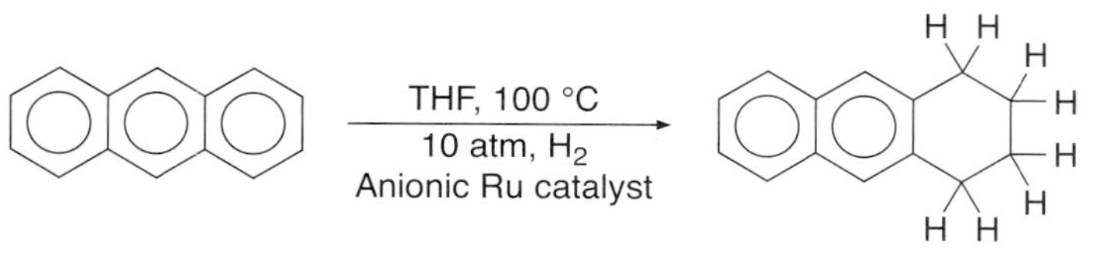

(15.96)

The coordination chemistry of this anionic, cyclometallated precatalyst and its derivatives, along with its stoichiometric reactions with anthracene, have been reported by Halpern.[394] On the basis of these studies, a plausible mechanism can be proposed for the hydrogenation of polycyclic arenes by this catalyst. Although kinetic studies of the complete catalytic cycle have not been reported, Halpern's investigation of the stoichiometric reactions of this system provides strong evidence that the reaction occurs by a series of steps involving soluble complexes.

The putative catalyst is an anthracene complex $[Ru(PPh_3)_2H(\eta^4\text{-anthracene})]^-$ generated by irreversible addition of H_2 to the cyclometallated species to form $[Ru(PPh_3)_3H_3]^-$ (Equation 15.97), followed by reaction of the trihydride with anthracene to form the anthracene complex in Equation 15.98. Addition of four moles of H_2 to the anthracene complex affords tetrahydroanthracene and the pentahydride complex in Equation 15.99. This reaction is rapid at 25 °C. The fluxional pentahydride then reacts with two moles of anthracene, regenerating the η^4-anthracene complex and forming another mole of tetrahydroanthracene (Equation 15.100). The reaction of the η^4-anthracene complex with hydrogen to form the pentahydride and reaction of the pentahydride to regenerate the η^4-anthracene complex creates a plausible catalytic cycle. The reaction of the pentahydride to regenerate $[Ru(PPh_3)_2H(\eta^4\text{-anthracene})]$ (Equation 15.100) is slower than the addition of hydrogen to the η^4-anthracene complex. The η^4 coordination of anthracene in $[Ru(PPh_3)_2H(\eta^4\text{-anthracene})]$ explains the regiochemistry of the reduction and the lack of reaction with monocyclic arenes, such as benzene, toluene, or tetrahydronaphthalene, which form less stable η^4-arene complexes.

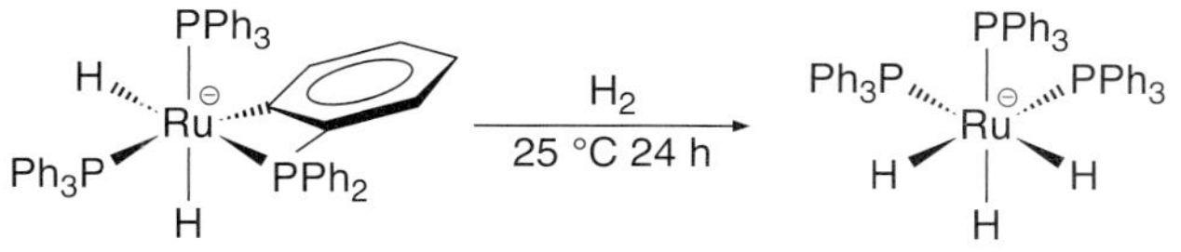

(15.97)

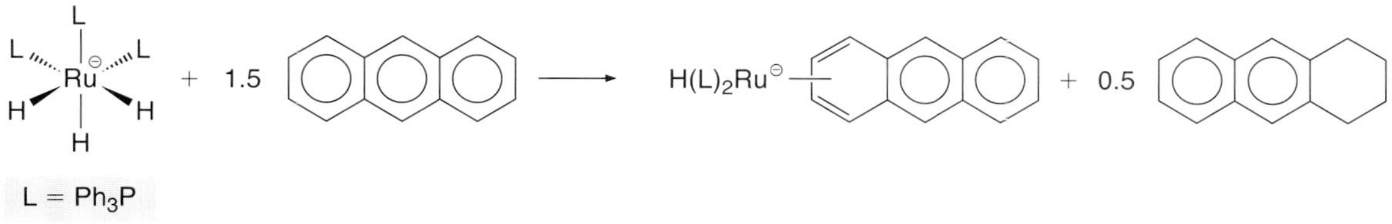

L = Ph$_3$P

(15.98)

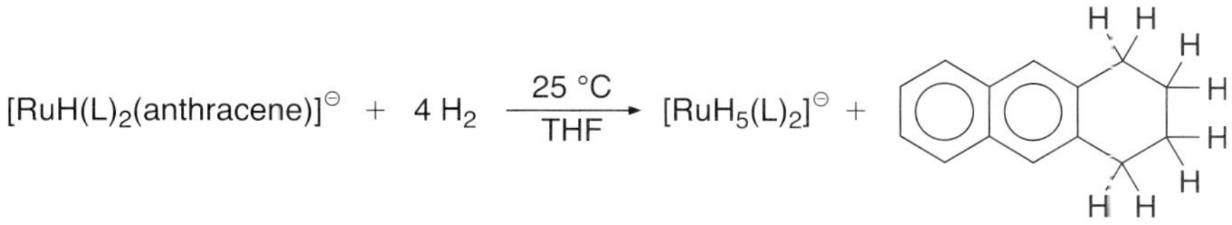

(15.99)

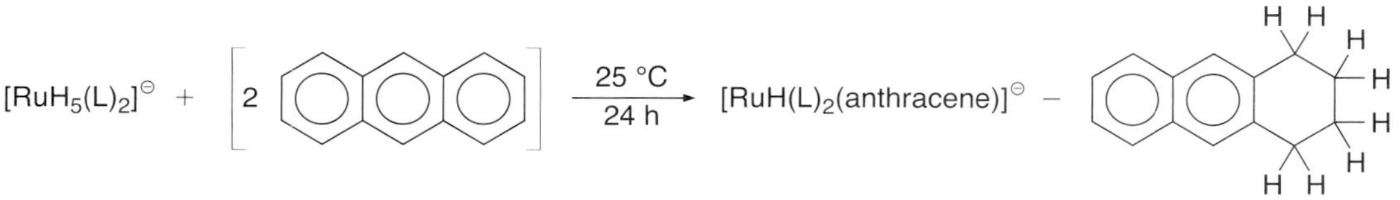

(15.100)

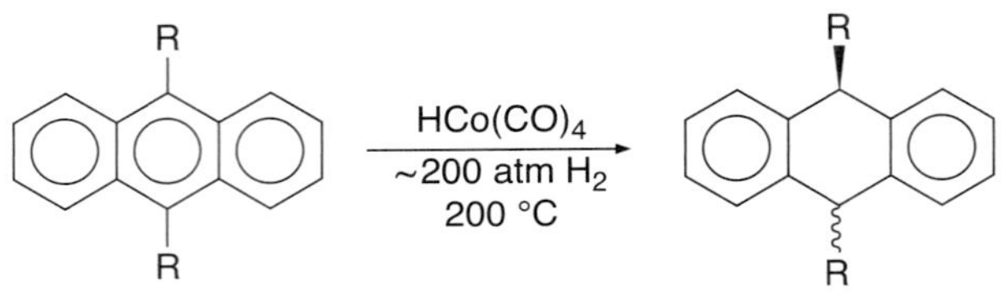

(15.101)

Polycyclic arenes are also hydrogenated in the presence of the monohydride catalyst, $HCo(CO)_4$ (Equation 15.101). The regioselectivity of this reduction is different from that of the hydrogenation of anthracene catalyzed by the ruthenium complex of Grey and Pez. This difference in regioselectivity results from a different mechanism. Halpern showed that the hydrogenation of polycyclic arenes catalyzed by $HCo(CO)_4$ occurs by hydrogen atom transfer.[395]

Several pieces of data suggest that the reaction catalyzed by $HCo(CO)_4$ occurs by hydrogen atom transfer, rather than by formation of a π-anthracenyl complex. First, added carbon monoxide does not inhibit the rate, nor does it result in carbonylated products. Second, only 9,10-dihydroanthracene products are formed, and they are formed as a mixture of cis and trans isomers. Third, exchange occurs between deuterium in the 9- and 10-positions and H_2.

These data are accommodated by the free-radical mechanism[395] shown stepwise in Scheme 15.23. The donation of a hydrogen atom from $HCo(CO)_4$ to the anthracene occurs reversibly. This step is then followed by fast hydrogen-atom abstraction from another $HCo(CO)_4$ to form the reduced product and $\bullet Co(CO)_4$. Dimerization of $\bullet Co(CO)_4$ and hydrogenolysis of $Co_2(CO)_8$ regenerates the starting $HCo(CO)_4$.

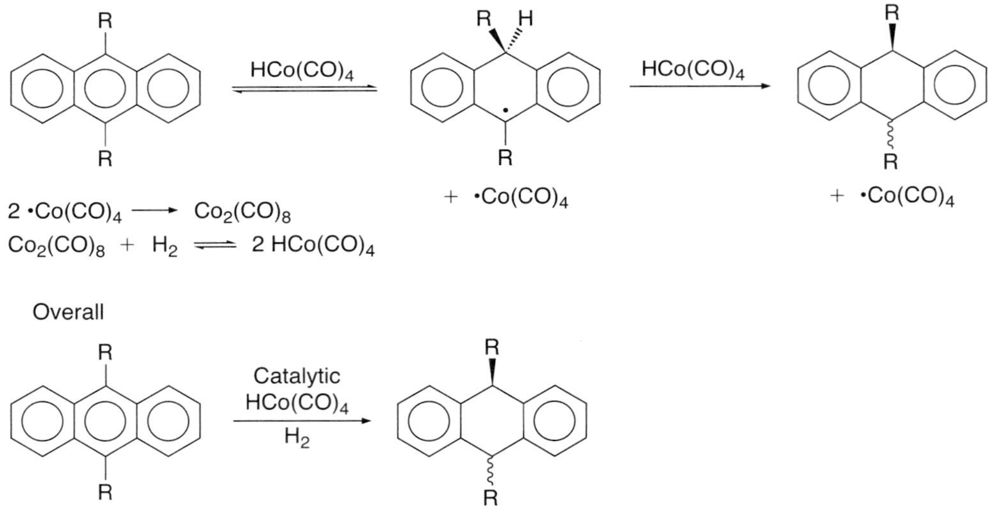

$$2 \bullet Co(CO)_4 \longrightarrow Co_2(CO)_8$$
$$Co_2(CO)_8 + H_2 \rightleftharpoons 2\, HCo(CO)_4$$

Overall

Scheme 15.23

The enthalpy of activation for the turnover-limiting hydrogen-atom transfer step depends upon the affinity of the anthracene substrate for free radicals. The 9,10-dihydroanthracene is formed as the product because the radical is most stable in these positions, and the mixture of stereoisomers is formed because the process does not involve migratory insertion. This mechanism also explains why simple benzene derivatives are unreactive. The 1,4-hexadienyl radical lies too far uphill of benzene and the cobalt hydride to be a product from a reaction of a catalytic cycle that occurs under mild conditions.

15.9.2. Hydrogenation of Monocyclic Arenes

Several soluble transition metal complexes, such as {Co(η^3-allyl)[P(OMe)$_3$]$_3$}, [Cp*RhCl$_2$]$_2$, [Ru(η^6-C$_6$H$_6$)(η^4-C$_6$H$_6$)], and {[Ru(η^6-C$_6$H$_6$)]$_2$(μ-H)(μ-Cl)}Cl, serve as precatalysts for the hydrogenation of benzene and its derivatives. Typically, these catalysts require a high pressure of hydrogen or generate catalysts that react slowly and with short lifetimes as a soluble species. From a practical point of view, "homogeneous" monocyclic arene hydrogenation catalysts have not proven to be simultaneously selective and reactive enough to supplant the well established heterogeneous catalysts.

Much effort has been spent to ascertain whether these soluble precatalysts generate active heterogeneous catalysts.[382–384] It is currently thought that most or all of these complexes actually do form heterogeneous arene hydrogenation catalysts. Finke has shown that the catalysts react with induction periods that are characteristic of the association of small particles to generate the true catalyst.[383,384]

One of the most sophisticated arene hydrogenation systems was designed by Angelici, and it involved rhodium pyridylphosphine and bipyridyl complexes tethered to a silica-supported heterogeneous palladium catalyst.[385,389] These systems are remarkably active for arene hydrogenation, reducing benzene with hydrogen under ambient conditions. Like other species, this system generates colloidal catalysts. In this case, however, the combination of rhodium and palladium generates more active catalysts than had been generated previously, and study of this bimetallic heterogeneous catalyst continues.[390,391]

15.9.3. Asymmetric Hydrogenation of Heteroarenes

Much progress has been made on the homogeneous hydrogenation of heteroarenes.[393,396,397] The major interest in this process is to access saturated heterocycles, such as piperidines, tetrahydroquinolines, pyrrolidines, dihydroindoles, furans, dihydrofurans, and related structures, with the control of stereochemistry. The two compounds noted by Glorius[392] to be potential targets for pyridine hydrogenation, shown in Figure 15.19, are the insect repellent Bayrepel and the well-known drug Ritalin for the treatment of attention-deficit hyperactivity disorder.[398]

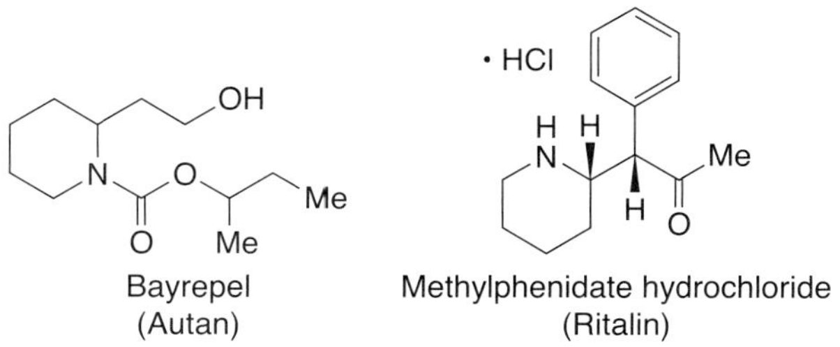

Bayrepel
(Autan)

Methylphenidate hydrochloride
(Ritalin)

Figure 15.19.
Two potential synthetic targets for the asymmetric hydrogenation of pyridine.

Several approaches have been followed for the generation of optically active heteroarenes by homogeneous asymmetric hydrogenation. One approach involves direct asymmetric hydrogenation of the unactivated heteroarene. This approach has been most successful for heteroarenes that are polycyclic, like quinolines, isoquinolines, or quinoxalines (benzopyrazines). Less success has been achieved on the direct asymmetric hydrogenation of pyridines.[399] To address this limitation, the hydrogenation of modified pyridines has been conducted. In one set of examples, pyridines and related benzo-fused heteroarenes were modified at the nitrogen by acylation or the installation of another auxiliary to make the pyridine more electron poor and to dock the catalyst. In a second set of examples, a chiral auxiliary was placed on the pyridine, and the product was formed diastereoselectively by an achiral catalyst.

A particular challenge in the reduction of six-membered ring heteroarenes is the propensity of these basic heteroarenes to bind to the metal to deactivate it or to serve as an achiral ligand. Thus, most success has been gained with heteroarenes containing a benzo-fused group next to the nitrogen or with the substituent and resulting stereocenter α to nitrogen. These substrates bind the metal less tightly for steric reasons. A second challenge results from the lower thermodynamic driving force for the reduction of the heteroarene and the smaller binding constant for coordination of the C–N π-bond to the metal than for coordination of an imine. Because of these challenges, much of the success with pyridine hydrogenations has been gained using iridium catalysts related to the most active imine hydrogenation catalysts or iridium catalysts related to Crabtree's system for the hydrogenation of highly substituted alkenes.

15.9.3.1. Asymmetric Hydrogenation of Six-Membered Ring Heteroarenes

Early studies on the asymmetric hydrogenation of heteroarenes, such as quinazolines, furans, and pyrimidines, were reported by Murata, Takaya, and researchers at Lonza,[400–402] but these reactions occurred with modest enantioselectivities. Subsequent systems for the enantioselective hydrogenation of the heteroaromatic ring of benzo-fused six-membered ring nitrogen heterocycles have been developed that react with high ee. Two of these early examples are shown in Equations 15.102 and 15.103. The first such hydrogenation (Equation 15.102) was the reduction of the quinazoline by Bianchini.[403] The reduction of substituted quinolines has more recently been shown to occur in high yield and enantioselectivity with an iridium catalyst containing the MeO–Biphep ligand[404] under conditions closely mirroring those for the reduction of the precursor to metolachlor (Equation 15.103).

$$(15.102)$$

$$(15.103)$$

The hydrogenations of isoquinoline and pyridine derivatives that exploit the activating ability of substituents at nitrogen are shown in Equations 15.104 and 15.105. The isoquinoline can more easily bind the metal than the 2-substitued benzoquinolines of Equation 15.103, and low conversions were observed with the unactivated substrate. However, Zhou showed that acylation, followed by hydrogenation, formed the reduced products with substantial enantioselectivities.[405] A different approach was used by Charette and co-workers to activate pyridines for hydrogenation. Conversion of the pyridine to a hydrazide derivative in the form of an N-benzoyliminopyridinium ylide makes the pyridine more electron poor and provides a potential second point of binding for the metal. Hydrogenation then forms the 2-substituted piperidine with substantial enantioselectivity after cleavage of the N–N bond.[406] Finally, Glorius has shown that heterogeneous catalysts can be used to generate 3-substituted pyridines with high enantioselectivities by incorporation of an oxazolidinone auxiliary at the 2-position of the pyridine (Equation 15.106). Removal of the auxiliary by hydrolysis generates the optically active 3-subsituted piperidine with high enantioselectivity.[407]

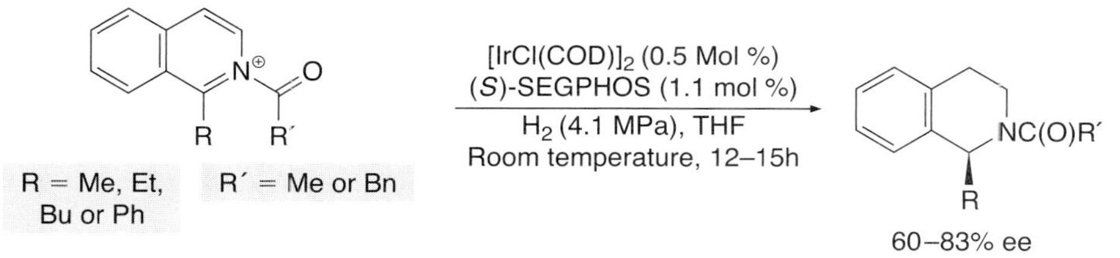
R = Me, Et, Bu or Ph
R′ = Me or Bn
[IrCl(COD)]2 (0.5 Mol %)
(S)-SEGPHOS (1.1 mol %)
H2 (4.1 MPa), THF
Room temperature, 12–15h
NC(O)R′
60–83% ee

(15.104)

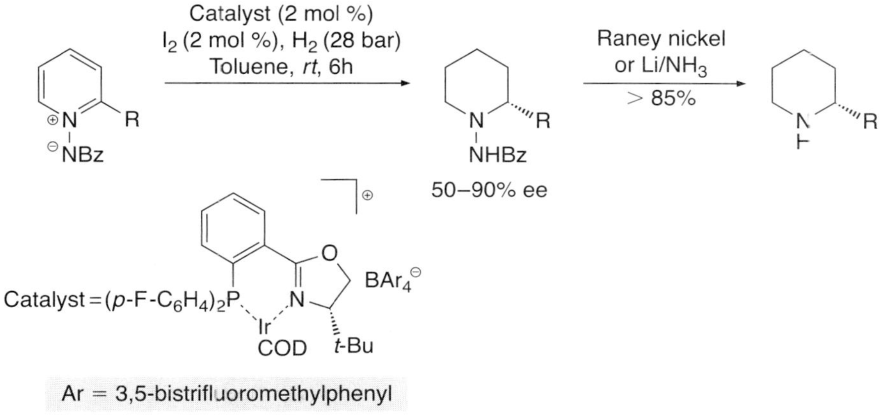
Catalyst (2 mol %)
I2 (2 mol %), H2 (28 bar)
Toluene, rt, 6h
NBz
NHBz
50–90% ee
Raney nickel
or Li/NH3
> 85%
Catalyst = (p-F-C6H4)2P
Ir
COD
t-Bu
BAr4
Ar = 3,5-bistrifluoromethylphenyl

(15.105)

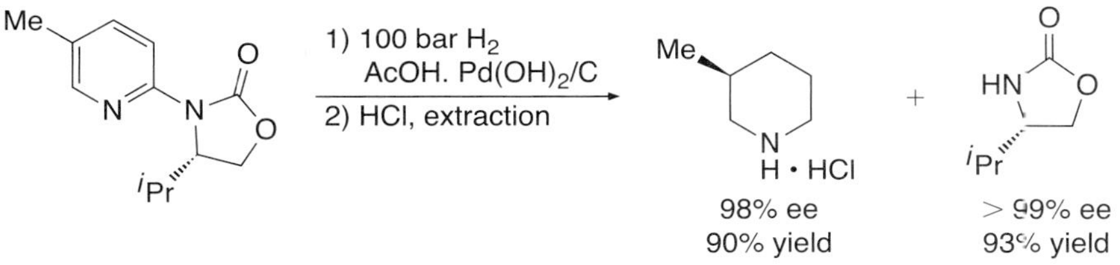
1) 100 bar H2
AcOH. Pd(OH)2/C
2) HCl, extraction
H • HCl
98% ee
90% yield
> 99% ee
93% yield

(15.106)

15.9.3.2. Asymmetric Hydrogenation of Five-Membered Ring Heteroarenes

Pyrroles, indoles, furans, and benzofurans possess more C=C double bond character than C=N or C=O double bond character. Thus, the hydrogenation of pyrroles, indoles, furans, and benzofurans is more closely related to the hydrogenation of enamides and enol ethers than imines or ketones. Like the hydrogenation of pyridines, pyrazines, and their benzo-fused analogs, enantioselective hydrogenation of pyrroles, indoles, furans, and benzofurans generate heterocycles with defined stereochemistries that are important substructures of biologically active molecules and natural products. Significant progress has been made by Kuwano on the enantioselective hydrogenation of these compounds with rhodium and ruthenium catalysts containing the TRAP ligand shown at the beginning of this chapter and iridium catalysts containing P,N ligands.

Some of the most enantioselective hydrogenations of five-membered ring heteroarenes have occurred with indoles. These indoles typically possess an acetyl, Boc, or tosyl group on the nitrogen. Two examples of such enantioselective hydrogenations are shown in Equations 15.107[408] and 15.108.[409] Both reactions, as well as reductions of structurally related indole derivatives, have been conducted with rhodium catalysts containing the trans-spanning bis-ferrocenyl phosphine "TRAP." Particularly high enantioselectivities were obtained for the hydrogenation of 3-substituted indoles. Reduction of *N*-Boc-substituted indoles has been conducted with a ruthenium catalyst containing the TRAP ligand.[410]

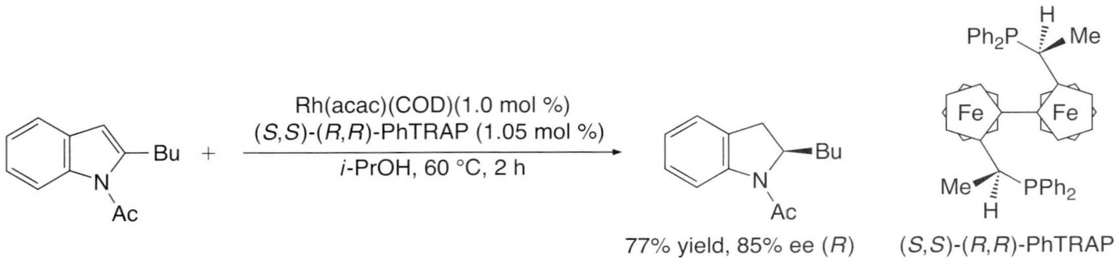

$$(15.107)$$

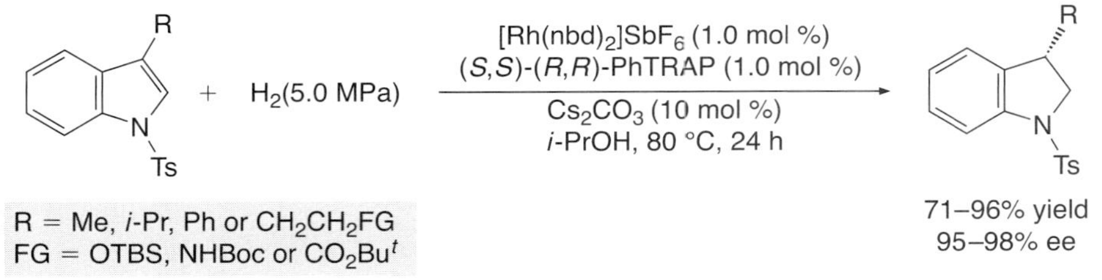

$$(15.108)$$

R = Me, *i*-Pr, Ph or CH$_2$CH$_2$FG
FG = OTBS, NHBoc or CO$_2$Bu^t

71–96% yield
95–98% ee

Enantioselective hydrogenation of pyrroles has also been conducted, and is somewhat more complex than that of indoles. These reactions were conducted with *N*-Boc-substituted pyrroles and a ruthenium–TRAP catalyst. The pyrroles can be mono-, di-, or tri-substituted. For example, the asymmetric reduction of 4,5-dimethylpyrrole-2-carboxylate gave the product from addition of hydrogen to the same face of both olefin units to form products containing three stereocenters in high enantioselectivity.[186] These reactions occur by initial reduction of the less-substituted carbon–carbon double bond.

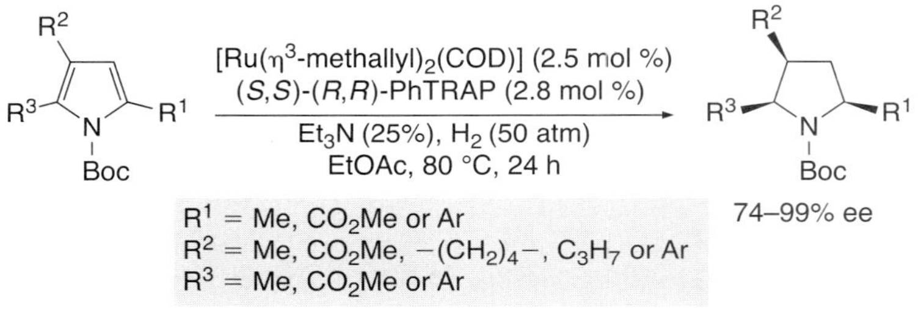

$$(15.109)$$

R^1 = Me, CO$_2$Me or Ar
R^2 = Me, CO$_2$Me, –(CH$_2$)$_4$–, C$_3$H$_7$ or Ar
R^3 = Me, CO$_2$Me or Ar

74–99% ee

Furans do not provide the opportunity to modulate the steric and electronic properties of the heteroarene by placing substituents on the heteroatom. When the substituents on the carbons of the furan lack functionality, hydrogenation of these heteroarenes is, therefore, most similar to the hydrogenation of unfunctionalized olefins. Thus, the asymmetric hydrogenation of furans has occurred with the highest selectivities using iridium catalysts possessing P,N ligands. Selected results published by Pfaltz with these types of catalysts are shown in Equation 15.110. Hydrogenation of 2-substituted furans and either 2- or 3-substituted benzofurans occurred with exceptionally high enantioselectivities.[411]

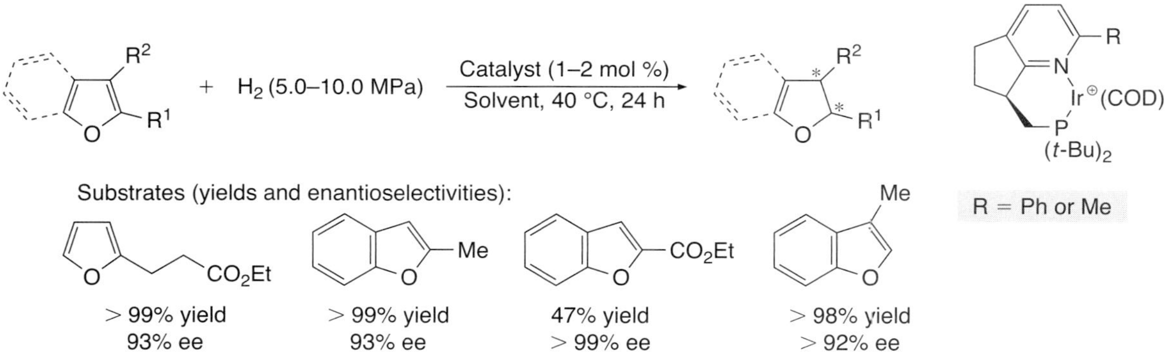

$$(15.110)$$

Substrates (yields and enantioselectivities):

> 99% yield
93% ee

> 99% yield
93% ee

47% yield
> 99% ee

> 98% yield
> 92% ee

15.10. Homogeneous Hydrogenation of Other Functional Groups (Written with Prof. Jing Zhao)

The hydrogenation of aldehydes, esters, nitriles, and nitro groups has received less attention than hydrogenations that lead to chiral products. However, the increased focus on processes that produce less waste is likely to increase progress in this area. Aldehydes, esters, and nitriles are often reduced with main group hydrides, and nitro groups are often reduced with procedures that involve stoichiometric amounts of metals, such as zinc. Thus, the catalytic hydrogenation of these functional groups would be a cleaner method to effect these reductions. This final section of the chapter provides selected examples of catalysts that lead to the hydrogenation of these types of functional groups.

15.10.1. Hydrogenation of Esters

Fewer complexes catalyze the hydrogenation of esters than the hydrogenation of ketones; even heterogeneous catalysts require relatively high temperatures (200–300 °C) and high hydrogen pressures (> 1000 psi). However, there has been considerable interest in developing the homogeneous catalytic reduction of carboxylic esters to alcohols with H_2 recently, and several significant advances have been reported.

In 1981, Grey and co-workers reported that the anionic ruthenium precatalyst in Equation 15.111 catalyzes the hydrogenation of activated esters (esters containing an electron-withdrawing group adjacent to the carboalkoxy group) under relatively mild conditions. The hydrogenation of unactivated esters occurred with significantly lower conversions.[412] Later, the active catalyst in this reaction was shown by Halpern and co-workers to be the neutral complex $RuH_4(PPh_3)_3$.[413] Mattedi and coworkers reported the hydrogenation of dicarboxylic esters catalyzed by the ruthenium cluster $[Ru_4H_4(CO)_8(PBu_3)_4]$ and the monomeric complex $[Ru(CO)_2(OC(O)Me)_2(PBu_3)_2]$, but these reactions are slow and require high pressures of hydrogen (130 bar H_2 pressure, 180 °C, 144 h).[414–416]

$$(15.111)$$

Ru-precatalyst $= [K(diglyme)]^{\oplus} [RuH_2(PPh_3)_2(PPh_2C_6H_4)]^{\ominus}$

More active catalysts have been generated from ruthenium and either polydentate phosphines or P,N ligands. Elsevier and co-workers reported the hydrogenation of dimethyl maleate (Equation 15.112), as well as the unactivated esters benzyl benzoate and methyl palmitate (Equation 15.113) in fluorinated solvents. The catalyst used in this work was generated from $Ru(acac)_3$ and the trisphosphine $MeC(CH_2PPh_2)_3$. Although milder than the previous hydrogenations of esters, these reactions still required higher temperatures and pressures (85 bar H_2 pressure, 120 °C) than the hydrogenation of ketones.[417,418] Related systems that include zinc to help reduce the Ru(III) have also been reported.[419]

$$(15.112)$$

Dimethyl maleate (DMM)   Butane-1,4-diol (BDO)

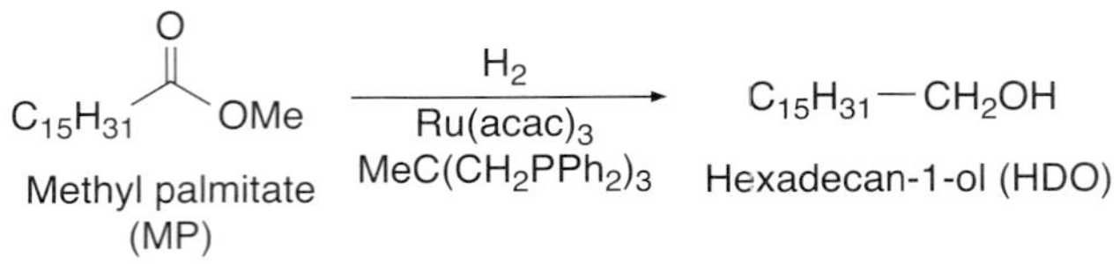

$$C_{15}H_{31}-\overset{\displaystyle O}{\overset{\displaystyle \|}{C}}-OMe \quad \xrightarrow[\substack{Ru(acac)_3 \\ MeC(CH_2PPh_2)_3}]{H_2} \quad C_{15}H_{31}-CH_2OH$$

Methyl palmitate   Hexadecan-1-ol (HDO)
(MP)

(15.113)

Significant improvements were made by Milstein,[420] Saudan,[421] and Ikariya[422] by using P,N ligands. Milstein reported the hydrogenation of non-activated esters catalyzed by a ruthenium pincer complex.[420] With Milstein's ruthenium pincer system, the hydrogenation can be conducted under relatively mild, neutral conditions without additives. Limited experimental data are available on the mechanism of this process, but the proposed pathway, summarized in Scheme 15.24, is not straightforward. The pincer ligand was shown to undergo oxidation and reduction, and the amino group was proposed to dissociate to provide a site for binding of the substrate. One cycle was proposed to lead to the reduction of the ester to the aldehyde and a second cycle for the reduction of the aldehyde. The proposed mechanism begins with the ruthenium dihydride at the top of the left cycle. This species is proposed to bind the ester and reduce it by hydride transfer to form a bound anion of an acetal. Reductive elimination would then generate the free acetal that converts to the aldehyde, and transfer of a hydride from the ligand to the metal, followed by hydrogenation of the resulting unsaturated ligand, would regenerate the starting complex. A second related cycle then reduces the resulting aldehyde to the alcohol.

More recently, Saudan and co-workers reported a series of ruthenium complexes containing P,N ligands that have connectivity reversed from that of Noyori's bisphosphine diamine catalysts. These catalysts containing P,N ligands are currently the most active homogeneous catalysts for ester hydrogenation (Equation 15.114).[421] These catalysts react with higher turnover numbers and turnover frequencies at lower temperatures and pressures than had been used previously (TON ≈ 2000, TOF = 800–2000 h⁻¹, 100 °C, 50 bar H$_2$). These reactions form alcohols with good to excellent chemoselectivity. Finally, catalysts containing a cyclopentadienyl and P,N ligand were reported by Ikariya to reduce benzoic esters and lactones to benzyl alcohol and diols.[422] The detailed mechanism of these reductions has not been revealed, but at least two features are needed for the catalytic activity. First, the ligand must be able to generate an intermediate with a hydride on the metal and proton on the ligand nitrogen atom. Second, the phosphorus and nitrogen must be present in the same ligand in the active catalysts.

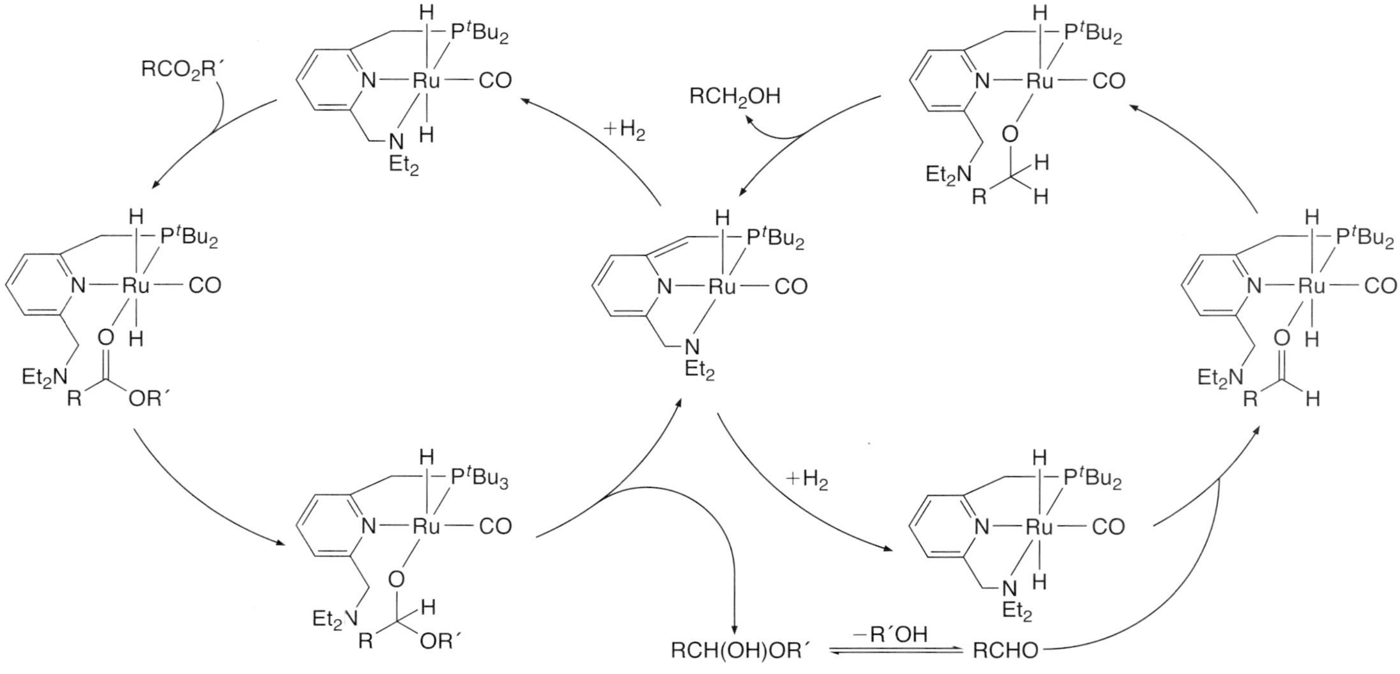

Scheme 15.24

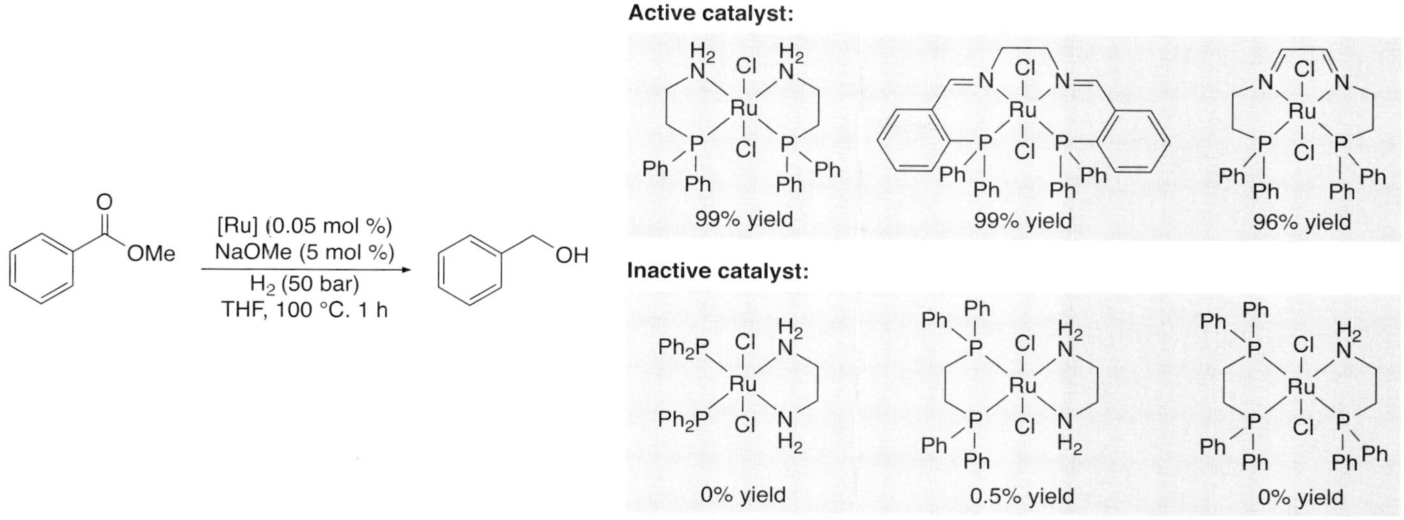

Active catalyst:

99% yield   99% yield   96% yield

Inactive catalyst:

0% yield   0.5% yield   0% yield

(15.114)

15.10.2. Hydrogenation of Carboxylic Anhydrides and Imides

The hydrogenation of carboxylic anhydrides is conducted with heterogeneous catalysts on a large scale. The hydrogenation of maleic anhydride produced by exhaustive oxidation of butane is one route to tetrahydrofuran.[423,424] Hydrogenation of these anhydrides can also be valuable on smaller scales with homogeneous catalysts. Partial hydrogenation of succinic anhydride produces lactones, and an early example conducted with the classic homogeneous [Ru(PPh$_3$)$_3$Cl$_2$] catalyst is shown in Equation 15.115.[425,426] The yield of 50% can be rationalized by the formation of water, which opens the anhydride to form the unreactive succinic acid. The hydrogenation of certain cyclic anhydrides can generate chiral products.[427]

More recently, the hydrogenation of imides has been reported.[422,428] These reactions were conducted with [Cp*Ru(PPh$_2$CH$_2$CH$_2$NH$_2$)Cl] or, for enantioselective hydrogenation of imides, a Cp*Ru catalyst containing a proline-derived P,N ligand. N-benzyl imides are opened to hydroxycarboxamides (Equation 15.116). This provides a method to deprotect phthalimide derivatives by hydrogenation. Hydrogenation of glutarimides substituted in the 3-position leads to chiral products by desymmetrization (Equation 15.117), and this reaction, conducted with the catalyst containing the proline-derived P,N ligand, occurs in high ee to form a precursor to the antidepressant paroxetine.

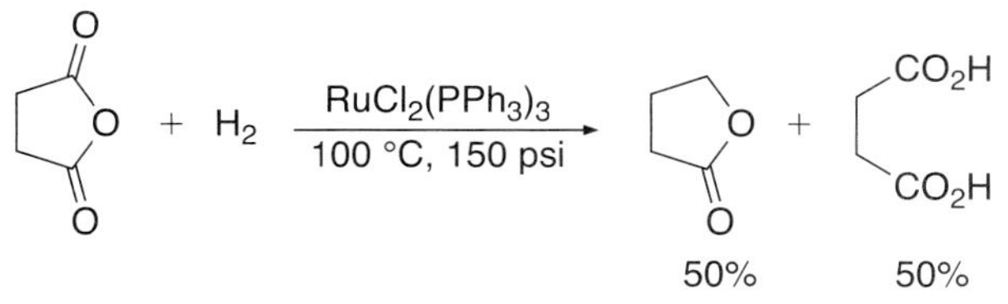

(15.115)

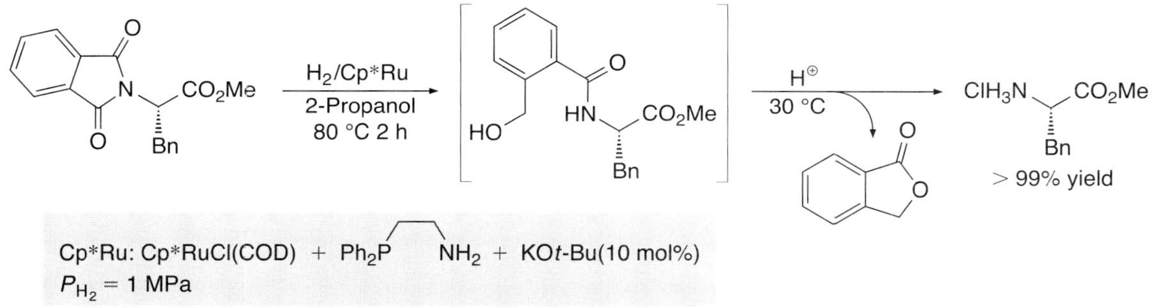

(15.116)

Cp*Ru: Cp*RuCl(COD) + Ph$_2$P⌒NH$_2$ + KOt-Bu(10 mol%)
P_{H_2} = 1 MPa

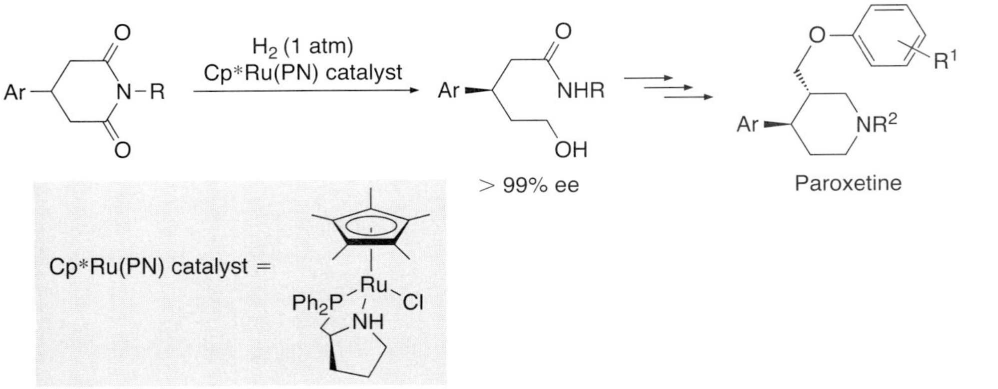

$$> 99\% \text{ ee} \qquad \text{Paroxetine}$$

(15.117)

Cp*Ru(PN) catalyst =

The conversion of carboxylic acids to aldehydes is typically conducted by reduction to the alcohol with main group metal hydride reagents, followed by oxidation to the aldehyde, or by reaction of the corresponding ester with stoichiometric amounts of the the aluminum hydride DIBAL. Thus, development of a selective hydrogenation of carboxylic esters to aldehydes would be valuable and would generate less waste. To overcome the typically lower reactivity of the acid reagent versus the aldehyde product, Yamamoto developed a system in which the acid is converted to an anhydride, and the anhydride is hydrogenated to the aldehyde (Equation 15.118).

$$\underset{\substack{\text{2 mmol}}}{RCO_2H} + \underset{\substack{\text{3.0 MPa}}}{H_2} \xrightarrow[\text{THF, 80 °C, 24 h}]{\substack{Pd(PPh_3)_4 \ 0.02 \ \text{mmol} \\ (^tBuCO)_2O \ 6 \ \text{mmol}}} RCHO$$

(15.118)

R = alkyl, aryl, heteroaryl

Yamamoto first found that the hydrogenation of acyclic anhydrides occurs in the presence of [Pd(PPh$_3$)$_4$] as catalyst.[429,430] The reaction was proposed to proceed by the mechanism in Scheme 15.25, involving oxidative addition of the anhydride, followed by hydrogenolysis of the Pd-acyl and Pd-carboxylato groups in {Pd(PPh$_3$)$_2$[C(O)R](OC(O)R)} to give an equal mixture of aldehyde and acid. To create a system that leads to the full conversion of a carboxylic acid to an aldehyde via the anhydride, mixed anhydrides were generated from the more-hindered and less-reactive pivaloyl anhydride. Thus, a combination of the transesterification of anhydrides and the selective hydrogenation of the mixed anhydride at the O–C(O)R bond of the less-hindered acyl group (Scheme 15.26) creates a direct, catalytic hydrogenation of carboxylic acids to aldehydes. This process can be conducted with aliphatic, aromatic, and heterocyclic carboxylic acids, as well as with di- and tribasic carboxylic acids.

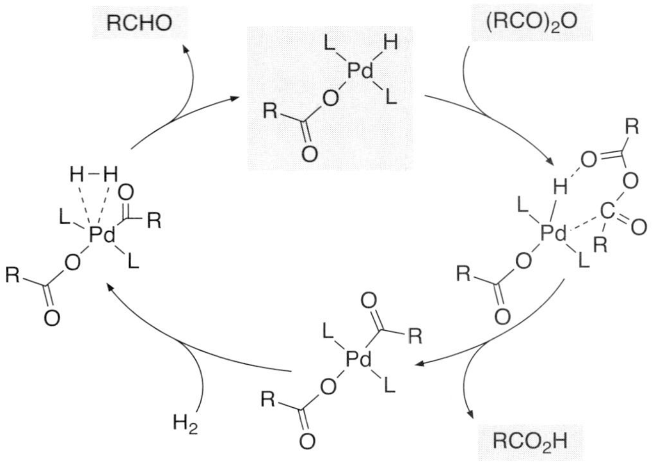

Scheme 15.25

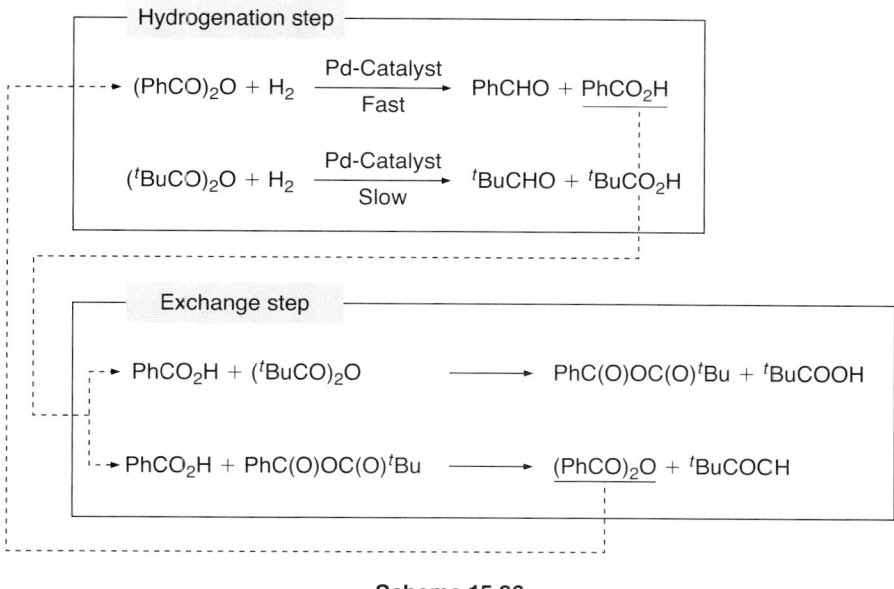

Scheme 15.26

15.10.3. Hydrogenation of Nitriles

The hydrogenation of nitriles is an important large-scale industrial process conducted with heterogeneous catalysts at high temperatures and pressures. The homogeneous hydrogenation of nitriles has received less attention, and some of the homogeneous catalysts could be truly heterogeneous systems.[431] Nevertheless, some very active catalysts for nitrile hydrogenation have been identified. In one case, the reaction most likely occurs by insertion of the nitrile and then the imine into a metal–hydride bond, whereas in a second case an outer-sphere mechanism characteristic of the metal–ligand bifunctional systems was proposed.

Complexes of ruthenium and rhodium have been identified that catalyze the hydrogenation of nitriles under mild conditions with very good selectivities to the primary amine. In particular, Otsuka and co-workers showed that $[RhH(P^iPr_3)_3]$ catalyzes the hydrogenation of a variety of nitriles to amines under mild conditions (1 atm H_2 pressure, 20 °C, 2 h; Equation 15.119).[432] The reaction is reversible; primary amines undergo dehydrogenation in the presence of $[RhH(P^iPr_3)_3]$ to form nitriles (Equation 15.120). Iridium catalysts for the dehydrogenation of nitriles have also been reported recently.[433]

$$R-C\equiv N \xrightarrow[\substack{20\ °C,\ THF \\ RhH(P^iPr_3)_3}]{1\ atm,\ H_2} R-CH_2NH_2 \quad 45-100\%$$
$$R = {}^nBu,\ {}^iPr,\ {}^tBu,\ Bn,\ CH_2{=}CHCH_2,\ or\ Ph \qquad (15.119)$$

$$PhCH_2NH_2 \xrightarrow[\substack{RhH(P^iPr_3)_3}]{110\ °C,\ 0.05\ torr} PhC\equiv N + H_2 \qquad (15.120)$$

More recently, Morris reported the hydrogenation of benzonitrile catalyzed by a ruthenium complex containing a P–NH–NH–P tetradentate ligand (Equation 15.121).[434] The presence of an amine N–H and metal hydride make it likely that the reaction occurs by an outer-sphere mechanism involving two sequential simultaneous transfers of the hydride and the N–H proton, first to the nitrile and second to the imine (Scheme 15.27). This proposal was supported by DFT calculations. The hydrogenation of aryl and heteroaryl nitriles catalyzed by a combination of $[Ru(COD)(2\text{-methylallyl})_2]$ and DPPF has also been reported to occur in high yields.[435]

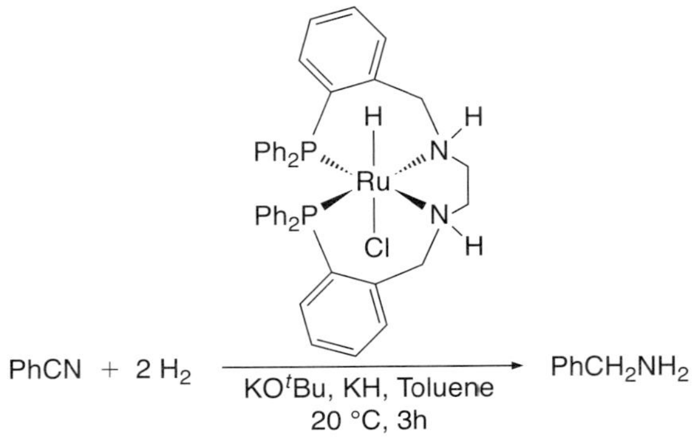

$$\text{PhCN} + 2\,\text{H}_2 \xrightarrow[\text{20 °C, 3h}]{\text{KO}^t\text{Bu, KH, Toluene}} \text{PhCH}_2\text{NH}_2$$

(15.121)

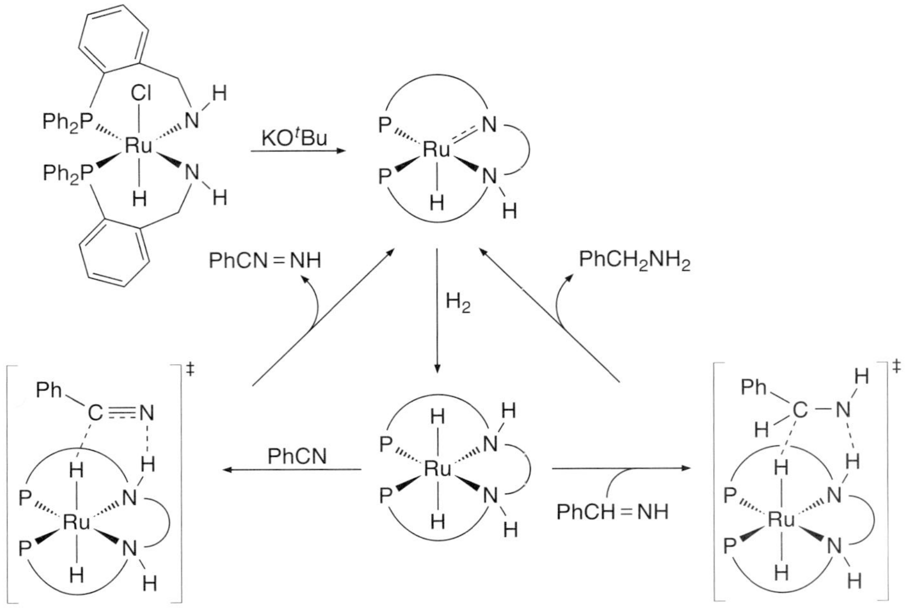

Scheme 15.27

15.11. Summary

Homogeneous hydrogenation is among the most utilized homogeneous catalytic reaction for the synthesis of fine chemicals on a production scale. Such reactions were first reported over 70 years ago and have been studied intensively for over 40 years. The first papers reporting the types of catalyst that are used today were published by Wilkinson and Coffey. This work was followed by the discovery of cationic rhodium catalysts, ruthenium catalysts, iridium catalysts, chromium catalysts, and even catalysts lacking d-electrons in the valence shell. The discovery of complexes that hydrogenate α-(acylamino)acrylates to form amino acids in enantioenriched form has ultimately inspired the development of hundreds of chiral, non-racemic catalysts for asymmetric hydrogenation. The most progress on asymmetric hydrogenation has been made since the discovery of the BINAP ligand in 1980, which generated catalysts for the hydrogenation of a wide range of unsaturated substrates. Alkylphosphines, aryl phosphines, P-chiral phosphines, mixed P,N ligands, phosphonites, phosphites, and phosphoramidites all generate catalysts that are highly selective for the hydrogenation of particular classes of olefins. The asymmetric hydrogenation of ketones has had an equal impact on the synthesis of optically active products, and the hydrogenation of imine derivatives shows promise for similar applications. Finally, the asymmetric hydrogenation of heteroarenes is beginning to be developed as a route to enantioenriched reduced heterocycles.

Hydrogenation with achiral catalysts has been widely used and is likely to experience increased development in the coming years. The concept of directed hydrogenation led to an increase in the ability to control diastereoselectivity, which is often as important to control in synthetic sequences as enantioselectivity. Furthermore, the drive for cleaner chemical processes is likely to make the hydrogenation of functional groups such as esters, acids, nitriles, and nitro groups important for the production of achiral products without the use of main group metal hydrides.

References and Notes

1. Birch, A. J.; Williamson, D. H. *Org. React.* **1976**, *24*, 1.
2. James, B. R. *Homogeneous Hydrogenation*; Wiley: New York, 1973.
3. James, B. R. In *Comprehensive Organometallic Chemistry*, 1st ed.; Wilkinson, G., Stone, F. G. A., Abel, E. W., Eds.; Pergamon Press: New York, 1982; Vol. 8, p 285.
4. James, B. R.; Stone, F. G. A.; West, R. *Adv. Organomet. Chem.* **1979**, *17*, 319.
5. Jardine, F. H. *Prog. Inorg. Chem.* **1981**, *28*, 63.
6. Tolman, C. A.; Faller, J. W. In *Homogeneous Catalysis with Metal Phosphine Complexes*; Pignolet, L., Ed.; Plenum Press: New York, 1983; p 13.
7. Calvin, M.; Polanyi, M. *Trans. Faraday Soc.* **1938**, *34*, 1181.
8. Calvin, M. *J. Am. Chem. Soc.* **1939**, *61*, 2230.
9. Iguchi, M. *J. Chem. Soc. Jpn.* **1939**, *60*, 1287.
10. Iguchi, M. *J. Chem. Soc. Jpn.* **1942**, *63*, 634.
11. Iguchi, M. *J. Chem. Soc. Jpn.* **1942**, *63*, 1752.
12. Jardine, F. H.; Osborn, J. A.; Wilkinson, G.; Young, J. F. *Chem. Ind. (London)* **1965**, 560.
13. Bennet, M. A.; Longstaff, P. A. *Chem. Ind. (London)* **1965**, 846.
14. Coffey, R. S. Br. Patent 1,121,642, 1965.
15. Evans, D.; Osborn, J. A.; Jardine, F. H.; Wilkinson, G. *Nature* **1965**, *208*, 1203.
16. Osborn, J. A.; Jardine, F. H.; Young, J. F.; Wilkinson, G. *J. Chem. Soc. A* **1966**, 1711.
17. Baird, M. C.; Mague, J. T.; Osborn, J. A.; Wilkinson, G. *J. Chem. Soc.* **1967**, 1347.
18. Hallman, P. S.; Evans, D.; Osborn, J. A.; Wilkinson, G. *J. Chem. Soc., Chem. Commun.* **1967**, 305.
19. Dawans, F.; Morel, D. *J. Mol. Catal.* **1978**, *3*, 403.
20. Halpern, J.; Okamoto, T.; Zakhariev, A. *J. Mol. Catal.* **1977**, *2*, 65.
21. Crabtree, R. H.; Felkin, H.; Morris, G. E. *J. Organomet. Chem.* **1977**, *141*, 205.
22. Crabtree, R. H.; Felkin, H.; Fillebeenkhan, T.; Morris, G. E. *J. Organomet. Chem.* **1979**, *168*, 183.
23. Horner, L.; Siegel, H.; Büthe, H. *Angew. Chem., Int. Ed. Engl.* **1968**, *7*, 942.
24. Knowles, W. S.; Sabacky, M. J. *J. Chem. Soc., Chem. Commun.* **1968**, 1445.
25. Dang, T. P.; Kagan, H. B. *J. Chem. Soc., Chem. Commun.* **1971**, 481.
26. Inoguchi, K.; Sakuraba, S.; Achiwa, K. *Synlett* **1992**, 169.
27. Knowles, W. S. *Angew. Chem. Int. Ed.* **2002**, *41*, 1999.
28. Miyashita, A.; Yasuda, A.; Takaya, H.; Toriumi, K.; Ito, T.; Souchi, T.; Noyori, R. *J. Am. Chem. Soc.* **1980**, *102*, 7932.
29. Ikariya, T.; Ishii, Y.; Kawano, H.; Arai, T.; Saburi, M.; Yoshikawa, S.; Akutagawa, S. *J. Chem. Soc., Chem. Commun.* **1985**, 922.
30. Kitamura, M.; Ohkuma, T.; Inoue, S.; Sayo, N.; Kumobayashi, H.; Akutagawa, S.; Ohta, T.; Takaya, H.; Noyori, R. *J. Am. Chem. Soc.* **1988**, *110*, 629.
31. Ohkuma, T.; Ooka, H.; Hashiguchi, S.; Ikariya, T.; Noyori, R. *J. Am. Chem. Soc.* **1995**, *117*, 2675.
32. Hashiguchi, S.; Fujii, A.; Takehara, J.; Ikariya, T.; Noyori, R. *J. Am. Chem. Soc.* **1995**, *117*, 7562.
33. Noyori, R. *Angew. Chem. Int. Ed.* **2002**, *41*, 2008.
34. Burk, M. J. *J. Am. Chem. Soc.* **1991**, *113*, 8518.
35. Burk, M. J.; Feaster, J. E.; Nugent, W. A.; Harlow, R. L. *J. Am. Chem. Soc.* **1993**, *115*, 10125.
36. Roseblade, S. J.; Pfaltz, A. *Acc. Chem. Res.* **2007**, *40*, 1402.
37. Blackmond, D. G.; Lightfoot, A.; Pfaltz, A.; Rosner, T.; Schnider, P.; Zimmermann, N. *Chirality* **2000**, *12*, 442.

38. Lightfoot, A.; Schnider, P.; Pfaltz, A. *Angew. Chem. Int. Ed.* **1998**, *37*, 2897.

39. van den Berg, M.; Minnaard, A. J.; Schudde, E. P.; van Esch, J.; de Vries, A. H. M.; de Vries, J. G.; Feringa, B. L. *J. Am. Chem. Soc.* **2000**, *122*, 11539.

40. Reetz, M. T.; Mehler, G. *Angew. Chem. Int. Ed.* **2000**, *39*, 3889.

41. Campbell, M. L. In *Ullmann's Encyclopedia of Industrial Chemistry (Online)*; Wiley-VCH: Weinheim, 2002; p DOI: 10.1002/14356007.a12_629.

42. Smiley, R. A. In *Ullmann's Encyclopedia of Industrial Chemistry (Online)*; Wiley-VCH: Weinheim, 2002; p DOI: 10.1002/14356007.a12_629.

43. Osborn, J. A.; Wilkinson, G. *Inorg. Synth.* **1990**, *28*, 77.

44. Koch, G. K.; Dalenber, J. W. *J. Labelled Compd.* **1970**, *6*, 395.

45. Collman, J. P.; Cooke, M. P.; Dolcetti, G.; Hegedus, L. S.; Marquardt, D. N.; Norton, J. R. *J. Am. Chem. Soc.* **1972**, *94*, 1789.

46. Schrock, R. R.; Osborn, J. A. *J. Am. Chem. Soc.* **1976**, *98*, 2134.

47. Schrock, R. R.; Osborn, J. A. *J. Am. Chem. Soc.* **1976**, *98*, 2143.

48. Schrock, R. R.; Osborn, J. A. *J. Am. Chem. Soc.* **1976**, *98*, 4450.

49. Brown, J. M. *Angew. Chem., Int. Ed. Engl.* **1987**, *26*, 190.

50. Evans, D. A.; Morrissey, M. M. *J. Am. Chem. Soc.* **1984**, *106*, 3866.

51. Schrock, R. R.; Osborn, J. A. *J. Chem. Soc., Chem. Commun.* **1970**, 567.

52. Crabtree, R. *Acc. Chem. Res.* **1979**, *12*, 331.

53. Wüstenberg, B.; Pfaltz, A. *Adv. Synth. Catal.* **2008**, *350*, 174

54. Smidt, S. P.; Zimmermann, N.; Studer, M.; Pfaltz, A. *Chem.—Eur. J.* **2004**, *10*, 4685

55. Suggs, J. W.; Cox, S. D.; Crabtree, R. H.; Quirk, J. M. *Tetrahedron Lett.* **1981**, *22*, 303.

56. Halpern, J.; Harrod, J. F.; James, B. R. *J. Am. Chem. Soc.* **1961**, *83*, 753.

57. Chatt, J.; Hayter, R. G. *J. Chem. Soc.* **1961**, 2605.

58. Grey, R. A.; Pez, G. P.; Wallo, A. *J. Am. Chem. Soc.* **1980**, *102*, 5948.

59. Noyori, R.; Ohkuma, T.; Kitamura, M.; Takaya, H.; Sayo, N.; Kumobayashi, H.; Akutagawa, S. *J. Am. Chem. Soc.* **1987**, *109*, 5856.

60. Jeske, G.; Lauke, H.; Mauermann, H.; Schumann, H.; Marks, T. J. *J. Am. Chem. Soc.* **1985**, *107*, 8111.

61. Giardello, M. A.; Conticello, V. P.; Brard, L.; Gagne, M. R.; Marks, T. J. *J. Am. Chem. Soc.* **1994**, *116*, 10241.

62. Crabtree, R. H.; Davis, M. W. *Organometallics* **1983**, *2*, 681.

63. Stork, G.; Kahne, D. E. *J. Am. Chem. Soc.* **1983**, *105*, 1072.

64. Hoveyda, A. H.; Evans, D. A.; Fu, G. C. *Chem. Rev.* **1993**, *93*, 1307.

65. Schultz, A. G.; McCloskey, P. J. *J. Org. Chem.* **1985**, *50*, 5905.

66. Chan, A. S. C.; Halpern, J. *J. Am. Chem. Soc.* **1980**, *102*, 838.

67. Halpern, J. *Inorg. Chim. Acta* **1981**, *50*, 11.

68. Halpern, J.; Riley, D. P.; Chan, A. S. C.; Pluth, J. J. *J. Am. Chem. Soc.* **1977**, *99*, 8055.

69. Halpern, J. *Science* **1982**, *217*, 401.

70. Gridnev, I. D.; Yasutake, M.; Higashi, N.; Imamoto, T. *J. Am. Chem. Soc.* **2001**, *123*, 5268.

71. Gridnev, I. D.; Imamoto, T.; Hoge, G.; Kouchi, M.; Takahashi, H. *J. Am. Chem. Soc.* **2008**, *130*, 2560.

72. Halpern, J.; Wong, C. S. *J. Chem. Soc., Chem. Commun.* **1973**, 629.

73. Wink, D. A.; Ford, P. C. *J. Am. Chem. Soc.* **1985**, *107*, 1794.

74. Wink, D. A.; Ford, P. C. *J. Am. Chem. Soc.* **1986**, *108*, 4838.

75. Brown, J. M.; Evans, P. L.; Lucy, A. R. *J. Chem. Soc., Perkin Trans. 2* **1987**, 1589.

76. Duckett, S. B.; Newell, C. L.; Eisenberg, R. *J. Am. Chem. Soc.* **1994**, *116*, 10548.

77. Meakin, P.; Tolman, C. A.; Jesson, J. P. *J. Am. Chem. Soc.* **1972**, *94*, 3240.

78. Halpern, J.; Okamoto, T. *Inorg. Chim. Acta* **1984**, *89*, L53.

79. Brown, J. M.; Canning, L. R.; Lucy, A. R. *J. Chem. Soc., Chem. Commun.* **1984**, 915.

80. Halpern, J. *Inorg. Chim. Acta* **1981**, *50*, 11.

81. Chan, A. S. C.; Pluth, J. J.; Halpern, J. *Inorg. Chim. Acta* **1979**, *37*, L477.

82. Brown, J. M.; Chaloner, P. A. *J. Chem. Soc., Chem. Commun.* **1980**, 344.

83. Brown, J. M.; Chaloner, P. A.; Morris, G. A. *J. Chem. Soc., Chem. Commun.* **1983**, 664.

84. Gridnev, I. D.; Imamoto, T. *Acc. Chem. Res.* **2004**, *37*, 633.

85. Gridnev, I. D.; Imamoto, T. *Organometallics* **2001**, *20*, 545.

86. Gridnev, I. D.; Yamanoi, Y.; Higashi, N.; Tsuruta, H.; Yasutake, M.; Imamoto, T. *Adv. Synth. Catal.* **2001**, *343*, 118.

87. Chodosh, D. F.; Crabtree, R. H.; Felkin, H.; Morehouse, S.; Morris, G. E. *Inorg. Chem.* **1982**, *21*, 1307.

88. Chodosh, D. F.; Crabtree, R. H.; Felkin, H.; Morris, G. E. *J. Organomet. Chem.* **1978**, *161*, C67.

89. Crabtree, R. H.; Felkin, H.; Morris, G. E.; King, T. J.; Richards, J. A. *J. Organomet. Chem.* **1976**, *113*, C7.

90. Cui, X. H.; Fan, Y. B.; Hall, M. B.; Burgess, K. *Chem.—Eur. J.* **2005**, *11*, 6859.

91. Brandt, P.; Hedberg, C.; Andersson, P. G. *Chem.—Eur. J.* **2003**, *9*, 339.

92. Dietiker, R.; Chen, P. *Angew. Chem. Int. Ed.* **2004**, *43*, 5513.

93. Crabtree, R. H. In *Handbook of Homogeneous Hydrogenation*; de Vries, J. G., Elsevier, C. J., Eds.; Wiley-VCH: Weinheim, 2007; Vol. 1, p 31.

94. Crabtree, R. H.; Lavin, M. *J. Chem. Soc., Chem. Commun.* **1985**, 1661.

95. Crabtree, R. H.; Davis, M. W. *J. Org. Chem.* **1986**, *51*, 2655.

96. Pfaltz, A.; Blankenstein, J.; Hilgraf, R.; Hormann, E.; McIntyre, S.; Menges, F.; Schonleber, M.; Smidt, S. P.; Wustenberg, B.; Zimmermann, N. *Adv. Synth. Catal.* **2003**, *345*, 33.

97. Perry, M. C.; Cui, X. H.; Powell, M. T.; Hou, D. R.; Reibenspies, J. H.; Burgess, K. *J. Am. Chem. Soc.* **2003**, *125*, 113.

98. Sanchez-Delgado, R. A.; DeOchoa, O. L. *J. Mol. Catal.* **1979**, *6*, 303.

99. O'Connor, C.; Yagupsky, G.; Evans, D.; Wilkinson, G. *J. Chem. Soc., Chem. Commun.* **1968**, 420.

100. O'Connor, C.; Wilkinson, G. *J. Chem. Soc. A* **1968**, 2665.

101. Hallman, P. S.; McGarvey, B. R.; Wilkinson, G. *J. Chem. Soc. A* **1968**, 3143.

102. Ashby, M. T.; Halpern, J. *J. Am. Chem. Soc.* **1991**, *113*, 589.

103. Morris, R. H. In *Handbook of Homogeneous Hydrogenation*; de Vries, J. G., Elsevier, C. J., Eds.; Wiley-VCH: Weinheim, 2007; Vol. 1, p 45.

104. Noyori, R.; Kitamura, M.; Ohkuma, T. *Proc. Natl. Acad. Sci. U.S.A.* **2004**, *101*, 5356.

105. Noyori, R.; Ohkuma, T. *Angew. Chem. Int. Ed.* **2001**, *40*, 40.

106. Daley, C. J. A.; Bergens, S. H. *J. Am. Chem. Soc.* **2002**, *124*, 3680.

107. King, S. A.; Thompson, A. S.; King, A. O.; Verhoeven, T. R. *J. Org. Chem.* **1992**, *57*, 6689.

108. Akotsi, O. M.; Metera, K.; Reid, R. D.; McDonald, R.; Bergens, S. H. *Chirality* **2000**, *12*, 514.

109. Sweany, R. L.; Halpern, J. *J. Am. Chem. Soc.* **1977**, *99*, 8335.

110. Persson, B. A.; Larsson, A. L. E.; Le Ray, M.; Bäckvall, J. E. *J. Am. Chem. Soc.* **1999**, *121*, 1645.

111. Martin-Matute, B.; Edin, M.; Bogar, K.; Bäckvall, J. E. *Angew. Chem. Int. Ed.* **2004**, *43*, 6535.

112. Martin-Matute, B.; Edin, M.; Bogar, K.; Kaynak, F. B.; Bäckvall, J. E. *J. Am. Chem. Soc.* **2005**, *127*, 8817.

113. Martin-Matute, B.; Bäckvall, J. E. *Curr. Opin. Chem. Biol.* **2007**, *11*, 226.

114. Pamies, O.; Bäckvall, J. E. *Trends Biotechnol.* **2004**, *22*, 130.

115. Pamies, O.; Bäckvall, J. E. *Chem. Rev.* **2003**, *103*, 3247.

116. Choi, J. H.; Kim, Y. H.; Nam, S. H.; Shin, S. T.; Kim, M.-J.; Park, J. *Angew. Chem. Int. Ed.* **2002**, *41*, 2373.

117. Kim, M. J.; Chung, Y. I.; Choi, Y. K.; Lee, H. K.; Kim, D.; Park, J. *J. Am. Chem. Soc.* **2003**, *125*, 11494.

118. Fujii, A.; Hashiguchi, S.; Uematsu, N.; Ikariya, T.; Noyori, R. *J. Am. Chem. Soc.* **1996**, *118*, 2521.

119. Doucet, H.; Ohkuma, T.; Murata, K.; Yokozawa, T.; Kozawa, M.; Katayama, E.; England, A. F.; Ikariya, T.; Noyori, R. *Angew. Chem. Int. Ed.* **1998**, *37*, 1703.

120. Noyori, R.; Hashiguchi, S. *Acc. Chem. Res.* **1997**, *30*, 97.

121. Ikariya, T.; Murata, K.; Noyori, R. *Org. Biomol. Chem.* **2006**, *4*, 393.

122. Tang, W.; Zhang, X. *Chem. Rev.* **2003**, *103*, 3029.

123. Gladiali, S.; Alberico, E. *Chem. Soc. Rev.* **2006**, *35*, 226.

124. Gladiali, S.; Alberico, E. In *Transition Metals for Organic Synthesis*, 2nd ed.; Beller, M., Bolm, C., Eds.; Wiley-VCH: Weinheim, 2004; Vol. 2, p 145.

125. Everaere, K.; Mortreux, A.; Carpentier, J. F. *Adv. Synth. Catal.* **2003**, *345*, 67.

126. Blacker, A. J.; Martin, J. In *Asymmetric Catalysis on Industrial Scale*; Blaser, H. U., Schmidt, E., Eds.; Wiley-VCH: Weinheim, 2004; p 201.

127. Carmona, D.; Lamata, M. P.; Oro, L. A. *Eur. J. Inorg. Chem.* **2002**, 2239.

128. Bäckvall, J. E. *J. Organomet. Chem.* **2002**, *652*, 105.

129. Wills, M.; Palmer, M.; Smith, A.; Kenny, J.; Walsgrove, T. *Molecules* **2000**, *5*, 4.

130. Palmer, M. J.; Wills, M. *Tetrahedron: Asymmetry* **1999**, *10*, 2045.

131. Naota, T.; Takaya, H.; Murahashi, S. I. *Chem. Rev.* **1998**, *98*, 2599.

132. Fehring, V.; Selke, R. *Angew. Chem. Int. Ed.* **1998**, *37*, 1827.

133. Hamilton, R. J.; Bergens, S. H. *J. Am. Chem. Soc.* **2008**, *130*, 11979.

134. Yamakawa, M.; Ito, H.; Noyori, R. *J. Am. Chem. Soc.* **2000**, *122*, 1466.

135. Abdur-Rashid, K.; Clapham, S. E.; Hadzovic, A.; Harvey, J. N.; Lough, A. J.; Morris, R. H. *J. Am. Chem. Soc.* **2002**, *124*, 15104.

136. Clapham, S. E.; Hadzovic, A.; Morris, R. H. *Coord. Chem. Rev.* **2004**, *248*, 2201.

137. Hamilton, R. J.; Bergens, S. H. *J. Am. Chem. Soc.* **2006**, *128*, 13700.

138. Hamilton, R. J.; Leong, C. G.; Bigam, G.; Miskolzie, M.; Bergens, S. H. *J. Am. Chem. Soc.* **2005**, *127*, 4152.

139. Sandoval, C. A.; Ohkuma, T.; Muniz, K.; Noyori, R. *J. Am. Chem. Soc.* **2003**, *125*, 13490.

140. Abdur-Rashid, K.; Faatz, M.; Lough, A. J.; Morris, R. H. *J. Am. Chem. Soc.* **2001**, *123*, 7474.

141. Abdur-Rashid, K.; Clapham, S. E.; Hadzovic, A.; Harvey, J. N.; Lough, A. J.; Morris, R. H. *J. Am. Chem. Soc.* **2002**, *124*, 15104.
142. Casey, C. P.; Singer, S. W.; Powell, D. R.; Hayashi, R. K.; Kavana, M. *J. Am. Chem. Soc.* **2001**, *123*, 1090.
143. Casey, C. P.; Johnson, J. B. *J. Org. Chem.* **2003**, *68*, 1998.
144. Casey, C. P.; Clark, T. B.; Guzei, I. A. *J. Am. Chem. Soc.* **2007**, *129*, 11821.
145. Casey, C. P.; Bikzhanova, G. A.; Cui, Q.; Guzei, I. A. *J. Am. Chem. Soc.* **2005**, *127*, 14062.
146. Yamakawa, M.; Ito, H.; Noyori, R. *J. Am. Chem. Soc.* **2000**, *122*, 1466.
147. Handgraaf, J. W.; Meijer, E. J. *J. Am. Chem. Soc.* **2007**, *129*, 3099.
148. Ito, M.; Hirakawa, M.; Murata, K.; Ikariya, T. *Organometallics* **2001**, *20*, 379.
149. Hedberg, C.; Kallstrom, K.; Arvidsson, P. I.; Brandt, P.; Andersson, P. G. *J. Am. Chem. Soc.* **2005**, *127*, 15083.
150. Di Tommaso, D.; French, S. A.; Zanotti-Gerosa, A.; Hancock, F.; Palin, E. J.; Catlow, C. R. A. *Inorg. Chem.* **2008**, *47*, 2674.
151. Leyssens, T.; Peeters, D.; Harvey, J. N. *Organometallics* **2008**, *27*, 1514.
152. Samec, J. S. M.; Ell, A. H.; Aberg, J. B.; Privalov, T.; Eriksson, L.; Bäckvall, J. E. *J. Am. Chem. Soc.* **2006**, *128*, 14293.
153. Johnson, J. B.; Bäckvall, J. E. *J. Org. Chem.* **2003**, *68*, 7681.
154. Koike, T.; Ikariya, T. *Adv. Synth. Catal.* **2004**, *346*, 37.
155. Bullock, R. M. *Chem.—Eur. J.* **2004**, *10*, 2366.
156. Bullock, R. M.; Voges, M. H. *J. Am. Chem. Soc.* **2000**, *122*, 12594.
157. Voges, M. H.; Bullock, R. M. *J. Chem. Soc., Dalton Trans.* **2002**, 759.
158. Kagan, H. B.; Fiaud, J. C. *Top. Stereochem.* **1978**, *10*, 175.
159. Morrison, J. D.; Masler, W. F.; Neuberg, M. K. *Adv. Catal.* **1976**, *25*, 81.
160. Kagan, H. B. *Pure. Appl. Chem.* **1975**, *43*, 401.
161. Marko, L.; Heil, B. *Catal. Rev.* **1973**, *8*, 269.
162. Valentine, D.; Scott, J. W. *Synthesis* **1978**, 329.
163. Knowles, W. S.; Sabacky, M. J.; Vineyard, B. D. *Adv. Chem. Ser.* **1974**, *132*, 274.
164. Lapporte, S. J.; Schuett, W. R. *J. Org. Chem.* **1963**, *28*, 1947.
165. Harada, K. In *Asymmetric Synthesis*; Morrison, J. D., Ed.; Academic Press: New York, 1985; Vol. 5, p 345.
166. Takaya, H.; Akutagawa, S.; Noyori, R. *Org. Synth.* **1989**, *67*, 20.
167. Kazlauskas, R. J. *Org. Synth.* **1998**, *Coll. Vol. 9*, 77.
168. Cai, D.; Hughes, D. L.; Verhoeven, T. R.; Reider, P. J. *Org. Synth.* **2004**, *Coll. Vol. 10*, 93.
169. Cai, D.; Payack, J. F.; Bender, D. R.; Hughes, D. L.; Verhoeven, T. R.; Reider, P. J. *Org. Synth.* **2004**, *10*, 112.
170. Saito, T.; Yokozawa, T.; Ishizaki, T.; Moroi, T.; Sayo, N.; Miura, T.; Kumobayashi, H. *Adv. Synth. Catal.* **2001**, *343*, 264.
171. Saito, T.; Yokozawa, T.; Zhang, X.; Sayo, N. U.S. Patent 5,872,273, 1999.
172. Wu, S.; Wang, W.; Tang, W.; Lin, M.; Zhang, X. *Org. Lett.* **2002**, *4*, 4495.
173. Dotta, P.; Kumar, P. G. A.; Pregosin, P. S.; Albinati, A.; Rizzato, S. *Organometallics* **2004**, *23*, 2295.
174. Ohkuma, T.; Koizumi, M.; Doucet, H.; Pham, T.; Kozawa, M.; Murata, K.; Katayama, E.; Yokozawa, T.; Ikariya, T.; Noyori, R. *J. Am. Chem. Soc.* **1998**, *120*, 13529.
175. Wu, J.; Ji, J. X.; Guo, R. W.; Yeung, C. H.; Chan, A. S. C. *Chem.—Eur. J.* **2003**, *9*, 2963.
176. Wu, J.; Pai, C. C.; Kwok, W. H.; Guo, R. W.; Au-Yeung, T. T. L.; Yeung, C. H.; Chan, A. S. C. *Tetrahedron: Asymmetry* **2003**, *14*, 987.
177. Wu, J.; Chen, H.; Kwok, W. H.; Lam, K. H.; Zhou, Z. Y.; Yeung, C. H.; Chan, A. S. C. *Tetrahedron Lett.* **2002**, *43*, 1539.
178. Pai, C. C.; Lin, C. W.; Lin, C. C.; Chen, C. C.; Chan, A. S. C. *J. Am. Chem. Soc.* **2000**, *122*, 11513.
179. Zhang, X. Y.; Mashima, K.; Koyano, K.; Sayo, N.; Kumobayashi, H.; Akutagawa, S.; Takaya, H. *J. Chem. Soc., Perkin Trans. 1* **1994**, 2309.
180. Zhang, X. Y.; Mashima, K.; Koyano, K.; Sayo, N.; Kumobayashi, H.; Akutagawa, S.; Takaya, H. *Tetrahedron Lett.* **1991**, *32*, 7283.
181. Marquarding, D.; Klusack, H.; Gokel, G.; Hoffman, P.; Ugi, I. *J. Am. Chem. Soc.* **1970**, *92*, 5389.
182. Togni, A.; Breutel, C.; Schnyder, A.; Spindler, F.; Landert, H.; Tijani, A. *J. Am. Chem. Soc.* **1994**, *116*, 4062.
133. McGarrity, J.; Spindler, F.; Fuchs, R.; Eyer, M. LONZA AG, EP-A 624587 A2, 1995.
184. Blaser, H.-U. *Adv. Synth. Catal.* **2002**, *344*, 17.
185. Blaser, H.-U.; Buser, H.-P.; Coers, K.; Hanreich, R.; Jalett, H.-P.; Jelsch, E.; Pugin, B.; Schneider, H.-D.; Spindler, F.; Wegmann, A. *Chimia* **1999**, *53*, 275.
186. Kuwano, R.; Kashiwabara, M.; Ohsumi, M.; Kusano, H. *J. Am. Chem. Soc.* **2008**, *130*, 808.
187. Berens, U.; Burk, M. J.; Gerlach, A.; Hems, W. *Angew. Chem. Int. Ed.* **2000**, *39*, 1981.

188. You, J. S.; Drexler, H. J.; Zhang, S. L.; Fischer, C.; Heller, D. *Angew. Chem. Int. Ed.* **2003**, *42*, 913.
189. Fryzuk, M. D.; Bosnich, B. *J. Am. Chem. Soc.* **1977**, *99*, 6262.
190. Kunin, A. J.; Farid, R.; Johnson, C. E.; Eisenberg, R. *J. Am. Chem. Soc.* **1985**, *107*, 5315.
191. Xie, J. H.; Wang, L. X.; Fu, Y.; Zhu, S. F.; Fan, B. M.; Duan, H. F.; Zhou, Q. L. *J. Am. Chem. Soc.* **2003**, *125*, 4404.
192. Pye, P. J.; Rossen, K.; Reamer, R. A.; Tsou, N. N.; Volante, R. P.; Reider, P. J. *J. Am. Chem. Soc.* **1997**, *119*, 6207.
193. For an exception see: Tani, K.; Suwa, K.; Tanigawa, E.; Ise, T.; Yamagata, T.; Tatsuno, Y.; Otsuka, S. *J. Organomet. Chem.* **1989**, *370*, 203.
194. Burk, M. J. *Acc. Chem. Res.* **2000**, *33*, 363.
195. Nugent, W. A.; Rajanbabu, T. V.; Burk, M. J. *Science* **1993**, *259*, 479.
196. Burk, M. J.; Feaster, J. E.; Harlow, R. L. *Organometallics* **1990**, *9*, 2653.
197. Xiao, D. M.; Zhang, Z. G.; Zhang, X. M. *Org. Lett.* **1999**, *1*, 1679.
198. Chi, Y. X.; Zhang, X. M. *Tetrahedron Lett.* **2002**, *43*, 4849.
199. Jiang, Q. Z.; Jiang, Y. T.; Xiao, D. M.; Cao, P.; Zhang, X. M. *Angew. Chem. Int. Ed.* **1998**, *37*, 1100.
200. Zhang, Z. G.; Zhu, G. X.; Jiang, Q. Z.; Xiao, D. M.; Zhang, X. M. *J. Org. Chem.* **1999**, *64*, 1774.
201. Vineyard, B. D.; Knowles, W. S.; Sabacky, M. J.; Bachman, G. L.; Weinkauff, D. J. *J. Am. Chem. Soc.* **1977**, *99*, 5946.
202. Knowles, W. S. *Acc. Chem. Res.* **1983**, *16*, 106.
203. Muci, A. R.; Campos, K. R.; Evans, D. A. *J. Am. Chem. Soc.,* **1995**, *117*, 9075.
204. Juge, S.; Stephan, M.; Laffitte, J. A.; Genet, J. P. *Tetrahedron Lett.* **1990**, *31*, 6357.
205. Imamoto, T.; Watanabe, J.; Wada, Y.; Masuda, H.; Yamada, H.; Tsuruta, H.; Matsukawa, S.; Yamaguchi, K. *J. Am. Chem. Soc.* **1998**, *120*, 1635.
206. Crépy, K. V. L.; Imamoto, T. *Adv. Synth. Catal.* **2003**, *345*, 79.
207. Yasutake, M.; Gridnev, I. D.; Higashi, N.; Imamoto, T. *Org. Lett.* **2001**, *3*, 1701.
208. Imamoto, T.; Watanabe, J.; Wada, Y.; Masuda, H.; Yamada, H.; Tsuruta, H.; Matsukawa, S.; Yamaguchi, K. *J. Am. Chem. Soc.* **1998**, *120*, 1635.
209. Ohashi, A.; Imamoto, T. *Org. Lett.* **2001**, *3*, 373.
210. Gridnev, I. D.; Higashi, N.; Asakura, K.; Imamoto, T. *J. Am. Chem. Soc.* **2000**, *122*, 7183.
211. Gridnev, I. D.; Yasutake, M.; Higashi, N.; Imamoto, T. *J. Am. Chem. Soc.* **2001**, *123*, 5268.
212. von Matt, P.; Loiseleur, O.; Koch, G.; Pfaltz, A.; Lefeber, C.; Feucht, T.; Helmchen, G. *Tetrahedron: Asymmetry* **1994**, *5*, 573.
213. Williams, J. M. J. *Synlett* **1996**, 705.
214. Helmchen, G.; Pfaltz, A. *Acc. Chem. Res.* **2000**, *33*, 336.
215. Pfaltz, A. *Acta Chem. Scand.* **1996**, *50*, 189.
216. Hilgraf, R.; Pfaltz, A. *Synlett* **1999**, 1814.
217. Cozzi, P. G.; Zimmermann, N.; Hilgraf, R.; Schaffner, S.; Pfaltz, A. *Adv. Synth. Catal.* **2001**, *343*, 450.
218. Blankenstein, J.; Pfaltz, A. *Angew. Chem. Int. Ed.* **2001**, *40*, 4445.
219. Menges, F.; Neuburger, M.; Pfaltz, A. *Org. Lett.* **2002**, *4*, 4713.
220. Menges, F.; Pfaltz, A. *Adv. Synth. Catal.* **2002**, *344*, 40.
221. Smidt, S. P.; Menges, F.; Pfaltz, A. *Org. Lett.* **2004**, *6*, 2023.
222. Hou, D.-R.; Reibenspies, J.; Colacot, T. J.; Burgess, K. *Chem.—Eur. J.* **2001**, *7*, 5391.
223. Duursma, A.; Lefort, L.; Boogers, J. A. F.; de Vries, A. H. M.; de Vries, J. G.; Minnaard, A. J.; Feringa, B. L. *Org. Biomol. Chem.* **2004**, *2*, 1682.
224. van Zijl, A. W.; Arnold, L. A.; Minnaard, A. J.; Feringa, B. L. *Adv. Synth. Catal.* **2004**, *346*, 413.
225. Reetz, M. T.; Sell, T.; Meiswinkel, A.; Mehler, G. *Angew. Chem. Int. Ed.* **2003**, *42*, 790.
226. Reetz, M. T. *Angew. Chem. Int. Ed.* **2008**, *47*, 2556.
227. Gergely, I.; Hegedus, C.; Gulyas, H.; Szollosy, A.; Monsees, A.; Riermeier, T.; Bakos, J. *Tetrahedron: Asymmetry* **2003**, *14*, 1087.
228. Reetz, M. T.; Mehler, G. *Angew. Chem. Int. Ed.* **2000**, *39*, 3889.
229. Reetz, M. T.; Mehler, G.; Meiswinkel, A.; Sell, T. *Tetrahedron Lett.* **2002**, *43*, 7941.
230. van den Berg, M.; Minnaard, A. J.; Schudde, E. P.; van Esch, J.; de Vries, A. H. M.; de Vries, J. G.; Feringa, B. L. *J. Am. Chem. Soc.* **2000**, *122*, 11539.
231. van den Berg, M.; Haak, R. M.; Minnaard, A. J.; de Vries, A. H. M.; de Vries, J. G.; Feringaa, B. L. *Adv. Synth. Catal.* **2002**, *344*, 1003.
232. Peña, D.; Minnaard, A. J.; de Vries, J. G.; Feringa, B. L. *J. Am. Chem. Soc.* **2002**, *124*, 14552.
233. Blaser, H. U.; Schmidt, E. *Asymmetric Catalysis on Industrial Scale*; Wiley-VCH: Weinheim, 2004.
234. Blaser, H.-U.; Spindler, F.; Thommen, M. In *The Handbook of Homogeneous Hydrogenation*; de Vries, J. G., Elsevier, C. J., Eds.; Wiley-VCH: Weinheim, 2007; Vol. 3, p 1279.
235. Burk, M. J.; Feaster, J. E.; Nugent, W. A.; Harlow, R. L. *J. Am. Chem. Soc.* **1993**, *115*, 10125.

236. Dobler, C.; Kreuzfeld, H. J.; Michalik, M.; Krause, H. W. *Tetrahedron: Asymmetry* **1996**, *7*, 117.
237. Burk, M. J.; Gross, M. F.; Martinez, J. P. *J. Am. Chem. Soc.* **1995**, *117*, 9375.
238. Cobley, C. J.; Johnson, N. B.; Lennon, I. C.; McCague, R.; Ramsden, J. A.; Zanotti-Gerosa, A. In *Asymmetric Catalysis on Industrial Scale*; Blaser, H. U., Schmidt, E., Eds.; Wiley-VCH: Weinheim, 2004; p 269.
239. Hiebl, J.; Kollmann, H.; Rovenszky, F.; Winkler, K. *J. Org. Chem.* **1999**, *64*, 1947.
240. Blaser, H. U.; Spindler, F. *Top. Catal.* **1997**, *4*, 275.
241. McGarrity, J. F.; Brieden, W.; Fuchs, R.; Mettler, H.-P.; Schmidt, B.; Werbitzky, O. In *Asymmetric Catalysis on Industrial Scale*; Blaser, H. U., Schmidt, E., Eds.; Wiley-VCH: Weinheim, 2004; p 283.
242. Imwinkelried, R. *Chimia* **1997**, *51*, 300.
243. Fuchs, R. Eur. Patent 803502, 1996.
244. Brieden, W. In *Proceedings of the Chiral USA '97 Symposium*; Spring Innovation: Stockport, UK, 1997, p 45.
245. Brieden, W. In *Proceedings of the ChiraSource '99 Symposium*; The Catalyst Group, Spring House: USA, 1999.
246. Lubell, W. D.; Kitamura, M.; Noyori, R. *Tetrahedron: Asymmetry* **1991**, *2*, 543.
247. Heller, D.; Holz, J.; Drexler, H. J.; Lang, J.; Drauz, K.; Krimmer, H. P.; Borner, A. *J. Org. Chem.* **2001**, *66*, 6816.
248. Zhu, G.; Chen, Z.; Zhang, X. *J. Org. Chem.* **1999**, *64*, 6907.
249. Lee, S. G.; Zhang, Y. J. *Org. Lett.* **2002**, *4*, 2429.
250. Tang, W.; Chi, Y.; Zhang, X. *Org. Lett.* **2002**, *4*, 1695.
251. Liu, D. A.; Zhang, X. M. *Eur. J. Org. Chem.* **2005**, 646.
252. Tang, W.; Zhang, X. *Org. Lett.* **2002**, *4*, 4159
253. Wu, H. P.; Hoge, G. *Org. Lett.* **2004**, *6*, 3645.
254. Fu, Y.; Hou, G. H.; Xie, J. H.; Xing, L.; Wang, L. X.; Zhou, Q. L. *J. Org. Chem.* **2004**, *69*, 8157.
255. Koenig, K. E.; Bachman, G. L.; Vineyard, B. D. *J. Org. Chem.* **1980**, *45*, 2362.
256. Schmidt, U.; Langner, J.; Kirschbaum, B.; Braun, C. *Synthesis* **1994**, 1138.
257. Burk, M. J.; Kalberg, C. S.; Pizzano, A. *J. Am. Chem. Soc.* **1998**, *120*, 4345.
258. Lotz, M.; Ireland, T.; Perea, J. J. A.; Knochel, P. *Tetrahedron: Asymmetry* **1999**, *10*, 1839.
259. Lotz, M.; Polborn, K.; Knochel, P. *Angew. Chem. Int. Ed.* **2002**, *41*, 4708.
260. Ohta, T.; Takaya, H.; Kitamura, M.; Nagai, K.; Noyori, R. *J. Org. Chem.* **1987**, *52*, 3174.
261. Zhang, X.; Uemura, T.; Matsumura, K.; Sayo, N.; Kumobayashi, H.; Takaya, H. *Synlett* **1994**, 501.
262. Uemura, T.; Zhang, X.; Matsumura, K.; Sayo, N.; Kumobayashi, H.; Ohta, T.; Nozaki, K.; Takaya, H. *J. Org. Chem.* **1996**, *61*, 5510.
263. Genet, J. P.; Pinel, C.; Ratovelomanana-Vidal, V.; Mallart, S.; Pfister, X.; Bischoff, L.; Deandrade, M. C. C.; Darses, S.; Galopin, C.; Laffitte, J. A. *Tetrahedron: Asymmetry* **1994**, *5*, 675.
264. Li, S.; Zhu, S.-F.; Zhang, C.-M.; Song, S.; Zhou, Q.-L. *J. Am. Chem. Soc.* **2008**, *130*, 8584.
265. Chiba, T.; Miyashita, A.; Nohira, H.; Takaya, H. *Tetrahedron Lett.* **1991**, *32*, 4745.
266. Burk, M. J.; Bienewald, F.; Harris, M.; Zanotti-Gerosa, A. *Angew. Chem. Int. Ed.* **1998**, *37*, 1931.
267. Pilkington, C. J.; Zanotti-Gerosa, A. *Org. Lett.* **2003**, *5*, 1273.
268. Tang, W.; Liu, D.; Zhang, X. *Org. Lett.* **2003**, *5*, 205.
269. Reetz, M. T.; Gosberg, A.; Goddard, R.; Kyung, S.-H. *Chem. Commun.* **1998**, 2077.
270. Reetz, M. T.; Neugebauer, T. *Angew. Chem., Int. Ed. Engl.* **1999**, *38*, 179.
271. Hu, X. P.; Zheng, Z. *Org. Lett.* **2004**, *6*, 3585.
272. Jia, X.; Guo, R.; Li, X.; Yao, X.; Chan, A. *Tetrahedron Lett.* **2002**, *43*, 5541.
273. Cui, X. H.; Burgess, K. *Chem. Rev.* **2005**, *105*, 3272.
274. Burk, M. J.; Casy, G.; Johnson, N. B. *J. Org. Chem.* **1998**, *63*, 6084.
275. Dobbs, D. A; Vanhessche, K. P. M.; Rautenstrauch, V. PCT/IB1998/000776, 1998.
276. Dobbs, D. A.; Vanhessche, K. P. M.; Brazi, E.; Rautenstrauch, V.; Lenoir, J. Y.; Genet, J. P.; Wiles, J.; Bergens, S. H. *Angew. Chem. Int. Ed.* **2000**, *39*, 1992.
277. Rautenstrauch, V. In *Proceedings of the Chiral USA '97 Symposium*; Spring Innovation: Stockport, UK, 1999; p 204.
278. Akutagawa, S. *Top. Catal.* **1997**, *4*, 271.
279. Takaya, H.; Ohta, T.; Sayo, N.; Kumobayashi, H.; Akutagawa, S.; Inoue, S.; Kasahara, I.; Noyori, R. *J. Am. Chem. Soc.* **1987**, *109*, 1596.
280. Källström, K.; Munslow, I.; Andersson, P. G. *Chem. Eur. J.* **2006**, *12*, 3194.
281. Wang, J.; Sun, Y. K.; Leblond, C.; Landau, R. N.; Blackmond, D. G. *J. Catal.* **1996**, *161*, 752.
282. Sun, Y. K.; LeBlond, C.; Wang, J.; Blackmond, D. G. *J. Am. Chem. Soc.* **1995**, *117*, 12647.
283. Schmid, R.; Scalone, M. In *Comprehensive Asymmetric Catalysis I-III*; Jacobsen, E. N., Pfaltz, A., Yamamoto, H., Eds.; Springer: Berlin, 1999; Vol. 3, p 1439.

284. Netscher, T.; Scalone, M.; Schmid, R. In *Asymmetric Catalysis on Industrial Scale*; Blaser, H. U., Schmidt, E., Eds.; Wiley-VCH: Weinheim, 2004; p 71.

285. Akutagawa, S. *Appl. Catal., A* **1995**, *128*, 171.

286. Broene, R. D.; Buchwald, S. L. *J. Am. Chem. Soc.* **1993**, *115*, 12569.

287. Troutman, M. V.; Appella, D. H.; Buchwald, S. L. *J. Am. Chem. Soc.* **1999**, *121*, 4916.

288. Schrems, M. G.; Neumann, E.; Pfaltz, A. *Angew. Chem. Int. Ed.* **2007**, *46*, 8274.

289. Bell, S.; Wustenberg, B.; Kaiser, S.; Menges, F.; Netscher, T.; Pfaltz, A. *Science* **2006**, *311*, 642.

290. Boaz, N. W.; Debenham, S. D.; Mackenzie, E. B.; Large, S. E. *Org. Lett.* **2002**, *4*, 2421.

291. Boaz, N. W. *Tetrahedron Lett.* **1998**, *39*, 5505.

292. Ager, D. J.; Laneman, S. A. *Tetrahedron: Asymmetry* **1997**, *8*, 3327.

293. Burk, M. J.; Harper, T. G. P.; Kalberg, C. S. *J. Am. Chem. Soc.* **1995**, *117*, 4423.

294. Yamano, T.; Taya, N.; Kawada, H.; Huang, T.; Imamoto, T. *Tetrahedron Lett.* **1999**, *40*, 2577.

295. Pye, P. J.; Rossen, K.; Reamer, R. A.; Volante, R. P.; Reider, P. J. *Tetrahedron Lett.* **1998**, *39*, 4441.

296. Genet, J. P.; Deandrade, M. C. C.; Ratovelomanana-Vidal, V. *Tetrahedron Lett.* **1995**, *36*, 2063.

297. Makino, K.; Goto, T.; Hiroki, Y.; Hamada, Y. *Angew. Chem. Int. Ed.* **2004**, *43*, 882.

298. Blaser, H. U.; Gamboni, R.; Pugin, B.; Rihs, G.; Sedelmeier, G.; Schaub, B.; Schmidt, E.; Schmitz, B.; Spindler, F.; Wetter, H. In *Process Chemistry in the Pharmaceutical Industry*; Gadamasetti, K. G., Ed.; Marcel Dekker: New York, 1999; p 189.

299. Kumobayashi, H. *Rec. Trav. Chim. Pays-Bas* **1996**, *115*, 201.

300. Noyori, R.; Tokunaga, M.; Kitamura, M. *Bull. Chem. Soc. Jpn.* **1995**, *68*, 36.

301. Kiely, A. F.; Jernelius, J. A.; Schrock, R. R.; Hoveyda, A. H. *J. Am. Chem. Soc.* **2002**, *124*, 2868.

302. Blaser, H.-U.; Spindler, F.; Thommen, M. In *The Handbook of Homogeneous Hydrogenation*; de Vries, J. G., Elsevier, C. J., Eds.; Wiley-VCH: Weinheim, 2007; Vol. 3, p 1306.

303. Rouhi, M. *Chem. Eng. News* **2004**, *82*, 47.

304. Mashima, K.; Kusano, K. H.; Sato, N.; Matsumura, Y.; Nozaki, K.; Kumobayashi, H.; Sayo, N.; Hori, Y.; Ishizaki, T.; Akutagawa, S.; Takaya, H. *J. Org. Chem.* **1994**, *59*, 3064.

305. Ohkuma, T.; Hattori, T.; Ooka, H.; Inoue, T.; Noyori, R. *Org. Lett.* **2004**, *6*, 2681.

306. Ohkuma, T.; Ishii, D.; Takeno, H.; Noyori, R. *J. Am. Chem. Soc.* **2000**, *122*, 6510.

307. Klinger, F. D.; Wolter, L.; Dietrich, W. Eur. Patent 1147075, 1999.

308. Noyori, R.; Ohkuma, T. *Angew. Chem. Int. Ed.* **2001**, *40*, 40.

309. Burk, M. J.; Hems, W.; Herzberg, D.; Malan, C.; Zanotti-Gerosa, A. *Org. Lett.* **2000**, *2*, 4173.

310. Wu, J.; Chen, H.; Kwok, W.; Guo, R. W.; Zhou, Z. Y.; Yeung, C.; Chan, A. S. C. *J. Org. Chem.* **2002**, *67*, 7908.

311. Kumobayashi, H.; Miura, T.; Sayo, N.; Saito, T.; Zhang, X. Y. *Synlett* **2001**, 1055.

312. Chaplin, D.; Harrison, P.; Henschke, J. P.; Lennon, I. C.; Meek, G.; Moran, P.; Pilkington, C. J.; Ramsden, J. A.; Watkins, S.; Zanotti-Gerosa, A. *Org. Process Res. Dev.* **2003**, *7*, 89.

313. Ohkuma, T.; Koizumi, M.; Muniz, K.; Hilt, G.; Kabuto, C.; Noyori, R. *J. Am. Chem. Soc.* **2002**, *124*, 6508.

314. Chen, C. Y.; Reamer, R. A.; Chilenski, J. R.; McWilliams, C. J. *Org. Lett.* **2003**, *5*, 5039.

315. Ohkuma, T.; Ooka, H.; Yamakawa, M.; Ikariya, T.; Noyori, R. *J. Org. Chem.* **1996**, *61*, 4872.

316. Bolm, C. *Angew. Chem., Int. Ed. Engl.* **1993**, *32*, 232.

317. Viso, A.; Lee, N. E.; Buchwald, S. L. *J. Am. Chem. Soc.* **1994**, *116*, 9373.

318. Willoughby, C. A.; Buchwald, S. L. *J. Am. Chem. Soc.* **1994**, *116*, 8952.

319. Willoughby, C. A.; Buchwald, S. L. *J. Org. Chem.* **1993**, *58*, 7627.

320. Ringwald, M.; Sturmer, R.; Brintzinger, H. H. *J. Am. Chem. Soc.* **1999**, *121*, 1524

321. Cobley, C. J.; Henschke, J. P.; Ramsden, J. A. WO 0208169, WO2001GB03271 20010720, 2002.

322. Cobley, C. J.; Henschke, J. P. *Adv. Synth. Catal.* **2003**, *345*, 195.

323. Abdur-Rashid, K.; Lough, A. J.; Morris, R. H. *Organometallics* **2000**, *19*, 2655.

324. Morimoto, T.; Achiwa, K. *Tetrahedron: Asymmetry* **1995**, *6*, 2661.

325. Rice, K. C. *J. Org. Chem.* **1980**, *45*, 3135.

326. Lie, T. S.; Maat, L.; Beyerman, H. C. *Rec. Trav. Chim. Pays-Bas* **1979**, *98*, 419.

327. Rice, K. C. In *The Chemistry and Biology of Isoquinoline Alkaloids*; Phillipson, J. D., Roberts, M. F., Zenk, M. H., Eds.; Springer-Verlag: Berlin, 1985; p 191.

328. Blakemore, P. R.; White, J. D. *Chem. Commun.* **2002**, 1159.

329. Tani, K.; Onouchi, J.; Yamagata, T.; Kataoka, Y. *Chem. Lett.* **1995**, 955.

330. Kitamura, M.; Hsiao, Y.; Noyori, R.; Takaya, H. *Tetrahedron Lett.* **1987**, *28*, 4829.

331. Noyori, R.; Ohta, M.; Hsiao, Y.; Kitamura, M.; Ohta, T.; Takaya, H. *J. Am. Chem. Soc.* **1986**, *108*, 7117.

332. Willoughby, C. A.; Buchwald, S. L. *J. Am. Chem. Soc.* **1992**, *114*, 7562.

333. Bakos, J.; Orosz, A.; Heil, B.; Laghmari, M.; Lhoste, P.; Sinou, D. *J. Chem. Soc., Chem. Commun.* **1991**, 1684.

334. Xiao, D. M.; Zhang, X. M. *Angew. Chem. Int. Ed.* **2001**, *40*, 3425.
335. Blaser, H.-U.; Hanreich, R.; Schneider, H.-D.; Spindler, F.; Steinacher, B. In *Asymmetric Catalysis on Industrial Scale*; Blaser, H. U., Schmidt, E., Eds.; Wiley-VCH: Weinheim, 2004; p 55.
336. Burk, M. J.; Feaster, J. E. *J. Am. Chem. Soc.* **1992**, *114*, 6266.
337. Spindler, F.; Blaser, H.-U. *Adv. Synth. Catal.* **2001**, *343*, 68.
338. Rouhi, M. *Chem. Eng. News* **2004**, *82*, 28.
339. Saito, T. In *19th NACS Meeting*; Philadelphia, 2005.
340. Blacker, J.; Martin, J. In *Asymmetric Catalysis on Industrial Scale*; Blaser, H. U., Schmidt, E., Eds.; Wiley-VCH: Weinheim, 2004; p 201.
341. Ikariya, T.; Blacker, A. J. *Acc. Chem. Res.* **2007**, *40*, 1300.
342. Uematsu, N.; Fujii, A.; Hashiguchi, S.; Ikariya, T.; Noyori, R. *J. Am. Chem. Soc.* **1996**, *118*, 4916.
343. Mashima, K.; Abe, T.; Tani, K. *Chem. Lett.* **1998**, 1199.
344. Mao, J. M.; Baker, D. C. *Org. Lett.* **1999**, *1*, 841.
345. Murata, K.; Ikariya, T.; Noyori, R. *J. Org. Chem.* **1999**, *64*, 2186.
346. Blacker, J.; Martin, J. In *Asymmetric Catalysis on Industrial Scale*; Blaser, H. U., Schmidt, E., Eds.; Wiley-VCH: Weinheim, 2004; p 201.
347. Moore, M. L. *Org. React.* **1949**, *5*, 301.
348. Kadyrov, R.; Riermeier, T. H. *Angew. Chem. Int. Ed.* **2003**, *42*, 5472.
349. Brown, J. M.; Chaloner, P. A. *J. Chem. Soc., Chem. Commun.* **1980**, 344.
350. Alcock, N. W.; Brown, J. M.; Derome, A. E.; Lucy, A. R. *J. Chem. Soc., Chem. Commun.* **1985**, 575.
351. Ojima, I.; Kogure, T.; Yoda, N. *J. Org. Chem.* **1980**, *45*, 4728.
352. Ojima, I.; Kogure, T.; Yoda, N. *Chem. Lett.* **1979**, 495.
353. Sinou, D. *Tetrahedron Lett.* **1981**, *22*, 2987.
354. Evans, D. A.; Campos, K. R.; Tedrow, J. S.; Michael, F. E.; Gagne, M. R. *J. Am. Chem. Soc.* **2000**, *122*, 7905.
355. Schmidt, T.; Baumann, W.; Drexler, H. J.; Arrieta, A.; Heller, D.; Buschmann, H. *Organometallics* **2005**, *24*, 3842.
356. Drexler, H. J.; Baumann, W.; Schmidt, T.; Zhang, S. L.; Sun, A. L.; Spannenberg, A.; Fischer, C.; Buschmann, H.; Heller, D. *Angew. Chem. Int. Ed.* **2005**, *44*, 1184.
357. Wiles, J. A.; Bergens, S. H.; Young, V. G. *J. Am. Chem. Soc.* **1997**, *119*, 2940.
358. Wiles, J. A.; Bergens, S. H. *Organometallics* **1998**, *17*, 2228.
359. Hutchins, R. O.; Hutchins, M. G. K. In *Chemistry of Triple-Bonded Functional Groups*; Rappoport, Z., Patai, S., Eds.; Wiley: 1983; Vol. 1, p 571.
360. Cui, X.; Burgess, K. *J. Am. Chem. Soc.* **2003**, *125*, 14212.
361. Muramatsu, H.; Kawano, H.; Ishii, Y.; Saburi, M.; Uchida, Y. *J. Chem. Soc., Chem. Commun.* **1989**, 769.
362. Burk, M. J.; Allen, J. G.; Kiesman, W. F. *J. Am. Chem. Soc.* **1998**, *120*, 657.
363. Beghetto, V.; Matteoli, U.; Scrivanti, A. *Chem. Commun.* **2000**, 155.
364. Heldal, J. A.; Frankel, E. N. *J. Am. Oil Chem. Soc.* **1985**, *62*, 1117.
365. Spencer, A. *J. Organomet. Chem.* **1975**, *93*, 389.
366. Speziali, M. G.; Moura, F. C. C.; Robles-Dutenhefner, P. A.; Araujo, M. H.; Gusevskaya, E. V.; dos Santos, E. N. *J. Mol. Catal. A* **2005**, *239*, 10.
367. Cais, M.; Frankel, E. N.; Rejoan, A. *Tetrahedron Lett.* **1968**, 1919.
368. Frankel, E. N.; Selke, E.; Glass, C. A. *J. Am. Chem. Soc.* **1968**, *90*, 2446.
369. Wrighton, M.; Schroeder, M. A. *J. Am. Chem. Soc.* **1973**, *95*, 5764.
370. Mirbach, M. J.; Tuyet, N. P.; Saus, A. *J. Organomet. Chem.* **1982**, *236*, 309.
371. Frankel, E. N.; Selke, E.; Glass, C. A. *J. Am. Chem. Soc.* **1968**, *90*, 2446.
372. Tucker, J. R.; Riley, D. P. *J. Organomet. Chem.* **1985**, *279*, 49.
373. Lemaux, P.; Jaouen, G.; Saillard, J. Y. *J. Organomet. Chem.* **1981**, *212*, 193.
374. Sodeoka, M.; Shibasaki, M. *J. Org. Chem.* **1985**, *50*, 1147.
375. Vasil'ev, A. A.; Serebryakov, E. P. *Russ. Chem. Bull.* **2002**, *51*, 1341.
376. Sodeoka, M.; Shibasaki, M. *Synthesis* **1993**, 643.
377. van Laren, M. W.; Elsevier, C. J. *Angew. Chem. Int. Ed.* **1999**, *38*, 3715.
378. Hauwert, P.; Maestri, G.; Sprengers, J. W.; Catellani, M.; Elsevier, C. J. *Angew. Chem. Int. Ed.* **2008**, *47*, 3223.
379. Sloan, M. F.; Matlack, A. S.; Breslow, D. S. *J. Am. Chem. Soc.* **1963**, *85*, 4014.
380. Kroll, W. R. *J. Catal.* **1969**, *15*, 281.
381. Lapporte, S. J. *Ann. N.Y. Acad. Sci.* **1969**, *158*, 510.
382. Dyson, P. J. *J. Chem. Soc., Dalton Trans.* **2003**, 2964.
383. Widegren, J. A.; Finke, R. G. *J. Mol. Catal. A* **2003**, *198*, 317.
384. Widegren, J. A.; Finke, R. G. *J. Mol. Catal. A* **2003**, *191*, 187.

385. Gao, H. R.; Angelici, R. J. *J. Am. Chem. Soc.* **1997**, *119*, 6937.

386. Gao, H. R.; Angelici, R. J. *J. Mol. Catal. A* **1999**, *149*, 63.

387. Gao, H. R.; Angelici, R. J. *Organometallics* **1999**, *18*, 989.

388. Perera, M.; Angelici, R. J. *J. Mol. Catal.* **1999**, *149*, 99.

389. Yang, H.; Gao, H. R.; Angelici, R. J. *Organometallics* **2000**, *19*, 622.

390. Abu-Reziq, R.; Avnir, D.; Miloslavski, I.; Schumann, H.; Blum, J. *J. Mol. Catal. A* **2002**, *185*, 179.

391. Barbaro, P.; Bianchini, C.; Dal Santo, V.; Meli, A.; Moneti, S.; Psaro, R.; Scaffidi, A.; Sordelli, L.; Vizza, F. *J. Am. Chem. Soc.* **2006**, *128*, 7065.

392. Glorius, F. *Org. Biomol. Chem.* **2005**, *3*, 4171.

393. Kuwano, R. *Heterocycles* **2008**, *76*, 909.

394. Wilczynski, R.; Fordyce, W. A.; Halpern, J. *J. Am. Chem. Soc.* **1983**, *105*, 2066.

395. Feder, H. M.; Halpern, J. *J. Am. Chem. Soc.* **1975**, *97*, 7186.

396. Zhou, Y.-G. *Acc. Chem. Res.* **2007**, *40*, 1357.

397. Glorius, F. *Org. Biomol. Chem.* **2005**, *3*, 4171.

398. Gilman, V. *Chem. Eng. News* **2005**, *83*, 108.

399. Studer, M.; Wedemeyer-Exl, C.; Spindler, F.; Blaser, H. U. *Monatsh. Chem.* **2000**, *131*, 1335.

400. Murata, S.; Sugimoto, T.; Matsuura, S. *Heterocycles* **1987**, *26*, 763.

401. Ohta, T.; Miyake, T.; Seido, N.; Kumobayashi, H.; Takaya, H. *J. Org. Chem.* **1995**, *60*, 357.

402. Fuchs, R. U.S. Patent 5,886,181, 1997.

403. Bianchini, C.; Barbaro, P.; Scapacci, G.; Farnetti, E.; Graziani, M. *Organometallics* **1998**, *17*, 3308.

404. Wang, W. B.; Lu, S. M.; Yang, P. Y.; Han, X. W.; Zhou, Y. G. *J. Am. Chem. Soc.* **2003**, *125*, 10536.

405. Lu, S. M.; Wang, Y. Q.; Han, X. W.; Zhou, Y. G. *Angew. Chem. Int. Ed.* **2006**, *45*, 2260.

406. Legault, C. Y.; Charette, A. B. *J. Am. Chem. Soc.* **2005**, *127*, 8966.

407. Glorius, F.; Spielkamp, N.; Holle, S.; Goddard, R.; Lehmann, C. W. *Angew. Chem. Int. Ed.* **2004**, *43*, 2850.

408. Kuwano, R.; Sato, K.; Kurokawa, T.; Karube, D.; Ito, Y. *J. Am. Chem. Soc.* **2000**, *122*, 7614.

409. Kuwano, R.; Kaneda, K.; Ito, T.; Sato, K.; Kurokawa, T.; Ito, Y. *Org. Lett.* **2004**, *6*, 2213.

410. Kuwano, R.; Kashiwabara, M. *Org. Lett.* **2006**, *8*, 2653.

411. Kaiser, S.; Smidt, S. P.; Pfaltz, A. *Angew. Chem. Int. Ed.* **2006**, *45*, 5194.

412. Grey, R. A.; Pez, G. P.; Wallo, A. *J. Am. Chem. Soc.* **1981**, *103*, 7536.

413. Linn, D. E.; Halpern, J. *J. Am. Chem. Soc.* **1987**, *109*, 2969.

414. Matteoli, U.; Bianchi, M.; Menchi, G.; Frediani, P.; Piacenti, F. *J. Mol. Catal.* **1984**, *22*, 353.

415. Matteoli, U.; Menchi, G.; Bianchi, M.; Piacenti, F. *J. Organomet. Chem.* **1986**, *299*, 233.

416. Matteoli, U.; Menchi, G.; Bianchi, M.; Piacenti, F.; Ianelli, S.; Nardelli, M. *J. Organomet. Chem.* **1995**, *498*, 177.

417. Teunissen, H. T.; Elsevier, C. J. *Chem. Commun.* **1997**, 667.

418. Teunissen, H. T.; Elsevier, C. J. *Chem. Commun.* **1998**, 1367.

419. Nomura, K.; Ogura, H.; Imanishi, Y. *J. Mol. Catal. A* **2002**, *178*, 105.

420. Zhang, J.; Leitus, G.; Ben-David, Y.; Milstein, D. *Angew. Chem. Int. Ed.* **2006**, *45*, 1113.

421. Saudan, L. A.; Saudan, C. M.; Debieux, C.; Wyss, P. *Angew. Chem. Int. Ed.* **2007**, *46*, 7473.

422. Ito, M.; Ikariya, T. *J. Syn. Org. Chem. Jpn.* **2008**, *66*, 1042.

423. Müller, H. In *Ullmann's Encyclopedia of Industrial Chemistry (Online)*; Wiley-VCH: Weinheim, 2002; p DOI: 10.1002/14356007.a26_221.

424. Kanetaka, J.; Asano, T.; Masumune, S. *Ind. Eng. Chem.* **1970**, *62*, 24

425. Lyons, J. E. *J. Chem. Soc., Chem. Commun.* **1975**, 412.

426. Morand, P.; Kayser, M. *J. Chem. Soc., Chem. Commun.* **1976**, 314.

427. Ishii, Y. *Kagaku to Kogyo* **1987**, *40*, 30.

428. Ito, M.; Sakaguchi, A.; Kobayashi, C.; Ikariya, T. *J. Am. Chem. Soc.* **2007**, *129*, 290.

429. Nagayama, K.; Shimizu, I.; Yamamoto, A. *Chem. Lett.* **1998**, 1143.

430. Nagayama, K.; Shimizu, I.; Yamamoto, A. *Bull. Chem. Soc. Jpn.* **2001**, *74*, 1803.

431. Debellefon, C.; Fouilloux, P. *Catal. Rev.* **1994**, *36*, 459.

432. Yoshida, T.; Okano, T.; Otsuka, S. *J. Chem. Soc., Chem. Commun.* **1979**, 870.

433. Bernskoetter, W. H.; Brookhart, M. *Organometallics* **2008**, *27*, 2036.

434. Li, T.; Bergner, I.; Haque, F. N.; Iuliis, M. Z. D.; Song, D.; Morris, R. H. *Organometallics* **2007**, *26*, 5940.

435. Enthaler, S.; Addis, D.; Junge, K.; Erre, G.; Beller, M. *Chem. Eur. J.* **2008**, *14*, 9491.

CHAPTER 16

Hydrofunctionalization and Oxidative Functionalization of Olefins

16.1. Introduction and Scope

The transition-metal-catalyzed addition of substrates containing H–X bonds across carbon–carbon multiple bonds (Equation 16.1) has been studied extensively for nearly 50 years, and studies of these types of reactions remain intense at this time. The addition of water and ammonia to alkenes with anti-Markovnikov regiochemistry has been one of the holy grails of catalysis and remains an unsolved problem. However, the addition of many other types of H–X bonds with anti-Markovnikov regioselectivity has been accomplished and includes reactions conducted industrially on the scale of millions of pounds per annum. The additions of C–H, N–H, O–H, B–H, Si–H, Al–H, S–H, and P–H bonds, as well as element–element bonds, such as Si–Si and B–B, B–Si, and B–Sn, with varying regioselectivity have all been reported. A recent monograph,[1] a compendium of catalysis,[2] and review articles[3] together provide a comprehensive treatment of these reactions. The factors controlling regioselectivity have been reviewed.[4]

The oxidative functionalization of olefins through π-olefin complexes of palladium also has a long history, including the industrial production of acetaldehyde and vinyl acetate. Related reactions, including the conversion of olefins to vinyl ethers and enamines, have been studied in more recent times for fine chemical synthesis. These oxidative C–O and C–N bond formations have been conducted with a variety of oxidants, including O_2, and have been studied as both intermolecular and intramolecular processes.

This chapter focuses on the subset of these reactions that have been studied most intensively and that draw from the stoichiometric reactions presented earlier in this text. Thus, the first sections of this chapter highlight certain aspects of hydrocyanation, hydrosilylation, disilylation, hydroboration, diboration, silylborations, and hydroamination. The last section presents aspects of palladium-catalyzed oxidation and metal-catalyzed oxidative amination of olefins.

$$R\diagup\!\!\!\!\diagdown + \text{ H-X} \xrightarrow{\text{ Catalyst }} \underset{R}{\overset{H}{|}}\!\!\diagdown\!\!\diagup X \text{ or } \underset{R}{\overset{X}{|}}\!\!\diagdown\!\!\diagup H \qquad (16.1)$$

16.2. Homogeneous Catalytic Hydrocyanation of Olefins and Alkynes

16.2.1. Introduction to Hydrocyanation

Hydrocyanation is the addition of HCN across carbon–carbon or carbon–heteroatom multiple bonds to form products containing a new C–C bond.[5–14] The majority of examples from organometallic chemistry involve the addition of HCN across carbon–carbon multiple bonds, as shown in Equations 16.2 and 16.3. Lewis acids and peptides have been used to catalyze the enantioselective addition of HCN to aldehydes and imines to form cyanohydrins and precursors to amino acids.[15–20] The addition of HCN to unactivated olefins requires a catalyst because HCN is not sufficiently acidic to add directly to an olefin, and the C–H bond is strong enough to make additions by radical pathways challenging. However, a large number of soluble transition metal compounds catalyze the addition of HCN to alkenes and alkynes.

$$R\diagup\diagdown + \text{HCN} \xrightarrow{\text{Catalyst}} \underset{R}{\overset{H}{\diagdown}}\diagup\diagdown\text{CN} \qquad (16.2)$$

$$\diagdown\diagup\diagdown + \text{HCN} \xrightarrow{\text{Catalyst}} \diagdown\diagup\diagdown\text{CN} \xrightarrow{\text{Catalyst}} \text{NC}\diagup\diagdown\diagup\diagdown\text{CN} \qquad (16.3)$$

The addition of HCN to olefins catalyzed by complexes of transition metals has been studied since about 1950. The first hydrocyanation by a homogeneous catalyst was reported by Arthur with cobalt carbonyl as catalyst.[21] These reactions gave the branched nitrile as the predominant product. Nickel complexes of phosphites are more active catalysts for hydrocyanation, and these catalysts give the anti-Markovnikov product with terminal alkenes. The first nickel-catalyzed hydrocyanations were disclosed by Drinkard[22–25] and by Brown and Rick.[6,26] The development of this nickel-catalyzed chemistry into the commercially important addition to butadiene (Equation 16.3) was conducted at DuPont. Taylor and Swift referred to hydrocyanation of butadiene,[27] and Drinkard exploited this chemistry for the synthesis of adiponitrile.[22–25] The mechanism of this process was pursued in depth by Tolman.[28] As a result of this work, butadiene hydrocyanation was commercialized in 1971.[7,13] The development of hydrocyanation is one of the early success stories in homogeneous catalysis. Significant improvements in catalysts have been made since that time, and many reviews have now been written on this subject.[5–14]

16.2.2. Examples of Alkene Hydrocyanation

Nickel-catalyzed hydrocyanation of α-olefins typically produces the terminal nitrile as the more abundant, but not exclusive isomeric product. In contrast, nickel-catalyzed hydrocyanation of vinylarenes typically generates the branched product.[29] This branched selectivity arises from the stability of η^3-phenethyl complexes, as is shown in more detail in Section 16.2.5 on asymmetric hydrocyanation. The relative rates for hydrocyanation follow the trend: ethylene > styrene > propene ≈ 1-hexene > disubstituted olefins.[21] Examples of these reactions and selectivities for formation of the linear and branched products are shown in Scheme 16.1.[30]

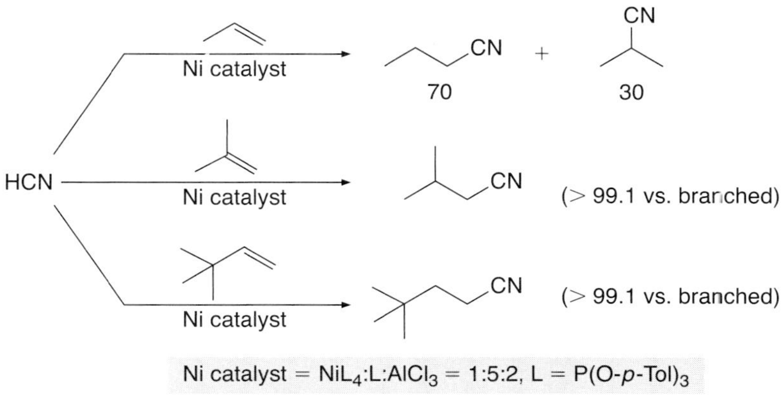

Ni catalyst $= NiL_4:L:AlCl_3 = 1:5:2$, L $= P(O\text{-}p\text{-Tol})_3$

Scheme 16.1

Reactions of internal olefins can even generate terminal alkylnitriles by a pathway that involves isomerization of intermediate cyanometal–alkyl complexes. This isomerization is similar to the isomerization that occurs during the hydroformylation of internal olefins discussed in Chapter 17. In fact, the nickel catalyst rapidly isomerizes hexene to the equilibrium ratio of olefins faster than it adds HCN to the C=C bond. Thus, internal hexenes generate the terminal alkane nitrile.

The toxicity of HCN has limited the use of hydrocyanation in routine synthesis and the development of this reaction as a synthetic method. However, acetone cyanohydrin can be used in some cases as a convenient surrogate for the more noxious HCN. One example of the use of acetone cyanohydrin is shown in Equation 16.4.[31,32]

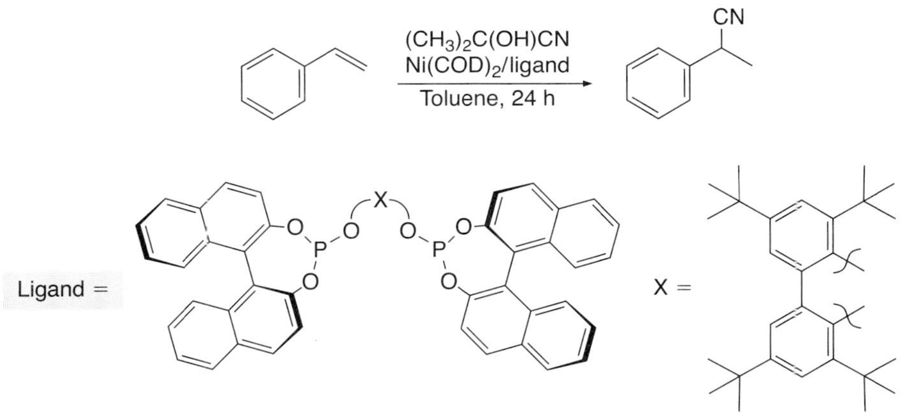

$$(16.4)$$

The use of a co-catalyst was crucial to the development of practical hydrocyanation.[33] The rate and catalyst lifetime for hydrocyanation of simple alkenes increases dramatically by conducting the reactions in the presence of a Lewis acid. As shown in Table 16.1, the reaction of propene occurs much faster in the presence of aluminum and zinc halides.[30] Lewis acid cocatalysts also promote isomerization and selective additions during some steps of the hydrocyanation of butadiene. This effect is presented later in this section.

Table 16.1. Relative rates for hydrocyanation of propylene in 75% toluene/25% CD_2Cl_2 in the presence of $HNi(CN)[P(O$-o-$Tol)_3]_3$ as catalyst.[30]

Lewis acid	Approximate $t_{1/2}$(min)			% Linear product
	−25°C	−0°C	+25°C	
$AlCl_3$	−10			72
$ZnCl_2$			<4	70
None		60		72
None			>7	70
BPh_3			>60	89

While the hydrocyanation of monoolefins remains under development, the hydrocyanation of butadiene shown in Equation 16.3 is one of the largest-scale homogeneous catalytic processes known. This reaction generates the 1,4-dinitrile called adiponitrile, which is the precursor to the diamine monomer in nylon 6,6. The formation of this regioisomer requires a set of reversible hydrocyanations and isomerizations that are discussed in Section 16.2.4 on hydrocyanation of dienes.

16.2.3. Mechanism of Hydrocyanation

16.2.3.1. Mechanism of the Hydrocyanation of Alkenes

The mechanism of hydrocyanation of alkenes catalyzed by soluble complexes is closely related to the mechanism of hydrogenation and hydrosilation. Hydrocyanation occurs by a sequence consisting of oxidative addition of HCN, olefin insertion into the M–H bond, and reductive elimination to form the new C–C bond. The mechanism of the original hydrocyanation catalyzed by cobalt carbonyl has not been studied in depth, but the mechanism of the reactions catalyzed by nickel complexes has been studied in depth and is better defined.

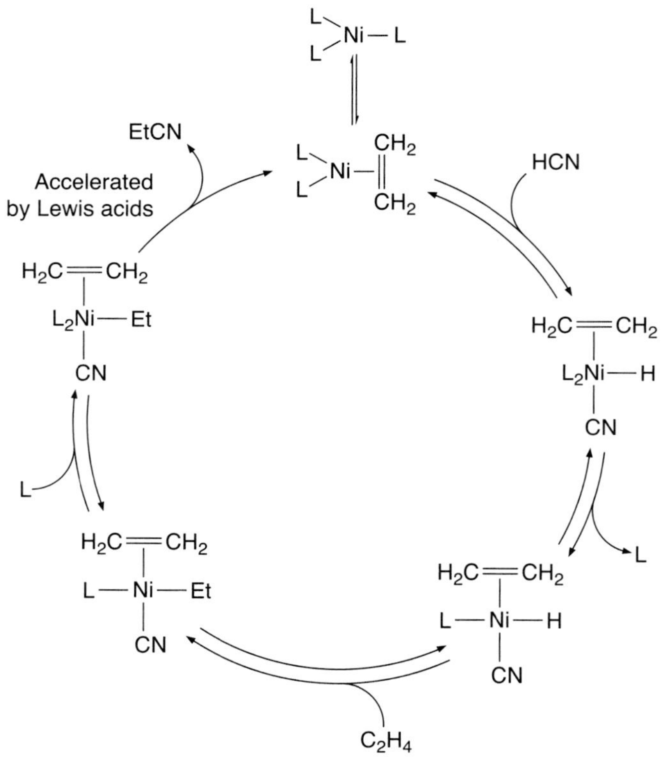

Scheme 16.2

The mechanism of hydrocyanation of ethylene catalyzed by the combination of Ni(0) and P(O-o-Tol)$_3$, as deduced by Tolman, is shown in Scheme 16.2.[12] The L$_2$Ni(ethylene) complex has been isolated and shown to add HCN. The Ni(0) complexes of higher olefins are less stable. In this mechanism, oxidative addition of HCN to a Ni(0) olefin complex forms a cyanometal–hydride complex. In the presence of ethylene, this complex contains olefin, but in the presence of higher olefins this complex has the composition L$_3$Ni(H)(CN). Insertion of an olefin into the metal hydride occurs by a migratory insertion mechanism initiated by coordination of olefin to the cyanometal hydride. Reductive elimination of the alkyl cyanide completes the cycle, and this step is accelerated by Lewis acids, as presented later in this chapter.

Reactions of internal olefins are more complex than reactions of terminal olefins (Scheme 16.3). As mentioned previously, terminal nitriles are often formed from reactions of internal olefins. The formation of terminal nitriles results from insertion of the internal olefin to form a branched alkylmetal intermediate (**A** in Scheme 16.3) that undergoes isomerization to the terminal alkyl intermediate **B** prior to reductive elimination of the final linear nitrile faster than it undergoes reductive elimination to form the branched nitrile. Internal olefins react more slowly than terminal olefins, and this relative rate can be traced to the slower insertion of internal olefins into metal hydrides. Lewis acids, such as ZnCl$_2$ and AlCl$_3$, promote these reactions of isolated alkenes.

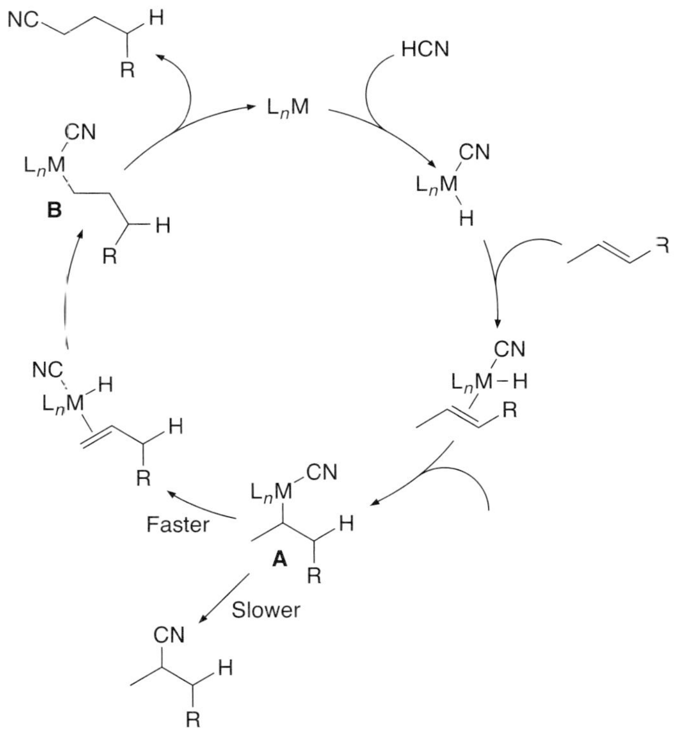

Scheme 16.3

The regioselectivity of the nickel-catalyzed hydrocyanation of terminal olefins results from the regioselectivity of reversible insertion and the relative rates for irreversible reductive elimination from the linear and branched cyanometal–alkyl complexes (Scheme 16.4). The regioselectivity of the insertion step is thought to override differences in the rate of reductive elimination. The regiochemistry is influenced both by the nature of the olefinic substrate and by the catalyst. The formation of terminal nitriles by addition of HCN to terminal olefins is typically observed with nickel catalysts when bulky ligands are present. This regioselectivity likely results from steric effects on the metal–hydride insertion step. These factors controlling regioselectivity are similar to the factors controlling regioselectivity in hydroformylation.

However, Lewis acids have a pronounced effect on the regioselectivity of olefin hydrocyanation. The Lewis acid is thought to coordinate to the cyano group, as noted in detail below. Coordination of the large BPh_3 increases steric constraints on the coordination sphere, and this steric effect is thought to further enhance the thermodynamic preference for formation of the terminal alkylnickel complex.[28]

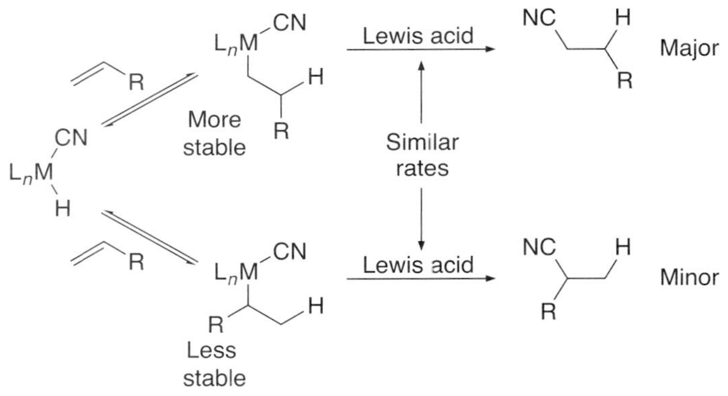

Scheme 16.4

Several experiments support the reaction sequences in Schemes 16.2–16.4. Jackson[11,34–36] and Bäckvall[37–39] studied the stereochemistry of the addition of HCN to alkenes catalyzed by Ni(0) and Pd(0). These studies showed that the addition occurs in a syn fashion. This mode of addition was shown for the reaction of DCN with trans, monodeurerio *tert*-butyl ethylene, as shown in Equation 16.5. The competitive formation of the trideuterated product shown in Equation 16.5 implies that olefin insertion is reversible. Insertion and β-hydrogen elimination will generate some dideuterated alkene that will form the trideuterated product upon addition of DCN.

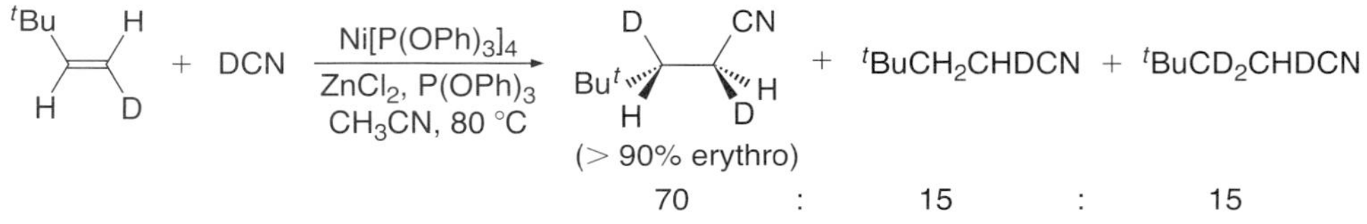

Certain reaction conditions and properties of the nickel complexes promote hydrocyanation. The oxidative addition of HCN requires an open coordination site. Catalysts containing $P(O\text{-}o\text{-}Tol)_3$ as ligand are particularly reactive because the unsaturated L_3Ni complex is the most stable form of the Ni(0) complex, whereas the saturated 18-electron L_4Ni complex is the most stable form of the catalysts containing smaller phosphite ligands. The need to understand the steric and electronic properties of the ligand on the dissociation of phosphine led to the classic work of Tolman on cone angles and electronic parameters.[40]

The slowest step of the hydrocyanation process is reductive elimination. Thus, turnover occurs more rapidly with catalysts containing phosphite ligands than with those containing phosphine ligands, which are more electron–donating than phosphite ligands. Moreover, additives that promote reductive elimination increase the rate of the overall process. It is thought that reductive elimination to form the C–C bond in the nitrile product is facilitated in certain cases by Lewis acid co-catalysts.[28] This acceleration of reductive elimination by Lewis acids is thought to result from coordination of the Lewis acid to the nitrile, which makes the cyanide ligand more electrophilic and promotes internal attack of the alkyl group on the cyanide carbon. This sequence is thought to parallel the promoting effect of Lewis acids on the migration of alkyl groups to carbonyl ligands.

16.2.3.2. Mechanism of Deactivation

Nickel catalysts for hydrocyanation are poisoned by the formation of dicyanide complexes, $L_2Ni(CN)_2$. The formation of this material is second order in HCN Thus, hydrocyanation reactions are typically run under conditions in which HCN is dilute. The presence of added phosphite also helps to minimize deactivation of the catalyst.

16.2.4. Hydrocyanation of Dienes

As noted in the introduction to hydrocyanation, the addition of two equivalents of HCN to 1,3-butadiene is conducted on a large commercial scale to generate adiponitrile, which is converted to hexamethylenediamine.[13,30] This reaction is complex and involves reversible steps. Overall, the process occurs in several stages, shown in Equations 16.6 and 16.7. Each step occurs only in the presence of a catalyst, and the most active and selective catalysts are the combination of a nickel–phosphite complex and a Lewis acid co-catalyst, often called a "promoter." In the first step, one equivalent of HCN adds to butadiene to form 3-pentenenitrile (3PN) and 2-methyl-3-butenenitrile (2M3BN). In the second step, the combination of nickel catalyst and Lewis acid promoter leads to the isomerization of 2M3BM to a roughly 93:7 equilibrium mixture of 3PN and 2M3BN. 3PN then undergoes isomerization to a mixture of pentene nitriles that includes 4-pentene nitrile (4PN). The addition of HCN to this mixture finally generates the α,ω-substituted adiponitrile product. In the presence of triarylboranes as Lewis acids, the isomerization process is sufficiently fast, and the selectivity for linear dinitrile is high. Some data on the selectivities with different Lewis acids are provided in Tables 16.2 and 16.3.[30] Many model studies of the isomerization process have been reported.[41–45]

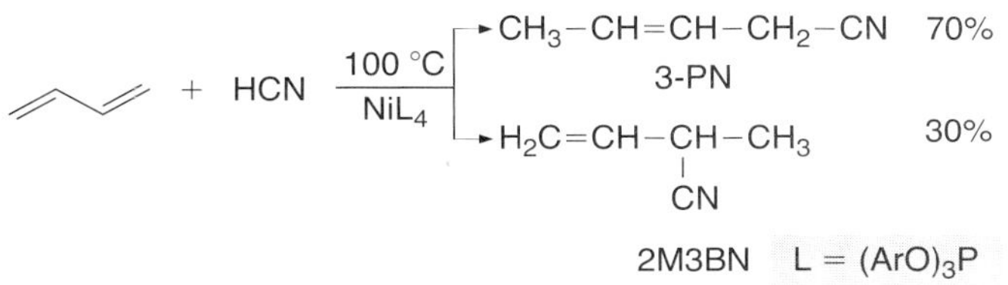

$$
\text{(16.6)}
$$

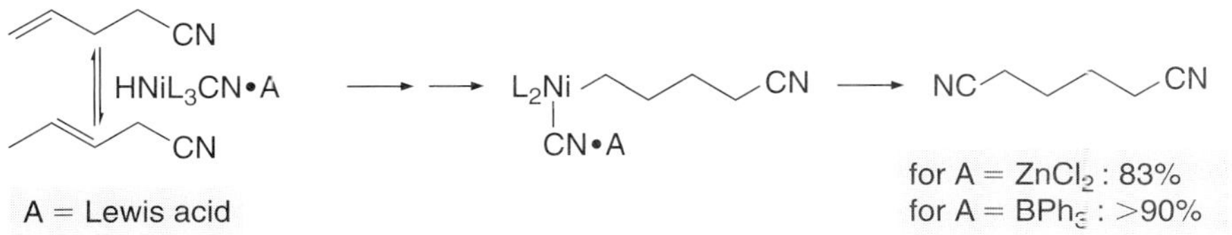

$$
\text{(16.7)}
$$

Table 16.2. Selectivities for the nickel-catalyzed hydrocyanation of propylene in the presence of a series of different classes of Lewis acids.

Lewis acid	% Linear product
$AlCl_3$	72
$ZnCl_2$	70
None	72
BPh_3	89

Table 16.3. Selectivities for formation of the linear dinitrile from nickel-catalyzed hydrocyanation of 4-pentene nitrile in the presence of a series of boron Lewis acids.

Lewis acid	% 1,4-butane dinitrile
B(p-tolyl)$_3$	99
BPh$_3$	98
B(CH$_2$Ph)$_3$	80
None	77
B(o-tolyl)$_3$	74
BCy$_3$	72
B(OPh)$_3$	70
B(O-o-tolyl)$_3$	66

The isomerization of 2-methyl-3-butenenitrile to 3-pentenenitrile is thought to occur by elimination and re-addition of HCN. A number of labeling experiments have been conducted to reveal the order and reversibility of the steps of hydrocyanation; an experiment that addresses the mechanism of isomerization of the branched nitrile is depicted in Equation 16.8.[30] As shown on the left of this equation, allylic transposition of the nitrile group without elimination would lead to 5-deuterio-3-pentenenitrile as the only isotopomer of 3-PN-d_1. However, elimination to form an H–Ni–CN complex and free 1-deuteriobutadiene would lead to a mixture of two labeled 3-PN-d_1 isotopomers after re-insertion of the labeled butadiene and reductive elimination of the free labeled 3-PN. A mixture of the two isotopomers was formed, and this result indicates that isomerization occurs by elimination and re-addition of HCN.

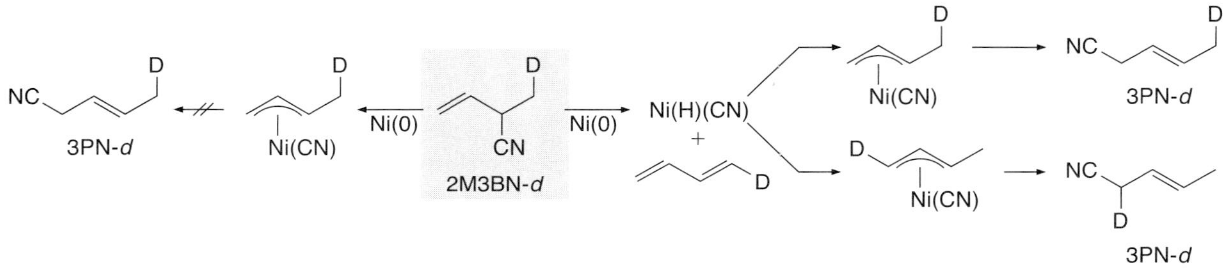

$$(16.8)$$

16.2.5. Asymmetric Hydrocyanation

The enantioselective hydrocyanation of alkenes has the potential to serve as an efficient method to generate optically active nitriles, as well as amides, esters, and amines after functional group interconversions of the nitrile group. As in asymmetric hydroformylation, asymmetric hydrocyanation requires control of both regiochemistry and stereochemistry because simple olefins tend to generate achiral terminal nitrile products. The hydrocyanation of norbornene will give a single constitutional isomer and was studied initially. However, modest enantioselectivities were obtained, and the synthetic value is limited.[46]

More success has been achieved with the enantioselective hydrocyanation of vinylarenes.[47] For reasons described below, the hydrocyanation of vinylarenes tends to generate the branched, chiral, α-aryl nitrile product, instead of the linear, achiral, β-aryl nitrile product. Much research has focused on the hydrocyanation of 6-methoxyvinylnaphthalene because hydrolysis of the nitrile product would lead to the profen drug Naproxen. As shown in Equation 16.9, the hydrocyanation of this vinylarene occurs with high enantioselectivity in the presence of a nickel catalyst containing a phosphinite derived from a carbohydrate. One key to obtaining high enantioselectivity was the use of weakly electron-donating groups on the phosphorus.

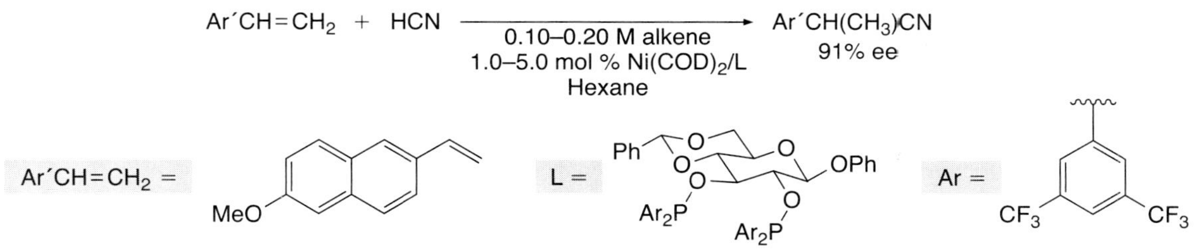

$$Ar'CH=CH_2 \;+\; HCN \xrightarrow[\substack{1.0-5.0 \text{ mol \% Ni(COD)}_2/L \\ \text{Hexane}}]{0.10-0.20 \text{ M alkene}} Ar'CH(CH_3)CN \qquad 91\% \text{ ee}$$

$$Ar'CH=CH_2 = \qquad L = \qquad Ar = \qquad (16.9)$$

The two pathways in Scheme 16.5 illustrate the origins of regioselectivity and the steps that are likely to control enantioselectivity.[47] The branched regioselectivity is almost certain to originate from an insertion of the vinylarene into the nickel hydride to generate an η^3-α-arylethyl complex. Insertion with the opposite regiochemistry would lead to a less stable η^1-β-arylethyl complex. Reductive elimination of alkyl cyanide from this complex, followed by either association of olefin and addition of HCN or addition of HCN followed by coordination of olefin, regenerates the precursor to the η^3-arylethyl complex. Isotopic labeling studies indicate that the olefin insertion is reversible. Thus, the stereochemistry of the process results from the relative stabilities of the two diastereomeric η^3-arylethyl complexes and the relative rates for reductive elimination from these complexes.

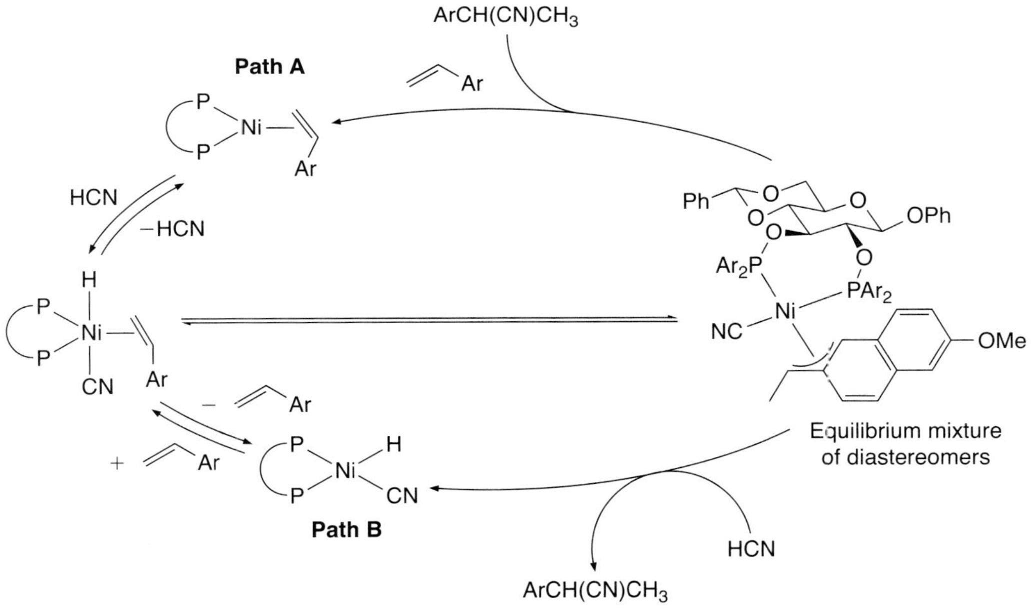

Scheme 16.5

The asymmetric hydrocyanation of dienes with substantial enantioselectivities has also been reported (Equation 16.10).[48,49] Like the reactions of vinylarenes, these reactions have been reported with catalysts containing carbohydrate-derived phosphinites. Reactions of aryl-substituted dienes occur to form the products from 1,2-hydrocyanation.[48] In addition to the reactions of purely acyclic dienes, such as 1-phenyl-1,3-butadiene, dienes containing an exocyclic vinyl group have been studied.[49] These are substrates for products possessing a chiral center at the carbon attached to the ring, and some of these reactions occurred with substantial enantioselectivity. The origin of enantioselection in the hydrocyanation of dienes is similar to that in the hydrocyanation of vinylarenes. The step that controls enantioselectivity was shown by Vogt to be reductive elimination that couples the allyl and cyanide ligands to form the allylnitrile product.[50]

$$(16.10)$$

16.2.6. Hydrocyanation of Alkynes

Although less studied, the hydrocyanation of alkynes in the presence of soluble transition metal complexes has also been reported.[36,51–53] The reactions conducted with nickel(0) catalysts occur with cis stereochemistry, high regioselectivity, and moderate-to-high yields. Again, both steric and electronic effects control the regioselectivity. These points are illustrated by the data in Equation 16.11. Terminal, straight-chain alkynes such as 1-hexyne react to form predominantly the branched nitrile, whereas *tert*-butyl acetylene reacts to form mostly the terminal nitrile. Reactions conducted with DCN have shown that the addition occurs in a syn fashion.[36]

$$(16.11)$$

16.2.7. Summary of Catalytic Hydrocyanation

Catalytic HCN addition has been demonstrated to occur with alkenes, vinylarenes, conjugated dienes, and alkynes. High regioselectivity and significant enantioselectivity have been obtained. Alkenes generally form the terminal nitriles, and vinylarenes generally form the branched nitriles. Conjugated dienes react more rapidly than alkenes or vinylarenes. These reactions occur through π-allyl intermediates, and high selectivity for formation of the α,ω-addition product requires isomerization of the minor branched nitrile product formed from the first addition of HCN to butadiene. The addition of HCN to acetylenes occurs with syn stereochemistry and is also catalyzed by nickel complexes.

16.3. Hydrosilylation and Disilylation

16.3.1. Introduction to Hydrosilylation and Disilylation

The hydrosilylation of alkenes (Equation 16.12) and alkynes (Equation 16.13), alternatively termed hydrosilation, is the addition of a silicon–hydrogen bond across the C–C π-bond to form a new alkylsilane or vinylsilane.[54–58] This reaction has been catalyzed by complexes containing many different metals, but is most commonly conducted with complexes of platinum, rhodium, and palladium. The hydrosilylation of alkenes typically forms terminal alkylsilanes as the major regioisomer, and the hydrosilylation of vinylarenes often generates the chiral branched alkylsilane. The hydrosilylation of alkynes has also been developed. As shown generally in Equation 16.13, these reactions can occur by either cis or trans addition, depending on the catalyst. In some cases, the reactions of silanes with olefins form vinylsilanes (called dehydrogenative silylation, Equation 16.14). The addition of an Si–Si bond of a disilane across an olefin has also been reported (Equation 16.15), and this reaction is called disilation of olefins.

$$R^1R^2C{=}CR^3R^4 + HSiR_3 \xrightarrow{\text{Catalyst}} \quad \quad \quad \quad (16.12)$$

$$R^1{-}{\equiv}{-}R^2 + HSiR_3 \xrightarrow{\text{Catalyst}} \quad \text{or} \quad \quad \quad (16.13)$$

$$2\ R^1R^2C{=}CR^3H + HSiR_3 \xrightarrow{\text{Catalyst}} \quad (+\ H_2 \text{ or } R^1R^3CH{-}CH_2R^2) \quad (16.14)$$

$$R^1R^2C{=}CR^3R^4 + R_3SiSiR_3 \xrightarrow{\text{Catalyst}} \quad \quad \quad (16.15)$$

$$R^1R^2C{=}X + HSiR_3 \xrightarrow{\text{Catalyst}} \quad \quad \quad (16.16)$$

X = O or NR

Much work has also been conducted on the hydrosilylation of ketones and imines (Equation 16.16).[56,57] The products from these reactions are silyl ethers and silylamines. These additions of silanes across C–X π-bonds have been conducted predominantly for the purpose of generating optically active alcohols and amines after hydrolysis. Because the mechanism of these reactions is less defined than the mechanism of alkene hydrosilylation, and this chemistry lies outside the theme of this chapter, the hydrosilylation of ketones and imines is presented only briefly. Instead, this chapter provides an overview of the scope and motivation for the hydrosilylation of alkenes and alkynes and provides details on the mechanisms of these reactions catalyzed by complexes of various metals. Several comprehensive reviews of the scope of these reactions have been published.[55–60]

16.3.2. Purpose for Hydrosilylation

Alkylsilanes are important monomers for industrial applications, and properties of the polysiloxanes generated from them are affected by the degree of crosslinking. Many alkylsilane monomers are produced by the hydrosilylation of olefins, and crosslinking of polysiloxanes has also been conducted by hydrosilylation. The uses of these siloxanes range from the production of common household silicon caulk to aerospace applications. Hydrosilylation also has potential utility in fine-chemical applications. Alkylsilanes undergo oxidation to alcohols under Tamao–Fleming conditions,[61-66] leading to applications of catalytic hydrosilylation that parallel applications of more classical hydroboration. The development of conditions for the cross coupling of vinylsilanes with aryl and alkenyl halides[67,68] has further increased the utility of the hydrosilylation of alkynes. The regioselectivity and stereoselectivity of olefin and alkyne hydrosilylation depends on the olefin or alkyne, but can be modulated by the structure of the catalyst. The catalytic asymmetric hydrosilylations of olefins that form branched products create a route to non-racemic chiral alcohols.

16.3.3. History and Types of Catalyst

Transition-metal-catalyzed hydrosilylation was first reported in the late 1950s with catalysts based on platinum, ruthenium, and iridium chlorides.[69-71] For industrial applications, chloroplatinic acid ($H_2PtCl_6 \bullet nH_2O$) has been used extensively and is highly active for this process. This catalyst has become known as Speier's catalyst. This catalyst is spectacularly reactive, as indicated by the low catalyst loading for the reaction in Equation 16.17.[55] A Pt(0) complex containing vinylsiloxane ligands (platinum divinyltetramethyldisiloxane) shown in Figure 16.1 has also been used frequently in industrial settings as a catalyst for hydrosilylation. This catalyst has become known as Karstedt's catalyst.[72]

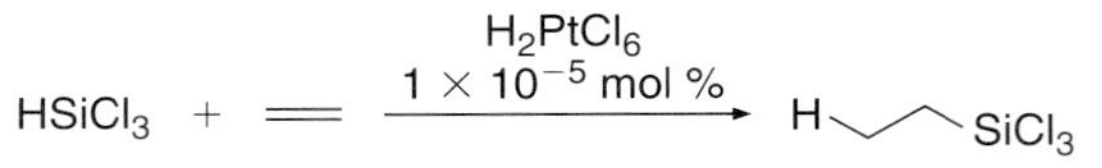

$$\text{HSiCl}_3 \quad + \quad == \quad \xrightarrow[\text{1} \times 10^{-5}\ \text{mol \%}]{\text{H}_2\text{PtCl}_6} \quad \text{H}\diagdown\diagup\text{SiCl}_3 \qquad\qquad (16.17)$$

A series of studies were conducted to determine if Karstedt's highly active Pt(0) catalyst was a homogeneous catalyst or a precursor to platinum colloids.[73-76] Some initial data indicated that catalytically active colloids were generated from the reduction of (COD)PtCl$_2$ with triethoxysilane.[73,74] Although colloidal platinum is also generated from Karstedt's catalyst at the end of reactions containing excess olefin, there is evidence from XAFS structural work that the active platinum catalyst is monomeric and contains a Pt–C bond during catalytic process. This last finding implies that the catalyst is homogeneous.[75] The mechanism of these reactions likely occurs by a process involving oxidative addition of silane, insertion of olefin into the M–H bond, and reductive elimination to form the C–Si bond.

Rhodium complexes also catalyze the hydrosilylation of olefins, and one of the earliest soluble catalysts used for hydrosilylation was Wilkinson's catalyst.[77,78] As is described in more detail below, the mechanisms for hydrosilylation catalyzed by rhodium complexes differ from those catalyzed by platinum complexes. Olefin insertion occurs into different ligands in the two mechanisms. The rhodium-catalyzed processes occur by a so-called modified Chalk–Harrod mechanism.[79-89] Rhodium complexes were also among the first complexes used for the asymmetric hydrosilylation of ketones.[90-92]

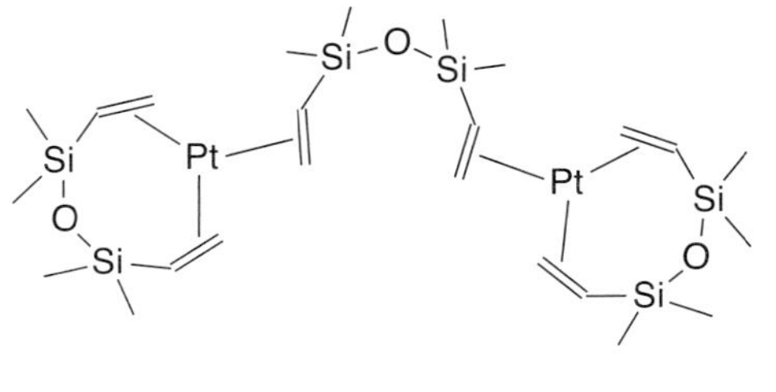

Figure 16.1.
Karstedt's Pt(0) catalyst containing tetramethyldivinylsiloxane ligands.

Palladium complexes also catalyze hydrosilylation, and particular emphasis has been placed on the use palladium catalysts for asymmetric hydrosilylation.[93] The most selective of these catalysts contains a binaphthyl monophosphine ligand.[93,94] Finally, lanthanides have also been used for hydrosilylation.[95] Lanthanide–metallocene catalysts can be highly active for the hydrosilylation of olefins, and lanthanides bearing chiral ligands catalyze asymmetric hydrosilylation with measurable enantiomeric excess.[95]

16.3.4. Examples of Hydrosilylations

16.3.4.1. Hydrosilylation of Olefins with Achiral Catalysts

The hydrosilylation of alkenes produces terminal alkylsilane products. Several examples of these reactions described in Speier's original paper are shown in Equations 16.18–16.22. These examples first show that the terminal anti-Markovnikov products are formed from α-olefins (Equation 16.18).[69] These results also show that linear products are formed from the hydrosilylation of α,β-unsaturated esters with Speier's catalyst (Equation 16.19). Reactions of internal olefins are more complex. Reactions of unsubstituted cyclic alkenes form a single symmetrical product (Equation 16.20). However, as shown in Equations 16.21a and 16.21b,[96] reactions of internal olefins form the same major product as reactions of terminal olefins. This result was confusing at the time, but the now well-known isomerization of secondary alkyl complexes to primary alkyl complexes accounts for this result. More details about this isomerization process are given in Section 16.3.5 that covers the mechanism of hydrosilylation. Finally, the silane can affect regioselectivity of the hydrosilylation of alkenes catalyzed by Speier's catalyst. Reaction of dichlorosilane with 2-hexene formed the 2- and 3-alkylsilanes without formation of the terminal alkylsilane (Equation 16.22).[97]

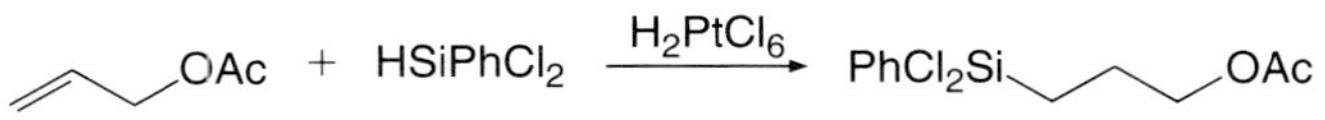

(16.18)

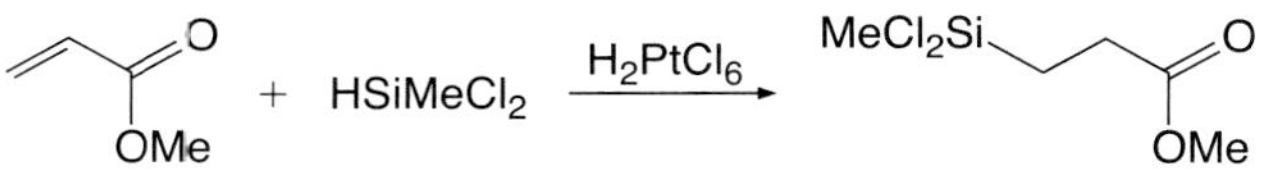

(16.19)

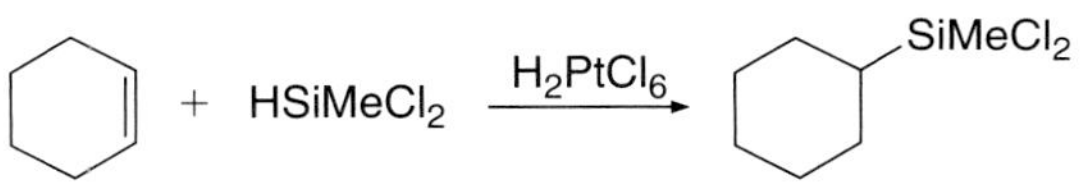

(16.20)

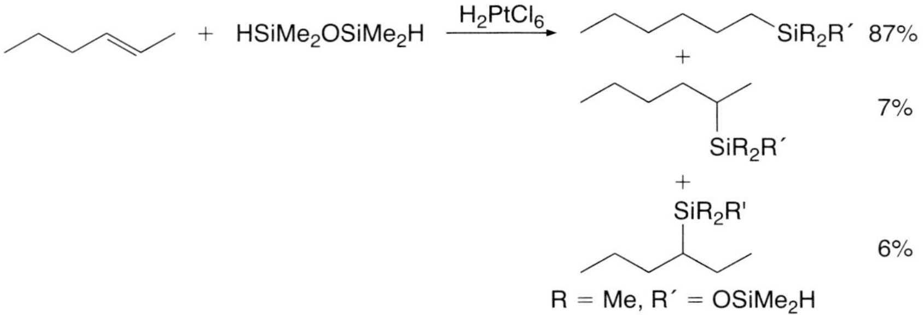

(16.21a)

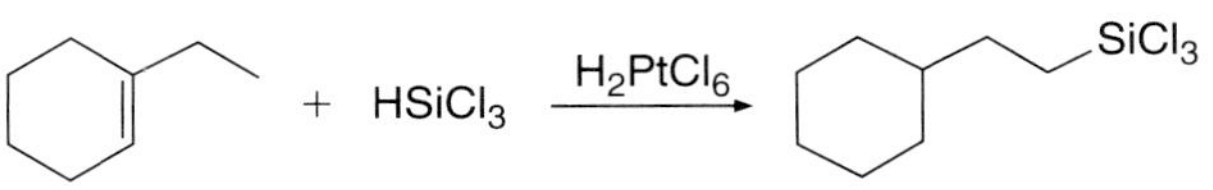

(16.21b)

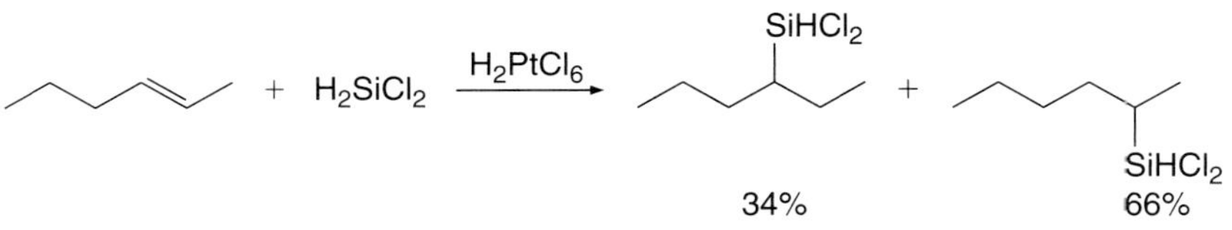

(16.22)

Many other catalysts based on cobalt, ruthenium, rhodium, and platinum are now known to catalyze the hydrosilylation of alkenes, and the types of products can be controlled by the choice of catalyst and silane. Rhodium complexes, such as Wilkinson's catalyst,[77] have been used frequently. A comprehensive treatment of selectivities is beyond the scope of this chapter, but several reviews provide information on the products formed from different catalysts and silanes.[98–101] As one example, crotononitrile undergoes hydrosilylation in the presence of Wilkinson's catalyst to form the α-silyl nitrile product (Equation 16.23),[102,103] and this regioselectivity contrasts with that for the reaction of the related acrylic acid ester in Equation 16.19 conducted with Speier's catalyst.

$$(16.23)$$

16.3.4.2. Hydrosilylation of Vinylarenes

Different catalysts also generate different products from the reactions of silanes with vinylarenes. Speier's catalyst generates a mixture of regioisomers,[104,105] whereas Wilkinson's catalyst generates predominantly the vinylsilane from dehydrogenative silylation discussed later in this chapter.[106] In contrast, palladium[94,107] and nickel[108] complexes tend to give clean conversion to the branched α-phenethylsilane (Equation 16.24). This selectivity is similar to that observed from the hydroformylation and hydrocyanation of vinylarenes discussed in previous chapters and sections of this chapter. This regioselectivity is thought to originate from interaction of the arene π-system with the metal in a structure akin to an η³-benzyl complex. This interaction stabilizes the branched alkyl intermediate and causes the product to form through this η³ binding mode of the aralkyl intermediate. The structure of such an intermediate formed by insertion of the vinylarene into the metal–hydride intermediate is shown schematically in Equation 16.24.

$$(16.24)$$

16.3.4.3. Hydrosilylation of Dienes

The hydrosilylation of dienes can also form different regioisomeric products, depending on the identity of the silane and catalyst. In principle, the hydrosilylation of dienes could be used as a route to form allylsilane reagents. However, these reactions are complex because allylic and homoallylic silanes can both be produced, and products from two hydrosilylations of the diene, or products from a combination of hydrosilylation and oligomerization of the diene, can be formed.

Specific examples of the hydrosilylation of dienes are shown in Equations 16.25–16.28. The hydrosilylation of butadiene with Speier's catalyst gives a mixture of mono- and di-addition products, but the hydrosilylation of isoprene forms predominantly 3-methyl-2-butenylsilane, as shown in Equation 16.25.[109] In contrast, triphenylphosphine complexes of palladium generate 2-butenylsilane from a single 1,4-addition of silane to butadiene[110,111] and 2-methyl-butenylsilane from a single 1,4-addition to isoprene (Equations 16.26 and 16.27).[112,113] The hydrosilylation of butadiene with triethoxysilane in the presence of Wilkinson's catalyst also forms the product from 1,4-addition (Equation 16.28),[114] but the hydrosilylation of isoprene with this reagent forms the same regioisomer as the reaction in the presence of Speier's catalyst.[113] Thus, many products can be formed from the hydrosilylation of dienes. Although conditions can be found to form single products, it is difficult to provide a simple picture of the factors that control regioselectivity.

$$\text{(16.25)}$$

$$\text{(16.26)}$$

$$\text{(16.27)}$$

$$\text{(16.28)}$$

16.3.4.4. Dehydrogenative Silylation of Olefins

The dehydrogenative silylation of olefins to form vinylsilanes is a process that competes with hydrosilylation. In most cases, the vinylsilane is a minor product, although the formation of this minor product can be important in applications of hydrosilylation. The mechanistic importance of the formation of this product is discussed later in this section. In some cases, however, the vinylsilane from dehydrogenative silylation can be the major product. As noted above, the reaction of styrene with triethylsilane catalyzed by Wilkinson's catalyst generates the vinylsilane as the major product.[106] The dehydrogenative silylation of styrene also occurs in high yield when catalyzed by $Ru_3(CO)_{12}$ (Equation 16.29).[115a] Hydrogen is transferred to the styrene to form ethylbenzene as a side product. The dehydrogenative silylation of ethylene catalyzed by a ruthenium polyhydride also occurs in good yield.[115b]

$$\text{(16.29)}$$

16.3.4.5. Hydrosilylation of Alkynes

Vinylsilanes are also formed by the hydrosilylation of alkynes. Similar to the selectivity of the hydrosilylation of alkenes, the selectivity of the hydrosilylation of alkynes can be controlled by both the catalyst and the silane. Alkyne hydrosilylation can form either of two regioisomeric vinylsilanes and can form products from either cis or trans addition of the silane across the alkyne unit.

The hydrosilylation of alkynes catalyzed by platinum complexes tends to generate products from cis addition of the silane across the alkyne unit. Thus, the product obtained from the reaction of a terminal acetylene is the trans (E)-vinylsilane, as shown in Equation 16.30. This example also shows that the simple Speier's catalyst does not adequately control the regioselectivity of alkyne hydrosilylation.

$$(16.30)$$

$$(16.31)$$

[Rh]	A	B	C
[Rh(COD)$_2$]BF$_4$/2PPh$_3$	5	95	0
[Rh(COD)Cl]$_2$	94	4	2

$$(16.32)$$

Takeuchi has shown that the stereoselectivity of alkyne hydrosilylation can be controlled by both the ligand and the solvent (Equation 16.31).[116] Syn addition of the Si–H bond across the alkyne has been observed for the hydrosilylation of hexyne by triethylsilane in ethanol solvent in the presence of a catalyst generated from the combination of [Rh(COD)$_2$]$^+$ and PPh$_3$. In contrast, trans addition occurs for the same reaction in the same medium with a neutral rhodium catalyst lacking added phosphine. Trans addition also occurs in the presence of Wilkinson's catalyst in toluene. (Equation 16.32). As discussed in Section 16.3.5.2.4 below, trans addition likely results from the isomerization of a vinyl intermediate.[57, 117, 118]

Regioselectivity and stereoselectivity can also be altered by using ruthenium catalysts.[119–123] Examples of the ruthenium-catalyzed hydrosilylation of alkynes are shown in Equations 16.33–16.35. Notice that the hydrosilylation of terminal alkynes catalyzed by [CpRu(NCMe)$_3$]$^+$ almost exclusively forms the product from addition of the silyl group to the internal position of the alkyne.[123, 124] This complex also catalyzes the hydrosilylation of internal alkynes. Most striking, this ruthenium complex and the Cp* analog, [Cp*Ru(NCMe)$_3$]$^+$, lead to products resulting from exclusive trans hydrosilylation of the internal alkynes.[119, 123] Likewise, [Ru(cymeme)Cl$_2$]$_2$[25] catalyzes the trans hydrosilylation of terminal alkynes, in this case to form the terminal vinylsilane. Extensive studies have shown that the ruthenium-catalyzed hydrosilylations are tolerant of the presence of various functional groups and have demonstrated the ability to use hydroxyl and carbonyl functionality to direct the regiochemistry of the hydrosilylation process. An example of this directing effect is shown in Equation 16.35.[126]

$$(16.33)$$

Ratio of (α:β)	Yield
> 20:1	89

$$(16.34)$$

$$(16.35)$$

16.3.4.6. Asymmetric Hydrosilylation of Olefins[127–131]

As noted earlier in this chapter, the enantioselective hydrosilylation of olefins could be a useful method to prepare chiral, non-racemic alcohols. Although the scope of highly enantioselective hydrosilylations is limited, high enantioselectivities have been obtained for the asymmetric hydrosilylation of alkenes and vinylarenes. A majority of the most selective chemistry has been conducted using a palladium catalyst containing an axially chiral monophosphine ligand.[128,129,131,132]

The formation of chiral products from hydrosilylation depends on the substitution pattern of the olefin and the regioselectivity of the hydrosilylation process. The products of the hydrosilylation of 1,1-disubstituted olefins are chiral if the two substituents on the alkene are different. Hydrosilylation of terminal olefins can generate chiral products if the regioselectivity of the hydrosilylation is reversed from that typically observed, and the hydrosilylation process forms branched products. Asymmetric hydrosilylation of geminally disubstituted alkenes has not generated products with high enantiomeric excess, but asymmetric hydrosilylation by additions to terminal olefins to form branched alkylsilanes has occurred with high ee.

	A/B	% ee of alcohol
R = n-C$_4$H$_9$:	89/11	94 (R)
R = CH$_2$CH$_2$Ph:	81/19	97 (S)
R = cyclo-C$_6$H$_{13}$:	97/3	95 (R)

$$(16.36)$$

Reactions of alkenes and vinylarenes catalyzed by palladium complexes of axially chiral biarylmonophosphines generate branched products, and these products are formed with high ee. As shown in Equation 16.36, the hydrosilylation of hexene, 4-phenyl-1-butene, and vinylcyclohexane predominantly form the branched hydrosilylation product, and the ee is between 94% and 97%.[132,133] Enantioselective reactions of simple α-olefins, such as hexene, are rare. Enantioselective reactions of norbornene are more common. Consistent with this trend, the hydrosilylation of norbornene occurs in the presence of this palladium catalyst to form the exo alkylsilane and subsequent exo norbornyl alcohol in 93–95% ee.[134]

R = H	% ee
Ar = 4-CF$_3$C$_6$H$_4$, R = H	96% ee
Ar = 4-MeC$_6$H$_4$, R = H	89% ee
Ar = 4-MeOC$_6$H$_4$, R = H	61% ee

Ar = Ph	% ee
R = Me	89% ee (at 20 °C)
R = n-Bu	92% ee (at 20 °C)

(16.37)

Examples of hydrosilylations of vinylarenes are shown in Equation 16.37. In this case, the catalyst containing the methoxy-substituted binaphthyl ligand gives rise to products with low enantioselectivity,[107] but the palladium catalyst containing the related unsubstituted 2-diphenylphosphino-2,2-binaphthyl ligand forms a hydrosilylation product with high selectivity.[94] These reactions occur with high regio- and stereoselectivity for styrene and β-alkyl styrenes. The hydrosilylation of dienes could provide a route to enantioenriched allylsilanes. However, these reactions have occurred so far with only modest ee's.[135,136]

16.3.4.7. Hydrosilylation of Ketones and Imines

Literature on the hydrosilylation of ketones and imines is extensive. This chemistry provides a method for the reduction of these unsaturated substrates with reagents that are inexpensive and either liquids or solids, rather than gases (H$_2$). As noted in the introduction, much of this literature focuses on the enantioselective hydrosilylation of ketones and imines to form non-racemic chiral alcohols and amines. In addition to the convenience of using a liquid reagent, the ability to vary the substituents on a silane allows for tuning of the stereoselectivity of the reductions.

Until recently, most enantioselective hydrosilylations of ketones were conducted with rhodium catalysts. In 1972, the hydrosilylation of carbonyl compounds catalyzed by the achiral Wilkinson's catalyst was reported.[137,138] Reactions catalyzed by ruthenium[139,140] and platinum[141] complexes were published about the same time. Among extensive studies with various chiral ligands, Brunner found that the hydrosilylation of acetophenone occurred in good yield with the thiazolidine-based ligand shown in Equation 16.38a.[90,91] Nishiyama showed that high selectivities were obtained in the presence of a combination of rhodium trichloride and the pybox ligand in Equation 16.38b.[142] The hydrosilylation of imines has also been conducted using late transition metal catalysts, but enantioselectivities have been low. Higher selectivities have been obtained using titanium(III) catalysts (Equation 16.38c).[143]

More recently, the asymmetric hydrosilylation of aryl ketones and aryl imines has been developed using copper catalysts.[144–146] In this case, axially chiral biaryl bisphosphine ligands bound to copper generate remarkably active catalysts for the hydrosilylation of ketones. These reactions occur with high selectivity using the hydrosilane polymer

(PMHS). One example of these reactions is shown in Equation 16.38d. These reactions have also been extended to highly enantioselective hydrosilylations of *N*-phosphoryl aryl ketimines (Equation 16.38e).[147]

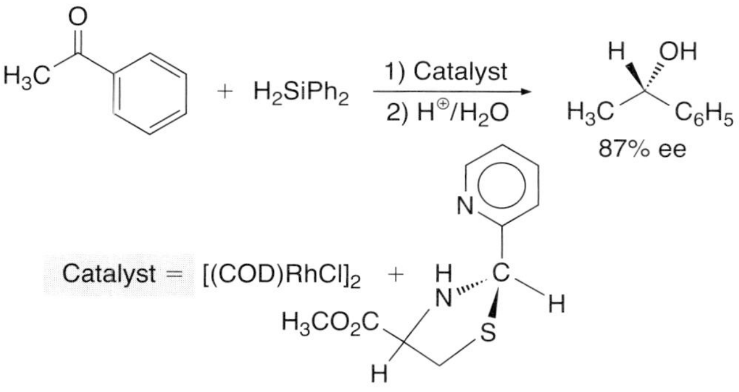

(16.38a)

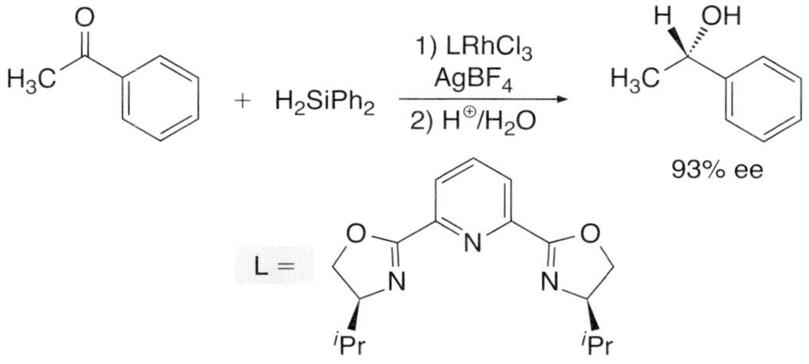

(16.38b)

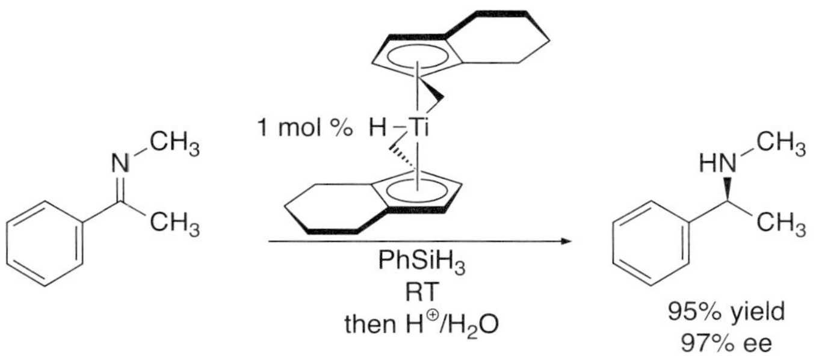

(16.38c)

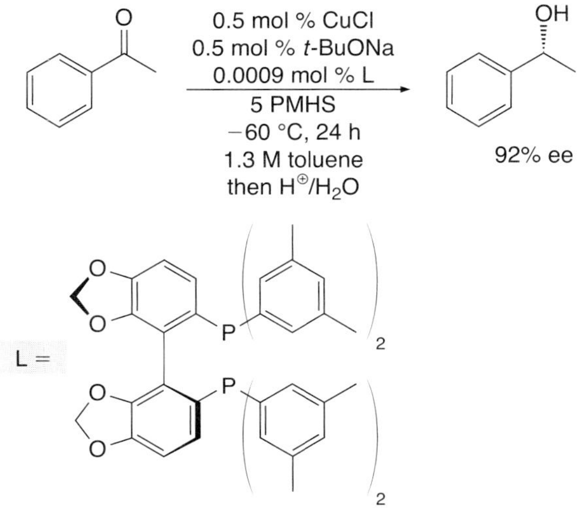

(16.38d)

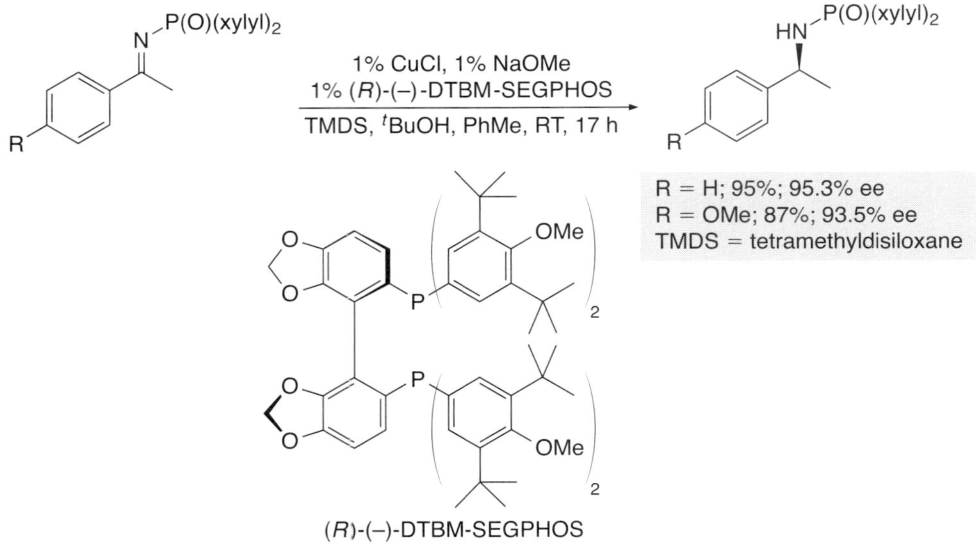

1% CuCl, 1% NaOMe
1% (R)-(–)-DTBM-SEGPHOS
TMDS, ᵗBuOH, PhMe, RT, 17 h
P(O)(xylyl)2
HN
R
OMe
P
(R)-(–)-DTBM-SEGPHOS
R = H; 95%; 95.3% ee
R = OMe; 87%; 93.5% ee
TMDS = tetramethyldisiloxane

(16.38e)

16.3.5. Mechanism of Hydrosilylation

The mechanism of hydrosilylation involves a sequence of elementary reactions described in the earlier chapters of the book. The most commonly cited mechanism for hydrosilylation was first described by Chalk and Harrod[148] and involves oxidative addition of the silane, insertion of an olefin into the metal–hydride bond, and reductive elimination to form the silicon–carbon bond in the organosilane product. More recently, a related but distinct mechanism involving insertion of the olefin into the silyl group has been recognized, and this mechanism is often called the modified Chalk–Harrod[79–89] mechanism. Before these steps are described, some of the mechanistic issues regarding the specific systems of Speier's catalyst and Karstedt's catalyst are described briefly.

16.3.5.1. Induction Periods and Phase of the Reactions Catalyzed by Speier's and Karstedt's Catalysts

Speier's "catalyst" ($H_2PtCl_6 \bullet nH_2O$) is a catalyst precursor, and Karstedt's "catalyst" [Pt(tetramethyldivinylsiloxane)$_3$] is also likely to be a catalyst precursor. Reactions conducted with Speier's catalyst occur with an induction period. This induction period is now assumed to result from reduction of the Pt(IV) species to a Pt(0) catalyst that is slower than the overall process catalyzed by the resulting Pt(0) complex.[99,149] This catalyst is typically dissolved in isopropanol, and this solvent can act as a reducing agent. Because Karstedt's catalyst is a Pt(0) species, the induction period of reactions conducted with this catalyst is not as pronounced as it is in reactions using Speier's catalyst. The current data described in Section 16.3.3 to determine if Karstedt's catalyst truly acts as a homogeneous catalyst or if platinum particles are the actual catalyst[73–75,149] imply that the catalyst is homogeneous.[75] The reactions conducted with rhodium and palladium complexes occur by homogeneous catalysis.

16.3.5.2. Overall Catalytic Cycles

The classic Chalk–Harrod mechanism[150] is shown in Scheme 16.6, and the modified Chalk–Harrod mechanism[79–89,100,117,151,152] is shown in Scheme 16.7. In the Chalk–Harrod mechanism, oxidative addition of silane occurs to form a silyl hydride complex. Migratory

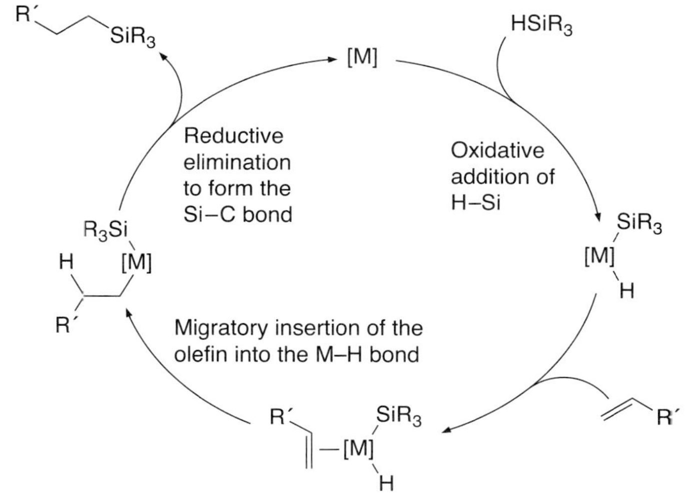

Scheme 16.6. Chalk–Harrod mechanism.

Pathway for isomerization of branched alkyl complexes formed from internal olefins:

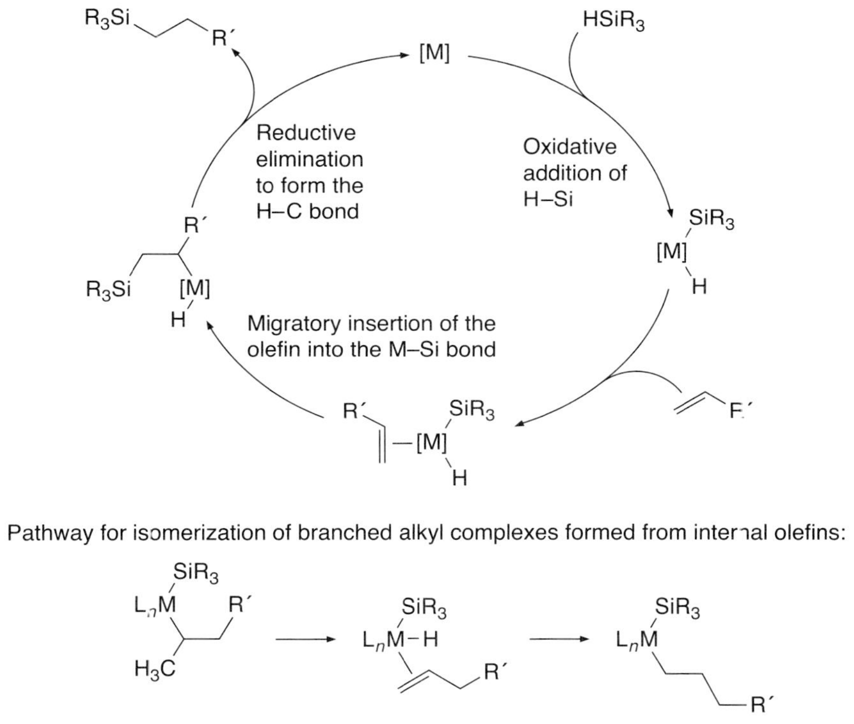

Scheme 16.7. Modified Chalk–Harrod mechanism.

insertion of olefin into the metal hydride generates an alkyl silyl complex that undergoes reductive elimination of the alkylsilane and regeneration of the starting low-valent metal complex. By the modified Chalk–Harrod mechanism, oxidative addition of the silane is followed by migratory insertion of the olefin into the metal–silicon bond. Reductive elimination then forms the new C–H bond in the organosilane product. Platinum catalysts, including both Speier's catalyst and Karstedt's catalyst, are believed to follow the Chalk–Harrod mechanism.[153,154] Rhodium catalysts are believed to follow the modified Chalk–Harrod mechanism.[152]

16.3.5.2.1. The Chalk–Harrod Mechanism

In the Chalk–Harrod mechanism, oxidative addition of silane occurs to a Pt(0) complex. The ancillary ligands on the active catalyst when reactions are initiated with "Speier's catalyst" are unknown, but are likely to be the olefin substrate. The ancillary ligands on the active form of Karstedt's catalyst or the related Pt(COD)$_2$ complex could be the original diene ligands or the olefin substrate. Details on the mechanism of oxidative additions of silanes are provided in Chapter 6. In brief, this reaction occurs by coordination of silane to an open site to form a silane σ-complex, followed by cleavage of the Si–H bond to form a silyl hydride species.

Following this oxidative addition, insertion of the olefin or alkyne occurs into the platinum–hydride bond to form an alkyl or vinyl hydride complex. The regioselectivity of the insertion step and the chemistry of the alkyl silyl complexes control the overall regioselectivity of the hydrosilylation of olefins. Recall that these platinum catalysts form terminal alkylsilane products. This regioselectivity indicates that insertions of terminal olefins occur in order to generate a linear alkylplatinum intermediate. Apparently, the insertions of styrene and acrylic acid derivatives into the platinum hydride in these catalysts occur with the same regiochemistry.

However, the observation that terminal alkylsilanes are produced from internal olefins indicates that isomerization of the alkylplatinum intermediate is faster than the final C–Si bond formation. This isomerization occurs by a series of elimination and insertion processes. This process was discussed in Chapter 9, the chapter on migratory insertions, and is shown at the bottom of Scheme 16.7. Following this insertion, or the combination of insertion and migration of the metal to the terminal carbon, reductive elimination of the alkylsilane occurs. Reductive elimination to form carbon–silicon bonds is slower than reductive elimination to form carbon–hydrogen bonds, but this reaction does have precedent. Examples of this reductive elimination process were provided in Chapter 8, the chapter on reductive elimination. This reaction is concerted and occurs with retention of configuration at the carbon and at the silicon.

16.3.5.2.2. Evidence for a Modified Chalk–Harrod Mechanism

Several observations led to the proposal that some of the catalysts containing metals other than platinum do not react by the Chalk–Harrod mechanism. First, carbon–silicon bond-forming reductive elimination occurs with a sufficiently small number of complexes to suggest that formation of the C–Si bond by insertion of olefin into the metal–silicon bond could be faster than formation of the C–Si by reductive elimination. Second, the formation of vinylsilane as side products[82,84] or as the major products[81,85,115] in some reactions of silanes with alkenes cannot be explained by the Chalk–Harrod mechanism. Instead, insertion of olefin into the M–Si bond, followed by β-hydrogen elimination from the resulting β-silylalkyl complex, would lead to vinylsilane products. This sequence is shown in Equation 16.39. Third, computational studies have indicated that the barrier for insertion of ethylene into the Rh–Si bond of the intermediate generated from a model of Wilkinson's catalyst is much lower than the barrier for reductive elimination to form a C–Si bond from the alkylrhodium–silyl complex.[152]

$$L_nRh\!-\!SiR_3 \;+\; {=\!\!=} \;\longrightarrow\; L_nRh\!\diagup^{\!\!\diagup^{SiR_3}} \;\longrightarrow\; L_nRh\!-\!H \;+\; {\diagup\!\!\diagup}^{SiR_3} \qquad (16.39)$$

It is now accepted that alkene hydrosilylation catalyzed by several systems occur by the modified Chalk–Harrod mechanism. Wrighton's studies on the insertion of ethylene into the M–Si bond of (CO)$_4$CoSiMe$_3$[87,88] led to the conclusion that this photocatalytic process occurs by the mechanism in Scheme 16.8. Conversion of the β-silylalkyl product of insertion to the free organosilane is thought to occur by a sequence of oxidative addition of silane and reductive elimination to form the C–H bond. Perutz's studies on the mechanism of hydrosilylation by CpRh(ethylene)$_2$[89] led the authors to propose that insertion of olefin occurs into the rhodium–silyl complex, and Brookhart's thorough study of the mechanism of hydrosilylation by [Cp(P(OMe)$_3$)CoEt]$^+$ complexes[155] provided additional strong evidence for this mechanism.

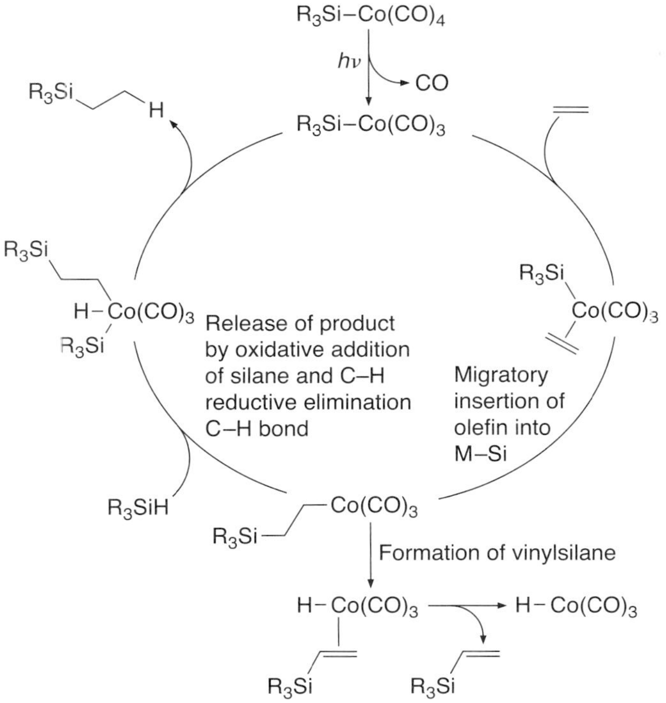

Scheme 16.8

16.3.5.2.3. Alkene Hydrosilylation by σ-Bond Metathesis

Because lanthanide catalysts do not have the appropriate accessible oxidation states for reaction by a sequence of oxidative addition and reductive elimination, an alternative mechanism is followed by these catalysts. These hydrosilylations are thought to occur by a version of a Chalk–Harrod mechanism in which the C–Si bond is formed by σ-bond metathesis instead of reductive elimination.[95] As shown in Scheme 16.9, conversion of the alkyl precursor to a metal hydride is followed by insertion of the alkene to form a yttrium alkyl complex. Reaction of this alkyl complex with the silane generates the alkylsilane and the metal hydride in a σ-bond metathesis process.

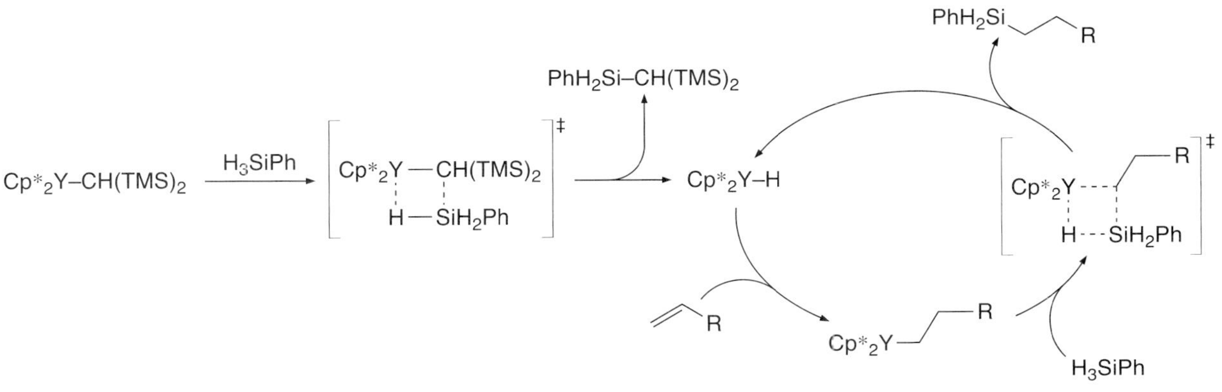

Scheme 16.9

16.3.5.2.4. Mechanism of Alkyne Hydrosilylation

The hydrosilylation of alkynes can also occur by Chalk–Harrod or modified Chalk–Harrod mechanisms, as shown in Scheme 16.10. Reactions forming products from cis addition of the silane across the alkyne would occur by one of these mechanisms. The origin of the regioselectivity of these hydrosilylations has not been well established experimentally. However, based on computational studies,[156] it has been argued that the unusual Markovnikov regiochemistry of the ruthenium-catalyzed process occurs by insertion of the alkyne into the M–Si bond to form an η^2-vinyl intermediate. These η^2-vinyl intermediates are shown in Scheme 16.10.

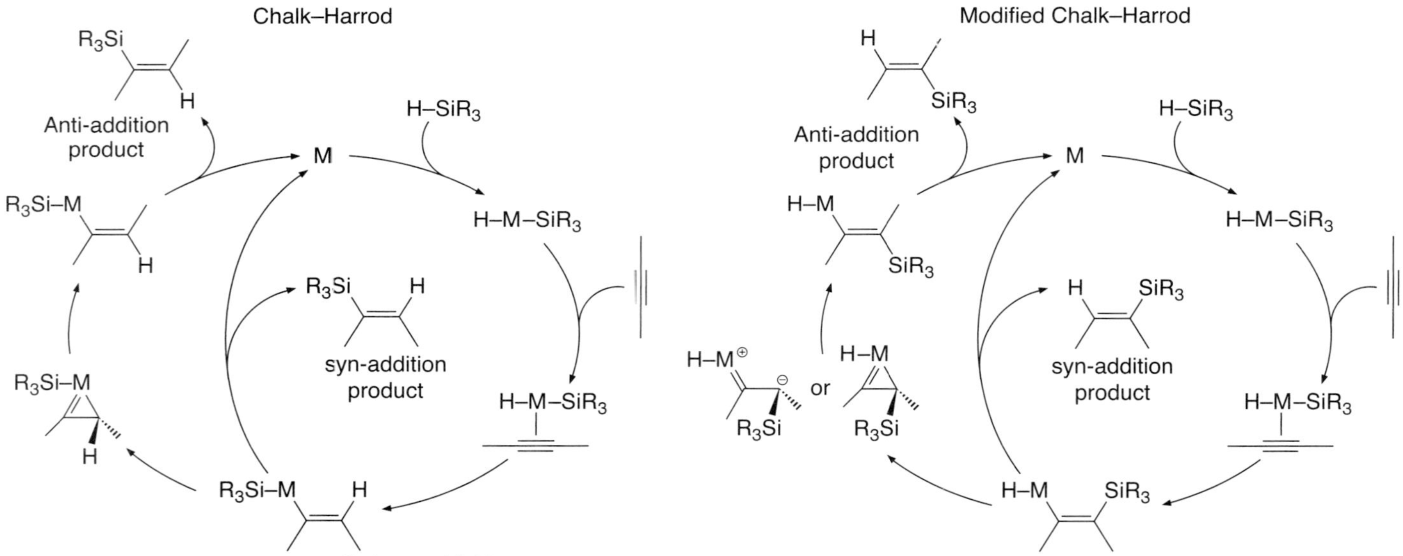

Scheme 16.10.
Trans hydrosilylation by a combination of Chalk–Harrod and modified Chalk–Harrod mechanisms for hydrosilylation and isomerization via η^2-vinyl intermediates.

Pathways to form trans addition products by Chalk–Harrod and modified Chalk–Harrod mechanisms are also shown in Scheme 16.10. The formation of trans addition products is rationalized by the dynamics of η^2-vinyl complexes or by a zwitterionic intermediate.[117] η^2-Vinyl complexes are known to exchange stereochemistry and are thought to do so by one of the mechanisms shown at the bottom left of the two cycles in Scheme 16.10. One pathway involves rotation of the C–C bond upon reopening of the η^2-vinyl complex and a second involves formation of a zwitterionic intermediate. The steric interactions in the initial cis η^1-vinyl intermediate make it less stable than the trans η^1-vinyl complex. Such trans insertions of alkynes were discussed in detail in Chapter 9 (migratory insertions).

16.3.6. Disilation[157]

The addition of the Si–Si bond across an alkyne or an olefin has also been studied in order to generate synthetic intermediates that can undergo cross coupling at two positions or that can undergo oxidations to form diols. This addition reaction is limited in scope at the current time, but certain palladium complexes catalyze the disilylation of alkenes and alkynes.[158–160] Palladium complexes of isonitrile ligands have proven to be the most active for this addition process. Examples of these disilations are shown in Equations 16.40 and 16.41. The mechanism of this process presumably involves oxidative addition of the disilane to a palladium(0) complex of the isonitrile ligand,[161,162] followed by migratory insertion of the olefin or alkyne and reductive elimination to form the second C–Si bond. The diboration of alkenes and alkynes has been developed more extensively and is discussed in the next section.

$$
\begin{array}{ccc}
\text{Me} \; \text{Me} & & \\
\text{Ph–Si–Si–Ph} + n\text{-hexyl–C}\equiv\text{C–H} & \xrightarrow[\substack{\text{Toluene}\\\text{Reflux}}]{\substack{2\text{ mol \%}\\ \text{Pd(OAc)}_2}} & \\
\text{Me} \; \text{Me} & &
\end{array}
\qquad (16.40)
$$

$$
(16.41)
$$

16.4. Transition-Metal-Catalyzed Hydroboration, Diboration, Silylboration, and Stannylboration

16.4.1. Overview of Hydroboration and Diboration

The hydroboration of olefins is a classic reaction in organic synthesis.[163,164] Dialkylboranes add rapidly to alkenes in the absence of catalyst. However, dialkoxyboranes, such as catecholborane and pinacolborane, add more slowly to olefins and alkynes. Thus, transition metal complexes could catalyze the addition of dialkoxyboranes to olefins and alkynes without interference from the background reaction. The potential to alter chemoselectivity, regioselectivity, enantioselectivity, and diastereoselectivity has led a number of groups to develop metal-catalyzed versions of hydroboration.[165–167] Enantioselective hydroboration would alleviate the need to use boranes containing stoichiometric amounts of chiral substituents to generate optically active alkylboranes.

This interest in catalytic hydroboration led to the development of the transition-metal-catalyzed diboration of alkenes and alkynes.[3,168] The diboration of alkenes and alkynes generates bifunctional products, and additions to alkenes have now been conducted with high enantioselectivity. The following sections describe the types of catalysts used for catalytic hydroboration and diboration of alkenes, alkynes, and dienes, as well as catalytic cycles that account for selectivities and side products formed during these processes.

16.4.2. History of Catalytic Hydroboration

The first examples of catalytic hydroborations were reported in the 1980s. Sneddon published a series of papers on the additions of the B–H bonds in boron clusters to alkynes catalyzed by transition metal complexes.[169–177] An example of these processes is shown in Equation 16.42. These reactions provided precursors to new boron cages and to boron–carbide materials. In 1985, Nöth published the hydroboration of olefins with catecholborane catalyzed by Wilkinson's catalyst.[178] One example of this process (Equation 16.43) shows the difference in chemoselectivity between the catalyzed and uncatalyzed processes. This report by Nöth led to the development of catalytic hydroboration as a method for organic synthesis. Studies on both early and late transition metal catalysts have been conducted, and these studies included experiments to probe for differences in selectivities between catalyzed and uncatalyzed processes.

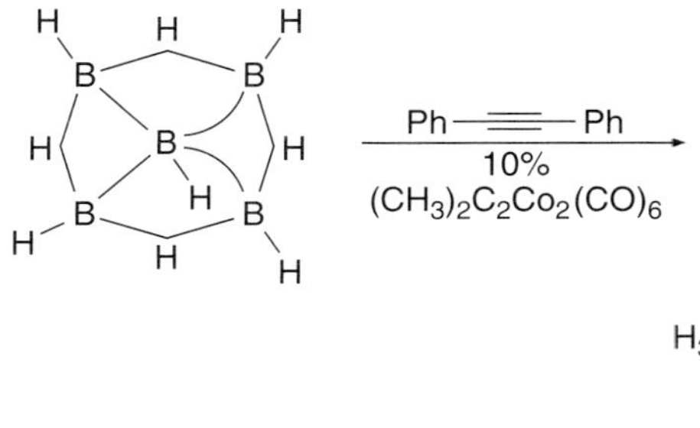

(16.42)

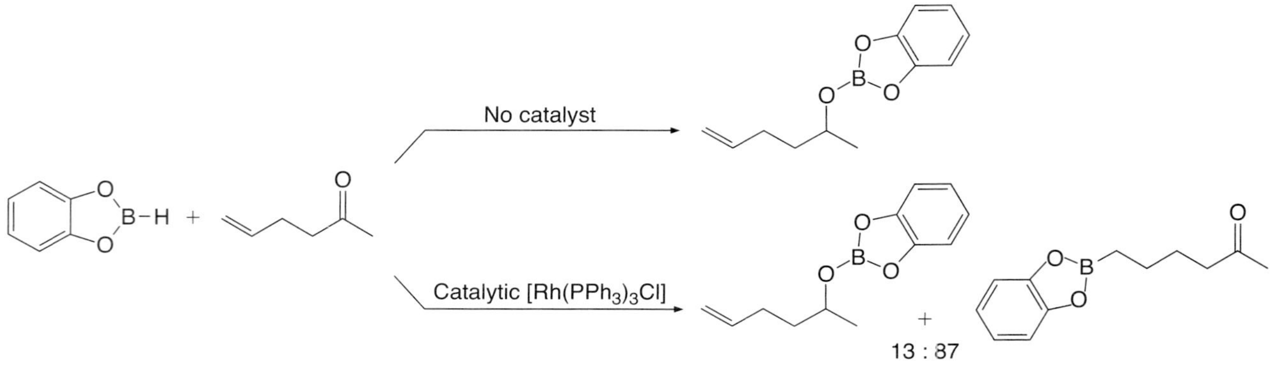

(16.43)

16.4.3. Examples of Metal-Catalyzed Hydroboration

Examples of metal-catalyzed hydroborations of alkenes and alkynes are shown in Equations 16.44–16.50. The examples of the hydroboration of vinylarenes in Equation 16.44[179–184] illustrate that the regiochemistry of the transition-metal-catalyzed hydroboration with catecholborane as reagent can be the opposite of that for the uncatalyzed process with dialkylboranes as reagent. For example, in the absence of air, the late metal catalysts preferentially form branched hydroboration products.[179,180] The regioselectivity parallels that for the hydrosilylation and hydroformylation of vinylarenes catalyzed by most late-metal complexes. The examples in Equation 16.44 also illustrate that catalysts based on different metals can form products with different regioselectivities[181,182] and that charge and number of ligands,[183,184] and even reaction conditions (the air in the second example presumably oxidizes the phosphine and changes the catalyst composition) can affect the selectivity.[183–185]

Catalyst		
RhCl(PPh$_3$)$_3$ (in argon)	> 99	< 1
RhCl(PPh$_3$)$_3$ (in air)	24	76
[Rh(COD)$_2$]BF$_4$/dppb	99	1
Cp$_2$TiMe$_2$ (in benzene)	0	100

(16.44)

The examples in Equation 16.45 show that the regiochemistry of the metal-catalyzed hydroboration of alkenes contrasts with that of the metal-catalyzed hydroboration of vinylarenes. The predominant products from the metal-catalyzed hydroborations of 1-alkenes are terminal boranes.[180,186–188] These terminal boranes can also be formed from internal alkenes, as shown in Equation 16.46. Similar to hydrosilylation with Speier's catalyst, the hydroboration of 4-octene with Wilkinson's catalyst can generate the terminal borane by a combination of isomerization and hydroboration.[189,190] The reaction of catecholborane conducted with Wilkinson's catalyst in THF generates the product from hydroboration without isomerization, but the reaction of pinacolborane conducted with the same catalyst in methylene chloride solvent forms the terminal alkylborane product (Equation 16.46).

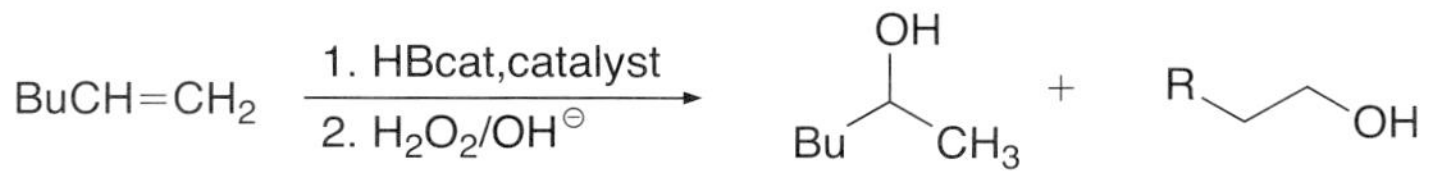

$$BuCH=CH_2 \xrightarrow[\text{2. H}_2\text{O}_2/\text{OH}^\ominus]{\text{1. HBcat, catalyst}}$$

(16.45)

Catalyst		
RhCl(PPh$_3$)$_3$	1	99
Cp*$_2$Sm(THF)	< 1	> 99

$$(E)\text{-C}_3\text{H}_7\text{CH}=\text{CHC}_3\text{H}_7 \xrightarrow[\text{2. H}_2\text{O}_2/\text{OH}^\ominus]{\text{1. HBX}_2/\text{catalyst}} \text{1-Octanol} + \text{2-Octanol} + \text{3-Octanol} + \text{4-Octanol}$$

Borane	Solvent	1-ol	2-ol	3-ol	4-ol	catalyst =	RhCl(PPh$_3$)$_3$	(16.46)
HBcat	THF	0	0	0	100			
HBpin	CH$_2$Cl$_2$	100	0	0	0			

Early transition metal and lanthanide complexes also catalyze the hydroboration of alkenes. As shown in Equation 16.44, titanocene complexes catalyze the hydroboration of vinylarenes,[181,182] and as shown in Equation 16.45, lanthanocene complexes catalyze the hydroboration of alkenes.[187] Ligand-less lanthanides and zirconocene complexes have also been reported to catalyze these reactions, but the apparent catalysis of the hydroboration in these cases more likely occurs by catalytic generation of BH$_3$ from catecholborane and uncatalyzed addition of BH$_3$ to the olefin.

The hydroboration of 1,3-dienes has also been reported, and these reactions generate Z-allylic boronic ester products. These reactions have been reported with palladium catalysts and are thought to occur through π-allyl intermediates. Reactions with butadiene and isoprene, followed by addition of the product to benzaldehyde, are shown in Equation 16.47.[191,192] This equation also shows the presumed mechanism that proceeds by generation of a palladium allyl from the combination of diene and palladium hydride formed by oxidative addition of the borane.

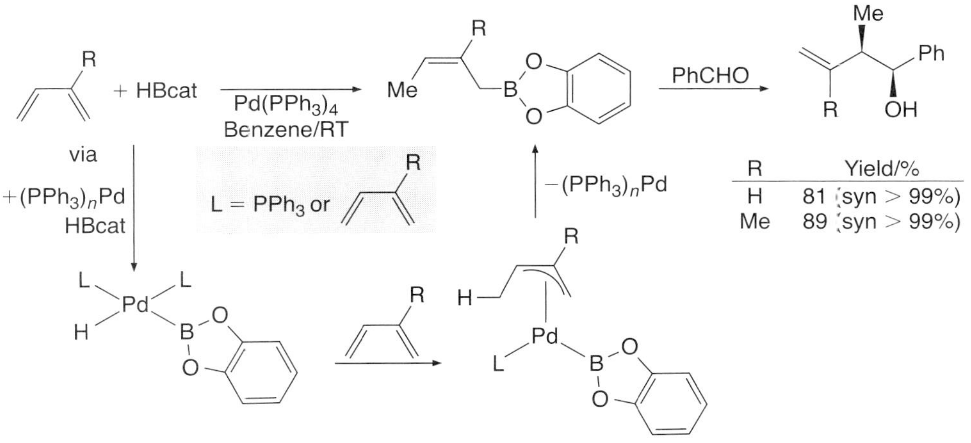

R	Yield/%
H	81 (syn > 99%)
Me	89 (syn > 99%)

(16.47)

The metal-catalyzed hydroboration of alkynes has also been studied, and this process provides a route to vinylboronate esters that are useful in palladium-catalyzed cross-coupling chemistry. In this case, the transition-metal-catalyst strongly affects the regioselectivity and stereoselectivity of the process. As shown in Equation 16.48, reactions catalyzed by both titanium[181,182] and rhodium[189] complexes favor formation of the terminal vinylboronate ester, while reactions catalyzed by the standard Wilkinson's catalyst are less selective. Reactions catalyzed by the rhodium(I) complex containing $PiPr_3$ as ligand even form products from trans hydroboration across the alkyne with high stereoselectivity.[193] The stereochemistry of this addition draws parallels to the trans hydrosilylation of alkynes catalyzed by some rhodium and ruthenium catalysts that were described in Section 16.3.4.5. Although the mechanism of this trans addition has not been determined, it presumably occurs by the same type of isomerization of a vinyl-rhodium complex as is thought to occur during the trans hydrosilylation of alkynes.

$$R^1-C\equiv C-H \xrightarrow[\text{Catalyst}]{HBX_2}$$

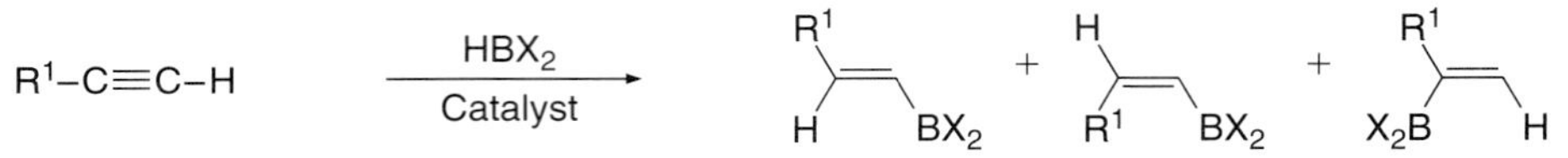

Alkyne	Borane	Catalyst			
p-Tol−≡−H	HBcat	Cp$_2$Ti(CO)$_2$	100	0	0
C$_4$H$_9$−≡−H	HBpin	Rh(CO)(PPh$_3$)$_2$Cl	99	0	1
p-Tol−≡−H	HBpin	RhCl(PPh$_3$)$_3$	48	0	52
Ph−≡−H	HBcat	[Rh(cod)Cl]$_2$/4P^{i}Pr$_3$	1	99	0

(16.48)

16.4.4. Asymmetric Hydroboration

Catalytic asymmetric hydroboration of olefins occurs without a chiral substituent on the stoichiometric reagent and could, therefore, be a more efficient alternative to uncatalyzed asymmetric hydroboration with stoichiometric chiral organoboron reagents. Similar to the asymmetric hydrosilylation of olefins, the asymmetric hydroboration of olefins occurs with the highest selectivities for additions to vinylarenes and for additions to norbornene. Enantioselective reactions of geminally disubstituted olefins to form products containing a new stereocenter at the reduced carbon have not occurred with high enantioselectivity. However, reactions to form products containing a new stereocenter attached to boron have occurred with high enantioselectivity. Rhodium complexes of the ligand QUINAP have been the most enantioselective for hydroboration of both vinylarenes and norbornene.[194,195] Some results with this ligand are shown in Equations 16.49 and 16.50.

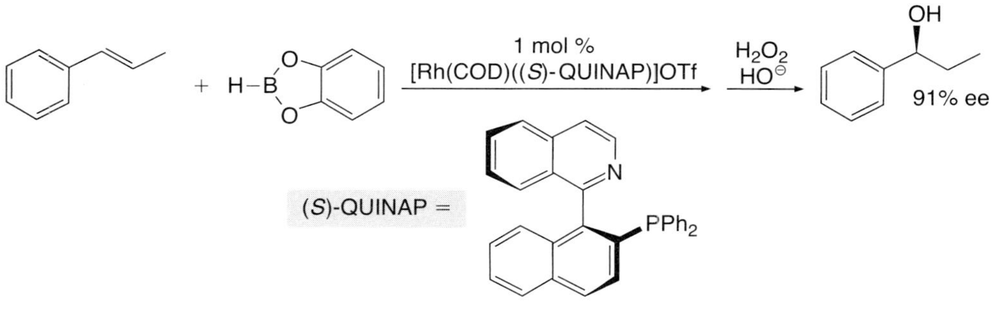

(16.49)

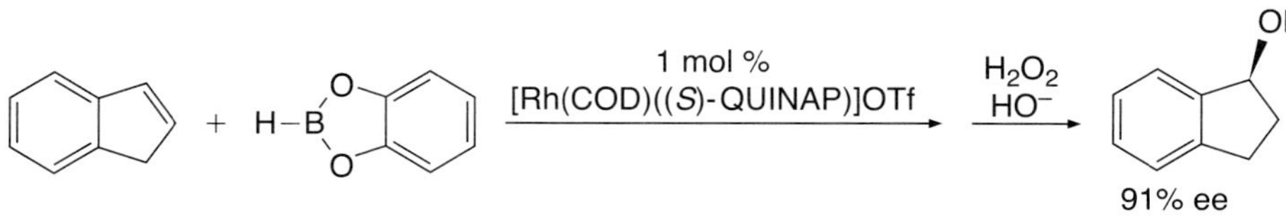

(16.50)

16.4.5. Mechanism of the Hydroboration of Olefins

Mechanisms of catalytic hydroborations of olefins are thought to fall into two categories, and each category can be further divided into two subsets.[167] The reactions catalyzed by Wilkinson's catalyst and by phosphine-free rhodium catalysts are thought to occur by oxidative additions, migratory insertions, and reductive eliminations through two distinct groups of intermediates. The mechanisms of the reactions catalyzed by early metals and lanthanides occur without oxidations and reductions, and two types of mechanisms for reactions catalyzed by early metals have been identified. These four mechanisms are shown in Schemes 16.11–16.14.

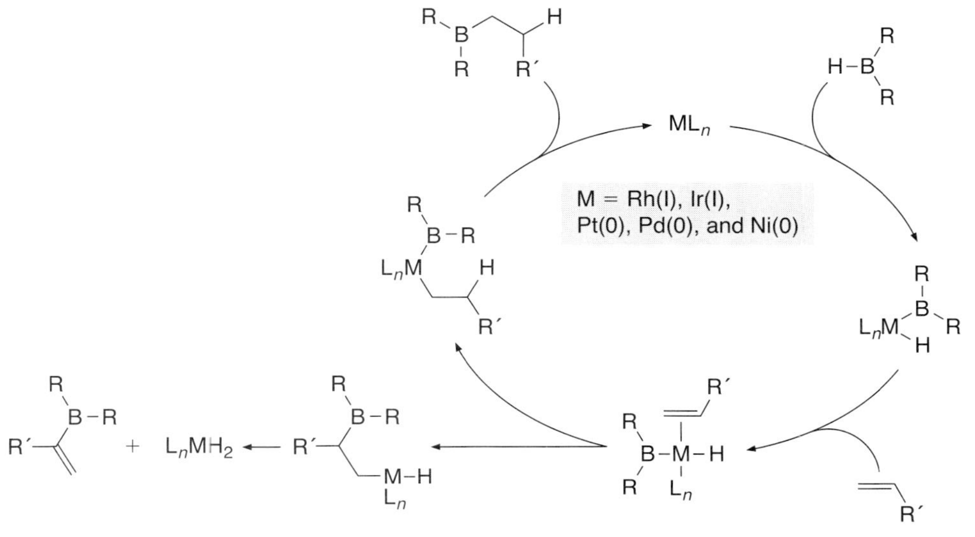

Scheme 16.11

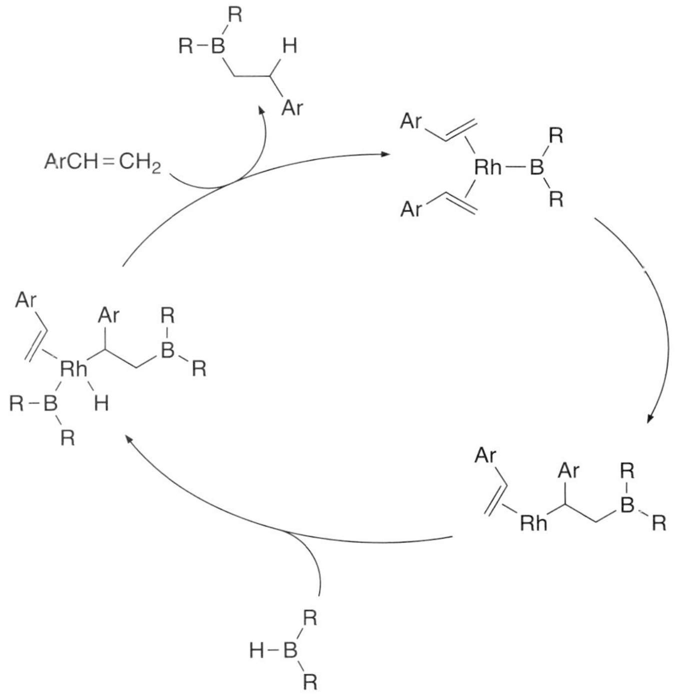

Scheme 16.12

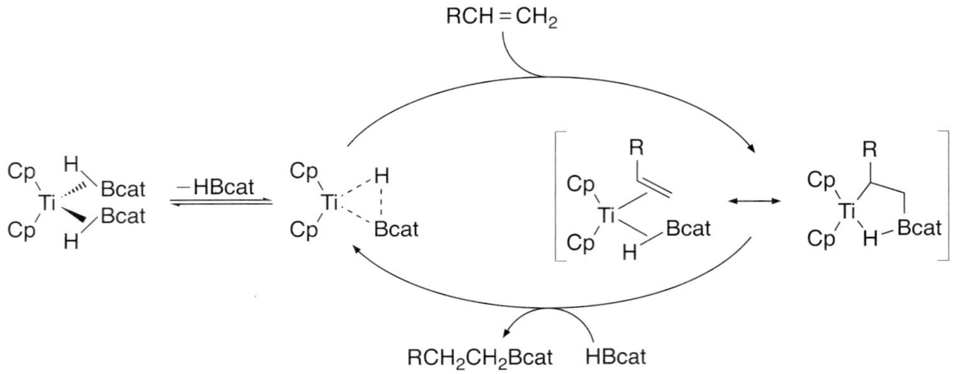

Scheme 16.13

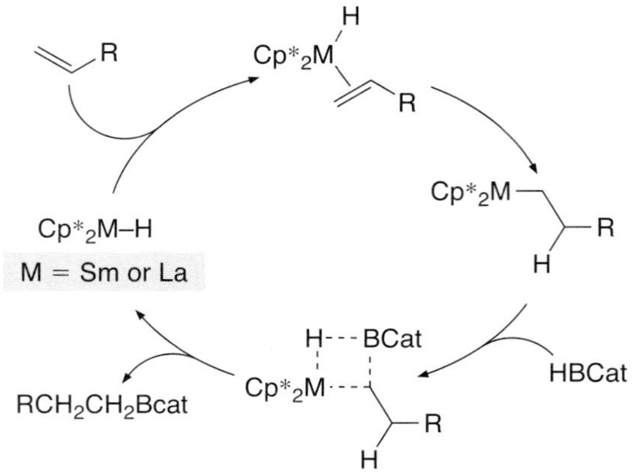

Scheme 16.14

The mechanism of the hydroboration with catecholborane conducted with Wilkinson's catalyst is the most studied,[180,196,197] but also the most complex. The major pathway is depicted in Scheme 16.11 and involves oxidative addition of borane to [Rh(PPh$_3$)$_3$Cl] to form the 16-electron [Rh(PPh$_3$)$_2$Cl(Bcat)(H)].[179] This complex then undergoes insertion of an alkene into the metal hydride to form the linear alkyl complex [Rh(PPh$_3$)$_2$Cl(Bcat)(CH$_2$CH$_2$R)]. Reductive elimination forms the terminal alkylborane product. However, additional events occur during the hydroboration of vinylarenes by Wilkinson's catalyst. First, oxidative addition does not occur to form a single product.[179] Second, vinylboronate esters form as side products.[198] Like the formation of vinylsilanes during hydrosilation, these products are thought to arise from competitive insertion of olefin into the metal–boron bond.[179] Furthermore, catecholborane is relatively unstable, and the borane can decompose to BH$_3$•PPh$_3$.[179,199]

The reactions conducted with ligand-less cationic systems are proposed to occur through the rhodium(I) boryl complexes shown in Scheme 16.12, but fewer data are available to support this mechanism. Theoretical studies[200] have implied that insertion of the olefin into the metal–boryl linkage occurs and that this step is followed by oxidative addition of the borane and reductive elimination to form the C–H bond of the product and the starting rhodium(I) boryl complex.

Reactions catalyzed by titanocene and lanthanocene complexes occur by different mechanisms (Schemes 16.13 and 16.14). The mechanism of the hydroboration of vinylarenes catalyzed by titanium complexes is shown in Scheme 16.13.[182] In this mechanism, a titanocene bis-borane complex dissociates borane to generate a 16-electron complex that coordinates the alkene or alkyne. Coupling of one carbon of the resulting metallacycle with the boron of the coordinated borane forms the final product. The mechanism of the lanthanocene-catalyzed reactions, shown in Scheme 16.14, relates to the mechanism of the lanthanocene-catalyzed hydrosilylation. In this mechanism, the olefin inserts into the metal hydride, and σ-bond metathesis occurs between the borane and the alkyl group to form the alkylboronate product and the starting hydride.[187]

16.4.6. Diboration, Silylboration, and Stannylboration

Diborane(4) reagents containing alkoxo groups are stable and commercially available reagents. Two of the most common diborane(4) reagents are shown in Figure 16.2 and contain pinacolate and catcholate groups at the two boron atoms (pinB–Bpin, catB–Bcat). Oxidative addition of the B–B bond in these diborane(4) reagents occurs to many transition metal complexes[201–206] and occurs more readily than oxidative addition of the Si–Si bond in disilane reagents. Thus, the diboration of alkenes and alkynes has been developed more extensively than has the disilation of these unsaturated substrates. The oxidative addition of silylboranes[207,208] and stannylboranes[208,209] also occurs more readily than the addition of disilanes, and the addition of silylboranes and stannylboranes to alkynes has been reported. The following sections focus on diboration processes.

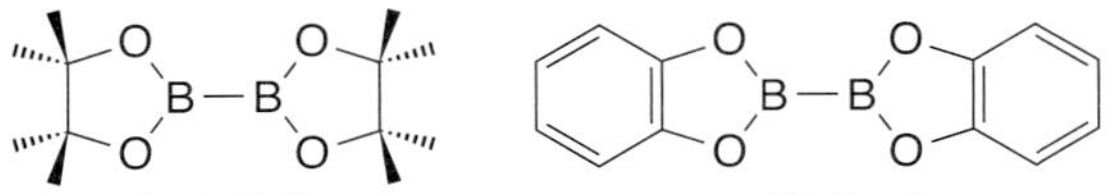

Figure 16.2.
Two tetraalkoxy diborane(4) reagents commonly used in the diboration of alkenes and alkynes.

pinB–Bpin
or B₂pin₂

catB–Bcat
or B₂cat₂

16.4.6.1. Diboration, Silylboration, and Stannylboration of Alkynes

Table 16.4, which is adapted from a comprehensive review by Miyaura from 2001,[167] illustrates the scope and types of catalysts that have been used for the diboration, silylboration, and stannylboration of alkynes. These data illustrate that the simple platinum(0) complex of PPh_3 is a suitable catalyst for diboration.[201,210,211] These data also show that both B_2pin_2 and B_2cat_2 add to terminal alkynes, but that the tetraaminodiboron reagent does not. Finally, these data illustrate that palladium complexes of isonitrile ligands catalyze the silylboration of alkynes[212,213] and that simple $Pd(PPh_3)_4$ catalyzes the stannylboration of terminal alkynes.[209]

Table 16.4. Examples of palladium- and platinum-catalyzed diboration, silylboration, and stannylboration of alkynes.

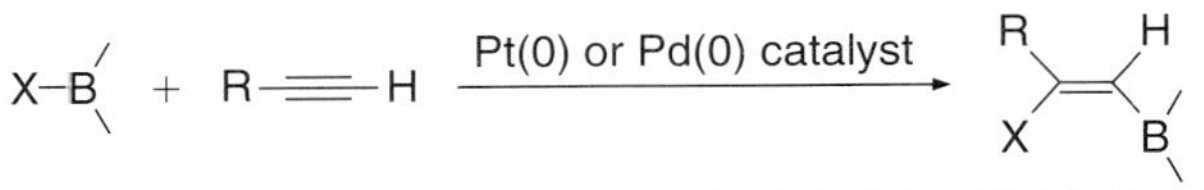

Entry	X–B	R	Catalyst	Yield/%
1	$(MeO)_2B-B(OMe)_2$	$n\text{-}C_6H_{13}$	$Pt(PPh_3)_4$/80 °C	89
2	catB–Bcat	$n\text{-}C_6H_{13}$	$Pt(PPh_3)_4$/80 °C	>95
3	pinB–Bpin	$n\text{-}C_6H_{13}$	$Pt(PPh_3)_4$/80 °C	92
4		Ph	$Pt(PPh_3)_4$/80 °C	79
5		$N\equiv C(CH_2)_3$	$Pt(PPh_3)_4$/80 °C	79
6		$MeO_2C(CH_2)_4$	$Pt(PPh_3)_4$/80 °C	89
7	$(Me_2N)_2B-B(NMe_2)_2$	$n\text{-}C_6H_{13}$	$Pt(PPh_3)_4$/120 °C	7
8	$R_3Si-B(NR_2)_2$	$n\text{-}C_6H_{13}$	$Pd_2(dba)_3$/4 etpo/80 °C[a]	92
9	R_3Si–Bpin	$n\text{-}C_6H_{13}$	$Pd(OAc)_2$/15 RNC/110 °C[b]	92
10	$Me_3Sn-B(NR_2)_2$	$n\text{-}C_6H_{13}$	$Pd(PPh_3)_4$/RT	83
11		Ph	$Pd(PPh_3)_4$/RT	73

[a]etpo = $P(OCH_2)_3CCH_2CH_3$
[b]RNC = $^tBuCH_2CMe_2NC$

16.4.6.2. Diboration of Alkenes

The diboration of alkenes has been more challenging to develop. Table 16.5, which is also adapted from the Miyaura review,[167] summarizes a series of diborations of alkenes. These examples show that complexes of group 10 metals lacking phosphine or isonitrile ligands have been most active for these additions.[214-216] Gold[217] and rhodium[218] complexes have also been shown to catalyze the diboration of vinylarenes.

Table 16.5. Examples of the diboration of alkenes.

(RO)$_2$BB(OR)$_2$	Alkene	Catalyst	Yield/%
pinB–Bpin	1-Decene	Pt(dba)$_2$/50 °C	82
	Styrene	Pt(dba)$_2$/50 °C	86
	Cyclopentene	Pt(dba)$_2$/50 °C	85
	Cyclohexene	Pt(dba)$_2$/50 °C	0
	Norbornene	Pt(dba)$_2$/50 °C	85
catB–Bcat	1-Hexene	Pt(cod)$_2$/RT	95
	Norbornene	Pt(cod)$_2$/RT	93
	4-MeOC$_6$H$_4$CH=CH$_2$	AuCl(PEt$_3$)/2dcpe^a/RT	High
	Styrene	[(acac)Rh(dppm)^b]B$_2$cat$_3$/RT	87
	PhCH=CHPh	[(acac)Rh(dppm)^b]B$_2$cat$_3$/RT	92–99

^{a}dcpe = 1,2-bis(dicyclohexylphosphino)ethane
^{b}1,1-bis(diphenylphosphino)methane

Initially, the absence of ligands other than dienes on the metal made the development of enantioselective versions of these additions challenging. Subsequently, however, systems have been identified that make diborations with ligated metal catalysts possible to conduct. Palladium complexes ligated by phosphines catalyze the diboration of allenes,[219] and enantioselective versions of these diborations have been reported.[220] One example of this process is shown in Equation 16.51. Moreover, rhodium complexes of bisphosphines have been found to catalyze the diborations of simple alkenes. With certain bisphosphine ligands, these reactions also occur with high enantioselectivity.[221] Enantioselective diborations of β-methyl styrene and of 5-decene, catalyzed by rhodium complexes of the QUINAP ligand, are shown in Equations 16.52 and 16.53.

(16.51)

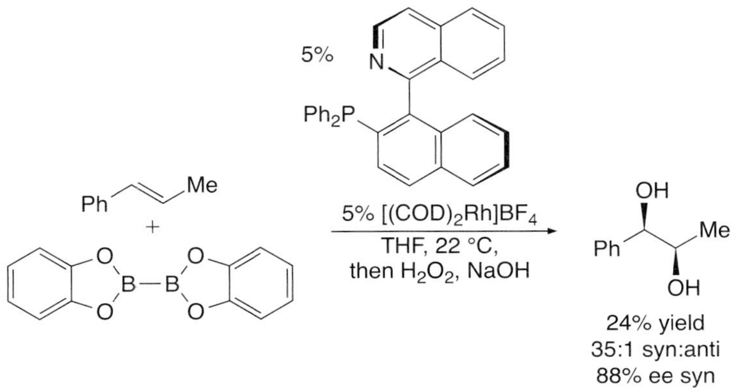

$$(16.52)$$

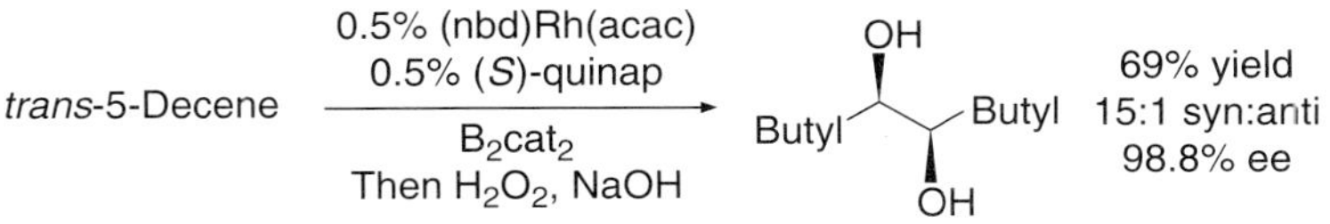

24% yield
35:1 syn:anti
88% ee syn

$$(16.53)$$

16.4.6.3. Mechanism of Diborations

The diborations of alkenes and alkynes are thought to occur by the general pathway shown in Scheme 16.15 and analogous silylborations and stannylborations would occur by related mechanisms. By this mechanism, the oxidative addition of the diborane(4), silylborane, or stannylborane reagent leads to a bisboryl, a silylboryl, or a stannylboryl complex. These oxidative addition processes are described in Chapter 6. In brief, Miyaura and Ishiyama, Marder, and Smith all published examples of the oxidative addition of diboron compounds to Pt(0),[201,222,223] and Marder has published examples of the oxidative addition of diboron reagents to Rh(I).[224] These reactions have also been studied computationally.[204,225] In addition, Tanaka has published an example of the oxidative addition of a stannylborane,[209] and Ozawa has published examples of the oxidative addition of a silylborane to Pt(0) that would initiate stannylboration and silylboration.[207] These oxidative addition processes would be followed by migratory insertion of the alkyne into the M–B bond in the case of diboration. Insertion of the alkyne could occur into the M–B, M–Si, or M–Sn bond in the case of silylboration and stannylboration. Reductive elimination is then thought to form the final product.

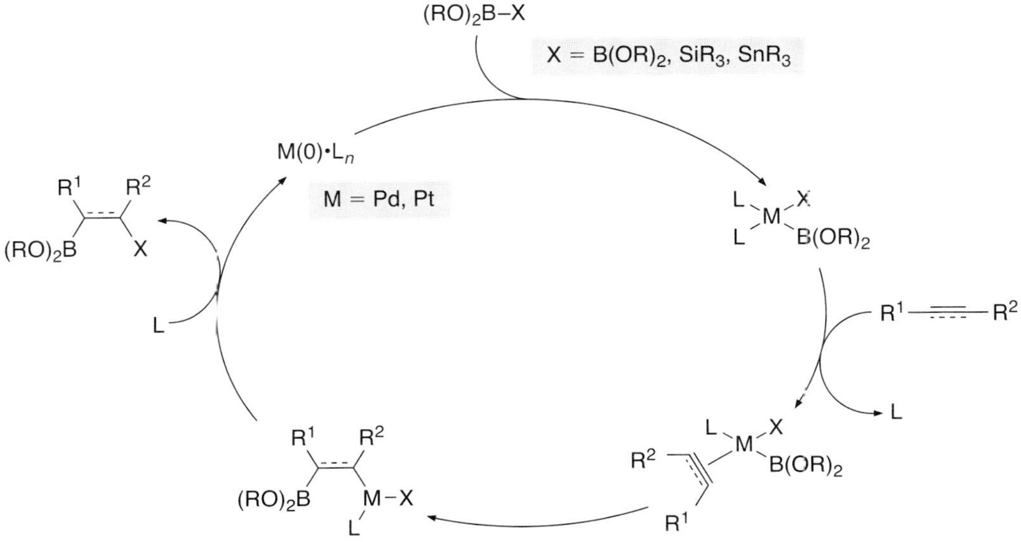

Scheme 16.15

16.5. Transition-Metal-Catalyzed Hydroamination of Olefins and Alkynes

16.5.1. Introduction and Fundamentals of Hydroamination

Hydroamination is the addition of the N–H bond of an amine or related compound across a C–C multiple bond (Equations 16.54 and 16.55). While these products can be obtained via classic organic methods, many of these methods involve the use of additional reagents and compounds derived from olefins. For these reasons, catalytic hydroamination has been the subject of studies for nearly 50 years, and the anti-Markovnikov addition of ammonia to alkenes has been included on top-ten lists of unsolved problems in catalysis.[226] At the same time, catalytic hydroamination is far from a general process, and catalytic anti-Markovnikov hydroamination of alkenes has not yet been published. Nevertheless, significant progress in transition-metal-catalyzed hydroamination has been made recently, and a large number of papers on this topic have been published during the late 1990s and in the first decade of the twenty-first century.

This section focuses on hydroamination catalyzed by transition metal complexes, but many studies on hydroamination catalyzed by acid,[227,228] base,[229,230] main group metals such as mercury and copper, and heterogeneous catalysts have been reported. Because the elementary steps of the mechanisms of these reactions lie outside the scope of this text, this chapter does not present details of the hydroaminations conducted with these types of catalysts. This material has been presented in many reviews.[3,231–241]

$$R^1 \!-\!\!=\!\!-\!H \; + \; HNR_2 \quad \xrightarrow{\text{Catalyst}} \quad \substack{R^1 \\ H} \!\!>\!\!=\!\!<\!\! \substack{H \\ NR_2} \quad \text{or} \quad \substack{R^1 \\ R_2N} \!\!>\!\!=\!\!<\!\! \substack{H \\ H} \tag{16.54}$$

$$R^1 \!-\!\!=\!\!-\!\!\bigcirc\!\! NHR \quad \xrightarrow{\text{Catalyst}} \quad \substack{R_1 \\ H}\!\!-\!\!\bigcirc\!\!N\!\!-\!\!R \quad \text{or} \quad \substack{R_1 \\ R_2N}\!\!-\!\!\bigcirc\!\!H \tag{16.55}$$

Computational studies of the intermolecular addition of ammonia to ethylene have shown that the reaction is enthalpically favored. The intermolecular reactions are entropically disfavored. The thermodynamics for additions of arylamines to vinylarenes have been measured recently in solution.[242] These studies are summarized in Equation 16.56, and have shown that additions to vinylarenes occur with free energies near zero. Additions of arylamines to vinylarenes are favored thermodynamically, but additions to β-substituted vinylarenes are unfavorable, and high concentrations are needed to obtain the addition products in acceptable yield.

$$
n = 0,1 \qquad R = \text{H or Me}
$$

2 mol % $L_2Pd(OTf)_2$
80 °C
2 mol % $L/CpPd(\eta^3\text{-allyl})$
2 mol% HOTf

$K = 0.16\text{–}155 \; M^{-1}$
$\Delta G = 1.3\text{–}(-3.5) \; \text{kcal/mol}$

$$\tag{16.56}$$

The scope of hydroamination now includes many types of compounds containing N–H bonds and many types of alkenes and alkynes. Because the scope of these reactions is rapidly changing and many reviews of these processes have been published elsewhere, this section of the chapter provides a broad overview of the scope of these reactions. It then provides information on the diverse mechanisms that have been identified for hydroamination. These mechanisms draw from the set of elementary reactions described in earlier chapters of this book.

16.5.2. Scope of Hydroamination

The scope of hydroamination can be subdivided into the types of compounds containing N–H bonds that add to olefin or alkyne, or it can be divided into types of reagents containing carbon–carbon π-bonds, such as alkenes, vinylarenes, dienes, allenes, and alkynes. Because the limitations of the catalysts and the rates of the reactions largely depend on the type of carbon–carbon π-bond involved in the reaction, this section of this chapter is divided into the reactions of alkenes, vinylarenes, allenes, dienes, and alkynes.

16.5.2.1. Hydroamination of Alkenes

Hydroaminations of alkenes have proven to be more challenging to develop than other hydroamination reactions, although some of the first hydroaminations were additions to alkenes. In 1971 Coulson reported the addition of secondary amines to ethylene catalyzed by RhCl$_3$.[243] Additions of amides to ethylene and propylene have been published more recently by Widenhoefer,[244] as shown in Equation 16.57, and the addition of aniline to norbornene was published by Milstein and Casalnuovo.[245] Although the turnover numbers for the addition of aniline to norbornene were low, several important mechanistic findings resulted from this work were presented in Chapter 9 and are reviewed in Section 16.5.3.3. Additions of amines to ethylene, propylene, and norbornene are less complicated than additions to higher olefins because these alkenes cannot undergo isomerization to a less reactive internal olefin. Nevertheless, Brunet has reported additions of arylamines to ethylene and hexene catalyzed by platinum halides with acid additive in an ionic liquid (Equation 16.58).[246,247]

$$(16.57)$$

70% (95:5 mixture)

$$(16.58)$$

Intermolecular additions of primary amines to alkenes have also been reported using lanthanide catalysts. These reactions, although slow, do occur to high conversion. Similar to hydroaminations catalyzed by late transition metal complexes, these reactions form the products from Markovnikov addition of the N–H bond across the olefin. One example of such a reaction is shown in Equation 16.59.[248,249]

$$(16.59)$$

90%, TOF = 0.4 h^{-1}

More examples of intramolecular additions of the N–H bonds of amines and related compounds to alkenes have been reported. The most active catalysts for these cyclizations contain lanthanide and group 4 metals. However, complexes of calcium have been reported to cyclize aminoalkenes containing geminal dimethyl groups in the linker (Equation 16.60),[250] and zirconocene and titanocene complexes catalyze the intramolecular hydroamination of aminoalkenes.[251]

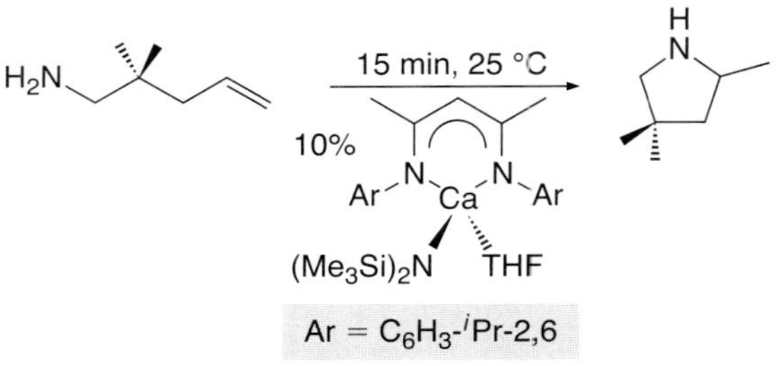

(16.60)

$$Ar = C_6H_3\text{-}^iPr\text{-}2,6$$

The bulk of the studies on intramolecular hydroamination of alkenes catalyzed by lanthanide complexes[248, 249, 252–269] have been conducted using lanthanocene complexes or half-sandwich lanthanide complexes. The prototypical cyclizations of aminoalkenes to form five- and six-membered rings are shown in Equation 16.61. These reactions occur with exclusive Markovnikov selectivity. These reactions have also been conducted using arylamines, as shown in Equation 16.62. The intramolecular reactions of amines catalyzed by certain lanthanide complexes occur with 1,1- and 1,2-disubstituted olefins (Equation 16.63), although such reactions require high temperatures.

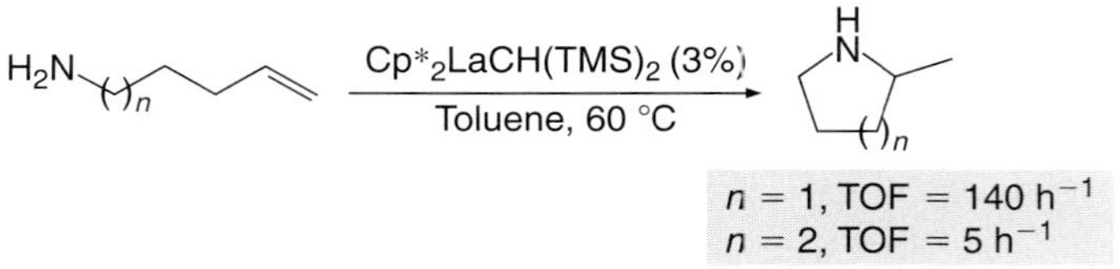

(16.61)

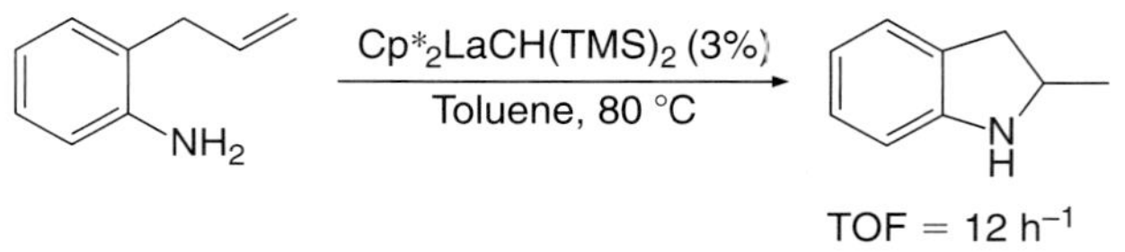

$$n = 1, TOF = 140\ h^{-1}$$
$$n = 2, TOF = 5\ h^{-1}$$

(16.62)

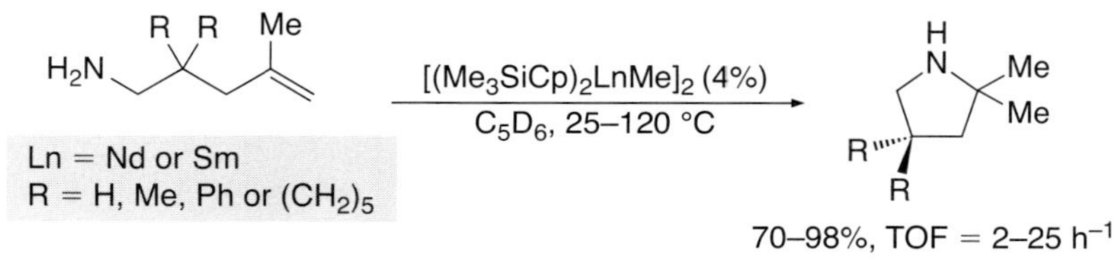

$$TOF = 12\ h^{-1}$$

(16.63)

Ln = Nd or Sm
R = H, Me, Ph or (CH$_2$)$_5$

70–98%, TOF = 2–25 h^{-1}

The size and type of cyclopentadienyl ligands affect the rates of these cyclizations (Tables 16.6–16.8). The cyclization in Equation 16.61 occurred fastest in the presence of the monocyclopentadienyl complexes containing samarium and neodymium (Table 16.6).[260] It also occurred when catalyzed by bis(pentamethylcyclopentadienyl) complexes, and the most active of these complexes were lanthanum and samarium derivatives. Reactions of 1,2-disubstituted olefins occurred significantly faster when catalyzed by complexes containing the larger lanthanum metal center than when catalyzed by complexes containing the smaller samarium (Table 16.7).[267] Moreover, reactions conducted with the lutecium (Table 16.6) or samarium (Table 16.8) catalyst containing a more open metal center, created by linking the cyclopentadienyl groups with the dimethylsilyl group, occurred faster than those catalyzed by the unlinked Cp$^*_2$M fragment.

Table 16.6. Effect of catalyst precursor on the intramolecular hydroamination in Equation 16.61.

Catalyst Precursor	Reaction Temp. (°C)	TOF^a (h^{-1})
[Me$_2$Si(C$_2$Me$_4$)(t-BuN)]SmN(TMS)$_2$	25	181
[Me$_2$Si(C$_2$Me$_4$)(t-BuN)]NdN(TMS)$_2$	25	200
[Me$_2$Si(C$_2$Me$_4$)(t-BuN)]YbN(TMS)$_2$	25	10
Me$_2$Si(C$_2$Me$_4$)$_2$LuCH(TMS)$_2$	25	90
(C$_5$Me$_5$)$_2$LaCH(TMS)$_2$	25	95
(C$_5$Me$_5$)$_2$SmCH(TMS)$_2$	60	48
(C$_5$Me$_5$)$_2$LuCH(TMS)$_2$	80	<1
(EBI)YbN(TMS)$_2$^{b}	25	0.7

^{a}TOF = turnover frequency
^{b}EBI = 1,2-bis(indenyl)ethane-diyl

Table 16.7. Effect of ionic radius on the intramolecular hydroamination of 1,2-disubstituted olefins.

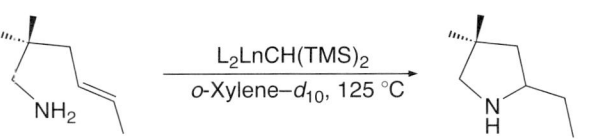

Entry	Cp′$_2$LnCH(TMS)$_2$^{b}	Ionic radii (A)	TOF^a(h^{-1})
1	Cp′$_2$LaCH(TMS)$_2$	1.160 (La^{3+})	8.54
2	Cp′$_2$SmCH(TMS)$_2$	1.079 (Sm^{3+})	0.50
3	Cp′$_2$YCH(TMS)$_2$	1.016 (Y^{3+})	No reaction

^{a}TOF = turnover frequency
^{b}Cp′ = C$_5$Me$_5$

Table 16.8. Effect of Cp–M–Cp bite angle on the intramolecular hydroamination of 1,2-disubstituted olefins.

Entry	L$_2$LnCH(TMS)$_2$^{b}	Bite angle[(c)-Ln-(c)]°	TOF^a(h^{-1})
1	Cp′$_2$SmCH(TMS)$_2$	~134°	0.50
2	Me$_2$Si(C$_5$Me$_4$)$_2$SmCH(TMS)$_2$	~122°	21.60

^{a}TOF = turnover frequency
^{b}Cp′ = C$_5$Me$_5$

These intramolecular hydroaminations lead to the formation of chiral products. Thus, several studies have been devoted to developing ligands for enantioselective, intramolecular hydroaminations of olefins catalyzed by lanthanide complexes.[237,270] Selectivities of these reactions with chiral cyclopentadienyl derivatives have been modest. Selectivities have been higher with catalysts containing non-cyclopentadienyl ligands. Some of the most selective catalysts are the yttrium complexes of the bis(thiolate) ligands reported by Livinghouse,[271] the scandium and lutetium complexes of 3,3′-disubstituted binaptholates reported by Hultzsch,[238] and the BINAM-based bisamidates of Schafer.[272] A representative cyclization catalyzed by a member of each of these classes of catalyst are shown in Equation 16.64.

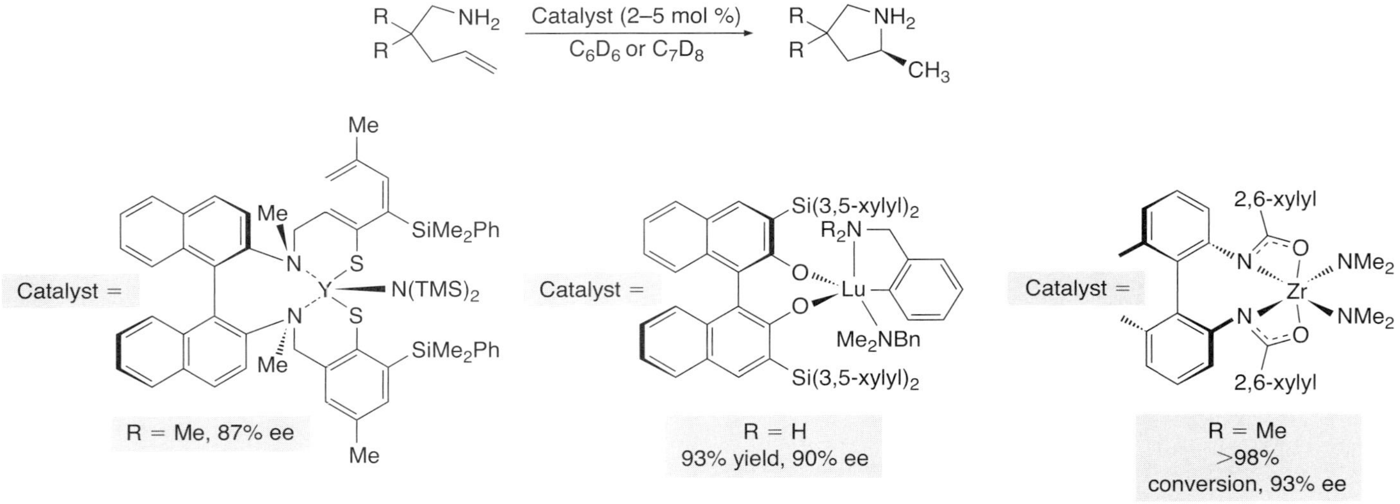

Catalyst (2–5 mol %)
C_6D_6 or C_7D_8

Catalyst =

R = Me, 87% ee

Catalyst =

R = H
93% yield, 90% ee

Catalyst =

2,6-xylyl

R = Me
>98%
conversion, 93% ee

(16.64)

Few examples of intramolecular additions of amines to alkenes catalyzed by late transition metals have been published; more examples of the additions of amides, carbamates, and tosylamides to alkenes catalyzed by this type of complex have been reported. Addition of a secondary amine across a tethered olefin catalyzed by a simple platinum–halide complex is shown in Equation 16.65a.[273] A more recent catalyst based on $[Rh(COD)_2]BF_4$ and a biaryldialkylphosphine leads to cyclizations of aminoalkenes with greater scope (Equation 16.65b).[274] These reactions occur to form five- and six-membered rings, with or without groups that bias the system toward cyclization. They also occur with both internal and terminal olefins and with both primary and secondary amines.

One of the best-characterized examples of intramolecular hydroamidation of an alkene with the tethered activated nitrogen of amide and carbamate groups is shown in Equations 16.66a and 16.66b. This reaction is catalyzed by a dicationic palladium complex ligated by a PNP pincer ligand.[275] Like the rhodium-catalyzed hydroamination, this process occurs to form five- and six-membered rings with or without substituents to bias the system toward cyclization.

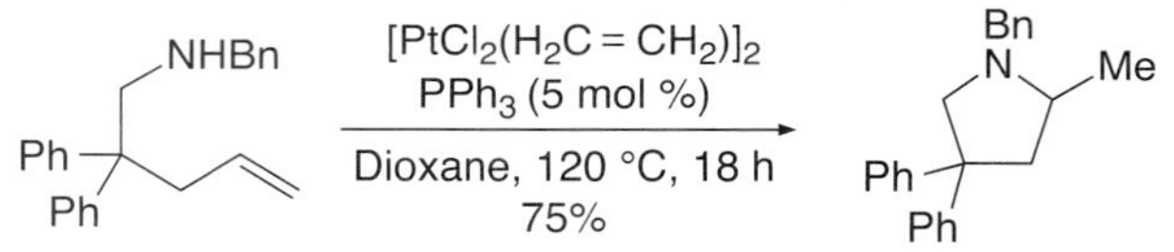

$[PtCl_2(H_2C=CH_2)]_2$
PPh_3 (5 mol %)
Dioxane, 120 °C, 18 h
75%

(16.65a)

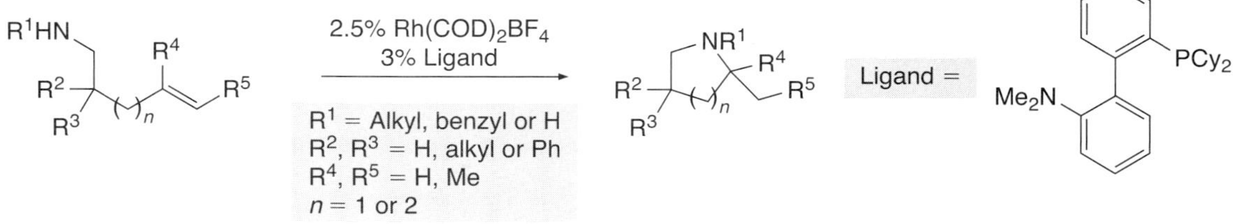

2.5% $Rh(COD)_2BF_4$
3% Ligand

R^1 = Alkyl, benzyl or H
R^2, R^3 = H, alkyl or Ph
R^4, R^5 = H, Me
n = 1 or 2

Ligand =

(16.65b)

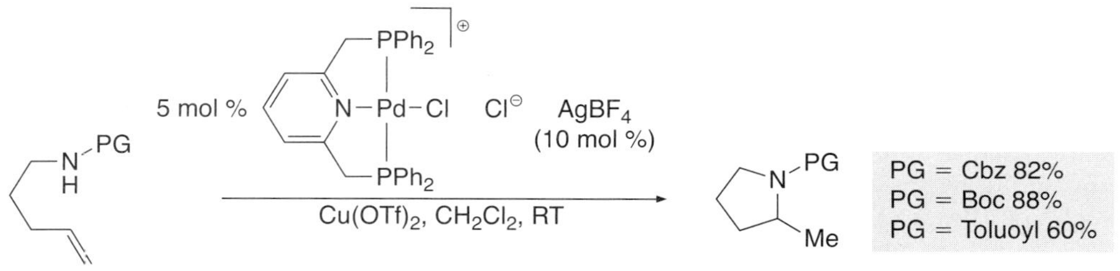

5 mol %

AgBF$_4$
(10 mol %)

$Cu(OTf)_2$, CH_2Cl_2, RT

PG = Cbz 82%
PG = Boc 88%
PG = Toluoyl 60%

(16.66a)

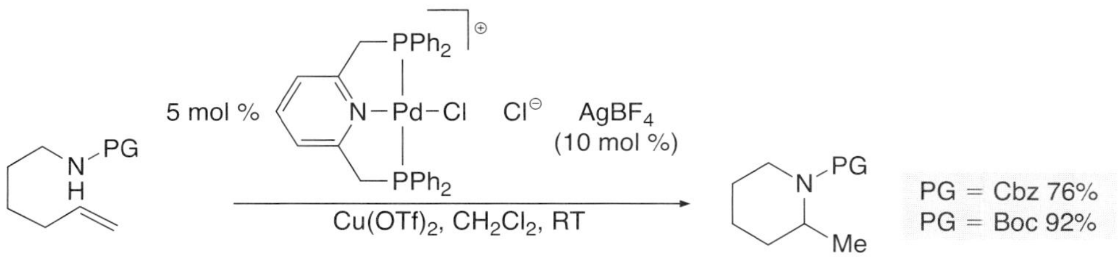
5 mol %
PPh2
N−Pd−Cl
PPh2
Cl⊖
AgBF4
(10 mol %)
N-PG
H
Cu(OTf)2, CH2Cl2, RT
N-PG
Me
PG = Cbz 76%
PG = Boc 92%

(16.66b)

16.5.2.2. Hydroamination of Vinylarenes

Catalysts for the additions of amines to vinylarenes have also been developed. These catalytic reactions include some of the first hydroaminations of unstrained olefins catalyzed by late transition metals, as well as examples catalyzed by lanthanide complexes. These additions occur with Markovnikov selectivity with one set of catalysts and with anti-Markovnikov selectivity with several others. These additions occur by several different mechanisms that are presented in Section 16.5.3.2.

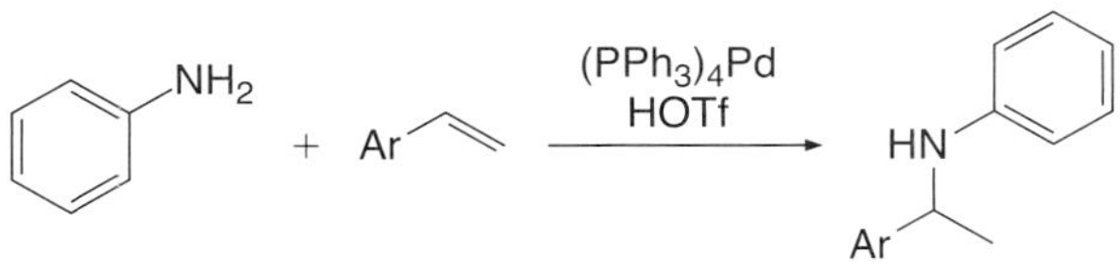
NH2
+ Ar
(PPh3)4Pd
HOTf
HN
Ar

(16.67)

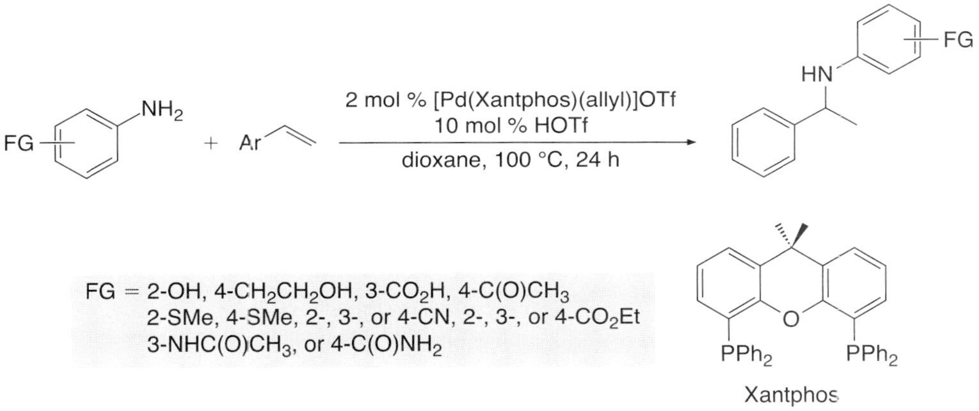
FG
NH2
+ Ar
2 mol % [Pd(Xantphos)(allyl)]OTf
10 mol % HOTf
dioxane, 100 °C, 24 h
HN
FG
FG = 2-OH, 4-CH2CH2OH, 3-CO2H, 4-C(O)CH3
2-SMe, 4-SMe, 2-, 3-, or 4-CN, 2-, 3-, or 4-CO2Et
3-NHC(O)CH3, or 4-C(O)NH2
O
PPh2
PPh2
Xantphos

(16.68)

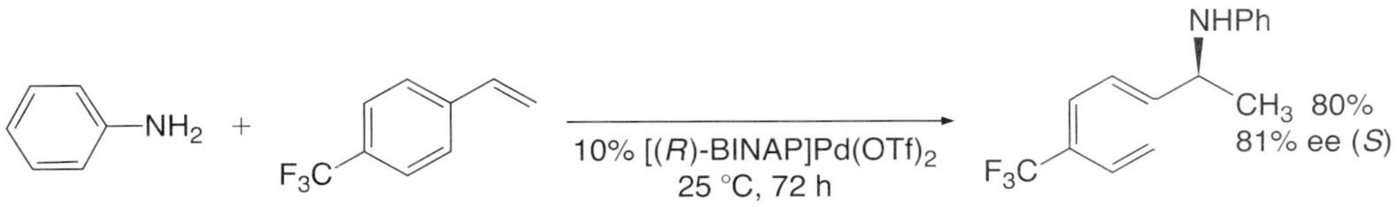
NH2 +
F3C
10% [(R)-BINAP]Pd(OTf)2
25 °C, 72 h
NHPh
CH3 80%
81% ee (S)
F3C

(16.69)

The hydroamination of vinylarenes was first conducted with a palladium catalyst generated either from $Pd(PPh_3)_4$ and triflic acid (Equation 16.67), or from Pd(tifluoroacetate)$_2$, 1,1′-bis(diphenylphosphino)ferrocene (DPPF), and triflic acid.[276] Related systems also catalyze the additions of alkylamines.[277] More active catalysts for the additions of arylamines to vinylarenes have been reported to be formed from [(Xantphos)Pd(allyl)]OTf.[278] The reactions catalyzed by the Xantphos complex have been shown to tolerate a range of functional groups, including ketones with enolizable hydrogens, free alcohols, free carboxylic acids, free amides, nitriles, and esters (Equation 16.68). Reactions also occurred with BINAP as ligand with substantial enantioselectivity.[276] One example is shown in Equation 16.69. These reactions have now been conducted with other atropisomeric biaryl ligands.[279]

Other late transition metal complexes catalyze the anti-Markovnikov hydroamination of vinylarenes, and lanthanide complexes catalyze the anti-Markovnikov hydroamination of vinylarenes when the reactions are conducted intermolecularly. Examples of these hydroaminations are shown in Equations 16.70–16.73.[249,280–282] In the first case,[280] the loosely chelating DPEphos generates a rhodium complex that catalyzes the anti-Markovnikov hydroamination between many vinylarenes and secondary amines in good yield (Equation 16.70). Enamines resulting from oxidative amination of the vinylarenes are the major side products. It has been previously reported that the enamine is the major product formed with catalysts containing monodentate ligands.[283] A rhodium catalyst containing the loosely chelating bisphosphine, 1,4-bis(diphenylphosphino)butane, induces cyclization of vinylarenes containing a pendant amino group to form 3-aryl-1-methyl piperidines (Equation 16.71).[281]

Ruthenium complexes also catalyze the anti-Markovnikov hydroamination of vinylarenes. In this case, the combination of 1,5-bis(diphenylphosphino)pentane (DPPPent), triflic acid, and a ruthenium(II) precursor generates a catalyst for the additions of secondary amines to vinylarenes (Equation 16.72).[282] This mixture of catalyst components has been shown to generate a cationic η^6-arene complex of a PCP pincer ligand generated from the DPPPent ligand.[284] The mechanism of this reaction involves nucleophilic attack of the amine on an η^6-vinylarene complex, as described in more detail in the section on the mechanisms of hydroamination.

Finally, a much different catalyst, a lanthanocene, generates β-phenethylamines from the anti-Markovnikov hydroamination of vinylarenes and primary alkylamines (Equation 16.73).[249] These reactions occur with vinylarenes containing a range of electronic properties. The reaction is thought to occur by insertion of styrene into a lanthanum–amido complex.

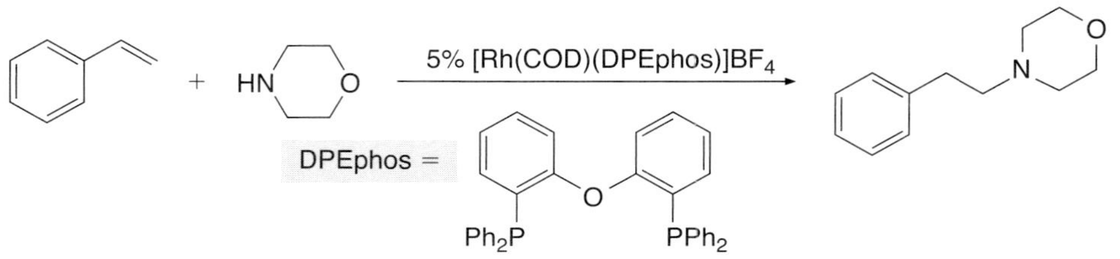

(16.70)

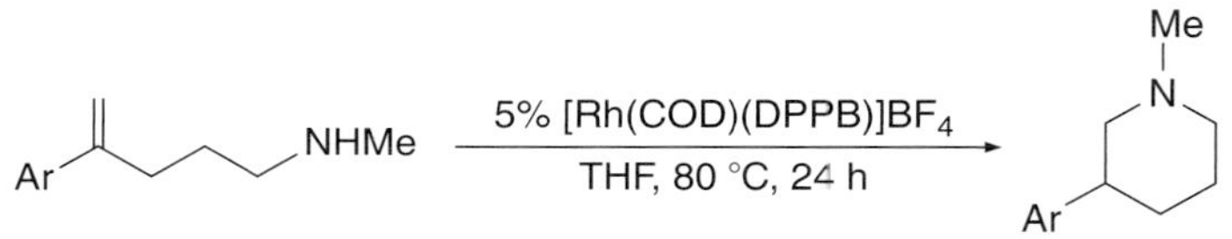

(16.71)

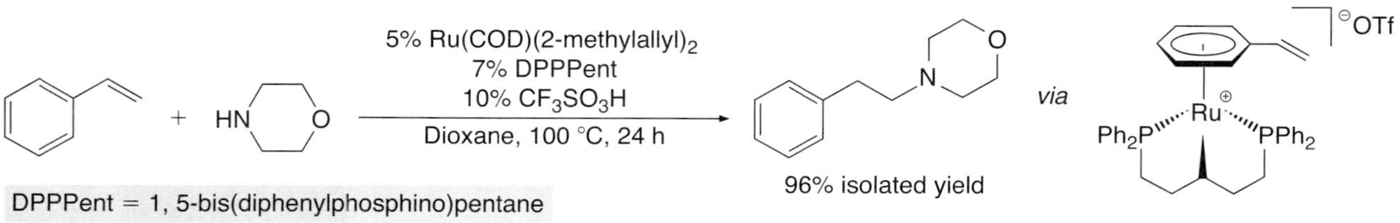

(16.72)

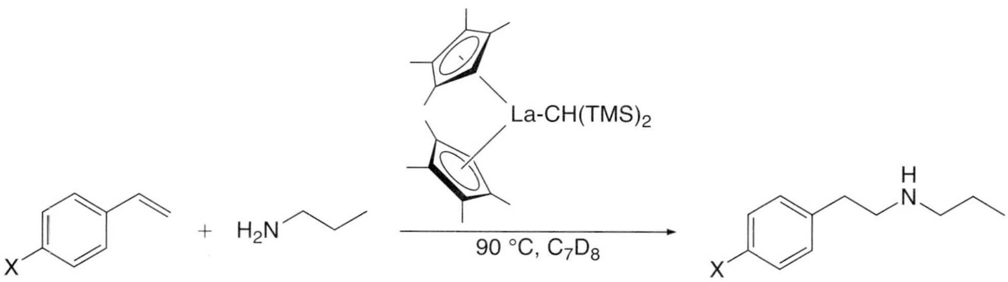

(16.73)

16.5.2.3. *Hydroamination of Allenes*

The hydroamination of allenes generates imines or allylamines, depending on the regioselectivity of the addition. The hydroamination of allenes has been reported with lanthanide complexes, group 4 metal complexes,[285–288] palladium complexes,[289,290] and gold complexes.[291,292] Many of the hydroaminations of allenes have been conducted intramolecularly to form nitrogen heterocycles. However, one of the original hydroaminations with group 4 metals was conducted as an intermolecular process with allene itself.

Selected examples of the hydroaminations of allenes are shown in Equations 16.74–16.78. The first hydroamination of an allene catalyzed by a group 4 metal complex is shown in Equation 16.74 and was conducted with $Cp_2Zr(NHAr)_2$ (Ar = 2,6-dimethylphenyl).[285] Later studies showed that bis-sulfonamide complexes of titanium are more reactive for this process (Equation 16.75).[288] As shown in Equation 16.75, the regioselectivity of the allene hydroaminations catalyzed by group 4 complexes of the bis-sulfonamide ligand depends on the identity of the metal. These hydroaminations are thought to occur by [2+2] cycloadditions of allenes with metal–imido complexes. Lanthanide complexes also catalyze cyclizations of allenes, and these reactions have been used to form naturally occurring alkaloids. One example of such a reaction is shown in Equation 16.76. $Cp^*_2SmN(SiMe_3)_2$ is one of the more reactive catalysts for this reaction. These hydroaminations of allenes are thought to occur by intramolecular insertion of the allene into a Sm–N bond.

$$\text{(16.74)}$$

	Yield	
M = Ti	> 98	—
M = Zr	11	78

$$\text{(16.75)}$$

$$\text{(16.76)}$$

$$
\text{(16.77)}
$$

$$
\text{(16.78)}
$$

The hydroamination of allenes catalyzed by late transition metal complexes has also been reported, and these reactions usually generate allylic amine derivatives. Intramolecular reactions with substrates containing electron-withdrawing groups on nitrogen are shown in Equations 16.77 and 16.78. Additions of amines to allenes catalyzed by a combination of a palladium salt and PPh$_3$ have been reported.[289] However, the scope of the reactions that occurred with high selectivity was limited. In contrast, the intramolecular (Equation 16.77)[293] or intermolecular[290] hydroamination of allenes with a tosylamide, catalyzed by an allylpalladium complex of DPPF in the presence of acid, occurred in high yield. This reaction is likely to occur through an allylpalladium intermediate generated from the acid and the allene. A related cyclization of an allene tethered to a carbamate occurs in the presence of a gold catalyst (Equation 16.78).[291]

16.5.2.4. Hydroamination of 1,3-Dienes

The hydroamination of 1,3-dienes has been known for many years, but these reactions have often generated mixtures of products. More recently, the intermolecular and intramolecular hydroamination of dienes has been reported with lanthanide and palladium catalysts to generate allylic amines in high yields. These reactions have also occurred with high enantioselectivity in some cases.

Nickel-catalyzed reactions between butadiene and alkylamines to generate mixtures of "telomeric" amines containing 4, 8, or 12 carbons (Equation 16.79) were studied extensively in the early 1970s.[294,295] Many examples of these reactions were also conducted with palladium catalysts.[294,296–299] This chemistry provides a method of preparing medium-length alkylamines after hydrogenation of the remaining double bonds. These telomerizations of dienes were typically conducted with a combination of nickel(II) or palladium(II) compounds and added phosphines. The activity of the catalyst was increased by adding alkylaluminum co-catalysts.[296,297] In many synthetic contexts, the 1:1 adduct would be desirable, and the selectivity toward formation of 1:1 adducts was greater in some cases when the ratio of ligand to metal was high.[300] The addition of an acid co-catalyst was also shown to increase selectivity for formation of the 1:1 adduct.[301]

$$
\text{(16.79)}
$$

Mixture of products with an additional C=C bond in the chain + analogous C$_{12}$ amine

More recently, the hydroamination of butadiene, isoprene, and cyclohexadiene to form 1:1 adducts with high selectivity catalyzed by nickel[302] and palladium[303] complexes has been reported. The nickel-catalyzed reactions were conducted with $Ni(COD)_2$ and DPPF as catalyst precursors with acid co-catalyst at or near room temperature. One example of such a reaction is shown in Equation 16.80. The scope of the reaction encompassed primary and secondary alkylamines and cyclic and acyclic dienes. Even the reaction of butadiene with a secondary alkylamine selectively formed the 1:1 adduct. These nickel-catalyzed hydroaminations of dienes were shown to be reversible, and this reversibility leads to racemization of chiral allylic amine products.

The palladium-catalyzed reactions occurred with the simple combination of [Pd(allyl) Cl]$_2$ and added phosphine, or with $Pd(PPh_3)_4$ and an acid co-catalyst. The scope of these reactions encompassed the additions of arylamines. As shown in Equation 16.81, reaction of various arylamines with cyclohexadiene occurred with high enantioselectivity using Trost's ligand. This ligand is discussed in more detail in Chapter 20. These reactions, and nickel-catalyzed reactions, occur by nucleophilic attack of amines on π-allyl intermediates generated by protonation of diene complexes or insertion of dienes into palladium hydrides.

Lanthanide complexes also catalyze the hydroamination of 1,3-dienes. The lanthanide catalysts originally developed for the intramolecular hydroamination of aminoalkenes are particularly active for the intramolecular additions of alkyl amines to dienes. The scope of this process is broad; an illustrative example showing the high diastereoselectivity of the cyclization of a chiral amine is shown in Equation 16.82. These reactions occur by insertion of the diene into a lanthanide–amide intermediate to form an allyl–metal intermediate.

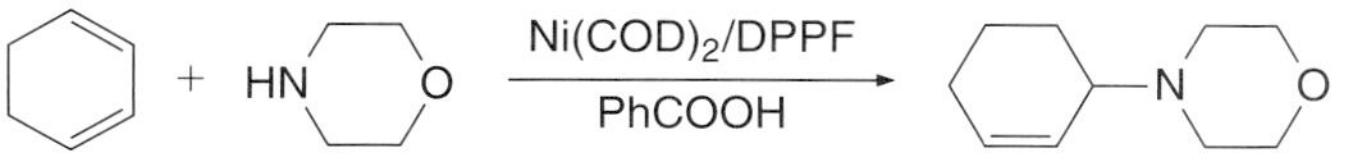

(16.80)

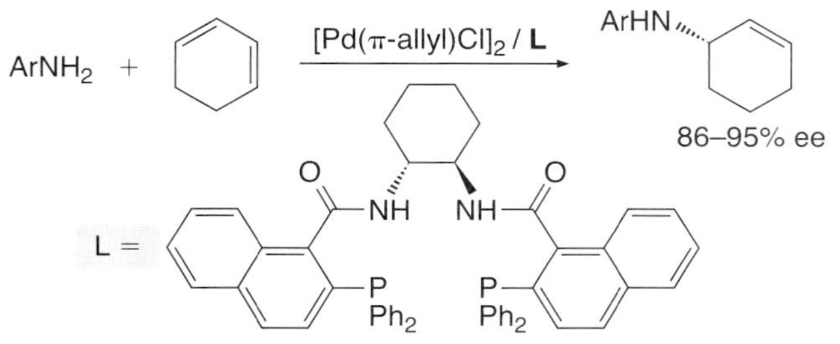

86–95% ee

(16.81)

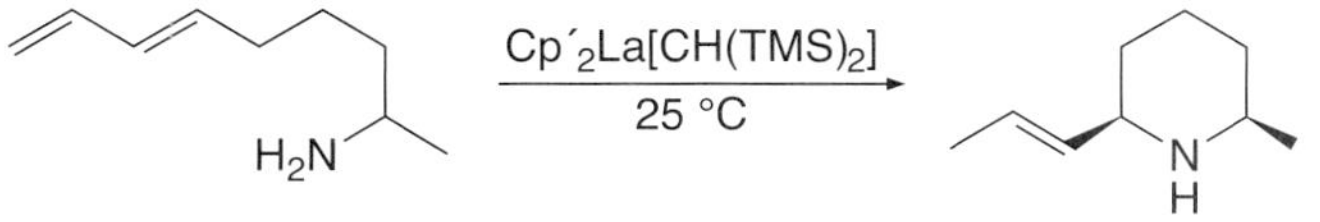

(16.82)

The hydroamination of cycloheptatriene has also been reported (Equation 16.83).[3C4] Although the mechanism of this reaction has not been revealed, it is likely to be related to the mechanism of the palladium-catalyzed hydroamination of dienes. The overall transformation must involve two hydroaminations: one intermolecular hydroamination of the triene unit and one intramolecular hydroamination of the resulting diene.

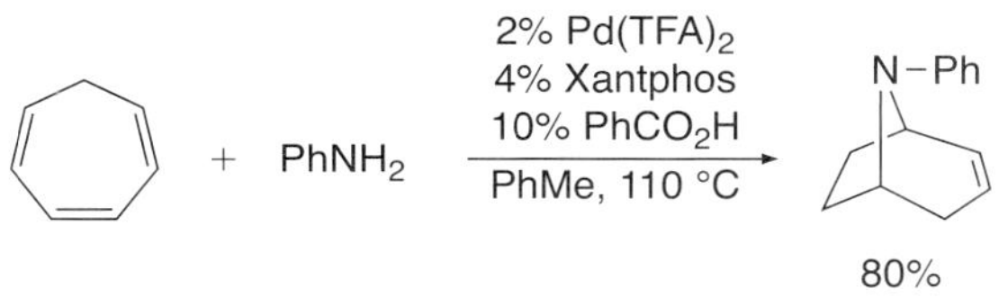

80%

(16.83)

16.5.2.5. Hydroamination of Alkynes[236]

The hydroamination of alkynes with secondary amines usually generates enamines, and the hydroamination of alkynes with primary amines usually generates imines. These hydroaminations of alkynes are more favorable thermodynamically than are the analogous hydroaminations of alkenes. Moreover, hydroaminations of alkynes are often faster than hydroaminations of alkenes, presumably because the coordination of alkynes to transition metals is more favorable than the coordination of alkenes. Also, reactions that can lead to the products of hydroamination, such as [2+2] cycloadditions and migratory insertions, are typically faster and more favored thermodynamically with alkynes than with alkenes. For these reasons, a broader range of hydroaminations of alkynes than of alkenes has been reported. These hydroaminations include examples catalyzed by complexes of group 4 metals, lanthanides, rhodium, and palladium. These reactions can occur with Markovnikov or anti-Markovnikov regiochemistry. In addition, the palladium-catalyzed reaction of an amine with an alkyne under one set of conditions containing added acid generates an allylic amine.

16.5.2.5.1. Hydroamination of Alkynes Catalyzed by Group 4 Metal Complexes

The hydroamination of alkynes catalyzed by group 4 complexes were some of the first transition-metal-catalyzed hydroamination reactions.[285,305] One example of these reactions is shown in Equation 16.84. The reactions only occur with hindered amines and are slow. Nevertheless, the reactions occur in high yield and, in the absence of air, the catalysts are stable indefinitely. An early intramolecular reaction catalyzed by CpTiCl$_3$ to form a cyclic enamine is shown in Equation 16.85. Reactions with internal alkynes occur to form products with Markovnikov regiochemistry.[306] As described in more detail below, these reactions occur by [2+2] additions of the alkyne to an intermediate metal–imido complex.

These results led to the use of other group 4 complexes for the hydroamination of alkynes. For example, Cp$_2$TiMe$_2$[307–309] and [Cp$_2$Ti(η^2-Me$_3$SiCCSiMe$_3$)][310] have been used as catalyst precursors for additions of amines and hydrazines to alkynes. Hydroaminations of alkynes catalyzed by non-metallocene group 4 metal complexes have also been reported.[311–315] Hydroaminations of alkynes catalyzed by group 4 metal complexes have been conducted in tandem with hydrosilylation to form the corresponding amine (Equation 16.85),[316] or in tandem with the addition of an alkyllithium reagent to the aldimine to form branched amines containing a new C–C bond.[317] Alkyne hydroaminations catalyzed by early metal complexes have also been conducted intramolecularly to form pyrroles[315] and indoles (Equation 16.86).[318]

$$R\!\!-\!\!\!\equiv\!\!\!-Me \; + \; H_2NAr \xrightarrow[\substack{R' = Me,\, NHAr \\ 110\,°C}]{Cp_2Zr(NHAr)R'} \xrightarrow{HCl/H_2O} R\!\!\overset{O}{\diagup\!\!\diagdown}\!\!Me \tag{16.84}$$

R = Ph, iPr

$$\tag{16.85}$$

R = Ph or n-Bu

94%, TOF ~9 h^{-1}

$$\tag{16.86}$$

Ph$-\!\!\equiv\!\!-$Me
+
H$_2$N$-$R

1. 10 mol % catalyst
Toluene, 105 °C, 24 h

2. PhSiH$_3$
40 mol % piperidine
40 mol % MeOH
105 °C, 24 h

NHR

Ph   Me

Catalyst: [Cp$_2$TiMe$_2$]

16.5.2.5.2. Hydroamination of Alkynes Catalyzed by Lanthanide and Actinide Complexes

Some of the most active catalysts for the hydroamination of alkynes are based on lanthanides and actinides. The turnover frequencies for the additions are higher than those for lanthanide-catalyzed additions to alkenes by one or two orders of magnitude. Thus, intermolecular addition occurs with acceptable rates. Examples of both intermolecular[248] and intramolecular[258,319] reactions have been reported (Equations 16.87 and 16.88). Tandem processes initiated by hydroamination have also been reported.[319,320] As shown in Equation 16.89, intramolecular hydroamination of an alkyne, followed by cyclization with the remaining olefin, generates a pyrrolizidine skeleton. Hydroaminations of aminoalkynes have also been conducted with the metallocenes of the actinides uranium and thorium.[321,322] These hydroaminations catalyzed by lanthanide and actinide complexes occur by insertion of the alkyne into a metal–amido intermediate.

$$H_3C\!\!=\!\!\equiv\!\!-R \;+\; \text{n-PrNH}_2 \xrightarrow[\text{C}_6\text{D}_6,\ 60\ °\text{C}]{\text{Me}_2\text{Si}(\text{C}_5\text{Me}_4)_2\text{NdCH(TMS)}_2\ (0.5\%)} \qquad (16.87)$$

R = Me or Ph

85–91%, TOF = 1–2 h^{-1}

$$\xrightarrow[\text{C}_6\text{H}_6]{\text{Cp*}_2\text{SmCH(TMS)}_2\ (2\%)} \qquad (16.88)$$

92% for R = Me$_3$Si; n = 1

$$\xrightarrow[\text{C}_6\text{H}_6,\ 21\ °\text{C}]{\text{Cp*}_2\text{SmCH(TMS)}_2\ (2\%)} \qquad (16.89)$$

R = Ph or Me

68–75%, TOF = 17–777 h^{-1}

16.5.2.5.3. Hydroamination of Alkynes Catalyzed by Rhodium and Palladium Complexes

Among the first hydroaminations of alkynes to be published were cyclizations catalyzed by palladium complexes to form indoles. Hydroaminations of alkynes catalyzed by low-valent, late transition metal complexes occur with a narrower scope than the reactions catalyzed by lanathanide complexes. More recently, examples of the hydroaminations of alkynes catalyzed by rhodium complexes have been reported.

Examples of palladium- and rhodium-catalyzed hydroaminations of alkynes are shown in Equations 16.90–16.92 and Table 16.9. The reaction in Equation 16.90[323] is one of many examples of intramolecular hydroaminations to form indoles that are catalyzed by palladium complexes.[323,324] The reaction in Equation 16.91 shows earlier versions of this transformation to form pyrroles by the intramolecular hydroamination of amino-substituted propargyl alcohols. More recently, intramolecular hydroaminations of alkynes catalyzed by complexes of rhodium and iridium containing nitrogen donor ligands have been reported,[325–327] and intermolecular hydroaminations of terminal alkynes at room temperature catalyzed by the combination of a cationic rhodium precursor and tricyclohexylphosphine are known.[328] The latter reaction forms the Markovnikov addition product, as shown in Equation 16.92 and Table 16.9. These reactions catalyzed by rhodium and iridium complexes are presumed to occur by nucleophilic attack on a coordinated alkyne.

$$(16.90)$$

$$(16.91)$$

$$(16.92)$$

Table 16.9. Scope of the hydroamination of alkynes catalyzed by the combination of [Rh(COD)$_2$]BF$_4$ and PCy$_3$.

Entry	Alkyne R	Amine R^1	Mol % catalyst	Yield [%]
1	n-hexyl	C$_6$H$_5$	1.5	79
2	n-butyl	C$_6$H$_5$	1.5	83
3[a]	n-butyl	2-Me-C$_6$H$_4$	1.5	55
4	n-hexyl	4-Me-C$_6$H$_4$	1.5	73
5	n-hexyl	4-MeO-C$_6$H$_4$	1.5	63
6	n-hexyl	3-F-C$_6$H$_4$	1.5	80
7	n-hexyl	4-Cl-C$_6$H$_4$	1.0	> 99
8	Ph	C$_6$H$_5$	2.5	10[b]

[a]44 h reaction time.
[b]Conversion of aniline; product identified by GC/MS analysis.

16.5.3. Mechanisms of Transition-Metal-Catalyzed Hydroamination

16.5.3.1. Overview of the Mechanisms of Transition-Metal-Catalyzed Hydroaminations

The mechanisms of hydroaminations of olefins and alkynes are varied. Many reviews divide these mechanisms into reactions that "activate" the N–H bond or "activate" the olefin. Because each mechanism must somehow cleave both the N–H bond and the C–C π-bond, and the term "activate" does not have a precise meaning, these mechanisms are divided in this discussion into three classes depending on the type of elementary reaction that forms the C–N bond. The first class of mechanism occurs by attack of an amine on a π-bound ligand. The second class of mechanism occurs by insertion of the olefin or alkyne into a metal–amido complex. The third class of mechanism occurs by [2+2] cycloaddition of the alkene or alkyne with a metal–imido complex. In each case, the mechanisms can be further subdivided. In particular, the mechanisms involving attack on a π-ligand can occur by attack on a coordinated alkene, a coordinated alkyne, a π-allyl or π-benzyl ligand, or even an η^6-arene complex. The reactions described in the previous sections all occur by one of the three general classes of mechanisms.

16.5.3.2. Hydroamination by Attack of Amines on π-Complexes

16.5.3.2.1. Hydroamination by Attack on π-Olefin and Alkyne Complexes

Hydroaminations occurring by nucleophilic attack on π-ligands are the oldest class of hydroamination and are discussed first. A mechanism for the hydroamination of alkenes and alkynes catalyzed by palladium(II) complexes is shown in Scheme 16.16. By this pathway, coordination of the alkene or alkyne through the π-system occurs to generate a cationic or electron-poor, neutral metal–olefin or metal–alkyne complex. Nucleophilic attack of the amine on the coordinated olefin or alkyne then occurs. Nucleophilic attack on coordinated olefins and alkynes is presented in detail in Chapter 11. As noted in Chapter 11, this nucleophilic attack occurs at the internal position of an alkene or alkyne.

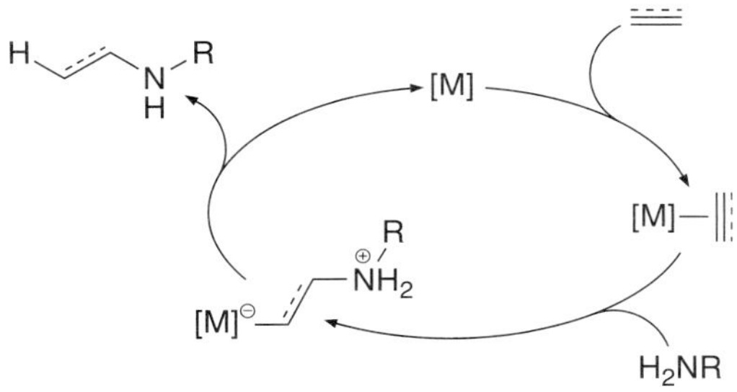

Scheme 16.16

The product of this nucleophilic attack is a metal–alkyl or metal–vinyl intermediate that can undergo protonolysis at the M–C bond to yield the organic product and regenerate the catalyst. This protonolysis can occur by direct proton transfer to the M–C bond, or an indirect sequence of proton transfer to the metal center, followed by reductive elimination to form the C–H bond. The N–H bond in the substrate must be sufficiently acidic to form a product that will undergo this proton transfer. The intramolecular hydroaminations of alkynes catalyzed by palladium(II) complexes occur by this mechanism. The intramolecular hydroaminations of olefins catalyzed by platinum(II) complexes[273,329,330] and the intramolecular hydroaminations of alkynes catalyzed by cationic rhodium and iridium complexes containing nitrogen ligands[325–327] have been proposed to occur by this general mechanism. The mechanism of the hydroamination of alkynes with amines catalyzed by rhodium complexes of PCy$_3$ is currently unknown.

16.5.3.2.2. Hydroamination by Attack of Amines on π-Allyl and π-Benzyl Complexes

Hydroaminations of 1,3-dienes and of vinylarenes occur by a different mode of nucleophilic attack. In this case, the attack occurs on a formally anionic π-allyl or π-benzyl complex. The hydroaminations of dienes catalyzed by nickel and palladium complexes and the hydroaminations of vinylarenes to form Markovnikov addition products catalyzed by palladium complexes occur by this mechanism. The mechanism of hydroamination of vinylarenes catalyzed by rhodium complexes to form the anti-Markovnikov addition product has not yet been established.

This catalytic cycle is shown in Scheme 16.17 for the reactions of cyclohexadiene catalyzed by palladium-phosphine complexes.[278] By this mechanism, the combination of a diene and either a Ni(0) or Pd(0) complex and acid or a cationic nickel(II) or palladium(II) hydride generates a cationic π-allyl complex. Similarly, the combination of a vinylarene and one of these combinations of species generates a cationic phenethylpalladium complex. These allyl and phenethyl complexes have been isolated using this procedure.[278,302,331] The amine then attacks the π-allyl or π-phenethyl complex to form the organic product and regenerate Ni(0) or Pd(0). This nucleophilic attack on π-allyl and π-benzyl complexes was discussed in detail in Chapter 11.

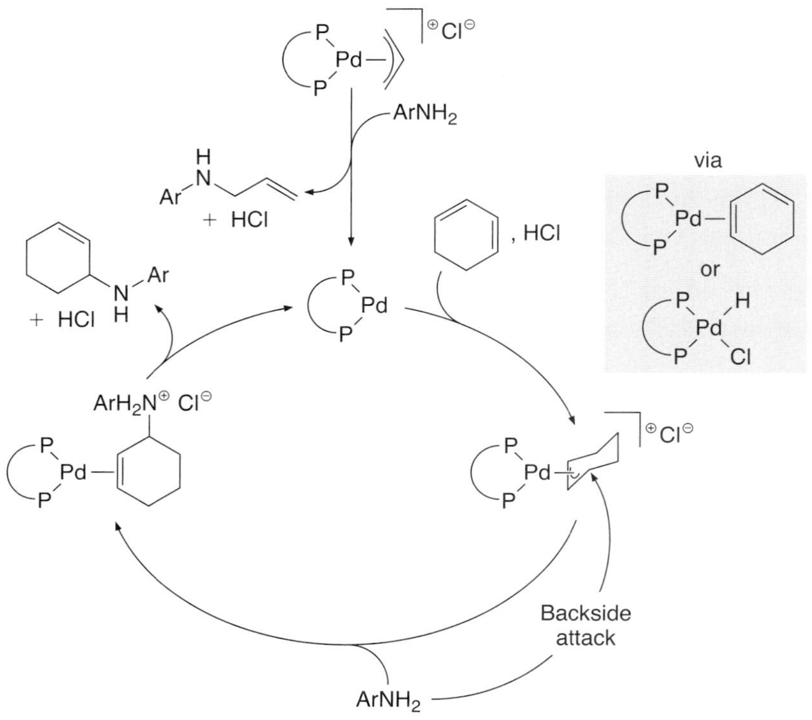

Scheme 16.17

16.5.3.2.3. Hydroamination by Attack of Amines on π-Arene Complexes

The mechanism of the hydroamination of vinylarenes catalyzed by the combination of [Ru(methallyl)$_2$(COD)], 1,5-bis(diphenylphosphinopentane), and triflic acid follows an unusual pathway involving attack on a π-arene complex.[284] This pathway is shown in Scheme 16.18. The combination of catalyst components and styrene generates a π-arene complex of the vinylarene. The ruthenium in this complex contains a PCP pincer ligand formed by activation of the center methylene of the bisphosphine ligand, as shown in Equation 16.93. This cationic η^6-vinylarene complex then undergoes attack by amine to generate an η^6-aminoalkylarene complex. With this ligand set at ruthenium, exchange of arenes occurs under relatively mild conditions to regenerate the starting η^6-vinylarene complex.

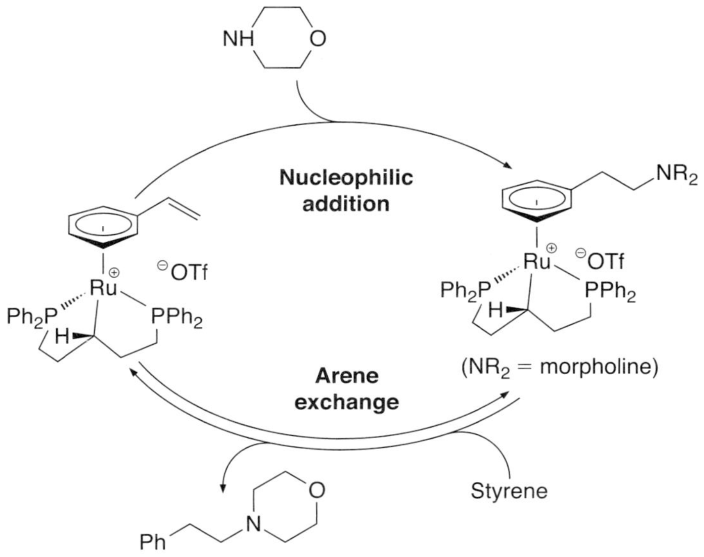

Scheme 16.18

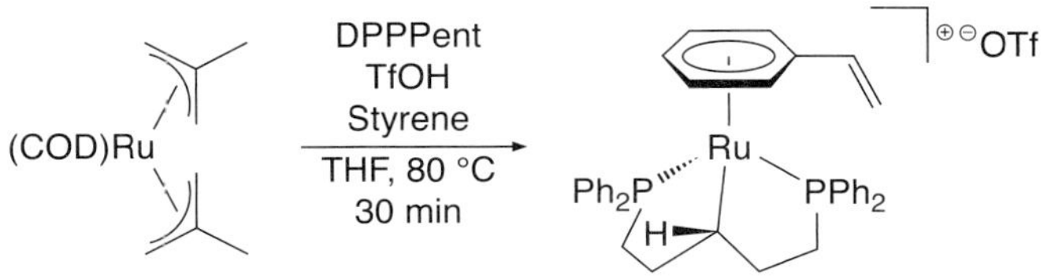

(16.93)

16.5.3.3. Hydroamination by Insertions of Olefins into Metal Amides

As discussed in Chapter 9, the insertion of olefins and alkynes into metal–amido complexes is limited to a few examples. Such insertion reactions are proposed to occur as part of the mechanism of the hydroamination of norbornene catalyzed by an iridium(I) complex and as part of the hydroamination of alkenes and alkynes catalyzed by lanthanide and actinide metal complexes. This reaction was clearly shown to occur with the iridium(I) amido complex formed by oxidative addition of aniline, and this insertion process is presented in Chapter 9.[245] The mechanism of the most active Ir(I) catalyst system for this process involving added fluoride[332] is unknown.

The mechanism of the lanthanide-catalyzed hydroamination of alkenes, dienes, and alkynes has been studied extensively by experimental and computational methods.[252,255,265,266,333–335] The mechanism in Scheme 16.19 provides a general pathway for lanthanide-catalyzed reactions of aminoalkenes. The reactions of aminoalkynes, aminoallenes, and aminodienes occur similarly. This pathway has also been proposed for the hydroamination of aminoalkenes catalyzed by cationic group 4 metal complexes[251,334] and for those catalyzed by complexes of calcium[250] and zinc.[336] The catalyst precursor undergoes protonolysis to generate the metal–amido complex. The tethered olefin or alkyne then coordinates to the metal center and undergoes migratory insertion to generate an aminoalkyl or vinyl intermediate. Protonolysis of this aminoalkyl complex by a second amine then regenerates the metal–amido complex. The migratory insertion step is irreversible and turnover limiting. As a result, the reaction is zero order in the aminoalkene or aminoalkyne reagent. When the cyclization of an aminodiene occurs, an allylmetal product forms, and protonolysis of this allyl intermediate is likely the turnover-limiting step.

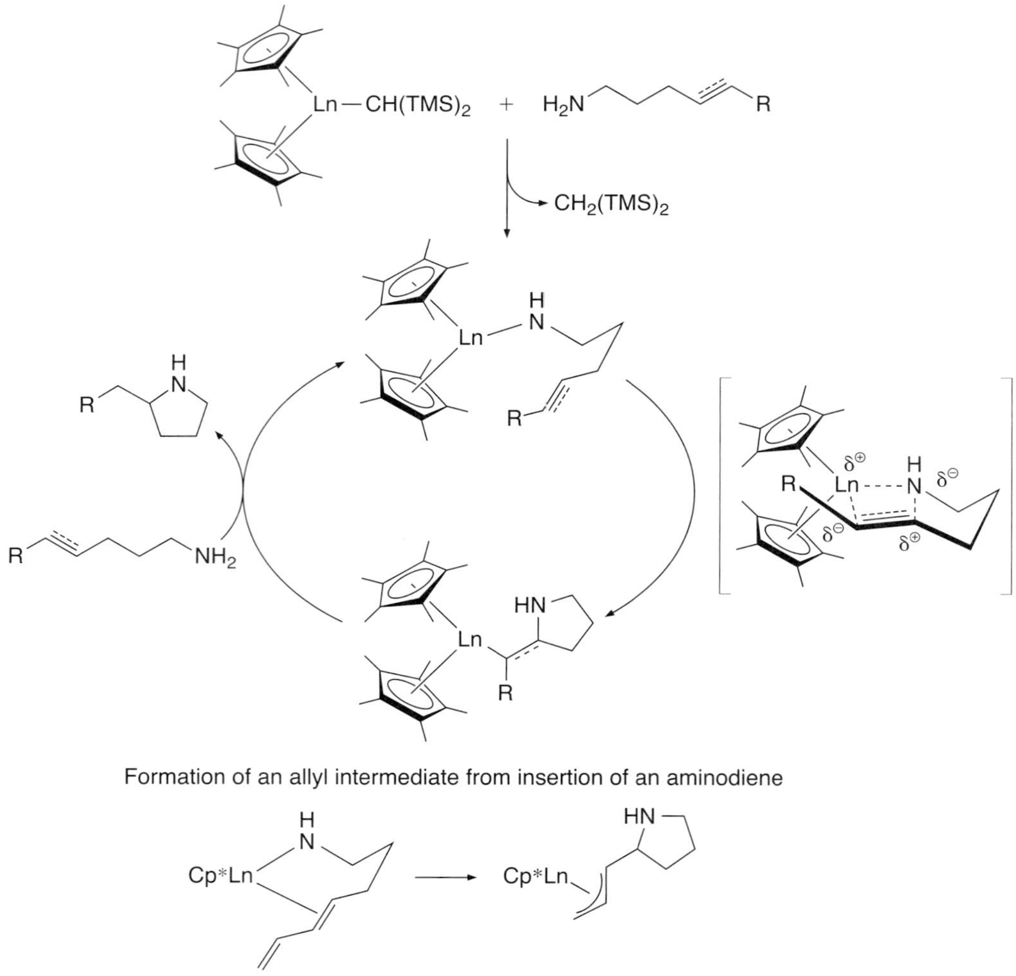
Ln—CH(TMS)2
H2N
R
CH2(TMS)2
Formation of an allyl intermediate from insertion of an aminodiene
Cp*Ln
HN

Scheme 16.19

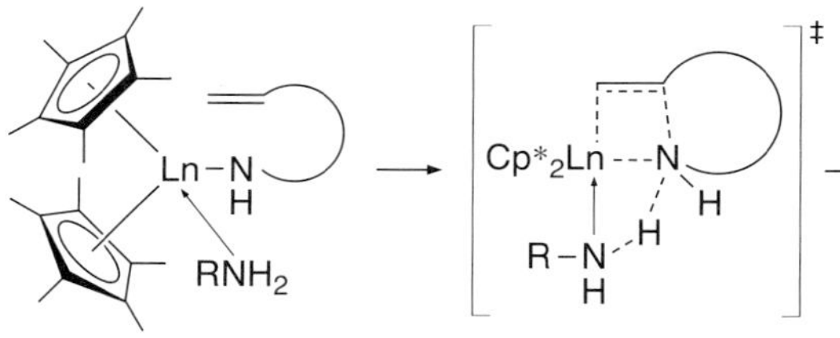
Ln-N
H
RNH2
Cp*2Ln
R-N
‡

(16.94)

Detailed studies of the stereochemistry of the cyclization process indicate that a second amine is likely coordinated to the metal center during the insertion process and primary KIE values for N–H and N–D suggest that protonolysis occurs in concert with insertion.[255] A possible mechanism that was proposed to account for this role of a second amine is shown in Equation 16.94. Hydrogen bonding between this coordinated amine and the amido complex is proposed to accelerate the insertion step.

16.5.3.4. Hydroamination by [2+2] Cycloadditions

The mechanism of the hydroaminations of alkynes and allenes catalyzed by neutral group 4 metal complexes has been studied in detail, and this reaction has been shown to occur through metal–imido complexes.[285,287,306] The basic steps of this mechanism are shown in Scheme 16.20. These reactions are initiated with either a bis(amido) complex, as

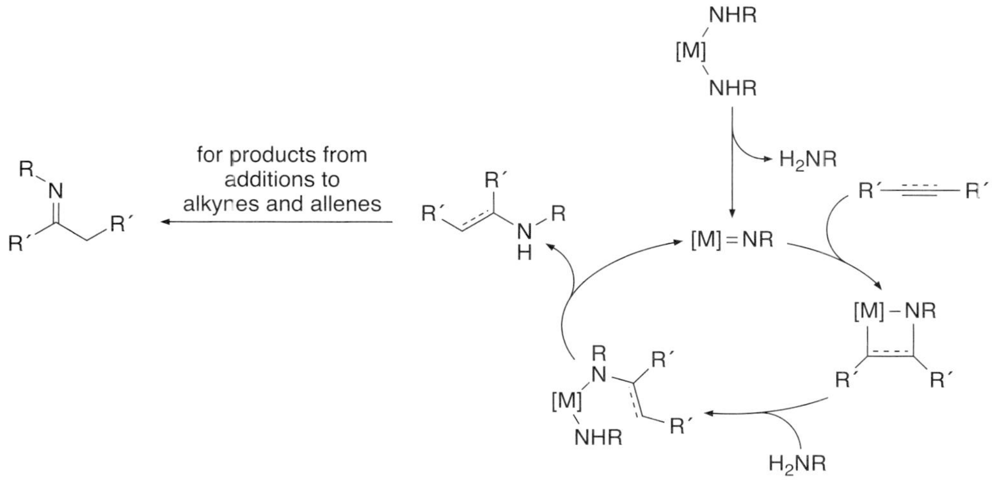

Scheme 16.20

shown, or a methyl amido complex. The former reversibly extrudes amine and the later irreversibly extrudes methane to generate a metal–imido complex. This imido complex then undergoes [2+2] addition of the alkyne, allene, or alkene to form an azametallacyclobutene or azametallacyclobutane. Amine protonates these azametallacycles to form the enamido amido complex. A second proton transfer from the amido group to the enamido ligand generates the enamine product of the hydroamination and the starting bis(imido) complex. The enamine usually tautomerizes to the imine.

Examples of the [2+2] cycloadditions and the mechanisms of these processes were presented in detail in Chapter 13 on complexes containing metal–ligand multiple bonds. In short, coordination of the alkyne or allene precedes the [2+2] cycloaddition.[337] This cycloaddition is thermodynamically favorable for alkynes and allenes, but is thermodynamically disfavorable for reactions of alkenes.[338] Studies on the regioselectivity of the stoichiometric [2+2] cycloaddition and of the regioselectivity of zirconocene-catalyzed hydroamination revealed that the [2+2] process is reversible during the hydroaminations catalyzed by zirconocene complexes.[306] Moreover, it has been shown that addition of an alkyne to an isolated zirconocene azametallacyclobutene leads to exchange.

Thus, the turnover-limiting step of the hydroamination cycle catalyzed by zirconocene complexes involves the protonolysis of the azametallacycle to form the enamine and the metal–imido complex. The lack of reactions with alkenes must result in some way from the unfavorable thermodynamics for the [2+2] cycloaddition. The mechanisms of the hydroaminations of alkynes and allenes with other neutral group 4 catalysts have all been thought to occur analogously, although detailed studies have not been conducted. It should be noted, however, that detailed studies on the catalytic hydroamination of alkynes by titanocene complexes showed that these reactions actually occur through monocyclopentadienyl complexes. Reaction of the metallocene complex with amine leads to protonolysis of one of the Cp ligands and formation of an active catalyst containing an ancillary amido group.

16.6. Oxidative Functionalization of Olefins[339]

16.6.1. Overview

The oxidative functionalization of olefins to generate aldehydes, ketones, vinyl ethers, vinyl acetates, allylic acetates, and allylic ethers, as well as enamides, allylic amines, and products containing new C–C bonds has been studied extensively in both industrial and academic laboratories.[340–344] Oxidations of olefins is a topic with a broad scope, and the oxidation of olefins through organometallic intermediates is a subset of these reactions. This section of this chapter focuses on the oxidations of olefins catalyzed by palladium complexes through organometallic intermediates. As discussed in Sections 16.6.2.2 and 16.6.4 on the mechanisms of organometallic oxidations of alkenes, these reactions are likely to involve nucleophilic attack on coordinated ligands (Chapter 11) and insertions of olefins into M–X bonds (Chapter 9), depending on the reaction conditions.

$$ \text{==} \ + \ \tfrac{1}{2} O_2 \ \xrightarrow[\text{and PdCl}_2]{\text{Catalytic CuCl}_2} \ \underset{H}{\overset{O}{\parallel}} \qquad\qquad (16.95) $$

$$ C_2H_4 \ + \ H_2O \ + \ PdCl_2 \ \longrightarrow \ \underset{H}{\overset{O}{\parallel}} \ + \ Pd(0) \ + \ 2\,HCl \qquad (16.96) $$

For many years a majority of the world's supply of acetaldehyde has been produced by palladium-catalyzed oxidations of ethylene with a copper cocatalyst through organometallic intermediates (Equation 16.95).[345, 346] This process was based on the stoichiometric reaction of palladium(II) with ethylene in water to form acetaldehyde, observed initially by Phillips in the 1890s (Equation 16.96).[344] The catalytic version of this process with palladium(II) and copper(II) was initially presented by the Consortium für Elektrochemische Industrie GmbH and developed into a technical process by Wacker Chemie.[340,345,347] This reaction is now commonly called the "Wacker process." At a similar time Moseev reported the oxidation of ethylene with acetic acid to form vinyl acetate.[348] This chemistry spawned efforts to develop related palladium-catalyzed oxidations, including the oxidations of dienes, vinylarenes, and α-olefins with a variety of oxygen and nitrogen nucleophiles. These reactions have proven to be valuable in the synthesis of commodity chemicals, fine chemicals, and natural products. This section of this chapter focuses on the most extensively developed reactions with heteroatom nucleophiles. Note, though, that oxidative functionalizations of olefins with carbon nucleophiles are also beginning to be developed,[349,350] as are oxidative difunctionalizations of olefins.[351–355] A majority of the principles and mechanisms that apply to the additions of heteroatom nucleophiles apply to these new C–C and C–X bond-forming processes.

16.6.2. The Wacker Process

16.6.2.1. Description of the Process

The oxidation of ethylene to acetaldehyde catalyzed by a combination of palladium and copper is called the Wacker process (Equation 16.96). This process and related reactions that form vinyl acetate from ethylene, acetic acid, and oxygen were discovered and disclosed in the late 1950s and early 1960s.[340,348,356,357] The Wacker process remains the major industrial route to acetaldehyde[345,346,358] because of the low cost of the ethylene feedstock.[346,358,359] These reactions were first studied as stoichiometric oxidations of olefins by palladium(II) complexes with water to form aldehydes and ketones. These reactions were stoichiometric because they formed palladium(0), rather than regenerating the starting palladium(II). It was the discovery that cupric chloride ($CuCl_2$), in the presence of chloride ions, mediates the efficient re-oxidation of palladium(0) to palladium(II) that caused this process to be catalytic in both palladium and copper.[358] The three reactions shown in Scheme 16.21 constitute the overall oxidation of ethylene to acetaldehyde.

While the industrial Wacker reaction is run with $CuCl_2$ as a catalyst for re-oxidation of the Pd(0) by O_2, other oxidants have been used in the laboratory without the need for copper cocatalysts. These oxidants include benzoquinone, heteropolyphosphates, many metal ions, and oxygen alone. The rate of the oxidation of ethylene (and other olefins) is

$$H_2C=CH_2 + PdCl_2 + H_2O \longrightarrow CH_3CHO + Pd(0) + 2HCl$$

$$Pd(0) + 2\,CuCl_2 \longrightarrow PdCl_2 + 2\,CuCl$$

$$2\,CuCl + 1/2\,O_2 + 2\,HCl \longrightarrow 2\,CuCl_2 + H_2O$$

$$\text{Overall:}\ \ H_2C=CH_2 + 1/2\,O_2 \xrightarrow[\text{PdCl}_2]{\text{CuCl}_2} CH_3-\overset{\overset{\displaystyle O}{\|}}{C}-H$$

Scheme 16.21

unaffected by the presence of benzoquinone or copper (although there are indirect effects if the presence of the re-oxidant changes the concentration of acid or Cl⁻ ions).[358,360] There are, however, reasons to believe that the oxidation of the olefin and the re-oxidation of palladium are not completely independent for reactions of higher alkenes. Palladium–copper clusters have been isolated from such reactions,[361] and Murahashi has suggested that they remain intact during olefin oxidation.[362]

16.6.2.2. Mechanism of the Wacker Process (Written with Prof. Jack R. Norton)

Many types of palladium-catalyzed oxidative functionalizations of olefins related to the Wacker process have been developed, and these reactions are presented later in this chapter. To understand the relationship between these reactions and the basic Wacker oxidation of ethylene to form acetaldehye, the mechanism of the Wacker process is discussed before the related oxidation processes.

The mechanism of the Wacker oxidation has been the subject of many mechanistic studies and much discussion for nearly 50 years.[348,363–367] At this point, the identity of the elementary steps of this process appears to depend on the reaction conditions. The majority of the mechanistic discussion has focused on whether the C–O bond is formed by nucleophilic attack of water on a coordinated olefin or by insertion of an olefin into a metal–hydroxo complex. These elementary reactions were discussed in Chapters 11 and 9, respectively. It appears that the mechanism involving nucleophilic attack occurs under conditions of high chloride concentration, and the mechanism involving olefin insertion occurs under conditions of low chloride concentration.[368]

Established mechanistic data have shown that the rate law for the stoichiometric reaction obeys the expression in Equation 16.97.[360] These data show that olefin reversibly replaces one of the chlorides in the starting palladium chloride and that water reversibly replaces a second. This equilibrium is shown in Equation 16.98.

$$\frac{d[CH_3CHO]}{dt} = -\frac{d[C_2H_4]}{dt} = \frac{[PdCl_4{}^{2-}][C_2H_4]}{[Cl^-]^2[H^+]} \tag{16.97}$$

$$[PdCl_4]^{2-} + C_2H_4 + H_2O \rightleftharpoons (C_2H_4)(H_2O)PdCl_2 + 2\,Cl^- \tag{16.98}$$

Three mechanisms that are consistent with this rate law have been proposed and are shown in Scheme 16.22. All three mechanisms generate a hydroxyethyl intermediate, but the three differ in the mechanism by which this intermediate is formed. The first mechanism involves formation of a palladium hydroxide, accompanied by isomerization of the trans isomer to the cis isomer.[346,360,369] The isomerization occurs via aquation of a chloride ligand and the formation of cis-[(C₂H₄)Pd(OH)₂Cl]⁻ as an intermediate, followed by insertion of olefin into the metal–hydroxide bond. The second involves rate-determining external attack of hydroxide on the coordinated olefin, and the third involves external attack of water on the coordinated olefin, followed by proton transfer and rate-determining dissociation of chloride.

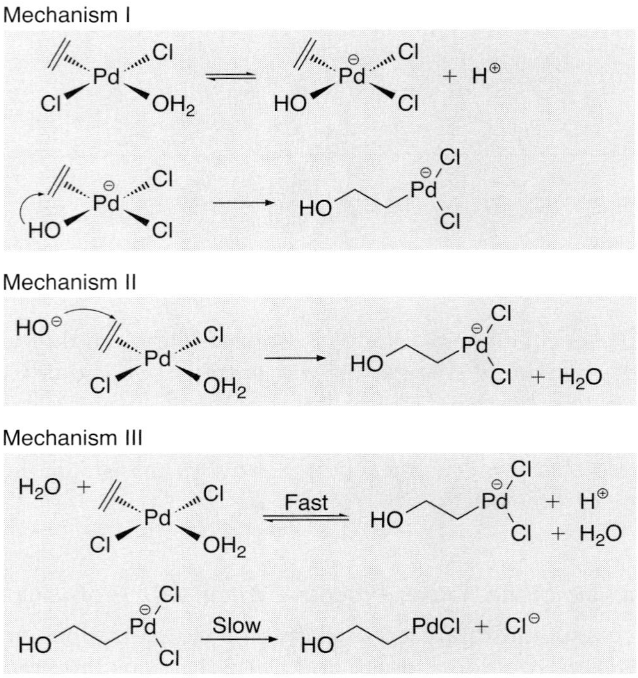

Mechanism I

Mechanism II

Mechanism III

Scheme 16.22

One classic mechanistic experiment showed that all four hydrogens in the acetaldehyde product were contained in the starting ethylene (Equation 16.99); none come from solvent water.[360,370] Thus, the formation of acetaldehyde from the hydroxyethyl species does not occur by formation of free vinyl alcohol that tautomerizes to the aldehyde; this tautomerization would lead to incorporation of deuterium from water into the final product. Instead, the 2-hydroxyethyl ligand in **A** of Scheme 16.23 must isomerize to the 1-hydroxyethyl ligand of **B**, either in two steps via a *coordinated* vinyl alcohol or in a single step. Both possibilities are shown in Scheme 16.23 for reactions in D_2O. β-Hydrogen elimination from the oxygen in **B** or loss of a proton to the solution then returns the hydroxyl proton to the solvent and liberates acetaldehyde.

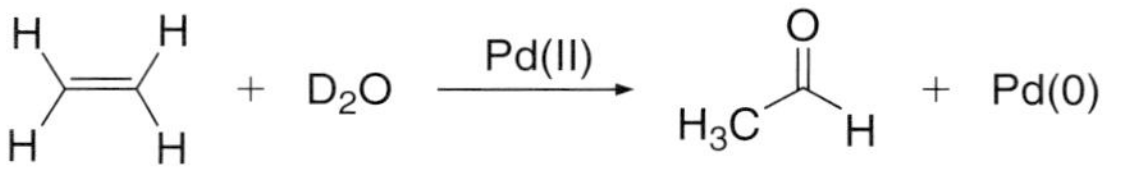

$$(16.99)$$

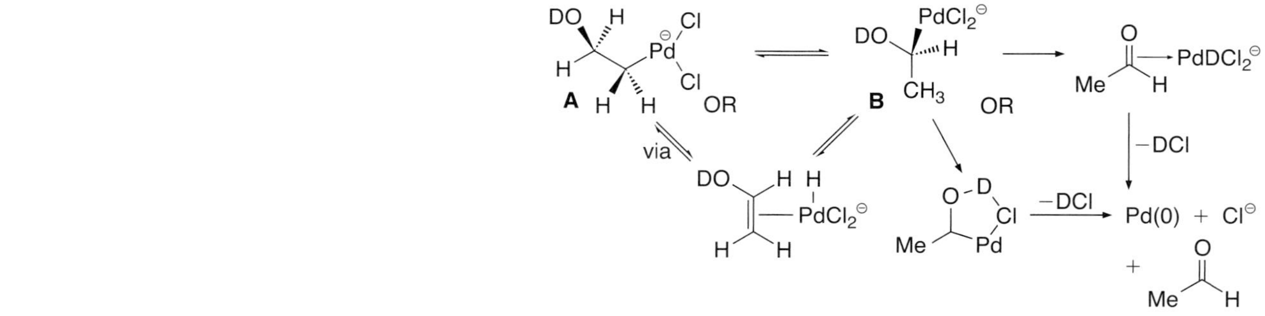

Scheme 16.23

To investigate the stereochemistry of the formation of the C–O bond in the hydroxyethyl intermediate of the stoichiometric Wacker reaction, Åkermark conducted the reaction of *trans*-ethylene-d_2 in the presence of CuCl$_2$ and added Cl⁻. The added chloride leads to the formation of ethylene chlorohydrin.[371,372] In a similar vein, Stille studied the reaction of *cis*-ethylene-d_2 in the presence of CO to form β-propiolactone (Equation 16.100).[373] Both experiments formed products from anti addition of the Pd and O across the double bond. This result is consistent with reaction by either Mechanism **II** or **III**, but not with reaction by Mechanism **I**.

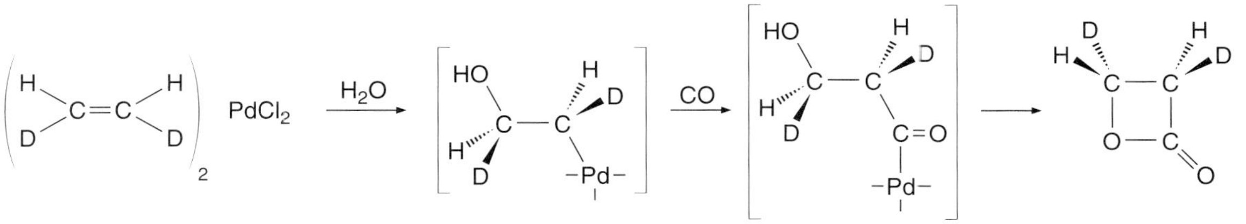

(16.100)

However, a series of papers by Henry[374–376] have established that Cl⁻ and CuCl$_2$ *do not trap the same intermediate that leads to the formation of acetaldehyde*. In the presence of high concentrations of chloride, the hydroxyalkyl intermediate contains chloride instead of hydroxide within the coordination sphere of the metal, and the hydroxyalkyl intermediate is stabilized toward β-hydrogen elimination. The formation of ethylene chlorohydrin is favored at high [Cl⁻] because its rate law is inverse first order in Cl⁻ (Equation 16.101), whereas oxidation to form aldehyde is inverse second order in Cl⁻.

$$\frac{d[\text{ClCH}_2\text{CH}_2\text{OH}]}{dt} = -\frac{d[\text{C}_2\text{H}_4]}{dt} = \frac{[\text{PdCl}_4^{2-}][\text{C}_2\text{H}_4]}{[\text{Cl}^-][\text{H}^+]} \qquad (16.101)$$

A stereochemical study reported by Henry illustrated that the formation of aldehyde and formation of chlorohydrin occur with different stereochemistry, and this result implies that one process occurs by syn addition and one by anti addition of water and palladium across the olefin.[375] This study is summarized in Scheme 16.24. Oxidation of the non-racemic, chiral allyl alcohol in the absence of added chloride forms the (*R*)-(*E*)-alcohol, whereas reaction of the allyl alcohol in the presence of added chloride forms the product with stereochemistry resulting from the opposite mode of attack. Because it is known that the allylic alcohol binds to palladium with hydrogen bonding between the hydroxyl group and the bound chloride, Henry concluded that the reaction conducted in the presence of high concentrations of added chloride occurs by external attack of the oxygen nucleophile, while the reaction with low concentrations of added chloride occurs by insertion of the olefin into a Pd–O bond.[375]

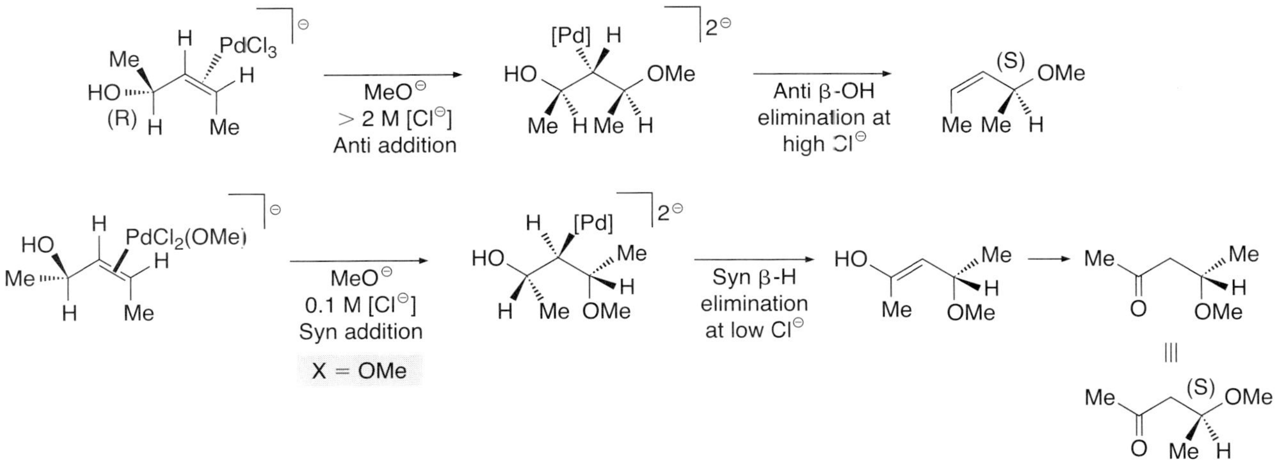

Scheme 16.24

16.6.2.3. Olefin Oxidations Related to the Wacker Process

The oxidation of higher olefins has also been studied, and these reactions form ketones (Equation 16.102). Thus, the C–O bond formation between water and the substituted olefin mediated by palladium occurs at the internal carbon. For example, palladium-catalyzed oxidation of propene forms acetone, and this reaction provides one industrial route to this material. Oxidations of substituted olefins to form ketones have also become a common method for the conversion of olefins to ketones during complex-molecule synthesis. Examples of the use of palladium-catalyzed oxidation in complex-molecule synthesis are described later in this chapter.

$$R\diagup\diagdown + \tfrac{1}{2}O_2 \xrightarrow{\text{Pd(II), CuCl}_2} R\overset{O}{\diagup\diagdown}CH_3 \qquad (16.102)$$

The oxidations of olefins with many oxygen nucleophiles other than water have also been reported. These reactions include the synthesis of vinylic and allylic ethers from reactions of olefins with alcohols and phenols, and vinylic and allylic esters from reactions of olefins with carboxylic acids. These reactions have been conducted with both monoenes and 1,3-dienes. Both intermolecular and intramolecular versions of each of these processes have been developed. Some discussion of these reactions was included in Chapter 11 because of their connection to the nucleophilic attack of oxygen nucleophiles on coordinated olefins and dienes.

16.6.2.3.1. Intermolecular Additions of Alcohols and Carboxylates

The intermolecular oxidations of olefins with alcohols as nucleophile typically generate ketals, whereas the palladium-catalyzed oxidations of olefins with carboxylic acids as nucleophile generates vinylic or allylic carboxylates.[377] As a result, many of the oxidations with alcohols have been conducted with diols to generate stable cyclic acetal products. Both types of oxidations have been conducted on large industrial scale,[378] and vinyl acetate is produced from the oxidative reaction of ethylene with acetic acid[348] in the gas phase over a supported palladium catalyst.[378]

This divergent oxidative reactivity of alcohols and carboxylic acids with olefins is illustrated in Equation 16.103. Both reactions generate products from β-hydrogen elimination of an alkoxyalkyl or acetoxyalkyl complex. However, the vinyl ether product generated by β-hydrogen elimination undergoes reaction with a second equivalent of alcohol to form the acetal in a process catalyzed by the acidic medium or the action of palladium(II) as Lewis acid.

$$ (16.103) $$

Simple examples of the reactions of olefins with diols are shown in Equations 16.104 and 16.105. Reactions of alkenes typically generate ketals, whereas reactions of olefins bearing electron-withdrawing groups, such as those in acrylates or acrylonitrile, tend to form acetals. This regioselectivity is shown by the reactions of butene and

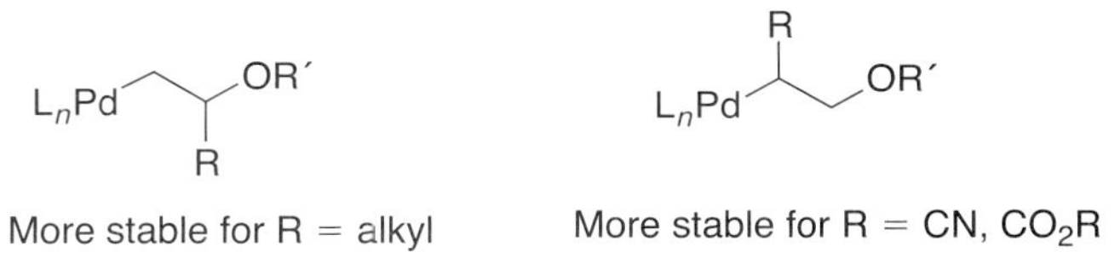

Figure 16.3.
Intermediates that control the selectivity of the Wacker-type oxidations of olefins.

acrylonitrile in Equations 16.104 and 16.105.[379] This difference in regioselectivity reflects the difference in the stability of the alkoxyalkyl intermediates shown in Figure 16.3. Although there are exceptions, an alkoxyalkyl complex containing an electron-withdrawing group tends to be more stable when this group is present in the α-position than when it is present in the β-position, but an alkoxyalkyl complex containing an electron-dontaining alkyl substituent tends to be more stable when the alkyl substituent is located in the β-position than when it is located in the α-position. One exception to the formation of acetal or ketal products is shown in Equation 16.106. In this case, the product from 1,2-addition across the alkene occurs.[355,380]

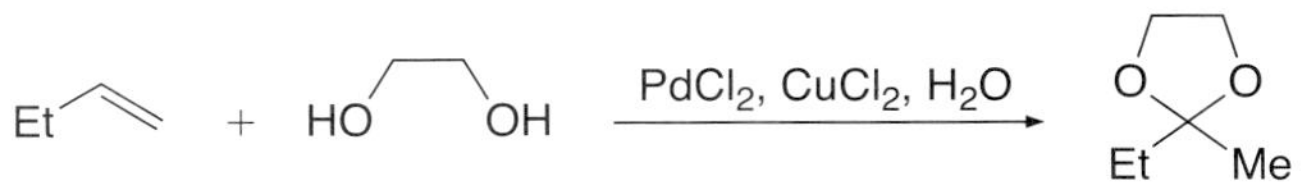

(16.104)

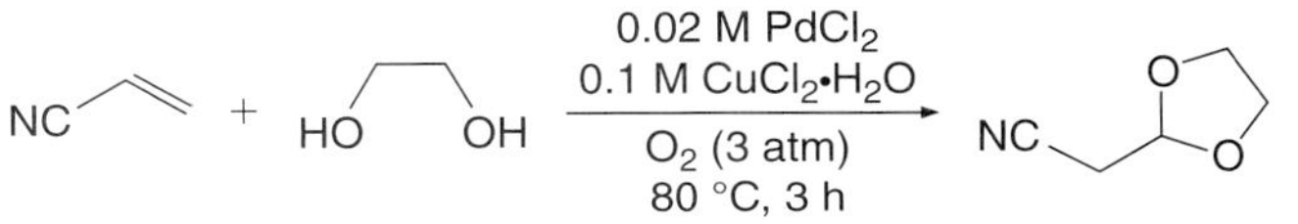

(16.105)

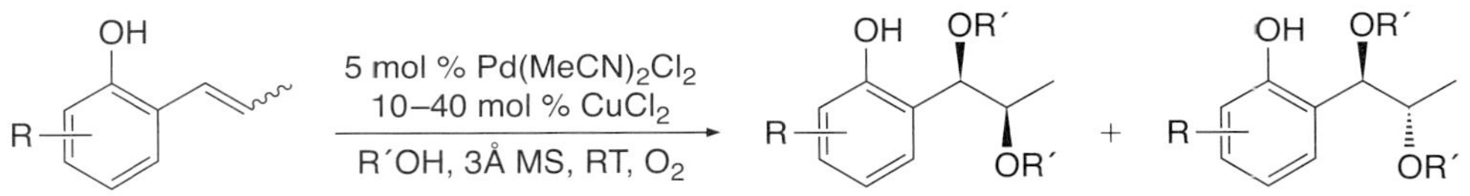

(16.106)

Examples of the oxidative reactions of olefins with carboxylic acids are shown in Equations 16.107–16.109. These examples illustrate the selectivities of the oxidations of ethylene, acylic alkenes, and cyclic alkenes. The reactions of alkenes with carboxylic acids generate either vinylic esters or allylic esters.

As noted above, the oxidative reaction of ethylene with acetic acid forms vinyl acetate (Equation 16.107), and this reaction has been conducted on industrial settings in both the liquid phase and in the gas phase with a supported palladium catalyst.[378] Allyl alcohol is also produced by palladium-catalyzed oxidations. The reaction of propene with acetic acid in the presence of a combination of palladium and copper forms allyl acetate (Equation 16.108), which is hydrolyzed to the alcohol. Cyclic olefins also undergo allylic oxidation, in this case to generate allylic esters. A variety of combinations of palladium acetate and co-oxidants and additives give rise to this allylic ester in good to excellent yields. Equation 16.109 summarizes results for the reactions of cyclohexene.[381–391]

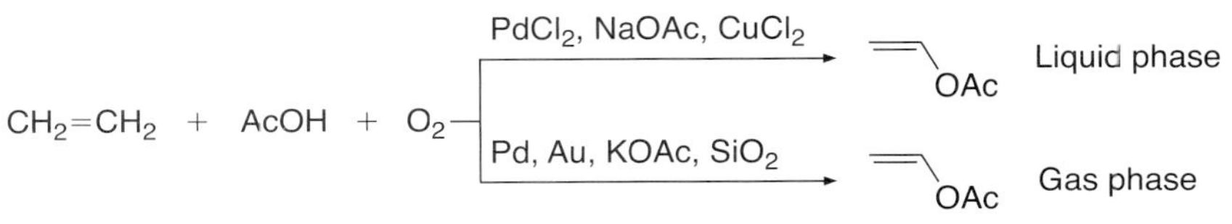

(16.107)

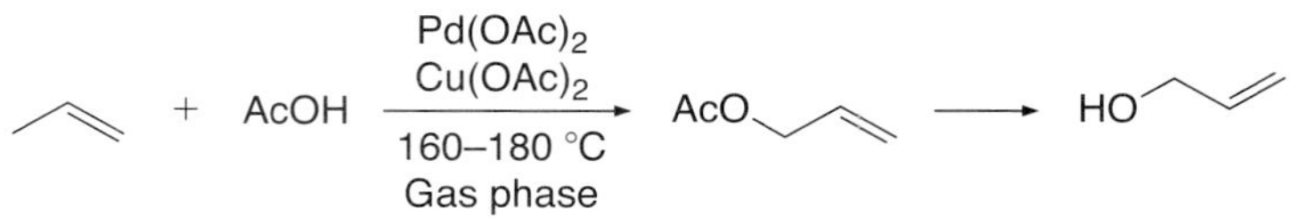

$$\text{\raisebox{0pt}{$\diagdown\!\!\diagup$}} + AcOH \xrightarrow[\substack{160-180\ ^\circ C \\ \text{Gas phase}}]{\substack{Pd(OAc)_2 \\ Cu(OAc)_2}} AcO\diagup\!\!\diagdown\!\!\diagup \longrightarrow HO\diagup\!\!\diagdown\!\!\diagup \qquad (16.108)$$

$$\text{(cyclohexene)} + HOAc \xrightarrow[\substack{\text{Co-oxidant} \\ \text{Salt additive}}]{\substack{0.2-5\% \\ Pd(OAc)_2}} \underset{75-100\%}{\text{(cyclohexenyl-OAc)}} \qquad (16.109)$$

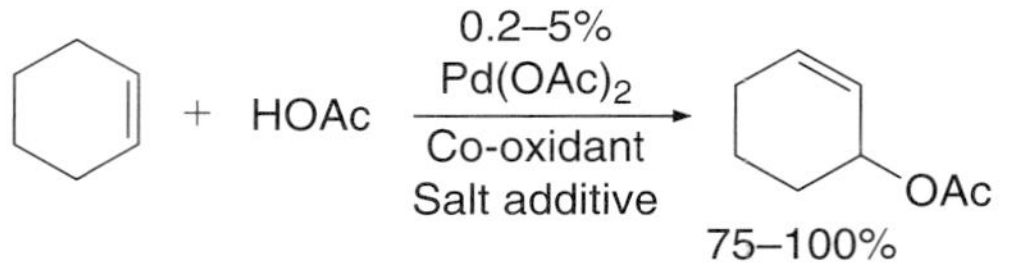

Co-oxidant (mol %)	Salt additive
MnO$_2$ (110)−benzoquinone (20)	
O$_2$−Fe(Pc) (5)−hydroquinone (20)	LiOAc
O$_2$−Co(TPP) (0.5)−hydroquinone (20)	LiOAc
O$_2$−Co(salophen) (5)−hydroquinone (20)	LiOAc
O$_2$−Cu(OAc)$_2$ (5)−hydroquinone (10)	
O$_2$−Cu(OAc)$_2$ (5)−Co(salen) (5)	
O$_2$−Fe(NO$_3$)$_3$ (5)	
O$_2$−hydroquinone (20)-NPMoV (25 mg)	Ac$_2$O
t-BuOOH (150)−benzoquinone (20)	Na$_2$CO$_3$
H$_2$O$_2$ (70%, 1.1 equiv.)−benzoquinone (20)	
Benzoquinone (100)−o-methoxyacetophenone (20)	

Palladium-catalyzed intermolecular oxidations of dienes with carboxylic acids and alcohols as donors give 1,4-addition products. This chemistry has been studied extensively by Bäckvall.[392,393] Early studies involved 1,4-additions of two acetoxy or alkoxy groups across a diene.[394] More recently, intermolecular additions of two different nucleophiles have been developed.[395] The ability to control the stereochemistry of the additions across cyclic dienes makes this procedure particularly valuable. As shown in Scheme 16.25, conditions for either cis or trans additions have been developed.[396] Reactions conducted in the absence of added chloride form products from trans 1,4-addition, while reactions conducted in the presence of added chloride form products from cis 1,4-addition.

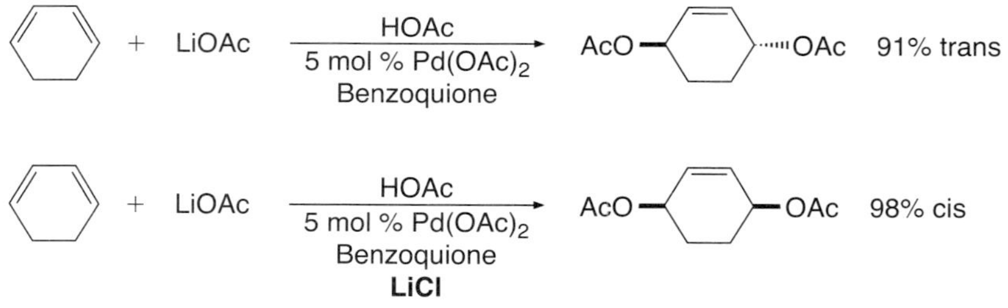

$$\text{(cyclohexadiene)} + LiOAc \xrightarrow[\substack{5 \text{ mol \% Pd(OAc)_2} \\ \text{Benzoquione}}]{HOAc} AcO\text{—(cyclohexene)—}{}^{\text{\tiny''''}}OAc \quad 91\% \text{ trans}$$

$$\text{(cyclohexadiene)} + LiOAc \xrightarrow[\substack{5 \text{ mol \% Pd(OAc)_2} \\ \text{Benzoquione} \\ \mathbf{LiCl}}]{HOAc} AcO\text{—(cyclohexene)—}OAc \quad 98\% \text{ cis}$$

Scheme 16.25

As noted in Chapter 11, the cis products are thought to form by external attack of acetate on a coordinated diene, followed by external attack of the second carboxylate on the intermediate allyl species. The trans products are thought to form by external attack of one nucleophile to a coordinated diene and subsequent internal migration of the second nucleophile from the metal to the resulting allyl intermediate.[396] These reaction mechanisms, including the coordination by benzoquinone to the intermediate to induce elimination, are shown in Scheme 16.26.

16.6.2.3.2. Intramolecular Additions of Alcohols and Carboxylates

Intramolecular oxidative reactions of alcohols, phenols, and carboxylic acids have also been studied extensively. These reactions have been exploited for the synthesis of natural

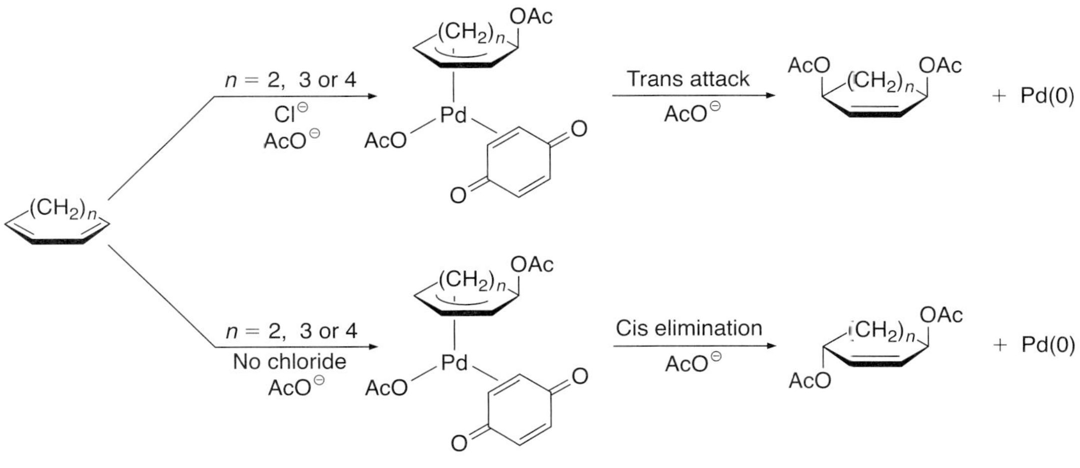
n = 2, 3 or 4
Cl⊖
AcO⊖
Pd
AcO
OAc
Trans attack
AcO⊖
+ Pd(0)
n = 2, 3 or 4
No chloride
AcO⊖
Cis elimination
+ Pd(0)

Scheme 16.26

products, in part because regioselectivity can be controlled by the design of the substrate. Murahashi, Bäckvall, and Larock reported some of the earliest examples of cyclizations of particular classes of substrates.

The cyclization of *ortho*-allyl phenols was reported by Murahashi in the late 1970s.[397–399] The reaction of the 2-(2-cyclohexenyl)phenol (Equation 16.110) was one of the early examples of Wacker-type reactions with alcohol nucleophiles and has been re-investigated in more recent years with chiral catalysts. Intramolecular reactions of alkene-ols and alkenoic acids form cyclic ethers and lactones. These reactions were reported by Larock and by Annby, Andersson, and co-workers, and examples are shown in Equations 16.111 and 16.112.[400,401] The use of DMSO as solvent was important to form the lactone products. More recently, reactions with alcohols were reported by Stoltz to form cyclic ethers by the use of pyridine and related ligands in toluene solvent.[402,403] The type of ligand, whether an additive or the solvent, is crucial to the development of these oxidative processes. However, the features of these ligands that lead to catalysis are not well understood at this time.

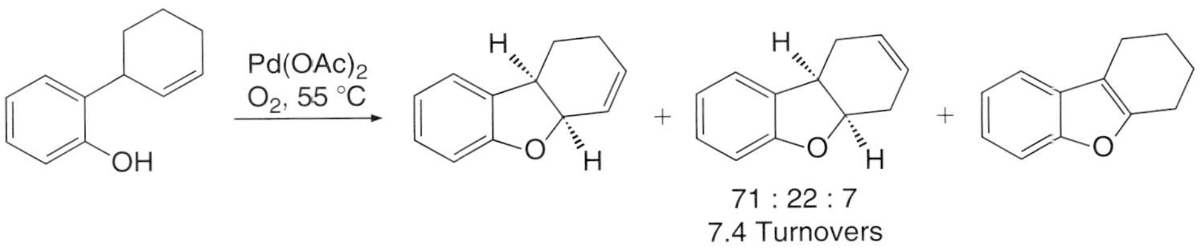
Pd(OAc)2
O2, 55 °C
71 : 22 : 7
7.4 Turnovers

(16.110)

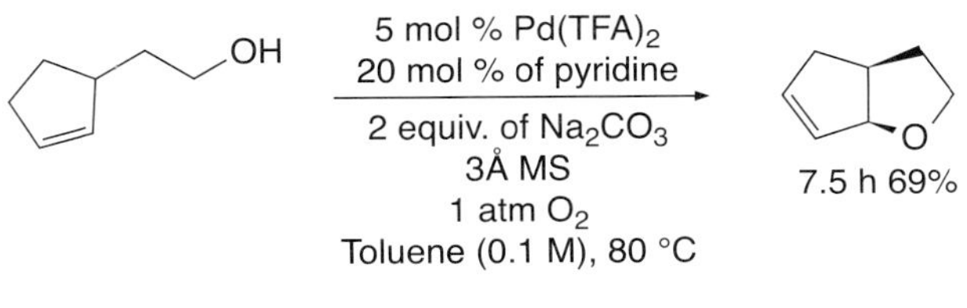
5 mol % Pd(TFA)2
20 mol % of pyridine
2 equiv. of Na2CO3
3Å MS
1 atm O2
Toluene (0.1 M), 80 °C
7.5 h 69%

(16.111)

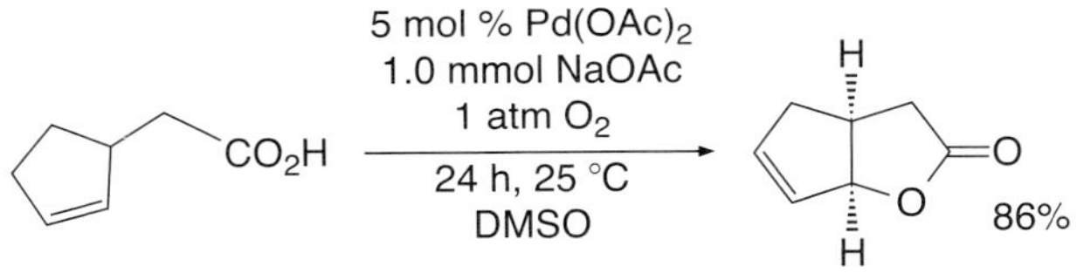
5 mol % Pd(OAc)2
1.0 mmol NaOAc
1 atm O2
24 h, 25 °C
DMSO
86%

(16.112)

In parallel with the development of these intramolecular palladium-catalyzed oxidations of alkenes with oxygen donors, the intramolecular oxidation of dienes to form 1,4-addition products was developed. Two examples of these reactions from the same starting diene-ol are shown in Equation 16.113,[404] and an example of the intramolecular diacetoxylation of a diene is shown in Equation 16.114.[405] As was observed for the intermolecular diacetoxylation of dienes, the relative stereochemistry of the 1,4-substituents depended on the reaction conditions. In the absence of chloride, the trans product was observed, and in the presence of chloride the cis product was observed.

$$(16.113)$$

$$(16.114)$$

16.6.2.3.3. Wacker-Type Oxidations in Natural Products Synthesis

Wacker-type oxidations have not only been used to prepare commodity chemicals, but have been used in complex molecule synthesis. Examples of the use of three classes of the oxidative functionalizations of olefins in natural products synthesis are shown in Schemes 16.27–16.30.

In one case, a Wacker oxidation of an olefin to form a ketone was used to prepare fragments for the synthesis of Calyculins.[406,407] The ketone was then converted via the vinyl triflate to the vinyl bromide. In a second example, Wacker oxidation was used in the synthesis of a taxol derivative.[408] In this case, the Wacker oxidation of an olefin was conducted at the later stage of a synthesis. Oxidation of a pendant olefin formed a ketone, which cyclized by condensation with a second keto functionality.

The third and fourth examples involve intramolecular oxidation of olefins with a protected alcohol and with a carboxylic acid. Cleavage of a silyl protected alcohol, followed by oxidative C–O bond formation at the β-position of an acrylate group was conducted at the late stage of the synthesis of the alkaloid alstophylline.[409] Intramolecular 1,4-oxidations of dienes have been used in the synthesis of a number of natural products. For example, the intramolecular oxidation shown in Scheme 16.30 led to the precursor to Paeonilactone B.[393,410]

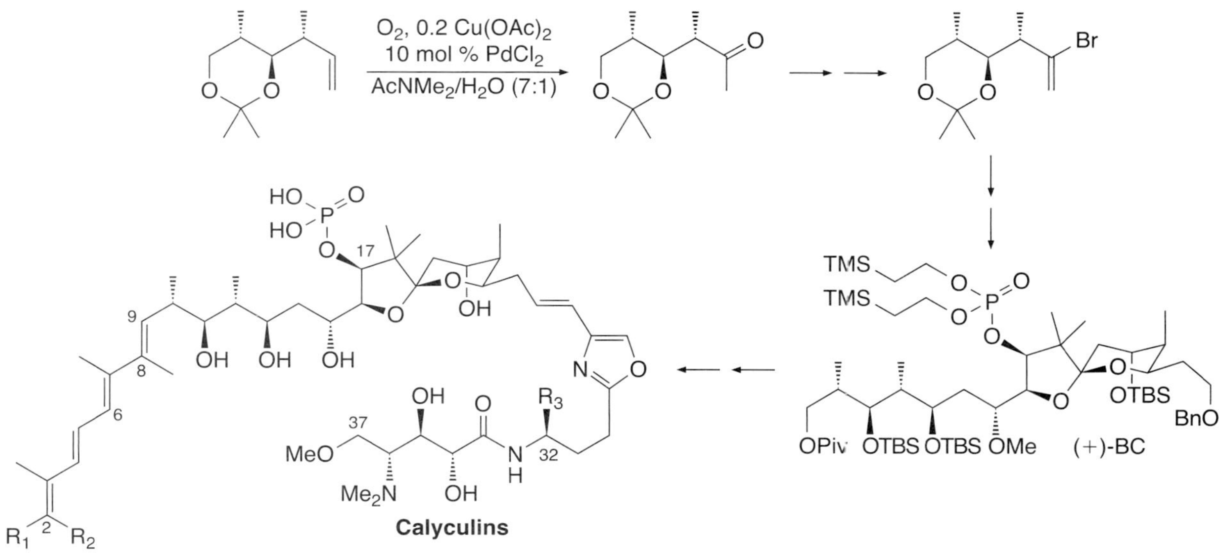

Scheme 16.27

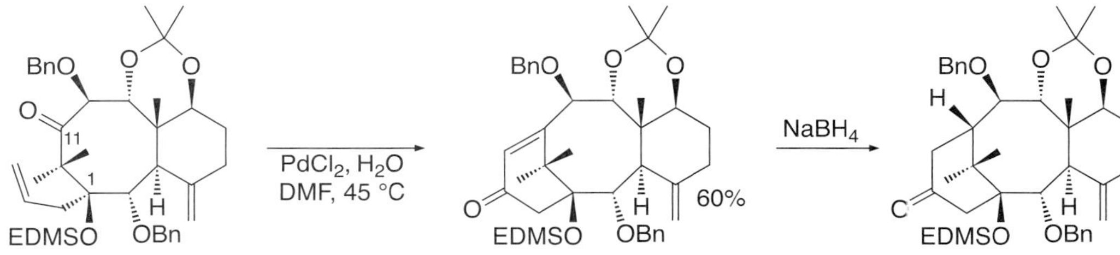

Scheme 16.28

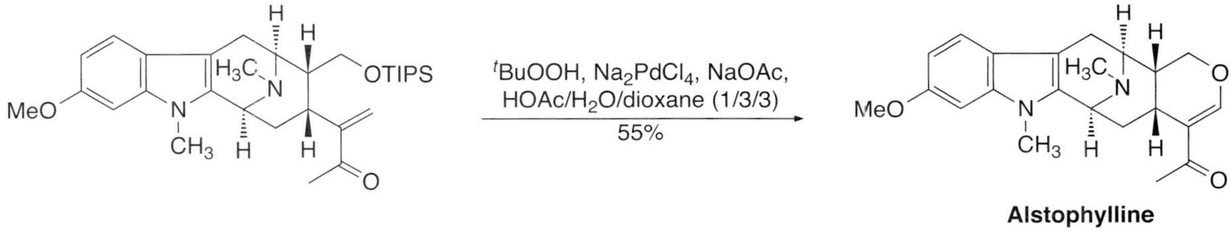

Scheme 16.29

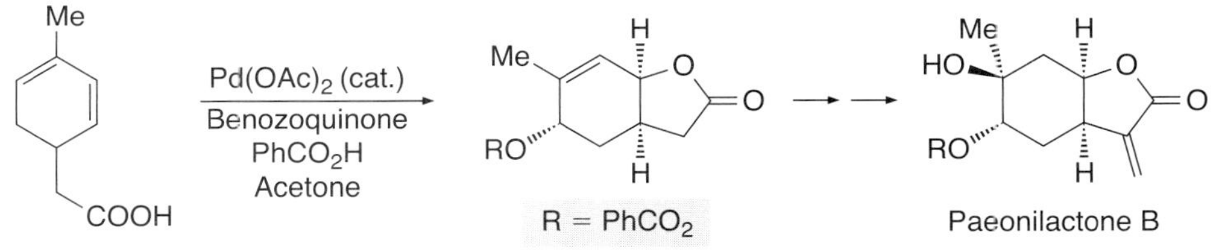

Scheme 16.30

16.6.3. Oxidative Aminations of Olefins

The oxidative aminations of olefins have been reported in parallel with the oxidations of olefins with alcohols, phenols, and carboxylic acids. These reactions are generally conducted with amides or imides; amines are thought to be protonated by the acidic medium or to bind the metal center too tightly to allow for the catalytic chemistry to occur. Some exceptions noted below have been observed with arylamines or with rhodium catalysts. These oxidations have been conducted both intermolecularly and intramolecularly. The oxidations have been conducted with benzoquinone, with copper, or with O_2 alone as the oxidant.

16.6.3.1. Intermolecular Oxidative Aminations

Some of the earliest oxidative aminations of olefins were reactions of lactams with electron-poor olefins reported by Murahashi. Examples of these reactions conducted with the combination of O_2 and CuCl with HMPA as additive are shown in Equation 16.115.[411] This type of oxidative amination of olefins can also be conducted with N-alkyl arylamines, as demonstrated by Hegedus (Equation 16.116).[412] Like the oxidations of acrylates with alcohols, these reactions occur at the β-carbon of the acrylate.

$$\text{EWG} \diagup + \text{HN}\underset{O}{\overset{}{\bigcirc}} \xrightarrow[\text{O}_2, \text{ HMPA (5\%), DME}]{\overset{\text{PdCl}_2(\text{MeCN})_2 \text{ (5\%)}}{\text{CuCl (5\%)}}} \text{EWG}\diagdown\diagup\text{N}\underset{O}{\overset{}{\bigcirc}} \qquad (16.115)$$

EWG = CO_2Me, C(O)Me, CHO, C(O)NEt$_2$

60–93%

$$\underset{\text{NHMe}}{\bigcirc} + \diagup\text{Z} \xrightarrow[\substack{\text{Benzoquinone (1 equiv.)} \\ \text{LiCl (10 equiv.), THF}}]{10 \text{ mol \% PdCl}_2(\text{MeCN})_2} \underset{\text{N}}{\bigcirc}\overset{\text{Me}}{\diagup}\diagdown\diagup\text{Z} \quad \begin{matrix} Z = CO_2Me & 73\% \\ Z = COMe & 63\% \\ Z = CN & 53\% \end{matrix} \qquad (16.116)$$

More recently, oxidative aminations of vinylarenes with amides and imides have been reported by Stahl.[413,414] In this case, the regioselectivity of the process depends on the presence or absence of a tertiary amine additive. As shown in Equation 16.117, the 1,1-disubstituted olefin is formed in the presence of amine, but a 1,2-disubstitued olefin is formed in the absence of amine.[414] This change in regioselectivity is proposed to arise from the degree of reversibility of the addition of the nucleophile to the coordinated vinylarene in the presence and absence of base. As shown in Scheme 16.31, the kinetic site of nucleophilic attack is proposed to be the internal carbon, but the thermodynamic site of attack is proposed to be the terminal carbon. The thermodynamic site of attack is proposed to be the terminal carbon because attack at this position forms an η^3-benzyl complex. In the presence of base, the kinetic product is rapidly deprotonated, and the 1,1-disubstituted olefin is formed. In the absence of base, the deprotonation step is slow enough that the thermodynamic product of attack is generated.

Oxidative aminations of vinylarenes are not limited to reactions catalyzed by palladium. Beller has reported oxidative aminations of vinylarenes with secondary alkylamines to form products from addition with "anti-Markovnikov" regioselectivity.[283,415,416] As shown in Equation 16.118, the enamine containing a new C–N bond at the terminal carbon is formed by this process. In these reactions, the vinylarene is the oxidant. Thus, two equivalents of vinylarene are consumed, one to form the enamine and one to accept the hydrogen and form ethylarene. The mechanism of this reaction and the role of the second equivalent of vinylarene in the catalytic cycle is poorly understood.

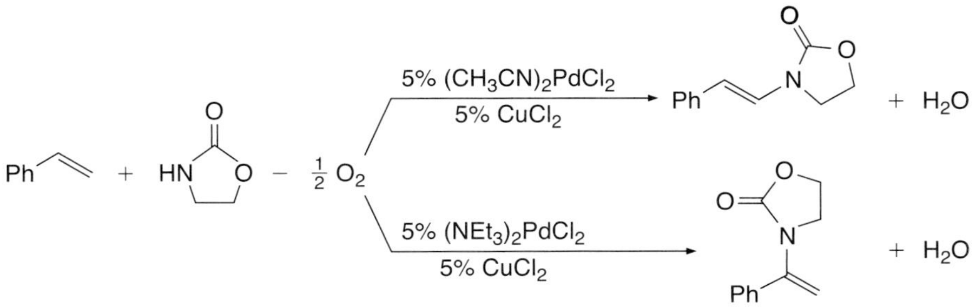

(16.117)

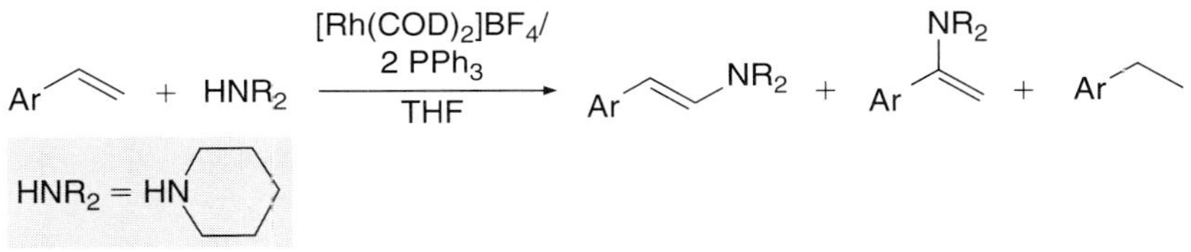

(16.118)

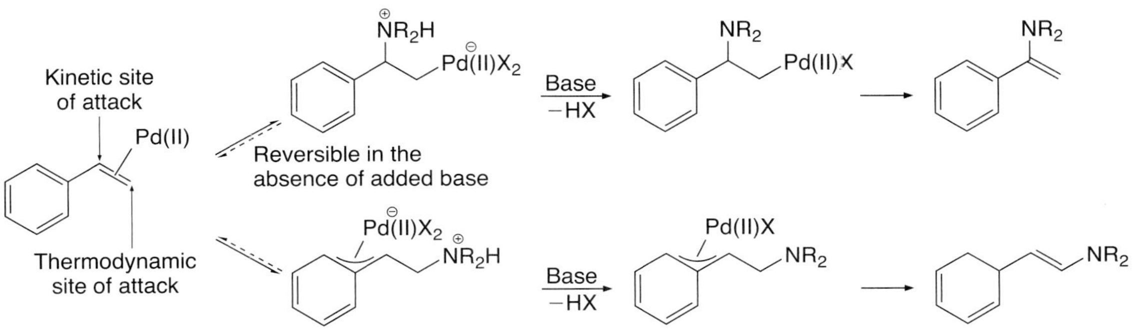

Scheme 16.31

Stahl has also shown that the product from oxidative amination of simple alkenes forms in the presence of a Pd(II) complex and dioxygen without copper.[417] The analogous reactions conducted in the presence of copper led to isomerization of the olefin by an unknown mechanism, and a mixture of products was formed. Thus, the reactions conducted with O_2 as oxidant in the absence of copper are more selective. An example of the reaction with phthalimide as nitrogen donor in the absence of copper is shown in Equation 16.119.

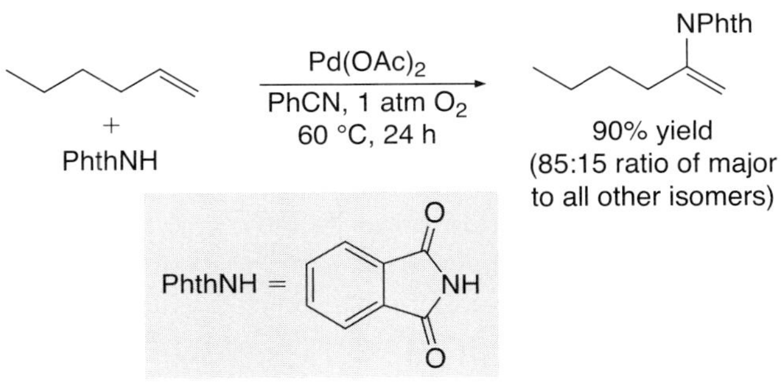

(16.119)

16.6.3.2. Intramolecular Oxidative Amination

Intramolecular oxidative aminations of olefins have also been studied, and many of these intramolecular processes were observed prior to the analogous intermolecular variants. The oxidative aminations of alkenes with arylamines and arylamine derivatives catalyzed by palladium complexes were shown by Hegedus to form indoles (Equation 16.120).[418,419] These reactions were conducted with *ortho*-allylaniline and *ortho*-allylaniline derivatives as substrate, Pd(NCMe)$_2$Cl$_2$ as catalyst, and benzoquinone as oxidant. Intramolecular reactions of *N*-tosylated aliphatic amines were reported by Larock.[420,421] For example, the tosylamide in Equation 16.121 undergoes cyclization in high yield in the presence of dioxygen with Pd(OAc)$_2$ as catalyst in DMSO.[421] A related reaction (Equation 16.122) was reported recently in toluene solvent with added pyridine.[422]

$$\text{(16.120)}$$

R = H	86%
R = Me	89%
R = Ac	71%

$$\text{(16.121)}$$

$$\text{(16.122)}$$

16.6.3.3. Palladium-Catalyzed Difunctionalizations of Olefins

The oxidative difunctionalization of alkenes through organometallic intermediates is less developed than the oxidative processes just discussed. However, several examples of this process catalyzed by palladium have been reported, and difunctionalizations of olefins is being investigated by many researchers at the time of publication of this text. Chlorohydrins have been known for many years to form from Wacker chemistry in the presence of chloride as additive. This product is thought to form by chlorination of the hydroxyalkyl species with a rate that exceeds that of β-hydrogen elimination. This synthesis of chlorohydrins has been developed into an asymmetric process with the dinuclear palladium catalyst in Equation 16.123.[423,424]

Cleavage of the product from initial addition of a nitrogen nucleophile can also lead to difunctionalized products. This cleavage has been conducted in some cases with an oxidant. For example, the carbamate in Equation 16.124 cyclizes in the presence of PhI(OAc)$_2$ to form the acetoxy oxazolidinone.[351] This strategy can also be followed in an intermolecular fashion, as shown in Equation 16.125.[425]

Other strategies have been followed to observe palladium-catalyzed diaminations of olefins. The diaminations of dienes have been conducted with ureas, and the use of this reagent leads to the 1,2-, rather than 1,4-, functionalization of the diene (Equation 16.126).[352] The use of an aziridinone as reagent in the presence of olefin and a palladium catalyst also leads to the diamination of conjugated dienes to form precursors to vicinal diamines (Equation 16.127).[426]

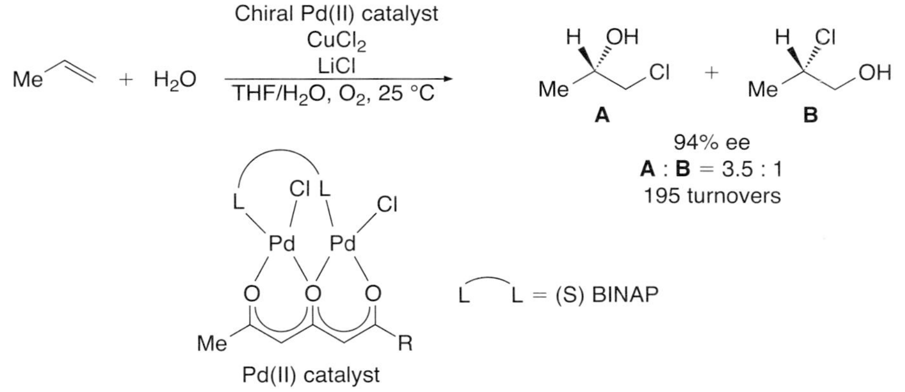
Chiral Pd(II) catalyst
CuCl2
LiCl
THF/H2O, O2, 25 °C
Me
+ H2O
H OH
Cl
A
+
H Cl
OH
B
94% ee
A : B = 3.5 : 1
195 turnovers
Cl L
Cl
Pd Pd
O O O
Me
R
Pd(II) catalyst
L = (S) BINAP

(16.123)

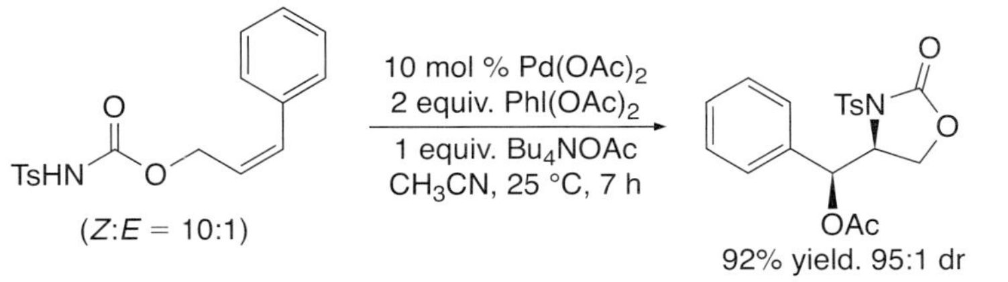
TsHN
O
(Z:E = 10:1)
10 mol % Pd(OAc)2
2 equiv. PhI(OAc)2
1 equiv. Bu4NOAc
CH3CN, 25 °C, 7 h
TsN
O
OAc
92% yield. 95:1 dr

(16.124)

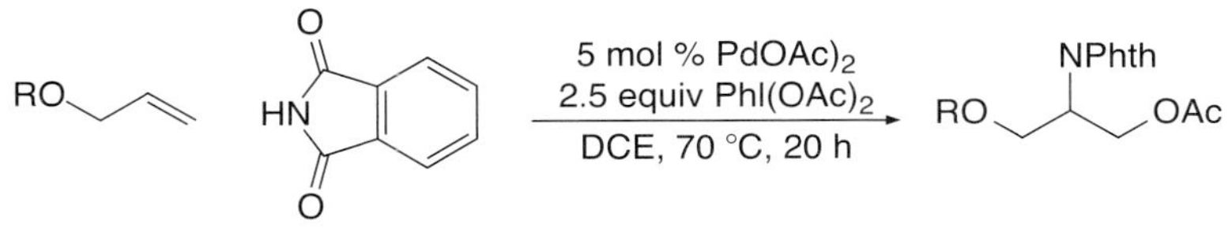
RO
HN
O
O
5 mol % PdOAc)2
2.5 equiv PhI(OAc)2
DCE, 70 °C, 20 h
NPhth
RO
OAc

R =	Yield
^{n}Pr	84
Bn	80
Bz	75
Ac	58

(16.125)

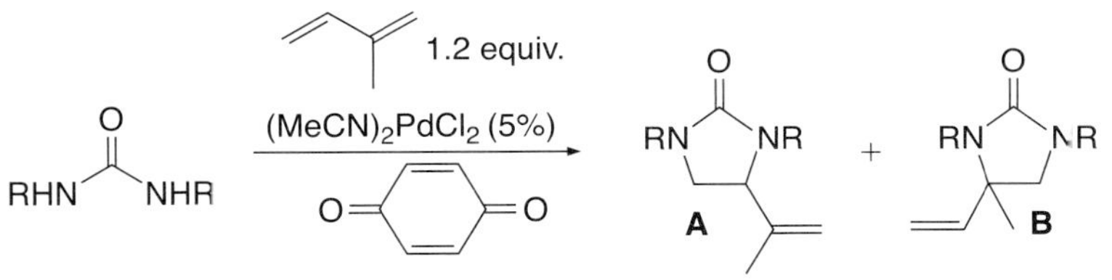
1.2 equiv.
(MeCN)2PdCl2 (5%)
O
RHN
NHR
O
O
RN
NR
A
+
B

R =	Yield (**7A** + **B**)	**A** / **B**
Et	81%	77 / 23
Bu	82%	90 / 10

(16.126)

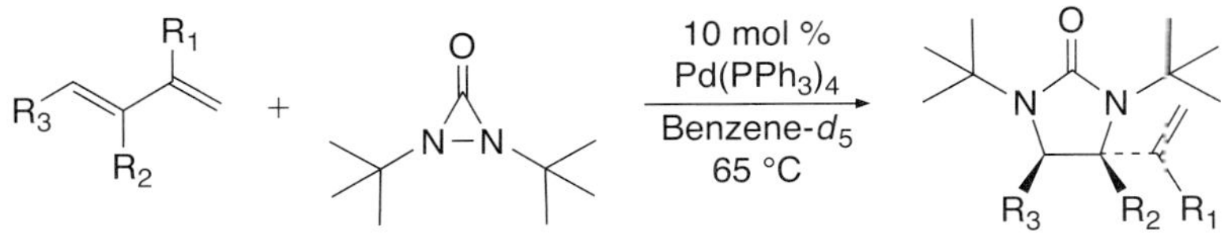
R1
R3
R2
+
O
N–N
10 mol %
Pd(PPh3)4
Benzene-d5
65 °C
O
N
N
R3
R2 R1

(16.127)

16.6.4. Mechanistic Studies on Wacker Oxidations with Alcohol, Phenol, and Amide Nucleophiles

16.6.4.1. Overview

Most of the mechanistic information that has been obtained on the reactions of hydroxide nucleophiles applies to reactions of alkoxide, phenoxide, and nitrogen nucleophiles. Henry has shown that the same stereochemistry is observed for the reactions of alcohols as is observed for the reaction of water.[365,427,403] In addition, studies on the mechanism of the oxidations with phenoxide nucleophiles have shown a similar dependence of the stereochemistry on the concentration of chloride,[366] and studies with amide nucleophiles have shown that the stereochemistry of addition depends on the dative ligand and nitrogen nucleophile.[428] Even the reactions of dienes described earlier in this section of the chapter show a reversal of stereochemistry with chloride additive.[396,404,405] Less studied is the mode of oxidation of the palladium catalyst. The mechanism of the C–X bond-forming step and the re-oxidation of the palladium are described briefly in the following two sections.

16.6.4.2. Mechanism of C–X Bond Formation

The C–O and C–N bond in the product of the oxidations with oxygen and nitrogen donors forms by either nucleophilic attack on the coordinated olefin or by insertion of the olefin into a palladium–oxygen or palladium–nitrogen bond. More detailed descriptions of the nucleophilic attack on coordinated ligands were provided in Chapter 11 and a more detailed description of migratory insertions was provided in Chapter 9. These reactions are discussed in the context of the effect of additives on the stereochemistry of the catalytic processes in several earlier sections on the Wacker process. Henry conducted the same stereochemical study for reactions of alcohols with the resolved allylic alcohol in Scheme 16.24 as was conducted for reactions of water. The results of these experiments were similar to those on the reactions of water.[365]

Other experiments have corroborated these results. It is appropriate to reiterate the results of Hayashi's study on the stereochemistry of the cyclization of the ortho-allyl phenol that was presented in Chapter 9.[366] As shown in Scheme 16.32, the reaction with *trans*-3-deuterio cyclohexenylphenol forms deuterated products, and the formation of these products demonstrates that the C–O bond formation occurs by migratory insertion. As noted in Chapter 9, migratory insertion to form a cis ring juncture places the palladium trans to deuterium, and this disposition of the deuterium and the palladium leads to β-hydrogen elimination instead of β-deuterium elimination.

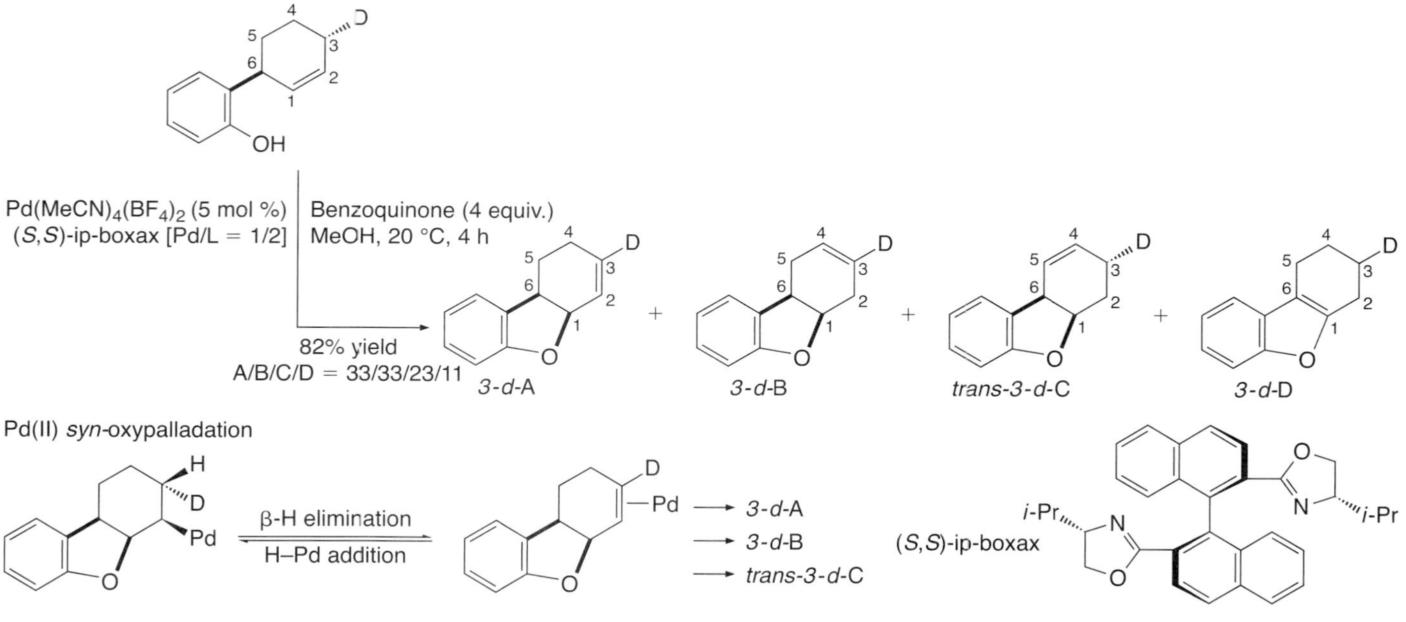

Scheme 16.32

Results that imply that some oxidative amination reactions occur by the insertions of olefins into palladium–nitrogen bonds have been gained. First, the stereochemistry of the oxidative coupling of two norbornenes with a tosylamide implied that the initial step of this reaction occurs by a cis aminopalladation, as shown in Scheme 16.33.[417] In addition, the formation of enamines from acyclic olefins but allylamines from cyclic olefins implies that the C–N bond in these products forms by a cis aminopalladation that places the metal in a position that cannot undergo β-hydrogen elimination to form an enamine product.[417] This stereochemistry is illustrated in Scheme 16.34. A similar formation of allylic esters from the palladium-catalyzed oxidations of cyclohexene in acetic acid could result from the same stereochemical requirements. Finally, more detailed analysis of the stereochemistry of olefin oxidative amination has shown a fine balance between pathways involving cis and trans aminopalladation. Reactions of stereochemically defined olefins show that cis aminopalladation occurs in many cases. However, reactions of olefins with pendant sulfonamides can undergo cyclization by cis or trans aminopalladation depending on the added ligand.[428] A clear understanding of the factors that dictate which pathway is followed awaits further study, but this work clearly shows that the activation energies of these two pathways are similar.

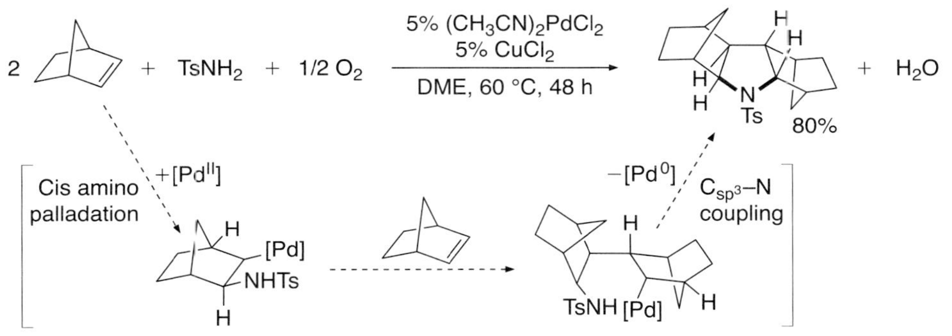

Scheme 16.33

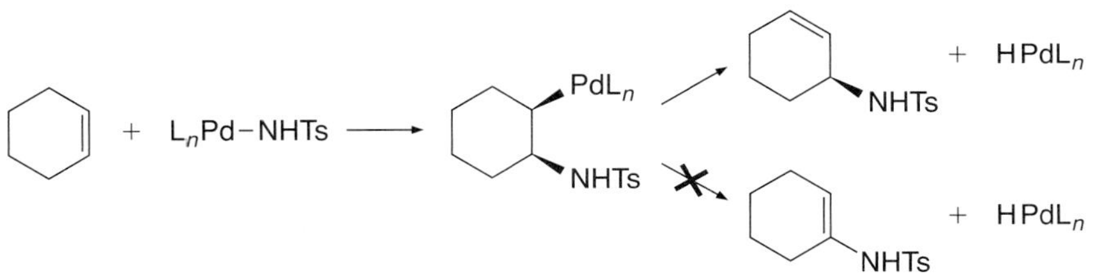

Scheme 16.34

16.6.4.3. Mechanism of Reoxidation[429]

As noted in Section 16.6.2.1, many reagents have been used to re-oxidize the Pd(0) product of the olefin oxidation to the starting Pd(II) species. Each oxidant reforms the Pd(II) by a different mechanism. The least understood is the re-oxidation of Pd(0) by the combination of CuCl$_2$ in acidic medium. This process formally occurs by the transfer of one Cl atom from two CuCl$_2$ to the Pd(0) to re-form PdCl$_2$. This re-oxidation could include Cu–Pd clusters possessing bridging oxo ligands.[361,362]

The re-oxidation with quinone can occur by several mechanisms. The re-oxidation of Pd(0) by quinone in acid has been studied by Bäckvall. This work showed that the well-known quinone complexes of Pd(0), such as the Pd(COD)(benzoquinone) complex in Equation 16.128, react with acid to protonate the oxygen of the quinone. A second protonation generates hydroquinone and a Pd(II) species. Attack of the conjugate base on the diolefin complex studied by Bäckvall occurred to generate the bridging acetate structure shown.[430]

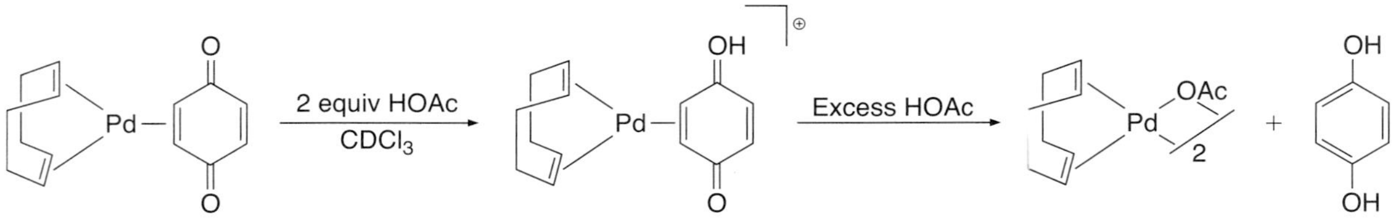

$$ \text{(16.128)} $$

One can also envision that the quinone could trap the hydridopalladium species resulting from the β-hydrogen elimination that releases the oxidized organic product (Equation 16.129). Insertion of quinone into the palladium hydride would form an enolate that would tautomerize to the phenoxide complex. Protonation with the reagent containing an O–H or N–H bond would generate the free hydroquinone and Pd(II). As noted in Chapter 17 on carbonylation, quinone has been used as an additive with this mechanism in mind to prevent the Pd(II) hydroesterification catalysts from undergoing reduction to palladium(0).[431]

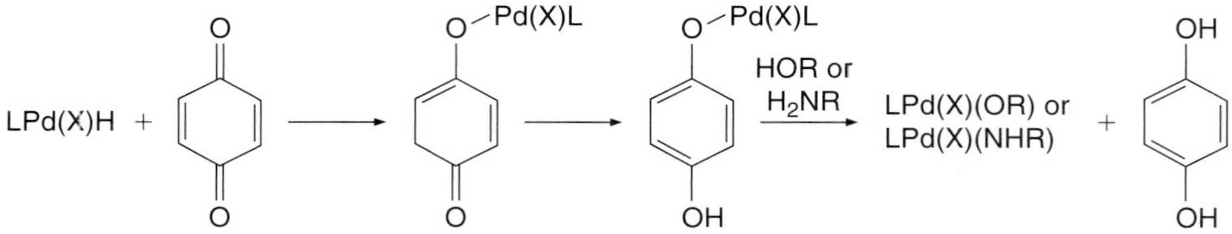

$$ \text{(16.129)} $$

Finally, re-oxidation with O_2 has been studied, in part because of the interest in using oxygen as a terminal oxidant for oxidation processes in general. This re-oxidation can occur by formation of a palladium–peroxo complex, $L_2Pd(O_2)$, followed by protonation with the acid to form peroxide and the starting palladium(II) species containing carboxylate, alkoxide, sulfonamidate, or imidate ligands. This sequence is shown generically in Equation 16.130.

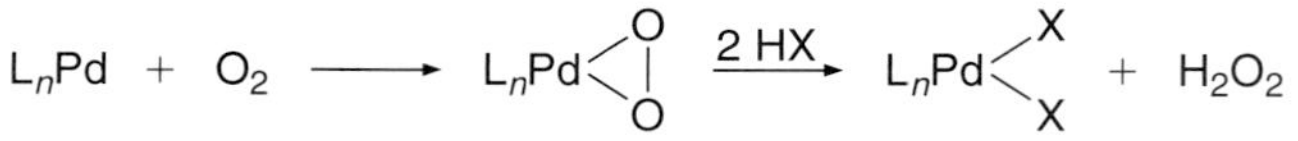

$$ L_nPd \; + \; O_2 \longrightarrow L_nPd\!\!\bigg\langle{}^{O}_{O} \xrightarrow{2\ HX} L_nPd\!\!\bigg\langle{}^{X}_{X} \; + \; H_2O_2 \qquad \text{(16.130)} $$

Each step of Equation 16.130 has been observed individually. The reaction of $Pd(PPh_3)_4$ with O_2 to form $(PPh_3)_2Pd(O_2)$ has been known for many years.[432–435] The reactions of Pd(0) complexes with O_2 to form diamagnetic Pd(II) peroxo complexes are spin forbidden, but recent computational studies show that the approach of dioxygen with an end-on trajectory, followed by triplet–singlet spin crossover, occurs with a low barrier. Thus, the spin-forbidden nature of the process does not create a large activation barrier.[436] The reaction of palladium–peroxo complexes with acid to form hydrogen peroxide shown in Equation 16.131 was observed many years ago.[437] Thus, the combination of the coordination of dioxygen and protonation of the peroxo ligand is a viable mechanism for the re-oxidation of palladium in the presence of dioxygen and acid without added copper.

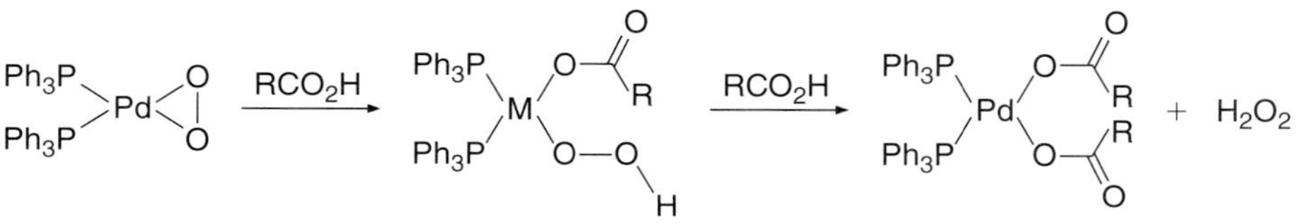

$$ \text{(16.131)} $$

16.7. Summary

As one can see from these sections on hydrofunctionalizations of C–C multiple bonds, the scope and level of development of these reactions varies widely. The hydrocyanation of the C–C double bonds in butadiene is highly refined and is conducted on a large industrial scale. Other types of hydrocyanations remain under development. Likewise, the hydrosilylation of olefins is conducted commercially, and highly reactive platinum catalysts have been developed. Control of the regioselectivity and stereoselectivity for other types of hydrosilylations has advanced significantly during the past few years. The addition of boron reagents and amines to olefins and alkynes is less developed, but has been studied intensively. The diboration of alkenes and alkynes has been developed recently. The hydroamination of alkenes and alkynes has been studied most extensively during the past decade, although the first examples with soluble catalysts were reported more than 35 years ago.

The mechanisms of these reactions are varied, but can still be categorized. The hydrocyanations, hydrosilylations, and many of the hydroborations, occur with late-metal catalysts. These reactions occur by oxidative addition of the H–X bond, followed by migratory insertion of the olefin into the M–H or M–X bond, and reductive elimination to form the final product. Hydrocyanation occurs by insertion of the unsaturated reagent into the M–H bond, while hydrosilylation and hydroboration have been shown to occur by insertion of the olefin into the M–H bond in some cases and into the M–X bond in others. Hydrosilylations and hydroborations of alkenes and alkynes catalyzed by d^0 transition metal complexes and by lanthanides follow a different pathway because these complexes cannot undergo oxidative addition. The mechanism of the reactions catalyzed by these complexes involves σ-bond metatheses.

The hydroamination of olefins has been shown to occur by the sequence of oxidative addition, migratory insertion, and reductive elimination in only one case. Because amines are nucleophilic, pathways are available for the additions of amines to olefins and alkynes that are unavailable for the additions of HCN, silanes, and boranes. For example, hydroaminations catalyzed by late transition metals are thought to occur in many cases by nucleophilic attack on coordinated alkenes and alkynes or by nucleophilic attack on π-allyl, π-benzyl, or π-arene complexes. Hydroaminations catalyzed by lanthanide and actinide complexes occur by insertion of an olefin into a metal–amide bond. Finally, hydroamination catalyzed by d^0 group 4 metals have been shown to occur through imido complexes. In this case, a [2+2] cycloaddition forms the C–N bond, and protonolysis of the resulting metallacycle releases the organic product.

Presumably, new pathways for all of these reactions, a clearer picture of the existing mechanisms, and catalysts with improved activities and selectivities will be revealed for each type of hydrofunctionalization process in the future.

References and Notes

1. *Catalytic Heterofunctionalization*; Togni, A., Grutzmacher, H., Eds.; Wiley-VCH: Weinheim, 2001.
2. *Encyclopedia of Catalysis*; Horváth, I. T., Ed.; Wiley-Interscience: Hoboken, 2003.
3. Han, L.-B.; Tanaka, M. *Chem. Commun.* **1999**, 395.
4. Beller, M.; Seayad, J.; Tilack, A.; Jiao, H. *Angew. Chem. Int. Ed.* **2004**, *43*, 3368.
5. Brown, E. S. In *Aspects of Homogeneous Catalysis*; Ugo, R., Ed.; D. Reidel: Dordrecht, The Netherlands, 1974; Vol. 2, p 57.
6. Brown, E. S. In *Organic Synthesis via Metal Carbonyls*; Wender, I., Pino, P., Eds.; Wiley-Interscience: New York, 1977; Vol. 2, p 743.
7. Parshall, G. W. *J. Mol. Catal.* **1978**, *4*, 244.
8. Parshall, G. W. In *Homogeneous Catalysis*, 1st ed.; Wiley: New York, 1980.
9. James, B. R. In *Comprehensive Organometallic Chemistry: The Synthesis, Reactions, and Structures of Organometallic Compounds*; Wilkinson, G., Stone, F. G. A., Abel, E. W., Eds.; Pergamon Press: Oxford, U.K., 1982; p 353.

10. Hubert, A. J.; Puentes, E. In *Catalysis in C1 Chemistry*; Keim, W., Ed.; D. Reidel: Dordrecht, The Netherlands, 1983; p 219.
11. Jackson, W. R.; Perlmutter, P.; Elmes, P. S.; Lovel, C. G.; Thompson, R. J.; Haarburger, D.; Probert, M. K. S.; Smallridge, A. J.; Campi, E. M.; Fitzmaurice, N. J.; Kertesz, M. A. In *5th IUPAC Symposium on Organic Synthesis*; Streith, J., Prinzbach, H., Schill, G., Eds.; Blackwell Scientific: Freiburg, 1984; p 55.
12. Tolman, C. A.; McKinney, R. J.; Seidel, W. C.; Druliner, J. D.; Stevens, W. R. In *Advances in Catalysis*; Eley, D. D., Pines, H., Weisz, P. B., Eds.; Academic Press: New York, 1985; Vol. 33, p 1.
13. Tolman, C. A. *J. Chem. Educ.* **1986**, *63*, 199.
14. Krill, S. In *Applied Homogeneous Catalysis with Organometallic Compounds*, 2nd ed.; Cornils, B., Herrmann, W. A., Eds.; Wiley-VCH: Weinheim, 2002; Vol. 1, p 468.
15. Yet, L. *Angew. Chem. Int. Ed.* **2001**, *40*, 875.
16. Krueger, C.; Kuntz, K.; Dzierba, C.; Wirschun, W.; Gleason, J.; Snapper, M.; Hoveyda, A. *J. Am. Chem. Soc.* **1999**, *121*, 4284.
17. Ishitani, H.; Komiyama, S.; Hasegawa, Y.; Kobayashi, S. *J. Am. Chem. Soc.* **2000**, *122*, 762.
18. Sigman, M. S.; Vachal, P.; Jacobsen, E. N. *Angew. Chem. Int. Ed.* **2000**, *39*, 1279.
19. Deng, H. B.; Isler, M. R.; Snapper, M. L.; Hoveyda, A. H. *Angew. Chem. Int. Ed.* **2002**, *41*, 1009.
20. Vachal, P.; Jacobsen, E. N. *J. Am. Chem. Soc.* **2002**, *124*, 10012.
21. Arthur, P.; England, D. C.; Pratt, B. C.; Whitman, G. M. *J. Am. Chem. Soc.* **1954**, *76*, 5364.
22. Drinkard, W. C., Jr.; Lindsey, R. V., Jr. U.S. Patent 3,496,217, 1968.
23. Drinkard, W. C., Jr. U.S. Patent 3,496,218, 1970.
24. Drinkard, W. C., Jr.; Lindsey, R. V., Jr. U.S. Patent 3,496,215, 1970.
25. Drinkard, W. C., Jr.; Kassal, R. J. U.S. Patent 3,496,217, 1970.
26. Brown, E. S.; Rick, E. A. *J. Chem. Soc. D* **1969**, 112b.
27. Taylor, B. W.; Swift, H. *J. Catal.* **1972**, *26*, 254.
28. Tolman, C. A.; Seidel, W. C.; Druliner, J. D.; Domaille, P. J. *Organometallics* **1984**, *3*, 33.
29. Nugent, W. A.; McKinney, R. J. *J. Org. Chem.* **1985**, *50*, 5370.
30. Tolman, C. A.; McKinney, R. J.; Seidel, W. C.; Druliner, J. D.; Stevens, W. R. *Adv. Catal.* **1985**, *33*, 1.
31. Yan, M.; Xu, Q.-Y.; Chan, A. S. C. *Tetrahedron: Asymmetry* **2000**, *11*, 845.
32. Baker, M. J.; Pringle, P. G. *J. Chem. Soc., Chem. Commun.* **1991**, 1292.
33. Tolman, C. A.; Seidel, W. C.; Druliner, J. D.; Domaille, P. J. *Organometallics* **1984**, *3*, 33.
34. Jackson, W. R.; Lovel, C. G. *Aust. J. Chem.* **1982**, *35*, 2052.
35. Jackson, W. R.; Lovel, C. G. *Tetrahedron Lett.* **1982**, *23*, 1621.
36. Jackson, W. R.; Lovel, C. G. *Aust. J. Chem.* **1983**, *36*, 1975.
37. Bäckvall, J. E.; Andell, O. S. *J. Chem. Soc., Chem. Commun.* **1981**, 1098.
38. Bäckvall, J. E.; Andell, O. S. *J. Chem. Soc., Chem. Commun.* **1984**, 260.
39. Bäckvall, J. E.; Andell, O. S. *Organometallics* **1986**, *5*, 2350.
40. Tolman, C. A. *Chem. Rev.* **1977**, *77*, 313.
41. Brunkan, N. M.; Jones, W. D. *J. Organomet. Chem.* **2003**, *683*, 77.
42. Brunkan, N. M.; Brestensky, D. M.; Jones, W. D. *J. Am. Chem. Soc.* **2004**, *126*, 3627.
43. Chaumonnot, A.; Lamy, F.; Sabo-Etienne, S.; Donnadieu, B.; Chaudret, B.; Barthelat, J. C.; Galland, J. C. *Organometallics* **2004**, *23*, 3363.
44. Acosta-Ramirez, A.; Munoz-Hernandez, M.; Jones, W. D.; Garcia, J. J. *J. Organomet. Chem.* **2006**, *691*, 3895.
45. Acosta-Ramirez, A.; Flores-Gaspar, A.; Munoz-Hernandez, M.; Arevalo, A.; Jones, W. D.; Garcia, J. J. *Organometallics* **2007**, *26*, 1712.
46. Horiuchi, T.; Shirakawa, E.; Nozaki, K.; Takaya, H. *Tetrahedron: Asymmetry* **1997**, *8*, 57.
47. Casalnuovo, A. L.; RajanBabu, T. V.; Ayers, T. A.; Warren, T. A. *J. Am. Chem. Soc.* **1994**, *116*, 9869.
48. (a) Saha, B.; RajanBabu, T. V. *Org. Lett.* **2006**, *8*, 4656; (b) Zhang, A. B.; RajanBabu, T. V. *J. Am. Chem. Soc.* **2006**, *128*, 54.
49. Zhang, A. B.; RajanBabu, T. V. *J. Am. Chem. Soc.* **2006**, *128*, 54.
50. Wilting, J.; Janssen, M.; Muller, C.; Vogt, D. *J. Am. Chem. Soc.* **2006**, *128*, 11374.
51. Funabiki, T.; Yamazaki, Y.; Sato, Y.; Yoshida, S. *J. Chem. Soc., Perkin Trans. 2* **1983**, 1915.
52. Funabiki, T.; Sato, H.; Tanaka, N.; Yamazaki, Y.; Yoshida, S. *J. Mol. Catal.* **1990**, *62*, 157.
53. Funabiki, T.; Tatsumi, K.; Yoshida, S. *J. Organomet. Chem.* **1990**, *384*, 199.
54. For reviews spanning four different decades, see references 55–58.
55. Speier, J. L. *Adv. Organomet. Chem.* **1979**, *17*, 407.
56. Ojima, I. In *The Chemistry of Organic Silicon Compounds*; Patai, S., Rappoport, Z., Eds.; Wiley: New York, 1989; p 1479.
57. Ojima, I.; Li, Z.; Zhu, J. In *The Chemistry of Organic Silicon Compounds*; Rappoport, Z., Apeloig, Y., Eds.; Wiley: New York, 1998; Vol. 2, p 1687.

58. Marciniec, B. *Coord. Chem. Rev.* **2005**, *249*, 2374.

59. Marciniec, B.; Gulinski, J. *J. Organomet. Chem.* **1993**, *446*, 15.

60. Marciniec, B. *Silicon Chem.* **2002**, *1*, 155.

61. Jones, G. R.; Landais, Y. *Tetrahedron* **1996**, *52*, 7599.

62. Smitrovich, J. H.; Woerpel, K. A. *J. Org. Chem.* **1996**, *61*, 6044.

63. Tamao, K.; Ishida, N.; Tanaka, T.; Kumada, M. *Organometallics* **1983**, *2*, 1694.

64. Tamao, K.; Kakui, T.; Akita, M.; Iwahara, T.; Kanatani, R.; Yoshida, J.; Kumada, M. *Tetrahedron* **1983**, *39*, 983.

65. Fleming, I.; Henning, R.; Plaut, H. *J. Chem. Soc., Chem. Commun.* **1984**, 29.

66. Fleming, I.; Henning, R.; Parker, D. C.; Plaut, H. E.; Sanderson, P. E. J. *J. Chem. Soc., Perkin Trans. 1* **1995**, 317.

67. Hatanaka, Y.; Hiyama, T. *Synlett* **1991**, 845.

68. Denmark, S. E.; Sweis, R. F. *Acc. Chem. Res.* **2002**, *35*, 835.

69. Speier, J. L.; Webster, J. A.; Barnes, G. H. *J. Am. Chem. Soc.* **1957**, *79*, 974.

70. For an initial report of hydrosilylation catalyzed by peroxides, see reference 71.

71. Sommer, L. H.; Pietrusza, E. W.; Whitmore, F. C. *J. Am. Chem. Soc.* **1947**, *69*, 188.

72. Karstedt, B. D. U.S. Patent 3,775,452, 1973.

73. Lewis, L. N.; Lewis, N. *J. Am. Chem. Soc.* **1986**, *108*, 7228.

74. Lewis, L. N.; Lewis, N.; Uriarte, R. J. *Adv. Chem. Ser.* **1992**, 541.

75. Stein, J.; Lewis, L. N.; Gao, Y.; Scott, R. A. *J. Am. Chem. Soc.* **1999**, *121*, 3693.

76. Lappert, M. F.; Scott, F. P. A. *J. Organomet. Chem.* **1995**, *492*, C11.

77. DeCharentenay, F.; Osborn, J. A.; Wilkinson, G. *J. Chem. Soc. A* **1968**, 787.

78. Rejhon, J.; Hetflejs, J. *Collect. Czech. Chem. Commun.* **1975**, *40*, 3680.

79. Schroeder, M. A.; Wrighton, M. S. *J. Organomet. Chem.* **1977**, *128*, 345.

80. Reichel, C. L.; Wrighton, M. S. *Inorg. Chem.* **1980**, *19*, 3858.

81. Millan, A.; Towns, E.; Maitlis, P. M. *J. Chem. Soc., Chem. Commun.* **1981**, 673.

82. Onopchenko, A.; Sabourin, E. T.; Beach, D. L. *J. Org. Chem.* **1983**, *48*, 5101.

83. Millan, A.; Fernandez, M. J.; Bentz, P.; Maitlis, P. M. *J. Mol. Catal.* **1984**, *26*, 89.

84. Ojima, I.; Fuchikami, T.; Yatabe, M. *J. Organomet. Chem.* **1984**, *260*, 335.

85. Onopchenko, A.; Sabourin, E. T.; Beach, D. L. *J. Org. Chem.* **1984**, *49*, 3389.

86. Oro, L. A.; Fernandez, M. J.; Esteruelas, M. A.; Jimenez, M. S. *J. Mol. Catal.* **1986**, *37*, 151.

87. Randolph, C. L.; Wrighton, M. S. *J. Am. Chem. Soc.* **1986**, *108*, 3366.

88. Seitz, F.; Wrighton, M. S. *Angew. Chem., Int. Ed. Engl.* **1988**, *27*, 289.

89. Duckett, S. B.; Perutz, R. N. *Organometallics* **1992**, *11*, 90.

90. Brunner, H.; Riepl, G.; Weitzer, H. *Angew. Chem., Int. Ed. Engl.* **1983**, *22*, 331.

91. Brunner, H.; Becker, R.; Riepl, G. *Organometallics* **1984**, *3*, 1354.

92. Riant, O.; Mostefaï, N.; Courmarcel, J. *Synthesis* **2004**, 2943.

93. Uozumi, Y.; Hayashi, T. *J. Am. Chem. Soc.* **1991**, *113*, 9887.

94. Kitayama, K.; Uozumi, Y.; Hayashi, T. *J. Chem. Soc., Chem. Commun.* **1995**, 1533.

95. Fu, P. F.; Brard, L.; Li, Y. W.; Marks, T. J. *J. Am. Chem. Soc.* **1995**, *117*, 7157.

96. Bank, H. M.; Saam, J. C.; Speier, J. L. *J. Org. Chem.* **1964**, *29*, 792.

97. Benkeser, R. A.; Muench, W. C. *J. Am. Chem. Soc.* **1973**, *95*, 285.

98. Harrod, J. F.; Chalk, A. J. In *Organic Synthesis via Metal Carbonyls*; Wender, I., Pino, P., Eds.; Wiley: New York, 1977; Vol. 2, p 673.

99. Speier, J. L. In *Advances in Organometallic Chemistry*; West, R., Stone, F., Eds.; Academic Press: New York, 1979; Vol. 17, p 407.

100. Ojima, I. In *The Chemistry of Organic Silicon Compounds*; Patai, S., Rappoport, Z., Eds.; Wiley: Chichester, U.K., 1989; Vol. 1, p 1479.

101. Trost, B. M.; Ball, Z. T. *Synthesis* **2005**, 853.

102. Ojima, I.; Kumagai, M.; Nagai, Y. *J. Organomet. Chem.* **1976**, *111*, 43.

103. Chalk, A. J. *J. Organomet. Chem.* **1970**, *21*, 207.

104. Musolf, M. C.; Speier, J. L. *J. Org. Chem.* **1964**, *29*, 2519.

105. Capka, M.; Svoboda, P.; Hetflejs, J. *Collect. Czech. Chem. Commun.* **1973**, *38*, 3830.

106. Wechsler, D.; Myers, A.; McDonald, R.; Ferguson M. J.; Stradiotto, M. *Inorg. Chem.* **2006**, *45*, 4562.

107. Uozumi, Y.; Kitayama, K.; Hayashi, T. *Tetrahedron: Asymmetry* **1993**, *4*, 2419.

108. Svoboda, P.; Sedlmayer, P.; Hetflejs, J. *Collect. Czech. Chem. Commun.* **1973**, *38*, 1783.

109. Benkeser, R. A.; Merritt, F. M., II; Roche, R. T. *J. Organomet. Chem.* **1978**, *156*, 235.

110. Belyakova, Z. V.; Pomerantseva, M. G.; Popkov, K. K.; Efremova, L. A.; Golubtsov, S. A. *Zh. Obshch. Khim.* **1972**, *42*, 889.

111. Takahashi, S.; Shibano, T.; Kojima, H.; Hagihara, N. *Oganomet. Chem. Synth.* **1970/1971**, *1*, 193.

112. Vaisarova, V.; Capka, M.; Hetflejs, J. *Synth. React. Inorg. Met-Org. Chem.* **1972**, *2*, 289.

113. Ojima, I.; Kumagai, M. *J. Organomet. Chem.* **1978**, *157*, 359.

114. Rejhon, J.; Hetflejs, J. *Collect. Czech. Chem. Commun.* **1975**, *40*, 3190.

115. (a) Seki, Y.; Takeshita, K.; Kawamoto, K.; Murai, S.; Sonoda, N. *Angew. Chem., Int. Ed. Engl.* **1980**, *19*, 928; (b) Delpech, F.; Mansas, J.; Leuser, H.; Sabo-Etienne, S.; Chaudret, B. *Organometallics* **2000**, *19*, 5750.

116. (a) Takeuchi, R.; Tanouchi, N. *J. Chem. Soc., Perkin Trans. 1* **1994**, 2909; (b) Takeuchi, R.; Tanouchi, N. *J. Chem. Soc., Chem. Commun.* **1993**, 1319.

117. Ojima, I.; Clos, N.; Donovan, R. J.; Ingallina, P. *Organometallics* **1990**, *9*, 3127.

118. Tanke, R. S.; Crabtree, R. H. *J. Am. Chem. Soc.* **1990**, *112*, 7984.

119. Trost, B. M.; Ball, Z. T.; Joge, T. *J. Am. Chem. Soc.* **2002**, *124*, 7922.

120. Chung, L. W.; Wu, Y. D.; Trost, B. M.; Ball, Z. T. *J. Am. Chem. Soc.* **2003**, *125*, 11578.

121. Trost, B. M.; Ball, Z. T. *J. Am. Chem. Soc.* **2003**, *125*, 30.

122. Trost, B. M.; Ball, Z. T.; Laemmerhold, K. M. *J. Am. Chem. Soc.* **2005**, *127*, 10028.

123. Trost, B. M.; Ball, Z. T. *J. Am. Chem. Soc.* **2005**, *127*, 17644.

124. Trost, B. M.; Fraisse, P. L.; Ball, Z. T. *Angew. Chem. Int. Ed.* **2002**, *41*, 1059.

125. Na, Y. G.; Chang, S. B. *Org. Lett.* **2000**, *2*, 1887.

126. Trost, B. M.; Ball, Z. T.; Joge, T. *Angew. Chem. Int. Ed.* **2003**, *42*, 3415.

127. Brunner, H.; Nishiyama, H.; Itoh, K. In *Catalytic Asymmetric Synthesis*; Ojima, I., Ed.; VCH Publishers: New York, 1993; p 303.

128. Hayashi, T. In *Comprehensive Asymmetric Catalysis I–III*; Jacobsen, E. N., Pfaltz, A., Yamamoto, H., Eds.; Springer: Berlin, 1999; Vol. 1, p 319.

129. Hayashi, T. *Acc. Chem. Res.* **2000**, *33*, 354.

130. Nishiyama, H.; Itoh, K. In *Catalytic Asymmetric Synthesis*, 2nd ed.; Ojima, I., Ed.; Wiley: New York, 2000; p 111.

131. Tang, J.; Hayashi, T. In *Catalytic Heterofunctionalization*; Togni, A., Grützmacher, H., Eds.; Wiley-VCH: Weinheim, 2001; p 73.

132. Uozumi, Y.; Hayashi, T. *J. Am. Chem. Soc.* **1991**, *113*, 9887.

133. Uozumi, Y.; Kitayama, K.; Hayashi, T.; Yanagi, K.; Fukuyo, E. *Bull. Chem. Soc. Jpn.* **1995**, *68*, 713.

134. Uozumi, Y.; Lee, S.-Y.; Hayashi, T. *Tetrahedron Lett.* **1992**, *33*, 7185.

135. Ohmura, H.; Matsuhashi, H.; Tanaka, M.; Kuroboshi, M.; Hiyama, T.; Hatanaka, Y.; Goda, K. *J. Organomet. Chem.* **1995**, *499*, 167.

136. Kitayama, K.; Tsuji, H.; Uozumi, Y.; Hayashi, T. *Tetrahedron Lett.* **1996**, *37*, 4169.

137. Ojima, I.; Kogure, T.; Nihonyanagi, M.; Nagai, Y. *Bull. Chem. Soc. Jpn.* **1972**, *45*, 3506.

138. Ojima, I.; Nihonyanagi, M.; Nagai, Y. *J. Chem. Soc., Chem. Commun.* **1972**, 938.

139. Corriu, R. J. P.; Moreau, J. J. E. *J. Chem. Soc., Chem. Commun.* **1973**, 38.

140. Eaborn, C.; Odell, K.; Pidcock, A. *J. Organomet. Chem.* **1973**, *63*, 93.

141. Yamamoto, K.; Hayashi, T.; Kumada, M. *J. Organomet. Chem.* **1972**, *46*, C65.

142. Nishiyama, H.; Sakaguchi, H.; Nakamura, T.; Horihata, M.; Kondo, M.; Itoh, K. *Organometallics* **1989**, *8*, 846.

143. Verdaguer, X.; Lange, U. E. W.; Reding, M. T.; Buchwald, S. L. *J. Am. Chem. Soc.* **1996**, *118*, 6784.

144. Lipshutz, B. H.; Frieman, B. A. *Angew. Chem. Int. Ed.* **2005**, *44*, 6345.

145. Lipshutz, B. H.; Noson, K.; Chrisman, W.; Lower, A. *J. Am. Chem. Soc.* **2003**, *125*, 8779.

146. Lipshutz, B. H.; Lower, A.; Noson, K. *Org. Lett.* **2002**, *4*, 4045.

147. Lipshutz, B. H.; Shimizu, H. *Angew. Chem. Int. Ed.* **2004**, *43*, 2228.

148. Chalk, A. J.; Harrod, J. F. *J. Am. Chem. Soc.* **1965**, *87*, 16.

149. Lappert, M. F.; Scott, F. P. A. *J. Organomet. Chem.* **1995**, *492*, C11.

150. Chalk, A. J.; Harrod, J. F. *J. Am. Chem. Soc.* **1965**, *87*, 16.

151. Onopchenko, A.; Sabourin, E. T. *J. Org. Chem.* **1987**, *52*, 4118.

152. Sakaki, S.; Sumimoto, M.; Fukuhara, M.; Sugimoto, M.; Fujimoto, H.; Matsuzaki, S. *Organometallics* **2002**, *21*, 3788.

153. Sakaki, S.; Mizoe, N.; Sugimoto, M. *Organometallics* **1998**, *17*, 2510.

154. Sakaki, S.; Mizoe, N.; Sugimoto, M.; Musashi, Y. *Coord. Chem. Rev.* **1999**, *192*, 933.

155. Brookhart, M.; Grant, B. E. *J. Am. Chem. Soc.* **1993**, *115*, 2151.

156. Chung, L. W.; Wu, Y. D.; Trost, B. M.; Ball, Z. T. *J. Am. Chem. Soc.* **2003**, *125*, 11578.

157. Suginome, M.; Ito, Y. *Chem. Rev.* **2000**, *100*, 3221.

158. Ito, Y.; Suginome, M.; Murakami, M. *J. Org. Chem.* **1991**, *56*, 1948.

159. Murakami, M.; Andersson, P. G.; Suginome, M.; Ito, Y. *J. Am. Chem. Soc.* **1991**, *113*, 3987.

160. Murakami, M.; Suginome, M.; Fujimoto, K.; Nakamura, H.; Andersson, P. G.; Ito, Y. *J. Am. Chem. Soc.* **1993**, *115*, 6487.

161. Yamashita, H.; Kobayashi, T.-a.; Hayashi, T.; Tanaka, M. *Chem. Lett.* **1990**, 1447.

162. Bottoni, A.; Higueruelo, A. P.; Miscione, G. P. *J. Am. Chem. Soc.* **2002**, *124*, 5506.
163. Brown, H. C. *Boranes in Organic Chemistry*; Cornell University Press: Ithaca, N.Y., 1972.
164. Brown, H. C. *Organic Synthesis via Boranes*; Wiley: New York, 1975.
165. Burgess, K.; Ohlmeyer, M. J. *Chem. Rev.* **1991**, *91*, 1179.
166. Beletskaya, I.; Pelter, A. *Tetrahedron* **1997**, *53*, 4957.
167. Miyaura, N. In *Catalytic Heterofunctionalization*; Togni, A., Grutzmacher, H., Eds.; Wiley-VCH: Weinheim, 2001; p 1.
168. Smith, M. R., III. *Prog. Organomet. Chem.* **1999**, *48*, 505.
169. Wilczynski, R.; Sneddon, L. G. *J. Am. Chem. Soc.* **1980**, *102*, 2857.
170. Wilczynski, R.; Sneddon, L. G. *Inorg. Chem.* **1981**, *20*, 3955.
171. Wilczynski, R.; Sneddon, L. G. *Inorg. Chem.* **1982**, *21*, 506.
172. Corcoran, E. W.; Sneddon, L. G. *J. Am. Chem. Soc.* **1985**, *107*, 7446.
173. Lynch, A. T.; Sneddon, L. G. *J. Am. Chem. Soc.* **1987**, *109*, 5867.
174. Sneddon, L. G. *Pure. Appl. Chem.* **1987**, *59*, 837.
175. Lynch, A. T.; Sneddon, L. G. *J. Am. Chem. Soc.* **1989**, *111*, 6201.
176. Pender, M. J.; Wideman, T.; Carroll, P. J.; Sneddon, L. G. *J. Am. Chem. Soc.* **1998**, *120*, 9108.
177. Pender, M. J.; Carroll, P. J.; Sneddon, L. G. *J. Am. Chem. Soc.* **2001**, *123*, 12222.
178. Mannig, D.; Nöth, H. *Angew. Chem., Int. Ed. Engl.* **1985**, *24*, 878.
179. Burgess, K.; van der Donk, W. A.; Westcott, S. A.; Marder, T. B.; Baker, R. T.; Calabrese, G. C. *J. Am. Chem. Soc.* **1992**, *114*, 9350.
180. Evans, D. A.; Fu, G. C.; Anderson, B. A. *J. Am. Chem. Soc.* **1992**, *114*, 6679.
181. He, X. M.; Hartwig, J. F. *J. Am. Chem. Soc.* **1996**, *118*, 1696.
182. Hartwig, J. F.; Muhoro, C. N. *Organometallics* **2000**, *19*, 30.
183. Hayashi, T.; Matsumoto, Y.; Ito, Y. *J. Am. Chem. Soc.* **1989**, *111*, 3426.
184. Hayashi, T.; Matsumoto, Y.; Ito, Y. *Tetrahedron: Asymmetry* **1991**, *2*, 601.
185. Burgess, K.; van der Donk, W. A.; Westcott, S. A.; Marder, T. B.; Baker, R. T.; Calabrese, J. C. *J. Am. Chem. Soc.* **1992**, *114*, 9350.
186. Evans, D. A.; Fu, G. C.; Hoveyda, A. H. *J. Am. Chem. Soc.* **1992**, *114*, 6671.
187. Harrison, K. N.; Marks, T. J. *J. Am. Chem. Soc.* **1992**, *114*, 9220.
188. Bijpost, E. A.; Duchateau, R.; Teuben, J. H. *J. Mol. Catal.* **1995**, *95*, 121.
189. Pereira, S.; Srebnik, M. *Tetrahedron Lett.* **1996**, *37*, 3283.
190. Pereira, S.; Srebnik, M. *J. Am. Chem. Soc.* **1996**, *118*, 909.
191. Matsumoto, Y.; Hayashi, T. *Tetrahedron Lett.* **1991**, *32*, 3387.
192. Satoh, M.; Nomoto, Y.; Miyaura, N.; Suzuki, A. *Tetrahedron Lett.* **1989**, *30*, 3789.
193. Ohmura, T.; Yamamoto, Y.; Miyaura, N. *J. Am. Chem. Soc.* **2000**, *122*, 4990.
194. Brown, J. M.; Hulmes, D. I.; Layzell, T. P. *J. Chem. Soc., Chem. Commun.* **1993**, 1673.
195. Valk, J. M.; Whitlock, G. A.; Layzell, T. P.; Brown, J. M. *Tetrahedron: Asymmetry* **1995**, *6*, 2593.
196. Burgess, K.; van der Donk, W. A.; Kook, A. M. *J. Org. Chem.* **1991**, *56*, 2949.
197. Brown, J. M.; Lloyd-Jones, G. C. *J. Am. Chem. Soc.* **1994**, *116*, 866.
198. Westcott, S. A.; Marder, T. B.; Baker, R. T. *Organometallics* **1993**, *12*, 975.
199. Westcott, S. A.; Blom, H. P.; Marder, T. B.; Baker, R. T.; Calabrese, J. C. *Inorg. Chem.* **1993**, *32*, 2175.
200. Musaev, D. G.; Mebel, A. M.; Morokuma, K. *J. Am. Chem. Soc.* **1994**, *116*, 10693.
201. Ishiyama, T.; Matsuda, N.; Murata, M.; Ozawa, F.; Suzuki, A.; Miyaura, N. *Organometallics* **1996**, *15*, 713.
202. Dai, C.; Stringer, G.; Marder, T. B. *Inorg. Chem.* **1997**, *36*, 272.
203. Marder, T. B.; Norman, N. C.; Rice, C. R.; Robins, E. G. *Chem. Commun.* **1997**, 53.
204. Sakaki, S.; Kikuno, T. *Inorg. Chem.* **1997**, *36*, 226.
205. Curtis, D.; Lesley, M. J. G.; Norman, N. C.; Orpen, A. G.; Starbuck, J. *J. Chem. Soc.* **1999**, 1687.
206. Hartwig, J. F.; He, X. *Angew. Chem., Int. Ed. Engl.* **1996**, *35*, 315.
207. Sagawa, T.; Asano, Y.; Ozawa, F. *Organometallics* **2002**, *21*, 5879.
208. Sakaki, S.; Kai, S.; Sugimoto, M. *Organometallics* **1999**, *18*, 4825.
209. Onozawa, S.; Hatanaka, Y.; Sakakura, T.; Shimada, S.; Tanaka, M. *Organometallics* **1996**, *15*, 5450.
210. Ishiyama, T.; Matsuda, N.; Miyaura, N.; Suzuki, A. *J. Am. Chem. Soc.* **1993**, *115*, 11018.
211. Lesley, G.; Nguyen, P.; Taylor, N. J.; Marder, T. B.; Scott, A. J.; Clegg, W.; Norman, N. C. *Organometallics* **1996**, *15*, 5137.
212. Onozawa, S.; Hatanaka, Y.; Tanaka, M. *Chem. Commun.* **1997**, 1229.
213. Suginome, M.; Matsuda, T.; Nakamura, H.; Ito, Y. *Tetrahedron* **1999**, *55*, 8787.
214. Ishiyama, T.; Yamamoto, M.; Miyaura, N. *J. Chem. Soc., Chem. Commun.* **1997**, 689.
215. Marder, T. B.; Norman, N. C.; Rice, C. R. *Tetrahedron Lett.* **1998**, *39*, 155.
216. Iverson, C. N.; Smith, M. R., III. *Organometallics* **1997**, *16*, 2757.

217. Baker, R. T.; Nguyen, P.; Marder, T. B.; Westcott, S. A. *Angew. Chem., Int. Ed. Engl.* **1995**, *34*, 1336.
218. Dai, C. Y.; Robins, E. G.; Scott, A. J.; Clegg, W.; Yufit, D. S.; Howard, J. A. K.; Marder, T. B. *Chem. Commun.* **1998**, 1983.
219. Ishiyama, T.; Kitano, T.; Miyaura, N. *Tetrahedron Lett.* **1998**, *39*, 2357.
220. Pelz, N. F.; Woodward, A. R.; Burks, H. E.; Sieber, J. D.; Morken, J. P. *J. Am. Chem. Soc.* **2004**, *126*, 16328.
221. Morgan, J. B.; Miller, S. P.; Morken, J. P. *J. Am. Chem. Soc.* **2003**, *125*, 8702.
222. Iverson, C. N.; Smith, M. R., III. *Organometallics* **1996**, *15*, 5155.
223. Clegg, W.; Lawlor, F. J.; Lesley, G.; Marder, T. B.; Norman, N. C.; Orpen, A. G.; Quayle, M. J.; Rice, C. R.; Scott, A. J.; Souza, F. E. S. *J. Organomet. Chem.* **1998**, *550*, 183.
224. Clegg, W.; Lawlor, F. J.; Marder, T. B.; Nguyen, P.; Norman, N. C.; Orpen, A. G.; Quayle, M. J.; Rice, C. R.; Robins, E. G.; Scott, A. J.; Souza, F. E. S.; Stringer, G.; Whittell, G. R. *J. Chem. Soc.* **1998**, 301.
225. Sakaki, S.; Kai, S.; Sugimoto, M. *Organometallics* **1999**, *18*, 4825.
226. Haggin, J. *C&EN* **1993**, *71, May 31*, 23.
227. Rosenfeld, D. C.; Shekhar, S.; Takemiya, A.; Utsunomiya, M.; Hartwig, J. F. *Org. Lett.* **2006**, *8*, 4179.
228. Li, Z. G.; Zhang, J. L.; Brouwer, C.; Yang, C. G.; Reich, N. W.; He, C. *Org. Lett.* **2006**, *8*, 4175.
229. Beller, M.; Breindl, C. *Tetrahedron* **1998**, *54*, 6359.
230. Seayad, J.; Tillack, A.; Hartung, C. G.; Beller, M. *Adv. Synth. Catal.* **2002**, *344*, 795.
231. Gasc, M. B.; Lattes, A.; Perie, J. J. *Tetrahedron* **1983**, *39*, 703.
232. Steinborn, D.; Taube, R. *Z. Anorg. Allem. Chem.* **1986**, *26*, 349.
233. Brunet, J.-J. *Gazz. Chim. Ital.* **1997**, *127*, 111.
234. (a) Muller, T. E.; Beller, M. *Chem. Rev.* **1998**, *98*, 675; (b) Müller, T. E.; Hultzsch, K. C.; Yus, M.; Foubelo, F.; Tada, M. *Chem. Rev.* **2008**, *108*, 3795.
235. Nobis, M.; Driessen-Holscher, B. *Angew. Chem., Int. Ed. Engl.* **2001**, *40*, 3983.
236. Pohlki, F.; Doye, S. *Chem. Soc. Rev.* **2003**, *32*, 104.
237. Roesky, P. W.; Müller, T. E. *Angew. Chem. Int. Ed.* **2003**, *42*, 2708.
238. Hultzsch, K. C. *Adv. Synth. Catal.* **2005**, *347*, 367.
239. Brunet, J. J.; Neibecker, D. In *Catalytic Heterofunctionalization*; Togni, A., Grutzmacher, H., Eds.; Wiley-VCH: Weinheim, 2001; p 91.
240. Beller, M.; Tillack, A.; Seayad, A. In *Transition Metals for Organic Synthesis*, 2nd ed.; Beller, M., Bolm, C., Eds.; Wiley-VCH: Weinheim, 2004; Vol. 2, p 403.
241. Müller, T. E. In *Encyclopedia of Catalysis*; Horváth, I. T., Ed.; Wiley-Interscience: Hoboken, 2003; Vol. 3, p 518.
242. Johns, A. M.; Sakai, N.; Ridder, A.; Hartwig, J. F. *J. Am. Chem. Soc.* **2006**, *128*, 9306.
243. Coulson, D. R. *Tetrahedron Lett.* **1971**, 429.
244. Wang, X.; Widenhoefer, R. A. *Organometallics* **2004**, *23*, 1649.
245. Casalnuovo, A. L.; Calabrese, J. C.; Milstein, D. *J. Am. Chem. Soc.* **1988**, *110*, 6738.
246. Brunet, J. J.; Cadena, M.; Chu, N. C.; Diallo, O.; Jacob, K.; Mothes, E. *Organometallics* **2004**, *23*, 1264.
247. Brunet, J. J.; Chu, N. C.; Diallo, O. *Organometallics* **2005**, *24*, 3104.
248. Li, Y.; Marks, T. J. *Organometallics* **1996**, *15*, 3770.
249. Ryu, J.-S.; Li, G. Y.; Marks, T. J. *J. Am. Chem. Soc.* **2003**, *125*, 12584.
250. Crimmin, M. R.; Casely, I. J.; Hill, M. S. *J. Am. Chem. Soc.* **2005**, *127*, 2042.
251. (a) Gribkov, D. V.; Hultzsch, K. C. *Angew. Chem. Int. Ed.* **2004**, *43*, 5542; (b) Wood, M. C.; Leitch, D. C.; Yeung, C. S.; Kozak, J. A.; Schaefer, L. L. *Agnew. Chem. Int. Ed.* **2007**, *46*, 354; (c) Gott, A. L.; Clarke, A. J.; Clarkson, G. J.; Scott, P. *Chem. Commun.* **2008**, 1422.
252. Hong, S.; Marks, T. J. *Acc. Chem. Res.* **2004**, *37*, 673.
253. Gagné, M. R.; Marks, T. J. *J. Am. Chem. Soc.* **1989**, *111*, 4108.
254. Gagné, M. R.; Nolan, S. P.; Marks, T. J. *Organometallics* **1990**, *9*, 1716.
255. Gagné, M. R.; Stern, C. L.; Marks, T. J. *J. Am. Chem. Soc.* **1992**, *114*, 275.
256. Gagné, M. R.; Brard, L.; Conticello, V. P.; Giardello, M. A.; Stern, C. L.; Marks, T. J. *Organometallics* **1992**, *11*, 2003.
257. Giardello, M. A.; Conticello, V. P.; Brard, L.; Gagné, M. R.; Marks, T. J. *J. Am. Chem. Soc.* **1994**, *116*, 10241.
258. Li, Y. W.; Marks, T. J. *J. Am. Chem. Soc.* **1996**, *118*, 9295.
259. Arredondo, V. M.; Tian, S.; McDonald, F. E.; Marks, T. J. *J. Am. Chem. Soc.* **1999**, *121*, 3633.
260. Tian, S.; Arredondo, V. M.; Stern, C. L.; Marks, T. J. *Organometallics* **1999**, *18*, 2568.
261. Ryu, J. S.; Marks, T. J.; McDonald, F. E. *Org. Lett.* **2001**, *3*, 3091.
262. Douglass, M. R.; Ogasawara, M.; Hong, S.; Metz, M. V.; Marks, T. J. *Organometallics* **2002**, *21*, 283.
263. Hong, S.; Marks, T. J. *J. Am. Chem. Soc.* **2002**, *124*, 7886.
264. Hong, S. W.; Tian, S.; Metz, M. V.; Marks, T. J. *J. Am. Chem. Soc.* **2003**, *125*, 14768.

265. Hong, S.; Kawaoka, A. M.; Marks, T. J. *J. Am. Chem. Soc.* **2003**, *125*, 15878.

266. Motta, A.; Lanza, G.; Fragala, I. L.; Marks, T. J. *Organometallics* **2004**, *23*, 4097.

267. Ryu, J. S.; Marks, T. J.; McDonald, F. E. *J. Org. Chem.* **2004**, *69*, 1038.

268. Molander, G. A.; Dowdy, E. D.; Pack, S. K. *J. Org. Chem.* **2001**, *66*, 4344.

269. Molander, G. A.; Pack, S. K. *J. Org. Chem.* **2003**, *68*, 9214.

270. Hultzsch, K. C. *Adv. Synth. Catal.* **2005**, *347*, 367.

271. Kim, J. Y.; Livinghouse, T. *Org. Lett.* **2005**, *7*, 1737.

272. Wood, M. C.; Leitch, D. C.; Yeung, C. S.; Kozak, J. A.; Schafer, L. L. *Angew. Chem. Int. Ed.* **2007**, *46*, 354.

273. Bender, C. F.; Widenhoefer, R. A. *J. Am. Chem. Soc.* **2005**, *127*, 1070.

274. Liu, Z.; Hartwig, J. F. *J. Am. Chem. Soc.* **2008**, *130*, 1570.

275. Michael, F. E.; Cochran, B. M. *J. Am. Chem. Soc.* **2006**, *128*, 4246.

276. Kawatsura, M.; Hartwig, J. F. *J. Am. Chem. Soc.* **2000**, *122*, 9546.

277. Utsunomiya, M.; Hartwig, J. F. *J. Am. Chem. Soc.* **2003**, *125*, 14286.

278. Johns, A. M.; Utsunomiya, M.; Incarvito, C. D.; Hartwig, J. F. *J. Am. Chem. Soc.* **2006**, *128*, 1828.

279. Hu, A. G.; Ogasawara, M.; Sakamoto, T.; Okada, A.; Nakajima, K.; Takahashi, T.; Lin, W. B. *Adv. Synth. Catal.* **2006**, *348*, 2051.

280. Utsunomiya, M.; Kuwano, R.; Kawatsura, M.; Hartwig, J. F. *J. Am. Chem. Soc.* **2003**, *125*, 5608.

281. Takemiya, A.; Hartwig, J. F. *J. Am. Chem. Soc.* **2006**, *128*, 6042.

282. Utsunomiya, M.; Hartwig, J. F. *J. Am. Chem. Soc.* **2004**, *126*, 2702.

283. Beller, M.; Eichberger, M.; Trauthwein, H. *Angew. Chem., Int. Ed. Engl.* **1997**, *36*, 2225.

284. Takaya, J.; Hartwig, J. F. *J. Am. Chem. Soc.* **2005**, *127*, 5756.

285. Walsh, P. J.; Baranger, A. M.; Bergman, R. G. *J. Am. Chem. Soc.* **1992**, *114*, 1708.

286. Johnson, J. S.; Bergman, R. G. *J. Am. Chem. Soc.* **2001**, *123*, 2923.

287. Straub, B. F.; Bergman, R. G. *Angew. Chem. Int. Ed.* **2001**, *40*, 4632.

288. Ackermann, L.; Bergman, R. G.; Loy, R. N. *J. Am. Chem. Soc.* **2003**, *125*, 11956.

289. Besson, L.; Gore, J.; Gazes, B. *Tetrahedron Lett.* **1995**, *36*, 3857.

290. Al-Masum, M.; Meguro, M.; Yamamoto, Y. *Tetrahedron Lett.* **1997**, *38*, 6071.

291. Zhang, Z.; Liu, C.; Kinder, R. E.; Han, X.; Qian, H.; Widenhoefer, R. A. *J. Am. Chem. Soc.* **2006**, *128*, 9066.

292. Nishina, N.; Yamamoto, Y. *Angew. Chem. Int. Ed.* **2006**, *45*, 3314.

293. Meguro, M.; Yamamoto, Y. *Tetrahedron Lett.* **1998**, *39*, 5421.

294. Dzhemilev, U. M.; Yakupova, A. Z.; Minsker, S. K.; Tolstikov, G. A. *Zh. Org. Khim.* **1979**, *15*, 1164.

295. Dzhemilev, U. M.; Yakupova, A. Z.; Tolstikov, G. A. *Izv. Akad. Nauk, Ser. Khim.* **1976**, 2346.

296. Dzhemilev, U. M.; Selimov, F. A.; Yakupova, A. Z.; Tolstikov, G. A. *Russ. Chem. Bull.* **1978**, *27*, 1230.

297. Dzhemilev, U. M.; Yakupova, A. Z.; Minsker, S. K.; Tolstikov, G. A. *Zh. Org. Khim.* **1979**, *15*, 1164.

298. Keim, W.; Roper, M.; Schieren, M. *J. Mol. Catal.* **1983**, *20*, 139.

299. Zakharkin, L. I.; Petrushkina, E. A.; Podvisotskaya, L. S. *Bull. Acad. Sci. USSR* **1983**, 805.

300. Rose, D. *Tetrahedron Lett.* **1972**, 4197.

301. Armbruster, R. W.; Morgan, M. M.; Schmidt, J. L.; Lau, C. M.; Riley, R. M.; Zabrowski, D. L.; Dieck, H. *Organometallics* **1986**, *5*, 234.

302. Pawlas, J.; Nakao, Y.; Kawatsura, M.; Hartwig, J. F. *J. Am. Chem. Soc.* **2002**, *124*, 3669.

303. Löber, O.; Kawatsura, M.; Hartwig, J. F. *J. Am. Chem. Soc.* **2001**, *123*, 4366.

304. Sakai, N.; Ridder, A.; Hartwig, J. F. *J. Am. Chem. Soc.* **2006**, *128*, 8134.

305. McGrane, P. L.; Jensen, M.; Livinghouse, T. *J. Am. Chem. Soc.* **1992**, *114*, 5459.

306. Baranger, A. M.; Walsh, P. J.; Bergman, R. G. *J. Am. Chem. Soc.* **1993**, *115*, 2753.

307. Haak, E.; Bytschkov, I.; Doye, S. *Angew. Chem., Int. Ed. Engl.* **1999**, *38*, 3389.

308. Pohlki, F.; Doye, S. *Angew. Chem. Int. Ed.* **2001**, *40*, 2305.

309. Heutling, A.; Doye, S. *J. Org. Chem.* **2002**, *67*, 1961.

310. Tillack, A.; Castro, I. G.; Hartung, C. G.; Beller, M. *Angew. Chem. Int. Ed.* **2002**, *41*, 2541.

311. Ong, T. G.; Yap, G. P. A.; Richeson, D. S. *Organometallics* **2002**, *21*, 2839.

312. Zhang, Z.; Schafer, L. L. *Org. Lett.* **2003**, *5*, 4733.

313. Bexrud, J. A.; Beard, J. D.; Leitch, D. C.; Schafer, L. L. *Org. Lett.* **2005**, *7*, 1959.

314. Li, Y. H.; Shi, Y. H.; Odom, A. L. *J. Am. Chem. Soc.* **2004**, *126*, 1794.

315. Ramanathan, B.; Keith, A. J.; Armstrong, D.; Odom, A. L. *Org. Lett.* **2004**, *6*, 2957.

316. Heutling, A.; Pohlki, F.; Bytschkov, I.; Doye, S. *Angew. Chem. Int. Ed.* **2005**, *44*, 2951.

317. Castro, I. G.; Tillack, A.; Hartung, C. G.; Beller, M. *Tetrahedron Lett.* **2003**, *44*, 3217.

318. Ackermann, L. *Org. Lett.* **2005**, *7*, 439.

319. Li, Y. W.; Fu, P. F.; Marks, T. J. *Organometallics* **1994**, *13*, 439.

320. Li, Y. W.; Marks, T. J. *J. Am. Chem. Soc.* **1996**, *118*, 707.

321. Haskel, A.; Straub, T.; Eisen, M. S. *Organometallics* **1996**, *15*, 3773.

322. Straub, T.; Haskel, A.; Neyroud, T. G.; Kapon, M.; Botoshansky, M.; Eisen, M. S. *Organometallics* **2001**, *20*, 5017.

323. Rudisill, D. E.; Stille, J. K. *J. Org. Chem.* **1989**, *54*, 5856.

324. Taylor, E. C.; Katz, A. H.; Salgadozamora, H.; McKillop, A. *Tetrahedron Lett.* **1985**, *26*, 5963.

325. Burling, S.; Field, l. D.; Messerle, B. A. *Organometallics* **2000**, *19*, 87.

326. Field, L. D.; Messerle, B. A.; Wren, S. L. *Organometallics* **2003**, *22*, 4393.

327. Burling, S.; Field, L. D.; Messerle, B. A.; Turner, P. *Organometallics* **2004**, *23*, 1714.

328. Hartung, C. G.; Tillack, A.; Trauthwein, H.; Beller, M. *J. Org. Chem.* **2001**, *66*, 6339.

329. Qian, H.; Han, X. Q.; Widenhoefer, R. A. *J. Am. Chem. Soc.* **2004**, *126*, 9536.

330. Qian, H.; Widenhoefer, R. A. *Org. Lett.* **2005**, *7*, 2635.

331. Nettekoven, U.; Hartwig, J. F. *J. Am. Chem. Soc.* **2002**, *124*, 1166.

332. Dorta, R.; Egli, P.; Zurcher, F.; Togni, A. *J. Am. Chem. Soc.* **1997**, *119*, 10857.

333. Tobisch, S. *J. Am. Chem. Soc.* **2005**, *127*, 11979.

334. Tobisch, S. *Dalton Trans.* **2006**, 4277.

335. Motta, A.; Fragala, I. L.; Marks, T. J. *Organometallics* **2006**, *25*, 5533.

336. Zulys, A.; Dochnahl, M.; Hollmann, D.; Lohnwitz, K.; Herrmann, J. S.; Roesky, P. W.; Blechert, S. *Angew. Chem. Int. Ed.* **2005**, *44*, 7794.

337. Walsh, P. J.; Hollander, F. J.; Bergman, R. G. *J. Am. Chem. Soc.* **1988**, *110*, 8729.

338. Polse, J. L.; Andersen, R. A.; Bergmann, R. G. *J. Am. Chem. Soc.* **1998**, *120*, 13405.

339. Beccalli, E. M.; Broggini, G.; Martinelli, M.; Sottocornola, S. *Chem. Rev.* **2007**, *107*, 5318.

340. Smidt, J.; Hafner, W.; Jira, R.; Sieber, R.; Sedlmeier, J.; Sabel, A. *Angew. Chem., Int. Ed. Engl.* **1962**, *1*, 80.

341. Hosokawa, T.; Murahashi, S.-I. *Acc. Chem. Res.* **1990**, *23*, 49.

342. Jira, R. In *Applied Homogeneous Catalysis with Organometallic Compounds: A Comprehensive Handbook in Two Volumes*; Cornils, B., Herrmann, W. A., Eds.; VCH Publishers: New York, 1996; p 394.

343. Jira, R. In *Applied Homogeneous Catalysis with Organometallic Compounds: A Comprehensive Handbook in Two Volumes*; Cornils, B., Herrmann, W. A., Eds.; VCH Publishers: New York, 1996; p 374.

344. (a) Phillips, F. C. *Am. Chem. J.* **1894**, *16*, 255; (b) Anderson, J. S. *J. Chem. Soc. II* **1934**, 971; (c) Kharasch, M. S.; Seyler, R. C.; Mayo, F. R. *J. Am. Chem. Soc.* **1938**, *60*, 882.

345. Smidt, J.; Hafner, W.; Jira, R.; Sedlmeier, J.; Sieber, R.; Ruttinger, R.; Kojer, H. *Angew. Chem.* **1959**, *71*, 176.

346. (a) Jira, R. In *Applied Homogeneous Catalysis with Organometallic Compounds: A Comprehensive Handbook in Two Volumes*; 2nd ed.; Cornils, B., Herrmann, W. A., Eds.; Wiley-VCH Publishers: Weinheim, 2002; p 386; (b) Henry, P. M. In *Handbook of Organopalladium Chemistry for Organic Synthesis*; Negishi, E.-i., de Meijere, A., Eds.; Wiley Interscience: New York, 2002; p 2189.

347. Smidt, J. *Chem. Ind. (London)* **1962**, 54.

348. Moiseev, I. I.; Vargaftik, M. N.; Syrkin, J. K. *Dok. Akad. Nauk* **1960**, *133*, 377.

349. Pei, T.; Wang, X.; Widenhoefer, F. A. *J. Am. Chem. Soc.* **2003**, *125*, 648.

350. Ferreira, E. M.; Stoltz, B. M. *J. Am. Chem. Soc.* **2003**, *125*, 9578.

351. Alexanian, E. J.; Lee, C.; Sorensen, E. J. *J. Am. Chem. Soc.* **2005**, *127*, 7690.

352. Bar, G. L. J.; Lloyd-Jones, G. C.; Booker-Milburn, K. I. *J. Am. Chem. Soc.* **2005**, *127*, 7308.

353. Streuff, J.; Hovelmann, C. H.; Nieger, M.; Muniz, K. *J. Am. Chem. Soc.* **2005**, *127*, 14586.

354. Zabawa, T. P.; Kasi, D.; Chemler, S. R. *J. Am. Chem. Soc.* **2005**, *127*, 11250.

355. Schultz, M. J.; Sigman, M. S. *J. Am. Chem. Soc.* **2006**, *128*, 1460.

356. Smidt, J.; Hafner, W.; Jira, R.; Sedlmeier, J.; Sieber, R.; Ruttinger, R.; Kojer, H. *Angew. Chem.* **1959**, *71*, 176.

357. Smidt, J.; Sedlmeier, J.; Hafner, W.; Sieber, R.; Sabel, A.; Jira, R. *Angew. Chem.* **1962**, *74*, 93.

358. Jira, R. In *Applied Homogeneous Catalysis with Organometallic Compounds*, 2nd ed.; Cornils B., Herrman, W. A., Eds.; Wiley-VCH: 2002; Vol. 1, p 386.

359. Fleischmann, G.; Jira, R. *Ullmann's Encyclopedia of Industrial Chemistry*, 7th ed.; Wiley-VCH: Weinheim, 2005.

360. Jira, R.; Freiesleben, W. In *Organometallic Reactions*; Becker, E. I., Tsutsui, M., Eds.; Wiley: New York, 1972; Vol. 3, p 1.

361. Hosokawa, T.; Takano, M.; Murahashi, S. I. *J. Am. Chem. Soc.* **1996**, *118*, 3990.

362. Hosokawa, T.; Murahashi, S. I. *Acc. Chem. Res.* **1990**, *23*, 49.

363. Stille, J. K.; Divakaruni, R. *J. Organomet. Chem.* **1979**, *169*, 239.

364. Bäckvall, J. E.; Åkermark, B.; Ljunggren, S. O. *J. Am. Chem. Soc.* **1979**, *101*, 2411.

365. Hamed, O.; Thompson, C.; Henry, P. M. *J. Org. Chem.* **1997**, *62*, 7082.

366. Hayashi, T.; Yamasaki, K.; Mimura, M.; Uozumi, Y. *J. Am. Chem. Soc.* **2004**, *126*, 3036.

367. Stacchiola, D.; Calaza, F.; Burkholder, L.; Schwabacher, A. W.; Neurock, M.; Tysoe, W. T. *Angew. Chem. Int. Ed.* **2005**, *44*, 4572.

368. Henry, P. M. In *Handbook of Organopalladium Chemistry for Organic Synthesis*; Negishi, E.-i., de Meijere, A., Eds.; Wiley-Interscience: New York, 2002.

369. Detailed kinetic studies are consistent with the suggestion. See references 358, 360, and 368.

370. Smidt, J.; Hafner, W.; Jira, R.; Sieber, R.; Sedlmeier, J.; Sabel, A. *Angew. Chem., Int. Ed. Engl.* **1962**, *1*, 80.

371. Bäckvall, J. E.; Åkermark, B.; Ljunggren, S. O. *J. Am. Chem. Soc.* **1979**, *101*, 2411.

372. Bäckvall, J. E. *Tetrahedron Lett.* **1977**, 467.

373. Stille, J. K.; Divakaruni, R. *J. Am. Chem. Soc.* **1978**, *100*, 1303.

374. Gregor, N.; Zaw, K.; Henry, P. M. *Organometallics* **1984**, *3*, 1251.

375. Hamed, O.; Thompson, C.; Henry, P. M. *J. Org. Chem.* **1997**, *62*, 7082.

376. Hamed, O.; Henry, P. M.; Thompson, C. *J. Org. Chem.* **1999**, *64*, 7745.

377. Hosokawa, T.; Murahashi, S.-I. In *Handbook of Organopalladium Chemistry for Organic Synthesis*; Negishi, E.-i., de Meijere, A., Eds.; Wiley-Interscience: New York, 2002; Vol. 2, p 2141.

378. Tsuji, J. *Synthesis* **1990**, 739.

379. Lloyd, W. G.; Luberoff, B. J. *J. Org. Chem.* **1969**, *34*, 3949.

380. Chevrin, C.; Le Bras, J.; Henin, F.; Muzart, J. *Synthesis* **2005**, 2615.

381. Heumann, A.; Åkermark, B. *Angew. Chem., Int. Ed. .Engl.* **1984**, *23*, 453.

382. Hansson, S.; Heumann, A.; Rein, T.; Åkermark, B. *J. Org. Chem.* **1990**, *55*, 975.

383. Heumann, A.; Åkermark, B.; Hansson, S.; Rein, T. *Org. Synth.* **1990**, *68*, 109.

384. Bäckvall, J. E.; Hopkins, R. B.; Grennberg, H.; Mader, M. M.; Awasthi, A. K. *J. Am. Chem. Soc.* **1990**, *112*, 5160.

385. Bystrom, S. E.; Larsson, E. M.; Åkermark, B. *J. Org. Chem.* **1990**, *55*, 5674.

386. Larsson, E. M.; Åkermark, B. *Tetrahedron Lett.* **1993**, *34*, 2523.

387. Yokota, T.; Fujibayashi, S.; Nishiyama, Y.; Sakaguchi, S.; Ishii, Y. *J. Mol. Catal.* **1996**, *114*, 113.

388. Åkermark, B.; Larsson, E. M.; Oslob, J. D. *J. Org. Chem.* **1994**, *59*, 5729.

389. Uemura, S.; Fukuzawa, S.; Toshimitsu, A.; Okano, M. *Tetrahedron Lett.* **1982**, *23*, 87.

390. Jia, C. G.; Muller, P.; Mimoun, H. *J. Mol. Catal.* **1995**, *101*, 127.

391. McMurry, J. E.; Kocovsky, P. *Tetrahedron Lett.* **1984**, *25*, 4187.

392. Bäckvall, J. E. *Acc. Chem. Res.* **1983**, *16*, 335.

393. Bäckvall, J. E. *Pure Appl. Chem.* **1999**, *71*, 1065.

394. Bäckvall, J. E. In *Metal-Catalyzed Cross-Coupling Reactions*; Stang, P. J., Diederich, F., Eds.; Wiley-VCH: Weinheim, 1998; p 339.

395. Aranyos, A.; Szabo, K. J.; Bäckvall, J. E. *J. Org. Chem.* **1998**, *63*, 2523.

396. Bäckvall, J. E.; Bystrom, S. E.; Nordberg, R. E. *J. Org. Chem.* **1984**, *49*, 4619.

397. Hosokawa, T.; Yamashia, S.; Murahashi, S. I.; Sonoda, A. *Bull. Chem. Soc. Jpn.* **1976**, *49*, 3662.

398. Hosokawa, T.; Miyagi, S.; Murahashi, S. I.; Sonoda, A. *J. Org. Chem.* **1978**, *43*, 2752.

399. Hosokawa, T.; Kono, T.; Uno, T.; Murahashi, S. I. *Bull. Chem. Soc. Jpn.* **1986**, *59*, 2191.

400. Larock, R. C.; Hightower, T. R. *J. Org. Chem.* **1993**, *58*, 5298.

401. Annby, U.; Stenkula, M.; Andersson, C.-M. *Tetrahedron Lett.* **1993**, *34*, 8545.

402. Trend, R. M.; Ramtohul, Y. K.; Ferreira, E. M.; Stoltz, B. M. *Angew. Chem. Int. Ed.* **2003**, *42*, 2892.

403. Trend, R. M.; Ramtohul, Y. K.; Stoltz, B. M. *J. Am. Chem. Soc.* **2005**, *127*, 17778.

404. Bäckvall, J. E.; Andersson, P. G. *J. Am. Chem. Soc.* **1992**, *114*, 6374.

405. Bäckvall, J. E.; Granberg, K. L.; Andersson, P. G.; Gatti, R.; Gogoll, A. *J. Org. Chem.* **1993**, *58*, 5445.

406. Smith, A. B.; Friestad, G. K.; Barbosa, J.; Bertounesque, E.; Hull, K. G.; Iwashima, M.; Qiu, Y. P.; Salvatore, B. A.; Spoors, P. G.; Duan, J. J. W. *J. Am. Chem. Soc.* **1999**, *121*, 10468.

407. Smith, A. B.; Friestad, G. K.; Barbosa, J.; Bertounesque, E.; Duan, J. J. W.; Hull, K. G.; Iwashima, M.; Qiu, Y. P.; Spoors, P. G.; Salvatore, B. A. *J. Am. Chem. Soc.* **1999**, *121*, 10478.

408. Iwadare, H.; Sakoh, H.; Arai, H.; Shiina, I.; Mukaiyama, T. *Chem. Lett.* **1999**, 817.

409. Liao, X. B.; Zhou, H.; Wearing, X. Z.; Ma, J.; Cook, J. M. *Org. Lett.* **2005**, *7*, 3501.

410. Jonasson, C.; Ronn, M.; Bäckvall, J. E. *J. Org. Chem.* **2000**, *65*, 2122.

411. Hosokawa, T.; Takano, M.; Kuroki, Y.; Murahashi, S. I. *Tetrahedron Lett.* **1992**, *33*, 6643.

412. Bozell, J. J.; Hegedus, L. S. *J. Org. Chem.* **1981**, *46*, 2561.

413. Timokhin, V. I.; Anastasi, N. R.; Stahl, S. S. *J. Am. Chem. Soc.* **2003**, *125*, 12996.

414. Timokhin, V. I.; Stahl, S. S. *J. Am. Chem. Soc.* **2005**, *127*, 17888.

415. Beller, M.; Trauthwein, H.; Eichberger, M.; Breindl, C.; Müller, T. E.; Zapf, A. *J. Organomet. Chem.* **1998**, *566*, 277.

416. Beller, M.; Trauthwein, H.; Eichberger, M.; Breindl, C.; Herwig, J.; Muller, T. E.; Thiel, O. R. *Chem. Eur. J.* **1999**, *5*, 1306.

417. Brice, J. L.; Harang, J. E.; Timokhin, V. I.; Anastasi, N. R.; Stahl, S. S. *J. Am. Chem. Soc.* **2005**, *127*, 2868.

418. Hegedus, L. S.; Allen, G. F.; Waterman, E. L. *J. Am. Chem. Soc.* **1976**, *98*, 2674.

419. Hegedus, L. S.; Allen, G. F.; Bozell, J. J.; Waterman, E. L. *J. Am. Chem. Soc.* **1978**, *100*, 5800.

420. Ronn, M.; Bäckvall, J. E.; Andersson, P. G. *Tetrahedron Lett.* **1995**, *36*, 7749.

421. Larock, R. C.; Hightower, T. R.; Hasvold, L. A.; Peterson, K. P. *J. Org. Chem.* **1996**, *61*, 3584.

422. Fix, S. R.; Brice, J. L.; Stahl, S. S. *Angew. Chem., Int. Ed. Engl.* **2002**, *41*, 164.

423. El-Qisairi, A.; Hamed, O.; Henry, P. M. *J. Org. Chem.* **1998**, *63*, 2790.

424. El-Qisairi, A. K.; Qaseer, H. A.; Henry, P. M. *J. Organomet. Chem.* **2002**, *656*, 168.

425. Liu, G. S.; Stahl, S. S. *J. Am. Chem. Soc.* **2006**, *128*, 7179.

426. Du, H. F.; Zhao, B. G.; Shi, Y. *J. Am. Chem. Soc.* **2007**, *129*, 762.

427. For the same stereochemistry from intramolecular cyclizations in the presence and absence of chloride, see the more recent study in reference 402.

428. Liu, G.; Stahl, S. S. *J. Am. Chem. Soc.* **2007**, *129*, 6328.

429. Popp, B. V.; Stahl, S. S. *Top. Organomet. Chem.* **2007**, *22*, 149.

430. Grennberg, H.; Gogoll, A.; Bäckvall, J. E. *Organometallics* **1993**, *12*, 1790.

431. Drent, E.; van Broekhoven, J. A. M.; Doyle, M. J. *J. Organomet. Chem.* **1991**, *417*, 235.

432. Takahashi, S.; Sonogashira, K.; Hagihara, N. *Nippon Kagaku Zaishi* **1966**, *87*, 610.

433. Wilke, G.; Schott, H.; Heimbach, P. *Angew. Chem., Int. Ed. Engl.* **1967**, *6*, 92.

434. Nyman, C. J.; Wymore, C. E.; Wilkinson, G. *J. Chem. Soc. A* **1968**, 561.

435. Aboelella, N. W.; York, J. T.; Reynolds, A. M.; Fujita, K.; Kinsinger, C. R.; Cramer, C. J.; Riordan, C. G.; Tolman, W. B. *Chem. Commun.* **2004**, 1716.

436. Landis, C. R.; Morales, C. M.; Stahl, S. S. *J. Am. Chem. Soc.* **2004**, *126*, 16302.

437. Muto, S.; Ogata, H.; Kamiya, Y. *Chem. Lett.* **1975**, 809.

CHAPTER 17

Catalytic Carbonylation

17.1. Overview

Carbonylation reactions were some of the first reactions catalyzed by soluble organotransition metal complexes. Today, many of these reactions are practiced industrially and are among the largest scale chemical processes. Others have recently led to new routes to important commodity chemicals. The large interest in carbonylation processes stems, in part, from the availability of CO and mixtures of CO and hydrogen from coal in large quantities. The ratio of CO and H_2 can be adjusted by converting CO and H_2O to CO_2 and H_2 by the "water gas shift" reaction. In the 1940s Walter Reppe at BASF discovered a series of carbonylations of olefins and acetylenes, and in the 1950s Richard Heck proposed valid mechanisms for these reactions. Many of these reactions were catalyzed by first row metals, such as cobalt. Today, most carbonylation processes are catalyzed by the "platinum-group" metals, in particular rhodium, iridium, and palladium.

Millions of tons of aldehydes and alcohols are produced each year by the hydroformylation of propene—the reaction of propene with CO and H_2—catalyzed by phosphine and phosphite complexes of rhodium. Esters and acids are produced by related reactions conducted with alcohols and water instead of hydrogen. In addition, a majority of acetic acid, again on the scale of millions of tons per year, is produced by the carbonylation of methanol by a process that has recently been updated with new catalysts based on iridium. Even a spectacularly efficient synthesis of polymers containing ketone functionality in the main chain has been discovered using just ethylene and CO as co-monomers. Other carbonylation reactions have been developed for the synthesis of fine chemicals and pharmaceutical intermediates. For example, the addition of amines to the hydroformylation process can generate alkylated amines, rather than alcohols, in a process called hydroaminomethylation, strained heterocyclic rings can be expanded or opened with CO to generate chemical building blocks, and the combination of an alkyne, an alkene and CO form cyclopentenones that are found in medicinally important compounds, such as prostaglandins.

At one time, it was thought that commodity chemicals could not be synthesized economically using homogeneous catalysts based on precious metals. However, the spectacularly high reactivity and longevity of these catalysts makes the synthesis of commodity chemicals with this type of catalyst feasible. The high selectivity of some of the reactions and the lack of waste generated by these addition processes also enables the efficient synthesis of higher value-added fine chemicals. This chapter describes the scope and mechanism of several of the most practiced carbonylation reactions in industrial and academic laboratories. For a more detailed description of these reactions, the reader is referred to the many reviews and monographs on catalytic carbonylation.[1]

17.2. Catalytic Carbonylation to form Acetic Acid and Acetic Anhydride (Written with Prof. Charles P. Casey)

17.2.1. Rhodium-Catalyzed Carbonylation of Methanol: Monsanto's Acetic Acid Process

The carbonylation of methanol to form acetic acid (Equation 17.1) is one of the largest and most successful industrial applications of homogeneous catalysis.[1-3] In 2004, it was estimated that the global manufacturing capacity of acetic acid was 17 billion pounds a year and that 80% of this capacity was based on methanol carbonylation technology. Methanol carbonylation is also one of the better understood catalytic processes, and the evolution of improved catalysts has been aided by mechanistic investigations. All of the group 9 and 10 metals, in combination with iodide, are active catalysts for methanol carbonylation. In 1965, BASF described a high-pressure process catalyzed by a combination of cobalt and iodide.[4-6] This process required severe conditions of 210 °C and 700 atm to achieve acceptable rates. In 1970, Monsanto commercialized a highly selective carbonylation of methanol to form acetic acid catalyzed by rhodium and iodide that operated at 10^2 lower catalyst loading, at lower temperature (180 °C), and at lower pressure (30–40 atm). Because Rh, and even iodide, is expensive, relative to acetic acid, the process is economical only because there is no appreciable loss of either catalyst component. Although the rhodium-catalyzed Monsanto process is still in use, BP introduced its Cativa™ process in 1996,[7] which is based on iridium, iodide and a promoter. This new technology is becoming dominant.

$$CH_3OH + CO \xrightarrow[\substack{30-40 \text{ atm} \\ 180 \,°C}]{[Rh(CO)_2I_2]^{\ominus}} H_3C-\overset{\overset{\displaystyle O}{\|}}{C}-OH \qquad (17.1)$$

The carbonylation of methanol requires catalysis of both organic and organometallic reactions. The catalytic process consists of five steps, which are shown in Scheme 17.1 for the reactions catalyzed by rhodium–carbonyl compounds: (1) the reaction of methanol

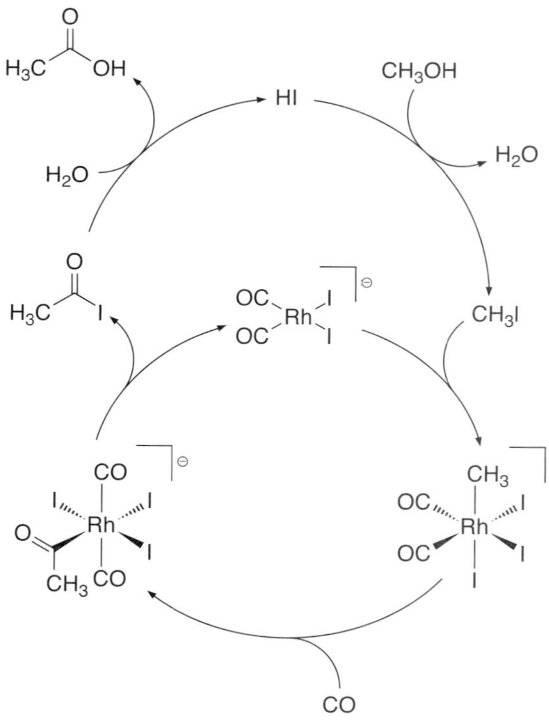

Scheme 17.1

or methyl acetate with HI to produce CH_3I; (2) the oxidative addition of CH_3I to an M(I) (M = Rh or Ir) species to give a $CH_3M(III)$ species; (3) methyl migration to coordinated CO to form a metal–acetyl complex; (4) nucleophilic attack on the acyl carbon by iodide or another nucleophile that leads to reductive elimination of acyl iodide or another acylated species and reformation of the starting M(I) species; and (5) reaction of acyl iodide or the acylated species with water to produce acetic acid and regenerate HI.

The Monsanto process has been thoroughly studied, and the relatively simple catalytic mechanism in Scheme 17.1, consisting of only anionic rhodium species, has been proposed.[3] Virtually any source of rhodium or of iodide may be introduced as a "precatalyst" to generate $[Rh(CO)_2I_2]^-$ and MeI under catalytic conditions. High-pressure IR spectroscopy has shown that the major rhodium species in the catalytic solutions is $[Rh(CO)_2I_2]^-$ (ν_{CO} = 2055 and 1985 cm^{-1}). The overall rate law for the formation of acetic acid contains a linear dependence on rhodium and methyl iodide and no dependence on the concentrations of reactants (CO and methanol) and product (acetic acid). Consistent with the rate law, the turnover-limiting step has been proposed to be the oxidative addition of MeI to $[Rh(CO)_2I_2]^-$.

Many of the individual steps in the catalytic process have been independently investigated, as summarized in Scheme 17.2. In early work, Forster reported that oxidative addition of CH_3I to $[Ph_4As][Rh(CO)_2I_2]$ occurred at room temperature, followed by migratory insertion of CO into the resulting rhodium–methyl bond to give an acetylrhodium complex that was isolated and characterized by X-ray crystallography as a dimer.[8] Reaction of this acetyl species with CO gave $[Rh(COCH_3)(CO)_2I_3]^-$, which was shown by subsequent IR and NMR experiments to be the mer,trans isomer.[9] Under vacuum, this isomer dissociated CO and reverted to the Rh dimer; upon warming above room temperature under CO, it reductively eliminated acetyl iodide to regenerate $[Rh(CO)_2I_2]^-$.

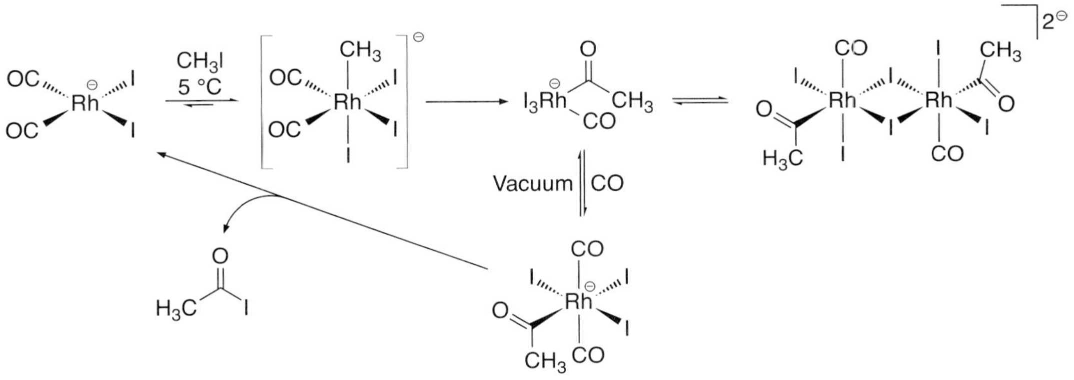

Scheme 17.2

Later, Haynes and Maitlis examined the reaction of $[Rh(CO)_2I_2]^-$ in neat CH_3I at 5 °C and observed small steady state amounts (1%) of the kinetically unstable methyl complex $[RhCH_3(CO)_2I_3]^-$. Under CO, this complex converts to the monomeric acetyl complex by insertion of CO into the methyl group about nine times faster that it eliminates CH_3I.[10] The anionic, monomeric acetyl complex $[Rh(COCH_3)(CO)_2I_3]^-$ was also prepared by oxidative addition of acetyl iodide to $[Rh(CO)_2I_2]^-$. Thus, all of the key steps in the catalytic cycle have been observed independently, and all except the conversion of the methyl complex to the monomeric acetyl complex are reversible.

The mechanisms of oxidative addition of CH_3I and acetyl iodide to $[Rh(CO)_2I_2]^-$ are depicted in Scheme 17.3. Oxidative additions of alkyl iodides were described in detail in Chapter 8. In brief, the addition of methyl iodide to $[Rh(CO)_2I_2]^-$ is proposed to proceed by S_N2 attack of $[Rh(CO)_2I_2]^-$ onto CH_3I. The high nucleophilicity of $[Rh(CO)_2I_2]^-$ is ascribed to the strong electron-donating properties of the iodide ligands and its net negative charge; it is much more reactive than the corresponding chloro and bromo complexes and is 10^5 times more reactive than neutral $[Rh(AsPh_3)_2(CO)I]$.[11] The reaction of acyl halides with $[Rh(CO)_2I_2]^-$ is proposed to occur by nucleophilic addition to the acyl carbonyl to give a tetrahedral intermediate that expels iodide. This mechanism follows the typical pathway for organic reactions of carboxylic acid derivatives. The reductive elimination of acetyl iodide from [Rh(COCH$_3$)(CO)$_2$I$_3$]$^-$ is proposed to occur by the microscopic reverse of the process. Alternatively, other nucleophiles present in the reaction medium, such as acetic acid, water, or methanol, could attack the acyl–rhodium complex to produce acetic acid derivatives.[12] It has been suggested that the *mer, trans*-[Rh(COCH$_3$)(CO)$_2$I$_3$]$^-$ isomer might rearrange to the fac isomer to facilitate reductive elimination of acetyl iodide,[13] but there are no data that require such an isomerization.

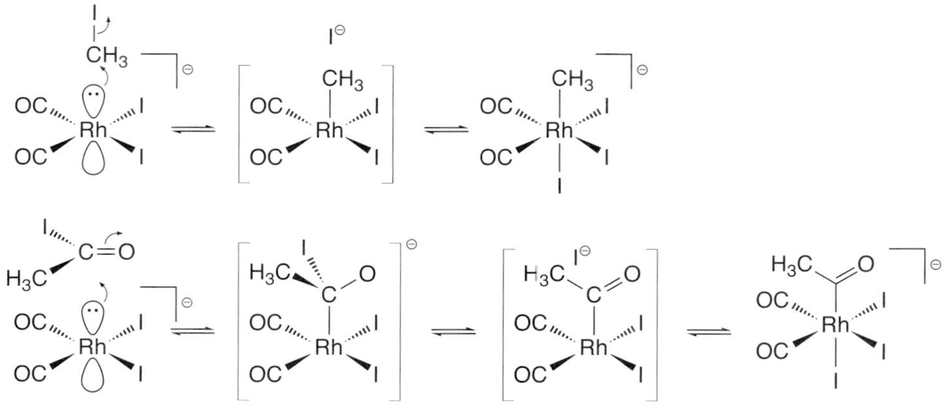

Scheme 17.3

A side reaction in the Monsanto process is oxidation of [Rh(CO)$_2$I$_2$]$^-$ by HI to less soluble RhI$_3$ and inactive [RhI$_4$(CO)$_2$]$^-$.[14] One way to circumvent this side reaction is to run the process with moderately high amounts of water (>10%) to increase the solubility of RhI$_3$ and to use the water gas shift reaction (CO + H$_2$O → CO$_2$ + H$_2$) to consume the Rh(III) (Scheme 17.4).[15] However, this added water must be removed in an energy intensive process if anhydrous acetic acid is to result. To ameliorate this problem, I$^-$ salts are used as promoters; researchers at Celanese found that the addition of LiI or LiOAc allowed the synthesis of acetic acid at low levels of water.[16] The possibility that high concentrations of added iodide create a more reactive dianionic complex [Rh(CO)$_2$I$_3$]$^{-2}$ was suggested,[17] but this species has not been directly observed, and the data pertaining to this hypothesis remain inconclusive.

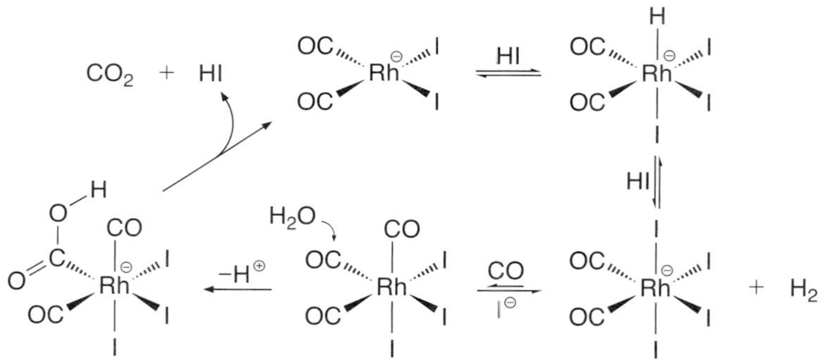

Scheme 17.4.
Reduction of Rh(III) to Rh(I) by the water gas shift reaction during acetic acid synthesis.

17.2.2. Carbonylation of Methyl Acetate: Eastman Chemical's Acetic Anhydride Process

Eastman Chemical's carbonylation of methyl acetate to produce acetic anhydride is closely related to the rhodium-catalyzed carbonylation of methanol to form acetic acid.[18,19] Eastman's carbonylation process was commercialized in 1983 and produces over

800 million pounds of acetic anhydride a year. The proposed mechanism for the acetic anhydride synthesis (Scheme 17.5) is closely related to that for acetic acid synthesis. The initial step is the formation of methyl iodide from methyl acetate and LiI or HI. Because this process necessarily operates under anhydrous conditions, it proved necessary to run reactions with added H_2 and lithium acetate to minimize the amount of inactive Rh(III) species present and to shift and accelerate the organic equilibrium between methanol and the MeOAc electrophile (generated from MeI and LiOAc).

The Eastman process was the first modern coal-to-chemicals facility. Coal is converted to synthesis gas (CO + H_2) by reaction with water in a Lurgi reactor. The synthesis gas is then purified and converted to methanol in a heterogeneous process. Reaction of methanol with acetic acid (which is recycled from applications of acetic anhydride) produces methyl acetate, which is the starting material for the Eastman carbonylation process.

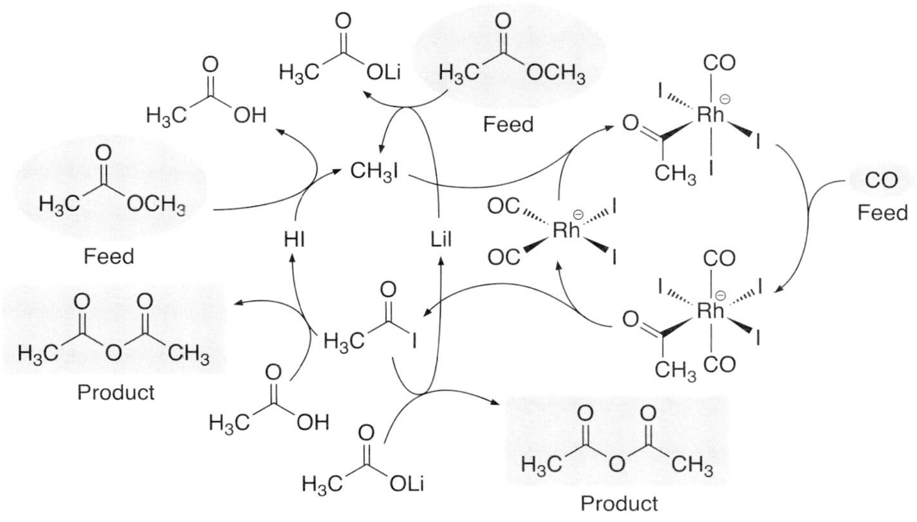

Scheme 17.5. Proposed mechanism of Eastman acetic anhydride process.

17.2.3. Iridium-Catalyzed Carbonylation of Methanol: BP's Cativa™ Process

Methanol carbonylation catalyzed by a combination of iridium–carbonyl compounds and iodide additives was first reported by Monsanto in the 1970s. The mechanism of this process was studied by Forster.[20] In the 1990s, BP reported an improved catalyst system based on iridium and iodide that included a "promoter," such as $[Ru(CO)_3I_2]_2$. These Ir-based Cativa™ catalysts are about five times more active than the Rh catalysts, more stable in the presence of low amounts of water (5 wt %), and more soluble.[7] In addition, Ir is usually less expensive than Rh. BP not only built new Cativa™ plants, but were able to convert existing plants containing rhodium catalysts to plants containing iridium Cativa™ catalysts because of the similarity of the Ir and Rh systems.

While, there are many similarities between the mechanism of the Ir-catalyzed process and that of the Rh-catalyzed stystem, there are important differences.[21] Unlike the dependence of the rate of the Monsanto process on only [Rh] and [CH$_3$I], the dependence of BP's iridium system on CO pressure, water, methyl acetate, methyl iodide, ruthenium promoter, and iridium are more complex and nonlinear. In situ IR spectroscopy of the iridium catalyst shows that the predominant species is the anionic Ir(III) methyl complex fac,cis-[Ir(CH$_3$)(CO)$_2$I$_3$]$^-$ (2100 and 2047 cm^{-1}).[22] Instead of occurring by turnover-limiting oxidative addition of MeI, the iridium-catalyzed process occurs by turnover-limiting insertion of CO into the metal–methyl complex.

Third-row transition metals are used less commonly than second-row transition metals in industrial homogeneous catalytic processes. Until recently, the third-row metals were considered less reactive and were used as models for the more active first- and second-row catalysts. The metal–ligand bonds to third-row metals are stronger, and the metal prefers higher oxidation states. These stronger metal–ligand bonds increase the stability of the catalyst (by reducing susceptibility to dissociation of CO and precipitation of IrI_3), but drastically decrease the rate of migratory insertion of CO. As a result, the insertion step becomes rate limiting, and commercial catalysts require the use of promoters to accelerate the carbonylation step.

Model studies explain why carbonylation is the turnover-limiting step of the reactions of the iridium catlaysts (Scheme 17.6). The rate of oxidative addition of CH_3I to $[Ir(CO)_2I_2]^-$ is 150 times faster than to $[Rh(CO)_2I_2]^-$. This faster rate can be attributed to the formation of stronger bonds to third-row metals than to second-row metals, to the higher nucleophilicity of third-row metals than of second-row metals, and to the preference of third-row metals for higher oxidation states.[23] In contrast, migratory insertion of CO into *fac,cis*-[Ir(CH$_3$)(CO)$_2$I$_3$]$^-$ occurs about 10^5 times more slowly than in the Rh system.[24] The stronger metal–ligand bonds in third-row metal complexes dramatically retard the migratory CO insertion step.

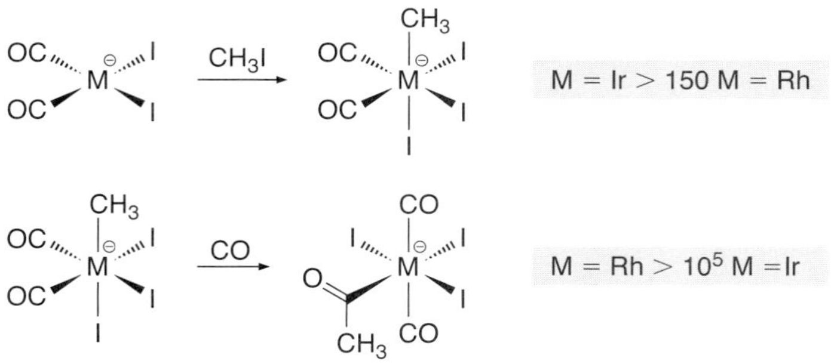

Scheme 17.6

The role of "promoters," such as $Ru(CO)_4I_2$, in the Cativa™ process has been clarified by model studies. Addition of substoichiometric amounts of $[Ru(CO)_3I_2]_2$ to the anionic methyliridium complex *fac,cis*-[Ir(CH$_3$)(CO)$_2$I$_3$]$^-$ accelerated carbonylation by a factor of 20. Added iodide negated the effect of the promotor (Scheme 17.7).[22] Thus, the promoter is concluded to accept iodide from the anionic iridium product from oxidative addition of MeI to form the neutral tricarbonyl species. This tricarbonyl species, $Ir(CH_3)(CO)_3I_2$, undergoes carbonylation 700 times faster than the anionic triiodide.

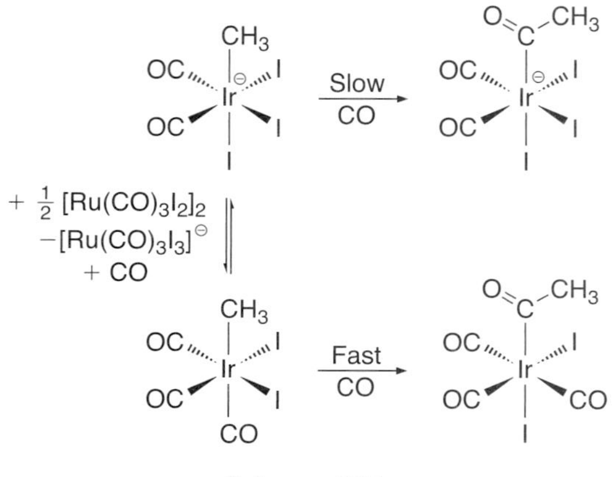

Scheme 17.7

Thus, the mechanism for the Cativa™ process involves both anionic and neutral iridium complexes (Scheme 17.8).[22] It begins with oxidative addition of CH_3I to $[Ir(CO)_2I_2]^-$ to produce a five-coordinate intermediate, which can either react reversibly with I^- to form the resting state *fac,cis*-$[Ir(CH_3)(CO)_2I_3]^-$ or can react directly with CO to form the neutral $Ir(CH_3)(CO)_3I_2$. This neutral methyliridium complex undergoes rapid, migratory insertion of CO to produce a five-coordinate intermediate, which either undergoes attack by water to produce acetic acid or reacts reversibly with CO or I^- to form stable species. The promoter binds iodide to form a $[promoter–I]^-H_3O^+$ adduct, which acts as a Brønsted acid catalyst for the reaction $HI + MeOAc \rightarrow MeI + HOAc$. The presence of the promoter moderates the standing concentration of ionic iodide, thereby aiding carbonylation via neutral intermediates.

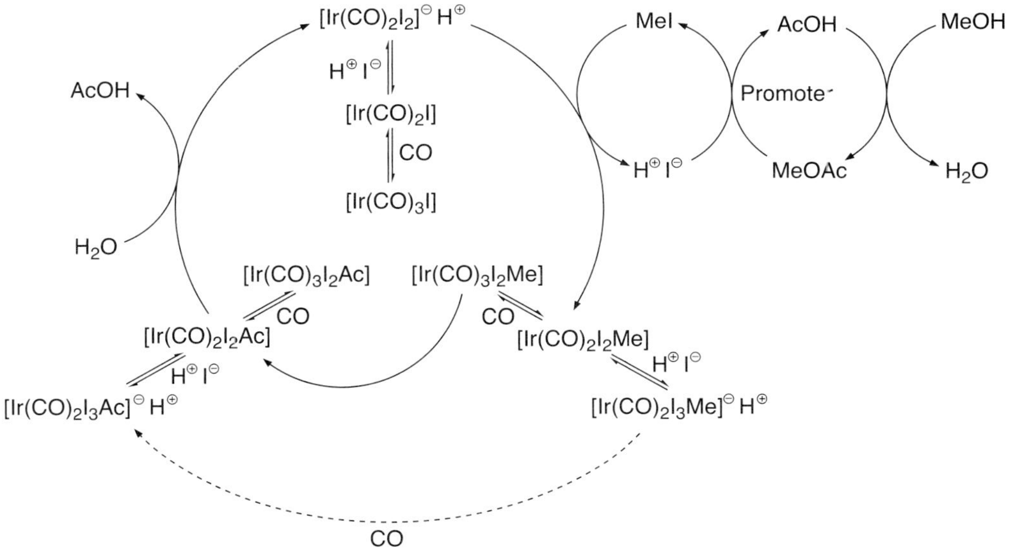

Scheme 17.8

17.3. Hydroformylation of Olefins (Written with Prof. Charles P. Casey)

17.3.1. Overview

Hydroformylation is a metal-catalyzed reaction in which an olefin, CO, and H_2 react to produce an aldehyde.[25,26] The reaction was discovered at BASF by Otto Roelen,[27] and was called hydroformylation by Adkins.[28] This transformation is also sometimes referred to as the "oxo" process. In a formal sense, the elements of formaldehyde are added across a C=C bond. Common side reactions include alkene hydrogenation, aldehyde hydrogenation, and alkene isomerization. Hydroformylation is one of the largest volume reactions conducted with homogeneous catalysts in the chemical industry. It is used to produce over 14 billion pounds of aldehydes per year (two pounds per year for every person on Earth!). The aldehydes are converted to alcohols, acids, and other materials as useful end products. One large-volume use of hydroformylation is the conversion of propene to a mixture of *n*-butyraldehyde and *i*-butyraldehyde (Equation 17.2). Since the desired product is *n*-butyraldehyde, a great deal of effort has been expended to maximize the *n:i* (normal to iso, or often also called *l/b* for linear to branched) ratio of aldehydes and to understand the factors that control it.

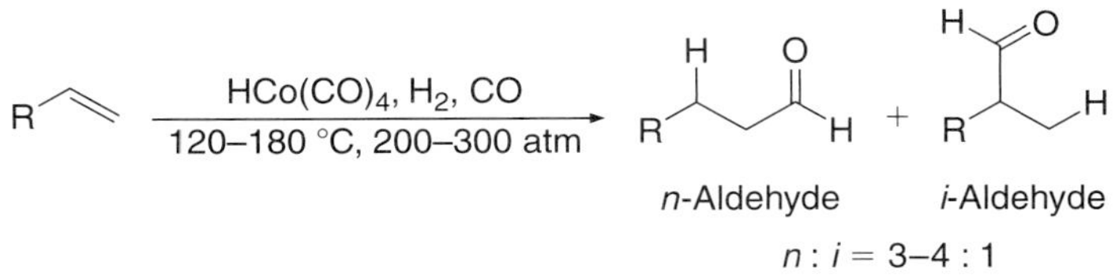

$$n\text{-Aldehyde} \qquad i\text{-Aldehyde}$$
$$n : i = 3\text{--}4 : 1$$

(17.2)

Equation 17.3 shows that aldol condensation of *n*-butanal followed by hydrogenation produces 2-ethyl-hexan-1-ol, which is then converted to di(2-ethylhexyl)phthalate (DEHP or "dioctyl phthalate"). DEHP is used as a plasticizer for polyvinyl chloride and is, therefore, ubiquitous.

(17.3)

The discovery of hydroformylation by Otto Roelen was made while investigating the influence of alkenes on the Fischer–Tropsch reaction using a heterogeneous cobalt oxide catalyst supported on silica. Later it was concluded that hydroformylation is actually a homogeneous process catalyzed by $HCo(CO)_4$ formed in situ.[29] Many metals catalyze hydroformylation, but the most active catalysts contain cobalt, rhodium, palladium, and platinum as the central metal. The discussion in this chapter centers on the most utilized catalysts: $HCo(CO)_4$, $HCo(CO)_3PR_3$, $HRh(CO)_2(PR_3)_2$, and $HRh(CO)_2$(diphosphine).

17.3.2. Hydroformylation Catalyzed by $HCo(CO)_4$

Hydroformylations catalyzed by $HCo(CO)_4$ are typically run at high temperature (120–170 °C) and high pressure of a 1:1 mixture of $CO:H_2$ (200–300 atm "synthesis gas").[30,31] These reactions are typically conducted with alkenes, such as propene, l-octene, or a mixture of internal C_{10} and C_{12} alkenes generated as part of the Shell higher olefin process (SHOP).

17.3.2.1. Mechanism of Hydroformylation Catalyzed by $HCo(CO)_4$

$HCo(CO)_4$, an exceptionally acidic transition metal "hydride", is formed by the reaction of H_2 and $Co_2(CO)_8$. $HCo(CO)_4$ is a trigonal bipyramidal d^8, 18-electron complex in which the hydride occupies an apical position. High temperature is required for industrially useful rates of hydroformylation catalyzed by $HCo(CO)_4$. Moreover, high pressure of CO is required to prevent formation of higher cobalt clusters and of metallic cobalt. The rate of hydroformylation catalyzed by cobalt carbonyl depends on $[H_2]$ and $[CO]^{-1}$.[32,33] Thus, increasing the pressure of a 1:1 mixture of $CO:H_2$ has little effect on the rate but prevents catalyst decomposition.

Although the mechanism of the cobalt-catalyzed oxo reaction has been the subject of many studies, it is not fully understood. Rate studies, product analyses, infrared spectroscopy under catalytic conditions, model reactions for various steps in the reaction cycle, and isotopic labeling have all been used to probe this multi-step process. The now generally accepted mechanism for cobalt-catalyzed hydroformylation (Scheme 17.9) was first proposed by Heck and Breslow on the basis of reactions of model organocobalt–carbonyl complexes.[34]

The catalytic cycle (inside the box in Scheme 17.9) involves a succession of 18-electron and 16-electron species. The precatalyst enters the cycle by the dissociation of CO from $HCo(CO)_4$ to generate the very reactive 16-electron complex $HCo(CO)_3$. Alkene coordination, followed by migratory insertion of the coordinated alkene into the cobalt hydride, produces either the linear (*n*-) or branched (*i*-) $RCo(CO)_3$ species. Coordination of CO then

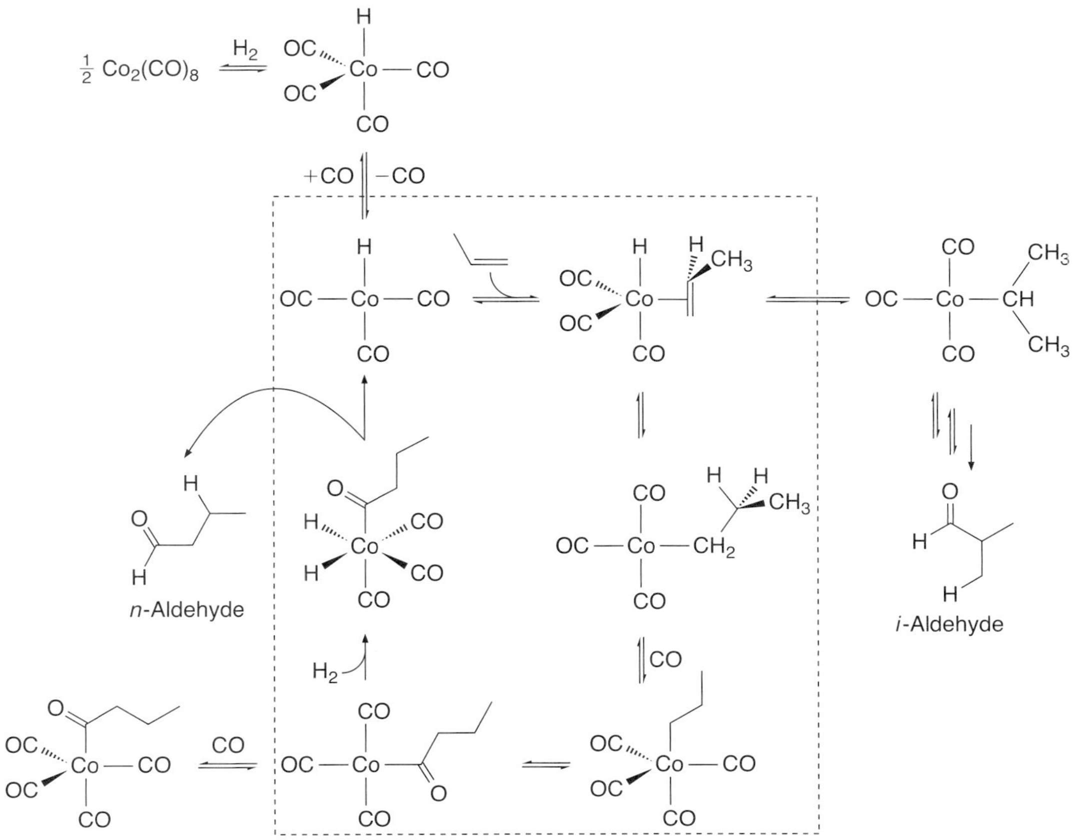

Scheme 17.9

forms the 18-electron RCo(CO)$_4$ species, which have been independently synthesized and characterized. Alkyl migration to CO then produces RC(O)Co(CO)$_3$, which can coordinate another CO to give RC(O)Co(CO)$_4$. These acylcobalt complexes have been independently synthesized and are the only cobalt species detected by infrared spectroscopy during hydroformylation of 1-octene at 150 °C and 250 atm of synthesis gas.[35] To this point of the catalytic cycle, all of the reactions have the potential to be reversible.

Two general pathways have been discussed for the formation of aldehyde from the acylcobalt intermediate. By one pathway, the coordinatively unsaturated RC(O)Co(CO)$_3$ undergoes oxidative addition of H$_2$, followed by reductive elimination of the aldehyde. The coordinatively unsaturated RC(O)Co(CO)$_3$ could form by dissociation of CO from the stable RC(O)Co(CO)$_4$, or H$_2$ could add before coordination of CO occurs to RC(O)Co(CO)$_3$. Alternatively, the aldehyde could form from the reaction of HCo(CO)$_4$ with the coordinatively unsaturated RC(O)Co(CO)$_3$ by an oxidative addition of the H–Co bond, followed by reductive elimination to form aldehyde and a dinuclear cobalt–carbonyl product.

Although the reaction of HCo(CO)$_4$ with RC(O)Co(CO)$_3$ has been observed in stoichiometric model studies,[36-38] the consensus is that the aldehyde is formed during catalytic hydroformylation by reaction of the acylcobalt species with H$_2$.[38-40] The observation of RC(O)Co(CO)$_4$ as the major species present in the system, along with the dependence of the rate on [H$_2$][CO]$^{-1}$, implies that hydrogenolysis of RC(O)Co(CO)$_4$ is turnover limiting and that it occurs by reversible dissociation of CO, followed by oxidative addition of H$_2$. This scenario is a classic case of the accumulation of the complex that is one of the reactants in the turnover-limiting step.

HCo(CO)$_4$ and Co$_2$(CO)$_8$ are observed during hydroformylation of less-reactive substrates, such as cyclohexene. These data are consistent with a different turnover-limiting step that involves insertion of the olefin. As is often the case in homogeneous catlaysis, the three species detected during catalysis [$RC(O)Co(CO)_4$, $HCo(CO)_4$, and $Co_2(CO)_8$] are not part of the catalytic cycle shown within the box in Scheme 17.9. However, they are intimately connected to the species on the catalytic cycle by equilibria, by ligand substitution or ligand dissociation.

17.3.2.2. Regioselectivity of Hydroformylation Catalyzed by HCo(CO)₄

Hydroformylation of terminal alkenes catalyzed by $HCo(CO)_4$ in the absence of any added ligands typically occurs to form a 3–4:1 ratio of straight-chain to branched aldehydes (*n:i* ratio). This regioselectivity might be established by irreversible insertion of the olefin into the cobalt hydride to form a combination of linear and branched cobalt–alkyl complexes. Alternatively, the *n:i* ratio might be controlled by reversible formation of the cobalt–alkyl species, followed by irreversible insertion of CO into the two alkyl complexes to form linear and branched cobalt–acyl complexes. Finally, the *n:i* ratio might be controlled by reversible formation of a cobalt–acyl complex from $HCo(CO)_4$, olefin, and CO, followed by irreversible hydrogenolysis of the isomeric acylcobalt species to yield the two aldehydes and regenerate the starting cobalt hydride.

Reversible formation of a cobalt–alkyl complex from $HCo(CO)_4$ and olefin has been established. The reversible formation of a cobalt alkyl is required to explain observations of alkene isomerization during hydroformylation by $HCo(CO)_4$ and scrambling of deuterium in substrates, such as $CD_3(CH_2)_3CH=CH_2$, under relatively low pressures of the mixture of CO and H_2 (Equation 17.4).[41] Little deuterium is lost during transformation of the alkene to product, but the deuterium atoms are distributed nearly statistically along the chain.[42]

$$D_3C \diagup\diagdown\diagup\diagdown\diagup \quad \xrightarrow[\text{2. Oxidation, MeOH, H}^\oplus]{\text{1. Co}_2\text{(CO)}_8\text{, H}_2\text{, CO}} \quad C_6H_{10}D_3CO_2Me \qquad (17.4)$$

In other cases, it has been shown that isomeric cobalt–alkyl complexes interconvert without dissociation of alkene from the hydridocobalt–alkene complex. For example, isomerization through a tertiary alkyl intermediate without alkene dissociation is required by retention of optical activity and deuterium labeling in the products of hydroformylation of the non-racemic alkene in Equation 17.5.[43,44] These results are consistent with extensive, rapid, intramolecular isomerization of the intermediate cobalt alkyl to place the metal at terminal and internal positions along the alkyl chain.

$$\xrightarrow[\substack{\text{2. Ag}_2\text{O} \\ \text{3. CH}_2\text{N}_2}]{\text{1. Co}_2\text{(CO)}_8\text{, H}_2\text{, CO}} \qquad (17.5)$$

92%   5%

17.3.3. Hydroformylation Catalyzed by HCo(CO)₃(PR₃)

17.3.3.1. Comparison of Rate, Selectivity, and Mechanism to Hydroformylation Catalyzed by HCo(CO)₄

The regioselectivities and rates of hydroformylation with cobalt catalysts were improved significantly by the addition of phosphines to $HCo(CO)_4$ (Equation 17.6). This hydroformylation catalyzed by $HCo(CO)_3(PR_3)$ complexes was developed by Slaugh and Mullineaux at

Shell.[45,46] In model studies, PBu_3 has typically been used as the added phosphine, but under industrial conditions, a more stable, high-molecular-weight phosphabicyclonane is used.[47-49] The electron-donating property of a phosphine ligand increases the strength of the Co–CO bond by increasing backbonding. This increased strength of the Co–CO bond allows the reaction to be conducted at lower CO pressures (50–100 atm) and at higher temperatures without deposition of cobalt metal. In addition, reactions of alkenes conducted with the phosphine-modified catalysts occur to form aldehydes with higher $n{:}i$ ratios (typically about 8:1).

$$n : i = 8 : 1$$

The electron-donating property of the phosphine also increases the rate of hydrogenation of the initially produced aldehydes to form alcohols. Thus, Shell employs the phosphine-ligated cobalt catalyst for conversion of internal linear alkenes to terminal linear alcohols that are ultimately used as precursors to detergents. However, these phosphine-ligated cobalt catalysts react about five times more slowly than the unmodified cobalt catalysts, and alkene hydrogenation is a significant side reaction (10–15%). Because the cost of raw materials is a major contributor to the overall cost of running this process, the competing hydrogenation of alkenes to undesired alkanes is a concern.

The major species in the catalytic system with added phosphine is different from that in the system lacking added phosphine. Infrared spectroscopic analyses under steady-state conditions for hydroformylation catalyzed by the phosphine-modified complexes provide no evidence for the accumulation of an acylcobalt complex; only phosphine-substituted cobalt carbonyl dimers and hydride complexes are observed.[50]

17.3.3.2. Hydroformylation of Internal Alkenes Catalyzed by $HCo(CO)_3(PR_3)$

One important process for the production of aldehydes and alcohols on commodity scale is the hydroformylation of internal olefins to terminal aldehydes. As noted in Chapters 21 and 22 on olefin oligomerization and olefin metathesis, a large portion of internal olefins generated by the Shell higher olefin process are used in hydroformylation. Hydroformylation of the internal olefin generates a linear aldehyde, and this aldehyde can then be converted to the linear alcohol. The formation of linear aldehydes from internal olefins occurs by insertion of the olefin into a cobalt hydride to form a branched alkyl species, followed by isomerization of the branched alkyl species to the linear alkyl. Insertion of CO, followed by hydrogenolysis of the resulting acyl complex, then generates the linear aldehyde. The phosphine-modified cobalt systems are reasonably good catalysts for alkene isomerization, and branched alkyl complexes rapidly undergo isomerization to linear alkyl complexes. Thus, the $n{:}i$ ratio is usually the same for reactions initiated with terminal or internal alkenes, and cobalt catalysts are principally used for the hydroformylation of internal alkanes. For example, hydroformylation of either 1- or 2-hexene catalyzed by $Co_2(CO)_8/2PBu_3$ at 170 °C and 30 atm of H_2 and CO gives the same 7:1 ratio of $n{:}i$ alcohols (Equation 17.7).[51]

17.3.4. Rhodium-Catalyzed Hydroformylation

17.3.4.1. Overview

The second generation of hydroformylation catalysts consists of rhodium complexes containing phosphorus ligands, and these catalysts were developed by researchers at Celanese Corporation and then Union Carbide Corporation in the 1970s.[31] These reactions occurred with lower pressures of CO and this class of hydroformylation was called the "low-pressure oxo" process. This process has been industrially attractive because the lower pressure of CO reduces capital and operating costs, and the presence of phosphorus ligands on rhodium has allowed for the fine-tuning of rates and selectivities.

In the absence of added phosphorus ligands, rhodium-catalyzed hydroformylation of alkenes is very rapid (10^3–10^4 faster than Co), but it is unselective and gives about a 1:1 ratio of linear and branched aldehydes. In 1965, Wilkinson discovered that $RhCl(PPh_3)_3$ could be used as a catalyst precursor for hydroformylation at room temperature and atmospheric pressure and that the hydroformylation was highly selective for n-aldehydes ($n:i$ >10) at high catalyst concentrations. Furthermore, these reactions occurred without competing formation of alcohol, hydrogenation of the alkene, or isomeriation of the alkene.[52] The active catalyst was soon discovered to be $HRh(CO)_2(PPh_3)_2$, and common catalyst precursors are now $HRh(CO)(PPh_3)_3$ and $Rh(acac)(CO)_2$ (acac = acetylacetonate), as shown in Equation 17.8. These phosphine-modified rhodium hydroformylation catalysts can be used in the synthetic laboratory without resorting to specialized high-pressure equipment.

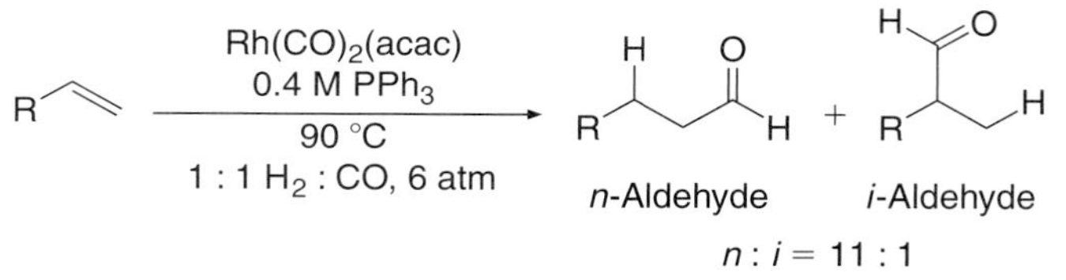

$$
R\diagup\!\!= \xrightarrow[\substack{90\ ^\circ C \\ 1:1\ H_2:CO,\ 6\ atm}]{\substack{Rh(CO)_2(acac) \\ 0.4\ M\ PPh_3}} \underset{n\text{-Aldehyde}}{R\diagdown\!\diagup\!\diagdown\!\overset{O}{\diagup}H} + \underset{i\text{-Aldehyde}}{R\diagdown\!\diagup\!\overset{\overset{H\diagdown\ O}{}}{}\diagdown H}
$$

$$n:i = 11:1$$

(17.8)

17.3.4.2. Rhodium Catalysts for Hydroformylation Containing Triarylphosphine Ligands

17.3.4.2.1 Discovery and Reactivity of the Original Catalyst

Pruett and coworkers at Union Carbide found that the combination of rhodium precursors and excess PPh_3 created an active, selective, and stable system that catalyzed hydroformylation with 5–10 atm of CO and H_2 at 90 °C. This system was developed into a commercial hydroformylation process in the early 1970s.[53,54] Excess PPh_3 was used to replenish PPh_3 that is lost by degradation through several different pathways and to favor formation of $HRh(CO)_2(PPh_3)_2$ over less selective rhodium complexes containing fewer PPh_3 ligands (Scheme 17.10). Hydroformylation reactions conducted with higher concentrations of PPh_3 occur more slowly, but with higher $n:i$ selectivity.

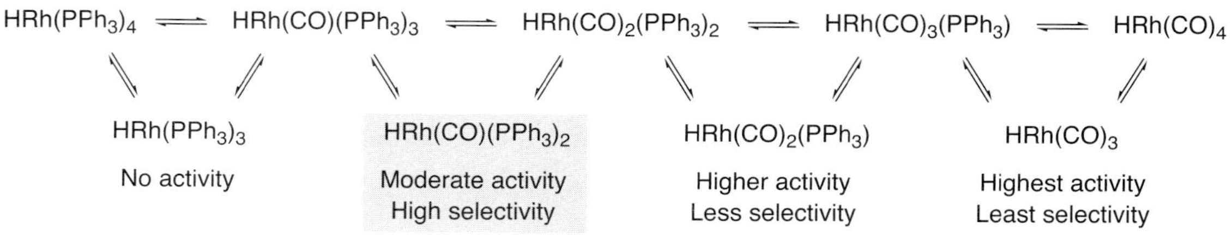

$$HRh(PPh_3)_4 \rightleftharpoons HRh(CO)(PPh_3)_3 \rightleftharpoons HRh(CO)_2(PPh_3)_2 \rightleftharpoons HRh(CO)_3(PPh_3) \rightleftharpoons HRh(CO)_4$$

$HRh(PPh_3)_3$	$HRh(CO)(PPh_3)_2$	$HRh(CO)_2(PPh_3)$	$HRh(CO)_3$
No activity	Moderate activity High selectivity	Higher activity Less selectivity	Highest activity Least selectivity

Scheme 17.10

The *n:i* ratio of aldehydes formed by the hydroformylation of a 1-alkene with rhodium catalysts containing PPh$_3$ varies from 3 to 12. The highest regioselectivity is obtained at high concentrations of PPh$_3$ and low pressures of CO. As depicted in Scheme 17.10, at high concentrations of PPh$_3$, *n*-aldehydes are formed selectively through diphosphine intermediates. At lower concentrations of PPh$_3$, more reactive and less selective monophosphine intermediates are responsible for the lower *n:i* ratios.

Steric interactions affect both the rate and regioselectivity of Rh/PPh$_3$-catalyzed hydroformylation. Terminal alkenes undergo hydroformylation one to two orders of magnitude more rapidly than internal alkenes and about three orders of magnitude more rapidly than tri-substituted alkenes.[55–57] *Cis*-2-alkenes react a few times faster than *trans*-2-alkenes. Among simple 1-alkenes containing methyl substituents at carbon-3 or carbon-4, greater amounts of linear product is formed from olefins containing a higher degree of substitution closer to the alkene bond, but the rates are not strongly affected by these substituents (Figure 17.1).[58]

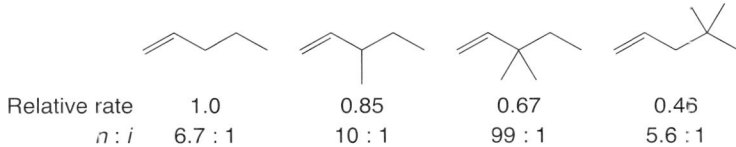

Relative rate	1.0	0.85	0.67	0.46
n : i	6.7 : 1	10 : 1	99 : 1	5.6 : 1

Figure 17.1.
Relative rates and *n:i* ratios for reaction of substituted alkenes.

17.3.4.2.2. Mechanism of Hydroformylation Catalyzed by HRh(CO)$_2$(PPh$_3$)$_2$

Many studies have been conducted on the mechanism of hydroformylation catalyzed by HRh(CO)$_2$(PPh$_3$)$_2$. Some of these studies have focused on identifying the complexes in the catalytic system, and others have focused on obtaining kinetic data. Together, these data have provided mechanistic insight. However, hydroformylation involves many different steps, the formation of at least two isomeric products, and the intermediacy of species that can exist in several isomeric forms. Thus, this catalytic system is very complex, and only broad conclusions are drawn here.

The structures of the hydridorhodium complexes that are present in the catalytic system have been deduced by NMR spectroscopy. Brown showed that HRh(CO)$_2$(PPh$_3$)$_2$ exists as an 85:15 mixture of diequatorial:apical–equatorial isomers of HRh(CO)$_2$(PPh$_3$)$_2$ (Scheme 17.11) that undergoes rapid equilibration at room temperature.[59] This rapid equilibration of trigonal bipyramidal complexes could occur by either a Berry pseudorotation mechanism or a turnstile mechanism. In situ IR transmission spectroscopy on the catalytic system demonstrated that these two isomers are the resting state of the catalyst and were by far the predominant species present during hydroformylation of 1-octene (60–100 °C, 5–20 atm, [Rh] = 1 mM, PPh$_3$/Rh = 5).[60]

The basic steps of the mechanism of hydroformylation catalyzed by the combination of Rh and PPh$_3$ are similar to those of hydroformylation catalyzed by HCo(CO)$_4$. However, the number of possible intermediates is much larger because many combinations of the number of phosphines and geometric orientation are possible in each intermediate complex. For example, phosphines in a five-coordinate complex can occupy the apical or equatorial positions, and these complexes are typically stereochemically nonrigid. Similarly, two phosphines in a four-coordinate intermediate can be located cis or trans to each other. A five-coordinate hydridorhodium–alkene complex containing diequatorial phosphines might be expected to convert to a four-coordinate Rh alkyl complex containing trans phosphines, but it could also lead to an alkyl complex containing cis phosphines.

In addition, many different combinations of steps can control the *n:i* ratio. The regiochemistry of aldehyde might be controlled by irreversible addition of the Rh–H unit across the alkene, irreversible insertion of CO into an equilibrating mixture of *n*-alkyl- and *i*-alkylrhodium complexes, or by irreversible hydrogenolysis of an equilibrating mixture

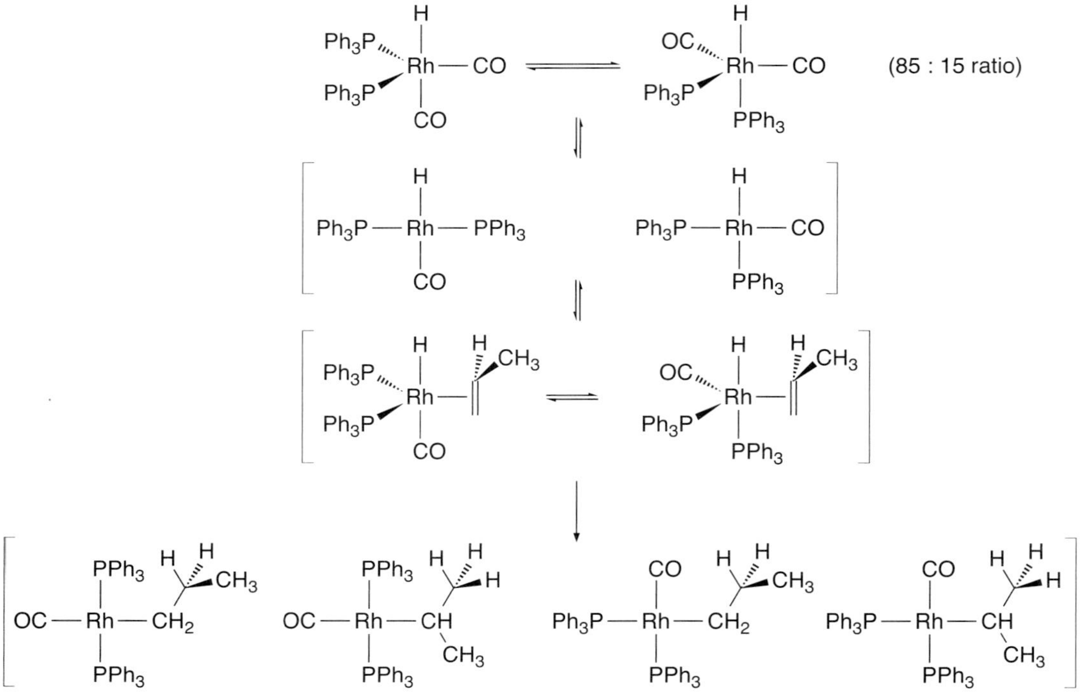

Scheme 17.11

of *n*-acyl- and *i*-acylrhodium complexes. Aldehyde formation is irreversible under hydro-formylation conditions. However, even formation of the aldehyde can be forced to be reversible. Brookhart has shown that Cp*Rh(CH$_2$=CHMe)$_2$ catalyzes the interconversion of *n*-butyraldehyde and *i*-butyraldehyde at 80 °C, and the only Rh species detected during catalysis is Cp*Rh(CH$_2$CH$_2$Me)$_2$(CO) (Equation 17.9).[61] The great complexity of the Rh/PPh$_3$ system and the sensitivity to reaction conditions has made it very difficult to disentangle the steps of the mechanism of this hydroformylation system.

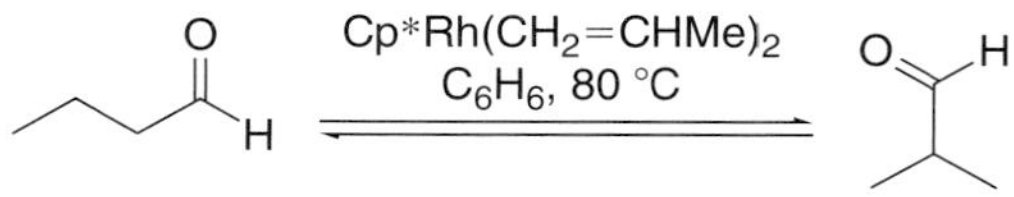

$$ (17.9) $$

The kinetics of the hydroformylation catalyzed by the combination of a rhodium precursor and PPh$_3$ depend strongly on reaction conditions. Under "standard industrial operating conditions" (T = 70–120 °C, CO = 5–25 atm, H$_2$ = 5–25 atm, Rh ≈ 1 mM, and alkene = 0.1–2 M), hydroformylation is first order in [alkene] and [Rh], zero order in [H$_2$], and negative order in the concentration of ligand (PR$_3$ or CO, or both).[62] In combination with the observation of HRh(CO)$_2$(PPh$_3$)$_2$ as the catalyst resting state, this kinetic behavior is consistent with a rate-limiting insertion of the coordinated alkene into the hydride ligand in RhH(PPh$_3$)$_2$(CO)(alkene), which is formed reversibly from RhH(PPh$_3$)$_2$(CO)(L) and the alkene.

17.3.4.3. Water-Soluble Rhodium Hydroformylation Catalysts

A serious problem encountered when practicing industrial hydroformylation is separation and recycling of the valuable rhodium catalyst. This problem is particularly severe when forming aldehydes that require temperatures for distillation that lead to decomposition of the rhodium catalyst. To address this issue, Kuntz at Rhone-Poulenc developed water-soluble Rh–triarylphosphine catalysts.[63,64] The reactions are then conducted in a biphasic system in which the catalyst resides in the aqueous phase and the product resides in the organic phase and can be removed from the catalyst without distillation.

The catalyst is generated from the product of sulfonation of PPh_3. This ligand is a mixture of species but largely consists of $P(C_6H_4\text{-}m\text{-}SO_3^-Na^+)_3$. Binding of this ligand to rhodium generates a highly water-soluble catalyst. A stirred tank reactor containing two immiscible phases is used. The aqueous phase contains the highly water soluble $HRh(CO)[P(Ph\text{-}m\text{-}SO_3^-Na^+)_3]_3$, and the organic phase contains the product aldehyde, the alkene reagent, and some products from condensation of the aldehyde. Propene is sufficiently water soluble that good rates are obtained, but longer chain alkenes are too insoluble in the aqueous phase to react at useful rates for commercial hydroformylation.

17.3.4.4. Rhodium Catalysts Containing Chelating Diphosphine Ligands

17.3.4.4.1. Early Studies with Less Selective Catalysts

As stated in the introduction to rhodium catalysts for hydroformylation, the ligands on rhodium have a large influence on rate and selectivity. Thus, catalysts containing ligands more sophisticated than PPh_3 have been developed. In particular, rhodium complexes containing chelating diphosphine ligands catalyze the formation of aldehydes with high *n:i* selectivity.[54] By limiting the number of potential isomers, complexes of these ligands have also led to considerable mechanistic insight.

Because $HRh(CO)_2(PPh_3)_2$ was believed to be the species that reacted with olefin to ultimately form aldehyde with a ~9:1 *n:i* ratio, it was thought that complexes containing diphosphine ligands (Figure 17.2) could form aldehydes with high regioselectivity. However, when hydroformylations were conducted with catalysts generated from $HRh(CO)$ $(PPh_3)_3$ and simple diphosphines, such as 1,2-bis(diphenylphosphino)ethane (dppe), 1,3-bis(diphenylphosphino)propane (dppp), or 1,2-bis(diphenylphosphino)butane (dppb), the regioselectivity was only about 4:1.[65–67] One possible explanation for this low selectivity is the requirement that the five-coordinate alkene hydride complex possess a structure in which the chelating ligand binds with one phosphorus in the apical position and one in the equatorial position and that this "apical–equatorial" chelation leads to low regioselectivity. Later, Unruh found that higher linear to branched (5–8:1 *n:i*) selectivity was observed when the catalyst was generated from $HRh(CO)(PPh_3)_3$ and diop or dppf.[68,69] These two ligands bind with wider P–M–P "bite angles" (see Chapter 2 for a description of bite angles) that might allow the bisphosphine to bind with both phosphorus atoms in equatorial positions. A higher population of this geometry could lead to higher regioselectivity.

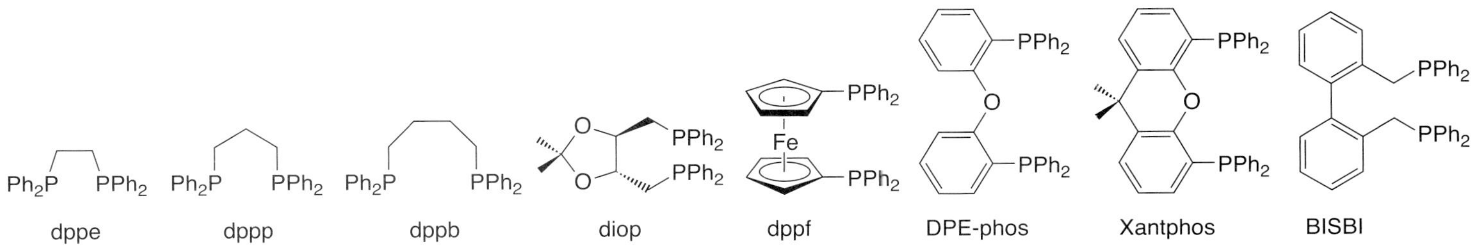

Figure 17.2.
Bisphoshine ligands tested for the generation of regioselective hydroformylation catalysts.

17.3.4.4.2. Catalysts Containing Wide-Bite-Angle Bisphosphines

In 1987, Devon at Texas Eastman reported that complexes generated from the bisphosphine BISBI catalyzed the formation of aldehyde with a very high (30:1) *n:i* regioselectivity at high temperatures.[70] A Shell patent in 1989[71,72] reported that reactions catalyzed by rhodium and ether-linked bisphosphines gave aldehydes with a l:b ratio of 10, and this result eventually led to the use of Xantphos.[73–75] Casey and Whiteker examined molecular models of the metal complexes of BISBI and concluded that this ligand prefers a chelate bite angle much greater than 90°.[76] These authors speculated that ligands possessing wider bite angles might lead to catalysts that react with high *n:i* regioselectivity in the hydroformylation of 1-alkenes by favoring diequatorial coordination over apical–equatorial coordination. This proposal for regioselectivity is depicted in Scheme 17.12.

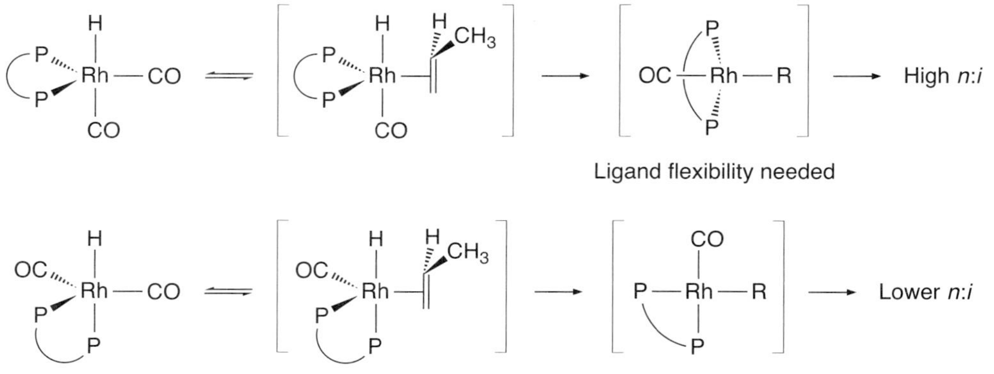

Scheme 17.12

To identify ligands possessing wide bite angles, Casey and Whiteker[77] introduced the concept of the "natural bite angle", which is the preferred L–M–L angle of a bidentate ligand determined only by ligand backbone constraints and not by metal valence angles. The natural bite angle is calculated using molecular mechanics. The natural bite angle is calculated by minimizing the strain energy of the diphosphine–M fragment with an L–M–L bending force constant of 0 kcal mol^{-1} rad^{-2} and M–P distances from X-ray structures of similar compounds. In addition, by fixing the bite angle at various angles by using a large bending force constant and then calculating the strain energy, potential diagrams can be constructed that allow an estimation of the flexibility of the ligand. The "flexibility range" was defined as the range of angles accessible with 3 kcal mol^{-1} or less additional strain. Flexibility is required for a wide bite angle diphosphine to span both diequatorial positions in five-coordinate complexes and trans positions in four-coordinate intermediates.

The structure of HRh(CO)(BISBI)(PPh$_3$) determined by X-ray diffraction demonstrated for the first time the unequivocal diequatorial coordination of a bidentate diphosphine.[76] The 124° bite angle measured for BISBI is similar to the calculated natural bite angle of 112°. When HRh(CO)(BISBI)(PPh$_3$) was placed under 1 atm of CO, PPh$_3$ was replaced by CO to give HRh(CO)$_2$(BISBI), which was shown by NMR and IR spectroscopy to possess a structure in which the two phosphorus atoms reside in the equatorial plane, and the hydrogen atom resides in an apical position.

Hydroformylation of 1-hexene under mild conditions (34 °C, 6 atm 1:1 CO:H$_2$, and 4 mM Rh) was carried out with a series of ligands possessing varying bite angles. Figure 17.3 shows that there is a strong correlation between the natural bite angle of the diphosphine and the *n:i* ratio of the aldehyde product, ranging from 66.5 for BISBI (bite angle 113°) to 2.1 for DPPE (bite angle 85°). Little alkene isomerization was observed. It was concluded that diequatorial coordination of a diphosphine led to high *n:i* ratios, although the reason for this relationship is not well understood.

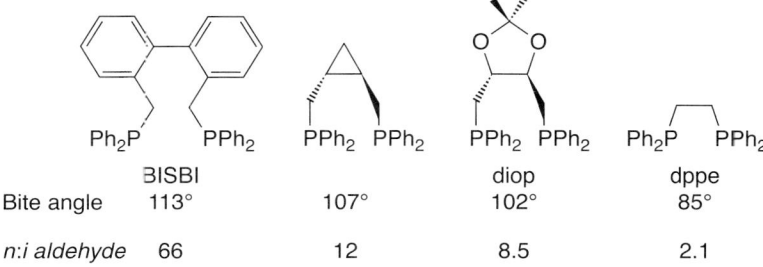

Figure 17.3.
Relationship between several ligand bite angles and *n:i* selectivity for
propene hydroformylation.

The origin of the high selectivity was evaluated by deuterium labeling studies. To test whether insertion of alkene into a rhodium hydride to produce 1-alkyl- and 2-alkylrhodium species was irreversible in the Rh/BISBI system, the deuterioformylation of 1-hexene was studied under mild conditions.[78] The potential equilibria and deuterium migrations are shown in Scheme 17.13. If formation of the rhodium alkyl were irreversible, then the product would contain deuterium only at the carbon β the carbonyl group; if formation of the rhodium alkyl were reversible, then a pathway would be available to place deuterium α to the carbonyl group and to incorporate deuterium into the 1-hexene. More than 96% of the initially formed deuterated n-alkylrhodium intermediate was converted to *n*-aldehyde that contained deuterium beta to the carbonyl moiety, establishing that formation of the *n*-alkylrhodium intermediate is irreversible. In contrast, only 25% of the initially formed deuterated *i*-alkylrhodium intermediate was converted to *i*-aldehyde that contained deuterium β to the carbonyl unit; the remaining 75% was converted to 1-deuterio-1-hexene. This result indicates that formation of the *i*-alkylrhodium species is partially reversible. The absence of significant amounts of deuterium on the carbon located α to the carbonyl of the branched aldehyde indicates that alkene dissociation occurs after β-hydrogen elimination. Thus, the high regioselectivity of the Rh/BISBI system appears to result from selective and irreversible insertion of the coordinated 1-alkene into the rhodium hydride to form an *n*-alkylrhodium intermediate and reversible insertion to form the *i*-alkylrhodium intermediate.

In contrast, deuterioformylation of 1-hexene with catalysts containing dppe and other bisphosphines under these mild conditions formed products containing deuterium only in the β-positions of the aldehdydes. This observation shows that formation of both *n*- and *i*-alkylrhodium complexes was irreversible with these less selective catalysts.

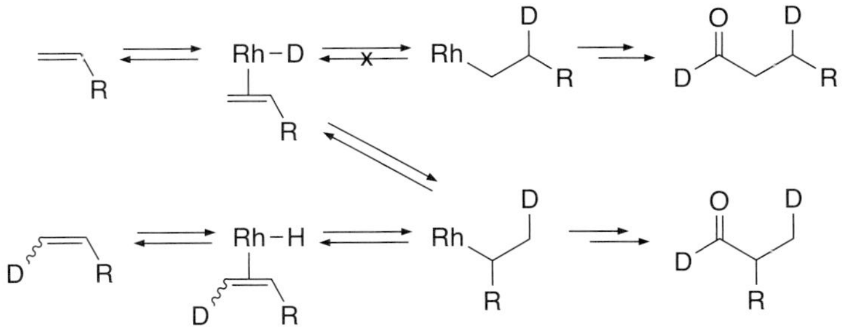

Scheme 17.13

The correlation between regioselectivity and chelation mode suggests that complexes containing diequatorial bisphosphines like BISBI and complexes containing apical–equatorial chelating diphosphines like DIPHOS have either significantly different steric or electronic properties. Molecular mechanics calculations were employed to explore whether the selectivity resulted from steric differences between the transition states leading to the primary and secondary alkylrhodium intermediates, but these calculations did not provide significant insight.[78]

17.3.4.4.3. Effect of Diphosphine Electronic Properties on Regioselectivity

Much work has been conducted to try to understand the relationship between ligand bite angle and ligand electronic properties on the *n:i* ratio. In one set of studies, the effect of ligand electronic properties on the isomeric ratio of aldehyde products was investigated by studying a series of symmetrically substituted diphosphines that adopt different chelation modes and possess varied electronic environments at the diarylphosphino groups.[79] These studies led to the hypothesis that an electron-withdrawing substituent on an equatorial phosphine increases the *n:i* ratio, while an electron-withdrawing substituent on an apical phosphine decreases the *n:i* ratio. To test this hypothesis, hydroformylation catalyzed by complexes containing the dissymmetric DIPHOS derivative [3,5-$(CF_3)_2$-$C_6H_3]_2PCH_2CH_2PPh_2$ was studied. NMR spectroscopy showed that the electron-withdrawing phosphino group bound in the equatorial position and the electron-donating phosphino group bound in the apical position. Reactions catalyzed by complexes of the dissymmetric diphosphine formed product with a higher *n:i* ratio than either of the related symmetric DIPHOS derivatives.[80] From these data, it was concluded that the most selective dppe catalysts contain the less electron-donating phosphino group in the equatorial position and the more electron-donating phosphino group in the apical position.

Van Leeuwen reported a range of Xantphos ligands possessing bite angles ranging from 102° to 121° that probe the effect of this ligand parameter on rhodium-catalyzed hydroformylation. Two of those ligands are shown in Figure 17.4. Crystallographic data on $HRh(CO)_2L_2$ complexes in which L_2 is a Xantphos ligand revealed distorted trigonal bipyramidal geometries in which the bidentate ligand occupies two equatorial sites.[81] ^{1}H and ^{31}P NMR spectra of these complexes showed that they exist as mixtures of diequatorial and apical–equatorial isomers that are in rapid, dynamic equilibrium. The percentage of diequatorial coordination varied from 30% for a complex containing a bite angle of 102° to 80% for a complex containing a bite angle of 114°. The same ratio of isomers was determined by IR spectroscopy in the catalytic hydroformylation system. The selectivity for linear aldehyde increased from 88 to 96% as the bite angle of the ligand increased, and the rate of hydroformylation increased nearly 10-fold.[82, 83]

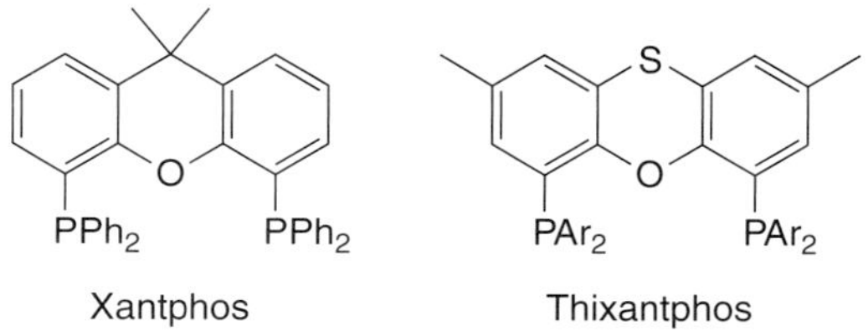

Figure 17.4.
Two Xantphos-type ligands used in rhodium-catalyzed hydroformylation.

Xantphos   Thixantphos

However, several studies by van Leeuwen and co-workers showed that the ratio of diequatorial and apical–equatorial isomers did not correlate well with the natural bite angle and that the ratio of diequatorial and apical–equatorial isomers did not correlate well with product selectivity. In one study, van Leeuwen and co-workers investigated electronic effects on the combination of the ratio of *n*- to *i*-aldehyde products and the ratio of diequatorial to apical–equatorial binding modes.[83] These studies were conducted with thixantphos ligands possessing electron-donating and electron-withdrawing groups on the aryl substituents of the diarylphosphino groups.

Three observations and conclusions were drawn from this study. First, hydroformylation of 1-octene conducted with catalysts containing electron-withdrawing substituents on the diarylphosphino groups occurred faster than reactions containing electron-donating substituents on the diarylphosphino groups. This observation was rationalized by the effect of ligand electronic properties on the equilibrium for dissociation of CO and binding of alkene. Because the binding of CO depends more on the degree of backbonding than the binding of alkenes, the equilibrium for dissociation of CO from complexes containing less electron-donating ligands is increased more than that for the dissociation of alkenes. Thus, the equilibrium for replacement of CO by olefins is larger for complexes containing less electron-donating ligands.

Second, the ratio of linear to branched aldehyde increased with decreasing electron donation by the ligand, but the percent of linear aldehyde versus all reaction products was nearly constant because the amount of 2-octene formed by olefin isomerization increased. This observation was explained by proposing that the ratio of n-alkyl to i-alkyl complex was constant and that the conversion of n-alkyl intermediate to n-aldehyde product was efficient, but that the conversion of the i-alkyl intermediate to 2-octene product was faster for catalysts containing the less electron-donating ligands.

Third, there was little relationship between the diequatorial versus equatorial–axial binding modes and the regioselectivity for the hydroformylation of 1-octene. The diequatorial versus equatorial–axial binding modes of the ligand depended on the electronic properties of the ligand (the more electron-poor the ligand, the larger the ratio of diequatorial to apical–equatorial binding modes), but the regioselectivity was similar for catalysts containing each of the electronically distinct ligands. This observation leaves the relationship between bite angle and selectivity a largely empirical one. Thus, van Leeuwen and co-workers have concluded that an increase in bite angle with selectivity results more from the effect of bite angle on alkene coordination to $HRhL_2(CO)$ and olefin insertion within the $HRhL_2(CO)(alkene)$ complex than on the structure of the $HRhL_2(CO)_2$ species.

17.3.4.5. Rhodium-Catalyzed Hydroformylation of Internal Alkenes

Rhodium complexes of certain phosphine ligands catalyze olefin isomerization and undergo rapid isomerization of branched alkyl complexes to linear alkyl complexes. These complexes catalyze the hydroformylation of internal alkenes to linear alkenes. For example, rhodium complexes of the dibenzophosphole-substituted Xantphos ligand catalyzes the hydroformylation of α-olefins with high activity and selectivity for formation of the linear product ($n{:}i > 60$). At the same time, rhodium complexes of this ligand exhibit high activity and selectivity for the hydroformylation of $trans$-2-octene and $trans$-4-octene to 86–90% linear nonanal (Equation 17.10).[84] These very bulky Xantphos ligands generate catalysts that are selective for formation of linear aldehydes from internal alkenes because the i-alkyl intermediate undergoes β-hydride elimination in competition with CO coordination.

$$\text{1.0 mM [Rh]} \quad \text{2 atm CO, H}_2 \quad 120\ °C$$

86%

(17.10)

17.3.4.6. Hydroformylation Catalyzed by Rhodium Complexes of Phosphites[85]

In addition to complexes of phosphine ligands, complexes of phosphite ligands have been shown to catalyze hydroformylation with high linear-to-branched selectivities, and these systems are used commercially. Early data were obtained on catalysts containing sterically hindered monophosphites,[53,86–91] and these catalysts reacted with fast rates. Subsequent studies focused on complexes of bisphosphite ligands, and a series of patents showed that catalysts containing this class of ligand were particularly reactive and selective for formation of the n-aldehyde from 1-alkenes without competing hydrogenation.[92,93]

The rates of hydroformylation catalyzed by complexes of bisphosphites were slower than those by complexes of monodentate phosphites, but they are faster than those catalyzed by complexes of PPh$_3$. The *n:i* regioselectivity of the aldehyde produced by reactions catalyzed by complexes of bidentate phosphites were high, particularly when bulky phosphites were joined by a relatively long 5–6 atom bridge. The bisphenol-bridged diphosphite shown in Equation 17.11 generated a catalyst that formed butyraldehyde from propene with over 50:1 *n:i* regioselectivity. Because these hydroformylations are relatively fast under easily achieved laboratory conditions (70 °C, 5 atm CO:H$_2$) and occur with excellent chemoselectivity, these reactions are beginning to be used in organic synthesis. For example, the Union Carbide catalyst has been used for selective hydroformylation of olefins in the presence of pendant ketones (Equation 17.12), esters, alcohols, nitriles, and halides.[94]

$$n : i = 53 : 1 \tag{17.11}$$

$$87\% \quad n : i > 40 : 1 \tag{17.12}$$

17.3.4.7. Rhodium-Catalyzed Hydroformylation of Functionalized Alkenes

The hydroformylation of alkenes bearing functional groups has been studied less extensively than the hydroformylation of unfunctionalized alkenes because of the importance of the products from the hydroformylation of alkenes as commodity chemicals. However, the use of hydroformylation for the preparation of fine chemicals has significant potential,[95] and the hydroformylation of geminally disubstituted alkenes, internal alkenes, and alkenes bearing functional groups have been studied in the context of organic synthesis. For example, the hydroformylation of geminally disubstituted alkenes has been shown to form linear products with high selectivity,[96–101] and reactions of cyclic alkenes of unsaturated heterocycles have been shown to form cyclic aldehydes. Examples of these reactions are shown in Equations 17.13–17.15.[102–105]

$$\tag{17.13}$$

$$\tag{17.14}$$

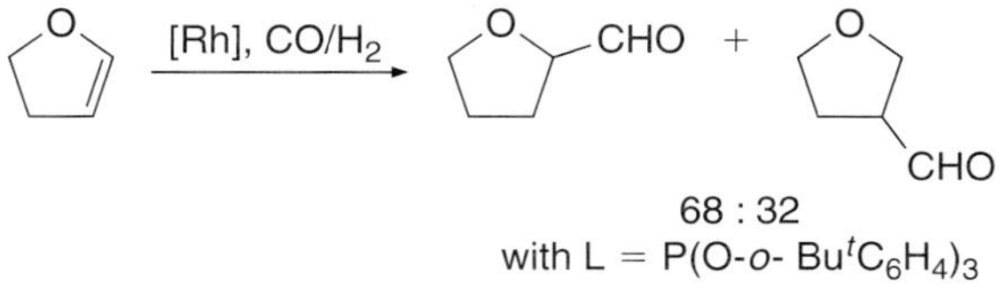

$$(17.15)$$

68 : 32
with L = P(O-o- Bu^{t}C$_6$H$_4$)$_3$

In addition, reactions of many olefins containing electron-withdrawing functional groups at the C=C bond react to form branched substitution products. Examples of these reactions catalyzed by rhodium–carbonyl complexes modified by PPh$_3$ are shown in Scheme 17.14.[95, 106–113] Directed hydroformylations, such as that in Equation 17.16, have also been studied.[114]

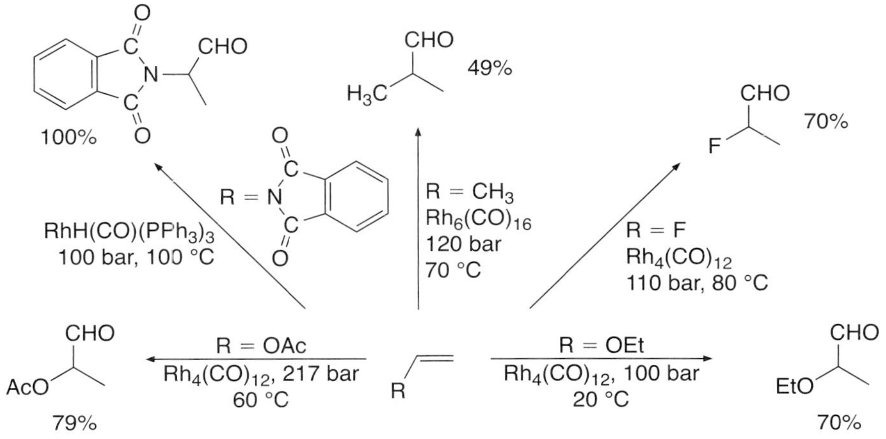

Scheme 17.14

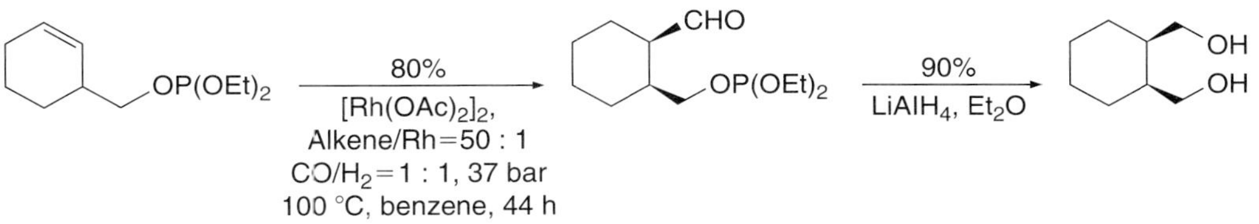

$$(17.16)$$

This branched regioselectivity in these cases results from the greater stability of the branched alkyl complex when the alkyl group bears an electron-withdrawing substituent on the α-carbon. This regioselectivity was discussed in Chapter 10 on insertion processes. The high selectivity for formation of branched products from the hydroformylation of vinylarenes results from the formation of an η^3-benzyl intermediate,[110, 1.1, 115, 116] as discussed in Chapter 10. The formation of these chiral, branched products has been a particular focal point for the development of enantioselective hydroformylation.

17.3.4.8. Enantioselective Hydroformylation[117–120]

Enantioselective hydroformylation has great synthetic potential as an efficient enantioselective carbon–carbon bond-forming process. This reaction has been studied with particular emphasis on the reactions of vinylarenes to form aldehydes that can be readily oxidized to analgesics, such as naproxen and ibuprofen (Figure 17.5). This process has been studied for more than 30 years. Early studies with catalysts containing bisphosphine ligands, such as DIOP, formed product in low ee. After nearly 20 years of research, a few bisphosphites and mixed phosphine–phosphite ligands have generated catalysts that form branched aldehydes with high enantioselectivities with certain classes of olefins. A new family of diazaphospholanes has also been discovered for this process. The following section describes selected systems that have been the focal point of research that has led to the most enantioselective hydroformylations.

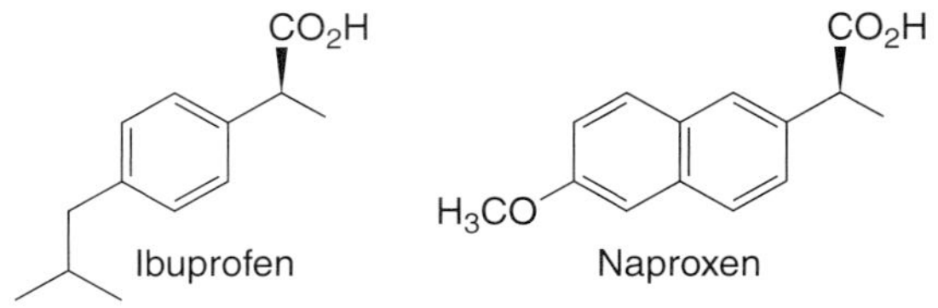

Figure 17.5.
Two non-steroidal analgesics that have been synthesized by hydroformylation, followed by oxidation.

The enantioselective hydroformylation process confronts a series of challenges. First, the hydroformylation of alkenes tends to form linear over branched products. Thus, enantioselective hydroformylation of alkenes must be conducted with unsymmetrical "vinylidenic" olefins (2,2-disubstituted) that establish a stereocenter at the carbon β to the carbonyl group, or with mono-substituted olefins that form the branched product. As noted above, vinylarenes, vinyl acetates, and other olefins bearing electron-withdrawing groups, form branched products. Thus, reactions of these olefins have been the greatest focus of enantioselective hydroformylations.

Second, simple alkenes bind solely through the π-system. Therefore, the two-point binding mode that leads to high enantioselectivities in many hydrogenations of enamides and related alkenes is absent in the hydroformylation of simple alkenes. Third, it is necessary that the hydroformylation catalyst be sufficiently reactive that the product is formed faster than it undergoes competing racemization at the enolizable α-stereocenter of the product. Finally, two of the ligands on the rhodium catalyst can create a chiral environment, but the two phosphines in a trigonal bipyramidal environment are farther from the olefin than in the square-planar geometry of many hydrogenation catalysts, and one of the ancillary ligands (carbon monoxide) is achiral.

Both platinum and rhodium catalysts have been studied for enantioselective hydroformylation. The asymmetric hydroformylation of an itaconate derivative occurs with the highest selectivity in the presence of platinum catalysts (Equation 17.17).[121] However, the chemical yields of branched aldehydes from reactions catalyzed by platinum catalysts are typically low. Therefore, enantioselective hydroformylation of mono-substituted olefins with platinum catalysts have generally been unsatisfactory, and most effort has been spent on developing rhodium catalysts for asymmetric hydroformylation.

$$PtCl(SnCl_3)[(R,R)\text{-DIOP}](0.05\%)$$
$$CO/H_2 = 1/1; 80 \text{ bar}; 100\ °C$$
$$\text{Conversion} = 80\% \text{ (45 h)}$$

$$ee = 82\%\ (R)$$

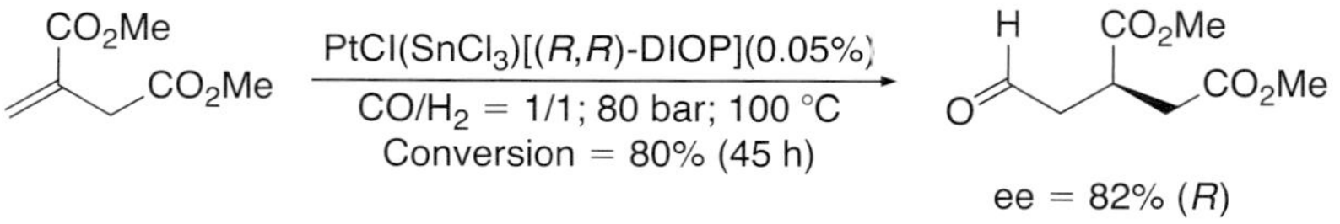

(17.17)

Van Leeuwen and co-workers first reported in the academic literature the enantioselective hydroformylation of styrene with high selectivity for formation of the branched product using 1,2- and 1,4-disposed bisphosphite ligands. However, enantioselectivities from this process were low.[122] Concurrent with this work, Babin and Whiteker described in a Union Carbide patent the most significant early work on asymmetric hydroformylation, in this case using 1,3-disposed bisphosphite ligands (Figure 17.6, left).[123] The Babin and Whiteker patent describes rhodium complexes of bisphosphite ligands derived from chiral 1,3-diols that catalyze the hydroformylation of styrene at moderate pressures with 98% selectivity for the branched product with 98% enantiomeric excess.

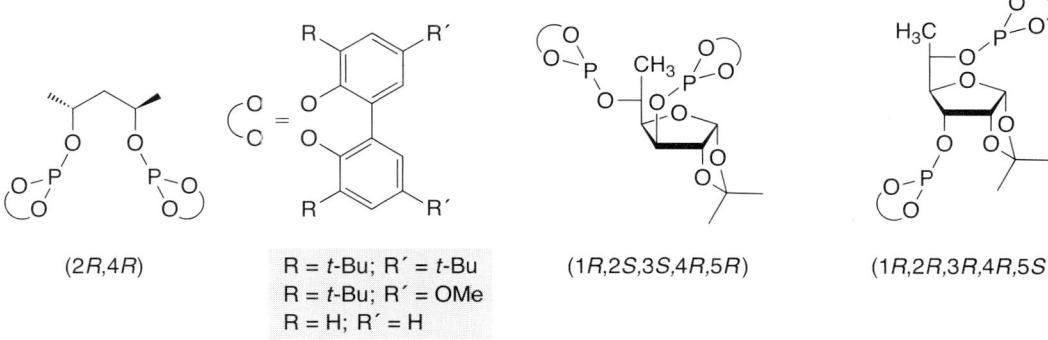

(2R,4R)

R = t-Bu; R′ = t-Bu
R = t-Bu; R′ = OMe
R = H; R′ = H

(1R,2S,3S,4R,5R)

(1R,2R,3R,4R,5S)

Figure 17.6. Chiral, non-racemic phosphite ligands for enantioselective hydroformylation.

Variation of this bisphosphite structure revealed the features of these ligands that are required for high enantioselectivity. The presence of a three-carbon linker in the diol backbone with stereocenters in the 1- and 3-positions is important for high selectivity, as is the presence of a sterically demanding ortho substituent on the biphenolate group on phosphorus.

A related series of tunable diphosphite ligands containing sugar backbones has been prepared and studied by van Leeuwen[124] and Claver.[125–128] These ligands also possess a 1,3-disposition of stereocenters in a three-carbon diol backbone, along with biphenolate substituents at phosphorus. In this case, the relative stereochemistry at the ring carbon and carbon-5 of the sugar backbone are important. Likewise, sterically demanding *tert*-butyl substituents in the ortho position and electron-donating *p*-methoxy substituents in the para position of the aryl groups create the most selective catalysts. These ligands are shown at the right of Figure 17.6 and generate catalysts that react with styrene to form > 98% branched product with 90% ee.

Most of the early bisphosphites used for asymmetric hydroformylation possessed C_2 symmetry. Thus, a major breakthrough was the use of mixed phosphite–phosphine ligands by Nozaki, Takaya, and co-workers.[128] The atropisomeric ligand (S,R)-BINAPHOS at the left of Figure 17.7 generates a rhodium complex that catalyzes the hydroformylation of vinylarenes with high branched-to-linear selectivities and high enantiomeric excess. With vinylarenes, vinyl acetate, and vinyl phthalimide, the percent of branched aldehyde was typically 85% or higher, and the ee of the product was between 83% and 97%. Many reactions, particularly those of vinylarenes, occurred in 92–95% ee.

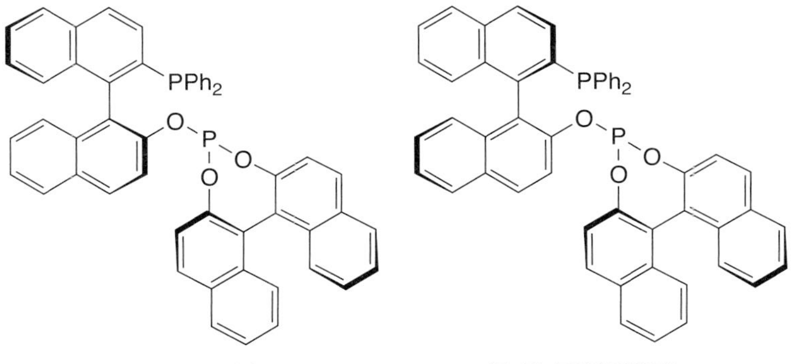

(S,R)-BINAPHOS
[95% (S)]

(R,R)-BINAPHOS
[25% (S)]

Figure 17.7.
Takaya's phosphine–phosphite ligands for asymmetric hydroformylation.

A series of studies have been conducted on phosphite–phosphine ligands containing varied substituents on the general structure of the Takaya ligand. Initial studies showed

Table 17.1. Scope of the enantioselective hydroformylation of olefins catalyzed by rhodium and (S, R)-BINAPHOS.[120]

Substrate	Product	% ee	Substrate	Product	% ee
allyl CN	CN / CHO	66	vinylcyclohexene	CHO	97
Ph	Ph * CHO	98.3	dihydrofuran	CHO	68
C₄H₉	C₄H₉ * CHO	90	OAc	OHC * OAc	92
	* CHO	89.9	SᵗBu	OHC * SᵗBu	90
Ph OH	Ph lactone	88	TBSO	TBSO CHO	89[a]

[a] Diastereomeric excess.

that the two diastereomeric catalysts containing the opposite relative stereochemistry of the two binaphtholate groups react with much different enantioselectivity. The (R,S)-BINAPHOS ligand (and its enantiomer) generate catalysts that react with high enantioselectivity, but the (R,R)-BINAPHOS ligand (Figure 17.7, right) generates a catalyst that reacts with low enantioselectivity. The aryl group on the phosphine also has an effect on enantioselectivity, and the complexes generated from ligands containing *meta*-methoxyphenyl substituents on phosphorus react with the highest enantioselectivity. An illustration of the scope of the enantioselective hydroformylation catalyzed by the complex generated from the Takaya ligand is shown in Table 17.1. Phosphoramidite–phosphine ligands related to the Takaya ligands containing an *N*-ethyl group in place of oxygen in the linker have also been shown recently to give rise to catalysts that react with high enantioselectivity.[133]

The geometry of the complex generated from these bisphosphite and phosphite–phosphine ligands is thought to influence enantioselectivity. The ligands of this group that bind in an equatorial–equatorial mode generate catalysts that react with high enantioselectivities. However, the ligands of this group that bind to form isomeric mixtures of complexes generate catalysts that react with low enantioselectivities.[129, 130] In contrast, the BINAPHOS system has been suggested by Nozaki to bind in an axial–equatorial mode in which the phosphine occupies the equatorial position.[131, 132] The phosphite–phosphine ligands generate catalysts that react with high enantioselectivity. As is the case for the bisphosphite systems, phosphite–phosphine ligands that give rise to mixtures of complexes react with lower enantioselectivities.

More recently, Landis has reported a series of modular 3,4-diazaphospholane ligands that form rhodium complexes that catalyze the hydroformylation of vinylarenes, allyl cyanide and vinyl acetate with good regioselectivity and enanatioselectivity (Scheme 17.15).[134] Most striking, catalysts generated from these ligands exhibit even faster rates of hydroformylation than those containing either (R,S)-BINAPHOS or Chiraphite. Moreover, the scope of the hydroformylation catalyzed by this ligand is unusually broad.

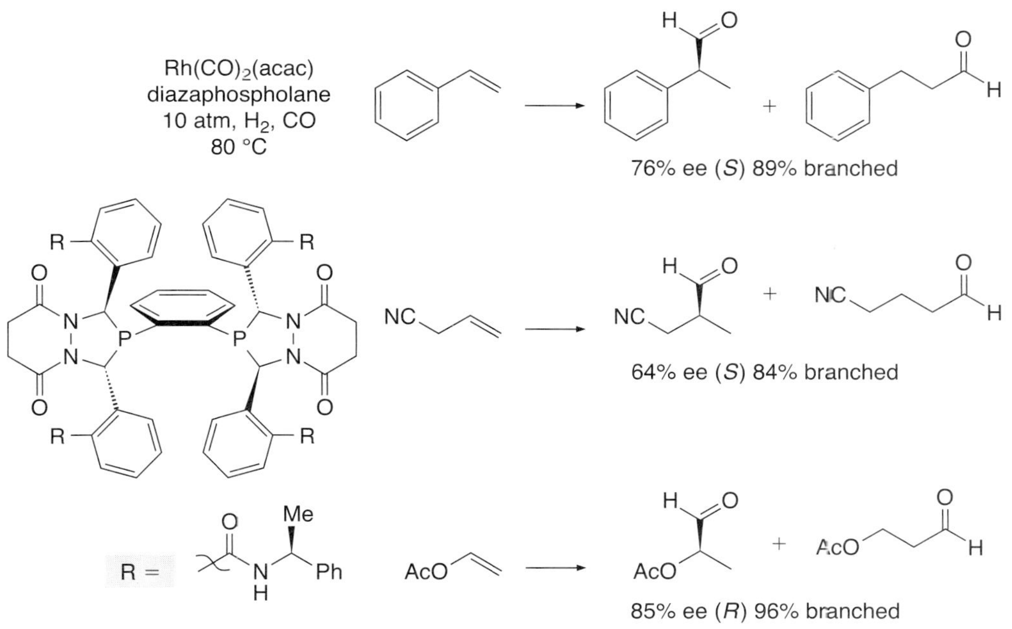

Scheme 17.15

Catalysts generated from the same ligand react with styrene, allyl cyanide, and vinyl acetate with good rates, regioselectivity, and enantioselectivity.

17.4. Hydroaminomethylation

Many reactions can be conducted in tandem with hydroformylation,[135] but the most extensively developed process is the combination of hydroformylation and reductive amination. The combination of these reactions catalyzed by a single complex is called hydroaminomethylation and is shown in Equation 17.18. These reactions formally add the C–H bond of an aminomethyl group across an olefin. Hydroaminomethylation reactions were discovered in the mid-twentieth century,[136,137] but have been developed into more practical processes in recent years. These reactions rely on a complex that can simultaneously catalyze the hydroformylation of an alkene and the reduction of an imine or enamine.

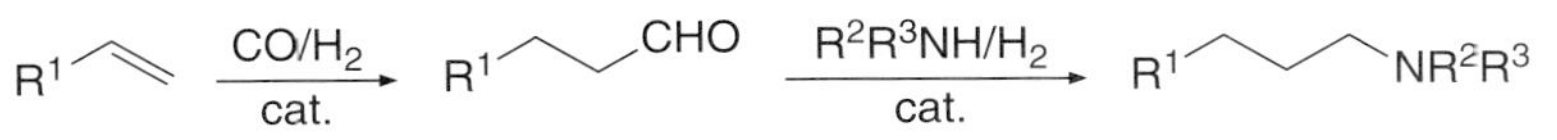

$$R^1 \diagdown \xrightarrow[\text{cat.}]{\text{CO/H}_2} R^1 \diagdown\diagup\text{CHO} \xrightarrow[\text{cat.}]{R^2R^3\text{NH/H}_2} R^1 \diagdown\diagup\text{NR}^2\text{R}^3 \qquad (17.18)$$

17.4.1. History and Overview of Recent Developments

Much of the early hydroaminomethylation chemistry was developed in industrial laboratories. Hydroaminomethylation was patented using heterogeneous catalysts in the 1940s, and the homogeneous version of this reaction was discovered in the mid-1950s by Walter Reppe at BASF. These reactions were conducted with iron pentacarbonyl.[136–138] Later, these reactions were published with cobalt catalysts.[139] Most of these early reactions were conducted with high pressures of carbon monoxide, and these high pressures limited the application of these processes in many synthetic environments. Moreover, these reactions catalyzed by cobalt and iron complexes did not occur with sufficiently high turnover numbers, selectivity for formation of amine, or regioselectivity for formation of terminal or branched amines to allow simple purification of the major product.

More recent studies have taken advantage of the highly active and regioselective rhodium catalysts for hydroformylation containing phosphine ligands. The phosphorus ligand makes the rhodium complex capable of hydrogenation of the enamine or imine formed by condensation of the hydoformylation product with the amine.[140,141] Catalysts for the formation of terminal aldehydes from terminal alkenes that were discussed in the previous sections of this chapter lead to the high terminal selectivity for formation of terminal amines by hydroaminomethylation.[141] Moreover, catalysts that form terminal aldehydes from internal alkenes by the combination of alkene isomerization and hydroformylation have been exploited to develop a hydroaminomethylation of internal alkenes that forms terminal amines.[140] Examples of these processes are described in the following section.

17.4.2. Scope of Hydroaminomethylation

The most important hydroaminomethylations for commodity chemicals would be hydroaminomethylations with ammonia, but this reaction remains a challenge. Hydroaminomethylation reactions have more often been conducted between alkenes and secondary aliphatic amines, and these reactions can be useful for fine-chemical synthesis. Intramolecular hydroaminomethylations can also be valuable, and many examples of intramolecular hydroaminomethylations to form nitrogen heterocycles have been reported.

Specific examples of the hydroaminomethylations of olefins with secondary amines are shown in Equations 17.19–17.21.[141] Cyclic and acyclic secondary amines occur in high yield with linear-to-branched ratios exceeding 50 to 1 in most cases when catalyzed by the rhodium complex generated from $[Rh(COD)_2]BF_4$ and xantphos. The reaction of pentene with piperidine is shown in Equation 17.19. These reactions are also compatible with alcohol (Equation 17.20) and acetal functional groups (Equation 17.21).

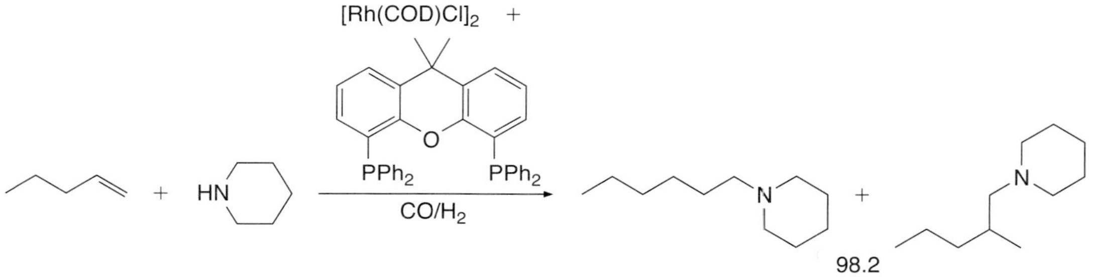

(17.19)

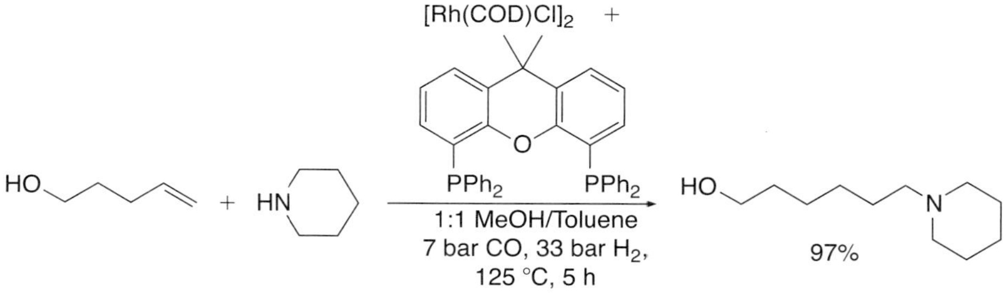

(17.20)

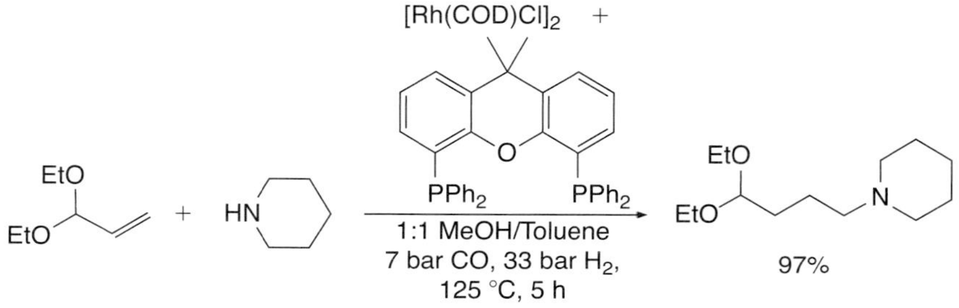

(17.21)

Hydroaminomethylations of vinylarenes also occur (Equations 17.22 and 17.23), and the major regioisomer depends on the catalyst. Reactions conducted with biarylphosphinomethylbinaphthyl ligands or Xantphos form predominantly the linear isomer.[141] A roughly 4:1 ratio versus the branched isomer was observed from the reaction of piperidine in the presence of the catalyst generated from Xantphos (Equation 17.22). In contrast, hydroaminomethylations of vinyl arenes catalyzed by the zwitterionic rhodium complex in Equation 17.23 generate predominately branched regioisomers.[142]

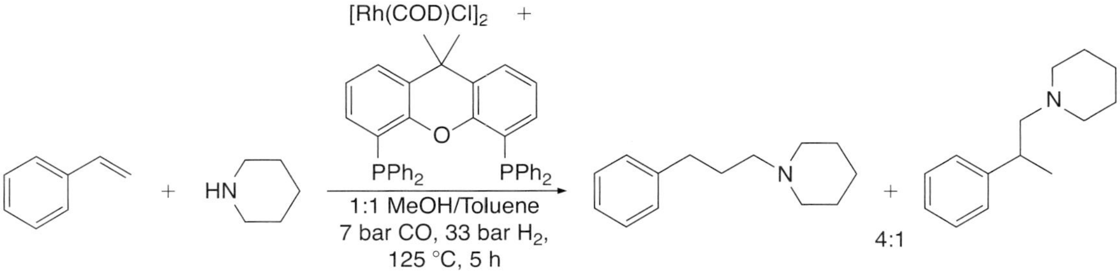

(17.22)

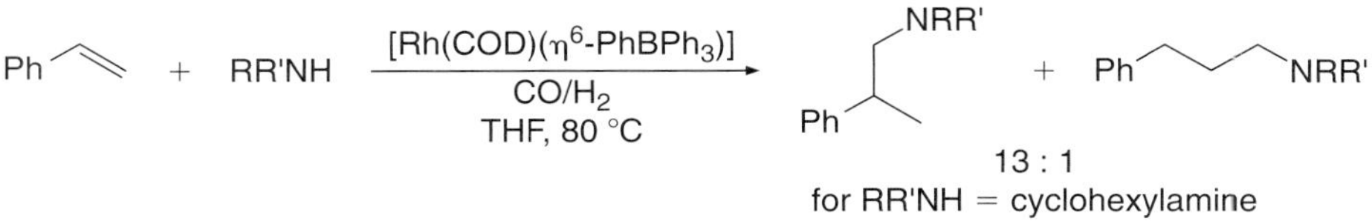

13 : 1
for RR'NH = cyclohexylamine

(17.23)

Reactions of internal alkenes can also form terminal amines, and examples of this process are shown in Equation 17.24.[140] These reactions occur in the presence of a rhodium catalyst containing the electron-poor bisphosphine ligand shown in these equations. For example, the reaction of 2-pentene (R = Me in Equation 17.24) with piperidine forms a 4:1 ratio of the linear to branched amines, and the same reaction with a excess of 2-butene (R = H in Equation 17.24) forms a 96:4 ratio of linear to branched products.

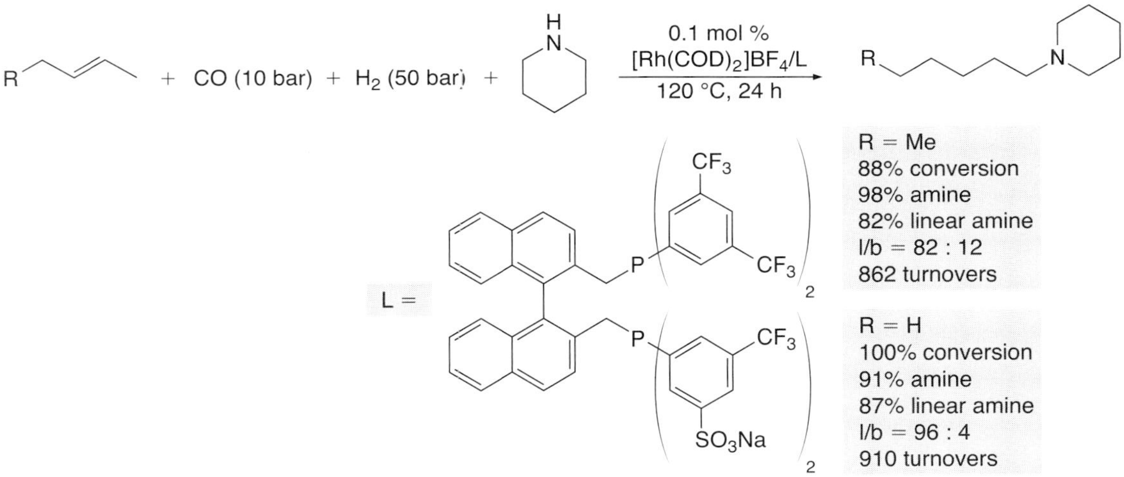

R = Me
88% conversion
98% amine
82% linear amine
l/b = 82 : 12
862 turnovers

R = H
100% conversion
91% amine
87% linear amine
l/b = 96 : 4
910 turnovers

(17.24)

Hydroaminomethylations with ammonia are difficult to develop because the product amine is more nucleophilic than ammonia and preferentially condenses with the aldehyde to form the precursor to the final amine product. Thus, hydroaminomethylations with ammonia tend to form dialkyl- and trialkylamines. To overcome this problem in selectivity, hydroaminomethylations of ammonia have been conducted with a water-soluble catalyst in a biphasic system.[143] Under these conditions, the reaction occurs preferentially with ammonia in the aqueous phase, because it is more soluble than the alkylamine product, and the reactions favor formation of the primary amine. The use of a water-soluble version of the biarylphosphinomethylbinaphthyl catalyst leads to high linear-to-branched ratios, as shown in Equation 17.25.

Yield (amine)	n:iso	prim:sec.
90	99:1	77:23

Ar = m-SO$_3$Na–C$_5$H$_4$ (17.25)

A variety of hydroaminomethylations have been reported that are more directly connected with fine-chemical or target-oriented synthesis. Several examples are shown in Equations 17.26–17.28. In the first case, hydroaminomethylation has been used to generate products with the structures of methadone analogs.[144] This reaction shows the selectivity for reaction at an alkene without interference from reductive amination of the ketone. Hydroaminomethylation has also been shown to be capable of producing pharmaceuticals. Briggs, Kosin, and Whiteker at Dow illustrated the utility of hydroaminomethylation by preparing Ibutilide and Aripiprazole (Equations 17.27 and 17.28).[145] These reactions show the tolerance of this chemistry to free sulfonamido and hydroxyl groups and the free N–H bond of a lactone. These reactions were conducted with the rhodium catalyst generated from Rh(CO)$_2$(acac) and the bisphosphite shown in the equation. At the same time, the results in Table 17.2 illustrate the fine balance between conditions that give high linear-to-branched ratios and conditions that give high yield of the amine product from hydrogenation of the intermediate imine.

In addition to linking two fragments, hydroaminomethylation has been used to generate nitrogen heterocycles by intramolecular chemistry. Two examples are shown in Equation 17.29.[146–148] Presumably these reactions would benefit from the use of catalysts that are highly selective for the formation of terminal aldehydes. Finally, hydroaminomethylation has been used in the field of materials chemistry, in this case to generate polymers containing amino groups in the side chains, as shown in Equation 17.30.[149–151]

4 : 1 to 8 : 1 l/b selectivity
63–96%

(17.26)

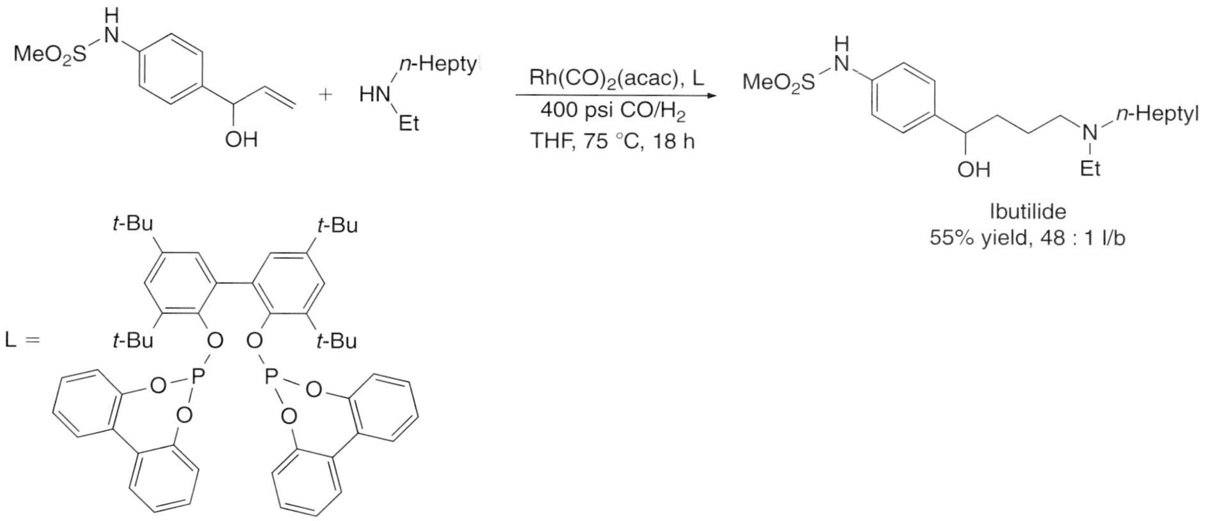

Ibutilide
55% yield, 48 : 1 l/b

$$(17.27)$$

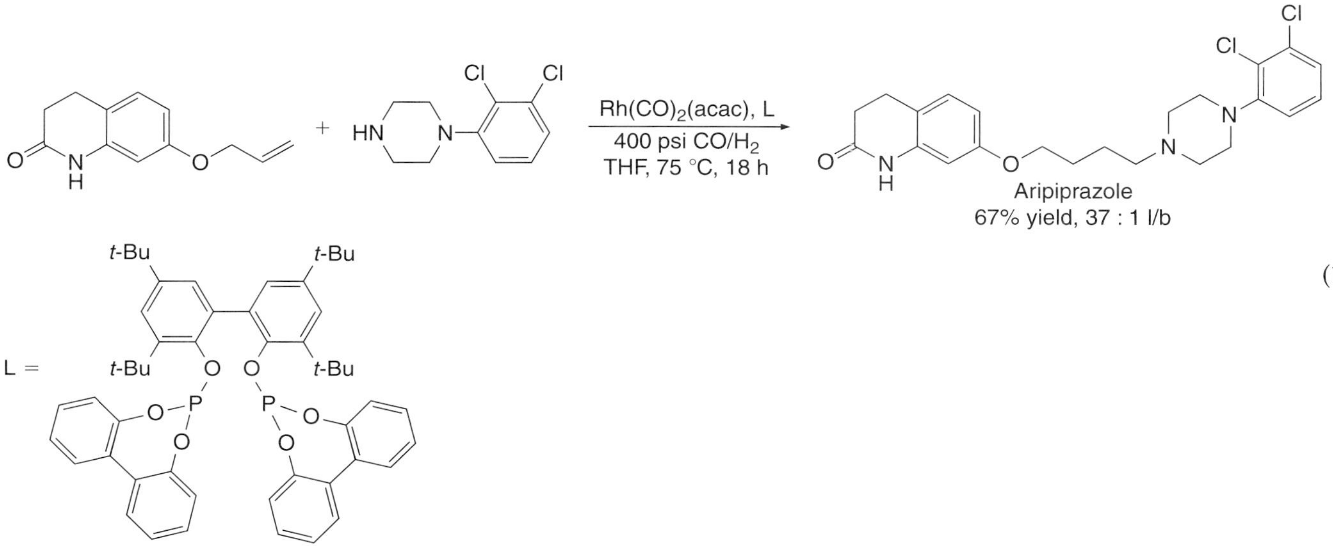

$L =$

Aripiprazole
67% yield, 37 : 1 l/b

$$(17.28)$$

$L =$

Table 17.2. Effect of L/Rh ratio on the hydroaminomethylation of 1-pentene with piperidine catalyzed by Rh(CO)$_2$(acac) and the ligand in Equations 17.27 and 17.28.

L/Rh	% conv	% amine	l/b
1.0	96.6	8	40.0
0.9	97.0	92	12.1
0.8	97.3	100	6.4

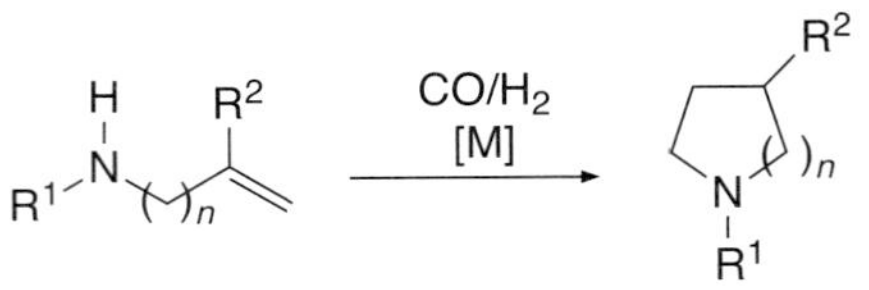

(17.29)

[M] = HRh(CO)$_3$PPh$_3$	R^1	R^2	n	Yield
[M] = [Co$_2$Rh$_2$(CO)$_{12}$]	CH$_2$Ph	H	3	83–85
	PhCO	CH$_3$	1	85

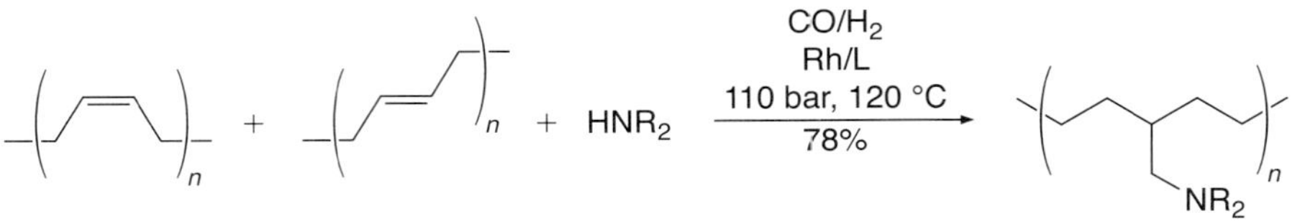

(17.30)

17.4.3. Mechanism of Hydroaminomethylation

Hydroaminomethylation can be considered to be a sequence of two catalytic processes linked by an uncatalyzed process, as summarized in Scheme 17.16. First, catalytic hydroformylation of an alkene occurs by the steps described in the previous section on hydroformylation. This process is then followed by an uncatalyzed condensation of the amine with the aldehyde. When this reaction is conducted with a secondary amine, an enamine results and when the reaction is conducted with a primary amine, an imine results. This condensation process is then followed by hydrogenation of the resulting enamine or imine by the mechanisms described in Chapter 15, the chapter on hydrogenation.

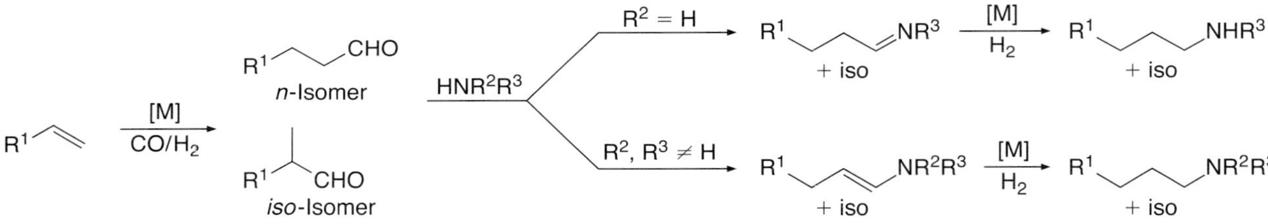

Scheme 17.16

Several factors are needed to realize the tandem process. First, the catalyst should not be poisioned by the high concentration of amine. The use of bidentate ligands helps to retard catalyst decomposition by this route. Second, the catalyst must be sufficiently electron rich to catalyze hydrogenation of the relatively electron-rich enamine, while not being too electron rich to prevent hydroformylation. These properties have been achieved with a number of systems, many of which have been outlined in a review on tandem processes initiated by hydroformylation.[135] Some of these catalysts lack phosphine ligands, while others contain PPh$_3$ or more modern bisphosphine or bisphosphite ligands shown by the examples in the equations above.

17.5. Hydrocarboxylation and Hydroesterification of Alkenes and Alkynes

17.5.1. Overview

The hydrocarboxylation and hydroesterification of alkenes and alkynes are additional classes of carbonylation processes that have been studied extensively and have commercial significance.[152,153] These reactions combine alkenes or alkynes with CO and water or an alcohol to form carboxylic acids or esters. These processes can form either branched or linear products, as summarized in Equations 17.31 and 17.32. The mechanism of this overall transformation is related to that of hydroformylation, with water or an alcohol taking the place of hydrogen. By one commonly invoked mechanism, olefin insertion into a hydride generates a transition metal alkyl, which undergoes insertion to generate a metal–acyl intermediate. Reaction of this acyl complex with water or alcohol releases the product and regenerates a metal hydride. This reaction is also closely related to the alternating copolymerization of CO and olefins described later in this chapter,[154] but the presence of the alcohol and type of catalyst leads to the formation of monomeric products. In contrast to hydroformylations that have been conducted most extensively with cobalt and rhodium catalysts, the hydrocarboxylation and hydroesterifications have been conducted most effectively with palladium catalysts and an acid co-catalyst.

$$\text{alkyl} \diagup\!\!= \; + \; CO \; + \; ROH \quad \xrightarrow[\text{Acid co-catalyst}]{\text{Pd/L}} \quad \text{alkyl} \diagdown\!\diagup\!\diagdown \overset{O}{\underset{OR}{\|}} \qquad (17.31)$$

R = H or alkyl

$$\text{Ar} \diagup\!\!= \; + \; CO \; + \; ROH \quad \xrightarrow[\text{Acid co-catalyst}]{\text{Pd/L}} \quad \underset{Ar}{\diagup}\!\!\overset{O}{\underset{OR}{\|}} \qquad (17.32)$$

R = H or alkyl

Like many carbonylation processes, the hydrocarboxylation and hydroesterification reactions were first reported by Reppe. These first reactions involved the hydrocarboxylation of alkynes.[155] These reactions were conducted with nickel carbonyl as catalyst and occurred with very low turnover numbers. Hydrocarboxylation and hydroesterification have now been studied extensively in both academic and industrial laboratories. As a result of these investigations, commercialization of this chemistry as part of new industrial processes has occurred, and the mechanism of these processes is now generally accepted. This section of Chapter 17 presents the scope and industrial applications of hydrocarboxylation and hydroesterification, the types of catalysts that have been used for these processes, and the elementary steps that constitute the catalytic cycle for olefin and alkyne hydroesterification.

17.5.2. Synthetic Targets for Hydroesterification and Hydrocarboxylation

The value of the hydroesterification of olefins originates from the common type of functionality created by the catalytic process and the fundamental types (and thereby low cost) of reagents that are used to generate the ester products. The ability to conduct the reaction with high selectivity for branched or linear isomers, depending on the olefin and the ligand on the metal, and the ability to conduct the reaction with alkenes, alkynes, and as cyclizations to form lactones gives the process significant synthetic value. Furthermore, much work has been spent to develop a process involving two sequential hydroesterifications of butadiene to form adipic acid,[156] which is one of the monomers of nylon. This process has not been conducted with high enough selectivity and activity to become a commercial process.

Several other types of hydrocarboxylations and hydroesterifications have been conducted with rates and selectivity that are appropriate for the synthesis of fine chemicals and commodity chemicals. One target for hydroesterification has been methyl methacrylate, the monomer of poly(methyl methacrylate), which is the polymer often called "acrylic". It is estimated that 2.1 million metric tons of methyl methacrylate was produced in 2005. Much of this material is produced from acetone cyanohydrin, but two alternative routes could involve catalytic carbonylation. The first route would involve the hydroesterification of methylacetylene, and this chemistry relates to the original route to methyl methacrylate by carbonylation of methylacetylene using nickel carbonyl as catalyst.[157] The second route involves the sequence of ethylene hydoesterification, aldol addition of the resulting ester to formaldehyde, and dehydration. This sequence comprises Lucite's new "Alpha" process and is shown in Equation 17.33.[158] The route to methyl methacrylate by hydrocarboxylation of ethylene produces water as the only byproduct.

$$+ \ CO \ + \ MeOH \quad \xrightarrow[\text{Acid co-catalyst}]{\text{Pd/L}} \quad \ldots \quad \xrightarrow[\text{base}]{H_2CO} \quad \ldots \quad \xrightarrow{-H_2O} \quad \ldots \qquad (17.33)$$

The hydrocarboxylation of vinylarenes has also been studied extensively as a simple, clean route to the α-aryl carboxylic acids that are common non-steroidal anti-inflamatory medicines, such as ibuprofen and naproxen.[159–161] By this process, a vinylarene undergoes hydrocarboxylation to form the branched α-aryl carboxylic acid. A series of patents and papers describe this hydrocarboxylation process[162–165] and the related hydroesterification.[160,161,166] Like the hydrosilylations and hydrocyanations presented in Chapter 16 and the hydroformylations described in this chapter, the regioselectivity for reactions of vinylarenes contrasts with that for reactions of alkenes. The reactions of vinylarenes form branched hydrocarboxylation products.

The selectivity for formation of the branched isomer and the activity for formation of the carboxylic acid or ester when the catalyst is generated from monodentate ligands, such as PPh₃[160,161] and neomenthyldiphenylphosphine,[162–166] is exceptionally high, as shown in Equation 17.34. The use of these ligands led to the development of a particularly efficient method to prepare naproxen by the Albemarle company shown in Equation 17.35.[159] Naproxen was synthesized from 2-bromo-6-methoxynapthalene by a sequence of a Heck reaction (Chapter 19) to form the vinylnapthalene, followed by palladium-catalyzed hydrocarboxylation. Unfortunately, changes in business focus have reduced the use of this process.

An enantioselective version of this reaction to form carboxylic acids and esters would certainly be valuable. Examples of asymmetric hydrocarboxylation of norbornene have been published with significant selectivities,[167,168] but asymmetric hydroesterification of vinylarenes have been less successful. No examples of these reactions that occur in high yields and with enantioselectivities over 90% have been reported.[166,169]

$$\xrightarrow[\text{MeOH–acetone, 50 °C}]{\substack{\text{PdCl}_2 \ (0.01 \text{ mol. amt.}) \\ \text{RPPh}_2 \ (0.02 \text{ mol. amt.}) \\ \text{CO (2.0 MPa) 24 h}}} \qquad (17.34)$$

	Branched	Linear
b : 1	100.0	
Yield	95%	

(17.35)

17.5.3. Catalysts for the Hydroesterification and Hydrocarboxylation of Olefins and Alkynes

The most active and selective catalysts for olefin hydrocarboxylation and hydroesterification are based on palladium, a phosphine ligand, and an acid co-catalyst. Most of the early catalysts for olefin hydroesterification contained monodentate ligands, but highly active and selective catalysts have recently been identified that contain chelating ligands. The most active and regioselective catalyst for the hydrocarboxylation of vinylnaphthalene contains neomenthyl diphenylphosphine,[159] and one of the most selective catalysts for the hydroesterification of butadiene is generated from a combination of palladium precursor, acid co-catalyst, pyridine to buffer the acid, and PPh$_3$ as ligand.[156]

However, the most active and selective catalysts for the hydroesterification of methylaceylene and ethylene contain bidentate ligands. The most active and selective catalyst for the hydroesterification of ethylene and higher olefins contains 1,2-bis(di-*tert*-butylphosphinomethyl)benzene.[170] This ligand was identified by Tooze at ICI Acrylics. (For earlier studies of hindered bidentate ligands, see reference 171). It is speculated that the flexibility and large bite angle of this ligand could allow the complex to adopt the same trans geometries that are adopted by complexes of monophosphines.[170] Prior to the identification of the palladium catalysts containing these ligands, reactions catalyzed by complexes of 1,4-bis(diphenylphosphino)butane were published by Alper.[172] The most active and selective catalysts for the hydroesterification of methylacetylene[157] contain 2-pyridyldiphenylphosphine as ligand, which could be envisioned to be a hemilabile bidentate ligand.[173]

Ruthenium catalysts for hydroesterification are also known, but these catalysts typically operate by the addition of formates, rather than by the use of the combination of CO and alcohol.[174] Nevertheless, some leading citations to this literature are provided here.[175–181] In some cases, milder conditions have been developed with pyridylmethyl formates (Equation 17.36). These substrates were designed so that the pyridyl group would bind to the catalyst.[179,182] The synthesis of amides from olefins, CO, and amines is much less developed, and the most active catalyst for this process and for the addition of formamides to olefins is simply Ru$_3$(CO)$_{12}$.[182] Ruthenium-catalyzed hydroesterification with pyridylmethylformates has been used to prepare the lactone fragment of an HIV integrase inhibitor shown in Scheme 17.17.

(17.36)

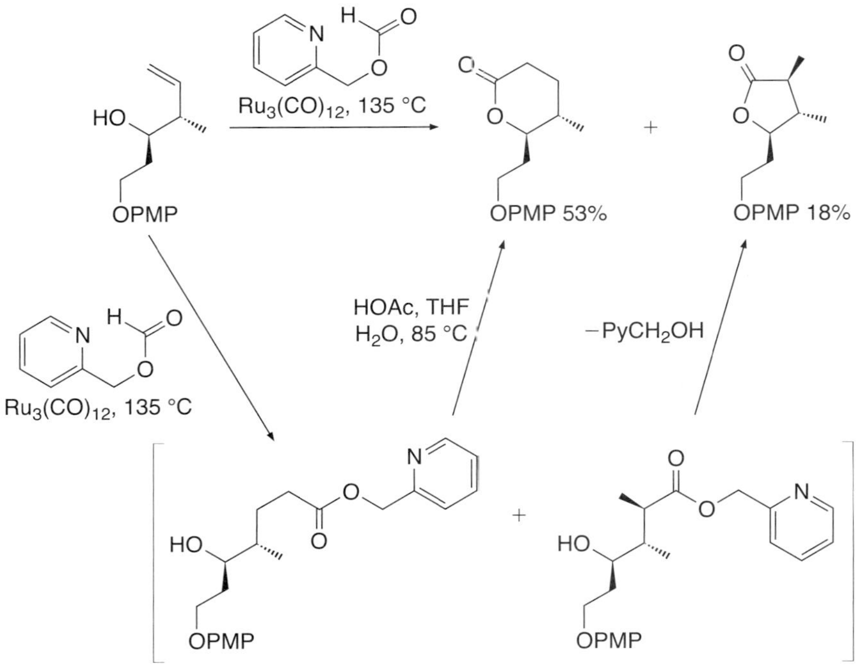

Scheme 17.17

17.5.4. Scope of Hydroesterification and Hydrocarboxylation

17.5.4.1. Hydroesterification and Hydrocarboxylation of Alkenes

17.5.4.1.1. Intermolecular Hydroesterification and Hydrocarboxylation of Alkenes

A majority of the development of hydroesterifications of alkenes and alkynes has focused on reactions of ethylene and propyne to form the precursor to methyl methacrylate or methyl methacrylate itself because of the importance of the products of these reactions in commodity chemical applications. However, these reactions have also been demonstrated with higher α-olefins, with internal olefins that produce terminal esters by a combination of isomerization and hydroesterification, with polymers containing pendant olefins, and with dienes, alkynes, and olefins containing auxiliary functional groups. These reactions have also been conducted with catalysts that generate optically active esters from prochiral olefins. The following paragraphs provide examples of each type of reaction and the complexes that catalyze these processes.

The reactions of ethylene with CO and methanol to form methyl propanoate catalyzed by complexes of various phosphinomethylbenzenes are shown in Equation 17.37 and Table 17.3. These data show the importance of the hindered *tert*-butyl groups on the 1,2-bis(di-*tert*-butylphosphinomethyl)benzene ligand (DTBMPB) and the relative insensitivity to the substituents on the arene ring of the backbone. Reactions catalyzed by complexes of this ligand form monomeric esters with high selectivity over formation of CO/ethylene copolymers and oligomers. Furthermore, the activity for formation of the ester is a remarkable 12,000 turnovers per hour.

Table 17.3. Effect of catalyst structure on the activity and selectivity for hydroesterification of ethylene.

	Complex	Activity[a]	Selectivity[b] (%)	P–Pd–P bite angle (°)
	$R^1 = R^2 = Bu^t$, X = H	12 000	99.9	103.9
	$R^1 = R^2 = Bu^t$, X = NO_2	11 500	99.9	
	$R^1 = R^2 = Bu^t$, X = OMe	11 800	99.9	
	$R^1 = R^2 = Pr^i$, X = H	200	20	104.3
	$R^1 = R^2 = Cy$, X = H	200	25	103.9
	$R^1 = R^2 = Ph$, X = H	400	20	104.6
dba = dibenzylidene acetone	$R^1 = Bu^t$, $R^2 = Cy$, X = H	500	30	

[a]Activity is given in turnovers per hour.
[b]Selectivity is for production of methyl propanoate vs. CO/ethylene copolymers and oligomers.

$$1{:}1 \; C_2H_4/CO \; (10 \; atm) + MeOH \xrightarrow[\substack{MeOH \\ MeSO_3H}]{\substack{Pd_2(dba)_3 \\ Ligand}} \text{(methyl propanoate)} \qquad (17.37)$$

This complex also catalyzes the reaction of higher alpha-olefins and internal olefins to form linear esters. The reaction of octane with 1 atm of CO in methanol with added methane sulfonic acid formed high yields of methyl nonanoate. The same reaction of 2-, 3-, or 4-octene with 4 atm of CO, shown in Equation 17.38, generated the same methyl nonanoate. The selectivity for the linear product ranged from 94% for 4-octene to 99% for 2-octene and 1-octene. The reactions of the branched olefins 2-, 3-, and 4-methyl-1-pentene formed the terminal ester, as did 4-methyl-2-pentene.[183]

$$\text{(olefins)} + 1\text{–}4 \; bar \; CO \xrightarrow[\substack{MeOH \\ MeSO_3H}]{\substack{Pd_2(dba)_3 \\ DTBPMB}} n\text{-}C_8H_{17}\text{-}CO\text{-}OMe \qquad (17.38)$$

DTBPMB = (structure with $P(^tBu)_2$, $P(^tBu)_2$)

The formation of linear carboxylic acids has also been reported under conditions in which formic acid is the source of the hydroxyl group. In this case, the reaction of olefins with CO and formic acid generates the carboxylic acid. These reactions have been conducted with palladium catalysts containing 1,4-bis(diphenylphosphino)butane as ligand at higher temperatures than the esterifications with palladium complexes of the DTBMPB ligand. The sequence that is proposed to generate the carboxylic acids is shown in Equation 17.39. Reactions with labeled CO form the labeled carboxylic acid shown. Thus, CO, not formic acid, is the source of the carbonyl group.

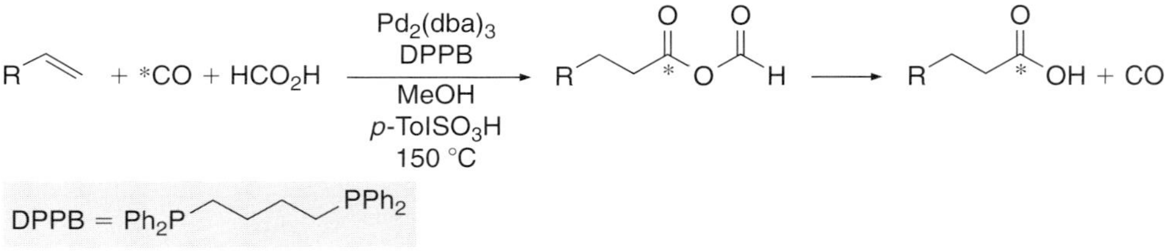

$$R\diagup\!\!\!\diagdown \;+\; {}^*CO \;+\; HCO_2H \xrightarrow[\substack{MeOH \\ p\text{-TolSO}_3H \\ 150\,°C}]{\substack{Pd_2(dba)_3 \\ DPPB}} R\diagdown\!\!\diagup\overset{O}{\underset{*}{\diagup}}\!O\overset{O}{\diagdown\!H} \longrightarrow R\diagdown\!\!\diagup\overset{O}{\underset{*}{\diagup}}OH \;+\; CO \tag{17.39}$$

DPPB = $Ph_2P\diagdown\!\!\diagup\!\!\diagdown\!\!\diagup\!PPh_2$

These hydrocarboxylation reactions have been conducted with terminal alkenes containing a series of functional groups.[172] As summarized in Equation 17.40, these reactions occur in the presence of keto, cyano, formyl, acetoxy, carboxylic acid, and amide functionality. In addition, these reactions have been conducted with polybutadienes containing pendant vinyl groups to form polymers possessing pendant carboxylic acid functionality (Equation 17.41).

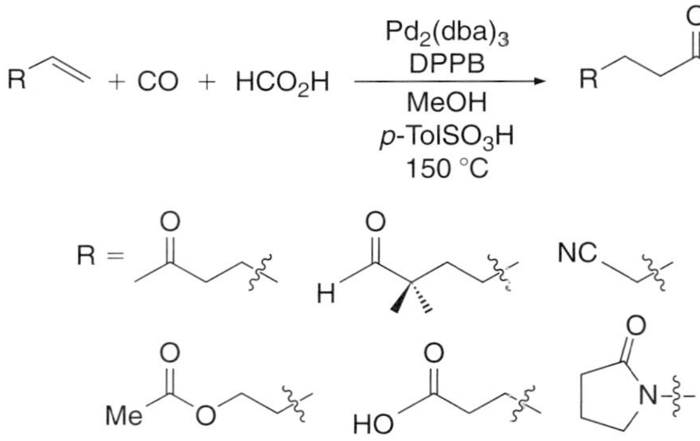

$$R\diagup\!\!\!\diagdown \;+\; CO \;+\; HCO_2H \xrightarrow[\substack{MeOH \\ p\text{-TolSO}_3H \\ 150\,°C}]{\substack{Pd_2(dba)_3 \\ DPPB}} R\diagdown\!\!\diagup\overset{O}{\diagdown}OH$$

$$R = $$

(17.40)

Pd(OAc)$_2$ (0.02 mmol), dppb (0.04 mmol)

HCO$_2$H (10 mmol), CO (100 psi)

DME, 150 °C, 21 hrs

$n \approx 45$

$\diagup\diagdown CO_2H$

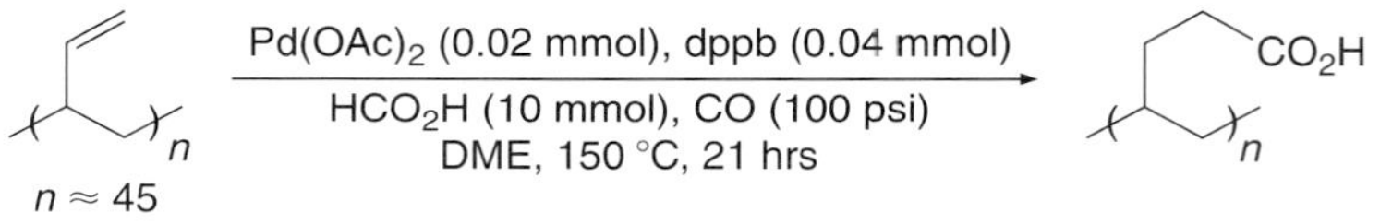

(17.41)

17.5.4.1.2. Intramolecular Hydroesterification of Olefins

The intramolecular hydroesterification of olefins provides a route to lactones. Although this reaction has not been applied extensively in the synthesis of complex molecules, several examples demonstrating the scope and potential utility of this reaction have been reported. As shown in Equations 17.42–17.44 this reaction can be used to prepare optically active lactones,[184] benzo-fused lactones,[185] and lactams.[185] The reaction in Equation 17.43 illustrates how the product can result from a combination of isomerization and carbonylation, and the reaction in Equation 17.44 shows how the ring size can be controlled by the composition of the catalyst.

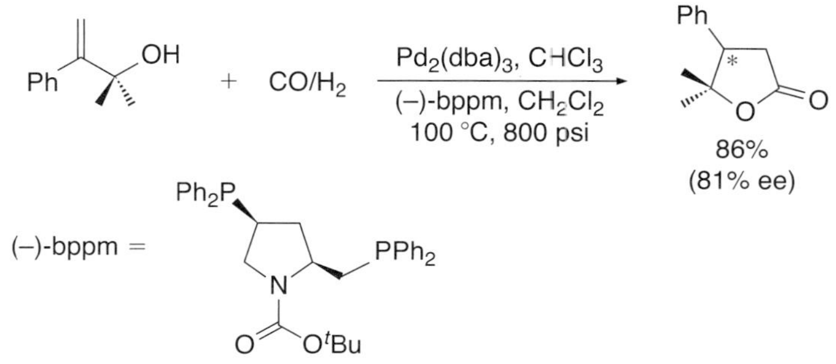

Pd$_2$(dba)$_3$, CHCl$_3$

(–)-bppm, CH$_2$Cl$_2$

100 °C, 800 psi

86%

(81% ee)

(17.42)

(–)-bppm =

$$(17.43)$$

$$(17.44)$$

| Ligand = PCy$_3$ | 95% yield | 100 | 0 |
| dppb | 75% yield | 0 | 100 |

17.5.4.2. Hydroesterification of Alkynes

The hydroesterification of alkynes (Equation 17.45) was first conducted with Ni(CO)$_4$ as catalyst[186] to produce acrylate esters as products. Much more active catalysts for this reaction have now been generated from a combination of palladium acetate and 2-pyridylphosphines.[173] The use of these ligands for alkyne hydroesterification was conducted by Drent at Shell.

R = H: methyl acrylate
R = Me: methyl methacrylate

$$(17.45)$$

The hydroesterification of propyne has been the focus of much research because this process forms methyl methacrylate (R = Me in Equation 17.45). The effect of the ligand on this reaction and the position of the nitrogen in the pyridyl group within the ligand can be seen by the results in Table 17.4. The rates are slower and the selectivity is lower for reactions catalyzed by complexes of PPh$_3$. The rates are improved by the use of pyridyldiphenylphosphines that cannot chelate, but they are dramatically improved by the use of the 2-pyridyldiphenylphosphine ligand. The rate and selectivity is further enhanced when the catalyst contains the more hindered 2-(6-methypyridyl)diphenylphosphine ligand. This ligand is proposed to chelate the metal, although a four-membered ring would result from this chelation. Further details on the origins of selectivity have been suggested,[173] but these proposals were based on a mechanism that has subsequently been ruled out.

Table 17.4. Effect of presence and position of the nitrogen atom in the ligand for propyne carbonylation.

Ligand type	Temperature (°C)	Average rate (mol/mol Pd h)	Selectivity (% MMA)
PPh$_3$	115	ca. 10	89
4-PyPPh$_2$	90	ca. 10	90
3-PyPPh$_2$	70	1000	99.2
2-PyPPh$_2$	45	40 000	98.9
2-(6-CH$_3$-Py)PPh$_2$	60	50 000	99.95

17.5.4.3. Hydroesterification of Butadiene

The hydroesterification of butadiene is not well developed, although much effort in many industrial laboratories has been spent to develop this process as a method to prepare adipic acid and caprolactam. Because little information on this process has been disclosed in the academic literature, this topic is not presented in detail here. Challenges encountered in developing this reaction include the regioselectivity for formation of terminal esters, the potential to form stable allylmetal intermediates, and the potential to form polymeric or oligomeric products containing multiple diene units.

17.5.5. Mechanism of Hydroesterification

Two mechanisms for the hydroesterification of alkenes have been considered. One pathway—the "alkoxide cycle"—begins with the insertion of CO into a metal alkoxide (Scheme 17.18) and the other—"the hydride cycle"—begins with the insertion of an alkene into a metal hydride (Scheme 17.19). The relative importance of the different pathways depends on the identity of the dative ligand. However, hydroesterification of ethylene with the bis(di-*tert*-butylphosphinomethyl)benzene ligand is now generally accepted to occur through a palladium hydride. The mechanism of the hydroesterification of alkynes is less established, but is likely to occur by a sequence that shares some steps with the mechanism for the hydroesterifcation of alkenes.

The alkoxide pathway occurs by initial insertion of CO into a palladium alkoxide, followed by insertion of the alkene into the bond between the metal and the alkoxycarbonyl group to form a palladium–alkyl complex (Scheme 17.18). Protonation of this metal alkyl by alcohol would form the free organic product and regenerate the palladium alkoxide. This mechanism has now been ruled out for the reactions of ethylene to form methyl propanoate. Although each of these steps has precedent, the absence of reduction products from the alkoxide argues against this pathway. Moreover, the alkyl generated from insertion of ethylene into the palladium–alkoxycarbonyl complex (Scheme 17.18) is chelated to the metal, and methanolysis of this species is slower than the steps of the alternative hydride mechanism.[187]

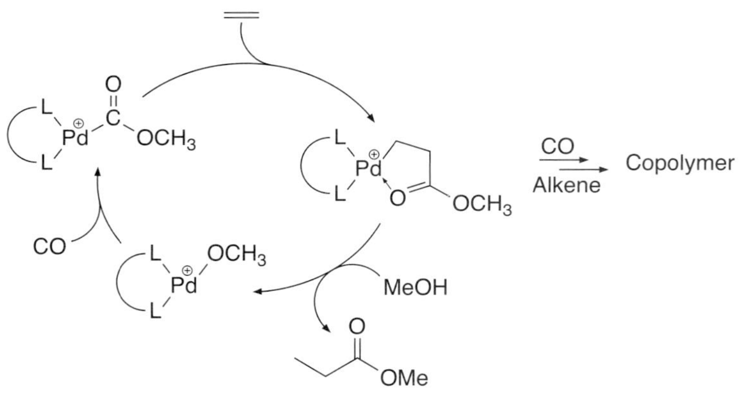

Scheme 17.18. "Alkoxide cycle" for olefin hydroesterification.

The more favored hydride mechanism (Scheme 17.19) involves generation of a palladium hydride by the acid co-catalyst.[187] This hydride complex contains a weakly coordinating sulfonate anion. Insertion of olefin into the palladium hydride then generates a palladium–alkyl complex. This insertion likely occurs by displacement of the sulfonate by olefin to generate a cationic olefin hydride complex. Subsequent insertion of CO into the metal alkyl then generates a palladium–acyl complex. Reaction of the acyl complex with alcohol then generates the ester and the starting palladium hydride. Like the alkoxide mechanism, each of these steps has precedent. Insertions of olefins into metal hydrides and insertion of CO into metal alkyl complexes were described in Chapter 9. Reactions of acyl complexes with nucleophiles, including water and alcohols, were described in Chapter 11.

Yamamoto has studied this reaction as part of the mechanism of the catalytic formation of esters from haloarenes, CO, and alcohols described later in this chapter.[188]

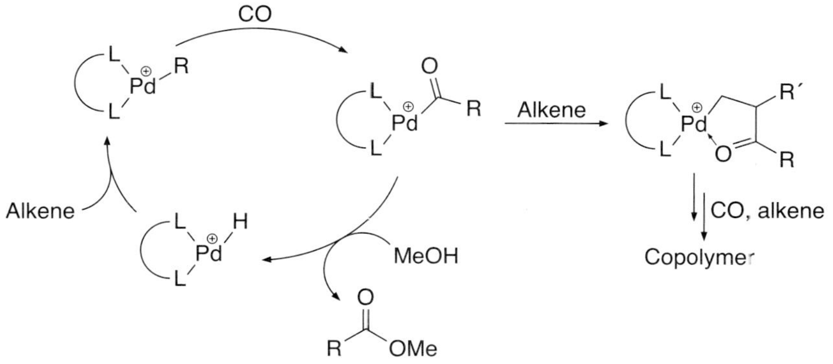

Scheme 17.19. Hydride cycle for olefin hydroesterification.

Each of the steps of the hydride mechanism has been observed directly in systems containing 1,2-bis(di-*tert*-butylphosphinomethyl)benzene (DTBMPB),[189–191] bis(diisobutyl-phosphino)propane,[187] and PPh$_3$.[192] These studies showed that the elementary steps of the hydride cycle in Scheme 17.19 occurred and showed that the turnover-limiting step was the methanolysis of the acylpalladium complex (Equation 17.46). The properties of the ligands that affect the rate of methanolysis have been studied. Bidentate ligands that allow access to a cis geometry are proposed to lead to faster rates than those that restrict the complex to a trans geometry. These data support a mechanism in which methanol coordinates cis to the acyl ligand, and the oxygen of the coordinated alcohol or alkoxide attacks the carbonyl of the acyl ligand.[193]

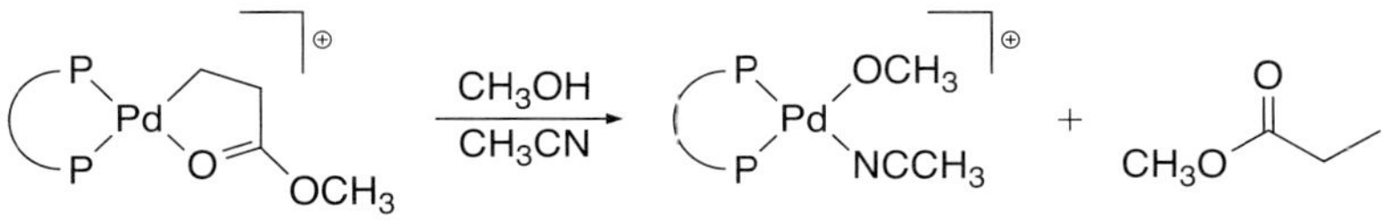

$$ (17.46) $$

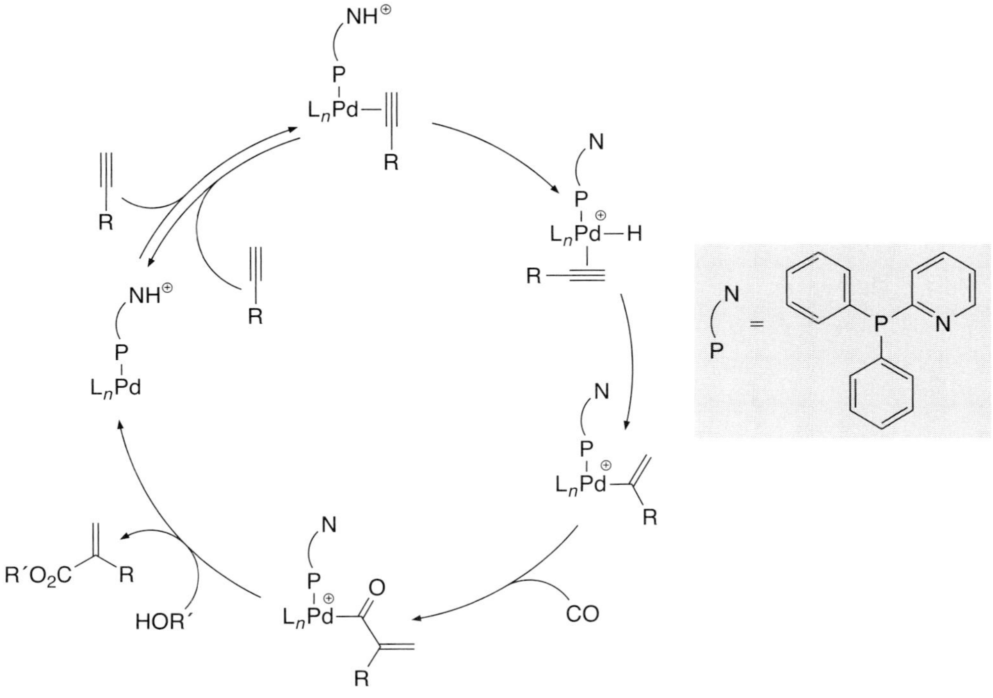

Scheme 17.20

The hydroesterification of alkynes also occurs by a mechanism involving CO insertion and alcoholysis of an intermediate acyl complex, but the steps by which proton transfers occur can be different from those in the hydroesterification of alkenes.[194] As noted above, a particularly active system for the hydroesterification of propyne to form methyl methacrylate contains 2-pyridyldiphenylphosphine.[173] The pyridyl nitrogen in this ligand could act as a base to assist in the addition of the alcohol to the acyl intermediate and transfer of the proton back to the alkyne to generate the intermediate vinyl complex. As shown in Scheme 17.20, one proposed mechanism for this process involves protonation of an alkyne complex to form a vinylpalladium intermediate. Insertion of CO into the palladium-vinyl linkage would then generate an acyl intermediate that undergoes reaction with methanol to form the organic product and a palladium(0) species containing a pyridinium group on the ligand. This reaction with methanol could be assisted by the 2-pyridyl group. Coordination of alkyne would then regenerate the starting species.

17.6. Carbonylation of Epoxides and Aziridines (Written with Prof. Geoffrey W. Coates)

The carbonylation of epoxides and aziridines has been studied for several decades, and two forms of this process are now well established.[195] Highly active catalysts for the ring expansion of epoxides and aziridines to β-lactones and β-lactams are now known. In addition, conditions have been developed for the hydroformylation of epoxides to α-hydroxy aldehydes (including protected α-hydroxyacetals), and similar conditions have been developed for the tandem hydroformylation and hydrogenation of epoxides to generate 1,3-diols. The catalysts for these reactions are either the neutral cobalt carbonyl [$Co_2(CO)_8$], anionic cobalt carbonyl [$Co(CO)_4^-$] with a Lewis acidic countercation, or a combination of cobalt and ruthenium carbonyl.[196]

These reactions are useful as processes to form building blocks for polymer synthesis and organic synthesis. The β-lactones and β-lactams formed by the carbonylation of epoxides and aziridines are monomers that can be used in ring-opening polymerization. The 1,3-diols—particularly propanediol—are monomers for large-scale syntheses of polyesters. At the same time, the α-hydroxy aldehydes from epoxide hydroformylation are the equivalent of products from difficult aldol condensations with aldehyde enolates. The advent of methods to prepare or to resolve optically active epoxides has increased the importance of epoxide carbonylation processes for fine-chemical synthesis.

17.6.1. Ring-Expansion Carbonylation of Epoxides and Aziridines

17.6.1.1. Overview

Strained heterocycles undergo carbonylation in the presence of several types of catalysts, including cobalt carbonyl hydride and cobalt carbonyl anion with a salen-ligated, Lewis acidic transition metal as cation. These carbonylation processes can be divided into three types: ring-opening carbonylation, ring-opening carbonylative polymerization, and ring-expansion carbonylation. The synthetic applications of these reactions include the large-scale commercial production of 1,3-propanediol used in polyester synthesis, the preparation of monomers for new polymerizations, the use of optically active epoxides to prepare synthetic building blocks, and the synthesis of products that are typical synthetic intermediates. As shown in Scheme 17.21, the products of these carbonylations generally include 1,3-diols, β-lactones, succinic anhydrides, β-lactams, poly(β-hydroxyalkanoate)s, poly(β-peptoid)s and β-hydroxyesters.[197,198]

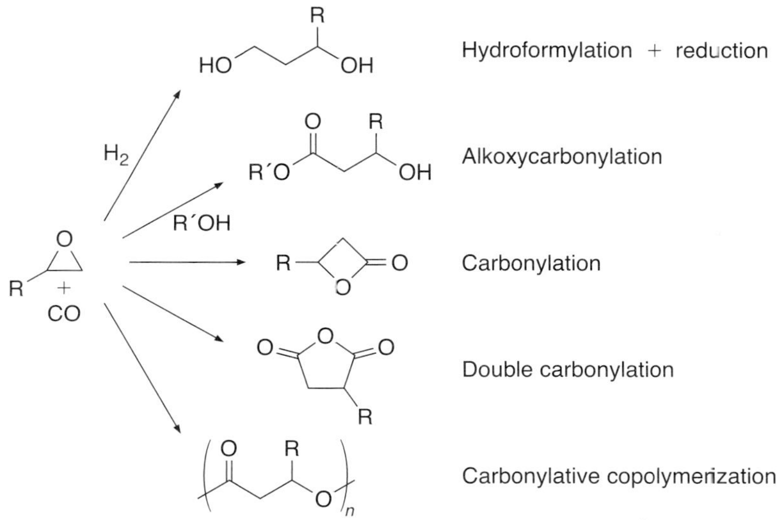

Scheme 17.21. Examples of the carbonylation of epoxides.

17.6.1.2. History of Epoxide and Aziridine Carbonylation

The carbonylation of epoxides was discovered in the 1960s, and advances have been made more recently. Analogous carbonylations of aziridines have also been studied extensively. The advent of well-defined catalysts creates the potential to bring this process from the synthesis of commodity materials to problems in complex molecule synthesis. This section focuses on modern catalysts for this reaction, the impact these processes can have on synthesis, and the mechanisms of these reactions.

In 1963, Heck reported the synthesis of cobalt–acyl compounds by the reactions of epoxides and oxetanes with $[Co(CO)_4]^-$ (Scheme 17.22).[199] Reactions of these acyl complexes with alcohol and amine bases generated β-hydroxyesters or γ-lactones. Catalytic ring expansions of various heterocycles were reported before and after this work,[200–206] and the first ring-expansive carbonylations of epoxides were reported between the late 1970s and early 1980s.[207–210] These reactions of epoxides were confined to vinyloxiranes and styrene oxides and involved high temperatures and pressures. Cobalt-catalyzed carbonylations of aziridines were reported in the mid-1990s.[211] In the 1990s, the carbonylation of epoxides was advanced by a patent[212] of Drent and Kragtwijk. This patent claimed the carbonylation of epoxides catalyzed by a cobalt source and a hydroxy-substituted pyridine, such as dicobalt octacarbonyl, $Co_2(CO)_8$, and 3-hydroxypyridine (3-HP), to form β-lactones. Subsequent investigations by other researchers have shown that the system makes a mixture of lactone and poly (hydroxyalkanoate) oligomers.[213–215]

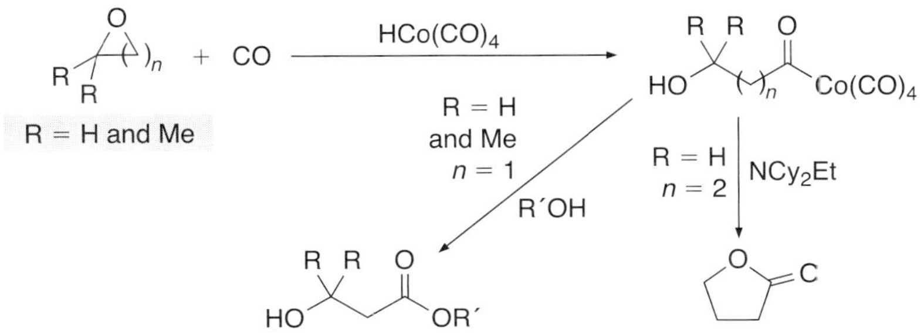

Scheme 17.22. Heck's carbonylation of epoxides and oxetanes.

17.6.1.3. Types of Catalysts and Scope of Substrates for Epoxide Carbonylation

Although many of the early examples of the carbonylation of heterocycles included reactions of tetrahydrofurans, oxetanes, and azetidines, the majority of recent work has focused on the reactions of epoxides and aziridines. At this point, the ring-expansive reactions of epoxides are more general than the reactions of aziridines and occur under milder conditions. Prior to 1994, ring-expansive carbonylation of epoxides was restricted to a few substrates. The patent by Drent and Kragtwijk's in 1994 inspired further work on these types of carbonylations, and this work led to dramatic improvements in reaction scope.

In 2001, Alper and co-workers reported that mixtures of neutral Lewis acids (such as $BF_3 \cdot OEt_2$) with $[PPN]^+[Co(CO)_4]^-$, in which $[PPN]^+$ = bis(triphenylphosphine)iminium, catalyzed the ring-expansion carbonylation of a range of epoxides (Scheme 17.23).[215] They showed that the reaction did not occur in the absence of a Lewis acid, thus demonstrating the value of a catalyst containing both a Lewis acid and a nucleophilic metal carbonyl. Although the reactions were slow (typically 24–48 h), they produced several β-lactones in 60–90% yield. Reactions of unsubstituted and 1,2-disubstituted epoxides occurred in lower yields. In 2002, Coates and co-workers described reactions catalyzed by [(salph) Al(THF)$_2$]$^+$[Co(CO)$_4$]$^-$ (in which salph = N,N'-o-phenylenebis(3,5-di-$tert$-butylsalicylideneimine) (Scheme 17.24), in which the Lewis acidic cation and the nucleophilic metal anion were present in a single complex.[213] Epoxides that undergo ring expansion carbonylation with these catalysts are shown in Figure 17.8. [213]

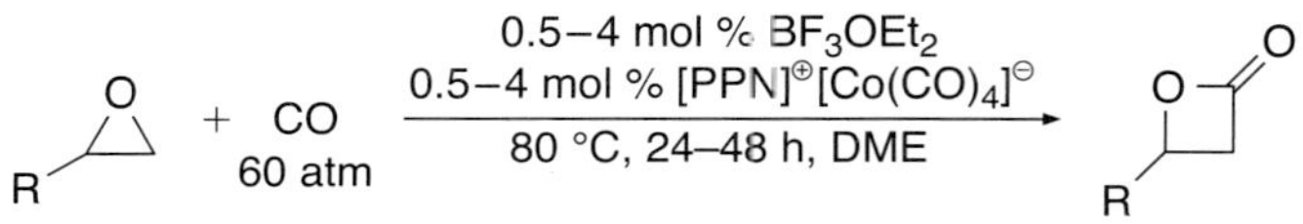

R = H, Me, ^{n}Bu, ^{n}Hex, CH$_2$Cl, CH$_2$O^{i}Pr, (CH$_2$)$_2$CH=CH$_2$ or (CH$_2$)$_4$CH=CH$_2$

Scheme 17.23. Alper's carbonylation of epoxides to β-lactones.

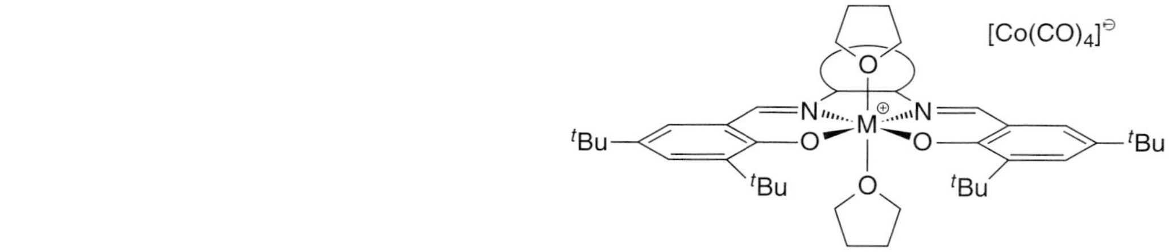

	M	
1	Al	1,2-C$_6$H$_4$
2	Al	(R,R)-1,2-C$_6$H$_{10}$
3	Cr	1,2-C$_6$H$_4$

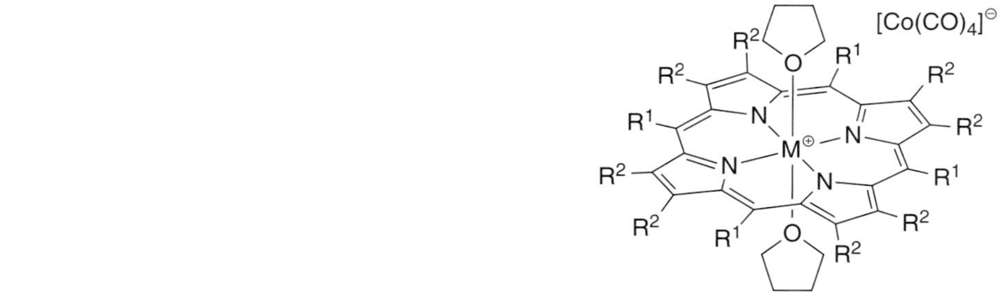

	M	R^1	R^2
4	Cr	Ph	H
5	Cr	H	Et
6	Al	p-Cl–C$_6$H$_4$	H

Scheme 17.24.
Well-defined [Lewis acid]$^+$[Co(CO)$_4$]$^-$ catalysts for the carbonylation of epoxides to β-lactones.

O
O
O
O
O
O

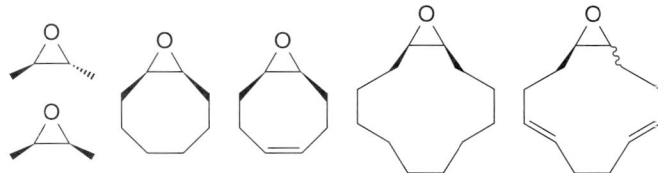

Figure 17.8.
Disubstituted epoxides undergoing ring-expansion carbonylation by catalysts **4** and **5** of Scheme 17.24.

Although the salen aluminum complex **1** of Scheme 17.24 and $[PPN]^+[Co(CO)_4]^-$/Lewis acid are active and selective catalysts for formation of β-lactones, further improvements in epoxide carbonylations were achieved by conducting reactions with $[(TPP)Cr(THF)_2]^+[Co(CO)_4]^-$ (**4**, Scheme 17.24; TPP = *meso*-tetraphenylporphyrinato) and $[(OEP)Cr(THF)_2]^+[Co(CO)_4]^-$ (**5**, Scheme 17.24; OEP = 2,3,7,8,12,13,17,18-octaethylporphyrinato) as catalyst. These complexes containing chromium-based Lewis acids catalyzed the carbonylation of epoxides with fast rates and broad substrate scope.[216,217] These complexes catalyzed the carbonylation of disubstituted and large-ring bicyclic epoxides (Figure 17.8) at lower pressures and catalyst loadings than were required with previous catalysts.[218] This complex also catalyzed the carbonylation of glycidyl ethers containing functionality that can give rise to the alcohol group in hydroxymethyl-substituted β-lactones[219] and catalyzed the ring-expansion carbonylation of epoxides containing pendant ester and secondary amide groups (Scheme 17.25).[216]

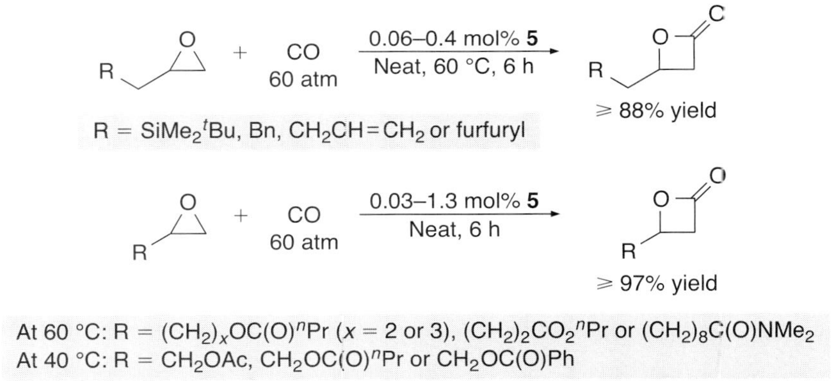

R
O
+
CO
60 atm
0.06–0.4 mol% 5
Neat, 60 °C, 6 h
O
C
R
≥ 88% yield
R = SiMe2tBu, Bn, CH2CH=CH2 or furfuryl
O
R
+
CO
60 atm
0.03–1.3 mol% 5
Neat, 6 h
O
O
R
≥ 97% yield
At 60 °C: R = (CH2)xOC(O)nPr (x = 2 or 3), (CH2)2CO2nPr or (CH2)8C(O)NMe2
At 40 °C: R = CH2OAc, CH2OC(O)nPr or CH2OC(O)Ph

Scheme 17.25.
Carbonylations of epoxides catalyzed by complex **5** of Scheme 17.24.

An even more active version of the catalysts containing Lewis acids and metal anions allows the reactions to be conducted at ambient temperatures with 1 atm of CO. This catalyst appears to suppress the competitive formation of ketones from isomerization of the epoxide[220] that possibly occurs through β-hydride elimination by the cobalt–alkyl intermediate generated prior to CO insertion.[221] Reactions of a range of epoxides catalyzed by $[(salph)Cr(THF)_2]^+[Co(CO)_4]^-$ occur with balloon pressures of carbon monoxide, even on the multigram scale.[222]

17.6.2. Carbonylation of Lactones and Epoxides to Succinic Anhydrides

The carbonylation of β-lactones to form succinic anhydrides also occurs with catalysts containing a combination of the salen-ligated aluminum Lewis acid and cobalt carbonyl anion (Equation 17.47).[223] These reactions occur with lactones bearing alkyl, alkenyl, and ether substituents, as well as with the unsubstituted β-propiolactone. All of these substrates

have been converted to succinic anhydrides in $\geq$ 90% yield. Using the enantiomerically pure (R)-β-butyrolactone, it was demonstrated that the reaction occurred with inversion of configuration at the β-position of the starting lactone.[224] Consistent with this result, carbonylation of the disubstituted cis-3,4-dimethyloxetan-2-one formed the trans product, implying that the configuration at the β-carbon is inverted, while the configuration at the α-carbon of the β-lactone is retained.

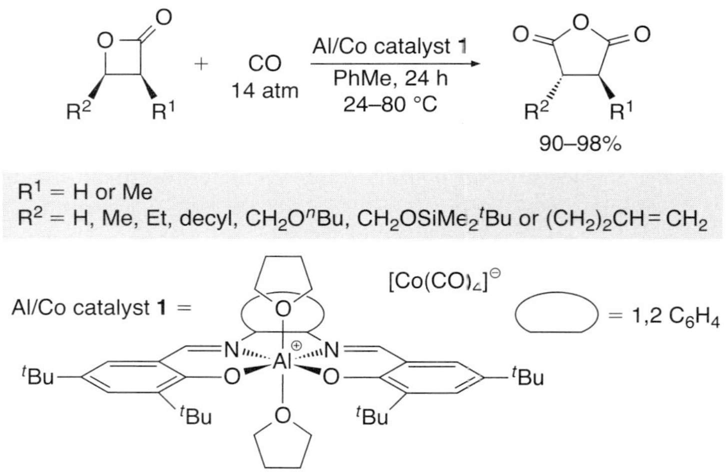

R¹ = H or Me
R² = H, Me, Et, decyl, CH₂OⁿBu, CH₂OSiMe₂ᵗBu or (CH₂)₂CH=CH₂

$$R^1 = \text{H or Me}$$
$$R^2 = \text{H, Me, Et, decyl, } CH_2O^nBu, CH_2OSiMe_2{}^tBu \text{ or } (CH_2)_2CH=CH_2$$

(17.47)

Direct synthesis of succinic anhydrides from epoxides has also been achieved by using the catalyst containing the aluminum Lewis acid and cobalt carbonyl anion (Equation 17.48).[225] The carbonylations of epoxide and β-lactone in the presence of this catalyst were found to occur separately and sequentially, with β-lactone carbonylation occurring only after all of the epoxide had undergone carbonylation. β-Lactone carbonylation was slow in polar or donor solvents. Because epoxide and β-lactone carbonylation occur with opposing solvent dependences, the carbonylation of the epoxide to the anhydride in a single pot depended upon the identification of a solvent (1,4-dioxane) that is sufficiently donating to accelerate epoxide carbonylation, but sufficiently weak in polarity to allow rapid β-lactone carbonylation.

R = p-Cl-C₆H₄

$$R = p\text{-Cl-}C_6H_4$$

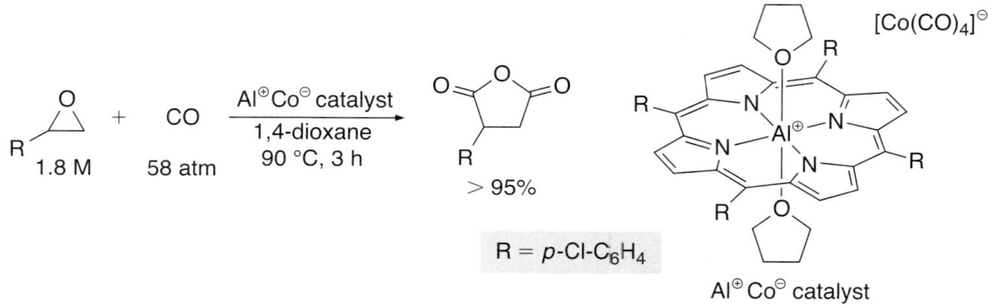

(17.48)

17.6.3. Ring-Opening Epoxide Carbonylation

Several catalytic processes have also been developed that lead to the opening of epoxides to form valuable products. These ring-opening processes include the hydroformylation of epoxides, the alkoxycarbonylation of epoxides, and the ring-opening carbonylative polymerization of epoxides. Although these reactions have been known for nearly 50 years, processes efficient enough for commercial production were developed in the late twentieth and early twenty-first century, largely at Shell.

The hydroformylation of epoxides—the reaction of an epoxide with CO and hydrogen—to form β-hydroxy aldehydes or to form 1,3-diols has been investigated extensively. This overall reaction is shown in Equation 17.49. The formation of 1,3-diols or β-hydroxyaldehydes depends on reaction conditions. Much of the information on this process is contained within patents,[226] but scattered journal publications have appeared over a long time period.[227,228] In short, diols are formed from reactions conducted with cobalt carbonyl and a co-catalyst, such as $Ru_3(CO)_{12}$, that catalyzes hydrogenation, or from reactions catalyzed by cobalt carbonyl modified with a diphosphine containing alkyl groups, such as a phospholane.[229-233] Reactions with cobalt carbonyl in the absence of a promoter of hydrogenation form mostly β-hydroxy aldehyde. The most important of these processes is the formation of 1,3-propanediol from ethylene oxide, CO, and H_2. 1,3-Propane diol is a monomer used in the production of polyesters on a large industrial scale.[234]

$$\text{(17.49)}$$

The alkoxycarbonylation of epoxides from reactions of epoxides with CO and alcohols generates β-hydroxy esters, and the aminocarbonylation of epoxides generates β-hydroxyamides. The products are analogous to the products of the difficult acetate aldol reaction. The methoxy carbonylation of epoxides was first reported with $Co_2(CO)_8$ as catalyst by Eisenmann and co-workers,[235a] and further studies with other alcohols[235b] and with epichlorohydrin followed.[236] The patent report of the cobalt-catalyzed ring-opening carbonylation of epoxides with 3-hydroxypyridine spawned efforts to use this catalyst system to convert enantiomerically enriched epoxides to enantioenriched β-hydroxy esters and amides. This transformation was accomplished using a 2:1 mixture of 3-hydroxypyridine and $Co_2(CO)_8$ (Equation 17.50).[237] The reaction was highly selective in many cases for the isomer shown,[238] and the absolute configuration and enantiomeric purity was typically retained. Subsequently, the methoxycarbonylation of propylene oxide (PO) under reduced CO pressures, giving higher yields and selectivities, were reported.[239] In addition, Lewis acids have been shown to accelerate the methoxycarbonylation of ethylene oxide.[240]

$$\text{(17.50)}$$

$$\text{(17.51)}$$

R = Me, Et, ^{n}Bu, CH_2Cl, $(CH_2)_2CH=CH_2$,
CH_2O^iPr, CH_2OBn, or $CH_2OC(O)^nPr$

Finally, the conversion of epoxides to β-hydroxy amides has been accomplished by the reaction of epoxides with CO and silylamines (Equation 17.51). This reaction was originally reported with $Co_2(CO)_8$ as catalyst to form the β-silyloxyamide products in 60 to 84% yields using 1 atm of CO at room temperature for 24–50 h. Optically active β-hydroxy morpholine amides, which can be converted to optically active β-hydroxy ketones, have been prepared more recently using the same $Co_2(CO)_8$ catalyst. Because optically active terminal epoxides can be accessed by Jacobsen's hydrolytic kinetic resolution[241] these carbonylations have gained importance for the generation of synthetic intermediates.

The ring-opening copolymerization of propylene oxide and carbon monoxide forms the polyester poly(β-hydroxybutyrate) (PHA, Equation 17.52). The physical and mechanical properties of some PHAs are similar to those of isotactic polypropylene. This polymerization was first reported by Furukawa and co-workers in 1965,[241] and more recent studies have been reported by Osakada,[242] Rieger,[243–250] and Alper.[251] These polymerizations have been conducted with $Co_2(CO)_8$ and additives. The combination of $Co_2(CO)_8$, a 1,10-phenanthroline derivative, and benzyl bromide afforded polyester with an M_n value of 19.4 kg/mol and M_w/M_n of 1.41. In addition to propylene oxide, 1,2-epoxybutane was successfully copolymerized with CO to yield the corresponding poly(β-hydroxypentanoate) with an M_n value of 16.7 kg/mol and a M_w/M_n of 1.28. The role of benzyl bromide is unclear. Related copolymerizations of aziridines and CO to form polyamides have also been reported.[252] Polymer M_n values as high as 27.5 kg/mol have been reported, and typical M_w/M_n values varied from 1.11 to 1.64.

$$\text{R}\overset{O}{\triangle} + \underset{\substack{40-87 \text{ atm}}}{CO} \xrightarrow[\text{R = H, Me, Et}]{\text{"Co"}} \text{[structure]}_n \tag{17.52}$$

These polymerizations could occur either by direct enchainment of the epoxide or by carbonylation of the epoxide to the β-lactone, followed by ring-opening polymerization of the lactone. Rieger and co-workers and Alper and co-workers showed that these polymerizations occur by direct enchainment of the epoxide.[214,251] A variety of mechanisms for epoxide enchainment have been proposed;[214,242,251] most involve reaction of the epoxide with an acylcobalt species and a nucleophile, either a nitrogen base or $[Co(CO)_4]^-$, in a transition state that is not well defined.

17.6.4. Types of Catalysts and Scope of Substrates for Aziridine Carbonylation

Efforts to develop aziridine carbonylation have occurred in parallel with efforts to develop epoxide carbonylation. The β-lactams that are formed by this process are important in medicinal,[253–255] organic,[256–261] and polymer chemistry.[262–265] Important contributions to aziridine carbonylation have been made by Alper, Davoli and Prati, and Coates.[266–271] A series of carbonylations of aziridines were developed using $[Rh(CO)_2Cl]_2$ and $Co_2(CO)_8$ as catalyst. More recently, faster rates and expanded scope have been found with catalysts that combine a Lewis acid cation with the $[Co(CO)_4]^-$ anion.

The carbonylation of several classes of aziridines have been reported. In general, the carbonylation of aziridines containing alkyl or benzyl substituents on the two carbon atoms occurs to insert the CO into the C–N bond containing the less-substituted carbon.[266,269] Equation 17.53 shows a series of these reactions catalyzed by $Co_2(CO)_8$. In contrast, the carbonylation of aziridines containing aryl groups on one of the two carbons occurs at the benzylic position. Examples of these reactions are shown in Equation 17.54.[266,269,271] Carbonylation of these phenyl-substituted aziridines catalyzed by $[Rh(CO)_2Cl]_2$ has also been reported by Alper, and a representative example is shown in Equation 17.55.[272]

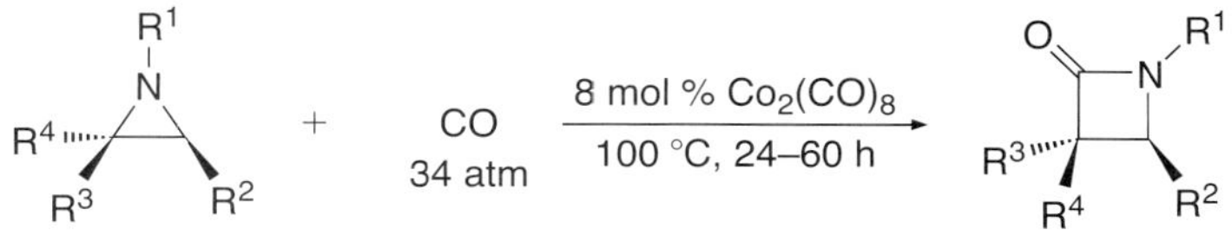

$$R^1, CO \xrightarrow[100\,°C,\ 24-60\ h]{8\ mol\%\ Co_2(CO)_8} \qquad (17.53)$$

34 atm

40–95%

$R^1 = p\text{-MeOC}_6\text{H}_4,\ (CH_2)_2\text{Ph, or Bn}$
$R^2 = {}^t\text{Bu, Me, }-(CH_2)_4-,\ CH_2\text{OSi}^t\text{BuMe}_2,\ \text{or } CH_2\text{OH}$
$R^3 = H,\ R^4 = H\ \text{or Me}$

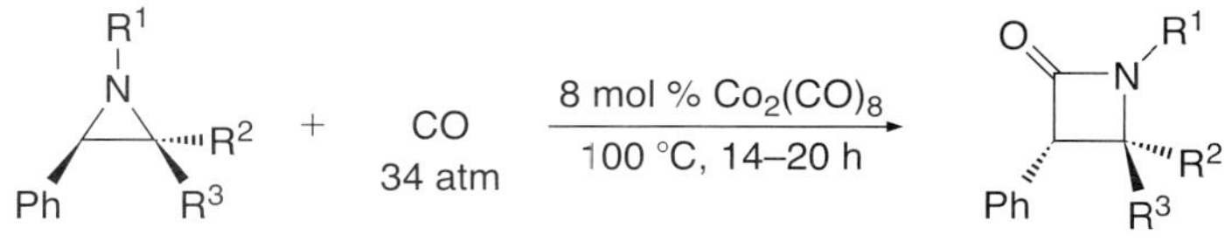

$$\xrightarrow[100\,°C,\ 14-20\ h]{8\ mol\%\ Co_2(CO)_8} \qquad (17.54)$$

34 atm

40–94%

$R^1 = \text{Bn, } CH_2CO_2\text{Et, }{}^i\text{Pr, or } CH_2CH=CH_2$
$R^2 = CH_2\text{OSi}^t\text{BuMe}_2\text{ or H}$
$R^3 = H,\ CH_2NH_2,\ CH_2\text{OSi}^t\text{BuMe}_2,\ \text{Me, or } CH(OH)CH_2CH=CH_2$

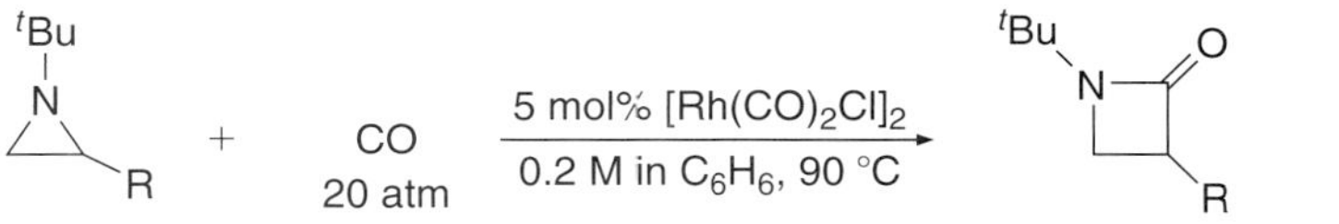

$$\xrightarrow[0.2\ M\ \text{in }C_6H_6,\ 90\,°C]{5\ mol\%\ [Rh(CO)_2Cl]_2} \qquad (17.55)$$

20 atm

R = Aryl, quantitative
R = Me, no reaction

The kinetic resolution of aziridines has also been reported. Alper showed that the carbonylation of *N-tert*-butyl- and *N*-adamantyl-2-arylaziridines[273] in the presence of catalytic amounts of the combination of [Rh(COD)Cl]$_2$ and enantiopure menthol produced the corresponding *N*-alkyl-3-phenylazetidin-2-ones in up to 99.5% optical yield, although the isolated yields were modest. Consistent with the cobalt-catalyzed carbonylation of epoxides and related substrates, inversion of configuration occurs at the site of carbonyl insertion.

The selectivity for reaction at the less hindered carbon results from steric effects. The selectivity for reaction at a phenyl-substituted carbon of an aziridine has been proposed to result from coordination of the catalyst to the phenyl ring,[274] but computational studies imply that the catalyst coordinates to the aziridine nitrogen.[275] In any case, oxidative additions that cleave benzylic C–X bonds are generally favored over oxidative additions that cleave aliphatic C–X bonds, and the preferential insertion of CO into the phenyl-substituted C–N bond of the aziridine is consistent with this general trend.

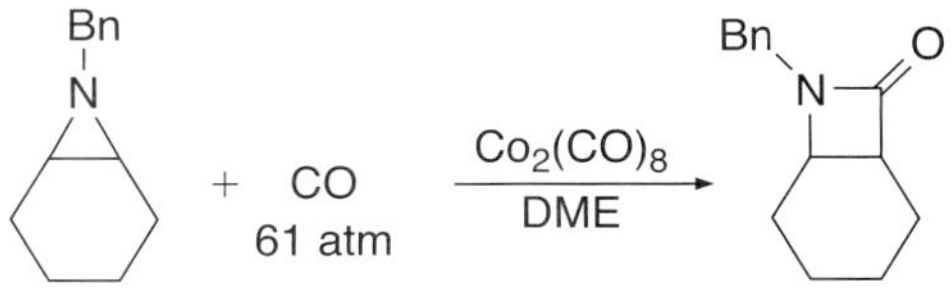

$$\xrightarrow[DME]{Co_2(CO)_8} \qquad (17.56)$$

61 atm

Catalyst	Mol%	T (°C)	t (h)	Yield (%)
$Cp_2Ti^{\oplus}/(CO)_4Co^{\ominus}$	5	80	18	80
Al/Co catalyst of Scheme 17.24	5	80	18	> 5
$Co_2(CO)_8$	8	100	48	28

Like the carbonylation of epoxides, the carbonylation of aziridines occurs with increased rates and scope in the presence of catalysts containing a Lewis acidic cation and $Co(CO)_4^-$ anion. A particularly active version of this catalyst for the carbonylation of aziridines is the uncommon species $[Cp_2Ti(THF)_2]^+[Co(CO)_4]^-$. As shown in Equation 17.56, this catalyst is substantially more active than $Co_2(CO)_8$ for the carbonylation of N-benzyl cyclohexene imine.[218] The carbonylation of N-tosyl-2-methylaziridine has also been accomplished (Equation 17.57), and this reaction is important because of the ability to prepare optically active N-tosyl aziridines.[276] Although the reaction catalyzed by the titanium and cobalt system occurred to only 35% conversion, the carbonylation of the N-tosyl-2-methylaziridine catalyzed by the aluminum and cobalt system occurred to completion under the same reaction conditions.

(17.57)

Catalyst	Mol%	T (°C)	t (h)	Yield (%)
$Cp_2Ti^{\oplus}/(CoCO)_4^{\ominus}$	5	90	6	35%
Al/Co catalyst of Scheme 17.24	5	90	6	99%

17.6.5. Mechanism of Epoxide Carbonylation

The mechanism of catalytic epoxide-expansion carbonylation has been the subject of several investigations. Because the catalytic formation of β-lactones from epoxides generates a new carbonyl compound, it is easily monitored by in situ IR spectroscopy.[277] In addition to kinetic and reactivity studies, computational studies[278,279] have been conducted to probe the mechanism of epoxide-expansion carbonylation. The mechanistic aspects of the carbonylation chemistry based on these results are presented in this section.

The accepted mechanism for the carbonylation of epoxides is shown in Scheme 17.26,[277] and the basic steps of this cycle are also thought to occur during the carbonylation of aziridines. Alper first proposed a catalytic cycle for the expansion carbonylation of aziridines by $[Co(CO)_4]^-$,[266,267] and Coates has proposed a similar cycle for epoxide carbonylation catalyzed by complexes containing both Lewis acids and cobalt-carbonyl anions (Scheme 17.26).[213,218] This mechanism consists of four steps: (1) the activation of substrate by coordination to a Lewis acid; (2) the S_N2 attack on the substrate by $[Co(CO)_4]^-$; (3) the insertion of CO into the new cobalt–carbon bond, and the subsequent uptake of CO; and (4) ring closing with extrusion of product and regeneration of the catalytic species.

This cycle is supported by several pieces of data. First, the carbonyl insertion generally occurs between the heteroatom and the least-hindered carbon. Attack of $[Co(CO)_4]^-$ on the activated epoxide at the least substituted carbon of the epoxide leads to CO insertion adjacent to this center and formation of the observed product. Second, the carbonylation of 1,2-disubstituted epoxides proceeds with inversion of configuration at the α-carbon (the carbon attached to the carbonyl group in the product). This observation is consistent with C–O or C–N bond cleavage by an S_N2 pathway. Also consistent with exclusive attack at the less substituted carbon, (R)-propylene oxide undergoes carbonylation with the cobalt catalyst containing the aluminum Lewis acid with complete retention of stereochemistry at the lactone β-carbon, yielding (R)-β-butyrolactone.[213] Although these trends are not without exceptions,[280] they provide strong support for this catalytic cycle.

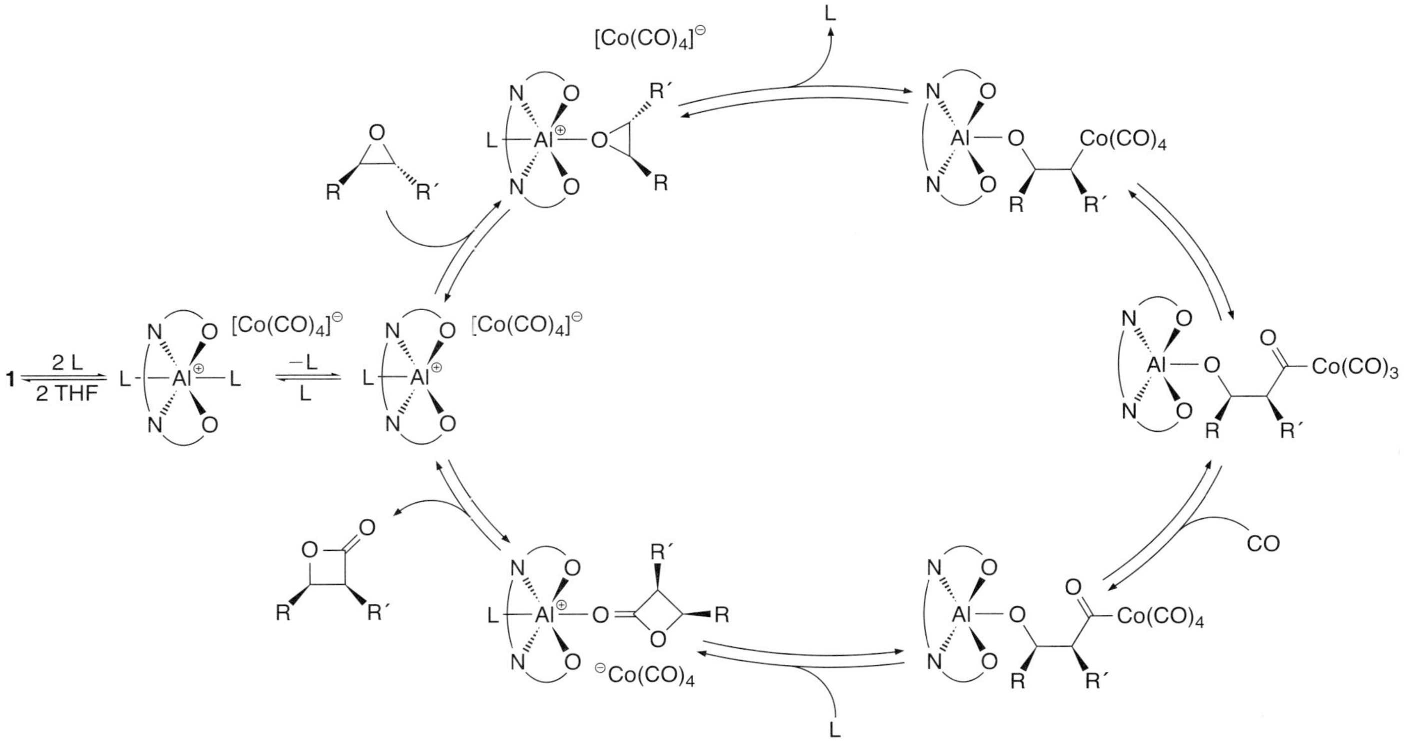

Scheme 17.26.
Catalytic cycle for the carbonylation of epoxides catalyzed by the aluminum/cobalt catalyst. L = Lewis
base (solvent, epoxide, lactone).

The mechanism for the formation of β-hydroxy aldehydes, β-hydroxy esters, and
β-hydroxy amides from the carbonylation of epoxides stems from the first steps of the
same cycle. As shown in Scheme 17.27, nucleophilic attack on the epoxide, followed by
carbonylation, forms a species that can react with hydrogen to form hydroxy aldehydes
(which can be further hydrogenated to diols), with alcohols to form esters, or with silylamines to form β-siloxy amides. The hydrogenolysis is similar to the hydrogenolysis during
the hydroformylation of olefins, and the resulting cobalt hydride is sufficiently acidic to
protonate the pendant alkoxide. The reaction of the acyl intermediate with alcohol is similar to the reaction that generates esters in the hydroesterification of olefins. The mechanism
of the reaction with silylamines is less studied.

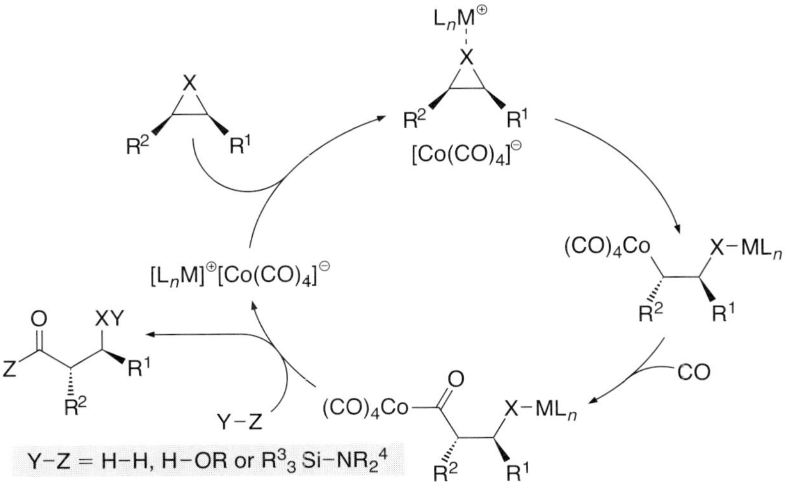

Scheme 17.27

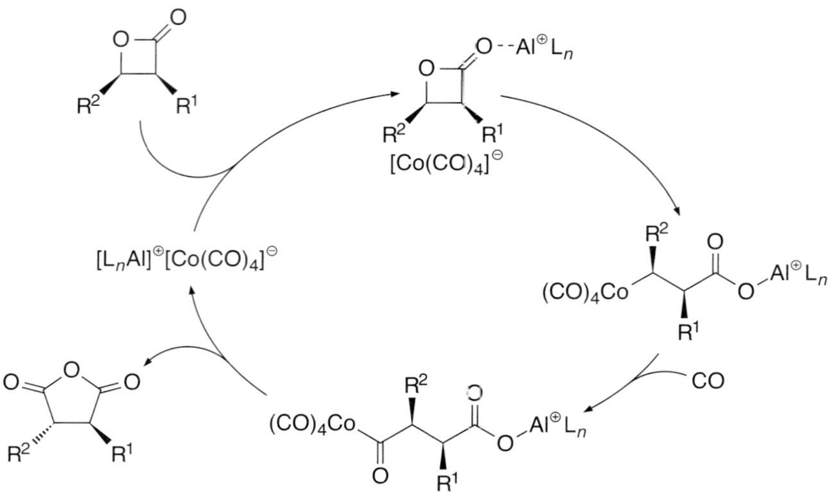

Scheme 17.28

The formation of succinic anhydride from the carbonylation of lactones likely results from a mechanism that is similar to that for the carbonylation of epoxides. This mechanism is shown in Scheme 17.28. In this case, the β-lactone would be activated through binding to the Lewis acid center and attacked in an S_N2 fashion by $[Co(CO)_4]^-$. Nucleophilic attack in this system occurs adjacent to the oxygen in the ring, with the lactone carboxylate functionality serving as a leaving group. Following insertion of CO into the cobalt–alkyl bond, the carboxylate group would then attack the cobalt acyl to form the five-membered anhydride ring and regenerate the catalytic species. Further investigation of β-lactone carbonylation by the catalyst containing the aluminum Lewis acid showed that it occurred by a pre-equilibrium involving coordination of β-lactone to the aluminum cation, followed by turnover-limiting opening of the β-lactone ring.

17.7. Carbonylations of Organic Halides

Palladium-catalyzed carbonylation of organic halides to form esters, amides, ketones, and aldehydes (Equation 17.58) has been studied and reviewed extensively.[281–288] These reactions are closely related to the palladium-catalyzed cross-coupling processes presented in Chapter 19. However, the addition of CO to these processes generates organic carbonyl compounds, rather than products from direct cross coupling.

$$R–X \ + \ CO \ + \ Nu^\ominus \xrightarrow{\text{Pd catalyst}} \underset{R}{\overset{O}{\underset{}{\|}}}\!\!-Nu \ + \ X^\ominus \qquad (17.58)$$

The carbonylations to form esters are typically conducted by combining an organic halide, carbon monoxide, an alcohol, and a base, and the carbonylations to form amides are typically conducted by combining an organic halide, carbon monoxide, and a primary or secondary amine. The synthesis of ketones requires a main group organometallic reagent to deliver the alkyl group and has been reported with the combination of organic halides, carbon monoxide, and either organoboronates, organostannanes, or organozinc reagents. The carbonylations of organic halides to form aldehydes are the least developed of these sets of reactions, but recent procedures involving the reaction of aromatic halides with the combination of carbon monoxide and hydrogen (synthesis gas) should make this version of organic halide carbonylation more practical.

A majority of the carbonylations of organic halides have been conducted with aryl and vinyl halides, although reactions have been developed with benzylic halides and even purely aliphatic halides. A majority of the reactions of aryl halides have been conducted with aryl iodides, although a few reactions have been reported with electron-poor aryl bromides. Few examples of these reactions have been reported with electron-rich aryl bromides or aryl chlorides. Most of these reactions have been conducted with palladium complexes containing phosphine ligands.

17.7.1. Carbonylations of Organic Halides to form Esters and Amides

17.7.1.1. Discovery and Scope

The first palladium-catalyzed carbonylations of aryl halides were published in 1974 and 1975 (Equation 17.59).[289,290] Heck first reported the synthesis of benzoates by the reaction of an aryl iodide, carbon monoxide, and an alcohol in the presence of a tertiary amine and catalytic amounts of the combination of palladium acetate and triphenylphosphine.[289] Concurrently, he reported the synthesis of benzamides from an aryl iodide, carbon monoxide, a primary amine, and a tertiary amine as base catalyzed by the same type of palladium complex.[291] The related reactions of vinyl halides were also reported, and these reactions occurred with retention of the olefin geometry.[289]

More recently, Osborn[292] and Milstein[293] reported the reactions of aryl chlorides with carbon monoxide and a base to form aromatic esters, and Alper reported related reactions of aryl chlorides in biphasic media to form carboxylic acids.[294] The reaction of Milstein is shown in Equation 17.60, and the reactions of Osborn were conducted under similar conditions with PCy_3 as ligand. These examples comprise some of the first coupling chemistry performed with aryl chlorides.

The carbonylations of haloarenes have also been conducted intramolecularly. Examples of these reactions to form lactams are represented generically in Equation 17.61[295-298] along with specific examples to form benzolactams.[295] The carbonylations of haloarenes has also been applied to the synthesis of polymers. As shown in Equation 17.62, these reactions can be conducted between dihaloarenes and diamines to form polyaramide materials.[299]

In addition to the carbonylation of aryl and vinyl halides, the carbonylation of aryl and vinyl triflates to form esters and amides has been conducted. An example of these reactions is shown in Equation 17.63.[300]

$$(17.59)$$

$$(17.60)$$

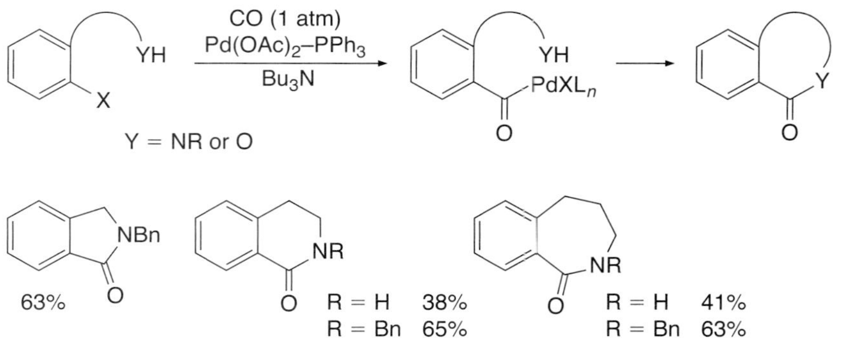

(17.61)

(17.62)

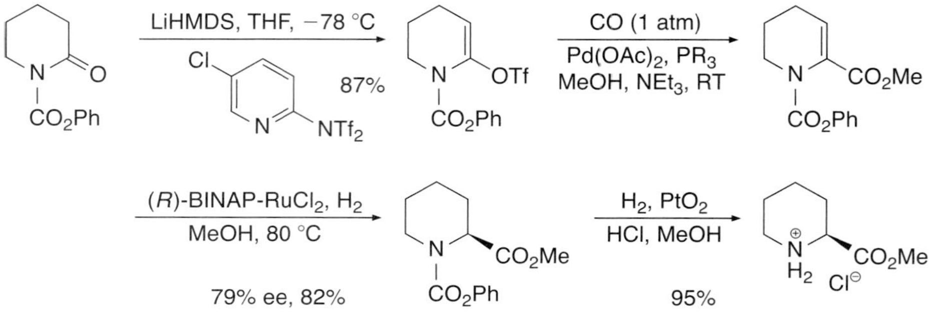

(17.63)

One application of the esterification of aryl halides is shown in Equation 17.64 and was reported by Tsuji.[301] The reaction of an electron-rich, hindered aryl iodide with a secondary alcohol generated the precursor to the macrocyclic lactone.

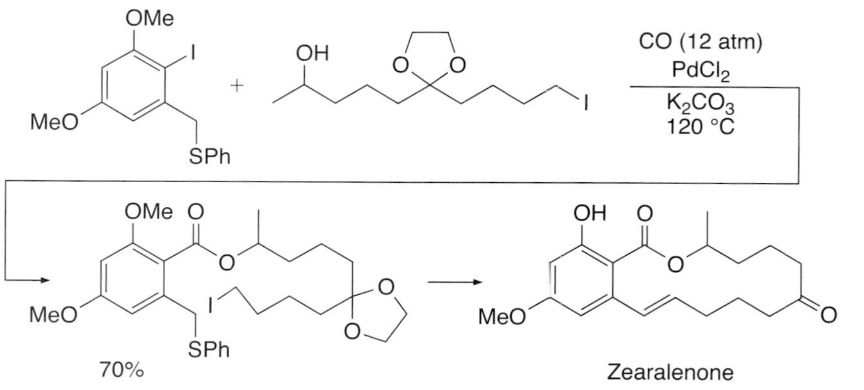

(17.64)

The reactions of benzylic halides with carbon monoxide and alcohols form esters in good yields.[302] However, the reactions of alkyl halides are more limited for two reasons. First, the oxidative addition of alkyl halides occurs less readily to palladium complexes than the oxidative addition of aryl halides. This difference was noted in Chapter 7. Second, the intermediate alkylpalladium halide can undergo β-hydrogen elimination. As noted in Chapters 9 and 10, these hurdles have been overcome in some cases, and cross-coupling reactions of alkyl halides are beginning to be developed. However, the carbonylation of purely aliphatic halides remains to be developed.

17.7.1.2. Mechanism of Aryl Halide Esterification and Amidation

The reactions of aryl halides with carbon monoxide and either alcohols or amines to form esters or amides occur by two related yet distinct mechanistic pathways. These pathways are related to the mechanism of cross coupling presented in Chapter 19, and the reader might find it helpful to refer to the discussion of mechanisms of cross coupling in that chapter. Scheme 17.29 outlines the pathways for the carbonylation of aryl halides to form esters and amides.

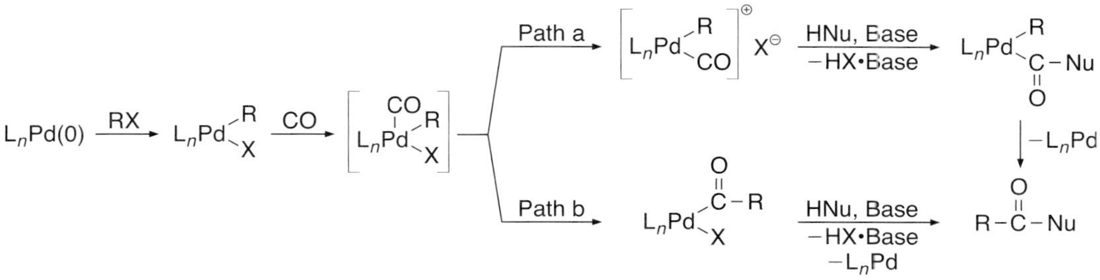

Scheme 17.29

The mechanism of the esterification and amidation of aryl halides begins with oxidative addition of the aryl halide to a Pd(0) complex containing carbon monoxide and phosphine ligands to form an arylpalladium halide intermediate. This complex can then react by two different pathways. It can undergo reaction with CO to displace halide and form a cationic carbonyl species, which then undergoes nucleophilic attack by the alcohol or amine to form an arylpalladium complex containing an alkoxycarbonyl or carbamoyl ligand. Reductive elimination then forms the ester or amide product. Alternatively, the arylpalladium halide complex can undergo migratory insertion of CO to form a benzoylpalladium halide complex. This complex can then react with the alcohol or amine to form the ester or amide product.

Careful mechanistic studies on the carbonylation process have been reported by Yamamoto, including studies with isolated model compounds.[188,303] These studies have revealed several reaction pathways, and the particular pathway depends on whether the electrophile is an aryl or benzyl halide and whether the nucleophile is an amine or an alkoxide. The existing experimental data suggest that the latter pathway b in Scheme 17.29 involving insertion of CO into the palladium–aryl bond to form a benzoylpalladium halide intermediate occurs.[188] These complexes have been isolated with PMe$_3$ and PPh$_3$ as ligand and have been shown to form the ester product upon reaction with an alcohol and amine base and to form the amide products upon reaction with amine alone.

These studies have also suggested that the formation of ester and amide from the benzoylpalladium complex occurs by two different mechanisms. Particularly revealing is a set of reactions of an acylpalladium complex with a mixture of amine and alcohol conducted with varied amounts of added PPh$_3$. The reactions with higher concentrations of PPh$_3$ formed more amine, and the reactions with lower concentrations of PPh$_3$ formed more ester. These results imply that the ester and amine form by two different pathways, such as the two in Scheme 17.30, in which the coordination number of the species undergoing reaction with the two nucleophiles is different. They imply that the formation of ester occurs by deprotonation of an alcohol complex generated by dissociation of phosphine to form a benzoylpalladium alkoxide complex that reductively eliminates ester and that formation of amide occurs by direct attack of amine on the benzoyl group without dissociation of a dative ligand.

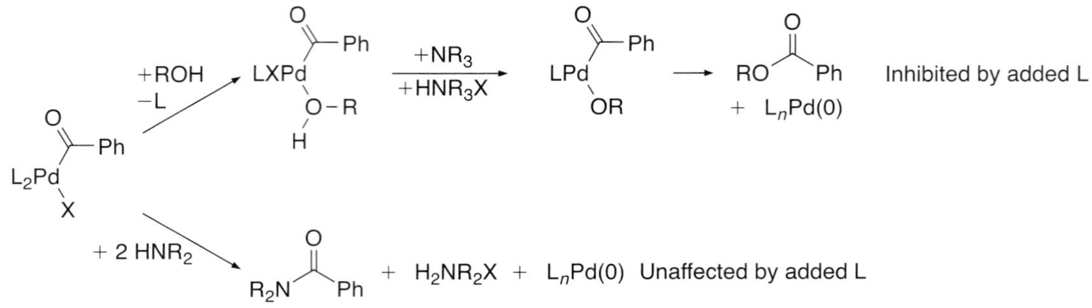

Scheme 17.30

17.8. Copolymerization of CO and Olefins

17.8.1. Overview of the Process and Polymer Properties

The copolymerization of carbon monoxide and olefins[304–308] forms the polyketone in Equation 17.65, and this polymerization is closely related to the hydroesterification and hydrocarboxylation of olefins. The rate of reaction of the acyl intermediate that was generated in the hydroesterification process with olefin or alcohol differentiates the formation of copolymer from the formation of monomeric esters. This difference in relative rates for reaction of the acyl intermediate with olefin versus alcohol results from a change in the ancillary ligand on the palladium, as described in this section.

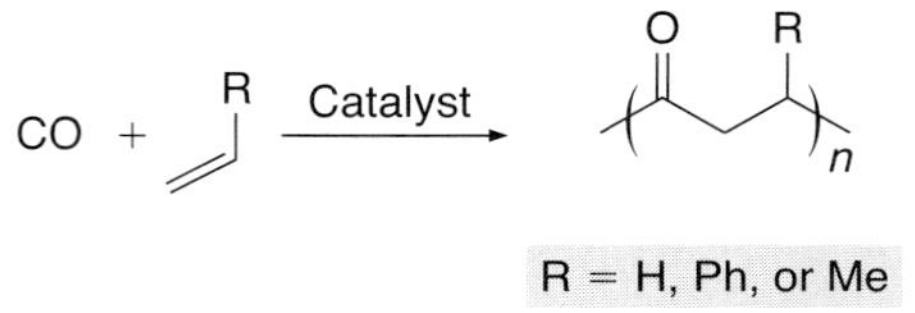

$$R = H, Ph, or Me$$

(17.65)

The polyketone containing alternating units of carbon monoxide and ethylene with a small percentage of incorporated propylene was developed at Shell during the 1980s and 1990s and had the trade name Carilon. According to a review of Sommazzi and Barbassi on the polymer properties of these materials,[308] "Carilon's properties are near to those of engineering polymers such as nylon and polyacetal. . . . They exhibit moduli, impact and thermal characteristics of amorphous polymers with the chemical resistance of crystalline polymers, while processing like a polyolefin. Their properties are at the border between commodity polymers (like PE and PVC) and engineering polymers of medium performance like polyamides and polyesters, etc." The invention of the palladium-catalyzed synthesis of these polymers by Drent and the blending of these materials to generate a commercial material was conducted at Shell Chemicals. A pilot plant for production of this material was constructed in the Netherlands. However, business issues led this technology to be abandoned by Shell, and the patents were donated to SRI International in 2002.

17.8.2. Development of Catalysts for the Synthesis of CO/Ethylene Copolymerization

The copolymer between ethylene and carbon monoxide has been known since the early 1940s. In 1941 this copolymer was produced at Bayer in a high-temperature, high-pressure process (230 °C, 200 atm).[309] Roughly 10 years later, this polymer was produced through free-radical chemistry by Brubaker at DuPont.[310] This material was not a perfectly alternating polymer, and for reasons described below, the regularity of the polymer is important for its stability. The first transition metal mediated process to generate such materials was reported by Reppe in 1948 with nickel cyanide.[311,312] Later, Gouch[313] patented the use of palladium chloride catalysts containing triphenylphosphine as ligand to form alternating copolymers of carbon monoxide and ethylene, and catalysts consisting of palladium cyanides,[314,315] combinations of palladium cyanides[316] or halides[317] and PPh_3, or combinations of Pd(0) phosphine complexes and acid[317] had all been reported in the patent literature for this process.

In the early 1980s, Sen reported the synthesis of CO–ethene oligomers using a combination of dicationic palladium complexes $[Pd(NCMe)_4](BF_4)_2$ and triphenylphosphine as precatalysts.[307,318] These catalysts induced polymerization under milder conditions than the previous catalysts for this process. Although the material reported by Sen possessed relatively low molecular weight, and the catalysts were less reactive than those known today, this finding showed that cationic palladium complexes were more reactive than the neutral systems studied previously.

The first perfectly alternating carbon monoxide/ethylene copolymer with high molecular weight was produced at Shell in the 1980s.[319–322] This material had a high melting temperature, high solvent resistance, and possessed the crystallinity and engineering properties mentioned above. Moreover, the polymers produced by the Shell catalysts were perfectly alternating. Perfect alternation of carbon monoxide and ethylene in the polymer chain is important for the stability of the polymer because α–diketones are photolytically unstable toward C–C bond cleavage, and ketones containing γ-hydrogens undergo photochemical abstraction from this γ-position (Norrish type II cleavage). Thus, the absence of γ-hydrogens that would be present from sequential incorporation of two ethylene units and the absence of α-diketone units from sequential incorporation of CO helps make these polyketones sufficiently stable for commercial applications.

These high-molecular-weight, perfectly alternating copolymers were produced by cationic palladium complexes containing bidentate ligands.[305] The presence of the bidentate ligands (at least with relatively short backbones) changed the selectivity of reactions in alcohol solvents from the formation of esters to the formation of copolymers. As described in more detail below, the polymers formed in alcohol solvents contain at least one ester end group,[305] and the presence of these end groups implies that one termination or initiation step of the polymerization processes is the reaction of the acylpalladium intermediate with alcohol to form ester and metal hydride.

The effect of the donor atom and backbone length on the catalyst activity and selectivity are more subtle. A chart showing the effect of ligand backbone is shown in Figure 17.9. The highest rates and molecular weights are obtained with the catalyst containing a three-carbon backbone. Because the most active current catalyst for the hydroesterification of olefins to form monomeric esters contains the bisphosphine 1,2-bis(di-*tert*-butylphosphino)-*o*-xylene, the factors that differentiate catalysts for hydroesterification and copolymerization are more subtle than simply the presence of a monophosphine (early hydroesterification catalysts) or a bisphosphine (most active copolymerization catalysts). Bidentate nitrogen ligands, such as phenanthroline, have also been studied extensively and generate active and selective catalysts for the copolymerization of CO and alkenes. To avoid competing hydroesterification and copolymerization processes, some of the copolymerizations have been conducted in weakly nucleophilic alcohol solvents, such as hexafluoroisopropanol, or aprotic solvents, such as methylene chloride.

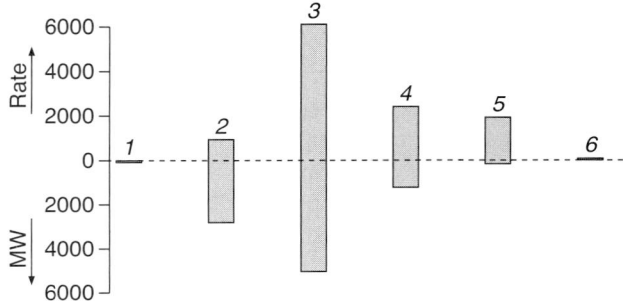

Figure 17.9.
Influence of chain length of ligand $Ph_2P(CH_2)_nPPh_2$ on the rate and molecular weight of polyketone from the copolymerization of CO and ethylene. Figure adapted from reference 304.

17.8.3. Mechanism of the Coplymerization of CO and Ethylene

17.8.3.1. Overall Cycle: The Steps of Chain Propagation

The basic steps of the copolymerization of carbon monoxide and ethylene catalyzed by dppp-ligated palladium are shown in Scheme 17.31.[323] The chain growth portion of this mechanism consists of a series of alternating migratory insertions of ethylene into a palladium–acyl complex and insertion of CO into a metallacyclic γ-ketoalkyl complex. CO binds sufficiently tightly to the metal center to displace the ketone carbonyl group in the metallacycle to generate an alkyl carbonyl species that precedes insertion to reform the acyl complex. These classes of insertion reactions were described in Chapter 9. Chain termination steps and catalyst deactivation steps are discussed later in this section.

The resting state of this catalyst (see Chapter 6 for definitions of catalytic species) is either the chelated γ-ketoalkyl complex or the acyl carbonyl complex. The γ-ketoalkyl complex is the resting state for reactions conducted with 1,3-bis(diphenylphosphino)propane at lower ratios of CO to ethylene, whereas the acyl carbonyl complex is the resting state at high ratios of CO to ethylene. The acyl carbonyl complex is the resting state of reactions conducted with phenanthroline as the dative ligand. At low ratios of CO to ethylene, all steps are reversible, and the insertion of ethylene to form the γ-ketoalkyl complex is turnover limiting.

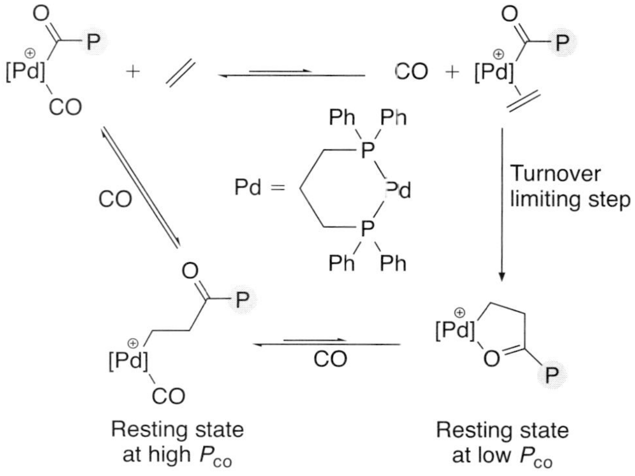

Scheme 17.31

The high activity of the catalysts containing bidentate ligands is thought to result from the favorable orientation of the growing polymer chain and the unsaturated reactants (CO and olefin).[323–325] The presence of a bidentate ligand on the palladium(II) catalysts ensures that coordination of the substrates occurs cis to the growing polymer chain. The migratory insertion reactions that generate the polymer require a cis disposition of the coordinated substrate and the covalent ligand into which the substrate inserts (see Chapter 9).

Vrieze and Elsevier[326, 327] and Brookhart[323–325, 328–330] have studied insertions involved in the polymerization process. Brookhart and co-workers have quantified the origin of the nearly perfect alternation between the incorporation of carbon monoxide and ethyelene. The activation barriers for the various insertion steps involving palladium complexes of a bidentate phosphine are provided in Figure 17.10.[323,330] The activation energies for the insertion of carbon monoxide into the alkylpalladium complex to form an acylpalladium species and the insertion of ethylene into a cationic palladium acyl to form a γ-keto alkylpalladium complex are both lower than that for insertion of ethylene into the palladium alkyl. These insertion reactions occur at −80 °C and −100 °C with barriers of 13.4 and 12.3 kcal/mol, respectively.

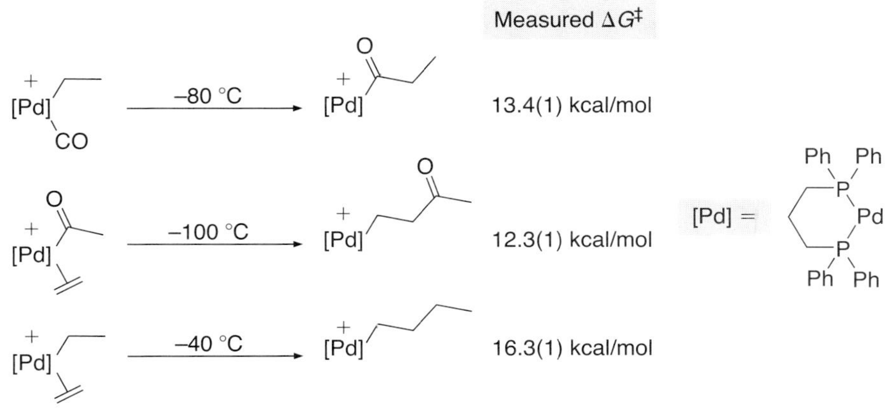

Figure 17.10.
Activation barriers for the insertion reactions of the copolymerization of CO and ethylene catalyzed by DPPP-ligated palladium and the competing insertion to form polyethylene.

One can compare these reactions to those that would form regio-irregularities in the polymer chain from sequential insertions of CO or ethylene. The ethylene-bound acyl complex formed by ligand substitution of ethylene for carbon monoxide lies uphill of the acyl carbonyl complex. Nevertheless, the ethylene acyl complex is the reactive complex for insertion into the acyl group because insertion of CO into a metal–acyl complex is thermodynamically disfavored (though not kinetically prohibited), as noted in Chapter 9. Cationic alkylpalladium complexes containing bidentate ligands polymerize ethylene with fast rates in the absence of CO. Thus, sequential insertion of ethylene might be expected to compete with the perfectly alternating insertion of CO and ethylene. The insertion of ethylene into the metal alkyl does occur at low temperature. However, it requires a temperature of −40 °C at comparable rates to the insertion of CO and has a much higher barrier of 16.3 kcal/mol.

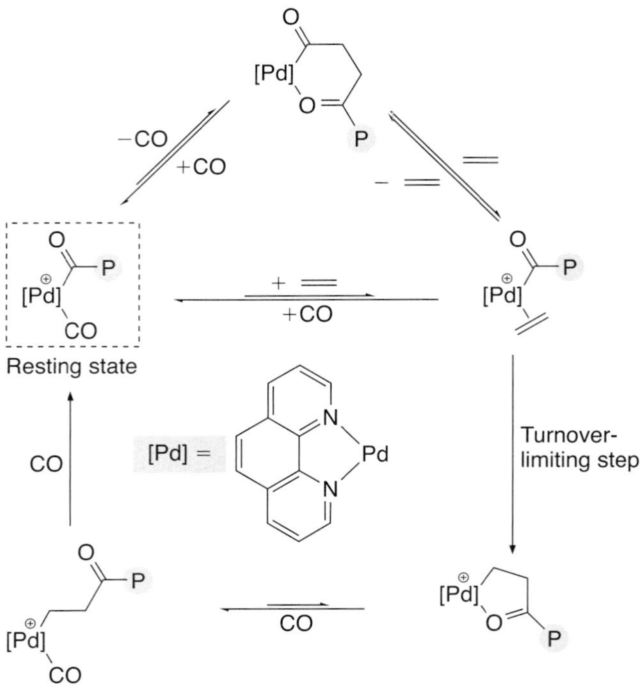

Scheme 17.32

These barriers have also been measured for the reaction of the cationic palladium species containing phenanthroline as a ligand.[324,328] A summary of the steps of this catalytic process is shown in Scheme 17.32. The barriers for insertion of CO and ethylene into the phenanthroline-ligated palladium–alkyl complex are shown as an energy diagram in Figure 17.11. In this case, the acyl carbonyl species is the catalyst resting state. The barrier for insertion of CO into the alkyl complex is 14.9 kcal/mol. The alkyl ethylene complex lies 2.7 kcal/mol uphill of the alkyl carbonyl species, and the barrier for insertion of ethylene lies a further 19.4 kcal/mol uphill from the alkyl carbonyl species. Thus, the composite barrier for insertion of ethylene in this system containing CO is 22.1 kcal/mol. Based on these barriers and the 14.9 kcal/mol barrier for insertion of CO into the alkyl complex, one double ethylene insertion occurs every 10^6 turnovers.

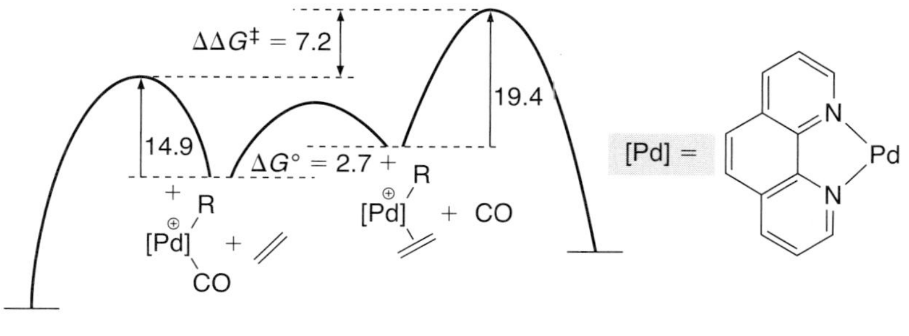

Figure 17.11.
Free energy diagram for the insertions of ethylene and CO during palladium-catalyzed copolymerization of ethylene and CO. All values are in kcal/mol.

17.8.3.2. Chain Termination and Catalyst Decomposition

The molecular weights of polymers are controlled by the relative rates for chain growth and chain transfer steps, as described in more detail in Chapter 22. Thus, the mechanism of chain transfer has also been studied. One mechanism for chain transfer is the reaction of the acyl group with alcohol to form ester and palladium hydride. This process is shown in Equation 17.66. The palladium hydride would then insert olefin to generate an alkyl group, and this alkyl group would insert carbon monoxide to generate an acyl group that begins chain growth. This mechanism leads to a polymer with one ester and one ketone end group, and these end groups have been observed on polymers that have been generated at low temperatures.[305]

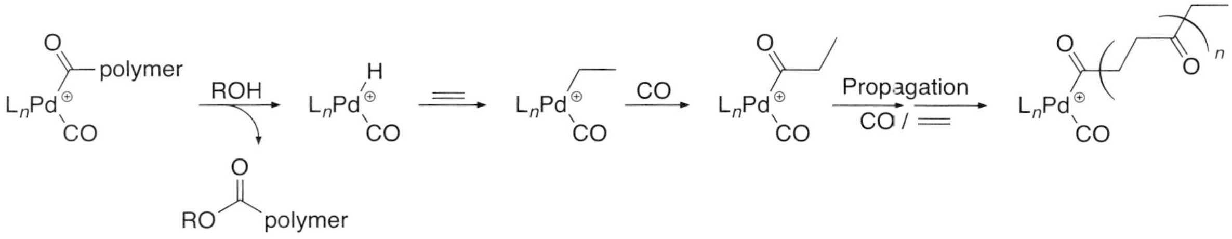

$$(17.66)$$

In the absence of an alcohol solvent, chain transfer can occur by β-hydrogen elimination, or more likely[331] by chain transfer to monomer in which the β-hydrogen is directly transferred to an incoming olefin (see Chapter 22 for details on this process in alkene polymerization). These mechanisms for chain transfer give rise to olefin end groups, as shown in Equation 17.67. The formation of olefin-terminated polymer is not observed when the copolymerization is conducted in alcohol solvent, but has been observed from reactions conducted in polar aprotic solvents.[332] Brookhart has shown that the copolymerization of CO and styrene (vide infra) in the absence of alcohol is a "living polymerization" in which the catalyst remains bound to the end of the growing polymer chain (for more details on living polymerizations, see Chapters 21 and 22).[333]

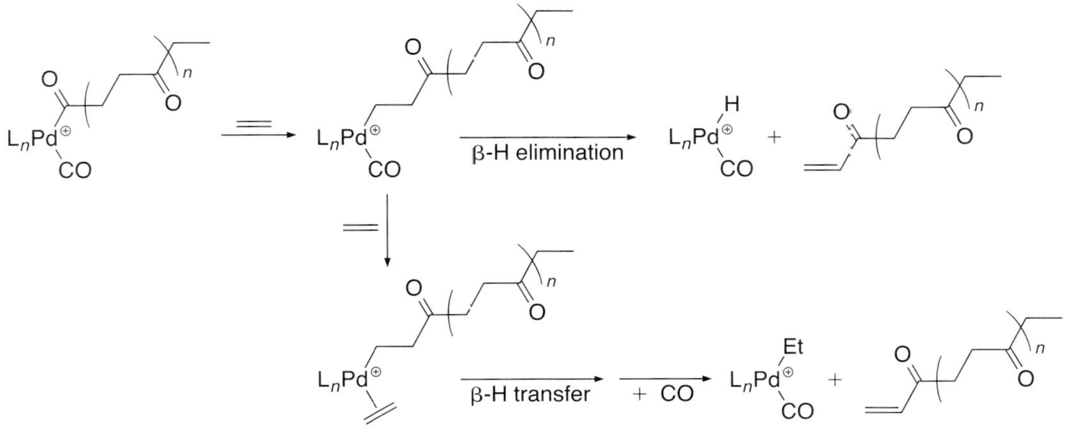

$$(17.67)$$

The pathway for catalyst decomposition is more challenging to determine because the percentage of events that lead to catalyst decomposition is much lower than the percentage of events that lead to polymer production. Nevertheless, the effect of quinone additives provides some indication of the pathways for catalyst deactivation. The chain termination processes in Equations 17.66 and 17.67 both generate palladium hydrides, which are often unstable toward reduction to palladium(0) complexes and palladium metal. Thus, the addition of quinone can intercept the palladium hydride to generate a palladium enolate species, or it can lead to oxidation of the palladium(0) to form a palladium(II) complex and hydroquinone. In addition, the unusual dimeric dicationic palladium species in Figure 17.12 containing bridging phosphine ligands (see Chapter 2 for a discussion of bridging phosphine ligands) has been identified as a potential catalyst decomposition product formed from the combination of a palladium(II) dication and a palladium(0) species.[334]

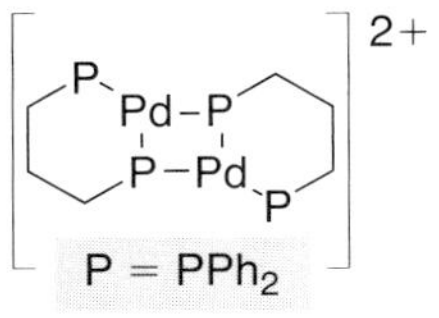

Figure 17.12.
One catalyst decomposition product identified in the copolymerization of CO and olefins.

17.8.4. Copolymerization of CO and α-Olefins[335]

17.8.4.1. Overview of the Copolymerization of CO and α-Olefins

Copolymers formed from carbon monoxide and α-olefins have also been studied intensively. These polymers have not reached the stage of commercialization, but have an unusual property that has led to efforts to develop catalysts for these copolymerizations and efforts to understand the factors that control selectivity pertinent to this type of polymerization. Unlike most other polymers possessing relative stereochemistry in the polymer chain, like polypropylene, the copolymer of carbon monoxide and propylene is chiral and optically active. The structures of the two stereo-regular isomers of this class of polymer are shown in Figure 17.13. These are among the few non-natural synthetic polymers that are chiral and contain the stereocenters within the main chain. Thus, enantioenriched polymers can be prepared with catalysts containing chiral non-racemic catalysts.

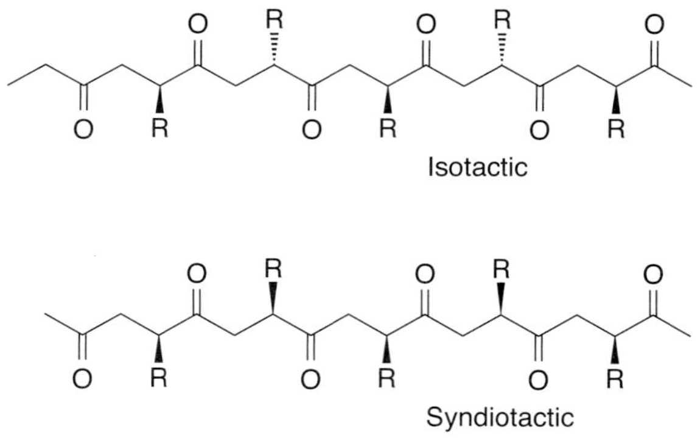

Figure 17.13.
The two stereoregular forms of copolymers generated from CO and substituted olefins.

The copolymerization of carbon monoxide and propylene confronts issues of both regiochemistry and stereochemistry. The catalyst must control the regioselectivity of the insertion of the α-olefin and the relative and absolute stereochemistry along the main chain. A majority of the studies on the copolymerization of carbon monoxide and substituted olefins have been conducted with styrene and propene as the olefin. The regiochemistry and stereochemistry of the copolymerization of carbon monoxide with styrene is distinct from those of the copolymerization of carbon monoxide with propene. These differences result from differences in the electronic properties of the olefins and its impact on the regiochemistry for the insertion. The copolymerization of carbon monoxide with styrene is presented first, and the copolymerization of carbon monoxide with propylene is presented second.

17.8.4.2. Copolymerization of Carbon Monoxide and Styrene

The copolymerization of carbon monoxide with styrene has been the most stereoselective of the copolymerizations of carbon monoxide with substituted olefins. Like the copolymerization of carbon monoxide with ethylene, the copolymerization of carbon monoxide with styrene is perfectly alternating. Moreover, the regioselectivity for insertion of styrene is high.

17.8.4.2.1. Overall Mechanism

The mechanism for the copolymerization of carbon monoxide and styrene is shown in Scheme 17.33. The regiochemistry of the olefin insertion step is noteworthy. This copolymerization occurs by 2,1-insertion of styrene into the metal–acyl linkage to place the aryl group α to the metal center. This regiochemistry for insertion has been demonstrated by the related hydroesterification and hydrocarboxylation of vinyl arenes. It has also been shown by direct observation of the product from insertion of styrene into the palladium–acyl intermediate of the polymerization, as shown in Equation 17.68.[333] As discussed previously in this text, the origin of this regioselectivity is due to the electronic properties of the aryl group and its ability to generate η^3-benzyl structures.

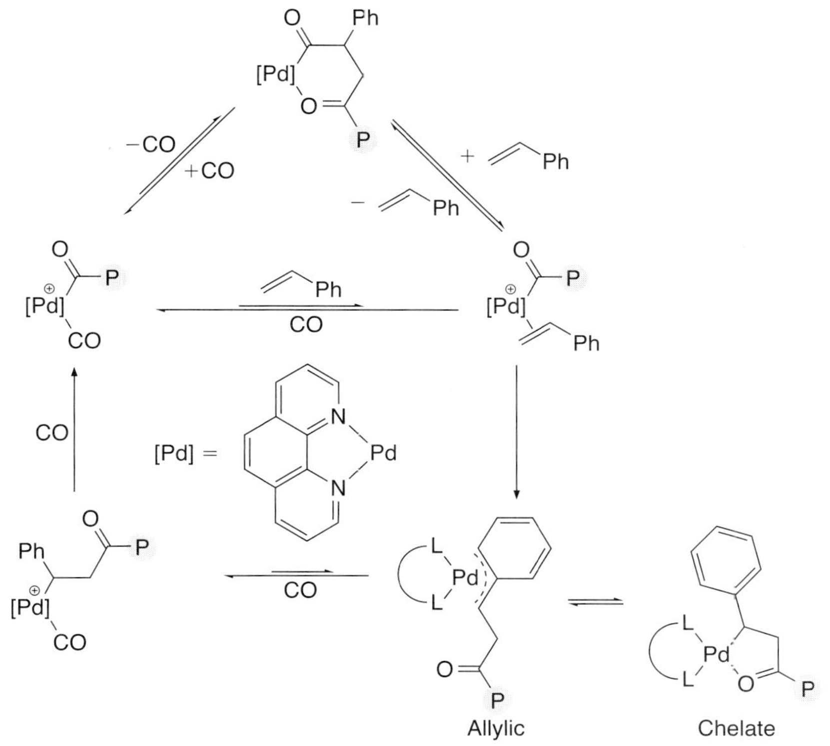

Scheme 17.33

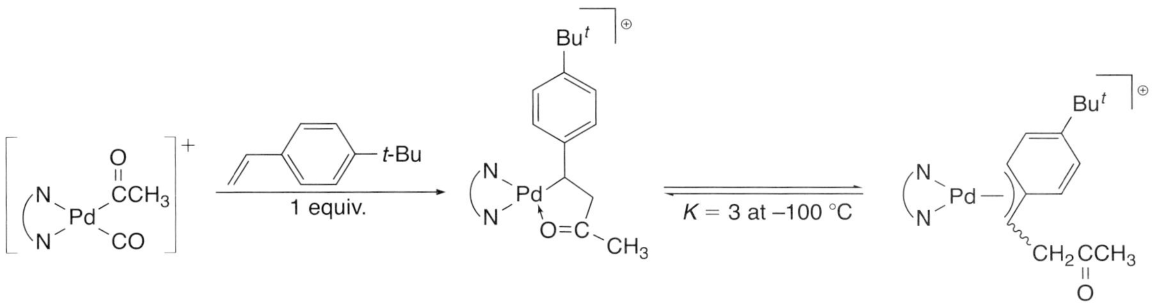

(17.68)

17.8.4.2.2. Control of Stereochemistry

The relative stereochemistry along the main chain can be controlled by the catalyst (catalyst-control or "enantiomorphic site-control") or by the growing polymer chain (chain-end control). The relative stereochemistry along the main chain is different for polymerizations conducted with achiral catalysts and with chiral catalysts and illustrates the contributions of these two effects.

The copolymerization of carbon monoxide and styrene catalyzed by cationic palladium complexes containing bipyridine[336] or 1,10-phenanthroline generates the "syndiotactic" polymer shown in Figure 17.13 The degree of stereoregularity depends on the ligand and conditions.[333,335,337,338] This polymer contains stereocenters with alternating configurations along the polymer chain. Because the catalyst contains an achiral ancillary ligand, the relative stereochemistry of the inserting monomer must be controlled by the growing polymer chain. The stereochemistry of these copolymers generated by catalysts containing other achiral ligands is less regular. One proposal for how the polymer chain could influence the coordination environment at palladium is through a structure such as the one in Figure 17.14.[304]

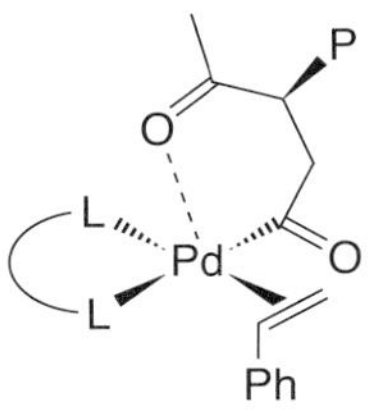

Figure 17.14.
Proposal for the transmission of polymer stereochemistry to the metal center.

In contrast to this relative stereochemistry generated from achiral catalysts, the "isot-actic" polymer shown in Figure 17.13 is generated from catalyst containing certain chiral ligands. For example, cationic palladium catalysts containing bisoxazoline ligands generate highly isotactic polymers,[339] and catalysts containing azabisoxazoline ligands have led to further improvements in activity.[340] This observation of isotactic polymers with these catalysts implies that the geometry of the stereoselectivity of the catalyst overrides the inherent stereoselectivity of the polymerization with achiral catalysts that resulted from chain-end control.

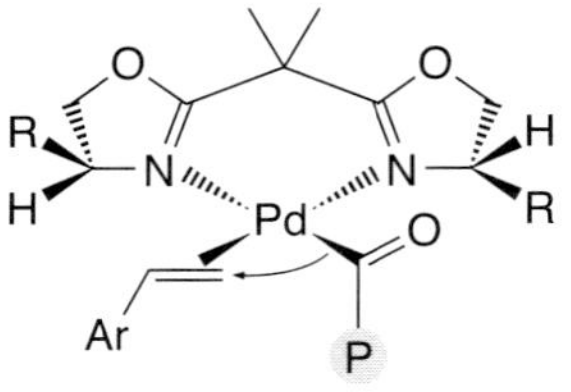

A stereochemical model for the high stereoregularity of the polymer formed from catalysts containing bisoxazoline ligands is shown in Figure 17.15.[339] In this model, the stereochemistry of the catalyst controls the stereochemistry of the insertion process by dictating that the olefin binds to the metal in a orientation that places the substituent away from the protruding substituent on the chiral ligand. Insertion of the olefin into the acyl group then leads to the syndiotactic stereochemistry in the main chain. Thus, each insertion occurs with the same stereochemistry, and syndiotactic polymer is formed.

The stereochemistry from chain-end control and catalyst-control are not necessarily different. More details on the stereochemistry of alkene polymerization are provided in Chapter 22. In brief, however, a C_2 symmetric catalyst like the ones used for the copolymerization of carbon monoxide with styrene will generate isotactic polymer, but polymerizations with stereochemistry dictated by chain-end control can form either syndiotactic or isotactic material. Catalysts displaying other symmetries have been designed for other types of polymerizations to generate polymer architectures beyond isotactic.

Figure 17.15.
Stereochemical model for catalyst-control of the stereochemistry of the copolymerization of carbon monoxide and styrene.

17.8.4.3. Copolymerization of Carbon Monoxide and Propene

The copolymerization of carbon monoxide and propene is related to both the copolymerization of carbon monoxide and ethylene and the copolymerization of carbon monoxide and styrene. The basic steps of chain growth of the two processes are similar and are shown in Scheme 17.34. However, there are significant differences in the regioselectivity of the insertion step. Moreover, the catalysts that give rise to regio- and stereo-regular polymers of CO and propene are distinct from those that give rise to regio- and stereo-regular copolymers of CO and styrene.

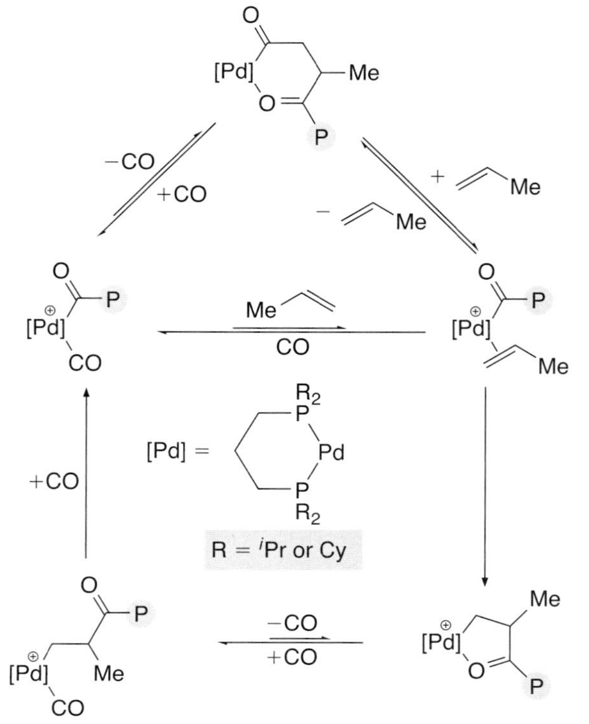

Scheme 17.34

17.8.4.3.1. Regiochemistry of Insertion

Like the copolymers generated from carbon monoxide and either ethylene or styrene, the coplymers generated from carbon monoxide and propene are perfectly alternating. Thus, the barriers for double insertions of propene and double insertions of CO are again higher than the barriers for insertion of propene into the acyl intermediate and CO into the alkyl intermediate. However, the polymers produced by these polymerizations with the ligands used in CO/ethylene copolymerization are not regio-regular.

The copolymer from carbon monoxide and propylene tends to result from a 1,2-insertion of the α-olefin into the metal–acyl linkage, not the 2,1 insertion into the metal–acyl linkage that occurs during copolymerizations of CO and styrene. Catalysts containing the 1,3-bis(diphenylphosphino)propane catalyst used for CO/ethylene copolymerization generate copolymers of carbon monoxide and propene possessing modest regio-regularity.[341] However, catalysts containing chelating alkylphosphine ligands give rise to more regio-regular polymers resulting from the 1,2-insertion of propene (vide infra). For example, Consiglio showed that the catalyst containing 1,3-bis(diisopropylphosphino)propane generates copolymers of carbon monoxode and propene that possess nearly perfect regio-regularity.

This regiochemistry is consistent with the regiochemistry of the hydroesterification and hydrocarboxylation of 1-alkenes that results from 1,2 insertion into a metal–hydride. 1,2-Insertion of propene has also been observed directly during studies on the individual steps of the copolymerization process. One example of the direct observation of this step is shown in Equation 17.69 for a catalyst containing BINAPHOS.[342,343]

(17.69)

(R,S)-BINAPHOS

17.8.4.3.2. Stereochemistry of Insertion

Like the stereochemistry of the copolymerization of carbon monoxide with styrene, the stereochemistry of the copolymerization of carbon monoxide and propene has been studied with catalysts containing both achiral and chiral ligands. Syndiotactic polymer is generated from catalysts containing the achiral 1,3-bis(diisopropylphosphino)propane and 1,3-bis(dicyclohexylphosphino)propane as ancillary ligand. Again, the absence of chirality in the catalyst makes it necessary that the growing polymer chain controls the relative stereochemistry of the olefin insertion to generate the syndiotactic structure.

The synthesis of isotactic copolymers from carbon monoxide and propene has been more difficult to achieve. The stereoselective copolymerization of carbon monoxide and propene has been most selective with catalysts containing alkyl-substituted bisphosphines. The catalyst containing the dialkylphosphino biaryl ligand in Figure 17.16 has been shown by Consiglio to generate highly isotactic CO/propylene copolymer.[344] Likewise, the palladium catalyst containing DuPhos (Figure 17.16) as ligand has been shown by Sen to generate highly isotactic CO/propylene copolymer.[345]

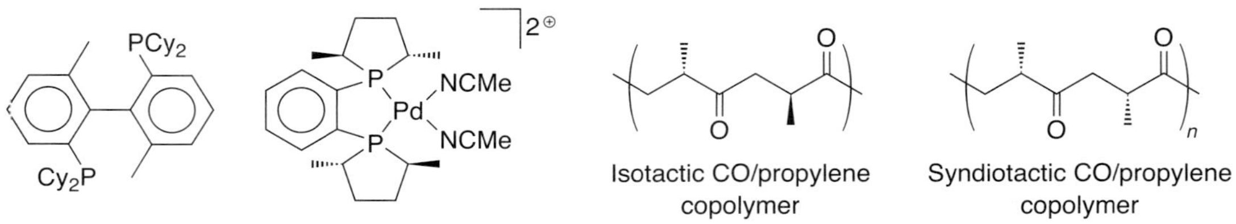

Figure 17.16.
Biaryl bisphosphine reported by Consiglio and catalyst reported by Sen for the copolymerization of carbon monoxide and propene to form syndiotactic CO/propylene copolymer.

Nozaki has shown that the less symmetric BINAPHOS ligand (Equation 17.70) containing one phosphine and one phosphite donor also gives rise to highly isotactic copolymers from carbon monoxide and propylene.[342, 343] NMR spectroscopic studies on this polymerization process have shown that the acyl group in one intermediate prefers to be located trans to the phosphine donor and the alkyl group of the other intermediate prefers to be located trans to the phosphite donor. Thus, olefin coordinates to the site trans to the phosphite prior to insertion, and CO coordinates trans to the phosphine donor prior to insertion.

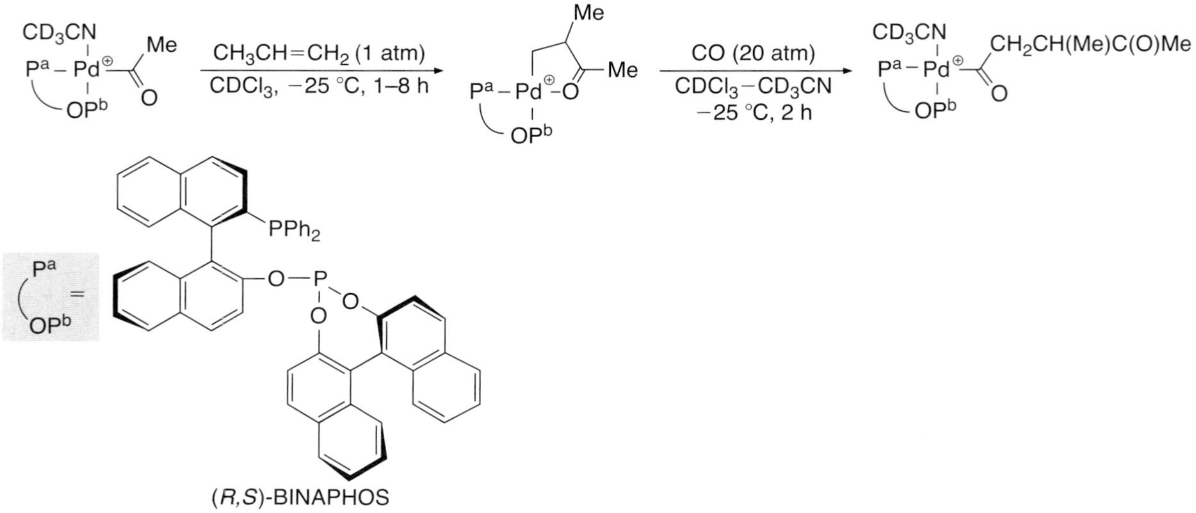

$$(17.70)$$

17.8.4.3.3. Polymer Structure from the Copolymerization of CO and Propene

The final difference in the copolymerization of carbon monoxide with propene or styrene is the overall connectivity of the initial polymer generated under some conditions. The polymer generated from the copolymerization of carbon monoxide and propene in protic solvents consists of the fused tetrahydrofuran ketal structure shown in Figure 17.17. This polymer reopens to the polymer shown in Figure 17.13 upon addition of acid in alcohol. Several mechanisms for formation of this product have been proposed, and the origin of the ketal structure remains unresolved. Polymers formed in aprotic solvents form the acylic polymer.

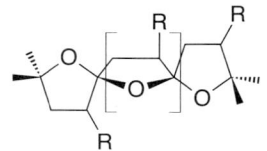

Figure 17.17.
Ketal structure of the copolymer of CO and propene formed in the absence of protic component.

17.9. Pauson–Khand Reactions (Written with Dr. Qilong Shen)

17.9.1. Overview

Chapter 17 closes with a brief presentation of the Pauson–Khand reaction.[346–351] The Pauson–Khand reaction (PKR) is a formal [2+2+1] cycloaddition reaction involving an alkyne, an alkene, and carbon monoxide to form a cyclopentanone shown generically in Equation 17.71. The Pauson–Khand reaction was initially reported as a stoichiometric reaction mediated by cobalt carbonyl, but it has been translated into a catalytic process in recent years. Most recently, it has developed into an enantioselective catalytic process. Complexes of Ti, Mo, W, Fe, Co, Ni, Ru, Rh, Ir, and Pd have all been shown to catalyze this reaction.

(17.71)

The Pauson–Khand reaction (PKR) was discovered by Pauson and Khand in the early 1970s.[352,353] Both dinuclear and mononuclear catalysts have now been developed for this process, and the mechanistic details are not established in most cases. However, the broad features of the mechanism proposed by Magnus in the 1980s is generally accepted.[354] The reaction is initiated by coupling of an alkyne and an alkene to form a metallacyclopentene complex. Insertion of CO into one of the M–C bonds and reductive elimination generates the cyclopentenone. The original PKR occurred with low conversion and with relatively narrow scope, but the modifications of this reaction to include several additives has dramatically increased the rate and broadened the scope of this process over the past two decades.

The Pauson–Khand reaction has been studied extensively in academic laboratories and has been used to prepare a number of natural products. The PKR is included in this text because of its extensive use in synthesis, but the presentation is brief because it has not experienced the same development into industrial scale synthesis as the other carbonylation processes described in this chapter. This section presents the original reaction, the modifications that increase the rate, allenic PKRs, asymmetric variants, and the application of the PKR in modern organic synthesis.

17.9.2. Origin of the Pauson–Khand Reaction

The PKR was first reported as a stoichiometric reaction between norbornadiene and a complex of acetylene bound to hexacarbonyl dicobalt (Equation 17.72).[352] Pauson and co-workers also reported the first catalytic intermolecular formation of a cyclopentene from an alkyne, an olefin, and CO (Equation 17.73).[353] The first intramolecular version of the PKR was reported almost 10 years later by Shore (Equation 17.74).[355] At this point, the intramolecular PKR has been studied in more detail than the intermolecular PKR.

(17.72)

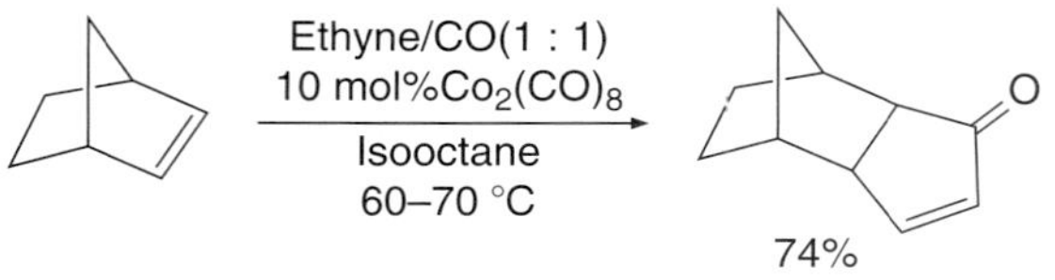

$$(17.73)$$

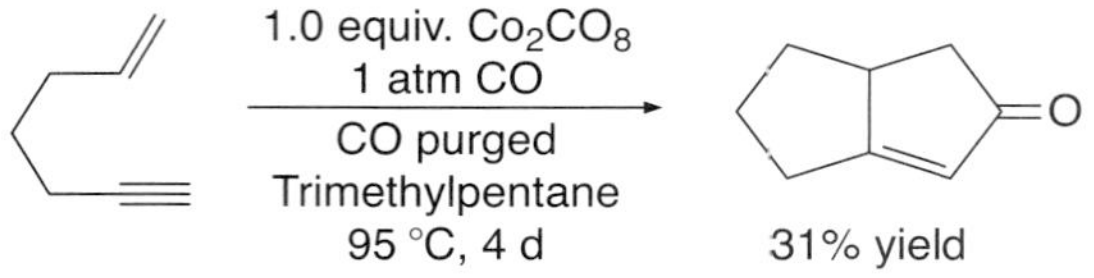

$$(17.74)$$

17.9.3. Effects of Additives

Many additives have been shown to accelerate the PKR. Smit reported that binding of the cobalt complex onto silica gel increased the rate and yield.[356] However, the addition of amine N-oxide, as reported by Schreiber[357] and Jeong,[358] is the most common method to promote the PKR (Equation 17.75). Amine N-oxide accelerates the rate because it leads to dissociation of CO, as presented in Chapter 5, and this dissociation allows coordination of the olefin. However, this approach had a smaller affect on intermolecular PKRs. Rapid decomposition of the alkyne–$Co_2(CO)_6$ complexes in the absence of alkene occurs when amine N-oxide is added.

$$(17.75)$$

Other additives, such as dialkyl sulfides, also improve the yields of intermolecular Pauson–Khand reactions.[359] One reaction that formed the Pauson–Khand product in 85% yield in the presence of methylbutyl sulfide, but which did not form the Pauson–Khand product in the presence of added N-methyl morpholine oxide, is shown in Scheme 17.35.

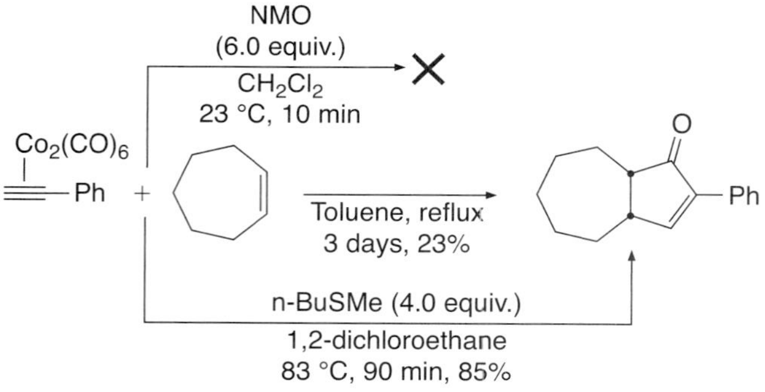

Scheme 17.35

17.9.4. Catalysts Other Than $Co_2(CO)_8$

Because of the instability of hexacarbonyl dicobalt complexes at elevated temperatures, other cobalt complexes have been used for the PKR. For example, Chung and Jeong reported the reaction of norbornadiene with phenylacetylene at 100 °C under 15 atm of CO in 93% yield in the presence of (indenyl)Co(COD)[360] (Equation 17.76). Chung also reported the combination of $Co(acac)_2$ and $NaBH_4$ as catalyst for the same transformation.[361]

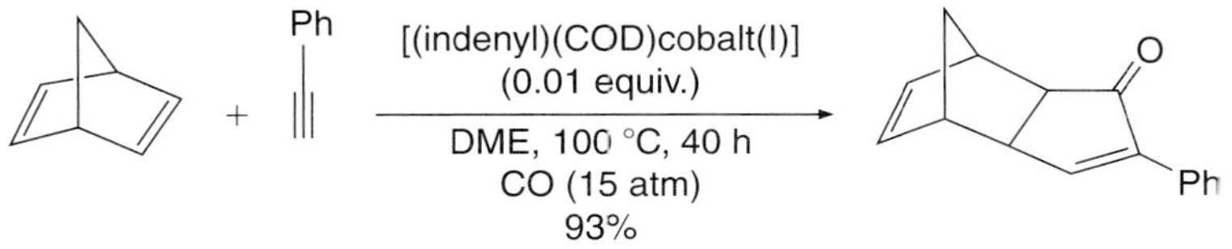

$$(17.76)$$

Complexes of other transition metals have been reported to catalyze Pauson–Khand reactions. Buchwald reported intramolecular PKRs with 1.2 atm of CO at 90 °C in the presence of $Cp_2Ti(CO)_2$.[362–367] However, most other catalytic Pauson–Khand reactions have been conducted with late transition metal catalysts. Murai[368] and Mitsudo[369] simultaneously reported intramolecular PKRs catalyzed by ruthenium carbonyl clusters in dioxane or DMAc at 140–160 °C under 10–15 atm of CO. The first Rh-catalyzed PKR was reported by Narasaka.[370] In this case, the reaction occurred with acceptable rates, even with CO pressures less than 1 atm. Shibata reported PKRs in refluxing xylenes under 1 atm of CO in the presence of catalytic amounts of PPh_3 and $[Ir(COD)Cl]_2$.[371] Adrio and Carretero showed that the solvated molybdenum carbonyl complex $Mo(DMF)_3(CO)_3$ catalyzed intramolecular PKRs with monosubstituted olefins, as well as with disubstituted electron-poor olefins,[372] and Hoye showed that $W(CO)_5(THF)$ catalyzes intramolecular PKRs.[373] Iron[374] and palladium[375] complexes have also been reported to catalyze the PKR.

17.9.5. Pauson–Khand Reactions with Allenes

Most Pauson–Khand reactions have been conducted with an alkene, an alkyne, and CO. However, PKRs have been developed with allenynes (Scheme 17.36).[347] Narasaka reported the first example of an intramolecular allenyne PKR catalyzed by an iron complex,[376] and Brummond has extensively studied the allenic Pauson–Khand reaction. These reactions have been catalyzed by the combination of $Mo(CO)_6$ and DMSO or by $[Rh(CO)_2Cl]_2$, and several examples are shown in Equations 17.77 and 17.78.[377–382] The reactions in Equations 17.77 and 17.78 illustrate the different regiochemistry of the products from reactions catalyzed by molybdenum and rhodium. Computational studies indicate that different geometries of octahedral Mo(0), and square-planar Rh(I) species account for the different regioselectivities.[382]

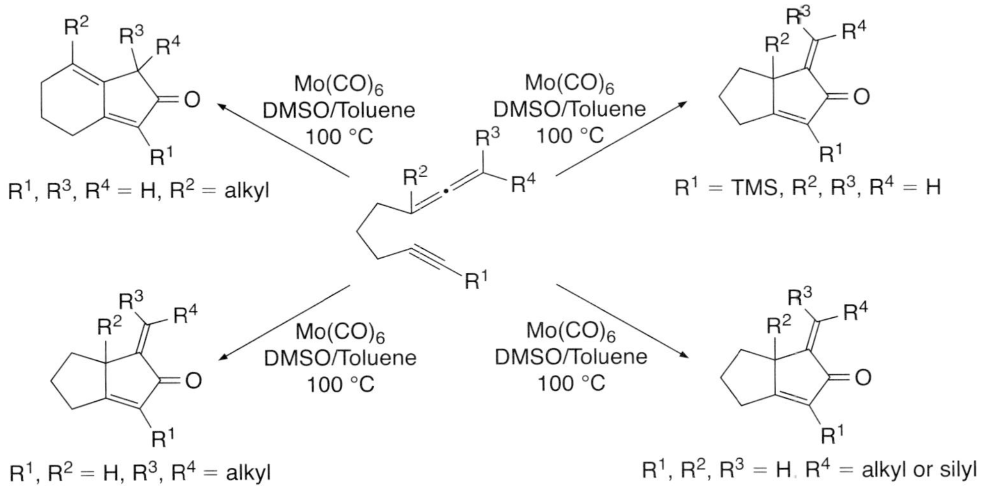

Scheme 17.36

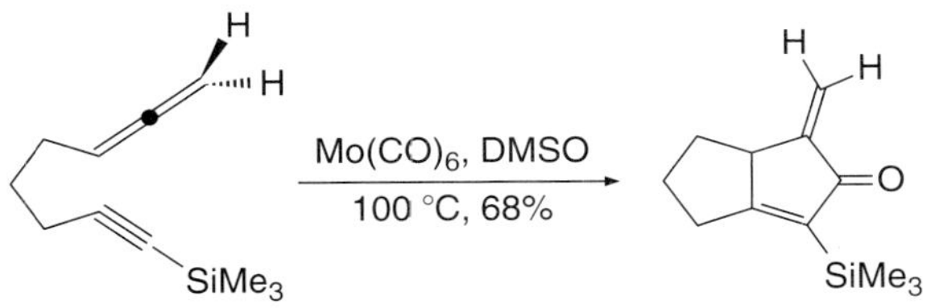

$$(17.77)$$

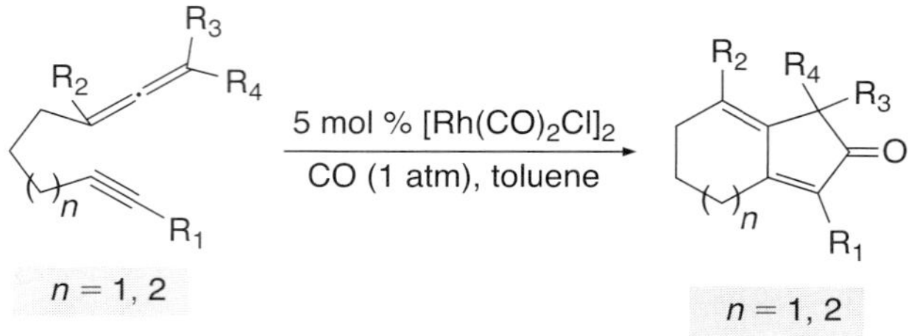

$$(17.78)$$

17.9.6. Catalytic Asymmetric Pauson–Khand Reactions

The first catalytic asymmetric PKR was reported by Buchwald and Hicks in 1996.[364] A chiral titanocene–carbonyl complex (Figure 17.18) containing an ansa-indenyl ligand described in Chapter 3 was used for the cyclization of an enyne to give a bicyclic cyclopentenone product in high yield with good enantioselectivity. Hiroi published the first example of an asymmetric cobalt-catalyzed PKR. These reactions occurred with BINAP as ligand and $Co_2(CO)_8$ as the cobalt component in moderate yield and with good enantioselectivity.[383]

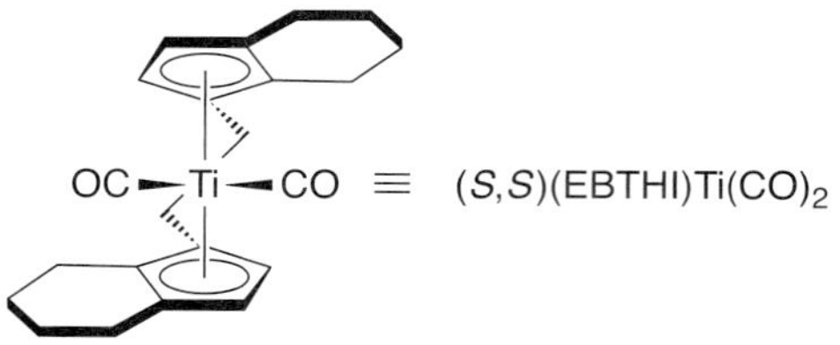

Figure 17.18.
Catalyst for the first catalytic asymmetric Pauson–Khand reaction.

More recent work on the asymmetric PKR has focused on reactions catalyzed by rhodium and iridium complexes. Jeong and co-workers reported reactions catalyzed by a combination of $[Rh(CO)_2Cl]_2$, (S)-BINAP, and AgOTf. Good to excellent ee's were obtained for a small range of substrates (Equation 17.79).[384] After the observation that phosphine ligands improve the yield of the Ir-catalyzed PKR, Shibata reported intramolecular PKRs catalyzed by the combination of Tol–BINAP and $[Ir(COD)Cl]_2$ in excellent yields and enantioselectivities (Equation 17.80).[371]

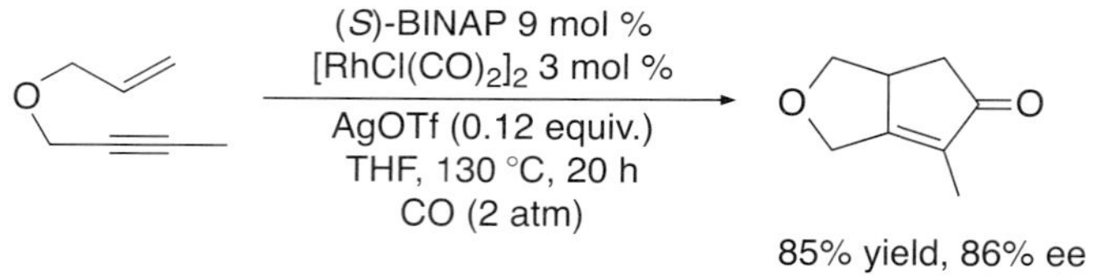

$$(17.79)$$

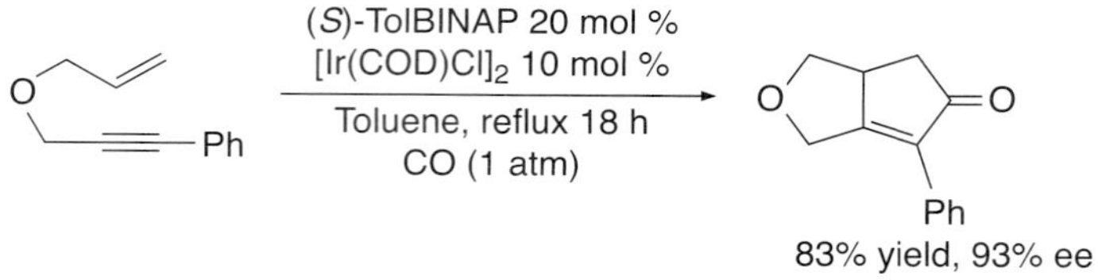

85% yield, 86% ee

$$(17.80)$$

83% yield, 93% ee

17.9.7. Intermolecular Pauson–Khand Reaction

Intermolecular PKRs have been less well studied, in part because of the difficulty in controlling regioselectivity.[350] Several strategies have been followed to develop intermolecular PKRs, including the incorporation of ligating groups on the alkene, and the use of strained alkenes.

Pre-association of the alkene and alkyne partners to the cobalt metal center usually leads to a highly ordered transition state. Following this strategy, the rate and selectivity of the process has been improved significantly. Examples of this strategy are shown in Equations 17.81 and 17.82. Krafft reported reactions of the alkene containing amino and thioether groups in the alkyl chain, and these reactions occurred with high regioselectivity.[385] More recently Yoshida reported the reactions with the alkenyl dimethyl-2-pyridylsilane as the alkene component (Equation 17.82).[386] These reactions occurred with symmetrical alkynes with high selectivity for formation of the product with the R^2 group of the alkene α to the carbonyl group. This substituent is directed to this position of the product because of the directing ability of the pyridyl group on silicon and the inherent silicon α-effect.

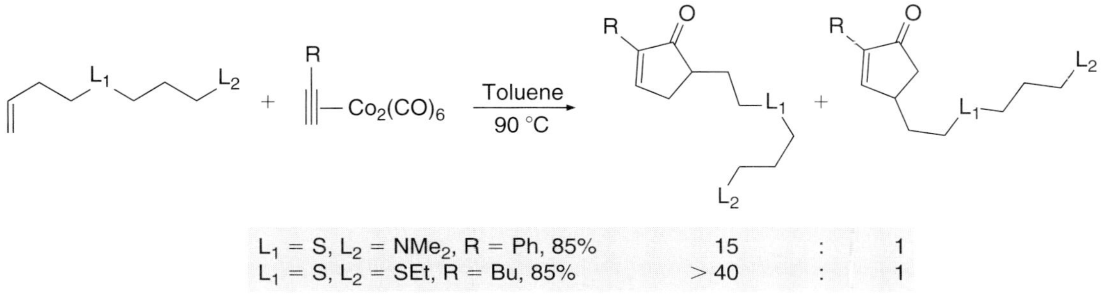

$$L_1 = S, L_2 = NMe_2, R = Ph, 85\% \qquad 15 \quad : \quad 1$$
$$L_1 = S, L_2 = SEt, R = Bu, 85\% \qquad > 40 \quad : \quad 1$$

(17.81)

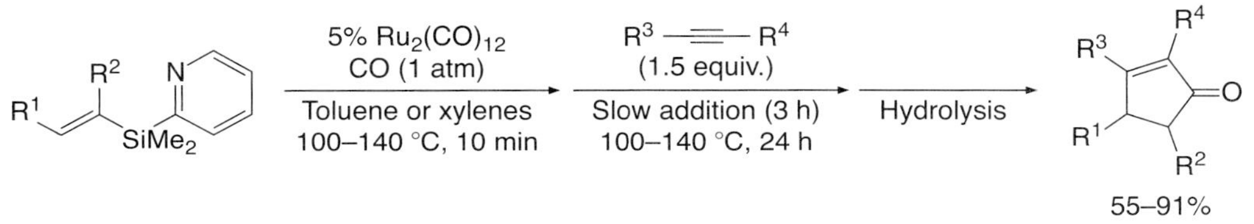

(17.82)

55–91%

The second strategy is illustrated in Equation 17.83. In this case, the LUMO of the olefin is lowered by using strained alkenes, such as cyclopropene and cyclobutene, and the binding constant of the olefin to the metal is higher because of relief of strain during coordination (see Chapter 3). These factors have been proposed to lead to faster reactions with these strained alkenes.[387]

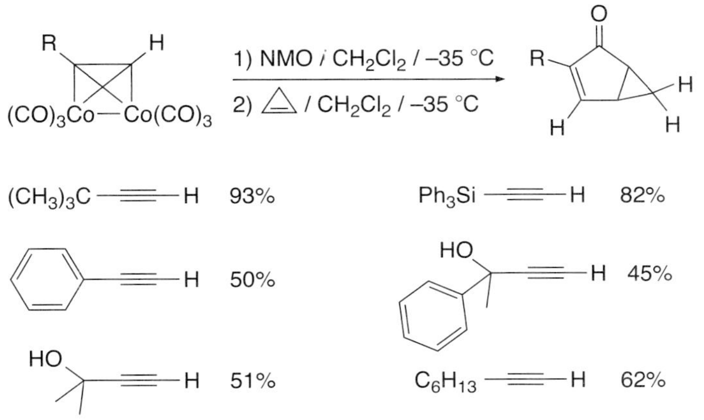

$(CH_3)_3C$ ═══ H   93% Ph_3Si ═══ H   82%

Ph ═══ H   50%

HO—C(Ph) ═══ H   45%

$(HO)C(CH_3)_2$ ═══ H   51% C_6H_{13} ═══ H   62%

(17.83)

17.9.8. Applications of the PKR

One of the most cited examples of the use of the PKR in a natural products synthesis is the synthesis of (+)-epoxydictymene reported by Schreiber and shown in Scheme 17.37. To do so, they first synthesized the eneyne in the center of the scheme.[388,389] This compound was then allowed to react with $Co_2(CO)_8$ to form the dicobalt–alkyne complex. The Nicolas reaction was then used to form a fused 5,8-ring system. This reaction was followed by an intramolecular PKR to give the polycyclic product in high yield.

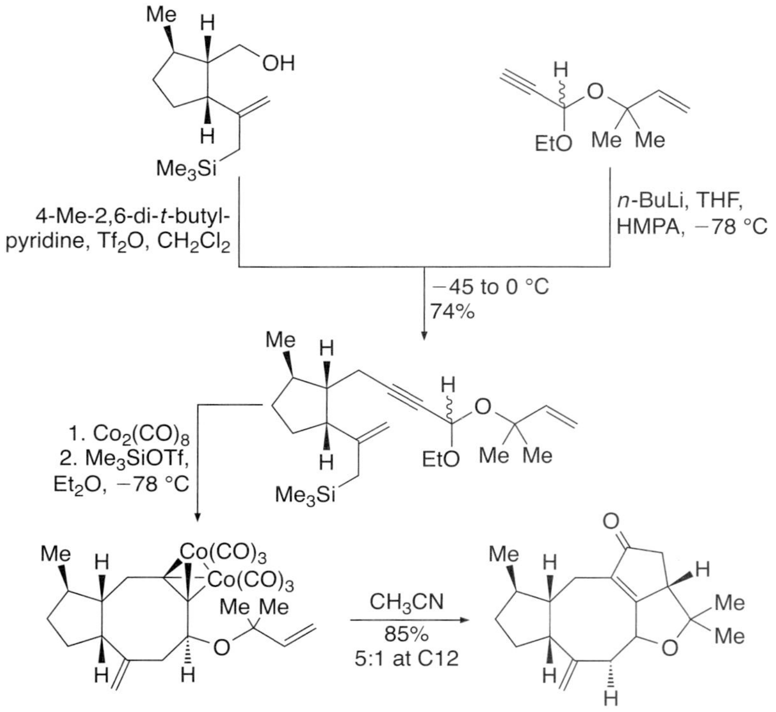

Scheme 17.37

An example of the allenic PKR in complex-molecule synthesis reported by Brummond is shown in Scheme 17.38. The alleneyne precursor to the PKR in this scheme was synthesized in five steps starting from 1,1-diacetylcyclopropane. Treatment of this alleneyne with 1.2 equivalents of $Mo(CO)_6$ and 10 equivalents of DMSO in toluene at 100 °C gave the desired product containing a 6,5-fused ring system. This material was then carried forward to prepare the hydroxymethylacylfulvene natural product.

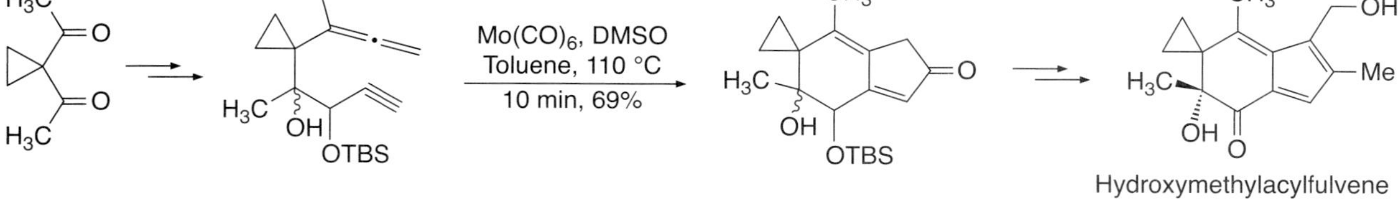

Scheme 17.38

17.9.9. Mechanism of the Pauson–Khand Reaction

The currently accepted mechanism of the Pauson–Khand reaction mediated by cobalt carbonyl systems was proposed by Magnus.[354] This mechanism is summarized in Scheme 17.39. The reaction is initiated by irreversible replacement of CO from the starting hexacarbonyl complex by an olefin. Following coordination of the olefin, reductive coupling of the alkyne and the olefin occurs. The resulting metallacycle undergoes insertion of CO, and reductive elimination to form the ketone product regenerates the starting dicobalt complex.

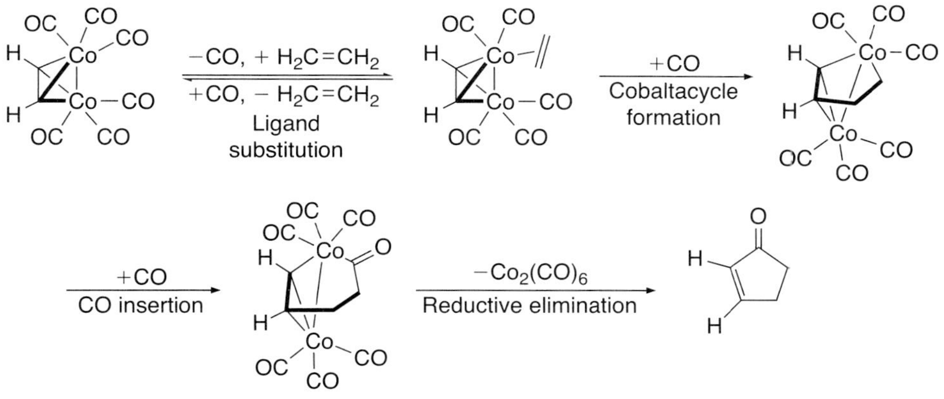

Scheme 17.39

Pericas has studied the mechanism using DFT calculations, and the results of these calculations are summarized in Figure 17.19.[387] These results imply that the dissociation of CO is endothermic by 33.5 kcal/mol, and the coordination of olefin is exothermic by 21.5 kcal/mol. Thus, the ligand substitution step is endothermic by 12 kcal/mol. The reductive coupling step is approximately thermo-neutral, but the binding of CO to the unsaturated product is strongly exothermic. Therefore, the overall formation of the cobaltacycle is strongly exothermic and irreversible when the reaction is performed in the presence of CO.

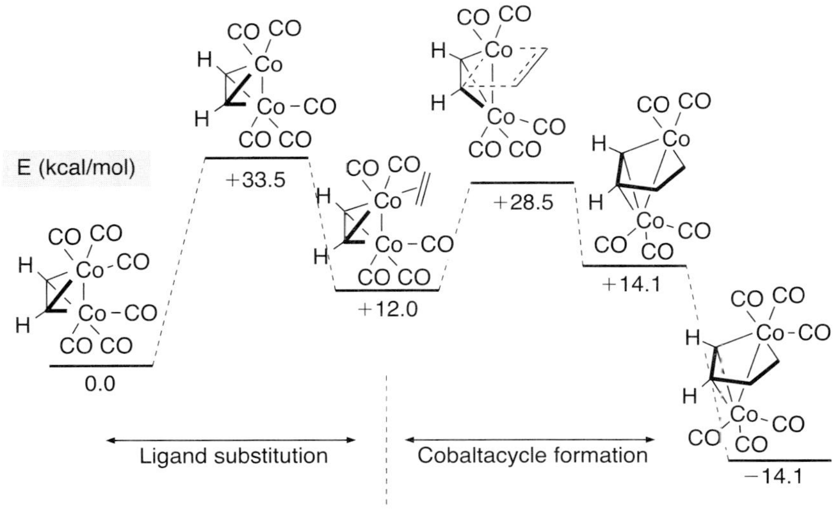

Figure 17.19.
Energies for the intermediates in the Pauson–Khand reaction mediated by cobalt carbonyl.

PKRs catalyzed by complexes other than $Co_2(CO)_8$ have not been the subject of detailed mechanistic studies. However, the catalytic cycle is assumed to follow the path in Scheme 17.40. This mechanism begins with the coordination of the olefin and the alkyne to a single metal center. Reductive coupling of the two groups generates a metal-lacyclopentene. Insertion of CO, followed by reductive elimination then forms the ketone product and the starting catalyst.

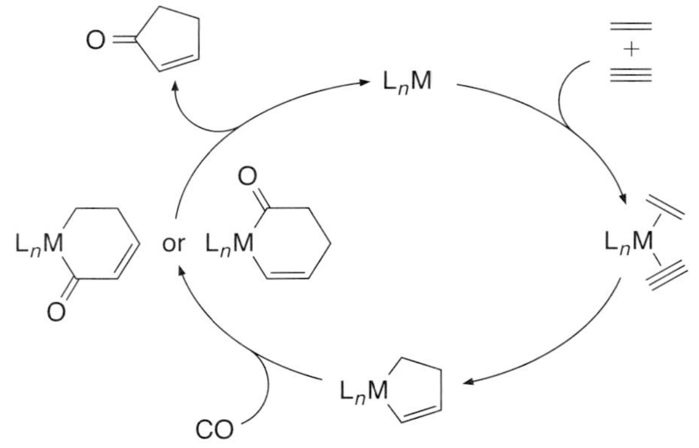

Scheme 17.40

References and Notes

1. For recent compendia and compilations of reviews, see (a) *Applied Homogeneous Catalysis with Organometallic Compounds*, 2nd ed.; Cornils, B., Herrmann, W. A., Eds.; Wiley-VCH: Weinheim, 2002; Vol. 1, Chapter 2.1; (b) *Catalytic Carbonylation Reactions (Topics in Organometallic Chemistry)*; Beller, M., Ed.; Springer: Berlin, 2006; (c) *Modern Carbonylation Methods*; Kollár, L., Ed.; Wiley-VCH: Weinheim, 2008.
2. (a) Roth, J. F.; Craddock, J. H.; Hershman, A.; Paulik, F. E. *Chemical Technology* **1971**, 600; (b) Grove, H. D. *Hydrocarbon Process.* **1972**, *51*, 76.
3. Forster, D. *Adv. Organomet. Chem.* **1979**, *17*, 255.
4. Ellwood, P. *Chem. Eng. News* **1969**, 148.
5. Falbe, J. *Carbon Monoxide in Organic Synthesis*; Springer-Verlag: New York, 1970.
6. *New Syntheses with Carbon Monoxide*; Falbe, J., Ed.; Springer-Verlag: New York, 1980.
7. Jones, J. H. *Platinum Metals Rev.* **2000**, *44*, 94.
8. Adamson, G. W.; Daly, J. J.; Forster, D. *J. Organomet. Chem.* **1974**, *71*, C17.
9. Adams, H.; Bailey, N. A.; Mann, B. E.; Manuel, C. P.; Spencer, C. M.; Kent, A. G. *J. Chem. Soc., Dalton Trans.* **1988**, 489.
10. Haynes, A.; Mann, B. E.; Morris, G. E.; Maitlis, P. M. *J. Am. Chem. Soc.* **1993**, *115*, 4093.
11. Forster, D. *J. Am. Chem. Soc.* **1975**, *97*, 951.
12. Parshall, G. *Homogeneous Catalysis*; Wiley-Interscience: New York, 1980; p 81.
13. Howe, L. A.; Bunel, E. E. *Polyhedron* **1995**, *14*, 167.
14. Forster, D.; Dekleva, T. W. *J. Chem. Educ.* **1986**, *63*, 204.
15. Baker, E. C.; Hendriksen, D. E.; Eisenberg, R. *J. Am. Chem. Soc.* **1980**, *102*, 1020.
16. Murphy, M. A.; Smith, B. L.; Torrence, G. P.; Aguilo, A. *J. Organomet. Chem.* **1986**, *303*, 257.
17. Hickey, C. E.; Maitlis, P. M. *J. Chem. Soc., Chem. Commun.* **1984**, 1609.
18. Polichnowski, S. W. *J. Chem. Ed.* **1986**, *63*, 206.
19. Larkins, T. H.; Polichnowski, S. W.; Tustin, G. C.; Young, D. A. U.S. Patent 4,374,070, 1983.
20. Forster, D. *J. Chem. Soc., Dalton Trans.* **1979**, 1639.
21. Haynes, A. *Top. Organomet. Chem.* **2006**, *18*, 179.
22. Haynes, A.; Maitlis, P. M.; Morris, G. E.; Sunley, G. J.; Adams, H.; Badger, P. W.; Bowers, C. M.; Cook, D. B.; Elliott, P. I. P.; Ghaffar, T.; Green, H.; Griffin, T. R.; Payne, M.; Pearson, J. M.; Taylor, M. J.; Vickers, P. W.; Watt, R. J. *J. Am. Chem. Soc.* **2004**, *126*, 2847.
23. Ellis, P. R.; Pearson, J. M.; Haynes, A.; Adams, H.; Bailey, N. A.; Maitlis, P. M. *Organometallics* **1994**, *13*, 3215.
24. Bassetti, M.; Monti, D.; Haynes, A.; Pearson, J. M.; Stanbridge, I. A.; Maitlis, P. M. *Gazz. Chim. Ital.* **1992**, *122*, 391.
25. Frohning, C. D.; Kohlpaintner, C. W.; Bohnen, H.-W. In *Applied Homogeneous Catalysis with Organometallic Compounds*, 2nd ed.; Cornils, B., Herrmann, W. A., Eds.; Wiley-VCH: Weinheim, 2002; Vol. 1, p 31.

26. For annual reviews of hydroformylation, see Ungvary, F. in *Coord. Chem. Rev.*

27. Roelen, O. DRP 849548, 1938.

28. Adkins, H.; Krsek, G. *J. Am. Chem. Soc.* **1948**, *70*, 383.

29. Wender, I.; Orchin, M.; Storch, H. H. *J. Am. Chem. Soc.* **1950**, *72*, 4842.

30. Cornils, B. In *New Syntheses with Carbon Monoxide*; Falbe, J., Ed.; Springer-Verlag: New York, **1980**; p 177.

31. Frohning, C. D.; Kohlpaintner, C. W.; Bohnen, H. W. In *Applied Homogeneous Catalysis with Organometallic Compounds*; 2nd ed.; Cornils, B., Hermann W. A., Eds.; Wiley-VCH: Weinheim, 2002; p 33.

32. Bálint Heil, L. M. *Chem. Ber.* **1968**, *101*, 2209.

33. Natta, G.; Ercoli, R.; Castellano, S.; Barbieri, F. H. *J. Am. Chem. Soc.* **1954**, *76*, 4049.

34. Heck, R. F.; Breslow, D. S. *J. Am. Chem. Soc.* **1961**, *83*, 4023.

35. Whyman, R. *J. Organomet. Chem.* **1974**, *81*, 97.

36. Ungvary, F.; Marko, L. *Organometallics* **1983**, *2*, 1608.

37. Azran, J.; Orchin, M. *Organometallics* **1984**, *3*, 197.

38. Kovacs, I.; Ungvary, F.; Marko, L. *Organometallics* **1986**, *5*, 209.

39. Mirbach, M. F. *Inorg. Chim. Acta* **1984**, *88*, 209.

40. Mirbach, M. F. *J. Organomet. Chem.* **1984**, *265*, 205.

41. Bianchi, M.; Piacenti, F.; Frediani, P.; Matteoli, U. *J. Organomet. Chem.* **1977**, *137*, 361.

42. The aldehyde product was converted into a carboxylic ester for analysis.

43. Piacenti, F.; Pucci, S.; Bianchi, M.; Lazzaroni, R.; Pino, P. *J. Am. Chem. Soc.* **1968**, *90*, 6847.

44. Casey, C. P.; Cyr, C. R. *J. Am. Chem. Soc.* **1973**, *95*, 2240.

45. Slaugh, L. H.; Mullineaux, R. D. *J. Organomet. Chem.* **1968**, *13*, 469.

46. Masters, C. In *Homogeneous Transition-Metal Catalysis, A Gentle Art*; Chapman and Hall: London, 1981.

47. Weise, K.-D.; Obst, D. *Top. Organomet. Chem.* **2006**, *18*, 1.

48. Crause, C.; Bennie, L.; Damoense, L.; Dwyer, C. L.; Grove, C.; Grimmer, N.; Rensburg, W. J. v.; Kirk, M. M.; Mokheseng, K. M.; Otto, S.; Steynberg, P. J. *J. Chem. Soc., Dalton Trans.* **2003**, 2036.

49. Winkle, J. L. v.; Lorenzo, S.; Morris, R. C.; Mason, R. F. U.S. Patent 3420898, 1969.

50. Piacenti, F.; Bianchi, M.; Bendetti, M. *Chim. Ind. Milan* **1969**, *49*, 245.

51. Hershman, A.; Craddock, J. H. *Ind. Eng. Chem. Prod. Res. Develop.* **1968**, *7*, 226.

52. Evans, D.; Yagupsky, G.; Wilkinson, G. J. *J. Chem. Soc. A* **1968**, 3133.

53. Pruett, R. L.; Smith, J. A. *J. Org. Chem.* **1969**, *34*, 327.

54. van Leeuwen, P. W. M. N.; Casey, C. P.; Whiteker, G. T. In *Rhodium Catalyzed Hydroformylation*; van Leeuwen, P. W. N. M., Claver, C., Eds.; Kluwer: Dordrecht, The Netherlands, 2000; Vol. 22, Chapter 4.

55. Evans, D.; Osborn, J. A.; Wilkinson, G. *J. Chem. Soc. A* **1968**, 3133.

56. Evans, D.; Yagupsky, G.; Wilkinson, G. J. *J. Chem. Soc. A* **1968**, 2660.

57. Brown, C. K.; Wilkinson, G. *J. Chem. Soc. A* **1970**, 2753.

58. VanRooy, A.; deBruijn, J. N. H.; Roobeek, K. F.; Kamer, P. C. J.; van Leeuwen, P. *J. Organomet. Chem.* **1996**, *507*, 69.

59. Brown, J. M.; Kent, A. G. *J. Chem. Soc., Perkin Trans. 2* **1987**, 1597.

60. Dieguez, M.; Claver, C.; Masdeu-Bulto, A. M.; Ruiz, A.; van Leeuwen, P. W. N. M.; Schoemaker, G. C. *Organometallics* **1999**, *18*, 2107.

61. Lenges, C. P.; Brookhart, M. *Angew. Chem. Int. Ed.* **1999**, *38*, 3533.

62. van Leeuwen, P. W. M. N.; Casey, C. P.; Whiteker, G. T. In *Rhodium Catalyzed Hydroformylation*; van Leeuwen, P. W. N. M., Claver, C., Eds.; Kluwer: Dordrecht, The Netherlands, 2000; Vol. 22, p 69.

63. Kuntz, E. French Patent 2,314,910, 1975.

64. In *Aqueous Phase Organometallic Catalysis*, 2nd ed.; Cornils, B., Herrmann, W. A., Eds.; Wiley: Wienheim, 2004; p 351.

65. Sanger, A. R.; Schallig, L. R. *J. Mol. Catal.* **1977**, *3*, 101.

66. Sanger, A. R. *J. Mol. Catal.* **1978**, *3*, 221.

67. Pittman, C. U.; Hirao, A. *J. Org. Chem.* **1978**, *43*, 640.

68. Hughes, O. R.; Unruh, J. D. *J. Mol. Catal.* **1981**, *12*, 71.

69. Unruh, J. D.; Christenson, J. R. *J. Mol. Catal.* **1982**, *14*, 19.

70. Devon, T. J.; Phillips, G. W.; Puckette, T. A.; Stavinoha, J. L.; Vanderbilt, J. J. U.S. Patent 4,694,109, 1987.

71. van Leeuwen, P. W. N. M.; Grotenhuis, P. A. M.; Goodall, B. L. Eur. Pat. Appl. 309056, 1989.

72. Thewissen, D.; Timmer, K.; Noltes, J. G.; Marsman, J. W.; Laine, R. M. *Inorg. Chim. Acta* **1985**, *97*, 143.

73. Kranenburg, M.; van der Burgt, Y. E. M.; Kamer, P. C. J.; van Leeuwen, P. W. N. M.; Goubitz, K.; Fraanje, J. *Organometallics* **1995**, *14*, 3081.

74. Bronger, R. P. J.; Kamer, P. C. J.; van Leeuwen, P. W. N. M. *Organometallics* **2003**, *22*, 5358.

75. van der Veen, L.; Keeven, P.; Schoemaker, G.; Reek, J.; Kamer, P.; van Leeuwen, P.; Lutz, M.; Spek, A. *Organometallics* **2000**, *19*, 872.

76. Casey, C. P.; Whiteker, G. T.; Melville, M. G.; Petrovich, L. M.; Gavney, J. A.; Powell, D. R. *J. Am. Chem. Soc.* **1992**, *114*, 5535.

77. Casey, C. P.; Whiteker, G. T. *Isr. J. Chem.* **1990**, *30*, 299.

78. Casey, C. P.; Petrovich, L. M. *J. Am. Chem. Soc.* **1995**, *117*, 6007.

79. Casey, C. P.; Paulsen, E. L.; Beuttenmueller, E. W.; Proft, B. R.; Petrovich, L. M.; Matter, B. A.; Powell, D. R. *J. Am. Chem. Soc.* **1997**, *119*, 11817.

80. Casey, C. P.; Paulsen, E. L.; Beuttenmueller, E. W.; Proft, B. R.; Matter, B. A.; Powell, D. R. *J. Am. Chem. Soc.* **1999**, *121*, 63.

81. van der Veen, L. A.; Boele, M. D. K.; Bregman, F. R.; Kamer, P. C. J.; van Leeuwen, P. W. N. M.; Goubitz, K.; Fraanje, J.; Schenk, H.; Bo, C. *J. Am. Chem. Soc.* **1998**, *120*, 11616.

82. Kranenburg, M.; van der Burgt, Y. E. M.; Kamer, P. C. J.; van Leeuwen, P. W. N. M.; Goubitz, K.; Fraanje, J. *Organometallics* **1995**, *14*, 3081.

83. van der Veen, L. A.; Boele, M. D. K.; Bregman, F. R.; Kamer, P. C. J.; van Leeuwen, P. W. N. M.; Goubitz, K.; Fraanje, J.; Schenk, H.; Bo, C. *J. Am. Chem. Soc.* **1998**, *120*, 11616.

84. van der Veen, L. A.; Kamer, P. C. J.; van Leeuwen, P. *Angew. Chem. Int. Ed.* **1999**, *38*, 336.

85. Kamer, P. C. J.; Reek, J. N. H.; van Leeuwen, P. W. N. M. In *Rhodium Catalyzed Hydroformylation*; van Leeuwen, P. W. N. M., Claver, C., Eds.; Kluwer: Dordrecht, The Netherlands, 2000; Vol. 22, Chapter 3.

86. Pruett, R. L.; Smith, J. A. African Patent 6,804,937, 1968.

87. van Leeuwen, P. W. N. M.; Roobeek, C. F. British Patent 2,068,377, 1980.

88. van Leeuwen, P.; Roobeek, C. F. *J. Organomet. Chem.* **1983**, *258*, 343.

89. Jongsma, T.; Challa, G.; van Leeuwen, P. *J. Organomet. Chem.* **1991**, *421*, 121.

90. Vanrooy, A.; Orij, E. N.; Kamer, P. C. J.; van Leeuwen, P. *Organometallics* **1995**, *14*, 34.

91. Billig, E.; Abatjoglou, A. G.; Bryant, D. R.; Murray, R. E.; Maher, J. M. U.S. Patent 4,599,206, 1989.

92. Billig, E.; Abatjoglou, A. G.; Bryant, D. R. U.S. Patent 4,668,651, 1987.

93. Billig, E.; Abatjoglou, A. G.; Bryant, D. R. U.S. Patent 4,769,498, 1988.

94. Cuny, G. D.; Buchwald, S. L. *J. Am. Chem. Soc.* **1993**, *115*, 2066.

95. Castillon, S.; Fernandez, E. In *Rhodium Catalyzed Hydroformylation*; van Leeuwen, P. W. N. M., Claver, C., Eds.; Kluwer: Dordrecht, The Netherlands, 2000; Vol. 22, p 284.

96. Matsui, Y.; Orchin, M. *J. Organomet. Chem.* **1983**, *246*, 57.

97. Botteghi, C.; Cazzolato, L.; Marchetti, M.; Paganelli, S. *J. Org. Chem.* **1995**, *60*, 6612.

98. Botteghi, C.; Marchetti, M.; Paganelli, S.; Sechi, B. *J. Mol. Catal.* **1997**, *118*, 173.

99. Kleemann, A.; Engel, J. *Pharmazeutische Wirkstoffe*; George Thieme Verlag: Stuttgart, 1987.

100. Elks, J.; Ganellin, G. R. *Dictionary of Drugs*; Chapman Hall: London, 1990.

101. Kleemann, A. *Chem.-Ztg.* **1977**, *101*, 389.

102. Leighton, J. L.; O'Neil, D. N. *J. Am. Chem. Soc.* **1997**, *119*, 11118.

103. Sarraf, S. T.; Leighton, J. L. *Tetrahedron Lett.* **1998**, *39*, 6423.

104. Botteghi, C.; Soccolini, F. *Synthesis* **1985**, 592.

105. Polo, A.; Claver, C.; Castillon, S.; Ruiz, A.; Bayon, J. C.; Real, J.; Mealli, C.; Masi, D. *Organometallics* **1992**, *11*, 3525.

106. Fell, B.; Barl, M. *J. Mol. Catal.* **1977**, *2*, 301.

107. Abatjoglou, A. G.; Bryant, D. R.; Desposito, L. C. *J. Mol. Catal.* **1983**, *18*, 381.

108. Becker, Y.; Eisenstadt, A.; Stille, J. K. *J. Org. Chem.* **1980**, *45*, 2145.

109. Delogu, G.; Faedda, G.; Gladiali, S. *J. Organomet. Chem.* **1984**, *268*, 167.

110. Ojima, I. *Chem. Rev.* **1988**, *88*, 1011.

111. Fuchikami, T.; Ojima, I. *J. Am. Chem. Soc.* **1982**, *104*, 3527.

112. Crameri, Y.; Ochsner, P. A.; Schudel, P. European Patent Appl. EP52775, 1982.

113. Lazzaroni, R.; Settambolo, R.; Uccellobarretta, G. *Organometallics* **1995**, *14*, 4644.

114. Jackson, W. R.; Perlmutter, P.; Tasdelen, E. E. *J. Chem. Soc.. Chem. Commun.* **1990**, 763.

115. Kawabata, Y.; Suzuki, T. M.; Ogata, I. *Chem. Lett.* **1978**, 361.

116. Hayashi, T.; Tanaka, M.; Ogata, I. *J. Mol. Catal.* **1981**, *13*, 323.

117. Claver, C.; van Leeuwen, P. W. M. N. In *Rhodium Catalyzed Hydroformylation*; van Leeuwen, P. W. N. M., Claver, C., Eds.; Kluwer: Dordrecht, The Netherlands, 2000; Vol. 22, Chapter 5.

118. Gladiali, S.; Bayon, J. C.; Claver, C. *Tetrahedron: Asymmetry* **1995**, *6*, 1453.

119. Agbossou, F.; Carpentier, J. F.; Mortreaux, A. *Chem. Rev.* **1995**, *95*, 2485.

120. Dieguez, M.; Pamies, O.; Claver, C. *Tetrahedron: Asymmetry* **2004**, *15*, 2113.

121. Kollar, L.; Consiglio, G.; Pino, P. *J. Organomet. Chem.* **1987**, *330*, 305.

122. Buisman, G. J. H.; Kamer, P. C. J.; van Leeuwen, P. *Tetrahedron: Asymmetry* **1993**, *4*, 1625.

123. Babin, J. E.; Whiteker, G. T. WO 93/03839, 1993.

124. Buisman, G. J. H.; Martin, M. E.; Vos, E. J.; Klootwijk, A.; Kamer, P. C. J.; van Leeuwen, P. *Tetrahedron: Asymmetry* **1995**, *6*, 719.

125. Dieguez, M.; Pamies, O.; Ruiz, A.; Castillon, S.; Claver, C. *Chem. Commun.* **2000**, 1607.

126. Dieguez, M.; Ruiz, A.; Claver, C. *J. Chem. Soc., Dalton Trans.* **2003**, 2957.

127. Pamies, O.; Net, G.; Ruiz, A.; Claver, C. *Tetrahedron: Asymmetry* **2000**, *11*, 1097.

128. (a) Nozaki, K.; Sakai, N.; Nanno, T.; Higashijima, T.; Mano, S.; Horiuchi, T.; Takaya, H. *J. Am. Chem. Soc.* **1997**, *119*, 4413; (b) Sakai, N.; Mano, S.; Nozaki, K.; Takaya, H. *J. Am. Chem. Soc.* **1993**, *115*, 7033; (c) Nozaki, K.; Matsuo, T.; Shibahara, F.; Hiyama, T. *Adv. Synth. Catal.* **2001**, *343*, 61.

129. Yan, Y.; Zhang, X. M. *J. Am. Chem. Soc.* **2006**, *128*, 7198.

130. Buisman, G. J. H.; van der Veen, L. A.; Klootwijk, A.; de Lange, W. G. J.; Kamer, P. C. J.; van Leeuwen, P.; Vogt, D. *Organometallics* **1997**, *16*, 2929.

131. Buisman, G. J. H.; van der Veen, L. A.; Kamer, P. C. J.; van Leeuwen, P. *Organometallics* **1997**, *16*, 5681.

132. Nozaki, K.; Matsuo, T.; Shibahara, F.; Hiyama, T. *Organometallics* **2003**, *22*, 594.

133. Yan, Y.; Zhang, X. M. *J. Am. Chem. Soc.* **2006**, *128*, 7198.

134. (a) Clark, T. P.; Landis, C. R.; Freed, S. L.; Klosin, J.; Abboud, K. A. *J. Am. Chem. Soc.* **2005**, *127*, 5040; (b) Klosin, J.; Landis, C. R. *Acc. Chem. Res.* **2007**, *40*, 1251.

135. Eilbracht, P.; Barfacker, L.; Buss, C.; Eilbracht, P.; Barfacker, L.; Buss, C.; Hollmann, C.; Kitsos-Rzychon, B. E.; Kranemann, C. L.; Rische, T.; Roggenbuck, R.; Schmidt, A. *Chem. Rev.* **1999**, *99*, 3329.

136. Reppe, V. W. *Experientia* **1949**, *5*, 93.

137. Olin, J. F.; Deger, T. E. U.S. Patent 2,422,631, 1947.

138. Brunet, J. J.; Neibecker, D.; Agbossou, F.; Srivastava, R. S. *J. Mol. Catal.* **1994**, *87*, 223.

139. Larson, A. T. U.S. Patent 2,497,310, 1950.

140. Seayad, A.; Ahmed, M.; Klein, H.; Jackstell, R.; Gross, T.; Beller, M. *Science* **2002**, *297*, 1676.

141. Ahmed, M.; Seayad, A. M.; Jackstell, R.; Beller, M. *J. Am. Chem. Soc.* **2003**, *125*, 10311.

142. Lin, Y. S.; El Ali, B.; Alper, H. *Tetrahedron Lett.* **2001**, *42*, 2423.

143. Zimmermann, B.; Herwig, J.; Beller, M. *Angew. Chem. Int. Ed.* **1999**, *38*, 2372.

144. Rische, T.; Eilbracht, P. *Synthesis* **1997**, 1331.

145. Briggs, J. R.; Klosin, J.; Whiteker, G. T. *Org. Lett.* **2005**, *7*, 4795.

146. Ojima, I.; Zhang, Z. *J. Org. Chem.* **1988**, *53*, 4422.

147. Zhang, Z.; Ojima, I. *J. Organomet. Chem.* **1993**, *454*, 281.

148. da Rosa, R. G.; de Campos, J. D. R.; Buffon, R. *J. Mol. Catal.* **1999**, *137*, 297.

149. Jachimowicz, F. GE Patent 3,106,139, 1981.

150. Jachimowicz, F.; Hansson, A. Can. Patent 1,231,199, 1984.

151. Jachimowicz, F.; Hansson, A. In *Catalysis of Organic Reactions*; Augustine, R. L., Ed.; Marcel Dekker: New York, 1985; p 381.

152. El Ali, B.; Alper, H. In *Handbook of Organopalladium Chemistry for Organic Synthesis*; Negishi, E.-i., Ed.; Wiley-Interscience: New York, 2002; Vol. II, p 2333.

153. El Ali, B.; Alper, H. In *Transition Metals for Organic Synthesis*, 2nd ed.; Beller, M., Bolm, C., Eds.; Wiley-VCH: Weinheim, 2004; Vol. I, p 113.

154. Drent, E.; Budzelaar, P. H. M. *Chem. Rev.* **1996**, *96*, 663.

155. Mullen, A. In *New Syntheses with Carbon Monoxide*; Falbe, J., Ed.; Springer-Verlag: New York, 1980; p 243.

156. D'Amore, M. B. U.S. Patent 5,026,901, 1991.

157. Happel, J.; Umemura, S.; Sakakibara, Y.; Blanck, H.; Kunichika, S. *Ind. Eng. Chem., Proc. Design Dev.* **1975**, *14*, 44.

158. http://pubs.acs.org/cen/news/83/i26/8326busc1a.html.

159. Chen, A.; Ren, L.; Crudden, C. M. *J. Org. Chem.* **1999**, *64*, 9704.

160. Seayad, A.; Jayasree, S.; Chaudhari, R. V. *Org. Lett.* **1999**, *1*, 459.

161. Seayad, A.; Jayasree, S.; Chaudhari, R. V. *Catal. Lett.* **1999**, *61*, 99.

162. Lin, R. W.; Herndon, R. C., Jr.; Allen, R. H.; Chockalingham, K. C.; Focht, G. D.; Roy, R. K. Wo, 9830529, 1998.

163. Wu, T.-C.; Chockalingham, K. C.; Klobucar, W. D.; Focht, G. D. Wo, 9830522, 1998.

164. Wu, T.-C. U.S. Patent 5,322,959, 1994.

165. Wu, T.-C. U.S. Patent 5,315,026, 1994.

166. Kawashima, Y.; Okano, K.; Nozaki, K.; Hiyama, T. *Bull. Chem. Soc. Jpn.* **2004**, *77*, 347.

167. Zhou, H.; Lu, S.; Hou, J.; Chen, J.; Fu, H.; Wang, H. *Chem. Lett.* **1996**, *25*, 339.

168. Zhou, H. Y.; Hou, J. G.; Cheng, J.; Lu, S. J.; Fu, H. X.; Wang, H. Q. *J. Organomet. Chem.* **1997**, *543*, 227.

169. Alper, H.; Hamel, N. *J. Am. Chem. Soc.* **1990**, *112*, 2803.

170. Clegg, W.; Eastham, G. R.; Elsegood, M. R. J.; Tooze, R. P.; Wang, X. L.; Whiston, K. *Chem. Commun.* **1999**, 1877.

171. Drent, E.; Kragtwijk, E. Eur. Pat. Appl. EP495548, 1992.

172. El Ali, B.; Alper, H. *J. Mol. Catal.* **1992**, *77*, 7.

173. Drent, E.; Arnoldy, P.; Budzelaar, P. H. M. *J. Organomet. Chem.* **1994**, *475*, 57.

174. Lugan, N.; Lavigne, G. *Organometallics* **1995**, *14*, 1712.

175. Keim, W.; Becker, J. *J. Mol. Catal.* **1989**, *54*, 95.

176. Nahmed, E. M.; Jenner, G. *J. Mol. Catal.* **1990**, *59*, L15.

177. Legrand, C.; Castanet, Y.; Mortreux, A.; Petit, F. *J. Chem. Soc., Chem. Commun.* **1994**, 1173.

178. Kondo, T.; Okada, T.; Mitsudo, T.-a. *Organometallics* **1999**, *18*, 4123.

179. Na, Y.; Ko, S.; Hwang, L. K.; Chang, S. *Tetrahedron Lett.* **2003**, *44*, 4475.

180. Ko, S.; Lee, C.; Choi, M.-G.; Na, Y.; Chang, S. *J. Org. Chem.* **2003**, *68*, 1607.

181. Park, E. J.; Lee, J. M.; Han, H.; Chang, S. *Org. Lett.* **2006**, *8*, 4355.

182. Ko, S.; Na, Y.; Chang, S. *J. Am. Chem. Soc.* **2002**, *124*, 750.

183. Rodriguez, C. J.; Foster, D. F.; Eastham, G. R.; Cole-Hamilton, D. J. *Chem. Commun.* **2004**, 1720.

184. Yu, W. Y.; Bensimon, C.; Alper, H. *Chem. Eur. J.* **1997**, *3*, 417.

185. ElAli, B.; Okuro, K.; Vasapollo, G.; Alper, H. *J. Am. Chem. Soc.* **1996**, *118*, 4264.

186. Reppe, W. *Ann. Chem.* **1953**, *582*, 1.

187. Liu, J. K.; Heaton, B. T.; Iggo, J. A.; Whyman, R.; Bickley, J. F.; Steiner, A. *Chem. Eur. J.* **2006**, *12*, 4417.

188. Lin, Y.-S.; Yamamoto, A. *Organometallics* **1998**, *17*, 3466.

189. Eastham, G. R.; Heaton, B. T.; Iggo, J. A.; Tooze, R. P.; Whyman, R.; Zacchini, S. *Chem. Commun.* **2000**, 609.

190. Clegg, W.; Eastham, G. R.; Elsegood, M. R. J.; Heaton, B. T.; Iggo, J. A.; Tooze, R. P.; Whyman, R.; Zacchini, S. *Organometallics* **2002**, *21*, 1832.

191. Clegg, W.; Eastham, G. R.; Elsegood, M. R. J.; Heaton, B. T.; Iggo, J. A.; Tooze, R. P.; Whyman, R.; Zacchini, S. *J. Chem. Soc., Dalton Trans.* **2002**, 3300.

192. Seayad, A.; Jayasree, S.; Damodaran, K.; Toniolo, L.; Chaudhari, R. V. *J. Organomet. Chem.* **2000**, *601*, 100.

193. van Leeuwen, P. W. N. M.; Zuideveld, M. A.; Swennenhuis, B. H. G.; Freixa, Z.; Kamer, P. C. J.; Goubitz, K.; Fraanje, J.; Lutz, M.; Spek, A. L. *J. Am. Chem. Soc.* **2003**, *125*, 5523.

194. Scrivanti, A.; Beghetto, V.; Campagna, E.; Zanato, M.; Matteoli, U. *Organometallics* **1998**, *17*, 630.

195. For reviews, see references 197a and 198.

196. Slaugh, L. H.; Knifton, J. F.; Weider, P. R.; Powell, J. B.; Allen, K. D.; James, T. G. U.S. Patent 6,576,802, 2003.

197. (a) Khumtaveeporn, K.; Alper, H. *Acc. Chem. Res.* **1995**, *28*, 414; (b) Vasapollo, G.; Mele, G. *Curr. Org. Chem.* **2006**, *10*, 1397.

198. Church, T. L.; Getzler, Y. D. Y. L.; Byrne, C. M.; Coates, G. W. *Chem. Commun.* **2007**, 657.

199. Heck, R. F. *J. Am. Chem. Soc.* **1963**, *85*, 1460.

200. For seminal work, see references 201–206.

201. Reppe, W.; Kröper, H.; V. Kutepow, N.; Pistor, H. J.; Weissbarth, O. *Liebigs Ann. Chem.* **1953**, *582*, 72.

202. Eisenmann, J. L.; Yamartino, R. L.; Howard, J. F. J. *J. Org. Chem.* **1961**, *26*, 2102.

203. Murahashi, S.; Horiie, S. *J. Am. Chem. Soc.* **1956**, *78*, 4816.

204. Reppe, W.; Kröper, H.; Pistor, H. J.; Weissbarth, O. *Liebigs Ann. Chem.* **1953**, *582*, 87.

205. Nienburg, H. J.; Elschnigg, G. German Patent 1,066,572, 1959.

206. Murahashi, S. *J. Am. Chem. Soc.* **1955**, *77*, 6403.

207. Aumann, R.; Ring, H. *Angew. Chem., Int. Ed. Engl.* **1977**, *50*, 16.

208. Aumann, R.; Ring, H.; Kruger, C.; Goddard, R. *Chem. Ber.* **1979**, *112*, 3644.

209. Kamiya, Y.; Kawato, K.; Ohta, H. *Chem. Lett.* **1980**, 154.

210. Alper, H.; Arzoumanian, H.; Petrignani, J.-F.; Maldonado, M. S. *J. Chem. Soc., Chem. Commun.* **1985**, 340.

211. Piotti, M. E.; Alper, H. *J. Am. Chem. Soc.* **1996**, *118*, 111.

212. Drent , E.; Kragtwijk, E. Eur. Pat. Appl. EP 577206; *Chem. Abstr.* **1994**, *120*, 191517c.

213. Getzler, Y. D. Y. L.; Mahadevan, V.; Lobkovsky, E. B.; Coates, G. W. *J. Am. Chem. Soc.* **2002**, *124*, 1174.

214. Allmendinger, M.; Eberhardt, R.; Luinstra, G.; Rieger, B. *J. Am. Chem. Soc.* **2002**, *124*, 5646.

215. Lee, J. T.; Thomas, P. J.; Alper, H. *J. Org. Chem.* **2001**, *66*, 5424.

216. Schmidt, J. A. R.; Lobkovsky, E. B.; Coates, G. W. *J. Am. Chem. Soc.* **2005**, *127*, 11426.

217. Schmidt, J. A. R.; Mahadevan, V.; Getzler, Y. D. Y. L.; Coates, G. W. *Org. Lett.* **2004**, *6*, 373.

218. Mahadevan, V.; Getzler, Y. D. Y. L.; Coates, G. W. *Angew. Chem. Int. Ed.* **2002**, *41*, 2781.

219. Greene, T. W.; Wuts, P. G. M. *Protective Groups in Organic Synthesis*, 2nd ed.; Wiley: New York, 1991.

220. Eisenmann, J. L. *J. Org. Chem.* **1962**, *27*, 2706.

221. Prandi, J.; Namy, J. L.; Menoret, G.; Kagan, H. B. *J. Organomet. Chem.* **1985**, *285*, 449.

222. Kramer, J. W.; Lobkovsky, E. B.; Coates, G. W. *Org. Lett.* **2006**, *8*, 3709.

223. Getzler, Y. D. Y. L.; Kundnani, V.; Lobkovsky, E. B.; Coates, G. W. *J. Am. Chem. Soc.* **2004**, *126*, 6842.

224. (*S*)-Methylsuccinic anhydride was produced with 99% ee when the carbonylation was performed at 55 °C. At 80 °C, some product racemization occurred.

225. Rowley, J. M.; Lobkovsky, E. B.; Coates, G. W. *J. A. Chem. Soc.* **2006**, *129*, 4948.

226. Allen, K. D.; James, T. G.; Knifton, J. F.; Powell, J. B.; Slaugh, L. H.; Weider, P. R. PCT Int. Appl. WO 098887; *Chem. Abstr.* **2002**, *138*, 14854.

227. Nakano, K.; Katayama, M.; Ishihara, S.; Hiyama, T.; Nozaki, K. *Synlett* **2004**, *8*, 1367.

228. Weber, R.; Englert, U.; Ganter, B.; Keim, W.; Möthrath, M. *Chem. Commun.* **2000**, 1419.

229. Arhancet, J. P.; Forschner, T. C.; Powell, J. B.; Semple, T. C.; Slaugh, L. H.; Thomason, T. B.; Weider, P. R.; Allen, K. D.; Eubanks, D. C. WO, 9,610,552, 1996.

230. Allen, K. D.; Knifton, J. F.; Powell, J. B.; Slaugh, L. H.; Talmadge, G. J.; Weider, P. R.; Williams, T. S. WO, 2,001,072,675, 2001.

231. Allen, K. D.; James, T. G.; Knifton, J. F.; Powell, J. B.; Slaugh, L. H.; Weider, P. R. WO, 2,002,098,887, 2002.

232. Allen, K. D.; Arhancet, J. P.; Knifton, J. F.; Powell, J. B.; Slaugh, L. H.; Weider, P. R. WO, 2,003,068,720, 2003.

233. Allen, K. D.; Powell, J. B.; Weider, P. R.; Knifton, J. F. U.S. Patent 2,003,040,647, 2003.

234. Sullivan, C. J. In *Ullman's Encyclopedia of Industrial Chemistry*; Elvers, B., Hawkins, S., Russey, W., Schulz, G., Eds.; Wiley: New York, 1993.

235. (a) Eisenmann, J. L.; Yamartino, R. L.; Howard, J. F. *J. Org. Chem.* **1961**, *26*, 2102; (b) Samain, H.; Carpentier, J. F.; Mortreux, A.; Petit, F. *New J. Chem.* **1991**, *15*, 367.

236. McClure, J. D. *J. Org. Chem.* **1967**, *32*, 3888.

237. Hinterding, K.; Jacobsen, E. N. *J. Org. Chem.* **1999**, *64*, 2164.

238. Though the structure of the branched product in methoxycarbonylation was not explicitly stated, we assume it to be the [α]-R,-[β]-hydroxyester.

239. Liu, J.; Chen, J.; Xia, C. *J. Mol. Catal.* **2006**, *250*, 232.

240. Kim, H. S.; Bae, J. Y.; Lee, J. S.; Jeong, C. I.; Choi, D. K.; Kang, S. O.; Cheong, M. *Appl. Catal., A* **2006**, *301*, 75.

241. Tokunaga, M.; Larrow, J. F.; Kakiuchi, F.; Jacobsen, E. N. *Science* **1997**, *277*, 936.

242. Furukawa, J.; Iseda, Y.; Saegusa, T.; Fujii, H. *Makromol. Chem.* **1965**, *89*, 263.

243. Takeuchi, D.; Sakaguchi, Y.; Osakada, K. *J. Polym. Sci., Part A: Polym. Chem.* **2002**, *40*, 4530.

244. Allmendinger, M.; Molnar, F.; Zintl, M.; Luinstra, G. A.; Preishuber-Pflügl, P.; Rieger, B. *Chem. Eur. J.* **2005**, *11*, 5327.

245. Allmendinger, M.; Zintl, M.; Eberhardt, R.; Luinstra, G. A.; Molnar, F.; Rieger, B. *J. Organomet. Chem.* **2004**, *689*, 971.

246. Allmendinger, M.; Eberhardt, R.; Luinstra, G. A.; Rieger, B. *Macromol. Chem. Phys.* **2003**, *204*, 564.

247. Allmendinger, M.; Eberhardt, R.; Luinstra, G. A.; Molnar, F.; Rieger, B. *Z. Anorg. Allg. Chem.* **2003**, *629*, 1347.

248. (a) Allmendinger, M.; Eberhardt, R.; Luinstra, G.; Rieger, B. *Polym. Mater. Sci. Eng.* **2002**, *87*, 223; (b) Allmendinger, M.; Eberhardt, R.; Luinstra, G.; Rieger, B. *Polym. Mater. Sci. Eng.* **2002**, *87*, 89.

249. Luinstra, G.; Molnar, F.; Rieger, B.; Allmendinger, M. WO 012860, 2004.

250. Zintl, M.; Hearley, A. K.; Rieger, B. In *Leading Edge Organometallic Chemistry Research*; Cato, M. A., Ed.; Nova Science Publishers: New York, 2006.

251. Lee, J. T.; Alper, H. *Macromolecules* **2004**, *37*, 2417.

252. Liu, G.; Jia, L. *Angew. Chem. Int. Ed.* **2006**, *45*, 129.

253. Kidwai, M.; Sapra, P.; Bhushan, K. R. *Curr. Med. Chem.* **1999**, *6*, 195.

254. Schofield, C. J.; Walter, M. W. In *Amino Acids, Peptides, and Proteins*, Davies, J. S., Ed; Royal Society of Chemistry: Cambridge, U.K., 1999; Vol. 30.

255. Konaklieva, M. I. *Curr. Med. Chem.: Anti-Infect. Agents* **2002**, *1*, 215.

256. Alcaide, B.; Almendros, P. *Curr. Org. Chem.* **2002**, *6*, 245.

257. Brown, A. G. *Pure Appl. Chem.* **1987**, *59*, 475.

258. Ojima, I. *Adv. Asym. Synth.* **1995**, *1*, 95.

259. Palomo, C.; Aizpurua, J. M.; Ganboa, I.; Oiarbide, M. *Amino Acids* **1999**, *16*, 321.

260. Alcaide, B.; Almendros, P. *Synlett* **2002**, 381.

261. Alcaide, B.; Almendros, P. *Curr. Med. Chem.* **2004**, *11*, 1921.

262. Hashimoto, K.; Hotta, K.; Okada, M.; Nagata, S. *J. Polym. Sci., Part A: Polym. Chem.* **1995**, *33*, 1995.

263. Hashimoto, K.; Oi, T.; Yasuda, J.; Hotta, K.; Okada, M. *J. Polym. Sci., Part A. Polym. Chem.* **1997**, *35*, 1831.

264. Hashimoto, K.; Yasuda, J.; Kobayashi, M. *J. Polym. Sci., Part A: Polym. Chem.* **1999**, *37*, 909.

265. Cheng, J. J.; Deming, T. J. *J. Am. Chem. Soc.* **2001**, *123*, 9457.

266. Piotti, M. E.; Alper, H. *J. Am. Chem. Soc.* **1996**, *118*, 111.

267. Davoli, P.; Moretti, I.; Prati, F.; Alper, H. *J. Org. Chem.* **1999**, *64*, 518.

268. Davoli, P.; Prati, F. *Heterocycles* **2000**, *53*, 2379.

269. Davoli, P.; Forni, A.; Moretti, I.; Prati, F.; Torre, G. *Tetrahedron* **2001**, *57*, 1801.

270. Lu, S.-M.; Alper, H. *J. Org. Chem.* **2004**, *69*, 3558.

271. Davoli, P.; Spaggiari, A.; Ciamaroni, E.; Forni, A.; Torre, G.; Prati, F. *Heterocycles* **2004**, *63*, 2499.

272. Alper, H.; Urso, F.; Smith, D. J. H. *J. Am. Chem. Soc.* **1983**, *105*, 6737.

273. Calet, S.; Urso, F.; Alper, H. *J. Am. Chem. Soc.* **1989**, *111*, 931.

274. (a) Calet, S.; Urso, F.; Alper, H.; *J. Am. Chem. Soc.* **1989**, *111*, 931; (b) Khumtaveeporn, K.; Alper, H.; *Acc. Chem. Res.* **1995**, *28*, 414.

275. Ardura, D.; López, R.; Sordo, T.; *J. Org. Chem.* **2006**, *71*, 7315.

276. (a) Leung, W. H.; Mak, W. L.; Chan, E. Y. Y.; Lam, T. C. H.; Lee, W. S.; Kwong, H. L.; Yeung, L. L. *Synlett* **2002**, 1688; (b) Osborn, H.; Sweeney, J. *Tetrahedron: Asymmetry* **1997**, *8*, 1693; (c) Tanner, D. *Angew. Chem., Int. Ed. Engl.* **1994**, *33*, 599.

277. Church, T. L.; Getzler, Y. D. Y. L.; Coates, G. W. *J. Am. Chem. Soc.* **2006**, *128*, 10125.

278. Molnar, F.; Luinstra, G. A.; Allmendinger, M.; Rieger, B. *Chem. Eur. J.* **2003**, *9*, 1273.

279. Stirling, A.; Iannuzzi, M.; Parrinello, M.; Molnar, F.; Bernhart, V.; Luinstra, G. A. *Organometallics* **2005**, *24*, 2533.

280. The carbonylation of cyclopentene oxide with retention of configuration is an exception to this trend. This exception presumably results from conformational constraints during the cyclization step. See reference 216.

281. Mori, M. In *Handbook of Organopalladium Chemistry for Organic Synthesis*; Negishi, E.-i., Ed.; Wiley: New York, 2002; Vol. 2, p 2313.

282. *Applied Homogeneous Catalysis with Organometallic Compounds: A Comprehensive Handbook in Two Volumes*; Cornils, B., Herrmann, W. A., Eds.; Wiley-VCH: Weinheim, 1996.

283. Tsuji, J. *Palladium Reagents and Catalysts: New Perspectives for the 21st Century*; Wiley: Chichester, U.K., 2004.

284. Beller, M.; Cornils, B.; Frohning, C. D.; Kohlpaintner, C. W. *J. Mol. Catal.* **1995**, *104*, 17.

285. Bates, R. W. In *Comprehensive Organometallic Chemistry, A Review of the Literature 1982–1994*; Abel, E. W., Stone, F. G. A., Wilkinson, G., Eds.; Pergamon Press: New York, 1995; Vol. 12, p 349.

286. Tkatchenko, I. In *Comprehensive Organometallic Chemistry*; Wilkinson, G., Stone, F. G. A., Abel, E. W., Eds.; Pergamon Press: Oxford, U.K., 1982; Vol. 8.

287. *Carbonylation: Direct Synthesis of Carbonyl Compounds*, Colquhoun, H. M., Thompson, D. J., Twigg, M. V., Eds.; Plenum: New York, 1991.

288. Heck, R. F. *Palladium Reagents in Organic Syntheses*; Academic Press: New York, 1985.

289. Schoenberg, A.; Bartoletti, I.; Heck, R. F. *J. Org. Chem.* **1974**, *39*, 3318.

290. Ito, T.; Mori, K.; Mizoroki, T.; Ozaki, A. *Bull. Chem. Soc. Jpn.* **1975**, *48*, 2091.

291. Shoenberg, A.; Heck, R. F. *J. Org. Chem.* **1974**, *39*, 3327.

292. Huser, M.; Youinou, M. T.; Osborn, J. A. *Angew. Chem., Int. Ed. Engl.* **1989**, *28*, 1427.

293. Ben-David, Y.; Portnoy, M.; Milstein, D. *J. Am. Chem. Soc.* **1989**, *111*, 8742.

294. Grushin, V. V.; Alper, H. *J. Chem. Soc., Chem. Commun.* **1992**, 611.

295. Mori, M.; Chiba, K.; Ban, Y. *J. Org. Chem.* **1978**, *43*, 1684.

296. Mori, M.; Chiba, K.; Inotsume, N.; Ban, Y. *Heterocycles* **1979**, *12*, 921.

297. Martin, L. D.; Stille, J. K. *J. Org. Chem.* **1982**, *47*, 3630.

298. Cowell, A.; Stille, J. K. *J. Am. Chem. Soc.* **1980**, *102*, 4193.

299. Kim, J. S.; Sen, A. *J. Mol. Catal.* **1999**, *143*, 197.

300. Foti, C. J.; Comins, D. L. *J. Org. Chem.* **1995**, *60*, 2656.

301. Takahashi, T.; Nagashima, T.; Tsuji, J. *Chem. Lett.* **1980**, *9*, 369.

302. Lin, Y.-S.; Yamamoto, A. *Tetrahedron Lett.* **1997**, *38*, 3747.

303. Komiya, S.; Akai, Y.; Tanaka, K.; Yamamoto, T.; Yamamoto, A. *Organometallics* **1985**, *4*, 1130.

304. Drent, E.; Budzelaar, P. H. M. *Chem. Rev.* **1996**, *96*, 663.

305. Drent, E.; Vanbroekhoven, J. A. M.; Doyle, M. J. *J. Organomet. Chem.* **1991**, *417*, 235.

306. Sen, A. *Adv. Polym. Sci.* **1986**, *73–4*, 125.

307. Sen, A. *Acc. Chem. Res.* **1993**, *26*, 303.

308. Sommazzi, A.; Garbassi, F. *Prog. Polym. Sci.* **1997**, *22*, 1547.

309. Ballauf, F.; Bayer, O.; Leichmann, L. German Patent 863,711, 1941.

310. Brubaker, M. M. U.S. Patent 2,495,286, 1950.

311. Reppe, W.; Mangini, A. U.S. Patent 880,297, 1948.

312. Reppe, W.; Mangini, A. U.S. Patent 2,577,208, 1951.

313. Gough, A. British Patent 1,081,304, 1967.

314. Fenton, D. M. U.S. Patent 3,530,109, 1970.

315. Fenton, D. M. U.S. Patent 4,076,911, 1978.

316. Nozaki, K. U.S. Patent 3,835,123, 1974.

317. Nozaki, K. U.S. Patent 3,689,460, 1972.

318. Sen, A.; Lai, T. W. *J. Am. Chem. Soc.* **1982**, *104*, 3520.

319. Drent, E. E.P. Appl. 121,965, 1984.

320. Drent, E. E.P. Appl. 181,014, 1986.

321. van Broekhoven, J. A. M.; Drent, E.; Klei, E. E.P. Appl. 213,671, 1987.

322. van Broekhoven, J. A. M.; Drent, E. E.P. Appl. 235,865, 1987.

323. Shultz, C. S.; Ledford, J.; DeSimone, J. M.; Brookhart, M. *J. Am. Chem. Soc.* **2000**, *122*, 6351.

324. Rix, F. C.; Brookhart, M. *J. Am. Chem. Soc.* **1995**, *117*, 1137.

325. Rix, F. C.; Brookhart, M.; White, P. S. *J. Am. Chem. Soc.* **1996**, *118*, 4746.

326. Dekker, G. P. C. N.; Elsevier, C. J.; Vrieze, K.; van Leeuwen, P. W. N. M. *Organometallics* **1992**, *11*, 1598.

327. Vanasselt, R.; Gielens, E.; Rulke, R. E.; Vrieze, K.; Elsevier, C. J. *J. Am. Chem. Soc.* **1994**, *116*, 977.

328. Rix, F. C.; Brookhart, M.; White, P. S. *J. Am. Chem. Soc.* **1996**, *118*, 4746.

329. Rix, F. C.; Brookhart, M.; White, P. S. *J. Am. Chem. Soc.* **1996**, *118*, 2436.

330. Ledford, J.; Shultz, C. S.; Gates, D. P.; White, P. S.; DeSimone, J. M.; Brookhart, M. *Organometallics* **2001**, *20*, 5266.

331. Maurice Brookhart, personal communication of unpublished material.

332. Drent, E. E.P. Appl. 317,003, 1989.

333. Brookhart, M.; Rix, F. C.; Desimone, J. M.; Barborak, J. C. *J. Am. Chem. Soc.* **1992**, *114*, 5894.

334. Budzelaar, P. H. M.; van Leeuwen, P.; Roobeek, C. F.; Orpen, A. G. *Organometallics* **1992**, *11*, 23.

335. Nozaki, K.; Hiyama, T. *J. Organomet. Chem.* **1999**, *576*, 248.

336. Corradini, P.; Derosa, C.; Panunzi, A.; Petrucci, G.; Pino, P. *Chimia* **1990**, *44*, 52.

337. Barsacchi, M.; Consiglio, G.; Medici, L.; Petrucci, G.; U.V. Suter *Angew. Chem., Int. Ed. Engl.* **1991**, *30*, 989.

338. Aeby, A.; Consiglio, G. *Helv. Chim. Acta* **1998**, *81*, 35.

339. Brookhart, M.; Wagner, M. I. *J. Am. Chem. Soc.* **1994**, *116*, 3641.

340. Schatz, A.; Scarel, A.; Zangrando, E.; Mosca, L.; Carfagna, C.; Gissibl, A.; Milani, B.; Reiser, O. *Organometallics* **2006**, *25*, 4065.

341. For a list of patents and theses containing these data, see references 41 and 42 of reference 154.

342. Nozaki, K.; Sato, N.; Takaya, H. *J. Am. Chem. Soc.* **1995**, *117*, 9911.

343. Nozaki, K.; Sato, N.; Tonomura, Y.; Yasutomi, M.; Takaya, H.; Hiyama, T.; Matsubara, T.; Koga, N. *J. Am. Chem. Soc.* **1997**, *119*, 12779.

344. Bronco, S.; Consiglio, G.; Hutter, R.; Batistini, A.; Suter, U. W. *Macromolecules* **1994**, *27*, 4436.

345. Jiang, Z. Z.; Sen, A. *J. Am. Chem. Soc.* **1995**, *117*, 4455.

346. Strübing, D.; Beller, M. *Top. Organomet. Chem.* **2006**, *18*, 165.

347. Alcaide, B.; Almendros, P. *Eur. J. Org. Chem.* **2004**, 3377.

348. Blanco-Urgoiti, J.; Añorbe, L.; Pérez-Serrano, L.; Domínguez, G.; Pérez-Castells. J. *Chem. Soc. Rev.* **2004**, *33*, 32.

349. (a) Gibson, S. E.; Stevenazzi, A. *Angew. Chem. Int. Ed.* **2003**, *42*, 1800; (b) Gibson, S. E.; Mainolfi, N. *Angew. Chem. Int. Ed.* **2005**, *44*, 3022.

350. Brummond, K. M.; Kent, J. L. *Tetrahedron* **2000**, 3263.

351. Rivero, M. R.; Adrio, J.; Carretero, J. C. *Eur. J. Org. Chem.* **2002**, 2881.

352. Khand, I. U.; Knox, G. R.; Pauson, P. L.; Watts, W. E. *J. Chem. Soc. D* **1971**, 36.

353. Khand, I. U.; Knox, G. R.; Pauson, P. L.; Watts, W. E.; Foreman, M. I. *J. Chem. Soc., Perkin Trans. 1* **1973**, 977.

354. Magnus, P.; Principe, L. M. *Tetrahedron Lett.* **1985**, *26*, 4851.

355. Schore, N. E.; Croudace, M. C. *J. Org. Chem.* **1984**, *46*, 5436.

356. Simonian, S. O.; Smit, W. A.; Gybin, A. S.; Shashkov, A. S.; Mikaelian, G. S.; Tarasov, V. A.; Ibragimov, I. I.; R., C.; Froen, D. E. *Tetrahedron Lett.* **1986**, *27*, 1245.

357. Shambayati, S.; Crowe, W. E.; Schreiber, S. L. *Tetrahedron Lett.* **1990**, *31*, 5289.

358. Jeong, N.; Chung, Y. K.; Lee, B. Y.; Lee, S. H.; Yoo, S. *Synlett* **1991**, 204.

359. Sugihara, T.; Yamadab, M.; Yamaguchi, M.; Nishizawa, M. *Synlett* **1999**, 771.

360. Lee, B. Y.; Chung, Y. K.; Jeong, N.; Lee, Y.; Hwang, S. H. *J. Am. Chem. Soc.* **1994**, *116*, 8793.

361. Lee, N. Y.; Chung, Y. K. *Tetrahedron Lett.* **1996**, *37*, 3145.

362. Hicks, F. A.; Buchwald, S. L. *J. Am. Chem. Soc.* **1999**, *121*, 7026.

363. Hicks, F. A.; Kablaoui, N. M.; Buchwald, S. L. *J. Am. Chem. Soc.* **1996**, *118*, 9450.

364. Hicks, F. A.; Buchwald, S. L. *J. Am. Chem. Soc.* **1996**, *118*, 11688.

365. Hicks, F. A.; Kablaoui, N. M.; Buchwald, S. L. *J. Am. Chem. Soc.* **1999**, *121*, 5881.

366. Kablaoui, N. M.; Hicks, F. A.; Buchwald, S. L. *J. Am. Chem. Soc.* **1996**, *118*, 5818.

367. Kablaoui, N. M.; Hicks, F. A.; Buchwald, S. L. *J. Am. Chem. Soc.* **1997**, *119*, 4424.

368. Morimoto, T.; Chatani, N.; Fukumoto, Y.; Murai, S. *J. Org. Chem.* **1997**, *62*, 3762.

369. Kondo, T.; Suzuki, N.; Okada, T.; Mitsudo, T. *J. Am. Chem. Soc.* **1997**, *119*, 6187.

370. Kobayashi, T.; Koga, Y.; Narasaka, K. *J. Organomet. Chem* **2001**, *624*, 73.

371. Shibata, T.; Takagi, K. *J. Am. Chem. Soc.* **2000**, *122*, 9852.

372. Adrio, J.; Rivero, M. R.; Carretero, J. C. *Org. Lett.* **2005**, *7*, 431.

373. Hoye, T.; Suriano, J. A. *J. Am. Chem. Soc.* **1993**, *115*, 1154.

374. Pearson, A. J.; Dubbert, R. A. *J. Chem. Soc., Chem. Commun.* **1991**, 202.

375. Tang, Y.; Deng, L.; Zhang, Y.; Dong, G.; Chen, J.; Yang, Z. *Org. Lett.* **2005**, *7*, 1657.

376. Narasaka, K.; Shibata, T. *Chem. Lett.* **1994**, 315.

377. Brummond, K. M.; Mitasev, B. *Org. Lett.* **2004**, *6*, 2245.

378. Kent, J. L.; Wan, H.; Brummond, K. M. *Tetrahedron Lett.* **1995**, *36*, 2407.

379. Brummond, K. M.; Chen, H.; Fisher, K. D.; Kerekes, A. D.; Rickards, B.; Sill, P. C.; Geib, S. J. *Org. Lett.* **2002**, *4*, 1931.

380. Brummond, K. M.; Wan, H.; Kent, J. L. *J. Org. Chem.* **1998**, *63*, 6535.

381. (a) Brummond, K. M.; Gao, D. *Org. Lett.* **2003**, *5*, 3491; (b) Brummond, K. M.; Curran, D. P.; Mitasev, B.; Fisher, S. *J. Org. Chem.* **2005**, *70*, 1745.

382. Bayden, A. S.; Brummond, K. M.; Jordan, K. D. *Organometallics* **2006**, *25*, 5204.

383. Hiroi, K.; Wanatabe, T.; Kawagishi, R.; Abe, I. *Tetrahedron Lett.* **2000**, *41*, 891.

384. Jeong, N.; Sung, B. K.; Choi, Y. K. *J. Am. Chem. Soc.* **2000**, *122*, 6771.

385. Krafft, M. E.; Juliano, C. A. *J. Org. Chem.* **1992**, *57*, 5106.

386. Itami, K.; Mitsudo, K.; Fujita, K.; Ohashi, Y.; Yoshida, J. *J. Am. Chem. Soc.* **2004**, *126*, 11058.

387. Pericàs, M. A.; Balsells, J.; Castro, J.; Marchueta, I.; Moyano, A.; Riera, A.; Vázquez, J.; Verdaguer, X. *Pure Appl. Chem.* **2002**, *74*, 167.

388. Jamison, T. F.; Shambayati, S.; Crowe, W. E.; Schreiber, S. L. *J. Am. Chem. Soc.* **1997**, *119*, 4353.

389. Jamison, T. F.; Shambayati, S.; Crowe, W. E.; Schreiber, S. L. *J. Am. Chem. Soc.* **1994**, *116*, 5505.

CHAPTER 18

Catalytic C–H Functionalization

18.1. Overview

The selective cleavage and functionalization of C–H bonds, shown most generally in Equation 18.1, has been a long-term goal in transition metal organometallic chemistry.[1–5] The term "C–H activation" is imprecise, but typically refers to catalytic or stoichiometric reactions of transition metal complexes with the unreactive C–H bonds of alkanes, arenes, or alkyl chains to form products containing a new metal–carbon bond. Thus, C–H activation by this definition would contrast in most cases with processes, such as Friedel–Crafts alkylations or aromatic nitrations, that cleave a C–H bond, but do so by initial electrophilic attack on the aromatic π-system and subsequent cleavage of the C–H bond by a second reaction component that acts as a base. By this definition, C–H activation would also differ from directed orthometallation of aromatic C–H bonds by reagents such as BuLi, which occur by the action of a strongly basic reagent.

$$R–H \ + \ A \ \xrightarrow{\text{Catalyst}} \ R–FG \ + \ B \qquad FG = \text{functional group} \qquad (18.1)$$

The interest in C–H activation is motivated by the desire to conduct chemical transformations by the cleavage of an unactivated C–H bond and subsequent installation of a functional group with high chemo- and regioselectivity under mild conditions. The cleavage of C–H bonds of alkanes and arenes is common: the oxidation of alkanes to generate carbon dioxide, water, and energy occurs at the burner of a common gas stove or in the engine in a car. Moreover, uncatalyzed halogenations of alkanes is typically taught in introductory organic chemistry, and alkyllithium reagents cleave C–H bonds by the directed metallation processes mentioned in the previous paragraph.[6] Even the common chemical reagent *tert*-butyl hydroperoxide is generated by a selective reaction of oxygen at the tertiary C–H bond of isobutane catalyzed by HBr.[7, 8] Thus, the challenge of "C–H activation" is not just to cleave a C–H bond, but to cleave one C–H bond in a selective fashion and to incorporate this C–H bond cleavage step into a catalytic process that leads to functionalization. The greatest challenge is to conduct this transformation at C–H bonds that are the least reactive toward radicals or strong electrophiles.

Two of the major goals of alkane functionalization have been the selective conversion of methane to methanol and the replacement of strong terminal alkyl C–H bonds with a functional group. These goals have also included the selective functionalization of a single aliphatic or aromatic C–H bond, directed by either a substituent attached to the arene or controlled by the particular steric or electronic properties of the substrate. Thus, many of the principles that have been developed for the functionalization of small molecules with the goal of producing chemical feedstocks have now been adopted for more intricate synthetic applications.

Functionalization of alkanes under mild conditions catalyzed by coordination complexes is known. For example, much of the cyclohexanol and cyclohexanone used in the synthesis of adipic acid is produced by the oxidation of cyclohexane with a cobalt catalyst.[9] Chapter 18, however, focuses on the types of alkane functionalization that occur by the reactions described in earlier chapters of this text involving cleavage of a C–H bond to form an organometallic product. Chapter 6 described oxidative addition and σ-bond metathesis of C–H bonds in aliphatic and aromatic reagents, and many of the catalytic processes described here in Chapter 18 stem from the cleavage of C–H bonds by these types of reactions under relatively mild conditions. This chapter also describes reactions in which an aliphatic C–H bond is cleaved by insertions of a carbene fragment from metal–carbenoid complexes described in Chapter 13.[10]

The selective functionalization of typically unreactive C–H bonds confronts several challenges. First, few reagents react with C–H bonds to generate products that lie thermodynamically downhill of the reactants. Many catalytic processes that could be envisioned with simple organometallic reactions, such as the carbonylation of alkanes and arenes (Equation 18.2), are endothermic ($\Delta H = 1.7$ kcal/mol for benzene).[11,12] Likewise, the dehydrogative coupling of alkanes with most reagents containing an X–H bond (Equation 18.3) are endothermic.[13–24] Similarly, the dehydrogenation of an alkane (Equation 18.4) to make an olefin is unfavorable thermodynamically, as one would expect from the common observation of catalytic olefin hydrogenation.[12] Therefore, conditions must be found to conduct these catalytic processes under non-equilibrium conditions, or reagents must be identified that react with alkanes by exothermic processes. Because the oxidation of alkanes by many oxidizing reagents is downhill, much effort has focused on this type of transformation.

$$\text{C}_6\text{H}_6 + \text{CO} \longrightarrow \text{C}_6\text{H}_5\text{CHO} \qquad \Delta H = 1.7 \text{ kcal/mol} \tag{18.2}$$

$$\text{R–H} + \text{HX} \longrightarrow \text{RX} + \text{H}_2 \qquad \begin{array}{l}\Delta H = 22 \text{ kcal/mol} \\ \text{for R} = \text{C}_5\text{H}_{11}, \text{X} = \text{OH}\end{array} \tag{18.3}$$

$$\text{RCH}_2\text{CH}_3 \longrightarrow \text{RCH=CH}_2 + \text{H}_2 \qquad \begin{array}{l}\Delta H = 30 \text{ kcal/mol} \\ \text{for R} = \text{C}_4\text{H}_9\end{array} \tag{18.4}$$

Second, the strongest C–H bonds are the most desirable to functionalize with metal catalysts. Because radical reactions can be used to functionalize alkanes or alkyl chains at secondary C–H bonds, one target of catalytic alkane functionalization is the development of reactions at terminal C–H bonds. Because these C–H bonds are stronger than secondary and tertiary C–H bonds[25] and because terminal carbons less readily support an accumulation of positive charge than internal carbons, many reactions of unfunctionalized C–H bonds occur less readily at primary C–H bonds than at secondary and tertiary C–H bonds.

Third, the products of C–H activation are typically more reactive than the reactants because the functional group in the product creates a reactive site. For example, the alcohols formed by oxidation typically undergo further oxidation to ketones, aldehydes, or acids. Likewise, aldehydes formed by alkane carbonylation undergo condensation reactions. One method to prevent subsequent functionalization is to install a functional group that is hindered or that makes the α C–H bonds less prone to oxidation. Products containing electron-withdrawing trifluoroacetate and sulfonate groups appear to be less prone to oxidation than the starting hydrocarbons.[26,27]

The following sections describe a selection of the types of C–H bond functionalization processes that involve organometallic intermediates. These reactions include platinum-catalyzed oxidations to alcohols and alkyl halides, oxidative and alkylative carbonylation of arenes and alkanes, oxidative olefination of arenes, hydroarylation of olefins, carbonylation of arenes and alkanes, dehydrogenation of alkanes, dehydrogenative coupling of arenes and alkanes with main group reagents, and insertions of carbenes generated from aryldiazoacetates into secondary C–H bonds. An additional class of reaction in which aryl and heteroaryl C–H bonds are cleaved and products containing the C–C bond in biaryl and heteroaryl compounds are formed is described in the last section of Chapter 19. In some cases, the mechanisms of the C–H bond functionalization reactions have been studied in detail; in these cases, mechanistic conclusions are presented.

18.2. Platinum-Catalyzed Alkane and Arene Oxidations via Organometallic Intermediates

18.2.1. Early Platinum-Catalyzed C–H Activation Processes

Shilov reported some of the earliest evidence that transition metal complexes could selectively cleave the C–H bonds of alkanes in a catalytic fashion.[28] Shilov showed that H/D exchange would occur between alkanes and deuterated acid in the presence of platinum complexes (Equation 18.5 and Table 18.1). In addition, Shilov showed that the oxidation of alkanes occurred in the presence of a platinum(II) catalyst, although a platinum(IV) complex was needed as the oxidant. These reactions led to a mixture of alkyl halides formed from the halide of the Pt(IV) oxidant (Equation 18.6) and trifluoroacetate from the trifluoroacetic acid solvent.[29–33] The cost of platinum(IV) as an oxidant makes this reaction impractical. However, these results provided hope that selective alkane functionalization could be developed because H/D exchange occurred faster at primary C–H bonds than at secondary C–H bonds (Table 18.1),[34] and some selectivity for oxidations of primary C–H bonds over secondary C–H bonds was observed.[28] As noted in Chapter 6, these results motivated a large number of groups to seek transition metal complexes that would insert into, or by other means selectively cleave, the C–H bond of alkanes and create products from this bond cleavage that could be observed directly.

Table 18.1. Regioselectivity of alkane H/D exchange with CH_3CO_2D/D_2O catalyzed by K_2PtCl_4.

| | | | D in[a] | | |
| | Time | D found | Me– | –CH$_2$– | –CH– |
Alkane	(h)	(%)	(%)	(%)	(%)
Methane	95	25	—	—	—
Ethane	137	91	91	—	—
Pentane	137	75	92	57	—
2-Methylbutane	137	69	83	37	9

[a] Full D incorporation at each position at equilibrium would be 92–96%.

$$R-H \xrightarrow[\text{DClO}_4,\ \text{CH}_3\text{CO}_2\text{D},\ \text{D}_2\text{O}]{\text{K}_2\text{PtCl}_4,\ 12\ \text{mol}\ \%} R-D \tag{18.5}$$

$$\text{(18.6)}$$

56 : 44

18.2.2. More Practical Platinum Catalysts for Alkane Functionalization

Several systems for selective catalytic reactions based on Shilov's system have been developed with oxidants more practical than platinum(IV). Periana reported two different systems for the oxidation of methane in sulfuric acid containing SO_3. One of the catalysts is a simple mercuric halide, and reactions catalyzed by this mercury compound generated methyl sulfate with turnover frequencies of 10^{-3} s^{-1}.[35] The second system is more reactive and is based on a platinum complex containing a bipyrimidine ligand (Equation 18.7).[27] In this case, methane is converted to methyl bisulfate with 81% selectivity, greater than 500 turnovers, and a turnover frequency of 10^{-2} s^{-1}. These reactions are selective for the functionalization of methane to this methanol derivative because the electron-withdrawing

$$CH_4 + 2 H_2SO_4 \xrightarrow{\text{Cat. (bpym)PtCl}_2} CH_3OSO_3H + 2 H_2O + SO_2$$

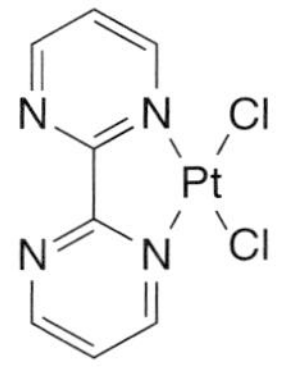

$$\text{TOF} = 10^{-2}\,s^{-1}$$
$$\text{TON} > 500$$

(bpym)PtCl$_2$ =

(18.7)

group attached to the oxygen protects the methanol from further oxidation. When the methane oxidation step is coupled to the hydrolysis and oxidation of SO$_2$, a closed process for oxidation of methane by O$_2$ results (Scheme 18.1). Unfortunately, removal of the product from sulfuric acid and the corrosive nature of this acid make the development of this chemistry into a commercially viable route to methanol challenging.

Step 1
$$CH_4 + HX + SO_3 \longrightarrow CH_3X + H_2O + SO_2$$

Step 2
$$CH_3X + H_2O \longrightarrow CH_3OH + HX$$

Step 3
$$SO_2 + 1/2\,O_2 \longrightarrow SO_3$$

Overall: $CH_4 + 1/2\,O_2 \longrightarrow CH_3OH$

Scheme 18.1.
Process scheme for the net oxidation of methane to methanol, in which X = OSO$_3$H.

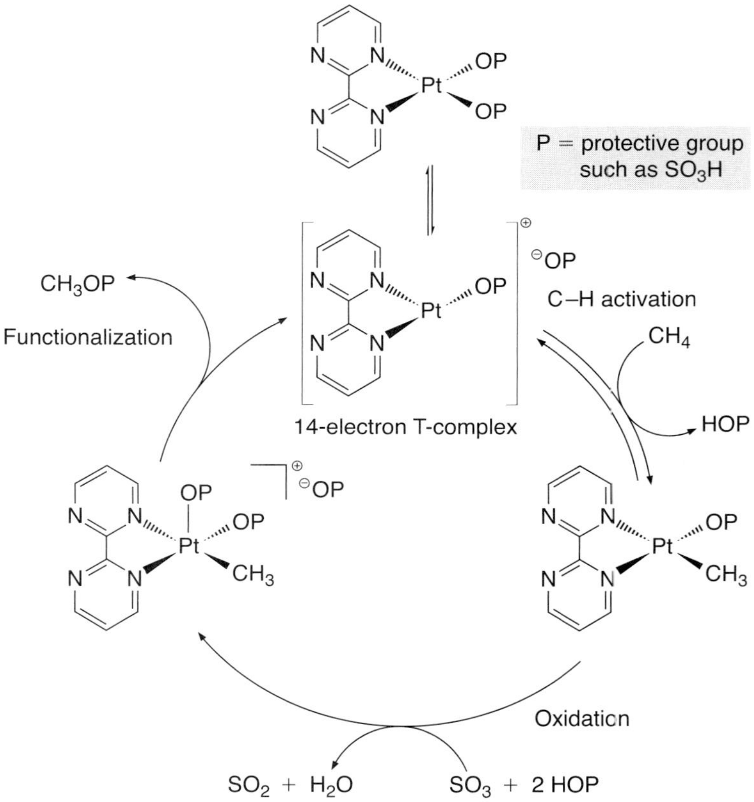

P = protective group
such as SO$_3$H

CH$_3$OP

Functionalization

14-electron T-complex

C–H activation

CH$_4$

HOP

Oxidation

$SO_2 + H_2O$ $SO_3 + 2$ HOP

Scheme 18.2

The proposed mechanism for this methane oxidation is shown in Scheme 18.2 and follows steps discussed in more detail in the next section. One remarkable feature of this platinum system in this highly acidic medium at high temperatures is the thermodynamic stability of platinum(II) bound by the bipyrimidine ligand. Addition of platinum metal to bipyrimidine in sulfuric acid leads to generation of the platinum(II) bipyrimidine complex (Equation 18.8).

$$Pt(0) + bpym + 3 H_2SO_4 \longrightarrow (bpym)Pr(HSO_4)_2 + SO_2 + 2 H_2O \qquad (18.8)$$

Thus, many groups have sought alternative oxidants. A polyoxometallate (POM) has been shown to act as a mediator of oxidation by O_2 (Equation 18.9).[36] In this case, the reaction of methane with O_2 in the presence of Periana's catalyst supported on $H_5PV_2Mo_{10}O_{40}$ as acid and mediator of oxidation has been reported to form a mixture of methanol and acetaldehyde. The mechanism of the formation of the C_2 acetaldehyde product from methane is not firm, but is proposed to occur by oxidative coupling of methane with formaldehyde, which would be generated from methanol. These reactions occur with modest turnover numbers of about 30, but the use of O_2 and a POM is a clear advance over the original Shilov process with platinum(IV) as the stoichiometric oxidant.

$$CH_4 + 1/2 O_2 \xrightarrow[50\ °C]{Catalyst} CH_3OH \xrightarrow[CH_4,O_2]{Catalyst} CH_3CHO$$

Catalyst

(18.9)

Aerobic oxidation of methane catalyzed by
$[Pt(Mebipym)Cl_2]^{\oplus}[H_4PV_2Mo_{10}O_{40}]^{\ominus}/SiO_2$

Products μmol				
CH$_3$OH	HCHO	CH$_3$CHO	Acid (μmol)	TON
3	1	13	None	6
30	12	48	H$_2$SO$_4$ (150)	31
24	19	49	H$_5$PV$_2$Mo$_{10}$O$_{40}$ (60)	32
46	9	41	H$_5$PV$_2$Mo$_{10}$O$_{40}$ (30)	33

18.2.3. Mechanism of the Pt-Catalyzed Oxidations

Many years after these initial results of Shilov and more concurrent with the developments of new catalysts and oxidants by Periana, detailed mechanistic studies have been conducted on this type of alkane functionalization.[37–41] The conclusion from these studies is summarized in Scheme 18.3. In brief, the Shilov oxidations, as well as related oxidations reported by Periana, are generally considered to occur by a C–H activation step in which a Pt(II) alkyl complex is generated from a Pt(II) halide or hydroxide complex, a subsequent step in which the Pt(II) alkyl complex is oxidized to a Pt(IV) alkyl complex by transfer of a halide from Pt(IV) to the Pt(II) alkyl, and a C–X bond-forming step by attack of water, hydroxide, sulfonate, or halide on the resulting Pt(IV) alkyl to generate the alcohol or alkyl halide product and to regenerate the starting Pt(II) catalyst.

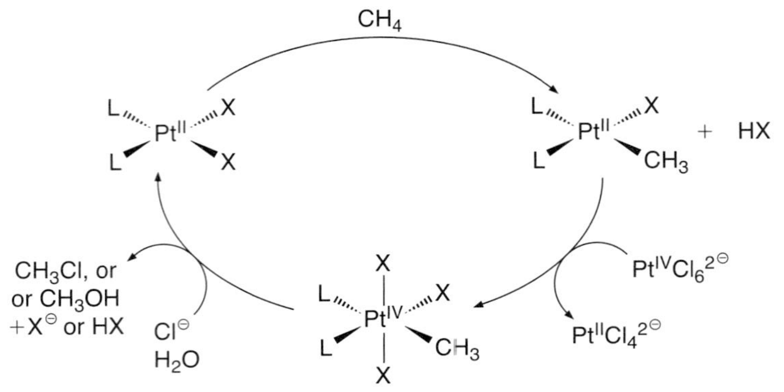

Scheme 18.3

The most extensive studies on the mechanism of this process have focused on how the C–H activation step occurs. Many of these studies were included in Chapter 6 on oxidative addition. Four reactions that occur via cationic Pt(II) intermediates are shown in Equations 18.10–18.13.[40, 42–48] The oxidative addition by these model compounds occurs by exchange of alkane or arene for a neutral ligand, such as trifluoroethanol, water, perfluoropyridine, or THF, in the starting cationic platinum complex, or dissociation of triflate from a neutral platinum complex. This exchange generates an alkane σ-complex. This σ-complex then undergoes oxidative cleavage of the C–H bond to form an alkylplatinum(IV) hydride intermediate that reductively eliminates alkane to form the final complex containing an aryl or labeled methyl group.

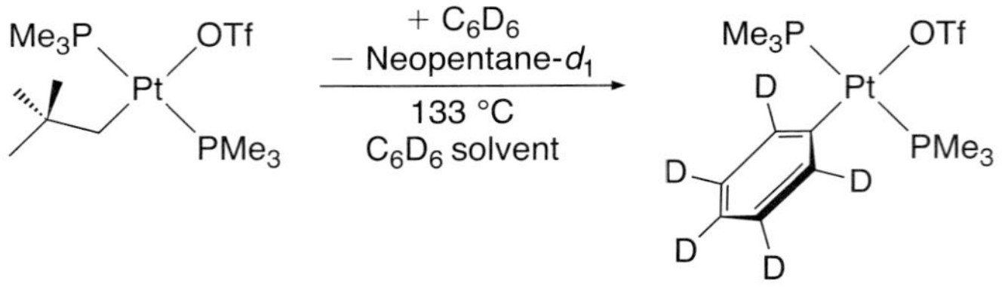

$$(18.10)$$

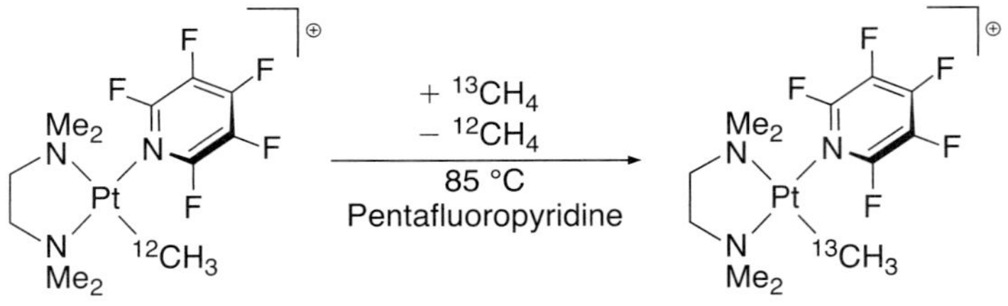

$$(18.11)$$

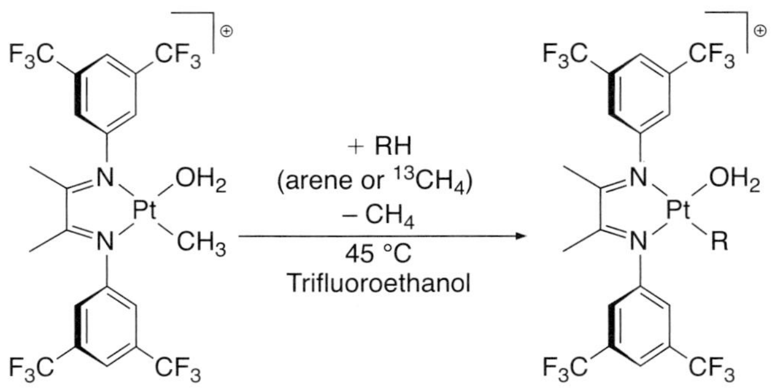

$$(18.12)$$

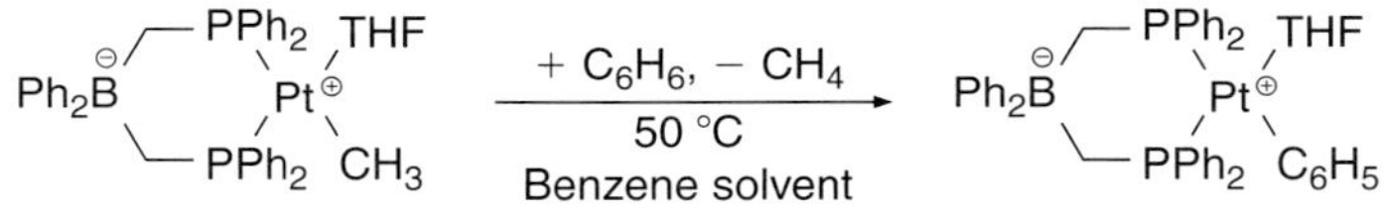

$$(18.13)$$

Much less information has been gained on the oxidation of the Pt(II) alkyl to Pt(IV). In one elegant study, oxidation of the chloroplatinum–methyl complex by ^{195}Pt-labeled Pt(IV) chloride formed a Pt(IV) methyl complex without ^{195}Pt enrichment (Equation 18.14). This result shows that oxidation of the Pt(II) alkyl complexes by Pt(IV) occurs by transfer of chlorides from the Pt(IV) to the platinum alkyl complex, not by transfer of the alkyl group to the Pt(IV) center.[49]

$$[MePtCl_3]^{2-} + [^{195}PtCl_6]^{2-} \xrightarrow{\text{H}_2\text{O}} [MePtCl_5]^{2-} + [^{195}PtCl_4]^{2-} \tag{18.14}$$

via

This finding led to the conclusion that oxidants other than Pt(IV) could be used to oxidize the Pt(II) alkyl species. During Periana's oxidation of methane described in the previous section, SO_3 appears to oxidize the Pt(II) alkyl complex in sulfuric acid to a Pt(IV) alkyl complex[27] that presumably contains $^-OSO_3H$ as ligand from the sulfuric acid and SO_3 oxidant. This oxidation step appears to be turnover limiting in the formation of methyl bisulfate from methane and SO_3. Likewise, the heteropolyacids can oxidize the Pt(II) alkyl to Pt(IV), although the coordination sphere of the resultant product is unknown.[36] Additional studies on the potential use of O_2 as oxidant in this step show that the oxidation of platinum–alkyl complexes in methanol generates platinum(IV) complexes containing hydroxo and alkoxo ligands (Equation 18.15).[50,51] These compounds have not undergone C–O bond-forming reductive elimination to form alcohol or ether products.

$$\tag{18.15}$$

$$\overset{N}{\underset{N}{\bigcap}} = \text{bipy, phen, or tmeda}$$

The mechanism of the final step of C–O bond formation by a formal reductive elimination was included as part of Chapter 11. In particular, two model systems have provided information on the mechanism of this reaction. First, the reaction orders, solvent effects, and electronic effects on the reductive elimination of methyl acetate and methyl aryl ethers from methylplatinum(IV) acetate and phenoxide complexes (Equation 18.16 and Scheme 18.4) indicated that these reductive eliminations occur by backside attack on a platinum methyl.[52,53] Second, a study of the stereochemistry of the attack of water on a Pt(IV) alkyl showed that the formation of alcohol occurred with the inversion of configuration that reflects a backside attack.[54]

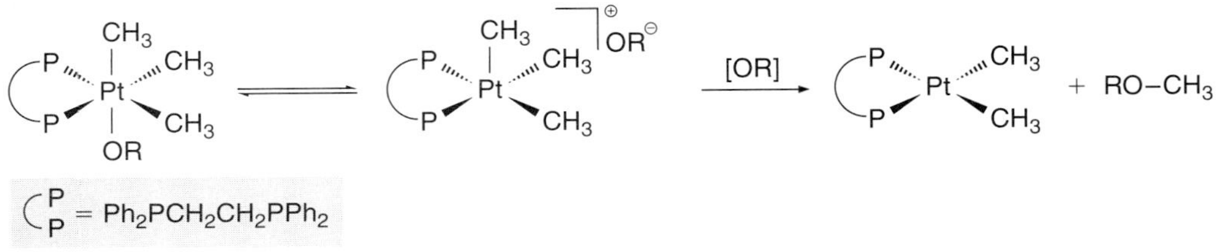

(18.16)

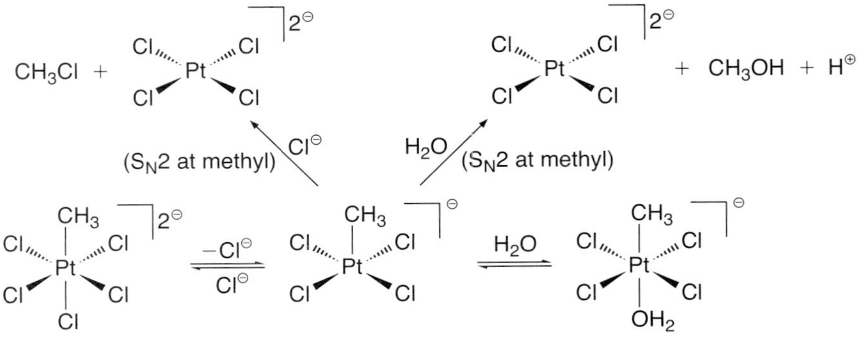

Scheme 18.4

18.3. Directed Oxidations, Aminations, and Halogenations of Alkanes and Arenes [55]

Many oxidative functionalization processes directed by groups attached to an alkyl chain or to an aryl group have been reported. The group on the alkyl or aryl chain is either a functional group or heteroaromatic system that can bind to the transition metal. This ability to bind to the transition metal serves two functions. First, the directing groups trigger C–H bond cleavage by metal centers that typically do not add C–H bonds in the absence of a directing group. Second, they allow for regioselective functionalization of a C–H bond by directing the chemistry to the position ortho to the ligating substituent. Much effort has been spent to develop these types of regioselective functionalization processes in the early twenty-first century. These reactions include acetoxylations, aminations, halogenations, and C–C bond formations. Dozens of papers based on this concept have been published during the writing of this book. A few examples that demonstrate this strategy applied to the functionalization of aliphatic and aromatic C–H bonds are described in this section.

Among the first examples of this directed C–H bond functionalization was the hydroxylation of amino acids reported by Sames. Sames showed that the Shilov system containing platinum(II) and a combination of oxygen and copper as the oxidant leads to the functionalization of the methyl groups of valine (Equation 18.17).[56] Sanford reported a range of oxidations of aryl and alkyl groups directed by a pendant imine (Equations 18.18a and b).[57,58] The readily accessible PhI(OAc)$_2$ is the oxidant in these reactions. A closely related strategy for directed amination of arene C–H bonds is the use of a persulfate oxidant and an amide as the nitrogenous reagent (Equation 18.18c).[59] The amination of alkyl[60] and allyl[61] C–H bonds using a tether to control regioselectivity has been developed by Du Bois. The aliphatic C–H amination occurs by the generation of rhodium–nitrene complexes that were discussed in Chapter 13. This type of amination is unlikely to occur through organometallic intermediates. The amination of allylic C–N bonds likely occurs by nucleophilic attack on an intermediate allyl species (shown in Equation 18.18d) formed by allylic C–H activation.

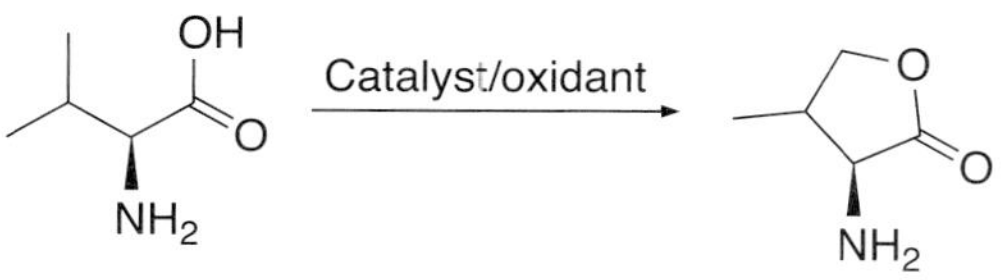

(18.17)

Catalyst/oxidant	Yield	Syn : anti
16% K_2PtCl_4/K_2PtCl_6	21%	5 : 1
10% $K_2PtCl_4/CuCl_2$	67	3 : 1
1% $K_2PtCl_4/5\%CuCl_2$	20	3 : 1

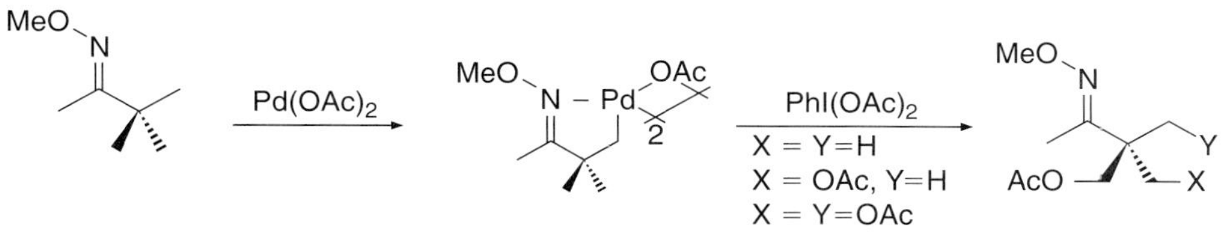

(18.18a)

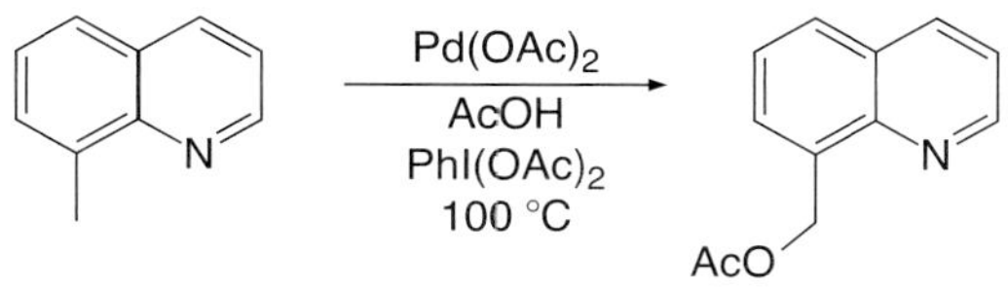

(18.18b)

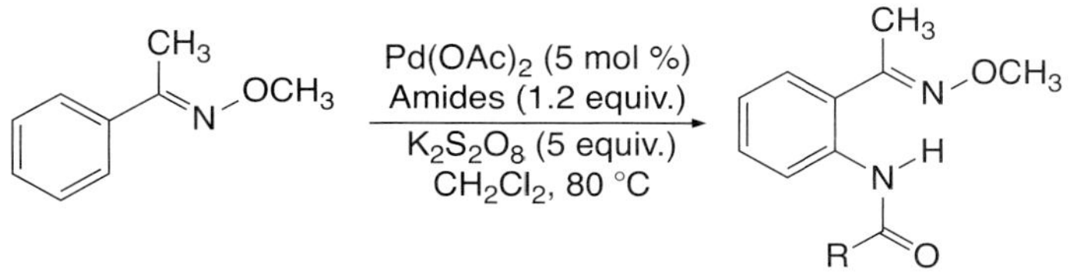

(18.18c)

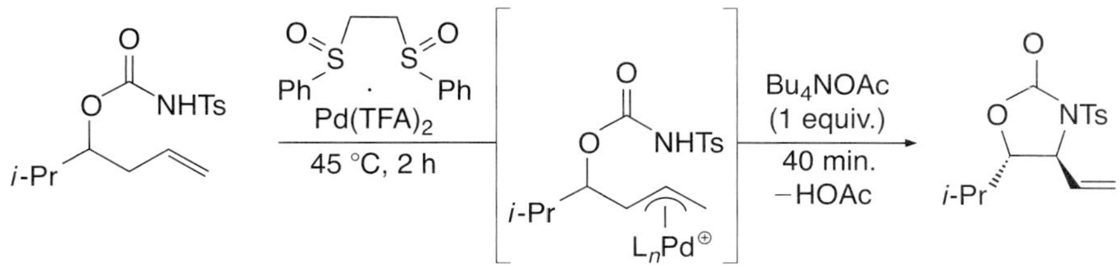

(18.18d)

Related strategies have been successfully applied to the halogenation of aliphatic and aromatic C–H bonds. Yu reported iodination of alkyl chains attached to an oxazoline that binds the palladium catalysts (Equation 18.19a).[62] In this iodination, the combination of PhI(OAc)$_2$ and I$_2$ are used as the reagents to effect iodination and oxidation. The use of directing groups on arenes has also made possible the regioselective halogenation of arenes.[63] As illustrated by the catalyzed and uncatalyzed processes in Equation 18.19b, the palladium-catalyzed process can occur faster than the uncatalyzed process to form a product directed by the catalyst, rather than the electronic properties of the arene. Under microwave conditions, directed fluorination of arene C–H bonds can also occur (Equation 18.19c).[64]

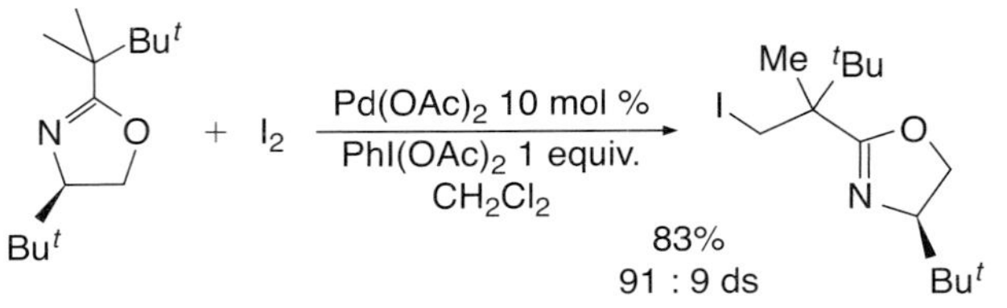

(18.19a)

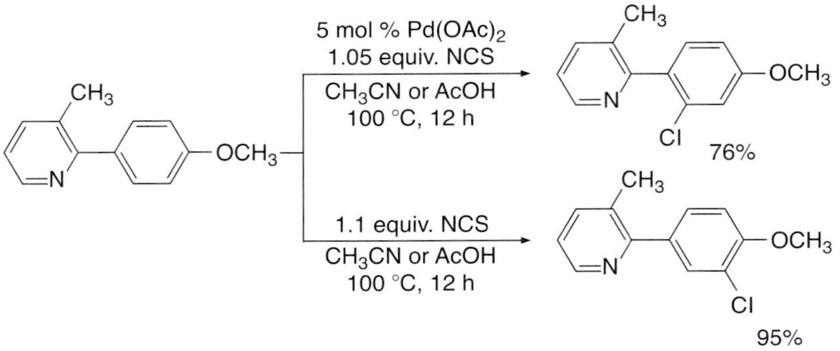

(18.19b)

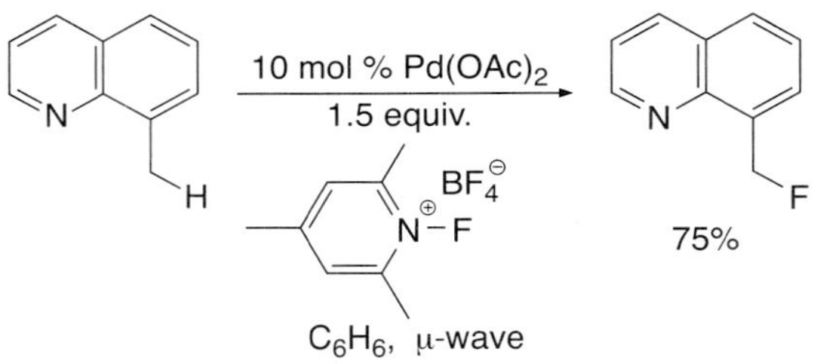

(18.19c)

These oxidative processes have been proposed to occur by formation of an arylpalladium(II) intermediate, followed by oxidation to a palladium(IV) intermediate, although dimeric Pd(III) species have been isolated and are a potential alternative to high-valent palladium intermediates.[65] In the case of the acetoxylation processes, the formation of the arylpalladium intermediate occurs by reaction of the arylpalladium complex with PhI(OAc)$_2$. In a model system, the Pd(IV) intermediate was formed by oxidation with this reagent and shown to undergo reductive elimination to form the Ar–OAc bond (Equation 18.20). Mechanistic data indicate that these reactions occur by direct reductive elimination or a less likely path involving rate-limiting dissociation of a pyridyl group from the chelate ring.[66] Such reductive eliminations to form carbon–heteroatom bonds were discussed in more detail in Chapter 8.

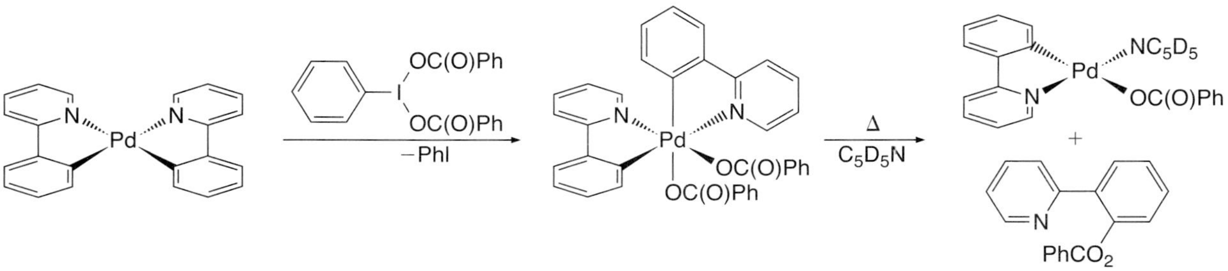

(18.20)

The use of directing groups has also allowed for regioselective coupling of arenes at alkyl and aryl C–H bonds. Daugulis initially reported the arylation of sp^3 C–H bonds (Equation 18.21a).[67] In Daugulis' arylation, an aryl iodide was used as reagent. More recently, approaches to couple two arenes to form biaryls without the need for halogenated reagents have been sought. One approach makes use of the distinct electronic properties of an electron-rich heteroarene and an arene. By this strategy, the electron-rich heteroarene would react with the electrophilic Pd(TFA)$_2$ by an electrophilic aromatic substitution path, while the arene would react by a σ-bond metathesis pathway with the more electron-rich arylpalladium intermediate.[68,69] This selectivity is illustrated in Equation 18.21b.[68] Alternatively, one substrate can be made more reactive by the presence of a directing group, and another more suitable for activation after the first arene has added because of steric effects. An example of this type of process is shown in Equation 18.21c.[70] The larger discussion of work on direct coupling of arenes with aryl halides is included in the Chapter 19.

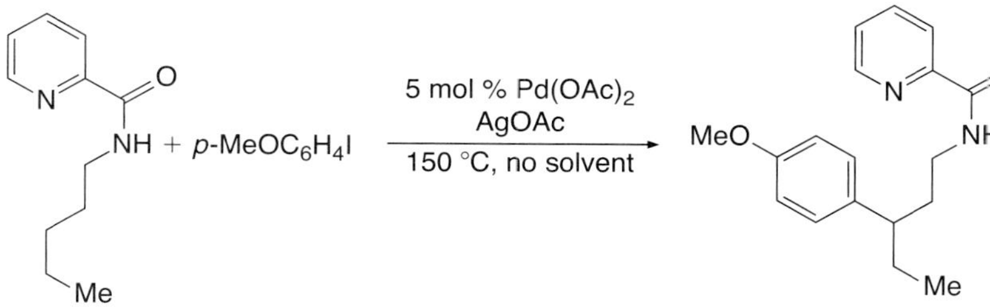

(18.21a)

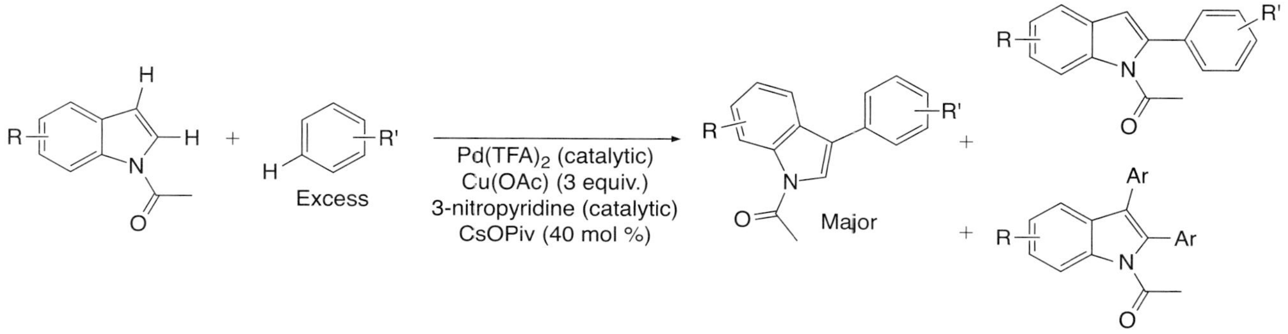

(18.21b)

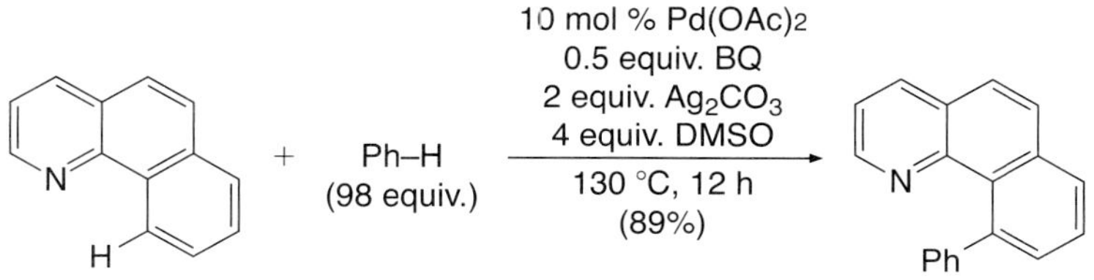

(18.21c)

18.4. Carbonylation of Arenes and Alkanes

18.4.1. Oxidative Carbonylation of Alkanes and Arenes

In parallel with progress on the oxidation of alkanes to alcohols and alkyl halides have been reports on the oxidative carbonylations of alkanes and arenes to form carboxylic acids.[71] Fujiwara showed that stoichiometric amounts of arylpalladium acetates formed from the reaction of Pd(OAc)$_2$ with arenes and that the resulting arylpalladium complex reacts with CO in acetic acid to form aromatic acids (Equation 18.22).[72] When O$_2$, ^{t}BuOOH, alkyl halides, or K$_2$S$_2$O$_8$ were added as oxidant, the reaction became catalytic in palladium, and benzoic acids were generated from benzene, CO, and the oxidant in the presence of palladium acetate (Equation 18.23).[73–76] The highest yields were obtained with K$_2$S$_2$O$_8$ as oxidant.[75,76] Sen has reported related oxidation reactions in acidic media.[77,78]

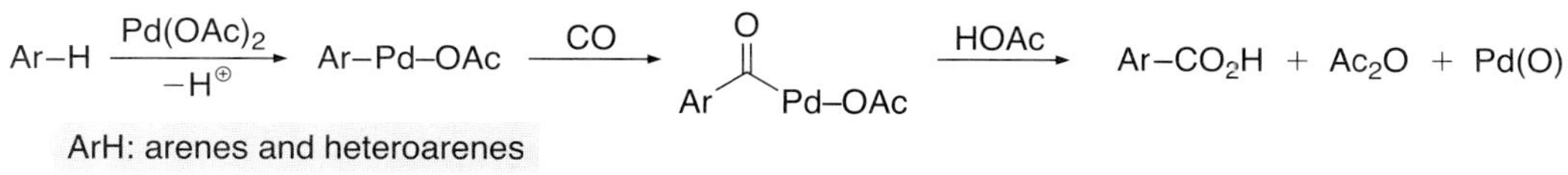

(18.22)

$$\text{C}_6\text{H}_6 \;+\; \text{CO} \quad \xrightarrow[\substack{\text{K}_2\text{S}_2\text{O}_8 \,/\, \text{TFA} \\ \text{RT, 20 h}}]{\text{10\% Pd(OAc)}_2} \quad \text{C}_6\text{H}_5\text{CO}_2\text{H} \qquad (18.23)$$

Yield: ~100%

These oxidative carbonylations also occur with alkanes, although with low conversions of alkane and relatively low turnover numbers. Fujiwara reported that cyclohexane reacts with CO and $\text{K}_2\text{S}_2\text{O}_8$ in the presence of palladium acetate and copper acetate in trifluoroacetic acid to form cyclohexane carboxylic acid with about 20 turnovers and about 4% yield based on alkane (Equation 18.24).[79,80] Fujiwara also reported the carboxylation of methane with $\text{K}_2\text{S}_2\text{O}_8$ as oxidant with V(O)(acac)_2 as catalyst (Equation 18.25).[81] This reaction occurred in 93% yield based on methane and with 18 turnovers. Sen reported a palladium(II)-catalyzed oxidative carbonylation of methane with hydrogen peroxide as the oxidant.[82] Subsequently, he showed that RhCl_3 catalyzes the conversion of methane, CO, and oxygen to acetic acid at 100 °C in water (Equation 18.26).[77] Periana has reported a somewhat related transformation of methane to acetic acid, although the reaction is conducted in the absence of CO, and both carbon atoms of the acetic acid arise from methane (Equation 18.27).[82] In this case, the CO appears to arise from oxidation of the methane, as shown in Scheme 18.5.

$$\text{C}_6\text{H}_{12} \;+\; \text{CO} \quad \xrightarrow[\substack{\text{K}_2\text{S}_2\text{O}_8/\text{TFA} \\ \text{80 °C, 20 h}}]{\text{Pd(OAc)}_2/\text{Cu(OAc)}_2} \quad \text{C}_6\text{H}_{11}\text{CO}_2\text{H} \qquad (18.24)$$

Yield: 4.3%
TON: 19.8

$$\text{CH}_4 \;+\; \text{CO} \quad \xrightarrow[\text{TFA, 80 °C, 20 h}]{\text{VO(acac)}_2/\text{K}_2\text{S}_2\text{O}_8} \quad \text{CH}_3\text{CO}_2\text{H} \qquad (18.25)$$

Yield: 93%
TON: 18

$$\text{CH}_4 \;+\; \text{CO} \;+\; \tfrac{1}{2}\text{O}_2 \quad \xrightarrow[\substack{\text{H}_2\text{O} \\ \text{100 °C}}]{\text{RhCl}_3} \quad \text{CH}_3\text{CO}_2\text{H} \qquad (18.26)$$

$$2\,\text{CH}_4 \;+\; 4\,\text{H}_2\text{SO}_4 \quad \xrightarrow{\text{Pd(II)}} \quad \text{CH}_3\text{CO}_2\text{H} \;+\; 4\,\text{SO}_2 \;+\; 6\,\text{H}_2\text{O} \qquad (18.27)$$

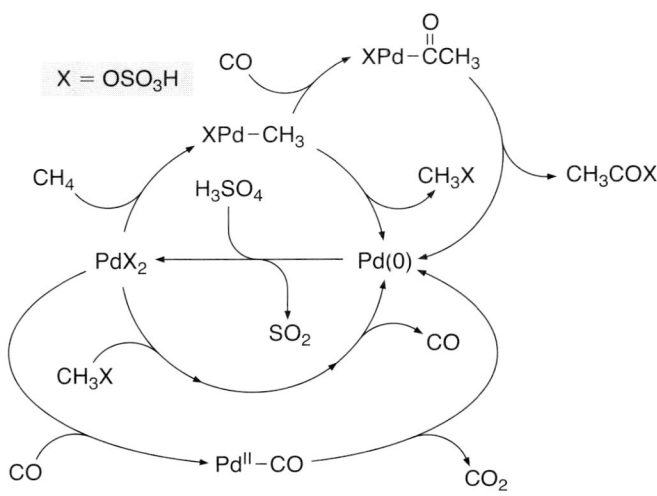

Scheme 18.5

18.4.2. Alkylative Carbonylation of Alkanes and Arenes

In parallel with the directed hydroarylation of olefins, a series of papers described the formation of ketones from heteroarenes, carbon monoxide, and an alkene. Moore first reported the reaction of CO and ethylene with pyridine at the position α to nitrogen to form a ketone (Equation 18.28).[83] Related reactions at the less-hindered C–H bond in the 4-position of an N-benzyl imidazole were also reported (Equation 18.29).[84,85] Reaction of CO and ethylene to form a ketone at the ortho C–H bond of a 2-arylpyridine[86] or an N-Bu^t aromatic aldimine[87] has also been reported (Equations 18.30 and 18.31). Reaction at an sp^3 C–H bond of an N-2-pyridylpiperazine results in both alkylative carbonylation and dehydrogenation of the piperazine to form an α,β-unsaturated ketone (Equation 18.32).[88] The proposed mechanism of the alkylative carbonylation reaction is shown in Scheme 18.6. This process is believed to occur by oxidative addition of the C–H bond, insertion of olefin into the metal–hydride bond, insertion of CO into the metal–heteroaryl or metal–alkyl linkage, and reductive elimination to form the new C–C bond in the product.

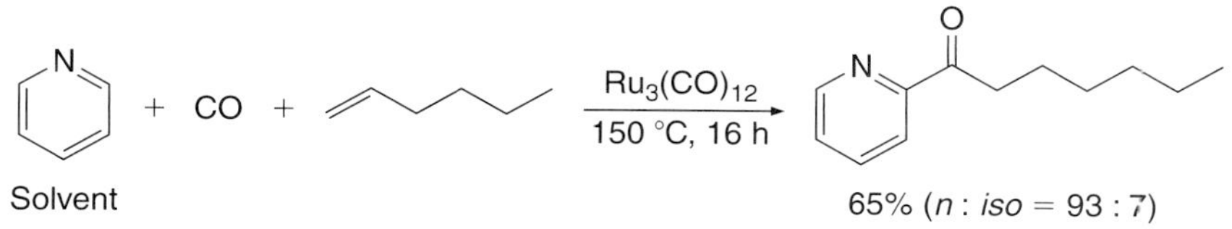

(18.28)

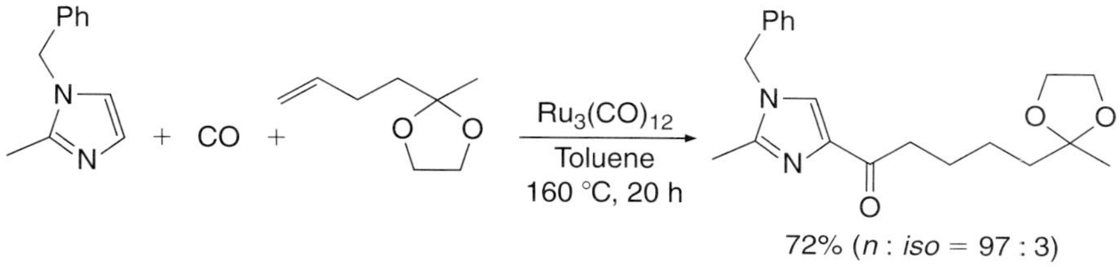

(18.29)

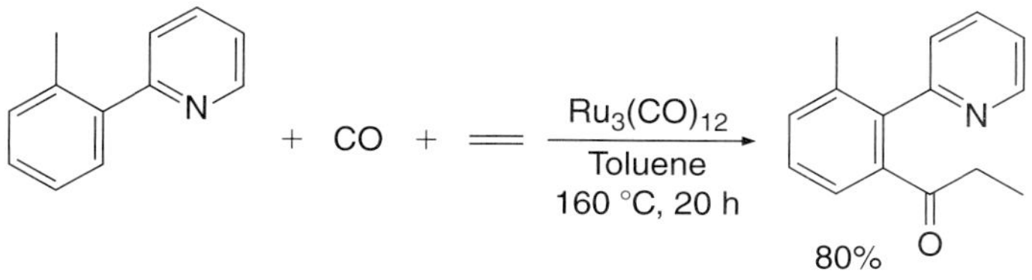

(18.30)

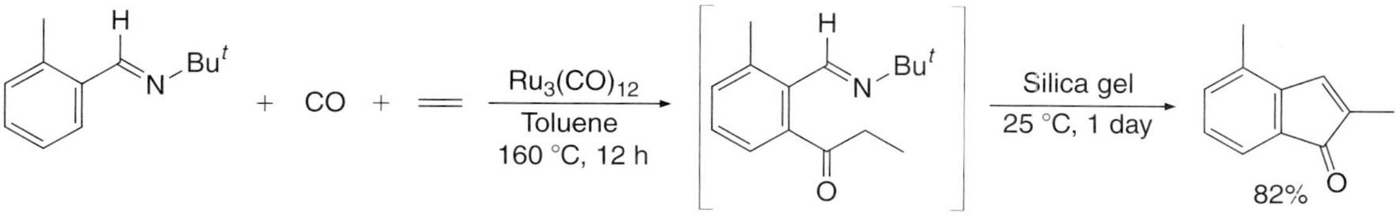

(18.31)

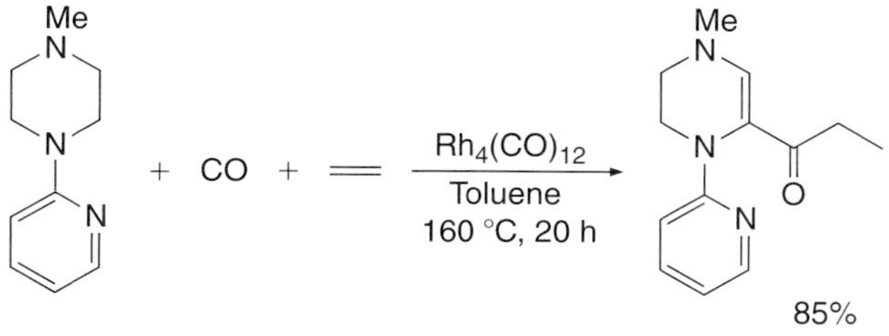

(18.32)

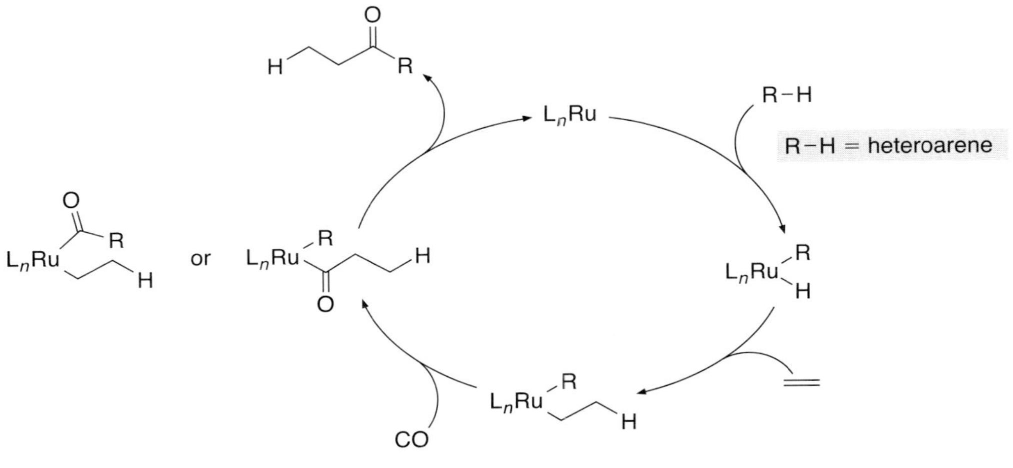

Scheme 18.6

18.4.3. Direct Carbonylation to Aldehydes

In contrast to oxidative carbonylation of alkanes and arenes to form carboxylic acids, the simple carbonylation of alkanes and arenes to form aldehydes is endothermic.[11] Nevertheless, this process can generate measurable concentrations of aromatic and aliphatic aldehydes by photochemical processes. The reactions of [Rh(CO)(PMe₃)₂Me] with alkanes and CO under photochemical conditions (Equation 18.33a) generate linear aldehydes as the major carbonylation product.[89–92] However, these aldehydes undergo Norrish type II photochemistry (Equation 18.33b), and a series of organic products, such as alkenes of one shorter carbon chain, are formed.[93] The carbonylation of arenes gives a higher yield of aldehyde product. However, this carbonylation process has been calculated to be endothermic by 1.7 kcal/mol.[11] These thermodynamics prevent these carbonylation reactions from producing high yields of aldehyde product under thermal conditions.

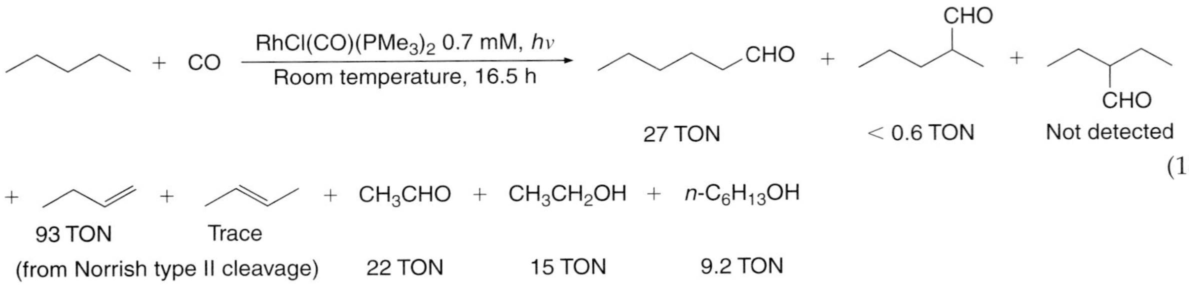

$$(18.33a)$$

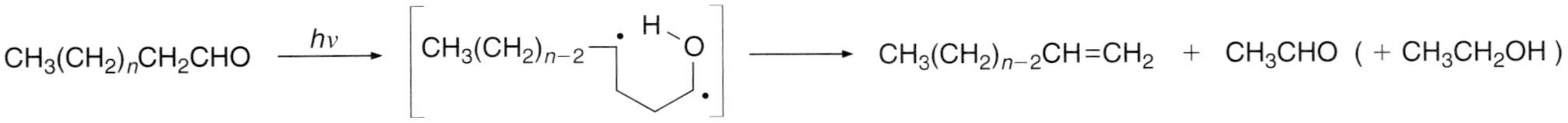

$$(18.33b)$$

The C–N multiple bond of an isocyanide is weaker than the C–O multiple bond in CO.[94] Thus, the insertion of an isocyanide into a C–H bond in a reaction that is analogous to the carbonylation of C–H bonds should be more favorable thermodynamically. An example of this reaction catalyzed by [RhCl(CO)(PMe₃)₂] under photochemical conditions is shown in Equation 18.34.[95] A more efficient, albeit more complex, reaction of an isocyanide involves intramolecular insertion into the *o*-methyl group in 2,6-dimethylphenylisocyanide, which leads to an indole product in high yield (Equation 18.35).[96,97]

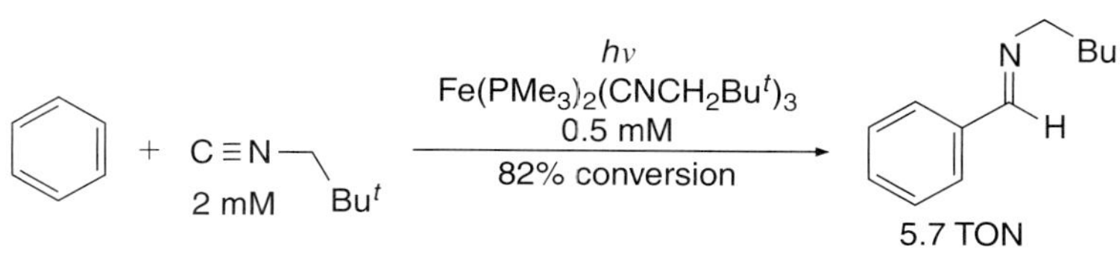

$$(18.34)$$

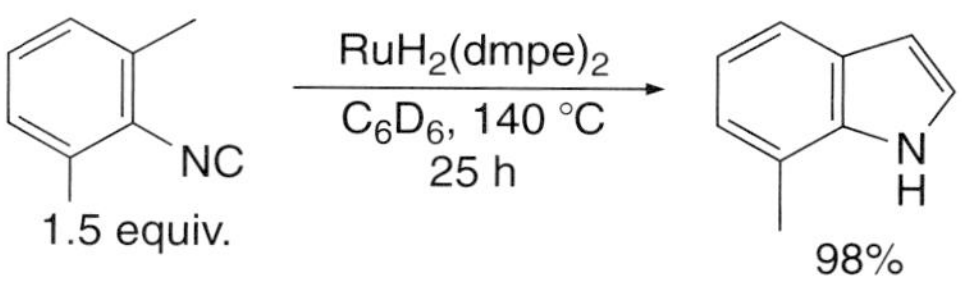

$$(18.35)$$

The mechanism of the photochemical carbonylation catalyzed by [Rh(CO)(PMe$_3$)$_2$Cl] is complex (Scheme 18.7).[92] Two pathways through a common intermediate formed by a common photochemical step appear to occur in parallel. One pathway appears to proceed by two photochemical events and leads to the linear aldehyde. A second pathway appears to occur by thermal steps after the first photochemical event. This first photochemical process is proposed to occur by the reaction of [Rh(CO)(PMe$_3$)$_2$Cl] with alkane to form an alkyl hydride intermediate by photochemical dissociation of CO, addition of alkane, and reassociation of CO. This step parallels the known reaction of arenes with the [Rh(PMe$_3$)$_2$Cl] fragment[98] generated by photodissociation of CO from [Rh(CO)(PMe$_3$)$_2$Cl].[92] The proposed alkyl hydride intermediate [Rh(CO)(PMe$_3$)$_2$Cl(alkyl)(H)] in the alkane carbonylation process was detected by NMR spectroscopy[92] and appeared to have a lifetime long enough to undergo a second photochemical process that would lead to the linear aldehyde. A radical trap inhibited the formation of the branched aldehyde, indicating that this aldehyde was formed by a second mechanism, perhaps by homolysis of the rhodium–carbon bond of a secondary alkyl intermediate that would form in parallel with the primary alkyl intermediate.

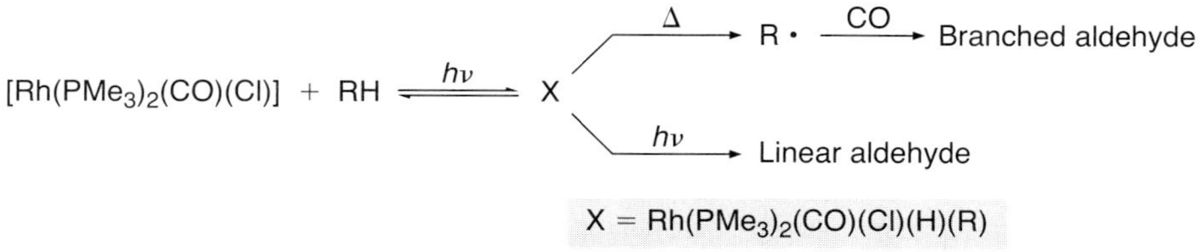

Scheme 18.7

18.5. Dehydrogenation

18.5.1. Early Studies

In the late 1970s, several groups focused on homogeneous dehydrogenation of alkanes. Crabtree showed that the stoichiometric reaction of cyclopentane with [IrH$_2$(acetone)$_2$ (PPh$_3$)$_2$]BF$_4$ generated the cyclopentadienyl complex [CpIrH(PPh$_3$)$_2$]BF$_4$ and that reaction with cyclooctane formed the cyclooctadiene complex [Ir(COD)(PPh$_3$)$_2$]BF$_4$ (Equations 18.36 and 18.37).[99–102] Felkin[103–105] reported the first homogeneous catalytic dehydrogenations (Equations 18.38 and 18.39), which were conducted with *tert*-butylethylene as the hydrogen acceptor. Felkin and co-workers showed, although with low turnover numbers, that linear alkanes generate α-olefins as the kinetic product (Equation 18.40).[105]

$$\text{(18.36)}$$

$$\text{(18.37)}$$

+ *tert*-Butylethane
71 vs. 29 (internal isomers)

$$\text{(18.38)}$$

+ *tert*-Butylethane
99 (+ other internal isomers) vs. 1 external

$$\text{(18.39)}$$

90% selectivity
but 0.3 TON

$$\text{(18.40)}$$

Crabtree and Felkin then developed several catalysts for the catalytic transfer dehydrogenation of cyclooctane to cyclooctene with a substantial number of turnovers (Equation 18.41).[102,104–106] These dehydrogenation reactions were conducted with the iridium complexes $[Ir(\kappa^2\text{-}O_2CR^1)_2(PR^2_3)_2(H_2)]^+$ ($R^1 = CF_3$ or C_2F_5, $R^2 = Cy$, Ph, or $p\text{-}C_6H_4CF_3$), $[Ir(i\text{-}Pr_3P)_2H_5]$, and $\{Ir[(p\text{-}FC_6H_4)_3P]_2H_5\}$, and with the ruthenium complex $\{Ru[(p\text{-}FC_6H_4)_3P]_3H_4\}$. Several other groups reported various catalysts for this type of thermal transfer dehydrogenation over the course of the subsequent decade.[107–109] Saito and co-workers and Crabtree and co-workers also showed that the dehydrogenation can be run in an open system.[106,110,111] In this case, the reaction occurs under non-equilibrium conditions because hydrogen is extruded, and this endothermic process can be run to accumulate a measurable concentration of olefins.

$$\text{(18.41)}$$

Crabtree: Catalyst = $[Ir(PR_3)_2(\kappa^2\text{-}O_2CC_2F_5)_2H_2]^+$ R = Cy or $C_6H_4CF_3$
35 turnovers with acceptor; 35 turnovers without acceptor in open reflux
Felkin: Catalyst = $(i\text{-}Pr_3P)_2IrH_5$, $[(p\text{-}FC_6H_4)_3P]_2IrH_5$, or $[(p\text{-}FC_6H_4)_3P]_3RuH_4$
45–70 turnovers with acceptor

18.5.2. Dehydrogenation Catalyzed by Complexes of Pincer Ligands

Because these dehydrogenation reactions occur at high temperatures, catalyst stability limits turnover numbers. The development of so-called pincer ligands, which are tridentate anionic ligands that give rise to thermally stable complexes, has led to the identification of complexes that dehydrogenate alkanes with remarkable turnover numbers (Equation 18.42).[112,113] Transfer dehydrations in which the hydrogen of the alkane is transferred to *tert*-butyl ethylene have been reported with these catalysts,[114] as have dehydrogenations in an open system to extrude hydrogen under non-equilibrium conditions.[115] The highest turnover numbers are observed for the transfer dehydrogenation of cyclododecane to cyclododecenes and concomitant reduction of *tert*-butylethylene to *tert*-butylethane (Equation 18.42). Lower turnover numbers are observed for "acceptorless dehydrogenation." The dehydrogenation of linear alkanes to α-olefins would be a more valuable process synthetically than the dehydrogenation of cyclooctanes. The linear α-olefin is formed as the kinetic product, but these olefins are converted to the isomeric internal olefins at high conversions with the current catalysts (Equation 18.43).[116]

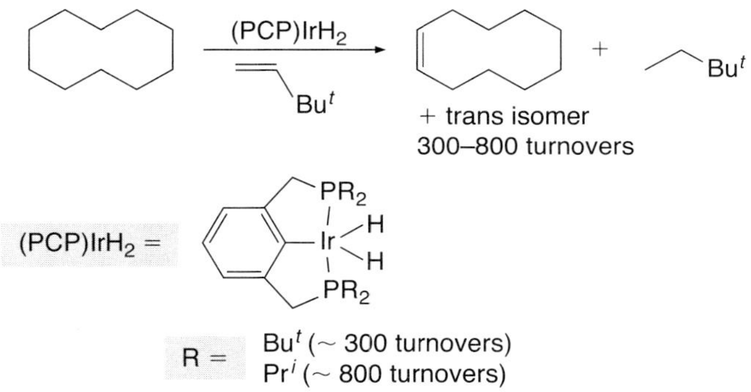

(18.42)

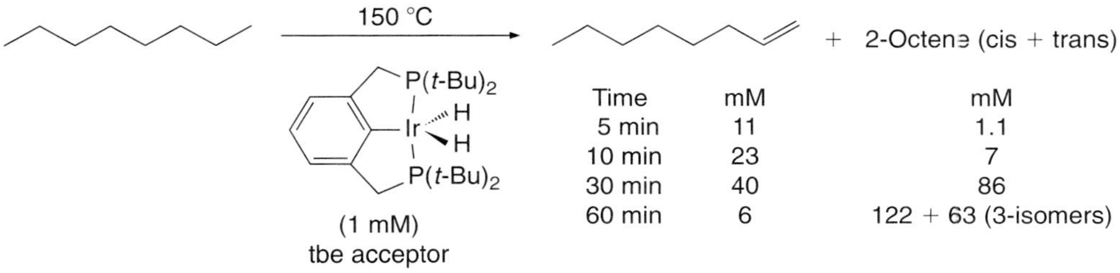

(18.43)

The most active catalysts for these dehydrogenations are the iridium complexes in Figure 18.1 containing a "PCP pincer" ligand in which the carbon donor is an aryl group and the two phosphorus donors are di-*tert*-butylphosphinomethyl groups bound to the 2- and 6-positions of the aryl ring or containing a "POCOP pincer" ligand in which Bu^t_2PO groups are bound at the 2- and 6-positions of the aryl ring. The most active of the phosphine pincer catalysts contains a methoxy group in the para position of the aryl group. The most active of these catalysts currently for the transfer dehydrogenation of cyclooctane is the "POCOP" catalyst.[117–119] As described in more detail below, the higher activity of this catalyst versus that containing phosphine pincer ligands results from changes in the resting state and rate of hydrogen transfer, and these differences result from the reduced electron density at the metal.

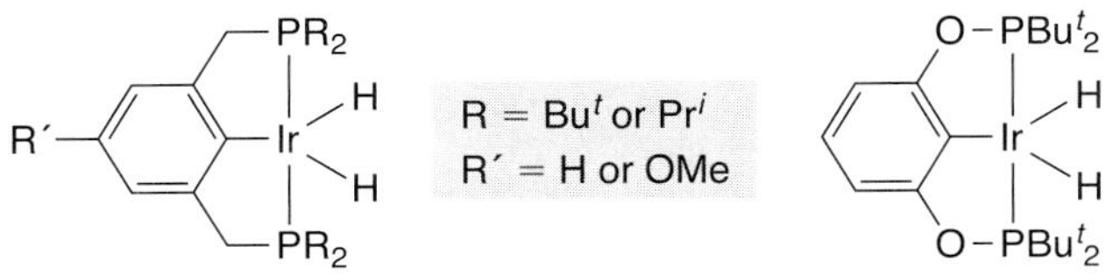

Figure 18.1.
Two catalysts for alkane dehydrogenation.

These pincer iridium complexes also have been applied to the dehydrogenation of polyolefins.[120] Although the activities for dehydrogenation of polymers are not as high as those for the dehydrogenation of low-molecular-weight alkanes, they have been shown to catalyze the dehydrogenation of octane–ethylene copolymers to generate side chains containing olefin functionality (Equation 18.44). In addition, the dehydrogenation chemistry has been applied to the dehydrogenation of alkylamines to generate imines (Equation 18.45)[121] and alkyl ethers to generate vinyl ethers.[114] Although the mechanism of this reaction is not understood in detail, it is likely that the reaction occurs by activation of the N–H or α-C–H bond, and not by dehydrogenation of an alkyl chain and isomerization to the imine.[121]

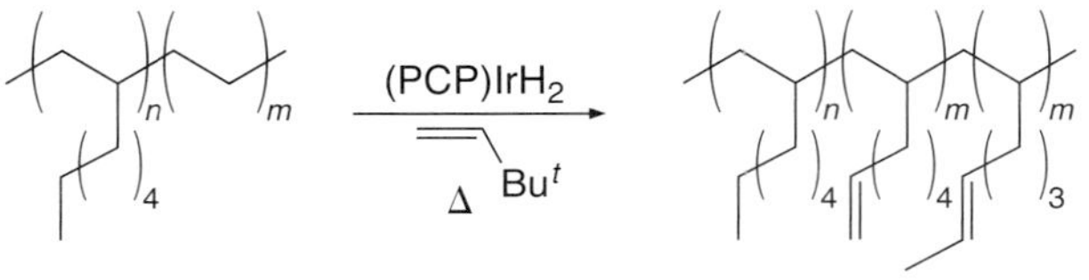

(18.44)

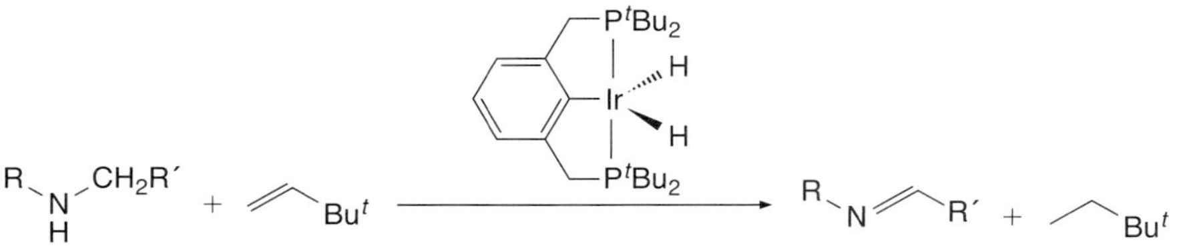

(18.45)

18.5.3. Alkane Metathesis via Dehydrogenation

Alkane metathesis is the formal cleavage of alkane C–C bonds and the re-formation of these bonds to form alkanes of different chain lengths (Equation 18.46), much like alkene metathesis (Chapter 21) cleaves and reforms C=C bonds to form new alkenes. In principle, this metathesis can form medium-weight alkanes from lighter and heavier alkanes or can form light and heavy alkanes from medium-weight alkanes. This alkane metathesis could ultimately be used, along with distillation, to convert the distribution of alkanes produced by a Fischer–Tropsch process into a set of alkanes possessing a narrower molecular weight distribution. The metathesis of alkanes by organometallic tantalum complexes was reported first, and remarkably active and selective homogeneous systems based on the combination of iridium and molybdenum complexes have been reported subsequently.

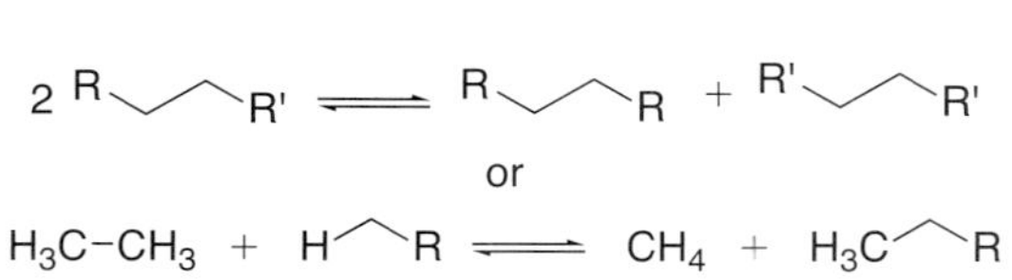

(18.46)

Scheme 18.8 shows the results of Basset, which involve two types of metathesis catalyzed by a tantalum hydride supported on silica.[122,123] In one case, an alkane forms methane and propane; in the other case, methane and an alkane combines to form ethane and a shorter alkane. In practice, the reaction of ethane (Scheme 18.8, top) has been shown to form methane and propane, and the reverse reaction of propane (Scheme 18.8, bottom) with methane has been reported to form two equivalents of ethane.

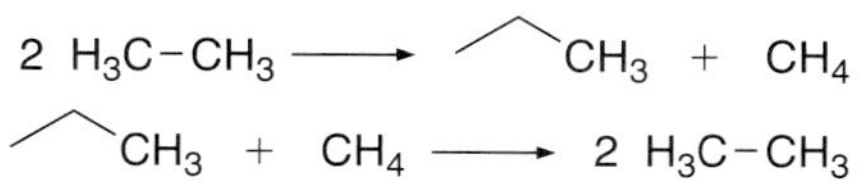

$$2 \ H_3C-CH_3 \longrightarrow \diagup\!\!\diagdown CH_3 \ + \ CH_4$$

$$\diagup\!\!\diagdown CH_3 \ + \ CH_4 \longrightarrow 2 \ H_3C-CH_3$$

Scheme 18.8

Although one can envision that alkane metathesis could occur by oxidative addition of an alkane C–C bond or σ-bond metathesis to cleave an alkane C–C bond, these stoichiometric alkane, metatheses reactions are unknown. Instead, these alkane metatheses reactions have been shown to occur via C–H bond cleavage by the supported Ta–H to form H_2 and a tantalum alkyl complex that undergoes β-hydrogen elimination to form an alkene in some cases and α-hydrogen elimination in others to form a tantalum alkylidene.[124] This combination of an alkylidene and an alkene leads to metathesis of the alkene. Hydrogenation of the resulting alkenes by the extruded hydrogen then forms the final alkanes. This network of reactions, as depicted by the authors, is shown in Scheme 18.9.

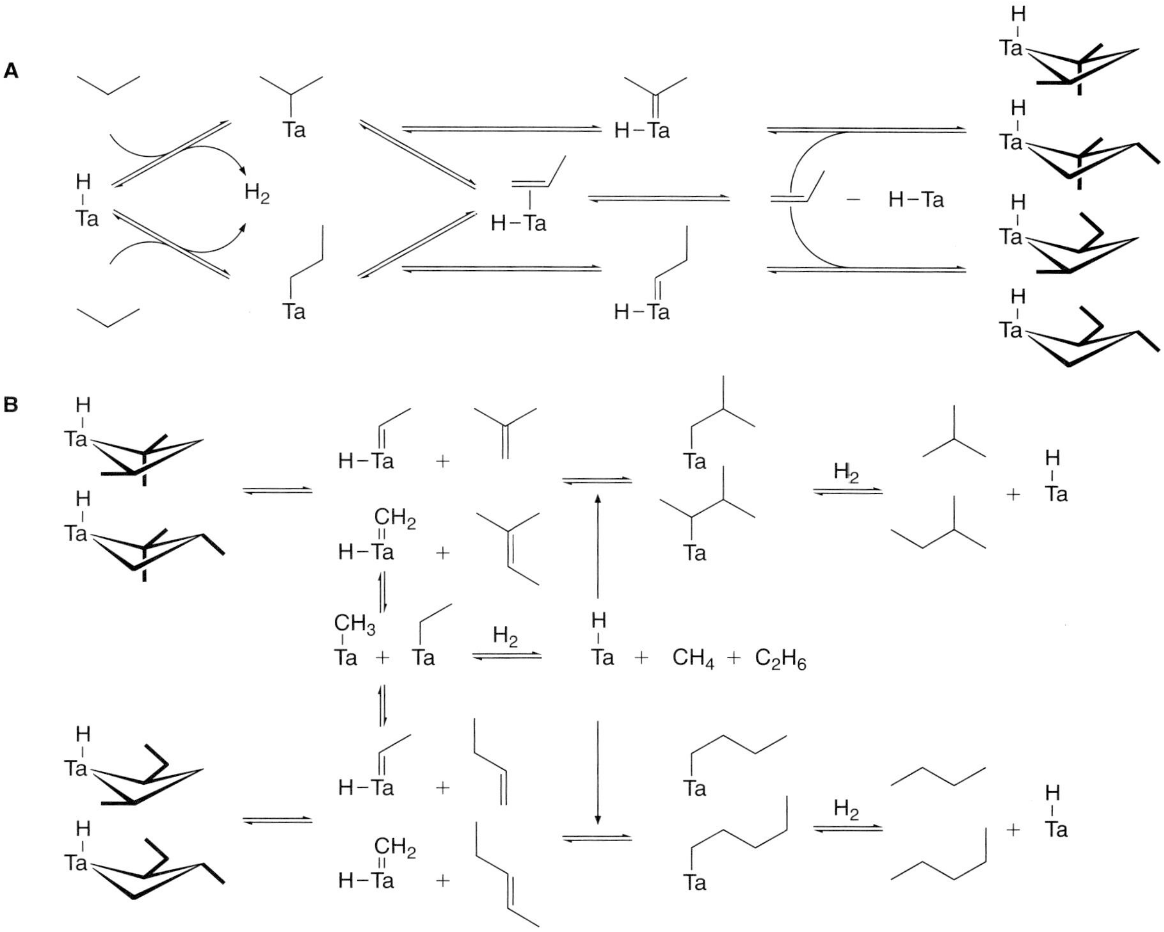

Scheme 18.9

More recently, the pincer iridium catalysts for alkane dehydrogenation, in combination with Schrock's molybdenum catalyst for olefin metathesis (Chapter 21), has been shown to catalyze alkane metathesis (Equation 18.47).[125] This combination of catalysts provides hundreds of turnovers and is relatively selective for generating ethane and the corresponding higher molecular weight alkane. For example, hexane is converted to ethane and decane as the major products. Equation 18.48 shows the distribution of products at various times from the metathesis of hexane catalyzed by the (PCP)IrH$_2$ and molybdenum carbene complexes shown. Alkanes other than ethane and decane are formed from isomerization of the initially formed α-olefin to an internal olefin and metathesis of the internal olefin.

$$(18.47)$$

Temperature (°C)	Time	Product concentration (mM)													Total product (M)
		C$_2$	C$_3$	C$_4$	C$_5$	C$_7$	C$_8$	C$_9$	C$_{10}$	C$_{11}$	C$_{12}$	C$_{13}$	C$_{14}$	C$_{15}$	
125	23 hours	(131)	176	127	306	155	37	49	232	18	4	4	10	2	1.25
	46 hours	(189)	255	193	399	208	61	81	343	31	9	9	22	7	1.81

$$(18.48)$$

18.5.4. Mechanism of Dehydrogenation

The mechanism of dehydrogenation involves the fundamental steps discussed in previous chapters of this book. The reaction mechanism includes oxidative addition of the C–H bond of the alkane, followed by β-hydrogen elimination of the resulting alkyl complex to generate the alkene product. Dissociation of the alkene and either reductive elimination of H$_2$ or transfer of the hydrogen to a hydrogen acceptor regenerates the species that adds the alkane (Scheme 18.8).

The mechanism of these reactions has been studied in detail by Goldman for the pincer ligand based on the xylene backbone[126, 127] and by Brookhart for the pincer ligand based on the resorcinol backbone.[117, 119, 128] These reactions have also been studied intensively by computational methods.[129, 130] The most revealing studies include entropy terms in the calculations of energies and include a treatment of the steric effects on the pincer ligands.[130]

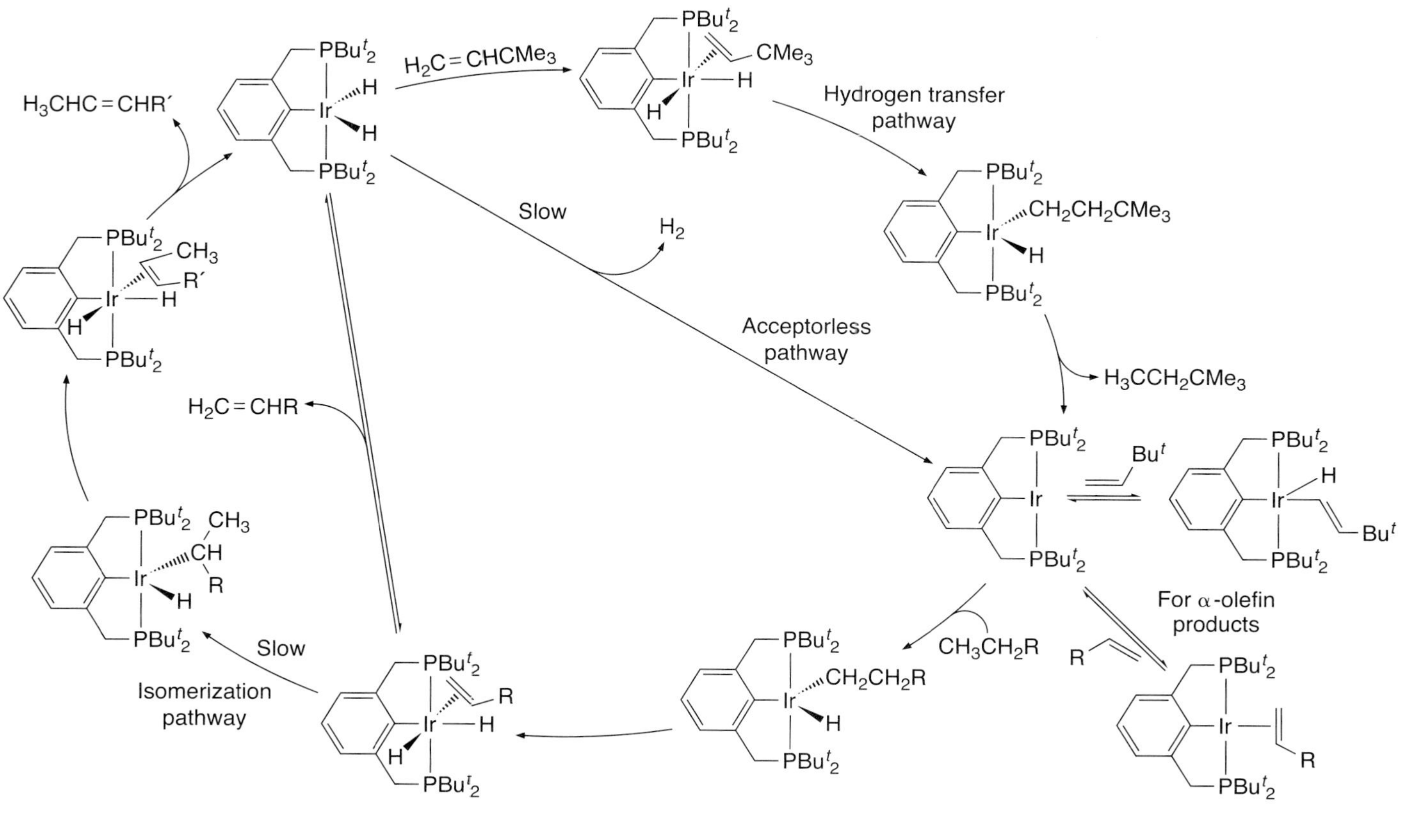

Scheme 18.10

These experimental and computational studies indicate that the reaction catalysed by the (PCP)Ir system follows the pathway summarized in Scheme 18.10.[113] First, the iridium(III) dihydride complex reacts with the *tert*-butylethylene hydrogen acceptor to generate an alkyl hydride complex that reductively eliminates alkane and generates the (PCP)Ir(I) complex. This Ir(I) complex undergoes several reactions with olefins and alkanes. It adds the vinylic C–H bond of *tert*-butylethylene to generate an alkenyl hydride complex that lies off of the cycle and is the most stable species in the catalytic dehydrogenation of cyclooctene. To re-enter the cycle, the alkenyl hydride complex reductively eliminates *tert*-butylethylene. The (PCP)Ir(I) complex also binds alkenes to form Ir(I) olefin complexes that lie off of the catalytic cycle. Finally, the (PCP)Ir(I) complex undergoes oxidative addition of the C–H bond of the reactant alkane in a productive step of the cycle. The resulting alkyl hydride complex undergoes β-hydrogen elimination and dissociation of alkene to regenerate the Ir(III) dihydride complex.

Differences in electron density at the metal in the PCP and POCCP catalysts lead to two significant differences in the relative energies of species in the catalytic system and faster rates for dehydrogenation of cyclooctene by the POCOP catalyst.[131] First, an Ir(III) vinyl hydride complex formed from addition of *tert*-butylethylene (tbe) is the resting state of the (PCP)Ir catalyst, whereas an η^2-olefin complex of cyclooctene is the resting state of the (POCOP)Ir catalyst. Second, the step in which the hydrides are transferred to the olefin hydrogen acceptor to generate *tert*-butylethane is faster for the POCOP complex. This faster rate occurs because the reduced electron density at the metal accelerates the rate of reductive elimination of alkane from the alkyl hydride intermediate formed by insertion of olefin into the metal hydride.

The same cycle is followed during the reactions of linear alkanes to form linear alkenes. Although the thermodynamics for dehydrogenation of cyclooctene are more favorable than those for the dehydrogenation of linear alkanes, primary C–H bonds typically undergo oxidative addition faster than secondary C–H bonds, as discussed in Chapter 6. Thus, linear alkanes react faster than cyclic alkanes. However, the accumulation of α-olefin inhibits the catalytic process. An η^2-olefin complex formed from the α-olefin becomes the resting state of the catalytic cycle for reactions catalyzed by the POCOP system, instead of the vinyl hydride complex that is the resting state of the PCP system. The accumulation of the olefin complex that lies off the cycle leads to a lower concentration of the iridium complexes within the cycle and slower reactions as the concentration of α-olefin product increases.

18.6. Hydroarylation

18.6.1. Directed Hydroarylation of Olefins

18.6.1.1. Overview

In 1993, Murai reported the reactions of aryl ketones with ethylene and vinylsilanes to form the product from the addition of the C–H bond ortho to the carbonyl group to the olefin (Equation 18.49).[132] This finding led to subsequent work on the addition of the C–H bonds of a variety of aromatic groups to a series of olefins. Catalysts that react under milder conditions than the original ones and extensions of the C–H activation to intramolecular cyclizations to form heterocyclic structures have made this reaction capable of being used in the synthesis of complex molecules. For example, alkylated diterpenoids[133, 134] and (+)-lithospermic acid (Equations 18.50 and 18.51) have been synthesized using directed hydroarylation.[135]

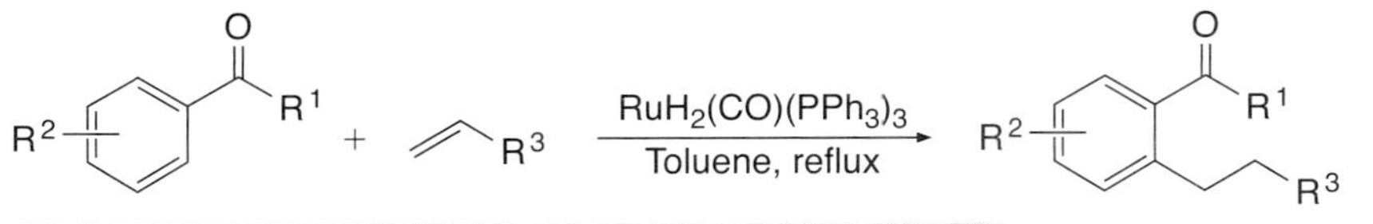

$$R^1 = \text{Me or Bu}^t; \quad R^2 = \text{Me}, \quad R^3 = \text{Si(OEt)}_3, \text{SiMe}_3, \text{Me or Ar}$$

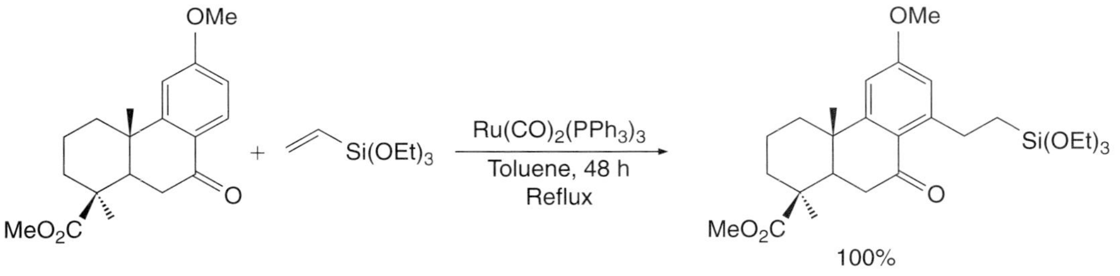

(18.49)

(18.50)

$$(18.51)$$

(+)-Lithospermic acid

18.6.1.2. Reaction Scope and Catalysts

The directed hydroarylation of olefins has developed into a reaction with broad scope. The directed hydroarylation has been conducted at the C–H bonds of aromatic ketones, aromatic imines (Equation 18.52),[136] 2-aryl pyridines (Equation 18.53),[137] and five-membered heterocycles, such as benzimidazoles (Equation 18.54).[138,139] Initially, the reactions required olefins, such as ethylene and vinylsilanes, that lack allylic hydrogens and cannot undergo isomerization. However, many intermolecular reactions of alkenes that could undergo isomerization now occur at the ortho C–H bonds of aryl imines without isomerization (Equation 18.55).[140,141] In many cases, dialkylation of an aryl ketone at the two aromatic C–H bonds is observed. Thus, substrates containing one ortho substituent on the aryl ring have also been used. These substrates can simply contain an ortho alkyl group or can be a bicyclic system like that of α-tetralone (Equation 18.56). This issue of selectivity is also eliminated by conducting intramolecular reactions, as illustrated by the reactions summarized in Equation 18.57.[142] Intramolecular hydroarylation has also been conducted with a rhodium catalyst containing an enantioenriched chiral phosphoramidite ligand to form the cyclization products with high enantiomeric excess (Equation 18.58).[143]

$$(18.52)$$

$$(18.53)$$

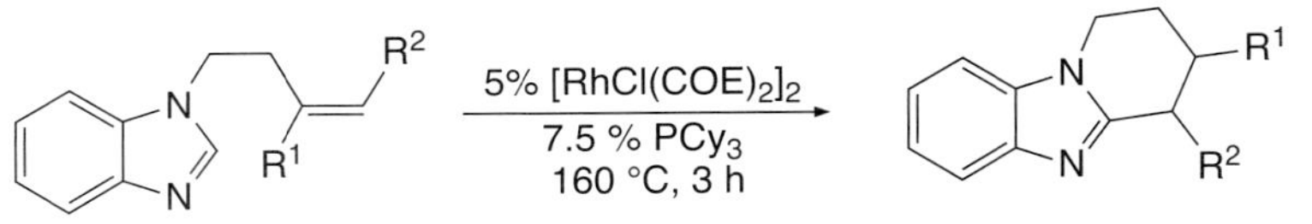

$$(18.54)$$

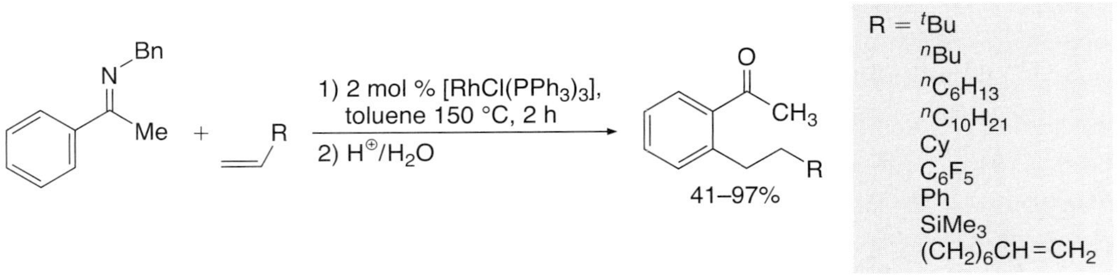

R = ^{t}Bu
^{n}Bu
^{n}C$_6$H$_{13}$
^{n}C$_{10}$H$_{21}$
Cy
C$_6$F$_5$
Ph
SiMe$_3$
(CH$_2$)$_6$CH=CH$_2$

$$(18.55)$$

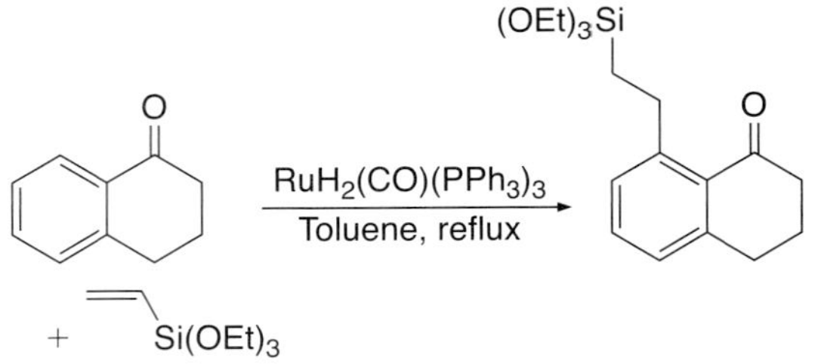

$$(18.56)$$

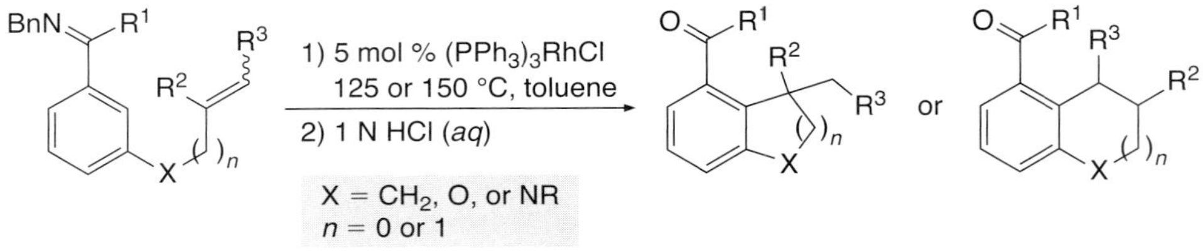

X = CH$_2$, O, or NR
n = 0 or 1

$$(18.57)$$

X = CH$_2$, or O
R = Me, SiMe$_2$Ph, or Ph

70–95% ee
90–95% yield

L =

R′ = N(i-Pr)$_2$
R = N[CH(CH$_3$)Ph]$_2$ (S, R, R)

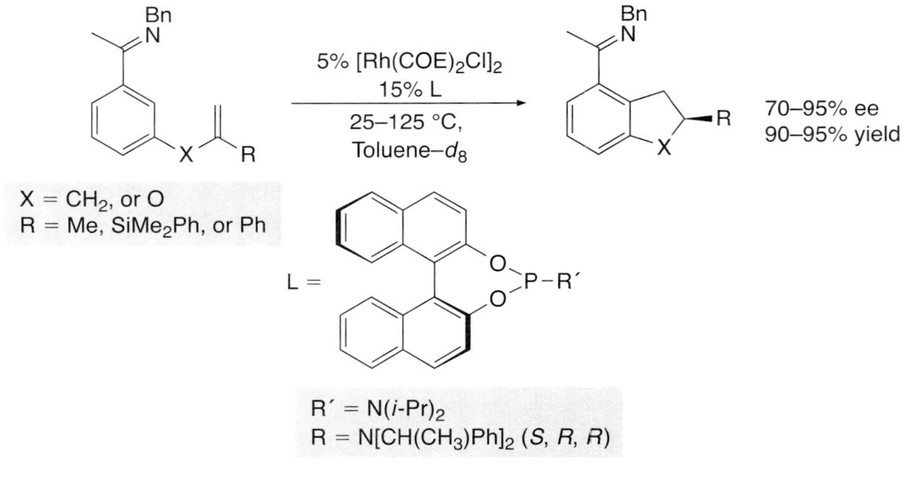

$$(18.58)$$

A variety of catalysts have been used for these hydroarylation reactions of the C–H bonds in aromatic ketones, imines, and 2-arylpyridines. The intermolecular additions of aromatic C–H bonds of aryl ketones to olefins reported by Murai were conducted with RuH$_2$(CO) (PPh$_3$)$_3$ as catalyst.[144, 145] [RuH$_2$(H$_2$)(PCy$_3$)$_2$] was shown subsequently to catalyze this process at room temperature[146] and even the much different Rh(I) complex [Cp*Rh(C$_2$H$_3$SiMe$_3$)$_2$] catalyzes this reaction.[147] Additions of the C–H bonds to N-Bu^t and N-Bn benzaldimines were conducted with $Ru_3(CO)_{12}$[136] or with [Rh(PPh$_3$)$_3$Cl] as catalyst,[142,148] and additions to 2-aryl pyridines were conducted with [RhCl(COE)]$_2$ and PCy$_3$ as catalyst.

18.6.1.3. Mechanisms of Directed Hydroarylation of Olefins

The mechanism of the directed hydroarylation has been studied by experiment[147,149] and theory.[150,151] The mechanism for the ruthenium-catalyzed reactions (Scheme 18.11) involves initial oxidative addition of the ortho C–H bond. This C–H activation step appears by labeling studies to be reversible.[147,149] This oxidative addition is then followed by insertion of olefin to generate an alkyl aryl intermediate. Reductive elimination of this alkyl aryl intermediate would form the carbon–carbon bond in the final product. The C–C bond-forming step is proposed to be accelerated by the presence of the electron-withdrawing group on the ortho-carbon. Computational studies have suggested that the reductive elimination resembles a migration of the alkyl group to the π-system of the arene, followed by elimination of the metal to re-aromatize the ring (Equation 18.59). This pathway was calculated to have a lower barrier than the more conventional one involving C–C bond-forming reductive elimination between two ligands bound through M–C σ-bonds.[150,151]

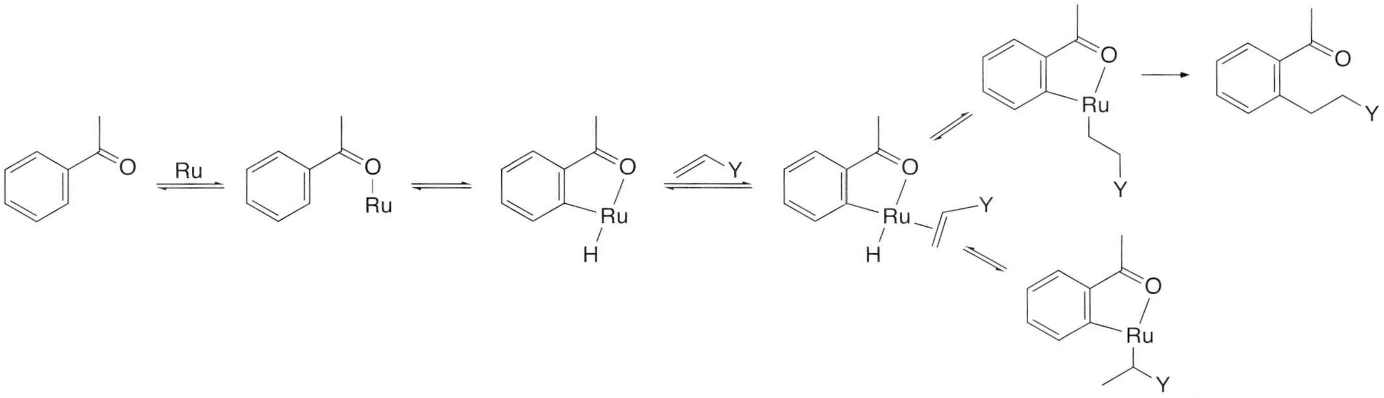

Scheme 18.11

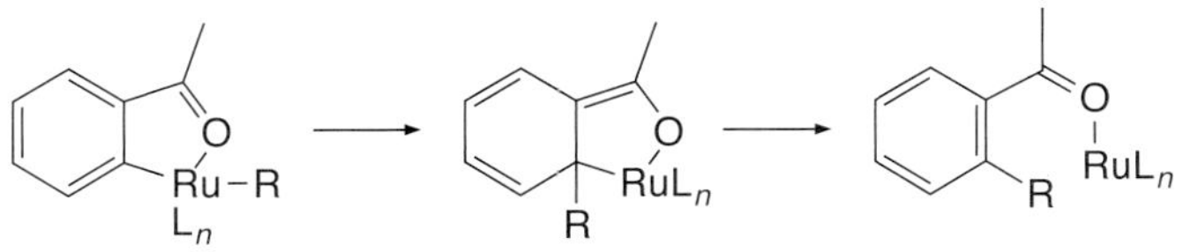

(18.59)

Other pathways are available for hydroarylation of heterocycles. A mechanistic study by Bergman and Ellman was conducted on intramolecular reactions of imidazole-type heterocycles.[139] These mechanistic studies have shown that the C–H activation of the imidazole is followed by isomerization to generate an N-heterocyclic carbene ligand (Equation 18.60). Following this isomerization, the olefin appears to couple with the carbene by a [2+2] process to generate the carbon–carbon bond, followed by proton transfer to generate a rhodium hydride and reductive elimination to form the C–H bond.

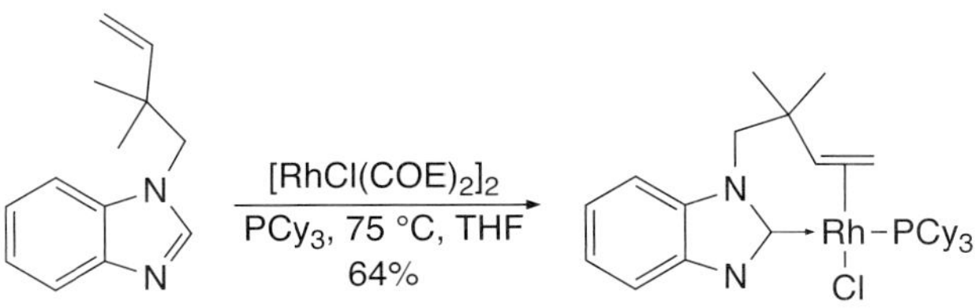

(18.60)

18.6.2. Directed Hydroarylation of Alkynes

Many examples of the intermolecular and intramolecular hydroarylation of alkynes have been reported.[152] Echavarrien has reported a series of gold-catalyzed hydroarylations of alkynes,[153] and Fujiwara has reported palladium-catalyzed hydroarylations of alkynes (Equations 18.61 and 18.62).[154] Although mechanistic data are not comprehensive on any system, most data imply that these reactions are more closely related to an electrophilic aromatic substitution than to a mechanism in which oxidative addition of an arene is followed by insertion of an alkyne into a metal hydride bond.[155] The electrophilic aromatic substitution pathway would occur by initial coordination of an electrophilic metal complex to the alkyne in strong acid medium to generate an intermediate that has accumulated positive charge on one carbon of the coordinated alkyne. After electrophilic addition to the arene, C–H bond cleavage and re-aromatization would occur.

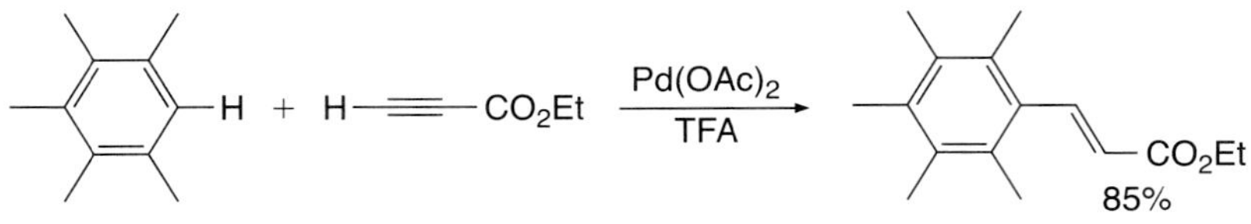

(18.61)

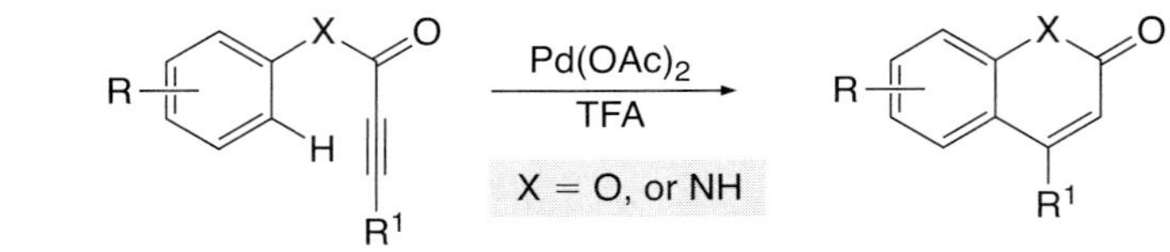

(18.62)

18.6.3. Undirected Hydroarylation and Oxidative Arylation of Olefins

A series of arylations of olefins by C–H bond cleavage without direction by an ortho functional group has also been reported, and these reactions can be divided into two sets. In one case, the C–H bond of an arene adds across an olefin to form an alkylarene product. This reaction has been called hydroarylation. In a second case, oxidative coupling of an arene with an olefin has been reported. This reaction forms an aryl-substituted olefin as product, and has been called an oxidative arylation of olefins. The first reaction forms the same types of products that are formed from Friedel–Crafts reactions, but with selectivity controlled by the metal catalyst. For example, the metal-catalyzed process can form products enriched in the isomer resulting from anti-Markovnikov addition, or it could form the products from Markovnikov addition with control of absolute stereochemistry. Examples of hydroarylation and oxidative arylation of olefins are shown in Equations 18.63[156,157] and 18.64.[154]

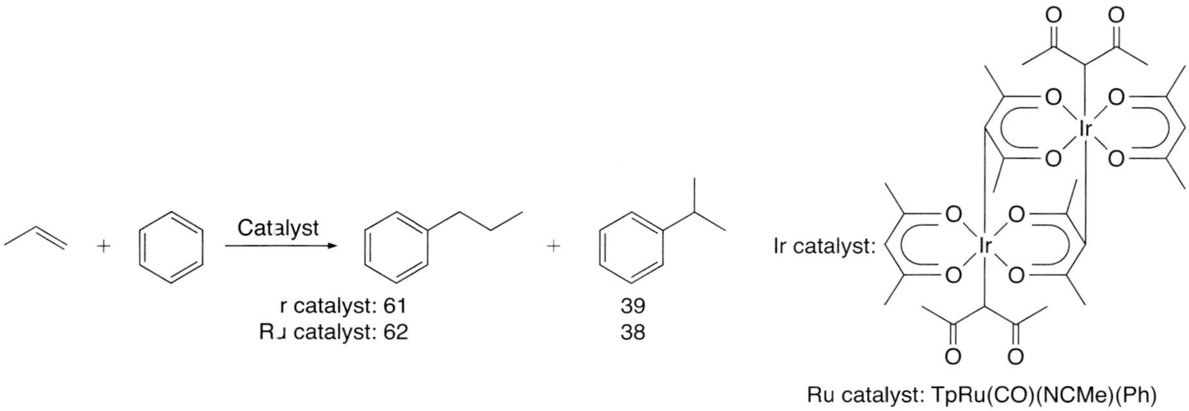

(18.63)

Ir catalyst: 61   39
Ru catalyst: 62   38

Ru catalyst: TpRu(CO)(NCMe)(Ph)

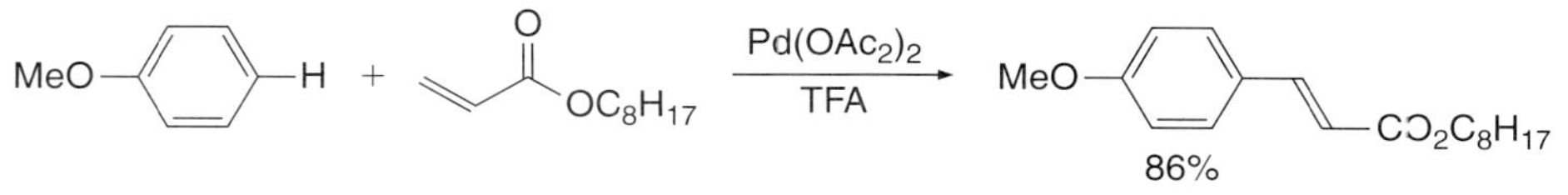

$$MeO-\langle\rangle-H \;+\; \overset{O}{\underset{OC_8H_{17}}{\diagdown}} \quad \xrightarrow[\text{TFA}]{\text{Pd(OAc}_2)_2} \quad MeO-\langle\rangle-\diagdown\diagdown CO_2C_8H_{17} \qquad (18.64)$$

86%

Progress on the addition of aromatic C–H bonds to olefins has been made by Matsumoto and Periana with iridium catalysts[156,158,159] and Gunnoe with ruthenium catalysts.[157,160] Both systems illustrate that the anti-Markovnikov addition products can be generated in larger quantities than the Markovnikov products,[156–158] although mixtures of regioisomers are still observed. Intramolecular additions of the C–H bonds of electron-rich heterocycles to electron-deficient alkenes have also been reported (Equation 18.65).[161] Most recently, Tilley has reported the addition of the C–H bond of methane across an olefin catalyzed by scandocene complexes.[162] This reaction occurs, albeit slowly, with Markovnikov regiochemistry.

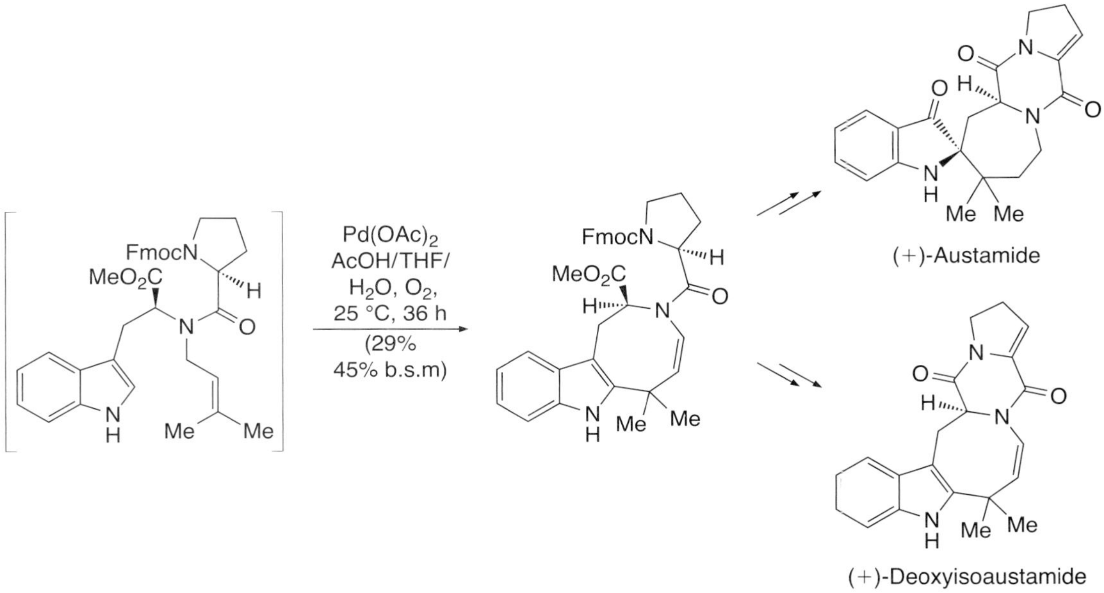

(18.65)

(+)-Austamide

(+)-Deoxyisoaustamide

The intermolecular oxidative arylation of olefins has been reported in most cases with acrylic acid derivatives. This process could be developed as an alternative to the Heck reaction, which occurs with aryl halides. Several groups have reported versions of this oxidative C–C bond formation. Fujiwara reported intermolecular examples of this reaction catalyzed by palladium and copper (Equation 18.64). Intermolecular versions of this reaction have also been reported with ruthenium catalysts and O_2 as the oxidant.[163] Other oxidative reactions in which electron-rich arenes add to olefins (Equation 18.66) have been reported as stoichiometric steps of natural products syntheses,[164,165] and later as a catalytic process.[157]

The mechanism of olefin hydroarylation has been studied by both experimental and theoretical methods.[160,166–168] These results are summarized in Scheme 18.12 for the ruthenium system, but similar results have been obtained on the iridium system. Experimental studies have shown that a metal–aryl complex reacts with the olefin to generate a β-arylalkyl intermediate. Computational studies by the two groups have suggested that this alkyl intermediate reacts with arenes by a pathway that lies in between the

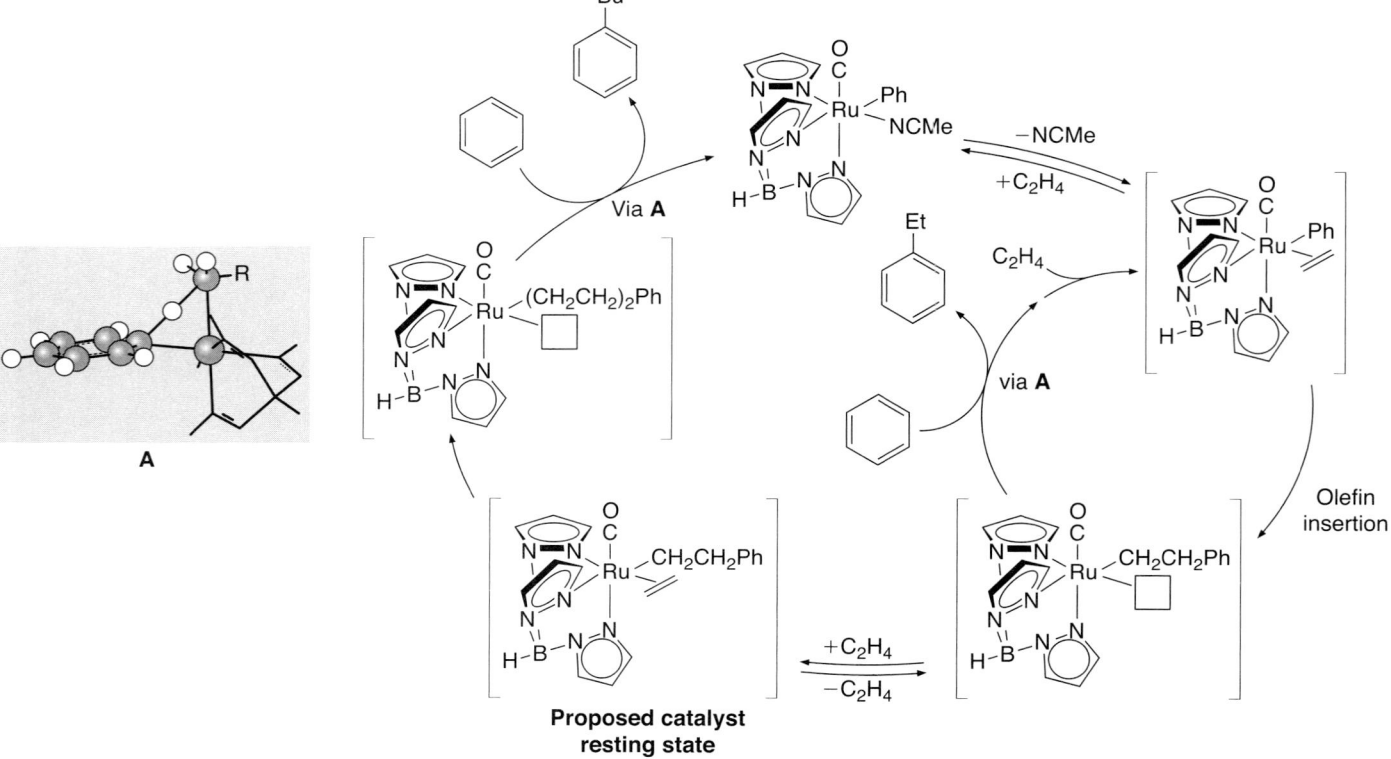

(18.66)

sequence of oxidative addition and reductive elimination and a pathway that involves σ-bond metathesis. This step of Scheme 18.12 has been called an oxidative hydrogen migration.[166–168]

Scheme 18.12

18.7. Functionalization of Alkanes and Arenes with Main Group Reagents

18.7.1. Borylation of Alkanes

The reactions of alkanes with borane reagents occurs in the presence of late transition metal catalysts to form products from the replacement of a terminal C–H bond by a boryl group, and these are the first reactions to form products with a functional group in the terminal position with such high selectivity. In 1999 and 2000, Hartwig and co-workers reported that alkanes reacted with diborane(4) reagents to generate alkyl boronate esters.[14,16] The first reactions with alkanes occurred under photochemical conditions with a rhenium catalyst,[14] but the reactions of boranes with boron reagents were later shown to occur under thermal conditions in the presence of a rhodium catalyst (Scheme 18.13).[16] The reactions of a series of alkanes and arenes with bis(pinacolato) diborane(4) formed the corresponding 1-alkylboronate esters catalyzed by $Cp*Rh(C_6Me_6)$ in high yields, with up to 144 turnovers.

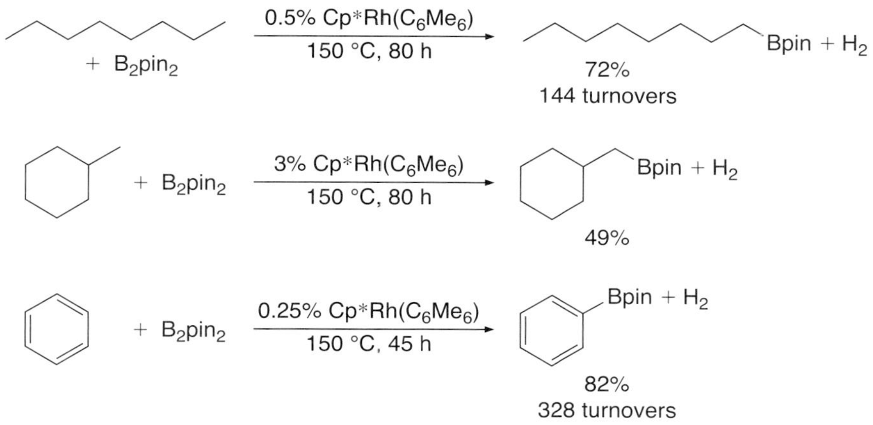

Scheme 18.13

This rhodium-catalyzed borylation of alkanes exclusively formed products from functionalization of the terminal C–H bonds. This selectivity had not been observed earlier with any catalyst. Even the oxidation of alkanes by enzymes typically occurs with variable selectivity for the "omega" (terminal) and omega-1 positions of an alkane.[169] Reactions of branched alkanes containing two different types of methyl groups occurred at the less hindered of the two methyl groups. Moreover, reactions of substrates containing heteroatoms occur at the least hindered methyl group, not at the C–H bond α to the heteroatom.[170] For example, the reactions of a pinacol acetal, a *tert*-butyl ether, a fluoroalkane, and a trialkylamine all occured at the least-hindered methyl group (Equation 18.67).

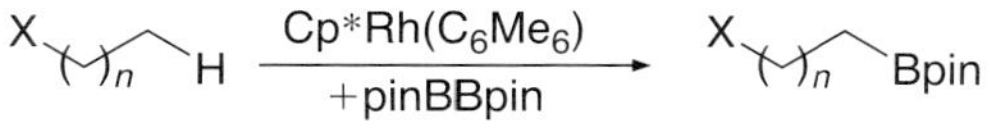

pin = Pinacolate
X = RO, RC(OR)$_2$, F, CF$_3$(CF$_2$)$_m$, or R$_2$N

(18.67)

The products from these reactions are common reagents in organic chemistry. The boronate esters can be oxidized to the corresponding alcohols,[171–173] oxidized with periodate to the boronic acid,[174] converted to the difluoroalkylborates,[168] or used in palladium-catalyzed cross-coupling reactions.[175,176]

18.7. 2. Borylation of Arenes

The borylation of arenes provides direct access to arylboronate esters that can be used in a number of subsequent metal-catalyzed processes. The first example of such a reaction is shown in Equation 18.68.[15] Although the initial system underwent only a few turnovers over long times, two subsequent types of catalyst are particularly effective for these reactions. The combination of [Ir(arene)(Bpin)$_3$] and diphenylphosphinoethane catalyzes the borylation of arenes with B$_2$pin$_2$ as reagent at the elevated temperature of 150 °C (Equation 18.69).[17] In contrast, the combination of [Ir(COD)(X)]$_2$ (X = Cl or OMe) and 4,4'-di-*tert*-butylbipyridine (d*t*bpy) catalyzes this reaction[19] in alkane or ether solvents with 1:1 ratios of arene to boron reagent at room temperature or slightly above (Equation 18.70).[18,177] These borylations of arenes occur with little electronic preference for the ortho, meta, or para positions of the substituted arenes. Instead, the regioselectivity of these reactions is controlled by steric effects. Thus, 1,3-disubstituted, 1,2,3-trisubstituted, or symmetric 1,2-disusbstituted arenes form a single functionalized product, as shown in Figure 18.2. In the absence of steric effects, the borylation of five-membered heteroarenes occurs regioselectively at the C–H bond ortho to the heteroatom (see the top of Figure 18.2).[178,179] These heteroarenes are more reactive than arenes, and reactions of benzofuran, benzothiophene, and indole occur at the C–H bond α to the heteroatom, not at the C–H bonds of the fused aryl ring. The C–H borylation of arenes is particularly valuable because it occurs without halogens or organolithium reagents, because it forms without the intermediacy of organomagnesium reagents, and because the regioselectivity complements that of electrophilic aromatic substitution.

$$C_6H_6 + HBPin \xrightarrow[150\,°C,\ 120\ h]{Ir\ cat,\ 17\ mol\ \%} Ph-Bpin + H_2$$
$$53\%$$
$$Ir\ cat = Cp^*Ir(PMe_3)(H)(Bpin)$$

(18.68)

$$Ar-H + HBpin \xrightarrow[100-150\,°C]{\substack{2\ mol\ \%\ (Indenyl)Ir(COD)\\ PMe_3,\ DPPE,\ or\ DMPE}} Ar-Bpin + H_2$$
$$pin = Pinacolate$$

(18.69)

With OMe precursor

$$B_2pin_2 + 2\ H-Ar \xrightarrow[\substack{Hexane\\ Room\ temperature}]{\substack{1/2[Ir(OMe)(COD)]_2/dtbpy\\ (3\ mol\ \%)}} 2\ Ar-Bpin + H_2$$
$$1.0\ equiv./B$$

(18.70)

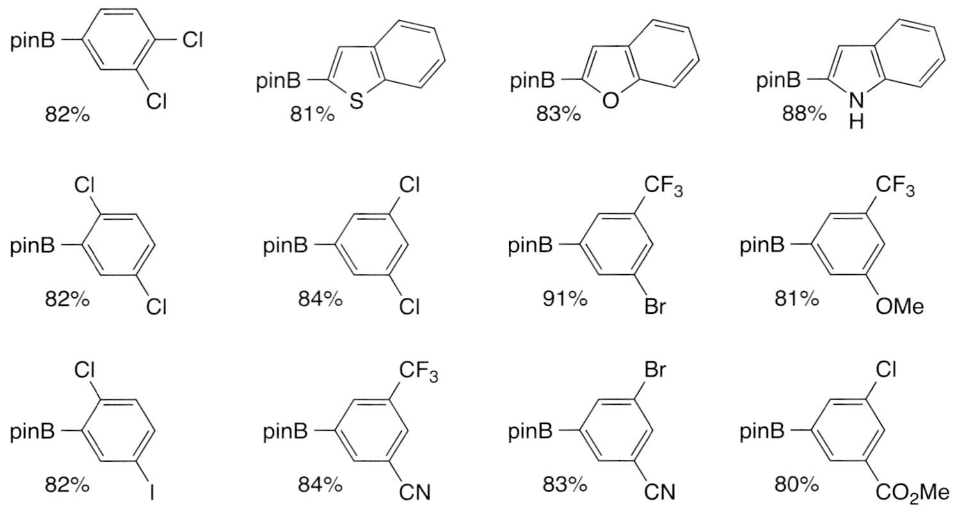

Figure 18.2. Products from the borylation of arenes with B$_2$pin$_2$ as reagent.

18.7.3. Borylation of Polyolefins

The borylation of polyolefins has also been conducted (Scheme 18.14).[180-182] Reactions of polybutene generate products containing boryl groups at the termini of the side chains. The borylation of polypropylene also occurs exclusively at the methyl side chains, although the efficiency of these reactions are lower because these methyl groups are more hindered than those on the ethyl side chains of polybutene. These reactions have also been conducted on copolymers of ethylene and octene. Again, these reactions occur regiospecifically at the methyl groups of the hexyl side chain of this material. Each of these borylated polymers has been oxidized with basic hydrogen peroxide solutions to generate polyolefins containing pendant polar hydroxyl functionalities.

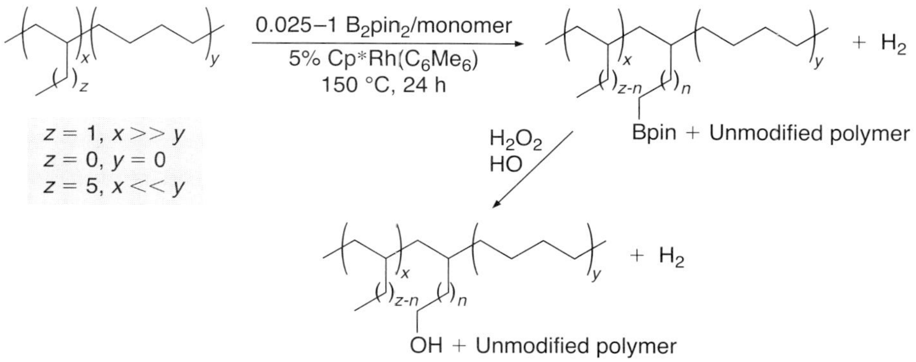

Scheme 18.14

18.7.4. Mechanism of the Alkane and Arene Borylation

Information on the mechanism of the borylation of alkanes and arenes has been reported by Hartwig.[183,184] Catalytic cycles proposed for the functionalization of alkanes and arenes by the Cp*Rh and (d*t*bpy)Ir complexes are shown in Schemes 18.15 and 18.16. Both cycles involve the formation of 16-electron metal–boryl intermediates and the reactions of these intermediates with C–H bonds, followed by the formation of the B–C bond in the final product. Details on the C–H bond cleavage step were gained by studies on isolated metal–boryl complexes.

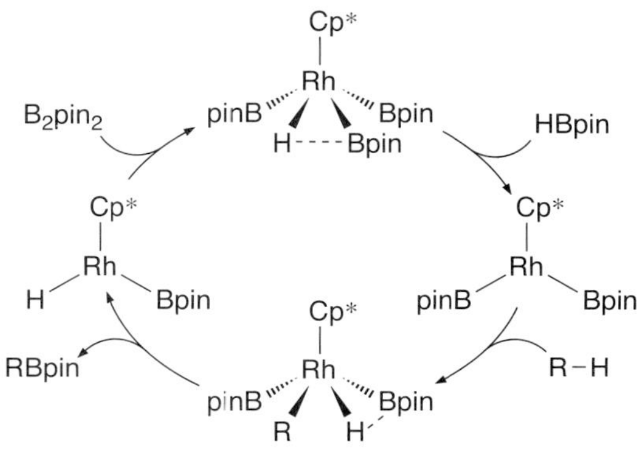

Scheme 18.15

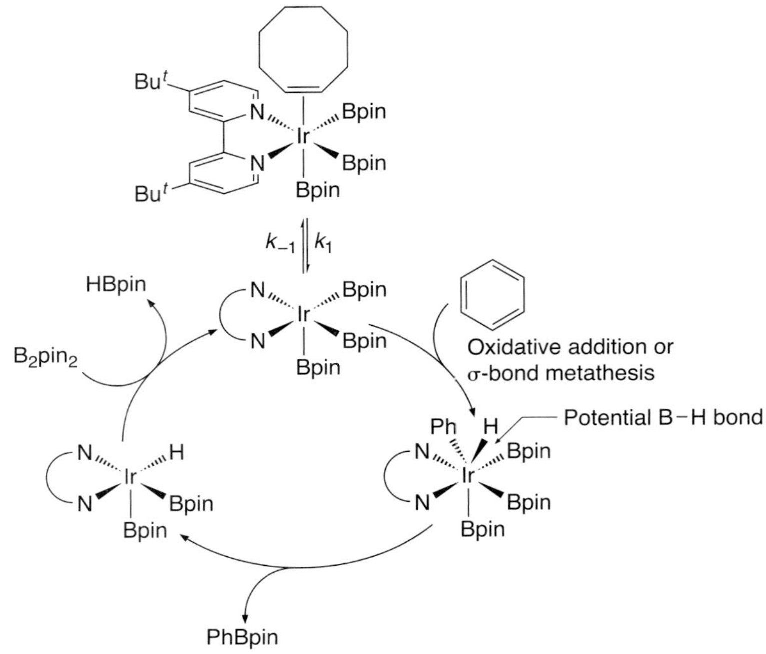

Scheme 18.16

Results from studies on the formation of rhodium–boryl complexes in the catalytic system and the reactions of these complexes with C–H bonds are depicted in Equations 18.71 and 18.72.[183] The rhodium bis(boryl) and tris(boryl) complexes have been observed by NMR spectroscopy in the catalytic system. Both purified complexes react with alkanes and arenes to generate the same products as are formed in the catalytic chemistry. Mechanistic studies on the reactions of these complexes with hydrocarbons show that borane first eliminates from these complexes to generate a 16-electron boryl complex that reacts with the alkane or arene. As presented in Chapter 6, theoretical studies have shown that the p-orbital on boron assists in the C–H bond cleavage step.[183, 185] The product from this C–H bond cleavage step is a metal–alkyl complex containig a coordinated borane, as shown for the computed pathway depicted in Equation 18.73. Subsequent formation of the B–C bond creates the alkylboronate ester product.[183, 186]

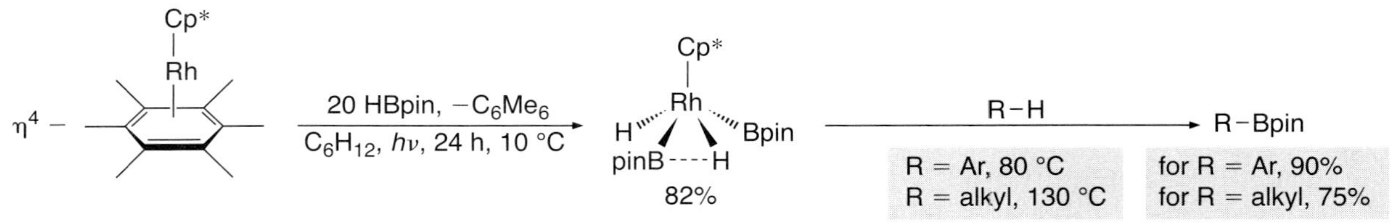

(18.71)

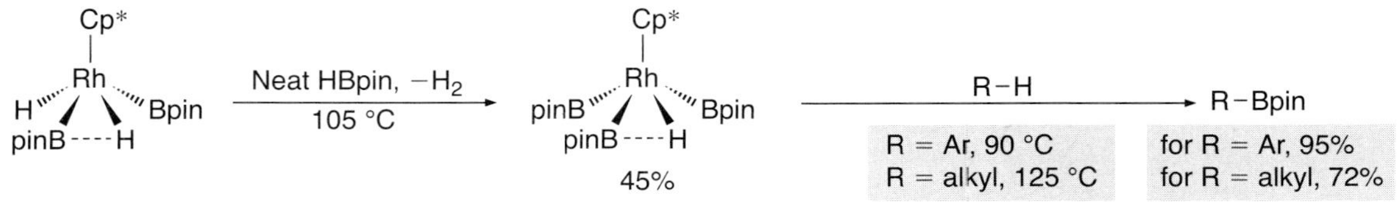

(18.72)

$$(18.73)$$

A tris(boryl)iridium complex ligated by di-*tert*-butylbipyridine (d*t*bpy) and cyclooctene was also observed directly in the borylation of arenes catalyzed by the combination of [Ir(COD)(OMe)]$_2$ and 4,4'-di-*tert*-butylbipyridine.[184] When prepared independently, this complex reacted with arenes below or slightly above room temperature to form the corresponding arylboronate ester (Equation 18.74). The COE-ligated tris(boryl) complex reacted with arenes after dissociation of the coordinated cyclooctene. Alternative pathways involving C–H bond cleavage by iridium(I) monoboryl complexes have been ruled out by labeling studies: the exchange of labeled diboron reagent with the boryl groups of the metal was slower than the rate of the C–H bond functionalization. Kinetic studies have shown that cleavage of the C–H bond by the tris(boryl) complex is the turnover-limiting step of the borylation of arenes catalyzed by the di-*tert*-butylbipyridine–iridium system.

$$(18.74)$$

18.7.5. Silylation of Aromatic and Aliphatic C–H Bonds

Although less developed, catalytic dehydrogenative couplings of silanes at the C–H bonds of aromatic and aliphatic groups have also been reported (Equation 18.75). The first of these reactions, although occurring in low yield, was observed prior to any of the functionalizations of alkanes or arenes with borane reagents.[187] Some of the silylations of arenes have been conducted with arenes containing a directing group to facilitate the reaction.[20,23] Silylations of these types of substrates are then best described as directed C–H functionalizations. Other silylations of arenes occur without any directing group, but require the presence of a hydrogen acceptor to make the reactions more favorable thermodynamically.[188,189] Most recently, a few silylations of aryl and alkyl C–H bonds have been reported in systems that lack a directing group or a hydrogen acceptor.[24]

$$\text{R–H} \ + \ \text{HSiR}'_3 \ \xrightarrow{\text{Catalyst}} \ \text{R–SiR}'_3 \ + \ \text{H}_2 \qquad (18.75)$$

More specifically, Chatani reported the orthosilylation of the aromatic imine in Equation 18.76[22] and the benzylic C–H bond of the methylquinoline in Equation 18.77.[23] The imine and pyridine groups facilitate the ortho silylation. Tanaka reported the silylation of arenes with the disilylarene reagent in Equation 18.78,[189] and Berry reported the dehydrogenative coupling of arenes with silanes catalyzed by silylrhodium cyclopentadienyl complexes, as shown in Equation 18.79.[188] The role of the second silyl group in the disilane reagent in Equation 18.78 is unclear. Most recently, Tilley reported the silylation of methane to form methylsilanes catalyzed by scandocene complexes without any hydrogen acceptor (Equation 18.80), although this reaction is slow.[190,191] Hartwig reported the platinum-catayzed silylation of arenes without a directing group or a hydrogen acceptor (Equation 18.81),[24] but these reactions required high temperatures. Intramolecular silylations of arenes and terminal methyl groups on an aliphatic chain with the same catalyst were also reported (Equation 18.82).[24]

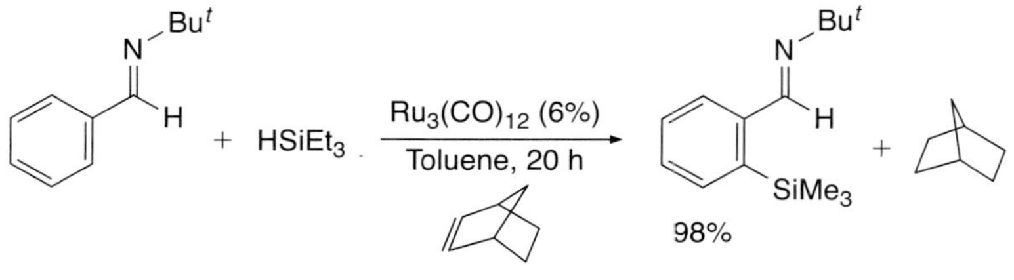

(18.76)

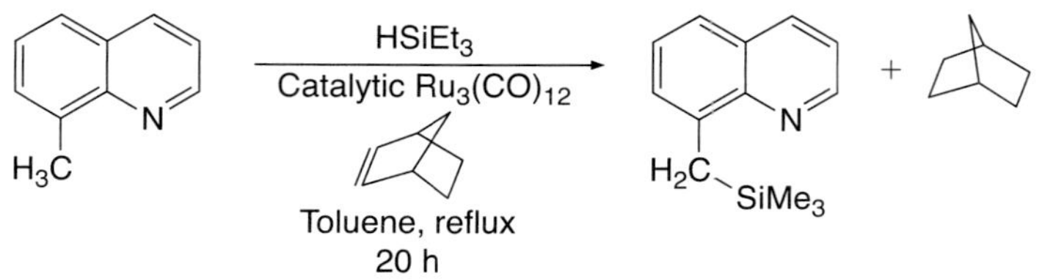

(18.77)

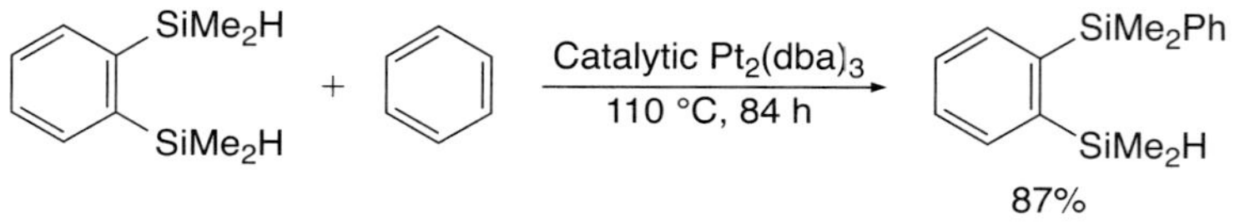

(18.78)

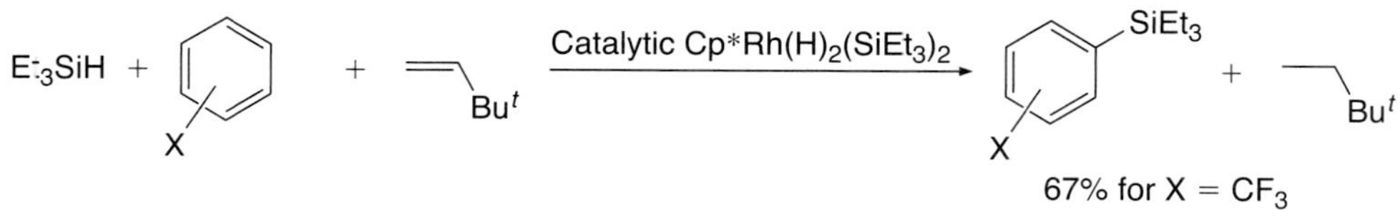

(18.79)

$$Ph_2SiH_2 + 150 \text{ atm } CH_4 \xrightarrow[\text{C}_6\text{H}_{12,} \ 80\,°C, \ 7 \text{ days}]{10\% \ Cp^*_2ScMe} Ph_2MeSiH + H_2 \tag{18.80}$$

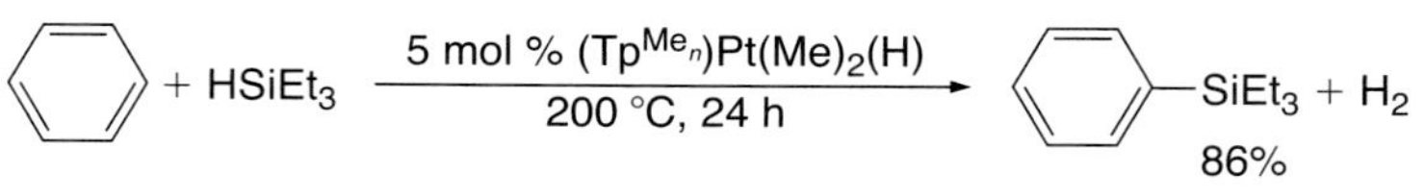

(18.81)

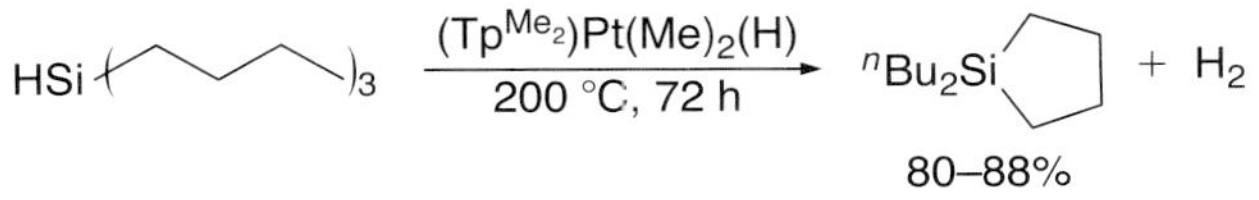

$$HSi \left(\diagdown \diagup \diagdown \right)_3 \xrightarrow[200\ °C,\ 72\ h]{(Tp^{Me_2})Pt(Me)_2(H)} {}^nBu_2Si \diagup\diagdown + H_2 \qquad (18.82)$$

80–88%

The mechanism of the silylation of arenes is not well documented at this point. However, the silylation of arenes by Cp*RhH$_2$(SiEt$_3$)$_2$ is proposed to occur through a disilyl complex Cp*Rh(SiEt$_3$)$_2$, which reacts with the arene to form Cp*Rh(SiEt$_3$)$_2$(H) (Ar).[188] This product from oxidative addition of the arene would then reductively eliminate the arylsilane and by some pathway regenerate the disilyl intermediate containing an additional silane and the hydrogen acceptor (Scheme 18.17). In contrast, the scandocene-catalyzed reactions occur by σ-bond metathesis chemistry because the catalyst contains a d^0 metal center.[190, 191] In this mechanism, the scandocene–methyl complex reacts with a silane to generate the methylsilane and a scandocene hydride, which reacts with methane to form dihydrogen and regenerate the scandocene methyl complex (Scheme 18.18).

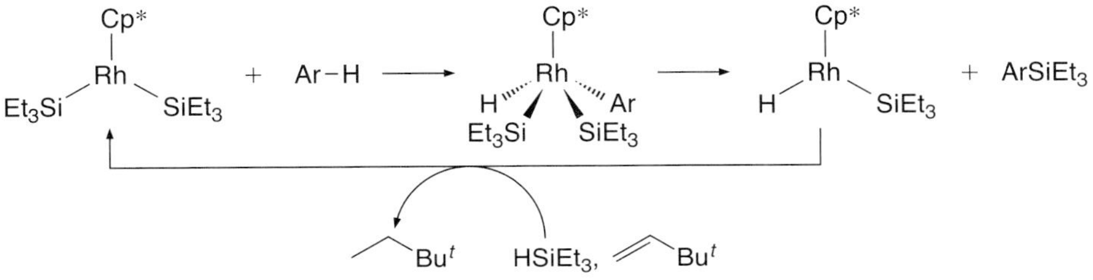

Scheme 18.17

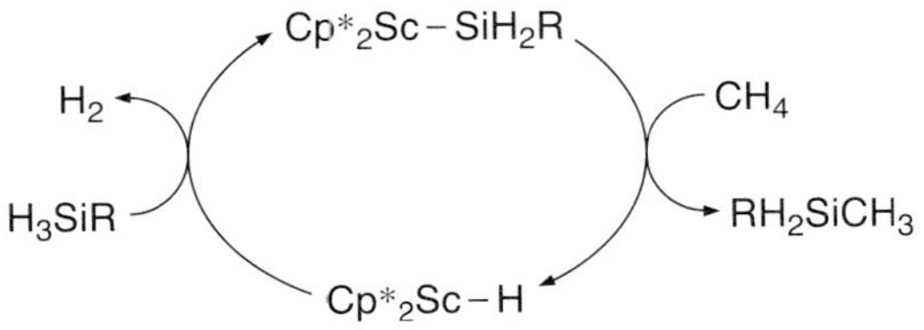

Scheme 18.18

18.8. Hydroacylation

18.8.1. Overview

The catalytic addition of a formyl C–H bond across an olefin to generate a ketone (Equation 18.83) is called "hydroacylation." Intramolecular hydroacylations were reported before intermolecular hydroacylations. The intramolecular reaction can be a valuable route to carbocycles, and it has been conducted enantioselectively. The intermolecular reaction could be a valuable route to acyclic ketones. However, both reactions are under development. The principal challenge in developing hydroacylation reactions is the need to prevent decarbonylation of the aldehyde. The decarbonylation of aldehydes to generate a product containing an aliphatic C–H bond was reported many years ago by Tsuji,[192,193] and has been improved as a synthetic methods since that time.[194] Aldol condensations of the product aldehydes can also reduce reaction yields.

$$R^1 \diagdown + H \diagup_{R^2}^{O} \xrightarrow{\text{Catalyst}} R^1 \diagdown\diagup_{R^2}^{O} \qquad (18.83)$$

18.8.2. Intermolecular Hydroacylation

Few mild, intermolecular hydroacylations that occur in high yield have been reported. The reaction of 4-hexenal with ethylene in the presence of [Rh(ethylene)(acac)] was reported by Miller to form the corresponding methyl ketone after 48 h at room temperature (Equation 18.84).[195] This reaction has now been reported with several rhodium catalysts.[196–198] Cobalt congeners also catalyze intermolecular hydroacylation. Brookhart reported that aromatic aldehydes add to vinylsiloxanes to form β-silyl ketones in the presence of Cp*Co(ethylene)$_2$ (Equation 18.85).[199,200] The hydroacylation of olefins has also been reported with Ru$_3$(CO)$_{12}$ as catalyst under CO pressure, but the temperatures are high and the reaction times are long.[201,202] The most active current catalyst is $(\eta^5\text{-C}_5\text{Me}_4\text{CF}_3)\text{Rh}(\text{CH}_2\text{CHSiMe}_2)_2$, and this complex catalyzes the addition of aromatic aldehydes to internal cyclic and terminal acyclic alkenes. The electron-withdrawing group on the Cp ligand accelerates the turnover-limiting reductive elimination of the final product.[203]

$$\diagup\diagdown\diagup_{H}^{O} + \ \| \quad \xrightarrow[\text{CHCl}_3,\ \text{RT, 48 h}]{\text{Cat. Rh(C}_2\text{H}_4)_2\text{(acac)}} \quad \diagdown\diagup\diagdown\diagup_{\quad}^{O} \qquad (18.84)$$
$$\text{84\% yield}$$

$$\text{MeO}\diagdown\diagup_{H}^{O} + \diagup\diagdown\text{SiMe}_3 \xrightarrow[\text{Toluene, RT, 24 h}]{2\text{-5 mol \% }(\eta^5\text{-C}_5\text{Me}_5)\text{Co(C}_2\text{H}_3\text{SiMe}_3)_2} \text{MeO}\diagdown\diagup_{\quad}^{O}\diagup\diagdown\text{SiMe}_3 \qquad (18.85)$$
$$\text{82\%}$$

18.8.3. Intramolecular Hydroacylation

The intramolecular hydroacylation of olefins typically occurs in higher yields than the intermolecular hydroacylation of olefins. In its simplest form, the reaction of 4-pentenone in the presence of stoichiometric amounts of Wilkinson's catalyst at room temperature in ethylene-saturated solvent formed the isomeric cyclopentanone in good yield after long reaction times (Equation 18.86).[204] Many different cyclopentanones were prepared by related methods.[205–210]

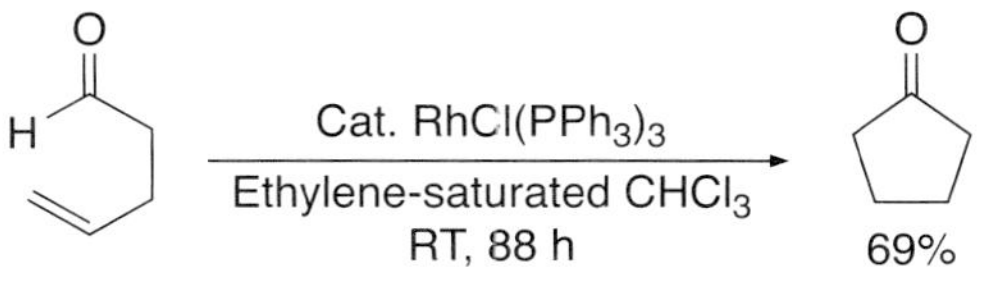

$$\text{(18.86)}$$

Because decarbonylation of the acyl intermediate competes with olefin insertion into the same species, and the carbonyl complexes are inactive as catalysts, the catalyst is poisoned by the competing decarbonylation. The use of a cationic rhodium complex containing a chelating ligand suppresses poisoning of the catalyst by decarbonylation, and reactions of alk-4-en-1-als catalyzed by rhodium complexes of bisphosphines formed the desired cyclopentanones faster than reactions catalyzed by neutral rhodium complexes.[211]

With procedures to generate cyclopentanones in the presence of catalytic amounts of rhodium complexes containing bisphosphines, several examples of enantioselective hydroacylations have been reported. Sakai reported cyclization of the substituted pentenal in Equation 18.87 with nearly perfect enantioselectivity in the presence of a catalyst containing (R)-BINAP as ligand.[212] Bosnich has published several papers on enantioselective hydroacylation with this system. High enantiomeric excess was observed for the cyclization of cyclopentenals with tertiary-alkyl or carboalkoxy groups at the 4-position when catalyzed by BINAP complexes of rhodium (Equation 18.88).[213–218]

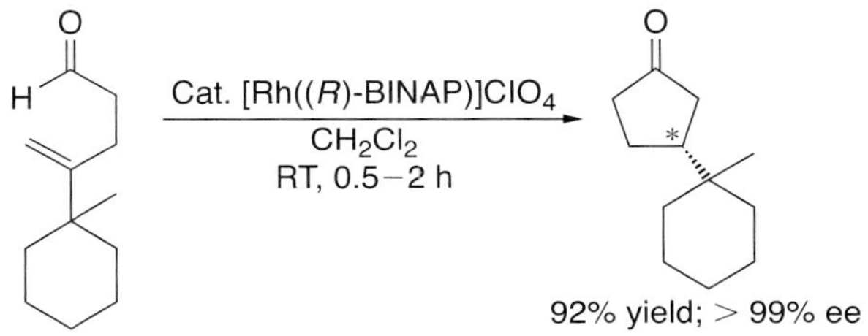

$$\text{(18.87)}$$

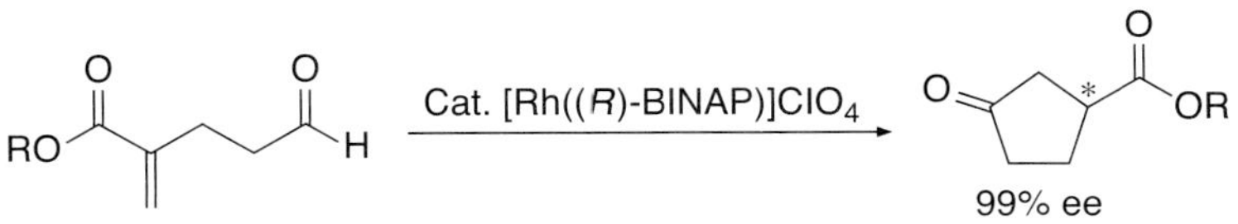

92% yield; > 99% ee

$$\text{(18.88)}$$

99% ee

18.8.4. Mechanism of Hydroacylation

A simple catalytic cycle for hydroacylation is shown in part A of Scheme 18.19.[218] Hydroacylation occurs by oxidative addition of the formyl C–H bond to generate an acyl hydride complex. Insertion of olefin into the metal hydride then generates an alkyl acyl intermediate. These complexes undergo reductive elimination, as described in Chapter 8. Although these basic steps constitute the catalytic cycle, many other processes occur outside of this cycle in the catalytic system.[218] Some of these steps lying off the cycle lead to poisoning of the catalyst and others are unproductive reversible processes that have been revealed by H/D exchange experiments. Part B of Scheme 18.19 shows a catalytic cycle that includes these side processes.

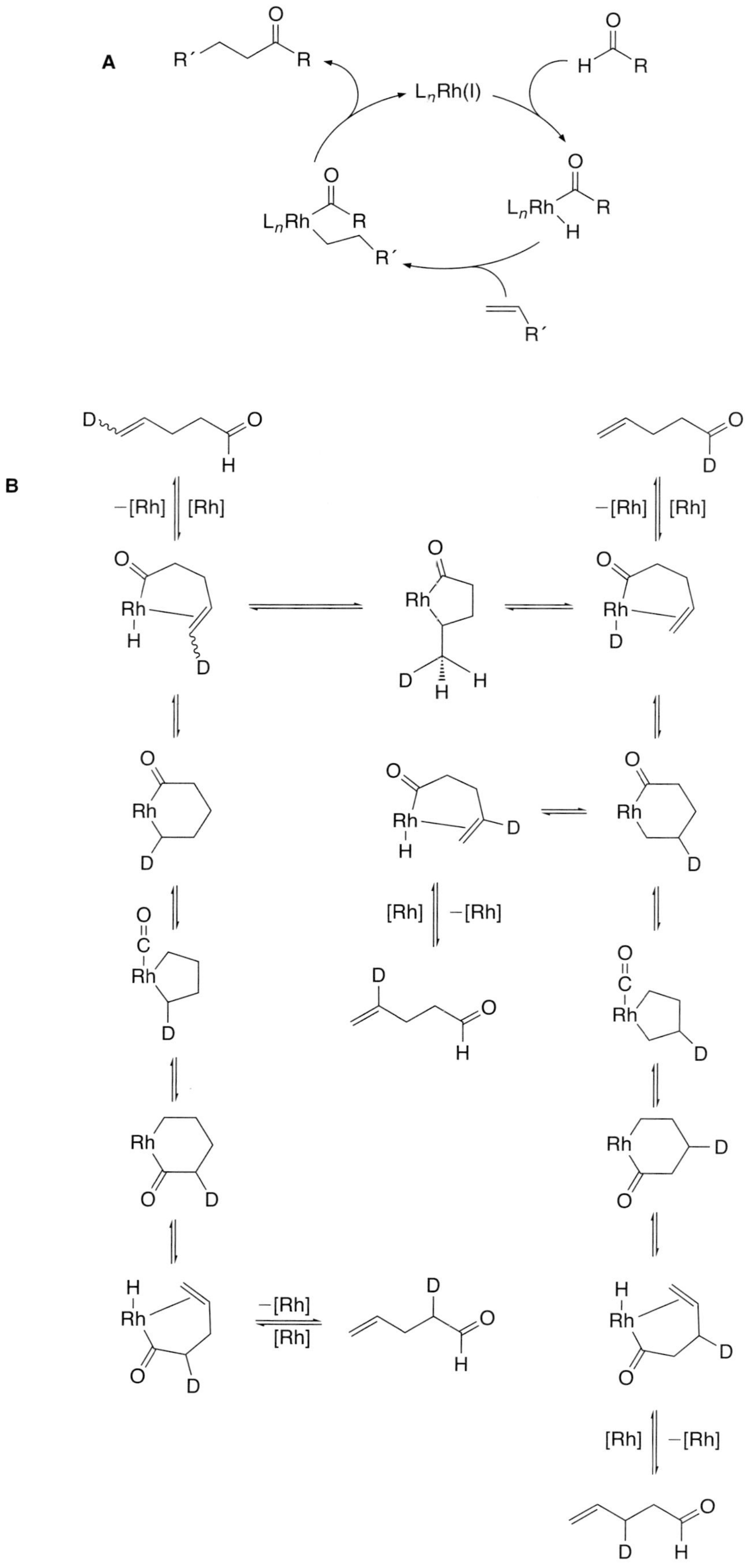

Scheme 18.19

One significant competing reaction is decarbonylation to generate a CO-ligated alkyl hydride intermediate that reductively eliminates alkane, as described above. However, studies on H/D exchange have revealed other side reactions. For example, reversible insertion of olefin leads to incorporation of the deuterium in the aldehyde into the olefin during intermolecular hydroacylation or into the olefinic portion of the enal during intermolecular hydrocylation. Insertion and de-insertion of the carbonyl group has been detected by the labeling studies on reactions conducted with catalysts that contain chelating ligands, which prevent irreversible de-insertion.

18.8.5. Directed Intermolecular Hydroacylation

Because the rates of hydroacylation are often slow, and decarbonylation can compete with hydroacylation, strategies to overcome these limitations have been devised. One strategy is to conduct reactions with substrates that will form stable metallacycles after cleavage of the aldehyde C–H bond. A second strategy is to conduct the hydroacylation with imine derivatives of the aldehyde because de-insertion of an isocyanide is less favorable than de-insertion of CO.

The first strategy of conducting reactions with aldehydes containing coordinating groups is illustrated by the reactions in Equations 18.89 and 18.90. During the hydroacylation of vinylcyclohexane by 8-quinolinylcarboxaldehyde, oxidative addition of the aldehydic C–H bond would lead to a five-membered metallacycle with the quinoline nitrogen coordinated to the metal.[219–221] Insertion of the olefin, followed by C–H bond-forming reductive elimination, would lead to the hydroacylation product. Apparently de-insertion of CO from the five-membered metallacycle to form a four-membered metallacycle is less favored than insertion of olefin to expand the size of the metallacycle. A related strategy was recently reported for reactions of acyclic aldehydes.[222,223] The reactions of aldehydes bearing 3-benzyloxy, methoxy, or methylthio substituents with acrylate derivatives catalyzed by Wilkinson's catalyst and related complexes containing chelating ligands were studied. Reactions with substrates containing a coordinating methylthio substituent occurred, and these reactions occurred in the highest yield in the presence of a catalyst containing dppe as ligand.

$$(18.89)$$

$$71\% \quad (18.90)$$

The second strategy of conducting hydroacylation with imine derivatives is illustrated by the reactions in Equation 18.91 and Scheme 18.20. Suggs reported hydroacylation of N-pyridylbenzaldimines in which the pyridine can assist oxidative addition of the aldimine C–H bond (Equation 18.91).[224] Hydrolysis of the imine would then form the ketone. This two-step process can be conducted in one pot with catalytic amounts of an aminopyridine, as shown in Scheme 18.20.[225–227] Addition of catalytic amounts of 2-amino-3-picoline to benzaldehyde generates an N-pyridylimine, which undergoes hydroacylation of an olefin. The extruded water then hydrolyzes the resulting imine to regenerate the aminopicoline and to form the final ketone product.

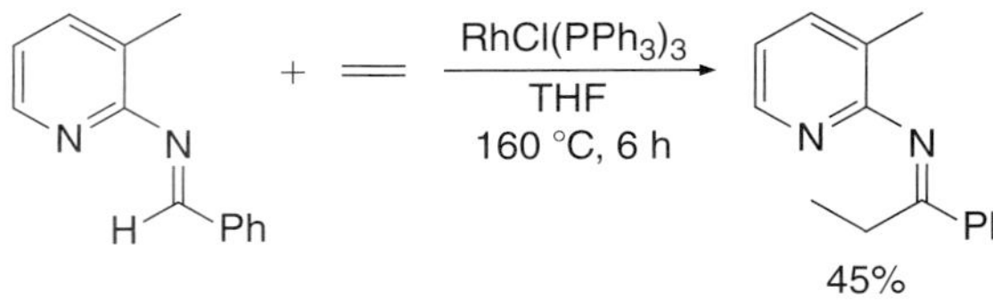

(18.91)

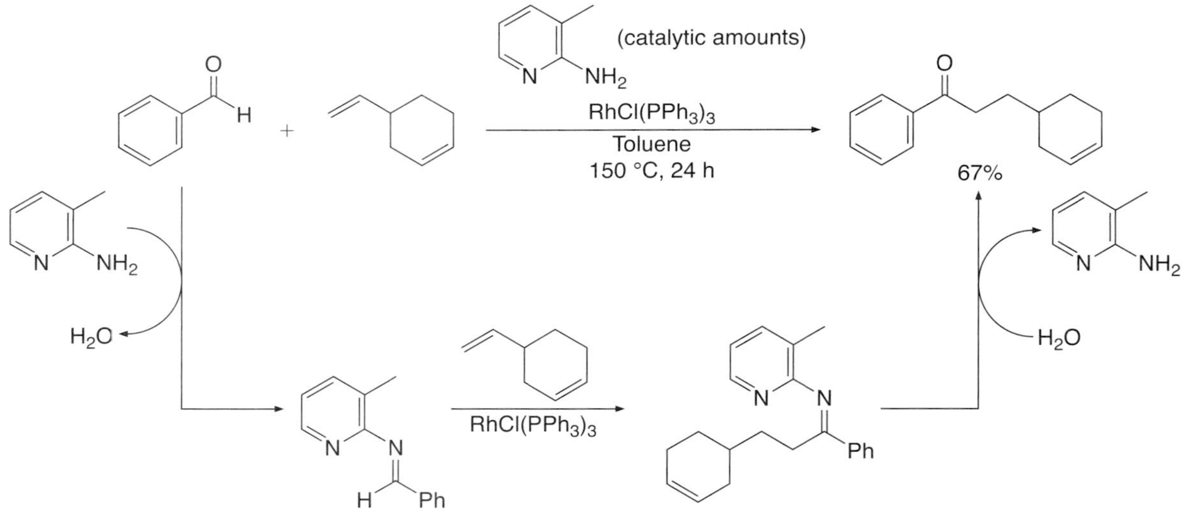

Scheme 18.20

18.9. Functionalization of C–H Bonds by Carbene Insertions

18.9.1. Overview

All of the reactions described in the preceding sections of this chapter involve functionalization of C–H bonds through an intermediate containing a new metal–carbon bond formed from reaction of the catalyst with the reagent undergoing C–H bond functionalization. A separate type of C–H bond functionalization with organometallic intermediates involves the reactions of metal carbene complexes containing carboxyl groups at the carbene carbon (a carbenoid complex). Many examples of these reactions have been conducted intramolecularly, and enantioselective versions of these reactions have been studied extensively. The use of rhodium–carbenoid complexes containing one π-donating (aryl or vinyl) and one π-accepting group (carboxy) on the carbene carbon has led to the development of intermolecular insertions of carbene units into unactivated C–H bonds. These reactions have been conducted enantio- and diastereoselectively. Most of the reactions are conducted with diazo compounds as the precursor to the carbenoid intermediate. Both the intramolecular and intermolecular insertion reactions have been used in the synthesis of a wide range of natural products.

The selectivity for different types of C–H bonds depends on both electronic and steric factors. The ring size also controls the regioselectivity of intramolecular carbene insertions into C–H bonds. Unlike the C–H bond functionalizations occurring by formation of metal–alkyl intermediates, the functionalization of C–H bonds by carbenoid complexes occurs preferentially at secondary over primary C–H bonds. For steric reasons, reactions tend to occur preferentially at secondary over tertiary C–H bonds.[228] The insertion reactions are also controlled by electronic factors. The reactions occur preferentially at sites that stabilize a buildup of positive charge at the carbon undergoing C–H cleavage.[229] Thus, these insertions are favored at the C–H bonds α to an alkoxo or amino group and are disfavored at C–H bonds α to ester or acetoxy groups. Also unlike the reactions described in the previous sections involving formation of metal–carbon bonds from C–H bond cleavage, the carbene insertion chemistry occurs preferentially at weaker aliphatic C–H bonds than at stronger aromatic C–H bonds.

18.9.2. Intramolecular C–H Functionalization by Carbene Insertion

The first work on intramolecular insertions of carbenes into C–H bonds was conducted with copper complexes.[230–231] Many of these reactions focused on developing enantioselective C–H insertions with optically active C_2-symmetric bisimine or bisoxazoline ligands. The highest enantioselectivities were observed with complexes containing the bisoxazoline ligand containing *tert*-butyl substituents.[232–238] Although these catalysts have been used almost exclusively for intramolecular reactions because intermolecular reactions occurred with low yields, reactions of ethyldiazoacetate at the alpha C–H bond of tetrahydrofuran led to high yields of the insertion product in the presence of a copper catalyst containing a bulky tris(pyrazolyl) borate ligand (Equation 18.92).[239]

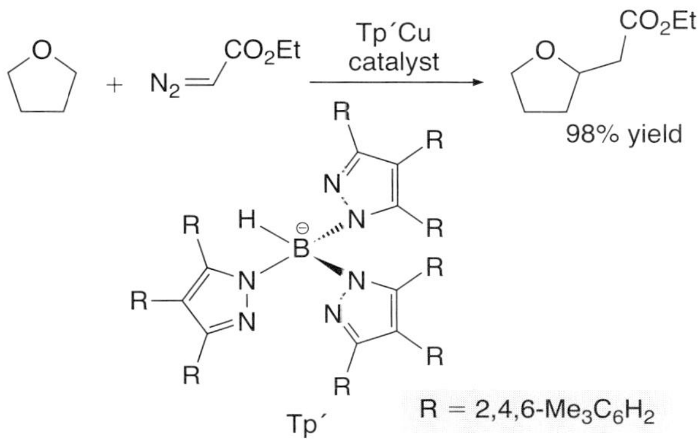

$$\text{(18.92)}$$

More recently, dirhodium complexes bridged by four carboxylate or carboxamide ligands have been studied for enantioselective carbene insertion. McKervey and co-workers showed that enantioselective, intramolecular C–H insertions of carbenes occurred with *N*-benzenesulfonyl prolinate groups bridging the two rhodium centers (Figure 18.3).[234,240–243] Davies showed that the combination of these catalysts and carbenes possessing π-donor (aryl or vinyl) and acceptor (carboalkoxy carbamoyl, or acyl) groups on the carbene led to reactions that occurred with high yield and enantiomeric excess.[10,244]

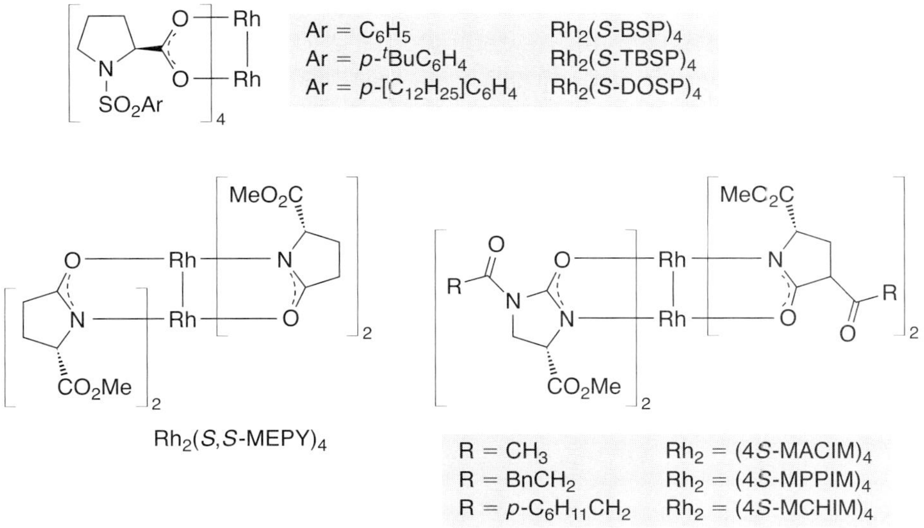

Figure 18.3. McKervey's and Doyle's catalysts for enantioselective carbene insertions.

The most active and selective catalysts for intramolecular reactions of acceptor carbenes derived from diazoesters or diazoacetamides were developed by Doyle and contain carboxamidate ligands bridging the two rhodium centers.[229,245–249] The MEPY catalyst shown in Figure 18.3 is a commonly used example of these rhodium catalysts, and the MPPIM complex has catalyzed C–H insertions with high enantiomeric excess. One example of such a reaction is shown in Equation 18.93.[250] Two examples of an intramolecular insertion of carboxamide-substituted carbenes catalyzed by MEPY-ligated dirhodium complexes are shown in Equations 18.94 and 18.95.[251,252] Many other examples occur with higher selectivity when conducted with other carboxamidate-substituted dirhodium complexes as catalyst. One of these is shown in Equation 18.96,[250,253–255] and many examples are presented in detail in Davies' review.[10] Examples of the synthesis of two small natural products by the use of this intramolecular insertion of diazoesters into C–H bonds are shown in Equations 18.97 and 18.98.[250,256]

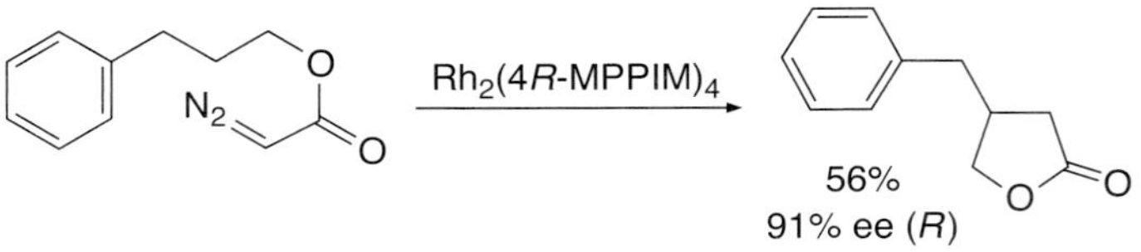

N2
O
O
Rh2(4R-MPPIM)4
O
O
56%
91% ee (R)

(18.93)

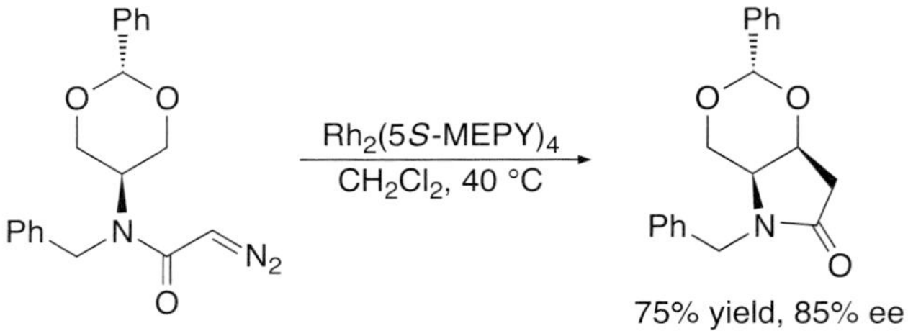

Ph
O
O
Ph
N
O
N2
Rh2(5S-MEPY)4
CH2Cl2, 40 °C
Ph
O
O
Ph
N
O
75% yield, 85% ee

(18.94)

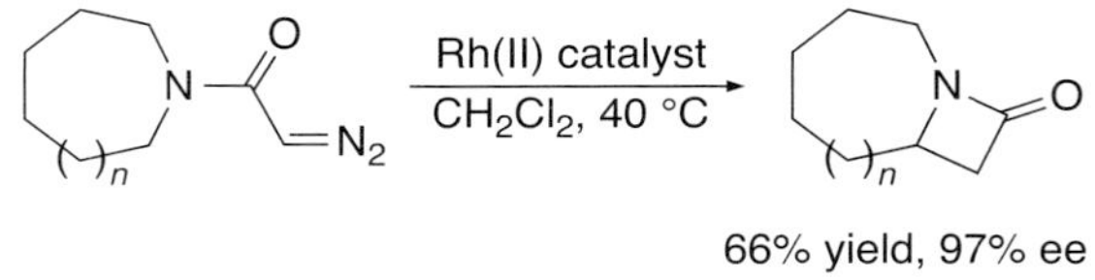

N
O
N2
n
Rh(II) catalyst
CH2Cl2, 40 °C
N
O
n
66% yield, 97% ee

(18.95)

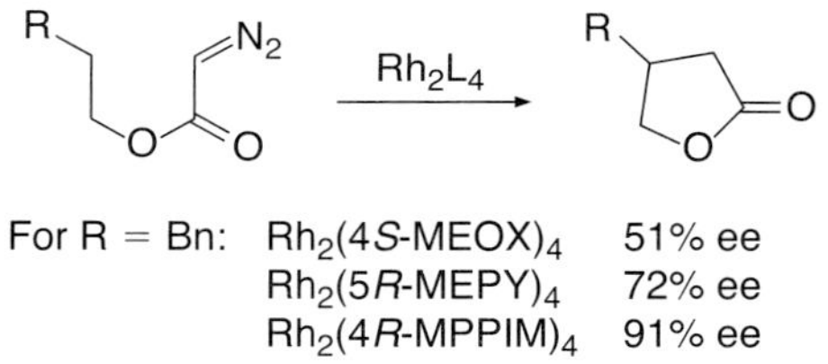

R
N2
O
O
Rh2L4
R
O
O
For R = Bn: Rh2(4S-MEOX)4 51% ee
Rh2(5R-MEPY)4 72% ee
Rh2(4R-MPPIM)4 91% ee

(18.96)

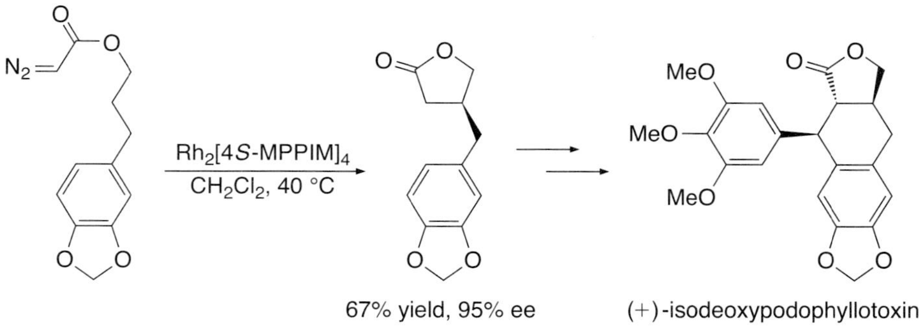

O
N2
O
O
O
Rh2[4S-MPPIM]4
CH2Cl2, 40 °C
O
O
O
O
67% yield, 95% ee
MeO
MeO
MeO
O
O
O
O
(+)-isodeoxypodophyllotoxin

(18.97)

61% yield, 95% ee   (R)-(−)-baclofen·HCl   (18.98)

18.9.3. Intermolecular C–H Functionalization by Carbene Insertion

Advances in the intermolecular insertion of carbenes fragments into C–H bonds have been made by Davies' studies on carbenes containing one π-donor and one π-acceptor substituent. Reactions of alkyl C–H bonds with rhodium carbenoid intermediates containing diazoacetate groups occur in low yields, in large part because these carbenes reacted with themselves to form olefins instead of undergoing intermolecular reactions.[229,257,258] However, the presence of a donor group on a carbenoid intermediate containing carboalkoxy acceptor groups causes the reactivity of carbenoid intermediates to be sufficiently attenuated to allow intermolecular insertion. Many intermolecular reactions of these carbenes with C–H bonds have been reported, and many of these reactions occur with high enantio- and diastereomeric excess. These reactions are typically conducted with the dirhodium complex containing the carboxamidate ligand (S)-DOSP as catalyst (DOSP = an N-arenesulfonylcarboxamidate with a p-dodecyl group on the arene for solubility; see Figure 18.3).

Many examples of enantioselective insertions of the donor/acceptor carbenoids into C–H bonds have now been reported, and these reactions are described in detail in reviews by Davies.[10,244] Some of the more striking examples are shown in Equations 18.99–18.102. Intermolecular examples of aryl-substituted diazoacetates into the C–H bonds of cyclopentane and cyclohexane occur in high yield in most cases and with high enantiomeric excess in the presence of the DOSP-ligated dirhodium catalysts.[259,260] Insertions into the C–H bonds α to oxygen and nitrogen also occur with high enantiomeric excess in cyclic ethers and amines, such as THF[259,260] and N-Boc pyrrolidine.[261,262] Reactions at the methyl groups of acyclic N-methyl N-Boc amines also have been reported.[263] Likewise, reactions of acylic trans-allylic silyl ethers occurs with high diasteromeric and enantiomeric excess at the allylic position (Equation 18.103).[264] Insertions into allylic C–H bonds often give modest diastereoselectivities, but insertions into a silyl enol ether[265] and into 1,4-cyclohexadiene[266,267] give high selectivities and little product from cyclopropanation (Equations 18.104 and 18.105). Insertions can also occur into benzylic C–H bonds with high stereoselectivity (Equation 18.106).[268] The regioselectivity of these carbene insertions into C–H bonds contrasts with that for most catalytic "C–H activations" because the C–H bond is cleaved by the carbene complex without formation of a new metal–carbon bond. Two examples of small natural products prepared by intermolecular enantioselective C–H insertion via enantioselective insertion products are shown in Scheme 18.21.[268]

23–81% yield   (18.99)
90–96% ee

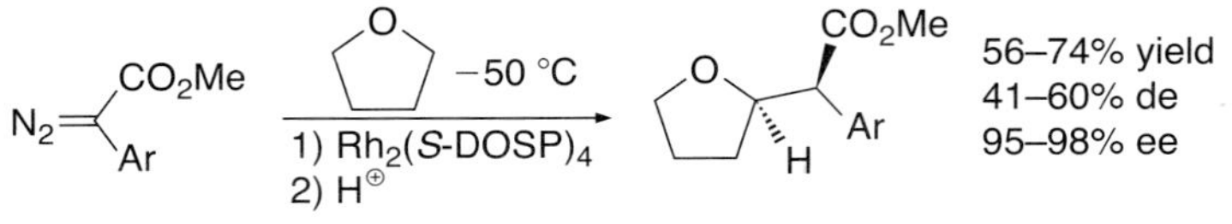

$$\text{(18.100)}$$

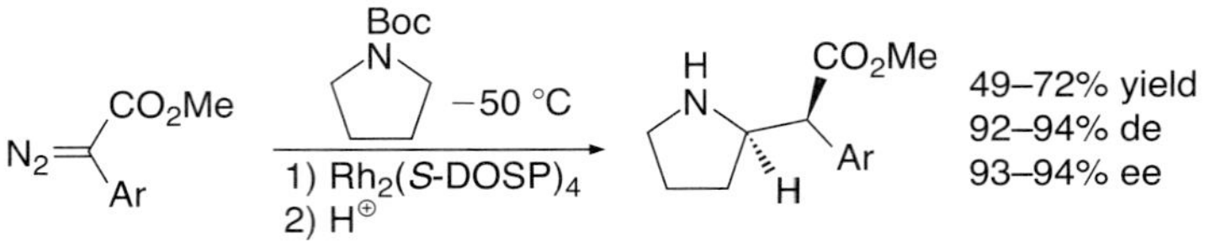

56–74% yield
41–60% de
95–98% ee

$$\text{(18.101)}$$

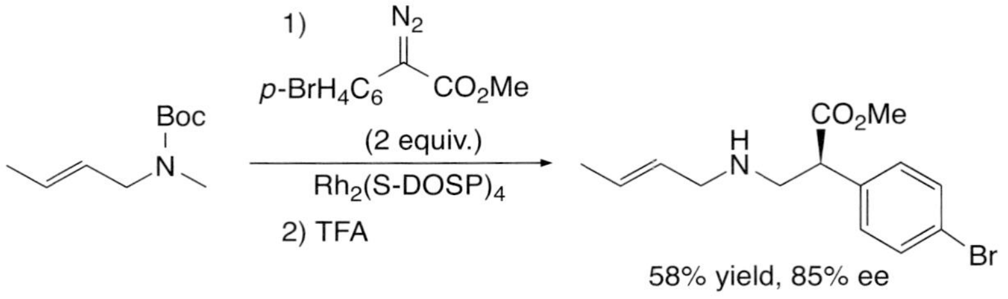

49–72% yield
92–94% de
93–94% ee

$$\text{(18.102)}$$

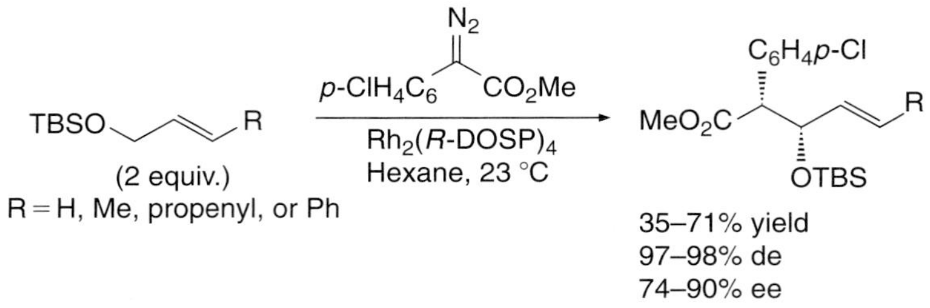

58% yield, 85% ee

$$\text{(18.103)}$$

35–71% yield
97–98% de
74–90% ee

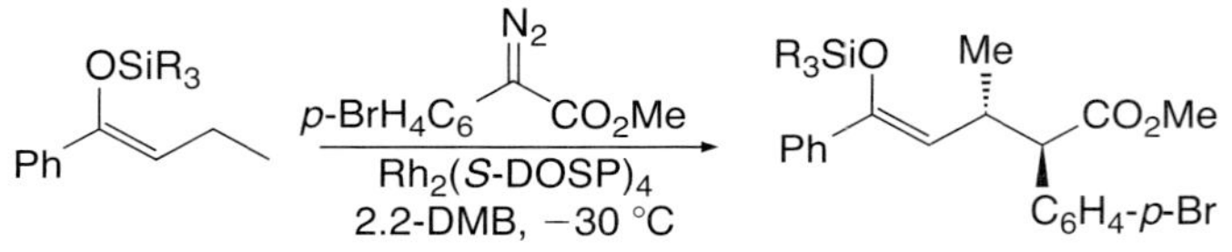

$$\text{(18.104)}$$

SiR$_3$	Yield, %	de, %	ee, %
TIPS	66	> 90	71
TBDPS	65	> 90	84

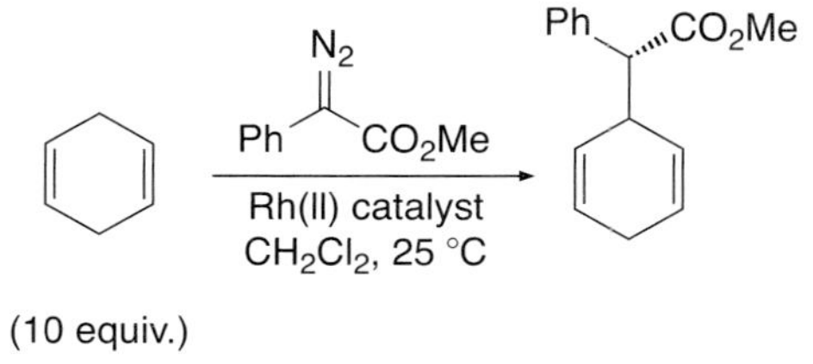

$$\text{(18.105)}$$

Catalyst	Yield, %		ee, %
Rh$_2$(S-DOSP)$_4$	98	>98:2	65 (R)
Rh$_2$(S-DOSP)$_4$	50	>98:2	71 (R)
Rh$_2$(S-DOSP)$_4$	37	>98:2	72 (R)
Rh$_2$(S-DOSP)$_4$	80	>98:2	91 (R)

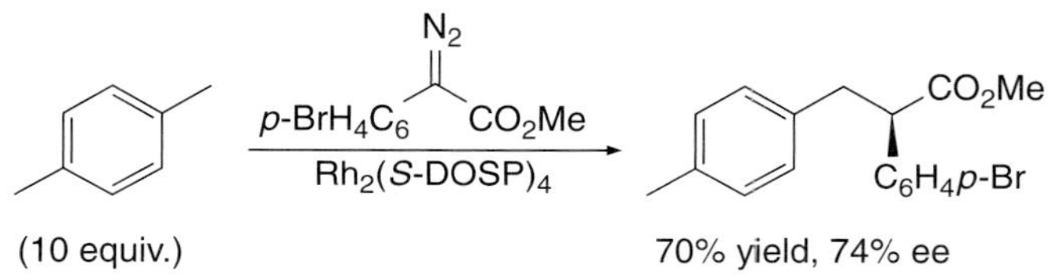

$$\text{(18.106)}$$

70% yield, 74% ee

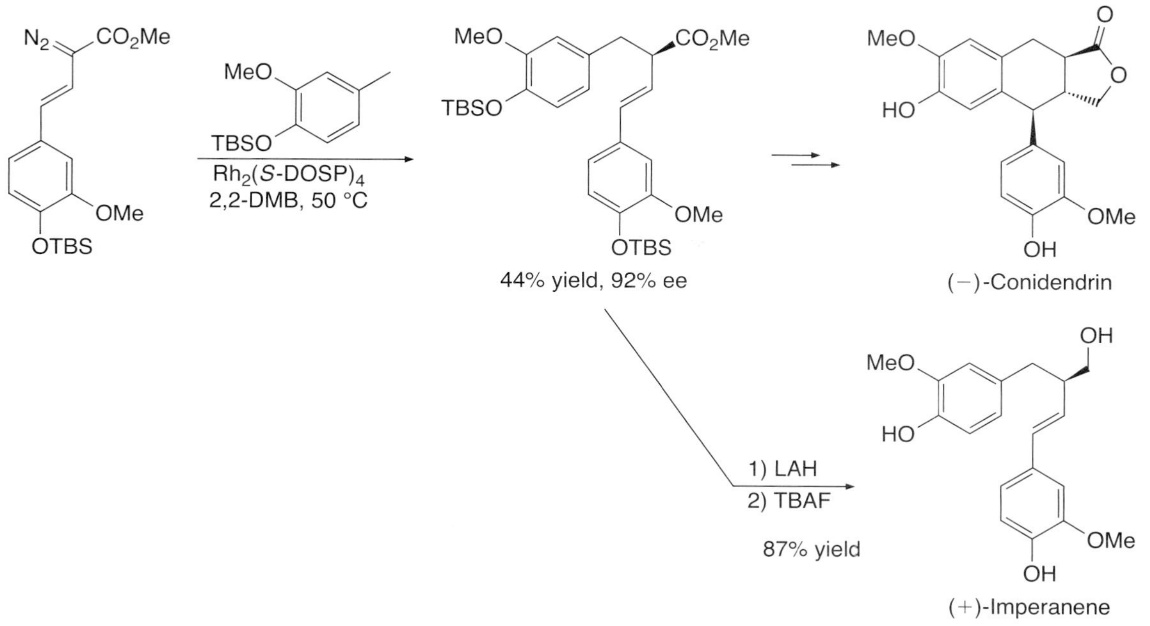

Scheme 18.21

18.10. H/D Exchange

Chapter 18 closes with a brief description of perhaps the most common type of C–H activation process: H/D exchange. Shilov reported H/D exchange almost 40 years ago.[33] Far too many examples of this reaction are now known to summarize in this forum. However, this reaction has been a test-bed for the capability of transition metal complexes to cleave aliphatic and aromatic C–H bonds. H/D exchange reactions have been conducted with homogeneous catalysts and either deuterium gas or a relatively inexpensive source of deuterium, such as benzene-d_6 or D_2O.[269,270] Seminal papers on the use of Crabtree's catalyst for H/D exchange in medicinally important compounds was published by Heys,[271] and this catalyst is frequently used for this application.

This reaction has become useful as a synthetic tool for radiolabeling (tritium, T) and the incorporation of stable isotopes (deuterium). This subject has recently been reviewed, and the topic of H/D exchange has experienced a renaissance in recent years.[272] This renaissance has resulted from the desire to label compounds directly from natural products or the final synthetic products for the analysis of drug metabolism and for the analysis of environmental and biological samples. For example, H/T exchange using cationic iridium catalysts related to Crabtree's hydrogenation catalyst has been used as a method to incorporate tritium labels into molecules used in biological assays.[271,273–276]

H/D exchange typically occurs by addition of an alkyl or aryl C–H bond to an intermediate containing at least one deuteride (Equation 18.107). Reductive elimination of alkane or arene regenerates the monodeuteride, but reductive elimination of labeled alkane or arene would lead to the H/D exchange. Depending on the geometry of the dihydride intermediate, site exchange of H and D might be necessary. Addition of hydrogen or H/D exchange with D_2O or C_6D_6 by an analogous mechanism would regenerate the metal deuteride.

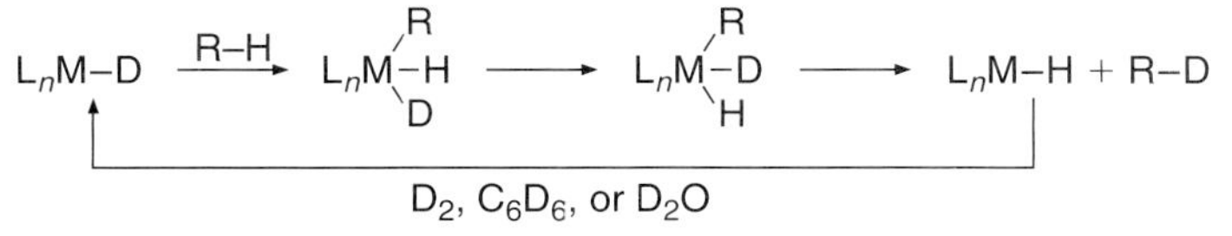

$$(18.107)$$

References and Notes

1. Bergman, R. G. *Science* **1984**, *223*, 902.
2. *Selective Hydrocarbon Activation*; Davies, J. A., Watson, P. L., Liebman, J. F., Greenberg, A., Eds.; VCH Publishers: New York, 1990.
3. Arndtsen, B. A.; Bergman, R. G.; Mobley, T. A.; Peterson, T. H. *Acc. Chem. Res.* **1995**, *28*, 154.
4. Shilov, A. E.; Shul'pin, G. B. *Chem. Rev.* **1997**, *97*, 2879.
5. Labinger, J. A.; Bercaw, J. E. *Nature* **2002**, *417*, 507.
6. Hartung, C. G.; Snieckus, V. *The Directed Ortho Metalation Reaction. A Point of Departure for New Synthetic Aromatic Chemistry*; Wiley-VCH: New York, 2002.
7. Shell, U.S. Patent 2,403,772, 1946.
8. Klenk, H.; Götz, P. H.; Siegmeier, R.; Mayr, W. In *Ullmanns Encyclopedia of Industrial Chemistry*; Wiley: New York, 2002; http://www.mrw.interscience.wiley.com/ueic/articles/a19_199/sect2.
9. Musser, M. T. In *Ullmanns Encyclopedia of Industrial Chemistry*; Wiley: New York, 2005; http://www.mrw.interscience.wiley.com/ueic/articles/a08_217/sect3.
10. Davies, H. M. L.; Beckwith, R. E. J. *Chem. Rev.* **2003**, *103*, 2861.
11. Kunin, A. J.; Eisenberg, R. *Organometallics* **1988**, *7*, 2124.
12. Jones, W. D. In *Selective Hydrocarbon Activation*; Davies, J. A., Watson, P. L., Liebman, J. F., Greenberg, A., Eds.; VCH Publishers: New York, 1990; p 113.
13. The borylation and silylation of arenes and alkanes with borane and silane reagents are exceptions.
14. Chen, H.; Hartwig, J. F. *Angew. Chem. Int. Ed.* **1999**, *38*, 3391.
15. Iverson, C. N.; Smith, M. R., III *J. Am. Chem. Soc.* **1999**, *121*, 7696.
16. Chen, H.; Schlecht, S.; Semple, T. C.; Hartwig, J. F. *Science* **2000**, *287*, 1995.
17. Cho, J. Y.; Tse, M. K.; Holmes, D.; Maleczka, R. E.; Smith, M. R. *Science* **2002**, *295*, 305.
18. Ishiyama, T.; Takagi, J.; Hartwig, J. F.; Miyaura, N. *Angew. Chem. Int. Ed.* **2002**, *41*, 3056.
19. Ishiyama, T.; Takagi, J.; Ishida, K.; Miyaura, N.; Anastasi, N.; Hartwig, J. F. *J. Am. Chem. Soc.* **2002**, *124*, 390.
20. Williams, N. A.; Uchimaru, Y.; Tanaka, M. *J. Chem. Soc., Chem. Commun.* **1995**, 1129.
21. Williams, N. A.; Uchimaru, Y.; Tanaka, M. *Chem. Commun.* **1997**, 461.
22. Kakiuchi, F.; Matsumoto, M.; Tsuchiya, K.; Igi, K.; Hayamizu, T.; Chatani, N.; Murai, S. *J. Organomet. Chem.* **2003**, *686*, 134.
23. Kakiuchi, F.; Tsuchiya, K.; Matsumoto, M.; Mizushima, B.; Chatani, N. *J. Am. Chem. Soc.* **2004**, *126*, 12792.
24. Tsukada, N.; Hartwig, J. F. *J. Am. Chem. Soc.* **2005**, *127*, 5022.
25. McMillen, D. F.; Golden, D. M. *Ann. Rev. Phys. Chem.* **1982**, *33*, 493.
26. Lin, M.; Shen, C.; Garcia-Zayas, E. A.; Sen, A. *J. Am. Chem. Soc.* **2001**, *123*, 1000.
27. Periana, R. A.; Taube, D. J.; Gamble, S.; Taube, H.; Satoh, T.; Fujii, H. *Science* **1998**, *280*, 560.
28. Shilov, A. E.; Shteinman, A. A. *Coord. Chem. Rev.* **1977**, *24*, 97.
29. Goldshleger, N. F.; Lavrushko, V. V.; Khrush, A. P.; Shteinman, A. A. *Bull. Acad. Sci. USSR Div. Chem. Sci.* **1976**, *25*, 2031.
30. Goldshleger, N. F.; Shteinman, A. A. *React. Kinet. Catal. Lett.* **1977**, *6*, 43.
31. Goldshleger, N. F.; Shteinman, A. A.; Shilov, A. E.; Eskova, V. V. *Russ. J. Phys. Chem.* **1972**, *46*, 785.
32. Goldshleger, N. F.; Shteinman, A. A.; Shilov, A. E.; Eskova, V. V. *Zh. Fiz. Khim.* **1972**, *46*, 1353.
33. Goldshleger, N. F.; Tyabin, M. B.; Shilov, A. E.; Shteinman, A. A. *Russ. J. Phys. Chem.* **1969**, *43*, 1222.
34. Hodges, R. J.; Webster, D. E.; Wells, P. B. *J. Chem. Soc., Chem. Commun.* **1971**, 462.
35. Periana, R. A.; Taube, D. J.; Evitt, E. R.; Loffler, D. G.; Wentrcek, P. R.; Voss, G.; Masuda, T. *Science* **1993**, *259*, 340.
36. Bar-Nahum, I.; Khenkin, A. M.; Neumann, R. *J. Am. Chem. Soc.* **2004**, *126*, 10236.
37. Stahl, S. S.; Labinger, J. A.; Bercaw, J. E. *Angew. Chem. Int. Ed.* **1998**, *37*, 2181.
38. Stahl, S. S.; Labinger, J. A.; Bercaw, J. E. *J. Am. Chem. Soc.* **1996**, *118*, 5961.
39. Luinstra, G. A.; Wang, L.; Stahl, S. S.; Labinger, J. A.; Bercaw, J. E. *J. Organomet. Chem.* **1995**, *504*, 75.
40. Johansson, L.; Ryan, O. B.; Tilset, M. *J. Am. Chem. Soc.* **1999**, *121*, 1974.
41. Johansson, L.; Ryan, O. B.; Rømming, C.; Tilset, M. *J. Am. Chem. Soc.* **2001**, *123*, 6579.
42. Brainard, R. L.; Nutt, W. R.; Lee, T. R.; Whitesides, G. M. *Organometallics* **1988**, *7*, 2379.
43. Holtcamp, M. W.; Labinger, J. A.; Bercaw, J. E. *J. Am. Chem. Soc.* **1997**, *119*, 848.
44. Johansson, L.; Tilset, M.; Labinger, J. A.; Bercaw, J. E. *J. Am. Chem. Soc.* **2000**, *122*, 10846.
45. Heiberg, H.; Johansson, L.; Gropen, O.; Ryan, O. B.; Swang, O.; Tilset, M. *J. Am. Chem. Soc.* **2000**, *122*, 10831.
46. Procelewska, J.; Zahl, A.; van Eldik, R.; Zhong, H. A.; Labinger, J. A.; Bercaw, J. E. *Inorg. Chem.* **2002**, *41*, 2808.

47. Zhong, H. A.; Labinger, J. A.; Bercaw, J. E. *J. Am. Chem. Soc.* **2002**, *124*, 1378.
48. Thomas, J. C.; Peters, J. C. *J. Am. Chem. Soc.* **2001**, *123*, 5100.
49. Wang, L.; Stahl, S. S.; Labinger, J. A.; Bercaw, J. E. *J. Mol. Cat. A* **1997**, *116*, 269.
50. Rostovtsev, V. V.; Labinger, J. A.; Bercaw, J. E.; Lasseter, T. L.; Goldberg, K. I. *Organometallics* **1998**, *17*, 4530.
51. Rostovtsev, V. V.; Henling, L. M.; Labinger, J. A.; Bercaw, J. E. *Inorg. Chem.* **2002**, *41*, 3608.
52. Williams, B. S.; Holland, A. W.; Goldberg, K. I. *J. Am. Chem. Soc.* **1999**, *121*, 252.
53. Williams, B. S.; Goldberg, K. I. *J. Am. Chem. Soc.* **2001**, *123*, 2576.
54. Luinstra, G. A.; Labinger, J. A.; Bercaw, J. E. *J. Am. Chem. Soc.* **1993**, *115*, 3004.
55. Dick, A. R.; Sanford, M. S. *Tetrahedron* **2006**, *62*, 2439.
56. Dangel, B. D.; Johnson, J. A.; Sames, D. *J. Am. Chem. Soc.* **2001**, *123*, 8149.
57. Desai, L. V.; Hull, K. L.; Sanford, M. S. *J. Am. Chem. Soc.* **2004**, *126*, 9542.
58. Dick, A. R.; Hull, K. L.; Sanford, M. S. *J. Am. Chem. Soc.* **2004**, *126*, 2300.
59. Thu, H. Y.; Yu, W. Y.; Che, C. M. *J. Am. Chem. Soc.* **2006**, *128*, 9048.
60. (a) Espino, C. G.; Wehn, P. M.; Chow, J.; Du Bois, J. *J. Am. Chem. Soc.* **2001**, *123*, 6935; (b) Espino, C. G.; Du Bois, J. *Angew. Chem. Int. Ed.* **2001**, *40*, 598.
61. Fraunhoffer, K. J.; White, M. C. *J. Am. Chem. Soc.* **2007**, *129*, 7274.
62. Giri, R.; Chen, X.; Yu, J. Q. *Angew. Chem. Int. Ed.* **2005**, *44*, 2112.
63. Kalyani, D.; Dick, A. R.; Anani, W. Q.; Sanford, M. S. *Org. Lett.* **2006**, *8*, 2523.
64. Hull, K. L.; Anani, W. Q.; Sanford, M. S. *J. Am. Chem. Soc.* **2006**, *128*, 7134.
65. Powers, D. C.; Ritter, T. *Nat. Chem.* **2009**, *1*, 302.
66. Dick, A. R.; Kampf, J. W.; Sanford, M. S. *J. Am. Chem. Soc.* **2005**, *127*, 12790.
67. Zaitsev, V. G.; Shabashov, D.; Daugulis, A. *J. Am. Chem. Soc.* **2005**, *127*, 13154.
68. Stuart, D. R.; Fagnou, K. *Science* **2007**, *316*, 1172.
69. Dwight, T. A.; Rue, N. R.; Charyk, D.; Josselyn, R.; DeBoef, B. *Org. Lett.* **2007**, *9*, 3137.
70. Hull, K. L.; Sanford, M. S. *J. Am. Chem. Soc.* **2007**, *129*, 11904.
71. Jia, C.; Kitamura, T.; Fujiwara, Y. *Acc. Chem. Res.* **2001**, *34*, 633.
72. Fujiwara, Y.; Kawauchi, T.; Taniguchi, H. *J. Chem. Soc., Chem. Commun.* **1980**, 220.
73. Jintoku, T.; Taniguchi, H.; Fujiwara, Y. *Chem. Lett.* **1987**, 1159.
74. Jintoku, T.; Fujiwara, Y.; Kawata, I.; Kawauchi, T.; Taniguchi, H. *J. Organomet. Chem.* **1990**, *385*, 297.
75. Taniguchi, Y.; Yamaoka, Y.; Nakata, K.; Takaki, K.; Fujiwara, Y. *Chem. Lett.* **1995**, 345.
76. Lu, W. J.; Yamaoka, Y.; Taniguchi, Y.; Kitamura, T.; Takaki, K.; Fujiwara, Y. *J. Organomet. Chem.* **1999**, *580*, 290.
77. Lin, M.; Sen, A. *Nature* **1994**, *368*, 613.
78. Sen, A. *Acc. Chem. Res.* **1998**, *31*, 550.
79. (a) Fujiwara, Y.; Jintoku, T.; Uchida, Y. *New J. Chem.* **1989**, *13*, 649; (b) Fujiwara, Y.; Takaki, K.; Watanabe, J.; Uchida, Y.; Taniguchi, H. *Chem. Lett.* **1989**, 1687.
80. Taniguchi, Y.; Hayashida, T.; Shibasaki, H.; Piao, D. G.; Kitamura, T.; Yamaji, T.; Fujiwara, Y. *Org. Lett.* **1999**, *1*, 557.
81. Kao, L.-C.; Hutson, A. C.; Sen, A. *J. Am. Chem. Soc.* **1991**, *113*, 700.
82. Periana, R. A.; Mironov, O.; Taube, D.; Bhalla, G.; Jones, C. J. *Science* **2003**, *301*, 814.
83. Moore, E. J.; Pretzer, W. R.; Oconnell, T. J.; Harris, J.; Labounty, L.; Chou, L.; Grimmer, S. S. *J. Am. Chem. Soc.* **1992**, *114*, 5888.
84. Chatani, N.; Fukuyama, T.; Kakiuchi, F.; Murai, S. *J. Am. Chem. Soc.* **1996**, *118*, 493.
85. Chatani, N.; Fukuyama, T.; Tatamidani, H.; Kakiuchi, F.; Murai, S. *J. Org. Chem.* **2000**, *65*, 4039.
86. Chatani, N.; Ie, Y.; Kakiuchi, F.; Murai, S. *J. Org. Chem.* **1997**, *62*, 2604.
87. Fukuyama, T.; Chatani, N.; Kakiuchi, F.; Murai, S. *J. Org. Chem.* **1997**, *62*, 5647.
88. Ishii, Y.; Chatani, N.; Kakiuchi, F.; Murai, S. *Organometallics* **1997**, *16*, 3615.
89. Sakakura, T.; Tanaka, M. *J. Chem. Soc., Chem. Commun.* **1987**, 758.
90. Boese, W. T.; Goldman, A. S. *Tetrahedron Lett.* **1992**, *33*, 2119.
91. Khannanov, N. K.; Menchikova, G. N.; Grigoryan, E. A. *Russ. Chem. Bull.* **1994**, *43*, 948.
92. Rosini, G. P.; Zhu, K. M.; Goldman, A. S. *J. Organomet. Chem.* **1995**, *504*, 115.
93. Sakakura, T.; Sodeyama, T.; Sasaki, K.; Wada, K.; Tanaka, M. *J. Am. Chem. Soc.* **1990**, *112*, 7221.
94. Lowry, T. H.; Richardson, T. H. *Mechanism and Theory in Organic Chemistry*, 3rd ed.; Harper & Row, Publishers: New York, 1987; p 162.
95. Tanaka, M.; Sakakura, T.; Tokunaga, Y.; Sodeyama, T. *Chem. Lett.* **1987**, 2373.
96. Hsu, G. C.; Kosar, W. P.; Jones, W. D. *Organometallics* **1994**, *13*, 385.
97. Jones, W. D.; Foster, G. P.; Putinas, J. M. *J. Am. Chem. Soc.* **1987**, *109*, 5047.
98. Choi, J. C.; Sakakura, T. *J. Am. Chem. Soc.* **2003**, *125*, 7762.
99. Crabtree, R. H.; Mihelcic, J. M.; Quirk, J. M. *J. Am. Chem. Soc.* **1979**, *101*, 7738.
100. Crabtree, R. H.; Mellea, M. F.; Mihelcic, J. M.; Quirk, J. M. *J. Am. Chem. Soc.* **1982**, *104*, 107.

101. Crabtree, R. H.; Demou, P. C.; Eden, D.; Mihelcic, J. M.; Parnell, C. A.; Quirk, J. M.; Morris, G. E. *J. Am. Chem. Soc.* **1982**, *104*, 6994.
102. Burk, M. J.; Crabtree, R. H. *J. Am. Chem. Soc.* **1987**, *109*, 8025.
103. Baudry, D.; Ephritikhine, M.; Felkin, H.; Holmes-Smith, R. *J. Chem. Soc., Chem. Commun.* **1983**, 788.
104. Felkin, H.; Fillebeen-Khan, T.; Gault, Y.; Holmes-Smith, R.; Zakrzewski, J. *Tetrahedron Lett.* **1984**, *25*, 1279.
105. Felkin, H.; Fillebeen-Khan, T.; Holmes-Smith, R.; Yingrui, L. *Tetrahedron Lett.* **1985**, *26*, 1999.
106. Aoki, T.; Crabtree, R. *Organometallics* **1993**, *12*, 294.
107. Braunstein, P.; Chauvin, Y.; Nahring, J.; DeCian, A.; Fischer, J.; Tiripicchio, A.; Ugozzoli, F. *Organometallics* **1996**, *15*, 5551.
108. Belli, J.; Jensen, C. M. *Organometallics* **1996**, *15*, 1532.
109. Miller, J. A.; Knox, L. K. *J. Chem. Soc., Chem. Commun.* **1994**, 1449.
110. Fujii, T.; Saito, Y. *J. Chem. Soc., Chem. Commun.* **1990**, 757.
111. Fujii, T.; Higashino, Y.; Saito, Y. *J. Chem. Soc., Dalton Trans.* **1993**, 517.
112. Jensen, C. M. *Chem. Commun.* **1999**, 2443.
113. Goldman, A. S.; Renkema, K. B.; Czerw, M.; Krogh-Jespersen, K. In *Activation and Functionalization of C–H Bonds*; Goldberg, K. I., Goldman, A. S., Eds.; American Chemical Society: Washington, DC, 2004; Vol. 885, p 440.
114. Gupta, M.; Kaska, W. C.; Jensen, C. M. *Chem. Commun.* **1997**, 461.
115. Xu, W. W.; Rosini, G. P.; Gupta, M.; Jensen, C. M.; Kaska, W. C.; Krogh-Jespersen, K.; Goldman, A. S. *Chem. Commun.* **1997**, 2273.
116. Liu, F.; Pak, E. B.; Singh, B.; Jensen, C. M.; Goldman, A. S. *J. Am. Chem. Soc.* **1999**, *121*, 4086.
117. Gottker-Schnetmann, I.; Brookhart, M. *J. Am. Chem. Soc.* **2004**, *126*, 9330.
118. Gottker-Schnetmann, I.; White, P.; Brookhart, M. *J. Am. Chem. Soc.* **2004**, *126*, 1804.
119. Gottker-Schnetmann, I.; White, P. S.; Brookhart, M. *Organometallics* **2004**, *23*, 1766.
120. Ray, A.; Zhu, K. M.; Kissin, Y. V.; Cherian, A. E.; Coates, G. W.; Goldman, A. S. *Chem. Commun.* **2005**, 3388.
121. Gu, X. Q.; Chen, W.; Morales-Morales, D.; Jensen, C. M. *J. Mol. Cat. A* **2002**, *189*, 119.
122. Vidal, V.; Theolier, A.; Thivolle-Cazat, J.; Basset, J. M. *Science* **1997**, *276*, 99.
123. Maury, O.; Lefort, L.; Vidal, V.; Thivolle-Cazat, J.; Basset, J. M. *Angew. Chem. Int. Ed.* **1999**, *38*, 1952.
124. Basset, J. M.; Coperet, C.; Lefort, L.; Maunders, B. M.; Maury, O.; Le Roux, E.; Saggio, G.; Soignier, S.; Soulivong, D.; Sunley, G. J.; Taoufik, M.; Thivolle-Cazat, J. *J. Am. Chem. Soc.* **2005**, *127*, 8604.
125. Goldman, A. S.; Roy, A. H.; Huang, Z.; Ahuja, R.; Schinski, W.; Brookhart, M. *Science* **2006**, *312*, 257.
126. Kanzelberger, M.; Singh, B.; Czerw, M.; Krogh-Jespersen, K.; Goldman, A. S. *J. Am. Chem. Soc.* **2000**, *122*, 11017.
127. Renkema, K. B.; Kissin, Y. V.; Goldman, A. S. *J. Am. Chem. Soc.* **2003**, *125*, 7770.
128. Gottker-Schnetmann, I.; White, P.; Brookhart, M. *J. Am. Chem. Soc.* **2004**, *126*, 1804.
129. Li, S.; Hall, M. B. *Organometallics* **2001**, *20*, 2153.
130. Zhu, K. M.; Achord, P. D.; Zhang, X. W.; Krogh-Jespersen, K.; Goldman, A. S. *J. Am. Chem. Soc.* **2004**, *126*, 13044.
131. Gottker-Schnetmann, I.; Brookhart, M. *J. Am. Chem. Soc.* **2004**, *126*, 9330.
132. Murai, S.; Kakiuchi, F.; Sekine, S.; Tanaka, Y.; Kamatani, A.; Sonoda, M.; Chatani, N. *Nature* **1993**, *366*, 529.
133. Harris, P. W. R.; Woodgate, P. D. *J. Organomet. Chem.* **1996**, *506*, 339.
134. Harris, P. W. R.; Woodgate, P. D. *J. Organomet. Chem.* **1997**, *530*, 211.
135. O'Malley, S. J.; Tan, K. L.; Watzke, A.; Bergman, R. G.; Ellman, J. A. *J. Am. Chem. Soc.* **2005**, *127*, 13496.
136. Kakiuchi, F.; Yamauchi, M.; Chatani, N.; Murai, S. *Chem. Lett.* **1996**, 111.
137. Lim, Y. G.; Kang, J. B.; Kim, Y. H. *J. Chem. Soc., Perkin Trans. 1* **1996**, 2201.
138. Tan, K. L.; Bergman, R. G.; Ellman, J. A. *J. Am. Chem. Soc.* **2001**, *123*, 2685.
139. Tan, K. L.; Bergman, R. G.; Ellman, J. A. *J. Am. Chem. Soc.* **2002**, *124*, 3202.
140. Jun, C.-H.; Hong, J.-B.; Kim, Y.-H.; Chung, K.-Y. *Angew. Chem. Int. Ed.* **2000**, *39*, 3440.
141. Jun, C.-H.; Moon, C. W.; Hong, J.-B.; Lim, S.-G.; Chung, K.-Y.; Kim, Y.-H. *Chem. Eur. J.* **2002**, 485.
142. Thalji, R. K.; Ahrendt, K. A.; Bergman, R. G.; Ellman, J. A. *J. Am. Chem. Soc.* **2001**, *123*, 9692.
143. Thalji, R. K.; Ellman, J. A.; Bergman, R. G. *J. Am. Chem. Soc.* **2004**, *126*, 7192.
144. Kakiuchi, F.; Murai, S. In *Activation of Unreactive Bonds and Organic Synthesis*; Murai, S., Ed.; Springer: Berlin, 1999; Vol. 3, p 47.
145. Kakiuchi, F.; Murai, S. *Acc. Chem. Res.* **2002**, *35*, 826.
146. Busch, S.; Leitner, L. *Adv. Synth. Catal.* **2001**, *343*, 192.
147. Lenges, C. P.; Brookhart, M. *J. Am. Chem. Soc.* **1999**, *121*, 6616.
148. Jun, C.-H.; Hong, J.-B.; Kim, Y.-H.; Chung, K.-Y. *Angew. Chem., Int. Ed. Engl.* **2000**, *39*, 3440.

149. Kakiuchi, F.; Ohtaki, H.; Sonoda, M.; Chatani, N.; Murai, S. *Chem. Lett.* **2001**, 918.

150. Matsubara, T.; Koga, N.; Musaev, D. G.; Morokuma, K. *J. Am. Chem. Soc.* **1998**, *120*, 12692.

151. Matsubara, T.; Koga, N.; Musaev, D.; Morokuma, K. *Organometallics* **2000**, *19*, 2318.

152. Nevado, C.; Echavarren, A. M. *Synthesis* **2005**, 167.

153. Nevado, C.; Echavarren, A. M. *Chem. Eur. J.* **2005**, *11*, 3155.

154. Jia, C. G.; Piao, D. G.; Oyamada, J. Z.; Lu, W. J.; Kitamura, T.; Fujiwara, Y. *Science* **2000**, *287*, 1992.

155. Li, K.; Foresee, L. N.; Tunge, J. A. *J. Org. Chem.* **2005**, *70*, 2881.

156. Matsumoto, T.; Taube, D. J.; Periana, R. A.; Taube, H.; Yoshida, H. *J. Am. Chem. Soc.* **2000**, *122*, 7414.

157. Lail, M.; Arrowood, B. N.; Gunnoe, T. B. *J. Am. Chem. Soc.* **2003**, *125*, 7506.

158. Matsumoto, T.; Periana, R. A.; Taube, D. J.; Yoshida, H. *J. Mol. Catal.* **2002**, 1.

159. Periana, R. A.; Liu, X. Y.; Bhalla, G. *Chem. Commun.* **2002**, 3000.

160. Lail, M.; Bell, C. M.; Conner, D.; Cundari, T. R.; Gunnoe, T. B.; Petersen, J. L. *Organometallics* **2004**, *23*, 5007.

161. Liu, C.; Han, X. Q.; Wang, X.; Widenhoefer, R. A. *J. Am. Chem. Soc.* **2004**, *126*, 3700.

162. Sadow, A. D.; Tilley, T. D. *J. Am. Chem. Soc.* **2003**, *125*, 7971.

163. Weissman, H.; Song, X.; Milstein, D. *J. Am. Chem. Soc.* **2001**, *123*, 337.

164. Baran, P. S.; Corey, E. J. *J. Am. Chem. Soc.* **2002**, *124*, 7904.

165. Baran, P. S.; Guerrero, C. A.; Corey, E. J. *J. Am. Chem. Soc.* **2003**, *125*, 5628.

166. Oxgaard, J.; Muller, R. P.; Goddard, W. A.; Periana, R. A. *J. Am. Chem. Soc.* **2004**, *126*, 352.

167. Oxgaard, J.; Goddard, W. A. *J. Am. Chem. Soc.* **2004**, *126*, 442.

168. Oxgaard, J.; Periana, R. A.; William A. Goddard, I. *J. Am. Chem. Soc.* **2004**, *126*, 11658.

169. Fisher, M. B.; Zheng, Y.-M.; Rettie, A. E. *Biochem. Biophys. Res. Commun.* **1998**, *248*, 352.

170. Lawrence, J. D.; Takahashi, M.; Bae, C.; Hartwig, J. F. *J. Am. Chem. Soc.* **2004**, *126*, 15334.

171. Zweifel, G. T.; Brown, H. C. *Org. React.* **1963**, *13*, 22.

172. Brown, H. C. *Boranes in Organic Chemistry*; Cornell University Press: Ithaca, N.Y., 1972.

173. Brown, H. C. *Organic Synthesis via Boranes*; Wiley: New York, 1975.

174. Falck, J. R.; Bondlela, M.; Venkataraman, S. K.; Srinivas, D. *J. Org. Chem.* **2001**, *66*, 7148.

175. Miyaura, N.; Suzuki, A. *Chem. Rev.* **1995**, *95*, 2457.

176. Miyaura, N. In *Metal-Catalyzed Cross-Coupling Reactions*; de Meijere, A., Diederich, F., Eds.; Wiley-VCH: Weinheim, 2004; Vol. 1, p 41.

177. Ishiyama, T.; Nobuta, Y.; Hartwig, J. F.; Miyaura, N. *Chem. Commun.* **2003**, 2924.

178. Takagi, J.; Sato, K.; Hartwig, J. F.; Ishiyama, T.; Miyaura, N. *Tetrahedron Lett.* **2002**, *43*, 5649.

179. Ishiyama, T.; Takagi, J.; Yonekawa, Y.; Hartwig, J. F.; Miyaura, N. *Adv. Synth. Catal.* **2003**, *345*, 1103.

180. Kondo, Y.; Garcia-Cuadrado, D.; Hartwig, J. F.; Boaen, N. K.; Wagner, N. L.; Hillmyer, M. A. *J. Am. Chem. Soc.* **2002**, *124*, 1164.

181. Bae, C.; Hartwig, J. F.; Boaen, N. K.; Long, R. O.; Anderson, K. S.; Hillmyer, M. A. *J. Am. Chem. Soc.* **2005**, *127*, 767.

182. Bae, C.; Hartwig, J. F.; Chung, H.; Harris, N. K.; Switek, K. A.; Hillmyer, M. A. *Angew. Chem. Int. Ed.* **2005**, *44*, 6410.

183. Hartwig, J. F.; Cook, K. S.; Hapke, M.; Incarvito, C.; Fan, Y.; Webster, C. E.; Hall, M. B. *J. Am. Chem. Soc.* **2005**, *127*, 2538.

184. Boller, T. M.; Murphy, J. M.; Hapke, M.; Ishiyama, T.; Miyaura, N.; Hartwig, J. F. *J. Am. Chem. Soc.* **2005**, *127*, 14263.

185. Webster, C. E.; Fan, Y.; Hall, M. B.; Kunz, D.; Hartwig, J. F. *J. Am. Chem. Soc.* **2003**, *125*, 858.

186. Kawamura, K.; Hartwig, J. F. *J. Am. Chem. Soc.* **2001**, *123*, 8422.

187. Gustavson, W. A.; Epstein, P. S.; Curtis, M. D. *Organometallics* **1982**, *1*, 884.

188. Ezbiansky, K.; Djurovich, P. I.; Laforest, M.; Sinning, D. J.; Zayes, R.; Berry, D. H. *Organometallics* **1998**, *17*, 1455.

189. Uchimaru, Y.; El Sayed, A. M. M.; Tanaka, M. *Organometallics* **1993**, *12*, 2065.

190. Sadow, A. D.; Tilley, T. D. *Angew. Chem. Int. Ed.* **2003**, *42*, 803.

191. Sadow, A. D.; Tilley, T. D. *J. Am. Chem. Soc.* **2005**, *127*, 643.

192. Tsuji, J.; Ohno, K. *Tetrahedron Lett.* **1965**, 3969.

193. Tsuji, J.; Ohno, K. *Synthesis* **1969**, *4*, 157.

194. O'Connor, J. M.; Ma, J. *J. Org. Chem.* **1992**, *57*, 5075.

195. Vora, K. P.; Lochow, C. F.; Miller, R. G. *J. Organomet. Chem.* **1980**, *192*, 257.

196. Okano, T.; Kobayashi, T.; Konishi, H.; Kiji, J. *Tetrahedron Lett.* **1982**, *23*, 4967.

197. Vora, K. P. *Synth. Commun.* **1983**, *13*, 99.

198. Marder, T. B.; Roe, D. C.; Milstein, D. *Organometallics* **1988**, *7*, 1451.

199. Lenges, C. P.; White, P. S.; Brookhart, M. *J. Am. Chem. Soc.* **1998**, *120*, 6965.

200. Lenges, C. P.; Brookhart, M. *J. Am. Chem. Soc.* **1997**, *119*, 3165.

201. Kondo, T.; Tsuji, Y.; Watanabe, Y. *Tetrahedron Lett.* **1987**, *28*, 6229.
202. Kondo, T.; Akazome, M.; Tsuji, Y.; Watanabe, Y. *J. Org. Chem.* **1990**, *55*, 1286.
203. Roy, A. H.; Lenges, C. P.; Brookhart, M. *J. Am. Chem. Soc.* **2007**, *129*, 2082.
204. Sakai, K.; Oda, O.; Nakamura, N.; Ide, J. *Tetrahedron Lett.* **1972**, 1287.
205. Campbell, R. E.; Lochow, C. F.; Vora, K. P.; Miller, R. G. *J. Am. Chem. Soc.* **1980**, *102*, 5824.
206. Campbell, R. E.; Miller, R. G. *J. Organomet. Chem.* **1980**, *186*, C27.
207. Larock, R. C.; Oertle, K.; Potter, G. F. *J. Am. Chem. Soc.* **1980**, *102*, 190.
208. Taura, Y.; Tanaka, M.; Funakoshi, K.; Sakai, K. *Tetrahedron Lett.* **1989**, *30*, 6349.
209. Taura, Y.; Tanaka, M.; Wu, X. M.; Funakoshi, K.; Sakai, K. *Tetrahedron* **1991**, *47*, 4879.
210. Sakai, K. *Bull. Chem. Soc. Jpn.* **1993**, *51*, 733.
211. Fairlie, D. P.; Bosnich, B. *Organometallics* **1988**, *7*, 936.
212. Wu, X.-M.; Funakoshi, K.; Sakai, K. *Tetrahedron Lett.* **1992**, *33*, 6331.
213. Barnhart, R. W.; Wang, X. Q.; Noheda, P.; Bergens, S. H.; Whelan, J.; Bosnich, B. *Tetrahedron* **1994**, *50*, 4335.
214. Barnhart, R. W.; Wang, X. Q.; Noheda, P.; Bergens, S. H.; Whelan, J.; Bosnich, B. *J. Am. Chem. Soc.* **1994**, *116*, 1821.
215. Barnhart, R. W.; Bosnich, B. *Organometallics* **1995**, *14*, 4343.
216. Barnhart, R. W.; McMorran, D. A.; Bosnich, B. *Inorg. Chim. Acta* **1997**, *263*, 1.
217. Barnhart, R. W.; McMorran, D. A.; Bosnich, B. *Chem. Commun.* **1997**, 589.
218. Bosnich, B. *Acc. Chem. Res.* **1998**, *31*, 667.
219. Jun, C. H.; Kang, J. B. *Bull. Korean Chem. Soc.* **1993**, *14*, 153.
220. Jun, C. H.; Han, J. S.; Kang, J. B.; Kim, S. I. *Bull. Korean Chem. Soc.* **1994**, *15*, 204.
221. Suggs, J. W. *J. Am. Chem. Soc.* **1978**, *100*, 640.
222. Willis, M. C.; McNally, S. J.; Beswick, P. J. *Angew. Chem. Int. Ed.* **2004**, *43*, 340.
223. Willis, M. C.; Randell-Sly, H. E.; Woodward, R. L.; Currie, G. S. *Org. Lett.* **2005**, *7*, 2249.
224. Suggs, J. W. *J. Am. Chem. Soc.* **1979**, *101*, 489.
225. Jun, C. H.; Lee, H.; Hong, J. B. *J. Org. Chem.***1997**, *62*, 1200.
226. Jun, C. H.; Lee, D. Y.; Hong, J. B. *Tetrahedron Lett.* **1997**, *38*, 6673.
227. Jun, C. H.; Huh, C. W.; Na, S. J. *Angew. Chem. Int. Ed.* **1998**, *37*, 145.
228. For a recently published example of sterically controlled preferential reaction at a primary over secondary C–H bond by a hindered carbenoid intermediate, see: Thu, H.-Y.; Glenna So-Ming, T.; Huang, J.-S.; Chan, S. L.-F.; Deng, Q.-H.; Che, C.-M. *Angew. Chem. Int. Ed.* **2008**, *47*, 9747.
229. Davies, H. M. L.; Hansen, T.; Churchill, M. R. *J. Am. Chem. Soc.* **2000**, *122*, 3063.
230. (a) Doyle, M.; McKervey, M.; Ye, T. *Modern Catalytic Methods for Organic Synthesis with Diazo Compounds: From Cyclopropanes to Ylides*; Wiley: New York, 1998; (b) Pfaltz, A. In *Comprehensive Asymmetric Catalysis*; Jacobsen, E. N., Pfaltz, A., Yamamoto, H., Eds.; Springer Publishing: Berlin, 1999; Vol. 2.
231. Maas, G. *Top. Curr. Chem.* **1987**, *137*, 75.
232. Muller, P.; Bolea, C. *Helv. Chim. Acta* **2002**, *85*, 483.
233. Lim, H. J.; Sulikowski, G. A. *J. Org. Chem.* **1995**, *60*, 2326.
234. Ye, T.; Garcia, C. F.; McKervey, M. A. *J. Chem. Soc., Perkin Trans. 1* **1995**, 1373.
235. Doyle, M. P.; Phillips, I. M. *Tetrahedron Lett.* **2001**, *42*, 3155.
236. Doyle, M. P.; Hu, W. H. *J. Org. Chem.* **2000**, *65*, 8839.
237. Doyle, M. P.; Kalinin, A. V. *Tetrahedron Lett.* **1996**, *37*, 1371.
238. Wee, A. G. H. *J. Org. Chem.* **2001**, *66*, 8513.
239. Diaz-Requejo, M. M.; Belderrain, T. R.; Nicasio, M. C.; Trofimenko, S.; Perez, P. J. *J. Am. Chem. Soc.* **2002**, *126*, 896.
240. Ye, T.; McKervey, M. A.; Brandes, B. D.; Doyle, M. P. *Tetrahedron Lett.* **1994**, *35*, 7269.
241. Kennedy, M.; McKervey, M. A.; Maguire, A. R.; Roos, G. H. P. *J. Chem. Soc., Chem. Commun.* **1990**, 361.
242. McKervey, M. A.; Ye, T. *J. Chem. Soc., Chem. Commun.* **1992**, 823.
243. Roos, G. H. P.; McKervey, M. A. *Synth. Commun.* **1992**, *22*, 1751.
244. Davies, H. M. L.; Antoulinakis, E. G. *J. Organomet. Chem.* **2001**, *617*, 47.
245. Doyle, M. P. In *Catalytic Asymmetric Synthesis*; Ojima, I., Ed; Wiley-VCH: New York, 2000; Chapter 5.
246. Timmons, D. J.; Doyle, M. P. *J. Organomet. Chem.* **2001**, *617*, 98.
247. Doyle, M. P.; Ren, T. *Prog. Inorg. Chem.* **2001**, *49*, 113.
248. Doyle, M. P. *Russ. Chem. Bull.* **1999**, *48*, 16.
249. Doyle, M. P. *Enantiomer* **1999**, *4*, 621.
250. Bode, J. W.; Doyle, M. P.; Protopopova, M. N.; Zhou, Q. L. *J. Org. Chem.* **1996**, *61*, 9146.
251. Doyle, M. P.; Yan, M.; Phillips, I. M.; Timmons, D. J. *Adv. Synth. Catal.* **2002**, *344*, 91.
252. Doyle, M. P.; Kalinin, A. V. *Synlett* **1995**, 1075.

253. Doyle, M. P.; Vanoeveren, A.; Westrum, L. J.; Protopopova, M. N.; Clayton, T. W. *J. Am. Chem. Soc.* **1991**, *113*, 8982.

254. Doyle, M. P.; Protopopova, M. N.; Zhou, Q. L.; Bode, J. W.; Simonsen, S. H.; Lyrch, V. *J. Org. Chem.* **1995**, *60*, 6654.

255. Doyle, M. P.; Hu, W. H.; Valenzuela, M. V. *J. Org. Chem.* **2002**, *67*, 2954.

256. Doyle, M. P.; Hu, W. H. *Chirality* **2002**, *14*, 169.

257. Ye, T.; McKervey, M. A. *Chem. Rev.* **1994**, *94*, 1091.

258. Spero, D. M.; Adams, J. *Tetrahedron Lett.* **1992**, *33*, 1143.

259. Davies, H. M. L.; Hansen, T. *J. Am. Chem. Soc.* **1997**, *119*, 9075.

260. Davies, H. M. L.; Hansen, T.; Churchill, M. R. *J. Am. Chem. Soc.* **2000**, *122*, 3063.

261. Davies, H. M. L.; Walji, A. M.; Townsend, R. J. *Tetrahedron Lett.* **2002**, *43*, 4981.

262. Davies, H. M. L.; Hansen, T.; Hopper, D. W.; Panaro, S. A. *J. Am. Chem. Soc.* **1999**, *121*, 6509.

263. Davies, H. M. L.; Venkataramani, C. *Angew. Chem. Int. Ed.* **2002**, *41*, 2197.

264. Davies, H. M. L.; Antoulinakis, E. G.; Hansen, T. *Org. Lett.* **1999**, *1*, 383.

265. Davies, H. M. L.; Ren, P. D. *J. Am. Chem. Soc.* **2001**, *123*, 2070.

266. Muller, P.; Tohill, S. *Tetrahedron* **2000**, *56*, 1725.

267. Davies, H. M. L.; Stafford, D. G.; Hansen, T. *Org. Lett.* **1999**, *1*, 233.

268. Davies, H. M. L.; Jin, Q. H. *Tetrahedron: Asymmetry* **2003**, *14*, 941.

269. Klei, S. R.; Golden, J. T.; Tilley, T. D.; Bergman, R. G. *J. Am. Chem. Soc.* **2002**, *124*, 2092.

270. Yung, C. M.; Skaddan, M. B.; Bergman, R. G. *J. Am. Chem. Soc.* **2004**, *126*, 13033.

271. Shu, A. Y. L.; Chen, W.; Heys, J. R. *J. Organomet. Chem.* **1996**, *524*, 87.

272. Atzrodt, J.; Derdau, V.; Fey, T.; Zimmermann, J. *Angew. Chem. Int. Ed.* **2007**, *46*, 7744.

273. Hesk, D.; Das, P. R.; Evans, B. *J. Labelled Compd. Radiopharm.* **1995**, *36*, 497.

274. Shu, A. Y. L.; Saunders, D.; Levinson, S. H.; Landvatter, S. W.; Mahoney, A.; Senderoff, S. G.; Mack, J. F.; Heys, J. R. *J. Labelled Compd. Radiopharm.* **1999**, *42*, 797.

275. Ellames, G. J.; Gibson, J. S.; Herbert, J. M.; Kerr, W. J.; McNeill, A. H. *J. Labelled Compd. Radiopharm.* **2004**, *47*, 1.

276. Johansen, S. K.; Sorenson, L.; Martiny, L. *J. Labelled Compd. Radiopharm.* **2005**, *48*, 569.

CHAPTER 19

Transition Metal-Catalyzed Coupling Reactions

19.1. Overview of Cross-Coupling

Catalytic nucleophilic substitution reactions comprise some of the most commonly used catalytic processes in synthetic organic chemistry. Substitutions at aromatic and vinylic halides and sulfonates, shown generically in Equation 19.1, are commonplace in the preparation of pharmaceutical candidates, have often been used in the syntheses of natural products, and have been used many times in the syntheses of sophisticated conjugated organic materials. These metal-catalyzed reactions are typically called cross-coupling reactions.[1–19]

$$ \text{Ar}-X \ + \ RM \ \xrightarrow{\text{Catalyst}} \ \text{Ar}-R \qquad (19.1) $$

The original cross-coupling reactions formed carbon–carbon bonds. However, cross-coupling reactions have now been developed that form carbon–heteroatom bonds. A majority of these reactions form carbon–nitrogen bonds in arylamines, but much effort has also been spent to develop catalysts for the formation of carbon–oxygen and carbon–sulfur bonds. Even catalysts and reaction conditions for the formation of carbon–phosphorus, carbon–silicon, and carbon–boron bonds have now been developeed. Cross-coupling reactions that form carbon–heteroatom bonds are presented after the discussion of cross-coupling reactions that form C–C bonds.

In addition to metal-catalyzed cross-coupling reactions to form the C–C bond in biaryls, homocouplings to form the C–C bond in biaryls have been developed. Homocouplings were traditionally conducted with stoichiometric amounts of copper (Equation 19.2),[20–23] but transition metal-catalyzed homocouplings (Equation 19.3) of aryl halides with a reducing agent have been developed. In other cases, transition metal-catalyzed homocouplings of main group organometallic compounds have been observed.[24–35]

$$ 2 \ \text{Ar}-I \ \xrightarrow{\text{Cu}} \ \text{Ar}-\text{Ar} \qquad (19.2) $$

$$ 2 \ \text{Ar}-X \ + \ M \ \xrightarrow{\text{Catalyst}} \ \text{Ar}-\text{Ar} \ + \ MX_2 \qquad (19.3) $$

In addition to coupling reactions that occur with aryl and vinyl nucleophiles and electrophiles, coupling reactions that occur with sp^3-hybridized nucleophiles or electrophiles have been developed. These reactions include those that form tertiary and quaternary stereocenters from racemic or prochiral nucleophiles, as shown in Equation 19.4. Substitution reactions at propargylic[36] and benzylic electrophiles[37] have also been reported, and several groups have reported in recent years progress in metal-catalyzed substitutions of alkyl electrophiles, including enantioselective substitutions of aliphatic organic halides.

$$
\begin{array}{c}
\underset{\substack{R^1 \quad R^2 \\ \text{+ enantiomer}}}{R^3 \diagdown M} \;+\; \underset{\text{or vinyl X}}{Ar–X} \xrightarrow{\text{Catalyst}} \underset{\substack{R^1 \quad R^2}}{R^3 \diagdown Ar} \;\text{or}\; \underset{\substack{R^1 \quad R^2 \\ \text{enantioenriched}}}{R^3 \diagdown \text{Vinyl}}
\end{array} \tag{19.4}
$$

Copper complexes also catalyze cross coupling. A renewed interest in copper-catalyzed coupling, principally by developing copper catalysts containing ancillary ligands, has led to significant improvements in these earliest classes of coupling reactions that were first reported nearly one hundred years ago. Copper-catalyzed coupling reactions are presented after the discussion of coupling reactions catalyzed by complexes of non-coinage transition metals.

Finally, coupling reactions that occur without a leaving group, a main group metal reagent, or both have been reported. These "direct arylation" reactions are a form of C–H bond activation, typically of aromatic or heteroaromatic C–H bonds. Due to the close relationship between these processes and cross-coupling reactions, however, direct arylations are described in the final section of this chapter after copper-catalyzed coupling processes.

Thus, this chapter focuses on the most investigated classes of metal-catalyzed coupling reactions—namely, reactions of aromatic and aliphatic electrophiles with main group carbon or nitrogen nucleophiles, reactions of aromatic halides with olefins, including enantioselective versions of these reactions, and direct coupling processes. The mechanisms of these reactions are presented with reference to the chapters on the stoichiometric steps of these catalytic processes.

19.2. The Classes of C–C Bond-Forming Coupling Reactions

19.2.1. Early Studies on Cross-Coupling: Coupling with Organomagnesium Reagents

In 1972, Kumada and Corriu independently reported the reaction of aryl halides with Grignard reagents in the presence of catalytic amounts of $Ni(dppb)Cl_2$ or the combination of $Ni(acac)_2$ and dppp (Equation 19.5).[38,39] At a similar time, Kochi reported reactions catalyzed by halide and acetylacetonate complexes of iron, but these complexes, until most recently, have generally been less active as catalysts for this chemistry than complexes of the nickel triad.[40] Nearly thirty years before these studies, Kharasch had shown that simple metal salts of cobalt and chromium catalyze the coupling of aryl Grignard reagents with aryl and vinyl halides.[41] Although cross coupling has been known for many years, developments in the past decade have made these reactions accessible and reliable for synthetic organic chemistry. Because the synthesis of unsymmetrical products is more general than the synthesis of symmetrical products, the greatest advances have been made in cross coupling.

$$
\begin{array}{c}
\underset{\substack{R^1 \diagdown X}}{\bigcirc\!\!\!-Br} \;+\; R^2MgX \xrightarrow[\substack{Ni(acac)_2, \\ (dppe)NiCl_2 \\ FeCl_3, \\ CoCl_2 \ or \ CrCl_2}]{Et_2O, \ 25\,°C} \underset{\substack{R^1 \diagdown R^2}}{\bigcirc\!\!\!-R^2} \;\text{or}\;
\end{array} \tag{19.5}
$$

19.2.2. Coupling of Organozinc Reagents

Since the reports of Kumada and Corriu, many advances have been made in the types of carbon nucleophiles that can be used in cross-coupling processes. Although Grignard reagents are easy to generate, and many are commercially available, the tolerance of these reagents for functional groups is lower that that of many other main-group organometallic reagents. Other main-group organometallic reagents, such as organozinc, organotin, organoboron, and organosilicon reagents are less basic and are more tolerant of functional groups. For this reason, much effort has been spent on using these types of organometallic reagents in cross-coupling chemistry.

Negishi first reported useful reactions of organozinc reagents with aryl halides[42] (Equation 19.6), and the coupling of organozinc reagents with organic halides has become known as Negishi coupling. Negishi also reported reactions of organozirconium and organo-aluminum reagents with aryl halides.[43] Although the first reactions reported by Negishi were catalyzed by nickel complexes, he, along with Cassar,[44] Heck,[45] Sonogashira,[46] and Murahashi,[47–49] demonstrated the advantages of palladium catalysts for cross coupling. Palladium catalysts tend to be less sensitive to oxygen, and are believed to be less toxic than nickel complexes. Furthermore, they tend to react without the intervention of radical intermediates that can lead to side products, such as those from homocoupling, and to products that have undergone racemization at a saturated carbon or loss of stereochemistry at a vinylic carbon.

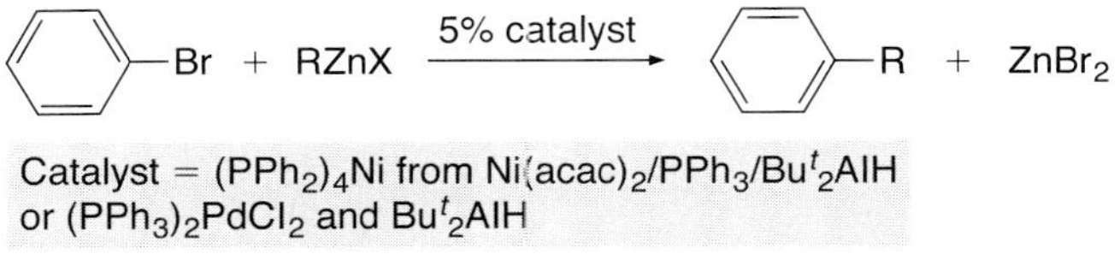

$$\text{Catalyst} = (PPh_2)_4Ni \text{ from } Ni(acac)_2/PPh_3/Bu^t_2AlH$$
$$\text{or } (PPh_3)_2PdCl_2 \text{ and } Bu^t_2AlH$$

(19.6)

19.2.3. Coupling of Organotin Reagents

Kosugi and Migita reported reactions of organotin reagents with acyl and aryl halides in 1977,[50] and Beletskaya also published early work on the coupling of aryl halides with organotin reagents in the early 1980s. Stille published some early work on the coupling of acyl and benzyl halides and demonstrated the synthetic utility of the reactions of organotin reagents with aryl halides (Equation 19.7).[51] The coupling of aryl halides with organotin reagents is now, most commonly, called the Stille reaction. The organotin reagents are stable to chromatography, water, and oxygen. Thus, they can be purified more readily than other main group organometallic reactions prior to the coupling process. However, organotin reagents are toxic, as are the tin halide byproducts of these reactions. As a result, the Stille reaction has been used many times for small-scale syntheses, but this reaction has rarely been used on large scale, particularly for the synthesis of pharmaceuticals.

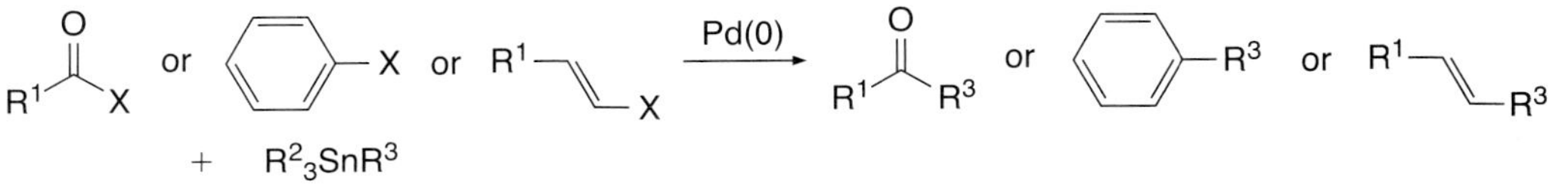

(19.7)

19.2.4. Coupling of Organosilicon Reagents

The silicon analogs of the tin reagents are less toxic, but are also less reactive. Yet, Hiyama showed that silicon reagents undergo cross-coupling reactions in the presence of added fluoride

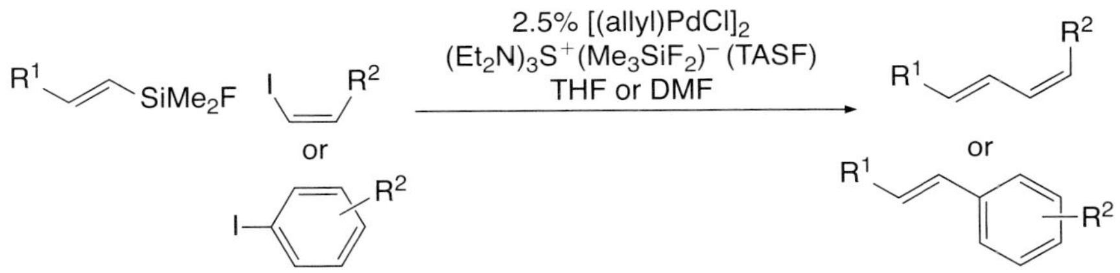

(19.8)

(Equation 19.8).[52–60] The addition of fluoride converts the four-coordinate silicon reagent into an anionic hypervalent silicon reagent. The anionic organosilicon reagent is more nucleophilic than the neutral silane and transfers the organic group from silicon to the palladium catalyst. The silicon reagent contains fluoride substituents in the work of Hiyama and alkoxo substituents in the work of Tamao and Itoh. These electronegative substituents make the silicon electrophilic enough to bind fluoride. These reactions are now called "Hiyama couplings" or "Hiyama–Tamao couplings." More recently Denmark has developed coupling reactions of organic silanolates with aryl and vinyl halides in the absence of fluoride additive.[60]

19.2.5. Coupling of Organoboron Reagents

The coupling of organoboron reagents has become the most commonly used cross-coupling process. Organoboron reagents are less toxic than organotin reagents and tend to undergo coupling reactions in the presence of a variety of functional groups. Like neutral organosilicon reagents, however, neutral organoboron reagents do not undergo metal-catalyzed cross coupling without an additive. Suzuki showed that addition of a hard base, such as hydroxide or fluoride, causes the organoboron reagent to undergo cross-coupling by generating a four-coordinate anionic organoboron reagent that transfers the organic group from boron to the metal catalyst (Equation 19.9).[61–63] This type of coupling reaction is now called "Suzuki coupling."[1,11,64]

$$(19.9)$$

The most common reagents for Suzuki reactions are boronic acids. These organoboron compounds are stable to water and oxygen and can be purified by crystallization. They are generally prepared by the reaction of a trialkylborate $[B(OR)_3]$ with a Grignard reagent, followed by acidic workup. Alternatively, air-stable organotrifluoroborates have been used in Suzuki couplings,[65–68] and these reagents are generated by the addition of KHF_2 to the boronic acid.[69] Trialkylboranes and alkenyldialkylboranes, along with added base, are also used in Suzuki couplings. These reagents are typically generated by hydroboration of an olefin or alkyne (Equation 19.10) to generate the trialkylborane or hydroboration of an alkyne to generate a vinyl borane. Most often, the hydroboration is conducted with 9-BBN (9-borabicyclononane) because the B–C bonds of the bicyclic BBN unit are not cleaved during the transmetallation process, and the alkyl or vinyl group is transferred selectively.

$$(19.10)$$

19.2.6. Coupling of Alkynes

Compounds containing reactive C–H bonds have been shown to undergo cross coupling in the presence of catalyst and base without initial formation and isolation of a main-group derivative. One classic version of this coupling is the reaction of an aryl halide with

$$(19.11)$$

a terminal alkyne to form an alkynylarene (Equation 19.11). Heck[45] and Cassar[44] independently reported this reaction in 1975; Sonogashira reported that reactions with added copper occur under milder conditions.[46] The copper additive is believed to generate a copper acetylide, which then reacts with the palladium catalyst during the coupling process. The coupling of aryl and vinyl triflates occurs under mild conditions in the absence of copper additive.[70] The coupling of an alkyne with an aryl halide is now called the "Sonogashira reaction."[71, 72]

19.2.7. Coupling of Enolates and Related Reagents

An additional set of palladium-catalyzed couplings of aryl and vinyl halides and sulfonates with compounds containing slightly acidic C–H bonds has been developed recently.[73, 74] These reactions couple aryl and vinyl groups with the enolates of ketones,[75–79] esters,[80–85] amides,[86–89] aldehydes,[90–92] nitriles,[93–97] malonates,[76, 98–102] cyanoacetates,[98, 103] cyclic 1,3-diketones,[79] and nitroalkanes[104] in the presence of a base of appropriate strength (Equation 19.12). In most cases, the reactions are conducted by combining the compound with the acidic C–H bond and an alkali metal base at room temperature in the same mixture as the aryl halide and catalyst, but in other cases the enolate is generated *in situ* prior to addition of the other reagents. Procedures for the addition of zinc and silicon enolates have also been developed to address the drawbacks of the strong basicity of alkali metal enolates (Equation 19.13).[82, 83, 105] The reactions of the zinc reagents occur without additives, but the reactions of silylketene acetals, silylketimine acetals and silylenol ethers require an activator, such as added fluoride. The reactions of aryl halides with ketones and base occur selectively at the less-hindered α-carbon, whereas the regiochemistry of the coupling of silicon enolates is controlled by the isomeric composition of the silyl enol ether or silyl ketene acetal. Reactions that generate quaternary carbons can be conducted with high enantioselectivity in some cases.[78, 85, 87, 106, 107]

$$(19.12)$$

$$(19.13)$$

19.2.8. Coupling at Aliphatic Electrophiles

Most couplings that join sp^2- with sp^3-hybridized centers are conducted with aryl or vinyl electrophiles and alkyl nucleophiles. The reverse coupling of an alkyl electrophile with an aryl nucleophile is more challenging because the alkylpalladium halide intermediate usually undergoes β-hydrogen elimination faster than it undergoes transmetallation. Yet, the proper choice of ligand and metal has led to successful couplings of sp^3-electrophiles. The range of complexes that catalyze the coupling of aliphatic electrophiles is growing rapidly.

Nickel complexes, in combination with an electron-poor olefin, led to some of the first couplings of alkyl halides in high yields.[108–110] Nickel complexes containing the tridentate nitrogen "pybox" ligands,[111] and palladium complexes containing trialkylphosphines with the appropriate steric properties have now been shown to catalyze the coupling of aryl nucleophiles with alkyl halides or tosylates (Equation 19.14a).[112–118] The coupling of secondary alkyl halides is particularly challenging because of the slow rate of oxidative addition of such alkyl electrophiles (see Chapter 7 on the oxidative addition of polar reagents). However, the combination of nickel halides and aminoalcohols or diamines, which are unusual ligands for coupling reactions catalyzed by late transition metals, have been shown to catalyze couplings of alkyl halides with arylboron[119] and arylsilicon[120] reagents.

These reactions are likely to occur by oxidative addition through a radical mechanism, as evidenced by the loss of stereochemistry of the starting alkyl halide during the coupling process (Equation 19.14b).[119] Despite the radical mechanism, some reactions of benzylic electrophiles have been conducted enantioselectively (Equation 19.14c).[121] Even reactions of alkylboron reagents with secondary alkyl halides catalyzed by nickel complexes have now been reported.[122] These reactions were conducted with nickel precursors in combination with *trans*-1,2-cyclohexanediamine (Equation 19.14d).[122]

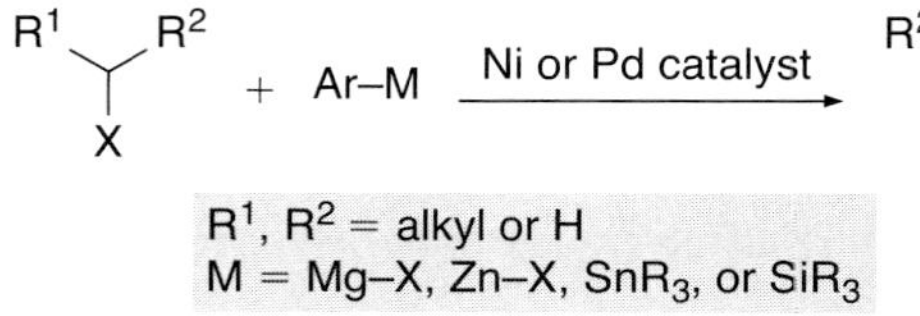

$$R^1, R^2 = \text{alkyl or H}$$
$$M = \text{Mg–X, Zn–X, SnR}_3\text{, or SiR}_3$$

(19.14a)

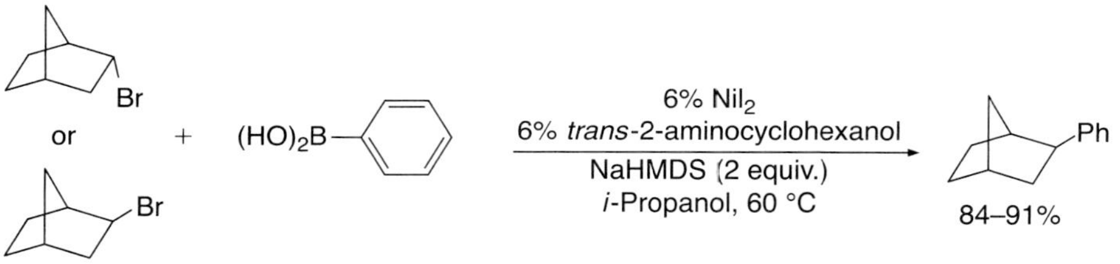

(19.14b)

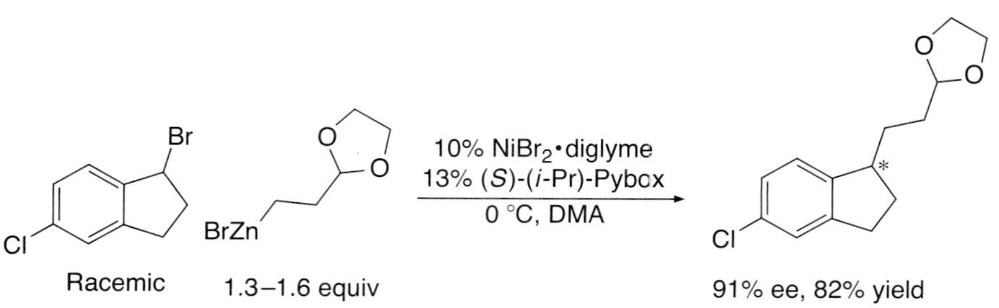

(19.14c)

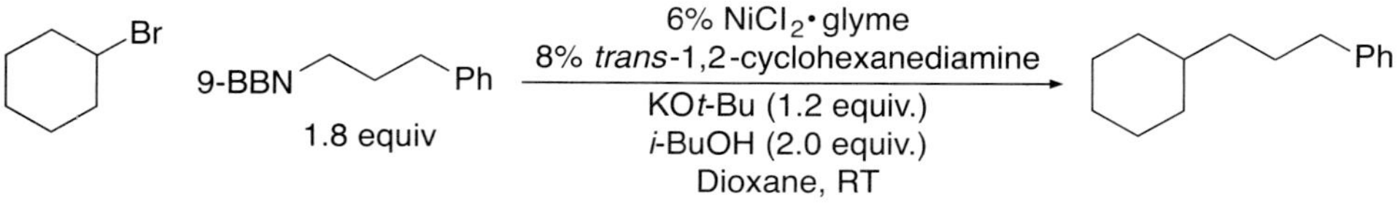

(19.14d)

A series of papers have also reported the coupling of alkyl and aryl electrophiles with aryl Grignard reagents catalyzed by iron (Equation 19.15) and cobalt complexes.[123–132] These reactions build upon Kochi's and Molander's early results on coupling reactions catalyzed by complexes of these metals.[40,133–137] The recent reactions have been conducted with simple metal salts in many cases and with discrete metal complexes as catalyst precursors in others. Although little recent mechanistic data is available on these reactions, they have earlier been shown to involve radical intermediates. Kochi concluded that the catalytic process occurs by an Fe(I)–Fe(III) couple; reactions of optically active alkyl halides generate racemic coupled products and reactions of diastereomerically pure alkyl halides generate equal ratios of diastereomeric products, as depicted in Equations 19.16 and 19.17.[123]

$$\text{EWG}-\!\!\!\diagdown\!\!\!\diagup\!\!\!-X \quad \text{or} \quad \diagup\!\!\diagdown_{N}\!\!-X \xrightarrow[\text{5\% Fe(acac)}_3]{\text{RMgX}} \text{EWG}-\!\!\!\diagdown\!\!\!\diagup\!\!\!-R \quad \text{or} \quad \diagup\!\!\diagdown_{N}\!\!-R \quad (19.15)$$

X = Cl, OTf, or OTs

$$\underset{R^1}{\overset{Br}{\diagup}}\!\!R^2 + \text{Ar–Mg–X} \xrightarrow[\text{5\% FeCl}_3]{0\,^{\circ}\text{C}} \underset{R^1}{\overset{Ar}{\diagup}}\!\!R^2 \qquad (19.16)$$
TMEDA

$$\text{Bu}^t\diagup\!\!\diagdown\!\!\diagup\!\!-Br \quad \text{or} \quad \text{Bu}^t\diagup\!\!\diagdown\!\!\diagup\!\!\overset{Br}{\diagdown} \xrightarrow[0\,^{\circ}\text{C}]{\text{FeCl}_3} \text{Bu}^t\diagup\!\!\diagdown\!\!\diagup\!\!-Ar \qquad (19.17)$$
+ ArMgX   trans/cis = 96 : 4

19.2.9. Coupling of Olefins

The final common class of coupling reactions to form C–C bonds described here is the coupling of an aryl halide with an olefin to cleave the C–H bond of the olefin and replace it with an aryl group. This reaction, which is shown generically in Equation 19.18, was first reported by Mizoroki;[138,139] the synthetic utility of this process and the most useful conditions for this process at the time were reported by Heck.[45,140,141] This process is often called the "Heck reaction," or more appropriately the "Mizoroki–Heck reaction."[14–17,19,142–144] The Heck reaction is most commonly conducted with electron-deficient olefins, such as styrene or acrylate derivatives. The electronic properties of these substrates tend to favor formation of the conjugated products. The reaction can also be conducted effectively with ethylene; a Heck reaction between 6-methoxy-2-bromonaphthalene and ethylene is one step of a short, catalytic commercial synthesis of naproxen.[145–148] In contrast, intermolecular reactions of internal olefins typically form mixtures of regioisomeric products.[149] Intramolecular Mizoroki–Heck reactions with internal olefins are more common.[14,15] Mizoroki–Heck reactions of aliphatic electrophiles have been reported, but remain rare.[150,151] Applications of the Mizoroki–Heck reaction have been reviewed.[16]

$$\diagup\!\!\diagdown\!\!-X + =\!\!\diagdown_{R} \xrightarrow[\text{Base}]{\text{Pd catalyst}} \diagup\!\!\diagdown\!\!-\!\!\diagup\!\!\diagdown_{R} \qquad (19.18)$$

19.2.10. Coupling of Cyanide

A less common, but developing, coupling process is the reaction of aryl halides with cyanide (Equation 19.19).[152–169] One might expect these reactions to be difficult to catalyze because cyanide tends to displace ligands from transition metals to form stable cyanide complexes. Although examples of the coupling of aryl halides with cyanides was reported a number of years ago, the catalysts did tend to be deactivated by the presence of high concentrations of cyanide. Most couplings occur with aryl bromides and iodides, but some couplings of aryl chlorides have been reported.[162,166]

$$(19.19)$$

Catalysts containing chelating ligands and conditions that limit the concentration of cyanide have allowed the development of the coupling of cyanide with aryl halides with improved catalyst lifetimes. Recent procedures have been devised that slowly release cyanide from a cyanohydrin[170] or trimethylsilylcyanide[163] or that include less reactive cyanide reagents, such as zinc cyanides. High-yield reactions have also been reported with potassium hexacyanoferrate(II) as a safe cyanide source (Equation 19.19).[167-169] This reagent releases little free cyanide, but does transfer cyanide in the catalytic process. It has such little toxicity that it is even used as a food additive.

19.3. Enantioselective Cross Coupling

Several forms of enantioselective cross coupling have been developed. In one form, a racemic phenethyl Grignard or zinc reagent couples with an aryl or vinyl halide. Because racemization of the phenethyl Grignard reagent is faster than the coupling process, the racemic Grignard reagent can form the coupled product in high enantiomeric excess (Equation 19.20).[171-175] In a second form, a prochiral enolate couples with an aryl or vinyl halide to generate a product containing a new stereocenter located α to the carbonyl group (Equation 19.21).[78,85,87,106,176]

$$(19.20)$$

$$(19.21)$$

In a third form, the cross-coupling process forms a biaryl compound that possesses axial chirality as a result of slow atropisomerism (Equations 19.22 and 19.23).[177–182] In most cases, this reaction has been studied for the formation of 1,1'-binaphthyls (Equation 19.22), which are found in BINAP, BINOL, and related ligands. In a fourth form, the Heck reaction has been conducted in a manner that forms a product containing a new stereocenter (Equations 19.24–19.26).[14,18,183] This type of enantioselective coupling has been the most widely used of the four in the synthesis of natural products.[14]

$$(19.22)$$

$$(19.23)$$

$$(19.24)$$

$$(19.25)$$

$$(19.26)$$

The coupling of phenethyl Grignard reagents was achieved most successfully with ligands developed by Hayashi and Kumada and by Kellogg that contain one phosphorus donor and an accompanying nitrogen, oxygen or sulfur substituent.[171–173,184–187] Several examples of these ligands are shown in Figure 19.1. The highest enantioselectivities were obtained for the coupling of β-bromostyrene with the combination of nickel catalyst and Grignard reagent[171,172] or palladium catalyst and zinc reagent[173,175] generated by transmetallation of the Grignard reagent to zinc halide (Equation 19.27). More recently, couplings that generate optically active compounds possessing a quaternary stereocenter have been prepared by desymmetrization of substrates containing two borane[188] or two triflate units (Equations 19.28 and 19.29).[189]

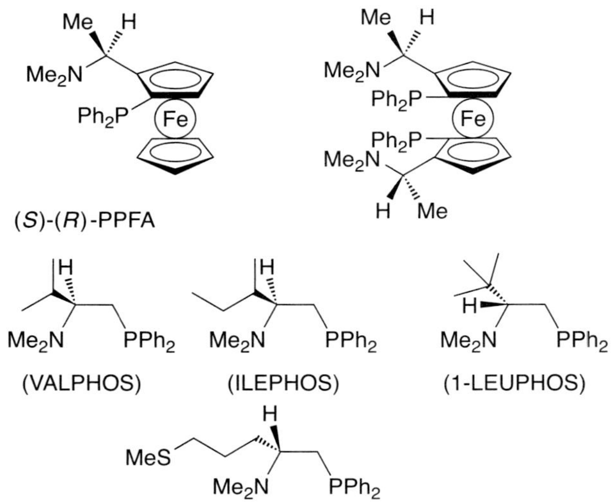

Figure 19.1.
Ligands used with nickel and palladium catalyst precursors for the coupling of phenethyl Grignard reagents with vinyl halides in > 80% ee.

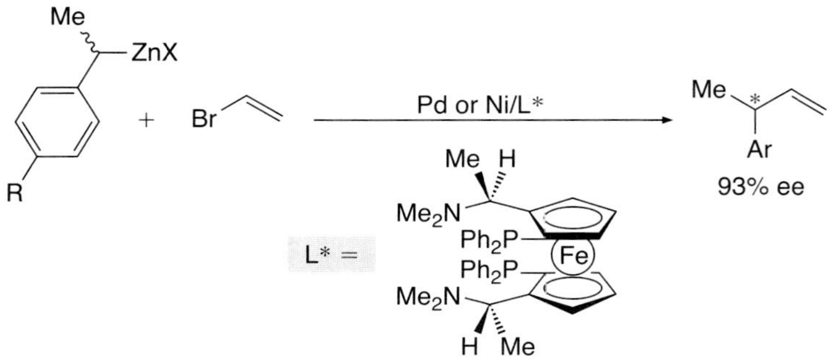

(19.27)

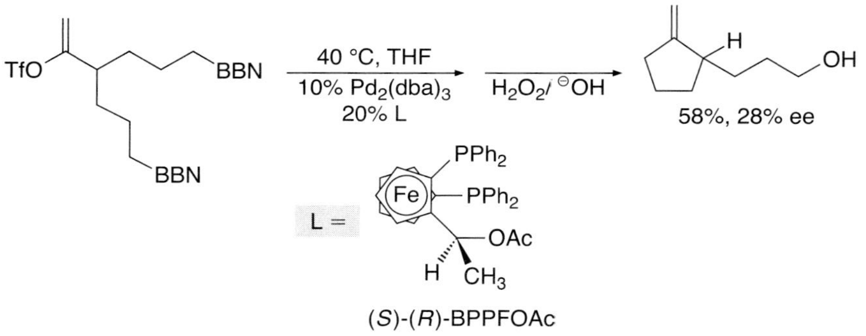

(19.28)

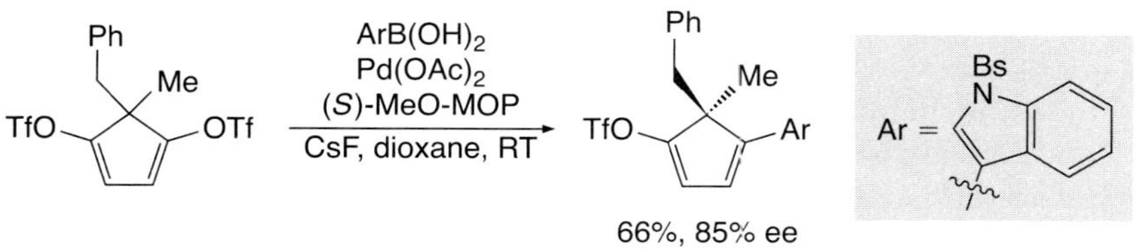

(19.29)

Asymmetric coupling of enolates with aryl halides and vinyl halides is an emerging enantioselective cross-coupling process. In this process, a ketone, ester, or amide with a single α-hydrogen is deprotonated to form the achiral enolate. This enolate then undergoes coupling with an aryl halide. Intermolecular enantioselective couplings of ketones with palladium or nickel catalysts containing bisphosphines possessing a biaryl backbone, or monophosphines possessing a biaryl substituent,[78,106,107,176] enantioselective couplings of lactones with nickel catalysts containing BINAP as ligand,[85] and intramolecular couplings of amides to form oxindoles with palladium catalysts bearing optically active carbenes[87] have been reported (Equations 19.30–19.32).

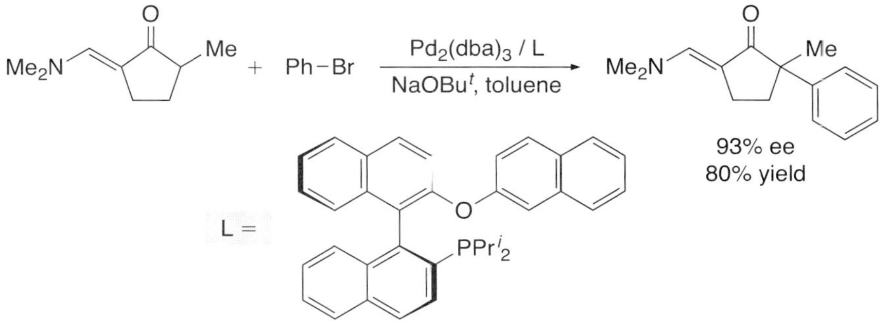

(19.30a)

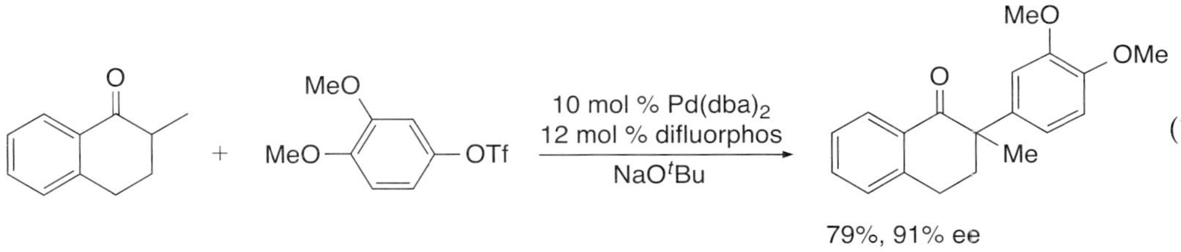

(19.30b)

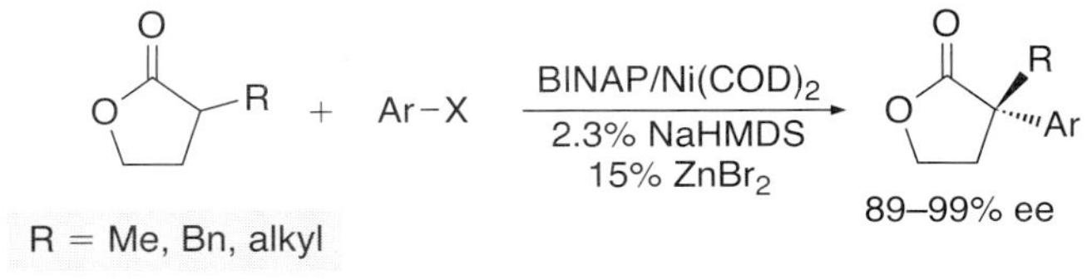

(19.31)

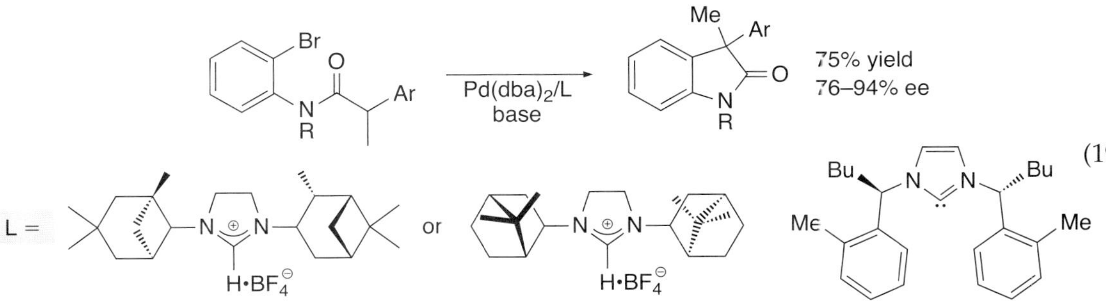

(19.32)

Several systems have now been developed for enantioselective couplings that form biaryls. The original results were obtained by Hayashi and Ito on couplings to form binaphthyls catalyzed by nickel complexes containing a methoxyethylferrocenylphosphine (Equation 19.33)[190] and on desymmetrization of a biarylditriflate (Equation 19.34).[177,178] More recently, the synthesis of biaryls by Suzuki and Kumada cross coupling of aryl halides has been investigated.[179–182] Several combinations of palladium and non-racemic chiral ligands have been tested for this reaction. High yields and enantioselectivities were observed with a catalyst containing a binaphthyl monophosphine that is a dicyclohexyl-phosphino version of Hayashi's MOP and Kocovsky's MAP ligands (Equation 19.35).[179]

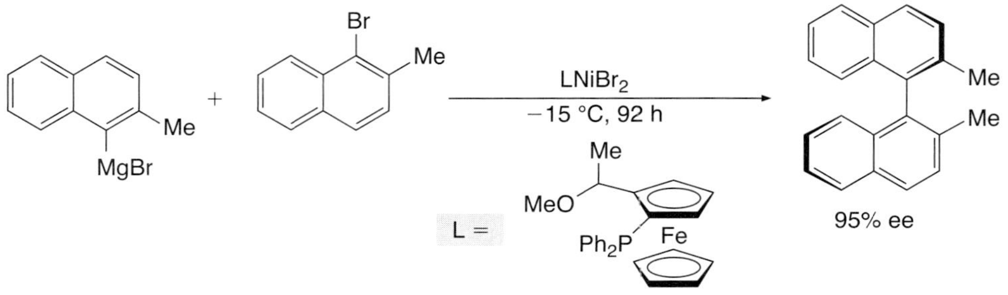

(19.33)

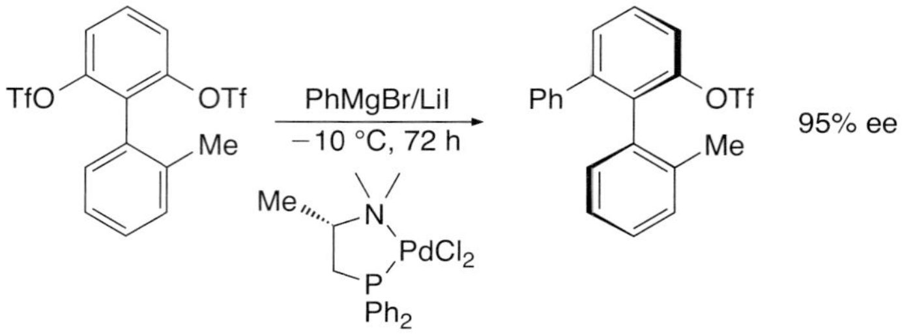

(19.34)

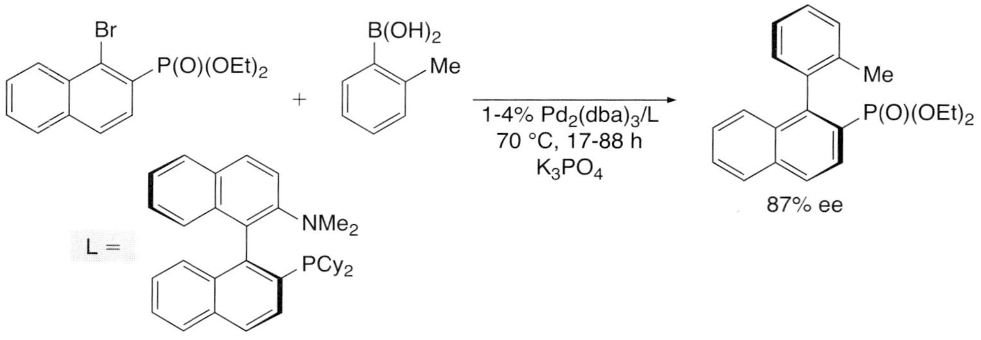

(19.35)

Shibasaki (Equation 19.36) and Overman first reported the enantioselective Heck reaction.[191,192] A few substrate combinations react intermolecularly with high enantiose-lectivity (Equation 19.37),[193,194] but most enantioselective Heck reactions used in synthetic applications have been conducted intramolecularly. The asymmetric Heck reaction in Equation 19.37[195] has begun to be a test-bed for new chiral ligands. Although most Heck reactions that occur with high enantioselectivity have been conducted with BINAP as ligand, a phosphinooxazoline ligand has been used successfully (vide infra).[196] Overman has applied an intramolecular version of the Heck reaction to form oxindoles (Equation 19.38)[197,198] in the synthesis of a variety of natural products.[14]

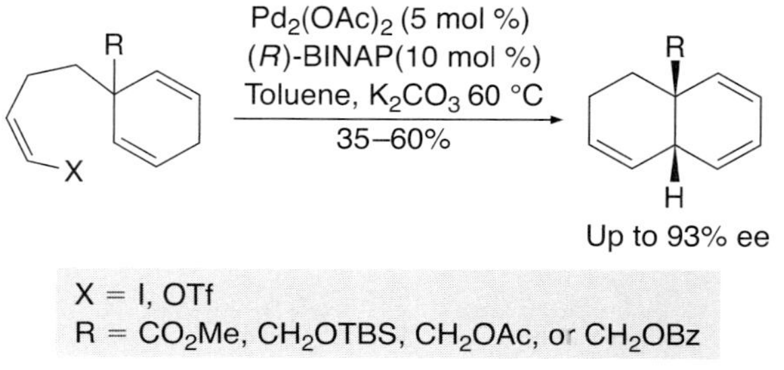

(19.36)

X = I, OTf
R = CO$_2$Me, CH$_2$OTBS, CH$_2$OAc, or CH$_2$OBz

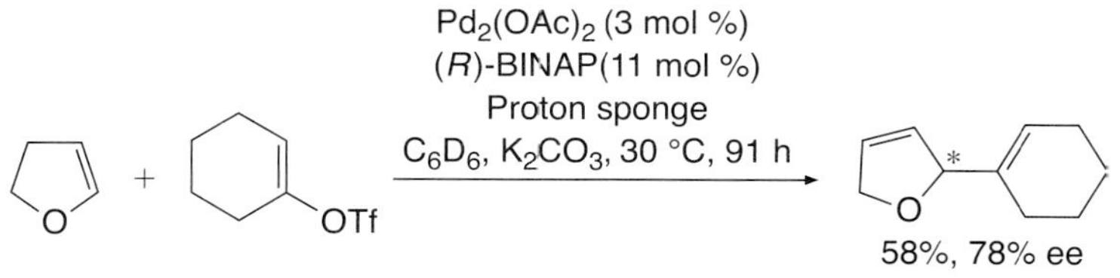

$$Pd_2(OAc)_2 \text{ (3 mol \%)}$$
$$(R)\text{-BINAP(11 mol \%)}$$
Proton sponge
$$C_6D_6, K_2CO_3, 30\ °C, 91\ h$$

58%, 78% ee

(19.37)

$$Pd_2(dba)_2 \text{ (5 mol \%)}$$
$$(R)\text{-BINAP(10 mol \%)}$$
DMA, Base, 60–80 °C

81–86%
70–71% ee

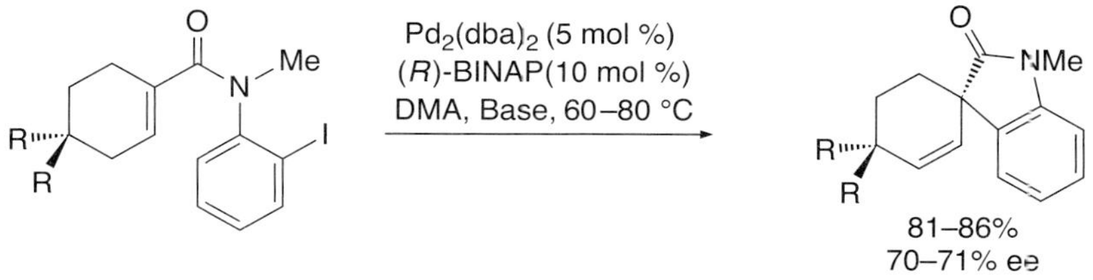

(19.38)

At first glance, one might think that the Heck reaction could not be conducted enantioselectively because it forms an olefin. However, Scheme 19.1 shows how the geometric requirements for β-hydrogen elimination can divert the reaction of a cyclic olefin to a product containing a new stereocenter. Recall from Chapter 10 that β-hydrogen elimination is most rapid when the complex can adopt a conformation with a syn coplanar arrangement of the metal, the hydrogen, and two carbons of the incipient olefin. Thus, reactions of a cyclic substrate or a substrate that generates a quaternary carbon upon insertion will occur by β-hydrogen elimination at the allylic carbon of the starting olefin instead of the hydrogen at the original olefin. In this case, a stereocenter forms at one of the carbons of the starting olefin. With an acyclic substrate, this geometrical constraint is not present, and no new stereocenter is formed unless a quaternary carbon is generated. This scenario is rare because intermolecular insertion of a 1,1-disbustituted olefin into a metal–carbon bond to form a quaternary carbon is slow. Thus, most successful applications of Heck reactions to form quaternary stereocenters are based on intramolecular reactions.

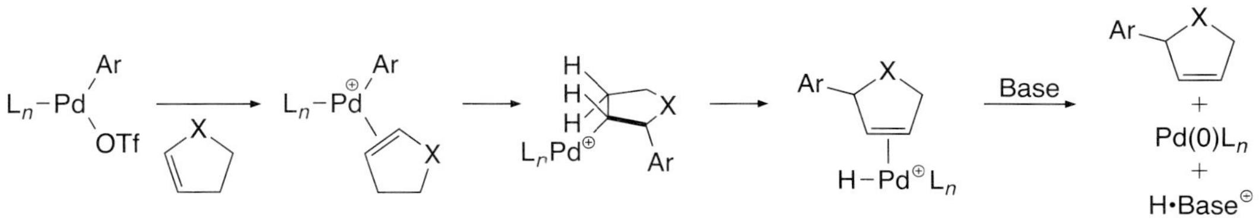

Scheme 19.1

The ligand most commonly used for the enantioselective Heck reaction is the chelating ligand BINAP. As noted in Section 19.4.3.1 on the effect of chelation, the Heck reactions of aryl triflates with catalysts containing chelating ligands are faster than those of aryl halides with the same catalysts because the olefin can more readily coordinate by displacement of the triflate than by displacement of the halide. However, intramolecular, enantioselective reactions with aryl halides catalyzed by complexes of BINAP have been reported. These reactions have been conducted in two ways. First, they have been conducted with a base containing a silver counterion to abstract the halide and generate a cationic palladium intermediate that binds olefin prior to the insertion step (Equation 19.38).[197] Second, they have been conducted with organic bases, and these reactions are thought to occur by the formation of neutral, five-coordinate olefin complexes prior to the insertion step.[198,199] Five-coordinate palladium–olefin complexes, although relatively unstable, are thought to form in these reactions because coordination of the olefin is intramolecular.

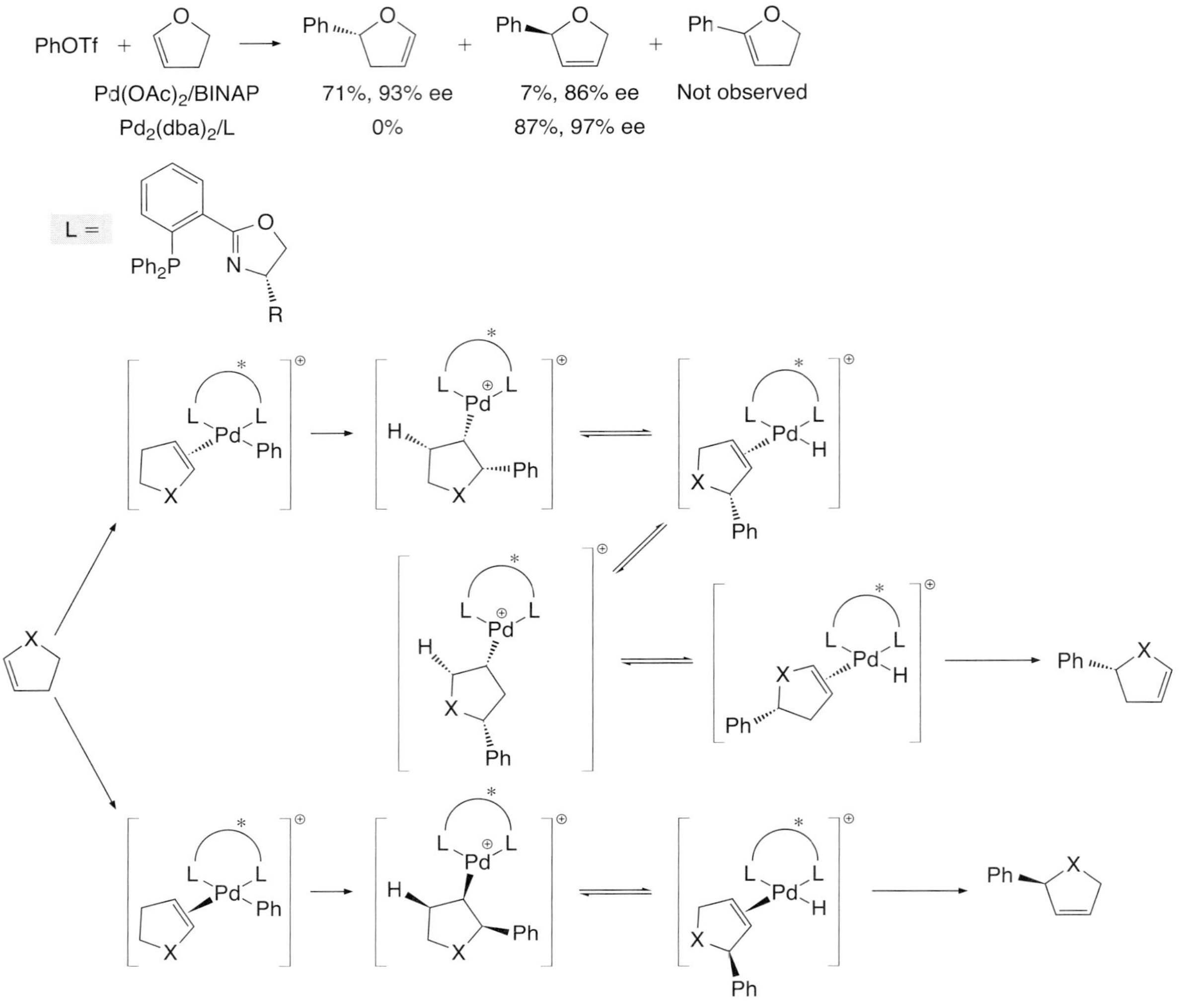

Scheme 19.2

Intermolecular reactions of aryl triflates with cyclic olefins have been studied with catalysts containing many ligands other than BINAP. The combination of phenyl triflate and 2,3-dihydrofuran has been studied most extensively, but enantioselective reactions of related 2,3-dihydropyrrole derivatives, as well as cyclopentene, have also been reported. This reaction generates the family of products shown at the top of Scheme 19.2 by the rearrangements displayed at the bottom of the same scheme.[18] Because the cyclic structure constrains the geometry of the intermediate that would undergo β-hydrogen elimination, the initial product is a 2,5-dihydrofuran. However, isomerization after the catalytic reaction or isomerization of an intermediate alkyl or olefin complex can occur to generate the vinyl ether product.[200] One ligand that generates catalysts that react with high yields and high enantioselectivities in these intermolecular reactions, including reactions of purely hydrocarbon substrates, is Pfaltz's phosphinooxazoline in Scheme 19.2.[196]

19.4. The Mechanisms of Cross Coupling

19.4.1. Mechanism of the Overall Catalytic Processes

19.4.1.1. Mechanism of Palladium-Catalyzed Cross Coupling with Main Group Organometallic Nucleophiles

The mechanism of the various cross-coupling reactions, with the exception of the Heck reaction, includes three stages: oxidative addition, transmetallation, and reductive

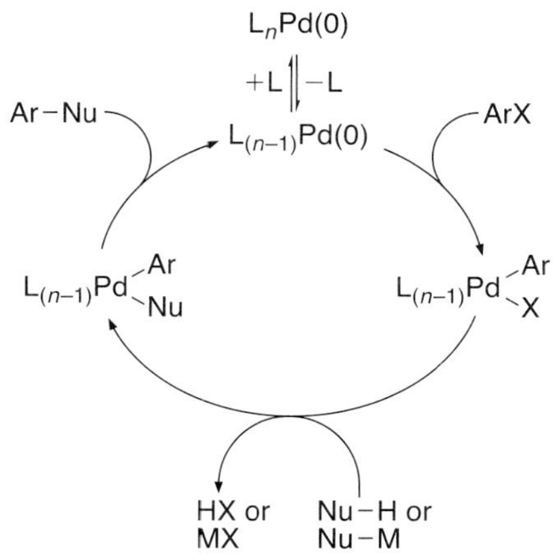

Scheme 19.3

elimination. A generic mechanism for these types of cross coupling is shown in Scheme 19.3. As noted in Chapters 6, 7, and 8 on oxidative addition and reductive elimination, these processes are typically multi-step because ligand association or dissociation usually occurs prior to the elementary bond-cleavage or bond-forming process. A low-valent metal complex containing a palladium(0) or nickel(0) metal center reacts in the first step with the aryl halide by oxidative addition to generate an arylnickel or arylpalladium halide complex. This complex then reacts with the Grignard reagent, organozinc reagent, organotin reagent, organosilicon reagent, or organoboron reagent, copper acetylide, or metal enolate to generate, after transmetallation, a species containing two metal–carbon bonds. Alternatively, in the coupling of alkynes without copper, the arylpalladium halide complex could coordinate the alkyne, and the coordinated alkyne could be deprotonated by a base to generate an arylpalladium–acetylide complex. The product of this transmetallation step then undergoes reductive elimination to form the final organic product that contains a new carbon–carbon bond and the starting nickel(0) or palladium(0) complex.

19.4.1.2. Mechanism of Homocoupling

Homocoupling likely occurs by a related catalytic cycle, but less information is known about this process. A nickel(0) or palladium(0) complex undergoes oxidative addition of an aryl halide. One could imagine that a second aryl halide adds to generate an intermediate in the M(IV) oxidation state, but this step is unlikely to occur with either palladium or nickel catalysts. One can also envision a mechanism involving disproportionation of the arylmetal–halide intermediate to form a biaryl complex and a dihalide complex. Reductive elimination would form the biaryl, and reduction of the dihalide with Zn or other terminal reductant would regenerate the catalyst.

Although these mechanisms could operate in some cases, Kochi showed many years ago that a more complex mechanism is likely to account for the homocoupling process that occurs with stoichiometric amounts of Ni(COD)$_2$ or with stoichiometric amounts of Zn and catalytic [Ni(PEt$_3$)$_2$Cl$_2$].[201] This mechanism is shown in Scheme 19.4 and involves several oxidation states of nickel. Oxidative addition of an aryl halide to a nickel(I) halide complex would form an arylnickel(III) complex, and this oxidative addition would be followed by transfer of an aryl group from an arylnickel(II) complex to the resulting Ni(III) intermediate. Reductive elimination of biaryl would then occur from the biaryl Ni(III) complex to regenerate Ni(I). Reduction of the Ni(II) dihalide complex resulting from the aryl transfer to Ni(III) would then generate a complex that adds aryl halide to regenerate the Ni(II) aryl halide complex. Since this early classic study by Kochi, little further mechanistic work has been conducted on homocoupling.

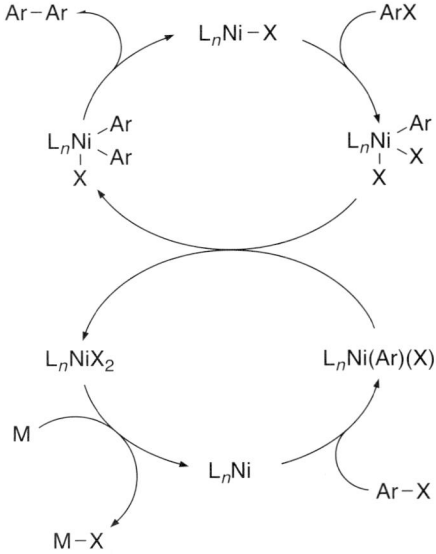

Scheme 19.4

19.4.1.3. Mechanism of the Olefination of Aryl Halides (Mizoroki–Heck Reaction)

The palladium-catalyzed reaction of olefins with aryl or vinyl halides or pseudohalides in the presence of base (the Heck reaction) follows a different course from the other cross-coupling reactions after the oxidative addition step. As shown in Scheme 19.5, the olefin coordinates to the palladium after oxidative addition of the aryl or vinyl halide. This coordination of olefin may occur by associative displacement of a monodentate ligand from the palladium, or it may occur by replacement of halide by the olefin to generate a cationic olefin complex. In some cases, these reactions are conducted with aryl or vinyl triflates. In this case, the olefin readily displaces the triflate to generate a cationic palladium–olefin

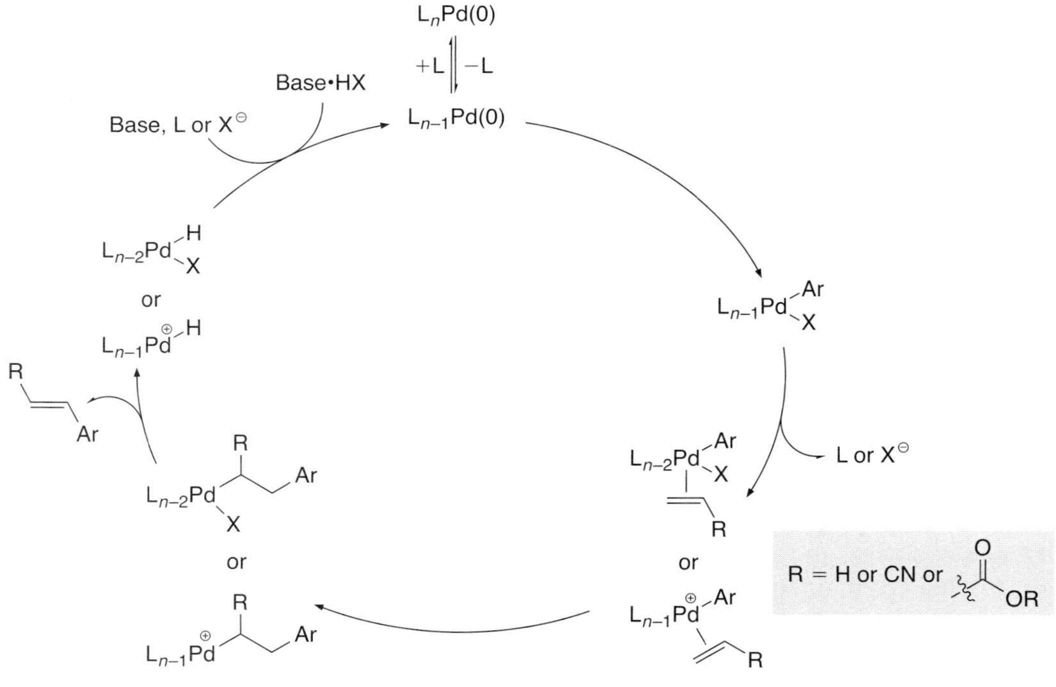

Scheme 19.5

complex. This olefin complex then undergoes insertion of the olefin into the palladium–aryl or –vinyl linkage to generate an alkylpalladium intermediate. This insertion process controls the regioselectivity of the Heck reaction. When the reaction is conducted with olefins containing an electron-withdrawing group, such as acrylates, acrylonitriles, or vinylarenes, the insertion occurs to form an alkylpalladium intermediate in which the alkoxycarbonyl, cyano, or aryl substituent is located α to the metal center. This alkylpalladium intermediate then undergoes β-hydrogen elimination to generate the olefin product and a hydridopalladium halide or triflate intermediate. This hydrido halide or pseudo-halide complex then reacts with base to regenerate the starting palladium(0) species and the protonated base. The mechanism of the final deprotonation of the hydrido halide complex has not been studied in detail, but it is likely that the base interacts directly with the palladium complex, rather than simply trapping the acid generated by a unimolecular reductive elimination.[202]

During the past decade, many metallacyclic palladium(II) precursors have been studied as catalysts for the Heck reaction.[203] Although some initial claims were made that these compounds must act as catalysts for the Heck reaction and other cross couplings by a new mechanism involving oxidative addition to palladium(II) and olefin insertion into an arylpalladium(IV) complex,[204, 205] subsequent studies have shown that these complexes are reduced to palladium(0) complexes by one of several processes that cleave the Pd–C bond of the metallacycle prior to the catalytic process.[17, 205–209] The turnover numbers of reactions initiated with palladacyclic catalysts are high, but the rates of reactions catalyzed by these complexes tend to be lower than those initiated with related palladium(0) complexes.[210, 211]

19.4.2. Mechanism of the Individual Steps of the Cross-Coupling Process

The fundamental information on the individual steps of these cross-coupling reactions was included in Chapters 6, 7, 8, and 9 on oxidative addition, reductive elimination, and migratory insertion. Details of the transmetallation step are less established, but some of the current data on this step are presented in this chapter in Section 19.4.2.2.

19.4.2.1. The Oxidative Addition Step

The oxidative addition of aryl and vinyl halides and triflates was presented in Chapter 7. A few features of this mechanism that are important for understanding the scope and limitations of the cross-coupling processes are presented here.

First, as shown in Scheme 19.6, oxidative addition occurs to a low-coordinate palladium(0) species, such as a 14-electron bisphosphine compound[212, 213] or a 12-electron monophosphine compound.[214] Complexes containing hindered monodentate ligands undergo rapid oxidative addition. It may seem counterintuitive that an addition reaction

$$(PPh_3)_4Pd \xrightarrow{-PPh_3} (PPh_3)_3Pd \underset{+PPh_3}{\overset{-PPh_3}{\rightleftharpoons}} (PPh_3)_2Pd \xrightarrow{Ar-X} (PPh_3)_2Pd\diagdown\substack{Ar \\ X}$$

$$(o\text{-tol})_3P-Pd-P(o\text{-tol})_3 \underset{+P(o\text{-tol})_3}{\overset{-P(o\text{-tol})_3}{\rightleftharpoons}} (o\text{-tol})_3P-Pd \xrightarrow{Ar-X} (o\text{-tol}_3P)Pd\diagdown\substack{Ar \\ X} \longrightarrow \substack{Ar \\ (o\text{-tol})_3P}Pd\substack{X \\ X}Pd\substack{P\text{-}(o\text{-tol})_3 \\ Ar}$$

$$L-Pd-L \underset{+L}{\overset{-L}{\rightleftharpoons}} LPd \xrightarrow{Ar-X} X-Pd-L \qquad L = (Ph_5Cp)FeP(t\text{-Bu})_2$$

Scheme 19.6

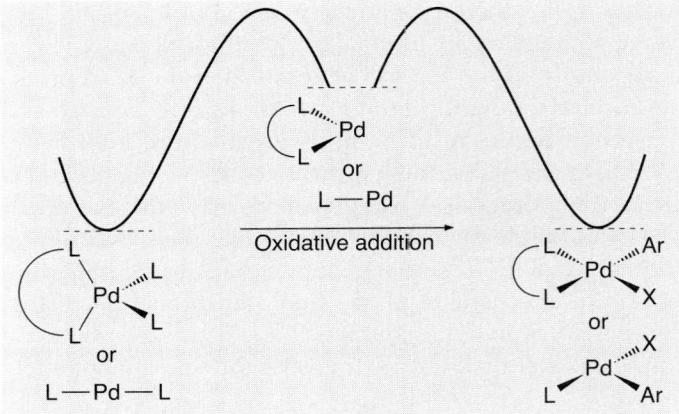

Figure 19.2.
Effect of steric hindrance on ligand dissociation and C–X bond cleavage.

will be faster to a complex containing more sterically hindered ligands than to a complex containing less-hindered ligands. However, this phenomenon is observed because the steric hindrance of the ligand causes it to dissociate more readily and to create a higher concentration of the unsaturated intermediate. This steric hindrance has a larger effect on the pre-equilibrium for dissociation of ligand than it has on the elementary carbon–halogen bond cleavage. These differences in energy are shown qualitatively in Figure 19.2.

The relative rates for oxidative addition of aryl halides follow the trend ArI > ArBr > ArCl, and the relative rates for reactions of aryl sulfonates follow the trend ArOTf > ArOTs. The relative rates for reaction of the halides and the sulfonates depend on the identity of the catalyst. Most systems follow the trend ArI > ArOTf > Br, but an exception to this trend—the preferential coupling of ArCl over ArOTf by complexes of $P(^tBu)_3$—has been observed.[215] These relative rates allow for selective reactions of one halogen over the other or sulfonates over halogens. Aryl iodides are reactive enough that coupling reactions of aryl iodides can be catalyzed by "ligandless" palladium complexes (those lacking phosphine or carbene donor ligands), such as $Pd_2(dba)_3$ or $Pd(OAc)_2$ (where dba = dibenzylidene acetone). The coupling of aryl bromides typically requires complexes formed from more electron donating ligands, such as phosphines, although procedures that generate colloidal palladium from palladium–dba and palladium–acetate complexes can lead to the coupling of aryl bromides.[216–219]

In addition, the choice of the proper ligand has led to the catalytic transformations of aryl chlorides and the direct observation of oxidative addition of aryl chlorides to palladium(0) complexes[220–222] under mild conditions. In general, the palladium-catalyzed couplings of aryl chlorides occur with complexes containing strongly electron-donating and sterically hindered ligands, such as trialkyl or biaryl dialkylphosphines or N-heterocyclic carbenes.[3] Complexes of these ligands oxidatively add aryl chlorides because they generate a higher concentration of a highly reactive, low-coordinate species than do complexes of less-hindered ligands and because the alkyl groups make this intermediate more electron rich than complexes of arylphosphine ligands. The reactions of aryl chlorides have been the target of catalyst development because they are less expensive than aryl bromides and more derivatives are commercially available.

The electronic properties of the aryl halides affect the rates of oxidative addition, and the rate of this elementary reaction, in turn, affects the scope of the coupling reactions. Aryl halides containing electron-withdrawing substituents on the aryl ring react faster than aryl halides containing electron-donating substituents on the aryl ring.[213] For example, 4-bromoacetophenone and 4-bromobenzonitrile undergo oxidative addition much faster than does 4-bromoanisole (4-bromo methoxybenzene). As a result, the scope of the cross-coupling chemistry and the turnover numbers obtained with most catalysts tend to be higher for reactions of the electron-poor aryl halides. Turnover numbers in the millions are obtained for reactions of electron-poor aryl halides with many catalysts; these turnover numbers tend to be much lower for reactions of electron-rich aryl halides.[223–225]

Nickel(0) complexes tend to undergo oxidative addition of aryl halides faster than palladium(0) complexes. There are some drawbacks to the use of the nickel complexes, described above, that can outweigh the lower cost of nickel. Nevertheless, the nickel complexes containing triarylphosphines do undergo oxidative addition of aryl chlorides (Equation 19.39)[226] and tosylates (Equation 19.40),[227] and some mild conditions have been developed for nickel-catalyzed cross couplings of aryl chlorides and aryl tosylates with ligands such as triphenylphosphine or tricyclohexylphosphine.[227–230]

$$(PPh_3)_4Ni \ + \ \langle C_6H_5 \rangle - Cl \ \xrightarrow[\substack{1\ day \\ -2\ PPh_3}]{Room\ temperature} \ Ph_3P-\underset{\underset{Cl}{|}}{\overset{\overset{Ph}{|}}{Ni}}-PPh_3 \qquad (19.39)$$

$$(COD)_2Ni \ + \ 2\ PCy_3 \ + \ MeO-\langle C_6H_4 \rangle-OTs \ \xrightarrow[\substack{15\ min \\ -2\ COD}]{Room\ temperature} \ Cy_3P-\underset{\underset{OTs}{|}}{\overset{\overset{C_6H_4OMe}{|}}{Ni}}-PCy_3 \qquad (19.40)$$

Because many of the precursors to catalysts for coupling contain halide or acetate ligands, because many protocols include addition of halide, and because many of the reactions generate halide byproducts, the effect of anions on the oxidative addition to palladium(0) have been studied.[231–239] These effects depend on the type of ligand on palladium and the solvent. A few trends can be established, however. First, complexes of relatively unhindered triarylphosphines coordinate halide in polar solvents to generate $[L_2PdX]^-$ complexes.[240] These complexes are less reactive than $L_2Pd(0)$ for the oxidative addition of aryl halides, but the reaction occurs through these species in the presence of halide in polar solvents because they are present in much higher concentration than the L_2Pd complexes. Second, many protocols involve addition of halogen to reactions of aryl sulfonates, and the effect of added halogen depends on the type of ligand. The oxidative addition of an aryl triflate to $Pd[P(o\text{-}tol)_3]_2$ in nonpolar solvents was faster in the presence of added halide, presumably because of the generation of a reactive, anionic, two-coordinate complex $Pd[P(o\text{-}tol)_3]X^-$. The reactions of aryl sulfonates with a complex of a bidentate alkylphosphine were also accelerated by added halides, but these reactions were accelerated even more strongly by salts containing weakly coordinating anions.[239] This effect of added salt was, therefore, attributed to an effect on the polarity of the medium, instead of the formation of a highly reactive anionic palladium(0) species.

19.4.2.2. Mechanism of Transmetallation

In the context of cross-coupling chemistry, the term "transmetallation" refers to the replacement of the halide or pseudohalide in a transition metal organometallic halide or pseudohalide complex with the organic group of a magnesium, zinc, tin, boron, or silicon reagent. In the cross-coupling mechanism, this reaction typically generates a transition metal species that contains two covalent, organic ligands and undergoes reductive elimination. The mechanism of the transmetallation with organomagnesium or zinc reagents has not been studied in any depth. One can imagine a mechanism by which the main group element would initially coordinate to the halogen and simultaneously assist dissociation of the halogen while delivering the carbon nucleophile to the palladium metal center (Equation 19.41). However, the mechanism by which less polar and less electrophilic organosilanes, stannanes, and boronates undergo transmetallation has been shown to be more complex. The majority of studies have been conducted on transmetallation of alkyl, aryl, and heteroaryl groups from tin to palladium, but an increasing number of studies have been conducted on transmetallation from boron and silicon.

$$L_nM-X \;+\; RMgX(THF)_2 \;\longrightarrow\; \underset{(THF)_n}{\overset{\displaystyle L_nM\cdots X}{R\cdots Mg-X}} \;\longrightarrow\; L_nM-R \;+\; MgX_2(THF)_2 \qquad (19.41)$$

Studies by Denmark and Echavarren have been conducted on the transmetallation of organosilicon reagents.[60,241] Recent studies have shown that reactions of organic silanols occur faster than those lacking the hydroxyl functionality.[60,242–244] Two pathways have been uncovered for this transmetallation, one that occurs in the absence of fluoride and one that occurs in the presence of added fluoride (Equations 19.42 and 19.43). In the absence of fluoride, an arylpalladium vinylsiloxide or arylsiloxide complex appears to be generated. This complex then decomposes to transfer the vinyl or aryl group from silicon to palladium, as shown in Equation 19.42, to generate the arylpalladium vinyl or aryl complex and $R_2Si=O$.[245] In the presence of fluoride, a hypervalent silicate is generated, and this complex transfers the aryl group to the arylpalladium–halide complex by a mechanism in which the aryl group presumably acts as a nucleophile.[246] Kinetic data imply the involvement of two silanols, in this case to generate the disiloxane shown in Equation 19.43 coordinated by fluoride.

$$(19.42)$$

$$(19.43)$$

The intimate details of the mechanism of transmetallation from organoboron reagents are also likely to be complex. A few studies have provided some information on how transmetallation from boron to palladium occurs. Soderquist and Woerpel have shown that the transmetallation of an aliphatic organoboron reagent occurs with retention of configuration of the carbon attached to boron (Equation 19.44).[247,248] Suzuki has suggested that transmetallation occurs after conversion of the arylpalladium–halide complex to a palladium-hydroxide complex and that the hydroxide complex reacts with the organoboron reagent (Equation 19.45).[63,249] However, the Suzuki cross-coupling reaction occurs under anhydrous conditions in many cases and with fluoride as the base in others. Thus, a palladium–hydroxide complex may be an intermediate in reactions in wet solvents or aqueous biphasic systems, but it is not required to achieve transmetallation.

$$(19.44)$$

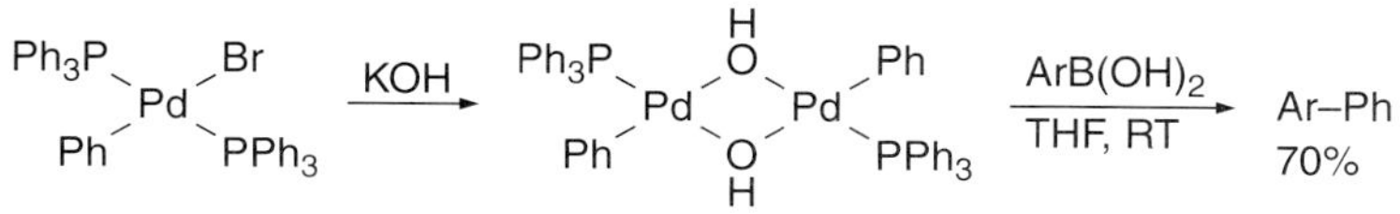

(19.45)

Many studies have been conducted on the transmetallation of organotin reagents. Studies with isolated arylpalladium–halide complexes have been conducted by Espinet,[250,251] Echavarren,[252,253] Ricci,[254] and Cotter (Equation 19.46).[255] These studies imply that unsaturated groups at tin that undergo transmetallation coordinate to palladium prior to the C–Sn bond cleavage. Farina, Hartwig, and Amatore have all concluded from kinetic studies that the transmetallation occurs to a three-coordinate palladium complex generated in many cases by dissociation of phosphine or arsine from a four-coordinate palladium product from the oxidative addition step (Equation 19.47).[256–258] Espinet has argued that the reversible dissociation of phosphine occurs after the metal has contacted the organotin reagent,[250] but Amatore provided further arguments against this proposal.[258]

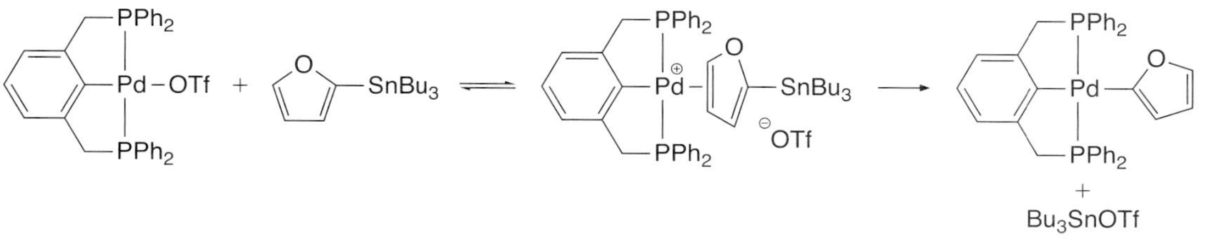

(19.46)

$$(PPh_3)_2Pd(Ar)Br \rightleftharpoons (PPh_3)Pd(Ar)Br \xrightarrow{R_3SnAr'} (PPh_3)Pd(Ar)Ar' \longrightarrow Ar-Ar' + (PPh_3)Pd(0)$$ (19.47)

Further clouding the mechanism of transmetallation from tin to palladium, the transfer of the organic group can occur with inversion or retention of configuration. The transmetallation occurs with retention of the stereochemistry of the olefinic portion of a vinylstannane (Equation 19.48).[259] However, the retention or inversion of configuration from the transfer of an alkylstannane appears to depend on solvent, reagent or both. In a polar solvent, the transmetallation of a benzyltin reagent occurs with *inversion* of configuration (Equation 19.49),[259] but in a less polar solvent, the reaction of an organcstannane containing an alkoxo substituent on the carbon α to tin occurs with *retention* of configuration (Equations 19.50 and 19.51).[260,261]

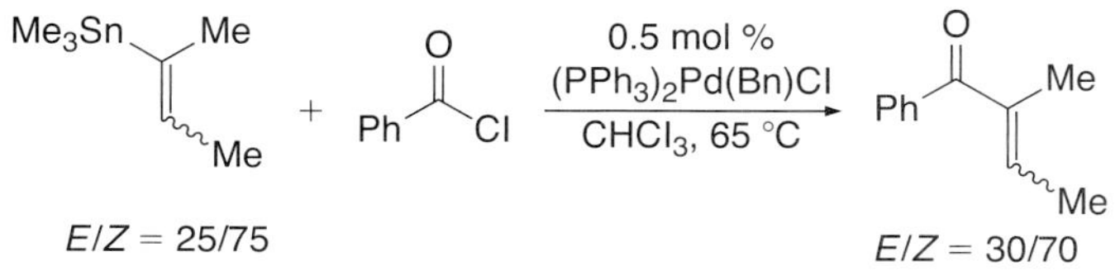

(19.48)

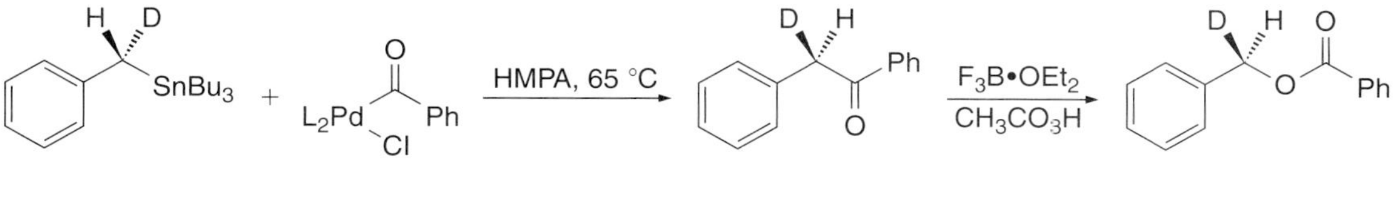

(19.49)

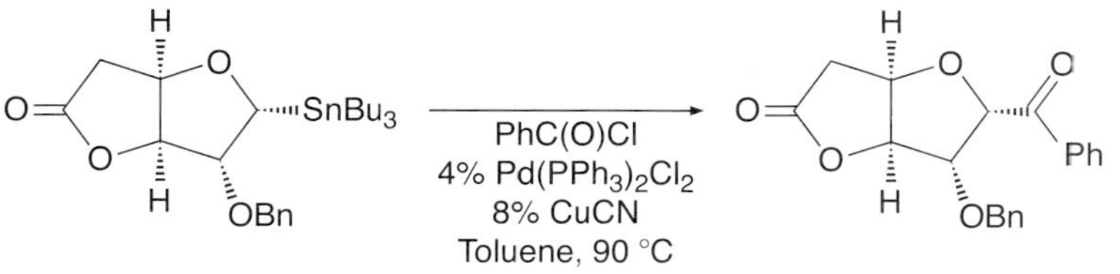

(19.50)

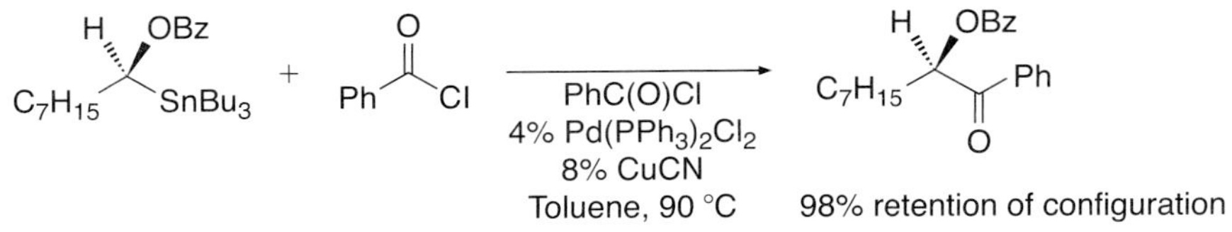

$$(19.51)$$

The electronic effects on the transmetallation are also counterintuitive. The reactions of benzylic stannanes are accelerated by electron-withdrawing groups on the benzyl group.[259] This electronic effect contradicts a view that the alkyl group is transferred to the transition metal as a nucleophile.

These data, particularly the effect of solvent on stereochemistry, have led to the suggestion that the reactions can occur through both open and closed (cyclic) transition states (Figure 19.3).[250,262] S_E-open transition states were proposed during the early development of the Stille reaction[259] because the inversion of configuration is akin to the stereochemistry from the cleavage of tin–carbon bonds by bromine.[263] However, the retention of configuration observed in less polar solvents and arguments based on changes in the geometry at the metal center[250,251,264] imply that the transmetallation can occur by a closed, cyclic transition state. This closed transition state may contain some oxidative addition character.[257]

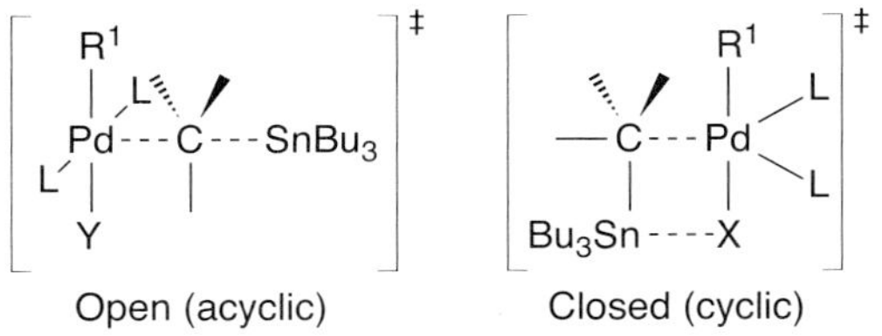

Figure 19.3.
Open and closed transition states proposed for transmetallation from organotin reagents to palladium(II).

The relative rates of transmetallation of various organic groups have been measured. Early studies showed that the rates of transmetallation followed the order PhCC > PrCC > PhCH=CH, CH_2=CH > Ph >PhCH$_2$ > CH_3OCH_2 > CH$_3$ > Bu.[259] The faster rates for the transfer of unsaturated groups may result from coordination of this group to the metal prior to the C–Sn bond cleavage.[250,255]

The rate of transmetallation is also affected by the ancillary groups on tin. In particular, transmetallation of organotin reagents can be accelerated by added base, just as the transmetallation of organosilicon reagents can; several papers report that Stille couplings are accelerated by added fluoride.[265,266] Furthermore, several studies have shown that transmetallation occurs faster with organostannanes containing an internal Lewis base than with organostannanes lacking an internal Lewis base. This phenomenon is illustrated by comparing the two reactions in Equations 19.52 and 19.53 and by comparing the relative rates for transmetallation of a phenyl group and an *ortho*-dimethylaminomethylphenyl group (Equation 19.54).[267-270]

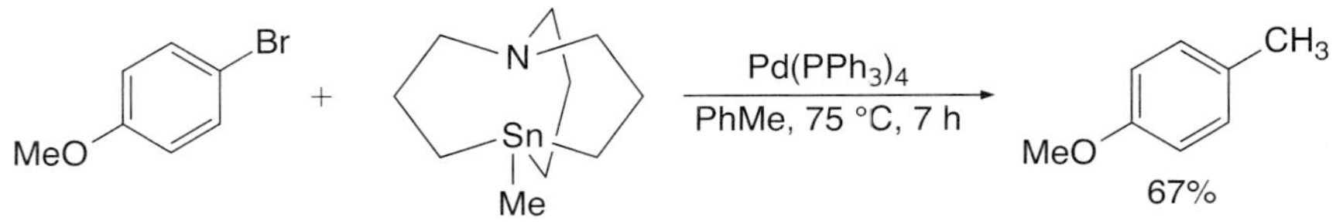

$$(19.52)$$

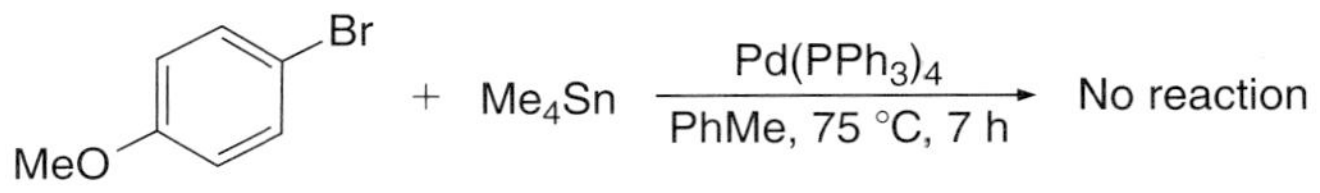

$$(19.53)$$

(19.54)

19.4.2.3. Mechanism of Reductive Elimination

The reductive elimination from organometallic palladium(II) complexes was discussed in depth in Chapter 8. In most cases, reductive elimination from complexes containing monodentate phosphine ligands occurs after dissociation from the more stable four-coordinate complex to form a three-coordinate palladium complex. Recall that a cis orientation of the two anionic ligands forming the new C–C bond is necessary for reductive elimination to occur. Also recall that reductive elimination is affected significantly by the steric and electronic properties of the ancillary and reacting ligands and by the hybridization of the reacting ligands. These points are addressed in relation to the cross-coupling process in this section.

Reductive eliminations to form C–C bonds tend to occur faster from palladium(II) aryl complexes and vinyl complexes than from palladium(II) alkyl complexes. This is one reason why cross-coupling reactions that form bonds between two sp^2-hybridized centers or one sp^2-hybridized center and one sp^3-hybridized center are more favorable than cross-coupling reactions that join two sp^3-hybridized carbons. The effect of the electronic properties of the alkyl or aryl groups on the rate of reductive elimination can influence the reaction scope. As noted in Chapter 8, reductive elimination is faster from symmetric complexes in which more donating groups participate in the reductive elimination,[271–273] but reductive elimination is faster from unsymmetrical complexes in which there are greater differences between the electron-donating and electron-withdrawing propensities of the two groups undergoing bond formation during the reductive elimination.[274]

19.4.3. Effects of Catalyst Structure on Cross Coupling

19.4.3.1. Effect of Chelation

Tremendous effort has been made to develop ligands that control and improve the selectivity, stability, and reactivity of palladium and nickel complexes for cross coupling. Initially, ligandless palladium complexes and triphenylphosphine–palladium complexes were used for these processes. Subsequently, bidentate ligands and monodentate ligands have been developed that possess particular electronic and steric properties that increase reaction rates, scope, and turnover numbers.

Hayashi conducted one of the early studies on ligand design to improve cross-coupling reactions.[275,276] Hayashi showed that the reactions of sec-butylmagnesium bromide with phenyl bromide generated sec-butyl benzene, n-butyl benzene, and butene (Table 19.1). The ratios of these products depended strongly on the type of phosphine bound to the palladium catalyst. When conducted with a catalyst containing triphenylphosphine as ligand, the reaction of sec-butylmagnesium bromide with phenyl bromide formed only 4% of the sec-butyl benzene product. Instead, 30% of the aryl halide remained, and 6% n-butyl benzene was formed. The reaction of sec-butylmagnesium bromide with trans-β-bromostyrene in the presence of the same catalyst formed the sec-butyl-substituted styrene in 33% yield and the n-butyl-substituted styrene in 36% yield. In contrast, the reaction of sec-butylmagnesium bromide with trans-β-bromostyrene in the presence of (dppf)PdCl₂ [dppf = bis(1,1′-diphenylphosphino)ferrocene], which contains a chelating phosphine, formed the sec-butylbenzene in 93% yield and sec-butyl-substituted styrene in 95% yield. No n-butyl-substituted styrene was formed. Reactions of other complexes

Table 19.1. Effect of chelation on isomerization during cross coupling.

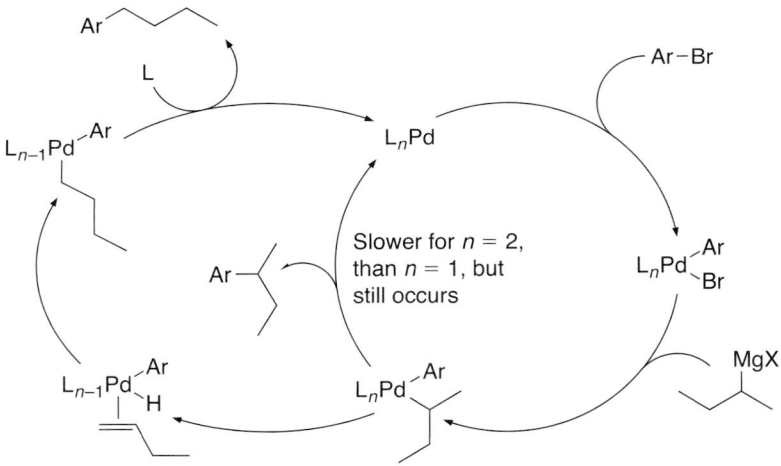

L = PPh₃	R = Ph	4	6	6	31
	R = Ph⌢⌢	33	36	4	—
L = dppf	R = Ph	95	0	—	0
	R = Ph⌢⌢	93	0	—	0

possessing bisphosphines containing flexible linkers, such as 1,4-bis(diphenylphosphino) butane formed mixtures of *sec*-butyl and *n*-butyl coupled products.

The origin of this effect of ligand structure on the product distribution is summarized in Scheme 19.7 and results from the effect of chelation on the relative rates for reductive elimination and for β-hydrogen elimination. After oxidative addition to form the arylpalladium–halide complex, transmetallation generates the corresponding *sec*-butylpalladium complex. With a monodentate or flexible bidentate ligand on palladium, dissociation of a phosphine can occur. This ligand dissociation generates a three-coordinate intermediate that can undergo β-hydrogen elimination to form a palladium–hydride complex containing a coordinated olefin.[277–279] This complex can then dissociate the olefin to generate butene and a hydride complex that reductively eliminates the arene or alkene product resulting from overall hydrodehalogenation. Alternatively, this complex can re-insert the olefin to generate a linear alkylpalladium complex. Reductive elimination from this complex would then generate the *n*-butyl-substituted product.

Scheme 19.7

In contrast, the *sec*-butylpalladium complex containing the chelating dppf ligand does not readily dissociate a phosphine donor. Because the complex of the chelating ligand does not readily generate an open coordination site, and β-hydrogen elimination is faster from complexes containing an open coordination site (see Chapter 10),[277–279] the alkylpalladium intermediate ligated by dppf undergoes slow β-hydrogen elimination. Reductive elimination can occur from a four-coordinate or a three-coordinate palladium complex.[273, 280–283]

Although reductive elimination from the four-coordinate species tends to be slower than from related three-coordiante species, the wide bite angle of dppf helps to accelerate the rate of reductive elimination.[284,285] Thus, reductive elimination occurs from the *sec*-butylpalladium complex to form the *sec*-butyl organic product without competing rearrangements to form *n*-butyl products or butene.

Thus, catalysts bound by monodentate and bidentate ligands can react with dramatically different rates and product distributions, and each has advantages for certain classes of substrates. Although catalysts containing bisphosphine ligands tend to retard β-hydrogen elimination that can lead to olefin side products, rearranged products, and arene products from net hydrodehalogenation, these catalysts have drawbacks. Complexes containing chelating ligands react slower with main group organometallic reagents that undergo transmetallation after dissociation of a ligand. Thus, Stille reactions, for which the transmetallation of the organotin reagent occurs via complexes containing open coordination sites, are conducted more commonly with catalysts containing monodentate ligands than with catalysts containing bidentate ligands. Furthermore, Heck reactions of aryl halides, which usually require an open coordination site for coordination and insertion of the olefin, are typically slower when they are catalyzed by complexes of bisphosphines than when they are catalyzed by complexes of monophosphines.[17] Finally, complexes containing chelating phosphines often undergo oxidative addition more slowly than those containing monodentate phosphines. An 18-electron, four-coordinate palladium(0) species containing two chelating ligands can form in the reaction mixture, and dissociation of the chelating ligand from such a complex to generate the 14-electron complex, which adds the aryl halide, often occurs slowly and lies far uphill of the 18-electron species.[286,287]

19.4.3.2. Effect of Steric Properties

More recently, a dramatic increase in the reactivity of catalysts for cross coupling has been obtained by using ligands that are sterically hindered and electron rich. For example, catalysts containing trialkyl phosphines, such as tri-*tert*-butylphosphine,[76,81–83,211,215,288–299] aryl dialkyl–phosphines, such as biaryl di-*tert*-butylphosphines,[104,223,225,300–305] ferrocenyl dialkyl phosphines,[306–308] constrained triaminophosphines,[93,94,309–313] and strongly electron-donating and sterically hindered N-heterocyclic carbenes[95,265,314–331] have been shown to be highly active for various types of cross coupling. Complexes containing these ligands undergo faster oxidative addition than complexes containing aryl phosphine ligands,[332–334] which were the most commonly used ligands for cross coupling during the early developments of these processes.

One might imagine that the steric and electronic properties of the ligand would accelerate one step but retard other steps of the catalytic cycle. However, steric hindrance appears to accelerate the rate of both oxidative addition and reductive elimination. The sterically hindered ligands accelerate oxidative addition because they favor generation of the unsaturated intermediate that reacts with the aryl halide (Figure 19.3; compare the rates of Equations 19.55 and 19.56), and they accelerate reductive elimination because reductive

$$\text{Pd(dba)}_2 \xrightarrow[\text{RT, minutes}]{\substack{\text{Ar–Br} \\ \text{P}(t\text{-Bu})_3}} \begin{array}{c} \text{Br} \\ | \\ \text{Ph–Pd} \\ | \\ \text{P}^t\text{Bu}_3 \end{array} \qquad (19.55)$$

$$\text{Pd(PPh}_3)_4 \xrightarrow[\text{50–80 °C, hours}]{\text{Ar–Br}} \begin{array}{c} \text{PPh}_3 \\ | \\ \text{Ph–Pd–Br} \\ | \\ \text{PPh}_3 \end{array} \qquad (19.56)$$

elimination reduces the steric congestion of the palladium(II) species created by the hindered ligands.[332,333,335] Moreover, steric hindrance can even accelerate transmetallation in some cases. Steric hindrance of a monophosphine ligand can accelerate transmetallation in the Stille reaction by increasing the concentration of the unsaturated palladium intermediate that reacts with the organotin reagent.[257,333] A similar effect on the rate of transmetallation may help increase the rate of other classes of cross couplings.

As a result of the greater reactivity of complexes containing sterically hindered, electron-donating ligands toward oxidative addition, the cross coupling of aryl bromides can be conducted at room temperature, and many cross couplings of aryl chlorides can be conducted between room temperature and 80 °C. For example, Suzuki reactions of organoboron reagents, Hiyama–Tamao couplings with organosilicon reagents and added base, Negishi couplings of organozinc reagents, Stille couplings of organotin reagents, and Heck reactions of olefins have all been reported with chloroarenes under mild conditions.[3]

Sterically hindered bisphosphine ligands can also generate highly active catalysts for cross coupling. Complexes of these ligands also appear to undergo fast oxidative addition. Milstein reported some of the earliest results on catalytic transformations of aryl chlorides using catalysts containing sterically hindered bisphosphines.[221,336–339] Although the conditions of the reactions reported by Milstein were not as mild as those reported more recently, these early examples did demonstrate the value of this class of ligand for increasing the scope of cross coupling to include chloroarenes. More recently, a bisphosphine containing a ferrocenyl backbone linking one di-*tert*-butylphosphino group and one dicyclohexylphosphino group has been shown to generate palladium complexes that are extremely reactive for oxidative addition (Equation 19.57). Chelation appears to decelerate the transmetallation of organotin and boron reagents, but Kumada couplings of aryl chlorides and aryl tosylates occur under mild conditions in the presence of catalysts containing this ligand, and, as discussed in section 19.62, the coupling of amines with aryl and heteroaryl chlorides also occurs when catalyzed by complexes of this ligand with high turnover numbers (Equation 19.58).[239,340]

$$\text{(19.57)}$$

$$\text{(19.58)}$$

19.4.3.3. Effect of Ligand Electronic Properties

Oxidative addition of aryl halides typically occurs faster to more electron-rich metal complexes than to closely related electron-poor complexes, and reductive elimination to form C–C bonds typically occurs faster from more electron-poor metal complexes than from closely related electron-rich complexes. Thus, when oxidative addition is the turnover-limiting step of the coupling cycle, complexes containing more electron-rich metal centers tend to react faster. For this reason, most couplings of the less reactive aryl bromides and of aryl chlorides occur faster when catalyzed by complexes of hindered alkylphosphines than by complexes of hindered arylphosphines. However, it is less straightforward to predict the effect of electronics when transmetallation or insertion is turnover limiting because less is known about the electronic effects of the metal center on these steps. Reductive elimination is not usually the turnover-limiting step, but the yields of coupling reactions can be controlled by the relative rates for reductive elimination versus side reactions, such as β-hydrogen elimination. Because reductive elimination is favored by complexes containing less electron-rich metal centers, ligands that are less electron donating tend to accelerate reductive elimination. However, steric effects on reductive elimination tend to be greater than electronic effects,[341] and complexes containing sterically hindered alkylphosphine ligands catalyze a broad range of coupling reactions.

19.5. Applications of C–C Cross Coupling

As noted in the introduction to this chapter, cross coupling has become one of the most common catalytic processes used for the generation of pharmaceutical candidates. In addition, many examples of cross couplings have been conducted as part of natural product syntheses.[342] Several examples of these reactions have also been conducted on large scales. Many reviews on the applications of cross-coupling chemistry have appeared.[4,342–344]

Losartan is one example of a molecule that has been synthesized on a large scale using cross-coupling methodology.[345,346] Reaction of the protected aryl tetrazole in Equation 19.59 with butyllithium and quenching of the organolithium reagent with a trialkylborate, followed by hydrolysis, generates the aryl boronic acid. Coupling of this species with the aryl bromide in Equation 19.59, followed by deprotection of the tetrazole generates losartan.[345] An alternative approach to the synthesis of materials with related structures has been conducted, as shown in Equation 19.60, by a nickel-catalyzed coupling of 2-bromobenzonitrile with an arylzinc reagent. The nitrile of the product of this reaction was then converted to the tetrazole, and the benzylic methyl group was brominated and used to alkylate imidazoles and benzimidazoles for drug development.[346]

(19.59)

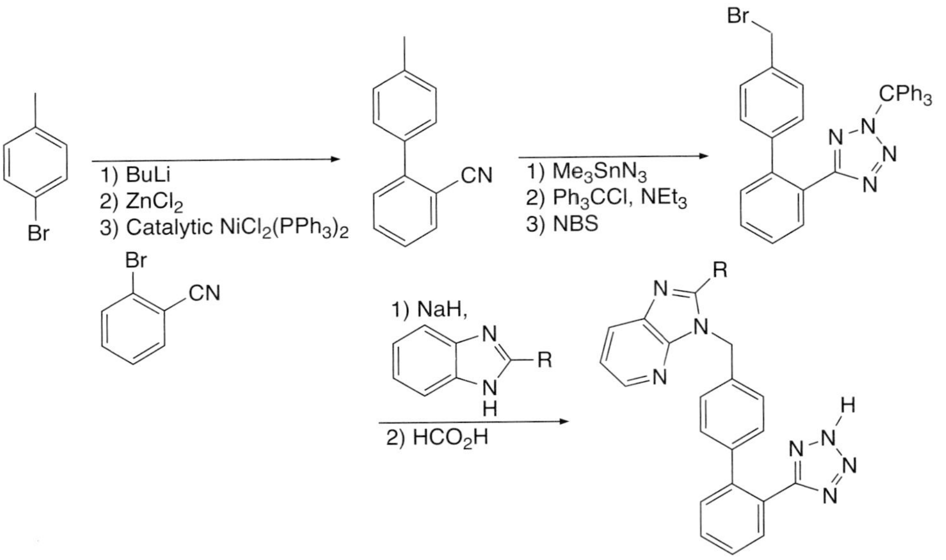

(19.60)

A second application of cross coupling in natural product synthesis followed by several groups is the preparation of members of the epothilone family. The natural products are macrocyclic structures containing a Z-olefin or an epoxide derived from a Z-olefin in the macrocyclic unit. The coupling of an alkylboron reagent with a geometrically pure haloalkene provides a method to generate geometrically pure alkenes. These natural products have been prepared by cyclization with olefin metathesis, but these reactions generate mixtures of E- and Z-olefins, as described in Chapter 21. As depicted in Equation 19.61, Danishefsky reported the synthesis of several epothilone derivatives by generating an alkylboron reagent from hydroboration of an olefin with 9-BBN, and generation of the precursor to ring closure by coupling of this alkylborane with the bromo alkene in the presence of a palladium catalyst.

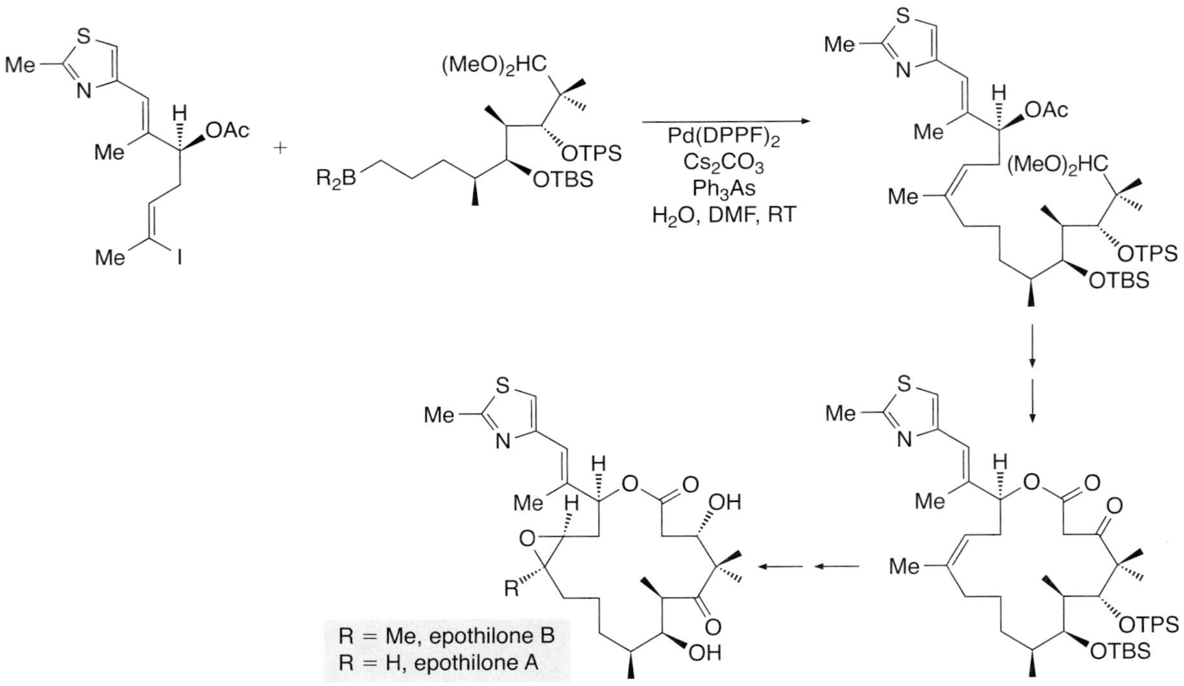

(19.61)

The Heck reaction has also been used in the preparation of pharmaceuticals, natural products, and materials. Compounds closely related to Merck's LTD$_4$ antagonist Singulair have been prepared by the Heck reaction.[347] In this process, an allylic alcohol is coupled with an electron-poor aryl iodide in the presence of a ligandless catalyst to generate the

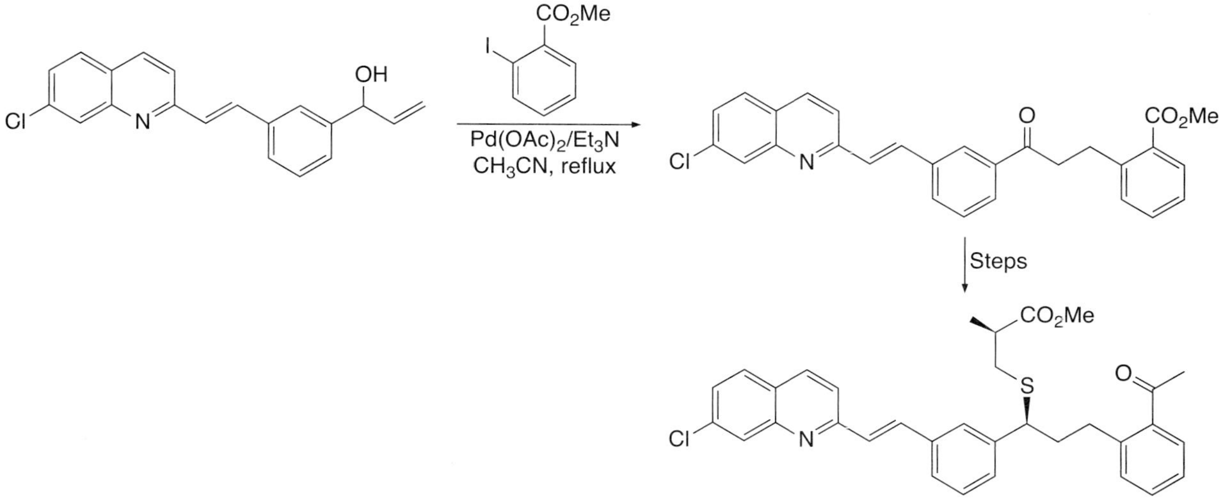

(19.62)

ketone product after isomerization (Equation 19.62). This product was then carried forward to L699,392. The Heck reaction has also been used in materials chemistry to prepare poly(phenylene vinylene)s[348,349] because of their electronic and photo-emissive properties. These materials include the Ru-tris(bpy) complexes in Equation 19.63. Dow has also been reported to produce a coating for electronic materials by ring-opening of the monomer produced by the Heck reaction (Equation 19.64).[343]

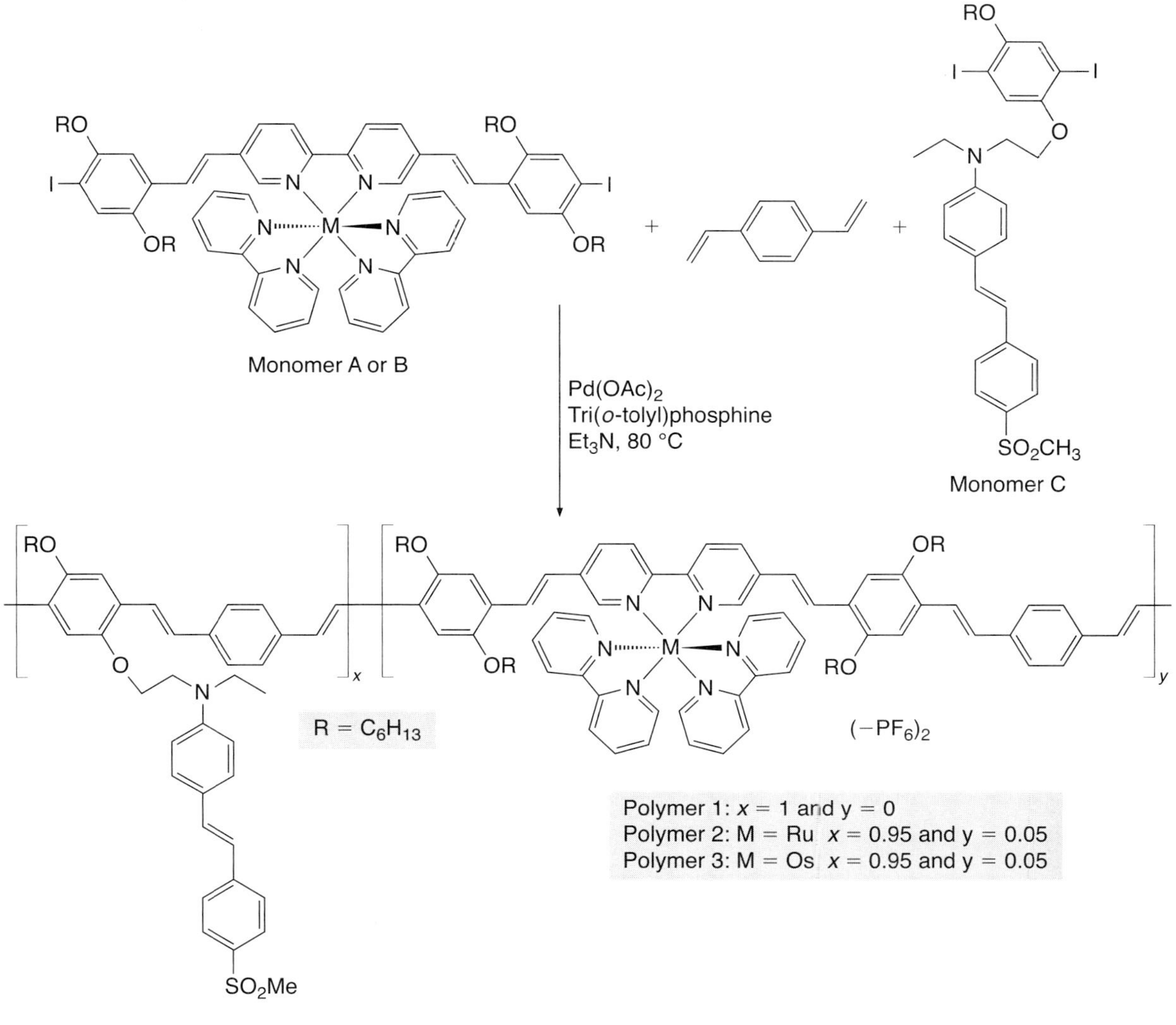

(19.63)

Polymer 1: x = 1 and y = 0
Polymer 2: M = Ru x = 0.95 and y = 0.05
Polymer 3: M = Os x = 0.95 and y = 0.05

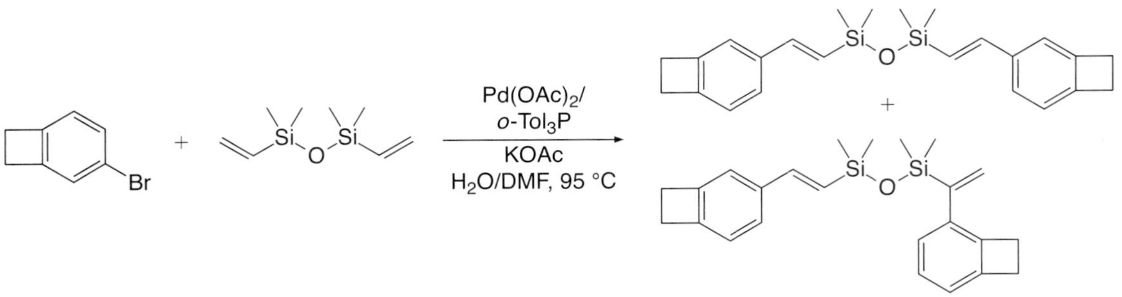

$$(19.64)$$

Many natural products have also been prepared with a Heck reaction as a key step. In many cases, these natural products contain tertiary or quaternary stereocenters created by enantioselective Heck reactions. The synthesis of many examples of these natural products have been completed by Overman and his co-workers. One particularly striking example is the synthesis of psycholeine, a member of the polypyrrolidinoindoline family of alkaloids (Scheme 19.8). During this synthesis, the meso intermediate shown in Scheme 19.8 is desymmetrized by an enantioselective intramolecular Heck reaction that sets the two biaryl quaternary stereocenters. The meso product is generated as a minor product, and the major optically active product is formed in 90% ee. This cyclization product was carried forward to quadrigemine C by reduction of the alkene, removal of the nitrogen protective groups, and cyclization. This material was converted to psycholeine by acid-catalyzed formation of the final two rings of the core.

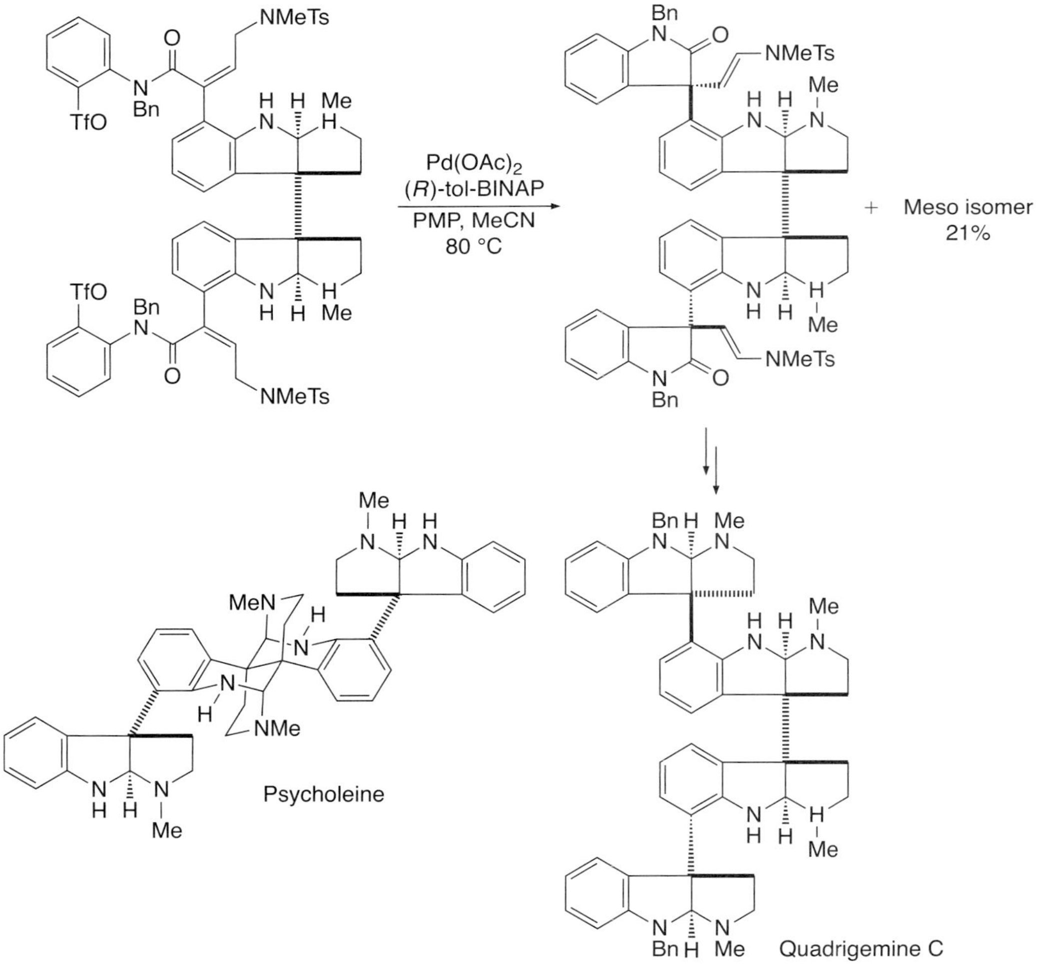

Scheme 19.8

19.6. Cross-Coupling Reactions that Form Carbon–Heteroatom Bonds

19.6.1. Overview

For many years, the reactions of aryl and vinyl halides with heteroatom nucleophiles[350] were not useful reactions. Scattered examples of these reactions had been published, but the processes had not been developed to the point of being useful for synthetic applications. Now, however, the coupling of aryl halides with amines has become a standard synthetic method for the preparation of aromatic amines.[203,351–360] The coupling of thiols with aryl halides[361–372] and the coupling of secondary phosphines and related phosphorus nucleophiles also occur with useful scope.[373–387] The reactions of alcohols with aryl halides catalyzed by transition metals has been challenging to develop, but intramolecular and some intermolecular reactions have approached the point of being reliable enough to use synthetically.[291,302,303,306,307,361,388–400] More recent work has again focused on copper-catalyzed coupling of aryl halides with heteroatom nucleophiles, and this work is presented as part of Section 19.8 on copper-catalyzed coupling reactions.

19.6.2. Coupling of Aryl Halides with Amines

19.6.2.1. Scope of the Reaction

The coupling of aryl halides with amines, commonly called the "Buchwald–Hartwig reaction," now occurs with broad scope, with fast rates, and with high turnover numbers in many cases (Equations 19.65 and 19.66). As a result, this reaction has been applied to the synthesis of biologically active molecules, as well as materials with potential electronic applications. This reaction has also been used to generate amido ligands that are now found in catalysts for other types of chemistry, including olefin polymerization and asymmetric catalysis. The scope of the coupling of amines with aryl halides has been reviewed several times.[203,351–359]

$$X = Cl, Br, I, OTf, or OTs$$
L = Hindered monodentate ligands:
$P(o\text{-tolyl})_3$, $P(t\text{-Bu})_3$, $Ph_5FcP(t\text{-Bu})_2$ (Q-phos), heterocyclic carbenes, $(Biaryl)PR_2$, $-OP(t\text{-Bu})_2$, and Verkade's proazaphosphatranes
L = Chelating bidentate ligands:
dppf, BINAP, Xantphos, and Josiphos ligands

(19.65)

$$X = Cl, Br, I, OTf, or OTs$$
Base = $NaO\text{-}t\text{-Bu}$, Cs_2CO_3, or K_3PO_4
L = ligands of Eq. 19.65

(19.66)

Aryl bromides are the most reactive aryl halide, and aryl chlorides are the next most reactive for this catalytic process. For reasons not well understood, aryl iodides tend to be less reactive toward palladium-catalyzed cross-couplings that form C–N bonds.[311,401–404] Aryl triflates also couple with amines,[301,405–410] and even the less-reactive aryl tosylates have been shown to couple with amines.[239,304,402,411]

Primary and secondary amines now react with broad scope, but the most effective catalyst for these two classes of amines tends to be different. The most efficient catalysts for reactions of secondary alkylamines possess hindered monodentate ligands.[293,294,301,304,308,311,328,412–415] The most active catalysts for the coupling of primary alkylamines contain hindered bisphosphine ligands.[340,416–418] Complexes of monodentate and bidentate ligands catalyze the couping of primary aromatic amines. Reactions of primary alkylamines catalyzed by palladium complexes of CyPF-t-Bu, a hindered ferrocenylbisphosphine of the Josiphos class, occur with remarkably high turnover numbers.[340] The combination of ammonia and $tert$-butoxide base (Equation 19.67) or lithium amide also couples with aryl halides in the presence of the catalyst containing this same ligand.[419]

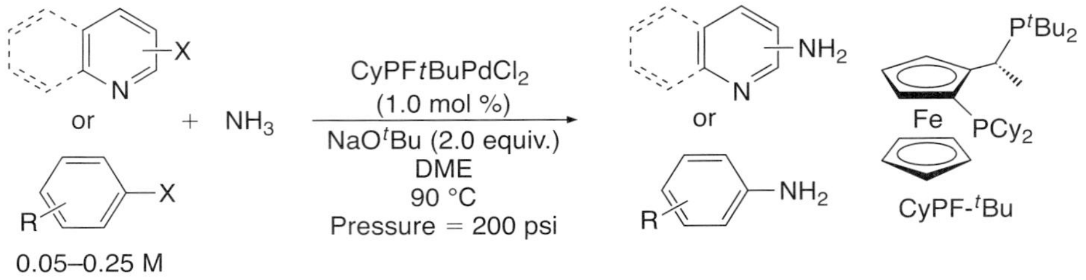

(19.67)

Other nitrogen nucleophiles that are synthetically valuable undergo this coupling process. Benzophenone imine reacts with a broad range of aryl bromides and chlorides (Equation 19.68).[340,408,420] Because the products of these coupling reactions can be converted to the primary arylamines, this reagent serves as an ammonia equivalent. Hexamethyldisilazide can also serve as an ammonia equivalent to generate primary arylamines after the combination of coupling and hydrolysis.[421–423] Both lithium and zinc hexamethyldisilazide undergo the coupling process with a broad range of meta- and para-substituted, but not ortho-substituted, aryl halides (Equation 19.69). Benzophenone hydrazone, which can serve as a precursor to nitrogen heterocycles, undergoes cross-coupling with aryl halides with broad scope as well (Equation 19.70).[340,424–426]

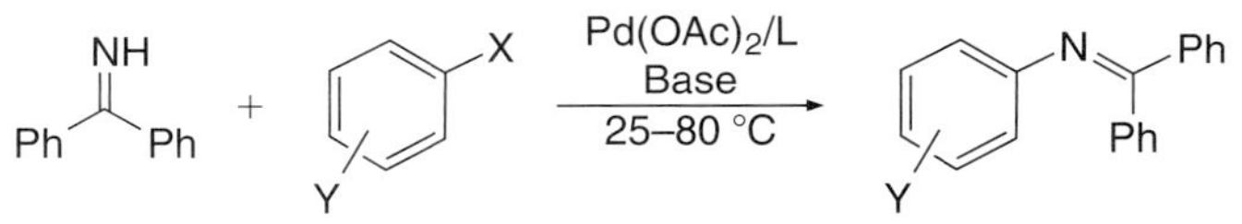

X = Cl, Br, I, or OTf
Base = NaO-t-Bu or Cs$_2$CO$_3$
L = dppf or BINAP, N-heterocyclic carbene
CyPPF-t-Bu (a hindered Josiphos ligand)

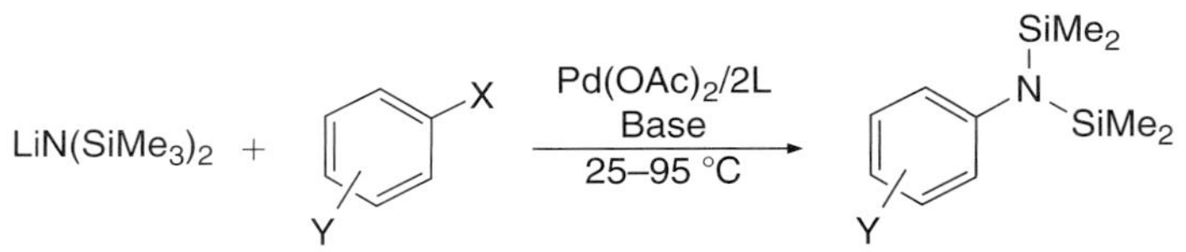

(19.68)

X = Cl or Br
L = P(t-Bu)$_3$ or (Biaryl)PR$_2$
Y ≠ ortho substituent

(19.69)

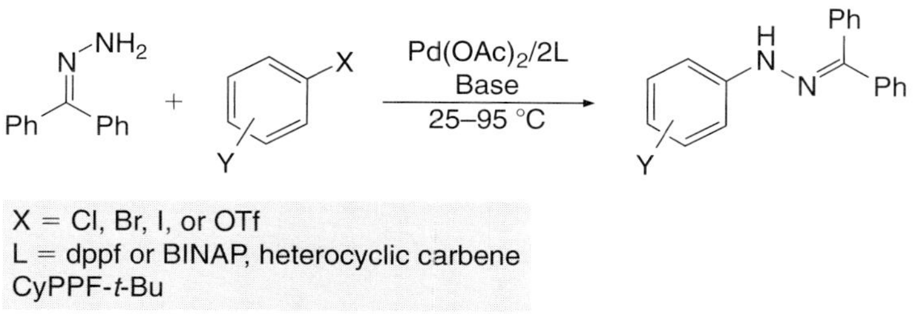
Pd(OAc)2/2L
Base
25–95 °C
X = Cl, Br, I, or OTf
L = dppf or BINAP, heterocyclic carbene
CyPPF-t-Bu

(19.70)

More recently, the direct coupling of ammonia or lithium amide with aryl bromides has been developed (Equation 19.71).[419,427] As noted above, this reaction was first conducted with a catalyst containing a strongly electron-donating and sterically hindered bisphosphine ligand.[419] Later work showed that complexes containing a monophosphine can also catalyze this type of coupling.[427] The factors that control selectivity for the monoarylation of ammonia were not determined, but the steric hindrance of the ligands is probably important to encourage reaction of the smaller ammonia over the larger arylamine product. In addition, the use of ligands that resist displacement by ammonia to form inactive amine complexes is required.

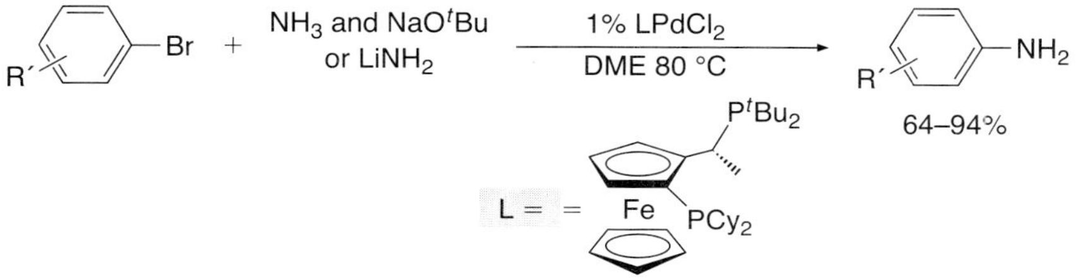
NH3 and NaOtBu
or LiNH2
1% LPdCl2
DME 80 °C
64–94%
L =

(19.71)

Amides,[304,428–431] carbamates,[294] sulfoximines,[432–434] and azoles also undergo C–N coupling (Equation 19.72).[292,435–438] However, the lower nucleophilicity of these substrates limits the scope of these reactions and leads to slower rates and the need for higher temperatures. The current most active and selective catalysts contain the biaryl dialkylphosphine ligand in Equation 19.72, but reactions catalyzed by even these systems require high loadings of both palladium and ligand. As discussed in Chapter 8, the weaker electron donation of the anions of amides, carbamates, sulfoximines, and azolyl anions, relative to the donation by formally anionic amide ligands, leads to slower rates for reductive eliminations from palladium(II) to form aryl carbon-nitrogen bonds. Moreover, amidates can bind in a κ^2-fashion, and complexes of such ligands undergo slower reductive elimination than those containing κ^1-amidates.[464]

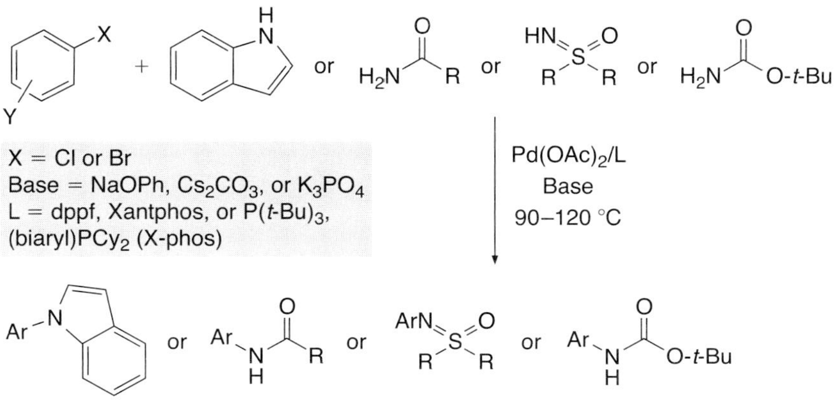
X = Cl or Br
Base = NaOPh, Cs2CO3, or K3PO4
L = dppf, Xantphos, or P(t-Bu)3,
(biaryl)PCy2 (X-phos)
Pd(OAc)2/L
Base
90–120 °C

(19.72)

19.6.2.2. Catalysts for C–N Coupling

Four classes of catalysts have been used for the coupling of amines and related nitrogen nucleophiles with aryl halides. Initially, complexes of the hindered monodentate aromatic tri-*ortho*-tolylphosphine catalyzed the reaction of aryl bromides with secondary amines in the presence of an alkoxide or silylamide base.[439–442] The use of this catalyst for this type of coupling to form C–N bonds was based on an earlier report of the reaction of aryl bromides with tin amides in the presence of a palladium complex of the same ligand as catalyst.[439,440] Later, complexes containing aromatic bisphosphines, such as BINAP[417,418] and dppf,[416] were shown to catalyze the coupling reaction with broader scope than complexes containing tri-*ortho*-tolylphosphine. These catalysts, particularly the complexes with BINAP, expanded the scope of the reaction to include the coupling of primary alkylamines with electron-neutral and electron-rich aryl bromides.[418]

Although these catalysts extended the scope of the amine reagent, the reactions of aryl chlorides did not occur, and reactions of acyclic secondary amines occurred in low yield in the presence of catalysts containing these aromatic phosphines. Because sterically hindered alkyl phosphines and *N*-heterocyclic carbenes generate complexes that oxidatively add aryl chlorides,[220,221,334,443] complexes of these ligands catalyze the coupling of amines with aryl chlorides. Several ligands that generate complexes that catalyze couplings of both aryl bromides and aryl chlorides under mild conditions are shown in Figure 19.4. Catalysts containing these types of ligands allow reactions of aryl bromides to occur at room temperature and the coupling of aryl chlorides to occur at room temperature to 110 °C.

Some of the most utilized and general catalysts for these transformations are generated from ligands shown in Figure 19.4. One set of catalysts contain biaryldialkylphosphines developed by Buchwald and co-workers,[225,304,359] as well as related heteroaryl analogs of these ligands developed by Beller and Singer.[444–447] The simple P(*t*-Bu)₃, in combination with palladium precursors, also catalyzes the coupling of aryl chlorides at high temperatures

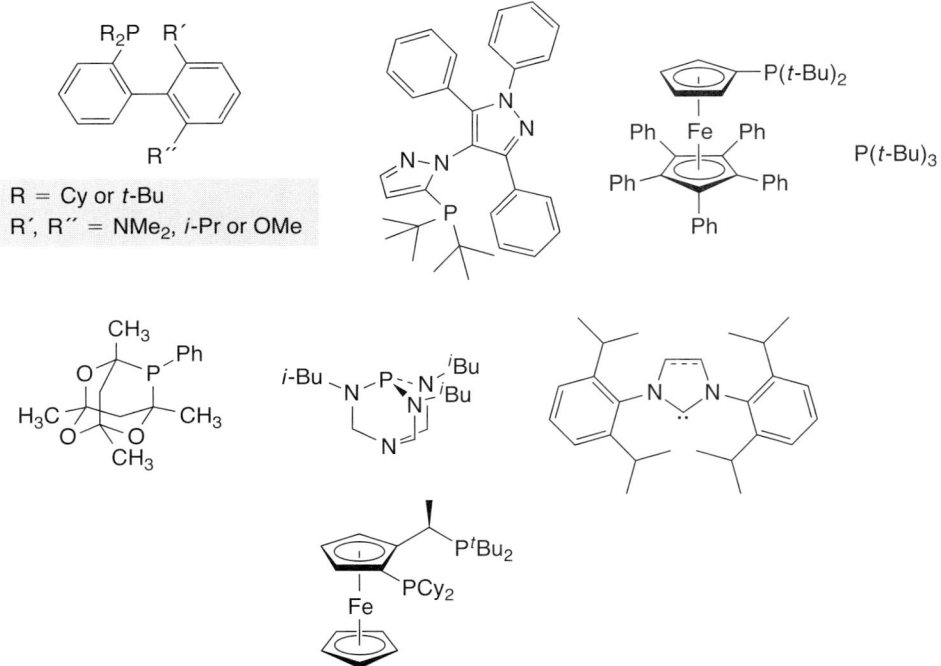

Figure 19.4.
A selection of ligands that generate catalysts for the coupling of amines with aryl chlorides.

with high turnovers[289] or at room temperature with higher loadings of catalyst.[293,294] In addition, complexes of the hindered ferrocenyl ligand couple primary and secondary amines with chloroarenes,[308] and complexes of the N-heterocylic carbenes catalyze the coupling of amines with chloroarenes under mild conditions.[412,448,449] Complexes of each of these types of ligands catalyze the reactions of both cyclic and acyclic secondary amines. Most likely, the faster reductive elimination, as a result of the steric properties of the hindered ligand and the secondary amine, causes C–N bond-forming reductive elimination to be faster than β-hydrogen elimination.[450]

In many cases, complexes of hindered monodentate alkylphosphines and N-heterocyclic carbenes are less stable as catalysts for reactions of primary amines than for reactions of secondary amines and are less stable for coupling of halogenated heterocycles, such as pyridines, than they are for coupling of haloarenes.[451–453] One factor that could lead to this reduced activity is the displacement of the ligand by the primary amine and the basic heterocycle. This point has been debated,[454] but turnover numbers for coupling of these two classes of substrate have tended to be higher for reactions catalyzed by complexes of bisphosphines.[455]

To prevent displacement of the phosphine by primary amines and basic heterocycles, complexes of sterically hindered bisphosphines were investigated for these C–N coupling reactions.[340,455] Reactions of primary alkylamines with aryl and heteroaryl chlorides catalyzed by palladium complexes of the ferrocenylbisphosphine in Equation 19.73 and Figure 19.4 occur under mild conditions and with extremely high turnover numbers. The steric hindrance of the bisphosphine causes reactions of secondary amines to be slower than reactions of primary amines. However, reactions of other primary nitrogen nucleophiles, such as benzophenone imine and benzophenone hydrazone, occur in high yield in the presence of catalytic amounts of palladium complexes of this ligand.

$$(19.73)$$

19.6.2.3. Mechanism of the C–N Coupling

The mechanism of the coupling of amines with aryl halides (Scheme 19.9) draws many analogies to the coupling of carbon nucleophiles with aryl halides. A palladium(0) complex undergoes oxidative addition to generate an aryl palladium halide complex. As noted in Chapter 7, this complex is typically generated, at least initially, by dissociation of ligand from a higher coordinate palladium(0) species and is either a 14-electron bent L_2Pd complex if the ligand is bidentate or a 12-electron monophosphine species if the ligand is a bulky monophosphine. This 12-electron species could be stabilized by interactions with ligand C–H bonds. The arylpalladium–halide complex then reacts with amine and base to generate an arylpalladium–amido complex by one of several possible mechanisms. This process probably occurs in a majority of cases by coordination of amine to the palladium and deprotonation of the coordinated amine by the base. In some cases, when alkoxide base is used, the base may generate an alkoxide intermediate that is protonated by the amine to generate the palladium amide. The resulting arylpalladium–amido complex then

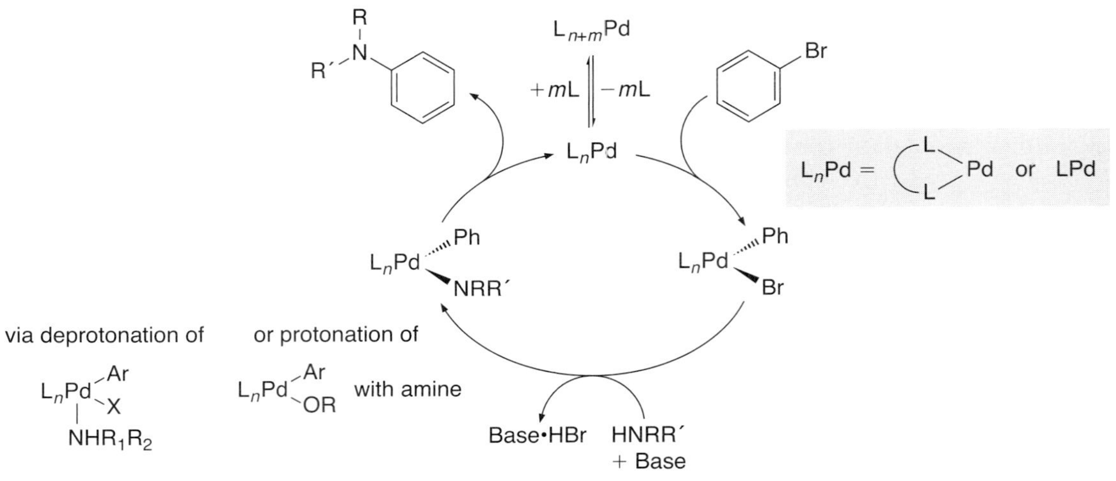

Scheme 19.9

undergoes reductive elimination to generate the arylamine product and to regenerate the starting palladium(0) species. These reactions were discussed in Chapter 8.

The mechanism of this process has been studied in detail. The identity of the palladium(0) species that lies on the catalytic cycle,[286,456] the effect of anions on the oxidative addition step,[457–459] the effect of amines in the dissociation of chelating ligands from the palladium(0) complex during the oxidative addition,[460] the mechanism of formation of the amido complex,[389] and the mechanism of reductive elimination of amine[283,416,461–464] have all been studied. The oxidative addition of aryl chlorides and bromides is generally the turnover-limiting step of the catalytic cycle.

In some cases, the halogen may coordinate to palladium(0) to accelerate the oxidative addition step. In other cases, the halogen might catalyze the displacement of the phosphine ligand to generate the unsaturated intermediate, or it might simply change the polarity of the medium. Although the halide has been shown to increase the rate of the oxidative addition process in polar solvents by generating an anionic palladium(0) species,[237,240] the rate constant for the elementary carbon–halogen bond cleavage step by the anionic palladium(0) compound is smaller than the rate constant for the analogous step by the lower-coordinate neutral palladium(0) complex. This situation noted in Section 19.4.2.1 is summarized in more detail in Scheme 19.10. The identity of the initial product from reaction of the anionic species is unclear, because recent computational work implies that oxidative addition to the three-coordinate anionic palladium(0) complex does not generate an anionic five-coordinate palladium(II) product.[465]

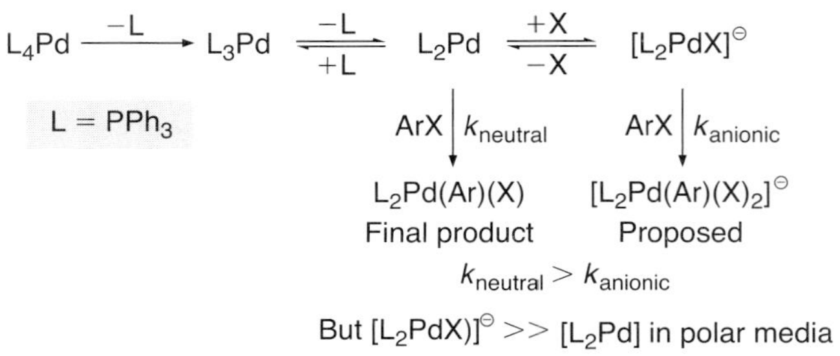

Scheme 19.10

Although the oxidative addition to form the arylpalladium–halide complex usually controls the rate of the C–N coupling process, the reactions of the arylpalladium–amido complexes tend to control the scope and yield. Studies on the reductive elimination of arylamines were presented in Chapter 8. This reaction may occur from a 14-electron, three-coordinate palladium species or a 16-electron, four-coordinate palladium species. If the palladium is ligated by a monodentate dative ligand, then the reaction occurs from the three-coordinate complex, but if the palladium is ligated by a bidentate dative ligand, then the reaction occurs from a four-coordinate complex.[283,352]

The major side reaction of the arylpalladium–amido complexes is β-hydrogen elimination of the amido group[450,466,467] to generate an arylpalladium–hydride complex and imine (Scheme 19.11). The arylpalladium–hydride complex then undergoes reductive elimination of arene, and the product from overall hydrodehalogenation of the arene is generated. Many different phosphine or carbene ligands are now known that favor reductive elimination over β-hydrogen elimination, as evidenced by the high yields of arylamine obtained with many different catalysts. Protonolysis of the amido complex by some unidentified proton source to generate free amine is also a competitive side reaction.

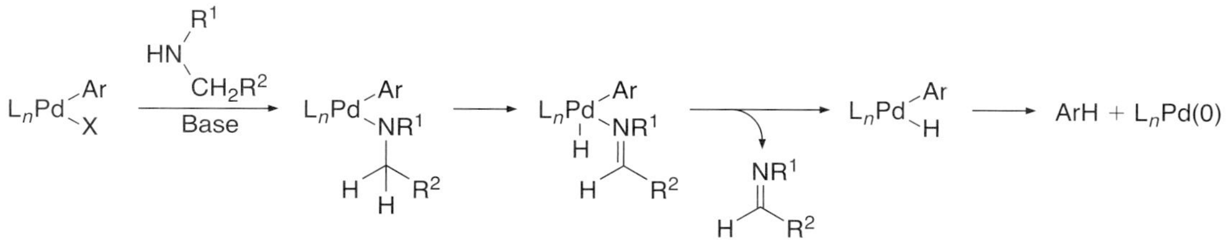

Scheme 19.11

The relative rates of β-hydrogen elimination and reductive elimination are controlled largely by the hapticity and electronic properties of the ligands. As noted in an earlier section of this chapter, Hayashi showed that chelation causes reductive elimination to be favored over β-hydrogen elimination. Because β-hydrogen elimination of amido complexes, like β-hydrogen elimination from alkyl complexes, appears to require an open coordination site,[466] chelating ligands tend to favor reductive elimination of amine over β-hydrogen elimination to form imine (Equation 19.74). As noted in Chapter 8 on reductive elimination, the bite angle of the phosphine also influences the rate of reductive elimination. Therefore, chelating ligands possessing large bite angles, such as dppf and Xantphos, tend to favor reductive elimination of amine over β-hydrogen elimination and generate effective catalysts for the C–N coupling. Finally, steric hindrance favors reductive elimination over β-hydrogen elimination.[450] Reductive elimination decreases the coordination number of the complex, while β-hydrogen elimination increases the coordination number of the complex (before reductive elimination of arene and dissociation of imine reduces the coordination number) (Scheme 19.11). For this reason, sterically hindered alkyl monophosphines generate catalysts that are selective for C–N coupling over hydrodehalogenation of the aryl halide, even though the arylpalladium–amido complex has an open coordination

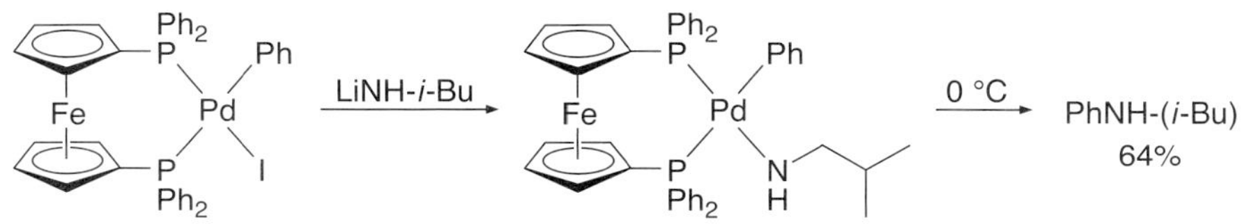

(19.74)

$$\text{(19.75)}$$

Ar=C$_6$H$_4$–p-OMe
(Compared to 80 °C, 2 h for dppf–ditolylamido complexes)

site available for β-hydrogen elimination. A large part of this selectivity is derived from the fast reductive elimination from unsaturated complexes containing hindered monodentate ligands. This fast rate of reductive elimination is illustrated by comparing the reactivity of the P(t-Bu)$_3$-ligated diamido complexes in Equation 19.75 to that of the dppf analogue.

19.7. Carbonylative Coupling Processes

The addition of CO to reactions of haloarenes and main group organometallic reagents, hydride reagents, or H$_2$ can generate organic carbonyl compounds, rather than products from direct cross coupling. The synthesis of ketones is typically conducted with a main group organometallic reagent to deliver the alkyl group and has been reported with the combination of organic halides, carbon monoxide, and either organoboronates, organostannanes, or organozinc reagents. The carbonylations of organic halides to form aldehydes are the least developed of these sets of reactions, but procedures involving the reaction of aromatic halides with the combination of carbon monoxide and hydrogen (synthesis gas) have made this version of organic halide carbonylation more practical.

19.7.1. Carbonylation of Organic Halides to Form Ketones

The conversion of aryl halides to aryl ketones in the presence of CO, a main group organometallic reagent, and a palladium catalyst has been studied for many years. A generic equation for this process is shown in Equation 19.76. This chemistry encompasses reactions of aryl and vinyl halides and triflates with organotin reagents, organozinc reagents, and organoboron reagents. As is described below, the selectivity of these reactions for the formation of ketone products depends on reaction conditions, such as CO pressure, and the identity of the halogen leaving group.

$$R-X \ + \ CO \ + \ R'-M \ \xrightarrow{\text{Pd catalyst}} \ \underset{R}{\overset{O}{\|}}R' \ + \ MX$$

R = Ar, vinyl, benzyl, or alkyl
M = SnR''$_3$, BR''$_2$, or ZnX

$$\text{(19.76)}$$

The reactions conducted with organotin reagents were among the first examples of carbonylative couplings to form ketones and were reported by Tanaka and by Beletskaya[468] (Equations 19.77 and 19.78). The relationship between this chemistry and what has been called "Stille coupling" has led these carbonylations to be called "carbonylative Stille

$$R-SnR''_3 \ + \ CO \ + \ X-R' \xrightarrow{[Pd]} R-C(O)R' \ + \ X-SnR''_3 \qquad (19.77)$$

$$Ph-I \ + \ Me_4Sn \ + \ CO \ (30 \ atm) \xrightarrow[120 \ °C]{Pd(PPh_3)_2(Ph)(I)} \underset{Ph}{\overset{O}{\|}} Me \ + \ Me_3SnI \qquad (19.78)$$

$$\underset{Ph}{\diagdown}SnBu_3 \ + \ I\diagdown Ph \xrightarrow[\substack{THF \\ 70\%}]{\substack{PdCl_2(PPh_3)_2 \\ 333 \ kPa \ CO}} Ph\diagdown\underset{O}{\overset{\|}{C}}\diagdown Ph \qquad (19.79)$$

$$Ar-I \ + \ RSnMe_3 \ + \ CO \ (1 \ bar) \xrightarrow[72 \ h]{[(allyl)PdCl]_2} \underset{Ar}{\overset{O}{\|}} R$$

Ar = p–C_6H_4X ; X = NO_2, CO_2Me, CN, Cl, I, or H

R = Me, CH=CH₂, (thiophene), p–C_6H_4X, C_6F_5, or C≡CPh

(X = OMe, Me, H, Cl, or NO_2)

$$(19.80)$$

$$(19.81)$$

$$(19.82)$$

reactions." Examples of these reactions are shown in the Equations 19.79–19.82. These examples illustrate the carbonylative coupling of vinyl halides with vinyltin reagents,[469] aryl halides with various organotin reagents,[470] and both aryl[471] and vinyl triflates[472] with organotin reagents.

Carbonylations to form ketones with less toxic main group organometallic reagents have also been developed. Although reactions with organoaluminum[473] and organozinc reagents[474] have been published, carbonylations to form ketones with organoboronates have been the most extensively studied. These reactions are now called "carbonylative Suzuki reactions." An early example of this process, shown in Equation 19.83, was reported by Kojima and co-workers.[475]

$$R-B\diagup \ + \ CO \ + \ ArX \xrightarrow[\substack{Zn(acac)_2 \\ THF-HMPA, \ 50 \ °C}]{PdCl_2(PPh_3)_2} \underset{R}{\overset{O}{\|}} Ar \qquad (19.83)$$

Several examples of these carbonylative Suzuki reactions are shown in Equations 19.84–19.86. The reactions of aryl halides[476,477] and vinyl halides[478] with aryl- and alkylboron reagents in the presence of carbon monoxide and a palladium catalyst ligated by triphenylphosphine forms the corresponding ketones in good yields. These reactions have also been conducted intramolecularly, and a representative example is shown in Equation 19.86.[478] The carbonylative Suzuki reactions require base to induce transmetallation of the organoboron reagent, and the identity of the base is important. Transmetallation of the organoboron reagent must be slower than incorporation of carbon monoxide, as described in more detail in the discussion of mechanism.

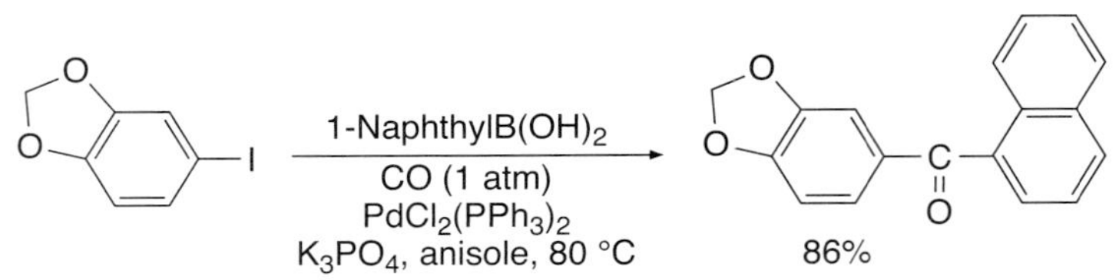

(19.84)

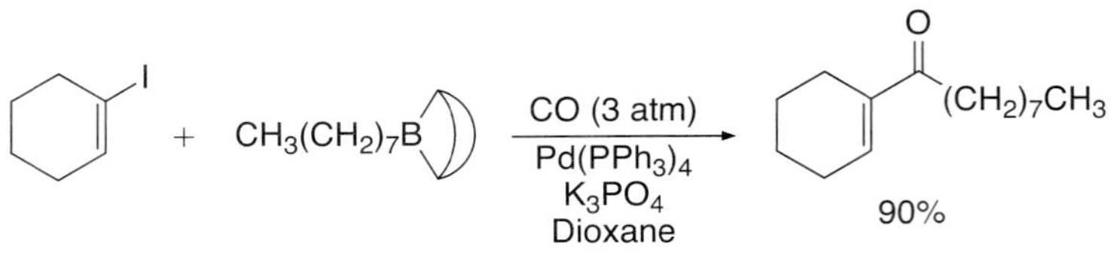

(19.85)

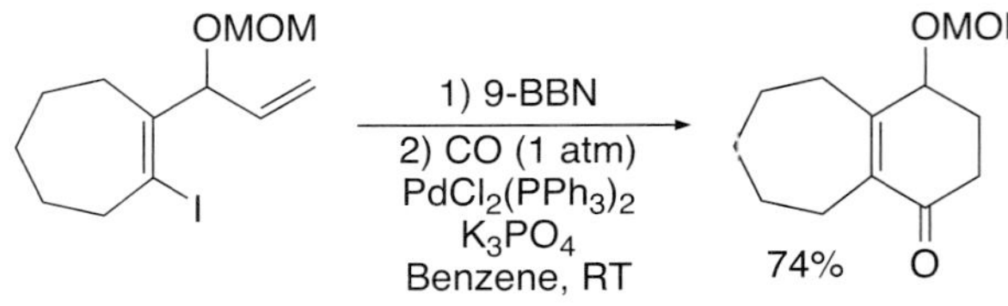

(19.86)

The carbonylation of aliphatic halides to form ketones is less common than the carbonylation of aryl and vinyl halides. However, examples of this process are known, and one example conducted under photochemical conditions to accelerate the oxidative addition of the alkyl halide via radical intermediates is shown in Equation 19.87.[479]

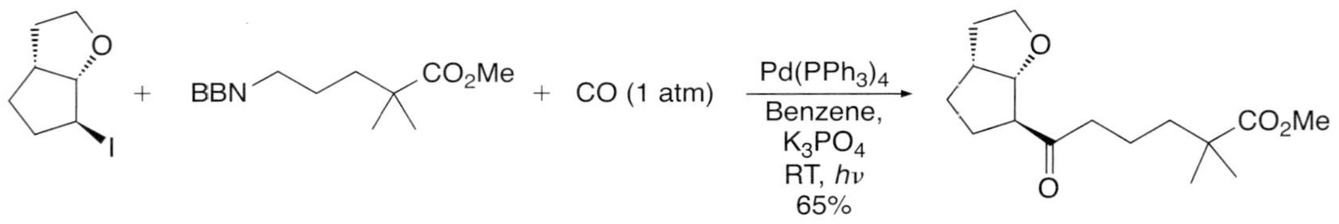

(19.87)

19.7.2. Mechanism of Carbonylative Coupling to form Ketones

Cross coupling of aryl and vinyl halides and triflates with organotin, -boron, -aluminum, and -zinc, reagents follows a mechanism related to the mechanism for the esterification of organic halides that was described in Chapter 17. This catalytic cycle is shown in Scheme 19.12. In this mechanism, a palladium(0) complex undergoes oxidative addition of the organic halide to form an organopalladium halide complex. This complex then reacts with carbon monoxide by a migratory insertion pathway to generate an acylpalladium–halide intermediate. This intermediate then undergoes transmetallation with a main group organometallic reagent to generate an organopalladium–acyl speices, which undergoes reductive elimination to form the free ketone and to regenerate the starting palladium(0) species. These steps of oxidative addition, migratory insertion of CO, and reductive elimination of ketone were described in Chapters 7, 9, and 8, respectively.

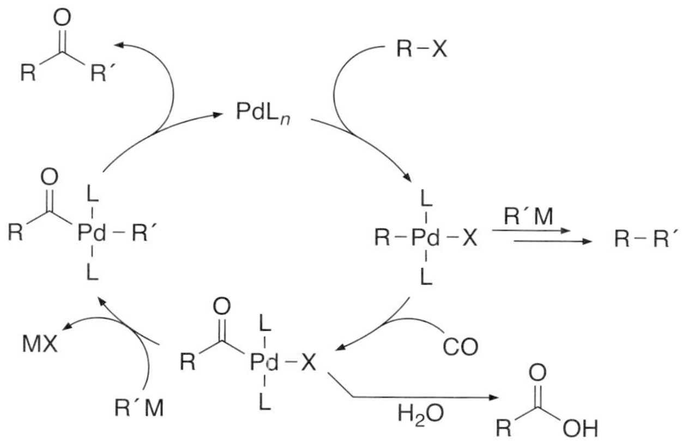

Scheme 19.12

The selectivity for formation of ketone requires the proper relative rates at several points in the catalytic cycle. For example, the organopalladium halide intermediate must react faster with carbon monoxide to generate the acyl intermediate than it reacts with the main group organometallic reagent to generate a diorganopalladium complex that would lead to the products of cross coupling. For this reason, reactions at higher pressure of carbon monoxide tend to form more ketone than reactions at lower pressures of CO. Modifications of reaction conditions, such as the addition of CuI or the use of AsPh$_3$ as ligand have also been shown to accelerate the CO insertion step versus the reductive elimination. Alternatively, the relative rates for CO insertion and transmetallation can be made more favorable by retarding the rate of transmetallation. The base and the halide can affect this rate. For example, reactions conducted with a weak base to activate the organoboron species should occur with slower rates of transmetallation than reactions conducted with a strong base. Moreover, the arylpalladium halide intermediates formed during the carbonylation of aryl iodides are thought to undergo slower transmetallation than those formed during reactions of aryl bromides. Thus higher yields of ketone have been observed from the reactions conducted with weaker bases than with stronger bases and with aryl iodides than with aryl bromides. Finally, the reductive elimination of the acylpalladium intermediate can occur to form carboxylic acids when the reactions are conducted in wet solvents. In this case, the acylpalladium–halide complex can convert to an acylpalladium–hydroxo complex, and reductive elimination from this species could occur to form a carboxylic acid, as shown at the bottom right of Scheme 19.12.[476,480] This step would be similar to the reductive elimination of esters described in Section 17.7.1 in the chapter on carbonylation of arylhalides to form esters.

19.7.3. Formylation of Organic Halides

The synthesis of aromatic aldehydes from aryl halides, carbon monoxide, and a hydride source has also been studied, but practical procedures for this process have been published only recently. Thus, the presentation of this process will be brief. The original formylations of aryl halides were conducted with a tin hydride, as shown in Equation 19.88,[481,482] or a silicon hydride.[483] More recently, reactions were reported with formate as the hydride source (Equation 19.89).[484] In fact, one of the earliest palladium-catalyzed reactions of aryl chlorides was the formylation of an aryl chloride to form aromatic aldehydes using

$$H-SnBu_3 \ + \quad \underset{I}{\overset{Bu}{\diagdown}} \quad \xrightarrow[\substack{THF, \ 50\ °C \\ 88\%}]{\substack{Pd(PPh_3)_4 \\ 100\ kPa\ CO}} \quad \underset{H}{\overset{O \qquad Bu}{\diagdown}} \tag{19.88}$$

(*Z*:*E* of the crude product = 85:15)

$$PhCl \ + \ CO \ + \ NaOC(O)H \quad \xrightarrow[\substack{80\ psi\ CO \\ 150\ °C \\ DMF}]{Pd(OAc)_2/dippp} \quad \underset{Ph \qquad H}{\overset{O}{\diagdown}} \quad 90\% \tag{19.89}$$

dippp = 1,3-*bis*(di-isopropylphosphino)propane

sodium formate as the hydride source.[337] Most recently, reactions have been conducted with synthesis gas as the source of carbon monoxide and a palladium hydride.

Examples of the formylation of aryl halides with synthesis gas catalyzed by palladium complexes are summarized in Equation 19.90.[485] These reactions relied upon the development of ligands with particular steric and electronic properties. The diadamantyl-*n*-butyl phosphine shown in the equation, in combination with palladium acetate, leads to the formation of aromatic aldehydes in high yields from electron-rich and electron-poor aryl bromides. Reactions of nitroarenes and 2-bromopyridine provided the aldehydes in low yield, but other examples occurred in satisfactory yield with only 0.1–0.75 mol % catalyst. The identity of the base is important in this process, and TMEDA was the most effective base. The mechanism of this process was not proposed in the initial work, but is likely to occur by oxidative addition of the aryl halide, insertion of the carbon monoxide into the palladium–aryl bond, and a combination of hydrogenolysis of the acyl intermediate and elimination of hydrogen halide to regenerate palladium(0). The base would then be involved in the hydrogenolysis and consumption of hydrogen halide.

$$R \overset{Br}{\diagup\!\!\!\diagdown} \ + \ CO/H_2 \ (1{:}1,\ 5\ bar) \quad \xrightarrow[\substack{TMEDA \\ 100-125\ °C}]{0.1-0.33\ mol\ \%\ Pd/L} \quad R \overset{O}{\diagup\!\!\!\diagdown}\!\!-\!\!\overset{H}{} \tag{19.90}$$

R = 4-OMe, 3-OMe, 2-OMe,
4-F, 3-F, 4-CF$_3$, 4-CN, 4-NMe$_2$,
4-C(O)Me or 4-Cl or
ArBr = 1- or 2-bromonaphthalene,
2- or 3- bromothiophene or
2- bromobenzothiophene

63–99%

L =

19.8. Copper-Mediated Cross-Coupling Reactions (Written with Dr. Shashank Shekhar)

At the beginning of 20th century, Fritz Ullmann[486–489] reported copper-mediated coupling reactions of aryl halides. The formation of biaryls from the reductive coupling of two aryl halides using stoichiometric amounts of copper was the first type of coupling reported by Ullmann's laboratory (Equation 19.91).[486] Two years later, Ullmann reported that stoichiometric amounts of copper compounds mediate the ipso-substitution of an aryl halide by heteroatom nucleophiles, such as phenoxides and arylamines

(Equation 19.92).[487] These reactions constituted the first metal-mediated nucleophilic aromatic substitutions of unactivated aryl halides. Previous reactions required electron-poor aryl halides, in combination with strong nucleophiles. In 1906, Irma Goldberg[490] extended the scope of the reactions that formed C(aryl)–N bonds to include the formation of C(aryl)–N bonds from the reactions of aryl halides with arylamines or amides, and these reactions were conducted with catalytic amounts of copper (Equation 19.93). Roughly 20 years later, Hurtley published related reactions in which the anions of 1, 3-dicarbonyl compounds act as nucleophile for aromatic substitutions mediated by copper (Equation 19.94),[491] and in the mid to late twentieth century, cuprates that react with alkyl halides to effect nucleophilic substitution of aliphatic halides were developed. Monoanionic "lower-order" cuprates with the formula R_2Cu^- were reported by Gillman,[492] and dianionic "higher-order" cuprates with the formula R_3Cu^{2-} were reported by House.[493] The widely used dianionic cuprates from CuCN having the empirical formula R_2CuCN^{2-} were developed by Lipshutz.[494] These "higher-order" cyanocuprates react with alkyl halides and pseudohalides to give substitution products, as shown in Equation 19.95. Catalytic versions of these reactions, in which the cuprate is generated from a Grignard reagent, are now known and utilized.

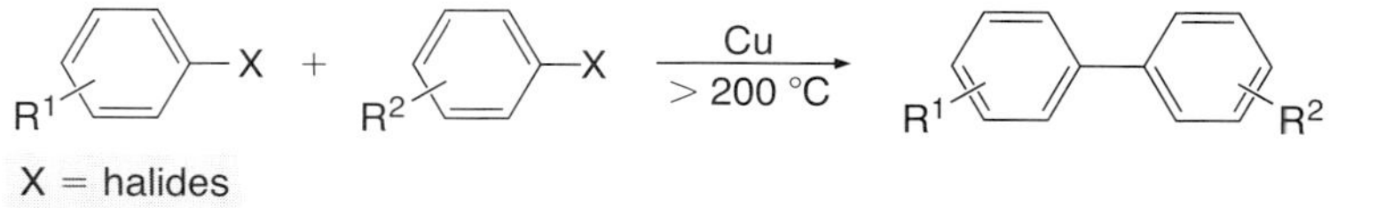
X
+
R1
R2
X
Cu
> 200 °C
R1
R2
X = halides

(19.91)

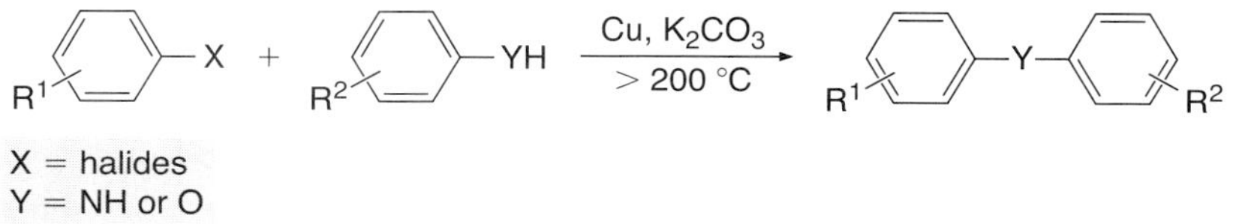
X
+
R1
R2
YH
Cu, K2CO3
> 200 °C
R1
Y
R2
X = halides
Y = NH or O

(19.92)

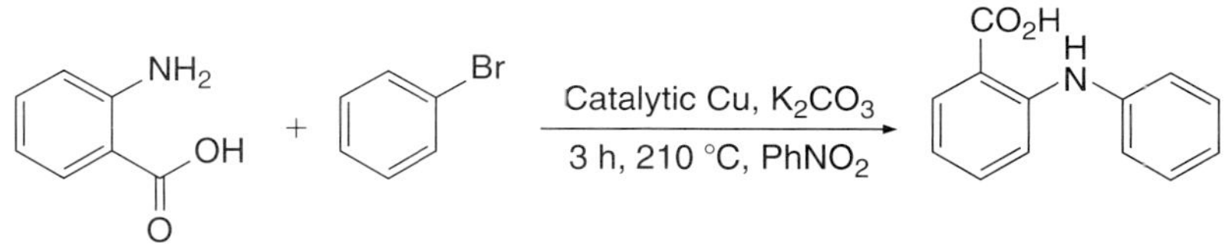
NH2
OH
O
+
Br
Catalytic Cu, K2CO3
3 h, 210 °C, PhNO2
CO2H
H
N

(19.93)

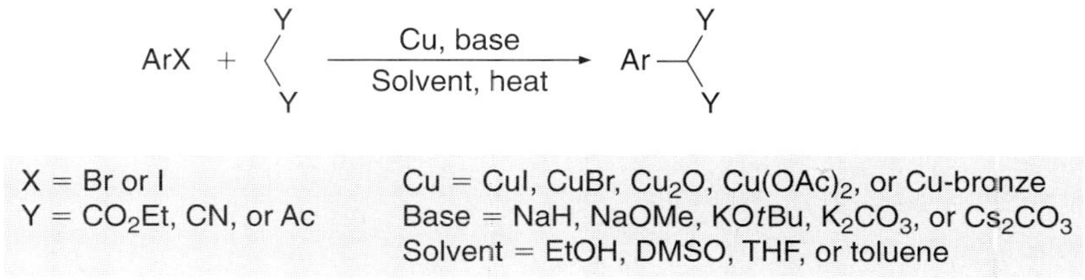
ArX
+
Y
Y
Cu, base
Solvent, heat
Ar
Y
Y
X = Br or I
Y = CO2Et, CN, or Ac
Cu = CuI, CuBr, Cu2O, Cu(OAc)2, or Cu-bronze
Base = NaH, NaOMe, KOtBu, K2CO3, or Cs2CO3
Solvent = EtOH, DMSO, THF, or toluene

(19.94)

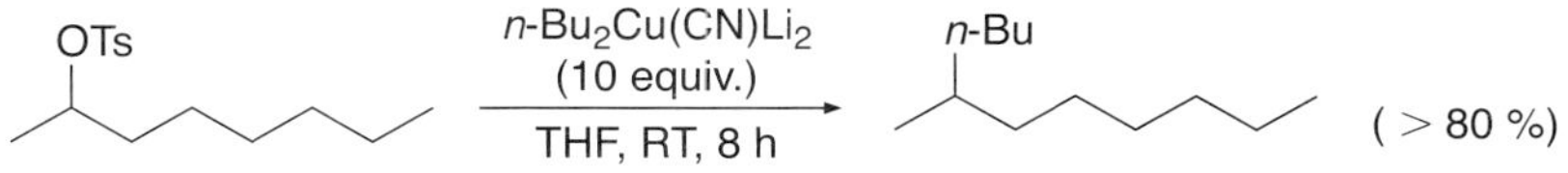
OTs
n-Bu2Cu(CN)Li2
(10 equiv.)
THF, RT, 8 h
n-Bu
(> 80 %)

(19.95)

In recent years, these copper-catalyzed reactions have been studied intensively, and a large number of review articles on the copper-mediated formation of C–X (X = C, N, or O) bonds have appeared.[495–502] This section of the chapter on cross-coupling provides an overview of the developments in copper-catalyzed coupling reactions. Mechanistic data on these reactions are far less extensive than those on cross-coupling reactions catalyzed by nickel and palladium, but recent experiments have provided an increased understanding of the mechanism, and these experiments, along with the mechanistic principles they reveal, are presented.

19.8.1. Copper-Mediated Cross Coupling to Form C(aryl)–N, C(aryl)–O and C(aryl)–S Bonds

The original Ullmann–Goldberg coupling reactions catalyzed by copper salts without added ligands typically required elevated temperatures (> 200 °C) and stoichiometric amounts of copper reagents. They often generated substantial amounts of products from the reductive dehalogenation of the aryl halides. More recent work has led to copper-catalyzed coupling reactions that occur under milder reaction conditions. The following sections provide a sampling of the coupling reactions of aryl halides with various carbon and heteroatom (N, O, or S) nucleophiles catalyzed by copper complexes. This portion of Chapter 19 is organized by the type of nucleophile. Before discussing the scope of copper-catalyzed coupling reactions, the different types of catalysts used for these reactions are described.

19.8.1.1. Classes of Copper Catalysts for Carbon–Heteroatom Bond-Forming Coupling Reactions

A wide variety of copper catalysts have been used for the coupling of aryl halides with nitrogen, oxygen, sulfur, and carbon nucleophiles. Many "ligandless" copper catalyst precursors were first used for these processes and continue to be useful for many substrate combinations. These "ligandless" systems include copper halides, copper triflates, copper nitrates, copper acetates, and copper metal itself. Whether copper metal, copper(I), or copper(II) catalyst precursors are used, the reactions are thought to proceed through copper(I) species.[503] These reactions are often conducted in basic, polar solvents, such as DMF or DMSO, which can act as a ligand for the metal.

As will be illustrated by the examples in the following sections on reaction scope, the copper-catalyzed coupling processes have been dramatically improved by combining these copper precursors with various ancillary ligands. The ligands that generate active copper catalysts are much different from the ligands that generate active palladium catalysts for cross coupling. Because copper is a first-row metal, it is harder than the second-row palladium, and it binds nitrogen and oxygen ligands more tightly than do palladium complexes. Moreover, studies described in the mechanistic discussion of these processes indicate that the key C–N bond-forming portion of the catalytic process occurs faster with nitrogen ligands than with softer and less electron-donating phosphorous ligands.

Many of the ligands that have been used in combination with copper precursors are shown in Figure 19.5. These ligands include those that are both neutral and formally anionic. They include classic ligands like phenanthrolines and α-dimines, as well as more recently studied 1,2-diamines. They also include classic 1,3-diketone ligand precursors that, presumably, generate 1,3-diketonate complexes. These ligands also include those containing a combination of nitrogen and oxygen donors, including salen derivatives, amino acid derivatives, and heterocyclic derivatives like hydroxyquinolines. At this point, a mechanistic foundation to explain the relative reactivities of complexes containing these different ligands is lacking. Much of the development of this chemistry has occurred through empirical evaluation of complexes of the various ligands.

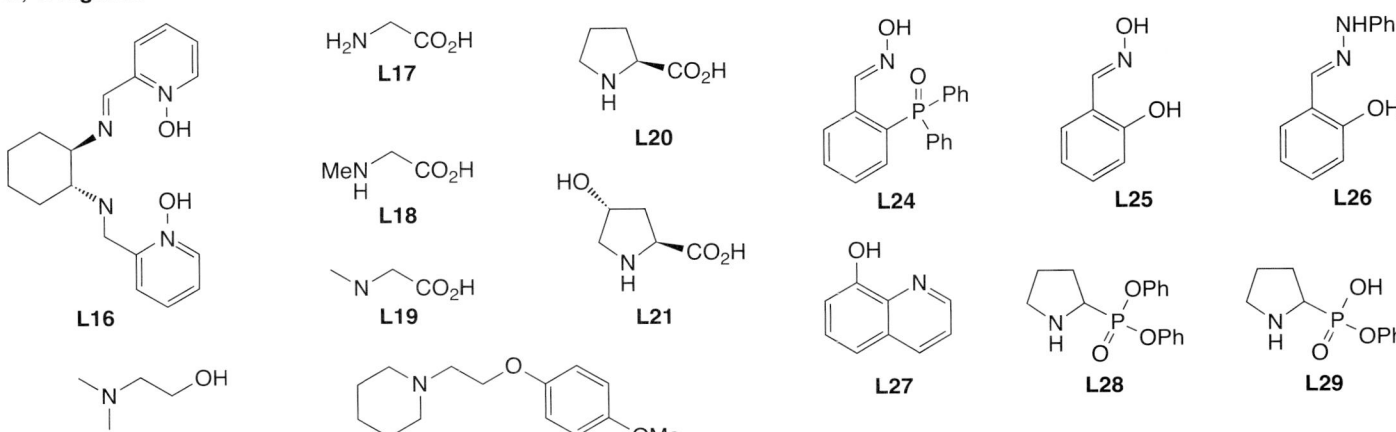

Figure 19.5.
Examples of the various classes of ligands used with copper precursors to generate catalysts for cross coupling. Adapted from Figure 1 of reference 502.

19.8.1.2. Copper-Catalyzed Carbon–Nitrogen Cross-Coupling Reactions

19.8.1.2.1. Copper-Catalyzed Coupling of Amines

19.8.1.2.1.1. Copper-Catalyzed Coupling of Arylamines

The first copper-mediated couplings to form C–N bonds were conducted between *ortho-*chlorobenzoic acid and aniline without solvent under refluxing conditions in the presence of a stoichiometric amount of copper metal (Equation 19.96).[487,490,504–507] A variety of anthranilic acids and diarylamines were prepared by this process, which was conducted in the presence of copper powder, copper salts, copper oxides, or copper alloys, along with a base, such as sodium or potassium carbonate or hydroxide, in a polar solvent, such as amyl alcohol or nitrobenzene, at high temperatures (> 150 °C). This chemistry has been used extensively for the synthesis of intermediates in pharmaceutical, agrochemical, specialty chemical and polymer synthesis.[496,508–521] For example, *N*-phenylanthranilic acid, an important intermediate in the synthesis of acridinones, was synthesized from reactions of *o*-halobenzoic acids with anilines in water or dimethylformamide with K_2CO_3 as base and Cu powder as catalyst (Equation 19.97).

$$(19.96)$$

$$(19.97)$$

Milder couplings of amines have been developed by using copper precursors in the presence of an added ligand. These conditions have significantly broadened the scope of this type of coupling. However, several significant limitations exist. First, the reactivity of aryl halides follows the order I > Br > Cl, and the differences in rates of reactions of substrates containing the three types of halides is often large. Most examples of the coupling of aryl halides with arylamines have been reported with aryl iodides, but some ligated copper systems couple aryl bromides. Only a few examples of copper-catalyzed couplings of aryl chlorides,[522,523] particularly unactivated aryl chlorides, have been reported.[524] Second, reactions of aromatic amines and aliphatic amines have both been reported, but reactions of acyclic secondary amines have not yet occurred in useful yields. Third, the copper-catalyzed reactions are strongly affected by the steric properties of the reagents. Reactions of ortho-substituted aryl halides have recently been reported with ligated catalysts, but many ligandless reactions occur in much lower yields with these more-hindered substrates than with unhindered substrates.

Some of the first mild, intermolecular couplings of amines with aryl halides were reported by Ma when studying the coupling of aryl halides with amino acids.[525] The *N*-arylation of α-amino acids with aryl halides catalyzed by CuI (10 mol %) with K_2CO_3 (1.5 equivalents) as base and DMA as solvent occurred at 90 °C, even with electron-rich aryl iodides and bromides (Equation 19.98a).[526] Systematic studies of a variety of amino acids, as well as different aryl halides and copper ion sources, led to the formation of a wide range of products.[527–529] β-Amino acids and β-amino acid esters underwent the coupling of aryl iodides and bromides catalyzed by CuI in the absence of any added ligand.[527] These studies, along with the known coordination of amino acids to copper, implied that the amino acid was serving as a ligand that made the copper a more active catalyst for Ullmann-type couplings. Amino acids containing secondary amino groups, such as proline, *N*-methylglycine, and pipecolinic acid, also generate active copper catalysts for the coupling of aryl iodides and bromides with aromatic amines,[530–534] as does pyrrole-2-carboxylic acid (Equation 19.98b).[535]

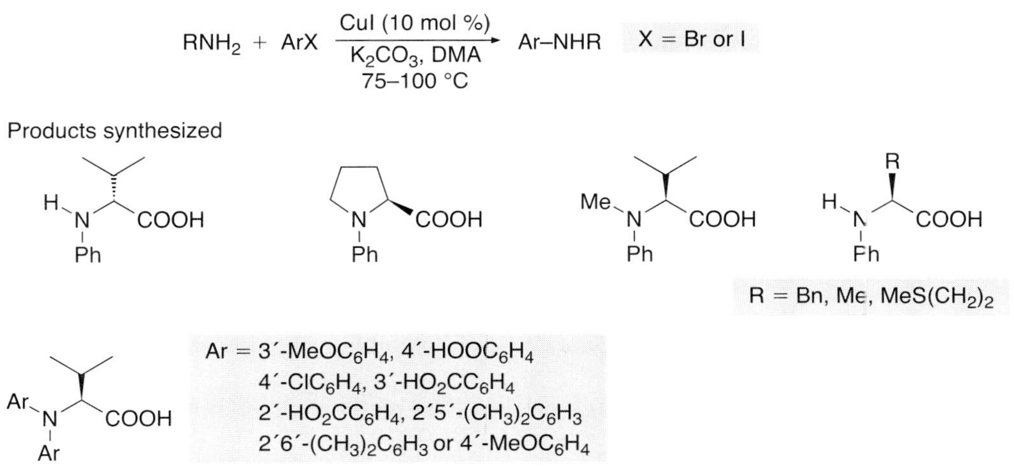
CuI (10 mol %)
RNH2 + ArX
K2CO3, DMA
75–100 °C
Ar–NHR
X = Br or I
Products synthesized
COOH
Ph
Me
R
Fh
R = Bn, Me, MeS(CH2)2
Ar = 3′-MeOC6H4, 4′-HOOC6H4
4′-ClC6H4, 3′-HO2CC6H4
2′-HO2CC6H4, 2′5′-(CH3)2C6H3
2′6′-(CH3)2C6H3 or 4′-MeOC6H4

(19.98a)

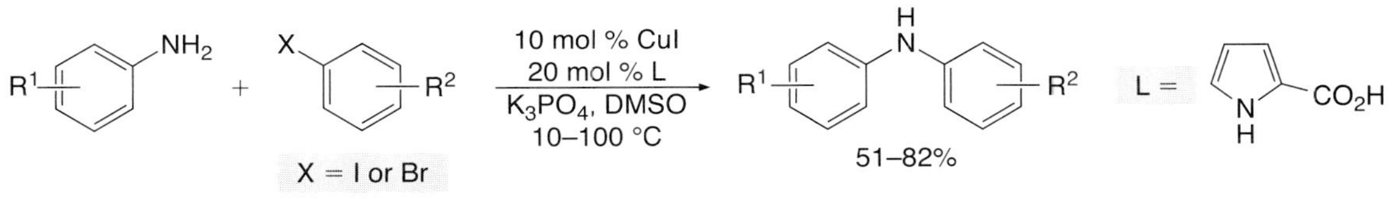
NH2
R1
X
R2
10 mol % CuI
20 mol % L
K3PO4, DMSO
10–100 °C
X = I or Br
51–82%
L =
CO2H

(19.98b)

Concurrent with this work, Goodbrand showed that the *N*-arylation of diarylamines in the presence of a catalyst consisting of CuCl and 1,10-phenanthroline occurred under milder conditions than had commonly been used previously for Ullmann chemistry (Equation 19.99).[536] In addition, reactions to form triarylamines from various aryl iodides and aromatic amines catalyzed by Cu(I) complexes, such as Cu(phen)(PPh_3)Br and Cu(neocup)(PPh_3)Br, which are soluble in organic solvents, were reported.[522,537]

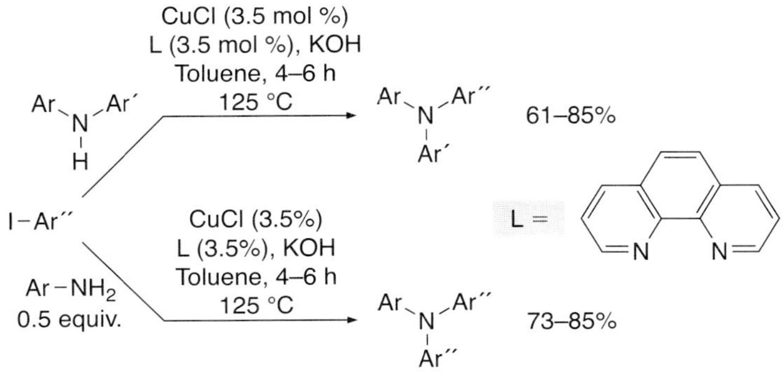
CuCl (3.5 mol %)
L (3.5 mol %), KOH
Toluene, 4–6 h
125 °C
61–85%
I–Ar″
CuCl (3.5%)
L (3.5%), KOH
Toluene, 4–6 h
125 °C
Ar–NH2
0.5 equiv.
73–85%
L =

(19.99)

A few reactions of aryl chlorides catalyzed by ligated copper complexes have been reported. The reaction of phenyl chloride with diphenylamine to form triarylamines catalyzed by the complexes of CuL(PPh_3)Br in which L is 1,10-phenanthroline, neocuproine, or 2,2′-bipyridine occurred in about 50% yield.[522] The favorable effect of potentially coordinating ortho substituents on the rates of Ullmann reactions has been exploited to couple *o*-chlorobenzoic acids with anilines to form *N*-arylanthranilic acids in aqueous or DMF solutions in the presence of pyridine.[538]

19.8.1.2.1.2. Copper-Catalyzed Coupling of Alkylamines

The classic Ullmann reactions catalyzed by ligandless copper do not occur with aliphatic amines, but the coupling of aryl halides with aliphatic amines does occur in the presence of some ligated copper catalysts. The scope of the coupling of aliphatic amines typically encompasses cyclic secondary amines, such as piperidine, piperazine, pyrrolidine, and morpholine, but not acyclic secondary amines. Like the reactions of aromatic amines, few reactions of aryl chlorides have been reported, and no reactions of aryl sulfonates have been reported.

A wide range of ligands has been used to generate copper catalysts for the coupling of aryl halides with alkylamines. Selected examples of such reactions are shown in Equations 19.100 and 19.101. Early examples of the copper-catalyzed coupling of aryl halides with aliphatic amines were conducted with copper complexes of α- and β-amino acids (Equation 19.100)[526,528,529] and β-amino alcohols.[539–543] The amination of aryl and heteroaryl iodides and bromides was also shown to occur in the presence of 0.5 mol % Cu_2O at 80 °C with ethylene glycol as additive.[544] Improvements in this system[545,546] led to a catalyst generated from a combination of CuI (5 mol %), ethylene glycol (2 equivalents), and K_3PO_4 as base in 2-propanol at 80 °C (Equation 19.101). These couplings of aliphatic amines also occurred with aryl bromides, although reactions conducted with the hindered phenols 2,6-dimethylphenol or 2-phenylphenol as additive occurred under milder conditions than those with ethylene glycol.[545] A large excess of amine was required in these reactions.

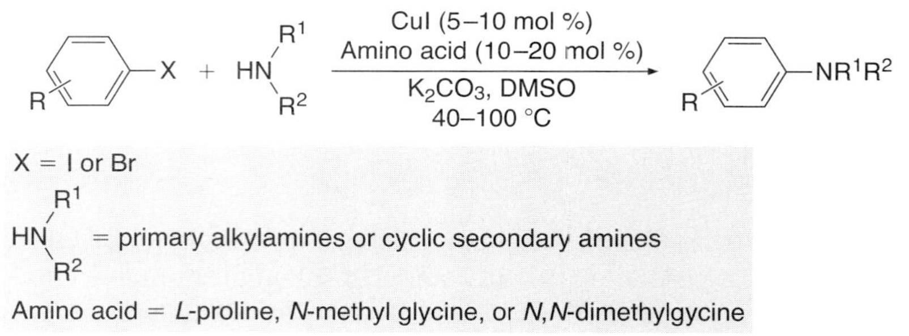

$$X = I \text{ or } Br$$ (19.100)

$$HN\overset{R^1}{\underset{R^2}{}} = \text{primary alkylamines or cyclic secondary amines}$$

Amino acid = L-proline, N-methyl glycine, or N,N-dimethylglycine

$$HN\overset{R^1}{\underset{R^2}{}} = \text{primary alkylamines or cyclic secondary amines}$$ (19.101)

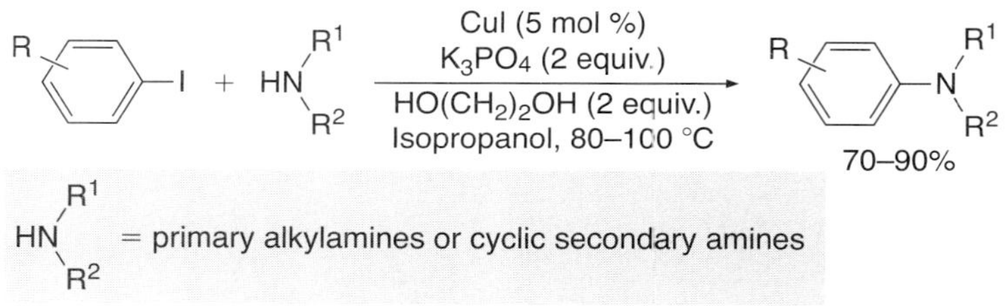

The use of copper catalysts containing bidentate O-donor ligands further improved the coupling of primary alkylamines with aryl bromides.[547] As shown in Equation 19.102, the combination of N,N-diethylsalicylamide (20 mol %) and CuI catalyzed the coupling of bromoarenes with alkylamines under relatively mild conditions. The scope of the reaction catalyzed by this system included aryl bromides containing thioether, hydroxy, nitrile, keto, nitro, and free amino groups. Buchwald and de Vries then reported the couplings of aryl iodides with primary amines in the presence of copper catalysts containing ligands derived from β-diketones (Equation 19.103).[548,549] Reactions of aryl iodides conducted with cyclic diketones occurred at room temperature, and reactions of aryl bromides occurred at 90–100 °C.

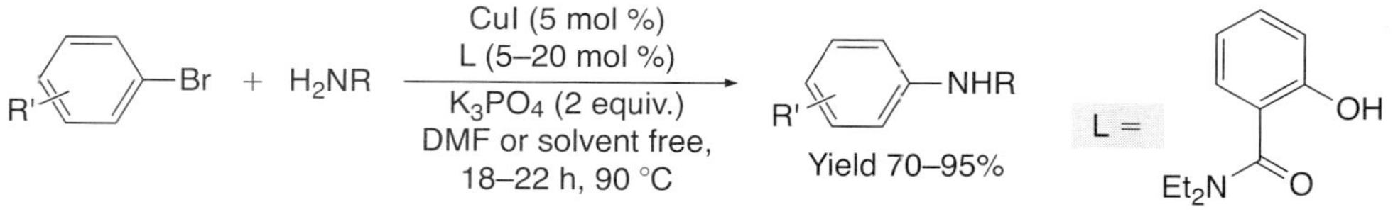

(19.102)

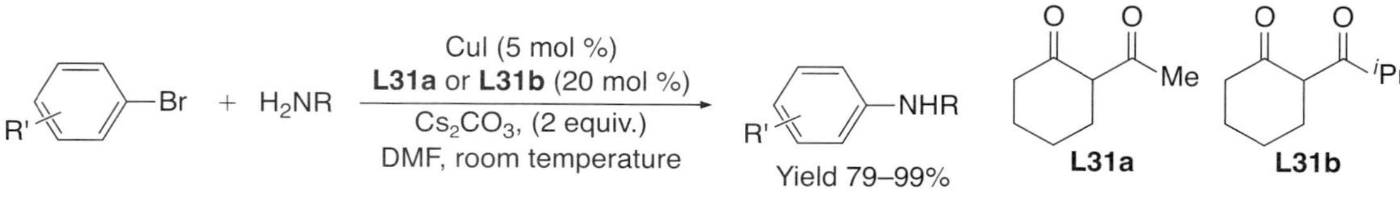

(19.103)

19.8.1.2.2. Copper-Catalyzed Coupling of Amides with Aryl Halides

The use of ligated copper catalysts has dramatically improved the coupling of aryl halides with amides—the Goldberg reaction.[550]. This reaction has been studied intensively because of the utility of the products and the limitations in scope and efficiency of the palladium-catalyzed coupling of aryl halides with amides. Moreover, it is the process that occurs with the broadest scope when conducted with certain ancillary ligands.

Buchwald showed that the combination of CuI and diamine ligands, along with phosphate or carbonate bases, catalyzes the coupling of aryl bromides and iodides with primary amides and cyclic secondary amides (Equation 19.104).[551,552] Reactions of chloroarenes occurred when conducted in neat chloroarene. The synthetically valuable and related carbamates were later shown to couple with aryl halides.[553] Copper complexes ligated by glycine,[554] 1,1,1-tris(hydroxymethyl)ethane,[555,556] or Chxn-Py-Al (see Figure 19.5 for the structure)[555,556] also catalyze the coupling of amides with aryl halides. Catalysts generated from β-ketoesters catalyzed the arylation of amides under particularly mild conditions; some of these reactions even occurred at room temperature.[557] Hydrazides also underwent coupling with aryl halides, in this case in the presence of the catalyst generated from copper(I) iodide and 1,10-phenanthroline and cesium carbonate as base (Equation 19.105). Unhindered aryl halides reacted at the amide nitrogen of the hydrazide, while the more-hindered ortho-substituted aryl halides reacted at the nitrogen lacking the carbonyl substituent.[558]

$$(19.104)$$

$$(19.105)$$

19.8.1.2.3. Copper-Catalyzed Reactions of Aryl Halides with Heterocyclic Amines

Several simple and inexpensive copper complexes catalyze the coupling of aryl iodides and bromides with a variety of nitrogen heterocycles under mild conditions. Complexes derived from CuI (5–10 mol %) and diamine ligands (10–20 mol %) catalyze the coupling of indoles, pyrroles, pyrazoles, indazoles, and triazoles to form the corresponding N-aryl azoles in high yields.[559,560] In addition to diamines, other chelating ligands generate active catalysts for the coupling of aryl iodides and bromides with heteroaromatic nucleophiles (Equation 19.106).[561,562] For example, copper complexes of salicylaldoxime derivatives,[561] amino acid derivatives,[531] 4,7-dichloro-1,10-phenanthroline,[563] 8-hydroxyquinoline,[564] and phosphoramidites,[565] among many other ligands,[566–572] catalyze the N-arylation of imidazoles with aryl iodides. Even ligandless Cu_2O (10 mol %) couples aryl iodides, bromides, and activated aryl chlorides with a variety of π-excessive heteroarenes in the presence of Cs_2CO_3 as base in DMF at 100–110 °C.[573] However, the coupling of imidazoles with aryl bromides or sterically hindered aryl iodides is more challenging and, so far, has been achieved in acceptable yields only when the copper catalyst contains 4,7-dimethoxy-1,10-phenanthroline as ancillary ligand.[574,575] The coupling of azoles with aryl chlorides is rare, but Choudary and co-workers reported a recyclable heterogeneous catalyst, copper-exchanged fluorapatite, for the coupling of aryl chlorides.[576]

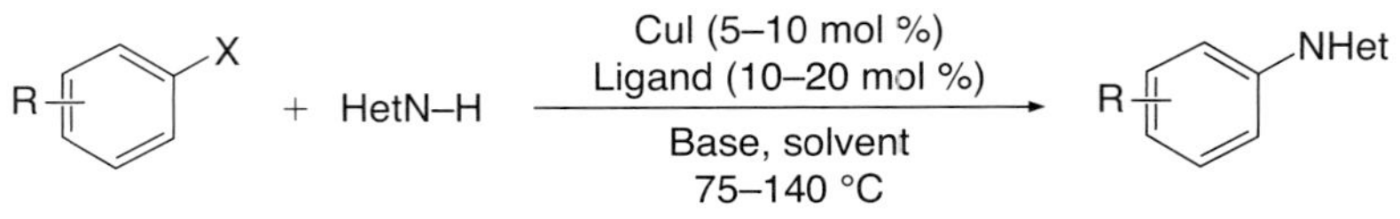

(19.106)

X = I or Br
Base = Cs_2CO_3, K_3PO_4, K_2CO_3
Ligand = **L4–L6**, **L10**, **L15**, **L16**, **L19**, **L20**, **L25**, **L27** of Figure 19.5

Solvent = DMF, dioxane, xylenes, CH_3CN, or DMSO
HetN–H = indoles, pyrroles, pyrazoles, indazoles, triazoles, imidizaloes, or benzimidazoles

19.8.1.3. Copper-Catalyzed Coupling of Aryl Halides with Alcohols and Thiols

19.8.1.3.1. Reactions of Aryl Halides with Phenols

As noted in the introduction to this section, Ullmann also reported the formation of biaryl ethers from phenol and phenyl bromide in the presence of copper and a base. This Ullmann ether synthesis has been used extensively to prepare biaryl ethers.[577–581] However, the original reaction conditions involved high temperatures (150–200 °C), neat phenol or highly polar aprotic solvents, and stoichiometric amounts of copper complexes. The yields for the reactions of unactivated aryl halides were often low.[582, 583] Conditions with catalytic amounts of copper at lower temperatures with broader scope have now been developed.

Examples of the coupling of aryl iodides and bromides with phenols in the presence of catalytic amounts of copper are shown in Equations 19.107–19.113. The reactions of aryl iodides and bromides with a variety of phenols occur in the presence of $(CuOTf)_2 \cdot PhH$ as catalyst and Cs_2CO_3 as base (Equation 19.107).[584] The air-stable $[Cu(MeCN)_4]PF_6$ (5 mol %) also was shown to catalyze the formation of biaryl ethers. As shown in Equation 19.108, o-halo tertiary and secondary benzamides and sulfonamides that were prepared by ortho-metallation procedures were suitable coupling partners.[585]

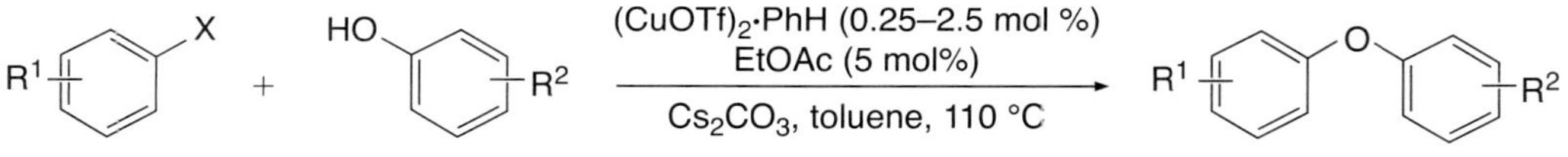

X = I or Br
R_1 = Me, OMe, CN, t-Bu, C(O)Me, or NMe_2
R_2 = Me, i-Pr, or Cl

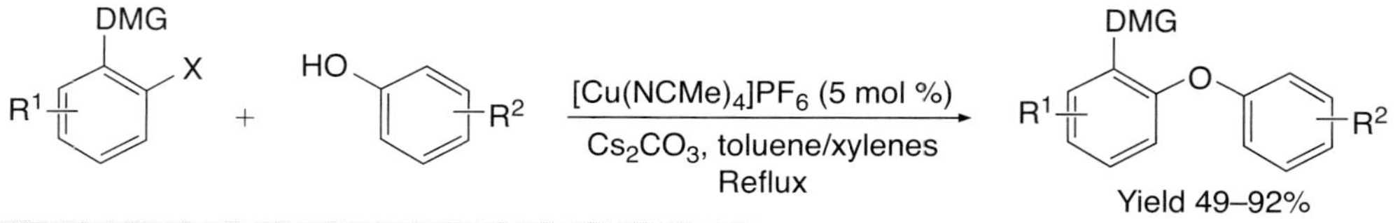

(19.107)

Yield 49–92%

X = I, Br, or Cl
DMG = C(O)NHEt, C(O)NEt_2, SO_2NHEt, or SO_2NEt_2
R_1 = H, Me, or OMe
R_2 = H, Me, F, OMe, or $CONEt_2$

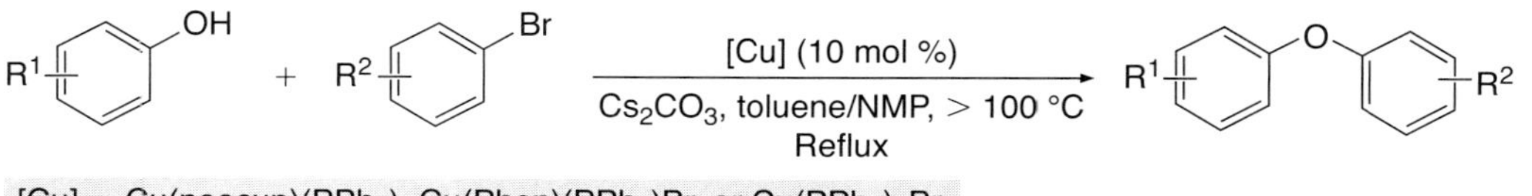

(19.108)

[Cu] = Cu(neocup)(PPh$_3$), Cu(Phen)(PPh$_3$)Br, or Cu(PPh$_3$)$_3$Br

(19.109)

Copper complexes containing ancillary ligands comprising nitrogen, oxygen, or phosphorus donors also catalyze the coupling of phenols with aryl halides. Some of the early examples of these couplings were conducted with complexes of phenanthroline or PPh_3 (Equation 19.109). The air-stable $[Cu(PPh_3)_3Br]$ (5 mol %), $Cu(neocup)(PPh_3)$ (10 mol %), and $Cu(Phen)(PPh_3)Br$ (10 mol %) catalyze the formation of diaryl ethers from the reactions of electron-rich aryl bromides and electron-rich phenols in the presence of Cs_2CO_3 as base in NMP as solvent.[586-588] The combination of CuCl (10 mol %) and the monodentate, neutral nitrogen donor 1-butylimidazole (50 mol %) also catalyzed the coupling of a variety of aryl bromides with phenols in the presence of K_2CO_3 as base in toluene in high yields.[589] Particularly mild couplings of aryl halides with phenols have also been conducted with CuI and N,N-dimethylglycine as ligand. These reactions occur with aryl iodides and bromides at only 90 °C (Equation 19.110), although reactions with hindered aryl bromides occurred in low yield.

$$\text{(19.110)}$$

Copper catalysts for the formation of biaryl ethers have also been developed that contain diketonate ligands. Researchers at Merck Laboratories reported the coupling of aryl halides with phenols catalyzed by the combination of CuCl and 2,2,6,6-tetramethylheptane-3,5-dione (TMHD), which generates a copper diketonate catalyst (Equation 19.111).[590] Catalysts containing related O-donor ligands have also been used. Two examples of reactions conducted with a polyol and with a phosphonic acid as ligand are shown in Equation 19.112[591] and Equation 19.113.[534]

$$\text{(19.111)}$$

$$\text{(19.112)}$$

$$\text{(19.113)}$$

19.8.1.3.2. Reactions of Aryl Halides with Aliphatic Alcohols

The synthesis of aryl alkyl ethers mediated by ligandless copper has limited utility because of the necessity to use super stoichiometric quantities of copper reagents, strong alkoxide bases, and very high temperatures (> 200 °C). The development of the copper-catalyzed coupling of aliphatic alcohols with aryl halides has been particularly challenging, but a few combinations of copper and added ligands are now known that catalyze these reactions under mild conditions with useful yields.[592-596] Among the most active current catalysts are those containing phenanthroline or amino acids as ligands.

The combination of CuI (10 mol %) and 1,10-phenanthroline has been shown to catalyze the reactions of aryl iodides with primary aliphatic alcohols to form aryl alkyl ethers in high yields.[592] Examples of these reactions are shown in Equation 19.114. Primary alcohols such as MeOH, EtOH, nBuOH, HeptOH, etc., react with aryl iodides to form aryl alkyl ethers. Reactions of alcohols with electron-rich aryl iodides and halopyridines occurred in high yield. Lower yields were observed for reactions of some secondary alcohols, due to the enhanced steric hindrance at the reaction center, but the reaction of an aryl halide with enantiopure phenethyl alcohol proceeded with retention of configuration. Vinyl iodides and bromides also couple with allylic, propargylic, benzylic, and aliphatic alcohols to form the corresponding ethers in moderate yields in the presence of the catalyst generated from tetramethyl-1,10-phenanthroline (20 mol %) and CuI (10 mol %) (Equation 19.115).[596]

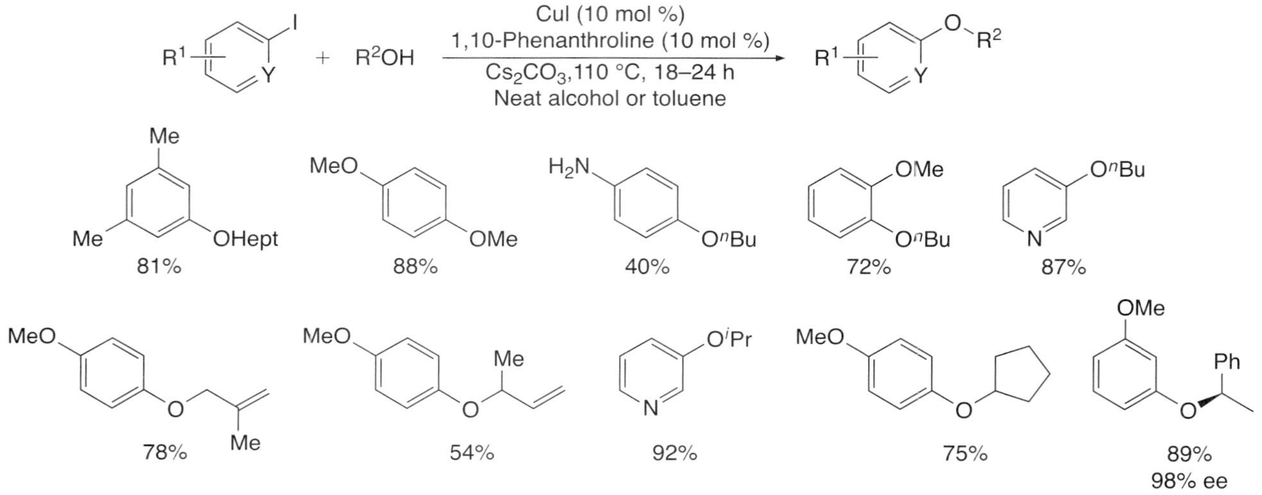

(19.114)

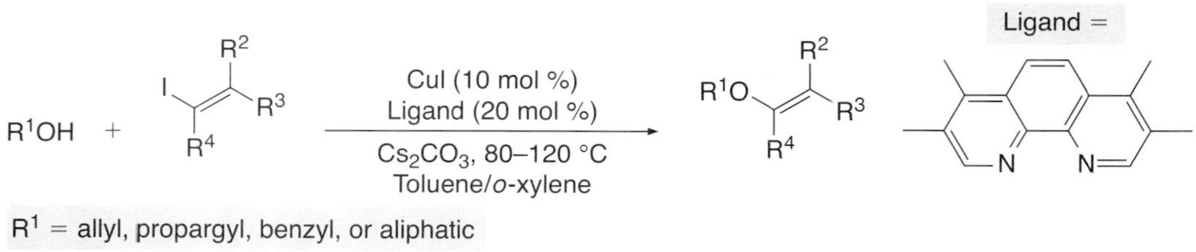

R^1 = allyl, propargyl, benzyl, or aliphatic

(19.115)

The combination of N,N-dimethylglycine and CuI also catalyzes the coupling of aryl halides with alcohols.[594] The scope of the reaction of aryl iodides with primary aliphatic alcohols in the presence of this catalyst was similar to that of the reaction catalyzed by copper complexes of 1,10-phenanthroline derivatives, but the scope of the reactions catalyzed by the complexes of tetramethylphenanthroline was somewhat broader.

19.8.1.3.3. Reactions of Aryl Halides with Amino Alcohols

The *N*-arylation of aminoalcohols reveals the contrasting reactivity of copper complexes containing neutral, bidentate nitrogen ligands and diketonate ligands. Hida[540] reported that β-amino alcohols are more reactive than amines in the classic Ullmann condensation with bromoanthroquinones. More recently, Buchwald[539] developed a practical protocol for the selective *N*-arylation of primary β-amino alcohols with limiting amine catalyzed by CuI (2.5 mol %) and NaOH in a mixture of DMSO and H_2O (2:1) or *i*PrOH at 90 °C. A comparison of the reactions in Equation 19.116 and Equation 19.117 reveals the importance of the ligand on copper.[597] The catalyst containing the diketonate ligand formed the *N*-arylation products in high yields and chemoselectivity (> 20:1) for aminoalcohol substrates in which the OH and NH_2 groups were separated by at least three methylene units. The catalyst containing tetramethylphenanthroline as the ligand afforded the *O*-arylation products in high yield and chemoselectivity (> 15:1) for the reactions of aminoalcohols in which OH and NH_2 groups were separated by at least four methylene units. The authors proposed that the Cu(I) diketonate complex is likely to be less electrophilic than the Cu(I) neocuproine complex and, therefore, to favor binding of the more basic amine over the less basic alcohol. The higher affinity of amine to the diketonate complex would then lead to the selectivity of this complex for *N*-arylation over *O*-arylation.

(19.116)

(19.117)

19.8.1.3.4. Copper-Catalyzed Reactions of Aryl Halides with Thiols

Copper-mediated couplings of aryl halides with thiols have not been explored in as much detail as the copper-mediated couplings of aryl halides with amines and alcohols. However, these reactions occur with aryl iodides in the presence of a variety of copper catalysts. One of the earliest studies showed that CuBr (20 mol %) in combination with a phosphazene base catalyzes the formation of diaryl sulfides from thiols and aryl iodies.[598] CuI (10 mol %) and neocuproine (10 mol %), in the presence of NaO*t*Bu in toluene, also catalyzes the formation of diaryl sulfides and aryl alkyl sulfides.[586] A general, efficient, and operationally simple protocol for the coupling of aryl iodides with thiols involves CuI (5 mol %) as catalyst in a medium consisting of ethylene glycol (2 equivalents) and K_2CO_3 (2 equivalents) in *i*PrOH at 80 °C (Equation 19.118).[599] Reactions under these conditions tolerate a variety of functional groups on the aryl iodides and lead to the coupling of thiols in the presence of aromatic and aliphatic amines, as well as aliphatic alcohols. Additional conditions for the formation of C–S bonds catalyzed by copper with additives,[591,600–602] as well as ligandless copper,[603,604] have been reported.

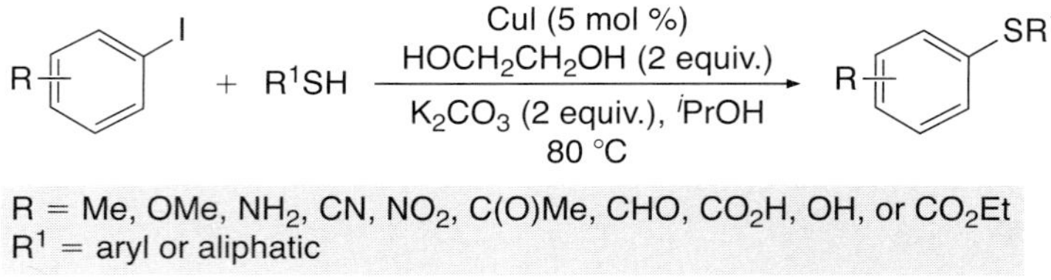

$$R \overset{\text{I}}{\bigcirc} + R^1SH \xrightarrow[\substack{K_2CO_3 \ (2 \ \text{equiv.}), \ ^i\text{PrOH} \\ 80 \ ^\circ C}]{\substack{\text{CuI} \ (5 \ \text{mol} \ \%) \\ \text{HOCH}_2\text{CH}_2\text{OH} \ (2 \ \text{equiv.})}} R \overset{\text{SR}^1}{\bigcirc} \tag{19.118}$$

R = Me, OMe, NH$_2$, CN, NO$_2$, C(O)Me, CHO, CO$_2$H, OH, or CO$_2$Et
R^1 = aryl or aliphatic

19.8.2. Mechanism of Copper-Catalyzed Coupling of Aryl Halides with Amines, Alcohols, and Thiols

Many mechanisms have been proposed for the copper-catalyzed coupling of aryl halides with nitrogen and oxygen nucleophiles, and related mechanisms are likely to be followed for the coupling with sulfur nucleophiles. Early studies focused on identifying the oxidation state of the active catalyst and on determining the composition of the species that react with aryl halides. Paine provided strong evidence that reactions catalyzed by or mediated by copper metal, copper(I), or copper(II) species all occurred through copper(I) complexes.[605] Whitesides,[606] van Koten,[607] and Cohen,[608] among others, studied reactions mediated by ligandless copper species. Mechanisms, including reactions of aryl halides with neutral copper(I) alkoxo[606] and amido complexes or anionic bis-alkoxide complexes,[607] have been proposed. Many mechanisms for how these complexes react with aryl halides have also been proposed, and a variety of these different mechanisms are summarized in Scheme 19.13. These mechanisms include single-electron transfer to generate aryl radicals,[609] nucleophilic aromatic substitution of bound aryl halides through Meisinheimer-type intermediates,[610,611] and the generation of copper(III) intermediates by oxidative addition of the aryl halide.[608] Studies by Cohen provided early evidence for the intermediacy of copper(III) intermediates.[608]

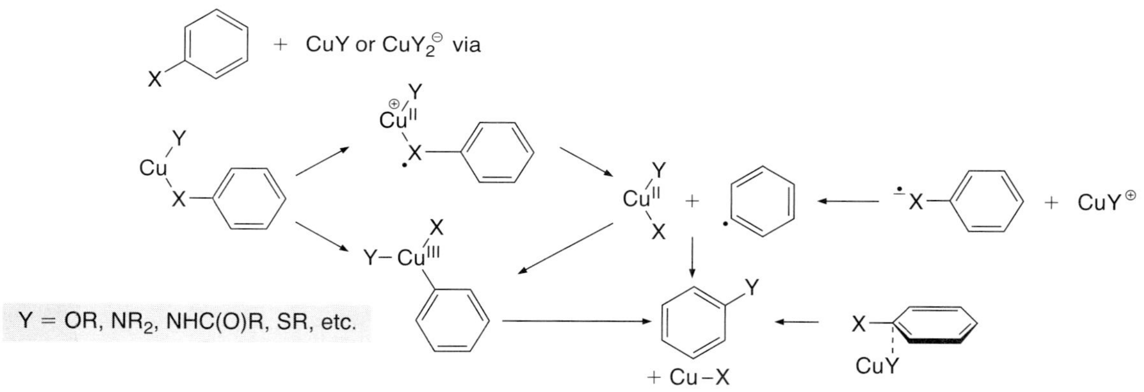

Scheme 19.13.
Network of various proposed mechanisms for the reaction of copper alkoxides, amides, and amides with aryl halides.

More detailed mechanistic studies have been conducted with isolated ligated copper complexes,[612] along with kinetic studies on reactions catalyzed by complexes of diamine ligands.[613] These studies have shown that copper(I) amidate and imidate complexes are competent to be intermediates in the catalytic coupling of aryl halide with amides and imides. These studies also implied that two-coordinate anionic cuprate complexes undergo oxidative addition of the aryl halide more slowly than do related three-coordinate, neutral copper complexes containing a bidentate dative ligand. This conclusion is shown clearly by the formation of coupled product from iodotoluene and the species that equilibrates between the ionic and three-coordinate neutral species (Equation 19.119) and the lack of reaction of iodotoluene with the two-coordinate anionic cuprate containing an ammonium counterion (Equation 19.120).[612] Finally, these studies showed that the neutral amidate compounds react with aryl halides to form the coupled product without the intermediacy of a free aryl radical formed by electron transfer. This conclusion was drawn from the two experiments summarized in Equations 19.121–19.123. First, a comparison between the reaction of an aryl bromide and a chloride, chosen such that the aryl bromide is more difficult to reduce than the aryl chloride, showed that the reaction occurred preferentially with the aryl bromide.[612] Second, reaction of an aryl iodide containing a tethered olefin that cyclizes with a rate constant on the order of 10^{10} occurred without forming cyclized product (Equation 19.123).[612] Although pathways for intramolecular electron transfer to generate a bound aryl halide radical anion are possible, these data support oxidative addition of the aryl halide to the copper amidate complex to form an aryl copper(III) amidate intermediate, followed by reductive elimination to form the carbon–nitrogen bond in the arylamine product.

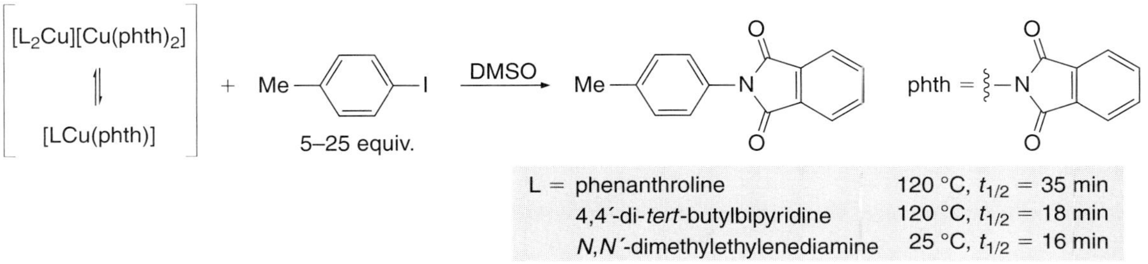

$$L = \text{phenanthroline} \qquad 120\ ^\circ C,\ t_{1/2} = 35\ \text{min}$$
$$4,4'\text{-di-}tert\text{-butylbipyridine} \qquad 120\ ^\circ C,\ t_{1/2} = 18\ \text{min}$$
$$N,N'\text{-dimethylethylenediamine} \qquad 25\ ^\circ C,\ t_{1/2} = 16\ \text{min}$$

(19.119)

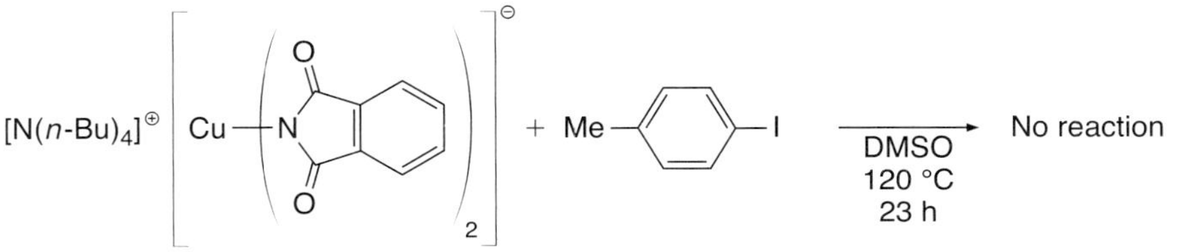

(19.120)

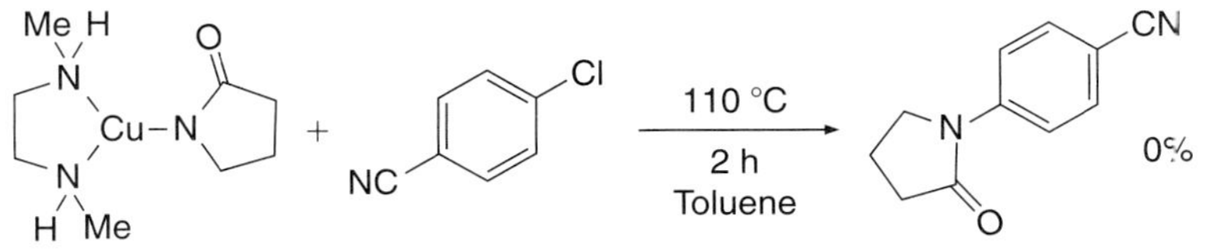

E_0 of 4-chlorobenzonitrile $= -2.03$ V (vs. SCE)

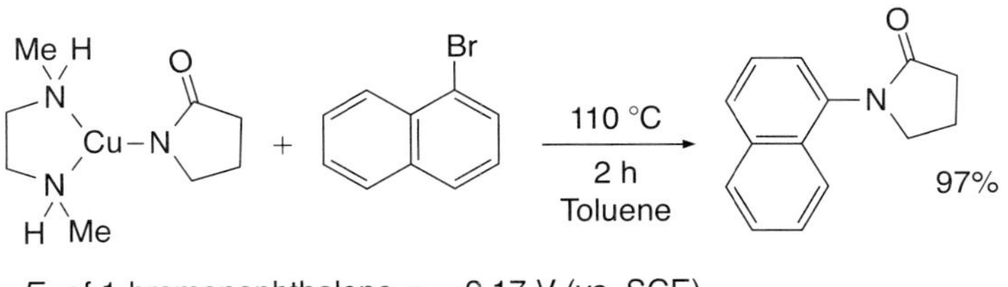

(19.121)

E_0 of 1-bromonaphthalene $= -2.17$ V (vs. SCE)

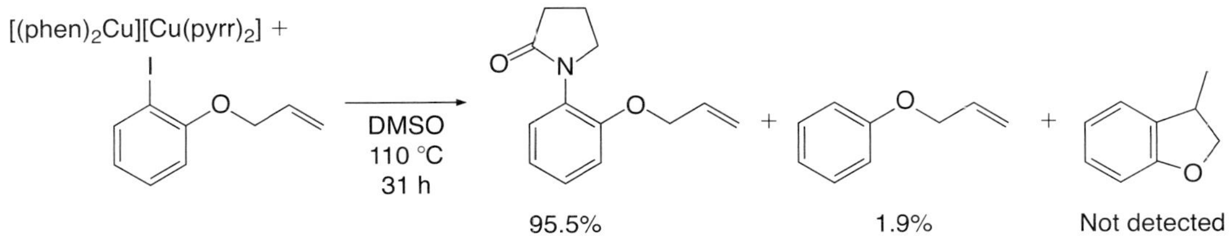

(19.122)

(19.123)

95.5%   1.9%   Not detected

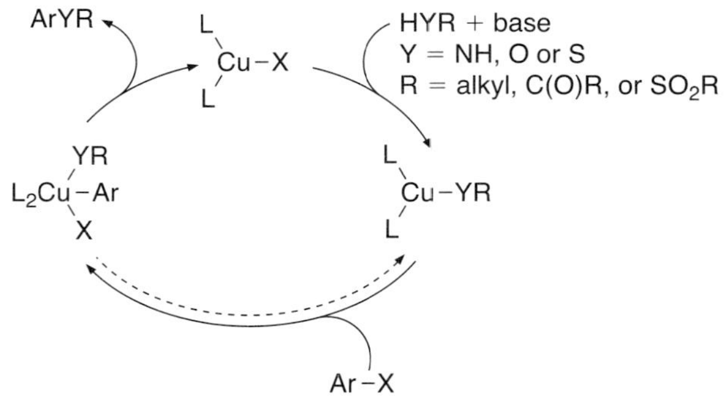

Scheme 19.14

Thus, the copper-catalyzed cross couplings to form aromatic C–N and C–O bonds appear to occur by the mechanism in Scheme 19.14. Like the mechanism for palladium-catalyzed cross coupling, this mechanism consists of oxidative addition, reductive elimination, and transmetallation processes. However, the transmetallation and oxidative addition steps occur in the reverse of the order that they occur in the palladium-catalyzed reactions. The copper-catalyzed coupling appears to occur by the conversion of a copper halide complex to a copper complex containing an anionic nitrogen or oxygen ligand, followed by oxidative addition of the aryl halide to form an arylcopper amide, amidate or alkoxide complex. Reductive elimination from this copper(III) intermediate then generates the organic product and regenerates the starting copper(I) halide complex.

19.8.3. Reactions of Aryl Boronic Acids with Amines and Alcohols (Chan–Evans–Lam Couplings)

Reports by the groups of Chan,[614] Evans,[615] and Lam[616] in 1998 revealed an alternative method to conduct copper-mediated couplings that form C(aryl)–O and C(aryl)–N bonds. In this process, arylboronic acids react with compounds containing N–H or O–H bonds in the presence of a Cu(II) reagent or catalyst. These reactions were initially conducted with stoichiometric amounts of copper reagents.[517-620] Amines, anilines, amides, ureas, carbamates, and sulfonamides underwent *N*-arylation in moderate to excellent yields by this process (Equation 19.124). The commercial availability of boronic acids and the ability to conduct these arylations in air under mild conditions has caused this method to be adopted quickly for synthetic applications on a small scale.

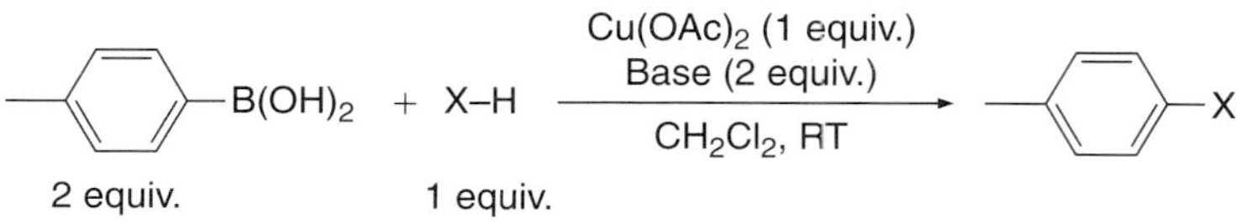

X–H = amine, amide, imide, urea, sulfonamide, or carbamate
Base = pyridine or triethylamine

(19.124)

Collman and Zhong reported the first catalytic, oxidative C–N cross coupling between an arylboronic acid and a substrate containing an N–H bond.[621] In the presence of 10 mol % of commercially available [Cu(OH)TMEDA]$_2$Cl$_2$ in dichloromethane, imidazole coupled with phenylboronic acid at room temperature. Lam and co-workers[622] and Antilla and Buchwald[623] expanded the scope of this type of catalytic coupling to encompass the reactions of amines. Lam and co-workers reported the first catalytic version of the reactions of arylboronic acids with phenols to yield aryl ethers in good yields.[624] The highest yields were obtained when O$_2$ was used as co-oxidant. After these early discoveries, several reports on copper-catalyzed couplings of boronic acids with aliphatic and aromatic amines, amides, heterocyclic amines, and alcohols have been published.[624–630]

Boronic acids have also been used as electrophiles for the copper-mediated formation of C–S bonds. Guy and co-workers reported the couplings of a wide variety of electronically and structurally diverse boronic acids (2 equivalents) with aliphatic and aromatic thiols using $Cu(OAc)_2$ (1.5 equivalents) to mediate the process in DMF at 155 °C.[631] Liebeskind and co-workers reported a milder version of the above reaction using a complex of 3-methylsalicylate with CuI (20–30 mol %) as the catalyst.[632]

19.8.4. Copper-Catalyzed Cross Coupling to Form C–C Bonds

Copper complexes have been used as reagents and as catalysts for the formation of carbon–carbon bonds. The most utilized reactions mediated by copper have been couplings of alkyl halides and sulfonates because copper complexes were unique for many years as reagents that would mediate the nucleophilic substitution of alkyl and aryl nucleophiles with alkyl halides. In recent years, work has been conducted to develop copper-catalyzed versions of cross couplings with aryl halides to address the issues of the cost of palladium catalysts. Although few examples of these processes currently rival those catalyzed by palladium complexes, they do illustrate the potential of copper complexes to catalyst these types of cross-coupling processes.

19.8.4.1. Cross Coupling to Form C(Alkyl)–C Bonds with Copper

Organometallic copper complexes have been important reagents for cross-coupling processes that form C–C bonds between Grignard reagents and alkyl halides. An extensive literature on organocuprate complexes as reagents for nucleophilic substitution of aliphatic halides has been published and reviewed.[633–636] As noted in the introduction to Section 19.8, the organocopper complexes that undergo these substitution reactions have been divided into two groups: "higher-order" cuprates and "lower-order" cuprates. The higher-order cuprates are dianionic triorganocopper complexes (R_2XCu^{2-}, X = alkyl, ^-CN, ^-SCN, etc.), and the lower order cuprates are monoanionic diorganocopper complexes (R_2Cu^-). The reactions of these reagents with alkyl halides have often been conducted with stoichiometric amounts of copper, but the reactions of cuprates have also been conducted with catalytic amounts of copper.[494,637,638] In this case, a copper catalyst is used in combination with the alkyl halide and a Grignard regent, and the combination of Grignard reagent and copper is thought to generate the higher-order cuprate in situ. Either Cu(II) or Cu(I) complexes can be used as the precatalyst, but the actual catalytic cycle starts with a Cu(I) species.

19.8.4.1.1. C(sp^3)–C(sp^3) Coupling Mediated by Copper Reagents

The coupling of alkyl halides with cuprates was originally conducted as shown generically in Equation 19.125 and more specifically in Equations 19.126 and 19.127.[639,640] These reactions are conducted by combining organolithium reagents with copper halides to

$$RCH_2I \xrightarrow[-78\ °C]{R'_2CuLi,\ THF} RCH_2R' \tag{19.125}$$

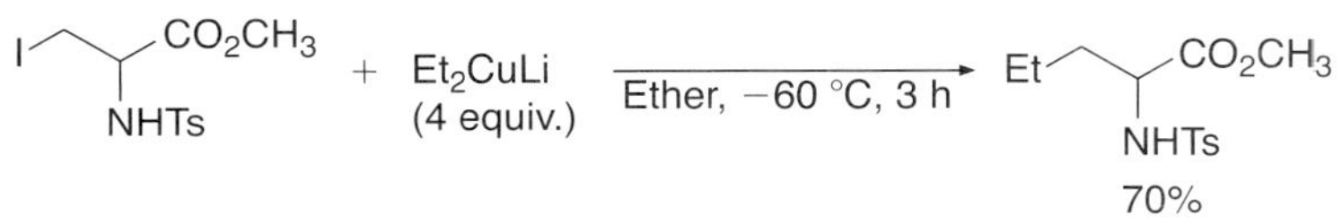

$$\tag{19.126}$$

70%

$$CH_3I \ + \ Li\left[\underset{Bu^t}{\overset{CH_3}{\bigwedge}}Cu{-}CH_3\right] \xrightarrow[\text{Ether, } -70 \text{ to } 0\,°C, \ 30 \text{ min}]{} t\text{-}C_4H_9CH(CH_3)_2 \qquad (19.127)$$

89%

$$\underset{Br}{\bigwedge\!\!\bigwedge} \xrightarrow[\text{THF, } 0\,°C - RT]{\substack{n\text{-}Bu_2Cu(CN)Li_2 \\ (10 \text{ equiv.})}} \underset{n\text{-}Bu}{\bigwedge\!\!\bigwedge} \qquad (19.128)$$

94%

form lower-order anionic "Gillman cuprates" that couple with primary alkyl iodides to form products containing a new $C(sp^3)$–$C(sp^3)$ bond. These reactions typically consume two equivalents of the organolithium reagent for every equivalent of product, but in some cases selective transfer of one group can be achieved (Equation 19.127). These reactions did not occur in high yield with secondary alkyl halides and typically required iodide as leaving group. The "higher-order" cyanocuprates developed by Lipshutz significantly improved the scope of these reactions. As shown in Equation 19.128, a secondary alkyl bromide reacts with this type of cuprate in high yield. These reactions have become even more valuable now that Knochel has developed methods to generate Grignard reagents at low temperatures from organic (aryl, alkenyl, and a very few examples of alkyl) halides containing auxiliary functional groups.[641–645]

19.8.4.1.2. Copper-Catalyzed $C(sp^3)$–$C(sp^3)$ Coupling

The coupling of alkyl halides and sulfonates with organolithium, organomagnesium, and even organozinc reagents can also be conducted with catalytic amounts of copper. When run as a catalytic process, Li_2CuCl_4 is the catalyst precursor most commonly used for the coupling of alkyl iodides and bromides with Grignard reagents (Equation 19.129).[646–649] Aryl and alkenyl Grignard reagents react with alkyl halides in the presence of this catalyst precursor. Recall that the active catalyst is accepted to be Cu(I); the Cu(II) center in Li_2CuCl_4 is apparently reduced to Cu(I) under the reaction conditions. The Cu(I) salt, $CuCl_2 \cdot LiCl$, has also been used as a catalyst, in some cases with better results.[635] Several reports describe the use of CuBr,[650,651] CuBr–HMPA[652,653] and CuBr–Me$_2$S–LiBr–PhSLi[654–656] to couple primary alkyl iodides, bromides, tosylates, or mesylates with a variety of Grignard reagents, including aliphatic, vinylic, and aromatic ones.

$$\underset{Br}{\bigwedge\!\!\bigwedge}\!Br \ + \ \underset{}{\bigvee}{-}MgCl \xrightarrow[\text{THF}]{1 \text{ mol } \% \ Li_2CuCl_4} \underset{(85\%)}{\bigwedge\!\!\bigwedge\!\!\bigwedge}Br \qquad (19.129)$$

A more recent development in copper catalysts has led to increased reaction scope. Kambe reported that 1,3-dienes and 1-phenylpropyne as additive creates a catalyst that couples Grignard reagents with unactivated alkyl fluorides[657] and chlorides (Equation 19.130),[658] respectively. One disadvantage of using Grignard reagents is the incompatibility with several functional groups in the second coupling partner. However, the addition of NMP allowed the coupling of Grignard reagents with alkyl halides and tosylates containing carbonyl, alkoxycarbonyl, and similar groups to occur.[659] The use of NMP also allows the cross-coupling reactions of *tert*-alkyl Grignard reagents to proceed in high yields.

$$\overset{n}{O}ct{-}Cl \ + \ \overset{s}{B}u{-}MgX \xrightarrow[\substack{2 \text{ mol } \% \ CuCl_2 \\ 10\% \ Ph{-}\!\!\!\equiv\!\!\!-Me \\ \text{THF, reflux, 6 h}}]{95\%} \overset{CH_3}{\underset{nOct}{\bigwedge}}CH_3 \qquad (19.130)$$

1 mol   1.5 mol

Two proposed mechanisms for the reaction of cuprates with alkyl halides are shown in Scheme 19.15. By one pathway, the cuprate reacts by an S_N2 path without formation of a new alkyl ligand from the alkyl halide. By a second pathway, the cuprate reacts with an alkyl halide to form a Cu(III) intermediate that undergoes reductive elimination to form the new carbon–carbon bond (Scheme 19.15).

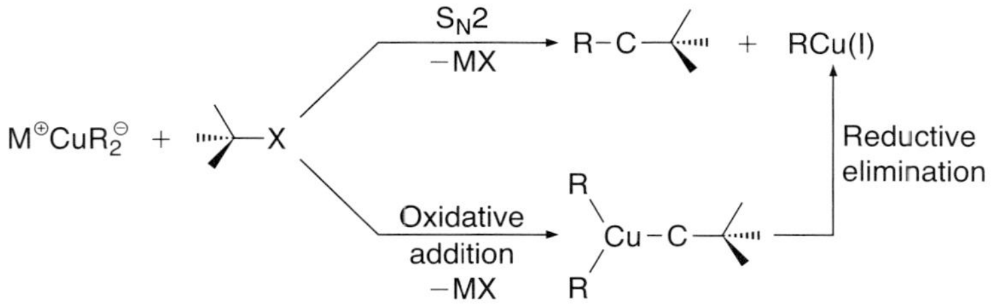

Scheme 19.15.
Mechanism for Cu-catalyzed alkyl halide coupling showing a Cu(III) intermediate.

Some data imply that these reactions occur through Cu(III) intermediates. For example, Cu(III) intermediates have now been observed directly by NMR spectroscopy.[660–663] Computational studies have led to the proposal that an alkali metal cation on the cuprate anion assists the cleavage of the halide from the alkyl group during formation of the Cu(III) intermediate.[664, 665] However, studies with radical clocks have shown that the reactions of alkyl halides can occur by S_N2 or electron tranfer processes. As shown in Equation 19.131, the substitutions of a secondary alkyl halide or sulfonate have generated products from either racemization or retention, depending on the leaving group. Consistent with the greater propensity of organic iodides to react by electron transfer mechanisms, the enantioenriched alkyl iodide reacted to give racemized product, while the enantioenriched alkyl bromide and tosylate formed products from inversion of configuration.[666, 667] Moreover, as shown in Equation 19.132, the reaction of a secondary alkyl iodide containing a pendant alkene predominantly formed products from ring closing of an alkyl radical, whereas the reaction of an analogous alkyl bromide formed little of this ring-closed product.[668, 669] Thus, reactions of alkyl iodides likely occur by electron-transfer, while reactions of alkyl bromides and sulfonates likely occur by S_N2 pathways.

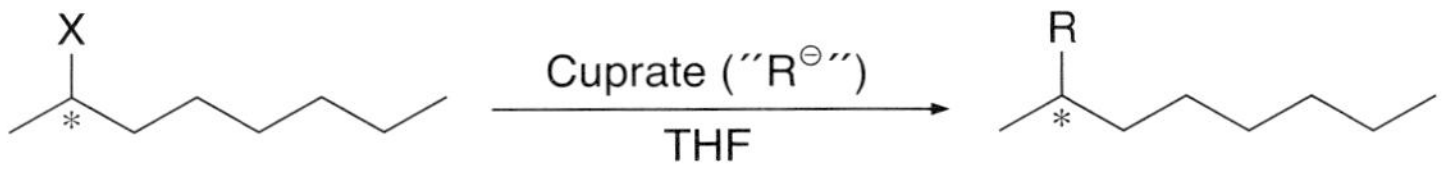

X	Outcome at C^+
I	Racemization
Br	Inversion
Cl	Unknown
OTs	Inversion

(19.131)

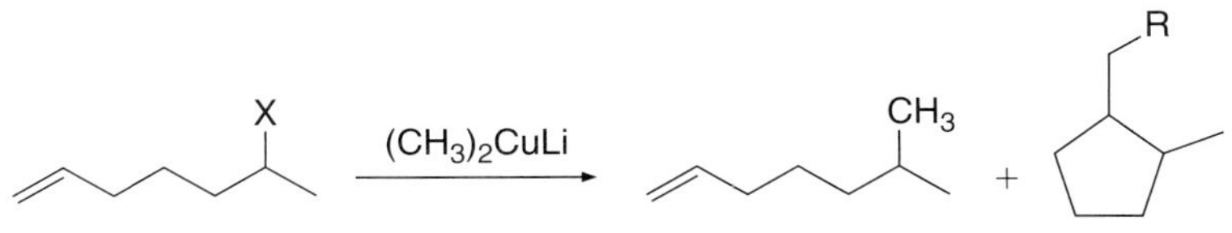

X	%	%
I	18	65 R=CH$_3$(+ 6% R=H)
Br	68	0 R=CH$_3$(+ 5% R=H)

(19.132)

19.8.4.2. Copper-Catalyzed Cross Coupling to Form Aromatic C–C Bonds

Like catalytic coupling reactions to form aromatic C–N bonds, catalytic coupling reactions to form aromatic C–C bonds were studied in the early twentieth century. However, this chemistry has not developed as substantially in more recent times as has the coupling to form carbon–heteroatom bonds. The reactions that occur under the mildest conditions in the highest yields couple aryl halides with reagents containing acidic C–H bonds, such as malonates, cyanoesters, and malononitrile (Equation 19.94). Copper-catalyzed versions of some of the classic palladium-catalyzed processes, such as Stille, Suzuki, and Sonogashira couplings, have been reported, but are not well developed.

19.8.4.2.1. Coupling of β-Diketones, Cyanoesters, and Malonates

The coupling of substrates containing acidic C–H bonds, such as β-diketones, cyanoesters, and malonates, was first reported by Hurtley in 1929,[491] and these couplings are often called "Hurtley reactions." Hurtley reported that 2-bromobenzoic acid coupled with the sodium enolate of acetylacetone in the presence of a catalytic amount of copper–bronze or copper acetate in refluxing EtOH. Later it was shown that 1,3-diketones, such as acetylacetone, couple with aryl halides containing ortho benzoate groups in the presence of a strong base, such as NaH, NaOMe, KO^{t}Bu, etc. and the copper catalyst.[670–672] However, it has been shown more recently that aryl halides lacking these ortho substituents also react with reagents possessing acidic C–H bonds, such as malonates, cyanoesters and malononitriles, in the presence of 10 mol % CuI, CuBr, Cu$_2$O, or Cu(OAc)$_2$, among other copper compounds, and K$_2$CO$_3$ as base in DMSO at 120 °C.[673]

Like the Ullmann and Goldberg reactions, these Hurtley reactions occur under milder conditions and with broader scope when catalyzed by copper precursors in combination with ligands. Buchwald showed that the combination of CuI (5 mol %) and 2-phenylphenol (10 mol %) as ligand catalyzes the coupling of aryl iodides with diethyl malonates under milder conditions than ligandless catalysts (70 °C).[674] However, these reactions also formed products from the coupling of the 2-phenylphenol ligand and from decarboxylation of the product malonate. Cristau, Taillefer, and their co-workers then showed that the combination of CuI (10 mol %) with a Schiff base (20 mol %) as ligand led to the coupling of iodobenzene with CH-acids without formation of these side products (Equation 19.133).[556] In 2005, Ma reported that the combination of CuI and L-proline as catalyst extended the scope of the Hurtley reaction to include reactions of aryl bromides lacking an ortho-directing group.[675] This catalyst also coupled acetylacetone or ethylcyanoacetate with aryl iodides and diketones with vinyl bromides. With CuI (5 mol %) and 2-picolinic acid (10 mol %), the Hurtley reaction can even occur at room temperature.[676] Finally, the Hurtley reaction has been conducted enantioselectively with the combination of CuI and (2S,4R)-2-hydroxyproline as ligand (Equation 19.134).[677] In this initial publication of such a process, however, the scope of the reaction was limited by the need for a chelating group in the ortho position of the aryl halide and for iodide as the halide leaving group.

Chxn-Py-Al

(19.133)

(19.134)

60–93% ee

19.8.4.2.2. Copper-Catalyzed Stille and Suzuki Couplings

Some of the classic coupling processes most commonly catalyzed by palladium or combinations of palladium and copper have now been reported with copper alone as catalyst. These reactions have not been used as widely as have other copper-catalyzed reactions, but a brief presentation of them illustrates the potential of these reactions for further development.

Both intramolecular[678] and intermolecular[679–682] versions of palladium-free, copper-catalyzed, Stille reactions have been reported. In most of these cases, the Cu(I) is added as a salt in the absence of added ligand. In these cases, CuBr or CuI is typically used as the catalyst with CsF as an additive in DMF or NMP solvent at 60–90 °C.[683] Li and Zhang reported a reusable copper catalyst for the Stille reaction of aryl iodides, bromides, and even activated chlorides based on 10% of cubic Cu_2O nanoparticles with 20 mol % of P(o-tol)$_3$, and the combination of KF•2H$_2$O and Bu$_4$NBr at 125–130 °C.[684]

Milder conditions have been achieved using a copper complex, albeit in stoichiometric amounts. Allred and Liebeskind reported the coupling of aryltin or alkenyltin compounds with alkenyl iodides at room temperature or below in the presence of a stoichiometric amount of Cu(I) thiophene-2-carboxylate (CuTC) (Equation 19.135).[685] The need for stoichiometric copper was attributed to the reversibility of the transmetallation step, which leads to inhibition of the cross-coupling reactions by the halide product. However, the mild reaction conditions, unparalleled rates, high tolerance of auxilary functionality, and lack of double bond migration for couplings of vinyl iodides has caused this reaction to be used in several synthetic applications.[686–690]

(19.135)

Copper catalysts have not been explored in detail for the coupling of aryl halides with organoboron reagents. However, some data suggesting that these reactions can occur without palladium have been gained. In one of the earliest examples, copper nanoparticles were found to couple PhB(OH)$_2$ with PhI in the presence of K$_2$CO$_3$ in DMF at 110 °C.[691] A system for a more general coupling of arylboronic acids with aryl iodides or bromides is based on the combination of CuI and DABCO. However, 100 mol % of CuI, 200 mol % of DABCO, and 200 mol % of TBAB were required for the coupling of electron-rich aryl bromides, and reactions of activated aryl chlorides yielded only trace amounts of the coupled product.[575,692,693] Catalytic coupling (10 mol % CuI) of aryl bromides was later achieved with CuI and TBAB in DMSO solvent, although yields were variable.[693] Copper powder with K$_2$CO$_3$ as base in the polyether PEG-400 has also been found to couple aryl iodides with boronic acids in high yields and aryl bromides and chlorides in moderate yields.[694] Thus, the efficiency of Suzuki couplings catalyzed by copper systems are much lower than that catalyzed by palladium systems, but it appears that such couplings can occur.

19.9. Direct Arylation (Written with Dr. Mark E. Scott, Dr. Dino Alberico, and Prof. Mark Lautens)

19.9.1. Introduction and Overview

An alternative to the formation of biaryl compounds by cross coupling between an aryl halide and a main group organometallic nucleophile is the reaction between an aryl halide or main group organometallic reagent with an unfunctionalized arene (Equation 19.136). This process is a form of C–H bond functionalization and, therefore, draws some parallels to the catalytic reactions presented in Chapter 18. However, it results in the same types of products as are formed from more classical cross couplings and is related to the couplings that occur by cleavage of the C–H bonds in carbonyl compounds and alkynes. Thus, the coupling of arenes with aromatic halides are presented here in Chapter 19. This overall process, called "direct arylation," began to be studied intensively roughly two decades after the first reports of this type of reaction.[695–698]

$$X = I, Br, Cl, OTf$$
$$B, Sn, Si \text{ or } Zn$$

(19.136)

19.9.2. Mechanisms of Direct Arylations

To date, many direct arylations have been reported between a wide range of activated arene partners and unactivated arene or heteroarene substrates. Thus, a presentation of the scope of this reaction is facilitated by an initial presentation of the mechanisms by which these reactions are proposed to occur. In general, the formation of biaryl products by direct arylation is proposed to occur by oxidative addition of the activated aryl halide to the transition metal to generate an arylmetal species, followed by reaction of the oxidative addition product with the arene reagent to form an intermediate that ultimately leads to the biaryl product. Several pathways for the reaction of the oxidative addition product with arene have been proposed and are shown in Scheme 19.16: (1) an attack of the metal center as an electrophile on the arene (S$_E$Ar),[699–706] (2) a concerted S$_E$3 process facilitated by base,[707] (3) σ-bond metathesis[702,708,709] or a concerted metallation–deprotonation[710,711] process, (4) a carbometallation of the arene, either through a formal anti β-hydride elimination or by isomerization followed by a syn β-hydride elimination,[701–704,706,712] (5) an oxidative addition of the arene C–H bond,[708,713,714] or (6) a process that generates a second metal–aryl species.[715] The mechanism of this step has been investigated in detail for some systems, but the mechanism for any given direct coupling can depend on the substrate, transition metal, solvent, base, and ligand.

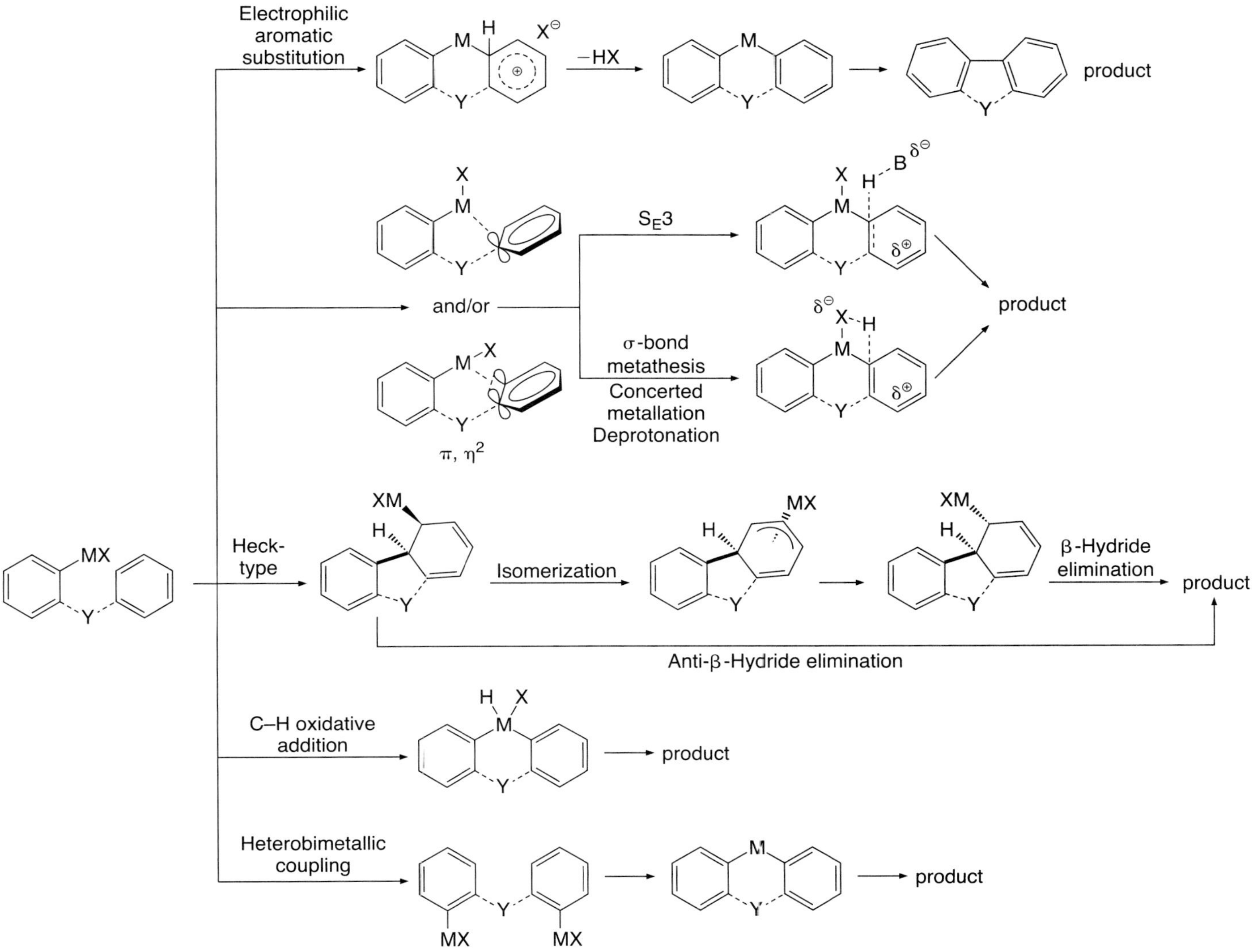

Scheme 19.16.
Six mechanisms for the reaction of an arylmetal halide complex with an arene by intermolecular and intramolecular processes.

19.9.3. Transition Metal Catalysts for Direct Arylation

Many different transition metal complexes catalyze direct arylation processes. The majority of such arylations reported so far have been conducted with second row transition metals, such as Pd, Ru, and Rh. Of these three metals, catalysts containing palladium have been the most versatile and widely used in direct arylation processes. Typically, the reactions catalyzed by palladium complexes occur by a mechanism involving Pd(0) and Pd(II) species. The scope of palladium-catalyzed direct arylations encompass intramolecular and intermolecular reactions of a variety of aryl and heteroaryl substrates (Equations 19.137–19.140).[716–719] The aryl electrophiles that undergo direct arylation through a mechanism involving palladium(0) and palladium(II) include aryl iodides, bromides, chlorides, and triflates, hypervalent aryl iodide reagents, as well as the combination of an oxidant and a main-group organometallic reagent, such as an aryl boronic ester.

$$(19.137)$$

$$(19.138)$$

$$(19.139)$$

$$(19.140)$$

Hypervalent aryl iodine reagents have also been used as the electrophile in direct arylation reactions. In these examples, the strongly oxidizing hypervalent iodine reagents are proposed to react with the intermediate Pd(II) species to generate a Pd(IV) intermediate. Subsequent reductive elimination of biaryl from this Pd(IV) intermediate is then believed to regenerate the active Pd(II) species (Equation 19.141).[720]

DG = directing group

$$(19.141)$$

Palladium catalysts for direct arylation reactions have been generated from many different ligands. The ability of the ligand to generate an active catalyst often depends on the nature of the aryl halide. Couplings of the more reactive aryl iodides are typically conducted with catalysts containing moderately electron-rich monodentate phosphines, such as PPh₃. Palladium complexes containing these same phosphines also catalyze the reactions of aryl bromides, although reactions conducted with more sterically bulky and electron-rich trialkylphosphine or biaryl dialkylphosphines occur in higher yields in many cases.[721–723] Recently, palladium-catalyzed direct arylations with aryl chlorides have been reported. Like the cross coupling of aryl halides with arylboronic acids,[724] the direct arylation of aryl chlorides occurs when the palladium catalyst contains electron-rich and sterically hindered trialkylphosphines, biphenylphosphines, or N-heterocyclic carbene ligands. These reactions have also been reported using "ligandless" conditions (Jeffery's conditions).[705,725,726]

Ruthenium complexes also catalyze the direct arylation of aryl halides. These reactions are thought to occur through mechanisms involving Ru(II) and Ru(IV) intermediates and can be conducted with substrates lacking protective groups. The higher oxidation state of these intermediates, relative to those in palladium-catalyzed direct arylations, makes the ruthenium catalysts well suited for conducting the direct arylation with arenes containing directing groups that coordinate to the metal through nitrogen or oxygen (Equations 19.142 and 19.143).[727] The aryl electrophiles that undergo direct arylation catalyzed by ruthenium complexes include aryl bromides, chlorides, and tosylates. These reactions also occur with arylboronic esters (Equation 19.143).[728,729] In this reaction conducted with arylboronic esters, the reaction required an aliphatic ketone (in this case, pinacolone) as a hydride scavenger. The proposed mechanism for this process shown in Scheme 19.17 begins with coordination of the ketone substrate to ruthenium, followed by C–H bond cleavage to afford a five-membered ruthenacycle. Insertion of pinacolone into the ruthenium-hydride bond affords an alkoxy–ruthenium intermediate, which subsequently undergoes transmetallation with the aryl boronic ester to generate the diaryl ruthenium intermediate and trialkoxyborane. Reductive elimination furnishes the biaryl product, thereby regenerating the active catalyst.

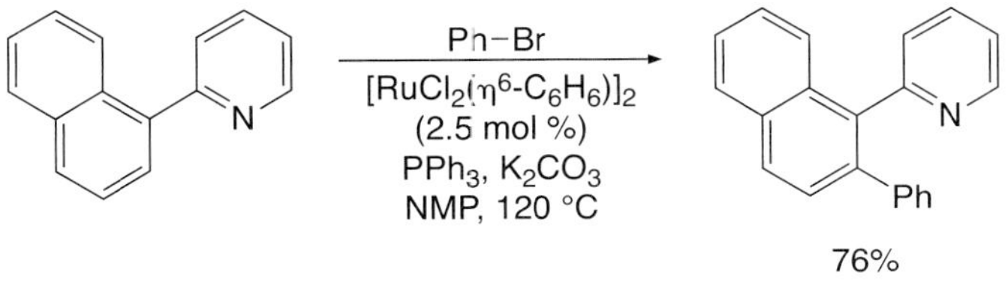

(19.142)

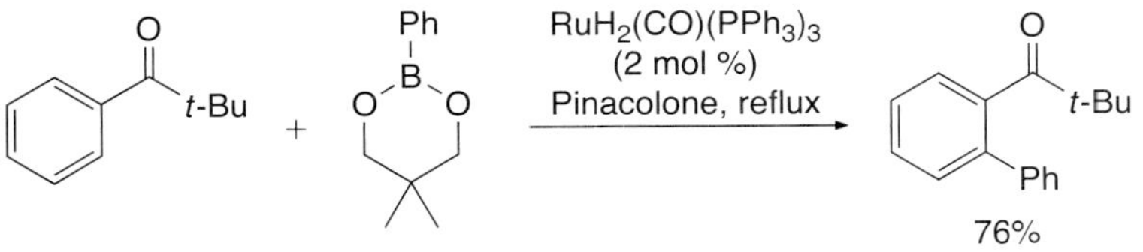

(19.143)

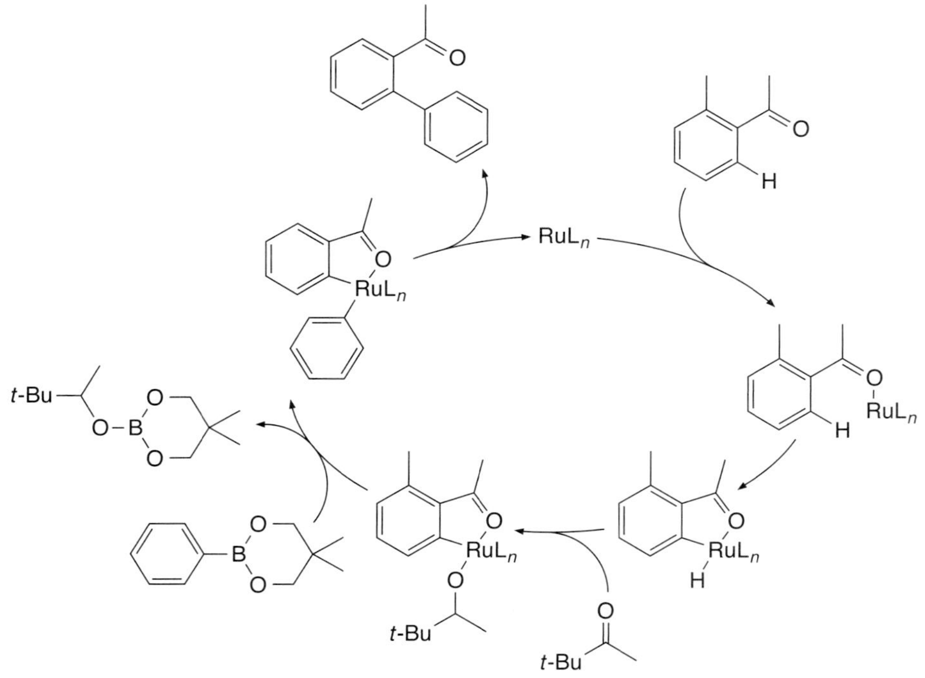

Scheme 19.17

Rhodium complexes catalyze the direct arylation of indoles,[730] pyrroles,[731] benzimidazoles, and benzoxazoles[732] (Equation 19.144), as well as arenes bearing imine[733] and phenolic[734,735] directing groups (Equation 19.145). These reactions have been conducted with aryl bromides and iodides as the electrophile. These coupling reactions are proposed to occur through a catalytic cycle involving a combination of Rh(I) and Rh(III) intermediates. In addition, rhodium-catalyzed direct arylations have been reported between arenes and a nucleophilic source of the second aryl group, such as aryl boronic acids,[733] sodium tetraphenyl borate,[733,736] and tetraphenyl tin.[737] In these instances, the reaction is proposed to occur by initial coordination of the directing group to a phenylrhodium species, followed by orthometallation to form a five-membered metallacyclic intermediate (Scheme 19.18). Subsequent reductive elimination generates the monophenylated product and a rhodium hydride species, which undergoes reduction by benzophenone imine[733] or α-chloroacetate[736] to regenerate the active catalyst.

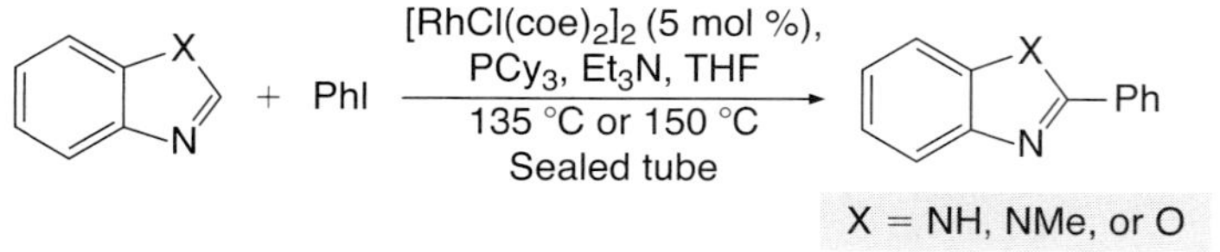

$$(19.144)$$

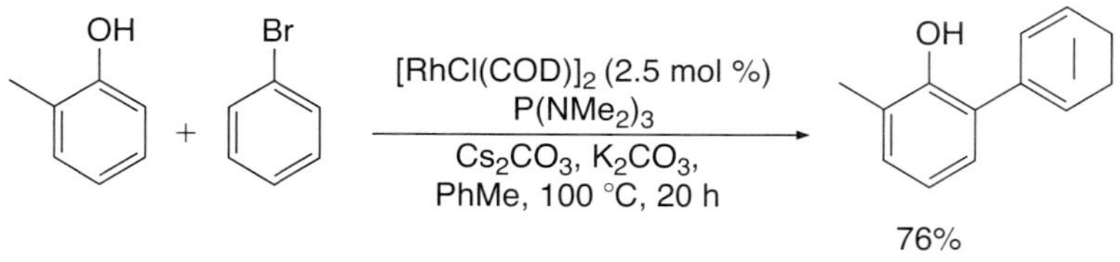

$$(19.145)$$

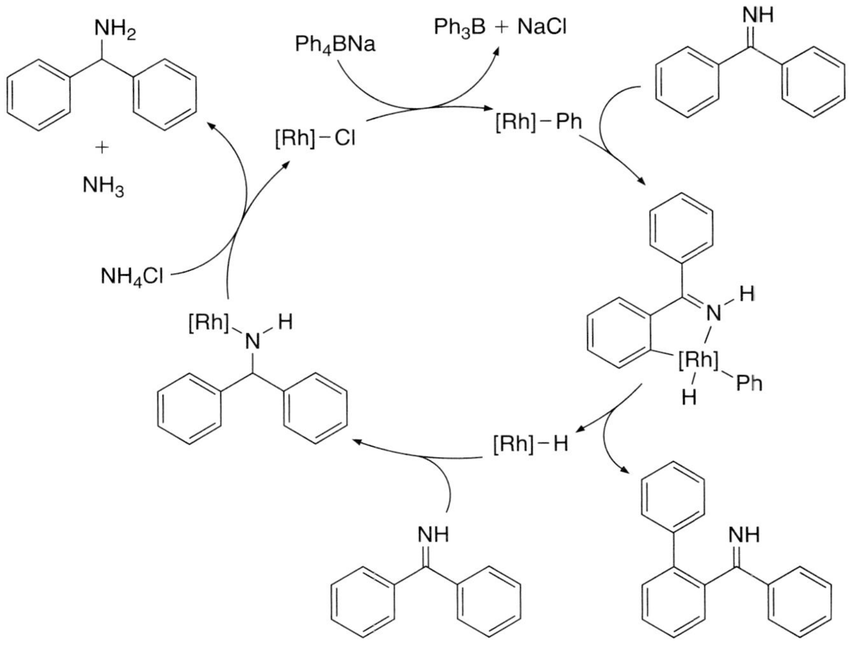

Scheme 19.18

Although the majority of direct arylations have been catalyzed by palladium, rhodium, and ruthenium, some additional studies have also focused on direct arylations catalyzed by first-row metals, such as iron and copper. For example, an iron-catalyzed direct arylation reaction between arylzinc reagents and 2-arylpyridine derivatives has been reported (Equation 19.146).[738] Several direct couplings of heteroarenes with aryl halides (Equation 19.147)[739] or hypervalent iodine reagents[740] catalyzed by copper halides have also been reported.

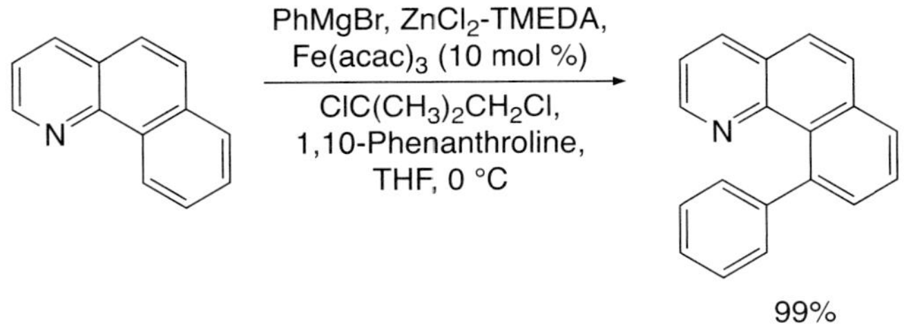

$$(19.146)$$

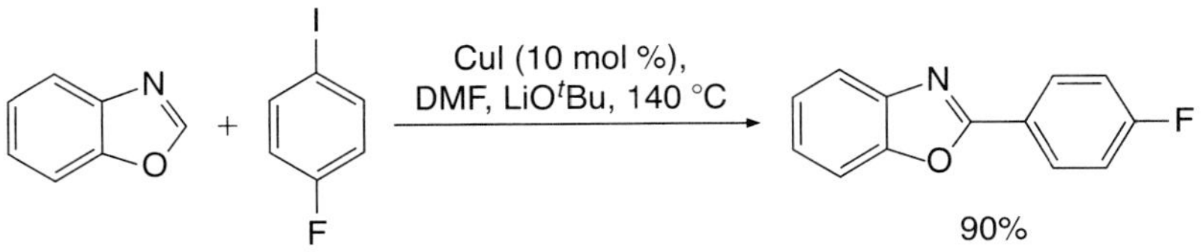

$$(19.147)$$

19.9.4. Regioselectivity of Direct Arylations

Regioselective direct arylations are difficult to achieve because the arene reagents often contain several inequivalent C–H bonds that can react with the metal center at similar rates. Several approaches have been investigated to address this challenge. In some cases the reactions have been conducted intramolecularly, while in other cases the reactions have been conducted intermolecularly with arenes that contain directing groups or that contain C–H bonds that have distinct steric or electronic properties.

Intramolecular reactions of the two arene components linked by a tether can occur regioselectively because of the geometric constraints. An example of an intramolecular direct arylation controlled by a temporary tether is shown in Equation 19.148.[741] In this synthesis of dioncophylline C, two important features of intramolecular direct arylation are illustrated. First, intramolecular arylation reactions are typically used for the construction of five-, six- (as shown in Equation 19.148), and seven-membered rings. Second, direct arylation tends to occur at the less-hindered aryl C–H bond.

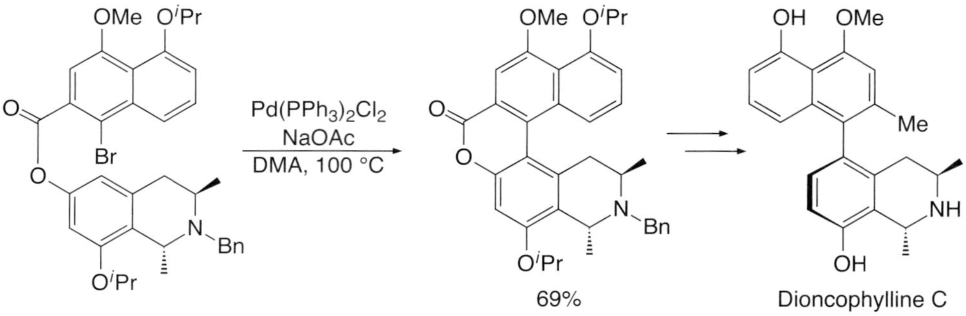

(19.148)

69%   Dioncophylline C

Regioselective, intermolecular direct arylation reactions are particularly challenging because of the lack of geometric preference for the cleavage of one C–H bond over another. However, several strategies have led to the development of regioselective intermolecular direct couplings (Figure 19.6). The most common strategy for conducting regioselective direct coupling of arenes involves the use of substrates containing directing groups. [c.f. Figure 19.6 (top), Equation 19.149, and Equation 19.150].[720,742] This strategy for conducting regioselective cleavage of a C–H bond was described in Chapter 18 on catalytic C–H bond functionalization. An arene containing a ligating substituent can direct the metal to cleave the ortho C–H bond to form a five- or six-membered metallacycle (c.f. Equations 19.149 and 19.150). Additionally, when two nonequivalent ortho C–H bonds are present, cleavage of the less-hindered ortho C–H bond is typically observed.

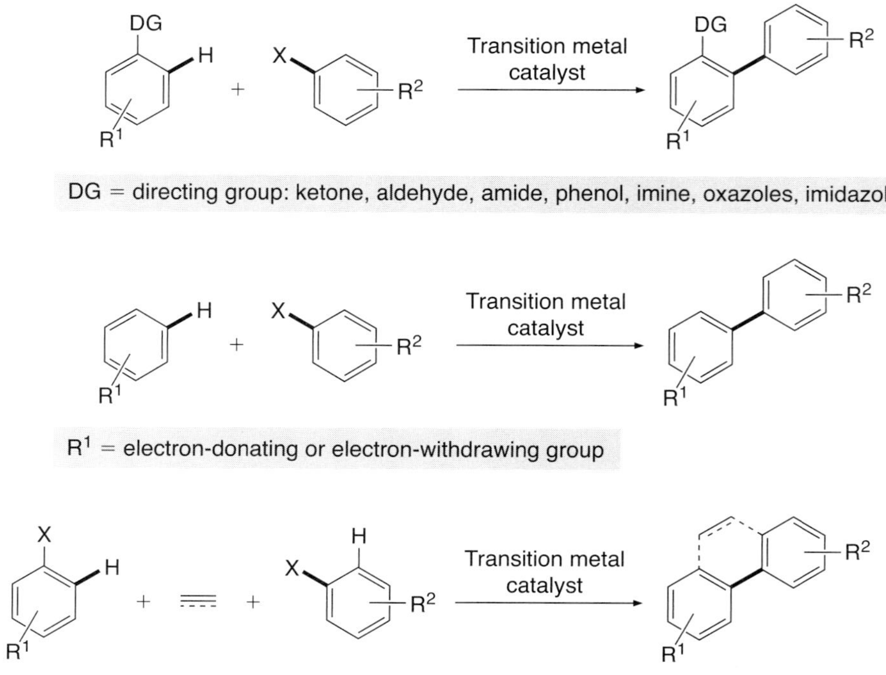

DG = directing group: ketone, aldehyde, amide, phenol, imine, oxazoles, imidazoles, etc.

R^1 = electron-donating or electron-withdrawing group

Figure 19.6. Methods to effect regioselective intermolecular direct arylation.

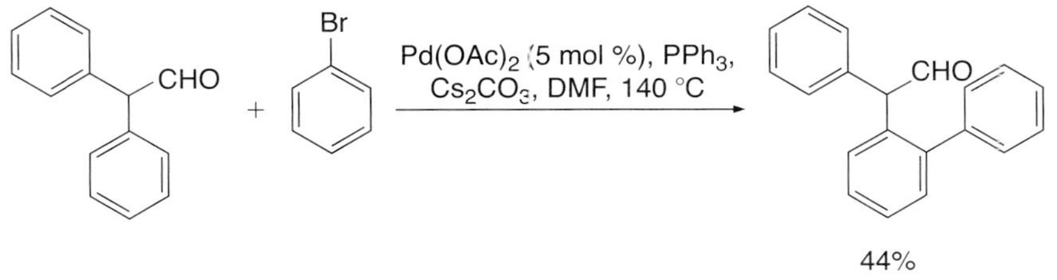

(19.149)

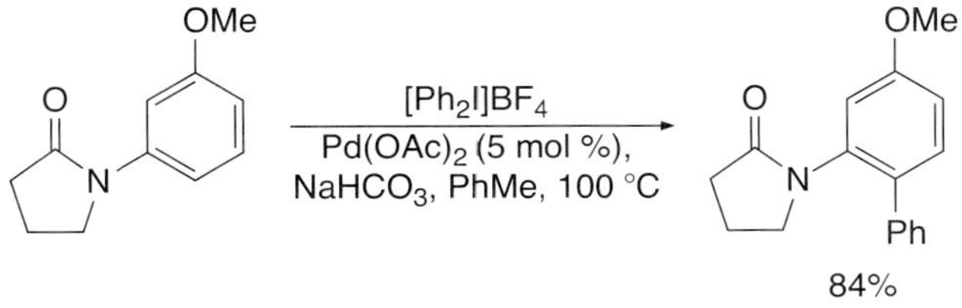

44%

(19.150)

84%

In the absence of directing groups, the electronic properties of the arene can control the position of C–H bond cleavage (c.f. Figure 19.6, middle). These electronic preferences for the cleavage of one C–H bond over another is particularly pronounced in heteroaromatic systems. The regioselectivity of the direct couplings of heteroarenes is highly dependent on the inherent electronic properties of the heterocycle. While these electronic properties can be difficult to override when the desired regioselectivity of the direct coupling differs from that predisposed by the inherent electronic bias, complementary or improved regioselectivities have been obtained in certain cases by changing the reaction solvent,[703] adding co-catalysts [i.e., Cu(I) salts],[715,743,744] or by conducting the reaction with more sterically hindered catalysts[706] or sterically encumbered substrates.[706,730,745] The positions of several heteroarenes that are most reactive toward direct coupling are shown in Figure 19.7.

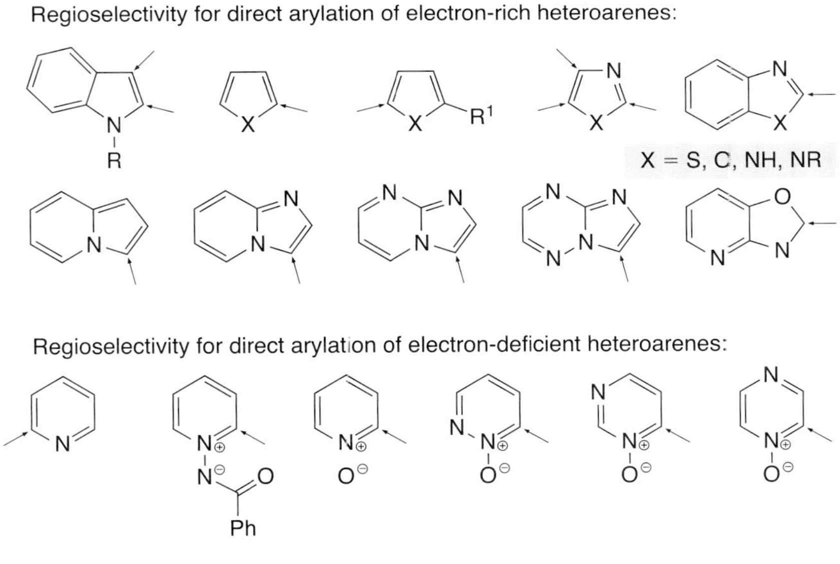

Regioselectivity for direct arylation of electron-rich heteroarenes:

X = S, C, NH, NR

Regioselectivity for direct arylation of electron-deficient heteroarenes:

Figure 19.7.
The positions of heteroarenes that are most reactive toward direct coupling.

In the absence of a directing group or an electronic bias for directing the coupling at one C–H bond over another, a cascade process involving external alkenes or alkynes can create a "directing" alkyl- or alkenylmetal species in situ (Figure 19.6, bottom). For example, bromobenzene reacts with norbornene in the presence of a palladium catalyst to generate a norbornane-diyl-containing biaryl compound, wherein the aryl–aryl bond is formed by direct arylation (Equation 19.151).[746,747] A mechanism that accounts for this transformation is depicted in Equation 19.152. This mechanism is initiated by oxidative addition of bromobenzene to palladium(0) to afford the phenylpalladium bromide complex. Syn insertion of the highly strained norbornene affords the cis, exo complex, which, in the absence of a syn β-hydrogen, undergoes insertion into an aryl C–H bond by electrophilic substitution to afford a five-membered palladacycle. The authors propose that this palladacycle undergoes an oxidative addition with bromobenzene to form a palladium(IV) intermediate, followed by reductive elimination to afford either intermediate **A** or **B**. Finally, cyclization of **A** or **B** occurs to give the desired product.[746] Various strained alkenes have been used in this process, including norbornene,[746,747] dicyclopentadiene, norbornenol, norbornenone,[748] indene,[749] and dihydronaphthalene tricarbonylchromium(0).[750] Reactions conducted with norbornene as the strained alkene are often referred to as the Catellani reaction because of the seminal work of Marta Catellani on its development.[751]

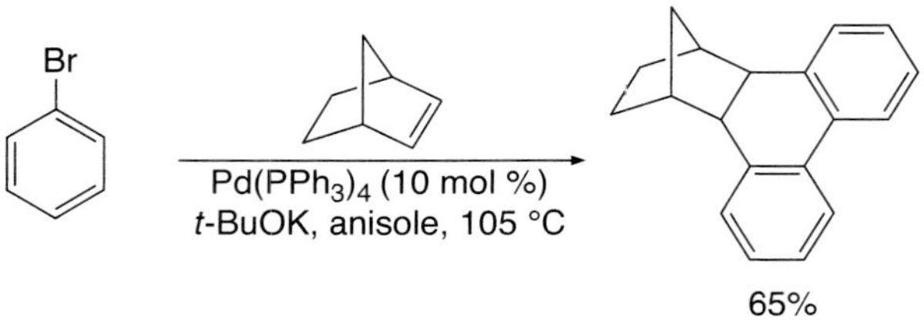

(19.151)

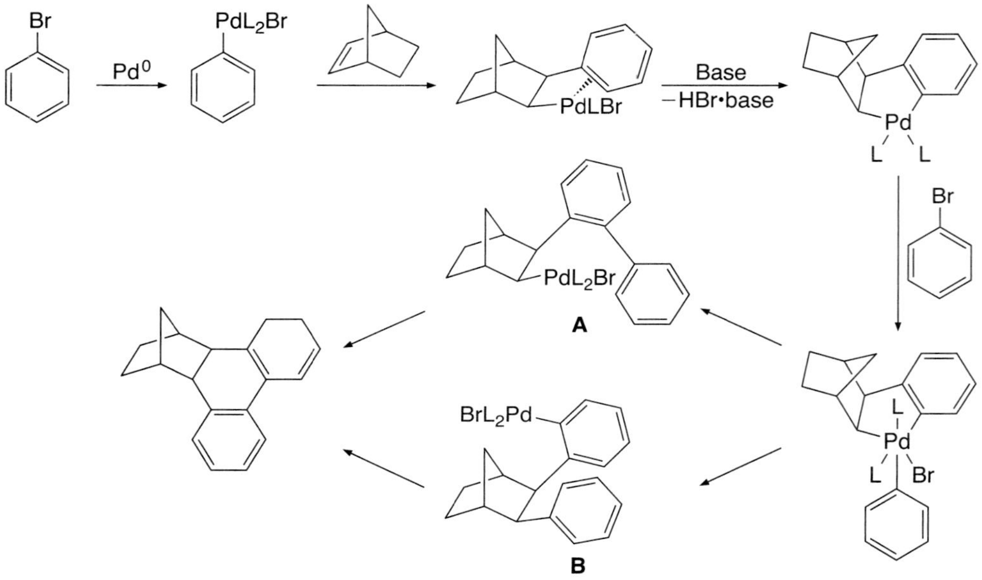

65%

(19.152)

Although the palladium(IV) species is often presumed to be an intermediate in this reaction, there is no experimental evidence for the oxidative addition of aryl halides to palladium(II) complexes. Echavarren and co-workers have found that this process may in fact proceed without the intermediacy of palladium(IV) complexes.[752] The authors conducted DFT calculations on model complexes to assess whether the reaction proceeds via oxidative addition of aryl halides to palladacycles to give palladium(IV) intermediates (path 1, Scheme 19.19), or by a transmetallation-type reaction of arene ligands between a palladacycle and a palladium(II) complex formed by the oxidative addition of an aryl halide to palladium(0) (path 2, Scheme 19.19). The results of this study suggest that the formation of $C(sp^2)$–$C(sp^2)$ bonds in this type of palladium-catalyzed reaction likely occurs without the intermediacy of palladium(IV) complexes.

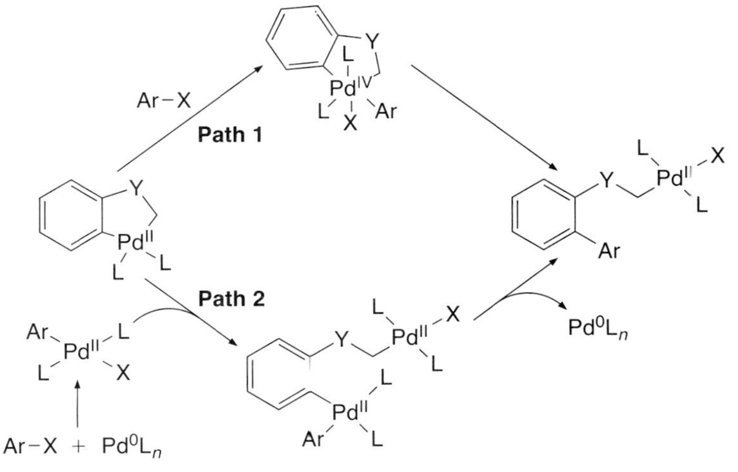

Scheme 19.19

Similar palladium-catalyzed cascade arylations[753] also occur with acyclic alkenes, including α,β-unsaturated sulfones (Equation 19.153), sulfonamides, phosphine oxides, and phosphonate esters. In contrast, typical conjugated olefins, such as α,β-unsaturated esters and enones almost exclusively react to form products from Heck reactions. Direct arylations have also been conducted with disubstituted alkynes containing a terminal arene and a large group, such as an aryl or *tert*-butyl group (Equation 19.154).[754]

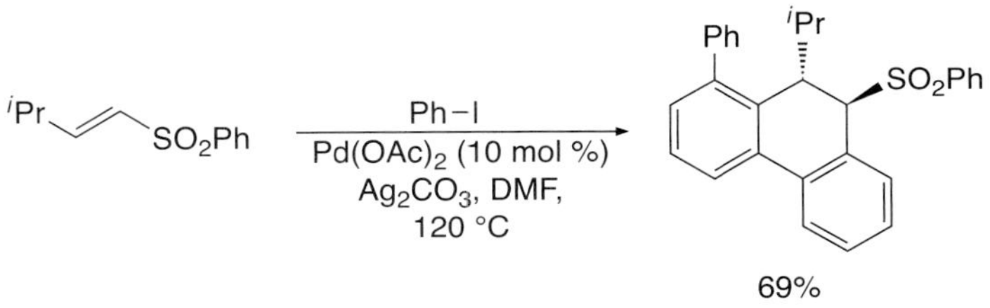

$$(19.153)$$

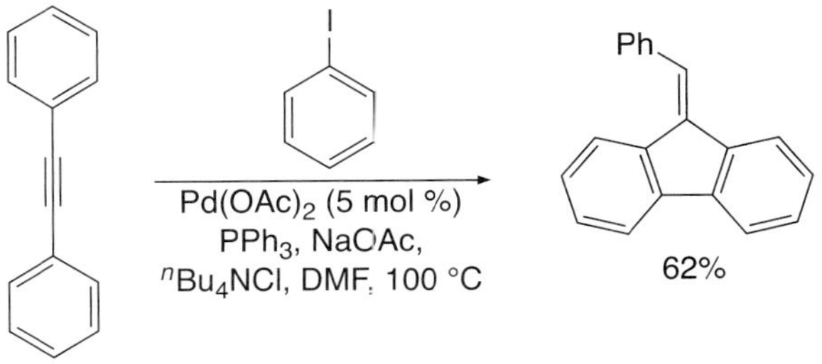

$$(19.154)$$

This palladium-catalyzed coupling mediated by norbornene has also been conducted with aryl halides containing an ortho substituent in the presence of a coupling partner, such as cyanide (Equation 19.155).[755] In this case, the product does not contain the norbornene additive. Various coupling partners include cyanide, olefins, allylic alcohols, diphenyl- and alkylphenylacetylenes, arylboronic acids, hydrogen donors, and amides.[751] The proposed mechanism (Scheme 19.20) begins with the oxidative addition of Pd(0)

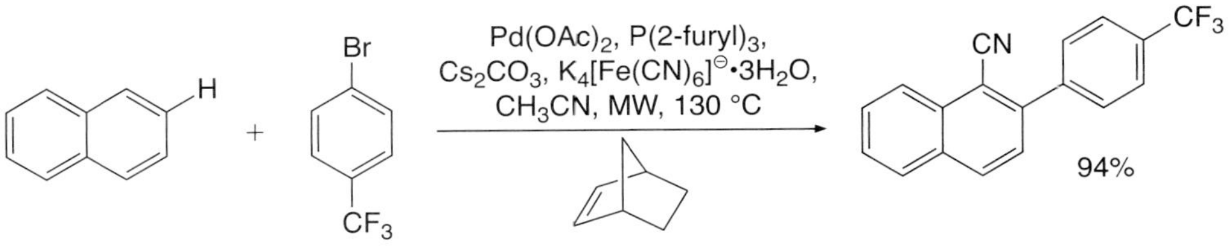

$$(19.155)$$

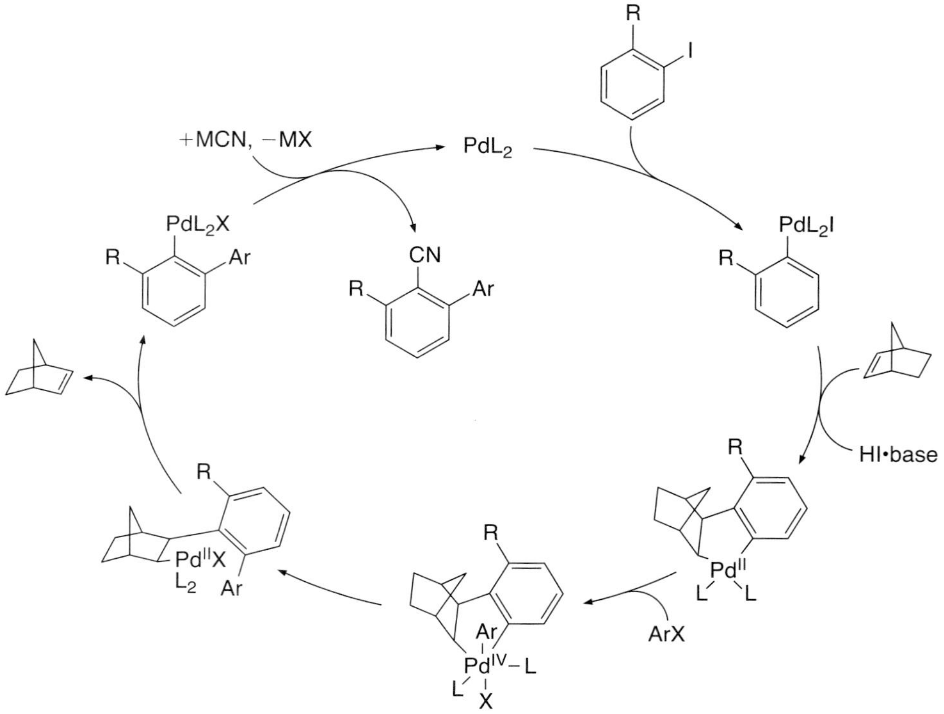

Scheme 19.20

to the aryl iodide to afford an arylpalladium iodide intermediate. Syn insertion of nor-bornene, followed by insertion into the ortho aryl C–H bond affords the five-membered palladacycle. This palladacycle has been proposed to undergo oxidative addition of the aryl bromide to form a Pd(IV) complex. Subsequent reductive elimination would then result in the formation of the aryl–aryl bond. Steric hindrance created by the ortho substituent would lead to the elimination of norbornene to generate an arylpalladium species that can undergo subsequent metal-catalyzed coupling processes (aryl–CN coupling in this example).

19.9.5. General Comments on Reaction Conditions for Direct Arylation

The base and the solvent for the direct coupling reactions are important. Because the net direct arylation process liberates acid as a byproduct, base is typically required.[756] Often, inorganic bases such as K_2CO_3, Cs_2CO_3, KOAc, KO^{t}Bu, and CsOPiv are used. The yields and rates depend on the identity of the base, and some data imply that the base participates in the reaction more intimately than simply to quench released acid.[757,758] In some systems the base may be involved in the formation of the penultimate diarylpalladium(II) species.[757,758] As discussed in Section 19.9.2, this process is believed to occur via either an S_E3 or a concerted process involving metalation and deprotonation (Scheme 19.16). Of the above carboxylate bases, Cs_2CO_3 and CsOPiv have proven to be the most effective and most commonly employed, due to their higher solubility in organic solvents than the other bases listed.

These reactions have typically been conducted in polar, aprotic solvents, such as DMF, DMA, CH_3CN, NMP, and DMSO. However, non-polar solvents, such as toluene and xylene, have also been used. These aromatic solvents have not been reported to participate in direct arylation reactions, primarily due to their lack of both a directing group and activating electronic properties.

Thus, direct arylation reactions are becoming a useful synthetic tool. However, one must be able to control the regioselective in a more general fashion. In addition, catalysts are needed that enable high yielding, direct arylation processes at lower temperatures.

19.10. Catalytic Direct Oxidative Cross Couplings (Written with Dr. Mark E. Scott, Dr. Dino Alberico, and Prof. Mark Lautens)

An even more direct approach to biaryl compounds than "direct arylation" is the coupling of two unactivated arenes by the cleavage of one C–H bond in each of two arenes. Oxidative dimerization reactions (homocoupling) have been reported for various aryl and heteroaryl compounds.[759,760] The coupling of two different unactivated arenes (Equation 19.156), called "oxidative arene cross-coupling," must address the challenge of obtaining products from cross coupling over homocoupling and the challenge of controlling the regioselectivity of C–H bond cleavage on each substrate.

$$(19.156)$$

Fagnou and co-workers overcame some of these challenges in a palladium-catalyzed oxidative cross-coupling reaction between N-pivaloyl indoles and substituted benzenes (Equation 19.157).[761] In this case, high regioselectivity was achieved for both the indole and benzene components. The high regioselectivity of the arene component results from cleavage of the more sterically accessible C–H bond. The origin of the selectivity at the heteroarene is currently speculative. Under the reaction conditions depicted in Equation 19.157, high C2-selectivity was achieved with N-pivaloyl indoles.[762] DeBoef and co-workers simultaneously reported related palladium-catalyzed oxidative cross-coupling reactions between various substituted arenes and either benzofuran or N-acetyl indoles at the 2-position.[763]

$$(19.157)$$

75%
30 : 1 selectivity

Sanford and co-workers have reported an oxidative cross coupling of benzo[h]quinoline with substituted arenes (Equation 19.158).[764] Once again, high regioselectivity was observed for both coupling partners. In this case, the regioselectivity results from coordination of the catalyst at the nitrogen of the benzo[h]quinoline and cleavage of the C–H bond at the 10-position, while the regioselectivity of the reaction at the arene component results from reaction at the less sterically hindered C–H bond. This process is believed to occur by C–H activation of the benzo[h]quinoline (Scheme 19.21) assisted by benzoquinone, followed by C–H bond activation of the arene, reductive elimination of the biaryl

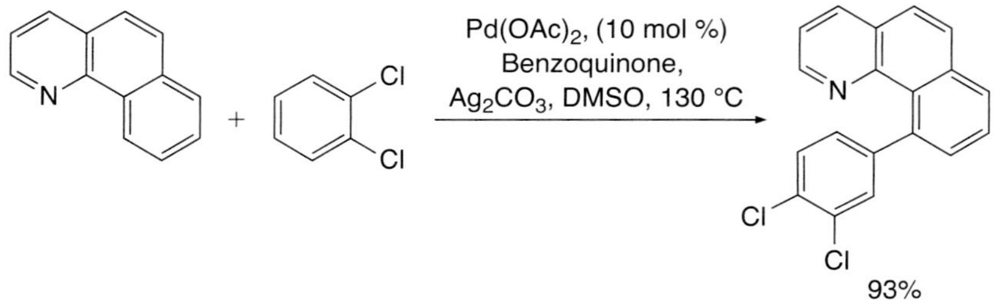

$$(19.158)$$

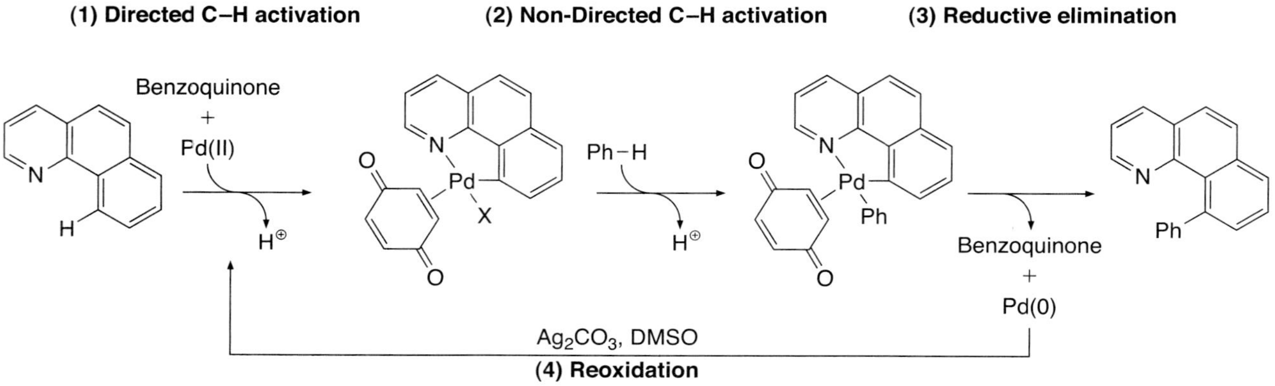

Scheme 19.21

product, and oxidation of Pd(0) to Pd(II) by Ag$_2$CO$_3$. Although these preliminary reports on oxidative arene cross-coupling reactions demonstrate the potential for general biaryl bond formation, the reactions are limited by the requirement for a large excess (30 to 100 equivalents) of the arene coupling partner, added silver for oxidation in some cases, and high loadings of palladium.

19.11. Summary

Classic cross-coupling reactions lead to the formation of carbon–carbon or carbon–heteroatom bonds by the reaction of aryl, vinyl, or alkyl halides or sulfonates with organomagnesium, zinc, tin, silicon, or boron reagents, enolates, cyanoalkyl anions, amines, alcohols, thiols, phosphines, or phosphine oxides and base. Reactions in the presence of CO form ketones and aldehydes, as well as the esters and amides presented in Chapter 17. Palladium or nickel complexes of phosphine ligands typically catalyze these reactions, but improvements in the classic copper-catalyzed versions of these reactions have been made, and couplings of alkyl halides catalyzed by iron and cobalt complexes have been developed beyond the initial results gained when cross-coupling was first discovered. Progress has been made on the direct coupling of aryl halides with arenes and the oxidative coupling of main group organometallic reagents with arenes. These direct coupling reactions alleviate the need to use two activated aromatic partners.

The basic steps of the palladium-catalyzed coupling reactions are well established, but the mechanism of other coupling reactions are less certain. The palladium-catalyzed reactions are initiated by oxidative addition of the organic halide or sulfonate. Transmetallation transfers the nucleophile from a main group reagent to the transition metal catalyst. Alternatively, a ligand substitution replaces the halide or sulfonate with an enolate, amide, alkoxide, thiolate, phosphide, or related ligand. Reductive elimination then forms the final product. The carbonylative couplings occur with an additional step of

CO insertion prior to the transmetallation step. The mechanism of nickel-catalyzed coupling reactions is less established. Early studies indicated that homocoupling processes occur by oxidative addition through radical intermediates and possible intermediacy of Ni(I) and Ni(III) complexes. The copper-catalyzed cross-coupling reactions likely occur by transmetallation prior to oxidiative addition of the aryl halide. Iron-catalyzed reactions likely occur by low-valent, even sub-valent, species.

These coupling and carbonylative coupling reactions have been applied to the synthesis of pharmaceuticals, natural products, organic materials, and even ligands for other catalytic processes. Catalysts with spectacular reactivity have been developed that allow these reactions to be conducted with aryl chlorides and tosylates, which have been unreactive toward transition metals for many years. These catalysts typically contain sterically hindered, electron-rich ligands. At the same time, simple complexes of the first-row elements, nickel, iron, and cobalt can catalyze a subset of reactions of aryl chlorides and tosylates under mild conditions, and these simple catalysts are likely to become equally practical for large-scale syntheses in the future. Overall, these reactions are among the most, if not the most, commonly used catalytic organometallic process in organic synthesis.

References and Notes

1. Bellina, F.; Carpita, A.; Rossi, R. *Synthesis* **2004**, 2419.
2. *Handbook of Organopalladium Chemistry for Organic Synthesis*; Negishi, E.-i., Ed.; Wiley-Interscience: New York, 2002; Vol. I, Chapter III.
3. Littke, A. F.; Fu, G. C. *Angew. Chem. Int. Ed.* **2002**, *41*, 4176.
4. Kotha, S.; Lahiri, K.; Kashinath, D. *Tetrahedron* **2002**, *58*, 9633.
5. Frost, C. G. In *Rodd's Chemistry of Carbon Compounds*, 2nd ed.; Elsevier: Amsterdam, 2001; Vol. 5, p 315.
6. Suzuki, A. *J. Organomet. Chem.* **1999**, *576*, 147.
7. Kingsbury, C. L.; Mehrman, S. J.; Takacs, J. M. *Curr. Org. Chem.* **1999**, *3*, 497.
8. Stanforth, S. P. *Tetrahedron* **1998**, *54*, 263.
9. *Metal-Catalyzed Cross-Coupling Reactions*; Diederich, F., Stang, P. J., Eds.; Wiley-VCH: Weinheim, 1998.
10. *Metal-Catalyzed Cross-Coupling Reactions*; de Meijere, A., Diederich, F., Eds.; Wiley-VCH: Weinheim, 2004.
11. Miyaura, N.; Suzuki, A. *Chem. Rev.* **1995**, *95*, 2457.
12. Stille, J. K. *Angew. Chem., Int. Ed. Engl.* **1986**, *25*, 508.
13. Tietze, L. F.; Ila, H.; Bell, H. P. *Chem. Rev.* **2004**, *104*, 3453.
14. Dounay, A. B.; Overman, L. E. *Chem. Rev.* **2003**, *103*, 2945.
15. Link, J. T. *Org. React.* **2002**, *60*, 157.
16. de Vries, J. G. *Can. J. Chem.* **2001**, *79*, 1086.
17. Beletskaya, I., P.; Cheprakov, A. V. *Chem. Rev.* **2000**, *100*, 3009.
18. Shibasaki, M.; Vogl, E. M. *J. Organomet. Chem.* **1999**, *576*, 1.
19. Demeijere, A.; Meyer, F. E. *Angew. Chem., Int. Ed. Engl.* **1995**, *33*, 2379.
20. Sakellarios, E.; Kyrimis, T. *Ber. Dtsch. Chem. Ges.* **1924**, *57B*, 322.
21. Kharasch, M. S.; Fields, E. K. *J. Am. Chem. Soc.* **1941**, *63*, 2316.
22. Rao, V. V. R.; Kumar, C. V.; Devaprabhakara, D. *J. Org. Chem.* **1979**, *179*, C7.
23. Hassan, J.; Sevignon, M.; Gozzi, C.; Schulz, E.; Lemaire, M. *Chem. Rev.* **2002**, *102*, 1359.
24. Song, Z. Z.; Wong, H. N. C. *J. Org. Chem.* **1994**, *59*, 33.
25. Moreno-Mañas, M.; Perez, M.; Pleixats, R. *J. Org. Chem.* **1996**, *61*, 2346.
26. Kang, S. K.; Kim, T. H.; Pyun, S. J. *J. Chem. Soc., Perkin Trans. 1* **1997**, 797.
27. Smith, K. A.; Campi, E. M.; Jackson, W. R.; Marcuccio, S.; Naeslund, C. G. M.; Deacon, G. B. *Synlett* **1997**, 131.
28. Yamaguchi, S.; Ohno, S.; Tamao, K. *Synlett* **1997**, 1199.
29. Ishikawa, T.; Ogawa, A.; Hirao, T. *Organometallics* **1998**, *17*, 5713.
30. Inoue, A.; Kitagawa, K.; Shinokubo, H.; Oshima, K. *Tetrahedron* **2000**, *56*, 9601.
31. Kabalka, G. W.; Wang, L. *Tetrahedron Lett.* **2002**, *43*, 3067.
32. Koza, D. J.; Carita, E. *Synthesis* **2002**, 2183.
33. Yoshida, H.; Yamaryo, Y.; Ohshita, J.; Kunai, A. *Tetrahedron Lett.* **2003**, *44*, 1541.

34. Punna, S.; Diaz, D. D.; Finn, M. G. *Synlett* **2004**, 2351.

35. Nagano, T.; Hayashi, T. *Org. Lett.* **2005**, 7.

36. Tsuji, J.; Mandai, T. *Angew. Chem., Int. Ed. Engl.* **1996**, *34*, 2589.

37. Kuwano, R.; Kondo, Y.; Matsuyama, Y. *J. Am. Chem. Soc.* **2003**, *125*, 12104.

38. Tamao, K.; Sumitani, K.; Kumada, M. *J. Am. Chem. Soc.* **1972**, *94*, 4374.

39. Corriu, R. J. P.; Masse, J. P. *J. Chem. Soc., Chem. Commun.* **1972**, 144.

40. (a) Tamura, M.; Kochi, J. *J. Organomet. Chem.* **1971**, *31*, 289; (b) Tamura, M.; Kochi, J. *J. Am. Chem. Soc.* **1971**, *93*, 1487.

41. (a) Kharasch, M. S.; Fields, E. K. *J. Am. Chem. Soc.* **1941**, *63*, 2316; (b) Kharasch, M. S.; Fuchs, C. F. *J. Am. Chem. Soc.* **1943**, *65*, 504.

42. Negishi, E.; King, A. O.; Okukado, N. *J. Org. Chem.* **1977**, *42*, 1821.

43. Negishi, E.-i.; Liu, F. In *Metal-Catalyzed Cross-Coupling Reactions*; Diederich, F., Stang, P. J., Eds.; Wiley-VCH: Weinheim, 1998; p 1.

44. Cassar, L. *J. Organomet. Chem.* **1975**, *93*, 253.

45. Dieck, H. A.; Heck, R. F. *J. Organomet. Chem.* **1975**, *93*, 259.

46. Sonogashira, K.; Tohda, Y.; Hagihara, N. *Tetrahedron Lett.* **1975**, *16*, 4467.

47. Yamamura, M.; Moritani, I.; Murahashi, S. I. *J. Organomet. Chem.* **1975**, *91*, C39.

48. Murahashi, S. I.; Yamamura, M.; Mita, N. *J. Org. Chem.* **1977**, *42*, 2870.

49. Murahashi, S. I.; Yamamura, M.; Yanagisawa, K.; Mita, N.; Kondo, K. *J. Org. Chem.* **1979**, *44*, 2408.

50. Kosugi, M.; Shimizu, Y.; Migita, T. *J. Organomet. Chem.* **1977**, *129*, C36.

51. Stille, J. K. *Pure Appl. Chem.* **1985**, *57*, 1771.

52. Hatanaka, Y.; Hiyama, T. *J. Org. Chem.* **1988**, *53*, 918.

53. Hatanaka, Y.; Hiyama, T. *J. Org. Chem.* **1988**, *53*, 918.

54. Tamao, K.; Kobayashi, K.; Ito, Y. *Tetrahedron Lett.* **1989**, *30*, 6051.

55. Hiyama, T. In *Metal-Catalyzed Cross-Coupling Reactions*; Diederich, F., Stang, P. J., Eds.; Wiley-VCH: Weinheim, 1998; Chapter 10.

56. Mowery, M. E.; DeShong, P. *Org. Lett.* **1999**, *1*, 217.

57. Mowery, M. E.; DeShong, P. *J. Org. Chem.* **1999**, *64*, 3266.

58. Mowery, M. E.; DeShong, P. *J. Org. Chem.* **1999**, *64*, 1684.

59. Lee, H. M.; Nolan, S. P. *Org. Lett.* **2000**, *2*, 2053.

60. Denmark, S. E.; Sweis, R. F. *Acc. Chem. Res.* **2002**, *35*, 835.

61. Miyaura, N.; Yamada, K.; Suzuki, A. *Tetrahedron Lett.* **1979**, *36*, 3437.

62. Miyaura, N.; Suginome, H.; Suzuki, A. *Tetrahedron Lett.* **1981**, *22*, 127.

63. Miyaura, N.; Yamada, K.; Sufinome, H.; Suzuki, A. *J. Am. Chem. Soc.* **1985**, *107*, 972.

64. Suzuki, A. In *Metal-Catalyzed Cross-Coupling Reactions*; Diederich, F., Stang, P. J., Eds.; Wiley-VCH: New York, 1998; Chapter 2.

65. Molander, G. A.; Bernardi, C. R. *J. Org. Chem.* **2002**, *67*, 8424.

66. Molander, G. A.; Katona, B. W.; Machrouhi, F. *J. Org. Chem.* **2002**, *67*, 8416.

67. Molander, G. A.; Yun, C. S. *Tetrahedron* **2002**, *58*, 1465.

68. Molander, G. A.; Ito, T. *Org. Lett.* **2001**, *3*, 393.

69. Vedejs, E.; Chapman, R. W.; Fields, S. C.; Lin, S.; Schrimpf, M. R. *J. Org. Chem.* **1995**, *60*, 3020.

70. Alami, M.; Ferri, F.; Linstrumelle, G. *Tetrahedron Lett.* **1993**, *34*, 6403.

71. Sonogashira, K. In *Handbook of Organopalladium Chemistry for Organic Synthesis*; Negishi, E.-i., Ed.; Wiley-Interscience: New York, 2002; Vol. 1, p 493.

72. Sonogashira, K. In *Metal-Catalyzed Cross-Coupling Reactions*; Diederich, F., Stang, P. J., Eds.; Wiley-VCH: New York, 1998; Chapter 5.

73. Culkin, D. A.; Hartwig, J. F. *Acc. Chem. Res.* **2003**, *36*, 234.

74. Miura, M.; Nomura, M. *Top. Curr. Chem.* **2002**, *219*, 211.

75. Hamann, B. C.; Hartwig, J. F. *J. Am. Chem. Soc.* **1997**, *119*, 12382.

76. Kawatsura, M.; Hartwig, J. F. *J. Am. Chem. Soc.* **1999**, *121*, 1473.

77. Palucki, M.; Buchwald, S. L. *J. Am. Chem. Soc.* **1997**, *119*, 11108.

78. Åhman, J.; Wolfe, J. P.; Troutman, M. V.; Palucki, M.; Buchwald, S. L. *J. Am. Chem. Soc.* **1998**, *120*, 1918.

79. Fox, J. M.; Huang, X. H.; Chieffi, A.; Buchwald, S. L. *J. Am. Chem. Soc.* **2000**, *122*, 1360.

80. Lee, S.; Beare, N. A.; Hartwig, J. F. *J. Am. Chem. Soc.* **2001**, *123*, 8410.

81. Jørgensen, M.; Liu, X.; Wolkowski, J. P.; Hartwig, J. F. *J. Am. Chem. Soc.* **2002**, *124*, 12557.

82. Hama, T.; Liu, X.; Culkin, D.; Hartwig, J. F. *J. Am. Chem. Soc.* **2003**, *125*, 11176.

83. Liu, X.; Hartwig, J. F. *J. Am. Chem. Soc.* **2004**, *126*, 5182.

84. Moradi, W. A.; Buchwald, S. L. *J. Am. Chem. Soc.* **2001**, *123*, 7996.

85. Spielvogel, D. J.; Buchwald, S. L. *J. Am. Chem. Soc.* **2002**, *124*, 3500.

86. Shaughnessy, K. H.; Hamann, B. C.; Hartwig, J. F. *J. Org. Chem.* **1998**, *63*, 6546.

87. Lee, S.; Hartwig, J. *J. Org. Chem.* **2001**, *66*, 3402.

88. Cossy, J.; de Filippis, A.; Pardo, D. G. *Org. Lett.* **2003**, *5*, 3037.

89. de Filippis, A.; Pardo, D. G.; Cossy, J. *Tetrahedron* **2004**, *60*, 9757.

90. Terao, Y.; Fukuoka, Y.; Satoh, T.; Miura, M.; Nomura, M. *Tetrahedron Lett.* **2002**, *43*, 101.

91. Martin, R.; Buchwald, S. L. *Angew. Chem. Int. Ed.* **2007**, *46*, 7236.

92. Vo, G. D.; Hartwig, J. F. *Angew. Chem. Int. Ed.* **2008**, *47*, 2127.

93. You, J. S.; Verkade, J. G. *Angew. Chem. Int. Ed.* **2003**, *42*, 5051.

94. You, J. S.; Verkade, J. G. *J. Org. Chem.* **2003**, *68*, 8003.

95. Gao, C. W.; Tao, X. C.; Qian, Y. L.; Huang, J. L. *Chem. Commun.* **2003**, 1444.

96. Culkin, D. A.; Hartwig, J. F. *J. Am. Chem. Soc.* **2002**, *124*, 234.

97. Satoh, T.; Inoh, J.; Kawamura, Y.; Kawamura, Y.; Miura, M.; Nomura, M. *Bull. Chem. Soc. Jpn.* **1998**, *71*, 2239.

98. Beare, N. A.; Hartwig, J. F. *J. Org. Chem.* **2002**, *67*, 541.

99. Aramendia, M. A.; Borau, V.; Jimenez, C.; Marinas, J. M.; Ruiz, J. R.; Urbano, F. J. *Tetrahedron Lett.* **2002**, *43*, 2847.

100. Kondo, Y.; Inamoto, K.; Uchiyama, M.; Sakamoto, T. *Chem. Commun.* **2001**, 2704.

101. Djakovitch, L.; Kohler, K. *J. Organomet. Chem.* **2000**, *606*, 101.

102. Soai, K.; Yokoyama, S.; Hayasaka, T.; Ebihara, K. *J. Org. Chem.* **1988**, *53*, 4149.

103. Stauffer, S. R.; Beare, N.; Stambuli, J. P.; Hartwig, J. F. *J. Am. Chem. Soc.* **2001**, *123*, 4641.

104. Vogl, E. M.; Buchwald, S. L. *J. Org. Chem.* **2002**, *67*, 106.

105. Chae, J.; Yun, J.; Buchwald, S. L. *Org. Lett.* **2004**, *6*, 4809.

106. Chieffi, A.; Kamikawa, K.; Åhman, J.; Fox, J. M.; Buchwald, S. L. *Org. Lett.* **2001**, *3*, 1897.

107. Liao, X.; Weng, Z.; Hartwig, J. F. *J. Am. Chem. Soc.* **2008**, *130*, 195.

108. Giovannini, R.; Studemann, T.; Dussin, G.; Knochel, P. *Angew. Chem. Int. Ed.* **1998**, *37*, 2387.

109. Giovannini, R.; Studemann, T.; Devasagayaraj, A.; Dussin, G.; Knochel, P. *J. Org. Chem.* **1999**, *64*, 3544.

110. Terao, J.; Watanabe, H.; Ikumi, A.; Kuniyasu, H.; Kambe, N. *J. Am Chem. Soc.* **2002**, *124*, 4222.

111. Zhou, J. R.; Fu, G. C. *J. Am. Chem. Soc.* **2003**, *125*, 14726.

112. Netherton, M. R.; Dai, C. Y.; Neuschutz, K.; Fu, G. C. *J. Am. Chem. Soc.* **2001**, *123*, 10099.

113. Kirchhoff, J. H.; Dai, C.; Fu, G. C. *Angew. Chem. Int. Ed.* **2002**, *41*, 1945.

114. Netherton, M. R.; Fu, G. C. *Angew. Chem. Int. Ed.* **2002**, *41*, 3910.

115. Kirchhoff, J.; Netherton, M. R.; Hills, I.; Fu, G. *J. Am. Chem. Soc.* **2002**, *124*, 13662.

116. Zhou, J. R.; Fu, G. C. *J. Am. Chem. Soc.* **2003**, *125*, 12527.

117. Wiskur, S. L.; Korte, A.; Fu, G. C. *J. Am. Chem. Soc.* **2004**, *126*, 82.

118. Menzel, K.; Fu, G. C. *J. Am. Chem. Soc.* **2003**, *125*, 3718.

119. Gonzalez-Bobes, F.; Fu, G. C. *J. Am. Chem. Soc.* **2006**, *128*, 5360.

120. Strotman, N. A.; Sommer, S.; Fu, G. C. *Angew. Chem. Int. Ed.* **2007**, *46*, 3556.

121. Arp, F. O.; Fu, G. C. *J. Am. Chem. Soc.* **2005**, *127*, 10482.

122. Saito, B.; Fu, G. C. *J. Am. Chem. Soc.* **2007**, *129*, 9602.

123. Nakamura, M.; Matsuo, K.; Ito, S.; Nakamura, B. *J. Am. Chem. Soc.* **2004**, *126*, 3686.

124. Martin, R.; Furstner, A. *Angew. Chem. Int. Ed.* **2004**, *43*, 3955.

125. Duplais, C.; Bures, F.; Sapountzis, I.; Korn, T. J.; Cahiez, G.; Knochel, P. *Angew. Chem. Int. Ed.* **2004**, *43*, 2968.

126. Furstner, A.; Mendez, M. *Angew. Chem. Int. Ed.* **2003**, *42*, 5355.

127. Quintin, J.; Franck, X.; Hocquemiller, R.; Figadere, B. *Tetrahedron Lett.* **2002**, *43*, 3547.

128. Furstner, A.; Leitner, A.; Mendez, M.; Krause, H. *J. Am. Chem. Soc.* **2002**, *124*, 13856.

129. Furstner, A.; Leitner, A. *Angew. Chem., Int. Ed. Engl.* **2002**, *41*, 609.

130. Cahiez, G.; Avedissian, H. *Synthesis* **1998**, 1199.

131. Dohle, W.; Kopp, F.; Cahiez, G.; Knochel. P. *Synlett* **2001**, 1901.

132. Cahiez, G.; Marquais, S. *Tetrahedron Lett.* **1996**, *37*, 1773.

133. Tamura, M.; Kochi, J. K. *J. Am. Chem. Soc.* **1971**, *93*, 1487.

134. Kochi, J. K. *Acc. Chem. Res.* **1974**, *7*, 351.

135. Neumann, S. M.; Kochi, J. K. *J. Org. Chem.* **1975**, *40*, 599.

136. Smith, R. S.; Kochi, J. K. *J. Org. Chem.* **1976**, *41*, 502.

137. Molander, G. A.; Rahn, B. J.; Shubert, D. C.; Bonde, S. E. *Tetrahedron Lett.* **1983**, *24*, 5449.

138. Mizoroki, T.; Mori, K.; Ozaki, A. *Bull. Chem. Soc. Jpn.* **1971**, *44*, 581.

139. Mori, K.; Mizoroki, T.; Ozaki, A. *Bull. Chem. Soc. Jpn.* **1973**, *46*, 1505.

140. Heck, R. F.; Nolley, J. P. *J. Org. Chem.* **1972**, *37*, 2320.

141. Heck, R. F. *Acc. Chem. Res.* **1979**, *12*, 146.

142. Tucker, C. E.; de Vries, J. G. *Top. Catal.* **2002**, *19*, 111.

143. Brase, S.; de Meijere, A. In *Handbook of Organopalladium Chemistry for Organic Synthesis*; Negishi, E.-i., Ed.; Wiley-Interscience: New York, 2002; Vol. 1, p 1123.

144. Brase, S.; de Meijere, A. In *Metal-Catalyzed Cross-Coupling Reactions*; Diederich, F., Stang, P. J., Eds.; Wiley-VCH: New York, 1998; Chapter 3.

145. McChesney, J. *Spec. Chem.* **1999**, *6*, 98.

146. Lin, R. W.; Herndon, R.; Allen, R. H.; Chockalingham, K. C.; Focht, G.D.; Roy, R. K. World Patent WO 98/30529, 1998.

147. Wu, T.-C. U.S. Patent 5,536,870, 1996.

148. Wu, T.-C. U.S. Patent 5,315,026, 1994.

149. Gurtler, C.; Buchwald, S. L. *Chem. Eur. J.* **1999**, *5*, 3107.

150. Firmansjah, L.; Fu, G. C. *J. Am. Chem. Soc.* **2007**, *129*, 11340.

151. Glorius, F. *Tetrahedron Lett.* **2003**, *44*, 5751.

152. Cassar, L. *J. Organomet. Chem.* **1973**, *54*, C57.

153. Takagi, K.; Okamoto, T.; Sakakiba.Y; Oka, S. *Chem. Lett.* **1973**, 471.

154. Cassar, L.; Ferrara, S.; Foa, M. *Adv. Chem. Ser.* **1974**, 252.

155. Cassar, L.; Foa, M.; Montanari, F.; Marinelli, G. P. *J. Organomet. Chem.* **1979**, *173*, 335.

156. Sakakibara, Y.; Okuda, F.; Shimobayashi, A.; Kirino, K.; Sakai, M.; Uchino, N.; Takagi, K. *Bull. Chem. Soc. Jpn.* **1988**, *61*, 1985.

157. Chambers, M. R. I.; Widdowson, D. A. *J. Chem. Soc., Perkin Trans. 1* **1989**, 1365.

158. Takagi, K.; Sakakibara, Y. *Chem. Lett.* **1989**, 1957.

159. Sakakibara, Y.; Ido, Y.; Sasaki, K.; Sakai, M.; Uchino, N. *Bull. Chem. Soc. Jpn.* **1993**, *66*, 2776.

160. Kubota, H.; Rice, K. C. *Tetrahedron Lett.* **1998**, *39*, 2907.

161. Sakamoto, T.; Ohsawa, K. *J. Chem. Soc., Perkin Trans. 1* **1999**, 2323.

162. Jin, F. Q.; Confalone, P. N. *Tetrahedron Lett.* **2000**, *41*, 3271.

163. Sundermeier, M.; Mutyala, S.; Zapf, A.; Spannenberg, A.; Beller, M. *J. Organomet. Chem.* **2003**, *684*, 50.

164. Sundermeier, M.; Zapf, A.; Beller, M. *Eur. J. Inorg. Chem.* **2003**, 3513.

165. Sundermeier, M.; Zapf, A.; Beller, M. *Angew. Chem. Int. Ed.* **2003**, *42*, 1661.

166. Sundermeier, M.; Zapf, A.; Mutyala, S.; Baumann, W.; Sans, J.; Weiss, S.; Beller, M. *Chem. Eur. J.* **2003**, *9*, 1828.

167. Schareina, T.; Zapf, A.; Beller, M. *J. Organomet. Chem.* **2004**, *689*, 4576.

168. Schareina, T.; Zapf, A.; Beller, M. *Chem. Commun.* **2004**, 1388.

169. Weissman, S. A.; Zewge, D.; Chen, C. *J. Org. Chem.* **2005**, *70*, 1508.

170. Sundermeier, M.; Zapf, A.; Beller, M. *Angew. Chem. Int. Ed.* **2003**, *42*, 1661.

171. Hayashi, T.; Tajika, M.; Tamao, K.; Kumada, M. *J. Am. Chem. Soc.* **1976**, *98*, 3718.

172. Hayashi, T.; Konishi, M.; Fukushima, M.; Mise, T.; Kagotani, M.; Tajika, M.; Kumada, M. *J. Am. Chem. Soc.* **1982**, *104*, 180.

173. Hayashi, T.; Hagihara, T.; Katsuro, Y.; Kumada, M. *Bull. Chem. Soc. Jpn.* **1983**, *56*, 363.

174. Hayashi, T.; Konishi, M.; Okamoto, Y.; Kabeta, K.; Kumanda, M. *J. Org. Chem.* **1986**, *51*, 3772.

175. Hayashi, T.; Yamamoto, A.; Hojo, M.; Ito, Y. *J. Chem. Soc., Chem. Commun.* **1989**, 495.

176. Hamada, T.; Chieffi, A.; Ahman, J.; Buchwald, S. L. *J. Am. Chem. Soc.* **2002**, *124*, 1261.

177. Kamikawa, T.; Hayashi, T. *Tetrahedron* **1999**, *55*, 3455.

178. Kamikawa, T.; Uozumi, Y.; Hayashi, T. *Tetrahedron Lett.* **1996**, *37*, 3161.

179. Yin, J. J.; Buchwald, S. L. *J. Am. Chem. Soc.* **2000**, *122*, 12051.

180. Cammidge, A. N.; Crepy, K. V. L. *Chem. Commun.* **2000**, 1723.

181. Shimada, T.; Cho, Y. H.; Hayashi, T. *J. Am. Chem. Soc.* **2002**, *124*, 13396.

182. Jensen, J. F.; Johannsen, M. *Org. Lett.* **2003**, *5*, 3025.

183. Shibasaki, M.; Boden, C. D. J.; Kojima, A. *Tetrahedron* **1997**, 7371.

184. Hayashi, T.; Konishi, M.; Fukushima, M.; Kanehira, K.; Hioki, T.; Kumada, M. *J. Org. Chem.* **1983**, *48*, 2195.

185. Hayashi, T.; Fukushima, M.; Konishi, M.; Kumada, M. *Tetrahedron Lett.* **1980**, *21*, 79.

186. Griffin, J. H.; Kellogg, R. M. *J. Org. Chem.* **1985**, *50*, 3261.

187. Vriesema, B. K.; Kellogg, R. M. *Tetrahedron Lett.* **1986**, *27*, 2049.

188. Cho, S. Y.; Shibasaki, M. *Tetrahedron: Asymmetry* **1998**, *9*, 3751.

189. Willis, M. C.; Powell, L. H. W.; Claverie, C. K.; Watson, S. J. *Angew. Chem. Int. Ed.* **2004**, *43*, 1249.

190. Hayashi, T.; Hayashizaki, K.; Kiyoi, T.; Ito, Y. *J. Am. Chem. Soc.* **1998**, *110*, 8153.

191. Sato, Y.; Sodeoka, M.; Shibasaki, M. *J. Org. Chem.* **1989**, *54*, 4738.

192. Carpenter, N. E.; Kucera, D. J.; Overman, L. E. *J. Org. Chem.* **1989**, *54*, 5846.

193. Loiseleur, O.; Hayashi, M.; Keenan, M.; Schmees, N.; Pfaltz, A. *J. Organomet. Chem.* **1999**, *576*, 16.

194. Loiseleur, O.; Hayashi, M.; Schmees, N.; Pfaltz, A. *Synthesis* **1997**, 1338.

195. Ozawa, F.; Kobatake, Y.; Hayashi, T. *Tetrahedron Lett.* **1993**, *34*, 2505.

196. Loiseleur, O.; Meier, P.; Pfaltz, A. *Angew. Chem., Int. Ed. Engl.* **1996**, *35*, 200.

197. Ashimori, A.; Bachand, B.; Overman, L. E.; Poon, D. J. *J. Am. Chem. Soc.* **1998**, *120*, 6477.

198. Ashimori, A.; Bachand, B.; Calter, M. A.; Govek, S. P.; Overman, L. E.; Poon, D. J. *J. Am. Chem. Soc.* **1998**, *120*, 6488.

199. Overman, L. E.; Poon, D. J. *Angew. Chem., Int. Ed. Engl.* **1997**, *36*, 518.

200. This isomerization can act as a kinetic resolution of the product. Therefore, the enantioselectivity of the intermolecular Heck reaction may not reflect a kinetic selectivity unless the yield of a single isomer and the enantioselectivity are both high.

201. Tsou, T. T.; Kochi, J. K. *J. Am. Chem. Soc.* **1979**, *101*, 7547.

202. Hills, I. D.; Fu, G. C. *J. Am. Chem. Soc.* **2004**, *126*, 13178.

203. Bedford, R. B. *Chem. Commun.* **2003**, 1787.

204. Herrmann, W. A.; Brossmer, C.; Öfele, K.; Reisinger, C.-P.; Priermeier, T.; Beller, M.; Fischer, H. *Angew. Chem., Int. Ed. Engl.* **1995**, *34*, 1844.

205. Beller, M.; Fischer, H.; Herrmann, W. A.; Öfele, K.; Brossmer, C. *Angew. Chem., Int. Ed. Engl.* **1995**, *34*, 1848.

206. Louie, J.; Hartwig, J. F. *Angew. Chem., Int. Ed. Engl.* **1996**, *35*, 2359.

207. Herrmann, W. A.; Brossmer, C.; Reisinger, C.-P.; Riermeier, T. H.; Öfele, K.; Beller, M. *Chem. Eur. J.* **1997**, *3*, 1357.

208. Rosner, T.; Le Bars, J.; Pfaltz, A.; Blackmond, D. G. *J. Am. Chem. Soc.* **2001**, *123*, 1848.

209. Ohff, M.; Ohff, A.; van der Boom, M. E.; Milstein, D. *J. Am. Chem. Soc.* **1997**, *119*, 11687.

210. Littke, A. F.; Fu, G. C. *J. Am. Chem. Soc.* **2001**, *123*, 6989.

211. Stambuli, J. P.; Stauffer, S. R.; Shaughnessy, K. H.; Hartwig, J. F. *J. Am. Chem. Soc* **2001**, *123*, 2677.

212. Fauvarque, J.-F.; Pflüger, F. *J. Organomet. Chem.* **1981**, *208*, 419.

213. Amatore, C.; Pfluger, F. *Organometallics* **1990**, *9*, 2276.

214. Hartwig, J. F.; Paul, F. *J. Am. Chem. Soc.* **1995**, *117*, 5373.

215. Littke, A.; Dai, C.; Fu, G. *J. Am. Chem. Soc.* **2000**, *122*, 4020.

216. Reetz, M. T.; Westermann, E. *Angew. Chem. Int. Ed.* **2000**, *39*, 165.

217. Farina, V. *Adv. Synth. Catal.* **2004**, *346*, 1553.

218. Beletskaya, I. P.; Cheprakov, A. V. *J. Organomet. Chem.* **2004**, *689*, 4055.

219. Reetz, M. T.; de Vries, J. G. *Chem. Commun.* **2004**, 1559.

220. Caddick, S.; Geoffrey, F.; Cloke, N.; Hitchcock, P. B.; Leonard, J.; Lewis, A. K. D.; McKerrecher, D.; Titcomb, L. R. *Organometallics* **2002**, *21*, 4318.

221. Portnoy, M.; Milstein, D. *Organometallics* **1993**, *12*, 1665.

222. Barios-Landeros, F.; Hartwig, J. F. *J. Am. Chem. Soc.* **2005**, *127*, 6944.

223. Yin, J. J.; Rainka, M. P.; Zhang, X. X.; Buchwald, S. L. *J. Am. Chem. Soc.* **2002**, *124*, 1162.

224. Bedford, R. B.; Welch, S. L. *Chem. Commun.* **2001**, 129.

225. Walker, S. D.; Barder, T. E.; Martinelli, J. R.; Buchwald, S. L. *Angew. Chem. Int. Ed.* **2004**, *43*, 1871.

226. Hidai, M.; Kashiwag.T; Ikeuchi, T.; Uchica, Y. *J. Organomet. Chem.* **1971**, *30*, 279.

227. Tang, Z. Y.; Hu, Q. S. *J. Am. Chem. Soc.* **2004**, *126*, 3058.

228. Percec, V.; Bae, J.; Hill, D. H. *J. Org. Chem.* **1995**, *60*, 1060.

229. Percec, V.; Golding, G. M.; Smidrkal, J.; Weichold, O. *J. Org. Chem.* **2004**, *69*, 7790.

230. Percec, V.; Golding, G. M.; Smidrkal, J.; Weichold, O. *J. Org. Chem.* **2004**, *69*, 3447.

231. Amatore, C.; Jutand, A.; Suarez, A. *J. Am. Chem. Soc.* **1993**, *115*, 9531.

232. Amatore, C.; Jutand, A.; Khalil, F.; Mbarki, M. A.; Mottier, L. *Organometallics* **1993**, *12*, 3168.

233. Amatore, C.; Carré, E.; Jutand, A.; M'Barki, M.; Meyer, G. *Organometallics* **1995**, *14*, 5605.

234. Amatore, C.; Carré, E.; Jutand, A.; Tanaka, H.; Ren, Q.; Torii, S. *Chem. Eur. J.* **1996**, *2*, 957.

235. Amatore, C.; Broeker, G.; Jutand, A.; Khalil, F. *J. Am. Chem. Soc.* **1997**, *119*, 5176.

236. Amatore, C.; Carré, E.; Jutand, A. *Acta Chem. Scand.* **1998**, *52*, 100.

237. Amatore, C.; Jutand, A.; Mottier, L. *Eur. J. Inorg. Chem.* **1999**, 1081.

238. Amatore, C.; Jutand, A.; M'Barki, M. A.; Meyer, G.; Mottier, L. *Eur. J. Inorg. Chem.* **2001**, 873.

239. Roy, A. H.; Hartwig, J. F. *Organometallics* **2004**, *23*, 194.

240. Amatore, C.; Jutand, A. *Acc. Chem. Res.* **2000**, *33*, 314.

241. Mateo, C.; Fernández-Rivas, C.; Echavarren, A. M.; Cárdenas, D. J. *Organometallics* **1997**, *16*, 1997.

242. Hagiwara, E.; Gouda, K.-i.; Hatanaka, Y.; Hiyama, T. *Tetrahedron Lett.* **1997**, *38*, 439.

243. Denmark, S. E.; Kallemeyn, J. M. *Org. Lett.* **2003**, *5*, 3483.

244. Denmark, S. E.; Sweis, R. F. *J. Am. Chem. Soc.* **2001**, *123*, 6439.

245. Denmark, S. E.; Sweis, R. F. *J. Am. Chem. Soc.* **2004**, *126*, 4876.

246. Denmark, S. E.; Sweis, R. F.; Wehrli, D. *J. Am. Chem. Soc.* **2004**, *126*, 4865.

247. Matos, K.; Soderquist, J. A. *J. Org. Chem.* **1998**, *63*, 461.

248. Ridgway, B. H.; Woerpel, K. A. *J. Org. Chem.* **1998**, *63*, 458.

249. Miyaura, N. *Top. Curr. Chem.* **2002**, *219*, 11.

250. Casado, A. L.; Espinet, P. *J. Am. Chem. Soc.* **1998**, *120*, 8978.

251. Casares, J. A.; Espinet, P.; Salas, G. *Chem. Eur. J.* **2002**, *8*, 4843.

252. Cárdenas, D. J.; Mateo, C.; Echavarren, A. M. *Angew. Chem., Int. Ed. Engl.* **1994**, *33*, 2445.

253. Mateo, C.; Cárdenas, D. J.; Fernández-Rivas, C.; Echavarren, A. M. *Chem. Eur. J.* **1996**, *2*, 1596.

254. Ricci, A.; Angelucci, F.; Bassetti, M.; Lo Sterzo, C. *J. Am. Chem. Soc.* **2002**, 1060.

255. Cotter, W. D.; Barbour, L.; McNamara, K. L.; Hechter, R.; Lachicotte, R. J. *J. Am. Chem. Soc.* **1998**, *120*, 11016.

256. Farina, V.; Krishnan, B. *J. Am. Chem. Soc.* **1991**, *113*, 9585.

257. Louie, J.; Hartwig, J. F. *J. Am. Chem. Soc.* **1995**, *117*, 11598.

258. Amatore, C.; Bahsoun, A. A.; Jutand, A.; Meyer, G.; Ntepe, A. N.; Ricard, L. *J. Am. Chem. Soc.* **2003**, *125*, 4212.

259. Labadie, J. W.; Stille, J. K. *J. Am. Chem. Soc.* **1983**, *105*, 6129.

260. Ye, J. H.; Bhatt, R. K.; Falck, J. R. *Tetrahedron Lett.* **1993**, *34*, 8007.

261. Ye, J. H.; Bhatt, R. K.; Falck, J. R. *J. Am. Chem. Soc.* **1994**, *116*, 1.

262. Casado, A. L.; Espinet, P.; Gallego, A. M. *J. Am. Chem. Soc.* **2000**, *122*, 11771.

263. McGahey, L. F.; Jensen, F. R. *J. Am. Chem. Soc.* **1979**, *101*, 4397.

264. Espinet, P.; Echavarren, A. M. *Angew. Chem. Int. Ed.* **2004**, *43*, 4704.

265. Grasa, G. A.; Nolan, S. P. *Org. Lett.* **2001**, *3*, 119.

266. Mee, S. P. H.; Lee, V.; Baldwin, J. E. *Angew. Chem. Int. Ed.* **2004**, *43*, 1132.

267. Vedejs, E.; Haight, A. R.; Moss, W. O. *J. Am. Chem. Soc.* **1992**, *114*, 6556.

268. Brown, J. M.; Pearson, M.; Jastrzebski, J.; van Koten, G. *J. Chem. Soc., Chem. Commun.* **1992**, 1440.

269. Brown, J. M.; Pearson, M.; Jastrzebski, J.; van Koten, G. *J. Chem. Soc., Chem. Commun.* **1992**, 1802.

270. Farina, V. *Pure. Appl. Chem.* **1996**, *68*, 73.

271. Culkin, D. A.; Hartwig, J. F. *Organometallics* **2004**, *23*, 3398.

272. Low, J. J.; Goddard, W. A. *J. Am. Chem. Soc.* **1986**, *108*, 6115.

273. Tatsumi, K.; Hoffman, R.; Yamamoto, A.; Stille, J. K. *Bull. Chem. Soc. Jpn.* **1981**, *54*, 1857.

274. Shekhar, S.; Hartwig, J. F. *J. Am. Chem. Soc.* **2004**, *126*, 13016.

275. Hayashi, T.; Knoishi, M.; Kumada, M. *Tetrahedron Lett.* **1979**, *21*, 1871.

276. Hayashi, T.; Konoshi, M.; Kobori, Y.; Kumada, M.; Higuchi, T.; Hirotsu, K. *J. Am. Chem. Soc.* **1984**, *106*, 158.

277. Whitesides, G. M.; Gaasch, J. F.; Stedronsky, E. R. *J. Am. Chem. Soc.* **1972**, *94*, 5258.

278. Miller, T. M.; Whitesides, G. M. *Organometallics* **1986**, *5*, 1473.

279. Cross, R. J. In *The Chemistry of the Metal–Carbon Bond*; Hartley, F. R., Patai, S., Eds.; Wiley: New York, 1985; Vol. 2, p 559.

280. Brown, J. M.; Cooley, N. A. *Chem. Rev.* **1988**, *88*, 1031.

281. Gillie, A.; Stille, J. K. *J. Am. Chem. Soc.* **1980**, *102*, 4933.

282. Moraviskiy, A.; Stille, J. K. *J. Am. Chem. Soc.* **1981**, *103*, 4182.

283. Driver, M. S.; Hartwig, J. F. *J. Am. Chem. Soc.* **1997**, *119*, 8232.

284. Brown, J. M.; Guiry, P. J. *Inorg. Chim. Acta* **1994**, *220*, 249.

285. See Table IV of reference 276.

286. Alcazar-Roman, L. M.; Hartwig, J. F.; Rheingold, A. L.; Liable-Sands, L. M.; Guzei, I. A. *J. Am. Chem. Soc.* **2000**, *122*, 4618.

287. Alcazar-Roman, L. M.; Hartwig, J. F. *Organometallics* **2002**, *21*, 491.

288. Brunel, J. M. *Mini-Rev. Org. Chem.* **2004**, *1*, 249.

289. Nishiyama, M.; Yamamoto, T.; Koie, Y. *Tetrahedron Lett.* **1998**, *39*, 617.

290. Yamamoto, T.; Nishiyama, M.; Koie, Y. *Tetrahedron Lett.* **1998**, *39*, 2367.

291. Watanabe, M.; Nishiyama, M.; Koie, Y. *Tetrahedron Lett.* **1999**, *40*, 8837.

292. Watanabe, M.; Nishiyama, M.; Yamamoto, T.; Koie, Y. *Tetrahedron Lett.* **2000**, *41*, 481.

293. Stambuli, J. P.; Kuwano, R.; Hartwig, J. F. *Angew. Chem. Int. Ed.* **2002**, *41*, 4746.

294. Hartwig, J. F.; Kawatsura, M.; Hauck, S. I.; Shaughnessy, K. H.; Alcazar-Roman, L. M. *J. Org. Chem.* **1999**, *64*, 5575.

295. Littke, A. F.; Fu, G. C. *Angew. Chem. Int. Ed.* **1998**, *37*, 3387.

296. Littke, A. F.; Fu, G. C. *J. Org. Chem.* **1999**, *64*, 10.

297. Littke, A. F.; Fu, G. C. *Angew. Chem. Int. Ed.* **1999**, *38*, 2411.

298. Dai, C.; Fu, G. C. *J. Am. Chem. Soc.* **2001**, *123*, 2719.

299. Littke, A. F.; Schwarz, L.; Fu, G. C. *J. Am. Chem. Soc.* **2002**, *124*, 6343.

300. Wolfe, J. P.; Singer, R. A.; Yang, B. H.; Buchwald, S. L. *J. Am. Chem. Soc.* **1999**, *121*, 9550.

301. Wolfe, J. P.; Tomori, H.; Sadighi, J. P.; Yin, J. J.; Buchwald, S. L. *J. Org. Chem.* **2000**, *65*, 1158.

302. Torraca, K.; Kuwabe, S.; Buchwald, S. *J. Am. Chem. Soc.* **2000**, *122*, 12907.

303. Torraca, K. E.; Huang, X. H.; Parrish, C. A.; Buchwald, S. L. *J. Am. Chem. Soc.* **2001**, *123*, 10770.

304. Huang, X. H.; Anderson, K. W.; Zim, D.; Jiang, L.; Klapars, A.; Buchwald, S. L. *J. Am. Chem. Soc.* **2003**, *125*, 6653.

305. Milne, J. E.; Buchwald, S. L. *J. Am. Chem. Soc.* **2004**, *126*, 13028.

306. Mann, G.; Incarvito, C.; Rheingold, A. L.; Hartwig, J. F. *J. Am. Chem. Soc.* **1999**, *121*, 3224.

307. Shelby, Q.; Kataoka, N.; Mann, G.; Hartwig, J. F. *J. Am. Chem. Soc.* **2000**, *122*, 10718.

308. Kataoka, N.; Shelby, Q.; Stambuli, J. P.; Hartwig, J. F. *J. Org. Chem.* **2002**, *67*, 5553.

309. Urgaonkar, S.; Nagarajan, M.; Verkade, J. G. *Tetrahedron Lett.* **2002**, *43*, 8921.

310. Urgaonkar, S.; Nagarajan, M.; Verkade, J. G. *J. Org. Chem.* **2003**, *68*, 452.

311. Urgaonkar, S.; Xu, J. H.; Verkade, J. G. *J. Org. Chem.* **2003**, *68*, 8416.

312. You, J.; Verkade, J. G. *J. Org. Chem.* **2003**, *68*, 8003.

313. Urgaonkar, S.; Verkade, J. G. *Adv. Synth. Catal.* **2004**, *346*, 611.

314. Hermann, W. A.; Reisinger, C.; Spiegler, M. *J. Organomet. Chem.* **1998**, *557*, 93.

315. Huang, J.; Nolan, S. *J. Am. Chem. Soc.* **1999**, *121*, 9889.

316. Weskamp, T.; Volker, P. W. B.; Herrman, W. A. *J. Organomet. Chem.* **1999**, *585*, 348.

317. Zhang, C.; Huang, J.; Trudell, M. L.; Nolan, S. P. *J. Org. Chem.* **1999**, *64*, 3804.

318. Bohm, V. P. W.; Gstöttmayr, C. W. K.; Weskamp, T.; Herrmann, W. A. *J. Organomet. Chem.* **2000**, *595*, 186.

319. Böhm, V. P. W.; Weskamp, T.; Gstöttmayr, C. W. K.; Herrmann, W. A. *Angew. Chem. Int. Ed.* **2000**, *39*, 1602.

320. Stauffer, S. R.; Lee, S.; Stambuli, J. P.; Hauck, S. I.; Hartwig, J. F. *Org. Lett.* **2000**, *2*, 1423.

321. Zhang, C.; Trudell, M. L. *Tetrahedron Lett.* **2000**, *41*, 595.

322. Andrus, M. B.; Song, C. *Org. Lett.* **2001**, *3*, 3761.

323. Caddick, S.; Cloke, F. G. N.; Clentsmith, G. K. B.; Hitchcock, P. B.; McKerrecher, D.; Titcomb, L. R.; Williams, M. R. V. *J. Organomet. Chem.* **2001**, *617–618*, 635.

324. Herrmann, W. A.; Weskamp, T.; Bohm, V. P. W. *Adv. Organomet. Chem.* **2001**, *48*, 1.

325. Yang, C.; Lee, H. M.; Nolan, S. P. *Org. Lett.* **2001**.

326. Desmarets, C.; Schneider, R.; Fort, Y. *J. Org. Chem.* **2002**, 3029.

327. Gstöttmayr, C. W. K.; Böhm, V. P. W.; Herdtweck, E.; Grosche, M.; Herrmann, W. A. *Angew. Chem. Int. Ed.* **2002**, *41*, 1363.

328. Viciu, M. S.; Germaneau, R. F.; Navarro-Fernandez, O.; Stevens, E. D.; Nolan, S. P. *Organometallics* **2002**, *21*, 5470.

329. Altenhoff, G.; Goddard, R.; Lehmann, C. W.; Glorius, F. *Angew. Chem. Int. Ed.* **2003**, *42*, 3690.

330. Navarro, O.; Kelly, R. A.; Nolan, S. P. *J. Am. Chem. Soc.* **2003**, *125*, 16194.

331. Viciu, M. S.; Kelly, R. A., III; Stevens, E. D.; Naud, F.; Studer, M.; Nolan, S. P. *Org. Lett.* **2003**, *5*, 1479.

332. Stambuli, J. P.; Incarvito, C. D.; Buhl, M.; Hartwig, J. F. *J. Am. Chem. Soc.* **2004**, *126*, 1184.

333. Stambuli, J. P.; Bühl, M.; Hartwig, J. F. *J. Am Chem. Soc.* **2002**, *124*, 9346.

334. (a) Barrios-Landeros, F.; Carrow, B. P.; Hartwig, J. F. *J. Am. Chem. Soc.* **2009**, *131*, 8141; (b) Barrios-Landeros, F.; Carrow, B. P.; Hartwig, J. F. *J. Am. Chem. Soc.* **2008**, *130*, 5842.

335. Fitton, P.; Rick, E. A. *J. Organomet. Chem.* **1971**, *28*, 287.

336. Ben-David, Y.; Portnoy, M.; Milstein, D. *J. Am. Chem. Soc.* **1989**, *111*, 8742.

337. Ben-David, Y.; Portnoy, M.; Milstein, D. *J. Chem. Soc., Chem. Commun.* **1989**, 1816.

338. Ben-David, Y.; Portnoy, M.; Gozin, M.; Milstein, D. *Organometallics* **1992**, *11*, 1995.

339. Ben-David, Y.; Gozin, M.; Portnoy, M.; Milstein, D. *J. Mol. Catal.* **1992**, *73*, 173.

340. Shen, Q.; Shekhar, S.; Stambuli, J. P.; Hartwig, J. F. *Angew. Chem. Int. Ed.* **2004**, *44*, 1371.

341. Mann, G.; Shelby, Q.; Roy, A. H.; Hartwig, J. F. *Organometallics* **2003**, 2775.

342. Nicolaou, K. C.; Bulger, P. G.; Sarlah, D. *Angew. Chem. Int. Ed.* **2005**, *44*, 4442.

343. de Vries, J. G. *Can. J. Chem.* **2001**, *79*, 1086.

344. Corbet, J. P.; Mignani, G. *Chem. Rev.* **2006**, *106*, 2651.

345. Larsen, R. D.; King, A. O.; Chen, C. Y.; Corley, E. G.; Foster, B. S.; Roberts, F. E.; Yang, C. H.; Lieberman, D. R.; Reamer, R. A.; Tschaen, D. M.; Verhoeven, T. R.; Reider, P. J. *J. Org. Chem.* **1994**, *59*, 6391.

346. Mantlo, N. B.; Chakravarty, P. K.; Ondeyka, D. L.; Siegl, P. K. S.; Chang, R. S.; Lotti, V. J.; Faust, K. A.; Schorn, T. W.; Chen, T. B.; Schorn, T. W.; Sweet, C. S.; Emmert, S. E.; Patchett, A. A.; Greenlee, W. J. *J. Med. Chem.* **1991**, *34*, 2919.

347. Shinkai, I.; King, A. O.; Larsen, R. D. *Pure. Appl. Chem.* **1994**, *66*, 1551.

348. Peng, Z.; Gharavi, A. R.; Yu, L. *J. Am. Chem. Soc.* **1997**, *119*, 4622.

349. Pan, M.; Bao, Z.; Yu, L. *Macromolecules* **1995**, *28*, 5151.

350. Baranano, D.; Mann, G.; Hartwig, J. F. *Curr. Org. Chem.* **1997**, *1*, 287.

351. Hartwig, J. F. *Angew. Chem. Int. Ed.* **1998**, *37*, 2046.

352. Hartwig, J. F. *Acc. Chem. Res.* **1998**, *31*, 852.

353. Wolfe, J. P.; Wagaw, S.; Marcoux, J.-F.; Buchwald, S. L. *Acc. Chem. Res.* **1998**, *31*, 805.

354. Belfield, A. J.; Brown, G. R.; Foubister, A. J. *Tetrahedron* **1999**, *55*, 11399.

355. Yang, B. H.; Buchwald, S. L. *J. Organomet. Chem.* **1999**, *576*, 125.

356. Hartwig, J. F. In *Modern Amination Methods*; Ricci, A., Ed.; Wiley-VCH: Weinheim, 2000; p 195.

357. Hartwig, J. F. In *Modern Arene Chemistry*; Astruc, C., Ed.; Wiley-VCH: Weinheim, 2002; p 107.
358. Hartwig, J. F. In *Handbook of Organopalladium Chemistry for Organic Synthesis*; Negishi, E.-i., Ed.; Wiley-Interscience: New York, 2002; Vol. 1, p 1051.
359. Muci, A. R.; Buchwald, S. L. *Top. Curr. Chem.* **2002**, *219*, 131.
360. Hartwig, J. F. *Acc. Chem. Res.* **2008**, *41*, 1534.
361. Hartwig, J. F. In *Handbook of Organopalladium Chemistry for Organic Synthesis*; Negishi, E.-i., Ed.; Wiley-Interscience: New York, 2002; Vol. 1, p 1097.
362. Cacchi, S.; Fabrizi, G.; Goggiamani, A.; Parisi, L. M. *Org. Lett.* **2002**, *4*, 4719.
363. Schopfer, U.; Schlapbach, A. *Tetrahedron* **2001**, *57*, 3069.
364. Li, G. Y.; Zheng, G.; Noonan, A. F. *J. Org. Chem.* **2001**, *66*, 8677.
365. Li, G. Y. *Angew. Chem. Int. Ed.* **2001**, *40*, 1513.
366. Hua, R.; Takeda, H.; Onozawa, S.-y.; Abe, Y.; Tanaka, M. *J. Am. Chem. Soc.* **2001**.
367. Zheng, N.; McWilliams, J. C.; Fleitz, F. J.; Armstrong, J. D.; Volante, R. P. *J. Org. Chem.* **1998**, *63*, 9606.
368. Rossi, R.; Bellina, F.; Mannina, L. *Tetrahedron* **1997**, *53*, 1025.
369. Ishiyama, T.; Mori, M.; Suzuki, A.; Miyaura, N. *J. Organomet. Chem.* **1996**, *525*, 225.
370. Carpita, A.; Rossi, R.; Scamuzzi, B. *Tetrahedron Lett.* **1989**, *30*, 2699.
371. Kosugi, M.; Ogata, T.; Terada, M.; Sano, H.; Migita, T. *Bull. Chem. Soc. Jpn.* **1985**, *58*, 3657.
372. Murahashi, S. I.; Yamamura, M.; Yanagisawa, K.; Mita, N.; Kondo, K. *J. Org. Chem.* **1979**, *44*, 2408.
373. Hirao, T.; Masunaga, T.; Yamada, N. *Bull.Chem. Soc. Jpn.* **1982**, *55*, 909.
374. Xu, Y.; Li, Z.; Xia, J.; Guo, H.; Huang, Y. *Synthesis* **1983**, 377.
375. Xu, Y.; Li, Z.; Xia, J.; Guo, H.; Huang, Y. *Synthesis* **1984**, 781.
376. Xu, Y.; Xia, J.; Guo, H. *Synthesis* **1986**, 691.
377. Xu, Y.; Zhang, J. *J. Chem. Soc., Chem. Commun.* **1986**, 1606.
378. Tunney, S. E.; Stille, J. K. *J. Org. Chem.* **1987**, *52*, 748.
379. Zhang, J.; Xu, Y.; Huang, G.; Guo, H. *Tetrahedron Lett.* **1988**, *29*, 1955.
380. Xu, H.; Wei, H.; Zhang, J.; Huang, G. *Tetrahedron Lett.* **1989**, *30*, 949.
381. Al-Masum, M.; Livinghouse, T. *Tetrahedron Lett.* **1999**, *40*, 7731.
382. Beletskaya, I. P.; Veits, Y. A.; Leksunkin, V. A.; Foss, V. L. *Bull. Russ. Acad. Sci.-Div. Chem. Sci.* **1992**, *41*, 1272.
383. Gilbertson, S. R.; Starkey, G. W. *J. Org. Chem.* **1996**, *61*, 2922.
384. Herd, O.; Hessler, A.; Hingst, M.; Tepper, M.; Stelzer, O. *J. Organomet. Chem.* **1996**, *522*, 69.
385. Song, Y.; Mok, K. F.; Leung, P. H.; Chan, S. H. *Inorg. Chem.* **1998**, *37*, 6399.
386. Vyskocil, S.; Smrcina, M.; Hanus, V.; Polasek, M.; Kocovsky, P. *J. Org. Chem.* **1998**, *63*, 7738.
387. Fernandez-Rodriguez, M. A.; Shen, Q. L.; Hartwig, J. F. *J. Am. Chem. Soc.* **2006**, *128*, 2180.
388. Keinan, E.; Sahai, M.; Poth, Z.; Nudelman, A.; Herzig, J. *J. Org. Chem.* **1985**, *50*, 3558.
389. Mann, G.; Hartwig, J. F. *J. Am. Chem. Soc.* **1996**, *118*, 13109.
390. Palucki, M.; Wolfe, J. P.; Buchwald, S. L. *J. Am. Chem. Soc.* **1996**, *118*, 10333.
391. Mann, G.; Hartwig, J. F. *J. Org. Chem.* **1997**, *62*, 5413.
392. Palucki, M.; Wolfe, J. P.; Buchwald, S. L. *J. Am. Chem. Soc.* **1997**, *119*, 3395.
393. Frost, C. G.; Mendonca, P. *J. Chem. Soc., Perkin Trans.* **1998**, 2615.
394. Olivera, R.; San Martin, R.; Dominguez, E. *Tetrahedron Lett.* **2000**, *41*, 4357.
395. Sawyer, J. S. *Tetrahedron* **2000**, *56*, 5045.
396. Kuwabe, S.; Torraca, K. E.; Buchwald, S. L. *J. Am. Chem. Soc.* **2001**, *123*, 12202.
397. Parrish, C. A.; Buchwald, S. L. *J. Org. Chem.* **2001**, *66*, 2498.
398. Ding, S.; Gray, N. S.; Wu, X.; Ding, Q.; Schultz, P. G. *J. Am. Chem. Soc.* **2002**, *124*, 1594.
399. Gao, G. Y.; Colvin, A. J.; Chen, Y.; Zhang, X. P. *Org. Lett.* **2003**, *5*, 3261.
400. Vorogushin, A. V.; Huang, X. H.; Buchwald, S. L. *J. Am. Chem. Soc.* **2005**, *127*, 8146.
401. Wolfe, J. P.; Buchwald, S. L. *J. Org. Chem.* **1996**, *61*, 1133.
402. Hamann, B. C.; Hartwig, J. F. *J. Am. Chem. Soc.* **1998**, *120*, 7369.
403. Ali, M. H.; Buchwald, S. L. *J. Org. Chem.* **2001**, *66*, 2560.
404. Meyers, C.; Maes, B. U. W.; Loones, K. T. J.; Bal, G.; Lemiere, G. L. F.; Dommisse, R. A. *J. Org. Chem.* **2004**, *69*, 6010.
405. Louie, J.; Driver, M. S.; Hamann, B. C.; Hartwig, J. F. *J. Org. Chem.* **1997**, *62*, 1268.
406. Wolfe, J. P.; Buchwald, S. L. *J. Org. Chem.* **1997**, *62*, 1264.
407. Åhman, J.; Buchwald, S. L. *Tetrahedron Lett.* **1997**, *38*, 6363.
408. Wolfe, J. P.; Åhman, J.; Sadighi, J. P.; Singer, R. A.; Buchwald, S. L. *Tetrahedron Lett.* **1997**, *38*, 6367.
409. Vyskocil, S.; Smrcina, M.; Kocovsky, P. *Tetrahedron Lett.* **1998**, *39*, 9289.
410. Anderson, K. W.; Mendez-Perez, M.; Priego, J.; Buchwald, S. L. *J. Org. Chem.* **2003**, *68*, 9563.
411. Ogata, T.; Hartwig, J. F. *J. Am. Chem. Soc.* **2008**, *130*, 13848.
412. Stauffer, S.; Hauck, S. I.; Lee, S.; Stambuli, J.; Hartwig, J. F. *Org. Lett.* **2000**, *2*, 1423.

413. Huang, J.; Grasa, G.; Nolan, S. P. *Org. Lett.* **1999**, *1*, 1307.

414. Grasa, G. A.; Viciu, M. S.; Huang, J. K.; Nolan, S. P. *J. Org. Chem.* **2001**, *66*, 7729.

415. Viciu, M. S.; Kissling, R. M.; Stevens, E. D.; Nolan, S. P. *Org. Lett.* **2002**, *4*, 2229.

416. Driver, M. S.; Hartwig, J. F. *J. Am. Chem. Soc.* **1996**, *118*, 7217.

417. Wolfe, J. P.; Wagaw, S.; Buchwald, S. L. *J. Am. Chem. Soc.* **1996**, *118*, 7215.

418. Wolfe, J. P.; Buchwald, S. L. *J. Org. Chem.* **2000**, *65*, 1144.

419. (a) Shen, Q.; Hartwig, J. F. *J. Am. Chem. Soc.* **2006**, *128*, 10028; (b) Vo, G.; Hartwig, J. F. *J. Am. Chem. Soc.* **2009**, *131*, 11049.

420. Mann, G.; Hartwig, J. F.; Driver, M. S.; Fernandez-Rivas, C. *J. Am. Chem. Soc.* **1998**, *120*, 827.

421. Huang, X. H.; Buchwald, S. L. *Org. Lett.* **2001**, *3*, 3417.

422. Lee, S.; Jorgensen, M.; Hartwig, J. F. *Org. Lett.* **2001**, *3*, 2729.

423. Lee, D.-Y.; Hartwig, J. F. *Org. Lett.* **2005**, *7*, 1169.

424. Wagaw, S.; Yang, B. H.; Buchwald, S. L. *J. Am. Chem. Soc.* **1998**, *120*, 6621.

425. Hartwig, J. F. *Angew. Chem. Int. Ed.* **1998**, *37*, 2090.

426. Wagaw, S.; Yang, B.; Buchwald, S. L. *J. Am. Chem. Soc.* **1999**, *121*, 10251.

427. Surry, D. S.; Buchwald, S. L. *J. Am. Chem. Soc.* **2007**, *129*, 10354.

428. Yang, B. H.; Buchwald, S. L. *Org. Lett.* **1999**, *1*, 35.

429. Edmondson, S. D.; Mastracchio, A.; Parmee, E. R. *Org. Lett.* **2000**, *2*, 1109.

430. Yin, J.; Buchwald, S. L. *Org. Lett.* **2000**, *2*, 1101.

431. Yin, J.; Buchwald, S. L. *J. Am. Chem. Soc.* **2002**, *124*, 6043.

432. Bolm, C.; Hildebrand, J. P. *Tetrahedron Lett.* **1998**, *39*, 5731.

433. Bolm, C.; Hildebrand, J. P. *J. Org. Chem.* **2000**, *65*, 169.

434. Bolm, C.; Martin, M.; Gibson, L. *Synlett* **2002**, *5*, 832.

435. Beletskaya, I. P.; Davydov, D. V.; Moreno-Mañas, M. *Tetrahedron Lett.* **1998**, *39*, 5617.

436. Old, D. W.; Harris, M. C.; Buchwald, S. L. *Org. Lett.* **2000**, *2*, 1403.

437. Beletskaya, I. P.; Davydov, D. V.; Gorovoy, M. S. *Tetrahedron Lett.* **2002**, *43*, 6221.

438. Lebedev, A. Y.; Izmer, V. V.; Kazyul'kin, D. N.; Beletskaya, I. P.; Voskoboynikov, A. Z. *Org. Lett.* **2002**, 623.

439. Kosugi, M.; Kameyama, M.; Migita, T. *Chem. Lett.* **1983**, 927.

440. Kosugi, M.; Kameyama, M.; Sano, H.; Migita, T. *Nippon Kagaku Kaishi* **1985**, *3*, 547.

441. Guram, A. S.; Rennels, R. A.; Buchwald, S. L. *Angew. Chem., Int. Ed. Engl.* **1995**, *34*, 1348.

442. Louie, J.; Hartwig, J. F. *Tetrahedron Lett.* **1995**, *36*, 3609.

443. Huser, M.; Youinou, M. T.; Osborn, J. A. *Angew. Chem., Int. Ed. Engl.* **1989**, *28*, 1427.

444. Harkal, S.; Rataboul, F.; Zapf, A.; Fuhrmann, C.; Riermeier, T.; Monsees, A.; Beller, M. *Adv. Synth. Catal.* **2004**, *346*, 1742.

445. Rataboul, F.; Zapf, A.; Jackstell, R.; Harkal, S.; Riermeier, T.; Monsees, A.; Dingerdissen, U.; Beller, M. *Chem. Eur. J.* **2004**, *10*, 2983.

446. Singer, R. A.; Tom, N. J.; Frost, H. N.; Simon, W. M. *Tetrahedron Lett.* **2004**, *45*, 4715.

447. Singer, R. A.; Dore, M. L.; Sieser, J. E.; Berliner, M. A. *Tetrahedron Lett.* **2006**, *47*, 3727.

448. Marion, N.; Navarro, O.; Mei, J. G.; Stevens, E. D.; Scott, N. M.; Nolan, S. P. *J. Am. Chem. Soc.* **2006**, *128*, 4101.

449. Marion, N.; Ecarnot, E. C.; Navarro, O.; Amoroso, D.; Bell, A.; Nolan, S. P. *J. Org. Chem.* **2006**, *71*, 3816.

450. Hartwig, J. F.; Richards, S.; Barañano, D.; Paul, F. *J. Am. Chem. Soc.* **1996**, *118*, 3626.

451. Paul, F.; Patt, J.; Hartwig, J. F. *Organometallics* **1995**, *14*, 3030.

452. Widenhoefer, R. A.; Buchwald, S. L. *Organometallics* **1996**, *15*, 3534.

453. Widenhoefer, R. A.; Buchwald, S. L. *Organometallics* **1996**, *15*, 2755.

454. Anderson, K. W.; Tundel, R. E.; Ikawa, T.; Altman, R. A.; Buchwald, S. L. *Angew. Chem. Int. Ed.* **2006**, *45*, 6523.

455. Shen, Q.; Ogata, T.; Hartwig, J. F. *J. Am. Chem. Soc.* **2008**, *130*, 6586.

456. Recall from Chapters 6 and 7 that ligand dissociation typically leads to the reactive Pd(0) species.

457. Alcazar-Roman, L. M.; Hartwig, J. F. *J. Am. Chem. Soc.* **2001**, *123*, 12905.

458. Shekhar, S.; Ryberg, P.; Hartwig, J. F. *Org. Lett.* **2006**, *8*, 851.

459. Shekhar, S.; Ryberg, P.; Hartwig, J. F.; Mathew, J. S.; Blackmond, D. G.; Strieter, E. R.; Buchwald, S. L. *J. Am. Chem. Soc.* **2006**, *128*, 3584.

460. For an analysis of this issue, see reference 458.

461. Driver, M. S.; Hartwig, J. F. *J. Am. Chem. Soc.* **1995**, *117*, 4708.

462. Yamashita, M.; Cuevas Vicario, J. V.; Hartwig, J. F. *J. Am. Chem. Soc.* **2003**, *125*, 16347.

463. Yamashita, M.; Hartwig, J. F. *J. Am. Chem. Soc.* **2004**, *126*, 5344.

464. Fujita, K. I.; Yamashita, M.; Puschmann, F.; Alvarez-Falcon, M. M.; Incarvito, C. D.; Hartwig, J. F. *J. Am. Chem. Soc.* **2006**, *128*, 9044.

465. Goossen, L. J.; Koley, D.; Hermann, H.; Thiel, W. *Chem. Commun.* **2004**, 2141.

466. Hartwig, J. F. *J. Am. Chem. Soc.* **1996**, *118*, 7010.

467. Hamann, B. C.; Hartwig, J. F. *J. Am. Chem. Soc.* **1998**, *120*, 3694.

468. Bumagin, N. A.; Bumagina, I. G.; Nashin, A. N.; Beletskaya, I. P. *Dokl. Akad. Nauk* **1981**, *261*, 532.

469. Goure, W. F.; Wright, M. E.; Davis, P. D.; Labadie, S. S.; Stille, J. K. *J. Am. Chem. Soc.* **1984**, *106*, 6417.

470. Bumagin, N. A.; Bumagina, G.; Kashin, A. N.; Beletskaya, P. *Dokl. Akad. Nauk* **1981**, *261*, 1141.

471. Echavarren, A. M.; Stille, J. K. *J. Am. Chem. Soc.* **1988**, *110*, 1557.

472. Crisp, G. T.; Scott, W. J.; Stille, J. K. *J. Am. Chem. Soc.* **1984**, *106*, 7500.

473. Bumagin, N. A.; Ponomaryov, A. B.; Beletskaya, I. P. *Tetrahedron Lett.* **1985**, *26*, 4819.

474. Tamaru, Y.; Ochiai, H.; Yamada, Y.; Yoshida, Z. *Tetrahedron Lett.* **1983**, *24*, 3869.

475. Wakita, Y.; Yasunaga, T.; Akita, M.; Kojima, M. *J. Organomet. Chem.* **1986**, *301*, C17.

476. Ishiyama, T.; Kizaki, H.; Miyaura, N.; Suzuki, A. *Tetrahedron Lett.* **1993**, *34*, 7595.

477. Ishiyama, T.; Kizaki, H.; Hayashi, T.; Suzuki, A.; Miyaura, N. *J. Org. Chem.* **1998**, *63*, 4726.

478. Ishiyama, T.; Miyaura, N.; Suzuki, A. *Bull. Chem. Soc. Jpn.* **1991**, *64*, 1999.

479. Ishiyama, T.; Miyaura, N.; Suzuki, A. *Tetrahedron Lett.* **1991**, *32*, 623.

480. Ishiyama, T.; Miyaura, N.; Suzuki, A. *Bull. Chem. Soc. Jpn.* **1991**, *64*, 1999.

481. Baillargeon, V. P.; Stille, J. K. *J. Am. Chem. Soc.* **1983**, *105*, 7175.

482. Baillargeon, V. P.; Stille, J. K. *J. Am. Chem. Soc.* **1986**, *108*, 452.

483. Kikukawa, K.; Totoki, T.; Wada, F.; Matsuda, T. *J. Organomet. Chem.* **1984**, *270*, 283.

484. Pribar, I.; Buchman, O. *J. Org. Chem.* **1984**, *49*, 4009.

485. Klaus, S.; Neumann, H.; Zapf, A.; Strubing, D.; Hubner, S.; Almena, J.; Riemeier, T.; Gross, P.; Sarich, M.; Krahnert, W. R.; Rossen, K.; Beller, M. *Angew. Chem. Int. Ed.* **2006**, *45*, 154.

486. Ullmann, F. *Ber. Dtsch. Chem. Ges.* **1901**, *34*, 2174.

487. Ullmann, F. *Ber. Dtsch. Chem. Ges.* **1903**, *36*, 2382.

488. Ullmann, F.; Sponagel, P. *Ber. Dtsch. Chem. Ges.* **1905**, *36*, 2211.

489. Ullmann, F.; Maag, R. *Ber. Dtsch. Chem. Ges.* **1906**, *39*, 1693.

490. Goldberg, I. *Ber. Dtsch. Chem. Ges.* **1906**, *39*, 1691.

491. Hurtley, W. R. H. *J. Chem. Soc.* **1929**, 1870.

492. Gilman, H.; Jones, R. G.; Woods, L. A. *J. Org. Chem.* **1952**, *17*, 1630.

493. House, H. O.; Koepsell, D. G.; Campbell, W. J. *J. Org. Chem.* **1972**, *37*, 1003.

494. Lipshutz, B. H.; Wilhelm, R. S.; Kozlowski, J. A. *Tetrahedron* **1984**, *40*, 5005.

495. Burnett, J. F. *Chem. Rev.* **1951**, *49*, 273.

496. Lindley, J. *Tetrahedron* **1984**, *40*, 1433.

497. Elliott, G. I.; Konopelski, J. P. *Tetrahedron* **2001**, *57*, 5683.

498. Finet, J.-P.; Fedorov, A. Y.; Combes, S.; Boyer, G. *Curr. Org. Chem.* **2002**, *6*, 597.

499. Kunz, K.; Scholz, U.; Ganzer, D. *Synlett* **2003**, 2428.

500. Thomas, A. W.; Ley, S. V. *Angew. Chem. Int. Ed.* **2003**, *42*, 5400.

501. (a) Soderberg, B. C. G. *Coord. Chem. Rev.* **2004**, *248*, 1085; (b) Beletskaya, I. P.; Cheprakov, A. V. *Coord. Chem. Rev.* **2004**, *248*, 2337.

502. Evano, G.; Blanchard, N.; Toumi, M. *Chem. Rev.* **2008**, *108*, 3054.

503. Paine, A. J. *J. Am. Chem. Soc.* **1987**, *109*, 1496.

504. Weston, P. E.; Adkins, H. *J. Am. Chem. Soc.* **1928**, *50*, 859.

505. Goldberg, I. *Ber. Dtsch. Chem. Ges.* **1907**, *40*, 4541.

506. Goldberg, I.; Nimerovsky, M. *Ber. Dtsch. Chem. Ges.* **1907**, *40*, 2448.

507. For a detailed procedure on the coupling of aniline without an activating acid group on the aryl halide, see Hager, F. D. *Org. Synth. Coll. Vol. 1*, **1941**, 544.

508. Acheson, R. M. *Acridines*; Interscience: New York, 1956.

509. Pellon, R. F.; Carrasco, R.; Rodes, L. *Synth. Commun.* **1993**, *23*, 1447.

510. Carrasco, R.; Pellon, R. F.; Elguero, J.; Paez, J. A. *Synth. Commun.* **1989**, *19*, 2077.

511. Hanoun, J.-P.; Galy, J.-P.; Tenaglia, A. *Synth. Commun.* **1995**, *25*, 2443.

512. Zeide, O. *Ann. Chim.* **1924**, *440*, 311.

513. Pellon, R. F.; Carrasco, R.; Rodes, L. *Synth. Commun.* **1996**, *26*, 3869.

514. Palacios, D. M. L.; Comdom, R. F. P. *Synth. Commun.* **2003**, *33*, 1777.

515. Stepanov, I.; Aingom, L. B. *Zh. Obshch. Khim.* **1959**, *29*, 3436.

516. Tuong, T. D.; Hida, M. *Bull. Chem. Soc. Jpn.* **1970**, *43*, 1763.

517. Iizuka, K.; Akahane, K.; Momose, D.; Nakazawa, M. *J. Med. Chem.* **1981**, *24*, 1139.

518. Lo, Y. S.; Nolan, J. C.; Maren, T. H.; Welstead, J., W. J.; Gripshover, D. F.; Shamblee, D. A. *J. Med. Chem.* **1992**, *35*, 4790.

519. Martinez, G. R.; Walker, K. A. M.; Hirshfeld, D. R.; Bruno, J. J.; Yang, D. S.; Maloney, P. J. *J. Med. Chem.* **1992**, *35*, 620.

520. Pavik, J. W.; Connors, R. E.; Burns, D. S.; Kurzweil, E. M. *J. Am. Chem. Soc.* **1993**, *115*, 7645.

521. Venuti, M. C.; Stephenson, R. A.; Alvarez, R.; Bruno, J. A. *J. Med. Chem.* **1987**, *30*, 2136.

522. Gujadhur, R.; Bates, C. G.; Venkataraman, D. *Org. Lett.* **2001**, *3*, 4315.

523. Haider, J.; Kunz, K.; Scholz, U. *Adv. Synth. Catal.* **2004**, *346*, 717.

524. Xia, N.; Taillefer, M. *Chem.—Eur. J.* **2008**, *14*, 6037.

525. Ma, D.; Cai, Q. *Acc. Chem. Res.* **2008**, *41*, 1450.

526. Cervetto, L.; Demontis, G. C.; Giannaccini, G.; Longoni, B.; Macchia, B.; Macchia, M.; Martinelli, A.; Orlandini, E. *J. Med. Chem.* **1998**, *41*, 4933.

527. Ma, D.; Xia, C.; Jiang, J.; Zhang, J.; Tang, W. *J. Org. Chem.* **2003**, *68*, 442.

528. Ma, D.; Xia, C. *Org. Lett.* **2001**, *3*, 2583.

529. Clement, J.-B.; Hayes, J. F.; Sheldrake, H. M.; Shledrake, P. W.; Wells, A. S. *Synlett* **2001**, 1423.

530. Ma, D.; Cai, Q.; Zhang, H. *Org. Lett.* **2003**, *5*, 2453.

531. Zhang, H.; Cai, Q.; Ma, D. *J. Org. Chem.* **2005**, *70*, 5164.

532. Rao, H.; Fu, H.; Jiang, J.; Zhao, Y. *J. Org. Chem.* **2005**, *70*, 8107.

533. Guo, X.; Rao, H.; Fu, H.; Jiang, Y.; Zhao, Y. *Adv. Synth. Catal.* **2006**, *348*, 2197.

534. Rao, H.; Jin, Y.; Fu, H.; Jiang, Y.; Zhao, Y. *Chem.—Eur. J.* **2006**, *12*, 3636.

535. Altman, R. A.; Anderson, K. W.; Buchwald, S. L. *J. Org. Chem.* **2008**, *73*, 5167.

536. Goodbrand, H. B.; Hu, N.-X. *J. Org. Chem.* **1999**, *64*, 670.

537. Gujadhur, R.; Venkataraman, D.; Kintigh, J. T. *Tetrahedron Lett.* **2001**, *42*, 4791.

538. Pellon, R. F.; Mamposo, T.; Carrasco, R.; Rodes, L. *Synth. Commun.* **1996**, *26*, 3877.

539. Job, G. E.; Buchwald, S. L. *Org. Lett.* **2002**, *4*, 3703.

540. Arai, S.; Yamagishi, T.; Ototake, S.; Hida, M. *Bull. Chem. Soc. Jpn.* **1977**, *50*, 547.

541. Kalinin, A. V.; Bower, J. F.; Riebel, P.; Snieckus, V. *J. Org. Chem.* **1999**, *64*, 2986.

542. Vedejs, E.; Trapencieris, P.; Suna, E. *J. Org. Chem.* **1999**, *64*, 6724.

543. Arterburn, J. B.; Pannala, M.; Gonzalez, A. M. *Tetrahedron Lett.* **2001**, *42*, 1475.

544. Lang, F.; Zewge, D.; Houpis, I. N.; Volante, R. P. *Tetrahedron Lett.* **2001**, *42*, 3251.

545. Kwong, F. Y.; Klapars, A.; Buchwald, S. L. *Org. Lett.* **2002**, *4*, 581.

546. Enguehard, C.; Allouchi, H.; Gueiffier, A.; Buchwald, S. L. *J. Org. Chem.* **2003**, *68*, 4367.

547. Kwong, F. Y.; Buchwald, S. L. *Org. Lett.* **2003**, *5*, 793.

548. Shafir, A.; Buchwald, S. L. *J. Am. Chem. Soc.* **2006**, *128*, 8742.

549. de Lange, B.; Lambers-Verstappen, M. H.; van de Vondervoort, L. S.; Sereinig, N.; de Rijk, R.; de Vries, A. H. M.; de Vries, J. G. *Synlett* **2006**, 3105.

550. Goldberg, I. *Ber. Dtsch. Chem. Ges.* **1906**, *39*, 1691.

551. Klapars, A.; Antilla, J. C.; Huang, X. H.; Buchwald, S. L. *J. Am. Chem. Soc.* **2001**, *123*, 7727.

552. Phillips, D. P.; Hudson, A. R.; Nguyen, B.; Lau, T. L.; McNeill, M. H.; Dalgard, J. E.; Chen, J. H.; Penuliar, R. J.; Miller, T. A.; Zhi, L. *Tetrahedron Lett.* **2006**, *47*, 7137.

553. Mallesham, B.; Rajesh, B. M.; Reddy, P. R.; Srinivas, D.; Trehan, S. *Org. Lett.* **2003**, *5*, 963.

554. Deng, W.; Wang, Y. F.; Zou, W.; Liu, L.; Guo, Q. X. *Tetrahedron Lett.* **2004**, *45*, 2311.

555. Taillefer, M.; Cristau, H. J.; Cellier, P. P.; Spindler, J. F.; Rhone Poulenc Chimie (FR), EP20030756038, France, **2005**.

556. Cristau, H. J.; Cellier, P. P.; Spindler, J. F.; Taillefer, M. *Chem.—Eur. J.* **2004**, *10*, 5607.

557. Lv, X.; Bao, W. L. *J. Org. Chem.* **2007**, *72*, 3863.

558. Wolter, M.; Klapars, A.; Buchwald, S. L. *Org. Lett.* **2001**, *3*, 3803.

559. Antilla, J. C.; Baskin, J. M.; Barder, T. E.; Buchwald, S. L. *J. Org. Chem.* **2004**, *69*, 5578.

560. Antilla, J. C.; Klapars, A.; Buchwald, S. L. *J. Am. Chem. Soc.* **2002**, *124*, 11684.

561. Cristau, H.-J.; Cellier, P. P.; Spindler, J.-F.; Taillefer, M. *Chem.—Eur. J.* **2004**, *10*, 5607.

562. Cristau, H.-J.; Cellier, P. P.; Spindler, J.-F.; Taillefer, M. *Eur. J. Org. Chem.* **2004**, 695.

563. Kull, M.; Bekedam, L.; Visser, G. M.; van den Hoogenband, A.; Terpstra, J. W.; Kamer, P. C. J.; van Leeuwen, P. W. N. M.; van Strijdonck, G. P. F. *Tetrahedron Lett.* **2005**, *46*, 2405.

564. Liu, L.; Frohn, M.; Xi, N.; Donminguez, C.; Hungate, R.; Reider, P. J. *J. Org. Chem.* **2005**, *70*, 10135.

565. Zhang, Z.; Mao, J.; Zhu, D.; Wu, F.; Chen, H.; Wan, B. *Tetrahedron* **2006**, *62*.

566. Alcalde, E.; Dinares, I.; Rodriguez, S.; Garcia de Miguel, C. *Eur. J. Org. Chem.* **2005**, 1637.

567. Jerphagnon, T.; van Klink, G. P. M.; de Vries, J. G.; van Koten, G. *Org. Lett.* **2005**, *7*, 5241.

568. Xu, L.; Zhu, D.; Wang, R.; Wan, B. *Tetrahedron* **2005**, *62*, 4435.

569. Hosseinzadeh, R.; Tajbakhsh, M.; Alikarami, M. *Tetrahedron Lett.* **2006**, *47*, 5203.

570. Kantam, M. L.; Venkanna, G. T.; Sridhar, C.; Kumar, K. B. *Tetrahedron Lett.* **2006**, *47*, 3897.

571. Xie, Y. X.; Pi, S. F.; Yin, D. L.; Li, J. H. *J. Org. Chem.* **2006**, *71*, 8324.

572. Yang, M.; Liu, F. *J. Org. Chem.* **2007**, *72*, 8969.

573. Arkaitz Correa, C. B. *Adv. Synth. Catal.* **2007**, *349*, 2673.

574. Altman, R. A.; Koval, E. D.; Buchwald, S. L. *J. Org. Chem.* **2007**, *72*, 6190.

575. Altman, R. A.; Buchwald, S. L. *Org. Lett.* **2006**, *8*, 2779.

576. Choudary, B. M.; Sridhar, C.; Kantam, M. L.; Venkanna, G. T.; Sreedhar, B. *J. Am. Chem. Soc.* **2005**, *127*, 9948.

577. Ley, S.V.; Thomas, A.W. *Angew. Chem. Int. Ed.* **2003**, *42*, 5400.

578. Lindley, J. *Tetrahedron* **1984**, *40*, 1433.

579. Evans, D. A.; Ellman, J. A. *J. Am. Chem. Soc.* **1989**, *111*, 1063.

580. Boger, D. L.; Yohannes, D. *J. Org. Chem.* **1990**, *55*, 6000.

581. Boger, D. L.; Patane, M. A.; Zhou, J. *J. Am. Chem. Soc.* **1994**, *116*, 8554.

582. Nicolaou, K. C.; Boddy, C. N. C.; Natarajan, S.; Yue, T. Y.; Li, H.; Brase, S.; Ramanjulu, J. M. *J. Am. Chem. Soc.* **1997**, *119*, 3421.

583. Pellon, R. F.; Carrasco, R.; Milian, V.; Rodes, L. *Synth. Commun.* **1995**, *25*, 1077.

584. Marcoux, J. F.; Doye, S.; Buchwald, S. L. *J. Am. Chem. Soc.* **1997**, *119*, 10539.

585. Kalinin, A. V.; Bower, J. F.; Riebel, P.; Snieckus, V. *J. Org. Chem.* **1999**, *64*, 2986.

586. Bates, C. G.; Gujadhur, R. K.; Venkataraman, D. *Org. Lett.* **2002**, *4*, 2803.

587. Gujadhur, R. K.; Venkataraman, D. *Synth. Commun.* **2001**, *31*, 2865.

588. Gujadhur, R. K.; Bates, C. G.; Venkataraman, D. *Org. Lett.* **2001**, *3*, 4315.

589. Schareina, T.; Zapf, A.; Cotté, A.; Müller, N.; Beller, M. *Tetrahedron Lett.* **2008**, *49*, 1851.

590. Buck, E.; Song, Z. J.; Tschaen, D.; Dormer, P. G.; Volante, R. P.; Reider, P. J. *Org. Lett.* **2002**, *4*, 1623.

591. Chen, Y. J.; Chen, H. H. *Org. Lett.* **2006**, *8*, 5609.

592. Wolter, M.; Nordmann, G.; Job, G. E.; Buchwald, S. L. *Org. Lett.* **2002**, *4*, 973.

593. Hosseinzadeh, R.; Tajbakhsh, M.; Mohadjerani, M.; Alikarami, M. *Synlett* **2005**, 1101.

594. Zhang, H.; Ma, D.; Cao, W. *Synlett* **2007**, 243.

595. Altman, R. A.; Shafir, A.; Choi, A.; Lichtor, P. A.; Buchwald, S. L. *J. Org. Chem.* **2008**, *73*, 284.

596. Nordmann, G.; Buchwald, S. L. *J. Am. Chem. Soc.* **2003**, *125*, 4978.

597. Shafir, A.; Lichtor, P. A.; Buchwald, S. L. *J. Am. Chem. Soc.* **2007**, *129*, 3490.

598. Palomo, C.; Oiabide, R.; Lopez, R.; Gomez-Bengoa, E. *Tetrahedron Lett.* **2000**, *41*, 1283.

599. Kwong, F. Y.; Buchwald, S. L. *Org. Lett.* **2002**, *4*, 3517.

600. Carril, M.; SanMartin, R.; Dominguez, E.; Tellitu, I. *Chem. Eur. J.* **2007**, *13*, 5100.

601. Verma, A. K.; Singh, J.; Chaudhary, R. *Tetrahedron Lett.* **2007**, *48*, 7199.

602. Zhang, H.; Cao, W.; Ma, D. *Synth. Commun.* **2007**, *37*, 25.

603. Sperotto, E.; van Klink, G. P. M.; de Vries, J. G.; van Koten, G. *J. Org. Chem.* **2008**, *73*, 5625.

604. Ranu, B. C.; Saha, A.; Ranjan, J. *Adv. Synth. Catal.* **2007**, *349*, 639.

605. Paine, A. J. *J. Am. Chem. Soc.* **1987**, *109*, 1496.

606. Whitesides, G. M.; Sadowski, J. S.; Lilburn, J. *J. Am. Chem. Soc.* **1974**, *96*, 2829.

607. Aalten, H. L.; Vankoten, G.; Grove, D. M.; Kuilman, T.; Piekstra, O. G.; Hulshof, L. A.; Sheldon, R. A. *Tetrahedron* **1989**, *45*, 5565.

608. Cohen, T.; Wood, J.; Dietz, A. G. *Tetrahedron Lett.* **1974**, 3555.

609. Arai, S.; Hida, M.; Yamagishi, T. *Bull. Chem. Soc. Jpn.* **1978**, *51*, 277.

610. Paine, A. J. *J. Am. Chem. Soc.* **1987**, *109*, 1496.

611. Couture, C.; Paine, A. J. *Can. J. Chem.* **1985**, *63*, 111.

612. Tye, J. W.; Weng, Z.; Johns, A. M.; Incarvito, C. D.; Hartwig, J. F. *J. Am. Chem. Soc.* **2008**, *130*, 9971.

613. Strieter, E. R.; Blackmond, D. G.; Buchwald, S. L. *J. Am. Chem. Soc.* **2005**, *127*, 4120.

614. Chan, D. M. T.; Monaco, K. L.; Wang, R.-P.; Winters, M. P. *Tetrahedron Lett.* **1998**, *39*, 2933.

615. Evans, D. A.; Katz, J. L.; West, T. R. *Tetrahedron Lett.* **1998**, *39*, 2937.

616. Lam, P. Y. S.; Clark, C. G.; Saubern, S.; Adams, J.; Winters, M. P.; Chan, D. M. T.; Combs, A. *Tetrahedron Lett.* **1998**, *39*, 2941.

617. Cundy, D. J.; Forsyth, S. A. *Tetrahedron Lett.* **1998**, *39*, 7979.

618. Lam, P. Y. S.; Bonne, D.; Vincent, G.; Clark, C. G.; Combs, A. P. *Tetrahedron Lett.* **2003**, *44*, 1691.

619. Combs, A. P.; Tadesse, S.; Rafalski, M.; Haque, T. S.; Lam, P. Y. S. *J. Comb. Chem* **2002**, *4*, 179.

620. Das, P.; Basu, B. *Synth. Commun.* **2004**, *34*, 2177.

621. Collman, J. P.; Zhong, M. *Org. Lett.* **2000**, *2*, 1233.

622. Lam, P. Y. S.; Vincent, G.; Clark, C. G.; Deudon, S.; Jadhav, P. K. *Tetrahedron Lett.* **2001**, *42*, 3415.

623. Antilla, J. C.; Buchwald, S. L. *Org. Lett.* **2001**, *3*, 2077.

624. Lam, P. Y. S.; Vincent, G.; Clark, C. G.; Deudon, S.; Jadhav, P. K. *Tetrahedron Lett.* **2001**, *42*, 3415.

625. Quach, T. D.; Batey, R. A. *Org. Lett.* **2003**, *5*, 4397.

626. Sasaki, M.; Dalili, S.; Yudin, A. K. *J. Org. Chem.* **2003**, *68*, 2045.

627. Lan, J.-B.; Zhang, G.-L.; Yu, X.-Q.; You, J.-S.; Chen, L.; Yan, M.; Xie, R.-G. *Synlett* **2004**, 1095.

628. Tzschucke, C. C.; Murphy, J. M.; Hartwig, J. F. *Org. Lett.* **2007**, *9*, 761.

629. Kantam, M. L.; Venkanna, G. T.; Sridhar, C.; Sreedhar, B.; Choudary, B. M. *J. Org. Chem.* **2006**, *71*, 9522.

630. Chiang, G. C. H.; Olsson, T. *Org. Lett.* **2004**, *6*, 3079.

631. Herradura, P. S.; Pendola, K. A.; Guy, R. K. *Org. Lett.* **2000**, *2*, 2019.

632. Savarin, C.; Srogl, J.; Liebeskind, L. S. *Org. Lett.* **2002**, *4*, 4309.

633. Tamura, M.; Kochi, J. *Synthesis* **1971**, 303.

634. Lipshutz, B. H. *Acc. Chem. Res.* **1997**, *30*, 277.

635. Johnson, D. K.; Cravarri, J. P.; Faoud, T. I.; Schillinger, K. J.; van Geel, T. A. P.; Stratton, S. M. *Tetrahedron Lett.* **1995**, *36*, 8565.

636. Beletskaya, I. P.; Cheprakov, A. V. *Coord. Chem. Rev.* **2004**, *248*, 2337.

637. Terao, J.; Todo, H.; Begum, S. A.; Kuniyasu, H.; Kambe, N. *Angew. Chem. Int. Ed.* **2007**, *46*, 2086

638. Lipshutz, B. H.; Sengupta, S. *Org. React.* **1992**, *41*, 149.

639. Bertz, S. H. *Tetrahedron Lett.* **1980**, *21*, 3151.

640. Bajgrowicz, J. A.; Elhallaoui, A.; Jacquier, R.; Pigiere, C.; Viallefont, P. *Tetrahedron* **1985**, *41*, 1833.

641. Knochel, P.; Dohle, W.; Gommermann, N.; Kneisel, F. F.; Kopp, F.; Korn, T.; Sapountzis, I.; Vu, V. A. *Angew. Chem. Int. Ed.* **2003**, *42*, 4302.

642. Yang, X.; Knochel, P. *Synlett* **2004**, 81.

643. Yang, X.; Althammer, A.; Knochel, P. *Org. Lett.* **2004**, *6*, 1665.

644. Fleming, F. F.; Zhang, Z.; Liu, W.; Knochel, P. *J. Org. Chem.* **2005**, *70*, 2200.

645. Stoll, A. H.; Krasovskiy, A.; Knochel, P. *Angew. Chem. Int. Ed.* **2006**, *45*, 606.

646. Beletskaya, I. P.; Cheprakov, A. V. *Coord. Chem. Rev.* **2004**, *248*, 2337.

647. Vyvyan, J. R.; Holst, C. L.; Johnson, A. J.; Schwenk, C. M. *J. Org. Chem.* **2002**, *67*, 2263.

648. Nunomoto, S.; Kawakami, Y.; Yamashita, Y. *Bull. Chem. Soc. Jpn.* **1981**, *54*, 2831.

649. Nivlet, A.; Dechoux, L.; Martel, J. P.; Proess, G.; Mannes, D.; Alcaraz, L.; Harrett, J. J.; Le Gall, T.; Mioskowski, C. *Eur. J. Org. Chem.* **1999**, 3241.

650. Fleming, F. F.; Jiang, T. *J. Org. Chem.* **1997**, *62*, 7890.

651. de Lang, R. J.; van Hooijdonk, M.; Brandsma, H.; Kramer, H.; Seinen, W. *Tetrahedron* **1998**, *54*, 2953.

652. Nishimura, J.; Yamada, N.; Horiuchi, Y.; Ueda, A.; Ohbayashi, A.; Oku, A. *Bull. Chem. Soc. Jpn.* **1986**, *59*, 2035.

653. Zhang, H. Y.; Blasko, A.; Yu, J. Q.; Bruice, T. C. *J. Am. Chem. Soc.* **1992**, *114*, 6621.

654. Burns, D. H.; Miller, J. D.; Chan, H. K.; Delaney, M. O. *J. Am. Chem. Soc.* **1997**, *119*, 2125.

655. Burns, D. H.; Chan, H. K.; Miller, J. D.; Jayne, C. L.; Eichhorn, D. M. *J. Org. Chem* **2000**, *65*, 5185.

656. Moreira, J. A.; Correa, A. G. *Tetrahedron: Asymmetry* **2003**, *14*, 3787.

657. Terao, J.; Ikumi, A.; Kuniyasu, H.; Kambe, N. *J. Am. Chem. Soc.* **2003**, *125*, 5646.

658. Terao, J.; Todo, H.; Begum, A. A.; Kuniyasu, H.; Kambe, N. *Angew. Chem. Int. Ed.* **2007**, *46*, 2086.

659. Cahiez, G.; Chaboche, C.; Jezequel, M. *Tetrahedron* **2000**, *56*, 2733.

660. Gschwind, R. M. *Chem. Rev.* **2008**, *108*, 3029.

661. Gartner, T.; Henze, W.; Gschwind, R. M. *J. Am. Chem. Soc.* **2007**, *129*, 11362.

662. Bartholomew, E. R.; Bertz, S. H.; Cope, S.; Dorton, D. C.; Murphy, M.; Ogle, C. A. *Chem. Commun.* **2008**, 1176.

663. Bertz, S. H.; Cope, S.; Dorton, D.; Murphy, M.; Ogle, C. A. *Angew. Chem. Int. Ed.* **2007**, *46*, 7082.

664. Mori, S.; Nakamura, E.; Morokuma, K. *J. Am. Chem. Soc.* **2000**, *122*, 7294.

665. Nakamura, E.; Mori, S.; Morokuma, K. *J. Am. Chem. Soc.* **1998**, *120*, 8273.

666. Lipshutz, B. H.; Wilhelm, R. S. *J. Am. Chem. Soc.* **1982**, *104*, 4696.

667. Hebert, E. *Tetrahedron Lett.* **1982**, *23*, 415.

668. Ashby, E. C.; Coleman, D. *J. Org. Chem.* **1987**, *52*, 4554.

669. Ashby, E. C.; Depriest, R. N.; Tuncay, A.; Srivastava, S. *Tetrahedron Lett.* **1982**, *23*, 5251.

670. Bruggink, A.; McKillop, A. *Tetrahedron* **1975**, *31*, 2607.

671. Quallich, G. J.; Makowski, T. W.; Sanders, A. F.; Urban, F. J.; Vasquez, E. *J. Org. Chem.* **1993**, *63*, 4116.

672. Ames, D. E.; Ribeiro, J. *J. Chem. Soc., Perkin Trans. 1* **1975**, 1390.

673. Okuro, K.; Furuune, M.; Miura, M.; Nomura, M. *J. Org. Chem.* **1993**, *58*, 7606.

674. Hennessy, E. J.; Buchwald, S. L. *Org. Lett.* **2002**, *4*, 269.

675. Xie, X. A.; Cai, G. R.; Ma, D. W. *Org. Lett.* **2005**, *7*, 4693.

676. Yip, S. F.; Cheung, H. Y.; Zhou, Z. Y.; Kwong, F. Y. *Org. Lett.* **2007**, *9*, 3469.

677. Xie, X.; Chen, Y.; Ma, D. *J. Am. Chem. Soc.* **2006**, *128*, 16050.

678. Piers, E.; Wong, J. *J. Org. Chem.* **1993**, *58*, 3609.

679. Takeda, T.; Matsunaga, K. I.; Kabasawa, Y.; Fujiwara, T. *Chem. Lett.* **1995**, 771.

680. Nudelman, N. S.; Carro, C. *Synlett* **1999**, 1942.

681. Tanaka, H.; Sumida, S.; Torii, S. *Tetrahedron Lett.* **1996**, *37*, 5967.

682. Falck, J. R.; Bhatt, R. K.; Ye, J. H. *J. Am. Chem. Soc.* **1995**, *117*, 5973.

683. Kang, S. K.; Kim, J. S.; Choi, S. C. *J. Org. Chem.* **1997**, *62*, 4208.

684. Marion, N.; Navarro, O.; Mei, J.; Stevens, E. D.; Scott, N. M.; Nolan, S. P. *J. Am. Chem. Soc.* **2006**, *128*, 4101.

685. Allred, G. D.; Liebeskind, L. S. *J. Am. Chem. Soc.* **1996**, *118*, 2748.

686. Savall, B. M.; Blanchard, N.; Roush, W. R. *Org. Lett.* **2003**, *5*, 377.

687. Armstrong, A.; Blench, T. J. *Tetrahedron* **2002**, *58*, 9321.

688. Dymock, B. W.; Kocienski, P. J.; Pons, J. M. *Synthesis* **1998**, 1655.

689. Maleczka, R. E.; Terrell, L. R.; Geng, F.; Ward, J. S. *Org. Lett.* **2002**, *4*, 2841.

690. Schuppan, J.; Wehlan, H.; Keiper, S.; Koert, U. *Angew. Chem. Int. Ed.* **2001**, *40*, 2063.

691. Thathagar, M. B.; Beckers, J.; Rothenberg, G. *J. Am. Chem. Soc.* **2002**, *124*, 11858.

692. Biscoe, M. R.; Barder, T. E.; Buchwald, S. L. *Angew. Chem. Int. Ed.* **2007**, *46*, 7232.

693. Alonso, F.; Beletskaya, I. P.; Yus, M. *Tetrahedron* **2008**, *64*, 3047.

694. Mao, J.; Guo, J.; Fang, F.; Ji, S.-J. *Tetrahedron* **2008**, *64*, 3905.

695. Alberico, D.; Scott, M. E.; Lautens, M. *Chem. Rev.* **2007**, *107*, 174.

696. Seregin, I. V.; Gevorgyan, V. *Chem. Soc. Rev.* **2007**, *36*, 1173.

697. *Handbook of C–H Transformations*; Dyker, G., Ed.; Wiley-VCH: Weinheim, 2005; Vols. 1 and 2.

698. Campeau, L.-C.; Stuart, D. R.; Fagnou, K. *Aldrichimica Acta* **2007**, *40*, 35.

699. Catellani, M.; Chiusoli, G. P. *J. Organomet. Chem.* **1992**, *425*, 151.

700. Martín-Matute, B.; Mateo, C.; Cárdenas, D. J.; Echavarren, A. M. *Chem.—Eur. J.* **2001**, *7*, 2341.

701. Hughes, C. C.; Trauner, D. *Angew. Chem. Int. Ed.* **2002**, *41*, 1569.

702. Hennessy, E. J.; Buchwald, S. L. *J. Am. Chem. Soc.* **2003**, *125*, 12084.

703. Glover, B.; Harvey, K. A.; Liu, B.; Sharp, M. J.; Tymoschenko, M. F. *Org. Lett.* **2003**, *5*, 301.

704. Park, C.-H.; Ryabova, V.; Seregin, I. V.; Sromek, A. W.; Gevorgyan, V. *Org. Lett.* **2004**, *6*, 1159.

705. Gómez-Lor, B.; Echavarren, A. M. *Org. Lett.* **2004**, *6*, 2993.

706. Lane, B. S.; Brown, M. A.; Sames, D. *J. Am. Chem. Soc.* **2005**, *127*, 8050.

707. Zollinger, H. *Adv. Phys. Org. Chem.* **1964**, *2*, 162.

708. Mota, A. J.; Dedieu, A.; Bour, C.; Suffert, J. *J. Am. Chem. Soc.* **2005**, *127*, 7171.

709. Davies, D. L.; Donald, S. A.; Macgregor, S. A. *J. Am. Chem. Soc.* **2005**, *127*, 13754.

710. Gorelsky, S. I.; Lapointe, D.; Fagnou, K. *J. Am. Chem. Soc.* **2008**, *130*, 10848.

711. Garcia-Cuadrado, D.; de Mendoza, P.; Braga, A. A. C.; Maseras, F.; Echavarren, A. M. *J. Am. Chem. Soc.* **2007**, *129*, 6880.

712. Toyota, M.; Ilangovan, A.; Okamoto, R.; Masaki, T.; Arakawa, M.; Ihara, M. *Org. Lett.* **2002**, *4*, 4293.

713. (a) Campo, M. A.; Huang, Q.; Yao, T.; Tian, Q.; Larock, R. C. *J. Am. Chem. Soc.* **2003**, *125*, 11506; (b) Capito, E.; Brown, J. M.; Ricci, A. *Chem. Commun.* **2005**, 1854.

714. Computational studies indicate that C–H insertion is higher in energy and less favorable than an alternative σ-bond metathesis which does not involve the formation of a palladium(IV) intermediate (reference 708).

715. Pivsa-Art, S.; Satoh, T.; Kawamura, Y.; Miura, M.; Nomura, M. *Bull. Chem. Soc. Jpn.* **1998**, *71*, 467.

716. Campeau, L.-C.; Parisien, M.; Fagnou, K. *J. Am. Chem. Soc.* **2004**, *126*, 9186.

717. Daugulis, O.; Zaitsev, V. G. *Angew. Chem. Int. Ed.* **2005**, *44*, 4046.

718. Lafrance, M.; Rowley, C. N.; Woo, T. K.; Fagnou, K. *J. Am. Chem. Soc.* **2006**, *128*, 8754.

719. Li, W.; Nelson, D. P.; Jensen, M. S.; Hoerrner, R. S.; Javadi, G. J.; Cai, D.; Larsen, R. D. *Org. Lett.* **2003**, *5*, 4835.

720. Kalyani, D.; Deprez, N. R.; Desai, L. V.; Sanford, M. S. *J. Am. Chem. Soc.* **2005**, *127*, 7330.

721. Wolfe, J. P.; Buchwald, S. L. *Angew. Chem. Int. Ed.* **1999**, *38*, 2413.

722. Wolfe, J. P.; Buchwald, S. L. *Angew. Chem. Int. Ed.* **1999**, *38*, 3415.

723. Wolfe, J. P.; Singer, R. A.; Yang, B. H.; Buchwald, S. L. *J. Am. Chem. Soc.* **1999**, *121*, 9550.

724. Littke, A. F.; Fu, G. C. *Angew. Chem. Int. Ed.* **2002**, *41*, 4176.

725. Hernández, S.; SanMartin, R.; Tellitu, I.; Domínguez, E. *Org. Lett.* **2003**, *5*, 1095.

726. Dyker, G. *Angew. Chem., Int. Ed. Engl.* **1994**, *33*, 103.

727. Oi, S.; Fukita, S.; Hirata, N.; Watanuki, N.; Miyano, S.; Inoue, Y. *Org. Lett.* **2001**, *3*, 2579.

728. Kakiuchi, F.; Kan, S.; Igi, K.; Chatani, N.; Murai, S. *J. Am. Chem. Soc.* **2003**, *125*, 1698.

729. Kakiuchi, F.; Matsuura, Y.; Kan, S.; Chatani, N. *J. Am. Chem. Soc.* **2005**, *127*, 5936.

730. Touré, B. B.; Lane, B. S.; Sames, D. *Org. Lett.* **2006**, *8*, 1979.

731. Wang, X.; Lane, B. S.; Sames, D. *J. Am. Chem. Soc.* **2005**, *127*, 4996.

732. Lewis, J. C.; Wiedemann, S. H.; Bergman, R. G.; Ellman, J. A. *Org. Lett.* **2004**, *6*, 35.

733. Ueura, K.; Satoh, T.; Miura, M. *Org. Lett.* **2005**, *7*, 2229.

734. Bedford, R. B.; Limmert, M. E. *J. Org. Chem.* **2003**, *68*, 8669.

735. Oi, S.; Watanabe, S.-I.; Fukita, S.; Inoue, Y. *Tetrahedron Lett.* **2003**, *44*, 8665.

736. Miyamura, S.; Tsurugi, H.; Satoh, T.; Miura, M. *J. Organomet. Chem.* **2008**, *693*, 2438.

737. Oi, S.; Fukita, S.; Inoue, Y. *Chem. Commun.* **1998**, 2439.

738. Norinder, J.; Matsumoto, A.; Yoshikai, N.; Nakamura, E. *J. Am. Chem. Soc.* **2008**, *130*, 5858.

739. Do, H.-Q.; Daugulis, O. *J. Am. Chem. Soc.* **2007**, *129*, 12404.

740. Phipps, R. J.; Grimster, N. P.; Gaunt, M. J. *J. Am. Chem. Soc.* **2008**, *130*, 8172.

741. Bringmann, G.; Holenz, J.; Weirich, R.; Rübenacker, M.; Funke, C.; Boyd, M. R.; Gulakowski, R. J.; François, G. *Tetrahedron* **1998**, *54*, 497.

742. Terao, Y.; Kametani, Y.; Wakui, H.; Satoh, T.; Miura, M.; Nomura, M. *Tetrahedron* **2001**, *57*, 5967.

743. Bellina, F.; Cauteruccio, S.; Mannina, L.; Rossi, R.; Viel, S. *J. Org. Chem.* **2005**, *70*, 3997.

744. Bellina, F.; Cauteruccio, S.; Mannina, L.; Rossi, R.; Viel, S. *Eur. J. Org. Chem.* **2006**, 693.

745. Akita, Y.; Itagaki, Y.; Takizawa, S.; Ohta, A. *Chem. Pharm. Bull.* **1989**, *37*, 1477.

746. Catellani, M. *Synlett* **2003**, 298.

747. Catellani, M. *Top. Organomet. Chem.* **2005**, *14*, 21.

748. Albrecht, K.; Reiser, O.; Weber, M.; Knieriem, B.; de Meijere, A. *Tetrahedron* **1994**, *50*, 383.

749. Reiser, O.; Weber, M.; de Meijere, A. *Angew. Chem., Int. Ed. Engl.* **1989**, *28*, 1037.

750. Dongol, K. G.; Matsubara, K.; Mataka, S.; Thiemann, T. *Chem. Commun.* **2002**, 3060.

751. Catellani, M.; Motti, E.; Della Ca', N. *Acc. Chem. Res.* **2008**, *41*, 1512.

752. Cárdenas, D. J.; Martín-Matute, B.; Echavarren, A. M. *J. Am. Chem. Soc.* **2006**, *128*, 5033.

753. Mauleón, P.; Núñez, A. A.; Alonso, I.; Carretero, J. C. *Chem.—Eur. J.* **2003**, *9*, 1511.

754. Larock, R. C.; Tian, Q. *J. Org. Chem.* **2001**, *66*, 7372.

755. Mariampillai, B.; Alliot, J.; Li, M.; Lautens, M. *J. Am. Chem. Soc.* **2007**, *129*, 15372.

756. Bellina, F.; Cauteruccio, S.; Rossi, R. *Eur. J. Org. Chem.* **2006**, 1379.

757. Campeau, L.-C.; Parisien, M.; Jean, A.; Fagnou, K. *J. Am. Chem. Soc.* **2006**, *128*, 581.

758. García-Cuadrado, D.; Braga, A. A. C.; Maseras, F.; Echavarren, A. M. *J. Am. Chem. Soc.* **2006**, *128*, 1066.

759. Bringmann, G.; Price Mortimer, A. J.; Keller, P. A.; Gresser, M. J.; Garner, J.; Breuning, M. *Angew. Chem. Int. Ed.* **2005**, *44*, 5384.

760. Takahashi, M.; Masui, K.; Sekiguchi, H.; Kobayashi, N.; Mori, A.; Funahashi, M.; Tamaoki, N. *J. Am. Chem. Soc.* **2006**, *128*, 10930.

761. Stuart, D. S.; Villemure, E.; Fagnou, K. *J. Am. Chem. Soc.* **2007**, *129*, 12072.

762. Stuart, D. S.; Fagnou, K. *Science* **2007**, *316*.

763. Dwight, T. A.; Rue, N. R.; Charyk, D.; Josselyn, R.; DeBoef, B. *Org. Lett.* **2007**, *9*, 3137.

764. Hull, K. L.; Sanford, M. S. *J. Am. Chem. Soc.* **2007**, *129*, 11904.

CHAPTER 20

Allylic Substitution

20.1. Overview

Catalytic reactions of allylic electrophiles with carbon or heteroatom nucleophiles to form the products of formal S_N2 or S_N2' substitutions (Equation 20.1) are called "catalytic allylic substitution reactions." These reactions have become classic processes catalyzed by transition metal complexes and are often conducted in an asymmetric fashion. The allylic electrophile is typically an allylic chloride, acetate, carbonate, or other type of ester derived from an allylic alcohol. The nucleophile is most commonly a so-called soft nucleophile, such as the anion of a β-dicarbonyl compound, or it is a heteroatom nucleophile, such as an amine or the anion of an imide. The reactions with carbon nucleophiles are often called allylic alkylations.

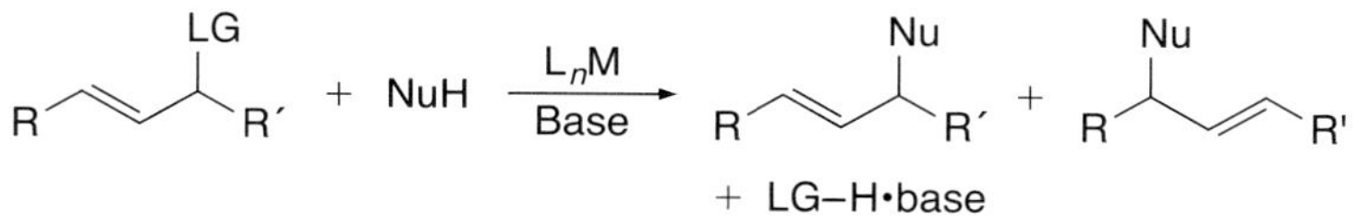

$$\text{(20.1)}$$

The mechanism most often invoked for these reactions (Scheme 20.1) involves oxidative addition of the allylic electrophile to a low-valent metal center to generate an η^3-allylmetal intermediate. In most catalytic allylic substitutions, the acetate, halide, or carbonate leaving group is a counterion for a cationic allyl complex and is not directly bound to the metal. The resulting cationic η^3-allylmetal intermediate then reacts with the nucleophile, in most cases, by attack on the opposite face of the allyl group to which the metal is bound.

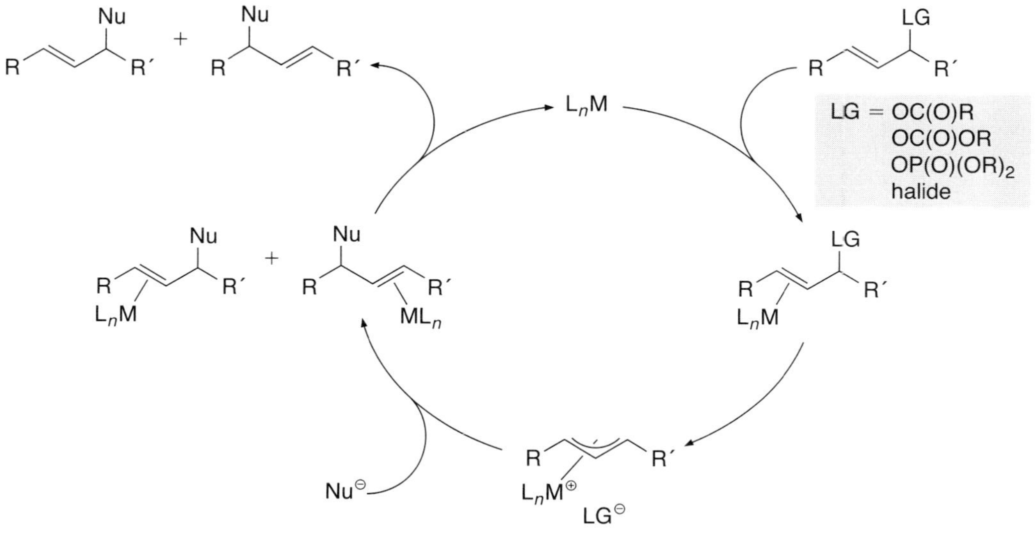

Scheme 20.1

967

Much effort has been devoted to developing catalysts that control the enantioselectivity of these substitution reactions, as well as the regioselectivity of reactions that proceed through unsymmetrical allylic intermediates. A majority of this effort has been spent on developing palladium complexes as catalysts. Increasingly, however, complexes of molybdenum, tungsten, ruthenium, rhodium, and iridium have been studied as catalysts for enantioselective and regioselective processes. In parallel with these studies of allylic substitution catalyzed by complexes of transition metals, studies on allylic substitution catalyzed by complexes of copper have been conducted. These reactions often occur to form products of S_N2' substitution. As catalytic allylic substitution has been developed, this process has been applied in many different ways to the synthesis of natural products.[1]

20.2. Early Developments Toward Enantioselective Allylic Substitution

20.2.1. Stoichiometric Attack on Palladium Allyl Complexes

In the early 1970s Tsuji demonstrated that palladium–allyl complexes bound in an η^3 fashion are electrophilic and undergo attack by stabilized carbon nucleophiles to generate products containing a new carbon–carbon bond (Equation 20.2).[2] This reactivity contrasts with that of most main group allyl compounds, such as allylMgBr, which are nucleophilic. The η^3 bonding mode leads to a net flow of electrons from the ligand to the metal and creates an electrophilic ligand. A positive charge on the complex further increases the electrophilicity of the allyl ligand. Therefore, most η^3-allyl complexes that react readily with nucleophiles are cationic.

$$\frac{1}{2}\left[\left(\text{–PdCl}\right)\right]_2 + CH_2(CO_2Et)_2 \xrightarrow[\text{DMSO}]{\text{NaH}} \diagup\diagdown CH(CO_2Et)_2 + Pd(0) + NaCl \qquad (20.2)$$

20.2.2. The First Catalytic Allylic Substitutions

After these initial results by Tsuji, this elementary step was incorporated into a catalytic process by Hata and co-workers at Toray Industries and by Atkins and co-workers at Union Carbide. These groups reported reactions of allylic phenyl ethers, allylic alcohols, and allylic acetates with carboxylates, alcohols, primary and secondary amines, and methyl acetoacetate catalyzed by Pd(0) complexes and precursors to Pd(0) complexes (Equation 20.3).[3,4] After these initial reports, early developments focused on reactions of "soft" carbanions derived from β-dicarbonyl compounds, cyanoesters, and related compounds containing two electron-withdrawing groups attached to the nucleophilic carbon. Although these reactions occur with allylic halides in the absence of a catalyst, these reactions are greatly accelerated by palladium catalysts. Thus, the palladium catalyst allows these reactions to occur under mild conditions with allylic acetates, which are more accessible than allylic halides, and with selectivities that are altered by the metal catalyst.

$$R\diagup\diagdown\diagup X + H–Nu \xrightarrow{L_nPd} R\diagup\diagdown\diagup Nu$$

X = OH, OAc, or OPh
Nu = OAr, NEt$_2$, or MeC(O)CHCO$_2$Et   (20.3)
L$_n$Pd = Pd(PPh$_3$)$_4$, or Pd(acac)$_2$/PPh$_3$

With this ability to control selectivity, palladium-catalyzed substitution chemistry was developed into an enantioselective process. In the early enantioselective reactions, a racemic or achiral allylic electrophile underwent substitution with a soft carbon nucleophile to generate a product containing a new stereocenter at one of the original allylic carbons. Trost and his co-workers were the first to report enantioselective allylic substitutions, in this case as a stoichiometric reaction (Equation 20.4).[5] Trost and co-workers have also developed some of the most useful catalysts for enantioselective allylic alkylation[6] and have reported many of the applications of this reaction to natural-products synthesis.[7] Palladium-catalyzed allylic substitution reactions are now often called "Tsuji–Trost reactions."[8–10]

$$\text{(20.4)}$$

20.2.3. The First Catalysts for Allylic Substitutions

Allylic substitution was initially conducted with palladium complexes lacking any added dative ligands or with palladium complexes containing classic aromatic phosphine ligands. More recently, catalyst development for this reaction has focused on the design and synthesis of new ligands for enantioselective allylic substitution processes. These ligands were presented as part of Section 18.6 on enantioselective allylic substitution.

Most of the early allylic substitution reactions[11] were conducted with precursors to palladium(0), such as [Pd(allyl)Cl]$_2$ or with palladium(0) complexes containing relatively labile ligands, such as [Pd$_2$(dba)$_3$] (in which dba = dibenzylideneacetone). Other reactions were catalyzed by complexes of phosphine ligands, such as PPh$_3$ or dppe [1,2-bis(diphenylphosphino)ethane].[12–18] The different rates of reaction with allylic acetates and carbonates (see below) imply that oxidative addition can be the turnover-limiting step, at least with some catalysts. However, the step that forms the new bond in the product involves nucleophilic attack on an electrophilic allyl complex, and this step may be turnover limiting with certain catalysts. Therefore, allylic substitutions must be catalyzed by complexes containing ligands that create a sufficiently electron-poor allyl complex for nucleophilic attack to occur. For this reason, complexes of alkylphosphines are rarely used for this reaction, although they have been effective for reactions of aryl halides presented in Chapter 19.

20.3. Substrate Scope and Catalysts

20.3.1. Scope of Electrophile

Most allylic substitution reactions have been conducted with derivatives of allylic alcohols, such as acetates, phosphates, and carbonates. These reactions occur with allyl electrophiles displaying a wide structural variation. The allylic electrophiles can be cyclic or acyclic, substituted with aliphatic or aromatic groups, substituted at one or both termini, and substituted or unsubstituted at the central carbon. As discussed in more detail below, these substituents affect the regioselectivity of the substitution process.

The relative reactivity of the derivatives follows the trend allyl carbonate > allyl phosphate > allyl acetate (see Equations 20.5–20.7).[11,19] This difference in reactivity allows for chemoselective substitutions of one allylic alcohol derivative over another. Other allylic electrophiles, such as allylic sulfonates, which undergo cleavage of the carbon–sulfur bond,[20–22] allylic nitro compounds, which undergo cleavage of the C–N bond,[23,24] and allylic amines,[25–29] which undergo carbon–nitrogen bond cleavage after protonation of the nitrogen, can be used to generate the allyl intermediates. Allylic ethers[30] can also be used in certain cases, and allylic acetals or geminal diacetates (Equations 20.8 and 20.9)[31,32] can be particularly useful.

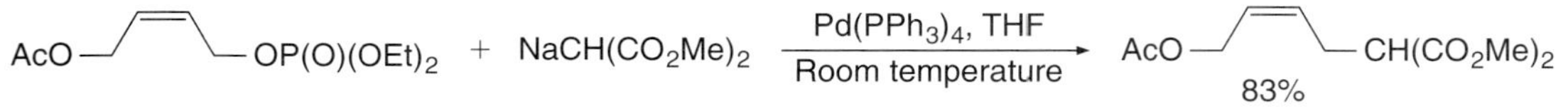
AcO
OP(O)(OEt)2
+ NaCH(CO2Me)2
Pd(PPh3)4, THF
Room temperature
AcO
CH(CO2Me)2
83%
(20.5)

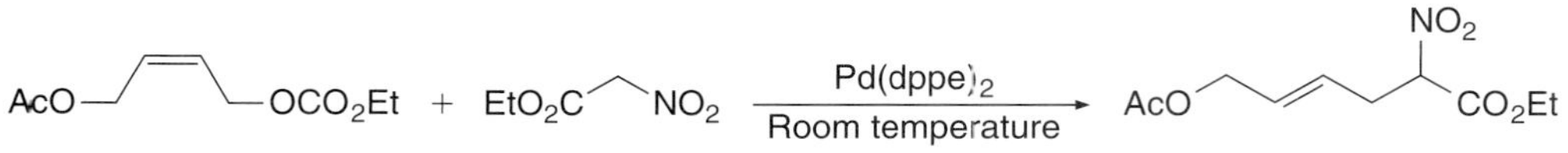
AcO
OCO2Et
+ EtO2C
NO2
Pd(dppe)2
Room temperature
AcO
NO2
CO2Et
(20.6)

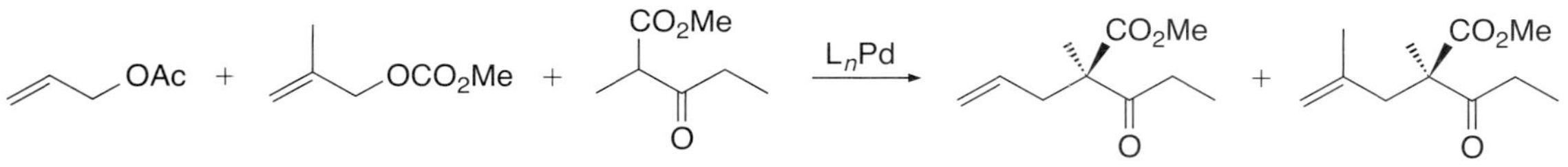
OAc
+
OCO2Me
+
CO2Me
O
LnPd
CO2Me
O
+
CO2Me
O

L		
PPh_3	16 : 84	
$P(OEt)_3$	3 : 97	

(20.7)

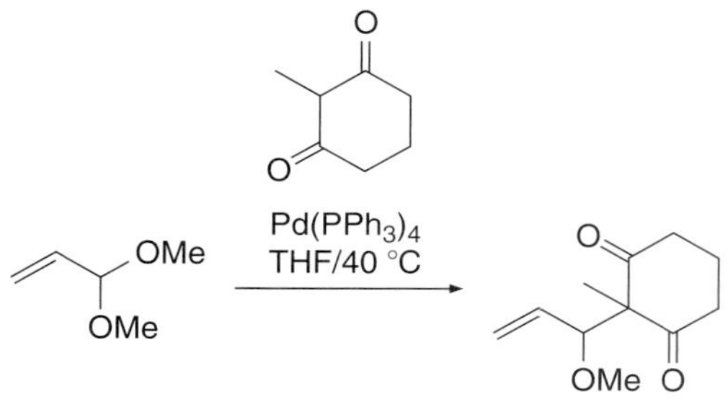
O
O
OMe
OMe
Pd(PPh3)4
THF/40 °C
O
OMe O
(20.8)

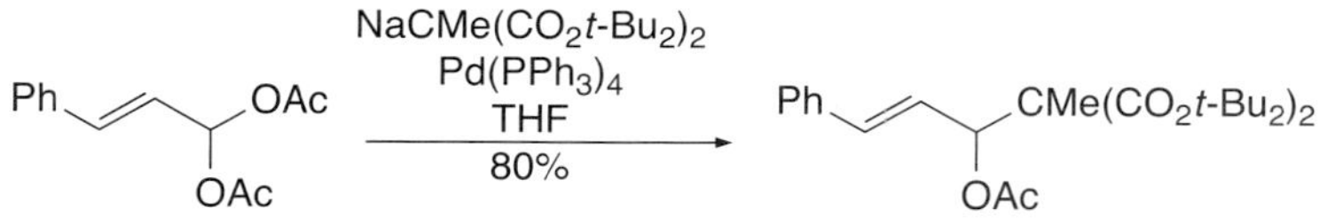
Ph
OAc
OAc
NaCMe(CO2t-Bu2)2
Pd(PPh3)4
THF
80%
Ph
CMe(CO2t-Bu2)2
OAc
(20.9)

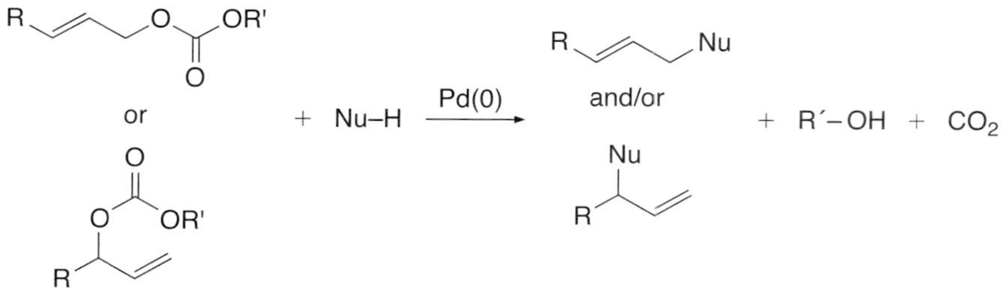
R
O
OR'
O
or
O
O
OR'
R
+ Nu–H
Pd(0)
R
Nu
and/or
Nu
R
+ R′–OH + CO2
(20.10)

Allylic carbonates are not only more reactive than allylic acetates, but they often undergo substitution by acidic pronucleophiles without addition of base (Equation 20.10). Scheme 20.2 shows the accepted mechanism for allylic substitution of allylic carbonates.[33,34] The oxidative addition of the allylic carbonate generates an allyl intermediate possessing an alkyl carbonate counterion. This alkyl carbonate anion extrudes carbon dioxide to form an alkoxide anion. This alkoxide anion then deprotonates the pronucleophile to form the anion that attacks the allyl group. Attack of the hard alkoxide is less favorable than attack by the soft carbanion, and deprotonation of the acidic C–H bond by the alkoxide is faster than attack by the alkoxide. These factors lead to preferential formation of a C–C bond over a C–O bond.

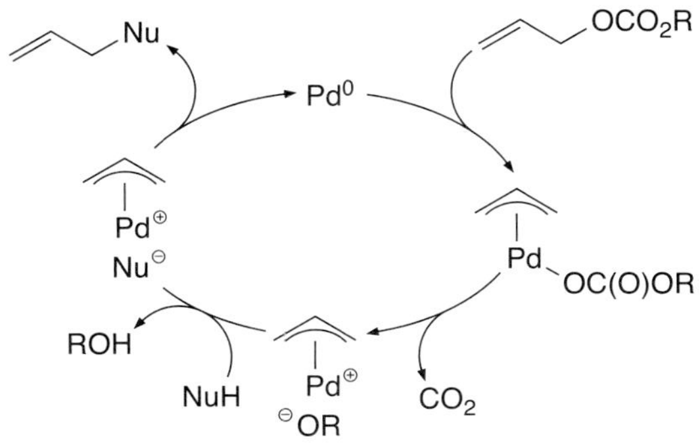

Scheme 20.2

Vinyl epoxides have also been used as electrophiles for allylic substitution.[35–46] Two examples of these reactions are provided in Equations 20.11[37] and 20.12.[36] The vinyl epoxide is a precursor to a palladium–allyl intermediate by coordination of the metal to the olefin and opening of the epoxide. Like allylic carbonates, the reactions of vinyl epoxides with neutral pronucleophiles do not need exogenous base. As shown in the mechanism of Scheme 20.3, the pendant alkoxide can deprotonate the pronucleophile to trigger nucleophilic attack on the metal–allyl intermediate. The same strategy has been followed to develop reactions of vinyl aziridines.[47] The pronucleophile for the additions to vinyl epoxides can be acidic compounds, such as malonates or β-ketoesters, or a preformed nucleophilic organometallic reagent, such as an organostannane.[38,39] In the latter case, the tin transfers to the alkoxide to generate a tin alkoxide as the initial product.

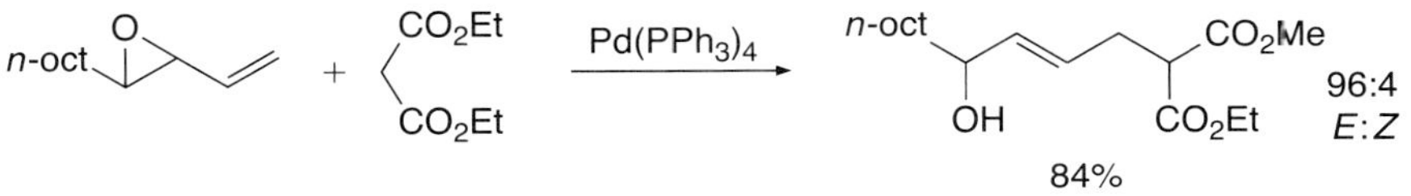

(20.11)

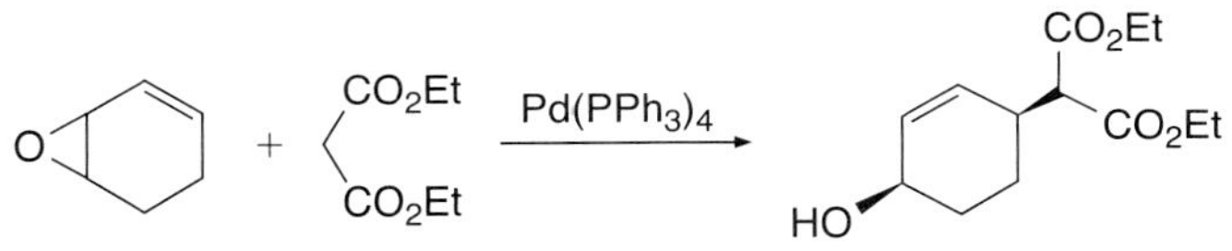

(20.12)

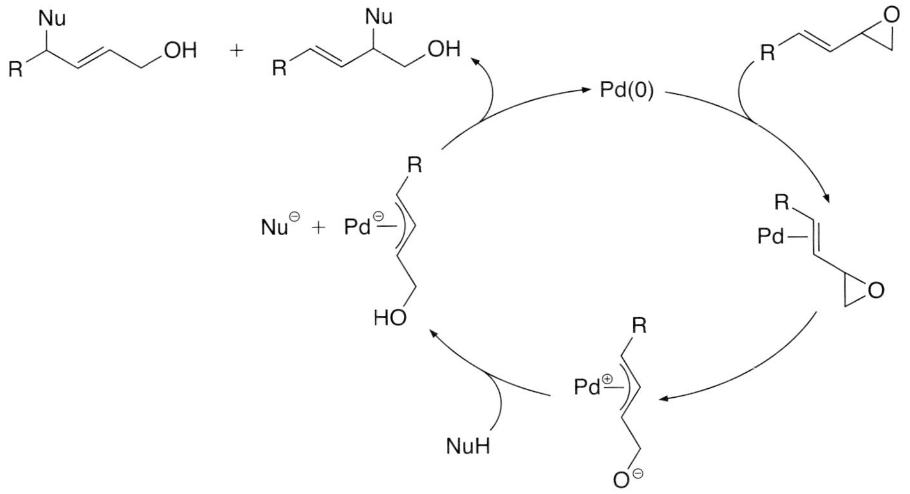

Scheme 20.3

When the nucleophile is a stabilized anion derived from a β-dicarbonyl compound, these reactions occur with retention of configuration from a sequence of two sequential inversions (see Section 20.4.1), but when the nucleophile is an unstabilized anion derived from a main-group organometallic reagent, such as an organotin, -zinc, or -magnesium reagent, they have been shown to occur with inversion of configuration. This inversion of configuration occurs by a sequence involving oxidative addition with inversion of configuration followed by C–C bond formation that occurs with retention of configuration.[38,48] More complex systems that generate non-racemic products from racemic epoxides are presented in the section on asymmetric allylic substitution.

20.3.2. Scope of Nucleophile

The scope of nucleophile that participates in allylic substitution catalyzed by transition metal complexes is broad, but some nucleophiles react in a more general fashion than others. Such allylic substitution reactions are most commonly conducted with soft, stabilized carbon nucleophiles, but allylic substitutions have also been reported with heteroatom nucleophiles, such as amines, imides, amides, and aryloxides. Although known, the reactions of hard, unstabilized carbon nucleophiles and of alkoxides are less common. The kinds of soft nucleophiles that undergo allylic substitution chemistry are shown in Figure 20.1 and include stabilized carbon nucleophiles formed by deprotonation of β-dicarbonyl compounds, cyanoesters, and malononitriles, as well as derivatives containing other electron-withdrawing groups, such as nitro, sulfonyl, and iminyl groups. Some reactions of alkali metal enolates have been reported, and specific cases have been reported to occur in high yield (Equation 20.13),[49–52] but these reactions most often occur in modest yields.[53,54] Thus, reactions of the harder nucleophiles of enolates have been sought with enolate derivatives that are milder than alkali metal enolates.[54] For example, reactions of boron (Equation 20.14),[53,55–57] silicon (Equation 20.15),[58,59] tin (Equation 20.16),[58,60,61] and zinc enolates (Equation 20.17)[62] have been reported.[51] Reactions of enamines, serving as ketone enolate equivalents, are also known.[63–71]

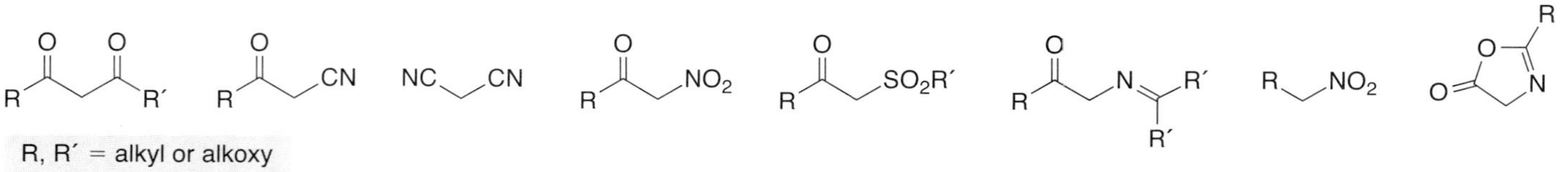

Figure 20.1. Common pronucleophiles for catalytic allylic substitution.

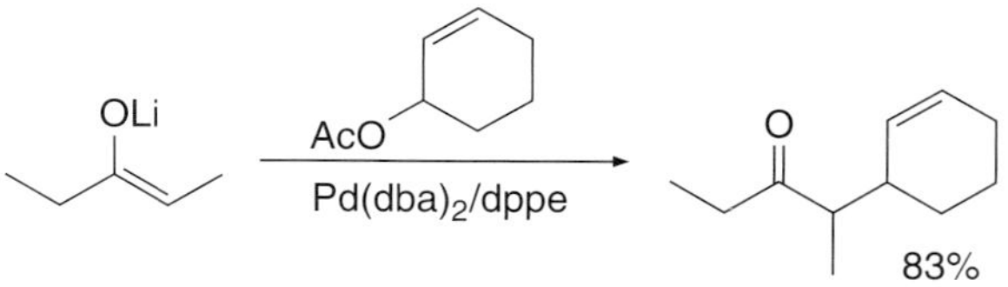

(20.13)

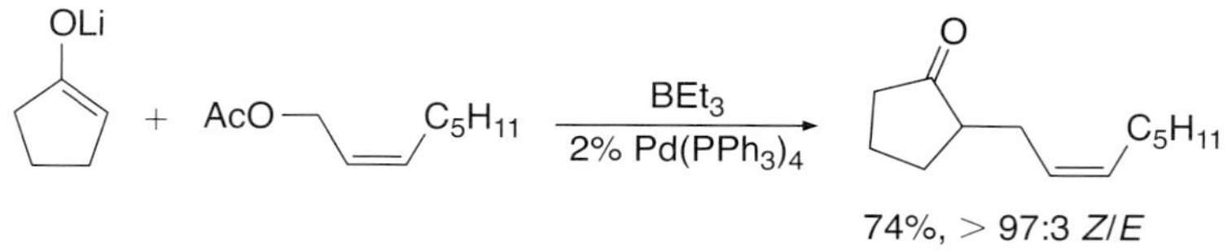

74%, > 97:3 Z/E

(20.14)

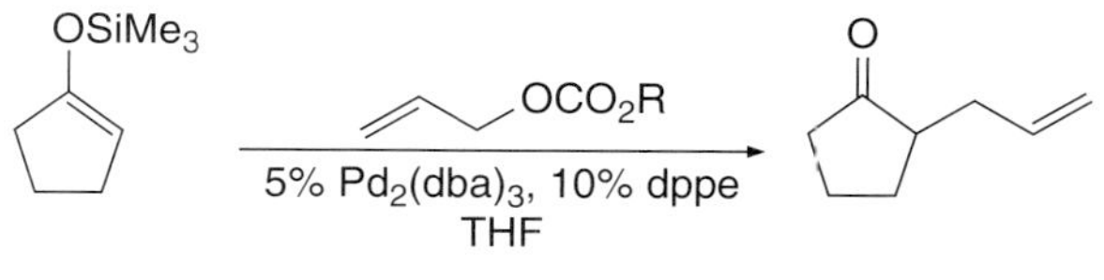

(20.15)

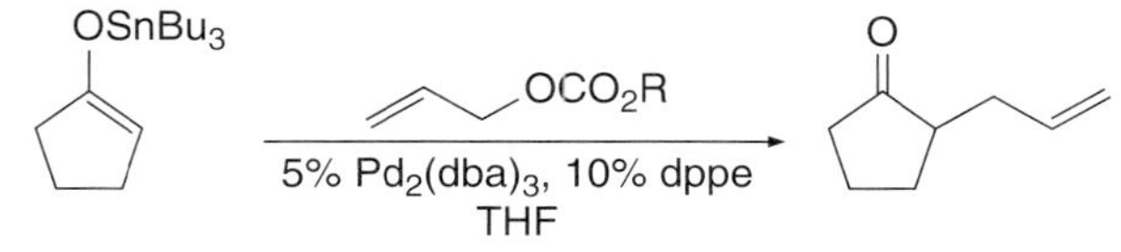

$$(20.16)$$

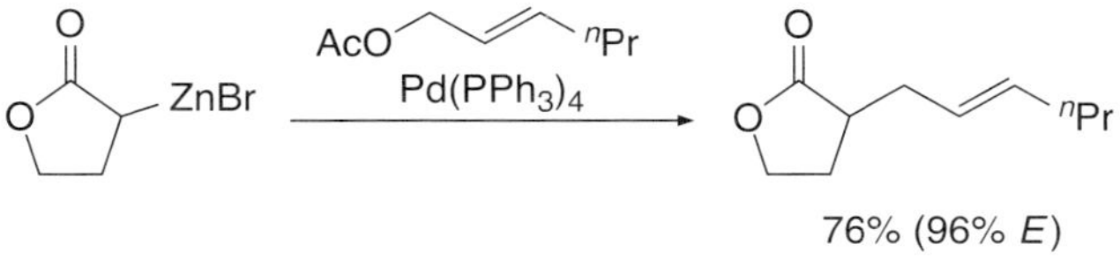

76% (96% E)

$$(20.17)$$

Alternatively, these nucleophiles can be tethered to the oxygen of the allylic electrophile, as shown for the allylic acetoester in Equation 20.18.[62,72–79] The mechanism for this transformation is shown in Scheme 20.4. Oxidative addition of the C–O bond of the allylic ester generates a carboxylate derivative of the enolate nucleophile, which undergoes decarboxylation. This decarboxylation has been proposed in some cases to occur within the coordination sphere of the metal[80,81] and in other cases to occur external, but nearby, to the allylmetal intermediate.[82,83] The resulting enolate nucleophile then attacks the allyl group to form the final substitution product.

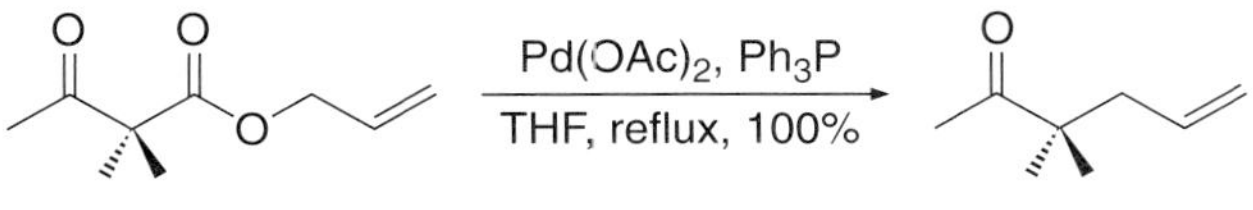

$$(20.18)$$

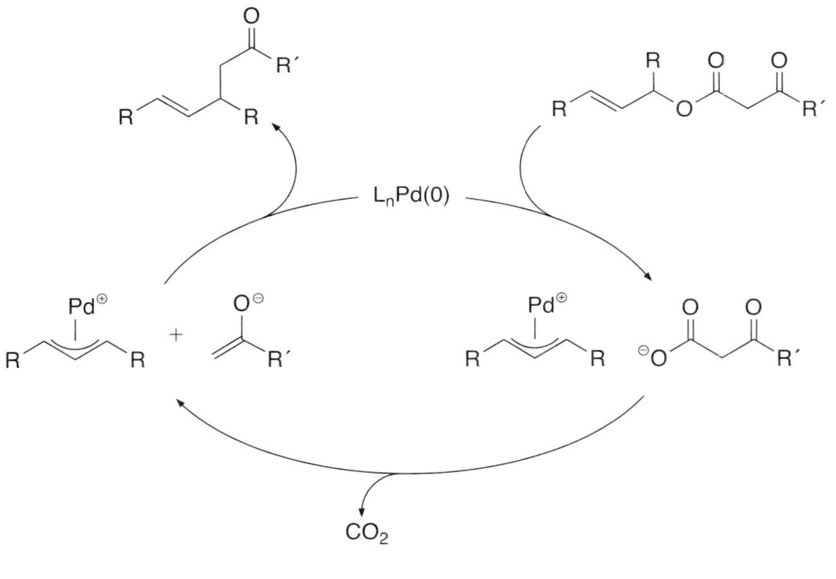

Scheme 20.4

20.3.3. Metals Used for Allylic Substitutions

A vast majority of the allylic substitution reactions have been reported with palladium catalysts. However, complexes of other metals also catalyze allylic substitution reactions. In particular, complexes of molybdenum,[41,45,46,84–96] tungsten,[97] ruthenium,[98–105] rhodium,[48,106–116] and iridium[71,117–134] have been shown to catalyze the reactions of a variety of carbon nucleophiles. In addition, complexes of ruthenium, rhodium, and iridium catalyze the reactions of phenoxides,[109,135] alkoxides,[136] amines,[99,100,137–139] and amine derivatives.[107,112,127–129,140] The regioselectivity of the allylic substitution process with these metals can often complement the regioselectivity of the reactions catalyzed by palladium complexes. The regioselectivity

$$R \diagdown \diagup O \diagup \overset{\overset{O}{\|}}{C} \diagdown R' \xrightarrow[\substack{-OC(O)R'}]{\substack{Nu-H,\ base \\ Catalytic,\ L_nM}} R \diagdown \overset{Nu}{\underset{\text{Usually major for}}{\diagup}} \diagup + R \diagdown \diagup \diagup Nu \qquad (20.19)$$

Usually major for
M = Mo,W, Ru, Rh, or Ir

Usually major for
M = Ni or Pd

for allylic substitution was discussed in more detail in Section 18.5. In broad terms, however, reactions of the unsymmetrical, monosubstituted allylic electrophiles shown in Equation 20.19 catalyzed by complexes of Mo, W, Ru, Rh, and Ir tend to form products from attack of the nucleophile at the more substituted terminus of the allylic electrophile. Based on the regioselectivity of these complexes, highly enantioselective reactions of amines, phenoxides, and alkoxides with terminal allylic carbonates to form branched allylic amines and ethers have been developed with iridium catalysts containing phosphoramidite ligands,[135,136,138,141–144] and enantioselective reactions of carbon nucleophiles catalyzed by complexes of molybdenum,[41,45,46,84–96,145] tungsten,[97] and iridium[71,117–134] have been developed. The regioselectivity of reactions of the same reagents catalyzed by palladium complexes depends on the ligand and the identity of the two reagents, but they tend to form more product from attack at the less-substituted terminus. In many cases of palladium-catalyzed reactions of carbon nucleophiles, the linear isomer is the major substitution product.[7,10,146,147]

The mechanism of the reactions catalyzed by complexes of these other metals is less well understood than the mechanism of the reactions catalyzed by complexes of palladium. This discrepancy in mechanistic understanding results, in part, from the accessibility of the allylpalladium intermediates and the detailed studies of the chemistry[148] and dynamics[149–152] of these complexes. Fewer allyl complexes relevant to the allylations catalyzed by other metals have been isolated.[104,105,153,154]

20.4. Mechanism of Allylic Substitution

The mechanism of the allylic substitution reaction was depicted in Scheme 20.1.[151] This mechanism occurs in two general steps, each of which is more accurately subdivided into two steps. First, a low-valent metal complex such as palladium(0), molybdenum(0), ruthenium(II), iridium(I), or rhodium(I) binds an allylic chloride or allylic alcohol derivative, and the resulting complex undergoes ionization to generate the intermediate allyl complex. This allyl complex then undergoes reaction with a nucleophile. This step usually occurs by external attack of the nucleophile on the allyl intermediate, and this nucleophilic attack on a coordinated allyl ligand was presented in Chapter 11. The product of this attack is an olefin complex, which is extruded from the metal by either a dissociative or associative process. The intermediates in these reactions can undergo rearrangements, and these rearrangements are important for controlling the regioselectivity and stereoselectivity of the catalytic process.

20.4.1. Mechanism of Palladium-Catalyzed Reactions

The oxidative addition of allylic esters to palladium(0) complexes is a common route to allylpalladium complexes. This reaction occurs to form cationic allylpalladium(II) complexes containing an acetate or other carboxylate derivative as the counterion. These reactions were described in Chapter 7. They likely occur by coordination of the olefinic unit of the allylic ester to the metal center, followed by ionization of the coordinated allylic ester to form the allylpalladium intermediate. These reactions occur with inversion of configuration, as shown explicitly by the reaction in Equation 20.20.[155,156]

The mechanism of the nucleophilic attack on the allyl intermediate has been studied in depth, and this process was discussed in Chapter 11. In general, stabilized carbon nucleophiles containing two electron-withdrawing groups on the methylene carbon, as well as

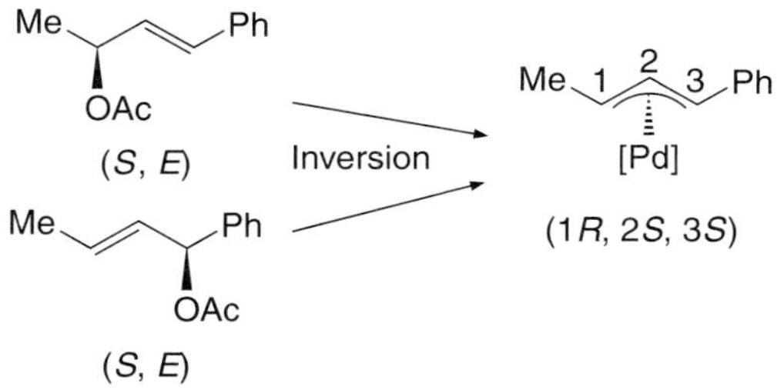

$$(20.20)$$

nitrogen nucleophiles, react by external attack on a cationic allylpalladium complex. This mechanism leads to inversion of configuration at the terminal carbon of the allyl ligand for this step of the cycle (Scheme 20.5). In contrast, the stereochemistry of reactions of unstabilized nucleophiles, such as borate, organozinc, organostannane, and Grignard reagents, indicate that these nucleophiles react at palladium to generate an intermediate that contains an alkyl ligand (Scheme 20.5). Although these intermediates have not been observed directly, and the binding mode of the allyl group is, therefore, unknown, it is reasonable to propose that the binding of the nucleophile to the metal causes the allyl to adopt an η^1 binding mode. Reductive elimination from the allyl complex containing the bound nucleophile would then form the final substitution product and regenerate the starting low-valent species. This mechanism leads to formation of the C–C bond with retention of configuration. As a result of the combination of the stereochemistry of the product-forming step and the inversion of configuration of the oxidative-addition step, the palladium-catalyzed allylic substitution by stabilized nucleophiles occurs with overall *retention* of configuration, while the palladium-catalyzed allylic substitution by hard nucleophiles occurs with overall *inversion* of configuration. This stereochemistry for the overall catalytic processes is revealed by reactions, such as those in Equations 20.21[156] and 20.22.[157, 158]

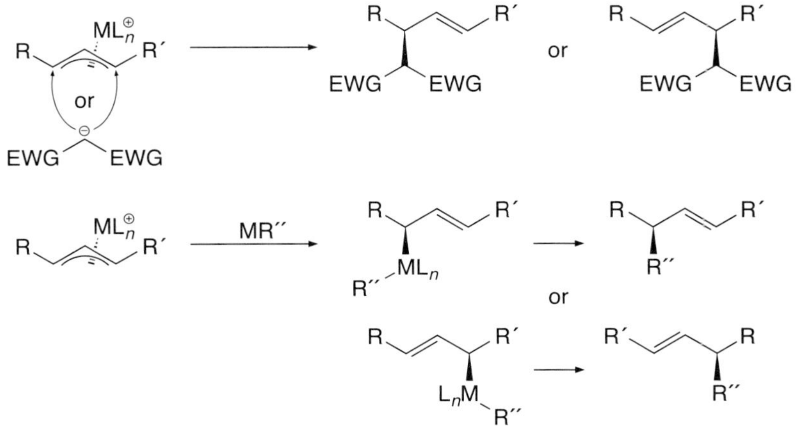

Scheme 20.5

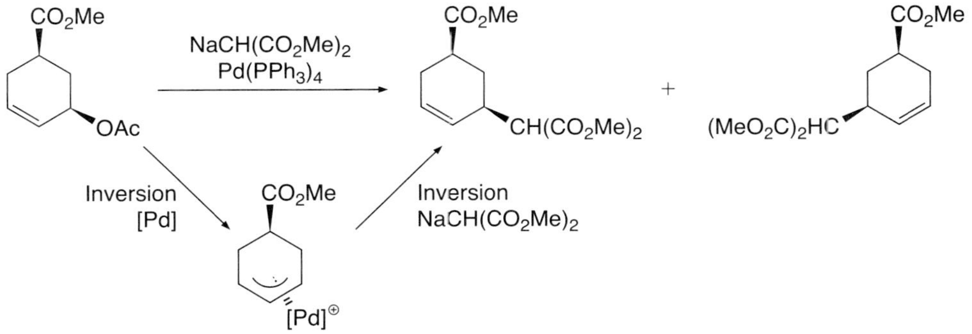

$$(20.21)$$

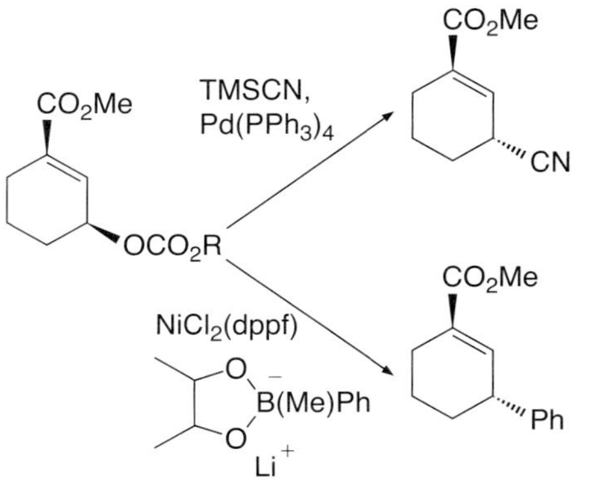

(20.22)

The dynamics of the allyl intermediates[149] can have a large effect on the regioselectivity and stereoselectivity of the catalytic reactions.[86,151,159,160] The dynamics of allyl complexes have been studied extensively by variable-temperature NMR spectroscopy.[149,150,152,159–161] Allyl complexes can undergo rearrangements by the combination of $\eta^3–\eta^1–\eta^3$ interconversions and exchange of the η^1-allyl group from one site in the complex to the other.[161–163] These intramolecular interconversions can lead to the formal changes in structure shown in Scheme 20.6. This intramolecular exchange of the metal from one face of the allyl group to the other occurs only when one terminus of the allyl group is symmetrically substituted. If both termini contain two different substituents, then the η^1 structure contains a stereogenic center bound to the metal, and the allyl group does not lose its stereochemistry through the same intramolecular processes as it does when the η^1 structure contains a non-stereogenic center bound to the metal (see Scheme 20.6).[164]

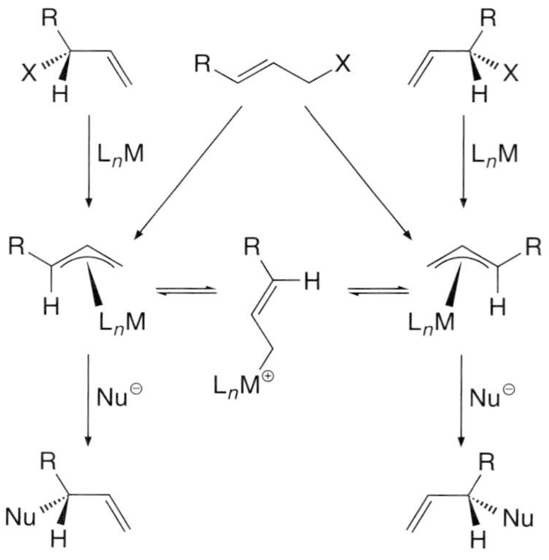

Scheme 20.6

However, Scheme 20.7 shows that inversion of configuration of allyl groups containing two different substituents at each allyl terminus can occur by an intermolecular process.[151,165,166] Transfer of the allyl group from one metal to the other, as shown in Scheme 20.7, inverts the stereochemistry and ultimately leads to racemization of the allyl complex.

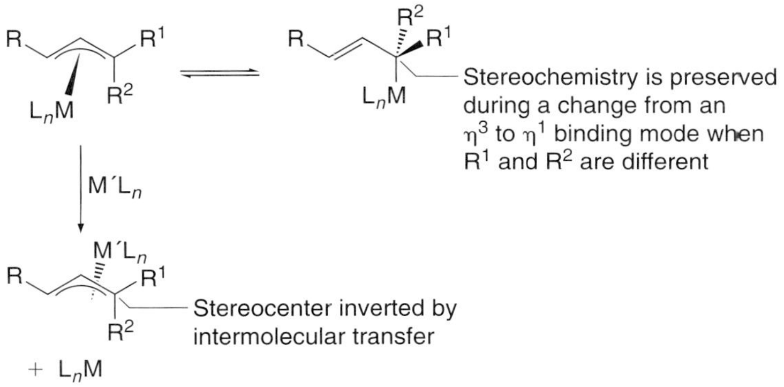

Stereochemistry is preserved during a change from an η^3 to η^1 binding mode when R^1 and R^2 are different

Stereocenter inverted by intermolecular transfer

Scheme 20.7

If the nucleophilic attack on the allyl intermediate is faster than any $\eta^3-\eta^1-\eta^3$ interconversions, then the reactions of soft nucleophiles occur with retention of the stereochemistry because the oxidative addition and nucleophilic attack steps both occur with inversion of configuration. However, when the attack is slower than the $\eta^3-\eta^1-\eta^3$ interconversions, and the allyl group is symmetrically substituted at one terminus, the interconversions eliminate the stereochemical content of the starting reagents. In this case, racemic substrates can form enantioenriched substitution products. These relative rates were revealed in a classic study by Bosnich.[151]

Added ligands, such as halide anions, phosphines, or coordinating solvents can accelerate the rearrangements of allyl complexes.[86,149,167] These ligands can bind to the metal center and encourage formation of η^1 structures and, therefore, accelerate $\eta^3-\eta^1-\eta^3$ interconversions. Alternatively, added ligands can induce rearrangements by generating five-coordinate intermediates that are stereochemically non-rigid (Chapter 1). The identity of the solvent can also affect the rate of these interconversions. More polar, coordinating solvents tend to accelerate rearrangements by binding to the metal. In addition to affecting the stereochemistry of the allylic substitutions, these rearrangements allow scrambling of the stereochemistry at the metal center. Because the position of the ancillary ligands relative to the two allyl termini can affect the position of nucleophilic attack, the scrambling of stereochemistry at the metal can help to form products with regiochemistry that is independent of the regiochemistry of the starting allylic ester reagent. These regiochemical issues are discussed in more detail in Section 20.5.

20.4.2. Mechanism of Reactions Catalyzed by Complexes Other Than Palladium

Few allyl intermediates in the reactions catalyzed by complexes of metals other than palladium have been identified. An allyl intermediate in the molybdenum chemistry was recently identified (Equation 20.23),[88,153] and studies of this intermediate revealed several surprising results. First, the allylic substitution occurs in a similar fashion with a catalyst generated from a bis(pyridinyl)amide ligand and from a monopyridinyl amide ligand. Thus, only one pyridine appears to be coordinated to molybdenum in the active catalyst. Second, as shown in Scheme 20.8, carbon monoxide dissociates and associates during the catalytic cycle. The malonate anion does not react with the isolated allyl complex alone. Instead, it reacts with the allyl complex and free CO or CO generated from the starting $Mo(CO)_6$.[88] Thus, the precise structure of the allyl complex that reacts with the malonate anion is unknown at this time.

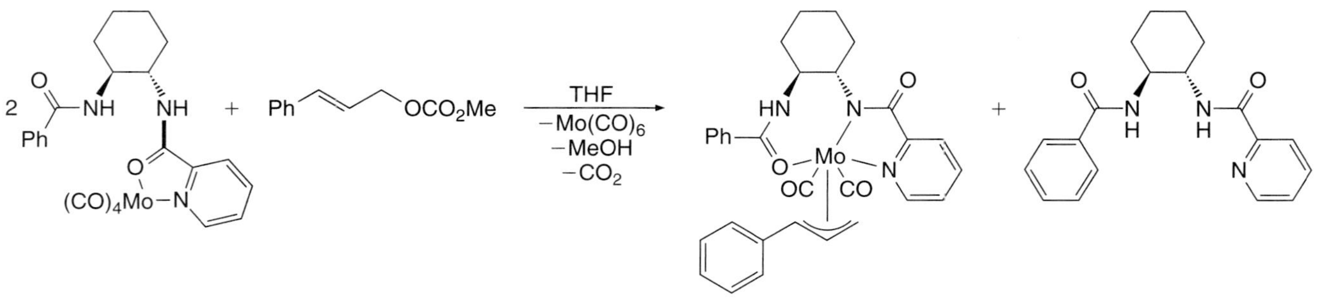

$$(20.23)$$

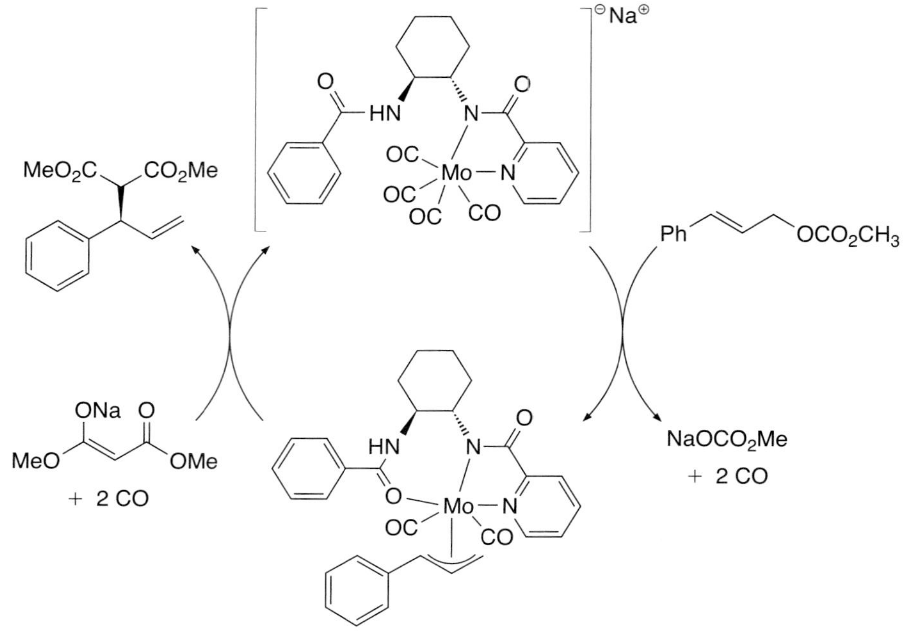

Scheme 20.8

Because of the dearth of allyl intermediates in allylic substitutions catalyzed by metals other than palladium, the stereochemistry of these allylic substitution reactions is less well studied. Evans has shown that rhodium complexes catalyze the reactions of various nucleophiles with allylic acetates with an overall retention of configuration.[106, 107, 109, 112, 113] Molybdenum complexes also catalyze the reactions of allylic acetates with malonate nucleophiles with retention of configuration.[153] The stereochemistry of the addition of the allyl acetate to molybdenum and the nucleophilic attack on the resulting allylmolybdenum intermediate have been studied by an elegant labeling experiment involving the diastereomerically defined, monodeuterio cinnamyl carbonate in Equation 20.24. This study showed that the addition and nucleophilic attack both occur with retention of configuration (Equation 20.24), instead of the inversion of configuration that occurs with palladium complexes.[153]

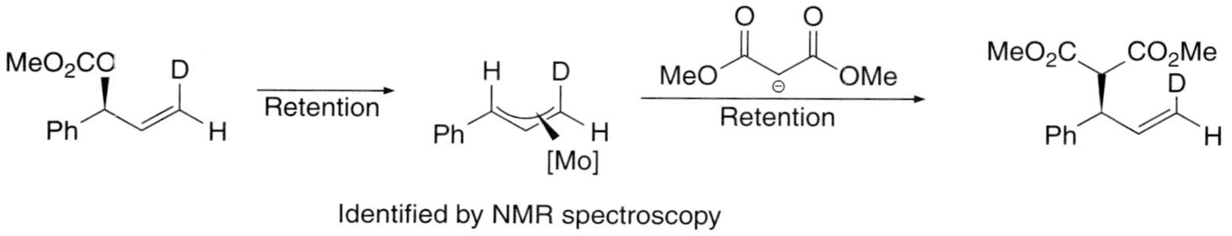

$$(20.24)$$

Identified by NMR spectroscopy

20.5. Regioselectivity of Allylic Substitutions

The regioselectivity of the allylic substitution process depends heavily on the identity of the metal and its coordination sphere. To bypass the issue of regioselectivity, most of the substrates used to demonstrate activity and enantioselectivity of catalysts for allylic substitution generate symmetrical allyl intermediates. However, many synthetic applications would require a regioselective substitution process involving unsymmetrical allyl intermediates. The regioselectivity of allylic substitution arises from the position of attack of the nucleophile on the allyl intermediate.

20.5.1. Trends and Origins of Regioselectivity of Palladium-Catalyzed Reactions

20.5.1.1. Reactions of Carbon Nucleophiles

The factors that control regioselectivity of palladium-catalyzed allylic substitutions have been the subject of numerous studies. Experiments that reveal the effects of both symmetric and unsymmetric ancillary ligands on this regioselectivity have been reported, along with experiments that have revealed effects of the type of nucleophile and the effects of basic additives. The following section describes some of this work on palladium-catalyzed processes.

Although exceptions have been reported,[86,168–171] palladium–allyl complexes tend to react with carbon nucleophiles at the less-hindered terminus of the allyl group when the electronic properties of the termini are similar. For example, reaction of dimethyl malonate with an allylic ester that generates an allyl intermediate containing one methyl and one phenyl or one methyl and one isopropyl group at the allyl termini occurs with roughly 12:1 selectivity for reaction at the less hindered position (Equation 20.25).[156,172] When related allylic substitutions are conducted with allylic esters that generate a monosubstituted allyl intermediate, mixtures of products that depend on the identity of the ligand have been known to form for many years but the achiral, linear product is typically the major one (Equation 20.26).[173]

$$ (20.25) $$

$$ (20.26) $$

Because attack at the more substituted position of the allyl ligand generates chiral products, much effort has been spent to direct attack at this position. Åkermark showed that the regioselectivity for attack at the less substituted versus the more substituted position of an allyl group on palladium depends on the electronic properties of the ancillary ligand.[174–176] He proposed that more accepting ligands lead to attack at the more substituted position because a greater degree of positive charge resides at the more substituted carbon. However, it is difficult to rank the electron-donating and electron-accepting properties of ligands possessing the widely varying structures used in palladium-catalyzed allylic substitution.

Particular ligands have been designed to induce attack at the more substituted positions of allylic electrophiles by orienting the nucleophile or rendering one end of the allyl moiety more electrophilic than the other by using unsymmetrical ancillary ligands. For example, reactions of branched allylic acetates with a palladium catalyst bound by a phosphine containing the ferrocenyl group and a pendant binding site for the nucleophile (Equation 20.27a)[177] form branched products, and reactions with phosphinooxazoline ligands containing one phosphorus and one nitrogen donor lead to preferential attack at the more substituted position of an allyl group (Equation 20.27b).[178] More recently, the reactions of amines and stabilized carbon nucleophiles with allylic acetates catalyzed by palladium complexes of particular ferrocenyl ligands predominantly generate the branched substitution product (Equation 20.28).[169]

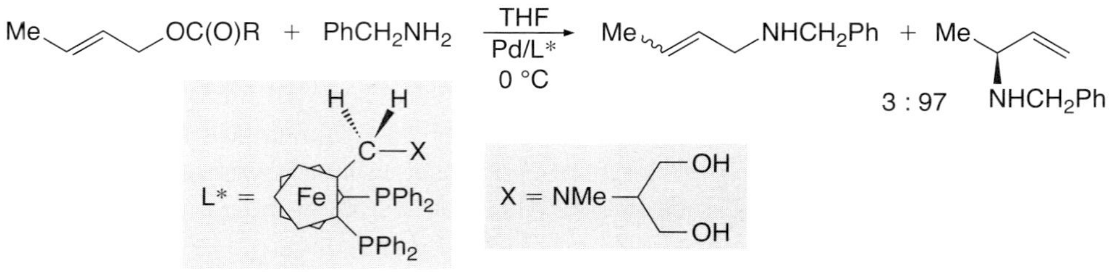

(20.27a)

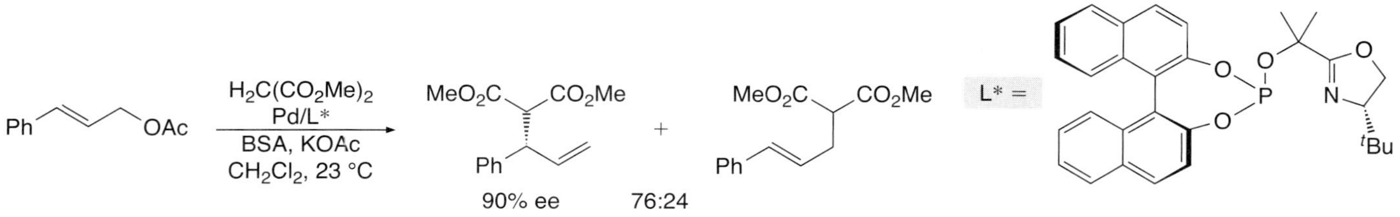

(20.27b)

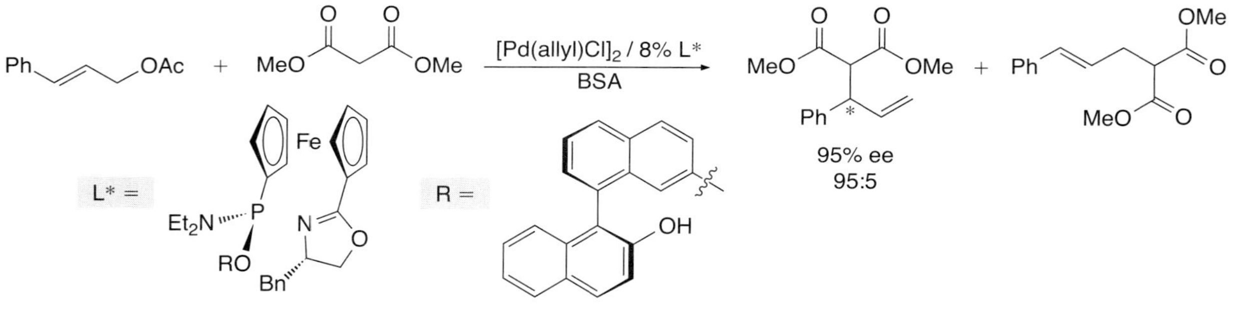

(20.28)

In addition to the steric properties of the allylic ester, the electronic properties of the allylic ester influence the regioselectivity of palladium-catalyzed reactions. Nucleophilic attack, somewhat counterintuitively, occurs away from electron-withdrawing groups. For example, reaction of an allylic electrophile with two different aryl groups at the termini occurred at the site attached to the more electron-rich aryl group (Equation 20.29).[179] Consistent with this trend, nucleophilic attack occurred at the methoxy-substituted allyl terminus of the substrate in Equation 20.30.[180, 181] Likewise, reactions of allylic electrophiles containing phosphonate or sulfonate groups at one terminus of the allyl ligand occur by nucleophilic attack at the terminus opposite to that bound to these electron-withdrawing groups (Equation 20.31).[182, 183] The regiochemistry of Equation 20.31 can be rationalized by the greater stability of products possessing conjugated olefins.

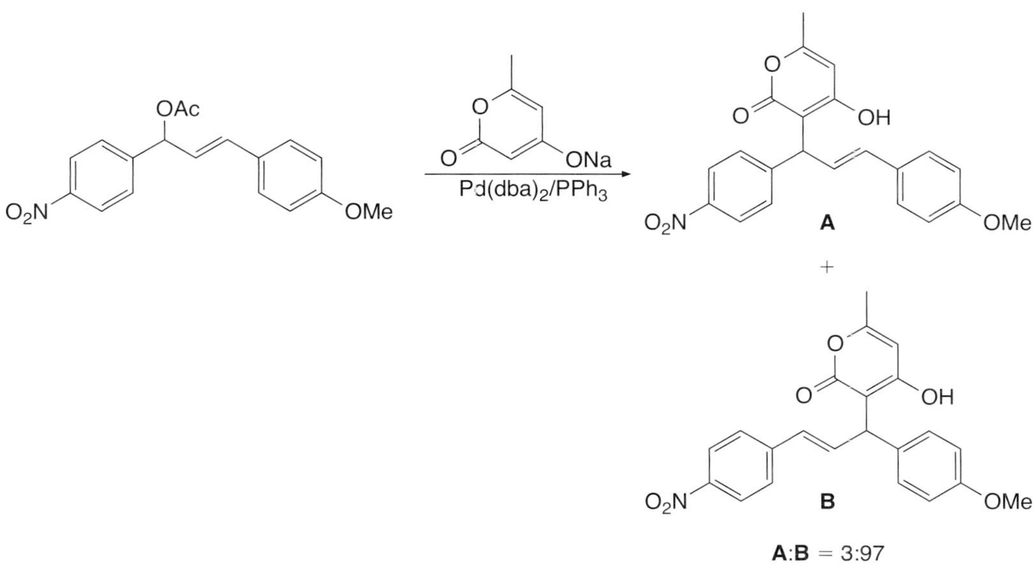

(20.29)

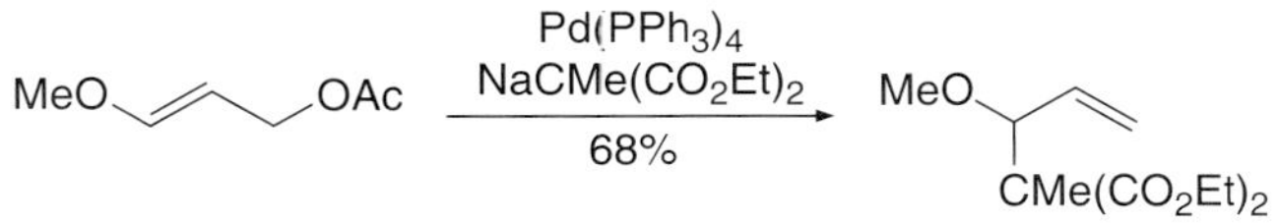

A:B = 3:97

(20.30)

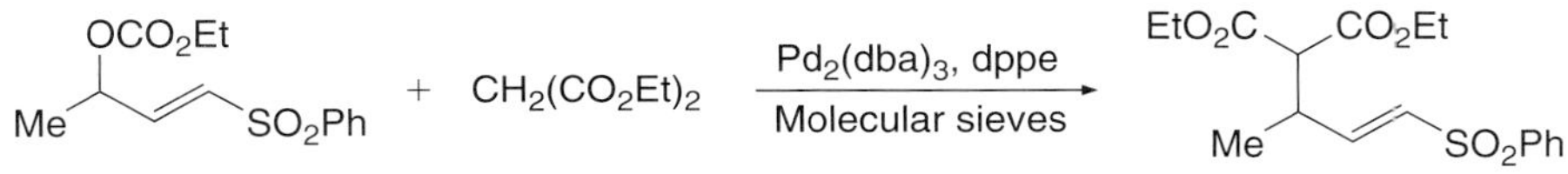

(20.31)

20.5.1.2. Reactions of Heteroatom Nucleophiles

The regioselectivity of palladium-catalyzed allylic substitutions with heteroatom nucleophiles is often much different from that of reactions with carbon nucleophiles. In general, reactions of oxygen and nitrogen nucleophiles form more branched product than do reactions of carbon nucleophiles, at least as the kinetic product of substitution. For example, reactions of aryloxide nucleophiles with mono-substituted allylic carbonates catalyzed by palladium complexes of Trost's ligand form more branched than linear product (Equation 20.32).[86] Reactions of geminally disubstituted allylic acetates with aziridines, hydroxylamine and hydrazine derivatives catalyzed by palladium complexes of bisphosphines form the branched prenyl product as the major isomer (Equations 20.33a and 20.33b).[185–187]

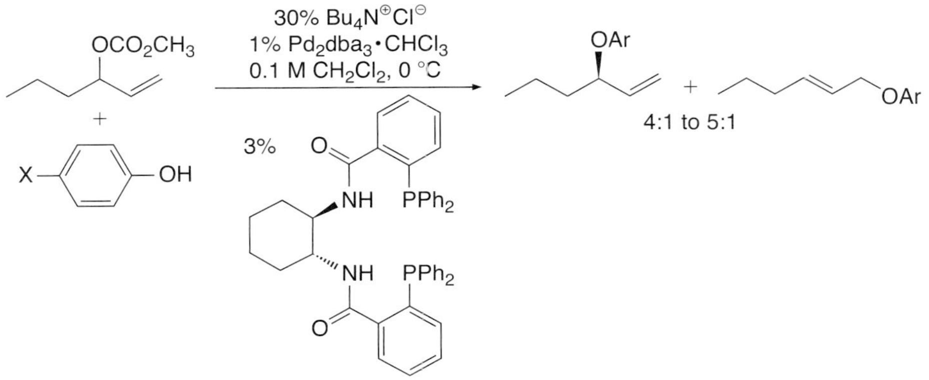

4:1 to 5:1

(20.32)

$$(20.33a)$$

X = OR or N=CPh₂   R = H or alkyl

$$(20.33b)$$

Studies on the origin of the regioselectivity of these reactions revealed that attack by amines occurred at the more-substituted position, but isomerization of the kinetic branched product to the thermodynamic linear product occurred faster than the catalytic process.[186] As a result, the linear isomer was observed as the final reaction product. However, isomerization of the products formed by reactions of aziridines, hydroxylamines, and hydrazone deriviatives was slower than the catalytic substitution process, and these different relative rates allowed isolation of the branched substitution products.[185–187] The isomerization process presumably occurs by protonation of the amine to form an ammonium salt that undergoes oxidative addition to palladium, as was observed in the initial allylic substitution processes that involved allylic ammonium salts as electrophile.[29] Thus, addition of a strong, non-nucleophilic base to the reactions of amine nucleophiles allowed isolation of the branched kinetic product.[188]

20.5.2. Memory Effect with Palladium

The regioselectivity and enantioselectivity of the palladium-catalyzed reactions can also be affected by the regiochemistry and stereochemistry of the starting allylic ester. In some cases, the branched allylic ester generates more branched substitution product than does the linear allylic ester. This effect has been called the regiochemical "memory effect."[184,189–191] One example of this phenomenon observed with a palladium catalyst containing the MOP ligand is shown in Equation 20.34.[184] An affect of the enantiomeric excess of the starting chiral allylic alcohol on the enantiomeric excess of the product has been termed a stereochemical "memory effect."

$$(20.34)$$

The origin of this memory effect has been studied in detail by Hayashi and by Lloyd-Jones and has many intricate features.[184,190–192] The proposed origin of this memory effect for reactions catalyzed by palladium bound by the MOP ligand is summarized in Scheme 20.9. If the leaving group is attached to the palladium–allyl intermediate, as originally proposed, or if the ligand is bound unsymmetrically through phosphorus and the biaryl π-system, as revealed by subsequent studies, then the regiochemistry of substitution can be explained by an electronic effect of the ligands located trans to the two termini of the allyl group, the principle of microscopic reversibility, and the dynamics of the allyl intermediate.

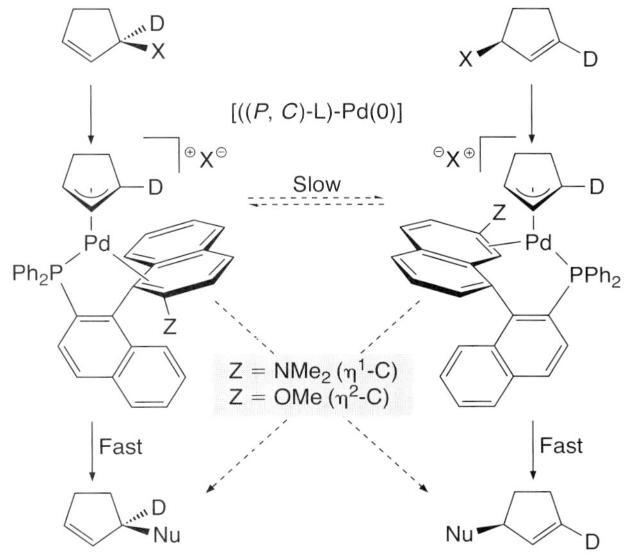

Scheme 20.9

As shown by the labeling experiment for reaction of the cyclopentenyl acetate in Scheme 20.9,[191] an oxidative addition of the allyl electrophile to generate the palladium–allyl intermediate places the aryl π-system of the MOP ligand cis to the terminus of the allyl group onto which the acetate was originally attached. Nucleophilic attack occurs trans to the more polarizable and π-accepting ligand,[193–198] which is thought to be the phosphine in this case. As a result, the nucleophile is attached to the same carbon that contained the starting acetate. This regiochemistry is observed if rearrangement of the allyl intermediate to scramble the two termini is slow. (If this rearrangement were fast, then a Curtin–Hammett situation would result, and the regiochemistry of the nucleophilic attack would not depend on the position of the leaving group in the reactant.) Thus, systems that undergo slow rearrangement of the allyl group, as well as those that generate intermediates in which the leaving group is attached to the metal in the allyl intermediate, can form products with retention of regiochemistry. Other mechanisms for retention of regiochemistry may also be followed by other catalysts, but the origin of this effect has been most firmly established for reactions of Hayashi's catalysts with MOP ligands and Kocovsky's catalysts with the aminophosphine analog MAP.

20.5.3. Regioselectivity of Reactions Catalyzed by Complexes of Other Metals

Allylic substitution reactions catalyzed by complexes of most metals besides palladium generate, as the major product, the species from substitution at the more hindered carbon. For example, the reactions of malonates with the branched or linear allylic carbonate or acetate derived from cinnamyl alcohol catalyzed by molybdenum, tungsten, ruthenium, rhodium, and iridium complexes containing nitrogen and phosphorus donor ligands generate the branched isomer as the major substitution product (Equation 20.35). The most thoroughly studied of these species are molybdenum and tungsten catalysts containing nitrogen-based ligands, such as pyridines and oxazolines,[41,90,97] rhodium catalysts that contain mixtures of achiral phosphine and phosphite ligands,[106,107,109,111–113,115,116,199] ruthenium

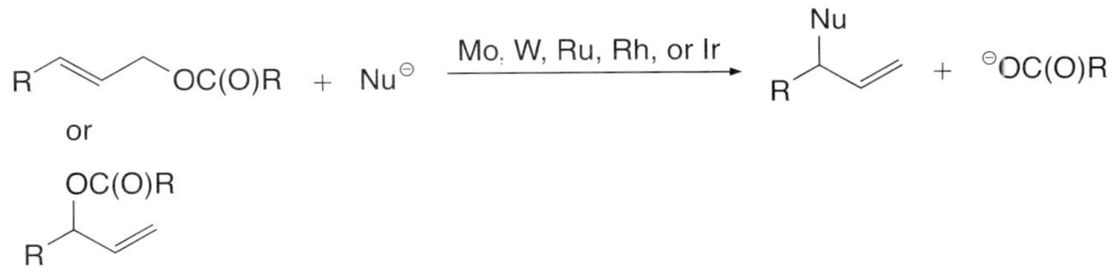

$$(20.35)$$

catalysts possessing cyclopentadienyl derivatives,[98–104,200] and iridium catalysts bound by phosphite and phosphoramidite ligands.[108,117–126,135–138,141–143,201–206] In other cases, reactions occur to form the product from substitution with retention of the regiochemistry of the starting allylic ester. This phenomenon has been observed with rhodium carbonyl chloride and Fe(CO)$_3$(NO)$^-$ as catalyst.[207–209] Because the complexes of the later transition metals catalyze the addition of heteroatom nucleophiles, these catalysts, particularly those based on iridium[135,136,138,141–144,206] and rhodium,[107,109,112,113,115] are useful for the synthesis of allylic amines and ethers.

20.6. Enantioselective Allylic Substitution

20.6.1. Overview of Enantioselective Allylic Substitution

20.6.1.1. Forms of Enantioselective Allylic Substitution[1,7,10,146,147,210]

The majority of recent efforts to develop catalytic allylic substitution reactions have focused on enantioselective transformations. Most enantioselective reactions form a stereocenter at one of the allylic carbons, and this new stereocenter is formed within the coordination sphere of the metal. Less common are enantioselective substitutions that form products containing a new stereocenter at the nucleophile. These reactions have been considered challenging because the stereochemistry is set at an atom that is outside of the direct coordination sphere of the metal.

Enantioselective allylic substitutions to form a stereocenter at one of the allylic carbons can take many forms, as summarized in Scheme 20.10. The diversity of methods by which

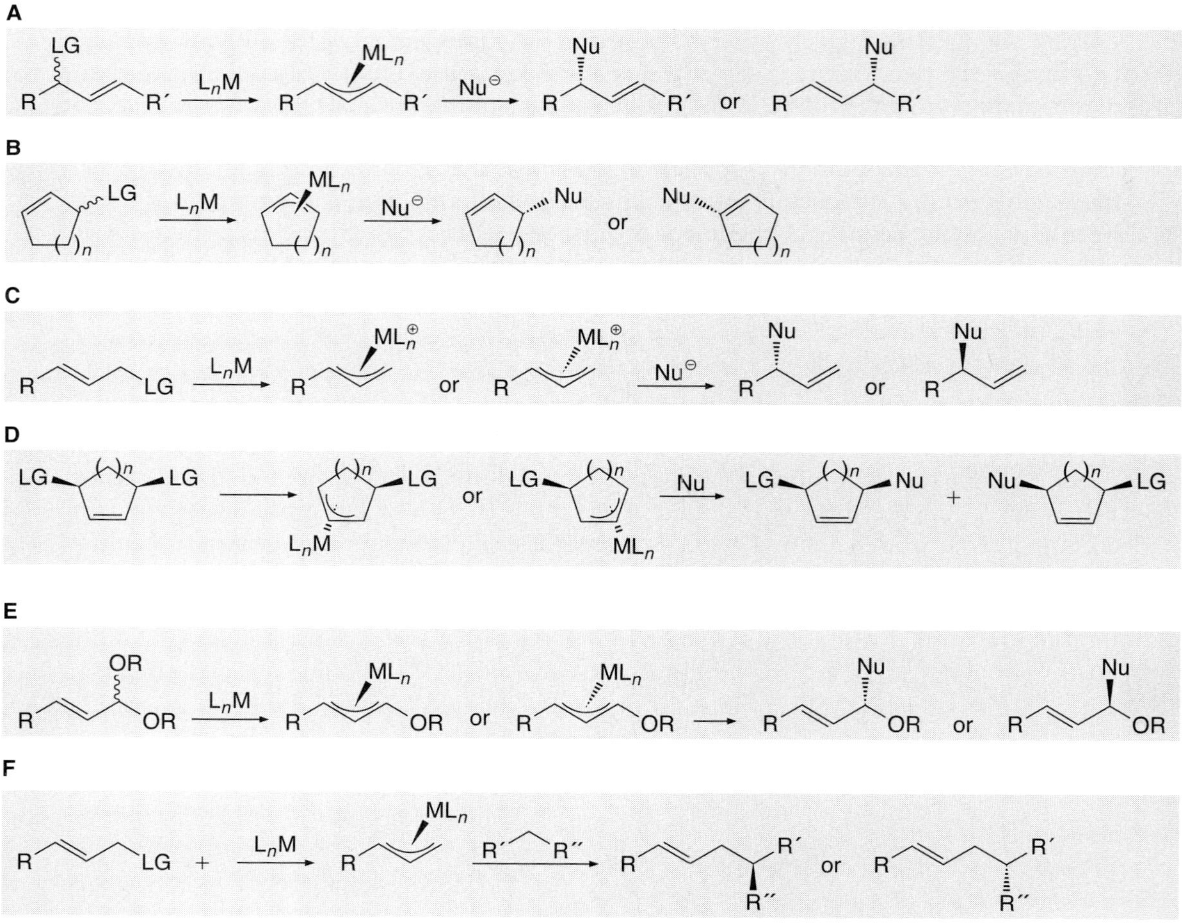

Scheme 20.10

enantioselective allylic substitution can be conducted has created opportunities to use this process in the synthesis of a variety of natural products.[1]

The reactions can occur as shown in Equations **A** and **B** of Scheme 20.10 to form a new stereocenter at a symmetric allylic electrophile, through control by the ancillary ligands of the catalyst, of the terminus that is attacked. These symmetric allylic electrophiles can be acyclic (**A**) or cyclic (**B**), and the geometries of the allyl intermediates are much different in these two cases. Therefore, different catalysts are optimal for these two classes of symmetric electrophiles.[6,8–10,211]

The enantioselective allylic substitution has also been conducted with unsymmetrical allylic electrophiles, as shown in Equation **C** of Scheme 20.10. In this case, enantioselectivity is dictated by the face to which the metal is bound and the resulting face of the allyl group that suffers nucleophilic attack. Complexes of many different metals have been studied for this type of reaction, and complexes of molybdenum and iridium are particularly selective for this class of of substitution.[41,90,135,136,138,141–144,206]

Enantioselective allylic substitution can also be achieved by selective cleavage of one of two enantiotopic leaving groups. This selectivity can occur within a cyclic substrate, as shown in Equation **D** of Scheme 20.10. Enantioselective allylic substitutions can also occur by replacement of one of two enantiotopic leaving groups on the same carbon, as shown for the acetal structure in Equation **E** of Scheme 20.10.

Finally, these reactions can be conducted with prochiral nucleophiles. In this case, shown as Equation **F** of Scheme 20.10, a stereocenter is generated at the nucleophile, instead of the electrophile. If the nucleophile and the electrophile are both prochiral, then control of both absolute and relative stereochemistry is necessary.

20.6.1.2. Catalysts for Enantioselective Substitutions

Several ligands that have been useful in enantioselective allylic substitution have been mentioned previously in this chapter. In this section, a more detailed presentation of ligands that generate complexes that catalyze enantioselective allylic substitution chemistry are described.

Palladium-catalyzed allylic substitution has been a test-bed for the evaluation of new ligands for enantioselective transformations. Thus, hundreds of ligands have been evaluated for these reactions, predominantly the reaction of 1,3-diphenylallyl acetate. A recent review[146] of enantioselective allylation cited more than 50 ligands that form palladium complexes that catalyze the reactions of 1,3-diphenylallyl acetate in greater than 90% ee. However, several ligands generate catalysts that are highly selective for substrates that undergo selective processes in fewer cases. Some of these ligands are shown in Figure 20.2.

Trost has developed two classes of ligands that create catalysts that are highly active and selective for allylic substitutions with a wide variety of allylic electrophiles and a variety of nucleophiles. First, Trost and Van Vranken[6] reported a bisphosphine ligand that was based on the principle that a wide bite angle would extend the chiral environment created by the ligand toward the incoming nucleophile and the developing chirality of the allylic substitution product. This ligand is generated by amide bond formation and has a relatively simple framework. A second class of ligands developed by Trost is based on a similar design principle. In this case, the ligand contains two pyridine nitrogen donors linked through a C_2 symmetric backbone and amide bonds.[41] This ligand was combined with molybdenum carbonyl to generate a catalyst for allylic substitution. As noted briefly in the prior discussion of the mechanism of Mo-catalyzed reactions, ligands containing a single pyridine donor were later found to be equally effective,[89] and the coordination sphere of the catalyst has been shown to contain one nitrogen and two oxygen donors from the amide linkages.

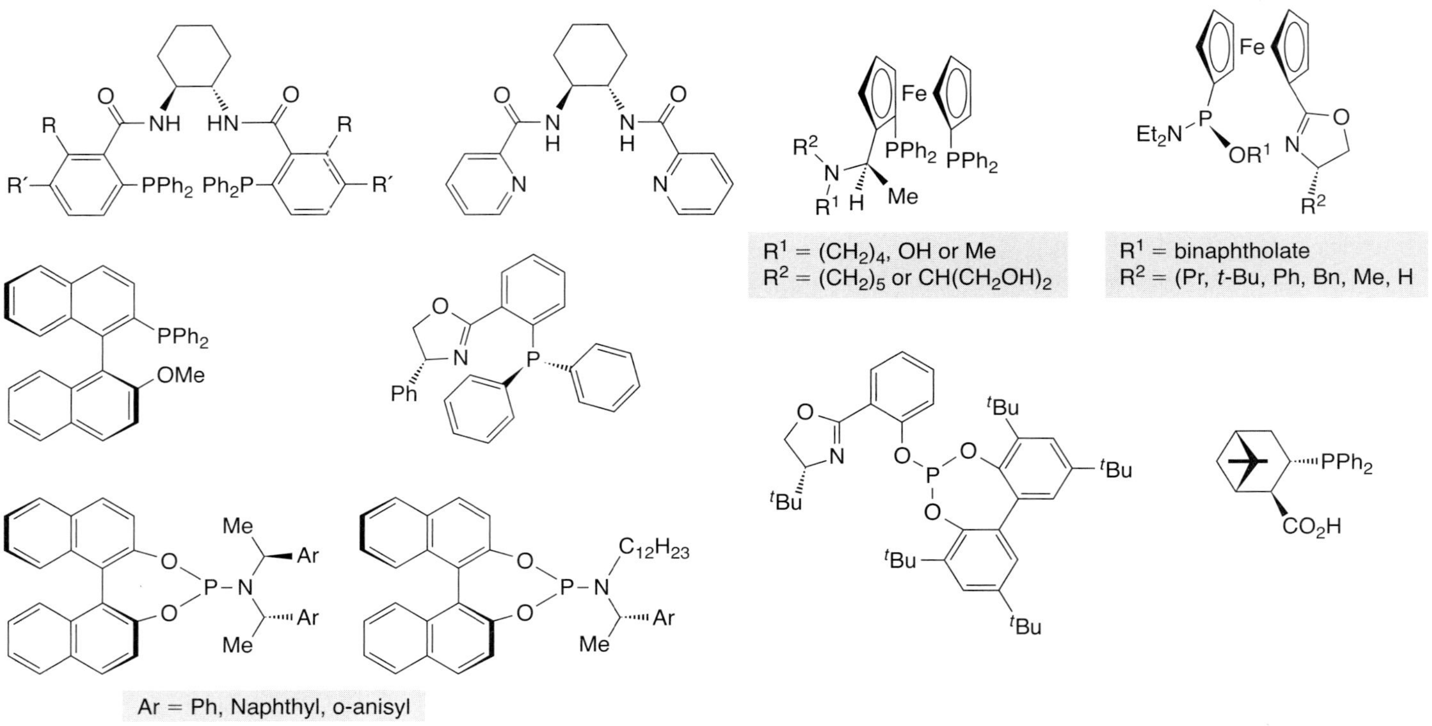

Figure 20.2.
Some of the ligands that generate highly enantioselective catalysts for allylic substitution with palladium, molybdenum, and iridium.

Hayashi and Ito reported a ferrocenyl ligand containing pendant heteroatom functionality that could direct the nucleophile to the more substituted site of a monosubstituted allyl intermediate.[168,212] A monodentate version of the binaphthyl phosphine MOP also leads to some highly enantioselective reactions of allylic acetates.[184] As mentioned in the sections covering regioselectivity, catalysts containing the MOP ligand lead to substantial retention of the regiochemistry of the starting allylic ester during the substitution process.

Other ligands contain a mixture of phosphorus and either nitrogen or oxygen donor groups. A well-known ligand developed independently by Pfaltz, Helmchen, and Williams[213,214] that has been employed for a number of catalytic processes is the phosphino oxazoline ligand in Figure 20.2. The high enantioselectivities obtained from reactions catalyzed by complexes of this ligand illustrated that C_2 symmetry in a ligand was not required to generate catalysts that reacted with high enantioselectivity. Claver and Pamies have studied a related ligand containing one oxazoline and one phosphite unit that is particularly reactive.[215]

Hou, Dai and co-workers reported a related oxazoline ligand that contains a pendant binaphtholate group that catalyzes the selective formation of branched allylic amines and allylic alkylation products from linear or branched allylic esters.[169] Helmchen developed the phosphine ligand with a pendant carboxylic acid that is particularly enantioselective for the reactions of aliphatic cyclic allylic acetates.[211]

Finally, phosphoramidite ligands have been shown to generate highly selective iridium catalysts for allylic substitution. The catalysts that are most selective are derived from the phosphoramidites in Figure 20.2 that contain one binaphtholate and one amino group possessing two phenethyl substituents or one phenethyl and one achiral alkyl group.[138,216] These ligands were first prepared for copper-mediated chemistry by Feringa and Alexakis[217–219] and were first used by Hartwig and co-workers to generate highly active and selective catalysts for allylic amination and etherification.[135,138,144] Alexakis has shown that an ortho-methoxy group on the phenethyl substituent enhances reactivity,[130] and Helmchen has then used this system for reactions of stabilized carbanions,[128,129,204,220] as well as some nitrogen nucleophiles.[140,221] The active iridium catalyst formed from these ligands contains a cyclom-etallated structure,[141] and it is this cyclometallated structure that led to the use of the unsym-metrical phosphoramidite.[142,206]

Finally, some optically active ruthenium complexes ligated by phosphines tethered to a cyclopentadienyl ring[100] or by bisoxazoline ligands and a cyclopentadienyl ligand[222] have been shown to catalyze enantioselective allylic substitution. The regioselectivity of unsymmetrical substrates tends to favor formation of chiral branched products, although the scope of reactions that occur with high enantiomeric excess have been limited to diphenylallylic esters.[102]

20.6.2. Enantioselective Allylic Substitution Classified by Electrophile

20.6.2.1. Enantioselective Allylic Substitution of Acyclic Electrophiles

20.6.2.1.1. Enantioselective Allylic Substitution of Symmetric Acyclic Allylic Esters

The enantioselectivity from reactions of substrates that generate symmetric allyl inter-mediates results from attack at one terminus over the other. The two termini of the allylic intermediate are enantiotopic, and the ligand on the metal controls the site of attack. As noted in the introduction to enantioselective allylic substitution, the most widely studied electro-phile for enantioselective allylic substitutions is 1,3-diphenylallyl acetate, and, as noted in the introduction to a symmetric allylic substitution, more than 50 ligands generate palladium complexes that catalyze this reaction in > 90% ee.[146] A much smaller number of catalysts lead to enantioselective reactions of the analogous aliphatic acyclic allylic esters. For example, reac-tions of pent-3-enyl-2-acetate with dialkyl malonates catalyzed by most complexes occur with low enantiomeric excess. The most selective reactions of this substrate include those catalyzed by complexes of the ligands developed by Trost (Equation 20.36, left), Helmchen (Equation 20.36, center), and Pamies and Claver (Equation 20.36, right).[215,223–225]

20.6.2.1.2. Enantioselective Opening of Vinyl Epoxides

Racemic vinyl epoxides have also been used as electrophiles for enantioselective allylic substitution chemistry.[34,226–228] These reactions occur by an intricate dynamic catalytic asym-metric sequence (Scheme 20.11). Reaction of the vinyl epoxide with a chiral catalyst generates diastereomeric allylic intermediates possessing two different allylic termini. This initial addi-tion presumably occurs with the same inversion of configuration as the oxidative addition of allylic esters. If reaction of the nucleophile with this intermediate at the more substituted carbon occurs, and if this addition is faster than epimerization of the allyl intermediate, then the reaction would generate racemic product resulting from retention or inversion of configuration of each component of the starting racemic mixture. However, if attack of the nucleophile on the allyl intermediate is slower than epimerization of the allyl group, then the product from the initial addition can equilibrate, and an enantioselective process can result when a non-racemic form of the catalyst is used. This equilibration of the allyl intermediate allows the establishment of a Curtin–Hammett situation, and the combination of the equilibrium concentrations of the two

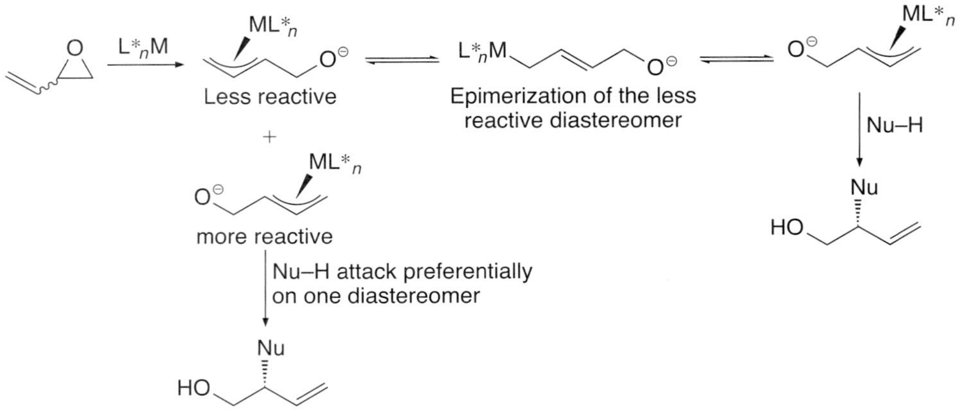

Scheme 20.11

intermediates and the rate constants for attack of the nucleophile on the two diastereomeric allyl intermediates controls the enantiomeric excess of the product. This dynamic catalytic process has been called "DYCAT" for dynamic catalytic asymmetric transformation. This DYCAT process with butadiene monoepoxide has been used in the synthesis of (+)-broussonetine G (Equation 20.37)[228] and has been extended to the use of vinyl aziridines as electrophiles.[47]

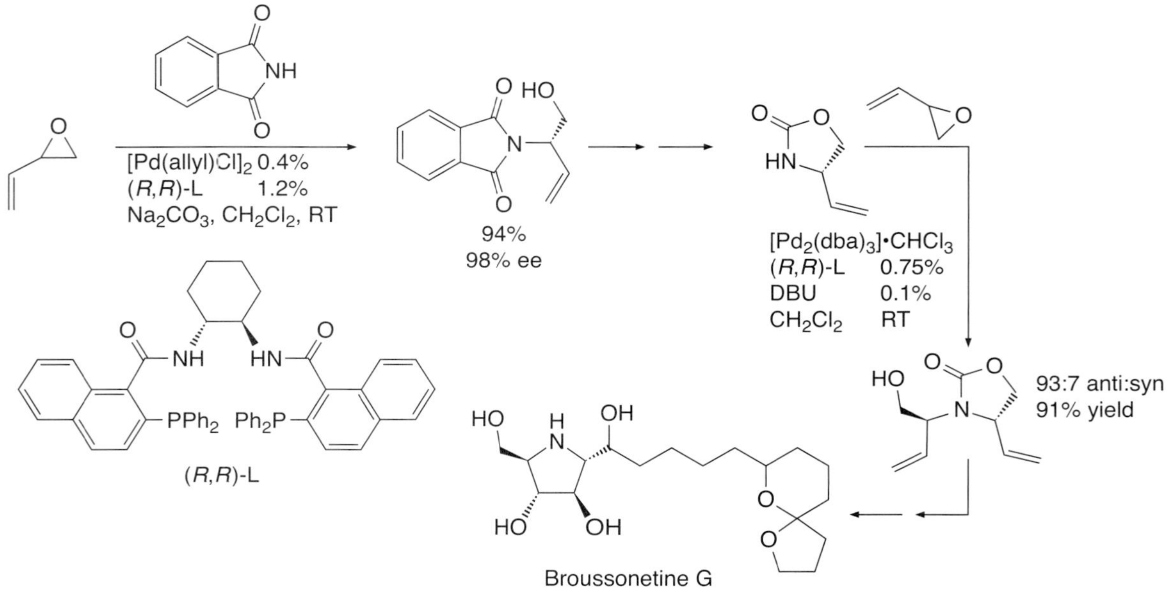

(20.37)

Broussonetine G

20.6.2.1.3. Enantioselective Reactions of Unsymmetrical Acyclic Substrates

20.6.2.1.3.1. Enantioselective Reactions of Unsymmetrical Acyclic Substrates Catalyzed by Palladium Complexes

Enantioselective and regioselective reactions of substrates that generate unsymmetrical allyl intermediates have been more challenging to develop with palladium catalysts than enantioselective reactions of substrates that generate symmetrical allyl intermediates. Two regioisomers can result from attack on the unsymmetrical allyl intermediates. As noted in Section 20.5.1, useful regioselectivities have been reported when the two allylic termini contain substituents of different size, when one carbon of the allyl group contains an alkoxo group, or when one carbon contains a conjugating group. When the allyl group contains a single substituent, most combinations of nucleophile and palladium catalyst generate the achiral product from substitution at the unsubstituted terminus, although exceptions to this general trend were noted in Section 20.5.1. In some cases, enantioselective and regioselective reactions have been developed with palladium catalysts, but many of the enantioselective reactions of unsymmetrical allylic electrophiles have been developed with metals other than palladium.

The reactions of acyclic geminal diacetates catalyzed by complexes of the Trost ligand occur with high regioselectivity and enantioselectivity.[229] These reactions can be viewed as carbonyl surrogates for allylic substitution. A variety of nucleophiles react with geminal diacetates derived from aldehydes to form monoacetate products with high enantiomeric excess. The scope of this chemistry is summarized in Equation 20.38. The high degree of functionality makes the products of these reactions synthetically valuable. For example, the enantioselective substitution of a geminal diacetate with the anion of an azlactone as the nucleophile shown in Equation 20.39 formed the quaternary stereocenter in products that were carried forward to prepare Sphingofungin F.[230, 231]

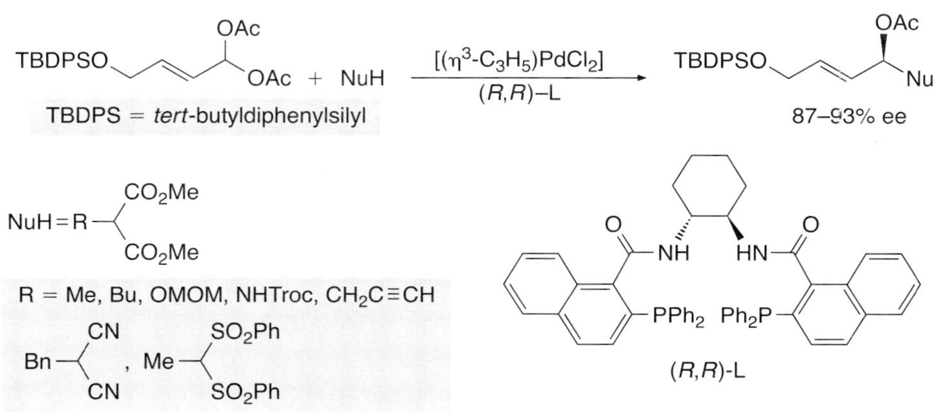

(20.38)

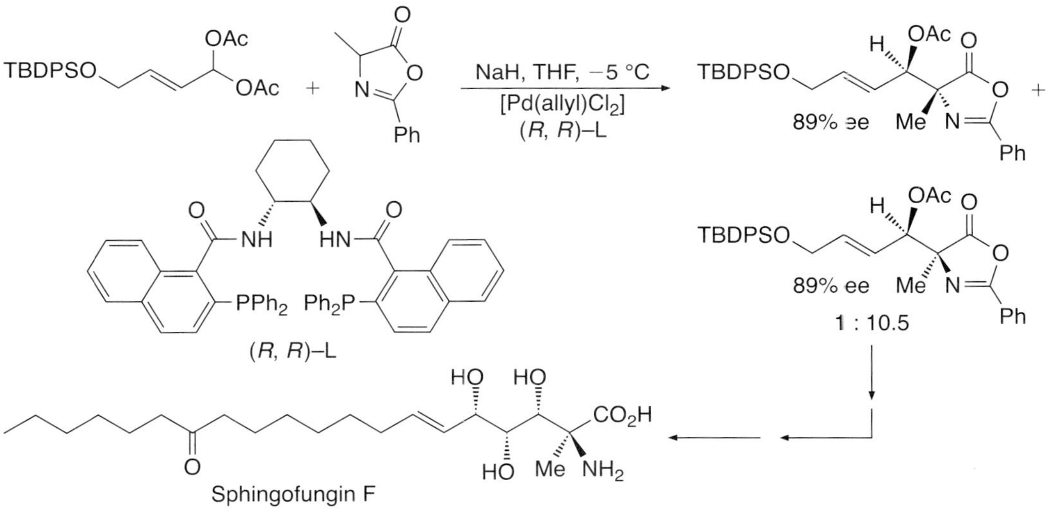

(20.39)

Sphingofungin F

20.6.2.1.3.2. Enantioselective Reactions of Unsymmetrical Allylic Esters Catalyzed by Molybdenum, Ruthenium, Rhodium, and Iridium

As noted in Section 20.5.3, catalysts generated from metals other than palladium tend to form the product from attack at the more substituted end of a monosubstituted allylic intermediate. Molybdenum and tungsten catalysts containing nitrogen-based ligands form, with high enantiomeric excess, branched substitution products from attack of sta-bilized carbon nucleophiles at the more substituted terminus of an allyl intermediate. Ruthenium catalysts for enantioselective allylic substitution with the same regioselectivity have also been reported,[100,222] but the enantiomeric excesses of the products formed by the ruthenium catalysts have been lower than those of the products formed by molybdenum and tungsten catalysts. However, the ruthenium complexes are able to catalyze reactions of nitrogen nucleophiles[100] and oxygen nucleophiles.[222]

Some of the most enantioselective catalysts for allylic alkylation to form branched products from terminal allylic carbonates or branched, racemic allylic carbonates are based on molybdenum and tungsten (Equation 20.26). These reactions are conducted with a com-bination of $M(CO)_3(NCMe)_3$ (M = Mo or W) or $Mo(CO)_6$ and the ligands of Figure 20.3. One might expect that the bipyridyldiamide ligands bind through both pyridines. However, as noted previously in this chapter, catalysts formed from the corresponding monopyridyl ligand (X = CH in Figure 20.3) reacted with similar enantioselectivities. This result suggests that the ligand is bound through nitrogen and the amide carbonyl group, as shown in the isolated allyl complex of Equation 20.23.

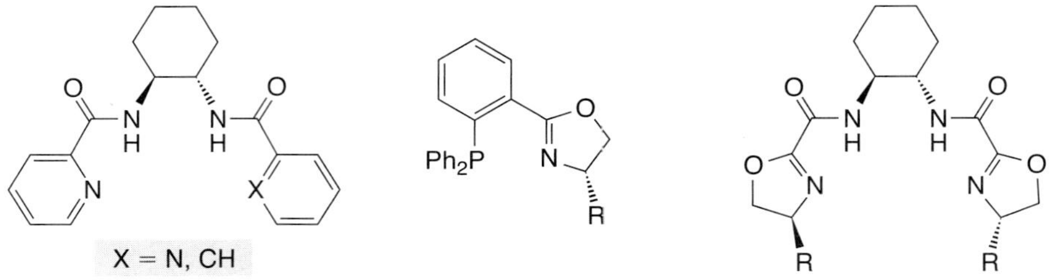

Figure 20.3. Ligands for enantioselective, molybdenum-catalyzed allylic substitution.

The molybdenum and tungsten complexes catalyze reactions of soft nucleophiles, such as malonates, related 1,3-dicarbonyl compounds, and nitroalkanes. Azlactones are also soft carbanions, and Trost has shown that complexes formed from molybdenum and the bis(pyridine) ligands catalyze enantioselective and diastereoselective allylation of azlactones with allylic phosphates to form quaternary amino acids (Equation 20.40). In these reactions, the nucleophile adds to the more substituted position of the ally-lic electrophile, and a stereocenter is formed at both the allyl carbon and the azlactone carbon. One route to the protease inhibitor tipranavir by the molybdenum-catalyzed allylation with 1,3-dicarbonyl compounds was demonstrated by Trost (Equation 20.41), and the Merck process group used related allylation chemistry with Trost's bis(pyridine) ligand to prepare the cyclopentanone precursor to various analogs of tipranavir (Equa-tion 20.42).

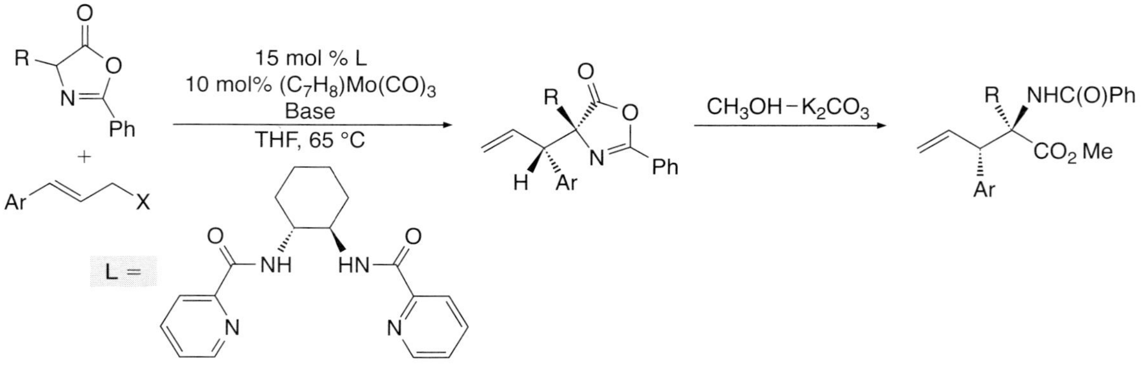

(20.40)

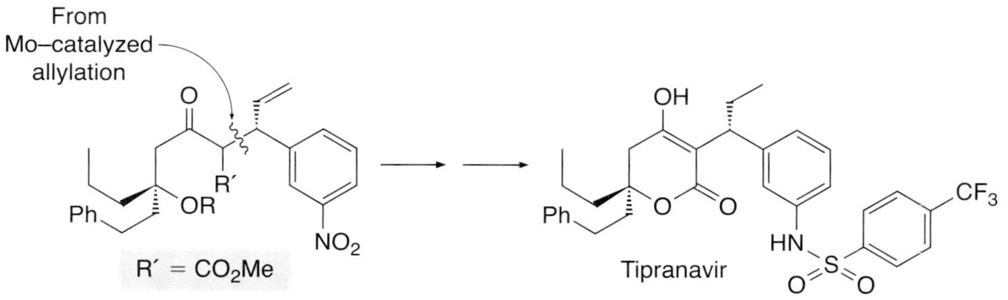

(20.41)

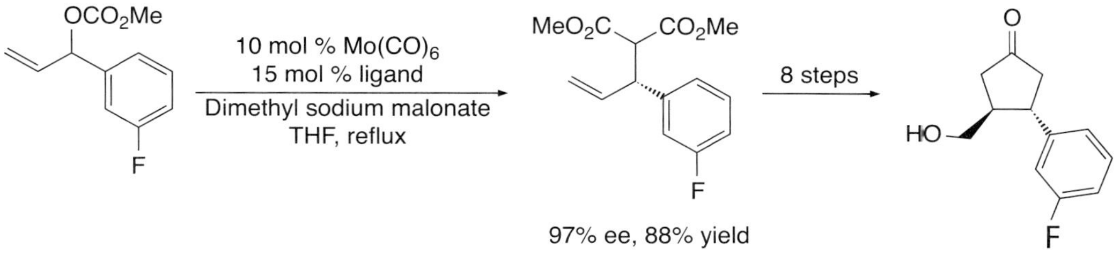

(20.42)

97% ee, 88% yield

Rhodium and iridium catalysts have generated the chiral, branched products with high enantiomeric excess from reactions of monosubstituted allylic electrophiles with, inter alia, stabilized or unstabilized carbon nucleophiles, amines, tosylamides, phenoxides, and alkoxides. Efforts to develop allylic substitutions with rhodium catalysts have focused on conducting reactions with conservation of the enantiomeric excess of an optically active branched allylic acetate or carbonate (Equation 20.43). A combination of Wilkinson's catalyst and trimethylphosphite generates a system that catlyzes the addition of stabilized carbon nucleophiles, such as malonates,[106,111] "unstabilized" copper enolates,[16] and heteroatom nucleophiles, such as tosylamides,[107,112] copper phenoxides,[109] and copper alkoxides,[113,115] to allylic carbonates with branched regioselectivity and predominant retention of configuration. In addition, an example of an enantioselective, rhodium-catalyzed addition of malonates to racemic branched allylic acetates[232] has been reported when the rhodium is bound by a phosphino–oxazoline ligand linked by an axial chiral binaphthyl unit (Equation 20.44).

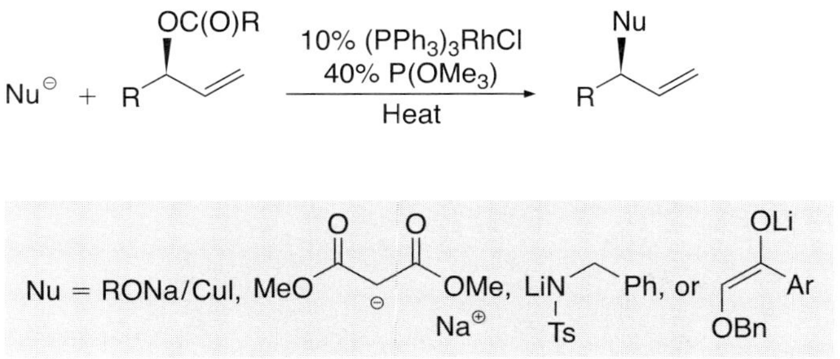

(20.43)

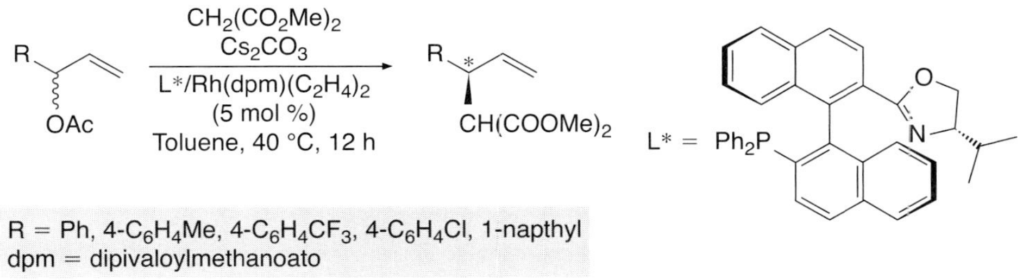

(20.44)

R = Ph, 4-C_6H_4Me, 4-$C_6H_4CF_3$, 4-C_6H_4Cl, 1-napthyl
dpm = dipivaloylmethanoato

In contrast, iridium complexes of phosphoramidite ligands catalyze the enantiose-lective formation of the branched allylic substitution products with high enantiomeric excess. Takeuchi[121,122,124,125,137] and Helmchen[117-120,129,139,204] reported that iridium complexes, like rhodium complexes, generate the chiral, branched product from reactions of mono-substituted allylic acetates and carbonates with carbon and nitrogen nucleophiles (Equations 20.45–20.47).

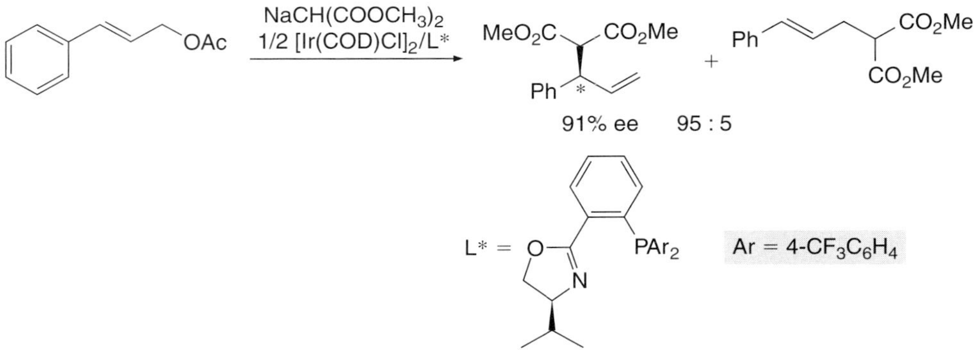

(20.45)

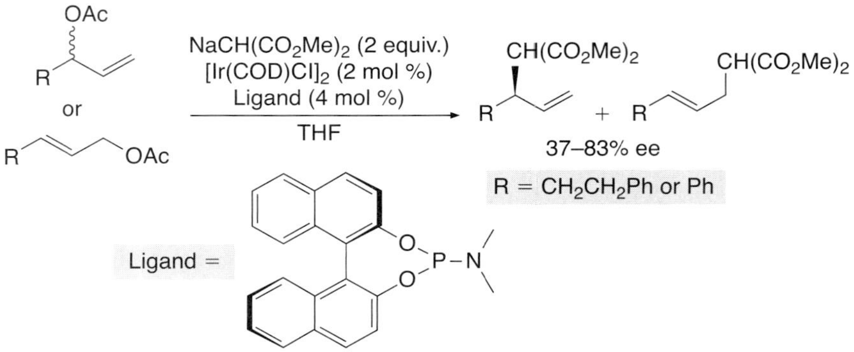

(20.46)

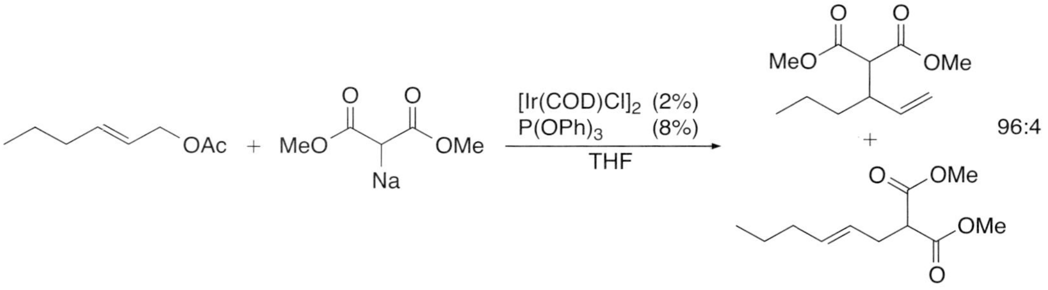

(20.47)

More recently, Hartwig and co-workers showed that iridium complexes of the phos-phoramidite ligand shown in Equation 20.48 catalyzes these reactions with amines, ary-loxides, the combination of alkali metal alkoxide and CuI, the combination of alcohol and base, trifluoroacetamide, azoles, and ketone enolate equivalents with enantiomeric excess near or above 95% in most cases.[71,134-136,138,141-144,233-235] They have also shown that this irid-ium system catalyzes the reaction of allylic carbonates with the combination of silyl enol ether and fluoride additive or with enamines of ketones to form products from attack of the enolate or enolate equivalent at the more substituted position of the allyl electrophile with high enantiomeric excess.[134]

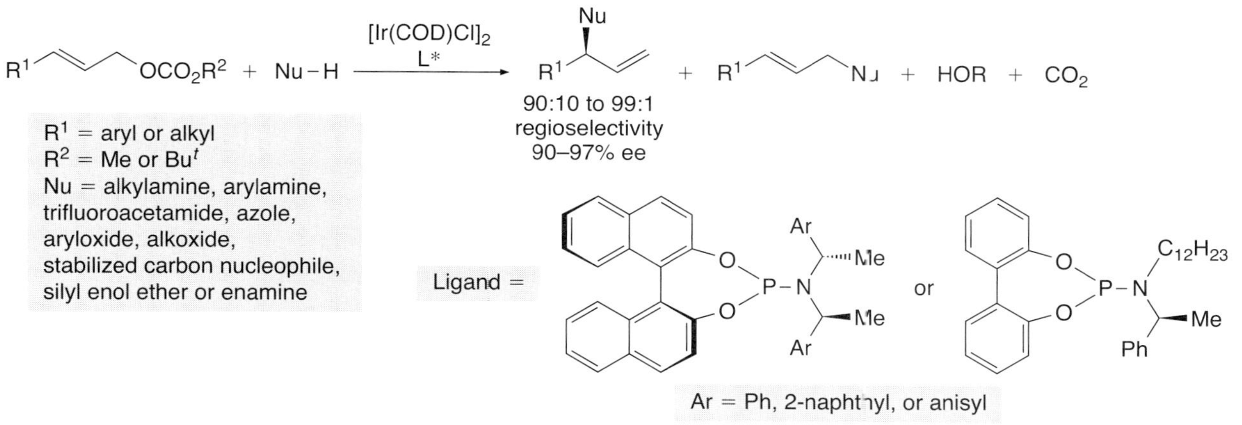

$$R^1\overset{\frown}{}OCO_2R^2 + Nu-H \xrightarrow[L^*]{[Ir(COD)Cl]_2} R^1\overset{Nu}{\underset{}{\overset{|}{\wedge}}}\overset{\frown}{} + R^1\overset{\frown}{}NJ + HOR + CO_2$$

90:10 to 99:1
regioselectivity
90–97% ee

R¹ = aryl or alkyl
R² = Me or Buᵗ
Nu = alkylamine, arylamine,
trifluoroacetamide, azole,
aryloxide, alkoxide,
stabilized carbon nucleophile,
silyl enol ether or enamine

Ligand =

or

(20.48)

Ar = Ph, 2-naphthyl, or anisyl

The active catalyst in the iridium chemistry is not a simple square-planar complex formed by coordination of the phosphoramidite to $[Ir(COD)Cl]_2$. Instead, the phosphoramidite ligand in $[Ir(COD)(L)(Cl)]$ (L = phosphoramidite) undergoes cyclometallation to generate the five-membered iridacycle in Equation 20.49,[141] and reactions initiated with this metallacycle and an additive to bind the κ^1-phosphoramidite after it dissociates occur much faster than those initiated with the combination of $[Ir(COD)Cl]_2$ and the free phosphoramidite. The knowledge of the structure of the active catalyst has led to the design of unsymmetrical phosphoramidite ligands that are much simpler in structure than the original ligands, but which form complexes that catalyze the substitution reactions with rates that are equivalent to those of the reactions catalyzed by complexes of the more complex ligands used originally. The simplest of these ligands is the one derived from the unsymmetrical N-cycloalkyl phenethylamine and biphenol shown at the right of Equation 20.48.[144]

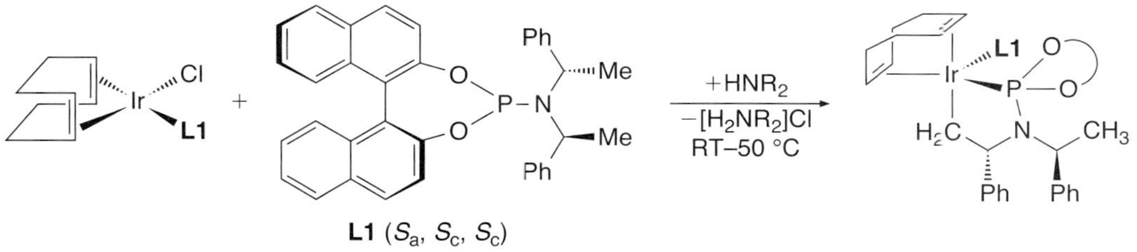

L1 (S_a, S_c, S_c)

$$\xrightarrow[-[H_2NR_2]Cl]{+HNR_2} \quad RT–50\ °C$$

(20.49)

20.6.2.2. Enantioselective Substitution of Cyclic Substrates

As noted in Section 20.6.1.1, the reactions of racemic cyclic allylic esters that generate symmetrical allylic intermediates and reactions of cyclic allylic diacetates in which the two acetate groups are enantiotopic have been developed. Examples of these reactions and catalysts that are most selective for these reactions are described in the next two sections.

20.6.2.2.1. Enantioselective Substitution of Cyclic Allylic Monoesters

Enantioselective allylic substitutions of cyclic allylic esters have been more challenging to develop than enantioselective reactions of symmetrical, acyclic allylic esters. In one set of reactions, racemic allylic esters react to form non-racemic products by addition of carbon or nitrogen nucleophiles in the presence of palladium catalysts. In these cases, attack at the two termini of the allylic intermediate generates the two enantiomers. Only a handful of ligands have generated catalysts that form products from the substitution of aliphatic,

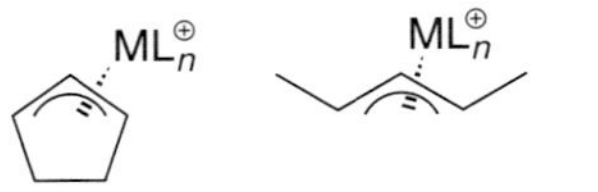

Figure 20.4.
Comparison of the structures of allyl complexes generated from cyclic and acyclic allylic esters.

cyclic, allylic carbonates in high enantiomeric excess (Equation 20.50). The ligands that have been shown to generate catalysts for highly enantioselective substitutions of five- and six-membered cyclic allylic acetates are shown in Equation 20.50.[211,236-238] These ligands include the bisphosphine ligands developed by Trost, a phosphinooxazoline ligand based on cymantrene developed by Helmchen, a phosphinocarboxylic acid ligand developed by Helmchen, a phosphinito sulfide developed by Evans, and a phosphite–oxazoline ligand developed by Pamies and Claver.[215] The difference in structures of the cyclic and acyclic metal–allyl fragments shown in Figure 20.4 illustrates why the optimal ligands for reactions of cyclic and acyclic substrates are often different.

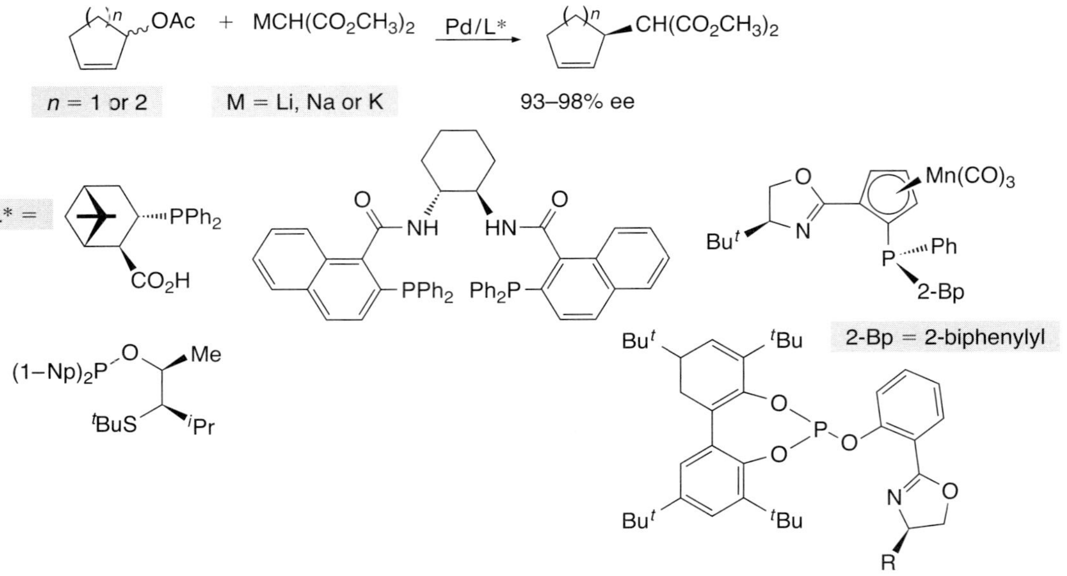

$$n = 1 \text{ or } 2 \qquad M = \text{Li, Na or K} \qquad 93\text{--}98\% \text{ ee}$$

(20.50)

2-Bp = 2-biphenylyl

20.6.2.2.2. Enantioselective Substitution of Meso Cyclic Diesters

In a second set of examples, cyclic substrates containing two enantiotopic allylic acetates undergo enantioselective substitution (Equations 20.51 and 20.52). These reactions have been conducted with palladium catalysts, particularly those of Trost.[239] In this case, the palladium catalyst selects for reaction at one of two ester groups of an achiral meso substrate. Reaction at the two esters generates enantiomeric products. This approach has been used to generate many natural products (for one example, see Equation 20.52),[239-241] carbonucleosides (for one example, see Equation 20.53),[242] and other optically active cyclic materials for further synthetic applications.

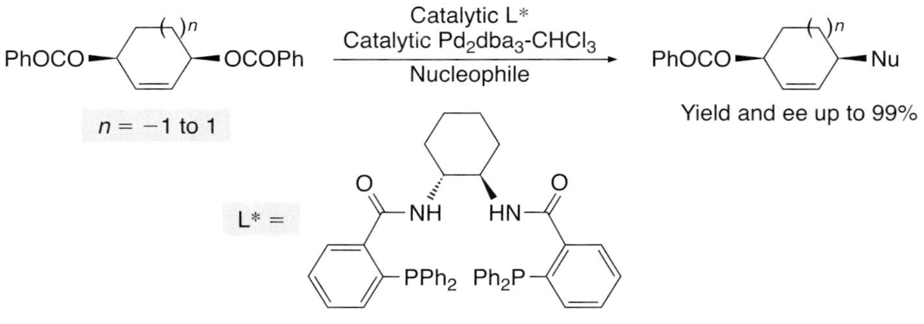

$$n = -1 \text{ to } 1 \qquad\qquad \text{Yield and ee up to } 99\%$$

(20.51)

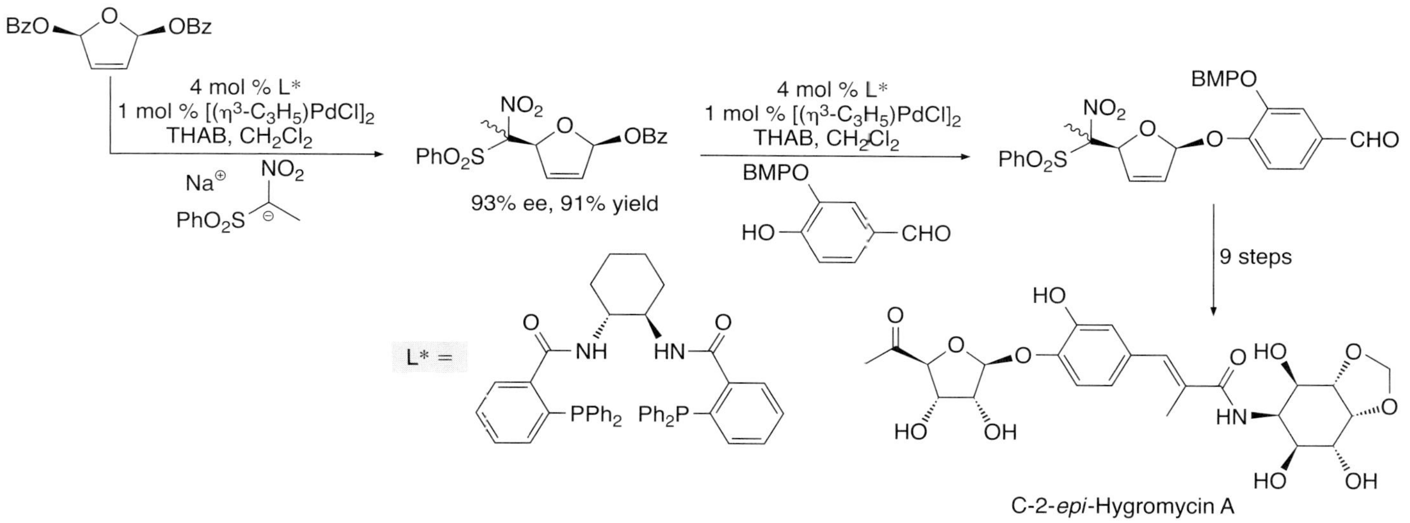

C-2-*epi*-Hygromycin A

(20.52)

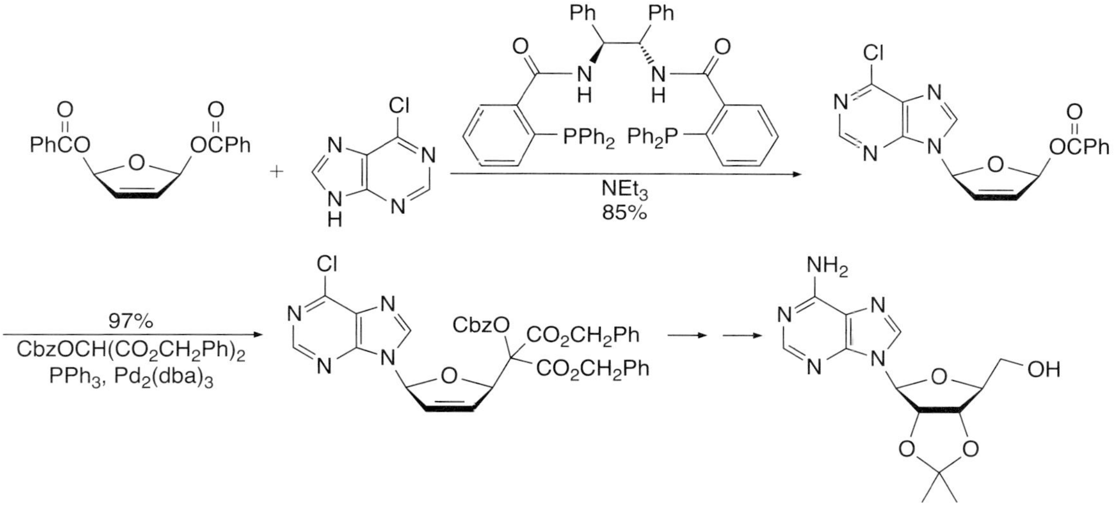

(20.53)

20.6.3. Kinetic Resolution

Kinetic resolutions of allylic esters have also been conducted. As noted in Chapter 14, in most cases, catalysts that are selective for kinetic resolution of substrates containing one type of functional group are also selective for reactions of meso substrates and vice versa.[243,244] The enantioselective reaction of one acetate of a meso substrate involves similar stereochemical recognition to reaction of two enantiomers of a racemic mixture of allylic esters. In one illustrative example, an allylic acetate underwent reaction with a pivalate nucleophile to form the allylic pivalate product, which is less reactive than the starting compound (Equation 20.54).[245] Reaction of one enantiomer of the racemic allylic acetate occurs preferentially, and the pivalate product is formed enantioselectively. The product of the resolution in Equation 20.54 was carried forward to form (+)-cyclophellitol.

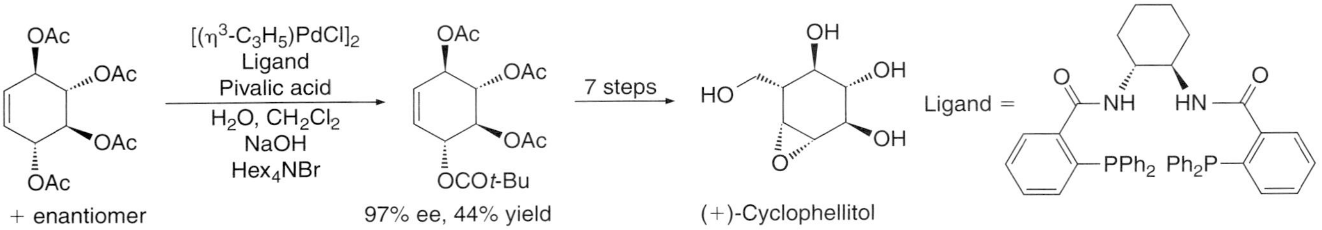

97% ee, 44% yield   (+)-Cyclophellitol   (20.54)

20.6.4. Enantioselective Allylation of Prochiral Nucleophiles

A final approach to enantioselective allylic substitution is the reaction of prochiral nucleophiles with allylic esters. In this case, the stereocenter is not generated on the allyl unit; it is generated at the nucleophilic carbon. This chemistry has been conducted with cyanoesters and related unsymmetrical stabilized carbon nucleophiles, including azlactones, which are a protected form of amino acids. This generation of a stereocenter in the nucleophile is thought to be particularly challenging because the position at which the stereocenter is formed is further from the metal than it is in reactions that form a stereocenter at the allyl group.

It was for this type of transformation that Trost designed the amide-linked bisphosphine ligands shown in Figure 20.2 possessing a wide bite angle. These ligands were envisioned to create a chiral pocket that extends from the metal and begins to encompass the prochiral nucleophile.[246,247] With these ligands, some of the first reactions of allylic esters with prochiral nucleophiles formed the substitution products with high enantiomeric excess (Equation 20.55). In addition to complexes of this ligand, complexes of a trans-spanning ligand called "TRAP" (for trans phosphine) reported by Ito also catalyzed enantioselective reactions with prochiral nucleophiles. In this case, the conditions were designed to generate the nucleophile on rhodium, and for the rhodium complex to transfer the enolate to the palladium allyl intermediate (Equation 20.56).[248] This property of exceptionally wide bite angle is not required for enantioselective versions of this reaction, however. Palladium complexes of BINAP also catalyze the allylation of prochiral nucleophiles with high enantioselectivity (Equation 20.57).[249]

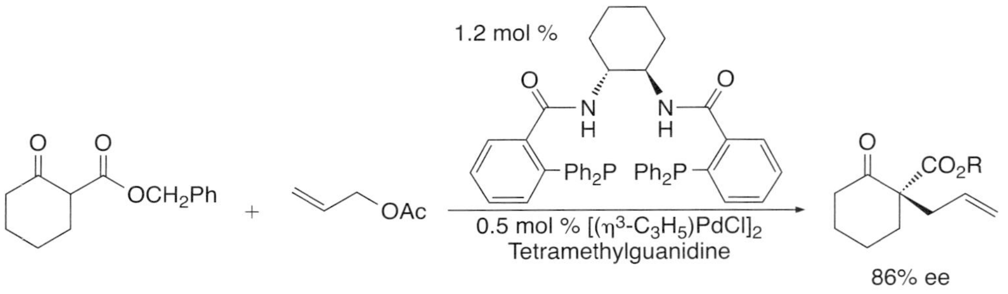

(20.55)

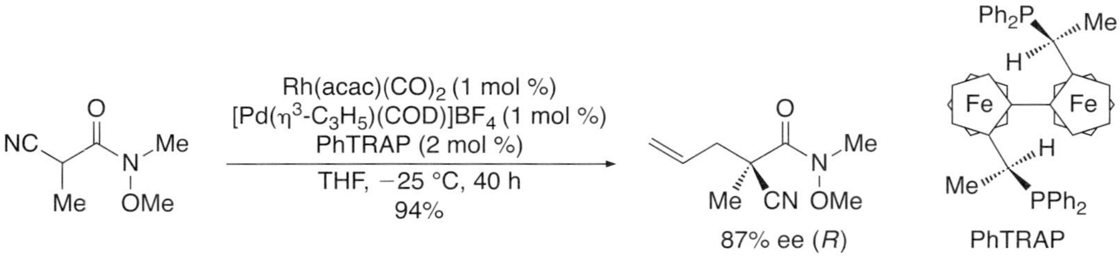

(20.56)

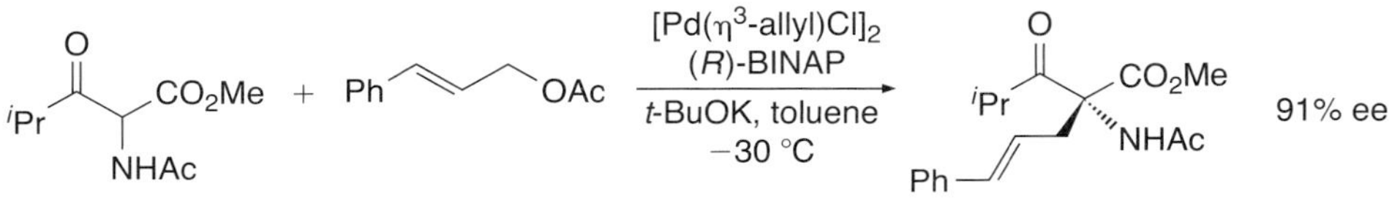

(20.57)

Many diastereoselective allylations form a new stereocenter at one of the allylic carbons and at the nucleophilic carbon. For example, an iridium complex containing a phosphite ligand catalyzes enantioselective and diastereoselective formation of products containing two stereocenters, one at the original nucleophile and one at the original allyl electrophile (Equation 20.58).[202,203] In another example shown in Equation 20.59, Trost's palladium catalyst leads to the reaction of allylic esters with chiral azlactone pronucleophiles with high diastereomeric and enantiomeric excess,[250] as does the related molybdenum catalyst.[90] In these cases, the metal appears to control the new stereocenter at the allyl group, as well as the relationship between this stereocenter and the new stereocenter formed at the nucleophile.

$$(20.58)$$

$$(20.59)$$

As a final set of examples, enantioselective allylic substitution of unstabilized enolates to form a new stereocenter at the enolate carbon have been developed through the decarboxylative reactions of allyl enol carbonates.[82,83,251–257] These reactions are enantioselective versions of reactions closely related to those in Equation 20.18 and Scheme 20.4, and two examples are shown in Equations 20.60 and 20.61. In these cases, a new stereocenter is formed at the α-carbon of the enolate nucleophile. Most of these reactions have been conducted with allyl enol carbonates that generate cyclic ketone enolates, but enantioselective reactions of acyclic allyl enol carbonates have also been reported.[257] Although allyl enol carbonates undergo decarboxylation faster than the β-keto ester isomers, the O-allyl β-keto esters are more difficult to prepare,[258] and enantioselective allylations starting with β-ketoesters have been reported.[255,258] Decarboxylative reactions of amines and α-amino acids have been conducted to form allylic and homoallylic amines (Equation 20.62), respectively, and enantioselective decarboxylative allylations of amides have been reported.[83,259] Iridium-catalyzed enantioselective decarboxylative allylation of amides starting with O-allyl imides has also been reported.[260]

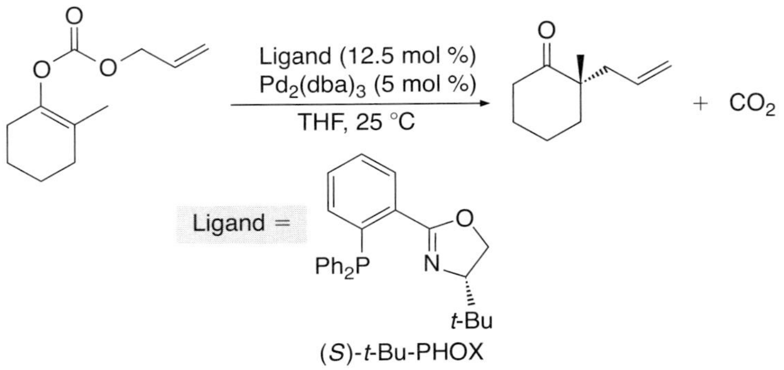

(20.60)

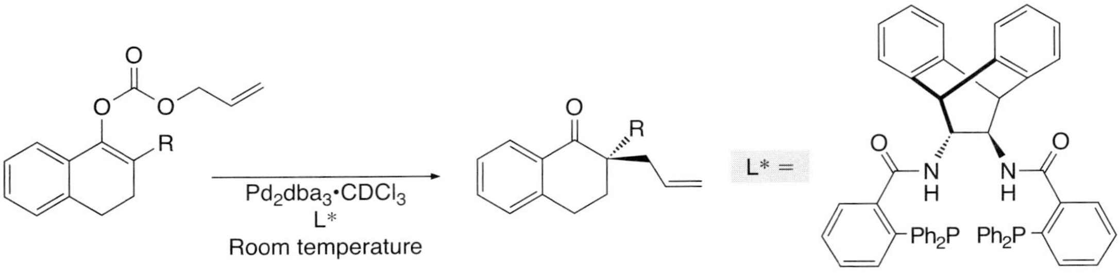

(20.61)

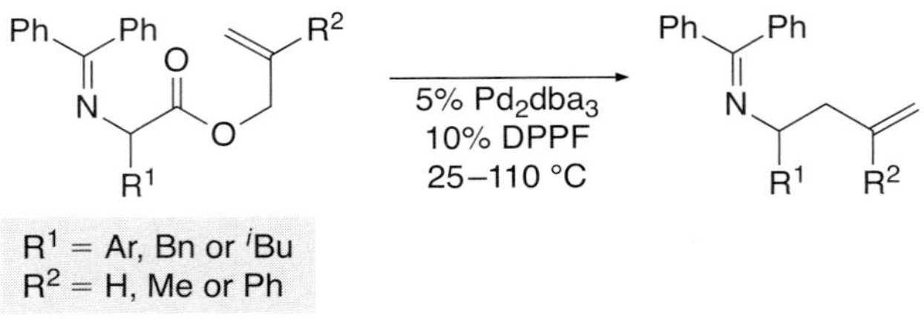

(20.62)

R^1 = Ar, Bn or ^{i}Bu
R^2 = H, Me or Ph

The mechanism of these reactions is not known in detail, but some data have been obtained. Considering the body of data described previously in this chapter on the reactions of allylic esters, the initial step of the reaction is almost certainly cleavage of the allyl–oxygen bond to form a metal–allyl intermediate. The resulting carbonate could be envisioned to extrude CO_2 spontaneously, but the anion of an α-imino acid and a β-keto acid require the palladium to extrude CO_2 under mild conditions.[83] After the enolate is generated, the stereochemistry is established. Whether the enolate adds after binding to the metal or by external attack is uncertain, but it has been observed that reactions conducted with formic acid as a proton source lead to enantioselective quenching of the enolate (Equation 20.62).[80] This observation implies that the enolate is chiral and, therefore, must be bound to the chiral ligand–palladium fragment at some point of the catalytic cycle, as depicted in Equation 20.63.

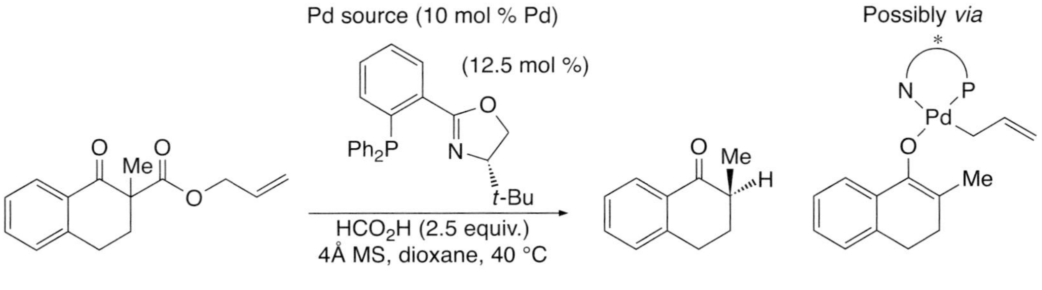

(20.63)

20.7. Copper-Catalyzed Allylic Substitution (Written with Levi Stanley)

20.7.1. Fundamentals

Allylic substitution catalyzed by copper[261–269] is a transformation that is related to allylic substitutions catalyzed by other transition metals discussed previously in this chapter, but several features of copper-catalyzed allylations make them worth differentiating. First, copper-catalyzed allylic substitutions are conducted with different types of nucleophiles than most allylic substitutions catalyzed by other metals. Second, the regioselectivity of the copper-catalyzed reactions is typically different from that of reactions catalyzed by complexes of other metals, particularly of reactions catalyzed by complexes of palladium. Thus, this last section of the chapter describes studies on allylic substitution catalyzed by copper, with an emphasis on enantioselective examples.

Copper-catalyzed allylic substitution is typically conducted with hard, nonstabilized carbon nucleophiles (Equation 20.64), instead of the "soft" stabilized carbon nucleophiles used most often in allylic substitutions catalyzed by other transition metal complexes. The most common nucleophiles used in copper-catalyzed allylic substitutions are diorganozinc reagents, Grignard reagents, organolithium reagents, and trialkylaluminum reagents. Common allylic electrophiles include allylic halides, acetates, carbononates, phosphates, and tosylates.

$$R^2 = \text{alkyl, aryl, vinyl, or allyl}$$
$$M = ZnX, MgX, AlX_2, Li, \text{etc.}$$
$$LG = Cl, Br, OC(O)R, OP(O)(OR_2)_2, OSO_2R, \text{etc.}$$

(20.64)

Copper-catalyzed allylic substitution typically occurs to form products from the addition of the nucleophile at the position γ to the leaving group. However, the tendency for copper-catalyzed allylic substitutions to form products from γ-substitution versus α-substitution depends on a number of reaction parameters, including the identity of the allylic electrophile, the reaction solvent, the reaction temperature, and the source of the organometallic nucleophile. Early examples of allylic substitutions with stoichiometric amounts of organocopper reagents demonstrated the importance of the identity of the organocopper species. For example, the reaction of the monoalkyl cuprate MeCu(CN)Li with a substituted cyclohexenyl acetate is regioselective for addition to the γ-position and occurs with anti stereoselectivity (Equation 20.65).[270] The corresponding reaction of the dialkyl cuprate Me$_2$CuLi also occurs with anti stereochemistry, but it gives a mixture of regioisomeric products.[271] Thus, copper-catalyzed allylic substitutions are often conducted under conditions that prevent the formation of symmetrical dialkyl cuprates.

MeCu(CN)Li	96	:	4
Me$_2$CuLi	50	:	50

(20.65)

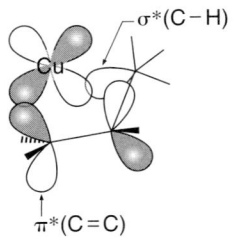

Figure 20.5.
Orbital interactions proposed to account for the anti diastereoselectivity. Figure adapted from reference 272.

The observed anti stereoselectivity is proposed to result from simultaneous interaction of a $3d$ copper orbital with the π^* (C=C) and σ^* (C–X) orbitals of the allylic electrophile (Figure 20.5).[272] Although anti diastereoselectivity is most often observed, additional factors, such as chelating leaving groups[273–277] or steric constraints[270,278,279] on the allylic electrophile, can override the inherent stereoselectivity.

Copper-mediated and copper-catalyzed allylic substitution reactions have been extensively developed over the past 40 years. Since the first report in 1969,[280] much of the early literature on allylic substitution with organocopper reagents describes reactions conducted with stoichiometric quantities of the metal.[270,271,278] However, allylic substitution reactions conducted with substoichiometric amounts of copper salts have been known since a report by Schlosser[281] in 1974, and most allylic substitutions with copper are now conducted with catalytic amounts of metal. Furthermore, the use of substoichiometric quantities of copper in combination with a chiral ligand has led to the development of catalytic, enantioselective copper-catalyzed allylic substitution reactions. A majority of this section of Chapter 20 summarizes factors that control regioselectivity and enantioselectivity, and illustrations of the reaction scope. However, mechanistic information is presented first to provide a foundation for understanding the synthetic methodology. This mechanistic information is still emerging and is less established than the mechanistic information on palladium-catalyzed allylic substitution.

20.7.2. Mechanism of Copper-Catalyzed Allylic Substitution

Bäckvall and Van Koten proposed the most commonly cited mechanism for copper-catalyzed allylic substitution (top of Equation 20.66).[282] This proposed mechanism involves initial coordination of the allylic electrophile to form an η^2,π-complex. Subsequent oxidative addition then sets the stereochemistry of the reaction and generates a σ-allyl intermediate in which the copper is bound to the position that is γ to the leaving group. In this mechanistic proposal, the regioselectivity of the reaction is dictated by the identity of the ligand X of the copper salt. When the ligand X is electron withdrawing, rapid reductive elimination occurs from the corresponding Cu(III) intermediate to form the γ-substitution product. When the ligand X is more electron donating, the σ-allyl intermediate in which the copper is bound at the γ-position is more stable. This greater stability is proposed to allow this intermediate to isomerize to the less sterically hindered α-substituted σ-allyl Cu(III) species through an η^3,π-allyl Cu(III) species. Subsequent reductive elimination would then form the α-substitution product.

Although many aspects of this mechanism have been accepted, computational studies by Nakamura[283,284] and experimental studies by Bertz and Ogle[285] suggest that modifications to this proposal are needed to fit the current data. Bertz and Ogle obtained the first spectroscopic data on the allylcopper(III) intermediates[285] that are commonly proposed in copper-catalyzed allylic substitution reactions. Their data imply that the structure of the allyl dimethylcopper(III) species contains a π-allyl unit.

Theoretical studies by Nakumura suggest that the regioselectivity of the allylic alkylation catalyzed by complexes of cyanocuprates can be rationalized by the unsymmetrical structure of the cuprate complex that undergoes oxidative addition of the allylic electrophile (bottom of Equation 20.66). These calculations imply that addition of MeCu(CN)Li to allyl acetate is γ-selective because the allyl complex contains two different ligands trans to the two termini of the η^3-allyl ligand. The transition state for oxidative addition to form the isomer in which the methyl group is located trans to the acetate leaving group (R^2=Me, LG=OAc in Equation 20.66) is calculated to be 3.2 kcal/mol lower in energy than the transition state that forms the isomer in which the cyanide ligand (X=CN in Equation 20.66) is located trans to the leaving acetate. Because the methyl group couples with the allyl terminus located cis to it, this sequence leads to coupling of the methyl group with the allyl terminus that was located γ to the leaving group in the starting allylic electrophile. A further understanding of the mechanism awaits experimental studies on allylcopper(III) complexes containing CN or other ligands in an unsymmetrical species that could be involved in the catalytic cycle for γ-selective coupling of hard nucleophiles with allylic electrophiles catalyzed by copper complexes.

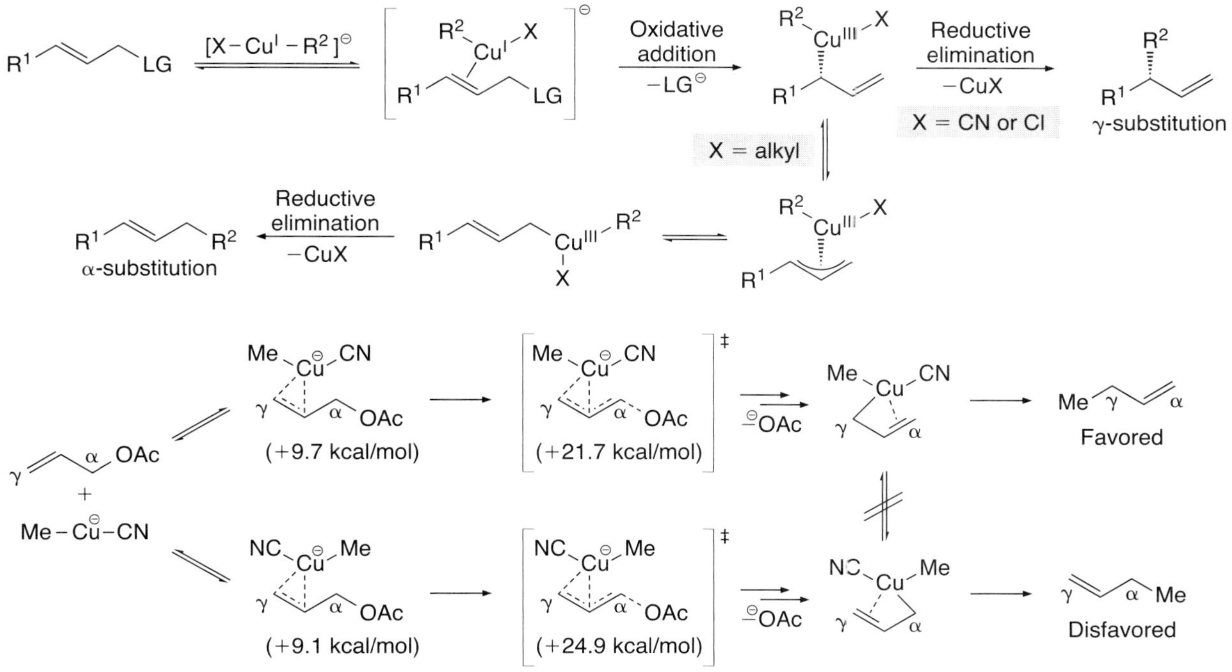

(20.66)

20.7.3. Enantioselective Copper-Catalyzed Allylic Substitution

The enantioselective copper-catalyzed addition of alkyl nucleophiles to allylic electrophiles is now a well-established reaction. Some of first examples of this process were reported by Bäckvall and Van Koten.[286] They showed that the addition of a Grignard reagent to an allylic acetate in the presence of a chiral arenethiolatocopper(I) complex proceeds with exclusive γ-selectivity and measurable enantioselectivity (Equation 20.67). Subsequently, Dübnar and Knochel published on the use of diorganozinc reagents as nucleophiles in enantioselective copper-catalyzed allylic substitution.[287,288] They reported additions of dialkylzinc reagents to allylic halides in the presence of a Cu(I) salt and ferrocenyl-based chiral amines (Equation 20.68). Pineschi, Feringa, and co-workers reported early examples of enantioselective copper-catalyzed additions of dialkylzinc reagents to related cyclic 1,3-diene monoepoxides.[289]

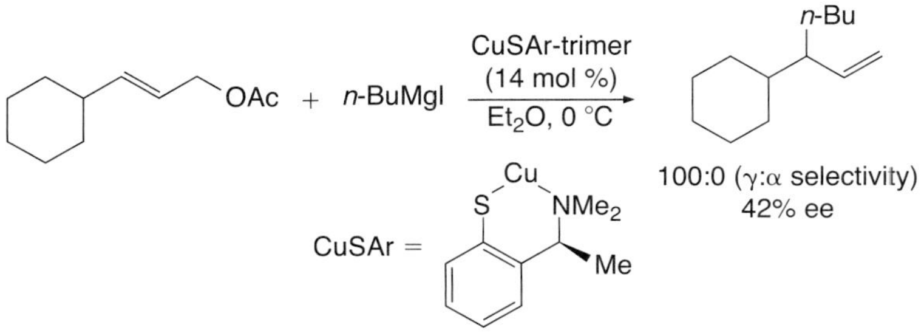

(20.67)

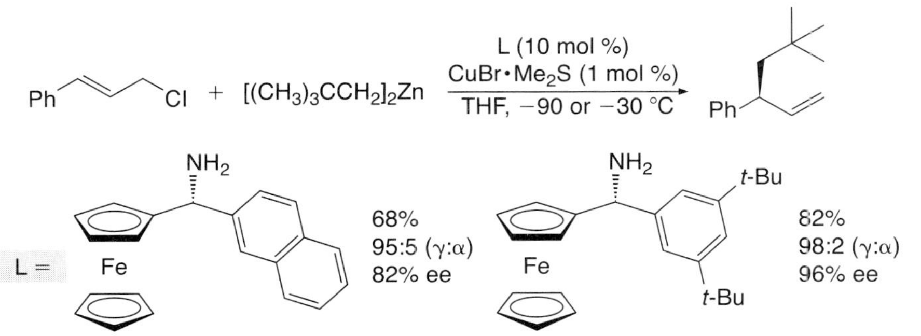

(20.68)

20.7.3.1. Diorganozinc Reagents as Nucleophiles

Dübnar and Knochel's initial work led to a number of additional reports on the use of diorganozinc reagents as nucleophiles in enantioselective copper-catalyzed additions to allyl halides. Zhou[290] and Feringa[291] reported that copper(I) complexes of phosphoramidite ligands catalyze additions of dialkylzinc reagents to allyl halides (Equations 20.69 and 20.70) with good to excellent regioselectivity and moderate to good enantioselectivity. Woodward subsequently showed that β,β-disubstituted α-methylene propionates formed with high enantioselectivities from the additions of a dialkylzinc reagent to electron-deficient allylic chlorides generated by Baylis–Hillman reactions (Equation 20.71).[292,293]

$$Ph\diagup\diagdown\diagup Br \; + \; i\text{-}Pr_2Zn \xrightarrow[\text{Digylme, } -30\,^\circ C]{\substack{(CuOTf)_2\cdot C_6H_6 \ (0.5\ mol\ \%) \\ L^* \ (2\ mol\%)}} $$

82%
91:9 (γ:α)
67% ee

(20.69)

$$Ph\diagup\diagdown\diagup Br \; + \; i\text{-}Pr_2Zn \xrightarrow[\text{THF, } -60\,^\circ C]{\substack{CuOTf \ (1\ mol\ \%) \\ L^* \ (2\ mol\%)}} $$

94%
97:3 (γ:α)
88% ee

(20.70)

53–95%
76–90% ee

(20.71)

L* =

MAO = Methylaluminoxane

(10 mol%)

The enantioselective copper-catalyzed addition of organozinc reagents to allylic electrophiles has been improved by Hoveyda and co-workers. They showed that a copper complex of a peptide-based ligand catalyzes enantioselective additions of dialkylzinc reagents to allylic phosphates in high yields and selectivities (Equation 20.72).[294,295] Furthermore, the methodology tolerates a remarkable variety of substituents on the allylic phosphate and can be used to generate both tertiary and quaternary carbon stereocenters. Related peptide-based ligands were also developed for the copper-catalyzed synthesis of α-alkyl-β,γ-unsaturated esters and α,α'-disubstituted-β,γ-unsaturated esters from reactions of diorganozinc nucleophiles with α,β-unsaturated carbonyl compounds bearing a leaving group at the γ-position.[296,297] Additional improvements in the amount of copper and chiral ligand required, as well as the selectivities for additions of dialkylzinc reagents to allylic phosphates, were realized by using a chiral, dimeric Ag(I)–N-heterocyclic carbene complex to generate a chiral copper(I)–NHC catalyst (Equation 20.73).[298,299] Related dimeric Ag(I)–NHC complexes have been used to generate copper catalysts for allylic alkylations of silicon-substituted allylic phosphates with diorganozinc reagents.[300] The resulting allylsilanes bearing silicon-substituted carbon stereocenters are formed with nearly perfect regioselectivity and excellent enantioselectivity (Equation 20.74).

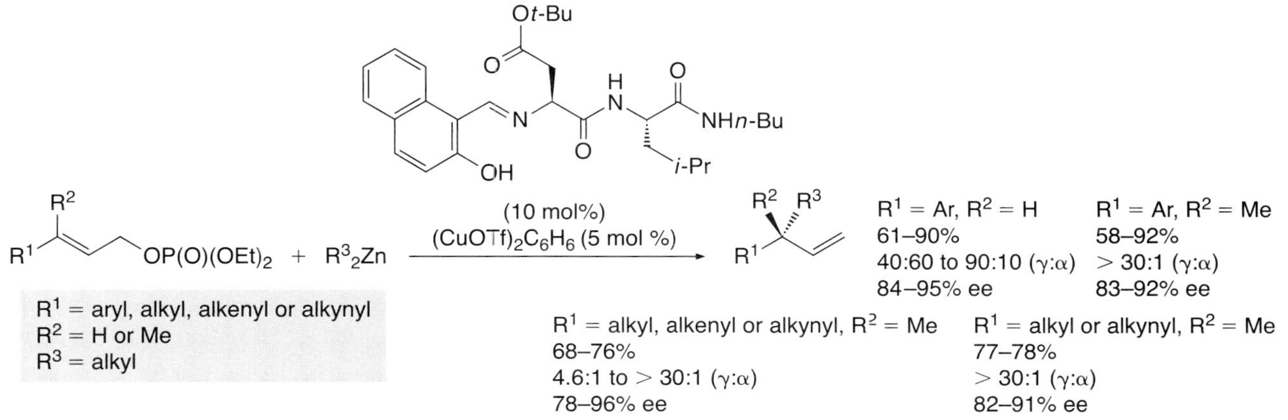
Ot-Bu
O
H
N
O
N
NHn-Bu
O
OH
i-Pr
R2
R1
OP(O)(OEt)2 + R32Zn
(10 mol%)
(CuOTf)2C6H6 (5 mol %)
R2 R3
R1
R1 = aryl, alkyl, alkenyl or alkynyl
R2 = H or Me
R3 = alkyl
R1 = Ar, R2 = H
61–90%
40:60 to 90:10 (γ:α)
84–95% ee
R1 = Ar, R2 = Me
58–92%
> 30:1 (γ:α)
83–92% ee
R1 = alkyl, alkenyl or alkynyl, R2 = Me
68–76%
4.6:1 to > 30:1 (γ:α)
78–96% ee
R1 = alkyl or alkynyl, R2 = Me
77–78%
> 30:1 (γ:α)
82–91% ee

(20.72)

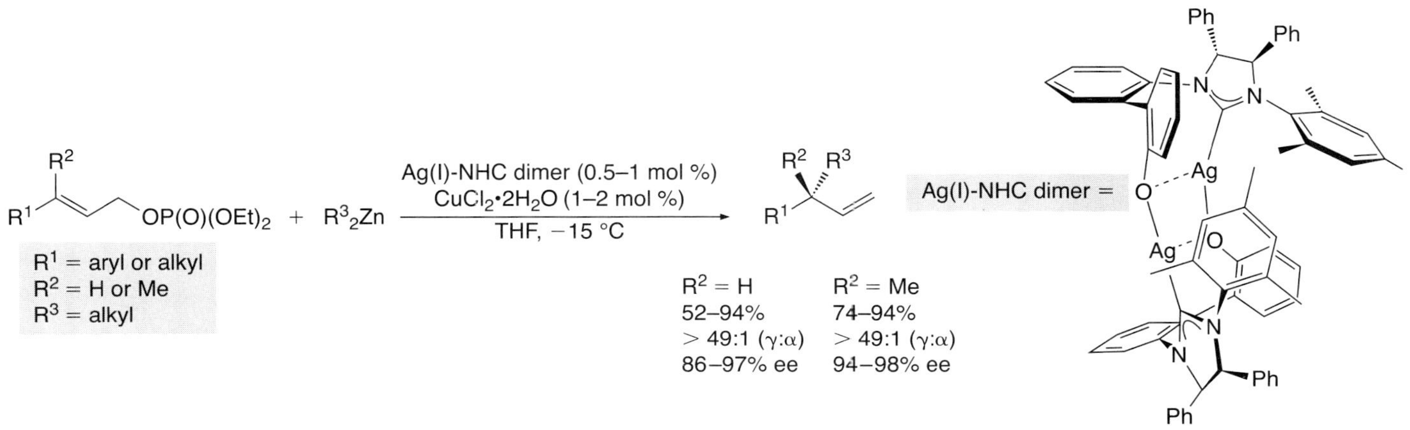
R2
R1
OP(O)(OEt)2 + R32Zn
Ag(I)-NHC dimer (0.5–1 mol %)
CuCl2•2H2O (1–2 mol %)
THF, −15 °C
R2 R3
R1
R1 = aryl or alkyl
R2 = H or Me
R3 = alkyl
R2 = H
52–94%
> 49:1 (γ:α)
86–97% ee
R2 = Me
74–94%
> 49:1 (γ:α)
94–98% ee
Ag(I)-NHC dimer =
Ph
Ph
N
N
O
Ag
Ag
O
N
N
Ph
Ph

(20.73)

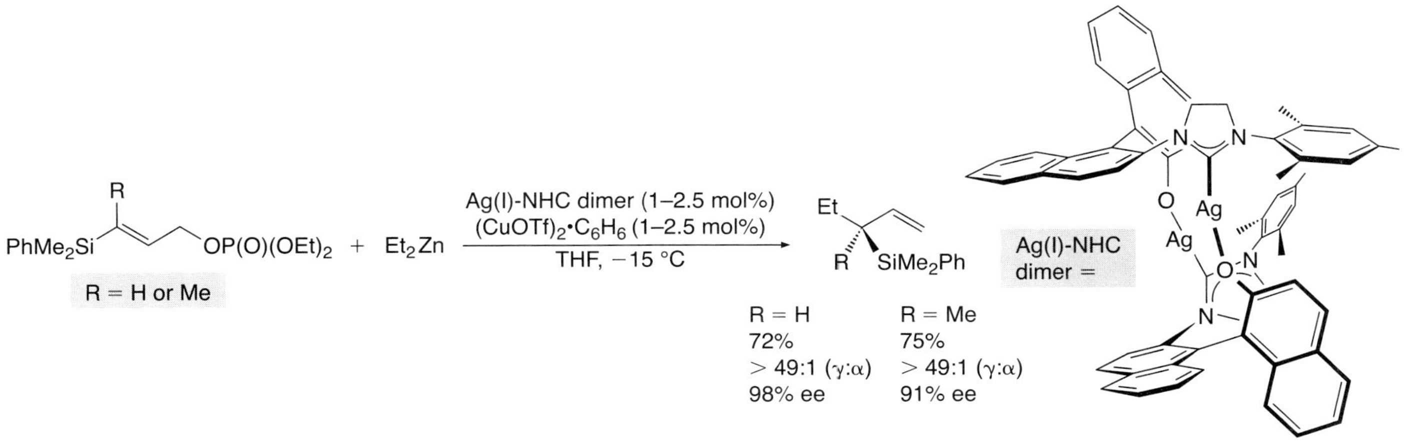
R
PhMe2Si
OP(O)(OEt)2 + Et2Zn
R = H or Me
Ag(I)-NHC dimer (1–2.5 mol%)
(CuOTf)2•C6H6 (1–2.5 mol%)
THF, −15 °C
Et
R
SiMe2Ph
R = H
72%
> 49:1 (γ:α)
98% ee
R = Me
75%
> 49:1 (γ:α)
91% ee
Ag(I)-NHC dimer =
N
N
O
Ag
Ag
O
N
N

(20.74)

Enantioselective copper-catalyzed desymmetrization of meso cyclic allylic bisdiethylphosphates has also been conducted with dialkylzinc reagents. Piarulli and Gennari initially showed that the copper(I) complex of a chiral Schiff base catalyzed the enantioselective desymmetrization of 4-cyclopentene-1,3-bis(diethylphosphate) with diethyzinc (Equation 20.75).[301] Piarulli, Gennari, and Feringa then improved upon the enantioselectivities

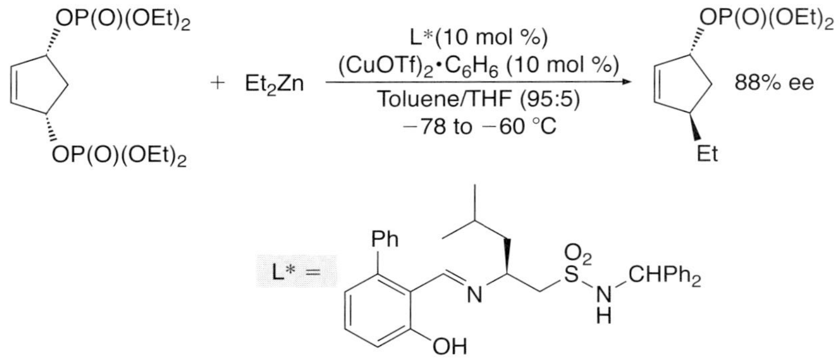

(20.75)

for desymmetrizations of cyclohexene and cycloheptene bisphosphates by using copper(I) complexes of chiral phosphoramidite ligands.[302,303]

20.7.3.2. Grignard Reagents as Nucleophiles

Following the initial reports of enantioselective copper-catalyzed allylic alkylation with Grignard nucleophiles by Bäckvall and Van Koten,[286] catalysts containing many chiral ligand architectures have been studied for these reactions. Bäckvall and co-workers showed that the reactions of an allylic acetate with n-BuMgBr occurred with improved enantioselectivity in the presence of a catalyst containing a chiral ferrocenyl thiolate ligand (Equation 20.76).[304,305] Okamoto and co-workers reported a chiral Cu(I)–NHC complex that catalyzes the allylic alkylation of a 4-siloxy-2-buten-1-ol derivative with moderate enantioselectivity (Equation 20.77).[306] However, synthetically useful levels of enantioselectivity in copper-catalyzed allylic substitution with Grignard reagents (Equation 20.78) were not achieved until Alexakis's use of first-, second-, and third-generation phosphoramidite ligands (Equation 20.78)[307–309] and Feringa's use of the Taniaphos ligand (Equation 20.79)[310] in these reactions.

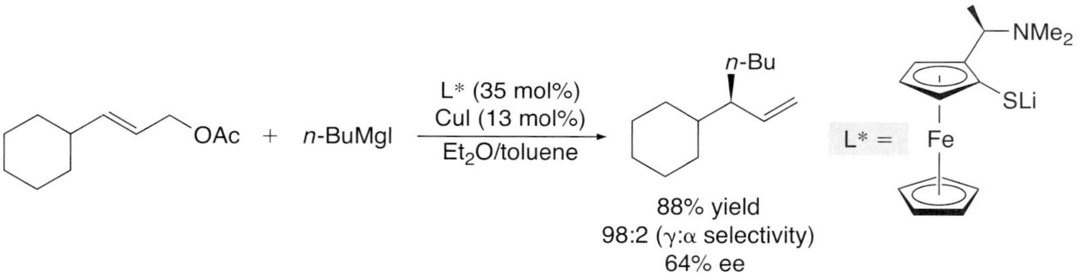

(20.76)

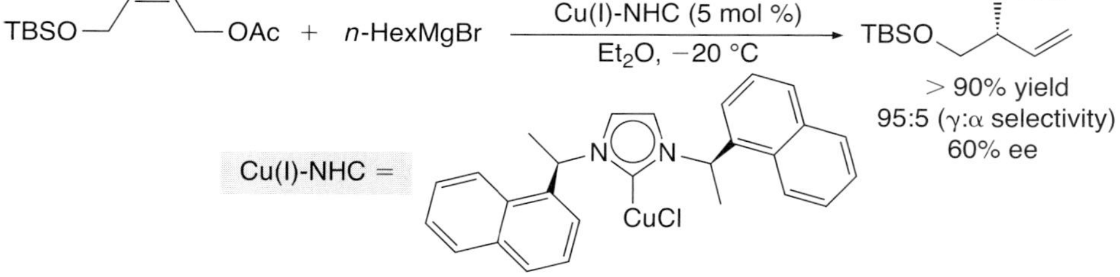

(20.77)

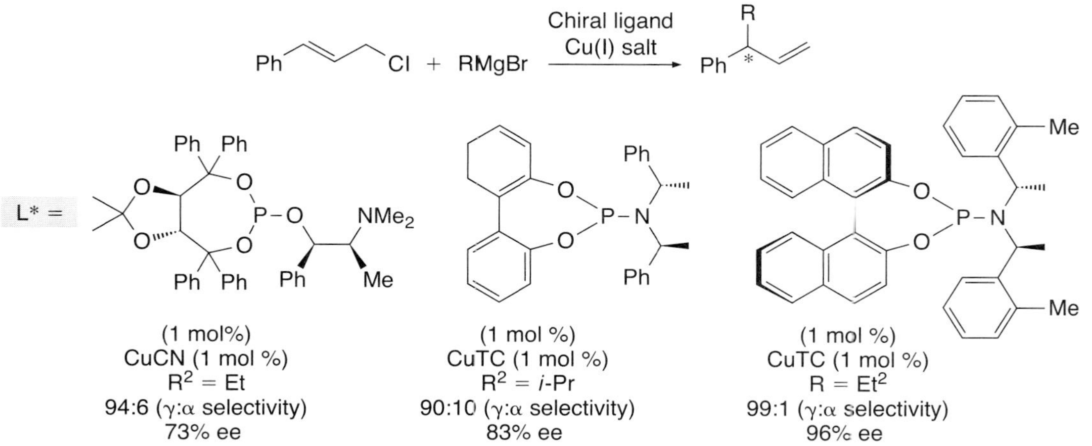

(20.78)

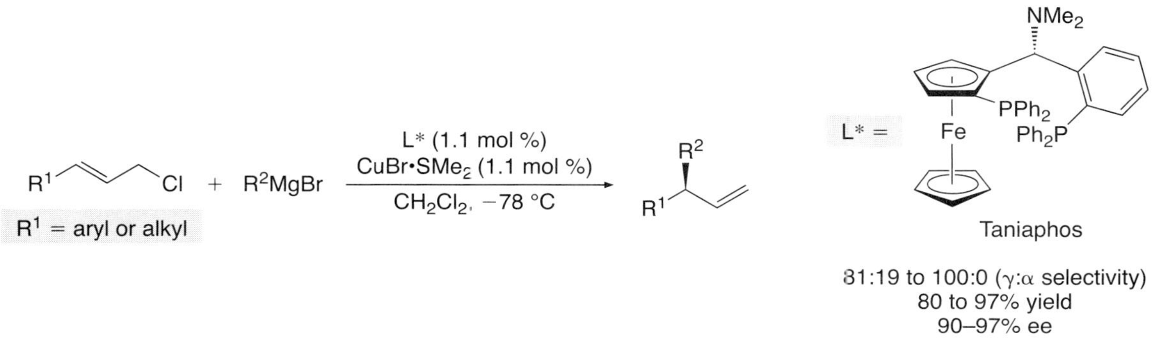

(20.79)

Alexakis and Feringa have since expanded the scope of allylic electrophiles that undergo copper-catalyzed allylic alkylations with Grignard reagents with high regio- and enantioselectivity. Alexakis showed that a copper complex of a phosphoramidite ligand is an efficient catalyst for additions of Grignard reagents to β,γ-disubstituted allylic chlorides (Equation 20.80) and endocyclic allylic chlorides (Equation 20.81).[311] Furthermore, Feringa reported that 3-bromopropenyl ester electrophiles undergo allylic alkylations in the presence of chiral allylic esters in high yield and enantioselectivity in the presence of the combination of Cu(I) and Taniaphos as catalyst (Equation 20.82).[312]

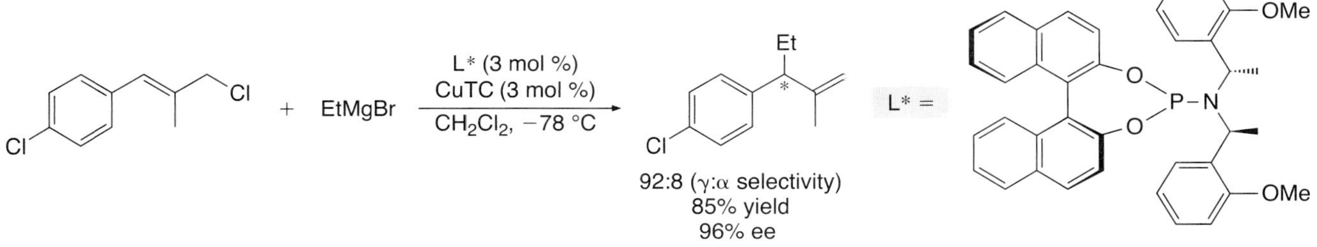

(20.80)

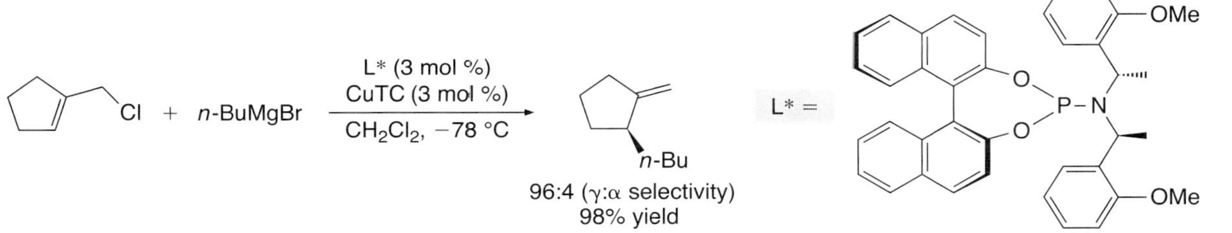

(20.81)

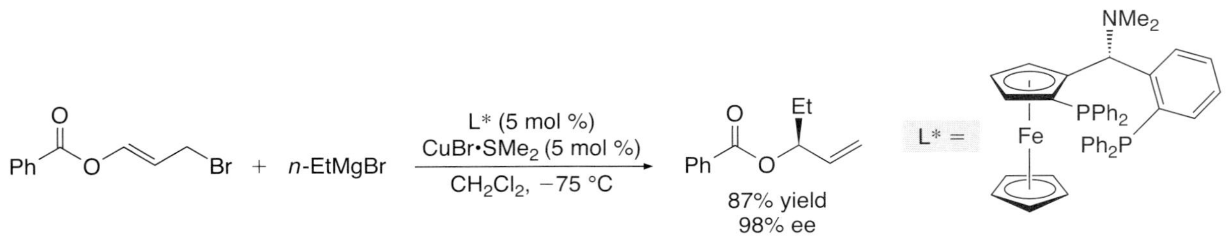

(20.82)

20.7.3.3. Organoaluminum Reagents as Nucleophiles

Organoaluminum reagents also serve as nucleophiles for enantioselective, copper-catalyzed allylic alkylation reactions, and reactions of these nucleophiles occur with faster rates than those of other nucleophiles in selected cases in which a sterically hindered allylic electrophile is used. Hoveyda and co-workers reported highly enantioselective, copper-catalyzed allylic alkylation of acyclic allylic electrophiles with trialkylaluminum reagents in 2007.[313] Low conversion (< 10%) was observed when Me_2Zn was used as the nucleophile in the allylic alkylation of a hindered allylic phosphate, but high (> 95%) conversion was observed for the identical reaction conducted with Me_3Al as the nucleophile. The method was applied to a double allylic alkylation reaction of a diene with Me_3Al to form the corresponding C_2-symmetric product of two alkylations with high enantioselectivity (Equation 20.83). A related protocol has since been developed for the enantioselective, copper-catalyzed addition of vinylaluminum reagents to β-disubstituted allylic phosphates generated in situ (Equation 20.84).[314]

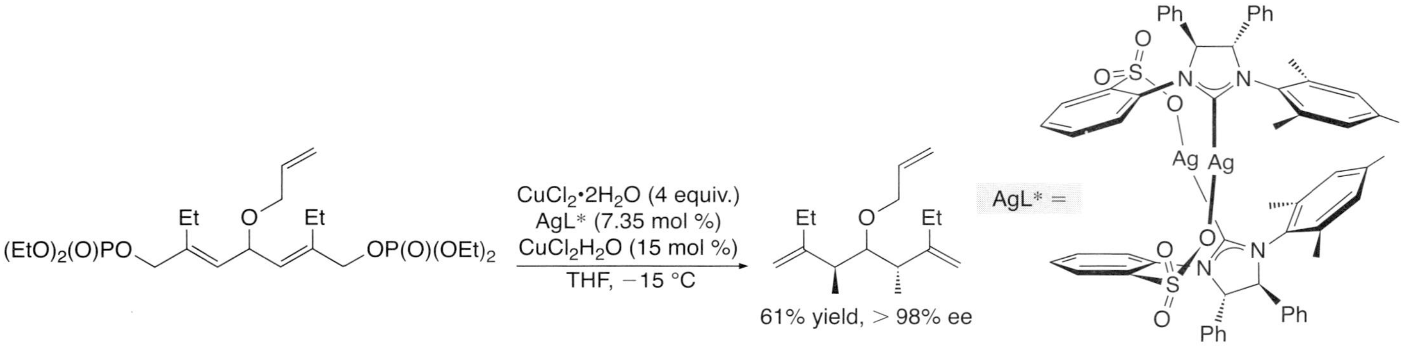

(20.83)

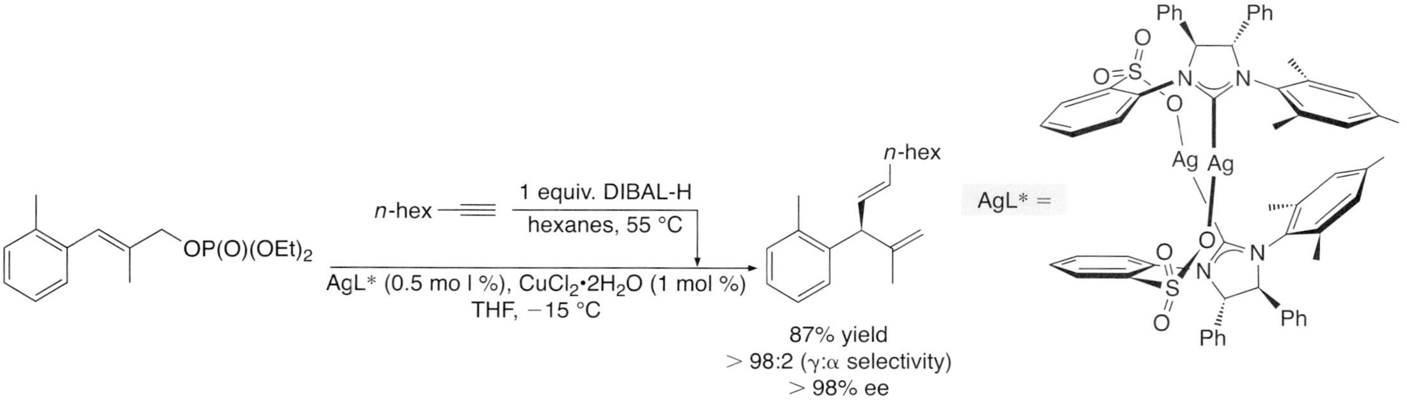

(20.84)

20.7.4. Miscellaneous Copper-Catalyzed Allylic Substitution Reactions

The scope of enantioselective, copper-catalyzed allylic substitution reactions is not limited to so-called hard carbon nucleophiles and achiral acyclic linear electrophiles. A recent report from Ito, Sawamura, and co-workers showed that a diboron reagent can serve as a pronucleophile for enantioselective, copper-catalyzed boronation of (Z)-allylic carbonates (Equation 20.85).[315] The corresponding chiral allylboronates were isolated in good yields with high enantioselectivities.

The asymmetric ring opening of racemic cyclic and meso bicyclic epoxides is related to allylic substitution and has also been conducted with copper catalysts. Pineschi, Feringa, and co-workers reported the kinetic resolution of racemic, cyclic 1,3-diene monoepoxides by ring opening of the epoxide unit with dialkylzinc reagents (Equation 20.86).[289] Later, Alexakis developed analogous kinetic resolutions with trialkylaluminum[316] and Grignard reagents (Equation 20.87).[317] Protocols for the desymmetrization of meso bicyclic substrates are also now well established. For example, oxabenzonorbornadiene derivatives react with dialkylzinc[318,319] and Gringard[320] reagents to form enantiomerically enriched 2-alkyl-1,2-dihydronaphth-1-ol derivatives (Equation 20.88). Finally, trialkylaluminum reagents serve as nucleophiles for enantioselective, copper-catalyzed desymmetrizations of meso bicyclic hydrazines (Equation 20.89).[321–323]

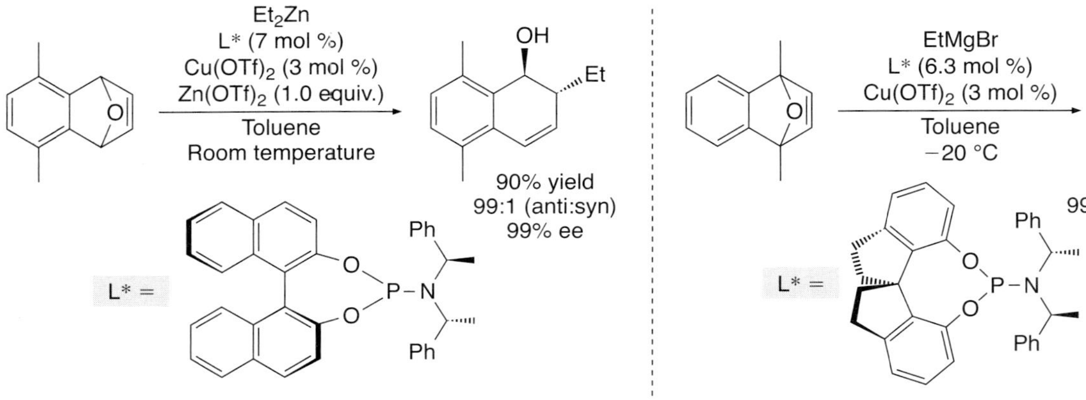

(20.88)

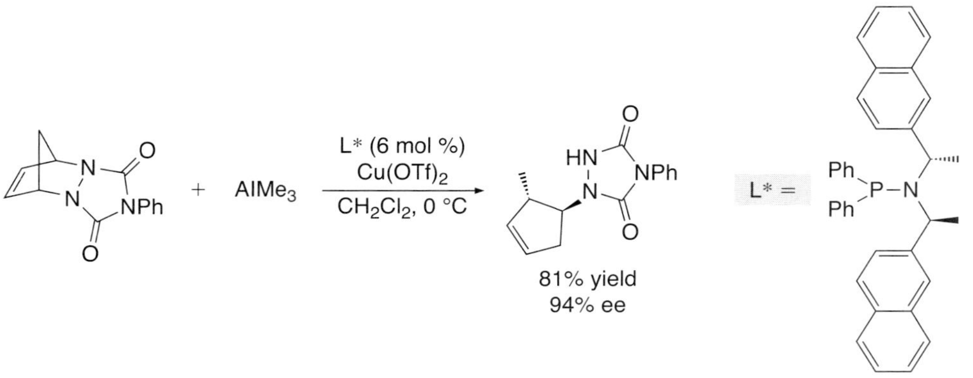

(20.89)

20.8. Summary

Enantioselective allyic substitution processes have been developed over the course of 30 years. Initial observations of the reactions of nucleophiles with palladium–allyl complexes led to the observation of catalytic substitutions of allylic ethers and esters, and then catalytic enantioselective allylic substitutions. The use of catalysts based on other metals has led to reactions that occur with complementary regiochemistry. Moreover, the scope of the reactions has expanded to include heteroatom and unstabilized carbon nucleophiles. Suitable electrophiles for these reactions include allylic esters of various types, allylic ethers, allylic alcohols, and allylic halides. Enantioselective reactions can be conducted with monoesters or by selection for cleavage of one of two equivalent esters. The mechanism of these reactions occurs by initial oxidative addition to form a metal–allyl complex. The second step involves nucleophilic attack on the allyl ligand for reaction of "soft" nucleophiles or inner-sphere reductive elimination for reactions of "hard" nucleophiles. The external nucleophilic attack typically occurs by reaction of the nucleophile with a cationic allyl complex at the face opposite to that to which the metal is bound. Exceptions include reactions of certain molybdenum–allyl complexes. Dissociation of product then regenerates the starting catalyst. Because of the diversity of the classes of these reactions, allylic substitution—in particular asymmetric allylic substitution—has been used to prepare a wide variety of natural products.

References and Notes

1. Trost, B. M.; Crawley, M. L. Chem. Rev. 2003, 103, 2921.
2. Tsuji, J.; Takahash.H; Morikawa, M. Tetrahedron Lett. 1965, 4387.
3. Hata, G.; Takahash.K; Miyake, A. J. Chem. Soc., Chem. Commun. 1970, 1392.
4. Atkins, K. E.; Walker, W. E.; Manyik, R. M. Tetrahedron Lett. 1970, 3821.
5. Trost, B. M.; Dietsch, T. J. J. Am. Chem. Soc. 1973, 95, 8200.

6. Trost, B. M.; Van Vranken, D. L. *Angew. Chem., Int. Ed. Engl.* **1992**, *31*, 228.

7. Trost, B. M. *J. Org. Chem.* **2004**, *69*, 5813.

8. Trost, B. M.; Van Vranken, D. L. *Chem. Rev.* **1996**, *96*, 395.

9. Pfaltz, A.; Lautens, M. In *Comprehensive Asymmetric Catalysis I–III*; Jacobsen, E. N., Pfaltz, A., Yamamoto, H., Eds.; Springer: Berlin, 1999; Vol. 2, p 833.

10. Trost, B. M.; Lee, C. In *Catalytic Asymmetric Synthesis*, 2nd ed.; Ojima, I., Ed.; Wiley-VCH: New York, 2000; p 593.

11. Tsuji, J. In *Handbook of Organopalladium Chemistry for Organic Synthesis*; Negishi, E.-i., Ed.; Wiley-Interscience: New York, 2002; Vol. 2, p 1669.

12. For the first demonstration that phosphines activate π-allylpalladium chloride dimers towards addition of nucleophiles, see Trost, B. M.; Fullerton, T. J. *J. Am. Chem. Soc.* **1973**, *95*, 292, and for the first use of a phosphine-ligated palladium complex in catalytic allylic substitution, see Trost, B. M.; Verhoeven, T. R. *J. Am. Chem. Soc.* **1976**, *98*, 630. For more recent uses of catalysts containing achiral phosphines, see references 13–18.

13. Bernocchi, E.; Cacchi, S.; Morera, E.; Ortar, G. *Synlett* **1992**, 161.

14. Shimizu, I.; Toyoda, M.; Terashima, T.; Oshima, M.; Hasegawa, H. *Synlett* **1992**, 301.

15. Sulsky, R.; Magnin, D. R. *Synlett* **1993**, 933.

16. Baldwin, I. C.; Williams, J. M. J.; Beckett, R. P. *Tetrahedron: Asymmetry* **1995**, *6*, 1515.

17. Baldwin, I. C.; Williams, J. M. J.; Beckett, R. P. *Tetrahedron: Asymmetry* **1995**, *6*, 679.

18. Zhang, D. Y.; Ghosh, A.; Suling, C.; Miller, M. J. *Tetrahedron Lett.* **1996**, *37*, 3799.

19. (a) Tanigawa, Y.; Nishimura, K.; Kawasaki, A.; Murahashi, S. I. *Tetrahedron Lett.* **1982**, *23*, 5549; (b) Tsuji, J.; Shimizu, I.; Minami, I.; Ohashi, Y.; Sugiura, T.; Takahashi, K. *J. Org. Chem.* **1985**, *50*, 1523; (c) Tsuji, J.; Shimizu, I.; Minami, I.; Ohashi, Y. *Tetrahedron Lett.* **1982**, *23*, 4809.

20. Cuvigny, T.; Julia, M.; Rolando, C. *J. Organomet. Chem.* **1985**, *285*, 395.

21. Kotake, H.; Yamamoto, T.; Kinoshita, H. *Chem. Lett.* **1982**, 1331.

22. Trost, B. M.; Schmuff, N. R.; Miller, M. J. *J. Am. Chem. Soc.* **1980**, *102*, 5979.

23. Ono, N.; Hamamoto, I.; Kaji, A. *J. Chem. Soc., Chem. Commun.* **1982**, 821.

24. Tamura, R.; Hegedus, L. S. *J. Am. Chem. Soc.* **1982**, *104*, 3727.

25. Aresta, M.; Quaranta, E.; Dibenedetto, A.; Giannoccaro, P.; Tommasi, I.; Lanfranchi, M.; Tiripicchio, A. *Organometallics* **1997**, *16*, 834.

26. Aresta, M.; Dibenedetto, A.; Quaranta, E.; Lanfranchi, M.; Tiripicchio, A. *Organometallics* **2000**, *19*, 4199.

27. Yamamoto, T.; Akimoto, M.; Saito, O.; Yamamoto, A. *Organometallics* **1986**, *5*, 1559.

28. Garrohelion, F.; Merzouk, A.; Guibe, F. *J. Org. Chem.* **1993**, *58*, 6109.

29. Hirao, T.; Yamada, N.; Ohshiro, Y.; Agawa, T. *J. Organomet. Chem.* **1982**, *236*, 409.

30. Iourtchenko, A.; Sinou, D. *J. Mol. Catal.* **1997**, *122*, 91.

31. van Heerden, F. R.; Huyser, J. J.; Williams, D. B. G.; Holzapfel, C. W. *Tetrahedron Lett.* **1998**, *39*, 5281.

32. Vicart, N.; Gore, J.; Gazes, B. *Tetrahedron* **1998**, *54*, 11063.

33. Moreno-Manas, M.; Pleixats, R. In *Handbook of Organopalladium Chemistry for Organic Synthesis*; Negishi, E.-i., Ed.; Wiley-Interscience: New York, 2002; Vol. 2, p 1707.

34. Tsuji, J. *Tetrahedron* **1986**, *42*, 4361.

35. Courillon, C.; Thorimbert, S.; Malacria, M. In *Handbook of Organopalladium Chemistry for Organic Synthesis*; Negishi, E.-i., Ed.; Wiley-Interscience: New York, 2002; Vol. 2, p 1795.

36. Trost, B. M.; Molander, G. A. *J. Am. Chem. Soc.* **1981**, *103*, 5969.

37. Tsuji, J.; Kataoka, H.; Kobayashi, Y. *Tetrahedron Lett.* **1981**, *22*, 2575.

38. Echavarren, A. M.; Tueting, D. R.; Stille, J. K. *J. Am. Chem. Soc.* **1988**, *110*, 4039.

39. Tueting, D. R.; Echavarren, A. M.; Stille, J. K. *Tetrahedron* **1989**, *45*, 979.

40. Furstner, A.; Weintritt, H. *J. Am. Chem. Soc.* **1998**, *120*, 2817.

41. Trost, B. M.; Hachiya, I. *J. Am. Chem. Soc.* **1998**, *120*, 1104.

42. Glorius, F.; Neuburger, M.; Pfaltz, A. *Helv. Chim. Acta* **2001**, *84*, 3178.

43. Hughes, D. L.; Palucki, M.; Yasuda, N.; Reamer, R. A.; Reider, P. J. *J. Org. Chem.* **2002**, *67*, 2762.

44. Krska, S. W.; Hughes, D. L.; Reamer, R. A.; Mathre, D. J.; Sun, Y.; Trost, B. M *J. Am. Chem. Soc.* **2002**, *124*, 12656.

45. Trost, B. M.; Dogra, K. *J. Am. Chem. Soc.* **2002**, *124*, 7256.

46. Trost, B. M.; Dogra, K.; Hachiya, I.; Emura, T.; Hughes, D. L.; Krska, S.; Reamer, R. A.; Palucki, M.; Yasuda, N.; Reider, P. J. *Angew. Chem. Int. Ed.* **2002**, *41*, 1929.

47. Trost, B. M.; Fandrick, D. R. *J. Am. Chem. Soc.* **2003**, *125*, 11836.

48. Evans, P. A.; Uraguchi, D. *J. Am. Chem. Soc.* **2003**, *125*, 7158.

49. Fiaud, J. C.; Malleron, J. L. *J. Chem. Soc., Chem. Commun.* **1981**, 1159.

50. Braun, M.; Meier, T. *Angew. Chem. Int. Ed.* **2006**, *45*, 6952.

51. Braun, M.; Meier, T. *Synlett* **2006**, 661.

52. Zheng, W.-H.; Zheng, B.-H.; Zhang, Y.; Hou, X.-L. *J. Am. Chem. Soc.* **2007**, *129*, 7718.

53. Luo, F. T.; Negishi, E. *Tetrahedron Lett.* **1985**, *26*, 2177.
54. Negishi, E.-i.; Liou, S.-Y. In *Handbook of Organopalladium Chemistry for Organic Synthesis*; Negishi, E.-i., Ed.; Wiley-Interscience: New York, 2002; Vol. 2, p 1769.
55. Negishi, E.; Matsushita, H.; Chatterjee, S.; John, R. A. *J. Org. Chem.* **1982**, *47*, 3188.
56. Negishi, E.; Luo, F. T.; Pecora, A. J.; Silveira, A. *J. Org. Chem.* **1983**, *48*, 2427.
57. Negishi, E.-i.; Chatterjee, S. *Tetrahedron Lett.* **1983**, *24*, 1341.
58. Tsuji, J.; Minami, I.; Shimizu, I. *Chem. Lett.* **1983**, 1325.
59. Baba, T.; Nakano, K.; Nishiyama, S.; Tsurya, S.; Masai, M. *J. Chem. Soc., Chem. Commun.* **1990**, 348.
60. Trost, B. M.; Keinan, E. *Tetrahedron Lett.* **1980**, *21*, 2591.
61. Tsuji, J.; Minami, I.; Shimizu, I. *Tetrahedron Lett.* **1983**, *24*, 4713.
62. Tsuda, T.; Chujo, Y.; Nishi, S.; Tawara, K.; Saegusa, T. *J. Am. Chem. Soc.* **1980**, *102*, 6381.
63. Hiroi, K.; Abe, J.; Suya, K.; Sato, S.; Koyama, T. *J. Org. Chem.* **1994**, *59*, 203.
64. Hiroi, K.; Abe, J. *Heterocycles* **1990**, *30*, 283.
65. Hiroi, K.; Abe, J.; Suya, K.; Sato, S. *Tetrahedron Lett.* **1989**, *30*, 1543.
66. Hiroi, K.; Suya, K.; Sato, S. *J. Chem. Soc., Chem. Commun.* **1986**, 469.
67. Hiroi, K.; Yamada, S. *Chem. Pharm. Bull.* **1973**, *21*, 47.
68. Onoue, H.; Moritani, I.; Murahashi, S. I. *Tetrahedron Lett.* **1973**, 121.
69. Hiroi, K.; Yamada, S.; Achiwa, K. *Chem. Pharm. Bull.* **1972**, *20*, 246.
70. Liu, D.; Xie, F.; Zhang, W. *Tetrahedron Lett.* **2007**, *48*, 7591.
71. Weix, D. J.; Hartwig, J. F. *J. Am. Chem. Soc.* **2007**, *129*, 7720.
72. Shimizu, I.; Yamada, T.; Tsuji, J. *Tetrahedron Lett.* **1980**, *21*, 3199.
73. Shimizu, I.; Minami, I.; Tsuji, J. *Tetrahedron Lett.* **1983**, *24*, 1797.
74. Tsuji, J.; Minami, I.; Shimizu, I. *Tetrahedron Lett.* **1983**, *24*, 1793.
75. Tanaka, T.; Okamura, N.; Bannai, K.; Hazato, A.; Sugiura, S.; Manabe, K.; Kurozumi, S. *Tetrahedron Lett.* **1985**, *26*, 5575.
76. Tsuda, T.; Tokai, M.; Ishida, T.; Saegusa, T. *J. Org. Chem.* **1986**, *51*, 5216.
77. Tsuda, T.; Okada, M.; Nishi, S.; Saegusa, T. *J. Org. Chem.* **1986**, *51*, 421.
78. Tsuji, J.; Yamada, T.; Minami, I.; Yuhara, M.; Nisar, M.; Shimizu, I. *J. Org. Chem.* **1987**, *52*, 2988.
79. Tsuji, J.; Ohashi, Y.; Minami, I. *Tetrahedron Lett.* **1987**, *28*, 2397.
80. Mohr, J. T.; Nishimata, T.; Behenna, D. C.; Stoltz, B. M. *J. Am. Chem. Soc.* **2006**, *128*, 11348.
81. Keith, J. A.; Behenna, D. C.; Mohr, J. T.; Ma, S.; Marinescu, S. C.; Oxgaard, J.; Stoltz, B. M.; Goddard, W. A. *J. Am. Chem. Soc.* **2007**, *129*, 11876.
82. Tunge, J. A.; Burger, E. C. *Eur. J. Org. Chem.* **2005**, 1715.
83. Burger, E. C.; Tunge, J. A. *J. Am. Chem. Soc.* **2006**, *128*, 10002.
84. Trost, B. M.; Lautens, M. *Tetrahedron* **1987**, *43*, 4817.
85. Trost, B. M.; Lautens, M. *J. Am. Chem. Soc.* **1987**, *109*, 1469.
86. Trost, B. M.; Toste, F. D. *J. Am. Chem. Soc.* **1999**, *121*, 4545.
87. Trost, B. M.; Hildbrand, S.; Dogra, K. *J. Am. Chem. Soc.* **1999**, *121*, 10416.
88. Krska, S. W.; Hughes, D. L.; Reamer, R. A.; Mathre, D. J.; Sun, Y.; Trost, B. M. *J. Am. Chem. Soc.* **2002**, *124*, 12656.
89. Trost, B. M.; Dogra, K.; Hachiya, I.; Emura, T.; Hughes, D. L.; Krska, S.; Reamer, R. A.; Palucki, M.; Yasuda, N.; Reider, P. J. *Angew. Chem. Int. Ed.* **2002**, *41*, 1929.
90. Trost, B. M.; Dogra, K. *J. Am. Chem. Soc.* **2002**, *124*, 7256.
91. Trost, B. M.; Andersen, N. G. *J. Am. Chem. Soc.* **2002**, *124*, 14320.
92. Krska, S. W.; Hughes, D. L.; Reamer, R. A.; Mathre, D. J.; Palucki, M.; Yasuda, N.; Sun, Y.; Trost, B. M. *Pure. Appl. Chem.* **2004**, *76*, 625.
93. Trost, B. M.; Zhang, Y. *J. Am. Chem. Soc.* **2006**, *128*, 4590.
94. Trost, B. M.; Dogra, K. *Org. Lett.* **2007**, *9*, 861.
95. Glorius, F.; Pfaltz, A. *Org. Lett.* **1999**, *1*, 141.
96. Malkov, A. V.; Baxendale, I. R.; Dvorak, D.; Mansfield, D. J.; Kocovsky, P. *J. Org. Chem.* **1999**, *64*, 2737.
97. Lloyd-Jones, G. C.; Pfaltz, A. *Angew. Chem., Int. Ed. Engl.* **1995**, *34*, 462.
98. Zhang, S. W.; Mitsudo, T.; Kondo, T.; Watanabe, Y. *J. Organomet. Chem.* **1993**, *450*, 197.
99. Morisaki, Y.; Kondo, T.; Mitsudo, T. A. *Organometallics* **1999**, *18*, 4742.
100. Matsushima, Y.; Onitsuka, K.; Kondo, T.; Mitsudo, T.; Takahashi, S. *J. Am. Chem. Soc.* **2001**, *123*, 10405.
101. Trost, B. M.; Fraisse, P. L.; Ball, Z. T. *Angew. Chem. Int. Ed.* **2002**, *41*, 1059.
102. Mbaye, M. D.; Demerseman, B.; Renaud, J.-L.; Toupet, L.; Bruneau, C. *Angew. Chem. Int. Ed.* **2003**, *42*, 5066.
103. Renaud, J.-L.; Bruneau, C.; Demerseman, B. *Synlett* **2003**, 408.
104. Hermatschweiler, R.; Fernandez, I.; Breher, F.; Pregosin, P. S.; Veiros, L. F.; Calhorda, M. J. *Angew. Chem. Int. Ed.* **2005**, *44*, 4397.

105. Hermatschweiler, R.; Fernandez, I.; Pregosin, P. S.; Watson, E. J.; Albinati, A.; Rizzato, S.; Veiros, L. F.; Calhorda, M. J. *Organometallics* **2005**, *24*, 1809.

106. Evans, P. A.; Nelson, J. D. *Tetrahedron Lett.* **1998**, *39*, 1725.

107. Evans, P. A.; Robinson, J. E.; Nelson, J. D. *J. Am. Chem. Soc.* **1999**, *121*, 6761.

108. Lavastre, O.; Morken, J. P. *Angew. Chem. Int. Ed.* **1999**, *38*, 3163.

109. Evans, P. A.; Leahy, D. K. *J. Am. Chem. Soc.* **2000**, *122*, 5012.

110. Muraoka, T.; Matsuda, I.; Itoh, K. *Tetrahedron Lett.* **2000**, *41*, 8807.

111. Evans, P. A.; Kennedy, L. J. *J. Am. Chem. Soc.* **2001**, *123*, 1234.

112. Evans, P. A.; Robinson, J. E.; Moffett, K. K. *Org. Lett.* **2001**, *3*, 3269.

113. Evans, P. A.; Leahy, D. K. *J. Am. Chem. Soc.* **2002**, *124*, 7882.

114. Evans, P. A.; Leahy, D. K.; Slieker, L. M. *Tetrahedron* **2003**, *14*, 3613.

115. Evans, P. A.; Leahy, D. K.; Andrews, W. J.; Uraguchi, D. *Angew. Chem. Int. Ed.* **2004**, *43*, 4788.

116. Evans, P. A.; Lawler, M. J. *J. Am. Chem. Soc.* **2004**, *126*, 8642.

117. Janssen, J. P.; Helmchen, G. *Tetrahedron Lett.* **1997**, *38*, 8025.

118. Bartels, B.; Helmchen, G. *Chem. Commun.* **1999**, 741.

119. Bartels, B.; Garcia-Yebra, C.; Rominger, F.; Helmchen, G. *Eur. J. Inorg. Chem.* **2002**, 2569.

120. Garcia-Yebra, C.; Janssen, J. P.; Rominger, F.; Helmchen, G. *Organometallics* **2004**, *23*, 5459.

121. Takeuchi, R.; Kashio, M. *Angew. Chem., Int. Ed. Engl.* **1997**, *36*, 263.

122. Takeuchi, R.; Kashio, M. *J. Am. Chem. Soc.* **1998**, *120*, 8647.

123. Takeuchi, R.; Shiga, N. *Org. Lett.* **1999**, *1*, 265.

124. Takeuchi, R. *Polyhedron* **2000**, *19*, 557.

125. Takeuchi, R.; Tanabe, K. *Angew. Chem. Int. Ed.* **2000**, *39*, 1975.

126. Takeuchi, R. *Synlett* **2002**, 1954.

127. Schelwies, M.; Dubon, P.; Helmchen, G. *Angew. Chem. Int. Ed.* **2006**, *45*, 2466.

128. Dahnz, A.; Helmchen, G. *Synlett* **2006**, 697.

129. Streiff, S.; Welter, C.; Schelwies, M.; Lipowsky, G.; Miller, N.; Helmchen, G. *Chem. Commun.* **2005**, 2957.

130. Tissot-Croset, K.; Polet, D.; Alexakis, A. *Angew. Chem. Int. Ed.* **2004**, *43*, 2426.

131. Polet, D.; Alexakis, A. *Org. Lett.* **2005**, *7*, 1621.

132. Gnamm, C.; Forster, S.; Miller, N.; Brodner, K.; Helmchen, G. *Synlett* **2007**, 790.

133. Alexakis, A.; Hajjaji, S. E.; Polet, D.; Rathgeb, X. *Org. Lett.* **2007**, *9*, 3393.

134. Graening, T.; Hartwig, J. F. *J. Am. Chem. Soc.* **2005**, *127*, 17192.

135. Lopez, F.; Ohmura, T.; Hartwig, J. F. *J. Am. Chem. Soc.* **2003**, *125*, 3426.

136. Shu, C. T.; Hartwig, J. F. *Angew. Chem. Int. Ed.* **2004**, *43*, 4794.

137. Takeuchi, R.; Ue, N.; Tanabe, K.; Yamashita, K.; Shiga, N. *J. Am. Chem. Soc.* **2001**, *123*, 9525.

138. Ohmura, T.; Hartwig, J. F. *J. Am. Chem. Soc.* **2002**, *124*, 15164.

139. Welter, C.; Dahnz, A.; Brunner, B.; Streiff, S.; Dubon, P.; Helmchen, G. *Org. Lett.* **2005**, *7*, 1239.

140. Weihofen, R.; Tverskoy, E.; Helmchen, G. *Angew. Chem. Int. Ed.* **2006**, *45*, 5546.

141. Kiener, C. A.; Shu, C.; Incarvito, C.; Hartwig, J. F. *J. Am. Chem. Soc.* **2003**, *125*, 14272.

142. Leitner, A.; Shu, C. T.; Hartwig, J. F. *Proc. Natl Acad. Sci. U.S.A.* **2004**, *101*, 5830.

143. Shu, C. T.; Leitner, A.; Hartwig, J. F. *Angew. Chem. Int. Ed.* **2004**, *43*, 4797.

144. Leitner, A.; Shu, C. T.; Hartwig, J. F. *Org. Lett.* **2005**, *7*, 1093.

145. Lautens, M.; Trost, B. M. *Abstr. Papers Am. Chem. Soc.* **1983**, *186*, 92.

146. Acemoglu, L.; Williams, J. M. J. In *Handbook of Organopalladium Chemistry for Organic Synthesis*; Negishi, E.-i., Ed.; Wiley-Interscience: New York, 2002; Vol. 2, p 1945.

147. Trost, B. M. *Chem. Pharm. Bull.* **2002**, *50*, 1.

148. Tsuji, J.; Minami, I. *Acc. Chem. Res.* **1987**, *20*, 140.

149. Vrieze, K. In *Dynamic Nuclear Magnetic Resonance Spectroscopy*; Jackman, L. M., Cotton, F. A., Eds.; Academic Press: New York, 1975.

150. Faller, J. W. *Adv. Organomet. Chem.* **1977**, *16*, 211.

151. Mackenzie, P. B.; Whelan, J.; Bosnich, B. *J. Am. Chem. Soc.* **1985**, *107*, 2046.

152. Faller, J. W.; Stokes-Huby, H. L.; Albrizzio, M. A. *Helv. Chim. Acta* **2001**, *84*, 3031.

153. Lloyd-Jones, G. C.; Krska, S. W.; Hughes, D. L.; Gouriou, L.; Bonnet, V. D.; Jack, K.; Sun, Y.; Reamer, R. A. *J. Am. Chem. Soc.* **2004**, *126*, 702.

154. (a) Markovic, D.; Hartwig, J. F. *J. Am. Chem. Soc.* **2007**, *129*, 11680; (b) Madrahimov, S.; Markovic, D.; Hartwig, J. F. *J. Am. Chem. Soc.* **2009**, *131*, 7228.

155. Hayashi, T.; Hagihara, T.; Konishi, M.; Kumada, M. *J. Am. Chem. Soc.* **1983**, *105*, 7767.

156. Hayashi, T.; Yamamoto, A.; Hagihara, T. *J. Org. Chem.* **1986**, *51*, 723.

157. Kobayashi, Y.; Mizojiri, R.; Ikeda, E. *J. Org. Chem.* **1996**, *61*, 5391.

158. Tsuji, Y.; Kusui, T.; Kojima, T.; Sugiura, Y.; Yamada, N.; Tanaka, S.; Ebihara, M.; Kawamura, T. *Organometallics* **1998**, *17*, 4835.

159. Consiglio, G.; Waymouth, R. M. *Chem. Rev.* **1989**, *89*, 257.

160. Pregosin, P. S.; Salzmann, R. *Coord. Chem. Rev.* **1996**, *155*, 35.

161. Faller, J. W.; Thomsen, M. E.; Mattina, M. J. *J. Am. Chem. Soc.* **1971**, *93*, 2642.

162. Corradini, P.; Maglio, G.; Musco, A.; Paiaro, G. *J. Chem. Soc., Chem. Commun.* **1966**, 618.

163. Tsutsui, M.; Courtney, A. *Adv. Organomet. Chem.* **1977**, *16*, 241.

164. Faller, J. W.; Tully, M. T. *J. Am. Chem. Soc.* **1972**, *94*, 2676.

165. Bäckvall, J. E.; Granberg, K. L.; Heumann, A. *Isr. J. Chem.* **1991**, *31*, 17.

166. Granberg, K. L.; Bäckvall, J. E. *J. Am. Chem. Soc.* **1992**, *114*, 6858.

167. Crociani, B.; Di Bianca, F.; Giovenco, A.; Boschi, T. *Inorg. Chim. Acta* **1987**, *127*, 169.

168. Hayashi, T.; Kishi, K.; Yamamoto, A.; Ito, Y. *Tetrahedron Lett.* **1990**, *31*, 1743.

169. You, S. L.; Zhu, X. Z.; Luo, Y. M.; Hou, X. L.; Dai, L. X. *J. Am. Chem. Soc.* **2001**, *123*, 7471.

170. Trost, B. M.; Toste, F. D. *J. Am. Chem. Soc.* **1998**, 9074.

171. Faller, J. W.; Wilt, J. C. *Org. Lett.* **2005**, *7*, 633.

172. Keinan, E.; Sahai, M. *J. Chem. Soc., Chem. Commun.* **1984**, 648.

173. Trost, B. M.; Weber, L.; Strege, P. E.; Fullerton, T. J.; Dietsche, T. J. *J. Am. Chem. Soc.* **1978**, *100*, 3416.

174. Åkermark, B.; Hansson, S.; Krakenberger, B.; Vitagliano, A.; Zetterberg, K. *Organometallics* **1984**, *3*, 679.

175. Åkermark, B.; Zetterberg, K.; Hansson, S.; Krakenberger, B.; Vitagliano, A. *J. Organomet. Chem.* **1987**, *335*, 133.

176. Åkermark, B.; Hansson, S.; Krakenberger, B.; Zetterberg, K.; Vitagliano, A. *Chem. Scr.* **1987**, *27*, 525.

177. Hayashi, T.; Kishi, K.; Yamamoto, A.; Ito, Y. *Tetrahedron Lett.* **1990**, *31*, 1743.

178. Pretot, R.; Pfaltz, A. *Angew. Chem. Int. Ed.* **1998**, *37*, 323.

179. Prat, M.; Ribas, J.; Morenomanas, M. *Tetrahedron* **1992**, *48*, 1695.

180. Vicart, N.; Gore, J.; Cazes, B. *Synlett* **1996**, 850.

181. Vicart, N.; Cazes, B.; Gore, J. *Tetrahedron Lett.* **1995**, *36*, 535.

182. Garrido, J. L.; Alonso, I.; Carretero, J. C. *J. Org. Chem.* **1998**, *63*, 9406.

183. Alonso, I.; Carretero, J. C.; Garrido, J. L.; Magro, V.; Pedregal, C. *J. Org. Chem.* **1997**, *62*, 5682.

184. Hayashi, T.; Kawatsura, M.; Uozumi, Y. *J. Am. Chem. Soc.* **1998**, *120*, 1681.

185. Watson, I. D. G.; Styler, S. A.; Yudin, A. K. *J. Am. Chem. Soc.* **2004**, *126*, 5086.

186. Watson, I. D. G.; Yudin, A. K. *J. Am. Chem. Soc.* **2005**, *127*, 17516.

187. Johns, A. M.; Liu, Z.; Hartwig, J. F. *Angew. Chem. Int. Ed.* **2007**, *46*, 7259.

188. Dubovyk, I.; Watson, I. D. G.; Yudin, A. K. *J. Am. Chem. Soc.* **2007**, *129*, 14172.

189. Lloyd-Jones, G. C.; Stephen, S. C. *Chem.—Eur. J.* **1998**, *4*, 2539.

190. Poli, G.; Scolastico, C. *Chemtracts* **1999**, *12*, 837.

191. Lloyd-Jones, G. C.; Stephen, S. C.; Murray, M.; Butts, C. P.; Vyskocil, S.; Kocovsky, P. *Chem.—Eur. J.* **2000**, *6*, 4348.

192. Lloyd-Jones, G. C.; Stephen, S. C. *Chem.—Eur. J.* **1998**, *4*, 2539.

193. Brown, J. M.; Hulmes, D. I.; Guiry, P. J. *Tetrahedron* **1994**, *50*, 4493.

194. Ward, T. R. *Organometallics* **1996**, *15*, 2836.

195. Blochl, P. E.; Togni, A. *Organometallics* **1996**, *15*, 4125.

196. Pfaltz, A. *Acta Chem. Scand.* **1996**, *50*, 189.

197. Steinhagen, H.; Reggelin, M.; Helmchen, G. *Angew. Chem., Int. Ed. Engl.* **1997**, *36*, 2108.

198. Jonasson, C.; Kritikos, M.; Bäckvall, J. E.; Szabo, K. J. *Chem.—Eur. J.* **2000**, *6*, 432.

199. Evans, P. A.; Nelson, J. D. *J. Am. Chem. Soc.* **1998**, *120*, 5581.

200. Kondo, T.; Morisaki, Y.; Uenoyama, S.-y.; Wada, K.; Mitsudo, T.-a. *J. Am. Chem. Soc.* **1999**, *121*, 8657.

201. Fuji, K.; Kinoshita, N.; Tanaka, K.; Kawabata, T. *Chem. Commun.* **1999**, 2289.

202. Kanayama, T.; Yoshida, K.; Miyabe, H.; Kimachi, T.; Takemoto, Y. *J. Org. Chem.* **2003**, *68*, 6197.

203. Kanayama, T.; Yoshida, K.; Miyabe, H.; Takemoto, Y. *Angew. Chem. Int. Ed.* **2003**, *42*, 2054.

204. Lipowsky, G.; Helmchen, G. *Chem. Commun.* **2004**, 116.

205. Miyabe, H.; Yoshida, K.; Yamauchi, M.; Takemoto, Y. *J. Org. Chem.* **2005**, *70*, 2148.

206. Leitner, A.; Shekhar, S.; Pouy, M.; Hartwig, J. F. *J. Am. Chem. Soc.* **2005**, *127*, 15506.

207. Ashfeld, B. L.; Miller, K. A.; Martin, S. F. *Org. Lett.* **2004**, *6*, 1321.

208. Plietker, B. *Angew. Chem. Int. Ed.* **2006**, *45*, 1469.

209. Ashfeld, B. L.; Miller, K. A.; Smith, A. J.; Tran, K.; Martin, S. F. *J. Org. Chem.* **2007**, *72*, 9018.

210. Sesay, S. J.; Williams, J. M. J. *Adv. Asym. Synth.* **1998**, *3*, 235.

211. Knuhl, G.; Sennhenn, P.; Helmchen, G. *J. Chem. Soc., Chem. Commun.* **1995**, 1845.

212. For an earlier conception of a directing substituent, see Hayashi, T.; Kanehira, K.; Tsuchiya, H.; Kumada, M. *J. Chem. Soc., Chem. Commun.* **1982**, 1162.

213. Koch, G.; Lloyd-Jones, G. C.; Loiseleur, O.; Pfaltz, A.; Pretot, R.; Schaffner, S.; Schnider, P.; Vonmatt, P. *Recl. Trav. Chim. Pays-Bas* **1995**, *114*, 206.

214. Helmchen, G.; Pfaltz, A. *Acc. Chem. Res.* **2000**, *33*, 336.

215. Pamies, O.; Dieguez, M.; Claver, C. *J. Am. Chem. Soc.* **2005**, *127*, 3646.

216. Leitner, A.; Shekhar, S.; Pouy, M. P.; Hartwig, J. F. *J. Am. Chem. Soc.* **2005**, *127*, 15506.

217. Feringa, B. L. *Acc. Chem. Res.* **2000**, *33*, 346.

218. Alexakis, A.; Vastra, J.; Burton, J.; Benhaim, C.; Mangeney, P. *Tetrahedron Lett.* **1998**, *39*, 7869.

219. Alexakis, A.; Burton, J.; Vastra, J.; Benhaim, C.; Fournioux, X.; Van den Heuvel, A.; Leveque, J.-M.; Maze, F.; Rosset, S. *Eur. J. Org. Chem.* **2000**, 4011.

220. Streiff, S.; Welter, C.; Schelwies, M.; Lipowsky, G.; Miller, N.; Helmchen, G. *Chem. Commun.* **2005**, 2957.

221. Weihofen, R.; Dahnz, A.; Tverskoy, O.; Helmchen, G. N. *Chem. Commun.* **2005**, 3541.

222. Mbaye, M. D.; Renaud, J.-L.; Demerseman, B.; Bruneau, C. *Chem. Commun.* **2004**, 1870.

223. Trost, B. M.; Krueger, A. C.; Bunt, R. C.; Zambrano, J. *J. Am. Chem. Soc.* **1996**, *118*, 6520.

224. Trost, B. M.; Breit, B.; Peukert, S.; Zambrano, J.; Ziller, J. W. *Angew. Chem., Int. Ed. Engl.* **1995**, *34*, 2386.

225. Wiese, B.; Helmchen, G. *Tetrahedron Lett.* **1998**, *39*, 5727.

226. Trost, B. M.; Tenaglia, A. *Tetrahedron Lett.* **1988**, *29*, 2931.

227. Trost, B. M.; McEachern, E. J. *J. Am. Chem. Soc.* **1999**, *121*, 8649.

228. Trost, B. M.; Horne, D. B.; Woltering, M. J. *Angew. Chem. Int. Ed.* **2003**, *42*, 5987.

229. Trost, B. M.; Lee, C. B. *J. Am. Chem. Soc.* **2001**, *123*, 3687.

230. Trost, B. M.; Lee, C. B. *J. Am. Chem. Soc.* **1998**, *120*, 6818.

231. Trost, B. M.; Lee, C. B. *J. Am. Chem. Soc.* **2001**, *123*, 12191.

232. Hayashi, T.; Okada, A.; Suzuka, T.; Kawatsura, M. *Org. Lett.* **2003**, *5*, 1713.

233. Pouy, M. J.; Leitner, A.; Weix, D. J.; Ueno, S.; Hartwig, J. F. *Org. Lett.* **2007**, *9*, 3949.

234. Ueno, S.; Hartwig, J. F. *Angew. Chem. Int. Ed.* **2008**, *47*, 1928.

235. Stanley, L. M.; Hartwig, J. F. *J. Am. Chem. Soc.* **2009**, *131*, 8971.

236. Trost, B. M.; Bunt, R. C. *J. Am. Chem. Soc.* **1994**, *116*, 4089.

237. Kudis, S.; Helmchen, G. *Angew. Chem. Int. Ed.* **1998**, *37*, 3047.

238. Evans, D. A.; Campos, K. R.; Tedrow, J. S.; Michael, F. E.; Gagne, M. R. *J. Am. Chem. Soc.* **2000**, *122*, 7905.

239. Trost, B. M.; Lee, C. B. In *Catalytic Asymmetric Synthesis*; 2nd ed.; Ojima, I., Ed.; Wiley-VCH: New York, 2000; p 503.

240. Trost, B. M.; Dudash, J.; Dirat, O. *Chem.—Eur. J.* **2002**, *8*, 259.

241. Trost, B. M.; Dirat, O.; Dudash, J.; E.J., H. *Angew. Chem. Int. Ed.* **2001**, *40*, 3658.

242. Trost, B. M.; Shi, Z. *J. Am. Chem. Soc.* **1996**, *118*, 3037.

243. Vedejs, E.; Jure, M. *Angew. Chem. Int. Ed.* **2005**, *44*, 3974.

244. Keith, J. M.; Larrow, J. F.; Jacobsen, E. N. *Adv. Synth. Catal.* **2001**, *343*, 5.

245. Trost, B. M.; Hembre, E. J. *Tetrahedron Lett.* **1999**, *40*, 219.

246. Trost, B. M.; Radinov, R.; Grenzer, E. M. *J. Am. Chem. Soc.* **1997**, *119*, 7879.

247. Trost, B. M.; Ariza, X. *Angew. Chem., Int. Ed. Engl.* **1997**, *36*, 2635.

248. Sawamura, M.; Sudoh, M.; Ito, Y. *J. Am. Chem. Soc.* **1996**, *118*, 3309.

249. Kuwano, R.; Ito, Y. *J. Am. Chem. Soc.* **1999**, *121*, 3236.

250. Trost, B. M.; Ariza, X. *Angew. Chem., Int. Ed. Engl.* **1997**, *36*, 2635.

251. Behenna, D. C.; Stoltz, B. M. *J. Am. Chem. Soc.* **2004**, *126*, 15044.

252. Burger, E. C.; Tunge, J. A. *Org. Lett.* **2004**, *6*, 4113.

253. Burger, E. C.; Tunge, J. A. *Org. Lett.* **2004**, *6*, 2603.

254. Burger, E. C.; Tunge, J. A. *Chem. Commun.* **2005**, 2835.

255. Mohr, J. T.; Behenna, D. C.; Harned, A. M.; Stoltz, B. M. *Angew. Chem. Int. Ed.* **2005**, *44*, 6924.

256. Trost, B. M.; Xu, J. Y. *J. Am. Chem. Soc.* **2005**, *127*, 2846.

257. Trost, B. M.; Xu, J. Y. *J. Am. Chem. Soc.* **2005**, *127*, 17180.

258. Trost, B. M.; Bream, R. N.; Xu, J. *Angew. Chem. Int. Ed.* **2006**, *45*, 3109.

259. Mellegaard-Waetzig, S. R.; Rayabarapu, D. K.; Tunge, J. A. *Synlett* **2005**, 2759.

260. Singh, O. V.; Han, H. *J. Am. Chem. Soc.* **2007**, *129*, 774.

261. Magid, R. M. *Tetrahedron* **1980**, *36*, 1901.

262. Lipshutz, B. H.; Sengupta, S. *Org. React.* **1992**, *41*, 135.

263. Karlstrom, A. S. E.; Bäckvall, J. E. In *Modern Organocopper Chemistry*; Krause, N., Ed.; Wiley-VCH: Weinheim, 2002; p 259.

264. Kar, A.; Argade, N. P. *Synthesis* **2005**, 2995.

265. Alexakis, A.; Malan, C.; Lea, L.; Tissot-Croset, K.; Polet, D.; Falciola, C. *Chimia* **2006**, *60*, 124.

266. Alexakis, A.; Bäckvall, J. E.; Krause, N.; Pàmies, O.; Diéguez, M. *Chem. Rev.* **2008**, *108*, 2796.

267. Harutyunyan, S. R.; den Hartog, T.; Geurts, K.; Minnaard, A. J.; Feringa, B. L. *Chem. Rev.* **2008**, *108*, 2824.

268. Falciola, C. A.; Alexakis, A. *Eur. J. Org. Chem.* **2008**, 3765.

269. Yorimitsu, H.; Oshima, K. *Angew. Chem. Int. Ed.* **2005**, *44*, 4435.

270. Goering, H. L.; Kantner, S. S. *J. Org. Chem.* **1984**, *49*, 422.

271. Goering, H. L.; Singleton, V. D. *J. Am. Chem. Soc.* **1976**, *98*, 7854.

272. Corey, E. J.; Boaz, N. W. *Tetrahedron Lett.* **1984**, *25*, 3063.

273. Gallina, C.; Ciattini, P. G. *J. Am. Chem. Soc.* **1979**, *101*, 1035.

274. Calò, V.; Lopez, L.; Carlucci, W. F. *J. Chem. Soc., Perkin Trans. 1* **1983**, 2953.

275. Greene, A. E.; Coelho, F.; Deprés, J.-P.; Brocksom, T. J. *Tetrahedron Lett.* **1988**, *29*, 5661.

276. Breit, B.; Demel, P. *Adv. Synth. Catal.* **2001**, *343*, 429.

277. Breit, B.; Herber, C. *Angew. Chem. Int. Ed.* **2004**, *43*, 3790.

278. Goering, H. L.; Singleton, V. D. *J. Org. Chem.* **1983**, *48*, 1531.

279. Chapleo, C. B.; Finch, A. W.; Lee, T. V.; Roberts, S. M. *J. Chem. Soc., Chem. Commun.* **1979**, 676.

280. Rona, P.; Tökes, L.; Tremble, J.; Crabbé, P. *J. Chem. Soc., Chem. Commun.* **1969**, 43.

281. Fouquet, G.; Schlosser, M. *Angew. Chem., Int. Ed. Engl.* **1974**, *13*, 82.

282. Persson, E. S. M.; van Klaveren, M.; Grove, D. M.; Bäckvall, J. E.; van Koten, G. *Chem.—Eur. J.* **1995**, *1*, 351.

283. Yamanaka, M.; Kato, S.; Nakamura, E. *J. Am. Chem. Soc.* **2004**, *126*, 6287.

284. Yoshikai, N.; Zhang, S.-L.; Nakamura, E. *J. Am. Chem. Soc.* **2008**.

285. Bartholomew, E. R.; Bertz, S. H.; Cope, S.; Murphy, M.; Ogle, C. A. *J. Am. Chem. Soc.* **2008**, *130*, 11244.

286. van Klaveren, M.; Persson, E. S. M.; del Villar, A.; Grove, D. M.; Bäckvall, J. E.; van Koten, G. *Tetrahedron Lett.* **1995**, *36*, 3059.

287. Dübner, F.; Knochel, P. *Angew. Chem. Int. Ed.* **1999**, *38*, 379.

288. Dübner, F.; Knochel, P. *Tetrahedron Lett.* **2000**, *41*, 9233.

289. Badalassi, F.; Crotti, P.; Macchia, F.; Pineschi, M.; Arnold, A.; Feringa, B. L. *Tetrahedron Lett.* **1998**, *39*, 7795.

290. Shi, W.-J.; Wang, L.-X.; Fu, Y.; Zhu, S.-F.; Zhou, Q.-L. *Tetrahedron: Asymmetry* **2003**, *14*, 3867.

291. van Zijl, A. W.; Arnold, L. A.; Minnaard, A. J.; Feringa, B. L. *Adv. Synth. Catal.* **2004**, *346*, 413.

292. Goldsmith, P. J.; Teat, S. J.; Woodward, S. *Angew. Chem. Int. Ed.* **2005**, *44*, 2235.

293. Börner, C.; Gimeno, J.; Gladiali, S.; Goldsmith, P. J.; Ramazzotti, D.; Woodward, S. *Chem. Commun.* **2000**, 2433.

294. Kacprzynski, M. A.; Hoveyda, A. H. *J. Am. Chem. Soc.* **2004**, *126*, 10676.

295. Luchaco-Cullis, C. A.; Mizutani, H.; Murphy, K. E.; Hoveyda, A. H. *Angew. Chem. Int. Ed.* **2001**, *40*, 1456.

296. Murphy, K. E.; Hoveyda, A. H. *J. Am. Chem. Soc.* **2003**, *125*, 4690.

297. Murphy, K. E.; Hoveyda, A. H. *Org. Lett.* **2005**, *7*, 1255.

298. Larsen, A. O.; Leu, W.; Oberhuber, C. N.; Campbell, J. E.; Hoveyda, A. H. *J. Am. Chem. Soc.* **2004**, *126*, 11130.

299. Van Veldhuizen, J. J.; Campbell, J. E.; Giudici, R. E.; Hoveyda, A. H. *J. Am. Chem. Soc.* **2005**, *127*, 6877.

300. Kacprzynski, M. A.; May, T. L.; Kazane, S. A.; Hoveyda, A. H. *Angew. Chem. Int. Ed.* **2007**, *46*, 4554.

301. Piarulli, U.; Daubos, P.; Claverie, C.; Roux, M.; Gennari, C. *Angew. Chem. Int. Ed.* **2003**, *42*, 234.

302. Piarulli, U.; Claverie, C.; Daubos, P.; Gennari, C.; Minnaard, A. J.; Feringa, B. L. *Org. Lett.* **2003**, *5*, 4493.

303. Piarulli, U.; Daubos, P.; Claverie, C.; Monti, C.; Gennari, C. *Eur. J. Org. Chem.* **2005**, 895.

304. Karlstrom, A. S. E.; Huerta, F. F.; Meuzelaar, G. J.; Bäckvall, J.-E. *Synlett* **2001**, 923.

305. Cotton, H. K.; Norinder, J.; Bäckvall, J.-E. *Tetrahedron* **2006**, *62*, 5632.

306. Tominaga, S.; Oi, Y.; Kato, T.; An, D. K.; Okamoto, S. *Tetrahedron Lett.* **2004**, *45*, 5585.

307. Alexakis, A.; Malan, C.; Lea, L.; Benhaim, C.; Fournioux, X. *Synlett* **2001**, 927.

308. Alexakis, A.; Croset, K. *Org. Lett.* **2002**, *4*, 4147.

309. Tissot-Croset, K.; Polet, D.; Alexakis, A. *Angew. Chem. Int. Ed.* **2004**, *43*, 2426.

310. López, F.; van Zijl, A. W.; Minnaard, A. J.; Feringa, B. L. *Chem. Commun.* **2006**, 409.

311. Falciola, C. A.; Tissot-Croset, K.; Alexakis, A. *Angew. Chem. Int. Ed.* **2006**, *45*, 5995.

312. Geurts, K.; Fletcher, S. P.; Feringa, B. L. *J. Am. Chem. Soc.* **2006**, *128*, 15572.

313. Gillingham, D. G.; Hoveyda, A. H. *Angew. Chem. Int. Ed.* **2007**, *46*, 3860.

314. Lee, Y.; Akiyama, K.; Gillingham, D. G.; Brown, M. K.; Hoveyda, A. H. *J. Am. Chem. Soc.* **2008**, *130*, 446.

315. Ito, H.; Ito, S.; Sasaki, Y.; Matsuura, K.; Sawamura, M. *J. Am. Chem. Soc.* **2007**, *129*, 14856.

316. Equey, O.; Alexakis, A. *Tetrahedron: Asymmetry* **2004**, *15*, 1531.

317. Millet, R.; Alexakis, A. *Synlett* **2007**, 435.

318. Pineschi, M.; Del Moro, F.; Crotti, P.; Di Bussolo, V.; Macchia, F. *Synthesis* **2005**, 334.

319. Bertozzi, F.; Pineschi, M.; Macchia, F.; Arnold, L. A.; Minnaard, A. J.; Feringa, B. L. *Org. Lett.* **2002**, *4*, 2703.

320. Zhang, W.; Wang, L.-X.; Shi, W.-J.; Zhou, Q.-L. *J. Org. Chem.* **2005**, *70*, 3734.

321. Pineschi, M.; Del Moro, F.; Crotti, P.; Macchia, F. *Org. Lett.* **2005**, *7*, 3605.

322. Bournaud, C.; Falciola, C.; Lecourt, T.; Rosset, S.; Alexakis, A.; Micouin, L. *Org. Lett.* **2006**, *8*, 3581.

323. Palais, L.; Mikhel, I. S.; Bournaud, C.; Micouin, L.; Falciola, C. A.; Vuagnoux-d'Augustin, M.; Rosset, S.; Bernardinelli, G.; Alexakis, A. *Angew. Chem. Int. Ed.* **2007**, *46*, 7462.

CHAPTER 21

Metathesis of Olefins and Alkynes

21.1. Introduction

21.1.1. Overview of the Catalytic Metathesis of Carbon–Carbon Multiple Bonds

Olefin metathesis reactions cleave carbon–carbon double bonds and reassemble them to generate products containing new carbon–carbon double bonds. This process requires a catalyst and is largely controlled by thermodynamics (Equation 21.1). Alkyne metathesis reactions cleave carbon–carbon triple bonds and reassemble them to form products containing new carbon–carbon triple bonds (Equation 21.2). The observation of complete cleavage of strong carbon–carbon multiple bonds by a catalytic process was remarkable when first discovered, but many transition metal complexes are now known that catalyze these reactions with fast rates. One might expect that the equilibrium control of this reaction would limit its use, but olefin metathesis has become one of the most useful reactions catalyzed by transition metal complexes.

Olefin metathesis was discovered in the 1950s, and initially was conducted with heterogeneous catalysts. Early studies focused on the mechanism by which the carbon–carbon bonds were cleaved. Well-defined soluble catalysts have now been developed, including those that tolerate a wide range of auxiliary functionality. These developments culminated in a Nobel Prize for Yves Chauvin, who first proposed the correct mechanism for this process, and Richard Schrock and Robert Grubbs, who developed highly active, well-defined catalysts for this process. Many review articles[1–6] and monographs[7–9] have now been published on olefin and alkyne metathesis, the most comprehensive being a three-volume compendium.[9] This chapter describes the major classes of metathesis processes, some of the most utilized catalysts, some of the classic and emerging applications, and information on the mechanism of this process.

$$2 \quad \underset{R^1 \quad \quad R^2}{\overset{R^1 \quad \quad R^2}{\diagup\!\!\!\diagdown}} \quad \underset{\text{Catalyst}}{\rightleftharpoons} \quad \underset{R^1 \quad \quad R^1}{\overset{R^1 \quad \quad R^1}{\diagup\!\!\!\diagdown}} \quad + \quad \underset{R^2 \quad \quad R^2}{\overset{R^2 \quad \quad R^2}{\diagup\!\!\!\diagdown}} \qquad (21.1)$$

$$2 \ R^1\!\!-\!\!\equiv\!\!-R^2 \quad \overset{\text{Catalyst}}{\rightleftharpoons} \quad R^1\!\!-\!\!\equiv\!\!-R^1 \ + \ R^2\!\!-\!\!\equiv\!\!-R^2 \qquad (21.2)$$

21.1.2. Overview of the Classes of Metathesis Processes

Six classes of olefin metathesis, shown in Scheme 21.1, have been studied most extensively. In addition, metathesis processes involving alkynes have been developed. The six classes of olefin metathesis reactions are discussed in detail in subsequent sections and are briefly described in the following paragraphs.

The first class of reaction shown in Scheme 21.1 is the simple equilibration of alkenes to broaden molecular weight distribution. This olefin metathesis reaction is conducted on large scale to produce commodity chemicals. Equilibration of the alkenes in conjunction with distillation of the fraction containing material with the desired molecular weight is part of the Shell higher olefin process (SHOP).[8] The olefin metathesis step of SHOP, summarized in Scheme 21.2, is conducted with alkenes generated by ethylene oligomerization. Another of the earliest olefin metathesis processes involved the formation of 2-butene from propylene by a similar redistribution of the molecular weight of a simple alkene.[10–12]

The second class of olefin metathesis in Scheme 21.1 is ring-opening metathesis polymerization (ROMP).[13] This class of olefin methathesis generates polymers by formally cleaving

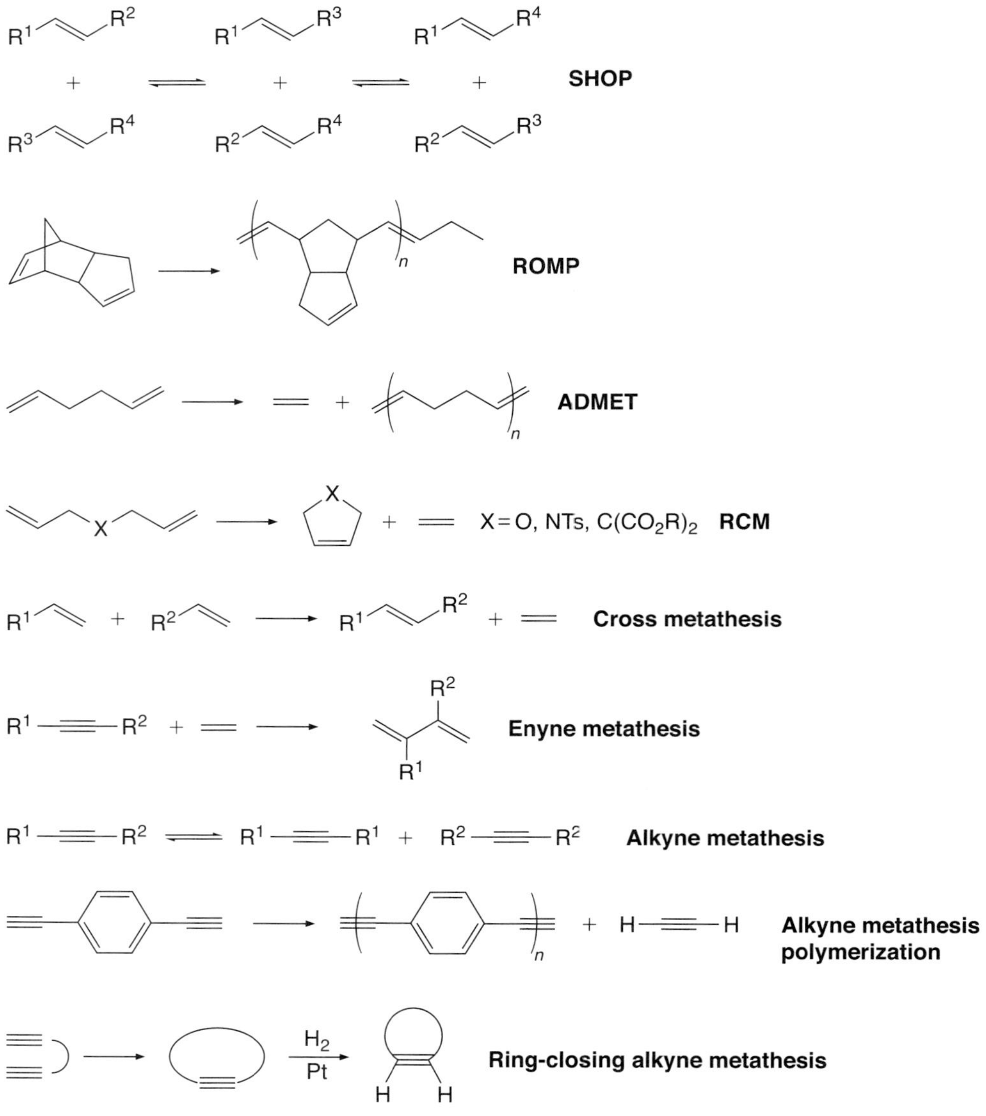

Scheme 21.1

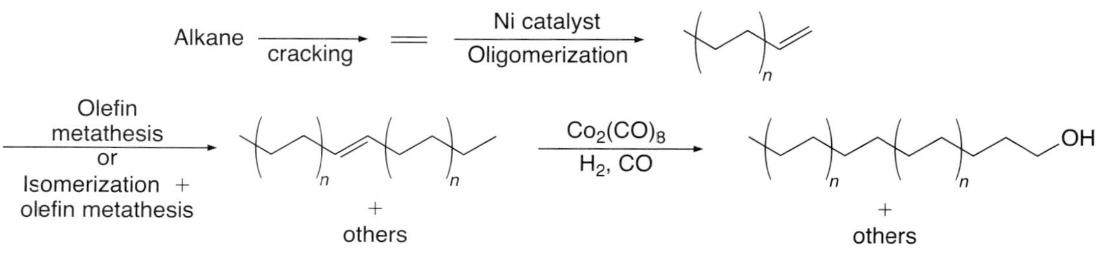

Scheme 21.2

the C=C bond in a strained cyclic olefin and forming new C=C bonds between monomer units. This process is used to prepare a material called Vestenamer® from the ROMP of cyclooctene, a material called Norsorex® from the ROMP of norbornene, and materials called Metton® and Telene® from the ROMP of dicyclopentadiene. The materials from this type of metathesis have been studied as precursors to elastomers after crosslinking of pendant olefins, such as those in the ROMP product of dicyclopentadiene shown in Scheme 21.1.

The third class of olefin methathesis in Scheme 21.1 is addition metathesis polymerization (ADMET).[14,15] This reaction is an alternative method to stitch together olefins into polymers, in this case by a combination of dienes with extrusion of ethylene. Control of molecular weight by the ADMET process is less precise than that by ROMP, but this reaction has been used to make polymers with precise architectures, such as polymers that would be perfectly alternating ethylene–propylene copolymers.[16]

The fourth class of olefin metathesis in Scheme 21.1 is ring-closing metathesis (RCM). RCM is now used commonly in complex molecule synthesis.[6] The thermodynamics for the ring-closing process are favored by the entropic benefit of generating two molecules from one. Moreover, these reactions are often conducted in an open system under non-equilibrium conditions that release gaseous ethylene. Some of the most favorable RCM reactions form five- and six-membered rings, but RCM has also been used to form macrocyclic units in natural products and pharmaceutical candidates.

The fifth class of olefin metathesis in Scheme 21.1 is cross-metathesis. Selective cross metathesis is a developing type of metathesis reaction.[17] Expulsion of ethylene usually provides the driving force for cross metathesis. In principle a statistical mixture of olefins would be formed by cross metathesis, as shown in the first reaction of Scheme 21.1. However, kinetic preferences for the reaction of a hindered carbene complex with an unhindered olefin or a kinetic preference for the reaction of an electron-rich carbene with an electron-poor olefin can provide the selectivity needed for the formation of one olefin over other potential homo- and cross-metathesis products.

In addition to the metathesis of olefins, metathesis between an olefin and an alkyne and metathesis between two alkynes are known and can be synthetically valuable. The metathesis between an olefin and an alkyne is called enyne metathesis,[18] and enyne methathesis is the final class of reaction involving an alkene shown in Scheme 21.1. This process combines an olefin with an alkyne to generate a diene. The thermodynamic driving force for this process is created by the generation of a new carbon–carbon single bond from the cleavage of one of the π-bonds in an alkyne.

Alkynes, like alkenes, can undergo cross or ring-closing metathesis, and these reactions are the final two processes shown in Scheme 21.1. Intermolecular alkyne metathesis has been investigated as a route to conjugated polymers.[19,20] Ring-closing alkyne methathesis has been used in natural products synthesis. Of course, ring-closing alkyne methathesis must form a macrocyclic product to accommodate the more linear structure of the alkyne, but this process, in combination with hydrogenation, has been developed as a method to generate Z-olefins in macrocyclic structures.[21]

21.2. Olefin Metathesis

21.2.1. Overview of Catalysts for Olefin Metathesis

Several of the important catalysts for olefin metathesis were described in Chapter 13 that covered complexes containing metal-ligand multiple bonds, and these structures are shown again in Figure 21.1. The molybdenum and tungsten catalysts have been developed by Schrock,[1] and the molybdenum complex is typically called "Schrock's Catalyst." The two ruthenium catalysts in the middle of the figure are commonly known as the first and second generation Grubbs catalysts.[22,23] The fourth complex in this figure, which is a modified version of the Grubbs catalyst lacking the second dative ligand, is often known as the Hoveyda–Grubbs catalyst.[24] Related cationic catalysts were developed by Hofmann.[25]

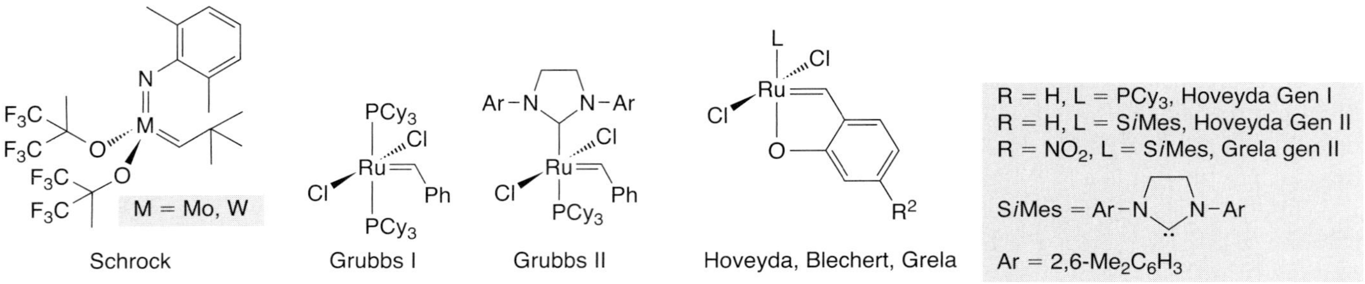

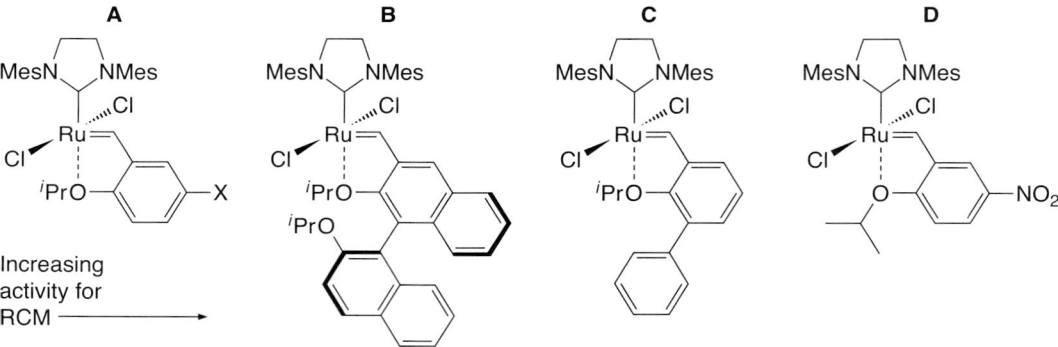

Figure 21.1. Current commonly used olefin metathesis catalysts.

Although the Hoveyda–Grubbs catalyst was originally designed to create a recyclable catalyst, it has proven to be useful for many classes of olefin metathesis processes for different reasons. This system is thought to be valuable, in part, because it lacks a phosphine which can ultimately lead to decomposition of the catalyst by attack at an intermediate methylene complex (vide infra). These complexes are also more stable thermally than the other ruthenium metathesis catalysts and, therefore, have been useful for reactions requiring higher temperatures. For these reasons, several groups including Blechert's and Grela's have modified the structure of this class of catalyst, and these modifications have led to systems that are more active for certain reactions than the original ones.

The relative rates of reaction of these catalysts for ring-closing metatheses to form five-membered rings are shown in Figure 21.2. Blechert has shown that an aryl group ortho to the alkoxo substituent destabilizes the metal–alkoxo linkage and leads to faster initiation and a more active catalyst.[26,27] Grela has shown that the electron-withdrawing property of a nitro group destabilizes the Ru–O linkage and leads to active catalysts.[28,29] However, combining the steric effects of a substituent ortho to the isopropoxide with the electronic properties of a nitro group led to catalysts that are overall less reactive, perhaps due to faster decomposition.

Figure 21.2. Relative activities of variants of the Hoveyda–Grubbs catalyst.

Pathways to deactivation of the Schrock-type catalysts are thought to proceed by association of two catalysts to form bridged structures. Thus, it might be possible to improve the catalyst activities and lifetimes by isolating the catalysts on a solid support, such that two catalysts do not interact with each other. Thus, Schrock, Coperet, and Basset have worked to support the Schrock tungsten and molybdenum catalysts on a partially dehydrated silica support.[30–34] This work has led to more active catalysts, partly from site isolation and partly from the unsymmetrical structure of the supported catalyst.

The overall scheme to generate the supported catalysts is shown in Equation 21.3a. Partially dehydrated silica is treated with a solution of Schrock's tungsten catalyst in pentane. The acidic hydroxyl groups protonate an alkyl group to form the siloxide-supported catalyst. In a similar fashion, partially dehydrated silica is treated with a solution of a bispyrrolyl version of Schrock's molybdenum catalyst to release pyrrole and generate a catalyst containing one siloxide linkage and one pyrrolide ligand, in addition to the imido ligand and the reactive carbene moiety (Equation 21.3b). A comparison of the reactivity of the supported and unsupported systems showed that the supported molybdenum complex catalyzes the self-metathesis of methyl oleate with turnover frequencies that are a factor of 10 higher than those of the unsupported catalyst.[31]

The faster rates of reaction of the supported molybdenum catalyst can be attributed to site isolation, but computational data also imply that an unsymmetical catalyst best balances the electronic requirements of the carbene and metallacycle structures and leads to low barriers for [2+2] and retro-[2+2] processes.[35]

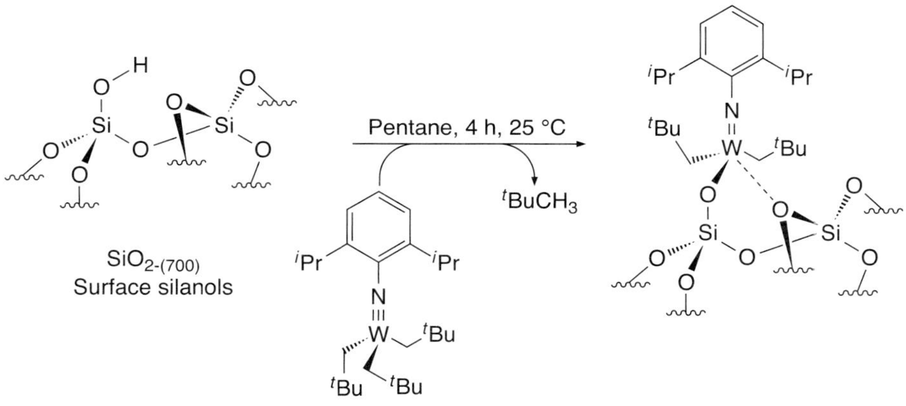

(21.3a)

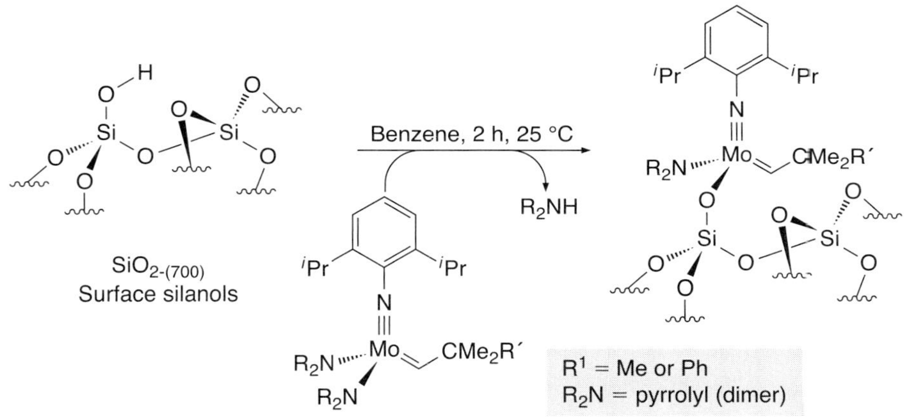

R^1 = Me or Ph
R_2N = pyrrolyl (dimer)

(21.3b)

21.2.2. History of Olefin Metathesis

A timeline for the development of olefin metathesis, adapted from a review by Grubbs,[23] is shown in Figure 21.3. Olefin metathesis is more than 50 years old.[10–12,36] It was first conducted with ill-defined rhenium, molybdenum, and tungsten systems generated from perrhenate, aluminum oxide,[37,38] and tetraethyl lead as additive,[39] from molybdenum oxide on β-TiO₂ and tetramethyltin as additive,[40] or from tungsten phenoxides supported on niobium oxide and silicon oxide activated with alkylaluminum reagents.[41] The temperatures for these processes are high, but the catalysts are relatively inexpensive and can be long lived. These are the types of catalysts that have been used for the synthesis of commodity chemicals by olefin metathesis.

In the 1960s, ring-opening metathesis polymerization was first developed, and simple ruthenium chloride was used for this reaction.[42–48] In the 1970s and 80s, the mechanism of this reaction involving sequential [2+2] steps was suggested by Chauvin[49] and Katz.[50] The first alkylidene complexes that underwent the [2+2] reactions of this proposed mechanism were prepared by Schrock,[51,52] Casey,[53] and Grubbs.[54,55] These mechanistic data led to the

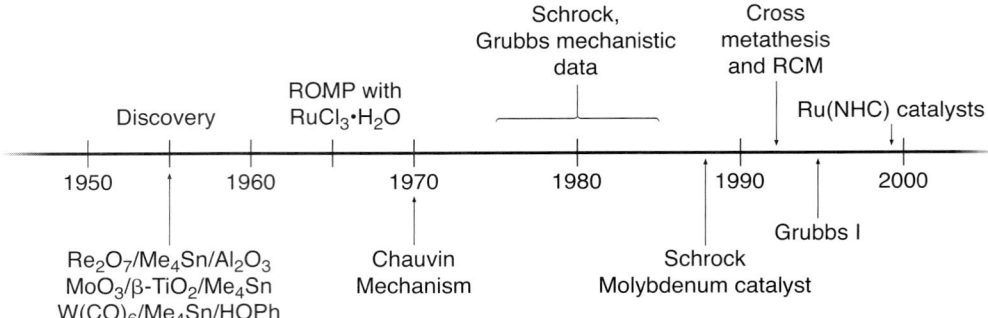

Figure 21.3.
Timeline of olefin metathesis (adapted from Trnka, T. M.; Grubbs, R. H. *Acc. Chem. Res.* **2001**, *34*, 18).

development of improved catalysts that are well defined. The benefits of the well-defined catalyst are high activities at moderate temperatures and in some cases higher functional group tolerance. Schrock's molybdenum catalyst was prepared in the late 1980s,[56] and the first-generation Grubbs catalyst was prepared in the mid-1990s.[57,58] The first cross metathesis with Schrock's catalyst was reported by Crowe in 1993,[59,60] and ring-closing metathesis emerged as a synthetically useful procedure in the work of Fu and Grubbs in the early to mid 1990s.[6,61] There is significant room to improve catalysts for the ring-closing process, but the current catalysts are effective enough that the reaction has gained widespread use.

21.2.3. Mechanism of Olefin Metathesis

Olefin metathesis occurs by a sequence of [2+2] cycloaddition and cycloreversion reactions shown as path A in Scheme 21.3. These [2+2] reactions were described in Chapter 13 on metal–ligand multiple bonds, and information on the effect of the ligand set at

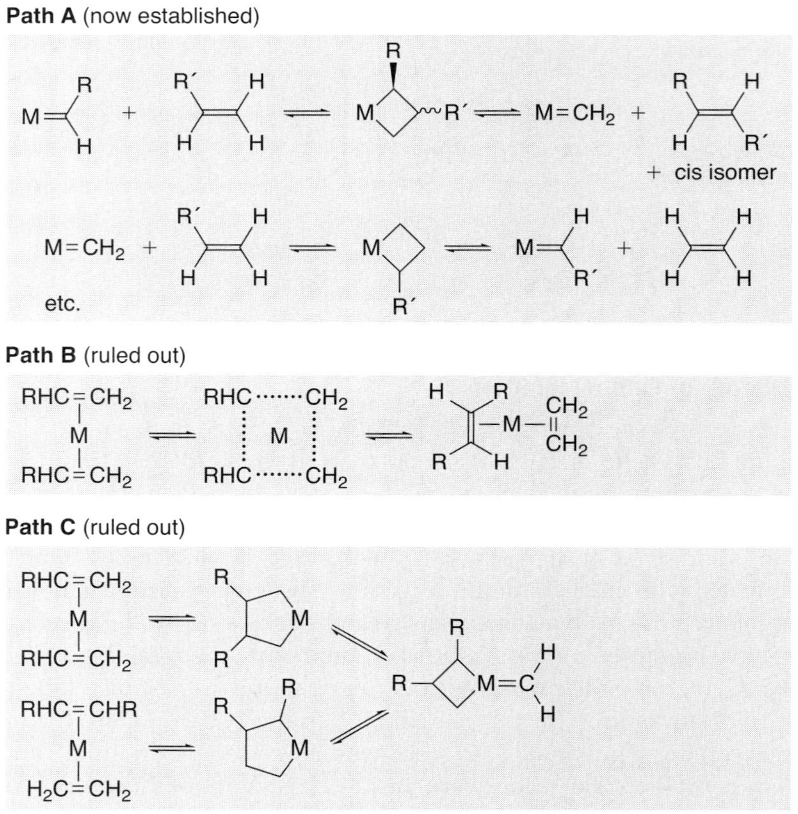

Path A (now established)

Path B (ruled out)

Path C (ruled out)

Scheme 21.3

the metal on the rate of these [2+2] reactions was included in that chapter. In brief, an open coordination site at the metal in the carbene complex is necessary for coordination of olefin prior to the [2+2] step. After coordination of the olefin, the complex containing the olefin and carbene generate a metallacyclobutane complex. The metallacyclobutane complex then reverts to a new olefin carbene complex and dissociates the new olefin.

This sequence of reactions, depicted as a catalytic cycle, is shown in Scheme 21.4. This cycle shows the dimerization of an olefin and extrusion of ethylene, but this cycle can be modified in a simple way to describe the mechanisms of all the types of olefin metatheses presented in this chapter. In this mechanism, the initial carbene catalyst reacts with the olefin reagent in an initiation step. Depending on the presence or absence of an open coordination site in the catalyst precursor, this step can involve ligand dissociation. This initiation is then followed by a [2+2] cycloaddition with a second olefin to form a metallacyclobutane. This [2+2] process can occur in a productive fashion, as shown in Scheme 21.4, or it can occur in a degenerate fashion, as shown at the bottom of this scheme. The degenerate [2+2] process is often faster than the productive one because of the 1,3-disposition of the two substituents. The metallacycle shown in the catalytic cycle then cleaves to generate the new olefin and a methylidene complex. The methylidene complex then reacts with a reagent olefin to extrude ethylene and form the starting alkylidene complex.

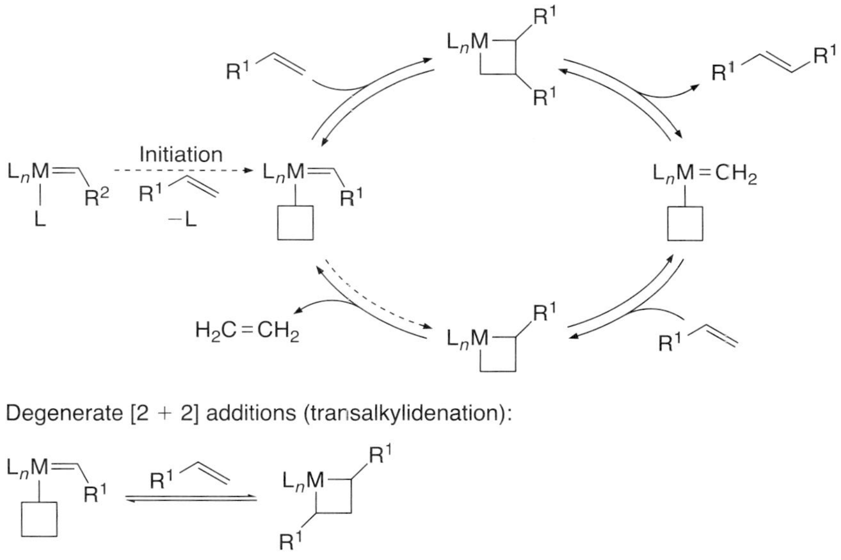

Degenerate [2 + 2] additions (transalkylidenation):

Scheme 21.4

Although it is now fully accepted that olefin metathesis ocurrs by this sequence of steps, the mechanism of olefin metathesis was not clear when it was first discovered. A series of papers were published with clever experiments to distinguish between "pairwise" and "non-pairwise" mechanisms. Pathway A of Scheme 21.3 is a "non-pairwise" mechanism because the olefin reacts with a metal complex containing only half of the second olefin reactant. In contrast, paths B and C contain all of both olefins within the coordination sphere of the metal. Several experiments to distinguish between these paths were reported by Grubbs,[54,55] Casey,[53,62] and Katz.[63-66]

The experiments by Katz depicted in Scheme 21.5 are perhaps the most intuitive to interpret. The first experiment involves the ring-opening of cyclooctene with a mixture of 2-butene and 4-octene. In the absence of any metathesis between acyclic olefins, the pairwise mechanism would only form products from the cyclooctene containing 12 and 16 carbons. The experiment is complicated by slower metathesis of the ring-opened products and metathesis of the 2-butene and 4-octene. However, extrapolation of the distribution

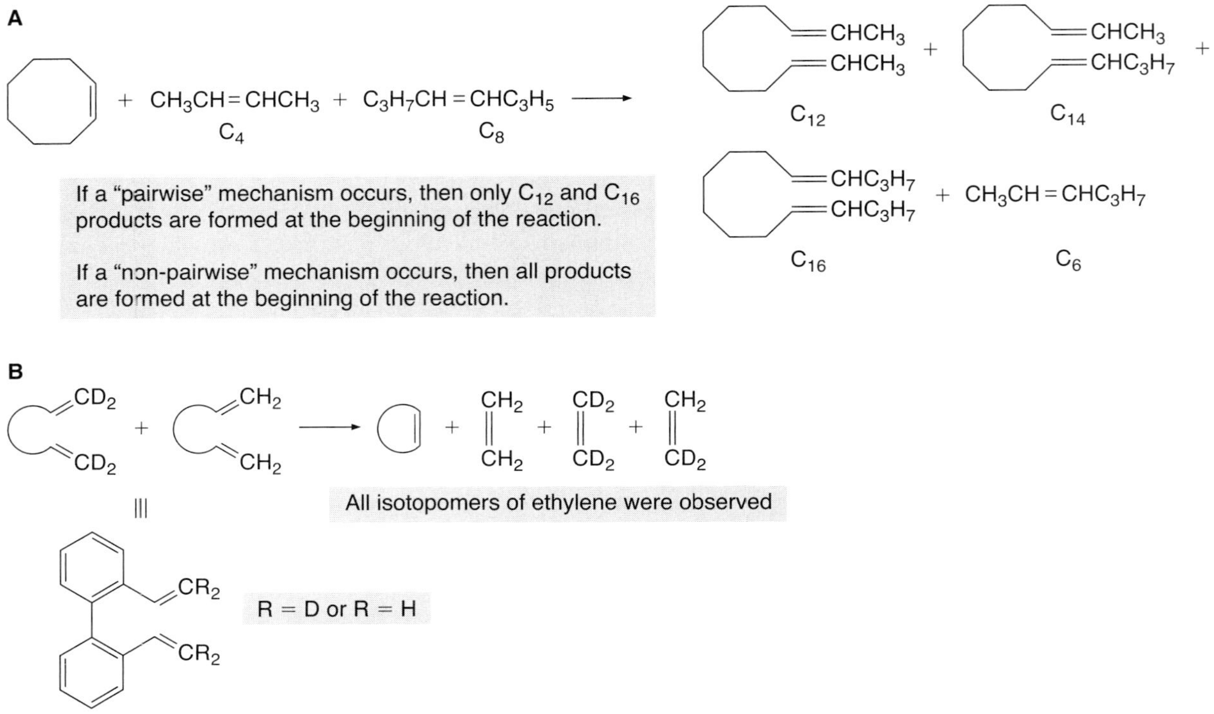

A

If a "pairwise" mechanism occurs, then only C_{12} and C_{16} products are formed at the beginning of the reaction.

If a "non-pairwise" mechanism occurs, then all products are formed at the beginning of the reaction.

B

All isotopomers of ethylene were observed

R = D or R = H

Scheme 21.5

of products to the beginning of the reaction can reveal whether any C_{14} products form initially or if only C_{12} and C_{16} products (along with reactants and lower molecular weight products) are present. This extrapolation showed that the C_{14} product was formed at the initial stage of the reaction. This experiment, despite its complications, ruled out pairwise mechanisms. A cleaner experiment is shown in part B of this scheme. In this experiment, a stable ring and ethylene are formed. The stability of the ring and the lack of activity of the catalyst toward ethylene prevents scrambling of the products. This reaction formed a statistical 1:2:1 ratio of C_2H_4, $C_2H_2D_2$, and C_2D_4, which is consistent with the "non-pairwise" mechanism and is inconsistent with the "pairwise" mechanisms.

21.2.4. Catalyst Decomposition

The convenience and functional group tolerance of the ruthenium catalysts has led these compounds to be used widely in synthesis, but the loading of catalyst needed for these applications is often high. Thus, several studies have been conducted to identify the reactions that lead to catalyst decomposition with the objective of designing systems that resist these decomposition pathways. Three reactions that lead to catalyst decomposition are shown in Equations 21.4a–c.

Grubbs and co-workers found that heating of the second-generation catalyst alone in benzene led to formation of a bridging carbide complex (Equation 21.4a). The proposed mechanism for formation of this species is speculative, but is thought to originate from attack of dissociated phosphine on the methylidene ligand to generate, at least formally, a 12-electron intermediate that reacts with another ruthenium complex to create a bridged species that eventually generates a carbide ligand.[67] Grubbs and co-workers have also found that heating the benzylidene complex with ethylene led to a complex resulting from cyclometallation at the ortho-methyl group of the aryl substitituent.[68] This process is shown in Equation 21.4b and has also been thought to originate from the attack of phosphine on a methylidene. The methylphosphonium salt is formed as a coproduct in quantitative yield. Finally, treatment of the methylidene complex with pyridine also leads to the formation of the methylphosphonium salt and the 18-electron trispyridine complex (Equation 21.4c).[32] Additional pathways for decomposition of metathesis catalysts are certain to continue to be uncovered.

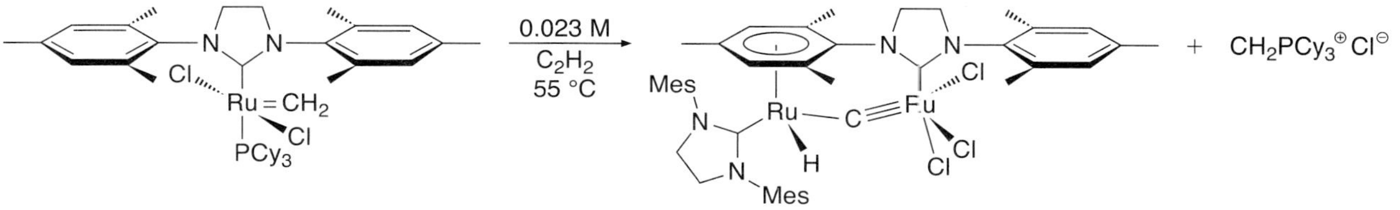

(21.4a)

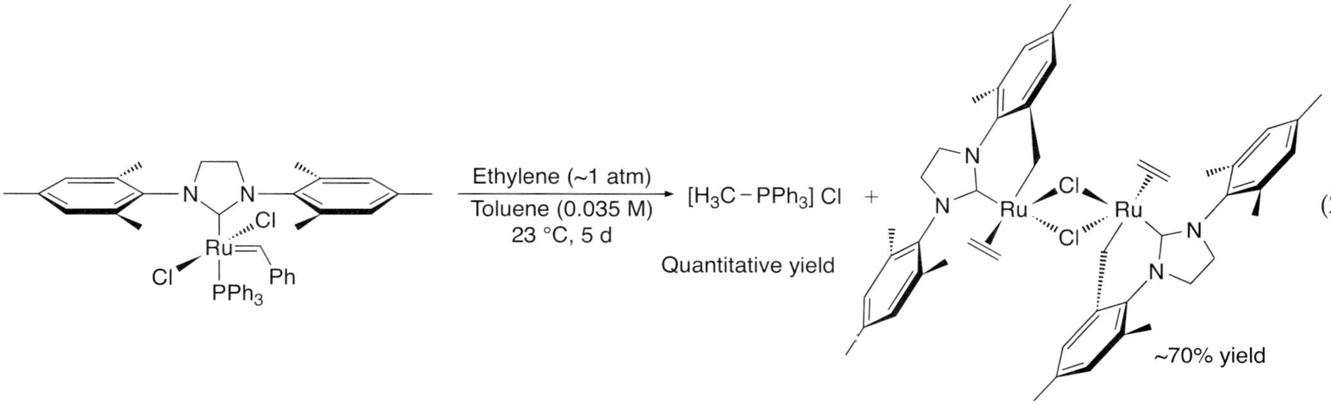

(21.4b)

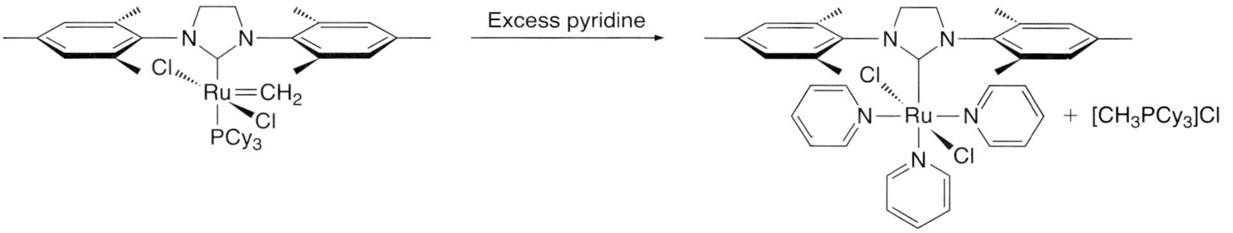

(21.4c)

21.2.5. Examples of Olefin Metathesis

21.2.5.1. Ring-Closing Olefin Metathesis[6]

Several simple prototypical examples of ring-closing metathesis are shown in Equation 21.5. In general, dienes undergo RCM to form five-, six-, and seven-membered cyclic alkenes containing oxygen[69,70] or a protected nitrogen.[70–72] Some of the most common substrates used to test catalyst activity are diallyl ethers, diallylamines containing protective groups on nitrogen, and diallyl malonates (Equations 21.5 and 21.6). However, the formation of large rings has been a particularly successful application of ring-closing metathesis to natural product synthesis. Such a macrocyclization was first reported by Tsuji[73] (Equation 21.7), but was developed in the context of natural products synthesis (vide infra) by Fürstner.[74–76] Although ADMET polymerization occurs with acyclic dienes under concentrated conditions, macrocyclic structures of natural products form under more dilute conditions or when the structure contains a conformational bias for cyclization. The formation of medium-sized rings remains a challenge,[77] and success depends largely on the conformation of the reactant diene.[77b] Moreover, the double bond in a macrocycle often forms as a mixture of E- and Z-isomers. The selective formation of E- or Z-olefins, particularly formation of the less thermodynamically stable Z-olefins by metathesis is challenging, but some progress has been made with molybdenum catalysts.[78] These catalysts that are thought to generate a metallacycle in which one very hindered and one less hindered ligand are bound to the metal on either side of the metallacycle. This distribution of ligands favors a structure in which both substituents of the product alkene would be located cis to each other on the side of the ring containing the less hindered ligand.

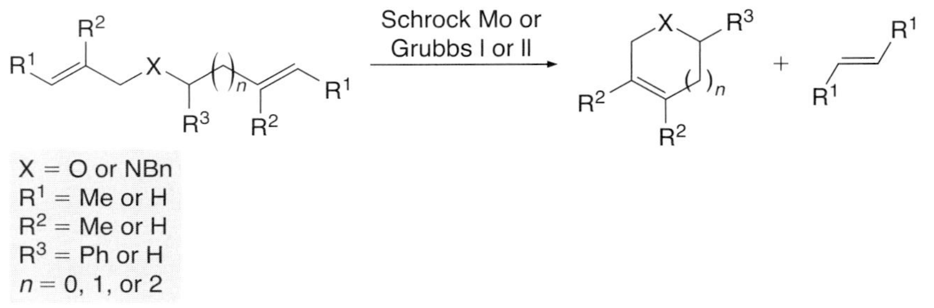

$$X = O \text{ or } NBn$$
$$R^1 = Me \text{ or } H$$
$$R^2 = Me \text{ or } H$$
$$R^3 = Ph \text{ or } H$$
$$n = 0, 1, \text{ or } 2$$

(21.5)

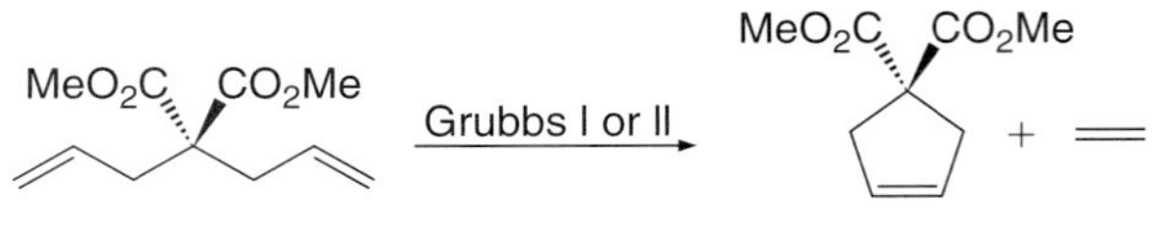

(21.6)

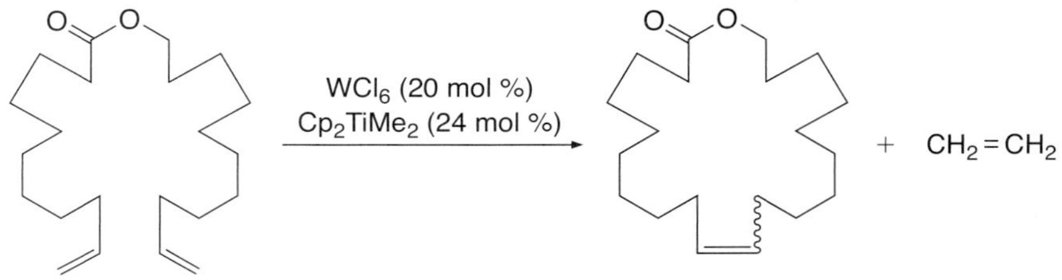

(21.7)

Manzamine A

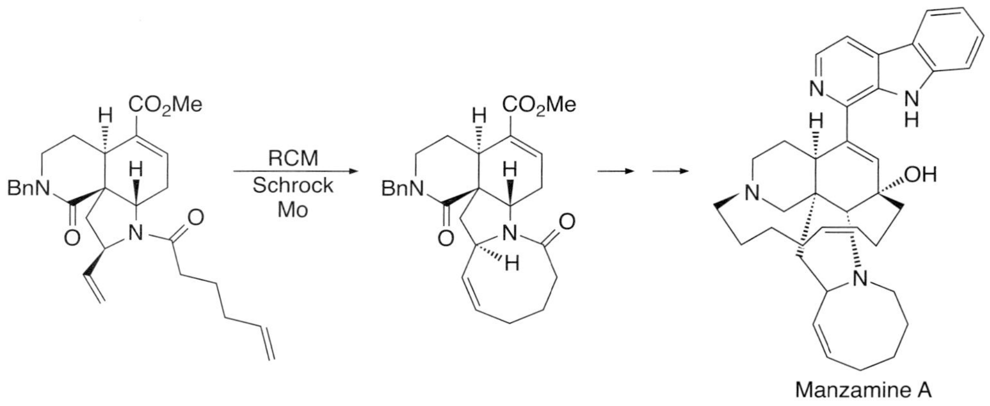

(21.8)

The synthesis of manzamine A by Martin and co-workers using olefin metathesis[79,80] was one of the earliest demonstrations[81,82] of how this reaction could be applied as a strategy in natural product synthesis. The ring-closing metathesis in this synthesis was conducted with Schrock's molybdenum catalyst, as shown in Equation 21.8.

The epothilones comprise an important set of natural products that were discovered at a time contemporary with the development of ring-closing olefin metathesis. These materials are 16-membered ring compounds containing an alkene in the macrocycle or an epoxide that can be generated from the alkene. A conventional retrosynthetic analysis

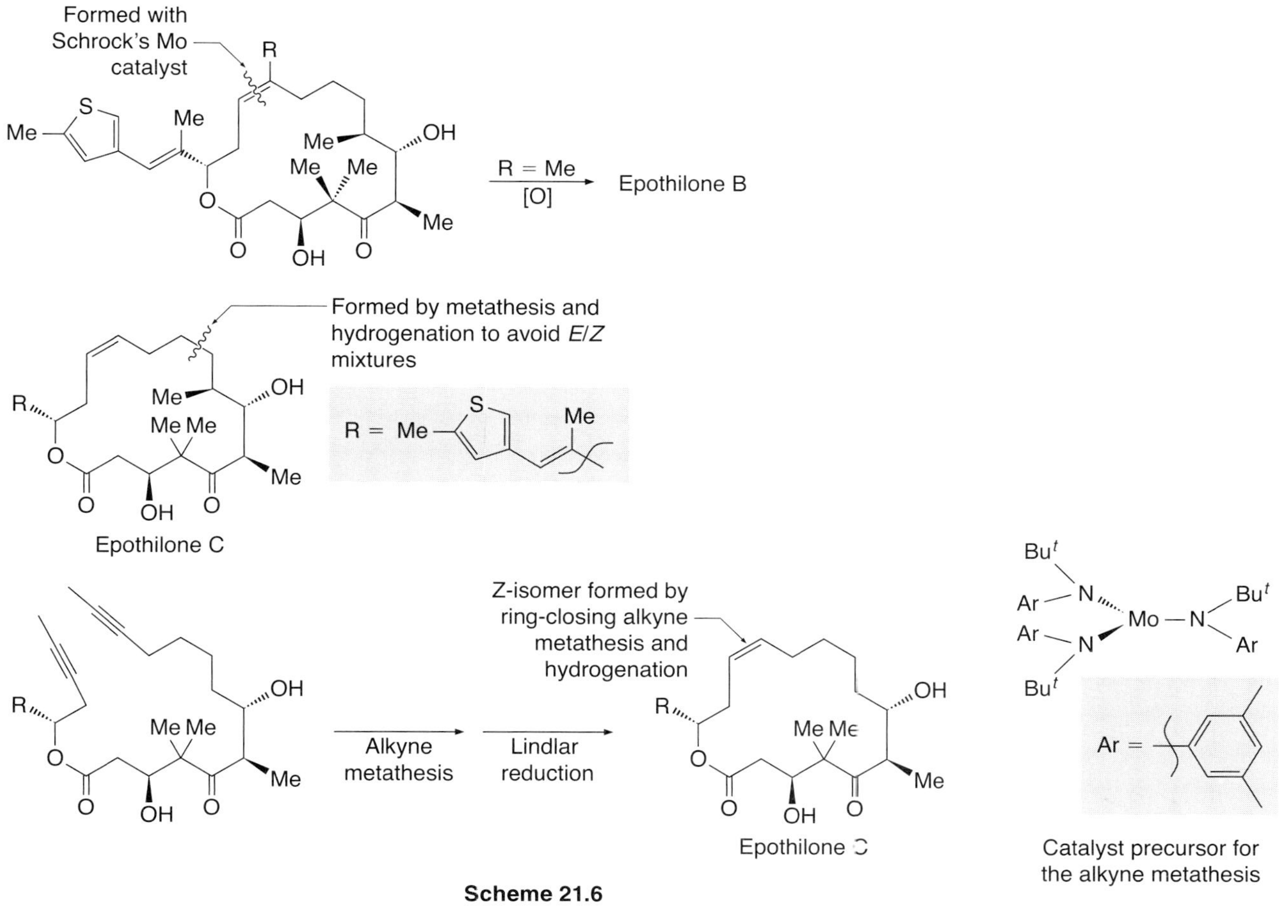

Scheme 21.6

of this compound would involve disconnection of the macrocycle at the ester. A less conventional approach at the time would involve disconnection at the alkene. One synthesis of epothilone B was conducted using the ring-closing metathesis of a diene in which the two alkenes were linked by the formation of an acyclic ester (top of Scheme 21.6).[83] This strategy generates a mixture of E- and Z-olefins.

Alternatively, the Z-olefin of epothioline C has been constructed by Fürstner and his co-workers by a sequence of alkyne methathesis and Lindar reduction of the resulting macrocyclic alkyne (the bottom of Scheme 21.6).[21,84] This approach cannot be used to generate trisubstituted olefins, such as that in the precursor to epothilone B, but it can be used to control the stereochemistry of disubstituted olefins. This group has shown that a molybdenum–trisamido complex of the type studied by Cummins for dinitrogen cleavage (Chapter 13)[85–87] generates a catalyst that is highly active for alkyne metathesis.[21,88]

In all of these examples, a disubstituted or trisubstituted alkene is formed. A major challenge for late transition metal olefin metathesis catalysts has been the construction of tetrasubstituted alkenes. This class of ring-closing metathesis has been most successfully conducted with molybdenum catalysts.[89] However, by reducing the steric hindrance of the N-heterocyclic carbene ligand in the Hoveyda–Grubbs catalyst, a ruthenium catalyst capable of forming tetrasubstituted olefins has been developed. As shown in Equation 21.9, these catalysts contain either meta substituents, one ortho-methyl substituent, or no substituents on the N-aryl group.[90,91] The complex containing ortho-methyl groups on the N-aryl ring catalyzes the formation of both five- and six-membered rings.

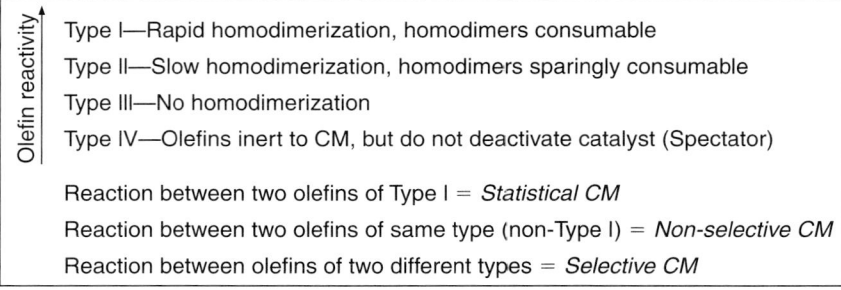

X = C(CO$_2$R)$_2$, n = 1 or 2
X = NTs, n = 1
X = O, n = 1

5- and 6- membered rings

(21.9)

21.2.5.2. Olefin Cross Metathesis

Cross metathesis is an emerging process that could have enormous synthetic value.[92] The utility of the cross metathesis reaction depends on the selectivity for formation of one olefin over the other possible olefin products. This selectivity can be obtained when one class of olefin is appropriately paired with a second class of olefin.

Guidelines about which pairs of olefins will give high yields of one cross-metathesis product have been published and are shown in Figure 21.4.[17] The reactivity of these classes of olefins is summarized in Table 21.1. These empirical guidelines state that two olefins that undergo fast homodimerization will generate a nearly statistical mixture of cross-metathesis products. They also imply that reactions between two olefins that undergo slow, nearly irreversible formation of homodimers (type II) or between two olefins that do not undergo homodimerization (type III) will lead either to low conversions or to product distributions that are non-statistical, but not sufficiently enriched in one product to be synthetically useful. In contrast, reactions between two olefins of two different classes lead to selective cross-metathesis reactions to form one major product. In general, unhindered, electron-rich olefins belong to type I, and more hindered, less-electron-rich olefins belong to types II–IV. The type of each olefin depends on the identity of the catalyst; to obtain details of these classes, consult the paper by Grubbs and co-workers[17] and subsequent literature. However, to illustrate these categories, examples of olefins of each type for reactions catalyzed by the second-generation Grubbs catalyst are shown in Table 21.1.

In early work,[59,60] Crowe showed that an α-olefin would undergo selective cross metathesis with acrylonitrile, styrene, and vinylsilanes in the presence of Schrock's molybdenum

Olefin reactivity

Type I—Rapid homodimerization, homodimers consumable

Type II—Slow homodimerization, homodimers sparingly consumable

Type III—No homodimerization

Type IV—Olefins inert to CM, but do not deactivate catalyst (Spectator)

Reaction between two olefins of Type I = *Statistical CM*

Reaction between two olefins of same type (non-Type I) = *Non-selective CM*

Reaction between olefins of two different types = *Selective CM*

Figure 21.4. The four types of olefin reactivities toward cross metathesis.

Table 21.1. Categorization of olefins by type for reactions with the second-generation Grubbs catalyst.

Type I (fast homodimerization)	Type II (slow homodimerization)	Type III (no homodimerization)	Type IV (spectators to CM)
terminal olefins, 1° allylic alcohols, esters, allyl boronate esters, allyl halides, styrenes (without large ortho substituents), allyl phosphonates, allyl silanes, allyl phosphine oxides, allyl sulfides, protected allylic amines	styrenes (with large ortho substituents) acrylates, acrylamides, acrylic acid, acrolein, vinyl ketones, unprotected 3° allylic alcohols, vinyl epoxides, 2° allylic alcohols, perfluoroalkyl olefins	1,1-disubstituted olefins non-bulky trisubstituted olefins, vinyl phosphonates, phenyl vinyl sulfone, alkenes with 4° allylic carbons (all alkyl substituents), 3° allylic alcohols (protected)	vinyl nitro olefins, trisubstituted allylic alcohols (protected)

catalyst (Scheme 21.7). Reaction with acrylonitrile formed α,β-unsaturated nitriles, which can then be hydrogenated to generate terminal amines. Since that time, cross-metathesis reactions with acrylates and allylic alcohols have been developed,[93] and reactions with these alkenes can lead to the formation of alkyl chains containing terminal functionality by the sequence of cross metathesis and hydrogenation of the resulting double bond (Scheme 21.7). This sequence has become a common synthetic strategy starting with acrylates and constitutes a formal hydroesterification of an olefin; related sequences with other olefins bearing functional groups can provide related formal hydrofunctionalizations of terminal olefins. Similar cross metathesis with a vinylborane generates higher vinylboranes for subsequent use in cross coupling or other processes.[94]

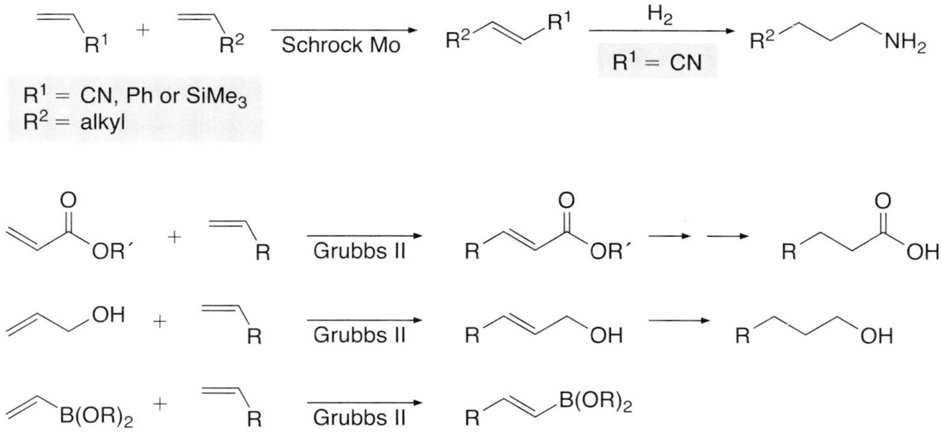

$R^1 = CN, Ph$ or $SiMe_3$
$R^2 = alkyl$

Scheme 21.7

Scheme 21.8 shows the complex network of equilibria that are possible when two different olefins are combined in a metathesis system. The relative rates and magnitudes of the individual steps are not well enough established to fully rationalize the trends in selectivities of cross-metathesis reactions. However, a few features of the [2+2] cycloadditions should be kept in mind when considering the equilibria involved in a cross metathesis. First, some of the [2+2] processes lead to metallacycles containing substituents located in positions 1,3 to each other, and these metallacycles do not lead to productive metatheses processes. Second, the [2+2] reactions are slow in some cases for steric or electronic reasons. For example, the formation of metallacycles containing two large substituents vicinal to each other are disfavored. Third, cross metathesis between two relatively unhindered olefins is the most favored sterically, but cleavage of the olefin product resulting from this process can also occur. Thus, metathesis reactions between two unhindered alkenes are reversible. An excess of the more-hindered olefin is often used in a cross metathesis between a hindered and unhindered olefin to favor formation of the carbene containing the hindered substituent and thereby to favor formation of the product from cross-metathesis.

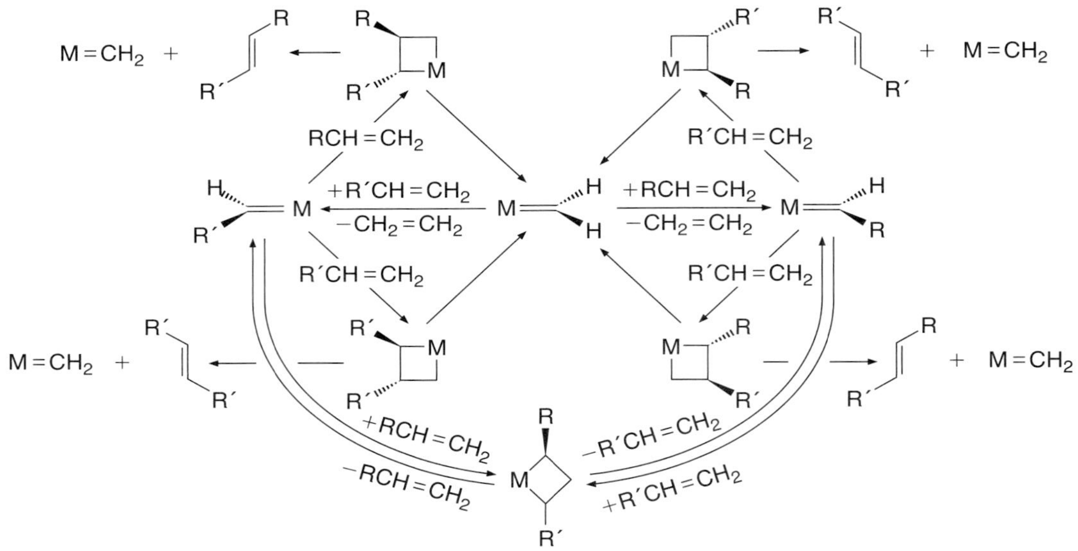

Scheme 21.8

21.2.6. Enantioselective Ring-Closing and Ring-Opening Metathesis

Ring-closing metathesis has also been developed by Hoveyda & Schrock and by Grubbs into an enantioselective process. Ring-closing metathesis does not form a new bond at a tetrahedral carbon and may, therefore, be considered an unusual candidate for the development of enantioselective chemistry. However, the metathesis reactions can be conducted as kinetic resolutions or desymmetrizations to generate optically active products containing a new stereocenter. This reaction was first conducted as a kinetic resolution of a racemic reactant by a ring-closing process, but subsequent studies have focused on desymmetrization of achiral reactants by ring closing reactions or sequences of ring opening and ring closing.

One particularly selective kinetic resolution is shown in Equation 21.10.[95] As discussed in Chapter 14, the selectivity of a kinetic resolution is best described by the relative rates (often called the selectivity factor, s) for reaction of the two enantiomers because the enantiomeric excess of the product and reactant changes over time.[96] The diene in Equation 21.10 containing a protected allylic alcohol undergoes ring-closing metathesis to generate optically active, cyclic products in the presence of a chiral, non-racemic molybdenum catalyst based on Schrock's original achiral catalysts. The selectivity factor is greater than 25 in this case, and an s-value of this magnitude allows for the product and remaining reactant to be isolated with high ee with conversions close to 50%.

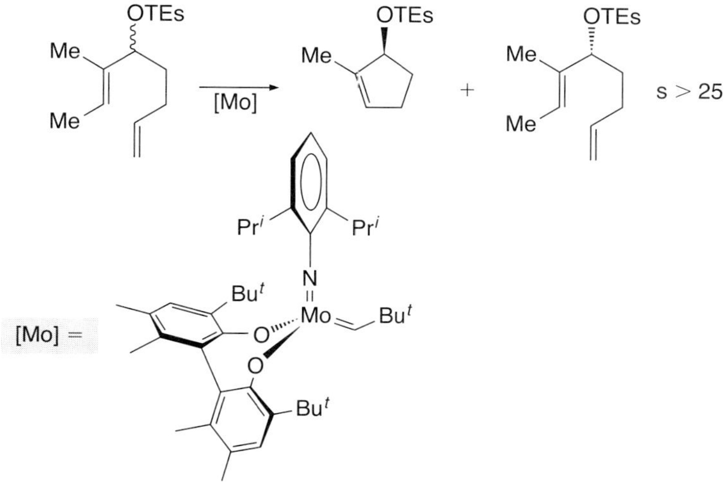

$$(21.10)$$

Alternatively, the reaction can be conducted as a desymmetrization. For example, the achiral triene in Equation 21.11 reacts to generate the product from ring closing metathesis in high yield and with high ee.[97] Such desymmetrizations can be conducted as a tandem sequence of ring opening and ring closing. The bisallyl ether in Equation 21.12 generates a new triene with a less strained five-membered ring with good enantiomeric excess.[98] Likewise, the achiral norbornyl allyl ether in Equation 21.13 undergoes ring opening, followed by ring closing, to generate the less strained, fused, bicyclic product in high enantiomeric excess.[99] In this case, the diallyl ether is added in substoichiometric amounts to help initiate the catalyst by generating the parent methylene compound from the starting neopentylidene.

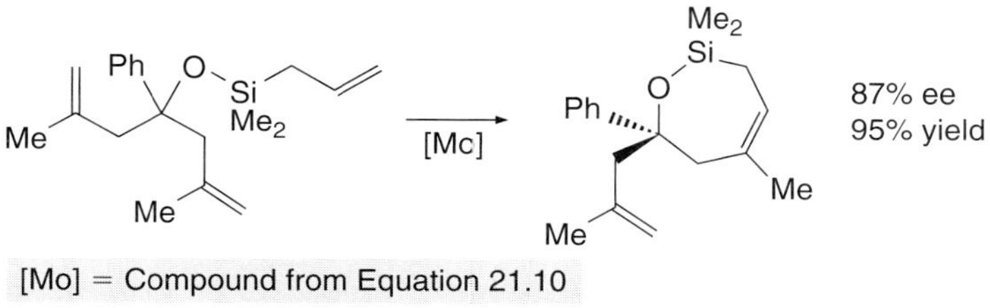

87% ee
95% yield

[Mo] = Compound from Equation 21.10

$$(21.11)$$

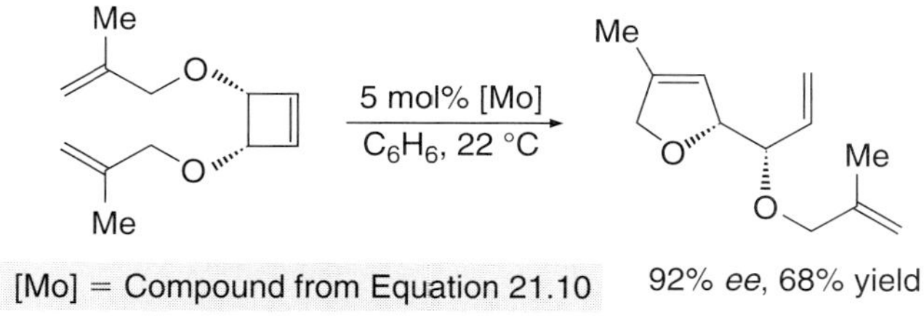

$$\frac{5 \text{ mol\% [Mo]}}{C_6H_6, 22 \degree C}$$

[Mo] = Compound from Equation 21.10   92% ee, 68% yield

$$(21.12)$$

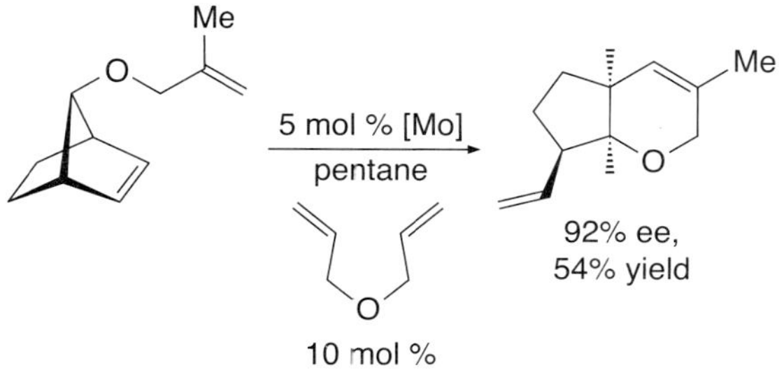

$$\frac{5 \text{ mol \% [Mo]}}{\text{pentane}}$$

10 mol %

92% ee,
54% yield

$$(21.13)$$

One valuable application of desymmetrization processes is the formation of quaternary stereocenters. In this case, a fully substituted carbon that lies in the plane of symmetry of the reactant becomes a stereocenter by reaction at one of the two substituents. This type of desymmetrization in the context of olefin metathesis is shown in Equation 21.14.[100] In this example, the achiral, symmetric triene is converted to the chiral, non-racemic diene with 87% ee.

$$(21.14)$$

Finally this type of reaction can be run as a sequence of ring opening and cross metathesis. For example, the norbornene derivative in Equation 21.15 undergoes reaction with the vinylsilane to generate the optically active cyclopentane in high enantiomeric excess.[101, 102]

$$(21.15)$$

The catalysts for these reactions are derived from both molybdenum and ruthenium, although products with higher enantiomeric excess have been obtained to date with molybdenum catalysts. As noted above, this molybdenum catalyst is based on the structure of Schrock's original bisalkoxide catalyst, but with an optically active bisnaphtholate in place of the two tertiary fluoroalkoxides. These catalysts can be used as isolated complexes or they can be generated in situ from the imido carbene molybdenum bistriflate DME complex and the diolate, as shown in Equation 21.16,[103] or from the bispyrrolyl complex, as shown in Equation 21.17.[104]

$$(21.16)$$

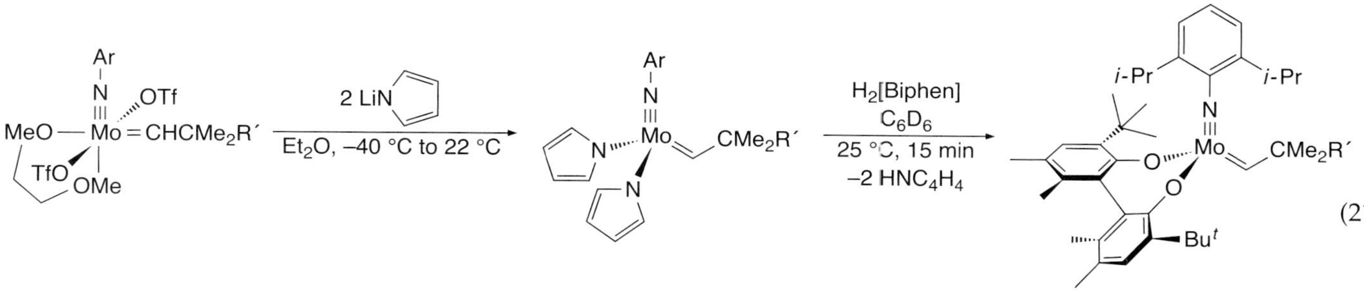

(21.17)

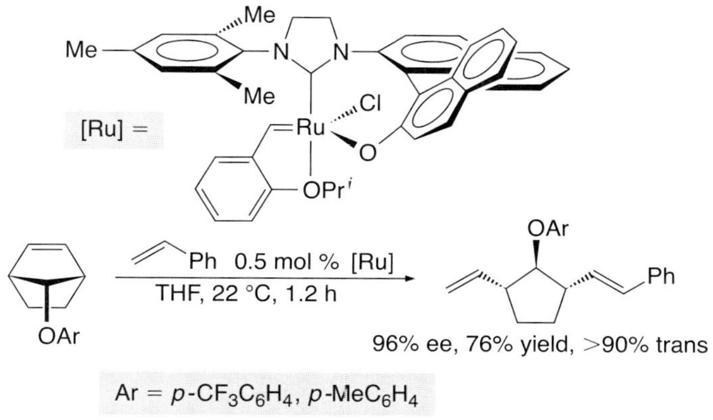

(21.18)

$Ar = p\text{-}CF_3C_6H_4, p\text{-}MeC_6H_4$

Because late metal catalysts can tolerate a broader, or at least complementary, range of functional groups and are easier to handle, efforts have also been made to generate optically active ruthenium catalysts for asymmetric olefin metathesis.[105] The ruthenium complex in Equation 21.18 forms products from a sequence of ring opening and cross metathesis with good enantiomeric excess.[106] Recent studies with this type of chiral, non-racemic ruthenium metathesis catalyst have shown the potential to obtain high enantioselectivities for certain classes of substrates with rates that are sufficiently fast to be able to use reasonably low loadings of catalyst.[107] This catalyst contains an unsymmetrically substituted aryl ring with a 1,2-diphenylethylene backbone, as shown in Figure 21.5.

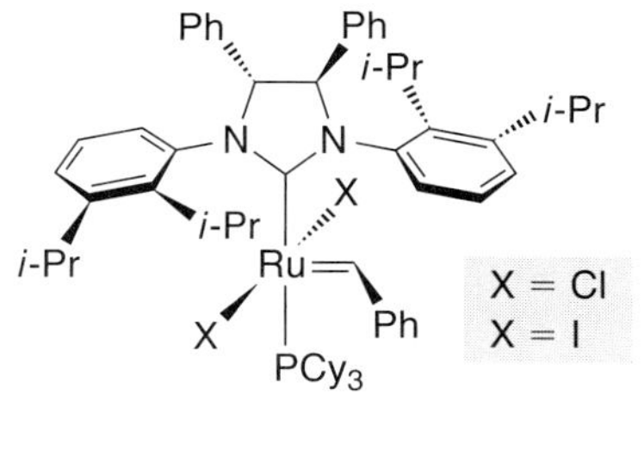

Figure 21.5.
A chiral, non-racemic ruthenium catalyst for asymmetric olefin metathesis.

21.2.7. Ring-Opening Metathesis Polymerization

21.2.7.1. Utility of Ring-Opening Metathesis Polymerization

Ring-opening metathesis is a reaction with enormous value for polymer synthesis. This reaction can be used to generate hydrocarbon polymers with favorable properties that are derived from the structures or functionalities of the monomers. Hydrocarbon polymers containing pendant olefins for subsequent crosslinking have been prepared with many different catalysts, and polymers containing pendant polar functionalities have been prepared with ruthenium catalysts that tolerate oxygen-containing functional groups. Cyclic olefins possessing ring strain are used as monomers for olefin metathesis because the thermodynamics will favor the ring-opened materials. Dicyclopentadiene is a particularly important monomer for ring-opening metathesis polymerization. The ring-opening metathesis polymerization of cyclooctene and norbornadiene have also been studied commonly. These polymerizations are shown in Scheme 21.9.

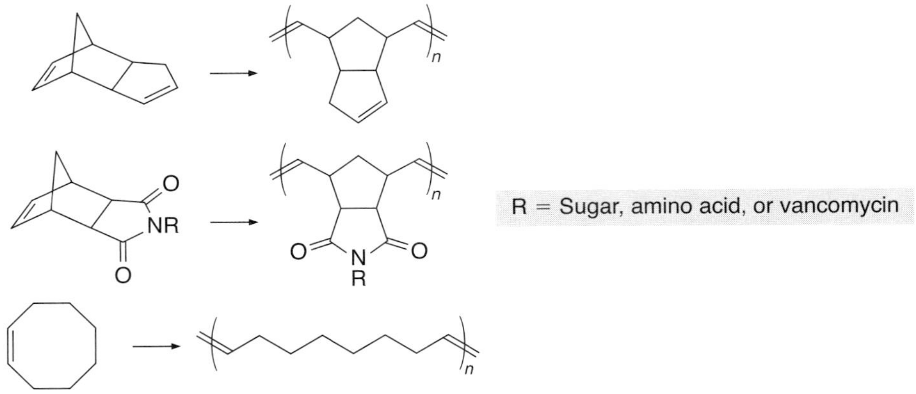

R = Sugar, amino acid, or vancomycin

Scheme 21.9

Many soluble complexes and heterogeneous systems noted in the introduction[37–41] to this chapter catalyze ring-opening metathesis polymerization (ROMP) of strained olefins. However, many practical aspects of the process distinguish one catalyst from another. Desirable properties of the catalyst include a high degree of control over molecular weight and molecular weight distribution, tolerance of the high temperatures used for polymerization in a blow-molding application, ability to polymerize monomers possessing various functional groups that modulate the properties of the resulting polymer, and activity after monomer is consumed to create a "living" polymerization system that makes it possible to generate block copolymers and materials with properties that take advantage of the phase separation of the two blocks. Polymerization processes that meet the requirements for technical applications of ROMP have been achieved through proper design or selection of the catalyst.

The properties of the polymer depend on the average molecular weight and the dispersity of molecular weights (polydispersity index, PDI),[108] and these properties of the polymer are controlled by the chemistry of the catalyst. Materials of low polydispersity (narrow molecular weight distribution) are generated when the rate for catalyst initiation (k_{init} in Equation 21.19) is much faster than the rate for chain propagation (k_{prop} in Equation 21.19). As an analogy of the importance of the rates of initiation and propagation, a series of people who knit at the same rate and begin knitting at the same time will produce a scarf of the same length at any single point in time. However, a series of people that knit at the same rate, but begin knitting at different times will each generate scarves of different lengths at a single point in time. Likewise, a catalyst that reacts with a rate of initiation that is faster than the rate of propagation will create a set of polymer chains that start growing at similar times and possess a narrow distribution of molecular weights, whereas a catalyst that reacts with a rate of initiation that is much slower than the rate of propagation will create a set of chains that begin growth at different times and possess a wide distribution of molecular weights at any point in time.

Because the key to control of molecular weight distribution depends on the relative rates for initiation and propagation, studies to control these relative rates have been conducted. Studies with the Grubbs-type ruthenium carbene complexes have shown that the ruthenium benzylidene complexes undergo faster initiation than vinyl alkylidene

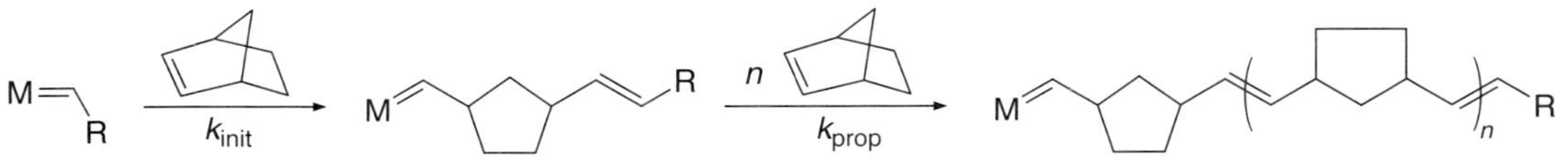

(21.19)

complexes studied previously.[58] Faster initiation than propagation is observed because the benzylidene unit is more reactive than the alkylidene unit contained in the catalyst that is present during polymerization. A further modification of this catalyst to replace PCy_3 by a pyridine derivative or by PPh_3 led to faster dissociation of the dative ligand to generate the reactive 14-electron carbene intermediate.[109,110] This faster dissociation of ligand creates a higher concentration of active catalyst and faster rates of initiation, relative to the rates of propagation. As a result, the polymers produced from these catalysts have narrower molecular weight distributions than those produced by previous catalysts.

21.2.7.2. Mechanism of Ring-Opening Metathesis Polymerization

The overall mechanism for ring-opening metathesis polymerization (Equation 21.20) is the reverse of the mechanism of ring-closing metathesis polymerization. During the ring-opening metathesis, the metal–carbene complex reacts with a cyclic olefin in a [2+2] process to generate a bicyclic or tricyclic, metallacyclobutane intermediate. Cleavage of this metallacyclobutane in the direction opposite to its formation opens the bicyclic system and generates a new carbene complex substituted with the growing polymer chain. Addition of a second cyclic olefin monomer to this carbene leads to a new metallacyclobutane intermediate, and cleavage of this intermediate further lengthens the chain.

(21.20)

A ring-opening metathesis polymerization in which the catalyst remains bound to the terminus of the growing polymer chain throughout chain growth and after the chain growth has ended[111] is a type of living polymerization. (A living polymerization is an addition polymerization that occurs without chain transfer or chain termination). A living ROMP system allows block copolymers to be prepared from two different alkenes, as shown in Equation 21.21.[112] The polymerization is typically conducted with one type of monomer first and, after consumption of this monomer, the length of the polymer chain is extended with a second monomer. This process can be repeated to create sophisticated tailor-made polymers containing alternating groups of monomers. For example, one group of monomer may generate a crystalline phase and the other an amorphous phase to give rise to an elastomeric material. Alternatively one could conduct the polymerization with a simple monomer and then with a monomer containing polar functionality, conductive properties, or stereochemical elements.

Block copolymer

(21.21)

Ring-opening metathesis polymerization has also been used to generate materials that are useful tools to address questions in polymer science. For example, cyclic analogs of linear polyethylene have been prepared by ring-opening metathesis polymerization, as depicted in in Equation 21.22, followed by hydrogenation of the initial unsaturated, cyclic material.[113] Polyacetylene has been prepared by ring-opening metathesis polymerization of cyclooctatetraene[114,115] (Equation 21.23) and a material containing purely Z-stilbene units has been prepared by ring-opening metathesis polymerization of the cyclophene in

Equation 21.24.[116,117] Furthermore, the high functional group compatibility of the recent ring-opening metathesis polymerization catalysts have created the opportunity to use olefin metathesis to study biological systems. As one of many examples of such applications, this reaction has been used to prepare polymers possessing pendant saccharides for studies on the effect of polyvalency on biochemical molecular recognition.[118–120]

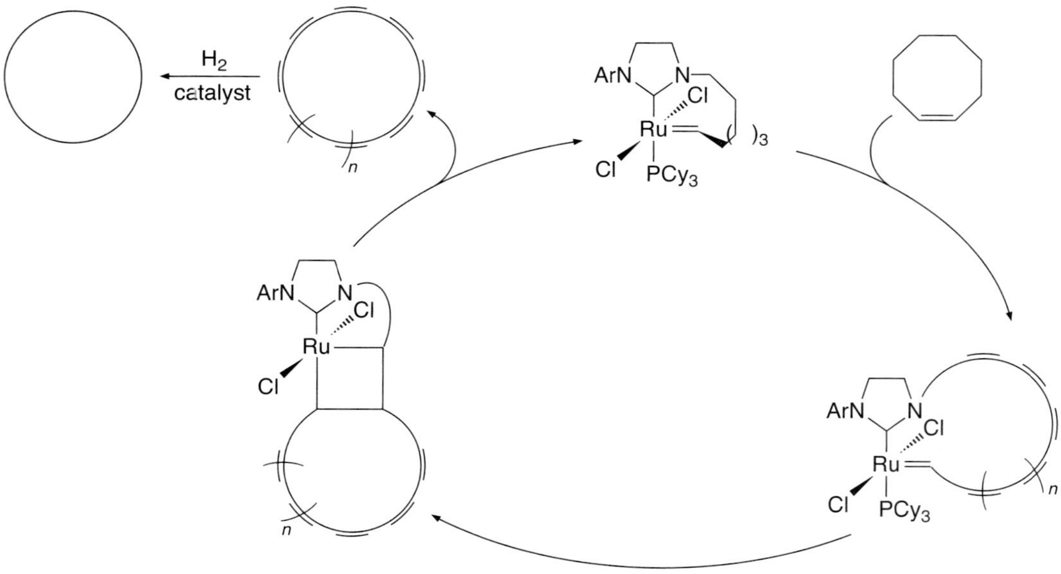

(21.22)

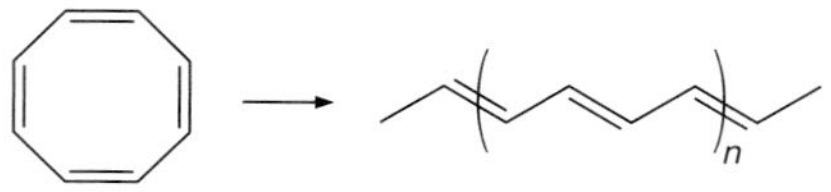

(21.23)

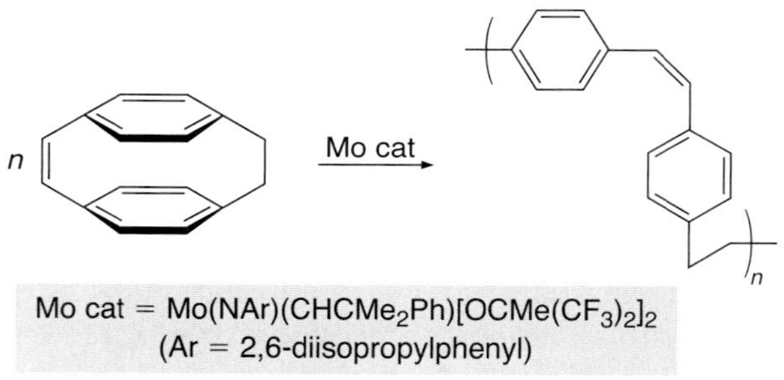

(21.24)

Mo cat = Mo(NAr)(CHCMe$_2$Ph)[OCMe(CF$_3$)$_2$]$_2$
(Ar = 2,6-diisopropylphenyl)

21.3. Alkyne Metathesis

21.3.1. Examples of Alkyne Metathesis

Alkyne metathesis (Equation 21.2 and Scheme 21.1)[121,122] is a process that is similar to alkene metathesis but cleaves the carbon–carbon triple bond in two alkynes and reforms the carbon–carbon triple bond in two new alkynes. Alkyne metathesis is less developed than alkene metathesis, and highly active catalysts for this process have been discovered only recently. Alkyne metathesis has been used in two synthetic contexts: the synthesis of conjugated polymers possessing interesting electronic properties and the synthesis of macrocycles containing alkene units possessing either E- or Z-geometry after reduction of the resulting alkyne.

The first homogeneous alkyne metathesis has been attributed to Mortreux and Blanchard.[123,124] They reported the high temperature reaction of 4-methyldiphenylacetylene in the presence of a catalyst generated from a mixture of Mo(CO)$_6$ and resorcinol (1,3-dihyroxybenzene) to form a mixture of the starting alkyne, 4,4'-dimethyldiphenylacetylene, and diphenylacetylene (Equation 21.25). Based on this observation, Villemin studied combinations of $Mo(CO)_6$ and other phenols and found that the combination of $Mo(CO)_6$ and 4-chlorophenol was most active for the metathesis of alkynes.[125] One proposal for the mechanism of these reactions involves the intermediacy of metal alkylidyne complexes (vide infra). After these studies with the combination of $Mo(CO)_6$ and phenols, Schrock studied the reactions of well-defined carbyne complexes with alkynes. These studies showed that metal alkylidynes catalyze alkyne metathesis (Equation 21.26).[126–131]

More recently, an alkyne metathesis catalyst generated from a molybdenum trisamido complex has been shown to be particularly reactive. Fürstner first showed that the catalyst generated from the reaction of the Cummins' trisamido complex[86] $Mo(NArBu^t)_3$ (Ar = 3,5-$Me_2C_6H_3$) with methylene chloride generates the alkylidyne complex $Mo(NArBu^t)_3(\equiv CH)$ (Equation 21.27).[88,132] This complex catalyzes the metathesis of alkynes, but this activity may result from some hydrolysis or alcoholysis of the trisamide. Regardless of the identity of the active catalyst in this system, the functional group tolerance of this catalyst and the ability to promote various types of alkyne metathesis has been improved.

Moore showed that a much more active alkyne metathesis catalyst than the trisamido species is generated from addition of electron-poor phenols to the trisamide complex containing a propylidyne ligand (Equation 21.27 and Table 21.2). Studies of various aryl alcohol and fluoro-alcohol additives showed that the most reactive catalyst was generated from p-nitrophenol.[133]

Table 21.2. Effect of phenol structure on the catalyst activity for alkyne metathesis.

Ligand									
$t_{1/2}$ (min)[a]	167	160	613[b]	20	12	< 8	86	12	38	< 8

[a]$t_{1/2}$ is the time required for the metathesis reaction to reach 50% of equilibrium conversion.
[b]Metathesis catalyzed by 3,5-bis(trifluoromethyl)phenol catalyst does not reach equilibrium within 2175 min.

21.3.2. Mechanism of Alkyne Metathesis

The mechanism of alkyne metathesis catalyzed by high-valent tungsten carbyne complexes has been studied by Schrock.[134,135] These reactions occur by a sequence of [2+2] and retro-[2+2] reactions of the carbyne complex with an alkyne to generate a metallacyclobutadiene complex. This reaction was directly observed with the reagents in Equation 21.28. Cycloreversion of the metallacycle then generates a new alkyne or regenerates the two starting materials. Thus, this mechanism parallels the accepted mechanism for alkene metathesis.

$$\tag{21.28}$$

The mechanism of the reaction of the alkyne metathesis catalyzed by the material generated from molybdenum carbonyl and a phenol derivative is less well defined. One possibility is that the complexes in this system are oxidized to a high-valent molybdenum species, which then catalyzes alkyne metathesis by the mechanism outlined by Schrock for isolated alkyne complexes. Alternatively, these complexes might react by a mechanism involving metallacyclopentadiene complexes (Equation 21.29).[136] By this mechanism, two alkynes and the low-valent metal undergo reductive coupling of the alkyne to generate the pentadienylmetallacyclopentadiene. Reductive elimination from this complex would generate a cyclobutadiene complex, which could then undergo oxidative addition to generate an isomeric metallacyclopentadiene. By this mechanism, the resulting metallacycle would then cleave to generate the two new alkynes. Little or no mechanistic data is available on these systems. The mechanism involving metallacyclopentadiene and butadiene complexes has been proposed because metallacyclopentadiene and butadiene complexes containing metal-carbonyl fragments are stable, and it is not clear how complexes in high oxidation states would be generated.

$$\tag{21.29}$$

21.3.3. Applications of Alkyne Metathesis

Alkyne metathesis has been used for the synthesis of polymers, electronically interesting macrocycles, and natural products. Early examples of the use of alkyne metathesis to prepare polymeric materials included ring-opening metathesis polymerizations. Two examples of these reactions are shown in Equations 21.30[137,138] and 21.31.[139] These two examples were conducted with Schrock's tungsten tris-*tert*-butoxy alkylidyne catalyst. Like ring-opening metathesis polymerizations of alkenes, these reactions are driven by the release of ring strain. Ring-opening alkyne metathesis polymerizations conducted with the most recently developed catalysts have not been published extensively.

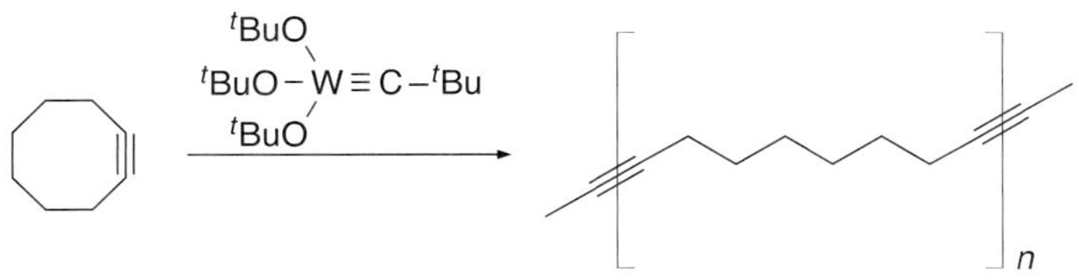

(21.30)

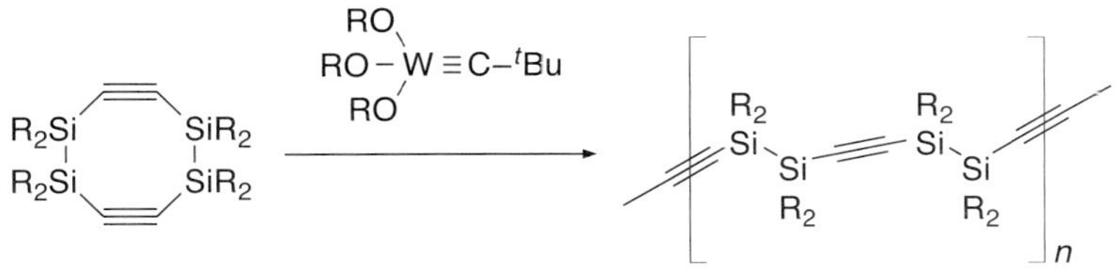

(21.31)

However, the new phenoxide-ligated molybdenum carbyne complexes have been used to form unsaturated macrocycles. In particular, the molybdenum catalyst containing phenoxide ligands catalyzes the reaction in Equation 21.32.[140] This reaction occurs in high yield because the macrocyclic product precipitates from solution.

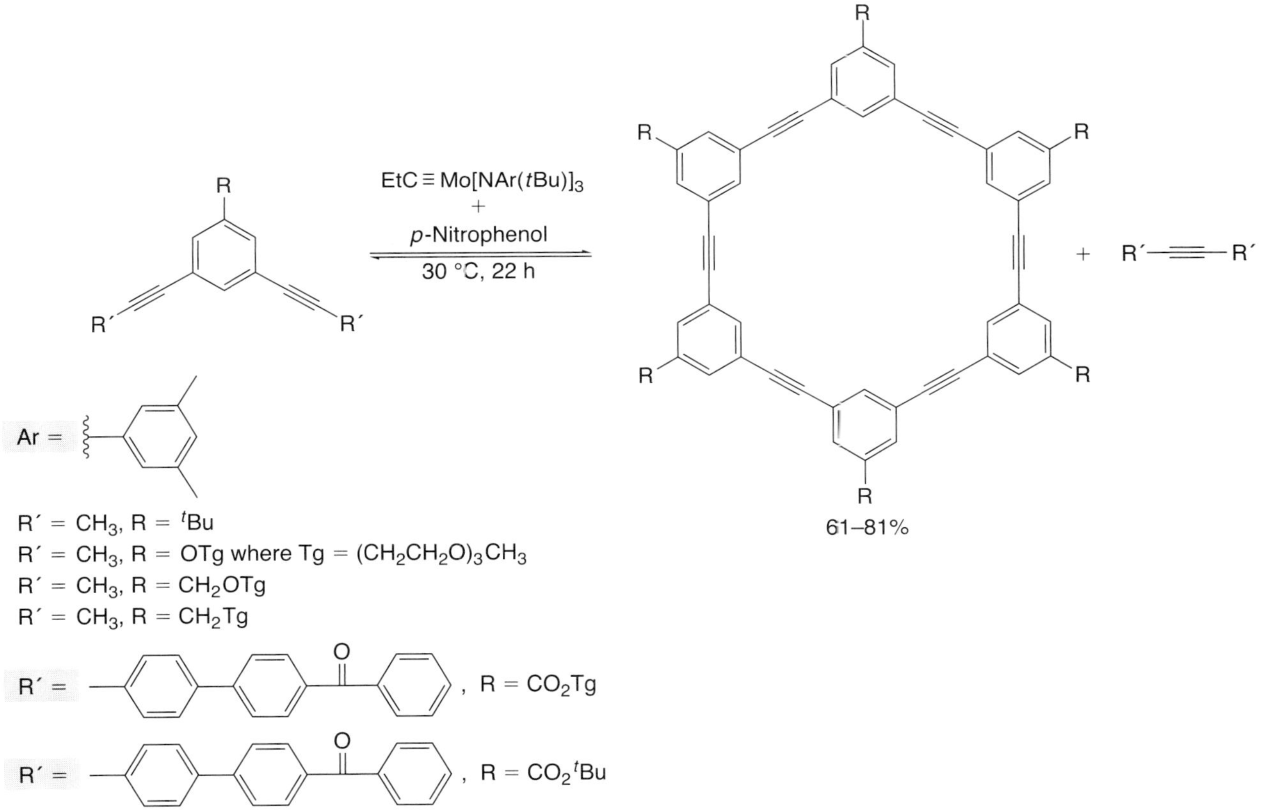

$$EtC \equiv Mo[NAr(tBu)]_3$$
$$+$$
$$p\text{-Nitrophenol}$$
$$\overline{\qquad 30\ °C,\ 22\ h \qquad}$$

$$+ \quad R' \!\!=\!\!\!=\!\! R'$$

61–81%

(21.32)

Ar =

R′ = CH₃, R = tBu
R′ = CH₃, R = OTg where Tg = (CH₂CH₂O)₃CH₃
R′ = CH₃, R = CH₂OTg
R′ = CH₃, R = CH₂Tg

R′ = [image], R = CO₂Tg

R′ = [image], R = CO₂tBu

More extensive use of alkyne metathesis in materials synthesis has exploited addition metathesis processes. Several examples of addition metathesis reactions are shown in Equations 21.33–21.36.[141] In Equations 21.33–21.35, dimethylacetylenes undergo alkyne metathesis in the presence of Schrock's alkylidyne catalyst to generate butyne and the polymeric acetylene. The reaction is driven by evaporation of the volatile butyne from the system. The Mo(VI) complex generated from the trisamido alkylidyne in Equation 21.35 and p-nitrophenol catalyzes such polymerizations at only 30 °C under vacuum.[142] Reactions catalyzed by the combination of molybdenum carbonyl and an aryl alcohol occur selectively at the alkyne units over the alkenes. Therefore, polymers containing alternating alkynyl and vinyl linkages can be prepared, as shown in Equation 21.36.[143–146]

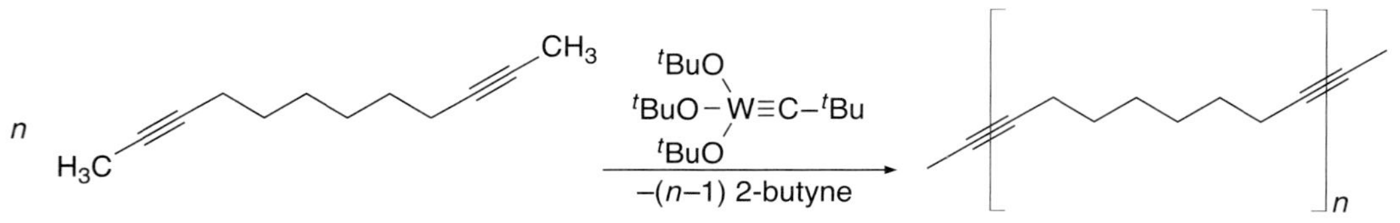

$$(21.33)$$

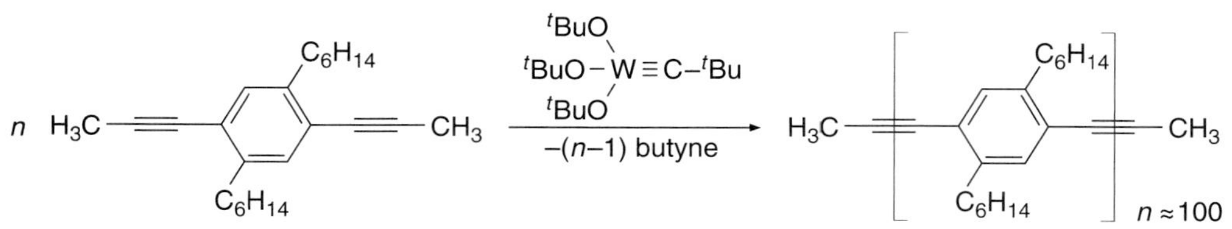

$$(21.34)$$

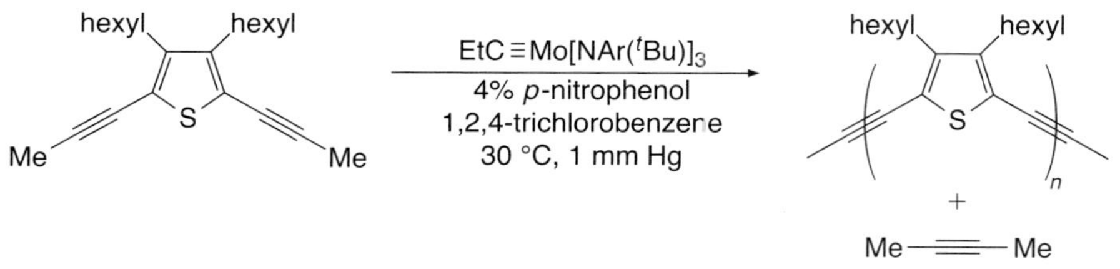

$$(21.35)$$

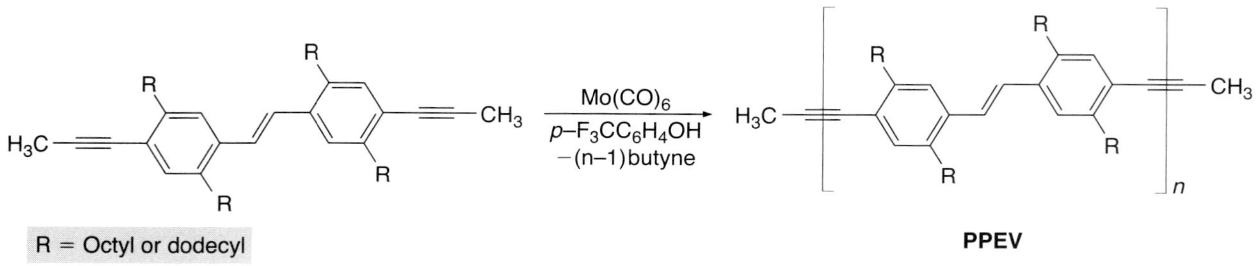

R = Octyl or dodecyl

PPEV

$$(21.36)$$

21.3.4. Alkyne Cross Metathesis

The use of alkyne cross metathesis for the synthesis of unsymmetrical alkynes has potential utility in synthesis, although it has been studied less intensively than alkene cross metathesis. Mori published some of the first examples of alkyne cross metathesis to generate unsymmetrical alkynes.[147] One example of this reaction conducted with the Montreux-type catalyst is shown in Equation 21.37. The selectivity for the cross-metathesis product was achieved by the use of an excess of the diphenylacetylene. A second example was conducted with the catalyst generated from Cummins' trisamido complex.[148] As shown in Equation 21.38, this cross metathesis was conducted in acceptable yields with an arylester that is sensitive to the Schrock-type tungsten alkylidyne catalysts. The application of such cross metathesis to the synthesis of prostaglandins is shown in Scheme 21.10.[149–151]

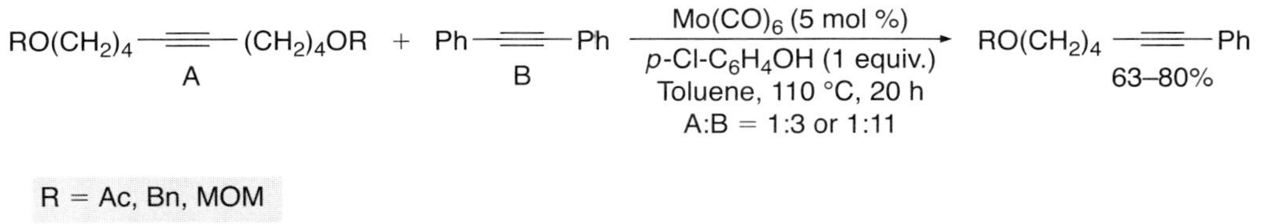
RO(CH2)4—≡—(CH2)4OR + Ph—≡—Ph
A
B
Mo(CO)6 (5 mol %)
p-Cl-C6H4OH (1 equiv.)
Toluene, 110 °C, 20 h
A:B = 1:3 or 1:11
RO(CH2)4—≡—Ph
63–80%
R = Ac, Bn, MOM

(21.37)

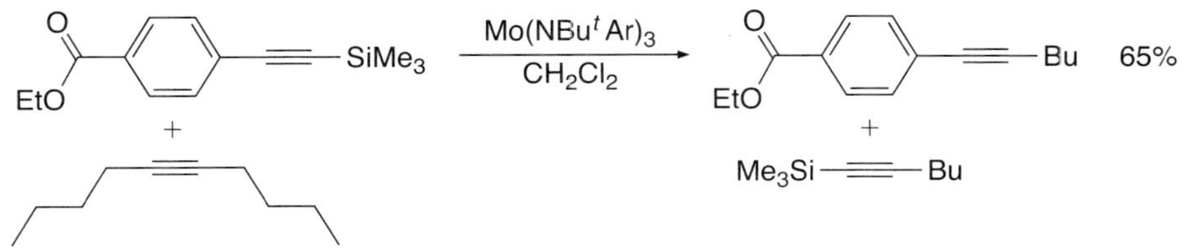
O
EtO
SiMe3
+
Mo(NBut Ar)3
CH2Cl2
Bu
65%
Me3Si—≡—Bu

(21.38)

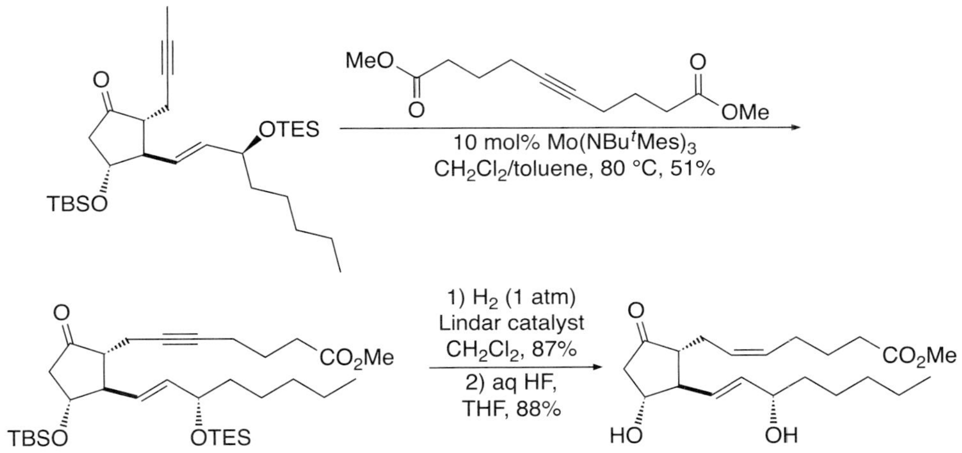
O
TBSO
OTES
MeO
O
OMe
10 mol% Mo(NBu^tMes)3
CH2Cl2/toluene, 80 °C, 51%
CO2Me
1) H2 (1 atm)
Lindar catalyst
CH2Cl2, 87%
2) aq HF,
THF, 88%
HO
OH

Scheme 21.10

21.3.5. Ring-Closing Alkyne Metathesis

Alkyne metathesis has also been used to generate macrocyclic natural products containing cis olefins. As noted in the section on olefin metathesis, control of the *E*/*Z*-geometry by olefin metathesis is challenging. Fürstner has developed an alternative approach to generating *Z*-olefins by a combination of alkyne metathesis and Lindlar reduction. This sequence has been used extensively in natural products synthesis, and many examples are summarized in a review article.[122] As noted in the introduction to catalytic metathesis, one particularly illustrative example of this challenge occurred during the synthesis of epothilone A and C. The *Z*-olefin in the macrocycle was generated in one synthesis by a ring-closing alkyne metathesis, as shown in Scheme 21.11. After this ring closure, the alkyne was reduced to the cis olefin by hydrogenation over platinum. Deprotection generated epothilone C and epoxidation generated epothilone A.

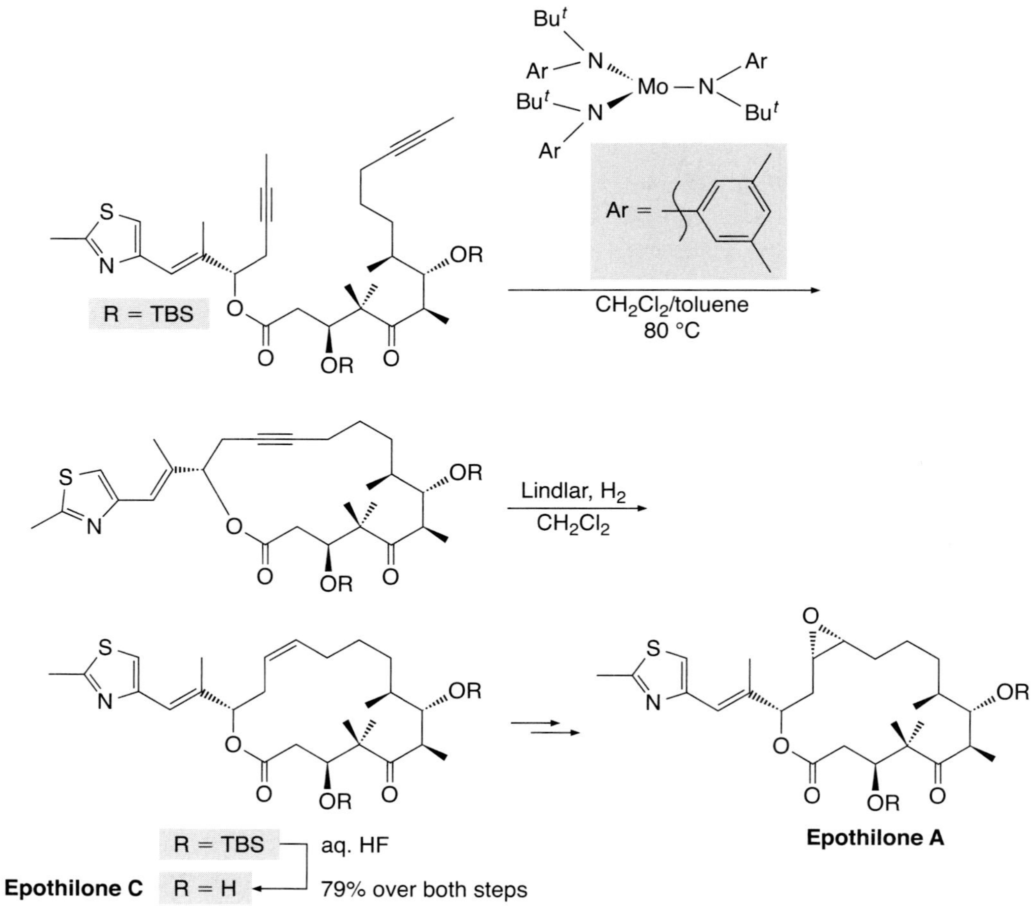

Scheme 21.11

21.4. Enyne Metathesis

21.4.1. Examples of Enyne Metathesis

Enyne metathesis, shown in Equation 21.39, generates dienes from alkenes and alkynes.[18,152,153,154] In this process, the alkene is formally split in half, and the two halves of the alkene are added across the alkyne. To date, this reaction has been studied most extensively with ruthenium carbene complexes, such as the Grubbs catalyst. Thus, the tolerance for heteroatom functionality is high. Four specific examples of this reaction are shown in Equations 21.40–21.43.[155–159] In many cases, ethylene has been used as the alkene reagent. However, enyne metathesis also occurs with terminal alkenes[157,158] and allylsilane.[156,159] Use of the combination of the second-generation Grubbs catalyst, along with added ethylene, improves the stereoselectivity. As shown in Equation 21.43,[160] a single isomer is formed from the enyne metathesis with octene in the presence of ethylene. This high selectivity in the presence of ethylene was rationalized by initial enyne metathesis with ethylene to form a mono-substituted diene and subsequent cross metathesis between the diene and 1-octene with selectivity for the *E*-olefin. The reaction in Equation 21.42 with allylsilane gives geometric isomers and appears to result from direct reaction of the allylsilane with the alkyne without intervention of the added ethylene.[160]

$$R\!-\!\!\!\equiv\!\!\!-H \;+\; \underset{H}{\overset{H}{\diagup}}\!\!=\!\!\underset{R'}{\overset{H}{\diagdown}} \;\xrightarrow{\text{Catalytic [Ru]}}\; R\!\!-\!\!\text{(diene)}\!\!-\!\!R' \tag{21.39}$$

$$(21.40) \quad 92\%$$

$$(21.41) \quad 91\%$$

$$79\% \quad E\!:\!Z = 1.3\!:\!1 \tag{21.42}$$

$$82\% \tag{21.43}$$

Although intermolecular enyne metathesis is the simplest to envision, intramolecular enyne metathesis was the major focus of the initial work. Two representative intramolecular enyne metatheses are shown in Equations 21.44 and 21.45. The reaction in Equation 21.44 shows the value of this chemistry to form heterocycles. The reaction in Equation 21.45 shows how enyne metathesis can be used in combination with olefin metathesis to form bicyclic products. The initial enyne metathesis process in Equation 21.45 terminates in a ruthenium carbene complex. The carbene complex is then trapped by the remaining olefin in a [2+2] and retro-[2+2] cycloaddition sequence to generate the bicyclic organic product and a ruthenium carbene complex that re-enters the catalytic cycle by reaction with the yne diene.

$$(21.44)$$

$$(21.45)$$

21.4.2. Mechanism of Enyne Metathesis

Enyne metathesis is proposed to occur by the sequence of [2+2] processes shown in Scheme 21.12.[161] The cycle is initiated by the formation of an alkylidene complex. This alkylidene complex then reacts with an alkyne to generate a metallacyclobutene complex.

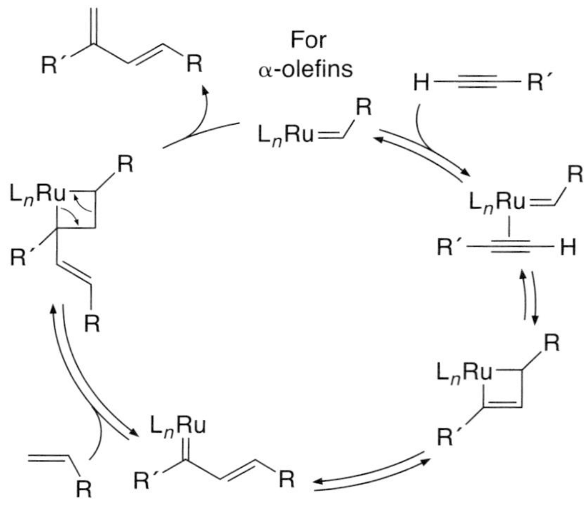

Scheme 21.12

The regiochemistry of this [2+2] process places the substituent from the alkylidene and the substituent from the alkyne 1,3 to each other. Ring opening generates the vinyl-substituted carbene. A [2+2] cycloaddition with the alkene generates a metallacycle in which the vinyl group and the substituent on the olefin are located 1,3 to each other. A retro-[2+2] reaction then generates the starting ruthenium alkylidene. The 1,3-disposition of the substituents on the metallacycles is required to account for the regiochemistry of the products. This regiochemistry presumably results from steric interactions that are weaker than those that would be present if the groups were located adjacent to each other.

21.5. Summary

Olefin metathesis is a reaction that is over fifty years old and has been developed over this time period from a process run at high temperatures with ill-defined catalysts by unknown mechanisms to a process that can be conducted under mild conditions with designed catalysts by mechanisms that occur by established steps. Olefin metathesis, and the related alkyne metathesis, fully cleaves carbon–carbon double and triple bonds and reforms these bonds to generate new alkenes and alkynes. The reaction is often under equilibrium control, but certain classes of reactions can be conducted in a selective fashion that is controlled by relative rates or thermodynamic preferences. This reaction can open strained rings to form polymers or small dienes. It can close small rings and macrocycles by a reaction that is driven by the expulsion of ethylene that makes the reaction favored entropically or by running in an open system under non-equilibrium conditions. It can also be run as a "cross metathesis" to form unsymmetrical alkenes when the steric or electronic properties of the two alkenes properly match.

The catalysts for this process include heterogeneous systems generated from main group alkyl complexes and high-valent tungsten compounds, catalysts generated from group 4 carbonyl compounds and added phenols, discrete molybdenum, as well as tungsten, alkoxides containing one imido and one alkylidene ligand, and ruthenium complexes containing two halides, a mixture of sterically hindered, electron-donating phosphines and N-heterocyclic carbenes, and one benzylidene ligand. Catalysts for alkyne metathesis include species generated from molybdenum carbonyl and electron-poor phenols, or molybdenum alkoxides containing an alkylidyne ligand.

The mechanism of alkene metathesis occurs by a series of [2+2] cycloadditions to form metallacyclobutanes from the metal-carbene and the alkene, followed by retro [2+2] additions. The mechanism of alkyne metathesis occurs by a series of [2+2] cycloadditions to form metallacyclobutenes from the metal-carbyne complex and the alkyne, followed by retro [2+2] additions. In cases of the reactions of well defined intermediates, these cycloadditions occur by initial coordination of the alkene, followed by formation of the metallacycle. High electrophilicity of the molybdenum complexes and modulation of the electronics by the ancillary imido ligand promote the [2+2] reaction, and strong electron donation by the N-heterocyclic carbene ligand is thought to promote [2+2] cycloaddition by the ruthenium catalysts.

Current goals for this process include further development of catalysts that form alkenes with high kinetic selectivity for E or Z isomers, catalysts that form hindered alkenes under mild conditions, and systems that react with higher selectivity for cross-metathesis products. Catalysts that are stable enough to react with complex substrates with high turnover numbers are also needed. Despite these current limitations, this reaction has been used to redistribute alkenes to form intermediates to alcohols produced on massive scale, to convert soybean oils into useful chemicals, to prepare many natural products, and to form medicinally important compounds. These developments, applications, and mechanistic insights were recognized by awarding the Nobel Prize in 2005 to Schrock, Grubbs, and Chauvin.

References and Notes

1. Schrock, R. R.; Hoveyda, A. H. *Angew. Chem. Int. Ed.* **2003**, *42*, 4592.
2. Hoveyda, A. H.; Schrock, R. R. *Chem.—Eur. J.* **2001**, *7*, 945.
3. Fürstner, A. *Angew. Chem. Int. Ed.* **2000**, *39*, 3013.
4. Grubbs, R. H.; Chang, S. *Tetrahedron* **1998**, *54*, 4413.
5. Schuster, M.; Blechert, S. *Angew. Chem., Int. Ed. Engl.* **1997**, *36*, 2037.
6. Grubbs, R. H.; Miller, S. J.; Fu, G. C. *Acc. Chem. Res.* **1995**, *28*, 446.
7. Ivin, K. J.; Mol, J. C. *Olefin Metathesis and Metathesis Polymerization*; Academic Press: London, 1997.
8. Dörwald, F. Z., Ed. *Metal Carbenes in Organic Synthesis*; Wiley-VCH: Weinheim, 1999.
9. Grubbs, R. H., Ed. *Handbook of Metathesis*; Wiley-VCH: Weinheim, 2003.
10. Peters, E. F.; Evering, B. L. (Standard Oil Company) **1960**, U.S. Patent 2 963 447.
11. Elleuterio, H. S. German Patent 1 072 811, 1960.
12. Banks, R. L.; Bailey, G. C. *Ind. Eng. Chem. Prod. Res. Dev.* **1964**, *3*, 170.
13. Novak, B. M.; Risse, W.; Grubbs, R. H. *Adv. Polym. Sci.* **1992**, *102*, 47.
14. Schwendeman, J. E.; Church, A. C.; Wagener, K. B. *Adv. Synth. Catal.* **2002**, *344*, 597.
15. Baughman, T. W.; Wagener, K. B. In *Advances in Polymer Science*; Buchmeiser, M., Ed.; Springer-Verlag GmbH: Berlin, 2005; Vol. 176, p 1.
16. Smith, J. A.; Brzezinska, K. R.; Valenti, D. J.; Wagener, K. B. *Macromolecules* **2000**, *33*, 3781.
17. Chatterjee, A. K.; Choi, T. L.; Sanders, D. P.; Grubbs, R. H. *J. Am. Chem. Soc.* **2003**, *125*, 11360.
18. Diver, S. T.; Giessert, A. J. *Chem. Rev.* **2004**, *104*, 1317. The term enyne metathesis originally referred to ring-closing metatheses of 1,n-enynes and ene-yne metathesis to intermolecular versions of this process. Enyne metathesis is now often used for both types of reactions, and is used in this way in this text.
19. Bunz, U. H. F. *Acc. Chem. Res.* **2001**, *34*, 998.
20. Bunz, U. H. F. *Chem. Rev.* **2000**, *100*, 1605.
21. Fürstner, A.; Mathes, C.; Lehmann, C. W. *Chem. Eur. J.* **2001**, *7*, 5299.
22. Trnka, T. M.; Morgan, J. P.; Sanford, M. S.; Wilhelm, T. E.; Scholl, M.; Choi, T. L.; Ding, S.; Day, M. W.; Grubbs, R. H. *J. Am. Chem. Soc.* **2003**, *125*, 2546.
23. Trnka, T. M.; Grubbs, R. H. *Acc. Chem. Res.* **2001**, *34*, 18.
24. Garber, S. B.; Kingsbury, J. S.; Gray, B. L.; Hoveyda, A. H. *J. Am. Chem. Soc.* **2000**, *122*, 8168.
25. Hansen, S. M.; Volland, M. A. O.; Rominger, F.; Eisentrager, F.; Hofmann, P. *Angew. Chem. Int. Ed.* **1999**, *39*, 1273.
26. Wakamatsu, H.; Blechert, S. *Angew. Chem. Int. Ed.* **2002**, *41*, 2403.
27. Wakamatsu, H.; Blechert, S. *Angew. Chem. Int. Ed.* **2002**, *41*, 794.
28. Grela, K.; Harutyunyan, S.; Michrowska, A. *Angew. Chem. Int. Ed.* **2002**, *41*, 4038.
29. Michrowska, A.; Bujok, R.; Harutyunyan, S.; Sashuk, V.; Dolgonos, G.; Grela, K. *J. Am. Chem. Soc.* **2004**, *126*, 9318.

30. Blanc, F.; Berthoud, R.; Salameh, A.; Basset, J. M.; Coperet, C.; Singh, R.; Schrock, R. R. *J. Am. Chem. Soc.* **2007**, *129*, 8434.

31. Blanc, F.; Thivolle-Cazat, J.; Basset, J. M.; Coperet, C.; Hock, A. S.; Tonzetich, Z. J.; Schrock, R. R. *J. Am. Chem. Soc.* **2007**, *129*, 1044.

32. Rhers, B.; Salameh, A.; Baudouin, A.; Quadrelli, E. A.; Taoufik, M.; Coperet, C.; Lefebvre, F.; Basset, J. M.; Solans-Monfort, X.; Eisenstein, O.; Lukens, W. W.; Lopez, L. P. H.; Sinha, A.; Schrock, R. R. *Organometallics* **2006**, *25*, 3554.

33. Blanc, F.; Coperet, C.; Thivolle-Cazat, J.; Basset, J. M.; Lesage, A.; Emsley, L.; Sinha, A.; Schrock, R. R. *Angew. Chem. Int. Ed.* **2006**, *45*, 1216.

34. Rhers, B.; Quadrelli, E. A.; Baudouin, A.; Taoufik, M.; Coperet, C.; Lefebvre, F.; Basset, J. M.; Fenet, B.; Sinha, A.; Schrock, R. R. *J. Organomet. Chem.* **2006**, *691*, 5448.

35. Poater, A.; Solans-Monfort, X.; Clot, E.; Coperet, C.; Eisenstein, O. *J. Am. Chem. Soc.* **2007**, *129*, 8207.

36. Eleuterio, H. S. *J. Mol. Catal.* **1991**, *65*, 55.

37. Sodesawa, T.; Ogata, E.; Kamiya, Y. *Bull. Chem. Soc. Jpn.* **1977**, *50*, 998.

38. Ogata, E.; Sodesawa, T.; Kamiya, Y. *Bull. Chem. Soc. Jpn.* **1976**, *49*, 1317.

39. Finkel'shtein, E. S.; Bykov, V. I.; Portnykh, E. B. *J. Mol. Catal.* **1992**, *76*, 33.

40. Tanaka, K. *J. Chem. Soc., Chem. Commun.* **1984**, 748.

41. Verpoort, F.; Bossuyt, A.; Verdonck, L. *Chem. Commun.* **1996**, 417.

42. Michelot, F. W.; Keaveney, W. P. *J. Polym. Sci.* **1965**, *3*, 895.

43. Rinehart, R. E.; Smith, H. P. *J. Polym. Sci.* **1965**, *3*, 1049.

44. Zenkl, E.; Stelzer, F. *J. Mol. Catal.* **1992**, *76*, 1.

45. Lu, S. Y.; Quayle, P.; Heatley, F.; Booth, C.; Yeates, S. G.; Padget, J. C. *Macromolecules* **1992**, *25*, 2692.

46. Feast, W. J.; Harrison, D. B. *J. Mol. Catal.* **1991**, *65*, 63.

47. Novak, B. M.; Grubbs, R. H. *J. Am. Chem. Soc.* **1988**, *110*, 7542.

48. Novak, B. M.; Grubbs, R. H. *J. Am. Chem. Soc.* **1988**, *110*, 960.

49. Hérisson, J.-L.; Chauvin, Y. *Makromol. Chem.* **1971**, *141*, 161.

50. Katz, T. J.; McGinis, J. *J. Am. Chem. Soc.* **1975**, *97*, 1592.

51. Schrock, R. R. *Science* **1983**, *219*, 13.

52. Schrock, R. R. *ACS Symp. Ser.* **1983**, *211*, 369.

53. Casey, C. P.; Tuinstra, H. E.; Saeman, M. C. *J. Am. Chem. Soc.* **1976**, *98*, 608.

54. Grubbs, R. H.; Carr, D. D.; Hoppin, C.; Burk, P. L. *J. Am. Chem. Soc.* **1976**, *98*, 3478.

55. Grubbs, R. H.; Burk, P. L.; Carr, D. D. *J. Am. Chem. Soc.* **1975**, *97*, 3265.

56. Schrock, R. R.; Murdzek, J. S.; Bazan, G. C.; Robbins, J.; Dimare, M.; O'Regan, M. *J. Am. Chem. Soc.* **1990**, *112*, 3875.

57. Nguyen, S. T.; Grubbs, R. H.; Ziller, J. W. *J. Am. Chem. Soc.* **1993**, *115*, 9858.

58. Schwab, P.; Grubbs, R. H.; Ziller, J. W. *J. Am. Chem. Soc.* **1996**, *118*, 100.

59. Crowe, W. E.; Zhang, Z. J. *J. Am. Chem. Soc.* **1993**, *115*, 10998.

60. Crowe, W. E.; Goldberg, D. R. *J. Am. Chem. Soc.* **1995**, *117*, 5162.

61. Fu, G. C.; Grubbs, R. H. *J. Am. Chem. Soc.* **1993**, *115*, 3800.

62. Casey, C. P.; Burkhardt, T. J. *J. Am. Chem. Soc.* **1974**, *96*, 7808.

63. Katz, T. J.; McGinnis, J. *J. Am. Chem. Soc.* **1975**, *97*, 1592.

64. Katz, T. J.; Rothchild, R. *J. Am. Chem. Soc.* **1976**, *98*, 2519.

65. McGinnis, J.; Katz, T. J.; Hurwitz, S. *J. Am. Chem. Soc.* **1976**, *98*, 605.

66. Katz, T. J.; McGinnis, J. *J. Am. Chem. Soc.* **1977**, *99*, 1903.

67. Hong, S. H.; Day, M. W.; Grubbs, R. H. *J. Am. Chem. Soc.* **2004**, *126*, 7414.

68. Hong, S. H.; Wenzel, A. G.; Salguero, T. T.; Day, M. W.; Grubbs, R. H. *J. Am. Chem. Soc.* **2007**, *129*, 7961.

69. Fu, G. C.; Grubbs, R. H. *J. Am. Chem. Soc.* **1992**, *114*, 5426.

70. Fu, G. C.; Nguyen, S. T.; Grubbs, R. H. *J. Am. Chem. Soc.* **1993**, *115*, 9856.

71. Fu, G. C.; Grubbs, R. H. *J. Am. Chem. Soc.* **1992**, *114*, 7324.

72. Buffat, M. G. P. *Tetrahedron* **2004**, *60*, 1701.

73. Tsuji, J.; Hashiguchi, S. *Tetrahedron Lett.* **1980**, *21*, 2955.

74. Fürstner, A.; Langemann, K. *J. Org. Chem.* **1996**, *61*, 3942.

75. Fürstner, A.; Langemann, K. *Synthesis* **1997**, 792.

76. Fürstner, A.; Seidel, G.; Kindler, N. *Tetrahedron* **1999**, *55*, 8215.

77. (a) Yet, L. *Chem. Rev.* **2000**, *100*, 2963; (b) Crimmins, M. T.; Powell, M.T. *J. Am. Chem. Soc.* **2003**, *125*, 7592

78. Ibrahem, I.; Yu, M.; Schrock, R. R.; Hoveyda, A. H. *J. Am. Chem. Soc.* **2009**, *131*, 3844.

79. Martin, S. F.; Humphrey, J. M.; Ali, A.; Hillier, M. C. *J. Am. Chem. Soc.* **1999**, *121*, 866.

80. Humphrey, J. M.; Liao, Y. S.; Ali, A.; Rein, T.; Wong, Y. L.; Chen, H. J.; Courtney, A. K.; Martin, S. F. *J. Am. Chem. Soc.* **2002**, *124*, 8584.

81. Martin, S. F.; Liao, Y. S.; Chen, H. J.; Patzel, M.; Ramser, M. N. *Tetrahedron Lett.* **1994**, *35*, 6005.

82. Martin, S. F.; Liao, Y. S.; Wong, Y. L.; Rein, T. *Tetrahedron Lett.* **1994**, *35*, 691.

83. Yang, Z.; He, Y.; Vourloumis, D.; Vallberg, H.; Nicolaou, K. C. *Angew. Chem. Int. Ed.* **1997**, *36*, 166.

84. Fürstner, A.; Mathes, C.; Grela, K. *Chem. Commun.* **2001**, 1057.

85. Laplaza, C. E.; Johnson, M. J. A.; Peters, J. C.; Odom, A. L.; Kim, E.; Cummins, C. C.; George, G. N.; Pickering, I. J. *J. Am. Chem. Soc.* **1996**, *118*, 8623.

86. Cummins, C. C. *Chem. Commun.* **1998**, 1777.

87. Cummins, C. C. *Prog. Inorg. Chem.* **1998**, *47*, 685.

88. Fürstner, A.; Mathes, C. *Org. Lett.* **2001**, *3*, 221.

89. Kirkland, T. A.; Grubbs, R. H. *J. Org. Chem.* **1997**, *62*, 7310.

90. Stewart, I. C.; Ung, T.; Pletnev, A. A.; Berlin, J. M.; Grubbs, R. H.; Schrodi, Y. *Org. Lett.* **2007**, *9*, 1589.

91. Berlin, J. M.; Campbell, K.; Ritter, T.; Funk, T. W.; Chlenov, A.; Grubbs, R. H. *Org. Lett.* **2007**, *9*, 1339.

92. Connon, S. J.; Blechert, S. *Angew. Chem. Int. Ed.* **2003**, *42*, 1900.

93. Chatterjee, A. K.; Grubbs, R. H. *Angew. Chem. Int. Ed.* **2002**, *41*, 3171.

94. Morrill, C.; Grubbs, R. H. *J. Org. Chem.* **2003**, *68*, 6031.

95. Alexander, J. B.; La, D. S.; Cefalo, D. R.; Hoveyda, A. H.; Schrock, R. R. *J. Am. Chem. Soc.* **1998**, *120*, 4041.

96. Keith, J. M.; Larrow, J. F.; Jacobsen, E. N. *Adv. Synth. Catal.* **2001**, *343*, 5.

97. Kiely, A. F.; Jernelius, J. A.; Schrock, R. R.; Hoveyda, A. H. *J. Am. Chem. Soc.* **2002**, *124*, 2868.

98. Weatherhead, G. S.; Ford, J. G.; Alexanian, E. J.; Schrock, R. R.; Hoveyda, A. H. *J. Am. Chem. Soc.* **2000**, *122*, 1828.

99. Harrity, J. P. A.; La, D. S.; Cefalo, D. R.; Visser, M. S.; Hoveyda, A. H. *J. Am. Chem. Soc.* **1998**, *120*, 2343.

100. Lee, A. L.; Malcolmson, S. J.; Puglisi, A.; Schrock, R. R.; Hoveyda, A. H. *J. Am. Chem. Soc.* **2006**, *128*, 5153.

101. La, D. S.; Ford, J. G.; Sattely, E. S.; Bonitatebus, P. J.; Schrock, R. R.; Hoveyda, A. H. *J. Am. Chem. Soc.* **1999**, *121*, 11603.

102. La, D. S.; Sattely, E. S.; Ford, J. G.; Schrock, R. R.; Hoveyda, A. H. *J. Am. Chem. Soc.* **2001**, *123*, 7767.

103. Aeilts, S. L.; Cefalo, D. R.; Bonitatebus, P. J.; Houser, J. H.; Hoveyda, A. H.; Schrock, R. R. *Angew. Chem. Int. Ed.* **2001**, *40*, 1452.

104. Hock, A. S.; Schrock, R. R.; Hoveyda, A. H. *J. Am. Chem. Soc.* **2006**, *128*, 16373.

105. Seiders, T. J.; Ward, D. W.; Grubbs, R. H. *Org. Lett.* **2001**, *3*, 3225.

106. (a) VanVeldhuizen, J. J.; Garber, S. B.; Kingsbury, J. S.; Hoveyda, A. H. *J. Am. Chem. Soc.* **2002**, 4954; (b) Van Veldhuizen, J. J.; Gillingham, D. G.; Garber, S. B.; Kataoka, O.; Hoveyda, A. H. *J. Am. Chem. Soc.* **2003**, *125*, 12502.

107. Funk, T. W.; Berlin, J. M.; Grubbs, R. H. *J. Am. Chem. Soc.* **2006**, *128*, 1840.

108. Odian, G. *Principles of Polymerization*, 3rd ed.; Wiley: New York, 1991.

109. Love, J. A.; Morgan, J. P.; Trnka, T. M.; Grubbs, R. H. *Angew. Chem. Int. Ed.* **2002**, *41*, 4035.

110. Love, J. A.; Sanford, M. S.; Day, M. W.; Grubbs, R. H. *J. Am. Chem. Soc.* **2003**, *125*, 10103.

111. Schrock, R. R.; Feldman, J.; Cannizzo, L. F.; Grubbs, R. H. *Macromolecules* **1987**, *20*, 1169.

112. Cannizzo, L. F.; Grubbs, R. H. *Macromolecules* **1988**, *21*, 1961.

113. Bielawski, C. W.; Benitez, D.; Grubbs, R. H. *Science* **2002**, *297*, 2041.

114. Scherman, O. A.; Grubbs, R. H. *Synth. Met.* **2001**, *124*, 431.

115. Scherman, O. A.; Rutenberg, I. M.; Grubbs, R. H. *J. Am. Chem. Soc.* **2003**, *125*, 8515.

116. Miao, Y. J.; Bazan, G. C. *Macromolecules* **1994**, *27*, 1063.

117. Miao, Y. J.; Bazan, G. C. *J. Am. Chem. Soc.* **1994**, *116*, 9379.

118. Gestwicki, J. E.; Cairo, C. W.; Strong, L. E.; Oetjen, K. A.; Kiessling, L. L. *J. Am. Chem. Soc.* **2002**, *124*, 14922.

119. Mortell, K. H.; Weatherman, R. V.; Kiessling, L. L. *J. Am. Chem. Soc.* **1996**, *118*, 2297.

120. Mortell, K. H.; Gingras, M.; Kiessling, L. L. *J. Am. Chem. Soc.* **1994**, *116*, 12053.

121. Bunz, U. H. F.; Kloppenburg, L. *Angew. Chem. Int. Ed.* **1999**, *38*, 478.

122. Fürstner, A.; Davies, P. W. *Chem. Commun.* **2005**, 2307.

123. Mortreux, A.; Blanchard, M. *J. Chem. Soc., Chem. Commun.* **1974**, 786.

124. Bray, A.; Mortreux, A.; Petit, F.; Petit, M.; Szymanskabuzar, T. *J. Chem. Soc., Chem. Commun.* **1993**, 197.

125. Villemin, D.; Cadiot, P. *Tetrahedron Lett.* **1982**, *23*, 5139.

126. Schrock, R. R.; Clark, D. N.; Sancho, J.; Wengrovius, J. H.; Rocklage, S. M.; Pedersen, S. F. *Organometallics* **1982**, *1*, 1645.

127. Listemann, M. L.; Schrock, R. R. *Organometallics* **1985**, *4*, 74.

128. Schrock, R. R. *Acc. Chem. Res.* **1990**, *23*, 158.

129. Feldman, J.; Schrock, R. R. *Prog. Inorg. Chem.* **1991**, *39*, 1.

130. Schrock, R. R. *J. Chem. Soc., Dalton Trans.* **2001**, 2541.

131. Schrock, R. R. *Chem. Rev.* **2002**, *102*, 145.

132. Fürstner, A.; Mathes, C.; Lehmann, C. W. *J. Am. Chem. Soc.* **1999**, *121*, 9453.

133. Zhang, W.; Kraft, S.; Moore, J. S. *J. Am. Chem. Soc.* **2004**, *126*, 329.

134. Wengrovius, J. H.; Sancho, J.; Schrock, R. R. *J. Am. Chem. Soc.* **1981**, *103*, 3932.

135. Pedersen, S. F.; Schrock, R. R.; Churchill, M. R.; Wasserman, H. J. *J. Am. Chem. Soc.* **1982**, *104*, 6808.

136. Nishida, M.; Shiga, H.; Mori, M. *J. Org. Chem.* **1998**, *63*, 8606.

137. Krouse, S. A.; Schrock, R. R. *Macromolecules* **1989**, *22*, 2569.

138. Krouse, S. A.; Schrock, R. R.; Cohen, R. E. *Macromolecules* **1987**, *20*, 903.

139. Zhang, X. P.; Bazan, G. C. *Macromolecules* **1994**, *27*, 4627.

140. Zhang, W.; Moore, J. S. *J. Am. Chem. Soc.* **2004**, *126*, 12796.

141. Weiss, K.; Michel, A.; Auth, E. M.; Bunz, U. H. F.; Mangel, T.; Mullen, K. *Angew. Chem., Int. Ed. Engl.* **1997**, *36*, 506.

142. Zhang, W.; Moore, J. S. *Macromolecules* **2004**, *37*, 3973.

143. Egbe, D. A. M.; Roll, C. P.; Birckner, E.; Grummt, U. W.; Stockmann, R.; Klemm, E. *Macromolecules* **2002**, *35*, 3825.

144. Pautzsch, T.; Klemm, E. *Macromolecules* **2002**, *35*, 1569.

145. Egbe, D. A. M.; Tillmann, H.; Birckner, E.; Klemm, E. *Macromol. Chem. Phys.* **2001**, *202*, 2712.

146. Brizius, G.; Pschirer, N. G.; Steffen, W.; Stitzer, K.; zur Loye, H. C.; Bunz, U. H. F. *J. Am. Chem. Soc.* **2000**, *122*, 12435.

147. Kaneta, N.; Hikichi, K.; Asaka, S.-i.; Uemura, M.; Mori, M. *Chem. Lett.* **1995**, 1055.

148. Fürstner, A.; Mathes, C. *Org. Lett.* **2001**, *3*, 221.

149. Fürstner, A.; Mathes, C. *Org. Lett.* **2001**, *3*, 221.

150. Fürstner, A.; Grela, K.; Mathes, C.; Lehmann, C. W. *J. Am. Chem. Soc.* **2000**, *122*, 11799.

151. Fürstner, A.; Grela, K. *Angew. Chem. Int. Ed.* **2000**, *39*, 1234.

152. For the first enyne metatheses, see references 153 and 154.

153. Katz, T. J.; Lee, S. J.; Nair, M.; Savage, E. B. *J. Am. Chem. Soc.* **1980**, *102*, 7940.

154. Katz, T. J.; Savage, E. B.; Lee, S. J.; Nair, M. *J. Am. Chem. Soc.* **1980**, *102*, 7942.

155. Smulik, J. A.; Diver, S. T. *Org. Lett.* **2000**, *2*, 2271.

156. Stragies, R.; Schuster, M.; Blechert, S. *Angew. Chem., Int. Ed. Engl.* **1997**, *36*, 2518.

157. Schurer, S. C.; Blechert, S. *Synlett* **1998**, 166.

158. Schurer, S. C.; Blechert, S. *Tetrahedron Lett.* **1999**, *40*, 1877.

159. Stragies, R.; Voigtmann, U.; Blechert, S. *Tetrahedron Lett.* **2000**, *41*, 5465.

160. Lee, H.-Y.; Kim, B. G.; Snapper, M. L. *Org. Lett.* **2003**, *5*, 1855.

161. Galan, B. R.; Giessert, A. J.; Keister, J. B.; Diver, S. T. *J. Am. Chem. Soc.* **2005**, *127*, 5762.

CHAPTER 22

Polymerization and Oligomerization of Olefins

22.1. Introduction

Polymerizations of ethylene and propylene (Equation 22.1) are some of the largest-scale chemical processes catalyzed by organometallic complexes. About 60% of the world's thermoplastics—deformable materials that melt when heated—are homopolymers or copolymers containing ethylene, propylene, or a combination of the two as the major monomer. In 2008, about 58.3 million metric tons of polyethylene was produced, and about 17 million metric tons of polypropylene was produced in the United States alone.[1] These materials are components of bags, packaging, water pipes, wire coatings, rope, carpeting, and textiles. Although some polyethylenes are made by polymerizations initiated by radicals, a majority of polyethylene and all polypropylene is made by coordination polymerization catalyzed by organometallic species. The metal coordination sphere and the relative rates of competing processes control the stereochemistry and degree of branching of the polymer. These and other properties of the microstructure affect the physical properties, which ultimately make the material suitable for specific applications. Thus, the organometallic chemistry of olefin polymerization dictates the ultimate utility of one of the world's most abundant chemical products.

$$\text{R} \xrightarrow{\text{Catalyst}} \left(\text{R} \right)_n \tag{22.1}$$

R = H or Me
Catalyst = Ti, Zr, Hf, Cr, Fe, Co, Ni or Pd-based

Two classes of catalysts are used commercially to produce polymers from ethylene and propylene. The first class of catalysts produces roughly 99% of the world's polyethylene and polypropylene.[2] They are heterogeneous catalysts activated by aluminum alkyls. This class of catalysts was discovered in the laboratories of Karl Ziegler at the Max Planck Institute in Mulheim, and was quickly developed for the formation of isotactic polypropylene by Giulio Natta at the Milan Polytechnic Institute.[3] These "Ziegler" catalysts were based on titanium halides. For their work, Ziegler and Natta were awarded the Nobel Prize in Chemistry in 1963. Heterogeneous catalysts formed from chromium precursors have also been used for many years to generate polyethylene.[4,5]

The second class of catalyst produces polymers with tailor-made architectures[6–9] and is based on the combination of ligated soluble metal complexes and alkylaluminum or borate activators. Most current commercial systems that fall into this class consist of a group 4 metal containing cyclopentadienyl or cyclopentadienyl analogs as ligands. The ligands contained in these catalysts control the polymer stereochemistry, the degree of branching, the length of branches, and the degree of incorporation of comonomers. However, catalysts based on the middle transition elements, such as iron and cobalt, and catalysts based on late transition metals, such as nickel and palladium, also have been studied extensively. The catalysts based on early and middle transition metals form linear polyethylene, while the catalysts based on late transition metals form linear to highly branched polyethylene. In addition to forming polymers with properties that are distinct from those formed by the early and middle transition metals, the late metal catalysts have been studied for their potential to form copolymers from monomers possessing polar functional groups. Catalysts based on vanadium, chromium, cobalt, nickel and palladium have been used for the industrially important polymerization, oligomerization, and cyclooligomerization of dienes (Equation 22.2).

$$\text{(22.2)}$$

In contrast to the state of the mechanistic understanding of ethylene and propylene polymerization twenty years ago when the predecessor to this text was published, detailed features of the mechanism of the polymerization of alkenes by the group 4 metal catalysts and of the mechanism of polymerization by nickel and palladium catalysts are now understood. Many of the features of this mechanism were alluded to during previous chapters of this text covering migratory insertion, β-hydrogen elimination, and electrophilic attack on coordinated ligands. In brief, polyethylene and polypropylene are produced by repetitive insertions of olefins into the metal–alkyl bond that links the catalyst to the growing polymer chain, and the polymerization is terminated by reactions with a chain transfer agent, β-hydrogen elimination, or direct transfer of the β-hydrogen of the growing chain to the incoming monomer.

Chapter 22 presents the different types of polymers produced from ethylene, propylene, and copolymers of these olefins with other monomers, along with some basic principles that apply to polyolefin synthesis and characterization. The basic features of the mechanism of the polymerization are presented first to provide a framework for the description of different catalysts and materials. After presenting different types of catalysts that have been used for the polymerization of these monomers, a more detailed description of several features of the mechanism is presented. Finally, an overview of ethylene oligomerization, as well as diene polymerization and oligomerization, is presented. The primary journal literature and patent literature on olefin polymerization is immense. Fortunately, many reviews of olefin polymerization have been published. The coverage of olefin polymerization and oligomerization in this chapter is selective, and the reader is directed to review articles for more comprehensive coverage of these topics.[10–18]

22.1.1. A Primer on Polyolefin Chemistry (Written with Prof. Geoffrey W. Coates and Prof. Gregory J. Domski)

A number of terms are used in this chapter that have not been used in earlier chapters of this book. Thus, it is appropriate at this point to define some of these terms and to describe some basic aspects of polymer chemistry as they apply to the synthesis of polyolefins. At the most basic level, a polymer consists of one or more repeating units, and a polymer chain possesses a molecular weight that depends on the number of monomers in the chain. Bulk samples of polyolefins consist of polymer chains of varying molecular weights. Polymer molecular weights are now most commonly reported as weight-averaged molecular weight values M_w and number-averaged molecular weight values M_n (Equation 22.3). Because the higher molecular weight chains contribute more to the weight-averaged molecular weight value than to the number-averaged molecular weight value, M_w is greater than M_n. The ratio of M_w to M_n is a measure of the molecular weight distribution (MWD) and is sometimes called the polydispersity index (PDI). For a pure material of single molecular weight, $M_w = M_n$.

$$M_n = \frac{\sum N_x M_x}{\sum N_x} \qquad M_w = \frac{\sum N_x M_x^2}{\sum N_x M_x}$$

(22.3)

in which N_x = the number of chains with molecular weight M_x.

The monomer units in a polyolefin synthesis are linked or "enchained" by the catalyst in a process called chain growth. In a chain-growth polymerization, monomer units are added to the chain end. In a perfect chain-growth process, in which no irreversible termination of the polymer chain occurs, the catalyst remains attached to the chain end in an active form. When this occurs and no irreversible transfer of the chain from one catalyst to another occurs, the molecular weight of the bulk sample increases linearly with conversion, and the catalyst is said to be "living." In the alternative type of polymer growth—step-growth or condensation polymerization—two polymer chains combine, and the molecular weight of the polymer chains grows exponentially with conversion. When initiation of a living polymerization is fast, relative to propagation, the polymer MWD is narrow and the PDI approaches 1.0.

Copolymers are produced by enchainment of two different monomers, and many types of copolymers have been produced. Copolymers discussed in this chapter include those produced by random enchainment of two monomers, by alternating enchainment of the two monomers, or by enchainment of a series of one monomer, followed by a series of another. These types of polymers are called random, alternating, and block copolymers, respectively. Stereoblock polymers are also known in which a catalyst forms two or more types of polymer segments, each containing a different stereochemical arrangement of the same monomer.

The stereochemistry of polymers can be more complex to describe than the stereochemistry of smaller organic molecules. As discussed in some detail in this chapter, the properties of polypropylene are controlled by the relative stereochemical configuration of the methyl groups along the polypropylene chain. Typically, one can measure the relative configuration of five contiguous propylene units or "pentads" by ^{13}C NMR spectroscopy. Thus, polymer chemists typically report the fraction or percentage of pentads that contain a certain stereochemical arrangement.

The relative configuration of the monomer units can be controlled by the structure of the catalyst or by the configuration of the last inserted unit. These two scenarios are called site control and chain-end control. The isotactic polypropylene generated by the types of stereodefined metallocene catalysts presented in this chapter results from site control. More sophisticated architectures are possible by site-control mechanisms than chain-end control mechanisms, as illustrated by the variety of polyolefins prepared by homogeneous catalysts.

The consequences of whether polymer stereochemistry results from site control or chain-end control are significant. As shown in Equation 22.4, when the catalyst controls the relative stereochemical configurations along the polymer chain, a monomer unit that inserts to generate the minor stereoisomeric relationship between these units (a "stereoerror") is followed by insertion of the next monomer with the same relative configuration as the bulk of the chain. However, when the relative configuration is controlled by the last monomer inserted, a "stereoerror" inverts the relative configuration of the subsequent enchained monomers until an additional stereoerror is made. These two modes of stereochemical control, therefore, lead to polymers with different overall relative stereochemistry, and are easily distinguished by the NMR signatures corresponding to these two types of stereoerrors.

$$(22.4)$$

22.2. Mechanism(s) of Monoene Polymerization and Oligomerization

The polymerization or oligomerization of simple monoenes involves three basic steps shown in Equation 22.5: (a) initiation, (b) propagation, and (c) termination. It is generally agreed that the catalytically active center at which polymerization or oligomerization occurs is a transition metal alkyl complex.

$$(22.5)$$

The initiation step (a) can be achieved in several ways,[19] and each of these methods generates a coordinatively unsaturated metal–alkyl complex. In one scenario, the unsaturated metal–alkyl complex is generated by the reaction of a transition metal dihalide with an alkylaluminum reagent. This reagent converts the dihalide to a dialkyl complex, and one of the alkyl groups is abstracted from the metal center by the alkylaluminum reagent to form the cationic monoalkyl species. By a second, related sequence, a dialkyl complex is treated with a strong boron Lewis acid, such as a perfluoroarylborane, and this reaction leads to abstraction of one of the alkyl groups to form a cationic alkyl complex paired with a borate anion. By a third process, a dialkyl complex is treated with an acid containing a noncoordinating conjugate base, such as BAr_4^- ($Ar = C_6F_5$). Such abstraction and protonation reactions were discussed in Chapter 12 (electrophilic attack on coordinated ligands).

Chain growth (step b in Equation 22.5) occurs by a combination of olefin coordination and migratory insertion. A vacant site is required for the olefin to coordinate before alkyl insertion can take place. Because coordination is required, fast insertions occur with less-hindered olefins, such as ethylene, propylene, linear α-olefins, and vinylarenes. The relative rates for the polymerization of alkenes typically follow the trend ethylene > propylene > α-olefin $\gg$ 1,2-disubstituted olefin $\approx$ 1,1-disubstituted olefin, and enchainment of tri-and tetra-substituted olefins is rare or unknown. Many polymerization catalysts are sensitive to poisoning by impurities that bind the open coordination site.

The second stage of the propagation process was the subject of some discussion, and the details of the interactions that control the stereochemical relationships between monomer units have been investigated extensively. These issues were discussed in Chapter 9 on olefin insertions. In brief, the carbon–carbon bond-forming step occurs by migratory insertion of the coordinated olefin into the metal–alkyl bond (the "Cossee mechanism").[20] The regioselectivity and stereoselectivity of this insertion control the distribution and relative configurations of the sidechains in the growing polymer chain.

Chain termination (step c in Equation 22.5) can occur by one of several processes.[21] In some cases, a "chain transfer" agent, such as hydrogen, is added to the polymerization process to control the molecular weight (Equation 22.6).[22,23] These reagents terminate the

$$L_nM-CH_2CH_2-P \xrightarrow{H_2} L_nM-H + CH_3CH_2P \qquad (22.6)$$

growth of a chain and leave a ligand on the metal, such as a hydride, that can be used to restart growth of a new chain. A second pathway for chain termination is β-hydrogen elimination. This step generates an olefin-terminated polymer and a metal hydride complex, which again can insert olefin to restart growth of a polymer chain (Equation 22.7). A third type of polymer termination process is related to the combination of β-hydrogen elimination and olefin insertion, but the olefin associates with the metal prior to cleavage of the β C–H bond. By this process, direct transfer of the β-hydride from the alkyl ligand to the coordinated olefin monomer is thought to occur without the intermediacy of a metal hydride complex (Equation 22.8).[21,24,25]

$$L_nM-CH_2CH_2-P \longrightarrow L_nM-H + CH_2=CH-P \qquad (22.7)$$

$$(22.8)$$

The relative rates of the propagation and the termination steps, k_p and k_t respectively, determine the molecular weight of the product. The polymerization can occur with three different sets of relative rates, and each set leads to a different class of product. When k_p is much greater than k_t, high-molecular-weight polymer is formed. When k_p and k_t are similar, a set of oligomers possessing a geometric molecular weight distribution ("Schultz/Flory" type[26-28]; see Section 22.9 on oligomerization) is formed. When k_t is much larger than k_p, dimers of the olefin are formed exclusively.

22.3. Ethylene-Based Polymers (Written with Prof. Geoffrey W. Coates and Prof. Gregory J. Domski)

It might seem at first glance that the only difference between polyethylene prepared from one source or another would be molecular weight and molecular weight distribution. However, polyethylenes produced from different catalysts can have different structures. Polyethylene materials can contain branches from side reactions or the incorporation of comonomers. This branching has a large effect on the crystallinity of the polymer, and this crystallinity has a large effect on the physical properties of the material. Two classes of polymers are prepared from ethylene alone, and one closely related class of polymers is prepared from ethylene and an α-olefin comonomer. Depictions of the structures of these polymers are shown in Figure 22.1. High-density polyethylene (HDPE) is strictly linear polyethylene. Low-density polyethylene (LDPE) contains extensive branching along the polymer chain and "branches on branches." Linear low-density polyethylene (LLDPE) is a linear polymer prepared from ethylene and an α-olefin that creates a material possessing a small number of branches.

Most high-density polyethylene (HDPE) is prepared by heterogeneous Ziegler catalysts containing titanium or by supported chromium oxide catalysts.[29,30] HDPE has a higher melting point and is much more crystalline than LDPE. Many transition metal catalysts polymerize ethylene to form HDPE, but few catalysts match the activity of the heterogeneous titanium systems, particularly considering the low cost of the Ziegler catalysts.

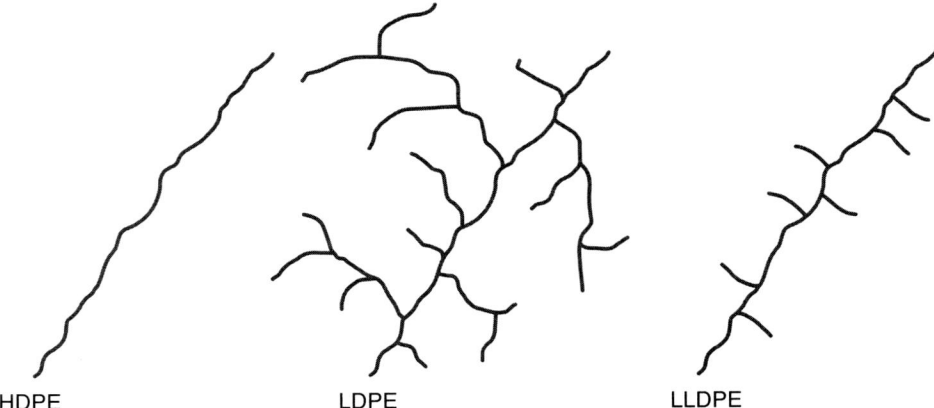

HDPE   LDPE   LLDPE

Figure 22.1.
Structures of common types of polyethylene: high-density polyethylene (HDPE), low-density polyethylene (LDPE), and linear low-density polyethylene (LLDPE).

Low-density polyethylene (LDPE) is industrially synthesized by a free-radical process using peroxide initiators at high temperatures and pressures. LDPE has a highly branched structure due to hydrogen atom shifts that convert a primary alkyl radical to a secondary alkyl radical. Plants that can accomodate this high-pressure process are expensive to build and operate. Therefore, LDPE formed by radical reactions is slowly being replaced by LLDPE prepared by reactions catalyzed by transition metal complexes.

The ability to form linear low-density polyethylene at lower pressures and temperatures in a controlled fashion has been an important advance in transition-metal-catalyzed polymerizations. LLDPE is typically formed from ethylene and short-chain α-olefins (typically about 10 weight percent of monomers containing 4–8 carbons). The resulting polymer exhibits good tensile strength, due to its long backbone, and toughness, due to the short-chain branches.

Other forms of polyethylenes include ultra-low-density polyethylene (ULDPE), high molecular weight (HMW) HDPE, and ultra-high molecular weight (UHMW) HDPE. ULDPE contains a higher percentage of α-olefin than LLDPE, and UHMW HDPE possesses a molecular weight exceeding three million.[31]

22.3.1. Catalysts for the Synthesis of HDPE

Because this chapter focuses on molecular transition metal complexes that catalyze the formation of polyolefins, an extensive description has not been included of the heterogeneous titanium systems of Ziegler and the supported chromium oxide catalysts that form HDPE. However, a brief description of these catalysts is warranted because of their commercial importance. The "Ziegler" catalysts are typically prepared by combining titanium chlorides with an aluminum-alkyl co-catalyst.[32–35] The structural features of these catalysts have been studied extensively, but it remains challenging to understand the details of how polymer architecture is controlled by the surface-bound titanium. This chapter does, however, include an extensive discussion of how group(IV) complexes that are soluble, molecular species polymerize alkenes to form many different types of polyolefins.

The second type of supported catalyst is based on chromium. Chromium catalysts for the polymerization of ethylene were discovered at Phillips by Hogan and Banks in the 1950s.[29,30] These catalysts are generated (as depicted in Equation 22.9) by depositing chromium complexes on silica, most often CrO_3 or a Cr(III) species, followed by calcining in oxygen to generate a supported chromate or dichromate species. These Cr(VI) species then react with ethylene to form oxidized olefin products, along with the lower-valent, supported chromium species that polymerize ethylene. A second class of chromium catalyst was discovered at Union Carbide and is generated by depositing chromocene onto silica.[36] The catalyst generated from chromocene is not currently being used in production, but the Phillips catalyst was stated in 2005 to be used to prepare roughly one third of the 40 million tons of polyethylene produced.[37]

The identity of the active form of the supported Phillips catalyst that polymerizes ethylene has been a center of controversy for more than fifty years.[37,38] Oxidation states ranging from Cr(II) to Cr(IV) have been claimed to be those of the active catalyst, but no one oxidation state has been accepted by the community to be that of the active catalyst. Cr(II) on silica has been generated from the Cr(VI) species and CO, and this material polymerizes ethylene, but it is unclear if the most reduced species or a site that has been oxidized from the Cr(II) state forms the polymer. Similar catalysts are generated by reducing Cr(IV) sites with H_2 or CO. In these catalysts a chromium hydride is speculated to be the initiation site.[5,39] After initiation, polymer growth is thought to occur by sequential migratory insertion reactions described earlier. The identity of the species generated from chromocene that polymerizes ethylene is equally ambiguous.

Homogeneous model compounds containing chromium have been prepared. In particular alkylchromium complexes ligated by Cp* and NacNac ligands have been shown by Theopold to polymerize ethylene (Equation 22.10).[39,40] Moroever, the NacNac compound reproduces the copolymerization of ethylene and α-olefins achieved with the Phillips catalyst. These compounds would then be models for an active, supported Cr(III) alkyl species in the Phillips catalyst system.

$$Cr(III) \;+\; silica \;\longrightarrow\; \xrightarrow{O_2/heat} \;\cdots\; \longrightarrow\; active\ Cr\ catalyst \tag{22.9}$$

$$LLDPE \;\longleftarrow\; \cdots \;\longrightarrow\; polyethylene \tag{22.10}$$

Most commercial polyethylene catalysts are either heterogeneous or colloidal.[22,41] Removal of the catalyst from the polymer is expensive, and the catalyst is usually left in the final product. It is possible to leave the catalyst in the final polymer because the catalysts are so active that $>10^4$ kg of polyethylene are typically formed per gram of catalyst.[21,22]

Homogeneous catalysts based on middle transition metals have been developed that polymerize ethylene with high activity to form HDPE. Bennett at DuPont, Brookhart at North Carolina, and Gibson at Imperial College discovered almost simultaneously cobalt and iron catalysts containing pyridyl diimine ligands for the polymerization of ethylene (Figure 22.2).[42–44] This class of catalysts forms polymers when the aryl groups on the imine nitrogen are sterically demanding. Related iron complexes lacking the steric hindrance on the N-aryl groups are highly active catalysts for the oligomerization of ethylene and form a Schultz–Flory distribution of α-olefins. The role of the steric bulk on the formation of polymers versus oligomers is discussed in more detail in Section 22.8.3. In brief, the steric bulk decreases the rate of chain transfer relative to the rate of propagation and thus leads to polymers possessing high molecular weights. It is thought that bulky ligands retard chain transfer either by preventing associative displacement of olefin formed reversibly by β-hydrogen elimination or by raising the transition state energy in the "chain transfer to monomer" process that was depicted in Equation 22.8.

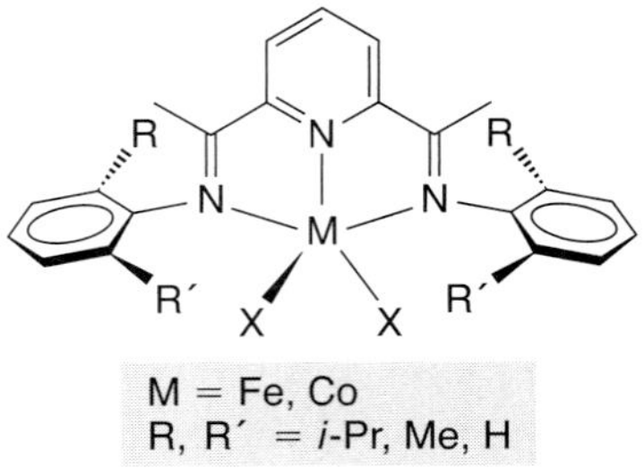

Figure 22.2.
Iron and cobalt pyridyl diimine catalysts for the synthesis of HDPE and the oligomerization of ethylene.

22.3.2. Catalysts for the Synthesis of LDPE Materials from Only Ethylene

Most low-density polyethylene synthesized with transition-metal catalysts is formed by copolymerization of ethylene and an α-olefin. This linear low-density polyethylene (LLDPE) is described in a later section on copolymerization. At the same time, some LDPE materials that are distinct from those formed by radical polymerizations can be produced by polymerization of ethylene with transition-metal catalysts.

The branching in the LDPE materials formed by metal-catalyzed polymerization originates from β-hydrogen elimination. Chain termination by β-hydrogen elimination to form an olefin-terminated polymer, followed by reassociation of the polymer with the catalysts and reinsertion, generates a material containing a long-chain branch (LCB). These long-chain branches give rise to a particular class of low-density polyethylene that is formed without adding α-olefin comonomers. Researchers at Dow and Exxon developed *ansa*-cyclopentadienyl–amido titanium catalysts for olefin polymerization (Figure 22.3)[45–49] that contain a particularly accessible metal center. These catalysts have been called "Constrained Geometry Catalysts" or CGCs. The accessibility of the metal center in these catalysts allows large α-olefin comonomers to be incorporated into the polymerization process (see also the later discussion of LLDPE). Thus, these catalysts are able to incorporate olefin-terminated polyethylene chains to form LDPE that contains long-chain branches.[50,51] High temperature, high conversion, and low ethylene or α-olefin concentrations favor formation of this type of polymer with the CGCs.[52,53] More recently, additional catalysts that re-insert long alkene-terminated polymer chains have been reported.[54–57] Systems containing two catalysts have also been identified in which one catalyst oligomerizes ethylene to 1-hexene or 1-octene, and a second catalyst incorporates these oligomers into the growing polyethylene chain to generate LLDPE from ethylene alone.[58–63] Dinuclear systems have also been developed as a means to promote long-chain branching by re-incorporation of olefin-terminated chains (Figure 22.4).[64] Heterobinuclear catalysts linked by covalent[65] or electrostatic[66] interactions generated more long-chain branches than mononuclear analogs.

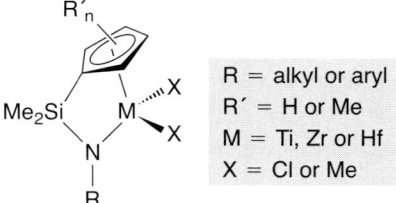

Figure 22.3.
ansa-Cyclopentadienyl–amido constrained geometry catalysts.

R = alkyl or aryl
R′ = H or Me
M = Ti, Zr or Hf
X = Cl or Me

22.3.3. Hyperbranched Polyethylenes from Late Metal Catalysts

Late metal nickel and palladium catalysts also generate a form of LDPE that is much different from any LDPE produced by the early transition metal catalysts.[67] These polyethylenes produced by nickel and palladium catalysts contain a much higher degree of

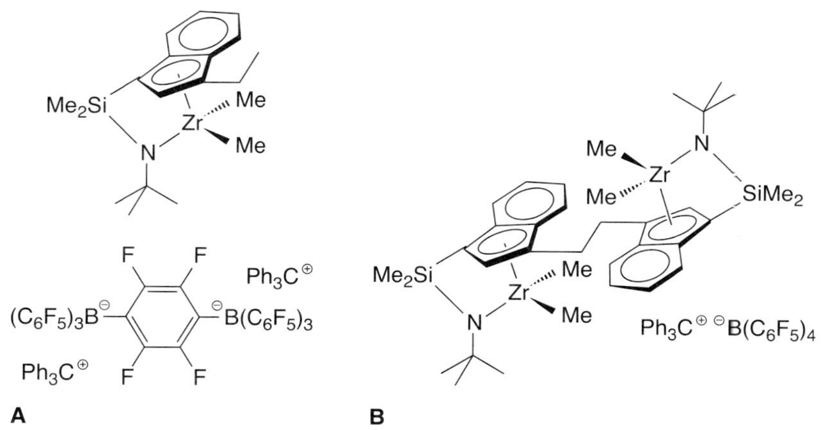

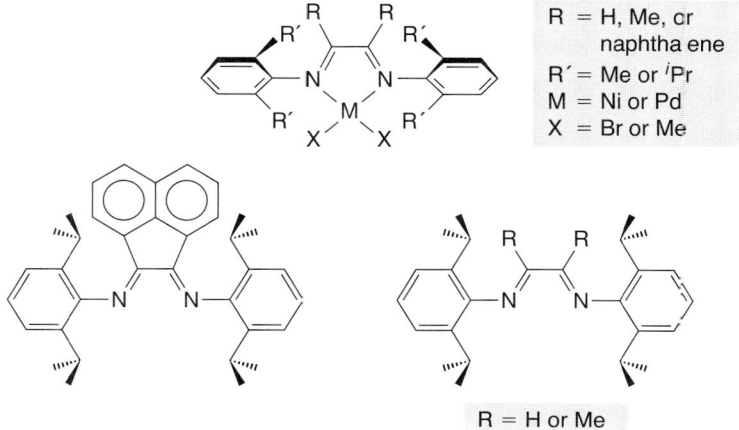

Figure 22.4.
Examples of dinuclear catalysts for the re-incorporation of olefinic polymers.
(A) A monomeric catalyst with a bis-borate activator and (B) a dinuclear catalyst
precursor with a monoborate activator.

branching than those generated by early metal catalysts and can even contain branches on
the branches. The polymers produced by these late metal catalysts have many more side
chains than LDPE produced by radical chemistry, and they are often oils.

The catalysts for these polymerizations contain α-diimine ligands. The key to
obtaining polymeric material with these catalysts was the use of α-diimines containing
sterically demanding aryl groups on the imine nitrogen. These catalysts were developed
by Brookhart and co-workers (Figure 22.5) and led to the iron and cobalt systems
mentioned in Section 22.3.1.[68] The degree of branching depends on the metal, ligand
structure, temperature, and pressure. In general, the nickel catalysts produce polyethylene
that can be nearly linear or moderately branched, while the palladium catalysts produce
polymers that are more highly branched.[69] Careful ^{13}C NMR analysis of the polymers[70] has
shown that the number of branch points (tertiary carbons) is on the order of 1 for every
10 carbons in the overall polymer. A revealing depiction of the structure of polyethylene
produced with a palladium catalyst containing an α-diimine ligand is shown in
Figure 22.6.

R = H, Me, or
    naphthalene
R′ = Me or ^{i}Pr
M = Ni or Pd
X = Br or Me

R = H or Me

Figure 22.5.
Catalysts and ligands used by Brookhart with palladium and nickel metal
centers to form highly branched polyethylene. The ligands shown provide
the highest molecular weights of polymer.

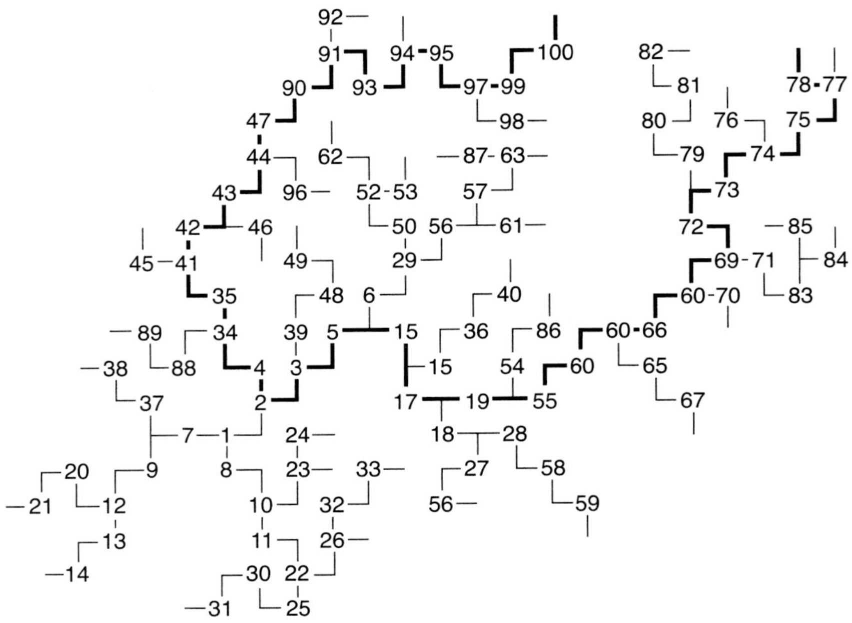

Figure 22.6.
Representation of the branching in a unit of 100 carbons within a polyethylene produced from the LPd(R)(ethylene) species with L = ligands in Figure 22.5. The dark line represents the longest linear segment. Adapted from *Macromolecules*, **2007**, *40*, 410.

The origin of extensive branching in the polyethylene produced by these late metal catalysts is different from that in LDPE produced by radical processes or the LLDPE produced by heterogeneous Ziegler-type catalysts. Branching from radical-induced polymerization occurs by hydrogen atom transfer, and branches containing various chain lengths result (Scheme 22.1). In LDPE produced by Ziegler catalysts, the branches consist of methyl groups from β-hydrogen elimination, rotation, re-insertion and continued polymer growth (Scheme 22.2). The late metal catalysts lead to branched polymers by a series of reactions involving more extensive β-hydrogen eliminations and reinsertions, as shown in Scheme 22.3.[71] During this process, the eliminated olefin remains bound to the catalyst. Free α-olefins, which would be formed by β-hydrogen elimination and dissociation or displacement of the resulting olefin, have not been observed in these polymerizations. Moreover, mechanistic studies have led to the conclusion that the isomerization occurs after dissociation of ethylene, not through five-coordinate hydride intermediates.

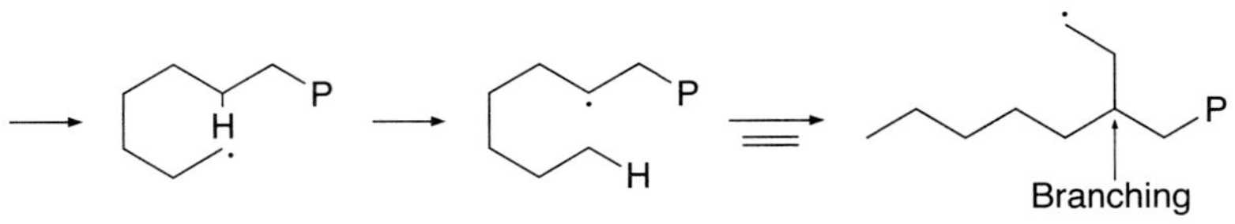

Scheme 22.1

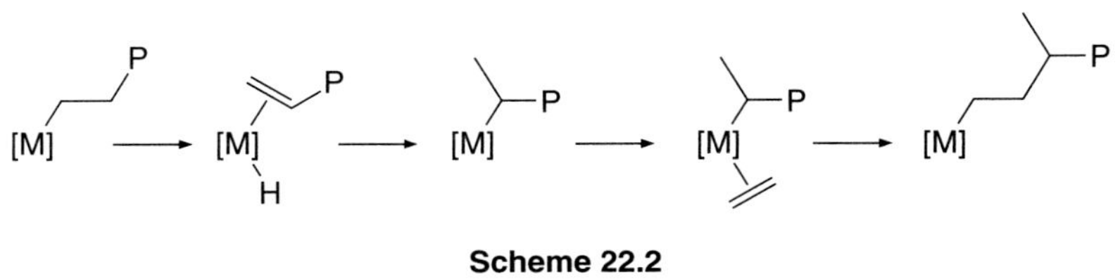

Scheme 22.2

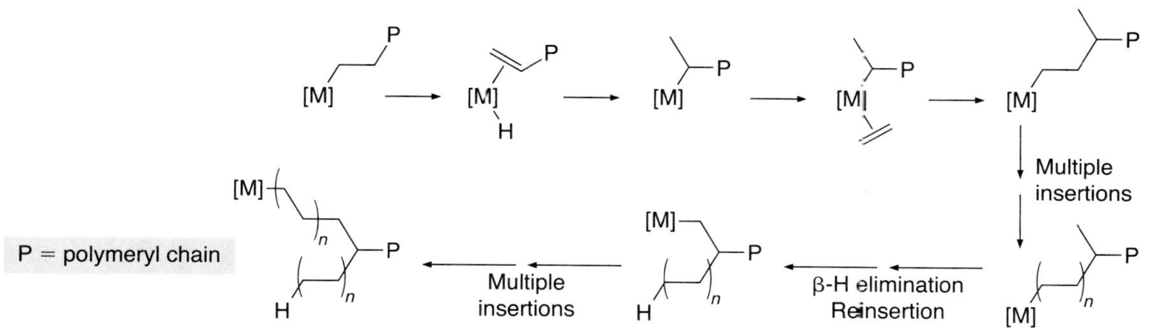

Scheme 22.3

22.4. Propylene-Based Polymers (Written with Prof. Geoffrey W. Coates and Prof. Gregory J. Domski)

Over 50 billion pounds of polypropylene (PP) were manufactured worldwide in 2004.[72] However, polypropylene encompasses many different materials (Figure 22.7), and the physical properties of polypropylene depend heavily on the relative configurations of the methyl groups in the polymer chain. Most of this polypropylene contains a microstructure or relative stereochemistry that is called isotactic. Isotatic polypropylene contains all syn 1,3-relationships between the adjacent methyl groups, and this material is highly crystalline. Syndiotactic polypropylene contains all anti relationships between the adjacent methyl groups and is also crystalline. Atactic polypropylene contains random relative stereochemistry along the chain and is amorphous. Materials containing more intricate stereochemical relationships also have been prepared. For example, hemiisotactic polypropylene contains two alternating units, one with the same relative configuration and one with a random configuration. Stereoblock polymers contain units of one microstructure, such as isotactic, and units of another microstructure, such as atactic. This particular combination of microstructures leads to an elastomer.[73]

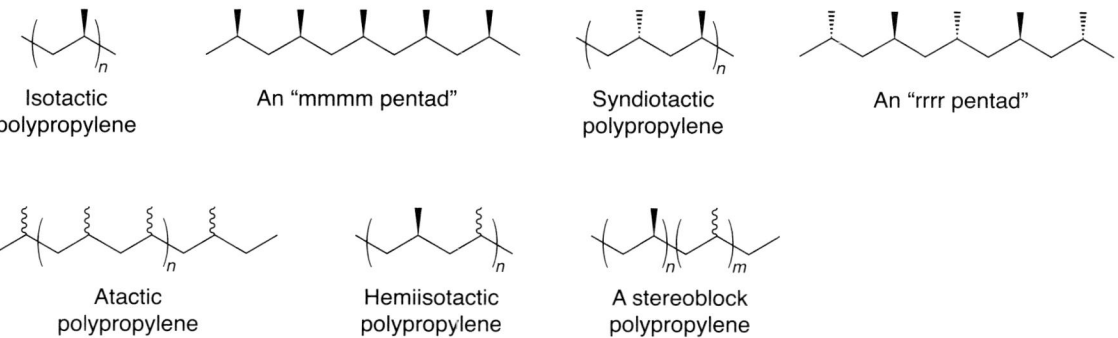

Figure 22.7. A selection of polypropylenes with various microstructures.

22.4.1. Mechanism of Stereocontrol in Isotactic Polypropylene Synthesis

Much research has been conducted to generate catalysts that form polypropylenes containing microstructures and architectures that can be rationally controlled. This catalyst development has involved some of the most elegant mechanism-based design of catalysts in any area of organometallic chemistry. This section describes some of the basic principles that explain how the stereochemistry of polyolefins is controlled.

The mechanisms for the stereoregular polymerization of α-olefins have been described in many reviews.[10,74–76] The basic principles that have been used to rationalize observed stereochemistry and to design catalysts that form polyolefins possessing particular stereochemical relationships between monomer units will be described here. Recall from the general discussion of polyolefin polymerization that the stereochemistry of the polymer involves the relative configuration of diad units and can be controlled by the catalyst or the last inserted monomer. This discussion focuses on control by the catalyst (site control), which allows for the design of systems that can produce more varied polymer microstructures than those dictated by chain-end control.

There are four possible combinations of regiochemistry and stereochemistry within diad units. The olefins can join in a head-to-head or head-to-tail fashion. Most polyolefins formed by early metal catalyts are formed by strict head-to-tail enchainment, but this head-to-tail enchainment can occur by a series of 1,2-insertions in which the α-olefin substituent is located β to the metal in the insertion product or 2,1-insertions in which the α-olefin substituent is located α to the metal in the insertion product. In addition, the olefins can join to give rise to a diad unit containing identical (meso, m) or opposite (racemo, r) stereochemical relationships to the last inserted monomer unit. Site control of polymerization to form isotactic polymer gives rise to rr-defects in the polymer from stereoerrors, but chain-end control of polymerization to form isotactic polymer gives rise to r-defects from stereoerrors.

Although the term olefin "insertion" is used in the literature and in this chapter, this "insertion" of the olefin is better considered to be migration of the alkyl chain to the olefin.[10,77] The polymer chain in this process occupies the binding site of the olefin after each enchainment step. This mechanism leads to an alternation of the polymer chain between two binding sites at the metal. This "windshield wiper" mechanism has consequences on the control of stereochemistry, as described later in this chapter. An event in which the polymer chain reverts to the original site faster than olefin inserts is called "back-skipping."

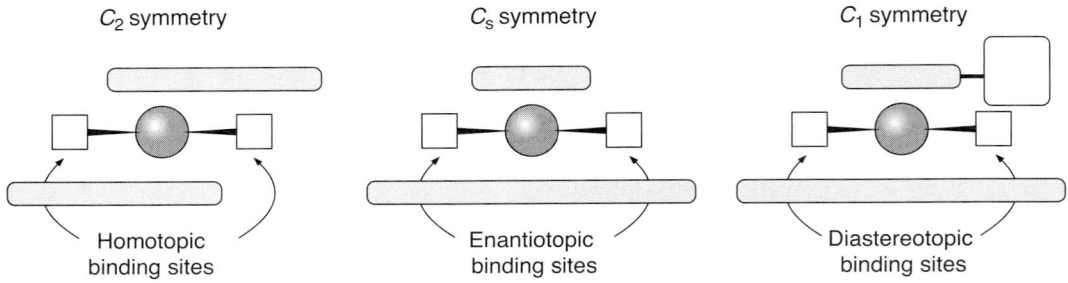

Figure 22.8.
Three general geometries for stereoselective olefin polymerization catalysts operating by site control.

Figure 22.8 shows three types of symmetries that can generate isotactic, syndiotactic, and hemiisotactic polypropylene, respectively. The metallocene on the left of Figure 22.8 contains C_2 symmetry (containing one C_2 axis) in the absence of the olefin and the polymer chain. The metallocene in the center contains C_s symmetry (containing one plane of symmetry) in the absence of the olefin and the polymer chain, and it contains two binding sites related by the mirror plane. The metallocene at the right of Figure 22.8 contains C_1 symmetry (containing no symmetry element higher than the identity element) in the absence of the olefin and the polymer chain. In this type of compound, the two binding sites are inequivalent.

The process of monomer enchainment for a metallocene containing each of these three symmetries is shown in Equations 22.11–22.13. The structure on the left of Figure 22.8 possessing C_2 symmetry contains two binding sites that are homotopic. Both binding sites favor binding to the same face of the olefin to generate two identical structures. These structures with olefin and polymer attached can be seen at the first, third, and fifth structures of

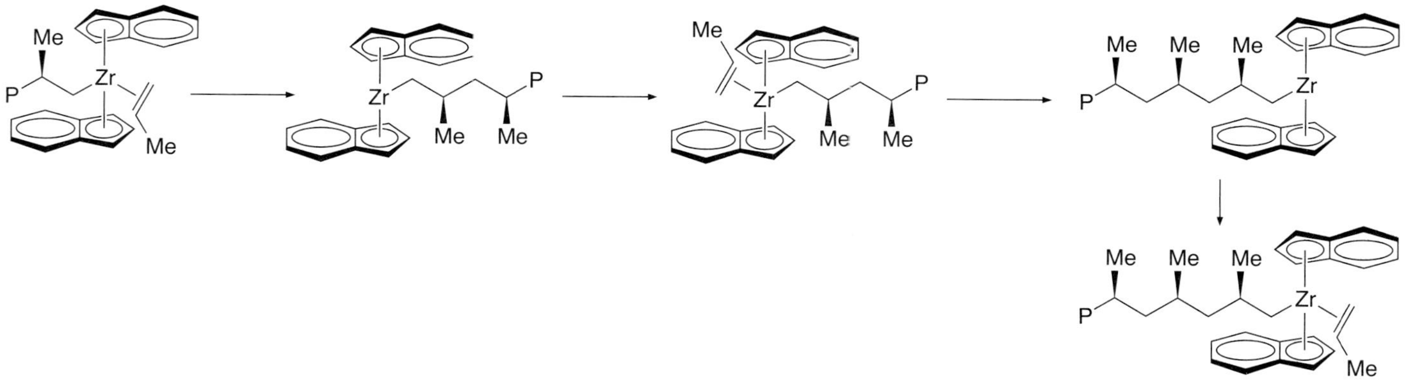

(22.11)

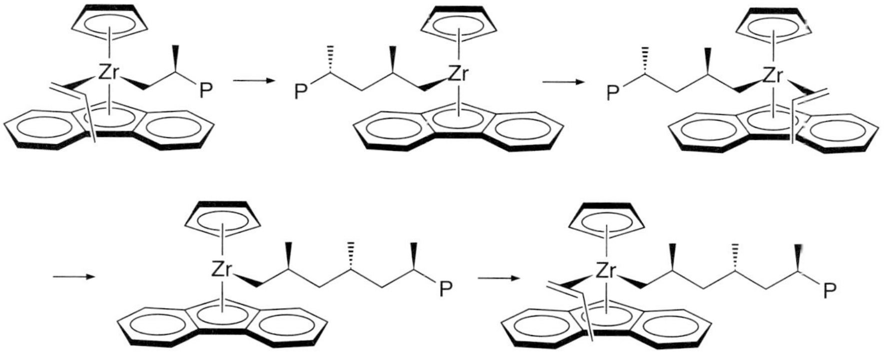

(22.12)

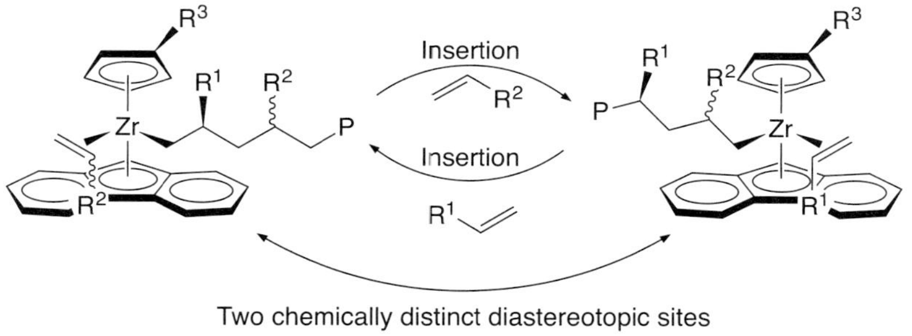

(22.13)

Two chemically distinct diastereotopic sites

Equation 22.11. As can be seen in this equation, insertion of the polymer during each of the two steps shown in this equation gives rise to a structure that is identical to the starting structure, except for the incorporation of an additional monomer unit. Thus, each inserted monomer has the same stereochemical relationship, and isotactic polypropylene is formed. This concept applies to catalysts other than metallocenes. The two binding sites in any catalyst possessing C_2 symmetry will be homotopic, and such catalysts should produce isotactic polypropylene if they follow these steps for insertion in this geometry.

The structure at the center of Figure 22.8 possessing C_s symmetry contains two binding sites that are enantiotopic. In this case, each binding site favors binding to the opposite face of the olefin to generate two enantiomeric structures. These structures can be seen as the first, third, and fifth structures of Equation 22.12. As can be seen in this equation, insertion of the polymer, followed by olefin coordination, gives rise to a structure that is the enantiomer of the preceding olefin complex, except for the incorporation of an additional monomer unit. Thus, each inserted monomer has the opposite stereochemical relationship, and syndiotactic polypropylene is formed. Again, this concept applies to catalysts other than metallocenes. For a C_s-symmetric catalyst whose active sites are reflected by a mirror plane of symmetry, and are thus homotopic, syndiotactic polypropylene will be produced if they follow these steps for insertion in this geometry.

The structure at the right of Figure 22.8 possessing C_1 symmetry contains two binding sites that are diastereotopic and, therefore, chemically inequivalent. These two chemically distinct binding sites can be seen in Equation 22.13, and these two sites will bind different olefins with different binding constants. The two resulting complexes insert the olefin at different rates. Thus, these structures can generate polymers in which the alternating olefins that are incorporated into the chain can have different stereochemical integrities and different degrees of incorporation of comonomer. As a result, catalysts possessing a C_1 structure are capable of producing hemiisotactic polypropylenes, as well as alternating copolymers of polyethylene and polypropylene.

22.4.2. Synthesis of Stereodefined Polypropylenes

The catalyst structures presented in the previous section have been used to prepare many types of polyolefins. The following several sections of this chapter provide an overview of the synthesis of polypropylenes containing various microstructures. These sections describe a selection of the catalysts that have been used to generate isotactic, syndiotactic, hemiisotactic, and stereoblock copolymers. Subsequent sections of this chapter describe catalysts that give rise to various copolymers of ethylene and propylene. Again, representative examples are provided, and the reader should consult other sources for comprehensive coverage of the synthesis and properties of each type of copolymer.[14]

22.4.2.1. Synthesis of Isotactic and Syndiotactic Polypropylene

Natta's first propylene polymerization conducted with heterogeneous catalysts produced a mixture of amorphous atactic PP and crystalline isotactic PP that was separated by fractionation. Current heterogeneous catalysts produce high molecular weight, highly isotactic PP. As noted in Section 22.1.1, the degree of stereoregularity is often measured in terms of five consecutive monomer units (pentads) because the relative configuration of the stereocenters in these units can be distinguished by routine ^{13}C NMR spectroscopy. For example, isotactic polypropylene contains mmmm pentads, as shown in Figure 22.7. Current state-of-the-art catalysts produce polymers containing greater than 99% (0.99) mmmm pentads. This leads to a melting point of the polymer that is around 165 °C.[78] This melting point is higher than that of polyethylene (135 °C for HDPE, 110–120 °C for LDPE). Therefore, isotactic PP is a better material than polyethylene for high-temperature applications.

Early homogeneous metallocene catalysts produced polypropylene that was either atactic or modestly isotactic. This situation changed when Ewen and Brintzinger reported the first stereoregular propylene polymerizations by conducting reactions with *ansa*-metallocene systems in which two indenyl groups are linked by a bridge (Figure 22.9).[79,80] Such *ansa*-metallocenes were presented in Chapter 3. The linker between the two indenyl ligands restricts ligand rotation and enforces a C_2 or C_s orientation of the indenyl groups. The meso C_s isomer of the catalyst in Figure 22.9 produced atactic polypropylene, while the racemic, C_2-symmetric isomer produced isotactic polypropylene ([mmmm] = 0.78).

Ewen later demonstrated that the zirconocene catalyst in Figure 22.9 containing a ligand in which one cyclopentadienyl and one fluorenyl ligand are connected by an *ansa*-bridge produces polypropylene that is predominantly syndiotactic. This material contained 86% [rrrr] pentads.[81] This syndiotactic polypropylene is less crystalline and melts at a lower temperature (145 °C) than isotactic PP, and current industrial production is limited. Nevertheless, heterogeneous catalysts had not been able to produce syndiotactic polypropylene, and the synthesis of this new form of polypropylene constituted a major breakthrough in the design of homogeneous catalysts for polyolefin synthesis.

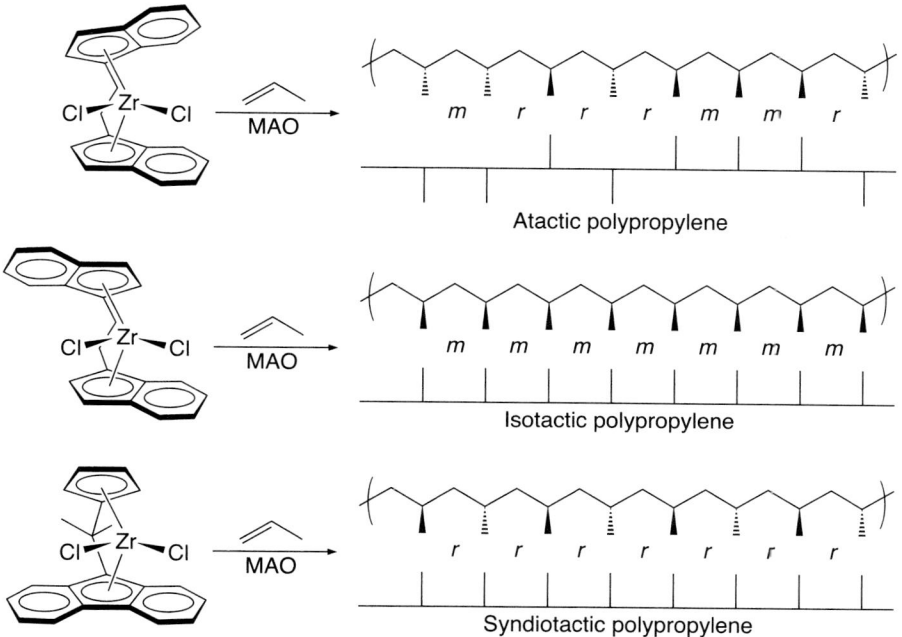

Figure 22.9.
Ansa-metallocene catalysts for the formation of atactic, isotactic, and syndiotactic polypropylene.

Unlike the bridged indenyl catalysts, the zirconocene fragment in this linked Cp-indenyl system contains a plane of symmetry along the M-centroid axes. When an alkyl group and an olefin are bound to the metal in the active form of the catalyst, this system possesses C_1 symmetry. This symmetry causes the catalyst to bind to opposite faces of the olefin prior to each of the consecutive insertions. Therefore, the insertions occur with alternating stereochemistry and syndiotactic polymer is formed.[7]

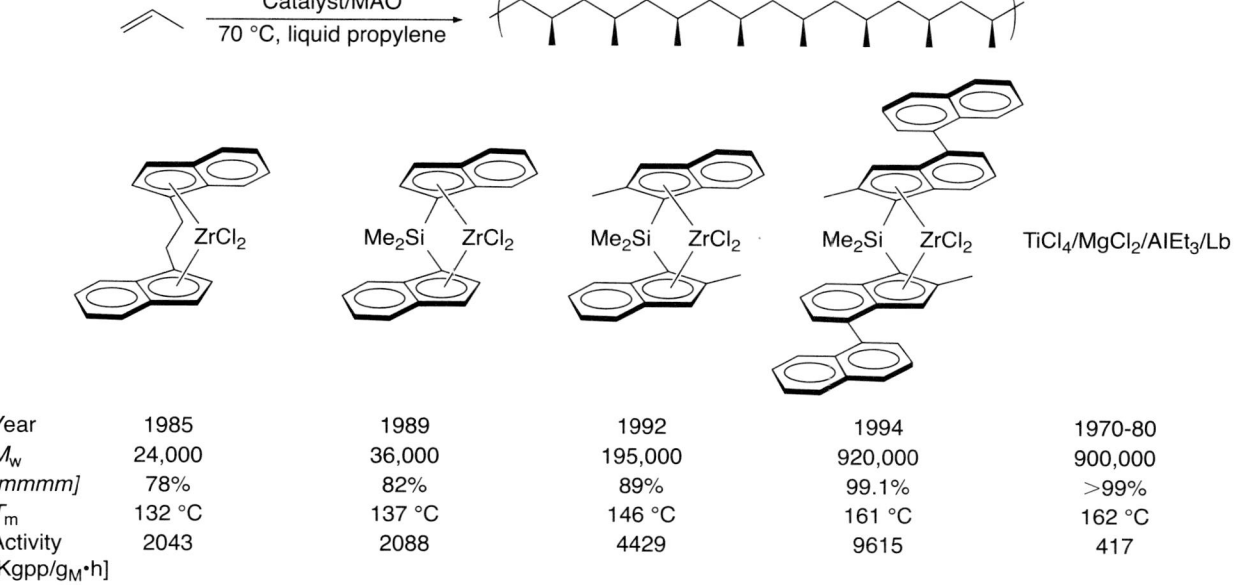

Year	1985	1989	1992	1994	1970-80
M_w	24,000	36,000	195,000	920,000	900,000
[mmmm]	78%	82%	89%	99.1%	>99%
T_m	132 °C	137 °C	146 °C	161 °C	162 °C
Activity [Kgpp/g$_M$·h]	2043	2088	4429	9615	417

Figure 22.10. Advances in isospecific metallocene catalysts.

Since these early discoveries, the relationship between the structure of ligands in metallocenes and the activity and selectivity of the resulting catalysts has been studied intensively.[7,82,83] Figure 22.10 shows a progression from the original, modestly selective systems studied by Brintzinger, to the more elaborate catalysts prepared at Hoechst.[10,84] Current isospecific metallocene catalysts react with rates and selectivities that equal those of heterogeneous α-TiCl$_3$ catalysts. Some of the most active and selective catalysts include those containing strapped bisindenyl ligands bearing bulky substituents on the benzo-fused ring of the indenyl groups. Because the industrial production of polypropylene is conducted at high temperatures due to the exothermocity of the process, catalysts that form polypropylene with high isotacticity at high temperatures are important. The naphthyl-substituted *ansa*-metallocene system in Figure 22.10 forms polypropylene with high isotacticity under these conditions.

22.4.2.2. Synthesis of Hemiisotactic Polypropylene

As noted in Section 22.1.1, hemiisotactic polypropylene is a polymer in which every other methyl group off the main chain has the same stereochemical configuration, and the intervening methyl groups have a random stereochemical configuration. The NMR spectrum of hemiisotactic PP matches that of the material formed independently by hydrogenation of poly(*trans*-2-methylpentadiene).[85]

The development of catalysts that form hemiisotactic polypropylene is a striking example of organometallic catalyst design. Because of the two types of stereocenters along the polymer chain, hemiisotactic polypropylene is generated from a catalyst that possesses two different sites at which insertion occurs, one that is selective for the generation of a single stereochemical configuration, and one that leads to nearly random configuration. Ewen developed the metallocene catalyst in Equation 22.14 in which the presence of a substituent on the cyclopentadienyl ring causes the metallocene fragment to possess C_1 symmetry and, therefore, to possess two distinct coordination sites. The presence of two distinct coordination sites, in combination with an exchange of the growing polymer chain from one site to the other, gives rise to hemiisotactic polypropylene.[86,87] Alternation of olefin insertion at an isospecific and an aspecific site produces the observed hemiisotactic PP. Stereoerrors appear to be the result of exchange of the polymer chain from one site to the other (back-skipping)[88] faster than insertion.[89–91] Since this work of Ewen, other catalysts for hemiisospecific polymerization have been reported.[92–94]

Hemiisotactic PP

(22.14)

22.4.2.3. Synthesis of Stereoblock Polypropylenes

In addition to polypropylenes in which the entire polymer chain consists of isotactic, syndiotactic, hemiisotactic, or atactic chains, polypropylenes in which the chain consists of alternating blocks of two microstructures have been prepared.[95] Perhaps the most interesting and useful of these stereoblock polymers consists of the combination of a crystalline block, such as a unit of isotactic or syndiotactic polypropylene, and an amorphous block, such as atactic polypropylene. Polymers containing this combination of microstructures often behave as a thermoplastic elastomer and have properties similar to natural rubber.[96] One significant advantage of the thermoplastic elastomers from polypropylene is that they can be melted and reprocessed, whereas natural rubber cannot.

Stereoblock polypropylene has been formed by reactions occurring through several different mechanisms with catalysts of varied structures. In one general case, the stereoblock polymer is generated by a mechanism in which polymerization occurs while changing the polymerization site within a catalyst. In this case, the catalyst contains at least two different sites for polymerization, one site that generates stereodefined polypropylene and one that generates atactic polypropylene. More specifically, the catalyst can contain two sites within the same coordination sphere, or it can contain one structure that interconverts with another, perhaps by coordination of the Lewis acid co-catalyst. In a second case, the stereoblock polymer is generated by a mechanism in which polymerization occurs by exchange of the growing polymer chain from a catalyst that generates stereodefined polypropylene to one that generates atactic polypropylene. In a third general case, the stereoblock polymer is generated by a living catalyst, and conditions, such as temperature, are changed to alter the stereospecificity of the catalyst. These stereoblock copolymers can be produced by both heterogeneous catalysts and homogeneous catalysts, and they have been produced by homogeneous catalysts possessing both metallocene and non-metallocene structures.

22.4.2.3.1. Isotactic–Atactic Stereoblock Polypropylene Generated from Heterogeneous Catalysts

The polypropylene originally synthesized by Natta using a heterogeneous titanium catalyst contained a small fraction of elastomeric polymer.[97] More recent heterogeneous systems have been developed that produce better yields of this elastomeric material. Because of the lack of detailed understanding of the structures of these systems and mechanisms by which they react, heterogeneous catalysts for the formation of stereoblock polypropylene are discussed only briefly. In general, these catalysts consist of tetraalkyl group 4 complexes containing bulky alkyl groups,[98,99] or bis(arene) complexes of group 4 metals,[100,101] both on an alumina support. In this system, the alternating sequences of isotactic and atactic PP are likely formed by chain transfer between different catalytic sites.[102] More recently, elastomeric PP was obtained using a titanium tetrachloride catalyst supported on magnesium chloride, with dibutylphthalate and aromatic ether additives.[103] Predominantly syndiotactic elastomers have been synthesized from a complicated catalyst mixture that includes a dialkoxymagnesium compound, a titanium tetrahalide, an organoaluminum activator, and an aromatic heterocycle.[104]

22.4.2.3.2. Isotactic–Atactic Stereoblock Polypropylene Generated from Homogeneous Catalysts

Chien and co-workers first reported the synthesis of elastomeric polypropylene with a homogeneous catalyst. These polymers were prepared with the unsymmetrical titanocenes at the left of Figure 22.11 activated with MAO at 25 or 50 °C.[105,106] These catalysts were relatively unstable and short-lived, and they produced polymer with a low molecular weight. However, they did produce elastomeric material with an [mmmm] pentad range of 0.30–0.40 and much narrower molecular weight distributions than heterogeneous systems ($M_w/M_n = 1.7$–1.9).

Collins and co-workers later developed a series of related C_1-symmetric hafnocene and zirconocene catalysts shown in Figure 22.11 that formed elastomeric polypropylene possessing narrow molecular weight distributions ($M_w/M_n = 1.7$–2.1) when activated with MAO at 25 °C.[107–109] The catalysts containing a silicon bridge produced relatively high

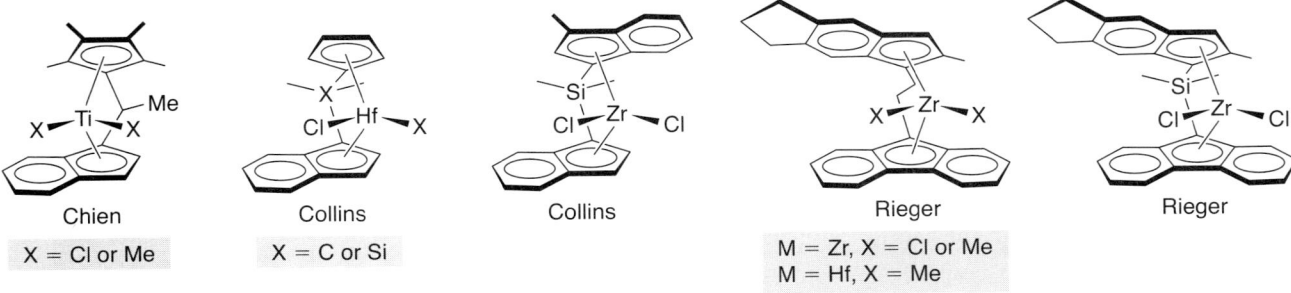

Chien

X = Cl or Me

Collins

X = C or Si

Collins

Rieger

M = Zr, X = Cl or Me
M = Hf, X = Me

Rieger

Figure 22.11. Unsymmetrical catalysts that form isotactic–atactic stereoblock polypropylene.

molecular weight polypropylene with [mmmm] ≈ 0.50 and elastomeric properties.[109] The bridged indenyl–fluorenyl metallocenes in Figure 22.11 containing a 5,6-fused cyclopentane ring fused to the indenyl ring were shown by Rieger to form elastomeric stereoblock polypropylene.[110–112] Additional catalysts that produce stereoblock copolymers have been described in the patent literature.[113,114]

Waymouth and Coates devised a strategy using unbridged metallocenes to form isotactic–atactic stereoblocks by allowing the catalyst to adopt two stereoisomeric conformations.[115,116] Without the bridge, the indenyl ligands in this catalyst can freely rotate between rac and meso structures. Coates and Waymouth envisioned that the C_2-symmetric rotamer could produce isotactic blocks, and the meso rotamer could produce atactic blocks. The structure of the unbridged metallocene in Figure 22.12 determined by X-ray diffraction contained both isomers in the unit cell. This unbridged metallocene was shown to form elastomeric polypropylene containing isotactic and atactic segments.[117,118]

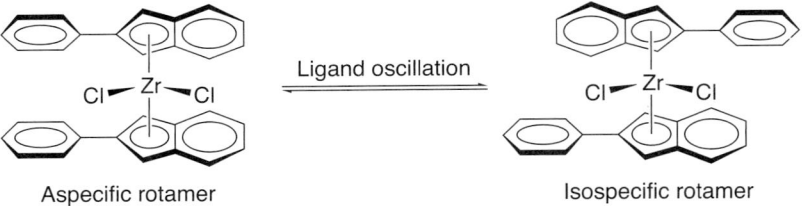

Ligand oscillation

Aspecific rotamer

Isospecific rotamer

Figure 22.12.
Oscillating metallocene catalyst for the formation of isotactic–atactic stereoblock polypropylene.

Busico and co-workers proposed that the two sites needed to generate stereoblocks can be generated by a change in the position of the counterion.[119,120] This conclusion was based on an analysis of stereoerrors and the absolute configuration of isotactic blocks produced by catalyst related to the ones studied by Coates and Waymouth. Busico suggests that the ion pair in this system locks the catalyst into the C_2 conformation, and isotactic polypropylene is formed in this state. Dissociation of the ion to generate a more separated ion pair would then allow the ligands to rotate freely, and atactic polypropylene would be formed in this state. By this theory, the event that controls the formation of atactic versus isotactic chain growth is a change in the type of ion pair, rather than rotation of the indenyl group.

In either case, the polymerization behavior of this catalyst would be expected to be sensitive to propylene pressure and reaction temperature. Indeed, higher pressures and lower temperatures favor formation of isotactic pentads. The effect of ligand structure on catalyst behavior and polymer properties for 2-arylindene metallocene catalysts has been extensively studied.[121,122]

22.4.2.3.3. Stereoblock Copolymers by Alternation of the Ligand Sphere

Stereoblock copolymers can also be formed if the catalyst alternates between two structures that contain different ligands. Eisen and co-workers have recently found that the octahedral, C_2-symmetric catalyst in Figure 22.13 forms isotactic–atactic stereoblock polypropylene when activated with MAO.[123] In this case, stereoblock formation is thought to result from a change to the identity of the ligands, rather than to a change in ligand conformation or a change in the site of the growing polymer chain. NMR studies of the combination of this complex and MAO indicate that the pyridine nitrogen donor can shift from coordination to the transition metal to coordination to an MAO counterion. As a result, the titanium can interconvert between two different species to generate two different polymer segments possessing different stereospecificity.

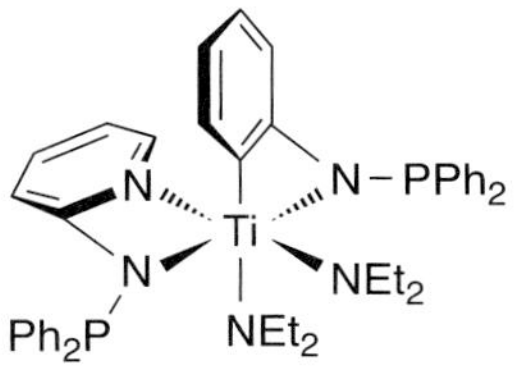

Figure 22.13.
A catalyst that forms stereoblock copolymers by reversible changes to the ligand structure.

22.4.2.3.4. Stereoblock Copolymers by Chain Transfer

In addition to the formation of stereoblock polypropylenes by exchange of the polymer chain from one site to another within one catalyst, stereoblock polypropylenes have been generated by the exchange of the polymer chain from one catalyst to another. Thus, mixtures of two different catalysts can be used to form stereoblock polypropylene if polymer chains can be exchanged between two different metal centers. Often trialkylaluminum additives are used as the chain transfer agents. The polymer chain transfers from the transition metal catalyst to aluminum to a second transition metal catalyst.

For example, Chien and co-workers combined the isospecific polymerization catalyst containing the bisindenyl ligand in Figure 22.14, the aspecific polymerization catalyst containing the bisfluorenyl ligand, the trialkylaluminun additive $^{i}Bu_3Al$, and the Lewis acid activator $[Ph_3C][B(C_6F_5)_4]$. This system[124] and related ones[125,126] formed a polymer mixture containing about 10% isotactic–atactic stereoblock PP. However, the majority of the product consisted of atactic and isotactic homopolymers.

Figure 22.14.
Ansa-metallocenes used as part of binary catalyst mixtures for the formation of stereoblock polypropylene.

22.4.2.3.5. Stereoblock Copolymers from Living Catalysts

Finally, stereoblock polypropylene has been prepared with living catalysts. In one case, a change in reaction medium was used to alter the stereospecificity. Shiono and co-workers synthesized syndiotactic–atactic diblock polypropylene using the living half-metallocene catalyst precursor in Equation 22.15.[127] In heptane solvent at 0 °C, the combination of this complex and modified MAO (MMAO) produces moderately syndiotactic polypropylene ([rr] = 0.72), and in chlorobenzene it produces atactic polypropylene ([rr] = 0.43). Thus, adding chlorobenzene to an ongoing polymerization in heptane converts the relative stereochemistry of the inserted monomers from syndiotactic to atactic, and a diblock polymer is formed.

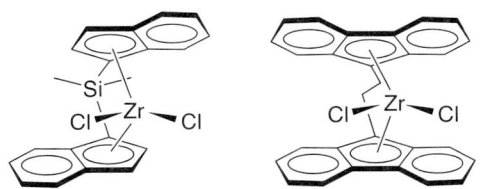

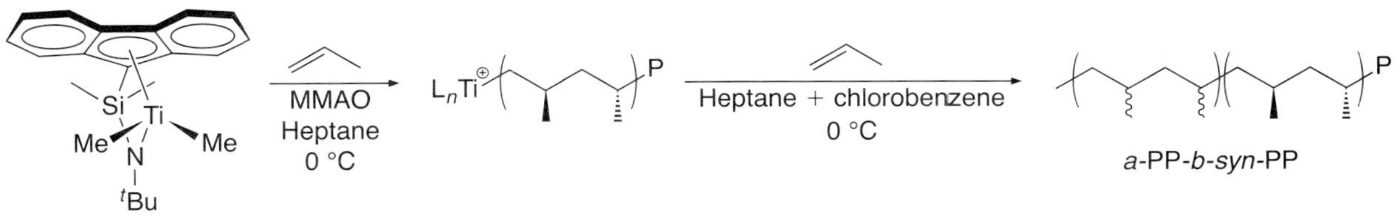

a-PP-b-syn-PP

(22.15)

22.5. Hyperbranched Polypropylenes

Like the polyethylene produced from nickel and palladium complexes coordinated by α-diimine ligands discussed earlier in this chapter, polymers of α-olefins produced by late transition metal catalysts coordinated by α-diimine ligands are much different from those formed by early-metal catalysts. These polymers generated from catalysts containing late transition metals are highly branched because of rapid β-hydrogen elimination and re-insertion of the resulting coordinated olefin.[70,128–130]

This rapid β-hydrogen elimination and re-insertion, along with less regioselective insertion, gives rise to several differences between the structural properties of these poly(α-olefins) and of those produced by early metals. First, the late metal catalysts produce polymers of α-olefins containing enchained units that are longer than the two carbons of the olefin and result from 2,ω– and 1,ω–enchainment (ω refers to the opposite chain end). These processes have been called "chain straightening."

The mechanism for 2,ω-enchainment is shown in Scheme 22.4. After insertion of the olefin, a series of β-hydrogen eliminations and re-insertions causes the metal to walk down the side-chain of the inserted α-olefin to its terminus, where insertion of the next olefin occurs. This sequence of β-hydrogen eliminations and re-insertions has been called "chain running." The mechanism for 1,ω-enchainment is shown in Scheme 22.5. In this case, the process is initiated by 2,1-insertion to place the metal at the internal carbon of the olefin. Studies of the structure of the polymer produced from α-olefins reveals the percentage of 1,2- versus 2,1-insertions, and these studies indicate that these two types of insertions occur with nearly equal rates. Thus, unlike the Ziegler catalysts for propylene polymerization, which enchain monomer predominantly by 1,2-insertions, the late metal catalysts incorporate monomer by a combination of 1,2- and 2,1-insertions.

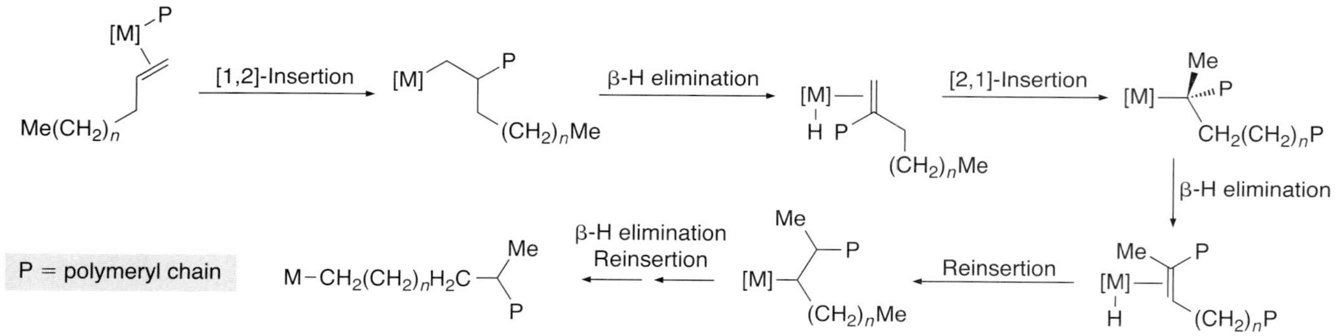

Scheme 22.4

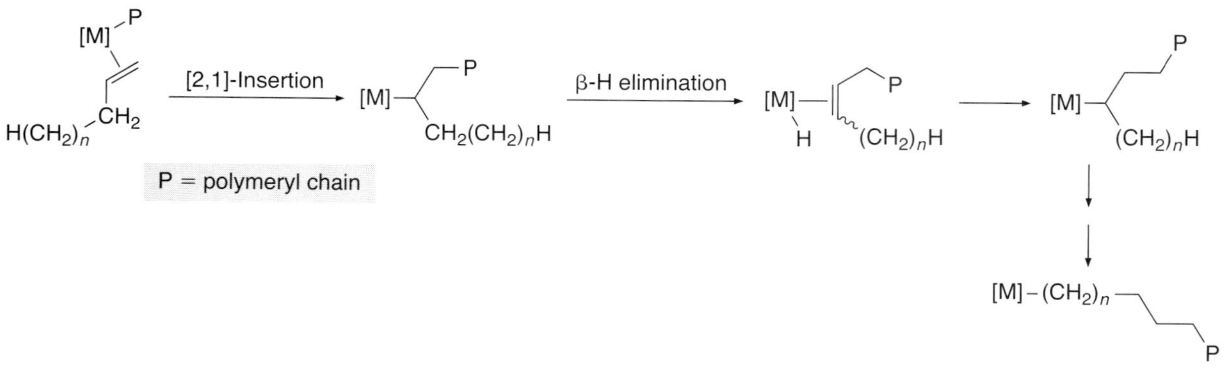

Scheme 22.5

Second, these catalysts give rise to polymers of α-olefins that possess branches on branches. These branches are similar to those formed from the polymerization of ethylene by the same catalysts. In both cases, these branches on branches are formed by chain-running events that allow insertions to occur at sites other than the M–C bond generated initially by olefin insertion.

22.6. Ethylene–α-Olefin Copolymers (Written with Prof. Geoffrey W. Coates and Prof. Gregory J. Domski)

As discussed in the section on ethylene polymerization, the copolymerization of ethylene with α-olefins generates a polymer called linear low-density polyethylene (LLDPE). LLDPE is less crystalline than strictly linear polypropylene and possesses a lower melting point. Thus, controlled synthesis of materials possessing varying melting points can be conducted by modulating the amount of α-olefin incorporated into the polymer chain. Much of this material is produced from a combination of ethylene and hexene or octene, but copolymers that are elastomeric are also produced from ethylene and propylene. In addition, catalysts that have sufficiently accessible coordination sites to re-incorporate a large olefin-terminated polymer chain can lead to LLDPE that contains long-chain branches. As noted in Section 22.1.1, these long-chain branches improve the toughness of the material.

The synthesis of copolymers of ethylene and α-olefins is one of the important commercial advances made possible by the development of homogeneous olefin polymerization catalysts. The difference in reactivity between ethylene and α-olefins with many homogeneous catalysts is often one or two orders of magnitude. Thus, the controlled synthesis of copolymers comprised of these simple molecules has been challenging.

Heterogeneous catalysts tend to produce polymers containing mixtures of copolymers and homopolymers.[131–133] In contrast, the defined homogeneous catalysts containing a single site produce polymers possessing a more uniform structure, with a more random distribution of comonomer, and with molecular weights that can be more easily controlled.

The group 4 complexes containing a Cp-type ligand linked to an amido ligand, which were developed in parallel by Dow[134] and Exxon[46,47] and were based on complexes of the general structure $[Me_2Si(Me_4C_5)(^tBuN)]MX_n$ published by Bercaw (M = Sc)[135] and Okuda (M = Ti)[136] (see Figure 22.3), are some of the most advanced catalysts for the synthesis of LLDPE. In these sterically less hindered systems, the relative reactivity of α-olefins and ethylene are sufficiently small to allow formation of copolymers in which roughly 10% of the α-olefin can be incorporated.[137] Perhaps most important for industrial use, these catalysts are stable enough to be used at temperatures as high as 170 °C. These catalysts and the process that is conducted with them has been trade-named Insite™ technology, and an ethylene–octene copolymer generated by this system is a tough material that has substantial chain entanglement and is called Engage™.

Catalysts that form ethylene–propylene copolymers can be used to produce a material known as ethylene–propylene rubber (EPR).[138] If an isospecific catalyst for the formation of polypropylene incorporates a small amount of ethylene, a crystalline copolymer is formed that has a lower melting point than isotactic PP. If an aspecific catalyst is used, or if more ethylene is incorporated into the polymer, an amorphous ethylene–propylene rubber (EPR) is formed. EPR generally has a lower glass transition temperature than atactic PP, and is a useful material for low-temperature applications. Catalysts that form isotactic or syndiotactic polypropylene can also generate polymers possessing defined stereochemistry within the propylene units in an EP copolymer.

22.6.1. Alternating Ethylene–Propylene Copolymers

Random EP copolymers containing equal molar amounts of ethylene and propylene can be prepared by using a catalyst suitable for copolymerization and the appropriate ratio of ethylene and propylene. However, catalysts for defined, *alternating*, incoporation of ethylene and propylene have also been reported. Soga and Waymouth developed the metallocene catalysts in Figure 22.15 for the formation of ethylene–propylene copolymers consisting of alternating monomer units (Equation 22.16).[139] Soga used the unsubstituted indenyl–fluorenyl catalyst in Figure 22.15 activated by MAO at –40 °C to form a polymer containing 93% alternating PEP and EPE sequences.[140] Waymouth used the substituted cyclopentadienyl–fluorenyl catalyst activated by MAO to form a similar polymer containing 70% alternating triads.[141] In both cases, a large excess of propylene monomer (about 15:1 or 20:1 [propylene]:[ethylene]) was used to generate the alternating microstructure. Other derivatives of the cyclopentadienyl–fluorenyl catalyst form isotactic alternating EP copolymer,[142] and other systems produce atactic alternating EP copolymer.[139–141]

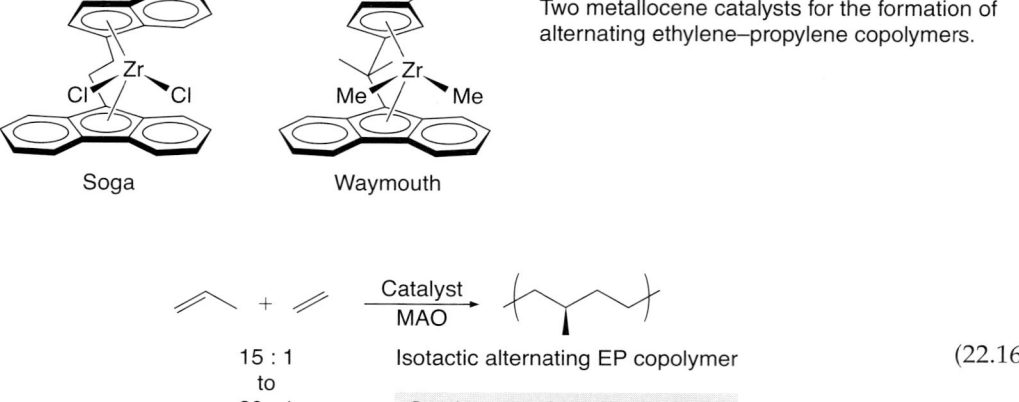

Figure 22.15.
Two metallocene catalysts for the formation of alternating ethylene–propylene copolymers.

Soga   Waymouth

$$15 : 1 \text{ to } 20 : 1$$

Isotactic alternating EP copolymer

Catalysts are from Figure 22.15

(22.16)

These catalysts are proposed to produce hemiisotactic PP by a mechanism involving two sites in the catalyst. Steric differences between the two coordination sites are thought to be the source of alternating enchainment (Equation 22.17). When the growing polymer chain is in the more open coordination site, higher propylene concentration is proposed to favor insertion to form isotactic polypropylene (analogous to hemiisotactic homopolymerization of propylene). When the growing chain is at the more crowded site, steric effects are proposed to favor insertion of ethylene over propylene. Since propylene insertion occurs at an isospecific site, an isotactic alternating polymer is formed. The small amount of contiguous propylene units results from "backskipping" of the chain from the more crowded site to the less crowded site and sequential enchainment

Insertion

Insertion

(22.17)

Aspecific site
ethylene insertion

Isospecific site
propylene insertion

of propylene units.[142] Kaminsky has analyzed these systems using statistical methods to fit experimental NMR spectroscopic data.[146]

22.6.2. Ethylene–Propylene Block Copolymers

Block copolymers made from ethylene and propylene are valuable industrial materials. They can be used as thermoplastic elastomers and as compatibilizing agents for homopolymer blends. The properties of this type of copolymer depend on the microstructure of the blocks, the relative lengths of the blocks, and the overall molecular weight. An ABA triblock copolymer structure containing crystalline A blocks and an amorphous B block can exhibit elastomeric behavior.[147] The crystalline "hard" blocks can consist of isotactic or syndiotactic polypropylene (iPP or sPP) units or linear polyethylene (PE). The amorphous "soft" blocks can consist of atactic polypropylene (aPP) or ethylene–propylene copolymer (ethylene–propylene rubber, EPR).

One challenge in producing these materials has been the production of true block copolymers, rather than a mixture of two homopolymers.[148–153] Therefore, living polymerization systems are more attractive for the synthesis of well-defined block copolymers because the long chain lifetimes allow the sequential addition of monomers. Block length can be controlled by the reaction time or monomer concentration. General depictions of the synthesis of sPP–EPR and iPP–EPR block copolymers with living catalysts are shown in Scheme 22.6.

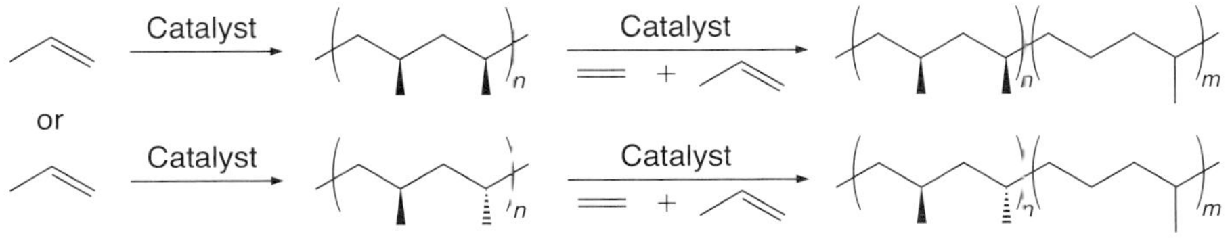

Scheme 22.6

Doi reported the first synthesis of a block copolymer of ethylene and propylene with a homogeneous, living polymerization system.[154] When activated with Et_2AlCl and anisole at $-78\ °C$, $V(acac)_3$ (acac = acetylacetonate) produced sPP–*block*–EPR-*block*–sPP triblock polymers. Turner and co-workers prepared PE–*block*–aPP–*block*–PE and PE–*block*–EPR–*block*–PE triblock copolymers using Cp_2HfMe_2 activated with $[HMe_2NC_6H_5][B(C_6F_5)_4]$ at $0\ °C$.[155] This system is not living under these conditions and probably produces a significant number of homopolymer chains, along with the triblock. At lower temperatures, simple metallocene complexes can exhibit living behavior. When activated with n-oct$_3$Al/B(C$_6$F$_5$)$_3$ at -75 or $-50\ °C$, Cp_2MMe_2 (M = Zr or Hf) forms aPP–*block*–EPR diblock copolymers with very narrow molecular weight distributions ($M_w/M_n = 1.07–1.16$).[156] All of these catalysts have low activities at such low temperatures.

Polyolefin block copolymers have been prepared at higher temperatures with non-metallocene systems. Fujita and Coates have independently developed bis(phenoxyimine) titanium complexes (Figure 22.16) for the living polymerization of propylene and ethylene.[157,158] These C_2-symmetric systems form highly syndiotactic polypropylene by a chain-end control mechanism.[159,160] Fujita and co-workers have used the mono-*tert*-butyl complex in Figure 22.16 with MAO as activator to synthesize a variety of block copolymers at $25\ °C$ possessing narrow molecular weight distributions.[158,161,162] sPP–*block*–EPR, PE–*block*–sPP, and PE–*block*–EPR diblock copolymers were formed, as well as PE–*block*–EPR–*block*–sPP and PE–*block*–EPR–*block*–PE triblock

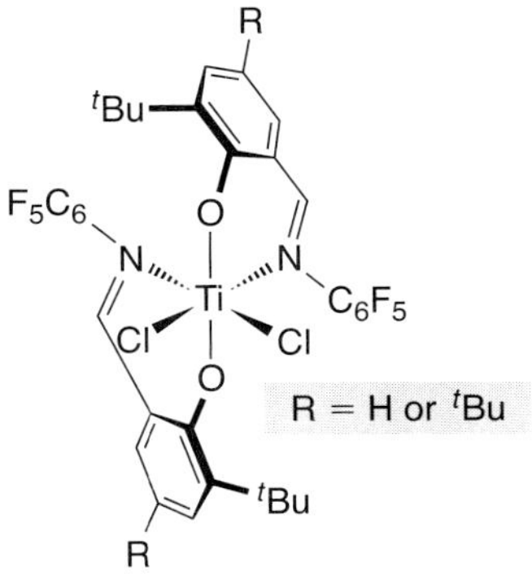

Figure 22.16.
Two non-metallocene catalysts for the formation of ethylene–propylene block copolymers.

copolymers. Coates and co-workers have used the di-*tert*-butyl complex in Figure 22.16 with MAO as activator at 0 °C to synthesize a number of sPP–*block*–EPR and sPP–*block*–PE diblock copolymers possessing varying block lengths and varying propylene content in the EPR block.[157,163] Related ligand frameworks have also given rise to living systems.[164–167]

22.7. Single-Site Catalysts for the Polymerization of Styrene (Written with Prof. Geoffrey W. Coates and Prof. Gregory J. Domski)

Most polystyrene is prepared by radical or cationic initiators.[168] These polymerizations occur with little stereocontrol (Scheme 22.7). Thus, the use of transition metal complexes as catalysts for the polymerization of styrene has offered opportunities to generate new materials with properties that differ from those prepared by traditional methods. The crystallinity imparted by the aryl group in polystyrene provides particular opportunities to generate high-melting polymers if the stereochemistry of the polymerization process could be controlled. Many metal centers can be used as radical initiators for styrene polymerization. However, a mechanism for polymerization involving coordination of the olefin and migratory insertion would be more likely to give rise to stereochemically defined polymer.

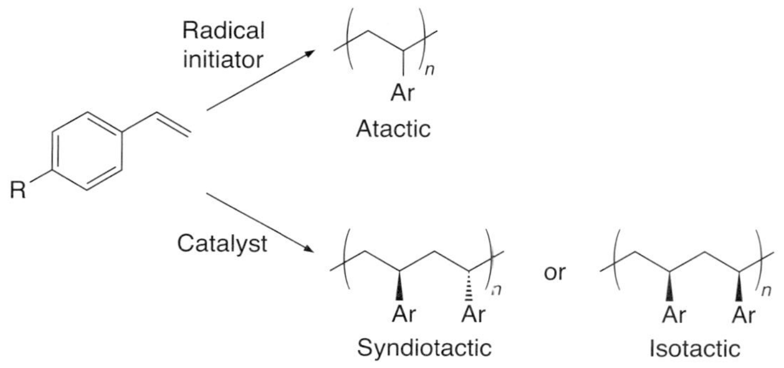

Scheme 22.7

22.7.1. Synthesis of Syndiotactic Polystyrene

The synthesis of polystyrene with a highly syndiotactic microstructure (Scheme 22.7) was achieved for the first time by Ishihara in 1986[165] with a catalyst consisting of a titanium precursor and aluminum activator. Roughly 98% of the polymer was insoluble in 2-butanone, and this material had an rrrr-pentad content greater that 98%. Many simple titanium catalysts activated by MAO were later shown to generate syndiotactic polystryrene. Tetrabenzyltitanium, when activated by MAO, catalyzes the polymerization of styrene to form a syndiotactic material with greater than 98% rr triads.[170–173] Tetrahalo and tetraalkoxo titanium compounds also form syndiotactic polystryrene when activated by MAO, but the activity of these catalysts was relatively low.

Some of the more active catalysts for the polymerization of styrene are shown in Figure 22.17. Monocyclopentadienyltitanium halides, such as $CpTiX_3$, $Cp*TiX_3$, and $CpTiX_2$, in combination with MAO, were more active catalysts than simple tetrahalides.[174,175] Of these Cp-ligated trihalide complexes, the fluoride complexes were most active,[176] followed by alkoxides[175,177] and then chlorides.[178] Moreover, cyclopentadienyl-ligated alkyltitanium compounds, such as $Cp'TiR_3$ (R = hydrocarbyl) activated by boron compounds $B(C_6F_5)_3$

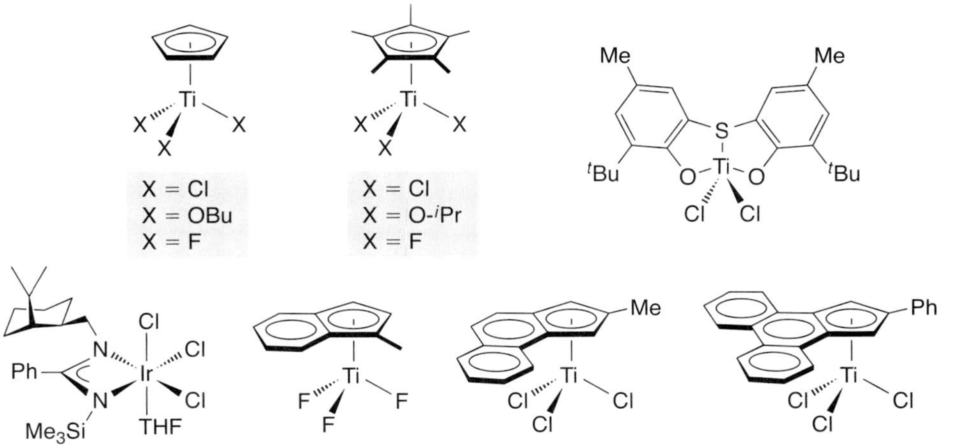

Figure 22.17. Catalysts for the synthesis of syndiotactic polystyrene.

or [PhNHMe$_2$][B(C$_6$F$_5$)$_4$] are active catalysts for styrene polymerization. Some complexes lacking cyclopentadienyl ligands also catalyzed the formation of syndiotactic polystyrene.[179–181] However, the compounds containing the sterically bulky, benzo-fused indenyl ligands form highly syndiotactic, high molecular weight polymer with the highest activities reported.[182–184] These complexes also catalyze the polymerization of styrenes containing alkyl and alkoxo substituents,[185,186] but they catalyze the polymerization of halostyrenes with lower activity. In contrast to the homoleptic alkyl, halide, and alkoxide complexes and the monocyclopentadienyl complexes, group 4 biscyclopentadienyl complexes that are highly active for ethylene and propylene polymerization form syndiotactic or even atactic polystyrene in low yields.[185,187]

Many mechanistic studies have been conducted on the polymerization of styrene, and these studies indicate that the polymerization process differs significantly from the stereoselective polymerization of propylene by group 4 metal complexes.[188] First, the active species is thought to be a cationic titanium(III) complex, rather than a titanium(IV) species. This cationic Ti(III) complex contains the Cp ligand of the precursor, along with the polymer chain, as the anionic ligands[189–191] (Equation 22.18). Studies by Pellecchia have shown that the polymerization occurs by the 2,1-insertion of styrene.[192] This regiochemistry for the insertion of vinylarenes was discussed in Chapter 9 on migratory insertions. It can result from the ability of the aryl group to stabilize the partial negative charge on the α-carbon or from stabilization through an η^3-benzyl complex. Isotopic labeling studies revealed that the polymerization of styrene by the combination of Ti(CH$_2$Ph)$_4$ and MAO occurs by syn addition of the metal–carbon bond across the olefin, implying that enchainment occurs by a migratory insertion pathway.[193] Chain transfer is thought to occur

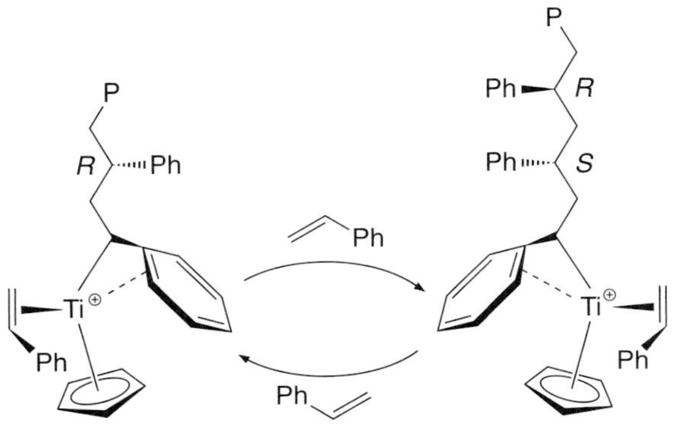

(22.18)

predominantly by β-hydrogen elimination, although chain transfer to aluminum also has been observed.[172,194]

The catalysts used to prepare syndiotactic polystyrene lack the stereochemically defined binding sites that are present in the catalysts for the polymerization of propylene. Thus, the stereochemistry is established by a chain-end control mechanism, rather than a site-control mechanism. Little is known about the structure of the transition state that leads to the site control. It is thought that the monomer could coordinate in an η^4 fashion through the olefin and a portion of the arene, and the polymer chain can be bound by an η^3-benzyl-type structure. Because the metal is highly unsaturated, many additional weak interactions are likely to lead to the observed stereochemistry. For example, interactions of the phenyl groups of the monomer and the growing polymer chain with the coordinatively unsaturated metal center are likely to be involved, giving rise to polymers with high stereoregularity.

22.7.2. Synthesis of Isotactic Polystyrene

Single-site catalysts for the synthesis of isotactic polystryrene have been more difficult to develop than single-site catalysts for the synthesis of syndiotactic polystyrene. Only a few soluble, single-site catalysts are known that form highly isotactic forms of this polymer.

Site control is needed to produce isotactic polystyrene because chain-end control typically produces syndiotactic polystyrene. Three classes of single-site catalysts are now known to produce isotactic polystyrene (Equation 22.19): phosphine-supported nickel cations,[195,196] isopropylidene-linked, Brintzinger-type zirconocenes,[197] and non-metallocene group 4 phenoxide catalysts. The mechanistic details of polymerizations by these catalysts are not known. However, it has been shown to be crucial to lock the bisphenoxide bis-thioether catalyst into a C_2 conformation by placing hindered groups on the phenoxide and to link the bisindenyl framework of the group 4 systems. The bis-thioether/bisphenoxide complexes that contain smaller substituents on the phenoxides possess more flexible geometries, as determined by dynamic NMR spectroscopy, and they give atactic polymer.

$$\text{Catalyst} = [\text{Ni(allyl)(COD)}] + \text{PCy}_3,$$

$$\text{MX}_2: \text{TiCl}_2, \text{Ti(O}^i\text{Pr)}_2, \text{Zr(CH}_2\text{Ph)}_2, \text{Hf(CH}_2\text{Ph)}_2,$$

(22.19)

22.8. Further Mechanistic Information on Alkene Polymerization

As noted in the introductory sections to this chapter, olefin polymerization occurs by catalyst initiation, chain propagation, and chain transfer. Catalyst initiation occurs by the generation of an alkyl or hydride ligand, or abstraction of one alkyl ligand from a dialkyl complex. Chain propagation occurs by olefin insertion into a metal–alkyl linkage, and chain transfer occurs by one of many paths. The following few sections provide more details on the chain propagation and chain transfer steps.

22.8.1. The Mechanism of the Chain Propagation Step

The chain propagation steps of alkene insertion consist of repetitive insertions of the olefin into the growing polymer chain. The mechanism of this olefin insertion process was discussed in detail in Chapter 9, and some of the first data in favor of a direct insertion mechanism for polymerization with early metals obtained by Watson is reiterated in Scheme 22.8.[198] These studies include spectroscopic observation of the insertions of propylene into lutetium alkyl linkages. As noted in this previous chapter, the mechanism of this insertion process is best described as a migratory insertion in which C–C bond formation occurs within an alkyl olefin complex.

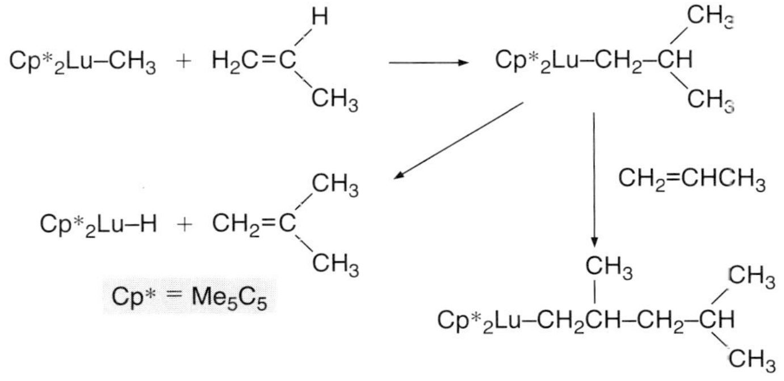

Scheme 22.8

This migratory insertion can occur with the aid of binding of the α-hydrogen of the alkyl group to the metal in an "α-agostic" interaction. As noted in Chapter 9, systems that are highly electrophilic and undergo insertion faster than dissociation of the olefin tend to show small effects from such an α-agostic interaction, most likely because olefin coordination is the first irreversible step of the overall migratory insertion process. However, olefin insertions involving less electrophilic systems that undergo insertion after reversible binding of the olefin can occur by a mechanism in which the olefin insertion assisted by an α-agostic interaction is the first irreversible (often considered the "rate-determining") step.[199]

As noted in Chapter 9, Jordan[200–204] and Casey[205–210] have generated and spectroscopically characterized alkyl and alkoxy olefin complexes of d^0 metal centers (Equations 22.20 and 22.21). Studies on the alkoxo olefin complexes that do not undergo insertion allow for equilibria to be established between the solvated species and the olefin complex.[203] In addition, studies on the alkyl olefin complexes have given rise to detailed information on the olefin coordination and insertion processes.[205,210] For example, the binding

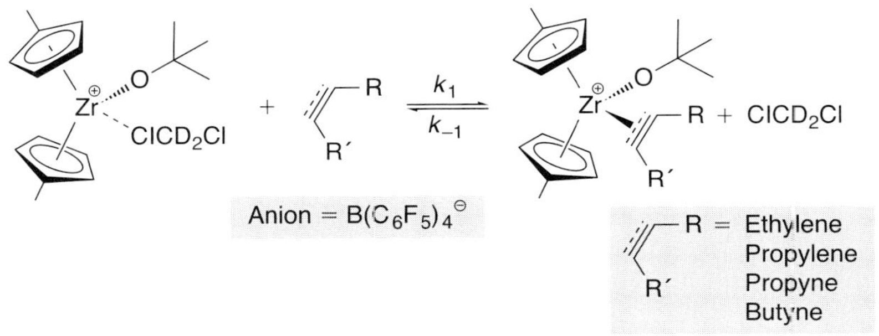

(22.20)

of propene to Cp_2YR has been shown to occur with the thermodynamic parameters $\Delta H° = 4.5$ kcal mol^{-1} and $\Delta S° = -30$ eu. The rate of insertion of propene from the bound species is fast. The rate of insertion was shown to depend significantly on the steric properties of the alkyl ligand. For example, the Cp^*_2YR complex in which the alkyl ligand possesses substituents on the β-carbon inserted olefins approximately 200 times more slowly than that containing substituents on the γ-carbon. Thus, the slower rate of propylene polymerization, relative to the rate of ethylene insertion, likely results as much or more from the difference in steric properties of the resulting alkyl chain as from the difference in steric properties of the two alkenes.

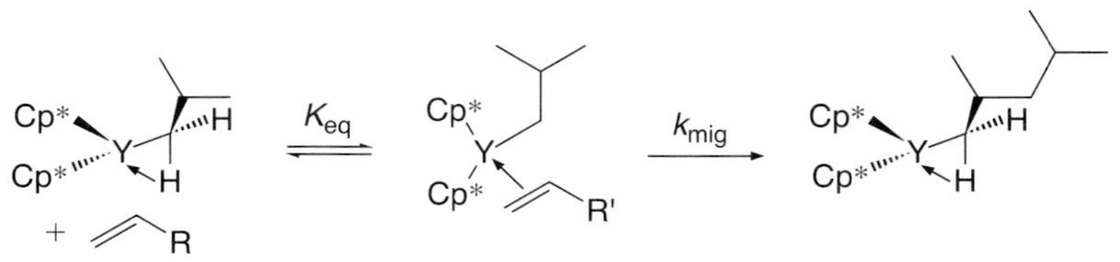

$$(22.21)$$

Landis has published studies in which ^{13}C-labeled polymerylzirconocene complexes have been used to monitor directly olefin insertion processes during the polymerizations of alkenes catalyzed by cationic zirconium complexes. The *ansa*-metallocene catalyst (EBI)ZrMe[MeB(C_6F_5)$_3$] reacts with excess 1-hexene at –40°C to form a ~3:1 mixture of propagating and uninitiated catalyst species upon complete consumption of 1-hexene.[211] Subsequent reaction of the polyhexenyl propagating species with ethene and propene provides soluble complexes in which polymeryl ligands are bound through polyethylene and polypropylene segments (Scheme 22.9).[211] The use of ^{13}C-labeled propylene leads to a species in which further reaction can be monitored by the decay of the ^{13}C NMR signal of the metal-bound carbon atom.

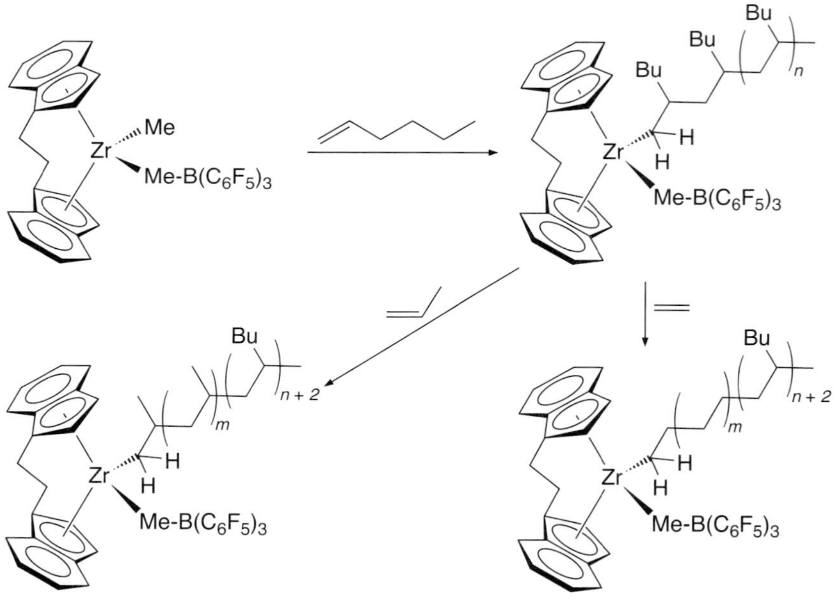

Scheme 22.9

Kinetic studies on the ^{13}C-labeled system have resolved two mechanistic issues. First, these studies addressed the question of whether a tight ion pair lies on the reaction pathway or serves as an inactive reservoir of the true catalyst by converting to a species that possesses a more loosely bound anion.[212] Data from these studies indicates that the cationic polymeryl complex associated with a borate anion through a tight ion pair does lie on the reaction pathway.

Second, data from these studies have provided insights into the relationship between epimerization of the chain end, chain termination, and isomerization.[213,214] These studies have shown that a mechanism proposed by Busico (Equation 22.22)[215] to account for chain epimerization is likely occurring. Landis' kinetic studies showed that epimerization, termination, and isomerization events occur at similar rates and are unaffected by the presence of excess 2-methyl-1-pentene (a model for the termination product).[216] Thus, epimerization, termination, and isomerization likely occur by a similar pathway, and all racemization and isomerization events must occur *without* dissociation of the coordinated alkene intermediate. Such a mechanism is shown in Equation 22.22.

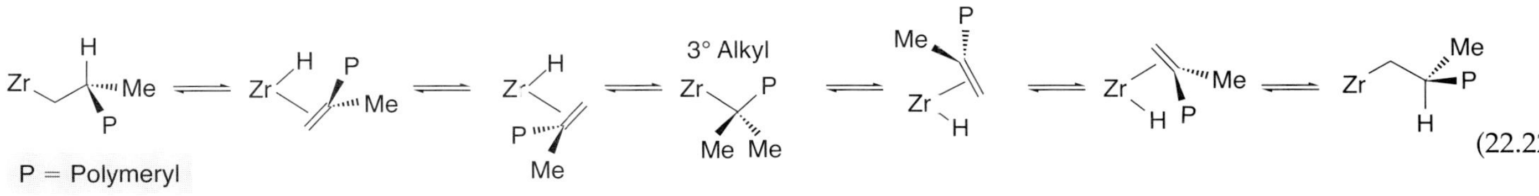

$$\text{(22.22)}$$

Also, as noted in Chapter 9, propagation during the polymerization of propylene occurs by 2,1- rather than 1,2-insertion in several systems studied recently. For example, 2,1-insertions occur during the polymerization of polypropylene catalyzed by titanium(IV) complexes of salicylaldiminato ligands (Equation 9.59 of Chapter 9).[217–222] In addition, end-group analysis shows that iron catalysts containing pyridine diimine ligands (Figure 9.3 of Chapter 9) react by 2,1-insertion.[223] Moreover, the regiochemistry of insertion during the palladium-catalyzed polymerization of α-olefins has a dramatic effect on polymer structure. This effect on polymer structure is illustrated in Scheme 9.8. By a series of insertion and elimination ("chain running") steps described in Section 22.5, the product of a 2,1-insertion of hexene transforms to a monomer unit in which the original side chain is contained in the main chain ("chain straightening"), whereas the product from 1,2-insertion, after chain running, transforms to a monomer unit containing a butyl branch or a methyl branch (Scheme 22.10).

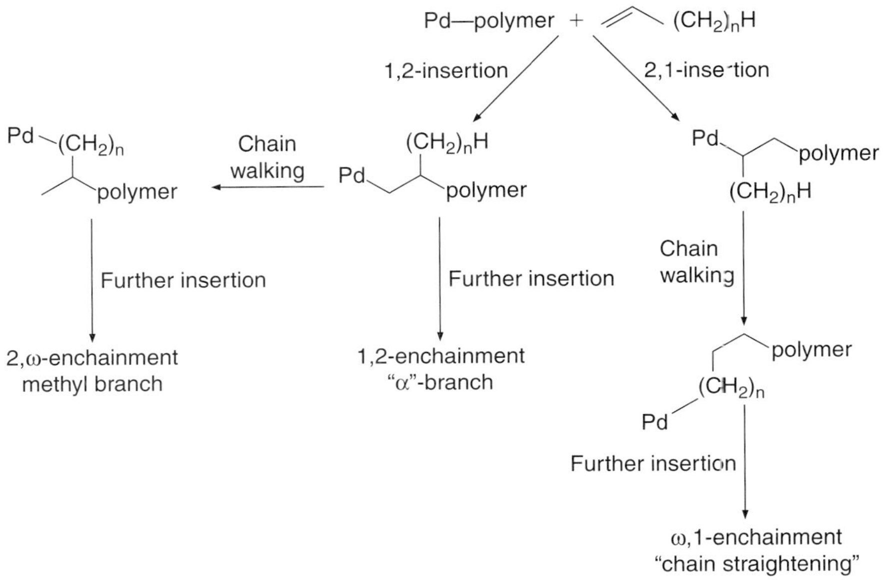

Scheme 22.10

22.8.2. Mechanism of Chain Transfer and Scope of Chain Transfer Agents

As noted in the general mechanistic description of Section 22.2, there are three general pathways for chain transfer. Chain transfer can occur by β-hydrogen elimination, followed by re-insertion of the olefin into the resulting hydride to restart the chain growth. It can occur by direct transfer of the hydride to the coordinated olefin, or it can occur by reaction of the catalysts attached to the growing polymer chain with an additional reagent (commonly H_2) that cleaves the polymer and regenerates a metal hydride. Although chain termination is often believed to occur by β-hydrogen elimination, followed by olefin insertion to restart growth of the polymer chain, substantial experimental and computational data imply that direct transfer of the β-hydrogen to monomer in a "chain transfer to monomer" process leads to chain termination.

For systems in which the resting state of the catalyst is the metal–alkyl complex, the molecular weight of the polymer will depend on the concentration of monomer if chain termination occurs by irreversible β-hydrogen elimination. This dependence is observed because the rate of β-hydrogen elimination is zero order in monomer, but the rate of chain growth is first order in monomer. Under these conditions, higher concentrations of olefin will lead to polymer with higher molecular weights. However, if chain transfer to monomer leads to termination of the polymer chain, then the rate of this process, like that of chain growth, is first order in monomer, and the molecular weight will be independent of polymer concentration. Thus, an independence of the molecular weight on monomer concentration supports a mechanism involving chain transfer to monomer, although other mechanisms for polymerization can also explain these data.[224] Such independence of molecular weight on monomer concentration has been observed.[225] In addition to these experimental data, several computational studies have suggested that direct chain transfer to monomer leads to termination of chain growth.[24, 226]

Initially, the scope of chain transfer agents was limited to hydrogen, which was used to control molecular weights. More recently, a number of additional reagents have been used that can lead to terminal-functionalized polymers. In some cases, this chain transfer occurs by addition of an X–H bond to generate an end-functionalized polymer and a metal hydride, and in other cases, addition of the X–H bond occurs to generate a catalyst containing the bound X-unit and a polymer containing hydrogen at the position formerly occupied by the metal center. Chain transfer agents that operate in these two fashions include silanes, boranes, phosphines, amines, vinylarenes, and heteroarenes.

Several groups have reported the use of silanes as as chain transfer agents. A system reported by Marks is depicted in Scheme 22.11.[227] Polymerizations catalyzed by Cp*$_2$LnR

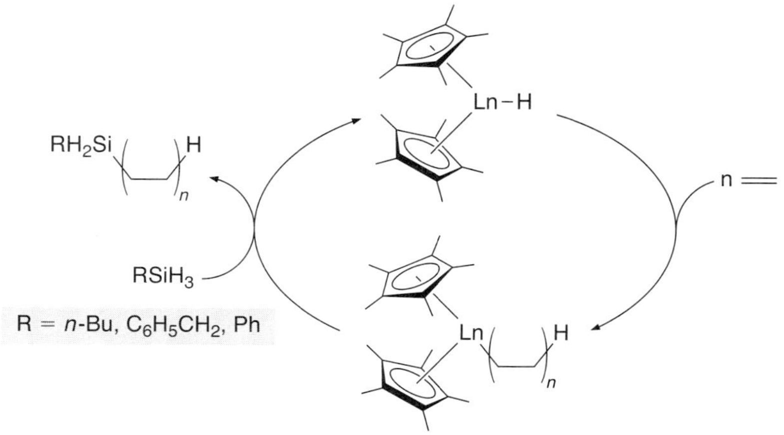

Scheme 22.11

complexes in the presence of the appropriate concentration of silane proceed with faster insertion than chain transfer and, therefore, form high molecular weight polymer. As shown in Scheme 22.11, silanes add to lanthanide catalysts with a regiochemistry that leads to an alkylsilane and metal hydride, and the metal hydride restarts polymer chain growth. Polymerizations conducted with dihydro or trihydrosilanes as the chain transfer agent formed polymers linked by the silyl group. Related studies have been reported on living ethylene polymerization catalyzed by $[Cp^*Co(L)R]^+$ in the presence of added silane to form end-capped polymers.[228] In this case, the silane added to release the polymer and form a metal–silyl species that inserts ethylene to restart the polymerization process. Chung showed that boranes can also be used as chain transfer agents.[229,230] The chain transfer process with this reagent yields a metal hydride and a functionalized polymer.

The addition of reagents containing X–H bonds in which X is more electronegative than H typically lead to addition across the M–C bond in the direction opposite to the addition of silane or borane to the early metal catalysts. Polymerization of ethylene with lanthanide catalysts in the presence of phosphines generates phosphine-terminated polymers (Scheme 22.12)[231,232] by a mechanism in which the alkyl chain is protonated, and a metal–phosphido complex is generated. This phosphido complex then inserts olefin to start the growth of a phosphine-functionalized polyolefin. Marks subsequently showed that a similar process can be conducted with amines.[233] In this case, the bulky dicyclohexylamine was needed to sufficiently retard the rate of protonation to allow chain growth. The steric bulk also makes the olefin insertion more favorable thermodynamically.

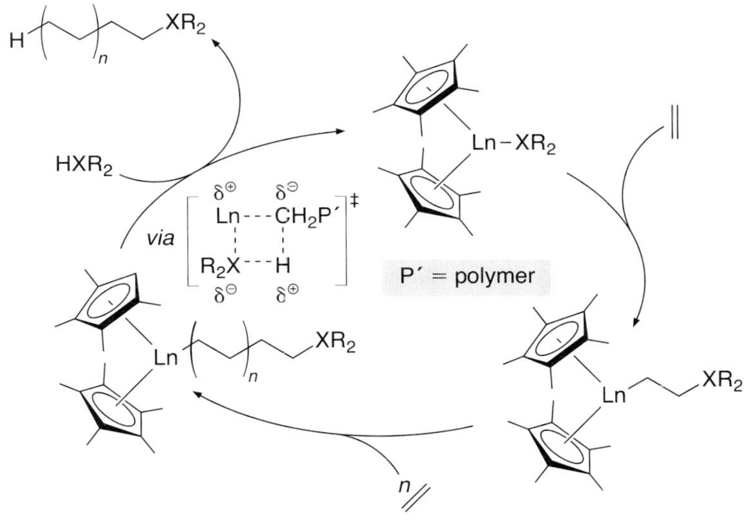

Scheme 22.12

In addition, Chung and Hessen have shown that olefins and thiophenes can be used as chain transfer agents. Chung showed that the combination of styrene and hydrogen can be used to terminate the growth of polyethylene (Scheme 22.13).[234–236] In this case, when styrene reacts, it undergoes a 2,1-insertion and generates a stable η^3-benzylic intermediate. This intermediate is stable until it undergoes hydrogenolysis to release the styrene-terminated polymer chain and form a metal hydride that restarts polymerization. This type of chain transfer has been conducted with vinylarenes containing a variety of functional groups to create polymers possessing terminal functionality. Hessen and Teuben reported chain transfer with thiophenes to form polyethylene containing a thienyl group at one terminus (Equation 22.23).[237]

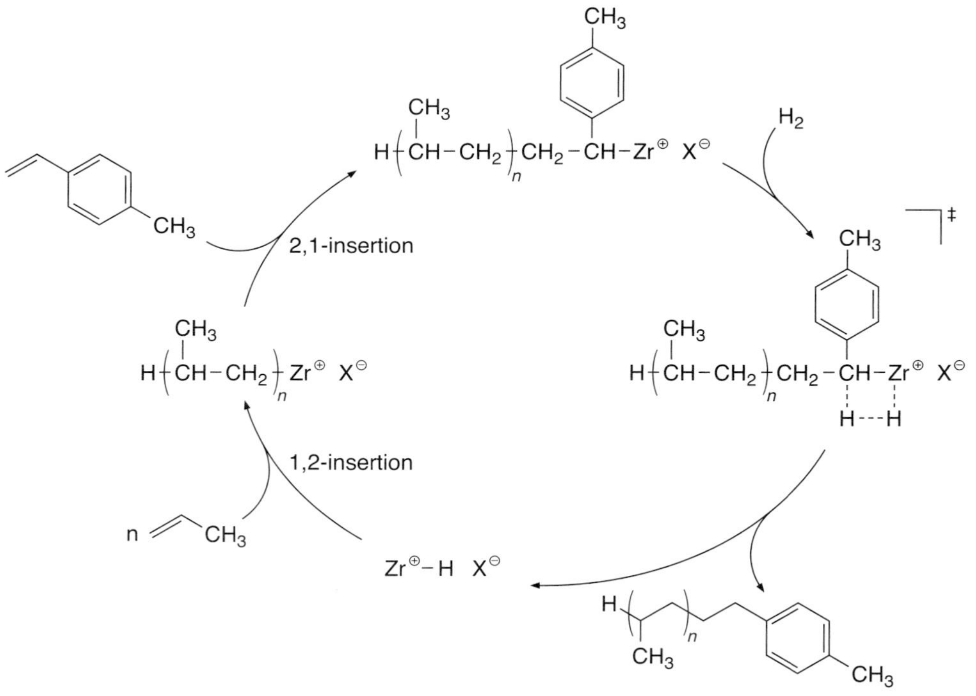

Scheme 22.13

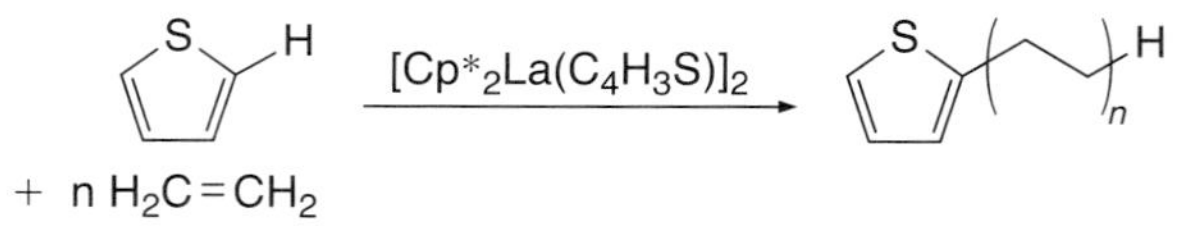

$$(22.23)$$

22.8.3. Effect of Catalyst Steric Properties on Chain Transfer

In several different systems, olefin oligomerization catalysts have been converted to olefin polymerization catalysts and vice versa through tuning of the steric properties of the ligand. Catalysts generated from sterically hindered ligands tend to react with larger relative rates for enchainment versus chain transfer, while catalysts generated from less sterically hindered ligands tend to react with smaller relative rates for enchainment versus chain transfer. As a result, the more hindered catalysts tend to form polymers while the less hindered catalysts tend to form oligomers. Among the most extensively developed of these relationships are those between the more and less hindered versions of middle and late transition metal catalysts containing diimine ligands studied by Brookhart.[238]

The steric properties of the ligand are thought to affect these relative rates because of the differences in the steric demands of olefin enchainment and chain transfer. The olefin insertion occurs by interconversion between three-coordinate alkyl complexes (containing an additional α- or β-agostic interaction) and four-coordinate olefin alkyl complexes. Chain transfer is thought to occur by either the combination of β-hydrogen elimination and associative displacement of the resulting olefin by monomer, or, more likely, by chain transfer to monomer through a transition state that closely resembles the intermediate in associative ligand substitution on an olefin hydride complex (Figure 22.18).[24,25] As a result, the catalyst in this figure containing 2,6-diisopropylphenyl groups on nitrogen forms polymer (as described in the section on ethylene polymerization to form HDPE,) while the catalyst containing 2-methylphenyl groups on nitrogen forms α-olefin oligomers from ethylene.

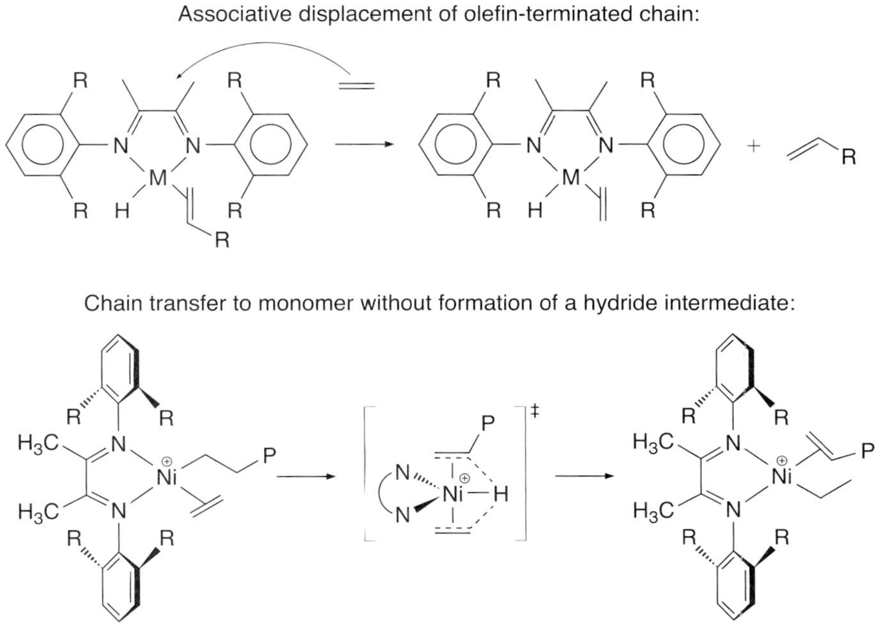

Associative displacement of olefin-terminated chain:

Chain transfer to monomer without formation of a hydride intermediate:

Figure 22.18.
Two possible mechanisms for chain transfer that involve increased steric hindrance, relative to that of the starting complex.

22.9. Oligomerization of Alkenes

The oligomerization of olefins[239-243] can form alkenes of a single length, or they can form a distribution of alkenes of various chain lengths. In the former case, dimers or trimers of olefins are the products that can typically be formed selectively. In the latter case, 1-alkenes (long-chain α-olefins) are produced that are significantly shorter than the material produced by olefin polymerization.

The distribution of olefin chain lengths produced from oligomerization depends on the relative rates for growth of the chain and termination of the chain. The distribution of molecular weights resulting from these relative rates is called a Schultz–Flory distribution[26-28] and obeys the quantitative relationship in Equation 22.24. Olefin polymerization occurs when chain transfer is much slower than chain growth. Olefin oligomerization occurs when the chain termination step occurs at rates that are close to those of chain growth. Olefin dimerization occurs when chain termination occurs after every insertion of an olefin into a metal alkyl. Olefin trimerization typically occurs by a mechanism that is distinct from dimerization, oligomerization, or polymerization. In this case, the chain length is controlled by the formation of metallacycles described later in this section.

$$K = [C_{n+2}]/[C_n] = 1/(1 + \beta)$$

in which β is the relative rates for chain transfer and propagation.

(22.24)

The relative reactivity of olefins for oligomerization follows the trend: ethylene $>>$ propene $>$ n-butene $>$ n-pentene. Only a few branched or internal olefins are readily oligomerized. The oligomerization of ethylene is conducted on the largest scale, but the oligomerization of long-chain α-olefins has also been conducted on large scale to produce certain lubricants.[242,243] The dimerization and oligomerization of dienes also has a long history, and the dimerization of butadiene has been investigated as a selective route to octane.[242, 243] The oligomerization of ethylene is discussed in the following section. The catalysts and the mechanism for the oligomerization of dienes are different from those typically used for the oligomerization of alkenes. Thus, the dimerization and oligomerization of dienes is presented in a separate section after the oligomerization of alkenes.

22.9.1. Ethylene Oligomerization

The synthesis of α-olefins from ethylene has been conducted by two general routes. By one route, triethylaluminum reacts with ethylene to form longer chain alkylaluminum compounds (Equation 22.25). On average, nine moles of ethylene are incorporated into one triethylaluminum to generate trioctylaluminum. These alkylaluminum products are then treated with ethylene at higher temperature to eliminate the olefin and regenerate triethylaluminum. This process is called "the ethyl process."[22, 244–248]

$$\text{(22.25)}$$

By a second route, ethylene is directly converted to a range of α-olefins (Equation 22.26). This process is more closely related to the polymerization of ethylene described previously in this chapter, and is described in the following section. To generate α-olefins from ethylene selectively, the oligomerization catalyst must be highly selective for reaction with ethylene over the α-olefin product.

$$n = \sim 0\text{–}18 \qquad \text{(22.26)}$$

22.9.1.1. The Shell Higher Olefin Process

The Shell Higher Olefin Process (SHOP)[249, 250] involves the oligomerization of ethylene by homogeneous nickel catalysts (Equation 22.27).[251–254] The α-olefins are separated into three molecular weight ranges by distillation: C_4–C_8, C_{10}–C_{18}, and > C_{19}. The lighter and heavier fractions are then isomerized to a mixture of internal olefins over a heterogeneous catalyst. The lighter and heavier fractions of the resulting internal olefins are then recombined over a heterogeneous olefin metathesis catalyst (MoO_3/Al_2O_3) (Equation 22.28). The resulting product consists of linear internal olefins containing 10–18 carbons. These internal olefins are then converted to linear aldehydes by hydroformylation and then reduced (Equation 22.29) using a cobalt catalyst that simultaneously isomerizes branched alkyl intermediates to terminal alkyl species prior to the carbonylation step of hydroformylation (see Chapter 17 for details). The resulting C_{11}–C_{19} primary alcohols are used as plasticizers ("fatty alcohols") and precursors to detergents. This remarkable overall process converts inexpensive ethylene into valuable α-olefins and primary alcohols without extruding any byproducts.

$$CH_2{=}CH_2 \xrightarrow{\text{"Ni–H"}} CH_2{=}CH(C_2H_4)_nH \quad
\begin{array}{ll}
C_4 - C_8 & 41\% \\
C_{10} - C_{18} & 40.5\% \\
C_{20+} & 18.5\%
\end{array} \qquad \text{(22.27)}$$

$$
\begin{array}{c}
R^1CH{=}CHR^2 \\
+ \\
R^3CH{=}CHR^4
\end{array}
\xrightarrow{MoO_3/Al_2O_3}
R^1HC{=}CHR^3 \ + \ R^2HC{=}CHR^4 \ + \ R^1HC{=}CHR^4 \ + \ R^2HC{=}CHR^3 \quad \text{(22.28)}
$$

$$C_{10}H_{21}CH=CHCH_3 \xrightarrow[CO/H_2]{HCo(CO)_4} C_{12}H_{25}CH_2\overset{\overset{\displaystyle O}{\|}}{C}-H \xrightarrow[HCo(CO)_4]{H_2} C_{13}H_{27}CH_2OH \qquad (22.29)$$

The ethylene oligomerization process (Equation 22.27) is initiated with a nickel catalyst generated from $Ni(COD)_2$ and precursors to a variety of P,O ligands. Several such systems are shown in Figure 22.19.[251–254] Some of these catalysts are generated from triaryl phosphinium salts of enolates or from phosphinoacetates, as shown in Equation 22.30. These catalysts have also been combined more recently with boranes to bind the carbonyl group and to make the nickel more electrophilic. This increased electrophilicity increases the rate of oligomerization[255,256] and has allowed the rates of nickel-catalyzed oligomerization to match the rates of titanium-catalyzed olefin polymerization. Used in tandem, the process with these two catalysts creates a route to linear low-density polyethylene from ethylene using a combination of nickel and titanium catalysts.[256] Related nickel catalysts containing monoanionic ligands based on a combination of imine and carboxylate donors have also been reported,[257] as have distinct nickel catalysts based on a relatively unhindered α-diimine ligand.[258]

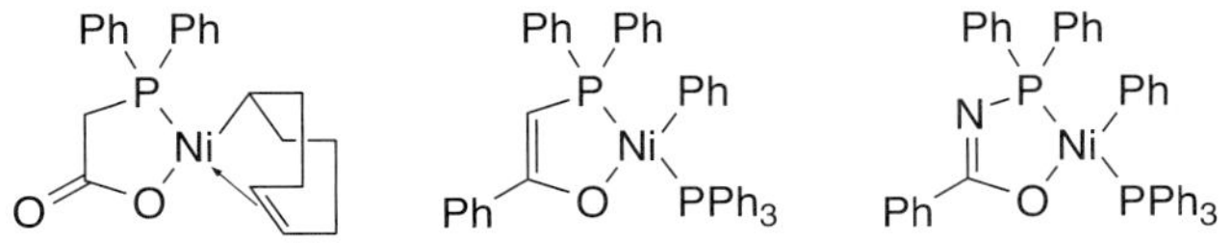

Figure 22.19.
Nickel catalysts for olefin oligomerization.

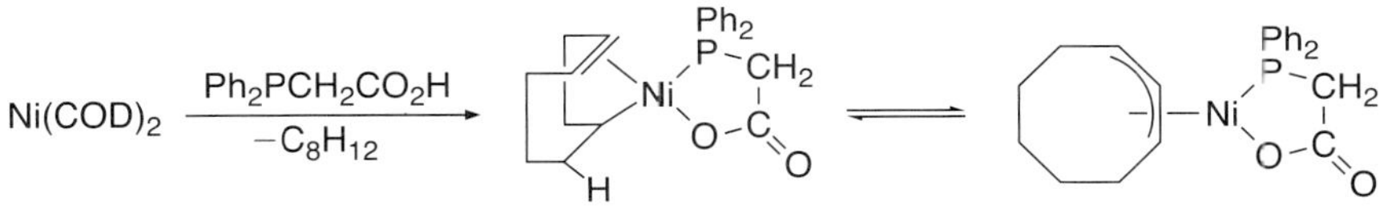

$$(22.30)$$

A plausible mechanism for this catalytic oligomerization is shown in Scheme 22.14. Chain growth occurs by olefin insertion, and chain termination occurs by β-hydride elimination or chain transfer to monomer. These mechanisms parallel the mechanisms for olefin polymerization, but the relative rate for chain growth versus chain transfer is smaller than that for polymerization. These relative rates, and therefore the average chain length, are influenced by added ligands such as tertiary phosphines.

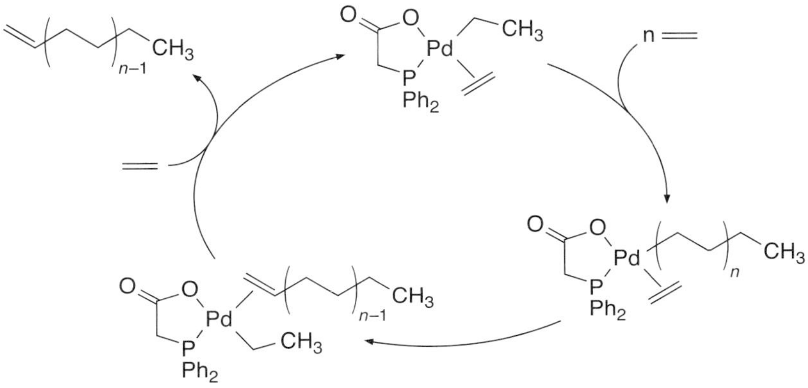

Scheme 22.14

$$\text{(22.31)}$$

22.9.1.2. Ethylene Oligomerization with Metals Other than Nickel

More recently, iron catalysts that display spectacular activity and selectivity for the oligomerization of ethylene to α-olefins have been reported (Equation 22.31).[259] These catalysts have the structure in Equation 22.31 that is related to the structure of the iron catalysts for the synthesis of strictly linear polyethylene. The polymerization catalyst is converted to an oligomerization catalyst by reducing the steric properties of the aryl groups on the imine nitrogen. The catalyst in which the nitrogen contains a 2-methylphenyl group catalyzes the oligomerization of ethylene with turnover frequencies between 1.0×10^5/h at 25 °C and 1 atm to an astonishing 1.8×10^8/h, which corresponds to 5.0×10^6 kg of oligomer/ (mol of Fe•h), at 90 °C and 600 psig. The origins of the remarkably high activity and the selectivity for enchainment of ethylene over α-olefins are not understood at this time, but studies to identify the active catalysts have begun to be conducted.[260–262]

Olefin oligomerization by group 4 metal catalysts has also been studied, and these catalysts are among the most used industrially.[263] These catalysts are typically generated from titanium alkoxides and aluminum alkyls, and are poorly defined.[264,265] Most published efforts to develop defined, single-site group 4 catalysts for the enchainment of ethylene have focused on developing catalysts for the polymerization of ethylene. However, complexes of boratabenzenes described in Chapter 3 have been shown to catalyze the oligomerization of ethylene to α-olefins. Apparently, the stronger electron donation by the boratabenzene ligand increases the rate of β-hydrogen elimination versus olefin insertion, and oligomers are formed instead of polymer.[266]

22.9.2. Olefin Dimerization by Insertion into Metal–Carbon Bonds

The dimerization of olefins can occur by two general mechanisms, one involving insertions of olefins, and the other involving a metallacyclic intermediate. The former involves the same catalytic cycle as olefin polymerization, except that the rate of chain termination by β-hydrogen elimination or β-hydrogen transfer is much greater than the rate of chain propagation. When these relative rates are observed, the process forms dimers nearly exclusively. This dimerization by coordination and insertion of the olefin is described in this section. The dimerization of olefins through metallacyclic intermediates is described in the next section. Early studies on the dimerization of ethylene have been reviewed.[239]

The dimerization of ethylene to form a mixture of butene isomers is not particularly useful in the field of commodity chemicals at this time because this mixture of butenes is usually cheaper than ethylene.[22] Selective dimerization of ethylene to 1-butene using a titanium catalyst is practiced, but this chemistry occurs through metallacycles and is described in the next section. The dimerization of propylene by migratory insertion chemistry typically produces the mixture of isomeric olefins shown in Equation 22.32. Four skeletal isomers of the intermediate metal alkyl can arise from the two different directions of M–H insertion, followed by two different modes of M–R insertion. The dimerization of ethylene is particularly fast when catalyzed by the combination of $NiBr(\eta^3\text{-}C_3H_5)(PCy_3)$ and $EtAlCl_2$; this dimerization in chlorobenzene at 25 °C occurs with turnover frequencies up to 60,000 per second.[267] The more selective dimerization of propene to 2,3-dimethylbutene is conducted on an industrial scale with titanium catalysts, again via metallacyclic intermediates described in the next section.

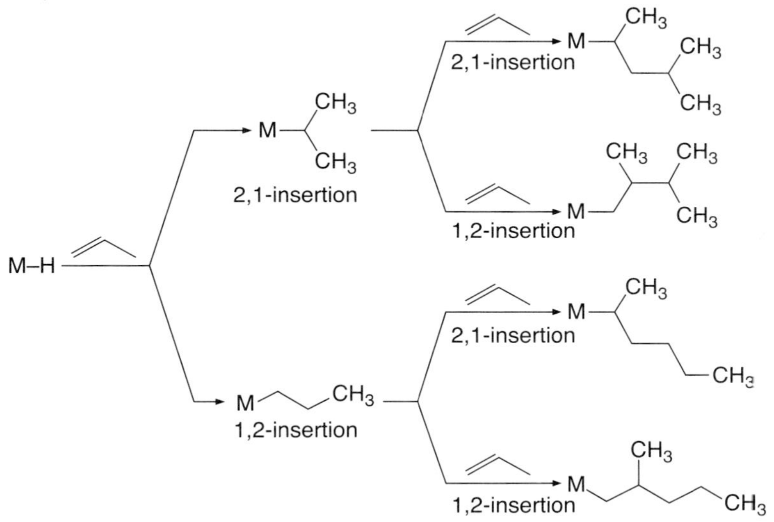

(22.32)

The catalytic dimerization of functionalized olefins has been studied intensively and occurs with high regioselectivity.[239] The dimerization of acrylates (Equation 22.33) has particular synthetic potential because the tail-to-tail dimerization can generate precursors to adipic acid, which is an intermediate in the synthesis of polyesters and polyamides, such as nylon 6,6. The coupling of acrylates to form predominantly linear dimers occurs in the presence of a variety of palladium,[268-277] ruthenium[278-281] and rhodium[275,282-284] catalysts.[270] However, the most active system reported is based on a Cp*Rh fragment containing acrylate ligands. A catalyst system generated from [Cp*Rh(ethylene)(ethyl)]+ was shown by Brookhart to be long lived for this dimerization under 1 atm of hydrogen. Total turnover numbers of up to 1.3×10^4 and full conversion of methyl acrylate were observed at room temperature to 60 °C. Selectivity for the tail-to-tail dimer was > 99%.[285,286]

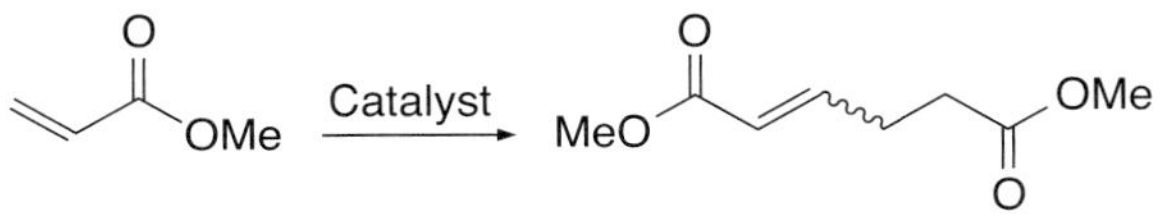

(22.33)

Scheme 22.15 shows a mechanism for this tail-to-tail dimerization of methyl acrylate catalyzed by [Cp*Rh(ethylene)(ethyl)]+.[285] Addition of acrylate converts the starting

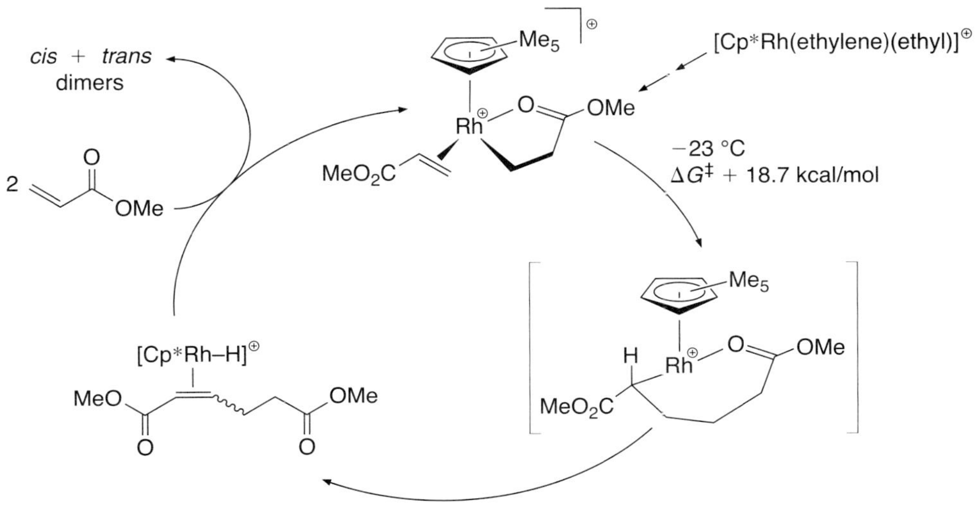

Scheme 22.15

$$\text{(norbornene)} + H_2C=CH_2 \xrightarrow{\text{Ni catalyst}} \text{(product)} \qquad (22.34)$$

$$\text{Ni catalyst} = [Ni(\eta^3\text{-}C_3H_5)Cl]_2 + Al_2Et_3Cl_3 + L^*$$

complex to the active catalyst in Scheme 22.15 containing a coordinated methyl acrylate and a homoenolate ligand. Insertion of the acrylate, followed by β-hydrogen elimination, generates a rhodium hydride that re-inserts the methyl acrylate and coordinates methyl acrylate at some stage of this sequence to regenerate the starting homoenolate complex.

The co-dimerization of ethylene with olefins has also been studied extensively and has been called hydrovinylation.[287] A seminal example, discovered by Bogdanovic and Wilke,[288,289] involved the co-dimerization of ethylene and norbornene catalyzed by a (π-allyl)nickel catalyst (Equation 22.34). This chemistry and more modern versions of these additions of one olefin C–H bond across another were presented in Chapter 16 on the hydrofunctionalization of olefins.

22.9.3. Olefin Oligomerization Through Metallacyclic Intermediates

22.9.3.1. Dimerization of Alkenes by a Metallacyclic Mechanism

Certain low-valent early transition metal complexes catalyze the dimerization of ethylene and propylene selectively to 1-butene and 2,3-dimethyl-1-butene.[290,291] The regioselectivity of this dimerization of propene signals a different mechanism than the insertion and elimination mechanism presented in the previous section. The formation of 1-butene occurs selectively because of the absence of a persistent metal hydride complex that isomerizes this olefin to the more stable 2-butene.

These regioselective dimerizations of ethylene or propylene occur through a metallacyclic intermediate (Equation 22.35).[292,293] By this mechanism, the two alkenes undergo "reductive coupling" (the alkenes are reduced and the metal is oxidized) to form a metallacyclopropane intermediate, and the dimerization of propylene occurs to generate a metallacycle containing two less-hindered primary metal-carbon bonds. The metallacycle then converts to the olefin by a two-step sequence involving β-hydride elimination, followed by reductive elimination. This process generates a metal–hydride complex as a fleeting intermediate, rather than an intermediate that can re-insert and isomerize the olefins.

$$\underset{CH_3}{\diagup\!\!\!\diagup} \xrightarrow{\text{Catalyst}} \left[\begin{array}{c} M \underset{CH_3}{\overset{CH_3}{\diagup\!\!\!\diagdown}} \end{array} \right] \longrightarrow \diagup\!\!\!\diagdown\!\!\!\diagup \qquad (22.35)$$

$$\text{Catalyst} = Zr(n\text{-}C_4H_6)_2(dmpe) \text{ or } TaCl_2(Cp)(RCH=CH_2)_2$$

22.9.3.2. Trimerization and Tetramerization of Alkenes by a Metallacyclic Mechanism

The selective trimerization and tetramerization of ethylene to form 1-hexene and 1-octene has become an important process to generate monomers for the synthesis of LLDPE. 1-Hexene was detected as a byproduct in the polymerization of ethylene catalyzed by homogeneous chromium complexes.[294] Chromium complexes have now been identified that catalyze this oligomerization to form 1-hexene with remarkably high selectivity.[295–297] Phillips patented the combination of 2,5-dimethylpyrrole, triethylaluminum, and diethylaluminum chloride with a chromium(III) salt,[298] and researchers at Union Carbide patented a catalyst generated from chromium(III) 2-ethylhexanoate and hydrolyzed triisobutylaluminum.[294,299,300] Researchers at BP have described a particularly active and selective catalyst based on a chromium(III) precursor, a bis(diphosphino)amine ligand, $(o\text{-MeO-C}_6H_4)_2PN(CH_3)P(o\text{-MeOC}_6H_4)_2$, and methylaluminoxane (MAO).[301,302]

Researchers at Sasol described chromium catalysts containing amido bisphosphine PNP[303] or amido bis-thioether SNS pincer ligands for ethylene trimerization.[304]

Most recently, studies at Sasol have revealed chromium systems containing Ph₂PNRPPh₂ ligands and very weakly coordinating anions that catalyze the tetramerization of ethylene to form 1-octene with >70% selectivity.[305–308] This work could have significant industrial importance because 1-octene is often olefin used in conjunction with ethylene for the formation of LLDPE. The catalyst for tetramerization of ethylene is generated in situ from $CrCl_3(THF)_3$ and the PNP ligand shown in Equation 22.36. Several factors were important for observing 1-octene over 1-hexene and for observing high activity for consumption of ethylene. First, more bulky groups on the phosphorus led to formation of C6 products, including 1-hexene, rather than C8 products.[306] Second, the anion must be stable.[307] Cleavage of the alkoxide from aluminum led to lower activities and lower selectivities for C8 product, perhaps from formation of chromium alkoxides. Finally, the anion must coordinate very weakly.[307] Catalysts containing more coordinating anions than the most electron-poor aluminates led to formation of C6 products. Thus, the selectivity for formation of octene over hexene appears to be sensitive to the steric properties of the metal center.

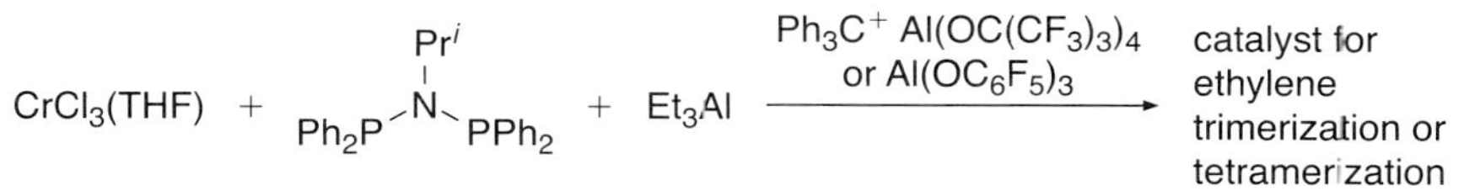

$$CrCl_3(THF) + \underset{Ph_2P \diagup N \diagdown PPh_2}{\overset{Pr^i}{|}} + Et_3Al \xrightarrow{\substack{Ph_3C^+ Al(OC(CF_3)_3)_4 \\ \text{or } Al(OC_6F_5)_3}} \begin{array}{l} \text{catalyst for} \\ \text{ethylene} \\ \text{trimerization or} \\ \text{tetramerization} \end{array} \quad (22.36)$$

anion	C8:C6 ratio
Ph₃C⁺ Al(OC(CF₃)₃)₄	3.3
for Al(OC₆F₅)₃	0.033

The mechanism of ethylene trimerization catalyzed by BP's chromium catalyst was studied by Bercaw.[309] Deuterium labeling confirmed the mechanism proposed by Briggs[295] involving metallacyclopentane and metallacycloheptane intermediates (Scheme 22.16). By this mechanism the low-valent chromium couples two ethylene molecules to form a metallacyclopentane intermediate. This metallacycle then inserts an additional ethylene to form a metallacycloheptane, which extrudes 1-hexene and regenerates the starting chromium complex. Trimers are formed because β-hydrogen elimination from the metallacyclopentane is slower than olefin insertion, and β-hydrogen elimination from the metallacycloheptane is faster than olefin insertion. The stability of the metallacyclopentane toward β-hydrogen elimination is consistent with the slow rate of β-hydrogen elimination from rigid metallacycles discussed in Chapter 10. The faster rate of β-hydrogen elimination from the metallacycloheptane, relative to ethylene insertion, is consistent with the combination of a more flexible metallacyclic structure that allows for the chromium to interact with the β-hydrogen and the rarity of nine-membered metallacycles like the one that would result from olefin insertion into the metallacycloheptane.

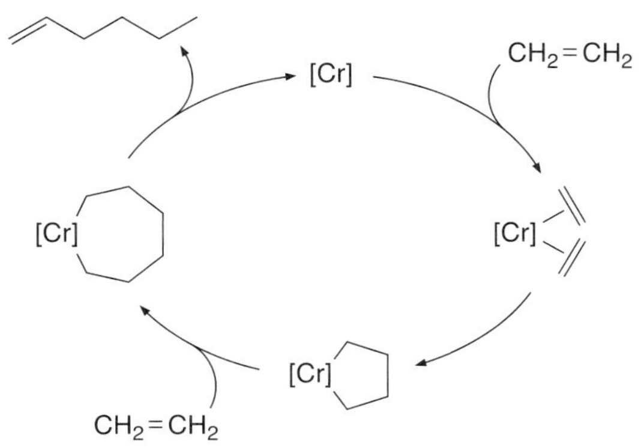

Scheme 22.16

A related mechanism for the tetramerization of ethylene has been deduced from studies on Sasol's chromium-catalyzed process that forms 1-octene.[305] The catalyst is, again, thought to operate by a Cr(I)/Cr(III) cycle such as the one in Scheme 22.17. This mechanism was supported by several studies. For example, Cr(0) complexes of the PNP ligand [Cr(CO)$_4$(PNP)] were shown to be catalytically inactive, but the complexes formed from these Cr(0) species and AgX species were found to be active.[308] Moreover, labeling studies[305] showed that the side products were formed from a metallacyclic intermediate, rather than a Cossee enchainment mechanism. This labeling study was analogous to that published by Bercaw discussed above. The octene presumably forms from an unusual nine-membered metallacyclic intermediate.

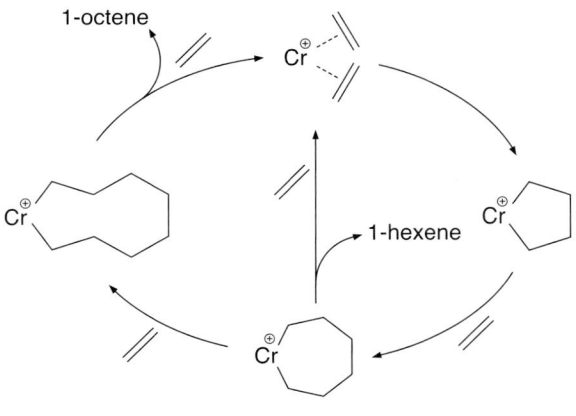

Scheme 22.17

22.10. Oligomerization and Polymerization of Conjugated Dienes [239,310]

The oligomerizations and polymerizations of conjugated dienes catalyzed by complexes of transition metals are industrially important processes for the preparation of materials ranging from synthetic rubbers (e.g., all-cis polybutadiene and polyisoprene) to raw materials for the production of certain nylons (e.g., cyclododecatriene). Catalysts based on cobalt and titanium catalysts (see above), along with catalysts based on (π-allyl)nickel and palladium complexes, have been used for diene oligomerization and polymerization. Although the early metal systems are used more extensively (at least for diene polymerization), the nickel and palladium systems are better understood mechanistically, and are, therefore, the focus of this section.

For all catalysts, the polymerizations and oligomerizations of conjugated dienes are thought to proceed through (η^3-allyl)metal complexes as intermediates. These η^3-allyl complexes can be formed from conjugated dienes in two ways. From catalyst precursors containing hydride or alkyl ligands, η^3-allyl complexes can form by insertion of the diene into the metal–hydride or metal–alkyl bond (Equation 22.37). The equilibrium between the η^3 and η^1 forms depends strongly on the nature of the other ligands and is a major factor in determining the further course of the reaction (see below).

$$\text{(22.37)}$$

Alternatively, many complexes containing low-valent metals, particularly Ni(0) and Pd(0) couple butadiene to form the corresponding bis-(η^3-allyl) complex (Equation 22.38).

$$L_nM(0) + 2 \quad \rightleftharpoons \quad \longrightarrow \qquad (22.38)$$

This process is important in many cyclooligomerization reactions of dienes described below. Further insertion reactions of these η^3-allyl intermediates with additional diene or monoene results in the oligomerization and polymerization reactions discussed in this section.

22.10.1. Polymerization of 1,3-Dienes

Ziegler-type catalysts are most commonly used for the commercial production of polybutadiene and polyisoprene. The selectivity of the process strongly depends on the identity of the transition metal catalyst (Table 22.1).[311] Although these differences in reactivity have not been rationalized, it is evident that (η^3-allyl)metal complexes are the active species.

Table 22.1. Stereoselective polymerization of butadiene.

	Structure of the polymer (%)		
Catalyst	*trans*-1,4	*cis*-1,4	1,2
$CoCl_2/AlEt_2Cl$	1	98	1
$MoO_2(OR)_2/AlEt_3$	1	3–6	92–96 (syndiotactic)
$Cr(acac)_3/AlEt_3$	1–2	0–3	97–99 (isotactic)
$VCl_3(THF)_3/AlEt_2Cl$	99	0	1

A general mechanism for the oligomerization of dienes by these high-valent early-metal systems is shown in Scheme 22.18.[312] In this mechanism, the allyl complex undergoes migratory insertion of the diene ligand to form a new allyl intermediate that re-coordinates diene. The insertion can occur from an η^1-allyl complex as shown or from an η^3-allyl complex. The cis or trans geometry of the 1,4-polybutadiene depends on the conformation of the allyl intermediate that participates in this migratory insertion. The formation of 1,2- or 1,4-polybutadiene depends on whether migratory insertion occurs at the terminal or internal carbon of the η^3-allyl ligand.

Extensive studies of the oligomerization of dienes by nickel catalysts were conducted by Wilke and co-workers.[313] More recently, Brookhart and co-workers reported the synthesis, isolation, and structural characterization of the active intermediate in nickel-catalyzed butadiene oligomerization (Equation 22.39). This species is formed from a cationic nickel–allyl species in the presence of butadiene. The coordination sphere of this complex contains one allyl and two olefin donors in a single hydrocarbon ligand that was formed from three butadiene ligands and the starting allyl group.[314]

$$(22.39)$$

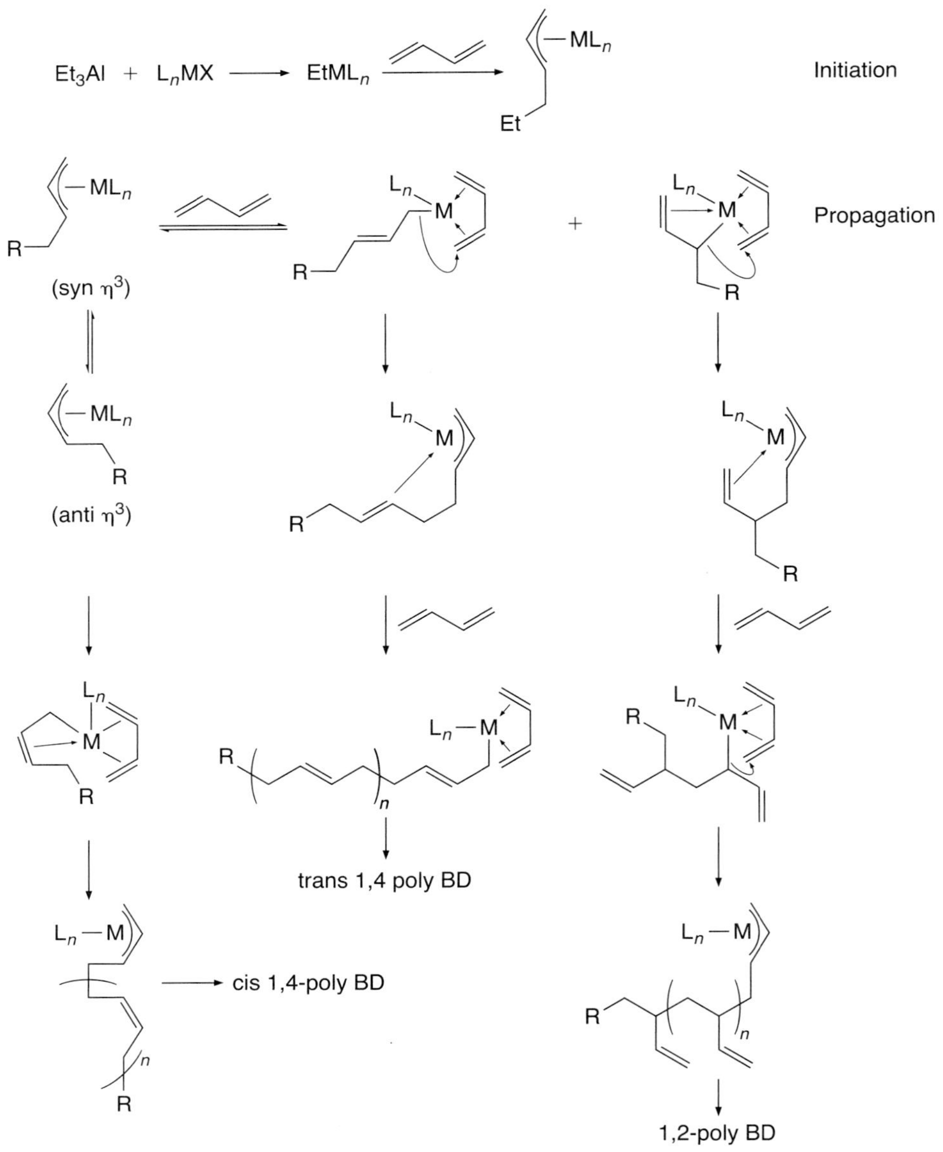

Scheme 22.18

22.10.2. Oligomerization and Telomerization of Conjugated Dienes

The catalytic dimerization and oligomerization of 1,3-dienes can lead to a large number of linear or cyclic products by mechanistically related paths. The major product depends strongly on both the metal and the auxiliary ligands, and much is known about controlling the product distribution (see below). In this section, a few mechanistically well-characterized systems are discussed. For more details, see reviews on the subject.[239]

22.10.2.1. Linear Oligomerization of Butadiene

Although the linear dimerization of butadiene has been conducted with many catalyst systems, one of the best understood and most specific systems is the nickel(0)–triethylphosphite–morpholine catalyst developed by Heimbach[315] (Scheme 22.19). Coupling of butadiene by Ni(0) produces a bis-(η^3-allyl) nickel complex. Protonation of an internal position of one of the η^3-allyl groups by morpholine, followed by proton abstraction α to the other η^3-allyl group, produces the observed octatrienes and regenerates the catalyst.

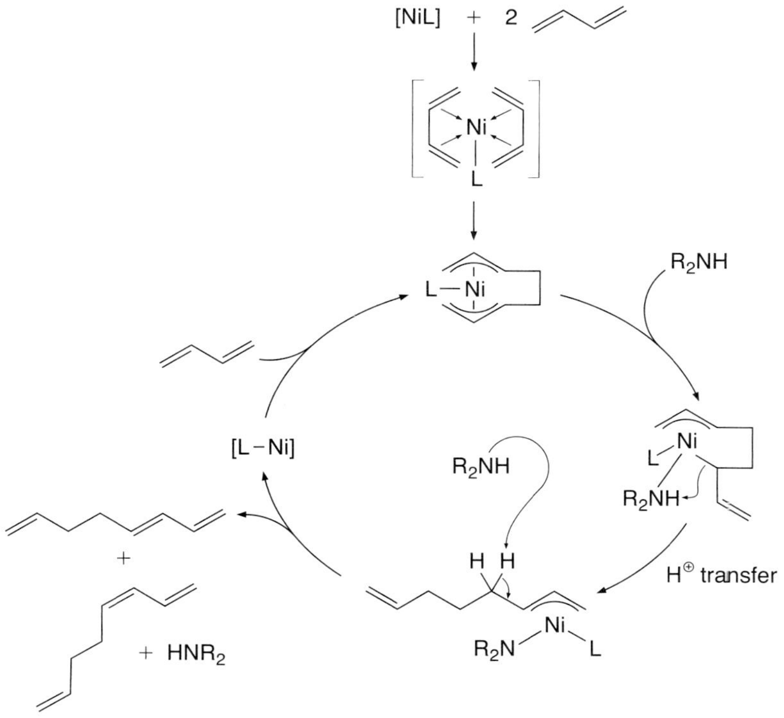

Scheme 22.19

The dimerization of butadiene has also been conducted in the presence of alcohols and amines in a fashion that incorporates these reagents into the product to produce ethers and amines containing eight-carbon chains. Such diene "telomerization" has been studied for many years[316–341] and is typically catalyzed by complexes of palladium and nickel. The challenge has been to obtain the dimeric products at the expense of longer chain oligomers and to produce linear dimers with the alkoxo or amino group at the chain end. Additional work has focused on using ammonia as the nucleophile and controlling the formation of primary versus secondary or tertiary amines. One approach to cause ammonia to be more reactive than the more nucleophilic primary and secondary amines has been to conduct the reactions in a biphasic system in which the catalyst and ammonia resides in the aqueous phase and the monoalkylation product is drawn into the organic phase.[357]

The telomerization of dienes with alcohols as the nucleophile has now been conducted in a practical fashion. A palladium catalyst containing an N-heterocyclic carbene ligand has been shown to form linear dimeric ethers selectively from alcohols and butadiene with remarkably high turnover numbers (Equation 22.40).[342–346] The activity of this catalyst is significantly higher than that of the more classical catalysts generated from $Pd(OAc)_2$ and PPh_3. Under optimized conditions involving some added carbene precursor, presumably to

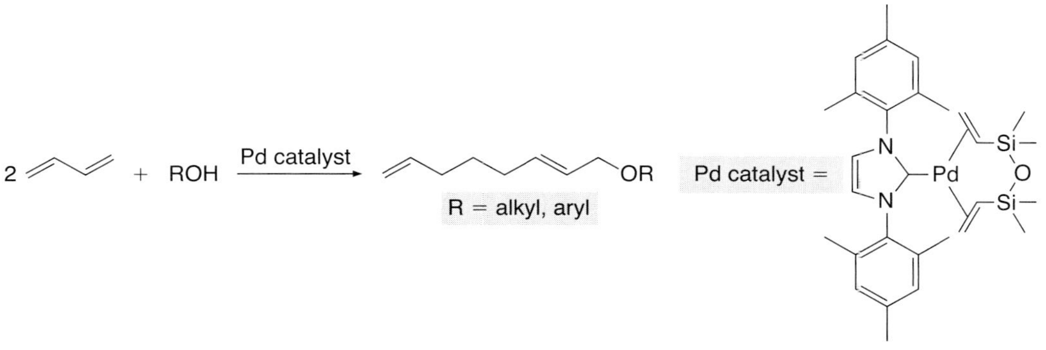

$$2 \quad + \quad ROH \quad \xrightarrow{\text{Pd catalyst}} \quad \text{OR} \qquad \text{Pd catalyst} = \qquad (22.40)$$

R = alkyl, aryl

prevent aggregation of Pd(0), turnover numbers of approximately 1.5 million are obtained for this process. Moreover, the selectivity of the carbene-ligated catalyst for formation of the octadienyl ether is higher than that with the PPh$_3$-ligated catalyst and exceeds 99:1. For reactions with other alcohols, the ratios of linear ether to branched ethers ranges from roughly 5:1 with phenols as nucleophile to between 40:1 and 100:1 with aliphatic alcohols. These telomerization reactions are important for the production of 1-octene. Hydrogenation of the dienyl ether, followed by acid-catalyzed elimination, forms octene and regenerates the alcohol nucleophile.

The telomerization reactions are thought to occur by the mechanism in Scheme 22.20. In this mechanism, the two dienes couple to form the tethered alkyl allyl complex. The isolation of this class of complex was reported by Jolly and Wilke[347–352] during mechanistic studies of diene oligomerization and later by others.[353,354] Reaction of this complex with methanol would then protonate the olefinic C-3 carbon, and the resulting methoxide would attack the terminal position of the coordinated allyl to generate the resulting diene complex. Replacement of the dienyl ether ligand by two equivalents of butadiene restarts the catalytic process.

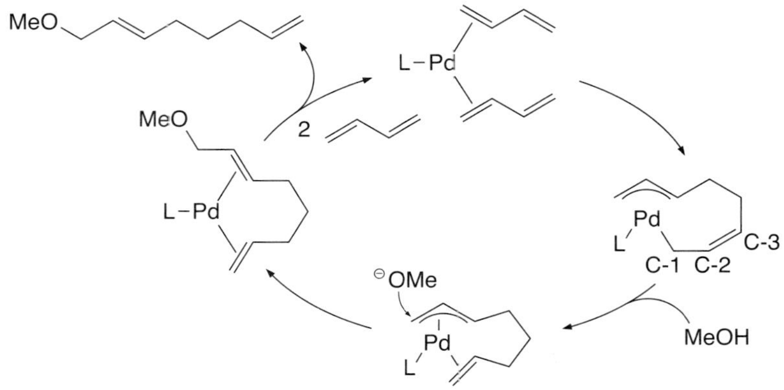

Scheme 22.20

In the absence of added dative ligands, many palladium catalysts convert butadiene into dodecatetraenes by the linear trimerization of butadiene (Scheme 22.21). Again, Jolly has reported the isolation of likely intermediates. The reaction of the palladium(0) precursor Pd$_2$(dba)$_3$ (dba = dibenzylidene acetone) generates the bis-η^3-allyl complex in Scheme 22.21 containing 12 carbons.[355] This complex was isolated as the bis-η^1-allyl complex after addition of bis(diphenylphosphino)ethane. Heating of the bis-η^3-allyl complex above −20 °C gave dodecatetraenes. In the presence of added ligands, only dimers of butadiene are obtained, probably because the ligand prevents coordination of the third diene necessary to produce the bis-η^3-allyl complex containing three butadiene units.

22.10.2.2. Cyclooligomerization of 1,3-Dienes

Nickel(0) complexes also catalyze a variety of cyclooligomerizations of butadiene. This process, studied intensively by Wilke and Jolly[313,356] and co-workers, can lead to a myriad of products that depend on the reaction conditions (Scheme 22.22). The cyclotrimerization of butadiene to cyclododecatriene was one of the first organometallic reactions subjected to careful mechanistic studies, and these studies led to the proposal that this process occurs by the mechanism shown in Scheme 22.23. Again, the bis-η^3-allyl complex containing 12 carbon atoms was isolated and characterized. It has been shown to enter the catalytic cycle as shown.

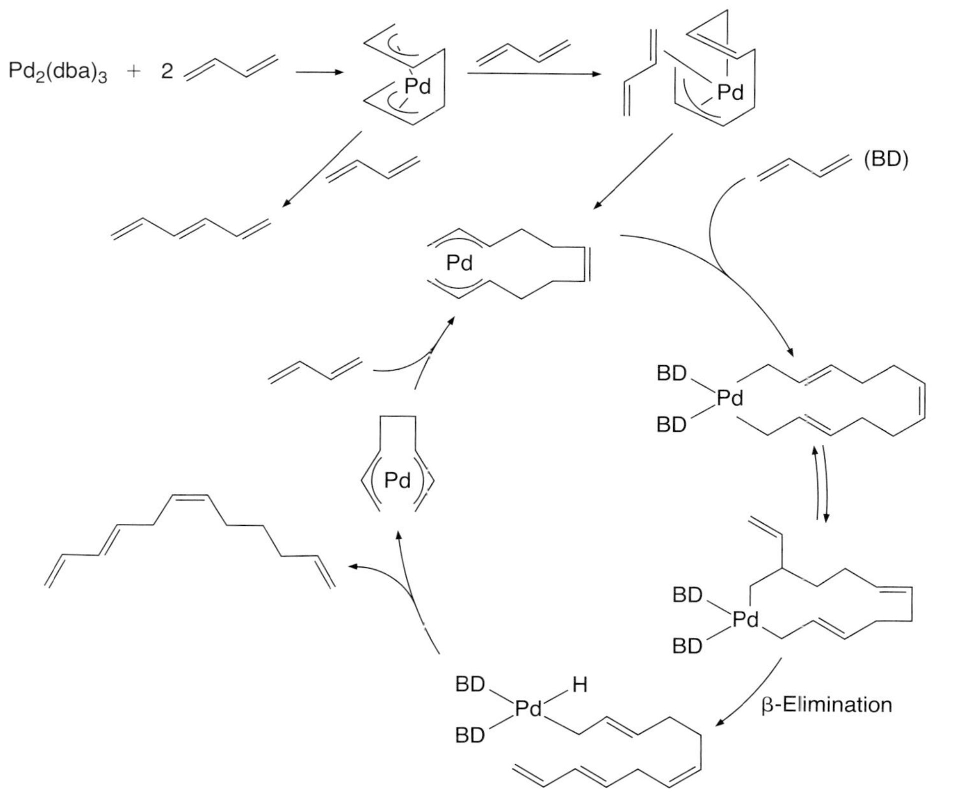

Scheme 22.21

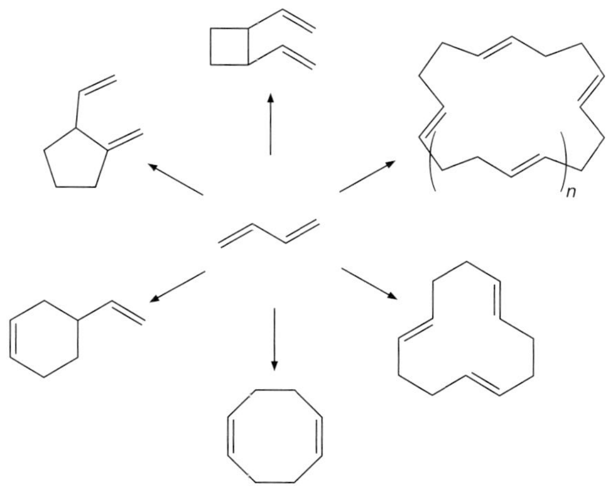

Scheme 22.22

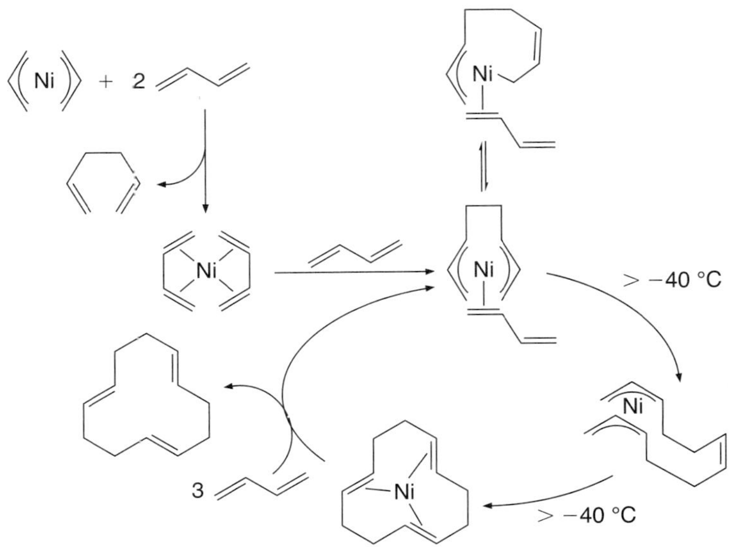

Scheme 22.23

22.11. Summary

The polymerization and oligomerization of alkenes has been one of the most successful applications of organometallic chemistry to the synthesis of organic products on a large scale. As noted in the introduction to this chapter, organometallic complexes are involved in the synthesis of close to, or in excess of, fifty to one hundred million metric tons of polyolefins and α-olefins per year. In most cases, these products are formed by a series of alkene insertions into metal alkyl complexes in competition with β-hydrogen elimination processes. In other cases, selective dimerization or trimerization of alkenes occurs by the intermediacy of metallacyclic intermediates.

The polymer properties are largely controlled by the microstructure of the polymer, which is controlled by the catalyst. This microstructure includes the degree of branching and the relative configuration of the polymer sidechains. In some cases, migration of the metal along the polymer chain leads to materials containing varying degrees of branching. When this occurs, polyethylenes are produced that consist of branched aliphatic structures, rather than a strictly linear chain, and polymers of α-olefins are produced that consist of structures in which the sidechains of the polymer have been "straightened." In addition to affecting the way in which monomers are linked, the catalyst affects the stereochemistry of the enchainment of the monomer units. This stereochemistry can be affected by the structure of the polymer chain or of the ancillary ligands on the metal. The most common structure for an aliphatic polyolefin is isotactic, but the ancillary ligands on the catalyst can create an environment that forms syndiotactic or hemisotactic structures, among others.

Olefin polymerization and oligomerization catalysts have been generated from all portions of the transition series. Group 4 catalysts are the most common for the polymerization of ethylene and propylene. Supported chromium catalysts are also used commercially, and many nickel and palladium catalysts have also been discovered that polymerize alkenes. The polymer produced from the palladium catalyst contains a high degree of branching due to facile rearrangements of the alkyl intermediate. Most of the catalysts are cationic complexes possessing a coordination site that is open or occupied by a weakly bound anion. The catalyst binds the incoming alkene at this site. The cationic catalyst is often generated by protonolysis or abstraction of an alkyl group from a neutral metal–alkyl precursor. Alternatively, metal halide complexes are treated with alkylaluminum species that simultaneously alkylate the dihalide complex and abstract one of the resulting alkyl ligands. Some of the most important systems for the initial work on controlled polymerization by soluble catalysts were based on cyclopentadienyl and C_2-symmetric *ansa*-metallocenes of titanium, zirconium and hafnium.

Many studies on the oligomerization of alkenes and dienes to form low molecular weight products have been conducted. In one case, α-olefins are the target products. These materials can be formed from ethylene with many types of metal complexes by a series of migratory insertions followed by β-hydrogen elimination or β-hydrogen transfer, with relative rates for insertions versus eliminations or transfers that are lower than those for catalysts that form high molecular weight polymer. Butene and hexene can also be formed from ethylene by selective dimerization and trimerization through metallacyclic intermediates. Related systems for the selective formation of octane have been discovered and are being developed. Ether and amine products containing eight carbons are also generated selectively, in this case by the selective dimerization of butadiene with added alcohol or amine. These reactions of butadiene have often been conducted with nickel and palladium catalysts, and these reactions occur via metal–allyl intermediates.

Thus, transition metal chemistry can form a bewildering array of valuable hydrocarbon products, each by the design of catalysts for the seemingly straightforward enchainment of alkenes. Some of the major future goals for the polymerization of alkenes include practical copolymerization of alkenes with polar monomers and the development of catalysts for the coordination insertion polymerization of polar monomer as a means to control the polymer architecture. A major goal for the oligomerization of alkenes includes the selective formation of oligomers longer than hexene.

References and Notes

1. *www.cen-online.org, 2007, July 2 issue*, page 55.
2. Kashiwa, N. *J. Polym. Sci. Part A: Polym. Chem.* **2004**, *42*, 1.
3. Natta, G.; Pino, P.; Corradini, P.; Danusso, F.; Mantica, E.; Mazzanti, G.; Moraglio, G. *J. Am. Chem. Soc.* **1955**, *77*, 1708.
4. Karol, F. J.; Wagner, B. E.; Levine, I. J.; Goeke, G. L.; Noshay, A. *Advances in Polyolefins;* Plenum, New York: 1987; p 337.
5. Karol, F. K.; Cann, K. J.; Wagner, B. E. In *Transition Metals and Organometallics as Catalysts for Olefin Polymerization;* Kaminsky, W., Sinn, H., Eds.; Springer-Verlag: Berlin, 1988; p 149.
6. Brintzinger, H. H.; Fischer, D.; Mülhaupt, R.; Rieger, B.; Waymouth, R. *Angew. Chem., Int. Ed. Engl.* **1995**, *34*, 1143.
7. Resconi, L.; Cavallo, L.; Fait, A.; Piemontesi, F. *Chem. Rev.* **2000**, *100*, 1253.
8. Fink, G.; Muelhaupt, R.; Brintzinger, H. H. *Ziegler Catalysts;* Springer: New York, 1995.
9. Kaminsky, W. *Metalorganic Catalysts for Synthesis and Polymerization;* Springer: New York, 1999.
10. Resconi, L.; Cavallo, L.; Fait, A.; Piemontesi, F. *Chem. Rev.* **2000**, *100*, 1253.
11. Gibson, V. C.; Spitzmesser, S. K. *Chem. Rev.* **2003**, *103*, 283.
12. Britovsek, G. J. P.; Gibson, V. C.; Wass, D. F. *Angew. Chem. Int. Ed.* **1999**, *38*, 428.
13. Ittel, S. D.; Johnson, L. K.; Brookhart, M. *Chem. Rev.* **2000**, *100*, 1169.
14. Coates, G. W. *Chem. Rev.* **2000**, *100*, 1223.
15. Coates, G. W.; Hustad, P. D.; Reinartz, S. *Angew. Chem. Int. Ed.* **2002**, *41*, 2236.
16. Domski, G. J.; Rose, J. M.; Coates, G. W.; Bolig, A. D.; Brookhart, M. *Progress in Polymer Science* **2007**, *32*, 30.
17. Chen, E. Y.-X.; Marks, T. J. *Chem. Rev.* **2000**, *100*, 1391.
18. An entire issue of Chemical Review was devoted to olefin polymerization: *Chem. Rev.,* **2000**, *100*, 1167.
19. Chen, E. Y. X.; Marks, T. J. *Chem. Rev.* **2000**, *100*, 1391.
20. Cossee, P. *J. Catal.* **1964**, *3*, 80.
21. Sinn, H.; Kaminsky, W. *Adv. Organomet. Chem.* **1980**, *18*, 99.

22. Parshall, G. In *Homogeneous Catalysis*; Wiley-Interscience: New York, 1980.
23. Kempe, R. *Chem. Eur. J.* **2007**, *13*, 2764.
24. Deng, L. Q.; Woo, T. K.; Cavallo, L.; Margl, P. M.; Ziegler, T. *J. Am. Chem. Soc.* **1997**, *119*, 6177.
25. Deng, L. Q.; Margl, P.; Ziegler, T. *J. Am. Chem. Soc.* **1997**, *119*, 1094.
26. Schulz, G. V. *Z. Phys. Chem. B* **1935**, *30*, 379.
27. Schulz, G. V. *Z. Phys. Chem. B* **1939**, *43*, 25.
28. Flory, P. J. *J. Am. Chem. Soc.* **1940**, *62*, 1561.
29. Hogan, J. P.; Banks, R. L. U.S. Patent 282 5721, 1958.
30. Hogan, J. P. *J. Polym. Sci., Part A: Polym. Chem.* **1970**, *8*, 2637.
31. Ulrich, H. *Introduction to Industrial Polymers*, 2nd ed.; Hanser: New York, 1993.
32. Kim, S. H.; Somorjai, G. A. *Proc. Natl. Acad. Sci. U. S. A.* **2006**, *103*, 15289.
33. Soga, K.; Shiono, T. *Prog. Polym. Sci.* **1997**, *22*, 1503.
34. Dusseault, J. J. A.; Hsu, C. C. *J. Macromol. Sci., Rev. Macromol. Chem. Phys.* **1993**, *C33*, 103.
35. Rodriguez, L. A. M.; van Looy, H. M. *J. Polym. Sci., Part A: Polym. Chem.* **1966**, *4*, 1951.
36. Karol, F. J.; Karapinka, G. J.; Johnson, R. N.; Wu, C.; Carrick, W. L.; Dow, A. W. *J. Polym Sci., Part A: Polym. Chem.* **1972**, *10*, 2621.
37. Groppo, E.; Lamberti, C.; Bordiga, S.; Spoto, G.; Zecchina, A. *Chem. Rev.* **2005**, *105*, 115.
38. McDaniel, M. P. *Adv. Catal.* **1985**, *33*, 47.
39. Theopold, K. H. *Eur. J. Inorg. Chem.* **1998**, 15.
40. MacAdams, L. A.; Buffone, G. P.; Incarvito, C. D.; Rheingold, A. L.; Theopold, K. H. *J. Am. Chem. Soc.* **2005**, *127*, 1082.
41. Gates, B. C.; Katzer, J. R.; Schuit, G. C. A. *Chemistry of Catalytic Processes*; McGraw-Hill: New York, 1979.
42. Britovsek, G. J. P.; Gibson, V. C.; Kimberley, B. S.; Maddox, P. J.; McTavish, S. J.; Solan, G. A.; White, A. J. P.; Williams, D. J. *Chem. Commun.* **1998**, 849.
43. Small, B. L.; Brookhart, M.; Bennett, A. M. A. *J. Am. Chem. Soc.* **1998**, *120*, 4049.
44. Bennett, A. M. A. *Chemtech* **1999**, *29*, 24.
45. McKnight, A. L.; Waymouth, R. *Chem. Rev.* **1998**, *98*, 2587.
46. Canich, J. A. M. U.S. Patent 5026798, 1991.
47. Canich, J. A. M.; Licciardi, G. F. U.S. Patent 5057475, 1991.
48. Stevens, J. C.; Neithamer, D. R. U.S. Patent 5064802, 1991.
49. Stevens, J. C.; Neithamer, D. R. U.S. Patent 5132380, 1992.
50. Lai, S.; Wilson, J. R.; Knight, G. W.; Stevens, J. C.; Chum, P. S. U.S. Patent 5272236, 1993.
51. Lai, S.; Wilson, J. R.; Knight, G. K.; Stevens, J. C. U.S. Patent 5278272, 1994.
52. Stevens, J. C. *Stud. Surf. Sci. Catal.* **1996**, *101*, 11.
53. Wang, W.; Yan, D.; Zhu, S.; Hamielec, A. E. *Macromolecules* **1998**, *31*, 8677.
54. Pellecchia, C.; Pappalardo, D.; Gruter, G. *Macromolecules* **1999**, *32*.
55. Izzo, L.; Caporaso, L.; Senatore, G.; Oliva, L. *Macromolecules* **1999**, *32*, 6913.
56. Kokko, E.; Malmberg, A.; Lehmus, P.; Löfgren, B.; Seppälä, J. V. *J. Polym. Sci., Part A: Polym. Chem.* **2000**, *38*, 376.
57. Kolodka, E.; Wang, W.; Charpentier, P. A.; Zhu, S.; Hamielec, A. E. *Polymer* **2000**, *41*, 3985.
58. Komon, Z. J. A.; Bazan, G. C. *Macromol. Rapid Commun.* **2001**, *22*, 467.
59. de Souza, R. F.; Casagrande, O. L., Jr. *Macromol. Rapid Commun.* **2001**, *22*, 1293.
60. Pettijohn, T. M.; Reagen, W. K.; Martin, S. J. U.S. Patent 5331070, 1994.
61. Barnhart, R. W.; Bazan, G. C. *J. Am. Chem. Soc.* **1998**, *120*, 1082.
62. Beigzadeh, D.; Soares, J. B. P.; Duever, T. A. *Macromol. Rapid Commun.* **1999**, *20*, 541.
63. Quijada, R.; Rojas, R.; Bazan, G.; Komon, Z. J. A.; Mauler, R. S.; Galland, G. B. *Macromolecules* **2001**, *34*, 2411.
64. Li, L. T.; Metz, M. V.; Li, H. B.; Chen, M. C.; Marks, T. J.; Liable-Sands, L.; Rheingold, A. L. *J. Am. Chem. Soc.* **2002**, *124*, 12725.
65. Wang, J.; Li, H.; Guo, N.; Li, L.; Stern, C. L.; Marks, T. J. *Organometallics* **2004**, *23*, 5112.
66. Abramo, G. P.; Li, L.; Marks, T. J. *J. Am. Chem. Soc.* **2002**, *124*, 13966.
67. Ittel, S. D.; Johnson, L. K.; Brookhart, M. *Chem. Rev.* **2000**, *100*, 1169.
68. Johnson, L. K.; Killian, C. M.; Arthur, S. D.; Feldman, J.; McCord, E. F.; McLain, S. J.; Kreutzer, K. A.; Bennett, M. A.; Coughlin, E. B.; Ittel, S. D.; Parthasarathy, A.; Tempel, D. J.; Brookhart, M. S. WO Patent 96/23010, 1996.
69. Johnson, L. K.; Killian, C. M.; Brookhart, M. *J. Am. Chem. Soc.* **1995**, *117*, 6414.
70. McCord, E. F.; McLain, S. J.; Nelson, L. T. J.; Ittel, S. D.; Tempel, D.; Killian, C. M.; Johnson, L. K.; Brookhart, M. *Macromolecules* **2007**, *40*, 410.
71. Shultz, L. H.; Tempel, D. J.; Brookhart, M. *J. Am. Chem. Soc.* **2001**, *123*, 11539.

72. *Chem. Eng. News* **2005**, *83(28)*, 67.
73. Natta, G. *J. Polym. Sci.* **1959**, *34*, 531.
74. Brintzinger, H. H.; Fischer, D.; Mulhaupt, R.; Rieger, B.; Waymouth, R. M. *Angew. Chem., Int. Ed. Engl.* **1995**, *34*, 1143.
75. Coates, G. W. *Chem. Rev.* **2000**, *100*, 1223.
76. Fujita, T.; Makio, H. In *Comprehensive Organometallic Chemistry III*; Crabtree, R., Mingos, M., Eds.; Elsevier: Amsterdam, 2007; p 691.
77. Cossee, P. *Tetrahedron Lett.* **1960**, *1*, 12.
78. Galli, P. *Macromol. Symp.* **1995**, *89*, 13.
79. Ewen, J. A. *J. Am. Chem. Soc.* **1984**, *106*, 6355.
80. Kaminsky, W.; Külper, K.; Brintzinger, H. H.; Wild, F. R. W. P. *Angew. Chem., Int. Ed. Engl.* **1985**, *24*, 507.
81. Ewen, J. A.; Jones, R. L.; Razavi, A.; Ferrara, J. D. *J. Am. Chem. Soc.* **1988**, *110*, 6256.
82. Spaleck, W.; Aulbach, M.; Bachmann, B.; Küber, F.; Winter, A. *Macromol. Symp.* **1995**, *89*, 237.
83. Spaleck, W.; Küber, F.; Winter, A.; Rohrmann, J.; Bachmann, B.; Antberg, M.; Dolle, V.; Paulus, E. F. *Organometallics* **1994**, *13*, 954.
84. Spaleck, W.; Küber, F.; Winter, A.; Rohrmann, J.; Bachmann, B.; Antberg, M.; Dolle, V.; Paulus, E. F. *Organometallics* **1994**, *13*, 954.
85. Farina, M.; Di Silvestro, G.; Sozzani, P. *Macromolecules* **1982**, *15*, 1451.
86. Ewen, J. A. U.S. Patent 5036034, 1991.
87. Ewen, J. A.; Elder, M. J.; Jones, R. L.; Haspeslagh, L.; Atwood, J. L.; Bott, S. G.; Robinson, K. *Macromol. Symp.* **1991**, *48/49*, 253.
88. Arlman, E. J.; Cossee, P. *J. Catal.* **1964**, *3*, 99.
89. Herfert, N.; Fink, G. *Macromol. Symp.* **1993**, *66*, 157.
90. Farina, M.; Di Silvestro, G.; Sozzani, P. *Macromolecules* **1993**, *26*, 946.
91. Guerra, G.; Cavallo, L.; Moscardi, G.; Vacatello, M.; Corradini, P. *Macromolecules* **1996**, *29*, 4834.
92. Razavi, A.; Peters, L.; Nafpliotis, L.; Vereecke, D.; Daw, K. D.; Atwood, J. L.; Thewald, U. *Macromol. Symp.* **1995**, *89*, 345.
93. Kleinschmidt, R.; Reffke, M.; Fink, G. *Macromol. Rapid Commun.* **1999**, *20*, 284.
94. Yano, A.; Kaneko, T.; Sato, M.; Akimoto, A. *Macromol. Chem. Phys.* **2000**, *200*, 2127.
95. Bravaya, N. M.; Nedorezova, P. M.; Tsvetkova, V. I. *Russ. Chem. Rev.* **2002**, *71*, 49.
96. Müller, G.; Rieger, B. *Prog. Polym. Sci.* **2002**, *27*, 815.
97. Natta, G. *J. Polym. Sci.* **1959**, *34*, 531.
98. Collette, J. W.; Tullock, C. W. U.S. Patent 4335225, 1982.
99. Collette, J. W.; Tullock, C. W.; MacDonald, R. N.; Buck, W. H.; Su, A. C. L.; Harrell, J. R.; Mülhaupt, R.; Anderson, B. C. *Macromolecules* **1989**, *22*, 3851.
100. Tullock, C. W.; Mülhaupt, R.; Ittel, S. D. *Makromol. Chem., Rapid Commun.* **1989**, *10*, 19.
101. Tullock, C. W.; Tebbe, F. N.; Mülhaupt, R.; Ovenall, D. W.; Setterquist, R. A.; Ittel, S. D. *J. Polym. Sci., Part A: Polym. Chem.* **1989**, *27*, 3063.
102. Collette, J. W.; Ovenall, D. W.; Buck, W. H.; Ferguson, R. C. *Macromolecules* **1989**, *22*, 3858.
103. Ohnishi, R.; Yukimasa, S.; Kanakazawa, T. *Macromol. Chem. Phys.* **2002**, *203*, 1003.
104. Job, R. C. U.S. Patent 5270410, 1993.
105. Mallin, D. T.; Rausch, M. D.; Lin, Y.; Dong, S.; Chien, J. C. W. *J. Am. Chem. Soc.* **1990**, *112*, 2030.
106. Chien, J. C. W.; Llinas, G. H.; Rausch, M. D.; Lin, G.; Winter, H. H. *J. Am. Chem. Soc.* **1991**, *113*, 8569.
107. Gauthier, W. J.; Corrigan, J. F.; Taylor, N. J.; Collins, S. *Macromolecules* **1995**, *28*, 3771.
108. Gauthier, W. J.; Collins, S. *Macromol. Symp.* **1995**, *98*, 223.
109. Bravakis, A. M.; Bailey, L. E.; Pigeon, M.; Collins, S. *Macromolecules* **1998**, *31*, 1000.
110. Dietrich, U.; Hackmann, M.; Rieger, B.; Klinga, M.; Leskelä, M. *J. Am. Chem. Soc.* **1999**, *121*, 4348.
111. Kukral, J.; Lehmus, P.; Feifel, T.; Troll, C.; Rieger, B. *Organometallics* **2000**, *19*, 3767
112. Rieger, B.; Troll, C.; Preuschen, J. *Macromolecules* **2002**, *35*, 5742.
113. Siedle, A. R.; Misemer, D. K.; Kolpe, V. V.; Duerr, B. F. U.S. Patent 6265512, 2001.
114. Meverden, C. C.; Nagy, S. U.S. Patent 6541583, 2003.
115. Coleman, B. D.; Fox, T. G. *J. Chem. Phys.* **1963**, *38*, 1065.
116. Erker, G.; Aulbach, M.; Knickmeier, M.; Wingbermuehle, D.; Krueger, C.; Nolte, M.; Werner, S. *J. Am. Chem. Soc.* **1993**, *115*, 4590.
117. Coates, G. W.; Waymouth, R. M. *Science* **1995**, *267*, 217.
118. Waymouth, R. M.; Hauptman, E.; Coates, G. W. U.S. Patent 5969070, 1999.
119. Busico, V.; Castelli, V. V. A.; Aprea, P.; Cipullo, R.; Segre, A.; Talarico, G.; Vacatello, M. *J. Am. Chem. Soc.* **2003**, *125*, 5451.

120. Busico, V.; Cipullo, R.; Kretschmer, W. P.; Talarico, G.; Vacatello, M.; Castelli, V. V. A. *Angew. Chem. Int. Ed.* **2002**, *41*, 505.
121. Lin, S.; Waymouth, R. M. *Acc. Chem. Res.* **2002**, *35*, 765.
122. Mansel, S.; Pérez, E.; Benavente, R.; Pereña, J. M.; Bello, A.; Röll, W.; Kirsten, R.; Beck, S.; Brintzinger, H. H. *Macromol. Chem. Phys.* **1999**, *200*, 1292.
123. Smolensky, E.; Kapon, M.; Woollins, J. D.; Eisen, M. S. *Organometallics* **2005**, *24*, 3255.
124. Chien, J. C. W.; Iwamoto, Y.; Rausch, M. D.; Wedler, W.; Winter, H. H. *Macromolecules* **1997**, *30*, 3447.
125. Chien, J. C. W. U.S. Patent 6121377, 2000.
126. Lieber, S.; Brintzinger, H. H. *Macromolecules* **2000**, *33*, 9192.
127. Nishii, K.; Shiono, T.; Ikeda, T. *Macromol. Rapid Commun.* **2004**, *25*, 1029.
128. McCord, E. F.; McLain, S. J.; Nelson, L. T. J.; Arthur, S. D.; Coughlin, E. B.; Ittel, S. D.; Johnson, L. K.; Tempel, D.; Killian, C. M.; Brookhart, M. *Macromolecules* **2001**, *34*, 362.
129. Gottfried, A. C.; Brookhart, M. *Macromolecules* **2003**, *36*, 3085.
130. Leatherman, M. D.; Svejda, S. A.; Johnson, L. K.; Brookhart, M. *J. Am. Chem. Soc.* **2003**, *125*, 3068.
131. Soga, K.; Sano, T.; Ohnishi, R.; Kawata, T.; Ishii, K.; Shiono, T.; Doi, Y. *Stud. Surf. Sci. Catal.* **1986**, *25*, 109.
132. Avella, M.; Martuscelli, E.; Volpe, G. D.; Segre, A.; Rossi, E.; Simonazzi, T. *Makromol. Chem.* **1986**, *187*, 1927.
133. Busico, V.; Corradini, P.; De Rosa, C.; Di Benedetto, E. *Eur. Polym. J.* **1985**, *21*, 239.
134. Stevens, J. C. *Stud. Surf. Sci. Catal.* **1996**, *101*, 11.
135. Shapiro, P. J.; Cotter, W. D.; Schaefer, W. P.; Labinger, J. A.; Bercaw, J. E. *J. Am. Chem. Soc.* **1994**, *116*, 4623.
136. Okuda, J. *Chem. Ber.* **1990**, *123*, 1649.
137. Okuda, J. *J. Chem. Soc., Dalton Trans.* **2003**, 2367.
138. Noordermeer, J. W. M. *Kirk-Othmer Encyclopedia of Chemical Technology Online;* Wiley: New York, online posting 2002, DOI: 10.1002/0471238961.0520082514151518.a01.pub2
138. Chien, H.; McIntyre, D.; Cheng, J.; Fone, M. *Polymer* **1995**, *36*, 2559.
140. Jin, J.; Uozumi, T.; Sano, T.; Teranishi, T.; Soga, K.; Shiono, T. *Macromol. Rapid Commun.* **1998**, *19*, 337.
141. Leclerc, M. K.; Waymouth, R. M. *Angew. Chem. Int. Ed.* **1998**, *37*, 922.
142. Fan, W.; Leclerc, M. K.; Waymouth, R. M. *J. Am. Chem. Soc.* **2001**, *123*, 9555.
143. Fan, W.; Waymouth, R. M. *Macromolecules* **2001**, *34*, 8619.
144. Fan, W.; Waymouth, R. M. *Macromolecules* **2003**, *36*, 3010.
145. Heuer, B.; Kaminsky, W. *Macromolecules* **2005**, *38*, 3054.
146. Arndt, M.; Kaminsky, W.; Schauwienold, A.; Weingarten, U. *Macromol. Chem. Phys.* **1998**, *199*, 1135.
147. Holden, G.; Kricheldorf, H. R.; Quirk, R. P. *Thermoplastic Elastaomers*, 3rd ed.; Hanser: Munich, 2004.
148. Ver Strate, G.; Cozewith, C.; West, R. K.; Davis, W. M.; Capone, G. A. *Macromolecules* **1999**, *32*, 3837.
149. Mori, H.; Yamahiro, M.; Tashino, K.; Ohnishi, K.; Nitta, K.; Terano, M. *Macromol. Rapid Commun.* **1995**, *16*, 247.
150. Yamahiro, M.; Mori, H.; Nitta, K.; Terano, M. *Macromol. Chem. Phys.* **1999**, *200*, 134.
151. Yamahiro, M.; Mori, H.; Nitta, K.; Terano, M. *Polymer* **1999**, *40*, 5265.
152. Mori, H.; Yamahiro, M.; Prokhorov, V. V.; Nitta, K.; Terano, M. *Macromolecules* **1999**, *32*, 6008.
153. Terano, M. European Patent 703253, 1998.
154. Doi, Y.; Ueki, S. *Makromol. Chem., Rapid Commun.* **1982**, *3*, 225.
155. Turner, H. W.; Hlatky, G. G.; Yang, H. W.; Gadkari, A. C.; Licciardi, G. F. U.S. Patent 5391629, 1995.
156. Fukui, Y.; Murata, M. *Appl. Catal. A* **2002**, *237*, 1.
157. Tian, J.; Hustad, P. D.; Coates, G. W. *J. Am. Chem. Soc.* **2001**, *123*, 5134.
158. Kojoh, S.; Matsugi, T.; Saito, J.; Mitani, M.; Fujita, T.; Kashiwa, N. *Chem. Lett.* **2001**, *30*, 822.
159. Tian, J.; Coates, G. W. *Angew. Chem. Int. Ed.* **2000**, *39*, 3626.
160. Milano, G.; Cavallo, L.; Guerra, G. *J. Am. Chem. Soc.* **2002**, *124*, 13368.
161. Saito, J.; Mitani, M.; Mohri, J.; Yoshida, Y.; Matsui, S.; Ishii, S.; Kojoh, S.; Kashiwa, N.; Fujita, T. *Angew. Chem. Int. Ed.* **2001**, *40*, 2918.
162. Mitani, M.; Mohri, J.; Yoshida, Y.; Saito, J.; Ishii, S.; Tsuru, K.; Matsui, S.; Furuyama, R.; Nakano, T.; Tanaka, H.; Kojoh, S.; Matsugi, T.; Kashiwa, N.; Fujita, T. *J. Am. Chem. Soc.* **2002**, *124*, 3327.
163. Ruokolainen, J.; Mezzenga, R.; Fredrickson, G. H.; Kramer, E. J.; Hustad, P. D.; Coates, G. W. *Macromolecules* **2005**, *38*, 851.
164. Matsugi, T.; Matsui, S.; Kojoh, S.; Takagi, Y.; Inoue, Y.; Nakano, T.; Fujita, T.; Kashiwa, N. *Macromolecules* **2002**, *35*, 4880.
165. Mason, A. F.; Coates, G. W. *J. Am. Chem. Soc.* **2004**, *126*, 16326.
166. Busico, V.; Cipullo, R.; Friederichs, N.; Ronca, S.; Togrou, M. *Macromolecules* **2003**, *36*, 3806.
167. Busico, V.; Cipullo, R.; Friederichs, N.; Ronca, S.; Talarico, G.; Togrou, M.; Wang, B. *Macromolecules* **2004**, *37*, 8201.

168. Maul, J.; Frushour, B. G.; Kontoff, J. R.; Eichenauer, H.; Ott, K.-H.; Schade, C. In *Ullmann's Encyclopedia of Industrial Chemistry Online*; Wiley-VCH: Weinheim, Online Posting Date: July 15, 2007; DOI: 10.1002/14356007.a21_615.pub2.

169. Ishihara, N.; Seimiya, T.; Kuramoto, M.; Uoi, M. *Macromolecules* **1986**, *19*, 2464.

170. Pellecchia, C.; Longo, P.; Grassi, A.; Ammendola, P.; Zambelli, A. *Makromol. Chem. Rapid Commun.* **1987**, *8*, 277.

171. Grassi, A.; Pellecchia, C.; Longo, P.; Zambelli, A. *Gazz. Chim. Ital.* **1987**, *117*, 249.

172. Zambelli, A.; Longo, P.; Pellecchia, C.; Grassi, A. *Macromolecules* **1987**, *20*, 2035.

173. Zambelli, A.; Oliva, L.; Pellecchia, C. *Macromolecules* **1989**, *22*, 2129.

174. Chien, J. C. W.; Salajka, Z. *J. Polym. Sci. Part A: Polym. Chem.* **1991**, *29*, 1243.

175. Chien, J. C. W.; Salajka, Z. *J. Polym. Sci. Part A: Polym. Chem.* **1991**, *29*, 1253.

176. Kaminsky, W.; Lenk, S. *Macromol. Symp.* **1997**, *118*, 45.

177. Kucht, A.; Kucht, H.; Barry, S.; Chien, J. C. W.; Rausch, M. D. *Organometallics* **1993**, *12*, 3075.

178. Kaminsky, W.; Lenk, S.; Scholz, V.; Roesky, H. W.; Herzog, A. *Macromolecules* **1997**, *30*, 7647.

179. Averbuj, C.; Tish, E.; Eisen, M. S. *J. Am. Chem. Soc.* **1998**, *120*, 8640.

180. Miyatake, T.; Mizunuma, K.; Kakugo, M. *Makromol. Chem. Macromol. Symp.* **1993**, *66*, 203.

181. Okuda, J.; Masoud, E. *Macromol. Chem. Phys.* **1998**, *199*, 543.

182. Xu, G. X.; Ruckenstein, E. *J. Polym. Sci. A* **1999**, *37*, 2481.

183. Foster, P.; Chien, J. C. W.; Rausch, M. D. *Organometallics* **1996**, *15*, 2404.

184. Schneider, N.; Prosenc, M. H.; Brintzinger, H. H. *J. Organomet. Chem.* **1997**, *546*, 291.

185. Ishihara, N.; Kuramoto, M.; Uoi, M. *Macromolecules* **1988**, *21*, 3356.

186. Grassi, A.; Longo, P.; Proto, A.; Zambelli, A. *Macromolecules* **1989**, *22*, 104.

187. Ricci, G.; Bosisio, C.; Porri, L. *Macromol. Rapid Commun.* **1996**, *17*, 781.

188. Zambelli, A.; Pellecchia, C.; Proto, A. *Macromol. Symp.* **1995**, *89*, 373.

189. Chien, J. C. W.; Salajka, Z.; Dong, S. *Macromolecules* **1992**, *25*, 3199.

190. Grassi, A.; Zambelli, A.; Laschi, F. *Organometallics* **1996**, *15*, 480.

191. Grassi, A.; Saccheo, S.; Zambelli, A.; Laschi, F. *Macromolecules* **1998**, *31*, 5588.

192. Pellecchia, C.; Pappalardo, D.; Oliva, L.; Zambelli, A. *J. Am. Chem. Soc.* **1995**, *117*, 6593.

193. Longo, P.; Grassi, A.; Proto, A.; Ammendola, P. *Macromolecules* **1988**, *21*, 24.

194. Duncalf, D. J.; Wade, H. J.; Waterson, C.; Derrick, P. J.; Haddleton, D. M.; McCamley, A. *Macromolecules* **1996**, *29*, 6399.

195. Ascenso, J. R.; Dias, A. R.; Gomez, P. T.; Romao, C. C.; Pham, Q.; Neibecker, D.; Tkatchenko, I. *Macromolecules* **1989**, *22*, 998.

196. Ascenso, J. R.; Dias, A. R.; Gomez, P. T.; Romao, C. C.; Tkatchenko, I.; Revillon, A.; Pham, Q. *Macromolecules* **1996**, *29*, 4172.

197. Arai, T.; Suzuki, S.; Ohtsu, T. In *Olefin Polymerization*; American Chemical Society: Washington, DC, 2000; Vol. ACS Symposium Series 749, p 66.

198. Watson, P. L. *J. Am. Chem. Soc.* **1982**, *104*, 337.

199. Grubbs, R. H.; Coates, G. W. *Acc. Chem. Res.* **1996**, *29*, 85.

200. Wu, Z.; Jordan, R. F.; Petersen, J. L. *J. Am. Chem. Soc.* **1995**, *117*, 5867.

201. Carpentier, J. F.; Wu, Z.; Lee, C. W.; Stromberg, S.; Christopher, J. N.; Jordan, R. F. *J. Am. Chem. Soc.* **2000**, *122*, 7750.

202. Carpentier, J. F.; Maryin, V. P.; Luci, J.; Jordan, R. F. *J. Am. Chem. Soc.* **2001**, *123*, 898.

203. Stoebenau, E. J.; Jordan, R. F. *J. Am. Chem. Soc.* **2003**, *125*, 3222.

204. Stoebenau, E. J.; Jordan, R. F. *J. Am. Chem. Soc.* **2004**, *126*, 11170.

205. Casey, C. P.; Lee, T. Y.; Tunge, J. A.; Carpenetti, D. W. *J. Am. Chem. Soc.* **2001**, *123*, 10762.

206. Casey, C. P.; Klein, J. F.; Fagan, M. A. *J. Am. Chem. Soc.* **2000**, *122*, 4320.

207. Casey, C. P.; Fagan, M. A.; Hallenbeck, S. L. *Organometallics* **1998**, *17*, 287.

208. Casey, C. P.; Hallenbeck, S. L.; Wright, J. M.; Landis, C. R. *J. Am. Chem. Soc.* **1997**, *119*, 9680.

209. Casey, C. P.; Hallenbeck, S. L.; Pollock, D. W.; Landis, C. R. *J. Am. Chem. Soc.* **1995**, *117*, 9770.

210. Casey, C. P.; Tunge, J. A.; Lee, T. Y.; Fagan, M. A. *J. Am. Chem. Soc.* **2003**, *125*, 2641.

211. Landis, C. R.; Rosaaen, K. A.; Sillars, D. R. *J. Am. Chem. Soc.* **2003**, *125*, 1710.

212. Schaper, F.; Geyer, A.; Brintzinger, H. H. *Organometallics* **2002**, *21*, 473.

213. Chirik, P. J.; Day, M. W.; Labinger, J. A.; Bercaw, J. E. *J. Am. Chem. Soc.* **1999**, *121*, 10308.

214. Harney, M. B.; Keaton, R. J.; Sita, L. R. *J. Am. Chem. Soc.* **2004**, *126*, 4536.

215. Busico, V.; Cipullo, R. *J. Am. Chem. Soc.* **1994**, *116*, 9329.

216. Sillars, D. R.; Landis, C. R. *J. Am. Chem. Soc.* **2003**, *125*, 9894.

217. Matsui, S.; Mitani, M.; Saito, J.; Tohi, Y.; Makio, H.; Matsukawa, N.; Takagi, Y.; Tsuru, K.; Nitabaru, M.; Nakano, T.; Tanaka, H.; Kashiwa, N.; Fujita, T. *J. Am. Chem. Soc.* **2001**, *123*, 6847.

218. Tian, J.; Hustad, P. D.; Coates, G. W. *J. Am. Chem. Soc.* **2001**, *123*, 5134.

219. Hustad, P. D.; Coates, G. W. *J. Am. Chem. Soc.* **2002**, *124*, 11578.
220. Hustad, P. D.; Tian, J.; Coates, G. W. *J. Am. Chem. Soc.* **2002**, *124*, 3614.
221. Lamberti, M.; Pappalardo, D.; Zambelli, A.; Pellecchia, C. *Macromolecules* **2002**, *35*, 658.
222. Makio, H.; Kashiwa, N.; Fujita, T. *Adv. Synth. Catal.* **2002**, *344*, 477.
223. Small, B. L.; Brookhart, M. *Macromolecules* **1999**, *32*, 2120.
224. Liu, Z.; Somsook, E.; White, C. B.; Rosaaen, K. A.; Landis, C. R. *J. Am. Chem. Soc.* **2001**, *123*, 11193.
225. Cherian, A. E.; Lobkovsky, E. B.; Coates, G. W. *Macromolecules* **2005**, *38*, 6259.
226. (a) Talarico, G.; Budzelaar, P. H. M. *J. Am. Chem. Soc.* **2006**, *128*, 4524; (b) Cavallo, L.; Guerra, G. *Macromolecules* **1996**, *29*, 2729; (c) Margl, P.; Deng, L.; Ziegler, T. *J. Am. Chem. Soc.* **1998**, *121*, 154.
227. (a) Fu, P. F.; Marks, T. J. *J. Am. Chem. Soc.* **1995**, *117*, 10747; (b) Koo, K.; Fu, P. F.; Marks, T. J. *Macromolecules* **1999**, *32*, 981.
228. Brookhart, M.; DeSimone, J. M.; Grant, B. E.; Tanner, M. J. *Macromolecules* **1995**, *28*, 5378.
229. Chung, T. C.; Xu, G.; Lu, Y. Y.; Hu, Y. L. *Macromolecules* **2001**, *34*, 8040.
230. Xu, G.; Chung, T. C. *J. Am. Chem. Soc.* **1999**, *121*, 6763.
231. Kawaoka, A. M.; Marks, T. J. *J. Am. Chem. Soc.* **2004**, *126*, 12764.
232. Kawaoka, A. M.; Marks, T. J. *J. Am. Chem. Soc.* **2005**, ASAP Article.
233. Amin, S. B.; Marks, T. J. *J. Am. Chem. Soc.* **2007**, *129*, 10102.
234. Chung, T. C.; Dong, J. Y. *J. Am. Chem. Soc.* **2001**, *123*, 4871.
235. Dong, J. Y.; Chung, T. C. *Macromolecules* **2002**, *35*, 1622.
236. Dong, J. Y.; Wang, Z. M.; Hong, H.; Chung, T. C. *Macromolecules* **2002**, *35*, 9352.
237. Ringelberg, S.; Meetsma, A.; Hessen, B.; Teuben, J. H. *J. Am. Chem. Soc.* **1999**, *121*, 6082.
238. Ittel, S. D.; Johnson, L. K.; Brookhart, M. *Chem. Rev.* **2000**, *100*, 1169.
239. Keim, W.; Bher, A.; Roper, M. In *Comprehensive Organometallic Chemistry*; Wilkinson, G., Stone, F. G. A., Abel, E. W., Eds.; Pergamon Press: New York, 1982; p 371.
240. Skupinska, J. *Chem. Rev.* **1991**, *91*, 613.
241. Mecking, S. *Coord. Chem. Rev.* **2000**, *203*, 325.
242. (a) Vogt, D. In *Applied Homogeneous Catalysis with Organometallic Compounds*; Cornils, B., Herrmann, W. A., Eds.; Wiley-VCH: Weinheim, 2002; Vol. 1, p 240; (b) Olivier-Bourbigou, H.; Saussine, L. In *Applied Homogeneous Catalysis with Organometallic Compounds*; 2nd ed.; Cornils, B., Herrmann, W. A., Eds.; Wiley-VCH: Weinheim, 2000; Vol. 1, p 253.
243. Clement, N. D.; Routaboul, L.; Grotevendt, A.; Jackstall, R.; Beller, M. *Chem. Eur. J.* **2008**, *14*, 7408.
244. Zietz, J. R.; Robinson, G. C.; Lindsay, K. L. In *Comprehensive Organometallic Chemistry*; Wilkinson, G., Stone, F. G. A., Abel, E. W., Eds.; Pergamon Press: New York, 1982; Vol. 7, p 365.
245. Lappin, G. R.; Nemec, L. H.; Sauer, J. D.; Wagner, J. D. In *Kirk-Othmer Encyclopedia of Chemical Technology*; Kroschwitz, J. I., Howe-Grant, M., Eds.; Wiley: New York, 1996; Vol. 17, p 839.
246. Parshall, G. W.; Ittel, S. D. *Homogeneous Catalysis*, 2nd ed.; Wiley: New York, 1992.
247. Freitas, E. R.; Gum, C. R. *Chem. Eng. Prog.* **1979**, *75*, 73.
248. Ziegler, K.; Gellert, H.-G.; Holzkamp, E.; Wilke, G.; Duck, E. W.; Kroll, W.-R. *Justus Liebigs Ann. Chem.* **1960**, *629*, 172.
249. Keim, W.; Kowaldt, F. H.; Goddard, R.; Krüger, C. *Angew. Chem., Int. Ed. Engl.* **1978**, *17*, 466.
250. Freitas, E. R.; Gum, C. R. *Chem. Eng. Prog.* **1979**, *75*, 73.
251. Keim, W.; Behr, A.; Kraus, G. *J. Organomet. Chem.* **1983**, *251*, 377.
252. Keim, W.; Behr, A.; Limbacker, B.; Kruger, C. *Angew. Chem., Int. Ed. Engl.* **1983**, *22*, 503.
253. Keim, W.; Kowaldt, F. H.; Goddard, R.; Kruger, C. *Angew. Chem., Int. Ed. Engl.* **1978**, *17*, 466.
254. Keim, W.; Schulz, R. P. *J. Mol. Catal.* **1994**, *92*, 21.
255. Komon, Z. J. A.; Bu, X. H.; Bazan, G. C. *J. Am. Chem. Soc.* **2000**, *122*, 12379.
256. Komon, Z. J. A.; Bu, X. H.; Bazan, G. C. *J. Am. Chem. Soc.* **2000**, *122*, 1830.
257. Lee, B. Y.; Bazan, G. C.; Vela, J.; Komon, Z. J. A.; Bu, X. H. *J. Am. Chem. Soc.* **2001**, *123*, 5352.
258. Killian, C. M.; Johnson, L. K.; Brookhart, M. *Organometallics* **1997**, *16*, 2005.
259. Small, B. L.; Brookhart, M. *J. Am. Chem. Soc.* **1998**, *120*, 7143.
260. Bouwkamp, M. W.; Lobkovsky, E.; Chirik, P. J. *J. Am. Chem. Soc.* **2005**, *127*, 9660.
261. Bart, S. C.; Chlopek, K.; Bill, E.; Bouwkamp, M. W.; Lobkovsky, E.; Neese, F.; Wieghardt, K.; Chirik, P. J. *J. Am. Chem. Soc.* **2006**, *128*, 13901.
262. Bart, S. C.; Lobkovsky, E.; Bill, E.; Wieghardt, K.; Chirik, P. J. *Inorg. Chem.* **2007**, *46*, 7055.
263. Skupinska, J. *Chem. Rev.* **1991**, *91*, 613.
264. Novaro, O.; Chow, S.; Magnouat, P. *J. Catal.* **1976**, *42*, 131.
265. Novaro, O.; Chow, S.; Magnouat, P. *J. Catal.* **1976**, *41*, 91.
266. Rogers, J. S.; Bazan, G. C.; Sperry, C. K. *J. Am. Chem. Soc.* **1997**, *119*, 9305.
267. Bogdanovic, B.; Spliethoff, B.; Wilke, G. *Angew. Chem., Int. Ed. Engl.* **1980**, *19*, 622.
268. Barlow, M. G.; Bryant, M. J.; Haszeldi.Rn; Mackie, A. G. *J. Organomet. Chem.* **1970**, *21*, 215.

269. Oehme, G.; Pracejus, H. *Tetrahedron Lett.* **1979**, 343.

270. Pracejus, H.; Krause, H. J.; Oehme, G. *Z. Chem.* **1980**, *20*, 24.

271. Nugent, W. A.; Hobbs, F. W. *J. Org. Chem.* **1983**, *48*, 5364.

272. Tkatchenko, I.; Neibecker, D.; Grenouillet, P. French Patent 2524341, 1983.

273. Grenouillet, P.; Neibecker, D.; Tkatchenko, I. *Organometallics* **1984**, *3*, 1130.

274. Nugent, W. A. U.S. Patent 4451665, 1984.

275. Nugent, W. A.; McKinney, R. J. *J. Mol. Catal.* **1985**, *29*, 65.

276. Grenouillet, P.; Neibecker, D.; Tkatchenko, I. French Patent 2596390, 1987.

277. Guibert, I.; Neibecker, D.; Tkatchenko, I. *J. Chem. Soc., Chem. Commun.* **1989**, 1850.

278. McKinney, R. J.; Colton, M. C. *Organometallics* **1986**, *5*, 1080.

279. McKinney, R. J. *Organometallics* **1986**, *5*, 1752.

280. McKinney, R. J. U.S. Patent 4485256, 1986.

281. Ren, C. Y.; Cheng, W. C.; Chan, W. C.; Yeung, C. H.; Lau, C. P. *J. Mol. Catal.* **1990**, *59*, L1.

282. Alderson, T. U.S. Patent 3013066, 1961.

283. Alderson, T.; Jenner, E. L.; Lindsey, R. V. *J. Am. Chem. Soc.* **1965**, *87*, 5638.

284. Singleton, D. M. U.S. Patent 4638084, 1987.

285. Brookhart, M.; Hauptman, E. *J. Am. Chem. Soc.* **1992**, *114*, 4437.

286. Hauptman, E.; Saboetienne, S.; White, P. S.; Brookhart, M.; Garner, J. M.; Fagan, P. J.; Calabrese, J. C. *J. Am. Chem. Soc.* **1994**, *116*, 8038.

287. Kagan, H. B. In *Comprehensive Organometallic Chemistry*; Wilkinson, G., Stone, F. G. A., Abel, E. W., Eds.; Pergamon Press: New York, 1982; Vol. 8.

288. Bogdanovic, B.; Henc, B.; Meister, B.; Pauling, H.; Wilke, G. *Angew. Chem., Int. Ed. Engl.* **1972**, *11*, 1023.

289. Bogdanovic, B.; Henc, B.; Lösler, A.; Meister, B.; Pauling, H.; Wilke, G. *Angew. Chem., Int. Ed. Engl.* **1973**, *12*, 954.

290. Ziegler, K.; Martin, H. U.S. Patent 2943125, 1954.

291. Al-Jarallah, A. M.; Anabtawi, J. A.; Siddiqui, M. A. B.; Aitani, A. M.; Al-Sa'doun, A. W. *Catal. Today* **1992**, *14*, 1.

292. Datta, S.; Fischer, M. B.; Wreford, S. S. *J. Organomet. Chem.* **1980**, *188*, 353.

293. Schrock, R.; McLain, S.; Sancho, J. *Pure Appl. Chem.* **1980**, *52*, 729.

294. Manyik, R. M.; Walker, W. E.; Wilson, T. P. *J. Catal.* **1977**, *47*, 197.

295. Briggs, J. R. *J. Chem. Soc., Chem. Commun.* **1989**, 674.

296. Emrich, R.; Heinemann, O.; Jolly, P. W.; Kruger, C.; Verhovnik, G. P. J. *Organometallics* **1997**, *16*, 1511.

297. Yang, Y.; Kim, H.; Lee, J.; Paik, H.; Jang, H. G. *Appl. Catal. A* **2000**, *193*, 29.

298. Reagen, W. K.; Conroy, B. K. U.S. Patent 5288823, 1994.

299. Manyik, R. M.; Walker, W. E.; Wilson, T. P.; Hurley, G. F. U.S. Patent 3231550, 1966.

300. Manyik, R. M.; Walker, W. E.; Wilson, T. P. U.S. Patent 3300458, 1967.

301. Carter, A.; Cohen, S. A.; Cooley, N. A.; Murphy, A.; Scutt, J.; Wass, D. F. *Chem. Commun.* **2002**, 858.

302. Wass, D. F. British Patent WO 2002004119, 2002.

303. McGuinness, D. S.; Wasserscheid, P.; Keim, W.; Hu, C. H.; Englert, U.; Dixon, J. T.; Grove, C. *Chem. Commun.* **2003**, 334.

304. McGuinness, D. S.; Wasserscheid, P.; Keim, W.; Morgan, D.; Dixon, J. T.; Bollmann, A.; Maumela, H.; Hess, F.; Englert, U. *J. Am. Chem. Soc.* **2003**, *125*, 5272.

305. Overett, M. J.; Blann, K.; Bollmann, A.; Dixon, J. T.; Haasbroek, D.; Killian, E.; Maumela, H.; McGuinness, D. S.; Morgan, D. H. *J. Am. Chem. Soc.* **2005**, *127*, 10723.

306. Bollmann, A.; Blann, K.; Dixon, J. T.; Hess, F. M.; Killian, E.; Maumela, H.; McGuinness, D. S.; Morgan, D. H.; Neveling, A.; Otto, S.; Overett, M.; Slawin, A. M. Z.; Wasserscheid, P.; Kuhlmann, S. *J. Am. Chem. Soc.* **2004**, *126*, 14712.

307. McGuinness, D. S.; Rucklidge, A. J.; Tooze, R. P.; Slawin, A. M. Z. *Organometallics* **2007**, *26*, 2561.

308. Rucklidge, A. J.; McGuinness, D. S.; Tooze, R. P.; Slawin, A. M. Z.; Pelletier, J. D. A.; Hanton, M. J.; Webb, P. B. *Organometallics* **2007**, *26*, 2782.

309. Agapie, T.; Schofer, S. J.; Labinger, J. A.; Bercaw, J. E. *J. Am. Chem. Soc.* **2004**, *126*, 1304.

310. Jolly, P. W. In *Comprehensive Organometallic Chemistry*; Wilkinson, G., Stone, F. G. A., Abel, E. W., Eds.; Pergamon Press: New York, 1982; Vol. 8, p 615.

311. Atlas, S. M.; Mark, H. F. *Catal. Rev. Sci. Eng.* **1976**, *13*, 1.

312. Nakamura, A.; Tsutsui, M. *Principles and Applications of Homogeneous Catalysis*; Wiley: New York, 1980.

313. Jolly, P. W.; Wilke, G. *The Organic Chemistry of Nickel*; Wiley: New York, 1975; Vol. 2.

314. O'Connor, A. R.; White, P. S.; Brookhart, M. *J. Am. Chem. Soc.* **2007**, *129*, 4142.

315. Heimbach, P. *Angew. Chem., Int. Ed. Engl.* **1968**, *7*, 882.

316. Estrine, B.; Soler, R.; Damez, C.; Bouquillon, S.; Henin, F.; Muzart, J. *Green Chem.* **2003**, *5*, 686.

317. Magna, L.; Chauvin, Y.; Niccolai, G. P.; Basset, J. M. *Organometallics* **2003**, *22*, 4418.

318. Drent, E.; Eberhard, M. R.; Made, R. H. V. d.; Pringle, P. G. WO Patent 2003 040 065, 2003.

319. Vollmuller, F.; Magerlein, W.; Klein, S.; Krause, J.; Beller, M. *Adv. Synth. Catal.* **2001**, *343*, 29.

320. Benvenuti, F.; Carlini, C.; Marchionna, M.; Patrini, R.; Galletti, A. M. R.; Sbrana, G. *J. Mol. Catal. A* **1999**, *140*, 139.

321. Basato, M.; Crociani, L.; Benvenuti, F.; Galletti, A. M. R.; Sbrana, G. *J. Mol. Catal. A* **1999**, *145*, 313.

322. Patrini, R.; Lami, M.; Marchionna, M.; Benvenuti, F.; Galletti, A. M. R.; Sbrana, G. *J. Mol. Catal. A* **1998**, *129*, 179.

323. Grenouillet, P.; Neibecker, D.; Poirier, J.; Tkatchenko, I. *Angew. Chem., Int. Ed. Eng.* **1982**, *21*, 767.

324. Perree-Fauvet, M.; Chauvin, Y. *Tetrahedron Lett.* **1975**, 4559.

325. Commereuc, D.; Chauvin, Y. *Bull. Soc. Chim. Fr.* **1974**, 652.

326. Beger, J.; Duschek, C.; Fullbier, H.; Gaube, W. *J. Prakt. Chem.* **1974**, *316*, 26.

327. Beger, J.; Reichel, H. *J. Prakt. Chem.* **1973**, *315*, 1067.

328. Beger, J.; Duschek, C.; Fullbier, H.; Gaube, W. *J. Prakt. Chem.* **1974**, *316*, 43.

329. Beger, J.; Meier, F. *J. Prakt. Chem.* **1980**, *322*, 69.

330. Takahashi, S.; Yamazaki, H.; Hagihara, N. *Bull. Chem. Soc. Jpn.* **1967**, *41*, 254.

331. Baker, R.; Cook, A. H.; Halliday, D. E.; Smith, T. N. *J. Chem. Soc., Perkin Trans. II* **1974**, 1511.

332. Green, M.; Scholes, G.; Stone, F. G. A. *J. Chem. Soc.* **1978**, 309.

333. Behr, A.; Keim, W. *Chem. Ber.* **1983**, *116*, 862.

334. Keim, W.; Roper, M.; Schieren, M. *J. Mol. Catal.* **1983**, *20*, 139.

335. Telin, A. G.; Fakhretdinov, R. N.; Dzhemilev, U. M. *B. Acad. Sci. USSR* **1986**, *35*, 2263.

336. Ahmad, M. U.; Hashem, M. A.; Khabiruddin, M.; Sarker, M. M. H.; Bäckvall, J. E. *Indian J. Chem., Sect. B* **1991**, *30*, 802.

337. Prinz, T.; Keim, W.; Driessen-Holscher, B. *Angew. Chem., Int. Ed. Engl.* **1996**, *35*, 1708.

338. Kiji, J.; Okano, T.; Nomura, T.; Saiki, K.; Sai, T.; Tsuji, J. *Bull. Chem. Soc. Jpn.* **2001**, *74*, 1939.

339. Zakharkin, L. I.; Petrushkina, E. A.; Podvisotskaya, L. S. *Bull. Acad. Sci. USSR* **1983**, 805.

340. Zakharkin, L. I.; Petrushkina, E. A. *Bull. Acad. Sci. USSR* **1986**, 1219.

341. Petrushkina, E. A.; Zakharkin, L. I. *Bull. Russ. Acad. Sci., Chem. Sci.* **1992**, *41*, 1392.

342. Beller, M.; Krotz, A.; Baumann, W. *Adv. Synth. Catal.* **2002**, *344*, 517.

343. Jackstell, R.; Andreu, M. G.; Frisch, A.; Selvakumar, K.; Zapf, A.; Klein, H.; Spannenberg, A.; Rottger, D.; Briel, O.; Karch, R.; Beller, M. *Angew. Chem. Int. Ed.* **2002**, *41*, 986.

344. Jackstell, R.; Frisch, A.; Beller, M.; Rottger, D.; Malaun, M.; Bildstein, B. *J. Mol. Catal. A* **2002**, *185*, 105.

345. Jackstell, R.; Harkal, S.; Jiao, H. J.; Spannenberg, A.; Borgmann, C.; Rottger, D.; Nierlich, F.; Elliot, M.; Niven, S.; Cavell, K.; Navarro, O.; Viciu, M. S.; Nolan, S. P.; Beller, M. *Chem.—Eur. J.* **2004**, *10*, 3891.

346. Harkal, S.; Jackstell, R.; Nierlich, F.; Ortmann, D.; Beller, M. *Org. Lett.* **2005**, *7*, 541.

347. Doehring, A.; Jolly, P. W.; Mynott, R.; Schick, K. P.; Wilke, G. *Z. Naturforsch., B: Chem. Sci.* **1981**, *36B*, 1198.

348. Goddard, R.; Jolly, P. W.; Krueger, C.; Schick, K. P.; Wilke, G. *Organometallics* **1982**, *1*, 1709.

349. Benn, R.; Jolly, P. W.; Mynott, R.; Raspel, B.; Schenker, G.; Schick, K. P.; Schroth, G. *Organometallics* **1985**, *4*, 1945.

350. Benn, R.; Gabor, G.; Jolly, P. W.; Mynott, R.; Raspel, B. *J. Organomet. Chem.* **1985**, *296*, 443.

351. Jolly, P. W.; Mynott, R.; Raspel, B.; Schick, K. P. *Organometallics* **1986**, *5*, 473.

352. Doehring, A.; Goddard, R.; Hopp, G.; Jolly, P. W.; Kokel, N.; Krueger, C. *Inorg. Chim. Acta* **1994**, *222*, 179.

353. Storzer, U.; Walter, O.; Zevaco, T.; Dinjus, E. *Organometallics* **2005**, *24*, 514.

354. Vollmuller, F.; Krause, J.; Klein, S.; Magerlein, W.; Beller, M. *Eur. J. Inorg. Chem.* **2000**, 1825.

355. Benn, R.; Jolly, P. W.; Mynott, R.; Schenker, G. *Organometallics* **1985**, *4*, 1136.

356. Jolly, P. W. In *Comprehensive Organometallic Chemistry*; Wilkinson, G., Stone, F. G. A., Abel, E. W., Eds.; Pergamon Press: New York, 1982; Vol. 8, p 371.

Contributor Listing

The following contributors made major contributions to the sections listed below in parentheses.

Dr. Dino Alberico (Sections 19.8, 19.9)
Alphora Research
2395 Speakman Drive
Suite 2001
Mississauga, Ontario
L5K 1B3 Canada

Professor Erik J. Alexanian (Section 3.3)
University of North Carolina
Department of Chemistry
Chapel Hill, NC 27599
USA

Dr. Elsa Alvaro (Section 4.6)
University of Illinois
Urbana, IL 61801
USA

Dr. Tim Boebel (Section 4.7)
Dow AgroSciences LLC
9330 Zionsville Road
Indianapolis, IN 46268
USA

Professor Charles P. Casey (Sections 17.2, 17.3)
University of Wisconsin
Department of Chemistry
1101 University Avenue
Madison, WI 53706
USA

Professor Geoffrey W. Coates (Sections 17.6, 22.1.1, 22.3, 22.4, 22.6, 22.7)
Cornell University
Department of Chemistry and Chemical Biology
Baker Laboratory
Ithaca, NY 14853
USA

Professor Gregory J. Domski (Sections 22.1.1, 22.3, 22.4, 22.6, 22.7)
Augustana College
Department of Chemistry
639 38th Street
Rock Island, IL 61201
USA

Professor Seth B. Herzon (Sections 4.2.1.2, 4.2.2)
Yale University
Department of Chemistry
350 Edwards Street
New Haven, CT 06511
USA

Professor Mark Lautens (Sections 19.8, 19.9)
University of Toronto
Department of Chemistry
Davenport Chemical Laboratories
80 St. George Street
University of Toronto
Toronto, Ontario
M5S 3H6 Canada

Dr. Jaclyn M. Murphy (Sections 4.2.5.3, 4.4)
Massachusetts Institute of Technology
Department of Chemistry
77 Massachusetts Avenue, Room 18-290
Cambridge, MA 02139
USA

Professor Jack R. Norton (Sections 3.2.1, 3.2.2, 3.2.3, 3.6, 3.8, 4.5, 16.6.2.2)
Columbia University
Department of Chemistry
3000 Broadway, MC 3102
New York, NY 10027
USA

Dr. Mark J. Pouy (Section 3.5)
Yale University
Department of Chemistry
225 Prospect Street
P.O. Box 208107
New Haven, CT 06520
USA

Dr. Devon C. Rosenfeld (Section 13.4.6)
The Dow Chemical Company
Hydrocarbons R&D, B-251
Freeport, TX 77541
USA

Dr. Mark E. Scott (Sections 19.8, 19.9)
Merck & Co., Inc.
BMB2-134
33 Avenue Louis Pasteur
Boston, MA 02115
USA

Dr. Shashank Shekhar (Section 19.8)
Abbott Laboratories
Process Research and Development
1401 Sheridan Road
North Chicago, IL 60064
USA

Dr. Qilong Shen (Section 17.9)
Shanghai Institute of Organic Chemistry
Chinese Academy of Sciences
345 Lingling Rd. Shanghai
200032 China

Dr. Levi M. Stanley (Section 20.7)
University of Illinois
Department of Chemistry
600 South Mathews Avenue
Urbana, IL 61801
USA

Professor Jesse W. Tye (Sections 3.4, 4.2.4)
Ball State University
Department of Chemistry
2000 W. University Avenue
Muncie, IN 47306
USA

Giang D. Vo (Section 4.2.5.1)
University of Illinois
Department of Chemistry
600 South Mathews Avenue
Urbana, IL 61801
USA

Professor Patrick J. Walsh (Section 4.2.5.2, Chapter 14)
University of Pennsylvania
Department of Chemistry
231 S. 34th Street
Philadelphia, PA 19104
USA

Professor Jing Zhao (Section 15.10)
Nanjing University
School of Life Sciences
Nanjing
210093 China

Professor Pinjing Zhao (Sections 4.2.1.1, 4.3)
North Dakota State University
Department of Chemistry and Molecular Biology
NDSU Dept. 2735
PO Box 6050
Fargo, ND 58108
USA

Professor Jianrong (Steve) Zhou (Section 4.2.3)
Nanyang Technological University
Division of Chemical and Biological Chemistry
School of Physical and Mathematical Sciences
21 Nanyang Link, SPMS-CBC-06-03
Singapore 637371

Index

A

"Acceptorless dehydrogenation," 841
Acenaphthyl-substituted
    cyclopentadienyl, 112
Acetaldehyde, 718
    See also Wacker process
(Z)-α-Acetamidocinnamate (MAC), 592
α-Acetamidocinnamic acid esters, 636–640
Acetic acid, 746–751
Acetic anhydride process, 748–749
Acetone cyanohydrin, 669
Acetoxylation, 834
Acetoxytubipofuran, 445
Acetylacetonate (acac), 185
Acetylene ligands, 25
Acetylenes
    insertions into metal–silicon and
        metal–boron bonds, 388–389
Acetylides, 487–488
    See also Alkynyl complexes; Alkynyl
        ligands
Achiral dienes, 570
Acid halides, 344
Acidities
    of hydride ligands, 129–131
    kinetic, 131
Acrylates, 1083–1084
"Acrylic," 776
Acrylic acids, 616–618
Acrylonitrile, 386
Actinide alkyl compounds, 133
Actinides, 87
    atom abstraction pathway of
        oxidative addition, 310
    hydroamination of alkynes and, 711
Activation enthalpy
    dissociative substitutions and,
        234–235
Active catalyst, 544
Acyclic dialkyl complex, 400
Acyclic electrophiles, 987–993
Acyclic N-alkyl imines, 630–631
Acyclic N-aryl imines, 631–632
α-(Acycloxy)acrylates, 616
Acyl complexes
    C–C bond-forming reductive
        elimination and, 333
    C–X bond-forming reductive
        elimination and, 344–345
    nucleophilic attack and, 426

α-(Acylamino)acrylic acids and esters,
    612–614
β-(Acylamino)acrylic acids and esters,
    614–615
Acylsilane complexes, 353
Addition metathesis polymerization
    (ADMET), 1016, 1017, 1037–1038
Adiponitrile, 668, 670
Adiponitrile hydrogenation, 577
ADMET. *See* Addition metathesis
    polymerization
"Agostic complexes," 279
Agostic interactions, 71–72, 1073
Alcohols
    aliphatic, 928
    allylic, 723, 969
    amino, 624, 929
    copper-catalyzed coupling
        with aryl boronic acids,
            932–933
        with aryl halides, 926–930
    intermolecular oxidation of olefins,
        722–724
    intramolecular oxidation of olefins,
        724–726
    racemic secondary, 566
    reductive elimination, 330–331
    unsaturated, 618
    unsaturated, asymmetric
        hydrogenation, 618
Aldehydes
    annual production, 751
    binding modes, 53
    direct carbonylation of alkanes and
        arenes to, 838–839
    formation by hydroformylation, 753
    insertions into metal–carbon
        bonds, 381
    nucleophilic attack, 435–436
Aliphatic alcohols, 928
Aliphatic amines, 923–924
Aliphatic backbones
    ligands containing, 607–608
Aliphatic bisphosphines, 608
Aliphatic C–H bonds, 71–72
    silylation, 857–859
Aliphatic electrophiles, 882–883
Aliphatic halides, 916
Aliphatic ketones, 628–629
Alkane complexes

dissociative substitutions, 237–238
    evidence for, 70–71
    intramolecular coordination of
        aliphatic C–H bonds, 71–72
    overview, 64, 65, 70
    stability, 70
Alkanes
    borylation, 852–853
        mechanism, 855–857
    carbonylation
        alkylative, 837–838
        direct carbonylation to
            aldehydes, 838–839
        oxidative, 835–836
    directed oxidations, aminations, and
        halogenations, 832–835
    examples of complexes that
        oxidatively add, 281–282
    functionalization
        major goals, 825
        overview, 825–826
    intermolecular oxidative addition of
        C–H bonds, 276–277
    metathesis via dehydrogenation,
        842–844
    oxidative addition
        mechanism, 279–281
        selectivity of, 278–279
    platinum-catalyzed oxidations
        catalysts, 827–829
        mechanisms, 829–832
    reaction of boryl complexes with,
        286
    reductive elimination, 325–326
        features of, 327–329
    σ-bond metathesis, 284, 286
"Alkene" hydrogenation mechanism,
    585, 586
Alkene oligomerization
    ethylene, 1080–1082
    olefin dimerization and trimerization
        through metallacyclic
        intermediates, 1084–1086
    olefin dimerization by insertion
        into metal–carbon bonds,
        1082–1084
    overview, 1079–1080
Alkene polymerization, 1048
    effect of catalyst steric properties on
        chain transfer, 1078, *1079*

Alkene polymerization (*Continued*)
  mechanism of chain transfer and
    scope of chain transfer agents,
    1076–1078
  mechanism of the chain propagation
    step, 1073–1075
  overview, 1072–1073
Alkenes
  diboration, 698–699
  hydroaminations, 526–527, 701–705
  hydroaminomethylations, 771
  hydroboration, 542–543, 692–693
  hydrocarboxylation and
    hydroesterification
      catalysts for, 777–778
      intermolecular, 778–780
      mechanism of
        hydroesterification, 782–784
      overview, 775
      scope of, 778–781
      synthetic targets, 775–777
  hydrocyanation, 668–670, 670–672
  hydroformylation, 755, 763,
    764–765
  hydrosilylation, 677, 689 (*see also*
    Hydrosilylation)
  insertions into metal–alkyl bonds,
    371–377
  overview, 47
  trimerization and tetramerization,
    1084–1086
  See also Internal alkenes; Olefin
    complexes
Alkenyl–metal bonds, 357
Alkoxide anions, 173
"Alkoxide cycle," 782
β-Alkoxide eliminations, 409, 410
Alkoxide ligands, 25
Alkoxides
  β-alkyl and β-aryl eliminations, 408
  early-transition-metal–alkoxides
    as ancillary ligands, 175–177
      bonding, 174
      catalytic reactions, 176–177
      overview, 174
      preparation, 174–175
      reactivity, 175
      steric and electronic
        properties, 175–176
  late-transition-metal–alkoxides
    as ancillary ligands, 180
      bonding of, 177–178
      catalytic reactions, 185
      overview, 173, 177
      preparation, 180–183
      reactivity, 183–184
      thermodynamics, 178–179
  metal alkoxides, 402–405
  nucleophilic attack on coordinated
    carbon monoxide, 421

Alkoxo complexes
  early-transition-metal–alkoxides,
    173, 174–177
  late-transition-metal–alkoxides, 173,
    177–185
  overview, 173
Alkoxo groups, 362
Alkoxycarbene complexes, 421
Alkoxycarbonyl complex, 421
Alkoxycarbonylation, *785*, 789
Alkyl C–H bonds, 284
Alkyl complexes
  β-alkyl eliminations, 349, 350,
    406–408
  alkyne insertions, 379–380
  carbon monoxide insertions, 351–364
  C–C bond-forming reductive
    elimination and, 333, 334
  electrophilic attack on
    at the β-position, 466–467
    with *d*-electrons absent,
      461–462
    with *d*-electrons present,
      457–460
  β-hydrogen eliminations, 398–402
  migratory insertion of other ligands,
    364–365
  selected reactions, 90–91
  synthesis, 87–89
  synthesis of hydride complexes
    from, 128
  thermodynamic properties of
    M–alkyl bonds, 86–87
Alkyl dihalides, 316
Alkyl electrophile, 302–303
β-Alkyl eliminations, 349, 350
  from alkoxide and amido
    complexes, 408
  from alkyl complexes, 406–408
Alkyl group
  abstraction of, 454–455, 457
  carbon monoxide insertions and,
    361, 362
  electrophilic attack at the α-position,
    465–466
  migration to oxo or imido ligand,
    524–525
  nucleophilic attack on, 423
  synthetic applications of C–H
    oxidative addition, 282
Alkyl halides
  atom abstraction pathway of
    oxidative addition, 309–310
  dinuclear oxidative additions and,
    316
  outer-sphere electron-transfer
    mechanism, 308–309
  oxidative additions, 301–302
  reductive elimination from
    platinum(IV) complexes, 423

Alkyl hydride complexes, 276–277
  isotopic scrambling, 327
  oxidative addition of C–H bonds
    and, 278–279
  reductive elimination of alkanes,
    327–329
Alkyl ligands
  bonds, 86
  common groups, 86
  history of, 86
  nucleophilic attack, 422
  stability, 132–133
Alkyl reagents, 420
Alkylamines, 923–924
Alkylarenes, 334
Alkylation, 88, 514
Alkylative carbonylation, 837–838
Alkylidene complexes, 20, 42
  addition of C–H and, 503–504
  reactivity, 498–504
  Schrock-type, 411
  synthesis, 488–491
Alkylidyne complexes, 45, 498–504
Alkylmetal–hydride complexes, 276–277
Alkylpalladium(II) complexes, 339
Alkylphosphines
  air sensitivity, 35
  in cationic iridium catalysts,
    594–595
  cationic rhodium catalysts
    containing, 592–593
  oxidative addition and, 312
  π-accepting ability, 37
Alkylplatinum(IV) complexes, 341–342
Alkylsilanes, 677–678
Alkyne complexes
  bonding, 51
  creation of vinyl complexes, 96
  hydroamination by attack of amines
    on, 713
  nucleophilic attack, 434–435
  origins of, 51
  physical and chemical properties, 52
  structural characteristics, 51–52
  synthesis of vinylidene complexes,
    486–487
Alkyne ligands, 22
Alkyne metathesis, 1016, 1017
  applications, 1036–1038
  cross metathesis, 1038–1039
  examples, 1034–1035
  introduction to, 1015
  mechanism, 1036
  ring-closing metathesis, 1039–1040
Alkynes
  coupling, 880–881
  diboration, silylboration, and
    stannylboration, 697
  directed hydroarylation, 850
  hydroamination, 526–527

by [2+2] cycloadditions, 716–717
catalyzed by group 4 metal complexes, 710
catalyzed by lanthaninde and actinide complexes, 711
catalyzed by rhodium and palladium complexes, 711–712
overview, 710
hydroboration, 693–694
hydrocarboxylation and hydroesterification
  catalysts for, 777–778
  overview, 775
  scope of, 781
  synthetic targets, 775–777
hydrocyanation, 676
hydroesterification, 781
hydrogenation
  chromium-catalyzed, 642–644
  overview, 640–641
  palladium-catalyzed, 642–644
  rhodium-catalyzed, 641–642
hydrosilylation, 677, 681–683, 690
  (see also Hydrosilylation)
insertions into metal hydrides, 368–370
insertions into metal–carbon bonds, 379–380
Alkynyl complexes
  C–C bond-forming reductive elimination and, 334
  overview, 92, 97–98
  stability, 92
Alkynyl copper complexes, 97–98
Alkynyl ligands
  electrophilic attack at the β-position, 468
  structure, 85
Alkynylarene, 881
Allene complexes
  nucleophilic attack on, 433–434
Allenes
  hydroamination, 707–708, 716–717
  Pauson–Khand reactions with, 811–812
Allyl complexes
  η¹-complexes, 105–106
    electrophilic attack at the γ-position, 469
    nucleophilic attack and, 426
  η³-complexes, 105, 106
    electrophilic attack at the γ-position, 469–470
    nucleophilic attacks on, 436–439
  π-complexes
    hydroamination by attack of amines on, 713–714
    synthesis, 107–108

dynamics, 106–107
overview, 104
reactions, 108
structures, 104–106
Allyl enol carbonates, 997
η¹-Allyl groups, 426
Allyl ligands
  classification, 5
  nucleophilic attack and, 426
  resonance structures, 5
Allylic acetates, 565, 566
Allylic alcohols, 723, 969
Allylic alkylations, 569, 967
  asymmetric, 557–558
Allylic carbonates, 970
Allylic monoesters, cyclic, 993–994
Allylic substitution
  copper-catalyzed
    enantioselective, 1001–1006
    fundamentals, 999–1000
    mechanism, 1000–1001
    miscellaneous reactions, 1007–1008
    organoaluminum reagents as nucleophiles, 1006
  enantioselective, 968–969
    of acyclic electrophiles, 987–993
    of cyclic substrates, 993–995
    kinetic resolution, 995–996
    overview, 984–987
    of prochiral nucleophiles, 996–998
  first catalysts for, 969
  mechanism, 974–978
  metals used for, 973–974
  overview, 967–968
  regioselectivity, 973–974
    memory effect with palladium, 982–983
    of palladium-catalyzed reactions, 979–982
    when catalyzed by complexes of other metals, 983–984
  scope of electrophile, 969–972
  scope of nucleophile, 972–973
(η³-Allyl)palladium chloride, 105
"Alpha" process, 776
α-Agostic complex, 375
Alstophylline, 726, 727
Alternating copolymers, 1049
Alternating ethylene–propylene copolymers, 1068–1069
Aluminohydride complexes, 127
Amidate complexes, 155
Amidation
  aryl halides, 797–798
Amides, 402–405
  binding modes, 53
  copper-catalyzed coupling with aryl halides, 925

formed by carbonylation of organic halides, 795–798
as π-donors, 25
reductive elimination of, 344
Amidinate complexes, 155
Amido complexes
  β-alkyl and β-aryl eliminations, 408
  early-transition-metal
    overview, 152–153
    reactivity, 154
    synthesis, 154
    thermodynamic properties, 153–154
  α-hydrogen elimination, 413
  late-transition-metal
    bonding, 148–149
    overview, 148
    reactivity, 151–152
    spectral properties, 150
    synthesis, 150–151
    thermodynamic properties, 149
  overview, 147–148
Amido groups, 362
Amidocyclopentadienyl ligand, 113
Amination
  alkanes and arenes, 832–833
  olefins, 431
Amine complexes, 57–58
Amine ligands
  cone angles, 57
Amine N-oxide, 810
Amines
  copper-catalyzed coupling, 922–924
    with aryl boronic acids, 932–933
    with aryl halides, 925–926
  cross-coupling with aryl halides
    catalysts for, 910–911
    mechanism, 911–914
    overview, 907–909
  hydroamination by attack on π-complexes, 713–715
  nucleophilic attack
    on coordinated carbon monoxide, 421
    on coordinated isonitrile, 421
  reductive elimination of, 330–331
  "telomeric," 708
Amino acids
  hydroxylation, 832–833
Amino alcohols, 624, 929
α-Amino ketones, 624
β-Amino ketones, 625
Aminoboratabenzenes, 119, 120
Aminocarbonyl complex, 421
Ammonia, 57
  creation from dinitrogen, 59, 61
  hydroaminomethylation and, 771–772

Ammonia (*Continued*)
 intermolecular addition to ethylene, 700
 oxidative additions and, 314
Ammonium hexafluorophosphate, 127
Ancillary ligands
 effect on C–H bond-forming reductive elimination, 329–330
Anilide ligand, 342
Aniline
 η^6-complex, 54
 oxidative additions and, 315
 reductive elimination of, 331
Anionic oxygen ligands
 alkoxo complexes, 173–185
 β-diketonate complexes, 185–186
Ansa bridges, 113, 116
Ansa metallocenes, 113–122
 See also Cyclopentadienyl ligands
Anthracene complex, 645–647
Anti-Markovnikov hydroamination, 700, 705–706
Anti-oxypalladation, 383
Antibonding orbitals, 18
"Apical–equatorial" chelation, 759
Arduengo carbenes, 43
Arene complexes
 as acceptors of electron density, 24
 η^2-complexes, 56
  electrophilic attack on, 475–476
  nucleophilic attack, 435
  oxidative addition and, 280
 η^4-complexes, 54–55
 η^6-complexes
  nucleophilic attack on, 442–444
  properties, 53–57
  representative complexes, 54
 π-complexes
  catalytic processes and, 542
  nucleophilic attack on, 444–446
 oxidative addition and, 279–280
 preparation, 54
Arene exchange reactions, 248–250
Arene hydrogenation, 577
Arenes
 borylation, 853–854
  mechanism, 855–857
 carbonylation
  alkylative, 837–838
  direct carbonylation to aldehydes, 838–839
  oxidative, 835–836
 directed oxidations, aminations, and halogenations, 832–835
 homogenous catalytic hydrogenation
  monocyclic arenes, 647
  overview, 644–645
  polycyclic arenes, 645–647

intermolecular oxidative addition of C–H bonds, 275–276
platinum-catalyzed oxidations
 catalysts, 827–829
 mechanism, 829–832
reductive elimination of, 329
σ-bond metathesis, 284
substitutions for, 248–250
synthesis of aryl complexes, 93
Aripiprazole, 772, *773*
Aromatic bisphosphines, 603–608
Aromatic C–C bonds, 936–938
Aromatic C–H bonds, 857–859
Aromatic phosphines, 590–592
Aroylhydrazones, 632–634
Arsine, 41
Ar–X bonds, 310
Aryl boronic acids, 932–933
Aryl bromides, 924
Aryl chlorides
 oxidative addition, 312–313, 894
Aryl complexes
 with bridging aryl ligands, 94
 C–C bond-forming reductive elimination and, 332–333, 334
 overview, 92
 properties, 95, *96*
 stability, 92
 synthesis of complexes with terminal ligands, 92–94
β-Aryl eliminations, 349, 350, 408
Aryl ethers, 932
Aryl groups
 carbon monoxide insertions and, 361, 362
 migration to oxo or imido ligand, 524–525
Aryl halides, 426
 amidation, 797–798
 carbonylation to form ketones, 914–915
  mechanism, 916–917
 copper-catalyzed coupling
  with alcohols and phenols, 926–930
  with alkylamines, 924
  with amides, 925
  with heterocyclic amines, 925–926
  mechanism, 930–932
 coupling, 880–881
  with cyanide, 883–884
  with organotin reagents, 879
 coupling, asymmetric, with enolates, 886–887
 cross-coupling with amines
  catalysts for, 910–911
  mechanism, 911–914
  overview, 907–909
 esterification, 796, 797–798

olefination, 892–893
oxidative addition, 894–895
 to palladium complexes, 310–313
phosphination, 569
reactions of nickel complexes with, 305
synthesis of aryl complexes, 93
Aryl heteroaryl ketones, 628
Aryl hydride complexes
 oxidative addition of C–H bonds and, 279
 reductive elimination of arenes, 329
N-Aryl imines, 631
Aryl iodides, 924
Aryl nitriles, 289
Aryl tosylates, 313
Aryl triflates, 313, 795
Arylamido complexes, 150
Arylamines
 addition to vinylarenes, 700
 copper-catalyzed coupling, 922–923
Arylation, 834–835
 direct (*see* Direct arylation)
 oxidative, of olefins, 850–852
Aryllithium reagents, 420
Aryloxide anions, 173
Aryloxide ligands, 173
Aryloxo complexes, 173
Arylpalladium alkoxides, 181
Arylpalladium diarylamido complexes, 344
Arylpalladium phenoxide complexes, 344
Arylpalladium(II) complexes
 C–X bond reductive elimination, 339–340
 reductive eliminations to form C–X bonds, 342–344
Arylphosphines, 35
Arylplatinum(IV) complexes
 C–X bond reductive elimination, 339, 341–342
Aryltungstenocene, 365
Assisted substitution reactions, 241–247
Associative interchange (I_A), 218
Associative substitutions
 versus dissociative substitutions of square-planar complexes, 229–231
 distinguishing by kinetic methods, 220
 factors affecting the rate of reaction, 226–228
 overview, 217–218
 rate law, 225–226
 17-electron complexes, 231–233
 square-planar, 16-electron complexes, 223–229
 stereochemistry, 224–225

trans and cis effects, 228–229
trends for, *219*
with and without solvent assistance, 223–224
Asymmetric allylic alkylation, 557–558
Asymmetric catalysis
    classes of, 550–551
    desymmetrization reactions, 569–571
    dynamic kinetic asymmetric transformations, 568–569
    dynamic kinetic resolution, 567–568
    energetics of stereoselectivity, 552–558
    importance of, 549–550
    kinetic resolutions, 563–567
    nomenclature, 551–552
    transmission of asymmetry
        effect of C_2 symmetry, 559
        overview, 559
        "privileged ligands," 561–563
        quadrant diagrams, 559–560
Asymmetric hydroboration, 694
Asymmetric hydrocyanation, 674–675
    dienes, 675–676
Asymmetric hydrogenation, 556–557
    of α-acetamidocinnamic acid esters, 636–640
    asymmetric transfer hydrogenation, 634–636
    heteroarenes, 648–651
    imines, 629–634
    ketones
        functionalized ketones, 621–625
        overview, 620–621
        unfunctionalized ketones, 626–629
    ligands used for
        aliphatic bisphosphines, 608
        aromatic bisphosphines, 603–608
        overview, 603
        P-chiral phosphines, 609
        phosphites and phosphoramidites, 610–611
        P,N ligands, 609–610
Asymmetric hydrosilylation, 683–684
Asymmetric Pauson–Khand reaction, 812
Asymmetric transfer hydrogenation, 634–636
Asymmetric transformation, 550–551
Atactic polypropylene, 1057, 1060, *1061*
Atom abstractions, 309–310
    ligand substitutions and, 243–244
Autan, *648*
Avecia, 634, 635
Axial chiral backbones, 603–605
Azaborindenyls, 118
1,2-Azaborolyls, 118
Azaferrocene, 155

Azetidinone, 567
Azide, 514
Aziridination, 521
Aziridines, 517
    carbonylation
        introduction to, 784
        mechanism, 792–794
        ring-expansion carbonylation, 784–787
        types of catalysts and scope of substrates, 790–792
Azlactones, 990
Azobenzene, 273
Azolyl complexes
    bonding, 155–156
    reactivity, 157–158
    synthesis, 156–157
Azolyl group, 155

B

Backbonding, 19–20
    in carbon monoxide binding, 29–30
    nucleophilic attack and, 419
    in terminal carbonyls, 30–31
"Backskipping," 1058, 1068
Bayrepel, 648
B–B bond, 291
Bent metallocenes
    defined, 113
    structure, 115–117
Benzene
    π-donation and, 22–24
    reductive elimination of, 329–330
    See also Arene complexes
η^6-Benzoic acid complex, 54
Benzonitrile, 656
Benzyl complexes
    η^3-complexes, 57
        nucleophilic attack and, 425
        overview, 108–109
    π-complexes, 713–714
    hydroamination by attack of amines on, 713–714
    nucleophilic attack and, 425
Benzyl ligands, *85, 86,* 361
Benzyltungstenocene complex, 366
Benzyne complexes, 52
Berry pseudorotation, 224, 225
Biaryl ethers, 344
Biarylplatinum complexes, 334
Biaryls, 334, 335, 887
Bidentate phosphines, 561
Bimetallic clusters
    ligand substitutions and, 253–254
BINAP, 561, 562, *563*, 577, 605
    asymmetric hydrogenation
        of α-(acycloxy)acrylates, 616
        of β-diketones, 624
        of β-keto esters, 622, 623

enantioselective Heck reaction and, 889
    hydroamination of vinylarenes and, 705
    synthesis, 603, 604
BINAP monoxide, 63
BINAPHANE ligands, 608
BINAPHOS ligand, 768
η^5-Binding mode, 155, 156
BINOL, 561
Biotin, 606
Biphemp, 624
Bipyridines, 58
Birch reduction, 473
Bis-sulfonamide complexes
    bonding, 165–166
    overview, 165
    synthesis, 166
    thermodynamics of metal–bis-sulfonamido bonds, 166–167
Bis(benzene)chromium, 53
    π-donation and, 22–24
    symmetry, 54
BISBI, 760
Bisimines, 58
Bisiminylpyridine complex, *373*
Bismuthines, 41
Bisoxazolines, 58
BisP* ligands, 609
Bisphosphine diamine catalysts, 626
Bisphosphine ligand, 985
Bisphosphines, *34,* 40
    aliphatic, 608
    wide-bite-angle, catalysts containing, 760–762
Bispyrazolylborate (Bp), 168
Bite angle
    bisphosphines-containing catalysts and, 760–762
    effect on C–C bond-forming reductive elimination, 335
    effect on rate of oxidative addition, 312
    "natural," 760
Block copolymers, 1049
    ethylene–propylene, 1069–1070
Bond dissociation energies (BDE)
    ligand substitution process, 220–223
    See also Bond strengths
σ-Bond metathesis, 689
Bond strengths
    in transition metals, 9–10
    See also Bond dissociation energies
Borane complexes, 64, 66, 186, 187
Boranes, 190
Boratabenzenes, 56, 118, 119
Borazines, 56
Borene ligand, 186
Borohydride, 169
Borohydride complexes, 127

Borollides, 118
Boronic acids, 880, 933
Boryl complexes
    bonding, 187
    intermolecular oxidative addition of
        C–H bonds, 278
    overview, 186–187
    reactivity, 190
    σ-bond metathesis, 286
    synthesis, 188–190
    thermodynamics, 188
Boryl group, 186
Borylation
    alkanes, 852–853
        mechanism, 855–857
    arenes, 853–854
        mechanism, 855–857
    polyolefins, 855
Borylene ligands, 187, 510
σ-Bound ligands
    nucleophilic attack, 422–427
BPE ligand, 608
Bridging thiolates, 194–195
Bromobenzene, 945–946
2-Bromoethylbenzene, 306
Brown's cationic rhodium system, 584
Buchwald–Hartwig reaction, 907–909
Bulk samples, 1048
Butadiene, 668, 670
    cyclooligomerization, 1090–1092
    hydroesterification, 782
    linear oligomerization, 1088–1090
t-Butyl, 128
sec-Butylmagnesium bromide, 899
sec-Butylpalladium complex, 900
Butyraldehyde, 751

C
C₂ symmetry, 559, 576
Caged radical-pairs, 305
Cahn–Ingold–Prelog convention, 551
C(alkyl)–C bonds, 933–935
Calvin, Melvin, 576
Calyculins, 726, 727
Carbapenem antibiotics, 567, 623
Carbazoles, 155
Carbene complexes
    agostic interactions, 71
    bonding, 44–45
    classes, 482
    defined, 481
    electronic properties of Fischer and
        Schrock carbenes, 483–484
    electrophilic attack on, 466
    first synthesis, 419
    α-hydrogen elimination, 412–413
    migratory insertions, 365–366
    nucleophilic attack on, 421–422
    π-bonding, 20
    properties, 41–44

reactivity
    alkylidene and alkylidyne
        complexes, 498–504
    Fisher complexes, 492–497
    vinylidene complexes, 498
spectroscopic characteristics, 45
synthesis
    alkylidene complexes, 488–491
    Fischer complexes, 484–486
    N-heterocyclic complexes,
        491–492
    vinylidene complexes, 486–488
[2+2] additions, 287
Carbene insertions
    C–H bond functionalization and
        intermolecular, 867–869
        intramolecular, 865–867
        overview, 864–865
    migratory, 365–366
Carbide complexes, 481
Carbon dioxide
    electrophilic insertion reactions,
        464–465
Carbon monosulfides, 31, 32
Carbon monoxide
    backbonding, 29–30
    coordinated, oxidation, 246
    copolymerization with olefins
        CO and α-olefins, 804–808
        development of catalysts for
            CO/ethylene, 798–799
        mechanism for CO/ethylene,
            800–803
        overview, 798
    isoelectronic analogs, 32–33
    nucleophilic attack, 419–421
    π-bonding, 19–20
    properties, 27–28
    See also Carbonyl complexes;
        Carbonyls
Carbon monoxide/ethylene
    copolymerization
    development of catalysts, 798–799
    mechanism
        chain propagation, 800–802
        chain termination and catalyst
            decomposition, 802–803
Carbon monoxide insertions
    catalysis of, 362–364
    into metal-hydrocarbyl complexes,
        351–352
    metal–alkyl complexes
        18-electron complexes, 354–355
        16-electron d⁸ complexes,
            355–356
        solvent effects, 359–360
        stereochemistry at carbon,
            356–357
        stereochemistry at the metal,
            357–358

    structure of the unsaturated
        intermediate, 358–359
    migratory aptitudes of R, 360–362
    into M–X bonds, 352–354
    overview, 351
Carbon nucleophiles
    attack on coordinated olefins, 433
Carbon–heteroatom coupling reactions,
    907–914, 920–921
Carbon–nitrogen cross coupling
    of amides with aryl halides, 925
    of amines, 922–924
    of aryl halides with heterocyclic
        amines, 925
Carbonyl clusters, 254
Carbonyl complexes, 28–29, 471–472
Carbonylation
    arenes and alkanes
        alkylative, 837–838
        direct carbonylation to
            aldehydes, 838–839
        oxidative, 835–836
    epoxides and aziridines, 784–794
    organic halides
        to form esters and amides,
            795–798
        to form ketones, 914–917
        overview, 794–795
    palladium-catalyzed, 426
    photochemical, 838–839
    See also Catalytic carbonylation
Carbonylative copolymerization, 785
Carbonylative coupling processes
    carbonylation of organic halides to
        form ketones, 914–917
    formylation of organic halides,
        917–918
    overview, 914
Carbonylative Stille reactions, 914–915
Carbonylative Suzuki reactions, 915
Carbonyls
    bridging, infrared and X-ray
        diffraction data, 31
    coordinated, oxidation of, 246
    ligand substitutions and, 253–254
    terminal, backbonding, 30–31
    thermodynamics of the M–CO
        bond, 31–32
Carboranyl ligands, 118
Carboxylates
    intermolecular oxidation of olefins,
        722–724
    intramolecular oxidation of olefins,
        724–726
Carboxylic acids, 779, 835
Carboxylic anhydrides, 654–656
Carbyne complexes
    bonding and structure, 45–46
    conversion of Fischer carbenes to,
        493–494

defined, 45, 481
electrophilic attack on, 466
nucleophilic attack on, 421–422
π-bonding, 20
preparation, 45
spectroscopic properties, 46
Carenoids, 482
Carilon, 798
Catalysis
asymmetric (*see* Asymmetric catalysis)
energetics of, 540
homogeneous and heterogeneous, 546–549
quantification of efficiency, 545–546
terminology, 543–546
See also Catalytic reactions
Catalyst deactivation, 545
Catalyst precursors, 544
"Catalyst resting state," 546
Catalysts
for C–N coupling, 910–911
definition of, 539
effects of structure on cross-coupling
chelation, 899–901
ligand electronic properties, 902–903
steric properties, 901–902
homogeneous and heterogeneous, 546–549
See also specific catalysts
Catalytic allylic substitution reactions, 967
See also Allylic substitution
Catalytic asymmetric hydrosilylation, 570
Catalytic asymmetric Pauson–Khand reaction, 812
Catalytic carbonylation
to form acetic acid, 746–751
hydroformylation of olefins, 751–769
overview, 745
palladium-catalyzed, 426
Catalytic cycle, 543–544
Catalytic direct oxidative cross couplings, 949–950
Catalytic enantioselective reactions, 553–558
Catalytic hydroboration, 691–692
Catalytic hydrocyanation. *See* Hydrocyanation
Catalytic metathesis. *See* Metathesis
Catalytic rates, 546
Catalytic reactions
kinetics of, 546
reaction coordinate diagrams of, 540–542
resting states, 546

transition-state stabilization, 542–543
See also Catalysis
Catalyzed ligand substitution reactions, 242–243
Catellani, Marta, 946
Catellani reaction, 946
CATHy catalysts, 634, 635, *636*
Cationic ligands, 6
Cationic rhodium catalysts, 580–581
hydrogenation by
with complexes containing alkylphosphines, 592–593
with complexes containing aromatic phosphines, 590–592
Cativa™ process, 746, 749–751
C–C bond-forming reductive elimination
effect of bite angle, 335
effect of coordination number, 334–335
effect of participating groups, 332–334
survey of, 336–338
trends and principles, 331–332
C–C bonds
copper-catalyzed cross coupling to form, 933–938
oxidative addition and, 264–266, 289–291
reductive elimination to form, 331–338
C–C multiple bonds
catalytic metathesis, 1015
Celanese Corporation, 756
CH and CH+
isolobal analogy, 17
C–H activation, 272, 825, 826
C–H bond-forming reductive elimination
effect of ancillary ligands on, 329–330
evidence for intermediate alkane and arene complexes, 327–329
examples, 326–327
overview and principles, 325–326
C–H bond functionalization
of alkanes and arenes with main group reagents, 852–859
by carbene insertions
intermolecular, 867–869
intramolecular, 865–867
overview, 864–865
carbonylation of arenes and alkanes, 835–839
dehydrogenation, 839–846
directed, 857
directed oxidations, aminations, and halogenations of alkanes and arenes, 832–835
H/D exchange, 869

hydroacylation, 859–864
hydroarylation, 846–852
overview, 825–826
platinum-catalyzed alkane and arene oxidations
catalysts, 827–829
mechanism, 829–832
C–H bond oxidative addition, 264–266
alkane additions
examples, 281–282
selectivity of, 278–279
dinuclear activation of hydrocarbons, 282–283
early history, 272–273
intermolecular, 275–278
intramolecular, 273–275
introduction to, 272
mechanism, 279–281
synthetic applications of addition of alkyl groups, 282
C–H bonds
2+2 additions, 287–289
aliphatic, 71–72
cleavage, direct arylation and, 944–945
oxidative addition, 271–283
reactions with imido compounds, 521
reactions with oxo compounds, 522–523
reductive elimination to form, 325–330
σ-bond metathesis, 283–286
C–H σ-bonds
[2+2] reactions with, 503–504
Chain-end control, 1049
Chain growth, 1049
"Chain running," 1066
"Chain straightening," 1066
Chain transfer
in alkene polymerization, 1076–1078
generation of stereoblock copolymers, 1065
Chain transfer agents, 1050–1051, 1076–1078
Chalk–Harrod mechanism
described, 688
evidence for the modified mechanism, 688–689
hydrosilylation of alkynes, 690
overview, 686–687
Chan–Evans–Lam couplings, 932–933
Charged ligands, 1
Chatt–Dewar–Duncanson bonding, 47
Chauvin, Yves, 1015
Chelating phosphines, 34–35
Chelation
cross-coupling reactions and, 899–901
Chien catalysts, 1063, *1064*

"Chiral-at-phosphorus" compound, 36
Chiral backbones
    axial, 603–605
    ferrocenyl, 606–607
Chiral lanthanide alkoxide catalysts, 176
Chiral lanthanide binaphtholate, 176
Chiral lanthanide biphenolate, 176
P-chiral phosphines, 609
Chiraphos complex, 560
β-Chloride eliminations, 409
Chlorides, 203–204
β-Chloroalkyl complexes, 409
Chlorohydrins, 730
Chloroplatinic acid, 678
Chromium catalysts
    ethylene polymerization, 1052
    hydrogenation of alkynes and
        conjugated dienes, 642–644
Chromocene, 114
Cis coordination, 331
"Cis effect," 203, 228–229, 238–240
Cis–trans isomerization, 224–225
Classical resolutions, 563
ClBcat, 189
C$_5$Me$_5$ ligand (Cp*), 112
C$_5$Me$_5$MgCl(THF), 112
C–N coupling
    catalysts for, 910–911
    mechanism, 911–914
Coal-to-chemicals, 749
Cobalt catalysts, 545
    phosphine-ligated, 755
Cobalt complexes
    C–C bond-forming reductive
        elimination and, 333
    coupling of aliphatic electrophiles
        and, 883
    insertions of olefins, 371
    for the Pauson–Khand reaction,
        810–811
Co–C bond, 87
Co(CO)$_4$–, 790, 792
(CO)$_4$Co-BH$_2$·X, 187
Co-dimerization, 1084
Coenzyme A, 621
Collins catalysts, 1063–1064
Collman's reagent, 189
η^1-Complexes
    C–O bond lengths, 53
η^2-Complexes
    C–O bond lengths, 53
Concerted oxidative additions
    reagents with C–X bonds of
        medium polarity, 310–313
    reagents with H–X bonds of
        medium polarity, 313–315
Condensation polymerization, 1049
"Cone angle," 38–39
    amine ligands, 57
Conjugated dienes

hydrogenation
    chromium-catalyzed, 642–644
    overview, 640–641
    palladium-catalyzed, 642–644
    rhodium-catalyzed, 641–642
oligomerization and polymerization
    cyclooligomerization of 1,3-
        dienes, 1090–1092
    linear oligomerization of
        butadiene, 1088–1090
    overview, 1086–1087
    polymerization of 1,3-dienes,
        1087–1088
[Co(NO)(NH$_3$)$_5$]$^{2+}$, 160
Constrained Geometry Catalysts (CGCs),
    1054
Coordinate bonds, 1
Coordinated CO, oxidation, 246
Coordinated dihydrogen complexes, 64
Coordination number
    C–X bond-forming reductive
        elimination and, 343
    defined, 14
    effect on C–C bond-forming
        reductive elimination, 334–335
    effect on reductive elimination rates,
        323–324
Copolymerization, 790
    carbon monoxide and olefins
        CO and α-olefins, 804–808
        development of catalysts for
            CO/ethylene, 798–799
        mechanism for CO/ethylene,
            800–803
        overview, 798
Copolymers
    ethylene–α-olefin, 1067–1070
    overview, 1049
Copper catalysts
    for carbon–heteroatom coupling
        reactions, 920–921
    C(sp^3)–C(sp^3) coupling, 934–935
    hydrosilylation of ketones and
        imines, 684–685
    insertions of carbenes into C–H
        bonds, 865
    "ligandless," 920
Copper-catalyzed allylic substitution
    enantioselective
        diorganozinc reagents as
            nucleophiles, 1002–1004
        Grignard reagents as
            nucleophiles, 1004–1006
        organoaluminum reagents as
            nucleophiles, 1006
        overview, 1001
    fundamentals, 999–1000
    mechanism, 1000–1001
    miscellaneous reactions, 1007–1008
Copper halides, 943

Copper-mediated cross-coupling
    reactions
        aryl boronic acids with amines and
            alcohols, 932–933
        aryl halides with alcohols and
            phenols, 926–930
        aryl halides with amines, alcohols
            and thiols, 930–932
        carbon–nitrogen reactions, 922–926
        classes of catalysts for carbon–
            heteroatom bond formation,
            920–921
        to form C(aryl)–N, C(aryl)–O and
            C(aryl)–S bonds, 920–930
        to form C–C bonds, 933–938
        introduction to, 918–920
        overview, 878
Copper–alkynyl species, 97–98
Corin complexes, 163
Corin ring system, 163
Cossee mechanism, 371, 1050
Cossee–Arlman mechanism, 375
Coupling reactions
    carbonylative, 914–918
    C–C bond forming
        at aliphatic electrophiles,
            882–883
        alkynes, 880–881
        applications, 903–906
        cyanide, 883–884
        enolates and related reagents,
            881
        olefins, 883
        organoboron reagents, 880
        organomagnesium reagents,
            878
        organosilicon reagents,
            879–880
        organotin reagents, 879
        organozinc reagents, 878–879
    See also Cross coupling reactions
Covalent ligands, 1
Cp. See Cyclopentadienyl ligands
Cp* ligands, 112, 115
(Cp*)$_2$Co$_2$(μ-H)$_3$, 123
Cp$_2$Fe, 114
CpFe(CO)$_2$, 304, 308
CpFe(CO)(Me)L, 358
Cp*Ir(PMe$_3$)(Ph)(OH), 383
Cp$_3$Ln, 114
Cp*$_3$Ln, 114
Cp$_2$M, 113, 114–115
Cp*$_2$M^{251}, 114
Cp*$_3$M, 114
Cp$_2$ML$_x$, 113, 115–117
CpML$_y$, 113, 117–118
Cp*$_2$Mn, 115
(Cp*)$_2$OsH$^+$, 136
Cp*$_2$Re, 115
[CPRu(NCMe)$_3$]$^+$, 682

Cp*$_3$U, 114
Cp$_2$ZrCl$_2$, 115
Cp$_2$Zr(H)Cl, 368
Cp$_2$Zr(II)L$_2$, 308–309
Cp$_2$Zr(SiPh$_3$)Cl, 198
Crabtree's catalyst, 548, 582–583, 594–595, 869
Cramer's mechanism, 371
[Cr(arene)(CO)$_3$], 642
Cr–benzyl bond, 86
Cr(CO)$_4$(butadiene), 247
Crixivan, 613
Cross coupling mechanisms
    effects of catalyst structure on
        chelation effects, 899–901
        ligand electronic properties and, 902–903
        steric properties and, 901–902
    overall catalytic processes, 890–893
    steps in
        oxidative addition, 893–895
        reductive elimination, 899
        transmetallation, 895–899
Cross coupling reactions
    applications, 903–906
    copper-mediated
        aryl boronic acids with amines and alcohols, 932–933
        aryl halides with alcohols and phenols, 926–930
        aryl halides with amines, alcohols and thiols, 930–932
        carbon–nitrogen reactions, 922–926
        classes of catalysts for carbon–heteroatom bond formation, 920–921
        to form C(aryl)–N, C(aryl)–O and C(aryl)–S bonds, 920–930
        to form C–C bonds, 933–938
        introduction to, 918–920
    direct oxidative, 949–950
    enantioselective, 884–890
    mechanisms (see Cross coupling mechanisms)
    with organomagensium reagents, 878
    overview, 877–878
    reactions that form carbon–heteroatom bonds, 907–914
    See also Coupling reactions
Cross metathesis, 1016, 1017
    alkyne, 1038–1039
    history, 1020
    olefin, 1026–1028
Crotononitrile, 680
C(sp^3)–C(sp^3) coupling
    copper-catalyzed, 934–935
    mediated by copper reagents, 933–934

Cu(acac)$_2$, 186
Cuprates
    "higher-order" and "lower-order," 933
Curtin–Hammett conditions
    described, 555
    examples of reactions under, 556–557
Curtin–Hammett principle, 638
    asymmetric allylic alkylation, 557–558
    asymmetric hydrogenation, 556–557
    theory of, 555–556
"Curtin–Hammett situation," 193
C–X bond-forming reductive elimination
    mechanisms of, 338–341
    overview, 338
    survey of, 341–345
C–X bonds
    concerted oxidative additions and, 310–313
    reductive elimination to form, 338–345
Cyanide
    coupling, 883–884
    properties, 102
    toxicity, 102
Cyanide complexes
    overview, 102
    spectral features, 103
    structures and electron counting, 102
    synthesis, 103–104
    thermodynamics of M–CN linkages, 102–103
Cyanoalkyl complexes, 100, 402
Cyanoesters
    coupling, 936
Cyclic allylic monoesters, 993–994
Cyclic imines, 629–630
Cycloadditions
    [2+2]
        hydroamination by, 716–717
        with imido and oxo groups, 515–518
    [3+2], 515–518
    with imido and oxo groups, 515–518
    nitrido complexes, 529
Cycloheptadienyl, 6
Cycloheptatriene, 709
Cycloheptatrienyl complexes, 442–446
Cycloheptyne, 52
Cyclohexadienes, 442
Cyclohexyne, 52
Cyclometallation, 183, 273–275
Cyclooligomerization, 1090–1092
Cyclopentadiene, 111, 112
η^1-Cyclopentadienyl complexes, 120
η^5-Cyclopentadienyl complexes, 111–112
Cyclopentadienyl ligands ("Cp")

bonding and thermodynamics, 111
electrophilic attack on, 474–475
isolobal relationship, 16–17
ligands electronically similar to, 118–120
modes of binding, 118
nucleophilic attack on, 441–442
overview, 111
reactions, 120–122
resonance structures, 5–6
similarity of trispyrazolylborate to, 168
substituted, 112–113
synthesis, 111–112
types of complexes, 113–118
Cyclopentanones, 860, 861
Cyclopentyltungsten(IV) alkyl complex, 412
Cyclopentyne, 52
Cyclopropanations, 495–496
Cy$_3$PH$^+$, 130
Cytochrome c oxidase, 102
Cytochrome P$_{450}$, 509

D
d^0 Complexes, 283–285
d-Electrons
    counting, 10
    geometries of transition metal complexes and, 15
    oxidation state and, 7, 10
d^8 Square-planar Pt(II), 307
Dative covalent bonds, 1
Dative ligands, 1
De-insertion reactions
    of CO (see Decarbonylation)
    overview, 397
    See also Elimination reactions
Decarbonylation
    overview, 349, 350
    stereochemistry at carbon, 356, 357
DEGPHOS, 608
Dehydro α-amino acids, 612–614
Dehydro β-amino acids, 614–615
Dehydrogenation
    alkane metathesis via, 842–844
    catalyzed by compounds of pincer ligands, 840–842
    early studies, 839–840
    mechanism, 844–846
Dehydrogenative silylation, 681
Denopamine hydrochloride, 624
Desymmetrization reactions
    examples of, 570–571
    overview, 569
Dialkoxyboranes, 691
Dialkyl ketones, 628
Dialkylamido ligands, 25
Diamagnetic complexes, 7
Diaminations, 730–731

Diamines, 57, 58
1,2-Diamines, 58
Diaminocarbene complex, 43
Diaryl ketones, 628
Diastereomeric transition states, 552
Diazoalkanes, 485
Dibenzo[a,e]cyclooctatetraene, 548
Diborane reagents, 189
Diboration
    alkenes, 698–699
    mechanisms, 699
    overview, 691, 697
Diboron reagents, 1007
1,3-Dicarbonyl compounds, 567–568
Dicyano metal complexes, 103
Dicyclopentadiene, 1032
Diels–Alder reactions, 542, 543
Diene complexes, 48
    achiral, desymmetrization, 570
    asymmetric hydrocyanation,
        675–676
    η⁴-complexes, 439–441
    hydride abstraction by electrophilic
        attack, 472–474
    hydroboration, 693
    hydrocyanation, 673–674
    hydrosilylation, 680–681
    intramolecular oxidation, 726
    nucleophilic attack on, 433–434
    substitutions for, 247–248
    synthesis of allyl complexes and,
        108
    See also 1,3-Dienes; Conjugated
        dienes
1,3-Dienes
    cyclooligomerization, 1090–1092
    hydroamination, 708–709, 713–714
    hydroboration, 693
    polymerization, 1087–1088
η⁵-Dienyl complexes, 441–442
Difunctionalizations
    of olefins, 730–731
Dihaloalkanes, 316
Dihydride complexes, 324
Dihydrogen
    heterolytic activation by late
        transition metal complexes, 285
    oxidative addition, 266–270
    photochemical reductive
        elimination of, 324, 325
    σ-bond metathesis, 284
"Dihydrogen bond," 136
Dihydrogen complexes
    Kubas compound, 65
    overview, 64
    oxidative addition and, 267
    reactivity, 68–70
    significance of, 66
    spectroscopic signatures, 67–68
    stability, 67

Dihydroimidazolium precursors, 491
α-Diimines, 58
Diiodorhodium complex, 303
Diiodoiridium–carbonyl complex, 303
β-Diketiminate complexes
    overview, 170
    structure and bonding of, 170–171
    synthesis, 171–172
β-Diketimines, 171, 172
β-Diketonate complexes, 185–186
Diketones, 171, 624, 936
β-Diketones, 624, 936
Dimerization. See Olefin dimerization
Dimethylpalladium complexes, 331
α-Dimine ligands, 1055
Dinitrogen complexes, 59–61
Dinuclear complexes, 289, 315–317
Dinuclear Rh(II) carbene complexes, 289
1,3-Diols, 789
DIOP ligand, 577
Diop* ligand, 607–608
Diorganozinc reagents, 11002–1004
Diphenylacetylene, 52
1,3-Diphenylallyl acetate, 987
DIPHOS catalyst, 590–591
Diphosphine ligands, 759–763
Direct arylation, 878
    comments on reaction conditions
        for, 948–949
    mechanisms, 938–939
    overview, 938
    regioselectivity, 943–948
    transition metal catalysts for,
        939–943
Direct oxidative cross couplings, 949–950
Directed C–H functionalizations, 857
Directed hydroarylation
    alkynes, 850
    olefins
        mechanisms, 849
        overview, 846–847
        reaction scope and catalysts,
            847–849
Directed hydrogenation, 584–585
Directed intermolecular hydroacylation,
    863–864
Directed oxidations, 832–835
Dirhodium complexes, 865, 866
Disilanes, 291, 292
Disilation, 690–691
Disilylation, 677, 690–691
Dissociation reactions
    photochemically induced, 244–246
Dissociative interchange (I_D), 218, 219, 242
Dissociative substitution reactions
    assisted reaction pathways, 241–247
    cis effect, 238–240
    distinguishing by kinetic
        methods, 220

18-electron complexes, 233–240
    kinetics, 233–235
    of Ni(CO)₄, 235
    stereochemistry of substitutions of
        octahedral compounds, 240
    steric effects on, 235–236
    trends for, 219
Dissociative substitutions
    versus associative substitutions
        of square-planar complexes,
        229–231
    overview, 217
    stereochemistry, 236–237
    of weakly bound ligands, 237–238
DME, 62
DMSO, 62, 63–64
    dissociative substitution reactions
        and, 229–230
    migratory insertion of carbon
        monoxide and, 359–360
L-Dopa, 556, 576, 636
Dormant states, 546
Dötz reaction, 496–497
Double carbonylation
    epoxides, 785
DuPhos ligand, 577, 608, 616
DuPhos monoxide, 63
Dynamic catalytic asymmetric
    transformation (DYCAT), 988
Dynamic kinetic asymmetric
    transformation (DyKAT), 568–569
Dynamic kinetic resolution (DKR),
    567–568, 623

E
Early metal–thiolate bond, 195
"Early–late heterobimetallic" complexes,
    315, 316
Eastman Chemical, 748–749
E–E bonds, 291–292
Effective atomic number rule, 13
18-electron complexes
    assisted reaction pathways for
        ligand substitution, 241–247
    dissociative substitution reactions,
        233–240
    ligand substitution mechanism, 219
    substitution of weakly bound
        ligands, 237–238
18-electron rule, 13, 219
Elastomeric polypropylene, 1063–1064
Elastomers, 1057
    thermoplastic, 1062–1063
"Electron count," 10
Electron counting, 10–12
    in polynuclear complexes, 13–14
Electron transfer, 242–243
Electron-transfer mechanism
    outer-sphere, 308–309
Electroneutrality principle, 6

Electrons
  donated in metal–ligand bonds, 3–4
Electrophiles
  reactions with imido and oxo
    complexes, 523–524
  unsaturated
    cleavage of metal–carbon and
      metal–hydride σ-bonds,
      454–462
    examples, 453
Electrophilic A–B, 315–317
Electrophilic attack
  on η²-arene and heteroarene
    complexes, 475–476
  cleavage of metal–carbon and
    metal–hydride σ-bonds, 454–462
  of coordinated ligands
    at the α-position, 465–466
    at the β-position, 466–468
    at the γ-position, 469–470
  on coordinated olefins and
    polyenes, 471–476
  insertion reactions with carbon
    dioxide, 464–465
  insertion reactions with sulfur
    dioxide, 462–464
  overview and basic principles, 453
  on π-polyenyl complexes, 474–475
Electrophilic cleavage, 454–462
Electrophilic insertion reactions, 462–465
Elimination reactions
  changes in geometry and electron
    count during, 350
  defined and described, 349
  overview, 397–398
  significance of, 350
  summary of, 413–414
  See also α-Eliminations;
    β-Eliminations; α-Hydrogen
    eliminations; β-Hydrogen
    eliminations
α-Eliminations
  formation of metal–imido
    complexes, 513
  formation of metal–oxo
    complexes, 514
  overview, 398
  See also α-Hydrogen eliminations
β-Eliminations, 349, 350
  β-alkyl eliminations, 406–408
  β-aryl eliminations, 408
  in chain termination, 1051
  β-hydrogen eliminations, 398–406,
    1054 (see also β-Hydrogen
    eliminations)
  overview, 397
γ-Eliminations, 398
Enamides, 612–616, 630
Enamines, 706
Enantiomeric ratio (er), 551, 552

Enantiomers, 551
Enantioselective allylic substitutions,
    968–969
  of acyclic electrophiles
    symmetric acyclic allylic
      esters, 987
    unsymmetric acyclic
      substrates, 988–993
    catalysts for, 985–987
  of cyclic substrates
    cyclic allylic monoesters,
      993–994
    meso cyclic diesters, 994–995
  forms of, 984–985
  kinetic resolution, 995–996
  of prochiral nucleophiles, 996–998
Enantioselective copper-catalyzed allylic
  substitution
    diorganozinc reagents as
      nucleophiles, 1002–1004
    Grignard reagents as nucleophiles,
      1004–1006
    organoaluminum reagents as
      nucleophiles, 1006
    overview, 1001
Enantioselective cross-coupling,
  884–890
Enantioselective Heck reaction, 888–889
Enantioselective hydroformylation,
  765–769
Enantioselective hydrogenation, 573
Enantioselective reactions
  reaction coordinates of, 553–558
Enantioselective transformations, 63,
  204
Enantioselectivity, 552–558
Enchainment, 1049, 1066
Encounter complex, 218–219
"End-on" binding mode, 29
Engage™, 1067
Enolate complexes
  overview, 98
  spectral features, 100–101
  structure, 98–100
  synthesis, 101
η³-Enolate complexes, 100
Enolates
  asymmetric coupling with aryl and
    vinyl halides, 886–887
  coupling, 881
  Fischer carbene complexes react
    like, 494
Enyne metathesis, 1016, 1017
  examples, 1040–1041
  mechanism, 1041–1042
Epimerization, 356
Epothilones, 904, 1024–1025
Epoxidations, 519–521
Epoxides
  alkoxycarbonylation, 789

carbonylation
  examples, 785
  introduction to, 784
  mechanism, 792–794
  ring-expansion carbonylation,
    784–787
  ring-opening carbonylation,
    788–792
  to succinic anhydrides,
    787–788
  types of catalysts and scope of
    substrates, 786–787
  hydroformylation, 789
Epoxydictymene, 814
Esterification, 796, 797–798
Esters
  binding modes, 53
  formed by carbonylation of organic
    halides, 795–798
  homogeneous hydrogenation, 651–653
  reductive elimination of, 344
Et₂Fe(bpy)₂, 337–338
Ethane, 264–266
Ether complexes, 62
Ethyl
  M–ethyl bond, 86
  "Ethyl process," 1080
Ethyl (Z)-α-acetamidocinnamate, 556
Ethylene
  co-dimerization, 1084
  CO/ethylene copolymerization
    development of catalysts,
      798–799
    mechanism, 800–803
  hydrocyanation, 671
  hydroesterification, 778–779
  insertion into metal–acyl bonds,
    377–379
  intermolecular addition of ammonia
    to, 700
  Shell Higher Olefin Process, 1080–1082
  trimerization and tetramerization,
    1084–1086
Ethylene-based polymers, 1051–1052
Ethylene bis(tetrahydroindenyl)
  ligand, 619
η²-Ethylene complexes, 280
Ethylene oligomerization, 1080–1082
Ethylene polymerization
  catalysts, 1047–1048
  significance of, 1047
Ethylene–α-olefin copolymers
  alternating ethylene–propylene,
    1068–1069
  ethylene–propylene block
    copolymers, 1069–1070
  overview, 1067
Ethylenebis(indenyl) (EBI), 116
Ethylenebis(tetrahydroindenyl)
  (EBTHI), 116

Ethylene–propylene block copolymers, 1069–1070

Ethylene–propylene rubber (EPR), 1067

F

Fe(CO)$_5$, 28

Fe(CO)$_4$2- ("supernucleophile"), 302

"Fences," 45

Ferrocene, 86, 111
　　reactions, 120–121
　　significance of, 114

Ferrocenyl backbones
　　chiral, 606–607

Ferrocenylphosphine, 63

"First-generation Grubbs catalyst," 492

Fischer, E. O., 42, 53, 482

Fischer carbenes, 42, 45
　　nucleophilic additions, 421
　　origin of the electrical properties of, 483
　　overview, 482
　　reactivity, 492–497
　　　　conversion to carbyne complexes, 493–494
　　　　cyclopropanations, 495–496
　　　　Dötz reaction, 496–497
　　　　reactions related to those of enolates, 494
　　　　reactions with nucleophiles, 493
　　synthesis, 484–486

"Fischer-type" carbyne complexes, 45

Fluorenyl, 112

β-Fluoride eliminations, 409, 410

Fluoride ligands, 201, 204

Fluorides, late-metal, 180

Fluoroalkyl complexes, 89

Fluouroalkane complexes, 237

Formate, 128

Formyl ligands, 360

Formylation, 917–918

Fredericamycin, 496

Friedel–Crafts acetylation, 474

Frontier orbitals, 15, 16

Functionalized alkenes, 764–765

Functionalized ketones
　　asymmetric hydrogenations, 621–625
　　defined, 620–621

Furands, 651

G

Gauche effect, 192

Geometries
　　of transition metal complexes, 14–15
　　valence bond theory and, 18

Geraniol, 618

Gibson, C. S., 86

Gillman cuprates, 934

Gilman, Henry, 87n

Gold compounds, 337

Gold(I) complex, 317

Gold(III) complex, 337

Goldberg, Irma, 919

Goldberg reaction, 925

Green–Rooney mechanism, 375

Grignard reagent
　　in copper-mediated cross-coupling, 919, 933, 934
　　in enantioselective copper-catalyzed allylic substitutions, 1004–1006
　　in enantioselective cross-coupling, 884, 885
　　formation, 261
　　synthesis of alkyl complexes and, 87

Group 4 metal complexes, 710

Grubbs, Robert, 1015

Grubbs catalyst, 1017, *1018*
　　enyne metathesis, 1041
　　history, 1020
　　second-generation, 43

Grubbs' cyclization, 377

H

H/D exchange, 869

H/T exchange, 869

Haber–Bosch process, 540

Hafnocene complex, 407

Hafnocene–olefin complexes, 471

"Half-sandwich" compounds
　　defined, 113
　　structure, 117–118

Halide complexes
　　organic, carbonylations, 794–798
　　reactivity of metal–halide complexes, 203–204
　　transmetallation, 895–899

β-Halide eliminations, 409–410

Halide ligands
　　cis effect, 239
　　overview, 200–201
　　as π-donors, 25
　　steric and electronic properties, 201–202

Halides
　　late metal, 180, 181

Haloarenes
　　carbonylations, 795
　　nucleophilic attack, 444

Halogenation
　　of alkanes and arenes, 833–834

Halogens
　　oxidative additions and, 313, 315

Hard soft acid base (HSAB) principle, 177

HCo(CO)$_4$, 130–131, 646–647, 752–754

HCo(CO)$_3$(PR$_3$), 754–755

HDPE. *See* High-density polyethylene

Heck, Richard, 745

Heck reaction
　　applications, 904–906
　　coupling of olefins and, 883
　　desymmetrization via, 570–571

enantioselective, 888–889

enantioselective coupling and, 885

mechanism, 892–893

overview, 374

Heck's arylpalladium complex, 399

Hemiisotactic polypropylene, 1057, 1058, 1062

Hermann's organyl trioxorhenium species, 509

Heteroarenes, 56
　　asymmetric hydrogenation
　　　　five-membered ring heteroarenes, 650–651
　　　　overview, 648
　　　　six-membered ring heteroarenes, 648–650
　　electrophilic attack on, 475–476
　　homogenous catalytic hydrogenation, 644–645

Heterobimetallic complexes
　　"early–late," 315, 316

Heterocyclic amines, 925–926

N-Heterocyclic carbenes (NHCs)
　　bonding, 44
　　overview, 482
　　properties, 41, 42, 43
　　symmetry, 45
　　synthesis, 491–492

Heterogeneous catalysts
　　distinguishing from homogeneous catalysts, 547–549
　　overview, 546–547

H–H bonds
　　σ-bond metathesis, 283–286

High-density polyethylene (HDPE)
　　catalysts for synthesis of, 1052–1054
　　overview, 1051, 1052

High molecular weight HDPE, 1052

Higher-nuclearity clusters, 253–254

Highly branched polyethylene, 1048

Hiyama–Tamao couplings, 880, 902

HMn(CO)$_5$, 122–123

Hofmann's catalyst, 1017

Homocouplings
　　mechanism, 891–892
　　overview, 877

Homogeneous catalysts
　　distinguishing from heterogeneous catalysts, 547–549
　　overview, 546–547

Homogeneous catalytic hydrocyanation. *See* Hydrocyanation

Homogeneous catalytic hydrogenation
　　arenes and heteroarenes
　　　　monocyclic arenes, 647
　　　　overview of catalysts, 644–645
　　　　polycyclic arenes, 645–647

Homogeneous hydrogenation
　　carboxylic anhydrides and imides, 654–656

esters, 651–653
nitriles, 656–657
olefins and ketones, 585–602
perspectives on, 576–578
significance of, 575–576
Homogeneous hydrogenation
catalysts
iridium, 582–583
lanthanide crystals, 584
rhodium, 578–581
ruthenium, 583
"Homoleptic" alkyl complexes, 86
Homoleptic metal cyanides [M(CN)$_n$],
104
Hoveyda–Grubbs catalyst, 1017, 1018
H$_5$Re(PMePh$_2$)$_3$, 123
HRh(CO)$_2$(PPh$_3$)$_2$, 757–758
HRh(PPh$_3$)$_4$, 123
Hurtley reactions, 936–937
HX
oxidative addition, 264
H–X bonds, 310, 313–315
Hydrazido complex, 61
Hydrazine complex, 61
Hydricities, 133–136
"Hydricity scale," 455, 456
Hydride abstraction, 472–474
Hydride complexes
carbon monoxide insertions,
360–361
hydrogen bonding, 66
insertions of alkynes, 368–370
ligand substitutions, 243–244
as a nucleophile, 135–136
polarization, 136
thermodynamic hydricity, 134
"Hydride cycle," 782–783
α-Hydride elimination, 412
"Hydride" hydrogenation mechanism,
585, 586
Hydride ligands
abstraction of, 455
acidities, 129–131
bridging, 123–124
defined, 122
hydricities, 133–136
hydrogen bonding, 136
metal–metal bonding in polynuclear
complexes, 13
M–H bond and bond strength,
131–133
migration to oxo or imido ligand,
524–525
nucleophilic attack and, 420
overview, 122
spectroscopic properties, 124
structural features, 122–124
synthesis, 124–128
terminal, 122–123
Hydrido carbene complex, 412

Hydridometal cyanide complexes, 103
Hydroacylation
intermolecular, 860
directed, 863–864
intramolecular, 860–861
mechanism, 861–863
overview, 859–860
Hydroamination
by [2+2] cycloadditions, 716–717
introduction and fundamentals,
700–701
mechanisms
attack of amines on
π-complexes, 713–715
insertions of olefins into metal
amines, 715–716
overview, 712–713
scope of, 701
1,3-dienes, 708–710
alkenes, 526–527, 701–705
alkynes, 526–527, 710–712
allenes, 707–708
vinylarenes, 705–706
Hydroaminomethylation
history and overview, 769–770
mechanism, 774
scope of, 770–774
Hydroarylation
directed
of alkynes, 850
of olefins, 846–849
undirected, of olefins, 850–852
Hydroboration
alkenes, 542–543
asymmetric, 694
catalytic, history of, 691–692
metal-catalyzed, 692–694
olefins, 695–697
overview, 691
Hydrocarbons
dinuclear activation, 282–283
Hydrocarboxylation
catalysts for, 777–778
overview, 775
scope of, 778–784
synthetic targets, 775–777
Hydrocarbyl complexes, 351–352
β-Hydrocarbyl eliminations, 406–408
Hydrocarbyl ligands
alkyl ligands, 86–91
alkynyl complexes and, 97–93
example structures, 85
significance of, 85
vinyl complexes and, 96–97
Hydrocyanation
alkenes
examples, 668–670
mechanism of, 670–672
alkynes, 676
asymmetric

dienes, 675–676
overview, 674–675
deactivation, 673
dienes, 673–674
asymmetric, 675–676
introduction to, 668
mechanism of, 670–673
summary of, 676
Hydroesterification
alkenes and alkynes, 781
catalysts for, 777–778
overview, 775
scope of, 778–784
synthetic targets, 775–777
butadiene, 782
intramolecular, of olefins, 780–781
mechanism, 782–784
Hydroformylation, 350, 352, 596
epoxides, 785, 789
of olefins
catalyzed by HCo(CO)$_4$, 752–754
catalyzed by HCo(CO)$_3$(PR$_3$),
754–755
catalyzed by rhodium, 756–769
overview, 751–752
rhodium-catalyzed
catalysts containing chelating
diphosphine ligands, 759–763
catalysts containing
triarylphosphine ligands,
756–758
enantioselective, 765–769
of functionalized alkenes,
764–765
of internal alkenes, 763
overview, 756
by rhodium complexes of
phosphites, 763–764
water-soluble catalysts,
758–759
Hydrogen
synthesis of hydride complexes
from, 124–126
Hydrogen bonding
compared to σ complexes, 66
hydride ligands, 136
metal hydrides, 66
β-Hydrogen elimination rate
effect of ancillary ligands, 402
effect of conformation and
coordination number, 399–400
effect of electronics, 400–401
α-Hydrogen eliminations, 90, 398, 410–413
β-Hydrogen eliminations, 90, 349, 350
branching of LDPEs and, 1054
C–N coupling and, 913
hyperbranched polypropylene and,
1066
insertions into metal–hydride bonds
and, 366–368

β-Hydrogen eliminations (*Continued*)
  late-metal alkoxides and, 183
  from metal alkoxides and amides,
    402-405
  from metal–alkyl complexes,
    398–402
  from metal–silyl complexes, 405–406
  overview, 397
Hydrogen-first mechanism, 593
Hydrogenation
  alkynes and conjugated dienes
    chromium-catalyzed, 642–644
    overview, 640–641
    palladium-catalyzed, 644
    rhodium-catalyzed, 641–642
  arene, 577
  catalysts in, 544
  directed, 584–585
  enantioselective, 575
  imine, 600–602
  ionic, 602
  "outer-sphere" mechanism, 586, 587
  "transfer," 575–576, 634–636
  See also Asymmetric hydrogenation;
    Homogeneous catalytic
    hydrogenation; Homogeneous
    hydrogenation; Olefin
    hydrogenation
Hydrogen–hydrogen bonds
  σ-bond metathesis, 283–286
Hydrosilances, 197
Hydrosilation. *See* Hydrosilylation
Hydrosilylation
  asymmetric, olefins, 683–684
  desymmetrization by, 570
  examples
    alkynes, 681–683
    dienes, 680–681
    ketones and imines, 684–686
    olefins with achiral catalysts,
      679–680
    vinylarenes, 680
  history and types of catalysts,
    678–679
  introduction to, 677
  mechanism, 686–690
  purpose for, 677–678
Hydrosulfido complexes, 197
Hydrovinylation, 1084
Hydroxide ion, 420–421
Hydroxide ligands, 173
Hydroxo complexes, 173
β-Hydroxy aldehydes, 789
α-Hydroxy ketones, 625
Hydroxycarbonyl, 128
Hydroxylation, 832
(Hydroxymethyl)acylfulvene, 814
Hydrozirconation, 368
Hyperbranched polyethylenes, 1054–1057
Hyperbranched polypropylene, 1066–1067

I

Ibuprofen, 617
Ibutilide, 772, *773*
Imides, 654–656
Imido complexes, 154
  bonding of, 510–512
  defined, 481
  formed from α-hydrogen
    elimination of amido
    complexes, 413
  reactions with electrophiles, 523–524
  synthesis, 512–514
Imido compounds
  atom transfer to olefins, 518–521
  catalytic reactions through
    organometallic intermediates,
    525–527
  cycloadditions, 515–518
  overview, 509
  reactions with C–H bonds, 521
Imido ligand, 524–525
Imine complexes, 58
  asymmetric hydrogenation
    acyclic *N*-alkyl imines, 630–631
    acyclic *N*-aryl imines, 631–631
    aroylhydrazones and
      phosphinylketimines,
      632–634
    classes of catalysts for, 629
    cyclic imines, 629–630
    overview, 629
  asymmetric transfer hydrogenation,
    634–636
  hydrosilylation, 677, 684–686 (*see
    also* Hydrosilylation)
  insertion into metal hydrides,
    370–371
  insertion into metal–carbon bonds,
    381–383
  nucleophilic attack, 435–436
Imine hydrogenation, 600–602
Imine ligand
  binding mode, *59*
  π-bonding, 22
η²-Iminoacyl ligand, 364
"Indenyl effect," 120
Indenyl ligands, 112
Indinavir, 613
Indoles, 155, 650
Inner-sphere electron transfer
    oxidative addition by, 305
1,2-Insertions, 373, 382, 1066
2,1-Insertions, 372, 373, 1066
Insite™, 1067
Interchange mechanism (I), 218
Intermolecular hydroacylation, 860,
    863–864
Intermolecular oxidative amination, 728–729
Intermolecular Pauson–Khand reaction,
    812–813

Internal alkenes
  hydroaminomethylations, 771
  hydroformylation by
    HCo(CO)$_3$(PR$_3$), 755
  rhodium-catalyzed
    hydroformylation, 763
Intramolecular hydroacylation, 860–861
Intramolecular hydroesterification,
    780–781
Intramolecular oxidative amination, 730
Iodide, 303
Iodide ligand, 201, 202
Ionic hydrogenation, 602
Ionization potentials, 8–9
[Ir(acetone)$_2$(PCy$_3$)(pyridine)], 576
[Ir(COD)(Py)PCy$_3$], 582
  See also Crabtree's catalyst
Iridium catalysts, 582–583
  cationic, containing
    alkylphosphines, 594–595
  enantioselective allylic substitutions
    and, 991, 992–993
Iridium-catalyzed carbonylation, 749–751
Iridium complexes
  intermolecular oxidative addition of
    C–H bonds, 276
  oxidative addition by the S$_N$2
    pathway and, 304
  reductive elimination of alcohols
    and amines from, 331
  See also individual complexes
Iridium–aminoalkyl complex, 386
Iridium–carbonyl bond, 32
Iridium(I) alkoxo complexes, 402–403
Iridium(I) amido complexes, 402–403
Iridium(I) complexes
  β-hydrogen elimination, 402–403
  oxidative addition and, 314
Iridium(III) alkoxide complexes, 404
Iridium(III) alkoxo complexes, 404
Iridium(III) complexes
  α-hydrogen elimination, 412
  β-hydrogen elimination, 404
Iron complexes
  as catalysts in direct arylation, 943
  coupling of aliphatic electrophiles
    and, 883
Iron–olefin complexes, 429
Ir–vinyl bond, 97
Isocarbonyl binding mode, 29
Isocyanides, 838
  migratory insertion and, 364–365
  overview, 32–33
  π-acceptor, 19
  preparation of Fischer carbenes
    and, 484
  See also Isonitriles
Isodityrosines, 445
Isoelectronic, 15, *16*
Isolobal, 15, 16–17

Isonitrile ligand
    migratory insertion and, 364–365
Isonitriles, 32
    nucleophilic attack, 419–420, 421
    See also Isocyanides
"Isotactic" polymer
    propene and CO, 807–808
    styrene and CO, 806
Isotactic polypropylene, 1057
    symmetry, 1058
    synthesis, 1060–1062
        mechanism of stereocontrol in,
            1057–1060
Isotactic polystyrene, 1072
Isotactic–atactic stereoblock
    polypropylene, 1063–1064
"Isotopic perturbation of degeneracy," 72
"Isotopic perturbation of equilibrium," 72
Isotopic scrambling, 327–328

J
Jacobsen hydrolytic kinetic resolution,
    565
Jahn–Teller distortion, 245
JM-Phos ligand, 609
"Josiphos" ligands, 606

K
Karstedt's catalyst, 678, 686
α-Keto esters, 621–622
β-Keto esters, 622–623
Ketone ligands
    binding modes, 53
    π-bonding, 22
Ketones
    asymmetric hydrogenation
        functionalized ketones, 621–625
        overview, 620–621
        unfunctionalized ketones,
           626–629
    asymmetric transfer hydrogenation,
        634–636
    carbonylation of organic halides to
        form, 914–916
        mechanism, 916–917
    hydrogenation
        ionic, 602
        outer-sphere mechanism,
           600–602
        by $RuL_2(\kappa^2\text{-OAc})_2$ and
           $[RuL_2Cl_2]_2$, 597–599
    hydrosilylation, 677, 684–686 (*see*
        also Hydrosilylation)
    insertion into metal hydrides, 370–371
Ketyl radical anion, 243
Kinetic acidities, 131
Kinetic hydricities, 134–135
Kinetic resolution (KR), 551
    described, 563–564
    desymmetrization reactions and, 569

    dynamic, 567–568
    energetics of selectivity in, 565
    examples, 565–567
    quantification of selectivity in,
        564–565
Knovenagel condensation, 494
Knowles, William S., 575
Kubas's dihydrogen complex, 65

L
L_2 ligands, 2
L-type ligands, 2
Lactones, 787–788
Lanthanide alkoxide catalysts,
    chiral, 176
Lanthanide binaphtholate, chiral, 176
Lanthanide biphenolate, chiral, 176
Lanthanide catalysts, chiral, 176, 619
Lanthanide-catalyzed hydroamination
    1,3-dienes, 709
    alkenes, 702
    alkynes, 711
    mechanism, 715–716
Lanthanide crystals, 584
Lanthanide–amido complex, 386
Lanthanides, 310
Lanthanocene catalyst, 706
Lanthanocene complexes, 696–697
Lanxess, 634
Late-metal-thiolate bond, 195
Late metal transition complexes
    σ-bond metathesis, 285–286
LDPE. *See* Low-density polyethylene
Le Châtelier's principle, 540
Leuckart–Wallach reaction, 635
Lewis acid catalysts, 176
Lewis acid co-catalysts, 673
Lewis acid–base interactions, 3
Lewis acids
    abstraction of alkyl groups, 454
    abstraction of hydrides, 455
    catalysis of carbon monoxide
        insertions and, 362–363
    examples, 453
    olefin hydrocyanation and, 672
    in ring expansion carbonylation,
        786–787
$Li[Cp_2TaH_2]$, 189
Ligand dissociation
    overview, 217
    oxidative addition and, 264
Ligand substitution mechanisms
    factors controlling, 219–220
    17-electron complexes, 231–233
    square-planar, 16-electron
        complexes, 223–231
Ligand substitutions
    catalyzed by electron transfer,
        242–243
    defined, 217

    involving polyhapto ligands,
        247–253
    in metal–metal bonded bimetallic
        and higher nuclearity clusters,
        253–254
    overview, 217
    by radical chains initiated by atom
        abstractions, 243–244
    thermochemical considerations,
        220–223
"Ligandless" copper catalysts, 920
Ligands
    classification, 1–2
Light
    effect on reductive elimination,
        324–325
Lindlar catalyst, 640
Linear low-density polyethylene
    (LLDPE), 1051, 1052, 1054, 1067, 1084
Linear polyethylene, 1048
Living catalysts
    block copolymers and, 1069
    generation of stereoblock
        copolymers, 1065
Living polymerizations, 1032, 1049
    block copolymers and, 1069
    copolymerization of CO and
        styrene, 803
L–M–L angle, 18, *19*
    See also Bite angle
Losartan, 903
Low-density polyethylene (LDPE)
    catalysts for synthesis of, 1054
    hyperbranched, from late metal
        catalysts, 1054–1057
    overview, 1051, 1052
"Low-pressure oxo" process, 756
L_2PtR_2, 275
Lurgi reactor, 749
Lutetium–isobutyl complex, 406
LX ligands, 2
L_2X ligands, 2

M
MAC, 592
Macrocyclic Co(I) complexes, 307
Malonates, coupling, 936
Manganese piano stool complex, 65
Manganese–oxo complexes, 519
Manganese–salen catalysts, 518–519
Manganocene, 114, 118
Manzamine A, 1024
Martin, Steve, 1024
$M(CO)_6$ complexes, 241–242, 245
$M(CO)_5$(amine) complexes, 241
McQuillin's catalyst, 548
M–E bonds
    carbon monoxide insertions, 353
Meerwein–Pondorf–Verlay reaction, 634
Meerwein's reagent, 420

MeMn(CO)₃, 357
Memory effect, 982–983
Me₃NO, 246
MeP, 36
MEPY catalyst, 866
Mercury, 548
Meso cyclic diesters, 994–995
Metal-catalyzed hydroboration, 692–694
Metal properties
    oxidation state, 6–7
Metal-to-ligand charge-transfer (MLCT)
    transition, 24
Metal–acyl bonds, 377–379
Metal–alkoxide bonds, 352–353
Metal–alkyl bonds, 371–377
Metal–amide bonds, 352, 353
Metal–bis-sulfonamido bonds, 166–167
Metal–boron bonds, 388–389
Metal–carbon bonds
    electrophilic attack on, 457–460
    electrophilic cleavage, 454–462
    insertions of aldehydes and imines,
        381–383
    insertions of alkynes, 379–380
    insertions of olefins, 371–383
    insertions of polyenes, 381
    protonolysis of, 460
Metal–carbon monoxide bond strength,
    221, 222
Metal–carbon σ–bonds
    nucleophilic cleavage, 422–427
Metal–heteroatom bonds
    insertions of olefins and acetylenes,
        383–389
Metal–heteroatom multiple bonds
    bonding of oxo and imido
        complexes, 510–512
    nitrido ligands, 527–530
    overview, 509–510
    reactions of imido and oxo
        compounds, 515–527
    synthesis of metal–imido
        complexes, 512–514
    synthesis of metal–oxo complexes,
        514–515
Metal–hydride bonds
    insertions, 366–371
    protonation, 461
Metal–hydride σ-bonds, 454–462
Metal–hydrocarbyl σ-bonds, 371–377
Metallacyclic complexes, 400
Metallacyclobutane complex, 490
Metallacyclopentadiene complex, 1036
Metallacyclopropenes, 52, 97
Metallaoxetanes, 181–182
"Metal–ligand bifunctional catalysis," 600
Metal–ligand bond strength
    steric properties of the ligand
        affect, 223
    trends in, 222

Metal–ligand bonds
    classification by electrons donated
        to the metal, 3–4
    covalent
        metal–carbon bonds, 371–383
        metal–hydride bonds, 366–371
    multiple
        carbene complexes, 482–505
        metal–heteroatom multiple
            bonds, 508–530
        overview, 481
        silylene complexes, 505–508
    trends in strength, 9–10
Metal–ligand complexes
    18-electron rule, 13
    electron counting, 10–12
    geometries, 14–15
    isoelectronic and isolobal analogies,
        15–17
    metal–metal bonds in polynuclear
        complexes, 13–14
    molecular orbitals, 17–19
    π-bonding in, 19–22
Metallocenes
    ansa, 113–122 (*see also*
        Cyclopentadienyl ligands)
    defined, 113
    significance of ferrocene, 114
    structure, 114–115
Metal–metal bonded bimetallic clusters,
    253–254
Metal–metal bonds, 13–14
Metal–nitrogen bonding
    complexes of anionic nitrogen
        heterocycles, 155–158
    early-metal–amidate and
        –amidinate complexes, 155
    metal–amido complexes, 147–154
    nitrosyl complexes, 158–162
    olefin insertions, 385–388
    polydentate nitrogen donor ligands,
        162–173
Metal–oxygen bonds, 383–385
    overview in late-metal alkoxides,
        177–178
    strengths for alkoxides, 179
Metal–phosphorus bond compounds. *See*
    Phosphido complexes
Metal–silicon bond compounds. *See* Silyl
    ligands
Metal–silicon bonds, 198, 388
Metal–silyl bonds, 352–353
Metal–sulfur bond compounds. *See*
    Thiolate complexes
Metathesis
    alkyne
        applications, 1036–1038
        cross metathesis, 1038–1039
        examples, 1034–1035
        mechanism, 1036

        ring-closing metathesis,
            1039–1040
    of C–C multiple bonds, 1015
    enyne
        examples, 1040–1041
        mechanism, 1041–1042
        overview, 1015–1017
    See also Olefin metathesis
Metathesis catalyst, 509
Metathesis reactions
    formation of metal–imido
        complexes, 513
    formation of metal–oxo complexes,
        514
    formation of nitrido complexes, 528
Metathetical substitution, 150
Methane
    oxidation, 827–829
    oxidative addition of the C–H bond,
        264–266
    σ-bond metathesis and, 284–285
Methanol carbonylation
    to form acetic acid, 746–748
    iridium-catalyzed, 749–751
Methoxycarbonylations, 789
Methyl acetate, 749
    carbonylation, 748–749
Methyl carbene complex, 366
β-Methyl elimination, 408
Methyl group
    alkyl ligand, 86
    carbon monoxide insertions and,
        361–362
    classification, 3
    M–methyl bond, 86, 87
    nucleophilic attack on, 424
    structure, *85*
Methyl iodide
    oxidative addition, 302, 303
    reductive elimination, 341
Methyl methacrylate, 776
Methylaluminoxane (MAO), 1065, 1070
Methylidene complex, 490
Methylphenidate hydrochloride, *648*
Methylplatinum(IV) complexes
    C–X bond reductive elimination,
        338–339
    nucleophilic attack of, 342
Metolachlor, 606, 629, 631–632
MeX, 264
Microscopic reversibility, 224, 240, 500
Migratory insertions
    into alkoxides, 184
    changes in geometry and electron
        count during, 350
    defined and described, 349
    of ligands bound by a single atom
        carbenes, 365–366
        carbon monoxide, 351–364
        other ligands, 364–365

into metal–heteroatom bonds, 383–389
of olefin (*see* Olefin insertions)
of polyhapto ligands into metal–ligand covalent bonds
metal–carbon bonds, 371–383
metal–hydride bonds, 366–371
significance of, 350
Mizoroki–Heck reaction, 371, 883
mechanism, 892–893
See also Heck reaction
Mμ(μ–H)M linkage, 123
Mn–benzyl bond, 86
Mn(CO)$_5$, 16
Mn$_2$(CO)$_{10}$, 253–254
Mn(NO)(CO)$_4$, 160
Mo(CO)$_6$, 1034–1035
Modified Green–Rooney mechanism, 375
Molecular orbitals
for transition metal complexes, 17–19
Molecular weight distribution (MWD), 1048–1049
Molecular weights
of polymers, 1048–1049
Molybdenum
dinitrogen complexes and, 60
Molybdenum catalysts, 989–990
Molybdenum complexes, 977–978
Molybdenum–allyl complexes, 437
Monocyclic arenes
hydrogenation, 647
Monodentate ligands, 610–611
Monoene polymerization and oligomerization, 1050–1051
Monohydride catalysts
d^0-, 600
in olefin hydrogenation, 599–600
"Monohydride mechanism," 596
Monoimines, 58
Mononuclear compounds, 261
Monophosphines, *34, 36*
Monsanto acetic acid process, 344, 746–748
Montreux-type catalyst, 1038
MOP ligand, 63, 982, 983, 986
Morphine alkaloids, 630
MPPIM catalyst, 866
M(PR$_3$)$_x$H$_y$, 123
M–SR bonds, 195–196
M–X bonds, 352–354

N

"NacNac" ligands, 170
Na$_2$[Fe(CO)$_4$], 189
Naproxen, 556, 617, 674, 776
Natta, Giulio, 1047
Natural bite angle, 35, 760

Neopentyl
alkyl ligand, 86
structure, *85*
Neopentyl zirconocene complex, 407
Neopentylcobalamin, 87
Neutral ligands, 1
Neutral nitrogen donor ligands
amine complexes, 57–58
dinitrogen complexes, 59–61
pyridine and imine complexes, 58
Neutral oxygen donors, 62–63
Neutral rhodium catalysts, 578–580
Neutral sulfur donors, 63–64
N–H bond
oxidative additions and, 314
reductive elimination and, 330
Nickel catalysts
hydroamination of 1,3-dienes, 708–709
mechanism of deactivation, 673
Nickel-catalyzed hydrocyanation, 668
Nickel complexes
C–C bond-forming reductive elimination and, 336
β-chloride elimination, 409
couplings of alkyl halides, 882
C–X bond reductive elimination, 340–341
homocoupling and, 891–892
oxidative addition of aryl halides and, 895
reactions with aryl halides, 305
steric effects on dissociative substitution, 235–236
Nickel phosphite complexes, 235–236
Nickel–hydrocarbyl complexes, 409
Nickel(II) complexes, 340–341
Nickelocene, 19
Ni(CO)$_4$, 28, 235
19-electron complexes, 231–232
Niobium complexes, 367
Nitric oxide (NO), 158–159
Nitride complexes, 481
Nitride ligands, 25
Nitrido complexes, 527–530
reactions, 529–530
structural and spectral features, 528
synthesis, 528–529
Nitrido ligands, 527
Nitrile ligands
π-bonding, 22
Nitriles
homogeneous hydrogenation, 656–657
Nitrogen. *See* Neutral nitrogen donor ligands
Nitrogen heterocycles, 772
Nitrogen heterocycles, anionic
metal–azolyl bonding, 155–156
overview, 155

reactivity of metal–azolyl complexes, 157–158
synthesis of metal–azolyl complexes, 156–157
Nitrosyl complexes
overview, 158–159
properties of nitric oxide, 159
reactivity, 162
spectral features, 161
structures and electron counting, 159–160
synthesis, 161–162
thermodynamics of M–NO linkages, 160–161
Nitrosyl ligands
classification, 6
migratory insertions, 365
π-acceptor, 19
NO complexes. *See* Nitrosyl complexes
Noble gases
dissociative substitutions, 237–238
Norbornene, 946, 948
Normal modes, 31
Norphos, 608
Noyori, Ryoji, 575
Nucleophilic attack
on carbene and carbyne complexes, 421–422
on complexes of carbon monoxide and isonitriles, 419–421
fundamental principles, 417–418
imine and aldehyde complexes, 435–436
metal–carbon σ–bonds, 422–427
methylplatinum(IV) complexes, 342
η^2-olefin complexes, 428–433
polyhapto ligands, 436–446
types of, 417
Nucleophilic cleavage
metal–carbon σ–bonds, 422–427
Nucleophilicity
of the M–H bond, 135–136
of pyramidal phosphide P, 192
"Number of d-electrons," 7

O

O-bound sulfinate, 462–463
Octahedral complexes, 238, 240
O–H bond
oxidative additions and, 314
reductive elimination and, 330
Olefin complexes
amination, 431
atom transfer from imido and oxo groups, 518–521
aziridination of, 521
bonding, 21–22, 47
catalysts for the hydrocarboxylation and hydroesterification of, 777–778

Olefin complexes (*Continued*)
  characteristics, 21
  copolymerization with CO
      CO and α-olefins, 804–808
      development of catalysts for
          CO/ethylene, 798–799
      mechanism for CO/ethylene,
          800–803
      overview, 798
  coupling, 883
  dehydrogenative silylation, 681
  directed hydroarylation
      mechanisms, 849
      overview, 846–847
      reaction scope and catalysts,
          847–849
  dissociative ligand substitutions, 230
  electrophilic attack of carbonyl
      compounds and protons, 471–472
  epoxidations, 519–521
  hydroamination by insertion in
      metal amides, 715–716
  hydroaminomethylations, 770
  hydroboration, 695–697
  hydroformylation
      catalyzed by HCo(CO)$_4$, 752–754
      catalyzed by HCo(CO)$_3$(PR$_3$),
          754–755
      overview, 751–752
      rhodium-catalyzed, 756–769
          (*see also* Rhodium-catalyzed
          hydroformylation)
  hydrogenation (*see* Olefin
      hydrogenation)
  hydrosilylation
      with achiral catalysts, 679–680
      asymmetric, 683–684
  induction of reductive elimination,
      336
  insertions (*see* Olefin insertions)
  intramolecular hydroesterification,
      780–781
  metathesis (*see* Olefin metathesis)
  nucleophilic attack, 427, 428–433
  orientation of coordinated olefins,
      49–51
  overview, 47
  oxidative amination, 387
      intermolecular, 728–729
      intramolecular, 730
      overview, 728
      palladium-catalyzed
          difunctionalizations,
          730–731
  oxidative functionalization
      mechanistic studies on Wacker
          oxidations, 731–734
      overview, 717–718
      oxidative aminations, 728–731
      Wacker process, 718–727

polymerization (*see* Olefin
    polymerization)
  spectral properties, 51
  stability, 47–48
  steric effects and, 48
  structural changes upon binding, 49
  synthesis of allyl complexes and, 108
  undirected hydroarylation and
      oxidative arylation, 850–852
  unfunctionalized, asymmetric
      hydrogenation, 618–620
  Wacker oxidation, 431, 731–734
π-Olefin complexes
  hydroamination by attack of amines
      on, 713
Olefin dimerization, 1079
  by insertion into M–C bonds,
      1082–1084
  through metallacyclic intermediates,
      1084
Olefin hydride complex, 366
Olefin hydrogenation
  asymmetric, classes of
      α-(acycloxy)acrylates, 616
      acrylic acids, 616–618
      enamides, 612–616
      imines, 629–634
      ketones, 620–629
      unfunctionalized olefins,
          618–620
      unsaturated alcohols, 618
  by insertions into dihydride
      complexes
      cationic iridium crystals
          containing alkylphosphines,
          594–595
      cationic rhodium crystals,
          590–593
      Wilkinson's catalyst, 588–590
  by insertions into monohydride
      intermediates
      monohydride catalysts
          reacting through radical
          pathways, 599–600
      monohydride catalysts
          reacting through σ-bond
          metathesis pathways, 600
      rhodium carbonyl hydride
          catalysts, 596
      RuL$_2$(κ^2-OAc)$_2$ and [RuL$_2$Cl$_2$]$_2$,
          597–599
      Ru(PPh$_3$)$_3$H(Cl), 597
      ruthenium catalysts, 597
  mechanisms
      background, 585
      overview, 585–602
  rhodium catalysts
      cationic, 580–581
      neutral, 578–580
  ruthenium catalysts, 583

Olefin insertions
  into alkoxides, 184
  in chain termination, 1051
  into dihydride complexes, 588–595
  metal–boron bonds, 388–389
  metal–carbon bonds, 371
      metal–acyl bonds, 377–379
      metal–hydrocarbyl σ-bonds,
          371–377
  metal–hydride bonds, 366–368
  metal–nitrogen bonds, 385–388
  metal–oxygen bonds, 383–385
  metal–silicon bonds, 388
  into monohydride intermediates,
      597–600
  olefin dimerization and, 1082–1084
  in polyolefin synthesis, 1058
  synthesis of zirconium alkyl, 89
Olefin metathesis, 43
  catalyst decomposition, 1022–1023
  chlorides as ancillary ligands,
      203–204
  enantioselective ring-closing and
      ring-opening, 1028–1031
  examples
      cross metathesis, 1026–1028
      ring-closing metathesis,
          1023–1026
  history of, 1019–1020
  introduction to, 1015
  mechanism, 1020–1022
  overview, 1015–1017
  overview of catalysts for, 1017–1019
  ring-opening metathesis
      polymerization
      mechanism, 1033–1034
      utility, 1031–1033
Olefin oligomerization
  alkene, 1079–1086
Olefin polymerization
  ethylene-based polymers, 1051–1057
  ethylene–α-olefin copolymers,
      1067–1070
  hyperbranched polypropylene,
      1066–1067
  introduction to, 1047–1048
  mechanism of stereocontrol in
      isotactic polypropylene synthesis,
      1057–1060
  monoene, 1050–1051
  a primer on the chemistry of,
      1048–1050
  styrene, 1070–1072
  synthesis of stereodefined
      polypropylenes, 1060–1065
  See also Alkene polymerization;
      Propylene-based polymers
Olefin trimerization, 1079, 1084–1086
Olefination
  of aryl halides, 892–893

α-Olefins
    copolymerization with CO
        CO and propene, 806–808
        CO and styrene, 804–806
        overview, 804
Z-Olefins, 1039
One-electron mechanisms
    of oxidative additions, 304–310
Open polyenes, 428
Optical rotations, 551, 552
Organic azide, 514
Organic carbonyl compounds, 53
Organic halides
    carbonylation
        to form esters and amides,
            795–798
        to form ketones, 914–917
        overview, 794–795
    formylation, 917–918
Organoaluminum, 87, 128
Organoaluminum reagents, 1006
Organoboron reagents, 880, 896
Organogold compounds, 86
Organolithium, 87, 128
Organomagnesium, 128
Organomagnesium reagents, 878
Organometal cyanide complexes, 103
Organosilicon reagents
    coupling, 879–880
    transmetallation, 896
Organotin reagents
    coupling, 879
    transmetallation, 898–899
Organotransition metal complexes,
    19–22
Organozinc, 87
Organozinc reagents, 878–879
Organyl trioxorhenium species, 509
Ortho fluorines, 95
Ortho metallations, 273
Ortholithiation, 121
Orthosilylation, 858
$Os_3(CO)_{10}(\mu\text{-H})_2$, 14
Osmium dimers, 222
Outer-sphere electron-transfer
    mechanism, 308–309
"Outer-sphere" hydrogenation
    mechanism, 586, 587
π-Oxaallyl, 99
Oxafloxazin, 625
Oxazoline ligand, 58, 986
Oxidation
    rate of carbon monoxide insertion
        and, 363–364
Oxidation state
    of metals, 6–7
    number of d-electrons and, 7
Oxidative addition
    aryl chlorides, 894
    aryl halides, 894–895

atom abstraction, 309–310
"bite angle" and, 312
C–C bonds, 289–291
C–H bonds, 271–283 (see also C–H
    bond oxidative addition)
concerted, 310–315
in cross coupling, 893–895
defined, 261
dihydrogen, 266–270
dinuclear, 315–317
E–E bonds, 291–292
formation of metal silyl complexes,
    200
generation of boryl complexes,
    189–190
groups of reagents that undergo, 262
by inner-sphere electron transfer
    and caged radical-pairs, 305
outer-sphere electron-transfer
    mechanism, 308–309
overview, 261–262
of polar reagents
    concerted additions, 310–315
    dinuclear additions of
        electrophilic A–B, 315–317
    one-electron mechanisms,
        304–310
    by S_N2 pathways, 301–304
qualitative trends for, 263–264
by radical pathways, 304–310
reductive elimination compared
    to, 321
salines, 270–272
thermodynamics, 264–266
Oxidative amination of olefins, 387
    intermolecular, 728–729
    intramolecular, 730
    overview, 728
    palladium-catalyzed
        difunctionalizations, 730–731
Oxidative arene cross-coupling, 949
Oxidative arylation of olefins, 850–852
Oxidative carbonylation, 835–836
Oxidative functionalization of olefins
    mechanistic studies on Wacker
        oxidations, 731–734
    overview, 717–718
    oxidative aminations, 727–731
    Wacker process, 718–727
Oxidative hydrogen migration, 286, 852
Oxidative ligation, 262, 264
Oxide ligands, 25
    See also Oxo ligands
Oxo complexes
    bonding of, 510–512
    defined, 481
    reactions with electrophiles, 523–524
    synthesis, 514–515
Oxo compounds
    atom transfer to olefins, 518–521

catalytic reactions through
    organometallic intermediates,
    525–527
cycloadditions, 515–518
overview, 509
reactions with C–H bonds, 522–523
Oxo ligands, 524–525
"Oxo" process, 751
Oxoallyl, 99
Oxygen
    complexes of neutral donors,
    62–63
Oxygen nucleophiles, 432
Oxypalladation, 383–384

P
P-chiral compound, 36
P-chiral ligands, 576
P-chiral phosphines, 609
P-Phos, 605
P-stereogenic phosphines, 192
Paeonilactone B, 726, 727
Palladium
    difunctionalizations of olefins and,
    730–731
Palladium carbonyl, 32
Palladium catalysts
    cross-coupling and, 879, 890–891
    in direct arylation, 939–941
    hydroamination
        of 1,3-dienes, 708–709
        of vinylarenes and, 705
    hydrogenation of alkynes and
        conjugated dienes, 644
Palladium-catalyzed allylic substitution,
    968–969, 985
    enantioselective
        with unsymmetrical acyclic
            substrates, 988–989
    mechanism, 974–977
    regioselectivity
        memory effect, 982–983
        reactions of carbon
            nucleophiles, 979–981
        reactions of heteroatom
            nucleophiles, 981–982
Palladium-catalyzed carbonylations, 426
Palladium-catalyzed cross coupling, 879,
    890–891
Palladium-catalyzed Heck reaction, 570–571
    See also Heck reaction
Palladium complexes
    bridging "cyclopentadienyl"
        ligands, 118
    catalysis of disilylation, 690
    to catalyze hydrosilylation, 679
    the Catellani reaction and, 946
    in direct arylation, 939–941
    hydroamination of alkynes and,
        711–712

Palladium complexes (*Continued*)
    oxidative addition of aryl halides to, 310–313
    steric effects on dissociative substitution, 235–236
    See also individual complexes
Palladium(0) complexes
    addition of alkyl halides, 301
    carbonylation of aryl halides and, 426
    C–N coupling and, 911, 912
Palladium enolate complex, 101
Palladium phosphine complexes, 235–236
Palladium–alkyl complexes, 409
Palladium–allyl complexes
    nucleophilic attacks on, 437–438
    stoichiometric attack on, 968
Palladium–amides, 387
Palladium–cyanoalkyl complexes, 100
Palladium(II) complexes, 337
    alkyl, β-hydrogen elimination, 401
    C–X bond reductive elimination, 339–340
    olefin, nucleophilic attack, 429–433
    reductive elimination, 899
    square planar
        allene, 433–434
        diene, 433–434
        olefin, 430–433
Pantothenic acid, 621
Paramagnetic complexes, 7, 19
Paroxetine, 654
Pauling's electroneutrality principle, 6
Pauson–Khand reaction (PKR)
    with allenes, 811–812
    applications, 814
    catalysts other than $Co_2(CO)_8$, 810–811
    catalytic asymmetric, 812
    effects of additives, 810
    intermolecular, 812–813
    mechanism, 814–816
    origin of, 809–810
    overview, 809
PCP pincer ligand, 841, 845, 846
Pd–N bonds, 387
Peachey, S. J., 86
PennPhos, 608
Pentadienyl ligand ring slip, 250–253
Pentadienyl systems, 109–110
Pentalenyl, 112
Pentamethylcyclopentadiene, 112
Pentamethylcyclopentadienyl, 5
Percentage enantiomeric excess (ee), 551, 552
Perchloric, 130
Perfluoroalkyl complexes, 89, 360
Permethylindenyl complexes, 112
Permethylscandocene–alkyl complex, 379
Pfaltz's phosphinooxazoline, 890

P–H bond
    oxidative additions and, 314
    reactivity, 40
PHA, 790
Phanephos, 608
Phenethyl Grignard reagent, 884, 885
Phenols
    copper-catalyzed coupling with aryl halides, 926–927
    reductive elimination of, 331
Phenyl iodide, 341
Phenyl ligands, *85*, 202
    See also Aryl complexes
N-Phenylanthranilic acid, 922
Phillips catalyst, 1053
Phosphide ligands, 190
Phosphido complexes
    dynamics of, 192
    overview, 190
    reactivity, 192–193
    structures, 191
    thermodynamic properties, 192
Phosphination
    of aryl halides, 569
Phosphine complexes
    bonding and electronic properties, 36–37
    chelating, 34–35
    cyclometallation, 274
    decomposition pathways, 39–40
    effect of steric and electronic properties on structure and reactivity, 39
    migratory insertions and, 365
    neutral oxygen donors and, 63
    NMR spectroscopic properties, 40–41
    oxidative additions and, 314
    properties when free, 35–36
    steric properties, 38–39
    tertiary, 33, *34*
Phosphine-ligated cobalt catalysts, 755
Phosphine oxides, 62, 246, 359
Phosphine–oxazoline, 58
Phosphine–phosphine oxide ligands, 63
Phosphines, 34–37
Phosphinidene ligands, 510
Phosphino oxazoline, *58*, 890, 986
Phosphinylketimines, 632–634
Phosphite ligands, 610–611
Phosphites, 35
    bonding properties, 36
    decomposition, 39–40
    hydroformylation catalyzed by rhodium complexes of, 763–764
Phosphoramide, 39
Phosphoramidite ligands, 610–611, 986–987, 993
Phosphorus ligands
    chelating phosphines, 34–35
    decomposition pathways, 39–40

heavier congeners, 41
    properties of free phosphines, 35–36
    properties of phosphine complexes, 36–39
    tertiary phosphines, 33, *34*
Photochemical carbonylation, 838–839
Photochemically induced reductive elimination, 324–325
Photochemically induced dissociation reactions, 244–246
Photolysis
    induction of ligand substitution, 244–246
    oxidative addition of alkanes and, 281
Phox ligand, 609
Piano stool complexes, 239
    manganese, 65
    pseudo-four-coordinate, 18-electron, 236–237
Pincer ligands, 148
    dehydrogenation catalyzed by, 840–842
    PCP, 841, 845, 846
    POCOP, 841, 845–846
    ruthenium pincer complex, 652
π-Acceptors, 19
    cis effect and, 238–239
    tertiary phosphines, 36–37
    trans effect and, 229
π-Bonded ligands
    in combination with σ-donors, 5–6
    overview, 4–5
π-Bonding
    with alkynes, 22
    of carbene and carbyne complexes, 20
    of carbon monoxide and analogs, 19–20
    with ketones, imines, and nitriles, 22
    in olefin complexes, 21–22
    with oxo and imido complexes, 510–512
π-Complexes
    hydroamination by attack of amines on, 713–715
π-Donation
    cis effect and, 239–240
π-Donor ligands, 22–25
    halides, 201–202
π-Ligands
    higher anionic complexes, 109–110
"Planar chirality," 606
Platinum
    η^1-allyl ligand and, 105–106
Platinum-catalyzed oxidations
    catalysts, 827–829
    mechanism, 829–832
Platinum complexes
    C–C bond-forming reductive elimination, 333, 337

nucleophilic attack, 423–424, 429–433, 434
oxidative additions and, 313
See also individual complexes
Platinum divinyltetramethyldisiloxane, 678
Platinum(II) complexes
   acetylene complexes, 434
   catalytic processes and, 542
   C–C bond-forming reductive elimination, 337
   d^8 square-planar, 307
   olefin complexes, 429–433, 542
Platinum(IV) complexes
   C–C bond-forming reductive elimination and, 333, 337
   C–X bond reductive elimination, 338–342
   nucleophilic attack and, 423–424
   as oxidants, 827
Platinum(IV) diphenyl didiodide complex, 341
Platinum(IV) methyl complexes, 338–339
P,N ligands, 577, 609–610, 652–653
POCOP pincer ligand, 841, 845–846
Polar solvents
   migratory insertion reactions and, 359
Polyacetylenes, 379, 1033
Polyamines, 57
Polycyclic arenes, 645–647
Polydentate nitrogen donor ligands
   bis-sulfonamide complexes, 165–167
   β-diketiminate complexes, 170–173
   overview, 162
   porphyrin and corin complexes, 162–165
   pyrazolylborate, 167–170
Polydispersity index (PDI), 1032, 1049
Polyenes
   insertions into metal–carbon bonds, 381
   open, nucleophilic attack, 428
π-Polyenyl complexes
   electrophilic attack on, 474–475
Polyethylenes, 1047, 1048
   forms, 1051–1052
   hyperbranched, from late-metal catalysts, 1054–1057
Polyhapto ligands, 109–110
   associative substitution by pentadienyl ligand ring slip, 250–253
   bond strength, 222
   nucleophilic attack on
      η^3-allyl complexes, 436–439
      η^6-arene and cycloheptatrienyl complexes, 442–446
      η^4-diene complexes, 439–441
      η^5-dienyl complexes, 441–442

substitutions for arenes and arene exchange reactions, 248–250
substitutions for dienes and trienes, 247–248
Polyhydride complexes, 67
Polyisocyanides, 364
Polyketones, 798
Polymer chain
   defined, 1048
Polymers
   defined, 1048
   molecular weights, 1048–1049
   stereochemistry, 1049
Polynuclear complexes
   metal–metal bonding and electron counting, 13–14
Polyolefins
   borylation, 855
   introduction to, 1047–1048
   mechanism of stereocontrol in isotacitc polypropylene synthesis, 1057–1060
   a primer on the chemistry of, 1048–1050
   See also Olefin polymerization
Poly(phenylene vinylene)s, 905
Polypropylene, 1047
   hyperbranched, 1066–1067
   overview, 1057
   stereochemistry, 1049
   See also Polypropylene synthesis; Propylene-based polymers
Polypropylene synthesis
   hemiisotactic polypropylene, 1062
   isotactic and syndiotactic polypropylenes, 1060–1062
   mechanism of stereocontrol in isotactic polypropylene, 1057–1060
   stereoblock polypropylenes
      by alternation of the ligand sphere, 1065
      by chain transfer, 1065
      isotactic–atactic, 1063–1064
      from living catalysts, 1065
      overview, 1062–1063
Polypyrazolylborate complexes, 170
Polypyrazolylborate ligands
   bonding, 168–169
   synthesis, 169–170
Polysiloxanes, 678
Polystyrene
   isotactic, 1072
   syndiotactic, 1070–1072
Pope, W. J., 86
Porphyrin complexes
   nucleophilic attack and, 423–424
   overview, 162–163
   reactivity, 164–165
   structures, 163–164
   synthesis, 164

PPh$_3$O complex, 62
PPyr$_3$, 250
"Precatalysts," 575
"Privileged ligands," 561–563
"Privileged structures," 561
Prochiral molecules, 550, 551
Prochiral nucleophiles
   enantioselective allylation, 996–998
Promoters, 545, 673
Propane diol, 789
Propargyl ligands, 470
Propene
   copolymerization with CO, 806–808
Propylene, 804
Propylene-based polymers
   overview, 1057
   stereocontrol in isotactic polypropylene synthesis, 1057–1060
   synthesis
      hemiisotactic polypropylene, 1062
      isotactic and syndiotactic polypropylenes, 1060–1062
      stereoblock polypropylenes, 1062–1065
   See also Polypropylene; Propylene polymerization
Propylene polymerization, 1047–1048
Propyne, 781
Protic acids, 363
Protonation
   of metal–hydride bonds, 461
   of olefin complexes, 471–472
   synthesis of hydride complexes from, 126–127
Protonolysis, 454
   of M–C bonds, 460
Protons
   electrophilic attack on olefin complexes, 471–472
Pseudo-four-coordinate, 18-electron "piano stool" complexes, 236–237
Pseudo-octahedral complexes
   cis effect and, 238, 240
   β-hydrogen elimination, 404
Pseudohalides, 895–899
Psycholeine, 906
trans-Pt(η^1-C$_3$H$_5$)Br(PEt$_3$)$_2$, 105–106
Pyrazole, 169
Pyrazolyl groups, 168, 169
Pyrazolylborate ligands, 167–168
Pyridine complexes, 56, 58, 648
Pyridyl bis(imine), *58*
Pyrroles, 155
   asymmetric hydrogenation, 650
   reductive elimination of, 331
η^1-Pyrrolyl complexes, 157
η^5-Pyrrolyl complexes, 155, 158
Pyrrolyl ligand, 157

Q

Quadrant diagrams, 559–560
Quadrigemine C, 906
Quinolines, 56

R

Racemic secondary alcohols, 566
Radical-chain pathway, 306–307
Radical pathways
    oxidative addition by, 304–310
Random copolymers, 1049
Rate constants, 546
RCM. *See* Ring-closing metathesis
RC(O)Co(CO)$_3$, 753
Re-oxidation
    in the Wacker process, 733–734
Reaction coordinate diagrams, 540–542
"Rebound mechanism," 522
Reduction
    activation of ligand substitution, 243
"Reductive coupling," 1084
Reductive elimination
    of alkyl halides from platinum(IV)
        complexes, 423–424
    changes in electron count and
        oxidation state, 321
    C–N coupling and, 913
    factors affecting the rates of,
        322–325
    to form C–C bonds, 331–338
    to form C–H bonds, 325–330
    to form C–X bonds, 338–345
    to form X–H bonds, 330–331
    general trends for, 263–264
    hydrocyanation and, 672
    mechanism, 899
    overview, 321–325
    oxidative addition compared to, 321
    photochemically induced, 324–325
Reductive elimination rates
    effect of coordination number,
        323–324
    effect of geometry, 324
    effect of light, 324–325
    effect of metal identity and electron
        density, 322
    effect of participating ligands, 323
    effect of steric properties, 322–323
Regiochemical "memory effect," 982–983
Regiochemistry
    of metal-alkyl addition, 372
Regioselective direct arylations, 943–948
Regioselectivity
    effect of diphosphine electronic
        properties on, 762–763
Reppe, Walter, 745, 769
Reppe-type catalysis, 421
Rhazinilam, 282
Rh–C bond, 87
RhClL$_3$, 589

RhCl(PPh$_3$)$_3$, 268
[Rh(CO)$_2$I$_2$]$^-$, 747–748
[Rh(DIPHOS)(MAC)]$^+$, 592
[Rh(DIPHOS)S$_2$]$^+$, 591, 592
Rh–DuPhos system, 613
Rhenium hydride complex, 7
Rhenium κ^1–"acylsilane" complex, 353
[Rh(H)(PPh$_3$)$_3$(CO)], 596
Rhodium arylmethoxide complex, 408
Rhodium carbene complexes, 289
Rhodium catalysts
    asymmetric, 612
    cationic, 580–581
        hydrogenation by, 590–593
    containing chelating disphosphine
        ligands
        early studies, 759
        regioselectivity effects, 762–763
        with wide-bite-angle
        bisphosphines, 760–762
    in direct arylation, 942–943
    enantioselective allylic substitutions
        and, 991
    hydrogenation of alkynes and
        conjugated dienes, 641–642
    in hydrosilylation, 678
    for olefin hydrogenation
        cationic, 580–581
        neutral, 578–580
    rhodium carbonyl hydride, 596
Rhodium-catalyzed borylation
    of alkanes, 853
Rhodium-catalyzed carbonylation
    to form acetic acid, 746–751
Rhodium-catalyzed hydroformylation
    catalysts containing chelating
        diphosphine ligands, 759–763
    catalysts containing
        triarylphosphine ligands, 756–758
    enantioselective, 765–769
    of functionalized alkenes, 764–765
    of internal alkenes, 763
    overview, 756
    by rhodium complexes of
        phosphites, 763–764
    water-soluble catalysts, 758–759
Rhodium complexes, 56
    hydroamination of alkynes and,
        711–712
    insertions of carbenes into C–H
        bonds and, 865–867
    insertions of olefins, 371–372
    intermolecular oxidative addition of
        C–H bonds, 276
    oxidative addition by the S$_N$2
        pathway and, 303–304
    of phosphites, hydroformylation
        catalyzed by, 763–764
    See also individual complexes
Rhodium enolate complex, 101

Rhodium hydrides, 381
    insertions of olefins, 367–368
Rhodium–alkoxo complex
    insertion of alkene, 384
Rhodium–amido complex
    olefin insertions, 386
Rhodium–aryl complexes
    insertion of aldehydes, 381
Rhodium–DuPhos system, 613
Rhodium(I) boryl complex, 696
Rhodium(II) complexes
    carbene complexes, dinuclear, 289
    dinuclear activation of
        hydrocarbons, 282–283
[Rh(PPh$_3$)$_3$Cl], 576
    See also Wilkinson's catalyst
[Rh(PPh$_3$)$_3$H$_2$Cl], 590
Rieger catalyst, 1064
Ring-closing metathesis (RCM), 1016, 1017
    alkyne, 1039–1040
    enantioselective, 1028–1031
    history, 1020
    olefin, 1023–1026
Ring-expansion carbonylation
    history of, 785
    overview, 784–785
    types of catalysts and scope of
        substrates, 786–787
Ring heteroarenes, asymmetric
    hydrogenation, 648–651
Ring-opening epoxide carbonylation,
    788–792
Ring-opening metathesis polymerization
    (ROMP)
    alkyne, 1036
    described, 1016–1017
    enantioselective, 1028–1031
    history, 1019–1020
    mechanism, 1033–1034
    utility of, 1031–1033
"Ring slip" reactions, 120
    associative substitution and, 250–253
Ritalin, 648
RNC, 419
Roelen, Otto, 751, 752
ROMP. *See* Ring-opening metathesis
    polymerization
Roscoe, Henry E., 86
ROSO$_2$F, 420
Ru-tris(bpy) complexes, 905
RuCl$_2$[(*S*)-Xyl-BINAP][(*S*)-DAIPEN], 627
RuH(Cl)(PPh$_3$)$_3$, 583
[RuL$_2$Cl$_2$]$_2$, 597–599
RuL$_2$(κ^2-OAc)$_2$, 597–599
Ru(PPh$_3$)$_3$H(Cl), 597
Ruthenacylobutane complex, 407
Ruthenium catalysts
    in direct arylation, 941–942
    enantioselective allylic substitutions
        and, 987, 990

for hydroesterification, 777
hydrogenation by, 597
    of esters, 651–653
hydrosilylation of alkynes, 682
for olefin hydrogenation, 583
$RuL_2(\kappa^2\text{-OAc})_2$ and $[RuL_2Cl_2]_2$, 597–599
Ruthenium complexes, 56
    anti-Markovnikov hydroamination
        of vinylarenes and, 706
    arene exchange, 249
    catalysis of enantioselective allylic
        substitutions, 987
    in olefin hydrogenations, 577
    pincer complex, 652
    See also individual complexes
Ruthenium dimers, *222*
Ruthenium hydride, 370–371
Ruthenium–alkyl complex, 285, 286
Ruthenium–amido complex, 353
Ruthenium–benzyne complex, 472
Ruthenium–BINAP, 562, *563*, 577
    hydrogenation by, 639–640
Ruthenium–carbene complex, 500
Ruthenium(II) aryloxide complex, 182

S

S-bound sulfinate, 462
Salicaldimine complex, 409
Salines
    oxidative addition, 270–272
Salt displacements, 199
Salt eliminations, 189
"Sandwich complexes," 114–115
"Saturated *N*-heterocyclic carbenes," 491
Saunders, Martin, 72
Scandocene β-arylethyl complex, 401
Scandocene–alkyl complexes, 406–407
Schrock, Richard, 42, 482, 1015
Schrock carbenes, 42, 45, 421–422
    [2+2] reactions, 502
    origin of the electrical properties
        of, 483
    overview, 482
    synthesis, 488
Schrock-type alkylidene complexes, 411,
    488–490
Schrock-type catalysts, 488–490, 509,
    1017, 1018–1019, 1020
Schultz–Flory distribution, 1079
Schwartz's reagent, 368, 410
SDP, 608
Second-generation Grubbs catalyst, 43
Segphos ligand, 605, 622
Selectivity factor, 1028
"Semibridging," 29
17-electron complexes
    associative substitutions, 231–233
    ligand substitutions, 220, 253
S–H bond, 314
Shapley, John, 72

Sharpless, K. Barry, 518
Sharpless asymmetric epoxidation
    process, 565
Shell Chemicals, 377, 798
Shell Higher Olefin Process (SHOP). 752,
    1016, 1080–1082
Si NMR spectroscopy, 199
σ-bond metathesis
    involving d^0 complexes, 283–285
    involving late transition metal
        complexes, 285–286
σ-bonded ligands
    in combination with π-donors, 5–6
σ complexes
    alkane and silane complexes, 70–72
    dihydrogen complexes, 66–67
    overview, 64–66
    reductive elimination and, 325–326
    σ-bond metathesis and, 284
σ donors
    cis effect and, 238
    trans effect and, 229
Si–H bond, 270–271
Silabenzene, 56
Silane complexes, 198
    η^2-complexes, 271
    overview, 64, 65, 70
    stability, 70
Silox, 175
Siloxanes, 678
Silyl complexes
    β-hydrogen elimination, 405–406
    intermolecular oxidative addition of
        C–H bonds, 278
    spectral properties, 198–199
    stability and reactivity, 200
    structures, 198
    synthesis, 199–200
Silyl ligands
    electronic properties, 197–198
    overview, 197
Silylation
    of aromatic and aliphatic C–H
        bonds, 857–859
    dehydrogenative, of olefins, 681
Silylboration
    alkynes, 697
    mechanism, 699
    overview, 697
Silylene complexes
    bonding of, 505
    examples, 506–507
    overview, 505
    reactivity, 507–508
Singulair, 904–905
Si–Si bond, 291
Site control, 1049
16-electron complexes, 219
    See also Square-planar, 16-electron
        complexes

16-electron Rh(I) macrocycle, 307
Sm(III) complex, 114
S_N2 reaction, 217–218, 301–304
Solid angle, 39
Solvents
    migratory insertion reactions and,
        359–360
Sonogashira reaction, 881
Speier's catalyst, 678, 680, 681, 686
Sphingofungin F., 989
Square-planar, 16-electron complexes
    associative substitutions, 223–229
    associative *versus* dissociative
        substitutions, 229–231
Square-planar geometry, 15
SRI International, 798
Stannylboration
    alkynes, 697
    mechanism, 699
    overview, 697
Step-growth polymerization, 1049
Stereoblock polymers, 1049, 1057
Stereoblock polypropylene synthesis
    isotactic–atactic
        generated from heterogeneous
            catalysts, 1063
        generated from homogeneous
            metallocene catalysts,
            1063–1064
    overview, 1062–1063
    stereoblock copolymers
        by alternation of the ligand
            sphere, 1065
        by chain transfer, 1065
        from living catalysts, 1065
Stereochemical "memory effect," 982
Stereodefined polypropylene synthesis
    hemiisotactic, 1062
    isotactic and syndiotactic, 1060–1062
    stereoblock polypropylenes,
        1062–1065
Stereoerror, 1049
Stereoselectivity
    description of, 551–552
    energetics, 552–558
    origin of, 552
Steric hindrance, 901–902
Steric properties
    cross-coupling reactions and,
        901–902
Z-Stilbene, 1033
Stille couplings (reaction), 879, 902,
    914, 937
Styrene
    copolymerization with CO
        control of stereochemistry,
            805–806
        overall mechanism, 804–805
    single-site catalysts for the
        polymerization of, 1070–1072

Succinic anhydrides, 787–788, 794
Sulfido ligands, 510
Sulfinates, 462–463
Sulfoxides, 63
Sulfur
    complexes of neutral donors, 63–64
Sulfur dioxide, 462–464
Sulfur ylides, 485–486
"Supernucleophile" Fe(CO)$_4$2-, 302
Suzuki couplings, 880, 896, 902, 937–938
Symmetric acyclic allylic esters, 987
Symmetry
    polypropylenes and, 1058
Symmetry-adapted linear combinations (SALCs), 17–18
Syn oxypalladation, 383, 384
Syndiotactic polymers
    propene and CO, 807
    styrene and CO, 805
Syndiotactic polypropylene, 1057
    symmetry, 1058
    synthesis, 1060–1062
Syndiotactic polystyrene, 1070–1072
Synthesis gas, 749

T

T_1, 68
Takasago Company, 623, 633
Takaya ligand, 768
Ta(V) hydrides, 410
Tebbe, Fred, 490
Tebbe's reagent, 490–491
"Telomeric" amines, 708
Telomerization, 1089–1090
Terminal olefins
    hydrocyanation, 671
Tertiary amine complexes, 405
Tertiary phosphines, 33, 34
    electronic properties, 36–37
Tetraalkyl ammonium cyanides, 104
"Tetrahedral enforcers," 168
Tetramerization
    of alkenes, 1084–1086
(Tetramethylethylenediamine)PdMe$_2$, 57
(Tetramethylethylenediamine)PdPh(I), 57
Thermolysis, 281
Thermoplastic elastomers, 1062–1063
Thermoplastics, 1047
THF, 62
κ^2-Thioacyl complexes, 365
Thiocarbonyl complexes, 32
Thiocarbonyl ligands, 365
Thioesters, 344
Thioethers, 63
Thiolate bridges, 194–195
Thiolate complexes
    bonding and structures, 194–195
    overview, 194
    reactivity, 197

synthesis, 196
    thermodynamics of M–SR bonds, 195–196
Thiols
    copper-catalyzed coupling with aryl halides, 929–930
    oxidative additions and, 314
Thiophene complexes, 56
Thiophenoxide ligand, 342
Three-phase test, 548
Tipranavir, 990
Titanium
    ansa metallocenes of, 116
Titanium–imido complex, 287
Titanium(IV) trimethylsilylmethyl benzylidene complex, 288
Titanium tetraisopropoxide, 176
Titanium–alkyne complexes, 471
Titanium–methylidene complex, 490
Titanocene, 116, 527
Titanocene complexes, 696
Titanocene dimethyl, 490–491
TMEDA, 918
TMM complexes, 110–111
Tolman, Chadwick, 37, 38
N-Tosyl-2-methylaziridine, 792
Tp. See Trispyrazolylborate (Tp) ligands
Tp' ligands, 169
Trans–1-bromo-2-fluorocyclohexane, 306
"Trans effect," 228–229
"Trans influence," 228
Transesterification, 218
"Transfer hydrogenations," 575–576, 634–636
Transition-metal-catalyzed hydroamination. See Hydroamination
Transition metal complexes
    geometries, 14–15
    molecular orbitals, 17–19
Transition metal properties
    bond strengths, 9–10
    ionization potentials, 8–9
    trends in size, 9
Transition state α-agostic mechanism, 375
Transition-state stabilization, 542–543
Transmetallation
    with alkynyl copper complexes, 97, 98
    generation of boryl complexes, 188
    mechanism, 895–899
    synthesis of alkyl complexes, 87–88
    synthesis of complexes with terminal aryl ligands, 92–93
TRAP ligand, 607, 650, 996
Tri-tert-butyl methoxide (tritox), 175
Trialkylphosphines, 35, 36
Triarylamines, 344
Triarylphosphine ligands, 756–758
Tridentate anionic ligands. See Pincer ligands

Trienes, 247–248
Trimerization
    of alkenes, 1084–1086
η^4-Trimethylenemethane (TMM) complexes, 110–111
Triphenylphosphine complex, 343
Tris(boryl)iridium complex, 857
Trispyrazolylborate (Tp) ligands
    bonding of, 168–169
    reactions of, 170
    similarity to Cp, 168
    synthesis, 169–170
Tritibutyl siloxide (silox), 175
Tritox, 175
Trost ligand, 996
Tsuji–Trost reactions, 969
TunePhos, 605
Tungsten catalysts, 989–990
Tungsten DME complex, 62
Tungsten–carbene complexes, 288
Tungstenocene, 329
Tunneling electron microscopy, 547
Turnover frequency, 545–546
"Turnover-limiting step," 546
Turnover number, 545
[2+2] additions
    across metal–ligand multiple bonds, 287–289
[2+2] reactions
    alkylidenes and alkylidynes, 499–503
    with C–H σ-bonds, 503–504
    N-heterocyclic carbene ligands, 503
    Schrock carbenes, 502
    vinylidenes, 498

U

U–C bond, 86
Ullmann, Fritz, 918
Ullmann ether synthesis, 918, 926
Ultra-high molecular weight HDPE, 1052
Ultra-low-density polyethylene (ULDPE), 1052
"Umbrella flip," 236
Undirected hydroarylation
    of olefins, 850–852
Unfunctionalized ketones, 621, 626–629
Unfunctionalized olefins, 618–620
Union Carbide Corporation, 756
Unsaturated alcohols, 618
Unsaturated electrophiles, 453
η^2-Unsaturated hydrocarbon ligands
    nucleophilic attack
        alkyne complexes, 434–435
        η^2-arene complexes, 435
        general trends, 427–428
        olefin complexes, 428–433
        square Pd(II) diene and allene complexes, 433–434
Uranium(III) complex, 114

V

Valence bond theory, 18
Valence electrons
 counting, 10–12
 18-electron rule, 13
Valence orbitals, 13
Vanadocene, 114
Vanomycin, 445
Vaska's complex, 150, 269, 306
Vinyl acetate, 723
Vinyl complexes
 C–C bond-forming reductive
  elimination and, 332–333, 334
 migratory insertions of alkynes and,
  369–370
 overview, 92, 96–97
 stability, 92
Vinyl epoxides, 971, 987–988
Vinyl halides
 asymmetric coupling with enolates,
  886–887
 carbonylation to form ketones,
  914–915
  mechanism, 916–917
Vinyl ligands, 97
 electrophilic attack at the β-position,
  467
 structure, 85
η^1-Vinyl ligands, 97
η^2-Vinyl ligands, 97
Vinyl lithium reagents, 96
Vinyl triflates, 795
Vinylarenes, 57
 addition of arylamines to, 700
 asymmetric hydrocyanation,
  674–675
 hydroamination, 705–706, 713–715
 hydroaminomethylations, 771
 hydroboration, 692
 hydrocarboxylation, 776

 hydrosilylation, 680, 684
 oxidative aminations, 728–729
Vinylcyclobutanes, 291
Vinylcyclopropanes, 291
Vinylidene complexes, 468
 origin of the electrical properties of,
  483–484
 overview, 482
 reactivity, 498
 synthesis of, 486–488
Vinylidenes, 43–44
"Vinylidenic" olefins, 766
Vinylsilanes, 681
Vitamin B$_{12}$, 163

W

Wacker Chemie, 718
Wacker oxidations, 401
 mechanistic studies
  C–X bond formation, 732–733
  overview, 731–732
  re-oxidation, 733–734
 of olefins, 431, 731–734
Wacker process
 description of, 718–719
 historical overview, 718
 mechanism, 719–721
 olefin oxidations related to
  intermolecular additions of
   alcohols and carboxylates,
   722–724
  intramolecular additions of
   alcohols and carboxylates,
   724–726
  natural products synthesis,
   726–727
  overview, 722
Water
 nucleophilic attack and, 342
Water-gas shift reaction, 419, 421

Water-soluble rhodium hydroformylation
 catalysts, 758–759
WH$_3$(dppe)$_2$+, 136
W(II) enolate complex, 101
Wilkinson's catalyst, 39, 268, 576, 678,
 680, 693
 olefin hydroboration and, 695, 696
 olefin hydrogenation and
  migratory insertion step, 590
  oxidative addition step, 589
 preparation of, 578
 reactivity, 579–580
"Windshield wiper" mechanism, 1058
WMe$_6$, 18

X

X-type ligands, 2
Xantphos complex, 705
Xantphos ligands, 762
X–H bonds, 330–331

Z

Zeiss's salt, 47, 49, 51
Ziegler, Karl, 1047
Ziegler catalysts, 1047, 1052, 1056
Ziegler–Natta olefin polymerization, 371
Zirconacyclopentadienes, 457
Zirconium
 ansa metallocenes of, 116, 117
 dinitrogen complexes and, 60, 61
 neutral oxygen donors and, 62
Zirconium alkyl, 89
Zirconium(IV)imido complexes, 287
Zirconocene, 1060, 1061
Zirconocene β-arylethyl complex, 401
Zirconocene–alkyl complexes, 409
Zirconocene–alkyne complexes, 471
Zirconocene–benzyne complexes, 472
Zirconocene–olefin complexes, 471